1700–1720
Fahrenheit constructs
mercury thermometer

Montague introduces
smallpox vaccination

1721–1740
Micheli publishes
work on fungi

Linnaeus describes
classification system
for living organisms

1741–1760
Osmosis described

Benjamin Franklin
shows lightning to be
electricity

1761–1780
Hydrogen, oxygen, and
nitrogen discovered

Spallanzani disputes
theory of spontaneous
generation

Watt perfects the
steam engine

1781–1799
Lavoisier produces
a table of chemical
elements

Jenner introduces
vaccination

1800–1820
Invention of battery

Discovery of uv rays

Dalton's atomic theory

Wave theory of light

Lamarck emphasizes
effects of environment
on species

Appert develops
canning

Avogadro proposes
molecular theory of
matter

1821–1840
Faraday's electric
motor

Ohm's law of current

Schwann, Kuntzing,
and de Latour report
that yeasts cause
fermentation

Morse invents
telegraph

Development of
centrifuge

1841–1860
Liebig studies on
biochemistry

Joule defines first law
of thermodynamics

Clausius outlines
second law of
thermodynamics

Bunsen invents
gas burner

Darwin's theory of
natural selection

Pasteur studies
fermentation

1348
DEC.31

1347
DEC

CITIES AND REGIONS
PARTIALLY SPARED
BY THE PLAGUE

February 4, 1988

Professor Robert Wheat
Department of Biology
Duke University Medical Center
Durham, NC 27706

Dear Professor Wheat:

I am very pleased to enclose a copy of the recently published Atlas,
Microbiology, Second Edition. As you can see, the book has been beautifully
produced and we expect good results in the marketplace.

Your careful reviewing has contributed much to the improvements incorporated
in this revision. Thanks again for your assistance.

Sincerely,

Robert L. Rogers
Senior Editor

RLR/at

MICROBIOLOGY

Fundamentals and Applications

Ronald M. Atlas

UNIVERSITY OF LOUISVILLE

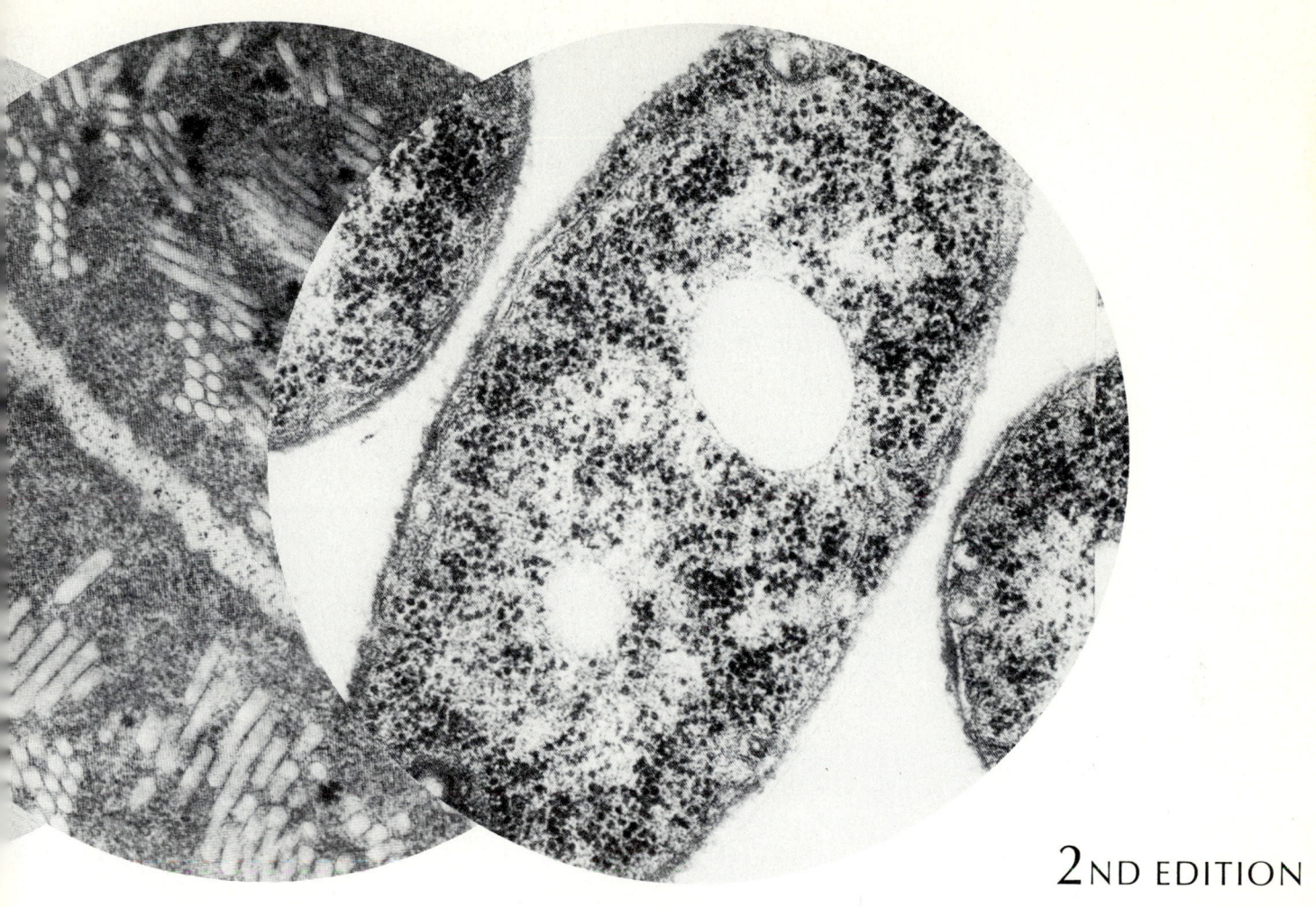

2ND EDITION

MICROBIOLOGY

Fundamentals and Applications

MACMILLAN PUBLISHING COMPANY
NEW YORK
COLLIER MACMILLAN PUBLISHERS
LONDON

Cover photo: Culture of *Pseudomonas* and *E. coli* on blood agar.
(© Carroll H. Weiss, 1988)

Macmillan Publishing Company
866 Third Avenue, New York, New York 10022

Collier Macmillan Canada, Inc.

LIBRARY OF CONGRESS CATALOGING IN PUBLICATION DATA

Atlas, Ronald M.
  Microbiology: fundamentals and applications / Ronald M. Atlas.—2nd ed.
      p. cm.
  Includes bibliographies and index.
  ISBN 0-02-304300-8
  1. Microbiology. I. Title.
QR41.2.A84 1988   87-23727
576--dc19   CIP

Printing: 1 2 3 4 5 6 7 8   Year: 8 9 0 1 2 3 4 5 6 7

ISBN 0-02-304300-8

Preface

Overview

Microbiology: Fundamentals and Applications covers the broad vistas of microbiology, providing students with a basis for understanding the various aspects of this field. It is intended for students who require a knowledge of microbiology for their future careers, as well as for students who desire a knowledge of microbiology for use in their daily lives. The depth of topic treatment is sufficient to satisfy and to stimulate student interest in the exciting field of microbiology.

The book is intended to be comprehensive and to meet the varied teaching needs of different institutions. Undoubtedly, there is more material in this text than can be reasonably handled in a one-semester course. Instructors may pick from the various chapters to meet their own needs. The text is designed to be flexible so that the order of chapters and topics covered are subject to the instructor's creativity. To aid in this endeavor, the first weight headings have been numbered for ease in assigning individual sections. Students can supplement the coursework by reading topics of interest that are omitted by the instructor from the formal class presentation.

There has been a general updating of material throughout this edition. Topics such as biotechnology and acquired immune deficiency syndrome have been expanded, as these are at the forefront of public attention in microbiology today. In these areas, both fundamental background information and applied aspects have been added to increase understanding.

General Organization of the Second Edition

Each chapter has the following general structure:

· Text of chapter
· Postlude
· Suggested supplementary readings
· Study questions
· Situational problems

The book is organized from the basic to the practical. Following an examination of the historical development of microbiology and the basic techniques of the microbiologist, the fundamental aspects of microbiology are examined in sections on structure, metabolism, genetics, and taxonomy. These sections build from the subcellular to the whole-organism level and establish the fundamental aspects of microbiology needed to understand the practical or applied aspects of microbiology covered in later chapters. Later chapters examine the practical ecological, industrial, and medical aspects of microbiology. There is a balance between the fundamental and applied aspects of microbiology.

The book includes coverage of both the eukaryotic and prokaryotic microorganisms, and, although the emphasis is undoubtedly placed on bacteria, eukaryotic organisms and viruses are also thoroughly covered. By comparing the similarities and differences among viruses, bacteria, and eukaryotic microorganisms, the student will develop an understanding of the basis for some of the applied aspects of microbiology, such as the use of antimicrobial agents in treating infectious diseases.

Several major changes have been made in the second edition of the text. There has been a major reorganization of chapters. The first chapter of the text still concerns the scope of microbiology and some of the early historical developments, but it has been greatly abbreviated to avoid overwhelming students with the history of all fields of microbiology and the associated terminology with these new subjects. Sections of the original historical discussion have been added to the postludes of the appropriate chapters.

The second chapter of the text covers microscopy and the third discusses the structures of prokaryotic and eukaryotic cells. The retention of a discussion of eukaryotic cells is critical for comparative purposes so that students understand the fundamental differences between prokaryotes and other organisms.

The growth and reproduction of bacteria, originally discussed in the chapter with viral replication and growth of eukaryotes, is moved towards the beginning of the book as the initial chapter on microbial metabolism. Bacterial reproduction is now combined with the discussion of nutrition. This chapter is followed by the discussion of microbial metabolism which, in turn, is followed by the discussion of microbial genetics.

The next section, entitled Survey of Microorganisms, begins with the chapter on virology. The consolidation of most of the material on virology into a single chapter on the viruses is perhaps the greatest organizational change between the first and second editions. Material on the structure of viruses, the replication of viruses, and viral taxonomy is included in this unified chapter on virology, which bridges the gap from the molecular-level discussion of the preceding chapters to the whole-organism level that takes place in the following chapters on bacterial taxonomy and the classification of eukaryotic microorganisms. Information on reproduction of eukaryotic microorganisms, previously found together with the discussion of bacterial and viral reproduction, has been included in the chapter on taxonomy of eukaryotes. This change should accommodate users of the book who do not wish to include eukaryotes in their discussions.

The survey of microorganisms section is now followed by coverage of environmental microbiology. The earlier placement of this section is aimed at tying this information more closely to the discussion of fundamental aspects of microbiology. The environmental section now has four chapters; the additional chapter is concerned with environmental effects on microorganisms and the distribution of microbes in nature. A chapter devoted to biogeochemical cycling and population interactions, one concerning agricultural microbiology, and one concerning water quality and pollutant biodegradation are retained in this section.

The environmental section is followed by the food and industrial microbiology section which in turn is followed by the medical section, which includes a discussion of immunology. Placement of medical microbiology at the end of the book allows instructors who do not cover this topic to easily omit it from their courses and students who want to use this material as a reference source can easily find the appropriate information. Within the medical section, the chapter on clinical microbiology that appeared in the first edition has been eliminated and the information incorporated into a unified chapter on prevention, diagnosis, and treatment of diseases caused by microorganisms.

Special Features

Analytical Process Boxes   Within most chapters where methodological processes are discussed, analytical boxes have been added. These analytical boxes set off material, describing the procedures used by microbiologists to gain the information on the topics being discussed.

Discovery Process Boxes   The discovery boxes, a popular feature of the first edition, have been retained and several new ones have been added. These boxes augment the figure captions, illustrating how discoveries were made. They highlight the paths scientists have followed in developing the knowledge that fills this text.

Illustrations   The book is elaborately illustrated, and each figure is accompanied by a clear and detailed figure legend. There has been a thorough review and upgrading of art, including additional use of color and use of overlay labels for many micrographs. These enhance the use of illustrations as teaching aids.

Key Terms   Within the text, key terms have been highlighted. These terms are defined in the glossary and in the text near where they are introduced. Succinct definitions of key terms permit students to continue reading and understanding the section without having to fully grasp every nuance of a term or to turn to the glossary if they are unfamiliar with a term. Every effort has been made to ensure that terms and concepts are clearly defined.

Postludes   Each chapter ends with a postlude that summarizes and places the information discussed in the chapter in proper perspective. As noted, historical information about the material covered in each chapter has been added to the postludes. The postludes remain more than simple summaries, attempting to place the information discussed in the chapters into a broader context of history and science.

Supplementary Readings List   There has been a thorough updating of the suggested reading lists to include the latest articles. This list is meant neither to be complete, nor of great depth, but to supplement the text for more advanced courses and to sustain the interest of the student who finds a particular topic relevant to his or her purpose for having enrolled in an introductory microbiology course.

Questions   Each chapter has a set of review questions intended to allow students to test their own comprehension of the material they have just examined.

Situational Problems   Situational problems have been added at the end of each chapter to cause students to think and to develop an in-depth understanding of microbiology. These problems should be challenging and interesting to students.

Glossary of Microbiological Terms   An extensive glossary has been included to help the student understand the information of the book. Key terms highlighted in the text when first introduced are included in the glossary.

Glossary of Bacterial Genera   There is an abridged glossary of bacterial genera that can be used as a handy reference for the descriptions and key features of the more common bacteria.

Appendices   The appendices include the metric system and a review of basic chemistry. The chemistry appendix can act as review or remedial material to ensure that students have ready access to the chemical principles they need to understand microbiology.

Supplements

Laboratory Manual   "Experimental Microbiology" by A. Brown, K. Dobra, L. Miller, and R. M. Atlas. This laboratory manual provides an extensive compilation of laboratory exercises that can be used for introductory laboratory courses.

Instructor's Manual   Outlines of text chapters and extensive listing of audiovisual material that can be used with each chapter.

Laboratory Instructor's Manual   Helpful hints for laboratory exercises and sources of needed materials.

Transparencies   Collection of transparencies from the text that can be used to illustrate lecture material.

Slides   Collection of 35 mm slides that can be used to illustrate lecture material.

Test Bank   Computerized collection of test questions compiled by H. Peery that can be used for preparing exams.

Computer Aided Instructional Package   "Micro-micro" by J. Snyder, G. Weinberg, and R. M. Atlas. This computer aided instructional package contains 42 lessons in general and medical microbiology. It can be run on Apple II series and IBM PC compatible microcomputers. The lessons can be used to review topics, such as enumeration of microorganisms, as well as to learn more advanced topics, such as specific areas of clinical microbiology.

Acknowledgments

Many individuals contributed to the writing of this book, and their efforts indeed are appreciated. Many of my colleagues, to whom I am indebted, contributed micrographs. Carl May, of the Biological Photo Service, found many of the photographs that help illustrate this text. I am especially grateful to the following reviewers for their comments and helpful suggestions that have aided greatly in the writing of this book and development of the second edition:

Richard Adler, University of Michigan, Dearborn
James P. Amon, Wright State University
Robert Bender, University of Michigan
Richard Blakemore, University of New Hampshire
John A. Breznak, Michigan State University
Alfred E. Brown, Auburn University
David Carlberg, California State University, Long Beach
Warren Cook, Georgia State University
Ronald L. Crawford, Gray Freshwater Biological Institute, University of Minnesota
Calvin Davenport, California State University, Fullerton
Arnold Damain, Massachusetts Institute of Technology
Kenneth W. Dobra, E. B. Sparrow Hospital
Loretta C. Ellias, Florida State University
Douglas Eveleigh, Rutgers University
Sharon File, University of Puerto Rico
Richard W. Fleming, California State University
James I. Frea, Ohio State University
James J. Gilroy, Boston College
Scott Graham, Upjohn Company
James Hauxhurst, University of Wisconsin, Fox Valley
Alice Helm, University of Illinois
John Holt, Iowa State University
Lewis Jacobson, University of Pittsburgh
D. C. Jordan, University of Guelph
David Kafkewitz, Rutgers University
Roger Lambert, University of Louisville
Hubert Lechevalier, Rutgers University
James MacMillan, Rutgers University
Thoyd Melton, North Carolina State University
Glenda Michaels, Western State College, Gunnison, Colorado
Thomas Montville, USDA, Philadelphia
William O'Dell, University of Nebraska, Omaha
James D. Oliver, University of North Carolina, Charlotte
Raymond Otero, Eastern Kentucky University
Harry E. Peery, Tompkins Cortland Community College
Morrison Rogosa, National Institutes of Health
Antonio Romano, University of Connecticut
Chester Roskey, Framingham State College
Ramon Seidler, Oregon State University
Rivers Singleton, University of Delaware
John Smith, Oklahoma State University
George Somkuti, USDA, Philadelphia
Frank X. Steiner, University of Massachusetts, Amherst
Daniel R. Tershak, Pennsylvania State University
James E. Urban, Kansas State University
John Wallace, University of Louisville
Robert W. Wheat, Duke University Medical Center
Roseann S. White, University of Central Florida
Gary R. Wilson, Texas A & M University
Rudy J. Wodzinski, University of Central Florida

I particularly want to thank the editors and production staff at Macmillan: Kate Aker, Greg Payne, Gary Carlson, Gary Ostedt, Bob Rogers, Bob Pirrung, Dora Rizzuto, and Bob Freese, for helping to develop and to produce this book. The many hours that they spent and their constant support made the writing of this text as painless as possible, allowing me to focus on the task of writing.

Finally, I want to thank my family who not only tolerated my obsession with writing this book, but who contributed to its writing. My wife Michel spent many hours researching parts of this book and editing numerous drafts; my children, Matt and Debbie, also made themselves an integral part of the work.

R. M. A.

Brief Contents

Detailed Contents

PART *3*

Microbial Growth and Metabolism   85

CHAPTER *4*

Culture, Nutrition, and Growth of Microorganisms   87

CHAPTER 11

Survey of Fungi, Algae, and Protozoa   313

PART 6

Environmental Microbiology   344

CHAPTER 12

Influence of Environmental Factors on the Growth and Distribution of Microorganisms   347

CHAPTER 13

Biogeochemical Cycling and Interactions Among Microbial Populations   371

CHAPTER *17*

Industrial Microbiology   495

PART *8*

Medical Microbiology   535

CHAPTER *18*

Interactions Between Microorganisms and Humans   537

CHAPTER *22*

Human Diseases Part 2: The Skin as a Portal of Entry and Diseases of Superficial Body Tissues   691

APPENDICES   719

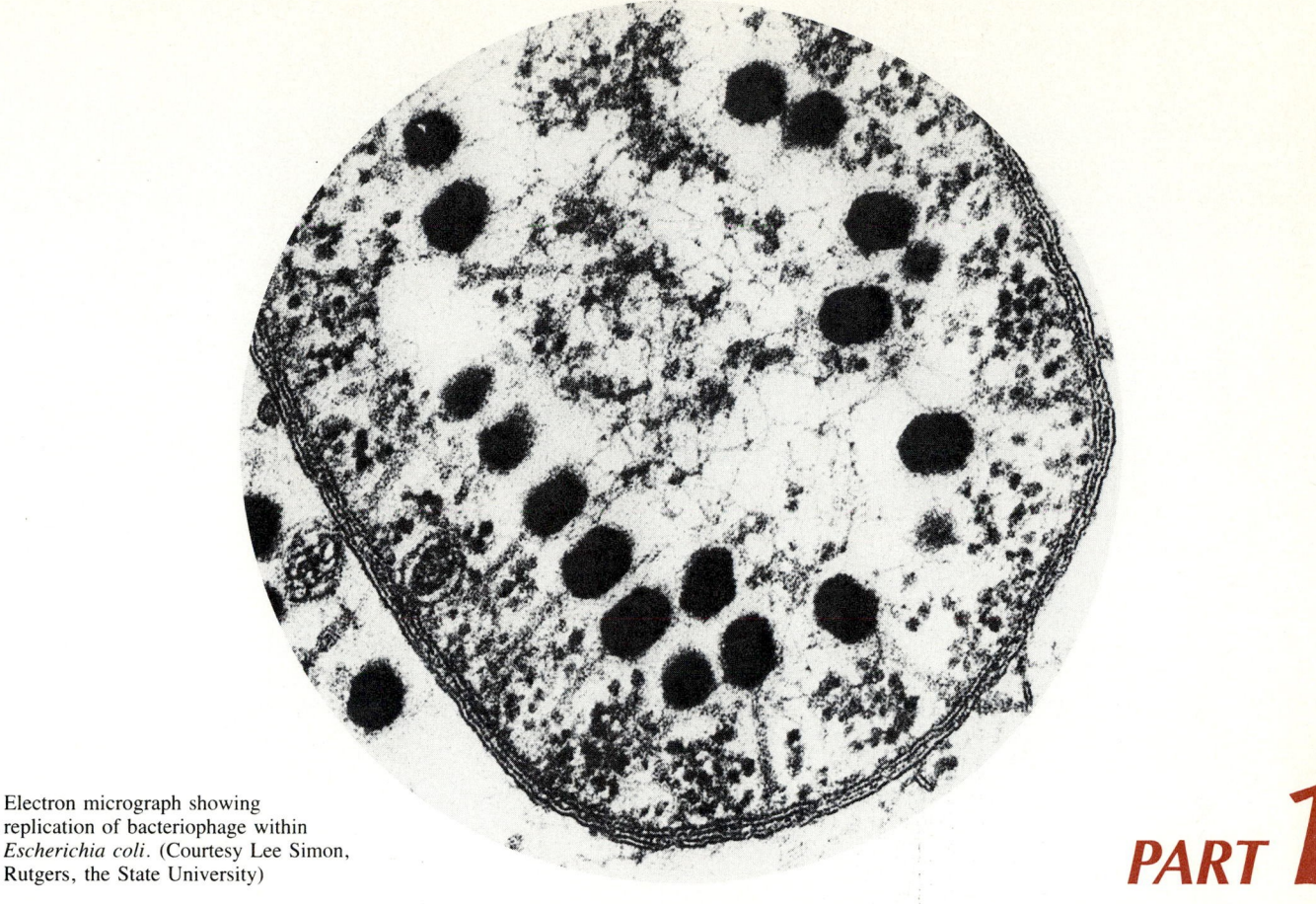

Electron micrograph showing
replication of bacteriophage within
Escherichia coli. (Courtesy Lee Simon,
Rutgers, the State University)

PART 1

Background to the Study
of Microbiology

CHAPTER 1   Evolution of Microbiology and Microorganisms

Evolution of Microbiology and Microorganisms

1.1 *Historical Evolution of Microbiology as a Scientific Discipline*

It has been only a little more than a century since we began to appreciate the roles of unseen microbes in processes such as decay and disease. This is, perhaps, not too surprising considering that **microbiology** is the field of study concerned with organisms—called **microorganisms**—so small that for the most part they were not observed prior to the advent of the microscope. The development of microbiology as a scientific discipline could occur only after the discovery of the physical principles needed to produce the optical instrument—the microscope—with which microorganisms could be observed, and the chemical principles needed to understand the fundamental biochemical and genetic properties of microorganisms.

Although microbiology did not exist as a true science before the latter part of the nineteenth century, humans have undoubtedly been practitioners of microbiology since prehistoric times. Early human beings saw the macroscopic structures formed by some microorganisms and made use of the fruiting bodies of certain fungi—mushrooms, for example—as food sources. Beer and wine were produced and enjoyed long before we understood the role of microorganisms in the transformation of sugar into alcohol (Figure 1.1). Long voyages of discovery, such as those to the Americas in the sixteenth century, depended on the ability to control microbial food spoilage by drying and salting foods, methods used even though no one understood why they worked. Even the eminent seventeenth-century English philosopher Francis Bacon had little understanding of the practical importance of microorganisms, and while testing his theory that freezing could retard the spoilage of chicken, he caught a cold and died.

For much of our history, we—like Bacon—have had to contend with disease. Our lack of scientific understanding has not dissuaded us from finding methods for treating disease. Early humans used various herbs and other plants to treat diseases caused by microorganisms. In the third century B.C., Theophrastus, the successor to Aristotle as head of the Greek Lyceum, compiled a list of medicinal properties of herbs, based largely on remedies imported from the Orient. Lacking any understanding of the underlying microbiological cause of plague or any other infectious disease, in the Middle Ages people even turned to such unscientific and ineffective practices as self-flagellation, beating themselves to drive out the force causing the ''black death'' during a severe outbreak of plague that left more than 25 million dead (Figure 1.2). A more scientifically relevant method of dealing with plague was to prevent those suspected of having the disease from coming into contact with the general public, a method known as **quarantining** that is still sometimes used today. The practice of isolating the sick dates back to biblical times, but the term *quarantine* (from the Italian *quarantenaria*, meaning forty) originated at the height of this fourteenth-century plague epidemic when sea voyagers coming into Sicily had to wait 40 days before being allowed to enter the city. Plague still occurs today, and the ''Western White House'' of the United States during the administration of Ronald Reagan, for example, was located within a posted plague-infested area. Fortunately, plague is now effectively controlled by sanitary measures that limit urban rodent populations—the primary reservoirs of plague-causing bacteria—and by the use of antibiotics to treat this disease.

FIGURE 1.1
The long history of human use of fermentative techniques is illustrated in this painted relief discovered on the wall of a Fifth Dynasty Egyptian tomb dating from about 2400 B.C. The top and middle panels show the production of bread. In the middle panel, leavening of dough and baking of "beer-bread" are shown. The bottom panel shows mash being strained into a fermenting vat resting on a stand that looks like a coiled rope. After being left to ferment for a few days, the finished beer was then poured into pottery jars, which were quickly capped and sealed with clay to be placed in storage. These early Egyptian brewers originally used yeasts in the air or on the skin or husk of fruits and cereals to carry out these fermentations but later developed an almost pure yeast. Their ancient breweries were able to produce different types of beer, some of which may have had an alcohol content as high as 12 percent. (Courtesy M. J. Vinkesteyn, National Museum of Antiquities, Leiden, The Netherlands)

FIGURE 1.2
The devastating effects of the plague in medieval Europe are shown in this painting. The plague was apparently introduced by seaborne rats from areas around the Black Sea. The spread of the disease generally followed trade routes from the Near East to Europe, reaching Scandinavia by 1350 and spreading perhaps as far as Iceland and Greenland; areas outside of the major trade routes remained virtually unaffected. (The Bettmann Archive, Inc.)

Two centuries after the major outbreaks of plague in Europe, Girolamo Fracastoro of Verona, a contemporary of the astronomer Copernicus, published a work on contagious diseases and their treatment in 1546. *De Contagione* was largely philosophical and Fracastoro did not recognize the true nature of microorganisms. Nevertheless, he did hypothesize that some diseases were caused by the passage of ''germs'' from one thing to another and described three processes for their transmission: direct contact, indirect transmission via inanimate objects such as clothing, and transmission from a distance via air. Fracastoro had the insight to recognize the similarity between **contagion** (disease processes) and **putrefaction** (spoilage of meat and other protein-containing foods with the production of foul-smelling decay products). He further recognized that disease-causing germs exhibit specificity, indicating that different diseases occur in different hosts and that different processes of transmission occur for different germs.

The Advent of the Microscope

Prior to the seventeenth century, the view of the importance of microorganisms in human health and other areas was, of necessity, mostly philosophical rather than scientific. The lack of an instrument with which to view these organisms and to confirm their existence precluded understanding the breadth of their activities. The **microscope**, an instrument used to magnify the images of objects too small to be seen with the naked eye, was not invented until the mid-sixteenth century; there is still debate as to whether the first useful microscope was made by the Dutch spectacle maker Hans Janssen or the Italian astronomer Galileo.

The advent of the microscope did not immediately reveal the existence of microorganisms. It was not until the mid-seventeenth century, when further development of the optical lens systems of the microscope permitted the visualization of microorganisms, that the great diversity of the microbial world began to be recognized. Robert Hooke's *Micrographia or Some Physiological Descriptions of Minute Bodies Made by Magnifying Glasses with Observations and Inquiries Thereupon*, published in 1665, contains drawings of the fruiting bodies of fungi (Figure 1.3). Hooke was an English experimental philosopher who made significant contributions to physics, architecture, and clockmaking, in addition to his microscopic observations of biological specimens.

The first recorded observations of bacteria, yeasts, and protozoa were made by the amateur Dutch microscope maker Antonie van Leeuwenhoek (Figure 1.4). Leeuwenhoek, a cloth maker and tailor by trade, was also a surveyor and the official wine taster of Delft, Holland. His interest in microscopes was probably related to the use of magnifying glasses by drapers to examine fabrics. Leeuwenhoek was a close friend of Christian Huygens, the noted Dutch physicist, mathematician, and astronomer, who undoubtedly encouraged his interest in science and lenses. Leeuwenhoek's findings were transmitted in a series of letters, from 1674 to 1723, to the Royal Society in London, through which his observations were widely disseminated; during part of this

FIGURE 1.3
Various species of fungi can clearly be identified in the drawings of Robert Hooke (1635–1703). This 1664 drawing shows the growth of blue mold on leather. Although macroscopic fungal structures had been described much earlier, Hooke's drawings represent some of the earliest descriptions of many of the common fungi. (Reprinted by permission of the Royal Society, London, from R. Hooke, 1665, *Micrographia*)

FIGURE 1.4
Antonie van Leeuwenhoek (1632–1723), here seen holding one of his microscopes, opened the door to the hidden world of microbes when he described bacteria. Although he was only an amateur scientist, Leeuwenhoek's keen interest in optics and diligence allowed him to make this important discovery. (Reprinted by permission of Russell and Russell, Dover Publications, and Mrs. Dobell, from C. E. Dobell, 1932, *Antony van Leeuwenhoek and his ''Little Animals''*)

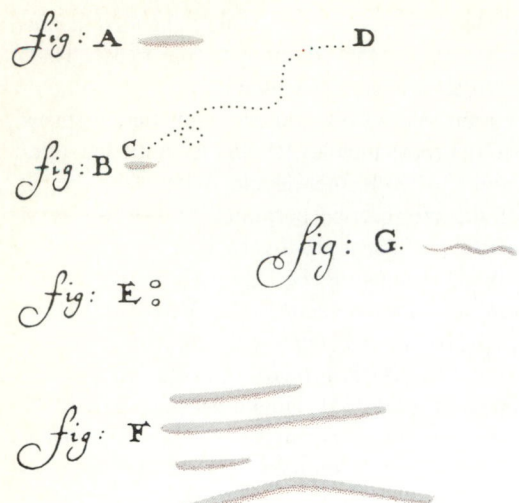

FIGURE 1.5

Leeuwenhoek's sketches of bacteria from the human mouth illustrate several common types of bacteria, including rods and cocci. (A) A motile *Bacillus*; (B) *Selenomonas sputigena*; (C) and (D) the path of *S. sputigena's* motion; (E) micrococci; (F) *Leptothrix buccalis*; (G) a spirochete, probably *Spirochaeta buccalis*. (Reprinted by permission of the Royal Society, London, from Letter 39, 17 September 1683)

period, Leeuwenhoek sent his communications to Robert Hooke, who was then the secretary of the Society. These letters contained detailed drawings, some of which clearly show microorganisms (Figure 1.5). Unfortunately, while Leeuwenhoek shared his observations with the scientific community, he never revealed the details of his methods of microscopy, nor how he constructed his hundreds of microscopes.

The early microscopes used by Hooke and Leeuwenhoek (Figure 1.6) permitted only a fuzzy view of the microorganisms they were observing. Many advances in the art of microscope making and the science of microscopy were needed before observations could be made on the fine structure of microorganisms. As microscopes were made that permitted higher magnifications and viewing of greater detail, scientists were able to perform increasingly complex studies on microorganisms; as the questions raised in these studies exceeded the capabilities of the instruments, new developments in the art of microscope making were mandated (Figure 1.7). The major advances made in microscopy in the late nineteenth century went hand in hand with a period of great advancement in microbiology. During this period, Ernst Abbe, a German physicist, developed microscope lenses that corrected for aberrations (distortions) inherent in magnifying lenses, which had limited the ability to view microorganisms.

FIGURE 1.6

Some early microscopes used in the late seventeenth century. (A) This microscope, used by Leeuwenhoek, consists of a single biconvex lens (l) inserted into a small hole on the left side of the leather base. The specimen was placed on the small pointed wire attached to the screw, which moved the specimen back and forth to bring it into the focus of the fixed glass, rather than focusing the lens on a fixed object, as is done on a modern microscope. Leeuwenhoek apparently kept the method for using his very tiny "magnifying-glasses" to himself. (Reprinted by permission of Russell and Russell, Dover Publications, and Mrs. Dobell, from C. E. Dobell, 1932, *Antony van Leeuwenhoek and his "Little Animals"*). (B) Hooke's compound microscope of the same period more closely resembles a modern microscope. Note the elaborate decoration of the body of the microscope, the candle illumination source, and the separation of the eyepiece from the objective lens. (Courtesy American Optical Co.)

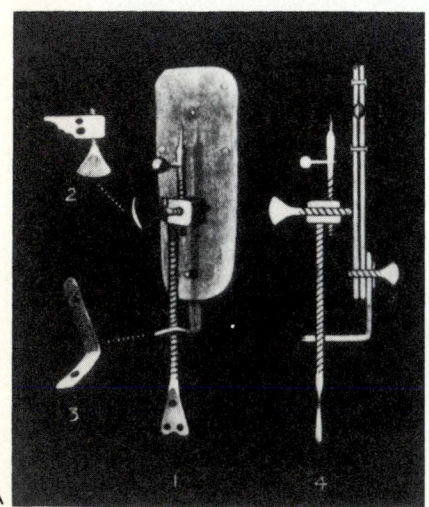

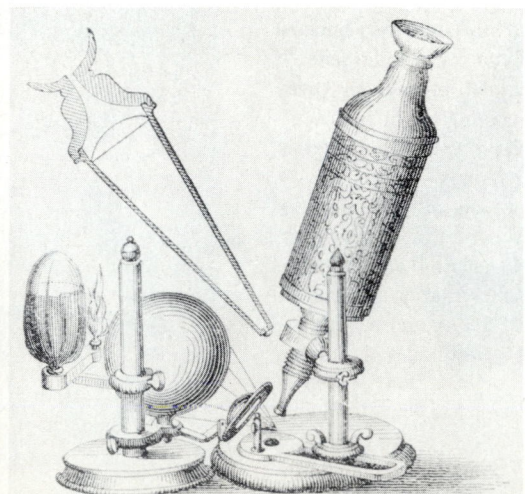

FIGURE 1.7

The development of advanced microscopes depended upon the elimination of optical defects that are natural properties of lenses. Simple magnifying glasses, like those used in the earliest microscopes, had two inherent problems: *spherical aberration* (distortion based upon the failure of light passing through the center of the lens to focus at the same point as light passing through the thinner edges of the lens) and *chromatic aberration* (distortion based upon the failure of light of differing colors to focus at the same point). To overcome these distortions, compound lenses were constructed with lenses of differing shapes and glass composition, as illustrated in this figure. The achromatic lens is a relatively inexpensive objective lens used on many microscopes intended for routine observations of microorganisms, including many of those used in introductory microbiology laboratory courses, that corrects for both spherical and chromatic aberrations. The apochromatic lens is a better and more expensive lens which is more finely corrected for spherical and chromatic aberrations; it produces very high-quality images that reveal the true colors of a specimen without distorting its shape and is excellent for photomicrography (photography through the microscope). Modern microscope lenses, called *flat field lenses*, also correct for curvature of field so that objects in the center and periphery of a field of view are simultaneously in focus.

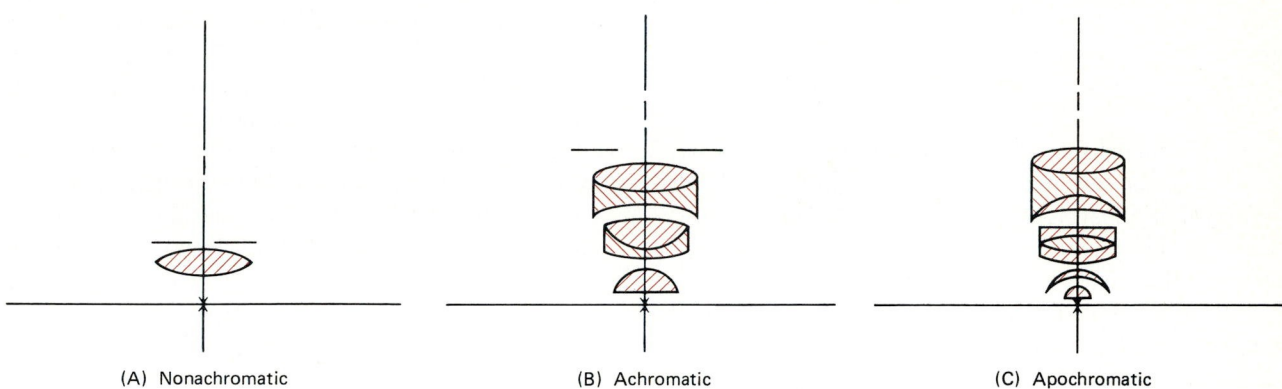

(A) Nonachromatic   (B) Achromatic   (C) Apochromatic

Discovery Process

Eighteenth-century microscopes were primitive instruments, limited in their usefulness by the chromatic and spherical aberration of their lenses. Chromatic aberration results from the failure of a simple double-convex lens to focus light rays of various wavelengths in the same plane and creates a blurred, multicolored image. The severity of blurring of the image due to chromatic aberration led many scientists in the late 1700s to believe that the microscope was useful only in creating artifacts. Chromatic aberration was the cause of a brisk argument between the famous English scientist Isaac Newton, who pessimistically thought that the aberration was an inherent property of light, and Leonhard Euler, a Swiss mathematician who reasoned that it would be corrected if lenses were made of two different materials. Euler thought that glass and water might be suitable, but this idea did not materialize. In 1759 an English artisan, John Dollond, successfully fashioned an achromatic objective lens for a telescope by combining two lenses made of glass with different indices of refraction; the different refractions of the two glasses canceled the aberration.

This practical invention made possible microscopy as we know it today. Despite the advances in the construction of microscope objective lenses, the manufacture of microscopes remained an art rather than a science until Carl Zeiss of Jena, Germany, formed a partnership with a young physicist, Ernst Abbe. From 1833 to 1895, Abbe solved the basic problems of light microscopy, particularly spherical aberration, which occurs because rays of light passing through the peripheral portions of a lens have a focal point different from those passing through the center. Abbe constructed objective lenses in which a concave lens was added to the basic convex lens system in order to bend the peripheral rays of light slightly to form an almost flat image. Schott's Optical Glass Works, another firm in Jena, furnished the glasses needed by the Zeiss factory. Jena became the mecca of microscopy, and it was there, in 1883, that the first lenses were made that corrected for both chromatic and spherical aberrations. By 1900 the light microscope had largely reached the state of technical development that characterizes light microscopes today.

Abbe also developed the oil immersion lens, which allowed for improved **resolution** (ability to see detail) in light microscopy and the use of higher magnifications for clearly viewing bacteria. Microscopic visualization of bacteria was greatly improved by the introduction in 1881 by Paul Ehrlich of vital staining of bacteria with methylene blue and the development of a differential staining method for bacteria by Hans Christian Gram in 1884.

The Gram stain technique is basic to the taxonomic description of bacteria even today, and the microscope continues to be the basic observational tool of the microbiologist because it allows the differentiation of microorganisms based upon fundamental structural differences. Modern microscopes and the current art of microscopy will be discussed in Chapter 2, and throughout this book micrographs illustrate the types of information that can be obtained about microorganisms by using different microscopic techniques.

Pasteur and the Refutation of the Theory of Spontaneous Generation

Although the microscope permitted microorganisms to be seen as early as the seventeenth century, it was not until the nineteenth century that microbiology began to develop as a true scientific discipline. The development of the scientific approach in microbiology, in which controlled experiments are used to prove or disprove hypotheses, is evident in the series of experiments that eventually discredited the **theory of spontaneous generation**, which held that living organisms could spring forth spontaneously from nonliving matter. In the seventeenth century, Francisco Redi had shown that flies do not spring forth from rotting meat. He experimentally covered one portion of meat with a loose-knit cloth, thereby preventing the flies from reaching the meat, and left a second portion of meat uncovered, so that flies could reach it. The observation of flies on the uncovered but not on the covered meat disproved the theory that *macroorganisms* arise spontaneously from decaying meat. However, the theory of spontaneous generation of *microorganisms* remained a viable idea throughout the eighteenth century and into the early nineteenth, subject to verification or refutation by experimental proof. In order to disprove the theory of spontaneous generation, it was necessary to begin to develop a true understanding of microbiology; several major advances in this field are tied to the disproval of this theory.

The relationship between the growth of **infusoria** (microorganisms) in organic broths and the onset of chemical changes that caused souring of wine and spoilage of meats, respectively known as *fermentation* and *putrefaction*, had frequently been observed. To demonstrate that the putrefaction of organic substances is caused by microorganisms that multiply by reproductive divisions and that do not arise by spontaneous generation, Lazzarro Spallanzani—an eighteenth-century priest who had an exceptionally inquiring mind and the daring to challenge the conventional wisdom of his time—sealed flasks containing meat broths that had been heated to destroy the microbes in the broth, thereby preventing spoilage indefinitely. Nineteenth-century advocates of spontaneous generation, though, claimed that the elimination of oxygen compromised these experiments because it eliminated the vital force needed for life to arise spontaneously. Noted chemists, such as Justus von Liebig, Jons Jakob Berzelius, and Friedrich Wöhler, lent support to this view, arguing that changes in organic chemicals, such as the putrefaction of proteins and the transformation of sugar into alcohol, occurred by strictly chemical processes without the intervention of living organisms. This premise was opposed in the 1830s by Charles Cagniard de Latour of France, Theodor Schwann, and Friedrich Kützing of Germany, each of whom separately proposed and conducted experiments to demonstrate that the products of fermentation—ethanol and carbon dioxide—were produced by microscopic forms of life; Schwann used a flame, and de Latour and Kützing used cotton plugs, to prevent microorganisms from entering the heat-sterilized broth. Each of these experiments was aimed at showing that living forms were responsible for fermentation; each was criticized by chemists for destroying or eliminating some essential component in air that was needed for the spontaneous generation of the fermentation products.

It took several more decades of debate and experimentation before **Louis Pasteur** (1822–1895) (Figure 1.8) succeeded in definitively discrediting the theory of spontaneous generation and establishing that living microorganisms are responsible for the chemical changes that occur during fermentation. Pasteur had been trained as a chemist, and this training had a marked influence on his approach to scientific questions. As a chemist, Pasteur was able to separate an optically inactive mixture of two optically active chemical compounds that differ only in spatial orientation such that the individual compounds bend light in opposite directions (called a *racemic mixture*) into its two optically active components, thus explaining the riddles concerning the optical activities of liquids and why light passing through a solution was bent sometimes to the left and sometimes to the right. Pasteur followed the same investigative approach throughout his long scientific career; he identified the problem, sought out all of the available information on the topic, formed a hypothesis, and devised experiments to test the validity of his theory.

Much of his work stemmed from the requests of local manufacturers to help solve the practical problems of their industrial processes; he loyally responded, attempting to solve these problems in order to improve the French economy and demonstrate French superiority. The problem of applied versus basic or pure research never bothered this aggressive, ambitious, argumentative, and highly patriotic Frenchman. He was concerned with problems such as why French beer was inferior to German beer. The answer to this practical question eventually led him to the basic discovery of the existence of **anaerobic life** (life in the absence of air).

In 1854 Pasteur was appointed dean of the Faculty of Science at the University of Lille. Following his appointment,

FIGURE 1.8

Louis Pasteur (1822–1895), seen working in his laboratory in this 1885 woodcut, began as a chemist but soon become a pioneer microbiologist. Pasteur's work encompassed both pure research and many areas of applied science that produced several important practical discoveries. Among his many accomplishments, Pasteur discredited the theory of spontaneous generation, introduced vaccination to treat rabies, and solved industrial problems related to the production and spoilage of foods. (From the Bettmann Archive)

one of the first problems Pasteur attacked was at the request of a local industrialist and concerned the souring of alcohol produced from beets; Pasteur's agreement to help solve this problem of the wine industry led his scientific career from chemistry to microbiology. By comparing, with the aid of a microscope, samples taken from productive and sour wine vats, Pasteur observed budding yeast cells in the productive vats and rod-shaped organisms in the nonproductive ones. He demonstrated that these two organisms determined the course of the chemical processes that result in different fermentations. The yeasts were responsible for the production of alcohol, and the rod-shaped bacteria produced the lactic acid that caused the wine to sour. The work of Pasteur led to an understanding of the role of microorganisms in food spoilage and the use of heat to destroy microorganisms in food products. In 1857 Pasteur demonstrated that the souring of milk was also caused by the action of microorganisms, and about 1860 he showed that heating could be used to kill microorganisms in wine and beer. This process of **pasteurization** (heating at moderate temperatures to reduce the number of living microorganisms), as it has come to be known, is based upon these experiments.

His work on microbial spoilage of wine and milk, and heat killing of microorganisms, led Pasteur in 1861 to report *On the Organized Bodies which Exist in the Atmosphere: an Examination of the Doctrine of Spontaneous Generation*; the series of experiments he described for the most part ended the controversy concerning spontaneous generation. In these experiments, Pasteur demonstrated that liquids subjected to boiling remain sterile, that is, free from any living microorganisms, as long as microorganisms in the air are not allowed to contaminate the liquid. By using a **swan-necked flask**, a flask with a long curved neck (Figure 1.9), Pasteur was able to leave a vessel containing a fermentable substrate open to the air and show that fermentation did not occur. The shape of the flask prevented airborne microorganisms from entering the liquid, and the fact that air could enter the

flask overcame the main argument that chemists had leveled against earlier studies using sealed flasks, namely, that oxygen was essential for the chemical reactions involved in the formation of alcohol.

Despite Pasteur's eloquent disproof of spontaneous generation, attempts to repeat his experiments occasionally failed. The problem was that boiling did not always kill all of the microorganisms in a broth. The English physicist John Tyndall (1820–1893), trying to confirm the results of Pasteur's experiments, determined that the variability of the results of heating was due to the capacity of bacteria to exist in two forms: a heat-labile form—likely to be changed or destroyed by exposure to heat—that was killed by exposure to elevated temperatures, and a heat-resistant form that could survive at such high temperatures. He found that intermittent heating could eliminate viable microorganisms and thus **sterilize** solutions, that is, completely eliminate living organisms from them, thereby validating Pasteur's disproof of spontaneous generation. Repeated heating on successive days, a process known as **tyndallization**, successfully sterilizes solutions containing bacteria that form heat-resistant structures known as *endospores*.[1] Alternatively, temperatures of 121°C, which can be achieved by heating water under a pressure of 15 pounds per square inch, can be used to sterilize solutions.

Koch and the Demonstration That Microorganisms Cause Disease

During the same period that Pasteur's studies were disproving spontaneous generation and contributing to an understanding of food spoilage and preservation, **Robert Koch**

[1] Tyndallization depends upon the germination and growth of the endospore-producing microorganism and the ability to kill the heat-sensitive growing microorganisms before they form new endospores. This method works only with solutions capable of supporting the growth of microorganisms.

FIGURE 1.9

Pasteur used a variety of shapes in the design of his swan-necked flasks.

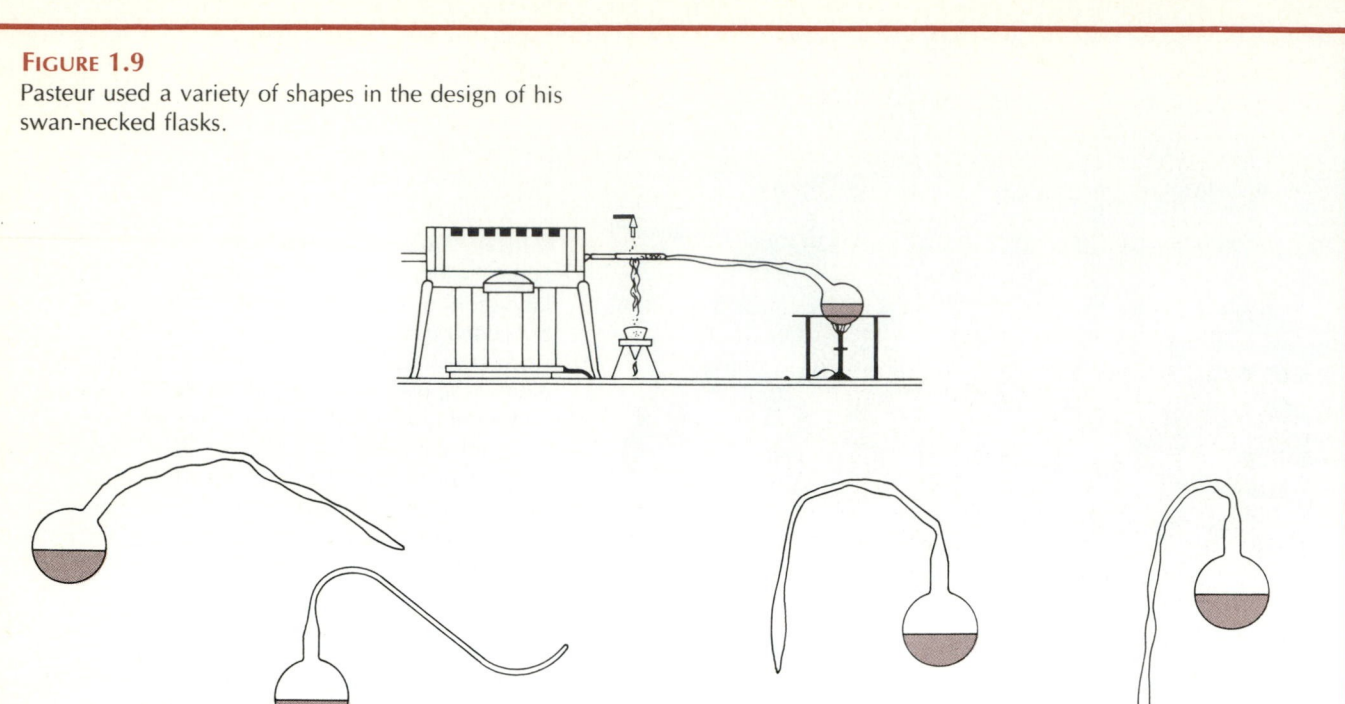

Discovery Process

Settling the question of spontaneous generation was one of Pasteur's great accomplishments. Pasteur began this classic experiment in 1861 by placing yeast water, sugared yeast water, urine, sugar beet juice, and pepper water into ordinary flasks. He then reshaped the necks of the flasks under a flame so that there were several curves in each neck, hence the term *swan-necked flasks*. Next, Pasteur boiled the liquids until they steamed through the necks. By using curved ends, Pasteur could leave the flasks open to the air, thus overcoming a major criticism of previous experiments aimed at disproving spontaneous generation where air, an "essential life force," had been excluded. Dust and microbes settled out in the curved neck of the flask; thus, while exposed to air, the broth did not become contaminated with microorganisms. Contrary to the opinion of those who believed in spontaneous generation, no change appeared in the liquid. The flasks were later sealed for exhibition and are displayed, still in their sterile state, at the Pasteur Institute in Paris.

(1843–1910) (Figure 1.10) was developing methods for growing pure cultures of microorganisms that permitted him to establish unequivocably the relationship between microbes and infectious diseases. Koch, a German country physician, began his studies isolated from the scientific community, working alone with primitive tools and materials. As a result of his medical practice, Koch was well aware of the diseases of humans and other animals. From 1873 to 1876, he studied the cattle disease anthrax and conducted experiments to show that the spores of anthrax bacilli isolated from pure cultures could infect animals. In his studies on anthrax, Koch demonstrated for the first time that germs grown outside a body could cause disease and that specific microorganisms caused specific diseases. Recognizing that to be of real use his findings would have to be published, Koch contacted Ferdinand Cohn, the esteemed director of the Botanical Institute at Breslau, Germany. Cohn quickly saw the significance of Koch's studies and arranged for their publication. This was the beginning of Koch's illustrious career. He went on to determine the causative organisms for several other diseases, including tuberculosis and cholera.

In his report on the **etiology** (cause) of tuberculosis, Koch reviewed his studies on anthrax and tuberculosis that permitted him to establish a cause-and-effect relationship between a given microorganism and a specific disease. Koch's studies were an extension of the ideas of Jacob Henle, a professor of anatomy and advocate of the **germ theory of disease**, who had been one of Koch's mentors at the University of Göttingen. Henle had proposed that contagion was due to living organized matter that could be transmitted through the air or by contact and that could multiply in the body. He reasoned that to establish the etiology of a specific disease, the agent would have to be found regularly in the host during the disease, the agent would have to be isolated,

FIGURE 1.10

Robert Koch (1843–1910), seen here viewing a specimen while a disciple looks on, pioneered studies in medical microbiology and developed many of the basic methods essential for the study of microbiology. Koch's postulates for establishing the etiology of infectious diseases and the methodological techniques he developed are still used today in scientific investigations. Many of Koch's students also made significant contributions to the development of the field of microbiology. (From the Bettmann Archive)

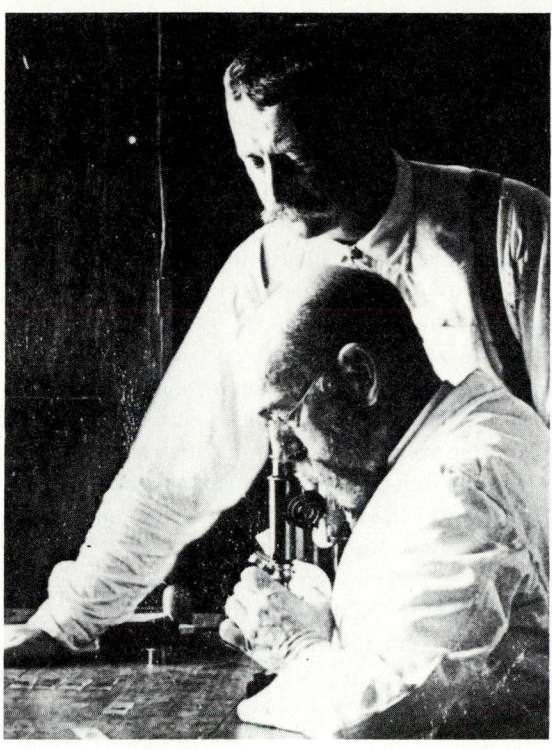

Discovery Process

"To obtain a complete proof of a causal relationship, rather than mere coexistence of a disease and a parasite, a complete sequence of proofs is necessary. This can only be accomplished by removing the parasites from the host, freeing them of all tissue elements to which a disease-inducing effect could be ascribed, and by introducing these isolated parasites into a healthy animal with the resulting reproduction of the disease with all its characteristic features. An example will clarify this type of approach. When one examines the blood of an animal that has died of anthrax one consistently observes countless colorless, non-motile, rod-like structures. . . . When minute amounts of blood containing such rods were injected into normal animals, these consistently died of anthrax, and their blood in turn contained rods. This demonstration did not prove that the injection of the rods transmitted the disease because all other elements of the blood were also injected. To prove that the bacilli, rather than other components of blood produce anthrax, the bacilli must be isolated from the blood and injected alone. This isolation can be achieved by serial cultivation. . . . The serial transfers can be continued for 3 or as many as 50 passages and in this manner the other blood components can be eliminated with certainty. Such pure bacilli produce fatal anthrax soon after injection into a healthy animal, and the course of the disease is the same as if produced with fresh anthrax blood or as in naturally occurring anthrax. These facts proved that anthrax bacilli are the unique cause of the disease." (Extracted from Robert Koch's memoir, 1884, *The Etiology of Tuberculosis*, in which he describes the basis for establishing a cause-and-effect relationship between a microorganism an a specific disease, discussing the etiology of both tuberculosis and anthrax)

and the isolated agent would have to be shown capable of producing the disease. Koch was able to fulfill this set of basic criteria experimentally, thus establishing their validity.

Historically Koch is credited with describing the steps that are necessary, known as **Koch's postulates**, for identifying the etiologic agent of a disease, which are as follows:

1. The organism should be present in all animals suffering from the disease and absent from all healthy animals.
2. The organism must be grown in pure culture outside the diseased animal host.
3. When such a culture is inoculated into a healthy susceptible host, the animal must develop the symptoms of the disease.
4. The organism must be reisolated from the experimentally infected animal and shown to be identical to the original isolate.

These four postulates, which are applicable to plant as well as animal diseases, still form the basic method for determining that a particular disease is caused by a given microorganism. For example, the search for the cause of Legionnaire's disease in 1976 followed Koch's 1890 postulates, resulting in the eventual identification of the bacterial etiologic agent. After many attempts, the bacterium *Legionella pneumophila*[2] was isolated from patients with this disease, grown in the laboratory, inoculated into test animals, causing

---

[2]Like other living organisms, microorganisms are named according to binomial system, so that the name of a bacterium, for example, *Legionella pneumophila*, includes the genus name *Legionella* and the species epithet *pneumophila*; both the genus and species names are italicized or underlined; the first letter of the genus name is capitalized, the first letter of the species name is not; if the name of the genus is clearly understood, it can be abbreviated as a single capitalized letter, for example, *L. pneumophila*.

the onset of disease symptomatology, and reisolated from the experimentally infected animals. Some modifications of Koch's postulates are required in some cases, such as the following:

1. The disease is caused by **opportunistic pathogens** (organisms that are normally associated with healthy animals and cause disease only under specific conditions).
2. The experimental host is **immune** (nonsusceptible due to host resistance) to the particular disease.

3. The disease process involves cooperation between multiple organisms.
4. The causative agent cannot be grown in **pure culture** (in the absence of any other organisms) outside of host cells, such as in the case of viruses.

In general, however, the philosophy of Koch's postulates for identifying the causes of infectious diseases remains intact.

1.2 Scope of Contemporary Microbiology

The discoveries of the fundamental nature of microorganisms and the applied aspects of microbiology, pioneered by the studies of Koch, Pasteur, Leeuwenhoek, and many others, continued into the twentieth century. With improved methods, such as the advent of the electron microscope, still smaller microbes, the viruses, were discovered. (See Figure 1.11.) The field of microbiology extended into areas such as basic molecular biology and biochemistry, largely because the nature of microorganisms makes them suitable for use in many basic scientific studies. Because microorganisms reproduce rapidly, large numbers of organisms with the same genetic composition can easily be grown, and because microorganisms are simpler than plants and animals, both genetically and biochemically, microbial systems are easier to deal with and understand than those of higher, more complex organisms. Accordingly, our fundamental understanding of molecular genetics and the relationship of genetics and metabolism has been developed using microorganisms. Besides, no one cares if bacteria are sacrificed for the advancement of science. Well, almost no one cares: some Japanese microbiologists, feeling a subtle sense of guilt for killing microorganisms as part of their studies, have established a memorial to microorganisms in which they buried a Buddhist prayer scroll and the ashen remains of *Bacillus subtilis.*

Besides its central position in basic biological studies, microbiology is an applied science at the center of biotechnology and genetic engineering. Many practical aspects of microbiology affect our daily lives and influence the overall quality of life. Applications of microbiology are important in medicine, industry, agriculture, and ecology. The health of humans and the existence of planet Earth depend on microorganisms, making microbiology a relevant and pragmatic science. **Microbial ecology**, the study of the environmental relationships of microorganisms, includes the examination of biogeochemical cycling reactions, which are important in maintaining air, water, and soil quality. The maintenance of soil fertility, the role of microorganisms in causing plant diseases, and the interactions of microorganisms with pesticides and fertilizers are important topics considered in **agri-** **cultural microbiology**, the study of the role of microorganisms in agriculture. The ability of microorganisms to degrade waste materials extends microbiology into the field of **sanitary engineering**, which is concerned with processes for waste removal.

Microbiology is an integral part of many industrial processes, and thus microorganisms have great economic importance. Quality control in many industries is concerned with preventing microbial contamination that could lead to the spoilage of products, reducing their economic value. The proper handling of food products is based on an understanding of microbiology. Various products, including many pharmaceuticals and food products such as beer, wine, and spirits, are produced by microbial fermentation.

Microbiology includes the study of pathogenic microorganisms that cause diseases in plants and animals, including human beings. Microbiology also includes the study of the prevention and treatment of diseases caused by microorganisms. **Medical microbiology** (the medical science relating to microorganisms) represents an important applied area in the study of microbiology. An understanding of the causative agents of disease has allowed us to develop preventive and

FIGURE 1.11 (*facing page*)
When microorganisms were discovered in the late seventeenth century, they were considered to be little plants and animals. Carl Linnaeus, who devised the first comprehensive classification system of living organisms in 1743, placed the microorganisms observed by Antonie van Leeuwenhoek into the genus *Chaos*, a name well chosen for this group, which at the time must have seemed to be a collection in disarray. Today we recognize that microorganisms, which include viruses, bacteria, fungi, algae, and protozoa are distinct from higher organisms, lacking the specialized organization in which cells (the structural and functional subunits of living systems) form differentiated tissues (aggregates of similar cells performing unified functions) that have specialized functions. True tissue differentiation is lacking in microorganisms, many of which are in fact unicellular (single-celled).

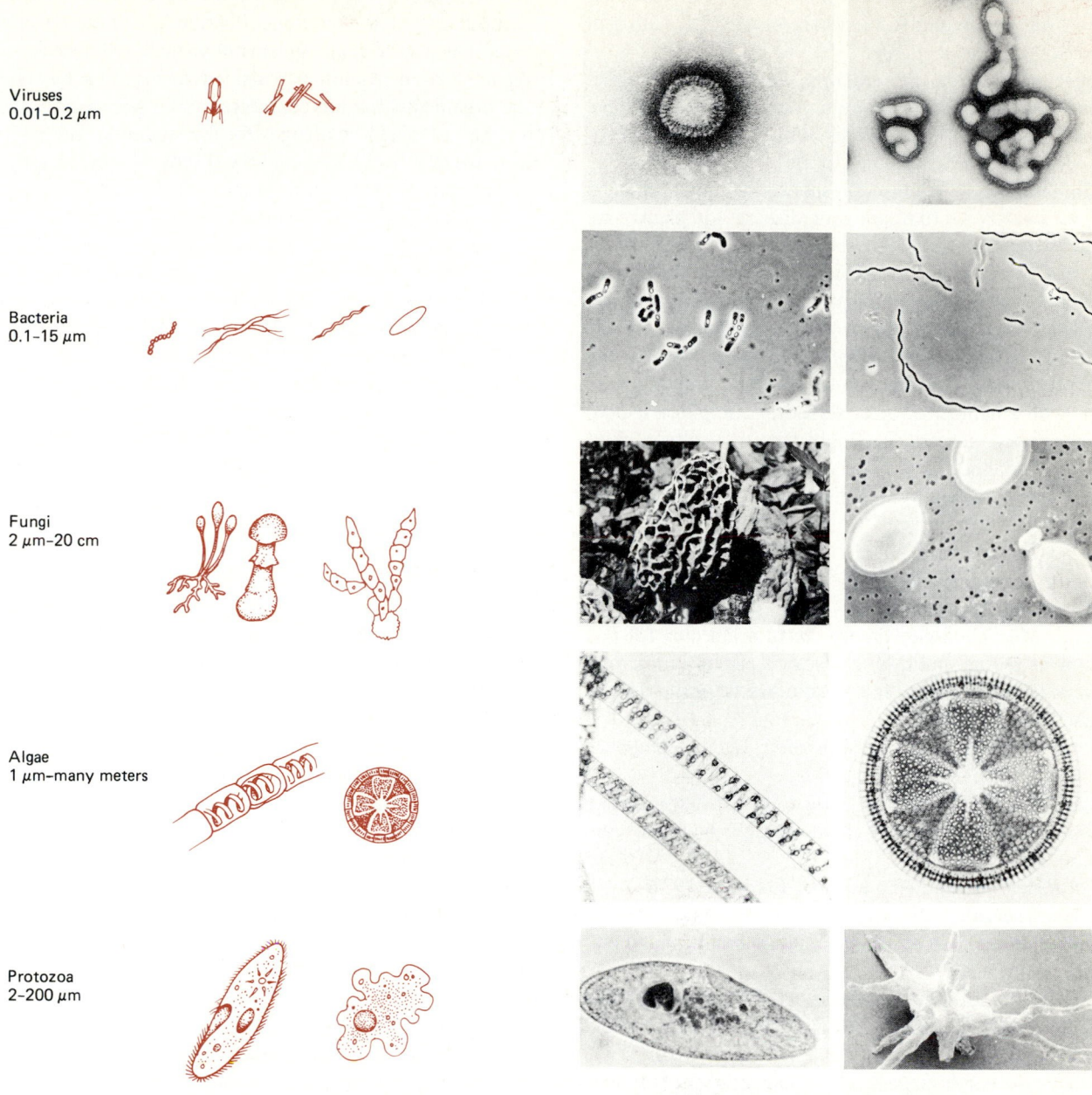

Viruses
0.01–0.2 μm

Bacteria
0.1–15 μm

Fungi
2 μm–20 cm

Algae
1 μm–many meters

Protozoa
2–200 μm

Microbiology, with its subdisciplines of virology (study of viruses), bacteriology (study of bacteria), mycology (study of fungi), phycology (study of algae), and protozoology (study of protozoa), is separated from botany (study of plants) and zoology (study of animals). In many universities and research institutes, departments of microbiology are independent of those of botany, zoology, and even biology. Microbiologists have their own scientific societies, such as the American Society for Microbiology, through which they communicate developments concerning microorganisms.

Despite their microscopic size, and despite the fact that their structural organization is simpler than that of higher plants and animals, microorganisms do exhibit great diversity of form and metabolic complexity that reflects their varying genetic compositions. Representative microorganisms are illustrated in this figure. The micrographs show VIRUSES—Respiratory syncytial virus (150,000×) (Courtesy Eugenie C. Ford, Georgetown Medical Center); Influenza A virus (200,000×) (Courtesy Erskine Palmer, Centers for Disease Control, Atlanta); BACTERIA—*Bacillus thuringiensis* (1,500×) (Courtesy B.N. Herbert, Shell Research, England); *Saprospira grandis* (700×) (Courtesy Hans Reichenbach, Institute for Biotechnical Research, Germany); FUNGI—*Morchella esculata* (Courtesy Orson Miller, Virginia Polytechnic Institute and State University); Baker's yeast *Saccharomyces cerevisiae* (Courtesy Robert Apkarian, Yerkes Primate Research Laboratory, Emory University); ALGAE—*Spirogyra elongans*; a marine diatom, *Actinoptychus* (Courtesy Varley Wiedeman, University of Louisville); PROTOZOA—*Paramecium* sp.; *Amoeba* sp. (Courtesy Eugene McArdle, Northeastern Illinois University).

treatment methods that have reduced **morbidity** (disease) and **mortality** (death) arising from certain diseases, resulting in an increase in life expectancy.

Because microorganisms are involved in causing infectious diseases, the field of microbiology logically has been extended to include the response of the diseased plant or animal. As such, the disciplines of **immunology** (the study of the immune—host defense—response of higher animals) and **plant pathology** (the study of diseases of plants) are included in the study of microbiology.

It is thus apparent that microbiology encompasses a very broad field. Some microbiologists are concerned with the basic sciences and the development of a fundamental understanding of living systems; others are concerned with the application of basic scientific knowledge. Microbiology overlaps with a number of other scientific disciplines including biochemistry, genetics, zoology, botany, ecology, pharmacology, medicine, food science, agricultural science, industrial science, and environmental science. The broad scope of microbiology attests to the diversity of microorganisms themselves, their ubiquitous distribution in nature, and the importance of microorganisms in virtually all aspects of life; the unity of microbiology rests with its central subject matter: the organisms that are considered to be microbes.

The vistas and challenges for the student of microbiology are virtually limitless, and new discoveries are made daily. Improvements in scientific communication have allowed microbiologists to capitalize rapidly on scientific advances, hastening the rate of development in the field of microbiology; today, there are numerous journals and publications through which worldwide distribution of microbiological information is made possible (see the list below). Additionally, the news media carry almost daily reports of microbiological interest ranging from outbreaks of disease to biotechnological patents. The field of microbiology promises to continue its rapid development for many years. It is an exciting and challenging field of science for students and professionals alike.

Some Journals Concerned Primarily with Microbiological Studies

Annales de Microbiologie. France. Institut Pasteur. 1887–.
Prominent journal publishing research papers in microbiology and immunology. Initial volumes from the time of Pasteur and Koch.

Antimicrobial Agents and Chemotherapy. U.S.A. American Society for Microbiology. 1972–.
Research papers covering all aspects of knowledge relating to antimicrobial agents and chemotherapy, including cancer chemotherapy.

Antiviral Research. The Netherlands. Elsevier. 1982–
Articles concerning effective control of viral infections of humans and other organisms, including treatment and prevention (vaccines).

Antonie van Leeuwenhoek Journal of Microbiology. The Netherlands. Stichting Antonie van Leeuwenhoek. 1935–.
Publishes research papers of investigations of limited scope devoted to microbiology and related fields of science and technology.

Applied and Environmental Microbiology. U.S.A. American Society for Microbiology. 1953–. (Formerly *Applied Microbiology*)
Dedicated to applied studies in microbiology including studies on clinical microbiology, virology, immunology, food microbiology, toxicology, environmental microbiology, ecology, taxonomy, metabolism, and fermentation products.

Archives of Microbiology. U.S.A. Springer-Verlag. 1930–.
Subtitled *Journal for the Investigation of Microorganisms*. Original papers and review articles covering the entire field of microbiology.

Biotechnology and Bioengineering. U.S.A. John Wiley & Sons, Inc. 1958–.
Publishes reports in both basic and applied aspects of biochemical and microbial technology.

British Phycological Journal. United Kingdom. British Phycological Society. 1953–.
Publishes papers on all aspects of the study of algae including original research reports, review articles on topics of phycological interest, field studies, and techniques relevant to algae.

Canadian Journal of Microbiology. Canada. National Research Council of Canada. 1954–.
Comprehensive, general microbiology journal, part of the Canadian Journal of Research series; includes research papers on microbial ecology, infection and immunity, physiology, virology, molecular biology, and genetics.

Comparative Immunology, Microbiology and Infectious Disease. U.S.A. Pergamon Press. 1978–.
Research papers related to medical microbiology, including works related to immunological responses to infection, pathogenic microorganisms, and infectious disease processes.

CRC Critical Reviews in Microbiology. U.S.A. CRC Press. 1971–.
Prominent scientific authorities present in-depth critical reviews, generally four per issue, centered on a single topic.

Current Microbiology. U.S.A. Springer-Verlag International. 1978–.
Publishes brief papers that deal concisely yet thoroughly with significant facts and ideas in all areas of microbiology. This journal is predicated on the idea that important benefits can be gained by substantially reducing the length of time required for scientific publication.

Diagnostic Microbiology and Infectious Disease. U.S.A. Elsevier Biomedical. 1983–.

Publishes articles, critical reviews, and commentaries oriented to the diagnostic microbiology laboratory and its role in the identification and treatment of infectious disease in the areas of bacteriology, immunology, immunoserology, infectious disease, mycology, parasitology, and virology.

FEMS—Microbiology Letters. The Netherlands. Elsevier Biomedical. 1977–.

Publishes short reports on new microbiological research.

Geomicrobiology Journal. U.S.A. Crane Russak & Co. 1978–.

Publishes reports on geomicrobiology and microbial biogeochemistry.

Infection and Immunity. U.S.A. American Society for Microbiology. 1970–.

Devoted to the advancement and dissemination of fundamental knowledge concerning pathogenic microorganisms and infections, and with ecology, epidemiology, and host factors, as well as immunology.

International Journal of Systematic Bacteriology. U.S.A. International Association of Microbiological Societies. 1951–.

Devoted to the advancement of the systematics of bacteria, yeasts, and yeast-like organisms.

Journal of Applied Bacteriology. United Kingdom. Society for Applied Bacteriology. 1954–. (Formerly *Proceedings of the Society for Applied Bacteriology*)

Review articles, original papers and short communications on microbiology and its application to agriculture and other industries.

Journal of Bacteriology. U.S.A. American Society for Microbiology. 1916–.

The leading American journal in the field, concerned primarily with general, fundamental reports on bacteria, their morphology and ultrastructure, genetics and molecular biology, physiology, and metabolism and enzymology. Papers also include studies on fungi and bacterial viruses.

Journal of Biotechnology. The Netherlands. Elsevier Biomedical. 1984–.

Publishes reports in the areas of industrial microbiology and biotechnology.

Journal of Clinical Microbiology. U.S.A. American Society for Microbiology. 1975–.

Disseminates new knowledge concerning the applied microbiological aspects of human and animal infections and infestations, particularly regarding their etiologic agents, diagnosis, and epidemiology.

Journal of General and Applied Microbiology. Japan. Microbiology Research Foundation. (English vol. 19) 1973–.

Research papers relating to applied aspects of microbiology and the underlying fundamentals.

Journal of General Microbiology. United Kingdom. Society for General Microbiology. 1947–.

Fundamental studies on bacteria, microfungi, microscopic algae, etc., discusses development and structure of microorganisms, their physiology and growth, biochemistry, genetics, and ecology.

Journal of General Virology. United Kingdom. Society for General Microbiology. 1967–.

Original works in general virology including studies of bacterial, plant, and animal viruses, their structure, genetics, systematics, and interactions with host cells, as well as investigations of pathogenesis.

Journal of Immunology. U.S.A. American Association of Immunologists. 1916–.

Covers all areas of immunology, including cellular, tumor, viral, and microbial immunology, immunochemistry, immunogenetics, and transplantation.

Journal of Industrial Microbiology. U.S.A. Society for Industrial Microbiology. 1986–.

Publishes original works in the areas of industrial microbiology and biotechnology.

Journal of Infectious Diseases. U.S.A. University of Chicago Press. 1904–.

Presents laboratory and clinical findings concerning infectious diseases, medical microbiology, and immunology, with emphasis on research.

Journal of Medical Virology. U.S.A. Alan R. Liss, Inc. 1977–.

Original reports of major research developments in the epidemiology, structure, and composition of viruses, diagnosis, pathology, and treatment.

Journal of Microbiological Methods. U.S.A. Elsevier. 1983–.

Original papers and reviews on new and revised methods in all areas of microbiology.

Journal of Phycology. Canada. Phycological Society of America. 1965–.

Covers all aspects of marine and freshwater algae including taxonomy, biochemistry, physiology, morphology, cytology, molecular biology, and ecology.

Journal of Protozoology. U.S.A. Society of Protozoologists. 1954–.

Original papers on both descriptive and experimental studies of protozoa. This journal provides a common medium for the taxonomist, physiologist, parasitologist, and biochemist working with protozoa.

Journal of Virology. U.S.A. American Society for Microbiology. 1967–.

Contains reports of original research in all areas of basic virology about the viruses of bacteria, plants, and animals.

Medical Microbiology and Immunology. Federal Republic of Germany. 1886–

Founded by Robert Koch, publishes original papers in virology, bacteriology, immunology, epidemiology, disease prevention, treatment of infectious diseases, and pathogenic mechanisms as underlying causes of the origin of infectious diseases as the core of its contents.

Microbial Ecology. U.S.A. Springer-Verlag. 1974–.

An international journal presenting research papers concerning those branches of ecology in which microorganisms are involved.

Microbiological Reviews. U.S.A., American Society for Microbiology. 1937–. (Formerly *Bacteriological Reviews*)
    Authoritative and critical review papers on the current state of research in all areas of microbiology, often including historical analyses.
Microbiology and Immunology. Japan. Japanese Society for Bacteriology, Society of Japanese Virologists, and Japanese Society for Immunology. 1957–. (Formerly *Japanese Journal of Microbiology*)
    Publishes original reports of research concerning significant findings in bacteriology, immunology, virology, and related fields.
Microbios. United Kingdom. Faculty Press. 1969–.
    Transworld biomedical research journal devoted to fundamental studies of viruses, fungi, microscopic algae, and protozoa, with emphasis on chemical microbiology.
Mikrobiologiya. Soviet Union. Akademiya Nauk S.S.R. (English translation) 1973–.
    Research papers on microbiological studies conducted primarily in the Soviet Union. Original papers published in

Russian, with volumes published in English after several years' delay.
Mycologia. U.S.A. Mycological Society of America. 1909–.
    Original research papers in all areas of mycology including works on the taxonomy, morphology, physiology, and ecology of fungi.
Phycologia. United Kingdom. International Phycological Society. 1961–.
    Review articles, reports, and notes concerning material of phycological interest.
Virology. U.S.A. Academic Press. 1955–.
    Articles on the biological, biochemical, and biophysical aspects of viral research, stressing contributions of a fundamental rather than applied nature.
Virus Research. The Netherlands. Elsevier. 1984–.
    Research papers in virology.
Zeitschrift für Allgemeine Mikrobiologie. German Democratic Republic. Akademie-Verlag. 1960–.
    Research papers on the morphology, physiology, genetics and ecology of microorganisms.

1.3 Evolution of Microorganisms

Having reviewed the evolution of the field of microbiology from its primitive state, through the golden age of discovery at the time of Pasteur and Koch, to its current multifaceted, highly technological state, let us consider how microorganisms themselves evolved. The lack of distinct shapes and structures that would permit the unequivocal recognition of most microbial species and the poor geological preservation of subcellular features make the fossil record of microbial evolution pitifully incomplete. Microorganisms were undoubtedly the first living inhabitants of Earth, but microbial evolution appears to have occurred very slowly during the 3 billion years that life has existed on Earth compared to the evolutionary time scale of multicellular organisms. The oldest macrofossils of plants and animals are only 0.6–0.7 billion years old, but we now have credible evidence that microbial life existed more than 3.5 billion years ago, that is, just 1 billion years after the formation of our planet and almost 3 billion years before the appearance of plants and animals. One is tempted to speculate that this long and seemingly uneventful period conceals a gradual evolution of the genetic basis for biochemical pathways and regulatory mechanisms that are poorly documented by the geological record, but that laid the groundwork for the subsequent explosive morphological diversification of multicellular life forms.

Before life developed on Earth, it appears that the atmosphere consisted largely of nitrogen, hydrogen, carbon dioxide, and water vapor, with smaller amounts of ammonia, carbon monoxide, and hydrogen sulfide; oxygen was absent or present only in trace amounts, and the atmosphere was thus **anaerobic** (oxygen free). Various types of organic chemicals were formed by **abiotic** (nonliving) reactions, and

this organic matter slowly evolved into larger and more complex organic molecules. The resulting **macromolecules** (very large organic molecules) had an inherent tendency to aggregate and form membrane-like interfaces with the surrounding liquid, foreshadowing a cellular organization. This scenario for the development of the organic chemicals that preceded life on Earth, which originally was proposed by A. I. Oparin and J. B. S. Haldane in the 1920s, based largely on theoretical considerations, received experimental support in the 1950s from the work of Stanley Miller and Harold C. Urey, who, using a relatively simple apparatus containing water and reducing gas mixtures and receiving energy input as heat, electric discharges, or ultraviolet (uv) radiation to simulate the conditions of the prebiotic atmosphere and sources of available energy, were able to produce a surprising array of organic molecules, including most of the amino acids and nucleic acid bases essential to life. Recent evidence indicates that complex carbon-, hydrogen-, oxygen-, and nitrogen-containing compounds exist in stellar gases, suggesting that organic compounds were available in the Earth's atmosphere when life evolved here.

Stanley Fox showed that moderate heating of amino acids could produce stable *protenoids* that possessed a self-ordering property, aggregating into microspheres (small, membrane-confined spheres). Oparin and his co-workers studied the properties of microspheres, finding that they develop semipermeable membranes and vacuoles, become capable of selectively acquiring substances from their environment, and mimic the growth and division of cells. Such microspheres form spontaneously when mixtures of certain organic substances are dried on a hot surface and then exposed to water.

ANALYTICAL PROCESS
RNA Fingerprinting

The method devised by Woese and co-workers provides a "fingerprint" of the rRNA of one organism that can be compared with the rRNA from another organism. In this method, cells are grown in the presence of phosphate containing the radioactive isotope ^{32}P so that the radiolabeled phosphate is incorporated into the nucleic acids, including 16S rRNA. The radioactive 16S rRNA is enzymatically cut into small fragments or *oligonucleotides* (Figure 1.12). Each nucleotide of RNA is composed of a sugar called *ribose*, a phosphate group, and one of four nitrogenous bases: adenine (A), uracil (U), guanine (G), or cytosine (C). The nucleotides are like letters in the alphabet and the oligonucleotides, which contain several nucleotides, are like words. The enzyme used to cut the RNA, T_1 ribonuclease, cuts at a specific site so that each oligonucleotide ends with a single G, as in UUUCCCAAAG.

Electrophoresis, a procedure in which an electric field is used to separate molecules based on size and electronic charge, is employed to separate the mixture of oligonucleotides. The sequence of nucleotide bases within each oligonucleotide is determined and the information catalogued. These catalogued sequences are analyzed by pairwise comparisons using a computer program to determine their similarities. Very small oligonucleotide sequences are ignored because they recur too frequently to be of value in comparing similarity. Oligonucleotides with six or more nucleotides are used because they are likely to appear only once in a particular 16S rRNA molecule. The typical 16S rRNA molecule has 25 useful oligonucleotides with at least six nucleotides; there are, incidentally, 243 possible six-nucleotide oligonucleotides that end in G. By restricting the analyses to "words" of six or more "letters," one can construct a "dictionary" characteristic of a given organism that can be compared with other such dictionaries to determine genetic and phylogenetic relatedness. These analyses have revealed signature sequences characteristic of closely related major groups of organisms. These signature sequences are important in allowing the determination of relationships when the organisms being compared are distantly related.

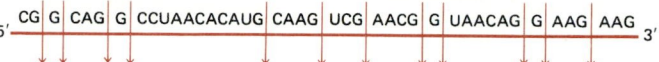

5′ CG | G | CAG | G | CCUAACACAUG | CAAG | UCG | AACG | G | UAACAG | G | AAG | AAG 3′

FIGURE 1.12
T_1 ribonuclease cuts 16S rRNA at specific sites where single guanosine (G) bases occur, producing a series of oligonucleotides, each ending with a single G. In this way, a typical 16S rRNA molecule is cut into oligonucleotides ("words") composed of 1 to 20 nucleotides ("letters"). The sites at which a typical bacterial 16S rRNA molecule is cut are illustrated in this figure. Words of six letters or more are compiled into a dictionary. The dictionaries of two organisms, A and B, are compared using a mathematical similarity coefficient to determine the relatedness of the two organisms. The similarity coefficient is equal to twice the number of letters in these words common to both A and B divided by the total number of letters in all such words in A and B.

The incorporation of enzymes, electron carriers, or chlorophyll into these microspheres permits metabolic reactions normally associated only with living cells. Although the model microspheres used in these experiments demonstrate the surprising self-organizing capacity of large organic molecules, they are not proposed as actual intermediates in the chemical evolution process. It is conceivable, though, that similar self-replicating, protein-containing microspheres represented the first step toward cellular organization. These postulated primitive cell-like structures are referred to as **progenotes** or **protobionts**. A further advance toward cellular organization was probably the acquisition and use of nucleic acids, first probably ribonucleic acid (RNA) and later both deoxyribonucleic acid (DNA) and RNA, as templates for protein synthesis. Along with further development of enzymatic capabilities and membrane organization, this led to the creation of **eugenotes** or primitive versions of the **prokaryotic cell** (bacterial cell).

The ability to reconstruct the process of evolutionary separation of species has been greatly bolstered by recent developments in molecular biology, which have provided the necessary tools for assessing the phylogenetic relationships among organisms at the molecular level. Because closely related organisms should have similar genetic compositions, analyzing the similarities of the nucleic acids (genetic informational macromolecules) is the most reliable and accurate method for determining the true relatedness of organisms. This approach was used by Carl Woese, who performed molecular-level genetic analyses of 16S ribosomal RNA (rRNA)[3] to determine the relatedness of organisms. Woese

[3]The designation 16S is a measure of nucleic acid size, where S stands for Svedberg units, a measure of the speed of movement of a molecule due to the gravitational field of a very high speed centrifuge; the rate of movement is influenced by the size, shape, and mass of the molecule. The 16S rRNA is one of the larger RNA (ribonucleic acid) molecules that make up ribosomes, the subcellular structures at which proteins are synthesized.

reasoned that rRNA is a good genetic marker because it is not transferred from one species to another, its function is relatively constant, and it is an appropriately sized macromolecule to permit an assessment of change over the evolutionary period being considered. The rRNA is like a clock that changes over the time of molecular evolution, acting as a good indicator of when one organism diverges as a distinct species from another.

Classification Based on Evolutionary Relatedness (Phylogeny)

Based on his rRNA analyses, Woese proposed a revolutionary new classification system for living organisms in 1980 with three primary kingdoms: **Archaebacteria**, **Eubacteria**, and **Eukaryotes** (Figure 1.13). This proposal represents the climax of the development of classification systems aimed at reflecting **phylogeny** (evolutionary relatedness).[4] Both eu-

[4]The system it replaces, and the one many general biologists still recognize, is the five-kingdom system proposed by Whittaker in 1969 based on phenotypic (observable) characteristics rather than genetic characteristics that reflect true relatedness. The five kingdoms in the Whittaker system, Monera (prokaryotic bacteria), Protista (eukaryotic and unicellular protozoa and most algae), Fungi, Plantae, and Animalia, were thought to have evolved along different lines of nutritional strategies—ingestion, adsorption, and photosynthesis.

bacteria and archaebacteria have **prokaryotic cells**, which are simply organized cells that lack a nucleus, and are structurally quite different from **eukaryotes**, which are organisms composed of cells that have a nucleus and various other organelles. Woese found that although the archaebacteria and eubacteria both have prokaryotic cells, they are genetically as distantly related to each other as each is to eukaryotes. Woese proposed that the **archaebacteria**, **eubacteria**, and an **urkaryote**, which was the original eukaryotic cell (German *ur* = prototype), all evolved from a common simple ancestor, the **progenote**.

According to Woese's theory, eukaryotes evolved after the urkaryote became a "host" for bacterial endosymbionts, that is, for bacteria living within the host cell in a mutually beneficial relationship. The theory of **endosymbiotic evolution**, which Woese's analyses support, is that bacteria living as endosymbionts within eukaryotic cells gradually evolved into the organelle structures called *mitochondria* and *chloroplasts*. The analyses of rRNA indicate that the chloroplasts and mitochondria of corn plants (*Zea mays*), for example, developed as separate lines of eubacterial evolution that were

FIGURE 1.13
The three-kingdom classification system proposed by Carl Woese was developed using the modern techniques of molecular biology and, in particular, the examination of the RNA macromolecules of ribosomes. Unlike previous classification systems, the analysis of conserved gene products permits a direct assessment of genetic and, thus, evolutionary relatedness. Based on rRNA analyses, Woese found that the archaebacteria are fundamentally different from all other organisms.

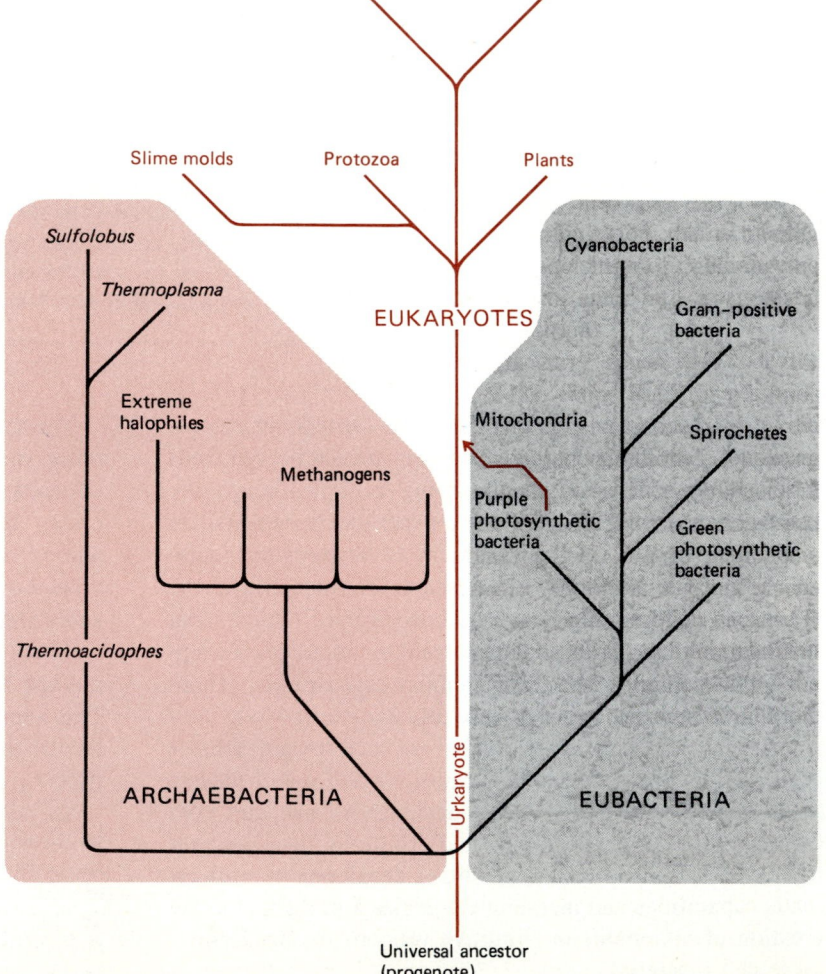

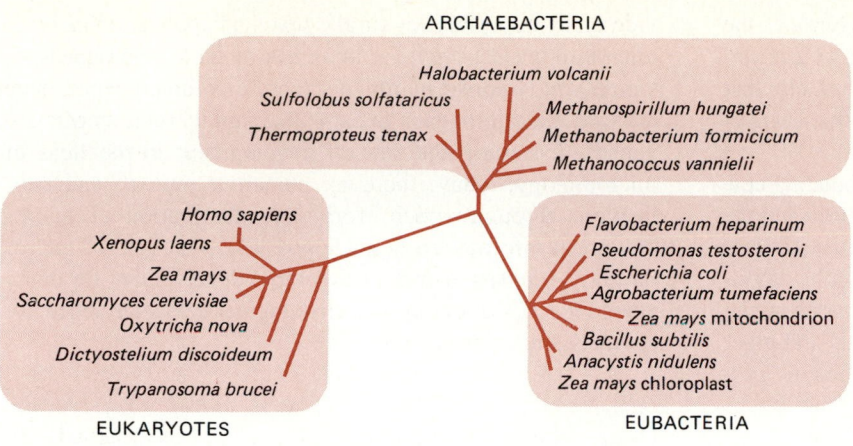

ARCHAEBACTERIA

Halobacterium volcanii
Sulfolobus solfataricus
Thermoproteus tenax
Methanospirillum hungatei
Methanobacterium formicicum
Methanococcus vannielii

Homo sapiens
Xenopus laens
Zea mays
Saccharomyces cerevisiae
Oxytricha nova
Dictyostelium discoideum
Trypanosoma brucei

Flavobacterium heparinum
Pseudomonas testosteroni
Escherichia coli
Agrobacterium tumefaciens
Zea mays mitochondrion
Bacillus subtilis
Anacystis nidulens
Zea mays chloroplast

EUKARYOTES   EUBACTERIA

FIGURE 1.14

Phylogenetic tree based on rRNA sequence analyses. These analyses establish that chloroplasts and mitochondria evolved along the eubacterial branch and therefore must have later been incorporated into eukaryotic cells, most likely by endosymbiotic evolution. (Courtesy N. Pace, Indiana University. Reprinted by permission of Cell Press from *Cell*:45#3)

very distant from the evolution of the eukaryotic corn plant (Figure 1.14). In fact, these analyses suggest that corn plants and humans are more closely related in terms of evolution than mitochondria and chloroplasts are to the plants in which they are found. The endosymbiotic theory of evolution of eukaryotic cells, strongly advocated by Lynn Margulis, is further supported by structural similarities between mitochondria, chloroplasts, and prokaryotic cells, which we will discuss further in Chapter 3, and the contemporary occurrence of many endosymbiotic bacteria in protozoa and other eukaryotes, which we will consider in Chapter 10.

Both the endosymbiotic theory and Woese's new classification system are receiving a great deal of attention from microbiologists and have begun to gain wider support from biologists in general who have been able to apply rRNA analyses in order to determine the phylogeny of higher organisms. As we have developed an understanding of the molecular-level events that led to genetic divergence and the evolution of new species, the systems used to classify organisms have progressively changed, from ones of supposition based on appearances to ones founded upon verifiable measurements of genetic relatedness.

Postlude

In just over 100 years, microbiology has developed as a major scientific discipline. Microbiology grew out of several other scientific disciplines, which undoubtedly contributes to its broad multidisciplinary nature. Without question the two greatest historical figures in microbiology were Louis Pasteur and Robert Koch. They set the stage for the future development of this field of science, emphasizing the relationships between germs and disease (including host resistance), and between microbes and fermentation (including metabolism and biochemistry). Louis Pasteur made major contributions to our understanding of fermentation; beginning as a chemist, he emerged as an outstanding microbiologist. He successfully refuted the theory of spontaneous generation. Robert Koch extended his training as a physician to tackle the problems of treating and preventing infectious diseases. He and his disciples discovered the etiologic agents of several diseases and developed methods for their cure and prevention. The pure culture techniques developed by Koch have dominated microbiological research up to the present day.

Pasteur's unique contribution to the science of microbiology was his conceptual ability, his ability to analyze and to bring together the pieces of the puzzle that form the grand scheme of microbiology. Koch's genius lay in his ability to devise the methods and techniques that allowed the science

to progress. Both men were excited by the process of scientific discovery and by solving the problems they had set for themselves. Their inquiring minds, sense of adventure, egos, and competitiveness caused them to strive ahead at a furious pace. They set a pattern for many great microbiologists who followed. The race to be first in making microbiological discoveries of basic and practical importance continues today.

Both Koch and Pasteur trained students to carry on their work and to publish extensively in order to make available the knowledge they had acquired. The science of microbiology is still young; advances in this discipline parallel progress in other fields of science. The development of appropriate methodologies for advancements in one field often depends on discoveries in other areas of scientific endeavor, and this is apparent in the historical development of microbiology. The science of microbiology has evolved from many fields—from the contributions of numerous amateur scientific observers who, like Leeuwenhoek, took pleasure in searching for the very small and discovering the wonders of the microbial world, to professional scientists who, like Pasteur and Koch, brought their training in other fields to the development of the new discipline of microbiology. Scientists continue to enter the field of microbiology from other disciplines. Francis Crick, for example, the codiscoverer of the nature of the genetic molecule DNA, was trained as a physicist, and it was his background in X-ray crystallography that provided the necessary expertise for unraveling the arrangement of DNA as a double helix.

Currently, studies in molecular genetics are producing new knowledge, such as the recognition of archaebacteria and the development of an understanding of evolutionary relatedness based upon analyses of informational macromolecules. Studies of rRNA by Carl Woese and his colleagues are enhancing our fundamental understanding of microorganisms and other living systems. Industrial and environmental problems pro-

vide many opportunities for the practical application of basic microbiological principles in an era of biotechnological advancement. Improvements in scientific communication have allowed microbiologists to capitalize rapidly on scientific advances, hastening the rate of development in the field of microbiology; today, there are numerous journals and publications through which worldwide distribution of microbiological information is made possible. The field of microbiology promises to continue its rapid development for many years. It is an exciting and challenging field of science for students and professionals alike.

Suggested Supplementary Readings

AINSWORTH, G. C. 1976. *Introduction to the History of Mycology.* Cambridge University Press, London.

ALLEN, G. E. 1978. *Life and Science in the Twentieth Century.* Cambridge University Press, London.

BROCK, T. D. (ed.). 1975. *Milestones in Microbiology.* American Society for Microbiology, Washington, D.C.

BULLOCH, W. 1938. *The History of Bacteriology.* Oxford University Press, London. (1979. Dover Publications, Inc., New York.)

CARLILE, M. J., J. F. COLLINS, and B. E. B. MOSLEY (eds.). *Molecular and Cellular Aspects of Microbial Evolution.* Cambridge University Press, Cambridge, England.

CHRISTENSEN, C. M. 1965. *The Molds and Man.* University of Minnesota Press, Minneapolis.

CLARK, P. F. 1961. *Pioneer Microbiologists of America.* University of Wisconsin Press, Madison.

COLE, F. J. 1926. *The History of Protozoology.* University of London Press, Ltd., London.

DE KRUIF, P. 1926. *Microbe Hunters.* Harcourt, Brace and Co., New York. (1966. Harcourt Brace Jovanovich, Inc., New York.)

DICKERSON, R. 1978. Chemical evolution and the origin of life. *Scientific American* 239(3):70–86.

DOBELL, C. (ed.). 1932. *Antony van Leeuwenhoek and his "Little Animals."* Constable and Co., Ltd., London. (1960. Dover Publications, Inc., New York.)

DOYLE, R. J., and N. C. LEE. 1986. Microbes, warfare, religion, and human institutions. *Canadian Journal of Microbiology* 32:193–200.

EGERTON, F. N. 1978. *The Select Works of Antony van Leeuwenhoek: His Microscopical Discoveries in Many Works of Nature.* Ayer Publishing, Salem, N.H.

ERON, C. 1981. *The Virus That Ate Cannibals: Six Great Medical Detective Stories.* Macmillan Publishing Co., Inc., New York.

FOX, S. W. 1965. *The Origins of Prebiological Systems and Their Molecular Matrices.* Academic Press, New York.

FOX, S. W., and K. DOSE. 1977. *Molecular Evolution and the Origin of Life.* Marcel Dekker, New York.

FOX, G. E., E. STACKEBRANDT, R. B. HESPELL, J. GIBSON, J. MANILOFF, T. A. DYER, R. S. WOLFE, W. E. BALCH, R. S. TANNER, L. MAGRUM, L. ZABLEN, R. BLAKEMORE, R. GUPTA, L. BONEN, B. J. LEWIS, D. A. STAHL, K. R.

LUEHRSEN, K. N. CHEN, and C. R. WOESE. 1980. The phylogeny of prokaryotes. *Science* 209:457–463.

GRAINGER, T. H. 1958. *A Guide to the History of Bacteriology.* Ronald Press Co., New York.

GRAY, M., and F. DOOLITTLE. 1982. Has the endosymbiont hypothesis been proven? *Microbiological Reviews* 46:1–42.

HALDANE, J. B. S. 1932. *The Causes of Evolution.* Harper & Row, New York.

HOOKE, R. 1665. *Micrographia.* Royal Society, London. (1961. Dover Publications, Inc., New York.)

JUDSON, H. F. 1979. *The Eighth Day of Creation: Makers of the Revolution in Biology.* Simon & Schuster, New York.

LECHEVALIER, H. A., and M. SOLOTOROVSKY. 1965. *Three Centuries of Microbiology.* McGraw-Hill Book Co., New York. (1974. Dover Publications, Inc., New York.)

LOVELOCK, J. E. 1979. *Gaia: A New Look at Life on Earth.* Oxford University Press, New York.

McNEIL, W. 1976. *Plagues and People.* Anchor Press, Garden City, N.Y.

MARGULIS, L. 1970. *Origin of Eukaryotic Cells.* Yale University Press, New Haven, Conn.

MARGULIS, L. 1971. The origin of plant and animal cells. *American Scientist* 59:230–235.

MARGULIS, L. 1981. *Symbiosis in Cell Evolution.* W. H. Freeman Co., San Francisco.

MARGULIS, L., and D. SAGAN. 1986. *Microcosmos.* Summit Books, New York.

MILLER, S. L., and L. E. ORGEL. 1974. *The Origins of Life on Earth.* Prentice-Hall, Inc., Englewood Cliffs, N.J.

OPARIN, A. I. 1938. *The Origin of Life.* Dover Publications, Inc., New York.

OPARIN, A. I. 1968. *Genesis and Evolutionary Development of Life.* Academic Press, New York.

PACE, N. R., G. J. OLSEN, and C. R. WOESE. 1986. Ribosomal RNA phylogeny and the primary lines of evolutionary descent. *Cell* 45:325–326.

REID, R. 1975. *Microbes and Men.* Saturday Review Press, New York.

SCHOPF, J. W. 1978. The evolution of the earliest cells. *Scientific American* 239(3):110–138.

TUNEVALL, G. (ed.). 1969. *Periodicals Relevant to Microbiology and Immunology.* John Wiley & Sons, Inc., New York.

VAN ITERSON, G., JR., L. E. DEN DOOREN DE JONG, and A. J.

KLUYVER. 1983. *Martinus Willem Beijerinck: His Life and His Work*. Science Tech, Inc.,Madison, Wis.

VIDAL, G. 1984. The oldest eukaryotic cells. *Scientific American* 250(2):48–57.

WAKSMAN, S. A. 1953. *Sergei N. Winogradsky—His Life and Work*. Rutgers University Press, New Brunswick, N.J.

WAKSMAN, S. A. 1954. *My Life with the Microbes*. Simon & Schuster, New York.

WATERSON, A. P., and L. WILKINSON. 1978. *An Introduction to the History of Virology*. Cambridge University Press, New York.

WATSON, J. D. 1968. *The Double Helix*. Atheneum Publishers, New York.

WHITTAKER, R. H. 1969. New concepts in kingdoms of organisms. *Science* 163:150–160.

WOESE, C. R. 1981. Archaebacteria. *Scientific American* 244(6):98–122.

WOESE, C. R. 1985. Why study evolutionary relationships among bacteria? *Evolution of Prokaryotes*. FEMS Symposium 29. (K. H. Scheifer and E. Stackebrandt, eds.), pp. 1–30. Academic Press, London.

Study Questions

1. What is a microorganism? What organisms do we consider as microbes?

2. What is the current taxonomic position of bacteria?

3. Why were Leeuwenhoek's observations the critical first step in the development of microbiology?

4. What major contributions to microbiology were made by Pasteur and Koch?

5. What was the theory of spontaneous generation? How was it disproved and what is the significance of having disproved this theory?

6. What are Koch's postulates? What is their significance to medical microbiology? Discuss a recent use of Koch's postulates to determine the cause of a disease outbreak.

Situational Problem

Searching the Literature

Progress in science occurs through a stepwise progression, building upon knowledge previously attained. Students of microbiology and professional microbiologists spend many hours of library research, reading journals such as those listed earlier in this chapter. Through these readings, they learn of methodological advances that can be applied to their own research and learn how their own activities mesh with those of others. The specialized journals in one's own field usually are read on a regular basis. Others are examined only during a search of the literature for specific information. The journal *Current Contents* often is used to review titles of recent articles that may be of interest.

All scientists must be able to retrieve information from the published literature. Such information gathering often is tedious, and only data in the accessible or open literature can be obtained. Proprietary secret data are increasingly common in this age of biotechnology, and access is restricted to a very few scientists. Data may be retrieved manually or with computer assistance.

Biological Abstracts is the main vehicle of literature searching for microbiologists and most others involved in the biological sciences. Abstracts are published every two weeks and are cumulated twice a year. When manual searches are carried out, each biweekly or semiannual volume must be consulted separately. Abstracts come from almost 10,000 serials. They are numbered and arranged by major subject headings, such as ecology, immunology, and medical and clinical microbiology. The abstract entry includes a full bibliographic citation of the article and a paragraph summary of the work described. Articles can be searched by author as well as subject. They are also indexed biosystematically by taxonomic category, including the genus and species names. The subject index employs a permutated keyword approach, with keywords arranged alphabetically and with additional keywords that describe the articles listed to the left and right. The reference number, which leads from the index volume to the abstract of the article, is listed at the right.

Because searching the literature is such a fundamental aspect of the work of a microbiologist, you should respond to the following situational problems by going to the library and retrieving the requested information. If you encounter problems, feel free to consult the reference librarian.

Let's start with the author index of *Biological Abstracts* by looking up the name of the author of this text (Atlas, R. M.) for the year 1984. If you do this correctly, you should find three entries in volume 77 for the first half of the year and one entry in volume 78 for the second half of the year. Look up the abstracts to learn about some of the research activities of Ronald Atlas and his collaborators.

Now try something more interesting. Search the author index for the name of the instructor of your course or any of the other instructors in your department, starting with the current year and working backward for one decade.[5] For each entry that you find, record the number and then go to the abstract indicated. Read the abstract to determine your instructor's area(s) of research. This will give you an idea of his or her area(s) of special expertise. See if you can find some of the articles listed so that you learn where the journals of microbiological interest in your institution's library are located.

To determine that you can also search the literature on a specified topic, find out what new studies are using RNA analyses to examine lines of microbial and higher-organism evolution. To do so, you will have to pick some keywords, such as *evolution* and *ribosomal RNA*. You should look at the other words in the index entry to try to determine whether it really applies to this topic, thereby limiting the number of abstract entries you actually consult. You may also wish to search this topic under the author entry for C. R. Woese in order to make sure that you are finding all of the relevant entries. You can extend this exercise to any other topics you wish.

If your library has computer search facilities, you may also be able to ask for a demonstration that will enable you to verify the thoroughness of your search.

---

[5]A word of caution about this exercise: Do not use this search to judge your instructor. Some of the best instructors are not currently active researchers, and some of the best researchers have difficulty communicating information to students.

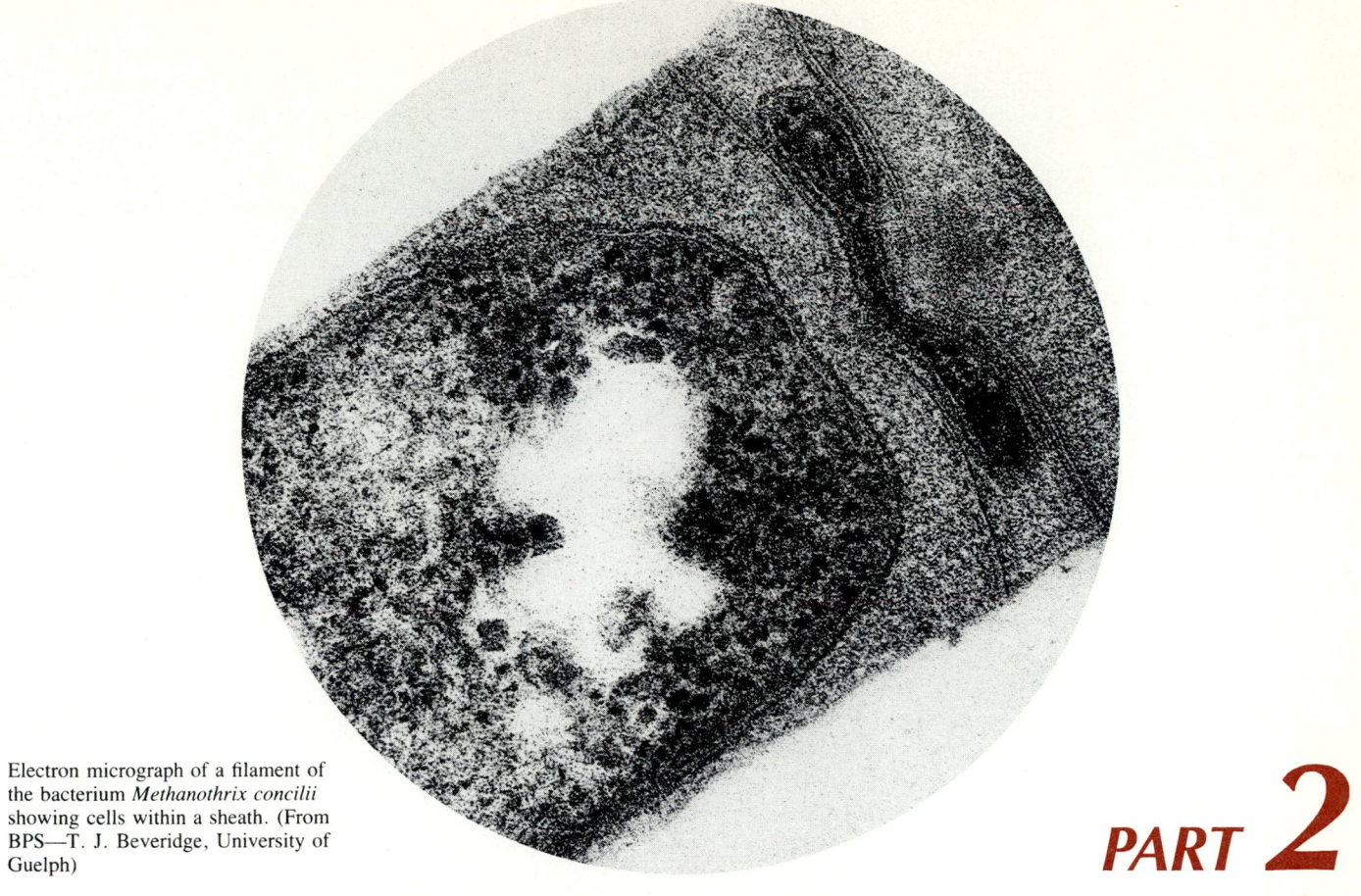

Electron micrograph of a filament of the bacterium *Methanothrix concilii* showing cells within a sheath. (From BPS—T. J. Beveridge, University of Guelph)

CHAPTER 2

Microscopic Methods for Observing Microorganisms

The **microscope**, which is an instrument used for viewing objects—such as microorganisms—too small to be visible with the naked eye, is the basic tool employed for the visualization of the microbial world. A microscope magnifies the size of the apparent image of an object, making it possible to see microorganisms. Through the microscope some microorganisms appear beautiful, diverse, and graceful in form, while others appear as small, pinhead-sized objects with little distinct form. The degree of magnification needed to see a microorganism depends upon the size of that microbe. Bacteria, fungi, algae, and protozoa can be viewed with the light microscope, that is, a microscope that uses visible light to illuminate the specimen. Smaller microorganisms, like viruses, as well as the internal structures of bacterial cells, require the use of more specialized electron microscopes in which an electron beam instead of light is used to reveal the detailed structure(s) of the specimen. The choice of a particular microscope, of which there are many, depends upon the size of the object, the degree of detail that must be viewed, the nature of the specimen, and the overall purpose of the microscopic observations. In this chapter we will examine some of the types of microscopes available for observing microorganisms, including their strengths, limitations, and applications (Table 2.1).

TABLE 2.1 Comparison of Various Types of Microscopes

Type of Microscope	Maximum Useful Magnification	Resolution	Comments
Bright-field	1,500×	100–200 nm	Extensively used for the visualization of microorganisms; usually necessary to stain specimens for viewing.
Dark-field	1,500×	100–200 nm	Used for viewing live microorganisms, particularly those with characteristic morphology; staining not required; specimen appears bright on a dark background.
Ultraviolet	2,500×	100 nm	Improved resolution over normal light microscope; largely replaced by electron microscopes.
Fluorescence	1,500×	100–200 nm	Uses fluorescent staining; useful in many diagnostic procedures for identifying microorganisms.
Phase contrast	1,500×	100–200 nm	Used to examine structures of living microorganisms; does not require staining.
Interference	1,500×	100–200 nm	Used to examine structures of microorganisms; produces sharp, multicolored image with three-dimensional appearance.
TEM	500,000–1,000,000×	1 nm	Used to view ultrastructure of microorganisms, including viruses; much greater resolving power and useful magnification than can be achieved with light microscopy.
SEM	10,000–1,000,000×	1–10 nm	Used for showing detailed surface structures of microorganisms; produces a three-dimensional image.

2.1 Light (Bright-Field) Microscopy

The most common type of microscope found in microbiology laboratories is the **light microscope** (also called the **bright-field microscope**), which is an optical microscope capable of magnifying the apparent size of an object in which light is transmitted through the object. The bright-field microscope used today for viewing bacteria and other cellular organisms is a compound microscope that uses multiple lens systems to magnify the object; it has a light source, a **condenser lens** that focuses the light on the specimen and two sets of lenses—objective and ocular—that contribute to the magnification of the image (Figure 2.1). Through the **refraction** or bending of light rays by the system of microscope lenses, an image of the specimen is formed that is larger than the object itself, permitting the structure(s) of the specimen to be seen.

Magnification

The magnifying capability of a compound microscope is the product of the individual magnifying powers of the **ocular lens** (the lens nearest the eye) and the **objective lens** (the lens nearest the specimen). At a magnification of $1,000\times$, bacteria and larger microorganisms can be visualized, but viruses and much of the fine structural detail of bacteria cannot be seen (Figure 2.2). A typical microscope used in bacteriology has objective lenses with powers of $10\times$, $40\times$, and $100\times$ and an ocular lens of $10\times$, and thus is capable of magnifying the image of a specimen 100, 400, and 1,000 times, respectively. If the various lenses are adjusted so that once the specimen is focused with one lens it remains in focus when one switches to other objective lenses, the mi-

croscope is said to be **parfocal**, meaning that the lenses are arranged so that they can be interchanged without refocusing.

Resolution

In addition to magnifying the image of a specimen so that it can be viewed, the usefulness of a microscope is dependent on its **resolving power**, that is, the degree to which the detail present in the specimen is retained in the magnified image. **Resolution** is defined as the closest spacing between two points at which the points can still be seen clearly as separate entities; the resolving power of a microscope, therefore, is the distance between two structural entities of a specimen at which the entities can still be seen as individual structures in the magnified image (Figure 2.3). The resolving power of a microscope is dependent on the wavelength of light (λ)[1] and on the **numerical aperture (NA)** of the lens.[2] The limit of resolution of a microscope is approximately equal to $0.5\lambda/NA$, which for a light microscope is approximately 200 nanometers (nm) or about the size of many bacterial cells. It follows that the shorter the wavelength of light and the higher

---

[1]Light is propagated through space as waves that are defined by specific amplitudes, frequencies, and wavelengths. The height (amplitude) of the peaks and troughs of the waves determines the intensity or brightness of the light. The frequency of the wave is the number of complete waves that occur per unit time. The wavelength is the distance between the apex of one peak and the apex of the next peak (or between one trough and the next). Visible light has wavelengths of approximately 400 to 700 nm; colors of light have specified ranges of wavelengths within this range.

[2]The NA is a property of a lens that describes the amount of light that can enter it.

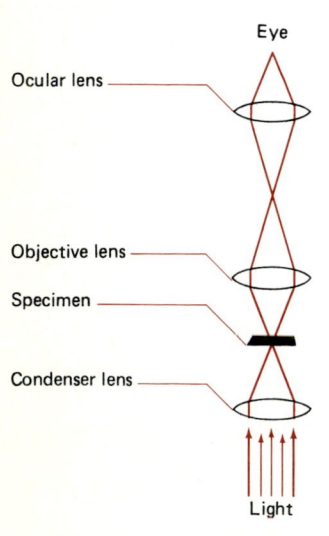

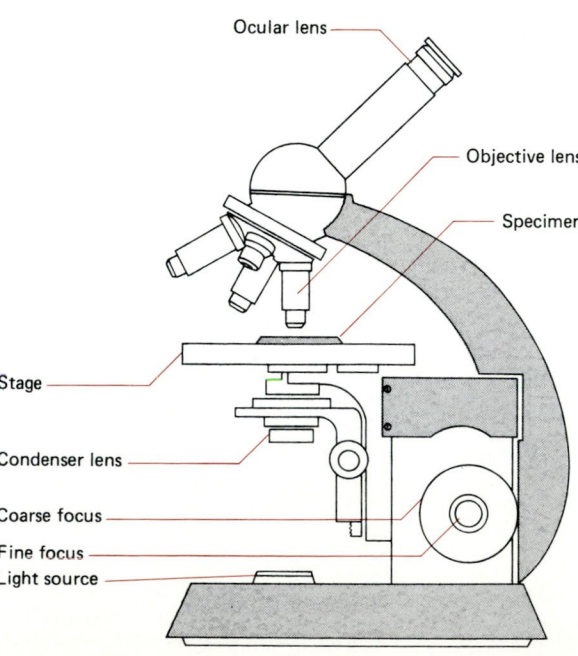

FIGURE 2.1
A light microscope and the path of light as it is modified by passage through the lenses.

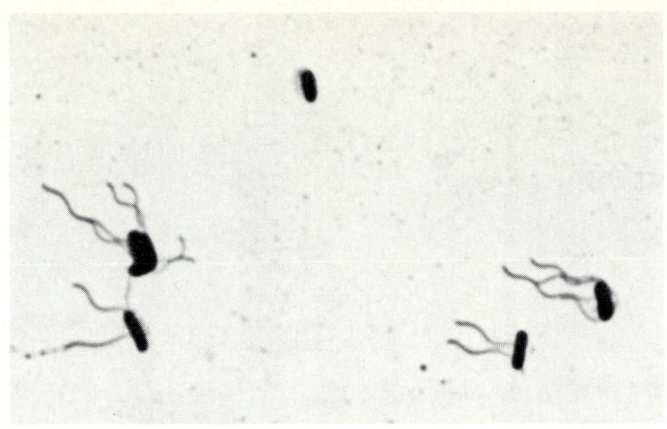

A

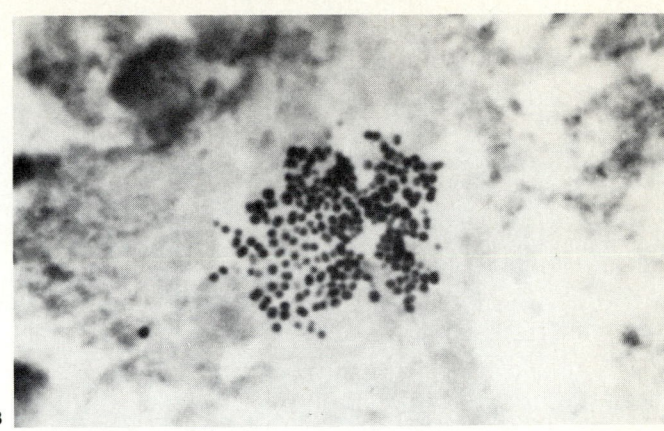

B

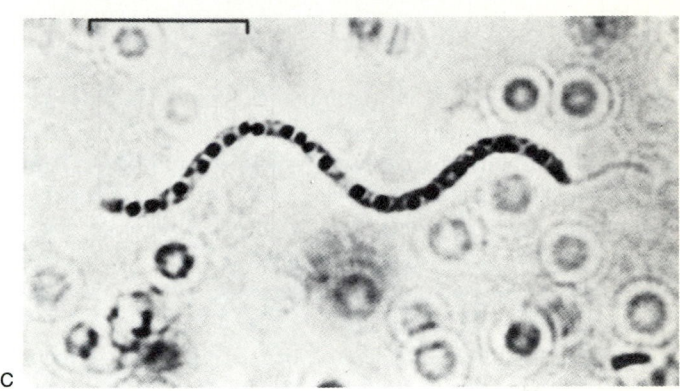

C

FIGURE 2.2

As revealed in these photomicrographs, by using bright-field
microscopy, the characteristic sizes, shapes, and arrangements
of microbial cells can be observed. These micrographs show
the commonly encountered rod-shaped, spherical (coccal),
and curved bacterial cells. (A) Light micrograph of rod-shaped
cells of *Bordetella bronchiseptica* with flagella emanating
from the cells. (Courtesy Centers for Disease Control, Atlanta).
(B) Micrograph of coccal-shaped cells (cocci) of *Streptococcus
viridans*. (Courtesy John J. Bochino, Norton's Kosair Children's
Hospital, Louisville, Kentucky). (C) Micrograph of spiral-
shaped cells of a *Thiovulum* species. (Courtesy J. W. M.
laRiviere, International Institute for Hydraulic and Environmental Engineering, Delft)

the NA of the lens, the better the resolving power of the
microscope. It should also be noted that the resolving power
of a light microscope is restricted by the obtainable NAs of
the lens systems and the wavelengths of the visible light
spectrum.

Because blue light has a shorter wavelength than red light,
greater resolution can be achieved by using a blue light
source to illuminate the specimen. Ultraviolet (uv) light,
which has a still shorter wavelength, is preferable to visible
light for increasing resolution, but because uv light will not
penetrate glass lenses well, and because viewing uv light
directly results in permanent eye damage and blindness, it is
normally not possible to capitalize on the improved resolving
power that could be achieved by using this shorter wavelength of light. Specialized uv microscopes—with approximately twice the resolving power of the normal light micro-

scope—can be constructed using quartz lenses and a screen
or photographic plate for indirectly viewing the image of the
specimen. However, the advent of the electron microscope,
which utilizes electrons, having even shorter electromagnetic
waves, has made such uv light microscopes obsolete.

Along with using blue light for illumination, using a lens
with a high NA permits high resolution. The NA of a lens
is dependent on the **refractive index (η)**[3] of the medium
filling the space between the specimen and the front of the
objective lens and on the angle of the most oblique rays of
light that can enter the objective lens (θ) (Figure 2.4). The
formula for calculating the NA is $NA = \eta \times \sin \theta$. Air has
a refractive index of 1, which limits the resolution that can
be achieved, but the NA can be increased by placing immersion oil between the specimen and the objective lens,
improving the resolving power of the microscope. Immersion
oil has a refractive index of 1.5, which considerably increases the NA and thus improves the resolving power of the
microscope. The observation of fungi, algae, and protozoa
can be achieved with dry objectives, that is, when air occupies the space between the specimen and the objective, but
the viewing of bacteria in sufficient detail to determine the
shape and arrangement of cells normally requires the use of

FIGURE 2.3

Resolution of two points. At low resolution, structures blur
together; the greater the resolution, the more detail that can
be observed.

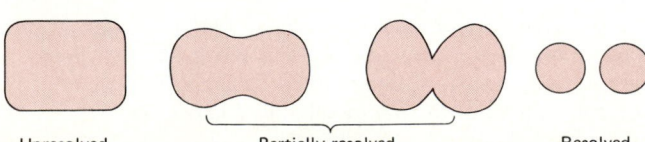

Unresolved   Partially resolved   Resolved

---

[3]The refractive index is the ratio of the speed of light in a given
medium to the speed of light in a vacuum.

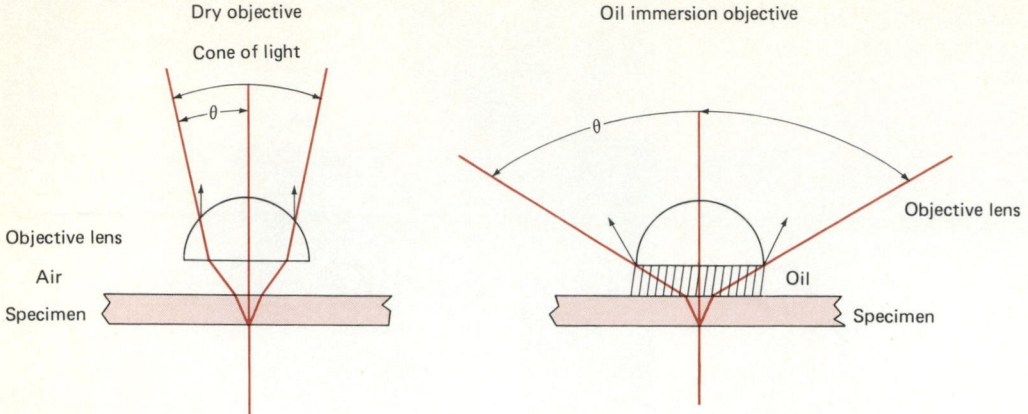

FIGURE 2.4
The numerical aperture is improved by the use of immersion oil to replace the air between the objective lens and the specimen, as shown by the wider cone of light obtained using an oil immersion compared with a dry objective. θ = the angle of peripheral light entering the lens, which is one-half the cone of light entering the front lens of the objective.

an **oil immersion lens**, in which immersion oil fills the space between the specimen and the object.

As a rule of thumb, the useful magnification of a microscope is 1,000 times the NA being used, and so it is possible with an oil immersion lens with an NA of 1.4 to achieve a useful magnification of approximately $1,400\times$. At higher magnifications the quality of the image deteriorates, and the magnification is therefore considered to be empty. A high-NA lens has a short focal length, that is, the plane of focus is near the lens, and therefore there is a short working distance between the lens and the object, that is, the lens and the specimen are very close to one another. Another consequence of the short focal length is a very shallow depth of field, and only a very thin section can be in focus at any one time. Because of the short working distance and shallow depth of field of oil immersion lenses, many students at first have great difficulty trying to focus the microscope on a specimen of bacteria without breaking the slide and scratching the objective lens, but with a little practice and proper instruction this problem is easily overcome.

Contrast and Staining

Another factor that must be considered in light microscopy is contrast, because without adequate contrast it is impossible to discern a structure from the surrounding background. Microorganisms are largely composed of water, as is the medium in which they are normally suspended, and simply viewing microorganisms with a light microscope without performing procedures to increase contrast can be likened to trying to see a white object on a white background. Microorganisms normally must be stained, that is, treated with dyes before they can be viewed using most microscopes.

ANALYTICAL PROCESS

Staining Microorganisms

Staining is used to increase the contrast between the specimen and the background, but unfortunately, staining generally precludes the observation of living microorganisms. In most staining procedures for light microscopy, a suspension of microorganisms is transferred to a glass microscope slide and allowed to dry (Figure 2.5). The slide is then quickly passed through a flame to heat fix the cells to the slide, that is, to affix the cells to the slide so that they cannot be washed off. A stain is then added to the slide, allowed to stand for a period of time, rinsed to remove excess stain, and viewed.

Simple Staining Procedures

In a simple staining procedure, a single stain reagent is used and all cells and structures generally stain in the same manner, regardless of type. The staining procedure may be positive, in which case the stain is attracted to the cells, or negative, in which case the stain is repelled by the cells (Figure 2.6). In positive staining procedures, the stain, which is basic (cationic), has a positively charged chromophore (colored portion of the stain molecule), that is attracted to the negatively charged outer surface of the microbial cell. A stain such as methylene blue has a blue-colored portion of its molecule, which is attracted to the cell's surfaces, resulting in positive staining of the microorganisms. In negative staining procedures, the stain, which is acidic (anionic), has a negatively charged chromophore that is repelled by the negatively charged microorganisms so that it fills the spaces between the microorganisms, resulting in the apparent negative or indirect staining of the microbial cell. Nigrosin and india ink are frequently used for negative staining of microbial cells, and this type of staining is particularly useful for viewing some structures, such as the capsules that surround some bacterial cells.

Differential Staining Procedures

In differential staining procedures, multiple staining reactions are employed such that specific types of microorganisms and/or particular structures of a microorganism exhibit different staining reactions that can be readily distinguished. The Gram stain procedure undoubtedly is the most widely used differential staining procedure in bacteriology (Figure 2.7). This staining procedure begins with primary staining with crystal violet, which stains all bacterial cells blue-purple, followed by application of Gram's iodine—a mordant (a substance that increases the affinity of the primary stain for the bacterial cells)—then rinsing with acetone-alcohol or some other decolorization agent (a substance that attempts to remove the primary stain), and finally application of the red counterstain safranin, which stains those bacteria that were decolorized in the previous step so that they can be easily visualized. Decolorization is the critical step that differentiates bacterial species. About half of all bacterial species are Gram positive and are not decolorized, and the other half are Gram negative and are decolorized. After treatment with safranin counterstain, the Gram-positive bacteria appear blue-purple and the remaining Gram-negative bacteria appear red-pink. The Gram stain procedure has great diagnostic value, because of its ability to differentiate among different bacterial species, and therefore is a key feature employed in many bacterial classification and identification systems.

Another differential staining procedure frequently used in bacteriology is acid-fast staining, a procedure in which decolorization with acid alcohol is the key differentiation step. Acid-fast bacteria do not decolorize and retain the red color of the primary stain, which is carbol fuchsin. The acid-fast stain procedure is especially useful in identifying members of the bacterial genus *Mycobacterium* and is important in identifying the causative organism of tuberculosis, *Mycobacterium tuberculosis*.

Yet another key differential staining procedure reveals the presence or absence of bacterial endospores. Endospores are produced by members of relatively few bacterial genera but are very important because of their resistance to high temperatures. In the endospore-staining procedure, the bacterial endospore typically is stained green and the rest of the bacterial cell pink, permitting differentiation of the endospore from the vegetative cell.

(1) Place a loopful of the culture on a clean slide

(2) Spread in a thin film over the slide

(3) Air dry

(4) Fix by passing the slide rapidly through the bunsen flame

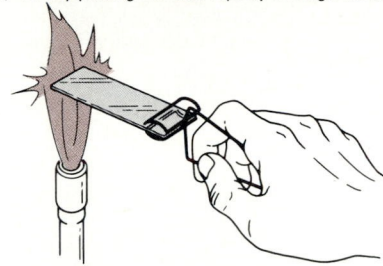

(5) Stain (e.g., with crystal violet)

(6) Wash off stain with water

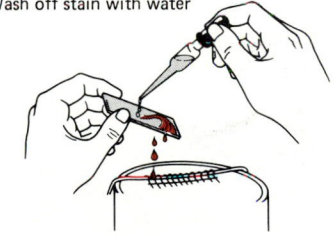

(7) Blot off excess water

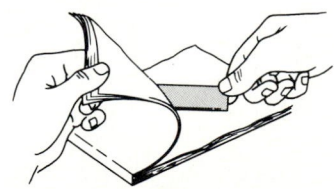

(8) View under microscope

FIGURE 2.5

In a simple staining procedure, microorganisms are affixed to a glass slide and stained with an appropriate dye to increase the contrast between the cells and the background so that they can easily be seen using a light microscope.

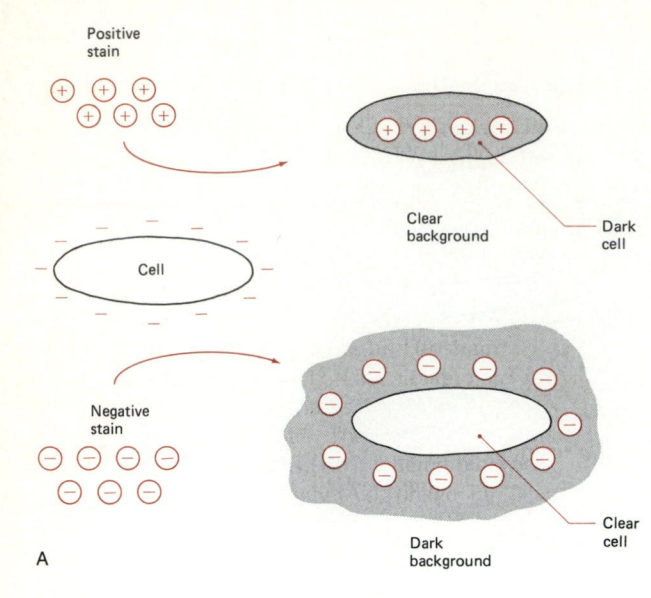

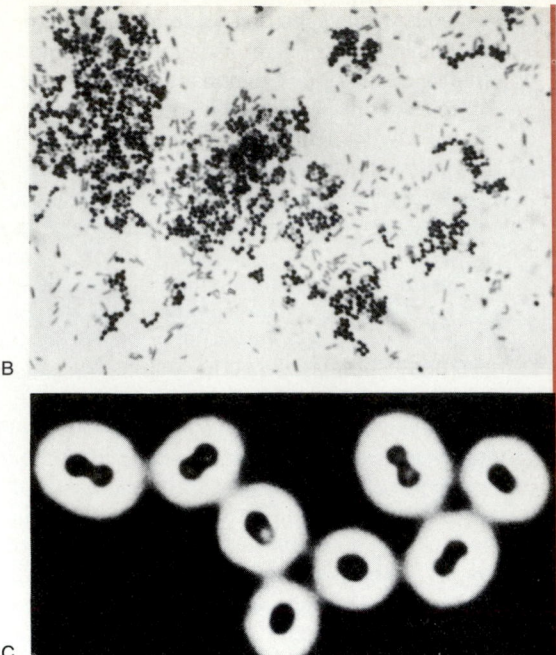

FIGURE 2.6

(A) The interaction of a cell with negative and positive stain reagents. Because the outer layer of a cell is negatively charged, a positive stain is attracted to the cell, whereas a negative stain is repelled. (B) This photomicrograph shows a mixture of the positively stained bacteria *Escherichia coli* and *Staphylococcus aureus* (100×). (From BPS—Leon J. LeBeau, University of Illinois Medical Center, Chicago). (C) The bacterium *Thiocapsa floridiana* following negative staining with india ink. The clear area around the cells is a capsule structure, which will be discussed in Chapter 3. (From BPS— Stanley C. Holt, University of Massachusetts)

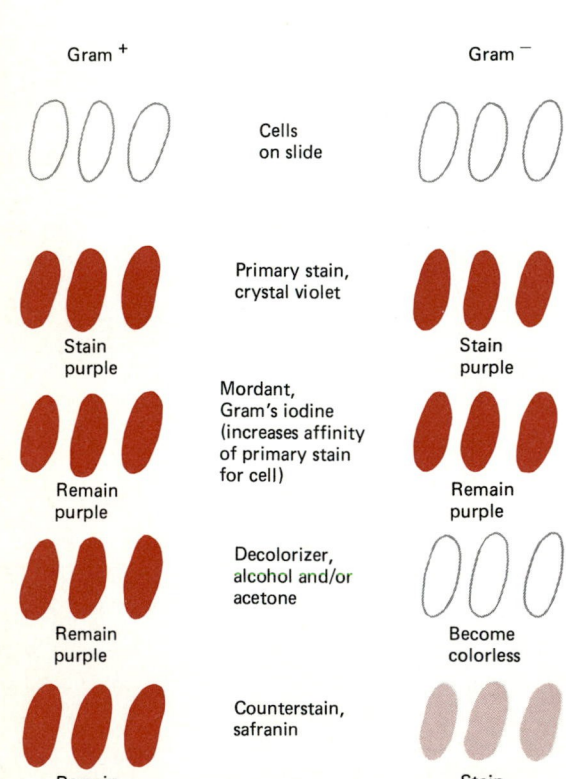

FIGURE 2.7

The Gram stain procedure is widely used to differentiate major groups of bacteria.

2.2 Fluorescence Microscopy

Microscopy that involves staining with fluorescent dyes is known as **fluorescence microscopy**, and the microscope used for viewing such specimens is referred to as a *fluorescence microscope*. When a fluorescent dye is illuminated by light of one wavelength—called the *excitation wavelength*—it gives off light at a different wavelength—called the *emission wavelength*. For example, when the fluorescent dye fluorescein isothiocyanate is illuminated with blue light, it emits green light. One of the reasons that fluorescence microscopy has become important in microbiology is that fluorescent dyes can be conjugated (linked) with antibodies (specific proteins produced as part of the immune response), providing great specificity in staining procedures.

The principles of magnifying power and resolution are no different for the fluorescence microscope than for the bright-field microscope (Figure 2.8). The excitation light may be transmitted either from below the specimen, in which case it is called **transmitted fluorescence**, or to the specimen through the objective lens, in which case the system is referred to as **epifluorescence** (*epi* from the Greek word meaning "upon"). The wavelength of the light used to excite the dye may be in the uv range, but the emitted light that is to be viewed must be in the visible range. Fluorescence microscopes are equipped with various excitation filters that permit the selection of the wavelength used to illuminate the specimen and barrier filters that prevent all but the emission wavelengths from reaching the ocular lens. When uv wavelengths are used, it is particularly important to illuminate the specimen so that the barrier filters preclude any uv light from reaching the eye to prevent blindness.

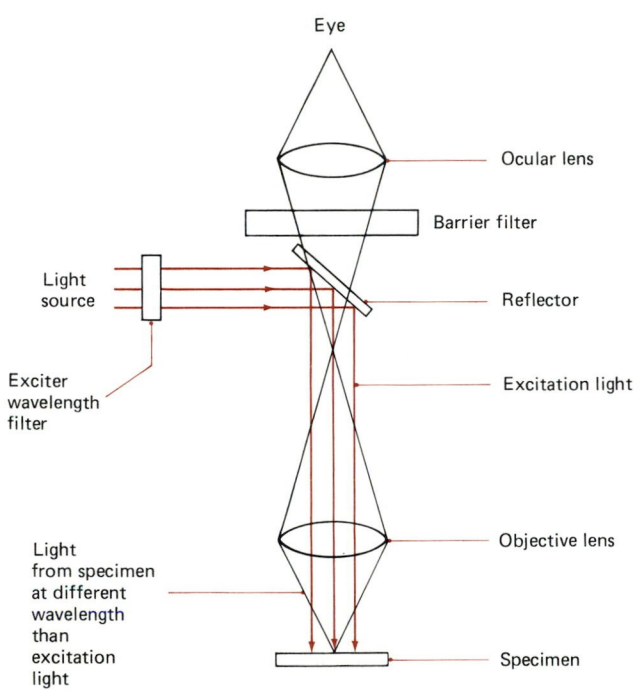

FIGURE 2.8
Diagram of an epifluorescence microscope showing the light path. This type of microscope does not use a condenser. Rather, the objective lens focuses light on the specimen. The illuminating light is at one wavelength, which can be in the ultraviolet range, and the light emitted from the fluorescent dye–stained specimen is at a different wavelength, which must be in the visible light range. A series of exciter and barrier filters are used to ensure that the specimen is illuminated with a particular wavelength and that the viewing is restricted to a different wavelength that corresponds to the emission wavelength of the fluorescent dye.

2.3 Dark-Field Microscopy

There are several alternative microscope designs that enhance the contrast between the specimen and the background without the use of staining, thus permitting the visualization of living specimens. In the simplest of these microscopes, the **dark-field microscope**, the normal condenser of the light microscope is replaced with a dark-field condenser that does not permit light to be transmitted directly through the specimen and into the objective lens (Figure 2.9). The dark-field condenser focuses light on the specimen at an oblique angle, such that light that does not reflect off an object does not enter the objective lens. Thus, only light that reflects off the specimen will be seen, and in the absence of a specimen the entire field will appear dark. Microorganisms viewed with a dark-field microscope appear very bright on a dark (black) background (Figure 2.10). The contrast between the specimen and the background is sufficient to permit the visualization of even small bacteria and large viruses, but it is not generally possible with dark-field microscopy to distinguish the internal structures of any bacteria being viewed.

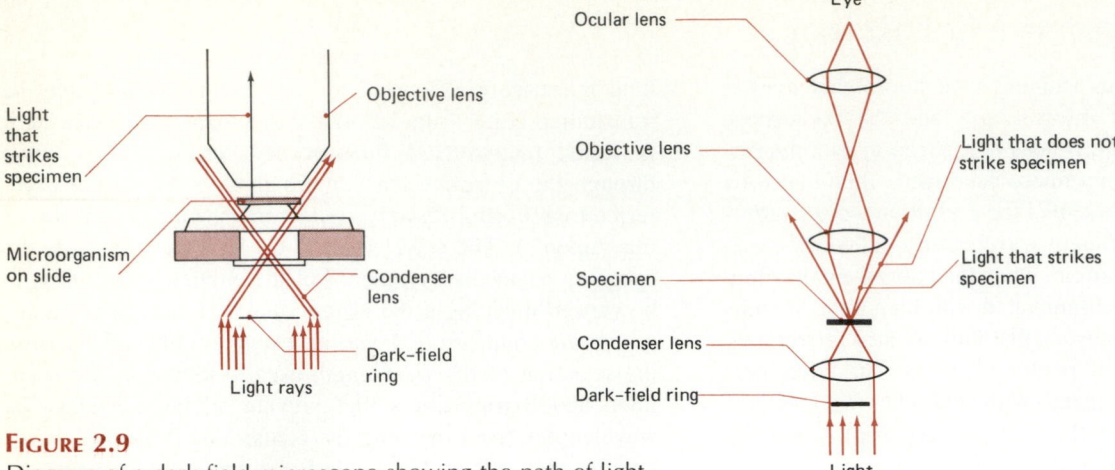

FIGURE 2.9
Diagram of a dark-field microscope showing the path of light
passing through the background and light striking the specimen.

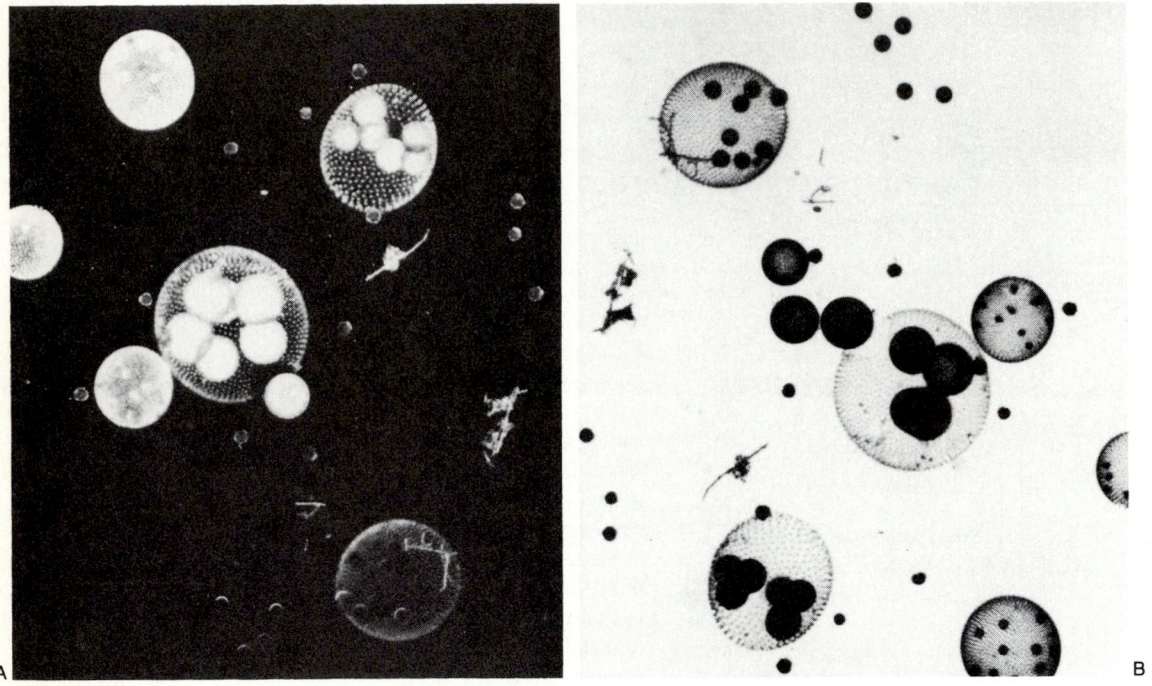

FIGURE 2.10
Dark-field microscopy is useful for visualizing microbial cells without the need for staining to
enhance contrast. (A) Micrograph of a colony of the green alga *Volvox* as it appears in dark-field
microscopy. (B) Micrograph of a colony of *Volvox* as it appears in bright-field microscopy is
shown for comparison. (Courtesy Gary B. Collins, USEPA, Cincinnati)

2.4 *Phase Contrast Microscopy*

The **phase contrast microscope** optically changes differences in the speed with which light passes through an object into differences in contrast that can be seen. Like the dark-field microscope, it is useful for visualizing living microorganisms and eliminates the necessity of staining to view microbial structures. This type of microscope relies upon the fact that light passing through a cell of **higher refractive index** (with a greater ability to change the direction of a ray of light) than the surrounding medium is slowed down relative to the light that passes directly through the less dense

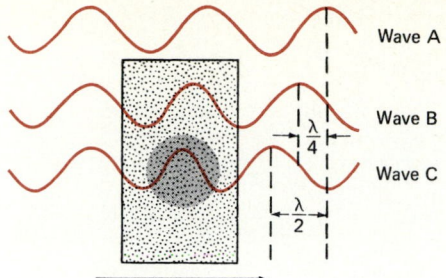

FIGURE 2.11

The retardation of phases of light waves as they pass through a transparent living cell mounted in saline. Compared to the waves that do not pass through the cell (A), the waves passing through the full thickness of the cytoplasm (B) have been retarded by $\frac{1}{4}\lambda$, and those passing through a more highly refractile inclusion (C) have been retarded by $\frac{1}{2}\lambda$.

background medium. The greater the refractive index of the cell or cellular structure, the greater the retardation of the light wave (Figure 2.11).

Thus, when light passes through a microorganism, there is a slight alteration in the phase of the light wave, that is, the point of advancement within the light wave cycle. The conversion of differences in the phase of the light waves to differences in light intensity is based on interference between

light waves that are out of phase with one another (Figure 2.12). The interaction of light waves that are in phase produces addition, which increases the intensity of the light, but the interaction of light waves that are 90° out of phase results in destruction of the propagating light wave, that is, a decrease in intensity. Thus, interference of light waves accomplishes the task of creating visible light intensity differences that are necessary for visualizing microorganisms.

In order to permit the observation of unstained microorganisms, the phase contrast microscope is designed to separate direct background light from the light passing through the object, causing these two different light waves to be approximately 90° out of phase with one another so that they interact destructively and cause changes in light intensity. To accomplish this, on the substage condenser the phase contrast microscope has a ring-shaped disk, called an **annulus**, which is solid except for a thin ring through which light can pass. An objective lens of a phase contrast microscope is constructed with a ring-shaped phase plate at its back focal plane that corresponds geometrically with the ring of direct light from the condenser, that is, the annulus ring is aligned with the ring of the phase plate. Normally, the ring-shaped area of the phase plate is thinner than the rest of the plate, so that the phase of the light wave passing through the ringed area is altered by 90°, or one-quarter of a wavelength. Light that passes directly through the object passes through the ring-

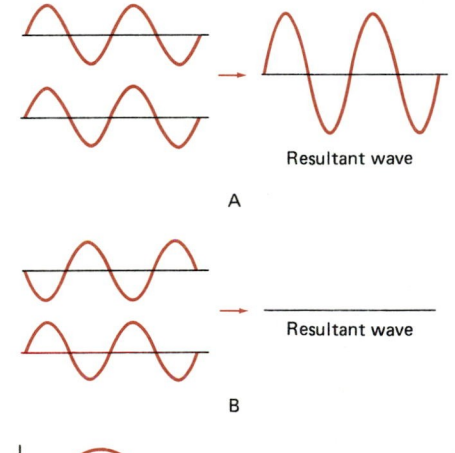

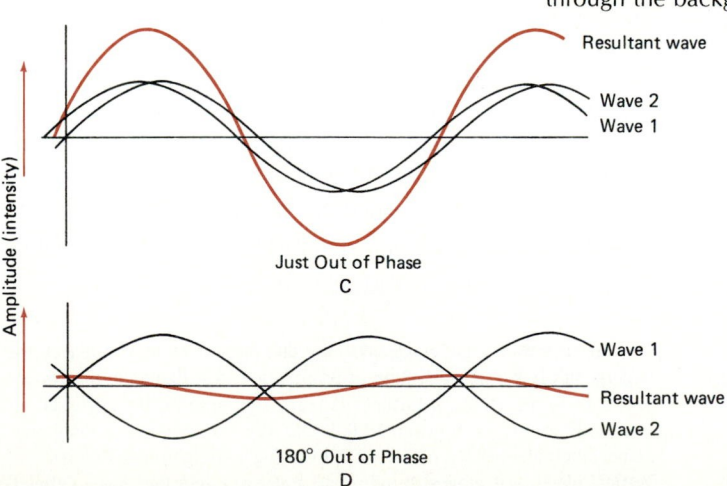

FIGURE 2.12

Additive and destructive interference of light waves. (A) Two waves in phase combining to produce a wave with twice the amplitude. (B) When the two waves are one-half wavelength out of phase, they combine to cancel each other out. (C) The superimposition of two waves of equal frequency (number of cycles of the light wave per unit time) and amplitude (height of the light wave) that are almost in phase results in a wave of almost twice the amplitude of either component. (D) The superimposition of two waves of equal frequency and amplitude, almost 180° out of phase, results in a wave whose amplitude is nearly zero. Note that in all of these cases the frequency is unchanged. The ability to alter the amplitude of a light wave by combining two different light waves forms the basis for phase contrast microscopy, allowing the visualization of different intensities of light passing through a specimen and light passing through the background.

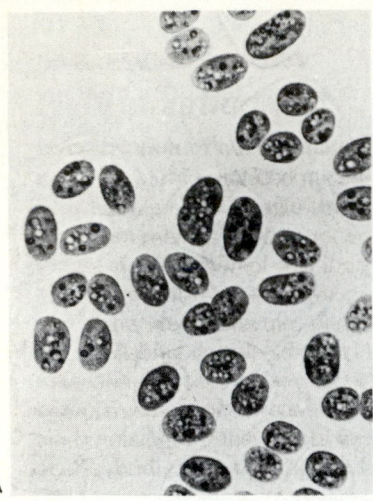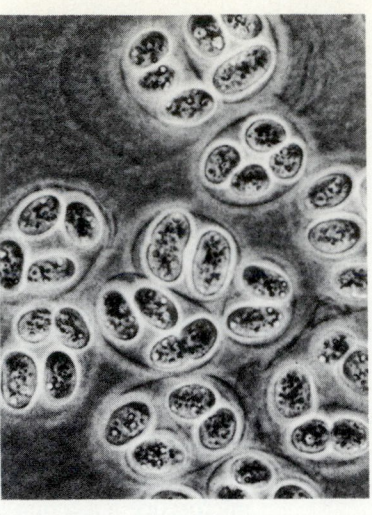

FIGURE 2.13

These photomicrographs of the cyanobacterium *Gleocapsa* compare its appearance by using (A) bright-field and (B) phase contrast microscopy. The cell structures are easier to see in the micrograph taken with the phase contrast microscope. (From BPS—J. Robert Waaland, University of Washington)

shaped area of the phase plate and is accordingly advanced by 90°. Light that is diffracted by the specimen passes through the thicker portion of the phase plate and is retarded by 90° relative to the light that is transmitted directly. When recombined, light waves that are out of phase interfere with each other, producing alterations in the amplitude of the light waves; these amplitude changes are seen as differences in light intensity.

If a specimen has the same refractive index as the surrounding medium, the direct light and the diffracted light will be 90° out of phase, causing destructive interference and producing a uniform decrease in intensity, that is, a uniform dark field. However, if the specimen has a higher refractive index than the surrounding medium, the light diffracted by the specimen will be retarded relative to the light passing directly through the surrounding background. The degree of

retardation depends on the difference in magnitude between the refractive indices of the specimen and its background and the thickness of the specimen. The greater the refractive index and thickness of the specimen, the greater the retardation. Similarly, differences in thickness and/or refractive index between structures within the specimen will produce differences in the retardation of the diffracted light, that is, phase changes that are convertible into visible differences in light intensity. Even difficult-to-stain structures often are conspicuous under a phase contrast microscope because small phase changes result in interference of light waves, giving rise to high-contrast images (Figure 2.13). Thus, with the phase contrast microscope, living organisms can be clearly observed in great detail without being stained, permitting the study of their movements in the medium in which they are growing.

2.5 *Interference Microscopy*

Besides the phase contrast microscope, there are several other types of microscopes that rely on light interference to produce a visible image. Both phase contrast and **interference microscopes** make use of the fact that light travels as waves and that the addition of light waves that are out of phase with each other produces interference that alters the amplitude of the light wave. Whereas the phase contrast microscope relies upon a single beam of light that passes through the specimen, interference microscopes have two beams of light that are combined after passing through the specimen to produce an interference pattern. Compared to the phase contrast microscope, interference microscopes can have higher NAs, can vary the degree of contrast to suit the subject, and can use colors to increase contrast. Like phase contrast microscopes, interference microscopes permit the visualization of living cells; also, as in the phase contrast microscope, phase changes that occur when light passes through the specimen are translated into changes in light intensity.

Nomarski Differential Interference Contrast Microscopy

Of the various types of interference microscopes, microbiologists most frequently employ **Nomarski differential interference contrast (NDIC) microscopes**, which produce high-contrast images of unstained, transparent specimens in what appear to be three dimensions (Figure 2.14). The NDIC microscope has three special features: a polarizing filter,[4] an interference contrast condenser, and a prism-analyzer plate. In the NDIC microscope, polarized light with its defined

[4]Light is a transverse wave, meaning that the direction of displacement is perpendicular to the direction of propagation, just like an ocean wave as it rolls to the shore. A linear polarizing filter absorbs the horizontal vibrations so that the transmitted light has vertical vibrations in only one plane, like reducing the ocean wave to a single straight up-and-down vertical plane as it moves to the shore. Polarized light then has a defined orientation, with the light wave vibrating in one plane.

A

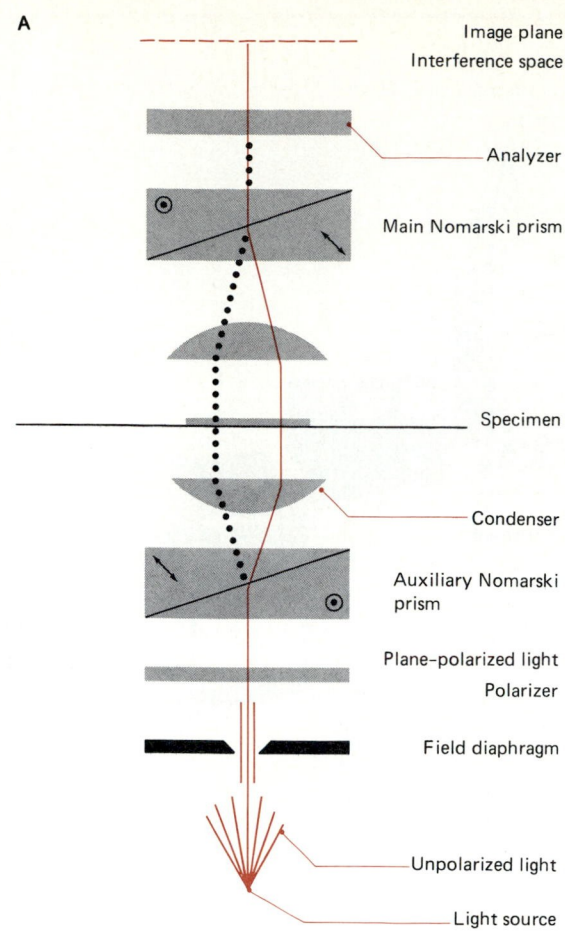

Image plane

Interference space

Analyzer

Main Nomarski prism

Specimen

Condenser

Auxiliary Nomarski prism

Plane-polarized light

Polarizer

Field diaphragm

Unpolarized light

Light source

B

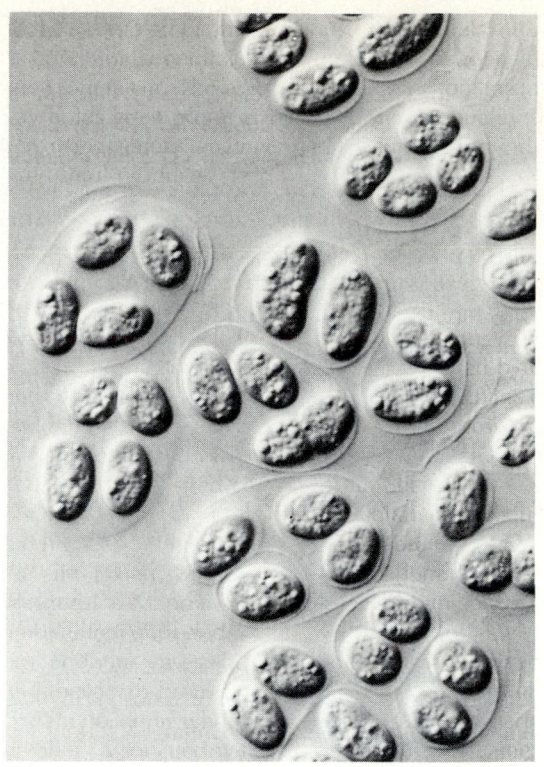

FIGURE 2.14

(A) Diagram of a Nomarski interference microscope, showing the path of light. (B) Photomicrograph of the cyanobacterium *Gleocapsa*, using Nomarski interference microscopy (2,800×). (From BPS—J. Robert Waaland, University of Washington).

pattern of aligned light waves is split into two beams at right angles to each other that travel closely parallel to each other through the specimen. The two beams of light are then combined and pass through an analyzer. When the two light rays that are differentially diffracted by the specimen are recombined, they produce an interference pattern.

The pseudo-three-dimensional image of NDIC is produced because the two beams of light traveling very close to each other through the specimen produce a stereoscopic effect. The degree of three-dimensional appearance is a function of the refractive index differences at the boundary surfaces of the specimen. Contrast in NDIC depends on the rate of change of the refractive index across a specimen; consequently, especially good contrast is produced at the edges of the specimen, where there is a large refractive index differential. Structures such as cell walls are very well defined when viewed with interference microscopes. Different structures of microorganisms appear in different colors that are related to the phase changes in the light passing through each structure, and images seen through the interference microscope are normally brilliantly colored. Interference microscopes are very useful for qualitative observations of unstained cells because they produce images with high contrast and striking topographic relief.

2.6 Electron Microscopy

The advent of the electron microscope marked a significant improvement over light microscopes for the visualization of microorganisms because electron microscopy permits better resolution, and therefore higher useful magnifications than can be achieved with light microscopy. Many of the fine structures of microorganisms have been elucidated through the use of electron microscopy. There are two basic types of electron microscopes: the transmission electron microscope and the scanning electron microscope.

Transmission Electron Microscope

The design of the **transmission electron microscope (TEM)**, in which an electron beam passes through the specimen, is similar to that of the compound light microscope, except that an electron beam is substituted for the light source and a series of electromagnets are substituted for the glass lenses (Figure 2.15). The source of the electrons is a hot tungsten filament in an electron gun. Electrons are drawn from the filament and accelerated as a fine electron beam

FIGURE 2.15

The TEM allows the visualization of the fine detail of the microbial cell. (A) Diagram of a TEM and (B) photograph of a high-resolution TEM. (Courtesy JEOL, Peabody, Mass.)

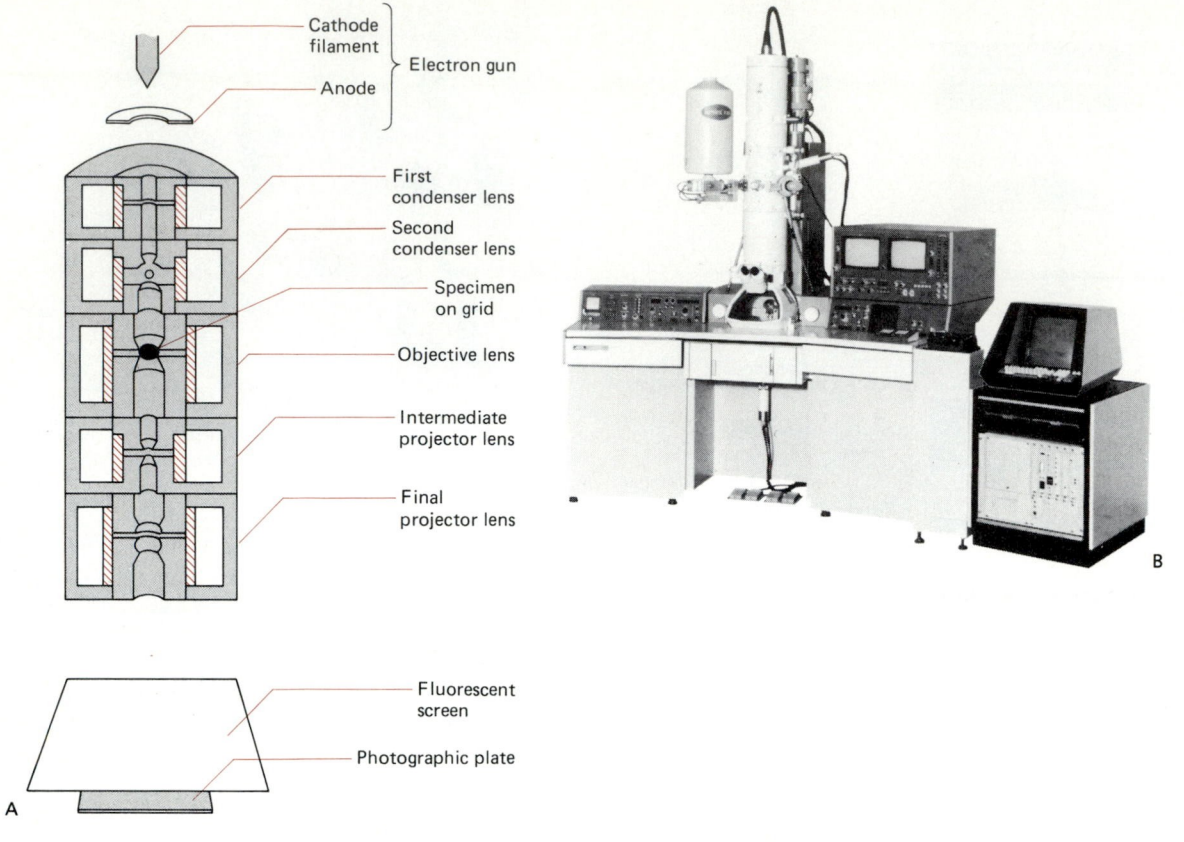

Discovery Process

The introduction of the wave theory to describe the nature of electrons by de Broglie in 1924 laid the foundation in the field of electronic optics needed for the development of electron microscopes. Magnetic coils were used as optical lenses, and a number of experimental microscopes were built to study the physical principles needed to develop the instrument. By 1931, Ernst Ruska—then 24 years old—had invented the electron microscope, for which he was corecipient of the 1986 Nobel Prize in physics. The first instruments that could be used in biological studies were used in about 1940. In a 1941 article entitled "The Electron Microscope: A New Tool for Bacteriological Research," L. Marton described the state of development of the electron microscope as follows: "Electron microscopy is based on the discovery that geometrical optics for electrons is quite similar to the optics of light. To understand the term 'geometrical optics' let us first consider the action of an electric or magnetic field on an electron beam. It is well known that an electron beam is deflected by such fields, and we can therefore compare their action on the beam to the action of a refractive medium on a light beam. A lens is nothing but a refractive medium of special symmetry—in this particular case of rotational symmetry. If we create

an electric or magnetic field of rotational symmetry, such a field acts on an electron beam as a lens, i.e., the electron beam is concentrated or made divergent in the same way that the light beam is acted upon by a glass lens. It has been proved mathematically that the laws of geometrical optics can be fully applied to such systems, and experimentally that we can obtain electronic images which can be made visible, for instance on a fluorescent screen. An image can be formed of any self-emitting object, as would be the case with light if we observed the image of the source itself, or we can illuminate an object in the same way as we illuminate one in a light microscope, and observe the image of the object with the help of the illuminating beam.

"The great advantage of an electron optical system lies in its highly increased resolving power. The practical microscopist knows that the resolving power, i.e., the smallest distance separately shown by an optical system, is about one-half of a wave-length for the best light optical systems. Since the fundamental discovery of de Broglie, we know that the electron behaves for some applications as a corpuscle, and for other applications as a wave, the wave-length of which depends on the speed of the electron; and, for the speeds generally used in electron microscopy, it is about one 100,000th of the wave-length of visible light."

Advances in electron microscopy have continued with the advent of the scanning electron microscope and, more recently, the scanning tunneling microscope. The inventors of the tunneling microscope shared the 1986 Nobel Prize in physics. This microscope uses magnets to draw electrons out of a specimen, rather than using an electron beam to scan the specimen surface. A scanning needle tip is used to explore the surface of the specimen. An electron cloud, which results from the indeterminancy of an electron because of its wave motion, occupies the space between the specimen surface and the needle tip. The intensity of the electron cloud decreases rapidly with the distance from the specimen surface, and therefore, a voltage-induced electron flow is very sensitive to the distance between the surface and the tip of the probe. As the tip is swept across the surface, this induced flow of electrons establishes a current—called the *tunneling current*—that is used to hold the tip at a uniform height above the surface of the specimen. The movement of the tip is detected and computer processed to produce an image on a screen that shows with high resolution the three-dimensional contours of the surface. This scanning tunneling microscope has been used to view the surface structures of viruses and the surface of DNA macromolecules.

past an anode by a high voltage established between the filament and the anode. The electron beam is focused on the specimen with an electromagnetic condenser lens by varying the current to the lens. Instead of placing the specimen on a glass slide, as is done in light microscopy, the specimen normally is placed on a copper mesh grid within the evacuated column of the electron microscope. The TEM has an objective lens, as does the light microscope, but instead of an ocular lens, the TEM has a projector lens that projects the image onto a fluorescent viewing screen or film plate, which is necessary because the electron beam cannot be viewed directly.

In an electron microscope, air is removed from the path of the electron beam to prevent collisions with gas molecules that would scatter the electron beam and make it impossible to resolve a high-magnification image. Additionally, the removal of air reduces the heat associated with the electron beam, reducing destruction of biological specimens in the path of the electron beam and of the filament that otherwise would deteriorate rapidly.

The TEM permits much greater resolution and thus much higher useful magnifications than the light microscope because the wavelength of an electron beam, generated at a high accelerating voltage, is much shorter than that of light in the visible range of the electromagnetic spectrum. The actual wavelength of the electron beam depends on the accelerating voltage of the microscope. At 60,000 volts, a typical accelerating voltage used in a TEM, the wavelength of the electron beam is approximately 0.005 nm, permitting a theoretical resolution of approximately 0.2 nm, which is about a thousand times better than can be achieved when using light microscopy. The useful magnification for an electron microscope, consequently, is in excess of $100,000\times$, which permits the visualization of all microorganisms, including viruses (Figure 2.16). In fact, the resolving power of the electron microscope permits visualization of molecules and atoms.

Contrast in electron microscopy occurs as a result of some electrons colliding with and being scattered by the specimen, while other electrons penetrate through the specimen without collision. The amount of electron scattering is proportional to the electronic density of the specimen, that is, the number of atoms per unit area and their atomic densities. Because the atoms composing biological cells have relatively low electron densities, biological specimens produce little scattering of the electron beam and, hence, low contrast. Staining is used to improve the contrast between the specimen and the background, but instead of the dyes used in light microscopy, the stains used for electron microscopy contain electron-dense heavy metal salts. The heavy metal stains scatter the electron beam and the stained areas thus appear dark, permitting visualization of the detailed ultrastructure of microorganisms.

FIGURE 2.16

To view viruses, an electron microscope is needed. This electron micrograph shows phage infecting a bacterium. Some viruses are outside of the cell, while new viruses are being made within the bacterium ($200,000\times$). (Courtesy Lee Simon, Rutgers, the State University)

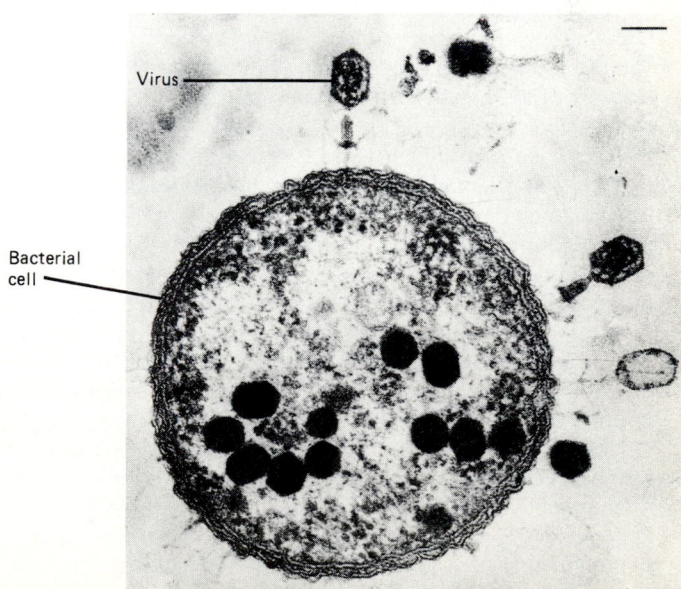

There are several problems in viewing biological specimens, including microorganisms, with the TEM. There is great potential for creating artifacts that could be mistakenly viewed as real structures in electron micrographs. An **artifact** is the appearance of something in an image or micrograph that is due to causes within the optical system or the preparation of the specimen and is not a true representation of the features of the specimen on view. This is a problem common to all microscopes but is particularly troublesome in electron microscopy because of the high magnifications that are used, the need to dehydrate (remove water from) the specimen, and the fact that the specimen must be placed in a high-vacuum chamber. Improper adjustment of the electron beam, excessive magnification, and improper sample preparation can all cause the formation of artifacts.

Scanning Electron Microscope

The **scanning electron microscope (SEM)** (Figure 2.19, page 41), in which an electron beam is scanned across the surface of a specimen, is used primarily for viewing surface details rather than the internal structures of microorganisms. The operational principles and design of the SEM are quite different from those of the TEM. In scanning electron microscopy, an accelerated electron beam is focused on the specimen by using a condenser lens. The lenses of the SEM are designed so that an extremely small electron beam is produced.

The primary electron beam knocks electrons out of the specimen surface, and the secondary electrons produced in this process are transmitted to a collector. Some of the primary electrons are also reflected or backscattered from the specimen surface, but the number of backscattered electrons is far fewer than the number of secondary electrons emitted from the specimen surface. Because the number of low-energy secondary electrons reaching the collector is far greater than the number of backscattered electrons, a more intense signal is developed by the secondary electrons than by the backscattered electrons. The electrons reaching the collector are transmitted to a detector consisting of a substance that emits light when struck by electrons. The emitted light is converted to an electrical current that is used to control the brightness of an image on a cathode ray tube (CRT) screen.

The secondary electrons emitted from each point on the specimen are characteristic of the surface at that point. The intensity of the image seen on the CRT screen thus reflects the composition and topography of the specimen surface. Contrast in the SEM is primarily determined by surface topography, which controls the number of secondary electrons reaching the detector. The shadowed image shown on the

---

ANALYTICAL PROCESS

Preparation of Specimens for the TEM

Before biological specimens—including microorganisms—can be viewed with a TEM, steps must be taken to maximize the amount of detail that can be visualized and to avoid the formation of artifacts. Biological specimens containing water cannot simply be placed under a high vacuum because the water would boil, destroying the integrity of the organism. Therefore, before viewing a microorganism with a TEM, it is necessary to dehydrate the specimen and to fix (preserve intact) the structures in their natural orientation. The fixation and dehydration process must be carried out carefully in several stages, because during the fixation process it is possible to shrink or otherwise distort the microorganisms. Additionally, microorganisms are too large (thick) to view with a TEM and to see the maximum amount of detail that is possible with its high resolving power. Therefore, it is normally necessary to slice the microorganisms into thin sections in order to view their ultrastructures (Figure 2.17).

The thin sectioning of microorganisms is achieved by using a microtome, which is a mechanical slicing instrument that advances a specimen incrementally across a knife surface, usually a diamond or glass knife. The microorganisms are normally embedded in a plastic resin to facilitate handling during thin sectioning. Once sec-

tioned, the specimens are stained with heavy metal-containing compounds, such as phosphotungstic acid that contains the heavy metal tungsten.

There are several special preparation procedures that are used in electron microscopy for revealing detailed structures of microorganisms. Freeze etching, for example, is used to reveal the various biochemically defined layers of a microorganism, including organelle structures. In this procedure, a specimen frozen in liquid nitrogen is fractured by striking it with a knife blade (freeze fractured); the fractured specimen is then etched, that is, some of the ice is allowed to evaporate, raising the surface layer of the specimen. The specimen is then exposed to vapors of a heavy metal while being held at a 45° angle to produce a shadow effect, after which it is rotated and exposed to vaporized carbon at a 90° angle to produce a replica of the surface. Any adhering biological material is removed, and the carbon replica is then viewed with an electron microscope (Figure 2.18). The freeze-etching method reveals much detail of both internal and external surface structures and also eliminates some problems with artifacts that arise through chemical fixation and sectioning of biological specimens.

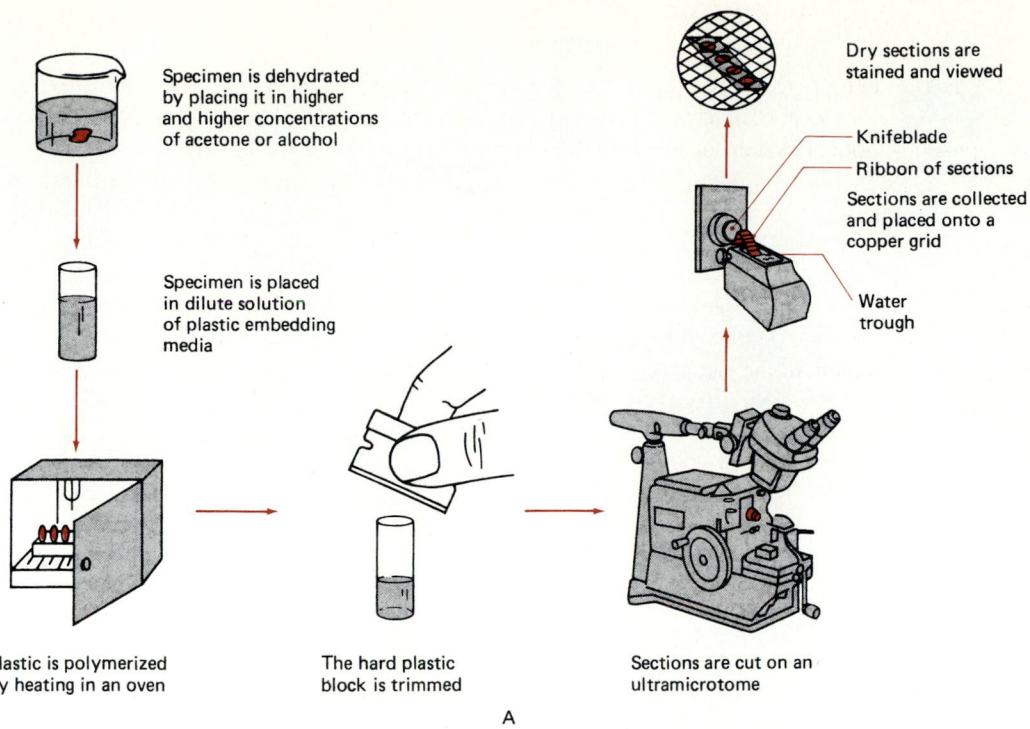

Specimen is dehydrated by placing it in higher and higher concentrations of acetone or alcohol

Specimen is placed in dilute solution of plastic embedding media

Dry sections are stained and viewed

Knifeblade

Ribbon of sections

Sections are collected and placed onto a copper grid

Water trough

Plastic is polymerized by heating in an oven

The hard plastic block is trimmed

Sections are cut on an ultramicrotome

A

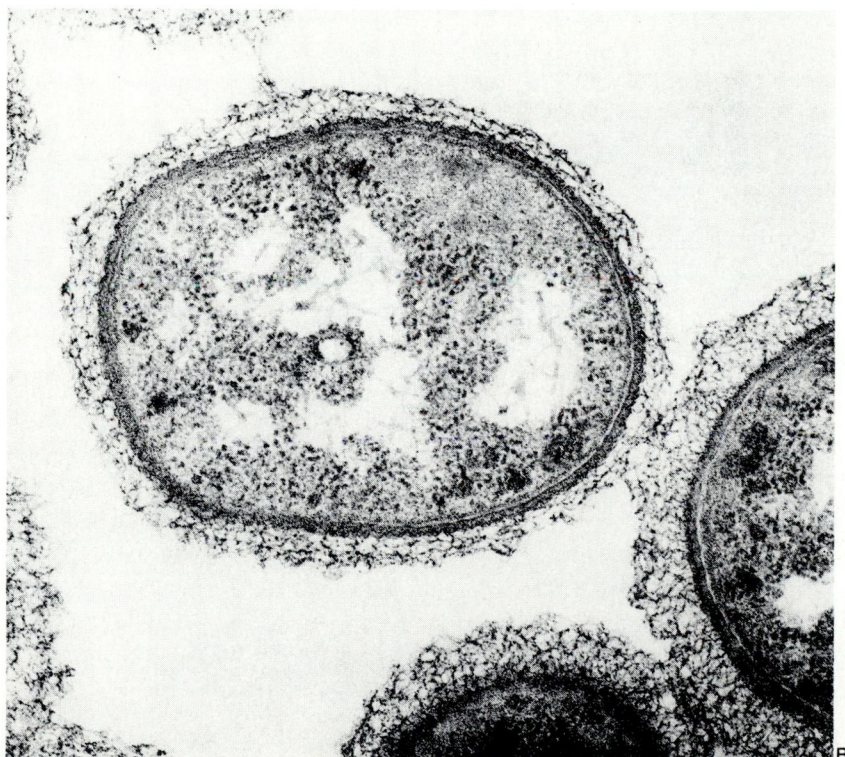

B

FIGURE 2.17

(A) Preparation of a specimen for viewing by transmission electron microscopy. The elaborate preparation requires far more time than the actual visualization of the specimen.
(B) The detailed structure of a bacterium is seen in this electron micrograph of a thin section of *Pseudomonas aeruginosa* (120,000×) viewed with a TEM (From BPS—John J. Cardamone, Jr., University of Pittsburgh)

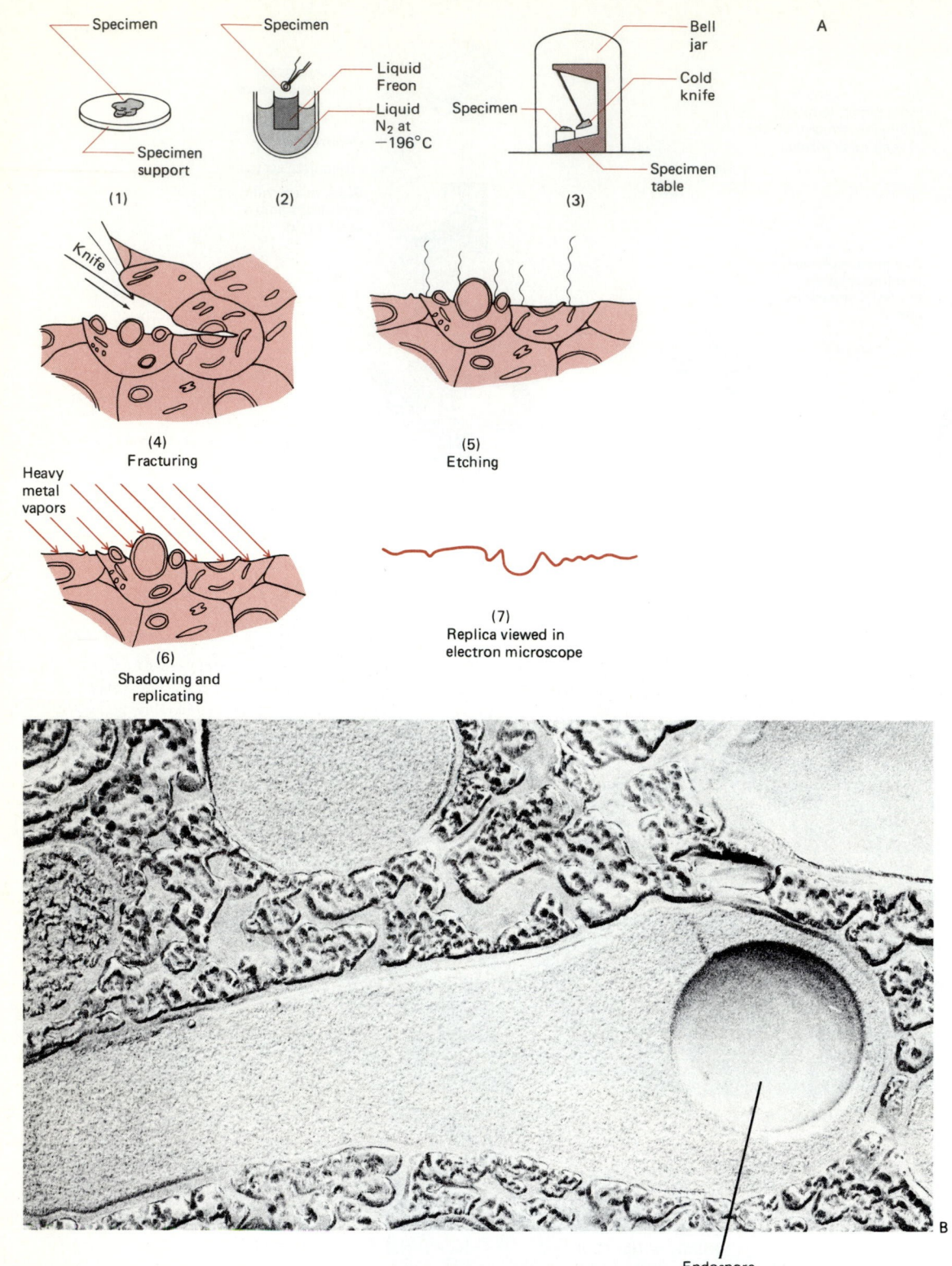

FIGURE 2.18

(A) Procedure for the formation of freeze-fracture replicas, used for visualizing surface structures in conjunction with transmission electron microscopy. (B) Electron micrograph of the endospore-forming bacterium *Clostridium botulinum* (type A) following freeze-etching as viewed under a TEM ($55,000 \times$).

Note the prominence and topographical relief of the terminal endospore. (From BPS—T.J. Beveridge, University of Guelph)

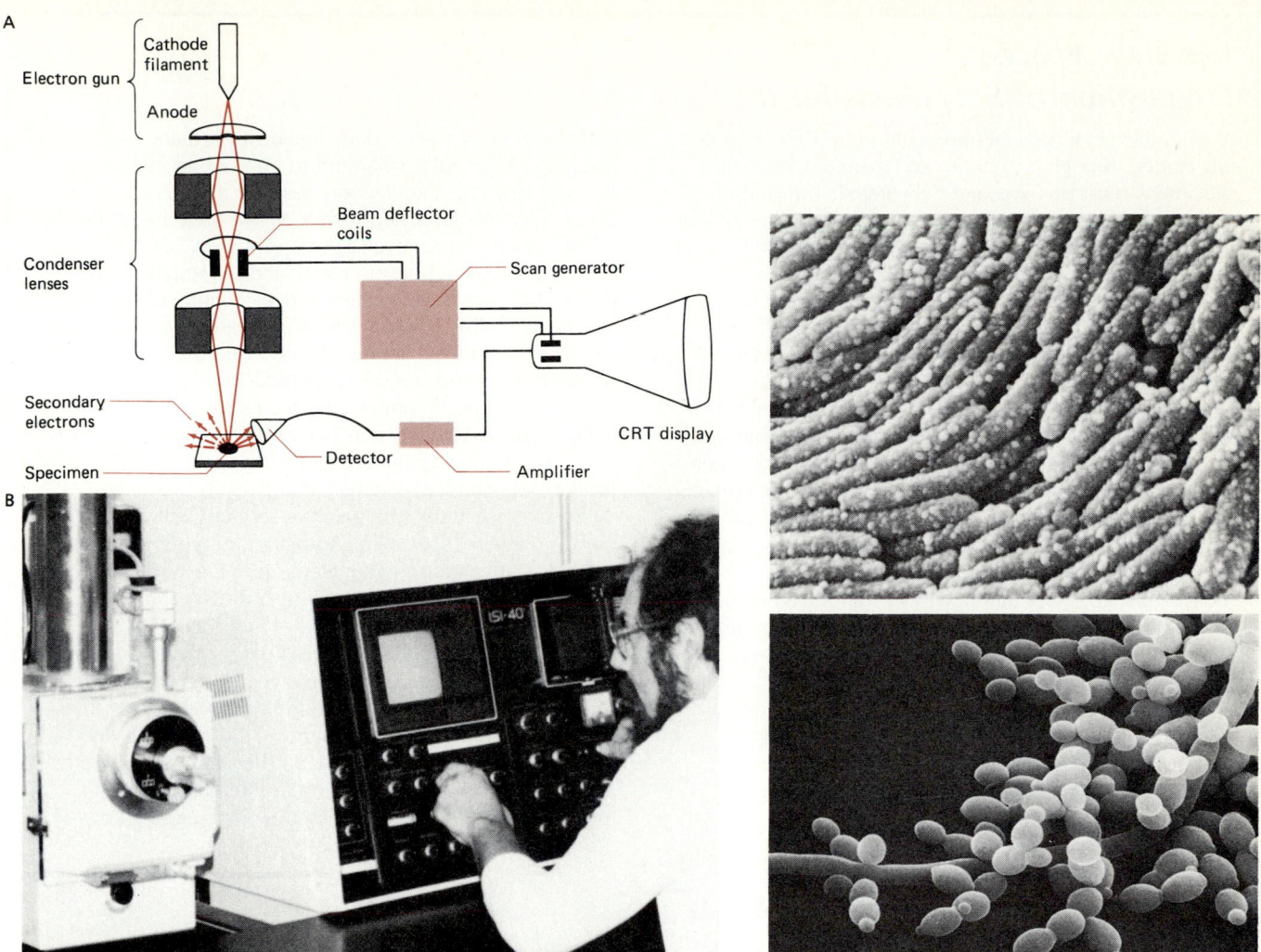

FIGURE 2.19
The SEM is used for viewing surface structures and their three-dimensional spatial relationships. (A) Diagram of SEM. (B) Photograph of SEM.

FIGURE 2.20
Scanning electron micrographs of microorganisms, showing their three-dimensional appearance. (A) Note the bumps on the outer membranes of these rod-shaped bacteria. (B) *Candida albicans* yeast and hyphal phases (3,400×). (A from BPS—Z. Skobe, Forsythe Dental Center; B from BPS—Garry T. Cole, University of Texas, Austin)

CRT screen gives a three-dimensional appearance, highlighting the topography of the specimen surface as seen with the SEM (Figure 2.20).

Magnification in a SEM is not achieved through lenses, as is the case in both the light microscope and the TEM. Rather, magnification in the SEM is determined by the ratio of the length of the scan across the specimen surface to the length of the scan of the CRT. If the electron beam scans 100 nm across the specimen and the corresponding image displayed on a CRT screen has a length of 100 mm, the magnification will be 100,000× (100 mm divided by 100 nm), whereas if the scan across the specimen is only 10 nm and the CRT screen is the same 100 mm, the magnification is 1,000,000× (100 mm divided by 10 nm). Thus, one de-termines the magnification using a single CRT screen by adjusting the scan distance across the specimen. The primary beam that scans across the specimen is synchronized with the scan across the CRT display so that the image on the CRT screen is an accurate reproduction of the scanning image.

As with other forms of microscopy, resolution is essential for achieving useful magnification. Resolution of the SEM depends on the size of the phosphorescent dots that are used to illuminate the CRT screen and the size (diameter) of the primary electron beam. Today SEMs are capable of achieving resolutions comparable to those of the TEM; commonly, a magnification of 10,000–100,000× with a resolution of 1–10 nm is achieved.

ANALYTICAL PROCESS
Preparation of Specimens for the SEM

As with the TEM, the electron beam in the SEM must be transmitted through a vacuum, and therefore, biological specimens must be fixed and dehydrated. The goal of specimen preparation is to retain the natural shape of the living organism when it is viewed in the SEM. Artifacts may be easily created when biological specimens are dehydrated during preparation for viewing in the SEM. Particularly destructive effects result from the surface tension properties of the air–water interface when specimens are simply air dried.

Critical point drying is normally utilized to minimize artifact creation because this technique avoids distortions caused by the effects of surface tension on the specimen when water is removed. This procedure takes advantage of the fact that there is a critical point, in terms of temperature and pressure, for a solvent where the liquid in a specimen and the gaseous phase over it become indistinguishable, permitting conversion of the liquid to a gas without any surface tension damage to the specimen. When all of the gas is removed, the specimen is dry. In critical point drying, the tissue water is replaced by a liquid with a low critical point, such as liquid carbon dioxide. The specimen is first immersed in ethanol or acetone to remove water, and then within a sealed chamber the dehydrating solvent is replaced by using pressurized liquid CO_2 as the transitional liquid. The temperature is raised above 32°C, the critical point of CO_2, and the fluid vaporizes without any surface tension. The pressure and gas are released slowly, leaving a dry, undistorted specimen.

Once the specimen is dehydrated, it is coated with a layer of gold or gold-palladium to improve its characteristics as a target for the electron beam of the SEM. Coating is normally accomplished with a device called a *sputter coater* which ensures even coverage with a thin film of metal by vaporizing the metal under vacuum and depositing it on the specimen. Proper coating of the specimen minimizes electronic charging of the surface, a serious problem in scanning electron microscopy that can produce an intense emission of electrons that precludes the achievement of contrast and the viewing of surface detail. Surface charge occurs because biological specimens have nonconducting surfaces. Coating with a metal produces a conductive surface that permits dissipation of the electrons without charging the surface. Following metallic coating, specimens normally are mounted and viewed. Unlike transmission electron microscopy, thin sectioning is unnecessary because only the surface structure is effectively viewed.

It is possible, though, to expose and then view subsurface layers. This can be accomplished by cryofracturing the specimen. In this technique the specimen is frozen at a very low temperature, usually in liquid nitrogen, and then fractured with a sharp blade. The specimen fractures along planes that correspond to the internal surfaces of the organism. The organism can then be coated with a suitable metal and viewed with the SEM.

Postlude

The development of various microscopes for viewing microorganisms, each with unique strengths and weaknesses, has gone hand in hand with the unraveling of the mysteries of the microbial world. The critical concerns of the microbiologist are that the image produced in the microscope is large enough to permit visualization of the organism being studied, that the detail of the specimen is resolved in the microscopic image, and that at the same time as the image is magnified, it is not distorted, nor are artifacts produced. Obviously, it is critical to know that what you see through the microscope is real. The ability to visualize the fine ultrastructure of microorganisms, as is accomplished with an electron microscope, must be matched with the critical intellectual ability of the microbiologist to interpret properly the significance of what is seen. Advances in microbiology are tied to developments in methodology and technology, but it is the creativity of the microbiologist using these tools that is essential for scientific discovery.

Suggested Supplementary Readings

ALDRICH, H. C., and W. J. TODD (eds.). 1986. *Ultrastructure Techniques for Microorganisms*. Plenum Press, New York.

BADBURY, S. 1984. *An Introduction to the Optical Microscope*. Oxford University Press, Oxford, England.

BERLYN, G. P., and J. P. M. MIKSCHE. 1976. *Botanical Microtechnique and Cytochemistry*. The Iowa State University Press, Ames, Iowa.

BURRELLS, W. 1978. *Microscope Techniques: A Comprehensive Handbook for General and Applied Microscopy*. Halsted Press, New York.

COLLINS, C. H., and P. M. LYNE. 1976. *Microbiological Methods.* Butterworth, Woburn, Mass.

GERHARDT, P. (ed.). 1981. *Manual of Methods for General Bacteriology.* American Society for Microbiology, Washington, D.C.

GOODHEW, P. J., and L. E. CARTWRIGHT. 1975. *Electron Microscopy and Analysis.* Crane, Rusack and Co., Inc., New York.

HAYAT, M. A. 1978. *Introduction to Biological Scanning Electron Microscopy.* University Park Press, Baltimore.

JAMES, J. 1976. *Light Microscopic Techniques in Biology and Medicine.* Kluwer Boston, Inc., Higham, Mass.

LASKIN, A., and H. A. LECHEVALIER. 1977–1981. *Handbook of Microbiology* (4 volumes). CRC Press, Boca Raton, Fla.

MEEK, G. A. 1976. *Practical Electron Microscopy for Biologists.* John Wiley & Sons, London.

NORRIS, J. R., and D. W. RIBBONS (eds.). 1969–. *Methods in Microbiology.* Academic Press, New York.

RASH, J., and C. S. HUDSON. 1982. *Electron Microscopy Methods and Applications.* Praeger Publishers, New York.

ROCHOW, T. G., and E. G. ROCHOW. 1979. *An Introduction to Microscopy: By Means of Light, Electrons, X-Rays or Ultrasound.* Plenum Publishing Co., New York.

SHIH, G., and R. KESSEL. 1982. *Living Images: Biological Microstructures Revealed by Scanning Electron Microscopy.* Jones and Bartlett, Boston.

WISCHNITZER, S. 1981. *Introduction to Electron Microscopy.* Pergamon Press, New York.

Study Questions

1. What is resolution, and why is it important in microscopy? What factors influence resolution, and why do we consider an electron microscope superior to a light microscope?

2. What organisms and structures can be seen with a light microscope? What organisms and structures can be seen with a TEM?

3. What is meant by the term *useful magnification*?

4. Why do we stain microorganisms before viewing them with a microscope?

5. What is the difference between a simple and a differential stain?

6. Name five types of microscopes, and discuss the advantages and disadvantages of each.

7. How does an SEM differ from a TEM? What are the different applications for each of these electron microscopes?

Situational Problem

Selecting a Microscope

At some time in your career, you may need to purchase a microscope. Microscopes have a wide price range, and there are many options. You must determine the applications for which the microscope will be used and the technical requirements for those applications. You will find that performance and cost depend largely on the objective lenses and any special applications, such as phase contrast capability. Once your own requirements for a microscope are established, you can begin to consult the microscope catalogs available from your departmental office, scientific buyer or purchasing department, and/or a scientific supply house or microscope company salesperson to obtain the necessary information concerning available options and costs.

Knowing the meaning of the descriptive terms used in microscope catalogs and the jargon used by microscope salespersons is critical to making the right choice. As many of us discovered when we first purchased microscopes, the terminology used in microscope catalogs and the buzzwords used by microscope salespersons often are not found or are inadequately described in introductory microbiology textbooks. Following are some examples of information provided in these catalogs. Do not be surprised if you have to consult a dictionary or specific reference volume on microscopy to understand some of the terms used.

By using the sample catalog information provided here and/or by consulting actual microscope catalogs, let us compare the capabilities and relative costs of the microscopes used in the introductory biology and microbiology laboratories. Try to determine whether the microscopes are adequate or excessive in their technical capabilities for their respective applications.

Next, assume that you have just been accepted by the medical school of your choice. In the letter of notification, you are informed that you will be required to purchase your own microscope. For some of you, this exercise may soon become reality. By using the sample catalog information provided here, by examining microscope catalogs, and by making inquiries of local microscope suppliers, decide what you will need and what your microscope will cost. Purchase of a microscope normally is "a la carte," that is, you must select each component separately to construct your personalized microscope. The basic microscope has a stand, observation tube, eyepieces, stage, condenser, nosepiece, and several objectives; additional options to consider are phase contrast and dark-field capabilities if you require them.

To determine your microscope requirements, assume that you will be taking courses in histology and microbiology. You may wish to make this exercise more realistic by consulting the catalog of a medical school to determine the courses you would actually be taking, which may be particularly fruitful if you are indeed planning to enter medical school in the near future. Based upon the sorts of microscopic observations you anticipate being required to make in these courses, you can determine the resolving and magnifying capabilities that you will need and whether you should add special options such as phase contrast. You will then be able to choose the lenses that you need and add up the costs. Pay careful attention to the extra cost needed to obtain increased resolution. Don't be shocked by the total cost. Some entering medical students purchase used or demonstrator microscopes at a substantial reduction.

Filling in the following table may prove helpful.

Item	Specification	Cost ($)
Stand	_____	_____
Observation tube	_____	_____
Eyepieces	_____	_____
Stage	_____	_____
Condenser(s)	_____	_____
	_____	_____
	_____	_____
Nosepiece	_____	_____
Objectives	_____	_____
	_____	_____
	_____	_____
	_____	_____
	_____	_____
Other	_____	_____
TOTAL	_____	_____

SAMPLE MICROSCOPE CATALOG

Stand

The standard microscope stand is very solid, with a stable frame and interchangeable dovetail nosepiece mount. The base has coaxial coarse and fine focus controls that operate over a range of 30 mm with fine focus knobs graduated in increments of 0.002 mm. The light source gives Koehler-type illumination, with a built-in precentered tungsten halogen light source and variable intensity control. There is a built-in, graduated, centerable field iris diaphragm and rack-and-pinion vertical condenser control with centering screws. The stage mount accepts interchangeable stages. The stand comes complete with dust cover, immersion oil, connecting cord, instruction manual, and warranty card. ... $600.00

Observation Tube

The binocular observation tube has parallel eyepiece tubes with an inclination of 30°, high transmission coated prisms, interpupillary distance adjustment 53mm-75mm, automatic tube length compensation and single diopter control ± 5 diopters that are rotatable 360° ... $400.00

Eyepieces

Widefield High Eyepoint Eyepiece 8×, paired ... $150.00
Widefield High Eyepoint Eyepiece 10×, paired ... $150.00
Widefield High Eyepoint Eyepiece 15×, paired ... $200.00

Stage

Both right-hand and left-hand square graduated mechanical stages are available in the 180 by 135-mm size. The stage will move 50 mm in the Y direction and 76 mm in the X direction. ... $200.00

Condensers

Available condensers include an achromatic/aplanatic condenser (NA 1.40) and an Abbe condenser (NA 1.25). The achromatic/aplanatic condenser is corrected for chromatic aberration, spherical aberration, and curvature of field, and is provided with a graduated aperture iris diaphragm, decenterable for oblique illumination. It is recommended for work with high-quality apochromats and plan apochromats. These condensers can be replaced with dark-field, phase contrast, or other special condensers.

Abbe Condenser, NA 1.25, with graduated variable aperture diaphragm ... $90.00
Achromat/aplanat Condenser, NA 1.30, with graduate aperture diaphragm ... $200.00
Dark-field immersion condenser, variable NA 1.2–1.4 ... $380.00
Phase contrast turret condenser, NA 1.25, with four centerable phase annuli, one bright-field position with aperture diaphragm, and built-in centering screws ... $450.00
Centerable phase annulus for phase 10× objectives ... $50.00
Centerable phase annulus for phase 20× objectives ... $50.00
Centerable phase annulus for phase 40× objectives ... $50.00
Centerable phase annulus for phase 100× objectives ... $50.00
Centering telescope CT-5 for aligning phase rings ... $80.00

Nosepieces

Five-position revolving ball-bearing nosepiece	$140.00
Six-position revolving ball-bearing nosepiece	$190.00

Objectives

Microscope objectives are categorized according to their optical performance and applications. The performance of an objective depends upon its resolving power, which is influenced by the NA and several aberrations. The design of an objective attempts to achieve high resolving power with few aberrations. Three main types of objectives are achromats, fluorites (semiapochromats), and apochromats. Achromats are corrected for chromatic aberration of two colors—red and blue. These lenses are adequate for microscopy that does not require especially high resolution and photomicrography in color. Achromatic objectives are also available as plan achromats, with special additional correction for curvature of field. Fluorites or semi-apochromats incorporate the mineral fluorite into the objectives. Fluorites make possible the attainment of a high order of correction for three colors chromatically, and are preferable to achromats in case more stringent resolution is required in the center of field. Apochromatic objectives are corrected chromatically for three colors and spherically for three colors. By limiting chromatic aberration, the resolving power can be enhanced by increasing the NA 30% to 50% over a chromatic lens. Therefore the apochromat is capable of forming a very sharp image with increased resolution of details that can support excellent color photomicrography. Plan (flat-field) objectives are capable of producing a flat image to the edge of the field. Plan achromats and plan apochromats are superior for the visual observation and photomicrography of flat objects such as stained smears and sections.

Plan achromat, bright-field objectives

4×, NA 0.13 dry	$160.00
10×, NA 0.30, dry	$300.00
20×, NA 0.46, dry, spring-loaded	$350.00
40×, NA 0.70, dry, spring-loaded	$400.00
100×, NA 1.25, oil, spring-loaded	$700.00

Plan apochromat, bright-field

4×, NA 0.16, dry	$400.00
10×, NA 0.40, dry, spring-loaded	$600.00
20×, NA 0.70, dry, spring-loaded	$700.00
40×, NA 0.95, dry, spring-loaded, correction collar	$1100.00
100×, NA 1.35, oil, spring-loaded, iris	$2000.00

Achromat flatfield

4×, NA 0.10, dry	$50.00
10×, NA 0.25, dry	$75.00
20×, NA 0.40, dry, spring-loaded	$100.00
40×, NA 0.65, dry, spring-loaded	$100.00
60×, NA 0.80, dry, correction collar	$100.00
100×, NA 1.30, oil, spring-loaded	$150.00
100×, NA 1.30, oil, spring-loaded, iris diaphram	$150.00

Phase contrast, positive low-contrast plan achromat

20×, NA 0.46, dry, spring-loaded	$500.00
40×, NA 0.70, dry, spring-loaded	$500.00
100×, NA 1.25, oil, spring-loaded	$900.00

Phase contrast, positive low-contrast achromat flatfield

10×, NA 0.25, dry	$100.00
20×, NA 0.40, dry, spring-loaded	$200.00
40×, NA 0.65, dry, spring-loaded	$200.00
100×, NA 1.30, oil, spring-loaded	$200.00

Nonfluorescing uv fluorite flatfield

10×, NA 0.40, dry, spring-loaded	$400.00
20×, NA 0.65, dry, spring-loaded	$600.00
40×, NA 0.85, dry, spring-loaded, correction collar	$700.00
100×, NA 1.30, iris, spring-loaded	$700.00

CHAPTER 3

Organization and Structure of Microorganisms

3.1 Cells: The Basic Organizational Units of Living Systems

A major characteristic of a living system is the ability to exist in a highly ordered state while interacting dynamically with its environment. From their surroundings, living organisms must acquire energy and materials, which they use to assemble their structural components and to maintain their structural integrity. When they fail to do so, they die and their highly organized state disappears. In this chapter we consider the properties of the functional and structural subunit of all living organisms—the **cell**—and how this basic unit of organization permits an organism to obtain and process the energy and materials it requires for life.

A cell is a self-contained unit separated from its surroundings by a membrane that serves as a limiting boundary. The membrane regulates the flow of materials into and out of the cell, allowing the critical maintenance of the cell's internal contents in a more highly organized state than the cell's external surroundings. The specific organizational pattern of the cell, including its metabolic capabilities and structural components, is determined by the genetic information contained in that cell. This genetic information permits each cell to reproduce its own organizational pattern by itself. The ability to reproduce is another essential attribute of a living organism.

Microorganisms can be classified on the basis of their cellular organization (Table 3.1). Although the cell is the basic unit of all living organisms, some microorganisms are acellular, that is, they lack cells. These microbes are the **viruses**, **viroids**, and **prions**, which will be discussed in Chapter 9. They do not have their own boundary membranes, and are therefore precluded from carrying out the organized biochemical reactions required for replication unless they enter the cell of another organism. Within the confines of a host cell, acellular microorganisms are capable of acting as living systems because the host cell provides the essential boundary

TABLE 3.1 Organizational Structure of the Major Groups of Microorganisms

Microbial Group	Structural Organization	Essential Macromolecules
Prions	Acellular; no protective coat around protein	Protein
Viroids	Acellular; no protective coat around nucleic acid	RNA
Viruses	Acellular; protein coat surrounds nucleic acid	DNA or RNA, protein
Bacteria	Prokaryotic cell	Nucleic acids (DNA + RNA), protein, lipid, carbohydrate
Fungi	Eukaryotic cell	Nucleic acids (DNA + RNA), protein, lipid, carbohydrate
Algae	Eukaryotic cell	Nucleic acids (DNA + RNA), protein, lipid, carbohydrate
Protozoa	Eukaryotic cell	Nucleic acids (DNA + RNA), protein, lipid, carbohydrate

membrane. Outside such host cells, acellular organisms are incapable of replication and act as nonliving entities. Because acellular organisms are so intimately tied to host cells, they can be thought of as extracellular genetic elements or degenerate portions of the very cells that act as their hosts for reproduction. They can also be the subject of unresolvable philosophical debate concerning whether they should be considered living, dead, or even organisms at all.

In contrast to acellular microbes, all other living organisms are composed of either prokaryotic or eukaryotic cells (Figure 3.1). A **prokaryotic cell** is a simply organized cell, lacking the specialized internal membrane–bound compartments known as **organelles** that characterize the more complex and larger eukaryotic cell. In particular, the genetic information of a prokaryotic cell is not contained within the specialized organelle called the *nucleus, prokaryotic* meaning "before the nucleus." A prokaryotic cell's **deoxyribonucleic acid (DNA)**, the double helical macromolecule that stores a cell's genetic information (genome),[1] is not separated by a membrane barrier from the rest of the cell constituents. Both archaebacteria and eubacteria have prokaryotic cells. Hence, all organisms with prokaryotic cells are bacteria, and conversely, all bacteria have prokaryotic cells.

In contrast to the bacteria,—plants, animals, protozoa, algae, and fungi all have eukaryotic cells. The genome of a

[1] The genome is a single complete set of a cell's genetic information.

eukaryotic cell is contained within a nucleus, and, by definition, all eukaryotic cells at some time have a nucleus, whereas prokaryotic cells never possess this organelle; *eukaryotic* means "true nucleus." Eukaryotic cells also have numerous other membrane-bound organelles that serve specialized functions (Table 3.2). Several cellular functions, including many of the metabolic reactions involved in the generation of energy, occur within membrane-bound organelles in eukaryotic cells, whereas comparable activities are not so structurally separated within prokaryotic cells.

The differences between prokaryotic and eukaryotic cells represent a fundamental division of living systems. Presumably, as cells evolved into more advanced forms, greater separation of function was needed to carry out the operations of the cell efficiently; hence, the eukaryotic cell acquired organelles, probably from prokaryotic cells by endosymbiotic evolution, as discussed in Chapter 1. Because of their differences in cellular organization, eukaryotic and prokaryotic microorganisms possess different structures and strategies for carrying out essentially the same physiological and reproductive functions, which include the generation of energy in the form of **adenosine triphosphate (ATP)**—the principal molecule used for transferring energy within cells—and the storage and expression of genetic information.

Besides their physiological ramifications, the differences between organisms with prokaryotic versus eukaryotic cells have great practical importance. For example, the ability to

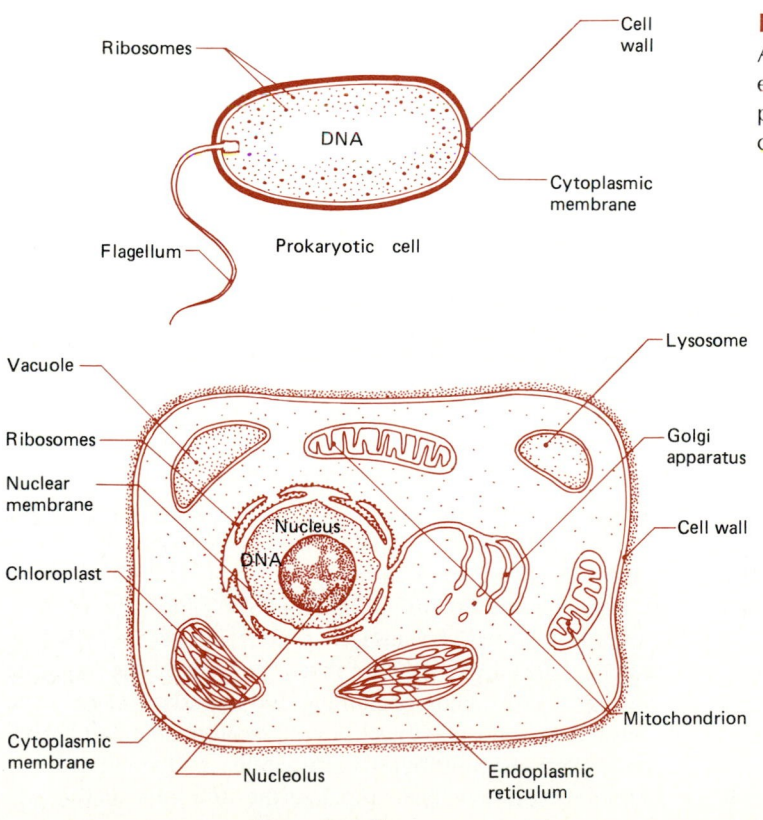

FIGURE 3.1

A comparison of structural organization reveals that the eukaryotic cell has far more internal organization than the prokaryotic cell; many of the organelles found in eukaryotic cells do not occur in prokaryotic cells.

TABLE 3.2   Descriptions of some Membrane-bound Organelles found in Eukaryotic Cells that do not occur in Prokaryotic Cells

Organelle	Description	Organelle	Description
Nucleus	Stores the genetic information (genome) of the eukaryotic cell; within the nucleus, genetic information is processed before it is sent out for use in directing protein synthesis.	Endoplasmic reticulum	Extensive membrane network used to coordinate the flow of material within the cell. Proteins made at ribosomes attached to the surface of the endoplasmic reticulum move through its tubular structure to other organelles.
Mitochondrion	Site of respiratory ATP generation. Hydrogen ions are pumped across the inner membrane of the mitochondrion to establish the electrochemical gradient needed for driving the generation of ATP.	Golgi apparatus	Associated with the endoplasmic reticulum, this organelle is involved with packaging of materials for export from the cell.
Chloroplast	Site of photosynthetic ATP generation. Chlorophyll and auxiliary pigments in the chloroplast trap light energy, which is used to generate ATP. In this process, hydrogen ions are pumped across the inner membrane of the chloroplast to establish the electrochemical gradient needed for driving the generation of ATP.	Vacuoles	Various types of vacuoles occur in different cells, where they serve different functions. Some vacuoles store reserve materials, others are involved with digestive functions, and one type—the contractile vacuole—pumps water out of the cell.
		Lysosomes	Organelles that contain digestive enzymes.
		Microbodies	Organelles that contain degradative enzymes that generate hydrogen peroxide during the metabolic transformations they catalyze.

use an antimicrobial agent such as penicillin to treat human bacterial diseases is dependent on the ability to target the mode of action of such agents against specific structures found in prokaryotic bacterial cells without killing eukaryotic human cells at the same time. It is more difficult to target agents selectively against disease-causing fungi and protozoa, which, like humans, have eukaryotic cells. Therefore, we have far fewer drugs of therapeutic value and more difficulty in treating infections when eukaryotic microbes invade the human body.

3.2 Cytoplasmic Membrane: Movement of Materials Into and Out of Cells

Because a cell depends upon its ability to exchange materials with its surroundings in order to meet its physiological needs, the functioning of the **cytoplasmic membrane**—the limiting boundary of the cell—is critical for the survival of the cell. In both prokaryotic and eukaryotic cells, the primary function of the cytoplasmic membrane is to regulate the flow of material into and out of the cell.[2] The cytoplasmic membrane is a **differentially permeable barrier**, meaning that the movement of molecules across the cytoplasmic membrane is selectively restricted. Some small molecules, such as water, move across the membrane quite readily, but larger molecules, such as glucose, do not move across the membrane freely, although they can cross it via specific transport systems. The membranes themselves mediate the selective transport process, and the biochemical structure of the membrane determines which molecules can enter and leave the cell.

Structure of the Cytoplasmic Membrane

The cytoplasmic membrane is composed largely of **phospholipid**, a molecule made up of two parts—a phosphate group and a lipid. The phosphate group is negatively charged and hence is hydrophilic (literally meaning ''water loving'' because such groups are attracted to water). The lipid portion is nonpolar and therefore hydrophobic (literally meaning ''afraid of water'' because such molecules are repelled by water). Typically, the hydrophobic portions of the molecule consist of fatty acids containing 16–18 carbon atoms. The fact that the phospholipid has both hydrophobic and hydrophilic portions contributes to the ability of the cytoplasmic membrane selectively to regulate the flow of material into and out of the cell.

[2]The available surface area of the cytoplasmic membrane that encloses the cell limits the rate of exchange between a cell and its surroundings. Because the volume of a sphere = $4.189 \times (radius)^3$ and the surface area of a sphere = $12.57 \times (radius)^2$, as the radius (size) of a spherical cell increases, its volume increases much more rapidly than its surface area. If a cell is too large, it does not possess sufficient surface area to permit adequate exchange across its limiting boundary to acquire its required nutrients and remove its waste products.

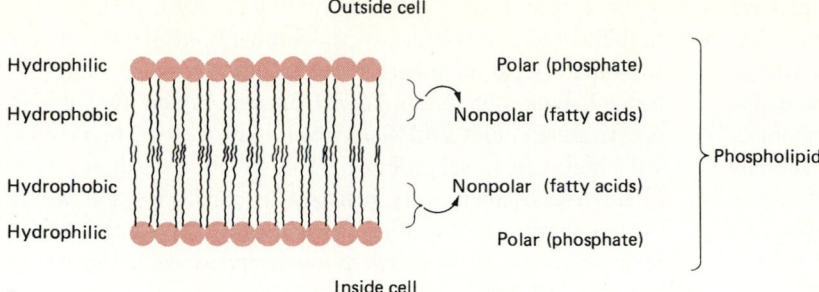

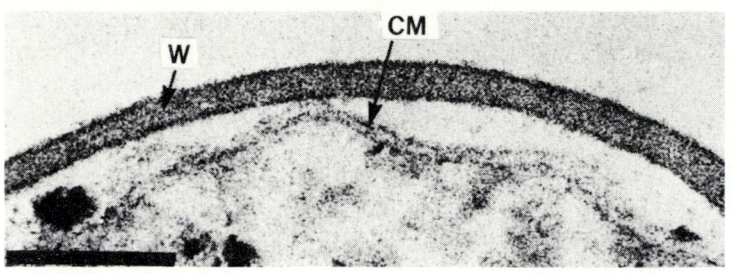

FIGURE 3.2

(A) The typical cytoplasmic membrane of both prokaryotic and eukaryotic cells is bilipid, as illustrated here showing the orientations of the hydrophilic and hydrophobic ends of the phospholipids that make up this structure. (B) The railroad track-like appearance of the cytoplasmic membrane of *Bacillus subtilus* is seen in this electron micrograph. Bar = 100 nm, W = cell wall outside of membrane, CM = cytoplasmic membrane. (From BPS— T. J. Beveridge, University of Guelph)

The cytoplasmic membranes of most eubacterial and eukaryotic cells have a bilipid structure; that is, there are two opposing layers of phospholipid that probably overlap slightly (Figure 3.2). The hydrophobic ends of the phospholipids orient toward each other and form the internal matrix of the membrane, and the hydrophilic ends point away from each other, with one layer of polar ends pointing away from the cell and the other pointing to the cell's **cytoplasm** (the fluid contents within the cell). When thin cross sections of cells are viewed with the transmission electron microscope (TEM), the cytoplasmic membrane has a railroad track appearance; the dark, rail-like portions of the membrane correspond to the electron-dense hydrophilic portions of the phospholipid molecule.

In addition to the lipid constituents of the cytoplasmic membrane, there are other biochemicals, including proteins, that are integrated into or associated with the basic membrane structure. Some of these protein molecules establish pores or channels through which substances can move across the membrane. The specific substituent groups of the amino

acids in the proteins that form these channels give the pores some degree of selectivity, and even molecules small enough to fit into and pass through the pores can be excluded on the basis of their interactions with the protein substituents. The distribution of proteins establishes a definite sidedness to the membrane so that the membrane has a distinguishable inside and outside with differing functional roles.

The Fluid Mosaic Model of the Cytoplasmic Membrane
Several models have been proposed to explain the relationship between membrane structure and function. The currently accepted **fluid mosaic model** reflects the biochemical and microscopic analyses of the membrane that have revealed the integral and dynamic relation between protein and phospholipid molecules in the membrane matrix and also explain the permeability properties of the cytoplasmic membrane. According to this model, the membrane is a bilipid layer, with the proteins associated with the membrane distributed in a mosaic-like pattern, both on the surfaces and within the membrane (Figure 3.3). Some of the proteins are confined

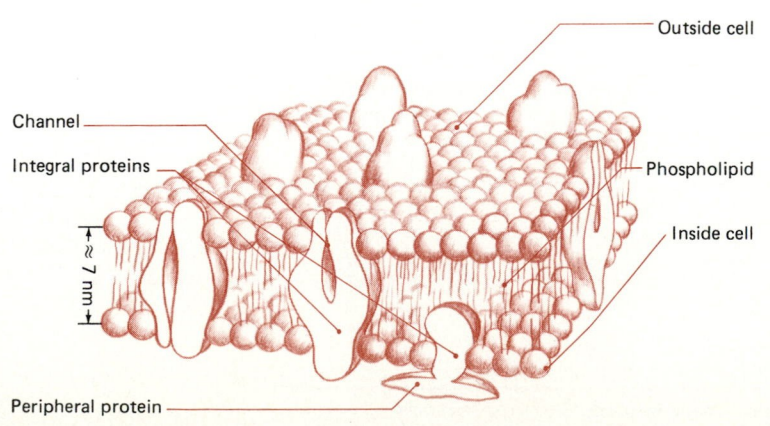

FIGURE 3.3

The fluid mosaic model of membrane structure accounts for the facts that proteins as well as phospholipids comprise an integral part of membranes and that the structure is dynamic as opposed to static.

to the membrane surfaces (extrinsic proteins), and others are partially or totally buried within the membrane matrix (intrinsic proteins). The structure of the membrane is not viewed as static. Rather, in the fluid mosaic model, both protein and lipids move within the phospholipid matrix of the membrane. Lipids held together by weak bonds can move laterally through the fluid membrane matrix. Proteins can also move laterally, but to a lesser extent than the phospholipid molecules.

Comparison of the Cytoplasmic Membranes of Eubacteria, Archaebacteria, and Eukaryotes   Although their structures are very similar, there are some differences in lipid composition between archaebacterial, eubacterial, and eukaryotic cytoplasmic membranes. Sterols, such as cholesterol, occur in eukaryotic cell membranes but generally are absent from those of prokaryotic cells (Table 3.3). The presence of sterols makes the cytoplasmic membranes of eukaryotic cells somewhat more rigid than those of prokaryotic cells. The rigidity conferred by the sterols makes the eukaryotic membrane stronger and permits eukaryotes to survive in some environments where eubacteria cannot unless they have other rigid protective structures.

Archaebacterial cytoplasmic membranes. The archaebacteria have cytoplasmic membranes with very different structures from either eubacteria or eukaryotes, and this is one of the key features that distinguishes them from all other organisms. Many of the archaebacteria live in extreme environments where unusual, physiologically specialized membranes are needed for survival. In contrast to the eubacterial and eukaryotic cytoplasmic membranes that contain straight chain fatty acids linked to glycerol by ester linkages, the cytoplasmic membranes of archaebacteria contain lipids that are branched and linked to glycerol by ether bonds (Figure 3.4). *Sulfolobus*—an archaebacterium that lives at high temperatures in acidic environments—has a cytoplasmic membrane that contains long chain branched hydrocarbons twice the length of the fatty acids found in the cytoplasmic membranes of eubacteria. The lipid chains are long enough to extend from one side of the membrane to the other, giving

it the appearance of a monolayer while concealing its true bilipid structure (Figure 3.5). Similar unusual membrane structures occur in other archaebacteria living in other extreme habitats, including *Thermoplasma*, which lives at high temperatures, and *Halobacterium*, which lives in habitats with high salt concentrations. The structure of these membranes makes them very resistant to conditions that would disrupt them and interrupt the function of a normal bilipid layer, enabling them to remain as semipermeable barriers in extreme habitats.

Transport Across the Cytoplasmic Membrane

Regardless of the specific structures of the cytoplasmic membranes of archaebacteria, eubacteria, and eukaryotes, the cytoplasmic membrane must permit selective movement across it. Organisms use different transport mechanisms for moving various biochemicals into and out of the cell and have different capabilities for transporting molecules across the cytoplasmic membrane. Some biochemicals can move across the cytoplasmic membrane by several different mechanisms.

Diffusion and Osmosis   If the concentrations of a substance are different on opposing sides of a membrane, there is a **concentration gradient** of that substance across the membrane. In such a case, the substance will attempt to move from the region of higher concentration to the one of lower concentration in order to maximize the degree of randomness that corresponds to the most stable energetic state. The movement of molecules by this process is known as **diffusion** (Figure 3.6). The cytoplasmic membrane, however, restricts free diffusion of many molecules. Because the movement of solutes (which in the case of a cell refers to substances dissolved in water) is restricted, water will move across the membrane from the region of lower to the region of higher concentration until the concentrations of the solute are equalized on both sides of the membrane or until a pressure force prevents further flow of the water. The process by which the water crosses the membrane in response to the concentration

TABLE 3.3   Differences in Membrane Lipid Composition (Shown as a percentage of membrane lipids) Between Prokaryotic and Eukaryotic Cells

	Eukaryotic Cell	Prokaryotic Cell	
Lipid	*Human erythrocyte*	Escherichia coli	Staphylococcus epidermidis
Cholesterol	25	0	0
Phosphatidyl ethanolamine	20	75	0
Phosphatidyl serine	11	0	0
Phosphatidyl choline	23	0	0
Phosphatidyl inositol	2	0	0
Phosphatidyl glycerol	0	15	67
Sphingomyelin	18	0	0
Diphosphatidyl glycerol (cardiolipin)	0	5	1
Other	2	5	32

Based on P. Komaratat and M. Kates. 1975. The lipid composition of a haloterant species of *Staphylococcus epidermidis. Biochimica et Biophysica Acta* 398:464–484.

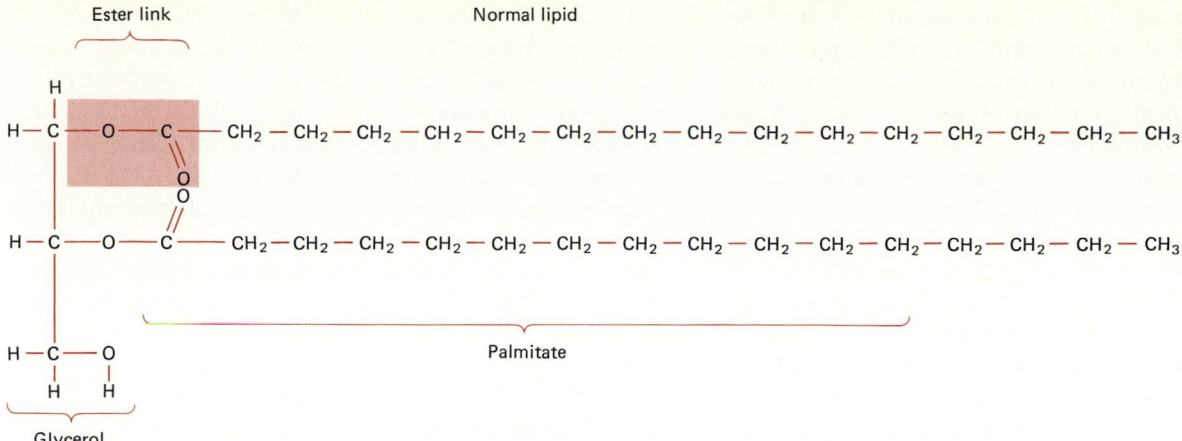

FIGURE 3.4

The membrane lipids of archaebacteria are different from the lipids found in other organisms.
The lipids of both eubacteria and eukaryotes are glycerol esters of straight chain fatty acids. Archaebacterial lipids are diethers in which a glycerol unit is connected by an ether link to phytanols,
branched chains in which carbon atoms at regular intervals carry a methyl group. Moreover,
glycerol has two optical isomers, distinguished by the configuration of the molecule around the
central carbon atom; the optical isomers rotate polarized light in opposite directions. The
configuration around the central atom of glycerol found in archaebacterial lipids is the mirror
image of the configuration found in both eubacterial and eukaryotic lipids. (Based on C. R.
Woese, 1981, *Scientific American*, 244(2):120)

FIGURE 3.5

The archaebacterial cytoplasmic membrane may be composed
of two completely overlapping lipid layers. In contrast, the
bilipid structure of the eubacterial cytoplasmic membrane
overlaps only slightly.

FIGURE 3.6

Diffusion across a membrane occurs when substances pass
through the pores of the membrane and when there is a
favorable concentration gradient; this type of transport represents the downhill flow of a substance along a concentration
gradient.

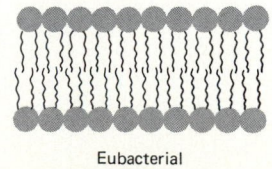

Eubacterial
bilipid
membrane

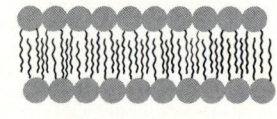

Archaebacterial
bilipid
membrane

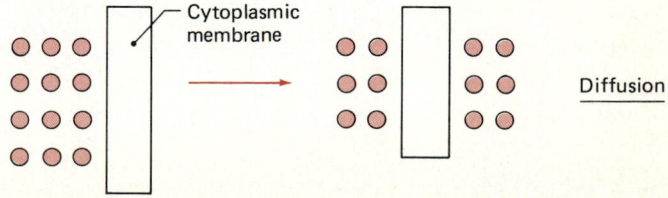

Cytoplasmic
membrane

Diffusion

gradient of the solute is known as **osmosis**. This process exerts a pressure on the membrane—known as **osmotic pressure**—that represents the force that must be exerted to maintain the concentration differences between solutions on opposite sides of the membrane.

Now let us examine the consequences of osmosis for a microbial cell. In a medium where the solute concentration inside the cell is equal to the solute concentration outside the cell, water will flow equally in both directions across the membrane (Figure 3.7). However, in a medium where the solute concentration is higher (**hypertonic**) outside than inside the cell, water will tend to flow out of the cell and the cell will tend to shrink, a process called *plasmolysis*. The reverse is true if the cell is in a medium where the solute concentration is lower (**hypotonic**) outside the membrane than inside the cell, in which case water will tend to flow into the cell, causing the cell to expand and—if unrestricted—the cytoplasmic membrane to burst (**cell lysis**). Microorganisms usually find themselves in this latter situation because the various biochemicals of the cell normally are in considerably higher concentrations than the solutes in the dilute aqueous medium in which most microorganisms exist. In order to survive, microorganisms have developed various strategies—to be discussed later in this chapter—for preventing such **osmotic shock**.

In addition to water, various small solute molecules can diffuse passively across the membrane. The rates of **passive diffusion** (unassisted movement from areas of high to areas of low concentration) are determined by the concentration gradient; the greater the concentration difference across a membrane, the more rapid the rate of passive diffusion. The rates of passive diffusion across the cytoplasmic membrane are not rapid enough for many of the exchanges of biochemicals with the environment that must be accomplished by a cell. Therefore, cells have evolved additional membrane transport mechanisms whereby molecules can move across a membrane at higher rates.

Facilitated diffusion is diffusion at an enhanced rate, that is, movement from a region of high concentration to one of low concentration that occurs more rapidly than it would based only on that concentration gradient (Figure 3.8). The reason for the enhanced rate of diffusion is that proteins, called **permeases**, within the cytoplasmic membrane selectively increase the permeability of the membrane for specific solutes, that is, the degree of ease with which that substance can cross the membrane. The mechanisms of permease action are not fully understood, but it is thought that permeases form channels through which facilitated diffusion occurs and that they act as carriers, picking up a molecule on one side of the membrane and transporting it across the membrane to the

FIGURE 3.7

Water flows into and out of cells because of osmosis in an attempt to balance the concentration of solute on each side of the membrane. In a hypertonic solution the cell shrinks, whereas in a hypotonic solution the cell swells and, if unprotected, bursts.

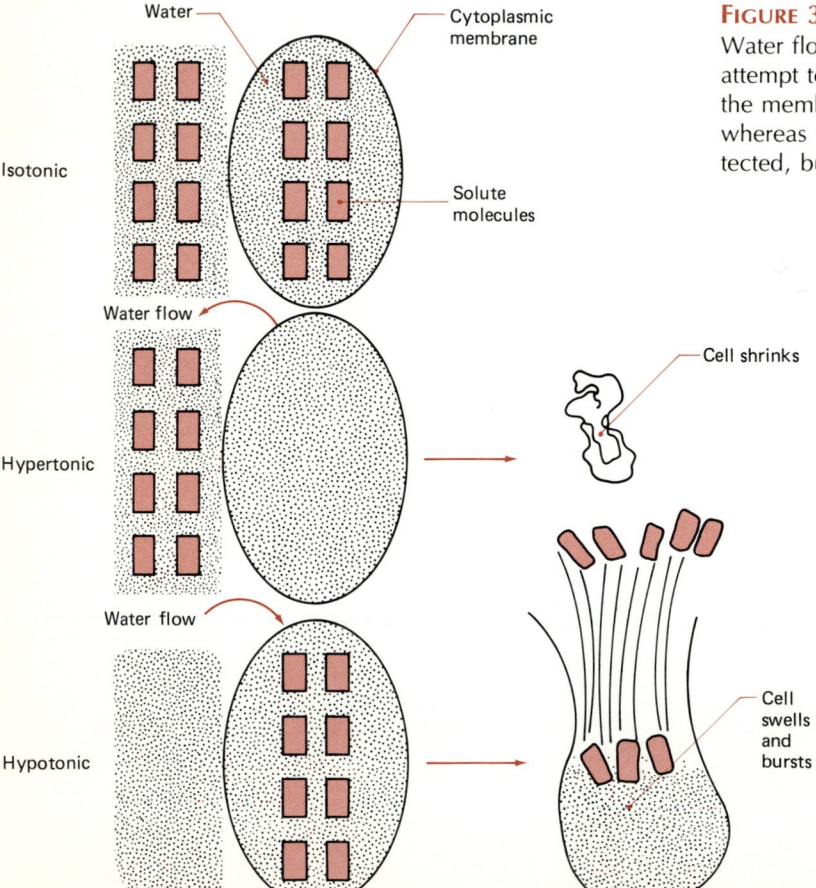

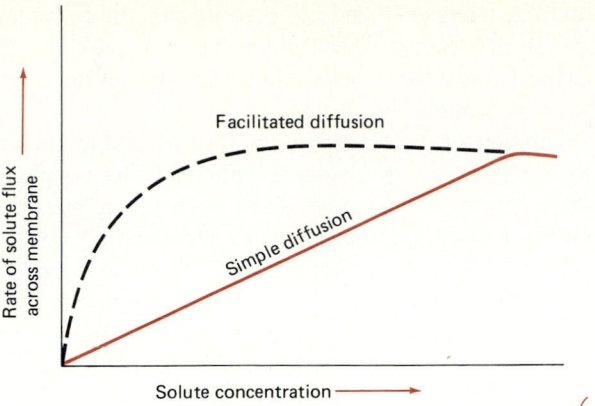

FIGURE 3.8

A comparison of the relative rates of passive and facilitated diffusion across a membrane shows that the rate of simple diffusion increases linearly with increasing solute concentration, whereas the rate of facilitated diffusion is far more rapid and exhibits saturation kinetics.

other side. It is likely that the binding of a solute to a permease alters the three-dimensional properties of the permease protein, and this change in conformation allows the solute to be moved or carried across the membrane.

Regardless of the mechanism, it should be emphasized that both passive and facilitated diffusion involve the movement of molecules down a concentration gradient that depends upon the relative concentrations of free solutes on opposing sides of the membrane. In many cases when a substance moves into a cell, it binds with other substances, forming

complexes, or is metabolically transformed. These processes prevent the buildup in concentration of the transported substance within the cell, allowing diffusion to continue along a favorable concentration gradient.

Active Transport   Not all substances move along a favorable concentration gradient. In fact, cells must often move substances across the membrane against a concentration gradient, that is, from a region of low concentration on one side of the membrane to a region of high concentration on the other. To do so requires metabolic energy. Membrane transport requiring the expenditure of energy in which the substance is not chemically modified as it is transported across the membrane is called **active transport**. This transport process can be likened to a pump that uses energy to move water uphill. In active transport, membrane proteins act as carriers to move substances across the membrane. The movement of the carrier–substrate complex across the membrane is driven by the energy of ATP or an electrochemical gradient caused by ion concentration differences across the membrane that act to change the conformational shape of the carrier so that the substrate can be moved across the membrane (Figure 3.9).

Group Translocation   In the diffusion and active transport processes discussed so far, substances are not altered by transport across the membrane. Although temporary binding with a protein may occur during passage through the membrane, the transported substance appears on both sides of the membrane in the same biochemical form. In contrast, in the transport process called **group translocation**, the substance

FIGURE 3.9

Active transport requires temporary binding of the substance with a carrier protein during transfer across the membrane and energy activation of the membrane using ATP or the electrochemical energy associated with an ionic concentration gradient across the membrane. The energy input may be used to change the shape of the carrier as it moves across the membrane.

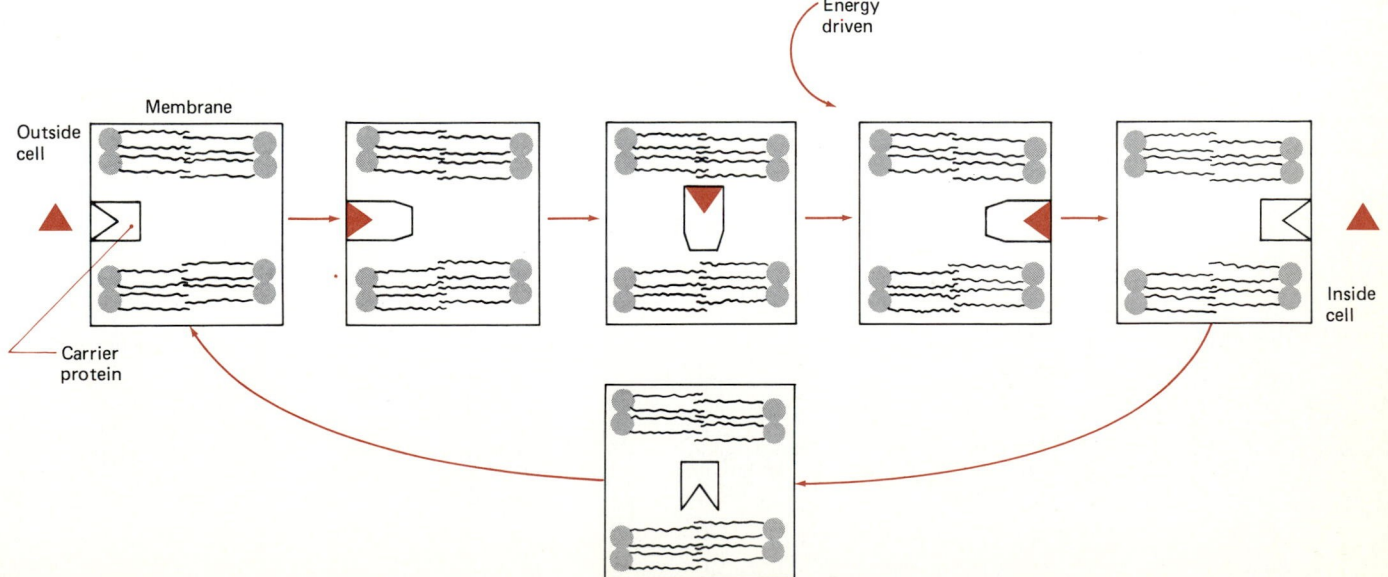

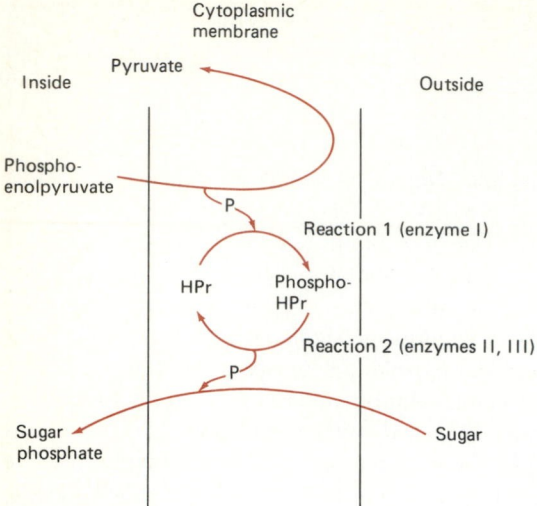

FIGURE 3.10

In the phosphoenolpyruvate-phosphotransferase (PEP-PTS) group transport system, the phosphate for the phosphorylation of the carbohydrate comes from PEP, which is a high-energy phosphate compound. There are three separate enzyme systems in this phosphotransferase system. The initial step in this process, reaction 1 catalyzed by enzyme I, involves the transfer of phosphate from PEP to a heat-stable protein molecule (HPr). This reaction occurs in the cytoplasm or at the inner surface of the membrane. Within the membrane, the phospho-HPr molecule then transfers phosphate to a carbohydrate in reaction 2. There are two enzymes, II and III, involved in the transfer of phosphate from the phospho-HPr molecule to the sugar. Different substrate-specific enzymes II and III are needed for the transport and initial phosphorylation of different carbohydrates.

is chemically altered during passage through the membrane (Figure 3.10). For example, in the **phosphotransferase system (PTS)** of the bacterium *Escherichia coli*, a sugar is phosphorylated—that is, a phosphate group is added—as it crosses the cytoplasmic membrane. Thus, when glucose is transported into a bacterial cell by PTS, it exists as glucose outside the cell but as glucose 6-phosphate within the cell because during transit across the membrane a phosphate group is added. Because carbohydrate substrates, such as glucose, are normally phosphorylated as part of their me-

tabolism to generate ATP and cell constituents, the chemical modification that occurs during group transport provides an efficient mechanism for initiating the metabolism of the compound as it is brought into the cell.

The evolution of this transport system by bacteria allows them to use their energy resources efficiently by coupling transport with the initiation of energy-generating metabolism. Group translocation seems to be restricted to prokaryotic cells; at least, it has not yet been found to occur in eukaryotes. The transport of substrates by this mechanism does not occur in all bacteria but is an important process in those bacteria that have the necessary enzymes associated with their membranes to phosphorylate substrates during their translocation across the membrane. By altering the chemical form of the substrate during transport, a more favorable concentration gradient is established. This is an important phenomenon because bacteria generally live in aqueous environments with very low concentrations of carbohydrate substrates.

Endocytosis and Exocytosis   Up to now, we have discussed mechanisms for moving substances through the membrane. Eukaryotic cells possess another mechanism, cytosis, by which substances may enter or leave a cell; prokaryotic cells do not have this capability. **Cytosis** does not involve transport through the membrane but rather relies upon the engulfment of a substance by the cytoplasmic membrane to form a **vesicle** (a membrane-bounded sphere) (Figure 3.11). This type of transport permits substances in a to sense to go around rather than through the membrane. Whether the substance is entering or leaving a cell by cytosis is designated by using the prefixes *endo-* (into) or *exo-* (out of). **Endocytosis** refers to the movement of materials into the cell and **exocytosis** denotes movement out of the cell. Both endo- and exocytosis require energy and are important in moving large substances into and out of cells where this could not be accomplished by transport through a membrane. **Phagocytosis** is a specific example of this mechanism in which one cell engulfs a smaller cell or particle. This transport mechanism is particularly important for some protozoa that feed on bacteria and for certain white blood cells that engulf and digest bacteria as part of the immune response of animals to infections.

FIGURE 3.11

In cytosis the substance is transported into the cell without actually passing through the membrane.

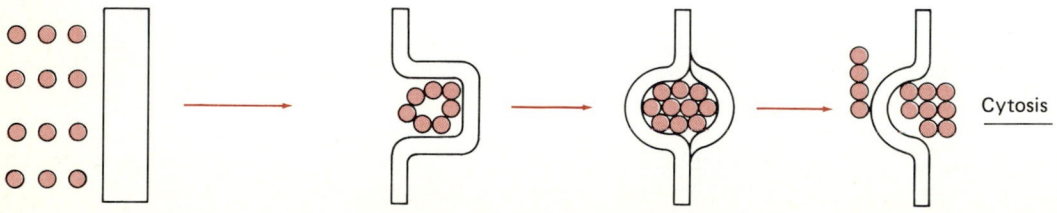

ANALYTICAL PROCESS

Mesosomes: Real Structures or Artifacts

When thin sections of bacteria are viewed by electron microscopy, extensive invaginations (infoldings) of the cytoplasmic membrane often are seen (Figure 3.12). These extensively invaginated portions of the cytoplasmic membrane have been called *mesosomes*. Careful observations indicate that mesosomes are continuous with the cytoplasmic membrane. They are frequently observed in the region where cell division occurs during bacterial reproduction. However, they are not always seen when bacterial cells divide, nor, when they are present, do they always occur near the site of cell division. Having observed mesosomes in many routine electron microscopic preparations, several investigators set out to find out their function(s), assuming that, as with other cell structures, a structure–function relationship would soon be elucidated.

Although many functions have been proposed for mesosomes, including a role in cell division, various metabolic processes, enzyme secretion, and the possibility that they provide additional necessary membrane surface within bacterial cells for functions accomplished by membranes of organelles in eukaryotic cells, the role of mesosomes has proven elusive. For each proposed function, cells were found with no apparent mesosomes that were not defective for that function. After many such studies, it became necessary to reevaluate the evidence for the existence of mesosomes, focusing on why they were only sometimes observed.

When mesosomes were observed, they were almost always found to be attached to or closely associated with DNA within the cell. Some investigators hypothesized that the DNA might be shrinking due to dehydration during preparation for electron microsopy, stretching an attached region of the cytoplasmic membrane as this occurred. To see if this was the case, cells were frozen in liquid nitrogen and then exposed to x-ray radiation to break up the DNA before the cells were dehydrated for electron microscopic viewing. When this procedure was followed, no mesosomes were observed. This suggests that these observed structures are artifacts of preparation for electron microscopic observation rather than real structures of the bacterial cell, formed by DNA pulling on the cytoplasmic membrane when the cells are dehydrated. The current view, therefore, is that mesosomes are artifacts rather than real cell structures with definable functions.

Mesosome

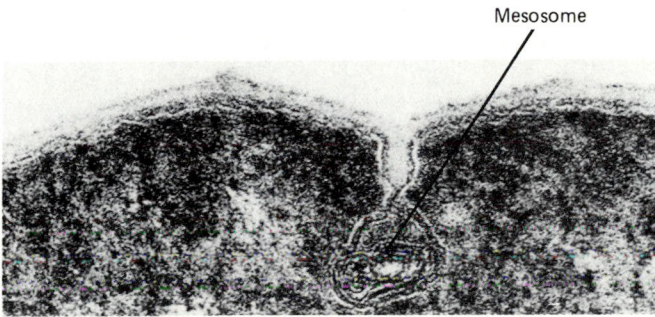

FIGURE 3.12

"Mesosomes" are seen as invaginations of the bacterial cytoplasmic membrane. These invaginations are now considered to be artifacts of specimen preparation rather than real structures. (Courtesy M. I. Higgins, Temple University; reprinted by permission of the American Society for Microbiology from M. I. Higgins and L. Daneo-Moore, 1972, *Journal of Bacteriology*, 109:1226)

3.3 External Structures that Protect the Cell

Cell Wall

Because the cytoplasmic membrane is essential for the maintenance of cell viability, it is important to protect this structure. The cytoplasmic membranes of many cells are protected by a rigid external structure called the **cell wall** that surrounds the cell. Generally, the cell wall is relatively porous so that it does not greatly restrict the flow of small molecules to or from the cytoplasmic membrane, although very large molecules usually are unable to pass across it.

The Eubacterial Cell Wall Because bacteria normally exist in dilute aqueous environments, most bacterial cells would burst from the osmotic pressure exerted on their cytoplasmic membranes were it not for their cell walls. In fact, this often occurs if the cell wall structure is disrupted, and it forms the basis for the action of some antibiotics such as penicillin. It is the rigidity of the cell wall surrounding the cytoplasmic membrane that prevents the fluid cytoplasmic membrane of the cell from expanding and bursting because of the pressure placed on it. The rigid wall also establishes the shape of a bacterial cell. Bacteria occur as spheres called **cocci**—cylinders called **rods** or **bacilli**—and spiral shapes called **spirilli**, as well as other diverse forms that typify different bac-

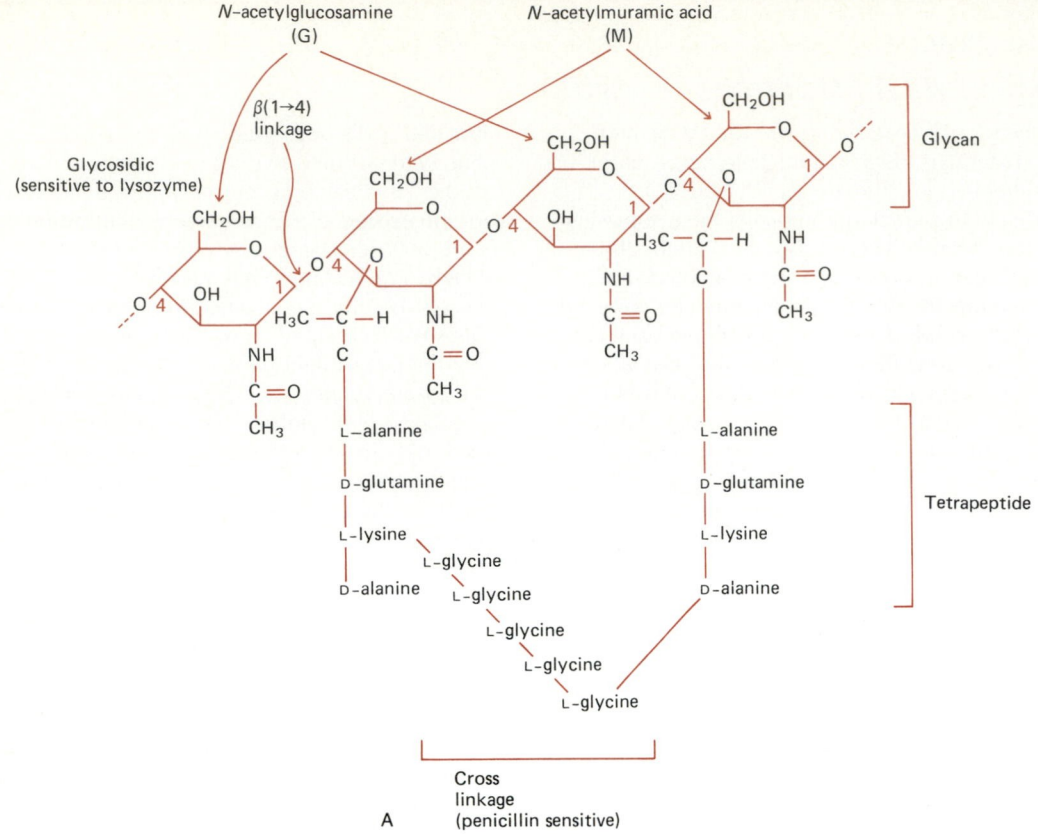

A

FIGURE 3.13
(A) Murein or peptidoglycan is the backbone biochemical of the bacterial cell wall; it is composed of repeating alternating units of N-acetylglucosamine and N-acetylmuramic acid and has cross-linked, short peptide chains, some of which have unusual amino acids. (B) The cross-links provide the needed structural support of the wall; if they are not present, as occurs in the cell walls of bacteria treated with penicillin, there is structural failure.

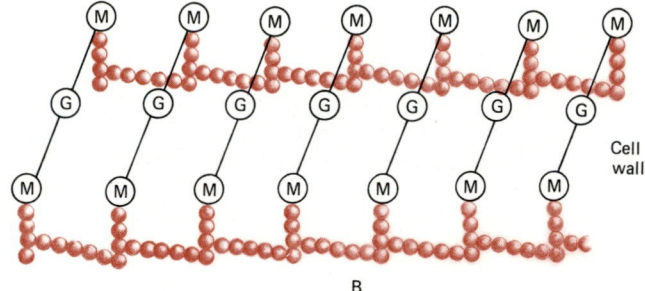

B

terial species; the diversity of bacterial forms can be seen in the micrographs throughout this book.

The cell wall of almost every eubacterial cell contains **peptidoglycan**, which is also known as **murein** or **mucopeptide**. This peptidoglycan layer is biochemically unique and is not found in archaebacteria or any eukaryotic cell. As the name implies, there are two parts to the peptidoglycan molecule—a peptide portion, which is composed of amino acids connected by peptide linkages, and a glycan or sugar portion. The glycan portion, which forms the backbone of the molecule, is composed of alternately repeating units of the amino sugars N-acetylglucosamine and N-acetylmuramic acid (Figure 3.13). Short peptide chains with four amino acids are attached to some of the N-acetylmuramic acid.

Some of the amino acids occurring in the peptide portion of the molecule are relatively unique in biological systems. These include D-amino acids and diaminopimelic acid, which

occur in peptidoglycan but not in proteins.[3] The specific amino acids comprising this peptide portion of the peptidoglycan molecule show limited variation between different bacterial species. All of the tetrapeptide sequences include L-alanine, D-glutamic acid (which may have a hydroxyl group added in some organisms), either L-lysine or diaminopimelic acid, and D-alanine. The major variation that occurs is the substitution of lysine and diaminopimelic acid, with lysine occurring in most Gram-positive bacteria and diaminopimelic acid occurring in all Gram-negative bacteria.

The tetrapeptide chains are interlinked by a peptide bridge between the carboxyl group of an amino acid in one tetra-

[3]Proteins in both eukaryotic and prokaryotic cells contain exclusively L-amino acids. The designations D- and L- for a molecule indicate that there are mirror images of that molecule; specifically, the D- and L-designations relate a structure to the D- and L- configurations of glyceraldehyde (see the appendix).

peptide chain and the amino group of an amino acid in another tetrapeptide chain. The greatest variation in the peptidoglycan occurs in these cross-linkages.[4] The cross-linkage can occur between tetrapeptides in different chains, as well as directly between adjacent tetrapeptides, so that the peptidoglycan forms a rigid, multilayered sheet. Without the cross-linkage of the peptide chains, the murein layer would not be rigid and would not protect the cell against osmotic shock.

When this cross-linkage is disrupted, the cell wall is defective and cannot adequately protect the bacterial cell against osmotic shock. It is for this reason that the antibiotic penicillin is effective in controlling bacterial infections. Penicillin prevents the formation of cross-linkages between the peptides, resulting in the production of defective cell walls and the death of growing bacteria. Penicillin, however, does not destroy preexisting cross-linkages of the peptidoglycan layer and thus has no effect on bacteria that are not growing. It should be emphasized that penicillin will work only on growing cells that contain peptidoglycan. One eubacterial genus, *Mycoplasma*, lacks a cell wall entirely. Since they lack a cell wall, members of the genus *Mycoplasma* will not be inhibited by penicillin because they lack the biochemical component that this antibiotic affects. This particular bacterial genus is also unusual in that it incorporates sterols, which it normally acquires from a eukaryotic host cell, as a component of its cytoplasmic membrane; the sterols add sufficient rigidity to the cytoplasmic membrane to permit the survival of this organism.

In contrast to penicillin, **lysozyme**—an enzyme that breaks the bonds of the backbone glycan portion of the peptidoglycan molecule—will degrade a preformed peptidoglycan molecule. This enzyme is produced by various organisms that consume bacterial cells, aiding in the digestion of the bacteria by such larger organisms. Lysozyme also occurs as part of various normal body secretions, such as tears and saliva, and is found in high concentration in egg white, providing protection against would-be bacterial invaders.

Lysozyme can destroy all or part of the cell wall structure. If a portion of the bacterial cell wall remains after lysozyme treatment, the remaining cell is called a *spheroplast*. If the cell wall is completely removed, the remaining intact bacterial cell is called a *protoplast*. Both protoplasts and spheroplasts can exist in a supporting medium of high solute concentration in which the osmotic pressure is high enough to prevent lysis, that is, bursting of the cell.

The difference between Gram-negative and Gram-positive bacteria is defined by the inherently different structure of the cell wall in these two groups of bacteria (Figure 3.14). This difference is critical because it distinguishes the major taxonomic groups of eubacteria and has practical importance with regard to the survival of bacteria in nature and the effectiveness of antibiotics against bacterial pathogens. Both Gram-positive and Gram-negative cell wall structures contain a peptidoglycan layer, but with different amounts of peptidoglycan. They further differ in which biochemicals form the cell wall complex in addition to peptidoglycan.

The Gram-positive Eubacterial Cell Wall. The **Gram-positive cell wall** has a peptidoglycan layer that is relatively thick and comprises approximately 90 percent of the cell wall (see Figure 3.14B). The rigid peptidoglycan acts to fulfill the primary protective function of the cell wall. In addition to peptidoglycan, Gram-positive cell walls generally have another component, **teichoic acids**, which are acidic polysaccharides. They contain a carbohydrate—such as glucose, phosphate, and an alcohol—either glycerol or ribitol. The teichoic acids are covalently bonded to the peptidoglycan, making them an integral part of the Gram-positive cell wall structure. Teichoic acids can bind protons, thereby maintaining the cell wall at a relatively low pH so that self-produced enzymes (autolysins) do not degrade the cell wall. They also have several other functions, such as binding metals and acting as receptor sites for some viruses.

The Gram-negative Eubacterial Cell Wall and Cell Envelope. The **Gram-negative cell wall** is biochemically far more complex (Figure 3.15) than its Gram-positive counterpart. The peptidoglycan layer of the Gram-negative cell wall is very thin and often comprises only 10 percent or less of the cell wall. Teichoic acids generally do not occur in Gram-negative bacteria, but lipoproteins (lipid linked to protein molecules) are bonded to the peptidoglycan, forming an integral part of the Gram-negative bacterial wall. Additionally, there are layers of lipopolysaccharide (lipid linked to carbohydrate molecules), phospholipids, and proteins outside the peptidoglycan layer. Although these layers have sometimes been considered to be part of the cell wall, it is now more common to view the peptidoglycan layer as the wall component of a larger, more complex structure called the **cell envelope** of the Gram-negative bacterial cell.

The cell envelope of the Gram-negative eubacterial cell extends outward from the cytoplasmic membrane to a second membrane—the **outer membrane**. The region between the two membranes is known as the **periplasmic space**.[5] Like the cytoplasmic membrane, the outer membrane is a bilipid layer containing phospholipids and proteins, but it also contains **lipopolysaccharide (LPS)**. LPSs are complex molecules composed of three distinct regions: lipid A, R core region, and O side chain (Figure 3.15B). Functionally, the **outer membrane** of the Gram-negative bacterial cell is a coarse molecular sieve that allows the diffusion of both hydrophilic and hydrophobic molecules up to a molecular weight of about 800 for *E. coli* and even higher—3,000 to

---

[4]In the Gram-positive bacterium *Staphylococcus aureus*, the cross-linkage involves a pentapeptide. In Gram-negative bacteria, cross-linkage almost always occurs by the direct bonding of diaminopimelic acid of one chain to the terminal D-alanine of another chain.

[5]Recent studies suggest that the term *periplasmic space* should be replaced with the term *periplasmic gel* to indicate that the peptidoglycan may actually fill the region between the cytoplasmic and outer membranes. The term *periplasmic gel* implies that the peptidoglycan is relatively porous and that proteins can migrate through its gel-like composition.

FIGURE 3.14
Gram-positive and Gram-negative bacteria differ in the structure of their cell walls. The Gram-positive cell wall has a relatively thick murein layer. The Gram-negative cell wall has a thin murein layer but also has an outer membrane and additional lipopolysaccharides and proteins not present in Gram-positive cell walls.

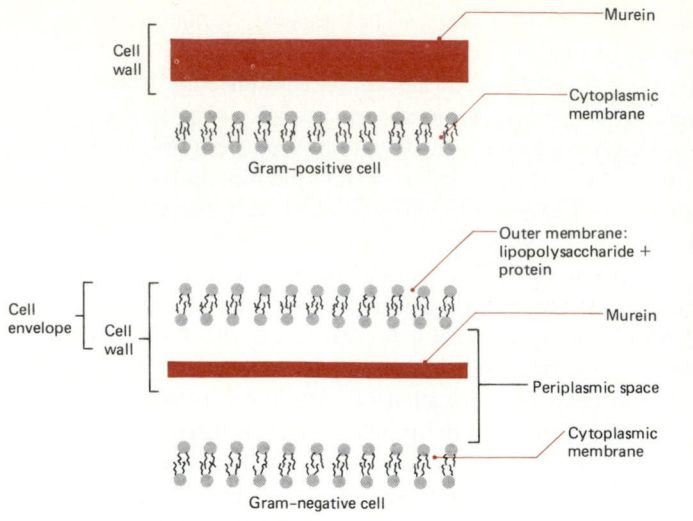

Discovery Process

The development of the Gram stain procedure by the Danish physician Hans Christian Gram remains one of the most important methodological contributions in bacteriology. At the time Gram published the description of his staining method, most bacteriologists were concerned with simply trying to see difficult-to-detect bacteria and with differentiating infecting bacteria from mammalian nuclei. Gram's paper describing the staining technique, "The Differential Staining of *Schizomycetes* in Sections and in Smear Preparations," was published in 1884. In this paper, Gram described primary staining with aniline-gentian violet, treatment with iodine-potassium iodide (Gram's iodine mordant), and decolorization with absolute alcohol followed by further decolorization with clove oil: "Bacteria are stained intense blue while the background tissues are light yellow" Gram observed that not all bacteria were stained in this procedure and suggested that counterstaining was possible. It was the detailed reporting of which bacteria were stained and which were not that was crucial to the recognition of the value of this staining procedure. "I. The following forms of schizomycetes retain the aniline-gentian violet after treatment with iodine followed by alcohol: (a) cocci of croupous pneumonia (19 cases) . . . (k) tubercle bacilli (5 cases). . . . II. The following schizomyetes are decolorized by alcohol subsequent to treatment with iodine: (a) encapsulated cocci from croupous pneumonia . . ."

Gram undoubtedly was disappointed with his procedure. He wanted to stain all bacteria, not the surrounding mammalian tissues. It is not known who thought of using the Gram stain procedure to differentiate bacteria, as routinely employed in bacteriology laboratories today. Gram died in 1935 without further developing his staining procedure but with the hope, stated in the conclusion to his paper, that "the method would be useful to other workers." Frequently in science, the discoverer of a method does not recognize the full potential of his or her discovery, and it remains for later scientists, working in an era of different concerns and enlightened by later discoveries, to realize the significance of the original finding.

Since the advent of the Gram stain, many investigations have been carried out to determine the basis for differential staining of bacteria in this procedure. The most recent investigations have used electron microscopy and heavy metal–labeled stains to see where the stains were going and what was happening to the cell. Based on such observations, it now appears that thick walls are able to trap the stain within the cell, whereas thin walls become porous when treated with a decolorizing agent, allowing the stain to wash out of the cell. Thus, a eubacterial cell with a Gram-positive wall with its thick peptidoglycan layer traps the stain, and one with a Gram-negative wall with its thin peptidoglycan layer does not.

10,000 for *Pseudomonas* and *Neisseria* species, whereas the cytoplasmic membrane excludes almost all hydrophilic substances except water. Permeability of the outer membrane to nutrients is provided in part by proteins collectively called **porins**, which, in aggregates usually of three, form cross-membrane channels through which certain small molecules can diffuse. Despite its permeability to small molecules, the outer membrane is less permeable than the cytoplasmic membrane to hydrophobic (nonpolar molecules) or amphipathic

molecules (molecules that have polar and nonpolar ends), such as phospholipids. For this reason, Gram-negative bacteria are less sensitive to antibiotics having these properties because such antibiotics cannot reach their targets in the cytoplasm, rendering the bacteria resistant.

Cell Walls of Archaebacteria   The archaebacteria do not contain murein in their cell walls: rather, their wall structures show great biochemical diversity. Some archaebacteria have

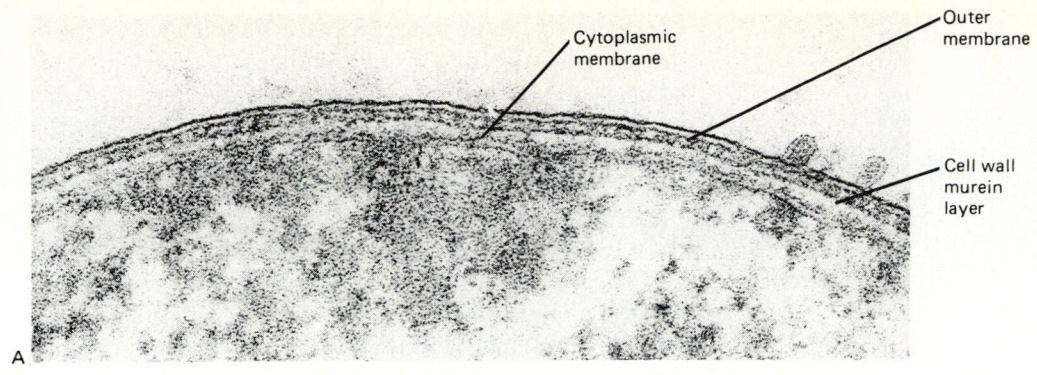

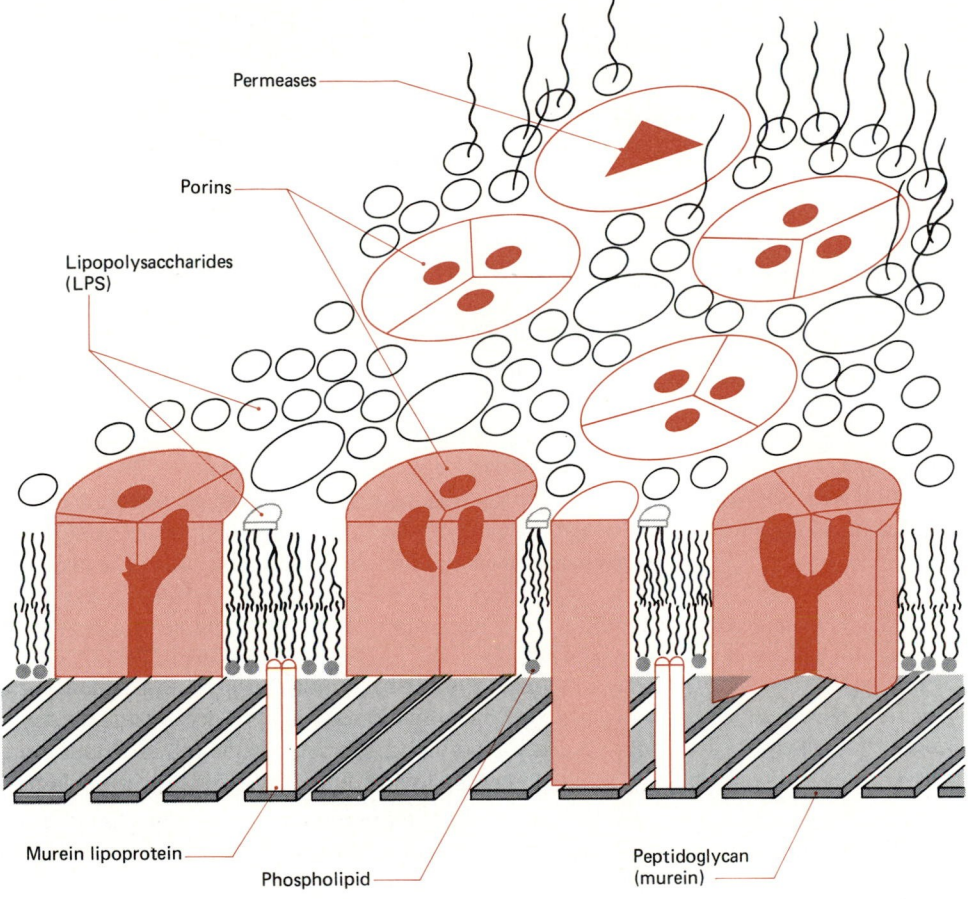

FIGURE 3.15
(A) Electron micrograph of the cell wall of *Escherichia coli*, a typical Gram-negative cell wall with a thin murein layer and an outer membrane (220,000×). (From BPS—T. J. Beveridge, University of Guelph). (B) Schematic model of the cell wall structure of *E. coli*, showing the presumed arrangement of the biochemical constituents. (Based on H. Nikaido and M. Varra, *Microbiological Reviews* 49:1–32)

walls composed of **pseudomurein**, which resembles the peptidoglycan of eubacteria but contains *N*-acetyltalosaminuronic acid instead of *N*-acetylmuramic acid and lacks the D-amino acids found in eubacterial cell walls. Other archaebacteria have walls composed of protein subunits (Figure 3.16); still others have walls with different biochemical compositions. Although there is no unifying structural composition, the walls are able to protect the cytoplasmic membranes even in the hot, acidic, and saline environments where many archaebacteria live.

Cell Walls of Eukaryotic Organisms   As with the archaebacteria, there is a great diversity of biochemical structures in the cell walls of eukaryotic microorganisms. In some cases, the cell walls of eukaryotic microorganisms protect the cell against osmotic shock, but they may also serve other functions, such as protection against physical damage. Many algae have cell walls composed primarily of cellulose, but various other polysaccharides are found as major components of some algae. Some algae have cell wall structures containing calcium or silicon, sometimes called the **test** or **frustule**.

FIGURE 3.16
Electron micrograph of the freeze-etched cell wall of the archaebacterium *Methanogenium marisuigri* (61,000×). S = surface layer composed of highly ordered protein subunits, F = flagella. (Courtesy F. Mayer, University of Gottingen)

The diatoms, for example, have frustules that are cell walls composed of silicon dioxide, protein, and polysaccharide. The frustule has two overlapping halves and distinctive markings that give these organisms their characteristically symmetrical and beautiful shapes (Figure 3.17). The coral algae deposit calcium carbonate in their wall structures, forming the basis of coral reefs. These structures protect the cell against physical damage rather than against osmotic shock, and the cell walls of these organisms are preserved long after the organisms die.

In fungi the cell walls are normally composed of microfibrils of cellulose or chitin or a combination of these biochemicals. Most fungi, including yeasts, have cell walls with distinctive biochemical characteristics. The slime molds are

FIGURE 3.17
Electron micrograph of the diatom frustule of *Xanthopyxis* (780×): Po = pore. (Reprinted by permission of Springer-Verlag, New York, from R. G. Kessel and C. Y. Shih, 1976, *Scanning Electron Microscopy for Biology*)

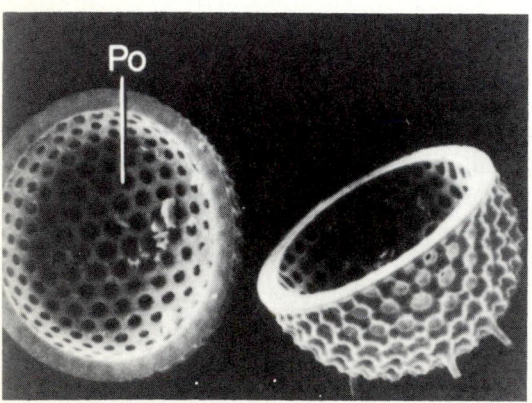

an exception; these fungi lack a cell wall and are pleomorphic, that is, they assume different forms rather than having a rigid shape. During the normal slime mold life cycle, however, the cells go through a stage in which they do have rigid cell walls. The biochemical composition of fungal cell walls is reflected in taxonomic relationships and is a useful criterion in fungal classification systems.

The protozoa usually do not have a true cell wall surrounding their membranes, and many protozoa have developed alternative mechanisms for protection against osmotic pressure. For example, *Paramecium* species have an internal structure called the **contractile vacuole** that actively pumps water out of the cell. Many protozoa do have a thin pellicle surrounding the cell that maintains the shape of the organisms. If the pellicle of a ciliate protozoan such as *Paramecium* is removed, the cell becomes spherical. Some protozoa form an outer wall or shell, composed of calcium carbonate in the foraminifera and silicon dioxide or strontium sulfate in the radiolaria that physically protects the organism. These shells are not a basis for protection against osmotic pressure, and, in fact, many foraminifera extend their cytoplasm beyond the shell.

Bacterial Capsules and Slime Layers

In addition to a cell wall, some bacteria form another protective structure called a **capsule**, which is generally composed of polysaccharides[6] that are attached externally to the cell wall and can be seen by negative staining (Figure 3.18). The capsule is especially important in protecting bacterial cells against phagocytosis by eukaryotic cells, such as by various protozoa and human white blood cells. The presence of a capsule can be a major factor in determining the **pathogenicity** of a bacterium, that is, the ability of a bacterium to cause disease in the organism that it infects. In some cases a bacterial species will have two variants, one that forms a capsule and is a virulent pathogen, and a nonencapsulated form that does not cause disease. The reason for this is that the nonencapsulated bacteria are subject to phagocytosis by blood cells involved in the immune response of the infected host organism. On the other hand, phagocytizing blood cells involved in the immune response are unable or less able to adhere to, engulf, and digest those bacteria that have capsules.

Although capsules and slime layers are often similar in composition, a distinction is made between them. **Slime layers** are not as tightly bound to the cell as capsules. These external layers may protect the cell against dehydration and a loss of nutrients. In some cases, they act as traps by restricting the flow of substrates away from the cell.

---

[6]The specific biochemical composition of the capsule varies among species of bacteria. Some *Bacillus* species produce capsules composed exclusively of glutamic acid, largely in the D form, rather than polysaccharide capsules.

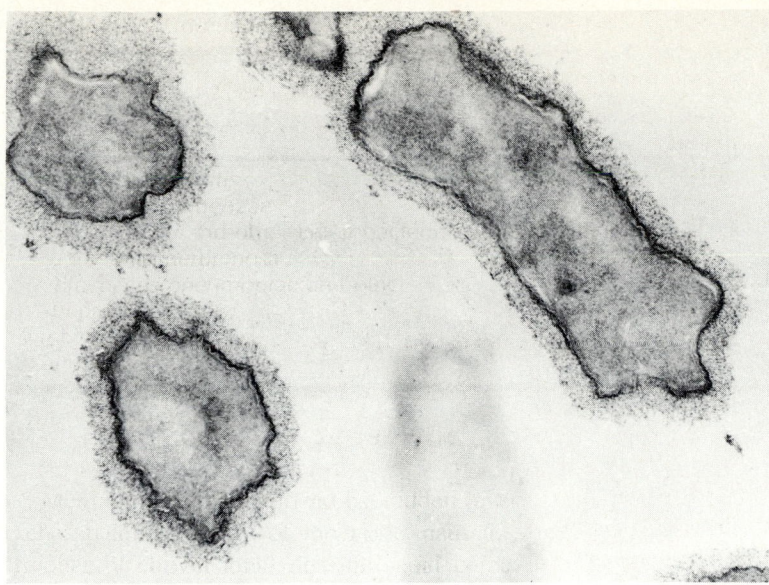

3.4 Cellular Storage of Genetic Information

By protecting cells, the cell wall and other structures described in the preceding section simultaneously protect the genetic information, which is essential for cellular survival and function. The genome of a cell, encoded within DNA macromolecules, determines the properties and structural characteristics of the cell. This is true for both eukaryotic and prokaryotic cells. As already indicated, the way in which the genetic information is stored is the prime distinction between prokaryotic and eukaryotic cells. In the eukaryotic cell, the genome is separated from the rest of the cell and is contained within the nucleus, whereas in the prokaryotic cell the DNA is not segregated from the rest of the cell constituents by a membrane barrier.

Bacterial Chromosome

The region occupied by the DNA is a prominent feature within prokaryotic cells (Figure 3.19). Most of the genetic information of the bacterial cell is contained within a **bacterial chromosome** composed of a single DNA macromolecule (Figure 3.20). The area occupied by the DNA is sometimes referred to as the **nucleoid region**, which is not a separate membrane-bound organelle. This DNA macromolecule is highly folded, but the nature of the forces maintaining this very condensed form of the bacterial chromosome is not fully understood and does not appear to be equivalent to the specific winding patterns observed in eukaryotic organisms.

The proteins normally associated with the bacterial chromosome are those involved in DNA replication, transcription of the information in the DNA molecule to RNA, and regulation of genetic expression. Although bacteria generally lack the basic histone proteins involved in the coiling of the DNA in eukaryotic chromosomes, histone-like proteins have been reported in some archaebacteria and eubacteria.

Plasmids

In addition to the bacterial chromosome, bacteria may contain one or more small, circular macromolecules of DNA known as **plasmids**. All bacterial cells contain a bacterial chromosome, but not all bacteria contain plasmids. Plasmids contain a limited amount of specific genetic information that supplements the essential genetic information contained in the bacterial chromosome. This supplemental information can be quite important, establishing such things as mating capabilities, resistance to antibiotics, and tolerance to toxic metals. Such supplemental genetic capability can permit the survival of the bacterium under conditions that are normally unfavorable for growth and survival. Although plasmids usually contain no more than 1–5 percent of the DNA found in the bacterial chromosome, the effect of this limited amount of DNA can mean 100 versus 0 percent survival if even 0.01 percent of the plasmid concerns an antibiotic resistance gene.

Pathogenic bacteria containing plasmids that code for multiple drug resistance have become a particular problem in treating some infectious diseases of humans. These bacteria are resistant to many antibiotics and can continue to grow in the body despite antibiotic treatment. On the other hand, plasmids can be quite useful and are employed in genetic engineering as carriers of genetic information from a variety of sources. Because of their relatively small size, plasmids are rather easy to manipulate. They can be isolated, genetic information from other sources can be spliced into them, and then they can be implanted into viable bacterial cells, permitting expression of the genetic information they contain in

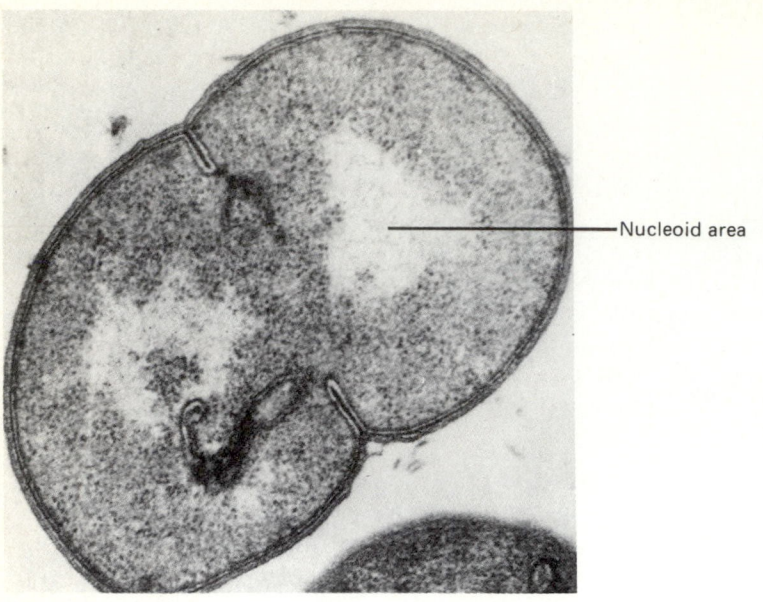

FIGURE 3.19

In this electron micrograph of dividing *Sporosarcina ureae*, the nucleoid region is seen as light areas within the cytoplasm (100,000×). (From BPS—T. J. Beveridge, University of Guelph)

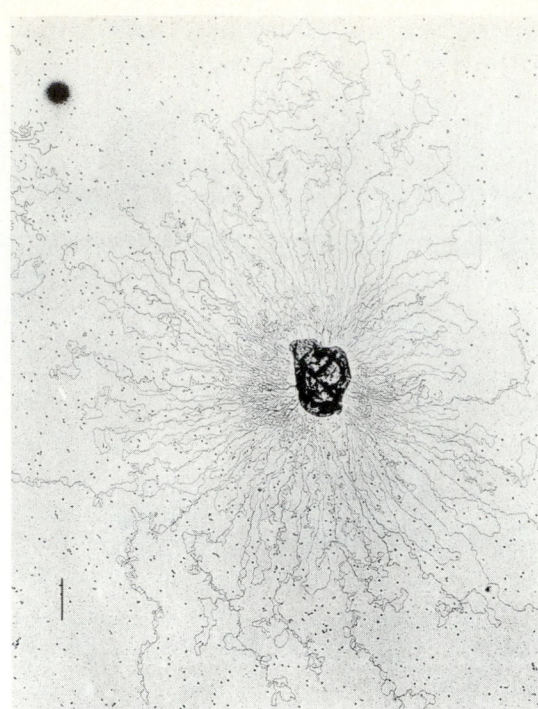

FIGURE 3.20

The mass of DNA contained within the circular loop of the bacterial chromosome is shown in this micrograph. Within a cell the DNA is maintained in a highly condensed, coiled form by interactions with proteins. When the proteins are removed, the DNA unwinds into the very long molecule shown in this micrograph. (Courtesy Ruth Kavenoff; reprinted by permission of Springer-Verlag from Kavenoff and Ryder, 1976, *Chromosoma*, 55:23)

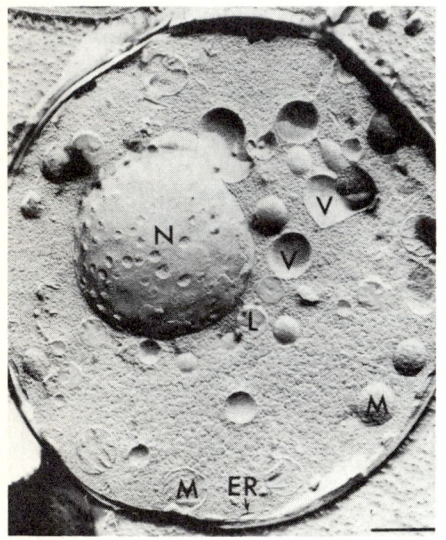

FIGURE 3.21

The double-membrane structure and pores of the nucleus are visible in these electron micrographs. An electron micrograph of a freeze-etched preparation of *Saccharomyces cerevisiae*, showing the surface of a nucleus (16,650×). Note the large number of pores in the membrane of the nucleus (N), making it look rather like a golf ball. ER = endoplasmic reticulum, L = lipid granules, M = mitochondrion, V = vacuole. (Courtesy Sam Conti, University of Massachusetts; reprinted by permission of the American Society for Microbiology from E. Guth, T. Hashimoto, and S. F. Conti, 1972, *Journal of Bacteriology*, 109:869–880).

the newly created organisms into which they are placed. Such genetic engineering has many industrial applications that will be discussed later.

Nucleus and Chromosomes of Eukaryotic Cells

In contrast to DNA within prokaryotic cells, the genome of eukaryotic cells is contained within the cell's **nucleus**, which

is separated from the rest of the cell by an inner and an outer membrane (Figure 3.21). The **nuclear membrane** is a double layer with a distinct space between the two membranes. This nuclear membrane has pores that permit the exchange of relatively large molecules between the nucleus and the cytoplasm of the cell. Nevertheless, the nuclear membrane is selective and controls which molecules pass into and out of the nucleus.

Within the nucleus the genetic information is stored in **chromosomes**, made up of **chromatin** consisting of DNA and protein.[7] All of the genetic information resides in the DNA. The more abundant protein component maintains the coiled structure of the chromatin. Unlike the bacterial chromosome, where the DNA forms a circular macromolecule, the chromosomes of eukaryotic cells contain linear DNA macromolecules arranged as a double helix. The chromatin proteins consist primarily of five cationic or basic proteins—called **histones**—that bind to the DNA by ionic interactions. The DNA coils around the histones to form subunits of the chromatin known as **nucleosomes** (Figure 3.22). Each nucleosome is composed of about 200 **nucleotides** (the structural units of nucleic acid) coiled around the histones. The resulting structures appear as spherical particles, looking like beads on a string when viewed by electron microscopy. The nucleosomes, which establish the structural configuration of eukaryotic chromosomes, are fundamental units of eukaryotic genetic material but are absent in bacterial chromosomes.[8]

The highly structured arrangement of DNA within the nucleus is critical because the genetic information coding for the synthesis of specific proteins can be widely separated in the chromosomes of eukaryotes and must be extensively processed before it can be properly interpreted. Separation from the rest of the cell, achieved by having DNA contained within the nucleus, is needed for processing of the genetic information to form a readable message.[9]

3.5 Information Flow in Cells

Ribosomes: Sites of Protein Synthesis

While the genome of an organism stores the genetic information, it is the enzymes of the cell that actually mediate the expression of that information. Enzymes are proteins, and it is, therefore, necessary for a cell to use the information stored in the DNA macromolecules to direct the synthesis of functional proteins. This requires that the information stored in the genome be sent to the **ribosomes**, the sites of protein synthesis in the cell. A typical prokaryotic cell may have 10,000 or more ribosomes, and eukaryotic cells contain considerably more. Ribosomes are intracellular particles composed of **ribosomal ribonucleic acid (rRNA)** and protein. In *E. coli*, about two-thirds of the ribosome is rRNA and the remainder is protein. During protein synthesis, the information stored in the DNA is transferred to an RNA molecule (mRNA) that acts as a messenger carrying the information to the ribosomes located in the cell's cytoplasm. There the information is translated in order to direct the synthesis of a protein.

There are significant differences between the ribosomes that occur in prokaryotic cells and those found in the cytoplasm of eukaryotic cells. The ribosomes of eukaryotic cells are larger and contain different-sized rRNA molecules than the ribosomes of prokaryotic cells (Figure 3.23). Prokaryotic cells have **70S ribosomes** composed of 50S and 30S subunits, whereas eukaryotic cells have **80S ribosomes** composed of 40S and 60S subunits.[10,11] Ribosomes are functional only when the two subunits are combined. The formation of functional ribosomes depends on the presence of magnesium ions for binding the subunits. In eukaryotic cells, the ribosomal subunits are synthesized within the nucleus in a region known as the **nucleolus** and are transported through the pores of the nuclear membrane to the cytoplasm, where assembly of the 80S ribosomes occurs.

The differences in the structural composition of prokaryotic 70S and eukaryotic 80S ribosomes forms an important basis for using antibiotics in the treatment of animal and plant diseases caused by bacteria. Protein synthesis that occurs at

---

[7]Chromosomes are visible with a light microscope only when the cell is undergoing division and the DNA is in a highly condensed form. Under these conditions the chromosomes appear as distinct thread-like structures in the nucleus; at other times the chromosomes are not condensed, and are therefore not visible.

[8]An exception to the usual eukaryotic chromosomal arrangement occurs in the dinoflagellate algae. These organisms are eukaryotic, and the DNA is contained within their nuclei, but the DNA is not associated with histones and is not supercoiled, as in the chromosomes of other eukaryotic organisms. The DNA within the nucleus of dinoflagellates resembles the nucleoid region of prokaryotic cells, but except for this feature the structure of dinoflagellates conforms to that of eukaryotic cells. Dinoflagellates may represent an evolutionary link between eubacteria and eukaryotic algae.

[9]In prokaryotic cells, where there is no nucleus, the information in the bacterial chromosome is not separated; rather, the information coding for specific proteins occurs as a contiguous sequence of nucleotides within the DNA macromolecule. Hence, the genetic information of prokaryotes does not have to be extensively processed before it can be used to code for the synthesis of proteins.

[10]As noted in Chapter 1, S stands for Svedberg units, which represent a measure of how rapidly particles or molecules sediment in an ultracentrifuge. Generally, the larger a substance the greater its S value. However, the rate of sedimentation in a centrifuge depends upon shape as well as size. Therefore, when the subunits of a ribosome combine to form the functional ribosome, the intact ribosome has a lower S value than would have been calculated based upon the S values of the individual subunits. This is why adding 50S and 30S subunits gives a 70S ribosome, and adding 60S and 40S subunits gives an 80S ribosome.

[11]In addition to their 80S ribosomes, eukaryotic cells have 70S ribosomes within their mitochondria and chloroplasts. The 70S ribosomes of these organelles are very similar to the ribosomes of prokaryotic cells, giving strong credence to the endosymbiotic theory that mitochondria and chloroplasts evolved from prokaryotic cells.

FIGURE 3.22

(A) This illustration of a nucleosome shows how histones establish the configuration of DNA in eukaryotic cells. It was originally proposed in 1974 that the structure of the 100-Å-wide chromatin fiber was composed of successive 200-pair stretches of DNA wound on a series of beads; octamers were composed of two molecules each of the four histones H2A, H2B, H3, and H4. At the time, neither the actual shape of the histone complex nor the path of the DNA was known. (Based on R. D. Kornberg and A. Klug, 1981, *Scientific American* 244(2):55). (B) In this electron micrograph of chromatin fibers, the nucleosomes look like beads on a string. The chromatin was stretched in the preparation process, increasing the distance between nucleosomes and making it easier to distinguish them. (Courtesy Aaron Klug, Medical Research Council, Cambridge, England)

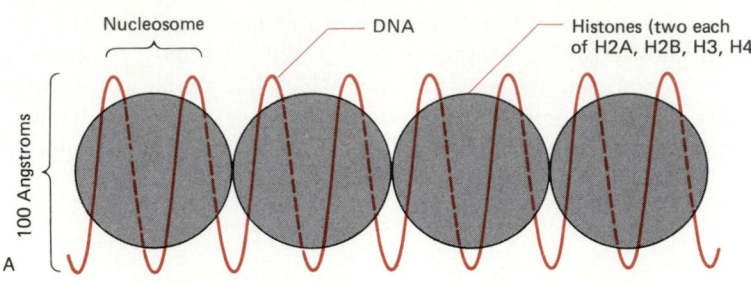

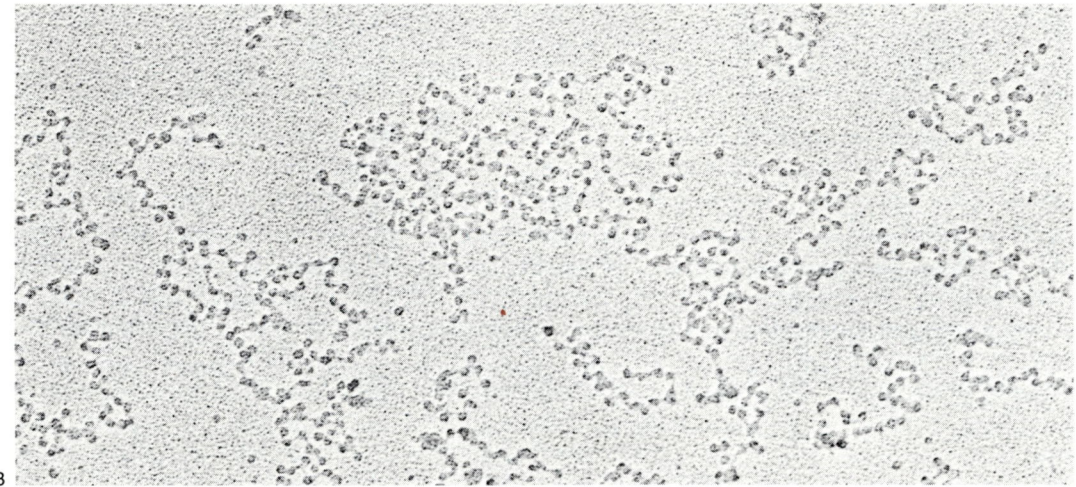

Discovery Process

The existence of the nucleosome has been established by two avenues of research: electron microscopy and studies on the enzymatic digestion of chromosomal DNA. By improving the methods for preparing chromatin fibers for electron microscopy to the point where clear and regular substructural patterns could be seen, L. Ada and Donald E. Olins at the Oak Ridge National Laboratory, C. L. F. Woodcock of the University of Massachusetts, Pierre Chambon at the Laboratoire de Genetique Moleculaire des Eucaryotes in Strasbourg, and Jack D. Griffith of Stanford University showed the chromatin fibers appearing as linear arrays of spherical particles, about 100 angstroms in diameter, connected by thin strands of apparently naked DNA. Micrococcal-nuclease digestion of chromatin and analysis of the DNA fragments by gel electrophoresis experiments showed that the cleavage of chromatin at sites spaced at regular intervals along the DNA is a property of chromatin structure. The nucleosome was isolated, free of DNA, following dissociation of the ionic bonds between histones and DNA in solutions of high salt concentrations. Combined cross-linking and nuclease-

digestion studies established the association of a histone octamer with a 200-nucleotide-pair unit of DNA.

Critical to the establishment of the ordered association of histones with chromosomal DNA was the work done by Aaron Klug and colleagues at the Medical Research Council Laboratory of Molecular Biology in Cambridge. These studies determined the three-dimensional structure of chromosomes by using high resolution electron microscopy and x-ray crystallographic analysis of the nuclease digests of chromosomes. Klug's work, which related the cellular and molecular levels of organization of chromatin, exemplifies his approach of chemically dissecting out parts of complex structures for detailed analysis by x-ray diffraction, the results of which are correlated with information about macromolecular organization of the intact assembly from electron microscopy. The elucidation of the histone–DNA organization of chromosomes was among several accomplishments cited in awarding the 1982 Nobel Prize in Chemistry to Aaron Klug.

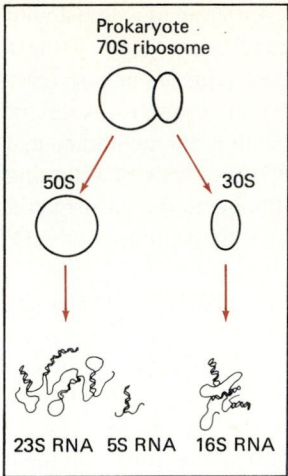 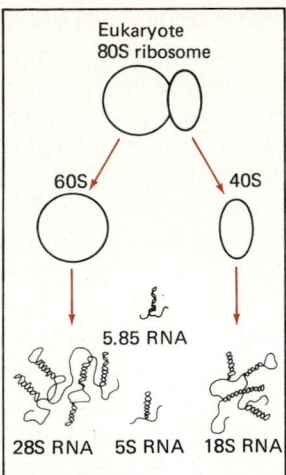

FIGURE 3.23
A basic difference between prokaryotic and eukaryotic cells is the nature of the ribosomes found in the cytoplasm. These differences form the basis for the specificity of action of some antibiotics that inhibit protein synthesis. The prokaryotic cell has 70S ribosomes composed of 30S and 50S subunits. The 30S subunit contains about 21 proteins and a 16S rRNA molecule, having approximately 1,540 nuleotides; the 50S subunit is composed of approximately 34 proteins, a 23S rRNA, having approximately 2,900 nucleotides, and a small 5S rRNA species having only about 120 nucleotides. A eukaryotic cell has 80S ribosomes in its cytoplasm composed of 60S and 40S subunits. The 40S subunit contains 18S rRNA, and the larger 60S subunit has 25-28S rRNA and 5S rRNA.

the ribosomes is essential for cells to carry out life-supporting metabolism, and any disruption of ribosomal conformation can disrupt this essential process. Many antibiotics, such as erythromycin and streptomycin, are effective because they bind to and alter the shape of 70S ribosomes. Such antibiotics are useful therapeutically because they selectively attach to 70S ribosomes and hence disrupt protein synthesis in bacteria, but do not exhibit any affinity for 80S ribosomes and therefore do not disrupt protein synthesis in eukaryotic human cells. Here we can see the practical application of a fundamental difference in the cellular structure of eukaryotes and prokaryotes.

3.6 Sites of Cellular Energy Transformations Where ATP Is Generated

Besides having mechanisms for storing and expressing genetic information, cells must have a means of obtaining energy to carry out life functions. The relationship between metabolism and cellular energy is the subject of Chapter 5. Here we will preface that discussion by considering the structures of a cell that are involved in the energy transformations that generate ATP.

Sites of ATP Generation in Prokaryotic Cells

Bacterial Cytoplasmic Membrane   The cytoplasmic membranes of some eubacteria and archaebacteria discussed in Section 3.2 are involved in energy transformations, namely, the generation of ATP. Hydrogen ions (protons) can be pumped across the cytoplasmic membrane to establish a concentration gradient, that is, a situation where the concentration of hydrogen ions is greater on the outside of the membrane than within the cell.[12] The more stable condition would be achieved with an equal concentration of hydrogen ions on both sides of the membrane; therefore, hydrogen ions attempt to move across the membrane to achieve this more stable state. Hydrogen ions cannot move freely across the cytoplasmic membrane, but they can move into pores in the membrane that are specifically associated with an enzyme system—adenosine triphosphatase (ATPase)—for generating chemical energy in the form of ATP. The generation of ATP by this mechanism is known as **chemiosmosis**.

Bacterial Internal Membranes   Although prokaryotic cells are generally characterized by a lack of internal membrane-bound organelles, certain specialized groups of bacteria do contain extensive internal membranes that are similarly involved with chemiosmotic generation of ATP. Such groups of bacteria include some nitrifying bacteria, which oxidize inorganic nitrogen-containing compounds to generate ATP, and the photosynthetic bacteria, which are able to use light energy to generate ATP.[13] The internal membranes of these bacteria, which may appear literally to fill the cell (Figure 3.24), are the sites of ATP generation. In photosynthetic bacteria, for example, these membranes are the anatomical sites where light is converted to chemical energy in the form of ATP during photosynthesis. These specialized **photosynthetic membranes** exhibit diversity among the different groups of photosynthetic bacteria. They can be simple extensions of the cytoplasmic membrane—as in the purple sulfur bacteria—cylindrically shaped vesicles—as in the green

---

[12]Some bacteria have more complex systems, pumping ions other than hydrogen to establish the gradient needed to generate ATP.

[13]Several other bacteria, including hydrocarbon oxidizers, form similar extensive internal-membrane networks that provide the necessary structure for synthesizing ATP.

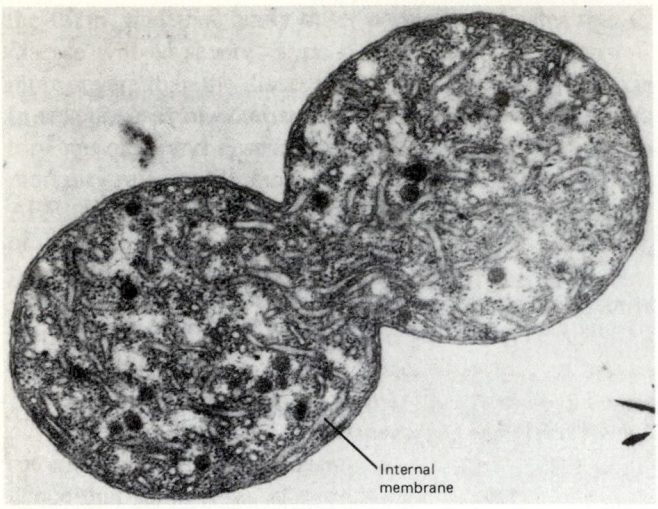

FIGURE 3.24

Some specialized bacteria have extensive internal membrane networks. This electron micrograph of the nitrifying bacterium *Nitrococcus mobilis* shows the extensive internal membranes (82,000×). (Courtesy Stan Watson, Woods Hole Oceanographic Institution, Woods Hole, Mass.)

photosynthetic bacteria—or extensive multilayered membrane structures— known as **thylakoids**—in the cyanobacteria. This diversity of structures suggests an evolutionary developmental sequence of photosynthetic membranes in these organisms.

Sites of ATP Generation in Eukaryotic Cells

Mitochondria   Unlike the prokaryotic cells, which lack internal membrane-bound organelles, the eukaryotic cell has various such organelles, some of which are involved with the generation of ATP. The **mitochondria** of eukaryotic cells are the sites of chemiosmotic ATP generation. As one would predict for an organelle involved with this function, mitochondria have a large membrane surface area. The mitochondrion has an independent interior membrane that is extensively folded and an outer membrane that acts as the boundary between it and the cell cytoplasm (Figure 3.25). The convolutions of the inner membrane that extend into the interior of the mitochondrion are called **cristae**. This inner

membrane has a higher proportion of protein associated with it than the outer mitochondrial membrane. Many of these proteins are involved in energy-transferring metabolic reactions. As a result of electron transport through a series of carriers embedded asymmetrically within the membrane and the resultant expulsion of protons, a gradient of returning hydrogen ions across the inner membrane to the inner matrix of the mitochondrion is used to drive the synthesis of ATP in eukaryotic cells.

In addition to considering the relationship between the structure and function of the mitochondrion, several aspects of the structure of this organelle are of interest in evolutionary theory. These organelles of eukaryotic cells show a marked resemblance to the prokaryotic cell and, as already noted, are believed to have evolved from an endosymbiotic relationship with a eubacterium related to the purple photosynthetic bacteria. Mitochondria have approximately the same size as a bacterial cell, a fact that probably explains why mitochondria never occur within prokaryotic cells. Within the mitochondrion there are 70S ribosomes, like the ribosomes of the prokaryotes, and a circular strand of DNA that resembles the bacterial chromosome. Mitochondrial membranes lack the sterols found in the cytoplasmic membranes of eukaryotic cells and thus resemble the prokaryotic cytoplasmic membrane. Further, the 16S rRNA of the 70S mitochondrial ribosomes shows a high degree of similarity in its nucleotide sequence to the 16S rRNA from prokaryotic cells.

These similarities indicate that mitochondria evolved from prokaryotic cellular organisms, supporting the endosymbiotic theory that mitochondria are descended from prokaryotic cells that became trapped in larger eukaryotic cells, eventually forming a stable relationship and evolving into the present mitochondrial structure. It is interesting to note, though, that the DNA contained in mitochondria is not sufficient to direct the synthesis of the entire mitochondrion structure and that mitochondria are in fact synthesized at the direction of the genome within the nucleus of the eukaryotic cell, indicating that contemporary mitochondria are no longer capable of independent existence.

Chloroplasts   **Chloroplasts**, which are the sites of photosynthetic generation of ATP, are quite similar in many ways to mitochondria. They also appear to have arisen as a result

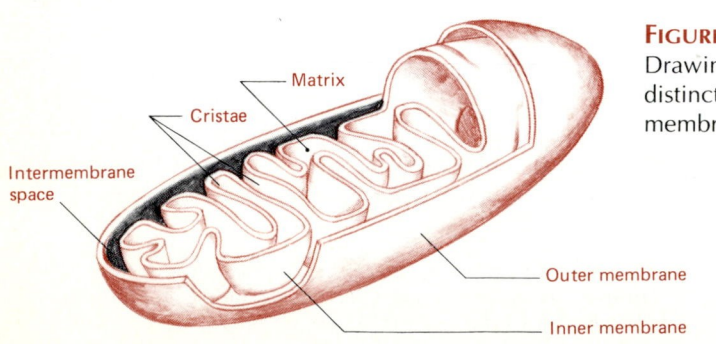

FIGURE 3.25

Drawing of the structure of a mitochondrion, showing its two distinct membranes and the extensive folding of the internal membrane.

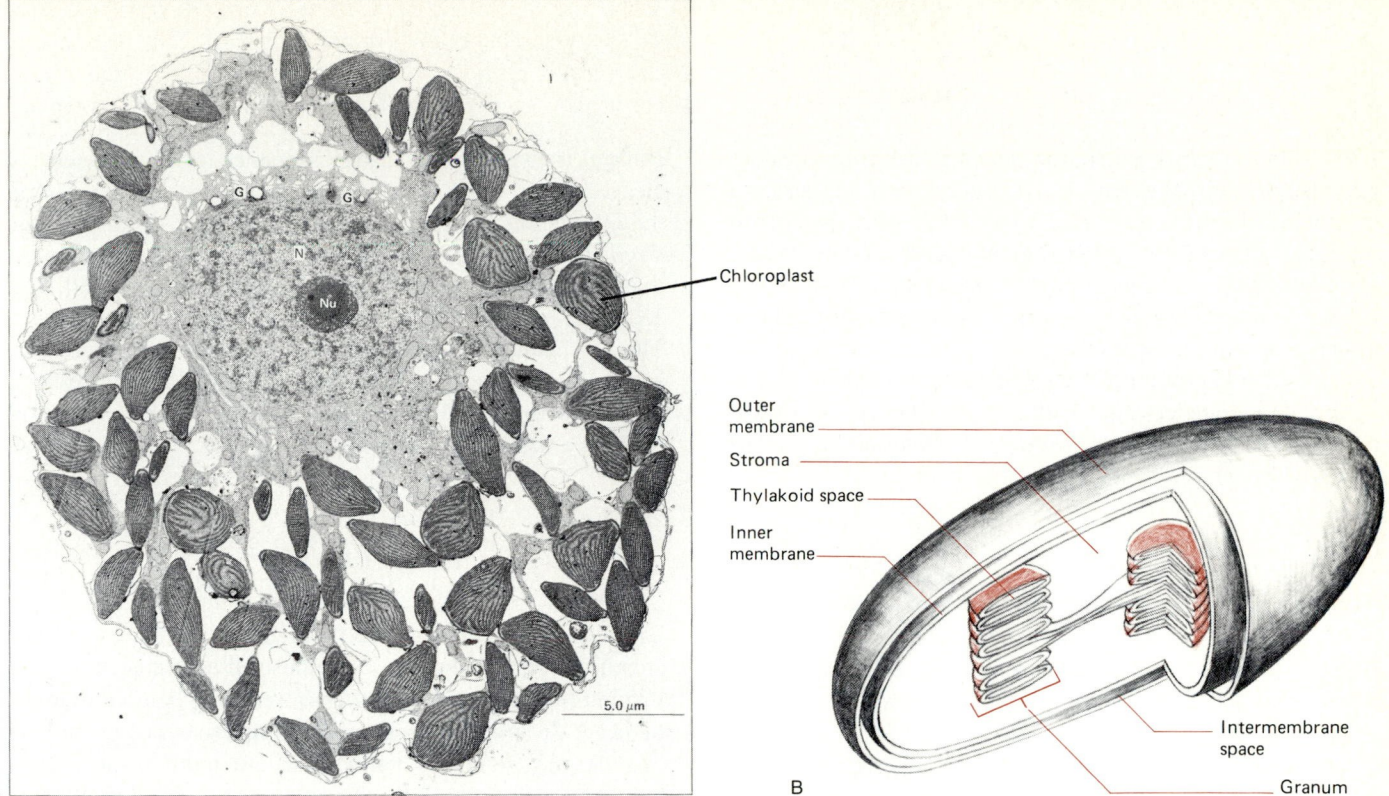

Chloroplast

Outer membrane

Stroma

Thylakoid space

Inner membrane

Intermembrane space

Granum

A

B

FIGURE 3.26

(A) Electron micrograph showing chloroplasts, the sites where light energy is converted to ATP in the alga *Vacuolaria virescens*. Note the large number of chloroplasts between the cell membrane and the dense cytoplasmic layer surrounding the nucleus (N): Nu = nucleolus, G = Golgi body. (Courtesy Peter Heywood, Brown University; reprinted by permission from P. Heywood, 1977, *Journal of Phycology* 13:69)

(B) Drawing of a chloroplast, showing its double membrane structure and the stacks of membranes (grana) within it.

of endosymbiotic evolution from a prokaryotic cell. They are composed of extensively invaginated membranes, contain 70S ribosomes, and have a circular DNA macromolecule, as do the mitochondria. Chloroplasts are one form of **plastid**, which are large cytoplasmic organelles occurring within the cytoplasm of photosynthetic eukaryotic organisms that contain pigments or other cellular products (Figure 3.26). Like the mitochondrion, the chloroplast contains an outer membrane that separates the organelle from the cytoplasm and an inner membrane, but the chloroplast also has an additional complex internal membranous system known as the **thylakoids**. Within the chloroplast, the thylakoids, which are sac-like membranous vesicles, may be stacked to form **grana** that normally are densely packed piles of individual thylakoids.[14]

Chloroplasts occur in algal and plant cells, where they are the sites of both photosynthetic ATP synthesis and carbon dioxide fixation. The interior compartment of the chloroplast, defined by the inner membrane and called the **stroma**, is where the fixation of carbon dioxide occurs during photosynthesis. The establishment of a hydrogen ion gradient across the thylakoid membranes drives the synthesis of ATP by chemiosmosis; this is analogous to the synthesis of ATP in the mitochondria, except that the flow of hydrogen ions is inward in the case of the chloroplast and outward in the case of mitochondria. The photosynthetic pigments found in the thylakoid membranes, including the chlorophylls, are responsible for trapping light energy and initiating the photosynthetic generation of ATP. The auxiliary pigments within the chloroplast confer characteristic colors on the algae and determine which wavelengths of light can be used for initiating photosynthetic ATP generation. The biochemical nature of the light-absorbing pigments is a basic taxonomic characteristic used in defining the major algal groups that are referred to by their common colors, such as the green, red, and brown algae.

[14]This is a somewhat oversimplified view, as there are great variations in the structures of chloroplasts in different algae. In general, the subunits of the chloroplast structure are less organized than the highly specialized structures characteristic of higher plants. The brown algae, for example, contain no grana and their thylakoid membranes are not stacked.

3.7 *Coordinated Material Movement and Storage in Cells*

In addition to energy generation, another task cells must accomplish is the movement of substances from one place to another within the cell. In prokaryotes this is relatively simple because the lack of internal membrane-bound organelles allows substances to mix freely within the cell's cytoplasm. In eukaryotic cells, however, materials must move from one organelle to another within the cell. The compartmentalization of the eukaryotic cell makes necessary a system of coordinated movement within such cells. In this section we will examine the movement and storage of materials in prokaryotic and eukaryotic cells.

Material Movement in Prokaryotic Cells

Secretion of Extracellular Enzymes   In some cases, it is necessary for bacterial cells to earmark materials for export from the cell. For example, because some large molecules, such as cellulose, are too big to transport through the pores of the cytoplasmic membrane, bacterial cells must secrete extracellular enzymes that can initiate an attack on such substances if they are to obtain energy and nutrients from them. These extracellular enzymes—called **exoenzymes**— degrade the large substance outside the cell, forming smaller molecules that can then be transported across the cell membrane and metabolized within the cell. In addition to their role in converting substances that cannot be transported through a membrane into usable substrates, exoenzymes are involved in destroying substances that are harmful to the cell. The secretion of extracellular enzymes represents an interesting regulatory mechanism whereby the cell recognizes which proteins to export (Figure 3.27).

Many enzymes that are designed to be secreted contain a leader segment at the amino terminal end of the molecule that acts as a signal to initiate the secretion process. This **signal sequence** contains about 20 predominantly hydrophobic amino acids that react with the membrane, initiating the translocation of the protein across the cell membrane barrier.

During transport, the leader sequence of the protein is cleaved by an enzyme within the membrane matrix, so that the exoenzyme released is smaller than when it was synthesized. In many cases, the secretion of the exoenzyme is initiated before the synthesis of the protein is completed, and secretion continues while protein synthesis is proceeding. The biochemical composition of the synthesized exoproteins and their interactions with the components of the cytoplasmic membrane provide the mechanism for the selective secretion of these extracellular enzymes.

Material Storage in Prokaryotic Cells: Inclusion Bodies

Under some conditions, microorganisms store various biochemicals, which they have synthesized and/or accumulated within the cell, to act as nutrient reserves to be used in times of need. In bacteria these reserve materials accumulate as cytoplasmic **inclusion bodies** that are not separated by a boundary membrane from the rest of the cytoplasm. Rather, the separation of the reserve inclusions is generally based on differential solubility.

The most common bacterial carbon reserve material is **poly-β-hydroxybutyric acid (PHB)**, a lipid-like molecule (Figure 3.28). Although PHB is not physically separated from the cytoplasm by a membrane, its nonpolar hydrophobic nature causes it to accumulate as a distinct inclusion body. PHB inclusions are surrounded by proteins thought to be involved with the metabolism of this carbon reserve.

Besides PHB, many microorganisms accumulate granules of **polyphosphate**, which are reserves of inorganic phosphate that can be used in the synthesis of ATP. Polyphosphate granules can be seen using light microscopy after staining and are sometimes called **volutin** or **metachromatic granules**. These accumulated reserves can be metabolized at a later time for the generation of ATP and cell constituents.

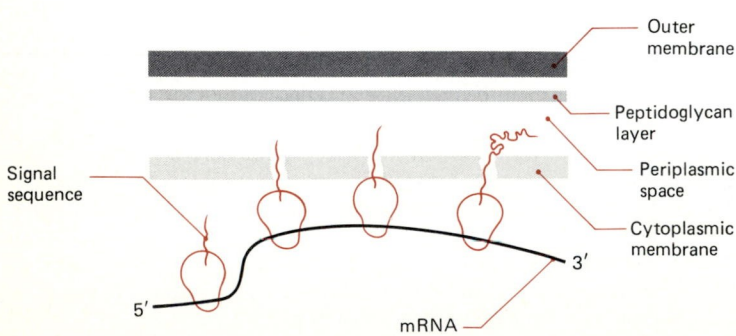

FIGURE 3.27

The utilization of high molecular weight compounds requires the secretion of exoenzymes that function outside the cell to break the substance down into molecules small enough to be transported across the membrane. Prokaryotic proteins destined for locations other than the cytosol are synthesized by ribosomes bound to the plasma membrane. A signal sequence on the peptide chain directs the ribosome to the plasma membrane and enables the protein to be translocated across it.

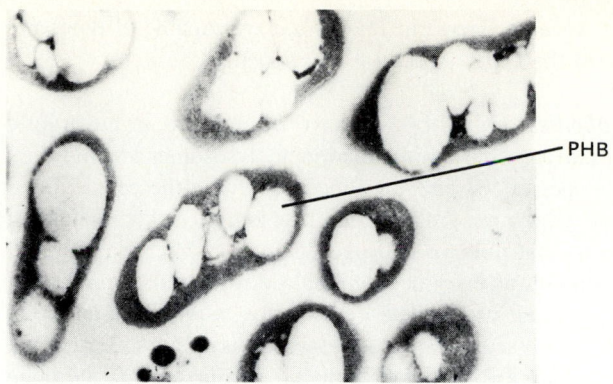

FIGURE 3.28

Electron micrograph of a nitrogen-starved *Alcaligenes eutrophus* (20,000×), showing accumulation of PHB, the white globular areas within the cell. (Courtesy Richard Bartha, Rutgers, the State University)

Network of Membrane-Bound Organelles in Eukaryotic Cells

In marked contrast to the prokaryotic cell, the eukaryotic cell is filled with membranous organelles involved with the processing and storage of materials within the cell. The extensive internal membrane system of eukaryotic microorganisms not only permits the efficient segregation of function, adding versatility to the metabolic functioning of the eukaryotic cell, but also increases the need to coordinate and manage the functions of the cell's subunit organelles. It is not surprising, then, that many of the organelles of the eukaryotic cell are linked together so that they can function in a coordinated manner.

Endoplasmic Reticulum   Eukaryotic cells contain an extensive membranous network known as the **endoplasmic reticulum**. The appearance of the endoplasmic reticulum varies greatly among different eukaryotic cells but always forms a system of fluid-filled sacs enclosed by the membrane network. There is evidence that the endoplasmic reticulum forms a continuum with the outer nuclear membrane and may provide a communication network for coordinating the metabolic activities of the cell. The endoplasmic reticulum shows two distinct morphologies when examined by electron microscopy. In one case, the endoplasmic reticulum looks rough and has attached ribosomes; and in the other case, it appears smooth and is not associated with ribosomes (Figure 3.29).

The attachment of ribosomes to the endoplasmic reticulum allows for coordinated activity, whereby proteins made at the ribosomes can immediately be sent through the channels of the endoplasmic reticulum to other organelles within the cell for use, packaging, or export.[15] In fact, many of the proteins

[15]No analogous membrane structure exists in prokaryotic cells to which ribosomes could attach, and bacteria have no system comparable to the endoplasmic reticulum for coordinated material movement within the cell.

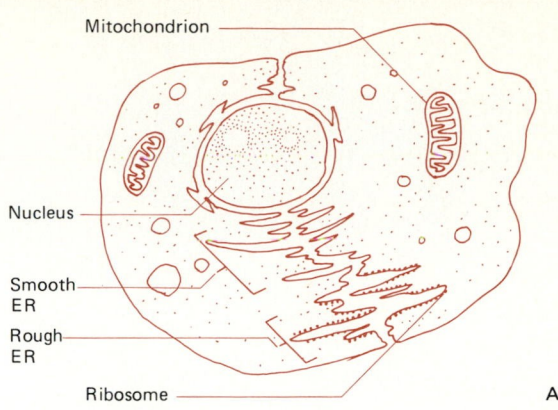

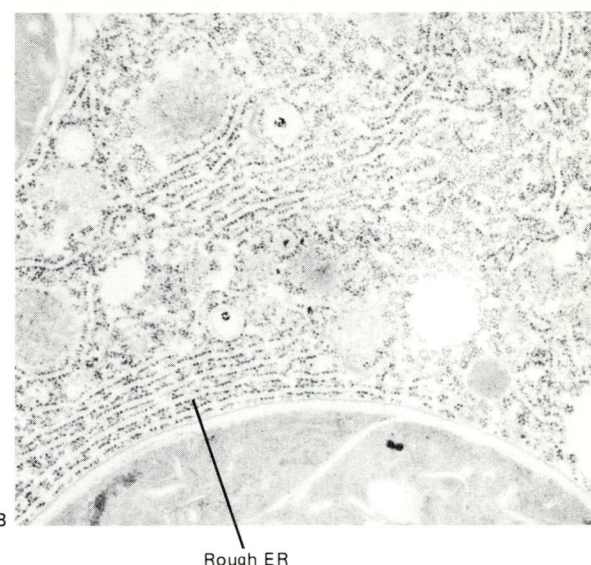

FIGURE 3.29

(A) Drawing of rough and smooth endoplasmic reticulum (ER). The terminology is derived from the fact that when viewed by electron microscopy, the endoplasmic reticulum with attached ribosomes appears bumpy or rough, whereas the endoplasmic reticulum lacking attached ribosomes appears smooth. (B) Electron micrograph of ribosomes arranged along the rough endoplasmic reticulum of the protozoan *Tetrahymena rostrata*. (Courtesy Eugene W. McArdle, Northeastern Illinois University)

synthesized by ribosomes attached to the endoplasmic reticulum are destined to be transported out of the cell or incorporated into membranes. Proteins synthesized on free ribosomes not associated with the endoplasmic reticulum are not transported through the channels of this membranous network and generally are used within the cytoplasm of the cell. Additionally, the endoplasmic reticulum provides a large surface for enzymatic activities, as well as a source of lipids and membranes for the other organelles of the eukaryotic cell.

The Golgi Apparatus   The **Golgi apparatus** forms a continuous network with the rough endoplasmic reticulum. The

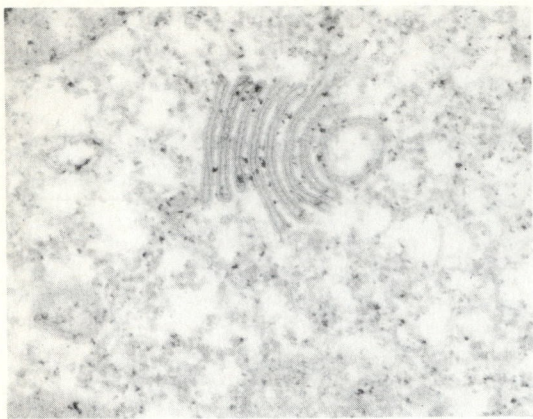

FIGURE 3.30

Electron micrograph of the Golgi apparatus of *Tetrahymena rostrata*, showing stacked Golgi bodies (60,000×). (Courtesy Eugene W. McArdle, Northeastern Illinois University)

proteins transported through the rough endoplasmic reticulum are transferred to the Golgi apparatus by a process of vesicular budding called *blebbing*. Normally, four to eight Golgi bodies, which are flattened membranous sacs, are stacked to form the Golgi apparatus (Figure 3.30). The Golgi apparatus is sometimes referred to as the *Golgi complex* and the individual stacks of membranes as **dictysomes**. Golgi bodies are the sites of various synthetic activities by which polysaccharides and lipids can be added to proteins to form lipoproteins, glycoproteins, and various polysaccharide derivatives that are essential for the synthesis of various cell constituents. Membrane sacs from the endoplasmic reticulum carry protein and lipids to the Golgi apparatus, where repackaging into secretory vesicles occurs. The secretory vesicles, which are formed by the Golgi apparatus and then move to the cytoplasmic membrane, where they release their contents through exocytosis. Such a process is important for the construction of structures external to the cytoplasmic membrane, such as the cell wall. Functioning together, the outer nuclear membrane, the endoplasmic reticulum, the Golgi bodies, and the secretory vesicles perform a sequence of coordinated activities that move and process biochemicals associated with protein synthesis through the cytoplasm of the cell to the exterior of the cytoplasmic membrane.

Lysosomes   **Lysosomes** are specialized membrane-bound organelles of eukaryotic cells, probably produced in the Golgi apparatus, that contain various digestive enzymes. Some of the digestive activities of eukaryotic cells occur within the lysosome. Indeed, one of the functions of the enzymes within the lysosomes is to digest prokaryotic cells that have been ingested by phagocytosis. The lysosome membrane is impermeable to the outward movement of digestive enzymes and is also resistant to their action. This segregation of certain enzymes within the lysosome is nec-

essary because these enzymes often are capable of digesting many of the cell's structural components.

Microbodies   Microbodies have been found in eukaryotic cells and are similar in function to lysosomes in that they isolate specialized enzyme functions within the cell. **Microbodies**, which are smaller than lysosomes, isolate metabolic reactions that involve hydrogen peroxide. Microbodies contain catalase, an enzyme that breaks down hydrogen peroxide to oxygen and water. The **peroxisome**, a type of microbody found in eukaryotic microorganisms, is a site where some amino acids are oxidized with the production of hydrogen peroxide. If the peroxides formed in these reactions were not contained or destroyed, they could oxidize a number of essential biochemicals within the cell, resulting in the death of the cell. Their isolation within the peroxisome protects the cell.

Vacuoles   Various types of membrane-bound organelles—called **vacuoles**—serve different purposes within the cells of eukaryotic microorganisms (Figure 3.31). Whereas reserve materials are not separated from the cytoplasm by a membrane barrier in prokaryotic cells, one type of vacuole, the **storage vacuole**, is involved in maintaining accumulated reserve materials segregated from the cytoplasm in eukaryotic cells. For example, yeasts can store polyphosphate, amino acids, and uric acid as reserve materials within storage vacuoles. Other organisms store other forms of organic carbon, nitrogen, and phosphate reserves for times of need. Other vacuoles are involved in the movement of materials out of the cell. These vacuoles can unite with the cytoplasmic membrane during endo- and exocytosis. In some cases, a vacuole formed when the cell engulfs a food source fuses with lysosomes, establishing a **digestive vacuole** that permits digestion of the contents.

Cytoskeleton   In addition to the numerous membrane-bound organelles that occur in eukaryotic but not prokaryotic cells, the eukaryotic cell has a **cytoskeletal network**, consisting of microtubules and microfilaments, that helps determine the ability of the cell to move and to maintain its shape. This cytoskeletal network links the various components of the cytoplasm into a unified structure, the **cytoplast**, providing the rigidity needed to hold the various structures in their appropriate locations (Figure 3.32). The microtubular-microfilament arrangement of the cytoskeleton runs throughout the eukaryotic cell, connecting membrane-bound organelles with the cytoplasmic membrane. This cytoskeletal structure appears to be involved in both the support and movement of membrane-bound structures, including the cytoplasmic membrane and the various organelles of the eukaryotic cell. It apparently provides the basis for membrane movement involved in transporting materials into and out of the cell by cytosis. The lack of a cytoskeleton in prokaryotic cells may explain why bacteria have not been found to be capable of cytosis.

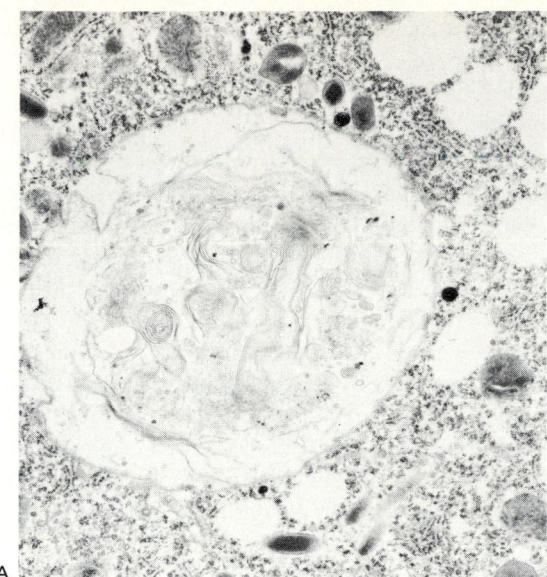

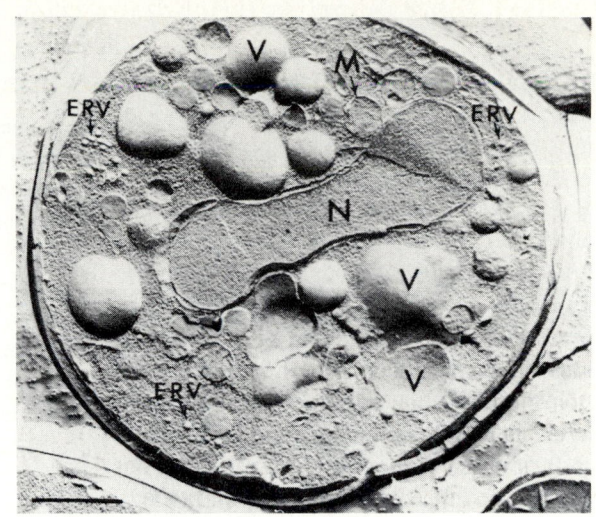

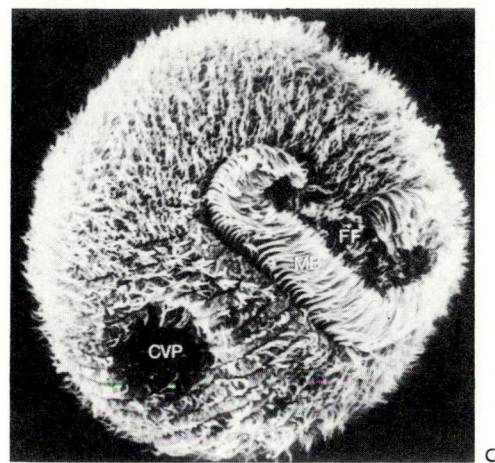

FIGURE 3.31

Electron micrographs showing various vacuoles found in eukaryotic cells. (A) A food vacuole of the ciliate protozoan *Tetrahymena rostrata* filled with membranes of bacteria that have been ingested. (Courtesy Eugene McArdle, Northeastern Illinois University). (B) A freeze-etched preparation of the yeast *Saccharomyces cerevisiae* in a late stage of meiosis, showing prominent storage vacuoles (V) and a dumbbell-shaped nucleus (N). ERV = endoplasmic reticulum vesicles. (Courtesy Sam Conti, University of Massachusetts; reprinted by permission of the American Society for Microbiology, from E. Guth, T. Hashimoto, and S. F. Conti, 1972, *Journal of Bacteriology*, 109:874). (C) Electron micrograph of a contractile vacuole pore in *Stentor coeruleus* (1,200×); CVP = contractile vacuole pore, MB = membranellar band of cilia, FF = frontal field of cilia. (Reprinted by permission of Springer-Verlag, New York, from R. G. Kessel and C. Y. Shih, 1976, *Scanning Electron Microscopy for Biology*)

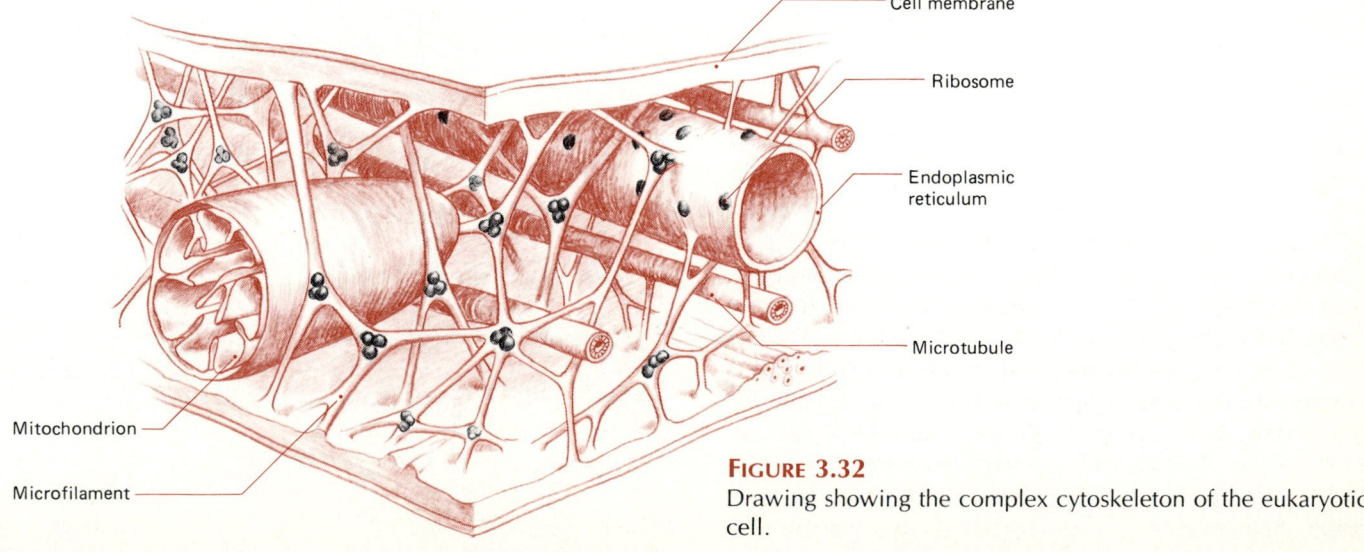

FIGURE 3.32

Drawing showing the complex cytoskeleton of the eukaryotic cell.

3.8 Structures Involved with Movement of Cells

In some cases, the cytoskeletal structure plays a role in the movement of an organism. For example, the movement of microtubules permits the extension of the cytoplasmic membrane to form the ''false feet'' (psuedopodia), used by some protozoa, like *Amoeba*, that move by extending their cytoplasm in a particular direction as they continuously change shape. More commonly, though, microorganisms move by means of flagella or cilia, which are specialized structures that project from the cell surface and propel the cell along. Although flagella serve the same function of locomotion in prokaryotes and eukaryotes, the flagella of bacteria and those of eukaryotic cells are markedly different in mechanism and structure. Motility is important because it allows microorganisms to move from place to place in order to obtain nutrition for growth and reproduction or to escape from noxious microenvironments.

Flagella and Cilia

Bacterial Flagella   Bacterial flagella are relatively long projections extending outward from the cytoplasmic membrane that propel bacteria from place to place (Figure 3.33). In some bacteria, such as *Pseudomonas*, the flagella are known as **polar flagella** because they originate from the end or pole of the cell (see Figure 3.33A). A bacterial cell may have one or more polar flagella, which they use to swim rapidly in what is generally described as a corkscrew motion. In contrast to polar flagella that emanate from an end of the cell, **peritrichous flagella**—such as those of the bacterial genus *Proteus*—surround the cell (see Figure 3.33B). The specific number of flagella varies, but there are always multiple peritrichous flagella emanating from lateral points around the cell. The arrangement of the flagella is characteristic of a bacterial genus and is an important diagnostic characteristic used in classifying bacteria.

The bacterial flagellum consists of a single filament composed of protein (flagellin) subunits. The bacterial flagellum is attached to the cell by a hook and a basal body, which has a set of rings that attach to the cytoplasmic membrane and a rod that passes through the rings to anchor the flagellum to the cell (Figure 3.34). In Gram-negative bacteria the basal body has a second set of rings that attaches to the outer membrane of the cell envelope, and in Gram-positive bacteria there is only one set of rings that attaches to the cytoplasmic membrane. The hook structure attaches the filament of the bacterial flagellum to the rod of the basal body. The structure of the bacterial flagellum allows it to spin like a propeller and thereby to propel the bacterial cell. Effectively, the structure allows the flagellum to spin like the shaft of an electric motor. Rotation of the flagellum requires energy, which is supplied by a hydrogen ion gradient across the cytoplasmic membrane.

Specialized Movement in Bacteria

Chemotaxis   The bacterial flagellum provides the bacterium with a mechanism for swimming toward or away from chemical stimuli, a behavior known as **chemotaxis** (Figure 3.35). Bacteria move toward certain chemicals, known as **attractants**, and away from others, known as **repellents**. Chemosensors in the cell envelope, called **binding proteins** because they bind specifically and tightly to substrates in the membrane transport process, detect certain chemicals and signal the flagella to respond. Bacterial cells, though, are too small to detect spatial chemical concentration differences directly; they do not have different sensors on the ends of the

FIGURE 3.33

(A) Electon micrograph of a marine vibrio showing a single polar flagellum (35,000×). (From BPS—Paul W. Johnson, University of Rhode Island). (B) Electron micrograph of negatively stained, peritrichously flagellated *Proteus* sp. (Courtesy Lee D. Simon, Rutgers, the State University)

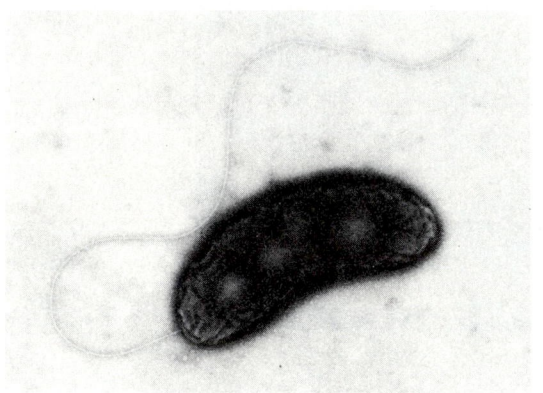

A

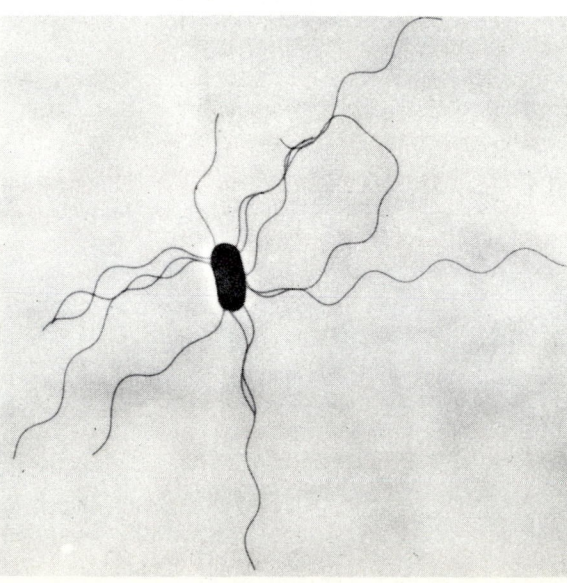

B

cell that indicate which way to move. Rather, bacteria have a memory system that allows them to compare the concentrations of chemicals as they swim along, so that they effectively detect chemical concentration differences over distances many times the length of a cell.

To understand how chemotaxis works, we need to recognize that when bacteria move, they periodically change direction rather than reaching their destination by swimming in a single straight line (Figure 3.36). The straight-line movements of bacteria are known as **runs**, and the turns—which occur when the bacteria stop—are called **tumbles** or **twiddles**. At least in bacteria with peritrichous flagella, the counterclockwise rotation of the flagella results in a run and the clockwise rotation in a twiddle. The direction of flagella rotation, and hence the length of a run—that is, the amount of time before the organism stops and tumbles—is determined by the interactions of the chemoreceptors in the cell's cytoplasmic membrane with attractants or repellents. An increasing concentration of attractant, for example, interacts with the chemoreceptors to decrease the frequency of tumbling, whereas a decreasing concentration of attractant causes increasing tumbling, and hence shorter runs. The same is true for the interactions with repellents. The net effect of this process, called a *random walk*, is a biased movement toward

an attractant or away from a repellent based upon the relative proportion of running and tumbling.

Magnetotaxis   Some motile bacteria contain structures that enable them to respond to environmental stimuli other than chemical concentration differences. One fascinating group of bacteria contains inclusions of iron granules, known as **magnetosomes**, that permit them to orient their movement in response to magnetic fields, a phenomenon known as **magnetotaxis** (Figure 3.37). Bacteria can use these granules to navigate along the Earth's magnetic field. Some bacteria move predominantly north, and others move south. Magnetotaxis allows some anaerobic bacteria to orient themselves so that they point downward into the sediment.

FIGURE 3.34

Drawing showing how the bacterial flagellum is anchored at the cell. The flexible hook serves as a universal joint coupling the rod to the filament. Torque is generated between the M ring, which is rigidly mounted on the rod and freely rotates in the cytoplasmic membrane, and the S ring, which is rigidly mounted on the cell wall. Torque is generated by the translocation through the cytoplasmic membrane and the M ring of ions that interact with charges on the surface of the S ring. The energy for rotation is derived from the transport of hydrogen ions across the cytoplasmic membrane, chemiosmosis, with transport of approximately 256 hydrogen ions required per revolution. The additional pair of outer rings (P and L) in this Gram-negative bacterium apparently serve as bearings for the rod's passage through the cell's more complex wall and minimize friction and leakage; the L and P rings do not occur in Gram-positive bacteria. The rings are about 0.2 μm in diameter.

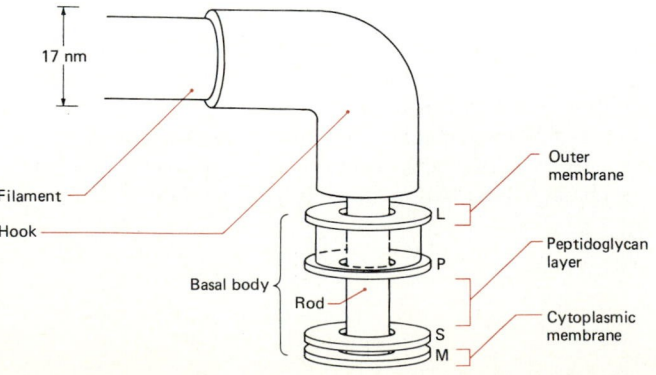

FIGURE 3.35

A demonstration of a chemotactic response to an attractant. That bacteria respond to their chemical surroundings by chemotaxis was an important finding, showing that even the simplest organisms show approach–avoidance behavior. Chemotactic behavior is readily demonstrated and measured by placing the tip of a thin capillary tube containing an attractant solution in a suspension of motile *E. coli* bacteria. The suspension is placed on a slide, in a chamber created by a U tube and a cover slip. (A) At first, the bacteria are distributed at random throughout the suspension; (B) after 20 minutes, they have congregated at the mouth of the capillary tube; and (C) after about an hour, many cells have moved up into the capillary tube. If the capillary tube had contained a repellant, few bacteria, if any, would have entered. Using this technique, it is possible to show which chemicals attract bacteria and which do not.

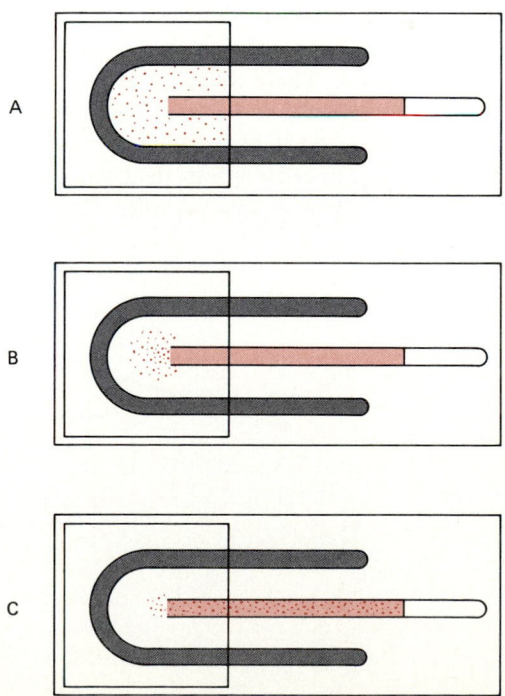

ANALYTICAL PROCESS
Rotary Movement of Bacterial Flagella

Through the use of immunological reactions, the flagella of bacteria can be attached to glass microscope slides, leaving the body of the cell free to move. While the flagella are difficult to see, the live bacterial cells can easily be observed by phase contrast microscopy. Because the bacterial cells are tethered to the microscope slides, the cells are not free to swim, but the motion of the flagella will move the cells and this movement can be observed. Just how bacteria move was revealed by photographing their movements using high-speed motion picture cameras. By tethering *E. coli* to a microscope slide, Howard C. Berg watched as the flagellum made the body of the bacteria rotate in a counterclockwise direction. Norbert Pfennig photographed *Chromatium okenii* at 600–1,000 frames per second and showed that it took about one-hundredth of a second for the flagellar fragments to complete one turn, during which time the cell body moved forward a fraction of its own length. The films of these experiments dramatically show the rotary motion of bacterial flagella. In later experiments, the flagella were coated with latex beads. The coated flagella were seen rotating as a line extending from the cell, occasionally stopping and changing direction. When combined with high-resolution electron microscopy, which revealed the structure of the flagellum, the mechanism of bacterial movement was shown to be similar to the working of an electric motor. These cleverly designed experiments and microscopic observations clearly show that bacterial flagella indeed rotate as they propel the cell.

Gas Vacuoles and Phototaxis   Some bacteria are able to detect and respond to differences in light intensity, a phenomenon called **phototaxis**. In some bacteria this process is similar to chemotaxis, and the bacteria use flagella to swim to areas of particular light intensities. Other bacteria form ''membrane-bound'' **gas vacuoles** that enable them to respond to light (Figure 3.38). Actually, the boundary layers of these vacuoles are not true membranes, but rather are composed exclusively of protein. The ''membrane'' layer appears to be only one protein molecule thick. The protein composing the boundary layer of these vacuoles has both hydrophilic and hydrophobic properties. The formation of gas vacuoles by aquatic bacteria provides a mechanism for adjusting the buoyancy of the cell and thus the height of the bacterium in the water column. Many aquatic cyanobacteria, for example, use their gas vacuoles to move up and down in the water column, depending on light irradiation levels, to achieve optimal conditions for carrying out their photosynthetic metabolism. The presence of gas vacuoles thus provides the necessary structural units for phototaxis in aquatic bacteria.

Flagella and Cilia of Eukaryotic Cells

Unlike the bacterial flagella that rotate, the cilia and flagella of eukaryotic cells undulate in a wave-like motion to propel the cell (Figure 3.39). Eukaryotic flagella emanate from the polar region of the cell, whereas **cilia**, which are somewhat shorter than flagella, surround the cell. The flagella and cilia of eukaryotic microorganisms are important taxonomic characteristics. Among the protozoa, the Ciliophora are grouped taxonomically because of the presence of cilia, and the Mastigophora are grouped on the basis of the presence of flagella. Both cilia and flagella are generally involved in cell locomotion, but cilia may also be involved in moving materials, such as food particles, past the cell surface while the organism or cell remains stationary.

In contrast to the rather simple structure of the bacterial

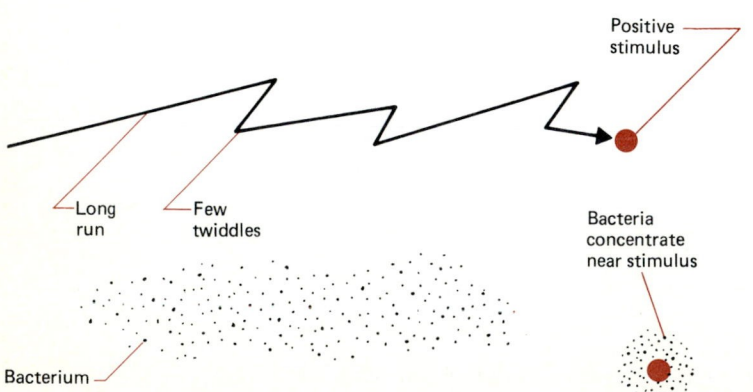

FIGURE 3.36

Illustration of patterns of bacterial movement during chemotaxis. When responding to chemotactic stimuli, bacteria move in characteristic straight lines (runs) and tumbling movements (twiddles). The runs and twiddles result from changes in the direction of the rotation of the flagella. Movement toward or away from a substance is characterized by long runs and few twiddles.

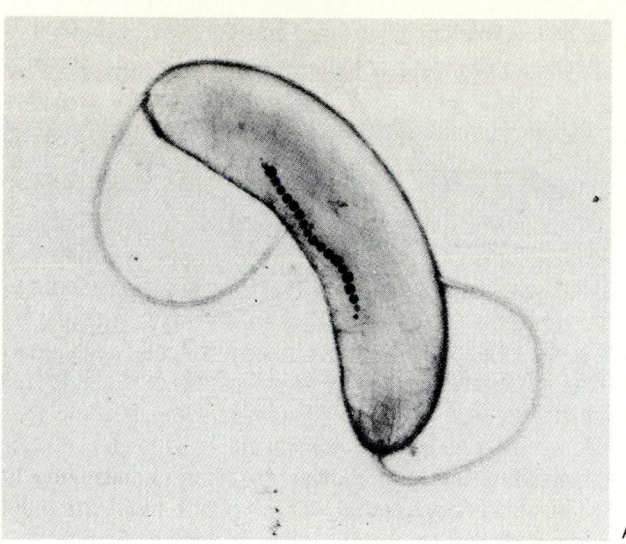
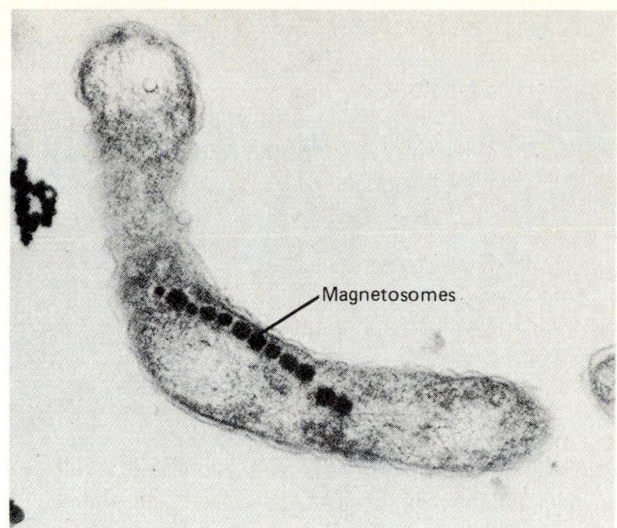

FIGURE 3.37

An interesting and relatively new finding is that some bacteria have internal magnetic structures that permit them to navigate. (A) In this electron micrograph, iron granules are prominent within a magnetotactic bacterium, which was found in a water treatment plant in Durham, New Hampshire (52,000×). (Courtesy R. Blakemore and N. Blakemore, University of New Hampshire). (B) This electron micrograph of *Aquaspirillum magnetotacticum* (108,000×), shown in thin section, illustrates the membrane around the magnetic iron particles. (Courtesy R. Blakemore and D. Maratea, University of New Hampshire)

FIGURE 3.38

This micrograph shows the gas vacuoles of *Nostoc muscorum*, a cyanobacterium that has been freeze-etched (57,600×). (From BPS—J. Robert Waaland, University of Washington)

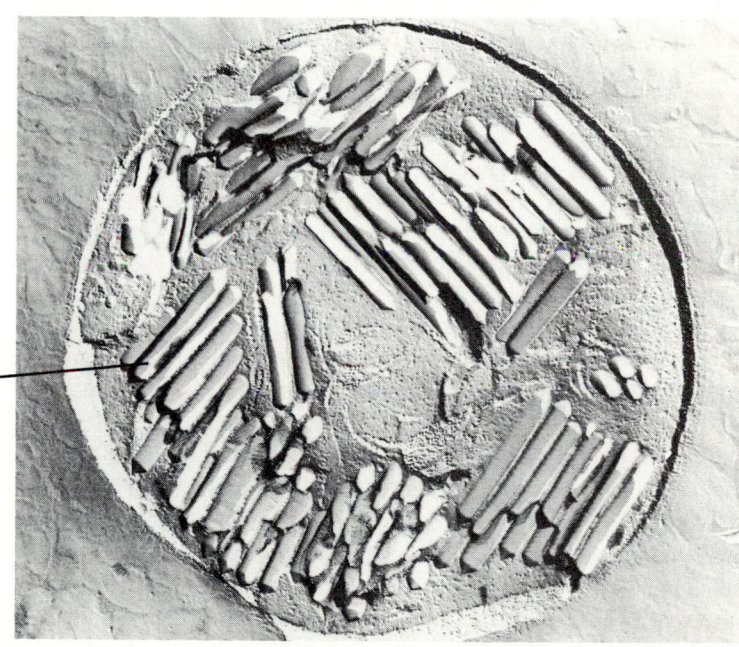

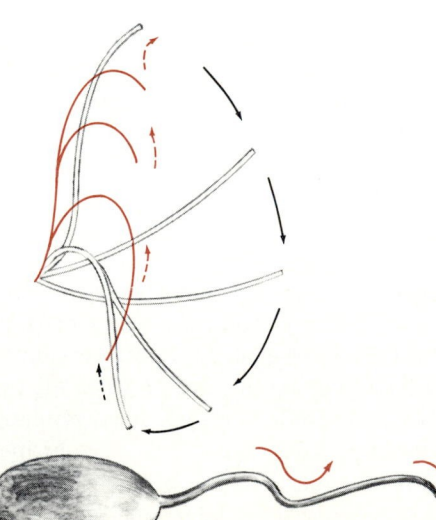

FIGURE 3.39

The movement of the eukaryotic flagellum occurs with a wave-like motion that propels the cell.

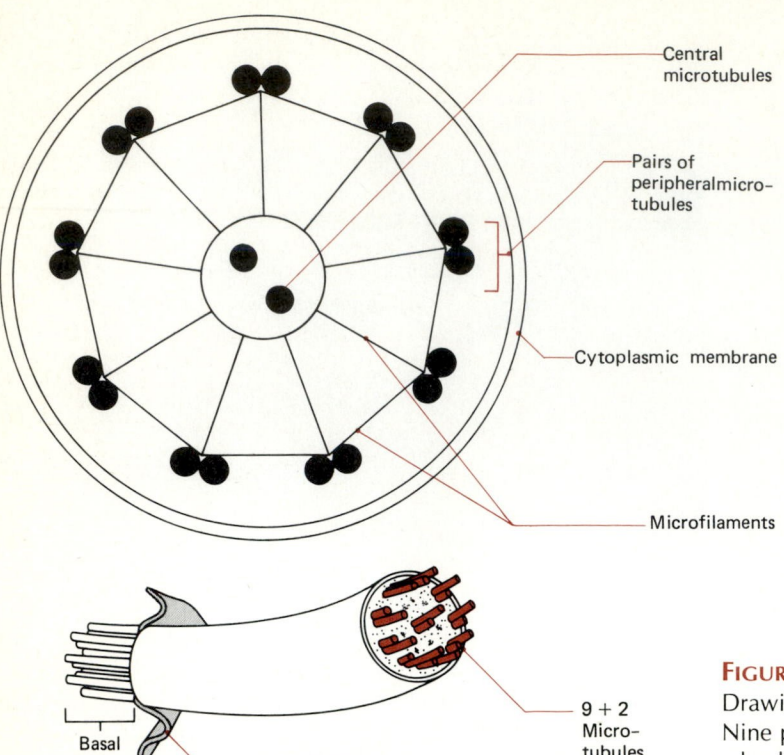

Central
microtubules

Pairs of
peripheralmicro-
tubules

Cytoplasmic membrane

Microfilaments

9 + 2
Micro-
tubules

Basal
body

Membrane

FIGURE 3.40
Drawing showing the structure of the eukaryotic flagellum.
Nine pairs of microtubules occur like the spokes of a bicycle
wheel around the central pair.

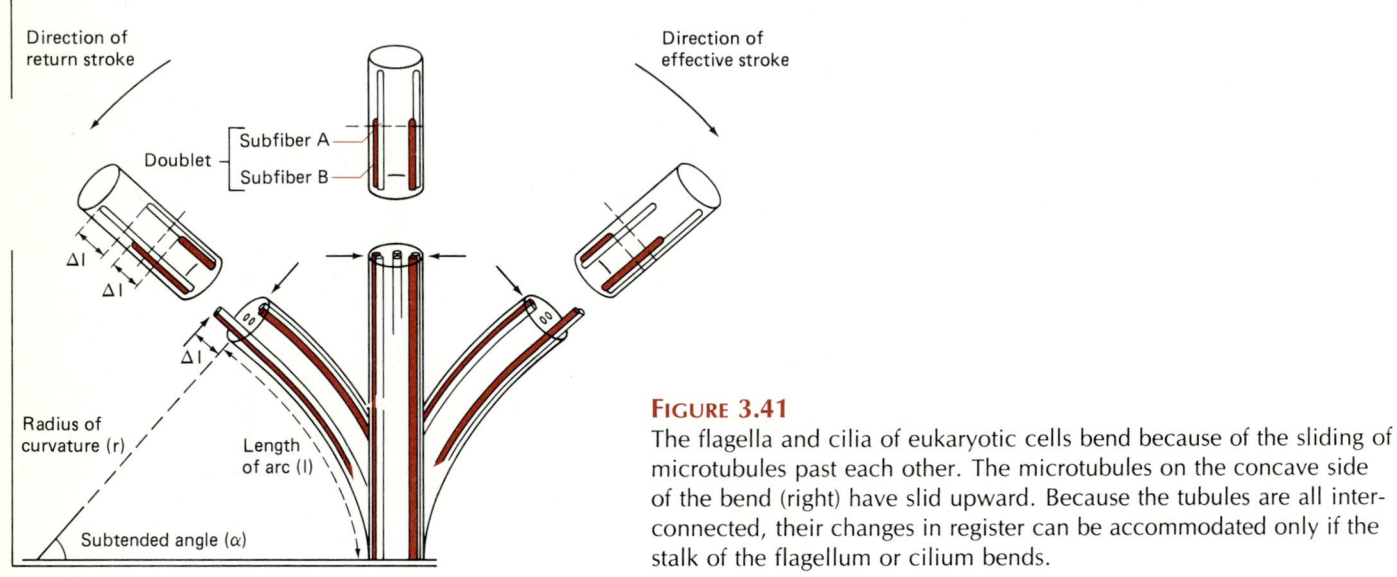

Direction of
return stroke

Direction of
effective stroke

Doublet ⎧ Subfiber A
        ⎩ Subfiber B

Δl
Δl
Δl

Radius of
curvature (r)

Length
of arc (l)

Subtended angle (α)

FIGURE 3.41
The flagella and cilia of eukaryotic cells bend because of the sliding of
microtubules past each other. The microtubules on the concave side
of the bend (right) have slid upward. Because the tubules are all inter-
connected, their changes in register can be accommodated only if the
stalk of the flagellum or cilium bends.

flagellum, the flagella and cilia of eukaryotic microorganisms
are far more complex and larger (Figure 3.40). The eukaryotic flagella and cilia both consist of a series of microtubules—hollow cylinders composed of proteins (tubulin)—
surrounded by a membrane. The arrangement of microtubules in eukaryotic flagella and cilia is known as the **9 + 2
system** because it consists of nine peripheral pairs of microtubules surrounding two single central microtubules. The
nine pairs of microtubules form a circle surrounding the central microtubules. The peripheral microtubule doublets are

linked to the central microtubules by radial spokes of protein
microfilaments; the peripheral microtubule doublets are also
similarly linked to each other to form a circular network
surrounding the spokes. The movement of flagella or cilia is
based on a sliding microtubule mechanism in which the peripheral doublet microtubules slide past each other, resulting
in bending of the flagella or cilia (Figure 3.41). The peripheral spokes of the microtubular network contain a protein
(dynein), which has ATPase activity and is involved in coupling ATP utilization to the movement of the flagella or cilia.

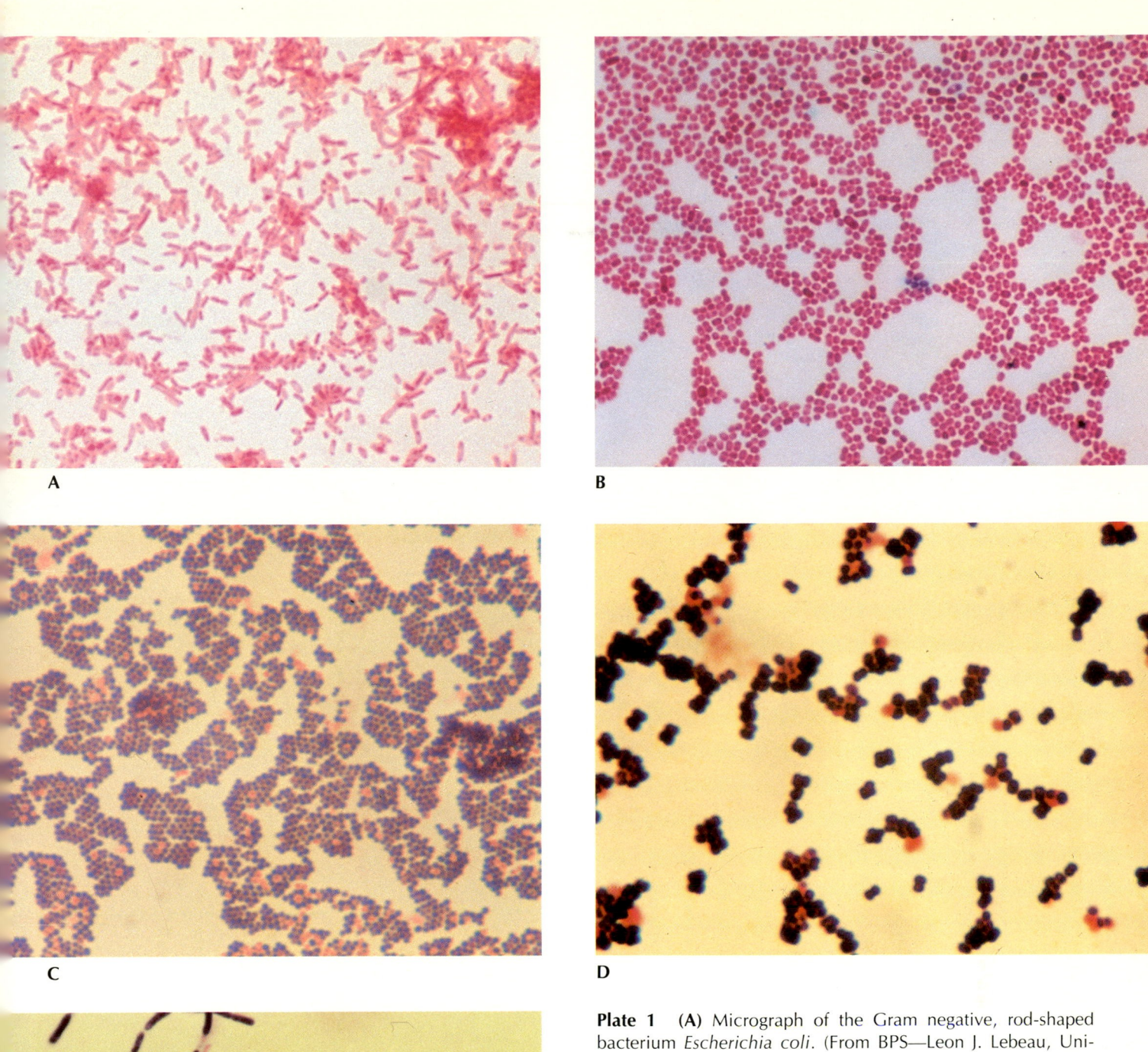

Plate 1 **(A)** Micrograph of the Gram negative, rod-shaped bacterium *Escherichia coli*. (From BPS—Leon J. Lebeau, University of Illinois Medical Center) **(B)** Micrograph of *Acinetobacter*, a Gram negative coccobacillus. (From BPS—Leon J. Lebeau, University of Illinois Medical Center) **(C)** Micrograph of *Staphylococcus*, a Gram positive coccal-shaped bacterium that forms grapelike clusters. (From BPS—Leon J. Lebeau, University of Illinois Medical Center) **(D)** Micrograph of a *Micrococcus* species that forms tetrads of coccal-shaped cells. (From BPS—Leon J. Lebeau, University of Illinois Medical Center) **(E)** Micrograph of *Bacillus subtilis*, a Gram positive rod-shaped bacterium that produces endospores. (From BPS—Paul Johnson, University of Rhode Island)

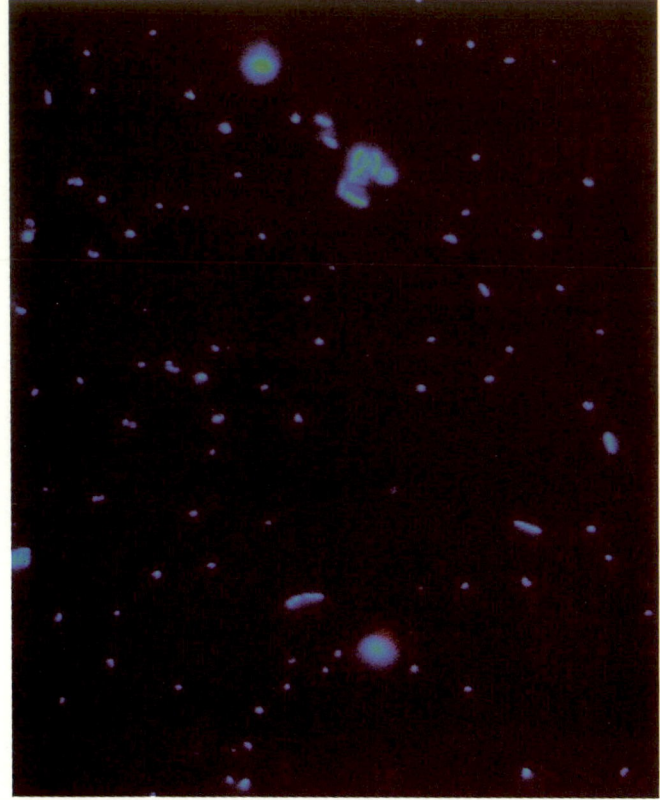

A

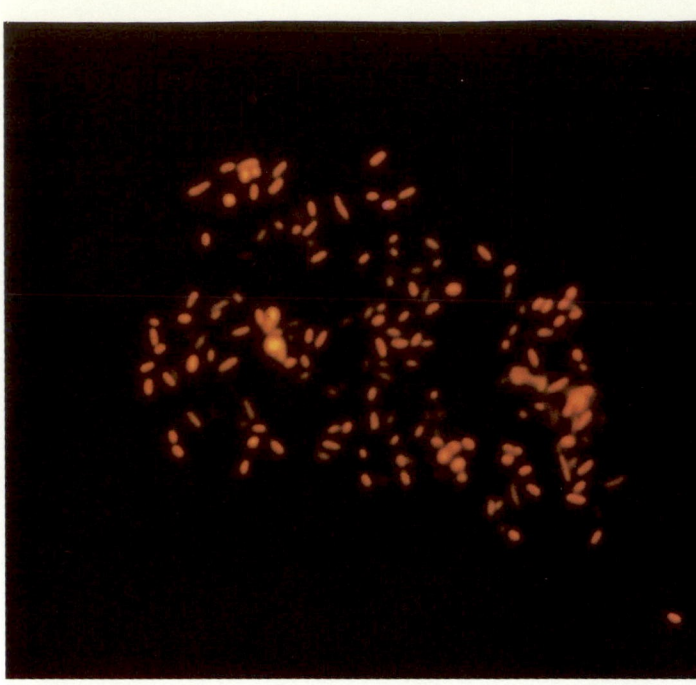

B

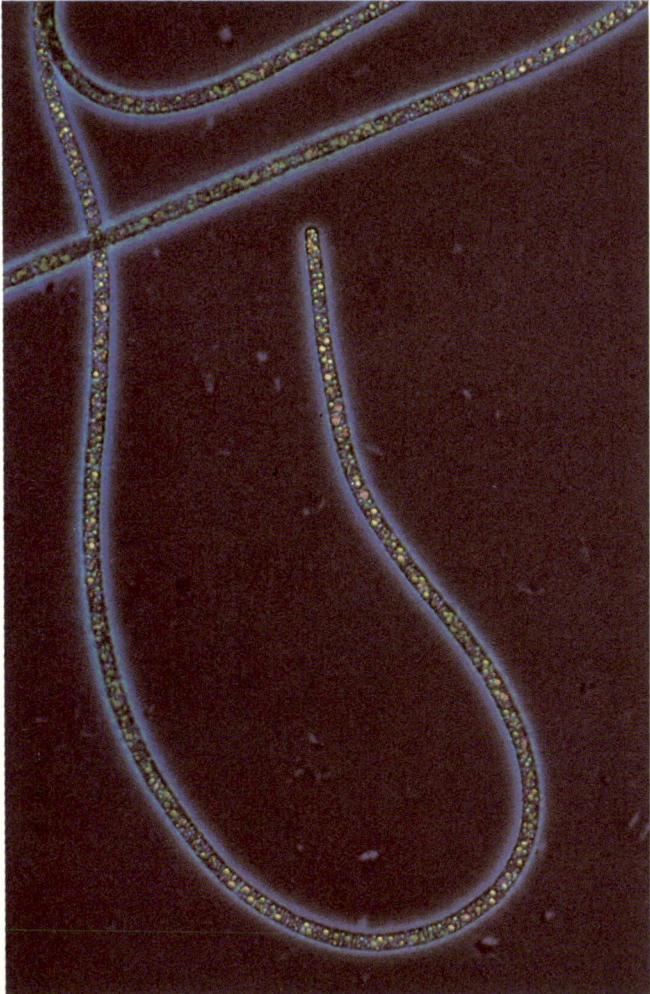

C

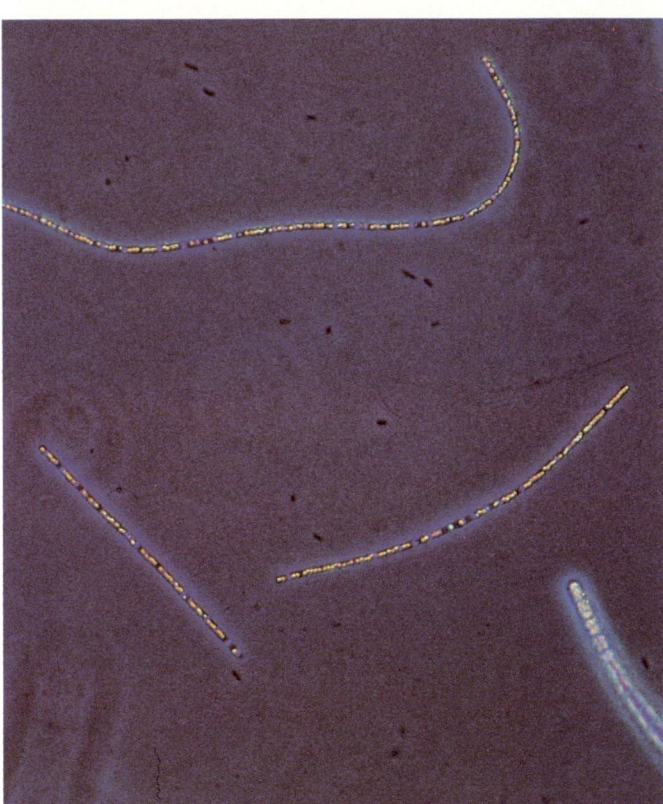

D

Plate 2 **(A)** Micrograph showing marine bacteria stained with DAPI and viewed by epifluorescence microscopy. (From BPS—Paul Johnson, University of Rhode Island) **(B)** Microcolony of bacteria stained with acridine orange and viewed by epifluorescence microscopy. (From BPS—Paul Johnson, University of Rhode Island) **(C)** Phase contrast micrograph of the bacterium *Beggiatoa*. (From BPS—Paul Johnson, University of Rhode Island) **(D)** Phase contrast micrograph of the bacterium *Thiothrix*. (From BPS—Paul Johnson, University of Rhode Island)

A

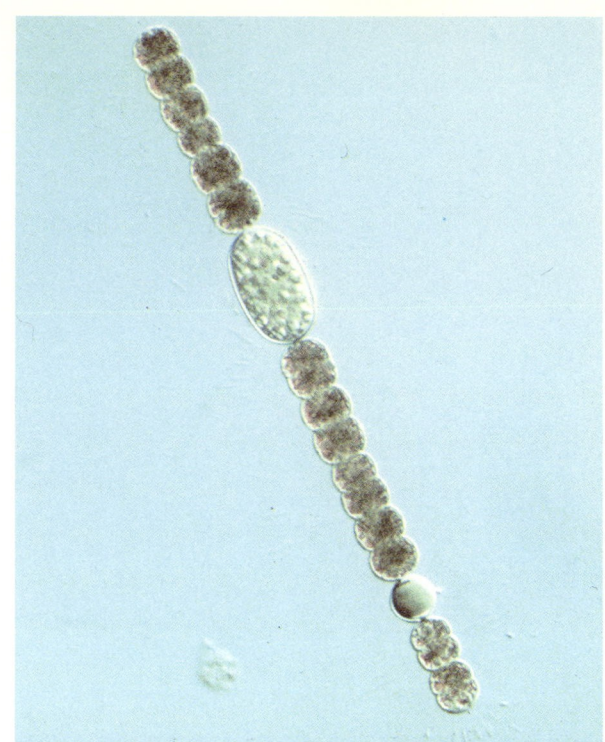

B

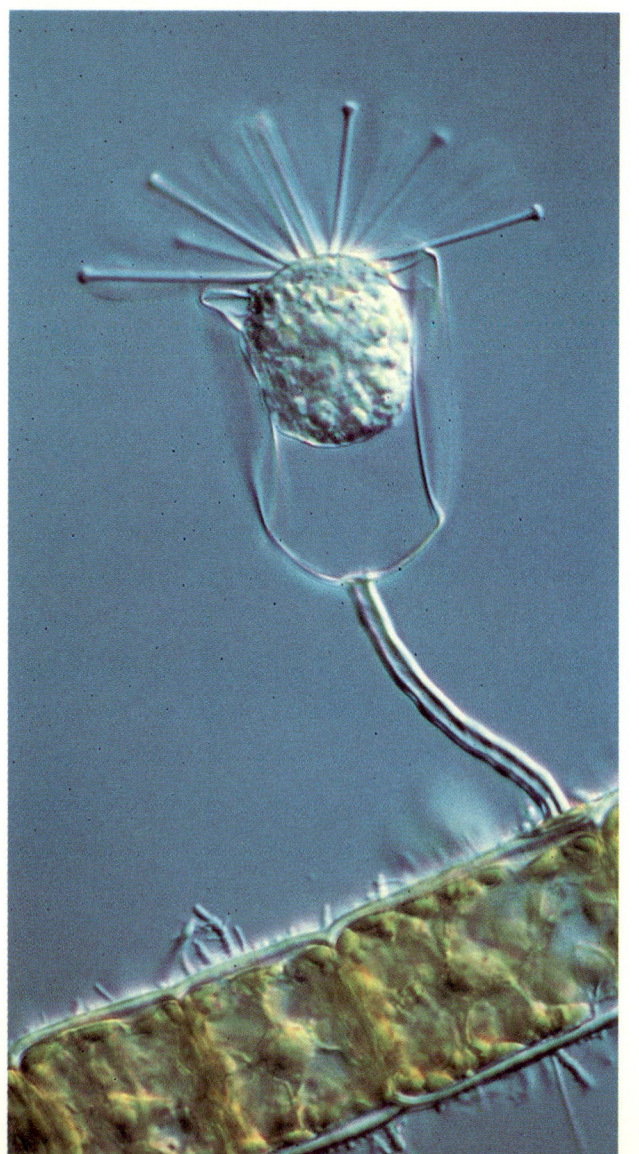

C

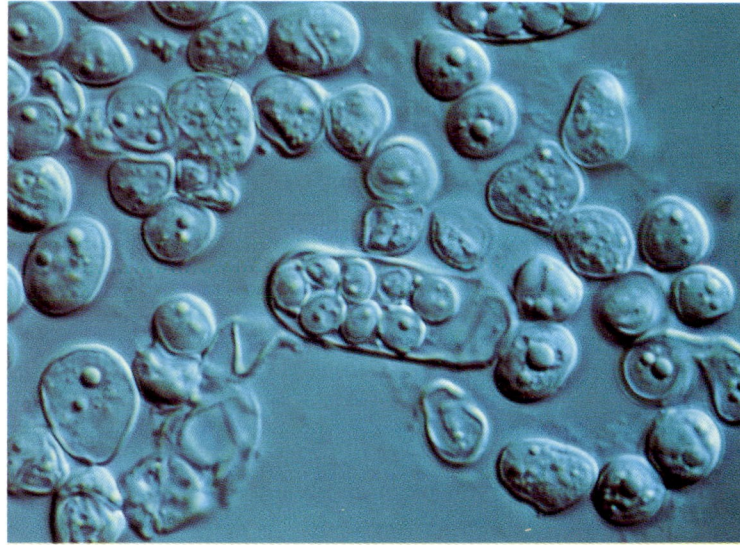

D

Plate 3 **(A)** Filaments of the cyanobacteria *Lyngbia* and *Oscillatoria* viewed by Nomarski interference microscopy. *Lyngbia* is larger than *Oscillatoria*. (From BPS—Paul Johnson, University of Rhode Island) **(B)** Micrograph of *Anabaena*, a filamentous cyanobacterium that forms heterocysts, viewed by Nomarski interference microscopy. (From BPS—Paul Johnson, University of Rhode Island) **(C)** Nomarski differential interference micrograph of the protozoan *Parcineta* on the alga *Spongomorpha*. (From BPS—Paul Johnson, University of Rhode Island) **(D)** Nomarski differential interference micrograph of the yeast *Schizosaccharomyces* containing ascospores. (From BPS—J. Robert Waaland, University of Washington)

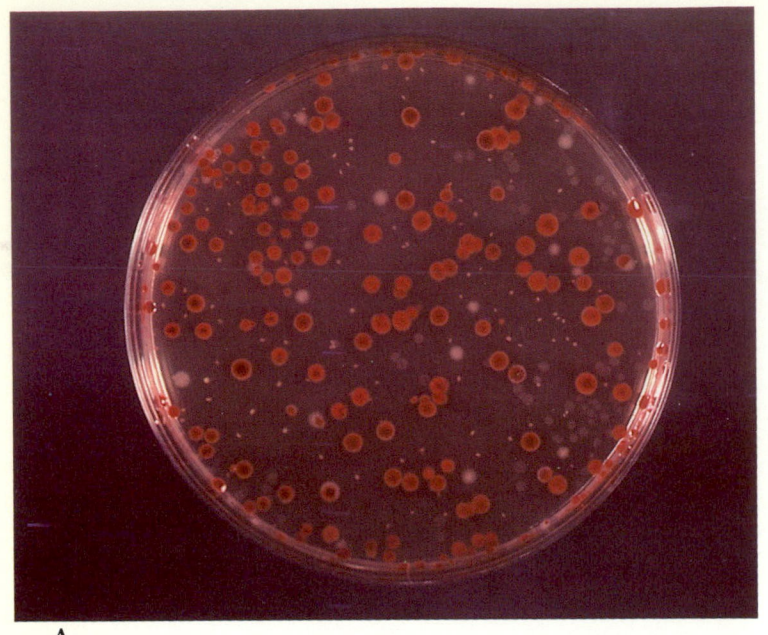

A

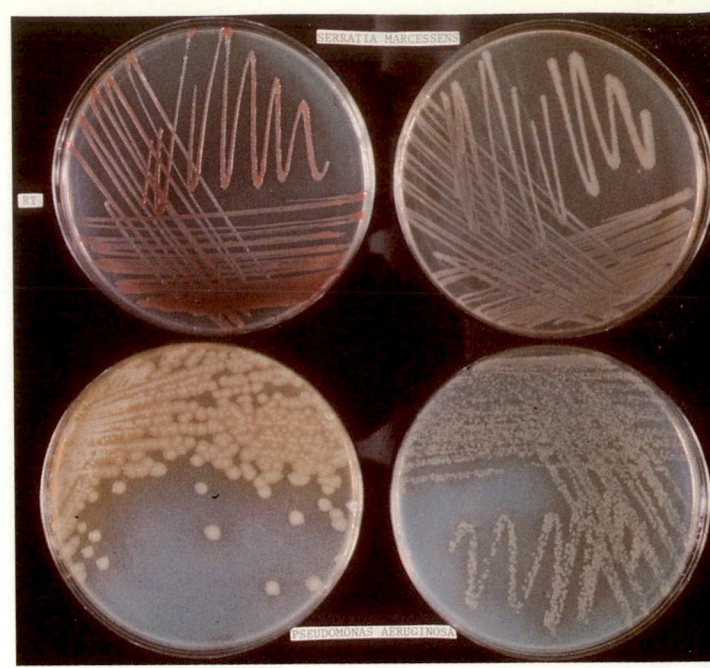

B

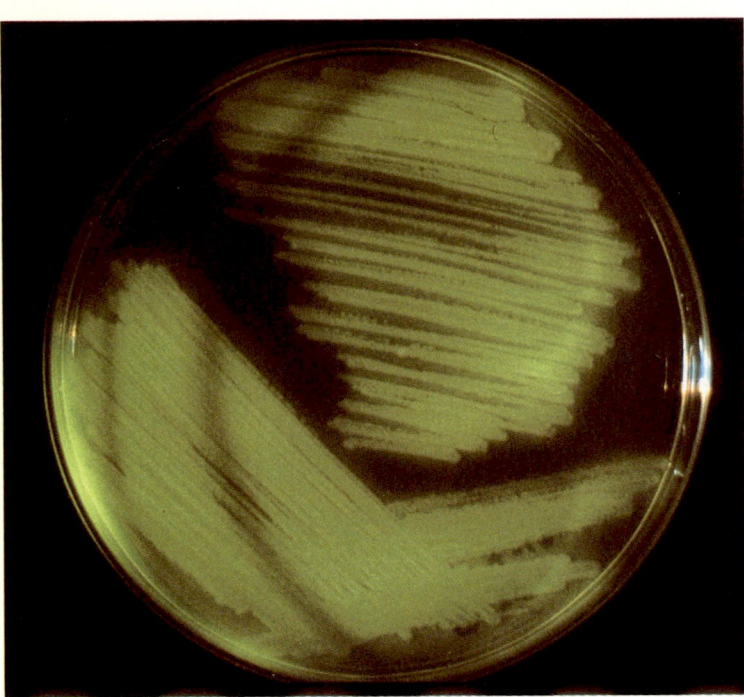

C

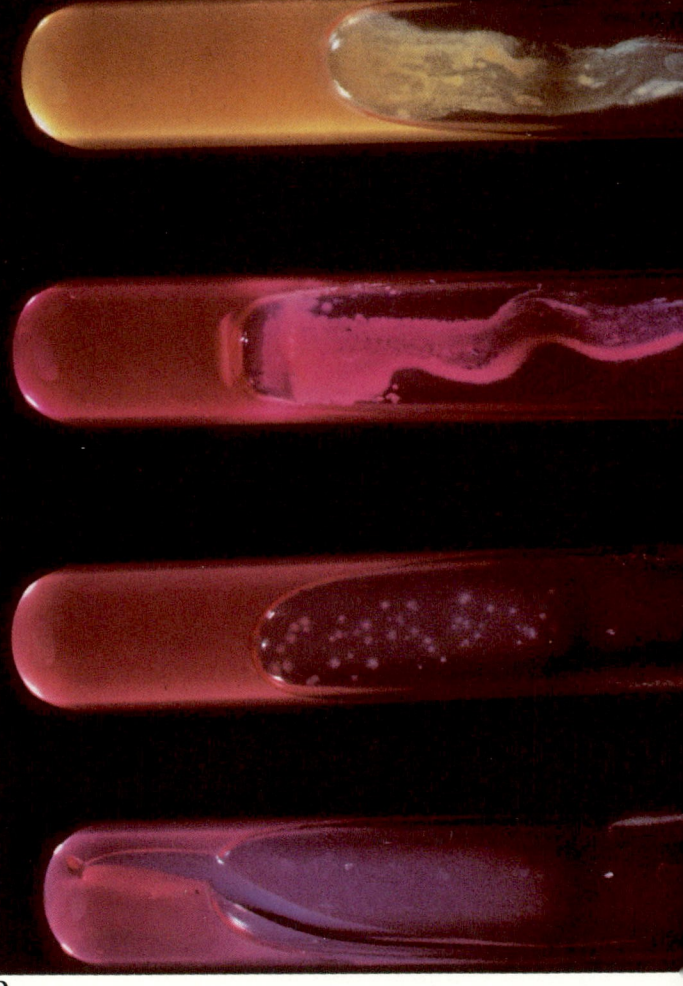

D

Plate 4 **(A)** Pour plate with colonies of *Serratia marcescens* (red) and *Escherichia coli* (grey). (From BPS—B. J. Miller) **(B)** Streak plates showing pigment production by *Serratia marcescens* (top two) and *Pseudomonas aeruginosa* (bottom two) grown at 25°C (left) and 37°C (right). (From BPS—R. L. Miller, BioTechniques Laboratories) **(C)** Plate showing fluorescent growth of *Pseudomonas aeruginosa*. (From BPS—Leon LeBeau, University of Illinois Medical Center) **(D)** Tubes with media containing different sugars inoculated with *Neisseria gonorrhoeae*. This bacterium ferments glucose (top tube that turns yellow due to acid production) but not maltose, sucrose, or lactose (remaining tubes that remain red indicating no acid production). (From BPS—Leon LeBeau, University of Illinois Medical Center)

3.9 Structures Involved in Attachment

Glycocalyx

Whereas some microorganisms move to meet their material and energy requirements, others have structures that enable them to attach to surfaces where they find favorable conditions for growth. Many bacterial cells are surrounded by a specialized structure called the **glycocalyx** that plays such a role in attachment processes. The glycocalyx is a mass of tangled fibers of polysaccharides or branching sugar molecules surrounding an individual cell or colony of cells. It may act to bind cells together, forming multicellular aggregates. Additionally, the glycocalyx of some bacteria are involved in attachment to solid surfaces (Figure 3.42). Some pathogenic bacteria, for example, adhere to the animal tissues they invade via a glycocalyx. Other bacteria in aquatic habitats seem to be held to rocks through the slime layers they secrete. Bacteria occurring in the oral cavity on the surfaces of teeth form an extensive polysaccharide slime, dental plaque, which enables them to adhere to the tooth (Figure 3.43). This adherence to the tooth surface is important in the formation of dental caries.

Pili

In addition to the glycocalyx and slime layers, **pili**, which are short hair-like projections composed primarily of protein subunits that emanate from the surface of some bacteria, are involved in attachment processes (Figure 3.44). Apparently there are several different types of pili associated with the bacterial surface, each serving a different function, but at-

FIGURE 3.42

The bacterial glycocalyx is involved in the attachment of bacteria to solid surfaces; in this electron micrograph the stabilized glycocalyx surrounds enteropathogenic *E. coli* cells. (Courtesy William Costerton, University of Calgary)

FIGURE 3.43

The slime layer of *Streptococcus* species permits the attachment of bacterial cells to tooth surfaces and the formation of dental plaque. This electron micrograph, showing bacteria in plaque and actinomycetes in the slime (plaque) layer, exemplifies the normal microbiota associated with human tooth enamel (13,600×). (From BPS—Max Listgarten, School of Dental Medicine, University of Pennsylvania)

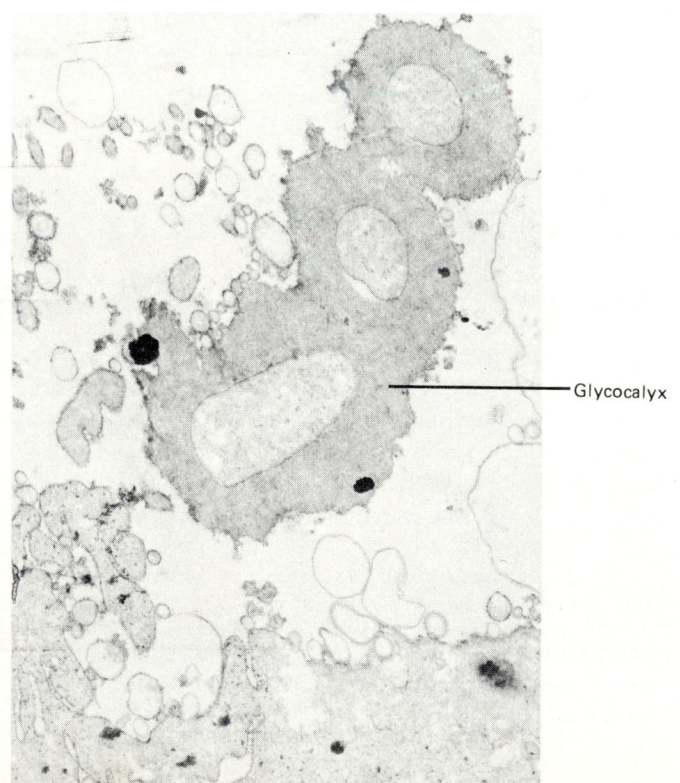

Glycocalyx

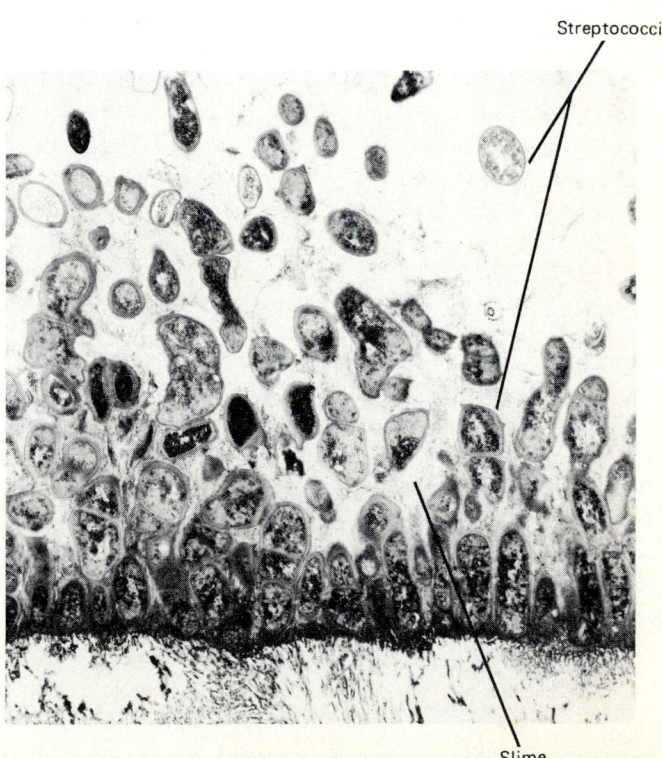

Streptococci

Slime

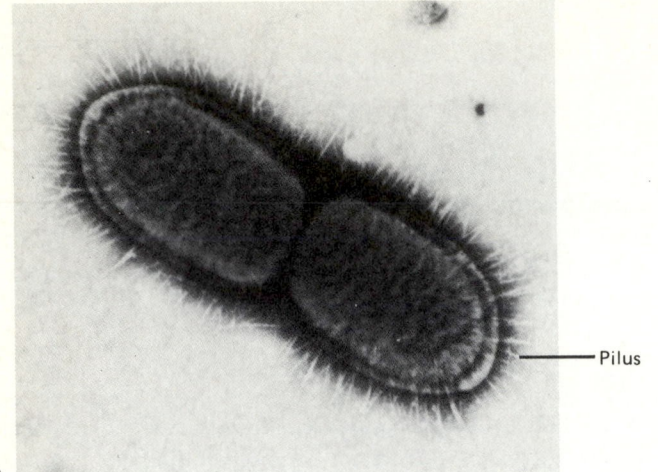

A

Pilus

FIGURE 3.44

(A) *Proteus mirabilis*, negatively stained with phosphotungstic acid and surrounded by pili (36,000×). (Courtesy J. F. M. Hoeniger, University of Toronto; reprinted by permission of the Society for General Microbiology, from J.F.M. Hoeniger, 1965, *Journal of General Microbiology*, 40:29–42). (B) Electron micrograph showing the F pilus acting as a conjugation tube. The bacterial cell covered with numerous appendages is a genetic donor connected to a recipient cell (without appendages) by the F pilus. The F pilus is necessary for the genetic donor to transfer bacterial genes to the recipient. In this micrograph, the F pilus has been labeled along its length by the use of phage that specifically attach to this pilus, permitting its recognition. The numerous other appendages on the donor are called *type I* pili and have no role in conjugation. (Courtesy Charles C. Brinton, Jr., and Judith Carnahan, University of Pittsburgh)

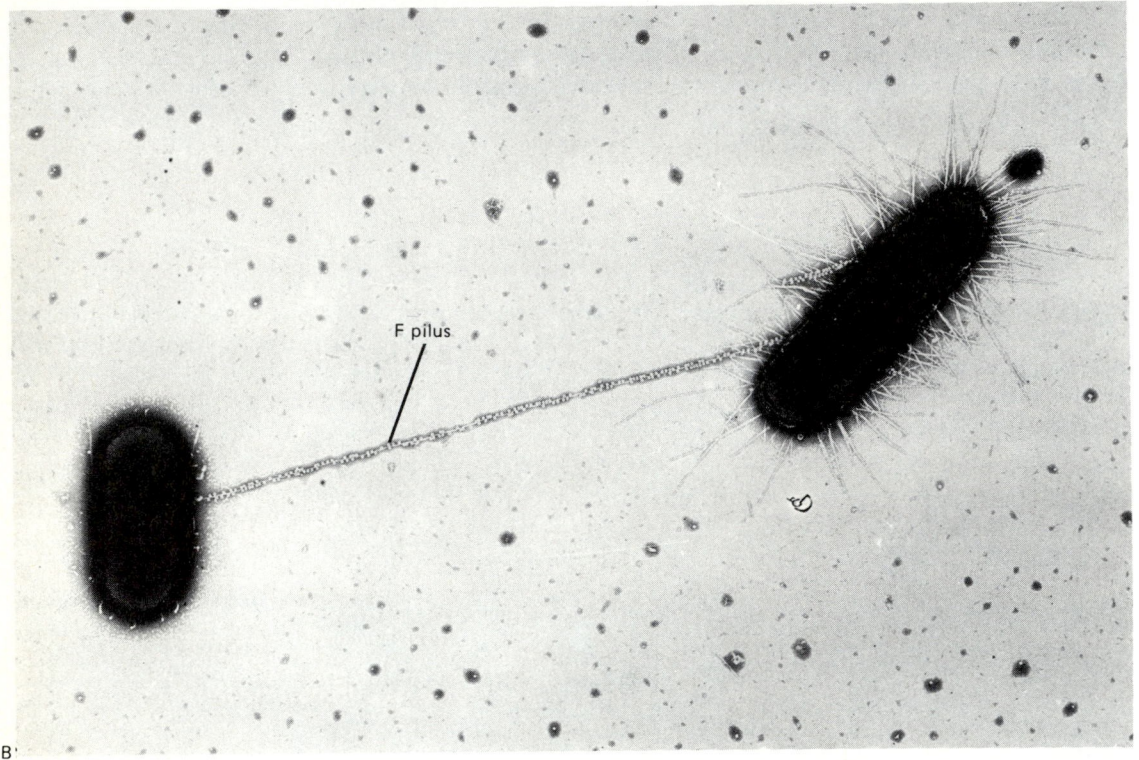

F pilus

B

tachment or adhesion is central to these various functions.[16] The **F or sex pilus** is involved in bacterial mating and is found exclusively on the cells that donate DNA during this process. Mating pairs cannot form in the absence of an F pilus or if the bridge established by the F pilus between the DNA donor and recipient cells is interrupted.[17]

Pili also act as receptor sites for some bacteriophages, providing these bacterial viruses with a site of attachment to the bacterial cell. The phage attach to the pili and subse-

quently transfer their genetic information to the bacterial cell. Pili have also been implicated in the ability of bacteria to recognize specific receptor sites that enable them to attach to the cytoplasmic membranes of eukaryotic cells. The pili seem to play the role of adhesins (substances involved in

[16]Sometimes a distinction is made between types of attachment processes, with the term *pilus* referring only to attachment between mating bacterial cells and the term *fimbriae* referring to all other attachment processes.

[17]F pili are phosphate–carbohydrate–protein complexes with a single peptide subunit called *pilin*. The cylindrical nature of the F pilus could permit the passage of single-stranded DNA, but once a mating pair is established through pilus wall contact, the cells rapidly establish wall–wall contact, and the DNA transfer therefore can occur directly through the contacting walls. Regardless of the exact mechanism, it appears that the contact of the F pilus of the donor cell with the wall of the recipient cell establishes the mating pair and initiates the transfer of DNA.

adhesion) by allowing bacteria to attach to and colonize host cells, sometimes leading to colonization and disease in the host organism. For example, *Neisseria gonorrhoeae*—the etiologic agent of gonorrhea—attaches to the surfaces of cells of the human genitourinary tract via its pili when it initiates colonization and the subsequent disease process. In such processes, the pili act as points of specific contact and attachment between the bacterial cell and another surface.

3.10 Survival Through the Production of Spores

Spores

Having discussed the various parts of a cell, let us consider the specialized resistant cells, called **spores**, that are produced by many microorganisms—especially eukaryotes—to enhance the survival potential of the organism. Spores typically are involved in reproduction, dispersal, or the ability of the organism to withstand adverse environmental conditions. Each of these functions is involved in the overall survival of the organism. The spores involved in reproduction are metabolically quite active, whereas those involved in dispersal or survival of the microorganism often are metabolically dormant. Spores involved in the dispersal of microorganisms usually are quite resistant to desiccation (drying), and the production of such spores is an important adaptive feature that permits the survival of microorganisms for long periods of time during transport in the air. The spread of many fungi depends on the successful transport of fungal spores from one place to another. Unfortunately, many of these fungi are plant pathogens and cause great agricultural damage as a result of their ability to move from field to field.

Bacterial Endospores   Of the many types of microbial spores, one specific type, the **bacterial endospore**—a heat-resistant spore formed within the cell—has special importance. The endospore is a complex, multilayered structure containing peptidoglycan within its complex spore coat and **calcium dipicolinate** within its core. The endospore is highly refractory and resistant to elevated temperatures and desiccation, retaining its viability over long periods of time under conditions that do not permit growth of the organism. Endospores can survive exposure to high temperatures for extended periods, whereas normal bacterial vegetative cells are killed by brief exposures to such high temperatures. The absence of water and the presence of calcium dipicolinate are involved in conferring heat resistance on the endospore. Cells growing in a medium lacking calcium and mutant strains that cannot form calcium dipicolinate produce endospores that are not particularly resistant to elevated temperatures.

Few bacterial genera are capable of forming endospores; the most important endospore producers are members of two genera, *Bacillus* and *Clostridium*. Members of both genera are defined as Gram-positive rods that form endospores; *Bacillus* is aerobic, growing in the presence of oxygen, and *Clostridium* is obligately anaerobic, growing only in the absence of oxygen. That members of these genera form endospores presents special problems for the food industry, which employs processes that rely on heat to prevent spoilage of products. Because some endospores can withstand boiling for more than 1 hour, in order to ensure that the endospores are killed, it is necessary to heat liquids to a temperature greater than 120°C and maintain that temperature for at least 15 minutes; dry materials require several hours at this temperature to ensure sterilization.

Endospores are formed when conditions are unfavorable for continued growth of the bacterium. **Sporulation**, that is, the formation of spores, can be initiated under conditions of starvation. Once started, the process at some point becomes irreversible, and sporulating bacteria continue to form spores even when starvation is relieved and conditions suitable for growth are restored. During the sporulation process, there is an invagination of the cytoplasmic membrane within the cell to establish the site of endospore formation (Figure 3.45). A copy of the bacterial chromosome is incorporated into the endospore, and the various layers of the endospore are then synthesized around the bacterial DNA. In this process, biochemicals are synthesized that are specifically related to endospore formation; these include dipicolinic acid, involved in conferring heat resistance to the spore, and polypeptides composed almost exclusively of single amino acids, such as cystine. The formation of the completed spore involves the synthesis of two cell wall layers and the formation of a spore cortex. Once the endospore is formed within the parent cell, it can be released by lysis of the parent cell. Endospores can retain viability for millennia, and viable endospores have been found in geological deposits where they must have been dormant for thousands of years.

Under favorable conditions, such as when water and nutrients are available and the temperature is permissive of growth, the endospore can germinate and give rise to an active vegetative cell of the bacterium (Figure 3.46). During **germination** the spore swells, breaks out of the spore coat, and elongates. One of the striking features of spore germination is the speed with which metabolism shifts from a state of dormancy to the high activity levels that characterize a germinating spore. This shift in metabolic activity can occur within minutes. The endospore is metabolically self-sufficient, and during germination ATP generation and protein synthesis can take place for at least 15 minutes, using the energy and substrates— principally phosphoglycerate—contained within the spore. After spore germination, the organism renews normal vegetative growth.

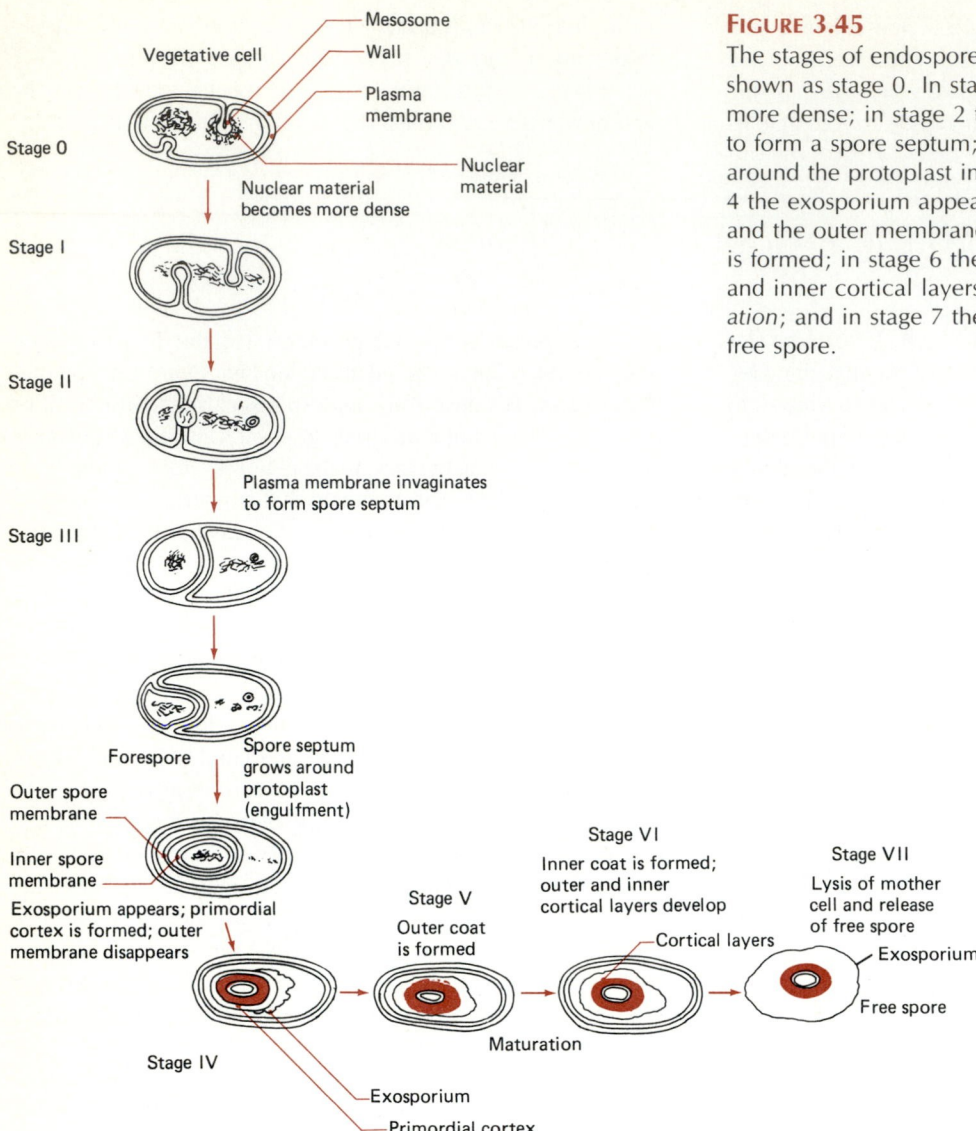

FIGURE 3.45

The stages of endospore formation. A typical vegetative cell is shown as stage 0. In stage 1 the nuclear material becomes more dense; in stage 2 the cytoplasmic membrane invaginates to form a spore septum; in stage 3 a spore septum grows around the protoplast in a process called *engulfment*; in stage 4 the exosporium appears, the primordial cortex is formed, and the outer membrane disappears; in stage 5 the outer coat is formed; in stage 6 the inner coat is formed as the outer and inner cortical layers develop in the process called *maturation*; and in stage 7 the mother cell lyses and releases a free spore.

Postlude

Even the smallest forms of life, the microorganisms, are highly organized living systems. They are composed of a large variety of biochemical constituents and contain a large number of specialized structures. The organizational patterns of viruses, bacteria, and other microorganisms are quite different. The structure of the cell allows a separation of the organism (the living system) from the surrounding environment, with the cytoplasmic membrane acting as a semipermeable barrier that controls the flow of materials into and out of the cell. Viruses, which lack a cytoplasmic membrane, are obligately dependent on host cells to provide a suitable environment for their replication. The membrane is critical in biological systems, as it permits the inside of the cell to remain structured and functional while the entropy of the surroundings increases. Transport of materials across this boundary layer must be carefully regulated to prevent the random diffusion of cellular constituents; the bilipid membrane, with its hydrophilic and hydrophobic portions, is well adapted for this function. The cell's boundary is an active layer, and the proteins associated with the cytoplasmic membrane play an active role in the transport of materials into and out of the cell. With the recognition of chemiosmosis as a major process in driving the synthesis of ATP, the importance of the membrane in the bioenergetics of the cell, as well as in transport, becomes apparent. It is now clear that the establishment of a hydrogen ion gradient across a membrane is essential for the generation of the majority of the ATP produced in many organisms.

The protection of the membrane against rupture due to osmotic shock is critical for the survival of microorganisms. Bacteria have evolved a distinctive, complex wall structure containing murein that acts as a rigid outer layer to protect

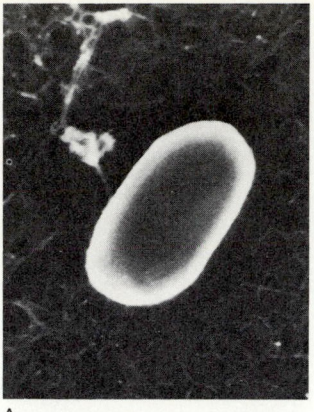
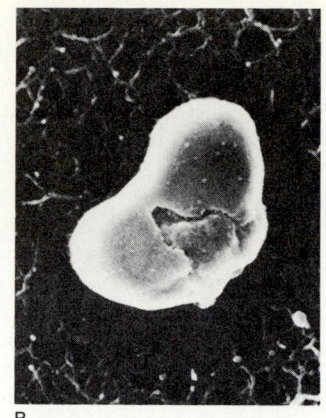
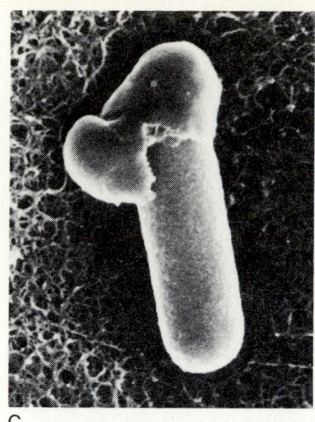
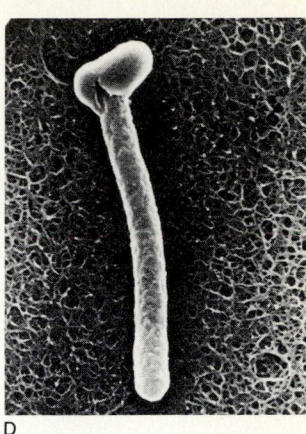

A   B   C   D

FIGURE 3.46

Micrograph showing spore germination. (A) Scanning electron micrograph of a spore of *Bacillus subtilis* (30,000×) 0 minutes following heat activation at 60°C for 1 hour; (B) after 90 minutes of incubation on nutrient agar at 37°C, a crack can be seen in an equatorial position (30,000×); (C) after 120 minutes on an agar surface, the vegetative bacterium has started to grow (22,500×); (D) after 150 minutes of incubation on agar, the vegetative bacterium has grown extensively, the bacterium retains an empty spore shell on one or both ends, and the cell is about 4–5 μm in size. (Courtesy of Akiko Umeda, Fukora University)

the cell. Other external structures, such as the bacterial capsule, also act to protect the cell against physical and chemical forces that could disrupt its organizational structure. Some cells also possess organelles of locomotion that permit them to move and find more favorable environments away from forces disruptive to growth and reproduction. Even bacteria can recognize and respond to their chemical surroundings, moving toward or away from chemical stimuli by chemotaxis.

Although both prokaryotic and eukaryotic cells are highly organized, the differences between them represent a true split in architectural organizational strategies. These different strategies give rise to the formation of distinct structures in prokaryotic and eukaryotic organisms, with some structures occurring exclusively in prokaryotes and others exclusively in eukaryotes (Table 3.4). Although the genetic information of both prokaryotic and eukaryotic cells is encoded within DNA, the storage of that information is inherently different. The genetic information of eukaryotes often is stored as split sections, whereas in prokaryotic cells functional units occur as contiguous segments of the genome. The nucleus of the eukaryotic cell allows the processing of the genetic infor-

TABLE 3.4   Comparison of Eukaryotic and Prokaryotic Cell Structures

Structure	Prokaryotic Cell	Eukaryotic Cell
Cytoplasmic membrane	+	+
Sterols in membranes	−	+
Nucleus	−	+
DNA arranged as true chromosomes with associated histone proteins	−	+
Ribosomes	70 S	80 S
Ribosomal subunits	50 S + 30 S	60 S + 40 S
Cell wall	Contains murein	Several types; none with murein
Internal membranous organelles	±	+
Chloroplasts	−	+
Mitochondria	−	+
Endoplasmic reticulum	−	+
Golgi apparatus	−	+
Vacuoles	±	+
Flagella	+	+
9 + 2 microtubular arrangement	−	+
Cytoskeleton	−	+

Note: ± denotes general absence, although membranous structures occur in some species of prokaryotes.

mation within a protected organelle before it is translated into proteins. In bacteria such processing cannot occur because the DNA is not segregated within a membrane-bound organelle. The membranous organelle substructures of eukaryotic organisms, including the nucleus, establish a compartmentalization of function and permit a large number of specialized activities to occur within the integrated framework of the eukaryotic cell. The eukaryotic cell also has an extensive support and communication network that permits the coordination of complex cellular activities.

Suggested Supplementary Readings

ADLER, J. 1976. The sensing of chemicals by bacteria. *Scientific American* 234(4):40–47.

ALLEN, M. M. 1984. Cyanobacterial cell inclusions. *Annual Review of Microbiology* 38:1–26.

AMES, G. F-L. 1986. Bacterial periplasmic transport systems: Structures, mechanism, and evolution. *Annual Review of Biochemistry* 55: 397–426.

BERG, H. C. 1975. How bacteria swim. *Scientific American* 233(2):36–44.

BLAKEMORE, R. P. 1982. Magnetotactic bacteria. *Annual Review of Microbiology* 36:217–238.

BLAKEMORE, R. P., and R. B. FRANKEL. 1981. Magnetic navigation in bacteria. *Scientific American* 245(6):58–67.

CAIRNS, J. 1963. The bacterial chromosome and its manner of replication as seen by autoradiography. *Journal of Molecular Biology* 6:208–213.

CAMPBELL, A. 1981. Evolutionary significance of accessory DNA elements in bacteria. *Annual Review of Microbiology* 35:55–84.

COSTERTON, J. W. 1979. The role of electron microscopy in the elucidation of bacterial structure and function. *Annual Review of Microbiology* 33:459–479.

COSTERTON, J. W., G. G. GEESEY, and K.-J. CHENG. 1978. How bacteria stick. *Scientific American* 238(1):86–95.

COSTERTON, J. W., R. T. IRWIN, and K.-J. CHENG. 1981. The bacterial glycocalyx in nature and disease. *Annual Review of Microbiology* 35:299–324.

DILLS, S. D., A. APPERSON, M. R. SCHMIDT, and M. H. SAIER, JR. 1980. Carbohydrate transport in bacteria. *Microbiological Reviews* 44:385–418.

DIRIENZO, J. M., K. NAKAMURA, and M. INOUYE. 1978. The outer membrane proteins of Gram-negative bacteria: Biosynthesis, assembly, and functions. *Annual Review of Biochemistry* 47:481–553.

DOETSCH, R. N., and R. D. SJOBLAD. 1980. Flagellar structure and function in eubacteria. *Annual Review of Microbiology* 34:69–108.

FINEAN, J. B., R. COLEMAN, and R. H. MITCHELL. 1978. *Membranes and Their Cellular Functions*. Blackwell Scientific Publications, Oxford, England.

GRAY, M. W., and F. DOOLITTLE. 1982. Has the endosymbiont hypothesis been proven? *Microbiological Reviews* 46:1–42.

GREENAWALT, J. W., and T. L. WHITESIDE. 1975. Mesosomes: Membranous bacterial organelles. *Bacteriological Reviews* 39:405–463.

HANCOCK, R. E. W. 1984. Alterations of outer membrane permeability. *Annual Review of Microbiology* 38:237–266.

HEPPEL, L. A. 1969. The effect of osmotic shock on release of bacterial proteins and active transport. *Journal of General Physiology* 54:95s–113s.

INOUYE, M. (ed.) 1979. *Bacterial Outer Membranes: Biogenesis and Functions*. John Wiley & Sons, Inc., New York.

JAIN, M. K., and R. C. WAGNER. 1980. *Introduction to Biological Membranes*. Wiley-Interscience, New York.

JENSEN, W. A., and R. B. PARK. 1967. *Cell Ultrastructure*. Wadsworth Publishing Co., Belmont, Calif.

KORNBERG, R. D., and A. KLUG. 1981. The nucleosome. *Scientific American* 244(2):55–72.

KOSHLAND, D. E., JR. 1980. *Bacterial Chemotaxis as a Model Behavioral System*. Raven Press, New York.

LAKE, J. A. 1981. The ribosome. *Scientific American* 245(2):84–97.

LOEWY, A. G., and P. SIEKEVITZ. 1970. *Cell Structure and Function*. Holt, Rinehart & Winston, Inc., New York.

MACNAB, R. W., and S. AIZAWA. 1984. Bacterial motility and the bacterial flagellar motor. *Annual Review of Biophysics and Bioengineering* 13:51–84.

Molecules to Living Cells: Readings from Scientific American. 1980. W.H. Freeman and Co., San Francisco.

NIKAIDO, H., and M. VAARA. 1985. Molecular basis of bacterial outer membrane permeability. *Microbiological Reviews* 49:1–32.

PETTIJOHN, D. E. 1976. Prokaryotic DNA in nucleoid structure. *CRC Critical Reviews of Biochemistry* 4:175–202.

PORTER, K. R., and J. B. TUCKER. 1981. The ground substance of the living cell. *Scientific American* 244:(3):57–67.

REUSCH, V. M. 1984. Lipopolymers, isoprenoids, and the assembly of the Gram-positive cell wall. *CRC Critical Reviews in Microbiology* 11:129–156.

ROGERS, H. J. 1983. *Bacterial Cell Structure*. American Society for Microbiology, Washington, D.C.

ROSEN, B. P. (ed.). 1978. *Bacterial Transport*. Marcel Dekker Inc., New York.

SALTON, M. R. J., and P. OWEN. 1976. Bacterial membrane structure. *Annual Review of Microbiology* 30:451–482.

SATIR, P. 1974. How cilia move. *Scientific American* 231(4):44–63.

SCHLEIFER, K. H., and O. KANDLER. 1972. Peptidoglycan types of bacterial cell walls and their taxonomic implications. *Bacteriological Reviews* 36:407–477.

SHIVELEY, J. M. 1974. Inclusion bodies of procaryotes. *Annual Review of Microbiology* 28:167–187.

SHOCKMAN, G. D., and J. F. BARRETT. 1983. Structure, function, and assembly of cell walls of Gram-positive bacteria. *Annual Review of Microbiology* 37:501–528.

SILVERMAN, M., and M. I. SIMON. 1977. Bacterial flagella. *Annual Review of Microbiology* 31:397–420.

SLEYTR, U. B., and P. MESSNER. 1983. Crystalline surface layers on bacteria. *Annual Review of Microbiology* 37:311–340.

SUTHERLAND, I. W. 1985. Biosynthesis and composition of

Gram-negative bacterial extracellular and wall polysaccharides. *Annual Review of Microbiology* 39:243–270.

WALSBY, A. E. 1977. The gas vacuoles of blue-green algae. *Scientific American* 237(2):90–97.

WITTMANN, H. G. 1983. Architecture of prokaryotic ribosomes. *Annual Review of Biochemistry* 52:35–66.

WOESE, C. R. 1981. Archaebacteria. *Scientific American* 244(6):98–122.

YATES, G. T. 1986. How microorganisms move through water. *American Scientist* 74:358–365.

Study Questions

1. What are the fundamental differences between prokaryotic and eukaryotic cells? Which groups of microorganisms are prokaryotic, and which are eukaryotic?

2. What structures occur in eukaryotic cells that are not found in prokaryotic cells? What structures are found in prokaryotic cells and not in eukaryotic cells? What structures occur in both prokaryotic and eukaryotic cells?

3. How does a virus differ from a bacterium?

4. What are the differences between bacteria and fungi in terms of how the genetic information is stored?

5. What is osmotic pressure, and what strategies have microorganisms evolved for protection against this force?

6. Describe the differences in cell wall structure between Gram-negative and Gram-positive bacteria.

7. What are the similarities between a mitochondrion and a bacterial cell?

8. What are the structural differences between archaebacteria and eubacteria?

9. What is a bacterial endospore, and what is the significance of this structure?

10. Discuss how materials move into and out of cells. What are the different transport mechanisms in prokaryotic and eukaryotic cells? How is the structure of the cytoplasmic membrane related to transport processes?

Situational Problem 1

Debating the Origins of Eukaryotic Flagella

The question "Has the endosymbiont hypothesis been proven?" appears to have been answered in the affirmative through the use of RNA analyses, at least as it relates to the evolution of mitochondria and chloroplasts. It now seems certain that chloroplasts and mitochondria evolved from an endosymbiotic relationship between a primitive eukaryotic cell and a eubacterial prokaryotic cell. Prior to molecular genetic-level analyses, however, this topic remained unresolved and was still the subject of argumentative debates even when the structural similarities between these eukaryotic organelles and the prokaryotic cell were known. The reason is that appearance alone cannot be the sole proof of a scientific hypothesis. It had been logically argued that even though mitochondria and chloroplasts appear to be more eubacterial than eukaryotic (nuclear), this was because the traits being considered were primitive ones and because mitochondria and chloroplast genomes changed more slowly than nuclear genomes after evolutionary divergence from a common ancestor occurred.

Even though the 16S rRNA analyses settled part of the debate on the endosymbiotic theory, it did not resolve the argument over the origins of eukaryotic flagella. Did eukaryotic flagella and cilia, with their 9 + 2 organization, arise from an endosymbiotic eubacterium? Let us consider the interesting organism *Mixotricha paradoxa*. This protozoan lacks its own mitochondria but harbors endosymbiotic eubacteria that carry out the essential metabolism that provides the protozoan with necessary ATP as an energy source. The protozoan swims along, apparently propelled by bacterial cells attached to the surface. These attached bacteria are spirochetes, which have a central axial filament connecting the two ends of the cell. The filament enables the spirochetes, which are approximately the same size as a eukaryotic flagellum, to contract and move with a creeping motion similar to that of a caterpillar. The filaments connecting the ends of the spirochete resemble those of the eukaryotic flagellum, although they do not have the typical 9 + 2 arrangement. There may be as many as 50,000 spirochetes attached to the surface of a single cell of *M. paradoxa*. There are also four eukaryotic flagella with a 9 + 2 arrangement that appear to steer rather than to propel the protozoan. Clearly, as its name implies, this organism is a biological paradox.

Now let us assume that you are a member of the university debating team and the topic for the next contest is the origin of the eukaryotic flagellum, a topic certain to attract all biology students. Choose a side in the debate and begin to prepare your arguments. Consider whether the evidence supports the view that flagella and cilia of eukaryotes originated from spirochetes, what additional lines of evidence you might require, and exactly how you could resolve this question. In preparing for this debate, you may want to read L. Margulis (1982), *Symbiosis in Cell Evolution*, W. H. Freeman, San Francisco; L. Margulis and D. Sagan (1986), *Micro-Cosmos: Four Billion Years of Microbial Evolution*, Summit Books, New York; M. W. Gray and W. F. Doolittle (1982), Has the endosymbiont hypothesis been proven?, *Microbiological Reviews* 46:1–42; M. A. Sleigh (1985), Origin and evolution of flagella movement, *Cell Motility* 5:137–173; and more recent relevant articles that you can find in the library. Be sure to consider the alternate views so that you are prepared for the opposition.

Situational Problem 2

Mission to Recognize Extraterrestrial Life Forms

Defining life and recognizing living systems is not always easy, as evidenced by the debate over whether viruses should be considered living organisms. Because life on Earth is so diverse, we have difficulty in recognizing the unifying structural and functional properties common to *all* living organisms. Living systems are highly organized, but so is an ice crystal; they exchange materials with their surroundings, but so does a mailbox; they transform energy, but so does a solar cell; they process information, but so does a computer; they reproduce themselves, but so does the robot that has been programmed to make more robots. No one would ever claim that the crystal, the mailbox, the solar cell, the computer, or the robot is alive, so we feel safe in proclaiming, "We know life when we see it," believing that we have little difficulty in recognizing living organisms, even microbes, when we see them.

Since we can't decide exactly what to tell a laymen to look for when searching for life, you—as a student of microbiology—have been chosen to travel on the next mission to planet X to search for life. You have been chosen with the confidence that you will know what to look for and will know how to discriminate between living organisms, inanimate objects, and extraplanetary artifacts. Previous space missions to planet X have failed to detect any visible forms of macroscopic life, but there is available water and environmental conditions are within the tolerance limits where living microorganisms are found on Earth. The chemical composition of the planet and the atmosphere, however, is quite different from that of Earth. Therefore, if microbial life exists on planet X, the microbes there need not have the same structures as microbes on Earth, both in terms of their biochemical composition and their physical arrangement. Your spacecraft is outfitted with the best light and electron microscopes. The ship has a computer available for your use, and you may bring up to 50 pounds of additional scientific equipment to assist you in your quest for life forms.

Before embarking, you should establish your observational and experimental plan and the criteria that you are going to use to define living organisms. Begin now by entering in your logbook what you are going to look for and how you are going to know if whatever you find is or is not a life form.

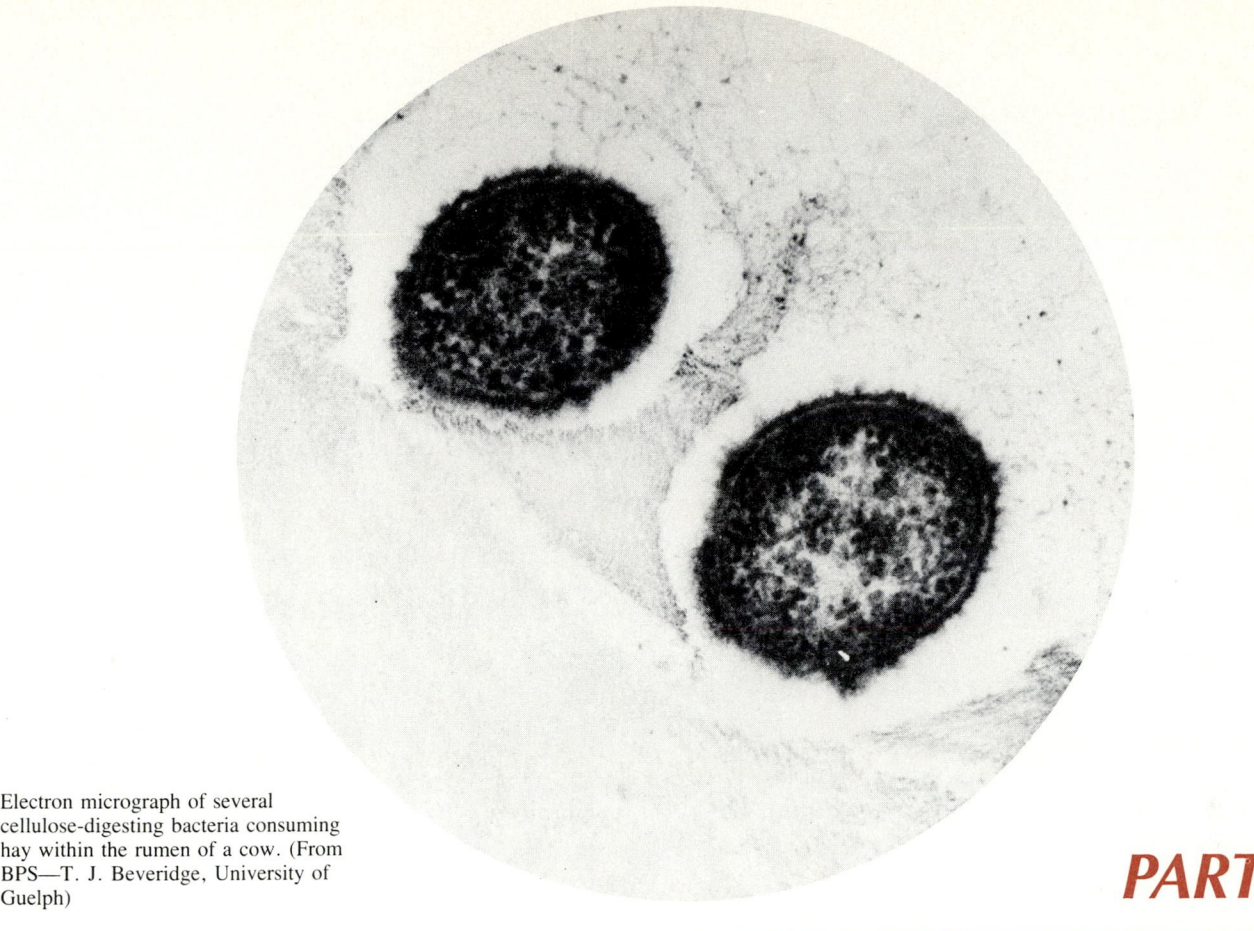

Electron micrograph of several cellulose-digesting bacteria consuming hay within the rumen of a cow. (From BPS—T. J. Beveridge, University of Guelph)

Microbial Growth and Metabolism

CHAPTER 4

Culture, Nutrition, and Growth of Microorganisms

4.1 *Culture of Microorganisms*

Much of the information we have just reviewed on the structures of microorganisms has come from work with **pure cultures**, that is, single types of microorganisms that are free from all other types. The ability to examine and study the characteristics of microorganisms, including obtaining organisms for microscopic visualization, depends in large part on being able to grow pure cultures of organisms in the laboratory, where they can be studied. In this section we will examine some of the procedures and their principles used for growing single species or types of microorganisms isolated from all others.

Sterilization

In order to establish a pure microbial culture, it is necessary to eliminate unwanted microorganisms. There are various ways of eliminating microorganisms from the liquids, containers, and instruments used in pure culture procedures, including exposure to elevated temperatures, toxic chemicals, or radiation to kill microorganisms and filtration to remove them.[1] The heat killing of microorganisms can be described by the **decimal reduction time (*D*)** (Figure 4.1). *D* is defined as the time required for a 10-fold reduction in the number of viable cells at a given temperature, that is, the time required for a log reduction in the number of microorganisms. As the temperature is increased above the maximal growth

[1]Removal of microorganisms by filtration generally is accomplished by passage through an 0.2- to 0.45-μm bacteriological filter; most bacteria are trapped on the filter, but viruses and some very small bacteria may pass through it.

temperature[2] for a microorganism, the decimal reduction time is shortened. In the food industry, where the decimal reduction time is important in establishing appropriate processing times for the safe production of canned products,

[2]The maximal growth temperature for a particular microorganism is the highest temperature at which that organism will reproduce. Above that temperature the organism usually dies. However, even though growth cannot occur at such a temperature, if the microbe produces endospores, the endospores may not be killed, permitting survival of the microorganism despite exposure at the elevated temperature.

FIGURE 4.1

This is a hypothetical survivor curve for a *D* value equal to 10 with an initial concentration of 10^4 cells per milliliter. After 10 minutes there is a log reduction in the number of surviving cells. Increased processing times are needed to sterilize highly contaminated materials. (Based on Frazier and Westhoff, *Food Microbiology*, 1978, McGraw-Hill Book Co.)

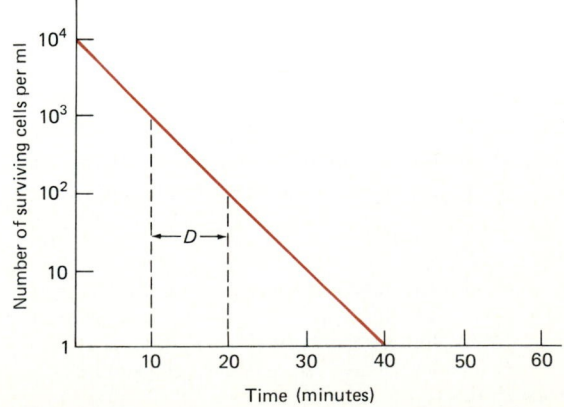

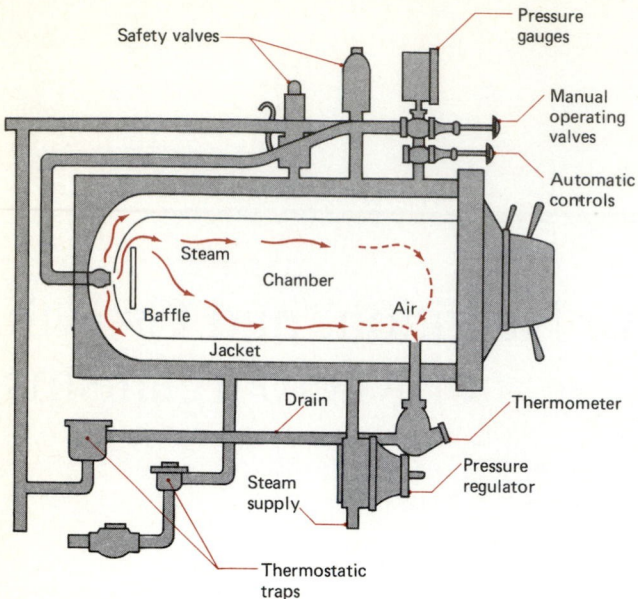

FIGURE 4.2

Diagram of an autoclave. This instrument is routinely used for sterilization of media and other items in the microbiology laboratory. In an autoclave, steam is introduced under pressure into a chamber containing the material to be sterilized. The pressure generally is adjusted to 15 lb/in.² so that a temperature of 121°C (250°F) is reached. The valving of the autoclave permits the rapid entry of steam from a preheated jacket into the chamber and the subsequent slow exhausting of steam from the chamber; this process permits rapid heating of the material and prevents liquids from boiling out of their containers, as would happen if the pressure was suddenly reduced.

an *8D* process is often used, that is, the food is exposed to a given temperature for a period equal to eight times the decimal reduction time.

Culture media in bacteriological laboratories are normally prepared by heat sterilization, which kills all microorganisms, including their endospores. Media preparation for the microbiology laboratory involves the use of an autoclave for sterilization (Figure 4.2). An **autoclave** is an instrument that exposes substances to steam at elevated temperatures; generally, exposure for 15 minutes at 121°C, achieved by using a pressure of 15 lb/in² (SI equivalent = 103.4 kPa), is used to sterilize microbiological media.[3] Much of the time spent in preparation for the bacteriology laboratory involves making the media for growing bacteria, that is, mixing and sterilizing the growth media in suitable culture vessels. The actual time used to sterilize materials may be considerably longer

because of the time needed to heat the material to 121°C and to allow it to cool.

There are several practical problems that must be taken into account when sterilizing materials. The general assumption is that the product will heat very rapidly. In determining the actual time for autoclaving, however, it is necessary to take into account the nature of the material being sterilized and its volume, as these factors can affect the ability of steam to penetrate and hence the time needed to ensure the sterility of a product. Also, while many autoclaves have automatic cycles, it is important to check the temperature recorders in order to ensure that the desired temperatures are reached for the proper length of time. Additionally, one must be certain that glassware is not tightly stoppered, lest it explode, and that when removing materials from the autoclave, superheated material is prevented from escaping from the containers and causing burns. It is also important to note that not all materials can be sterilized in the autoclave. Some materials, such as most plastics, cannot tolerate elevated temperatures. They must be sterilized by alternative means such as exposure to ethylene oxide or gamma radiation from cobalt-60 (^{60}Co), as is done by the commercial producers of the presterilized plasticware used in most microbiology laboratories.

Aseptic Transfer Technique

While many of the petri dishes and tissue culture plates now used for growing pure cultures of microorganisms are plastic and come presterilized from the manufacturer, the filling of these vessels with a sterile microbiological medium requires the use of a method that prevents contamination with microorganisms. This procedure, known as **aseptic technique** (aseptic means free from sepsis and not liable to microbial putrefaction), is also required to transfer the pure culture from one vessel to another. Besides preventing contamination of pure cultures with unwanted microorganisms, proper aseptic transfer technique also protects the microbiologist from contamination with the culture, which should always be treated as a potential pathogen.

Aseptic technique involves avoiding any contact of the pure culture, sterile medium, and sterile surfaces of the growth vessel with contaminating microorganisms. To accomplish this task, (1) the work area is cleansed with a disinfectant to reduce the number of potential contaminants; (2) the transfer instruments are sterilized, for example, by heating a transfer loop in a bunsen burner flame before and after transferring; and (3) the work is accomplished quickly and efficiently to minimize the time of exposure during which contamination of the culture or laboratory worker can occur. The normal steps for **transferring a culture** from one vessel to another are as follows: (1) flame the transfer loop; (2) open and flame the mouths of the culture tubes; (3) pick up some of the culture growth and transfer it to the fresh medium; (4) flame the mouths of the culture vessels and reseal them; and (5) reflame the inoculating loop (Figure

---

[3]Dry heat sterilization requires higher temperatures and much longer exposure periods, on the order of several hours, in order to kill all of the microorganisms in a sample. Exposure in an oven for 2 hours at 170°C (328°F) is generally used for the dry heat sterilization of glassware and other laboratory items.

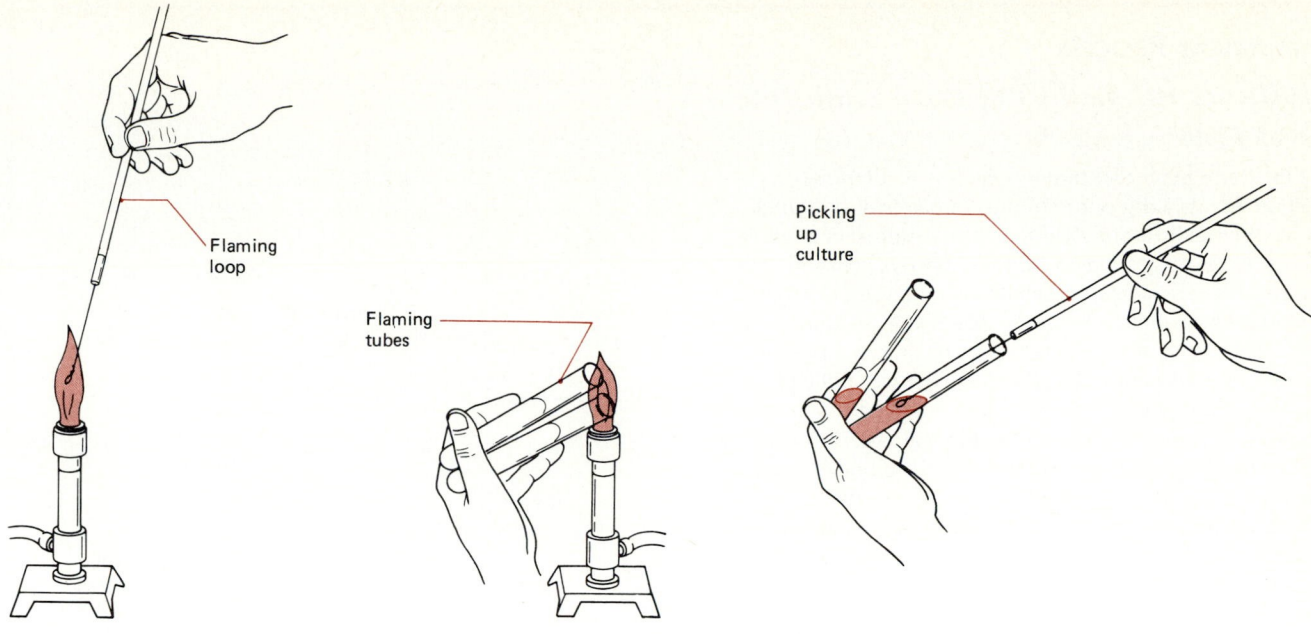

FIGURE 4.3
Steps in the aseptic transfer of bacteria. Aseptic transfer procedures are essential for preventing contamination of cultures and for ensuring that the microorganisms being cultured do not escape into the laboratory.

4.3). Essentially the same technique is used for inoculating petri dishes, except that the dish is not flamed, and for transferring microorganisms from a culture vessel to a microscope slide.

Isolation

Several different methods are used for the establishment of pure cultures of microorganisms. These **isolation methods** usually involve separating microorganisms on a solid medium into individual cells that are then allowed to reproduce to form clones of single microorganisms. Each clone represents a pure culuture. Isolation is achieved by the phys-

ical separation of the microorganisms, but the success of an isolation method also involves the maintenance of the viability and growth of a pure culture of the microorganism. Care must be taken to ensure that the microorganisms are not killed during the isolation procedure, which can easily occur by exposing the microorganisms to conditions they cannot tolerate, such as air in the case of obligately anaerobic microorganisms that are sensitive to oxygen. The success of an isolation method also depends on being able to grow the microorganism, that is, to define the growth medium and to establish the appropriate incubation conditions that permit the growth of the microorganism.

4.2 *Microbial Nutrition*

In order to grow cultures of microorganisms, it is necessary to provide the nutrients that are required for microbial metabolism and reproduction. The specific **nutritional requirements**, that is, the essential growth substances, for different microorganisms, vary greatly. In nature the distribution of microorganisms is determined in part by the availability in a given environment of the specific nutrients needed to support the growth of particular microbial species. The nutritional requirements of microorganisms are reflected in the chemical composition of the cell. Water accounts for some 80 to 90 percent of the total weight of cells and is always, therefore, a major essential nutrient. Hydrogen, oxygen, carbon, nitro-

gen, phosphorus, and sulfur constitute 95 percent of the cellular dry weight; potassium, magnesium, calcium, iron, manganese, cobalt, copper, molybdenum, zinc, and various anions, such as chloride, are also required in trace amounts for growth (Table 4.1, page 92).

Some microorganisms have simple nutritional requirements and are able to reproduce by using a single carbon source, such as glucose, as their growth substrate. Such organisms have the metabolic capacity to convert the single substrate into all of their required constituents. At the other extreme, some microorganisms have complex specific growth factor requirements and are able to reproduce only

ANALYTICAL PROCESS

Methods for Isolating Pure Cultures

Streak Plate

In the **streak plate technique**, which is a commonly used method for isolating pure cultures of bacteria, a loopful of bacterial cells is streaked across the surface of a sterile solidified agar plate that contains a nutrient medium. Several different streaking patterns can be used to separate individual bacterial cells on the agar surface (Figure 4.4). The plates are then incubated under favorable conditions to permit the growth of the bacteria. The key principle of this method is that, by streaking, a dilution gradient is established across the face of the plate as bacterial cells are deposited on the agar surface. Because of this dilution gradient, confluent growth occurs on part of the plate where the bacterial cells are not sufficiently separated and individual isolated colonies develop in other regions of the plate where few enough bacteria are deposited to form separate macroscopic colonies that can easily be seen with the naked eye. Each well-isolated colony is assumed to arise from a single bacterium and therefore to represent a clone of a pure culture. If this important premise is not sustained, for example, because two bacterial cells are deposited at the same location on the plate, the method fails to produce a pure culture. Assuming that each colony comes from a single cell, the isolated colonies can be picked up, using a sterile inoculating loop, and restreaked onto a fresh medium to ensure purity. A new colony is then picked up and transferred to an agar slant or other suitable medium for maintenance of the pure culture.

Spread Plate

In the **spread plate method** a drop of a liquid containing a suspension of microorganisms is placed on the center of an agar plate and spread over the surface of the agar, using a sterile glass rod (Figure 4.5). The glass rod is normally sterilized by being dipped in alcohol and flamed to burn off the alcohol. When the suspension is spread over the plate, individual microorganisms are separated from others in the suspension and are deposited at discrete locations. To accomplish this separation, it is often necessary to dilute the suspension before application to the agar plate to prevent overcrowding and the formation of confluent growth rather than the desired development of isolated colonies. After incubation, isolated colonies are picked up and streaked onto a fresh medium to ensure purity.

Pour Plate

In the **pour plate technique**, suspensions of microorganisms are added to melted agar tubes that have been cooled to approximately 42–45°C (Figure 4.6). The bacteria and agar medium are mixed well, and the suspensions are poured into sterile petri dishes using aseptic technique. The agar is allowed to solidify, trapping the bacteria at separate discrete positions within the matrix

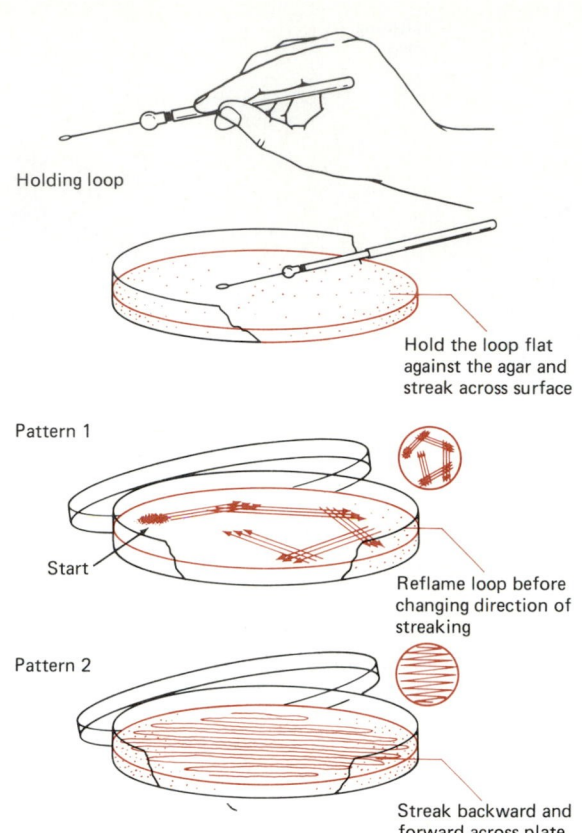

Holding loop

Hold the loop flat against the agar and streak across surface

Pattern 1

Start

Reflame loop before changing direction of streaking

Pattern 2

Streak backward and forward across plate

FIGURE 4.4
Streaking for the isolation of pure cultures, showing two different streaking patterns. In this procedure, a culture is diluted by drawing a loopful of the organism across a medium until only single cells are deposited at a given location. The growth of each isolated cell results in the formation of a discrete colony.

of the medium. While the medium holds bacteria in place, it is still soft enough to permit the growth of bacteria and the formation of discrete isolated colonies both within the fluid and on the surface of the agar. As with the other isolation methods, individual colonies are then picked up and streaked onto another plate for purification. In addition to its use in isolating pure cultures, the pour plate technique is used for the quantification of numbers of viable bacteria. The facts that agar solidifies below 42°C and that many bacteria survive at these temperatures ensure the success of this isolation technique. But because in some cases significant numbers of bacteria are killed under these conditions, this method cannot always be used.

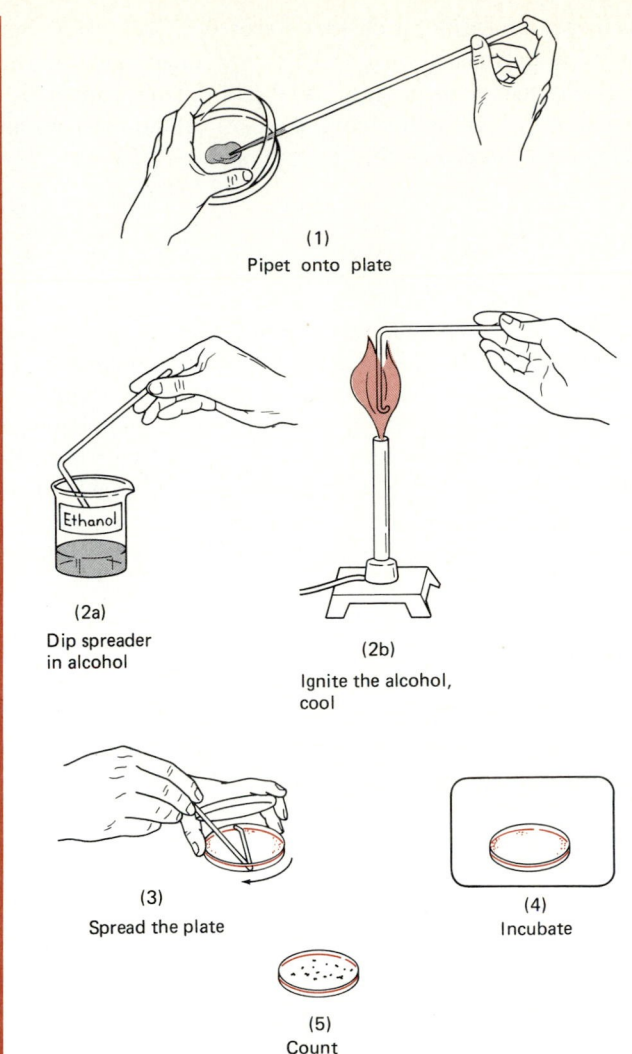

(1)
Pipet onto plate

(2a)
Dip spreader
in alcohol

(2b)
Ignite the alcohol,
cool

(3)
Spread the plate

(4)
Incubate

(5)
Count
colonies

FIGURE 4.5 *(LEFT)*
The spread plate technique for isolating and enumerating microorganisms. (1) Aseptically apply a known volume of a suspension to a suitable solid medium; (2) sterilize a spreading rod by dipping in alcohol and flaming; (3) use a sterile rod to spread suspension over the surface of the medium; (4) incubate; (5) count the colonies and calculate the number of microorganisms in the original suspension.

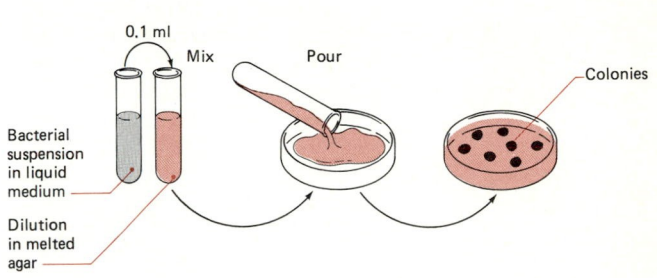

Bacterial
suspension
in liquid
medium

Dilution
in melted
agar

0.1 ml   Mix   Pour   Colonies

FIGURE 4.6 *(ABOVE)*
The pour plate technique for isolating and enumerating microorganisms. A known volume of a microbial suspension is mixed with a liquefied agar medium and poured into a petri plate. After incubation the numbers of colonies that develop are counted, and the concentration of microorganisms in the original suspension is calculated.

under greatly restricted conditions; such microorganisms are termed **fastidious**. The substrates that a microorganism can utilize and the specific growth factors it requires for reproduction are determined by the enzymatic capabilities of that microorganism as specified by its genome. Even slight changes in the genetic composition of an organism can greatly alter the metabolic capabilities of that organism and hence its nutritional requirements.

Autotrophic and Heterotrophic Modes of Nutrition

Microorganisms have developed several strategies of metabolism for meeting their nutritional needs (Table 4.2). Two distinct modes of microbial metabolism have evolved for accomplishing these tasks: **autotrophy** and **heterotrophy**.[4] These contrasting nutritional modes differ with respect to how the organism obtains energy and carbon, which is the backbone of the macromolecules of all living systems. **Au-**

totrophic (literally, self-feeding) microorganisms do not require preformed organic matter to generate cellular energy as adenosine triphosphate (ATP) or as a source of carbon for the biosynthesis of the macromolecules[5] of the cell. Rather, some autotrophic microorganisms are able to generate ATP from the oxidation of inorganic compounds; these organisms are known as **chemoautotrophs** or **chemolithotrophs**. The remaining autotrophs convert light energy to chemical energy and are known as **photoautotrophs** or **photolithotrophs**. The carbon for the macromolecules of autotrophs originates

[4]Autotrophs and heterotrophs are designated by a two-part term, the first part describing the source of energy (*chemo* if the energy for generating ATP is derived from chemical energy and *photo* if it comes from light energy) and the second part indicating the source of carbon used to synthesize the components of the organism (*organotroph* if the carbon comes from organic matter and *lithotroph* if it comes from inorganic carbon dioxide).

[5]Macromolecules are large molecules and include proteins, polysaccharides, and nucleic acids (RNA and DNA) that make up the structural and information-containing units of a cell.

from inorganic carbon dioxide or another inorganic carbon source. **Heterotrophic** microorganisms, on the other hand, require preformed organic matter. If the microorganism uses the organic matter both for the generation of ATP and for the synthesis of cellular macromolecules, it is called a **chemoorganotroph**, a **chemoheterotroph**, or, more commonly, simply a *heterotroph*. Those few microorganisms that use light energy to generate ATP and organic carbon as a source of the carbon for biosynthesis of cell components are called **photoheterotrophs**.

Nitrogen, Sulfur, and Phosphorus Requirements

Besides carbon, all cells require sources of nitrogen, sulfur, and phosphorus. Within cells, nitrogen occurs mainly in the amino form ($R-NH_2$), sulfur principally as sulfhydryl ($R-SH$) or disulfide ($R-S-S-R$) groups, and phosphate ($R-PO_4^{3-}$) and phosphate esters ($R_2-PO_4^{2-}$) are the main forms of phosphorus.[6] Some microorganisms are able to obtain their nitrogen and sulfur nutrients as nitrates and sulfates, but others do not have the metabolic capability to assimilate these oxidized forms and require reduced forms such as ammonium salts for nitrogen and cysteine for sulfur. The nitrogen and sulfur requirements can often also be met by organic nutrients such as amino acids or by more complex mixtures, such as peptones, which are derived by digesting proteins. Such compounds may simultaneously meet the cellular requirements for carbon, nitrogen, sulfur, and energy.

Influence of Oxygen on Microbial Growth

Another factor that has a great influence on microbial growth is the availability of molecular oxygen. Microorganisms vary with respect to their requirements and tolerances of molecular oxygen (Figure 4.7). Microorganisms that require molecular oxygen (O_2) because they are dependent on aerobic respiration to meet their energy requirements are called **obligate aerobes**. Some microorganisms, known as **microaerophiles**, grow only over a very narrow range of oxygen concentrations because they require oxygen for metabolism but are sensitive to oxygen at high concentrations. Still other microorganisms are **facultative anaerobes**, able to grow either in the presence or in the absence of molecular oxygen. Some facultatively anaerobic microorganisms have alternative metabolic capabilities that can shift from respiration, in which molecular oxygen is used, to fermentation, in which molecular oxygen is not used, or to anaerobic respiration, in which a compound, such as nitrate, substitutes for oxygen. Still other microorganisms cannot use molecular oxygen as a nutrient at all and are therefore called **obligate anaerobes**. Some of these anaerobic microorganisms, called **strict anaerobes**, cannot tolerate molecular oxygen and are inhibited or killed by exposure to oxygen.

TABLE 4.1 Principal Elements of the Cell and their Physiological Functions

Element	Percentage of Cell Dry Weight	Physiological Functions
Carbon	50	Constituent of all organic cell components
Oxygen	20	Constituent of cellular water and most organic cell components; molecular oxygen serves as an electron acceptor in aerobic respiration
Nitrogen	14	Constituent of proteins, nucleic acids, coenzymes
Hydrogen	8	Constituent of cellular water and organic cell constituents
Phosphorus	3	Constituent of nucleic acids, phospholipids, coenzymes
Sulfur	1	Constituent of some amino acids in proteins and some coenzymes
Potassium	1	Important inorganic cation and cofactor for some enzymatic reactions
Sodium	1	One of the principal inorganic cations in cells and important in membrane transport
Calcium	0.5	Important inorganic cation and cofactor for some enzymatic reactions
Magnesium	0.5	Important inorganic cation and cofactor for many enzymatic reactions
Chlorine	0.5	Important inorganic anion
Iron	0.2	Constituent of cytochromes and some proteins
All others	~0.3	

TABLE 4.2 Nutritional Types of Organisms

Type	Description
Photoautotrophs (photolithotrohs)	Use light as an energy source and CO_2 as the principal carbon source
Photoheterotrophs (photoorganotrophs)	Use light as the energy source and an organic compound as the principal carbon source
Chemoautotrophs (chemolithotrophs)	Use CO_2 as the principal carbon source; energy is obtained by the oxidation of inorganic compounds, such as NH_4^+, NO_2^-, H_2, H_2S, S, and Fe^{2+}
Chemoheterotrophs (chemoorganotrophs)	Use organic compounds for energy and as the principal carbon source

---

[6] Amino groups occur in all amino acids and, as such, are major groups found in proteins. Nitrogen also occurs in the ring structures that make up nucleic acid (RNA and DNA) macromolecules. Sulfhydryl and disulfide groups occur in some amino acids and are important functional groups of proteins. Phosphate occurs in the phospholipids of membranes, and phosphate diesters form the backbone linkages of nucleic acid macromolecules.

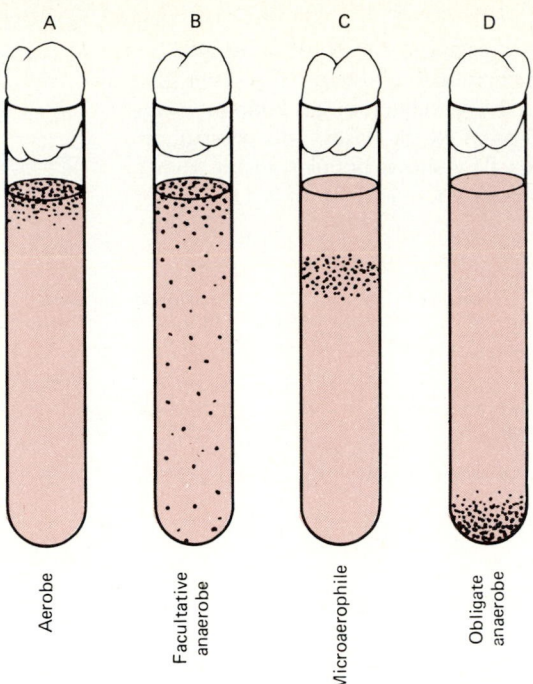

FIGURE 4.7

These test tubes show the relationship of the growth of various types of microorganisms to the presence of oxygen. (A) Aerobes grow in the presence of oxygen on the surface; (B) facultative anaerobes grow throughout the tube; (C) microaerophiles grow in a narrow band where the oxygen tension is reduced; (D) obligate anaerobes grow at the bottom of the tube where there is no free oxygen.

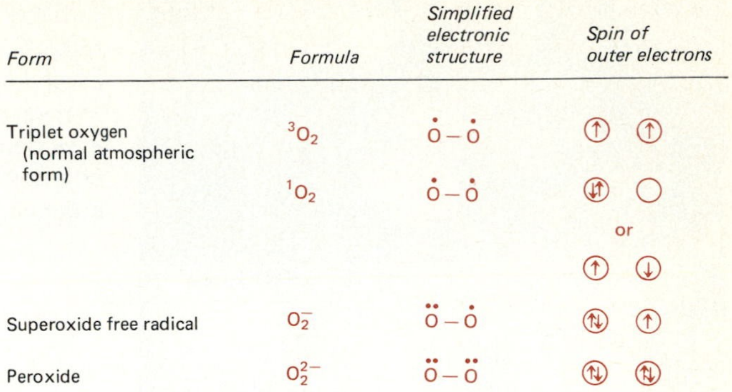

Form	Formula	Simplified electronic structure	Spin of outer electrons	
Triplet oxygen (normal atmospheric form)	$^3O_2$	Ȯ — Ȯ	↑	↑
	$^1O_2$	Ö — Ö	↑↓	○
			or	
			↑	↓
Superoxide free radical	O_2^-	:Ö — Ȯ	↑↓	↑
Peroxide	O_2^{2-}	:Ö — Ö:	↑↓	↑↓

FIGURE 4.8

Oxygen occurs in a variety of electronic states. Some forms of oxygen are particularly toxic to microorganisms, and microorganisms have evolved various enzymes for the removal of such toxic forms of oxygen. The singlet oxygen form is chemically reactive and is extremely toxic to living organisms. Phospholipids can be oxidized by singlet oxygen, leading to a disruption of membrane function and the death of microorganisms. Peroxidases in saliva and phagocyte cells (blood cells involved in the defense of the human body against invading microorganisms) generate singlet oxygen, accounting in part for the antibacterial activity of saliva and the ability of phagocytic blood cells to kill invading microorganisms.

As a consequence of the toxicity of oxygen to microorganisms, even anaerobic microorganisms generally possess enzyme systems for detoxifying various forms of oxygen. Oxygen can exist in a number of energetic states, some of which are more toxic than others (Figure 4.8).[7] The common state of oxygen is the triplet form in which two of the electrons in the valence shell are unpaired and spin in parallel directions. In singlet oxygen, which has a higher energy level than the triplet form, these two electrons have antiparallel spins. Various enzymatic reactions in a cell convert triplet oxygen to singlet oxygen. For example, the conversion of oxygen to water, which occurs during respiration, forms singlet oxygen. This process also produces an intermediary form of oxygen known as the *superoxide anion* (O_2^-) and hydrogen peroxide. These are extremely reactive oxidative chemicals that can irreversibly denature many biochemicals.

Microorganisms generally contain enzymes that protect them against the toxicity of these forms of oxygen. **Catalases**, for example, convert hydrogen peroxide to water and triplet-state oxygen (Figure 4.9). Additionally, the superoxide radical is converted to hydrogen peroxide and water by the action of the enzyme **superoxide dismutase**.

Both obligate aerobes and facultative anaerobes usually produce both catalase and superoxide dismutase enzymes (Figure 4.10). These enzymes permit such microorganisms to use oxygen and continue growing without accumulating toxic forms of oxygen that would kill the organism. In contrast, strict anaerobes generally lack both catalase and superoxide dismutase enzymes. The inability of these organisms to remove toxic forms of oxygen enzymatically probably accounts for the fact that they are strictly anaerobic and sensitive to oxygen. The obligate anaerobes, which can tolerate some exposure to oxygen but cannot grow in its presence, generally produce superoxide dismutase, which permits them to survive, but not catalase, which would allow them to grow in the presence of oxygen.

Defining Culture Media to Support Microbial Growth

In order to grow, microorganisms require a suitable environment, including a growth medium that can support their nutritional needs. Additionally, the culturing of microorganisms requires careful control of various environmental factors, including temperature, which normally is maintained

[7]In addition to the various electron states of normal molecular oxygen (O_2), oxygen can also occur as ozone (O_3). Ozone is an extremely strong oxidizing agent and consequently is very toxic to microorganisms. Ozone is sometimes utilized as a disinfecting agent, for instance in some water purification systems, to kill microorganisms. Ethylene oxide is another strong oxidizing agent that is used to sterilize many products that cannot tolerate high temperatures. Most hospitals have ethylene oxide sterilizers for sterilizing various medical items. Caution must be used when handling both ozone and ethylene oxide, as these compounds are quite toxic to humans as well as to microbes.

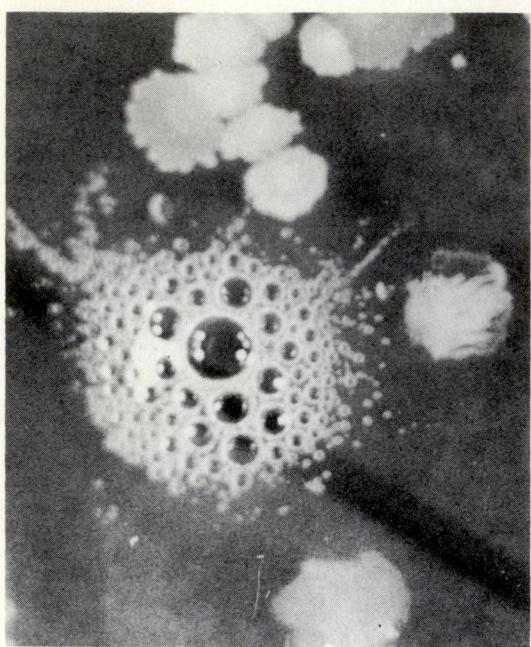

FIGURE 4.9 *(LEFT)*
Microbial production of catalases can be demonstrated by adding a loopful of a microbial culture to a 3 percent solution of hydrogen peroxide. The evolution of gas bubbles (oxygen) is evidence of the action of the catalases. This photograph of a positive test for catalase shows bubbles arising from a *Bacillus* bacterial colony upon addition of H_2O_2.

	Catalase $(2 H_2O_2 \rightarrow 2 H_2O + O_2)$	Superoxide dismutase $(2 O_2^- + 2 H^+ \rightarrow H_2O_2 + O_2)$
Aerobe	+	+
Facultative anaerobe	+	+
Oxyduric anaerobe	−	+
Oxylabile anaerobe	−	−

FIGURE 4.10 *(ABOVE)*
Enzymes produced by aerobes and anaerobes that remove toxic forms of oxygen.

ANALYTICAL PROCESS

Growing Cultures of Aerobic and Anaerobic Microorganisms

Under controlled laboratory conditions, it is possible to adjust oxygen concentrations to maximize the growth rate of a particular microorganism. Because oxygen diffuses only slowly into liquid, the concentration of oxygen frequently limits the rate of microbial growth of aerobic and facultatively anaerobic microorganisms in liquid culture. In order to supply oxygen for the growth of aerobic microorganisms and overcome the growth rate limitations caused by low oxygen concentrations, liquid cultures can be agitated at high speed on a shaker table or by an impeller within the culture vessel, or oxygen can be supplied to the culture vessel through forced aeration (Figure 4.11). Interrupting the supply of oxygen to an actively growing culture for even a brief period of time can lead to anaerobic conditions, in some cases causing a rapid die-off of the microorganisms. Some microbial populations can lose viability if a rotary shaker is turned off for only a few minutes, such as may occur when changing flasks on the shaker table.

Whereas aeration is used to enhance the rate of aerobic growth, oxygen must be excluded from the growth medium in order to permit the growth of obligate and strict anaerobes. This can be accomplished by adding chemicals that react with and remove molecular oxygen from the growth medium. For example, sodium thioglycollate is frequently added to liquid culture media for the growth of anaerobes because it reacts with molecular oxygen, removing free oxygen from solution. Similarly,

the amino acid cysteine and other compounds containing sulfhydryl groups can also be used to scavenge molecular oxygen from a growth medium. For liquid cultures, nitrogen may be bubbled through the medium to remove air and traces of oxygen, and then the culture vessel is sealed tightly to prevent oxygen from reentering.

Additionally, there are many types of **anaerobic culture chambers** that can be employed to exclude oxygen from the atmosphere (Figure 4.12). Common forms of anaerobic chambers, such as the Gas Pak system, generate hydrogen, which reacts with the oxygen as a catalyst within the chamber to produce water. Carbon dioxide is also generated in this system to replace the volume of gas depleted by the conversion of oxygen to water. It is also possible to combine several approaches to ensure absolutely anaerobic conditions. In the **roll tube method** after sterilization of a prereduced medium (a medium from which oxygen is excluded by the incorporation of a chemical that scavenges the free oxygen) within a sealed test tube, the medium is rolled during cooling so that it covers the inside of a test tube; the medium is then inoculated with a microorganism under a stream of carbon dioxide or nitrogen and tightly sealed; the development of microbial colonies can be seen on the tube surface, and individual cultures can be observed without disturbing other cultures, as must be done when a large anaerobic incubator is used (Figure 4.13).

FIGURE 4.11

(A) A three-tiered, variable-speed rotary shaker that can hold a variety of flasks for growth of aerobic micoorganisms. (Courtesy New Brunswick Scientific); (B) a benchtop fermenter in which aeration, temperature, agitation, and other factors can be carefully regulated to optimize conditions for microbial growth. (From BPS—David Cross)

FIGURE 4.12

Anaerobic jars are used to maintain cultures under strict anaerobic conditions.

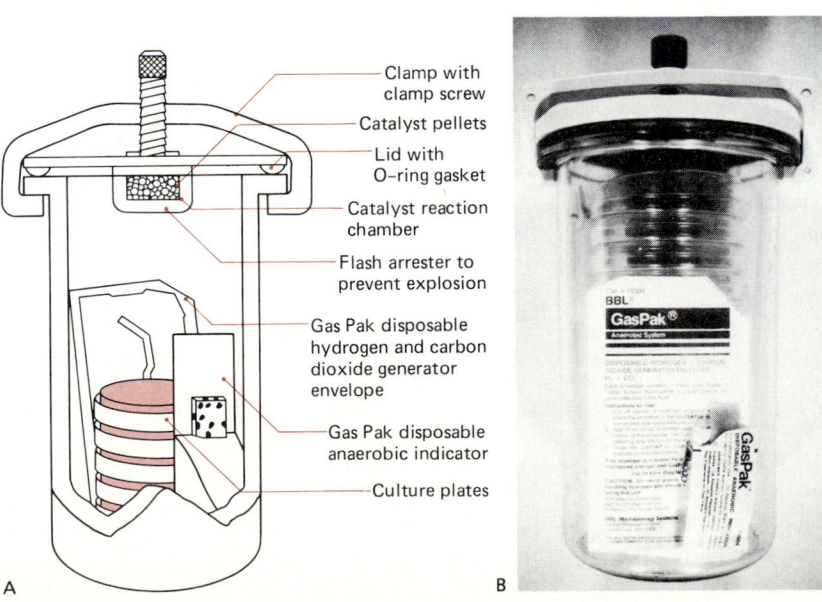

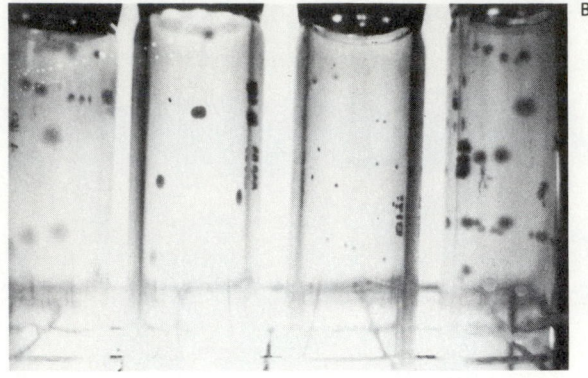

FIGURE 4.13

The culture of obligate and strict anaerobes requires special inoculation and culture procedures. (A) The VPI anaerobic culture transfer system. (B) Roll tube cultures showing colony development of anaerobes within the tubes. (Courtesy Donald E. Hash, Anaerobe Laboratory, Virginia Polytechnical Institute and State University)

within narrow limits by using a temperature-controlled incubator. By understanding the growth requirements of a given microbial species, it is possible to establish the necessary conditions *in vitro* to support the optimal growth of that microorganism.

Many bacterial species can be grown in the laboratory on a **defined medium,** that is, on a medium in which all components are known. Such a medium usually includes an organic carbon growth substrate—such as glucose, protein, or mineral nutrients—including a source of nitrogen and phosphorus, and water. Some microorganisms require a **complex medium** for growth, that is, a medium made with constituents whose composition is not totally known and may in fact vary from one batch to another. There are many different types of media used for growing bacteria and fungi in pure culture. Commonly used laboratory culture media contain proteins and/or carbohydrates as growth substrates (Table 4.3).

For the growth of particular heterotrophic microorganisms, specific growth substrates may be included in the culture medium. For the growth of autotrophic microorganisms, the organic carbon source is omitted from the growth medium and an inorganic source of carbon, carbon dioxide or carbonate, is supplied. A culture medium must also include a variety of required inorganic chemicals to support the growth of either heterotrophic or autotrophic microorganisms (Table 4.4). A typical growth medium normally contains a source of nitrogen (such as ammonium nitrate), phosphate, sulfate, iron, magnesium, sodium, potassium, and chloride ions. These inorganic chemicals are required for the biosynthesis of a variety of cellular biochemicals and for the maintenance of transport activities across the cytoplasmic membrane.

Microorganisms generally have many other specific inorganic nutritional requirements. For example, various metals, like zinc, manganese, and copper, among others, are generally required as trace elements. Some growth factors, such as vitamins and amino acids, may also be included in a medium designed to support the growth of microorganisms.

TABLE 4.3 The Composition of Some Culture Media for the Growth of Heterotrophic Bacteria

Nutrient broth (general medium for cultivation of heterotrophic bacteria)	
Beef extract	3 g
Peptone	5 g
Water	1,000 g

Glucose-inorganic salts medium (for growth of *E. coli*)	
Glucose	5 g
K_2HPO_4	7 g
KH_2PO_4	2 g
$MgSO_4$	0.1 g
$(NH_4)_2SO_4$	1 g
Water	1,000 g

Glucose-yeast extract medium (for growth of bacteria with growth factor requirements)	
Glucose	5 g
K_2HPO_4	7 g
KH_2PO_4	2 g
$MgSO_4$	0.1 g
$(NH_4)_2SO_4$	1 g
$FeSO_4$	0.1 mg
Yeast extract (contains numerous vitamins, amino acids, and other growth factors)	5 g
Water	1,000 g

TABLE 4.4   Some Inorganic Nutrients Required for Microbial Growth and Their Concentrations in a Typical Bacteriological Medium

Nutrient	Quantity
Water	1,000 g
NH_4Cl	1 g
K_2HPO_4	1 g
$MgSO_4 \cdot 7H_2O$	200 mg
$FeSO_4 \cdot 7H_2O$	10 mg
$CaCl_2$	10 mg
Trace elements (manganese, molybdenum, copper, cobalt, zinc) as inorganic salts	0.02–0.5 mg of each

Not all microorganisms can be grown on defined media. The nutritional requirements of many microorganisms are simply not known. These microorganisms are able to reproduce in nature, where their nutritional needs are met, but we do not understand their growth requirements well enough to define the appropriate laboratory conditions needed for their growth. From a natural soil or water sample, we are typically able to culture less than 1 percent of the microorganisms that are present. In some cases, the natural environment in which the microorganism reproduces can be established in the laboratory without actually defining the growth requirements. For example, soil extract is often added to growth media to support the reproduction of soil microor-

ANALYTICAL PROCESS
Enriching for Specific Microorganisms

By taking into account the metabolic capabilities of specific microorganisms, it is possible to design growth media that will favor the growth of particular microorganisms based on their nutritional requirements. This is the basis of the **enrichment culture technique,** a method used to isolate specific groups of microorganisms based upon a design of culture medium and incubation conditions that preferentially support the growth of a particular microorganism (Table 4.5). Liquid enrichment media tend to select the microorganisms that are able to grow best among all of the microbes introduced into the media. For example, in order to isolate microorganisms capable of metabolizing petroleum hydrocarbons, one can design a culture medium containing a hydrocarbon as the sole source of carbon and energy. By doing so, one establishes conditions whereby only microorganisms that are capable of metabolizing hydrocarbons can grow. Because other microorganisms cannot reproduce in this medium, one thereby selects for hydrocarbon-utilizing microorganisms. Similarly, a culture medium that favors the growth of

autotrophic microorganisms that derive their energy from the oxidation of ammonium ions and their carbon from inorganic carbon could be designed by providing ammonium ions and carbonate in the medium.

The design of an enrichment procedure takes into account not only the composition of the medium but also environmental factors such as temperature, aeration, pH, and so forth. For example, the temperature can be adjusted to 5°C in order to favor the growth of microorganisms that live at low refrigerator temperatures, or to 37°C in order to enrich for microbes capable of growth at the temperature of the human body. Cultures may be aerated by shaking or by sparging with air to favor the growth of aerobes, or oxygen may be totally excluded to enrich for anaerobes. The enrichment culture technique mimics many natural situations in which the growth of a particular microbial population is favored by the chemical composition of the system and by environmental conditions.

TABLE 4.5   Some Culture Conditions Employed for Enrichment Cultures

Carbon Source	Oxygen	Nitrogen Source	Inorganic Sulfur Source	Light	Organism Enriched For
Nonfermentable organic carbon	+	N_2	−	−	Azotobacter
Nonfermentable organic carbon	−	NO_3^-	−	−	Denitrifying bacteria
Nonfermentable organic carbon	−	−	$SO_4^=$	−	Desulfovibrio
Hydrocarbon	+	NH_4^+	$SO_4^=$	−	Hydrocarbon oxidizers
−	+	NH_4^+	$SO_4^=$	−	Nitrosomonas
−	+	NO_2^-	$SO_4^=$	−	Nitrobacter
−	+	−	S	−	Thiobacillus
−	+	N_2	$SO_4^=$	+	Cyanobacteria
−	−	NH_4^+	H_2S	+	Green sulfur bacteria

Note: − = none

ANALYTICAL PROCESS
Maintenance and Preservation of Pure Cultures

Once a microorganism has been isolated and grown in pure culture, it is necessary to maintain the viable culture, free of contamination, for some period of time. There are several methods available for maintaining and preserving pure cultures. The organisms may simply be subcultured periodically onto or into a fresh medium to permit continued growth and to ensure the viability of a stock culture. Proper aseptic technique must be used each time the organism is transferred, and there is always a risk of contamination. Furthermore, repeated subculturing is extremely time-consuming, making it difficult to maintain large numbers of pure cultures successfully for indefinite periods of time. Additionally, genetic changes are likely to occur when cultures are repeatedly transferred.

Therefore, a variety of methods have been developed for preserving pure cultures of microorganisms that do not require frequent subculturing into fresh media to maintain viability. These methods include refrigeration at 0–5°C for short storage times, freezing in liquid nitrogen at −196°C for prolonged storage, and **lyophilization** (also known as **freeze-drying**)[8] to dehydrate the cells (Figure 4.14). By sufficiently lowering the temperature or by removing water, microbial growth is precluded but viability in a dormant state is maintained, permitting preservation of microorganisms for extended periods.

Often valuable cultures are deposited in centralized **type culture collections**, such as the American Type Culture Collection in Rockville, Maryland, where they are preserved. It is especially important that all new microbial species be deposited in such culture collections to ensure their indefinite preservation and to make them available for scientific study. The choice of the preservation method depends on the nature of the culture and the facilities that are available. When freezing is used to preserve microorganisms, the rates of freezing and thawing must be carefully controlled to ensure the survival of the microorganisms because ice crystals formed during freezing can disrupt membranes. Glycerol is often employed as an "antifreeze" agent to prevent damage due to ice crystals and to ensure the ability to recover viable microorganisms when frozen cultures are thawed.

FIGURE 4.14
(A) This freeze dryer has a 6-L capacity and is mobile. (Courtesy Virtis Co). (B) This photograph shows the storage of freeze-dried cultures in a culture collection. (Courtesy American Type Culture Collection)

---

[8]In lyophilization or freeze-drying, the culture is frozen at a very low temperature and placed under a high vacuum. Under these conditions, the water in the culture and microbial cells goes directly from the frozen solid state to the gaseous state, thereby drying the cells without disrupting them.

ganisms. Often the task of defining the proper medium for growing microorganisms with rigorous growth requirements (fastidious microorganisms) is tedious and taxes the creativity of the microbiologist.

In many cases, rather than defining the specific constituents, complex media are used. For example, many media contain beef extract obtained by extracting the water-soluble components from beef tissue (a complex mixture of proteins, carbohydrates, lipids, and other biochemical constituents), peptones (an enzymatic digest of protein-containing amino acids and other nitrogen-containing compounds, as well as vitamins and other compounds), and yeast extract (an aqueous extract of yeast cells containing vitamins and other growth factors). Such complex culture media will support the growth of many different types of heterotrophs but will exclude fastidious organisms and those organisms that cannot tolerate particular components in the undefined medium.

Selective and Differential Culture Media

In contrast to establishing conditions that favor the growth of certain microorganisms, inhibitory substances may intentionally be added to a culture medium to prevent the growth of particular microorganisms. **Selective media** contain components that select for the growth of particular microorganisms, often by inhibiting the growth of others. For example, the stain methylene blue is more toxic to Gram-positive than to Gram-negative bacteria. By incorporating appropriate concentrations of methylene blue (about 0.5 percent) into a culture medium, the growth of Gram-positive bacteria may be inhibited, while not interfering with the growth of Gram-negative bacteria. This principle is exemplified by the medium eosin methylene blue (EMB) agar, which is frequently used for the selective culture of Gram-negative bacteria such as *Escherichia coli*. Methylene blue in this medium inhibits the growth of Gram-positive bacteria, and the other constituents of the medium favor the growth of microorganisms found in association with animal intestinal tracts. By incorporating the right inhibitory compound into a growth medium, the selectivity of that medium for a particular type of microorganism is established.

Selective media are used in many areas of microbiology, including clinical microbiology, where they are employed for the isolation of pathogenic microorganisms. Because clinical specimens often contain large numbers of indigenous non-pathogenic microorganisms, it is usually necessary to employ differential media in addition to selective media for the isolation of pathogens. **Differential media** contain indicators that permit the recognition of microorganisms with particular metabolic activities. For example, pH indicators are often incorporated into media for the detection of acidic metabolic products. Common selective and differential media that are employed for the isolation of intestinal tract pathogens include *Salmonella-Shigella* (SS), Hektoen enteric (HE), xylose-lysine-desoxycholate (XLD), brilliant green, eosin methylene blue (EMB), Endo, and MacConkey agars (Table 4.6). A combination of a differential medium, such as MacConkey agar, and a selective medium, such as HE agar, for example, is often used for the isolation of intestinal tract pathogens.

4.3 Bacterial Reproduction

Having considered the ways in which microorganisms can be cultured and maintained, let us examine some aspects of microbial reproduction. At the microbial level, growth is essentially synonymous with reproduction. Although individual cells may increase in size, growth in biological systems normally occurs through cell multiplication, during which genetic information is transmitted to the next generation. In the case of unicellular microorganisms, growth is equivalent to reproduction because reproducing the cell also accomplishes the reproduction of the entire organism. Even in multicellular eukaryotic microorganisms, the growth of the whole organism can be viewed in most cases as an increase in the number of individual reproductive units within a biological assemblage.

Binary Fission

Bacteria normally reproduce by **binary fission**, a process in which a cell divides to produce two equal-sized progeny cells (Figure 4.15). In binary fission the inward movement of the cytoplasmic membrane and cell wall, **septa formation**, pinches off and separates the two complete bacterial chromosomes, providing each of the progeny cells with a complete genome. The formation of **septae** or **cross walls** physically cuts apart the bacterial chromosomes and distributes them to the two daughter cells. Upon completion of cross wall formation, there are two equal-sized cells that can separate. Repeating the process results in the multiplication of the bacterial population.[9]

Reproduction by binary fission requires the replication of the bacterial chromosome so that each daughter cell receives a complete genome, that is, a complete set of genetic information; hence, cell division must be synchronized with rep-

---

[9]Although binary fission is the most common mode of bacterial reproduction, some bacteria reproduce by other means. These include budding, a type of division characterized by an unequal division of cellular material, and the formation of a relatively long, and generally branched filament or hypha (plural, hyphae) that may be composed of individual cells separated by cross walls.

TABLE 4.6   Media for Isolation of Enterobacteriaceae

MacConkey Agar

MacConkey agar is a differential plating medium for the selection and recovery of Enterobacteriaceae and related enteric Gram-negative rods. Bile salts and crystal violet are included to inhibit the growth of Gram-positive bacteria and some fastidious Gram-negative bacteria. Lactose is the sole carbohydrate. Lactose-fermenting bacteria produce colonies that are varying shades of red because of the conversion of the neutral red indicator dye (red below pH 6.8) from the production of mixed acids. Colonies of nonlactose-fermenting bacteria appear colorless or transparent.

Eosin Methylene Blue (EMB) Agar

EMB agar is a differential plating medium that can be used in place of MacConkey agar in the isolation and detection of the Enterobacteriaceae and related coliform rods from specimens with mixed bacteria. The aniline dyes (eosin and methylene blue) in this medium inhibit Gram-positive and fastidious Gram-negative bacteria. They also combine to form a precipitate at acid pH, thus also serving as indicators of acid production.

Desoxycholate-citrate (DCA) Agar

DCA agar is a differential plating medium used for the isolation of members of the Enterobacteriaceae from mixed cultures. The medium contains about three times the concentration of bile salts (sodium desoxycholate) of MacConkey agar, making it most useful in selecting species of *Salmonella* from specimens overgrown or heavily contaminated with coliform bacteria or Gram-positive organisms. Sodium and ferric citrate salts in the medium retard the growth of *E. coli*. Lactose is the sole carbohydrate, and neutral red is the pH indicator and detector of acid production.

Endo Agar

Endo agar is a solid plating medium used to recover coliform and other enteric organisms from clinical specimens. The medium contains sodium sulfite and basic fuchsin, which serve to inhibit the growth of Gram-positive bacteria. Acid production from lactose is not detected by a pH change, but rather from the reaction of the intermediate product, acetaldehyde, which is fixed by the sodium sulfite.

Salmonella-Shigella (SS) Agar

SS agar is a highly selective medium formulated to inhibit the growth of most coliform organisms and to permit the growth of species of *Salmonella* and *Shigella* from clinical specimens. The medium contains high bile salts concentration and sodium citrate, which inhibit all Gram-positive bacteria and many Gram-negative organisms, including coliforms. Lactose is the sole carbohydrate and neutral red the is indicator for acid detection. Sodium thiosulfate is a source of sulfur, and any bacteria that produce H_2S gas are detected by the black precipitate formed with ferric citrate.

Hektoen (HE) Enteric Agar

HE agar is devised as a direct plating medium for fecal specimens to increase the yield of species of *Salmonella* and *Shigella* from the heavy numbers of normal microbiota. The high bile salt concentration of this medium inhibits the growth of all Gram-positive bacteria and retards the growth of many strains of coliforms. Acids may be produced from three carbohydrates, and acid fuchsin reacting with thymol blue produces a yellow color when the pH is lowered. Sodium thiosulfate is a sulfur source, and H_2S gas is detected by ferric ammonium citrate, producing a black precipitate.

Xylose Lysine Desoxycholate (XLD) Agar

XLD agar is less inhibitory to the growth of coliform bacteria than HE and was designed to detect *Shigella* species in feces after enrichment in Gram-negative broth. Bile salts in relatively low concentration make this medium less selective than the other media included in this table. Three carbohydrates are available for acid production, and phenol red is the pH indicator. Lysine-positive organisms, such as most *Salmonella enteriditis* strains, produce initial yellow colonies from xylose utilization and delayed red colonies from lysine decarboxylation. The H_2S detection system is similar to that of HE agar.

---

lication of the bacterial chromosome. Because it can take longer to duplicate the bacterial chromosome, the bacterium initiates a new round of DNA replication every time the cell divides, even though the previously initiated replication of the DNA has not been completed. By initiating a new round of DNA replication every time the cells divide, the bacteria are able to produce completely duplicated genomes in time for cell division, with DNA replication occurring at the same frequency as cell reproduction. The regulatory mechanism is such that cell division takes place at a specified time after the completion of the replication of the bacterial chromosome.[10]

[10]Completion of the replication of the bacterial chromosome is a prerequisite for cell division; if the termination of DNA replication is blocked, the cell division that normally would occur 20 minutes later is prevented. The expression of specific genes that are required for cell division occurs at or immediately after the termination of replication of the bacterial chromosome.

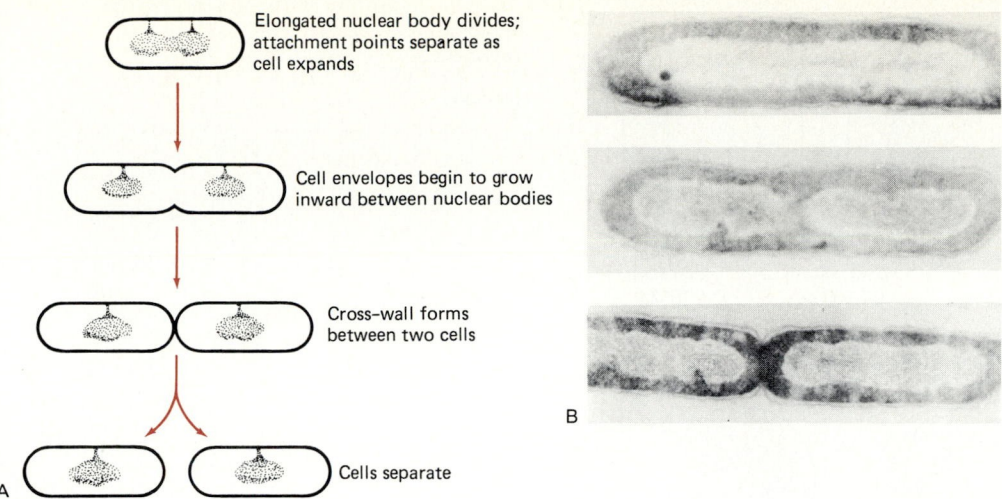

FIGURE 4.15

In binary fission, inward growth of the cytoplasmic membrane and the new wall material (septa formation) progresses to separate the two daughter cells, ensuring the equal distribution of genetic information between the cells. At each cell doubling, two processes are initiated almost simultaneously: the initiation of DNA replication and the initiation of the sequence of events leading to division. For a bacterium such as *E. coli*, the first 40 minutes of the division sequence involve protein synthesis and replication of the genome. Termination of genome replication at 40 minutes induces the synthesis of termination protein. After an additional 15–20 minutes, an interaction between some component of the septa and the termination protein leads to cell division. Thus, the entire cell cycle for *E. coli* would be 60 minutes, 40 minutes for replication of the bacterial chromosome and synthesis of various proteins and a 20-minute interval after the completion of genome replication before cell division occurs. However, because the cell initiates a new round of DNA replication every time it divides, it can synchronize the divisional process and continue to multiply every 20 minutes. (A) Schematic of the binary fission process. (B) Micrographs of the stages of cell division for synchronously grown *Erwinia*, showing invagination of cytoplasmic membrane and formation of the cross wall septa. (Reprinted by permission of the American Society for Microbiology, Washington, D.C., from E. A. Grula, and G. L. Smith, 1965, *Journal of Bacteriology*, 90: 1054–1058)

ANALYTICAL PROCESS

Enumeration of Bacteria

In order to assess the rate of microbial reproduction, it is necessary to determine the numbers of microorganisms present. There are a variety of methods that can be employed for **enumerating bacteria**. These include viable plate count, direct count, and most probable number (MPN) determinations.

Viable Count Procedures

The **viable plate count** method is one of the most common procedures for the enumeration of bacteria (Figure 4.16). In this procedure, serial dilutions of a suspension of bacteria are plated onto a suitable solid growth medium. The dilution procedure influences the overall counting process. Rapid chilling during dilution can cause the death of a significant portion of the population in certain cases, a phenomenon known as *cold shock*; in other cases, the cells die unless the dilution blanks and even the pipettes used in transferring are chilled. Because viable counts require viable cells, specific experimental protocols must be followed to enumerate specific microbial populations.

Normally the suspension is spread over the surface of an agar plate containing growth nutrients (surface spread technique) or mixed with the agar prior to allowing it to solidify and poured into the plate (pour plate technique). The plates are incubated to allow the bacteria to grow and form colonies. Relatively high numbers of colonies must be counted because, statistically, the standard error is approximately equal to the square root of the number of colonies counted. Preferably two or three plates, each with several hundred colonies, are counted. In some cases, when bacterial numbers in a sample are low, it is necessary to filter the suspension in order to concentrate

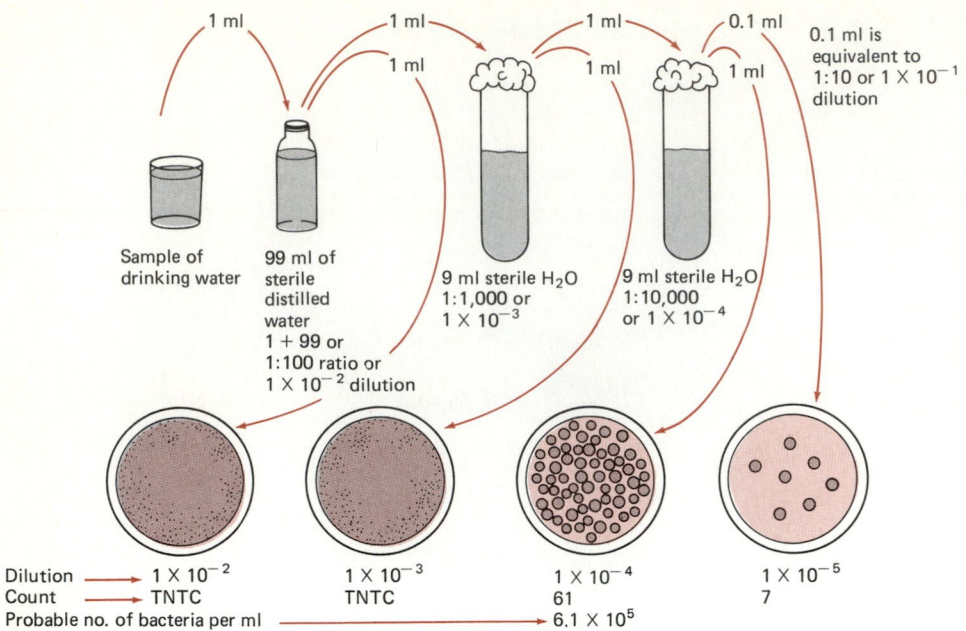

FIGURE 4.16

The viable plate count procedure is used to determine the viable population in a bacterial culture. Here plating is accomplished using the pour plate technique. There are several important points to note in this example. First, dilutions are usually achieved by diluting 1:100, as in the first dilution step, where 1 mL of sample is added to 99 mL sterile water (this could also have been achieved by adding an 0.1 mL sample to 9.9 mL sterile water), or by diluting 1:10, as in the second step, where 1 mL from the first dilution bottle is added to 9 mL sterile water. Second, greater dilutions are achieved by sequentially diluting the sample in series, as shown by the dilutions of 1:100, 1:1,000, and 1:10,000 that are achieved in the dilution tubes in this example. Third, if 1 mL is plated from a dilution tube, the concentration in the plate is the same as that in the tube, but if only 0.1 mL is plated, the dilution factor is increased by 10. Thus, in this example, both a 1:10,000 plate (using 1 mL) and a 1:100,000 plate (using 0.1 mL) are obtained from the third dilution tube. Fourth, counts on the plates must be in the range of 30–300 colonies to be counted and used in the calculation. In this example, there are too many colonies on the plates from the first two dilutions—recorded with the standard notation "TNTC" meaning too numerous to count—and too few on the the final plate. The only countable plate in this series is the 1:10,000 one, and this count gives the estimate of 6.1×10^5 bacteria per milliliter in the original sample.

the bacterial cells. The membrane filter can then be placed on a suitable medium, and the colonies that develop on the filter can be counted. The agar plates are incubated under conditions suitable for bacterial growth.

Multiplication of a bacterium on solid media results in the formation of a macroscopic colony visible to the naked eye. The formation of visible colonies generally takes 16–24 hours. It is assumed that each colony arises from an individual bacterial cell. Therefore, by counting the number of colonies that develop **colony-forming units (CFUs)** and by taking into account the dilution factors, the concentration of bacteria in the original sample can be determined. Countable plates are those having between 30 and 300 colonies. Fewer than 30 colonies are not acceptable for statistical reasons, and more than 300 colonies on a plate are likely to produce colonies too close to each other to be distinguished as individual CFUs.

A major limitation of the viable plate count procedure is its selectivity. There is no set of incubation conditions and medium composition that permits the growth of all bacterial types. The nature of the growth medium and the incubation conditions determine which bacteria can grow and thus be counted. Viable counting measures only those cells that are capable of growth on the given plating medium under the set of incubation conditions that are used. Sometimes cells are viable but nonculturable unless rigorous steps are taken to acclimate the microorganisms to laboratory culture conditions.

Direct Count Procedures

Bacteria can also be enumerated by **direct counting procedures**, that is, counting without the need to first grow the cells in culture. In one direct counting procedure, dilutions of samples are observed under a microscope, and the number of bacterial cells in a given volume of sample is counted and used to calculate the concentration of bacteria in the original sample. Special counting chambers, such as a hemocytometer or Petroff-Hausser chamber, are sometimes employed to determine the number of bacteria (Figure 4.17). These chambers are ruled with squares of known area and are so constructed that a film of liquid of known depth can be introduced between the slide and the cover slip. Consequently, the volume of liquid overlying each square is known. In order to help visualize bacterial cells, it is often desirable to stain the cells. Alternatively, a known volume of a sample containing a suspension of bacteria is filtered through a filter, such as a Nuclepore 0.2-μm pore size filter. The bacteria are stained on this filter and counted under a microscope. Fluorescent dyes are often used to stain bacteria in direct counting procedures. Such dyes stain all cells, making it impossible to differentiate living from dead bacteria. The difficulty in establishing the

metabolic status of the observed bacteria, that is, whether the cells are living or dead, is a major limitation of this procedure.

Another approach to direct counting is to use a particle counter such as a Coulter counter. The instrument can register the magnitude and duration of the changes in conductivity of a suspension of bacterial cells as it passes through a small orifice, and thus can register and record both the number and distribution of the size of a cellular population. Such instruments permit the discrimination of particles based on size so that particles the size of bacteria can be counted automatically. As long as there are no nonliving interfering particles in the same size range of bacteria, this is a good rapid counting method.

Turbidometric Procedures

Because particles within certain size limits scatter light in proportion to their concentration, when a beam of light passes through a suspension of bacteria, the amount of light transmitted is reduced due to the turbidity of the solution. Measuring the amount of light that passes through a suspension of microorganisms with a **spectrophotometer** (Figure 4.18) or other optical measuring device can be used for estimating cell mass, since the amount of light absorbed or scattered by the microorganisms is proportional to the cell density.[11] Spectrophotometers measure absorbancy units (A), which are defined as follows: $A = \log (I_0/I)$ where I_0 is the intensity of light striking a suspension and I is the intensity of light transmitted by the suspension.[12] When calibrated against bacterial suspensions of known concentration, a requirement for estimating cell concentrations, spectrophotometers provide an accurate and rapid way to estimate the dry weight (mass) of bacteria per unit volume of culture (Figure 4.19). An increase in cell mass, which can be equated with increases in the number of bacterial cells, is useful for establishing a growth curve for a bacterium.

Because bacteria are in suspension, not in solution, a measure of the absorbancy of a bacterial suspension is not a direct measure of the bacterial cell concentration. In fact, because light scattering also contributes so significantly to the determination, the measured value of A depends on the precise geometry of the instrument used. The value of A of a bacterial suspension measured on one instrument will not be the same as that measured on a different instrument. The instrument must be calibrated for the particular bacterium and media being studied by directly determining the number of bacteria in a dense suspension and by measuring its absorbance A, as well as

FIGURE 4.17

The direct counting procedure, using the Petroff-Hauser counting procedure. The sample is added to a counting chamber of known volume. The slide is viewed and the number of cells determined in an area delimited by a grid. In the counting chamber shown, the whole grid has 25 large squares for a total area of 1 mm² and a total volume of 0.02 mm. There are 12 cells within the single large grid (composed of 16 smaller boxes) shown in this example. Assuming that the number of cells in this single grid is representative of all the grids, the number of cells within the total area under the grids is 12 × 25 or 300 cells. The concentration of cells is therefore 300/0.02 mm³ or 1.5×10^7/ml.

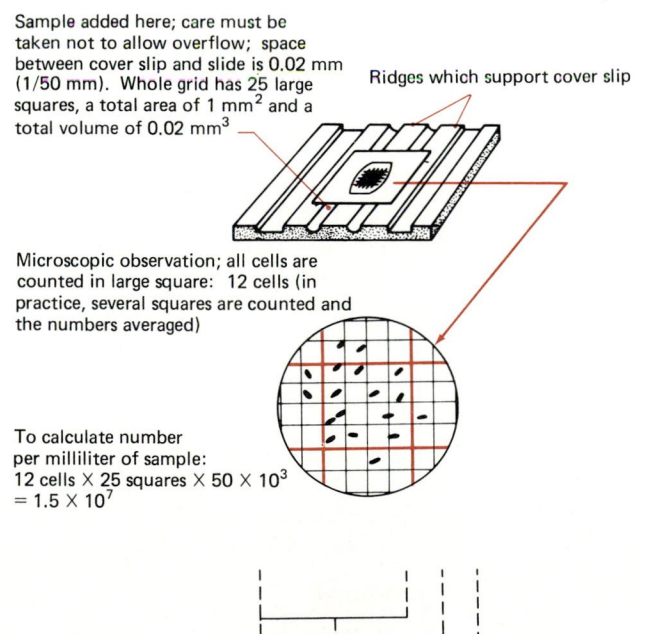

Sample added here; care must be taken not to allow overflow; space between cover slip and slide is 0.02 mm (1/50 mm). Whole grid has 25 large squares, a total area of 1 mm² and a total volume of 0.02 mm³

Ridges which support cover slip

Microscopic observation; all cells are counted in large square: 12 cells (in practice, several squares are counted and the numbers averaged)

To calculate number per milliliter of sample:
12 cells × 25 squares × 50 × 10³ = 1.5 × 10⁷

Number/sq mm

Number/cubic mm

Number /cubic cm (ml)

---

[11]Because light scattering is inversely proportional to the fourth power of the wavelength of the light being scattered, the sensitivity of the measurements increases sharply if shorter-wavelength light is used, but even when using relatively short-wavelength light, the lower limit of sensitivity is about 10^6 bacteria per milliliter. Nephelometers, which have the light-sensing device arranged at right angles to the incident beam of light, are somewhat more sensitive than spectrophotometers because they measure the scattered light directly.

[12]The absorbency or optical density (OD) of a solution is related to the percentage of light transmitted (%T) through the solution according to the formula OD = $\log 100 - \log \%T$.

the absorbance of known dilutions of the suspension. At low densities A is roughly proportional to the cell number, but at higher densities there is a significant deviation from linearity. Such deviation is a consequence of double scattering, that is, the probability of a scattered ray of light being scattered back at high culture densities so that it strikes the photodetection system is increased.

MPN Procedures

Another approach to bacterial enumeration, determination of the **most probable number (MPN)**, is a statistical method based on probability theory. In an MPN enumeration procedure, multiple serial dilutions are performed to reach a point of extinction, that is, a dilution level at which not even a single cell is deposited into one or

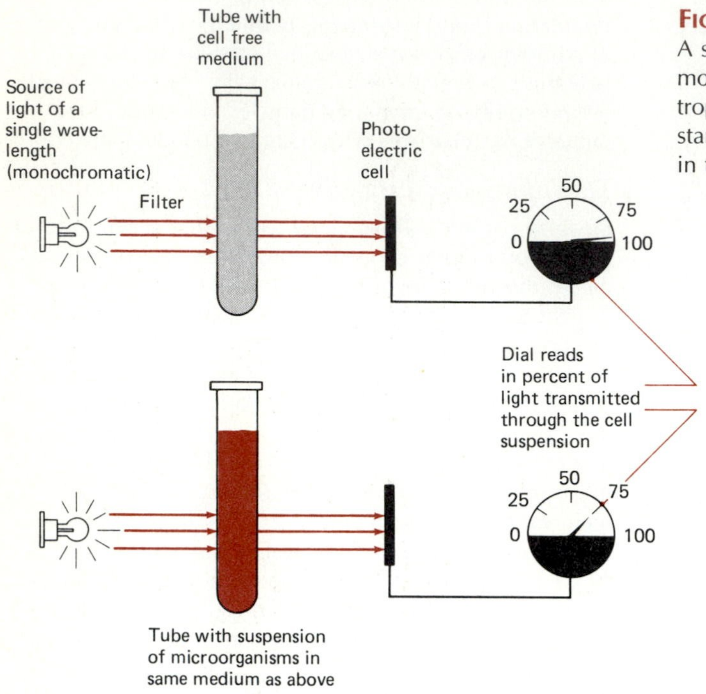

FIGURE 4.18

A spectrophotometer is used to measure cell mass by monitoring light transmission. The reading on the spectrophotometer is used to estimate the mass of cells, and standard curves are used to estimate the number of cells in the microbial suspension.

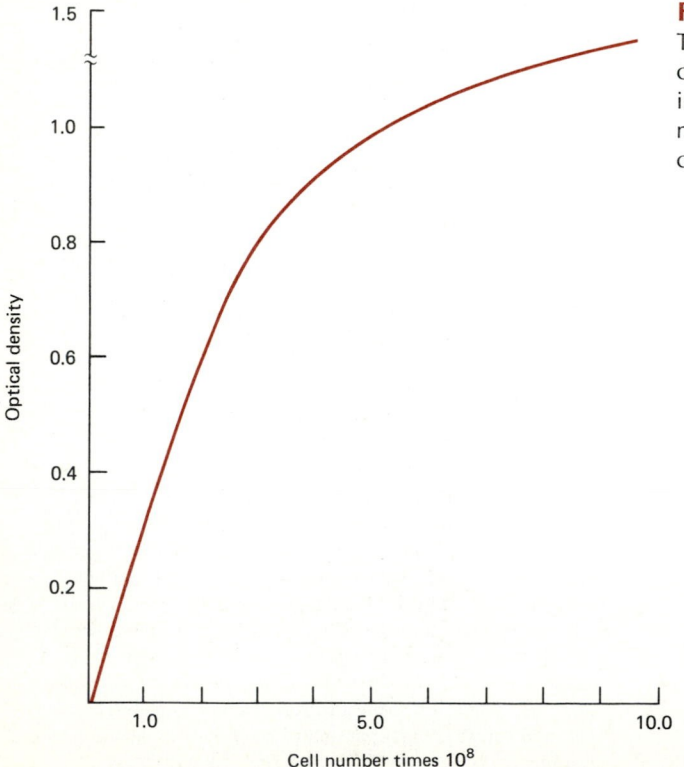

FIGURE 4.19

This graph shows the relationship of optical density to cell number. Here, an optical density reading of 0.6 indicates a cell count of 2.0×10^8. Calibration curves must be made for each instrument, culture, and set of culture conditions.

more of the multiple tubes at that dilution level. A criterion, such as the development of cloudiness or turbidity in a liquid growth medium, is established for indicating whether a particular dilution tube contains bacteria. The pattern of positive and negative test results is then used to estimate the concentration of bacteria in the original sample, that is, the MPN of bacteria, by comparing the observed pattern of results with a table of statistical probabilities for obtaining those results (Figure 4.20).

Procedures Based on Quantifying Specific Biochemical Constituents

Various other procedures, based on the detection of specific microbial macromolecules or metabolic products, can be used to estimate the number of microorganisms. For example, murein can be quantified, and because this biochemical occurs exclusively in the cell walls of bacteria, the concentration of murein can be used to estimate bacterial numbers. Similarly, lipopolysaccharide concentrations can be utilized to estimate the number of Gram-negative bacteria, as this biochemical occurs primarily in the outer cell wall structure of Gram-negative bacteria. More general estimates of microbial numbers can be obtained indirectly by measuring protein, ATP, and DNA concentrations. All of these biochemical approaches for determining bacterial numbers depend on the development of analytical chemical procedures for quantifying the particular biochemical and determining what proportion of a bacterial cell is composed of the specific biochemical constituent. Problems with these approaches involve developing appropriate conversion factors for equating the concentration of the particular biochemical with the actual number of bacterial cells and establishing that the particular biochemical is exclusively of bacterial origin.

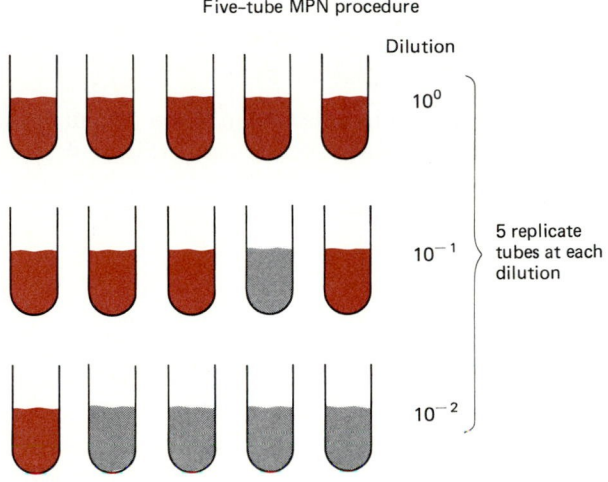

Five-tube MPN procedure

Dilution

10^0

10^{-1}

10^{-2}

5 replicate tubes at each dilution

1 — Inoculate replicate tubes
2 — Incubate
3 — Record number of positive tubes at each dilution
4 — Consult MPN table
5 — Record MPN of bacteria

FIGURE 4.20

In the five-tube MPN procedure, five replicate tubes are used for each dilution. First, inoculate replicate tubes; incubate for an appropriate period of time; record the number of positive tubes at each dilution; consult an MPN table; and finally, record the MPN of bacteria. In this example, all tubes are positive at the 10^0 dilution, four of the five tubes are positive at the 10^{-1} dilution, and one of the five tubes is positive at the 10^{-2} dilution. Thus, in this example, the MPN of bacteria is 17/100 ml.

Number of Positive Tubes at Stated Dilution

10^0	10^{-1}	10^{-2}	MPN/100 ml	10^0	10^{-1}	10^{-2}	MPN
0	1	0	0.18	5	0	0	2.3
1	0	0	0.20	5	0	1	3.1
1	1	0	0.40	5	1	0	3.3
2	0	0	0.45	5	1	1	4.6
2	0	1	0.68	5	2	0	4.9
2	1	0	0.68	5	2	1	7.0
2	2	0	0.93	5	2	2	9.5
3	0	0	0.78	5	3	0	7.9
3	0	1	1.1	5	3	1	11.0
3	1	0	1.1	5	3	2	14.0
3	2	0	1.4	5	4	0	13.0
4	0	0	1.3	5	4	1	17.0
4	0	1	1.7	5	4	2	22.0
4	1	0	1.7	5	4	3	28.0
4	1	1	2.1	5	5	0	24.0
4	2	0	2.2	5	5	1	35.0
4	2	1	2.6	5	5	2	54.0
4	3	0	2.7	5	5	3	92.0
				5	5	4	160.0

4.4 Bacterial Growth

Kinetics of Bacterial Growth

The measurement of bacterial growth reveals that once cell division begins, it proceeds as a geometric progression, with one cell dividing to form two, each of these cells dividing to form four, and so forth in a geometric progression (Figure 4.21). This is because most bacteria reproduce by binary fission, which results in doubling of the number of viable bacterial cells. Therefore, during active bacterial growth, the size of the microbial population is continuously doubling. The time required to achieve a doubling of the population size, known as the **doubling time** or **generation time**, is the unit of measure of the microbial growth rate (Table 4.7). The generation time for bacteria can be expressed mathematically as

$$g = t/3.3(\log B_t - \log B_0)$$

where g is the generation time, $\log B_t$ is the logarithm to the base 10 of the number of bacteria at time t, $\log B_0$ is the logarithm to the base 10 of the number of bacteria at the starting time, and t is the time period of growth.[13] This math-

[13]Although the use of the natural logarithm is preferable conceptually because the unit of growth is then equal to one generation or doubling of cell number ($B_t = B_0 \times 2^n$, where n is the number of generations, B_0 is the number of bacteria at the starting time, B_t is the number of bacteria at time t, and t is the time period of growth), this equation usually is expressed using logarithm to the base 10. The conversion from natural logarithms to base 10 logarithms uses the relationship $\ln B = 3.3 \log_{10} B$.

ematical formula for the bacterial growth rate is based on the premises that the rate of increase is proportional to the number or mass of cells present at any given time and that the doubling time is constant during a period of growth.

By determining cell numbers during the period of active cell division, the generation time can be estimated (Figure 4.22). In comparing generation times, one finds that bacteria reproduce more rapidly than higher organisms. A bacterium such as *E. coli* can have a generation time as short as 20 minutes under optimal conditions, although in nature many bacteria have generation times of several hours. In a bacterium with a 20-minute generation time, one cell would multiply to 1,000 cells in 3.3 hours and to 1,000,000 cells in 6.6 hours.

Growth Curve of Bacteria

When a bacterium is inoculated into a new culture medium, it exhibits a characteristic growth curve (Figure 4.23). The

FIGURE 4.21
Geometric progression in the number of cells resulting from binary fission.

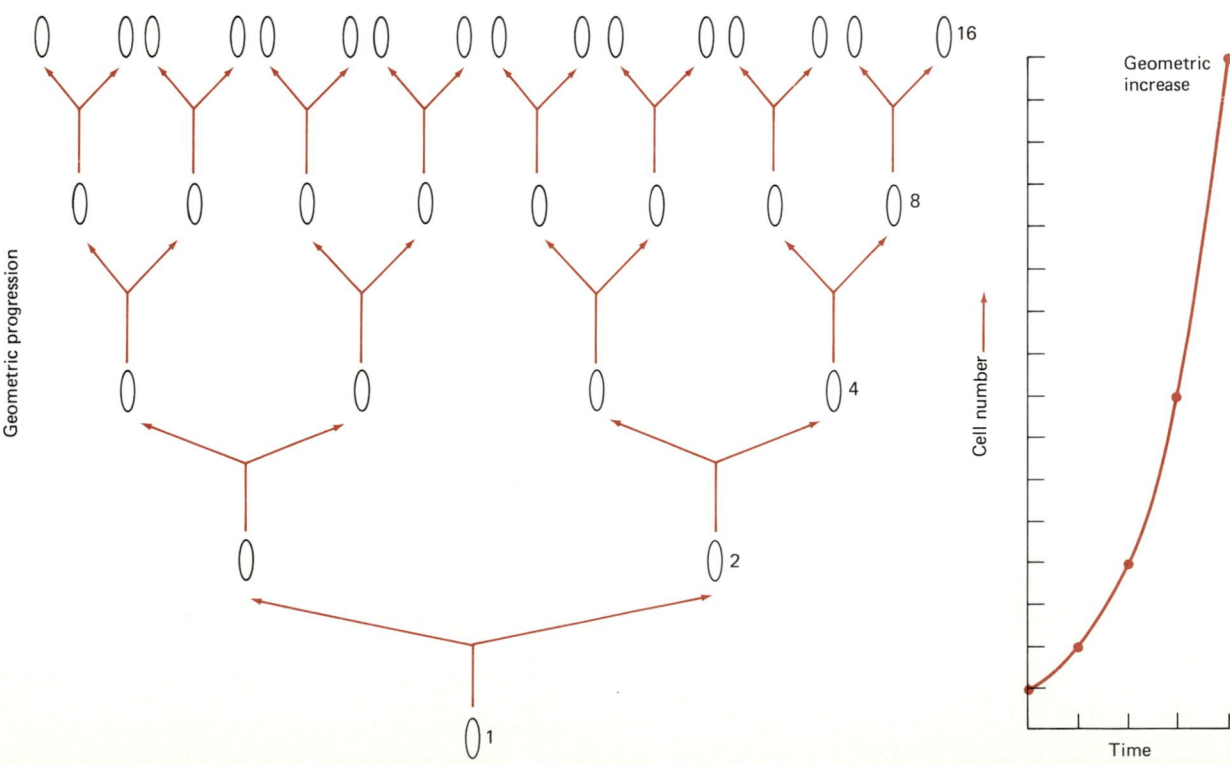

TABLE 4.7   Growth Rates for Some Representative Bacteria under Optimal Conditions

Organism	Temperature (°C)	Generation Time (min)
Bacillus stearothermophilus	60	11
Escherichia coli	37	20
Bacillus subtilis	37	27
Bacillus mycoides	37	28
Staphylococcus aureus	37	28
Streptococcus lactis	37	30
Pseudomonas putida	30	45
Lactobacillus acidophilus	37	75
Vibrio marinus	15	80
Mycobacterium tuberculosis	37	360
Bradyrhizobium japonicum	25	400
Nostoc japonicum	25	570
Anabaena cylindrica	25	840
Treponema pallidum	37	1980

typical **growth curve** of a bacterial culture has four phases: the lag phase, the log or exponential growth phase, the stationary phase, and the death phase. During the **lag phase** there is no increase in cell numbers. Rather, during this phase the bacteria are preparing for reproduction, synthesizing DNA and various inducible enzymes needed for cell division. They may increase in size during this process.

During the **log phase** of growth, so named because the logarithm of the bacterial biomass increases linearly with time, bacterial reproduction occurs at a maximal rate for the specific set of growth conditions. This growth phase is also

called the **exponential growth phase** because the number of cells is increasing as an exponential function of 2^n. Growth during much of the log phase is said to be balanced, that is, all properties of the cell are increasing at the same rate and, therefore, the average composition of the cells remains constant. During the log phase of the growth curve, the growth rate of a bacterium is proportional to the biomass of bacteria that is present.

$$dB/dt = \alpha B$$

where B is the bacterial biomass, t is time, and α is the instantaneous growth rate constant. It is during this period that the generation time of the bacterium is determined. If a bacterial culture in the exponential growth phase is inoculated into an identical fresh medium, the lag phase is usually bypassed and exponential growth continues. This occurs because bacteria are already actively carrying out the metabolism necessary for continued growth. If, however, the chemical composition of the new medium differs significantly from that of the original growth medium, the bacteria do go through a lag phase wherein they synthesize the enzymes needed for growth in the new medium before entering the logarithmic growth phase.

If, however, the bacterium is not transferred to a new medium and no fresh nutrients are added, the **stationary growth phase** eventually is reached, and there is no further net increase in bacterial cell numbers. The transition between the exponential and stationary phases involves a period of unbalanced growth during which the various cellular components are synthesized at unequal rates. Consequently, cells in the stationary phase have a different chemical composition from cells in the exponential phase. During the stationary phase, the growth rate is exactly equal to the death rate. A bacterial population may reach stationary growth when a required nutrient is exhausted, when inhibitory end products accumulate, or when physical conditions are appropriate. In all cases, there is a feedback mechanism that regulates the bacterial enzymes involved in key metabolic steps. The duration of the stationary phase varies; some bacteria exhibit a

FIGURE 4.22

The estimation of the generation time of a bacterium based on observed cell numbers during a period of active division. Note that the Y axis of this graph is ploted on a log scale.

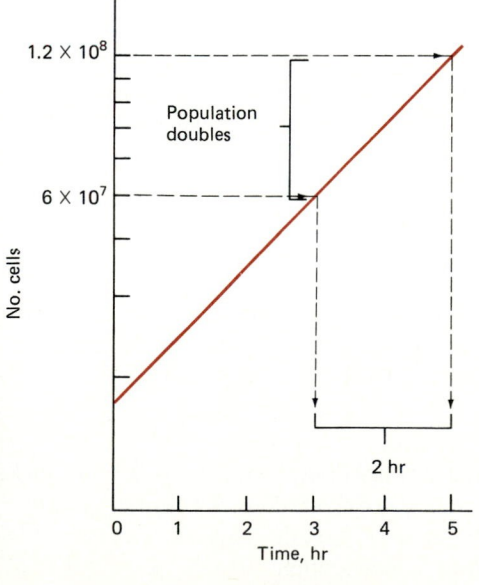

FIGURE 4.23

The normal bacterial growth curve has four stages: lag, exponential, stationary, and death.

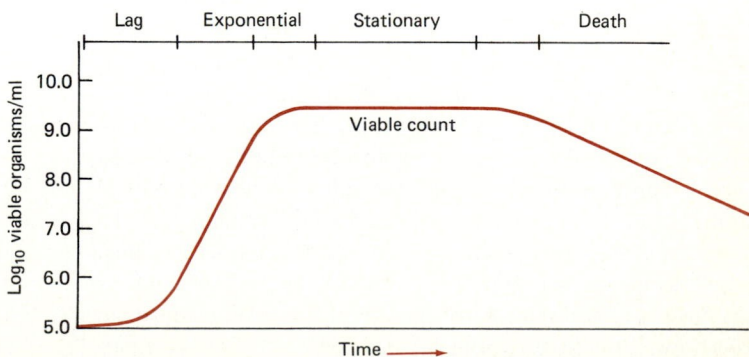

very long stationary phase. Eventually, however, the number of viable bacterial cells begins to decline, signaling the onset of the **death phase**.

The kinetics of bacterial death, like those of growth, are exponential because the death phase really represents the result of the inability of the bacteria to carry out further reproduction. The rate of the death phase need not, however, be equal to the rate of growth during the exponential phase. The rate of death is proportional to the number of survivors. Modifying environmental conditions can alter both the rate of exponential growth and the death rate of a bacterium.

Synchronous Growth of Bacteria

Synchronous cultures of bacteria, that is, cultures in which all cells divide at the same time, can be obtained by a number of techniques. **Synchrony** can be induced by manipulating environmental conditions, such as by repeatedly changing the temperature or by adding fresh nutrients to cultures as soon as they enter the stationary phase. Alternatively, a synchronous population can be obtained by physical separation procedures. For example, an unsynchronized culture of bacteria can be filtered through a membrane filter. The loosely associated bacteria are washed from the filter, leaving some cells tightly adsorped to it. The filter is inverted and fresh medium allowed to flow through it. New bacterial cells that arise through cell division are not tightly bound to the membrane and are washed into the effluent. All of the cells in the effluent are newly formed and are therefore at the same stage of the cell cycle.

Batch and Continuous Culture of Bacteria

The normal bacterial growth curve is characteristic of bacteria in **batch culture**, that is, under conditions in which a fresh medium is simply inoculated with a bacterium. A flask containing a liquid nutrient medium inoculated with a bacterium such as *E. coli* is an example of a batch culture. In batch culture growth nutrients are expended, and metabolic products accumulate in the closed environment. The batch culture models situations such as occur when a canned food product is contaminated with a bacterium.

Bacteria may also be grown in **continuous culture** in which nutrients are supplied and end products continuously removed so that the exponential growth phase is maintained. Continuous culture systems can be operated as chemostats or as turbidostats. In a **turbidostat** the system includes an optical sensing device that measures the turbidity of the culture in the growth vessel and generates an electrical signal that is used to regulate the flow of fresh medium into the vessel and the flow of spent medium and cells out it. Thus, in a turbidostat the number of cells in the culture controls the flow rate and the rate of growth of the culture adjusts to this flow rate. In a **chemostat** the flow rate is set at a par-

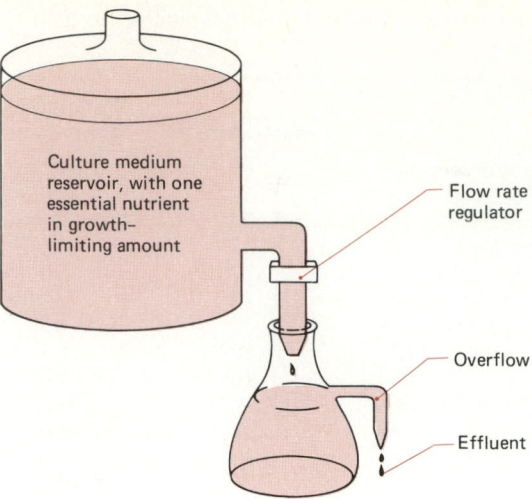

FIGURE 4.24
A chemostat is a continuous culture device. In such a device, the population density is controlled by the concentration of the limiting nutrient and the growth rate is controlled by the rate, which can be arbitrarily set.

ticular value and the rate of growth of the culture adjusts to this flow rate.

Because end products do not accumulate and nutrients are not completely expended, the bacteria never reach stationary phase in a chemostat. Continuous growth of bacteria can be accomplished in this device by continuously feeding a liquid medium into the bacterial culture (Figure 4.24). The liquid medium contains some nutrient in growth-limiting concentrations, and the concentration of the limiting nutrient in the growth medium determines the rate of bacterial growth. In steady-state operation of a continuous culture device, the concentration of the limiting nutrient remains constant because the rate of addition of the nutrient equals the rate at which it is used by the culture plus that lost through overflow. Even though bacteria are continuously reproducing, a number of bacterial cells are continuously being washed out and removed from the culture vessel.

Because the rate of cell washout is equal to the growth rate, the dilution rate is equal to the growth rate of a bacterium growing in a chemostat, which can be expressed as

$$\mu = 0.693k$$

where μ is the growth rate of the culture and k is the cell growth rate constant ($1/k$ = cell doubling or generation time). The relationship between the culture generation time and the concentration of the limiting substrate is

$$\mu = \mu_{max} \times s/(k_s + s)$$

where μ is the culture generation time, μ_{max} is the maximal growth rate at saturating concentrations of substrate, s is the substrate concentration, and k_s is the saturation constant defined as the substrate concentration at $\frac{1}{2}\mu_{max}$ (Figure 4.25). Cell numbers and the concentration of the limiting nutrient

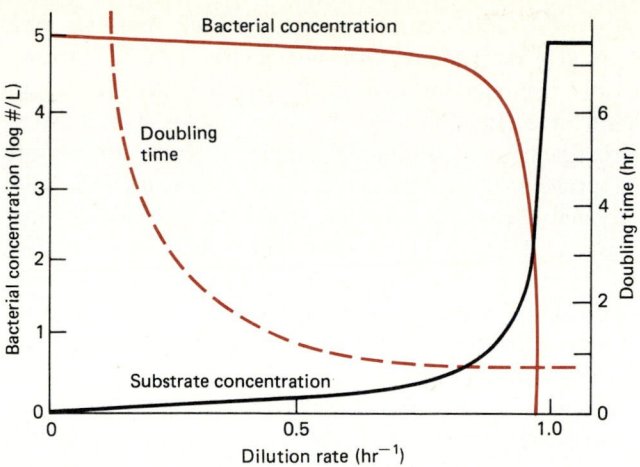

FIGURE 4.25

This graph shows the relationship between dilution rate, growth rate (doubling time), and bacterial numbers in a chemostat. The bacterial concentration tends to remain constant as the dilution rate increases because of the simultaneous increase in growth rate evidenced by the shorter doubling times. The excess cells are washed out of the chemostat, which is why even though cells are being produced more rapidly with the input of additional nutrients as the as the dilution rate increases the number of cells in the chemostat remains constant. Eventually, however, the dilution rate exceeds the reproductive capacity and the cells are washed out of the chemostat. With this loss of cells, the concentration of substrate increases because no cells are metabolizing the incoming nutrients.

change little at low dilution rates. As the dilution rate approaches k_{max}, the cell concentration drops rapidly to zero, and the concentration of the limiting nutrient approaches its concentration in the reservoir. A chemostat is probably a good model for bacterial growth in open systems, such as rivers and oceans, and by using chemostats and the appropriate mathematical calculations, the growth rates of bacteria in nature can be estimated.

Bacterial Growth on Solid Media

The development of bacterial **colonies** on solid growth media follows the basic normal growth curve. However, the dividing cells do not disperse, and hence, the population is densely packed. Under these conditions, nutrients rapidly become limiting at the center of the colony, and microorganisms in this area rapidly reach stationary phase. At the periphery of the colony, cells can continue to grow exponentially even while those at the center of the colony are in the death phase. Bacterial colonies generally do not extend indefinitely across the surface of the media but have a well-defined edge, the shape of which is characteristic for a given bacterial species. Therefore, individual well-isolated colonies develop from the growth of individual bacterial cells. The fact that the bacteria have reproduced asexually by binary fission means that, barring mutation, all of the bacteria in the colony should be genetically identical—each colony being a clone of cells derived from a single parental cell.

Postlude

Many basic microbiological methods developed by Robert Koch are still in use today, with only minor modifications. It was Koch who, between 1881 and 1883, developed simple methods for the isolation and maintenance of pure cultures of microorganisms on chemically defined solid media. Prior to that time, microorganisms were grown in liquid broths. At first, Koch cultured bacteria on solid fruits and vegetables, such as slices of boiled potato, but many bacteria cannot grow on such substrates. Koch developed a way of solidifying liquid broths that could support the growth of a greater variety of microorganisms, initially using gelatin and later agar (an algal extract) as the solidifying agent. Because gelatin is not solid above 30°C, it is not useful for growing the many microorganisms that are grown at 37°C. Agar, on the other hand, remains solid well above 37°C and therefore is very useful for this purpose. The suggestion for using agar originated with the New Jersey-born wife of one of the investigators at Koch's Institute, Mrs. Hess, who had seen her mother using agar to make jellies. The use of these solidified media permitted the isolation and unequivocal identification of disease-causing microorganisms. The isolation and growth

of microorganisms in pure culture have dominated most microbiological studies since the necessary techniques were developed by Koch. One of the modifications in Koch's original method was made by Richard J. Petri, who in 1887 described the use of a new type of culture dish for growing bacteria on semisolid media. The basic design has become known as the **petri dish** and is used in virtually all microbiological laboratories in essentially the same design described by Petri.

The ability to grow a microorganism in pure culture is basic to the development of our understanding of the fundamentals of microbiology. Establishing and maintaining pure cultures is fundamental to most microbiological studies. The establishment of a pure culture requires that the microorganisms be isolated from all other types of microorganisms. This mandates the creation of a sterile environment in which the pure culture can be grown, the isolation and aseptic handling of the pure culture to avoid contamination, and the use of media and incubation conditions favorable to the physiological requirements of the isolated microorganism. The culture medium used must meet the nutritional requirements of the particular microbial species, and these requirements vary greatly because of the great diversity of metabolic capabilities found in the microbial world.

Reproduction ensures that the genetic information of a microorganism is passed from one generation to the next. As a rule, reproduction in microorganisms is synonymous with microbial growth. Much of the metabolic activity of a microorganism is aimed at producing the macromolecules and structures needed to form progeny. Most bacteria reproduce by binary fission, forming two equal-sized cells with identical genomes. As a consequence of this reproductive mode, bacterial numbers double at regular time intervals. When inoculated into a fresh medium, bacteria exhibit a characteristic growth cycle consisting of a lag phase, a log or exponential phase, a stationary phase, and a death phase. The mathematics of the exponential growth and death phases reflect the binary fission mode of bacterial replication.

The particular reproductive strategy of any organism must be adaptive or that organism will perish. Among microorganisms, reproduction is normally characterized by the formation of a large number of progeny. It is the ability of microorganisms to proliferate rapidly when conditions are appropriate that ensures survival of the species. The high reproductive capacity of microorganisms also allows rapid evolution. Undoubtedly, many nonadaptive evolutionary lines rapidly become extinct. The many microbial species that continue to flourish represent the survivors whose reproductive capacities are high enough to overcome those factors that can lead to the extinction of a species. The important thing is that the particular reproductive strategy of the microorganism ensures the passage of its genome to progeny that are able to continue reproducing, establishing the basis for the survival of the species.

Suggested Supplementary Readings

ATLAS, R. M. 1982. Enumeration and estimation of biomass of microbial components in the biosphere. In: *Experimental Microbial Ecology* (R. G. Burns and J. H. Slater, eds.), pp. 84–102. Blackwell Scientific Publications, Oxford, England.

BAZIN, M. J. (ed.). 1982. *Microbial Population Dynamics*. CRC Press, Inc., Boca Raton, Fla.

COLLINS, C. H., and P. M. LYNE. 1976. *Microbiological Methods*. Butterworth, Woburn, Mass.

DAWSON, P. S. S. (ed.). 1975. *Microbial Growth* (Benchmark Papers in Microbiology). Academic Press, New York.

Difco Manual of Dehydrated Culture Media and Reagents for Microbiological and Clinical Laboratory Procedures. 1984. Difco Laboratories, Detroit, Mich.

DONACHIE, W. D., N. C. JONES, and R. TEATHER. 1973. The bacterial cell cycle. *Symposium of the Society for General Microbiology* 23:9–44.

EDWARDS, C. 1981. *The Microbial Cell Cycle*. American Society for Microbiology, Washington, D.C.

GERHARDT, P. (ed.). 1981. *Manual of Methods for General Bacteriology*. American Society for Microbiology, Washington, D.C.

HIGGINS, M. L., and G. D. SHOCKMAN. 1971. Procaryotic cell division with respect to wall and membrane. *CRC Critical Reviews in Microbiology* 1:29–72.

HUGO, W. B. (ed.). 1971. *Inhibition and Destruction of the Microbial Cell*. Academic Press, London.

JOHN, P. C. L. (ed.). 1981. *The Cell Cycle*. Cambridge University Press, New York.

LASKIN, A., and H. A. LECHEVALIER. 1977–1981. *Handbook of Microbiology* (4 volumes). CRC Press, Inc., Boca Raton, Fla.

MANDELSTAM, J., K. McQUILLEN, and I. W. DAWES. 1982. *Biochemistry of Bacterial Growth*. Blackwell Scientific Publications, Oxford, England.

MAZIA, D. 1974. The cell cycle. *Scientific American* 230(1): 55–64.

NORRIS, J. R., and D. W. RIBBONS (eds.). 1969–. *Methods in Microbiology*. Academic Press, New York.

RUSSELL, A. D., W. B. HUGO, and G. A. J. AYLIFFE. 1982. *Principles and Practice of Disinfection, Preservation and Sterilisation*. Blackwell Scientific Publications, Oxford, England.

SIROCKIN, G., and S. CULLIMORE. 1969. *Practical Microbiology*. McGraw-Hill Publishing Co., Ltd., London.

SKERMAN, V. D. B. 1969. *Abstracts of Microbiological Methods*. John Wiley & Sons, Inc., New York.

SLATER, M., and M. SCHAECHTER. 1974. Control of cell division in bacteria. *Bacteriological Reviews* 38:199–221.

Study Questions

1. What is a pure culture? Why do we place such importance on obtaining and maintaining pure cultures?

2. What is aseptic transfer technique? Why must you master this technique to work in a microbiology laboratory?

3. Discuss three methods for isolating pure cultures of microorganisms.

4. Why are so many types of microbiological media employed in laboratories for the culture of microorganisms?

5. How are selective and differential media used in the clinical laboratory?

6. How do bacteria normally reproduce? Why are bacterial growth and reproduction considered synonymous?

7. How do we measure microbial growth? What units do we use to express it?

8. Discuss three approaches to the enumeration of bacteria. What are the advantages and disadvantages of each?

9. Describe the typical bacterial growth curve. What is occurring during each of the growth phases?

10. Compare batch and continuous culture methods for growing microorganisms.

11. Calculate the growth rate for a bacterium at 37°C based on the following data. At the time of inoculation the bacterial concentration is 10/ml; after 1 hour it is still 10/ml; after 2 hours it is 30/ml; after 3 hours it is 480/ml; and after 4 hours it is 7,680/ml.

12. Calculate the decimal reduction time for a bacterium at 70°C based on the following data. The initial concentration is 1,800/ml. After 10 minutes of exposure at 70°C it is reduced to 18/ml.

Situational Problem

Ensuring Drinking Water Safety

Based on your expertise in bacteriology, you have gotten a part-time summer job working with the municipal board of health to perform routine tests on the bacteriological quality of food and water. Your job involves performing tests to determine the number of bacteria in samples sent to the health department and reviewing test results from independent laboratories. Your main concern is with enteric pathogens, which are bacteria that cause disease when they enter the gastrointestinal tract and which tend to be shed with fecal matter. To test for the presence of such bacteria, you look for coliform bacteria (*E. coli*), which are found in high numbers in human fecal matter. By using this indicator organism, you provide a margin of safety because the actual enteric pathogens generally are present in much lower numbers than coliforms and hence might be missed.

To detect coliforms, your laboratory uses EMB agar because it is selective for Gram-negative bacteria and because coliforms produce colonies with a green metallic sheen due to their ability to use lactose in the medium. The standard that you are using for determining the safety of the water supply is a maximal permissible coliform count of 4/100 ml. To detect coliforms in this concentration, you filter a water sample to collect the bacteria on the filter and place the filter on the surface of an agar plate. The nutrients diffuse through the filter and colonies develop directly on the surface of the filter. The procedure that you use is as follows. You filter 1-liter, 100-ml, and 10-ml water samples through separate 0.45-μm nuclepore filters and place them on EMB agar plates. After they incubate for 24 hours, you examine the filters and count only the colonies with a green metallic sheen.

1. During your first week of work, you fail to detect more than five colonies on any of the plates on samples from the municipal water supply. On one well water sample from a rural farm, the 1-liter filter is completely overgrown, the 100-ml filter has 80 colonies with a green metallic sheen, and the 10-ml filter has 13 colonies with a green metallic sheen. What recommendations would you make?

2. You receive a report from an independent laboratory on a water sample from a cattle ranch within the jurisdiction of your health department. This laboratory uses a five-tube MPN procedure with dilutions of 10^0, 10^{-1}, and 10^{-2}. The tubes contain a medium that permits the selective detection of coliforms based on the production of acid and gas from lactose. They report five positive reactions at the 10^0 dilution, two positive reactions at the 10^{-1} dilution, and one positive reaction at the 10^{-2} dilution. What recommendations would you make?

3. During your second week of work, there is a water main break in the west end of town and the water supply is contaminated with sludge that may contain enteric pathogens. You immediately issue a warning to residents in the affected area that they boil their water before usage. Immediate tests on the water indicate that the municipal water supply reaching that end of town may contain 500 coliforms per milliliter. A resident calls to ask how long to boil the water. Assuming that the *D* value for all enteric pathogens at 100°C is 0.1 minute, what would you recommend?

CHAPTER 5

Microbial Energetics: The Generation of ATP

5.1 Enzymes and Microbial Metabolism

The transformation of nutrients into the constituents of microbial cells occurs through a complex integrated network of biochemical reactions that make up the metabolism of the organism. Metabolic reactions proceed via various defined series of small, discrete steps that establish **metabolic pathways**, the sequential steps between the starting growth substances (**substrate molecules**), and the **end product(s)** constituting the particular pathway. The metabolic pathways are interconnected to form a complex network through which energy and materials flow and are transformed.

All metabolic reactions of microorganisms are mediated by enzymes. Thousands of different enzymes are needed for a microorganism to generate and use adenosine triphosphate (ATP) and to reproduce. **Enzymes** are biological catalysts, that is, they are molecules that lower the energy of activation required for a biochemical reaction to occur. The **activation energy** is the amount of energy required in a collision between two molecules to bring about that reaction. Enzymes permit essential metabolic steps to proceed rapidly at the temperatures at which living systems exist by lowering the activation energy (Figure 5.1). To lower the activation energy, enzymes combine with their specified substrates to form enzyme–substrate complexes. After the substrate reacts to form products, the enzyme is released in its original state. Because the enzyme is not consumed or modified in the overall reaction, it can be used over and over again.

The rate of an enzymatically catalyzed reaction is dependent on the temperature, the concentrations of the enzyme and substrate, and the affinity of the enzyme for the substrate. The kinetics of enzymatic reactions exhibit the phenomenon of **saturation**, in which raising the concentration of the sub-

strate does not continue to increase the rate of the reaction. The maximal rate of the reaction is termed V_{max} and the substrate concentration at $\frac{1}{2}V_{max}$ is termed the K_m. The K_m, known as the *Michaelis constant*, is a measure of the affinity of the enzyme for the particular substrate; the greater the affinity the lower the K_m. The relation of V_{max} and K_m and the kinetics of enzyme reactions are described by the **Michaelis-Menten equation**:

$$v = V_{max}/(K_m + [S])$$

where v is the velocity (rate) of the reaction, [S] is the sub-

FIGURE 5.1
This graph illustrates the energy of activation required for a chemical reaction to start and shows that an enzyme effectively lowers this energy of activation.

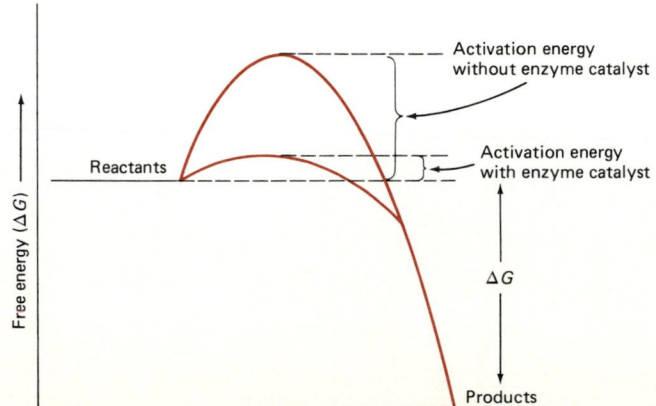

112

strate concentration, V_{max} is the maximal rate of the reaction, and K_m is the substrate concentration at $\frac{1}{2}V_{max}$ (Figure 5.2).

The degree of **substrate specificity** exhibited by enzymes reflects the fact that the enzyme and the substrate must fit together in a specific way for the enzyme to lower the activation energy. The binding of the substrate molecule to the enzyme actually alters the three-dimensional configuration of the enzyme, inducing the proper fit (Figure 5.3). The enzyme has specific active sites at which the substrate binds. When all the active sites of all the molecules of a particular enzyme of an organism are occupied, saturation occurs and the reaction proceeds at the maximal rate. The binding of the enzyme to the substrate involves the formation of weak bonds that are sufficient to place a strain on the substrate molecule, which lowers the activation energy and permits the reaction to proceed. In the case of multiple reactive molecules, which

bind at the active sites of an enzyme, the enzyme effectively positions the reactants in space so that they are brought together with the correct orientation to produce the biochemical reaction. The precision of fit and the structural complementarity between enzyme and substrate molecule permit the establishment of exactly the right spatial orientation so that the numerous biochemical reactions of a microorganism can occur with great rapidity.

The ability to modify the three-dimensional shape of the enzyme molecule provides a basis for microorganisms to regulate the rates of their enzymatic activities at the biochemical level. Substances called **allosteric effectors** may bind to the enzyme at locations some distance from the active site that reacts with a substrate, thereby altering the properties of the enzyme (Figure 5.4). Reaction with an allosteric effector may either increase (activate) or decrease (inhibit)

FIGURE 5.2

(A) Graph showing the relationship between the velocity of an enzymatic reaction and the substrate concentration. (B) To estimate K_m and V_{max}, data on reaction kinetics are usually graphed as a Lineweaver Burk plot, with $1/v$ plotted on the Y axis and $1/[s]$ on the X axis. Using this graphic approach, the Y intercept is $1/V_{max}$ and the X intercept is $-1/K_m$. The ability to determine K_m and V_{max} in this manner is based upon the conversion of the Michaelis-Menten equation for enzyme kinetics to the form $1/v = (K_m/V_{max} \times 1/[s]) + 1/V_{max}$, which is in the form of an equtaion for a straight line.

FIGURE 5.3

The fit between the enzyme and the substrate has been likened to that of a lock and key. Actually, this interaction modifies the three-dimensional structure of the enzyme. The precision of fit is responsible for the high degree of specificity of enzymes for particular substrates.

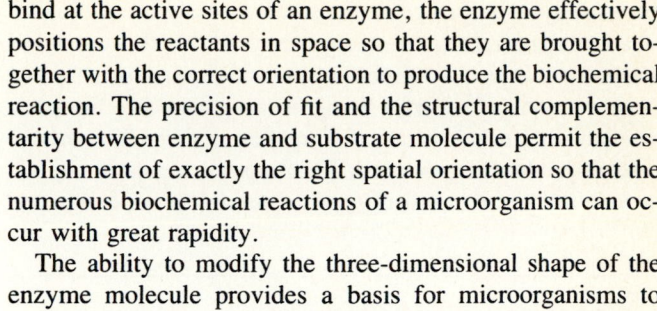

FIGURE 5.4 *(below)*

The activities of enzymes can be increased or decreased by the binding of a substance other than the substrate to allosteric effector sites.

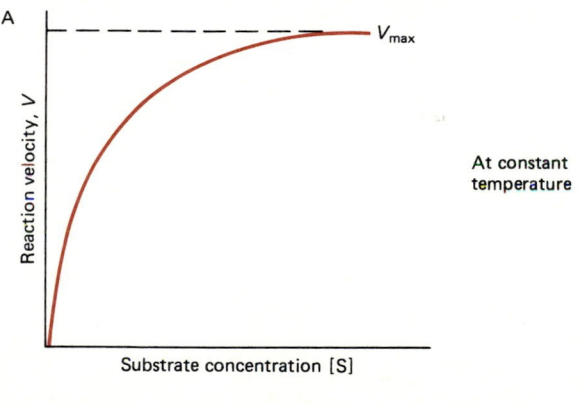

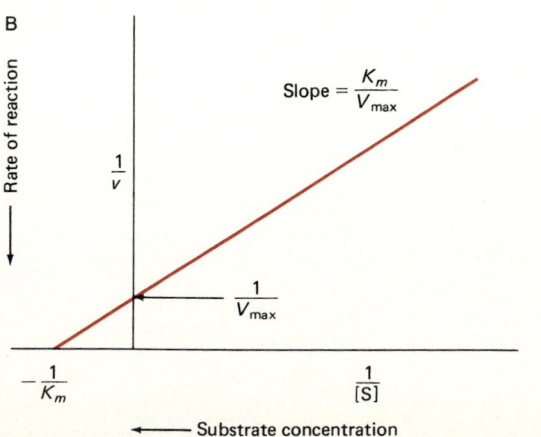

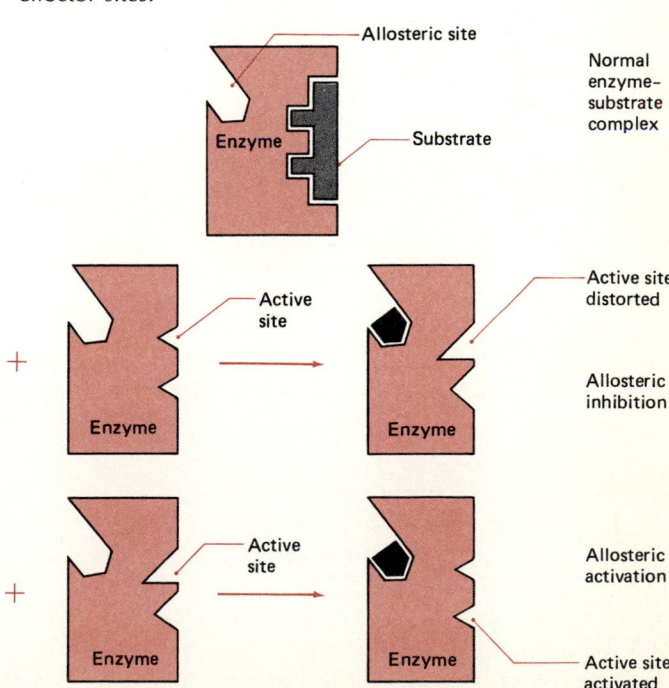

the activity of the enzyme. In some cases, the end product of an enzyme reaction sequence may bind with an enzyme within that sequence, inhibiting the activity of that enzyme. Such a system is self-regulating because there is a feedback mechanism through which the reaction is shut off when excessive product is produced. This type of regulation by allosteric inhibition is called **feedback inhibition** or **end prod-** **uct inhibition**. In other cases, the binding of a substance to an allosteric site may activate the enzyme, increasing its activity (**feedback activation**). Both feedback inhibition and feedback activation are important processes that regulate the activities of enzymes and thus the rates of metabolic reactions carried out by microorganisms.

5.2 Bioenergetics

Although they do not determine the rates of enzymatic reactions, the laws of **thermodynamics** govern the chemical reactions that occur in a cell and prescribe the flow of energy through the microorganism. According to the first law of thermodynamics, energy is conserved; that is, chemical reactions neither create nor destroy energy. Energy, however, can be transformed, and thus chemical, heat, light, electrical, and mechanical energy can be interchanged. In a chemical reaction there is a net balance between the energy required to break chemical bonds, the energy released by the new bonds that are formed, and the energy—such as heat energy—that is exchanged with the surrounding environment. The change in the stored energy between the amount contained in the bonds of the reactants and those of the products of a chemical reaction is described by the ΔH (**enthalpy**) of the reaction, the change in heat content of the molecules. Chemical reactions that absorb heat have a positive ΔH and are known as **endothermic reactions**; those that release heat have a negative ΔH and are called **exothermic reactions** (Figure 5.5).

Of more importance than the change in enthalpy in predicting whether a biological reaction can occur is the change in **free energy**, ΔG, which describes the change in the energy of the system that is available for doing work. The concept of free energy takes into consideration the degree of order of the system as well as the stored energy. The change in the free energy of a reaction (ΔG) is a function of the change of the heat of reaction (ΔH), the temperature (T, absolute temperature in degrees Kelvin[1]), and the change in the state of order or **entropy** (ΔS) between products and reactants:

$$\Delta G = \Delta H - T\Delta S.$$

According to the second law of thermodynamics, all processes proceed in the direction that increases the total entropy of the system and the surroundings, that is, in the direction of maximum randomness or disorder. Order can increase within a biological system only if disorder increases elsewhere in the universe. Reactions that release free energy have a negative ΔG and are called **exergonic**; these reactions can proceed spontaneously. Reactions that require the addition of free energy from another source in order to proceed have a positive ΔG and are called **endergonic** (Figure 5.6). Viewed in another way, some chemical reactions require free energy to drive them uphill (endergonic reactions), and others release free energy as they run downhill (exergonic re-

[1]Degrees Kelvin (°K) are related to degrees Celsius (°C) according to the equation °K = °C + 273°; therefore 0°K = −273°C.

FIGURE 5.5
Energy diagrams showing ΔH for exothermic and endothermic reactions.

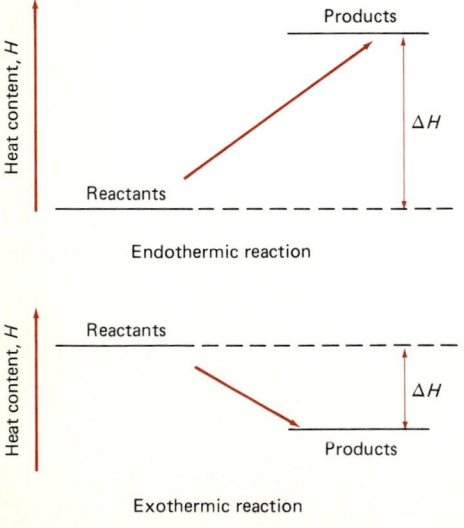

FIGURE 5.6
This graph shows the changes in free energy during exergonic and endergonic reactions.

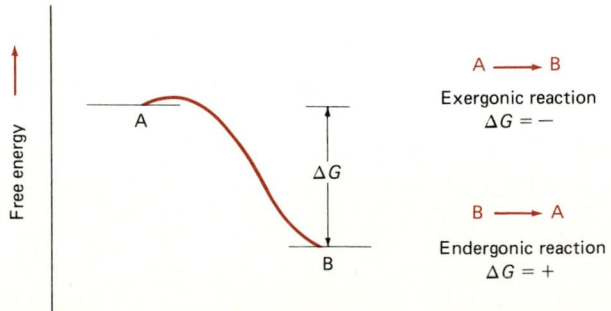

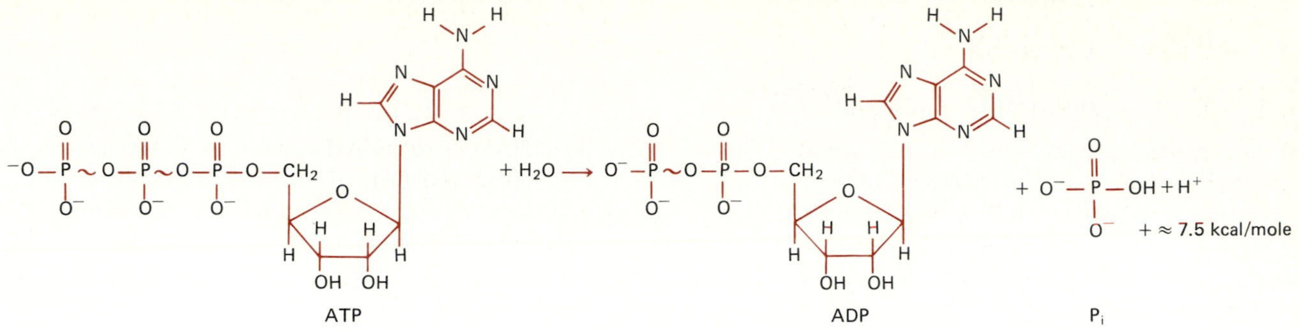

FIGURE 5.7
The conversion of ATP to ADP is an exergonic reaction that can be used to drive endergonic reactions.

actions). By knowing whether the ΔG for a reaction is positive or negative, one can tell whether or not that reaction requires an input of energy to proceed.

The actual ΔG of a reaction is a function of the relative concentrations of reactants and products and the standard free energy ($\Delta G°$) change of the reaction at 1 atmosphere (atm) pressure and 1 molar concentrations (mole) of reactants. The $\Delta G°$ is a thermodynamic constant for a given chemical reaction and is related to ΔG by the formula:

$$\Delta G = \Delta G° + RT \ln K_{eq}$$

where R is the gas constant (1.99 cal/mole/degree), T is the absolute temperature in °K, and $\ln K_{eq}$ is the natural logarithm of the equilibrium constant.[2] The $\Delta G°$ is the difference between the sum of the standard free energies of all of the products and the sum of the standard free energies of all of the reactants.[3] For any given biochemical reaction, the $\Delta G°$ of the reaction indicates whether the formation of products or reactants is favored. The $\Delta G°$ for an overall reaction will be the same regardless of the number of steps required to go from reactants to products. Consequently, it is possible to couple individual reactions to achieve a favorable overall $\Delta G°$ that will allow the complete process to occur, such that the energy released by an exergonic reaction is used to drive an endergonic reaction.

Many of the metabolic pathways of microorganisms are involved with coupling thermodynamically favorable reactions with the endergonic conversion of ADP + inorganic phosphate (P_i) to ATP; others use ATP and, in doing so, generate adenosine diphosphate (ADP) as a product. This cycling of ADP and ATP within the cell is fundamental to the bioenergetics of microorganisms, and the cell must continuously form and consume ATP.[4] A growing cell of the bacterium *Escherichia coli*, as a typical example, must synthesize approximately 2.5 million molecules of ATP per second to support its energy needs. The coupling of the free energy released when ATP is converted to ADP with thermodynamically unfavorable (endergonic) reactions shifts the ratio of products to reactants in such endergonic reactions by a huge factor, on the order of 10^8. When ATP is converted to ADP, the electrostatic repulsion between the negatively charged phosphate groups is reduced, and it is this fact that accounts for the relatively large release of free energy associated with this reaction (Figure 5.7). ATP is particularly useful because of its intermediate position in terms of stored energy, making it possible for microorganisms to generate as well as to use this molecule as a currency of free energy (Table 5.1).

---

[4]Although more work could be accomplished using a biochemical with a higher free energy potential than ATP for storage of cellular energy, it would be inefficient to do so considering the number of individual endergonic reactions of the cell that must be driven by coupling with the central energy storage molecule.

TABLE 5.1   Free Energies of Hydrolysis of Some Phosphorylated Compounds

Compound	$\Delta G°$(kcal/mole)
Phosphoenolpyruvate	− 14.8
Carbamyl phosphate	− 12.3
Acetyl phosphate	− 10.3
Creatine phosphate	− 10.3
Pyrophosphate	− 8.0
ATP to ADP	− 7.3
Glucose 1-phosphate	− 5.0
Glucose 6-phosphate	− 3.3
Glycerol 3-phosphate	− 2.2

---

[2]For the generalized reaction $aA + bB \rightleftarrows cC + dD$ the equilibrium constant (K) is $K_{eq} = [C]^c[D]^d/[A]^a[B]^b$.

[3]$\Delta G°$ reaction = $\Sigma\Delta G°$ products − $\Sigma\Delta G°$ reactants. At equilibrium the $\Delta G°$ of a reaction is given by the equation $\Delta G° = -RT \ln K_{eq}$, where R is the gas constant (1.99 cal/mole/degree), T is the absolute temperature, and $\ln K_{eq}$ is the natural logarithm of the equilibrium constant, which is equal to the concentration(s) of the products divided by the concentration(s) of the reactants.

5.3 Coenzymes and Oxidation-Reduction Reactions

Many of the metabolic reactions of microorganisms, including those involved in energy transfers, are oxidation-reduction reactions[5] that require the use of coenzymes. **Coenzymes**, small, nonprotein organic compounds that are not tightly bound to the enzymes with which they function, frequently accept a chemical group produced in one enzymatic reaction—such as an electron—hold that group for a short period of time, and donate it to another biochemical reaction.[6] In enzymatic reactions in which a substrate is oxidized, the electron is often transferred to a coenzyme, which becomes reduced in this process. An example of such a reaction is the conversion of the coenzyme nicotinamide adenine dinucleotide (NAD^+) to its reduced form, nicotinamide adenine dinucleotide (NADH) (Figure 5.8). The reverse is true for an enzymatic reaction in which the substrate is reduced. In such reactions, a reduced coenzyme can donate an electron, thus becoming oxidized, as when the reduced coenzyme nicotinamide adenine dinucleotide phosphate (NADPH) is oxidized to nicotinamide adenine dinucleotide phosphate ($NADP^+$) in a reaction that is coupled with the reduction of a substrate (Figure 5.9). Because biosynthetic reactions involve a net conversion of relatively oxidized carbon-containing molecules—such as carbon dioxide—to relatively reduced carbon-containing molecules—such as carbohydrates ($C_nH_{2n}O_n$)—a cell must generate reduced coenzymes (**reducing power**) that can be coupled with the reductive biosynthetic reactions.

The relative susceptibility of a substance to oxidation or reduction can be described quantitatively by its **reduction potential**. The reduction potential is the relative voltage required to remove an electron from a given compound compared to the voltage required to remove an electron from H_2

---

[5]Oxidation represents a loss of electrons, reduction a gain of electrons. Oxidation-reduction reactions always occur together, and the electrons gained by the molecule that is reduced must balance those given up by the substance that is oxidized.

[6]Coenzymes differ from cofactors, which are inorganic substances—such as minerals—required for enzymatic activity. A coenzyme is also distinct from a prosthetic group, which is a nonprotein organic substance that binds tightly to an enzyme, forming a permanent part of that enzyme.

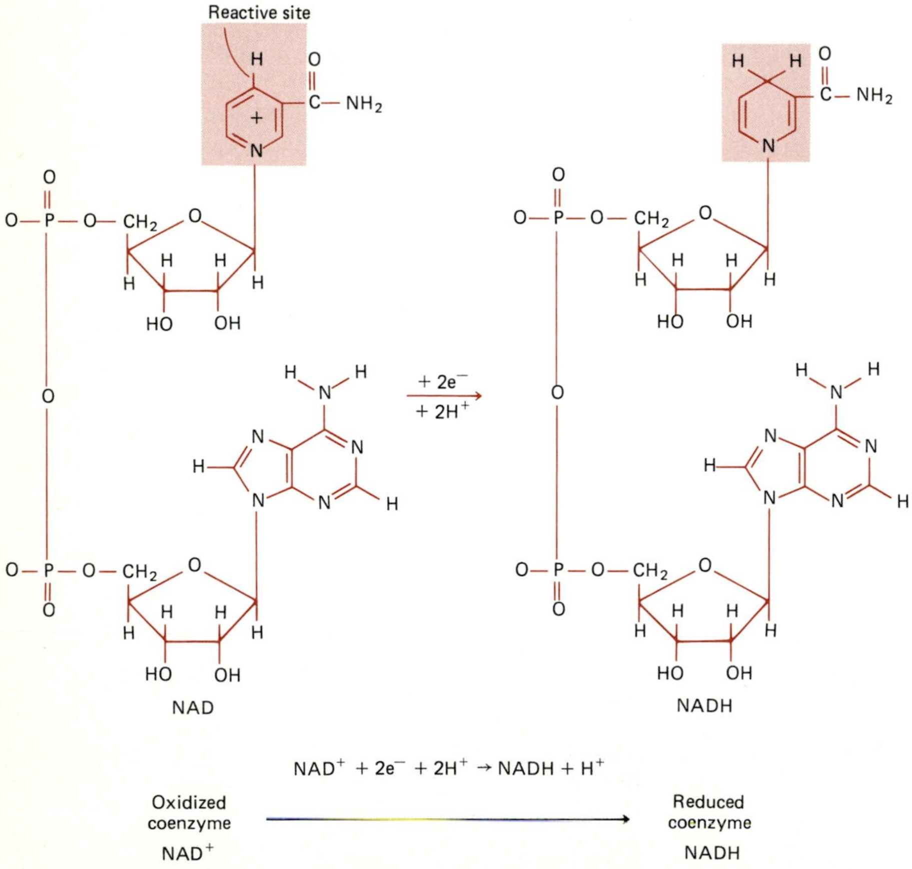

FIGURE 5.8
The reduction of NAD^+ to $NADH + H^+$ is a critical reaction that often is coupled with the oxidation of substrates within a cell. This reaction can be written several ways; the form $NAD^+ \rightarrow NADH$ is employed throughout this book.

$$NAD^+ + 2e^- + 2H^+ \rightarrow NADH + H^+$$

Oxidized coenzyme NAD^+ ⟶ Reduced coenzyme NADH

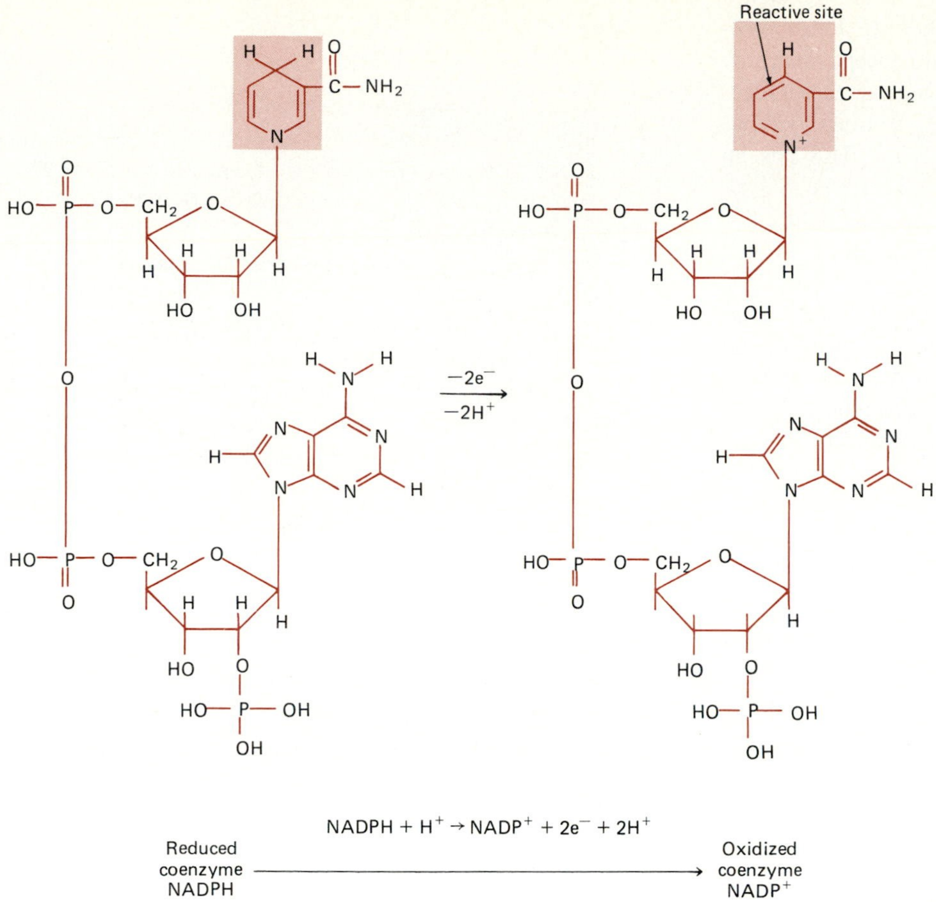

$$NADPH + H^+ \rightarrow NADP^+ + 2e^- + 2H^+$$

Reduced coenzyme NADPH $\longrightarrow$ Oxidized coenzyme NADP$^+$

FIGURE 5.9

The reduced coenzyme NADPH is important in biosynthetic reactions. It can be formed by the reduction of NADP$^+$ to NADPH + H$^+$ and can donate electrons in reactions when it is reoxidized.

under standard conditions.[7] Most molecules can be reduced (accept electrons) or oxidized (donate electrons) at different times, depending on the other molecules with which they react. Reducing agents donate electrons and become oxidized in the process, whereas oxidizing agents accept electrons and are reduced in the process. In an oxidation-reduction reaction the reducing agent has the more negative reduction potential, and therefore donates electrons, while the oxidizing agent has the more positive reduction potential and accepts electrons. The overall oxidation-reduction reaction consists of two half-reactions, one in which electrons are added (reduction) and the other in which electrons are donated (oxidation). By convention, equations for the half-reactions are both written as reductions even though one of the two half-

reactions must be an oxidation that proceeds in the reverse direction from which it is written. Thus, even if hydrogen acts in a reaction as an electron donor, the half-reaction is written as $2H^+ + 2e^- \rightarrow H_2$ with a reduction potential of -0.42 volt.

As an example of the reduction potentials in an oxidation-reduction reaction, let us consider the reaction of hydrogen with oxygen that forms water:

$$H_2 + O_2 \rightarrow H_2O.$$

The components of this reaction are

$$2H^+ + 2e^- \rightarrow H_2$$
$$\tfrac{1}{2}O_2 + 2e^- \rightarrow O^-$$
$$2H^+ + O^- \rightarrow H_2O$$

The reduction potentials for the half-reactions are $2H^+/H_2 = -0.42$ volt and $\tfrac{1}{2}O_2/H_2O = +0.82$. Therefore, in this reaction, hydrogen donates electrons because it has the more negative reduction potential and oxygen accepts electrons because it has the more positive reduction potential.

One can arrange the possible half-reactions on a scale, with the most negative reduction potentials at the top and the

[7]The standard reduction potential of the hydrogen electrode, $\tfrac{1}{2}H_2 = H^+ e^-$, is given an arbitrary value of 0.0 volt when all reactants and products are at 1 molar or 1 atm and the pH is 0.0. At pH 7.0, which is more typical of biological systems, the potential of the hydrogen electrode is -0.42 volts. The symbol E'_0 is used to indicate the reduction potentials measured at 25°C, with all reactants and products at 1.0 molar or 1 atm and pH 7.0. Thus, the E'_0 for the hydrogen electrode is -0.42.

most positive ones at the bottom (Figure 5.10). On such a scale, the half-reactions at the top are likely to donate electrons and those at the bottom are likely to accept electrons. The half-reactions in the middle of the scale can accept electrons from those above them or donate electrons to those below. For example, the half-reaction SO_4^{2-}/H_2S has an intermediate reduction potential of -0.22. Therefore, SO_4^{2-} can accept electrons from hydrogen and become reduced to H_2S, or alternatively, H_2S can donate electrons to oxygen and become oxidized to SO_4^{2-}.

An important aspect of reduction potentials is their relationship to the free energies of reactions. The greater the difference in voltage between the half-reactions, the greater the energy potential of the reaction. Reduction potential (E'_0) is related to free energy according to the equation

$$\Delta G^{\circ\prime} = -nF\Delta E'_0$$

where $\Delta G^{\circ\prime}$ is the free energy change at pH 7.0, n is the number of electrons transferred, F is the faraday constant (a physical constant equal to 23,000 cal/volt), and $\Delta E'_0$ is the difference between the potentials of the two half-reactions involved in an oxidation-reduction reaction. For example, based upon E'_0 values of -0.32 volt for the half-reaction $NAD^+/NADH$ and $+0.82$ for the half-reaction $\frac{1}{2}O_2/H_2O$, the oxygen-linked oxidation of NADH to NAD^+ has an $\Delta E'_0 = 1.14$, which converts to a free energy for this reaction of -52.4 kcal. This particular reaction is very important in the generation of cellular energy by respiration. Other oxidation reactions can also be coupled with the generation of ATP, the molecule primarily used to transfer energy within the cell.

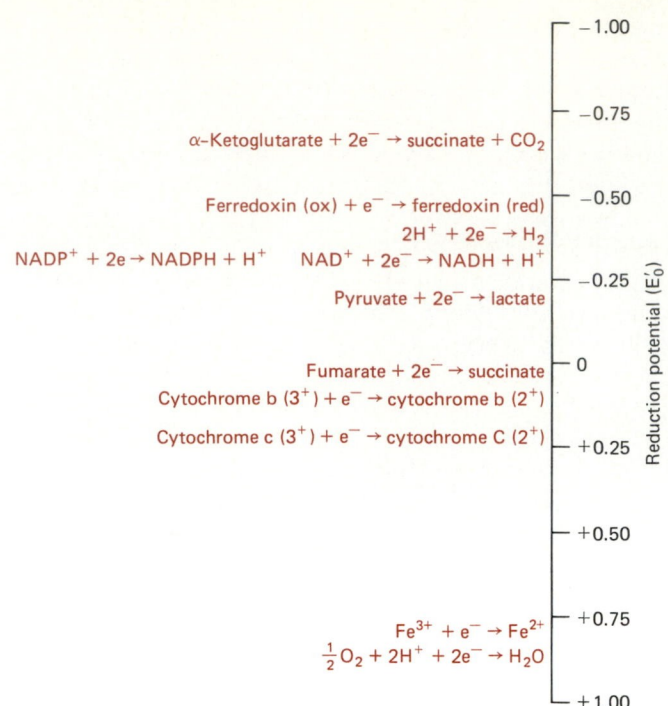

FIGURE 5.10
The reduction potentials for half-reactions can be shown as an electron tower, with the reactions most likely to donate electrons (most negative E'_0) at the top and those most likely to accept electrons (most positive E'_0) at the bottom. In an oxidation-reduction reaction, the difference in E'_0 values between the electron donor and the electron acceptor determines the free energy of the reaction.

5.4 *Heterotrophic Generation of ATP*

The heterotrophic generation of ATP involves the conversion of an organic substrate molecule to end products via a metabolic pathway that has sufficient free energy so that it can be coupled with the synthesis of ATP. This process involves the breakdown of an organic molecule to smaller molecules and is called **catabolism**. In such a pathway, oxidation reactions provide sufficient free energy to drive the conversion of ADP to ATP. These oxidation reactions are coupled with a simultaneous reduction reaction that balances the transfer of electrons, often the reduction of the coenzyme NAD^+ to its reduced form, NADH. To sustain its metabolism, the cell must reoxidize the reduced coenzyme in subsequent biochemical reactions; the reoxidation of NADH ensures the continuous supply of NAD^+ required for use as an oxidizing agent (electron acceptor) in metabolic pathways aimed at generating ATP. Thus, the ability to generate energy in the form of ATP is integrally tied to the cell's ability to balance its oxidation-reduction reactions.

Fermentation versus Respiration

For oxidizing organic compounds to release the free energy that can drive the formation of ATP, while still maintaining the required balance between oxidation and reduction reactions, heterotrophic microorganisms exhibit two basic strategies, fermentation and respiration. In **fermentation pathways** the organic substrate acts as an electron donor (reducing agent) and a product of that substrate acts as an electron acceptor (oxidizing agent). Therefore the both the electron donor and the acceptor are internal to the organic substrate. There is no net change in the oxidation state of the products relative to the starting substrate molecule in fermentation pathways. The oxidized products are exactly counterbalanced by the reduced products, and thus the required oxidation-reduction balance is achieved. The coenzymes that are reduced during this pathway are reoxidized by its end, so that they are in fact not consumed in the process. Fermentation pathways can occur in the absence of air because there is no requirement for oxygen or another electron acceptor to achieve a balance in the oxidation-reduction reaction; the organic substrate provides both the electron donor and the acceptor needed to achieve this balance.

In contrast to fermentation, a **respiration pathway** requires an external electron acceptor; that is, a molecule other than one derived from the electron donor must act as the oxidizing agent to achieve a balance of oxidation-reduction reactions. By reducing the terminal electron acceptor, the cell is able to balance the change in the oxidation state of the metabolic products relative to the starting substrate, that is, to achieve a balance of oxidation-reduction reactions without consuming coenzymes. The most common external electron acceptor in respiration pathways is molecular oxygen. When molecular oxygen serves as the terminal electron acceptor, the pathway is known as **aerobic respiration**. When another molecule, such as nitrate or sulfate, serves as the terminal electron acceptor, the metabolic pathway is called **anaerobic respiration** (literally, respiration not requiring the presence of air).[8]

Considerably different energy yields can be achieved by microorganisms using fermentation and respiration pathways, with fermentation yielding far less ATP per substrate molecule than respiration. This is because the organic substrate molecule must serve as both the internal electron donor and the electron acceptor during a fermentation pathway, and thus the substrate cannot be completely oxidized to carbon dioxide. The $\Delta G°$ for the complete oxidation of glucose to carbon dioxide and water is 686 kcal/mole, compared to only 58 kcal/mole when glucose is partially oxidized to two molecules of the fermentation product lactic acid. As a result, not as much energy can be released by fermentation to drive the synthesis of ATP as can be obtained during a respiration pathway in which glucose is oxidized completely to carbon dioxide and water. Because more ATP can be generated per molecule of substrate by a respiration pathway than by a fermentation pathway, fewer substrate molecules must be metabolized during respiration than during fermentation to achieve equivalent growth, that is, to support the metabolic requirements of an equivalent number of cells. From the viewpoints of both bioenergetics and conservation of available organic nutrient resources, respiration is more efficient and favorable than fermentation, and organisms that have the metabolic capability to carry out both types of metabolism will generally use the energetically more favorable respiration pathway when conditions permit and will rely on fermentation only when there is no available external electron acceptor that can be used by the organism.

---

[8]An organism such as *Pseudomonas denitrificans* can use either oxygen or nitrate as the terminal electron acceptor and thus can carry out both aerobic and anaerobic respiration. Other bacterial species are restricted to one or the other form of respiration.

Respiratory Metabolism

The overall aerobic respiration pathway results in the formation of carbon dioxide from the organic substrate molecule and water from the reduction of oxygen. In the process, a substantial amount of ATP is also formed. The classic equation for aerobic respiration of glucose is

$$C_6H_{12}O_6 + 6O_2 \rightarrow 6CO_2 + 6H_2O.$$

The overall pathway can be divided into three distinct phases (Figure 5.11), each of which we will examine in some detail:

1. A catabolic pathway, during which the organic molecule is broken down into smaller molecules so that acetyl coenzyme A (CoA) is formed, usually with the generation of some ATP and reduced coenzymes,
2. The Krebs cycle, during which organic carbon is oxidized to inorganic carbon dioxide and reduced coenzyme is generated,
3. Oxidative phosphorylation, during which reduced coenzyme molecules are reoxidized, electrons are transported through a series of membrane-bound carriers to establish a hydrogen ion gradient across a membrane, the terminal electron acceptor is reduced, and ATP is synthesized.

Catabolism of Organic Substrate to Form Acetyl CoA

Carbohydrate Catabolism. In the case of carbohydrates, a substrate molecule such as glucose is initially broken down to pyruvate via a **catabolic pathway**, that is a degradative pathway called **glycolysis**. In the most common of these pathways—the **Embden-Meyerhof pathway**, also known as the **Embden-Meyerhof-Parnas pathway** (Figure 5.12)—one molecule of glucose is converted to two molecules of pyruvate; this is accompanied by the formation of two reduced coenzyme (NADH) molecules and the release of sufficient free energy to permit a net synthesis of two ATP molecules (Table 5.2).

During the initial three reactions of the Embden-Meyerhof pathway of glycolysis, glucose is converted to fructose 1,6-bisphosphate. This beginning sequence of reactions of the glycolytic pathway involves endergonic reactions that require energy to drive them, and the pathway starts with the use, rather than the synthesis, of ATP. This series of reactions requires the input of the energy equivalent of two ATP molecules, one for the conversion of glucose to glucose 6-phosphate and one for the conversion of fructose 6-phosphate to fructose 1,6-bisphosphate.[9] Subsequent to the formation of fructose 1,6-bisphosphate, the individual stepwise reactions are exergonic and, thus, further utilization of ATP is not required. In fact, sufficient ATP is synthesized in two of the steps to yield a net gain of ATP.

---

[9]Bacteria can initiate the glycolytic pathway during the transport of the substrate across the membrane in the group transport system. Although the phosphorylation of the carbohydrate in this system is coupled with the conversion of phosphoenolpyruvate to pyruvate rather than the conversion of ATP to ADP, the energy balance is roughly equivalent.

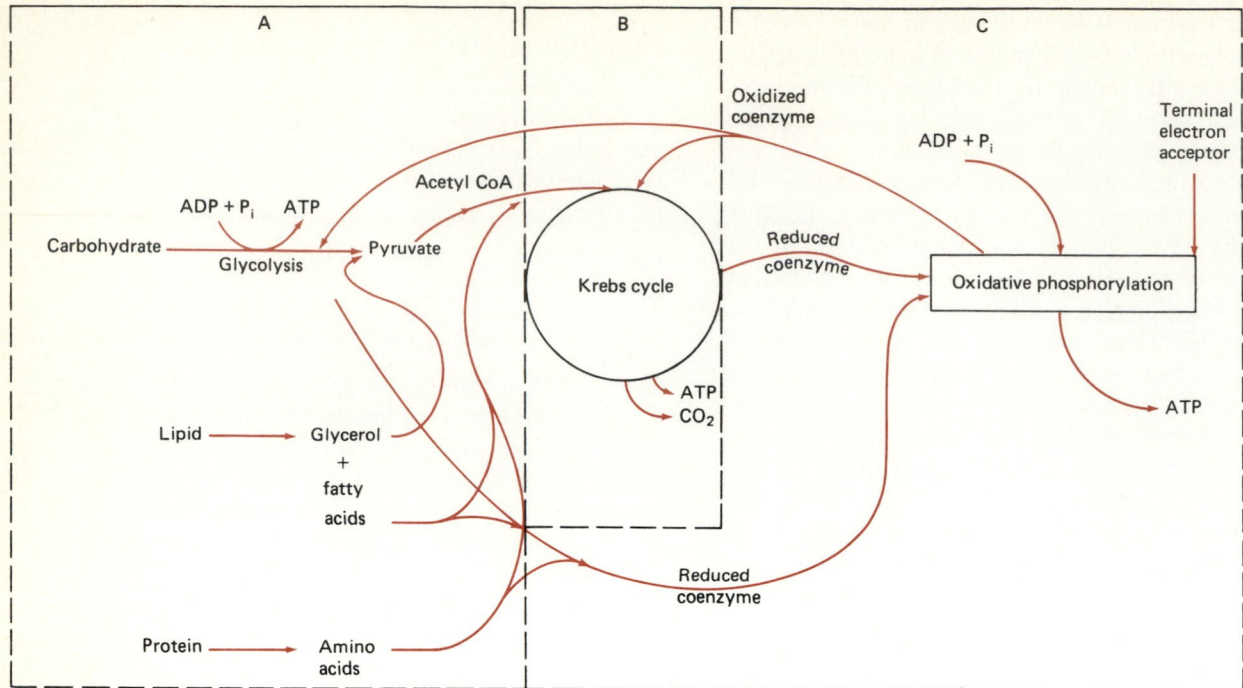

FIGURE 5.11

The metabolic pathways of a cell form a complex, interrelated network. This drawing shows a simplified view of the interrelationships of several major pathways: (A) catabolic pathways, including glycolysis and β-oxidation, that form small molecules that can be fed into the Krebs cycle via acetyl CoA; (B) the Krebs cycle; and (C) oxidative phosphorylation. Using these pathways, cells can generate needed cellular energy and carbon constituents from various classes of substrates, including carbohydrates, lipids, and proteins.

TABLE 5.2   Free Energies of Glycolysis Reactions

Reaction	Enzyme	ΔG°	ΔG
Glucose + ATP → glucose 6-phosphate + ADP + P_i	Hexokinase	−4.0	−8.0
Glucose 6-phosphate → fructose 6-phosphate	Phosphoglucose isomerase	+0.4	−0.6
Fructose 6-phosphate + ATP → fructose 1,6-diphosphate + ADP + P_i	Phosphofructo-kinase	−3.4	−5.3
Fructose 1,6-diphosphate → dihydroxyacetone phosphate + glyceraldehyde 3-phosphate	Aldolase	+5.7	−0.3
Dihydroxyacetone phosphate → glyceraldehyde 3-phosphate	Triose phosphate isomerase	+1.8	+0.6
Glyceraldehyde 3-phosphate + P_i + NAD → 1,3-diphosphoglycerate + NADH	Glyceraldehyde 3-phosphate dehydrogenase	+1.5	−0.4
1,3-Diphosphoglycerate + ADP → 3-phosphoglycerate + ATP	Phosphoglycerate kinase	−4.5	+0.3
3-Phosphoglycerate → 2-phosphoglycerate	Phosphoglycero-mutase	+1.1	+0.2
2-Phosphoglycerate → phosphoenolpyruvate	Enolase	+0.4	−0.8
Phosphoenolpyruvate + ADP → pyruvate + ATP	Pyruvate kinase	−7.5	−4.0

The ΔG° is based on the standard free energy; the ΔG represents the approximate actual free energy change based on measured concentrations of reactants under typical conditions.

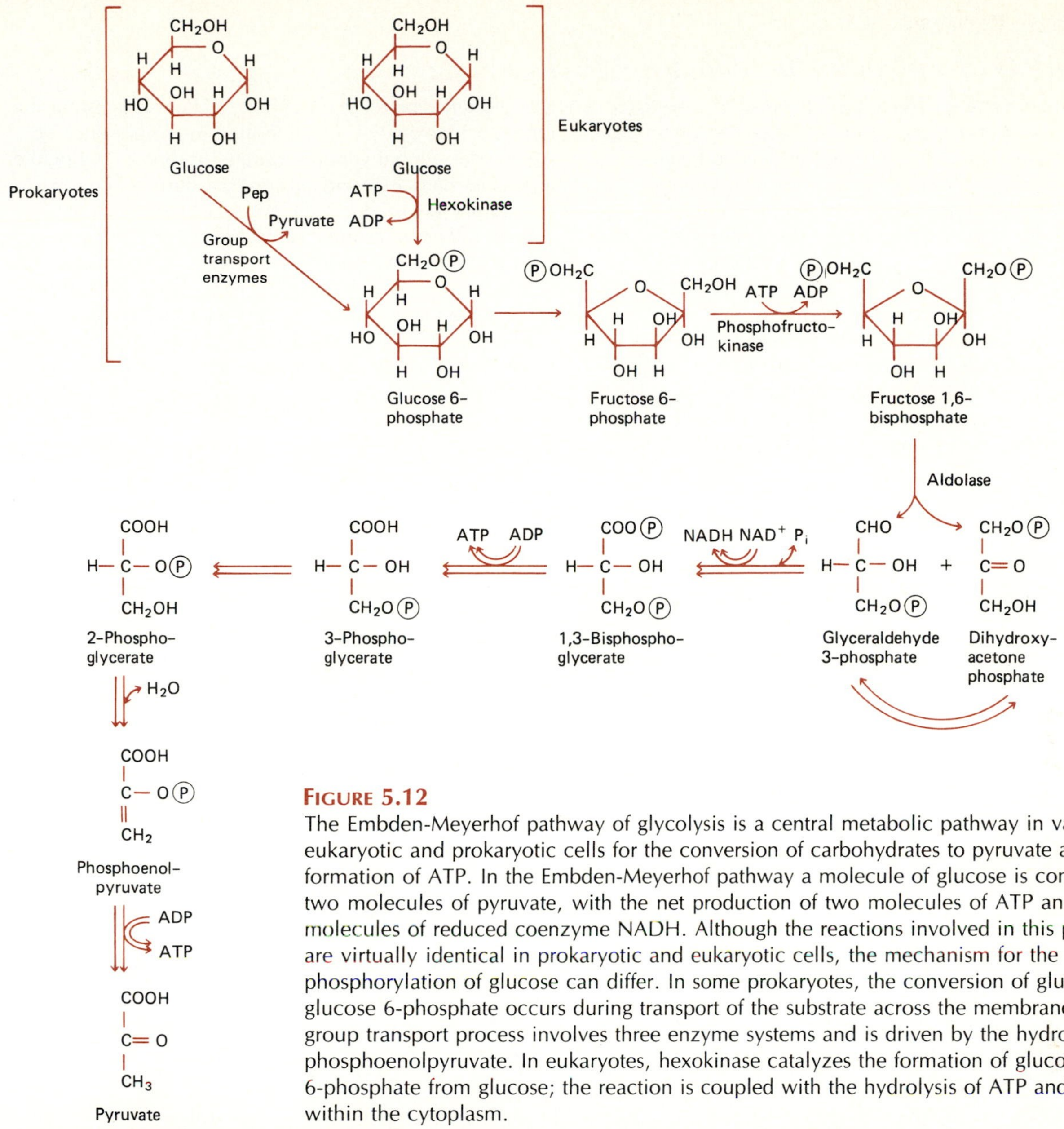

FIGURE 5.12

The Embden-Meyerhof pathway of glycolysis is a central metabolic pathway in various eukaryotic and prokaryotic cells for the conversion of carbohydrates to pyruvate and the formation of ATP. In the Embden-Meyerhof pathway a molecule of glucose is converted to two molecules of pyruvate, with the net production of two molecules of ATP and two molecules of reduced coenzyme NADH. Although the reactions involved in this pathway are virtually identical in prokaryotic and eukaryotic cells, the mechanism for the initial phosphorylation of glucose can differ. In some prokaryotes, the conversion of glucose to glucose 6-phosphate occurs during transport of the substrate across the membrane; this group transport process involves three enzyme systems and is driven by the hydrolysis of phosphoenolpyruvate. In eukaryotes, hexokinase catalyzes the formation of glucose 6-phosphate from glucose; the reaction is coupled with the hydrolysis of ATP and occurs within the cytoplasm.

Discovery Process

The delineation of this central metabolic pathway—the Embden-Meyerhof pathway—was central to the development of the field of biochemistry. The process of discovery began accidentally in 1897 when Hans and Eduard Buchner, while trying to manufacture cell-free extracts of yeast for therapeutic use, attempted to preserve the extracts without using antiseptics, such as phenol, and tried sucrose. They were surprised to find that the sucrose was rapidly fermented to alcohol by the yeast juice, thus demonstrating for the first time that fermentation could occur outside of living cells. This contradicted the accepted dogma of the time, held even by Pasteur, that fermentation was inextricably tied to living cells and allowed the study of metabolism to proceed along biochemical paths.

Beginning in 1905, Arthur Harden and William Young studied what happened when yeast juice was added to a glucose solution. They found that fermentation began almost immediately, that its rate decreased unless organic phosphate was added, and that the added inorganic phosphate disappeared during fermentation. From these findings, they inferred that the inorganic phosphate was incorporated to form a sugar phosphate; later, it was shown that fructose 1,6-diphosphate and other phosphorylated carbohydrates are key intermediates in the metabolism of carbohydrates. Later studies of muscle extracts showed that many of the reactions of lactic fermentation were the same as those of alcoholic fermentation, thus revealing the underlying unity in biochemistry.

ANALYTICAL PROCESS

Regulation of Phosphofructokinase Activity

The flow of substrates through glycolysis, and hence the amount of ATP generated are carefully regulated within the cell. The conversion of fructose 6-phosphate to fructose 1,6-bisphosphate, catalyzed by the enzyme **phosphofructokinase**, is an important regulatory step in the pathway. Even though ATP is required for the conversion of fructose 6-phosphate to fructose 1,6-bisphosphate, phosphofructokinase is inhibited by excess ATP because ATP is an allosteric inhibitor of the enzyme. If the cell has a sufficient supply of ATP, the inhibition of this enzyme stops the glycolytic pathway near its beginning, preventing further ATP synthesis. When ATP is depleted, the cell has a relatively high concentration of **adenosine monophosphate (AMP)**—the monophosphate formed by the hydrolysis of ADP. AMP is an allosteric activator for phosphofructokinase; thus, when the cell really needs to generate ATP, the key rate-limiting reaction of glycolysis is stimulated, leading to increased synthesis of ATP.

The allosteric control of phosphofructokinase is responsible for the paradoxical observation that, in the presence of oxygen, less carbohydrate substrate disappears during the growth of many microorganisms than during the growth of the same organisms in the absence of air. This phenomenon, known as the **Pasteur effect**, occurs because during aerobic respiration a high level of ATP accumulates within the cell and inhibits phosphofructokinase, greatly slowing the rate of substrate conversion. In the absence of oxygen, when the cell is using fermentative metabolism, less ATP is produced and glycolysis proceeds without inhibition. The allosteric regulation of phosphofructokinase activity is key to controlling the flow of carbon through the metabolic pathways of a cell, directing the cell toward ATP-generating or ATP-utilizing pathways.

The result of these first steps of glycolysis is the formation of a compound that can be broken down into two phosphorylated 3-carbon units without loss of energy. The fructose 1,6-bisphosphate molecule contains six carbon atoms; it is split into two 3-carbon pieces by the action of the enzyme aldolase. This splitting of the 1,6-bisphosphate molecule into two 3-carbon units is sometimes called the *aldolytic reaction*. One of the 3-carbon molecules formed is 3-phosphoglyceraldehyde; the other 3-carbon molecule, dihydroxyacetone phosphate, may be converted to 3-phosphoglyceraldehyde. The equilibrium constant for the conversion of dihydroxyacetone phosphate and 3-phosphoglyceraldehyde favors the formation of dihydroxyacetone phosphate, which is not in the direct glycolytic pathway. However, the constant removal of 3-phosphoglyceraldehyde, which is in the direct glycolytic pathway, shifts the balance of reactants and products so that the dihydroxyacetone is converted to 3-phosphoglycerate. Thus, for each 6-carbon glucose substrate molecule, two molecules of 3-phosphoglyceraldehyde are formed.

After forming two molecules of 3-phosphoglyceraldehyde, the next portion of the glycolytic pathway is concerned with using the energy stored in this compound to drive the synthesis of ATP. It is important to keep in mind that two phosphorylated 3-carbon molecules are formed for each 6-carbon carbohydrate substrate molecule in order to keep track of the net yield of ATP and reduced coenzyme (NADH) molecules formed during the overall pathway. Each of the steps subsequent to the formation of 3-phosphoglyceraldehyde occurs twice for each 6-carbon carbohydrate molecule.

The 3-phosphoglyceraldehyde molecule is converted to 1,3-bisphosphoglycerate by the incorporation of inorganic phosphate (P_i) into the molecule during this exergonic reaction. The oxidative conversion of 3-phosphoglyceraldehyde to 1,3-bisphosphoglycerate is coupled with the conversion of oxidized NAD^+ to the reduced coenzyme NADH. Because there are two molecules of 3-phosphoglyceraldehyde generated from each glucose molecule, there is a net production of two NADH molecules per molecule of glucose. The 1,3-bisphosphoglycerate is converted to 3-phosphoglycerate, an exergonic reaction that can be coupled with the synthesis of ATP. The formation of ATP in this coupled reaction is called a **substrate-level phosphorylation** reaction, so designated because ATP is formed from ADP by the direct transfer of a high-energy phosphate group from an intermediate substrate in the pathway.[10] Because this reaction occurs for each of the two 3-carbon molecules generated from glucose, two molecules of ATP are generated per glucose molecule, and therefore, the synthesis and utilization of ATP balance at this point in the metabolic pathway; that is, the net production of ATP equals zero at this stage of the metabolic pathway because two ATP molecules had been consumed in earlier steps.

The 3-phosphoglycerate is further converted to phosphoenolpyruvate and finally to pyruvate. The conversion of

---

[10]Substrate-level phosphorylation is contrasted to oxidative phosphorylation, a process in which ATP is generated by chemiosmosis based upon a hydrogen ion gradient across a membrane established through the reoxidation of reduced coenzymes.

phosphoenolpyruvate to pyruvate is coupled with the synthesis of ATP. Thus, this glycolytic pathway results in the conversion of the 6-carbon molecule glucose to two molecules of the 3-carbon molecule pyruvate, with the net production of two molecules of reduced coenzyme, NADH, and the net synthesis of two ATP molecules. The overall equation for glycolysis by the Embden-Meyerhof pathway can be written as follows:

Glucose $+$ 2 ADP $+$ 2 P$_i$ $+$ 2 NAD$^+$ $\rightarrow$
$$2 \text{ pyruvate} + 2 \text{ NADH} + 2 \text{ ATP.}$$

To enter the Krebs cycle, the next phase of a respiration pathway, the pyruvate molecules generated during glycolysis react with CoA in a reaction catalyzed by the pyruvate de-

hydrogenase complex. This reaction, which is coupled with the conversion of the coenzyme NAD$^+$ to reduced NADH, forms acetyl CoA and carbon dioxide.

The Embden-Meyerhof pathway is only one of several glycolytic pathways utilized by different microorganisms. The **Entner-Doudoroff pathway** of glycolysis, for example, is an alternative pathway possessed by some microorganisms that results in the net production of only one ATP molecule per molecule of glucose substrate metabolized (Figure 5.13). Only one ATP molecule is produced because the glucose 6-phosphate is oxidized before aldolytic cleavage and the substrate-level phosphorylation that usually accompanies oxidation is lost. The Entner-Doudoroff pathway is utilized by various *Pseudomonas* species and other bacteria. The net

FIGURE 5.13

The Entner-Doudoroff pathway is one of several types of glycolysis. Compared to the Embden-Myerhof pathway, less ATP is generated when this metabolic pathway is used.

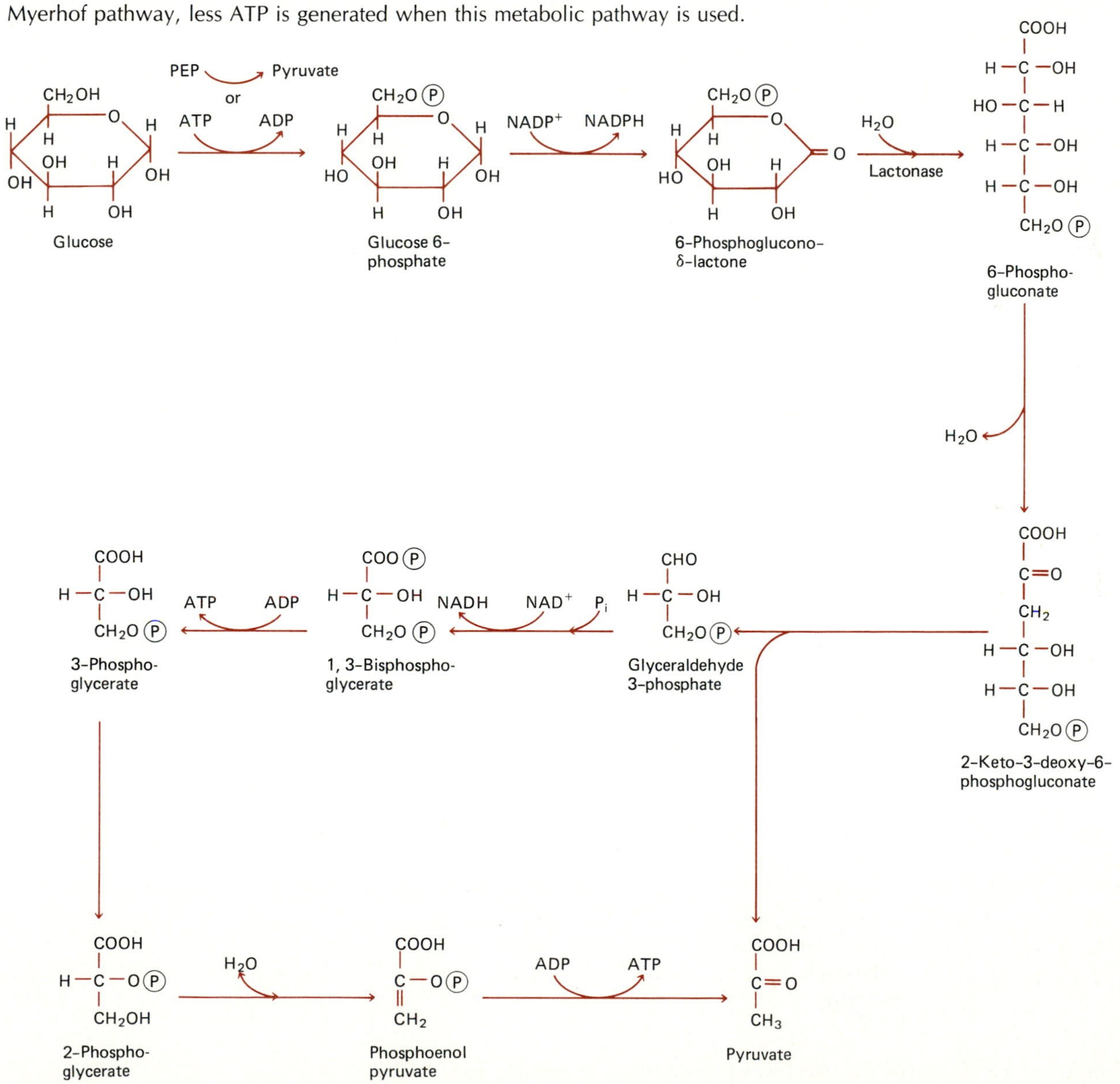

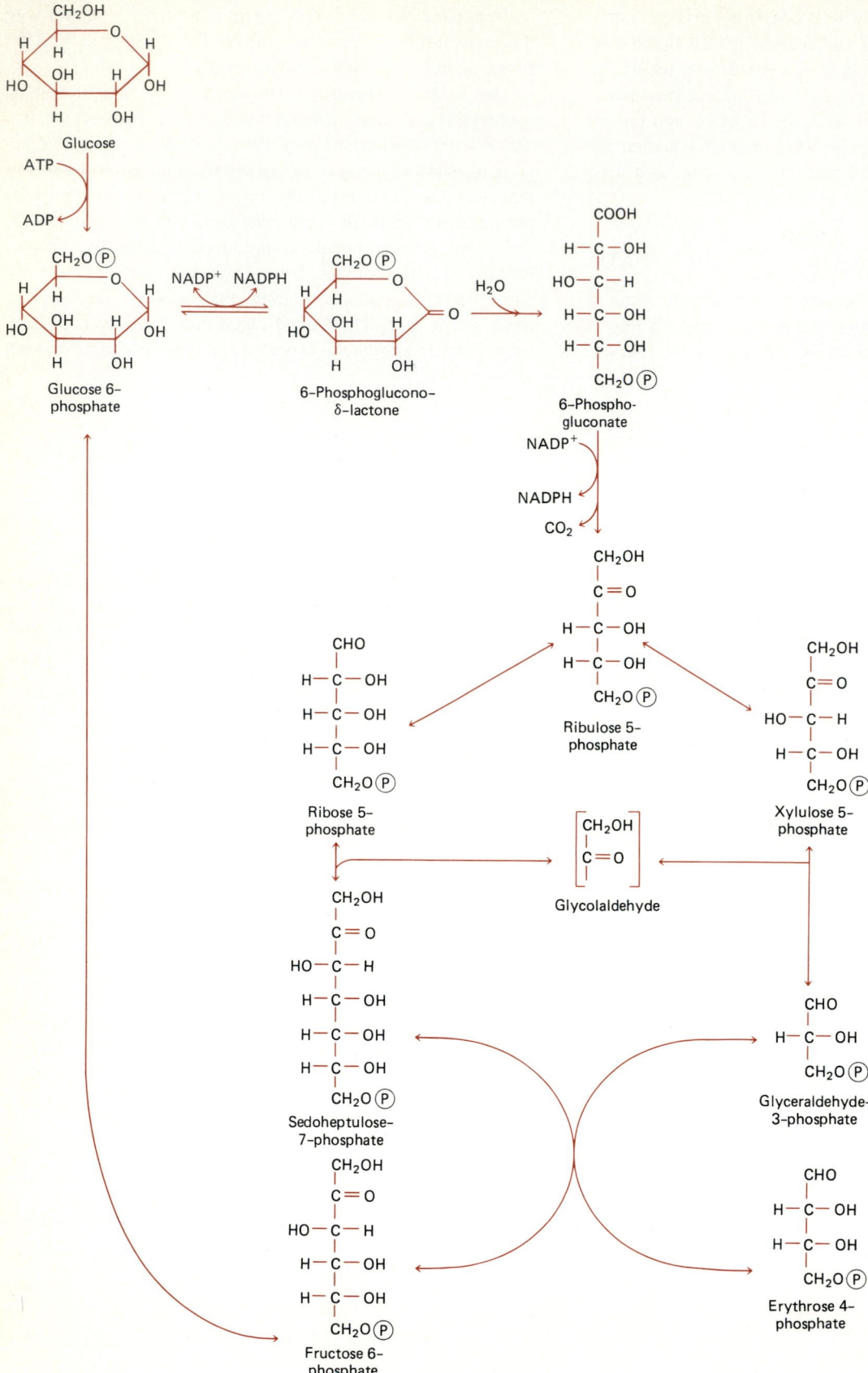

FIGURE 5.14

There are several variations of the pentose phosphate pathway that connects the metabolism of 6-carbon and 5-carbon carbohydrates.

equation for the Entner-Doudoroff pathway of glycolysis can be written as follows:

$$\text{Glucose} + 2\,\text{NADP}^+ + \text{ADP} + \text{P}_i \rightarrow$$
$$2\text{pyruvate} + 2\,\text{NADPH} + \text{ATP}.$$

This pathway provides an important mechanism for producing reduced coenzyme and 3-carbon building blocks that can be used in biosynthetic reactions when such a need is greater than that for ATP. Generally, the coenzyme NAD$^+$ and its reduced form, NADH, are used in metabolic reactions that generate ATP, whereas NADP$^+$ and NADPH are used in biosynthetic reactions. The two forms of coenzyme can be interconverted by the reaction

$$\text{NADPH} + \text{NAD}^+ \rightleftharpoons \text{NADP}^+ + \text{NADH}.$$

Another pathway—the **pentose phosphate pathway**—is especially important for generating reducing power in the form of NADPH needed for biosynthesis. Several different variations in the pentose phosphate pathway are possible depending on the need for NADPH, ATP, and small precursor molecules for incorporation into macromolecules. In one version of the pentose phosphate pathway, glucose is converted into ribulose 5-phosphate and carbon dioxide, a process that requires the hydrolysis of one ATP molecule and results in the generation of two NADPH molecules (Figure 5.14). When a large amount of reducing power is required, the glucose molecule can be completely metabolized to carbon dioxide, with the production of 12 molecules of reduced coenzyme NADPH (Figure 5.15). This series of reactions really involves a cyclic pathway in which glucose 6-phosphate is broken down and resynthesized, providing a large amount of reducing power needed by microorganisms during times of active growth. When the cell requires both NADPH and ATP, the phosphoglycerate molecule can be converted to pyruvate, with NADPH being generated during the initial steps of the pentose phosphate pathway and ATP being generated as a result of the oxidation of the pyruvate (Figure 5.16).

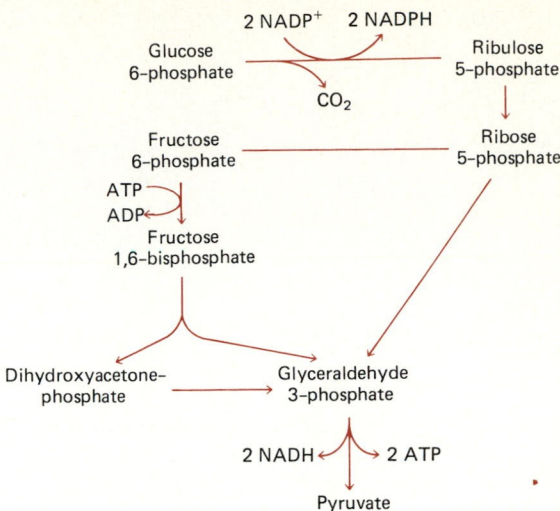

FIGURE 5.16

This version of the pentose phosphate pathway produces limited amounts of ATP, pyruvate, and reduced coenzymes.

Glucose is not the only carbohydrate that can be converted to pyruvate by glycolysis. In addition to monosaccharides, microorganisms can use disaccharides and polysaccharides as substrates for growth. Common disaccharides that can be used by microorganisms are maltose, which can be broken down by maltase to form glucose; sucrose, which can form glucose and fructose by the action of the enzyme sucrase; and lactose, which can form galactose and glucose by the action of the enzyme β-galactosidase. The monosaccharides formed from these disaccharides can enter the Embden-Meyerhof or pentose phosphate pathways (Figure 5.17). For example, the galactose derived from lactose can be converted to glucose 1-phosphate, which can then be transformed to glucose 6-phosphate, an intermediate in the glycolytic pathway. The glucose derived from lactose similarly can react to form glucose 6-phosphate. When sucrose or glycogen is used, the glucose that is formed reacts with inorganic phosphate to produce glucose 1-phosphate. The glucose 1-phosphate is then transformed, by the action of phosphoglucomutase, to glucose 6-phosphate, which enters the Embden-Meyerhof pathway. Because of the initiation of glycolysis without the need for ATP to form the phosphorylated carbohydrate, there is an increase in the net production of ATP.

Lipid Catabolism. Like carbohydrates, lipid molecules can serve as substrates supporting the growth of microorganisms. Lipases are enzymes that can cleave the fatty acids from the glycerol portion of a triglyceride lipid molecule. The glycerol molecule can be metabolized to form dihydroxyacetone phosphate and then 3-phosphoglycerate, thereby entering the metabolic pathways that have already been discussed for carbohydrate metabolism (Figure 5.18). In the case of a phospholipid, a phospholipase can cleave both the fatty acid and phosphate groups from the glycerol molecule, similarly

FIGURE 5.15

This version of the pentose phosphate pathway leads to the formation of 12 NADPH molecules and is important when reducing power is needed for biosynthesis.

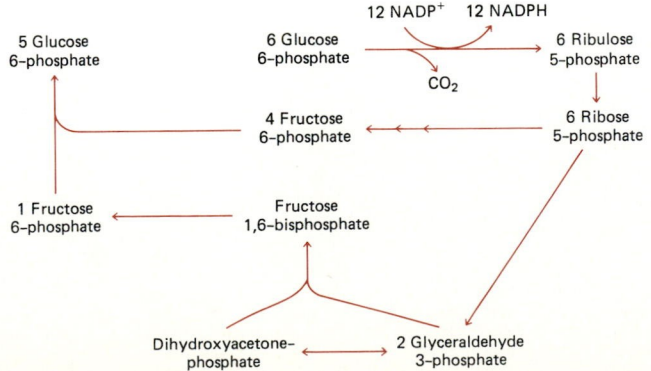

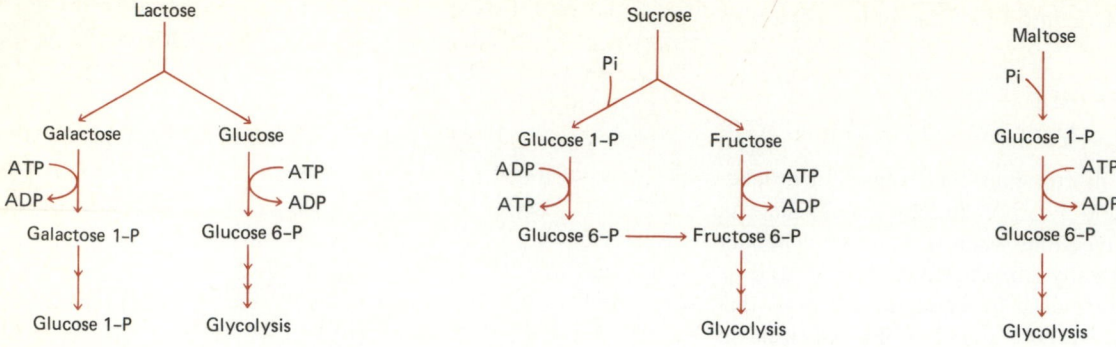

FIGURE 5.17

To enter the normal glycolytic pathways, disaccharides must be hydrolyzed and phosphorylated. In some cases, such as the metabolism of sucrose, the initial reaction is a substrate-level phosphorylation and thus more ATP is generated compared to the amount produced during the metabolism of other disaccharides, such as lactose. It should be noted that the removal of one or even several monosaccharide units from a polysaccharide, such as cellulose, has little effect on the overall size of the very large polysaccharide; that is, many glucose units can be fed into a glycolytic pathway from a polysaccharide.

converting the glycerol portion of the molecule to intermediate metabolites of the glycolytic pathways.

The metabolism of the fatty acid portions of lipid molecules proceeds by a different pathway. Fatty acids can be broken down into small 2-carbon acetyl CoA units in the process of **β-oxidation** (Figure 5.19). The fatty acid molecule initially reacts with CoA to form a fatty acid–CoA molecule. The activation of the fatty acid with CoA is coupled with the hydrolysis of an ATP molecule. Oxidation of this molecule releases acetyl CoA and forms a fatty acid–CoA complex that is two carbon atoms shorter than the parent fatty acid molecule. The formation of fatty acids that are successively two carbon atoms shorter than the parent molecule is characteristic of the β-oxidation process. The release of acetyl CoA from the fatty acid is coupled with the for-

mation of reduced coenzyme, as one molecule of reduced flavin adenine dinucleotide ($FADH_2$) and one molecule of NADH. The process is repeated continuously, forming fatty acid molecules, which are successively two carbon atoms shorter, with the production each time of acetyl CoA, NADH, and $FADH_2$.

Krebs Cycle   The acetyl CoA formed by β-oxidation of a fatty acid or from pyruvate produced from a carbohydrate by one of the glycolytic pathways can feed into the **Krebs cycle**, the next stage of respiratory metabolism (Figure 5.20).[11] The acetyl group enters the Krebs cycle by reacting with oxalo-

---

[11]The Krebs cycle is also known as the *tricarboxylic acid cycle* or the *citric acid cycle*.

FIGURE 5.18

When triglycerides are metabolized, the glycerol can enter a glycolytic pathway via the ATP-driven formation of glycerol phosphate and the NADH-coupled reduction to form dihydroxyacetone phosphate.

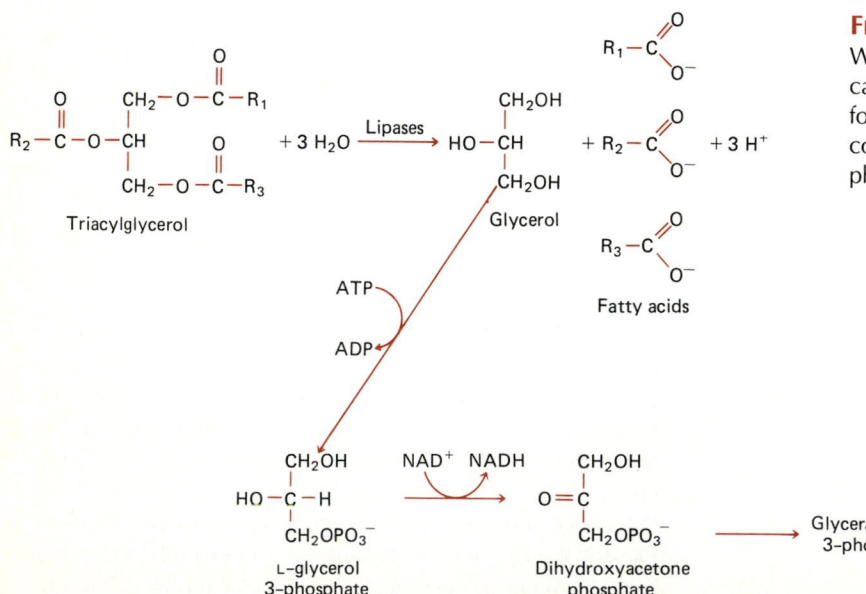

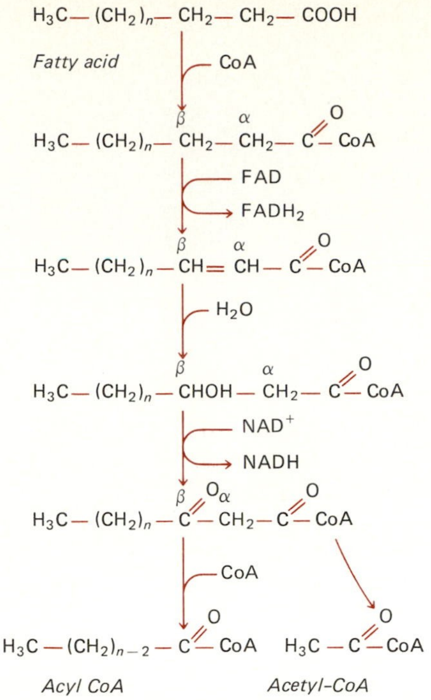

FIGURE 5.19
The metabolism of fatty acids occurs via the β-oxidation pathway, in which acetate and fatty acids that are progressively two carbon atoms shorter than the parent fatty acid are produced.

acetic acid to form citric acid. During the Krebs cycle there are two reactions that liberate carbon dioxide—the conversion of the 6-carbon compound isocitrate to the 5-carbon compound α-ketoglutarate, and the next step that converts the α-ketoglutarate, through an intermediate of succinyl CoA, to the 4-carbon compound succinate. Reduced coenzyme NADH is generated during three reactions of the Krebs cycle. Additionally, another coenzyme, flavin adenine dinucleotide (FAD), is reduced to FADH$_2$ during the conversion of succinate to fumarate.

Only one of the exergonic reactions of this cycle, the conversion of α-ketoglutarate to succinate, is directly coupled with the generation of a high-energy phosphate-containing compound. In this reaction, instead of the normal generation of ATP, guanosine triphosphate (GTP) is synthesized from guanosine diphosphate (GDP) and P$_i$. The GTP can be converted to ATP, and for accounting purposes, the GTP generated in this reaction will be treated as equivalent to ATP in determining the net synthesis of ATP during respiration. The Krebs cycle is completed with the regeneration of oxaloacetate, and thus, the intermediary carboxylic acids of the Krebs cycle can be continuously reformed.

The net reaction of the Krebs cycle, starting with the pyruvate generated from glucose, can be summarized as follows:

$$2 \text{ pyruvate} + 2 \text{ ADP} + 2 \text{ FAD} + 8 \text{ NAD}^+ \rightarrow$$
$$6 \text{ CO}_2 + 2 \text{ ATP} + 2 \text{ FADH}_2 + 8 \text{ NADH}.$$

At the end of the Krebs cycle, the cell has managed to convert all of the substrate carbon of the glucose molecule to carbon dioxide. There has also been a net synthesis of 4 ATP molecules (2 from the Embden-Meyerhof pathway and 2 from the Krebs cycle), the production of 10 reduced coenzyme molecules as NADH (2 from the Embden-Meyerhof pathway and 8 from the Krebs cycle), and the generation of 2 reduced coenzyme molecules as FADH$_2$ (both from the Krebs cycle).

Several of the enzymes mediating the reactions of the Krebs cycle are inhibited by ATP. In particular, ATP is an allosteric inhibitor of the enzyme citrate synthetase, which catalyzes the first step of the Krebs cycle—the synthesis of citrate from acetyl CoA and oxaloacetate. This is an important control point, and as the need for ATP decreases, the rate of the Krebs cycle slows through the allosteric inhibition of the enzyme catalyzing the first step of the pathway.

In addition to its role in the overall respiratory generation of ATP, the Krebs cycle occupies a central place in the flow of carbon through the cell, supplying precursors to many biosynthetic pathways, as will be discussed in the next chapter. Because some of the intermediates are siphoned off out of the Krebs cycle for biosynthetic reactions, some of the intermediary metabolites of this pathway must be resynthesized to maintain Krebs cycle activity. Many microorganisms oxidize only part of the substrate for driving the synthesis of ATP, and the remainder is used for biosynthesis. Similarly, the reduced coenzymes generated in this pathway can be used for generating ATP or synthesizing the reduced coenzyme NADPH for use in biosynthesis.

Oxidative Phosphorylation The reoxidation of the reduced coenzyme molecules generated during both glycolysis and the Krebs cycle can be used to generate ATP in the process of **oxidative phosphorylation**. During oxidative phosphorylation, electrons from the reduced coenzymes NADH and FADH$_2$ are transferred through a series of steps known as the **electron transport chain**. This transfer of electrons involves a series of oxidation-reduction reactions of membrane-bound carrier molecules and the eventual reduction of a terminal electron acceptor (Figure 5.21, page 129). The carriers in the electron transport chain participate in a series of reactions of increasing E_0' values between that of the primary electron donor and the terminal electron acceptor. As discussed earlier, the oxygen-linked reoxidation of NADH has a $\Delta G^{\circ\prime}$ of -52.4 kcal. The other carriers in the electron chain have smaller differences in E_0' values with the terminal electron acceptor and hence lower free energy values.

Although different microorganisms can have different carriers in their electron transport chains, **cytochrome molecules** play a key role in all electron transport chains. These molecules contain a central iron atom, which can be cycled between the oxidized ferric state (Fe^{3+}) and the reduced ferrous state (Fe^{2+}). Through a series of oxidation-reduction reactions, the electron is passed to a terminal electron acceptor. The terminal electron acceptor during aerobic respi-

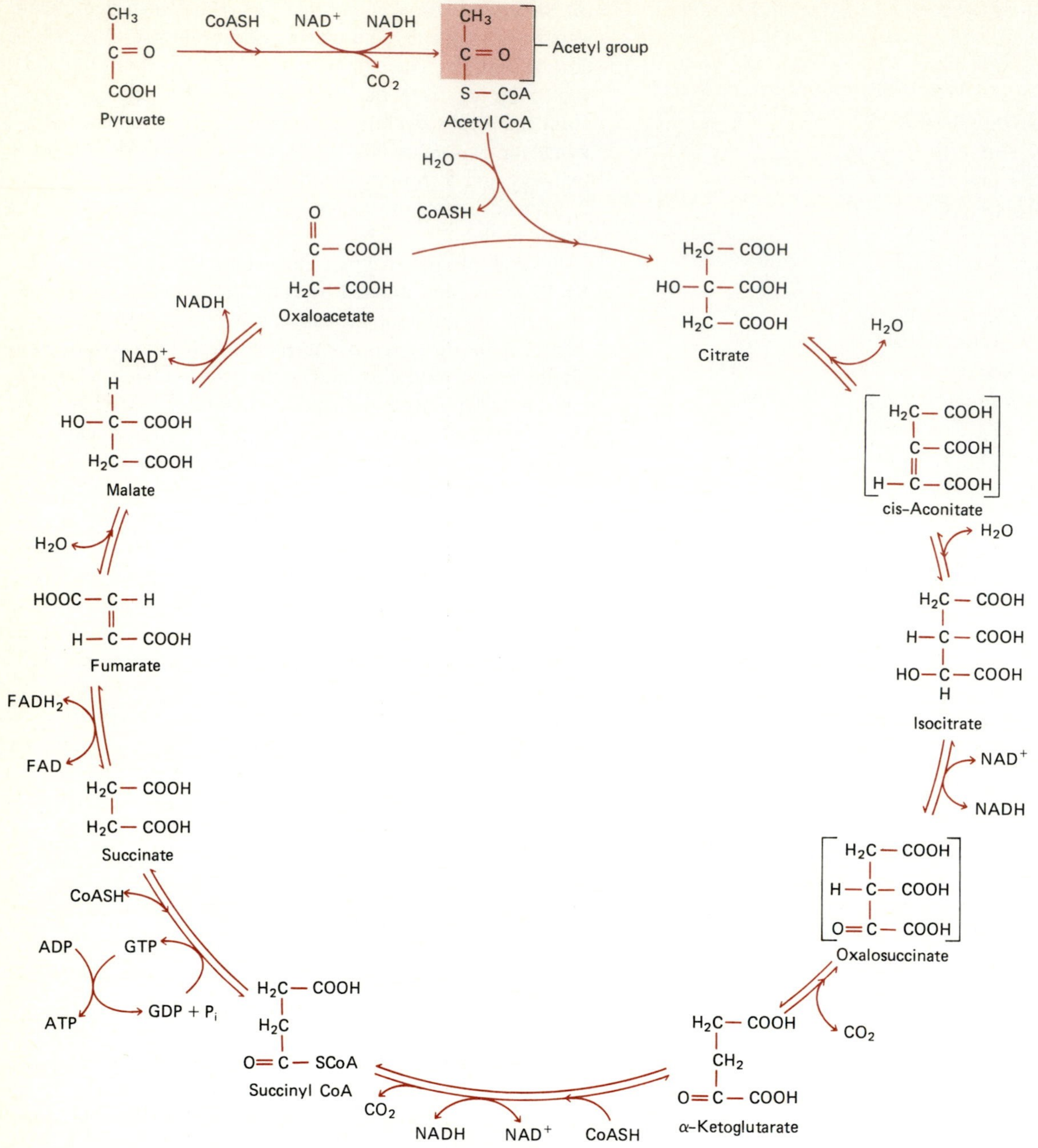

FIGURE 5.20

The Krebs cycle is a metabolic pathway central to respiratory metabolism and provides a critical link between the metabolism of the different classes of macromolecules. The metabolism of pyruvate through the complete Krebs cycle results in the generation of ATP and reduced coenzymes and the formation of CO_2. When the pathway is completed, the intermediate carboxylic acids are regenerated and continue to cycle throughout the reactions. The Krebs cycle begins when oxaloacetate reacts with acetyle CoA and H_2O to yield citrate and CoA. Citrate is next isomerized into isocitrate to enable the 6-carbon unit to undergo oxidative decarboxylation. The next step, which is the first of four oxidation-reduction reactions in the Krebs cycle, results in the conversion of isocitrate to α-ketoglutarate and CO_2 with the coupled generation of NADH. The conversion of isocitrate into α-ketoglutarate is followed by a second oxidative decarboxylation reaction that results in the formation of succinyl CoA, CO_2, and NADH. The cleavage of the thioester bond of succinyl CoA produces succinate and is coupled to the phosphorylation of guanosine diphosphate (GDP) to form guanosine triphosphate (GTP). Succinate is converted into oxaloacetate in three steps—an oxidation, a hydration, and a second oxidation reaction—thereby regenerating the oxaloacetate for another round of the cycle and simultaneously generating $FADH_2$ and NADH.

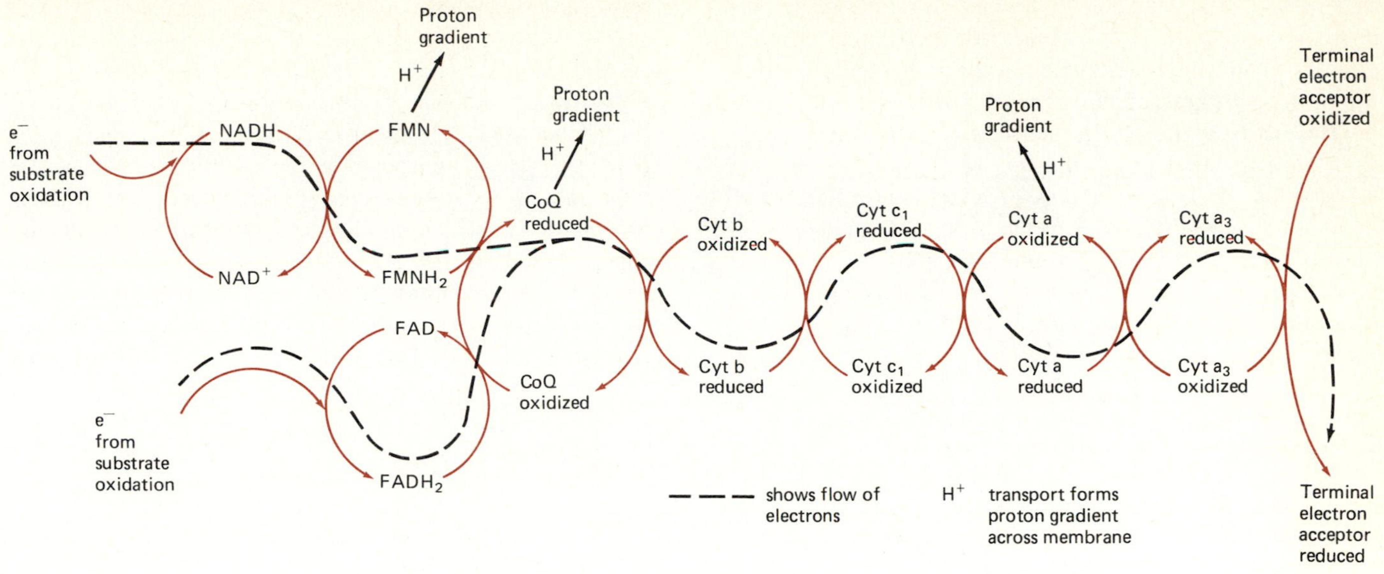

FIGURE 5.21
The electron transport chain is a membrane-embedded series of reactions that results in the reoxidation of reduced coenzymes. The transport of electrons through the cytochrome chain of this pathway results in the pumping of hydrogen ions across the membrane, and the return flow of hydrogen ions resulting from this proton gradient drives the generation of ATP. The requirement for such a terminal electron acceptor differentiates respiration from fermentation pathways. In aerobic respiration, oxygen is the terminal electron acceptor; in the absence of oxygen some, microorganisms can carry out anaerobic respiration by using the reduction of nitrate or sulfate to terminate the reaction sequence of this pathway. The actual carriers of electrons involved in this transport system vary among microorganisms. What is critical is the establishment of a sequence of oxidation-reduction reactions that establish a link between the electron donor and the terminal electron acceptor. The flow of electrons is shown as a dotted line. In this example, electrons that enter the system from NADH are transported through flavin mononulceotide (FMN) to coenzyme Q; those that enter from $FADH_2$ go directly to coenzyme Q. Electrons then flow through a series of cytochromes, designated cyt b, c_1, a, and a_3, to the terminal electron acceptor. As the electron is transported through each carrier, there is an oxidation-reduction reaction, so that in the case of the cytochromes, for example, iron within the cytochrome alternates between the oxidized Fe^{3+} and reduced Fe^{2+} states. At three locations, hydrogen ions are transported out of the cell.

ration is oxygen that is reduced to form water. When nitrate serves as the terminal electron acceptor during anaerobic respiration, the products of its reduction are also capable of serving as terminal electron acceptors, thereby establishing a series of anaerobic respirations: nitrate to produce nitrite, then nitrite to produce nitrous oxide, and finally nitrous oxide to form molecular nitrogen. Similarly, when sulfate acts as terminal electron acceptor, a series of reactions can produce hydrogen sulfide and water.

The reactions of the electron transport chain form the basis for generating ATP. However, the generation of ATP during oxidative phosphorylation is not directly coupled to the specific biochemical reactions that occur during electron transport. Within the electron transport chain, some carriers transport hydrogen atoms (an electron plus a proton), whereas others transport only electrons. The transfer of electrons, from the reduced coenzyme to the terminal electron acceptor, establishes a hydrogen ion gradient across the membrane that is critical for the generation of ATP (Figure 5.22). The inter-

mediate electron carriers of the electron transport chain are asymmetrically distributed through a membrane, and the movement of hydrogen ions across the membrane, as a result of electron transport, forms an electrochemical gradient that establishes the mechanism for generating ATP. In the case of eukaryotes this electrochemical gradient occurs across the inner mitochondrial membrane, whereas for bacteria it is formed across the cytoplasmic membrane.

The orientation of the carriers in the bacterial cytoplasmic membrane is such that hydrogen carriers transport toward the outside of the cell and electron carriers transport toward the inside. At each conjunction in the chain of a hydrogen carrier and an electron carrier, a proton (hydrogen ion) is transported out of the cell. The cytoplasmic membrane is otherwise impermeable to protons. Consequently, a portion of the chemical energy released by the net reaction of the electron transport chain (oxidation of the primary electron donor by the terminal electron acceptor) is trapped in the form of a gradient of hydrogen ions (ΔpH) and electrical potential ($\Delta\psi$)

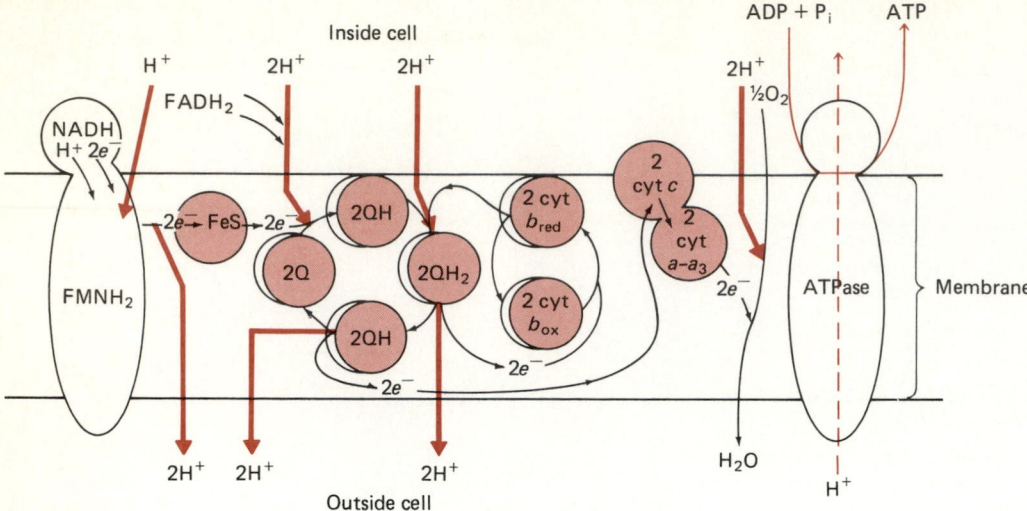

FIGURE 5.22

The generation of the hydrogen ion gradient across a membrane that is needed to generate ATP by chemiosmosis is depicted in this drawing. The carriers of electrons are asymmetrically arranged within the membrane. As a result of the reactions of the electron carriers, protons are transported across the membrane, establishing a hydrogen ion (proton) gradient across it. The force that results from this concentration and electrical difference, the protonmotive force, is used to generate ATP and to do other work, such as the active transport of some substrates and the rotation of flagella.

across the membrane. This potential energy can be expressed as a **protonmotive force (Δp)** that can be used to generate ATP or perform other work according to the formula

$$\Delta p = \Delta\psi - (2.3 \, RT/F)\Delta pH.$$

The protonmotive force measured across the cytoplasmic membrane of *E. coli* is sufficient to generate a high-energy phosphate bond of ATP.

The process of synthesizing ATP at the expense of the protonmotive force is called **chemiosmosis**. The generation of ATP by chemiosmosis depends on the fact that protons cannot simply diffuse back across the membrane but can enter the membrane via a proton channel established by the membrane-bound adenosine triphosphatase (ATPase). ATPase is a multicomponent enzyme system containing two major polypeptide complexes called F_0 and F_1 (Figure 5.23). The F_0F_1 complex couples the synthesis of ATP with proton diffusion. The F_1 polypeptides sit on the surface of the membrane, whereas the F_0 polypeptides are embedded in it. F_0 forms a channel across the membrane through which protons flow to F_1 and F_1 catalyzes the synthesis of ATP from ADP.[12]

As a consequence of the protonmotive force established by the electron transport chain, three ATP molecules can be synthesized for each NADH molecule oxidized to NAD$^+$ and two ATP molecules can be made for each FADH$_2$ molecule oxidized to FAD during oxidative phosphorylation by chemiosmosis.[13] The 10 NADH molecules generated during gly-

[12]F_1 is also capable of catalyzing the conversion of ATP to ADP.

[13]The difference between the amount of ATP that can be generated from NADH compared to FADH$_2$ occurs because the oxygen-linked oxidation of NADH liberates 52.4 kcal/mole compared to only 42 kcal/mole for the oxygen-linked oxidation of FADH$_2$.

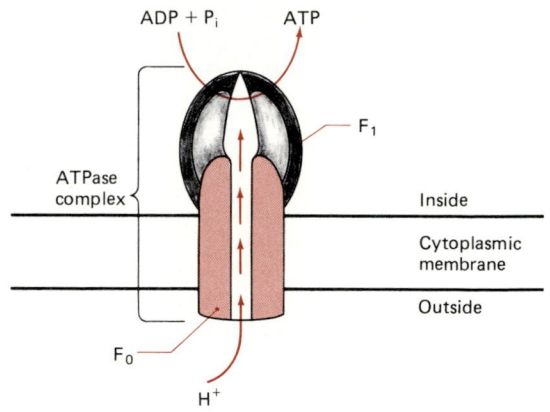

FIGURE 5.23

ATPase is a multicomponent enzyme system composed of two major complexes designated F_0 and F_1. The F_0F_1 components link the protonmotive force with ATP synthesis. Protons are channeled by the F_0 component of the system to F_1, where the conversion of ADP to ATP is catalyzed.

colysis and the Krebs cycle, therefore, can be converted to 30 ATP molecules during oxidative phosphorylation; the 2 FADH$_2$ molecules generated during the Krebs cycle can generate 4 ATP molecules. This synthesis of ATP is in addition to the ATP formed during glycolysis and the Krebs cycle. Thus, the overall reaction for the respiratory metabolism of glucose (using the Embden-Meyerhof pathway of glycolysis) can be expressed as follows:

$$\text{Glucose} + 6 \, O_2 + 38 \, \text{ADP} + 38 \, P_i \rightarrow$$
$$6 \, CO_2 + 6 \, H_2O + 38 \, \text{ATP}.$$

Fermentation

Unlike respiration, oxidative phosphorylation and chemiosmotic generation of ATP do not occur in fermentation.[14] Rather, the synthesis of ATP in fermentation is due to substrate-level phosphorylation and is largely restricted to the amount formed during glycolysis. Because they do not require oxygen, all fermentation pathways are anaerobic, and microorganisms that generate their energy by fermentation are carrying out anaerobic metabolism, regardless of whether or not the organism is growing in the presence of molecular oxygen.

Although obligately fermentative bacteria, such as *Streptococcus* species, do not use oxidative phosphorylation to generate ATP, they have an F_0F_1–ATPase system in their cytoplasmic membranes. In such bacteria, ATP is used to pump protons through the F_0F_1 complex in the reverse direction (Figure 5.24). ATP generated by substrate-level phosphorylation in a fermentation pathway is converted to ADP by ATPase, and the energy of this reaction is coupled to the export of protons from the cell. The F_0F_1–ATPase system thereby generates a protonmotive force across the cytoplasmic membrane, maintaining the intracellular pH at the appropriate value and providing a mechanism for driving processes that depend upon the protonmotive force across the membrane, such as active uptake of sugars and other substances, export of Na^+ and Ca^{2+}, and rotation of flagella.

Fermentation Pathways In considering the actual way in which ATP is generated in fermentative bacteria, we find that the initial metabolic steps of the fermentation pathway are identical to those of a respiration pathway. The metabolic pathway for carbohydrate fermentation, for example, begins with glycolysis. If the microorganism follows the Embden-Meyerhof glycolytic pathway, it generates two pyruvate molecules, two reduced coenzyme NADH molecules, and two ATP molecules for each molecule of glucose that goes through glycolysis. In general, the two ATP molecules formed during glycolysis represent the total energy conversion of the fermentation pathway, as measured by the number of ATP molecules synthesized. The remainder of the fermentation pathway is concerned with reoxidizing the coenzyme.

In fermentation, the reoxidation of NADH to NAD^+ depends on the reduction of the pyruvate molecules formed during glycolysis to balance the oxidation-reduction reactions. Different microorganisms have developed different pathways for utilizing the pyruvate for reoxidizing the reduced coenzyme with the different terminal sequences of the various fermentation pathways, resulting in the formation of various end products (Figure 5.25). The different fermenta-

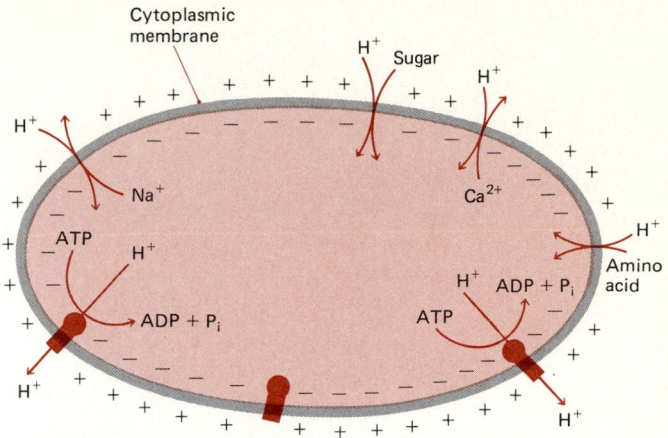

FIGURE 5.24

All bacteria, including fermentative bacteria, have a protonmotive force across the membrane. In the fermentative bacteria, ATP— all of which is derived from substrate-level phosphorylation—is used to establish the proton gradient across the membrane rather than having ATP synthesis based upon such a gradient. To establish the protonmotive force, the F_0F_1-ATPase system uses ATP, generated during fermentation, to form a proton concentration gradient across the cytoplasmic membrane. The protonmotive force is used to power flagella rotation, the export of Ca^{2+} and Na^+, and the uptake of sugars and amino acids into the cell.

tion pathways generally are named for the characteristic end products that are formed.

Ethanolic Fermentation. The **ethanolic fermentation pathway** derives its name from the fact that ethanol is one of the end products. In this fermentation pathway, pyruvate is converted to ethanol and carbon dioxide (Figure 5.26). This reaction is coupled with the conversion of NADH to NAD^+. The equation for ethanolic fermentation when glucose is used as a substrate and the Embden-Meyerhof pathway is followed is:

$$Glucose + 2\ ADP + 2\ P_i \rightarrow$$
$$2\ ethanol + 2\ CO_2 + 2\ ATP.$$

The complete fermentation pathway begins with the substrate, includes glycolysis, and terminates with the formation of end products There is no net change in the oxidative state of the coenzymes during the overall fermentation pathway, and thus, the coenzymes do not appear in the overall equation for the fermentation.

Ethanolic fermentation is carried out by many yeasts, such as *Saccharomyces cerevisiae*, but by relatively few bacteria. This fermentation pathway is quite important in food and industrial microbiology and is used to produce beer, wine, and distilled spirits. Besides its importance in alcoholic beverages, the ethanol produced by *S. cerevisiae* in this fermentation is used as a fuel in gasohol. In addition to its use for ethanol formation, *Saccharomyces cerevisiae* (Baker's yeast) is used in the production of bread; the carbon dioxide released by the ethanolic fermentation causes the bread to rise.

---

[14]Remember that in fermentation there is no external electron acceptor. The organic substrate acts as electron donor, and an organic molecule derived from that substrate acts as an electron acceptor to balance the oxidation-reduction reactions.

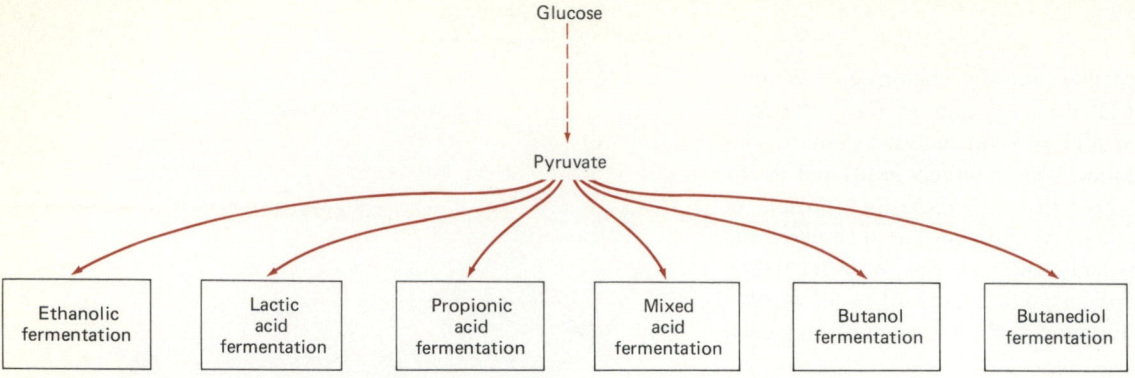

FIGURE 5.25

Various fermentation pathways, each branching from pyruvate, are carried out by different microorganisms. Compared to respiration, these fermentation pathways are energetically less favorable. In each complete fermentaion there is a balance of oxidation-reduction reactions, so that there is no net production of reduced coenzyme. Each pathway results in the formation of characteristic products, many of which are of industrial importance.

All of these uses of the ethanolic fermentation pathway have considerable economic importance.

Lactic Acid Fermentation. Another fermentation pathway of great economic importance because of its use in making cheeses and various other products is the **lactic acid fermentation pathway**. This fermentation pathway is carried

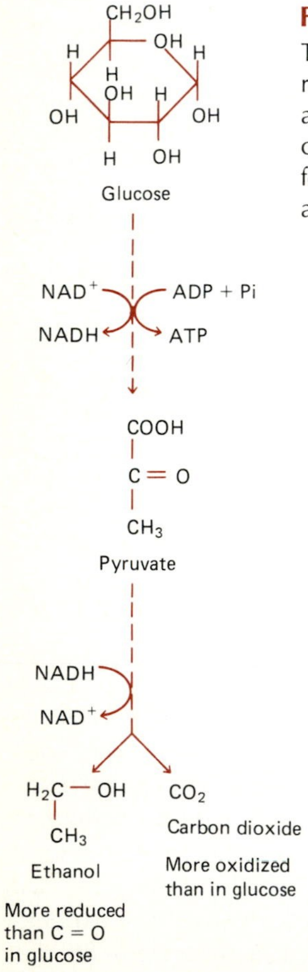

FIGURE 5.26

The ethanolic fermentation pathway results in the formation of ethanol and CO_2. The fermentation of carbohydrates to these end products forms the basis of the beer, wine, and spirits industries.

out by bacteria, which by virtue of their metabolic reactions are classified as lactic acid bacteria. Two important genera of lactic acid bacteria are *Streptococcus* (Gram-positive cocci that tend to form chains) and *Lactobacillus* (Gram-positive rods that tend to form chains). In the lactic acid fermentation pathway, pyruvate is reduced to lactic acid, with the coupled reoxidation of NADH to NAD^+ (Figure 5.27). The overall lactic acid fermentation pathway can be expressed as follows:

Glucose + 2 ADP + 2 P_i → 2 lactic acid + 2 ATP.

When the Embden-Meyerhof scheme of glycolysis is used in the lactic acid fermentation pathway, the overall pathway is a **homolactic fermentation** because the only end product formed is lactic acid. Homolactic fermentation is carried out by *Streptococcus*, *Pediococcus*, and various *Lactobacillus* species.

The homolactic acid fermentation pathway is quite important in the dairy industry. It is the pathway responsible for souring milk and is used in the production of numerous types of cheese, yogurt, and various other dairy products. Streptococci living on tooth surfaces in the oral cavity (mouth) produce lactic acid that is held against the tooth by dental plaque and gradually eats through the enamel of the tooth, creating cavities. Even though they can grow in the mouth, *Streptococcus* species are metabolically obligate anaerobes. *Lactobacillus* species occur in the human digestive tract and aid in the digestion of milk. These species are the initial colonizers of the intestinal tract. Adults often lack the ability to digest the carbohydrates in milk and suffer disease symptoms if they consume milk. *Lactobacillus acidophilus* is added to various commercial milk products (acidophilus milk) to aid individuals who are unable to digest milk products adequately. The enzymes produced by *L. acidophilus* convert milk sugars to products that do not accumulate and cause gastrointestinal problems.

In contrast to the homolactic acid pathway, some microorganisms carry out a **heterolactic acid fermentation**, using

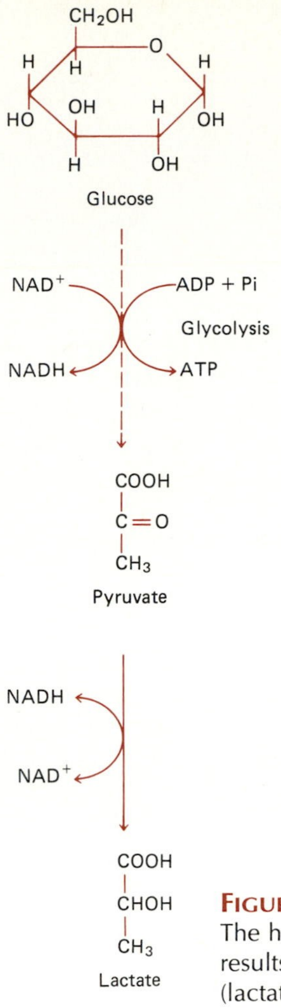

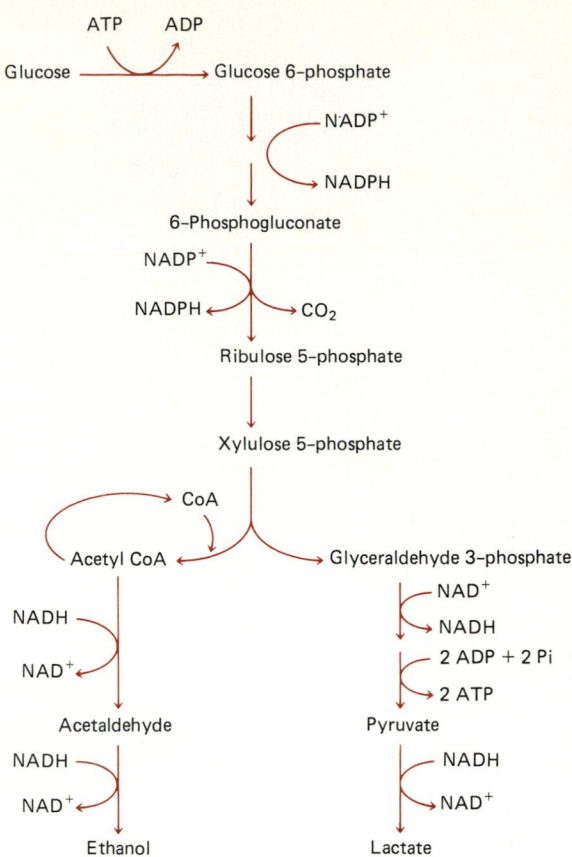

FIGURE 5.27
The homolactic acid fermentation pathway results in the production of lactic acid (lactate).

FIGURE 5.28
The heterolactic fermentation pathway results in the production of lactic acid (lactate), ethanol, and CO_2.

the pentose phosphate pathway rather than the Embden-Myerhof pathway of glycolysis. Heterolactic acid fermentation is so named because ethanol and carbon dioxide are produced in addition to lactic acid (Figure 5.28). The ethanol and carbon dioxide come from the glycolytic portion of the pathway. The overall reaction for the heterolactic fermentation can be expressed as follows:

Glucose + ADP + P_i →

lactic acid + ethanol + CO_2 + ATP.

The heterolactic fermentation pathway produces only one molecule of ATP per molecule of glucose substrate metabolized. This fermentative pathway is carried out by *Leuconostoc* and various *Lactobacillus* species. The heterolactic fermentation pathway carried out by *Leuconostoc* species is responsible for the production of sauerkraut.

Propionic Acid Fermentation. Another fermentation pathway of some interest is the **propionic acid fermentation pathway**. As the name indicates, the end product of this fermentation pathway is propionic acid. This metabolic sequence is carried out by the propionic acid bacteria, which are especially interesting because they have the ability to carry out a fermentation pathway beginning with lactic acid as the substrate. The bacterial genus *Propionibacterium* is

defined as a group of Gram-positive rods that produce propionic acid from the metabolism of carbohydrates. The ability to utilize the end product of another fermentation pathway is quite unusual, and it permits species of *Propionibacterium* to carry out late fermentation during the production of cheese. The lactic acid bacteria convert the initial substrates in the milk to lactic acid. The propionic acid bacteria subsequently convert the lactic acid to propionic acid and carbon dioxide. Only after the cheese curd has formed, through the action of lactic acid bacteria, do the propionic acid bacteria begin their fermentation. The release of carbon dioxide during this late fermentation forms gas bubbles in the semisolid cheese curd, which we recognize as Swiss cheese holes. The propionic acid formed during this fermentation also gives Swiss cheese its characteristic flavor.

Mixed Acid Fermentation. In contrast to propionic acid fermentation, which is restricted to a few bacterial species, the **mixed acid fermentation pathway** is relatively common. Mixed acid fermentation is so named because of the mixture of end products that are formed (Figure 5.29). It is carried out by members of the family Enterobacteriaceae, a large family of bacteria that includes *E. coli*, as well as hundreds of other bacterial species. In this metabolic pathway the pyruvate formed during glycolysis is converted to a

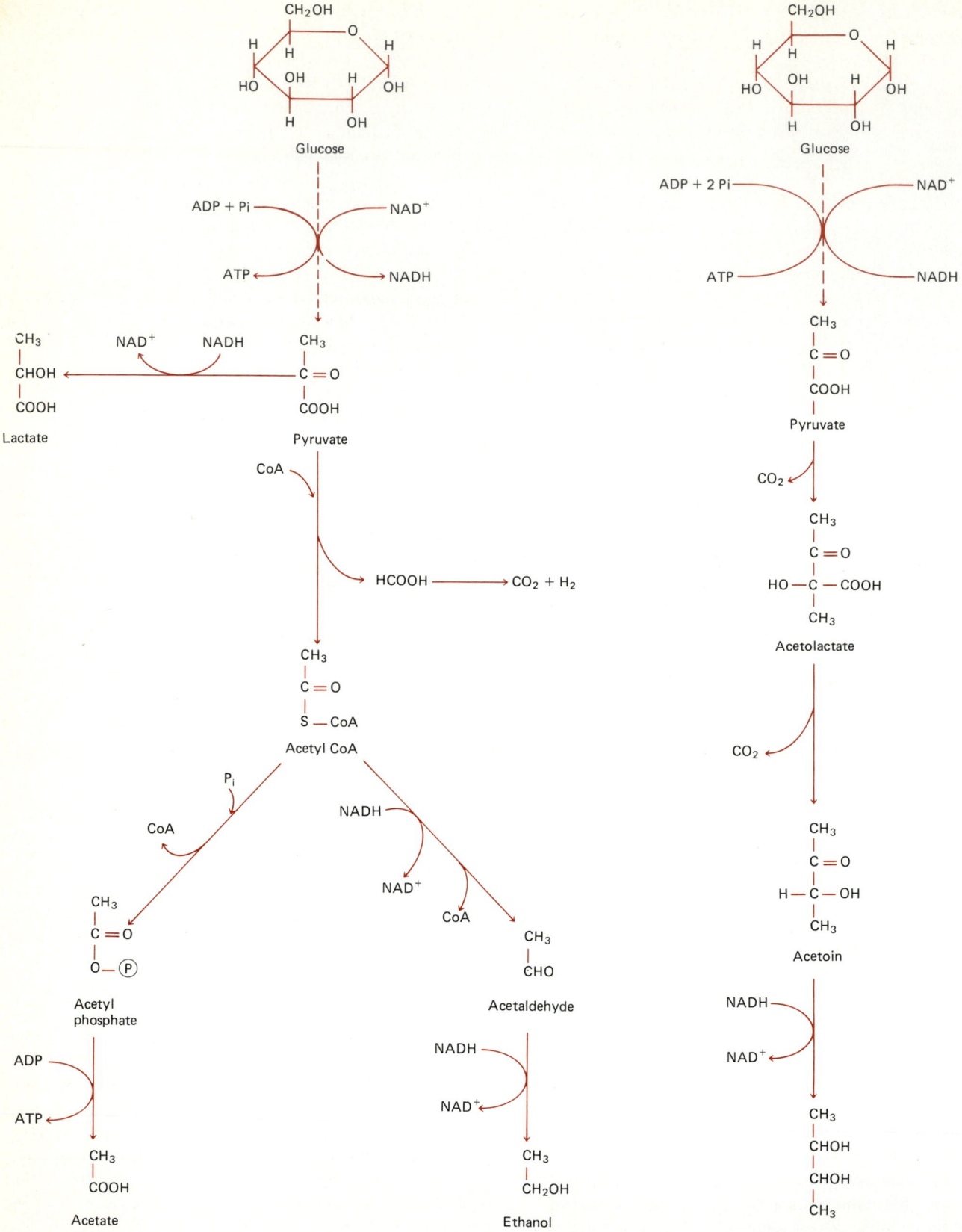

FIGURE 5.29
The mixed acid fermentation pathway, which is carried out by enteric bacteria such as *E. coli*, results in the production of carbon dioxide, hydrogen gas, acetic acid (acetate), lactic acid (lactate), and ethanol.

FIGURE 5.30
The butanediol fermentation pathway results in the production of the neutral product 2,3-butanediol. The production of acetoin, an intermediary metabolite, is diagnostic of this pathway.

variety of products, including ethanol, acetate, formate, molecular hydrogen, and carbon dioxide. The proportions of the products vary, depending on the bacterial species. During formation of these various products, the reduced NADH is reoxidized to NAD. The formation of acetate is also accompanied by a substrate-level phosphorylation that forms additional ATP. The mixed acid fermentation can be detected by the methyl red (MR) test, which is based on the color reaction of the pH indicator methyl red. This test is one of several typically used in identification systems, including miniaturized commercial identification systems used in clinical laboratories for the identification of bacteria, such as *E. coli*, that can cause urinary tract and other infections.

Butanediol Fermentation. Another test used in many such systems is the Voges-Proskauer (VP) test, which detects yet another fermentation pathway—the **butanediol fermentation pathway** (Figure 5.30). The VP test has classically been used together with the MR test to distinguish between *Enterobacter aerogenes*, which is VP+ and MR−, and *E. coli*, which is VP− and MR+. Among other reasons, this distinction is important because *E. coli* is used as an indicator of human fecal contamination in assessing the safety of water supplies. The VP test actually detects acetoin (acetyl methyl carbinol), which is an intermediary product in the butanediol fermentation pathway. The end product of the pathway, though, is butanediol. During this fermentation pathway, carbon dioxide is released and NADH is reoxidized to NAD^+, but no additional ATP is generated. Some bacteria, such as members of the genus *Klebsiella*, carry out both the butanediol and mixed acid fermentations. Such bacteria show a positive VP test but a negative MR test because they are producing a neutral product—butanediol—and not enough

acid to effect a color change for the indicator methyl red.

Butanol Fermentation. In yet another pathway, members of the genus *Clostridium* carry out a **butanol fermentation pathway**, which is also known as the **butyric acid pathway**. Different *Clostridium* species form a variety of end products via this fermentation pathway, with pyruvate being converted to either acetone and carbon dioxide, isopropanol and carbon dioxide, butyrate, or butanol. Many of these fermentation products are good organic solvents that have commercial applications, such as the use of acetone for nail polish remover. Today the choice of using microbial or organic-synthetic means of producing solvents is based on economics, but when the butanol fermentation pathway was first discovered by Chaim Weitzman, it was particularly important because it allowed Britain to produce acetone for use in the manufacture of munitions during World War I, which was instrumental in determining the outcome of that war.

Mixed Amino Acid Fermentation. In addition to various pathways for the fermentative metabolism of carbohydrates, amino acids can be metabolized by a **mixed amino acid fermentation pathway**—the **Strickland reaction**. This fermentation occurs when there is extensive protein degradation, and involves one amino acid serving as an electron donor and another amino acid acting as an electron acceptor to achieve the necessary oxidation-reduction balance. The Strickland reaction results in the deamination and decarboxylation of the amino acids. For example, a mixture of alanine and glycine can yield the end products acetate, carbon dioxide, and ammonia. The mixed amino acid fermentation contributes to the pleasant odors of some wines and cheeses but is also responsible in part for the horrible smell of a gangrenous wound.

5.5 *Autotrophic Generation of ATP*

In contrast to the heterotrophic metabolism of both fermentation and respiration pathways, the metabolism of autotrophic microorganisms does not require preformed organic matter to drive the conversion of ADP + P_i to ATP. Instead, **photoautotrophic (photosynthetic) microorganisms** use light energy and **chemoautotrophic (chemolithotrophic) microorganisms** use the energy derived from the oxidation of inorganic compounds to supply the energy needed for the synthesis of ATP.

Chemolithotrophy

Chemolithotrophic bacteria are very important because their metabolic activities form critical links in the biogeochemical cycling of various elements, such as nitrogen and sulfur. Chemolithotrophs carry out respiration based on inorganic compound metabolism, coupling the oxidation of an inorganic compound with the reduction of a suitable coenzyme (Figure 5.31). The transfer of electrons from the reduced coenzyme molecules through an electron transport chain of

membrane-bound carriers establishes a hydrogen ion gradient that can drive the synthesis of ATP by chemiosmosis. The terminal electron acceptor for chemolithotrophs most frequently is oxygen, but some chemolithotrophs are able to carry out anaerobic respiration. The chemiosmotic mechanism for the synthesis of ATP is equivalent to the mechanism for ATP generation during oxidative phosphorylation discussed earlier.

Some chemolithotrophs produce the enzyme hydrogenase and are able to metabolize molecular hydrogen, forming water and the reduced coenzyme NADH from the oxidation of hydrogen. The oxidation of reduced compounds containing either iron, sulfur, or nitrogen, however, does not produce a sufficient change in the oxidation state to drive the reduction of NAD^+ to NADH. Therefore, the electrons derived from the oxidation of reduced iron, sulfur, and nitrogen compounds enter the electron transport chain via one of the cytochrome carrier molecules that are at a lower energy level than the reduced coenzyme NADH. Some of the electrons flow toward the terminal electron acceptor oxygen, estab-

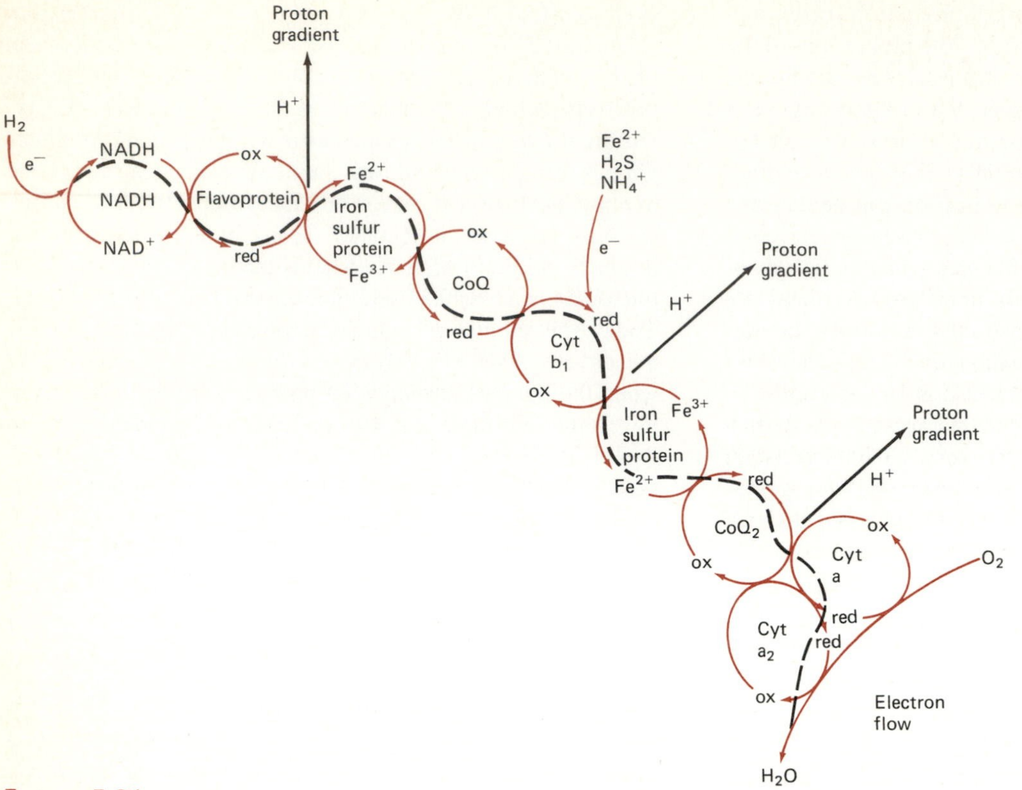

FIGURE 5.31

The chemolithotrophic oxidation of substrates, such as molecular hydrogen, ammonium ions, ferrous iron, and hydrogen sulfide, can be coupled with the reduction of an electron carrier. The resulting flow of electrons through a series of membrane-bound carriers (shown as a dotted line) is used to establish a proton gradient across the membrane that can drive the formation of ATP by chemiosmosis. In some of the chemolithotrophs, both forward and reverse electron transport chains occur. For the nitrifying bacteria, for example, electrons derived from the oxidation of nitrite (NO_2^-) to nitrate (NO_3^-) flow into a branched electron transport chain composed of various cytochromes and a flavoprotein. Some of the electrons flow in a forward direction and are used to generate the protonmotive force needed to generate ATP. The other electrons move in the reverse direction to reduce NAD, forming NADH.

lishing a proton gradient across the membrane that is used to generate ATP. Other electrons flow in the reverse direction to produce reduced coenzymes needed for biosynthesis. The amount of ATP that can be synthesized by this mechanism is thus less than the three molecules of ATP per molecule of NADH that can be generated by hydrogen-oxidizing chemolithotrophs. Consequently, in order to generate the amount of ATP required for growth and reproduction, these chemolithotrophic microorganisms must oxidize very large amounts of reduced nitrogen, sulfur, or iron-containing compounds and, therefore, normally exhibit very high turnover rates of inorganic substrates. This accounts for their importance in large-scale global biogeochemical cycling reactions, including those that transfer nitrogen and sulfur compounds between air, water, and soil so that they circulate around the Earth.

Sulfur Oxidation The most common reduced sulfur compounds, which are oxidized by chemolithotrophs, are hydrogen sulfide, elemental sulfur, and thiosulfate. The chemo-

lithotrophic activities of sulfur-oxidizing microorganisms have received considerable attention as a result of the finding that a highly productive submarine area off the Galapagos Islands is supported by the productivity of chemolithotrophs growing on reduced sulfur released from thermal vents in the ocean floor. It is quite unusual to find an ecological system driven by chemolithotrophic metabolism; most ecosystems are dependent on the primary productivity of photoautotrophs, either higher plants, algae, or photosynthetic bacteria.

When hydrogen sulfide is utilized as the substrate for chemolithotrophic metabolism, elemental sulfur is often deposited by the bacteria, with some such bacteria depositing sulfur granules within their cells and others depositing elemental sulfur extracellularly. Some sulfur-oxidizing bacteria, such as *Thiobacillus thiooxidans*, can oxidize large amounts of reduced sulfur compounds with the formation of sulfate. The sulfur-oxidizing activities of this bacterium are important because of their involvement in the formation of acid mine drainage problems and their use in mineral recovery processes. *T. ferrooxidans*, which oxidizes both reduced sulfur and reduced iron for generating ATP, often is found in acid mine drainage streams, where it has an available source of reduced sulfur and reduced iron that it can utilize for the chemolithotrophic generation of ATP. Various other bacteria also have the ability to oxidize ferrous iron to ferric iron, often forming a deposit of insoluble ferric hydroxide.

Nitrification   Nitrifying bacteria oxidize either ammonium or nitrite ions. Bacteria such as *Nitrosomonas* oxidize ammonia to nitrite; other bacteria, such as *Nitrobacter*, oxidize nitrite to nitrate. Again, because the chemolithotrophic oxidation of reduced nitrogen compounds yields relatively little energy, chemolithotrophic bacteria carry out extensive transformations of nitrogen in soil and aquatic habitats in order to synthesize their required ATP. The activities of these bacteria are quite important in soil because the alteration of the oxidation state radically changes the mobility of these nitrogen compounds in the soil column. Nitrifying bacteria have a marked influence on soil fertility because positively charged ammonium ions bind to negatively charged soil clay particles, whereas the negatively charged nitrite and nitrate ions do not bind and are leached from soils by rainwater.

Methanogenesis   Some chemolithotrophic archaebacteria, the **methanogens**, are able to convert CO_2 to methane (CH_4), using electrons from molecular hydrogen to reduce CO_2 and form CH_4 in a specialized anaerobic respiration pathway (Figure 5.32). Methanogenic bacteria may have been among the earliest living organisms on Earth because they can grow autotrophically on hydrogen and carbon dioxide under anaerobic conditions, and it is presumed that the early atmosphere of the Earth would have permitted the growth of just such organisms. The conversion of carbon dioxide to methane, using molecular hydrogen as an electron donor, is an exergonic reaction with a $\Delta G°$ of -31 kcal/mole and an actual ΔG at cellular concentrations of reactants and products of about -15 kcal/mole. The synthesis of ATP in methanogens is based on an oxidative phosphorylation mechanism rather than direct substrate-level phosphorylation. Several cofactors unique to this system have been identified. About one molecule of ATP can be synthesized for every molecule of CO_2 converted to CH_4. During the conversion of carbon dioxide to methane, NADPH is also generated. This NADPH is used for the incorporation of carbon dioxide into the macromolecules of the cell. Approximately 90–95 percent of the carbon dioxide used by methanogens is converted to methane, presumably supporting ATP and NADPH synthesis, and the remainder is incorporated into cellular carbon.

Photosynthesis (Oxidative Photophosphorylation)

The other group of autotrophic microorganisms, the photoautotrophs (algae and photosynthetic bacteria), are able to convert light energy to chemical energy in a process known

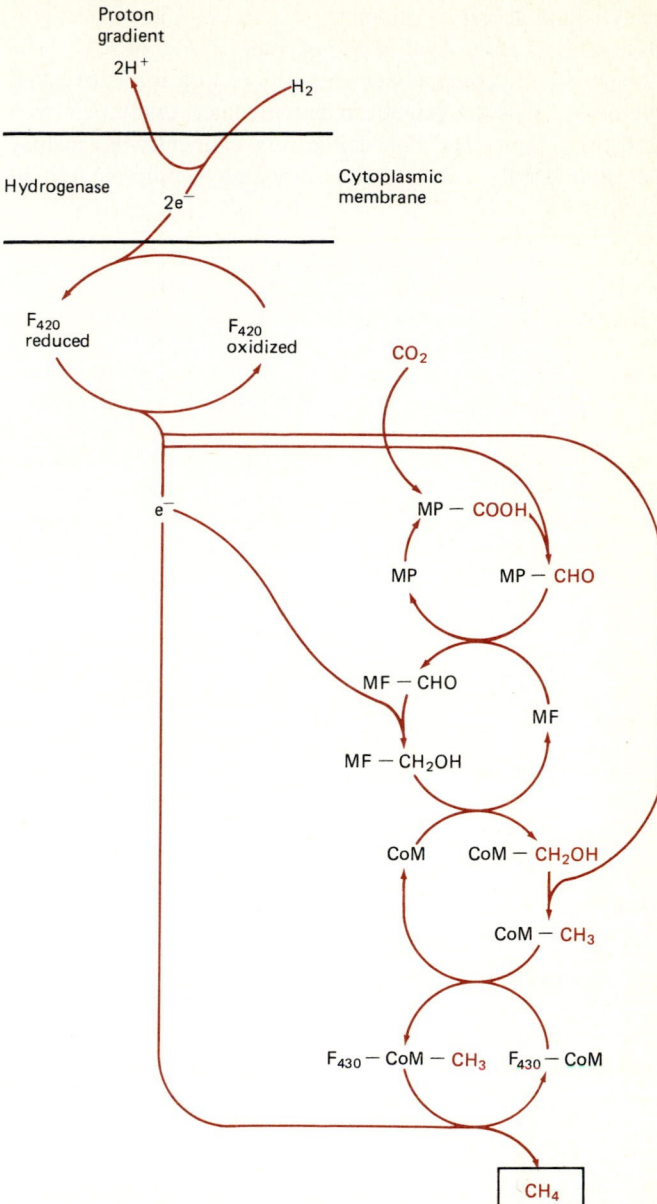

FIGURE 5.32
The conversion of CO_2 to CH_4 is carried out by a specialized group of archaebacteria. This is a strictly anaerobic pathway involving the flow of electrons from a hydrogen donor. Several unique electron carriers are involved in the transfer of electrons in this pathway, including factor 420 (F_{420}), factor 430 (F_{430}), coenzyme M (CoM), methanopterin (MP), and methanofuran (MF). The oxidation of hydrogen, which occurs outside of the cell, produces hydrogen ions and supplies electrons for the reduction of F_{420}, which occurs inside the cell. Because the reduction of F_{420} inside the cell also consumes protons, whereas the oxidation of hydrogen produces protons outside the cell, the net result is the establishment of a proton gradient (protonmotive force) across the membrane.

as **oxidative photophosphorylation**. Photosynthetic microorganisms have chlorophylls and accessory pigments that allow them to absorb light energy and to initiate a process that results in the generation of ATP. The particular set of chlorophylls and accessory pigments determines the wavelengths of light that can be used in photosynthesis (Figure 5.33). The photosynthetic apparatus consists of a set of light-harvesting pigments, a photosynthetic reaction center, and an electron transport chain. The light-harvesting pigments, which may be chlorophylls, carotenoids, and/or phycobiliproteins, ab-

FIGURE 5.33

Various chlorophylls and auxiliary pigments are involved in the capture of light energy and its conversion to ATP. The various photosynthetic pigments absorb light energy of differing wavelengths. (A) Chlorophyll a is the primary photosynthetic pigment in cyanobacteria and algae. (B) Various bacteriochlorophylls, such as bacteriochlorophyll a shown here, are the primary photosynthetic pigments in noncyanobacterial photosynthetic bacteria. (C) Carotenoids, such as as the β-carotene shown here, are widely found accessory pigments in many photosynthetic microorganisms; these molecules, which have long hydrocarbon chains with repeating double bonds and are typically yellow in color, are able to absorb light energy, usually blue light, and transfer it to chlorophyll molecules. (D) Cyanobacteria and some algae conatin biliproteins (phycobilins), such as phycoerythrin (red color) and phycocyanin (blue color). The structure shown here is part of the phycocyanin molecule.

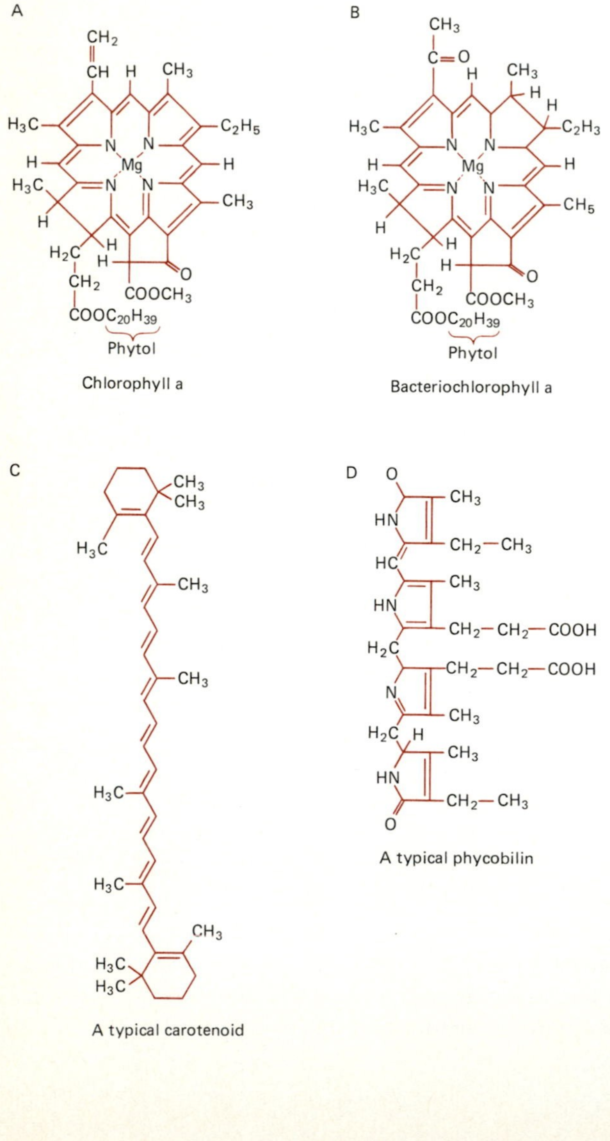

Chlorophyll a

Bacteriochlorophyll a

A typical carotenoid

A typical phycobilin

sorb light energy of particular wavelengths and transmit the energy to the photosynthetic reaction center, which has a chlorophyll molecule that can initiate a flow of electrons.

When a chlorophyll molecule absorbs light energy, it becomes energetically excited and emits an electron that is transferred through an electron transport chain analogous to the electron transport system of oxidative phosphorylation. The electron transfer reactions of oxidative photophosphorylation involve a series of membrane-bound carriers collectively known as a **photosystem**, that is, a pathway of electron transfer. The transfer of electrons through the carriers of the electron transport chain establishes a hydrogen ion gradient across the membrane. In accordance with the chemiosmotic mechanism for ATP generation, the extrusion of hydrogen ions to the inside of the thylakoid membrane establishes the protonmotive force that drives the synthesis of ATP (Figure 5.34). In chloroplasts the flow of hydrogen ions is the opposite of that in mitochondria; the ATPase is located on the outer side of the thylakoid membrane, whereas the ATPase spheres occur on the inner surface of the inner mitochondrial membrane.

Although the basic chemiosmotic mechanism driving the synthesis of ATP during oxidative photophosphorylation is the same in all photoautotrophs, there is a major difference between the algae and the anaerobic photosynthetic bacteria in the photosystems that they employ to transfer the energy absorbed from light irradiation to the synthesis of ATP. In the case of the anaerobic green and purple photosynthetic sulfur bacteria, there is only one photosystem, known as **photosystem I** or **cyclic oxidative photophosphorylation**. The cyanobacteria and the algae have two photosystems, referred to as **photosystem I and photosystem II**, or **cyclic and noncyclic oxidative photophosphorylation**.

The Z Pathway of Oxidative Photophosphorylation Photosystems I and II are normally linked in a **Z pathway** of oxidative photophosphorylation in cyanobacteria and algae (Figure 5.35). The combined photosystems require two separate photo-acts, that is, the absorption of light energy at two different photo-activation centers. The light energy actually is often absorbed by accessory pigments that transfer the energy to the primary photosynthetic pigment of the photoactivation center. In the algae and cyanobacteria, the primary photosynthetic pigment is **chlorophyll** a. In photosystem I, chlorophyll a absorbs light at a wavelength of 700 nm, and therefore is designated a P_{700} molecule. In photosystem II the photoreceptor is designated P_{680}, representing a molecule of chlorophyll a that has been modified to absorb light at a wavelength of 680 nm. When the P_{680} chlorophyll reaction center absorbs light energy from pigments in the photosynthetic unit, it becomes energetically excited and emits an electron initiating a noncyclic pathway in which the electron is transferred through a series of membrane-bound electron carriers.

In the combined Z pathway the electrons from photosystem II are transferred through a series of membrane-bound

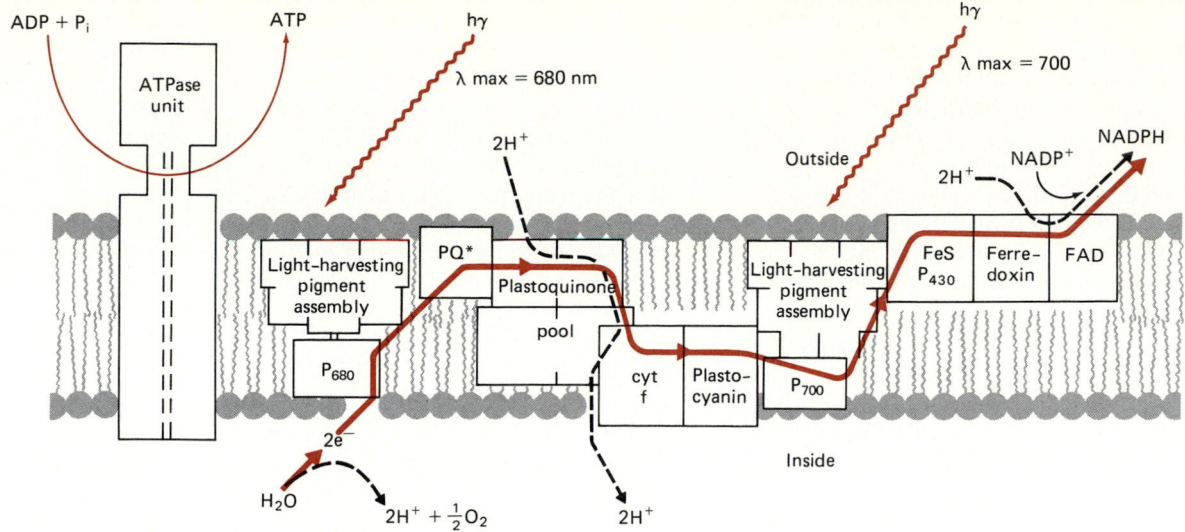

FIGURE 5.34

In photosynthetic organisms, light activation of a photoreceptor molecule, e.g., chlorophyll, initiates the flow of electrons through an electron transport chain. The proton gradient across the membrane is used to drive the formation of ATP by chemiosmosis in this process of oxidative photophosphorylation.

carriers to the P_{700} chlorophyll reaction center molecule of photosystem I. The transfer of an electron from an excited P_{680} chorophyll molecule of photosystem II to the P_{700} chlorophyll molecule of photosystem I establishes a sufficient hydrogen ion gradient across the membrane to drive the synthesis of one ATP molecule. The electron transport chain is continued when a molecule of P_{700} chlorophyll absorbs light energy, initiating the electron transfer sequence of photosystem I. The electron that is transferred from photosystem II is used to balance the oxidation state of the P_{700} chlorophyll molecule. The electrons transferred through photosystem I normally are used to reduce the coenzyme $NADP^+$ to NADPH, providing an essential source of reducing power for biosynthetic metabolic reactions.

The unidirectional flow of electrons to reduce $NADP^+$ to NADPH leaves a charged P_{680} chlorophyll molecule that must be reduced to balance the oxidation-reduction state of this pathway. In order to accomplish the needed balance of oxidation-reduction molecules, a water molecule is split, forming oxygen and transferring electrons to the P_{680} chlorophyll of photosystem II. The oxygen produced in the pathway is essential for supporting the aerobic respiratory metabolism of heterotrophic microorganisms and animals. It also appears that a second ATP molecule may be generated as a consequence of the electron transport from water to the P_{680} chlorophyll molecule and thus creates a hydrogen ion gradient across the membrane.

Anoxygenic Photosynthesis Although the movement of the electrons through the entire Z pathway is normally noncyclic, with the electron flow from the electron donor, H_2O, to the electron acceptor, $NADP^+$, electrons may also flow cyclically through photosystem I. When this occurs, reduced coenzyme NADPH is not generated, but ATP is synthesized; that is, photosystem I can act as a cyclic oxidative photophosphorylation system, providing the organism with an increased yield of ATP. At low light intensities the cyanobacteria can carry out **anoxygenic photosynthesis** (nonoxygen-evolving photosynthesis), during which photosystem I follows a cyclic oxidative photophosphorylation pathway. In anoxygenic photosynthesis, photosystem II is inoperative, and thus, water is not split to yield oxygen. While carrying out anoxygenic photosynthesis, the cyanobacteria derive their reducing power from the coupled oxidation of hydrogen sulfide with coenzyme reduction. When cyanobacteria utilize hydrogen sulfide as the reducing agent in this process, they form elemental sulfur granules that are deposited outside of the cells.

The process of anaerobic photosynthesis carried out by photoautotrophic bacteria other than the cyanobacteria is similar in many ways to cyanobacterial anoxygenic photosynthesis. The **anaerobic photosynthetic bacteria** only carry out the reactions of photosystem I (Figure 5.36). In the green and purple anaerobic photosynthetic bacteria the primary photosynthetic pigment is **bacterial chlorophyll**,[15] which absorbs light energy of longer wavelengths than does chlorophyll a. As with the cyanobacteria and algae, the incoming light energy may initially be absorbed by accessory pigments and then transferred to the primary photosynthetic pigments.

The excitation of the bacterial chlorophyll P_{870} causes it to emit an electron that is transported through an electron

---

[15]The primary photoreaction center in the anaerobic bacteria is labeled P_{870}. The P_{870} refers to bacterial chlorophyll that absorbs light at 870 nm, which is in the infrared wavelength region.

FIGURE 5.35

The Z pathway of oxidative photophosphorylation combines two separate photosystems into a unified pathway. Two separate photoactivation steps are needed to complete this pathway. These are the excitation of P_{680} to P_{680}^* and the excitation of P_{700} to P_{700}^*. The P_{680} has a sufficiently positive reduction potential to use H_2O as an electron donor. The resulting P_{680}^* is at a considerably more negative reduction potential, such that the resulting electrons can "fall down" a potential gradient in which protons are pumped across the membrane. The electrons are passed to the P_{700} complex, which when excited is at a potential more negative than that of the NADP/NADPH redox pair and is thereby capable of reducing $NADP^+$ to NADPH. The pathway is called the Z *pathway* because, when these reactions are plotted as a function of reduction potential, the resulting figure resembles a Z.

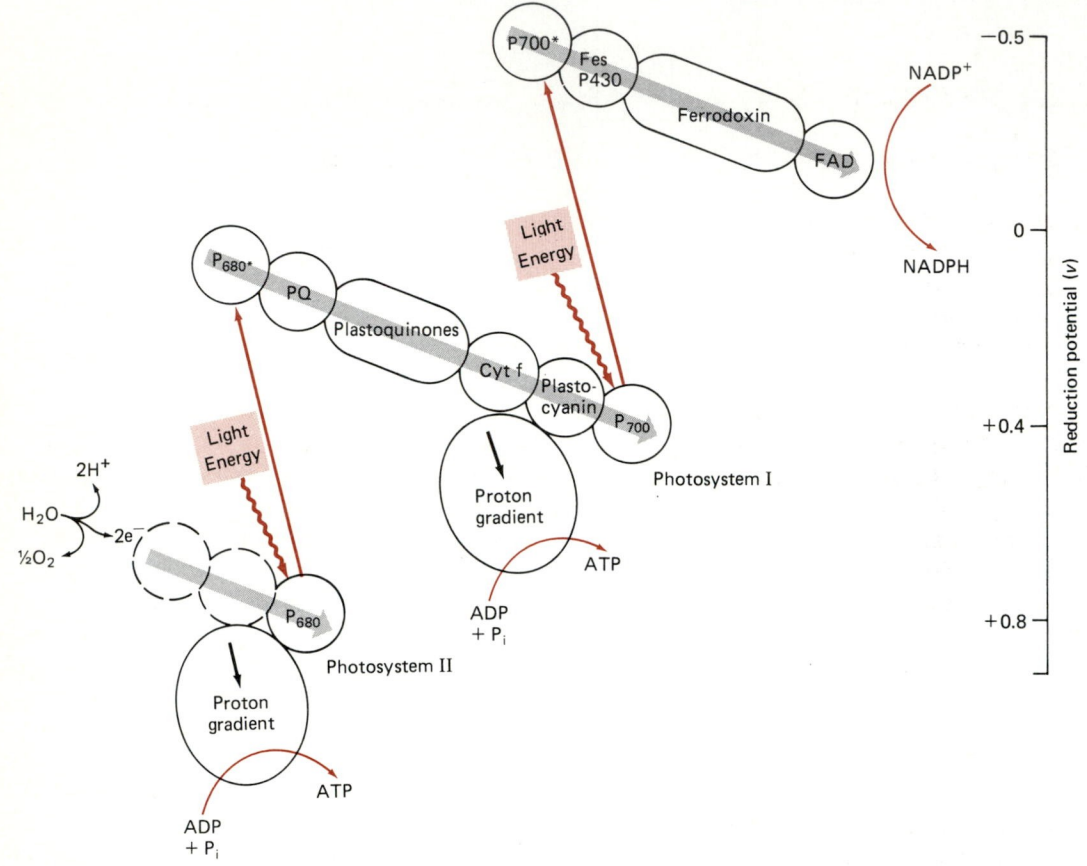

Discovery Process

Daniel Arnon and co-workers, in 1954 at the University of California, Berkeley, first demonstrated complete photosynthesis by isolated chloroplasts, showing that these chloroplasts could apply absorbed light purely to the formation of ATP. They discovered this reaction by depriving illuminated chloroplasts of carbon dioxide and pyridine nucleotide (coenzyme) while giving them large doses of ADP and inorganic phosphate. With no raw material (substrate) for making carbohydrates or NADPH, the chloroplasts used light energy to force the third phosphate on ADP to form ATP, which accumulated in substantial quantities. By 1958, Arnon and associates had constructed a theoretical model to fit the experimental facts of photosynthetic phosphorylation based on another photochemical reaction worked out by Gilbert Lewis and David Lipkin, also at the University of California.

They proposed that in the primary photochemical act a photon of light strikes a chlorophyll molecule, exciting one of the electrons to an energy level sufficient to remove it from the molecule. After losing this electron, the molecule can act as an electron acceptor. If the chlorophyll molecule took back the electron directly, it would not be useful in generating ATP, but if the excited electrons are transferred to carriers, the electrons are forced to return to chlorophyll in a series of graded steps. Such a chain of electron transfers can lead to the generation of ATP and, in some cases, to the production of reduced coenzyme. Now we know that the series of electron transfers initiated by the light excitation of chlorophyll leads to the establishment of an electrochemical gradient across the photosynthetic membrane and the chemiosmotic generation of ATP.

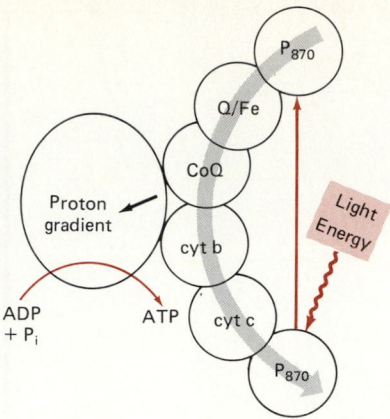

FIGURE 5.36

Cyclic oxidative photophosphorylation of anaerobic photosynthetic bacteria uses P_{870}. This pathway generates the proton gradient needed to drive the formation of ATP and does not produce reduced coenzyme.

transport chain. Instead of using an external electron donor to reduce the oxidized chlorophyll molecule, the electron is transferred back to the chlorophyll molecule, forming a cyclic pathway. In this system, the chlorophyll molecule acts as an internal electron donor and acceptor for photosystem I. In the course of returning the excited electron to the oxidized chlorophyll molecule, a sufficient energy gradient is developed to permit the synthesis of one ATP molecule. Because the electron is returned to the chlorophyll molecule, completing a cycle, and because ATP is synthesized as a consequence of light having excited the chlorophyll molecule, the pathway is referred to as cyclic oxidative photophosphorylation.

Cyclic oxidative photophosphorylation generates ATP without generating reduced coenzyme. The anaerobic photosynthetic bacteria, however, require the reduced coenzyme NADPH for biosynthetic reactions. In order to generate NADPH, these organisms utilize reduced compounds, such as H_2S, as an electron donor. The flow of electrons from the external electron donor to the reduced coenzyme may pass through photosystem I in a noncyclic pathway. Insufficient energy, however, is released from hydrogen sulfide, or other sulfur-containing compounds, to drive the formation of NADPH directly. In the purple photosynthetic bacteria, the electrochemical potential established by chemiosmosis across a membrane provides the energy to drive the formation of reduced coenzyme.

ANALYTICAL PROCESS

The Role of the Purple Membrane of Halobacterium in Generating ATP

Halobacterium is an archaebacterium that can generate ATP both by respiration using an organic substrate and by using light energy (Figure 5.37). In the presence of oxygen, halobacteria use aerobic respiration to generate ATP, including oxidative phosphorylation for chemiosmotic ATP synthesis. The mediators of the electron transport chain in this process are located in a portion of the cytoplasmic membrane that is red in color and hence is called the *red membrane*.

In the absence of oxygen, these bacteria can turn to a form of oxidative photophosphorylation for synthesizing ATP. Light energy is converted to chemical energy by a mechanism different from the ones already discussed for photoautotrophic microorganisms. It is based upon a **purple membrane** portion of the cytoplasmic membrane that contains bacteriorhodopsin, a protein that has a chemical structure similar to that of the rhodopsin pigment of the human eye. When illuminated by light, bacteriorhodopsin pumps protons to the outside of the membrane, establishing a hydrogen ion gradient. The counterflow of hydrogen ions drives the synthesis of ATP by chemiosmosis. The ATPase enzyme used for converting ADP + P_i to ATP is contained in a separate red membrane fraction so that the same ATP-synthesizing system used for oxidative photophosphorylation is used for oxidative phosphorylation. The halobacterial membrane system for using light energy to establish a hydrogen ion gradient that can drive the synthesis of ATP provides firm evidence for the essential role of chemiosmosis in the synthesis of ATP.

Establishing the role of the purple membrane in the light-coupled generation of ATP was accomplished largely through the work of Walter Stoeckenius and co-workers. In studying *Halobacterium* they had observed that, when exposed to light, the medium surrounding cell suspensions of *Halobacterium* became acidic and that exposure of the intact cells to light slowed their respiration. Aware of the chemiosmotic hypothesis of Peter Mitchell, they postulated that in the intact cell the purple membrane acts as a proton pump, and that under light illumination protons move from the inner side of the cytoplasmic membrane to the outer side and into the medium. To show that light was coupled with ATP formation, they suspended the bacteria in the dark in a salt solution without nutrients and bubbled nitrogen through the medium. Under these conditions, the intracellular ATP concentration decreased to about 30 percent of the original level because the cells could not carry out either

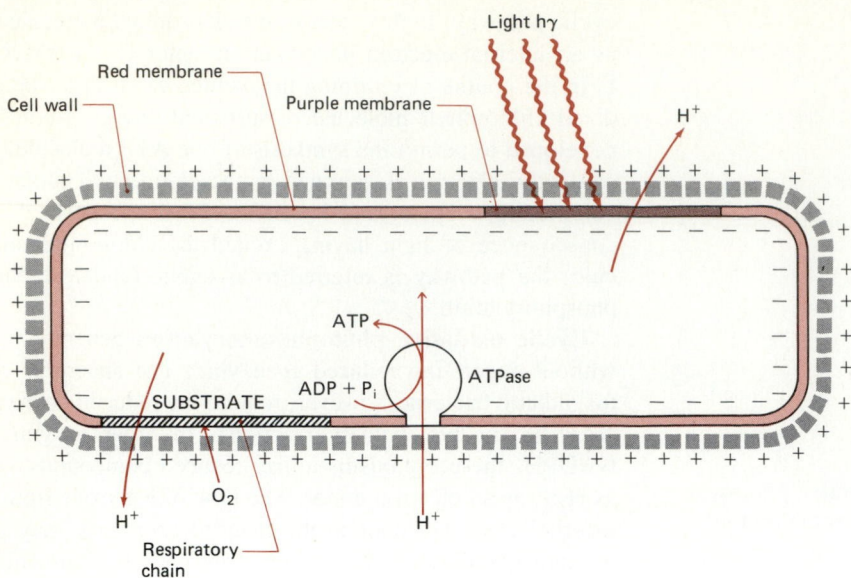

FIGURE 5.37

The relationship of proton transport to ATP formation in *Halobacterium* is illustrated here. Regions of the cytoplasmic membrane contain bacterio-rhodopsin and have a purple color. The purple membrane is involved with light-coupled generation of ATP. A larger portion of the cytoplasmic membrane is red in color and contains ATPase and mediators of the respiratory electron transport chain. When illuminated with light or supplied with an organic substrate in the presence of oxygen, protons are pumped across the cytoplasmic membrane, establishing an electrochemical and hydrogen ion gradient across the membrane. The protonmotive force and backflow of protons through the enzyme ATPase result in the formation of ATP.

aerobic respiration or light-coupled ATP generation. Adding oxygen in the dark led to a reestablishment of the original ATP level, as did illuminating the cells in the absence of oxygen, indicating that both aerobic respiration and light-coupled photophosphorylation could be used by *Halobacterium* to form ATP. Examining the effect of the wavelength of light revealed that only light of wavelengths absorbed by the purple membrane was effective in ATP synthesis.

These observations were consistent with the hypothesis that respiration and light generate an electrochemical proton gradient across the cell membrane and that the gradient drives the synthesis of ATP by membrane ATPase. To verify this hypothesis, Stoeckenius and co-workers used inhibitors of membrane ATPase, showing that they prevented the accumulation of ATP driven by either light or respiration. They also used electron transport uncouplers, substances that permit electron transport to proceed but make the membrane very permeable to protons, preventing the generation of an electrochemical gradient. The uncouplers inhibited ATP accumulation driven by light or respiration and blocked acidification of the surrounding medium. Finally, they used substances that interfere specifically with the respiratory electron transport

chain, all of which blocked respiratory but not light-driven ATP formation.

To establish definitively that it is the bacteriorhodopsin in the purple membrane that is responsible for the light-coupled protonmotive force, Stoeckenius and co-workers made artificial vesicles from the halobacterial membrane. When the vesicles, which were free of substances from the cytoplasm of the arachaebacterial cell, were exposed to light, an electrochemical proton gradient formed across the membrane. That excluded the possibility that enzyme systems in the cytoplasm were responsible for the formation of the proton gradient. Working with Efraim Racker, Stoeckenius next removed the red portion of the cytoplasmic membrane and replaced it with artificial phospholipid vesicles, leaving the bacteriorhodopsin-containing purple membrane as the only source of protein. When the resulting preparation was exposed to light, the medium became acidic, indicating that protons were being pumped across the membrane. Because bacterio-rhodopsin was the only protein in these preparations and because it worked in vesicles made of various lipids, there was no longer any doubt that bacteriorhodopsin converts light energy to generate a proton gradient.

Postlude

During the early twentieth century, major advances were made in the understanding of biochemistry and microbial metabolism. Albert Kluyver followed in the footsteps of Martinius Beijerinck, both as director of the Delft school in Holland and in the biochemical direction of his study, continuing the great tradition of microbiological study that had been begun by Leeuwenhoek 300 years earlier in that small Dutch town. Kluyver examined the unity and diversity in metabolism of microorganisms, emphasizing the unifying features of microbial metabolism, correctly recognizing the nature of intermediary metabolism, and establishing that hydrogen transfer (oxidation-reduction reactions) is a basic feature of all metabolic processes. The flow of carbon and energy through a series of metabolic transformations is an essential feature of living microorganisms. Kluyver was instrumental in developing an understanding of the role of central metabolic pathways in microbial metabolism. C. B. van Niel, a student of Kluyver, continued the tradition of advancing microbial physiology. He made important contri-

butions to our understanding of photosynthesis, recognizing the similarity between H_2S and H_2O in the photosynthetic processes of the anaerobic photosynthetic sulfur bacteria and in higher plants.

Important later discoveries, which emphasized the unity of intermediary metabolism, included the elucidation of the citric acid cycle, for which Sir Hans Krebs received the Nobel Prize in 1953. In 1978 the Nobel Prize was awarded to Peter Mitchell for the development of the chemiosmotic theory to explain how the biochemical reactions occurring at membranes generate energy in the form of ATP. While these Nobel Prizes were awarded in chemistry, they represent fundamental advances in our understanding of microbial metabolism.

Clearly, microorganisms exhibit a variety of strategies for converting chemical and light energy into the energy stored within ATP, the central currency of energy of the cell (Table 5.3). The synthesis of ATP can be achieved autotrophically—either through the oxidation of inorganic substrates or

TABLE 5.3   Types of Microbial Metabolism Used to Generate ATP

Autotrophic	Does not require preformed organic matter to generate ATP
Photoautotrophic (photolithotropic)	Uses light energy to drive synthesis of ATP
Oxygenic photosynthesis	Uses two connected photosystems and results in evolution of oxygen, as well as generation of ATP; carried out by algae and cyanobacteria
Anoxygenic photosynthesis	Uses one photosystem and does not result in evolution of oxygen; carried out by anaerobic photosynthetic bacteria, e.g., green and purple sulfur bacteria, and under some conditions by cyanobacteria
Chemoautotrophic (chemolithotrophic)	Uses oxidation of inorganic compounds such as sulfur, nitrite, nitrate, and hydrogen to establish an electrochemical gradient across a membrane that results in generation of ATP by chemiosmosis
Heterotrophic	Requires preformed organic matter to form ATP
Respiration	Uses complete oxidation of organic compounds, requiring an external electron acceptor to balance oxidation-reduction reactions used to generate ATP; much of the ATP is formed as a result of chemiosmosis based upon the establishment of a proton gradient across a membrane
Aerobic	Uses oxygen as the terminal electron acceptor in the membrane-bound pathway that establishes the proton gradient for chemiosmotic ATP generation
Anaerboic	Uses compounds other than oxygen, e.g., nitrate or sulfate as the terminal electron acceptor in the membrane-bound pathway that establishes the proton gradient for chemiosmotic ATP generation
Fermentation	Does not require an external electron acceptor, achieving a balance of oxidation-reduction reactions using the organic substrate molecule; various fermentation pathways produce different end products

through the conversion of light energy to chemical energy—or heterotrophically through the utilization of organic substrates. The amount of ATP that can be synthesized in these processes varies greatly, and microorganisms accordingly show great variation in the efficiency with which they synthesize sufficient ATP to meet their energy requirements. Chemolithotrophic metabolism, for example, is relatively inefficient energetically. Thus, chemolithotrophic microorganisms must metabolize a large number of inorganic substrate molecules to synthesize sufficient ATP for their metabolic needs during growth and reproduction.

Similarly, respiration is energetically far more favorable than fermentation. To take a representative number, many microorganisms are capable of generating 38 ATP molecules per molecule of glucose by respiration, compared to only 2 molecules of ATP per molecule of glucose by fermentation. As a consequence of the inefficiency of the fermentation pathway, more substrate must be utilized during fermentation than during respiration to achieve similar amounts of growth. Although fermentation does not yield as much ATP per molecule of substrate as respiration, the end products of fermentation are of more practical interest than those of respiration. Respiration normally results in the formation of carbon dioxide and water. The various fermentation pathways carried out by different organisms produce numerous end products of economic value. For example, the ethanolic and lactic acid fermentation pathways are of particular importance in the food industry because they are directly involved in the production of bread, beer, wine, and cheese, among other foods and beverages.

Although microorganisms show great metabolic versatility in generating ATP and reducing power, several central pathways play key roles in the metabolism of microbial cells. The metabolic pathways involved in ATP generation involve various intermediary metabolites, linked together in a series of small steps to form unified biochemical pathways. These metabolic pathways are carefully regulated so that when the cell has sufficient ATP, pathways that result in the synthesis of ATP are shut off through feedback inhibition by an allosteric inhibitor of a critical enzyme in the pathway. For example, phosphofructokinase, a key enzyme in the glycolysis of carbohydrates, is inhibited by ATP; therefore, when the cell has enough ATP, further synthesis of ATP is restricted.

Considering the variety of ways in which a microorganism can drive the synthesis of ATP, heterotrophic metabolism of an organic substrate may occur either by respiration or by fermentation, and autotrophic metabolism may involve the oxidation of inorganic compounds or the conversion of light energy. In respiration an external electron acceptor is required to complete the metabolic pathway; oxygen serves as the terminal electron acceptor in aerobic respiration, and nitrate or sulfate act as possible terminal electron acceptors in anaerobic respiration. In fermentation the organic substrate molecules serve as internal electron donors, and products of their metabolism serve as acceptors. Chemolithotrophic microorganisms are able to oxidize inorganic compounds to generate ATP. Photoautotrophic microorganisms are able to trap light energy, in some cases involving one cyclic photosystem and in others two different photosystems linked in a Z pathway.

With the exception of fermentation, all of these pathways, which are aimed at the synthesis of ATP, involve electron transport through a series of membrane-bound carriers and the establishment of a hydrogen ion gradient across a membrane. For example, in photosynthetic microorganisms the flow of electrons—initiated when a chlorophyll molecule is energetically excited by absorbing light energy—establishes an electrochemical gradient across a membrane during the process of oxidative photophosphorylation. Similarly, a hydrogen ion gradient across a membrane is formed during oxidative phosphorylation in both respiration and chemolithotrophic metabolism. The counterflow of hydrogen ions, channeled through membrane-bound ATPase, drives the synthesis of ATP. This process of chemiosmosis is central to many of the ATP-synthesizing pathways of microorganisms.

As a consequence of their metabolic activities, microorganisms are capable of channeling energy into the synthesis of ATP. They are also capable of channeling reducing power into the synthesis of the reduced coenzymes NADH and NADPH. The reducing power and the ATP generated by microbial metabolism are used by microorganisms for their growth and reproduction. The particular pathways that a microorganism can use for generating ATP and reducing power depend on the enzymes that the organism possesses. The additional end products of metabolism that are produced also are a function of the cell's enzymatic potentialities, which are genetically determined. Regardless of the mode of metabolism, the strategies are the same: synthesize ATP, reduce coenzyme (NADPH) and make small precursor molecules to serve as the building blocks of macromolecules, and then use the energy, reducing power, and precursor molecules to synthesize the macromolecular constituents of the organism.

Suggested Supplementary Readings

ANDERSON, J. M., and B. ANDERSON. 1982. The architecture of photosynthetic membranes: Lateral and transverse organization. *Trends in Biochemical Sciences* 7:288–292.

ATKINS, P. W. 1984. *The Second Law*. W. H. Freeman Co., New York.

BATTLEY, E. H. 1987. *Energetics of Microbial Growth*. John Wiley & Sons, New York.

BECKER, W. M. 1977. *Energy and the Living Cell*. Harper & Row, Publishers, Inc., New York.

BRONK, J. R. 1973. *Chemical Biology: An Introduction to*

Biochemistry. Macmillan Publishing Co., Inc., New York.

CHANCE, B. 1977. Electron transfer: Pathways, mechanisms, and controls. *Annual Review of Biochemistry* 46:967–980.

CLAYTON, R. K. 1980. *Photosynthesis: Physical Mechanisms and Chemical Patterns*. Cambridge University Press, Cambridge, England.

DARNELL, J., H. LODISH, and D. BALTIMORE. 1986. *Molecular Cell Biology*. Scientific American Books, New York.

DAWES, I. W., and I. W. SUTHERLAND. 1976. *Microbial Physiology*. Blackwell Scientific Publications, Oxford, England.

FERGUSON, S. J., and M. C. SORGATO. 1982. Proton electrochemical gradients and energy transduction processes. *Annual Review of Biochemistry* 51:185–218.

GOTTSCHALK, G. 1986. *Bacterial Metabolism*. Springer-Verlag, New York.

GOVINDJEE. 1975. *Bioenergetics of Photosynthesis*. Academic Press, New York.

GOVINDJEE (ED.) 1982. *Photosynthesis: Energy Conversion by Plants and Bacteria*. Academic Press, New York.

GOVINDJEE, and R. GOVINDJEE. 1974. The absorption of light in photosynthesis. *Scientific American* 231(6):68–82.

HINKLE, P. C., and R. E. McCARTHY. 1978. How cells make ATP. *Scientific American* 238(3):104–123.

INGRAHAM, J. L., O. MAALE, and F. C. NEIDHARDT. 1983. *Growth of the Bacterial Cell*. Sinauer Associates, Inc., Sunderland, Mass.

JONES, C. W. 1982. *Bacterial Respiration and Photosynthesis*. American Society for Microbiology, Washington, D.C.

KASHKET, E. R. 1985. The proton motive force in bacteria: A critical assessment of methods. *Annual Review of Microbiology* 39:219–242.

KREBS, H. A. 1970. The history of the tricarboxylic acid cycle. *Perspectives in Biology and Medicine* 14:154–170.

LEHNINGER, A. L. 1971. *Bioenergetics*. Benjamin/Cummings Publishing Co., Menlo Park, Calif.

LEHNINGER, A. L. 1982. *Principles of Biochemistry*. Worth

Publishers, Inc., New York.

LESSIE, T. G., and P. V. PHIBBS, JR. 1984. Alternative pathways of carbohydrate utilization in *Pseudomonas*. *Annual Review of Microbiology* 38:359–388.

MANDELSTAM, J., K. McQUILLEN, and I. W. DAWES. 1982. *Biochemistry of Bacterial Growth*. Blackwell Scientific Publications, Oxford, England.

MEYER, H., O. KAPPELI, and A. FIECHTER. 1985. Growth control in microbial cultures. *Annual Review of Microbiology* 39:299–319.

MITCHELL, R. 1979. Keilin's respiratory chain concept and its chemiosmotic consequences. *Science* 206:1148–1159.

NICHOLLS, D. G. 1982. *Bioenergetics: An Introduction to the Chemiosmotic Theory*. Academic Press, New York.

ORMEROD, J. G. (ED). 1983. *The Phototrophic Bacteria: Anaerobic Life in the Light*. University of California Press, Berkeley, Calif.

PARSON, W. W. 1974. Bacterial photosynthesis. *Annual Review of Microbiology* 28:41–59.

RACKER, E. 1980. From Pasteur to Mitchell: A hundred years of bioenergetics. *Federation Proceedings* 39:210–215.

SCHLEGEL, H. G., and B. BOWIEN. 1981. Physiology and biochemistry of aerobic hydrogen-oxidizing bacteria. *Annual Review of Microbiology* 35:405–452.

SOKATCH, J. A. 1969. *Energy Metabolism in Bacteria*. Academic Press, New York.

STOECKENIUS, W. 1976. The purple membrane of salt loving bacteria. *Scientific American* 234(6):38–46.

STOECKENIUS, W., and R. A. BOGOMOLNI. 1982. Bacteriorhodopsin and related pigments of halobacteria. *Annual Review of Biochemistry* 51:587–616.

STRYER, L. 1987. *Biochemistry*. W. H. Freeman Co., San Francisco.

ZUBAY, G. 1983. *Biochemistry*. Addison-Wesley Publishing Co., Reading, Mass.

Study Questions

1. What is autotrophic metabolism? Discuss the different ways in which autotrophs generate ATP. Consider both photoautotrophs and chemolithotrophs.
2. What is heterotrophic metabolism?
3. What is the difference between fermentation and respiration?
4. What individual pathways are involved in respiratory metabolism?
5. What is a terminal electron acceptor? What is the difference between aerobic and anaerobic respiration?
6. Name five different fermentation pathways. For each pathway, what are the metabolic end products, and what organisms characteristically carry out the pathway?
7. What is the difference between oxygenic and anoxygenic photosynthesis?
8. Based only on their metabolism, should the blue-greens be considered bacteria or algae?
9. What is chemiosmosis, and how does it explain the generation of ATP both in oxidative phosphorylation and in oxidative photophosphorylation?
10. Where do the following processes occur?
    a. oxidative phosphorylation in bacteria
    b. oxidative phosphorylation in fungi
    c. oxidative photophosphorylation in algae
    d. oxidative photophosphorylation in cyanobacteria
    e. the Krebs cycle in bacteria

Situational Problem 1

Determining the Pathways of ATP Generation in Newly Discovered Bacteria

You have been working in a research laboratory that received sample sediments from the deep ocean thermal vent regions. Working as part of a team, you have isolated three distinct bacteria that appear to be new species. They all grow at elevated temperatures. Two of the three will grow on a medium with glucose as the sole source of carbon and energy, provided that mineral nutrients are also added. One of these bacteria will grow in the presence or absence of oxygen, and produces acid and gas when growing on glucose. Another bacterium will also grow in the presence or absence of oxygen; it produces gas only when growing in the absence of oxygen and does not produce acid in either case. The third bacterium will not grow with added glucose but will grow if bicarbonate and thiosulfate are added.

Because these are newly discovered bacteria from a unique and very interesting ecosystem, you decide to determine how they are generating their required ATP. Specifically, you want to know whether they are autotrophs, heterotrophs, or mixotrophs (organisms capable of both autotrophic and heterotrophic growth). You also want to know about the specific pathways that are involved in ATP generation, for example, whether the bacteria carry out one or more fermentation and/or respiration pathways. Before initiating actual experimental studies, you, like all competent scientists, must design an experimental protocol that will provide unequivocal answers. The laboratory in which you are working is well equipped with pH meters to measure hydrogen ion concentrations, oxygen meters to measure oxygen concentrations, balances, spectrophotometers, and other routine equipment. There are also nonspecialty chemicals, such as glucose, inorganic salts, and the like. You also have a budget of $500 with which to purchase additional supplies. You could buy substrates, metabolic inhibitors such as uncoupling reagents, analytical standards, and so forth.

Design an experiment that would reveal the metabolic pathway(s) used by each of these organisms to generate ATP. Be specific in your design. If appropriate, consult chemical supply catalogs to determine the costs of the specific reagents you intend to use and make sure that their cost is within your allotted budget.

Situational Problem 2

Determining the Pathways of ATP Generation Based on Measurements of ATP Concentrations Under Different Conditions

ATP concentrations can be measured using a relatively simple assay known as the *luciferin-luciferase assay*. It is based on the reaction that produces light in the tail of a firefly and uses the intensity of light as a quantitative measure of ATP concentration. In this assay, ATP is extracted from cells by boiling in Tris buffer, pH 7.75. The heating disrupts the cytoplasmic membranes of the cells, permitting ATP to diffuse out of the cells. After cooling, the extracted ATP is added to a mixture of reduced luciferin and luciferase in a buffer containing magnesium ions. Under these conditions, the reduced luciferin reacts with oxygen in a luciferase-catalyzed reaction to produce oxidized luciferin. In this reaction light is emitted, and its intensity is directly proportional to the concentration of ATP, which is the limiting factor in this reaction. The amount of light emitted can be measured with a commercial instrument that has a photodetector and a photomultiplier. Using a standard curve, it is thus easy to measure the ATP in the bacterial cells.

Using the luciferin-luciferase assay, you have measured the ATP concentration for samples of a bacterial culture grown under several different conditions. When the culture is illuminated in the presence of oxygen without any added organic carbon source, the measured ATP concentration is 200 μg per milliliter of culture. When the light source is turned off, the ATP concentration drops to 5 μg/mL. After glucose is added in the dark, the concentration of ATP goes up to 100 μg/mL. Sparging the culture with nitrogen again lowers the ATP concentration to 5 μg/mL. Illuminating the culture again raises the ATP concentration to 200 μg/mL.

Based upon these data, what conclusions can you draw about the way(s) in which this bacterium generates ATP?

CHAPTER 6

Carbon Flow: Biosynthesis of Macromolecules

In addition to requiring energy in the form of adenosine triphosphate (ATP), microorganisms must carry out metabolic activities that transform substrate molecule(s)—that is, growth nutrients—into many different cell constituents. Among the macromolecules that both autotrophs and heterotrophs must make are proteins for enzymes, lipids for membranes, carbohydrates for various structures such as cell walls, and nucleic acids for the storage and expression of genetic information. The basic strategy of the cell in terms of carbon flow—the movement of carbon from substrate to cellular components—is to form relatively small biochemical molecules that can act as central building blocks for establishing the carbon skeletons of a variety of large macromolecules. When a heterotrophic microorganism starts with an organic substrate, like glucose, it carries out a catabolic pathway to break down that molecule into smaller compounds

that then act as precursors for the biosynthesis of macromolecules (**catabolism** = degradative process). Then the microorganism uses an anabolic pathway in which small molecules are transformed into larger molecules (**anabolism** = biosynthetic process).

As a general rule, the macromolecules of microorganisms are in a more reduced oxidation state than the starting substrate molecules, and therefore, biosynthetic pathways require reducing power. Some of the metabolic activities of a microorganism are aimed at generating the reduced coenzyme nicotinamide adenine dinucleotide phosphate (NADPH) (reducing power) required for biosynthesis.[1] Biosynthesis also requires ATP because anabolic pathways are endergonic and therefore must be coupled with exergonic reactions, generally the conversion of ATP to adenosine diphosphate (ADP).

6.1 Amphibolic Pathways

Besides ATP and NADPH, biosynthetic pathways require small precursors for the molecules being synthesized, that is, the substrates that can be transformed into macromolecules in the anabolic pathway. Many of the intermediary metabolites of the catabolic pathways serve as these substrates for biosynthesis. These compounds can serve in the generation of ATP and the biosynthesis of macromolecules, and often are siphoned off from catabolic, ATP-generating pathways into anabolic, ATP-utilizing pathways. In fact, there are several **central metabolic pathways** (pathways that play key roles in both catabolism and biosynthesis), and the flow of carbon can be reversed through these so-called **amphibolic**

pathways (dual-purpose pathways). The Krebs cycle, for example, serves a dual role; in catabolism it forms reduced coenzymes for ATP generation, and in anabolism it provides the small precursors needed to form macromolecules. The central pathways permit the interconversion of all four major

[1]There is an interesting and critical difference in the coenzymes used in catabolic and anabolic pathways; in catabolism NAD$^+$ serves as an oxidizing agent when it is reduced to reduced nicotinamide adenine dinucleotide (NADH), whereas in anabolism the coenzyme NADPH serves as the reducing agent and is converted to its oxidized form, NADP$^+$.

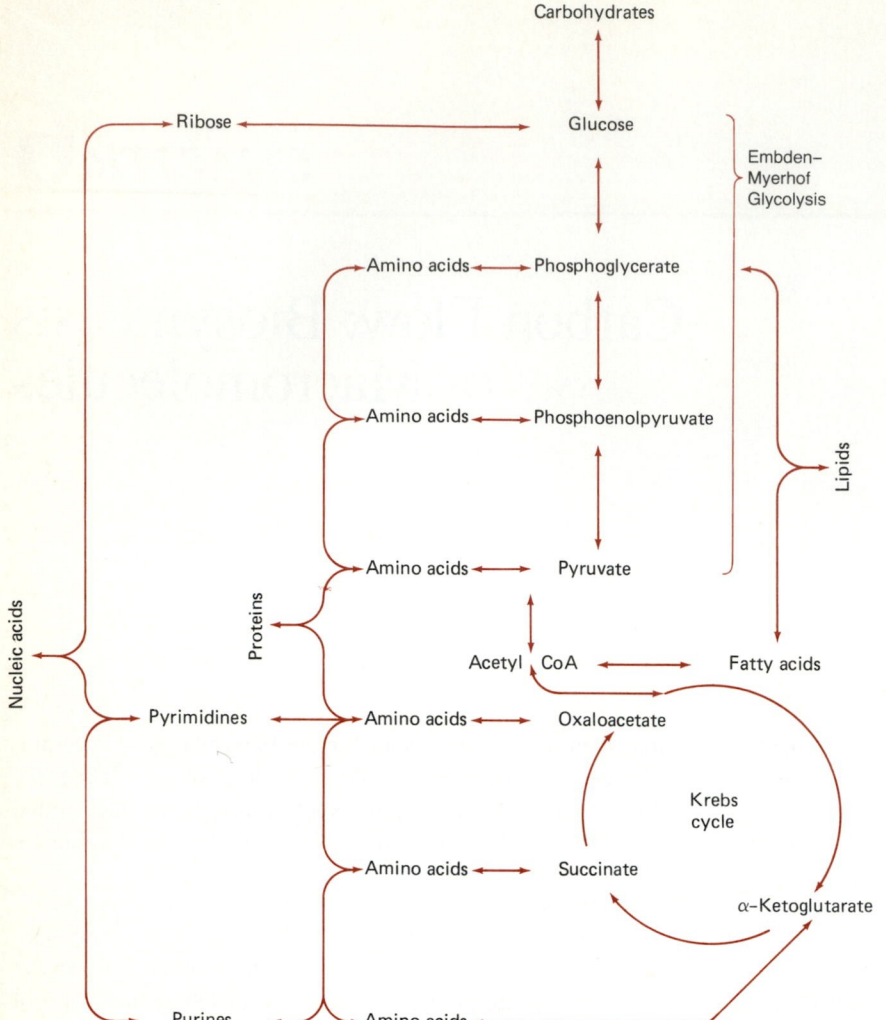

FIGURE 6.1
The central metabolic pathways are inter-connected so that carbon flows through the cell via a unified network of metabolic reactions. This network connects all of the major classes of macromolecules. Small carbon molecules formed from the breakdown of one macromolecular class can be used for the biosynthesis of compounds of the same or other classes of macromolecules.

classes of microbial biochemicals, namely, carbohydrates, lipids, proteins, and nucleic acids. They permit the necessary metabolic connections between the degradation of one class of compound, serving as substrate, and the biosynthesis of another class of compound, composing a necessary macro-molecular product (Figure 6.1).

6.2 The Conversion of Inorganic Carbon Dioxide to Organic Carbon

The formation of organic compounds from carbon dioxide is fundamental to the central metabolic pathways of the cell that involve interconversions of organic compounds. Autotrophic microorganisms use the organic molecules generated from the reduction of CO_2 as precursors for the rest of the macromolecules that they require; these organic molecules synthesized by autotrophs, both microbial autotrophs and plants, form the substrate organic molecules required for the metabolism of heterotrophic organisms. Because organic carbon is in a more reduced oxidation state than carbon dioxide, the autotrophic conversion of carbon dioxide to organic carbon requires reducing power in the form of reduced coenzyme NADPH, and because the pathway of carbon dioxide

fixation is endergonic, carbon dioxide fixation also requires ATP. We have already seen that photoautotrophic microorganisms generate NADPH and ATP as part of their photophosphorylation systems, and that chemoautotrophs generate reduced coenzyme and ATP through the oxidation of inorganic compounds. We shall now examine how such autotrophic microorganisms use NADPH and ATP to convert inorganic carbon dioxide to organic carbohydrate molecules.

The Calvin Cycle

The **fixation of CO_2** (conversion of inorganic carbon dioxide to organic compounds) by autotrophic microorganisms oc-

FIGURE 6.2

The Calvin or carbon reduction cycle is the main metabolic pathway used by autotrophs for the conversion of carbon dioxide to organic carbohydrates. The pathway, which is active in both photoautotrophs and chemolithotrophs, requires the input of carbon dioxide, ATP (energy), and NADPH (reducing power).

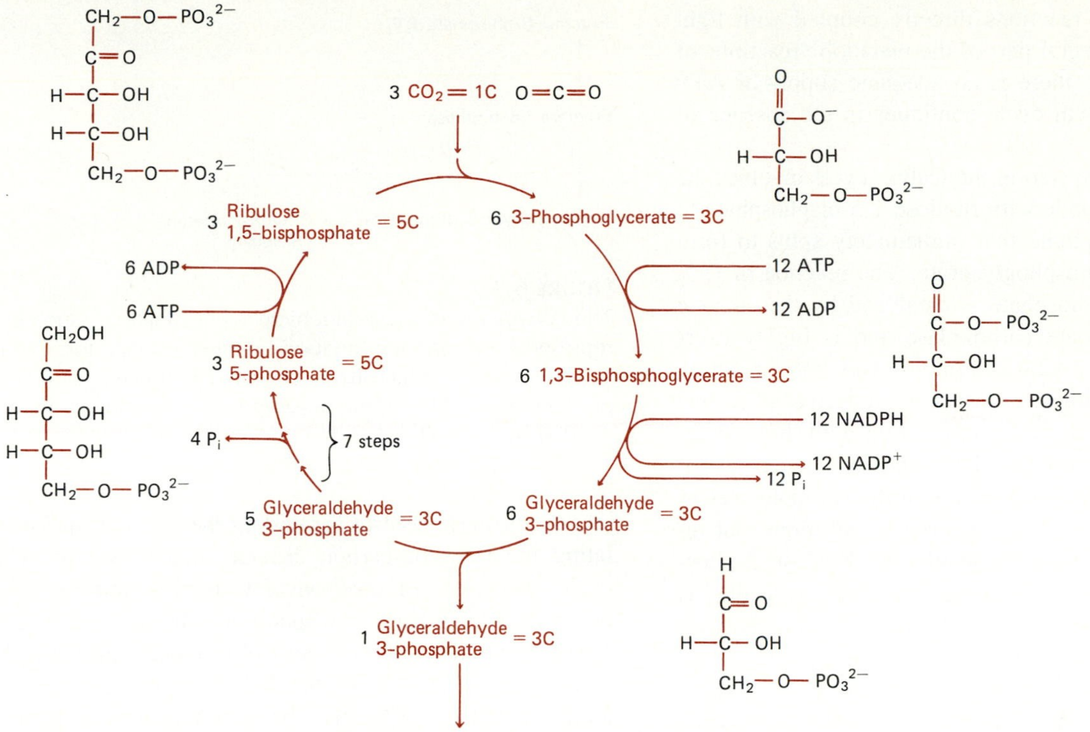

Discovery Process

In 1945 Melvin Calvin and his co-workers began the work that determined the dark reaction of photosynthesis and that also won him a Nobel Prize in 1961. The unicellular green alga *Chlorella* was their test organism because it is easy to culture in a highly reproducible way. In order to elucidate the pathway by which CO_2 becomes fixed in carbohydrate, Calvin and colleagues centered their experimental strategy on the use of radioactive ^{14}C to trace the fate of the CO_2. Radioactive $^{14}CO_2$ was injected into an illuminated suspension of algae that had been carrying out photosynthesis with normal CO_2. The algae were killed at selected intervals by pouring the suspension into alcohol, which stopped the enzymatic reactions. The radioactive compounds in the algae were then separated and identified by using two-dimensional paper chromatography. The resulting paper chromatogram was pressed against photographic film to produce an autoradiogram in which black spots indicate the locations of the labeled products arising from CO_2 fixation; the next 10 years were devoted to identifying these products positively, thus establishing the CO_2 reduction pathway and all of the intermediary metabolites in what later became known as the *Calvin cycle*.

---

curs in a metabolic pathway known as the **Calvin cycle** (Figure 6.2). The Calvin cycle is a complex series of reactions that actually represents three slightly different but fully integrated metabolic sequences. It effectively takes three turns of the Calvin cycle to synthesize one molecule of glyceraldehyde 3-phosphate, the net organic product of this metabolic pathway. Because glyceraldehyde 3-phosphate contains three carbon atoms, the Calvin cycle is sometimes referred to as a C_3 *pathway*. The formation of one molecule of glyceraldehyde 3-phosphate requires three molecules of carbon dioxide, nine molecules of ATP, and six molecules of NADPH:

$$3 \ CO_2 + 9 \ ATP + 6 \ NADPH \rightarrow$$
$$\text{glyceraldehyde 3-phosphate} + 9 \ ADP + 6 \ NADP^+.$$

It is thus apparent that the conversion of carbon dioxide to glyceraldehyde 3-phosphate requires a great deal of energy and reducing power. In photoautotrophs the ATP and NADPH, the energy and reducing power required to drive the Calvin cycle, come from the light reactions of photosynthesis and, in chemolithotrophs, from the oxidation of inorganic compounds. The Calvin cycle itself is known as a *dark reaction* because, although it requires ATP and NADPH, it does not require any reactions directly coupled with light energy input as an integral part of the metabolic reactions of this cycle. As long as there is an adequate supply of ATP and NADPH, the Calvin cycle continues in the absence of light.

The initial metabolic step in the Calvin cycle involves the reaction of carbon dioxide with ribulose 1,5-bisphosphate to form a 6-carbon compound that immediately splits to form two molecules of 3-phosphoglycerate. The reaction of CO_2 with ribulose 1,5-bisphosphate is catalyzed by the enzyme ribulose 1,5-bisphosphate carboxylase and is highly exergonic, with a $\Delta G°$ of -12.4 kcal/mole. This initial reaction occurs three separate times, so that three molecules of carbon dioxide enter the Calvin cycle by reacting with three molecules of ribulose 1,5-bisphosphate and six molecules of 3-phosphoglycerate are formed. Five of the six molecules of 3-phosphoglycerate go through a series of reactions that regenerate the original three molecules of ribulose 5-phosphate. The remaining 3-phosphoglycerate molecule is reduced to form the net product of the cycle, the glyceraldehyde 3-phosphate molecule. Because the ribulose 1,5-bisphosphate is regenerated by these reactions, the metabolic pathway forms a cycle, and the only net carbon flow is the consumption of three carbon dioxide molecules and the production of one molecule of glyceraldehyde 3-phosphate.

The glyceraldehyde 3-phosphate molecules that are formed during the Calvin cycle can further react to form glucose and polysaccharides of glucose, such as starch and cellulose. The conversion of glyceraldehyde 3-phosphate to glucose occurs by a pathway that effectively reverses the carbon flow of the Embden-Meyerhof pathway (Figure 6.3). It takes six turns of the Calvin cycle to form a 6-carbon carbohydrate, such as glucose. The net input of energy, as ATP, and reducing power, as NADPH, required for the conversion of carbon dioxide to glucose is 18 ATP and 12 NADPH molecules. The overall conversion of carbon dioxide to glucose is highly endergonic and requires 114 kcal/mole. In algae and cyanobacteria, in order to meet the ATP and NADPH requirements of this process, eight photo-acts, four each in photosystems I and II, are needed. Because 1 mole of photons is equivalent to approximately 47 kcal, the efficiency of photosynthesis is about $114/(47 \times 8)$, or approximately 30 percent.

The key enzyme that determines the rates of the Calvin cycle is **ribulose 1,5-bisphosphate carboxylase**, and thus, the key metabolic step is the reaction of carbon dioxide with ribulose 1,5-bisphosphate. This enzyme occurs on the surfaces of the thylakoid membranes of photosynthetic microorganisms. In the chloroplasts of some eukaryotic microorganisms, it constitutes approximately 15 percent of the total protein. Ribulose 1,5-bisphosphate carboxylase is sub-

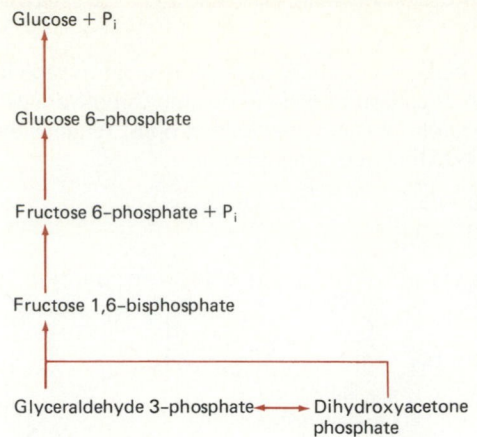

FIGURE 6.3

The conversion of glyceraldehyde 3-phosphate to glucose represents an important anabolic pathway. Once formed, the glucose can be metabolized via other pathways for the production of ATP and the biosynthesis of the other compounds needed by the cell.

ject to allosteric control, providing the mechanism for regulating the rates of carbon dioxide fixation via the Calvin cycle. The rates of reaction of CO_2 plus ribulose 1,5-bisphosphate to form 3-phosphoglycerate are increased by NADPH, an allosteric activator of ribulose 1,5-bisphosphate carboxylase. Thus, when the light reactions are generating large amounts of NADPH, the dark reactions of photosynthesis are stimulated. When reducing power is not available, the Calvin cycle ceases to function. Ribulose 1,5-bisphosphate carboxylase also has higher activities at alkaline than at neutral pH values, and the pumping of hydrogen ions across the photosynthetic membrane that occurs during the light reactions of photosynthesis raises the pH in the stroma of chloroplasts where the Calvin cycle occurs within eukaryotic photosynthetic organisms. Similar hydrogen ion pumping in photosynthetic and chemolithotrophic prokaryotes likewise increases the rate of carbon dioxide fixation by raising the pH so that ribulose 1,5-bisphosphate carboxylase functions optimally. By having the regulation of the Calvin cycle occur at the first metabolic step, the cell efficiently conserves its metabolic energy and reducing power.

The C_4 Pathway for the Fixation of Carbon Dioxide

While the Calvin cycle is the primary pathway for carbon dioxide fixation in most autotrophic microorganisms, there are other metabolic pathways in which carbon dioxide is converted to organic molecules. In fact, carbon dioxide fixation is a universal reaction of all living systems. A common pathway for the fixation of CO_2 in both heterotrophs and autotrophs is the **C_4 pathway**, so designated because the product formed via this pathway, oxaloacetate, is a 4-carbon molecule. In this metabolic pathway, pyruvate or phospho-

FIGURE 6.4

The reaction of pyruvate or phosphoenolpyruvate with CO_2 forms oxaloacetate. This reaction is critical for replenishing the intermediates of the Krebs cycle that are used up in biosynthesis.

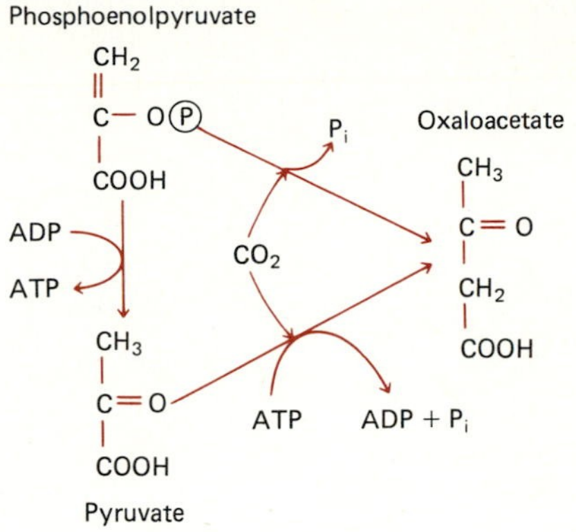

Discovery Process

Winogradsky discovered chemoautotrophs in the 1880s; since then, it has been known that autotrophic bacteria assimilate CO_2. However, not until 1935 was it accepted that heterotropic forms of life assimilate CO_2 or that CO_2 plays an important role as a chemical reactant in all heterotrophic organisms. The first experimental evidence of heterotrophic assimilation of CO_2 to form carbon-to-carbon linkages was presented by C. H. Werkman and H. G. Wood. In working with the propionic acid bacterium *Propionibacterium*, they found that the CO_2 present in the calcium carbonate, which had been added to neutralize acidity, was transformed into organic carbon. These results showed that nonphotosynthetic, typically heterotrophic organisms were able to assimilate CO_2. The initial findings were based on carbon balances determined by meticulous gravimetric analyses. Conclusive proof was provided in 1941, when the radioisotope technique showed the role CO_2 plays in heterotrophic metabolism. Werkman demonstrated that many heterotrophic bacteria, including anaerobes and aerobes, protozoa, fungi, yeasts, and animal tissues, assimilate CO_2. It is unlikely that life could be possible in the absence of available CO_2 because bacteria, as shown by Werkman and S. J. Ajl in 1949, and protozoa, as shown by O. Rahn in 1941, die when deprived of CO_2 for several hours. It appears that all forms of life assimilate CO_2 and that this assimilation is an essential physiological function providing for the synthesis of indispensable metabolic intermediates. This principle has been used in the search for life on other planets; the Viking-Mars lander was equipped with a system for detecting the conversion of $^{14}CO_2$ to organic compounds.

enolpyruvate—metabolites of the glycolytic pathway—react with carbon dioxide to form oxaloacetate, an intermediate metabolite of the Krebs cycle (Figure 6.4). The oxaloacetate formed in this pathway can go into amino acid and nucleic acid biosynthesis, which will be discussed later in this chapter. Whereas all organisms fix carbon dioxide as part of their metabolism, heterotrophic organisms are unable to form a significant portion of their carbon skeleton from CO_2 fixation; therefore, they still depend on organic compounds as substrates for forming macromolecules.

6.3 *Carbohydrate Biosynthesis*

Gluconeogenesis

Having seen that autotrophs can synthesize carbohydrates from carbon dioxide, we can now examine the other routes by which carbohydrates are made. The biosynthesis of carbohydrates is essential because carbohydrates comprise portions of the macromolecules of the cell, including DNA and RNA. The biosynthesis of glucose from noncarbohydrate molecules, **gluconeogenesis**, involves the conversion of a substrate to pyruvate, which is then converted to glucose (Figure 6.5). Gluconeogenesis is important because it allows microorganisms to convert noncarbohydrates to essential carbohydrate molecules. Amino acids derived from proteins, for example, can be converted to pyruvate or phosphoenolpy-

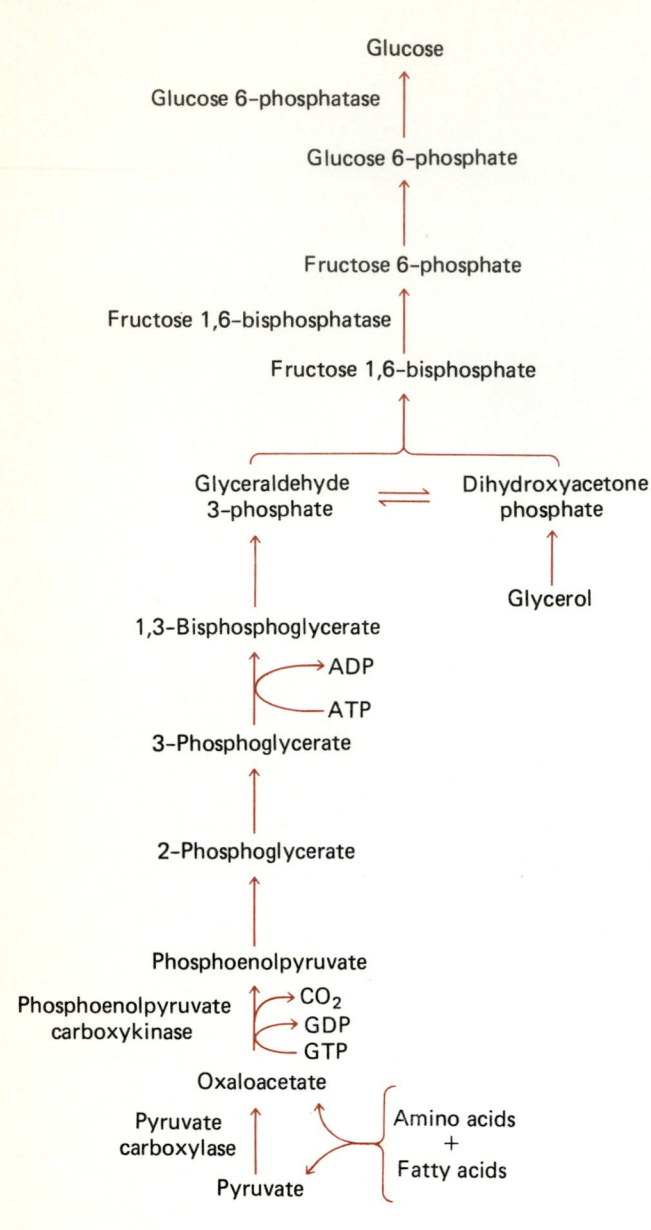

FIGURE 6.5

The conversion of noncarbohydrate substrates, e.g., amino acids, to carbohydrates, e.g., glucose, is accomplished via a gluconeogenic pathway.

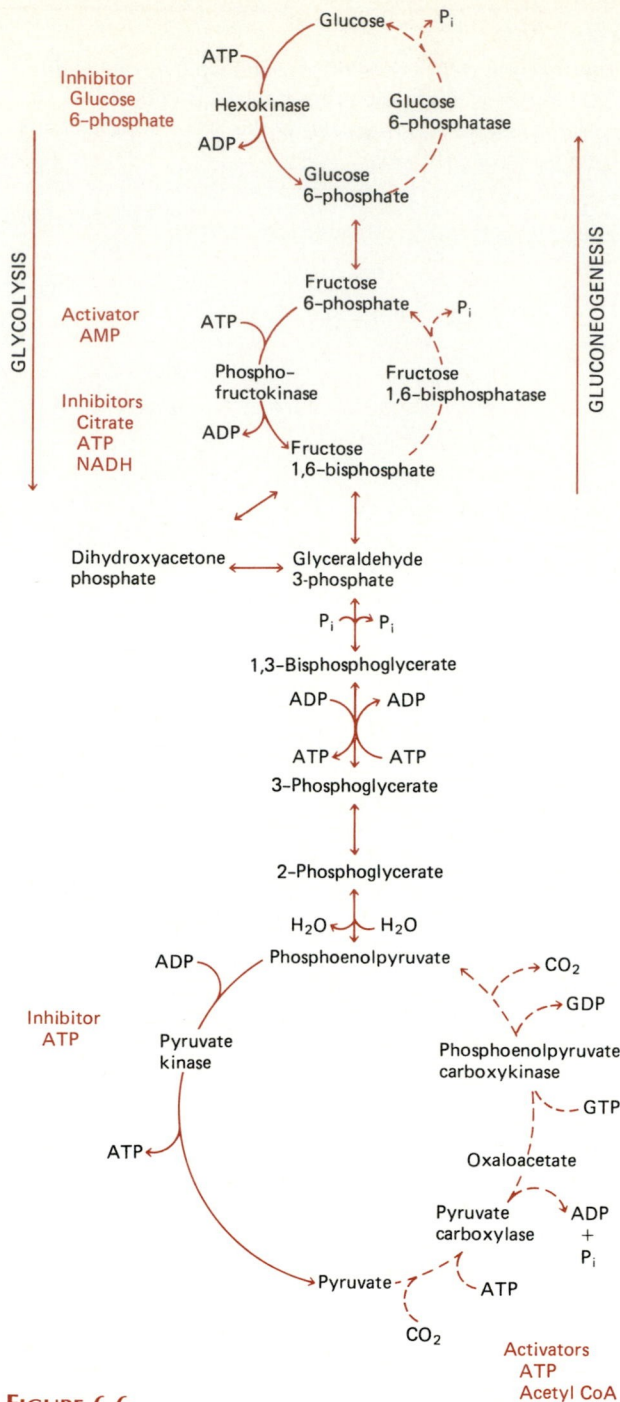

FIGURE 6.6

The gluconeogenic and glycolytic pathways have the same intermediates but represent two distinct pathways because of irreversible reactions in each that are catalyzed by different enzymes.

ruvate, which are intermediary metabolites of the gluconeogenic pathway. Similarly, lipids in some organisms can be broken down into the 3-carbon intermediates of the gluconeogenic pathway. The glucose synthesized in this pathway can be fed into the pentose phosphate shunt to supply 5-carbon carbohydrate molecules for the synthesis of molecules, such as nucleic acid nucleotides.

The pathway of gluconeogenesis effectively reverses the flow of carbon occurring during glycolysis, and the intermediary metabolites of gluconeogenesis are identical to those of the glycolytic pathway (Figure 6.6). While the intermediates are the same, the pathways are actually different. In each direction there is a critical enzymatic step that is irreversible so that a particular enzyme catalyzes the unidirectional flow of carbon. For example, during Embden-Meyerhof glycolysis the conversion of fructose 6-phosphate to fructose 1,6-bisphosphate, which is an irreversible reaction, is catalyzed by the enzyme phosphofructokinase; during gluconeogenesis, the conversion of fructose 1,6-bisphosphate to fructose 6-phosphate, also an irreversible reaction, is catalyzed by the enzyme fructose 1,6-bisphosphatase. These enzymes have different allosteric inhibitors that regulate the direction of carbon flow. Biosynthesis of carbohydrates is favored when the cell has an adequate supply of ATP, whereas catabolism of carbohydrates is favored when ATP concentrations are relatively low.

Because gluconeogenesis requires the input of carbon skeletons with three carbon atoms, such as occurs in a pyruvate molecule, fatty acids cannot be converted directly into carbohydrates via the pathways that have already been discussed. This is because fatty acid metabolism forms acetyl groups that contain only two carbon atoms. In order to permit the flow of carbon from fatty acids to carbohydrates, some microorganisms have an additional pathway, the glyoxylate cycle (Figure 6.7). The **glyoxylate cycle** is actually a shunt across the Krebs cycle. In the glyoxylate cycle, malate (a C-4 compound) is formed from the reaction of glyoxylate (C-2) and acetyl coenzyme A (CoA). The malate is converted via oxaloacetate to phosphoenolpyruvate—an intermediary metabolite of the gluconeogenic pathway—thereby linking the pathway of fatty acid metabolism with the pathway of carbohydrate metabolism. The formation of a 6-carbon carbohydrate from a fatty acid requires the use of four acetyl CoA molecules, two carbons of which are released as carbon dioxide when oxaloacetate is transformed into two phosphoenolpyruvate ($2 \times C$-3) molecules during the synthesis of the carbohydrate.

Like lipids, proteins can be converted to carbohydrates. Proteins are normally broken down extracellularly by proteases, enzymes that catalyze the formation of small peptides and individual amino acids that can enter the cell. The amino acids that are formed can be deaminated by removal of the amino group to form a carboxylic acid. After deamination, the organic acids formed from the various families of amino acids can enter the Krebs cycle or glycolytic pathways, and thus the breakdown of proteins can establish a flow of carbon to form carbohydrate molecules.

Biosynthesis of Polysaccharides

The carbohydrates formed by gluconeogenesis may also be converted to larger polysaccharide molecules. Some of these polysaccharide molecules, such as starch or glycogen, may store carbon and energy within the cell. These polysaccharides can later be broken down to form glucose 1-phosphate or glucose 6-phosphate, and thus enter the metabolic pathways of glycolysis and the pentose phosphate shunt. The glucose formed in gluconeogenesis can also be used to synthesize the polysaccharides that make up the cell walls of microorganisms.

Glucose is a precursor for *N*-acetylglucosamine and *N*-acetylmuramic acid, the repeating polysaccharide units of the glycan portion of the murein (peptidoglycan) layer of the bacterial cell wall. The formation of *N*-acetylglucosamine involves the conversion of glucose to fructose 6-phosphate and the subsequent reactions with glutamic acid and acetyl CoA. To form *N*-acetylmuramic acid, *N*-acetylglucosamine reacts with uridine triphosphate (UDP), to form *N*-acetylglucosamine-uridine diphosphate (UDP), which then reacts with phosphoenolpyruvate to form *N*-acetylmuramic acid-UDP. These reactions form the necessary units of the glycan portion of the bacterial wall.

The actual synthesis of the peptidoglycan layer of the bacterial cell wall occurs in several stages (Figure 6.9, page 155). Initially, a peptide chain is added to *N*-acetylmuramic acid to form a monosaccharide peptide unit. A molecule of *N*-acetylglucosamine is then added to the *N*-acetylmuramic acid peptide unit to form a disaccharide peptide. An additional peptide unit, which forms the cross-linkage of the cell wall, may also be added to the disaccharide unit. The disaccharides are then transferred to a growing peptidoglycan molecule. The details of the actual biosynthetic pathways

FIGURE 6.7
The glyoxylate cycle is a shunt across the normal Krebs cycle. This pathway is needed to maintain the Krebs cycle intermediates when lipids are metabolized. It also provides a way of converting lipids into carbohydrates.

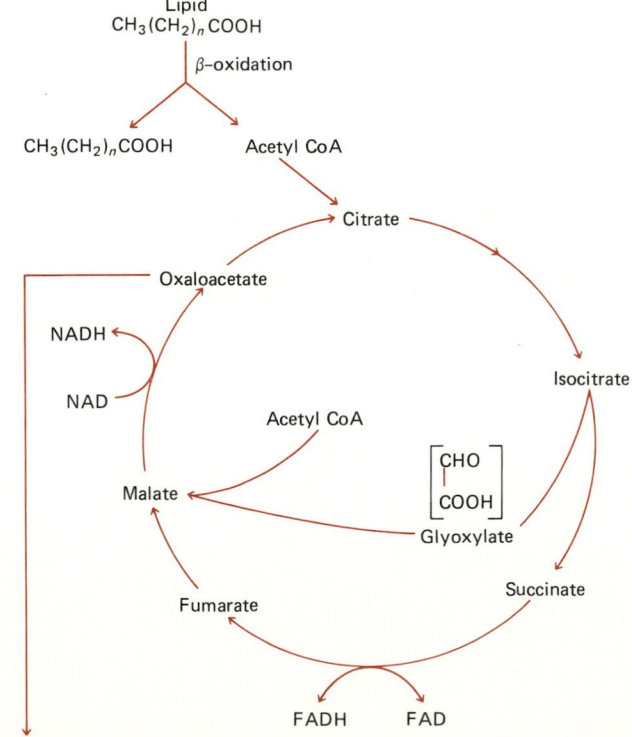

ANALYTICAL PROCESS

Regulating the Direction of Carbon Flow: Catabolism versus Anabolism

Because the same intermediary metabolites in these amphibolic pathways serve for both catabolism and anabolism, there must be mechanisms to regulate the direction of carbon flow, that is, some major differences between the anabolic and catabolic functions of these pathways that determine whether the flow of carbon is from large molecules to small ones or whether larger molecules are synthesized from smaller ones. Indeed, while the intermediates are the same, the catabolic and anabolic functions of amphibolic pathways are distinct. Elucidating the controlling mechanisms has provided insight into the fundamental functioning of a cell at the molecular level.

One difference is the separation of coenzymes that are involved in catabolic and anabolic pathways. Nicotinamide adenine dinucleotide (NAD⁺) is involved in catabolic pathways, and the related coenzyme nicotinamide adenine dinucleotide phosphate (NADP⁺) is involved in anabolic pathways. The use of different coenzymes enables the cell to differentiate the direction of carbon flow through the same intermediary metabolites, making it possible for reductive biosynthesis to occur at the same time that the cell is generating ATP through glycolysis.

Of greater importance in regulating the flow of carbon is the fact that in both the anabolic and catabolic pathways there are irreversible steps that are catalyzed by different enzymes (Table 6.1). Thus, carbon cannot simply flow in both directions. These enzymes catalyzing the key reactions generally have allosteric effectors that inhibit or accelerate the rates of the metabolic reactions catalyzed by the enzymes. In this manner the cell is able to fine tune the flow of carbon, balancing catabolic activities involved in generating ATP and anabolic activities involved in the synthesis of essential macromolecules. In particular, feedback inhibition, whereby an end product of a metabolic pathway acts as an allosteric inhibitor of an enzyme catalyzing a reaction near the beginning of that pathway, often plays a critical role in the regulation of a cell's metabolism.

A major factor controlling the direction of carbon flow through amphibolic pathways is the energy status of the cell, a measure of which is known as the **energy charge**. The energy charge describes the relative proportions of ATP, ADP, and adenine monophosphate (AMP) in the cell and is defined as follows:

$$\text{Energy charge} = \frac{[\text{ATP}] + \frac{1}{2}[\text{ADP}]}{[\text{ATP}] + [\text{ADP}] + [\text{AMP}]}$$

As a rule, ATP-generating pathways are inhibited by a high energy charge, whereas ATP-utilizing pathways are stimulated by a high energy charge (Figure 6.8). The reason for this is that ATP generally acts as an allosteric inhibitor of a key enzyme in an energy-generating pathway, as we saw for the enzyme phosphofructokinase, which catalyzes the conversion of fructose 6-phosphate to fructose 1,6-bisphosphate, when we considered the regulation of the rates of the Embden-Meyerhof pathway in Chapter 5. In contrast, AMP is an allosteric inhibitor of fructose 1,6-bisphosphatase, which catalyzes the reverse reaction that converts fructose 1,6-bisphosphate to fructose 6-phosphate in the energy-requiring gluconeogenic pathway. Thus, when the catabolic ATP-generating pathway is active, the reverse anabolic ATP-requiring pathway is inhibited, and vice versa, so that the flow of carbon through an amphibolic pathway at any point in time is unidirectional, regulated by differences in enzymes, their allosteric inhibitors, and the energy charge of the cell.

FIGURE 6.8

The effects of energy charge on rates of ATP-generating (catabolic) and ATP-utilizing (anabolic) pathways. By having ATP and cyclic AMP serve as allosteric effectors for key enzymes, the flow of carbon through identical intermediates of biosynthetic and catabolic pathways is controlled. When more ATP is needed, the energy charge is low and the rate of ATP-generating pathways is accelerated. When the cell has sufficient ATP, the rates of biosynthetic pathways are relatively high and the rates of catabolic pathways relatively low.

TABLE 6.1 Some Key Reactions in the Glycolytic and Gluconeogenic Pathways and the Enzymes that Catalyze Those Reactions

Glycolysis Reaction (enzyme)	Gluconeogenesis Reaction (enzyme)
Glucose → glucose 6-phosphate (hexokinase)	Glucose 6-phosphate → glucose (glucose 6-phosphatase)
Fructose 6-phosphate → fructose 1,6-bisphosphate (phosphofructokinase)	Fructose 1,6-bisphosphate → fructose 6-phosphate (fructose 1,6-bisphosphatase)
Phosphoenolpyruvate → pyruvate (pyruvate kinase)	Pyruvate → phosphoenolpyruvate (pyruvate carboxylase/ phosphoenolpyruvate carboxykinase)

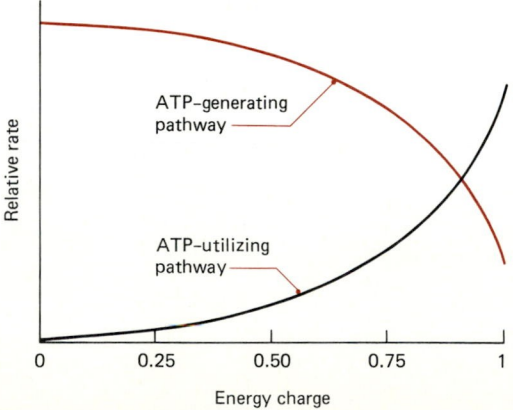

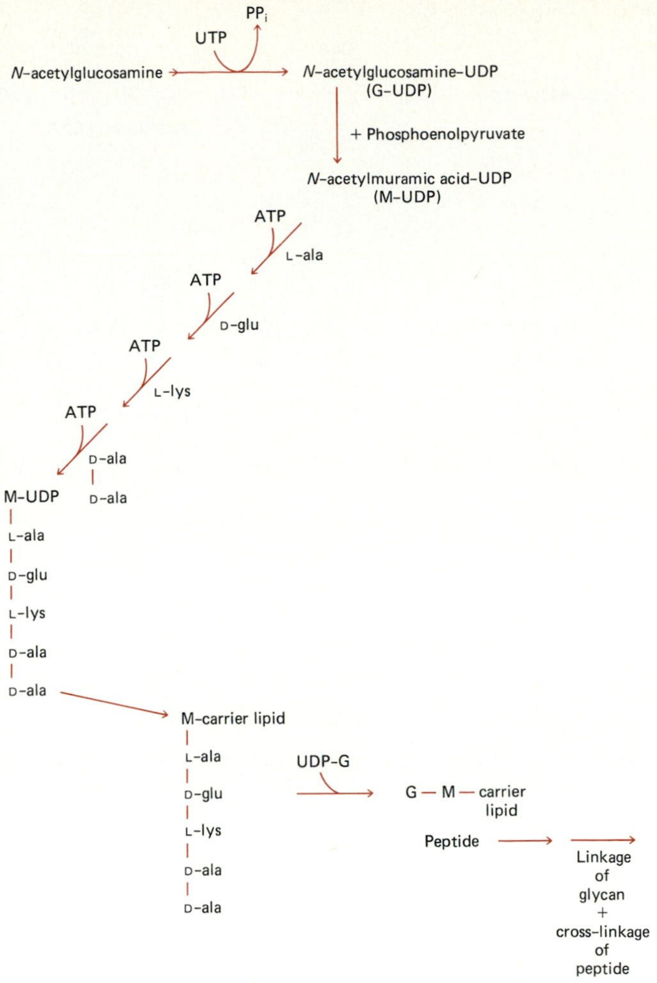

FIGURE 6.9
The synthesis of peptidoglycan by bacteria is important to provide the backbone material of the cell wall. To form murein, amino acids must be added to N-acetyl muramic acid, a repeating and alternating glycan layer of N-acetyl muramic acid and N-acetyl glucosamine must be formed and the peptide chains must be cross-linked.

involved in the synthesis of the peptidoglycan layer are quite complex, involving the activation of the carbohydrate molecules by UTP and the use of a membrane phospholipid (undecaprenol phosphate) as a carrier during biosynthesis of the wall. The synthesis of the bacterial cell wall can be viewed as occurring in four stages: (1) synthesis of water-soluble carbohydrate precursors; (2) attachment to a membrane lipid; (3) formation of linear polymers outside the cytoplasmic membrane; and (4) cross-linking of the polymers.

Because the assembly steps occur outside the cell, it is important that the reactions be exergonic, as ATP is not available outside the cell to drive endergonic reactions. The activation of the cell wall subunits, using UTP, provides a mechanism for ensuring that the assembly reactions are exergonic and that ATP is not needed. In the case of the Gram-negative cell wall, the synthesis and addition of lipopolysaccharide (LPS) also involves undecaprenol-phosphate as a carrier. The synthesis of the LPS component of the cell wall occurs at the cytoplasmic membrane and involves the initial addition of fatty acids to a glucosamine disaccharide phosphate, followed by successive sugar additions. The assembled LPS component is then transferred to the cell wall structure.

6.4 Lipid Biosynthesis

Fatty Acid Biosynthesis

Besides carbohydrates, all cells require lipids for incorporation in membranes. These lipids contain fatty acids of varying composition. The synthesis of fatty acids proceeds by the sequential addition of two carbon units derived from acetyl CoA (Figure 6.10). During fatty acid biosynthesis, the reactants are bound to a protein known as the *acyl carrier protein*. A key intermediate in the synthesis of fatty acids is malonyl CoA, which is formed from the reaction of acetyl CoA with carbon dioxide that requires ATP and biotin.[2] It is the 3-carbon malonyl CoA unit that successively contributes two carbon units for the elongation of the fatty acid. Carbon dioxide is released when the 3-carbon malonyl-acyl carrier protein molecule contributes a 2-carbon unit for the biosynthesis of fatty acids. The synthesis of a C_{16} saturated

fatty acid, a common component of membrane phospholipids, can be described according to the equation

$$8 \text{ acetyl CoA} + 7 \text{ ATP} + 14 \text{ NADPH} \rightarrow$$
$$\text{palmitate } (C_{16} \text{ fatty acid}) + 14 \text{ NADP}^+$$
$$+ 8 \text{ CoA} + 6 \text{ H}_2\text{O} + 7 \text{ ADP} + 7 \text{ P}_i.$$

It is thus apparent that the synthesis of fatty acids requires both energy in the form of ATP and reducing power in the form of NADPH. Some of the required NADPH for this reaction comes from the pentose phosphate pathway. Additionally, NADPH can be formed by the reaction

$$\text{NADP}^+ + \text{NADH} \rightarrow \text{NADPH} + \text{NAD}^+.$$

Biosynthesis of Poly-β-hydroxybutyric Acid

The pathway for the synthesis of poly-β-hydroxybutyric acid, a common storage product of bacteria, is similar to the

---

[2]The requirement for biotin in this reaction is one reason that many organisms need biotin in trace quantities as a growth factor.

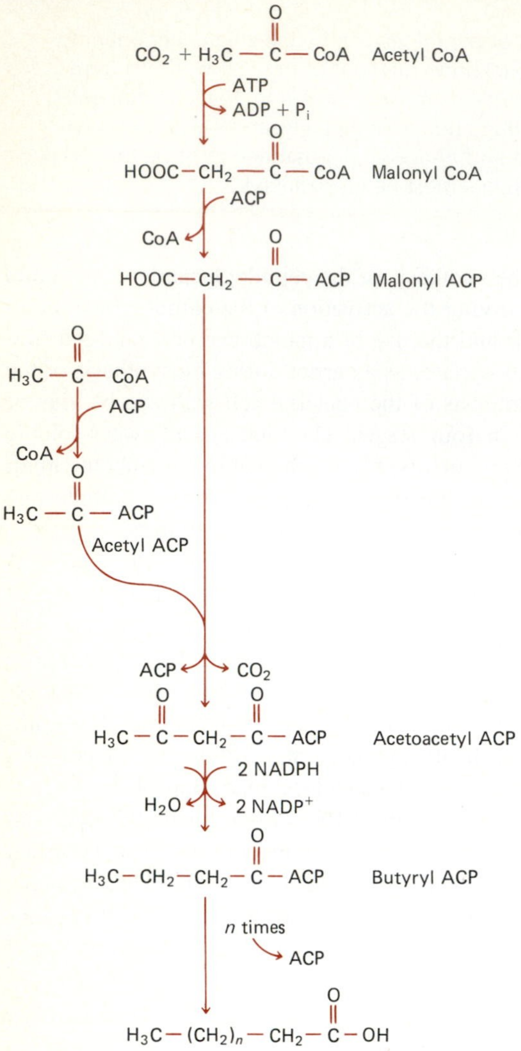

FIGURE 6.10

The synthesis of fatty acids involves two carbon additions from acetyl CoA and an acyl carrier protein. Fatty acid biosynthesis is not a simple reversal of β-oxidation used for the catabolism of fatty acids.

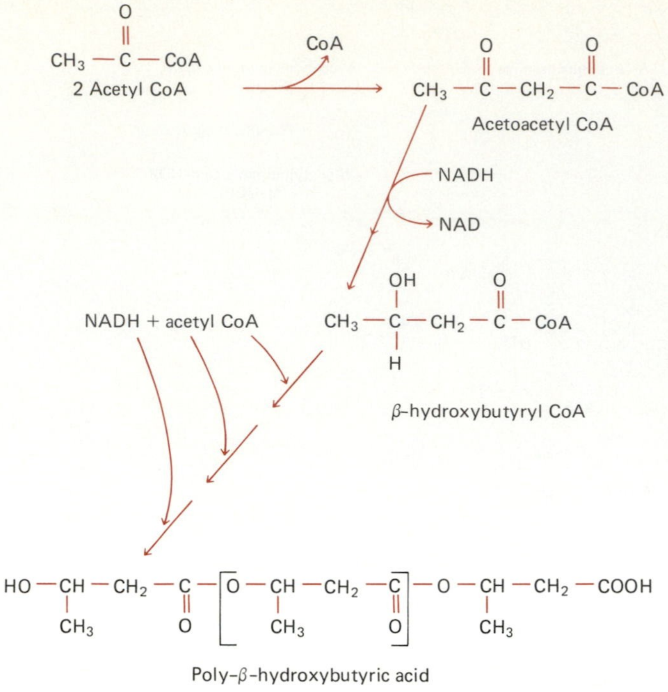

FIGURE 6.11

The synthesis of poly-β-hydroxybutyrate is used by bacteria to store carbon and energy reserves. This is an unusual biosynthetic pathway in that NADH, rather than NADPH, is used as a source of reducing power.

pathway for fatty acid biosynthesis (Figure 6.11). Acetyl CoA reacts to form acetoacetyl CoA (a 4-carbon derivative of CoA that can be reduced with NADH to form β-hydroxybutyryl-CoA). Repetitive sequential addition of acetyl CoA results in chain length elongation, and subsequent removal of the CoA portion of the molecule forms the poly-β-hydroxybutyric acid, which can accumulate in large amounts in bacteria. Interestingly, unlike other biosynthetic reactions, the formation of poly-β-hydroxybutyrate uses the coenzyme NADH rather than NADPH.

Biosynthesis of Phospholipids

The synthesis of phospholipids is required by both prokaryotic and eukaryotic organisms for incorporation into membranes. The formation of phospholipids involves the addition of fatty acids to glycerol phosphate. The glycerol 3-phosphate reacts with acyl CoA to form phosphatidate, a common intermediary metabolite in the synthesis of phospholipids and triglycerides (Figure 6.12). The fatty acid-acyl carrier protein molecules react with the phosphatidate to form elongated fatty acid derivatives.

Biosynthesis of Sterols

Besides phospholipids, the membranes of eukaryotic cells contain sterols, such as cholesterol, which are made up of repeating units of the unsaturated hydrocarbon isoprene (Figure 6.13). Isoprenoid hydrocarbons are synthesized from activated acetyl CoA molecules. The synthesis of isoprenoid hydrocarbons, which differs from fatty acid biosynthesis in the mechanism of chain elongation, comes from reactions that initially form a 6-carbon compound that is then decarboxylated. The synthesis of mevalonic acid from 3-hydroxy-3-methylglutaryl-CoA derived from the reaction of acetyl CoA with acetoacetyl CoA is the key step in the formation of cholesterol. The activity of the enzyme 3-hydroxy-3-methylglutaryl-CoA reductase regulates the rates of cholesterol biosynthesis.

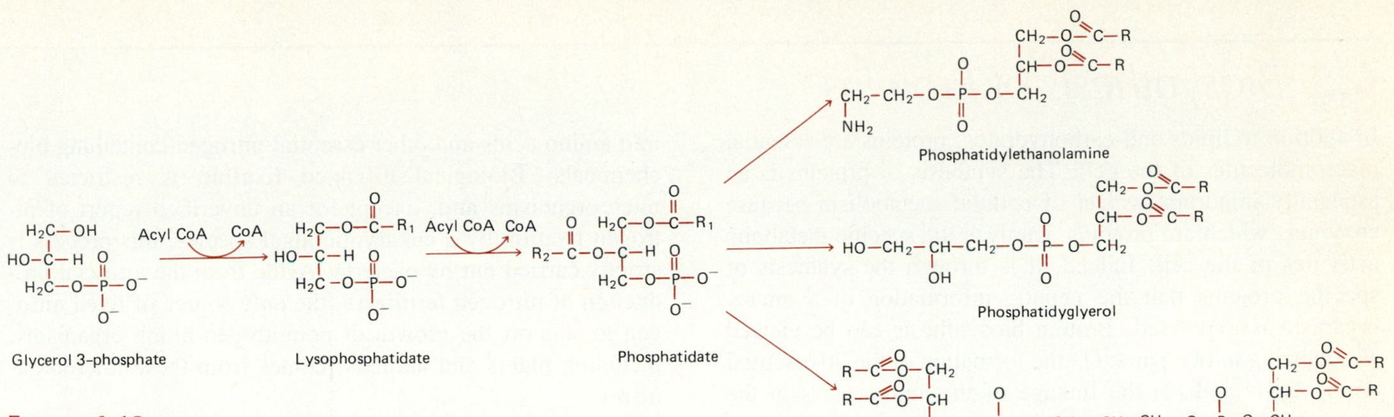

FIGURE 6.12

The formation of phosphatidate and lipids is necessary for the formation of cellular membranes.

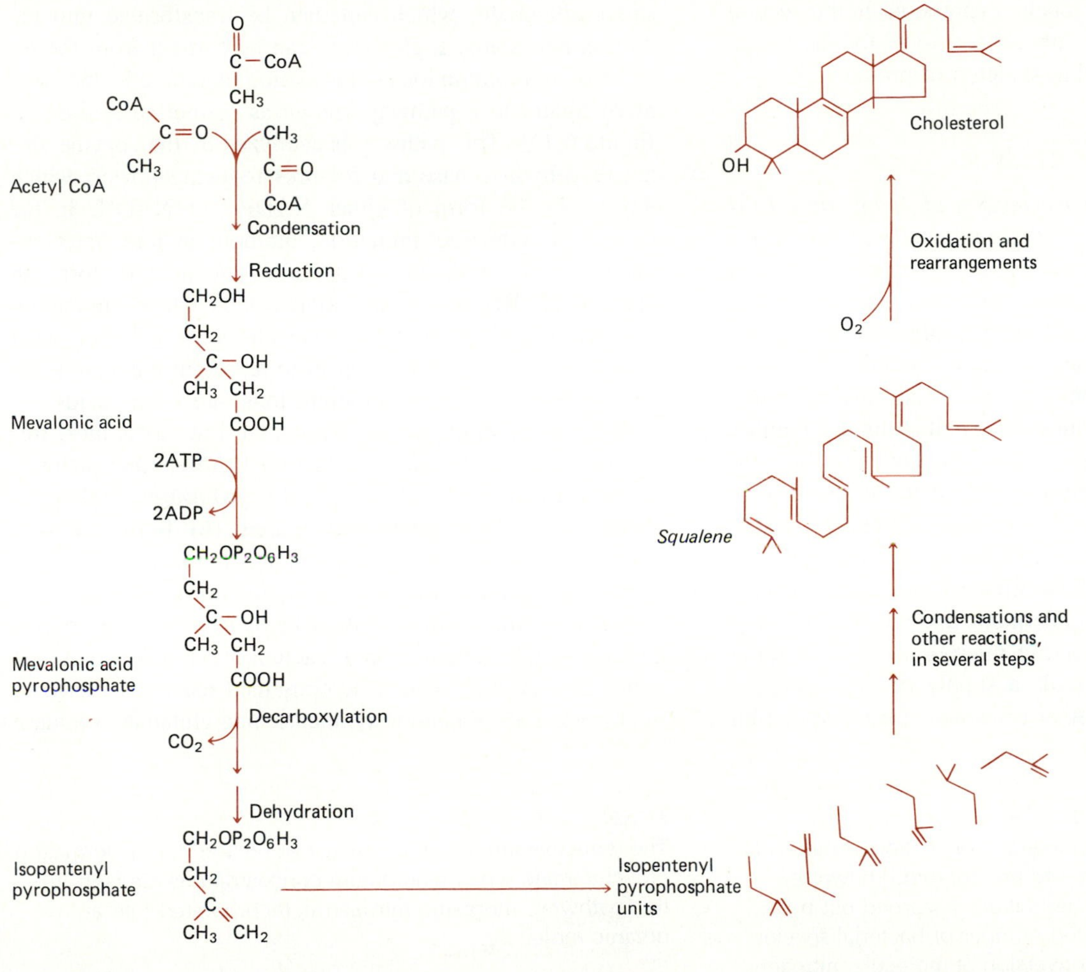

FIGURE 6.13

The synthesis of cholesterol involves the formation and subsequent condensation of isopentenyl pyrophosphate units.

The biosynthesis of cholesterol is exemplary of the fundamental mechanisms of long chain carbon skeleton formation from 5-carbon isoprenoid units. In addition to forming the backbone of sterols, isoprenoid hydrocarbons form the backbone of carotenoids, the brightly colored red, orange, and yellow pigment compounds of some microorganisms, and the phytyl portion of the chlorophyll molecules, which are essential for photosynthesis.

6.5 Biosynthesis of Proteins

In addtion to lipids and carbohydrates, proteins are essential macromolecules of the cell. The synthesis of proteins is an especially important aspect of cellular metabolism because enzymes, which are proteins, catalyze the specific metabolic activities of the cell. Indeed, it is through the synthesis of specific proteins that the genetic information of a microorganism is expressed. Protein biosynthesis can be viewed as occurring in two parts: (1) the formation of the 20 essential amino acids and (2) the linkage of the amino acids in the proper sequence to establish the primary structure of the protein molecule. The sequential ordering of amino acids to form the primary protein structure is under the direct control of the genetic informational macromolecules; as such, that aspect of protein biosynthesis will be covered in Chapter 7 as part of the discussion of genetic expression. In this section we shall consider the biosynthetic pathways for the amino acids that establish the carbon skeleton of proteins.

Biosynthesis of Amino Acids

Nitrogen Fixation and the Formation of Ammonium Ions

Unlike most lipid and carbohydrate molecules, amino acids contain nitrogen. The incorporation of inorganic nitrogen into organic molecules is needed for the synthesis of amino acids. Although molecular nitrogen is abundant in the atmosphere, most organisms are unable to use this as a source of nitrogen for incorporation into amino acids and the nitrogen-containing macromolecules of the cell. A limited number of prokaryotic microorganisms, however, have the ability to transform molecular nitrogen into ammonium nitrogen (Figure 6.14). These microorganisms possess the enzyme **nitrogenase**, which enables them to carry out nitrogen-fixing metabolic activities. The transformation of molecular nitrogen into ammonium nitrogen is endergonic and requires reducing power and ATP. Nitrogen-fixing microorganisms are quite important because they provide a supply of fixed nitrogen that can be assimilated by other organisms for incorporation

into amino acids and other essential nitrogen-containing biochemicals. Biological **nitrogen fixation** is restricted to microorganisms and, except for an unverified report of nitrogen fixation by a eukaryotic algal species, this process is strictly carried out by bacteria. Aside from the artificial production of nitrogen fertilizers, the only source of fixed nitrogen to support the growth of nonnitrogen-fixing organisms, including plants and animals, comes from these microorganisms.

Incorporation of Ammonium Ions into Amino Acids

There are two pathways for the incorporation of inorganic ammonium nitrogen into organic amino acids. In both of these pathways, ammonium ions are assimilated to form the amino acid L-glutamate, which can then be transformed into the other amino acids. L-glutamate can be formed from the reaction of ammonium ions with α-ketoglutarate, a Krebs cycle intermediate, in a pathway known as **reductive amination** (Figure 6.15). This pathway is catalyzed by the enzyme glutamate dehydrogenase and requires reducing power, which may be in the form of either NADPH or NADH. In the process of reductive amination, ammonium ions are combined directly with an α-ketocarboxylic acid to form an amino acid. Reductive amination occurs when ammonium ions are in relatively high concentrations and, for most microorganisms, does not appear to represent the main pathway for incorporating ammonium ions into amino acids.

When ammonium concentrations are low, as is most frequently the case, microorganisms resort to another pathway for the formation of L-glutamate, the **glutamine synthetase/glutamate synthase pathway** (Figure 6.16). In this pathway L-glutamate reacts with ammonium ions to form L-glutamine, a reaction that requires energy in the form of ATP. The L-glutamine then reacts with α-ketoglutarate to form two molecules of L-glutamate in a reaction that requires reducing power as NADPH. Thus, the combined reactions catalyzed by the enzymes glutamine synthetase and glutamate synthase

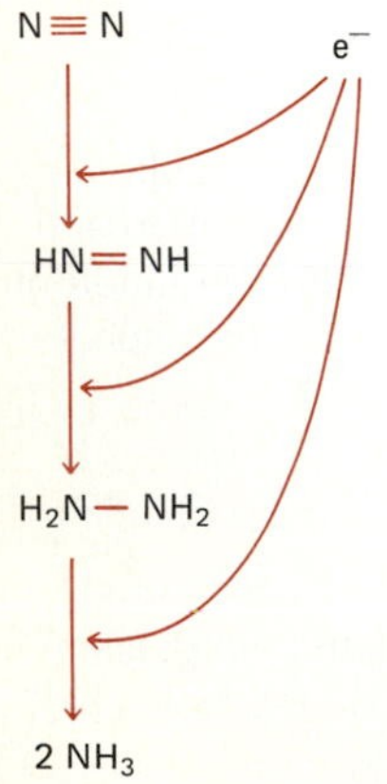

FIGURE 6.14

The biological conversion of molecular nitrogen to ammonium, known as *nitrogen fixation*, is carried out by a restricted number of bacterial species. This conversion of molecular nitrogen to a fixed form that can be assimilated by other organisms often governs rates of productivity of plants, animals, and other microorganisms.

FIGURE 6.15

The reductive amination pathway can be used for the formation of L-glutamate when ammonium concentrations are high. In this pathway, inorganic nitrogen is incorporated into an organic molecule.

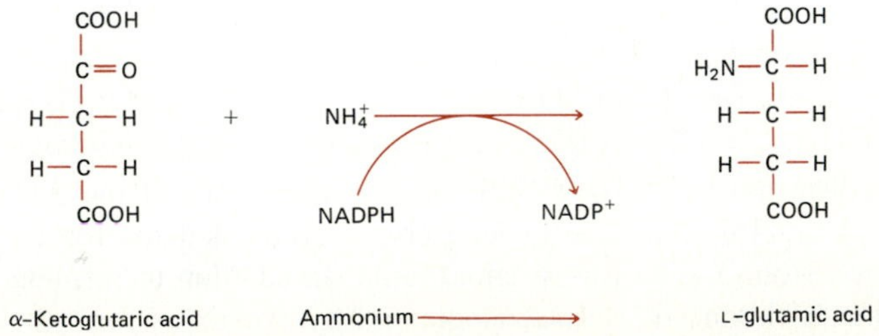

α-Ketoglutaric acid   Ammonium ──────────►   L-glutamic acid

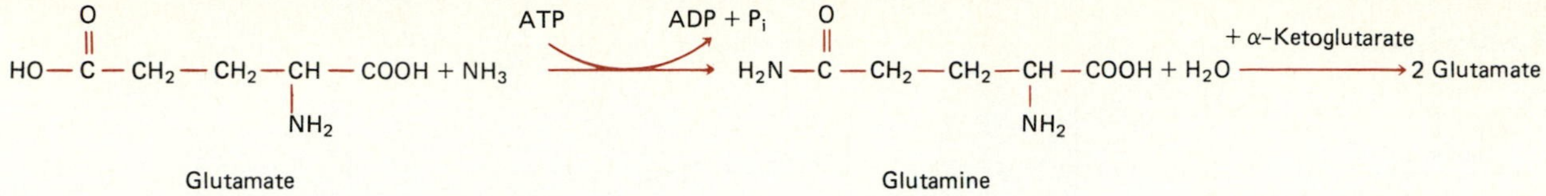

$$\underset{\text{Glutamate}}{HO - \overset{\overset{\displaystyle O}{\|}}{C} - CH_2 - CH_2 - \underset{\overset{\displaystyle |}{NH_2}}{CH} - COOH} + NH_3 \xrightarrow[\quad]{\overset{\displaystyle ATP \quad ADP + P_i}{\quad}} \underset{\text{Glutamine}}{H_2N - \overset{\overset{\displaystyle O}{\|}}{C} - CH_2 - CH_2 - \underset{\overset{\displaystyle |}{NH_2}}{CH} - COOH} + H_2O \xrightarrow[\quad]{+ \alpha\text{-Ketoglutarate}} 2 \text{ Glutamate}$$

FIGURE 6.16

When the ammonium concentration is low, the conversion of inorganic ammonium ions to the amino acid L-glutamate occurs in a two-step process, the glutamine synthetase/glutamate synthase pathway. Glutamine synthase is sometimes called *GOGAT*, an acronym for its alternative name, *glutamine-oxoglutarate aminotransferase*. This is the principal pathway used by bacteria for the formation of L-glutamate from the Krebs cycle intermediate α-ketoglutarate.

are described by the equation

$$\alpha\text{-ketoglutarate} + NH_4^+ + NADPH + ATP \rightarrow$$
$$L\text{-glutamate} + NADP^+ + ADP + P_i.$$

The enzyme glutamine synthetase plays a key role in regulating the rates of intermediary metabolism because of the regulatory control it exerts on the flow of nitrogen into amino acids and consequently into proteins and nucleic acids. Glutamine synthetase is subject to cumulative feedback inhibition by each of the products of glutamine metabolism; that is, a series of different inhibitors can act additively to reduce the activity of this enzyme. Inhibitors of glutamine synthetase include tryptophan, histidine, alanine, glycine, carbamoyl phosphate, glucosamine 6-phosphate, cytidine triphosphate (CTP), and AMP. Each of these allosteric inhibitors appears to have its own binding site on the enzyme, and when all eight inhibitors are bound to the enzyme, the activity of glutamine synthetase is virtually shut off. AMP in particular affects the sensitivity of this enzyme to feedback inhibition because the adenylated enzyme—that is, the enzyme with AMP attached—is more susceptible to cumulative feedback inhibition than the deadenylated form of the enzyme.

Biosynthesis of the Major Families of Amino Acids
The assimilation of nitrogen to form the amino acid L-glutamate establishes the basis for the biosynthesis of the other essential amino acids found in protein macromolecules, as well as for the biosynthesis of other essential nitrogen-containing compounds. The 20 amino acids originate from 6 different non-amino acid precursors, and as a result, there are only six biosynthetic families of amino acids (Figure 6.17). L-Glutamate is the parent molecule of one of the amino acid families and can be further metabolized to form the amino acids L-glutamine, L-proline, and L-arginine. L-Glutamate also serves as the nitrogen source for the other amino acids; that is, the amino group of the other amino acids is derived from L-glutamate. The carbon skeletons for the various amino acids come from glycolytic, pentose phosphate, or Krebs cycle pathway intermediate metabolites.

The ability to transfer the amino group of one amino acid to form another amino acid, a process known as **transami-**nation, is essential for the synthesis of all of the amino acids. This process involves specific transaminase enzymes and also requires the coenzyme pyridoxal phosphate (a derivative of vitamin B_6), which binds with the amino group being transferred during the transamination reaction. Glutamate transaminase is the most important of the transaminase enzymes because it catalyzes the transfer of the amino group from L-glutamate to form the parental amino acids of the various amino acid families. L-Glutamate can transfer its amino group to an α-ketocarboxylic acid to form a new amino acid and α-ketoglutarate, the carboxylic acid precursor of L-glutamate. For example, L-glutamate can react with oxaloacetate to form the amino acid L-aspartate (Figure 6.18).

L-Aspartate, made by transamination with L-glutamate, is the parent molecule of another family of amino acids and can be further metabolized to form L-asparagine, L-methionine, L-threonine, L-lysine, or L-isoleucine. Two intermediary metabolites in the conversion of L-aspartate to L-lysine, dihydrodipicolinic acid, and meso-diaminopimelic acid (Figure 6.19) are used in prokaryotic organisms but do not form part of the structures of eukaryotic microorganisms. Diaminopimelic acid is one of the unusual amino acids that forms part of the peptide portion of the peptidolglycan molecule that makes up the bacterial cell wall. Dipicolinic acid occurs uniquely in bacterial endospores.

Transamination reactions can also be utilized to generate the amino acids L-alanine, L-valine, and L-leucine from reactions with pyruvate. The formation of L-alanine involves a single-step transamination reaction that converts pyruvate to L-alanine. Similarly, L-glutamate can react with 3-phosphoglycerate, an intermediate of the glycolytic pathway, to form the amino acid L-serine. Because 3-phosphoglycerate does not have a keto group that can react with the amino group of L-glutamate, the 3-phosphoglycerate must first be oxidized to 3-phosphohydroxypyruvate, a reaction that is coupled with the reduction of NAD^+ to NADH. The initial product formed by the transamination reaction is 3-phosphoserine, and the phosphate group is subsequently removed to yield the final amino acid product of L-serine.

L-Serine is a precursor for the biosynthesis of the amino acids L-glycine and L-cysteine. L-Cysteine is one of the sulfur-containing amino acids, and the transformation of

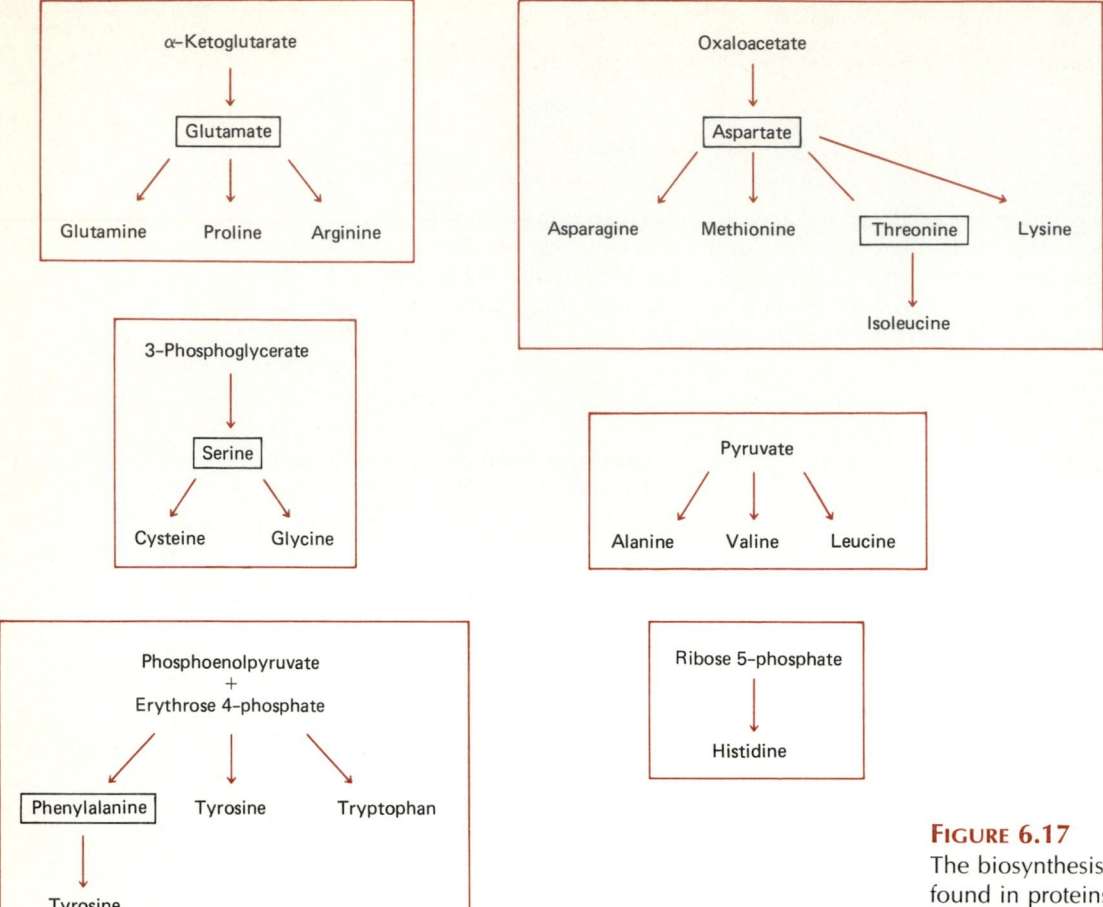

FIGURE 6.17
The biosynthesis of the 20 L-amino acids found in proteins represents the formation of six families of related amino acids.

FIGURE 6.18
Once nitrogen is incorporated into an amino acid, the amino group can be transferred to an α-ketocarboxylic acid to form a new amino acid via transamination. This illustration demonstrates transamination to form L-aspartate.

L-serine to L-cysteine involves a reaction with hydrogen sulfide, which can be derived from the reduction of sulfate (Figure 6.20). The sulfur-containing amino acids are important in determining the secondary structure of a protein. The sulfur-containing amino acids also are often involved in establishing the active site of enzyme molecules.

The formation of the aromatic amino acids L-phenylalanine, L-tyrosine, and L-tryptophan originates with phosphoenolpyruvate (a glycolytic pathway intermediary metabolite) and erythrose 4-phosphate, which is formed via the pentose phosphate cycle. The details of the formation of the aromatic ring structure are relatively complex, involving the intermediate metabolite shikimic acid (Figure 6.21). The

amino group of the amino acids L-phenylalanine and L-tyrosine arise through transamination reactions with L-glutamate; L-tryptophan derives its amino group from a transamination reaction with L-serine.

The formation of L-histidine also is quite complex (Figure 6.22). Histidine and nucleic acid purines arise from a common precursor molecule, ribose 5-phosphate, which is formed by the pentose phosphate cycle. The adenine unit of the molecule ATP provides one nitrogen and one carbon atom for the ring of L-histidine; the other nitrogen atom of the ring comes from the side chain of the amino acid L-glutamine. The amino group of L-histidine comes from a transamination reaction with L-glutamate.

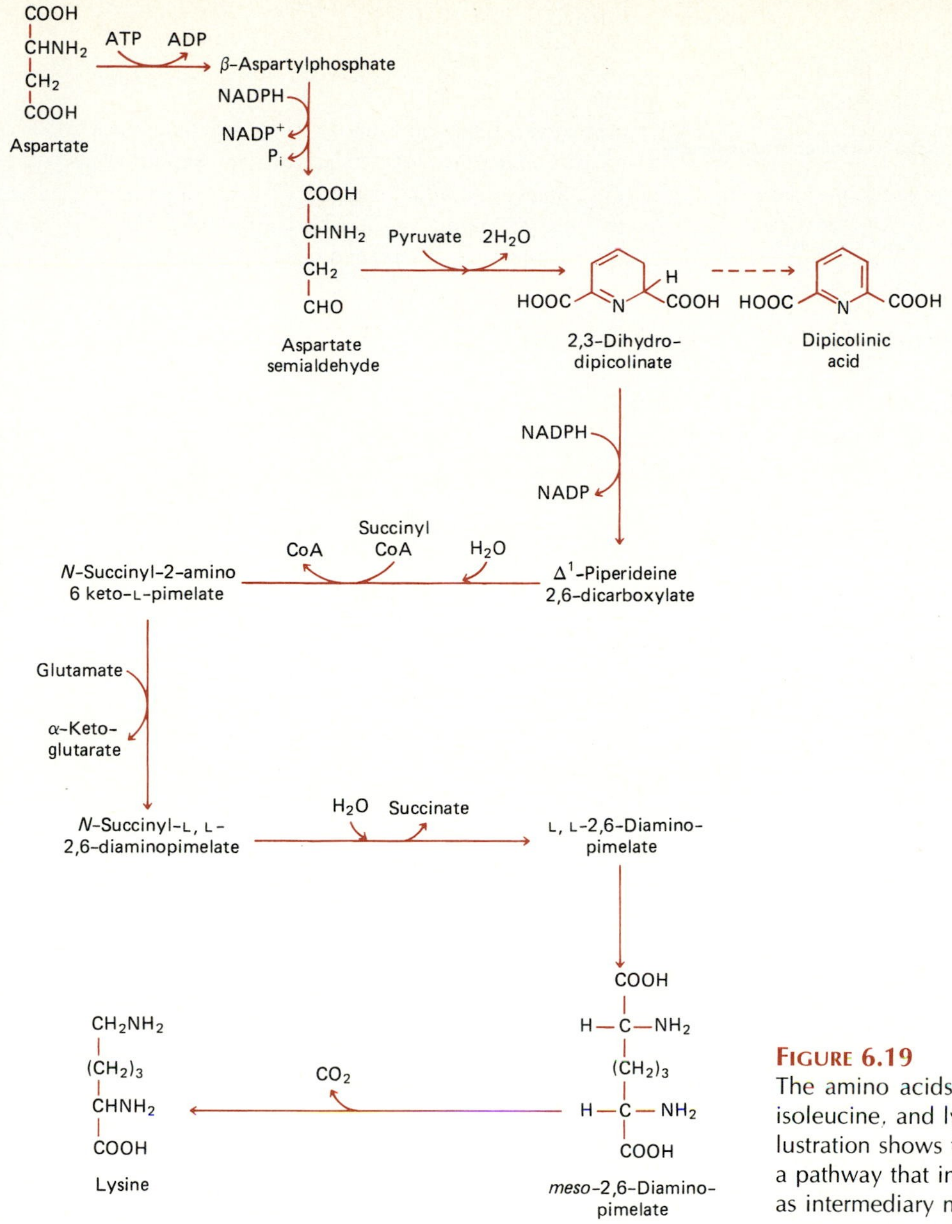

FIGURE 6.19

The amino acids asparagine, methionine, threonine, isoleucine, and lysine are derived from aspartate. This illustration shows the conversion of L-aspartate to L-lysine, a pathway that involves dipicolinate and diaminopimelate as intermediary metabolites.

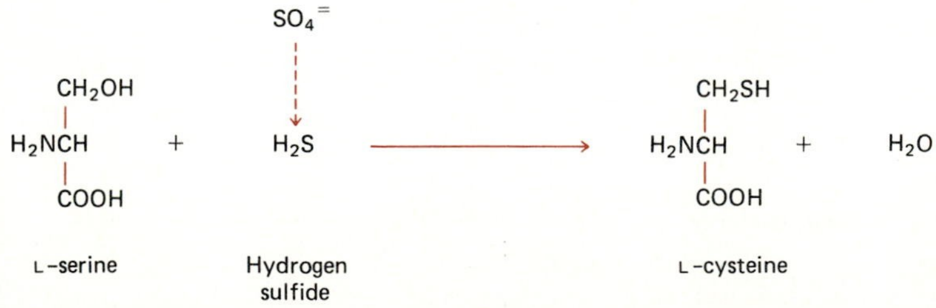

FIGURE 6.20

The formation of L-cysteine, a sulfur-containing amino acid, involves the reaction of L-serine with hydrogen sulfide.

Through these metabolic reactions, the 20 essential amino acids of proteins can be synthesized. The fact that the precursors for the biosynthesis of amino acids are intermediary metabolites of the glycolytic, pentose phosphate, and Krebs cycle pathways establishes a unity of metabolic pathways that permits the flow of carbon between the major biochemical classes. The carbon from carbohydrates and lipids, for example, can be used to generate the carbon skeletons of the 20 essential amino acids and thus can form the backbone of protein molecules. The rates of amino acid biosynthesis are regulated in large part by feedback inhibition. Usually there is a major regulatory step in each of the amino acid biosyn-

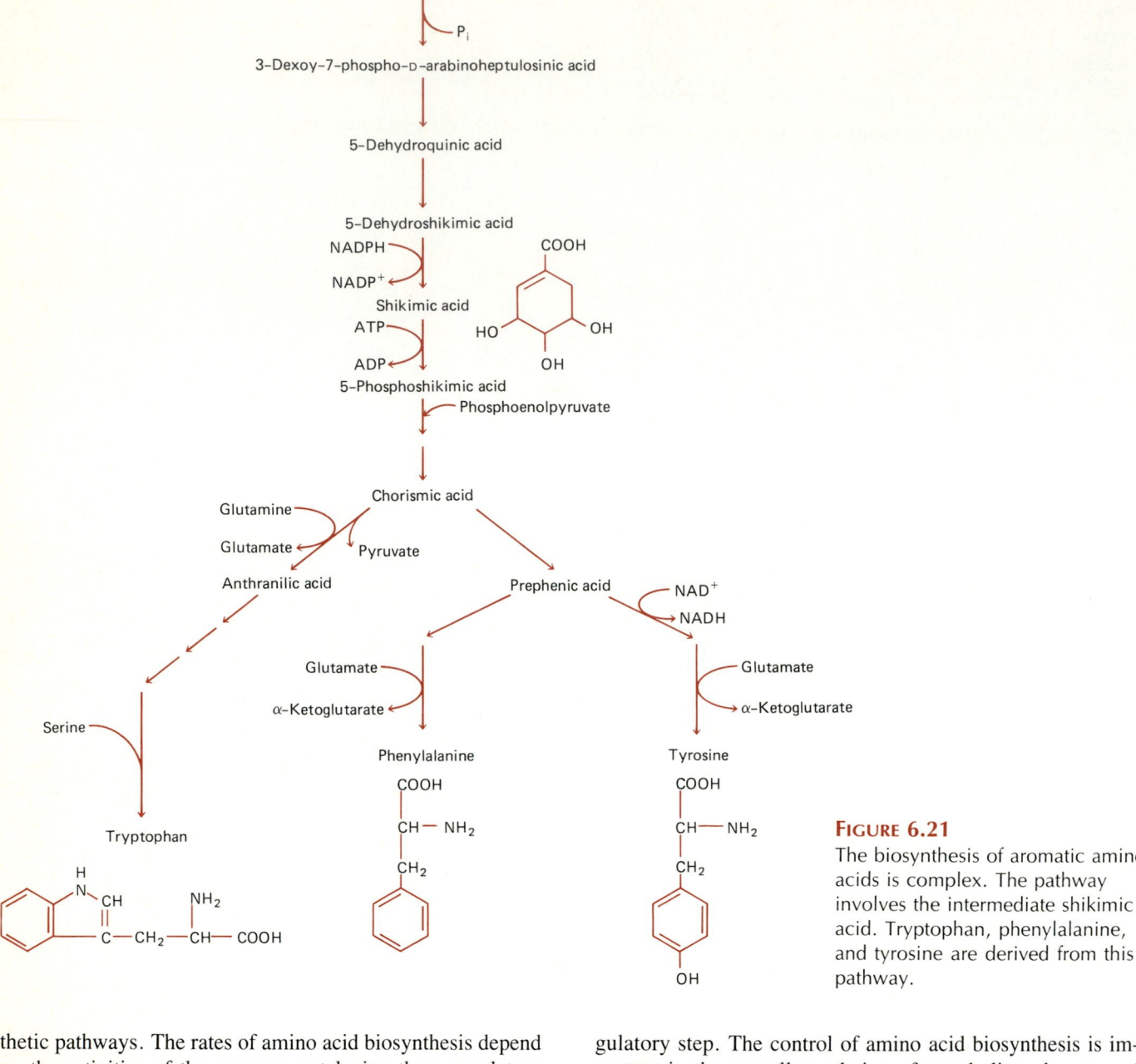

FIGURE 6.21
The biosynthesis of aromatic amino acids is complex. The pathway involves the intermediate shikimic acid. Tryptophan, phenylalanine, and tyrosine are derived from this pathway.

thetic pathways. The rates of amino acid biosynthesis depend on the activities of the enzymes catalyzing these regulatory steps, and the final product of a pathway often acts as an allosteric inhibitor of the enzyme catalyzing the critical regulatory step. The control of amino acid biosynthesis is important in the overall regulation of metabolism, because of the central role of enzymes that are composed of amino acids, in catalyzing metabolic reactions.

6.6 Nucleic Acid Biosynthesis

As with proteins, the biosynthesis of nucleic acid molecules can be examined from the viewpoint of synthesizing the nucleotides that go into the biosynthesis of DNA and RNA or considering the sequencing of the nucleotides that make up the macromolecules DNA and RNA. In this chapter, we will consider only briefly the biosynthesis of the nucleic acid bases, and in the next chapter we will examine how the nucleotides are aligned and linked together to form the DNA and RNA macromolecules.

There are two classes of nucleic acid bases, purines and

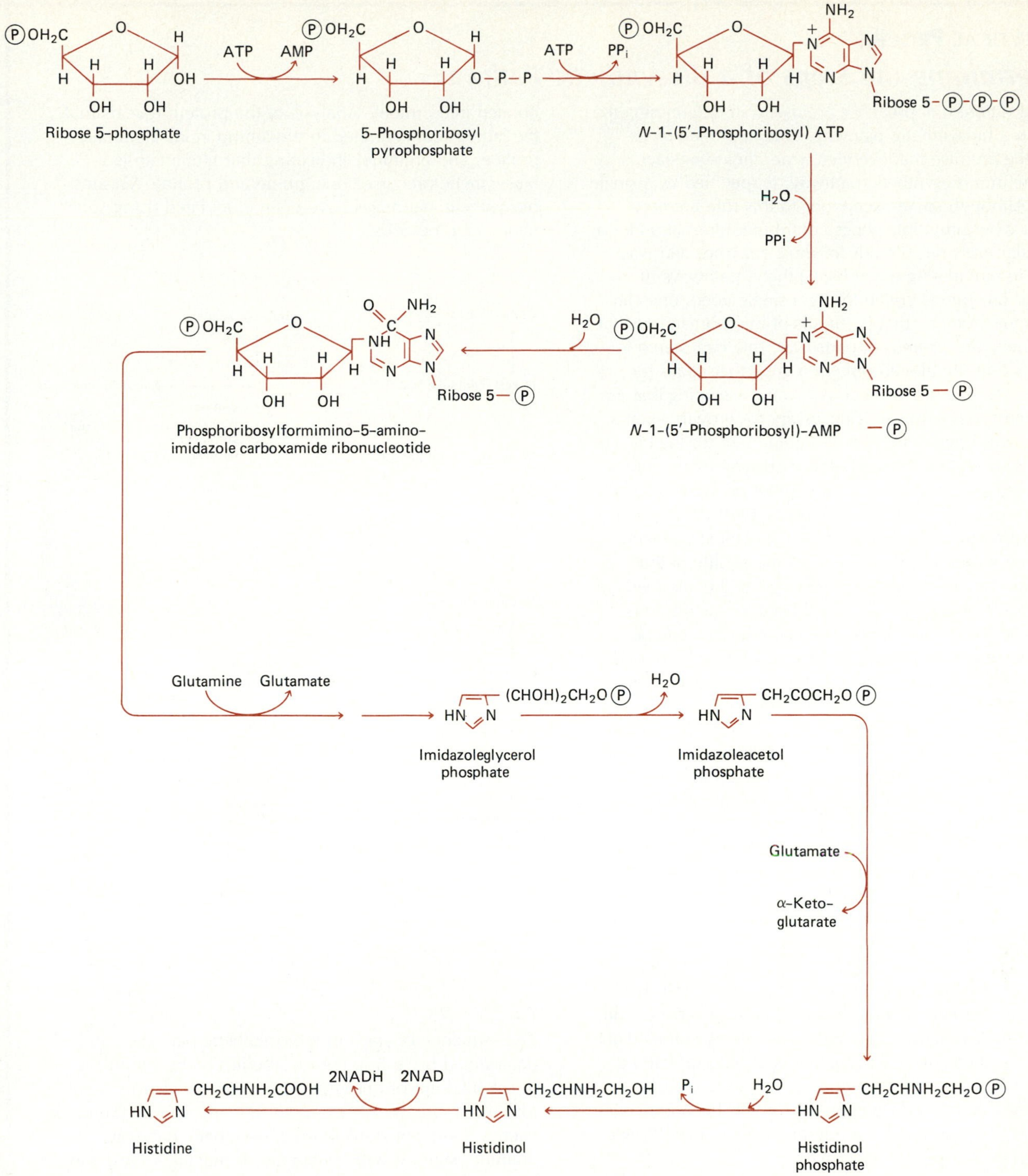

FIGURE 6.22

The synthesis of histidine, an amino acid with a nonaromatic ring structure.

pyrimidines. The pathways for the biosynthesis of the purine and pyrimidine ring structures of nucleotides are quite complex, and only a cursory overview of the process will be given here. The biosynthesis of purine and pyrimidine nucleotides is important because these compounds are involved in many biochemical processes; they are the precursors of DNA and RNA. ATP is an adenine nucleotide, as are the major coenzymes, NAD$^+$, NADP$^+$, and CoA, and nucleotides are important activators and inhibitors that regulate the rates of metabolic reactions within the cell.

ANALYTICAL PROCESS
Determining the Steps of a Biosynthetic Pathway

Several different approaches are useful in determining the steps in a biosynthetic pathway. One approach relies upon the premise that every enzyme, including each enzyme in a biosynthetic pathway, is specified by a single gene. Although some exceptions to this rule are now known where multiple genes contribute information for a particular enzyme, it holds for most enzymes and was useful in elucidating many biosynthetic pathways. If a single enzyme is coded for by a single gene, one can search for mutants, that is, strains of microorganisms that are genetically altered. In particular, one can search for mutants that are altered only in a gene that codes for an enzyme in a biosynthetic pathway. Assuming that the product of that pathway is necessary for growth, such a mutant will have a nutritional requirement for the biosynthetic product that it would otherwise synthesize. Supplying the specific compound that cannot be synthesized because of the loss of the enzyme as a growth factor will permit growth of the organism and specifically identify that compound as an intermediary metabolite of that pathway. By repeating this process, all of the intermediates in the pathway can be identified because each step is catalyzed by a different enzyme and a mutant can be found for each. For example, using mutants of *Salmonella typhimurium*, the eight steps in the biosynthetic pathway of arginine from glutamate were identified.

Fortunately, the search for possible intermediary metabolites is not random. The blockage of a pathway tends to cause metabolic intermediates produced in the pathway prior to the blocked step to accumulate and to be excreted into the surrounding medium. Using the metabolites that accumulated from one mutant as growth factors in the media used to grow other mutant strains can permit the growth of those other strains if they are in fact blocked at an earlier step in the metabolic pathway but not if they are blocked at the same or a later step in the pathway (Figure 6.23). Hence the order of the enzymes missing in a series of mutant strains can be determined. Alternatively, the abilities of substances with biochemical structures that make them likely intermediates in a pathway to overcome metabolic blockages can be tested. Empirical testing of compounds with probable structures, that is, educated guessing, has been frequently employed in the search for intermediates in biosynthetic pathways.

A quite different approach is the use of radiolabeled tracers to identify the intermediates of a pathway. A radiolabeled compound is supplied, either continuously in the growth medium or for only a brief period. After allowing the organism to grow and incorporate that compound into its biochemicals, the cells are chemically fractionated and the compounds containing the radiolabel are identified. For example, when ^{14}C-radiolabeled glutamate is added to growing bacterial cells, it is incor-

porated into protein. Analysis of the protein reveals that the label is concentrated in the amino acids glutamate, proline, and arginine, indicating that glutamate is a biosynthetic precursor of arginine and proline. Various biosynthetic pathways have been elucidated using such radiotracer methods.

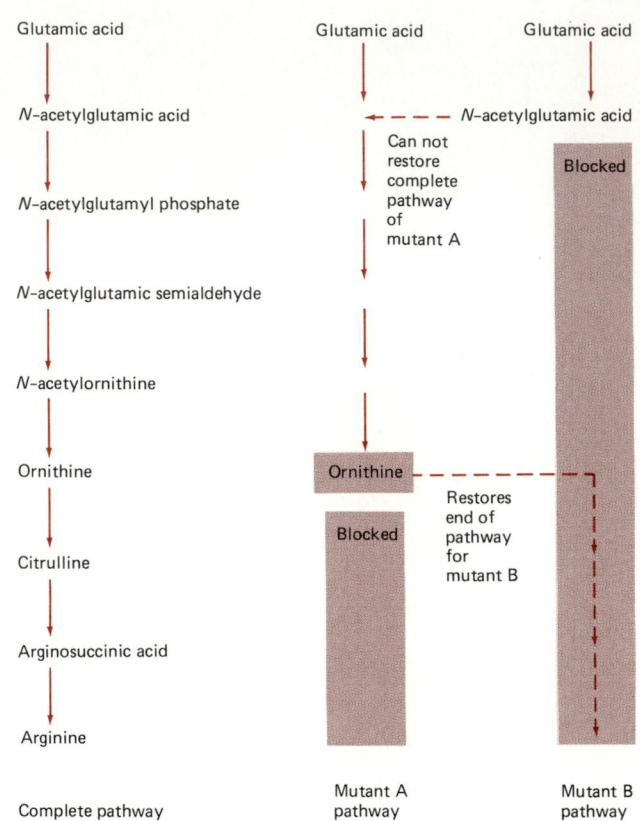

FIGURE 6.23
The sequence of steps in a biosynthetic pathway can be determined by testing the metabolites that accumulate in the medium of one mutant as potential growth factors for other mutants. For example, mutants can be found for each of the eight steps in the biosynthetic pathway for arginine, starting with glutamate. If mutant A lacks the enzyme needed to convert ornithine to citrulline, it accumulates ornithine and other metabolites that occur earlier in the pathway. If mutant B lacks the enzyme needed to convert *N*-acetylglutamic acid to *N*-acetylglutamyl phosphate, it accumulates *N*-acetylglutamic acid. Because of the relative positions of the steps in the pathway that are blocked, the accumulated metabolites from mutant A are able to support the growth of mutant B, but those from B are unable to permit the growth of A.

Biosynthesis of Pyrimidines

The precursors of the atoms of the **pyrimidine ring** are ammonia, carbon dioxide, and L-aspartate (Figure 6.24). The synthesis of the pyrimidine ring begins with the formation of carbamoylaspartate from the reaction of carbamoyl phosphate and aspartate, catalyzed by the enzyme aspartate transcarbamoylase. This is the key regulatory step in the synthesis of the pyrimidine ring, and the enzyme aspartate transcarbamoylase is subject to allosteric feedback inhibition by the products of the reaction and to allosteric activation by ATP. The succeeding steps in the formation of the pyrimidine ring

FIGURE 6.24

The formation of a pyrimidine ring is necessary for the biosynthesis of nucleic acids. This pathway results in the formation of uridine triphosphate from aspartate and carbamoyl phosphate. The conversion of uridine triphosphate to cytidine triphosphate forms one of the nucleotides found in both DNA and RNA. The conversion of ribose to deoxyribose produces the sugar backbone molecule of DNA and uracil to thymine forms the thymidine found in DNA.

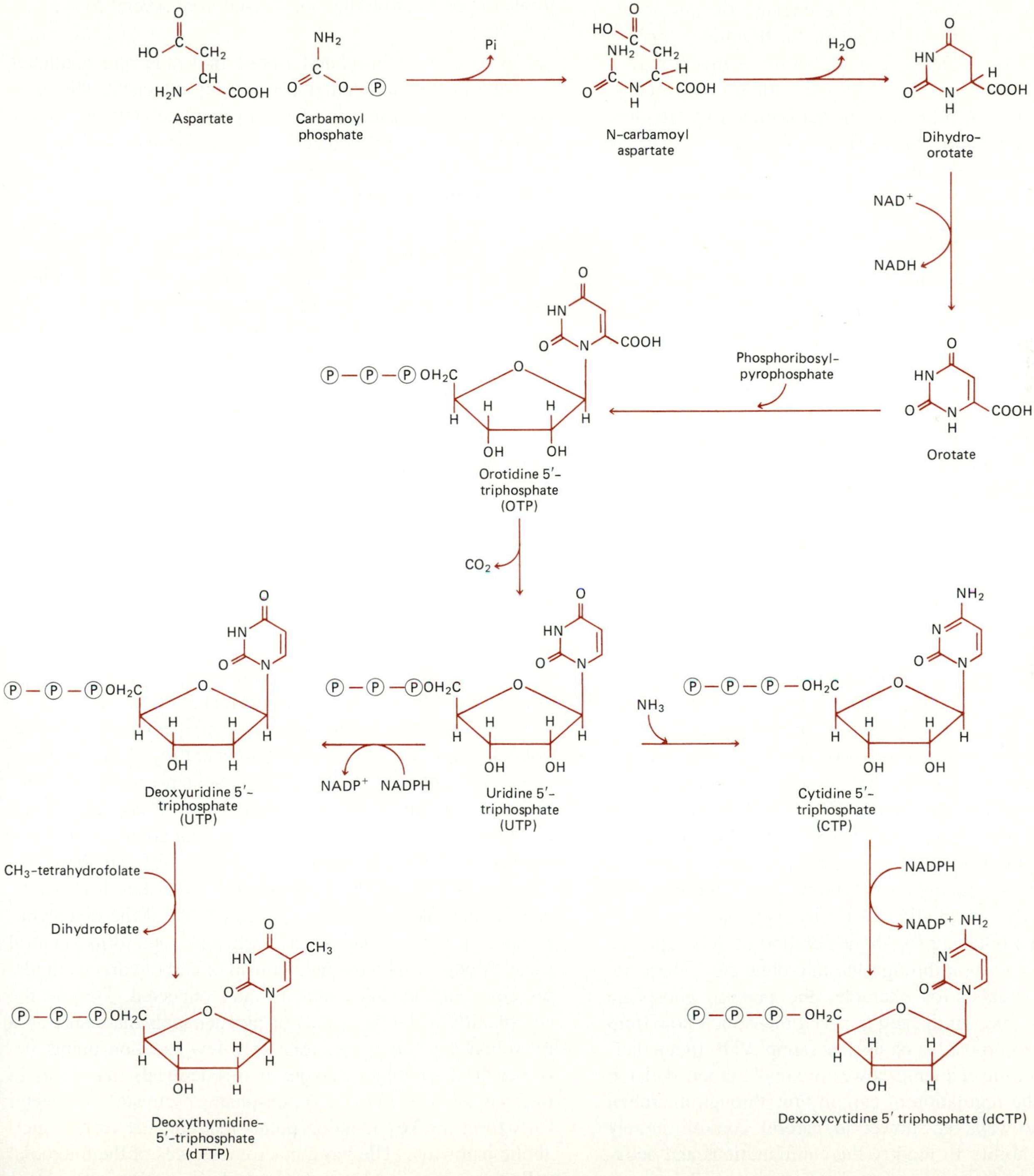

involve dehydration, cyclic ring formation, and oxidation to form orotate. After the pyrimidine ring is synthesized, ribose and phosphate are added to the molecule using 5-phosphoribosyl-1-pyrophosphate (PRPP). A carboxyl group is subsequently removed as carbon dioxide, forming uridine monophosphate (UMP) which is then phosphorylated to form UTP, one of the nucleotides of RNA.

The UTP can be modified to form cytidine triphosphate by the replacement of a keto group with an amino group in the pyrimidine ring. Cytidine triphosphate is a nucleotide in both RNA and DNA. The ribose sugar is reduced to the deoxyribose form of the nucleotide for DNA, using the reducing power of NADPH and the enzyme ribonucleoside reductase. UTP is also the precursor for thymidine triphosphate that occurs in DNA. The formation of thymidine for incorporation into DNA involves the addition of a methyl group, derived from folic acid, to the uridine ring structure to form thymine and the reduction of ribose using the NADPH to form deoxyribose.

Biosynthesis of Purines

The formation of the **purine nucleotides**, which have two rings, is somewhat more complex than that of the pyrimidines. Ten metabolic steps are involved in the formation of the basic purine ring structure. It is synthesized from a variety of amino acids, including L-aspartate, L-glycine, and L-glutamine (Figure 6.25). Additionally, carbon dioxide and a methyl group donated by folic acid are essential for the formation of the purine ring skeleton. The initial step in the synthesis of the purine ring involves the addition of an amino

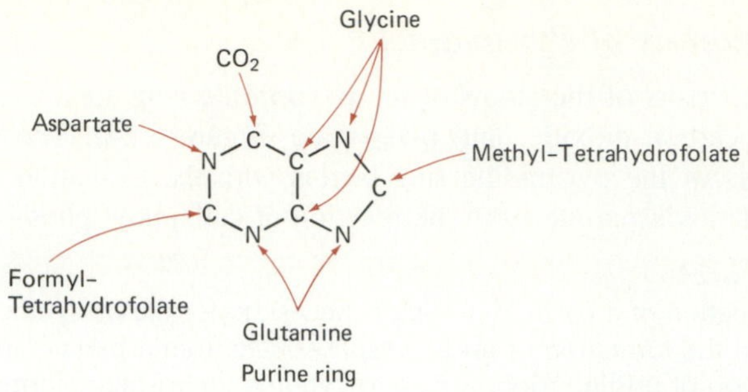

FIGURE 6.25

As shown in this illustration, the carbons involved in the formation of a purine ring are derived from several sources.

group to a phosphorylated ribose molecule and continues with the phosphorylated ribose already attached. This is in contrast to the synthesis of the pyrimidine ring, where the carbohydrate moiety is added after the formation of the ring structure. Once the basic purine ring is completed, it is modified to form the adenine and guanine nucleotides (Figure 6.26). The biosynthesis of the adenine ring involves the substitution of an amino group for a keto group.

The biosynthesis of the guanine ring involves the addition of an amino group without the removal of a keto group. The adenosine-phosphate molecule that is formed serves not only as a nucleotide base in DNA and RNA but also in the ATP, NAD^+, $NADP^+$, CoA, and FAD molecules. Thus, the synthesis of purine nucleotides is critical for the biosynthesis of many of the important molecules that act as energy carriers and coenzymes in the biochemical reactions of cellular metabolism.

Postlude

The metabolic pathways of microorganisms form an interlocking network through which carbon flows. Some of the pathways involve the breakdown of organic molecules (catabolism), and other pathways involve the synthesis of the macromolecules (anabolism) that constitute microorganisms. ATP and the reduced and oxidized forms of the coenzymes NAD^+ and $NADP^+$ play key roles in the flow of carbon through microorganisms because they supply energy and act as oxidizing and reducing agents, respectively, for microbial metabolism. The glycolytic and Krebs cycle pathways form a central focus not only for the generation of ATP but also for the flow of carbon through the microbial cell. There are several key shunts—for example, the pentose phosphate pathway— that permit the cell to divert the carbon flow from one metabolic purpose, such as generating ATP, to another, such as generating reducing power or small carbon skeleton molecules. The regulation of carbon flow through microbial metabolic pathways is subject to careful control, largely based on the ability to modify the configurations and activities of key regulatory enzymes. In many cases, the energy

charge of the cell determines the activities of these regulatory enzymes and thus the direction of metabolic activity.

The central metabolic pathways serve both for the catabolic breakdown of substrates—generating ATP, reduced coenzyme, and the precursors of anabolism—and for the biosynthesis of the cell's macromolecules. These amphibolic pathways use the same intermediary metabolites, and the regulation of key reactions, catalyzed by different enzymes, determines whether the pathway at a given time is catabolic or anabolic. The intermediary metabolites of the glycolytic, pentose phosphate, and Krebs cycle pathways form a central focus through which the metabolism of carbohydrates, lipids, proteins, and nucleic acids is interconnected. Despite the complexity of the thousands of biochemicals that make up a microbial cell, there are relatively few junction points between the four major classes of biochemicals that occur in microorganisms; glucose 6-phosphate, pyruvate, and acetyl CoA form the key junction points that connect various metabolic pathways. The large macromolecules of the microbial cell are synthesized from relatively few, small organic com-

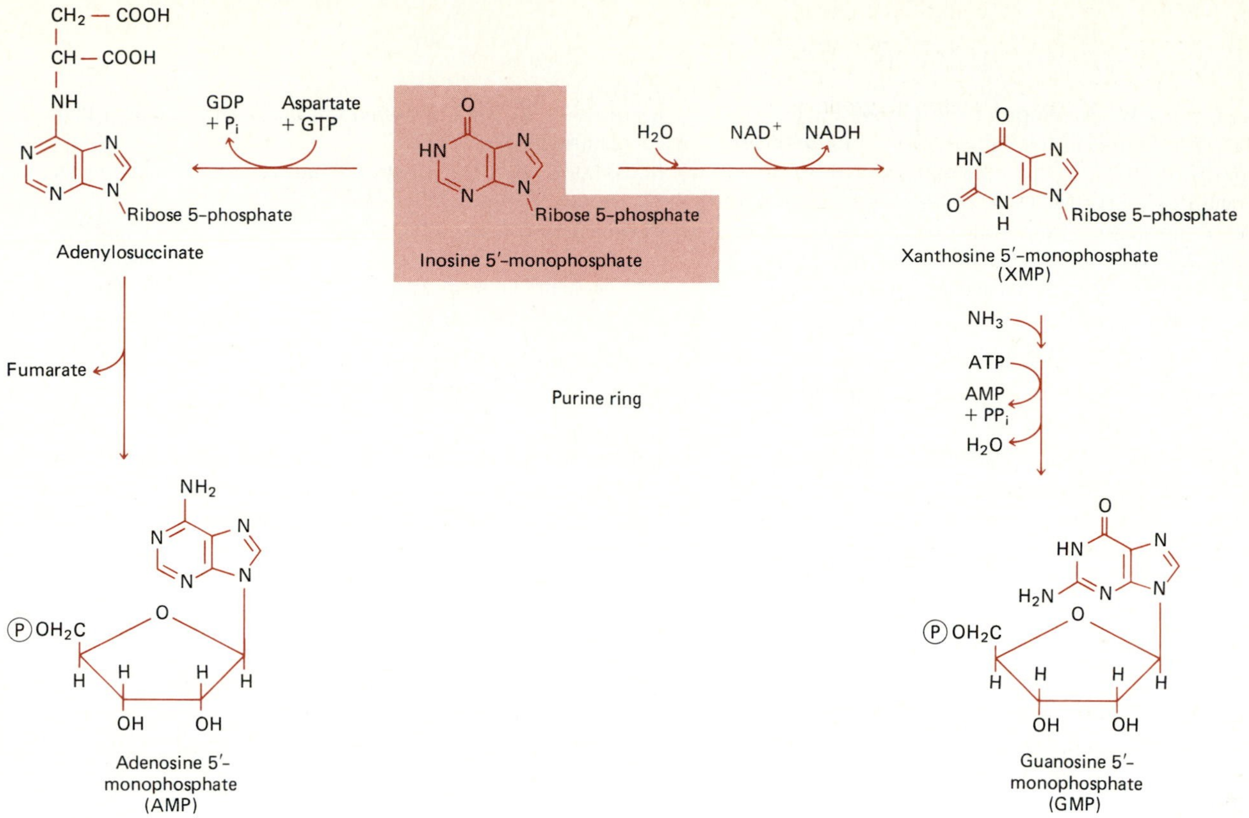

FIGURE 6.26
The conversion of an inosine purine ring to adenine and guanine is needed to form the purine nucleotides found in both DNA and RNA. The AMP formed via this pathway also is used for the synthesis of ATP.

pounds. A general theme of microbial metabolism is the conversion of a substrate molecule into a series of small organic compounds, which can then be utilized for the biosynthesis of many different large macromolecules. The metabolic pathways that generate the ATP and NADPH required for biosynthesis also generate the small organic molecules that establish the carbon skeletons of larger molecules.

Whereas the metabolites of the central metabolic pathways are involved in both anabolism and catabolism, there is a real distinction between carbon flow in biosynthetic and catabolic pathways. Part of the distinction is maintained by the use of different coenzymes in these opposing pathways. As a rule, NAD^+ and NADH are involved in catabolic pathways aimed at the generation of ATP, and NADPH is generally involved in biosynthetic reactions. Biosynthetic pathways almost always require reducing power in the form of NADPH and energy in the form of ATP. Additionally, while the intermediary metabolites are identical, the pathways are distinct because of the presence of irreversible steps in each direction, requiring different key enzymes to proceed in the biosynthetic versus degradative pathways. The activities of these key enzymes regulate the rates and direction of carbon flow through the central amphibolic pathways. The key enzymes are usually subject to allosteric control so that the flow of carbon through a particular pathway can be accelerated or inhibited, depending on the metabolic needs of the cell. The allosteric activators and inhibitors act as feedback indicators that regulate the direction of carbon flow.

The ability of microorganisms to move carbon from one class of compounds to another enables some microorganisms to synthesize all of their required biochemicals from a single starting substrate molecule. In the case of autotrophic microorganisms, all of the biochemicals of the cell can be synthesized from carbon dioxide and water or other hydrogen donors. The Calvin cycle is the key metabolic pathway for the conversion of carbon dioxide to organic carbon by autotrophic microorganisms. The organic carbon formed by this pathway supports not only the autotrophic microorganisms that possess this pathway but also heterotrophic microorganisms that require preformed organic compounds for their existence. Regardless of their starting substrates, all microorganisms must synthesize or acquire the precursors for all four major classes of macromolecules. The biosynthesis of 20 essential amino acids forms the building blocks for proteins, and the biosynthesis of 8 nucleotides forms the basic building blocks of nucleic acids. In the case of proteins and nucleic acids, the ordering of these building blocks within the macromolecule (see the discussion in Chapter 7) establishes its configuration and determines the macromolecule's functional properties. The informational nucleic acid macromolecules, and the functional enzymes they encode, determine the metabolic functioning of microorganisms.

Suggested Supplementary Readings

ANDERSON, R. L., and W. A. WOOD. 1969. Carbohydrate metabolism in microorganisms. *Annual Review of Microbiology* 33:539–575.

CONN, E. E., and P. K. STUMPF. 1976. *Outlines of Biochemistry*. John Wiley & Sons, Inc., New York.

DAGLEY, S., and D. E. NICHOLSON. 1970. *An Introduction to Metabolic Pathways*. Blackwell Scientific Publications, Oxford, England.

DOELLE, H. W. 1975. *Bacterial Metabolism*. Academic Press, New York.

GOTTSCHALK, G. 1986. *Bacterial Metabolism*. Springer-Verlag, New York.

INGRAHAM, J. L., O. MAALØE, and F. C. NEIDHARDT. 1983. *Growth of the Bacterial Cell*. Sinauer Associates, Sunderland, Mass.

LEHNINGER, L. 1982. *Principles of Biochemistry*. Worth Publishers, Inc., New York.

MAGASANIK, B. 1982. Control of nitrogen assimilation in bacteria. *Annual Review of Genetics* 16:135–168.

MANDELSTAM, J., K. MCQUILLEN, and I. W. DAWES. 1982. *Biochemistry of Bacterial Growth*. Blackwell Scientific Publications, Oxford, England.

STEWART, W. D. P. 1980. Some aspects of structure and function in N₂-fixing cyanobacteria. *Annual Review of Microbiology* 34:497–536.

STRYER, L. 1987. *Biochemistry*. W. H. Freeman Co., San Francisco.

UMBARGER, H. E. 1978. Amino acid biosynthesis and its regulation. *Annual Review of Biochemistry* 47:533–606.

ZUBAY, G. 1983. *Biochemistry*. Addison-Wesley Publishing Co., Reading, Mass.

Study Questions

1. What is the Calvin cycle, and what is its function? Why is it called a *dark reaction cycle*?

2. What is an amphibolic pathway? What is a catabolic pathway? What is an anabolic pathway?

3. What is gluconeogenesis? How would a cell make glucose starting with a protein?

4. What is the problem with metabolizing compounds containing only two carbons, such as acetate?

5. How is nitrogen incorporated into organic compounds to form amino acids?

6. How are allosteric effectors involved in regulating the flow of carbon through a cell?

7. What is the general strategy of a cell in terms of carbon flow?

Situational Problem

Determining What Is Needed to Synthesize a Cell

We have seen that given the necessary enzymes, cells can transform a starting substrate molecule into all of its macromolecular constituents. Let us consider the composition of a simple hypothetical bacterial cell. The cell has 2,000,000 protein molecules, 200,000 RNA molecules, 1 DNA molecule, and 20,000,000 lipid molecules. It has no cell wall and hence no peptidoglycan. Besides lacking a cell wall, this bacterial cell is very unusual in several other ways. Its proteins are composed only of the amino acids proline (an amino acid with five carbon atoms) and glycine (an amino acid with two carbon atoms), which occur in equal concentrations. There are 10 different types of proteins. Each protein has 300 amino acids. The DNA and RNA contain only cytidine (a nucleotide containing 9 carbon atoms) and guanidine (a nucleotide containing 10 carbon atoms), which also occur in equal concentrations. The DNA molecule contains 1,000 nucleotide kilobase pairs. There are 30 types of RNA molecules. The RNA molecules each have 1,000 nucleotides. The lipids are all phospholipids containing glycerol and two attached chains of palmitic acid, which is a straight chain saturated fatty acid containing 16 carbon atoms. The bacterium can grow on glucose as its sole source of carbon and energy.

1. Considering only the flow of carbon (ignore the need for ATP), draw the pathways needed to convert glucose into the macromolecules of this cell. (You may wish to consult one of the biochemistry texts listed in the Suggested Supplementary Readings.)

2. Based on the composition of this cell, and again considering only the flow of carbon, determine how many glucose molecules are needed to synthesize the macromolecules required for this cell to divide and produce an exact replicate.

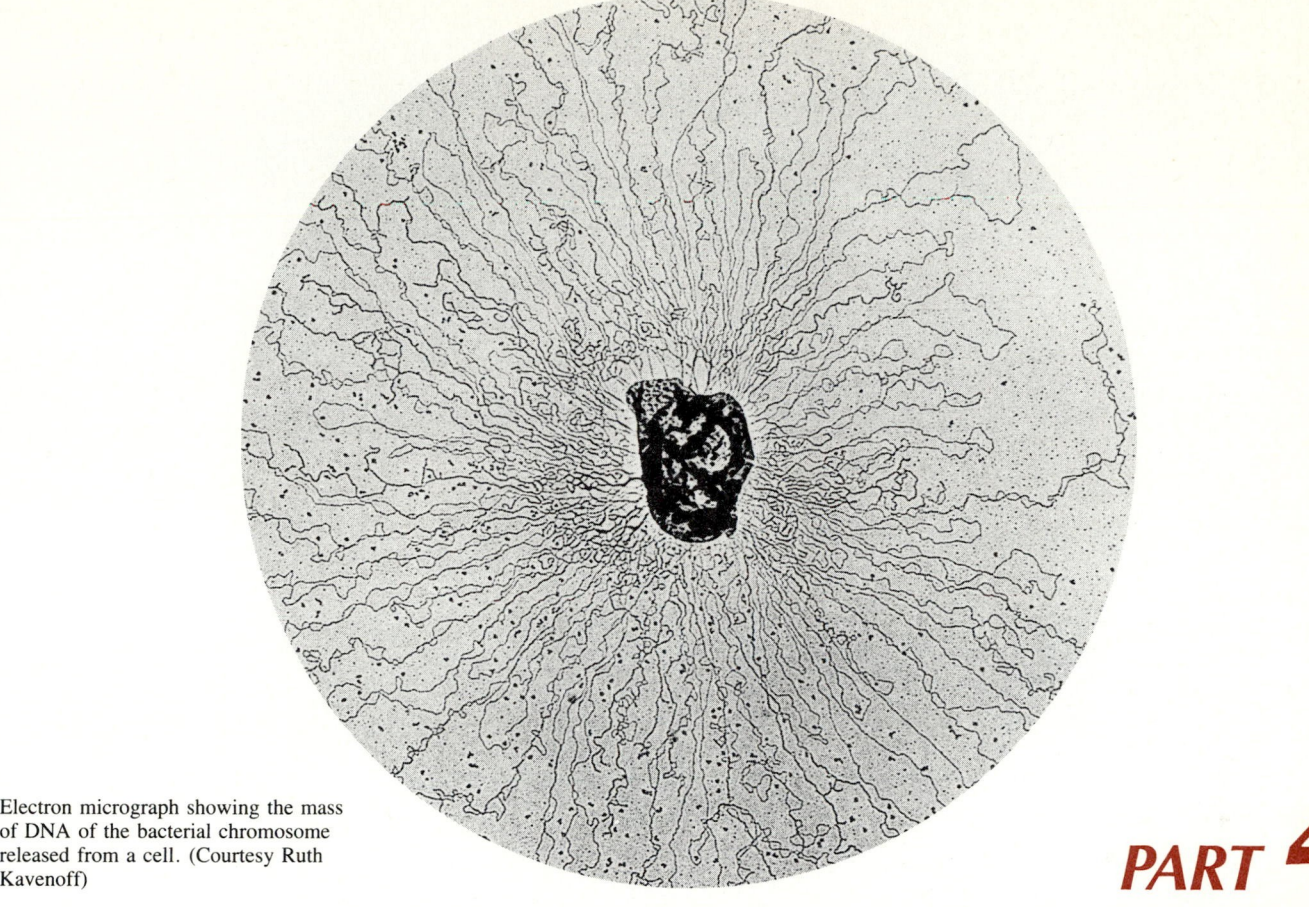

Electron micrograph showing the mass of DNA of the bacterial chromosome released from a cell. (Courtesy Ruth Kavenoff)

PART **4**

Microbial Genetics

DNA, RNA, and Protein Synthesis: The Replication and Expression of Genetic Information

As we have seen, cellular metabolism is essential for the functional activities of an organism. The expression of this functional activity is largely mediated by proteins, which—acting as enzymes—determine the metabolic capacity and hence the phenotype (appearance) of the cell. In addition, cells have informational functions, collectively known as the *genetics* of the cell, which encompass the storage and expression of hereditary information. The genetic information of prokaryotic and eukaryotic cells, which determines the metabolic and structural nature of the organism, is encoded in the DNA molecule(s) of the cell. The transmission of hereditary information necessitates the precise replication of the DNA macromolecules containing this genetic information. How an organism accurately replicates its genetic information so that it can be passed on to its progeny is one of the topics we will explore in this chapter.

7.1 Nucleic Acid Macromolecules

Genetic information is stored and transmitted by two types of nucleic acid macromolecules, **deoxyribonucleic acid (DNA)** and **ribonucleic acid (RNA)**. These informational macromolecules are made up of **nucleotides**; each nucleotide consists of a nucleic acid base, a sugar, and a phosphate group (Figure 7.1). DNA stores the genetic information of the cell, while the RNA, of which there are several types, is involved in the conversion of stored genetic information into proteins that carry out the biological functions of the organism. DNA and RNA differ with respect to the sugars and specific nucleic acid bases found in the nucleotides. In DNA the sugar that occurs is deoxyribose, whereas ribose is the sugar found in RNA. The extra hydroxyl group that occurs in ribose as compared to deoxyribose makes RNA a less stable structure than DNA. It is more advantageous to have greater stability in the macromolecule used for storage of genetic information, the DNA, than in the temporary carrier of that information, the RNA.

With respect to the nucleic acid bases that occur in DNA and RNA, each has four repetitive nitrogenous bases. In DNA these nucleic acid bases are adenine, guanine, cytosine, and thymine. **Adenine (A)** and **guanine (G)** are **purines**, which are two-ringed structures, whereas **cytosine (C)** and **thymine (T)** are **pyrimidines**, which have only one ring. In RNA, adenine, guanine, cytosine, and another pyrimidine—**uracil (U)**—occur.

In both RNA and DNA the nucleotides are linked by $3'$-$5'$ phosphate diester bonds (Figure 7.2). Consequently, at one end of the nucleic acid molecule, there is no phosphate diester bond to the 3-carbon of the monosaccharide, and thus there is an unattached or free hydroxyl group at the 3-carbon position (**3'OH free end**); at the other end of the molecule, the 5-carbon is not involved in forming a phosphate diester linkage, and there is a free phosphate ester group at the 5-carbon position (**5'P free end**). The fact that the ends of the nucleic acid macromolecule differ is extremely important

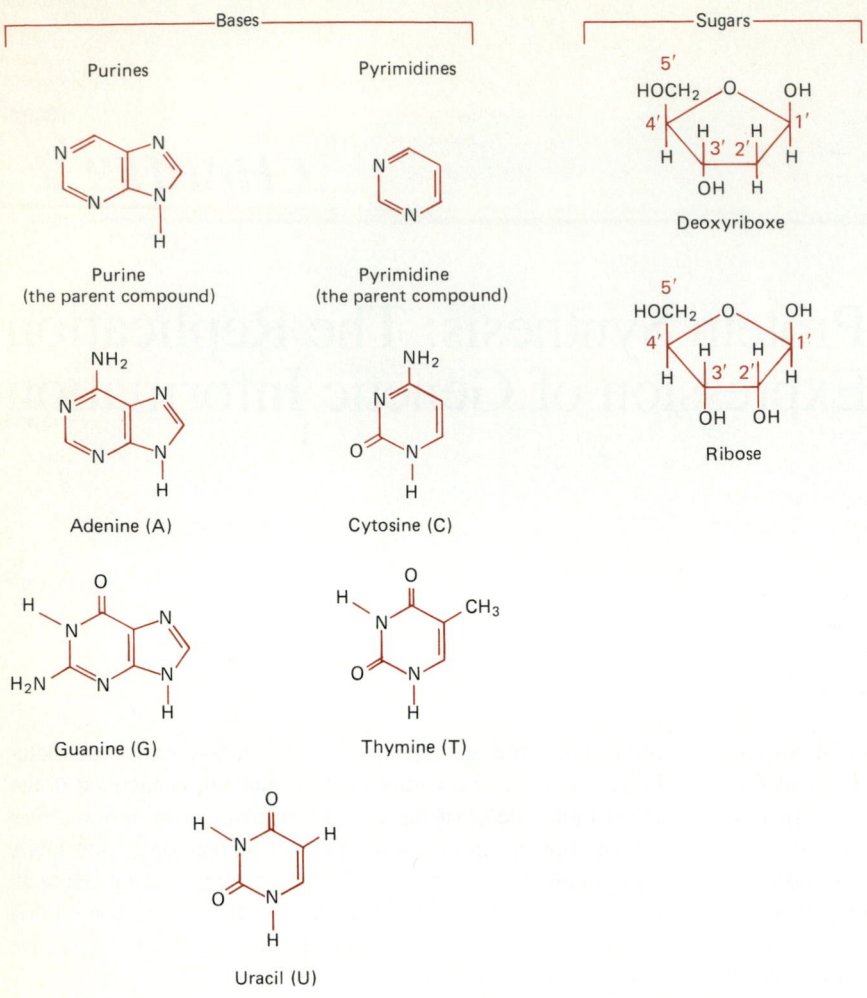

Bases

Purines

Purine
(the parent compound)

Adenine (A)

Guanine (G)

Pyrimidines

Pyrimidine
(the parent compound)

Cytosine (C)

Thymine (T)

Uracil (U)

Sugars

Deoxyriboxe

Ribose

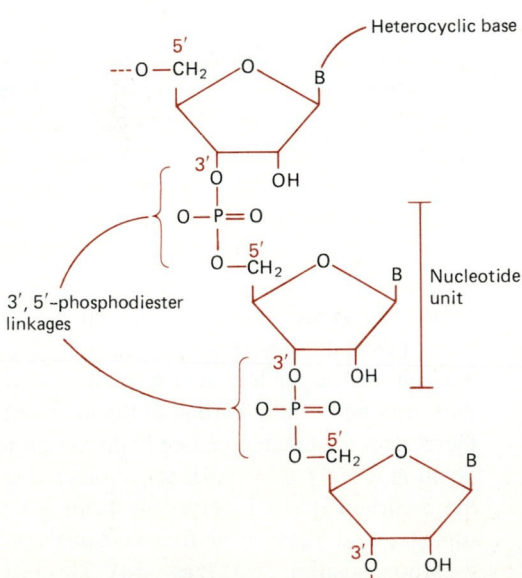

Heterocyclic base

3′, 5′-phosphodiester
linkages

Nucleotide
unit

FIGURE 7.1

DNA and RNA, the informational macromolecules of the cell, are composed of nucleotides. Each nucleotide contains a nucleic acid base attached to a sugar—ribose in the case of RNA and deoxyribose in the case of DNA—and a phosphate group. The nucleic acid bases attached to a sugar without the phosphate are called *nucleosides*. The structural formulas of the various bases and sugars in DNA and RNA are shown in this figure.

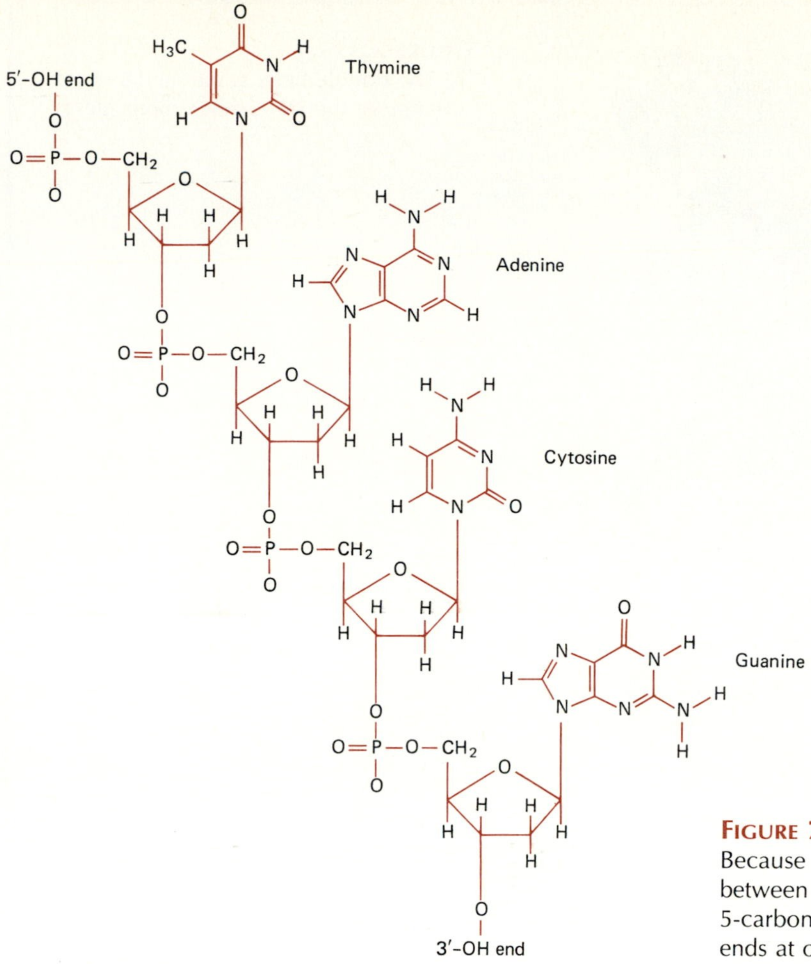

FIGURE 7.2
Because nucleic acids are linked by phosphodiester bonds between the 3-carbon of one nucleic acid base and the 5-carbon of the other, nucleic acids have 3'OH and 5'P free ends at opposite ends of polymeric molecules.

because this permits directional recognition at the biochemical level in the same sense that we can recognize left and right. This is essential for the nucleic acids to perform their principal function in biological systems, which is to store and transmit the genetic information of the cell.

The **DNA double helix** is composed of two primary polynucleotide chains held together by **hydrogen bonding**[1] between complementary nucleotide bases (Figure 7.3). Within the double helical DNA, adenine always pairs with thymine, and guanine always pairs with cytosine. There are two hydrogen bonds established between the adenine-thymine base pairs and three hydrogen bonds established between the guanine-cytosine base pairs. A representative bacterial chromosome, that of *Escherichia coli*, contains 3.8×10^3 kilobase pairs (kb) and has a molecular weight of 2.6×10^9 daltons. An average gene of 1.1 kb is needed to encode the information for a typical protein, and hence, the bacterial chromosome of *E. coli* theoretically can code for 3,500 different proteins.

[1]A hydrogen bond is a weak linkage based upon charge separations within the molecules that occurs when a hydrogen atom that is covalently bonded to an oxygen or nitrogen atom within a molecule is simultaneously attracted to a nitrogen or oxygen molecule of a neighboring molecule.

Proteins mediate functional activity seen as the phenotype, while nucleic acids mediate the informational capacity of the cell, that is, the **genotype**. A single copy of the genetic information of a cell constitutes its **genome**. The genome is divided into segments of DNA, known as **genes**, that have specific functions. Some genes code for the synthesis of RNA and proteins, determining, respectively, the sequences of the subunit nucleotide bases and amino acids in these macromolecules. Such genes are known as **structural genes** or **cistrons**. Other genes have regulatory functions (**regulatory genes**) and act to control the activities of the cell. Together the structural and regulatory genes constitute the genotype and determine the phenotype, that is, the actual appearance and activities of the organism.

The sequence of nucleotides within the DNA molecule, which establishes the sequence of bases in RNA, ultimately determines the sequence of amino acids in a particular protein. This is like saying that the letters of a word create that word and hence its meaning. Similarly, just as we must have a convention for the order of reading the letters of words (left to right in the English language), reading the sequence of nucleotides in the appropriate order is essential for converting stored genetic information into the functional activ-

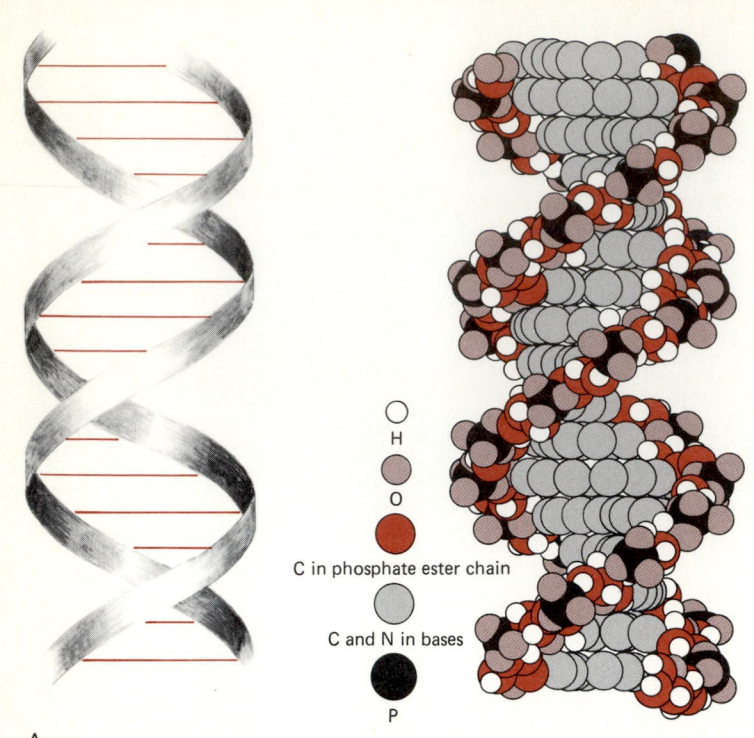

FIGURE 7.3
(A) The double helix is the fundamental structure of the DNA macromolecule.

H
O
C in phosphate ester chain
C and N in bases
P

(B) The two strands are held together by hydrogen bonding between complementary base pairs.

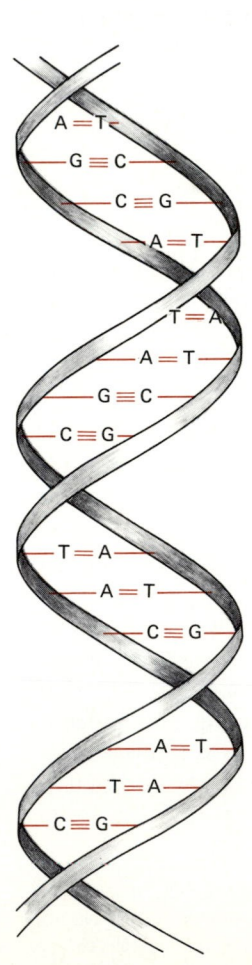

A=T
G≡C
C≡G
A=T

T=A
A=T
G≡C
C≡G

T=A
A=T
C≡G

A=T
T=A
C≡G

CH₃   O---H—NH

Thymine (T)   Adenine (A)
To strand of helix   To strand of helix
T=A

H
N—H---O

Cytosine (C)   Guanine (G)
To strand of helix   To strand of helix
C≡G

A

B

C

FIGURE 7.3 *(CONT'D)*
(C) James Watson (left), age 23, and Francis Crick, age 34, posed with their DNA model in the Cavendish Laboratory at Cambridge University, England, in 1953 when they announced their discovery of the molecular structure of DNA. They shared the Nobel Prize for Medicine in 1962 with Maurice Wilkins. (Photograph by Barr-Brown Camera Press, Central Office of Information, London)

Discovery Process

This particular scientific breakthrough was not accomplished through the research we often think of as involving test tubes and petri plates, guinea pigs, and bubbling chemical retorts, but rather was the product of a great deal of thought, discussion, examination of evidence already available, and intuition, as exemplified by the decision to build two-chain models simply because of the "repeated finding of twoness in biological systems." Watson and Crick knew that they had to rely on the simple laws of structural chemistry. "The essential trick was to ask which atoms like to sit next to each other. In place of pencil and paper, the main working tools were a set of molecular models superficially resembling the toys of preschool children." So, after getting the tautometric forms of guanine and thymine serendipitously corrected by a visiting scientist, who pointed out that the structures in the organic chemistry books were wrong, Watson and Crick began "shifting the bases in and out of various pairing possibilities." In the next step, their tools were a plumb line and a measuring stick to determine the relative positions of all of the atoms in a single nucleotide. By assuming a helical symmetry, it became clear that the locations of the atoms in one nucleotide would automatically generate the other position.

ities of the organism. The directional nature of nucleic acid molecules is critical for establishing the necessary direction of reading the genetic information. Within the double helical DNA macromolecule the two polynucleotide chains run in opposite directions; that is, one chain runs from the 3'OH to the 5'P free end and the complementary chain runs in an antiparallel direction from the 5'P to the 3'OH free end.

A consequence of the antiparallel nature of the DNA molecule is that different information is stored in each of the chains. This further means that for a given region of stored information, there must be some mechanism for designating which of the complementary chains is running in the appropriate manner for extracting the appropriate information coding for a particular function. As will be discussed later, there are indeed recognition sequences encoded within the DNA molecule that designate which chain to read, where to begin, and where to end. The genetic code, based on only the "few letters" (nucleotides) in its "alphabet," provides the necessary biochemical basis for encoding the genetic information of the great diversity of living organisms.

The genetic information of a microorganism is specified by the order or sequence of the nucleotides in its DNA macromolecule(s). Replication of the genome involves synthesizing new DNA molecules that have the same nucleotide sequence as the genome of the parental organism, a process that requires great precision. (Replication = DNA → DNA.) Any change in the sequence of nucleotides can greatly alter the characteristics of that organism. A change of even a single nucleotide can prove lethal for the progeny of a microorganism. The genome of the progeny must contain the appropriate information to permit the survival and growth of the organism.

Expression of genetic information involves using information encoded within the DNA to effect the synthesis of proteins. By specifying and regulating protein synthesis, the genetic informational macromolecules define and control the metabolic capabilities of microorganisms. The sequence of nucleotides within the genome determines the sequence of amino acids within protein molecules and thus the functional properties of microbial enzymes. The expression of the genetic information of the organisms through the activities of the enzymes that are produced is reflected in the **phenotypic**

features (observable characteristics) that distinguish one microorganism from another.

Transferring the information contained in DNA to form a functional enzyme occurs through protein synthesis, a process accomplished in two stages— transcription and translation. Briefly, these processes—details of which will be covered later in this chapter—are structured so that the order of nucleotides in the DNA is used to specify the order of amino acids in a protein. The information in the DNA molecule is initially transferred to RNA molecules in a process called *transcription*. (Transcription = DNA → RNA.) Several different types of RNA molecules are involved in protein synthesis. One type of RNA molecule, **messenger RNA (mRNA)**, carries the information from the DNA molecule to the ribosomes, the cellular sites of protein synthesis. The information encoded in the mRNA molecule is then translated into the sequence of amino acids that comprise the protein. (Translation = RNA → protein.) The specific polypeptide sequences are specified by the sequences of nucleotide bases within the mRNA. In addition to mRNA, the process of translation requires **transfer RNA molecules (tRNA)** that help align the amino acids during the translational process. Further, the ribosomes themselves are largely comprised of **structural (ribosomal) RNA (rRNA)** molecules. Thus, in most microorganisms the nucleic acids of RNA act as informational mediators between the DNA where the genetic information is stored and the proteins that functionally express that information.

7.2 DNA Replication

Semiconservative Nature of DNA Replication

Having established the importance of DNA for determining the properties of microorganisms, let us examine the replication of this macromolecule in more detail. The replication of the double helical DNA molecule is a semiconservative process, so-called because when a DNA molecule is replicated to form two double helical DNA molecules, each of the new daughter DNA molecules consists of one intact (conserved) strand from the parental double helical DNA and one newly synthesized complementary strand (Figure 7.4).

The actual replication of the DNA molecule involves a complex series of coordinated enzymatic reactions. Strand separation, the first step in the process of replication, requires adenosine triphosphate (ATP) and several proteins. Energy is required to break the hydrogen bonds. The energy needed for separation is reduced by the action of helix-destabilizing protein (HDP) that binds to the separated single strands. Other proteins, including DNA-unwinding enzyme I (topoisomerase I) and Rep protein, known collectively as the *unwinding proteins*, actively separate the DNA strands. Another protein, DNA gyrase, prevents the formation of twists by periodically breaking a phosphodiester bond in one of the strands, thereby allowing free rotation of the opposite strand. Later this enzyme reforms the same bond.

Although **semiconservative replication** requires separation of the two strands of the parental DNA, the entire DNA molecule is not unwound prior to replication. Rather, there is a localized unwinding of the DNA double helix, mediated by unwinding enzymes, that establishes a **replication fork** in the DNA molecule at the site of DNA synthesis (Figure 7.5). The parental DNA molecule acts as a template that codes for the synthesis of DNA. At the replication fork, free nucleotide bases are aligned opposite their base pairs in the parental DNA molecules, A opposite T and G opposite C. At the replication forks there are four strands of DNA, two of which are conserved and two of which are newly synthesized.

DNA Polymerases

The newly synthesized strands of DNA are established by linking the nucleotide bases together with phosphodiester bonds by the action of **DNA polymerases** (Figure 7.6). The action of DNA polymerase results in the elongation of the nucleotide chain of the synthesized DNA molecule and can be likened to a zipper where the teeth of the zipper are initially aligned and progressively linked together in a continuous motion. DNA polymerases have several interesting properties. All DNA polymerases add deoxynucleotides only to the free 3′OH end of an existing nucleic acid polymer. These enzymes require an RNA primer molecule with a 3′OH free end for DNA synthesis. The DNA polymerases are also DNA dependent and require an existing DNA molecule to act as a template; such an enzymatic dependence on an existing template molecule is exceptional among biochemical reactions.

Several different DNA-dependent DNA polymerases have been isolated, each serving somewhat different functions during DNA synthesis. The DNA polymerases of bacteria can remove the RNA primers from the DNA strand. In eukaryotes the removal of the RNA primer is accomplished by a ribonuclease. In bacteria the DNA polymerases also exhibit exonuclease activity, that is, the ability to degrade or depolymerize a nucleic acid chain. Exonuclease activity is not associated with the DNA polymerases of eukaryotic organisms. The exonuclease activity of bacterial DNA polymerases allows them to correct errors in the DNA molecule, and if an inappropriate sequence of nucleotide bases has been inserted, the DNA polymerase can reverse direction, removing nucleotide bases from the free end of the DNA molecule

FIGURE 7.4

The semiconservative nature of DNA replication was demonstrated by labeling DNA in one generation by the incorporation of heavy nitrogen (^{15}N) and following the fate of this tagged DNA from one generation to the next, using density gradient ultracentrifugation. The location of the bands obtained by ultracentrifugaton, i.e., the distance that the DNA moves, which is a function of the molecular weight of the DNA, permitted the tracking of the fate of the heavy DNA when the cells were grown in the presence of normal light (^{14}N). The banding pattern obtained in these experiments, which is illustrated in the figure, proved that DNA replication occurs by a semiconservative method.

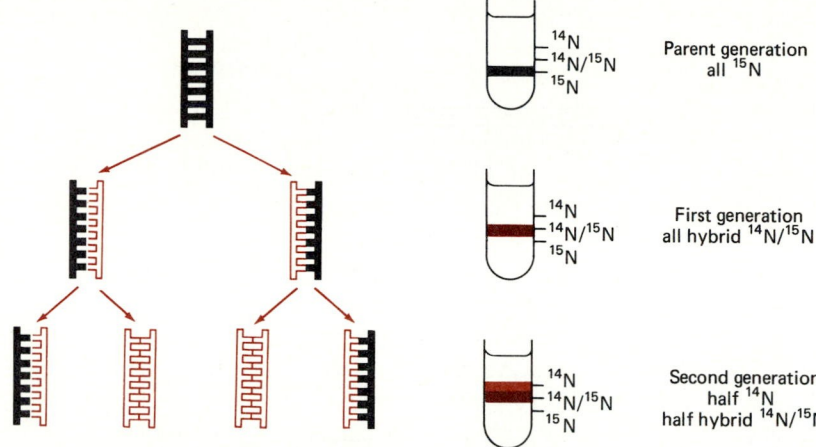

^{14}N
$^{14}N/^{15}N$ Parent generation
^{15}N all ^{15}N

^{14}N
$^{14}N/^{15}N$ First generation
^{15}N all hybrid $^{14}N/^{15}N$

^{14}N
$^{14}N/^{15}N$ Second generation
^{15}N half ^{14}N
                     half hybrid $^{14}N/^{15}N$

Discovery Process

The semiconservative nature of DNA replication was elegantly demonstrated in 1958 by Matthew Meselson and Franklin Stahl in a series of experiments that used the heavy isotope of nitrogen (^{15}N). In these experiments, *Escherichia coli* was initially grown in a medium with a sole nitrogen source of ^{15}N ammonium ions. The bacteria incorporated the heavy nitrogen into their nucleic acids. The bacterial culture was then transferred to a medium with a nitrogen source of ^{14}N ammonium ions. During incubation the bacterial DNA was replicated and the bacterial cells were reproduced. Cells were collected for analysis of the DNA after they had been allowed to grow for different generation times, and the DNA was then analyzed for the presence of ^{15}N and ^{14}N using density gradient ultracentrifugation. In this analytical method heavy molecules move farther than lighter molecules, and thus, DNA containing ^{15}N moves a greater distance than DNA containing only ^{14}N. When Meselson and

Stahl performed this experiment, all of the initial DNA formed as a single band corresponding to heavy DNA. After one generation, a single sedimentation band occurred corresponding to a hybrid DNA molecule of a mixture of ^{14}N- and ^{15}N-labeled DNA. After two generations, two bands occurred, one corresponding to light DNA (containing only ^{14}N) and the other corresponding to the $^{14}N-^{15}N$ hybrid DNA.

These results are consistent with our understanding of a semiconservative mode of DNA replication. In the first generation, the *E. coli* cells each had one parental strand of DNA containing ^{15}N and one newly synthesized strand of DNA containing ^{14}N. In the second generation, some of the cells contained the ^{15}N-labeled parental strand of DNA and a newly synthesized ^{14}N strand, and other cells contained the parental ^{14}N strand and a newly synthesized ^{14}N-containing strand.

and then renewing its polymerization activity. Some bacterial DNA polymerases can remove bases from the 3'OH end; others have exonuclease activity from the 5'P end. Because the same enzyme involved in the synthesis of the DNA molecule also removes and corrects errors in the DNA molecule, bacteria have an added safeguard for ensuring the accuracy of DNA replication.

Because the DNA polymerases can add nucleotides only to a 3'OH free end, the direction of DNA synthesis is

5'P →3'OH (see Figure 7.6). The fact that DNA polymerases can add deoxynucleotides only to the 3'OH free end of a nucleic acid primer creates a paradox for our understanding of the replication of DNA. The two strands of the double helical DNA molecule are antiparallel, one strand running from the 5'P to the 3'OH free end and the other complementary strand running from the 3'OH to the 5'P free end. Therefore, synthesis of complementary strands requires that DNA synthesis proceed in opposite directions, while the double helix is progressively unwinding and replicating in only one direction. One of the DNA strands can be continuously synthesized because it runs in the appropriate direction for the continuous addition of new free nucleotide bases to the free 3'OH end of the primer molecule (Figure 7.7). Synthesis of this **continuous** or **leading strand of DNA** occurs simultaneously with the unwinding of the double helical molecule and progresses toward the replication fork. The other strand of DNA, however, must be synthesized discontinuously. The initiation of the synthesis of the **discontinuous strand of**

DNA begins only after some unwinding of the double helix has occurred and therefore lags behind the synthesis of the continuous strand; it is referred to as the **lagging strand**.

Short segments of DNA, known as **Okazaki fragments**, are formed by the DNA polymerase running opposite to the direction of unwinding of the parental DNA molecule. The Okazaki fragments are joined together by the action of **ligases**, which establish phosphodiester bonds between the 3'OH and 5'P ends of chains of nucleotides. Ligases are not involved in chain elongation; rather, they act as repair en-

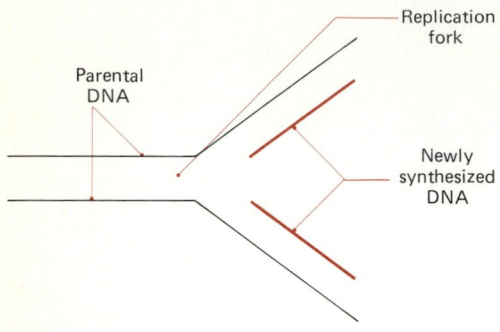

FIGURE 7.5
Replication of the double-stranded DNA molecule requires localized unwinding and the establishment of a replication fork.

FIGURE 7.6
The action of DNA polymerase results in the formation of a diester linkage and the elongation of the DNA chain. This reaction is pulled in the direction of DNA synthesis by the splitting of PP_i to P_i, an essentially irreversible reaction that results in the removal of the products of the polymerase reaction. The formation of the phosphate diester linkage between two sugar molecules occurs between the 3- and 5-carbon atoms. Thus the polymerized molecule has a free 3'OH group at one end and a free 5'P group at the other end.

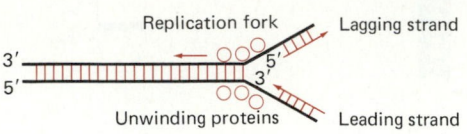
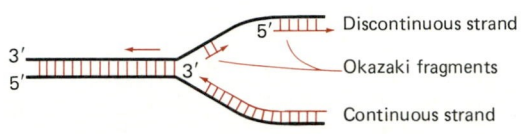
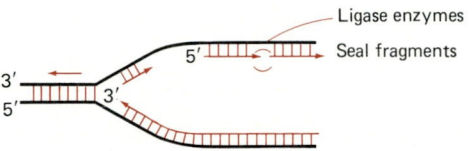

FIGURE 7.7 (above)
The unidirectional nature of DNA polymerase and the antiparallel nature of the double-stranded DNA molecule indicate that one strand of DNA is synthesized continuously and the other discontinuously.

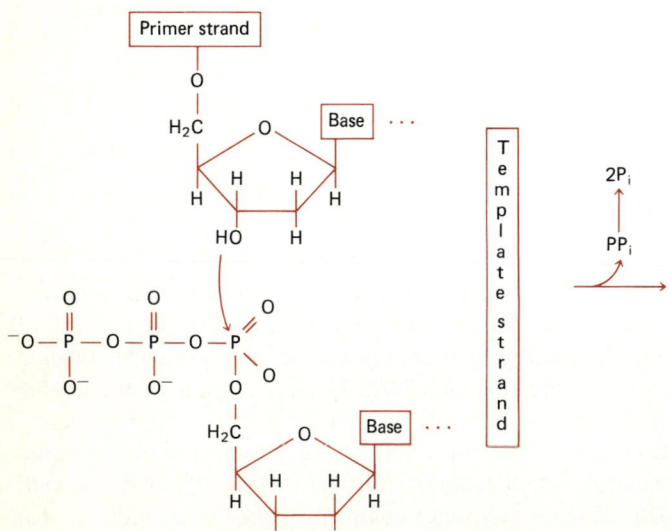
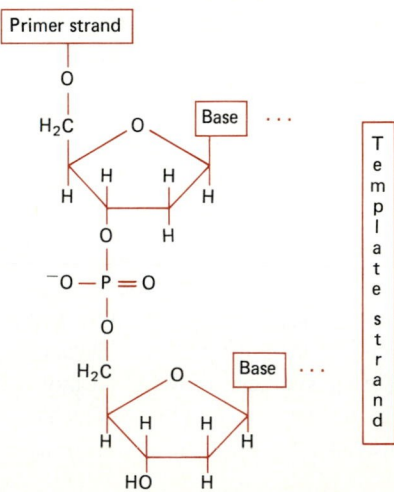

zymes for sealing "nicks" within the DNA molecule. Thus, through the combined actions of DNA polymerase and DNA ligase enzymes, both complementary strands of the DNA can be synthesized during DNA replication.

Differences in DNA Replication in Prokaryotes and Eukaryotes

Although the basic mechanism of DNA polymerization is identical in all microorganisms, there are some differences between the actual replication of the DNA of prokaryotic and eukaryotic microorganisms. The bacterial chromosome occurs as a circular macromolecule, whereas the DNA macromolecules of eukaryotic chromosomes are linear. The replication of a circular DNA molecule must necessarily be different from that of a linear DNA molecule. Whereas DNA

is the universal macromolecule for storing genetic information, and DNA replication is semiconservative in both eukaryotic and prokaryotic cells, it is now becoming apparent that there are major differences in how the information is stored within the DNA and how it is processed and replicated in prokaryotic and eukaryotic organisms.

One major difference between prokaryotic and eukaryotic DNA replication concerns the number of sites at which replication can begin. The bacteria that have been examined all have a single point of origin of DNA replication. Polymerases move bidirectionally from the origin to the terminus of DNA replication, so that there are two replicating forks moving in opposite directions. The bidirectional movements of the replication forks produce a loop of DNA that extends out of the plane of the parental bacterial chromosome (Figure 7.8). In *E. coli* and presumably in other bacteria, the ter-

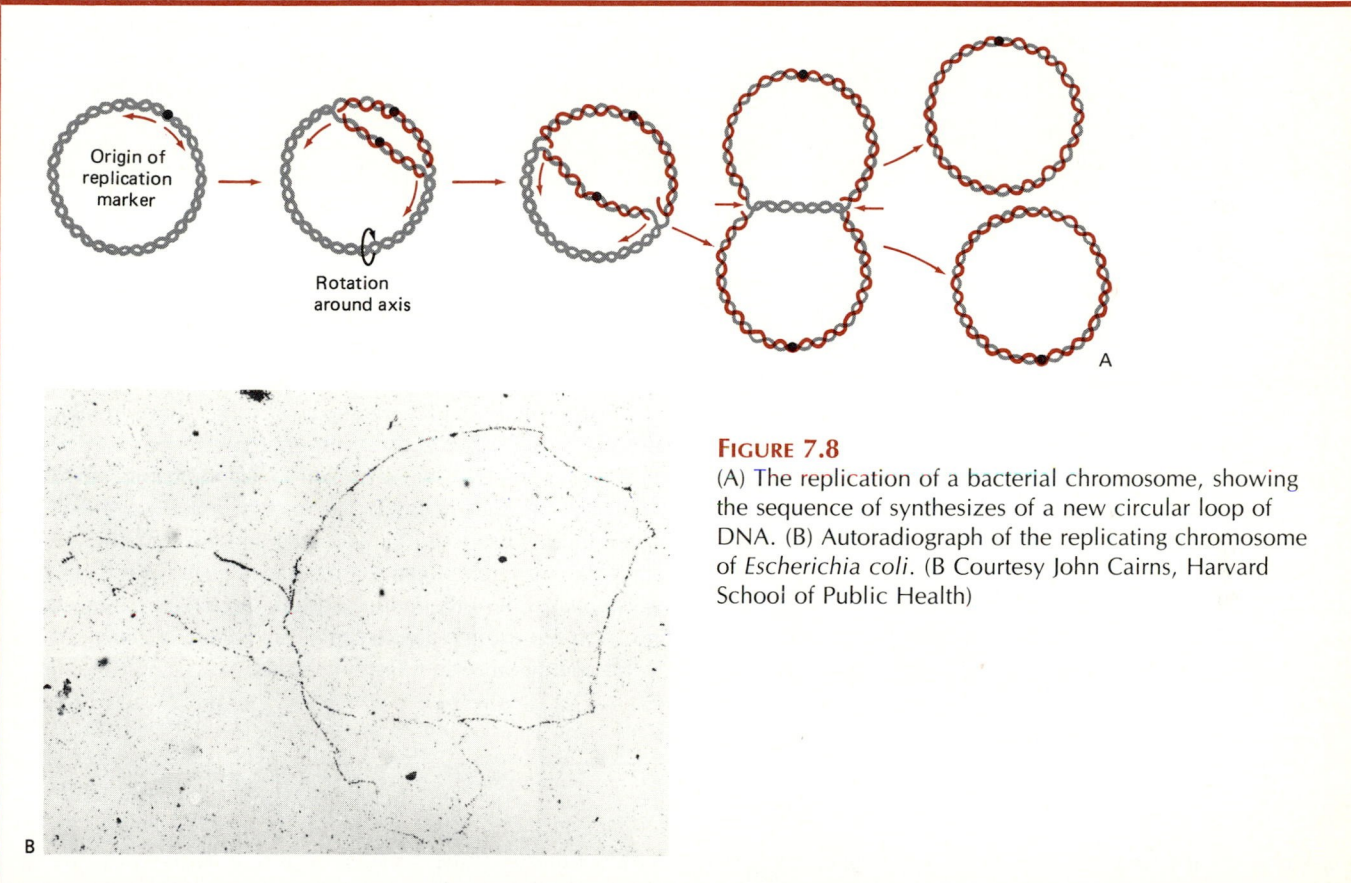

Figure 7.8

(A) The replication of a bacterial chromosome, showing the sequence of synthesizes of a new circular loop of DNA. (B) Autoradiograph of the replicating chromosome of *Escherichia coli*. (B Courtesy John Cairns, Harvard School of Public Health)

Discovery Process

The circular nature of the bacterial chromosome and the fact that during synthesis the new circular loop DNA grows out of the plane of the parental DNA were demonstrated by John Cairns using autoradiography. In this method, bacteria are grown in the presence of a radioactive compound, e.g., tritiated thymidine, which is incorporated into the DNA. The bacterial cells are lysed, releasing the radioactive DNA molecules. A fine-grained photographic emulsion is placed over the bacterial cells and incubated; the areas of radioactivity are detected when the film is developed, as seen in this autoradiograph of the replicating chromosome of *E. coli*. The visualization of the bacterial chromosome by autoradiography leaves no doubt as to its circular nature.

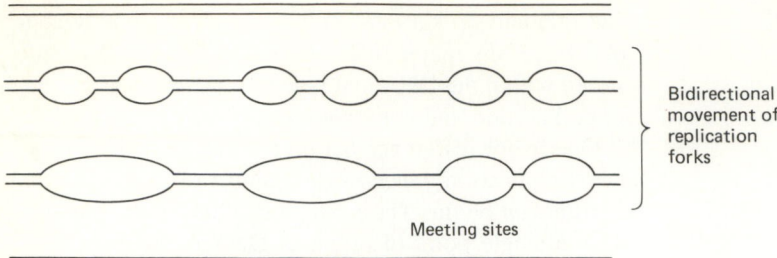

Initiation sites

Bidirectional
movement of
replication
forks

Meeting sites

A

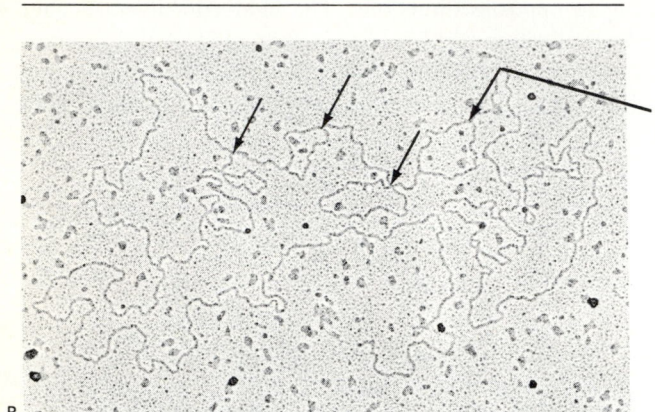

Replication
forks

B

FIGURE 7.9
(A) As shown in this illustration, DNA replication begins at multiple points in eukaryotes, producing multiple replication forks. (B) The multiple replication forks of a eukaryotic chromosome are visible in this electron micrograph of a replicating DNA molecule. (B Courtesy D. R. Wolstenholme, University of Utah; reprinted by permission of Springer-Verlag, from D. R. Wolstenholme, 1973, *Chromosoma*, 43:1–18)

minus for DNA replication is exactly opposite the origin in the circular bacterial chromosome, and the bidirectional replication of the DNA proceeds at identical speeds, meeting precisely at the termination site.

In contrast, the replication of eukaryotic DNA begins at multiple points of origin (Figure 7.9). In this process, DNA replication proceeds bidirectionally, utilizing polymerase and ligase enzymes that are analogous to those of prokaryotes. The rate of synthesis of DNA along a replicating fork may be slower in eukaryotic than in prokaryotic microorganisms, and although the replication of DNA in prokaryotes proceeds at a uniform rate, the rate of DNA synthesis can vary in a eukaryote. Despite these potential differences in the rates of

DNA synthesis within a particular region of DNA, the overall rate of DNA replication is higher in eukaryotes than in prokaryotes. This is because the DNA of eukaryotes has multiple **replicons** (segments of a DNA macromolecule having their own origin and termini) compared to the single replicon of the bacterial chromosome. Consequently, even though there is much more DNA in a eukaryotic chromosome than in a bacterial chromosome, the eukaryotic genome can be replicated much faster than the bacterial genome because of these multiple initiation points for DNA synthesis. This distinction between single and multiple origins of DNA synthesis appears to represent a fundamental difference between prokaryotic and eukaryotic microorganisms.

7.3 *RNA Synthesis: Transcription*

Regardless of the mechanisms used for replicating the genome, organisms depend on their ability to use genetic information in a functional manner. As indicated earlier, this expression of genetic information depends on a two-stage process in which the genome of an organism is used to direct the synthesis of RNA and thence proteins. **Transcription** is the process in which the information stored in the DNA is used to code for the synthesis of RNA. In transcription DNA serves as a template for synthesis of RNA, accomplishing a critical transfer of information for the eventual expression of genetic information. The transcription of DNA results in the

production of three classes of RNA: rRNA, tRNA, and mRNA. All three classes of RNA are required for the expression of genetic information.

Transcription is similar in several ways to DNA replication (Figure 7.10). The molecule of RNA that is synthesized is antiparallel to the strand of DNA that serves as a template. The process of transcription involves unwinding of the double helical DNA molecule for a short sequence of nucleotide bases, alignment of complementary ribonucleotides by base pairing opposite the nucleotides of the DNA strand being transcribed, and linkage of these nucleotides with phospho-

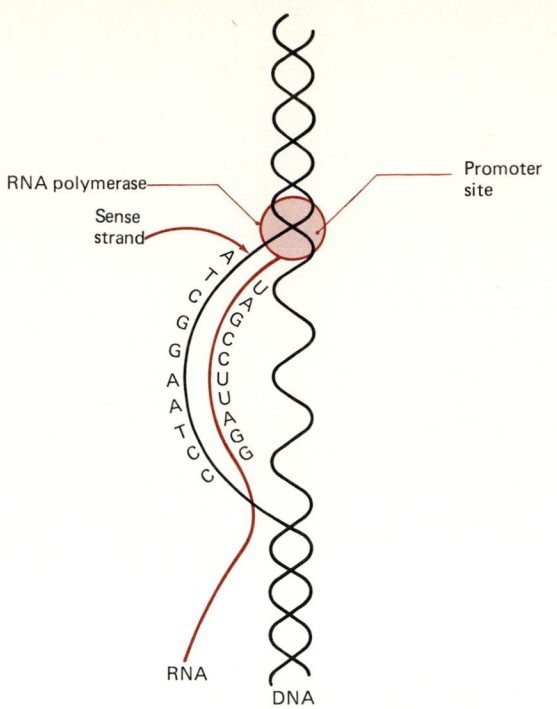

FIGURE 7.10
During transcription one strand of the DNA codes for the synthesis of RNA. The process of transcription begins at the promoter site, where RNA polymerase begins its job of linking the aligned nucleotides of the RNA molecule being synthesized.

FIGURE 7.11
This autoradiograph shows transcription occurring at multiple sites along a region of the genome of *E. coli* (165,000×). The double-stranded DNA is visible as thin parallel tracks and the RNA as dark chains growing away from the DNA. (Courtesy of O. L. Miller, Jr., University of Virginia; reprinted by permission of the American Association for the Advancement of Science, from O. L. Miller, Jr., B. A. Hamkalo, and C. A. Thomas, Jr., 1970, *Science* 169:392–397)

diester bonds by a DNA-dependent RNA polymerase.[2] RNA polymerases are able to link nucleotides only to the 3'OH free end of the polymer; thus, the synthesis of RNA, like that of DNA, occurs in a 5'P → 3'OH direction.

Although transcription is similar in several ways to DNA replication, there are some major differences between RNA and DNA synthesis. The RNA that is synthesized is single-stranded. Thus, for a given region, only one strand of DNA serves as a template; the strand of DNA coding for the synthesis of RNA is known as the **sense strand**.[3] Another difference is that the RNA polymerase is capable of linking two nucleotides only as long as they are aligned opposite the complementary DNA template nucleotides; therefore, unlike DNA replication, RNA synthesis does not require a primer. Bacteria have one basic type of RNA polymerase that produces all three classes of RNA molecules. In *E. coli* there is only one form, although other bacteria may possess several variants of the basic type of RNA polymerase. The evidence suggests that eukaryotic microorganisms have three distinct polymerase enzymes that are responsible for the synthesis of the three different classes of RNA.

Initiation and Termination Sites of Transcription

The transfer of information from DNA to RNA requires that transcription begin at precise locations. There are multiple initiation sites for transcription along the DNA molecule in both prokaryotes and eukaryotes (Figure 7.11). Different initiation sites are needed to begin the synthesis of different classes of RNA and the synthesis of RNA for different polypeptide sequences. DNA contains specific sequences of nu-

[2]Binding of RNA polymerase to DNA may also be involved in localized unwinding and proper alignment of complementary RNA nucleotides.

[3]Both strands of the DNA can serve as sense strands in different regions. The term *sense strand* is applied only to the specific region of the DNA that is being transcribed.

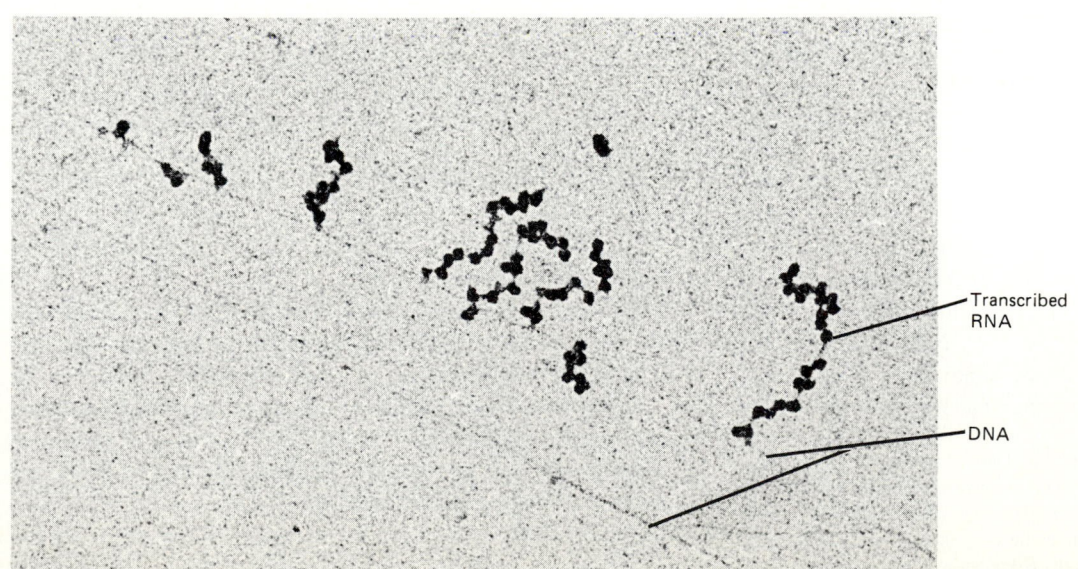

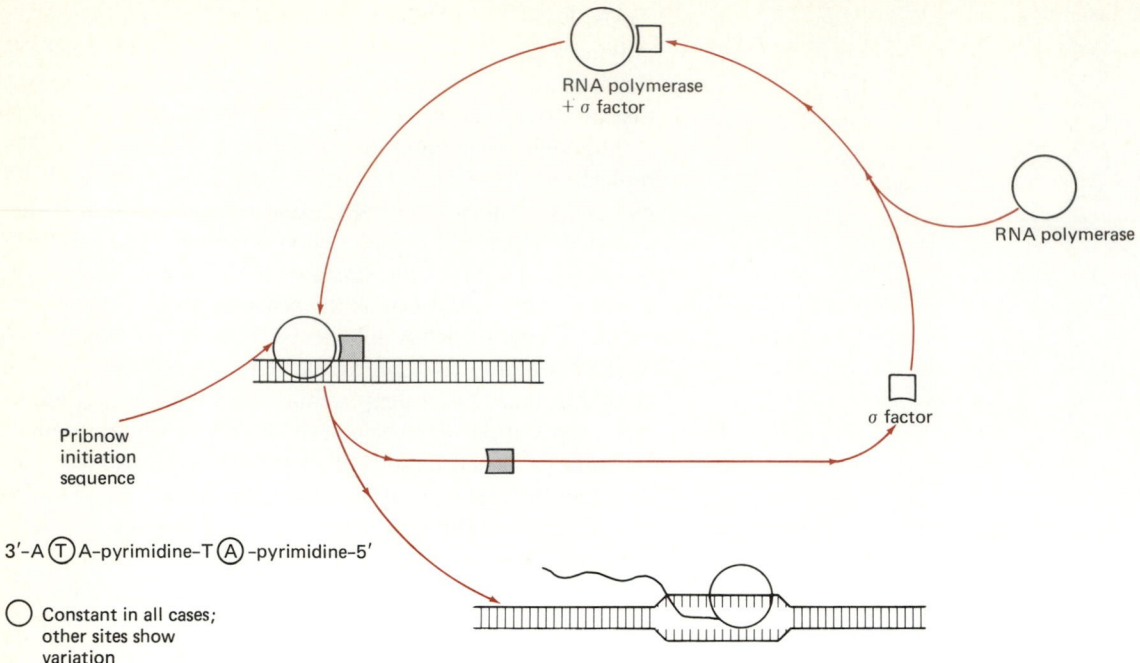

FIGURE 7.12

The site of binding of RNA polymerase to the promoter region is specified by the Pribnow sequence. Binding requires the addition of the sigma unit (σ factor) to the RNA polymerase.

cleotides, known as **promoter regions**, that serve as signals for the initiation of transcription. The promoter region of DNA is the site where RNA polymerase binds for transcription. The presence of the promoter region specifies both the site of transcription initiation and which of the two DNA strands is to serve as the sense strand for transcription in that region.

The promoter regions in the DNA of bacteria that have been examined consist of about 40 nucleotides and contain a seven-nucleotide sequence, known as the **Pribnow sequence**, which appears to be a key part of the recognition signal. A second recognition site, located about 35 bases from the start of the mRNA, has also been implicated in the initial binding of the RNA polymerase needed for the initiation of transcription. The binding of **RNA polymerase** to the promoter region is dependent on the presence of a subunit of the total RNA polymerase, known as the **sigma (σ) factor** (Figure 7.12).[4] Without the sigma subunit, the RNA polymerase fails to exhibit the necessary specificity for recognizing the initiation sites for transcription. The sigma factor ensures that RNA synthesis begins at the correct site. After the binding of the RNA polymerase molecule to the DNA, the sigma subunit dissociates from the RNA polymerase. The sigma subunit is then free to associate with another RNA polymerase molecule, completing that molecule and estab-

---

[4]The entire RNA polymerase has two identical alpha units, two very similar beta units, and the sigma unit; it is designated $\alpha_2\beta\beta'\sigma$ to specify the five subunits that compose the complete enzyme. The RNA polymerase without the sigma subunit is known as the *core enzyme*, and the complete RNA polymerase (core + sigma unit) is known as the *holoenzyme*.

```
             C
          /     \
       U         C
                  \
       U           G
          \       /
           G—C
           A—U
           C—G
           C—G
           G—C
           C—G
           C—G
           G—C
  U A A U C C C A C A G        A U U U U —OH
  5'                                      3'
```

FIGURE 7.13

The stem-and-loop structure of mRNA results from the transcription termination sequence. The sequence of nucleotides at the 3'OH end of the mRNA transcript allows a stable hairpin structure to form.

lishing the necessary specificity for the recognition of the transcriptional initiation site.

In addition to the specific initiation sites, there are specific **termination sites** that act as a signal to stop transcription. The termination sequence of nucleotides in the DNA contains a region with an abundance of GC bases followed by a region with an abundance of AT bases. The GC-rich region exhibits a symmetry that enables the synthesized RNA to fold back on itself, forming a stem and loop (Figure 7.13). The A bases in the AT-rich region code for a terminal sequence of several U nucleotides. These termination sequences cause the RNA polymerase to pause. In some cases, once the RNA polymerase pauses, transcription is terminated, but in other cases a protein, referred to as the **rho protein**, is required to interrupt transcription.

Differences in the Synthesis of mRNA in Prokaryotes and Eukaryotes

It is the mRNA molecules formed by transcription that actually encode the information for protein synthesis. In prokaryotes, there is usually only one DNA sequence coding for a particular mRNA, whereas in eukaryotes there often are multiple copies of genes coding for the same mRNA molecules. In both prokaryotic and eukaryotic microorganisms, however, there can be multiple copies of a particular mRNA macromolecule, enabling establishment of multiple sites of synthesis for identical proteins using the multiple copies of the mRNA molecule.

There is a drastic difference between prokaryotic and eukaryotic cells in the longevity of their mRNA molecules. In prokaryotic cells the mRNA molecules last for only a few minutes, whereas in eukaryotic cells an mRNA molecule can remain functional for hours or days. The long period of activity of an mRNA molecule in a eukaryotic cell imparts relative stability to the protein complement, compared to the changing situation in a prokaryotic cell, where the mRNA molecules are quickly degraded. The bacterial cell, as a result, can rapidly alter its metabolism in response to changing environmental conditions, whereas eukaryotic microorganisms are better adapted for continuous metabolism in a stable environment.

In addition to the differences in the half-lives of mRNA molecules in prokaryotic and eukaryotic cells, there is a fundamental difference in the way the mRNA is formed and whether transcription leads to the direct formation of mRNA or whether additional processing is necessary after transcription to form functional mRNA. The mRNA molecule in bacteria is not modified much between the time it is synthesized, through transcription of the DNA, and the time it is translated into the amino acid sequence of a polypeptide at the ribosome. In bacteria the mRNA molecule is transcribed directly from DNA, and the DNA nucleotides coding for the mRNA occur as a contiguous sequence (Figure 7.14). Often bacterial mRNA is **polycistronic**; that is, the mRNA contains the information for several proteins—often with related functions—coded for by a continuous region of the DNA. The mRNA molecule of bacteria contains a sequence of nucleo-

tides at the beginning and end of the molecule that do not code for specific amino acids in a polypeptide sequence. Rather, the beginning or **leader sequence** of nucleotides in the mRNA molecule is involved in the initiation of protein synthesis at the ribosomes. The nontranslated leader promotes binding of mRNA to the ribosomes. The attachment of mRNA to the ribosomes in prokaryotic cells often occurs before transcription of the mRNA is complete, indicating the close proximity of the transcriptional and translational processes.

In contrast to the formation of mRNA in prokaryotic cells, the RNA molecules of eukaryotic cells are generally modified extensively after transcription from DNA to form mRNA. The mRNA that is formed is **monocistronic** (contains only the information for one polypeptide sequence), and the transcriptional and translational processes are spatially and temporally separated. The precursor of mRNA in eukaryotes, known as **heterogeneous nuclear RNA (hnRNA)**, is several times larger than the final mRNA molecule and is subjected to substantial **post transcriptional modification** within the nucleus to form the mRNA. The processing of the hnRNA involves removing, adding, and rearranging sequences of nucleotides. The 5'P end of the hnRNA is capped with an inverted guanosine triphosphate (GTP) residue, some of the terminal adenine bases are methylated, and the 3'OH end of the hnRNA molecule is modified by the addition of a sequence of adenosine nucleotides (polyA tail).

Perhaps the most surprising discovery concerning the transcription of DNA and the processing of hnRNA is the occurrence of intervening sequences or **introns** (Figure 7.15). (The regions that code for amino acid sequences are known as **exons**.) Introns do not code for amino acid sequences, and the reason for their existence is unknown. In one case in yeast mitochondria, an intron within the gene for one protein has been found to code for the removal of introns

FIGURE 7.15

In eukaryotes extensive posttranscriptional modification of the primary RNA transcript, including the removal of introns, is needed to produce the mRNA. As a result of these modifications, eukaryotic mRNA is not colinear with DNA. In this example, gene G is split into three regions, a, b, and, c, designated in the DNA as Ga, Gb, and Gc.

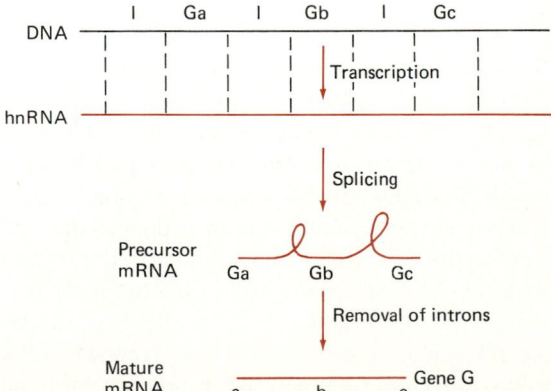

FIGURE 7.14

In prokaryotes mRNA is colinear with the region of the DNA that is transcribed.

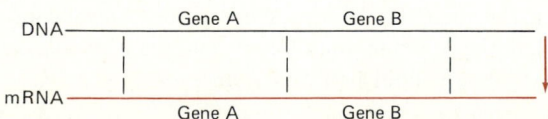

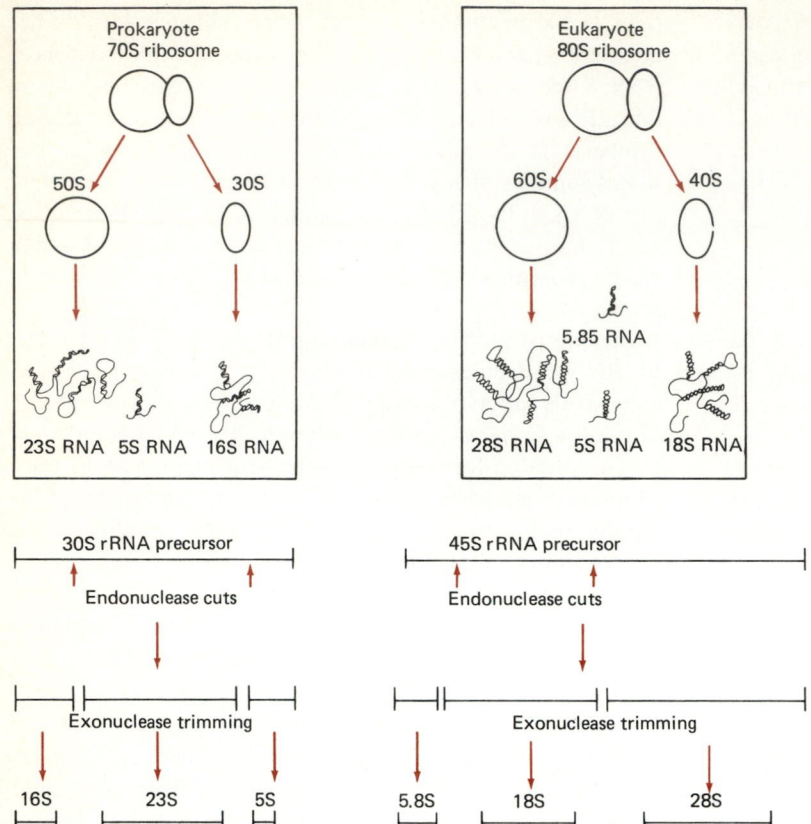

FIGURE 7.16

Posttranscriptional modification of RNA is needed to produce the rRNA molecules of the 70S and 80S ribosomes.

from genes that are subsequently transcribed, suggesting that introns are important regulators of gene function. Part of the processing of hnRNA involves the excision of introns to form the mature mRNA molecule. The rearrangement of RNA molecules transcribed from DNA to form a mature mRNA molecule involves cutting out and splicing together pieces of RNA. Thus, the eukaryotes are said to have **split genes** because the nucleotides that comprise the gene that codes for a specific protein are separated in the DNA. As a result, mRNA molecules of eukaryotic microorganisms generally are not colinear with the DNA molecule; that is, the sequence of nucleotide bases in an mRNA is not complementary to the specific contiguous linear sequence of bases in a DNA molecule.

Synthesis of tRNA and rRNA

Unlike mRNA, which is not modified after transcription in bacteria, both rRNA and tRNA molecules are substantially modified in both prokaryotic and eukaryotic cells. Although RNA molecules are single-stranded, they can fold back on themselves, establishing double-stranded regions. Both tRNA and rRNA molecules have extensive double-stranded regions that arise from the folding of the primary RNA chain. In both prokaryotic and eukaryotic cells, the RNA molecules transcribed from DNA are larger than the rRNA molecules found in the ribosomes (Figure 7.16). The precursor rRNA molecules must therefore be processed in order to form the

rRNA molecules. The ribosomes of prokaryotes contain 5S, 16S, and 23S RNA. The initial transcript from the DNA, however, is a large 30S molecule that can be cleaved by nuclease enzymes to form these different-sized RNA molecules. In eukaryotic microorganisms a large precursor RNA molecule is cleaved to form the 28S, 18S, and 5S rRNA molecules that make up the RNA portions of the ribosomes. (In some cases, a separate large precursor is used for the production of the 5S rRNA molecules.)

The tRNA molecules are similarly synthesized as high molecular weight precursors that are then processed to produce the mature tRNA molecules (Figure 7.17). tRNA molecules have a multilobed structure that is formed because of hydrogen binding between complementary regions of the molecule. All tRNA molecules have a 3'OH terminus with the nucleotide sequence CCA 3' that may be encoded in the primary nucleotide sequence or added enzymatically as a cap after transcription from the DNA template. Within the tRNA molecule, several of the nucleotides are modified to form unusual nucleotide bases through the posttranscriptional modification of the normal RNA nucleotides. Some of the unusual nucleotide bases found in tRNA are pseudouridine, dihydrouridine, ribothymidine, and inosine. The specific functions of the unusual bases have not been established, but their hydrophobic nature may be important in the interactions of tRNA and ribosomes. The presence of the unusual nucleotides and the specific multilobed configuration of the tRNA molecule distinguish it from other classes of RNA molecules.

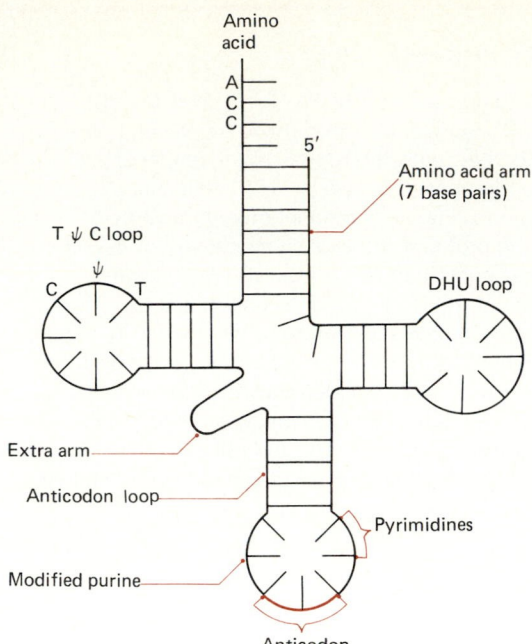

FIGURE 7.17
All tRNA molecules have a characteristic four-lobe structure that results from internal base pairing of some of the nucleotides. Each lobe of the tRNA molecule has a distinct function. Several of the lobes are characterized by the inclusion of unusual nucleotides. These nucleotides are formed by enzymatic modification of the nucleotides directly coded for by the DNA; that is, the DNA does not have additional nucleotides that directly call for the insertion of nucleic acid bases other than adenine, uracil, cytosine, and guanine into the RNA. One of the lobes, designated the *DHU* or *D loop*, contains dihydrouracil (DHU). This lobe binds to the enzyme involved in forming the peptide during translation. The TψC loop contains the sequence ribothymine (T), pseudouracil (ψ), and cytosine (C). This lobe binds to the ribosome. A third loop, which also contains modified purines, is designated the *anticodon loop* because it is complementary to the region of the mRNA, the codon, that specifies the amino acid to be incorporated during protein synthesis. The final end always has the terminal sequence 3'OH ACC, which is where the amino acid binds. This terminal sequence is usually referred to as the *CCA end*, reading from the 5'P end of the tRNA molecule.

7.4 *Protein Synthesis: The Translation of the Genetic Code*

All three types of RNA (rRNA, tRNA, and mRNA) are involved in transferring the information obtained from the sequence of nucleotides in DNA to the sequence of amino acids in the polypeptide chain of a protein. Translation of the genetic information into protein molecules, which can functionally express genetic information, occurs on the ribosomes. Ribosomes provide the spatial framework and structural support for aligning the translational process of protein synthesis. Distortion of the proper configuration of the ribosome can prevent the proper informational exchange and expression of the genetic information, forming the basis for the action of many antibiotics. tRNA molecules carry the amino acids to the site of protein synthesis and properly align amino acids for incorporation into the polypeptide chain. It is the mRNA molecule, however, that actually contains the coded information that specifies the sequence of amino acids in the polypeptide chain.

The Genetic Code

Within the mRNA three sequential nucleotides are used to code for a given amino acid; the **genetic code**, therefore, is called a **triplet code** (Figure 7.18). Each triplet nucleotide sequence is known as a **codon**. Because there are 4 different nucleotides, there are 64 possible codons; that is, there are 64 possible three-base combinations of the 4 different nucleotides. The genetic code, which is almost universal, can therefore be said to have 4 letters in the alphabet and 64 words in the dictionary, each word containing 3 letters. Although there are 64 possible codons, proteins in biological systems normally contain only 20 L-amino acids.[5] Thus, there are many more codons than are needed for the translation of genetic information into functional proteins. More than one codon can code for the same amino acid, and therefore, the genetic code is said to be **degenerate**. Stated another way, the genetic code is redundant, with several codons coding for the insertion of the same amino acid into the polypeptide chain. Additionally, there are three **nonsense codons**, so named because they do not code for any amino acid. Nonsense codons do, though, serve a very important function, acting as punctuators that signal the termination of the synthesis of a polypeptide chain.

Translating the Information in mRNA

Translation of the information in the mRNA molecule, that is, reading of the codons, is a directional process (Figure 7.19). mRNA is read in a 5'P → 3'OH direction, and the polypeptide is synthesized from the amino terminal to the carboxyl terminal end. The mRNA molecule is read one codon (three nucleotides) at a time. There are no spaces between the codons. Therefore, establishing a reading frame,

[5]Some proteins do contain other amino acids, but in such cases the unusual amino acids are usually formed by posttranslational enzymatic modification of the chemical structure rather than by coding sequences that specify the unusual amino acids.

FIGURE 7.18

The codons of the genetic code. Each three-base codon calls for a specific amino acid or acts as a stop signal.

Second letter

First letter	U	C	A	G	Third letter
U	UUU ⎤ Phe UUC ⎦ UUA ⎤ Leu UUG ⎦	UCU ⎤ UCC ⎥ Ser UCA ⎥ UCG ⎦	UAU ⎤ Tyr UAC ⎦ UAA Stop UAG Stop	UGU ⎤ Cys UGC ⎦ UGA Stop UGG Trp	U C A G
C	CUU ⎤ CUC ⎥ Leu CUA ⎥ CUG ⎦	CCU ⎤ CCC ⎥ Pro CCA ⎥ CCG ⎦	CAU ⎤ His CAC ⎦ CAA ⎤ Gln CAG ⎦	CGU ⎤ CGC ⎥ Arg CGA ⎥ CGG ⎦	U C A G
A	AUU ⎤ AUC ⎥ Ile AUA ⎦ AUG Met	ACU ⎤ ACC ⎥ Thr ACA ⎥ ACG ⎦	AAU ⎤ Asn AAC ⎦ AAA ⎤ Lys AAG ⎦	AGU ⎤ Ser AGC ⎦ AGA ⎤ Arg AGG ⎦	U C A G
G	GUU ⎤ GUC ⎥ Val GUA ⎥ GUG ⎦	GCU ⎤ GCC ⎥ Ala GCA ⎥ GCG ⎦	GAU ⎤ Asp GAC ⎦ GAA ⎤ Glu GAG ⎦	GGU ⎤ GGC ⎥ Gly GGA ⎥ GGG ⎦	U C A G

Discovery Process

Polyuridylic acid (. . . –U–U–U–U . . .) was the first synthetic mRNA created. It was made by reacting only uracil nucleotides with the RNA-synthesizing enzyme polynucleotide phosphorylase. In 1961 Marshall W. Nirenberg and J. Heinrich Matthei mixed this poly-U *in vitro* with the protein-synthesizing machinery of *E. coli* in a system that contained ribosomes, tRNAs, enzymes, amino acids, and ATP and GTP as energy donors. They succeeded in synthesizing a protein, which was identified as polyphenylalanine, a string of phenylalanine molecules attached to form a polypeptide. They concluded that the triplet U–U–U must code for phenylalanine. Additional mRNAs of varying compositions and complexities were used, and by 1966 all 64 code words had been identified.

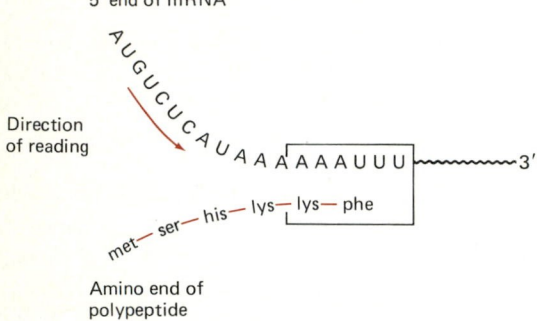

Direction of reading

Amino end of polypeptide

FIGURE 7.19

Protein synthesis (translation) is a directional process. The mRNA is read in the 5′P to 3′OH direction, and the peptide chain is synthesized from the amino end to the carboxylic acid end.

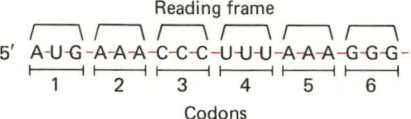

FIGURE 7.20

The initiating codon of mRNA establishes the reading frame. The contiguous bases are read three at a time, with no spacing between adjacent codons.

that is, determining which nucleotide is used to initiate the reading of the three-nucleotide sequences, is critical for extracting the proper information. Within the mRNA molecule there are sequences of nucleotides that define the beginning and end of each encoded polypeptide chain. Here we see the importance of having a mechanism for recognizing direction in the informational macromolecules. Just as we establish a convention for reading the English language from left to right, the correct interpretation of the information stored in the mRNA molecule requires that it be read from the 5′P to the 3′OH free end.

Initiating the Translation of mRNA   The reading of an mRNA molecule begins from a fixed starting point. The first codon normally read is AUG, which codes for the amino acid methionine. In eukaryotic microorganisms, methionine is always the first amino acid of the polypeptide sequence.

In prokaryotic cells the codon AUG codes for *N*-formylmethionine to initiate the polypeptide chain, although the same codon (AUG) codes for methionine when it occurs elsewhere in the mRNA molecule. In some bacteria the codon GUG acts as the initiator codon that specifies *N*-formylmethionine to initiate the polypeptide chain, although when this codon occurs at an internal position in the mRNA molecule, it codes for valine. Whereas either methionine or *N*-formylmethionine is the initial amino acid in all peptide chains, these terminal amino acids can later be enzymatically removed, so that not all polypeptides found in microorganisms have methionine at their amino terminal ends.

The initiating codon establishes the **reading frame** (three-nucleotide sequences) for the rest of the mRNA molecule (Figure 7.20). Because the nucleotide bases are read three at a time along a continuous chain of nucleotides, changing the reading frame by a single nucleotide base can completely alter the informational content of the mRNA molecule. Several proteins are required as initiating factors to begin the translational process. The actual sequence for the initiation of translation may depend on the base pairing of a region of the leader sequence of the mRNA to the 16S rRNA in prokaryotes or the 18S rRNA in eukaryotes. The initiation of polypeptide synthesis begins with the association of the 30S or 40S ribosomal subunit with the mRNA, formylmethionyl tRNA, or methionyl tRNA and GTP. The 50S or 60S ribosomal subunits then join to establish the 70S or 80S ribosomes in prokaryotes and eukaryotes, respectively. The association of the ribosomal subunits, which is essential for protein synthesis, requires magnesium ions.

The Role of tRNA   Before going on, we need to consider the role of tRNA in the translation process. (See Figure 7.17 for the structure of tRNA.) A tRNA molecule contains approximately 80 nucleotides, about half of which exhibit base pairing. The tRNA molecule forms four lobes, some of which are characterized by the presence of unusual nucleotides. tRNA brings the amino acids to the ribosomes and properly aligns them during translation. The binding of the amino acid to the tRNA molecule occurs at the 3′OH end of the tRNA molecule. Attachment of an amino acid to its specific tRNA molecule is called **charging** (Figure 7.21); a tRNA molecule with its attached amino acid is said to be *charged*. Charging tRNA molecules requires ATP and an aminoacyl synthetase enzyme. There is at least one aminoacyl synthetase enzyme for each of the 20 amino acids. There are at least 20 different types of tRNA molecules, with each of the 20 amino acids that occur in proteins binding with a different tRNA molecule. The structure of the tRNA molecule plays a role in establishing the proper alignment of molecules during translation. One of the four lobes is attached to the amino acid; one contains a nucleotide sequence that interacts with the rRNA, establishing the proper orientation to the ribosome; one interacts with aminoacyl synthetase; and one contains a region, the anticodon, that interacts with mRNA.

The **anticodon** has three nucleotides that are complementary to the three-based nucleotide sequence of the codon. It is the pairing of the codons of mRNA molecules and the anticodons of tRNA molecules that determines the order of the amino acid sequence in the polypeptide chain. The third base of the anticodon does not always properly recognize the third base of the mRNA codon. As a result, the tRNA molecules may **wobble**, resulting in the recognition by a tRNA anticodon of more than one codon.

Forming the Polypeptide   Returning to the sequence of events during translation, there are two sites on the ribosome involved in protein synthesis, the **peptidyl site** and the **aminoacyl site**. The aminoacyl site is the location where tRNA molecules bring individual amino acids to be sequentially inserted into the polypeptide chain. The peptidyl site is the location where the growing peptide chain is aligned.[6] When tRNA molecules move to the aminoacyl site, the proper anticodon pairs with its matching codon (Figure 7.22). The polypeptide chain is then transferred from the tRNA occupying the peptidyl site to the aminoacyl site. A peptide bond is formed between the amino group of the amino acid attached at the aminoacyl site and the carboxylic acid group of the last amino acid that had been added to the peptide chain. At the initiation of protein synthesis, *N*-formylmethionine in prokaryotes and methionine in eukaryotes are transferred to the aminoacyl site, forming a peptide bond with the second amino acid coded for by the mRNA molecule. The mRNA then moves along the ribosome by three nucleotides.

The movement of the mRNA, tRNA, and growing polypeptide chain along the ribosome, a process known as **translocation**, requires the input of energy from GTP. This is one of the rare biochemical reactions that requires a specific energy carrier other than ATP. Translocation moves the tRNA molecule with an attached peptide chain to the peptidyl site, leaving the aminoacyl site open for the anticodon of the next charged tRNA molecule to pair with the next codon of the mRNA. During translocation the tRNA molecule that has transferred its attached amino acids, and is thus no longer charged, returns to the cytoplasm, where it can be recycled, that is, where it can once again be charged with an amino acid and returned to the aminoacyl site. The process is repeated over and over, resulting in the elongation of the polypeptide chain. Eventually, one of the nonsense codons appears at the aminoacyl site. Because no charged tRNA molecule pairs with the nonsense codon, the aminoacyl site is empty when the peptide chain tries to transfer to that site. Therefore, the translational process is terminated and the polypeptide is released to the cytoplasm, where it can play a functional role in mediating the metabolism and structure of the organism.

[6]At the initiation of protein synthesis, the formylmethionyl tRNA or methionyl tRNA occupies the peptidyl site on the ribosome rather than the aminoacyl site, which is normally where the charged tRNA molecules associate with the ribosome.

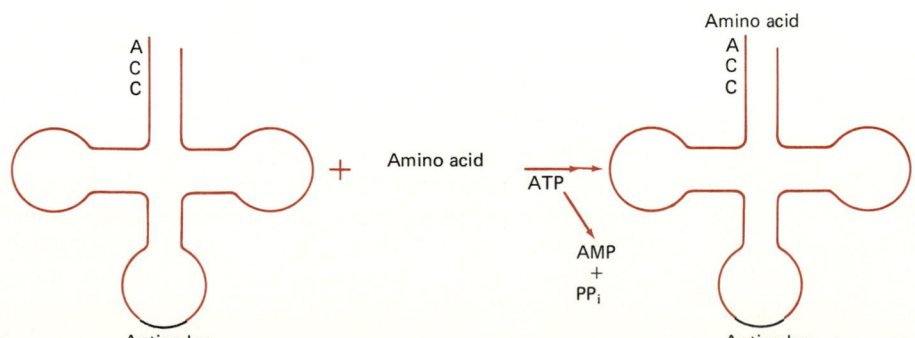

FIGURE 7.21

Charging of tRNA with its specific amino acid is necessary so that the tRNA can properly align the amino acid during translation and ensure that the correct amino acid will be placed in its proper sequence in the polypeptide chain.

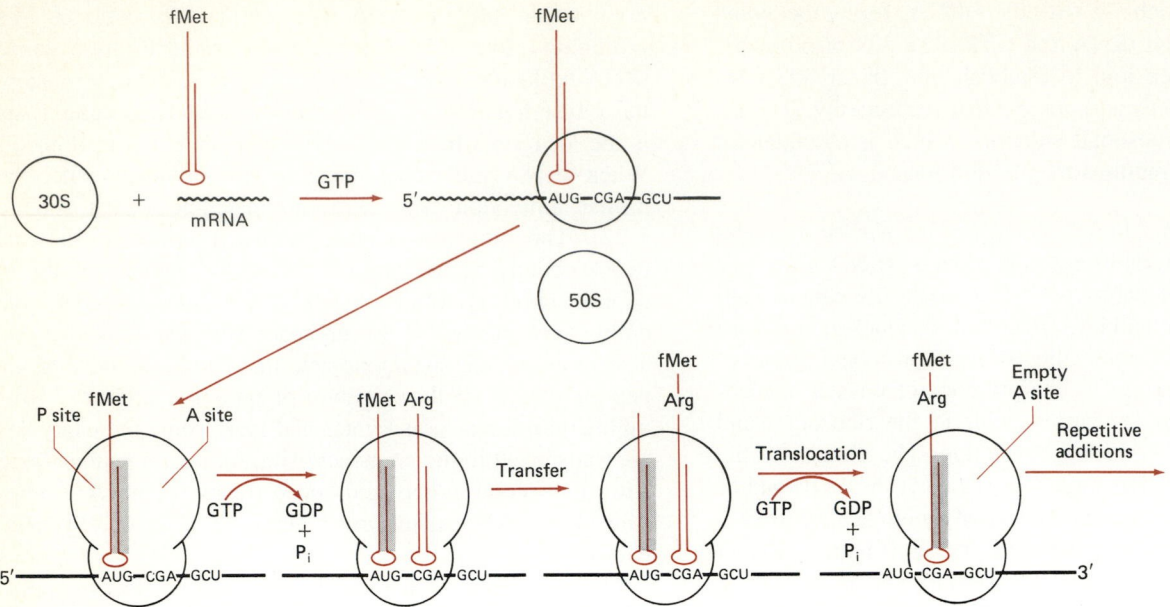

FIGURE 7.22

Translation is a multistep process in which the information in the mRNA is used to produce a polypeptide. This process occurs at the ribosomes that provide structural support for the proper alignment of the macromolecules involved in the translational process. During translation the anticodon of the charged tRNA pairs with the codon of the mRNA. Amino acids are transferred to the growing peptide chain as the mRNA is translocated along the ribosome, moving three bases at a time.

7.5 Regulation of Gene Expression

After considering the processes involved in converting the information stored in the DNA to functional proteins, we can now examine the regulatory mechanisms that control transcription and translation. The expression of the genetic information of microorganisms is subject to regulation at several levels. As we have already seen in Chapters 5 and 6, enzymes are subject to allosteric control; thus, once synthesized, the activities of enzymes are subject to careful regulation. Microorganisms are also able to control their metabolic activities by regulating protein synthesis. In addition to encoding the information for the specific polypeptide sequences of proteins, the genome of the cell codes the information that regulates its own expression. By controlling which of the genes of the organism are to be translated into functional enzymes, the cell is able to regulate its metabolic activities.

Some regions of DNA are specifically involved in regulating transcription, and these regulatory genes can control the synthesis of specific enzymes. In some cases gene expression is not subject to specific genetic regulatory control, and the enzymes coded for by such regions of the DNA are **constitutive**, that is, they are continuously synthesized. In contrast to constitutive enzymes, some enzymes are synthesized only when the cell requires them. Such enzymes are either **inducible**, that is, made only in response to a specific inducer substance, or **repressible**, that is, made unless stopped by

the presence of a specific repression substance (Figure 7.23). Often several enzymes that have related functions are controlled by the same regulatory genes. Induction and repression are based upon whether or not a regulatory repressor protein binds at a regulatory gene of the DNA, blocking transcription of the succeeding structural genes (Figure 7.24).

Regulating the Metabolism of Lactose: The lac Operon

One of the best-studied regulatory systems concerns the enzymes produced by *E. coli* strain K-12 for the metabolism of lactose. Three enzymes are specifically synthesized by *E. coli* for the metabolism of lactose: β-galactosidase, galactoside permease, and thiogalactoside transacetylase. The β-galactosidase cleaves the disaccharide lactose into the monosaccharides galactose and glucose. Permease is required for the transport of lactose across the bacterial cytoplasmic membrane, and the role of transacetylase is not yet established. The structural genes that code for the production of these three enzymes occur in a contiguous segment of the DNA, that is, a polycistron. Although there is some basal level of gene expression in the absence of an inducer, these structural genes are appreciably transcribed only in the presence of an inducer.

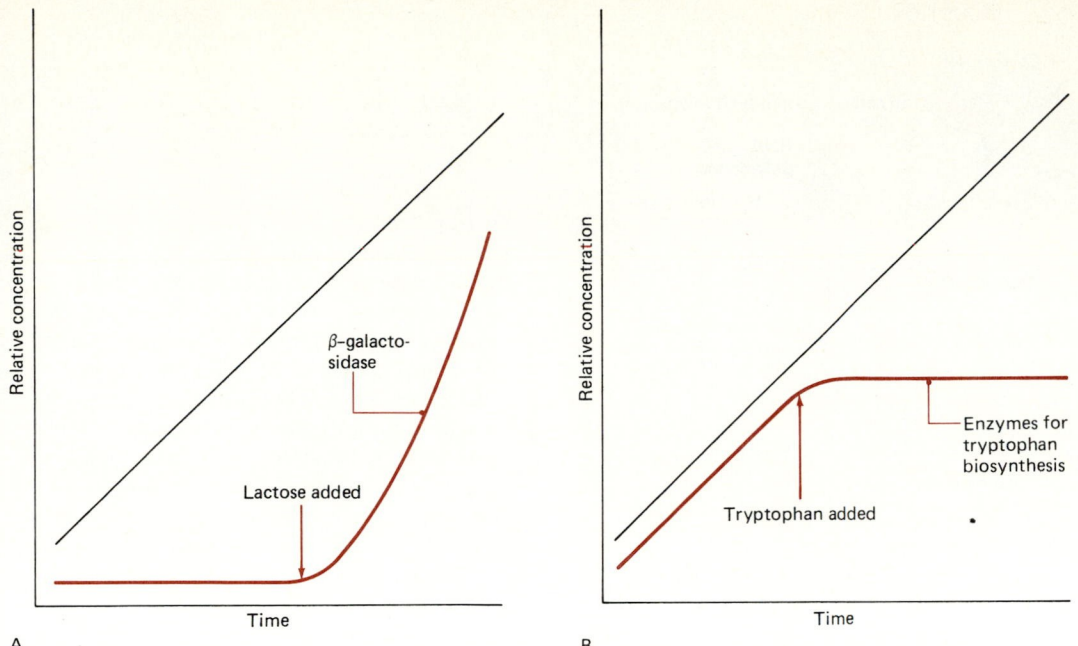

FIGURE 7.23

(A) Induction occurs when a specific inducer substance is present. In this example, the enzyme β-galactosidase is made in response to the addition of lactose. Actually, the inducer of this system is allolactose, a derivative of lactose. β-Galactosidase is one of the enzymes needed for the metabolism of lactose. Induction leads to the synthesis of one or several specific proteins. (B) Repression occurs when a specific repressor substance is present. In this example, the addition of tryptophan causes the cell to cease producing the enzymes involved in tryptophan biosynthesis. As in the case of induction, repression specifically affects the synthesis of a limited number of enzymes and overall protein synthesis continues.

Control of the expression of these structural genes for lactose metabolism is explained in part by the **operon model**, which demonstrates how the transcription of mRNA directing the synthesis of these enzymes is regulated. The operon for lactose metabolism *lac* **operon** includes a **promoter region** (*p*) where RNA polymerase binds, a **regulatory gene** (*i*) that codes for the synthesis of a repressor protein, and an **operator region** (*o*) that occurs between the promoter and the structural genes involved in lactose metabolism (Figure 7.25). The regulatory gene codes for a **repressor protein**, which in the absence of lactose binds to the operator region of the DNA. The binding of the repressor protein at the operator region blocks the transcription of the structural genes under the control of that operator region.

In the case of the *lac* operon, this means that in the absence of lactose, the three structural *lac* genes, which code for the three enzymes involved in lactose metabolism, are not transcribed. The operator region is adjacent to or overlaps the promoter region. The binding of the repressor protein at the operator region interferes with the binding of RNA polymerase at the promoter region. An **inducer** such as allolactose, which is a derivative of lactose, binds to the repressor protein and alters the conformation of the repressor protein; that is, it acts as an allosteric effector, so that it is unable to interact with and bind at the operator region. Thus, in the presence of an inducer that binds with the repressor protein, transcription of the *lac* operon is derepressed, and the synthesis of the three structural proteins needed for lactose metabolism proceeds. As the lactose is metabolized and its concentration diminishes, the concentration of the derivative allolactose, produced from lactose by low levels of β-galactosidase, also declines, making it unavailable for binding with the repressor protein; therefore, active repressor protein molecules are again available for binding at the operator region, and the transcription of the *lac* operon is repressed, ceasing further production of the enzymes involved in lactose metabolism that are controlled by this regulatory region of the DNA. The *lac* operon is typical of operons that control catabolic pathways; in the presence of an appropriate inducer, the system is derepressed.

Catabolite Repression

In addition to the specific operator-mediated regulation of transcription, there is a generalized type of repression known as **catabolite repression**. In the presence of an adequate concentration of glucose, a number of catabolic pathways are repressed, including those involved in the metabolism of lac-

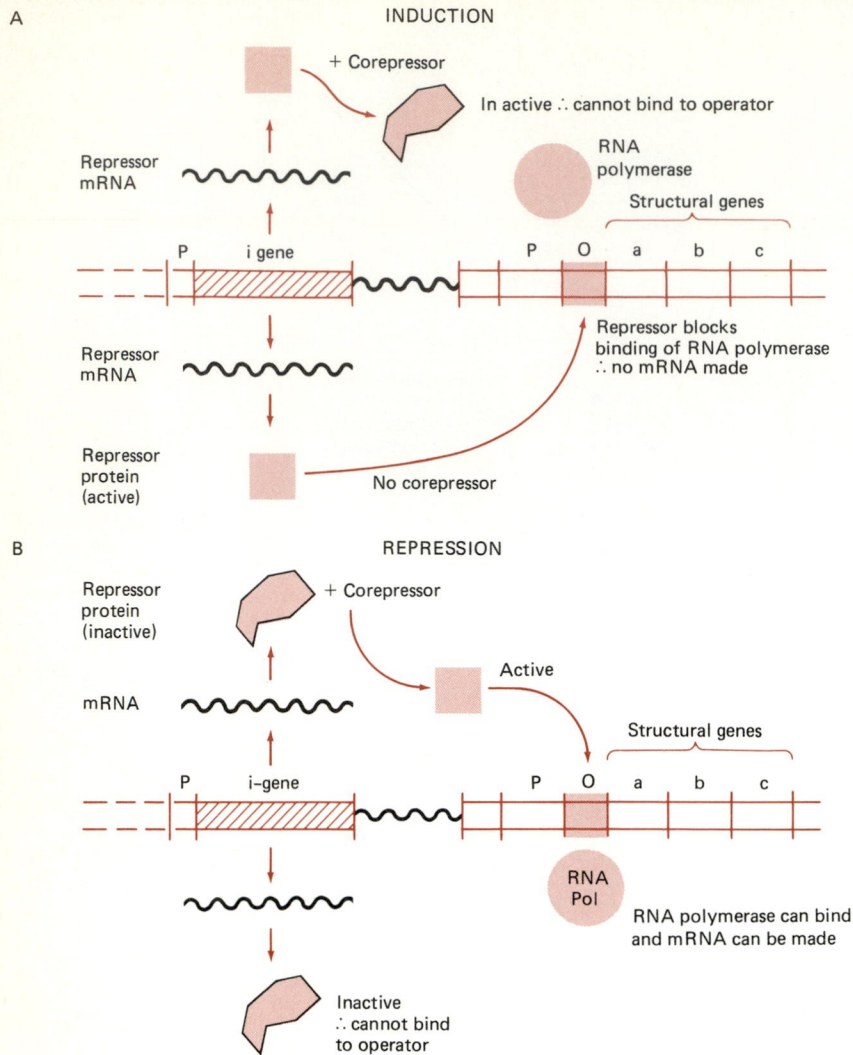

FIGURE 7.24

Induction and repression occur because of the interactions of repressor proteins with regulatory genes. The binding of a repressor protein at a regulator gene blocks the synthesis of proteins coded for by the adjacent structural genes. The ability of the repressor protein to bind to the regulator gene depends upon its conformation, which is affected by its interactions with inducer or repressor substances. (A) In the case of induction, the repressor protein normally is active and binds to the regulatory gene, blocking transcription of the subsequent structural genes. An inducer substance binds to the repressor protein, modifying its conformation and rendering it inactive so that it cannot bind to the regulatory gene. When this happens, transcription of the subsequent structural genes can proceed. (B) In the case of repression, the repressor protein normally is inactive and cannot bind to the regulatory gene so that transcription of the subsequent structural genes is not prevented. A repressor substance binds to the repressor protein, modifying its conformation and rendering it active so that it can bind to the regulatory gene. When this happens, transcription of the subsequent structural genes is blocked.

tose and galactose. When glucose is available for catabolism in the glycolytic pathway, disaccharides and polysaccharides need not be hydrolyzed to supply monosaccharides for the catabolic activities of the cell, and by blocking the metabolism of these more complex carbohydrates, the cell is able to conserve its metabolic resources.

Catabolite repression acts via the promoter region of DNA, and by doing so it complements the control exerted by the operator region. The efficient binding of RNA polymerase to promoter regions subject to catabolite repression requires the presence of a catabolite activator protein (Figure 7.26). In the absence of the catabolite activator protein, the RNA polymerase enzyme has a greatly decreased affinity to bind to the promoter region. The catabolite activator protein, in turn, cannot bind to the promoter region unless it is bound to cyclic adenosine monophosphate (AMP). There is an inverse relationship between the concentrations of cyclic AMP and ATP, and levels of cyclic AMP respond to the state of cellular metabolism. Intracellular levels of cyclic AMP are low when rapidly metabolizable substrates, such as glucose, are used. In the presence of glucose, cyclic AMP levels are

greatly reduced, and thus, the catabolite-activator protein is unable to bind at the promoter region. Consequently, RNA polymerase enzymes are unable to bind to catabolite repressible promoters, and transcription at a number of regulated structural genes ceases in a coordinated manner. In the absence of glucose, cyclic AMP is synthesized from ATP by the action of adenylate cyclase, and there is an adequate supply of cyclic AMP to permit the binding of RNA polymerase to the promoter region. Thus, when glucose levels are low, cyclic AMP stimulates the initiation of many inducible enzymes.

The combination of catabolite repression and the operon control mechanisms results in a **biphasic growth curve** rather than a normal growth curve under some conditions (Figure 7.27). The biphasic growth curve reflects a sequential utilization of substrates, a phenomenon known as **diauxie**. For example, cultures of *E. coli* exhibit biphasic growth when inoculated into a medium containing both glucose and lactose, indicative of the preferential metabolism of glucose before lactose as a substrate. While growing on glucose, *E. coli* exhibits the normal lag, log, and stationary phases of

FIGURE 7.25

The operation of the *lac* operon permits initiation of the synthesis of the enzymes needed for lactose utilization when lactose is present.

Absence of lactose

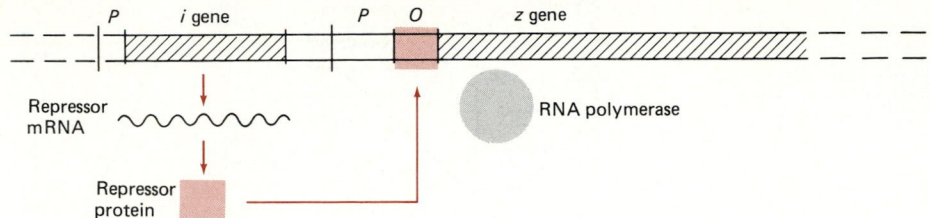

Presence of lactose

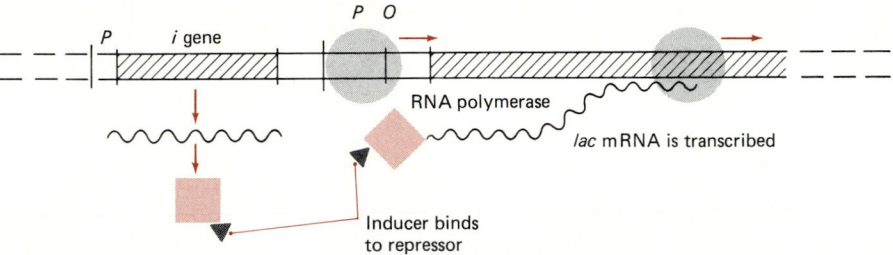

Discovery Process

Biochemical studies on enzyme synthesis have added much to our knowledge of induction and repression, but it was the development of the techniques of bacterial genetics that permitted a more thorough understanding. Francois Jacob and Jacques Monod, working at the Pasteur Institute in the late 1950s, isolated and genetically analyzed many mutants in the lactose pathway. Their results led to the conclusion that induction and repression were under the control of specific proteins, which were coded for by regulatory genes. They proposed that regulatory genes are closely associated with the structural genes coding for specific enzyme proteins but are distinct from them. At Harvard University, Walter Gilbert and Benn Müller-Hill in 1966 isolated and purified the Lac repressor molecule. *In vitro* studies showed that the repressor protein binds to DNA that contains the *lac* operon but does not bind when the DNA carries a deletion of the *lac* genes. Gilbert used the enzyme DNase to break apart DNA bound to repressor protein and recovered short DNA strands shielded from enzymatic activity by the repressor molecule; these strands were presumed to be the operator sequence. The sequence was determined, and each operator mutation was shown to involve a change in the sequence. The results of Gilbert and Müller-Hill confirmed the identity of the operator sequence and showed the specificity of repressor-operator recognition, which can be disrupted by a single base substitution. Later, when the sequence of bases in the mRNA transcript of the *lac* operon was determined, the first 21 bases on the 5′P initiation end were shown to be complementary to the operator sequence Gilbert had determined, confirming the model of the system; the repressor protein binds to the operator, preventing initiation of transcription by the RNA polymerase bound at the promoter site.

growth. Rather than exhibiting a prolonged stationary phase, *E. coli* enters a secondary lag phase when the glucose is no longer readily available in concentrations that suppress disaccharide utilization by catabolite repression. During this second lag phase, allolactose acts as an inducer to derepress the *lac* operon system. The enzymes that are necessary for lactose metabolism are synthesized, and the bacteria begin to grow exponentially by using the lactose substrate. When the lactose is used up or inhibitory products accumulate, the bacteria enter a second stationary phase.

Regulating the Biosynthesis of Tryptophan: The trp Operon

In contrast to the inducible *lac* operon, which is involved in regulating the catabolism of lactose, some regulatory genes that can be shut off under specific conditions, called **repressible operons**, have been found that control specific biosynthetic pathways. For example, the *trp* **operon**, which contains the genes that code for the enzymes required for the biosynthesis of the amino acid tryptophan, is repressible

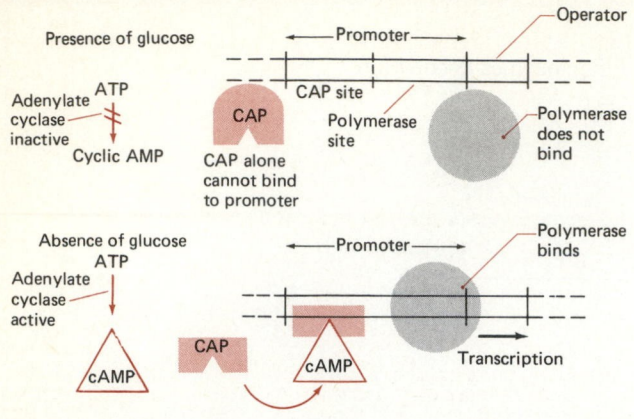

FIGURE 7.26
Catabolite repression explains why, in the presence of glucose, several catabolic pathways are shut off. As shown in this illustration, cyclic AMP plays a critical role in permitting RNA polymerase to bind to some promoters. Thus, the concentration of cyclic AMP can regulate metabolism by controlling the synthesis of enzymes in operon regions adjacent to such promoters.

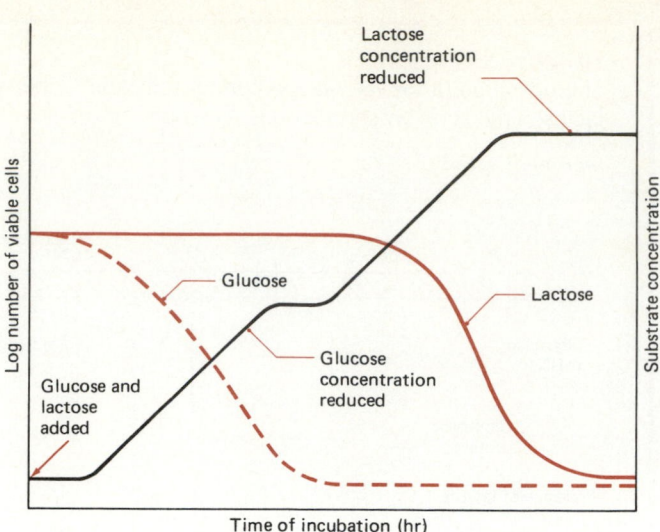

FIGURE 7.27
A biphasic growth curve reflects the preferential utilization of substrates and the phenomenon of diauxie.

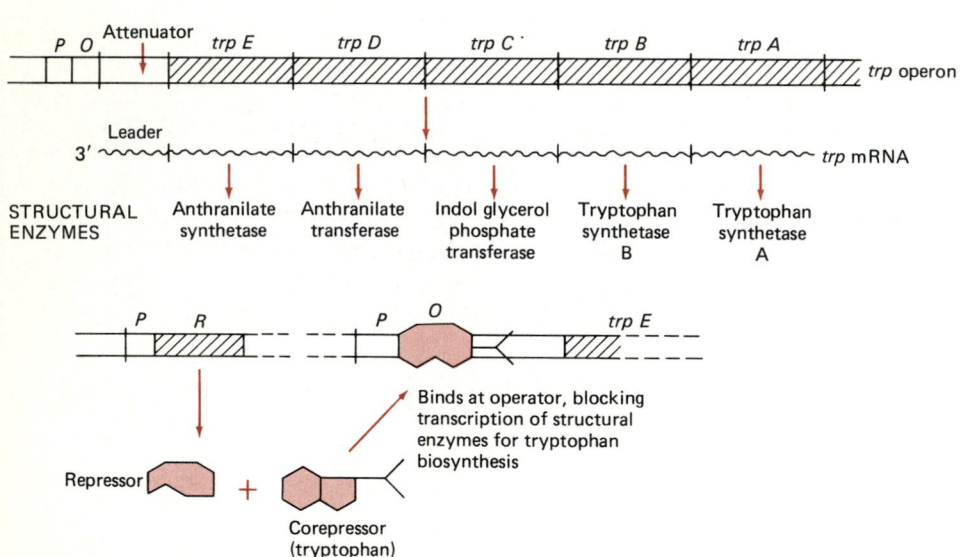

FIGURE 7.28
The operation of the *trp* operon permits the cell to stop synthesizing the enzymes involved in the biosynthesis of tryptophan when there is a sufficient concentration of this amino acid. Tryptophan interacts with a repressor protein, altering its conformation so that it can bind to the operator gene controlling the synthesis of several enzymes needed for the biosynthesis of tryptophan. Thus, when there is enough tryptophan, the enzymes needed for tryptophan biosynthesis are not made, thereby regulating this pathway.

(Figure 7.28). There are five structural genes in the *trp* operon that are responsible for the synthesis of five enzymes, and as with other operons there is also an operator region, a promoter region, and a gene that codes for a regulator protein. The trp repressor protein is normally inactive and unable to bind at the operator region, but tryptophan can act as an allosteric effector or corepressor. In the presence of excess tryptophan, the trp repressor protein binds with tryptophan and, as a result, is also then able to bind to the *trp* operator region. When the trp repressor protein—the tryptophan complex—binds at the *trp* operator region, the transcription of the enzymes involved in the biosynthesis of tryptophan is repressed. In the case of the *trp* operon, then, tryptophan acts as the repressor substance that shuts off the biosynthetic pathway for its own synthesis when there is a sufficient supply of tryptophan.

This is an example of **end product repression**, that is, the process of shutting off transcription when a by-product of the metabolism coded for by the genes in that transcription region accumulates. When the level of tryptophan in the cell declines, there is insufficient tryptophan to act as corepressor, and the transcription of the genes for the biosynthesis of tryptophan, therefore, resumes. The *trp* operon is typical of anabolic pathways, where in the presence of a sufficient supply of the biosynthetic product of the pathway, the system is repressed.

Attenuation: The Regulation of Transcription by Translation

As described earlier, when tryptophan is plentiful, it acts as corepressor and the operator region is blocked, but when its concentration diminishes, the repressor protein cannot bind at the operator region and transcription of the *trp* operon begins. A second mechanism exists for regulating expression of the genes involved in tryptophan metabolism. This mechanism functions through the regulation of the transcriptional process by interaction with the translational process. In this mode of regulating gene expression—called **attenuation**—the events that occur during translation affect the transcription of an operon region of the DNA. As indicated earlier, translation in prokaryotes normally begins before transcription of the mRNA is completed, with the leader portion of the mRNA attaching to the ribosome and translation beginning before the complete mRNA is transcribed. The leader sequence in prokaryotes occurs between the operator and the first structural genes. In the case of the *trp* operon, there is an **attenuator site** between the operator region and the first structural gene of the operon (at or near the end of the leader sequence) where transcription can be interrupted (Figure 7.29).

Whether or not termination occurs is determined by the secondary structure of the mRNA at the attenuator region. There are two possible structures of mRNA. In one form, mRNA folds to establish a double-stranded region, known as the *terminator hairpin*, that causes termination of transcription. The particular structure that forms depends upon the availability of tryptophan because tryptophan is one of the amino acids coded for in the leader sequence. When tryptophan is available for incorporation into proteins, the peptide sequence coded for by this leader sequence can be successfully translated. When tryptophan reaches very low concentrations, however, translation is delayed at the leader codons that code for the insertion of tryptophan into the polypeptide because sufficient tryptophan is not available. When the translational process is slowed, the mRNA is in the form that permits transcription to proceed through the entire sequence of the *trp* operon. However, if there is sufficient tryptophan present to permit rapid translation in order to proceed through the attenuator site, the mRNA forms the ter-

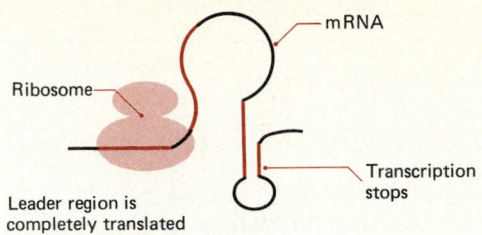

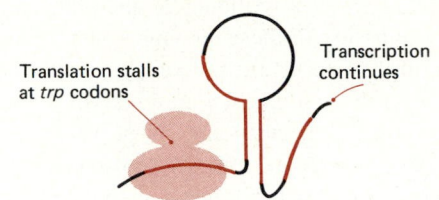

FIGURE 7.29

Attenuation permits translational control of transcription of the *trp* operon. The phenomenon of attenuation depends upon the intramolecular interactions within the leader sequence of the mRNA. The occurrence of base pairing results in the formation of a double-stranded segment that forms a hairpin loop. When sufficient tryptophan is present, translation of the leader sequence, which calls for a high proportion of tryptophan insertion, occurs rapidly and the portion of the leader sequence that has not yet been translated folds itself into a specific stem loop structure shown in the top portion of the figure. This structure results in termination of transcription of the *trp* genes. If, on the other hand, sufficient tryptophan is not present, translation of the leader sequence does not occur rapidly and a different stem loop structure—shown in the lower portion of the figure—occurs. This structure does not expose the terminator sequence for the *trp* operon genes, and hence, transcription continues.

minator hairpin structure and further transcription of the *trp* operon ceases, so that none of the structural genes for tryptophan metabolism are transcribed.

In addition to the tryptophan operon, the histidine and phenylalanine operons in *E. coli* also contain attenuator regions. In the case of histidine, there is a sequence of 7 contiguous codons in the leader sequence; in the case of phenylalanine, there are 15. Only when the concentrations of these amino acids are very low does translation stall, allowing transcription to proceed through the attenuator site. The attenuator complements the regulation of gene expression by the operator gene; thus, there is a redundancy in the control mechanisms for the biosynthesis of amino acids such as tryptophan. The leader sequence and the associated attenuator site thus provide a mechanism for even finer control of transcription and the expression of genetic information than does the operator region.

7.6 Gene Expression in Eukaryotic Microorganisms

The regulation of genetic expression in eukaryotic microorganisms is more complex than in bacteria, and many of the mechanisms described may not operate in the same manner in prokaryotic and eukaryotic cells. Some enzymes in eukaryotic microorganisms are clearly inducible and others are repressible, but the expression of these enzyme systems may not be under the control of mechanisms comparable to the *lac* operon in *E. coli*. mRNA molecules in eukaryotes normally are not polycistronic, as they are in prokaryotes.[7] Thus, control over several different mRNA molecules may be required to achieve coordinated control in eukaryotes, whereas control of a single mRNA molecule, carrying the information for the expression of several contiguous and sequential genes in a prokaryotic cell, can regulate the synthesis of several enzymes with related functions. Additionally, in almost all cases, the sequence of codons of a given gene in eukaryotic microorganisms is not colinear with the mRNA molecule or with the polypeptide sequence of the protein. It is, therefore, unlikely that in eukaryotic microorganisms an operator region could regulate the transcription of a series of genes that are contiguous with the region of the regulator gene. This effectively precludes the existence of a unified operon region, such as occurs in prokaryotic microorganisms, although an analogous type of operon control over a gene cluster can still exist in eukaryotic cells.

A series of different control mechanisms may be present in eukaryotic microorganisms to control genetic expression. These include the loss of genes, gene amplification, rearrangement of genes, differential transcription of genes, posttranscriptional modification of RNA, and translational control (Figure 7.30). For example, in the protozoan *Oxytricha*,

[7] Remember: *polycistronic* means that the mRNA contains the information for several proteins. Each mRNA of a eukaryotic cell generally contains the information for only one protein, and hence is moncistronic rather than polycistronic.

FIGURE 7.30
Several types of genetic control mechanisms appear to control metabolism in eukaryotes.

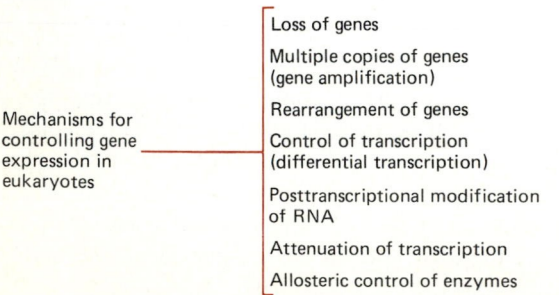

Mechanisms for controlling gene expression in eukaryotes
- Loss of genes
- Multiple copies of genes (gene amplification)
- Rearrangement of genes
- Control of transcription (differential transcription)
- Posttranscriptional modification of RNA
- Attenuation of transcription
- Allosteric control of enzymes

elimination of most of the DNA (gene loss) occurs in the vegetative cells. The elimination of some of the genetic information restricts the number of genes that can be expressed, and this mechanism allows for specialization in vegetative cells, analogous to differentiation of somatic cells in higher organisms.[8] In contrast to gene loss, some eukaryotic microorganisms are capable of **amplifying gene expression**, thereby producing large amounts of the enzyme coded for by a given gene. Eukaryotes do so by increasing the amount of rRNA and thus the number of ribosomes that can be used to translate the information in a stable mRNA molecule. The ribosomes line up along the same mRNA so that translation results in multiple copies of the synthesized protein. In the protozoan *Tetrahymena*, for example, there are hundreds of copies of the genes for rRNA in the vegetative cell that provide a mechanism for gene amplification. Within the eukaryotic genome, the position of some genes can be changed, thereby rearranging the location of genes within the eukaryotic chromosome. The **rearrangement of genes** (change in relative position within the chromosome) can alter the expression of the information contained in those genes. For example, the rearrangement of genes has been found to result in altered mating types in yeasts and the production of altered surface biochemicals in protozoa. This phenomenon has also been observed in prokaryotes, as for example in the case of flagellin protein synthesis in *Salmonella*.

In addition to these mechanisms, promoter regions in eukaryotic organisms are involved in the binding of RNA polymerase enzymes, and these may be sites for genetic regulation of the type exhibited in catabolite repression by prokaryotes. It is also likely that posttranscriptional modification of hnRNA is involved in the control of genetic expression in eukaryotes. The leader sequences, introns, and trailer sequences of RNA molecules in eukaryotes probably affect the expression of genetic information in eukaryotic microorganisms. Different strains of *Tetrahymena* have been found to have different introns in the genes that code for their rRNA. Further, the regulation of translation may be more important in controlling gene expression in eukaryotes than in prokaryotes because of the relative longevity of mRNA in eukaryotic cells.

It is thus clear that there are a number of mechanisms for regulating gene expression in eukaryotic microorganisms that do not exist in prokaryotes. Unraveling the complexity of gene regulation in eukaryotes and developing a better understanding of genetic regulation in prokaryotes remain prime challenges for the microbial geneticist.

[8] Most higher eukaryotic organisms do not seem to use gene elimination, although several insects use it as a means of differentiation.

Postlude

With the establishment of the structural and biochemical nature of the genetic material of the cell, the time was right for a unification of concepts in microbial genetics and biochemical metabolism. A major breakthrough in our understanding of genetics occurred in 1953, when James Watson and Francis Crick proposed the double helical structure of DNA. The revelation of DNA opened the field of molecular genetics for major new investigations. The establishment of the structure of the molecule housing the genetic information of living organisms permitted unraveling of the ways in which genetic information is stored and expressed. The structural model proposed by Watson and Crick provides the basis for understanding how the vast amount of genetic information of a cell can be encoded within a macromolecule and used to direct the synthesis of other macromolecules involved in cellular functions.

Genetic studies during the early 1960s established the basis for understanding how genetic information is stored in the DNA molecule and how that information is transcribed and translated into proteins that act as enzymes in determining the metabolic activities of microorganisms. In 1960 Francois Jacob and Jacques Monod proposed the operon theory, which explains how genetic information controls protein synthesis, that is, the nature of the control regions of the DNA molecule that act as switches, turning on and off the synthesis of enzymes. Francis Crick and co-workers correctly proposed that three nucleic acid bases in DNA coded for one amino acid. Several research groups determined which triplet base sequences specified which amino acids, and in 1968 the Nobel Prize was awarded to Robert Holley, H. G. Khorana, and Marshall Nirenberg for their work on breaking the genetic code. These researchers had examined different facets of the translation of the genetic code into protein structure. A number of other investigators also contributed to the basic understanding of the genetic code. Frederick Sanger and Walter Gilbert shared the 1980 Nobel Prize for their development of methods for sequencing the bases in DNA; determining the correspondence of the nucleic acid base sequence in DNA and the amino acid sequence in proteins was a major accomplishment.

In this chapter, we have considered the mechanisms for the storage and expression of genetic information. In prokaryotic and eukaryotic microorganisms, DNA is the molecule that stores the genetic information; in some viruses, this function is performed by RNA. The replication of the genome of microorganisms is essential for the transmission of the hereditary information from one generation to the next, and the process of DNA replication is designed to ensure the fidelity of the DNA that is transmitted to the progeny. The replication of the double helical DNA molecule uses the parental DNA as a template for directing the sequencing of the nucleotides in the synthesis of new DNA molecules. DNA replication is semiconservative; the progeny receive one parental and one newly synthesized strand of DNA in their double helical DNA molecules. DNA-dependent DNA polymerase enzymes, which catalyze the replication of DNA, add nucleotide bases only to the 3'OH free end of a nucleotide chain. DNA polymerases form phosphodiester linkages only when the free nucleotides are correctly aligned opposite their base pairs. Because of the directional nature of the DNA polymerase enzymes, only one strand of the DNA can be continuously synthesized, and the synthesis of the opposing strand lags behind and involves the formation of discontinuous segments of DNA, which are then linked together by ligases. The directional nature of DNA is essential for its proper functioning, and the replication process reflects and ensures the continuance of directionality in this macromolecule.

The replication of DNA requires an unwinding of the double helical DNA molecule to form replication forks where complementary strands of DNA can be synthesized. Replication of the bacterial chromosome has a single initiation site, and DNA replication in the organisms so far examined proceeds bidirectionally from this origin, with the replication forks moving at identical rates in opposite directions around the circular bacterial chromosome until they meet at the termination point. In eukaryotic microorganisms DNA synthesis also occurs bidirectionally, but initiation occurs at multiple origin sites along the chromosome. Because there are multiple initiation sites for DNA replication, eukaryotes are able to replicate their genome more rapidly than prokaryotes. These differences represent fundamental distinctions between eukaryotic and prokaryotic cells.

The replication of the genome provides the mechanism for passing the genetic information from one generation to the next. For the genetic information to determine the structure and function of the organism, it must also be expressed and used to control the metabolism of the cell. The expression of genetic information involves transferring the information encoded in the genome to functional enzymes. This transfer is accomplished by having the genome direct the process of protein synthesis, a two-stage process in which information encoded in the DNA is first transcribed to RNA molecules and then translated into the amino acid sequence of the polypeptide. Three types of RNA molecules are involved in protein synthesis: rRNA, tRNA, and mRNA. Transcription to form RNA uses DNA as a template, with only one strand of the DNA, the sense strand, serving as the template for the transcription of a given RNA molecule. Specific initiation and termination sites for transcription are encoded in the DNA molecule. Transcription requires that RNA polymerase bind to the promoter region of the DNA to initiate RNA synthesis; the promoter region contains a sequence of nucleotides recognized by the RNA polymerase enzyme as the site of transcriptional initiation.

A critical function of transcription is the formation of

mRNA because the information for sequencing amino acids in a polypeptide is encoded within mRNA molecules. In prokaryotic microorganisms, mRNA is colinear with the DNA molecule. In contrast, eukaryotes have split genes, and an hnRNA molecule is initially synthesized and then extensively processed after transcription to produce mRNA molecules.

The translation of the mRNA molecule occurs at the ribosomes. During translation the genetic code, consisting of 64 triplet nucleotide sequences known as *codons*, specifies the amino acids that are to be inserted into a polypeptide chain. Three of the codons do not specify any of the 20 amino acids that are found in proteins; rather, these nonsense codons act as termination signals for the translational process. Because more than one codon can specify the same amino acid, the genetic code is said to be *degenerate*. The genetic code appears to be almost universal, with the same code operating in both prokaryotic and eukaryotic cells. The direction of reading of the mRNA molecule is from the 5′P to the 3′OH free end, and the direction of synthesis of the polypeptide chain is from the amino to the carboxyl end. In the process of translation, the ribosome supplies the structural framework for the proper alignment required for protein synthesis, the mRNA specifies the order of the amino acid sequence for the polypeptide chain, and the tRNA molecule brings the amino acid to the ribosome and aligns it in its proper sequence.

Both prokaryotic and eukaryotic microorganisms can regulate the expression of genetic information in ways that permit them to carry out specialized activities during specific periods of time. The ability to direct the metabolic activities of the cell to the areas where they are needed is a critical conservation mechanism for microorganisms. The expression of genetic information can be controlled at several levels. In prokaryotic microorganisms some genes are clustered together, forming a polycistron, and their transcription is under the control of a regulatory operator region. The operator acts as a switch that regulates transcription and thus turns protein synthesis on and off, permitting the coordinated expression of genetic information. Some operons, such as the *trp* operon, are repressible, and others, such as the *lac* operon, are inducible. The functioning of the operon can be likened to a building with multiple light fixtures and electrical switches. When a light switch is turned on, the light bulbs function and emit light in a specified area of the building. In some rooms the light switch is turned on only when someone enters the room, a case that is analogous to the functioning of *lac* operon, where the system is induced (derepressed) only in the presence of an inducer. In other areas of the building, such as the entrances, the lights are normally left on but may be shut off when they are not needed, a situation that is analogous to the functioning of the *trp* operon, which is repressed only when the enzymes coded for by the *trp* structural genes are truly not needed.

Thus, the genome of the cell codes for its own replication, establishing a mechanism for hereditary transmission of genetic information; directs the synthesis of proteins, determining the metabolic capabilities and potential phenotype of the organism; and contains various mechanisms for controlling genetic expression that allow microorganisms to finely regulate their metabolic activities. The elucidation of the molecular-level events involved in the transmission and expression of genetic information has revolutionized our understanding of microbial genetics. These advances in molecular genetics permit the understanding of many fundamental processes in microorganisms and the use of this understanding in many applied areas of microbiology.

Suggested Supplementary Readings

BIRGE, E. A. 1981. *Bacterial and Bacteriophage Genetics.* Springer-Verlag, New York.

CAMPBELL, J. L. 1986. Eukaryotic DNA replication. *Annual Review of Biochemistry* 55:733–772.

CECH, T. R. 1983. RNA splicing: Three themes with variations. *Cell* 34:713–716.

CHAMBON, P. 1975. Eucaryotic nuclear RNA polymerases. *Annual Review of Biochemistry* 44:613–638.

CHAMBON, P. 1981. Split genes. *Scientific American* 244(5):60–71.

CHASE, J. W., and K. R. WILLIAMS. 1986. Single-stranded DNA binding proteins required for DNA replication. *Annual Review of Biochemistry* 55:103–136.

CRICK, F. H. C. 1966. The genetic code: III. *Scientific American* 215(4):55–62.

DARNELL, J. E. 1982. Variety in the level of gene control in eukaryotic cells. *Nature* 297:365–371.

DARNELL, J. E. 1983. The processing of RNA. *Scientific American* 249:90–100.

DARNELL, J., H. LODISH, and D. BALTIMORE. 1986. *Molecular Cell Biology.* Scientific American Books, New York.

DYNAN, W. S., and R. TIJAN. 1985. Control of eukaryotic messenger RNA synthesis by sequence-specific DNA-binding proteins. *Nature* 316:774–778.

FREIFELDER, D. 1978. *The DNA Molecule: Structure and Properties.* W. H. Freeman Co., San Francisco.

FRISTROM, J. W., and P. T. SPIETH. 1980. *Principles of Genetics.* Chiron Press, Inc., Portland, Ore.

Genetics: Readings from Scientific American. 1981. W. H. Freeman Co., San Francisco.

GOLD, L., D. PRIBNOW, T. SCHNEIDER, S. SHINEDLING, B. S. SINGER, and G. STORMO. 1981. Translational initiation in prokaryotes. *Annual Review of Microbiology* 35:365–403.

GOLDSTEIN, L., and D. M. PRESCOTT (eds.). 1980. *Gene Expression: The Production of RNA's.* Academic Press, New York.

HERSHEY, J. W. B. 1980. The translational machinery: Components and mechanisms. In *Cell Biology: A Comprehensive Treatise* (D. M. Prescott and L. Goldstein, eds.). Academic Press, New York.

JACOB, R., and J. MONOD. 1959. Genetic regulatory mechanisms in the synthesis of proteins. *Journal of Molecular Biology* 3:318–356.

KAPLAN, A. S. (ed.). 1982. *Organization and Replication of Viral DNA*. CRC Press, Inc., Boca Raton, Fla.

KORNBERG, A. 1980. *DNA Replication*. W. H. Freeman and Co., San Francisco.

LEWIN, B. 1974. *Gene Expression—1: Bacterial Genomes*. John Wiley & Sons, Inc., New York.

LEWIN, B. 1977. *Gene Expression—3: Plasmids and Phages*. John Wiley & Sons, Inc., New York.

LEWIN, B. 1980. *Gene Expression—2: Eukaryotic Chromosomes*. John Wiley & Sons, Inc., New York.

LEWIN, B. 1985. *Genes II*. John Wiley & Sons, Inc., New York.

McCLURE, W. R. 1985. Mechanism and control of transcription initiation in prokaryotes. *Annual Review of Biochemistry* 54:171–204.

MILLER, J. H., and W. S. REZNIKOFF (eds.). 1980. *The Operon*. Cold Spring Harbor Laboratory, Cold Spring Harbor, N.Y.

PABO, C. T., and R. T. SAUER. 1984. Protein–DNA interactions. *Annual Review of Biochemistry* 53:293–321.

PADGETT, R. A., P. J. GRABOWSKII, M. M. KONARSKA, S. SEILER, and P. A. SHARP. 1986. Splicing of message RNA precursors. *Annual Review of Biochemistry* 55:1119–1150.

PLATT, T. 1986. Transcription termination and the regulation of gene expression. *Annual Review of Biochemistry* 55:339–372.

RAIBAUD, O., and M. SCHWARTZ. 1984. Positive control of transcription in bacteria. *Annual Review of Genetics* 18:173–206.

REZNIKOFF, W. S., D. W. COWING, and C. GROSS. 1985. The regulation of transcription initiation in bacteria. *Annual Review of Genetics* 19:355–388.

RICH, A., and S. H. KIM. 1978. The three-dimensional structure of transfer RNA. *Scientific American* 238(1):52–62.

STENT, G. S., and R. CALENDAR. 1978. *Molecular Genetics: An Introductory Narrative*. W. H. Freeman Co., San Francisco.

STEWART, P. R., and D. S. LETHAM (eds.). 1977. *The Ribonucleic Acids*. Springer-Verlag, New York.

STRICKBERGER, M. W. 1985. *Genetics*. Macmillan Publishing Co., Inc., New York.

STRYER, L. 1987. *Biochemistry*. W. H. Freeman Co., San Francisco.

SUZUKI, D. T., A. J. F. GRIFFITHS, J. H. MILLER, and R. C. LEWONTIN. 1986. *An Introduction to Genetic Analysis*. W. H. Freeman Co., San Francisco.

WATSON, J. D., N. HOPKINS, J. ROBERTS, J. STEITZ, and A. WEINER. 1987. *Molecular Biology of the Gene*. Benjamin/Cummings Publishing Co., Menlo Park, Calif.

YANOFSKY, C. 1981. Attenuation in the control of expression of bacterial operons. *Nature* 289:751–758.

ZIPSER, D., and J. BECKWITH (eds.). 1977. *The lac Operon*. Cold Spring Harbor Laboratory, Cold Spring Harbor, N.Y.

ZUBAY, G. 1987. *Genetics*. Benjamin/Cummings Publishing Co., Menlo Park, Calif.

Study Questions

1. How was the semiconservative nature of DNA replication demonstrated?

2. What is a DNA polymerase enzyme? An RNA polymerase enzyme? How are these enzymes different from the enzymes involved in microbial metabolism?

3. What are the similarities and differences between DNA replication in prokaryotes and eukaryotes?

4. What are the steps in protein synthesis? What roles do different nucleic acids play in the expression of genetic information?

5. How is information encoded within nucleic acid molecules? What are the essential properties of the genetic code that permit the storage and extraction of genetic information?

6. Why is it important that DNA and RNA have distinct 3′OH and 5′P ends?

7. What signals the initiation and termination of transcription? Of translation?

8. How is the expression of DNA regulated at the level of transcription? How are operons and promoters involved in gene expression? Discuss the functioning of the *lac* and *trp* operons.

9. What is the glucose effect, and how does catabolite repression help explain its molecular basis?

10. What is a split gene, and why does it represent a fundamental difference between prokaryotes and eukaryotes?

Situational Problem 1

Constructing a DNA Sequence for a Polypeptide

Today automated systems are available for determining the sequence of amino acids in a polypeptide and for synthesizing DNA molecules with a specified order of nucleotides.

You are assisting in a laboratory that has instruments for both amino acid sequencing and DNA synthesis. You have isolated a bacterium that produces a surface polypeptide that acts to initiate ice crystal formation. The bacterium, a strain of *Pseudomonas syringae*, was isolated from the surface of a leaf. In this entrepreneurial era, you decide that there may be commercial value in producing this polypeptide; for example, it might be useful in increasing the efficiency of artificial snow production at ski resorts. You have been reading about the future of biotechnology and the power of genetic engineering, and you decide to produce a sequence of nucleotides that codes for this ice-nucleation polypeptide with the expectation of later introducing it into a bacterium, such as *E. coli*, and producing it in commercial quantities.

The first thing you do is to isolate and purify the polypeptide. Then you slip into the laboratory after hours and run it through the amino acid analyzer, with the following results for a segment of the polypeptide: amino terminal-phe-phe-his-trp-lys-lys-lys-lys-asp-arg-lys-ser-ser-trp-his-ile-phe -met-asp-glu-glu-glu-gly-gly-pro-gly-gly-val-leu. Based on these data, you program the DNA synthesizer for the desired sequence of nucleotides. Using a table of mRNA codons (see Figure 7.18), write an appropriate sequence of DNA nucleotides that would code for this polypeptide.

Situational Problem 2

Determining the Regulatory Mechanisms of a Bacterial Isolate

As part of an undergraduate research project, you have been given a bacterial strain that can use carbohydrates and hydrocarbons as growth substrates. You discover that the strain uses glucose following a minimal lag period after culture inoculation, regardless of the other carbohydrates and hydrocarbons present in the growth medium. Lactose, however, is not used until much later if glucose is present. In the absence of glucose, lactose is used after a lag period about three times as long as the lag period for glucose utilization, but well before it would have been used if glucose had been present. The presence of hydrocarbons does not affect the lag period for the utilization of lactose. The utilization pattern for all hydrocarbons is similar to that of lactose, that is, there is an

intermediate lag period in the absence of glucose and a long delay before any utilization occurs if glucose is initially present. The presence of lactose does not affect the lag period before hydrocarbon utilization occurs. Also, it is observed that branched hydrocarbons are not immediately used if straight chain hydrocarbons are initially present, but they are used much sooner in the absence of straight chain hydrocarbons.

As part of your research project report, you have been asked to explain these data. What regulatory mechanisms would be consistent with the observed patterns of carbohydrate and hydrocarbon utilization?

CHAPTER 8

Genetic Variation: The Introduction and Maintenance of Heterogeneity Within the Gene Pool

Despite the biochemical mechanisms that are designed to ensure the fidelity of DNA replication, alterations in genetic information can and do occur. The genomes of microorganisms are subject to change by a variety of mechanisms. Heritable changes in the sequence of nucleotide bases of an organism, known as **mutations**, alter the genome of the organism and introduce variability into the gene pool of microbial populations. Variability within the genomes of microorganisms provides a mechanism for selection and evolution. Genetic variability also provides the heterogeneity needed within a population to adapt to environmental variability.

Once variability is introduced into the gene pool of a microbial population, there are several ways in which it is maintained. Genetic information can be transferred between microorganisms, with the exchange of such information ensuring the passage of genetic variation from one generation to another. Recombination of genetic information from two different microbes produces progeny that contain genetic information derived from two potentially different genomes.

In the case of eukaryotic microorganisms, genetic exchange during sexual reproduction affords a mechanism for gene reassortment within the population and maintenance of genetic heterogeneity. Even in prokaryotic microorganisms, where reproduction is asexual or parasexual (not involving gamete formation or a long-lasting diploid state), there are genetic exchange processes that lead to the recombination of genetic information. The natural evolution of microorganisms is based on the selection of the genetic variants best suited for survival, and artificial manipulation of genetic exchange between organisms can be used to create new microorganisms. The development of an understanding of the molecular basis of genetics has spawned the new and exciting field of genetic engineering, and applications of genetic engineering promise to revolutionize the industrial applications of microorganisms. In this chapter, we will examine how variability is introduced into and maintained within the genome both naturally and by genetic engineering.

8.1 Mutation

We have previously emphasized that the mechanisms inherent in the DNA replication processes are designed to ensure that the process is error free; we will now see that despite all of the safeguards some errors do occur, resulting in mutations. Any heritable change in the sequence of nucleotide bases in the DNA can be reflected in the expression of the genetic information. Because the genetic code consists of triplet nucleotide sequences that determine the biochemical structure of proteins, changing the order of bases even

slightly can alter the ability of the cell to produce enzymes that function properly. Similarly, because part of the microbial genome regulates the synthesis of proteins, mutations may alter the capacity of the cell to control its metabolic activity properly. Mutations always change the **genotype** of the microorganism, that is, its genetic composition specified by the ordered nucleotides of the DNA, often changing the functions of regulatory or structural genes. However, in part because the genetic code is degenerate, mutations do not

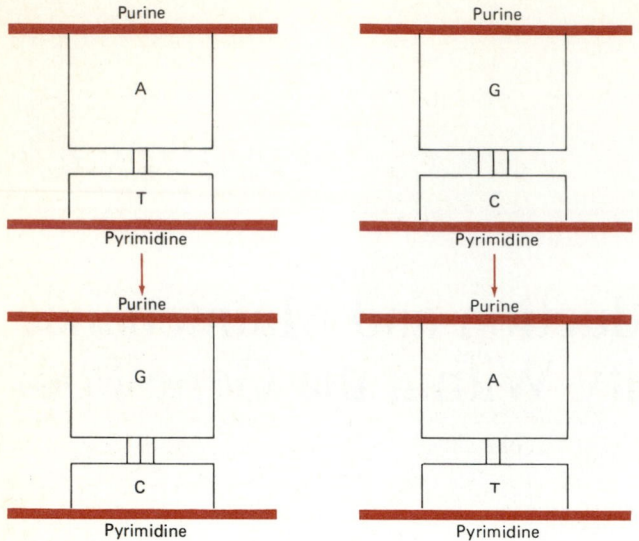

Transition mutation

FIGURE 8.1

In a transition mutation, one purine is replaced by the other purine or one pyrimidine is replaced by the other pyrimidine found in DNA.

always alter the **phenotype**, that is, the appearance of the organism based upon expression of genes under the given set of conditions, since different codons can specify the same amino acids.

Types of Mutations

There are various classes of mutations. One type of mutation, **base substitution**, occurs when one pair of nucleotide bases in the DNA is replaced by another. There are two general types of base substitution: transitions and transversions. **Transitions** involve the replacement of a purine on one strand by a different purine and the replacement of a pyrimidine on the other strand by a different pyrimidine, that is, the replacement of an adenine-thymine (AT) pair by a guanine-cytosine (GC) pair or vice versa (Figure 8.1). **Transver-**

sions, on the other hand, are base substitutions in which purines replace pyrimidines and pyrimidines replace purines (Figure 8.2). The conversion of an AT pair to a TA or CG pair represents a transversion mutation. Similarly, the change of a GC pair to a CG or TA pair also establishes a transversion mutation. Because only one strand of the DNA acts as the sense strand, a transversion, such as from a GC to a CG pair, changes the sequence of nucleotides in the mRNA molecule that is transcribed.

Most base substitutions are **missense mutations**, so named because they result in a change in the amino acid inserted into the polypeptide chain specified by the gene in which the mutation occurs. Missense mutations can result in the production of an inactive enzyme. Changes in a single amino acid within a polypeptide, though, often do not drastically reduce the activity of an enzyme and are rarely fatal to the microorganism. Because the genetic code is degenerate, the substitution of one nucleotide base for another may not even change the amino acid specified by the codon. Such **silent mutations** do not alter the phenotype of the organism and go undetected. Changes in the nucleotide sequence that alter the third base of codon are most likely to produce such silent mutations because this is where most of the redundancy in the genetic code occurs. This phenomenon is described by the **wobble hypothesis**, which states that changes in the third position of the codon often do not alter the amino acid sequence of the polypeptide. As discussed in Chapter 7, the tRNA molecules sometimes match only the first two nucleotides of the codon and thus are said to *wobble* because of the variability in the third base position (Figure 8.3).

Whereas base substitutions sometimes have little effect on the phenotype of the organism, changing even a single nucleotide base within the large DNA macromolecule can radically change the genetic information of a cell. One special type of mutation that often has a major effect on the expression of the genetic information occurs when the alteration in the base sequence of the DNA results in the formation of a nonsense codon (Figure 8.4). Because nonsense codons act as terminator signals during protein synthesis, the formation

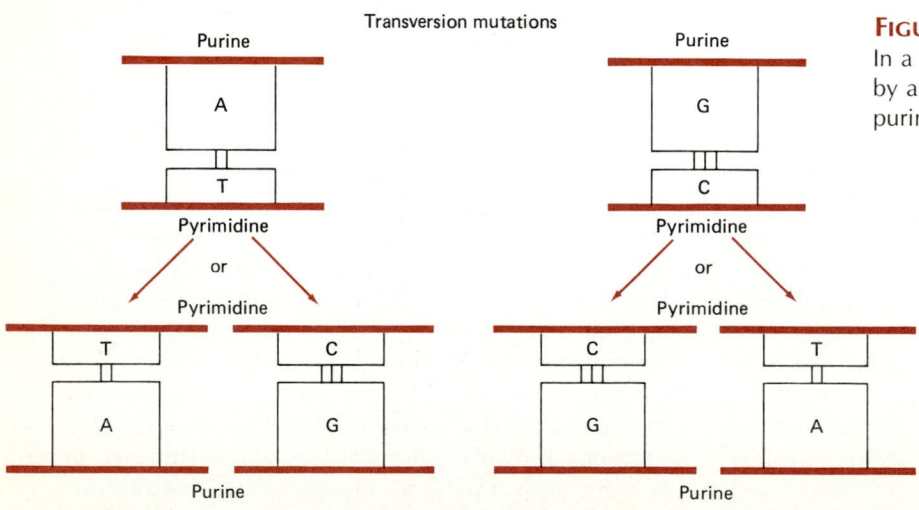

Transversion mutations

FIGURE 8.2

In a transversion mutation, a purine is replaced by a pyrimidine or a pyrimidine is replaced by a purine.

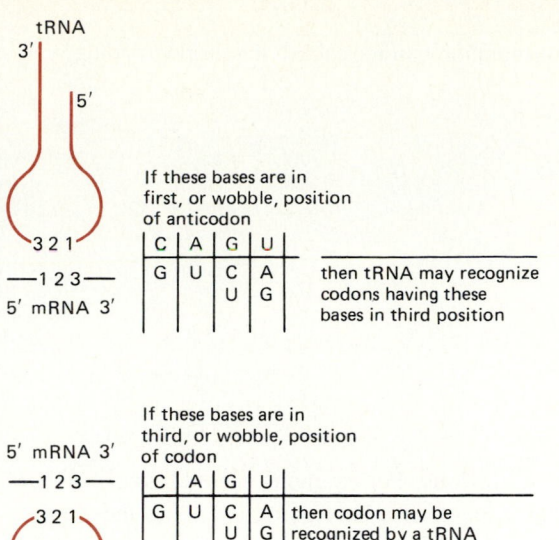

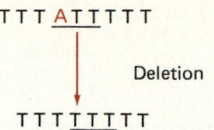

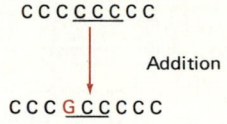

FIGURE 8.3

The wobble hypothesis is based upon the observation that sometimes only the first two nucleotides match between the codon and the anticodon. However, because the genetic code is degenerate—with codons varying in the third nucleotide often coding for the same amino acid—such "wobbling" is often of no consequence.

of a nonsense codon often signals the premature termination of a polypeptide chain, preventing the formation of a functional enzyme molecule. In bacteria where the mRNA molecule often is polycistronic, a **nonsense mutation** (a mutation that produces a nonsense codon) can affect the synthesis of several polypeptides; such mutations that prevent the translation of subsequent polypeptides coded for in the same mRNA molecule are said to be **polar mutations**. The degree to which a mutation within one gene affects the expression

FIGURE 8.5

A frame shift mutation can be accomplished by deletion or addition.

of other genes depends upon the degree of polarity, which is highly dependent on the relative locations of the nucleotide sequences involved in initiation and termination of the specific genes. Mutations may also alter normal termination codons, and failure to terminate translation properly results in coding for an elongated polypeptide.

In addition to base substitutions, mutations can result from the insertion or deletion of nucleotide bases (Figure 8.5). A **deletion mutation** involves the removal of one or more nucleotide base pairs from the DNA, and an **insertion mutation** involves the addition of one or more base pairs. Deletions of large numbers of base pairs, called **deficiencies**, can result in the loss of genetic information for one or more complete genes. Perhaps the greatest effect of deleting or adding a base occurs because the nature of the translation process is dependent on the establishment of the proper reading frame in order for the codons to specify the proper amino acids to be inserted into the polypeptide chain. Because adding or deleting a single base pair changes the reading frame of the transcribed mRNA, the deletion or addition of a single base pair can have as great an effect as a large deficiency. Such **frame shift mutations** can result in the misreading of large numbers of codons.

To understand how such mutations can alter the informational content of a message, let us consider what would happen if the English language contained only three-letter words; if we did not use spaces between words; if we used the three-

FIGURE 8.4

A nonsense mutation can result in the premature termination of the synthesis of a polypeptide.

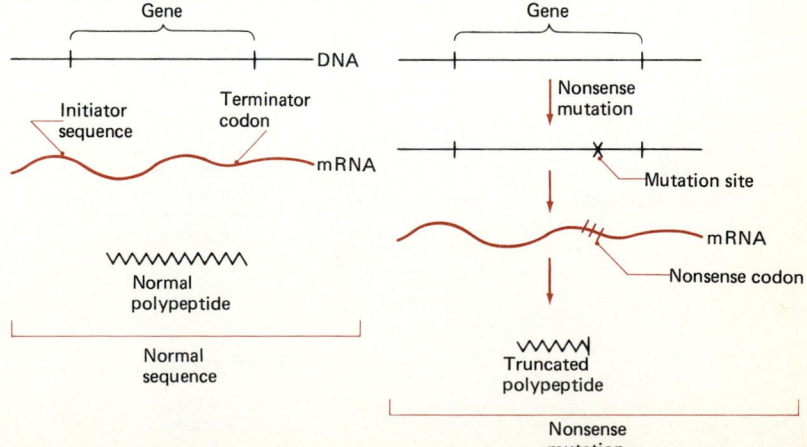

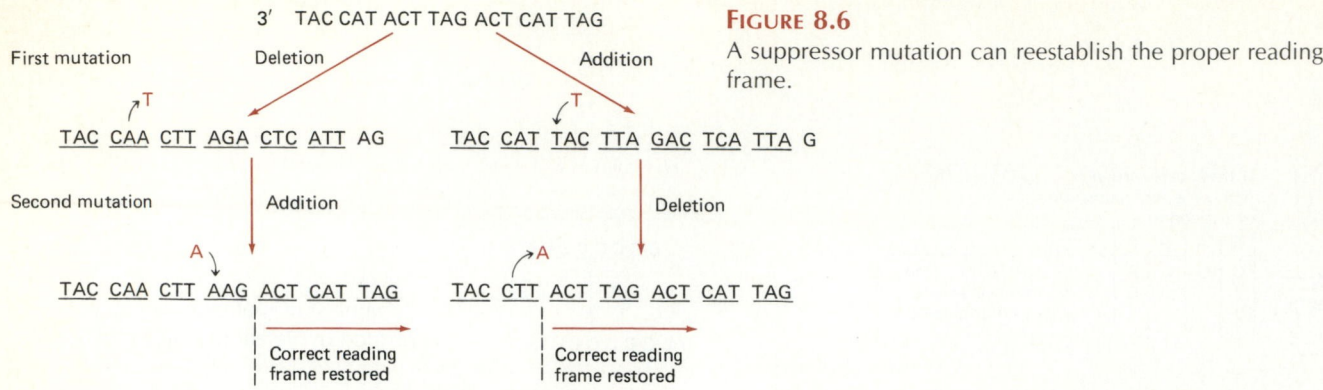

FIGURE 8.6

A suppressor mutation can reestablish the proper reading frame.

letter sequence XXX instead of a period to indicate the end of a sentence; and if we changed, deleted, or added a letter. We could understand the simple sentence "THECATA-TETHERATXXX" as "The cat ate the rat." Changing a single letter can alter the meaning but still convey information; for example, "THECATATETHEBATXXX," translated as the "The cat ate the bat," still conveys meaning, although a somewhat different meal for the cat. However, deleting a letter, for example, "THEATATETHER-ATXXX" alters the reading frame and renders the message nonsensical. In this case, we recognize only the words "THE" and "HER" in the sentence "The ata tet her atx." Similarly, adding a letter can alter the reading frame and greatly change the informational content.

In some cases, a second mutation can occur that reestablishes the reading frame (Figure 8.6). Single base pair additions and deletions can be reversed or reverted by such **suppressor mutations**. For example, the addition of a base pair after the deletion of a base pair can restore the reading frame, suppressing the expression of the first mutation. An **intragenic suppressor mutation** (a mutation within one gene), such as one that reestablishes a reading frame, permits the successful synthesis of the polypeptide specified by the gene in which the mutation occurs. Returning to our English language analogy to understand this concept, we saw that deleting a letter that resulted in "THEATATETHER-ATXXX" formed a nonsensical sentence. An addition after the first deletion, such as "THEANTATETHERATXXX," could have created a new but interpretable informational content, in this case telling us that "The ant ate the rat."

A suppressor mutation may also be an **intergenic mutation**, which is a mutation within one gene that affects another gene. Mutations that alter the anticodon region of tRNA molecules can be involved in such intergenic suppression of

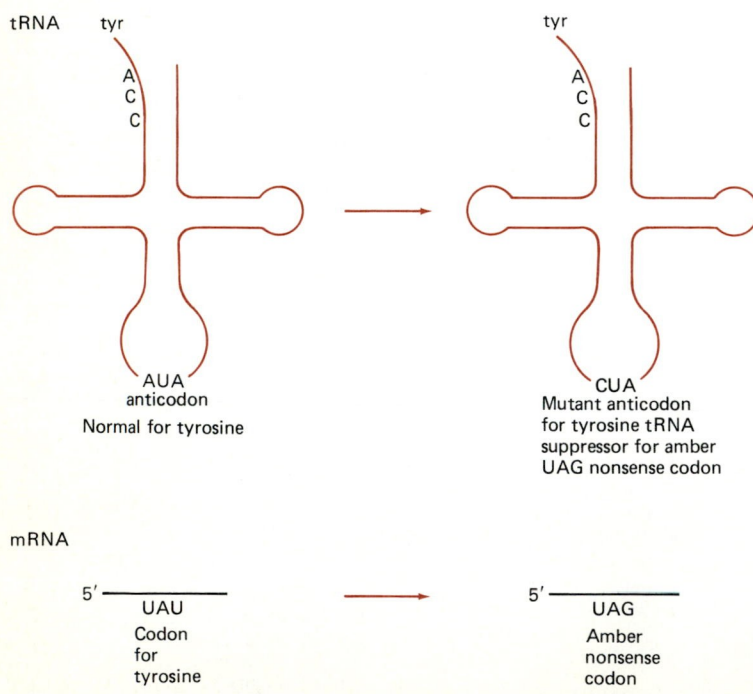

FIGURE 8.7

An intergenic suppression mutation changes the anticodon of tRNA and thereby negates a nonsense mutation. In the example shown here, the normal codon for tyrosine is UAU and the corresponding anticodon is GUA. A mutation results in the production of mRNA with a codon of UAG that would not normally pair with the anticodon of the tRNA carrying tyrosine. However, a second mutation in a gene for the tRNA that produces a tyrosine-carrying tRNA with an anticodon of CUA, which will pair with the mutant UAG codon, negates the mutation and the peptide has a normal insertion of tyrosine.

ANALYTICAL PROCESS

Detection of Mutants

Several approaches are used to detect mutant microorganisms. In some cases, a colony growing on an agar plate can easily be seen to be different from the normal parental type. For example, if the parental strain is pigmented, the observation of nonpigmented colonies may indicate the presence of mutants. Indicators can also be incorporated into the medium to detect organisms with and without particular metabolic capabilities, or various incubation conditions can be used. For example, pH indicators can be incorporated into the medium to detect the production of acidic products. The indication of acid production by one strain and not another of the same organism growing under identical conditions would show the presence of a mutant.

Nutritional mutants, as well as various other types of mutants, are often detected using the **replica plating technique**, a method that allows the observation of microorganisms under a series of growth conditions (Figure 8.8). In replica plating, a piece of sterile velvet is touched to the surface of an agar plate containing surface bacterial colonies. The fibers of the velvet act as fine inoculating needles, picking up bacterial cells from the surface of this master plate. The velvet with its attached microorganisms is then touched to the surface of a sterile agar plate, inoculating it. In this manner, microorganisms can be repeatedly stamped onto media of differing composition. The distribution of microbial colonies should be exactly replicated on each plate unless the colonies represent strains of differing genetic composition. Should a colony that develops on a complete medium fail to develop on a minimal medium that lacks a specific growth factor, the occurrence of a nutritional mutation is indicated. The microorganisms that do not grow on the minimal medium represent auxotrophic strains. By determining which biochemicals permit the growth of the auxotroph, the step in the metabolic pathway and the genetic site of the metabolic blockage can be determined.

FIGURE 8.8

Replica plating is used to identify mutants by transferring identical colonies to different types of media and comparing the colonies that develop on the respective plates. This method is critical in identfying auxotrophic mutants. All colonies develop on a complete medium that satisfies the nutritional needs of both the parental and mutant strains. Colonies of the auxotrophic mutant fail to develop on a medium lacking the specific nutritional growth factors required by the mutant.

The replica plating technique was developed by Joshua and Esther Lederberg in 1952 in order to provide direct evidence for the existence of preexisting mutations. Their actual experiment consisted of replicating master plates of sensitive cells to two or more plates containing either streptomycin or T1 phage. When the replicas were grown, they were compared to the locations of colonies on the master plate, and any resistant colonies that appeared at the same position on all of the replica plates were marked. The area of the master plate corresponding to the marked areas was cut out, and the bacteria on it were resuspended in a liquid medium. If the hypothesis of preexisting mutations was correct, the culture derived from these cells would be enriched for resistant mutants because only a very few cells from a small area in the agar were removed from the master plate. The enriched culture could then be used to prepare a new master plate and the whole process repeated. The final result was a master plate containing nothing but resistant bacteria, even though the cells and their progenitors had never been directly subjected to selection.

The replica plating method has been applied in numerous experiments to identify the occurrence of mutations; the method permits both the detection of mutations and the retention of viable colonies of mutant strains that can be readily identified and cultured for further study. Many of the biochemical pathways described in Chapters 5 and 6 were elucidated in this way by using nutritional mutations; by examining the growth requirements of auxotrophic strains, it is possible to determine the sequential order of intermediary metabolites in a metabolic pathway.

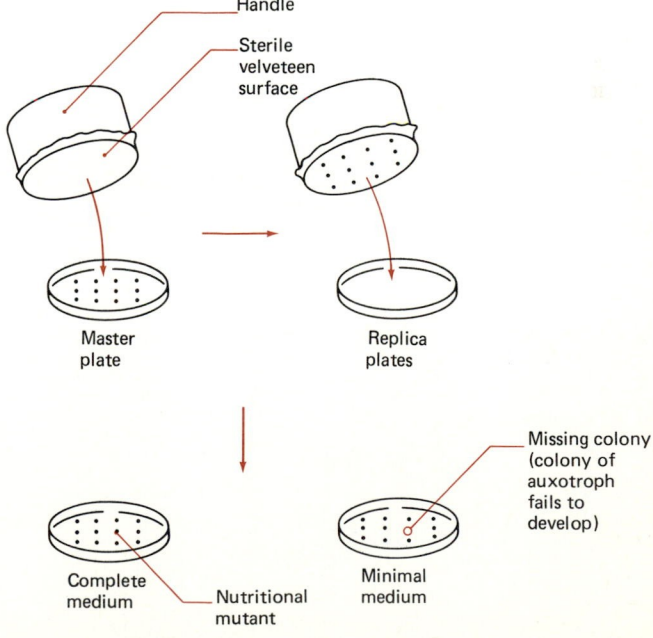

mutations (Figure 8.7). For example, a mutation in the anticodon of the tRNA molecule can suppress a nonsense mutation if the change in the anticodon results in the insertion of an amino acid where the mutant nonsense codon normally causes premature termination of the polypeptide chain.

In addition to altering structural genes, mutations may affect regulatory genes. Various types of mutations can alter the sequence of bases in the promoter or operator regions of the DNA, changing the ability of the organism to regulate protein synthesis at the level of transcription. Thus, for example, a mutation may change an inducible or repressible enzyme system to a constitutive enzyme system and vice versa.

Another way of viewing mutations is to consider their effect on the phenotype of the organism. When the mutation results in the death of the microorganism or its inability to reproduce, it is said to be a **lethal mutation**. Such mutations may be conditional, causing the death of the organism only under certain environmental conditions, or unconditional, being lethal to the organism regardless of the environmental conditions. A **conditionally lethal mutation** causes a loss of viability only under some specified conditions where the organism would normally survive. **Temperature-sensitive mutations**, for example, alter the range of temperatures over which the microorganism may grow when using specific substrates. For instance, a temperature-sensitive mutation of *Escherichia coli* that alters the stability of the enzymes involved in lactose utilization can prevent that strain of *E. coli* from growing on lactose at elevated temperatures, while not altering its ability to grow on lactose at a lower temperature or its ability to grow on glucose at the temperature at which it can no longer use lactose.

Nutritional mutations occur when a mutation alters the nutritional requirements for the progeny of a microorganism. Often nutritional mutants are unable to synthesize essential biochemicals, such as amino acids. Nutritional mutants that require growth factors, such as specific amino acids, that are not needed by the parental or wild-type (**prototroph**) strain are called **auxotrophs**.

Factors That Influence Mutation Rates

The mutation rate is the probability that any one cell will mutate during a period of time defined as the time required by a cell to divide to form two cells. The time required for a bacterial cell to divide is called the **generation time**. The **mutation rate** is equal to the average number of mutations per cell generation, described by the equation

$$a = (0.69) \, m \div (n - n_0),$$

where a stands for the mutation rate and m for the average number of mutations occurring when n_0 cells increase in number to n cells. The mutation rate of a given culture can be determined by growing a population of cells on a solid medium where each mutation gives rise to a mutant clone that can be detected as a single colony. Mutations occur spontaneously only at relatively low rates. In *E. coli*, for example, the spontaneous mutation rate is approximately one change per billion nucleotide pairs replicated. This spontaneous rate varies among different bacterial species. Additionally, a variety of chemical and physical agents can increase the incidence of mutation.

Chemical Mutagens   Various chemicals can modify the nucleotide bases and act as **chemical mutagens** (Figure 8.9). Hydroxylamine, for example, chemically modifies cytosine to uracil so that it pairs with adenine instead of its normal complementary base guanine. After one generation this change results in the replacement of a GC pair with an AT pair, that is, in a transition. Nitrosoguanidine and several other chemical mutagens can alkylate nucleotide bases, causing transitions that result in the substitution of GC for AT in

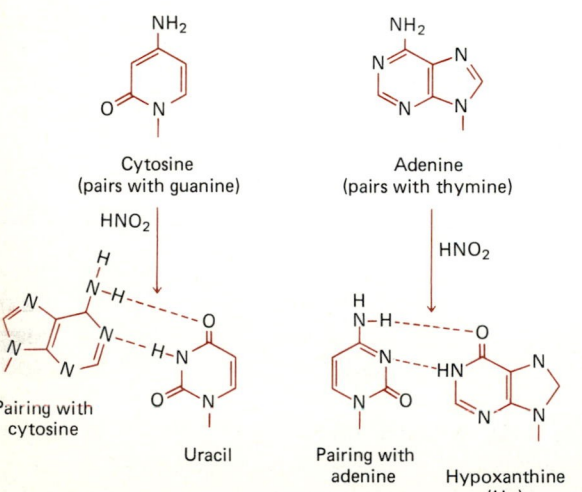

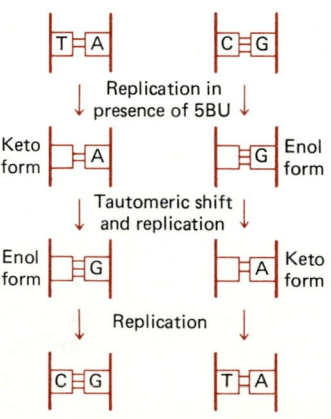

FIGURE 8.9
Exposure to mutagenic biochemicals, such as nitrous acid, causes changes within the DNA.

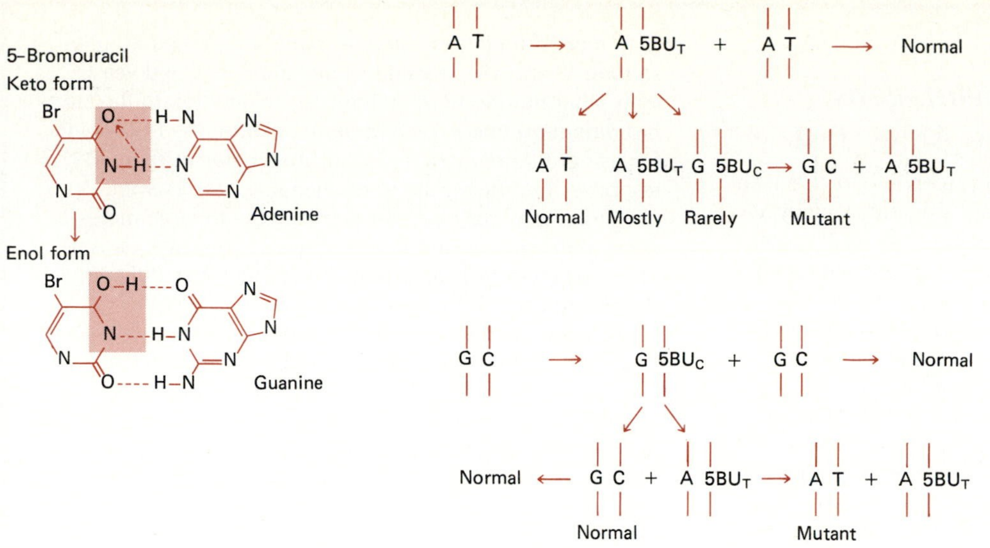

5-Bromouracil
Keto form

Adenine

Enol form

Guanine

A T → A 5BU_T + A T → Normal

A T   A 5BU_T G 5BU_C → G C + A 5BU_T

Normal Mostly Rarely   Mutant

G C → G 5BU_C + G C → Normal

Normal ← G C + A 5BU_T → A T + A 5BU_T

Normal   Mutant

FIGURE 8.10

Exposure to a base analog, such as 5-bromouracil, results in mutation.

the second generation. Nitrous acid oxidizes the amino group of cytosine or adenine, forming keto (-CO-) groups, converting cytosine to uracil and adenine to hypoxanthine, and resulting in a base substitution. There are several biochemicals that resemble DNA nucleotides, and such chemicals are able to act as **base analogs** (chemicals that structurally resemble a DNA nucleotide, and therefore may substitute for it, but do not function in the same manner) (Figure 8.10). For example, 5-bromouracil can replace thymine and pair with adenine or replace cytosine and pair with guanine, thus producing base substitutions in the DNA. Several chemical mutagens, such as acridine, result in base deletion or base addition mutations and cause frame shift mutations. Other mutagens, such as mitomycin C, form covalent cross-linkages between base pairs, preventing the replication of the DNA molecule. Thus, exposure to a variety of chemicals that act in different ways can greatly increase mutation rates.

Radiation   In addition to exposure to chemical mutagens, exposure to radiation can increase the rate of mutation. Exposure to high-energy radiation such as x-rays can cause mutations because such high-energy ionizing radiation produces breaks in the DNA molecule. Exposure to gamma radiation, such as that emitted by ^{60}Co, can be used to sterilize objects, including plastic petri plates, because sufficient exposure to ionizing radiation results in lethal mutations and the death of all exposed microorganisms. The time and intensity of exposure determine the number of lethal mutations that occur (Figure 8.11) and thus establishes the required exposures when ionizing radiation is employed in sterilization processes.

Exposure to **ultraviolet light** also can result in base substitutions by creating covalent linkages (dimers) between pyrimidine bases on the same strand of the DNA (Figure 8.13). A **thymine dimer** cannot act as a template for DNA polymerase, and the occurrence of such dimers therefore prevents

the proper functioning of polymerases. Exposure to ultraviolet light can cause lethal mutations and is sometimes used to kill microorganisms in sterilization procedures.

Microorganisms, though, have several mechanisms for repairing DNA containing thymine dimers. In one repair mechanism, exonucleases remove the dimers, polymerases fill in the gap, and ligases attach the newly inserted bases. The enzymes involved in this repair process are believed to be the same ones that carry out normal DNA replication. Thymine dimers can also be removed by a **photoreactivation** mechanism that breaks the covalent linkages between the thymine bases. The photoreactivation mechanism depends on an enzyme that functions only in the presence of light. The

FIGURE 8.11

Exposure to x-rays greatly increases the rate of mutation.

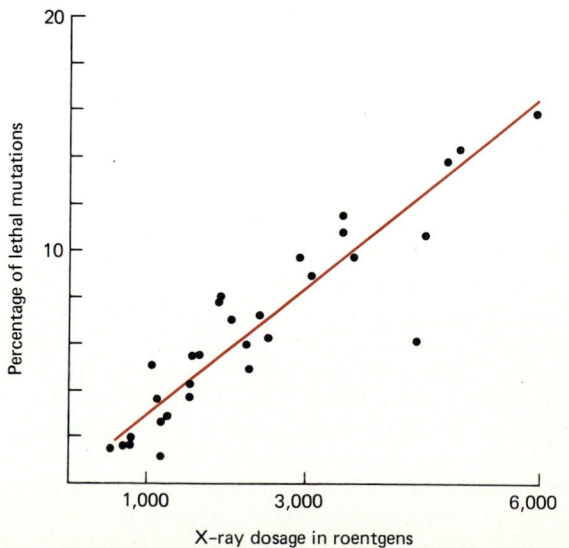

ANALYTICAL PROCESS

Detecting Chemical Mutagens and Carcinogens: The Ames Test

The fact that microorganisms are susceptible to chemical mutagens can be used to determine the mutagenicity of various chemicals. Microorganisms are especially useful in such testing because it is easy to expose huge numbers of microorganisms, perhaps a billion, to the chemical in a single petri plate. To expose this number of macroorganisms, say rabbits or mice, to the chemical for mutagenicity testing would be impossible.

In the **Ames test procedure** strains of the bacterium *Salmonella typhimurium* typically are used as test organisms for determining chemical mutagenicity (Figure 8.12). The strains employed in the Ames test procedure are auxotrophs that require the amino acid histidine for growth. Several different strains are used, each specific

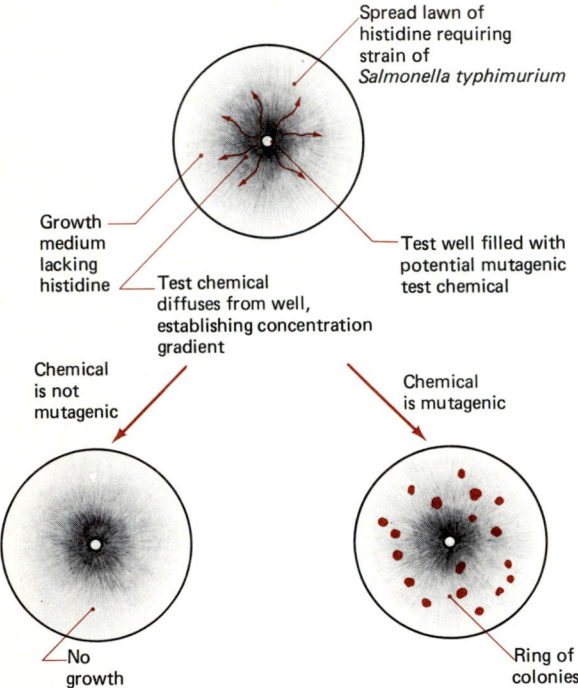

Spread lawn of histidine requiring strain of *Salmonella typhimurium*

Growth medium lacking histidine

Test well filled with potential mutagenic test chemical

Test chemical diffuses from well, establishing concentration gradient

Chemical is not mutagenic

Chemical is mutagenic

No growth

Ring of colonies

FIGURE 8.12
The Ames test procedure is used to screen for mutagens and potential carcinogens.

for a type of mutation, such as frame shift, deletion, and so forth. Some test procedures include *E. coli* and yeasts with other nutritional requirements, in addition to the histidine-requiring *S. typhimurium* auxotrophs. The reason for using different strains is that they differ in their responses to different types of chemicals; for example, one strain may have greater permeability to large molecules than another and hence may be a better organism to use when testing large molecules. Often five different strains are used in the test protocol.

The organisms are exposed to a gradient of the chemical being tested on a solid growth medium that contains only a trace of histidine, only enough to support the auxotrophs long enough for the potential mutagenic chemicals to act. Normally, the bacteria cannot grow sufficiently to form visible colonies because of the lack of histamine, and therefore, in the absence of a chemical mutagen, no colonies can develop. If the chemical is a mutagen, lethal mutations will occur in the areas of high chemical concentration, and no growth will occur in these areas. At lower chemical concentrations along the concentration gradient, however, fewer mutations will occur, and some of the cells may revert because of mutagenic action, producing histidine prototrophs (able to synthesize their own histidine) that will be able to grow and produce visible bacterial colonies on the medium. The appearance of bacterial colonies demonstrates that the chemical has mutagenic properties.

The Ames test is also useful in determining if a chemical is a potential carcinogen (cancer-causing agent). There is a high correlation between mutagenicity and carcinogenicity. Thus, determining whether a chemical has mutagenic activity is useful in screening large numbers of chemicals for potential mutagenicity, even though the Ames test does not actually establish whether a chemical causes cancer. In testing for potential carcinogenicity, the chemical is incubated with a preparation of rat liver enzymes to simulate what normally occurs in the liver, where many chemicals are inadvertently transformed into carcinogens in an apparent effort by the body to detoxify the chemical. Following this activation step, various concentrations of the transformed chemical are incubated with the *Salmonella* auxotroph to determine whether it causes mutations and is a potential carcinogen. Further testing for carcinogenicity is done on those chemicals that have tested positive for mutagenicity.

Discovery Process

The theoretical basis of the Ames test is that nearly all proven carcinogens that act directly, that is, by attacking the DNA as opposed to indirect hormonal action, are also mutagens. So, rather than screening chemicals directly for carcinogenicity, Bruce Ames and his coworkers thought it would be simpler to test them first for mutagenicity. Those chemicals not shown to be mutagenic would be assumed to be noncarcinogenic or indirect acting, and those shown to be mutagenic would be subjected to further testing. The investigators wanted a convenient bacterial genetic system that involved a positive selection for the desired event, so they chose a reversion mutation, the reacquisition of a genetically controlled trait—specifically, the reversion of the histidine mutations in *Salmonella typhimurium*. They assembled a collection of 13 mutant strains that revert to the prototype very rarely and that represent different types of mutagenic events, including frame shift, base substitution, and nonsense mutations. Today the use of this bacterial assay greatly simplifies the task of screening many potentially dangerous chemicals, allowing us to recognize potentially carcinogenic compounds.

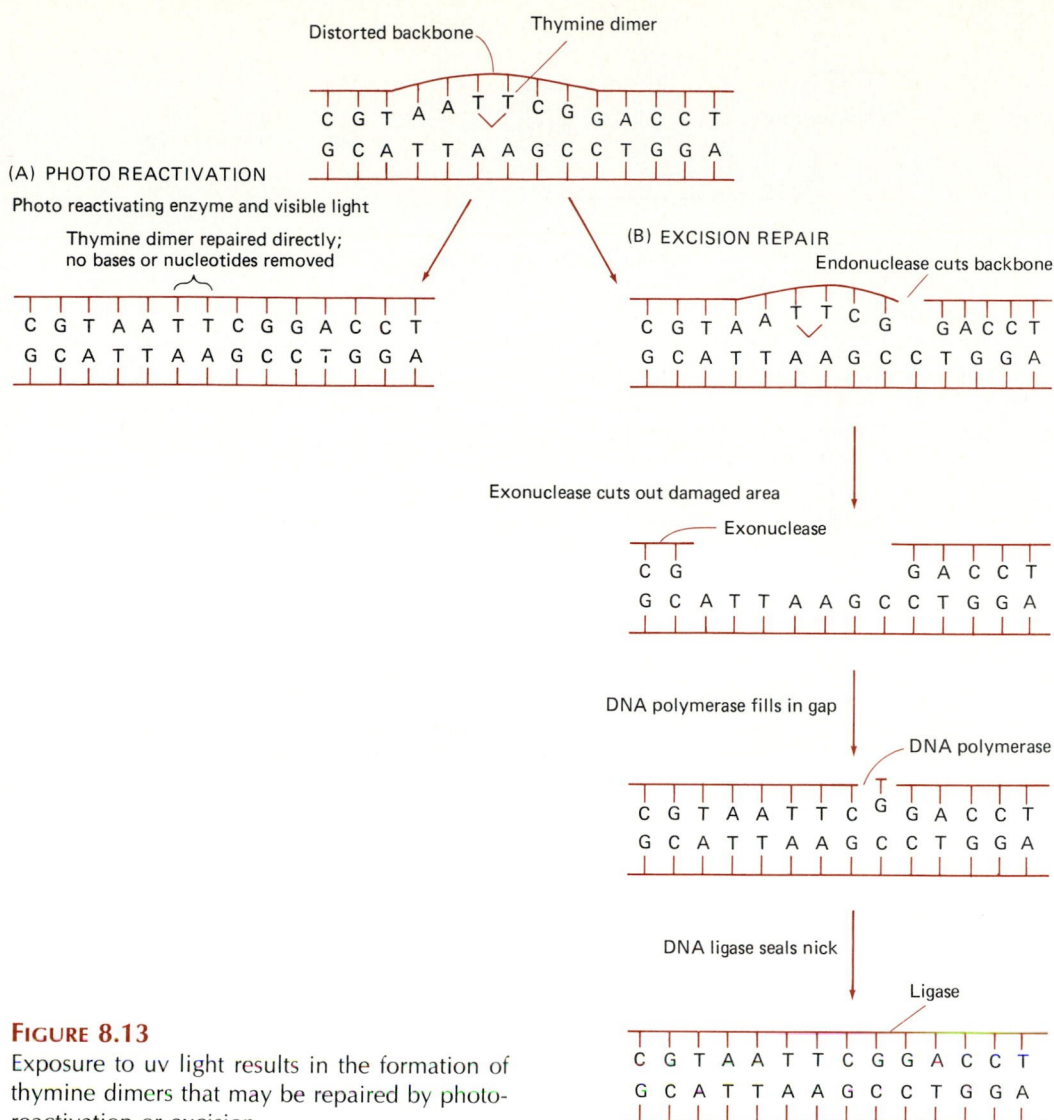

FIGURE 8.13

Exposure to uv light results in the formation of thymine dimers that may be repaired by photoreactivation or excision.

ability to remove thymine dimers is a particularly important adaptation in microorganisms that are normally exposed to high levels of solar radiation.

Another mechanism for repair of damaged DNA is the **SOS system**, which is a complex multifunctional process. This mechanism of DNA repair is not restricted to damage induced by exposure to ultraviolet radiation. Rather, it appears to be a radical repair system designed to save the cell when there is persistent DNA damage and is not restricted to the excision of thymine dimers. When the SOS system is activated, cell division ceases, resulting in filamentous growth. The induction of this system occurs only after a delay during which incomplete replication of DNA has occurred.

Activation of the SOS system occurs when the RecA protein is altered, probably by interaction with oligonucleotides formed as a result of DNA damage (Figure 8.14). The activated RecA protein, which has proteolytic activity, attacks

several DNA-binding proteins that function as repressors of transcription. Proteolytic cleavage of these repressor proteins turns on the SOS system. Specifically, the SOS system is normally repressed by the LexA protein, but LexA—a protein product of the regulatory gene *lex*A—is inactivated by the proteolytic action of the RecA protein. Both the *rec*A gene itself and the genes for uvrA nuclease, which excises thymine dimers, are expressed when the LexA protein is digested. The increase in RecA production increases recombination, and other functions of the system increase rates of mutation, so that there is a great increase in the rate of DNA modification. Once the SOS system is activated, DNA repair occurs in the absence of template direction. The damage to the DNA activates repair mechanisms that fill in gaps in the DNA without copying the template so that errors are not promulgated. Once the DNA damage has been repaired, the SOS system is switched off and further modifications of DNA cease.

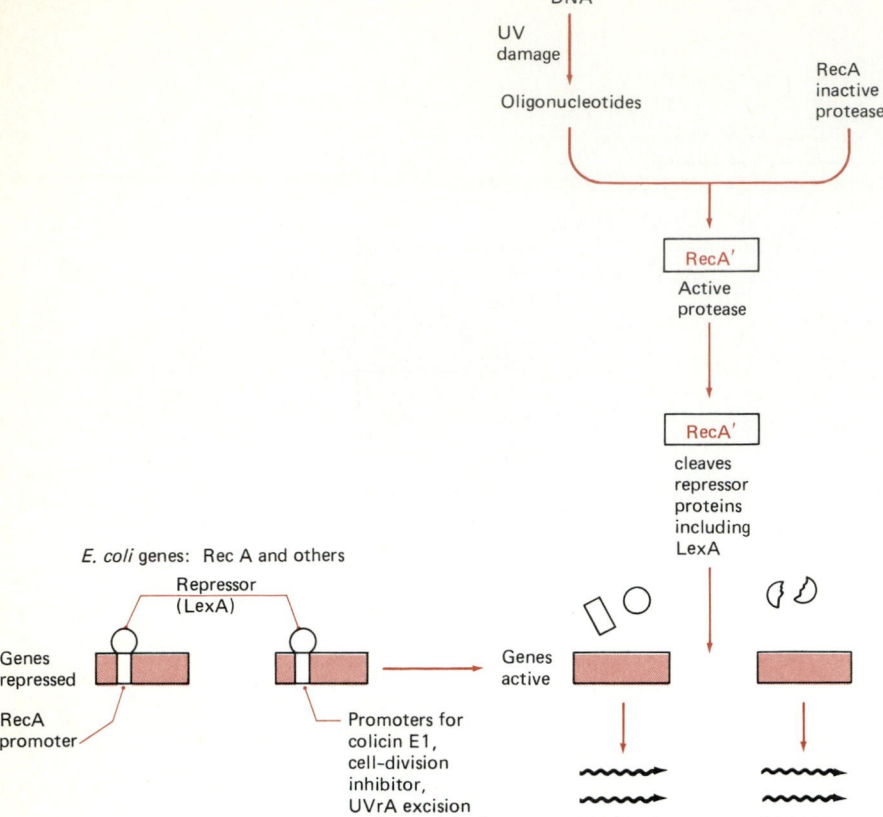

FIGURE 8.14
The SOS system is an extensive, integrated repair system. It is initiated when *rec*A is activated and acts as a proteolytic enzyme, cleaving LexA protein. This process, in turn, activates other genes, as well as initiating additional synthesis of RecA.

8.2 Genome Structure

Haploid and Diploid Cells

As previously emphasized, prokaryotic and eukaryotic cells differ in their genome structure, and this difference has an important effect on how genetic variability introduced into the gene pool through mutation is maintained and transmitted. Bacterial cells have a single genome contained in the bacterial chromosome. Because there is a single set of genes, bacteria are **haploid**. In contrast, eukaryotic microorganisms generally have pairs of matching chromosomes and are therefore **diploid**. Some eukaryotic microorganisms are haploid during much of their life cycle, whereas others are predominantly diploid. For eukaryotic microorganisms, which normally contain pairs of chromosomes and thus a duplicate set of genetic information, the variability introduced into the **gene pool** (set of genes) through mutation can result in differences in the information encoded in the corresponding genes (**alleles**) of the pairs of chromosomes. Because diploid microorganisms contain two copies of each gene, a mutation in one of the copies may not be expressed immediately.

Different forms of the same gene (alleles) arise through mutational events (Figure 8.15). When both copies of the gene are identical, the microorganism is **homozygous**, and when the corresponding copies of the gene differ, the micro-

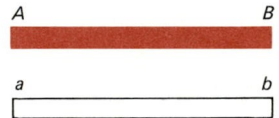

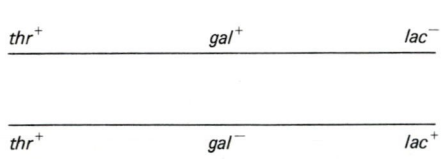

FIGURE 8.15
Alleles are different forms of the same genes.

organism is **heterozygous**. Allelic forms of a gene often code for different amino acid sequences in polypeptides. The information encoded in one of the alleles may dominate over the information in the other allele; that is, one gene may be dominant and the other recessive. For example, a **recessive** allele may code for an inactive enzyme, whereas the **dominant allele** codes for a fully active enzyme. In such an organism, production of an enzyme could continue even though a mutation had altered the genetic content encoded in one of the alleles. In some cases, alleles may exhibit **codominance**, producing a hybrid state with an intermediate phenotype. For example, microorganisms with homozygous alleles may appear red or colorless, and those with heterozygous alleles may appear pink.

Even though a single diploid organism can have no more than two alleles, there may be more than two allelic forms of a given gene in the entire population. The number of alleles determines the number of potential genotypes for a given gene locus. If there are two alleles, there are three possible genotypes—two homozygotes and one heterozygote. If there are 4 alleles, however, there are 10 possible genotypes, 4 homozygotes and 6 heterozygotes. The number of different genotypes that can arise from multiple alleles raises the potential degree of heterogeneity within the gene pool of a microbial population.

Plasmids

In addition to the genetic variability that can occur within the genome of the cell contained within the bacterial chromosomes of prokaryotes and of eukaryotes, some microorganisms contain plasmids (Figure 8.16). **Plasmids** are small

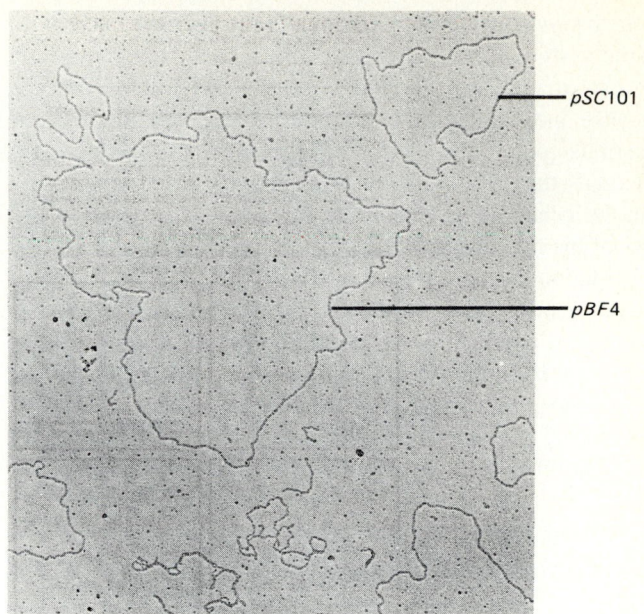

FIGURE 8.16
This electron micrograph shows two different plasmid DNA molecules. The larger loop is plasmid pBF4 and was isolated from *Bacteroides fragilis*; it encodes resistance to the antibiotic clindamycin. The smaller DNA loop is plasmid pSC101 from *Escherichia coli*; it encodes resistance to tetracycline (30,000×). (From BPS—Rod Welch, School of Medicine, Stanford University)

ANALYTICAL PROCESS
Detection of Plasmids

Plasmids can be isolated and detected using gel electrophoresis. Cells are lysed, generally using lysozyme and detergent; the DNA is extracted and placed on a gel, such as agarose. An electric charge is applied across the gel that causes the negatively charged DNA molecules to migrate toward the anode. Conditions are adjusted using buffers so that the rate of migration of the DNA depends on the size and shape of the DNA molecule. The locations of the separated bands of DNA in the gel can be stained by ethidium bromide, a fluorescent dye that becomes concentrated in the DNA-containing regions of the gel and visualized under ultraviolet light (Figure 8.17).

FIGURE 8.17
Plasmids can be detected following electrophoretic separation using ethidium bromide. The position of the plasmid is a function of its size, and, with appropriate standards, it is possible using this procedure, to determine the approximate number of nucleotides in the plasmid. In this photograph the large chromosomal DNA (Chr) is separated from plasmid DNA (P) of *Escherichia coli* strains containing various different plasmids. The distance of migration of the plasmid DNA in these six separate prepartions is a function of plasmid size. (Courtesy Michael Perlin, University of Louisville)

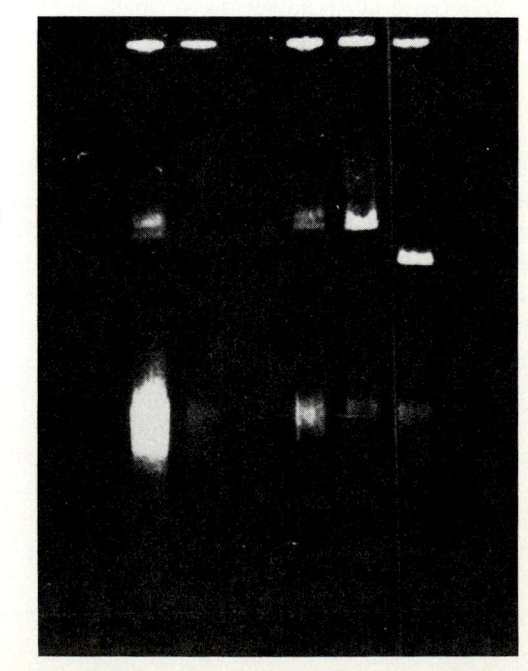

extrachromosomal genetic elements that permit microorganisms to store additional genetic information. The acquisition or loss of plasmids alters the genomes of microorganisms because individuals of a microbial population that possess plasmids contain different genetic information than those individuals that lack plasmids. Plasmids do not normally contain the genetic information for the essential metabolic activities of the microorganism, but they generally contain genetic information for specialized features, such as resistance to heavy metals and antibiotics. Plasmids are capable of self-replication and can be exchanged between bacteria, permitting the transfer of information encoded within the plasmid DNA from one microbial population to another. The transmission of plasmids is a mechanism by which specialized information can be transferred from one bacterium to another. The fact that plasmids are self-reproducing, relatively small, and readily transmitted makes them useful tools in genetic engineering. This topic will be discussed later in this chapter.

Bacterial cells may contain more than one plasmid, but certain pairs of plasmids cannot be stably replicated in the same bacterial cell. Such **incompatible plasmids** do not coexist in the same cell and are said to belong to an **incompatibility group**. The incompatibility group of a plasmid is designated *Inc* followed by a capital letter and sometimes also a number, for example, *Inc*P1. It appears that plasmids of the same incompatibility group are closely related and have similar properties. The property of incompatibility is encoded within the genes of the plasmid.

Functions of Plasmids   Although plasmids contain only a very small portion of the microbial genome, they are important. Plasmids can contain, among other things, the information that determines the ability of bacteria to mate and whether a bacterial strain acts as a ''male,'' donating DNA during mating; the information that codes for resistance to antibiotics and other chemicals, such as heavy metals, which are normally toxic to microorganisms; the information for the degradation of various complex organic compounds, such as the aromatic hydrocarbons found in petroleum; the information for toxin production that renders some bacteria pathogenic to humans; and the genes responsible for nitrogen fixation and the formation of root nodules on leguminous plants.

There are several different types of plasmids that serve different functions. The **F (fertility) plasmid**, for example, codes for mating behavior in *E. coli*. Strains of *E. coli* that have the F plasmid are donor strains, and those that lack the F plasmid are recipient strains. Bacteria that have the F plas-

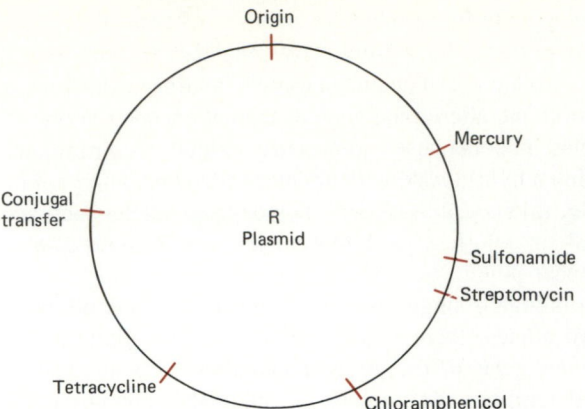

FIGURE 8.18
An R plasmid encodes the information for antibiotic and heavy metal resistance.

mid are able to form pili of the F type that are involved in establishing mating pairs. The F plasmid may exist as an independent circular molecule of double helical DNA or may become incorporated into the bacterial chromosome. (The F plasmid will be discussed further in connection with the process of bacterial mating later in this chapter.)

Colicinogenic plasmids carry the genes for a protein that is toxic only to closely related bacteria. For example, when strains of *E. coli* containing colicinogenic plasmids are mixed with other strains of *E. coli*, only one strain can survive. The toxins produced by *E. coli* are called *colicins*. In addition to the genes for toxin production, colicinogenic plasmids have genes that protect the host cell; they also may carry the information necessary for bacterial conjugation, that is, mating involving cell–cell contact. The acquisition of colicinogenic plasmids enables bacterial strains to enter into antagonistic relationships with other bacterial strains.

The **R (resistance) plasmids** carry genes that code for antibiotic resistance (Figure 8.18). Some R factor plasmids also carry genes for mating. The enzymes coded for by the genes of R plasmids are able to degrade antibiotics, rendering them inactive and thus conferring resistance on bacterial strains that possess such R plasmids. R factors can be passed among bacteria, for example from *E. coli* to pathogenic strains of *Shigella* or *Salmonella*. Antibiotic-resistant strains of bacteria have become a serious health problem because R plasmids can occur in pathogenic bacteria, and the treatment of human bacterial diseases has been complicated by the occurrence of these pathogens that are resistant to multiple antibiotics.

8.3 Recombination

One of the ways that genetic variability is maintained within a population is through **recombination**, a process involving the exchange of genetic information among different DNA molecules that results in a reshuffling of genes. Recombi-

nation provides a mechanism for redistributing the informational changes that occur in DNA as a result of mutation and can produce numerous new combinations of genetic information.

Regardless of the mechanism, recombination results in an exchange of allelic forms of genes that can produce new combinations of alleles. Recombination thus provides a mechanism for generating diversity within the gene pool of a microbial population. The generation of heterogeneity, however, is restricted by the fact that the exchanges must be between alleles of the same gene or between genes that have large sequences of corresponding nucleotides. As a result, reciprocal exchanges occur within species, but this process does not generally permit genetic exchange between different species.

Types of Recombinational Processes

There are two different types of recombinational exchange processes (Figure 8.19). The classic type of exchange occurs between homologous regions of DNA molecules, that is, between regions containing the same or nearly the same nucleotide sequences, as seen in the crossing over of chromosomes where pairs of chromosomes containing the same gene loci pair and exchange allelic portions of the same chromosomes. The term *homologous* indicates that the exchange is between alleles of the same gene and is not meant to imply that the exchanged DNA segments have exactly the same nucleotide sequences. **Homologous recombination** can also be considered a general or reciprocal exchange of DNA. Genetic recombination can occur as well between nonhomologous segments of DNA molecules. **Nonhomologous recombination** does not mean that there is no nucleotide homology in the segments of DNA that are exchanged, but it does imply that the extent of such homologous regions is limited. The difference between homologous and nonhomologous recombination is a matter of degree, and the distinction is not always clear.

Nonhomologous recombination, which also is called **nonreciprocal recombination**, can be a **site-specific** exchange process, that is, a process in which DNA exchange occurs only at a given location within the genome. Nonhomologous recombinations permit the joining together of DNA molecules from different sources. For example, viral DNA may become incorporated into a bacterial chromosome, plasmids may enter into bacterial chromosomes, and the locations of DNA segments may be transposed within chromosomes. Both general and site-specific recombination alter the arrangement of the genome and introduce variation into the gene pool of the population.

In homologous recombination, there is relatively good base pairing of corresponding regions of the DNA, and the aligned chromosomes may establish duplexes between homologous DNA regions. In eukaryotic microorganisms this often occurs during meiosis, the process whereby homologous chromosome pairs are separated and one member of each pair is distributed to each of the two daughter cells. Meiosis results in the conversion of a diploid cell into a haploid cell. A similar homologous alignment of DNA molecules can occur when a bacterial chromosome or portion thereof is transferred from a donor to a recipient bacterium.

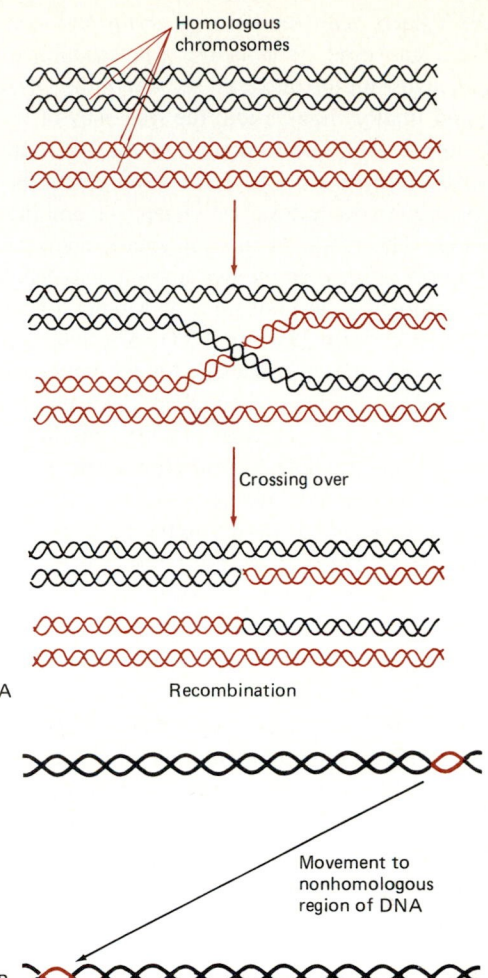

FIGURE 8.19

There are two different processes of genetic recombination, one with extensive homology between the nucleotide sequences that are recombining (homologous recombination) and another with relatively little homology (nonhomologous recombination). Each process is important in redistributing genetic information and maintaining genetic diversity. (A) Chromosomal crossing-over, which results in the recombination of genes, is a classic example of homologous recombination. (B) The movement of transposons, which are sometimes called *jumping genes,* is an example of nonhomologous recombination.

Events That Occur During Recombination

Several models have been developed to describe the events that occur during recombination. The experimental evidence indicates that single-stranded regions of DNA are involved in genetic recombination and that enzymes are intimately involved in the process. In one model, recombination is initiated by an **endonuclease**—an enzyme that cleaves a DNA macromolecule by breaking bonds within the molecule—that nicks one of the strands of DNA; the free 3'OH end of the DNA molecule that results from endonuclease activity acts as a primer for DNA synthesis involving a DNA polymerase that produces a single strand of DNA; the newly synthesized region of the DNA pairs with the corresponding region of the homologous chromosome, establishing a heteroduplex;[1] an endonuclease cleaves out the unpaired section of the DNA macromolecule; and finally, ligases join the free ends of the DNA strands (Figure 8.20). The formation of the heteroduplex is catalyzed by enzymes coded for by *rec* (**recombination**) **genes**; in organisms lacking *rec* genes, recombination does not occur. The result of this enzymatic action is the formation of a bridge between the two homologous DNA strands, and the joining of chromosomes at a homologous region establishes a chi form (Figure 8.21). The chromosomes then rotate so that the two strands no longer cross each other but are still held together by covalent linkages. An endonuclease cleaves the DNA strands, breaking the heteroduplex and establishing two independent chromosomes. In some cases, cleavage by the endonuclease results only in the exchange of DNA in the short region where the heteroduplex had formed; this type of exchange is not considered to establish recombinant DNA molecules. In other cases, the action of the endonuclease results in the formation of chromosomes that have exchanged large portions of homologous DNA regions.

In *E. coli*, homologous recombination depends on the enzymes coded for by three recombination genes—designated *recA*, *recB*, and *recC*. The enzymes coded for by *recB* and *recC* are involved in unwinding double helical DNA and breaking the chains into small fragments. It appears that these enzymes generate single-stranded DNA that can then interact with a double-stranded DNA macromolecule. The RecA protein catalyzes an ATP-driven assimilation reaction in which the single-stranded DNA hybridizes with a duplex (double-stranded) DNA macromolecule. In this process the single-stranded DNA locates a region of homology within the duplex, pairs with the complementary strand, and displaces the other strand.

[1] A heteroduplex forms when two strands of DNA that are complementary over only part of their lengths join together. The homologous regions form a duplex (double-stranded complementary segment) and the noncomplementary segments remain single-stranded. Because there are duplex and nonduplex regions, the term *heteroduplex* is used.

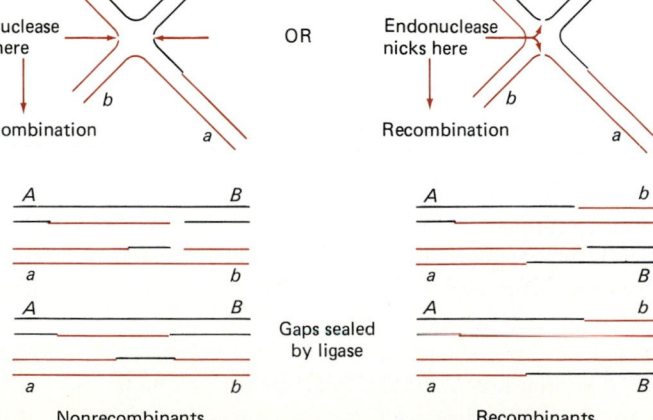

Endonuclease nicks one strand of the lower helix.

DNA polymerase begins synthesis on the available 3′ end, displacing a 5′ tail, which then pairs with the upper helix to form a heteroduplex.

Endonucleases remove the unpaired section of the upper helix. Adjacent ends are joined by ligase.

The upper heteroduplex extends, displacing a 5′ tail that forms a heteroduplex with the lower helix.

Adjacent ends are sealed by ligase.

A different configuration of the helices with the ends of the helices bent away from each other.

Isomerization occurs by rotating the bottom helices 180° with respect to the top ones.

Endonuclease nicks here   No recombination

OR

Endonuclease nicks here   Recombination

Gaps sealed by ligase

Nonrecombinants   Recombinants

FIGURE 8.20
The steps of homologous recombination.

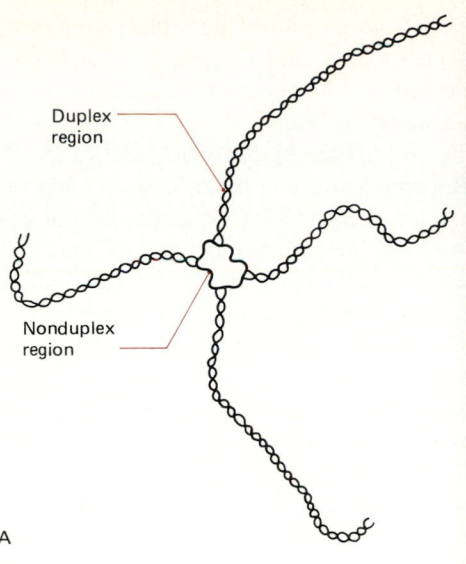

Duplex region

Nonduplex region

A

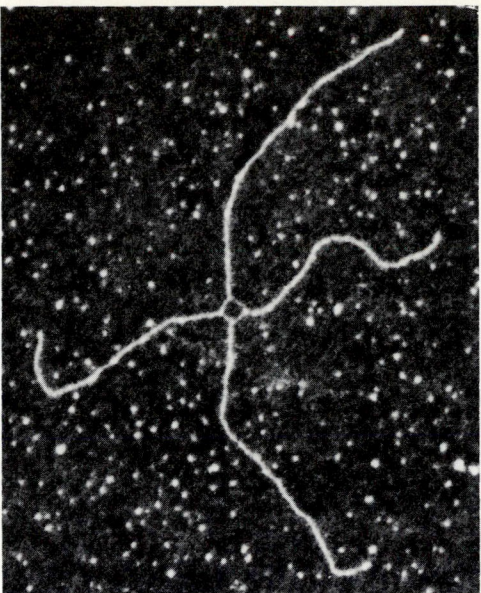

B

FIGURE 8.21
The chi form of DNA occurs during homologous recombination. (A) Diagrammatic representation. (B) Electron micrograph. (Courtesy of Huntington Potter and David Dressler, Harvard Medical School)

In contrast to the general recombination process involving homologous segments of DNA, **nonreciprocal recombination** can occur between very different DNA segments and provides a mechanism for genetic exchange between species. Nonreciprocal recombination permits the incorporation of viral DNA into a bacterial or eukaryotic chromosome; the insertion of plasmids, such as the F plasmid of *E. coli*, into the bacterial chromosome; and the translocation of antibiotic resistance genes from one plasmid location to another. There probably are several different mechanisms that can bring about nonreciprocal recombination. A number of enzymes that are different from those involved in reciprocal recombination appear to be involved in the site-specific transposition of genetic information. The *rec* genes are not involved in nonreciprocal recombination, and microorganisms deficient in the enzymes needed for homologous recombination still carry out nonreciprocal transpositions.

Transposable Genetic Elements

The ability of specific segments of DNA to undergo nonreciprocal recombination, moving from one location to another, appears to depend in some cases on the nucleotide sequences at the ends of the transposable elements. The ends of the **transposable genetic elements** often contain inverted repetitive sequences of nucleotide bases that permit the folding of DNA stabilized by hydrogen bonding between complementary bases within the DNA macromolecule (Figure 8.22). The occurrence of these inverted sequences appears to be important in establishing the ability of genetic elements to exhibit nonreciprocal recombination. In nonreciprocal exchanges of DNA, the lack of large regions of DNA homology suggests that DNA–protein interactions, rather than base pairing, play a particularly important role in the insertion process.

There are several different types of transposable genetic elements. One type, an **insertion sequence (IS)**, can move around bacterial chromosomes, occurring at different locations on the chromosome. ISs are small, transposable genetic elements with about 1,000 nucleotide bases. An IS is not homologous with the regions of the plasmids or the chromosomes into which it is inserted. The IS has an identical nucleotide sequence repeated at each end. The nucleotide bases in the IS regions do not appear to code for structural proteins but may have a regulatory function and may be involved in specifying the locations at which site-specific recombination occurs. ISs, for example, can alter a promoter site and thereby affect the expression of nearby genes.

In contrast to IS elements, **transposons** are larger transposable genetic elements that contain genetic information for the production of structural proteins.[2] Many of the transposons that have been studied code for antibiotic resistance.

---

[2]The distinction between transposons and ISs is not always clear, and some ISs may code for proteins whose functions have yet to be recognized.

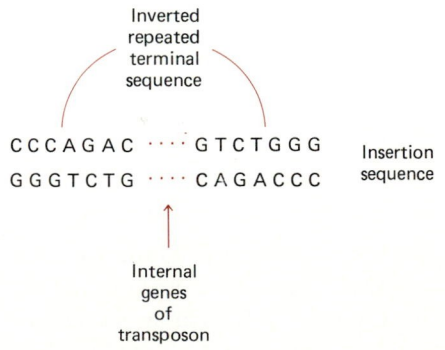

Inverted repeated terminal sequence

CCCAGAC ····· GTCTGGG
GGGTCTG ····· CAGACCC

Insertion sequence

Internal genes of transposon

FIGURE 8.22
Transposable genetic elements contain inverted, repeated nucleotide sequences.

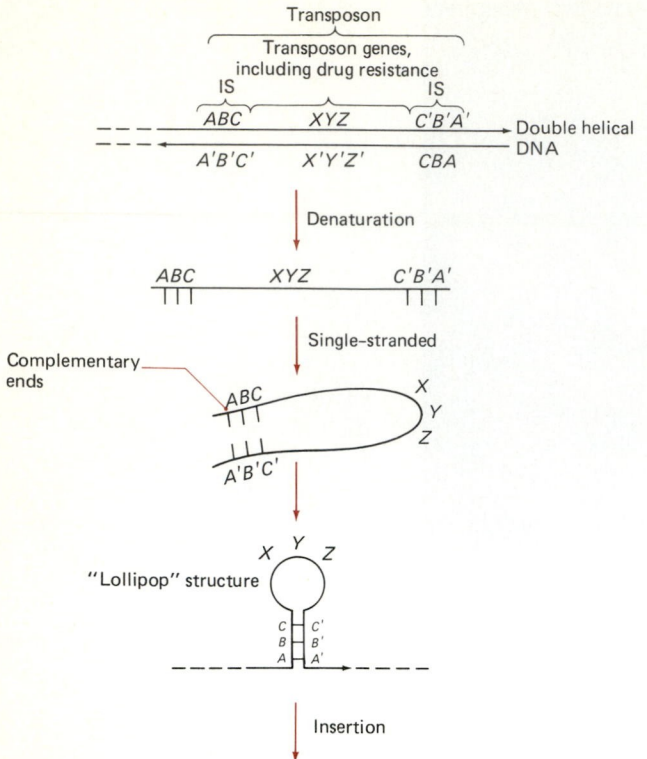

FIGURE 8.23
The steps in the nonreciprocal recombination for the insertion of a transposon.

Transposons, like ISs, have identical nucleotide sequences that are repeated at each end of the DNA molecule, and these terminal nucleotide sequences appear to establish the basis for their enzymatic insertion (Figure 8.23). In some cases, transposons may be constructed by the attachment of ISs to structural genes. Recombination can occur between an IS on a bacterial chromosome and an IS on a transposon. For example, the F plasmid may be incorporated into the bacterial chromosome by recombination between the IS on the F plasmid and a homologous IS on the bacterial chromosome.

Like transposons and ISs, a viral genome may be incorporated into a bacterial chromosome in the process called *lysogeny*. (Lysogeny and the lytic life cycles of bacteriophage will be discussed in Chapter 9.) The genes of temperate phage (viruses capable of lysogeny) can be expressed by the bacterial host, with the bacterium producing proteins that are coded for by the viral genes, a process called *lysogenic conversion*. In one bacterium, *Corynebacterium diphtheriae*, the presence of a temperate phage renders the bacterium pathogenic; strains of *C. diphtheriae* that contain the viral genome produce proteins that are toxins and cause the disease symptoms of diphtheria, whereas strains of *C. diphtheriae* that lack the incorporated viral genome are harmless nonpathogens. Similarly, plasmids may be inserted into the bacterial chromosome. Plasmids or other transposable elements that become incorporated into a bacterial chromosome can later be excised. For example, bacteriophage that establish lysogeny may later reestablish an independent viral element that directs the normal lytic life cycle of the virus.

8.4 *DNA Transfer in Prokaryotes*

Before recombination can occur, DNA from different sources must come together. This can occur in various ways. In bacteria, genetic exchange followed by recombination occurs principally by three mechanisms that differ in the way DNA is transferred between a donor and a recipient cell (Figure 8.24):

Transformation, in which free or naked DNA moves from a donor to a recipient cell.
Transduction, in which DNA is carried from a donor to a recipient cell by a virus.
Conjugation, in which cell–cell contact is needed for transfer of DNA from a donor to a recipient cell.

Transformation

In **transformation** a free DNA molecule is transferred from a donor to a recipient bacterium; that is, the donor bacterium leaks its DNA, generally as a result of lysis of the bacterium, and the recipient bacterium is able to take up the DNA. In order to take up DNA, a recipient cell must be **competent**, that is, its membrane must be in a state so that free DNA can pass across it. The competency of a cell depends upon

its growth phase and environmental conditions. Relatively few bacterial genera have been demonstrated to be capable of taking up naked DNA. When the DNA is taken up by a competent bacterium, one strand of the double helical DNA is enzymatically degraded[3] (Figure 8.25). The intact strand of DNA forms a heteroduplex with the bacterial chromosome of the recipient bacterium. A nuclease degrades the corresponding region of DNA in one of the strands of the recipient cell, and ligases join the donor DNA with the DNA of the recipient bacterial chromosome. This is an example of a reciprocal recombinational event, and if the allelic forms of the donor and recipient genes are not identical, the progeny of the recipient cell may have a composite (hybrid) genome different from that of either the donor or the recipient strain.

A classic example of transformation involves the bacterium *Streptococcus pneumoniae* (formerly referred to as *Diplococcus pneumoniae*) (Figure 8.26). One strain of this bacterium produces a capsule and is a virulent pathogen (disease-causing microbe), whereas another strain of the same

---

[3]In some species of *Bacillus* and *Streptococcus* it appears that only single-stranded DNA enters the recipient cell.

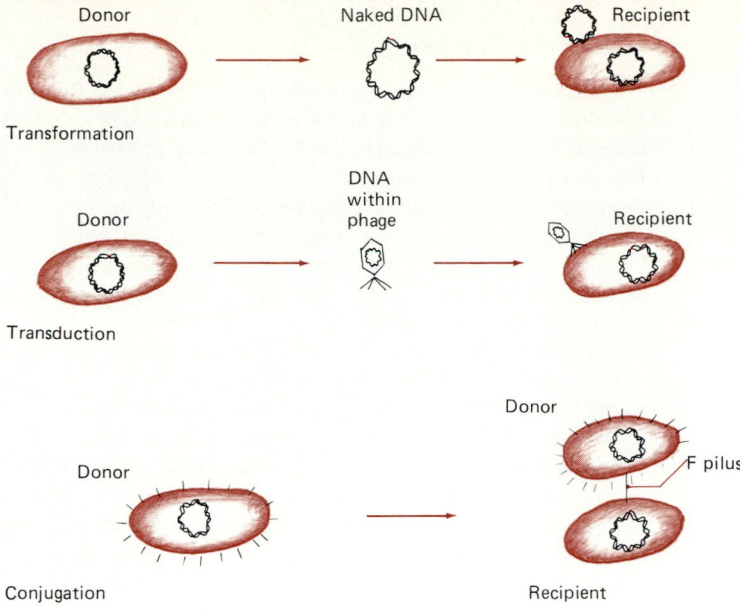

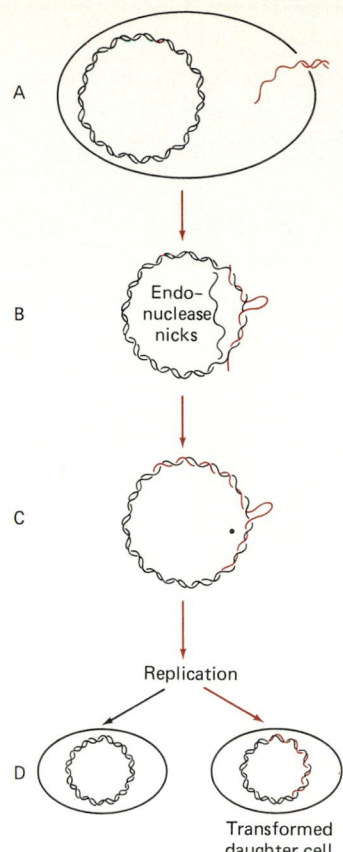

FIGURE 8.24

Genetic transfer in bacteria occurs by three distinct mechanisms: transformation, involving transfer of naked DNA; transduction, involving transfer of DNA by a phage; and conjugation, involving direct contact of bacteria during transfer of DNA.

FIGURE 8.25

Diagram showing the steps of bacterial transformation. (A) A piece of double-stranded DNA from a wild-type strain enters the bacterial cell. Exonuclease converts the transforming DNA to a single strand as it passes into the cell. (B) A heteroduplex forms as transforming DNA pairs with complementary host DNA. (C) Ligase completes the integration of transforming DNA. (D) One daughter cell is a mutant like the recipient cell and one is transformed.

bacterium lacks the genetic information for capsule production and is avirulent (non-disease-causing microbe). When dead cells of the virulent strain are mixed with avirulent live bacteria, transformation can occur, producing a mixture of avirulent and virulent bacteria; that is, the DNA containing the genes for capsule production can leak out of the dead bacteria and be taken up by the living bacteria that normally lack the genetic information for capsule production. Recombination can occur and the progeny of the transformed bacteria become capable of producing capsules; in this manner, nonpathogenic strains can be transformed into deadly pathogens.

Transduction

Whereas in transformation free (naked) DNA is transferred, in **transduction** the DNA is transferred from a donor to a recipient cell by a viral carrier. For transduction a virus must acquire a portion of the genome of the host cell in which it reproduces. A bacteriophage, for example, can acquire bacterial DNA when it infects a bacterial host cell and can then transfer this acquired bacterial DNA to another bacterial cell. There are two different types of transduction—generalized and specialized—which, as the names imply, differ in whether they bring about the general transfer of genes or only the transfer of specific genes.

Generalized Transduction **Generalized transduction** can result in the exchange of any of the homologous alleles. In generalized transduction, pieces of bacterial DNA are accidentally acquired by developing phage during their normal lytic growth cycle[4] (Figure 8.27). If a phage carries bacterial instead of viral DNA, it cannot reestablish a lytic growth cycle and cannot cause lysis in a recipient bacterium. Such phage, however, can attach to and inject DNA into a recipient bacterium, permitting it to carry bacterial DNA from a donor cell and inject it into a recipient bacterial cell. Once inside the recipient cell, the DNA may be degraded by nucleases, in which case genetic exchange does not occur. The injected DNA, however, may undergo homologous recombination. If recombination occurs, the transduced recipients may possess new combinations of genes.

[4]The normal lytic growth cycle of phage, which will be discussed in detail in Chapter 9, involves the invasion of a host cell, the reproduction of the virus within the host, and the lysis or bursting of the host cell to release the newly formed phage.

FIGURE 8.26

The transformation of *Streptococcus pneumoniae* shows how the properties of a bacterial strain can be altered by this recombination process.

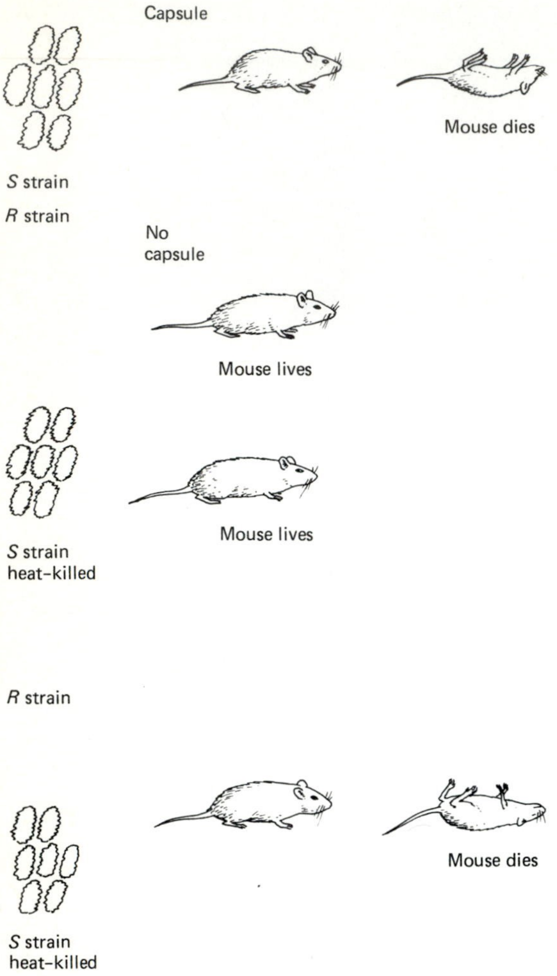

Capsule

S strain

R strain

Mouse dies

No capsule

Mouse lives

S strain heat-killed

Mouse lives

R strain

S strain heat-killed

Mouse dies

Discovery Process

The discovery of the transformation process provided the first direct proof that DNA is the genetic material. In the late 1920s a British scientist, Fred Griffith, was working with *Streptococcus pneumoniae*, a causative organism of pneumonia that produces a polysaccharide capsule. Griffith also isolated avirulent mutants, called *R strains* because they appeared as rough colonies on agar; the R strains lack a capsule, and Griffith demonstrated that they are unable to cause disease. Mice infected with even minimal doses of S (nonmutant, capsulated, smooth, colony-producing) strains died a few days after exposure to the bacteria, whereas injection of even massive doses of R strains did not cause death. Griffith injected mixtures of heat-killed S strains and live R strains into mice; surprisingly, the mice died, and Griffith was able to isolate live S strains from their corpses. The R strains had been transformed into a new encapsulated, pathogenic strain by what appeared to be a genetic process, later called *transformation*.

Oswald Avery, C. M. McCarty, and M. MacLeod provided the molecular explanation for this event in 1941 by separating the classes of molecules in the debris of the dead S cells and testing each one for its ability to cause transformation. First, they showed that it was not the polysaccharides in the capsules themselves that transform the R cells. They found that only one class of molecules, DNA, induces transformation. They deduced that DNA is the agent that determines the polysaccharide character, and hence the pathogenic characteristic, and that providing R cells with S DNA was the same as providing them with S genes. The unavoidable conclusion of this classic work is that the genetic information of the cell is contained within its DNA.

FIGURE 8.27

Diagram showing how phage can acquire bacterial DNA for generalized transduction.

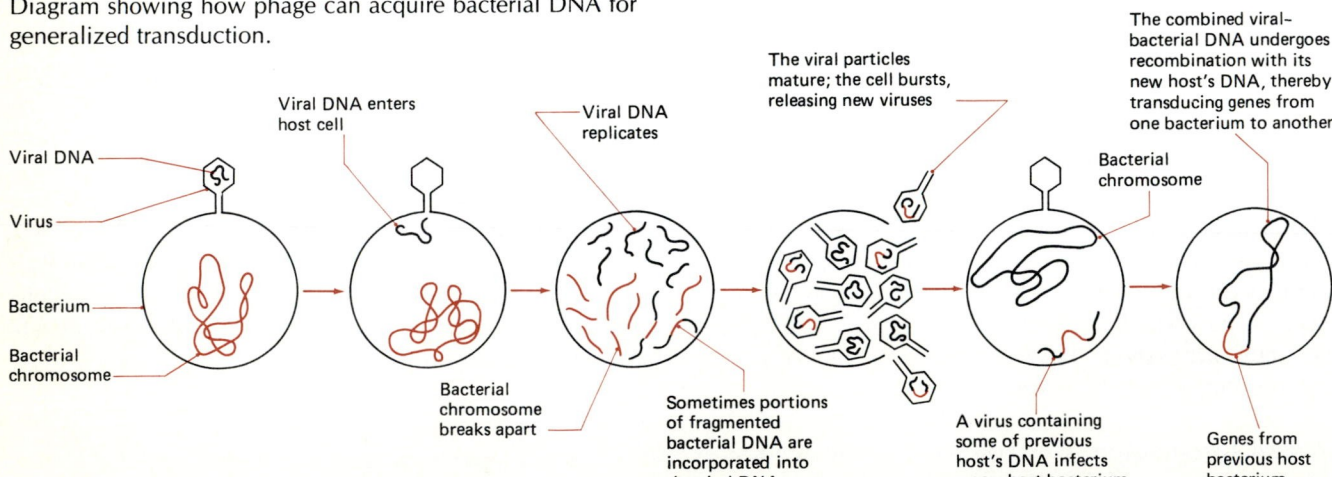

Viral DNA

Virus

Bacterium

Bacterial chromosome

Viral DNA enters host cell

Bacterial chromosome breaks apart

Viral DNA replicates

Sometimes portions of fragmented bacterial DNA are incorporated into the viral DNA

The viral particles mature; the cell bursts, releasing new viruses

A virus containing some of previous host's DNA infects a new host bacterium

Bacterial chromosome

The combined viral–bacterial DNA undergoes recombination with its new host's DNA, thereby transducing genes from one bacterium to another

Genes from previous host bacterium

Specialized Transduction In contrast to generalized transduction, which is mediated by a phage that causes lysis of the cell, **specialized transduction** depends on the transmission of bacterial DNA from a donor to a recipient cell by a temperate phage. Such temperate phage may incorporate their DNA into the bacterial chromosome rather than initiating replication that ends with lysis of the host cell. The incorporation of the phage DNA into the bacterial chromosome is an example of nonhomologous recombination and is regulated by an operon type of control mechanism. Lambda (λ) and mu (μ) bacteriophage are capable of establishing lysogeny, a state in which the phage genome is incorporated into the bacterial genome. The viral genome can be passed on in this state from one bacterial generation to the next. At some point, such as when stressed by exposure to ultraviolet light, the phage genome can be excised from the bacterial genome. When this occurs, the phage DNA occasionally carries with it adjacent bacterial genes and leaves behind some of the viral genes, making the phage defective in some viral genes.

The formation of such a defective phage establishes a viral carrier of bacterial DNA. Because there is a specific site for incorporation of the phage DNA into the bacterial chromosome, only the genes that are adjacent to the site of insertion of the viral genome may be transferred by specialized transduction (Figure 8.28). As an example, the λ phage may acquire from its host *E. coli* either the genes for galactose utilization or the genes for biotin synthesis, because these are the two genes that flank the site where this phage inserts its DNA into the bacterial chromosome. Phage containing *E. coli* DNA may infect new host cells but are unable to insert DNA into the bacterial chromosome in the normal manner because they lack the necessary IS. However, successful insertion can occur in the presence of a second normal λ phage DNA molecule (Figure 8.29). This type of recombination produces a bacterial chromosome with a section containing both a normal and a defective phage DNA molecule and two loci for either the galactose or biotin genes. In this manner, the recipient bacterium becomes diploid for either galactose or biotin. If these genes are of different allelic forms, the organism produced by specialized transduction is heterozygous.

Conjugation

Unlike the genetic exchange mechanisms already discussed, **conjugation (mating)** requires the establishment of physical contact between the donor and recipient bacterial cells. To demonstrate that contact is needed for mating, two different auxotrophic strains can be placed in a U-shaped tube separated by only a glass disk barrier that is impermeable to cells (Figure 8.30). The barrier blocks physical contact, and thus conjugation; free DNA and phages, however, can pass through the membrane, and thus, the other recombinational processes of transduction and transformation can still occur. The physical contact between mating cells of *E. coli* is es-

tablished through the F pilus that is coded for by the F plasmid genes (Figure 8.31).

The F plasmid encodes transfer of itself to other cells that lack an F plasmid. Thus, when cells lacking the F plasmid (F$^-$) are mixed with cells possessing the F plasmid (F$^+$), they become F$^+$ because the transfer of the F plasmid occurs at a high frequency. Chromosomal genes are transferred to the F$^-$ cell along with the F plasmid, but at a lower frequency. The F plasmid contains certain genes (*rep* genes) that allow it to be replicated by the host cell. The F plasmid belongs to the *Inc*F1 incompatibility group; that is, the F plasmid is not replicated when another plasmid of this incompatibility group is present.

The F plasmid and a number of others, collectively known as **conjugative plasmids**, encode for self-transfer. In the case of the F plasmid, transfer is encoded by 13 genes; some encode the synthesis of F pili and others encode a special type of replication called *transfer replication* of the F plasmid that occurs during transfer and mediates it. In addition to mediating its own self-transfer, the F plasmid is capable of mediating the transfer of other plasmids that are incapable of self-transfer. The ability of the F plasmid to mobilize chromosomal DNA rests with its complement of insertion sequences. The capacity of *E. coli* to exchange genetic material by conjugation and the way in which this occurs are determined by the presence of the F plasmid in certain strains. This plasmid, which has the capacity to transfer itself from one cell to another, occasionally interacts with the chromosome or another plasmid, causing some or all of this other element also to be transferred. The F plasmid can replicate only in Gram-negative enteric bacteria, but other conjugative plasmids can self-transfer in other Gram-negative bacteria. Some conjugative plasmids have a broad host range, replicating in most Gram-negative bacteria. Not all conjugative plasmids, however, readily mobilize the transfer of chromosomal DNA.

The F plasmid confers the ability to produce the F pilus needed to form mating pairs of *E. coli* and other Gram-negative bacteria. Gram-negative bacterial strains that produce F pili act as donors during conjugation, whereas those lacking the F plasmid are recipient strains. Recipient bacterial strains are designated F$^-$ and donor strains are designated either F$^+$ if the F plasmid is independent or **Hfr (high frequency recombination strain)** if the F plasmid DNA is incorporated into the bacterial chromosome. The F plasmid may incorporate at different specified sites within the bacterial chromosome. It may also later become detached from the bacterial chromosome, reestablishing itself and sometimes carrying genes from the chromosome, in which case the plasmid is called an F′ plasmid. For example, the *F′lac* plasmid contains genes from the *lac* locus (genes involved in lactose utilization) of the bacterial chromosome.

During bacterial mating, DNA from the donor strain is replicated by a rolling circle mechanism (see Figure 9.10) that produces a single strand of DNA that is transferred into the recipient cell. Normally, only a portion of the donor

A

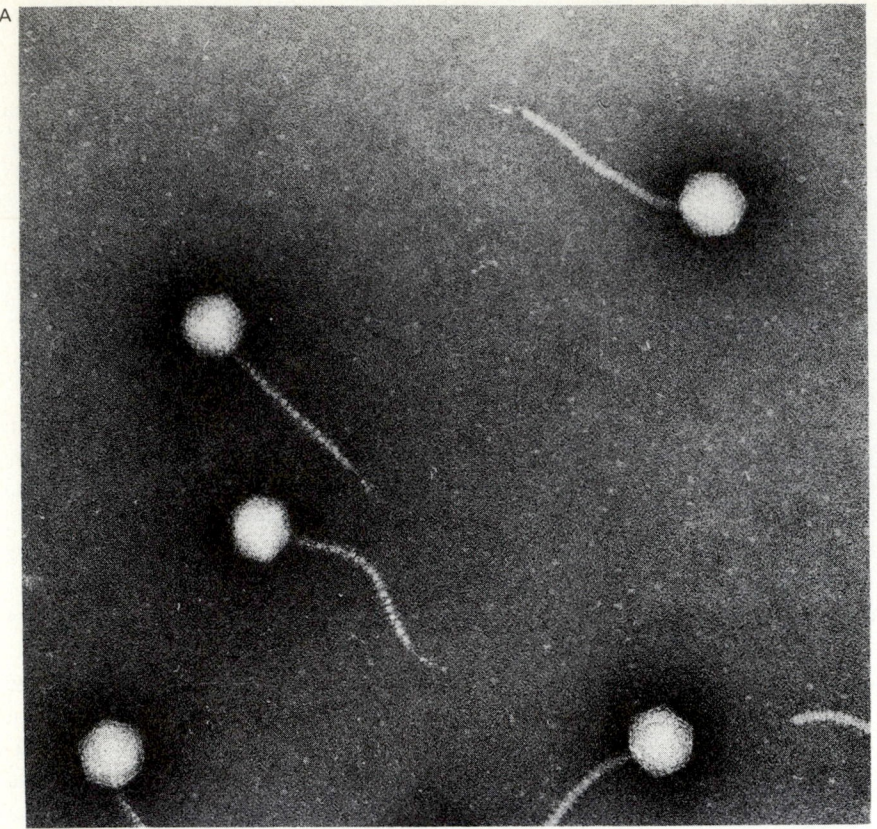

Integration

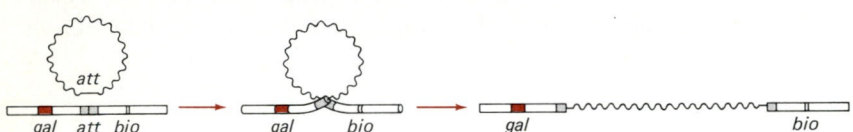

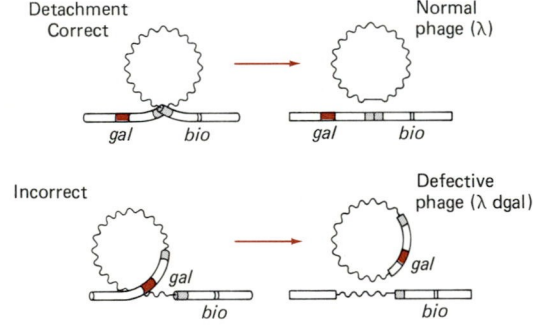

B

FIGURE 8.28

The temperate bacteriophage λ is incorporated into the bacterial genome at a specific site adjacent to the genes for galactose and biotin. When this process occurs, λ can be carried as a prophage, that is, the phage genome can be replicated during replication of the bacterial chromosome. At a later, time the phage genome can be excised from the bacterial chromosome. Depending on the manner of excision, the phage can become complete or defective, that is, temperate phage containing bacterial genes and lacking some portion of the phage genome. (A) Electron micrograph of λ (250,000×). (Courtesy R. C. Williams, University of California, Berkeley); (B) map showing site of insertion of λ into the chromosome of *E. coli*.

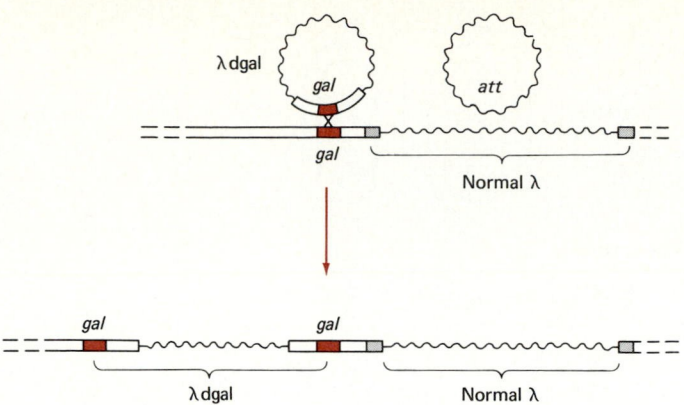

FIGURE 8.29
Specialized transduction by λ phage results in the transfer of a limited amount of genetic information.

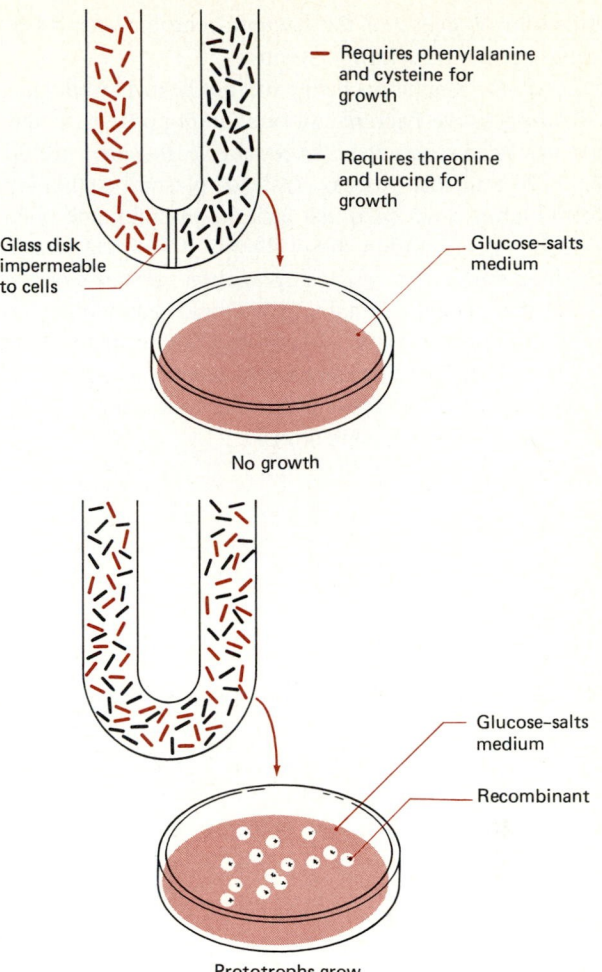

FIGURE 8.30
When direct contact between bacterial strains in a U-shaped tube is prevented by a glass disk barrier separating the two arms, genetic exchange by conjugation cannot occur.

bacterial chromosome is transferred during bacterial mating. The precise portion of the DNA that is transferred depends on the time of mating, that is, on how long the F pilus maintains contact between the mating cells.

When an F^+ cell is mated with an F^- cell, the F plasmid DNA is usually transferred from the donor to the recipient (Figure 8.32). The F plasmid does not normally recombine with the bacterial chromosome of the recipient bacterium; instead, the single-stranded linear DNA molecule that is transferred acts as a template for the synthesis of a complementary strand of DNA, and the double-stranded DNA then reestablishes a circular form. The independent circular F plasmid confers the genetic information for acting as a donor strain, and the offspring of the recipient of such a cross, therefore, are mostly donor strains. When an Hfr strain is mated with an F^- strain, the bacterial chromosome with the integrated F plasmid begins rolling circle DNA replication in response to attachment of the F pilus, that is, replication of the DNA in which a single strand is copied and "rolls" off the circular chromosome (Figure 8.33). A single strand of DNA is transferred from the donor Hfr cell to the F^- bacterium, and the DNA that is transferred may undergo homologous recombination with the recipient DNA. Only part of the F plasmid DNA normally is transferred in this

type of mating cross, and as a result, the recipient cell normally remains F^-. However, a relatively large portion of the bacterial chromosome is transferred from the donor to the recipient, and this results in a relatively high frequency of

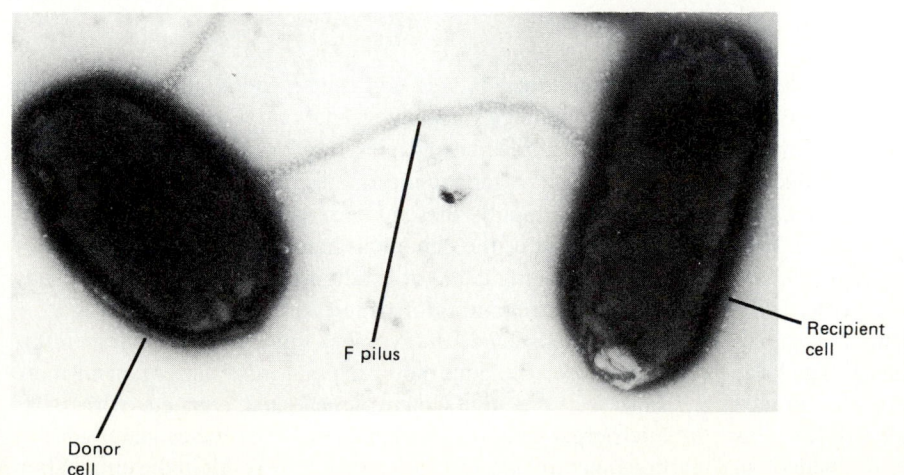

FIGURE 8.31
Electron micrograph showing the attachment of mating *E. coli* cells by an F pilus. (34,000×) (From BPS—D. P. Allison, Oak Ridge National Laboratory)

recombination of genes of the bacterial chromosome when Hfr strains are mated with F⁻ strains.

Conjugational genetic exchange has not been as well studied in Gram-positive bacteria. In one Gram-positive bacterium that has been examined, *Streptococcus faecalis*, pili do not play a role in this process. Instead, plasmid-containing cells form clumps with cells that lack the plasmid, and plasmid transfer occurs within these clumps. Clumping results from the interaction between an aggregation substance on the surface of the plasmid-containing cell and a binding substance on the surface of a plasmid-lacking recipient. The aggregation substance is produced only when a plasmid-containing (donor) cell is in close proximity to a cell that lacks that particular plasmid (recipient cell).

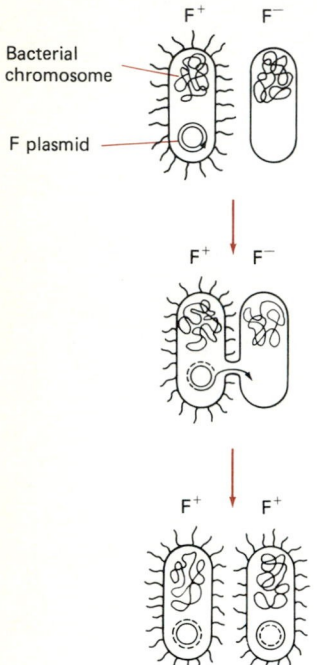

FIGURE 8.32
Diagram showing conjugation of F⁺ and F⁻ strains.

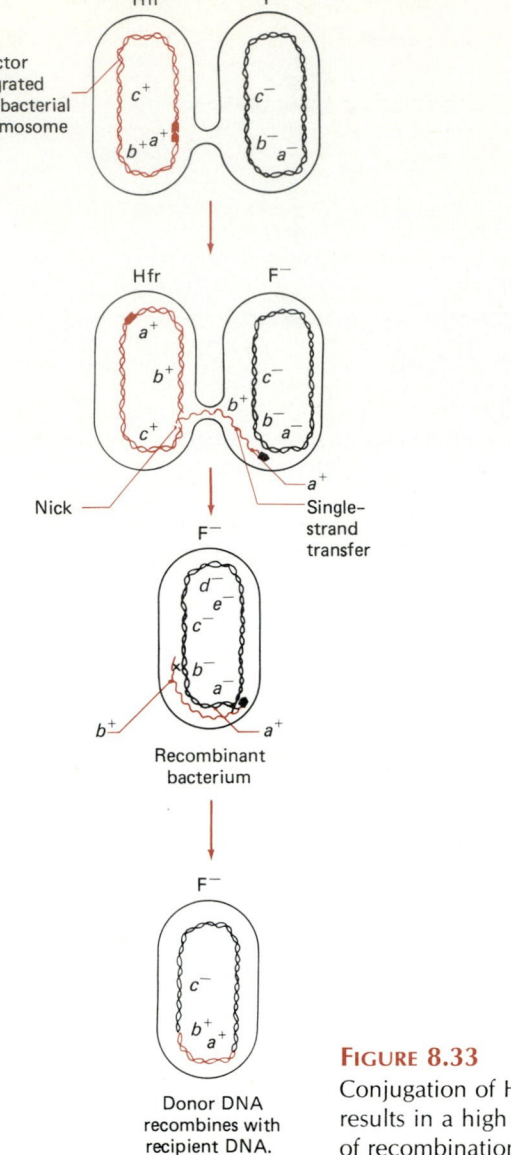

FIGURE 8.33
Conjugation of Hfr × F⁻ results in a high frequency of recombination.

8.5 *Genetic Modification and Microbial Evolution*

In addition to mutation, which is the basis for introducing variability into genetic information, genetic exchange and recombination play critical roles in redistributing genetic information and establishing new genotypes. The introduction of diversity into the gene pools of microbial populations establishes a basis for selection and evolution, that is, a basis for the better adaptation of some organisms for survival under a given set of conditions, which are therefore favored (selected) over other less well-adapted organisms. There are numerous theories to explain evolutionary change. The basis of evolution lies in the ability to change the gene pool and to maintain favorable new combinations of genes. Mutations introduce variability into genomes, resulting in changes in the enzymes that the organism synthesizes. The variations in

the genome are passed from one generation to another and are disseminated through populations. Because of their relatively short generation times compared to higher organisms, changes in the genetic information of microorganisms can be widely and rapidly disseminated. Although in some cases the modification of the genome is harmful to the organism, some mutations being lethal or conditionally lethal, a mutation may change the genetic information in a favorable way.

The occurrence of favorable mutations introduces information into the gene pool that can make an organism more fit to survive in its environment and compete with other microorganisms for available resources. Over many generations, natural selective pressures may result in the elimination of unfit variants and the continued survival of organisms possessing favorable genetic information. Change toward more favorable variants in a particular environment is the essence of evolution. Recombination creates new allelic combinations that may be adaptive. Altering the organization of genetic information within populations provides a basis for directional evolutionary change. The exchange of genetic information can produce individuals with multiple attributes that favor the survival of a microbial population. The long-term stability of a population depends on its incorporating adaptive genetic information into its chromosomes. Mutation and general recombination appear to provide a basis for the gradual selection of adaptive features. In particular, reciprocal recombination are expected to produce an evolutionary link between closely related organisms, and nonreciprocal recombinational events appear to provide a mechanism for rapid, stepwise evolutionary changes. The fact that unrelated genomes can recombine suggests that different lines of evolution can suddenly come together.

Plasmids and other transposable genetic elements may contribute to rapid changes in the genetic composition of a population, but the evolutionary stability of such changes is not clear. For example, the incidence of bacteria containing plasmids that code for antibiotic resistance has certainly increased since the widespread introduction of antibiotic use in medicine, particularly in hospital settings where antibiotic use is extensive (Figure 8.34). Possession of the genetic information that encodes for antibiotic resistance is adaptive for microorganisms trying to survive in the presence of a

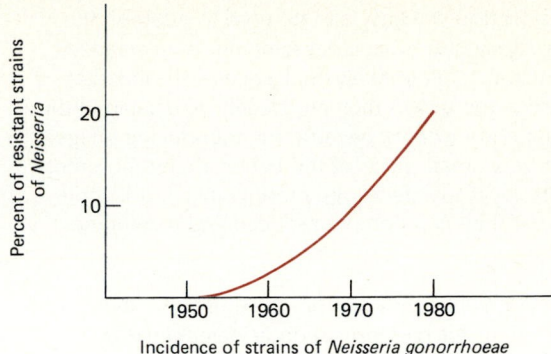

FIGURE 8.34

The incidence of antibiotic-resistant strains has increased in recent years as a result of selective pressure. This graph shows the increase in the number of strains of *Neisseria gonorrhoeae* that are resistant to 0.5 unit of penicillin per milliliter. New drugs must continually be developed to keep up with the rapid evolution of antibiotic-resistant strains.

variety of antibiotics. The information contained in plasmids, however, can be readily lost, especially if selective pressure diminishes. It is too early to say whether possessing the information for antibiotic resistance will be of long-term evolutionary advantage to bacteria and, if so, whether this information will be permanently incorporated into the bacterial chromosome.

ANALYTICAL PROCESS
Genetic Mapping

Mating of different strains of microorganisms that have different allelic forms of multiple genes can be used for **genetic mapping** of microbial genomes, that is, for determining the relative positions of genes. The occurrence of recombinants that result from mating often is used to map the order and, thus, to determine the relative locations (loci) of genes. In the case of *E. coli*, which has been well studied, mating between Hfr and F⁻ strains has been used to map large sections of the bacterial chromosome (Figure 8.35). By vibrating a culture of mating bacteria, one can interrupt mating by breaking the F pilus, which stops further transfer of DNA. Such interruption of mating can be done at various times after conjugational cell–cell contact begins. The order of genes on the bacterial chromosome can be determined by examining the times at which recombinants for given genes are found (Figure 8.36). In mating experiments aimed at mapping the order of genes, the recovery of recombinants of marker genes is normally used as a reference point for establishing the fine structure of the genome (Figure 8.37). If a gene of unknown location shows a high frequency of recombination along with the marker gene, it is likely that the marker and unknown genes are closely associated in the chromosome. If, however, the genes are far apart, it is unlikely that recombinants of both the marker gene and the gene of unknown location will occur in the progeny.

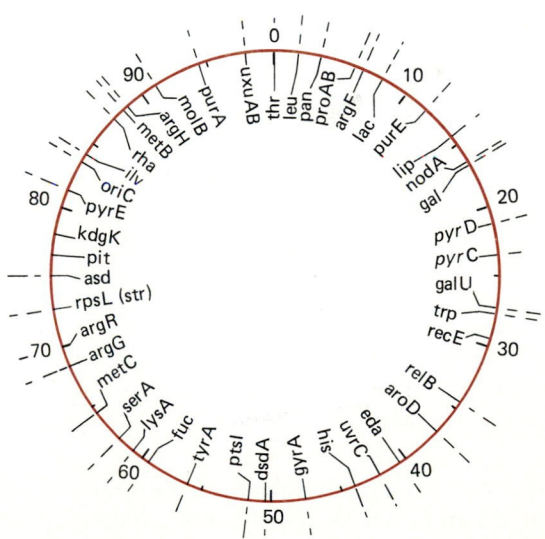

FIGURE 8.35

Circular gene map of *E. coli*, showing only a few of the genes that have been mapped. The numbers in the figure refer to the time (in minutes) of transfer of the markers by conjugation. (Based on B. J. Bachman and K. B. Low. 1980, *Microbiological Reviews* 44:1–56)

Transduction similarly can be used to establish the fine structure of the bacterial genome. In generalized transduction, it is unlikely that genes will undergo cotransduction unless they are closely associated in the bacterial chromosome because the transducing phage carry a very small piece of the bacterial chromosome. Conversely, if two genes are closely linked, it is more likely that they will be cotransduced and recombine

together than if they are not located adjacently on the bacterial chromosome. Cotransformation can similarly be used to map the microbial genome. Thus, using a variety of processes to achieve genetic exchange, the rates of recombination can be measured and the relative locations of the genes deduced, thus producing a detailed genetic map.

FIGURE 8.36

The locations of genes can be determined by the transfer times for recombination as determined by interrupted mating. In this case, an Hfr strain that is Thr$^+$, Leu$^+$, Az^S, T1^S, Lac$^+$, Gal$^+$, Str^S is mated with an F$^-$ strain that is Thr$^-$, Leu$^-$, Az^R, T1^R, Lac$^-$, Gal$^-$, Str^R. These genetic markers are threonine biosynthesis (Thr), leucine biosynthesis (Leu), azide sensitivity (Az), phage T1 sensitivity (T1), lactose utilization (Lac), galactose utilization (Gal), and streptomycin sensitivity (Str). The superscript + indicates that the organism has the genes for biosynthesis or utilization, whereas the superscript − indicates that the organism lacks these genes; the superscript R indicates resistance, whereas the superscript S indicates sensitivity. Mating of the Hfr and F$^-$ strains is initiated by mixing the two cultures at time $t = 0$. After mating for 10, 15, 20, 25, 30, 40, 50, and 60 minutes, a portion of the mixed culture is removed and agitated in blender to interrupt mating, and the cells are then plated on a medium containing glucose and streptomycin. On this medium, Thr$^+$ Leu$^+$ Str^R recombinants are selected, with the respective selected markers of thr^+, leu^+, and str^R. Azide sensitivity (azi), phage T1 sensitivity (ton), lactose utilization (lac), and galactose utilization (gal) are unselected markers because this medium does not specifically detect the different alleles of these genes. The Thr$^+$ Leu$^+$ Str^R recombinants that form colonies are scored for the alleles of the unselected markers that are present in the selected recombinants by replica plating on media that individually select for azi, ton, lac, and gal. As shown in this figure, the frequencies of unselected markers among Thr$^+$ Leu$^+$ Str^R selected recombinants are plotted as a function of time until mating is physically interrupted. Extrapolation of the frequency of each unselected marker to zero indicates the earliest time at which markers become available for recombination with the chromosome of the F$^-$ cell. These times permit the ordering of genes, that is, construction of a genetic map, with the assignment of distances between genes based upon the time (in minutes) elapsed from the initiation of conjugation until the earliest time at which a marker from the Hfr strain is detected as a recombinant with the F$^-$ strain.

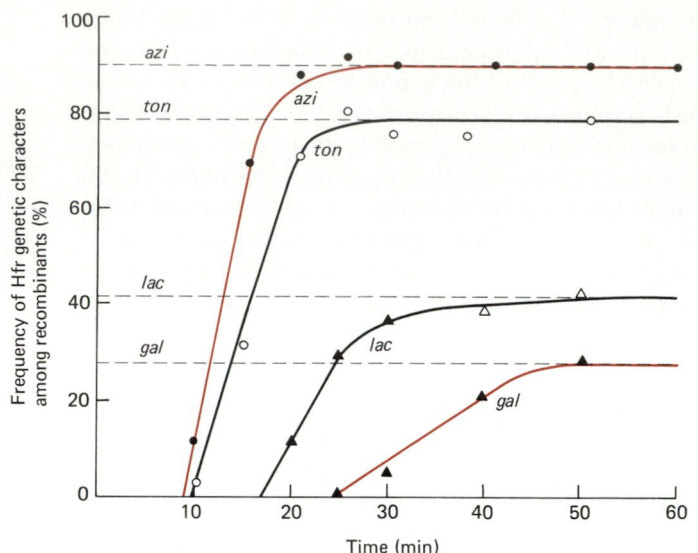

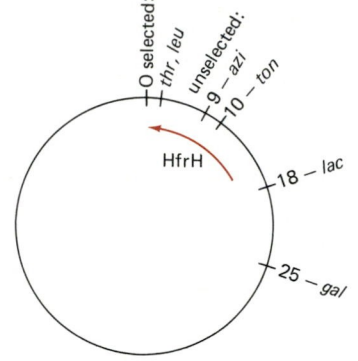

FIGURE 8.37

The frequency of corecombination with marker genes permits determination of the relative positions of genes along the chromosome and the establishment of the genetic map.

	Percentage of each genotype			
Cross	$r^- h^+$	$r^+ h^-$	$r^+ h^+$	$r^- h^-$
$r_a^- h^+ \times r_a^+ h^-$	34.0	42.0	12.0	12.0
$r_b^- h^+ \times r_b^+ h^-$	32.0	56.0	5.9	6.4
$r_c^- h^+ \times r_c^+ h^-$	39.0	59.0	0.7	0.9

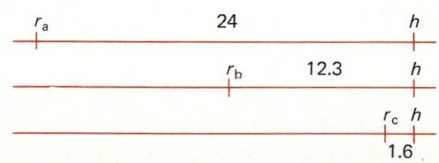

8.6 Genetic Engineering

Besides the potential role of genetic recombination in natural evolutionary changes, it is possible to employ DNA recombination to direct or engineer the evolution of new microorganisms. The intentional recombination of genes from different sources by artificial means (**recombinant DNA technology**) is the basis for **genetic engineering**, that is, the creation of new genetic varieties of organisms using recombinant DNA technology. Recombinant DNA technology promises to be a powerful tool for understanding basic genetic processes, and genetic engineering has tremendous potential industrial applications.

Enzymes Involved in Recombinant DNA Technology

The ability to manipulate DNA for genetic engineering depends largely upon enzymes (**endonucleases** commonly called **restriction enzymes**) that can cut double-stranded DNA at specified locations and enzymes (**ligases**) that can attach segments of double-stranded DNA. Restriction endonucleases normally function to prevent the incorporation of foreign DNA into a microbial genome by cutting both strands of a foreign DNA molecule. Endonucleases are named using a system whereby the first letter indicates the genus from which it was isolated, the next two letters indicate the species, the fourth letter—when needed—indicates the strain, and the number indicates the order of discovery of endonucleases from that strain. For example, the restriction enzyme *EcoRI* was isolated from the genus *Escherichia* (E), species *E. coli* (co), strain R, and it was the first (I) endonuclease isolated from that strain.

Different types of endonucleases vary with respect to the site at which they cut DNA. Type I restriction endonucleases cleave DNA at a random distance from a recognition site in the DNA nucleotide sequence. Type II restriction endonucleases cleave the DNA at the recognition site, and type III endonucleases cut the DNA at some precise distance from the recognition site. Types II and III restriction endonucleases are useful in genetic engineering. A type II restriction endonuclease cuts the DNA at a **palindromic sequence** of bases, which is a sequence of nucleotide bases that can be read identically in both the $3'OH \rightarrow 5'P$ and $5'P \rightarrow 3'OH$ directions; type II restriction endonucleases frequently produce DNA with staggered single-stranded ends (Figure 8.38). Those ends of the cut DNA that are staggered can act as cohesive or sticky ends during recombination, making them suitable for splicing with segments of foreign source DNA that has been excised using the same endonuclease.[5]

[5]Bacteria normally protect themselves against their own endonucleases by modifying DNA bases at the recognition sites where the endonucleases act, using specific modification enzymes for this purpose.

When two different endonucleases are used, one to open the plasmid ring and another to form a segment of donor DNA, an artificial homology at the terminal ends of the donor and plasmid DNA molecules can be synthesized (Figure 8.39). This can be accomplished by adding polyA (polyadenine) tails to the plasmid and polyT (poly-thymine) tails to the donor DNA. A transferase is employed to add polyA tails to the $3'OH$ ends of the DNA molecule. Pairing occurs between homologous regions of complementary bases, and ligases are used to seal the circular plasmid. The tails left by the action of the endonuclease are cleaved *in vitro*, using

FIGURE 8.38

Illustration of the palindromic sequences at the sites of gene cutting by several type II endonucleases.

Enzyme	Source organism	Restriction site
*Eco*RI	*Escherichia coli*	–C–T–T–A–A–G–5' 5'–G–A–A–T–T–C–
*Eco*RII	*E. coli*	–C–G–G–A–C–C–G–5' 5'–G–C–C–T–G–G–C–
*Hind*II	*Hemophilus influenzae*	–C–A–Pu–Py–T–G–5' 5'–G–T–Py–Pu–A–C–
*Hind*III	*H. influenzae*	–T–T–C–G–A–A–5' 5'–A–A–G–C–T–T–
*Hae*III	*H. aegyptius*	–C–C–G–G–5' 5'–G–G–C–C–
*Hpa*II	*H. parainfluenzae*	–G–G–C–C–5' 5'–C–C–G–G–
*Pst*I	*Providencia stuartii*	–G–A–C–G–T–C–5' 5'–C–T–G–C–A–G–
*Sma*I	*Serratia marcescens*	–G–G–G–C–C–C–5' 5'–C–C–C–G–G–G–
*Bam*I	*Bacillus amyloliquefaciens*	–C–C–T–A–G–G–5' 5'–G–G–A–T–C–C–
*Bgl*II	*B. globiggi*	–T–C–T–A–G–A–5' 5'–A–G–A–T–C–T–

exonucleases. Virtually any source of DNA can be used as a donor, including human DNA. By adding a polyT tail to the donor DNA after its excision with an endonuclease, the donor DNA can be made complementary to the polyA tails of the plasmid DNA, permitting the formation of a circular plasmid molecule. If the same restriction endonuclease is used to cut both the donor and recipient plasmid DNA, the strands will have homologous ends and it will be unnecessary to add polyA and polyT tails (Figure 8.40). The sealing of the ends of the DNA molecules is accomplished by using ligases, which can form a circular loop of DNA that contains a foreign segment of DNA that can be replicated (**cloned**) within a suitable host cell.

Protoplast Fusion

In addition to methods using isolated DNA and restriction enzymes, induced protoplast fusion is a useful process for establishing genetic exchange among prokaryotic and eukaryotic microorganisms. **Protoplasts** are cells that have had their walls removed by enzymatic and/or detergent treatment. They are protected against lysis due to osmotic shock by suspension in a buffer containing a high solute (such as sucrose) concentration. Protoplast fusion is a particularly useful technique for achieving gene transfer and genetic recombination in organisms with no efficient, natural gene transfer mechanism. Interestingly, more than two strains can be com-

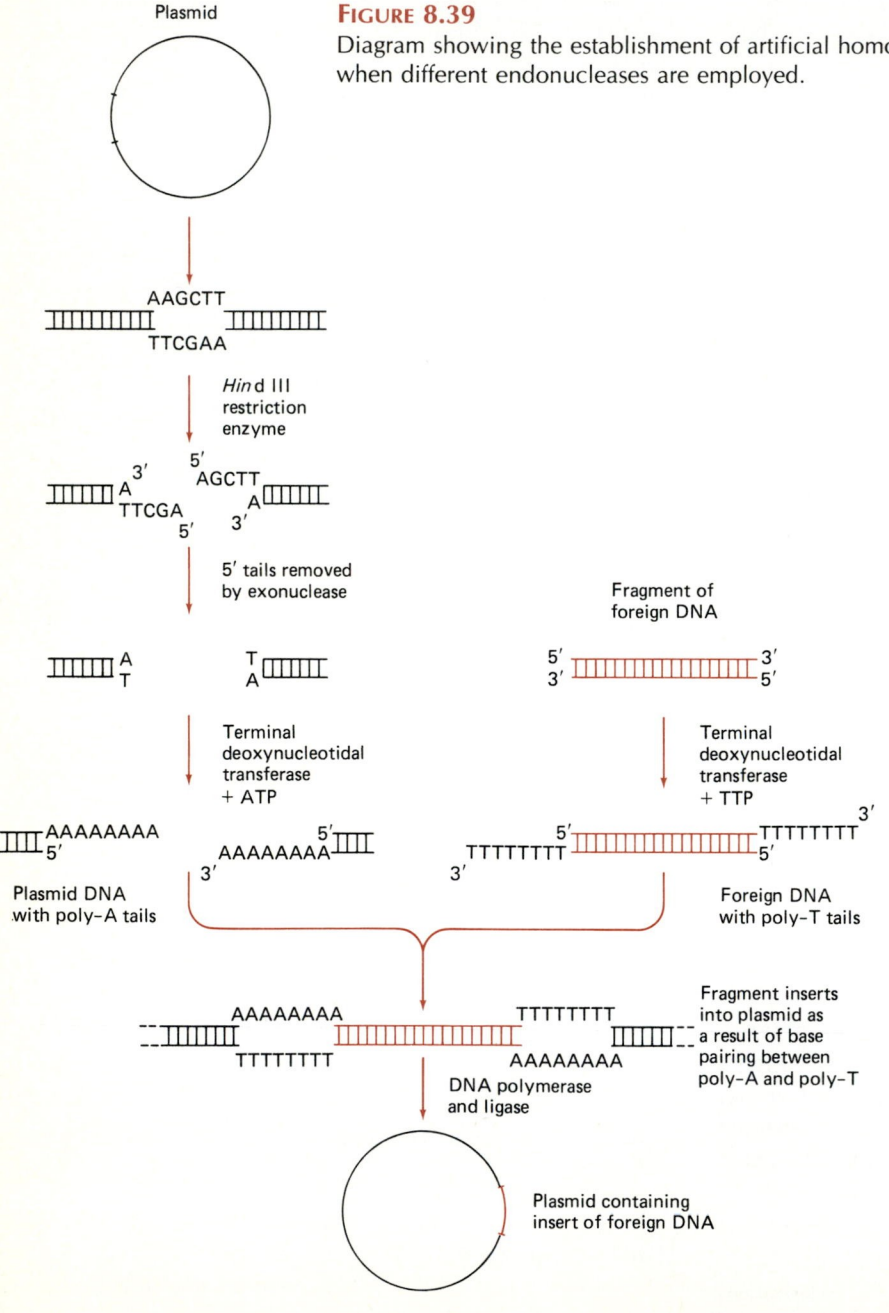

FIGURE 8.39

Diagram showing the establishment of artificial homology when different endonucleases are employed.

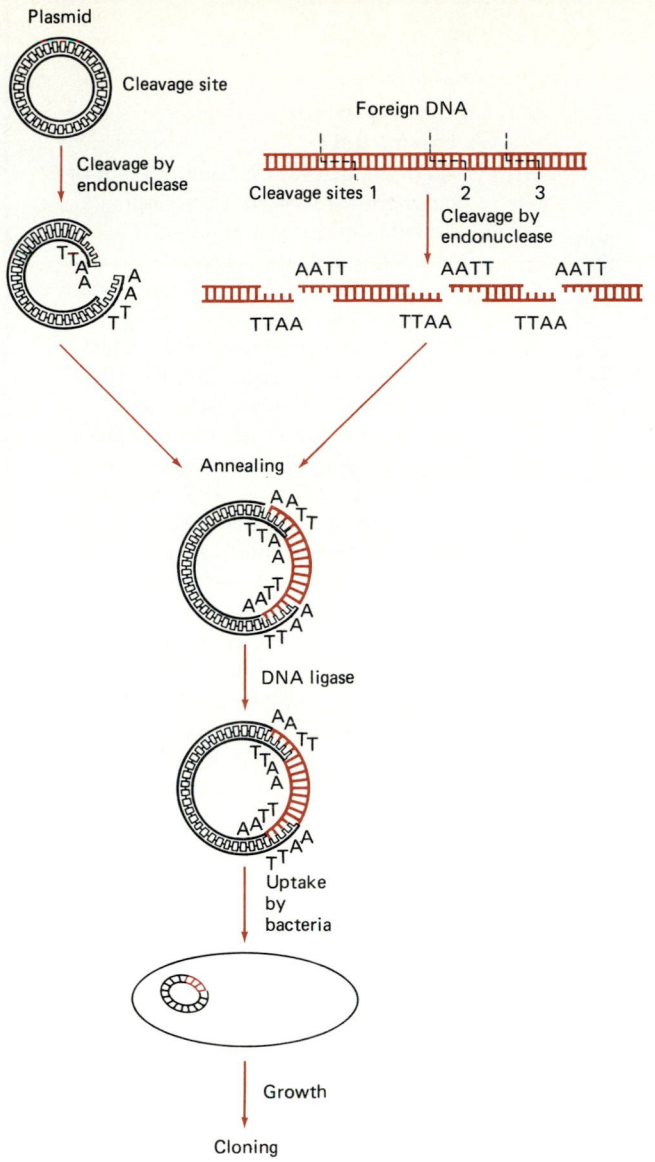

FIGURE 8.40

Diagram showing the formation of a recombinant DNA molecule when the same endonuclease is used for both the donor and recipient DNA.

bined in one fusion, generating recombinants that have inherited genes from all parents in the fusion. The basic procedure involves polyethylene glycol–induced fusion of protoplasts followed by the regeneration of normal cells. An important feature of bacterial protoplast fusion is that it enables establishment of a transient quasi-diploid state during fusion. This permits recombination between complete bacterial chromosomes, as opposed to fragments of the donor bacterial chromosome and the recipient bacterial chromosome. The fusion-induced recombinants are normally haploid.

Gene Cloning

The ability to introduce a segment of foreign DNA by recombinant DNA technology is typically used in conjunction with methods for replicating that DNA. Such replication of the DNA is accomplished by **gene cloning**, a process that involves (1) isolation and fragmentation of the source DNA and incorporation of the fragments obtained into a **cloning vector** (segment of DNA used for replication of foreign DNA fragments), with the use of restriction endonuclease to cut and ligase to rejoin DNA molecules; (2) incorporation of the genetically transformed DNA into a recipient organism that can replicate the cloning vector; (3) detection of the newly transformed cells containing DNA and isolation of a pure culture therefrom; and (4) growth of cultures of cells containing the cloned DNA fragment. A cloning vector used for replicating foreign DNA fragments should replicate autonomously in a suitable host; should be readily separable from the host DNA so that it can be purified; and should have regions, which are not essential for self-replication, that can be substituted for a fragment of foreign source DNA. It should also be able to enter a host cell and replicate to a **high copy number** (large number of repetitive copies of the genes, such as can be achieved by having multiple identical plasmids) and should be stable in the host. Plasmids, bacteriophage (such as λ phage), and phage–plasmid artificial hybrids (**cosmids**) that have these properties are frequently used as cloning vectors. Another key property of a good cloning vector is selectibility, such as antibiotic resistance, to allow selection as well as detection of those host cells that carry the vector with the donor fragment of interest.

Genetic engineering often uses plasmids as carriers of unrelated DNA and the enzymes involved in the normal recombination and replication of DNA for splicing foreign DNA into the plasmid carriers (Figure 8.42). Site-specific endonucleases are used to create sites where foreign donor DNA can be inserted. Once the plasmid containing the desired additional DNA segments is formed, it can be added to a culture of a suitable recipient bacterium that will incorporate the plasmid. The plasmid is taken up by the bacterium, and then, regardless of its source, it comprises part of the bacterial genome. The plasmid DNA, including the foreign DNA segments, can be replicated and passed from one generation to another. It is possible to add several copies of the same plasmid to a single bacterium so that the presence of the foreign DNA is amplified. When this approach is used, genes can be **cloned** (asexually reproduced), with bacteria acting as factories to produce multiple copies of identical genes.

Plasmid pBR322, which was created specifically for use in cloning DNA, is a frequently used cloning vector that can replicate in *E. coli*. It contains genes for ampicillin and tetracycline resistance and has multiple specific restriction enzyme sites (Figure 8.43). The presence of antibiotic resistance markers permits its detection using selective media containing these antibiotics. Because plasmid pBR322 con-

ANALYTICAL PROCESS
Hybridization of Nucleic Acids

Hybridization of nucleic acids, that is, artificial construction of a double-stranded nucleic acid by complementary base pairing of two single-stranded nucleic acids, is an important method used in recombinant DNA technology for the identification of specific genes. DNA–DNA hybrids, as well as DNA–RNA hybrids, can be made by reacting single-stranded molecules under appropriate **reanealing** (reestablishing double-stranded DNA) conditions. For reanealing to occur, there must be a high degree of complementarity between two single-stranded nucleic acid molecules if they are to form a stable hybrid. Nucleic acid hybridization may be used to detect specific cloned DNA sequences in recombinants, provided that a suitable labeled single-stranded DNA or RNA probe is available.

Probe DNA can be labeled by the incorporation of radiolabeled ^{32}P-nucleoside triphosphates, and the presence of the radioactive DNA can then be detected by autoradiography. The nick-translation reaction of *E. coli* DNA polymerase I has been widely exploited for labeling DNA probes. This enzyme will add deoxyribonucleotides in a 5' → 3' direction to a DNA or RNA primer with a 3'OH terminus. The substrate for nick translation is duplex DNA that has been nicked at random. The 5' → 3' exonuclease activity of polymerase I removes nucleotides from the exposed 5'P terminus generated at a nick. This, in turn, exposes the template strand. The degraded strand can be resynthesized by the polymerizing activity of polymerase I. Incorporation of one or more radiolabeled nucleotides ensures that the newly synthesized DNA is labeled uniformly to high specific activity. An alternative to radiolabeling is the use of biotinylated (biotin-labeled) nucleotides. The presence of biotinylated DNA can be detected by exploiting the affinity of the egg white glycoprotein, avidin, for protein. Avidin can be linked to enzymes, such as horseradish peroxidase or alkaline phosphatase, whose activity can be detected by colorimetric assays. Such nonradioactive probes are increasingly being used as clinical diagnostic tools.

Often probes are used to detect specific DNA fragments that have been separated using gel electrophoresis. Gel electrophoresis uses an electric field to separate molecules that have been placed in a slab of agarose or polyacrylamide gel. DNA fragments up to 50 kb pairs in length can be separated in this way if they differ in size by at least 1 percent. Visualization of DNA fragments on gels can be accomplished either by staining with ethidium bromide, a fluorescent compound that binds strongly to nucleic acids, or by autoradiography on x-ray films if the

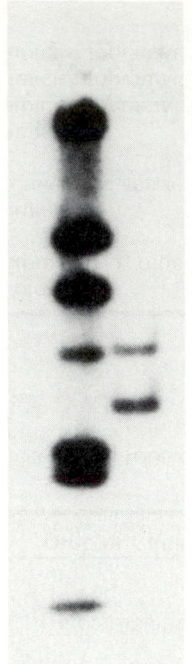

FIGURE 8.41
Autoradiograph of a Southern blot of fragments of DNA of bacteriophage lambda (left lane) and a bacterial plasmid (right lane). The DNA is first cut with the restriction endonuclease *Hind*III. The DNA fragments are then separated by electrophoresis. After electrophoretic separation, the DNA is transferred to filter paper and incubated with a radiolabelled probe. Hybridization occurs if there is extensive homology between the probe DNA and the DNA affixed to the filter. A sheet of film is exposed to the filter and developed. Regions where the radiolabelled probe has hybridized to the digested DNA appear as black bands. In this case the bands of lambda DNA serve as size markers for the fragments from the bacterial plasmid. (Courtesy Scott W. Hooper, University of Tennessee)

fragments are radioactive. Separated DNA fragments also can be denatured by alkali treatment and transferred to a sheet of nitrocellulose filter by blotting. The pattern of DNA fragments on the gel is thus preserved on the membrane filter.

In the Southern transfer or Southern blot procedure, named for its developer, E. M. Southern, the DNA immobilized on the filter is then hybridized to radiolabeled probe DNA or RNA. Autoradiography is used to locate the position(s) of DNA bands that are complementary to the probe (Figure 8.41). Nucleic acids that have been separated on agarose gels can also be detected by direct hybridization of the dried gel with an appropriate probe. This does not involve transfer to a support and is therefore more rapid than conventional Southern blot transfers. Bacterial colonies can also be lysed, and the released DNA transferred to filters and detected using labeled probes. After hybridization to radioactive probes, autoradiography is done to detect which DNA fragments have hybridized with the RNA. This procedure is most commonly used after the DNA has been digested with restriction enzymes, thus permitting localization of genes or partial gene sequences on the DNA sequence.

tains single restriction sites for several endonucleases, including sites within the genes coding for antibiotic resistance, fragments of foreign DNA can be inserted into specific sites. If the insertion occurs at the antibiotic resistance site, resist-

ance is lost because the nucleotide sequence of the antibiotic resistance gene is disrupted, an event known as **insertional inactivation** that is useful for detecting the presence of foreign DNA within a plasmid.

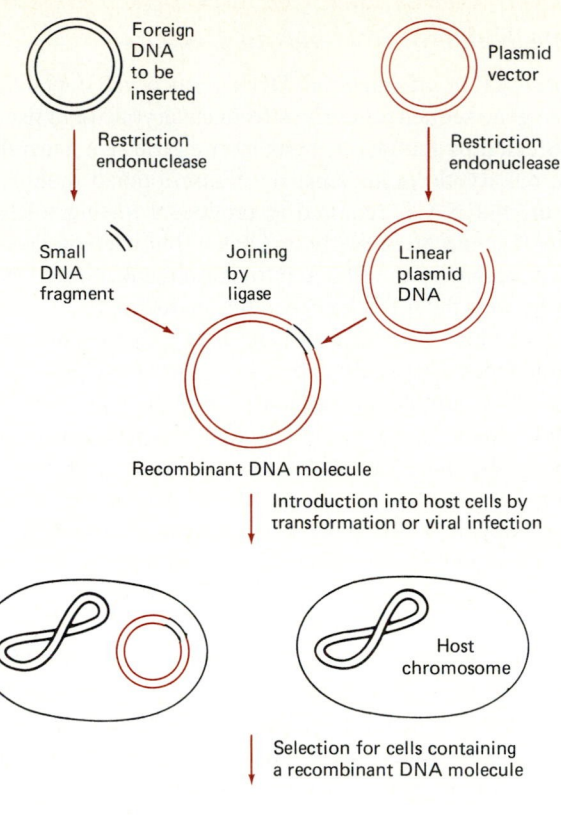

FIGURE 8.42
Diagram showing a generalized outline for genetical engineering of a new organism. The ability to create new organisms in this way holds great promise for improving the quality of living by using genetic engineering in medicine and industry to produce useful products.

FIGURE 8.43
Restriction endonuclease cleavage maps, constructed by analysis of the sizes of subfragments produced by complete- and/or partial-cleavage DNA with single and pairs of restriction endonucleases, are necessary for strategies designed to locate the minimum part of the nucleotide sequence that actually codes for the desired characteristic. Such maps are required to obtain fragments of DNA that are suitable for sequencing of the nucleotides and cloning. This map of pBR322 shows the sites where various restriction endonucleases cut the plasmid DNA and also the regions containing the genes for ampicillin (Amp) and tetracycline (Tet) resistance. The restriction endonucleases shown in boldface cleave at only one site, and these can be used to open the circular plasmid at a specific location where foreign DNA can be inserted. The numbers associated with the enzymes refers to the 5′-base at the position of cleavage within the recognition sequence, using the first T in the unique *Eco*RI recognition sequence (GAATTC) as nucleotide position number 1.

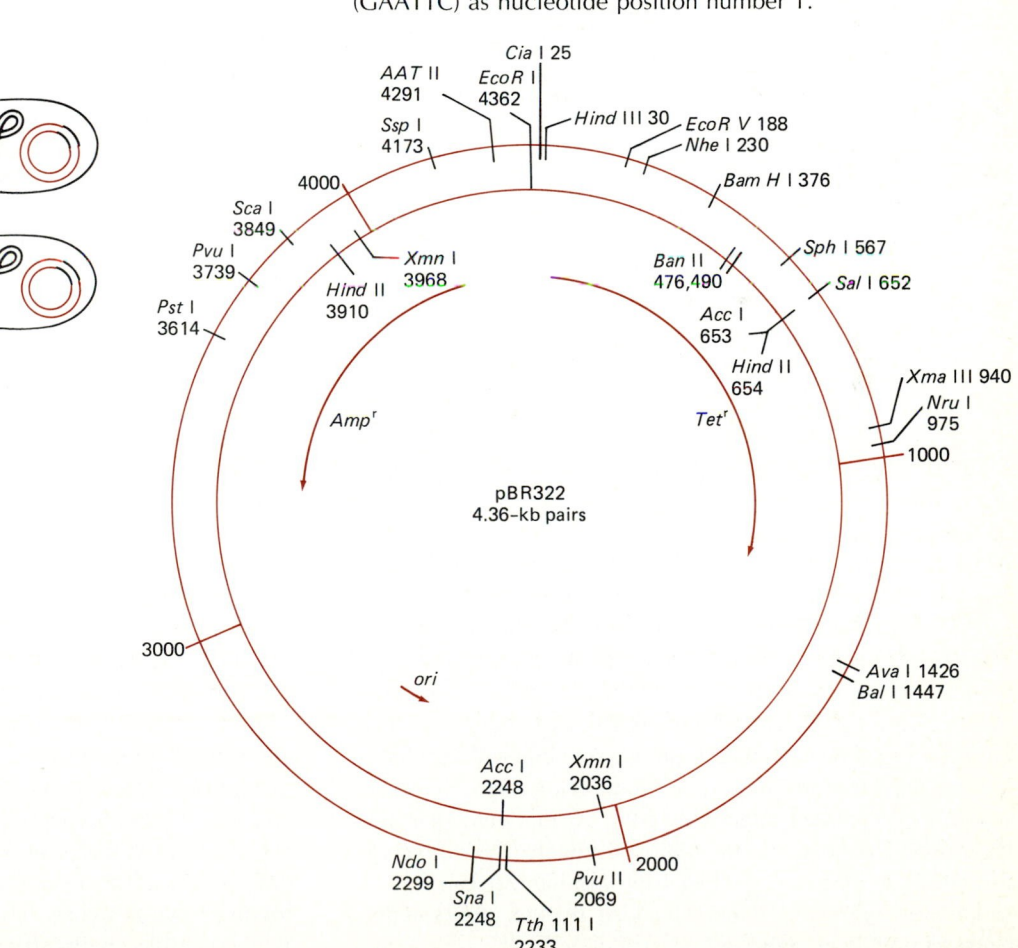

Another useful cloning vector is λ phage because there are specified restriction enzyme sites where the phage DNA can be cut for insertion of foreign DNA and because a portion of its genome can be replaced with foreign DNA without eliminating its replication capability, making it suitable for cloning large fragments of foreign DNA. Cloning with λ replacement vectors uses fragments of the genome that are separated by centrifugation or electrophoresis after isolated phage DNA has been treated with appropriate restriction endonucleases. Foreign DNA is added to the purified λ fragments using ligases to form a molecule that can be packaged into phage particles, a process accomplished by mixing the hybrid DNA with the structural proteins of the phage obtained by extracting infected cells. *E. coli* cells infected with the phage clones are isolated and checked for the presence of the desired foreign DNA sequence by nucleic acid hybridization procedures or observation of genetic properties.

Gene Expression

The ability to clone a gene does not necessarily ensure the production of a useful product. Production requires that a cloned gene be expressed. Because the genetic code is essentially universal, the information encoded in the DNA sequence theoretically can be expressed, and the polypeptide chain specified by the foreign DNA segment can be transcribed and translated to form an active protein molecule. The expression of the foreign genetic information, however, requires that the appropriate reading frame be established and the transcriptional and translational control mechanisms turned on to permit the expression of the DNA. Often the genes produced using a cloning vector must be transferred to an **expression vector** that contains not only the desired gene but also the necessary regulatory sequences that permit control of the expression of that gene. Several factors influence the level of expression of a gene. In general, more product is made if multiple copies of the gene are present, and expression vectors should be able to obtain a high copy number. Additionally, there should be a strong promoter to ensure binding of RNA polymerase. Strong *E. coli* promoters used in the construction of expression vectors include *lacuv5*, which normally controls β-galactosidase expression, *trp*, which normally controls λ virus production, and *ompF*, which regulates production of outer membrane protein. Besides a strong promoter, it is important that the early part of the RNA transcript contain a ribosome-binding site that establishes the appropriate reading frame. The ideal condition would allow the culture containing the expression vector to grow until a large population of cells is obtained, each containing a large copy number of the vector; then all copies would be expressed simultaneously. To this end, plasmids have been engineered with the lac promoter, ribosome-binding site, and operator such that the *lac* inducer can initiate production of the protein(s) encoded by the gene(s) engineered into the organism.

Cloning Eukaryotic Genes in Bacteria

The construction of bacterial DNA sequences, containing complete gene sequences derived from eukaryotic organisms, is complicated by the fact that eukaryotic genes are generally split. In eukaryotic organisms, posttranscriptional modification of the hnRNA is required to produce a mature mRNA molecule that can be properly translated, but bacteria do not possess the capacity to remove introns from eukaryotic DNA in order to form the mRNA needed for producing a functional protein molecule. Therefore, it is necessary to cut and splice eukaryotic DNA artificially or to use an alternative procedure to establish a contiguous sequence of nucleotide bases to define the protein that is to be expressed (Figure 8.44). The problem of the discontinuity of the eukaryotic gene can be overcome by using an mRNA molecule and a reverse transcriptase enzyme to produce a DNA molecule that has a contiguous sequence of nucleotide bases containing the complete gene. A major advantage of using mRNA is that the noncoding information present in the DNA (introns) has been removed. The single-stranded DNA molecule that is formed in this procedure is complementary to the complete mRNA molecule and is therefore called **complementary DNA (cDNA)**. The RNA can be removed using ribonuclease and the second complementary strand of DNA synthesized. The double-stranded DNA molecule formed in this manner can then be inserted into a carrier plasmid.

In addition to overcoming problems with sequencing the gene itself, when cloning eukaryotic genes in prokaryotic cells, it is necessary to take steps to ensure proper expression and stability of the product. One method of providing a ribosome-binding site in the proper reading frame when a mammalian DNA sequence is added to a bacterial host cell is to establish a nucleotide sequence that produces a fusion protein that contains a short prokaryotic sequence at the amino end and the desired eukaryotic sequence at the carboxyl end. Fusion proteins are often more stable in bacteria than unmodified eukaryotic proteins. Also, the bacterial portion can contain the bacterial sequence coding for the signal peptide that enables transport of the protein across the cell membrane, making possible the development of a bacterial system that not only synthesizes the mammalian protein but actually excretes it.

Using these methods, many eukaryotic genes can be cloned in prokaryotic cells. Several human proteins, including insulin, are now being commercially produced using genetically engineered bacteria. Diabetes, a disease resulting from an insulin deficiency, can be treated by injection of insulin extracted from cattle pancreas, which occasionally elicits an allergic reaction, or by injection of humulin, human insulin produced by genetically engineered bacteria. Insulin consists of two polypeptides, labeled A and B, that are coded for by separate parts of a single insulin gene. The genetic engineering of bacteria for the production of human insulin involved synthesizing the DNA sequence coding for A and B polypeptide chains that were elucidated by analyzing the

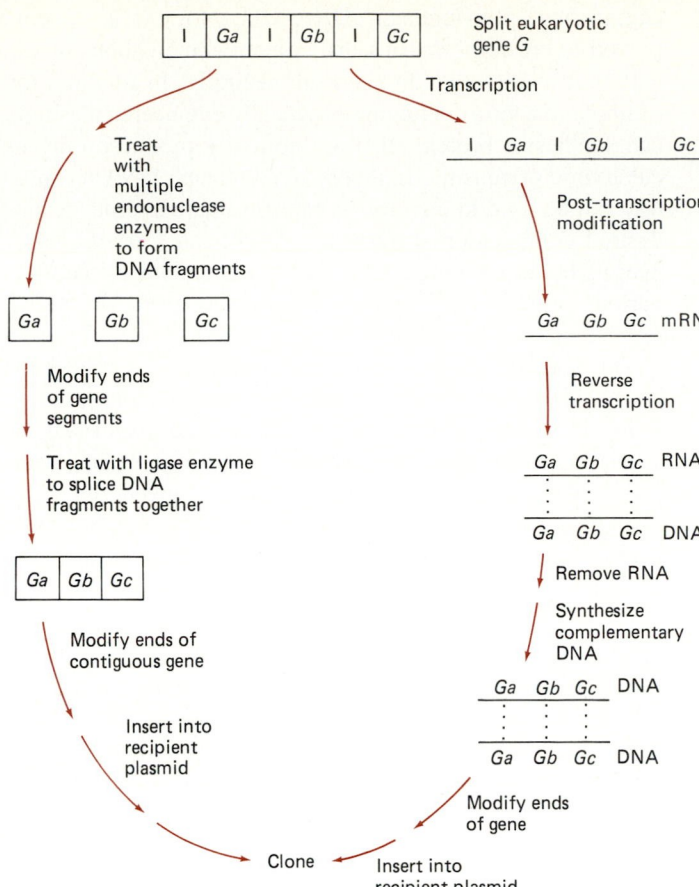

FIGURE 8.44
The establishment of a contiguous sequence of nucleotides is needed for the cloning of eukaryotic genes in bacterial cells. The need to remove intervening sequences is an unexpected complication in employing bacteria to produce eukaryotic gene products. This can be accomplished by isolating the natural mRNA following posttranscriptional processing and then synthesizing a complementary DNA for insertion into the cloning vehicle or by removing the intervening sequences with endonucleases and reassembling fragments into functional DNA segments.

amino acid sequences of the insulin molecule and determining the corresponding codons. Because the insulin protein is fairly small, it was more convenient to synthesize the proper DNA sequence chemically rather than to isolate the insulin gene from human tissue. Each DNA sequence was added separately to an expression vector, plasmid pBR322, containing the *trp* promoter and the bacterial ribosome-binding

site, which was cloned in separate cultures of *E. coli*. The two bacterial cultures were grown in large-scale fermentors, where they produced commercial quantities of A and B chains fused to a trp protein. The fused A and B proteins were purified and the A and B portions separated from *trp* by cleavage using cyanogen bromide. Finally, the A and B peptides were chemically joined to produce human insulin.

Postlude

Microbial genetics is probably the most rapidly expanding field in microbiology today. Our understanding of microbial genetics did not start until the middle of the twentieth century. In 1941 George Beadle and Edward Tatum published their studies on the genetic control of biochemical reactions in the fungus *Neurospora*. Tatum also showed, in 1945, that exposure to x-rays increased the mutation rate in the bacterium *E. coli*. In 1944 Oswald Avery, Colin MacLeod, and Maclyn McCarty published their studies on the nature of the substance that induces transformation of pneumococcal types and concluded that a nucleic acid of the deoxyribose type is the fundamental unit of the transforming principle of *Streptococcus pneumoniae*. Joshua Lederberg and co-workers, between 1946 and 1956, studied genetic exchange processes in

bacteria and made the first reports on conjugation and transduction. In 1958 George Beadle, Edward Tatum, and Joshua Lederberg shared the Nobel Prize for their studies on microbial genetics. Their pioneering investigations led to many other studies aimed at using genetic recombination processes to map the genomes of microorganisms.

With our growing understanding of molecular biology, it has become quite evident that despite the great precision of the DNA replication process, the genetic information of microorganisms is continuously changing. Various types of mutational events introduce modifications into the DNA molecule with varying degrees of frequency, producing multiple allelic forms of the same gene, and recombinational processes permit further redistribution of genetic information. Heterogeneity within the gene pool of a microbial population is necessary for maintaining stability under fluctuating environmental conditions; genetic variability essentially provides protection against environmental uncertainty for the survival of the species. Whereas asexual reproduction could lead to rapid divergence toward a gene pool with temporarily advantageous restricted information, sexual reproduction ensures continued genetic heterogeneity that has long term advantages when environmental conditions vary. The diversity introduced into the gene pool through mutation and recombination also establishes the basis for the selective evolution of microorganisms.

Recombination involves a restructuring of DNA molecules so that new genomes are formed containing information from different DNA sources. General recombination results in the exchange of corresponding genes, and nonreciprocal recombination permits the interchange of unrelated segments of DNA. A variety of mechanisms permit the transfer of genetic information from a donor to a recipient microorganism so that recombination can occur. In eukaryotes, sexual reproduction results in the exchange of genetic information between closely related organisms. In bacteria, DNA may be transferred by conjugation, which involves direct contact between the donor and recipient strains; transduction, which utilizes a virus as a vector for carrying the DNA from the donor to the recipient strain; and transformation, which involves the uptake of naked DNA by a competent recipient strain. DNA exchanges between microorganisms of differing allelic forms and the frequency of recombination can be used to map the microbial genome.

By the early 1970s, the genetics of bacteria was sufficiently understood to permit the development of experiments that could create new organisms. In 1978 the Nobel Prize was awarded to Werner Arber, Hamilton O. Smith, and Daniel Nathans for the development of restriction endonucleases that can be used to study genetic organization (genetic mapping) and to manipulate DNA for genetic engineering. Along with Frederick Sanger and Walter Gilbert, the 1980 Nobel Prize was shared by Paul Berg for his work on constructing recombinant DNA molecules. Pioneering efforts in genetic engineering also were made in the early 1970s by Harold Boyer and Stanley Cohen, who used plasmids and recombinant DNA methodology to clone genes, opening up an entirely new area of applied microbiology and spurring the formation of many genetic engineering companies. Genetic engineering, using recombinant DNA technology, holds the promise of solving many problems of mankind, including the ability to produce drugs for treating currently incurable diseases, to produce sufficient food to feed the world's population, and to solve problems of environmental pollution.

Recombination involving plasmids, as well as other vectors, provides a mechanism for the particularly rapid dissemination of genes through a population. Plasmids are quite useful as carriers of foreign genetic information in genetic engineering. Recombinant DNA technology can be employed to create organisms that contain combinations of genetic information that do not occur naturally. In addition, for example, bacteria containing genetically engineered plasmids can synthesize proteins that are normally produced only in eukaryotic organisms. In theory, recombinant DNA technology can be used to engineer organisms that can produce any desired combination of proteins. As a result, genetic engineering holds great promise in industry and medicine because various proteins of economic importance or use in curing disease may be produced.

The potential of genetic engineering to short-circuit evolution has raised numerous ethical, legal, and safety questions. The Supreme Court of the United States has ruled, in a landmark decision, that genetic engineering can create novel living systems that can be patented as inventions. This ruling has established the precedent for future genetic engineering efforts. The problem of safety has been temporarily solved by using mutation and selection procedures to develop a fail-safe strain of *E. coli* that is unable to grow outside a carefully defined culture medium. A committee was established in the United States by the National Institutes of Health—the Recombinant DNA Advisory Committee (RAC)—to oversee proposed government-sponsored research using recombinant DNA technology. Gradually the stringent guidelines established by this committee have been relaxed as it has become apparent that many of the initial concerns with regard to the unintentional escape of genetically engineered organisms from the laboratory were exaggerated.

A new concern, however, has focused on the development of genetically engineered organisms designed for deliberate release into the environment. Environmentalists have questioned the safety of such a procedure, and new governmental regulations have been developed amid a series of court cases concerning safety and regulatory issues. The questions of whether novel genomes will survive in the environment, whether they will transfer to other microorganisms and spread, and whether this dissemination could represent a serious biological hazard are being actively debated. The debate also continues as to whether it is ethical to clone DNA in all cases and whether recombinant DNA technology will permit the cloning of higher eukaryotic organisms, including plants and animals.

One must weigh these ethical questions against the benefits that can be derived through genetic engineering. There is little question that through genetic engineering the quality of human life can be improved. However, scientists have a responsibility to see that the public is informed about genetic engineering; to see that research remains within acceptable guidelines; to use this powerful technique for examining basic questions about molecular-level genetics; and to develop genetically engineered organisms of agricultural, ecological, medical, and economic importance that can be safely used.

Suggested Supplementary Readings

AYALA, F. J., and J. A. KIGER, JR. 1984. *Modern Genetics.* Benjamin/Cummings Publishing Co., Menlo Park, Calif.

BIRGE, E. A. 1981. *Bacterial and Bacteriophage Genetics.* Springer-Verlag, New York.

CAMPBELL, A. 1981. Evolutionary significance of accessory DNA elements in bacteria. *Annual Review of Microbiology* 35:55–83.

CHAKRABARTY, A. M. 1978. *Genetic Engineering.* CRC Press, Inc., Boca Raton, Fla.

COHEN, S. N., and J. A. SHAPIRO. 1980. Transposable genetic elements. *Scientific American* 242(2):40–49.

DAVIS, R. W., D. BOTSTEIN, and J. R. ROTH. 1980. *Advanced Bacterial Genetics: A Manual for Genetic Engineering.* Cold Spring Harbor Laboratory, Cold Spring Harbor, N.Y.

FRISTROM, J. W., and M. T. CLEGG. 1986. *Principles of Genetics.* Chiron Press, Inc., Portland, Ore.

Genetics: Readings from Scientific American. 1981. W. H. Freeman & Co., San Francisco.

GILBERT, W., and L. VILLA-KOMANOFF. 1980. Useful proteins from recombinant bacteria. *Scientific American* 242(4):74–94.

GLASS, R. E. 1982. *Gene Function:* E. coli *and Its Heritable Elements.* University of California Press, Berkeley, Calif.

GRINDLEY, N. D. F., and R. R. REED. 1985. Transpositional recombination in prokaryotes. *Annual Review of Biochemistry* 54:863–890.

HIGGINS, I. J., D. J. BEST, and J. JONES. 1985. *Biotechnology: Principles and Practice.* Blackwell Scientific Publications, Oxford, England.

HOPWOOD, D. A. 1981. Genetic studies with bacteria protoplasts. *Annual Review of Microbiology* 35:237–272.

KLECKNER, N. 1977. Translocatable elements in prokaryotes: Review. *Cell* 11:11–23.

KOLODNY, G. M. (ed.). 1981. *Eukaryotic Gene Regulation.* CRC Press, Inc., Boca Raton, Fla.

LEWONTIN, R. C. 1974. *The Genetic Basis of Evolutionary Change.* Columbia University Press, New York.

MANIATIS, T., E. F. FRITSCH, and J. SAMBROOK. 1982. *Molecular Cloning: A Laboratory Manual.* Cold Spring Harbor Laboratory, Cold Spring Harbor, N.Y.

MARGULIS, L., and D. SAGAN. 1986. *Origin of Sex: Three Billion Years of Genetic Recombination.* Yale University Press, New Haven, Conn.

MAYS, L. L. 1981. *Genetics: A Molecular Approach.* Macmillan Publishing Co., Inc., New York.

NOVICK, R. 1980. Plasmids. *Scientific American* 243(6):103–123.

OLD, R. M., and S. B. PRIMROSE. 1985. *Principles of Gene Manipulation: An Introduction to Genetic Engineering.* Blackwell Scientific Publications, Oxford, England.

OLIVER, S. G., and J. M. WARD. 1985. *A Dictionary of Genetic Engineering.* Cambridge University Press, Cambridge, England.

ROUGHGARDEN, J. 1979. *Theory of Population Genetics and Evolutionary Ecology: An Introduction.* Macmillan Publishing Co., Inc., New York.

SCAIFE, J., D. LEACH, and A. GALIZZI (eds.). 1985. *Genetics of Bacteria.* Academic Press, New York.

SHIMKE, R. T. 1980. Gene amplification and drug resistance. *Scientific American* 243(5):60–69.

SMITH, H. O., D. B. DANNER, and R. A. DEICH. 1981. Genetic transformation. *Annual Review of Biochemistry* 50:41–68.

STAHL, F. W. 1979. *Genetic Recombination: Thinking About It in Phage and Fungi.* W. H. Freeman & Co., San Francisco.

STRICKBERGER, M. W. 1985. *Genetics.* Macmillan Publishing Co., Inc., New York.

SUZUKI, D. T., A. J. F. GRIFFITHS, J. H. MILLER, and R. C. LEWONTIN. 1986. *An Introduction to Genetic Analysis.* W. H. Freeman & Co., San Francisco.

WALKER, G. C. 1985. Inducible DNA repair systems. *Annual Review of Biochemistry* 54:425–458.

WALKER, G. C., C. MARSCH, and L. A. DODSON. 1985. Genetic analyses of DNA repair: Inference and extrapolation. *Annual Review of Genetics* 19:103–126.

WATSON, J. D., N. HOPKINS, J. ROBERTS, J. STEITZ, and A. WEINER. 1987. *Molecular Biology of the Gene.* Benjamin/Cummings Publishing Co., Menlo Park, Calif.

ZUBAY, G. 1987. *Genetics.* Benjamin/Cummings Publishing Co., Menlo Park, Calif.

Study Questions

1. What is a mutation? How do mutations occur?

2. What effect does exposure to ionizing radiation or chemical mutagens have on rates of mutation?

3. What is a plasmid, and what are some functions associated with plasmids?

4. What is recombination, and how is this process involved in maintaining heterogeneity within the gene pool?

5. Compare homologous and nonhomologous recombination.

6. What is a transposable genetic element?

7. How is DNA exchanged in prokaryotes?

8. Compare general and specialized transduction.

9. How is genetic variability related to the evolution of microorganisms?

10. What is recombinant DNA technology? How is genetic engineering able to create new organisms?

Situational Problem 1

Testing Potential Carcinogenicity of Water Supplies

You have just begun working part-time for an analytical laboratory that services the municipal water company. Most of the work concerns chemical analyses for heavy metals and toxic chemicals. The water company has requested that tests be added to determine the mutagenicity and potential carcinogenicity of the water before and after disinfection. Because of your expertise in bacteriology, you have been asked to perform these tests. You decide to employ the Ames test as a routine screening procedure, using two histidine-requiring, auxotrophic strains of *Salmonella typhimurium*.

You collect 100-L water samples and pass the water through an ion exchange column to concentrate organic chemicals. You then elute the organics with a solvent, which is then evaporated. The concentrated organics are dissolved

in 10 mL dimethyl sulfoxide (DMSO) for use in the Ames test procedure. You add a series of volumes—2 to 20 μL—of the concentrated organics to suspensions of the *Salmonella* strains suspended in liquefied agar. A control suspension with only DMSO and no organic concentrate is also prepared. A microsomal preparation of rat liver homegenate is added to the suspensions to activate potential mutagens. The agar suspensions are then poured onto minimal media (lacking histidine) and incubated for 48 hours. A replicate sample is also streaked onto a complete medium to ensure viability of the bacteria in the suspension.

Having done this you observe the following numbers of colonies:

DMSO alone	2 μL organic concentrate	10 μL organic concentrate	20 μL organic concentrate	DMSO alone	2 μL organic concentrate	10 μL organic concentrate	20 μL organic concentrate
Water Sample A				Water Sample C			
Minimal Medium—*S. typhimurium* strain 1				Minimal Medium—*S. typhimurium* strain 1			
75	140	230	380	65	70	67	72
Complete Medium—*S. typhimurium* strain 1				Complete Medium—*S. typhimurium* strain 1			
400	400	400	400	400	400	400	400
Minimal Medium—*S. typhimurium* strain 2				Minimal Medium—*S. typhimurium* strain 2			
10	75	150	300	25	30	30	25
Complete Medium—*S. typhimurium* strain 2				Complete Medium—*S. typhimurium* strain 2			
300	300	300	300	300	300	300	300
Water Sample B				Water Sample D			
Minimal Medium—*S. typhimurium* strain 1				Minimal Medium—*S. typhimurium* strain 1			
65	80	35	0	60	120	260	375
Complete Medium—*S. typhimurium* strain 1				Complete Medium—*S. typhimurium* strain 1			
400	400	350	150	400	400	400	400
Minimal Medium—*S. typhimurium* strain 2				Minimal Medium—*S. typhimurium* strain 2			
10	25	0	0	10	15	12	10
Complete Medium—*S. typhimurium* strain 2				Complete Medium—*S. typhimurium* strain 2			
200	200	180	0	300	300	300	300

Based on these data, what specific conclusions can you reach, and based on these conclusions, what recommendations would you make to the water company in your report?

Situational Problem 2

Mapping a Genome Based Upon Interrupted Mating Data

You are preparing for the next bacteriology exam, which you have been informed will contain a problem on genetic mapping. To make sure that you can handle this problem, you construct the following hypothetical data. Suppose you mated an Hfr strain that is $Leu^+Gal^+Trp^+His^+Str^S$ and an F^- strain that is $Leu^-Gal^-Trp^-His^-Str^R$ and used a medium that selected for Leu^+Str^R. You then screened the isolates for Gal, Trp, and His, with the following results:

Gal^+Str^R		Trp^+Str^R		His^+Str^R	
Time (min)	Recombinants per 100 Hfr	Time (min)	Recombinants per 100 Hfr	Time (min)	Recombinants per 100 Hfr
0	0	0	0	0	0
5	2	6	0	16	0
7	9	8	0	18	0
9	15	10	0	20	0
11	21	12	0	22	0
13	27	14	2	24	0
15	33	16	3	26	0
17	39	18	7	28	0
19	45	20	11	30	0
21	46	22	14	32	2
23	46	24	14	34	15
25	46	26	18	36	24
27	46	28	22	40	46
		30	27		
		32	30		
		34	38		

Using these data, graph the results and construct the genetic map for these genes.

Situational Problem 3

Engineering a Bacterial Strain to Degrade the Herbicide 2,4,5-T

2,4,5-T is an herbicide that is persistent in soil because of the limited capacity of microorganisms to degrade this synthetic organic compound. You have isolated two bacterial strains that will partially degrade 2,4,5-T, but neither strain alone will completely detoxify it. Therefore, you have decided to use recombinant DNA technology to engineer genetically a single bacterial strain that will accomplish this environmentally important task.

Complete degradation of 2,4,5-T involves sequential cleavage of the acetic acid residue, removal of the three chlorine atoms, and cleavage of the phenol ring. The number 5 chlorine residue is the most difficult to remove and is cleaved only after all other substitutions are removed. Of the two organisms that you have isolated, organism A is capable of cleaving the acetic acid residue and the number 4 chlorine. Organism B is capable of cleaving the remaining chlorines and the phenol ring. Electron microscopic analysis of DNA preparations from organism A reveals the presence of two plasmids 4.5kb (plasmid A1) and 6.5kb (plasmid A2) in size. Organism B contains only one plasmid (B1) of 10.5 kb. When removed from the two organisms, the plasmids no longer have the ability to degrade 2,4,5-T.

The restriction fragments produced by digesting the three plasmids with the restriction endonuclease enzymes *Bam*HI and *Bgl*II are as follows:

*Bam*HI			*Bgl*II		
A1	A2	B1	A1	A2	B1
2.5	3.0	5.0	3.3	4.1	6.9
2.0	3.5	4.0	1.2	2.4	3.6
		1.5			

By incorporating the individual restriction fragments into the multiple cloning site of the 2.9kb *Pseudomonas cepacia* cloning vector pRS101, you can demonstrate that the 3.0kb fragment from A2 codes for cleavage of the acetic acid residue of 2,4,5-T. None of the other fragments from A1 and A2 produced changes in the herbicide. When the 6.9kb fragment of B1 was cloned into pRS101 containing the 3.0kb fragment from A2, the organism was capable of degrading benzene and removing the acetic acid residue and number 2 chlorine from 2,4,5-T. Cloning the 1.5kb fragment from B1 into pRS101 also resulted in degradation of phenol.

Based on these results, how would you complete the construction of a plasmid that would enable the common soil bacterium *P. cepacia* to degrade 2,4,5-T completely?

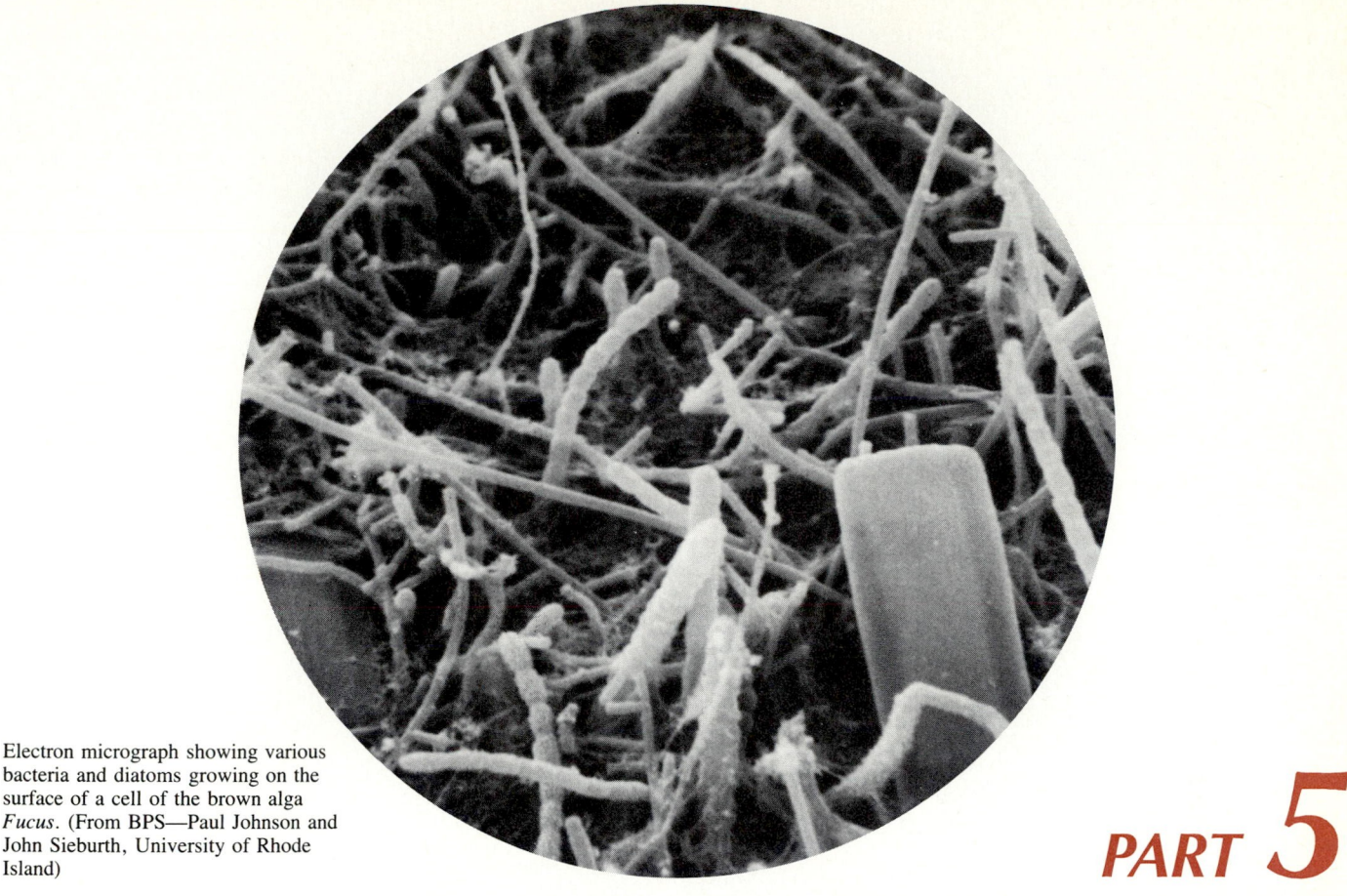

Electron micrograph showing various bacteria and diatoms growing on the surface of a cell of the brown alga *Fucus*. (From BPS—Paul Johnson and John Sieburth, University of Rhode Island)

PART 5

Survey of Microorganisms

CHAPTER 9

Viruses and Other Acellular Microbes

Lacking a cytoplasmic membrane to separate them from their surroundings, the acellular microbes are obligate intracellular inhabitants that totally depend on compatible host cells for their replication. Outside of host cells they are essentially nonliving organic molecules with varying degrees of complexity, whereas within host cells they exhibit various functions that are characteristic of living systems. The acellular microbes raise a profound question of what is life, the answer to which is largely philosophical and based upon individual perspectives.

9.1 Structure of Viruses and Other Acellular Organisms

Unlike prokaryotic and eukaryotic cells, viruses, viroids, and prions are acellular. A **viroid** is composed exclusively of RNA and, like other acellular microbes, can direct its own replication within a host cell. **Prions** have been described as proteins capable of self-replication inside a host cell.[1] **Viruses** have more complex structures, with a nucleic acid core surrounded by a proteinaceous coat and in some cases a lipid-containing envelope.

Viruses

All **viruses** have a central genetic nucleic acid core surrounded by a protein coat. The viral coat structure surrounding the nucleic acid, known as the **capsid**, is composed of protein subunits called **capsomers**. Whereas the capsid of some viruses is composed of a single type of protein, many other viruses have more complex capsids containing several different proteins. There are two basic types of capsids: helical and isometric. The capsid of a virus, such as tobacco mosaic virus, forms a helical coil around the nucleic acid (Figure 9.1). Isometric viruses have a capsid with a geometric stucture, often a pyramid-like structure known as an **icosahedron** (Figure 9.2). The capsomers of polyhedral viruses typically are proteins composed of five, six, or more subunits. In the simplest cases, capsid proteins aggregate into pentamers, 12 of which form the capsid, but in large isometric viruses the capsomers may bind into pentamerous and hexamerous arrangements to form more complex structures. For example, adenoviruses have four hexamers on each edge, six hexamers on each face, and one pentamer at each vertex. Because this virus has an icosahedral capsid with 20 faces, there are, therefore, 30 edges and 12 vertices so that adenoviruses have 240 ($6 \times 20 + 4 \times 30$) hexamers in addition to 12 pentamers. In some viruses, such as the T-even bacteriophage, the capsid structures are even more complex. T-even phage have a head structure that is isometric and a tail structure that is helical (Figure 9.3). This combined structure is termed *binal* and is quite common among the bacteriophage.

Unlike prokaryotic and eukaryotic cells, viruses contain

---

[1]This description may be erroneous since prions are not yet fully defined; they have been found as proteins in animals that may have had prior undetected viral infections.

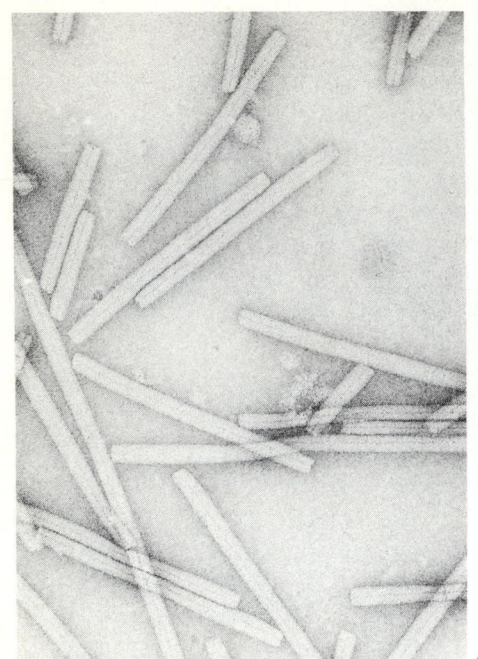

A

FIGURE 9.1

In TMV the subunits of the capsid are helically coiled around the virus's RNA nucleic acid. (A) An electron micrograph of TMV. (Courtesy Lee Simon, Rutgers, the State University). (B) A drawing showing the detailed structure of TMV.

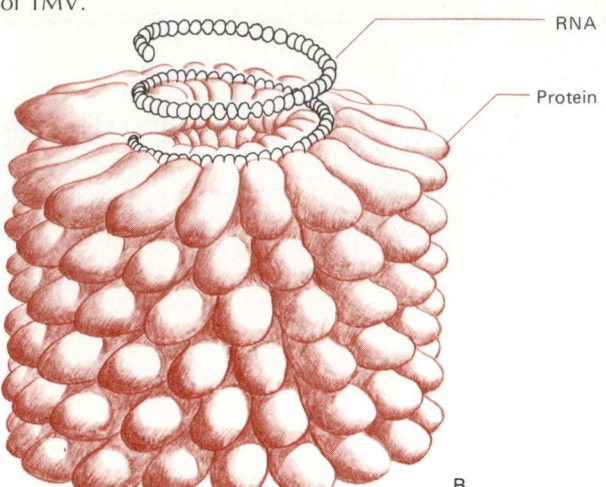

RNA

Protein

B

Discovery Process

The structure of TMV was elucidated in large part by workers at the Medical Research Council's Laboratory of Molecular Biology in Cambridge, England, under the direction of Aaron Klug. It took 20 years to gain a complete picture of the structure and assembly of TMV. Klug developed electron microscopic methods for the visualization of three-dimensional structures; his development of crystallographic electron microscopy permitted the elucidation of nucleic acid–protein complexes that are essential to the construction of viruses, such as TMV. Understanding structure is an important part of learning how biological entities function. Specifically, Klug's work showed how the components of TMV are put together; starting with one disk of protein surrounding a strand of RNA, the initial complex serves as a center for the assembly of a stack of 100 such disks composed of a total of 2,200 identical protein molecules that make up the complete virus.

only one type of nucleic acid, either RNA or DNA. The genome of the virus may consist of double-stranded DNA, single-stranded DNA, single-stranded RNA, or double-stranded RNA. The ability of nucleic acid molecules other than double-stranded DNA to store the genetic information of the organism is unique among biological systems.

By appropriately interpreting recombination frequencies and by using genetic markers, it has been possible to develop detailed, complete genetic maps for a variety of viruses. The genetic maps of several viruses show that genes are clustered according to their function. In some viruses, such as bacteriophage T7, the viral genome is linear (Figure 9.5); in others, such as bacteriophage T4, it is circular (Figure 9.6). A particularly interesting finding is that some viruses maximize the amount of information that is stored within the genome by using overlapping genes (Figure 9.7) and transcription of both strands of the DNA in opposite directions to code for different protein products (Figure 9.8). This is particularly important because the viral genome must be small and compact but must still code for a large number of gene products and regulatory functions.

FIGURE 9.2

Electron micrograph of Herpesvirus, an isometric virus, the capsid of which has been negatively stained. There is no envelope surrounding this virus. (From BPS—B. Roizman, University of Chicago)

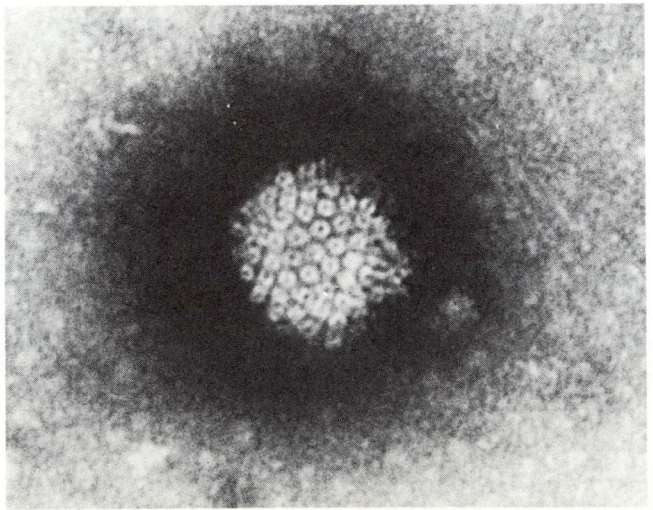

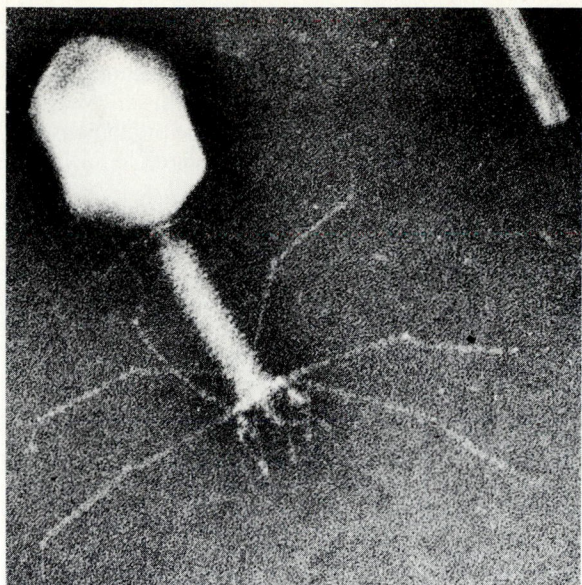
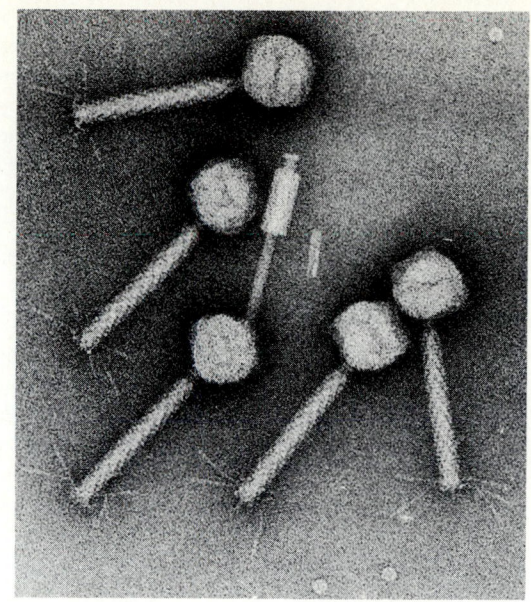

FIGURE 9.3
(A) Electron micrographs of bacteriophage T4 (280,000×). (B) Electron micrographs of bacteriophage P2 phage (240,000×). (Courtesy Robley Williams, University of California, Berkeley)

ANALYTICAL PROCESS
Mapping of Viral Genomes

In addition to the detailed genetic maps of several genomes of prokaryotic and eukaryotic cells, the genetic maps of several viruses have been determined. Like the study of genetics of cellular organisms, the study of viral genetics is based on the ability to recognize mutants and to determine rates of recombination. Four primary types of mutations have been employed in viral genetic studies: host range; plaque type; and two conditional mutations, temperature sensitive and nonsense. A **host-range** mutation alters the host that a virus can infect. For example, the wild type of bacteriophage T2 (T2h$^+$) can reproduce within *Escherichia coli* strain B but not within *E. coli* strain B/2, whereas a mutant of this virus (T2h)[2] can reproduce within both *E. coli* strains B and B/2. Host-range mutants played an important part in the early development of our understanding of phage genetics. These mutants are based on whether or not the phage can successfully attach to a bacterial host cell and initiate replication within it.

When bacteriophage replicate within bacterial cells growing on a solid surface, they produce zones of clearing, known as **plaques**, where bacterial cells are lysed when viruses are released from the host cell. The characteristics of the plaque, such as size and clearness or turbidity, are under the control of the viral genome.

For example, wild-type r$^+$ bacteriophage exhibit lysis inhibition and thus produce small plaques surrounded by a halo of increased turbidity,[3] whereas r mutants exhibit rapid lysis and produce large, clear plaques with sharp edges.

Conditional mutations occur under specific conditions; under other conditions the phage is able to replicate. **Temperature-sensitive mutants** are conditionally lethal because they permit a virus to replicate at one temperature but not at another within the normal growth range of the wild-type virus. For example, a temperature-sensitive mutant may replicate at 35°C but not at 43°C. Nonsense mutations occur when the formation of a nonsense codon within the viral genome causes premature termination of transcription of polypeptide synthesis and thus the inability of the virus to replicate. These mutants can multiply, though, in bacterial strains carrying a suppressor mutation that enables completion of the synthesis of the viral polypeptide.

When two mutant viral strains are mixed in the same host cell, recombination can occur, and the frequency of recombinant types can be used to map the viral genome. Genetic recombination in viruses was discovered in phage T$_2$ by Max Delbrück and Alfred Hershey in 1946, at the same time that Joshua Lederberg demonstrated mating in bacteria. Delbrück and Hershey performed genetic crosses by infecting bacteria with a mixture of a

[2]In viral genetics the wild type is designated +, but unlike other areas of genetics, the mutant is not labelled −. Rather, the mutant is designated only by the type of mutation, e.g., h for host range mutant or r for rapid-lysis mutant.

[3]The halo is formed by bacteria infected more than once because multiple infections lead to a delay in lysis known as *lysis inhibition*.

host range (h) and a rapid-lysis (r) mutant of a T-even phage (Figure 9.4). Four distinguishable viral types were found after incubation: the two parental types and two recombinant types with a marker from each parent. Determining viral genetic maps by such experiments is complicated because DNA molecules continue to recombine throughout the period of viral multiplication within a host cell. Because bacteriophage undergo several rounds of mating during an infection cycle, that is, before the phenotypes of the recombinants can be determined, the consequences of a mating between two phage DNA molecules cannot be inferred directly from the proportions of recombinants. Statistical analyses are required to account for the effect of repeated matings.

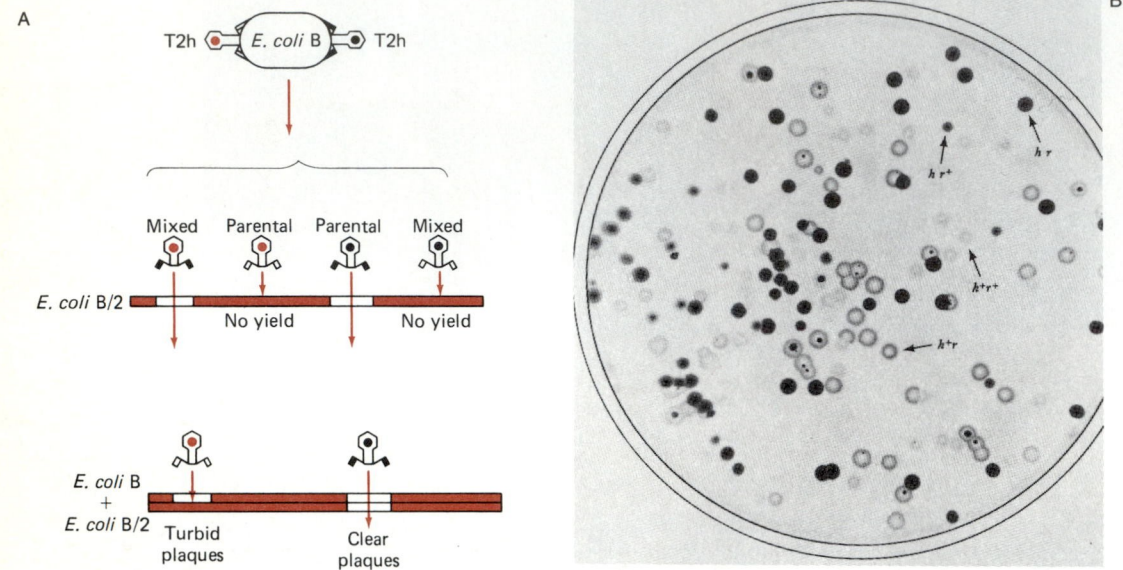

FIGURE 9.4

(A) Mixing of wild-type and h mutant bacteriophage in *E. coli* B produces four types of progeny— hr, hr$^+$, h$^+$r, and h$^+$r$^+$. The designation h refers to host cell range and r refers to plaque type, with the superscript + designating the wild-type mutant form. In this figure, the red dot in the head of the phage indicates the r$^+$ genotype, whereas the black dot indicates the r. The black tail structure indicates h and the white tail h$^+$. After infecting *E. coli* B/2, only phenotypically h phage produce progeny. Plating on a mixture of *E. coli* B and *E. coli* B/2 yields clear and turbid plaques. The turbid plaques are from genotypically h$^+$ phage that do not infect strain B/2 cells, and clear plaques are from genotypically h phage that infect both types of cells. When similar experiments are performed with phage carrying mutations at both the r and h loci, four phenotypically distinguishable plaques are produced. (B) Plaques formed by a mixture of T2 phage carrying mutations at the h and r loci. Phage with the h$^+$ allele produce turbid plaques (gray); phage with the h allele produce clear plaques (black); phage with the r allele are larger than those with the r$^+$ allele. (Reprinted by permission from Gunther Stent, 1963, *Molecular Biology of Bacterial Viruses*, Copyright © 1963, W. H. Freeman & Co.)

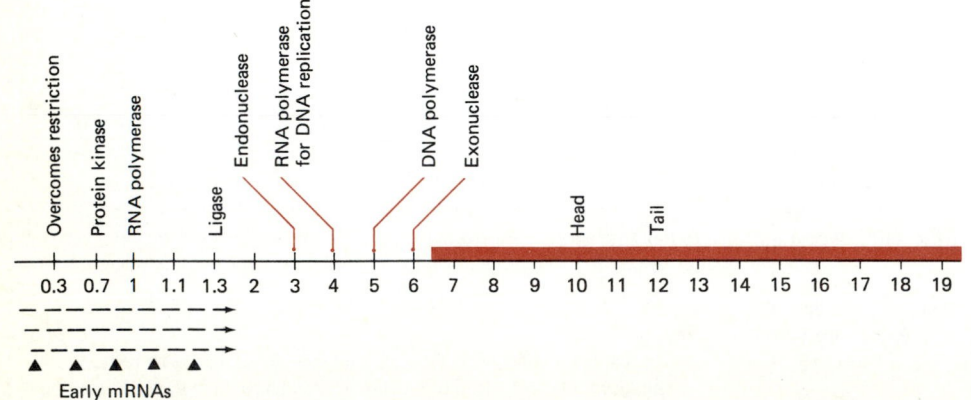

FIGURE 9.5

The linear genetic map of phage T7. The numbers represent the time (in minutes) after penetration of the host cell at which each component appears in the replication process.

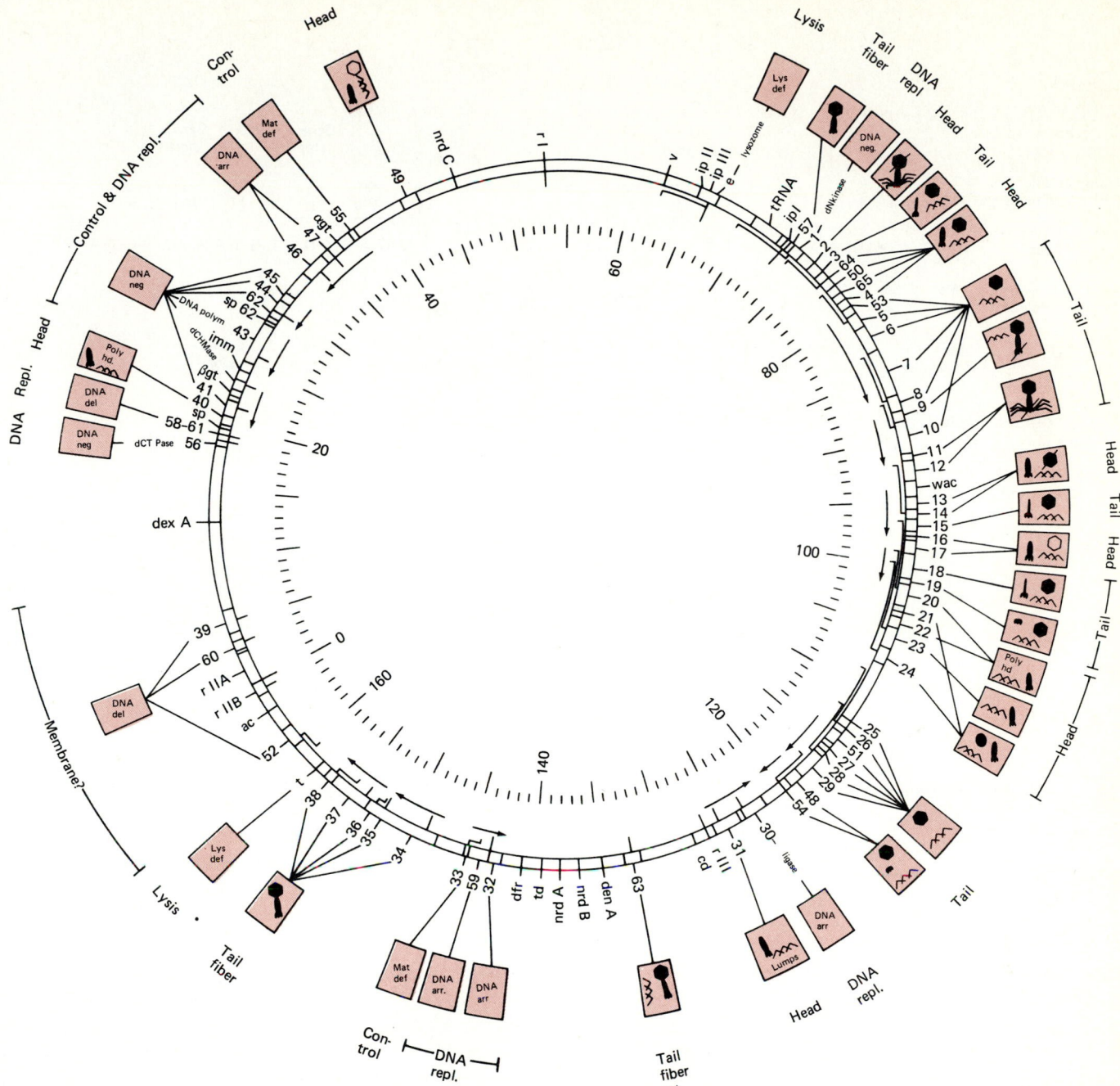

FIGURE 9.6

Map of the known genes of bacteriophage T4. The numbers on the inner dial are map distances in recombination units. The arrows indicate the direction of transcription. The symbols in the rectangles indicate the functions altered by mutations in the corresponding genes. NEG = negative, DEL = delayed, ARR = arrested, HD = head, LYS = lysis, MAT = maturation, DEF = defective. (After Wood, 1974, *Handbook of Genetics*, R. C. King, ed., Vol. 1, p. 327)

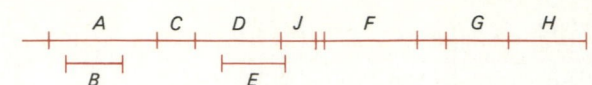

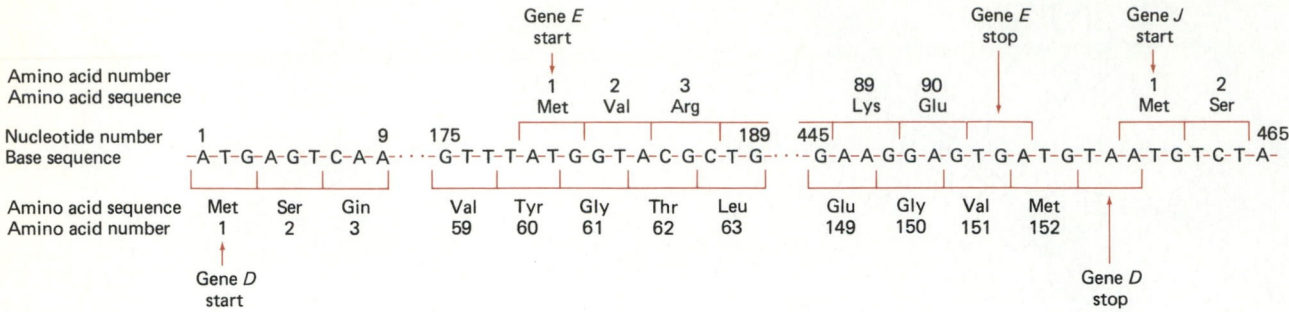

FIGURE 9.7

The genome of bacteriophage φ, showing overlapping genes. The reading frame of gene E, which begins in the middle of the nucleotide sequence for gene D, is offset by one base from that of gene D. The stop codon for gene D also overlaps with the start codon for gene J. Similar overlapping genes occur in animal viruses such as SV40.

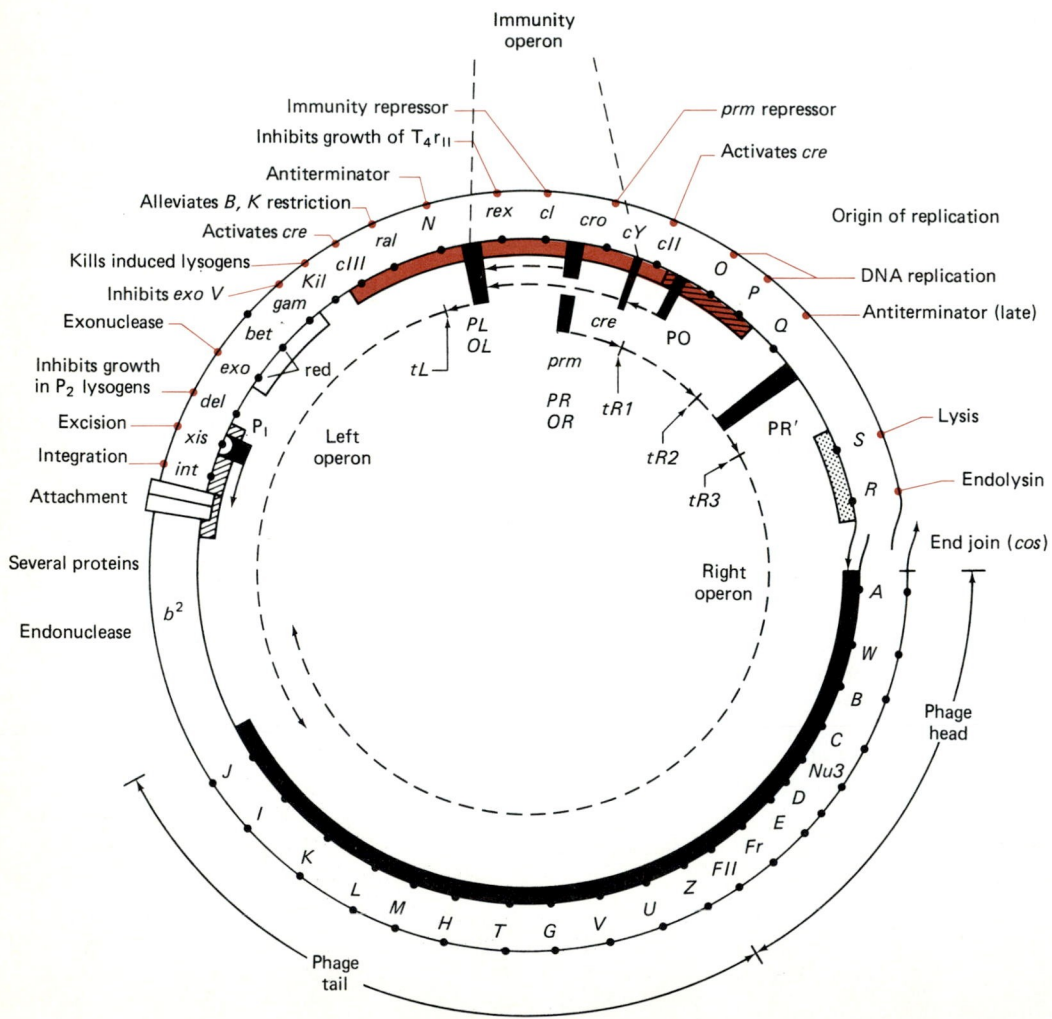

FIGURE 9.8

The genetic map of bacteriophage lambda shows that both left-hand and right-hand transcription are used for production of viral gene products. The directions for transcription are shown by the dashed lines and arrow within the circle. Regulation of gene expression determines which genes are expressed at any given time and controls the development of this bacteriophage.

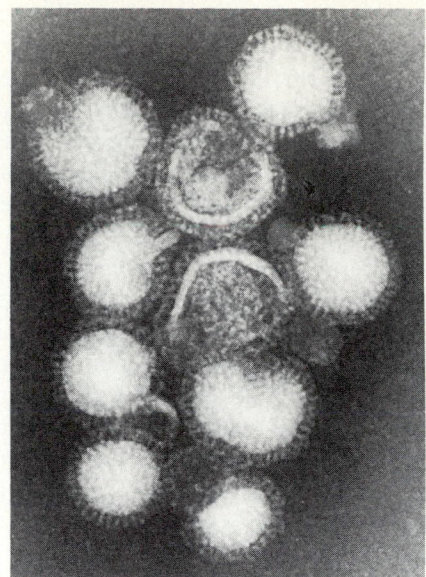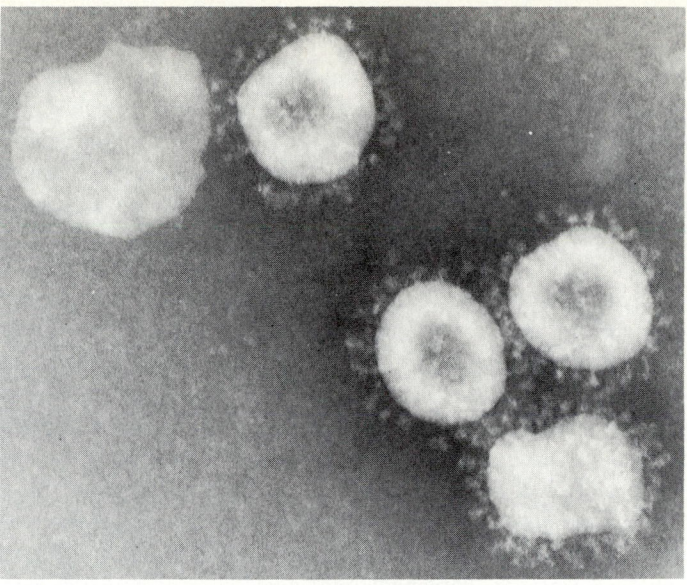

FIGURE 9.9

Electron micrographs showing enveloped viruses. (A) Electron micrograph of the influenza A_2 Hong Kong virus (170,000×), showing protruding spikes of the envelope structure. (B) A Coronavirus that causes respiratory tract infections. (A and B from BPS—F. A. Murphy, Centers for Disease Control, Atlanta)

In addition to the nucleic acid and capsid components, some viruses have an **envelope** that surrounds the virus particle. Like cellular membranes, these envelopes possess lipid bilayers and proteins with specialized functions.[4] Some of the envelope proteins are glycosylated, and the hydrophilic carbohydrate ends of such proteins may protrude from the viral particle. Such glycoproteins often occur as spikes on the outer surface of the virus (Figure 9.9). Some of these proteins are involved in binding the virus to a host cell, and others cause **cell lysis** (rupture of the cell). In addition to the glycoproteins, other proteins of the viral envelope form a matrix layer that attaches the envelope to the capsid. The proteins of the envelope are specified by the viral genome, but the carbohydrate moieties of the glycoproteins and the lipid components of the viral envelope are obtained from the host cell. When the virus leaves the host cell, it picks up a portion of the nuclear or cytoplasmic membrane and that piece of host cell membrane can surround the viral capsid, forming the lipid portion of the envelope. The presence of host cell membranes surrounding a viral particle can help the virus evade the normal host defense mechanisms designed to recognize and destroy foreign substances.

Viroids

Whereas viruses contain both protein and nucleic acid, **viroids** have only very small RNA genomes and no other struc-

tures. In essence, a viroid is simply an RNA macromolecule that can be preserved and transmitted to cells, where it is reproduced. Such macromolecules contain the information needed to direct their own replication but are totally dependent on the metabolic activities of a host cell for accomplishing this task. Inside a suitable host cell, however, the RNA is capable of initiating its own replication. Compared to viruses, viroids introduce far less genetic information. The presence of viroids sometimes manifests itself as disease symptoms in the host organism, and certain plant diseases have been identified as caused by viroids. Diseases caused by viroids include, among others, potato spindle tuber, citrus exocortis, chrysanthemum stunt, cucumber pale fruit, avocado sunblotch, and coconut cadang-cadang.

It is believed that viroid RNA is replicated within host cells, using host cell enzymes and the viroid RNA as a template, but that the viroid RNA is not translated into viroid-specified polypeptides. It is not yet clear how these molecules survive outside of host cells or how they are transported. That such macromolecules can be transmitted and cause infectious diseases of higher organisms is a relatively new finding, the ramifications of which have yet to be fully appreciated. While the origin of viroids is unclear, some have been found to have nucleotide sequences in common, suggesting a relationship or common ancestry. Recently viroid-like entities have been discovered in which the viroid RNA is surrounded by a protein coat, as occurs in viruses. Such viroid-like entities are incapable of independent replication and depend upon viral RNA to reproduce. Much work remains to be done before we fully understand the nature of viroids and their replication in biological systems.

---

[4]The viral envelope does not function to regulate the flow of materials, nor does it exhibit any of the other physiological activities associated with the membranes of living cells.

Prions

The most recently discovered and least understood microbes are the *prions*. The discovery of prions in 1983 was quite unexpected, and their nature is very controversial. What is so unusual is that some analyses of prions indicate that they may be composed only of protein. If these ''organisms'' are nothing more than specific infectious protein molecules that contain the information that codes for their own replication, they are unique entities that violate the commonly accepted dogma that nucleic acids are the conveyers of information.

In the discovery of prions, we have found an exception to what appeared to be a universal characteristic of living systems—the storage of genetic information in nucleic acid molecules. All other organisms store their genetic information in nucleic acids, DNA or, less commonly, RNA. We know how the genetic information in these nucleic acid molecules is replicated so that it can be passed from one generation to the next. We also have broken the genetic code and know how the information contained within nucleic acid molecules is expressed. What we do not know is how a protein can direct its own replication; thus, we do not understand how prions replicate. Prions actually could be encoded by cellular genes and their action could result from their binding to a cell membrane. They could also be the product of an earlier undetected viral infection.

Although we do not know exactly how prions reproduce or how prevalent they are, we do recognize their potential importance. Prions were discovered during the search for the cause of scrapie. Scrapie is an infectious and usually fatal disease of sheep. It is a neurological disease characterized by wild facial expressions, nervousness, twitching of the neck and head, grinding of the teeth, and scraping of portions of the skin against rocks with subsequent loss of wool. This disease was known to be caused by an agent that could pass through a bacteriological filter and, therefore, was believed to be a virus designated as a *slow virus* because of the slow development of the disease. However, no virus could be found and eventually the cause of scrapie was attributed to a prion. It is likely that other diseases, including some human diseases, such as kuru, result from the reproduction of prions within host cells. Kuru is a degenerative neurological disease associated with cannibal rituals in New Guinea; ingestion of infected brains results in transmission of the prions and kuru to the cannibal. Other diseases that have been ascribed to prion infections include transmissible mink encepalopathy, chronic wasting disease, Creutzfeldt-Jakob disease, and Gerstmann-Straüssler syndrome. Several of these diseases are characterized by a slow degeneration of the nervous system.

Prions and the consequences of their reproduction within the cells of the organisms they infect may explain several other diseases that still have no known cause. Some scientists have suggested that Alzheimer's disease, a degenerative disease of the nervous system that afflicts a large number of people over 40 years old, is caused by a prion. This is one of several hypotheses to explain the etiology of degenerative nervous disorders. Much more research is needed to reveal the importance of the discovery of prions.

9.2 Viral Replication

Viral Replication Cycle

Unlike the reproduction of all other organisms, the replication of viruses does not occur through cellular multiplication. Rather, viruses are replicated in parts and assembled within a host cell; then the completed viruses, sometimes called **virion particles** or **virions**, are released. To replicate, a virus must invade a suitable host cell because only within such a cell do conditions permit viral replication. While some microbiologists view viruses as parasites of the host cells within which they reproduce, others consider viruses to be genetic extensions of the host cells in which they replicate. Supportive of both viewpoints is the high degree of specificity between a particular virus and the specific host cell within which that virus can reproduce. Viruses that replicate only within specific host bacterial cells are known as **bacteriophage** or simply **phage**; others that replicate within plant cells are known as **plant viruses**; and those that replicate only within animal cells are known as **animal viruses**. The specificity of viruses and their host cells is based upon compatibility in terms of the ability of the virus to attach physically to the host cell and the ability of the viral genome to direct viral replication within that host cell.

For a specific virus to replicate within a host cell, (1) the host cell must be permissive, and the virus must be compatible with the host cell; (2) the host cell must not degrade the virus; (3) the viral genome must possess the information for modifying the normal metabolism of the host cell; and (4) the virus must be able to use the metabolic capabilities of the host cell to produce new virus particles containing replicated copies of the viral genome.

Although the specific details of viral replication vary from one virus to another, the general strategy for reproduction is the same for most viruses. The virus initially attaches to the outer surface of a suitable host cell. Generally, the **adsorption** of a virus to a host cell involves specific binding sites on the cell surface, which explains in part the high degree of specificity between the virus and the host cell. The nucleic acid of the virus then penetrates the cytoplasmic membrane.[5] Within the host cell, the viral genome achieves control of the cell's metabolic activities; in many cases, the viral ge-

---

[5]At this stage of viral replication the virus has lost its identity, that is, there are no complete viruses within the host cell. This loss of identity distinguishes viral replication from the reproduction of cellular organisms, including these cellular organisms that reproduce within host cells.

nome actually codes for the shutdown of these metabolic activities normally involved in host cell reproduction. The virus then uses the needed biochemical components and anatomical structures of the host cell for the production of new viruses. In particular, the virus employs the host cell's ribosomes for producing viral proteins and the cell's ATP and reduced coenzymes for carrying out biosynthesis. The nucleic acid and protein capsid structures of the virus (the essential structural components of a virus) are synthesized separately and then assembled prior to release from the host

cell. Generally, many virus particles are produced within a single host cell and are released together. The replication of a virus results in changes in the host cell, often causing the death of that cell. The cycle of viral adsorption to a host cell, invasion, synthesis of viral nucleic acid and proteins, assembly of the virus structure, and release of viral progeny is repeated when a virus encounters another suitable host cell.

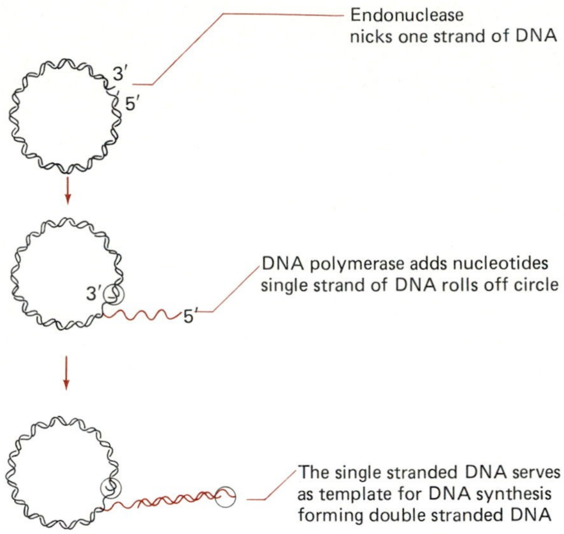

Replication of Viral Genomes

The replication of the viral genome can occur in a variety of ways, depending on the nature of the nucleic acid macromolecule in which the genetic information is stored. Some viruses can act like *E. coli*, exhibiting bidirectional DNA replication from a single point of origin, although in some cases, the terminus for DNA replication is offset from the origin. Some linear viruses exhibit multiple initiation points for DNA synthesis and thus resemble eukaryotes. In other cases, the replication of viral DNA follows a **rolling circle model** in which a circular DNA molecule is used to spin off unidirectionally a linear DNA molecule (Figure 9.10). The rolling circle replication of a DNA molecule requires an endonuclease enzyme that can nick the circular DNA molecule, establishing a free end of the nucleotide chain that can "roll off" the circle. A variation of the rolling circle model of DNA replication appears in the synthesis of single-stranded DNA for those DNA viruses that lack the normal double helical DNA molecule.

Several methods are employed by different RNA viruses for the replication of their genome (Figure 9.11). Some RNA viruses code for RNA-dependent RNA polymerase enzymes, and within a host cell these viruses are able to use their RNA genome as a template for RNA synthesis. The replication of genetic information in these RNA viruses is analogous to DNA replication in prokaryotic and eukaryotic organisms. In the case of some single-stranded RNA viruses, such as the bacteriophage Qβ, the genome of the virus (designated as a

FIGURE 9.10

The rolling circle model provides an explanation for how circular DNA replication occurs. According to this mechanism, an endonuclease nicks one strand of DNA, DNA polymerase adds nucleotides, and a single strand of DNA falls off the circle. The single-stranded DNA then serves as a template for DNA synthesis to form double-stranded DNA.

FIGURE 9.11

There are two different modes of replication used by different single-stranded RNA viruses for the replication of their genomes. (A) In some viruses, a double-stranded RNA replicating form is produced that acts as a template for the synthesis of new RNA macromolecules. In single-stranded RNA viruses, one of the strands then degenerates. (B) In other viruses, the viral RNA genome acts as a template for the synthesis of a DNA molecule by reverse transcription. Normal transcription of the DNA then produces new copies of the RNA genome.

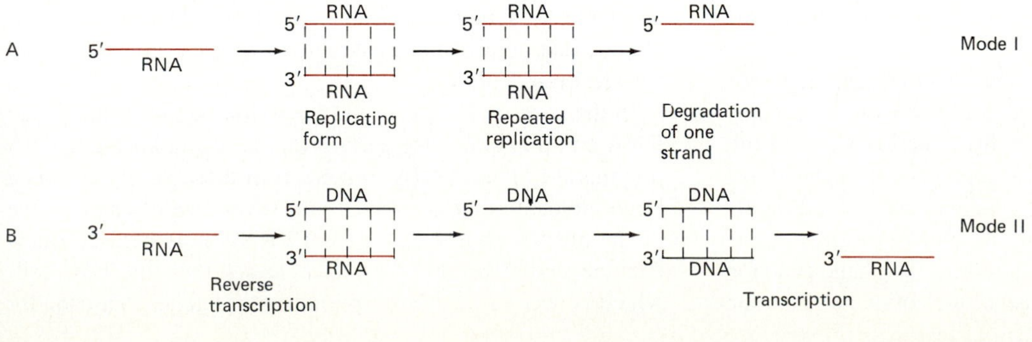

plus RNA strand) is used as a template for forming a replicative double-stranded form that has both plus and complementary minus RNA strands. When the RNA-dependent RNA polymerase enzyme uses the replicative form as a template, the product includes mostly plus strands of RNA, which go into the genomes of the viral progeny.

Another mechanism for RNA synthesis in some animal virus–infected cells employs reverse transcription. In **reverse transcription** the RNA viruses use their RNA genome as a template for an RNA-directed DNA polymerase. These viruses produce replicates of DNA instead of RNA. The DNA is then integrated into the host gene. The information in the DNA molecule is then used to direct the synthesis of RNA, which is accomplished by transcription. Some of the RNA acts as mRNA for synthesizing proteins, and some of the RNA is put into the RNA genomes of the viral progeny. Thus, there are a variety of mechanisms used by different viruses for replicating their genetic information. Several of these mechanisms are distinct from those used in prokaryotic and eukaryotic cells for the replication of the genome.

Lytic Phage

Reproductive Cycle   The complete reproductive cycle of bacteriophage results in the lysis of the host cell at the completion of the viral replication cycle, and so, these bacteriophage are referred to as **lytic phage**. The lytic phage, of which there are numerous types, exhibit characteristic developmental sequences or stages in their reproductive cycles. The specific details of reproduction vary among the different types of lytic phage, but the general developmental sequence is exemplified by the reproduction of the T-even phage, such as T2, T4, and T6. T-even phages are DNA viruses that have a complex symmetry with distinct head and tail structures (Figure 9.12). The general sequence of events in the lytic reproductive cycle of a T-even phage is depicted in Figure 9.13. Reproduction of T-even phages within cells of bacteria such as *E. coli* begins with the attachment of a T-even phage to the bacterial cell. There appear to be specific receptor sites on the bacterial cell surface where the phage may attach (Figure 9.14). The entire phage particle does not penetrate the bacterial cell; rather, the phage injects its DNA into the bacterium, using its tail structure like a syringe. The phage tail penetrates the cell wall, but not the cytoplasmic membrane, and the contraction of the tail structure forces the phage DNA into the periplasmic space. The phage DNA subsequently migrates across the cytoplasmic membrane and into the cell.

When the phage DNA enters the bacterial cell, it is not degraded by the exonucleases and endonucleases of a compatible host cell. The reasons for this situation vary. In the case of T4 phage, the phage DNA contains glucosylated hydroxymethyl cytosine instead of cytosine, a chemical modification of the DNA that prevents the nucleases of the bacterium from degrading the phage genome. T4 phage code for enzymes that synthesize this unusual nucleotide and also for

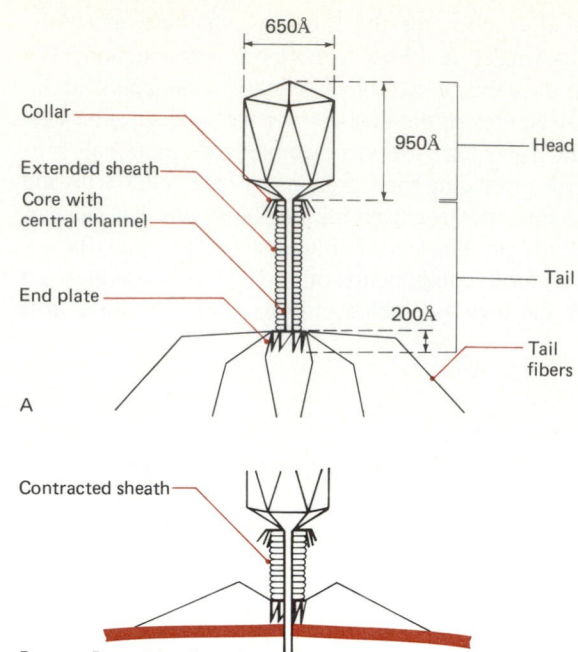

FIGURE 9.12
Drawing of the T-even phage, T4, showing its components. The phage tail fibers attach to a bacterial cell, and the contraction of the sheath injects the viral DNA through the bacterial envelope. (A) Virion with extended tail fibers. (B) Virion with the tail sheath contracted and the spikes of the tail plate pointing against the bacterial cell wall. (After Simon and Anderson, 1967, *Virology*, 32:279)

enzymes that break down the normal precursor for normal DNA, which is deoxycytidine triphosphate. The T-even phage also codes for a nuclease that degrades the host cell DNA, and the deoxynucleotides released by the degradation of the bacterial chromosome can be used as precursors for the synthesis of viral DNA. The T4 phage genome codes for DNA polymerases and other enzymes; these enzymes are made in large amounts so that synthesis of T4 DNA occurs rapidly. The viral DNA is initially transcribed by a bacterial RNA polymerase, and among the first proteins coded for are ones for the modification of the bacterial RNA polymerase. The phage-coded polypeptides replace or modify the sigma subunits of the bacterial RNA polymerase and, by doing so, alter the recognition sequence so that the RNA polymerase no longer binds at the normal Pribnow sequences of the bacterial DNA. When this occurs, the RNA polymerase binds at sites that control the transcription of other genes required for viral replication.

In the case of T7 phage, transcription begins at the phage DNA end that codes for early proteins, using host cell RNA polymerase. The mRNA formed from these early genes is a large molecule that is cleaved into several short chains. One of the mRNA chains codes for T7 RNA polymerase, a new enzyme coded for by the phage rather than the host cell, which is needed for transcription of other genes that code for

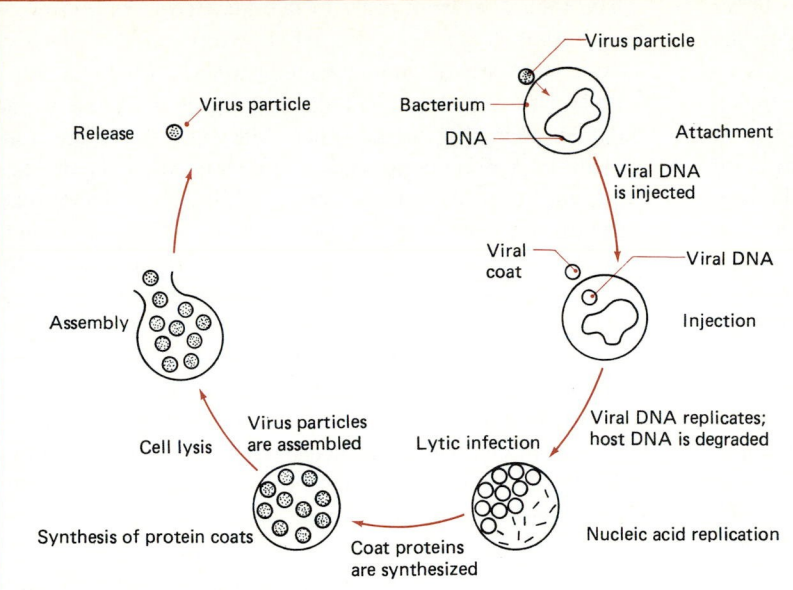

FIGURE 9.13

In the lytic phage reproduction cycle, the bacteriophage attaches to specific chemical sites on the surface of a bacterium and the viral nucleic acid penetrates the host cell. Once the viral nucleic acid enters the bacterium, bacterial metabolism is halted by the destruction of the host genome. First viral DNA and then viral coat proteins appear, which then self-assemble and the host cell lyses. The free viruses may infect other susceptible host bacteria.

Discovery Process

In spite of the large variety of viruses and their differences in structure and genetic information, the use of bacteriophage as a model system has been important in the advance of the field of virology because all viruses are similar in their mode of multiplication. Burnet in Australia, Schlesinger in Hungary, Lwoff in France, and Delbruck, Luria, and Cohen in the United States all used bacteriophage as model systems, even though it is now known that they are among the most complex of all viruses. It was this very complexity, however, that allowed the discovery of aspects of viral multiplication that might have been overlooked with simpler viruses. For example, the presence of 5-hydroxymethylcytosine, instead of cytosine, in the DNA of the bacteriophage made it possible to study the intracellular multiplication of viral DNA. Also, the presence in the infected cells of deoxycytidylate hydroxymethylase, which synthesizes that unusual base and its absence in uninfected cells, gave the first indication that virus-specified enzymes are made in the infected cells. Additionally, using these markers, the replication of bacteriophage could be followed, permitting the elucidation of the lytic growth cycle.

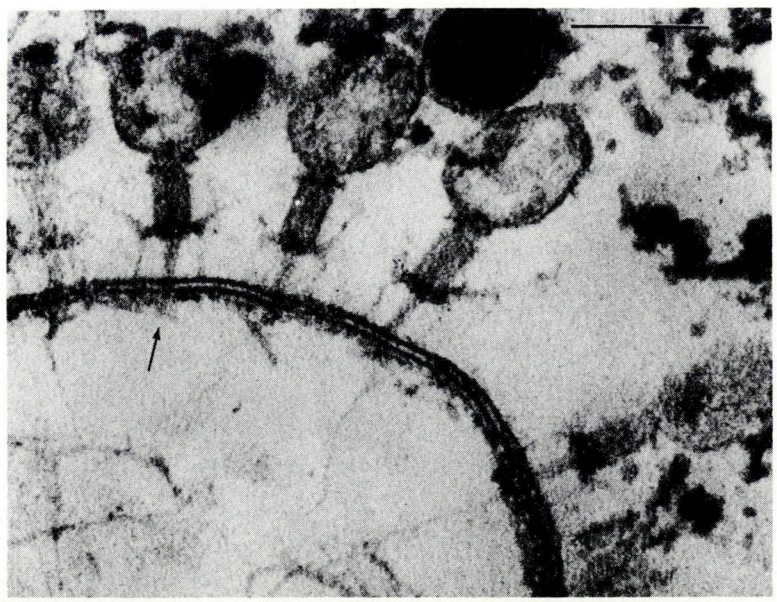

FIGURE 9.14

Electron micrograph showing the attachment of T-even phage to *E. coli*. Different viruses attach at different receptor sites on the host cell surface. The tail structure penetrates the bacterial cell (arrow) and the phage DNA is injected into the bacterium as the sheath contracts. (Courtesy Lee Simon, Rutgers, the State University)

proteins involved in the formation of the T7 viruses. Other proteins formed before the T7 RNA polymerase stop the action of host cell RNA polymerase, ending the early events in viral replication and stopping transcription of host cell genes.

The entire sequence of penetration, shutting off host cell transcription and translation, and the degradation of the bacterial chromosome takes only a few minutes. The penetration of phage DNA into the bacterial cell and the stoppage of host cell macromolecular synthesis, involving several enzymes referred to as **early proteins** (proteins that are made early in viral replication), represent the early developmental steps of viral replication. After these early events in the phage reproduction cycle, there is a further modification of the RNA polymerase, resulting in the cessation of further synthesis of the early phage proteins. This shift in the recognition site of the RNA polymerase coincides with the beginning of the synthesis of **late proteins**, proteins coded for late in the developmental sequence of the phage. The early genes coded for by the phage genome include some enzymes involved in the replication of the phage DNA. The late phage genes also code for the various proteins that make up the capsid structure of the phage. The tail, tail fiber, and head structures of the phage are made up of proteins coded for by different phage genes, with at least 32 genes involved in the formation of the tail structure and at least 55 genes involved in the formation of the head structure.

The transition from early to late gene transcription in T4 phage involves an interesting shift with regard to which of the two DNA strands of the phage genome serves as the sense strand, with one strand initially serving as the sense strand and then switching to the other. The early genes are transcribed in a counterclockwise direction, whereas the late genes are transcribed in a clockwise direction. Because transcription proceeds in the 5′P to 3′OH direction, the change from counterclockwise to clockwise must mean that the opposing strands of the viral DNA code for the early and late proteins. The alteration of the recognitional subunit of the RNA polymerase enzyme accounts for the change in the base sequences of the DNA recognized as promoter sites for transcription of the viral genome and permits changes in the DNA strand that acts as the sense strand. By altering reading frames and by changing the DNA strand that serves as the sense strand, the phage genome can encode the almost 150 genes involved in the replication of T4 phage.

After the production of its individual components the virus is assembled, with packaging of the nucleic acid genome within the protein capsid. The assembly of the T-even phage capsid is a complex process (Figure 9.15), but one that follows a sequential pathway. **Assembly** of the head and tail structures requires several enzymes that are coded for by the phage genome. The head, tail, and tail fiber units of the T-even phage capsid are assembled separately, and the tail fibers are added after the head and tail structures are com-

FIGURE 9.15

The assembly of phage head and tail structures within the bacterial host cell. Three separate subassembly lines are illustrated here: the head, sheath, hollow spike, and tail fiber components. The assembly of the individual components is enzymatically mediated on the basis of activities directed by the viral genome. Assembly of the viral head and tail structure occurs spontaneously. (Based on Wood et al., 1968, *Federation Proceedings*, 27:1160)

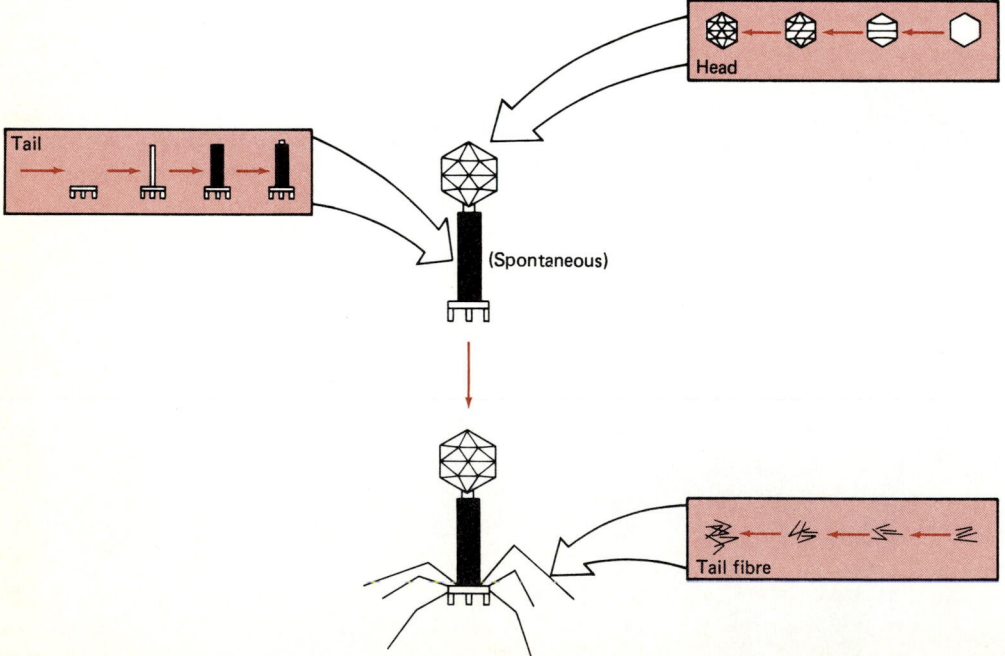

ANALYTICAL PROCESS
Assaying for Lytic Bacteriophage

The growth curve for a bacteriophage can be determined by inoculating a suspension of host bacterial cells with the phage and assaying for the number of infective phage particles at various time intervals. In one assay for infective bacteriophage, a lawn of bacteria is prepared on a suitable solid nutrient medium, and dilutions of the phage suspension are then spread over the same surface (Figure 9.16). In the absence of lytic bacteriophage, the bacteria form a confluent lawn of growth. Lysis by bacteriophage is indicated by the formation of a zone of clearing or **plaque** within the lawn of bacteria. Each plaque corresponds to the site where a single bacteriophage acted as an infectious unit and initiated its lytic reproductive cycle. The spread of infectious phage from the initially infected bacterial cell to the surrounding cells results in the lysis of the bacteria in the vicinity of the initial phage particle and hence this zone of clearing. The number of plaques that develop and the appropriate dilution factors can be used to calculate the number of bacteriophages in a sample. The medium used in these assays has a relatively low percentage of agar and therefore is called *soft agar*; it permits diffusion of phage to nearby uninfected cells but does not permit the viruses that are produced to move to remote parts of the plate. Plaques do not continue to spread indefinitely. With T4 phage the plaque size is limited because when a cell is heavily reinfected with phage before the time of normal lysis, the lysis of the cell is inhibited, a phenomenon known as *lysis inhibition*, which is actually an extension of the period of viral synthesis. In other phage, plaques are

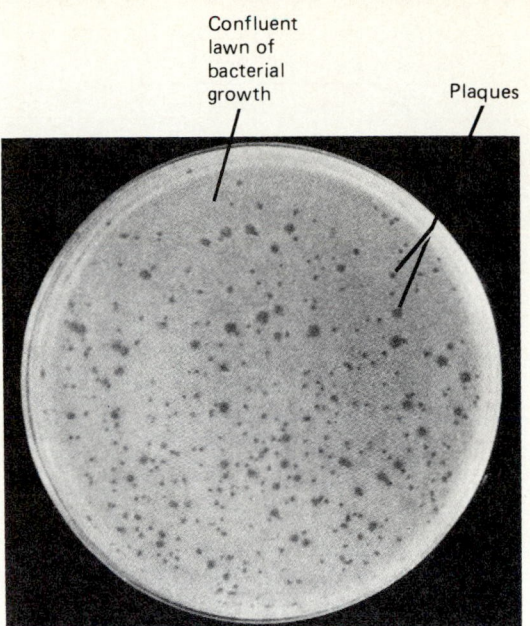

FIGURE 9.16
The lytic replication of phage on a lawn of bacteria growing on a solid medium produces zones of clearing known as *plaques*. Here the plaques of bacteriophage lambda growing on *E. coli* are clearly visible. (From BPS— Richard Humbert, Stanford University)

limited in size because the host bacterial cells are no longer in a growth phase in which viruses can be produced.

---

bined. Because of the small size of the viral particle, the DNA must be tightly packed within the phage head assembly. Packaging DNA into the head structure involves stuffing the head with DNA and cutting away the excess. When the head structure is completely filled with DNA, any extra DNA is cleaved by a nuclease. The specific mechanism that is responsible for filling the phage head with DNA and folding the DNA within the assembled phage particle has not been totally elucidated.

Once the phage are assembled, they must be released from the host cell and encounter another host cell for further reproduction. One of the late proteins coded for by the phage genome is **lysozyme**, which catalyzes the breakdown of the peptidoglycan wall structure of bacteria. The action of the lysozyme results in sufficient damage to the cell wall so that the wall is unable to protect the cell against osmotic shock, which results in the lysis of the bacterial cell and the release of the phage particles into the surrounding medium. Lysozyme activity appears to be subject to phage-directed regulation, which ensures that the wall is not degraded prematurely before a sufficient number of phage particles has been completely assembled.

Growth Curve of Lytic Phage

The lysis of the bacterial cell releases a large number of phage simultaneously, and consequently, the lytic reproduction cycle exhibits a **one-step growth curve** (Figure 9.17). The growth curve begins with an **eclipse period** during which there are no complete infective phage particles, and the naked DNA within the host cell is unable to infect other cells in order to initiate a new reproductive cycle (Figure 9.18). The end of the eclipse period is taken as the time at which an average of one infectious unit has been produced for each productive cell. The eclipse period, thus, is the time between entry of the viral DNA and formation of the first complete virus within the host cell. The eclipse period is part of a longer period, the **latent period**, which, like the eclipse period, begins when the phage injects DNA into a host cell, but which does not end until the first assembled virus from the infected cells appear extracellularly. The latent period for a T-even phage typically is about 15 minutes. During the time between the end of the eclipse period and the end of the latent period, assembled phage accumulate within the bacterial cell. Completely assembled phage accumulate

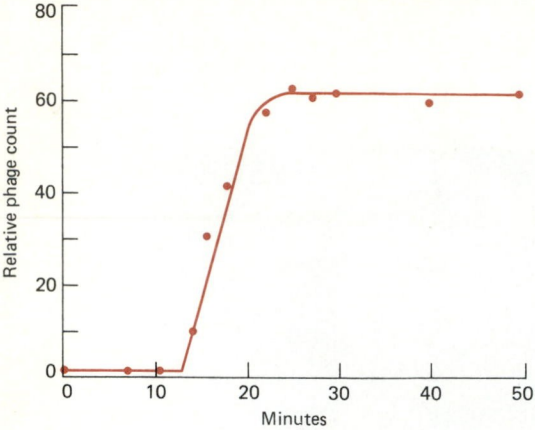

FIGURE 9.17

The one-step growth curve for the lytic bacteriophage T1 on *E. coli* strain B grown in nutrient broth at 37°C. Phage and bacteria were mixed in a ratio of 1:10 at time 0. After 4 minutes, 45 percent of the phage had been adsorbed, at which time the mixture was diluted. Assays were made at intervals after dilution. Average phage yield per infected cell = 138.

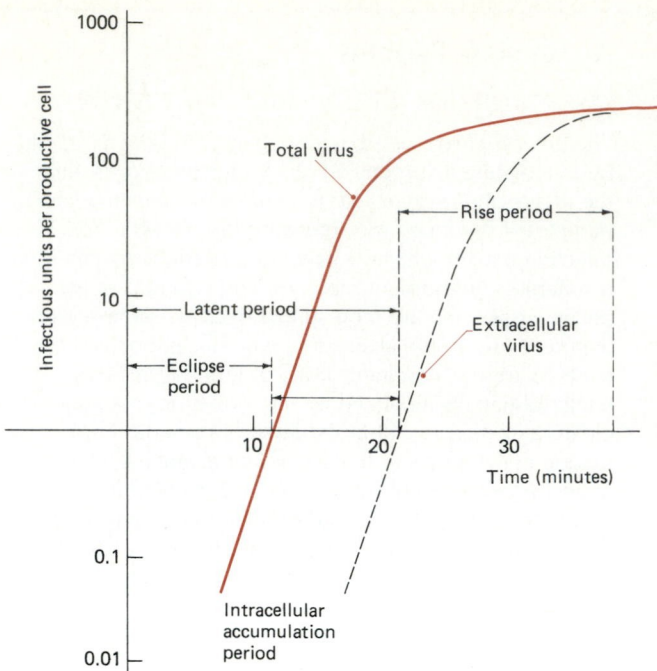

FIGURE 9.18

Detailed growth curve for bacteriophage T2, showing the eclipse, latent, and rise periods. To determine this viral growth curve, bacteria and phage were mixed for 2 minutes and then samples were periodically collected to determine the total number of bacteriophage and the number of extracellular phage. To determine the total number of bacteriophage, the host bacterial cells were lysed with chloroform and the lysate was then analyzed for phage. The presence of extracellular phage was determined by centrifuging the host bacterial cells and assaying the supernate for bacteriophage that could initiate infection (infectious units). The eclipse period is the time when there are no completed virus particles inside or outside the host cell. This period is followed by the intracellular accumulation period, when complete virus particles can be found inside but not outside the host cell. The end of the latent period marks the time when complete viral particles begin to appear extracellularly. During the rise period that follows, the number of extracellular viruses increases.

within the bacterial cell until the cell lyses, releasing the viruses into the extracellular fluid (Figure 9.19). The **burst size**, which varies from cell to cell, represents the average number of infectious viral units produced per cell; a typical burst size for a T-even phage may be as high as 200. As a result of the simultaneous release of a number of infective phage, the number of phages that can initiate a lytic reproduction cycle increases greatly in a single step. The entire lytic growth cycle for some T-even phage can occur in less than 20 minutes under optimal conditions.

In addition to the T-even DNA phage, there are also some RNA phage, such as Qβ and f2, that carry out a lytic reproductive cycle. RNA phage tend to be quite small. The RNA genome of these phage can serve as mRNA molecules for the production of viral proteins, but the replication of the viral genome requires the synthesis of a complementary replication strand of RNA that can then serve as a template for the production of the RNA genome. The complete assembly of the phage is followed by lysis of the host cell, catalyzed by a viral-coded lytic enzyme.

Lysogeny

In addition to eliciting a lytic cycle, some bacteriophage are capable of lysogeny in which only the integrated genome of the virus is replicated. Such phage are called **temperate bacteriophage** because only the genome is replicated, rather than producing complete viral particles when the viral genome is incorporated; also, the host cell is not killed by lysis in this mode of viral replication (Figure 9.20). In lysogeny the phage genome is usually incorporated into the bacterial chromosome. In some cases, exemplified by the phage P1, the phage genome exists as a circular, self-replicating, double-stranded DNA element called a **plasmid** rather than being

integrated into the bacterial chromosome. Incorporation of the phage genome into a bacterial chromosome is an example of nonreciprocal recombination. Once incorporated into the bacterial chromosome, the viral genome (referred to as a **prophage**) is replicated along with the bacterial DNA during normal host cell DNA replication. At a later time the prophage can be excised from the bacterial chromosome or plasmid DNA, reestablishing a normal lytic reproductive cycle.

The bacteriophage lambda (λ), for example, is a temperate phage that can alternate between the lytic reproductive and lysogenic reproductive cycles. The regulation of lambda reproduction, which determines whether the reproductive cycle is lytic or lysogenic, represent an interesting example of the ramifications of molecular-level control of gene expression. During the lytic reproduction cycle of lambda, transcription

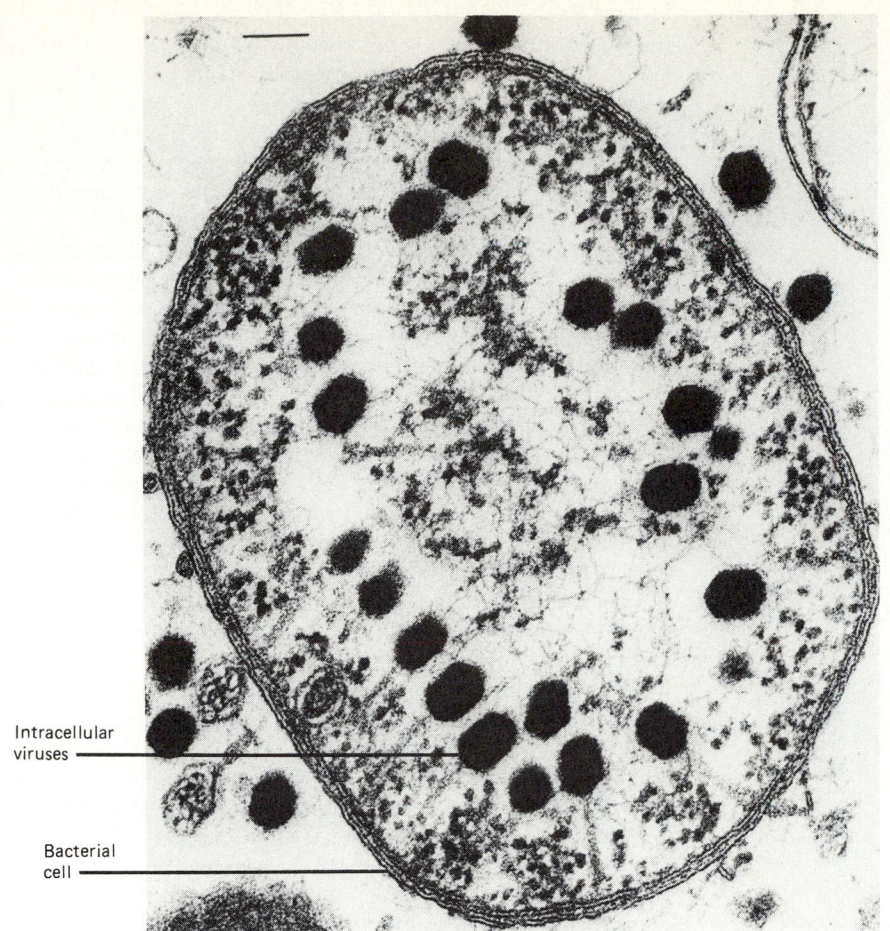

FIGURE 9.19

Assembled phage T2 within the bacterial cell of *E. coli* strain B. Note the cell envelopes of the wall and cytoplasmic membrane; the clear area of the phage DNA pool contains many condensed phage DNA cores. Three phages attached to the cell surface; one is empty, and two are still partially filled. The phage at the top shows the long tail fibers and the spikes of the tail plate in contact with the cell wall. The tail sheath is contracted, and the tail core has apparently reached the cell surface but has not penetrated it. (Courtesy Lee Simon, Rutgers, the State University)

Intracellular viruses

Bacterial cell

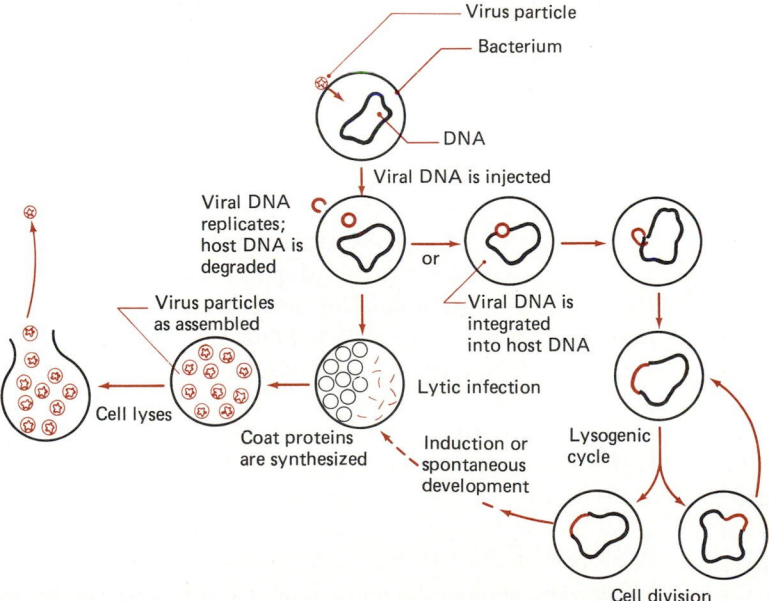

FIGURE 9.20

When a temperate bacteriophage infects, the result is usually a lysogenic cycle. The viral genome and the bacterial genome are fused enzymatically. The lysogenic cell continues growth and division normally through binary fission. When the cell is induced, the viral DNA de-integrates, and a productive or lytic infection occurs.

begins at two promoter sites during the early phase of reproduction (Figure 9.21). One of the promoter sites initiates clockwise transcription, and the other initiates counterclockwise transcription. The completion of the clockwise transcription of the phage genome requires the expression of a *Q* gene, which codes for a Q protein required for late gene expression. The complete counterclockwise transcription of the lambda genome requires an N protein. In the absence of N protein synthesis, transcription of the genes involved in the delayed early stage of phage reproduction cannot occur,

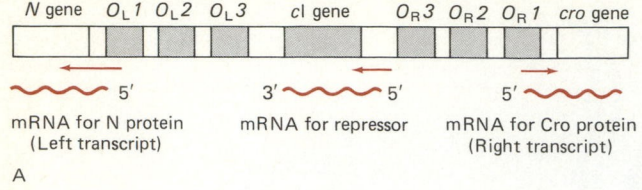

mRNA for N protein
(Left transcript)

mRNA for repressor

mRNA for Cro protein
(Right transcript)

A

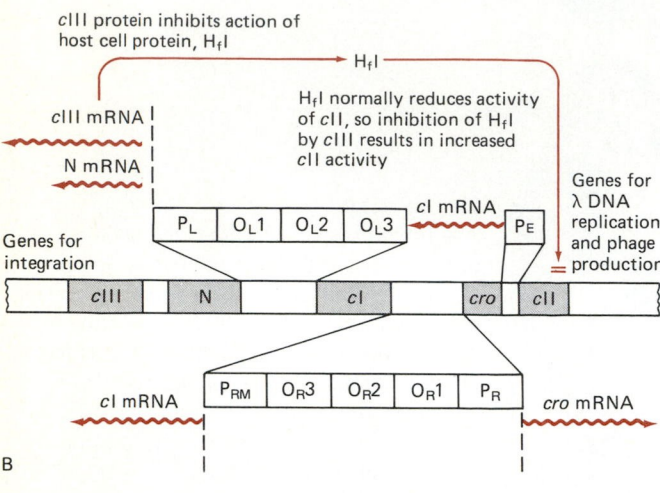

B

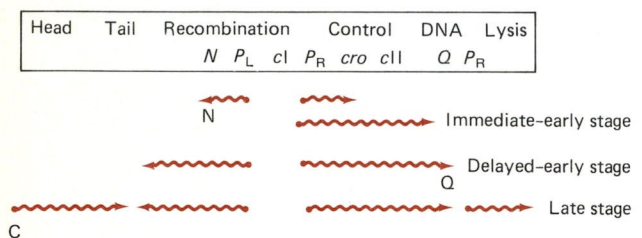

C

FIGURE 9.21

(A) Diagram of lambda genome, showing promoter sites. The left (O_L) and right (O_R) operator regions and the adjacent genes O_L1 and O_R1 have the highest affinity for the λ repressor. The repressor gene is cl. The cl gene product is a repressor protein that regulates transcription at two key promoters (P_L and P_R), and in so doing plays the central role in suppressing lytic growth of bacteriophage λ. The left transcript starts with the N gene, whereas the right transcript starts with the cro gene. When λ infects a host cell, no cl repressor is present, and therefore unregulated transcription begins at both the P_L and P_R promoter sites. The products of this early transcription are mRNAs for the N and Cro proteins. If enough cl repressor is synthesized just after λ enters a host cell, the cell is pushed into lysogeny; if this synthesis is interrupted and the supply of repressor diminishes, lytic growth can begin. There are three stages of transcription of the λ phage. These stages either can lead to lysogeny or to lytic growth. (B) In the first stage that leads to lysogeny, transcription of both the left and right promoters occurs, leading to an increase in the production of N and Cro proteins. The most important effect is that the antiterminator N protein allows the transcription of the cII and cIII genes, as well as genes to the left of cIII and genes to the right of cII. This leads to the second stage, in which concentrations of cl (λ repressor), cII, and cIII increase. The promoter P_E, whose function is not understood, is stimulated by cII. Host cell protein (H_fL) is inhibited by cIII. Additionally, there is an increase in the concentrations of integration enzymes. Thus, the critical aspects of the second stage are the marked increase in cl and integration enzymes. In the third stage—when lysogeny occurs—cl dominates, leading to decreases in N, integration enzymes, and Cro. (C) The lytic growth of λ phage also involves three stages of development based on differential regulation of transcription. In this process, cl does not dominate. The N protein formed in the immediate-early stage activates the delayed-early stage, in which protein Q is formed. In turn, the Q protein activates the late stage, in which head and tail proteins are made and lysis eventually is activated.

and therefore, the replication of lambda phage DNA cannot proceed. Thus, both the N and Q genes must be expressed for the transcription of the complete gene.

The expression of both the N and Q proteins, however, can be repressed. The lambda phage genome contains a cl gene that codes for a repressor protein that binds to the operator region of the phage genome, which then controls the expression of the N protein, blocking the lytic reproductive cycle. The repressor protein also binds to another operator region, blocking the clockwise transcription of the lambda phage DNA and, thus, the production of the Q protein. In lysogeny only the repressor protein preventing the lytic reproduction of the phage is produced; consequently, only the cl gene of the prophage is expressed. The expression of the cl gene is itself subject to regulation. Transcription of cl increases when the lambda repressor is present in low concentrations and high concentrations of repressor protein inhibit further transcription. If the concentration of lambda repressor protein declines sufficiently to permit further tran-

scription of the phage genome, a protein, coded for by the cro gene, is produced. The Cro protein represses further transcription of the cl gene, thus stopping synthesis of the repressor protein responsible for preventing complete expression of the phage genome. When this occurs, the phage can carry out a lytic reproduction cycle. The reproductive cycle of bacteriophage is one of the few processes for which details of the molecular-level events controlling development have been determined.

Replication of Plant Viruses

Many plant viruses exhibit a reproductive cycle similar to the lytic reproduction cycle of bacteriophage, involving penetration by the virus of a susceptible plant cell (generally through abrasions or aphid bites); release of the viral nucleic acid within the plant cell; assumption by the viral genome of control of the synthetic activities of the host cell; coding by the viral genome for the synthesis of viral nucleic acid and capsid components; assembly of the viral particles within the host cell; and finally, release of the complete viral particles from the host plant cell. In contrast to bacteriophage, both the capsid and the nucleic acid core of viruses infecting eukaryotic cells may cross the cytoplasmic membrane by endocytosis, with release of the nucleic acid from the capsid occurring within the host cell.

As an example of the reproductive cycle of a plant virus, let us examine tobacco mosaic virus (TMV). TMV has a single-stranded RNA genome contained within a helical array of protein subunits that comprise the viral capsid. Replication of TMV occurs within the cytoplasm of the infected cell. The RNA genome of TMV codes for an RNA-dependent RNA replicase that is used for the synthesis of a complementary RNA (minus strand) to serve as a template for the synthesis of the RNA genome (plus strand) of TMV. The complementary RNA (minus strand) also acts as a template for the synthesis of mRNA, which is subsequently translated at the plant cell ribosomes for the production of the protein coat subunits. Once the RNA and protein components of TMV are synthesized, the assembly of the protein coat around the central RNA core can proceed spontaneously; that is, TMV is self-assembled.

The initiation of TMV assembly involves the attachment of the viral RNA to a protein disk subassembly of the core structure (Figure 9.22). The TMV RNA is capable of binding with amino acids to initiate the assembly of the virus. The RNA molecule forms a loop, and the protein disk subunits are added continuously to the looped end of the RNA. The RNA overcomes the electrostatic forces that would prevent binding of the protein subunits, and without the RNA the protein subunits will not bind together at physiological pH and low ionic strength. Thus, in the absence of the RNA core, the protein units will not assemble, whereas with the RNA, self-assembly occurs.

Within infected plant cells, the replicated TMV particles form cytoplasmic inclusions (Figure 9.23). These viral inclusions are crystalline in nature. The chloroplast of a TMV-infected leaf becomes chlorotic (yellow due to loss of chlorophyll), leading to the death of the cell. The death of the plant cell releases completely assembled TMV particles and viral nucleic acid that has not been packaged with the protein subunits. Within plants, both completely assembled viral particles and viral RNA can move from one cell to another, establishing new sites of infection. As a consequence of the viral reproduction within the plant cells, the plant develops characteristic disease symptoms including the appearance of a mosaic pattern of chlorotic spots on the leaves that gives both the disease and the virus their names.

Replication of Animal Viruses

There are many types of animal viruses and many variations in the details of their reproduction. In some cases, the reproductive cycle of animal viruses closely resembles that of lytic bacteriophage, and in such instances there is a stepwise

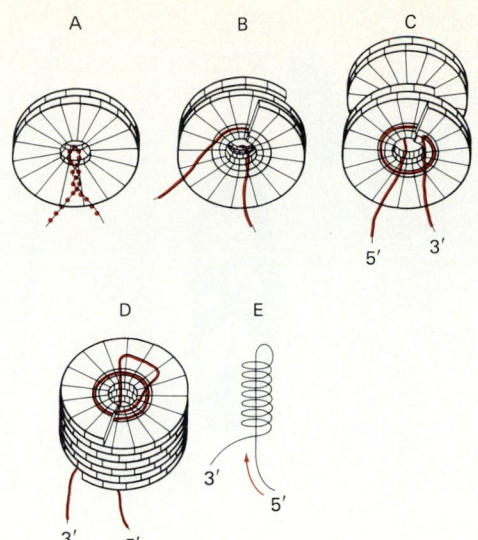

FIGURE 9.22

Diagram showing the self-assembly of TMV. (A) The initiation region of the RNA loops into the central hole of the protein disc and transforms it into (B) the helical lock-washer form; (C) additional discs add to the looped end of the RNA; and (D) one of the RNA tails is continually pulled through the central hole to interact with incoming discs. (E) A schematic diagram of RNA in a partially assembled virus. (After P. J. G. Butler and A. Klug, 1978, *Scientific American*, 239(5):68–69)

growth curve, with a burst of a large number of viruses released simultaneously. Unlike bacteriophage, however, the single-step growth curve for viruses occurs within hours rather than minutes (Figure 9.24). Though many viruses exhibit single-step growth curves characterized by the death of the host cell and the simultaneous release of a large number of viruses, some animal viruses characteristically do not kill the host cell and instead reproduce with a gradual, slow release of intact viruses. Additionally, some animal viruses transform the host cells, resulting in tumor formation rather than death of the host cells.

The essential steps of the reproductive cycle for animal viruses are (1) attachment, the adsorption of the virus to the surface of the animal cell; (2) penetration, the entry of the intact virus or viral genome into the host cell; (3) uncoating, the release of the viral genome from the capsid; (4) transcription to form viral mRNA; (5) translation using viral-coded mRNA to form early proteins; (6) replication of viral nucleic acid to form new viral genomes; (7) translation of mRNA to form late proteins needed for structural and other functions; (8) assembly of complete viral particles; and (9) release of new viruses (Figure 9.25). Viruses appear to adsorb onto specific receptor sites on animal cell surfaces, and as a rule, the entire viral particle enters the cell, often by endocytosis. Within the host cell, uncoating of animal viruses varies from one virus to another (Figure 9.26). The viral nucleic acids may be released at the cytoplasmic membrane, as occurs in enteroviruses, which are single-stranded

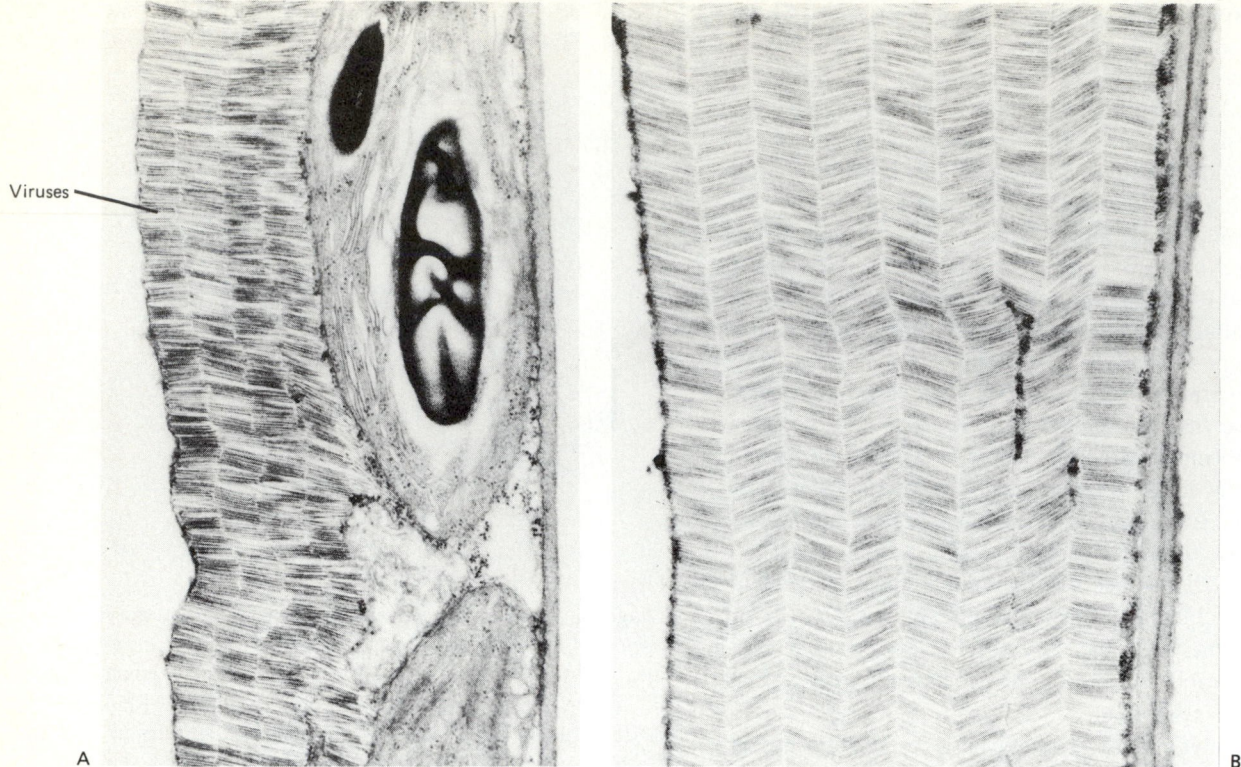

FIGURE 9.23

Electron micrograph of a thin section of a tobacco leaf infected with TMV. (A) Note the chloroplast at the bottom and the section of the viral crystal showing alignment of multiple viral particles. (B) Note the herringbone pattern of the three-dimensional array of viral particles. (Courtesy Harry E. Warmke, University of Florida)

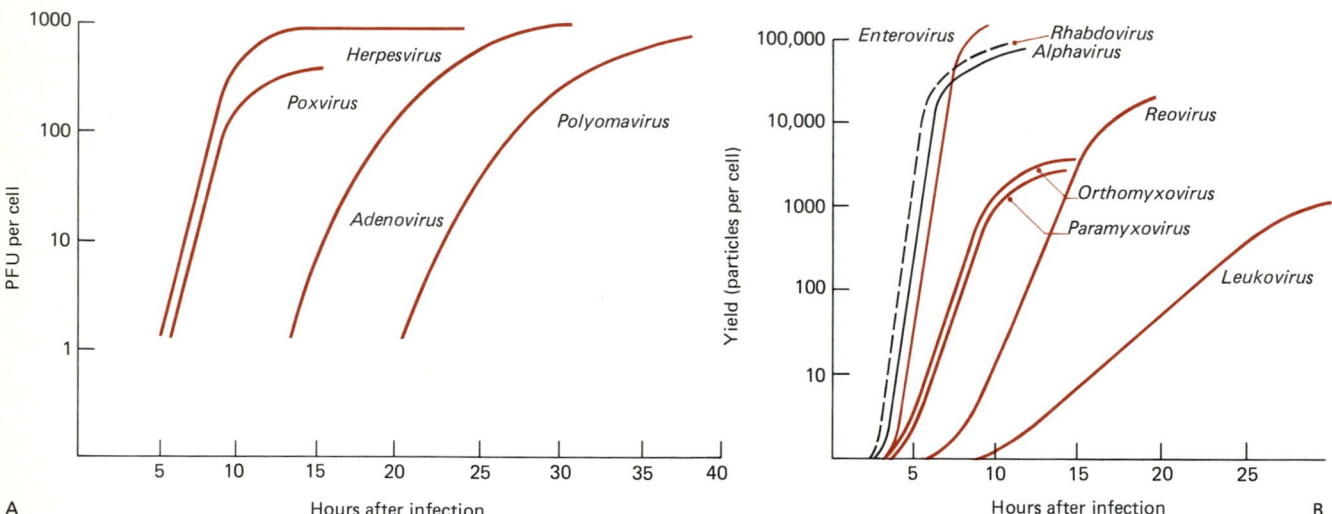

FIGURE 9.24

Growth curves for various DNA and RNA viruses. The number of viruses is determined by counting the number of plaques (areas of no cells because of lysis) that form in a culture of animal cells; the number of viruses is recorded as plaque-forming units (PFUs). (A) Idealized multiplication curves (total infectious viruses per cell at intervals after infection at high multiplicity) of viruses representing the major genera of DNA viruses. (B) Idealized multiplication curves (total virus particles per cell at intervals after infection at high multiplicity) of viruses representing the major genera of RNA viruses (Enterovirus: poliovirus type 1; Alphavirus: Sindbis virus; Rhabdovirus: vesicular stomatitis virus; Orthomyxovirus: influenza type A; Paramyxovirus: Newcastle disease virus; Reovirus: reovirus type 3; Leukovirus: RAV-1). Latent periods, multiplication rates, and final yields are all affected by the species and strain of the virus and by the cell strain and the kind of culture; the latent period and multiplication rate, but not the yield, are affected by the multiplicity of infection.

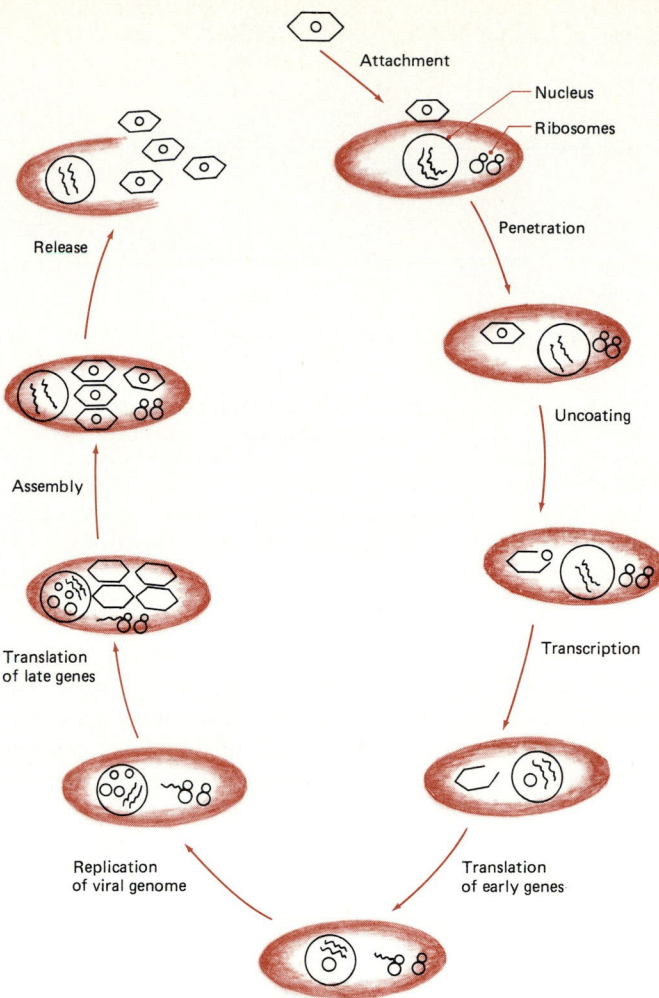

FIGURE 9.25

Animal viral reproduction involves attachment, penetration, uncoating, transcription of viral nucleic acid, translation of early genes, replication of the viral genome, translation of late genes, assembly, and release.

RNA viruses; the virus may be uncoated in a series of complex steps within the host cell, as occurs in poxviruses, which are large, double-stranded DNA viruses; or the virus may never be completely uncoated, as occurs in reoviruses, which are large, double-stranded RNA viruses. After uncoating, the genome of a DNA animal virus generally enters the nucleus, where it is replicated, whereas the genome of most RNA animal viruses need only enter the cytoplasm of the animal cell to be replicated. The diversity and complexity of animal virus reproduction make it impossible to go into further detail. Instead, a few representative examples of the reproduction of different types of animal viruses will be discussed.

Replication of DNA Animal Viruses   In the replication of Adenovirus, a representative DNA animal virus, the host cell continues its normal metabolic activities for a short period of time after the virus enters it. Uncoating of the virus takes several hours, and during this period the viral nucleic acid is released from the capsid, entering the nucleus possibly through a nuclear pore. Within the nucleus, the viral genome codes for the inhibition of normal host cell synthesis of macromolecules. The viral genome also acts as a template for its own replication. Viral genes are transcribed, the resulting mRNA is translated, and the proteins and enzymes needed for the assembly of the viral capsid are produced, with synthesis of viral proteins occurring at the ribosomes within the cytoplasm. The assembly of the Adenovirus particles occurs within the nucleus, and therefore, the nucleus of an infected animal cell contains inclusion bodies consisting of crystalline arrays of densely packed Adenovirus particles (Figure 9.27). Upon lysis of the host cell, numerous Adenovirus particles are released.

Replication of RNA Animal Viruses   RNA animal viruses exhibit many diverse strategies for reproduction. In the case of poliovirus, the RNA genome of the virus acts as an mRNA

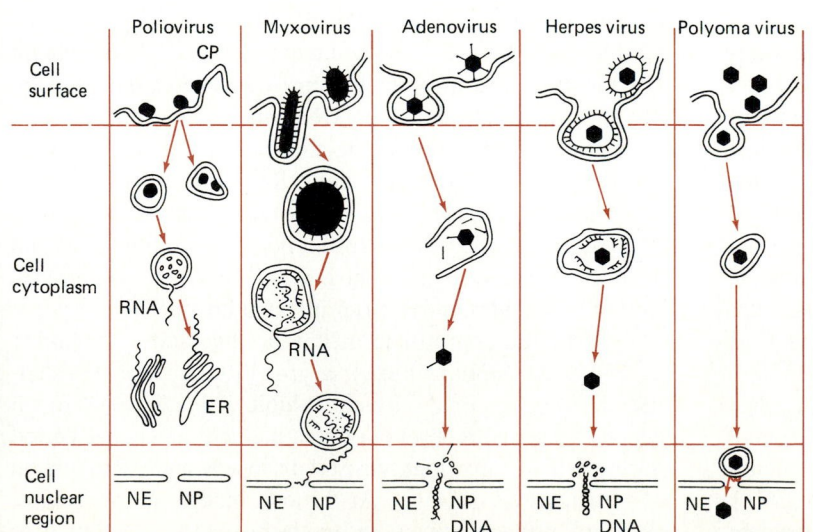

FIGURE 9.26

Some of the mechanisms associated with the attachment, penetration, and uncoating of certain animal viruses, as observed in the electron microscope from thin sections of infected cells sampled over known time courses. NE = nuclear envelope, NP = nuclear pore, CP = cell plasmalemma, ER = endoplasmic reticulum. (Based on S. Dales, 1973, *Bacteriological Reviews*, 37:103–135)

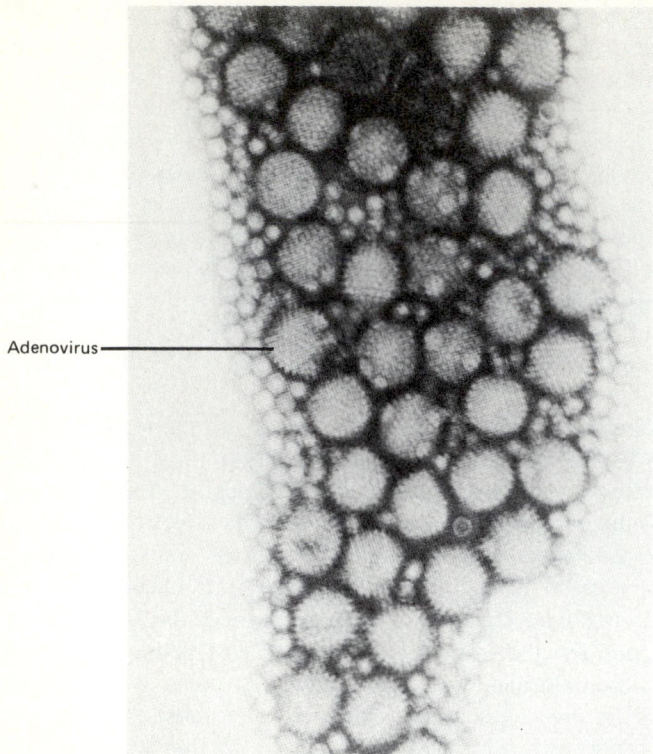

FIGURE 9.27
Micrograph of Adenovirus type 18 within the host cell
(300,000×). (From BPS—C. Garon and J. Rose, National
Institutes of Health)

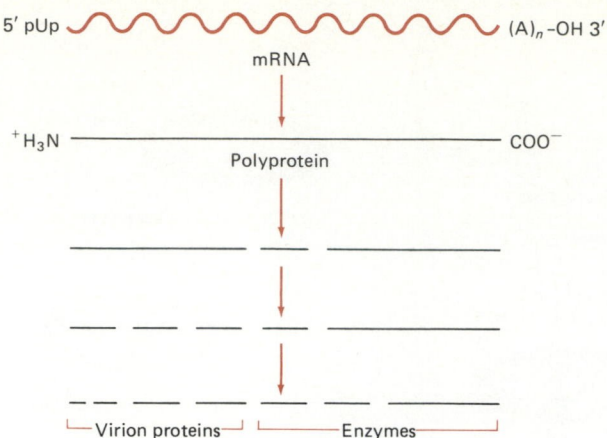

FIGURE 9.28
Large precursor poliovirus proteins are processed following
synthesis by multiple cleavages to produce the poliovirus
protein.

upon entering the host cell, coding for the production of
capsid proteins and an RNA-dependent RNA polymerase.
Viral RNA that can serve as mRNA is called *plus-strand
RNA*. Interestingly, the poliovirus RNA codes for a very
large polypeptide chain, which is cleaved by proteases, encoded both by the virus and the host cell, to form many
different proteins, including the RNA-dependent RNA
polymerase and four proteins of the viral capsid (Figure
9.28). The RNA polymerase is used to produce a complementary **replicative RNA strand** that can act as a template
for the synthesis of new viral genomes. The morphogenesis
of the viral particle, that is, the assembly of the capsid and
insertion of the RNA genome, is followed by the release of
a large number of viral particles. Release of the poliovirus
occurs because blockage of cellular protein synthesis by the
poliovirus leads to breakdown of lysosomes, with the release
of digestive enzymes from the lysosomes causing cell lysis.

In the case of influenza viruses, the viral RNA genome
serves as a template for the transcription of mRNA molecules
rather than acting as the mRNA, as occurs in poliovirus. The
genome of the influenza viruses is a **segmented genome**,
that is, it is composed of several separate RNA molecules.
Actually, the influenza genome consists of eight different
RNA molecules, each of which codes for a different monocistronic mRNA molecule. For example, one of the RNA
genome segments specifically codes for the RNA-dependent

RNA polymerase required for transcription of the viral genome. The replication of the viral RNA genome involves the
production of a complementary RNA strand that then serves
as a template for the synthesis of new viral RNA genome
molecules. The maturation of some viruses, such as influenza
virus, occurs by **budding**, a process in which the viral particles are wrapped within a piece of the cytoplasmic membrane of the host cell (Figure 9.29). As a result, budding
slowly releases encapsulated (lipid-enveloped) influenza viruses from infected host cells.

Reovirus, a double-stranded RNA virus, carries an RNA
polymerase that is used for the synthesis of new viral genome
molecules. The reovirus genome is segmented, containing 10
different double-stranded RNA genome molecules. Each of
these molecules codes for the production of a different protein. The proteins are then assembled into the viral capsid,
and the RNA genome is inserted prior to release of the completed reoviruses.

Retroviruses are RNA viruses that use a reverse transcriptase to produce a DNA molecule within the host cell. The
production of the DNA molecule requires an RNA-dependent
DNA polymerase in order to carry out reverse transcription
of the viral RNA. It is the DNA molecule "transcribed"
from the viral RNA genome that actually codes for viral
replication within the host cell. Retroviruses are released
from host cells by budding, and viral replication is nondes-
tructive to the host cell because it does not result in lysis and
death of the infected cell. Thus, these viruses can be released
slowly and continuously from infected host cells.

By using an appropriate method to quantitate the number
of infectious animal viruses, a growth curve can be established. Many animal viruses exhibit a single-step growth
curve for normal reproduction that includes an eclipse period
during which infectious viruses disappear and reproduction
of viral particles occurs. At the end of the eclipse period,
new viral progeny appear within the host cell, but often there

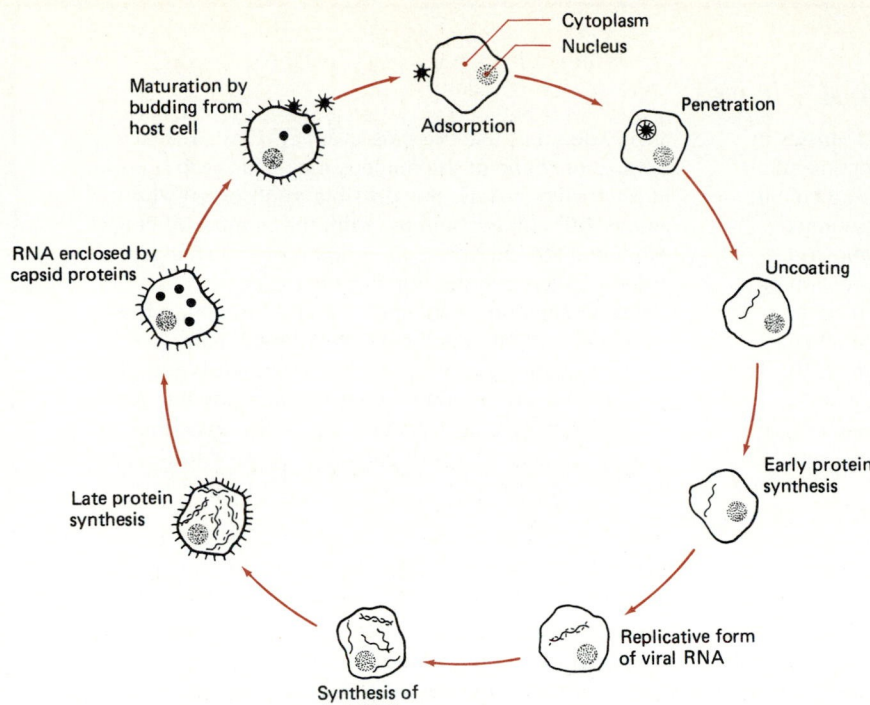

FIGURE 9.29
The life cycle of the influenza virus, showing budding release. The replication of the influenza virus illustrates many of the features that characterize animal virus interaction with host cells. Replication of this virus does not cause lysis of the cell. The virus is taken into the cell intact; it is then uncoated, releasing the single-stranded RNA genome in eight segments (only one of which is shown here). A replicative form (Rf = shaded area) is made for each segment. A replicative intermediate, in turn, serves as the template for the synthesis of new genomes. As the late proteins are synthesized, they come to lie just beneath the cytoplasmic membrane of the host cell. Virions are released by budding through the membrane. A segment of the host cell's membrane becomes their envelope.

is a further delay before they are released, except for viruses released by budding. Thus, the latent period, the time from the adsorption of the virus onto the host cell until the release of new viruses, generally exceeds the eclipse period.

Transformation of Animal Cells   The DNA produced during the replication of retroviruses, as well as the DNA of some other viruses—such as herpes viruses, can also be incorporated into the host cell's chromosomes.[6] Within the chromosomes of the host cell the viral genome can be transcribed, resulting in the production of virus-specific RNA and viral proteins. The DNA coded for by the virus, which is incorporated into the host cell genome, can be passed from one generation of animal cells to another. It is, therefore, possible for animals to inherit a viral genome. The presence of virus-derived DNA within the host cell can transform the animal cell.

Transformed cells, which are produced *in vitro*, have altered surface properties and continue to grow even when they contact a neighboring cell. *In vitro* infections that result in virus-derived DNA can result in the formation of a tumor. Viruses that transform cells and cause cancerous growth are called **oncogenic viruses**. These viruses are able to replicate in permissive hosts, but not in nonpermissive host cells; rather, in nonpermissive host cells, part of the viral genome, is incorporated into the host cell genome resulting in the transformation of the host cell. Several different retroviruses produce malignancies within infected cells. Rous sarcoma virus, for example, is an RNA tumor retrovirus that causes malignancies in chickens. Some DNA viruses, such as Simian virus 40 (SV40) and Polyomavirus, also are capable of transforming host cells and producing malignant tumors. At least one not uncommon form of malignancy, cervical cancer, may result from transformation of cells by certain papillomaviruses.

9.3 *Classification of Viruses*

There are several problems in classifying viruses. Viruses are not included in the recognized kingdoms of living organisms. Phylogenetic relationships among the viruses are unclear, and establishing a hierarchy of taxonomic relationships adds little conceptually to viral systematics.

Viruses have generally been classified nonsystematically, and their names generally are based on the disease they cause. Formal systems for the classification and nomencla-

ture of viruses are relatively new, and an International Committee on Nomenclature of Viruses is still working to establish an acceptable formal classification system that will gain wide acceptance. Most viruses are still referred to by their common names, rather than by the proper name assigned in a formal classification system. For example, we commonly refer to the measles virus, the poliovirus, and others, even though these viruses have been classified and given other formal names.

---

[6]This process is analogous to lysogeny in bacterial cells.

ANALYTICAL PROCESS

Quantitative Assays for Animal Viruses

It is possible to assay quantitatively for animal viruses in a method analogous to the plaque assay for enumeration of bacteriophage. In a typical procedure, a tissue culture monolayer of animal cells growing on a plate surface is inoculated with dilutions of a viral suspension and incubated for various periods of time. Viral infection of the animal tissue culture cells may result in plaque formation, indicative of localized death of animal cells, which can be observed microscopically (Figure 9.30), or, more commonly, with the naked eye. The number of plaques that form and the dilution factors are employed to determine the concentration of viruses in the sample.

Additionally, virus-infected animal cells often develop abnormally, which is visible as a change in their appear-ance, known as the **cytopathic effect (CPE)**. For example, the size or shape of the nucleus may change in a cell infected with a virus. It is possible to observe CPE in animal cell cultures and to count the number of cells exhibiting the characteristic morphological changes in order to determine the number of viruses. Hemagglutination, involving a reaction in which red blood cells clump together or agglutinate, can also be used for quantitating viruses. By mixing red blood cells with dilutions of a viral suspension, it is possible to determine the greatest dilution that still brings about hemagglutination. This dilution can be used to determine the concentration of viruses in the viral suspension.

FIGURE 9.30

Micrographs showing CPE in animal cells growing in tissue culture due to viral infection. (A) Normal, unstained bovine embryonic kidney cells (BEK, 40×); (B) CPE of infectious bovine rhinotracheitis virus (bovine herpesvirus 1) in monolayers of BEK cells 48 hours after inoculation (unstained, 40×). (Reprinted by permission of Lea & Febiger, Philadelphia, from S. B. Mohanty and S. K. Dutta, 1981, *Veterinary Virology*)

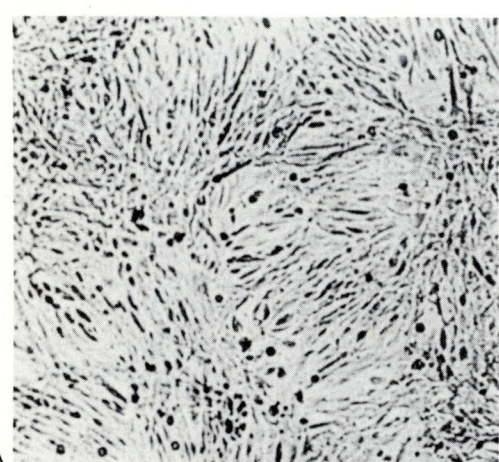
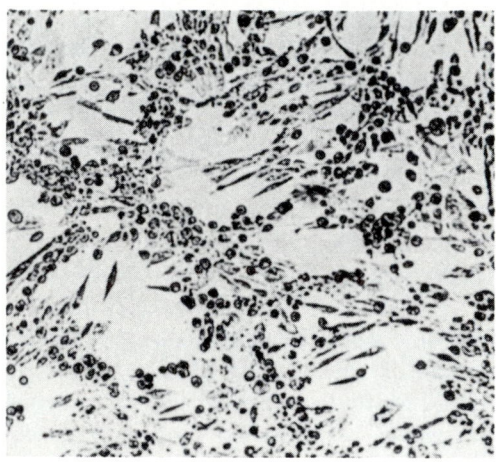

Viruses can be separated into groups on the basis of the type and form of the nucleic acid genome and the size, shape, structure, and mode of replication of the virus particle. Within each group, subgroupings can be based on immunological properties, that is, on the specific biochemicals (antigens) associated with the virus that evoke an immune response. The fact that viruses exhibit great host-cell specificity probably indicates that most viruses infect only one type of cell. As such, it is probably legitimate to consider the host cell in which the virus is capable of reproducing as a pertinent taxonomic criterion for classifying viruses. Viruses that infect different types of cells are usually classified separately. Three large groups of viruses have usually been recognized: **animal viruses** (viruses that replicate within animal cells), **plant viruses** (viruses that replicate within plant cells), and **bacteriophage** (viruses that replicate within bacterial cells).

Animal viruses traditionally refer to those viruses that infect only vertebrate animals. Viruses that infect invertebrate animals and microorganisms other than bacteria are not covered in the classification systems that rely on these three lines of division. As a result, separate taxonomic consideration has been given to the insect viruses and to the viruses that infect fungi, protozoa, and algae. Viral taxonomists working with viruses that infect different types of host cells have developed separate taxonomic schemes for classifying the viruses. Various subcommittees of the International Committee on Nomenclature of Viruses have been working to develop classification systems for each of these viral groups, with separate subcommittees developing criteria for classifying animal viruses, plant viruses, and viruses of microorganisms. The result is that a unified system has yet to be developed.

While not unified, the classification systems applied to plant, animal, and microbial viruses are based largely on the nature of the nucleic acid molecule and the arrangement of the capsid (Table 9.1). Critical questions include whether the virus is a DNA or RNA virus, whether the nucleic acid is single-stranded or double-stranded, and whether the virus is naked or enveloped. Important features used in classifying viruses include the molecular weight of the nucleic acid, the symmetry of the capsid, the number of capsomers, and the site of capsid assembly. Nomenclature of animal viruses is based largely on molecular considerations. Similarly, bacteriophage are characterized by their molecular structure. Plant viruses are still grouped largely on the basis of the host cells infected by the virus.

In the following section, brief descriptions of the major groups of viruses are given.

Animal Viruses

DNA Viruses   The DNA animal viruses are divided into six major families that replicate in vertebrate animal cells (Figure 9.31) and one family that replicates in invertebrate animal cells, according to the criteria established by the International Committee on Taxonomy of Viruses.

Parvoviridae. The parvoviruses are small viruses that contain single-stranded DNA. They have cubical symmetry with 32 capsomers and have no envelope. Replication and capsid assembly take place in the nucleus of the host cell. Members of the genus *Parvovirus* are resistant to heat inactivation and can withstand exposure to 60°C for 30 minutes. This group includes some interesting ''defective'' viruses that cannot multiply alone but can reproduce with the aid of another virus. For example, *Adeno-associated* satellite viruses can multiply only in the presence of replicating adenoviruses.

Papovaviridae. The papovaviruses are small viruses containing double-stranded, circular DNA. They have cubical symmetry with 72 capsomers. Papovaviruses are assembled within the nucleus of the host cell. This group has been recognized in the viral family Papovaviridae and causes tumors in vertebrate animals. One genus included in this group is *Papillomavirus* (wart virus), which causes human warts and may be associated with cervical cancer. The other genus in this group, *Polyomavirus*, contains some viruses that cause tumor development.

Adenoviridae. Adenoviruses are medium-sized viruses containing double-stranded DNA. They exhibit cubic symmetry and have 252 capsomers. Adenoviruses normally have

TABLE 9.1   The Characteristics of Various Families of Viruses

Viridae	Nucleic Acid	Symmetry	Nucleocapsid	Virions with Cubic Symmetry		Diameter of the Envelope (nm)	Virions with Helical Symmetry		Mol wt (× 10⁶) of Nucleic Acid	Numer of Nucleic Acid Strands
				Number of Capsomers	Diameter of the Nucleocapsid (nm)		Diameter and Length of the Nucleocapsid (nm)	Dimensions of the Enveloped Viruses (nm)		
Ino-	D	H	N				0.5×85		1.7–3	1
Pox-	D	H?	E				?	250×160 300×230	160–240	2
Micro-	D	C	N	12	25				1.7	1
Parvo-	D	C	N	32	22				1.8	1
Denso-	D	C	N	42	20				160–240	1
Papilloma- (papova-)	D	C	N	72	45–55				3–5	2
Adeno-	D	C	N	252	70				20–25	2
Irido-	D	C	N	812	130				126	2
Herpes	D	C	E	162	77	150–200			54–92	2
Uro-	D	BC	N							2
Rhabdo-	R	H	N				2×13 1×125			1
Myxo-	R	H	E				$9 \times ?$	100	2–3	1
Paramyxo-	R	H	E				$18 \times ?$	120	7.5	1
Stomato- (rhabdo-)	R	H	E				$18 \times ?$	175×68	6	1
Thylaxo-	R	H	E				?	1000	10	1
Napo-	R	C	N	32	22–27				1.1–2	1
Reo-	R	C	N	92	70				10	2
Cyano-	R	C	N	32 or 42	54					2
Encephalo-	R	C	E	?	?		60–80		2–3	1

Based on A. Lwoff and P. Tournier. 1971. Remarks on the classification of viruses. In *Comparative Virology*, K. Maramorosch and E. Kurstak, eds., Academic Press, New York.

Code: D = DNA; R = RNA; H = helical; C = cubic; B = binal; N = naked; E = enveloped.

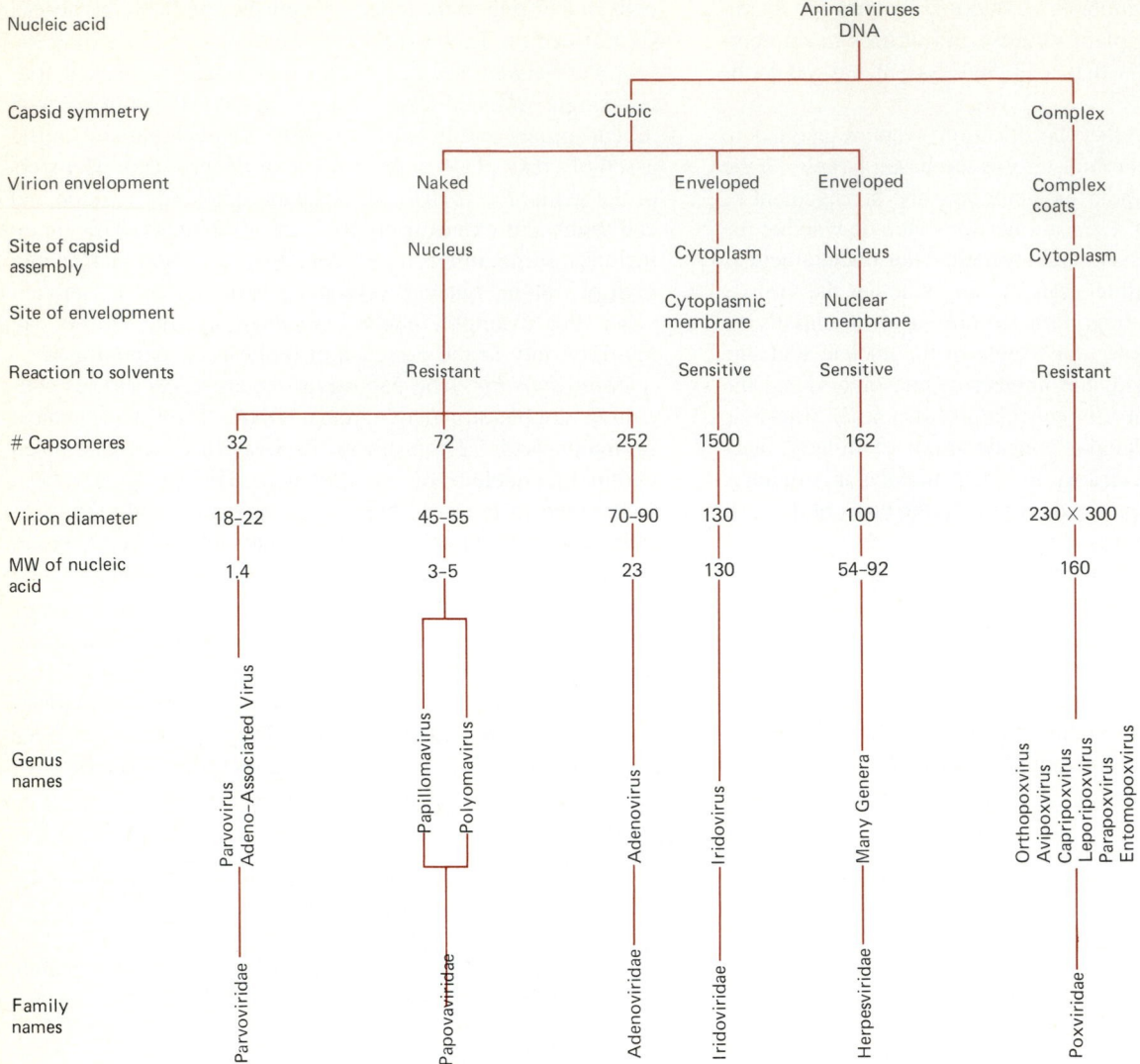

FIGURE 9.31

The families and genera of the DNA viruses of vertebrates. (Reprinted by permission of CRC Press, Inc., Cleveland, Ohio, from A. I. Laskin and H. A. Lechevalier, eds., 1978, *Handbook of Microbiology*, p. 625)

spikes projecting from the capsid that give these viruses a characteristic shape (Figure 9.32). The spikes are involved in the adsorption of the virus to the host cell. The adenoviruses multiply within the nuclei of infected cells, where they produce an array of crystalline particles. Several viruses in this group are able to reproduce within human cells, causing disease. Members of the genus *Adenovirus*, for example, are associated with acute respiratory tract infections.

Iridoviridae. The iridoviruses may be divided into mammalian, fish-amphibian, and insect genera. Members of the genus *Iridovirus* contain double-stranded DNA. The viruses are isometric, and some are enveloped. These are among the largest viruses.

Herpesviridae. The herpesviruses are medium-sized viruses containing linear, double-stranded DNA. The nucleocapsid is about 190 nm in diameter. The capsid has cubical symmetry with 162 capsomers. Herpesviruses are assembled in the nucleus and are surrounded by a lipid envelope. They acquire membrane lipids by budding through the nuclear membrane. Herpesviruses establish latent infections within host animals that can last for the entire life of the host.

Herpesviruses, including *Herpes simplex* types 1 and 2, cause a number of diseases in human beings. The *Herpes simplex* viruses can cause inflammations of the mouth and gums, known as *gingivostomatitis*, and one form of sexually transmitted disease is caused by a strain of *Herpes simplex* virus. The Epstein-Barr virus, the causative agent in infectious mononucleosis, and the *Varicella-Zoster* virus, the causative agent in chicken pox and shingles, are also included among the herpesviruses.

Poxviridae. Poxviruses are large viruses containing double-stranded DNA. Poxviruses may also include enzymes, such as RNA polymerase, within the viral particle. The inclusion of enzymes within the virus particle appears to be

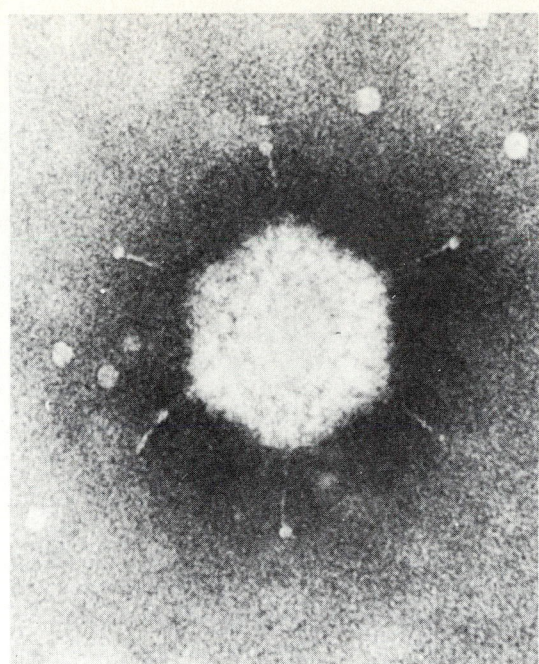

FIGURE 9.32
Micrograph of an Adenovirus, showing projecting spikes (Courtesy H. C. Pereira; reprinted by permission of Academic Press from R. C. Valentine and H. C. Pereira, 1965, *Journal of Molecular Biology*, 13:13–20)

quite unusual. The poxviruses reproduce within the cytoplasm of host cells. Many poxviruses are human pathogens that reproduce primarily within skin tissues. The formation of vesicular lesions in superficial body tissues is symptomatic of many diseases caused by poxviruses. Smallpox, cowpox, monkeypox, and fowlpox are examples of diseases caused by members of the poxvirus group.

Baculoviridae. The baculoviruses are rod-shaped viruses with lipid bilayer envelopes that frequently occur in bundles occluded in ''crystalline'' protein bodies. The genomes of these viruses are circular, double-stranded DNA macromolecules. These viruses occur in insects, spiders, and crustaceans. The members of the nuclear polyhedrosis group are frequently multiply enveloped and always have many virions in each occlusion body; the granulosis virions occur singly.

RNA Viruses The RNA animal viruses are divided into 11 major families that replicate in vertebrate animal cells (Figure 9.33); there is an additional family of viruses that replicate within insect cells and a newly proposed family of fish viruses.

Picornaviridae. The picornaviruses are small, single-stranded RNA viruses. The nucleocapsid has cubical symmetry and is nonencapsulated. Maturation of the picornaviruses occurs in the cytoplasm of the host cell. Two genera within the Picornaviridae, *Enterovirus* and *Rhinovirus*, have members that infect humans. Species of Rhinovirus cause 25 percent of all common colds in adults. The enteroviruses include *Poliovirus*, *Echovirus*, *Hepatitisvirus*, and *Coxsack-*

ievirus. Diseases caused by members of this group include poliomyelitis, infectious hepatitis, and foot and mouth disease. Picornaviruses also cause mild infections of the gastrointestinal and respiratory tracts.

Caliciviridae. The members of the genus *Calicivirus* are isometric, single-stranded RNA viruses. This group is closely related to the Picornaviridae and is sometimes included as part of that family. Diseases caused by members of this group include respiratory disease of cats and vesicular exanthema of swine and sea lions.

Reoviridae. Members of the Reoviridae contain segmented, double-stranded RNA genomes. Three genera are included in this group: *Reovirus*, *Orbivirus*, and *Rotavirus*. Rotaviruses have been identified as an important causative agent of human infantile diarrhea. Several orbiviruses also cause important diseases, including bluetongue of sheep and cattle, epizootic hemorrhagic disease of deer, African horse sickness, and human Colorado tick fever.

Togaviridae. Many viruses in the family Togaviridae were formerly known as *arboviruses* because they are transmitted by arthropod vectors. The members of the Togavirus group are single-stranded RNA viruses that have an enveloped viral particle. Some of the diseases caused by the togaviruses include equine encephalitis, St. Louis encephalitis, dengue fever, rubella, and yellow fever.

Orthomyxoviridae. The orthomyxoviruses are single-stranded RNA viruses that exhibit helical symmetry. The viral particles are enveloped (Figure 9.34). Most orthomyxovirus particles have surface projections or spikes as part of their outer wall. The viruses that cause influenza are members of this group. Many influenza viruses are referred to by common names that indicate their geographic origins, such as *Hong Kong flu virus*.

Paramyxoviridae. The paramyxoviruses are similar to the orthomyxoviruses but generally are somewhat larger. The paramyxovirus group includes several viruses that cause diseases in humans, including mumps virus and measles virus.

Rhabdoviridae. Rhabdoviruses are single-stranded RNA viruses that have a characteristic rod shape, resembling a bullet (Figure 9.35). Members of the rhabdovirus group are known to infect vertebrate animals, insects, and plants. Rabies is probably the best-known example of a disease caused by a member of the rhabdovirus group.

Retroviridae. The common feature of the retroviruses is that they contain RNA that must be transcribed to DNA. The retroviruses carry a reverse transcriptase. They also contain tRNAs, one of which serves as the primer for the reverse transcriptase. Three subfamilies have been distinguished. One subfamily—Oncorvirinae—causes animal cell transformations and its members were formerly called *RNA tumor viruses*; they are known to be oncogenic, that is, to cause malignancies in birds and mammals, including humans. Another subfamily—the Lentivirinae—causes slowly progressive lethal diseases. This subfamily contains the human immunodeficiency viruses (HIVs), including the virus that causes acquired immunodeficiency syndrome (AIDS).

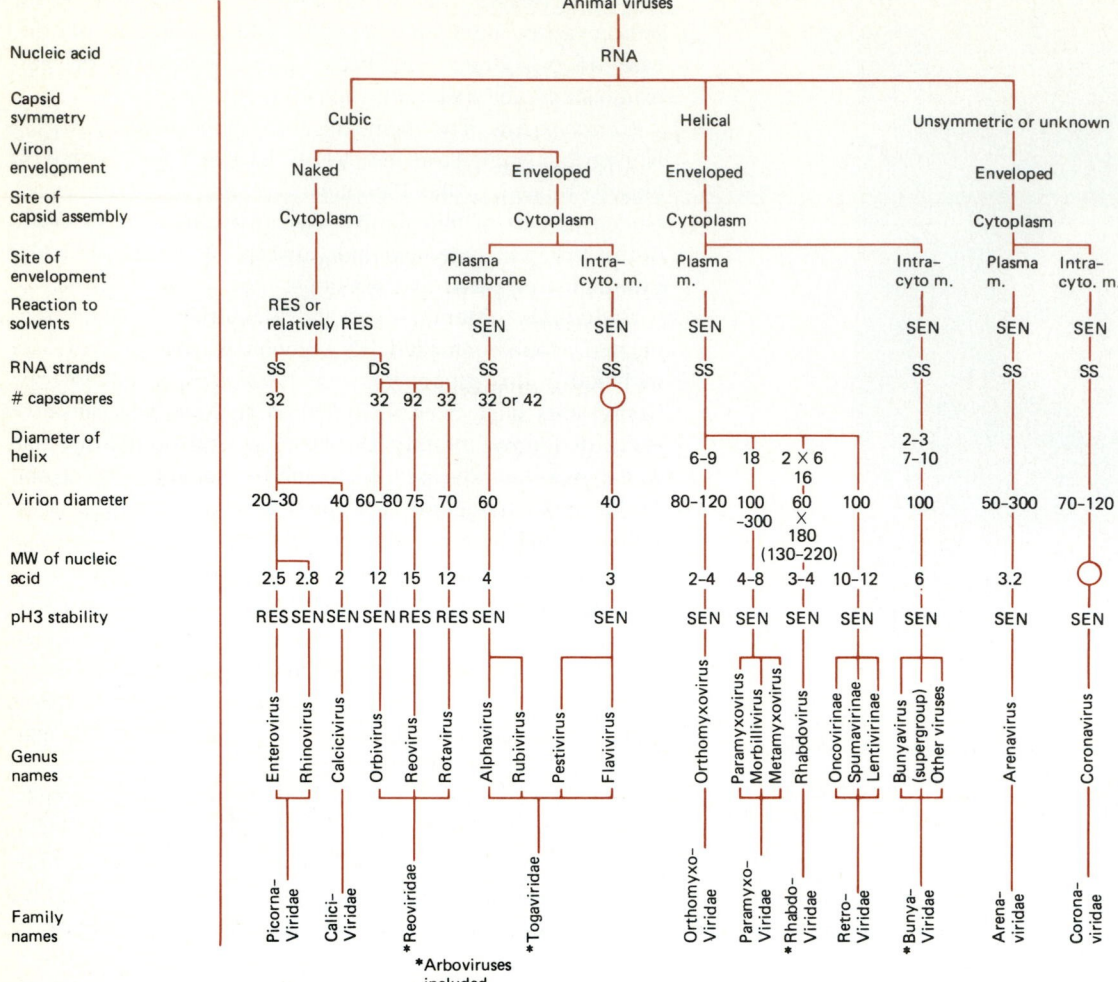

FIGURE 9.33

The families and genera of the RNA viruses of vertebrates. (Reprinted by permission of CRC Press, Inc., Cleveland, Ohio, from A. I. Laskin and H. A. Lechevalier, eds., 1978, *Handbook of Microbiology*, p. 626)

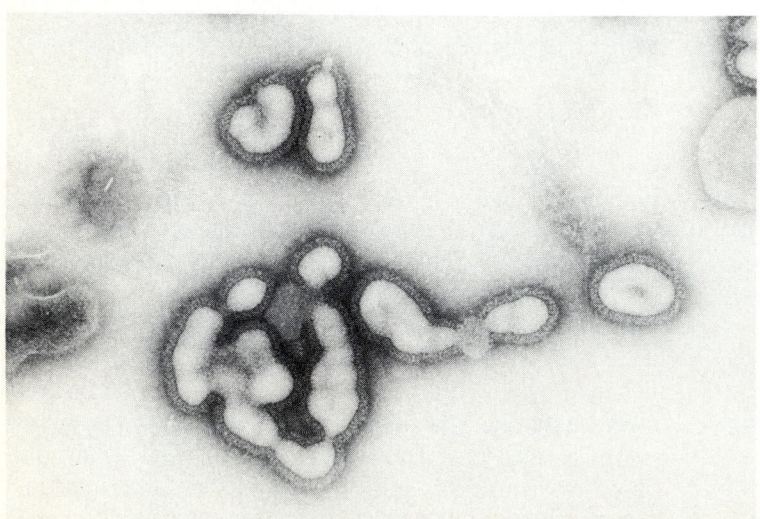

FIGURE 9.34

Micrograph of influenza A virus, strain USSR. (Courtesy E. Palmer, Centers for Disease Control, Atlanta)

ease and diarrheal diseases of humans and other animals are caused by viruses in this family.

Birnaviridae. A newly proposed family, the birnaviruses have two linear, double-stranded RNA molecules. These viruses cause diseases of fishes—including infectious pancreatic necrosis of trout, diseases of molluscs, and bursal disease of chickens.

Insect Viruses The insect viruses can be classified into a number of groups or genera. Some of these genera appear to be coincident with those assigned to viruses that infect vertebrate animal cells. For example, the entomopoxviruses are poxviruses of invertebrates. The insect viruses include the family Iridoviridae (Figure 9.36). These are very large, double-stranded DNA viruses and include *Aedes* and *Chironomus* iridescent viruses. It appears likely that some members of the iridovirus group are also able to infect vertebrate animal cells. The nodaviruses, which occur in the family Nodaviridae, have a wide host range among insects. These viruses have two plus-strand RNAs, both of which are required for infectivity.

Plant Viruses

The formal systematics of plant viruses have attempted to retain information concerning the infected host cells within the classification system. As a result, plant viruses are still primarily grouped according to the type of disease they cause. The classification of plant viruses has tended to maintain smaller groups of viruses.

The recognized groups or genera of plant viruses include *Cauliflower mosaic* virus (DNA virus of higher plants), *Tobravirus* (tobacco rattle virus), *Nepovirus* (tobacco ringspot virus), *Tobamovirus* (tobacco mosaic virus), *Tombusvirus* (tomato bushy stunt virus), *Tobacco necrosis* virus, *Bromovirus* (brome mosaic virus), *Cucumovirus* (cucumber mosaic virus), *Potyvirus* (potato virus Y), *Watermelon mosaic* virus, *Potexvirus* (potato virus X), *Tymovirus* (turnip yellow mosaic virus), *Luteovirus* (barley yellow dwarf virus), and *Ilarvirus* (necrotic ringspot). Some major properties of each of the plant virus groups are listed in Table 9.2. No attempt will be made here to describe the groups of plant viruses further. The names of the viral groups themselves are a key to the types of diseases caused by the viruses in these groups, including the nature of the host plant and the symptoms of the disease. It is obvious from this listing that the taxonomy of plant viruses has not yet established a unified, systematic scheme comparable to that of the families of animal viruses.

Viruses of Microorganisms

The viruses that multiply within the host cells of microorganisms are divided initially according to the group of microorganisms within whose cells they can reproduce, that is, into **bacteriophage** (infect bacteria), **mycoviruses** (infect

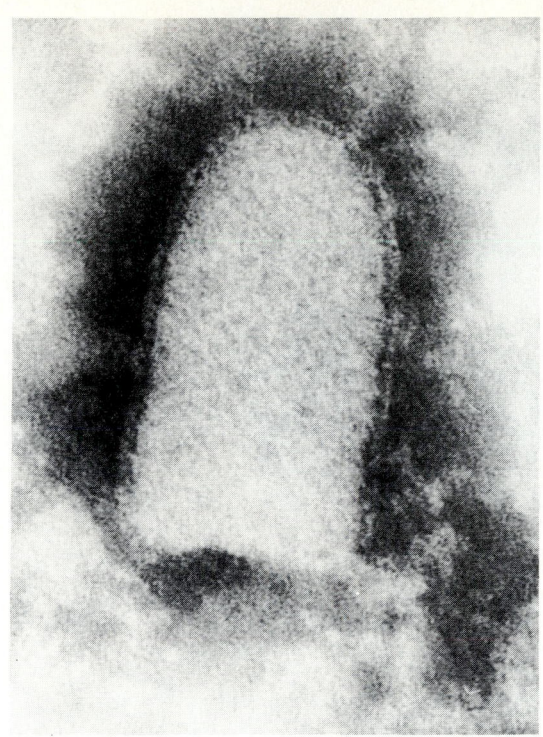

FIGURE 9.35
Micrograph of a rabies virus particle (280,000×). (From BPS— Alyne K. Harrison, Centers for Disease Control, Atlanta)

There has been considerable controversy about the proper nomenclature for the virus that causes AIDS. Researchers at the Institute Pasteur in Paris originally proposed the name *lymphadenopathy virus (LAV)*, and investigators at the United States National Institutes of Health named the virus *human T-cell lymphotrophic virus type III (HTLV-III)*. To solve the problem, the International Committee on the Taxonomy of Viruses approved the unified term *human immunodeficiency virus (HIV)* for those retroviruses that cause human immunodeficiency diseases. They also proposed that other related retroviruses with nonhuman hosts be similarly named, for example—*SIV* for *simian immunodeficiency virus* and *AGMIV* for *African green monkey immunodeficiency virus*. This newly adopted nomenclature is now being used in many scientific publications, but—at least for the moment—the older nomenclature of HTLV-III is still widely used.

Bunyaviridae. Viruses in the family Bunyaviridae contain single-stranded RNA in three linear segments. The viruses are formed by budding from organelles within the host cells, primarily from intracytoplasmic membranes of the Golgi apparatus. Rift Valley fever and Nairobi sheep disease are among the diseases caused by bunyaviruses.

Arenaviridae. The arenaviruses are single-stranded RNA viruses with the genome arranged in segments. Lassa fever is caused by members of the genus *Arenavirus*.

Coronaviridae. Members of the genus *Coronavirus* are pleomorphic, with bulbous projections from the membrane envelope that surrounds the viruses. Acute respiratory dis-

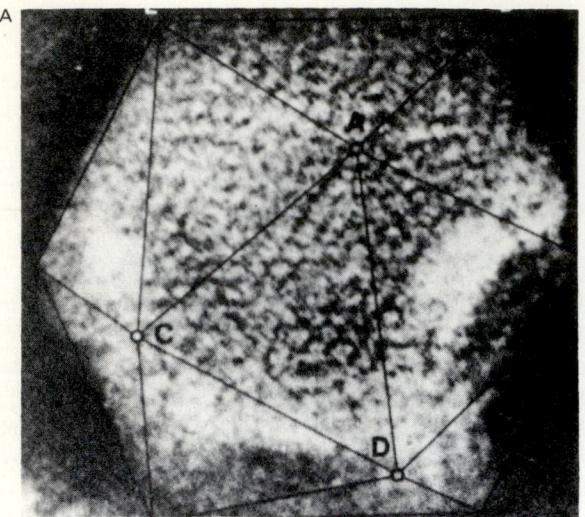
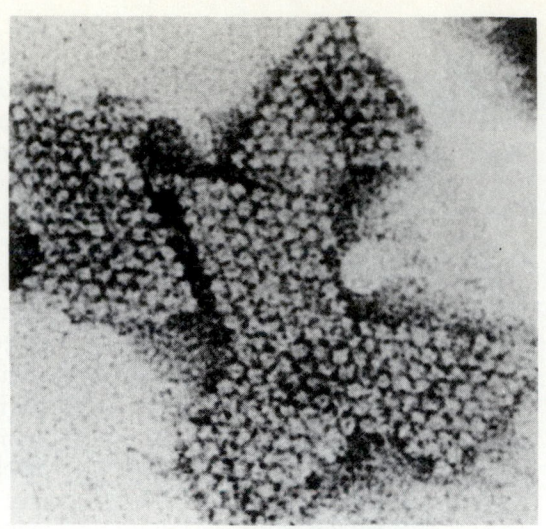

9.36
(A) The icosahedral structure of an iridovirus. (Reprinted by permission of the Society for General Microbiology, from N. G. Wrigley, 1969, *Journal of General Virology*, 5:123).
(B) The triangular faces of an iridovirus. (Reprinted by permission of the Society for General Microbiology, from N. G. Wrigley, 1970, *Journal of General Virology*, 6:169)

fungi), **phycoviruses** (infect algae), and viruses of protozoa. The viral groups are largely defined and separated on the basis of molecular considerations. Of the microbial viruses, the bacteriophage have been most extensively studied. The characteristics of some of the groups of bacteriophages are listed in Table 9.3. These viruses exhibit great morphological diversity (Figure 9.37).

TABLE 9.2   The Groups of Plant Viruses

Bromovirus (brome mosaic virus)
  Small, icosahedral RNA viruses
Cauliflower mosaic virus (DNA virus of higher plants)
  Double-stranded DNA; reproduces in cytoplasm
Cucumovirus (cucumber mosaic virus)
  Naked, icosahedral, RNA viruses
Luteovirus (barley yellow dwarf virus)
  Small isometric virus, RNA genome
Nepovirus (tobacco ringspot virus)
  Polyhedral, nematode transmitted, RNA viruses
Potexvirus (potato virus X)
  Flexous rods, 480–580 nm, RNA genome
Potyvirus (potato virus Y)
  Flexous, rod-shaped, helical symmetry, single-stranded RNA
Tobacco necrosis virus
  Isometric RNA viruses
Tobamovirus (tobacco mosaic virus)
  Rod-shaped, helical symmetry, single-stranded RNA
Tobravirus (tobacco rattle virus)
  Rod-shaped, nematode transmitted, plus-stranded RNA viruses, segmented genome
Tombusvirus (tomato bushy stunt virus)
  Small RNA viruses, cubic symmetry, resistant to elevated temperatures and organic solvents
Tymovirus (turnip yellow mosaic virus)
  Icosahedral virus, RNA genome, transmitted by flea beetles
Watermelon mosaic virus
  Flexous rods, 700–950 nm, RNA genome

The viruses of microorganisms exhibit great host cell specificity that may be due in part to the nature of the receptor sites that are required for viral adsorption. Bacteriophage, therefore, are often named according to the species of bacteria that they infect, for example, coliphage for viruses that infect *E. coli*. An even finer division of the coliphage can be made because some viruses infect only donor strains. Such viruses adsorb onto the F pili of bacteria, which explains the host cell's specificity. These viruses can be classified separately, and the International Committee on Nomenclature of Viruses at one point recommended the establishment of the viral group Masculovirus for viruses that infect male or donor bacteria.

TABLE 9.3   Characteristics of Some of the Groups of Viruses that Infect Bacteria

Tailed bacteriophage.
  Genome: DNA, double-stranded. Virion: complex shape, binary symmetry, variable number of capsomers. The tails of the phage are long and contractile in group A (Myoviridae), long and noncontractile in group B (Styloviridae), and very short in group C (Pedoviridae). Example: T-even coliphages.
Cubic bacteriophage.
  Group 1 (Microviridae)—Genome: DNA, single-stranded. Virion: icosahedral, cubic symmetry, 12 capsomers. Example: φx174
  Group 2 (Corticoviridae)—Genome: DNA, double-stranded. Virion: cubic symmetry, enveloped. Example: PM-2
  Group 3 (Leviviridae)—Genome: RNA, single-stranded. Virion: icosahedral, cubic symmetry, 32 capsomers. Example: f_2
  Group 4 (Cystoviridae)—Genome: RNA, double-stranded. Virion: cubic symmetry, enveloped. Example: φ6
Filamentous bacteriophage (Inoviridae). Genome: DNA, single-stranded.
  Virion: rod-shaped, helical symmetry. Example: fd

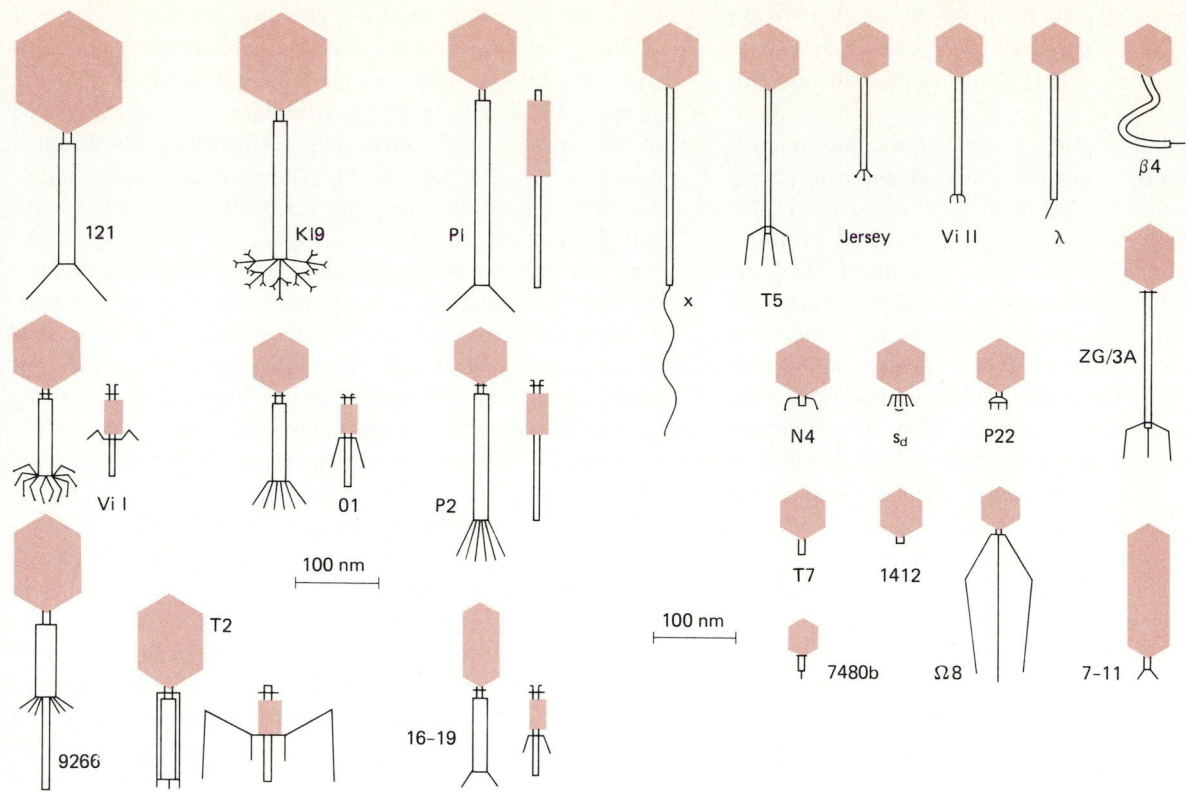

FIGURE 9.37
Schematic representation of the morphologies of phage of enterobacteria.

Postlude

The word *virus* was used by the ancient Romans to mean poison, venom, and/or secretion—all unpleasant-sounding meanings. When the field of bacteriology began to develop, the medical use of the term *virus* implied any microscopic etiologic agent of disease. It was not until the invention of bacterial filters in 1884 by Charles Chamberland, Pasteur's co-worker, who also invented the autoclave, that the field of virology really began to develop. In 1882, when Dmitrii Ivanowski reported that the agent responsible for tobacco mosaic disease could pass through a bacteriological filter, it became apparent that the microbial world contained even smaller members, the viruses, than had been previously recognized.

While the observation of most viruses awaited a further advance in microscopy, that is, the development of the electron microscope in the 1940s, the field of virology continued to progress. In 1898 Friedrich Loeffler and Paul Frosch reported that foot and mouth disease was caused by an agent that passed through a bacteriological filter, and they suggested that the causal agents of many other infectious diseases, including smallpox, cowpox, and measles, might be similar filterable agents (viruses). Their discoveries opened up the field of animal virology. Also in 1898, Martinus Bei-

jerinck, unaware of Ivanowski's work, ascribed tobacco mosaic disease to a "contagious living liquid." In 1911 Peyton Rous demonstrated that a cell-free filtrate could cause malignant growths in animals. The work of Beijerinck and Rous established the basis for tumor virology. The work of Walter Reed and others on yellow fever in the early 1900s showed that this disease was caused by a filterable agent (virus) that could be transmitted by a mosquito carrier (vector); this work demonstrated that viruses could infect more than one animal species and that viral diseases could be transmitted to humans by biting arthropods. Bacteriophage, viruses that infect bacteria, were discovered by Frederick Twort and Felix d'Herelle separately. Consequently, it was known by 1915 that viruses could infect even the smallest organisms that had been seen to that date.

In the mid-1920s, F. Parker and R. N. Nye successfully cultivated viruses using tissue culture techniques. Further advances were made by others during the following three decades, including the culture of viruses by Ernest Goodpasture and colleagues, using chick embryos (1931), and the establishment of the HeLa cell line (isolated from a cervical carcinoma of Henrietta Lacks) by G. O. Gay and co-workers (1952) that could be used for cultivating viruses. The ability to grow viruses in culture to facilitate the study of these organisms and to permit various experiments with viruses was a significant milestone in the advancement of microbiology.

Although it was not until the 1940s that the development of electron microscopy was sufficiently advanced to permit microbiologists to observe viruses in detail and discover that even these smallest microbes exhibit a diversity of form, Wendell Stanley in 1935 successfully crystallized TMV and showed that it was composed largely of protein. Stanley was a pioneer in molecular biology, establishing the importance of studying the molecules involved in replication and genetic transmission. In 1937, Frederick Bawden and Norman Pirie showed that TMV contains 6 percent RNA. Other viruses subsequently have been shown to contain RNA or DNA in addition to protein. The only essential components of a virus are a protective protein coat and a nucleic acid, either DNA or RNA, containing the genetic information of the virus.

The latter half of the twentieth century has been marked by great advances in our understanding of molecular biology. With respect to viral genetics, Heinz Fraenkel-Conrat and Robley Williams reported in 1955 on the reconstitution of TMV from its inactive protein and nucleic acid components. That same year, Seymour Benzer published a paper on the fine structure of a genetic region in bacteriophage. In 1969 Max Delbrück, Alfred Hershey, and Salvator Luria shared the Nobel Prize for their studies on viral genetics. These researchers had performed studies on the reproduction of viruses, with the finding that the viral nucleic acid coded for the sequential production of viral components within a host cell. Viruses are well suited for use in genetic studies because of their relatively small genomes.

For viruses, reproduction can occur only intracellularly. These organisms lack the necessary metabolic capability and cellular structures to accomplish independent reproduction and therefore are dependent on their compatibility with host cells for continued existence. The key steps in the reproduction of viruses are (1) adsorption of the virus onto the host cell; (2) penetration of the viral nucleic acid across the host cell's cytoplasmic membrane; (3) separation of the nucleic acid from the viral capsid; (4) shutdown of normal host cell synthesis of macromolecules; (5) production of proteins coded for by the viral genome; (6) replication of the viral genome; (7) assembly of the viral particles; and (8) release of the viral progeny. RNA and DNA viruses use different modes of reproduction for the expression and replication of the viral genome. In many viruses, different reading frames and different directions of reading are used during transcription, enabling a small viral genome to code efficiently for the variety of proteins that are required for viral reproduction.

Often the reproduction of a virus within a host cell results in the death of that cell. The lytic reproductive cycle of bacteriophage, for example, can be illustrated as a one-step growth curve, with a large number of viruses released simultaneously from the lysed bacterial cell. Similarly, some plant and animal viruses reproduce with the simultaneous release of a large number of progeny from the host cell. Other viruses reproduce without killing the host cell, exhibiting a slow, continuous release of progeny from it as occurs in budding. Some bacteriophage are temperate and can establish a prophage state that does not result in the death of the host cell. During lysogeny the viral DNA is usually incorporated into the bacterial chromosome and is replicated along with the bacterial DNA. At a later time, the viral DNA can be excised from the bacterial chromosome, and a lytic reproduction cycle can occur.

The viruses are classified both systematically, based largely on their molecular properties (modern approach), and nonsystematically, based largely on the host cells within which they reproduce and the diseases they cause (classical approach). Different classification systems are used for vertebrate animal viruses, insect viruses, plant viruses, bacteriophage, and viruses that infect other microorganisms. The nature of the genome and the capsid structure are important characteristics used in the formal classification of viruses. In the case of the viruses, it seems appropriate to consider the host cell as a primary basis for classification because viruses may have evolved as genetic extensions of prokaryotic and eukaryotic cells rather than as a direct evolutionary line from one viral species to another.

Suggested Supplementary Readings

Birge, E. A. 1981. *Bacterial and Bacteriophage Genetics*. Springer-Verlag, New York.

Buck, K. W. 1986. *Fungal Viology*. CRC Press, Boca Raton, Fla.

Butler, P. J. G., and A. Klug. 1978. The assembly of a virus. *Scientific American* 239(5):62–69.

Calendar, R. 1970. The regulation of phage development. *Annual Review of Microbiology* 24:241–296.

Casjens, S. 1985. *Virus Structure and Assembly*. Jones and Bartlett, Boston.

Crowell, R. L., and K. Lonberg-Holm. 1986. *Virus Attachment and Entry into Cells*. American Society for Microbiology, Washington, D.C.

Dalton, A. J., and F. Haguenau (eds.). 1973. *Ultrastructure of Animal Viruses and Bacteriophages: An Atlas*. Academic Press, New York.

Diener, T. O. 1981. Viroids. *Scientific American* 224:66–73.

Diener, T. O. 1982. Viroids and their interactions with host cells. *Annual Review of Microbiology* 36:239–258.

Diener, T. O., M. P. McKinley, and S. B. Prusiner. 1982. Viroids and prions. *Proceedings of the National Academy of Sciences, U.S.A.* 79:5220–5224.

Dulbecco, P. 1980. *Virology*. Harper & Row Publishers, Hagerstown, Md.

Fenner, F. 1976. The classification and nomenclature of viruses. *Journal of General Virology* 31:463–470

FENNER, F., B. R. McAUSLAN, C. A. MIMS, J. SAMBROOK, and D. O. WHITE. 1974. *The Biology of Animal Viruses*. Academic Press, New York.

FIELD, B. 1986. *Fundamentals of Virology*. Raven Press, New York.

FRAENKEL-CONRAT, H. 1985. *The Viruses: Catalogue, Characterization, and Classification*. Plenum Press, New York.

FRAENKEL-CONRAT, H., and P. C. KIMBALL. 1982. *Virology*. Prentice-Hall, Inc., Englewood Cliffs, N.J.

FRAENKEL-CONRAT, H., and R. R. WAGNER (eds.). 1977. *Comprehensive Virology: Regulation and Genetics—Plant Viruses*. Plenum Press, New York.

FRAENKEL-CONRAT, H., and R. R. WAGNER (eds.). 1978. *Comprehensive Virology: Newly Characterized Protist and Invertebrate Viruses*. Plenum Press, New York.

FRANCKI, R. I. B, and R. G. MILNE. 1985. *Atlas of Plant Viruses*. CRC Press, Inc., Boca Raton, Fla.

HOLLINGS, M. 1978. Mycoviruses: Viruses that infect fungi. *Advances in Virus Research* 22:2–54.

HUGHES, S. S. 1977. *The Virus: A History of the Concept*. Heinemann, London.

KAPLAN, A. S. (ed.). 1982. *Organization and Replication of Viral DNA*. CRC Press, Inc., Boca Raton, Fla.

LASKIN, A., and H. A. LECHEVALIER. 1979. *Handbook of Microbiology: Fungi, Algae, Protozoa, and Viruses*. CRC Press, Inc., Boca Raton, Fa.

LEMKE, P. A., and C. H. NASH. 1974. Fungal viruses. *Bacteriological Reviews* 38:29–56.

LIN, E. C. C., R. GOLDSTEIN, and M. SYVANEN. 1984. *Bacteria, Plasmids, and Phages*. Harvard University Press, Cambridge, Mass.

LURIA, S. E., J. E. DARNELL, D. BALTIMORE, and A. CAMPBELL. 1978. *General Virology*. John Wiley & Sons, Inc., New York.

MARAMOROSCH, K. (ed.). 1977. *Insect and Plant Viruses: An Atlas*. Academic Press, New York.

MARAMOROSCH, K., and J. J. McKELVEY, JR. (eds.). 1985. *Subviral Pathogens of Plants and Animals: Viroids and Prions*. Academic Press, Inc., Orlando, Fla.

MATHEWS, C., E. KUTTER, G. MOSIG, and P. BERGET (eds.). 1983. *Bacteriophage T$_4$*. American Society for Microbiology, Washington, D.C.

MATTHEWS, R. 1985. Viral taxonomy for the nonvirologist. *Annual Review of Microbiology* 39:451–474.

PALMER, E. L., and M. L. MARTIN. 1982. *An Atlas of Mammalian Viruses*. CRC Press, Inc., Boca Raton, Fla.

PRINCE, A. M. 1983. Non-A, non-B hepatitis viruses. *Annual Review of Microbiology* 37:217–232.

PRUSINER, S. B. 1984. Prions. *Scientific American* 251(4):50–60.

PTASHNE, M., A. D. JOHNSON, and C. O. PABO. 1982. A genetic switch in a bacterial virus. *Scientific American* 247:128–140.

RAO, V. C., and J. L. MELNICK. 1986. *Environmental Virology*. American Society for Microbiology, Washington, D.C.

REANNEY, D. C. 1982. The evolution of RNA viruses. *Annual Review of Microbiology* 36:47–73.

RIESNER, D., and H. J. GROSS. 1985. Viroids. *Annual Review of Biochemistry* 54:531–561.

WHITE, D. O., and F. J. FENNER. 1986. *Medical Virology*. Academic Press, Orlando, Fla.

WILDY, P. 1971. Classification and nomenclature of viruses. *Monographs in Virology* 5:1–81.

WILLIAMS, R. C., and H. W. FISHER. 1974. *An Electron Micrographic Atlas of Viruses*. Charles C Thomas, Publisher, Springfield, Ill.

WOOD, W. B. 1978. Bacteriophage T4 assembly and the morphogenesis of subcellular structures. *Harvey Lectures* 73:203–223.

Study Questions

1. Compare the lytic and lysogenic viral reproductive cycles. What are the similarities and differences in these two modes of replicating viral genomes?

2. How do we assay for lytic bacteriophage?

3. Discuss two ways that an RNA virus can replicate its genome.

4. How can viruses transform animal cells?

5. Is there a situation in animal viruses that is analogous to lysogeny in bacteria?

6. How does the structure of a viroid differ from that of an RNA virus?

7. How do viroids and RNA viruses compare with respect to their mode of replication?

8. How does a prion differ from a virus? How could a prion reproduce in a host cell?

9. How do the approaches used to classify animal, plant, and bacterial viruses differ? How could viral classification be unified?

10. How do retroviruses differ from other viruses?

Situational Problem 1

Debating Whether Viruses Are Alive

Microbiologists often argue about whether or not acellular microbes should be considered as living entities. Geneticists may consider that information flow is the principal life function and accordingly view viruses as alive because they are able to store and use genetic information to specify their own replication. Physiologists, on the other hand, may consider viruses as nonliving entities because of their inability to carry out physiological functions, such as ATP generation. Thus, the perspective of an individual can bias the opinion on whether viruses should be viewed as living or not. A philosophical and/or scientific discussion of whether viruses are alive requires a fundamental understanding of the meaning of life and the ability to examine the processes that distinguish living from nonliving entities.

Consider that you are a member of the university debate team and that the topic for the next debate is "Viruses: Are They Alive?" As in any formal debate, you must be prepared to argue either side of this issue. Prepare notes for use in this debate, making sure that you consider both sides. Think about how the properties of viruses can be viewed as evidence for determining whether they are alive or not. Also, decide what experiments could be designed to help resolve this issue.

Situational Problem 2

Determining the True Nature of Prions

Prions are the most recently discovered major group of microorganisms, and their nature is extremely controversial. You have been asked to review a journal article on prions that proposes that (1) prions are composed exclusively of proteins and (2) a prion interacts with its host cell genome via a regulator gene such that a structural gene controlled by that regulator gene is turned on by the presence of the prion protein, with that structural gene coding for the production of a protein identical to the prion protein.

1. In order to accept the arguments regarding these structural and reproductive properties of prions, what lines of evidence would you consider necessary? Consider structural data obtained by electron microscopic observation, chemical data gathered by enzymatic analyses, and genetic data determined by recombinant methods and the examination of mutants.

2. Having established the criteria that you would use to accept or reject the validity of an article on prion structure and mode of reproduction, go to the library and find a recent article on prions, perhaps one by Stanley Prusiner and his colleagues. Read the article and, using the criteria that you have developed, see whether the information in that article tends to support or refute the previously discussed properties of prions.

3. Next, go to the library and read the "The Game of the Name Is Fame. But Is It Science?" by Gary Taubes in the December 1986 issue of *Discover* magazine, which describes the controversy about the evidence concerning the nature of prions. See whether the criteria that you have developed would permit resolution of the issues raised in that article and whether, in fact, recent articles on prions have helped resolve this controversy.

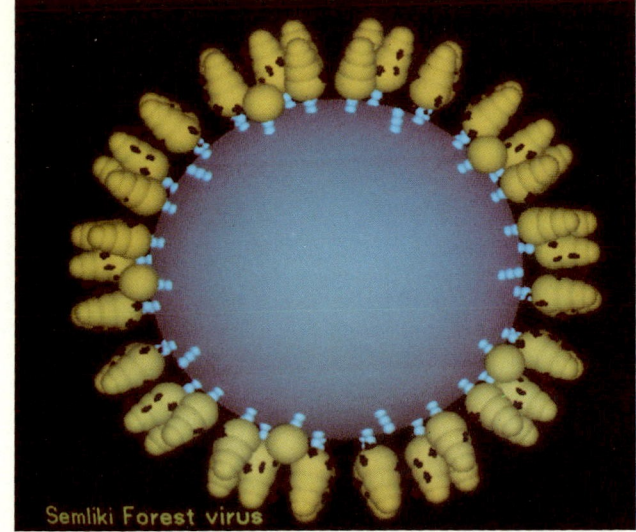

A

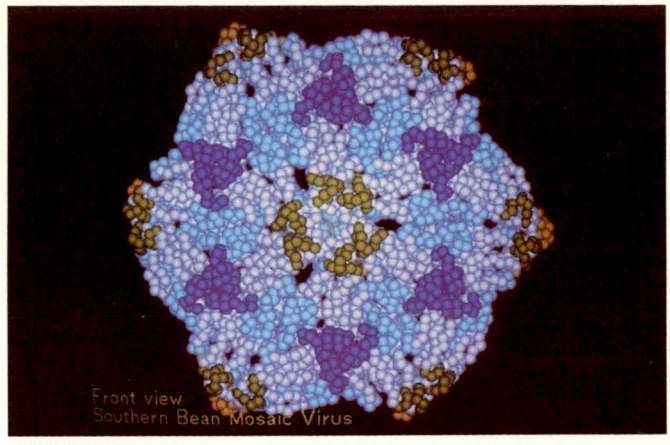

B

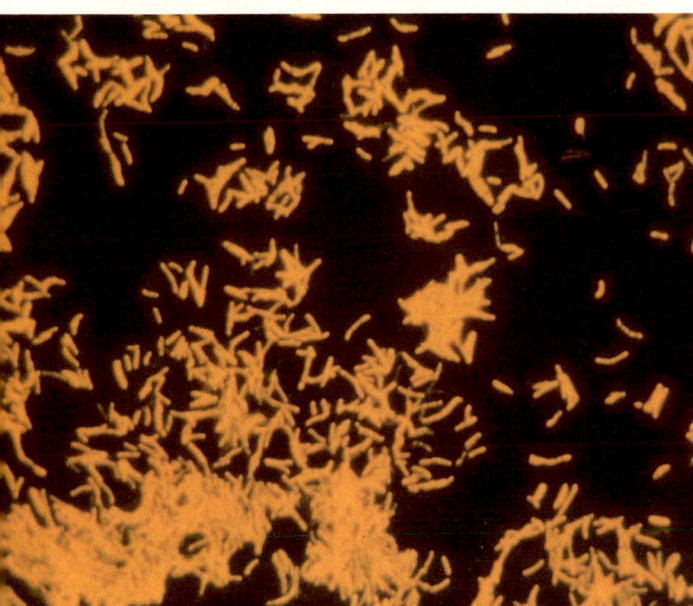

C

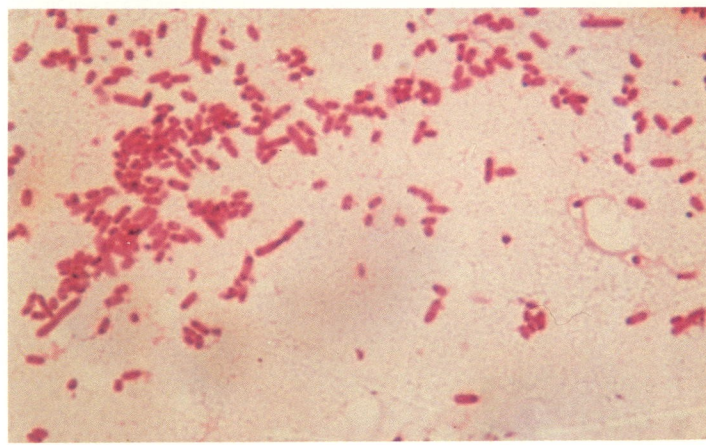

D

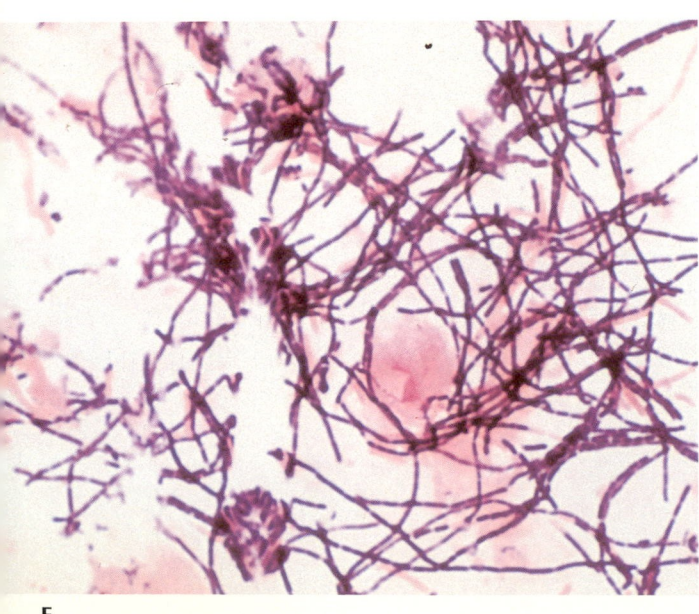

E

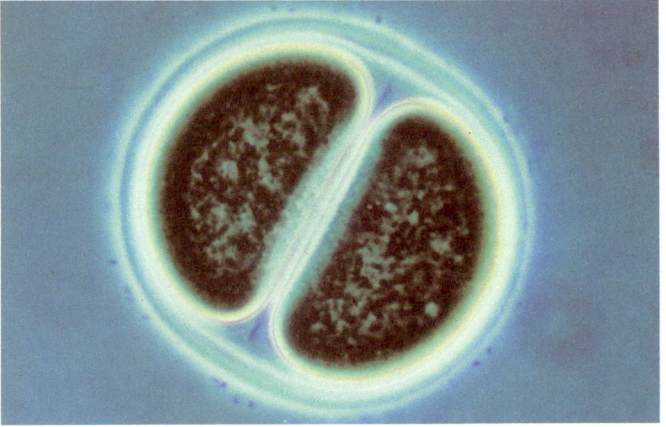

F

Plate 5 **(A)** Color-enhanced computer image of a cross section of Semliki Forest virus. (R. Feldmann, National Institutes of Health) **(B)** Color-enhanced computer image of a front view of Southern bean mosaic virus. (R. Feldmann, National Institutes of Health) **(C)** Micrograph of *Mycobacterium tuberculosis* stained with auramine 0 and viewed by fluorescence microscopy. (From BPS—R. L. Moore, BioTechniques Laboratories) **(D)** Gram stain of *Bacteroides fragilis*. (From BPS—Leon LeBeau, University of Illinois Medical Center) **(E)** Gram stain of *Lactobacillus* sp. (From BPS—Leon LeBeau, University of Illinois Medical Center) **(F)** The cyanobacterium *Chroococcus turgidus*. (From BPS—J. Robert Waaland, University of Washington)

A

B

C

D

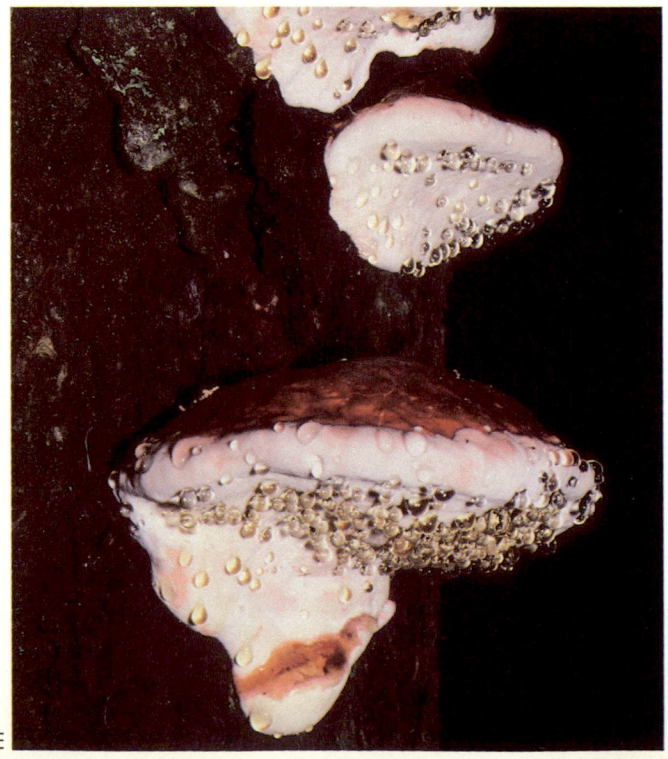

E

Plate 6 (A) The slime mold *Physarum polycephalum* growing in Acadia National Park, Maine. (From BPS—B. J. Miller) (B) The white slime mold *Brefeldia maxima* growing in Acadia National Park, Maine. (From BPS—B. J. Miller) (C) Yellow morel *Morchella esculenta*, family Morchellaceae, from Prince William Forest Park, Virginia. (From BPS—B. J. Miller) (D) ''Splash cup'' bird's nest fungi (*Cyathus striatus*); when raindrops strike the fungus, spores splash out of the sup shaped fruiting body. (From BPS—B. J. Miller) (E) The fungus *Ischnoderma resinosum* growing in Acadia National Park, Maine. (From BPS—B. J. Miller)

CHAPTER 10

Systematics of Bacteria

Bacterial systematics is concerned with the ordered relationships that exist among the bacteria. **Systematics** is the study of organisms, their diversity and interrelationships, with the aim of arranging them in an orderly manner. Accordingly, bacterial systematics involves many areas of bacteriology, including the taxonomy, ecology, biochemistry, genetics, and morphology of bacteria. Although *systematics* and *taxonomy* are sometimes used as synonymous terms, taxonomy is a subdivision of systematics. **Taxonomy** is the process, based on established procedures and rules, of describing the groups of organisms, their interrelationships, and the boundaries between microbial groups. Two functions of tax-

onomy are, first, to identify and describe the basic taxonomic units and, second, to devise an appropriate way of arranging and cataloging these units. The cardinal principles of taxonomy are that organisms exist as real, separate groups and that there is a natural ordering of these groups. A microbial taxonomist is concerned with **classification** (ordering or placing organisms into groups based on their relationships), **nomenclature** (assigning names to the units described in a classification system), and **identification** (applying the system of classification and nomenclature in order to assign the proper name to an unknown organism and to place it in its proper position within the classification system).

10.1 Nomenclature

Because one function of a taxonomic system is to establish unambiguous names for organisms, a logical system of nomenclature is needed. Organisms are normally named according to a binomial system in which the organism is identified by its genus and species. Bacteria, like other microorganisms, are referred to by their unique **binomial name**, consisting of the **genus** and **species** names of each organism.[1] The rules of nomenclature for microorganisms are

established by international committees. Different codes of nomenclature are used for different microbial groups: The code of nomenclature of bacteria applies to all bacteria; fungi and algae are covered by the botanical code; protozoa are named according to the zoological code; and viruses are named according to the viralogical code. In general, the codes of nomenclature are aimed at avoiding ambiguity and ensuring that the name of a microorganism specifically and unambiguously designates that organism. In the field of bacteriology, a summary list of the approved names of bacteria was published in the January 1980 issue of the *International Journal of Systematic Bacteriology*. Only names published in that listing and those validated and published as supplements to that list in that journal are considered to be valid. Similar listings are published periodically in other journals, establishing the validity of revised and new names of other microorganisms.

---

[1]The names of bacteria and all other organisms are given in Latin, because Latin was the classical language of science when early classification systems were developed and formal names were first given to organisms on a systematic basis. When typed or handwritten, bacterial genus and species names are underlined to indicate that they are in Latin. In print the genus and species names are italicized. The first letter of the genus name is capitalized and the species name is written in all lowercase letters. For example, we have already made frequent reference to the bacterial species *Escherichia coli*.

10.2 Classification of Microorganisms

The second objective of taxonomy, classification, attempts to differentiate microbial taxa into structured groups so that the members of a group are more closely related to each other than they are to members of any other group. Classification is a coherent scheme by which a collection of organisms are arranged so as to reflect the relationships between individuals and groups. The ordering of organisms into groups is based on an assessment of their similarities. Ideally, the classification of microorganisms should follow the natural ordering established by evolutionary processes, and therefore, taxonomic systems should be based on the genetic interrelationships among groups of microorganisms. However, many classification systems are artificial rather than natural, being based upon observable phenotypic features and not on evolutionary (genetic) relatedness.

Taxonomists frequently debate the validity of a taxonomic system, questioning whether it truly reflects the evolutionary relationships of the organisms and sometimes arguing vehemently whether the structure of the system contains the proper taxonomic units and reflects the appropriate ordering of those units. As a result, taxonomic systems are frequently revised. In Chapter 1 we considered the taxonomic position of microorganisms and examined three different classification systems that at one time or another have been considered to define properly the primary kingdoms and reflect the evolutionary relationships of living organisms. Obviously, the classification systems proposed by Whittaker and Woese show the evolution of our understanding of **phylogenetic (evolutionary) relationships**. The use of genetic analyses for assessing similarity holds great promise for the development of a valid natural classification system.

Although microbial classification systems should reflect genetic similarities, the classification of bacteria has been traditionally based on phenotypic characteristics. Phenotypic characteristics are readily determined by observing bacteria growing in pure culture, whereas, until recently, methodology for directly analyzing the genome did not exist. Taxa based on observed phenotypic characteristics may not accurately reflect genetic similarities, and such a classification may not correspond to the evolutionary flow of events. It is possible for genetically dissimilar microorganisms (**homologously dissimilar**) to resemble each other phenotypically (**analogously similar**). For example, many genetically dissimilar bacteria produce yellow pigments, and a classification scheme based on such a phenotypic characteristic could produce a taxonomic group of genetically unrelated bacteria. In fact, classification systems are filled with errors made by using such phenotypic characteristics. Various groups of bacteria that have been defined on the basis of their apparent phenotypic relationship are now considered to be "groups of uncertain taxonomic affinity" because the taxonomic group may not be homologously similar and therefore may not accurately represent genetic similarities.

Hierarchical Organization of Classification Systems

Although the hierarchical organization of a classification system should reflect the hierarchical branching of groups that is a natural consequence of evolution, evolutionary affinities among microorganisms are difficult to discern because there is no fossil record of most microorganisms. The lack of a fossil record makes examination of the remains of ancient microorganisms impossible. Additionally, the examination of bacteria requires their culture in the laboratory, and many bacteria that may prove to be critical evolutionary links have yet to be grown on defined media. As a result, bacterial taxonomy generally requires many subjective decisions, resulting in an artificial systematic classification scheme. Some taxonomists are "lumpers," tending to lump together many similar organisms into large taxonomic units. In contrast, other taxonomists are "splitters," favoring small taxonomic groups that emphasize even minor differences between organisms. Arguments over the proper taxonomic position of a microorganism can be quite heated, and these debates enhance the interest of microbiologists in the field of microbial taxonomy.

When classifying microorganisms, taxonomists use a hierarchy consisting of different organizational levels. The usual levels of a **taxonomic hierarchy**, from the highest to the lowest, are kingdoms, phyla, classes, orders, families, genera, and species (Table 10.1). Ideally, each level represents a different degree of homology, that is, of genetic and evolutionary similarity. In reality, many levels represent varying degrees of analogous (phenotypic) similarity. The higher the taxonomic level, the greater the diversity of the organisms classified as belonging to that group.

The **species** is considered to be the basic taxonomic unit of a classification system. Whereas species of higher organisms are readily recognized as a result of their reproductive

TABLE 10.1   Hierarchy of Taxonomic Organization

Level	Description	Example
Kingdom	A group of related phyla	Prokaryotate
Phylum	A group of related classes or divisions	—
Division or class	A group of related orders	Eubacteria
Order	A group of related families	Spirochaetales
Family	A group of related tribes or genera	Spirochaetaceae
Tribe	A group of related genera	—
Genus	A group of related species	*Treponema*
Species	A group of organisms of the same kind	*T. pallidum*
Subspecies or type	Variants of a species	—

isolation, a microbial species is difficult to define objectively.[2] A microbial species can be considered as a group of isolated strains that have an overall similarity and are significantly different from other fundamental groups. In defining a new microbial species, the organism must be described, named, and shown to be different from previously described species. International committees have been established to rule on the validity of defining new species. Once an organism or group of organisms has been defined as representing a new species, a culture should be deposited in an appropriate culture collection. That type culture and its description become the reference for future identification.

Although species are the basic taxonomic units, the genetic variability of microorganisms permits a further division into **subspecies** or **types** that describe the specific clone of cells. It is often important to identify the subspecies, or even a specific **strain**, of a given microorganism. For example, one strain of a bacterial species may produce a toxin and be a virulent pathogen, and other strains of the same species may be nonpathogenic. The ability to distinguish correctly between such strains and subspecies of a particular microbial species is of obvious importance in medical and industrial microbiology.

Approaches to the Classification of Microorganisms

A number of different approaches are used in microbiology for developing classification systems and establishing hierarchical relationships. Early approaches grouped microorganisms on the basis of the similarity of their morphological appearance. Later, physiological properties dominated microbial classification. Modern microbial classification systems still include morphological and physiological characteristics, but molecular-level similarity, that is, genetic relatedness, is given greater emphasis. Modern classification systems may be **phenetic** (assessing similarity) or **phylogenetic** (assessing true evolutionary relationships). Because organisms could have evolved quite differently and still have developed similar characteristics, microorganisms that are distantly related in evolutionary terms may still be similar and grouped together in a phenetic but not a phylogenetic classification system. Phenetic approaches require quantification and explicitness and do not permit a priori weighting of features. In contrast, phylogenetic approaches may consider evolutionary evidence and give added importance to those key features viewed as indicative of different evolutionary paths.

[2]Implicit in the definition of a species for higher organisms is a similar and shared gene pool. The asexual means of reproduction typically exhibited by bacteria that gives rise to clones of genetically identical cells limits the diversity of the gene pool, but closely related bacteria do exchange genetic information by several mechanisms. Recombination among closely related bacteria gives rise to a limited degree of genetic diversity within the population of even a bacterial species.

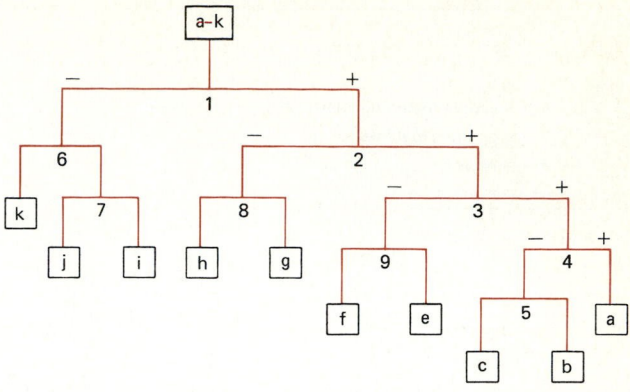

FIGURE 10.1

A dichotomous key for classification, showing branching based on critical features. The letters a–k designate individual groups of organisms. The numbers represent individual features (tests) that are used to separate the groups. In a bacterial classification system, the features used might be: 1—Gram stain reaction; 2—rod-shaped morphology of cells; 3—ability to form endospores; and so forth.

In a conventional or classical approach to classification, a number of features of a group of organisms are examined, and organisms are grouped on the basis of their similarity with respect to selected features. In the development of such classification systems, certain features are generally considered more important than others for defining taxonomic units. These key features are those considered to be of primary importance in the separation of species. For example, the Gram stain reaction is considered a key test in many bacterial classification systems. A taxonomic system based on this approach employs a sequential series of hierarchical decisions to separate the taxonomic units called **taxa** or **taxons**. This type of classification system emphasizes the branch points between groups that are presumed to represent fundamental differences between taxa. Often the classification system is represented as a tree diagram showing the various key branch points and the individual taxa at the the ends of the branches (Figure 10.1).

An alternative approach, numerical taxonomy, does not emphasize points of branching but rather uses overall degrees of similarity between organisms to establish a taxon. In **numerical taxonomy** a single characteristic does not determine the taxonomic position of an organism. Instead, overall similarity and the definition of the taxa are based on several characteristics; a large number of characteristics are examined to ensure that no single one achieves undue weighting. Various measures can be applied to assess similarity. These measures are often expressed as indices of similarity or similarity coefficients (Figure 10.2). The most common indices of similarity employed in microbiology are the simple matching (S_{SM}) and Jaccard (S_J) coefficients. The **simple matching coefficient** is based on all of the measured characteristics. In contrast, the calculation of the **Jaccard coefficient** does not use a characteristic when the organisms being compared are

$$S_m = \frac{(++) + (--)}{(++) + (--) + (+-) + (-+)}$$

Where

S_m = simple matching coefficient
$++$ = positive matches
$--$ = negative
$(+-) + (-+)$ = mismatches

$$S_j = \frac{(++)}{(++) + (+-) + (-+)}$$

Where

S_j = Jaccard coefficient
$++$ = positive matches
$(+-) + (-+)$ = mismatches

FIGURE 10.2

Formulas for simple matching and Jaccard coefficients for assessing similarity. Using these equations, the similarities between microorganisms can be assessed on the basis of phenotypic or genetic characteristics.

both negative for that feature; the assumption is made that such a feature may be an inappropriate description for the group under consideration. For example, a weight of over 1 ton may be an appropriate descriptive characteristic for elephants but is clearly inappropriate for microorganisms. When the simple matching coefficient is used, the inclusion of such an irrelevant feature in a classification system would artifi-

cially, even nonsensically, make organisms appear more similar than they really are; the Jaccard coefficient eliminates this problem.

The classification of organisms by the use of numerical taxonomy is normally represented graphically by a similarity matrix or a dendrogram (Figure 10.3). Organisms are grouped on the basis of the overall similarity of all of the tested characteristics by using a type of statistical analysis known as *cluster analysis*. In the graphic representations of these analyses, organisms of high similarity occur in close geometric proximity, whereas organisms of low similarity are separated.

Criteria for Classifying Microorganisms

Phenotypic Characteristics   Regardless of the classification approach employed, it is necessary to characterize organisms in order to assess their similarity. To characterize a microorganism, it is generally necessary to isolate and maintain pure cultures of the organism. Otherwise one might examine the combined features of several interactive organisms or an anomaly based on environmental variations. Determining the characteristics of a pure culture of a microorganism permits the description of that organism so that it can be classified. To exemplify the types of characteristics employed as microbial descriptors, some of the criteria frequently used in bacterial classification are listed in Table 10.2. The various criteria used in classifying microorganisms represent a continuum from the analysis of the genome itself to the pheno-

FIGURE 10.3

The similarities between microorganisms can be represented in several ways. In each, strains of the highest similarity are grouped together. (A) Dendrogram showing the hierarchical relationships between organisms; the horizontal lines show the levels of similarity, and the vertical lines represent individual strains, (B) A similarity triangle does not show as much detail as a dendrogram, but groups with high similarity (taxa) are readily distinguished in this graphic representation.

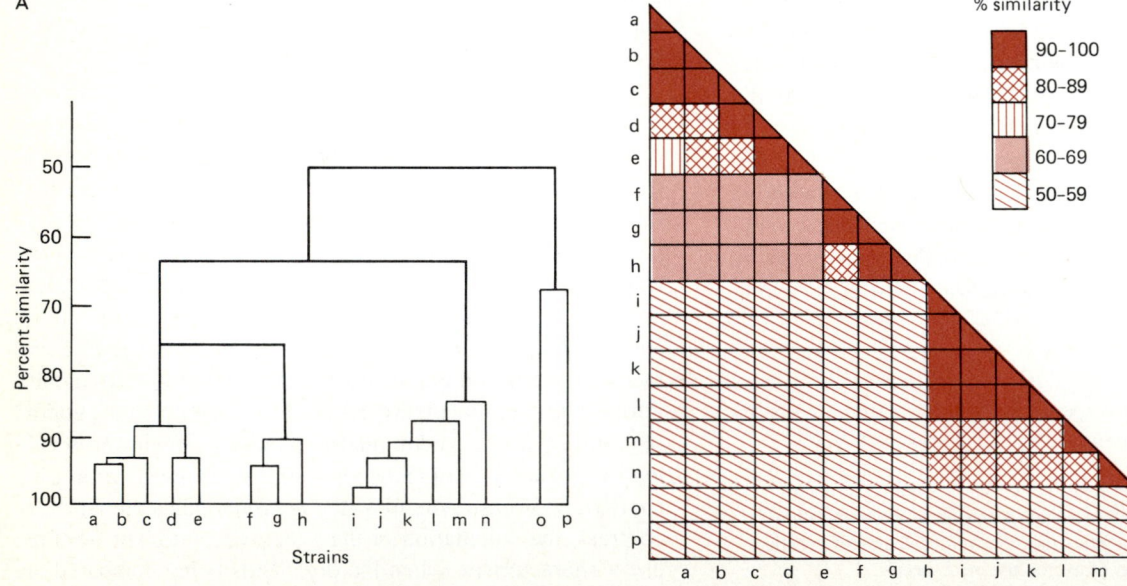

TABLE 10.2   Some Criteria for Classifying Bacteria

Microscopic characteristics
    Morphology
        Cell shape
        Cell size
        Arrangement of cells
        Arrangement of flagella
        Capsule
        Endospores
    Staining reactions
        Gram stain
        Acid-fast stain
Growth characteristics
    Appearance in liquid culture
    Colonial morphology
    Pigmentation
Biochemical characteristics
    Cell wall constituents
    Pigment biochemicals
    Storage inclusions
    Antigens
    RNA molecules
Physiological characteristics
    Temperature range and optimum
    Oxygen relationships
    pH tolerance range
    Osmotic tolerance
    Salt requirement and tolerance
    Antibiotic sensitivity
Nutritional characteristics
    Energy sources
    Carbon sources
    Nitrogen sources
    Fermentation products
    Modes of metabolism (autotrophic, heterotrophic,
        fermentative, respiratory)
Genetic characteristics
    DNA %G + C
    DNA hybridization

typic expression of genetic information. Presumably, classification systems based on genetic analyses reflect a natural evolutionary hierarchy more accurately than other artificial classification systems based on phenotypic characteristics. For practical reasons, however, phenotypic characteristics are employed in most microbial classification and identification systems. The morphological, physiological, biochemical, and nutritional features normally examined include mode of reproduction, morphology, staining reactions, ability to form spores, motility, specific metabolic activities, and life cycle stages. Analysis of the biochemical constituents of the cell and particular structures, such as cell walls and membranes, among others, may also be valuable in revealing major differentiating features of groups of microorganisms.

Genetic Characteristics   In addition to phenotypic analyses, it is possible to include genetic analyses in the development of classification schemes for microorganisms. One of the genetic analyses used for classifying microorganisms is the

determination of the relative proportion of guanine (G) and cytosine (C) compared to the total number of nucleotide base pairs in the DNA. Because there is pairing between complementary bases in DNA, specifying G + C also specifies adenosine + thymine (A + T). Therefore, the composition of the DNA is normally described by specifying the **mole% G + C**, that is the proportion of the DNA macromolecule composed of G + C base pairs. Measuring the proportion of G + C in the DNA is a crude way of analyzing the microbial genome. Closely related organisms should have similar proportions of G + C in their DNA, and organisms with vastly differing proportions of G + C in their DNA can safely be said to be unrelated. Microorganisms that are otherwise similar, but differ by 2 to 3% G + C content in their genomes, represent different species. However, a similarity of G + C ratios does not necessarily establish the relatedness of organisms because the sequence of genes may be different even when the mole% G + C content is the same.

Epigenetic Analyses   Genetic analyses can also be extended to the analysis of gene products, namely, RNA molecules and proteins. The subunits of these macromolecules can be described explicitly. Proteins can be separated and analyzed by using electrophoresis and other analytical methods. The nucleotide sequences of RNA molecules can be determined and the similarity between these molecules assessed. In performing these analyses, a particular type of RNA molecule, such as 16S rRNA, is broken into oligonucleotides by using a nuclease. The oligonucleotides are then separated and their specific nucleotide sequences recorded to establish a catalog or library of oligonucleotides that characterize that organism. By comparing the RNA oligonucleotide catalogs of two different organisms, the similarity between those two organisms (S_{AB}) can be expressed as follows: S_{AB} = 2 × the number of common oligonucleotides in organisms A and B / the total number of oligonucleotides in A and B. Because the order of nucleotides in RNA and amino acids in proteins is directly controlled by the genome, analyses of these gene products should permit the combining of phenetic and phylogenetic approaches to microbial taxonomy.

Such comparisons of gene products represent an intermediary stage between the analysis of genotype and phenotype. The examination of rRNA molecules is particularly useful for establishing the higher levels of taxonomic organization, for example, at the kingdom level. As has been discussed in previous chapters, analyses of rRNA molecules have been used to establish a new classification system composed of three primary kingdoms: Archaebacteria, Eubacteria, and Eukaryotes. Similar analyses of rRNA molecules are now used to assess the relatedness of microbial genera and species. As more is learned about the molecular-level composition of microorganisms and theories about the course of evolution are revised, phylogenetic classification systems will continue to change in order to reflect the prevailing evolutionary concepts.

ANALYTICAL PROCESS

Analysis of DNA Composition

To determine the mole% G + C, DNA can be slowly heated and its ability to absorb ultraviolet (UV) light at a wavelength of 260 nm measured spectrophotometrically. When double-stranded DNA melts, that is, when it is converted into two single strands or denatured, there is a change in its absorption characteristics. UV absorption by the two single strands of DNA is about 40 percent greater than it is by the same strands in the double helical form. This change in uv absorption is known as **hyperchromatic shift** (Figure 10.4). The midpoint temperature of the denaturation curve is known as the **melting temperature** (T_m) of the DNA. Because certain ions stabilize the hydrogen bonds of DNA, the melting temperature is influenced by the chemical composition of the solution in which the DNA is suspended. The empirical relationship between the melting temperature and the mole% G + C is given by the equation $T_m = 0.41\%$ (mole% G + C) + $16.6 \log M + 81.5$, where M is the concentration of monovalent cations within the range 0.0001–0.2 molar. Because G + C base pairs form three hydrogen bonds, compared to only two such bonds for A + T pairs, the melting point of the DNA molecule is proportional to the mole% G + C of that molecule because it takes more heat to break the three hydrogen bonds between G + C base pairs than it does to break the two hydrogen bonds between A + T base pairs.

The mole% G + C of the DNA may also be determined by measuring the buoyant density of the DNA. The density of the DNA can be determined by using $CsCl_2$ density gradient ultracentrifugation. The higher the mole% G + C, the greater the density of the DNA because the molecular weight of G + C is higher than that of A + T. The buoyant density of the DNA is therefore a measure of the proportion of G + C in the DNA. The relationship between buoyant density (σ) and mole% G + C is given by the equation $\sigma = 1.66 + 0.00098$ (mole% G + C). Thus, by two different approaches, the mole% G + C can be determined to assess the genetic relatedness of microorganisms.

Although the mole% G + C is a gross measure of the genome, a more precise measure of relatedness is the **DNA homology** between two organisms. A measure of the DNA homology accounts for the order of nucleotides as well as the overall composition of the genome. DNA homology measurements provide a powerful tool for classifying microorganisms, especially for establishing the affinities of strains within species and species within genera. In one method for determining DNA homology, one of the organisms to be compared is grown in a medium containing tritiated thymine so that the radio-labeled thymine is incorporated into the DNA (Figure 10.5). The other organism is grown in a medium without the radiolabeled nucleotide so that its DNA is cold (nonradioactive). The DNA from the organisms being compared is then melted (converted to single-stranded DNA), mixed, and allowed to reanneal (reestablish double-stranded regions). Only homologous regions of

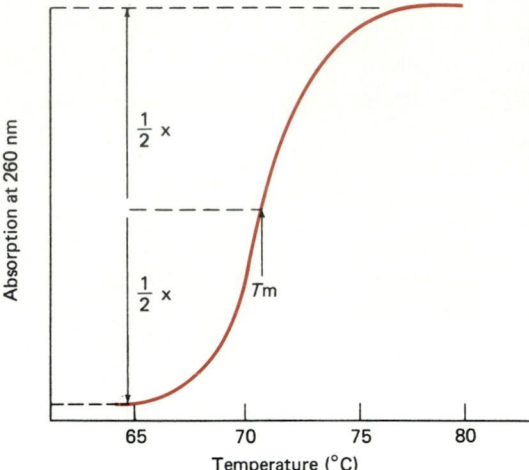

FIGURE 10.4
Melting curve for DNA, showing hyperchromatic shift and T_m. The hyperchromatic shift occurs because of separation of the two strands of DNA (melting). The temperature at which melting occurs is proportional to the mole% G + C. The mole% G + C is an important characteristic in assessing the genetic relatedness of two organisms. The two strands of a DNA helix readily come apart when the hydrogen bonds that unite the base pairs are disrupted. Heating a solution of DNA or adding acid or alkali to ionize the bases will unwind the helix. The melting point of DNA (T_m) is defined as the temperature at which half of the helical structure is lost. The melting of DNA is easily followed by measuring its absorbance at 260 nm. The unstacking of the base pairs, that is, the melting of the double helix into single strands, results in increased absorbance. This effect is called *hyperchromism*. The T_m of a DNA molecule depends heavily on its base composition. DNA molecules rich in GC base pairs have a higher melting temperature than those with an abundance of AT base pairs. GC base pairs are more stable than AT pairs because their bases are held together by three, rather than two, hydrogen bonds. Therefore, the AT-rich regions of the DNA are the first to melt, and the T_m of the DNA from many species varies linearly with the GC content, rising from 77°C to 100°C as the fraction of GC pairs increases from 20 to 78 percent. Separated complementary strands of DNA spontaneously reassociate to form a double helix when the temperature is lowered in a process called *annealing*.

the DNA (those regions having identical or nearly identical nucleotide sequences) will reanneal. Single-stranded DNA is removed chromatographically or by other procedures, and the degree of reannealing is then determined by measuring the amount of radioactivity in the double-stranded DNA form.

DNA homology may also be determined by density gradient ultracentrifugation. One organism is grown in the presence of a heavy isotope of an element, such as

[15]N, which is incorporated into the DNA, and another organism is grown in a medium containing the normal isotope, such as light [14]N. Thus, one of the organisms has heavy DNA and the other has light DNA. After the heavy DNA is broken into short strands, the DNA of both organisms is melted, mixed, and allowed to reanneal so that the hybrid DNA formed contains one strand of light DNA and segments of the strands of heavy DNA. The hybrid DNA is separated from the normal DNA by density gradient ultracentrifugation and quantitatively measured.

Data from such analyses of DNA homology permit the determination of genetic relatedness and should facilitate the development of classification systems that truly reflect the natural phylogeny of microorganisms.

Determining the sequence of nucleotides within the genome is yet another way of determining the relatedness of organisms. With the development of automated nucleotide sequencers, future classification systems may well be based upon direct determinations of nucleotide sequences.

FIGURE 10.5

The genetic (true) similarity of two organisms can be determined by DNA hybridization. In this method, the DNA of one organism is radiolabeled and the DNA of the other is not. After melting the DNA of both organisms, that is, converting it to single-stranded DNA molecules, the radiolabeled and nonlabeled DNA molecules are mixed. Reannealing (formation of hybrid double-stranded DNA) occurs in regions of high nucleotide sequence similarity. The hybrid DNA is separated from the single-stranded DNA by column chromatography using a hydroxyapatite column and differential phosphate buffer elution. The extent of hybridization, determined by the amount of radioactivity in the double-stranded (hybrid) DNA fraction, is a measure of the genetic similarity between the organisms.

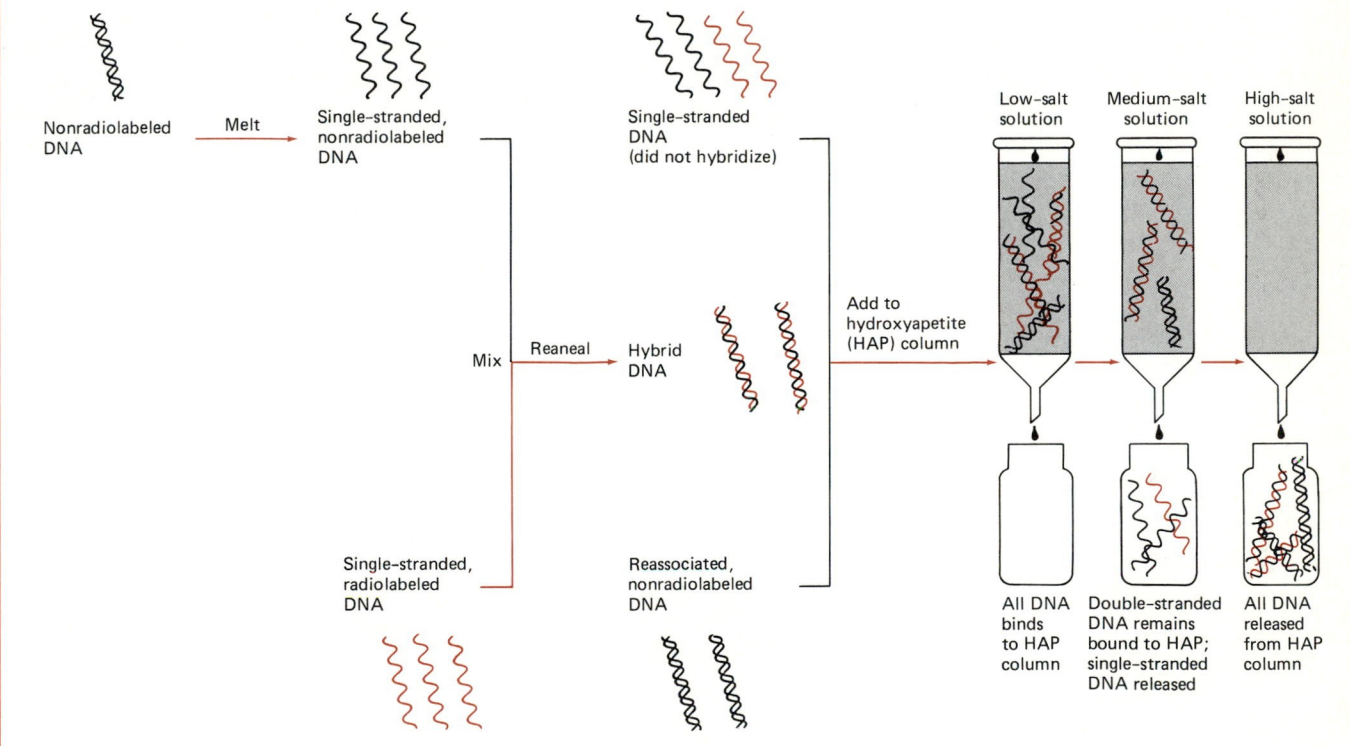

10.3 *Identification of Microorganisms*

While classification attempts to establish an orderly arrangement of taxonomic groups, **identification** attempts to compare an unknown organism with previously established groups in order to determine whether it should be considered as a member of one of these groups. The two processes, classification and identification, are not synonymous. Whereas classification defines taxonomic groups, identifica- tion assigns an organism of unknown taxonomic affinity to the correct group. One can view identification as a practical application of classification, permitting the correct recognition of an unknown microorganism as a member of a preexisting taxonomic group. An identification system should permit the efficient and reliable distinguishment of microorganisms. For identification purposes, any artificial classification system will suffice, as long as it permits the organism to be identified with others that have previously been placed in the system.

Although classification and identification of microorganisms are distinct processes, both generally require that pure cultures be examined and characterized. The characteristics used in an identification system are limited to those significant features that distinguish one group from another. As a rule, identification systems try to use the minimal number of characteristics that will produce a reliable identification. Because the purpose of these systems is to make a correct iden-

tification efficiently, the features included in the identification system are often different from those used in the classification system.

Identification Keys and Diagnostic Tables

The classical approach to microbial identification involves the development of keys or diagnostic tables. An **identification key** consists of a series of questions that lead through a classification system to the determination of the identity of the organism. In a **dichotomous key** a series of yes-no questions is asked that leads through the branches of a flow chart to the identification of a microorganism as a member of a specified microbial group (Table 10.3; see also Figure 10.1). The path to an identification in a true dichotomous key is unidirectional, and a single atypical feature or error in determining a feature will result in a misidentification. Most students in introductory microbiology laboratory courses use dichotomous keys for the identification of an unknown organism. Although dichotomous keys are frequently used in identification systems, other keys based on multiple-choice questions also can be used. Regardless of whether the choices are dichotomous or not, the characteristics used in establishing an identification key must be constant for the particular group. For example, if the Gram stain is employed as a key feature in a dichotomous key, the groups separated by this characteristic must be either Gram-positive or Gram-negative; a group cannot contain both Gram-positive and Gram-negative members.

Besides identification keys, **diagnostic tables** can be developed to aid in microbial identification (Table 10.4). Such tables summarize the characteristics of the taxonomic groups but do not indicate a hierarchical separation of the taxa. Diagnostic tables generally appear to be far more complicated than keys for the identification of microorganisms because

TABLE 10.3 Dichotomous Key for the Diagnosis of Species of the Genus *Pseudomonas*

1. Oxidase negative (Go to 2)
   Oxidase positive (Go to 4)
2. Lysine positive   *P. maltophilia*
   Lysine negative (Go to 3)
3. Motile   *P. paucimobilis*
   Nonmotile   *P. malleii*
4. Fluorescent (Go to 5)
   Nonfluorescent (Go to 8)
5. Pyocyanin and pyorubin positive   *P. aeruginosa*
   Pyocyanin and pyorubin negative (Go to 6)
6. Growth at 42°C   *P. aeruginosa*
   No growth at 42°C (Go to 7)
7. Gelatinase positive   *P. fluorescens*
   Gelatinase negative   *P. putida*
8. Nonmotile   (probably not a pseudomonad)
   Motile (Go to 9)
9. Peritrichous   (not a pseudomonad)
   Polar (Go to 10)
10. Glucose oxidation negative (Go to 11)
    Glucose oxidation positive (Go to 13)
11. Two or more flagella   *P. diminuta*
    One flagellum (Go to 12)
12. PHB positive   *P. testosteroni*
    PHB negative   *P. alcaligenes*
13. Mannose negative (Go to 14)
    Mannose positive (Go to 16)
14. Ornithine positive   *P. putrefaciens*
    Ornithine negative (Go to 15)
15. Mannitol positive   *P. acidovorans*
    Mannitol negative   *P. pseudoalcaligenes*
16. Arginine positive (Go to 17)
    Arginine negative (Go to 19)
17. 6.5% NaCl positive   *P. mendocina*
    6.5% NaCl negative (Go to 18)
18. Gelatinase positive   *P. fluorescens*
    Gelatinase negative   *P. putida*
19. Galactose negative (Go to 20)
    Galactose positive (Go to 21)
20. Mannose positive   *P. maltophilia*
    Mannose negative   *P. vesicularis*
21. Lactose negative (Go to 22)
    Lactose positive (Go to 23)
22. 6.5% NaCl positive   *P. stutzeri*
    6.5% NaCl negative   *P. pickettii*
23. Nitrogen production   *P. pseudomallei*
    No nitrogen production (Go to 24)
24. Citrate positive   *P. cepacia*
    Citrate negative   *P. paucimobilis*

TABLE 10.4 Diagnostic Table for Differentiating the Species *Pseudomonas aeruginosa*, *P. fluorescens*, and *P. putida*

Characteristic	P. aeruginosa	P. fluorescens	P. putida
Monotrichous polar flagella	+	−	−
Pyocyanin	+	−	−
Growth at 4°C	−	+	+
Growth at 42°C	+	−	−
Denitrification	d	−	−
Lecithinase	−	+	−
Gelatinase	d	+	−
Utilization			
Acetamide	+	−	
Creatinine	−		+
Benzylamine	−	−	+
Geranitol	+	−	−
Hippurate	−	−	+
Inositol	−	+	−
Phenylacetate	−		+
Trehalose	−	+	−

+, over 90%; −, less than 10%; d, 10–90% of strains tested are positive.

they contain more information. However, in cases where some features are variable for different groups, diagnostic tables are better than keys for the successful identification of an unknown microorganism.

Computer-Assisted Identification

Computers greatly facilitate the identification of microorganisms. When computers are used, the data gathered on an unknown microorganism can rapidly be compared to a data bank containing information on the characteristics of defined taxa. Keys and diagnostic tables can readily be programmed for computer-assisted identification of unknown isolates. Because computers rapidly perform large numbers of calculations and comparisons, computerized identification systems have also been developed to assess the statistical probability of correctly identifying a microorganism. In these methods the results of a series of phenotypic tests are scored and compared to the test results of organisms that have been classified as belonging to a particular taxonomic group. Unlike keys in which individual tests are critical in achieving proper diagnostic identification, these identification systems assess the statistical likelihood of obtaining a particular pattern of test results.

Such identification systems often involve the development and use of **probabilistic identification matrices**, which are compilations of the frequencies of occurrence of individual features within separate taxonomic groups (Figure 10.6).

FIGURE 10.6

The probabilistic matrix approach to organism identification allows organisms of unknown affiliation to be identified as members of established taxa.

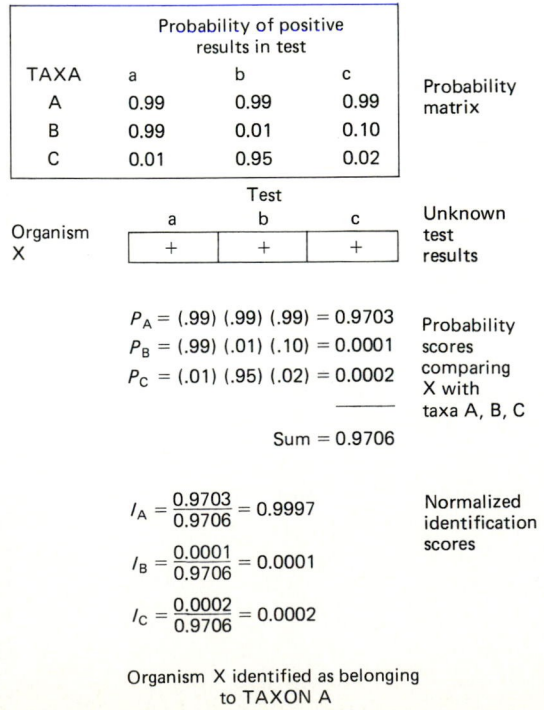

TAXA	Probability of positive results in test			
	a	b	c	Probability matrix
A	0.99	0.99	0.99	
B	0.99	0.01	0.10	
C	0.01	0.95	0.02	

Organism X	Test			Unknown test results
	a	b	c	
	+	+	+	

$P_A = (.99)(.99)(.99) = 0.9703$ — Probability scores comparing X with taxa A, B, C

$P_B = (.99)(.01)(.10) = 0.0001$

$P_C = (.01)(.95)(.02) = 0.0002$

Sum = 0.9706

$I_A = \dfrac{0.9703}{0.9706} = 0.9997$ — Normalized identification scores

$I_B = \dfrac{0.0001}{0.9706} = 0.0001$

$I_C = \dfrac{0.0002}{0.9706} = 0.0002$

Organism X identified as belonging to TAXON A

Digit 1, Digit 2, Digit 3 profile

	Digit 1			Digit 2			Digit 3			
	A	B	C	D	E	F	G	H	I	Test results
	1	2	4	1	2	4	1	2	4	
Unknown	+	−	−	−	−	+	+	−	−	

Profile 141

A — ornithine decarboxylase (ODC)
B — citrate utilization (CIT)
C — H₂S production (H₂S)
D — Voges-Proskauer test (VP)
E — gelatin hydrolysis (GEL)
F — glucose utilization (GLU)
G — mannose utilization (MAN)
H — inositol utilization (INO)
I — sorbose utilization (SOR)

Probability matrix showing percent positive for all strains that have been classified

		ODC	CIT	H₂S	VP	GEL	GLU	MAN	INO	SOR	Profile
Escherichieae	*Escherichia coli*	75.7	0	1.0	0	0	99.9	99.0	0.5	94.4	145
	Shigella dysenteriae	0	0	0	0	0	94.1	2.0	0	18.6	040
	Sh. flexneri	0.9	0	0	0	0	99.0	91.8	0	24.0	041
	Sh. boydii	3.4	0	0	0	0	99.1	94.1	0	55.9	041
	Sh. sonnei	92.9	0	0	0	0	99.9	99.0	0	2.2	141

Unknown matches profile of *Shigella sonnei*

FIGURE 10.7

A numerical profile can be created to identify an unknown organism. This approach is employed in several widely used commercial systems for the identification of clinical isolates.

These probability matrices are developed by characterizing large numbers of strains belonging to each taxonomic group. In this way, the variability of the group for a particular feature can be determined. Many of the commercial identification systems used in clinical laboratories for diagnosing infectious disease agents are based on such probabilistic identification matrices. In many of these systems, the test results of an isolated organism are rapidly compared with a previously developed probability matrix. Such multiple comparisons require the use of a computer for rapid identification and generally compute the statistical probability that the identification is correct. Because of the critical need to make correct identifications in medical microbiology, a positive identification in a clinical identification system generally requires that the unknown organism be far more similar to the group to which it is identified as belonging than to any other group. For example, in some identification systems, an unknown microorganism must be a thousand times more similar to one group than to all other groups in the system for a positive identification to be established.

Several of the commercial systems simplify the process of identification by calculating a numerical profile to describe unambiguously the pattern of test results (Figure 10.7). The numerical profile is simply a way of compressing the data so that they can easily be compared to the data collected on other organisms. The numerical profile of an unknown organism can be compared with the test pattern of a defined group to determine the probability that the test results represent a member of that taxon.

ANALYTICAL PROCESS

Miniaturized Commercial Identification Systems

Several commercial systems have been developed for the identification of members of the family Enterobacteriaceae and other pathogenic microorganisms. These systems are widely used in clinical microbiology laboratories because of the high frequency of isolation of Gram-negative rods indistinguishable except for characteristics determined by detailed biochemical and/or serological testing. Systems commonly used in clinical laboratories include the Enterotube, API 20-E, Minitek, Micro-ID, Enteric Tek, and r/b enteric systems. The pattern of test results obtained in these systems is converted to a numerical code that can be used to calculate the identity of the isolate. The numerical code describing the test results obtained for a clinical isolate is compared with results in a data bank describing test reactions of known organisms. Some of the commercial systems list a series of possible identifications indicating the statistical probability that a given organism (biotype) could yield the observed test results. All of the commercial systems employ miniaturized reaction vessels, and some are designed for automated reading and computerized processing of test results. The systems differ in how many and which specific biochemical tests are included. They also differ in whether they are restricted to identifying members of the family Entero-

bacteriaceae or whether they can be utilized for identifying other Gram-negative rods. The test results obtained with all of these systems show excellent correlation with conventional test procedures, and these package systems yield reliable identifications as long as the isolate is one of the organisms that the system is designed to identify.

The Enterotube system contains eight solid media from which 11 different features can be determined (Figure 10.8). This system has a self-contained inoculating needle that is touched to a colony on the isolation plate and drawn through the tube. The characteristics determined in the Enterotube are used to generate a four-digit biotype number from which bacterial identifications can be made. The identification is made by comparing the biotype number of an unknown organism with those of previously identified organisms.

The **API 20-E system,** as the name implies, utilizes 20 miniature capsule reaction chambers (Figure 10.9). A suspension of bacteria is used to inoculate each of the reaction chambers. The results of the API 20-E test system yield a seven-digit biotype number from which a computer-assisted identification can be made. The results of the API system can also be used in the identification of some nonfermentative, Gram-negative rods and for the

FIGURE 10.8

Conversion of Enterotube test results to a numerical code. The speed and ease of identification make these systems very popular in clinical laboratories.

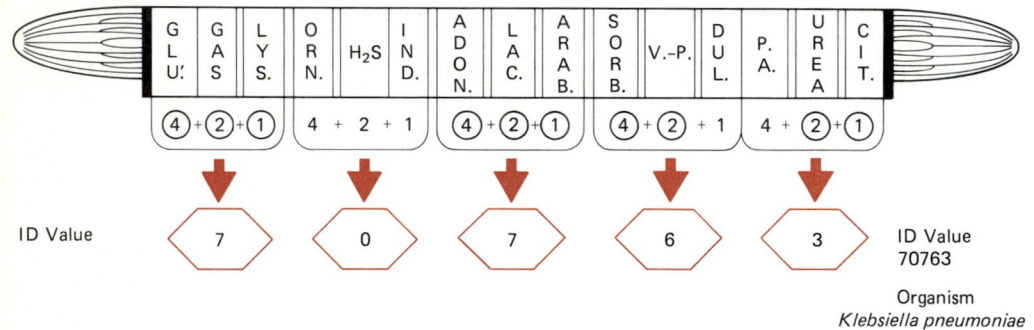

Features Examined in the Enterotube I System

| | Visual reactions | |
Characteristics	Positive	Negative
Dextrose	Yellow	Red
Gas in dextrose chamber	Bubbles	No bubbles
Lysine decarboxylase	Purple–blue	Yellow
Ornithine decarboxylase	Purple–blue	Yellow
Hydrogen sulfide	Black media	No blackening
Indole (add Kovac's reagent to H_2S chamber)	Red ring	No red ring
Lactose	Yellow	Red
Phenylalanine (add 10% $FeCl_3$)	Brown	Light green
Dulcitol (read in phenylalanine chamber)	Yellow	Light green
Urease	Red	Light yellow
Citrate	Deep blue	Light green

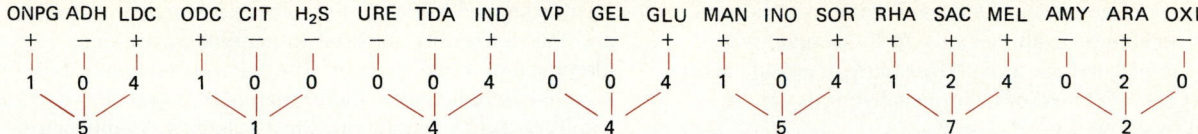

Normal 7 digit code 5144572 = *E. coli.*

Construction of a 9 digit profile

To the seven–digit profile described above, two digits are added corresponding to the following characteristics:

NO_2: Reduction of nitrate to nitrite only
N_2 GAS: Complete reduction of nitrate to N_2 gas or amines
MOT: Observation of motility
MAC: Growth on MacConkey medium
OF/O: Oxidative utilization of glucose (OF open)
OF/F: Fermentative utilization of glucose (OF closed)

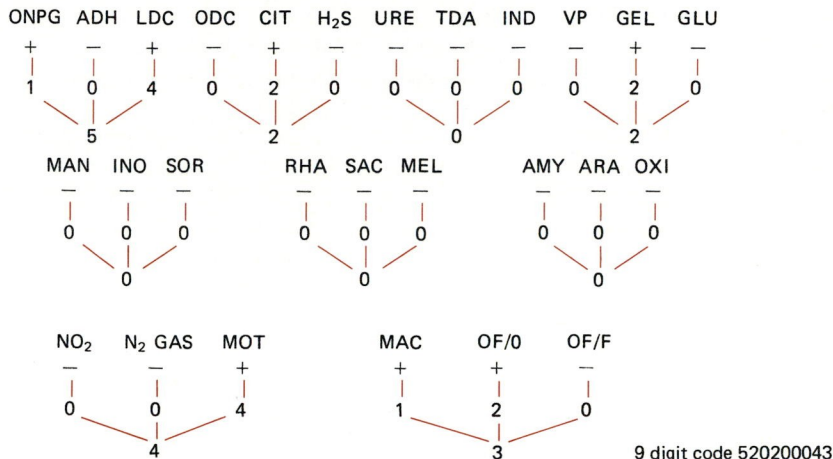

9 digit code 520200043 = *Pseudomonas maltophilia*

Features Examined in the API 20-E System

Characteristics	Visual reactions	
	Positive	Negative
ONPG (β–galactosidase)	Yellow	Colorless
Arginine dihydrolase	Red-orange	Yellow
Lysine decarboxylase	Red-orange	Yellow
Ornithine decarboxylase	Red-orange	Yellow
Citrate	Dark blue	Light green
Hydrogen sulfide	Blackening	Colorless
Urease	Cherry red	Yellow
Tryptophan deaminase (add 10% $FeCl_3$)	Red-brown	Yellow
Indole	Red ring	Yellow
Voges–Proskauer (add KOH plus α–naphthol)	Red	Colorless
Gelatin	Pigment diffusion	No pigment diffusion
Glucose	Yellow	Blue-green
Mannitol	Yellow	Blue-green
Inositol	Yellow	Blue-green
Sorbitol	Yellow	Blue-green
Rhamnose	Yellow	Blue-green
Sucrose	Yellow	Blue-green
Melibiose	Yellow	Blue-green
Amygdalin	Yellow	Blue-green
Arabinose	Yellow	Blue-green

FIGURE 10.9

Conversion of API 20-E test results to a numerical code for both fermentative and nonfermentative bacteria.

identification of anaerobic bacteria. For the identification of nonfermentative, Gram negative rods, six additional tests are run to generate a nine-digit biotype identification number. Over 100 taxa of Gram-negative rods can be identified by using the API-20E system.

The **Minitek system** consists of reagent-impregnated paper disks to which a broth suspension is added for the determination of characteristic reactions. The number of disks and tests employed is variable; generally, 17 tests at a time are used to differentiate clinical isolates. The profile of test results obtained with the Minitek system is useful in identifying obligate anaerobes as well as facultative enteric bacteria.

In contrast, the **Micro-ID system**, which is based upon constitutive enzymes, is designed primarily for identifying members of the Enterobacteriaceae, although it can also be used for biotyping *Haemophilus influenzae* and *H. parainfluenzae*. Bacteria must be screened for oxidase activity, and only oxidase-negative strains are tested with this system. All commercial ID systems actually recommend performing the oxidase test before attempting an

identification. This system employs 15 reaction chambers, and the test results are used to generate a five-digit identification code number. The Micro-ID system lists possible identifications and probabilities based on the results of the 15 biochemical test reactions. Identifications with the Micro-ID system can be accomplished in as little as 4 hours.

The API 20-E and Minitek systems have been expanded to identify obligate anaerobes. In order to use these systems to identify anaerobes, the isolates are grown in a liquid culture medium suitable for the growth of anaerobes, and the tests are carried out under anaerobic conditions. There is also a 20 test-API-20C system and a Uni-Yeast-Tek system with 11 test chambers that can be used in clinical laboratories for yeast identification. Several other systems are also available for the identification of nonfermentative, Gram-negative rods, such as the Automicrobic (AMS), Oxi-Ferm, and Oxitech systems. The Oxi-Ferm system is similar in design to the Enterotube, but eight different media are used in the tube for the identification of nonfermentative bacteria.

10.4 *Bacterial Systematics*

Having discussed the general aspects of microbial nomenclature, classification, and identification, we will now consider bacterial systematics. Extremely diverse groups of organisms exhibiting widely differing morphological, ecological, and physiological properties are found among the bacteria. The unifying feature of the bacteria is the fact that they all have prokaryotic cells. Bacterial systematics is always in a state of flux, with descriptions of new bacterial genera and species, and revisions of older taxonomic classifications frequently appear in the *International Journal of Systematic Bacteriology*. Periodically, the status of bacterial taxonomy has been summarized in a comprehensive volume, *Bergey's Manual of Determinative Bacteriology*. The eighth edition of *Bergey's Manual*, published in 1974, was the last comprehensive volume produced.

Starting with the latest edition, *Bergey's Manual*—now called *Bergey's Manual of Systematic Bacteriology*—will appear in separate parts describing the taxonomic status of particular groups of microorganisms; Volume 1 deals only with the Gram-negative bacteria of general, medical, and industrial importance; Volume 2 deals with the Gram-positive bacteria other than the filamentous actinomycetes; Volume 3 deals with the archaebacteria, the cyanobacteria, and the remaining Gram-negative bacteria including the chemolithotrophs; and Volume 4 covers the Gram-positive, filamentous bacteria.

Aside from its being the last unified volume, several very

significant features characterized the eighth edition of *Bergey's Manual*. The cyanobacteria were included as a division of the kingdom Prokaryotae, recognizing that the blue-greens are properly considered bacteria and not algae. The bacteria were placed within a single kingdom (the Prokaryotae) that is separated into two divisions (division I, the Cyanobacteria; division II, the Bacteria). In the latest edition of *Bergey's Manual*, four divisions of the kingdom Prokaryotae are recognized: the Gracilicutes (bacteria with a typical Gram-negative cell wall); Firmicutes (bacteria with a Gram-positive cell wall); Tenericutes (bacteria lacking a cell wall); and Mendosicutes (bacteria lacking peptidoglycan in their cell walls).

Perhaps of even greater significance, the newest editions of *Bergey's Manual* do not use a hierarchical ordering of bacteria; instead, they simply organize the bacteria into groups that have no official taxonomic status. To quote the editors: "The manual is meant to assist in the identification of bacteria. No attempt has been made to provide a complete hierarchy, as in previous editions, because a complete and meaningful hierarchy is impossible." The significance of this philosophical change in *Bergey's Manual* has yet to be appreciated by most microbiologists, who still turn to *Bergey's* as the official registry of the taxonomic status of bacteria. The publication of the first edition of *Bergey's Manual of Systematic Bacteriology* as multiple volumes published over several years means that there is no longer a single reference source that describes bacterial taxonomy.

10.5 *The Major Groups of Bacteria*

Following the lead of *Bergey's Manual*, this section will discuss the major groups of bacteria without attempting to place them into a hierarchical classification system (Table 10.5). These groups of bacteria are defined primarily on the basis of physiological and morphological criteria. In the remainder of this chapter, we will consider the diversity of these organisms, for the most part in the order in which they are described in *Bergey's Manual*, giving a very brief description of each group.

The Eubacteria

The Spirochetes   The **spirochetes** are helically coiled rods, with the cell wound around one or more central axial fibrils (Figure 10.10). The cell length varies in different genera from 3 to 500 μm. In addition to their characteristic morphology, the spirochetes exhibit a unique mode of motility. These bacteria move by a flexing motion of the cell and exhibit their greatest velocities in very viscous solutions where motility by bacteria with external flagella is slowest. Five genera are recognized in the family Spirochaetaceae, and several more have been proposed (Table 10.6). Members of the genus *Spirochaeta* are nonpathogens, occurring in aquatic environments, in hydrogen sulfide–containing mud, in sewage, and in polluted waters. Many spirochetes, though, are human pathogens. Several members of the genus *Treponema*, for example, are human pathogens, with *T. pallidum* causing syphilis and *T. pertenue* causing yaws. Like-

TABLE 10.5   Major Groups of Bacteria

Eubacteria
  The spirochetes
  Aerobic or microaerophilic, Gram-negative, spiral and
     curved bacteria
  Gram-negative aerobic rods and cocci
  Gram-negative, facultatively anaerobic rods
  Gram-negative, anaerobic bacteria
  Dissimilatory sulfate- or sulfur-reducing bacteria
  Gram-negative anaerobic cocci
  The rickettsia and chlamydias
  The mycoplasmas
  Endosymbionts
  Gram-positive cocci
  Endospore-forming rods and cocci
  Gram-positive, asporogenous rods of regular shape
  Gram-positive rods of irregular shape
  Anaerobic, nonfilamentous or filamentous rods
  Mycobacteria and nocardioforms
  Phototrophic bacteria
    Anoxyphotobacteria
    Oxyphotobacteria
  Gliding bacteria
  Sheathed bacteria
  Budding and/or appendaged bacteria
  Chemolithotrophic bacteria
  Actinomycetes
Archaebacteria
  Methane-producing bacteria
  Halobacteriaceae
  Sulfolobus

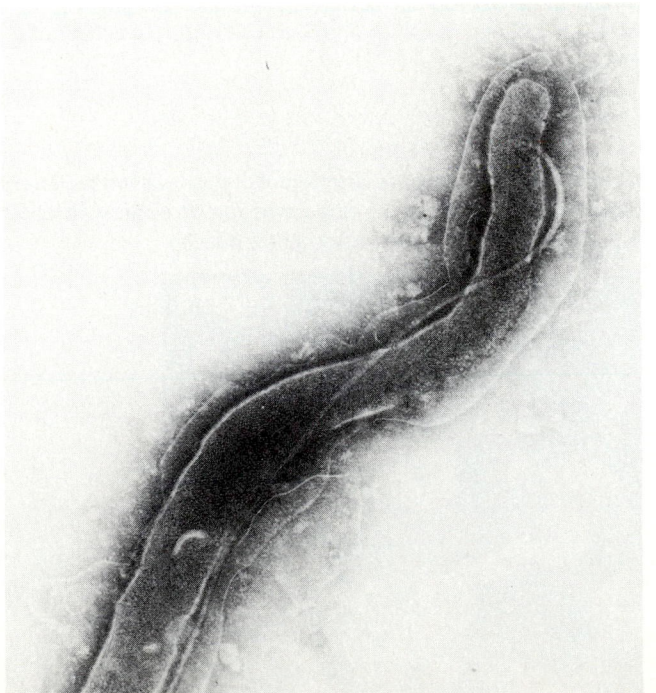

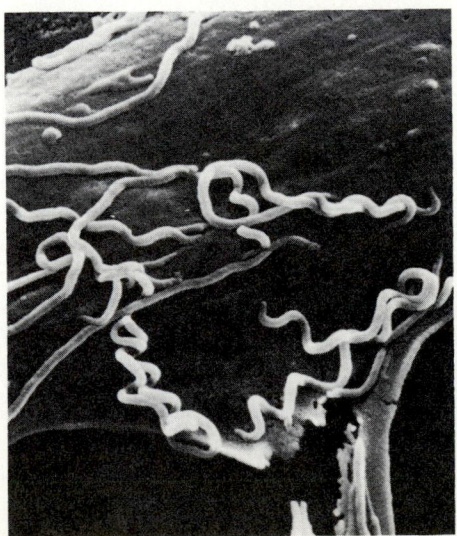

FIGURE 10.10

Micrographs of spirochetes. (A) Spirochete showing the central axial filament (99,560×). (B) *Treponema pallidum* (8,000×). (From BPS— Stanley C. Holt, University of Massachusetts)

wise, some members of the genera *Cristispira*, *Borrelia*, and *Leptospira* are animal and human pathogens.

Aerobic or Microaerophilic, Gram-Negative, Spiral and Curved Bacteria   Members of the heterogeneous group of aerobic or microaerophilic, Gram-negative, spiral and curved bacteria are helically curved rods that may have less than one complete turn (vibroid or comma shaped) or many turns (helical), but unlike the spirochetes, the cells are not wound around a central axial filament. The cells of members of this group are motile by means of polar flagella. Members of *Spirillum*, for example, have multiple polar flagella, usually at both ends (Figure 10.11). The original genus *Spirillum* has recently been divided into several genera—*Aquaspirillum*, *Spirillum*, *Azospirillum*, and *Oceanospirillum*—based upon genetic analyses that showed the mole% G + C for these genera to be 38, 70, 42–51, and 30–38, respectively.

Some members of *Campylobacter*, another genus affiliated with this group, are important pathogens of humans and other animals. *Bdellovibrio*, a genus of uncertain affiliation within this group, has the outstanding characteristic of being able to penetrate and reproduce within prokaryotic cells (Figure 10.12). All naturally occurring strains of *Bdellovibrio* have been found to be bacterial parasites. Unlike viruses, *Bdellovibrio* do not lose their integrity when they reproduce within host cells. Within host cells, bacteria in the genus

Bdellovibrio reproduce by binary fission. After reproduction of *Bdellovibrio* within a host cell, the host cell lyses, releasing the *Bdellovibrio* progeny. Another curved bacterium, *Microcyclus*, forms unique closed, ring-shaped cells (Figure 10.13); these bacteria are nonmotile and are considered in the latest edition of *Bergey's Manual* to be part of a separate group that also includes *Spiromonas*, *Runella*, *Flectobacillus*, *Meniscus*, *Brachyarcus*, and *Pelosigma*.

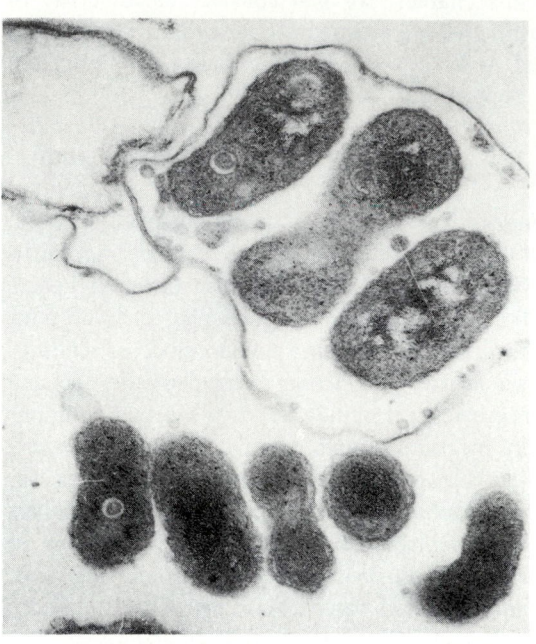

FIGURE 10.12
Bdellovibrio stolppii, an elongated, coiled organism, is shown here within the cell wall of its host *E. coli* and growing saprophytically adjacent. Note the curved shape of the cell to which the arrow points. (Courtesy Jeffrey C. Burnham, Medical College of Ohio, and Sam F. Conti, University of Massachusetts)

TABLE 10.6   The Spirochetes

Order Spirochaetales	
Family Spirochaetaceae	
Spirochaeta	5–500 μm long, 0.2–0.75 μm wide, free-living; anaerobic or facultative
Cristispira	30–150 μm long, 0.5–3.0 μm wide with 3 to 10 complete turns; not free-living
Treponema	5–15 μm long, 0.09–0.5 μm wide; not free-living; anaerobic
Borrelia	3–15 μm long, 0.2–0.5 μm wide; not free-living; anaerobic
Leptospira	6–20 μm long, 0.1 μm wide; aerobic

FIGURE 10.13
Micrographs of *Microcyclus marinus* Raj (= *Cyanobacterium marinus* Raj), showing the ring formation. (Courtesy H. D. Raj, California State University, Long Beach)

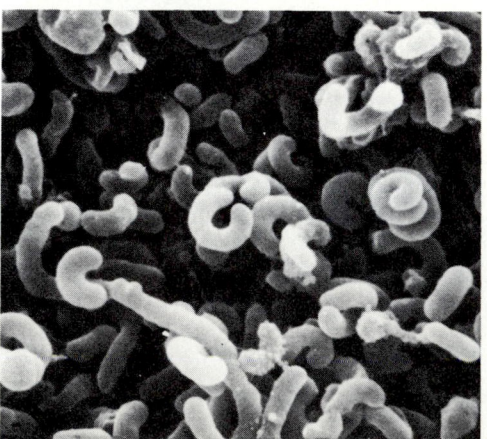

FIGURE 10.11
Micrograph of *Aquaspirillum bengal* showing its helical shape. The arrows point to tufts of flagella that emanate from the ends of this bacterial cell. (From BPS—Terry Beveridge, University of Guelph)

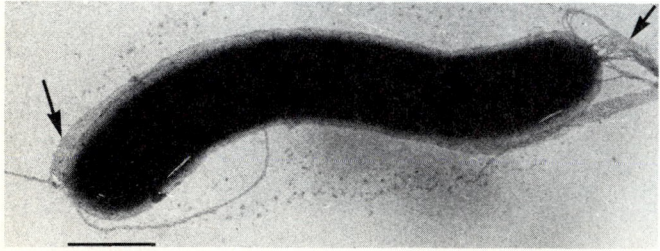

TABLE 10.7   Gram-Negative, Aerobic Rods and Cocci

Taxonomic Group	Flagella	Carbon Source	N₂ Fixation	Salt Requirement
Pseudomonadaceae	Polar	Numerous	No	3% NaCl for marine species
Azotobacteraceae	Peritrichous or polar	Numerous	Yes	None
Rhizobiaceae	Peritrichous or polar	Numerous	Yes	None
Methylococcaceae	Polar or none	1-carbon compounds only; methane oxidized	Yes (some)	None
Halobacteriaceae	Polar	Various	No	12–15% NaCl
Acetobacteraceae	Peritrichous or polar	Various; ethanol oxidized to acetic acid	No	None
Legionellaceae	Polar and lateral	Various; requires growth factors	No	None
Neisseriaceae	None	Various	No	None

Gram-Negative Aerobic Rods and Cocci   The Gram-negative aerobic rods and cocci encompass a large number of taxonomic units. Several major families are included in this group: Pseudomonadaceae, Azotobacteraceae, Rhizobiaceae, Methylomonadaceae, Halobacteriaceae, Acetobacteriaceae, Legionellaceae, and Neisseriaceae (Table 10.7). Members of the family Pseudomonadaceae are Gram-negative, straight or curved rods that are motile by means of polar flagella. Because their metabolism is respiratory, members of this family cannot carry out fermentative metabolism. Whereas the metabolism of the **Pseudomonadaceae** is obligately aerobic, some strains are able to carry out anaerobic respiration. For example, some pseudomonads, such as *Pseudomonas denitrificans*, are able to use nitrate as a terminal electron acceptor in anaerobic respiration, forming molecular nitrogen in a process called *denitrification*. The Pseudomonadaceae are unable to fix atmospheric nitrogen. Many *Pseudomonas* species are nutritionally versatile and are capable of degrading many natural and synthetic organic compounds. Some *Pseudomonas* species produce characteristic fluorescent pigments, but others do not. For example, *P. aeruginosa* produces yellow-green diffusible pigments that fluoresce when excited at a wavelength of less than 260 nm. *Pseudomonas* species are widely distributed in soil and aquatic ecosystems, occurring as free-living bacteria or in association with plants and animals, and some species are plant and animal pathogens. *P. aeruginosa*, for example, can be a human pathogen and is commonly isolated from wound, burn, and urinary tract infections. All recognized species of *Xanthomonas*, a genus included in the Pseudomonadaceae, are plant pathogens. *Xanthomonas* species are Gram-negative rods that are motile by means of polar flagella and in most cases produce yellow pigments.

Whereas nitrogen fixation does not occur in the Pseudomonadaceae, the family **Azotobacteraceae** is characterized by its capacity to fix molecular nitrogen. This family consists of Gram-negative rods exhibiting pleomorphic morphology. The genera *Azotobacter* and *Beijerinckia* are particularly im-

portant free-living, nitrogen-fixing bacteria. The practical importance of these bacteria will be considered later in the section on environmental microbiology. The **Rhizobiaceae** are also capable of fixing atmospheric nitrogen. *Rhizobium* and slower-growing *Bradyrhizobium* species are able to infect leguminous plant roots, causing the formation of tumorous growths called *nodules*. Free-living cells of these bacteria are rod-shaped, but within the nodules they occur as pleomorphic (irregularly shaped) cells termed **bacteroids**. *Rhizobium* can fix atmospheric nitrogen only within root nodules and thus is considered an obligately symbiotic nitrogen fixer, whereas some strains of *Bradyrhizobium* can fix nitrogen nonsymbiotically under defined laboratory conditions. Unlike *Rhizobium* and *Bradyrhizobium*, *Agrobacterium* species do not fix molecular nitrogen. *Agrobacterium* species, however, produce tumorous growths on infected plants known as **galls**. *Agrobacterium tumefaciens* causes galls of many different plants and is an extremely important plant pathogen causing large economic losses in agriculture.

The family **Methylomonadaceae** are bacteria that can utilize carbon monoxide, methane, or methanol as their sole source of carbon. The ability to use 1-carbon-containing (C_1) organic compounds as the sole source of carbon and energy requires a special metabolic capability. Some of these bacteria are restricted to growth on C_1 compounds. The metabolism of these organisms is respiratory, using molecular oxygen as the terminal electron acceptor. Members of this group are of interest to industry as a potential source of protein for animal feed or as a human dietary supplement. The C_1 compounds, such as methane and methanol, are considered prime candidates as substrates for industrial processes aimed at growing microorganisms as a source of protein.

The Legionellaceae includes only one genus, *Legionella*. Species of *Legionella* have unique physiological properties; they are Gram negative, nonfermentative rods that exhibit a requirement for iron and cysteine as growth factors, and form predominantly branched chain fatty acids. *Legionella pneumophila*, which causes Legionnaire's disease, frequently is

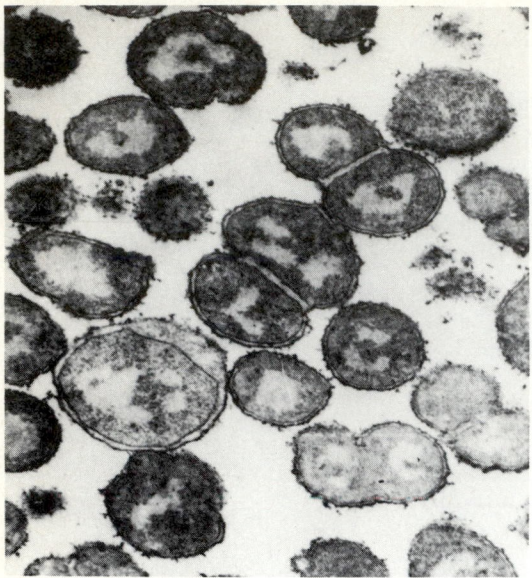

FIGURE 10.14
Electron micrograph of *Neisseria gonorrhoeae*, showing diplococci (100,000×). (From BPS—Centers for Disease Control, Atlanta)

associated with rapidly evaporating water bodies, including air-conditioning cooling towers.

Only one family of Gram-negative cocci and coccobacilli, **Neisseriaceae**, is recognized (Figure 10.14). The family Neisseriaceae includes the genera *Neisseria*, *Branhamella*, *Moraxella*, and *Acinetobacter*. The cells of *Neisseria* and *Branhamella* are cocci, whereas those of *Moraxella* and *Acinetobacter* are coccobacilli (oval-shaped) (Table 10.8). Members of the genera *Neisseria*, *Branhamella*, and *Moraxella* are parasitic, and some are important human pathogens. *Acinetobacter* species are saprophytic, although some are opportunistic pathogens. Members of the genus *Acinetobacter* are nutritionally versatile and can utilize a variety

TABLE 10.8   Descriptions of Some Genera of Gram-Negative Cocci and Coccobacilli

Genus	Description
Neisseria	Cocci; divide in two planes; sensitive to penicillin; oxidase positive; G + C% 47–52
Branhamella	Cocci; divide in two places; oxidase positive; reduce nitrates; G + C% 40–45
Moraxella	Coccobacilli; divide in one plane; oxidase positive; sensitive to penicillin; G + C% 40–46
Acinetobacter	Coccobacilli; divide in one plane; oxidase negative; penicillin resistant; G + C% 39–47
Paracoccus	Cocci; divide in one plane; aerobic; G + C% 64–67
Lampropedia	Cells rounded or cubical; aerobic; divide in two planes; G + C% 61

of organic compounds as their sole source of carbon and energy. Several species of the genus *Neisseria* are important human pathogens. For example, *N. gonorrhoeae* causes gonorrhea and *N. meningitidis* causes meningitis.

Several genera of uncertain affiliation are Gram-negative, aerobic rods. These include *Brucella*, *Bordetella*, and *Francisella*. Some members of these three genera are important human pathogens; *Bordetella pertussis* is the causative agent of whooping cough and *Francisella tularensis* is the causative agent of tularemia. Other genera of uncertain affiliation in this group are *Alcaligenes*, *Acetobacter*, and *Thermus*. *Thermus* is an ecologically interesting genus that grows well at temperatures over 70°C. Strains of this organism have been isolated from hot springs and the hot water tanks of laundromats.

Gram-Negative Facultatively Anaerobic Rods   There are two major families of Gram-negative, facultatively anaerobic rods: the **Enterobacteriaceae**, which are motile by means of peritrichous flagella, and the **Vibrionaceae**, which are motile by means of polar flagella (Table 10.9). The Enterobacteriaceae are divided further into five tribes: the Escherichieae, Klebsielleae, Proteeae, Yersinieae, and Erwinieae. These tribes are distinguished on the basis of their metabolic pathways. Many of the bacteria studied in introductory microbiology laboratory courses belong to the Gram-negative, facultatively anaerobic rods.

The family Enterobacteriaceae includes the genera *Escherichia*, *Edwardsiella*, *Citrobacter*, *Salmonella*, *Shigella*, *Klebsiella*, *Enterobacter*, *Hafnia*, *Serratia*, *Proteus*, *Yersinia*, and *Erwinia* (Figure 10.15). Members of the genus *Escherichia* occur in the human intestinal tract. *E. coli* has achieved a special place in microbiology. It has been used as the test organism in many metabolic and genetic studies, and much of what we know about bacterial metabolism and genetics has been elucidated in studies using *E. coli*. In addition, *E. coli* is employed as an indicator of fecal contamination in environmental microbiology. The genera *Salmonella* and *Shigella* contain numerous species, many of which are important human pathogens. In particular, typhoid fever and various gastrointestinal upsets are caused by *Salmonella* species, and bacterial dysentery is caused by *Shigella*. *Serratia marcesens*, once thought to be a nonpathogen, is now recognized as causing insect diseases and as an opportunistic human pathogen. *Serratia* strains can produce a red pigment known as *prodigiosin*. All members of the genus *Erwinia* are plant pathogens. *Erwinia amylovora*, for example, causes fire blight of pears and apples.

The family **Vibrionaceae** includes the genera *Vibrio*, *Aeromonas*, *Plesiomonas*, and *Photobacterium*. Many of the *Vibrio* have curved, rod-shaped cells. The habitat of *Vibrio* species is generally aquatic. *V. cholerae* is an important human pathogen that causes cholera. *Photobacterium* species are interesting because of their ability to luminesce. The mechanism of luminescence involves an ATP-driven reaction, an electron transport system, and a key reaction me-

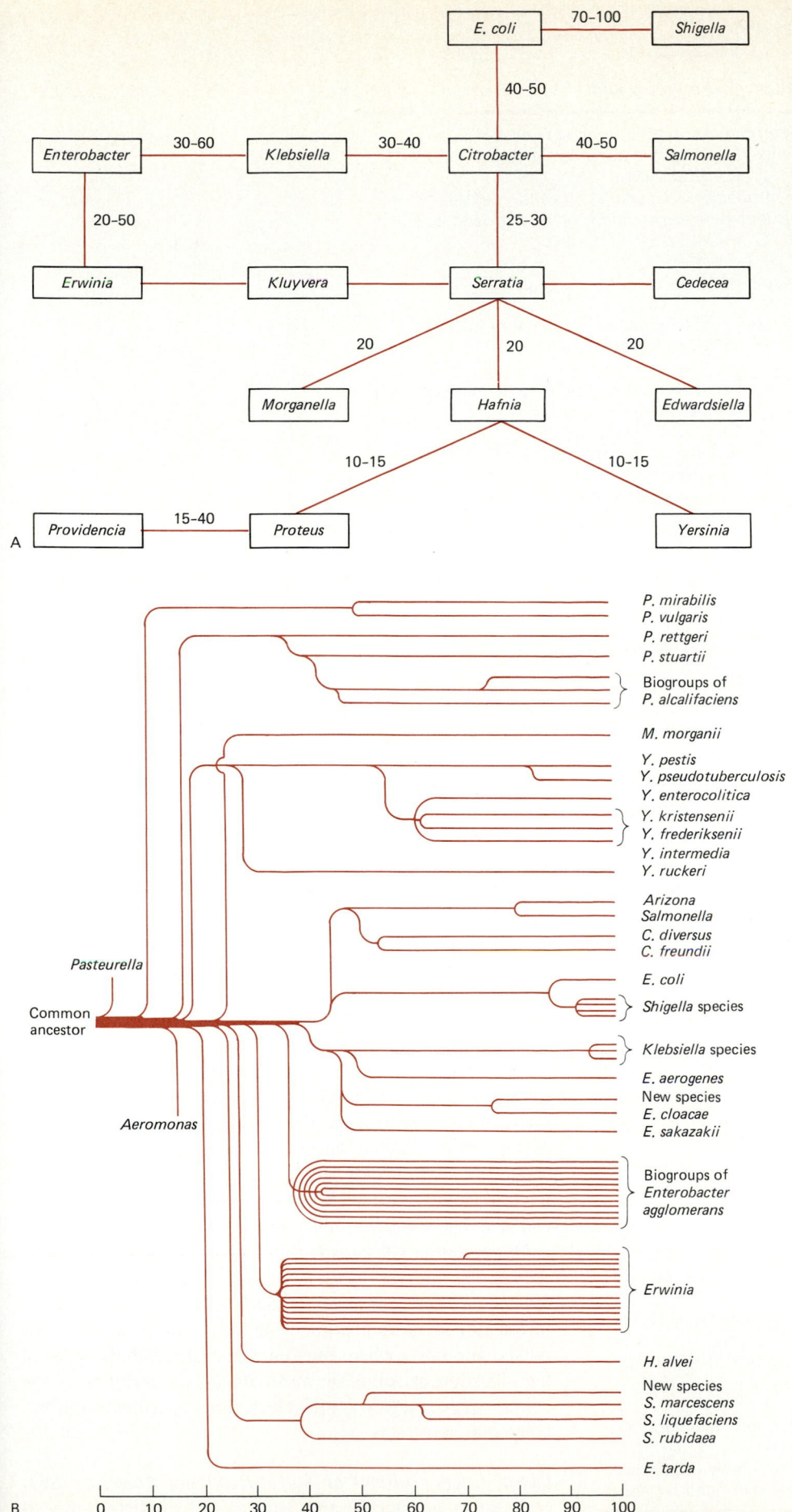

FIGURE 10.15
(A) Genetic relationships among the genera of Enterobacteriaceae. The numbers represent the approximate percentage of relatedness based upon DNA hybridization analyses.
(B) Dendrogram showing the divergence of Enterobacteriaceae based upon the assumption that all of the organisms diverged from a common ancestor. The vertical axis shows the percentage of relatedness between the organisms.

285

TABLE 10.9   The Gram-Negative, Facultatively Anaerobic Rods

Family Enterobacteriaceae	Peritrichous flagella; nonpigmented
Tribe Escherichiae	Mixed acid fermentation; optimal growth at 37°C; G + C% 50–53
Genera: *Escherichia, Edwardsiella, Citrobacter, Salmonella, Shigella*	
Tribe Klebsielleae	Butanediol fermentation; optimal growth at 37°C; G + C% 52–59
Genera: *Klebsiella, Enterobacter, Hafnia, Serratia*	
Tribe Proteeae	Optimal growth at 39°C; G + C% 39–42
Genus: *Proteus*	
Tribe Yersiniae	Mixed acid fermentation; optimal growth at 30–37°C; G + C% 45–47
Genus: *Yersinia*	
Tribe Erwineae	Mixed acid and butanediol fermentation; optimal growth at 27–30°C; G + C% 50–58
Genus: *Erwinia*	
Family Vibrionaceae	Polar flagella; nonpigmented; G + C% 39–63
Genera: *Vibrio, Aeromonas, Pleisomonas, Photobacterium*	
Family Pasteurellaceae	Nonmotile; nonpigmented; fermentative; G + C% 38–47
Genera: *Pasteurella, Actinobacillus, Haemophilus,*	
Genera of uncertain affiliation	
Chromobacterium, Flavobacterium, Zygomonas, Cardiobacterium, Streptobacillus Calymmatobacterium	

diated by the enzyme luciferase. Some species of luminescent bacteria occur in association with fish; some of these fish are known as *flashlight fish* because of the light emitted by these bacteria. The association of luminescent bacteria with these fish is important in various behavioral aspects of these fish, including mating activities.

Several genera of uncertain affiliation are Gram-negative, facultatively anaerobic rods. These include the genus *Chromobacterium*, which produces violet pigments. *Flavobacterium*, which produces a variety of insoluble yellow, orange,

red, or brown pigments, has been associated with this group, but, based on its mole% G + C, *Flavobacterium* clearly represents more than one group; in group I it is 30–42 and in group II it is 63–70. Also included in this group are the genera *Zymomonas*, *Cardiobacterium*, *Streptobacillus*, and *Calymmatobacterium*. Various species in these genera are important human pathogens.

Gram-Negative Anaerobic Bacteria   There is only one family, **Bacteroidaceae**, and relatively few genera in the Gram-negative, anaerobic bacteria group (Table 10.10). The genera include *Bacteroides*, *Fusobacterium*, and *Leptotrichia*. *Bacteroides* are important members of the normal microbiota of humans and may be the dominant microorganisms in the intestinal tract. *Bacteroides* species characteristically form pleomorphic rods (Figure 10.16, page 287). Some members of the family Bacteroidaceae are human pathogens. Some species of *Leptotrichia*, for example, produce abscesses in the oral cavity. Various *Bacteroides* and *Fusobacterium* species also cause human infections.

Other genera affiliated with the Gram-negative, anaerobic rods include *Butyrovibrio*, *Lachnospira*, *Succinovibrio*, *Succinomonas*, and *Selenomonas*. Species of these genera occur in the rumen (a compartment of the stomach of cows and related animals), where they play critical metabolic roles in the digestion of cellulosic materials and the nutrition of the animal, producing low molecular weight fermentation carbohydrate substrates.

Dissimilatory Sulfate- or Sulfur-Reducing Bacteria   Several genera of bacteria are characterized by their ability to

TABLE 10.10   The Gram-Negative Anaerobic Bacteria

Family Bacteriodaceae	
Bacteroides	Produce mixtures of acids including succinic, acetic, formic, lactic, and propionic
Fusobacterium	Produce butyric acid as the major product
Leptotrichia	Produce lactic acid as the only major fermentation acid
Genera of uncertain affiliation	
Butyrovibrio	Curved rods; fermentative; produce butyrate
Succinovibrio	Curved rods; fermentative; produce succinate and acetate
Succinomonas	Straight rods; fermentative; produce large amounts of succinate and some acetate
Lachnospira	Curved rods; fermentative; produce mixture of acids
Selenomonas	Curved rods; fermentative; produce acetate, propionate, and lactate

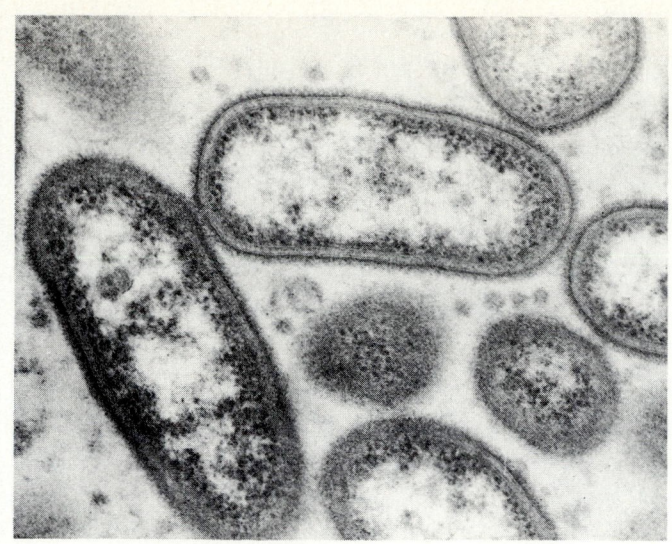

FIGURE 10.16

Bacteroides asaccharolyticus is a Gram-negative rod isolated from the oral cavity, where it causes gingivitis. (From BPS—Stanley C. Holt, University of Massachusetts)

reduce sulfate, other oxidized sulfur-containing compounds, or elemental sulfur (Table 10.11). Members of the genus *Desulfovibrio* are curved, Gram-negative rods capable of reducing sulfates, or other reducible sulfur compounds, to hydrogen sulfide. Sulfate-reducing bacteria, including *D. desulfuricans*, are normally found in anaerobic soils and sediments—such as those found in bogs and marshes—where they play an important role in the biogeochemical cycling of sulfur.

Gram-Negative Anaerobic Cocci   The Gram-negative, anaerobic cocci include the genera *Veillonella*, *Acidaminococcus*, *Megasphaera*, and *Gemmiger*. Each of these genera contains very few species. The cells of species in all of these genera typically occur as diplococci (pairs of cocci). *Veillonella* species have complex nutritional requirements and are unable to grow on individual organic substrates. They

TABLE 10.11   The Sulfate- and Sulfur-Reducing Bacteria

Genus	Description
Desulfovibrio	Curved rods; motile; anaerobic respiration using sulfate or other sulfur compounds as terminal electron acceptors and producing hydrogen sulfide; G + C% 46.1–61.2
Desulfuromonas	Acetate completely oxidized to CO_2 using elemental sulfur as the electron acceptor; sulfate never reduced; dissimilatory reduction of sulfur; straight or slightly curved rods; motile; G + C% 50–63
Desulfomonas	Straight rods; nonmotile; reduce sulfate to hydrogen sulfide; incomplete oxidation of pyruvate to acetate and CO_2; G + C% 66–67
Desulfococcus	Spherical cells; nonmotile; completely oxidize fatty acids or benzoate; sulfate and other oxidized sulfur compounds serve as terminal electron acceptors and are reduced to hydrogen sulfide; ferment lactate or pyruvate to acetate and propionate; G + C% 57.4
Desulfobacter	Ellipsoidal to rod-shaped cells with rounded ends; nonmotile or motile by polar flagellum; oxidize acetate to CO_2; sulfate or other oxidized sulfur compounds serve as terminal electron acceptors and are reduced to hydrogen sulfide; G + C% 45.9
Desulfobulbus	Ellipsoidal cells with pointed ends; motile by polar flagellum; incompletely oxidize propionate or lactate to acetate and CO_2; in absence of an external electron acceptor, ferment pyruvate or lactate to propionate and acetate; sulfate or other oxidized sulfur compounds serve as terminal electron acceptors and are reduced to hydrogen sulfide; G + C% 59.9
Desulfosarcina	Irregularly shaped cells that tend to be coccoid after extended growth; usually nonmotile; benzoate can serve as electron donor for anaerobic respiration; chemoautotrophic growth using molecular hydrogen as electron donor and CO_2 as carbon source; sulfate or other oxidized sulfur compounds serve as terminal electron acceptors and are reduced to hydrogen sulfide; ferment pyruvate or lactate to propionate and acetate; G + C% 51.2
Desulfotomaculatum	Anaerobic endospore former; sulfate or other oxidized sulfur compounds serve as terminal electron acceptors and are reduced to hydrogen sulfide
Desulfonema	Rod shaped; gliding motility; sulfate or other oxidized sulfur compounds serve as terminal electron acceptors and are reduced to hydrogen sulfide

ANALYTICAL PROCESS

Identification of Anaerobes

A novel approach to the identification of obligate anaerobes used in clinical laboratories involves the gas liquid chromatographic (GLC) detection of metabolic products. The anaerobes are grown in a suitable medium, and the short chain volatile fatty acids produced are extracted in ether. Fatty acids detected in this procedure include acetic, propionic, isobutyric, butyric, isovaleric, valeric, isocaproic, and caproic acids (Figure 10.17). The pattern of fatty acid production can be used to differentiate and identify various anaerobes. When coupled with observations of colony and cell morphology and a limited number of biochemical tests, the common anaerobes isolated from clinical specimens can be identified.

FIGURE 10.17

Gas chromatographic tracing, showing detection of the fatty acids produced by anaerobes:
(A) *Propionibacterium* sp.; (B) *Bacteroidies fragilis*.

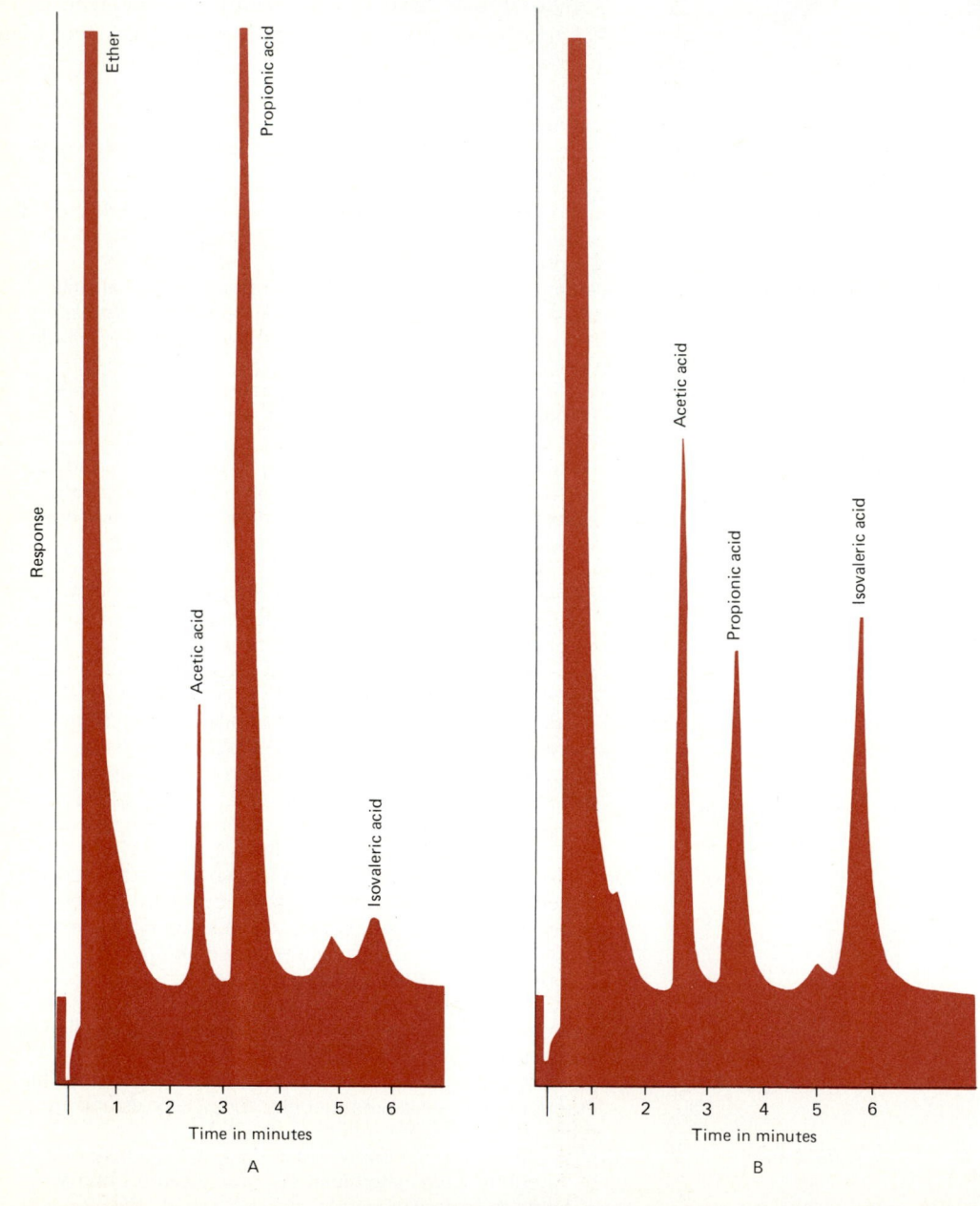

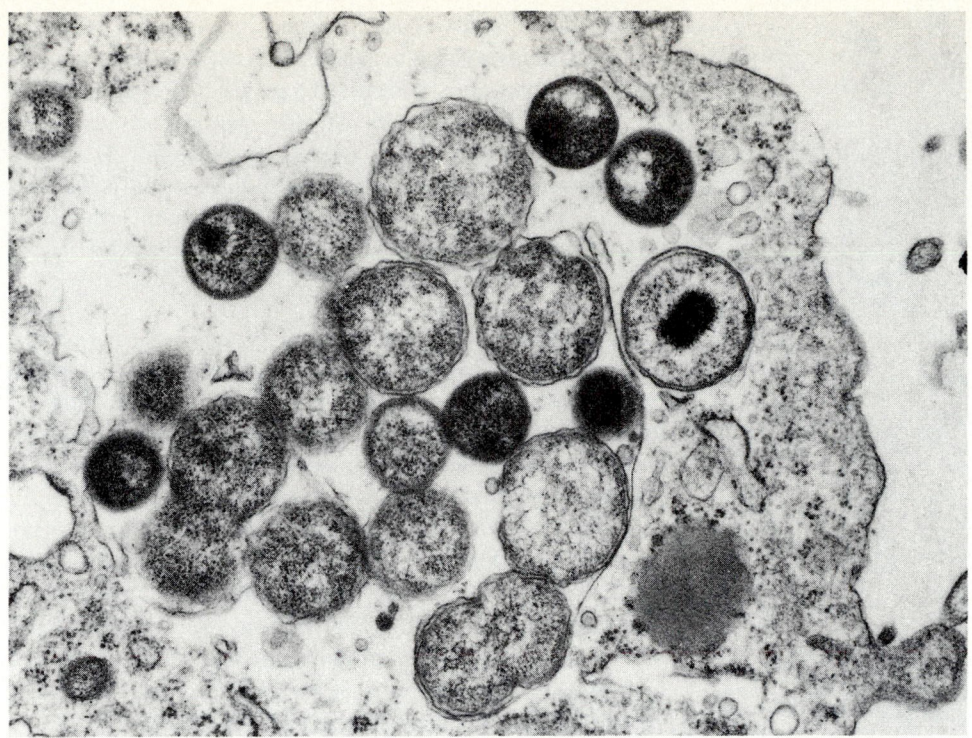

FIGURE 10.18
Micrograph of *Chlamydia psittaci* in the cytoplasm of a cell (47,500×). (From BPS—R. C. Cutlip, National Animal Disease Center, Ames, Iowa)

also require carbon dioxide for growth. Although these organisms are fastidious in their nutritional requirements, they comprise part of the normal human microbiota, representing, for example, 5–16 percent of the bacteria found in the oral cavity. Some *Veillonella* species are human pathogens, causing infections in the oral cavity and the intestinal and respiratory tracts.

The Rickettsias and Chlamydias   The **rickettsias** are intracellular parasites. The majority of members of the Rickettsiales are Gram-negative and multiply only within host cells. Within host cells the rickettsias reproduce by binary fission. They lack the enzymatic capability to produce sufficient amounts of ATP to support their reproduction; they are able to obtain the ATP from the host cells in which they grow.

Many species of Rickettsiales cause disease in humans and other animals. Some members of the family Rickettsiaceae are adapted to existence in arthropods but are capable of infecting vertebrate hosts, including humans. For example, many members of the genus *Rickettsia* are carried by insect vectors and cause diseases in humans; *R. rickettsii* is transmitted by ticks and causes Rocky Mountain spotted fever.

The **chlamydias** are obligate intracellular parasites whose reproduction is characterized by the change of the small, rigid-walled infectious form of the organism (**elementary body**) into a larger, thin-walled, noninfectious form (**initial body**) that divides by fission (Figure 10.18). Members of the order Chlamydiaceae are metabolically limited. These organisms are unable to generate sufficient ATP to support their reproduction. The chlamydias have sometimes been referred to as *large viruses*, but they are truly bacteria. Members of the genus *Chlamydia* are Gram-negative. The reproductive cycle for these organisms takes about 40 hours. *Chlamydia* cause human respiratory and genitourinary tract diseases, and in birds they cause respiratory diseases and generalized infections. For example, the disease psittacosis (parrot fever) is caused by *C. psittaci*.

The Mycoplasmas   The **mycoplasmas** are classified as mollicutes (soft skin) because they lack a cell wall. They are bacteria that are bounded by a single triple-layered membrane. Although it is difficult to define the mycoplasmas in a way that clearly distinguishes them from cell wall–deficient mutants of other bacterial groups, the mycoplasmas appear to be fundamentally different from other bacterial groups. They are the smallest organisms capable of self-reproduction. When growing on artificial media, mycoplasmas form small colonies that have a characteristic "fried-egg" appearance (Figure 10.19). Members of the genus *Mycoplasma* require sterols for growth. Several of them cause diseases in humans. For example, some forms of pneumonia are caused by *Mycoplasma* species.

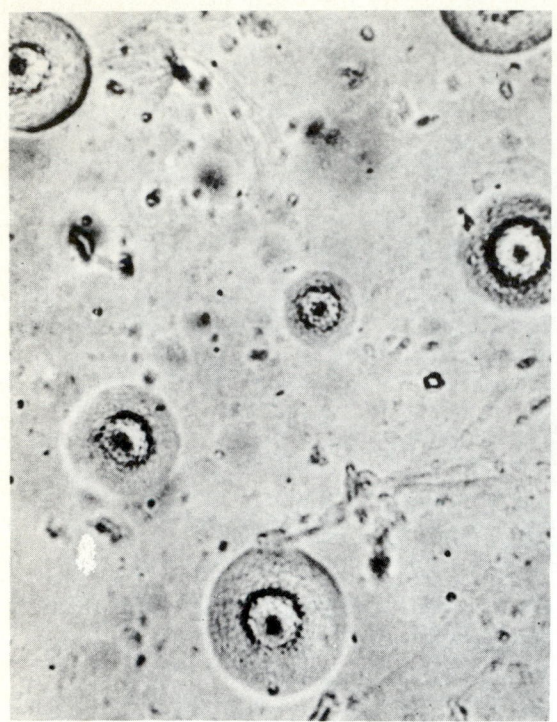

Like the colonies of *Mycoplasma*, the colonies of *Spiroplasma* species exhibit a typical biphasic fried-egg appearance. *Spiroplasma* is considered a genus of uncertain affiliation in the class Mollicutes. *Spiroplasma* species lack a cell wall, have a triple-layered cytoplasmic membrane, and require sterols for growth. They cause diseases in plants and animals. For example, *S. citri* causes "stubborn" disease of citrus plants. Suckling mouse cataract disease is also caused by a *Spiroplasma* species.

Endosymbionts   Several bacterial genera are obligate endosymbionts of invertebrates; that is, they live within the cells of invertebrate animals without adversely affecting the host. For example, the protozoan *Paramecium aurelia* can harbor a variety of endosymbiotic bacteria (Table 10.12). Additionally, new genera of endosymbionts have recently been described for other protozoa, insects, and various other invertebrates. Although these bacteria can be readily seen within host cells, difficulties in culturing them outside of the host have hampered efforts to determine their proper taxonomic status. Developing an understanding of the nutritional requirements of these bacteria has permitted the creation of complex media for their culture and identification.

FIGURE 10.19
Micrograph of colonies of *Mycoplasma*, showing the fried egg appearance. (Courtesy Leonard Hayflick, University of Florida)

TABLE 10.12   Symbionts of *Paramecium aurelia*

Genus	Common Names	Description
Caedibacter	Kappa	Varying in size and distinguished by the presence of a 0.5 μm-diameter inclusion within the host cell; exhibits killing of sensitive strains
Pseudocaedibacter	Pi	Slender rod until recently considered as a mutant of kappa; nonkilling symbiont
	Nu	A nonkilling symbiont similar in appearance to pi and mu
	Mu	Slender rod, often elongated; distinguished because its killing action is wholly dependent on cell–cell contact between mating paramecia
	Gamma	A diminutive bacterium, frequently appearing as doublets; strong killing of other strains is shown by gamma bearers
Tectibacter	Delta	Rod distinguished by an electron-dense material surrounding the outer of its two membranes
Lyticum	Lambda	Appears as a typical motile bacterium with peritrichous flagella, although its movement within the cytoplasm has not been observed
	Sigma	Largest of all endosymbionts of *Paramecium aurelia*; curved, flagellated rod resembling lambda
Hotospora	Omega	Present in the micronucleus existing in two forms, a short reproductive form and a long infective form with spiral tapered ends
	Iota	Present in the macronucleus existing in two forms, a short reproductive form and a long infective form with rounded ends
	Alpha	Present in the macronucleus existing in two forms, a short reproductive form and a long infective form with spiral tapered ends

Gram-Positive Cocci   The Gram-positive cocci include the families Micrococcaceae, Streptococcaceae, and Peptococcaceae (Table 10.13). The coccoid cells of the **Micrococcaceae** may occur singly or as irregular clusters (Figure 10.20). For example, the genus *Staphylococcus* typically forms grape-like clusters. Most strains of Staphylococcus can grow in the presence of 15 percent NaCl. Species of *Staphylococcus* commonly occur on skin surfaces. *S. aureus* is a potential human pathogen, infecting wounds and causing food poisoning.

In the family **Streptococcaceae**, Gram-positive cocci occur as pairs or chains (Figure 10.21). The metabolism of the Streptococcaceae is fermentative. Even though their metabolism is anaerobic, the Streptococcaceae are listed as being facultatively anaerobic because they grow in the presence of air. Indeed, many species *Streptococcus* occur in the oral cavity, where—although continuously exposed to air—they grow and where some cause formation of dental caries. Other streptococci cause more serious diseases, such as pneumonia due to *S. pneumoniae* infections and rheumatic fever associated with *S. pyogenes* infections.

The **Peptococcaceae** have complex nutritional requirements. Cells of organisms of this family may occur singly, in pairs, or in regular or irregular masses. They are obligately anaerobic and produce low molecular weight volatile fatty acids, carbon dioxide, hydrogen, and ammonia as the main products of amino acid metabolism.

Endospore-Forming Rods and Cocci   The endospore-forming rods and cocci are extremely important because of the heat resistance of the endospore structure. The genera *Bacillus*, *Sporolactobacillus*, *Clostridium*, *Desulfotomaculum*, *Sporosarcina*, and *Thermoactinomyces* are all characterized

FIGURE 10.20
Scanning electron micrograph of *Staphylococcus aureus*, showing grape-like clusters. (6,000×). (Courtesy Robert Apkarian, Primate Research Laboratory, Emory University)

TABLE 10.13   The Gram-Positive Cocci

Family Micrococcaceae	
Micrococcus	Irregular clusters; G + C% 60–75
Staphylococcus	Irregular clusters; G + C% 30–40
Planococcus	Tetrads; G + C% 39–52
Family Streptococcaceae	
Streptococcus	Pairs or chains; homofermentative; G + C% 33–42
Leuconostoc	Pairs or chains; heterofermentative; G + C% 38–44
Pediococcus	Pairs or tetrads; homofermentative; G + C% 34–44
Aerococcus	Pairs or tetrads; homofermentative; G + C% 36–40
Gemella	Single or pairs; G + C% 31–35
Family Peptococcaceae	
Peptococcus	Irregular clusters; no growth at pH 2.5; cellulose not degraded; G + C% 36–37
Peptostreptococcus	Chains; no growth at pH 2.5; cellulose not degraded; G + C% 33–34
Ruminococcus	Irregular clusters or chains; no growth at pH 2.5; cellulose degraded; G + C% 40–46
Sarcina	Tetrads or octads; growth at pH 2.5; cellulose not degraded; G + C% 28–31

FIGURE 10.21
Micrograph of *Streptococcus*. Note chains of coccus-shaped bacteria. This particular species, *S. mutans*, occurs on tooth surfaces, where it produces acid that erodes the enamel, leading to dental caries. (From BPS—Z. Skobe, Forsythe Dental Center)

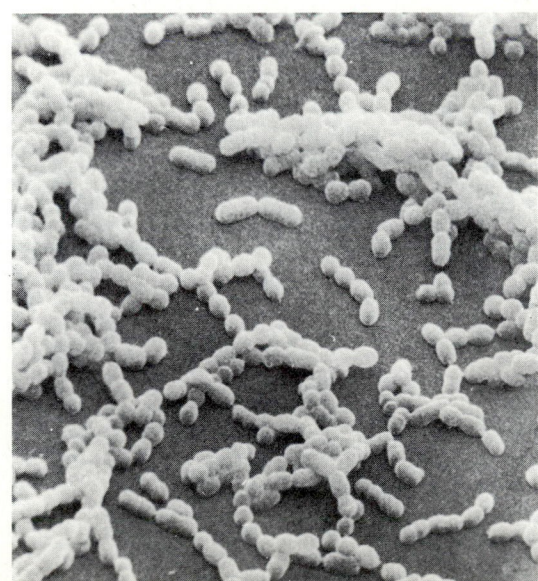

TABLE 10.14 Endospore Producers

Family Bacillaceae	
Bacillus	Rods; aerobic or facultative; catalase usually produced
Sporolactobacillus	Rods; microaerophilic; catalase not produced
Clostridium	Rods; anaerobic; sulfate not reduced to sulfide
Desulfotomaculum	Rods; anaerobic; sulfate reduced to sulfide
Sporosarcina	Cocci; tetrads or octads
Family Micromonosporaceae	
Thermoactinomyces	Filamentous thermophile

by the formation of endospores (Table 10.14). These genera occur in the family **Bacillaceae**. Most endospore formers are Gram-positive rods. Only members of the genus *Desulfotomaculum* are Gram-negative endospore formers. With the exception of members of the genus *Sporosarcina*, the other endospore formers are rod-shaped.

The two most important genera of endospore-forming bacteria are the genera *Bacillus* and *Clostridium*. *Bacillus* species are strict aerobes or facultative anaerobes. *Clostridium* species are obligately anaerobic. The endospore-forming bacteria are extremely important in food, industrial, and medical microbiology. Food spoilage by *Bacillus* and *Clostridium* species is of great economic importance. Several *Clostridium* species are important human pathogens. For example, *C. botulinum* is the causative agent of botulism, *C. tetani* causes tetanus, and *C. perfringens* causes gas gangrene.

Gram-Positive, Asporogenous Rods of Regular Shape The Gram-positive, asporogenous (nonsporulating), rod-shaped bacteria include the family **Lactobacillaceae**. These are Gram-positive rods that produce lactic acid as the major fermentation product; they occur in fermenting plant and animal

products that have available carbohydrate substrates; they are also found as part of the normal human microbiota in the oral cavity, vaginal tract, and intestinal tract. *Lactobacillus*, the only genus in the family Lactobacillaceae, is extremely important in the dairy industry. Cheese, yogurt, and many other fermented products are made by the metabolic activities of *Lactobacillus* species.

There are several genera of uncertain affiliation that are Gram-positive, nonspore-forming, rod-shaped bacteria. These include *Listeria*, *Erysipelothrix*, and *Caryophanon*. The *Listeria*, which are Gram-positive rods that tend to produce chains, include several species that are animal pathogens. Some recent outbreaks of human foodborne infections have been caused by *Listeria*-contaminated milk and cheeses. *C. latum*, another bacterium in this group, produces large rods or filaments up to 3 μm in diameter. This bacterium is normally found on animal fecal matter. The filaments of *C. latum* are divided by closely spaced cross walls into numerous disc-shaped cells less than 1 μm long, giving them an unusual morphology that is quite striking (Figure 10.22).

Gram-Positive Rods of Irregular Shape

Coryneform Bacteria. The **coryneform group** of bacteria is a heterogeneous group defined by the characteristic irregular morphology of the cells and their tendency to show incomplete separation following cell division. The coryneform bacteria exhibit pleomorphic morphology. They do not form true filaments. The irregular morphology and the association of the cells after division, however, indicate a relationship to the filament-forming actinomycetes. This group includes the genera *Corynebacterium*, *Arthrobacter*, *Brevibacterium*, *Cellulomonas*, and *Kurthia* (Table 10.15). Many species of *Corynebacterium* are plant or animal pathogens. For example, *Corynebacterium diphtheriae* is the causative agent of diphtheria. As noted in an earlier chapter, *C. diphtheriae* causes diphtheria only when it is infected with a temperate phage. Cells of *Corynebacterium* exhibit **snapping**

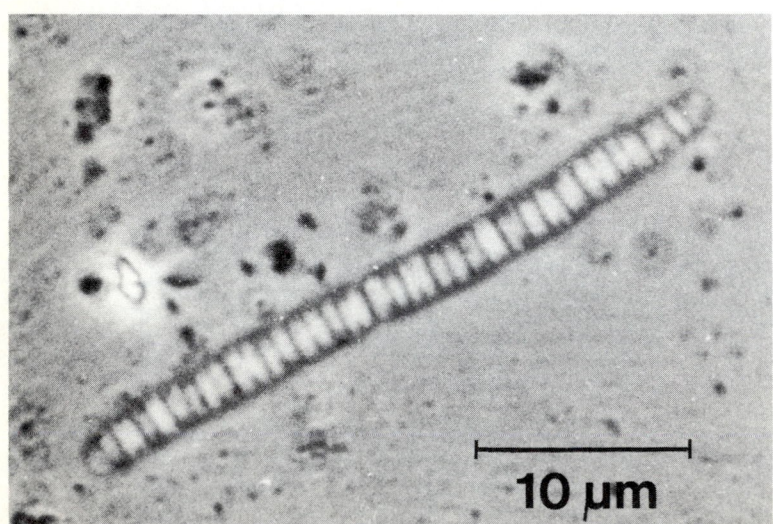

FIGURE 10.22
This phase contrast micrograph of *Caryophanon latum*, found in cow dung enrichment cultures, shows the multicellularity of the trichome. (Courtesy William Trentini, Mount Allison University)

division; that is, after binary fission the cells do not completely separate (Figure 10.23) and appear to form groups resembling "Chinese letters" when viewed under the microscope.

The genus *Arthrobacter*, which is widely distributed in soils, is interesting because it exhibits a simple life cycle (Figure 10.24) in which there is a change from rod-shaped cells to coccoid cells. The sequence of morphological changes in the growth cycle distinguishes *Arthrobacter* from other genera. The coccoid cells present during the stationary growth phase are sometimes referred to as *arthrospores* and *cystites*. The formation of arthrospores represents the beginning of a regular life cycle that is characteristic of eukaryotic microorganisms but is rare among the prokaryotes.

TABLE 10.15   The Coryneform Bacteria

Corynebacterium	Gram-positive rods frequently showing club-shaped swellings; snapping division produces angular arrangement of cells; G + C% 57–60
Arthrobacter	Gram-positive rods showing a marked change in form; exhibit a rudimentary life cycle; G + C% 60–72
Cellulomonas	Gram-positive rods that attack cellulose; G + C% 71–73
Kurthia	Gram-positive rods in young culture, cocci in old culture

Anaerobic Nonfilamentous or Filamentous Rods   The family **Propionibacteriaceae** contains Gram-positive rods that produce propionic acid, acetic acid, or mixtures of organic acids by fermentation; lactic acid is not a major fermentation

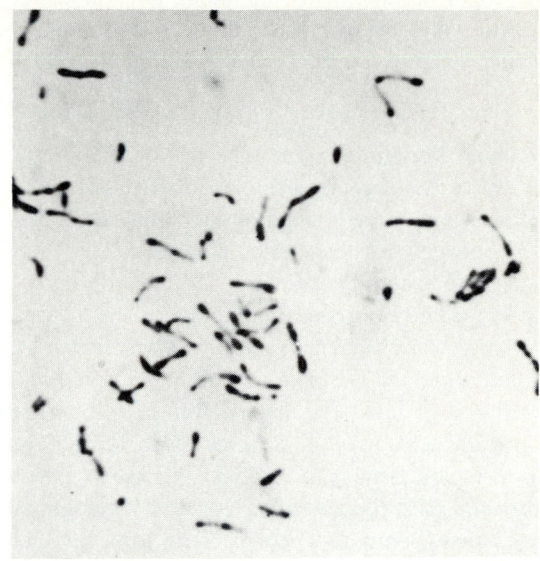

FIGURE 10.23
This micrograph of *Corynebacterium diphtheriae* shows the snapping division typical of the species. (From BPS—Centers for Disease Control, Atlanta)

FIGURE 10.24
Micrographs showing the life cycle of *Arthrobacter*. A medium was inoculated with coccoid cells and incubated at 25°C. (A) After 6 hours there is outgrowth of rods from coccoid cells; (B) after 12 hours the rods are longer; (C) after 24 hours the rods predominate; (D) after 3 days the rods have given rise to coccoid cells. (Reprinted by permission of Springer-Verlag, from R. M. Keddie and D. Jones, 1981, Saprophytic, aerobic coryneform bacteria, in *The Prokaryotes*, M. P. Starr, H. Stolp, H. G. Troper, A. Balows, and H. G. Schlegel, eds., p. 1848)

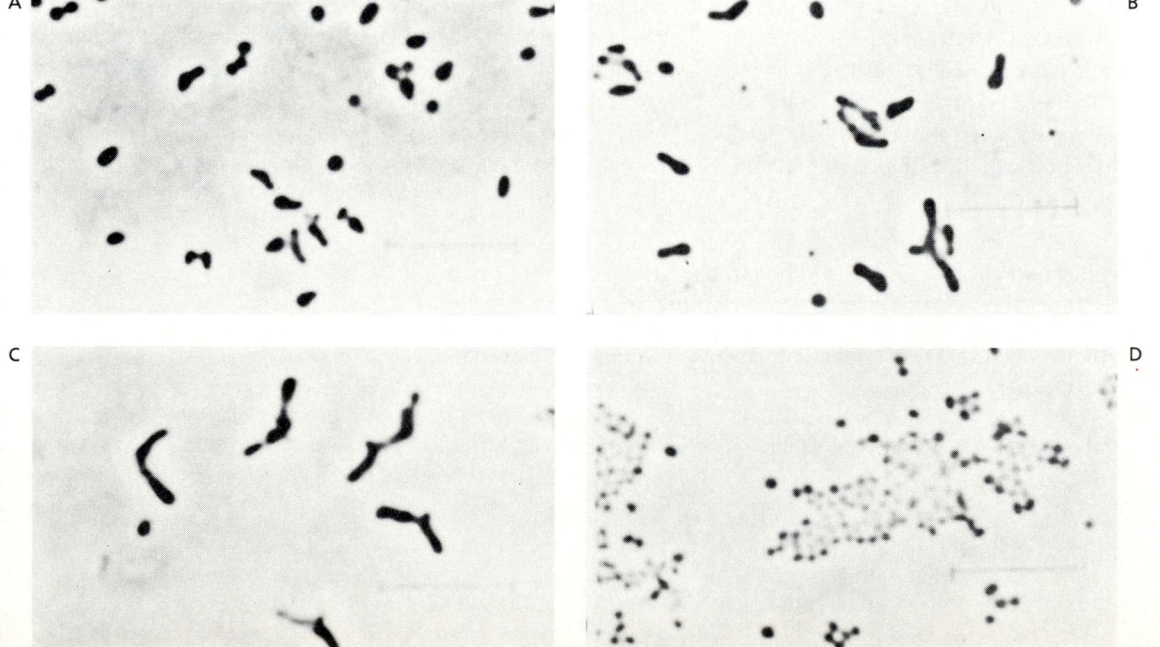

product but is itself used as a fermentation substrate. There are two genera in the family Propionibacteriaceae: *Propionibacterium* and *Eubacterium*. Species of *Propionibacterium* are important in the dairy industry; they normally carry out propionic acid fermentation. Species of *Eubacterium* usually produce mixtures of organic acids, including large amounts of butyric, acetic, and formic acids; they do not produce propionic, lactic, succinic, or acetic acids as major fermentation products.

Mycobacteria and Nocardioforms   The mycobacteria and nocardioforms have traditionally been considered as related to the coryneforms and/or the filamentous actinomycetes, and the separation between mycobacteria, corynebacteria, and various pleomorphic bacteria is not easy. Different observers may classify the same strain as belonging to the genera *Corynebacterium*, *Arthrobacter*, *Nocardia*, or *Mycobacterium*. *Mycobacterium*, for example, sometimes forms rudimentary mycelia. It is acid fast; that is, stained cells resist decolorization with acid alcohol. Several mycobacteria are important human pathogens that establish persistent infections. *M. tuberculosis* is the causative agent of tuberculosis and *M. leprae* causes leprosy. These bacteria form unusual acids, called *mycolic acids*, that are associated with their cell walls and contribute to their resistance to phagocytosis.

The nocardioforms typically produce filaments that fragment into nonmotile cells. Some bacteria in this group form mycelia (filaments) that do not readily fragment. Often the nocardioforms produce brightly colored pigments. *Nocardia* is a representative genus of this group.

Phototrophic Bacteria   The phototrophic bacteria (Table 10.16) are distinguished from other bacterial groups by their ability to use light energy to drive the synthesis of ATP. Most of the organisms included in this group are autotrophs capable of using carbon dioxide as the source of cellular carbon. Some of the phototrophic bacteria use water as an electron donor and liberate oxygen. These bacteria can be considered as belonging to the Oxyphotobacteria. The remainder of the photobacteria do not produce oxygen and, with one exception, can be classified as belonging to the Anoxyphotobacteria. The exception, *Halobacterium*, belongs to the Archaebacteria and has a unique mode of phototrophic metabolism. *Halobacterium* will be considered together with the Archaebacteria later in this chapter.

Anoxyphotobacteria. The **Anoxyphotobacteria** are photosynthetic but do not produce oxygen. They require an electron donor other than water and carry out only one form of oxidative photophosphorylation. Physiologically, these bacteria carry out photosynthesis anaerobically. The anaerobic photosynthetic bacteria typically occur in aquatic habitats, often growing at the sediment–water interface of shallow lakes where there is sufficient light penetration to permit photosynthetic activity; anaerobic conditions are sufficient to permit the existence of these organisms, and there is a source of reduced sulfur or organic compounds to act as electron donors for the generation of reduced coenzymes. The phototrophic bacteria include the Rhodospirillaceae (purple nonsulfur bacteria), Chromatiaceae (purple sulfur bacteria), Chlorobiaceae (green sulfur bacteria), and Cloroflexaceae (green flexibacteria) (Table 10.17). The green and purple sulfur bacteria utilize reduced sulfur compounds, such as hydrogen sulfide, as electron donors for generating reducing power. Most of the purple nonsulfur bacteria and green flexibacteria are unable to use reduced sulfur compounds; rather, these organisms utilize organic compounds to support photosynthetic growth.

Members of the family **Chromatiaceae** produce carotenoid pigments and may appear orange-brown, red-brown, purple-red, or purple-violet. They deposit elemental sulfur as a consequence of their utilization of reduced sulfur compounds as electron donors for generating reducing power (Figure 10.25). Because of their color and sulfur metabolism, the Chromatiaceae are called the *purple sulfur bacteria*. In all but one genus of organisms within this large family, the sulfur accumulates intracellularly. Members of the Chromatiaceae are potentially **mixotrophic**, that is, they are capable of both photoautotrophic and heterotrophic growth, and all strains are capable of photoassimilating simple organic substrates such as acetate. The cells of *Chromatium*, *Thiocystis*, *Thiosarcina*, *Thiospirillum*, and *Thiocapsa* do not contain gas vacuoles, but some genera of the family Chro-

TABLE 10.16   Characteristics of the Major Groups of Phototrophic Bacteria

Metabolism	Taxonomic Group	Photosynthetic Pigments	Electron Donors	Carbon Source
Anoxygenic photosynthesis	Purple bacteria	Bacteriochlorophyll a or b, carotenoids	H_2, H_2S, S	Organic C or CO_2
Anoxygenic photosynthesis	Green bacteria	Bacteriochlorophyll a or b, carotenoids	H_2, H_2S, S	CO_2
Oxygenic photosynthesis*	Cyanobacteria	Chlorophyll a, phycobiliproteins	H_2O	CO_2
Oxygenic photosynthesis	Prochlorobacteria	Chlorophyll a + b, β-carotenes	H_2O	CO_2
Purple membrane mediated	*Halobacterium*	Bacteriorhodopsin	—	Organic C

*Under some conditions, photosynthesis is anoxygenic, and H_2S serves as the electron donor.

TABLE 10.17   The Phototrophic Bacteria

Order Rhodospiralles	Cells capable of photosynthetic anaerobic metabolism
Suborder Rhodospirillineae (purple bacteria)	Cells contain bacteriochlorophyll a or b
Family Rhodospirillaceae (purple nonsulfur bacteria)	Cells photoassimilate simple organic substrates; most species unable to grow with sulfide as the sole electron donor
Genera: *Rhodospirillum, Rhodopseudomonas, Rhodomicrobium*	
Family Chromatiaceae (purple sulfur bacteria)	Cells able to grow with sulfide and sulfur as the sole electron donor; sulfur deposited inside or outside of cell
Genera: *Chromatium, Thiocystis, Thiosarcina, Thiospirillum, Thiocapsa, Lamprocystis, Thiodictyon, Thiopedia, Amoebobacter, Ectothiorhodospira*	
Suborder Chlorobiineae (green bacteria)	Cells contain bacteriochlorophyll a, b, or c
Family Chlorobiaceae (green sulfur bacteria)	Cells able to grow with sulfide and sulfur as the sole electron donor; sulfur deposited only outside of cell
Genera: *Chlorobium, Prosthecochloris, Chloropseudomonas, Pelodictyon, Clathrochloris*	
Family Chloroflexaceae (green flexibacteria)	Cells have flexible walls, gliding motility; form filaments and utilize organic C sources
Genera: *Chloroflexus, Chloronema, Oscillochloris*	

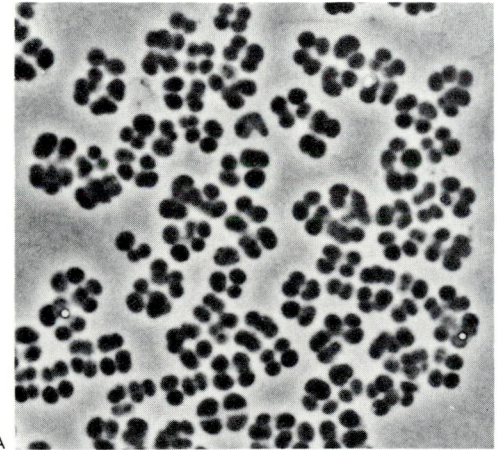

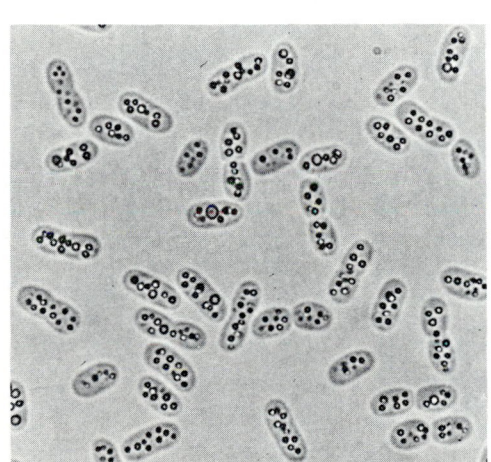

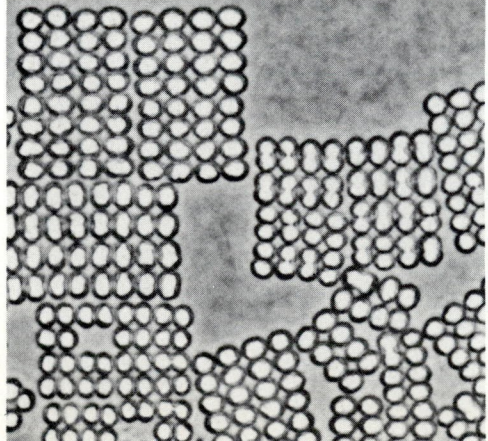

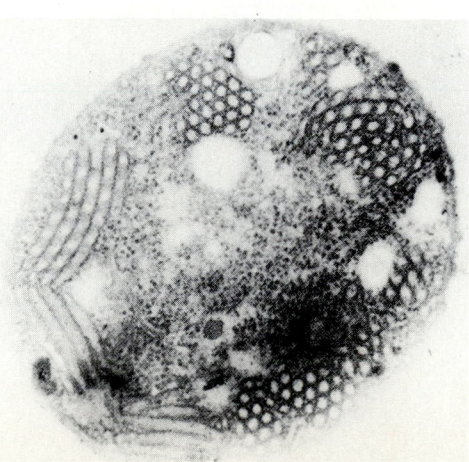

A   B   C   D

FIGURE 10.25
Morphologies of the purple sulfur bacteria. (A) *Thiocapsa roseopersicina* (2,000×); (B) *Chromatium vinosum*— note the intracellular sulfur globules (2,000×); (C) *Thiopedia rosea* (2000×)—note the gas vacuoles; (D) *Thiocapsa pfennigui* (60,000×)— note the bundled tube type of intracytoplasmic membrane system. (Reprinted by permission of the Bergey's Trust, John Holt, executor, from *Bergey's Manual of Determinative Bacteriology*, Williams & Wilkins Co., Baltimore)

matiaceae, such as *Thiodictyon* and *Thiopedia*, do contain gas vacuoles that permit an adjustment of cell buoyancy in a water column to a depth appropriate for light penetration and oxygen concentration, making anaerobic photosynthetic metabolism possible.

The **Chlorobiaceae** produce green or green-brown carotenoid pigments and are obligately phototrophic. They assimilate carbon dioxide, utilizing sulfide or elemental sulfur as electron donors, and they deposit sulfur granules extracellularly. Because of their color and sulfur metabolism, the Chlorobiaceae are called the *green sulfur bacteria*. Some genera of Chlorobiaceae, such as *Pelodictyon*, have gas vacuoles, but others, such as *Chlorobium* and *Chloropseudomonas*, do not (Figure 10.26). All members of the Chlorobiaceae are nonmotile. These bacteria often occur in ecological situations similar to those of the Chromatiaceae.

The **Rhodospirillaceae** generally produce red-purple carotenoid pigments. Three genera are recognized: *Rhodospirillum* has spiral-shaped cells; *Rhodopseudomonas* has spher-

ical or rod-shaped cells that do not form filaments; and *Rhodomicrobium* has oval cells that do form filaments and exhibit budding division (Figure 10.27). Members of the genera *Rhodospirillum* and *Rhodopseudomonas* are motile by means of polar flagella, whereas those of the genus *Rhodomicrobium* are peritrichously flagellated. Their photosynthetic development depends on the ability of the cells to photoassimilate simple organic compounds. When sulfide or thiosulfate is utilized as an electron donor, elemental sulfur is not deposited within the cell. The organic substrates utilized by the Rhodospirillaceae may serve as electron donors for generating reducing power or may be photoassimilated. Because they generally require preformed organic matter for growth and are able to utilize light energy for generating ATP, the type of metabolism they carry out is sometimes referred to as *photoheterotrophic metabolism* and the organisms are called the *purple nonsulfur bacteria*.

The Rhodospirillaceae convert carbon dioxide to organic matter by means of the Calvin cycle pathway. Some typical

FIGURE 10.26

Morphologies of the green sulfur bacteria. (A) *Chlorobium limicola* 2,000×)—note the extracellular sulfur globules; (B) *Pelodictyon clathratiforme* (2,000×)—note the gas vacuoles; (C) *Chlorobium phaeovibrioides* (3,000×); (D) *Pelodictyon clathratiforme* (105,000×). (Reprinted by permission of the Bergey's Trust, John Holt, executor, from *Bergey's Manual of Determinative Bacteriology*, Williams & Wilkins Co., Baltimore)

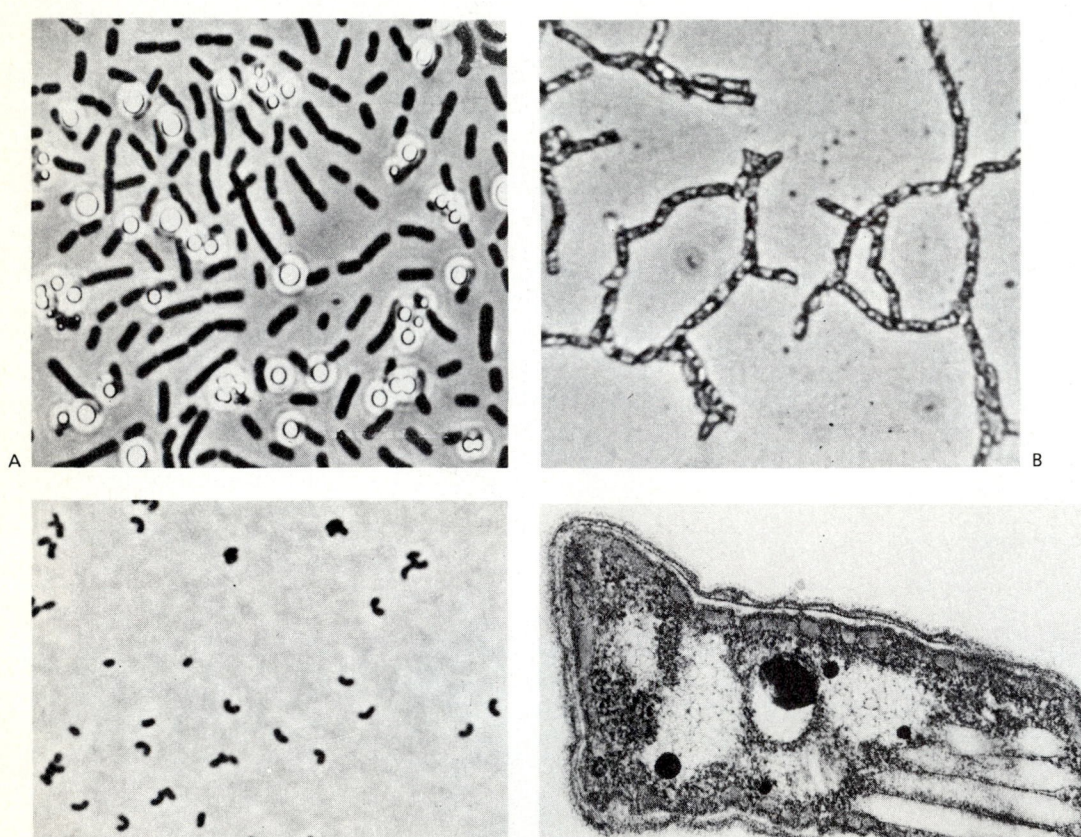

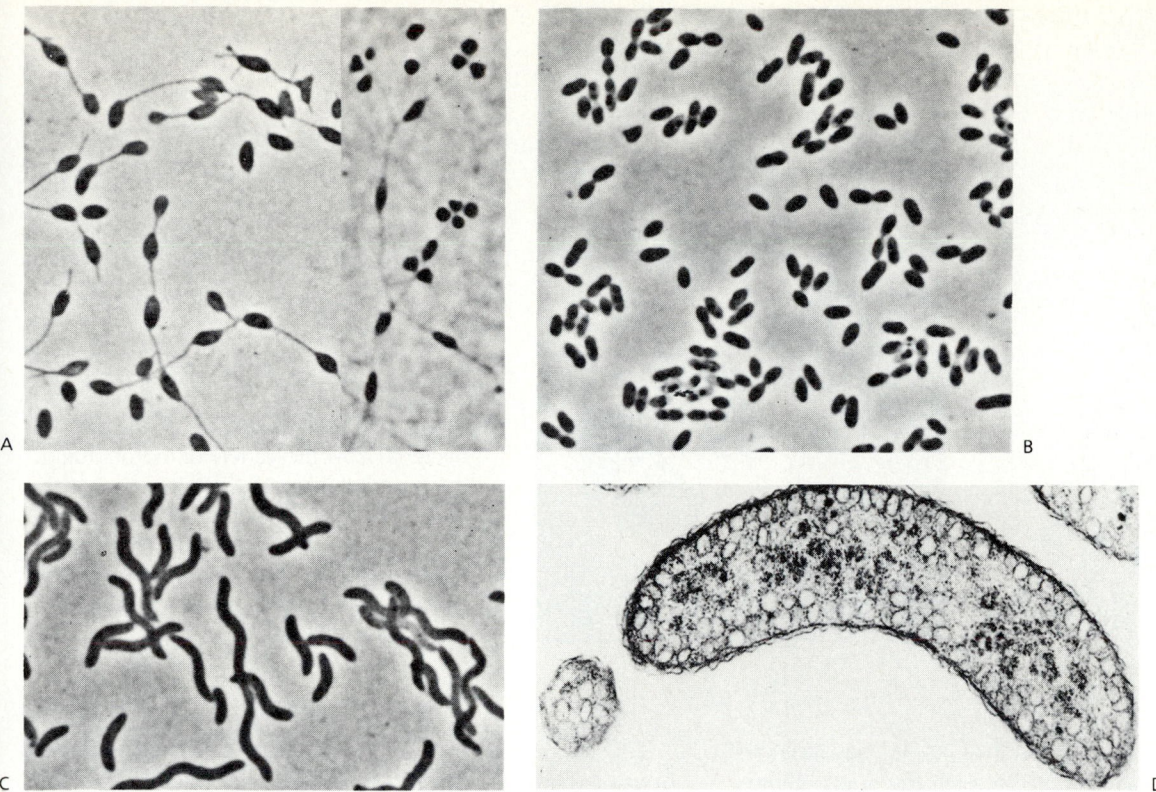

FIGURE 10.27

Morphologies of the purple nonsulfur bacteria. (A) *Rhodomicrobium vannielii* (2,000×);
(B) *Rhodopseudomonas acidophila* (2,000×); (C) *Rhodospirillum molischianum* (2,000×); and
(D) *Rhodospirillum rubrum* (51,000×). (Reprinted by permission of the Bergey's Trust, John Holt,
executor, from *Bergey's Manual of Determinative Bacteriology*, Williams & Wilkins Co.,
Baltimore)

members of the Rhodospirillaceae use molecular hydrogen
or sulfide as an electron donor and can grow without organic
compounds. As such, it may be best to consider these organisms as photoautotrophs, generally requiring organic
growth factor compounds. Indeed, most strains in the Rhodospirillaceae require one or more vitamins. Clearly, the
Rhodospirillaceae occupy a boundary position between autotrophs and heterotrophs. The basic metabolic pathways of
the Rhodospirillaceae are the same as those of other autotrophic microorganisms. Their ability to assimilate organic
compounds and the requirement of many members of the
Rhodospirillaceae for such compounds establish the resemblance of these organisms to heterotrophs.

The **Chloroflexaceae** represent a relatively newly discovered family of anaerobic phototrophic bacteria. In terms of
both physiology and morphology, the Chloroflexaceae exhibit unique combinations of characteristic features of other
phototrophic bacteria. These bacteria have flexible walls,
form filaments, and exhibit gliding motility. Among phototrophic bacteria, gliding motility and formation of filaments
were previously thought to be restricted to the cyanobacteria.
Photosynthesis is anoxygenic, and some organic compounds
are needed to achieve optimal growth. *Chloroflexus*, which

resembles a green sulfur bacterium in cell ultrastructure and
photosynthetic pigments but resembles a nonsulfur purple
bacterium in its photosynthetic and catabolic metabolism, is
the only genus in this family that has been characterized in
pure culture, although other genera have been assigned to
the Chlorofexaceae based on field observations. *C. aurantia-
cus* has been isolated from alkaline hot springs in various
parts of the world.

Oxyphotobacteria. The **Oxyphotobacteria** are capable of
splitting water to form oxygen as part of their photosynthetic
metabolism. Two orders are contained in this subclass of the
Photobacteria: **Cyanobacteriales** and **Prochlorales**. Both of
these orders occupy intermediary positions between the phototrophic bacteria and the eukaryotic algae, indicating a probable evolutionary link to these higher photosynthetic organisms. The primary photosynthetic pigment in both cases is
chlorophyll a, but the prochlorophytes also possess chlorophyll b, making them very similar to the green algae. Presumably, the prochlorobacteria are more closely related to
the green algae than are the cyanobacteria. Some cyanobacteria, on the other hand, are capable of anoxygenic photosynthesis, making them closely related to the Anoxyphotobacteria. Clearly, there is a phylogenetic relationship among

A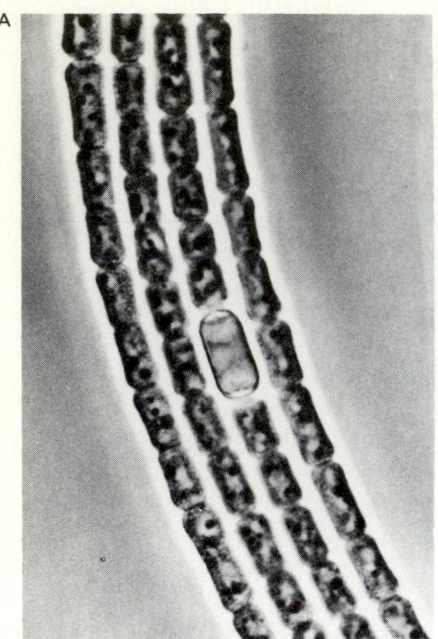
B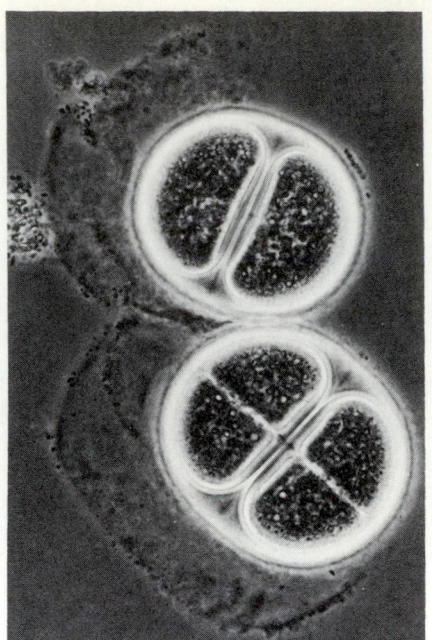

FIGURE 10.28
These micrographs of different cyanobacteria show the variety of morphologies they can exhibit. (A) An *Anabaena* sp. (4,000×); (B) *Chroococcus turgidus* (1,280×). Both organisms were viewed under the phase contrast microscope. (From BPS—J. Robert Waaland, University of Washington)

the photosynthetic organisms, with the Oxyphotobacteria occupying an intermediate position between the Anoxyphotobacteria and the algae.

The **cyanobacteria**, or blue-green bacteria, are the most diverse and widely distributed group of photosynthetic bacteria. Over 1,000 species of cyanobacteria have been reported, based largely on field observations. Field observations, however, leave many uncertainties about the variability of particular features and ambiguities concerning the separation of taxa. Examination of pure cultures has indicated that by eliminating ambiguous features, the 170 genera described on the basis of field observations can be reduced to 22 genera. Among the cyanobacteria some genera characteristically are unicellular, whereas others are filamentous (Figure 10.28). The cell wall structures of cyanobacteria are of the Gram-negative type. The cytoplasm of cyanobacteria is filled with photosynthetic membranes (thylakoids) containing the primary photosynthetic pigment—chlorophyll a. The outer surfaces of the photosynthetic membranes have associated granules known as **phycobilisomes**, which are composed of auxiliary photosynthetic pigments.

There are four major subgroups of cyanobacteria (Table 10.18). The **chroococcacean cyanobacteria** are unicellular rods or cocci. They reproduce either by binary fission (family Chroococcaceae) or by budding (family Chamesiphonaceae). Chroococcacean cyanobacteria are generally nonmotile. *Synechococcus*, *Synechocystis*, and *Chamaesiphon* are representative genera. One interesting genus in this group, *Gloeobacter*, lacks thylakoids and is purple in color. *Gloeobacter* can easily be confused with the anaerobic phototrophs, but in pure culture studies its biochemistry and metabolism have been shown to be typical of those of cyanobacteria.

The **pleurocapsalean cyanobacteria** are distinguished from the chroococcacean cyanobacteria by the fact that they exhibit multiple fission to produce small coccoid reproductive cells. In the phycological literature these reproductive cells are referred to as *endospores*, but to avoid confusion with endospore-forming bacteria, it has been proposed that the term **baeocyte** be used to describe the reproductive cells of the pleurocapsalean cyanobacteria. These cyanobacteria are unicellular, but the cells generally fail to separate completely following binary fission. Because binary fission does not result in complete separation of the cells, the pleurocapsalean cyanobacteria form multicellular aggregates.

The **oscillatorian cyanobacteria** form filamentous structures, composed exclusively of vegetative cells, known as **trichomes**. In some cases the trichomes are straight, and in others they are helical. *Spirulina*, *Oscillatoria*, and *Pseudanabaena* are representative genera of oscillatorian cyanobacteria.

Like the oscillatorian cyanobacteria, the **heterocystous cyanobacteria** are filamentous. Unlike the oscillatorian cyanobacteria, however, the heterocystous cyanobacteria form dif-

TABLE 10.18   The Subgroups of the Cyanobacteria

Group	Description
Chroococcaean	Unicellular rods or cocci reproducing by binary fission or budding
Pleurocapsalean	Single cells enclosed in a fibrous layer; reproduce by multiple fission, producing baeocytes
Oscillatorian	Cells form trichomes but not heterocysts
Heterocystous	Form trichomes with both vegetable cells and heterocysts

ferentiated cells known as heterocysts when growing in the absence of fixed forms of nitrogen. **Heterocysts** are nonreproductive cells that are distinguished from the adjoining vegetative cells by the presence of refractory polar granules

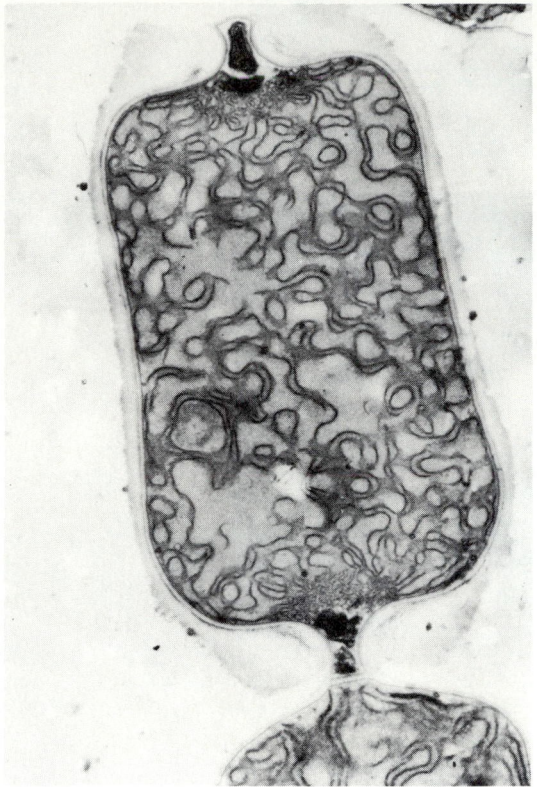

FIGURE 10.29

Heterocyst of the cyanobacterium *Anabaena cylindrica* (14,000×). (From BPS—Norma Lang, University of California, Davis)

and a thick outer wall (Figure 10.29). The ability to form heterocysts is associated with the physiological capability of fixing atmospheric nitrogen. The physiologically specialized heterocyst cells appear to be the anatomical site of nitrogen fixation in heterocystous cyanobacteria. *Nostoc* and *Anabaena* are probably the best-known genera among the heterocystous cyanobacteria. The ability to carry out both oxygen-yielding photosynthesis and nitrogen fixation is a unique characteristic of cyanobacteria principally found among the heterocystous cyanobacteria. The heterocystous cyanobacteria are ecologically important because they can form both organic carbon and fixed forms of nitrogen that can support the nutritional requirements of other organisms.

The **prochlorales** are similar to the cyanobacteria except that they also synthesize chlorophyll b. Although they originally were considered to be cyanobacteria, their unique ability as prokaryotes to produce chlorophyll b is now considered significant enough to separate them into their own order. The only known genus, *Prochloron*, occurs as single-celled, extracellular symbionts of marine invertebrates. These bacteria appear bright green on the surfaces of the animals with which they are associated. Various species of *Prochloron* have been recognized in field studies, but until the organisms are grown in pure culture, the validity of these species remains ambiguous.

The Gliding Bacteria   We have already seen that some phototrophic bacteria are capable of gliding motility. In addition to the cyanobacteria and Chloroflexaceae, the **Myxobacterales** (fruiting myxobacteria) and the **Cytophagales** are grouped together on the basis of their **gliding motility** on solid surfaces (Table 10.19). The mole% G + C of the gliding bacteria covers the entire range from 30 to 70 percent, and in all likelihood the gliding bacteria represent a phylogenetically heterogeneous group.

TABLE 10.19   The Gliding Bacteria

Order Myxobacterales	Produce fruiting bodies
Family Myxococcaceae	Vegetative cells tapered; microcysts spherical or oval
Genus: *Myxococcus*	
Family Archangiaceae	Vegetative cells tapered; microcysts rod-shaped, not in sporangia
Genus: *Archangium*	
Family Cystobacteraceae	Vegetative cells tapered; microcysts rod-shaped, in sporangia
Genera: *Cystobacter, Melittangium, Stigmatella*	
Family Polyangiaceae	Myxospores resemble vegetative cells
Genera: *Polyangium, Nannocystis, Chondromyces*	
Order Cytophagales	Fruiting bodies not produced
Family Cytophagaceae	Pigmented; filaments not attached
Genera: *Cytophaga, Flexibacter, Herpetosiphon, Flexithrix, Saprospira, Sporocytophaga*	
Family Beggiatoaceae	Nonpigmented; filaments not attached; cells in cylindrical filaments
Genera: *Beggiatoa, Vitreoscilla, Thioploca*	
Family Simonsiellaceae	Nonpigmented; filaments attached; cells in flat filaments
Genera: *Simonsiella, Alysiella*	
Family Leucotrichaceae	Filaments attached at one end
Genera: *Leucothrix, Thiothrix*	

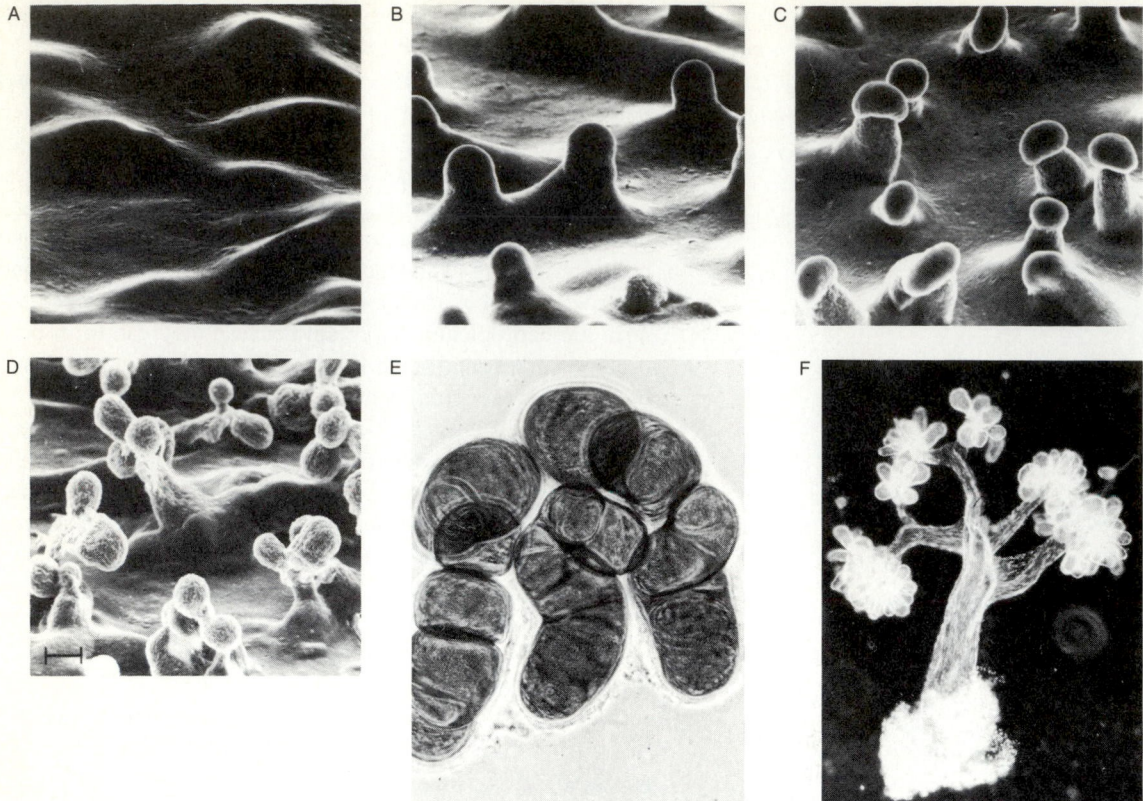

FIGURE 10.30
Micrograph showing morphologies and fruiting of myxobacteria. The stages of fruiting body
formation of the myxobacterium *Stigmatella aurantiaca* are shown in (A) early aggregates;
(B) early stalks; (C) late stalks; and (D) mature fruiting body; (E) *Cystobacter fucus* fruiting bodies
(170×); (F) *Chondromyces croatus* (125×). (A–D From BPS—Karen Stephens, Stanford Medical
Center; E, F Courtesy Hans Reichenbach, Gesellschaft fur Biotechnologische Forschung, West
Germany)

Myxobacteriales. The **myxobacteria** are small rods that
are normally embedded in a slime layer. They lack flagella
but are capable of gliding movement. A unique feature of
the myxobacteria is that under appropriate conditions they
aggregate to form **fruiting bodies** (specialized structures
bearing spores) (Figure 10.30). The taxonomy of the myxobacteria is based largely on the fruiting body structures,
which are often brightly colored and visible without the aid
of a microscope. Frequently, the fruiting bodies of myxobacteria occur on decaying plant material, on the bark of
living trees, or on animal dung, appearing as highly colored,
slimy growths that may extend above the surface of the substrate. Within the fruiting body, the cells of the myxobacteria
are dormant and are called **myxospores**. In some genera of
myxobacteria, the myxospores cannot be distinguished from
vegetative cells, but in others the myxospores are refractile
and encapsulated, in which case they are known as **microcysts**. Most of the myxobacteria produce a variety of hydrolytic enzymes, such as cellulases, and many are capable of
lysing other microorganisms.

Cytophagales. In contrast to the Myxobacterales, the **Cytophagales** do not produce fruiting bodies. Members of both

groups, however, do exhibit gliding motion. Genera in the
order Cytophagales exhibit widely differing morphological
forms (Figure 10.31) and modes of metabolism. They are
unified only by the presence of gliding motion and lack of
fruiting body formation. Some Cytophagales form filaments,
and others do not. Some *Flexibacter* species, for example,
may form filaments measuring as much as 100 μm in length.
Some of the Cytophagales are chemolithotrophs. For example, *Beggiatoa* forms filaments, oxidizes hydrogen sulfide,
and deposits sulfur intracellularly when growing on hydrogen
sulfide. *Cytophaga* species, on the other hand, do not form
filaments. Cells of *Cytophaga* contain deep yellow-orange or
red pigments and hydrolyze agar, cellulose, and chitin. As
a consequence of their hydrolytic activities, these gliding
bacteria play an important ecological role in the decomposition of organic matter.

The Sheathed Bacteria   The **sheathed bacteria** comprise
those bacteria whose cells occur within a filamentous structure known as a *sheath* (Figure 10.32). The sheathed bacteria
include the genera *Sphaerotilus*, *Leptothrix*, *Haliscomeno-
bacter*, *Lieskeella*, *Phragmidiothrix*, *Crenothrix*, and *Clon-*

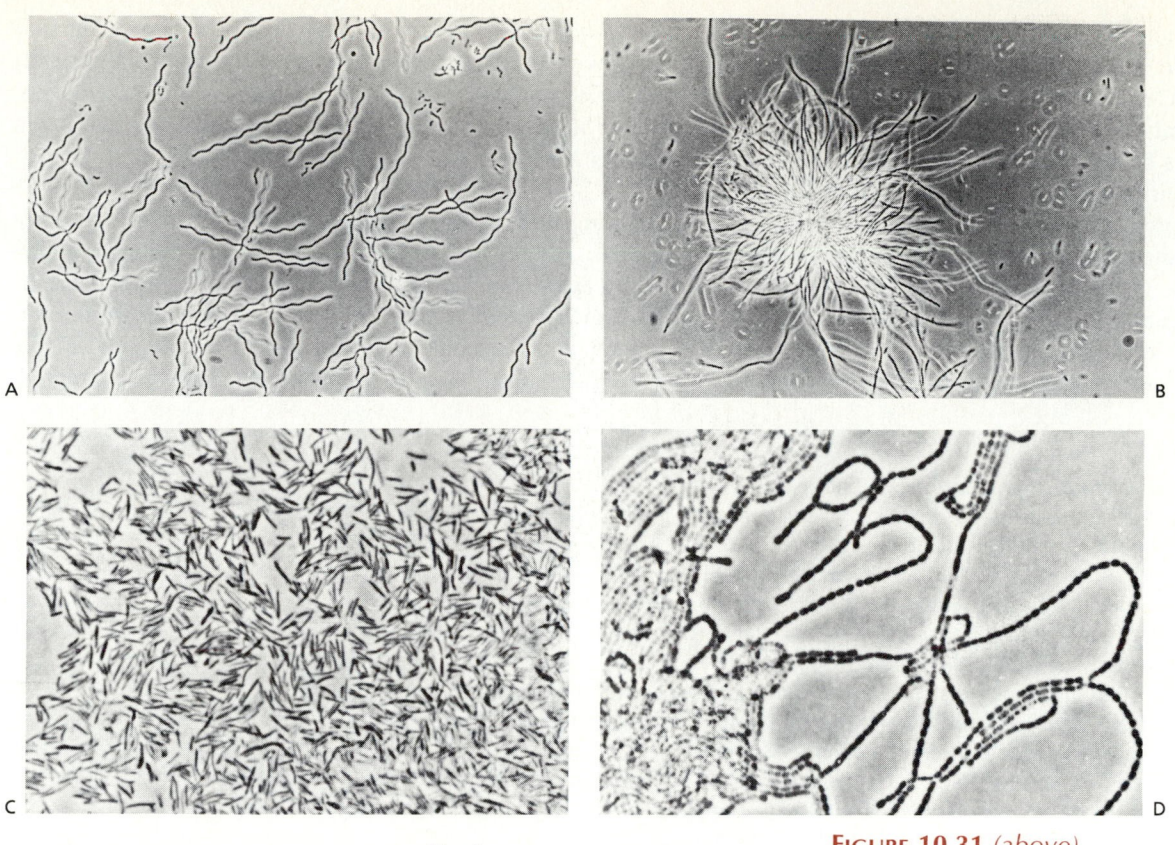

FIGURE 10.31 *(above)*

Phase contrast micrographs showing morphologies of the Cytophagales. (A) *Saprospira grandis* showing helical filaments (350×); (B) *Leucothrix mucor* showing a cluster of filaments from liquid culture and numerous cells released from their ends (470×); (C) *Cytophaga hutchinsonii* vegetative cells (1,500×); (D) *Vitreoscilla stercoraria* showing filaments (1,200×). (Courtesy Hans Reichenbach, Gesellschaft für Biotechnologische Forschung, West Germany)

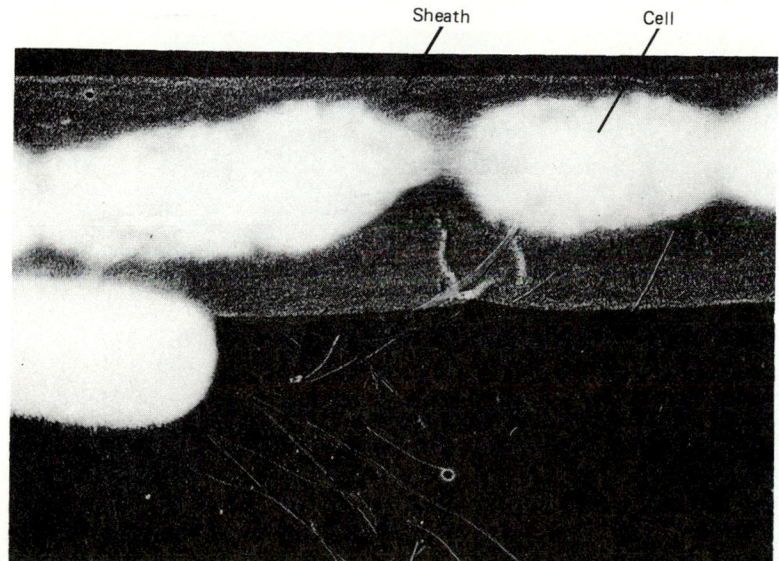

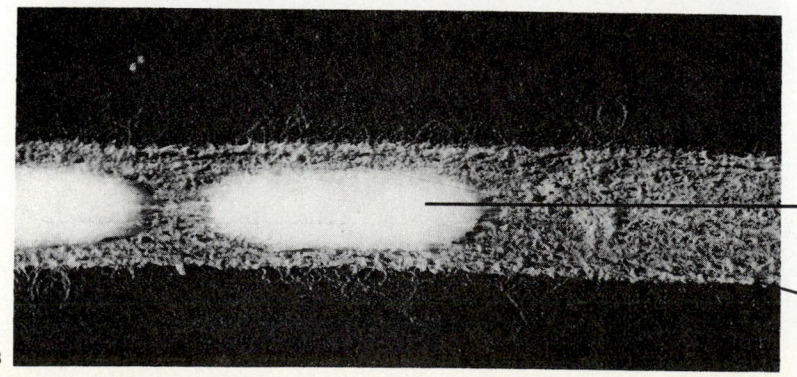

FIGURE 10.32 *(left)*

Micrographs showing morphologies of sheathed bacteria. (A) *Sphaerotilus natans* (19,000×). (B) *Leptothrix cholodnii* (19,000×). (Courtesy M. H. Deinema, Landbouwhogeschool Wageningen, the Netherlands: (A) reprinted by permission of the American Society for Microbiology, Washington, D.C., from W. L. vanVeen, E. G. Mulder, and M. H. Deinema, 1978, *Bacteriological Reviews*, 42:329–356; (B) reprinted by permission of Springer-Verlag, New York, from E. G. Mulder and M. H. Deinema, 1981, The sheathed bacteria, In *The Prokaryotes*, M. P. Starr, H. Stolp, H. G. Truper, A. Balows, and H. G. Schlegel, eds.)

othrix. The formation of a sheath enables these bacteria to attach themselves to solid surfaces. This is important to the ecology of these bacteria because many sheathed bacteria live in low-nutrient aquatic habitats. By absorbing nutrients from the water that flows by the attached cells, these bacteria are able to conserve their limited energy resources. Additionally, the sheaths afford protection against predators and parasites. In some cases, the sheaths may be covered with metal oxides. For example, in the genus *Leptothrix*, sheaths are encrusted with iron or manganese oxides. In other genera, such as *Heliscomenobacter*, this does not occur. In the genus *Sphaerotilus* the sheath is sometimes encrusted with iron oxides.

S. natans, often referred to as the *sewage fungus*, is the only species in the genus *Sphaerotilus*. This organism normally occurs in polluted flowing waters, such as sewage effluents, where it may be present in high concentrations just below sewage outfalls.

Budding and/or Appendaged Bacteria   Like the sheathed bacteria, the **budding and/or appendaged bacteria** represent a heterogeneous group on the basis of a particular morphological feature. These bacteria have in common the formation of extensions or protrusions from the cell (Figure 10.33). In some cases the cellular extensions have a reprod-

FIGURE 10.33

Micrographs of budding, stalked, and prothecate bacteria.
(A) *Planctomyces bekefil* rosette from pondwater, a negative-contrast TEM micrograph; b = bud, S = stalk, and sp = spore appendages. (Courtesy Jean M. Schmidt, Arizona State University; reprinted by permission of Springer-Verlag, New York, from J. M. Schmidt, and M. P. Starr, 1980, *Current Microbiology*, Vol. 4, pp. 183–188), (B) Interference light micrograph showing rosette formation by *Hyphomicrobium* at the air–water interface, (C) Scanning electron micrograph showing typical morphology of *Hyphomicrobium*. (Courtesy Richard L. Moore, University of Calgary), (D) Electron micrograph of the stalked bacterium *Caulobacter* (Courtesy Jeanne Poindexter, City of New York Public Health Laboratories), (E) Electron micrograph of *Prothecomicrobium pneumaticum*, showing multiple prothecae (Courtesy J. T. Staley, University of Washington), (F) Micrograph of the flat, six-pointed bacterium *Stella* (Reprinted by permission of Springer-Verlag, New York, from P. Hirsch and H. Schlesner, 1981, The genus *Stella*, in *The Prokaryotes*, M. P. Starr, H. Stolp, H. G. Truper, A. Balows, and H. G. Schlegel, eds.)

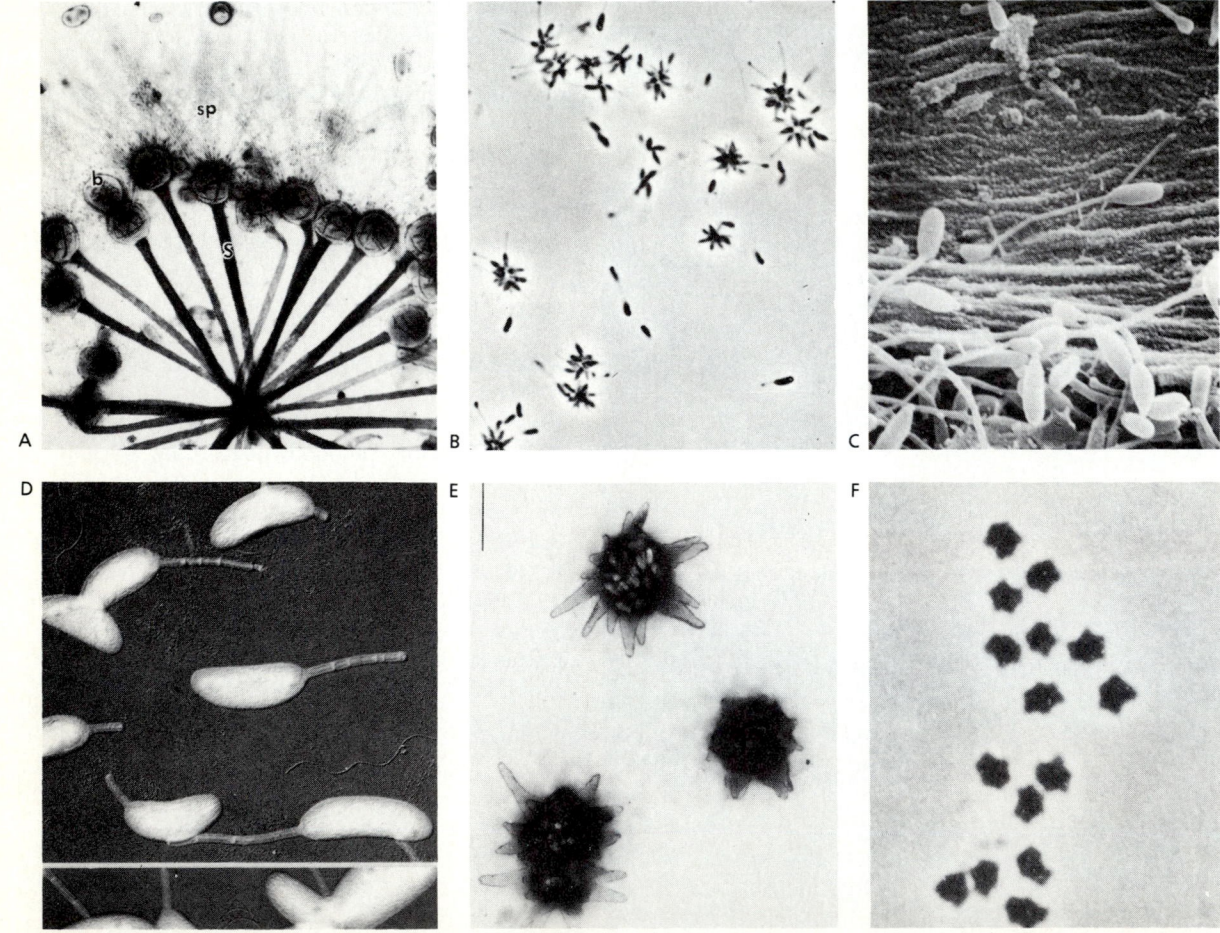

uctive function, but in others they have a physiological purpose. Some bacteria in this group reproduce by budding, others by binary fission (Table 10.20). Several budding bacteria, such as *Rhodomicrobium*, are associated primarily with other groups on the basis of their characteristic modes of metabolism. Budding and/or appendaged bacteria occur in all nutritional categories. Members of the genus *Gallionella* are capable of chemolithotrophic metabolism; they are probably facultatively chemolithotrophic because they are able to oxidize ferrous to ferric iron and fix carbon dioxide. *Gallionella* are sometimes considered to be sheathed bacteria because their "stalks" may be covered with iron hydroxide. The growth of *Gallionella* species often causes problems in iron pipes of water delivery systems.

The cell appendages of the bacteria in this group, known as **prosthecae**, provide greater efficiency in concentrating available nutrients. Many of the appendaged bacteria grow well at low nutrient concentrations. The appendages provide sufficient membrane surface to transport adequate nutrients into the cell to support the metabolic requirements of the organism. Many of the bacteria in this group primarily occur in aquatic habitats where concentrations of organic matter typically are low. *Caulobacter*, for example, grows in very dilute concentrations of organic matter in lakes and even in distilled water. Its appendages are referred to as **stalks**. In some cases, the stalks of individual cells provide a **holdfast** by which the organisms can attach to a substrate. In other cases, stalks do not function in attachment but may permit cells to adhere to each other, forming rosettes. Some of the appendaged bacteria form bizarre-looking cells. For example, members of the genus *Prosthecomicrobium* form prosthecae extending in all directions from the cell. *Seliberia* form radial clusters (star-like aggregates) of rod-shaped bacteria with a screw-like twisting of the rod surface and the formation of round reproductive cells by budding. At low nutrient concentrations, *Stella* forms flat cells resembling six-pronged stars. The isolation of various new types of appendaged bacteria has greatly increased our knowledge of the morphological diversity among the bacteria and the relationship between morphology and nutritional status. Many of the varied morphological forms of these bacteria are observed only at very low nutrient concentrations.

Chemolithotrophic Bacteria   The metabolic activities of the **chemolithotrophic bacteria** are extremely important in biogeochemical cycling reactions (Table 10.21). These bacteria oxidize inorganic compounds in order to generate ATP. Because their ATP-generating metabolism is inefficient, they must metabolize large amounts of substrate to meet their energy requirements. The metabolic transformations of inorganic compounds mediated by these organisms cause global-scale cycling of various elements among the air, water, and soil.

The family **Nitrobacteraceae** oxidizes ammonia or nitrite in order to generate ATP. Organisms in this family, commonly referred to as **nitrifying bacteria**, are commonly

TABLE 10.20   The Budding and/or Appendaged Bacteria

Prosthecate bacteria
  Prosthecae have a reproductive function; form new cells by budding
    Hyphomicrobium
    Hyphomonas
    Pedomicrobium
    Thiodendron
  Prosthecae have no reproductive function
    Caulobacter
    Asticcacaulis
    Ancalomicrobium
    Prosthecobacter
    Prosthecomicrobium
    Stella
Nonprosthecate bacteria; reproduce by budding
    Pasteuria
    Blastobacter
    Seliberia
Bacteria with excreted appendages and holdfasts
  Reproduce by binary fission only
    Gallionella
    Nevskia
  Reproduce by budding
    Planctomyces
Genera of uncertain affiliation
    Metallogenium
    Caulococcus
    Kusnezovia

TABLE 10.21   The Chemolithotrophic Bacteria

Family Nitrobacteraceae	Oxidize ammonia or nitrite
Nitrobacter	Oxidize nitrite to nitrate
Nitrospina	
Nitrococcus	
Nitrosomonas	Oxidize ammonia to nitrite
Nitrosospira	
Nitrosococcus	
Nitrosolobus	
Thiobacillus	Oxidize sulfur and sulfur compounds
Sulfolobus	
Thiobacterium	
Macromonas	
Thiovulum	
Thiospira	
Family Siderocapsaceae	Oxidize iron or manganese
Siderocapsa	Iron or manganese oxides deposited
Naumanniella	
Ochrobium	
Siderococcus	Iron but not manganese deposited

found in soil, fresh water, and seawater. Many of the nitrifying bacteria have extensive internal membrane systems (Figure 10.34). There are two physiological groups in the family Nitrobacteraceae; the first group oxidizes nitrite to nitrate, and the second group oxidizes ammonia to nitrite. Most members of this family are obligate chemolithotrophs. There are seven genera: *Nitrobacter*, *Nitrospina*, *Nitrococcus*, *Nitrosomonas*, *Nitrosospira*, *Nitrosococcus*, and *Nitro-*

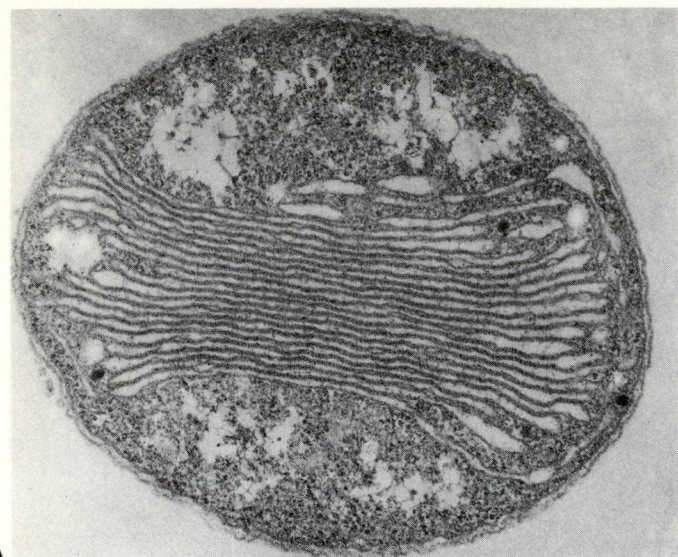

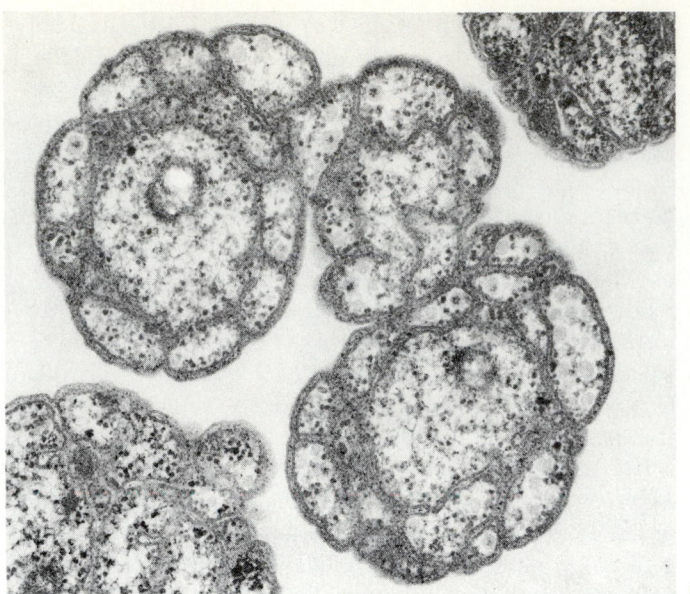

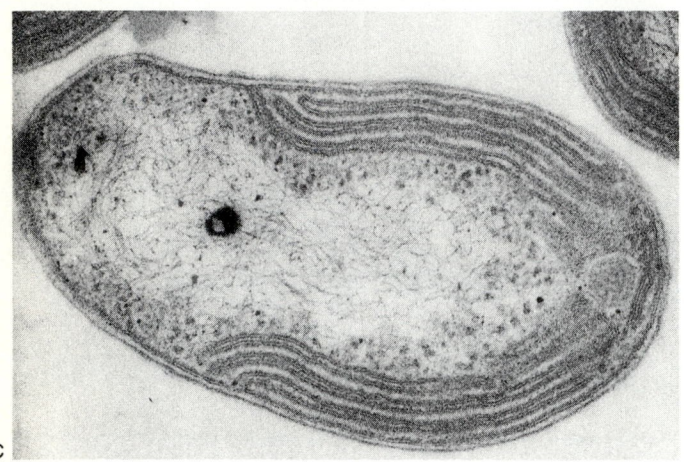

FIGURE 10.34

Micrographs showing morphologies and extensive internal membrane systems of nitrifying bacteria. (A) *Nitrosococcus oceanus* (123,700 ×); (B) *Nitrobacter multiformis* (88,200 ×); and (C) *Nitrobacter winogradsky* (165,759 ×). (Courtesy Stan Watson, Woods Hole Oceanographic Institute, Woods Hole, Mass. (A) reprinted by permission of Academic Press, New York, from S. W. Watson, and C. C. Remsen, 1970, *Journal of Ultrastructure Research*, 33:148–160; (B) reprinted by permission of Springer-Verlag, New York, from S. W. Watson, L. B. Graham, C. C. Remsen, and F. W. Valois, 1971, *Archiv für Mikrobiologie*, 76:183–203; (C) reprinted by permission of the American Society for Microbiology, Washington, D.C., from S. W. Watson, 1971, *International Journal of Systematic Bacteriology*, 21:254–270)

solobus. The first three genera, whose names begin with the prefix *nitro-*, oxidize nitrite; the remaining four genera, whose names begin with the prefix *nitroso-*, oxidize ammonia. *Nitrobacter* species are extremely important nitrifiers in soil, oxidizing nitrite to nitrate. *Nitrosomonas* species, likewise, are important nitrifiers in soil, oxidizing ammonia to nitrite. The combined actions of the members of the genera *Nitrosomonas* and *Nitrobacter* permit the conversion of ammonia to nitrate. The change in electronic charge between NH_4^+ and NO_3^- alters the mobility of these nitrogenous ions in soil and has a major influence on soil fertility.

Several genera of chemolithotrophic bacteria metabolize sulfur and sulfur-containing inorganic compounds. *Thiobacillus* derives energy from the oxidation of reduced sulfur compounds; its members are Gram-negative rods, motile by means of polar flagella. Some members of the genus *Thiobacillus* oxidize only sulfur compounds, whereas others, such as *T. ferrooxidans*, also oxidize ferrous to ferric iron in order to generate ATP. *Thiobacillus* species can be used in the recovery of minerals, including uranium, and their oxidation of reduced iron and sulfur compounds mobilizes various metals so that they can be extracted from even low-grade

ores. *T. thiooxidans*, frequently used in biological metal recovery, is an acidophile, with optimum growth occurring in the pH range of 1–3.5. The metabolic activities of *T. thiooxidans*, which is often found in association with waste coal heaps, produce acid mine drainage, a serious ecological problem associated with some coal mining operations.

Members in the family **Siderocapsaceae** oxidize iron or manganese, depositing iron and/or manganese oxides in capsules or extracellular material. Members of the genus *Siderocapsa*, for example, have spherical cells embedded in a common capsule partially encrusted with iron and/or manganese oxides. The taxonomic status of the entire family, and of the genus *Siderocapsa* in particular, has been questioned frequently. The description of these bacteria as unicellular, non-thread-forming or non-stalk-forming iron and/or manganese bacteria that under natural conditions deposit metal oxides on or in extracellular mucoid material is taxonomically imperfect and undoubtedly the source of the controversy. Although their proper taxonomic position is in doubt, these bacteria are ecologically important. They are widely distributed in nature, and their metabolic activities are of geological importance. Members of this family are found in iron-bearing waters, forming high concentrations in the lower portions of some lakes.

The Actinomycetes The order **Actinomycetales** contains bacteria characterized by the formation of branching filaments. Many of the more evolved actinomycetes resemble the fungi in appearance, but their cells are prokaryotic, and they are clearly eubacteria. The families in the order Actinomycetales include Frankiaceae, Actinoplanaceae, Dermatophilaceae, Streptomycetaceae, and Micromonosporaceae (Table 10.22). The various families of the order Actinomycetales are distinguished from one another by the nature of their mycelia and spores (Figure 10.35). Spore production is an important diagnostic characteristic for the identification of actinomycetes. Various types of spores are produced by actinomycete species, many of which are involved in the dispersal of actinomycetes; only *Thermoactinomyces* produces endospores.

The actinomycetes are widely distributed in nature. The oxidative forms are numerous and occur primarily in soils. The main ecological role of actinomycetes is in the decomposition of organic matter in soil. Fermentative types are primarily found in association with humans and other animals. Many of the Actinomycetaceae, such as *Streptomyces griseus*, produce antibiotics that are extremely important in the pharmaceutical industry. The availability of such antibiotics has revolutionized medical practice, and many previously fatal diseases are now easily controlled by the use of antibiotics produced by *Streptomyces* and other actinomycetes.

The Archaebacteria

As discussed in Chapter 1, members of the archaebacteria have been shown to be phylogenetically related to each other, on the basis of analysis of their 16S rRNA molecules, and to be distinct from eubacterial species. The archaebacterial species share several morphological and physiological features that make them distinct from other bacteria, including the lack of murein in their cell walls and the unusual ether linkage that occurs in their phospholipid molecules. They appear to be primitive bacteria and, as members of the kingdom Archaebacteria, are considered to be distantly related to other prokaryotes.

Methane-Producing Bacteria The methane-producing bacteria or **methanogens** represent a highly specialized physiological group. Members of the family Methanobacteriaceae form methane by the reduction of carbon dioxide. The methanogenic bacteria are very strict obligate anaerobes. In order to produce methane, these organisms utilize electrons generated in the oxidation of hydrogen or simple organic compounds, such as acetate and methanol. Methanogenic bacteria are unable to use carbohydrates, proteins, or other complex organic substrates. Improvements in anaerobic isolation techniques have permitted the isolation of many new species of methanogens. The methanogens often form consortia in association with other microorganisms. The microorganisms associated with the methanogens maintain the low oxygen tensions and provide the carbon dioxide and fatty acids required by the methanogenic bacteria. Such associa-

---

TABLE 10.22 The Actinomycetes

Order Actinomycetales	
Family Frankiaceae	Mycelium formed; symbionts in plant nodules with free stage in soil
Genus: *Frankia*	
Family Actinoplanaceae	Mycelium formed; saprophytes or facultative parasites; spores borne inside sporangia
Genus: *Actinoplanes, Spirillospora, Streptosporangium, Amorphosporangium, Ampullariella, Pilimelia, Planomonospora, Planobispora, Dactylosporangium, Kitasatoa*	
Family Dermatophilaceae	Mycelium divides transversely to form motile cocci; saprophytes or faculative parasites; spores not borne in sporangia
Genera: *Dermatophilus, Geodermatophilus*	
Family Streptomycetaceae	Mycelium tends to remain intact; saprophytes or facultative parasites; spores not borne in sporangia; usually abundant aerial mycelia and long spore chains
Genera: *Streptomyces, Streptoverticillium, Sporichthya, Microellobosporia*	
Family Micromonosporaceae	Mycelium remains intact; saprophytes or facultative parasites; spores not borne in sporangia; spores formed singly or in short chains
Genera: *Micromonospora, Thermoactinomyces, Actinobifida, Thermomonospora, Microbispora, Nicropolyspora*	

A 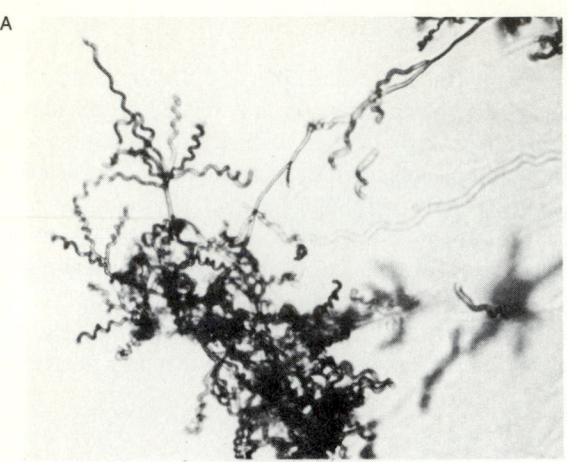 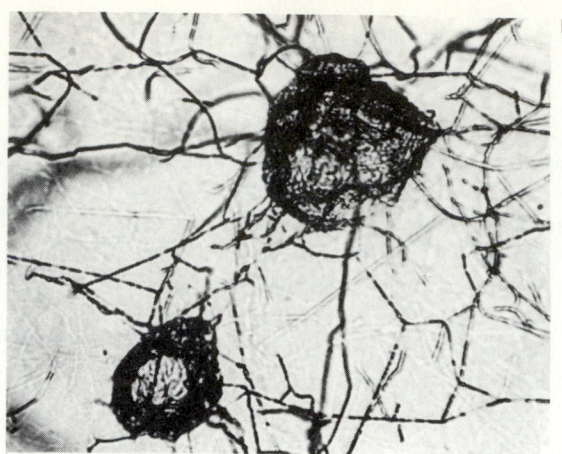 B

FIGURE 10.35

Micrographs of actinomycetes, showing hyphae and spores. (A) Aerial hyphae of a *Streptomyces* sp., showing the formation of coils of chains of conidia (1,000×); (B) the aerial hyphae of this *Streptomyces* sp., from sclerotia (1,000×). (Courtesy Hubert Lechevalier, Rutgers, the State University)

tions are extremely important in the rumen of animals such as cows. A major source of atmospheric methane is the rumen of such animals. The United States Environmental Protection Agency once issued an indelicate report stating that the burping cow was the major source of atmospheric hydrocarbon pollutants. Clearly, though, in urban areas, hydrocarbon pollutants originate primarily from automotive exhausts and not cows.

The family **Methanobacteriaceae** has been extensively revised recently to accommodate several new taxa. These

genera of methane-generating bacteria can be differentiated on the basis of relatively few morphological and physiological features (Table 10.23). Among these methanogens, *Methanobacterium* consists of rods or lancet-shaped cocci and *Methanosarcina* of large cocci occurring in packets; *Methanococcus* species occur as cocci arranged singly or in irregular clusters (Figure 10.36).

Halobacteriaceae   The genera *Halobacterium* and *Halococcus*, included in the family Halobacteriaceae, have the char-

TABLE 10.23   The Methanogenic Bacteria

Family Methanobacteriaceae	
Order Methanobacteriales	Cells short, lancet-shaped cocci to long filaments; cells appear Gram-positive but lack murein; strict anaerobes; oxidize hydrogen; reduce CO_2 to methane
Family Methanobacteriaceae	
Genera *Methanobacterium*	Slender rods often forming filaments
Methanobrevibacter	Short rods or lancet-shaped cocci often in pairs or chains
Order Methanococcales	Cells cocci; oxidize hydrogen or formate; reduce CO_2 to methane; Gram-negative, with protein units external to cytoplasmic membrane
Family Methanococcaceae	
Genus *Methanococcus*	Cocci single, in pairs or clumps
Order Methanomicrobiales	Cells cocci to rods; Gram-negative or Gram-positive; motile or nonmotile; strict anaerobes; oxidize hydrogen or formate; reduce CO_2 to methane or form methane via fermentation of methanol and related compounds
Family Methanomicrobiaceae	Gram-negative cocci to rods
Genera: *Methanomicrobium*	Short motile rods
Methanogenium	Irregular coccoid cells
Methanospirillum	Curved slender motile rods often forming filaments
Family Methanosarcinaceae	Gram-positive coccoid cells occurring in packets
Genus: *Methanosarcina*	

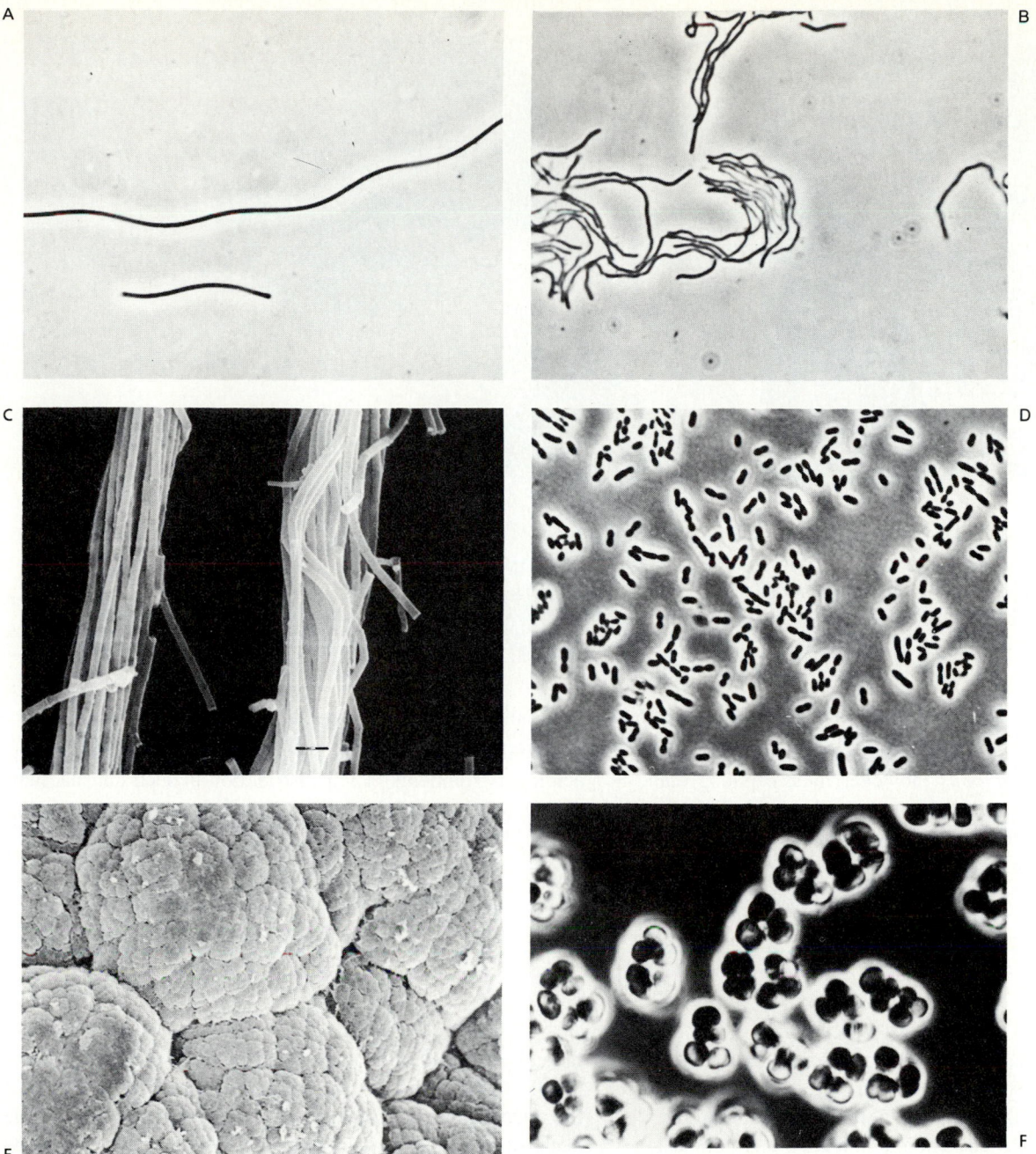

FIGURE 10.36

Micrographs of methanogens. (A) Phase contrast micrograph of *Methanospirillum hungatei*;
(B) phase contrast micrograph of *Methanobacterium thermoautotrophicum*; (C) scanning electron
micrograph of *Methanobacterium soehngenii*; (D) phase contrast micrograph of *Methanobrevi-
bacter ruminatium*; (E) scanning electron micrograph of *Methanococcus mazei*; (F) phase contrast
micrograph of *Methanosarcina* sp. (A–F courtesy Robert Mah, C with Robert Sleat; E with Jack
Pangborn, University of California, Los Angeles)

acteristic properties of the archaebacteria. The unusual
bacteriorhodopsin-mediated phototrophic metabolism of *Halobacterium* was discussed earlier. In addition to this unique
feature, all members of the family Halobacteriaceae are obligate halophiles, growing only in media containing at least
15 percent NaCl. Members of this family are found in ecosystems that have extremely high sodium chloride concentrations, such as some salt lakes, the Dead Sea, and foods
preserved by salting.

Sulfolobus   On the basis of their staining reactions, the bacteria in the sulfur-oxidizing genus *Sulfolobus* have been char-

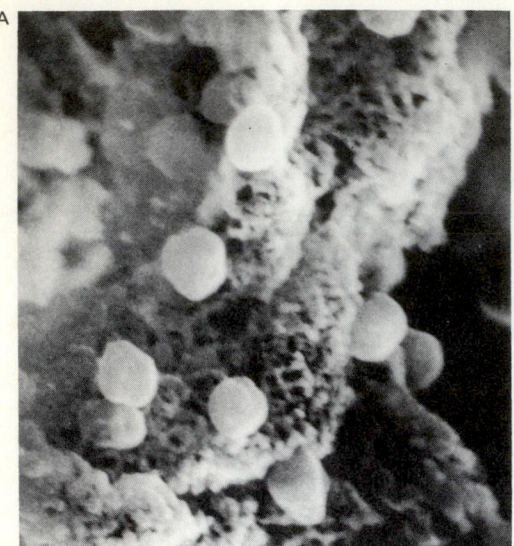

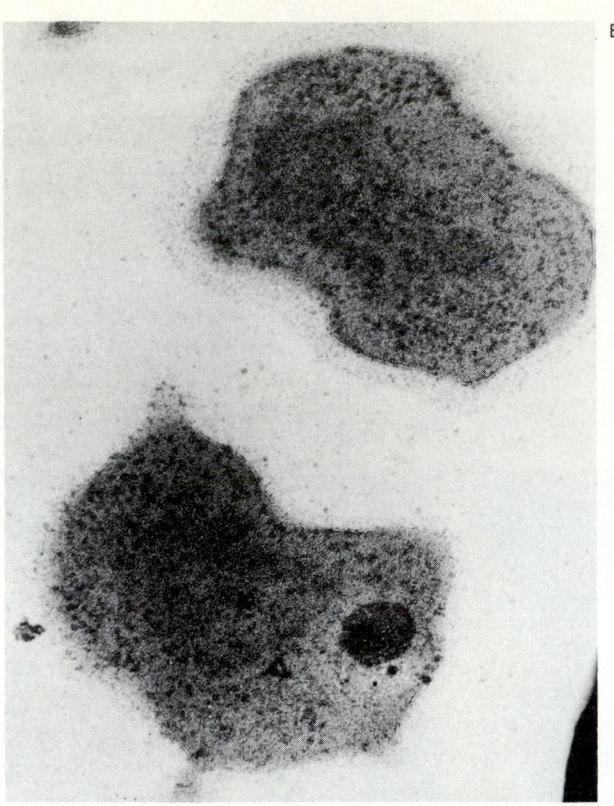

FIGURE 10.37

Micrographs of *Sulfolobus*. (A) Scanning electron micrograph showing *Sulfolobus* on sulfur particles (10,000×); (B) transmission electron micrograph of *S. brierleyi* (133,000). (Courtesy of C. L. and J. A. Brierley, New Mexico Technical University)

acterized as Gram-negative spherical cells (Figure 10.37). The cell walls of *Sulfolobus*, however, lack peptidoglycan. They also share other properties with the Archaebacteria, including rRNA homology and membrane structure. Members of this genus are thermophiles, with an optimum growth temperature of 70–75°C. *Sulfolobus* species occur in hot, acidic environments. Their distribution in such unusual habitats appears to be another common characteristic of many archaebacteria.

Postlude

Microbial systematics is a very broad area of microbiology. The more we know about an organism, the better we are able to understand its relationships to other organisms, so that it can be properly classified. The classification and nomenclature of microorganisms is important in all fields of microbiology. One purpose of a taxonomic system is to provide a basis for nomenclature. Microorganisms are named by using a binomial system in which they are specified by their genus and species names. The naming of microorganisms permits one to refer to a particular organism unambiguously, which is essential for communication regarding the properties of a particular microorganism. In this chapter and throughout the rest of this book, you will find the names and descriptions of microbial taxa, including the practical importance of particular species. Associating the proper name with an organism is critical for its classification.

Microbial classification systems attempt to describe the natural organization of the microbial world, showing the similarities among organisms in one taxonomic group and the diversity of microbial taxa. Ideally, classification systems should reflect the evolutionary separation of organisms. At the microbial level, however, this has been nearly impossible because of the lack of a fossil record. Arbitrary decisions traditionally have had to be made in classifying microorganisms, with some taxonomists emphasizing differences between microbial groups and others emphasizing similarities. All microbial classification systems developed so far are artificial, and are subject to constant review and revision. Although some microbiologists consider taxonomy dull, the field is marked by constant strife and differing opinions as to whether particular systems of classification assign microorganisms to their proper taxonomic place. Taxonomy is the only area of microbiology where it has been necessary to establish a formal judicial system to arbitrate and rule on the validity of differing opinions.

A variety of approaches are used to classify microorganisms, including phenetic approaches aimed at assessing group similarities and phylogenetic approaches aimed at reflecting natural evolutionary patterns. Most classification systems for microorganisms are based on phenotypic characteristics, sometimes leading to the formation of genetically heterogeneous groups of uncertain affiliation. Many different phenotypic features are used in classifying microorganisms. To determine the characteristics of a microorganism so that it can be unambiguously classified, the organism must be grown in pure culture. Modern culture techniques, including the development of methods for isolating strict anaerobes in pure culture, have permitted the proper classification of many microorganisms that were previously unknown.

The classical approach to microbial classification emphasizes the points of separation between microbial groups. In contrast, the numerical taxonomic approach examines overall similarity. The numerical taxonomic approach is less sensitive to individual features than are approaches that establish key features for distinguishing microbial taxa. Each system has its own merits. Modern approaches to microbial taxonomy include genetic analyses, and such analyses promise to revolutionize microbial classification systems. Already we have seen the recognition of the Archaebacteria, and as genetic and epigenetic approaches are used to examine other groups, a clearer picture of the phylogenetic relationships among microorganisms will certainly develop. In describing microbial taxa, it is now routine to determine the mole% G + C of the DNA. Studies of DNA and RNA homologies to determine the true relatedness of species of particular taxa are clarifying the taxonomic status of many groups of microorganisms. Although genetic analyses are emphasized in modern microbial classification systems, physiological and morphological features are still extremely important. The inclusion of the cyanobacteria with other groups of bacteria is an example of a major recent taxonomic revision based on the recognition of the fundamental difference between organisms having prokaryotic and eukaryotic cells.

In addition to establishing the relationships among microorganisms, a taxonomic system provides a basis for identifying them. Classification and identification are distinct processes. Identification involves the comparison of an unknown organism with previously described taxa. A microbial identification system should permit the unambiguous identification of an unknown microorganism by using the minimal number of tests; the system should also be accurate and efficient. Computers have greatly facilitated the identification of microorganisms by allowing the rapid comparison of numerous characteristics and the assessment of the statistical probability of making a correct identification. Computer-assisted identification systems have become an integral part of the clinical microbiology laboratory, where rapid and accurate identification of microorganisms is essential for the proper diagnosis and treatment of infectious diseases.

In addition to considering the general aspects of microbial systematics, the features of the major groups of bacteria have been reviewed. The major groups, organized on the basis of unifying morphological and/or physiological properties, represent the great diversity of microbial morphological forms and physiological functions. Only the highlights of each group have been considered here. The discussion of various applied topics in many of the following chapters will refer to and add information about these bacteria. The suggested supplementary readings should be consulted for more detailed consideration of a microbial group of interest.

Suggested Supplementary Readings

BALCH, W. E., G. E. FOX, L. J. MAGRUM, C. R. WOESE, and R. S. WOLFE. 1979. Methanogens: Reevaluation of a unique biological group. *Microbiological Reviews* 43:260–296.

BARKSDALE, L., and K. S. KIM. 1977. *Mycobacterium. Bacteriological Reviews* 41:217–372.

BAUMANN, P., L. BAUMANN, M. J. WOOLKALIS, and S. S. BANG. 1983. Evolutionary relationships in *Vibrio* and *Photobacterium*: A basis for a natural classification. *Annual Review of Microbiology* 37:369–398.

BECKER, Y. 1978. The chlamydia: Molecular biology of procaryotic obligate parasites of eucaryocytes. *Microbiological Reviews* 42:274–306.

BUCHANAN, R. E., and N. E. GIBBONS (eds.). 1974. *Bergey's Manual of Determinative Bacteriology*, 8th ed. Williams & Wilkins Co., Baltimore.

COLWELL, R. R. 1973. Genetic and phenetic classification of bacteria. *Advances in Applied Microbiology* 16:137–176.

COWAN, S. T., and L. R. HILL (eds.). 1978. *A Dictionary of Microbial Taxonomy*. Cambridge University Press, New York.

CROSS, T., and M. GOODFELLOW. 1973. Taxonomy and classification of the actinomycetes. In *Actinomycetales: Characteristics and Practical Importance* (G. Sykes and F. A. Skinner, eds.), pp. 11–112. Academic Press, London.

DELWICHE, E. A., J. J. PESTKA, and M. L. TORTORELLO. 1985. The Veillonellae: Gram-negative cocci with a unique physiology. *Annual Review of Microbiology* 39:175–193.

GERHARDT, P. (ed.). 1981. *Manual of Methods for General Bacteriology*. American Society for Microbiology, Washington, D.C.

GHIORSE, W. C. 1984. Biology of iron- and manganese-depositing bacteria. *Annual Review of Microbiology* 38:515–550.

GIBBONS, N. E., and R. G. E. MURRAY. 1978. Proposals concerning the higher taxa of bacteria. *International Journal of Systematic Bacteriology* 28:1–6.

GOODFELLOW, M., and D. E. MINNIKIN. 1977. Nocardioform bacteria. *Annual Review of Microbiology* 31:159–180.

HARWOOD, C. S., and E. CANALE-PAROLA. 1984. Ecology of spirochetes. *Annual Review of Microbiology* 38:161–192.

HASTINGS, J. W., and K. H. NEALSON. 1977. Bacterial bioluminescence. *Annual Review of Microbiology* 31:549–595.

HENRIKSEN, S. D. 1976. *Moraxella, Branhamella*, and *Acinetobacter. Annual Review of Microbiology* 30:63–83.

HOLT, S. C. 1978. Anatomy and chemistry of spirochetes. *Microbiological Reviews* 42:114–160.

JOHNSON, R. C. 1976. The spirochetes. *Annual Review of Microbiology* 31:39–61.

KRIEG, N. R. 1976. Biology of the chemoheterotrophic spirilla. *Bacteriological Reviews* 40:55–115.

KRIEG, N. R., and J. HOLT (eds.). 1984. *Bergey's Manual of Systematic Bacteriology*, Volume 1. Williams & Wilkins Co., Baltimore.

LAPAGE, S. P., P. H. A. SNEATH, E. F. LESSEL, V. B. D. SKERMAN, H. P. R. SEELIGER, and W. A. CLARK (eds.). 1975. *International Code of Nomenclature of Bacteria*. American Society for Microbiology, Washington, D.C.

LARKIN, J. M., and W. R. STROHL. 1983. *Beggiatoa, Thiothrix, and Thioploca. Annual Review of Microbiology* 37:341–367.

LASKIN, A. I., and H. A. LECHEVALIER. 1977. *Handbook of Microbiology: Bacteria*. CRC Press, Inc., Boca Raton, Fla.

MACY, J. M., and L. PROBST. 1979. The biology of gastrointestinal bacteriodes. *Annual Review of Microbiology* 33:561–594.

MOORE, R. L. 1983. The biology of *Hyphomicrobium* and other prosthecate, budding bacteria. *Annual Review of Microbiology* 37:567–594.

PFENNIG, N. 1977. Phototrophic green and purple bacteria: A comparative systematic survey. *Annual Review of Microbiology* 31:275–290.

POINDEXTER, J. S. 1981. The caulobacters: Ubiquitous unusual bacteria. *Microbiological Reviews* 45:123–179.

RAZIN, S. 1978. The mycoplasmas. *Microbiological Reviews* 42:414–470.

REICHENBACH, H. 1983. Taxonomy of the gliding bacteria. *Annual Review of Microbiology* 37:339–364.

SANDERSON, K. E. 1976. Genetic relatedness in the family Enterobacteriaceae. *Annual Review of Microbiology* 30:327–349.

SCHLEIFER, K. H., and E. STACKEBRANDT. 1983. Molecular systematics of prokaryotes. *Annual Review of Microbiology* 37:143–187.

SHAPIRO, L. 1976. Differentiation in the *Caulobacter* cell cycle. *Annual Review of Microbiology* 30:377–408.

SHEWAN, J. M., and T. A. McMEEKIN. 1983. Taxonomy (and ecology) of *Flavobacterium* and related genera. *Annual Review of Microbiology* 37:233–252.

SKERMAN, V. B. D. 1967. *A Guide to the Identification of the Genera of Bacteria*. Williams & Wilkins Co., Baltimore.

SKERMAN, V. B. D., V. McGOWAN, and P. H. A. SNEATH (eds.). 1980. Approved lists of bacterial names. *International Journal of Systematic Bacteriology* 36:225–420.

SKINNER, F. A., and D. W. LOVELOCK. 1980. *Identification Methods for Microbiologists*. Academic Press, New York.

SNEATH, P. H. A. 1978. Classification of microorganisms. In *Essays in Microbiology* (J. R. Norris, ed.), chapter 9. John Wiley & Sons, Inc., Chichester, England.

SNEATH, P. H. A. 1978. Identification of microorganisms. In *Essays in Microbiology* (J. R. Norris, ed.), chapter 10. John Wiley & Sons, Inc., Chichester, England.

SNEATH, P. H. A., and R. R. SOKAL. 1973. *Numerical Taxonomy: The Principles and Practice of Numerical Classification*. W. H. Freeman and Co., San Francisco.

SOKATCH, J. R. (ed.). 1986. *The Biology of* Pseudomonas. Academic Press, Orlando, Fla.

STANIER, R. Y., and G. COHEN-BAZIRE. 1977. Phototrophic prokaryotes: The cyanobacteria. *Annual Review of Microbiology* 31:225–274.

STARR, M. P., H. STOLP, H. G. TRUPER, A. BALLOWS, and H. G. SCHLEGEL (eds.). 1981. *The Prokaryotes: A Handbook on Habits, Isolation, and Identification of Bacteria*. Springer-Verlag, Berlin.

THORNSBERRY, C., A. BALOWS, J. C. FEELEY, and W. JAKUBOWSKI. 1984. *Legionella*. American Society for Microbiology, Washington, D.C.

vanVEEN, W. L., E. G. MULDER, and M. H. DEINEMA. 1978. The *Sphaerotilus-Leptothrix* group of bacteria. *Microbiological Reviews* 42:329–356.

WEISS, E. 1984. The biology of rickettsiae. *Annual Review of Microbiology* 36:345–370.

WHITCOMB, R. F. 1980. The genus *Spiroplasma. Annual Review of Microbiology* 34:677–709.

WOESE, C. R. 1981. Archaebacteria. *Scientific American* 244(6):98–122.

WOESE, C. R. 1985. Why study evolutionary relationships among bacteria? *Evolution of Prokaryotes*. FEMS Symposium 29. (K. H. Scheifer and E. Stackebrandt, eds.), pp. 1–30. Academic Press, London.

WOESE, C. R., and R. S. WOLFE (eds.). 1985. *The Archaebacteria*. Academic Press, Orlando, Fla.

ZEIKUS, J. G. 1977. The biology of methanogenic bacteria. *Bacteriological Reviews* 41:514–541.

Study Questions

1. How are bacteria named?

2. What is the difference between classification and identification?

3. Compare phenetic and phylogenetic approaches to microbial classification.

4. Why has the use of DNA homology caused the reclassification of many bacterial species? Why is DNA homology a better measure of relatedness than phenotypic characteristics for developing classification systems?

5. Why does comparison of the mole percent G + C permit the assessment of genetic relatedness? Why does DNA homology better describe genetic relatedness than the proportion of G + C in the DNA?

6. What is an identification key? How is a dichotomous key used in identifying an unknown bacterium?

7. How do computers aid in the identification of bacteria?

8. Describe the similarities and differences between cyanobacteria and other phototrophic bacteria.

9. How are the archaebacteria different from other bacterial groups? How has the use of rRNA analysis helped define the relationships among these organisms?

10. Which bacterial genera are characterized by endospore formation?

11. What phenotypic characteristics are used to distinguish major groups of bacteria?

12. Where would you find a description of a bacterial genus that had been described 10 years ago? A description of one that had been described for the first time this year?

Situational Problem

Designing an Identification Scheme

One of the simplest and most efficent methods of identifying a bacterial isolate that is one of a very limited number of possible species is the use of a dichotomous key. The dichotomous, or two-branching, key facilitates the separation of bacteria into increasingly smaller groups until only one bacterium remains (see Figure 10.1). This process of separation and identification is based on determining a series of physical and biochemical characteristics of the bacterium including its morphology, growth requirements, and enzymatic processes. Suppose you are asked to help design an identification scheme for a limited number of common bacterial isolates that are expected to be found in the specimens to be examined. The positive identification of particular organisms in the specimens is critical. Because of limited funds, you decide to design a dichotomous key rather than to advise the selection of a miniaturized commercial identification system that would be far more costly per specimen.

The first step in the development of a dichotomous key involves the characterization of all of the possible bacterial isolates. By referring to a taxonomic guide such as *Bergey's Manual*, you can find the characteristics of most known bacteria. Once the characteristics of the bacteria have been determined, the construction of the key can begin. The most efficient dichotomous keys utilize tests that divide a group of bacteria into two groups of equal size. Each group, in turn, should be divided into two smaller groups—again preferably of equal size—and the process of division repeated until only one organism remains in each of the subgroups. The tests used should be selected so that there are only two possible results, one positive and the other negative. Ideally, the dichotomous key should separate bacteria into individual families, which are subsequently divided into one or more genera, which in turn are divided into species.

The first test used in many such keys is the Gram stain. This divides most bacteria into two groups, which may be further divided into individual families according to shape, endospore production, or some other characteristic. Each family is divided into its respective genera by other tests that depict more specific differences between the organisms, and each genus is divided by still more specific characteristics until only one species remains at the end of each branch. Thus, with the use of a dichotomous key, the identity of

most common bacterial isolates can be determined quickly, accurately, and efficiently.

1. Given the results for the following 10 Gram-positive bacteria, design a dichotomous key that would permit their identification using the minimum number of tests.

Bacterial Species

0 *Bacillus anthracis*	5 *Mycobacterium tuberculosis*
1 *Bacillus cereus*	6 *Staphylococcus aureus*
2 *Clostridium perfringens*	7 *Staphylococcus epidermidis*
3 *Clostridium tetani*	8 *Streptococcus pneumoniae*
4 *Corynebacterium diphtheriae*	9 *Streptococcus pyogenes*

Characteristics Tested

A. Rods	L. Acid from mannitol
B. Cocci	M. Catalase
C. Endospores	N. Oxidase
D. Endospores central	O. Nitrate reduced
E. Endospores subterminal	P. Motile
F. Endospores terminal	Q. Snapping division
G. Acid fast	R. Coagulase
H. Growth in presence of air	S. Beta hemolysis
I. Growth in absence of air	T. Growth at 45°C
J. Acid from glucose	U. Growth at 6.5 percent NaCl
K. Gas from glucose	

Test Results

Organism	A	B	C	D	E	F	G	H	I	J	K	L	M	N	O	P	Q	R	S	T	U
0	+	−	+	+	−	−	−	+	+	+	−	−	+	+	+	−	−	−	−	−	−
1	+	−	+	+	−	−	−	+	+	+	−	−	+	+	+	−	−	−	−	−	+
2	+	−	+	−	+	−	−	+	+	+	−	−	−	−	−	−	−	−	−	−	+
3	+	−	+	−	−		+	−	−	+	−	−	−	−	−	−	+	−	−	+	−
4	+	−	−	−	−	−	+	+	+	−	−	+	+	+	−	+	−	+	−	−	
5	+	−	−	−	−	−	+	+	+	−	−	+	+	+	−	−	−	−	−	−	−
6	−	+	−	−	−	−	+	+	+	−	+	+	+	+	−	−	+	+	−	+	
7	−	+	−	−	−	−	+	+	+	−	−	+	+	+	−	−	−	−	+	+	
8	−	+	−	−	−	−	+	+	+	−	−	−	−	−	−	−	−	−	+	−	
9	−	+	−	−	−	−	+	+	+	−	−	−	−	−	−	−	−	−	+	−	−

+	=	positive test result
−	=	negative test result
blank	=	indeterminate test result

2. Given the results for the following 10 Gram-negative bacteria, design a dichotomous key that would permit their identification using the minimum number of tests.

Bacterial Species

0 *Citrobacter freundii*	5 *Pseudomonas aeruginosa*
1 *Enterobacter aerogenes*	6 *Salmonella typhimurium*
2 *Escherichia coli*	7 *Serratia marcescens*
3 *Klebsiella pneumoniae*	8 *Shigella flexnerii*
4 *Proteus vulgaris*	9 *Yersinia enterocolitica*

Characteristics Tested

A. Rods	L. Hydrogen sulfide
B. Acid from glucose	M. Urease
C. Gas from glucose	N. Gelatinase
D. Catalase	O. Indole
E. Oxidase	P. Lysine decarboxylase
F. Nitrates reduced	Q. Ornithine decarboxylase
G. Methyl red	R. Phenylalanine deaminase
H. Voges-Proskauer reaction	S. Motile
I. Citrate	T. Red pigments
J. β-Galactosidase	U. Fluorescent pigments
K. Acid from lactose	V. Growth at 37°C

```
Organ-                          Test
ism   A B C D E F G H I J K L M N O P Q R S T U V
 0    + + + + − + + − + + + + + − − − − − + − − +
 1    + + + + − + − + + + + + − − + + + + − + − − +
 2    + + + + − + + − + + + + − − − + + + − − + − − +
 3    + + + + − + − + + + + + − + − − + − − − − − +
 4    + + + + − + + − + − − + + + + − − + + − − +
 5    + − − + + + − − − − − − − + + − − − + − + +
 6    + + + + − + + + − + − − + − − + + + + − − +
 7    + + − + − + − + + + + − − + − − + + + + + − +
 8    + + − + − + + − − − − − − − − − − − − − − +
 9    + + − + − + + − − + − − + − − + − − + − + − − +
```

```
    + = positive test result
    − = negative test result
blank = indeterminate test result
```

3. Create a key for separating the following bacterial genera: *Bacillus, Clostridium, Escherichia, Enterobacter, Streptococcus,* and *Staphylococcus.*

4. Consider that the following 30 bacterial species represent all the possible bacteria that may occur in the specimens received by a laboratory for identification.

Bacillus cereus	*Clostridium sporogenes*
Bacillus firmus	*Enterobacter aerogenes*
Bacillus licheniformis	*Enterobacter cloacae*
Bacillus stearothermophilus	*Escherichia coli*
Bacillus subtilis	*Micrococcus luteus*
Citrobacter freundii	*Micrococcus roseus*
Clostridium perfringens	*Mycobacterium gastri*

Mycobacterium phlei	*Pseudomonas fluorescens*
Mycobacterium smegmatis	*Salmonella typhimurium*
Mycobacterium terrae	*Serratia marcescens*
Planococcus citreus	*Shigella flexneri*
Proteus mirabilis	*Staphylococcus aureus*
Proteus vulgaris	*Staphylococcus epidermidis*
Pseudomonas aeruginosa	*Streptococcus faecalis*
Pseudomonas flava	*Streptococcus lactis*

Using any edition of *Bergey's Manual*, look up the characteristic features of these species. The tests that you probably want to consider are as follows:

Morphological features
   Colony color
   Cell shape
   Cell arrangement
   Gram-stain reaction
   Acid-fast stain reaction
   Endospore stain reaction
   Capsule stain
   Motility
Physiological features
   Aerobic growth
   Anaerobic growth
   Catalase production
   Cytochrome oxidase production
   Growth temperature range and optimum
pH growth range
   Glucose fermentation (acid and gas)
   Galactose fermentation (acid and gas)
   Lactose fermentation (acid and gas)
   Maltose fermentation (acid and gas)
   Sucrose fermentation (acid and gas)
   Amylase production (starch hydrolysis)
   Citrate utilization
   DNase
   Gelatinase
   Blood hemolysis reactions (alpha and beta hemolysis)
   Indole production
   Lysine decarboxylase
   Methyl red reaction
   Voges-Proskauer reaction
   Ornithine decarboxylase
   Phenylalanine deaminase
   Hydrogen sulfide production

Starting with the Gram-stain reaction, select the features that will enable you to separate the species in the most efficient manner, that is, using the fewest number of tests, and construct an unambiguous dichotomous key that can be used to identify any of these bacteria that may occur in the specimen.

CHAPTER 11

Survey of Fungi, Algae, and Protozoa

11.1 Fungi

Reproduction of Fungi

Fungi exhibit a wide variety of reproductive strategies. The most common mode of reproduction for **yeasts**, which are fungi that exist predominantly as unicellular organisms, is **budding**—a process in which a daughter cell is formed by pinching off a segment of the mother cell (Figure 11.1). Budding involves the formation of a cross wall that separates the bud from the mother cell. The cross wall of *Saccharomyces*, for example, consists of chitin, which does not occur elsewhere in the cell wall. Budding follows mitotic division, so that both the progeny and the parent cell contain a complete genome. Budding can occur all around the mother cell (**multilateral budding**) or may be restricted to the end (**polar budding**) (Figure 11.2). The budding process leaves a **bud scar** on the mother cell, and consequently, only a limited number of progeny may be derived from an individual yeast cell. Although budding is the most common form of reproduction, various other reproductive strategies exist among the yeasts, including sexual reproduction and fission.

In contrast to yeasts, the **filamentous fungi** normally develop multicellular structures known as **hyphae**. The growth of the fungus involves the elongation of the hyphae, generally with the formation of branches and cross walls separating individual cells. Some fungi, however, form multinucleate **coenocytic mycelia** that lack cross walls; therefore, these fungi are actually one-celled, multinucleate organisms (Figure 11.3). **Dimorphism** is characteristic of some fungi; this is a phenomenon in which the fungus reproduces as a unicellular organism and appears yeast-like under some conditions, and grows as a filamentous form under other conditions. Because filamentous fungi form elongated filaments, growth can occur in the absence of reproduction. Growth requires an increase in biomass, but unlike reproduction, cellular division does not always occur. In filamentous fungi, growth normally originates from the hyphal tip (Figure 11.4) The apical growth of a fungal hypha requires that the necessary polymers be transported to the area of new cell wall synthesis, and in eukaryotes this area generally contains a large number of microfibrils that are involved in cell wall synthesis.

During the asexual phase of fungal reproduction a variety of spores may be produced, depending on the species (Figure 11.5). These include various **conidia**, asexual spores borne externally on hyphae, or specialized **conidiophore** structures. The conidia can be separated from the fungal hypha as single cells. One type of conidium, the **arthrospore**, represents fragmented hyphae. The fragmentation of multicellular eukaryotic microorganisms constitutes a form of reproduction because the individual fragments are each capable of reproducing the original organism. Other asexual fungal spores include **sporangiospores**, which are produced within a specialized structure known as the **sporangium**, and **chlamydospores**, which are thick-walled spores that occur within hyphal segments. These fungal spores can be dispersed from the fungal hyphae and later germinate to form new vegetative structures.

Fungi also can produce various types of sexual reproductive spores. Some sexual spores of fungi, **ascospores**, are formed within a specialized structure known as the **ascus**

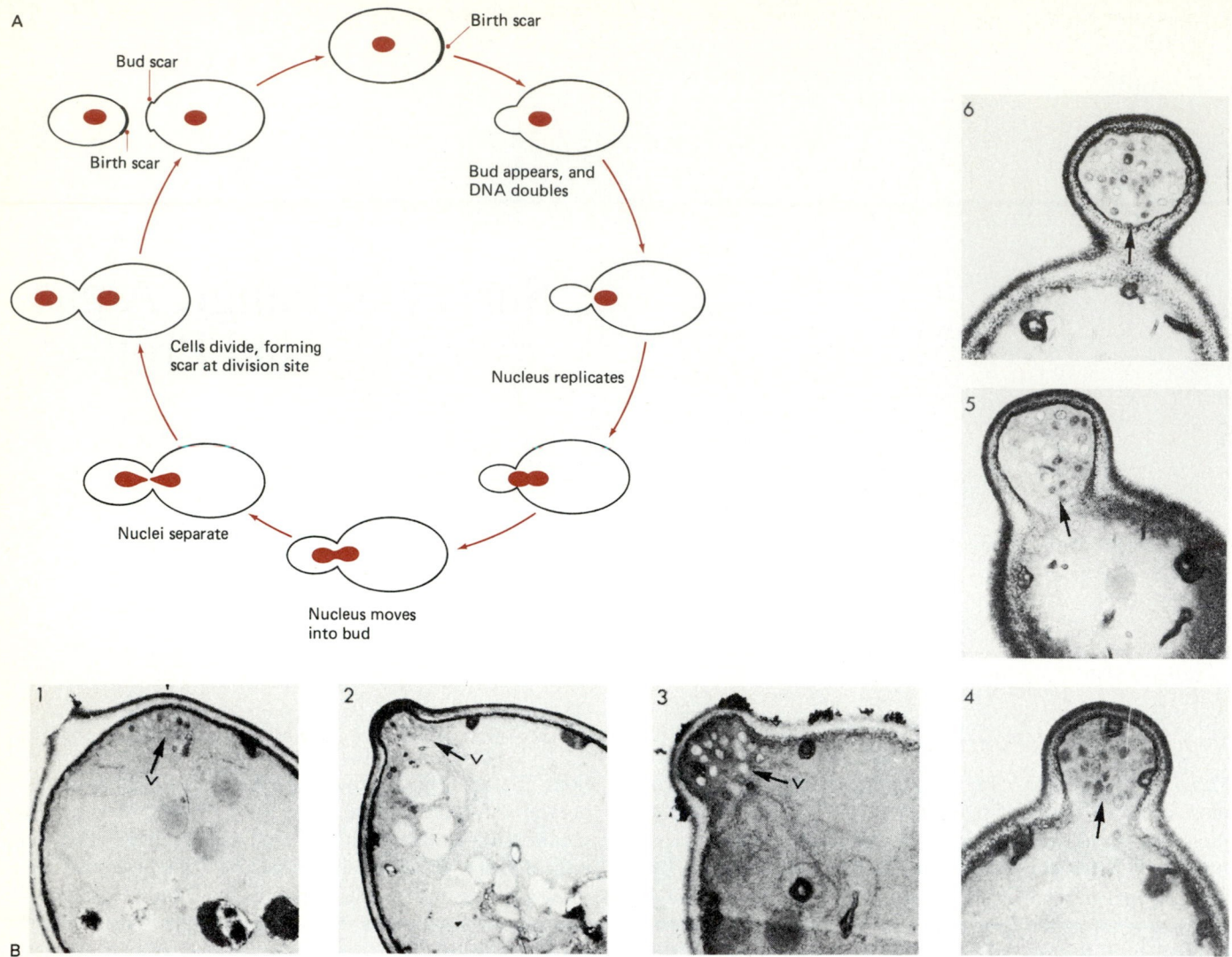

FIGURE 11.1

(A) Diagram illustrating reproduction by budding by a yeast. (B) Series of electron micrographs of *Saccharomyces cerevesiae* at different stages of budding; as shown in these micrographs, vesicles (v) of uniform size accumulate preferentially at a cytoplasmic location where the bud grows. (Courtesy Bijan Ghosh, Rutgers, the State University)

(Figure 11.6, page 316). The ability to produce ascospores distinguishes the ascomycete fungal group, which includes both yeasts and filamentous fungi, from other major fungal groups. In another major group of fungi, the **basidiomycetes**, the sexual spores are produced on a specialized structure known as the **basidium** (Figure 11.7, page 316). The sexual reproduction of basidiomycetes usually involves the fusion of hyphal cells. In several fungal groups, the Mastigomycota and Zygomycotina, sexual reproduction generally occurs by fusion of specialized **gametes** (Figure 11.8, page 317). These gametes are haploid, and their fusion reestablishes a diploid state. In many of the Gymnomycota, the gametes are motile and are thus known as **zoospores**, but Zygomycotina species, such as bread molds, produce only nonmotile reproductive spores. The deuteromycetes, or fungi

imperfecti, have no known sexual reproductive phase, and as far as we know, they are restricted to asexual means of reproduction. If a sexual stage is discovered for a fungus that has been classified among the fungi imperfecti, it is reclassified into one of the other major fungal groups on the basis of the types of sexual spores that are produced.

Classification of Fungi

As with other microorganisms, the classification of the fungi is difficult, sometimes ambiguous, vehemently debated, and subject to constant revision. The diversity of the fungi and the interrelationships among the fungi and other groups of eukaryotic microorganisms make fungal taxonomy particularly difficult. The fungi are eukaryotic heterotrophic micro-

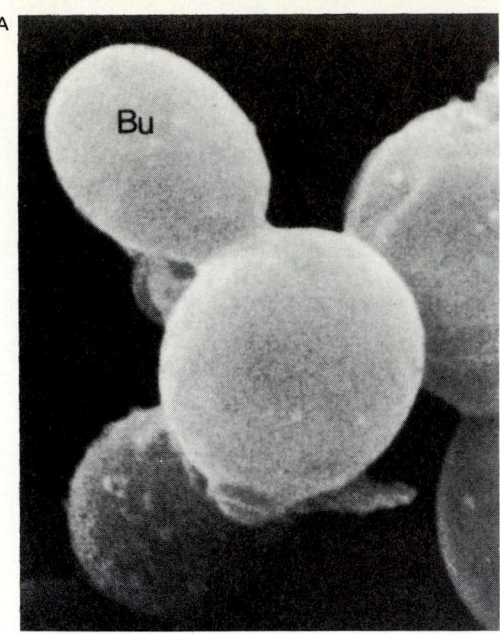

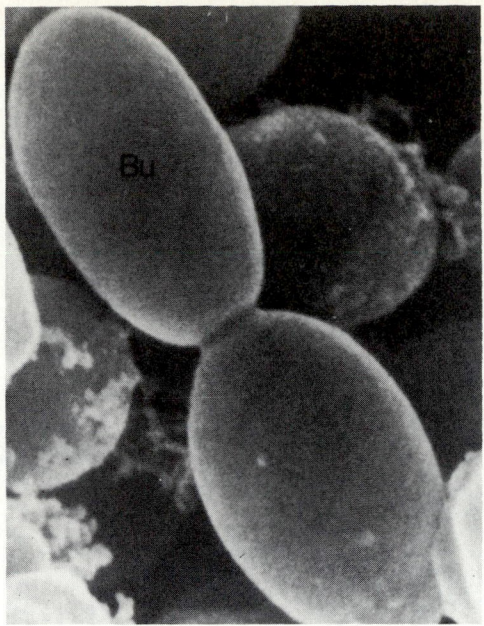

FIGURE 11.2
Budding in yeasts occurs only at the pole (polar budding) or all around the cell (multi-lateral budding). These electron micrographs illustrate the appearance of different yeasts exhibiting these two forms of reproduction. Multilateral (A) and polar (B) budding in a yeast (13,125× and 15,000×). Bu = bud. (Reprinted by permission of Springer-Verlag, New York, from R. G. Kessel and C. Y. Shih, 1976, *Scanning Electron Microscopy for Biology*)

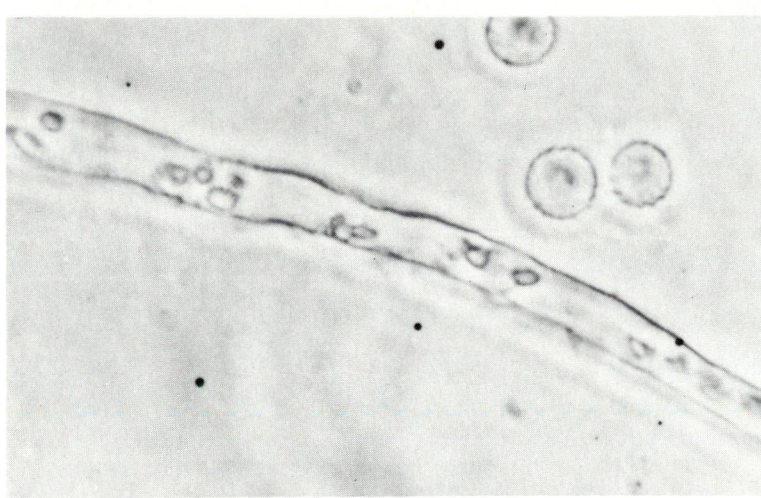

FIGURE 11.3 (*left*)
Micrograph of the coenocytic hypha of a *Rhizopus* sp., showing multiple nuclei within one elongated cell and the lack of cross walls.

FIGURE 11.4 (*above*)
Filamentous fungi reproduce by extension of hyphae. Here the growth of a hyphal tip of *Gelasinospora autosteira* is illustrated at 30-minute intervals. (Based on C. J. Alexopolus and C. M. Mims, *Introductory Mycology*, John Wiley & Sons, Inc.)

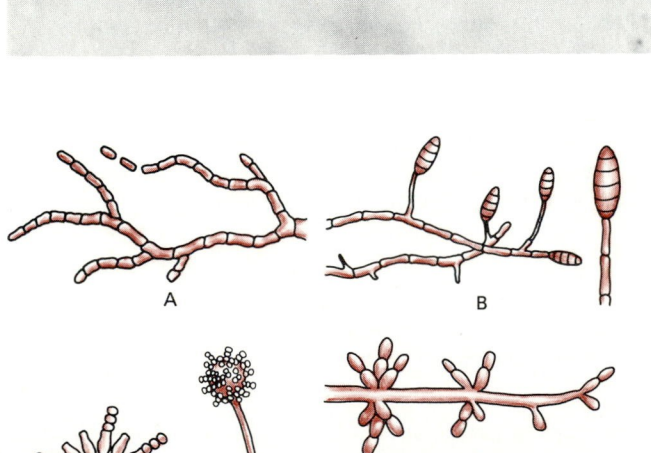

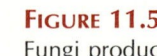

FIGURE 11.5
Fungi produce a variety of asexual spores. (A) Arthrospores; (B) aleurospores; (C) conidiospores and conidia; (D) blastospores; and (E) sporangia and sporangiospores.

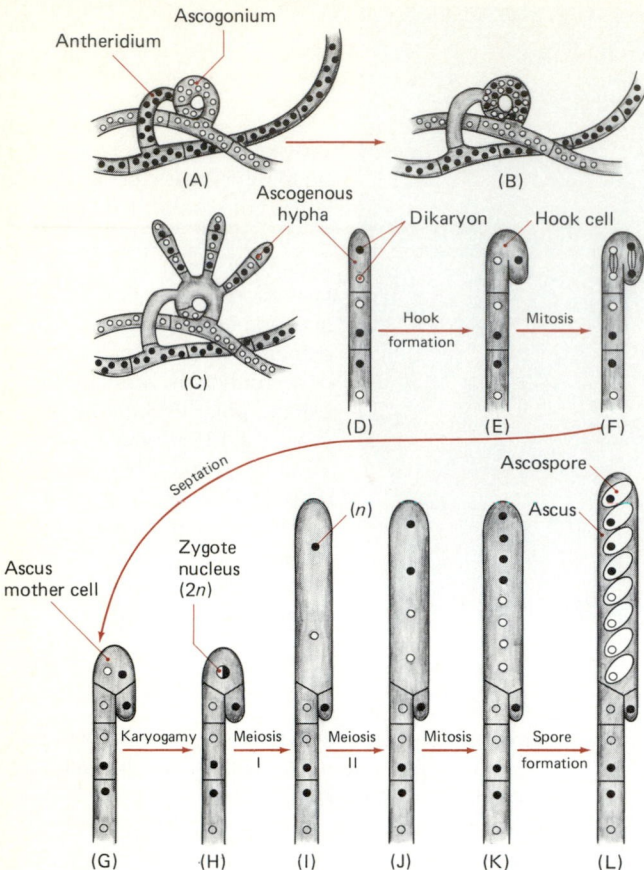

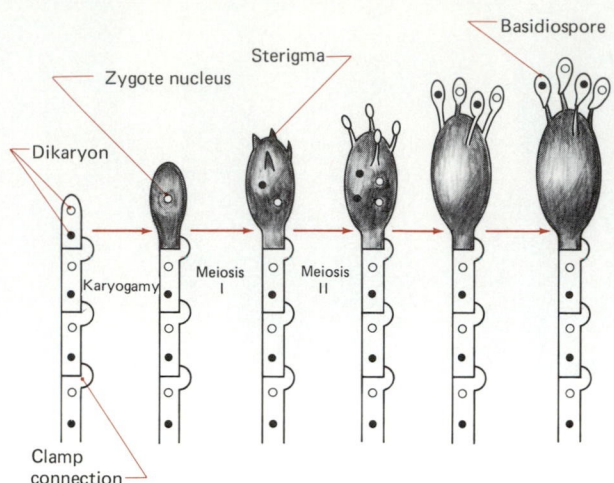

FIGURE 11.6 (*left*)
Sexual reproduction of one of the major groups of fungi is characterized by the formation of ascospores. This diagram shows the process of ascospore formation in a herothallic ascomycete. Filled and unfilled nuclei represent the two compatible mating types.

FIGURE 11.7 (*above*)
Basidiospore formation, without crossing over, in a basidiomycete. Filled and unfilled nuclei represent two compatible mating types.

ANALYTICAL PROCESS

Enumeration of Fungi

Defining the number of fungi is often a difficult task because individual organisms frequently represent multicellular aggregations that can be considered as one or many individuals. The task is simplest if it involves yeasts that can be enumerated by using viable count or direct count procedures analogous to the procedures used for the enumeration of bacteria. Enumeration of the filamentous fungi is far more difficult. Plate count enumeration procedures are biased toward fungal spores and underestimate the number of cells in a hyphal filament. The enumeration of filamentous fungi can be accomplished by determining the length of hyphae, which is considered a measure of fungal biomass, rather than the number of individual cells. Direct microscopic observations with the aid of a micrometer can be used to measure the length of hyphae. This approach, however, has some limitations. Fungi growing in an aqueous solution lacking growth nutrients can exhibit rapid growth of individual hyphae but minimal change in total biomass because the density of the hyphae is sparse. It is probably best, therefore, to determine the biomass of filamentous fungi by measuring the dry weight or a specific biochemical component of the cell walls, such as chitin. In the case of filamentous fungi, a change in biomass is the appropriate measure of growth, rather than a change in cell numbers.

organisms. They are nonphotosynthetic and typically form reproductive spores. Many fungi exhibit both sexual and asexual forms of reproduction. Some fungi are unicellular, but many form filaments of vegetative cells known as **hyphae**. Integrated masses of hyphae are called **mycelia**. The hyphae usually exhibit branching and are typically surrounded by cell walls containing chitin and/or cellulose. By a concise definition, the fungi are achlorophyllous, sapro-phytic, or parasitic, with unicellular or more typically filamentous vegetative structures usually surrounded by cell walls composed of chitin or other polysaccharides, propagating with spores, and normally exhibiting both asexual and sexual reproduction. If this definition seems complex and vague, it is; a broad definition is necessary to accommodate all of the morphological and physiological anomalies that occur among the fungi.

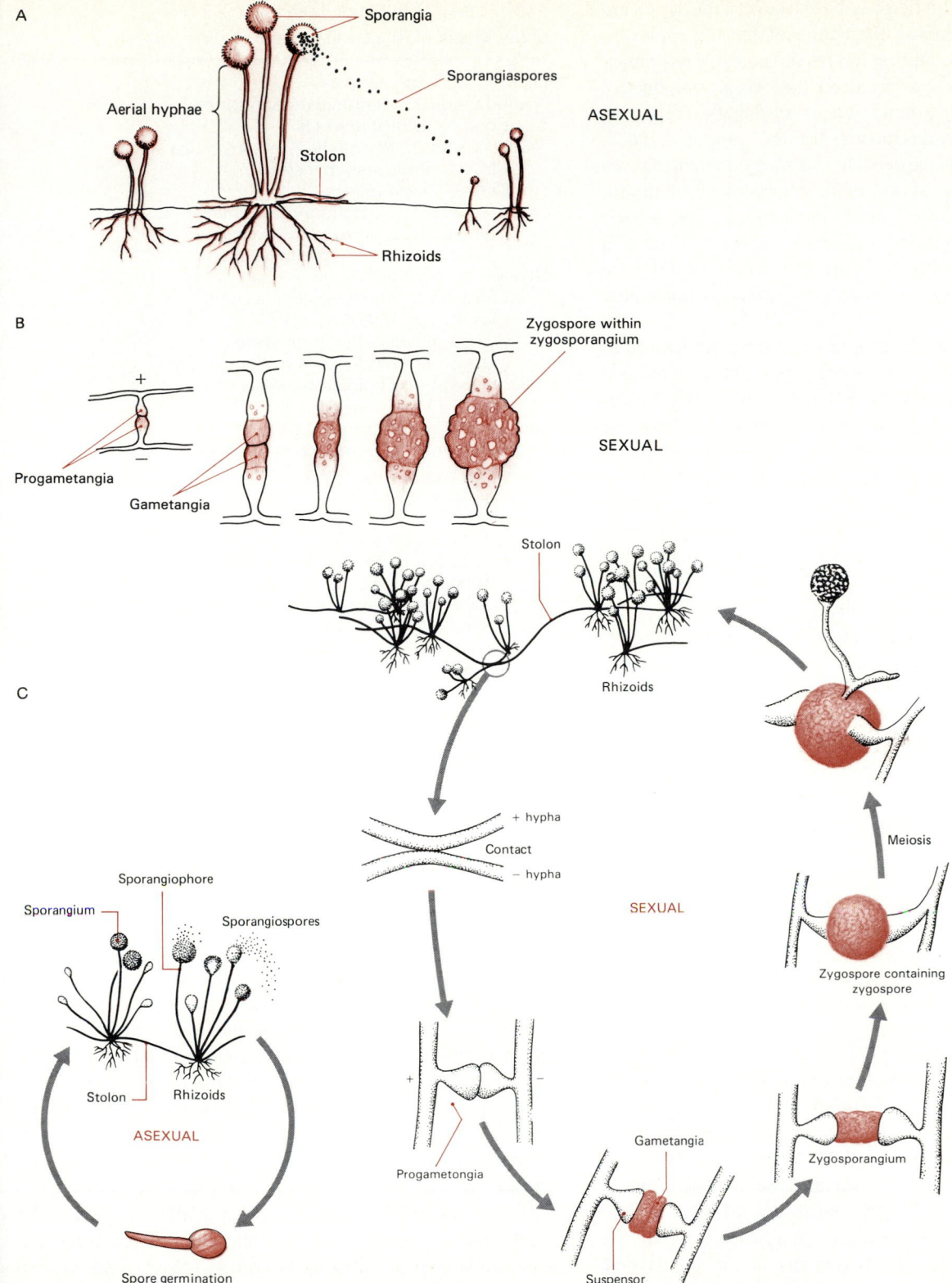

FIGURE 11.8

Zygospore formation in *Rhizopus*. (A) In asexual reproduction, a mycelium arises from the outgrowth of a fungal spore. Once the mycelium reaches a certain size, aerial hyphae arise, bearing sporangia filled with haploid spores. On dissemination, these spores give rise to new mycelial nets. (B) In sexual reproduction, when the subsurface rhizoids of a plus and minus type meet, they fuse and the result is zygospore formation. (C) The life cycle of the bread mold fungus *Rhizopus stolonifer*, showing asexual and sexual reproduction.

The classification of fungi is based largely on the means of reproduction, including the nature of the life cycle, reproductive structures, and reproductive spores. The primary taxonomic groupings are based on the sexual reproductive spores. To a lesser extent, fungal systematics relies on the morphological characteristics of the vegetative cells. Whereas most classical approaches to fungal systematics are based largely on observation of the morphology of the reproductive forms, physiological, biochemical, and genetic characteristics are included in some modern classification systems. Physiological features are particularly important in the classification of yeasts, which are primarily unicellular fungi.

In a formal systematic sense, yeasts are not recognized as being separate from the rest of the fungi and are classified along with their filamentous counterparts. In practice, however, the yeasts are typically treated separately from the filamentous fungi in both classification and identification systems. Separate classification and identification systems, for example, have been developed that include only the yeasts. Revisions of yeast systematics are published in the *International Journal of Systematic Bacteriology* rather than in the mycological literature. Commercial clinical systems are also available for identifying pathogenic yeasts. Similarly, separate classification and identification systems have been developed for other groups of fungi, such as the mushrooms. Such systems are very important because of their functional utility for identification purposes, even if they do not follow a formal taxonomic scheme based on phylogenetic relationships.

In formal classification systems, the fungi have traditionally been split into two large groups: the **slime molds** and the **true fungi**. The slime molds represent a borderline case between the fungi and the protozoa. They can just as well be classified with the protozoa, but traditionally they have been studied by mycologists. The vegetative cells of the slime molds are amoeboid and lack a cell wall, making them similar to protozoa. The slime molds and the true fungi are similar because they both produce spores that are surrounded by wall structures.

Gymnomycota   In the classification system described by Alexopoulos and Mims, the slime molds are placed in the division **Gymnomycota** of the kingdom Myceteae (fungi) (Table 11.1). The vegetative cells of organisms in the Gymnomycota lack cell walls and their nutrition is phagotrophic; that is, they engulf and ingest nutrients. All slime molds exhibit characteristic life cycles, a feature used to subdivide this division (Figure 11.9). Organisms in the subdivision **Acrasiogymnomycotina (Acrasiales)**, which often feed largely on bacteria, are known as *cellular slime molds*. The Acrasiomycetes, the single class within the subdivision Acrasiogymnomycotina, form a fruiting (spore-bearing) body known as a **sporocarp**, which is a special type of fruiting body that bears a mucoid droplet at the tip of each branch, containing spores with cell walls. The sporocarps of Acra-

TABLE 11.1   The Major Taxonomic Divisions of the Kingdom Myceteae (Fungi)

Division I. Gymnomycota
  Subdivision 1. Acrasiogymnomycotina
    Class 1. Acrasiomycetes
  Subdivision 2. Plasmodiogymnomycotina
    Class 1. Protosteliomycetes
    Class 2. Myxomycetes
      Subclass 1. Ceratiomyxomycetidae
      Subclass 2. Myxogastromycetidae
      Subclass 3. Stemonitomycetidae
Division II. Mastigomycota
  Subdivision 1. Haplomastigomycotina
    Class 1. Chytridiomycetes
    Class 2. Hyphochytridiomycetes
    Class 3. Plasmodiophoromycetes
  Subdivision 2. Diplomastigomycotina
    Class 1. Oomycetes
Division III. Amastigomycota
  Subdivision 1. Zygomycotina
    Class 1. Zygomycetes
Class 2. Trichomycetes
  Subdivision 2. Ascomycotina
    Class 1. Ascomycetes
      Subclass 1. Hemiascomycetidae
      Subclass 2. Plectomycetidae
      Subclass 3. Hymenoascomycetidae
      Subclass 4. Laboulbeniomycetidae
      Subclass 5. Loculoascomycetidae
  Subdivision 3. Basidiomycotina
    Class 1. Basidiomycetes
      Subclass 1. Holobasidiomycetidae
      Subclass 2. Phragmobasidiomycetidae
      Subclass 3. Teliomycetidae
  Subdivision 4. Deuteromycotina
    Class 1. Deuteromycetes
      Subclass 1. Blastomycetidae
      Subclass 2. Coclomycetidae
      Subclass 3. Hyphomycetidae

siomycetes are generally stalked structures. The stalks normally consist of walled cells, and this characteristic forms the basis for designating these organisms as the cellular slime molds. The sporocarp releases spores that germinate, forming **myxamebae**, which are amoeboid cells that form pseudopodia. The myxamebae swarm together or aggregate to form a **pseudoplasmodium**. Within the pseudoplasmodium, the cells of Acrasiomycetes do not lose their integrity. The pseudoplasmodium undergoes a developmental sequence (differentiation), culminating in the formation of a sporocarp.

The pseudoplasmodium formation of the Acrasiomycetes is of special interest because of the biochemical communication involved in initiating swarming activity. The swarming behavior of *Dictyostelium discoideum* has been extensively studied. Under appropriate conditions, when food sources become limiting, the myxamoebae cease their feeding activity and swarm to an aggregation center. The swarming activity is initiated when one or more cells at the aggregation center release cyclic AMP (**acrasin**). At the biochemical level, cyclic AMP is responsible for communi-

FIGURE 11.9

The life cycle of the slime mold *Dictyostelium discoideum*.
(A) Germination of a spore with a single myxamoeba
issuing from it; (B) myxamoebae; (C) streams of aggregating
myxamoebae; (D) pseudoplasmodium, grex, slug;
(E) beginning of culmination, myxamoebae at the front
end of the grex pushing down the middle to the stalk
cylinder; (F) later stage, showing stalk formation;
(G) mature sorocarp with cellular stalk; (H) young macrocyst in which keryogamy occurs; and (I) mature macrocyst
where meiosis takes place. Myxamoebae form on
germination of the mature macrocyst.

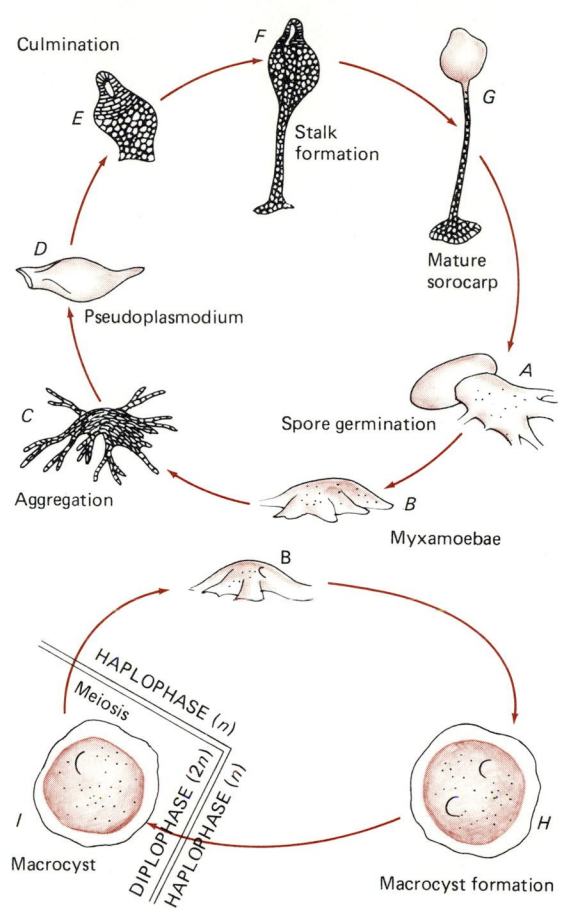

Discovery Process

The observation that slime molds aggregate under certain
conditions led to a long series of fascinating experiments
to elucidate the mechanism of developmental behavior.
In using quite different methods, such as the attraction of
amebae by aggregation centers located across semipermeable barriers and the interruption of aggregation
patterns by currents of water, it was argued that the
gathering of amebae at a central collection point occurred
by chemotaxis. The chemotactic behavior was postulated
by J. T. Bonner in the late 1940s to be the response to
a gradient of a substance, which he called acrasin, that
somehow orients the cells toward areas of higher concentration.

In 1953 B. M. Schaffer devised an ingenious test for
acrasin in which sensitive *Dictyostelium discoideum* were
sandwiched under a small agar block and arcrasin water
was applied to the outside meniscus. When this was
done at a rate of about once every 10 seconds, the amebae
were attracted to the outside edge; lengthening the
interval produced no effect. The interpretation of this
reaction was that the acrasin disappeared rapidly, thus
requiring frequent repeat applications. Schaffer also
showed that the acrasin was degraded enzymatically. A
quantitative assay was needed, and two were developed.
T. M. Konijn put sensitive *D. discoideum* in small drops
in washed agar; drops of test solution were placed nearby,
and if acrasin was present, the amebae burst out of the
original drop. This test was made quantitative by diluting
the test solutions and by varying the distance between the
test solution drops and the amebae. In the cellophane
square test, amebae were placed on a small square
of dialysis membrane, which was then placed on the
surface of unwashed agar containing the test substance.
The rate at which the amebae moved away from the
square was measured; the faster the rate, the greater the
power of the attractant, for the amebae do not actually
move faster, but their paths are more perfectly directed
away from the square. This test has shown both vegetative
and aggregating amebae to be subject to chemotaxis.
Now that it was clear that bacterial extracts attract slime
mold amebae, Konijn and T. J. Barkley tested many biochemicals for their ability to mediate aggregation, including 3',5' cyclic AMP, which they found to be extraordinarily active with *D. discoideum*. This was found to be
the component secreted by *E. coli*, which affects aggregating amebae. The cells of *D. discoideum* also secrete
large quantities of a specific phosphodiesterase that
converts cyclic AMP to 3',5' cyclic AMP. By using a
variety of techniques to prevent this phosphodiesterase
from immediately converting the slime mold's own cyclic
AMP, it was possible to show that *D. discoideum* does
synthesize its own cyclic AMP. From this finding, it is
deduced that acrasin is cyclic AMP; the production
of 3',5' cyclic AMP by an amebae that finds a food source
acts as a chemotactic signal, communicating to other
amebae to migrate to that aggregation center.

cation between the myxamoebae. The myxamoebae move along the concentration gradient of cyclic AMP until they reach the center of aggregation. They then mass together to form a pseudoplasmodium. Swarming occurs as a pulsating wave motion in which the chemical stimulus, cyclic AMP, is transmitted from cells that are proximal to the aggregation center to distant cells. Different species of Acrasiomycetes exhibit different waveforms in their swarming behavior, some moving in linear wave-like motion and others exhibiting spiral wave motion.

The second subdivision of the Gymnomycota, the **Plasmodiogymnomycotina**, includes two classes: the **Protosteliomycetes** and the **Myxomycetes**. The Myxomycetes are known as the **true slime molds**. Some species of Myxomycetes form myxamoebae, and others form flagellated cells known as **swarm cells**. The myxamoebae or swarm cells fuse together to form a true plasmodium. The **plasmodium** of the Myxomycetes is a multinucleate protoplasmic mass that is devoid of cell walls and is enveloped in a gelatinous slime sheath. The plasmodium gives rise to brilliantly colored fruiting bodies. The classification of the Myxomycetes is based largely on the structure of the fruiting body. In many species of Myxomycetes, spores are formed inside the fruiting body. These spores are sometimes referred to as *endospores*, but they do not bear any resemblance to the endospores of bacteria. The spores of myxomycetes generally have a definite thick wall. In the life cycle of Myxomycetes, the spores are released from the sporangia and disseminated. At a later time they germinate, producing myxamoebae and/or swarm cells (Figure 11.10). These structures later unite by sexual fusion to initiate formation of the plasmodium. The brightly colored fruiting bodies of Myxomycetes are often seen on decaying logs or other moist areas of decaying organic matter (Figure 11.11). Myxomycetes are often conspicuous on grass lawns, appearing as large blue-green colonies. To remove these unsightly blemishes from an otherwise luxuriant lawn, simply mow the grass.

Mastigomycota   The **Mastigomycota** comprise the second major division of the kingdom Myceteae. Some Mastigomycota are unicellular, but most form extensive filamentous, coenocytic mycelia. Mastigomycota typically produce motile cells with flagella during part of their life cycle. Asexual reproduction by species in this division normally involves motile spores called **zoospores**. In contrast to the phagotrophic mode of nutrition exhibited by members of the Gymnomycota, nutrition is accomplished in the Mastigomycota by absorption of nutrients. The division Mastigomycota includes four classes: Chytridiomycetes, Hyphochytridiomycetes, Plasmodiophoromycetes, and Oomycetes (Table 11.2).

The **Chytridiomycetes** are differentiated from all other fungi by the production of zoospores, which are motile with a single posterior flagellum of the whiplash type. Members of the order Chytridiale are referred to as *chytrids*. Many

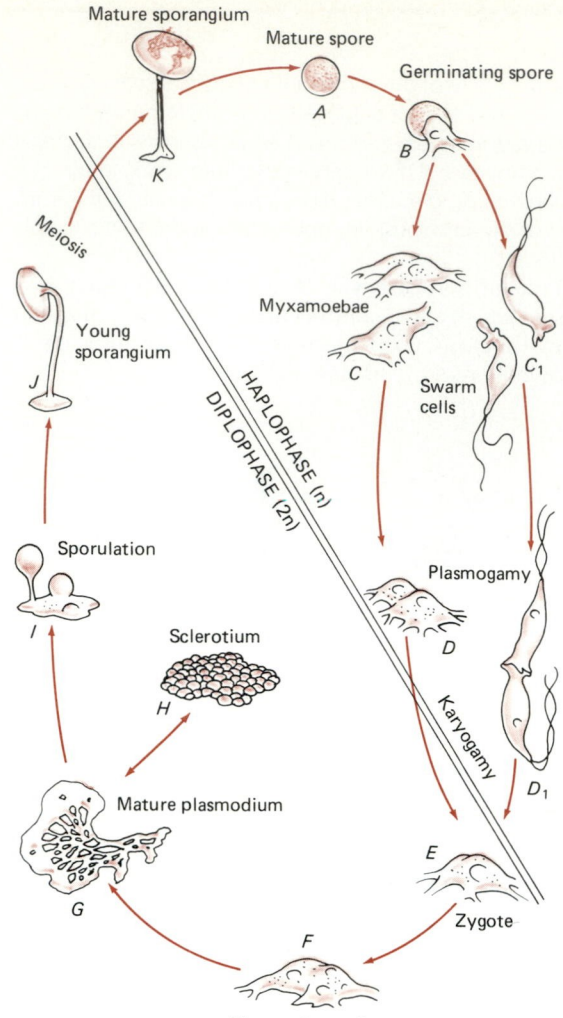

FIGURE 11.10

The life cycle of a Myxomycete. (A) Mature haploid spore; (B) germinating spore; (C) myxamoeba; (C₁) swarm cells; (D) fusing myxamoebae; (D₁) fusing swarm cells; (E) young zygote; (F) young plasmodium; (G) mature plasmodium; (H) sclerotium; (I) sporulation, sporangial initials; (J) young pre-meiotic spongium with spores; (K) mature postmeiotic sporangium.

chytrids are parasitic on other fungi, algae, and plants. A few, such as *Olpidium brassicae* which infects cabbage roots, are responsible for plant diseases. The **Hyphochytridiomycetes**, of which only about 15 species are known, produce uniflagellate zoospores of the tinsel type. The **tinsel-type flagellum** has many short projections extending down its length. The **Plasmodiophoromycetes** are known as the **endoparasitic slime molds** and are obligate parasites of other fungi, algae, and plants. They cause an abnormal enlargement of the host cells, called *hypertrophy*, and often an abnormal multiplication of the host cells called *hyperplasia*. The Plasmodiophoromycetes form a plasmodium that devel-

FIGURE 11.11
The slime mold *Lycogula epidendrum* growing on a log.
(Courtesy Varley Weideman, University of Louisville)

TABLE 11.2   The Four Classes in the Division Mastigomycota

Class 1. *Chytridiomycetes*   Varied vegetative form; produces
posteriorly uniflagellate motile cells with whiplash flagella

Class 2. *Hyphochytridiomycetes*   A small group of fungi;
produces motile, anteriorly uniflagellate cells with tinsel
flagella

Class 3. *Plasmodiophoromycetes*   Parasitic fungi with
multinucleate thalli (plasmodia) within the cells of their
hosts; resting cells (cysts) produced in masses but not in
distinct sporophores; motile cells with two anterior whiplash
flagella

Class 4. *Oomycetes*   Varied vegetative form, usually
filamentous, with a coenocytic, walled mycelium; hyphal
wall; produces zoospores, each with one whiplash and one
tinsel flagellum; sexual reproduction is oogamous, resulting
in the formation of oospores

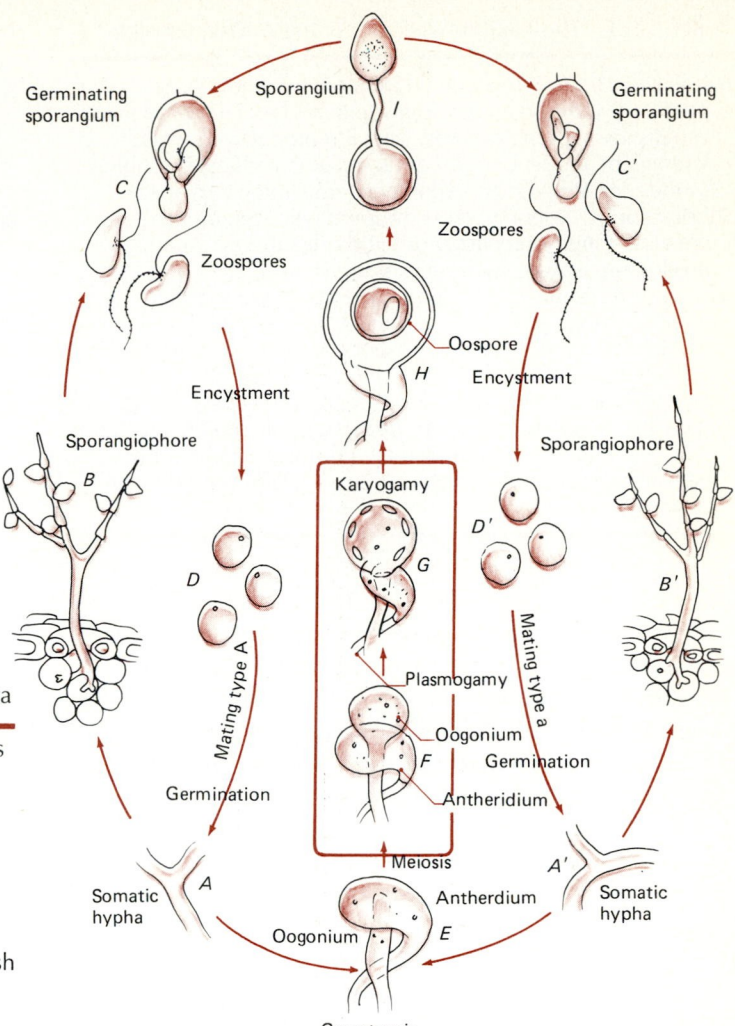

FIGURE 11.12
The life cycle of an Oomycete, *Phytophthora infestans*.

ops within the host cells. Various species of Plasmodiophoromycetes cause diseases of plants that are of economic significance. For example, *Plasmodiophora brassicae* causes
clubfoot of cabbage and *Spongospora subterranea* causes
powdery scab of potatoes.

The **Oomycetes**, known as the **water molds**, reproduce
by using flagellated zoospores. The zoospores typically have
two flagella, one of the tinsel type and the other of the whiplash type. Sexual reproduction in the Oomycetes typically
involves the formation of **oospores**, which are thick-walled
spores that develop by contact with specialized **gametangia**
(structures containing differentiated cells involved in sexual
reproduction). The gametangia of Oomycetes normally occur
at the terminal ends of the mycelia. During gametangial contact, male gametes pass through a fertilization tube into the
female gametangia (Figure 11.12). The male gametangium
is referred to as an *antheridium* and the female gametangium

as an *oogonium*. Members of the Saprolegniales, an order of
the Oomycetes, are abundant in aquatic ecosystems. Several
species in this order are animal and plant pathogens. For
example, *Phytophthora infestans* causes potato blight and
was responsible for the great Irish potato famine of 1845 and
1846, which resulted in the great wave of immigration from
Ireland to the United States.

Amastigomycota   The vegetative cells of the **Amastigomycota**, the third division of the fungi, vary from single cells
to mycelia that may be coenocytic (multinucleate) or may
have extensive septation (cross walls separating cells within
the mycelia). Unlike the Gymnomycota and Mastigomycota,
the Amastigomycota do not produce motile cells. There are
four subdivisions in the Amastigomycota: Zygomycotina,
Ascomycotina,   Basidiomycotina,   and   Deuteromycotina
(Table 11.3).

TABLE 11.3   The Four Subdivisions of the Amastigomycota

Subdivision I. *Zygomycotina*   Saprophytic, parasitic or predatory fungi, coenocytic mycelium; asexual reproduction usually by sporangiospores; sexual reproduction, where known, by fusion of equal or unequal gametangia resulting in the formation of zygosporangia containing zygospores

Subdivision II. *Ascomycotina*   Saprophytic, symbiotic or parasitic fungi; unicellular or with a septate mycelium, producing ascospores in sac-like cells (asci)

Subdivision III. *Basidiomycotina*   Saprophytic, symbiotic or parasitic fungi; unicellular or, more typically, with a septate mycelium, producing basidiospores on the surface of various types of basidia

Subdivision IV. *Deuteromycotina*   Saprophytic, symbiotic, parasitic, or predatory fungi; unicellular or, more typically, with a septate mycelium, usually producing conidia from various types of conidiogenous cells; sexual reproduction unknown

Zygomycotina. The **Zygomycotina** typically have coenocytic mycelia and are characterized by the formation of a **zygospore**, a sexual spore that results from the fusion of gametangia. For sexual reproduction, some species require gametangia of two different mating types, whereas others are homothallic, requiring only one type for zygote formation. In addition to sexual reproduction, the Zygomycotina characteristically produce asexual sporangiospores within a sporangium. Many members of the Zygomycotina are plant or animal pathogens. There are two classes within the subdivision Zygomycotina: Zygomycetes and Trichomycetes. Species of the Trichomycetes are obligately associated with arthropods and normally grow within the guts of these animals, where they attach to the chitinous lining of the gut by

means of a specialized structure known as a *holdfast*. The order Mucorales (the bread molds) occurs in the class Zygomycetes. *Rhizopus stolonifer* is the common bread mold (Figure 11.13). Some *Rhizopus* species are important in the food industry. For example, *R. oryzae* is used for the production of fermented Oriental foods, such as tempeh. Other *Rhizopus* species are important causes of food spoilage.

Several species of Mucorales exhibit an interesting morphological change known as *dimorphism*, which occurs under some conditions as filamentous mycelia and under other conditions as yeast-like, unicellular forms. For example, *Mucor rouxii* grows in a yeast-like form in atmospheres with a high percentage of carbon dioxide but produces filamentous mycelia at normal atmospheric concentrations of CO_2. *Mucor* species are important opportunistic human pathogens and can cause serious infections in burn wounds. The genus *Pilobolus* also occurs in the order Mucorales. *Pilobolus* species are interesting because of their forceful mode of spore discharge, shooting their spores several meters into the air, with the entire spore cluster ejected in the direction of highest light intensity (Figure 11.14). In this way, the spores of *Pilobolus* are released to an area of open air where air currents are likely to disperse them further.

Ascomycotina. Members of the subdivision **Ascomycotina** produce sexual spores within a specialized sac-like structure known as the **ascus**. The ascomycetes, which are sometimes called the **sac fungi**, normally produce a definite number of ascospores within the ascus. The mycelia of ascomycetes is composed of septate hyphae, and the cell walls of the hyphae of most ascomycetes contain chitin. Asexual reproduction in the ascomycetes may be carried out by fission, fragmentation of the hyphae, formation of **chlamydospores** (thick-walled spores within the hyphal filaments), and production of con-

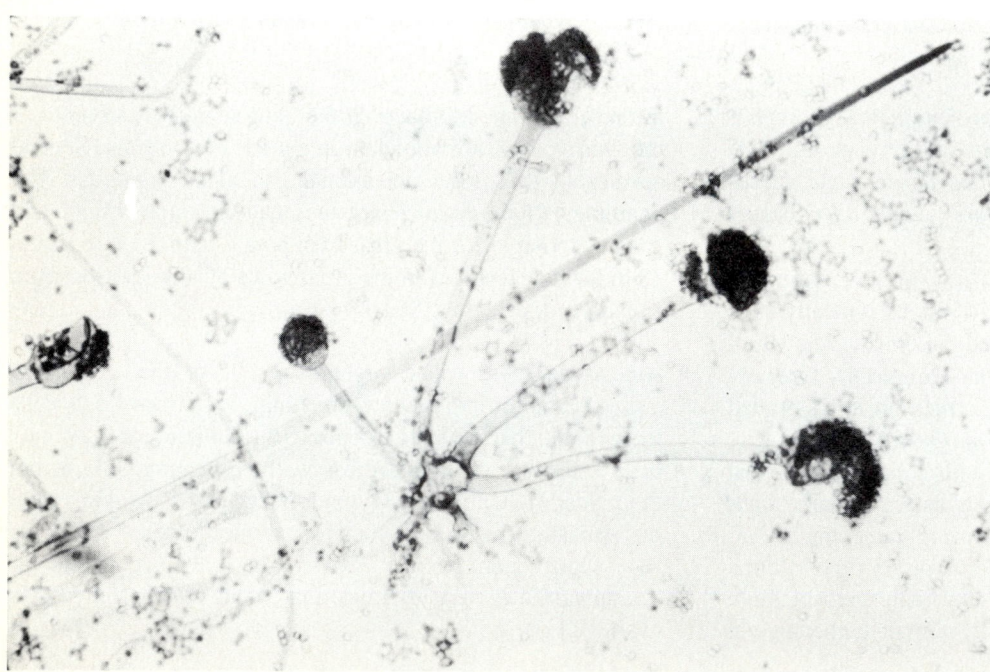

FIGURE 11.13
Micrograph of the bread mold *Rhizopus arrhizus*. (From BPS— C. Emmons, Phoenix, Ariz.)

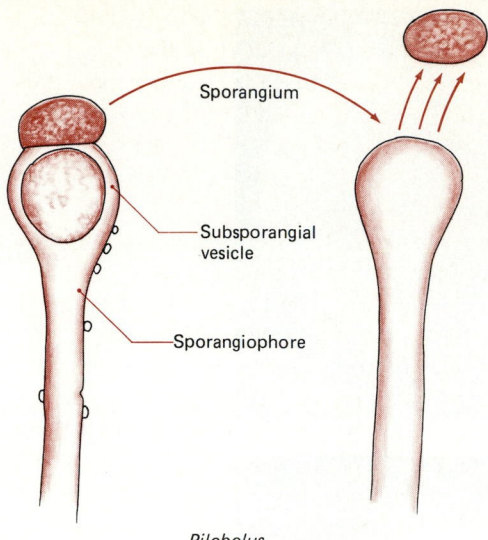

Pilobolus

FIGURE 11.14

In the spore discharge by *Pilobolus*, the entire spore mass is shot toward an area of high light intensity.

idia (nonmotile spores produced on a specialized spore-bearing cell). During sexual reproduction the ascomycetes normally exhibit a short-lived dikaryotic stage (having cells containing two nuclei) between the time of fusion of gametes (**plasmogamy**) and the time of fusion of the two nucleii (**karyogamy**).

The members of the subclass Hemiascomycetidae, in which the ascomycetous yeasts occur, are morphologically simple ascomycetes that generally lack hyphae. Many yeasts are ascomycetes (Figure 11.15). The morphology of the ascospore is a critical taxonomic feature in classifying yeasts at the genus level (Table 11.4). Classification of the yeasts at the species level normally employs numerous biochemical and physiological characteristics, as well as morphological features. The metabolic activities of the ascosporogenous yeasts have many industrial applications. *Saccharomyces cerevisiae* is used as baker's yeast. Many fermented beverages are also produced by using members of the genus *Saccharomyces*. Most commonly, *S. carlsbergensis* and *S. cerevisiae* are used for the production of beer, wine, and spirits.

In addition to the ascosporogenous yeasts, the subclass Hemiascomycetidae includes the order Taphrinales, the members of which resemble yeasts in that they reproduce asexually by budding and sexually by producing ascospores but differ from the yeasts in that they produce a definite true mycelium. Members of the Taphrinales are parasitic on plants. For example, *T. deformans* causes peach leaf curl, *T. cerasi* causes witches' broom of cherries, *T. pruni* causes prune pockets, and *T. coerulescens* causes puckering of oak leaves.

The **Euascomycetidae (true ascomycetes)** produce **asci** that normally develop from dikaryotic hyphae. The asci are produced in or on a specific structure known as the **ascocarp**.

TABLE 11.4 Descriptions of Ascospores Found in Different Genera of Yeasts

Genus	Number of Ascospores	Shape of Ascospores
Citeromyces	1	Spheroidal
Coccidiascus	8	Fusiform
Debaryomyces	1–4	Spheroidal, ovoidal, warty
Dekkera	1–4	Hat-shaped
Endomycopsis	1–4	Spheroidal, hat-shaped, saturn-shaped, sickle-shaped
Hanseniaspora	1–4	Hat-shaped, helmet-shaped, walnut-shaped
Hansenula	1–4	Hat-shaped, spheroidal, hemispheroidal, saturn-shaped
Kluyveromyces	1–many	Crescentiform, reniform, spheroidal, ellipsoidal
Lipomyces	1–16	Ellipsoidal, lenticular
Lodderomyces	1–2	Oblong, ellipsoidal
Metschnikowia	1–2	Needle-shaped
Nadsonia	1–2	Spheroidal, warty
Nematospora	8	Spindle-shaped
Pachysolen	4	Hemispheroidal
Pichia	1–4	Spheroidal, hat-shaped, saturn-shaped, warty
Saccharomyces	1–4	Spheroidal, ellipsoidal
Saccharomycodes	4	Spheroidal
Saccharomycopsis	1–4	Ovoidal, double-walled
Schizosaccharomyces	4–8	Spheroidal, ovoidal
Schwanniomyces	1–2	Walnut-shaped, warty
Wickerhamia	1–16	Cap-shaped
Wingea	1–4	Lens-shaped

The euascomycetes are divided according to the structure of the ascocarp into the Plectomycetes, in which the ascocarp has no special opening; the Pyrenomycetes, in which the ascocarp is shaped like a flask; and the Discomycetes, in which the ascocarp is cup-shaped (Figure 11.16). Species of Pyrenomyctes are important for their roles in basic scientific investigations. For example, studies on the genetics of *Neurospora*, a pyrenomete, have greatly added to our understanding of recombinational processes. *Neurospora* is useful in genetic studies because the spores can be isolated from the ascus and the genotypes readily determined.

Besides their usefulness in basic microbiological studies, Pyrenomycete species are of practical importance. Some Pyrenomycetes are important plant and animal pathogens, with the powdery and black mildews, for example, occurring in this taxonomic group. *Claviceps purpurea* causes ergot of rye; cattle and other such animals are poisoned when grazing

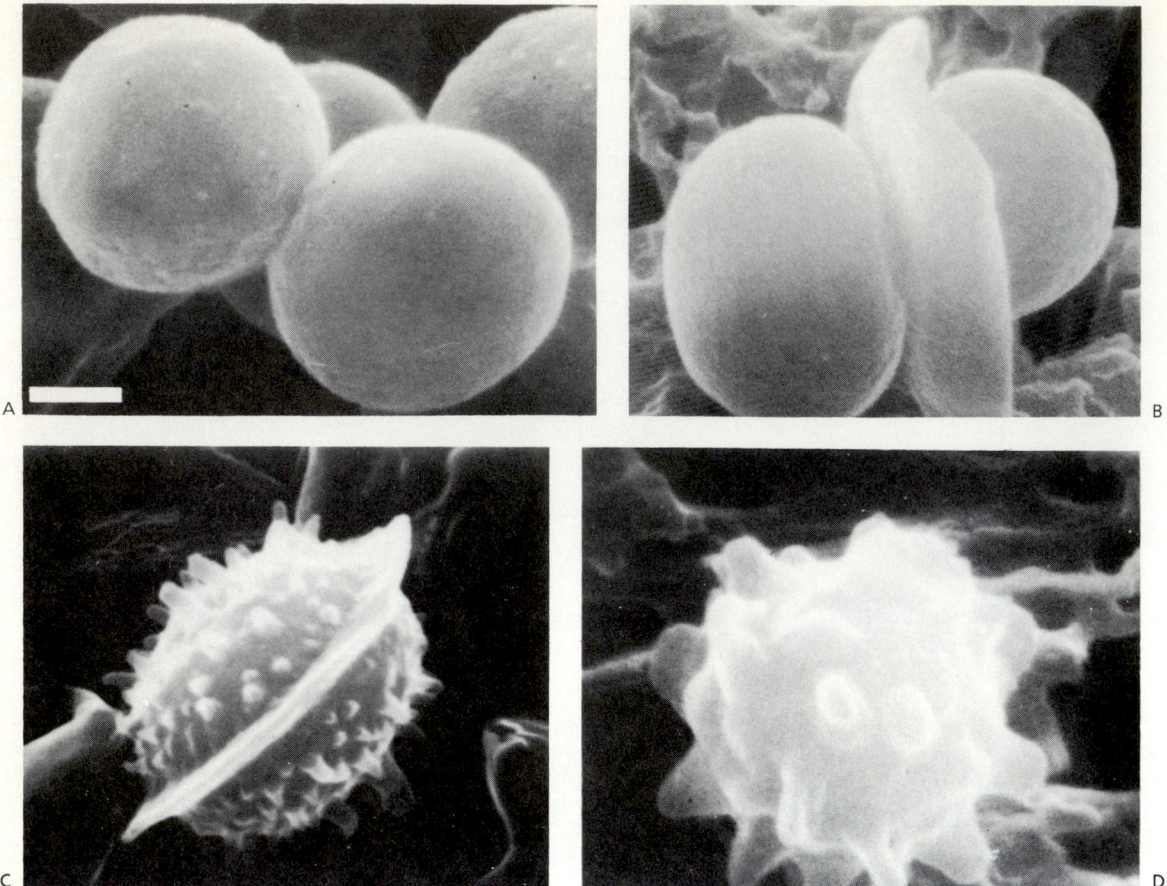

FIGURE 11.15

Micrograph of ascospores of various yeasts, showing characteristic morphologies. (A) Smooth spheroidal ascospores of *Pichia carsonii*; (B) hat-shaped ascospores of *Pichia spartinae*; (C) saturn-shaped ascospore of *Schwanniomyces occidentalis*; (D) ascospore of *Saccharomyces kloeckerianus* (Courtesy C. P. Kurtzman, United States Department of Agriculture, Northern Regional Center, Peoria, Ill.)

on grasses contaminated with the resting bodies (**sclerotia**)[1] of the fungus. Various alkaloid biochemicals are produced by *C. purpurea*, some of which have hallucinogenic properties and others of which are useful medicinals, such as those used to induce labor for childbirth. *Endothia parasitica*, another Pyrenomycete, is the causative agent of chestnut blight. This organism was introduced into North America from eastern Asia in the early twentieth century and quickly devastated the chestnut trees of the United States and Canada.

The Plectomycetes also include some important plant and animal pathogens, such as the black molds, blue molds, and ringworms. Several species of *Ceratocystis* are responsible for blue stain, which reduces the commercial value of lumber. *C. ulmi* is the causative agent of Dutch elm disease, a great threat to elm trees in North America. The fungus that causes the human disease histoplasmosis, *Emmonsiella cap-*

sulata,[2] also belongs to the subclass Plectomycetes. This fungus commonly occurs in soils that are contaminated with fecal droppings from birds. Other fungi known by the names of their imperfect forms associated with the Plectomycetes include the well-known genera *Penicillium* and *Aspergillus*.

The Discomycetes include the cup fungi, morels, and truffles. Species of the genus *Morchella* (morels) are gastronomical delights (Figure 11.17). All morels are edible and delicious. Truffles occur in the order Tuberales and, like morels, are considered edible delicacies.

Basidiomycotina. The subdivision **Basidiomycotina**, with its single class, Basidiomycetes, contains the most complex fungi, including the smuts, rust, jelly fungi, shelf fungi, stinkhorns, bird's nest fungi, puffballs, and mushrooms. The Basidiomycetes are distinguished from other classes of fungi by the fact that they produce sexual spores, known as **basi-**

---

[1]Sclerotia are hard resting bodies that are resistant to unfavorable conditions and may remain dormant for prolonged periods.

[2]*E. capsulata* was formerly known as *Histoplasma capsulatum* before the sexual reproductive stage of the organism was known, and some medical mycologists still retain the name of the imperfect (nonsexual) form when referring to this organism.

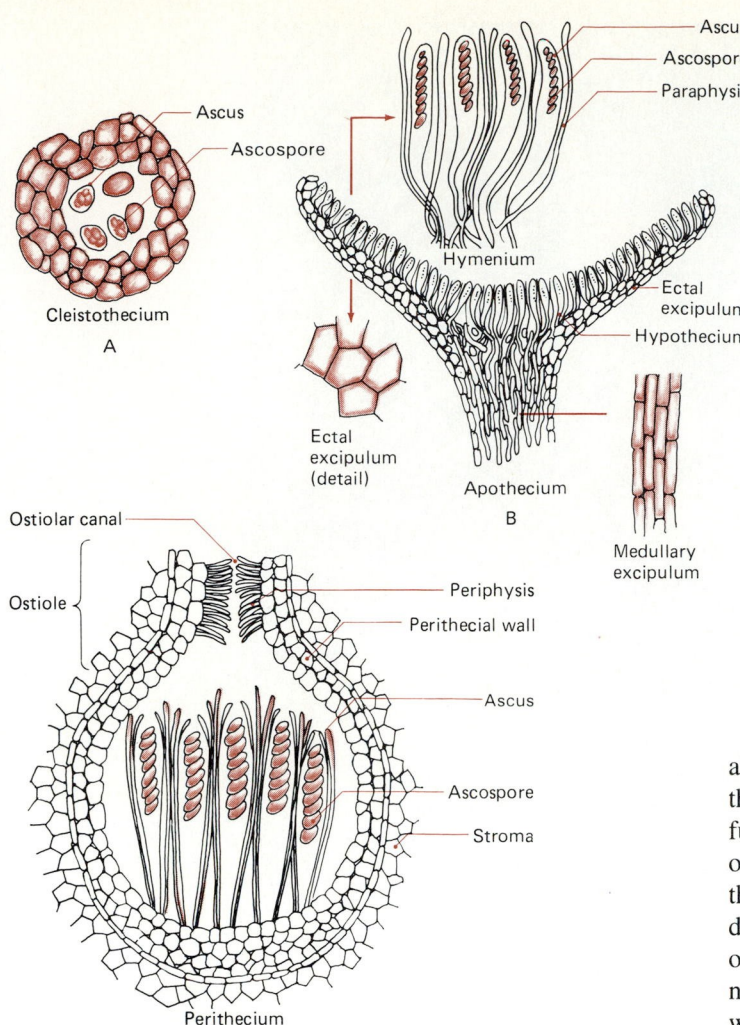

FIGURE 11.16

Asci occur in various characteristic shapes. (A) *Eurotium* sp.; (B) cross section of an apothecium; (C) an ascocarp.

FIGURE 11.17

Photograph of a morel, *Morchella esculenta*. These fungi are considered gastronomic delights. (Courtesy Orson K. Miller, Jr., Virginia Polytechnical Institute and State University)

diospores, on the surfaces of specialized spore-producing structures, known as **basidia** (Figure 11.18). The Basidiomycetes are also known as the **club fungi** because of the typical shape of the basidia. The mycelia of Basidiomycetes typically form **clamp connections** between cells (Figure 11.19). The clamp cell connections are generally indicative of a dikaryotic mycelium. Additionally, the mycelia of many Basidiomycetes are characterized by specialized cross walls between connecting cells, known as **dolipore septa** (Figure 11.20). The dolipore septum has a central pore surrounded by a barrel-shaped swelling of the cross wall. Effectively, the clamp cell connections and the dolipore septa permit enhanced chemical communication through the mycelia of the organism.

The Basidiomycete order **Aphyllophorales** contains the **shelf or bracket fungi**. These are some of the most conspicuous fungi, often seen growing on trees (Figure 11.21). The fruiting bodies of these fungi are tough and leathery. In addition to the bracket fungi, the Aphyllophorales includes the cantharelles, the coral fungi, the tooth fungi, and the pore fungi. The majority of these fungi are saprophytic, growing on dead and living plant materials. The growth of fungi in this group on wood results in two characteristic types of decay, called *brown rots* and *white rots* because of the color of the rotted wood. In brown rot only the cellulose component of wood is decomposed, leaving the brown lignins. In white rot both cellulose and lignin are degraded, producing white-colored wood.

The Basidiomycete order **Agaricales** includes the **mushrooms**. The mushrooms that we see are the fruiting bodies (**basidiocarps**) of basidiomycetes. The spore-bearing structures (basidia) of mushrooms are borne on the surface of the gills of the basidiocarp (Figure 11.22). In the boletes, which also are included in the Agaricales, the basidia are not borne on gills but rather within tubes. Some mushrooms are edible, but others are extremely poisonous—making the proper identification of mushrooms critical, lest one become the victim of mushroom poisoning. It is sometimes easy to confuse an edible species with one that is deadly.

The order Agaricales is divided into several families (Table 11.5). The family Russulaceae includes two genera, *Russula* and *Lactarius*. *Russula* species produce white spores and brittle fruiting bodies, and many species produce brilliantly colored, beautiful caps. Some of these attractive *Russula* species are poisonous. The family Amanitaceae includes the genus *Amanita*, which is characterized by free gills and the presence of an annulus and a volva (Figure 11.23). Many *Amanita* species are quite beautiful, but most are deadly. *Amanita phalloides* is known as the *death cap* because most deaths due to mushroom poisoning have been attributed to its ingestion. *Agaricus* species, which occur in the family

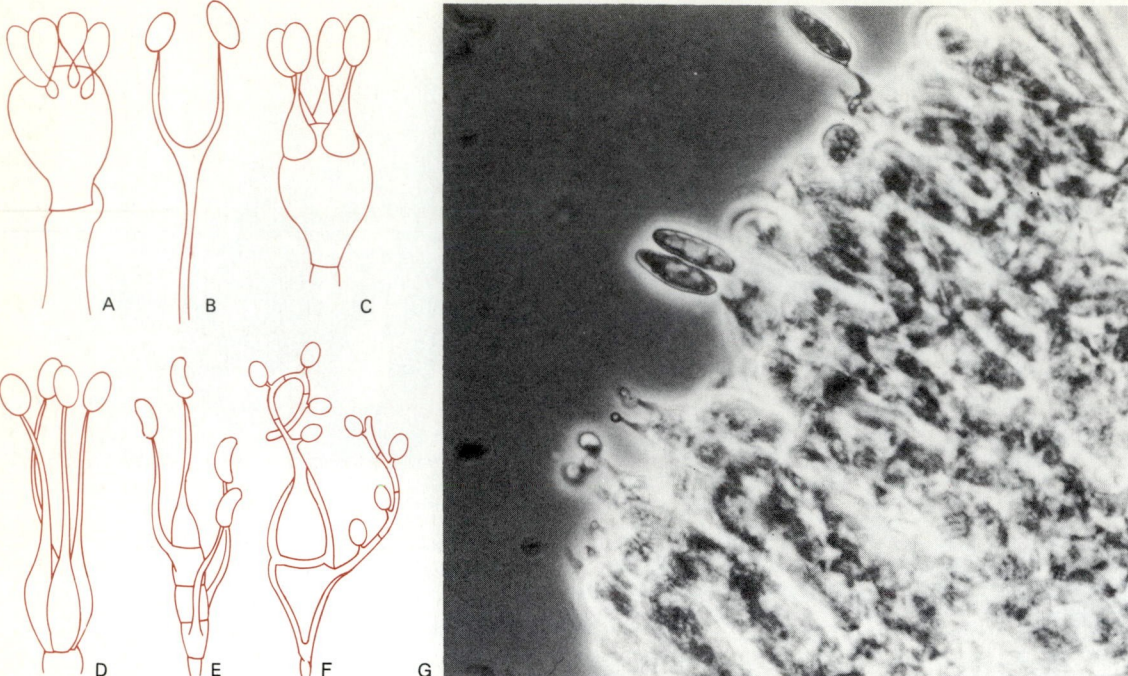

FIGURE 11.18

There are various types of basidia. (A) Typical holobasidium; (B) tuning fork basidium of *Dacrymyces*; (C) basidium of *Tulasnella*; (D) basidium of *Tremella*; (E) basidium of *Auricularia*; and (F) basidium of *Puccinia*. (G) Micrograph of mature basidia of *Gomphidius glutinosus* in all stages (400×). (Courtesy Orson K. Miller, Jr., Virginia Polytechnical Institute and State University)

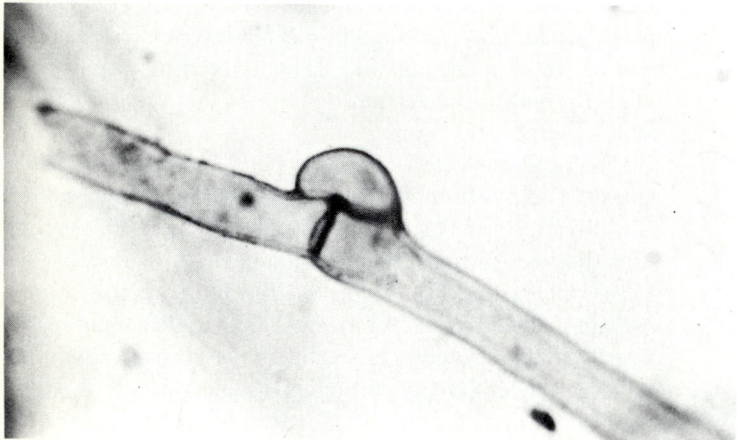

FIGURE 11.19

Basidiomycetes characteristically exhibit clamp cell connections. This micrograph of *Chroogomphus loculatus* shows clamped vegetative hyphae. (Courtesy Orson K. Miller, Jr., Virginia Polytechnical Institute and State University)

Agaricaceae, produce white to brown mushrooms, with free gills and an annulus but no volva. Several mushrooms of this genus, such as *Agaricus bisporus*, are grown commercially for human consumption. The family Coprinaceae contains the genus *Coprinus*, which is known as the *ink cap* mushroom because autodigestion (self-decomposition) causes it to dissolve into a black, ink-like liquid.

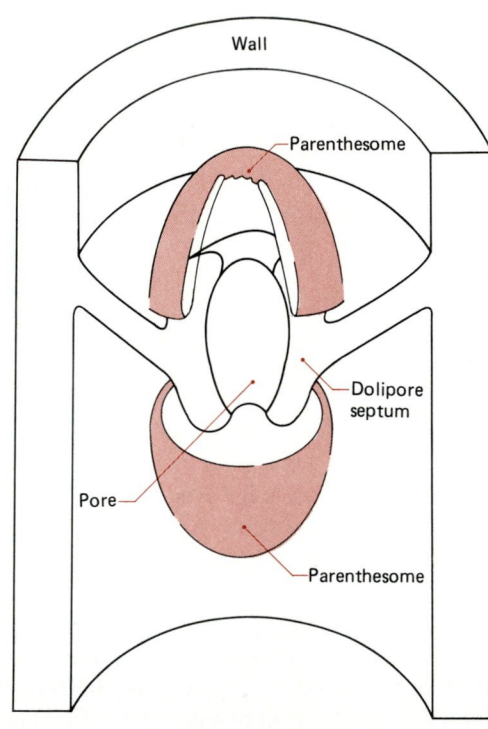

FIGURE 11.20

Dolipore septum. Septa of this form have been reported from basidiocarps of *Auricularia*, *Calvatia*, *Dacrymyces*, *Exidia*, *Merilius*, and *Polyporus*, and are suggested in the mature *Calvatia gigantea* mycelium.

FIGURE 11.21
The bracket fungi often are conspicuous, growing on trees. (From BPS—J. Robert Waaland, University of Washington)

A

B

TABLE 11.5   The Families in the Order Agaricales

Family Agariaceae	Possess free gills and an annulus, but no volva; produce brown spores
Family Amanitaceae	Possess free gills, an annulus, and a volva
Family Boletaceae	Possess vertically arranged tubes instead of gills; the basidia occur within the tubes
Family Coprinaceae	Possess gills that dissolve into a black, inky liquid at maturity
Family Cortinariaceae	Possess attached gills; may or may not have an annulus; produce light brown spores
Family Lepiotaceae	Some members resemble the amanitas but lack a volva
Family Rhodophyllaceae	Possess attached gills; produce annular spores
Family Russulaceae	Possess attached gills; produce white spores and are distinguished by the presence of specialized spherical cells known as spherocysts
Family Strophariaceae	Possess attached gills; produce dark spores
Family Tricholomataceae	Possess attached gills; produce white spores
Family Volvariaceae	Produce salmon- to pink-colored basidiospores

C

FIGURE 11.22
Mushrooms are visible macroscopic fruiting bodies of basidiomycetes. Some mushrooms are edible, but others are quite poisonous. (A) This *Pleurotus ostreutus* is edible and good. (B) This *Amanita virosa* is deadly. (Courtesy Orson K. Miller, Jr., Virginia Polytechnical Institute and State University) (C) This *Amanita pantherina* is also poisonous. (From BPS—J. Robert Waaland, University of Washington)

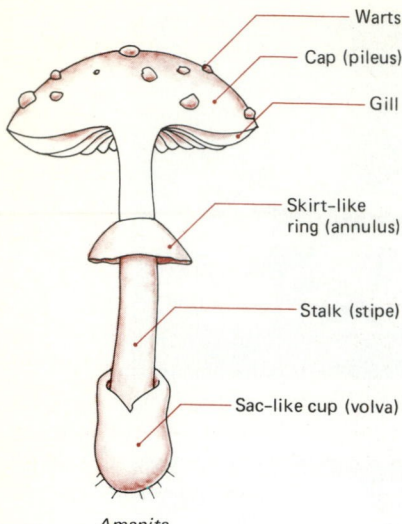

FIGURE 11.23

Diagram of an *Amanita* mushroom, showing the annulus and volva. Recognizing members of this genus is important because they are among the most deadly and can be mistaken for edible forms.

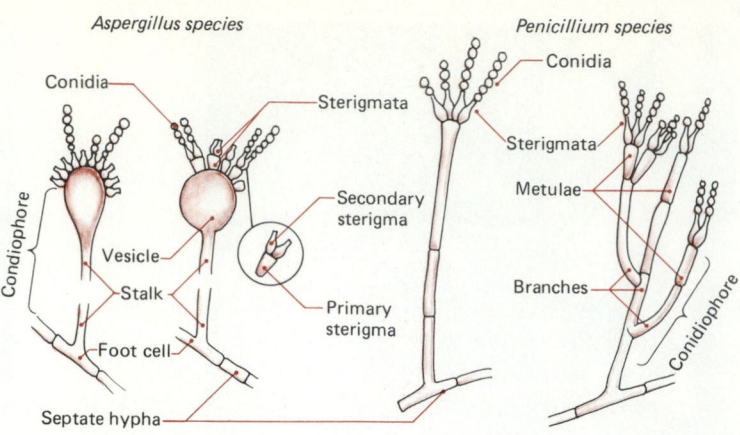

FIGURE 11.24

Diagram of asexual spores in *Penicillium* and *Aspergillus* spp. These asexual spores are important diagnostic characterisitics of the Fungi Imperfecti.

Several orders comprise the **gastromycetes**, another basidiomycete group, including the puffballs, earthstars, stinkhorns, and bird's nest fungi. Unlike other Basidiomycetes, the spores of the gastromycetes are not forcefully discharged. The order Phallales (stinkhorns) produces a green gelatinous ooze and a foul smell when the basidiocarp undergoes autodigestion, releasing the basidiospores. Although humans find the odor quite offensive, flies are attracted by it. Some of the ooze containing the basidiospores adheres to the flies, providing a mechanism for the dissemination of the basidiospores of these fungi.

The subclass Teliomycetidae includes the very few yeasts that are Basidiomycetes, along with the rusts and smuts. *Rhodosporidium* and *Leucosporidium*, which have been placed in this subclass, are basidiomycetous yeasts found in marine environments. The numerous species of rust and smut fungi are the most serious fungal plant pathogens. There are over 20,000 species of rust fungi and over 1,000 species of smut fungi. Rusts and smuts are characterized by the production of a resting spore known as a **teliospore**, which is thick-walled and binucleate. The rusts, all of which are plant pathogens, occur in the order *Uredinales*. The rust fungi require two unrelated hosts for the completion of their normal life cycle. For example, white pine blister rust uses gooseberry bushes as its alternate host. Important plant diseases are caused by rust fungi, and these fungal plant pathogens cause great economic losses in agriculture. Rust of cereals is caused by members of the genus *Puccinia*, rust of beans by *Uromyces* species, and pine blister rust by a *Cronartium* species. The smuts occur within the order Ustilaginales. The smut fungi cause serious economic losses in agriculture.

Members of the genus *Ustilago* cause smut of corn, wheat, and other plants; *Tilletia* species cause stinking smut of wheat; *Sphacelotheca* species cause loose smut of sorghum; and *Urocystis* species cause smut of onion. These are but a few examples of the common plant diseases caused by smut fungi. Many of the diseases caused by rusts and smuts will be considered in Chapter 14.

Deuteromycotina. The final subdivision of the fungi is the **Deuteromycotina**, also known as the **Deuteromycetes** or **Fungi Imperfecti**. There are about 15,000 species in the Fungi Imperfecti. The vegetative structures of most of these fungi resemble Ascomycetes, although the vegetative structures of a few members resemble Basidiomycetes. Sexual forms of reproduction in the Deuteromycetes, however, do not occur or have not been detected, and it is the lack of observed sexual spores that places fungal species in this group.

Without observation of the sexual reproductive stage, it is impossible to place members of the Deuteromycetes into either the Ascomycetes or Basidiomycetes, although many Fungi Imperfecti are clearly closely related to the Ascomycetes or Basidiomycetes. Whereas the sexual stages of several fungi have now been detected, a wholesale reclassification of fungi that have traditionally been placed in the Deuteromycetes has fortunately been avoided. For example, the sexual or perfect stages for several members of the genera *Penicillium* and *Aspergillus* have been found to involve the production of ascospores, but the members of these genera are still classified among the Fungi Imperfecti. It is likely, though, that many genera of Deuteromycetes will eventually be reclassified into other subdivisions.

The Fungi Imperfecti, or Deuteromycetes, are classified largely on the basis of the morphological structure of the vegetative phase and the types of asexual spores produced (Figure 11.24). The Deuteromycetes include many important

TABLE 11.6 Some Representative Genera of Deuteromycetes

Alternaria Soil saprophytes and plant pathogens; muriform spores fit together like bricks of a wall	*Coccidioides* *C. imminitis* causes mycotic infections in humans and animals
Arthrobotrys Soil saprophytes; some form organelles for the capture of nematodes	*Cryptococcus* Yeasts; saprophytic in soil, but some may cause mycoses in animals and humans
Aspergillus Common molds; radically arranged; colored, often black, conidiospores	*Geotrichum* Common soil fungus; older mycelial filaments break up into arthrospores
Aureobasidium (Pulullaria) Shore mycelial filaments, lateral blastospores; often damage painted surfaces	*Helminthosporium* Cylindrical, multiseptate spores; many are economically significant plant pathogens
Candida Common yeast; some cause mycoses; some species grow in concentrated sugar solutions; others grow on hydrocarbons	*Penicillium* Common mold with colored, often green, conidiospores arranged in a brush shape
	Trichoderma Common soil saprophyte with highly branched conidiophores

genera of filamentous fungi, such as *Penicillium*, and yeasts, such as *Candida*. Representative genera of the Deuteromycetes are listed in Table 11.6. Some species of Fungi Imperfecti are important in food and industrial microbiology.

The antibiotic penicillin, for example, is produced by *Penicillium* species, and both *Aspergillus* and *Penicillium* species are used in the production of various foods, such as blue cheese and soy sauce.

11.2 Algae

Classification of Algae

Whereas the fungi are nonphotosynthetic, the **algae** are eukaryotic photosynthetic organisms. As such, they are differentiated from all other microorganisms. The algae are separated from the plants by their lack of tissue differentiation. In Whittaker's five-kingdom classification system, some of the algae are placed in the kingdom Protista along with the protozoa, and other algae exhibiting more extensive organizational development are placed in the kingdom Plantae. Indeed, some organisms that are classified as algae are borderline cases with higher plants, and others are borderline cases with protozoa. There are some algae that lose their ability to carry out photosynthetic metabolism, rendering them indistinguishable from the protozoa. Some motile, unicellular algae have traditionally been studied by both protozoologists and phycologists. This situation has led to an inevitable confusion in the literature because zoologists and botanists typically use different features and criteria for establishing classification systems. Most traditional algal classification systems include the blue-green algae, but these organisms are properly considered as cyanobacteria because of their prokaryotic cells. The reclassification of the blue-greens as cyanobacteria is still considered controversial and is opposed by many phycologists.

The algae are classified into seven major divisions largely on the basis of the types of photosynthetic pigments that are produced, the types of reserve materials that are stored intracellularly, and the morphological characteristics of the cell (Table 11.7). The relative concentrations of photosynthetic pigments give the algae their characteristic colors. Many of

TABLE 11.7 The Major Divisions of Algae

Chlorophycophyta (green algae) Photosynthetic pigments: chlorophylls a and b, carotenes, several xanthophylls. Storage product: starch. Cell wall: cellulose, xylans, mannans, absent in some, calcified in others. Flagella: 1, 2–8, many, equal, apical.

Chrysophycophyta (golden and yellow-green algae, including diatoms) Photosynthetic pigments: chlorophylls a and c, carotenes, fucoxanthin, and several other xanthophylls. Storage product: chrysolaminaran. Cell wall: cellulose, silica, calcium carbonate. Flagella: 1–2, unequal or equal, apical.

Cryptophycophyta (cryptomonads) Photosynthetic pigments: chlorophylls a and c, carotenes, xanthophylls (alloxanthin, crocoxanthin, monadoxanthin), phycobilins. Storage product: starch. Cell wall: absent. Flagella: 2, unequal, subapical.

Euglenophycophyta (euglenoids) Photosynthetic pigments: chlorophylls a and b, carotenes, several xanthophylls. Storage product: paramylon. Cell wall: absent. Flagella: 1–3, apical, subapical.

Phaeophycophyta (brown algae) Photosynthetic pigments: chlorophylls a and c, carotenes, fucoxanthin, and several other xanthophylls. Storage product: laminaran. Cell wall: cellulose, alginic acid, sulfated mucopolysaccharides. Flagella: 2, unequal, lateral.

Pyrrophycophyta (dinoflagellates) Photosynthetic pigments: chlorophylls a and c, carotenes, several xanthophylls. Storage product: starch. Cell wall: cellulose or absent. Flagella: 2, one trailing, one girdling.

Rhodophycophyta (red algae) Photosynthetic pigments: chlorophyll a (also d in some), phycocyanin, phycoerythrin, carotenes, several xanthophylls. Storage product: Floridean starch. Cell wall: cellulose, xylans, galactans. Flagella: absent.

the major algal divisions have common names based on these characteristic colors, such as the green algae, red algae, and brown algae.

Chlorophycophyta   The **Chlorophycophyta**, or **green algae**, are widely distributed in aquatic ecosystems. The cellular organization in different species of Chlorophycophyta may be unicellular, colonial, or filamentous. Most cells are uninucleate, but several orders of green algae are characterized by the formation of coenocytic filaments. The chloroplasts of many unicellular green algae contain a pigmented region known as the **stigma** or *red eyespot*. Some green algae contain contractile vacuoles that serve an osmoregulatory function, protecting the cell against osmotic shock. The Chlorophycophyta normally store starch as a reserve material. The cell walls of different species of Chlorophycophyta are composed of cellulose, mannans, or xylans, but a high proportion of the cell wall also may be composed of protein.

Unicellular and Colonial Green Algae. The green algae can be divided into a number of orders. The order Volvocales includes the unicellular green algae, which are normally motile by means of flagella. Several well-known genera of algae occur in this order, including *Chlamydomonas* and *Volvox*. *Chlamydomonas* contains several hundred species. Members of this genus are unicellular and biflagellate.

Members of the genus *Volvox* form sphaeroidal colonies (Figure 11.25). The colonies contain many small vegetative cells and relatively few reproductive cells. The reproductive cells lack flagella and are called **gonidia**. The cells within a colony of *Volvox* act in a cooperative fashion so that the entire colony behaves as a superorganism. The flagella of the vegetative cells face outward and are able to move the entire colony in a unified manner. Colonies act as reproductive individuals, some producing male gametes and others female gametes. The colonies of some species thus exhibit sexual differentiation. Reproduction is dependent on the intact colony. The colonies of *Volvox* approach the level of tissue differentiation. It would appear that algae having such complex levels of organization represent an evolutionary link between microorganisms and higher plants. Based on ultrastructural analyses of the microtubules involved in mitotic division, however, *Volvox* does not appear to represent the missing evolutionary link between the green algae and green plants.

Filamentous Green Algae. There are several filamentous types of green algae, including members of the genera *Ulothrix* and *Spirogyra*. Probably the best-known genus of green algae, *Spirogyra*, occurs in the order Zygnemales. *Spirogyra* is an example of a filamentous green alga. The walls of the filament are continuous. The chloroplasts of *Spirogyra* form a spiral within the filaments (Figure 11.26). Another well-known algal genus, *Ulva*, occurs in the order Ulvales. Species of *Ulva*, commonly known as *sea lettuce*, are restricted to marine habitats. *Ulva* grows in marine habitats attached to rocks and other surfaces. Another marine form, the genus *Acetabularia*, is a tubular green alga (Figure

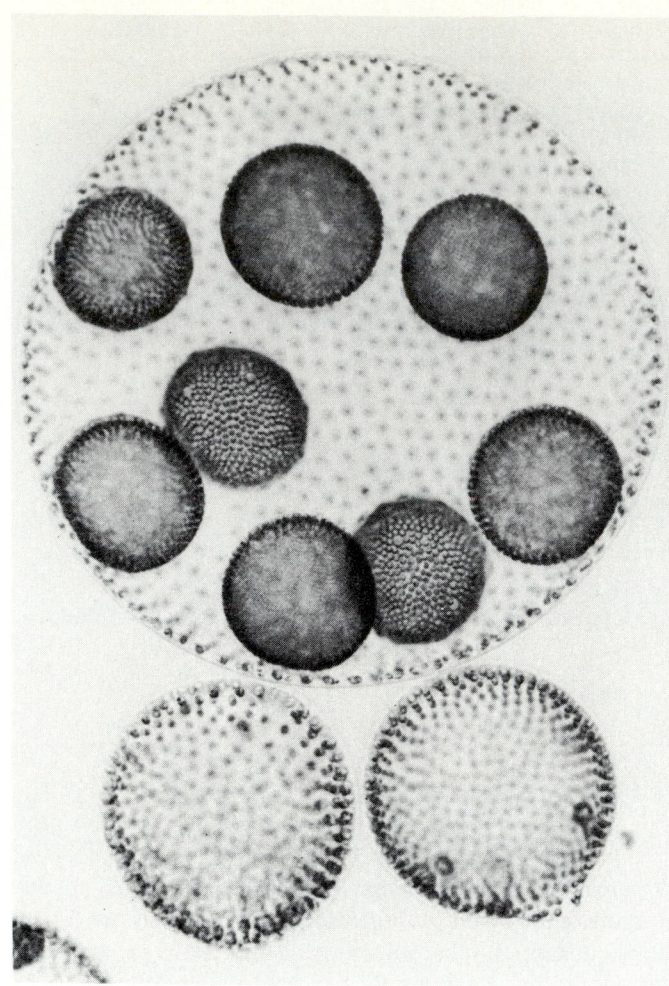

FIGURE 11.25
Colony of *Volvox aureus*. (Courtesy Carolina Biological Supply Co.)

11.27). This organism is known as the *mermaid's wine goblet*. There are several nonmotile coccoid types of green algae. These include members of the genera *Chlorococcum* and *Chlorella*, which are unicellular types, and members of the genus *Scenedesmus*, which form colonies of four to eight laterally united cells.

Euglenophycophyta   The **Euglenophycophyta** are similar to the Chlorophycophyta in that they contain chlorophylls a and b and typically appear green in color. The Euglenophycophyta differ from the Chlorophycophyta with respect to their cellular organization and their intracellular reserve storage products. The Euglenophycophyta are unicellular. They lack a cell wall but normally are surrounded by an outer layer, known as a **pellicle**, composed of lipid and protein. The Euglenophycophyta do not store starch but rather paramylon, a β-1,3-glucose polymer. The Euglenophycophyta appear to be closely related to the protozoa (Figure 11.28). Members of this division that lose their photosynthetic ap-

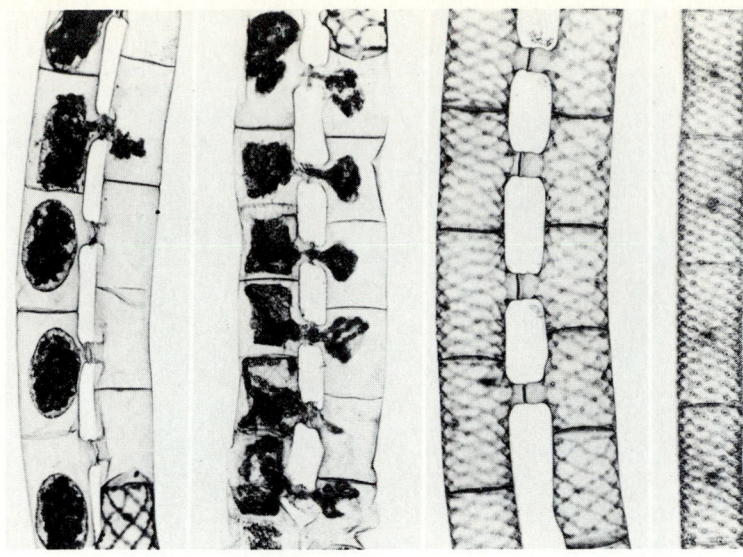

FIGURE 11.26
Micrograph of *Spirogyra* in conjugation. (Courtesy Carolina Biological Supply Co.)

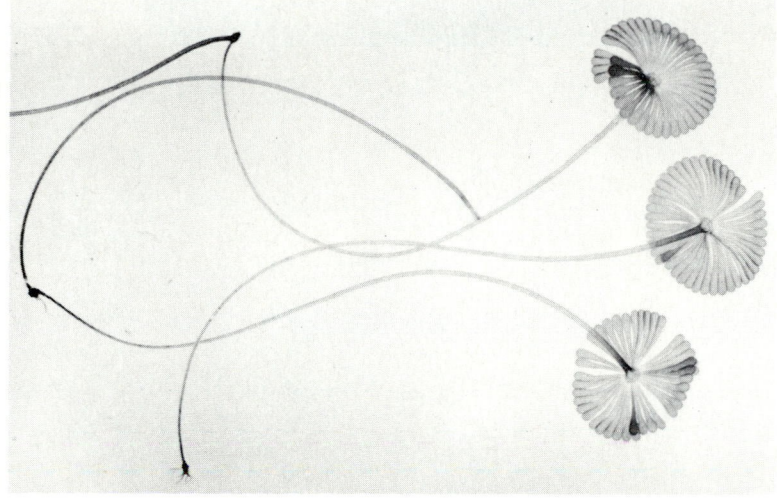

FIGURE 11.27
Photograph of *Acetabularia*, showing its delicate shape. (Courtesy Carolina Biological Supply Co.)

paratus are indistinguishable from protozoa. Reproduction in the Euglenophycophyta is normally by longitudinal division. These algae are widely distributed in aquatic and soil habitats. *Euglena* is the best-known genus of Euglenophycophyta. *Euglena* species have two flagella for locomotion and normally contain a contractile vacuole to protect it against osmotic shock.

Chrysophycophyta   The division **Chrysophycophyta** includes the classes Xanthophyceae (yellow-green algae), Chrysophyceae (golden algae), and Bacillariophyceae (the diatoms). Chrysophycophyta species produce a diversity of pigment biochemicals, cell wall biochemicals, and cell types. The Chrysophycophyta are unified by the production of the same reserve storage material, chrysolaminarin. Chrysolaminarin is a β-linked polymer of glucose. Chrysophycophyta species produce carotenoid and xanthophyll pigments that tend to dominate over the chlorophyll pigments; this confers golden-brown hues on members of this division. Most mem-

bers of the Chrysophycophyta are unicellular, although some are colonial.

Xanthophyceae and Chrysophyceae. Genera in the family Xanthophyceae include *Botrydioposis*, a unicellular form; *Tribonema*, a filamentous form; and *Vaucheria*, a coenocytic tubular form. *Vaucheria* is known as the *water felt* and is widely distributed in moist soils and aquatic habitats. The Xanthophyceae generally reproduce asexually by zoospores (motile spores) or aplanospores (nonmotile spores). Members of the class Chrysophyceae are typically motile by means of flagella. Asexual reproduction in the Chrysophyceae is by cell division, zoospores, or statospores (silicified cysts). Many members of the Chrysophyceae form siliceous or carbonaceous walls.

Bacillariophyceae (diatoms). The Bacillariophyceae or **diatoms** produce distinctive cell walls known as **frustules**. There are approximately 200 genera in the Bacillariophyceae. The diatoms typically are golden brown in color. The frustules of diatoms, which are also known as **valves**, have

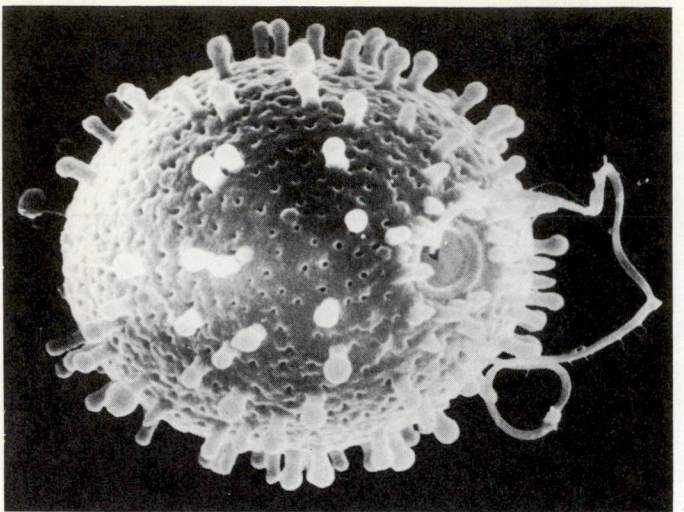

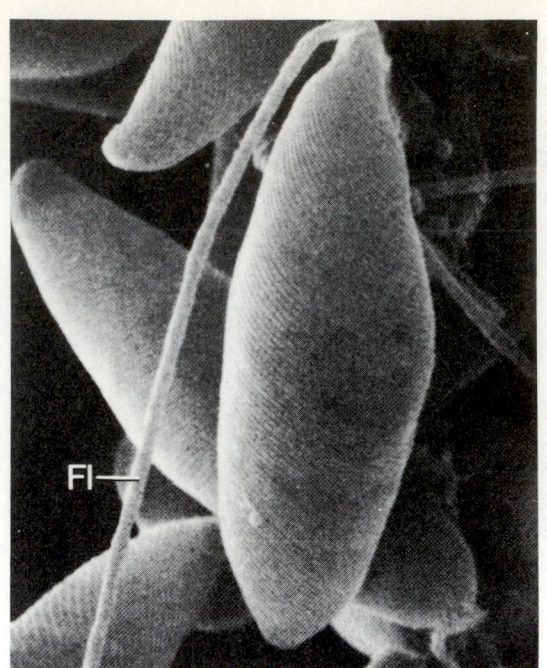

FIGURE 11.28

(A) Scanning electron micrograph of a *Trachelomonas* sp., a freshwater euglenoid flagellate. Based on their morphological appearance, these euglenoid organisms have been classified by both phycologists and protozoologists (5,000×) (From BPS—G. A. Antipa and E. B. Small). (B) Micrograph of *Euglena*; FL = flagellum (4,900×). (Reprinted by permission of Springer-Verlag, New York, from R. G. Kessel and C. Y. Shih, 1976, *Scanning Electron Microscopy for Biology*)

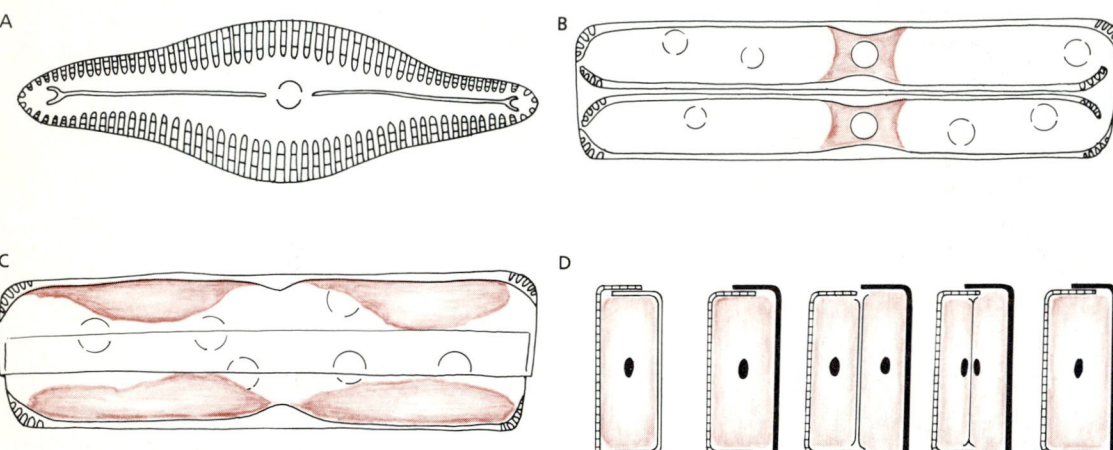

FIGURE 11.29

Diagrams of diatom frustules. (A–C) Cellular organization of *Pinnularia*: (A) surface view of the frustule—note the raphe and central and polar nodules; (B) living cell of *Pinnularia* in the girdle view—note the massive chloroplasts, central nucleus, and overlapping valves; (C) girdle view of a recently divided cell. (D) During the division process of a diatom, the progeny synthesize the inner frustula, leading to a progressive decrease in the size of one cell line.

two overlapping halves; the larger portion is referred to as the **epitheca** and the smaller one as the **hypotheca** (Figure 11.29). The halves of the frustule fit together like a petri dish. The geometric appearance of diatoms renders them aesthetically attractive (Figure 11.30). Pennate diatoms have bilateral symmetry, whereas centric diatoms have radial symmetry. Some diatoms are benthic, living at the bottom of aquatic ecosystems at the sediment layer, and other diatoms are planktonic, living suspended in open water bodies. The growth of diatoms is dependent on the concentrations of available silica because the cell walls of diatoms are impregnated with silica. Holes in the silica walls, called *puntae*, allow exchange of nutrients and metabolic wastes between the cell and its surroundings.

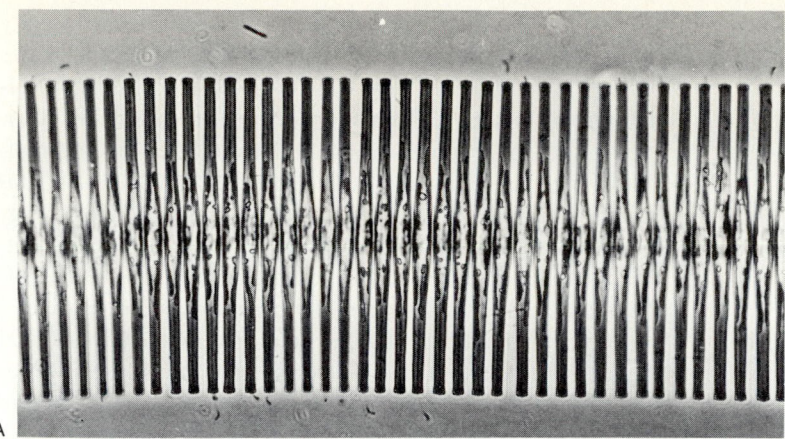

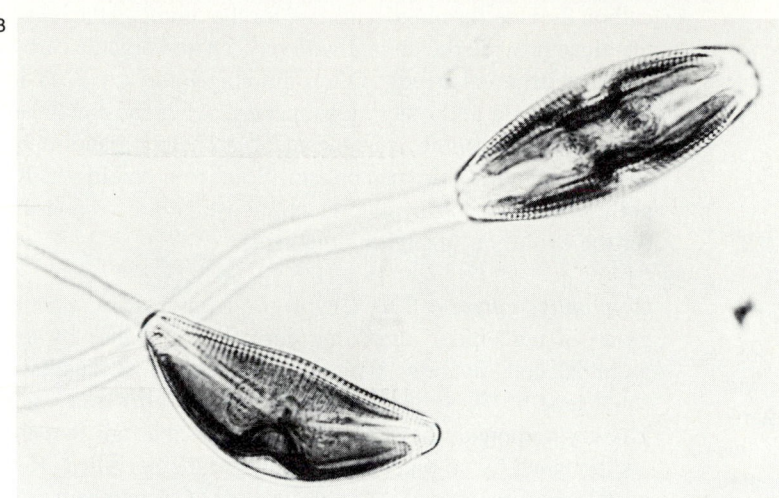

FIGURE 11.30 (*left*)
Micrographs of a few diatoms. The distinctive morphologies of the diatoms make their identification possible based only on microscopic observation. (A) *Fragilaria*, a colonial diatom (1,120×); (B) front and side views of *Cymbella*, a stalked diatom (1,280×). (From BPS—J. Robert Waaland, University of Washington)

FIGURE 11.31 (*above*)
Micrograph showing the raphe (Ra) and central node (No) structure of the diatom *Navicula* (2,300×). (Reprinted by permission of Springer-Verlag, New York, from R. G. Kessel and C. Y. Shih, 1976, *Scanning Electron Microscopy for Biology*)

The frustules of diatoms are resistant to natural degradation and accumulate over geologic periods. As a result, diatoms are preserved in fossil records dating back to the Cretaceous period 65 million years ago. There are significant deposits of diatom frustules in the world. Such deposits are known as *diatomaceous earth* and are mined for a number of commercial uses. Diatomaceous earth is sometimes used as an abrasive in toothpaste and metal polish. The most extensive use of diatomaceous earth is in the filtration of liquids, especially those from sugar refineries.

Reproduction in diatoms is normally by the formation of uneven-sized cells. Asexual reproduction involves the synthesis of a new cell wall structure in which each daughter cell reconstructs the smaller segment of the frustule, regardless of which segment of the parent frustule it receives. Therefore, continued asexual reproduction tends to result in diatoms of progressively smaller size. Occasionally, environmental conditions or the severe reduction in cell size leads to sexual reproduction with the production of auxospores, which are larger cells that act to reestablish larger diatoms. The valves of some diatoms have an opening along the apical axis known as the **raphe** (Figure 11.31). Diatoms that have a raphe exhibit gliding motility, with the direction of movement depending on the shape of the raphe. For example, due

to differences in the shapes of their raphes, *Navicula* species exhibit straight movement and those of *Nitzschia* exhibit curved movement. The gliding motility of diatoms permits these organisms to exhibit phototaxis, allowing them to move toward or away from light.

Pyrrophycophyta   The **Pyrrophycophyta** or **fire algae** are generally brown or red because of the presence of xanthophyll pigments. The Pyrrophycophyta are unicellular and biflagellate, and store starch or oils as their reserve material. The **dinoflagellates**, which are members of the Pyrrophycophyta, are characterized by the presence of a transverse groove that divides the cell into two semicells (Figure 11.32). The two flagella of the dinoflagellates emerge from an opening in the groove. Reproduction in the Pyrrophycophyta is primarily by cell division. The cell walls of Pyrrophycophyta contain cellulose and sometimes form structured plates, called **theca**. Because of their cell wall structures, some dinoflagellates that produce thecal plates are referred to as *armored dinoflagellates* (Figure 11.33).

Several dinoflagellates exhibit **bioluminescence**, the characteristic on which the designation *fire algae* is based. Some species also exhibit regular 24-hour behavioral patterns, known as *circadian rhythms*. For example, *Gonyaulax po-*

FIGURE 11.32
Scanning electron micrograph of the dinoflagellate *Gonyaulax tamerensis*, the organism that causes red tide (4,100×) (From BPS—J. Sieburth, University of Rhode Island)

lyedra exhibits a cyclic expression of luminescence, with peak luminescence occurring in the middle of the dark period. The luminescent capacity of *G. polyedra* allows this organism to glow at night. The glow rhythm is associated with a nightly increase in the level of luciferin and luciferase, the same enzyme substrate system that is operative in fire flies. *G. polyedra* also exhibits circadian rhythms with respect to its photosynthetic activities and cell division, with maximal cell divisions occurring at dawn and maximal photosynthetic activities occurring at midday.

Species of *Gonyaulax* and other dinoflagellates are extremely important because they produce the toxic blooms

FIGURE 11.33
Diagrams of armored dinoflagellates, so-named because of their thecal plates.

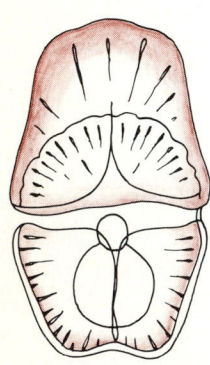
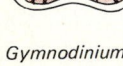
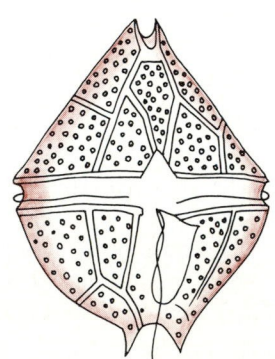

Gymnodinium   *Peridinium*

known as **red tides** that tend to color water in the vicinity of the bloom red or red-brown. The toxins of dinoflagellates during such blooms may kill invertebrate organisms. Although the blooms kill relatively few marine organisms, their toxins are concentrated in the tissues of filter-feeding molluscs, such as clams and oysters. Ingestion of shellfish containing dinoflagellates results in paralytic shellfish poisoning, a serious form of food poisoning. In order to prevent outbreaks of paralytic shellfish poisoning in humans, the collection of shellfish is banned during outbreaks of red tide.

Some dinoflagellates tend to enter into mutually beneficial (symbiotic) relationships with a variety of marine invertebrates. Such associations are termed *zooxanthellae*. Within such associations, the animal cell provides protection and carbon dioxide for photosynthesis for the dinoflagellates, and the algae provide the animal with oxygen and organic carbon for its nutritional needs. Often dinoflagellates grow on ingested bacteria and other algal species. As such, dinoflagellates are mixotrophic, capable of both chemoorganotrophic and photolithotrophic metabolism. Some microbiologists hypothesize that such symbiotic relationships are responsible for the evolution of higher organisms.

Cryptophycophyta   The **Cryptophycophyta** are a small group of unicellular algae that normally reproduce by longitudinal cell division, typically producing two flagella of equal length. These algae normally appear brown in color. The cryptomonads have asymmetric cells that are flattened and bounded by an outer covering called the **periplast**. Representative genera of this group are *Cryptomonas* and *Chroomonas*.

Rhodophycophyta   The **Rhodophycophyta** or **red algae** contain **phycocyanin** and **phycoerythrin** in addition to chlorophyll pigments. The red color of the Rhodophycophyta is due to the phycoerythrin. The primary reserve material in the Rhodophycophyta is **Floridean starch**, a polysaccharide similar to amylopectin found in higher plants. Rhodophycophyta exhibit a specialized type of oogamous sexual reproduction involving specialized female cells called **carpogonia** and specialized male cells called **spermatia**. Tetraspores, which are spores produced in a tetrasporangium, are formed during the life cycle of some red algae, and the tetraspores eventually differentiate into the male and female gametes. *Nemalion, Callithamnion, Delesserica, Anthithamnion, Callophyllis,* and *Porphyridium* are representative genera of red algae (Figure 11.34). Most red algae occur in marine habitats. They typically have a bilayered cell wall with an inner microfibrillar, rigid layer and an outer mucilaginous layer. Various biochemicals, including agar and carrageenin, occur in the cell walls of red algae. The agar and carrageenin of red algae are widely used as thickening agents and binders in various food products. Agar is also used as a solidifying agent in culture media, upon which the cultivation of bacteria largely depends. The carrageenin of *Chondrus*

A

B

FIGURE 11.34
Micrographs of some red algae. (A) *Callophyllis flabellulata*; (B) *Delesseria decipiens*
(From BPS—J. Robert Waaland, University of Washington)

crispus is utilized in puddings. The red alga *Porphyra* is cultivated and harvested by the Japanese as a source of food.

Phaeophycophyta   The division **Phaeophycophyta** or **brown algae** includes over 200 genera and 1,500 species. The Phaeophycophyta produce **xanthophylls** that dominate over the carotenoid and chlorophyll pigments and impart a brown color to these organisms. The main reserve materials for the brown algae are laminarin and mannitol. The cell walls of the Phaeophycophyta are generally composed of two layers, an inner cellulosic layer and an outer mucilaginous layer. Alginic acid is normally found as a biochemical constituent of the cell wall.

The Phaeophycophyta are almost exclusively marine organisms and are found primarily in coastal zones. The **kelps**, which are brown algae, can form macroscopic structures up to 50 m in length (Figure 11.35). It is difficult to consider organisms of that size as members of the microbial world. Most kelps have vegetative structures consisting of a holdfast, stem, and blade (Figure 11.36). These are histologically complex organisms that exhibit a degree of cellular differentiation. The genera *Fucus* and *Sargassum* are important representatives of the Phaeophycophyta. Large populations of *S. natans* occur in the Atlantic Ocean in the region known as the Sargasso Sea. Species of *Fucus* commonly occur along rocky shores, attached to the rocks by disc-like holdfasts. These brown algae clearly are the most complex organisms classified as algae, or for that matter, as microorganisms, representing a borderline case between algae and plants. Green plants, however, do not appear to have evolved directly from brown algae. Rather, parallel evolution appears to have occurred in which organisms in different evolutionary lines developed similar adaptive, organized structures.

FIGURE 11.35
Photograph of the kelp, *Lamanaria saccharina*. (From BPS—J. Robert Waaland, University of Washington)

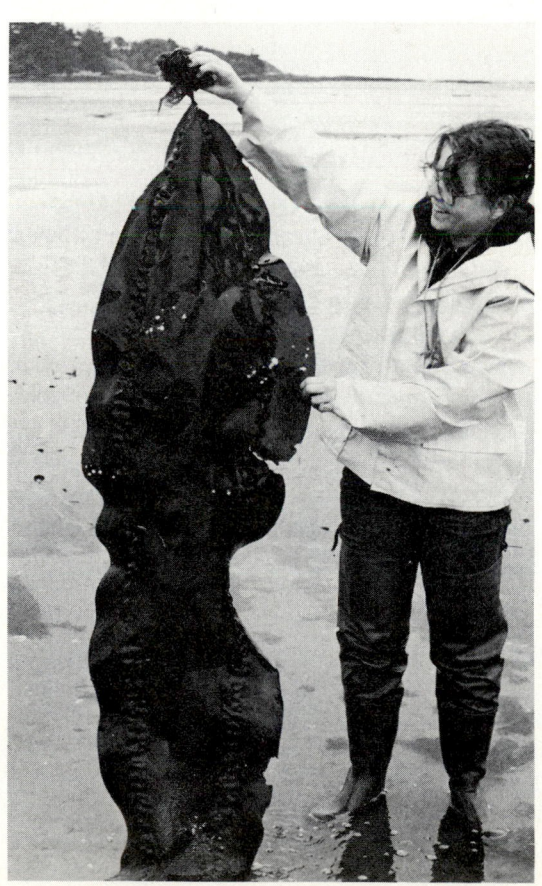

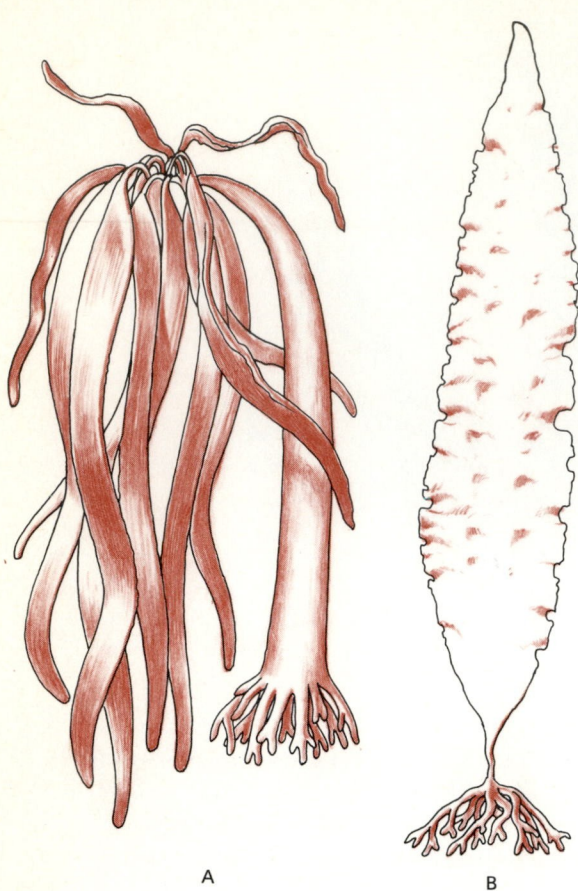

A   B

FIGURE 11.36
Drawings showing the complex structure of brown algae.
(A) *Postelsia palmaeformis*, the sea palm; (B) *Laminaria agardhii*.

11.3 Protozoa

Classification of Protozoa

Protozoa occur in the kingdom Protista. This subkingdom includes over 65,000 species. According to the revised 1980 classification of the protozoa reported by the Committee on Systematics and Evolution of the Society of Protozoologists, the protozoa, which are essentially unicellular organisms, are divided into seven phyla (Table 11.8). This new classification system departs from earlier systems that recognized four phyla based primarily on their means of motility: the **Mastigophora** (flagellates), **Sarcodina** (pseudopodia formers), **Ciliophora** (cilliates), and **Sporozoa** (spore formers). The new system encompasses the classification of several groups claimed by other disciplines, including most of the algae and the lower fungi, and for purposes of overall classification of microorganisms with this book, the slime molds are still considered as fungi and the photosynthetic microorganisms as algae. There are several noteworthy aspects of the new protozoa classification system. First, the Sarcodina and the Mastigophora are included in one phylum, the Sarcomastigophora. Within this phylum, the Mastigophora and Sarcodina are treated as separate subphyla. The other major change in this classification system is the division of the Sporozoa into four separate phyla: the Apicomplexa, Microspora, Acetospora, and Myxospora. The last three have spores, but many

members of the Apicomplexa do not; the Apicomplexa have an apical complex visible by electron microscopy.

Sarcomastigophora

Sarcodina. The **Sarcodina** are motile by means of pseudopodia. **Pseudopodia** are cytoplasmic extensions, sometimes referred to as *false feet* or *rhizopods*. The false feet may occur in a variety of forms, including extensions of the ectoplasm that encompass the flow of endoplasm (**lobopodia**); filamentous projections composed entirely of ectoplasm (**philopodia**); filamentous projections with branching **rhizopodia**; and axial rods within a cytoplasmic envelope (**axopodia**) (Figure 11.37). The pseudopodia are used for engulfing and ingesting food as well as for locomotion. Members of the Sarcodina may move at rates of 2–3 cm per hour under optimal conditions. Some members of the Sarcodina are important because they cause human diseases. For example, *Entamoeba histolytica* causes amoebic dysentery, a serious, debilitating disease.

The **Amoebida** form lobopodia and lack a skeletal structure. Members of the genus *Amoeba* have no distinct shape because the flow of cytoplasm continuously changes the shape of true amoebae. The giant amoeba, *A. proteus*, is

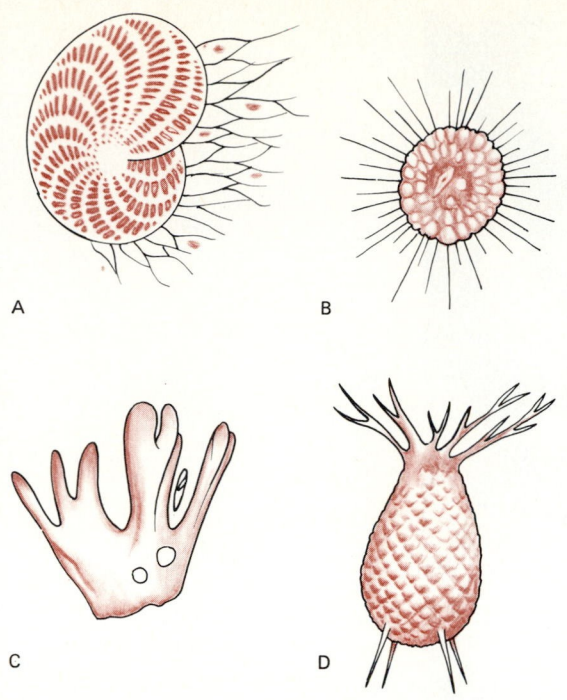

FIGURE 11.37

Illustrations of members of the *Sarcodina* with different types of false feet. (A) *Elphidium crispa* with rhizopodia; (B) *Actinospaerium eichhorni* with axopodia; (C) *Chaos carolinensis* with lobopodia; (D) *Euglypha alveolata* with filopodia.

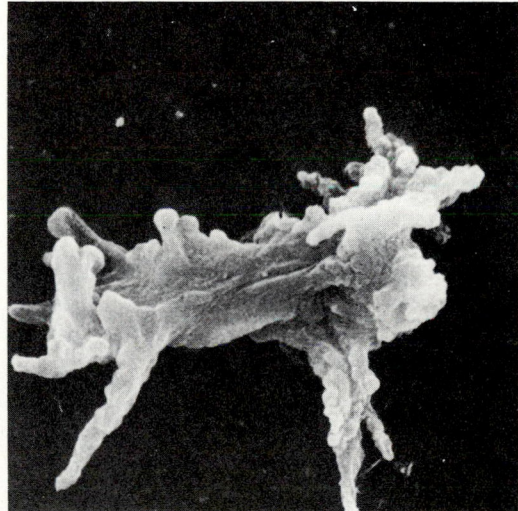

FIGURE 11.38

Micrograph of *Amoeba proteus* (Courtesy Robert Apkarian, Yerkes Primate Research Center, Emory University)

normally about 250 μm in length (Figure 11.38). This organism is readily found and visualized microscopically in pond water samples. *Amoeba* feed on a number of smaller organisms, including bacteria and other protozoa. For example, *A. proteus* can ingest the protozoa *Tetrahymena* and *Paramecium*.

The **Heliozoida** typically produce numerous radiating axopodia. The heliozoans are freshwater forms. The **Radiolaria**,

TABLE 11.8   The Classification of the Protozoa

Old System

Ciliophora   Locomotion: cilia. Reproduction: asexual, transverse fission; sexual, conjugation. Nutrition: ingestive.
Mastigophora.   Locomotion; usually paired flagella. Reproduction: asexual, longitudinal fission. Nutrition: heterotrophic, absorptive.
Sarcodina   Locomotion: pseudopodia (false feet). Reproduction: asexual, binary fission. Nutrition: phagocytic.
Sporozoa   Locomotion: usually none; some stages with flagella. Reproduction: asexual, multiple fission; sexual, within host; spores formed. Nutrition: absorptive.

New System

Sarcomastigophora   Locomotion: flagella, pseudopodia, or both. Reproduction: when sexual, essentially syngamy. Representative genera: *Monosiga, Bodo, Leishmania, Trypanosoma, Giardia, Opalina, Amoeba, Entamoeba, Difflugia*
Labyrinthomorpha   Synonymous with the net slime molds. Produce an ectoplasmic network with spindle-shaped or spherical nonamoeboid cells; in some genera, amoeboid cells move within a network by gliding.
Apicomplexa   Produce an apical complex visible with the electron microscope; all species parasitic. Representative genera: *Eimeria, Toxoplasma, Babesia, Theileria, Plasmodium*
Microsporal   Unicellular spores, each with an imperforate wall; obligate intracellular parasites. Representative genus: *Metchnikovella*
Ascetospora   Multicellular spore; no polar capsules or polar filaments; all species parasitic. Representative genus: *Paramyxa*
Myxospora   Spores of multicellular origin with one or more polar capsules; all species parasitic. Representative genera: *Myxidium, Kudoa*
Ciliophora   Cilia produced at some stage in the life cycle. Reproduce by binary transverse fission; budding and multiple fission also occur. Sexuality involving conjugation, autogamy, and cytogamy; most are free-living heterotrophs. Representative genera: *Didinium, Tetrahymena, Paramecium, Stentor*

like the heliozoans, are members of the Actinopodea. Radiolarians typically have axopodia, with a skeleton of silicon or strontium sulfate (Figure 11.39). They occur in marine ecosystems. The silica-containing exoskeletons of radiolarians are quite attractive when viewed microscopically.

The **Foraminiferida** are also marine members of the Sarcodina. Foraminiferans form one or many chambers composed of siliceous or calcareous tests (Figure 11.40). A **test** is a skeletal or shell-like structure. Tests of the Foraminiferida accumulate in marine sediments and are preserved in the geological record. The white cliffs of Dover are composed largely of the test structures of foraminiferans. Many of the foraminiferans are recognized in fossil records, whereas there is no fossil record for many microorganisms.

Mastigophora. The **Mastigophora** are the flagellate protozoa. Because some members of the Mastigophora are able to produce pseudopodia in addition to flagella, these organisms are now classified together with the Sarcodina. It is in this subphylum that protozoologists place the dinoflagellates,

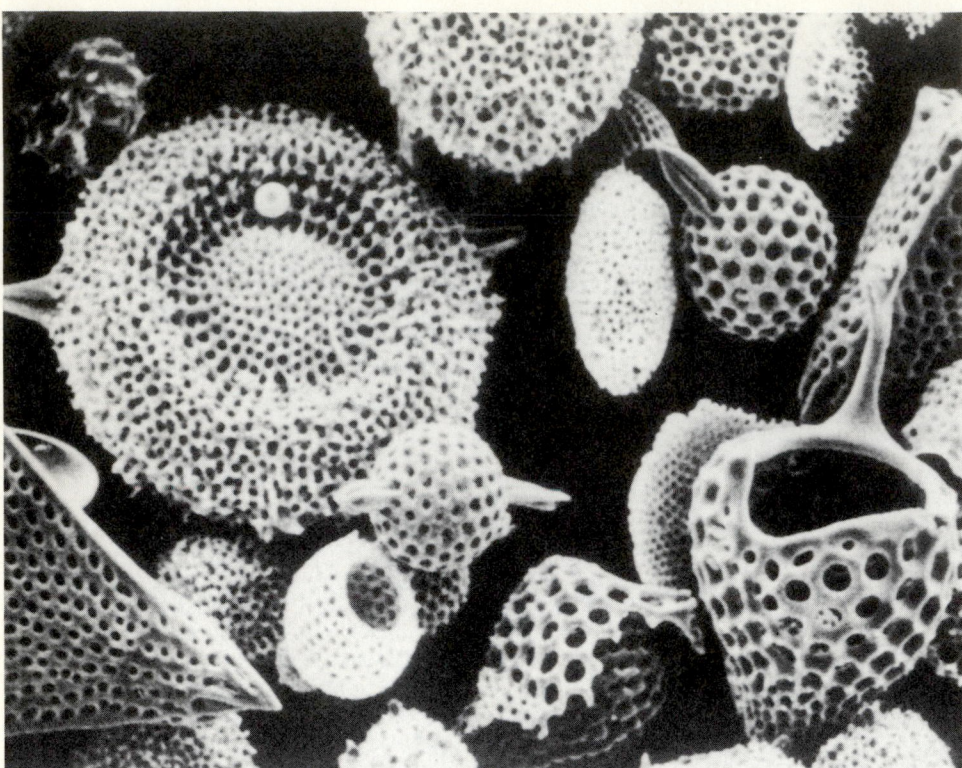

FIGURE 11.39
Micrograph of radiolarian shells (400 ×). (Reprinted by permission of Springer-Verlag, New York, from R. G. Kessel and C. Y. Shih, 1976, *Scanning Electron Microscopy for Biology*)

FIGURE 11.40
A larger foraminiferan, *Heterostegina depressa*, showing the chambers of the test (test size = 2 mm). The calcareous test consists of many chambers coiled in a plane. Each chamber is subdivided into chamberlets. As in all larger foraminiferans, it harbors unicellular symbiotic algae in its protoplasm, which give a yellow color to the living specimen. These larger foraminiferans occur in tropical and subtropical shallow seas. (Courtesy Rudolf Rottger, Universität Kiel)

FIGURE 11.41
Micrograph of *Trypanosoma cruzi*, the causative agent of Chagas' disease. (From BPS—Stephen Baum, Albert Einstein College of Medicine, New York)

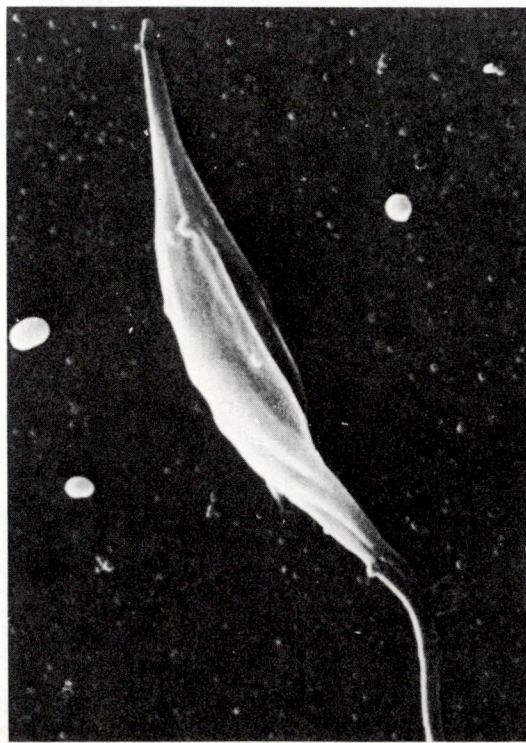

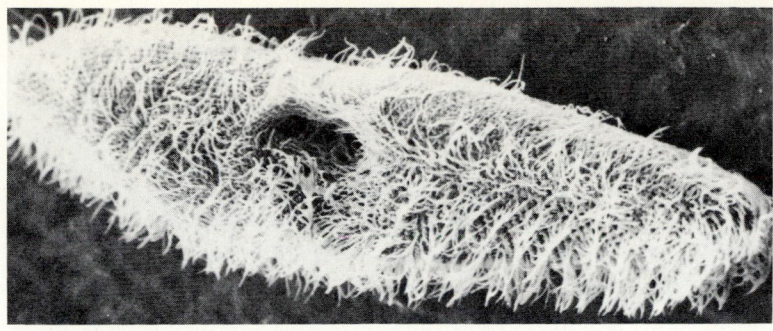

A

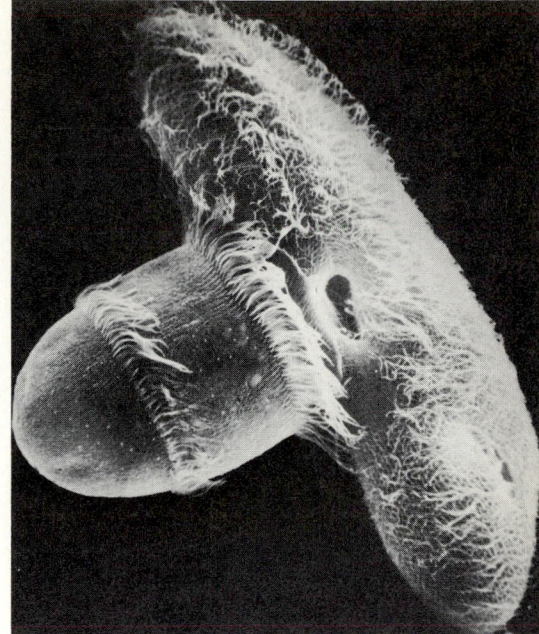

FIGURE 11.43
Micrograph of one ciliate protozoan, *Didinium nasutum*, consuming another ciliate, *Paramecium multimicronucleatum*, at an early stage of ingestion (1,400×). (From BPS—H. S. Wessenberg and G. A. Antipa, 1970, *Journal of Protozoology*, 17:250–270).

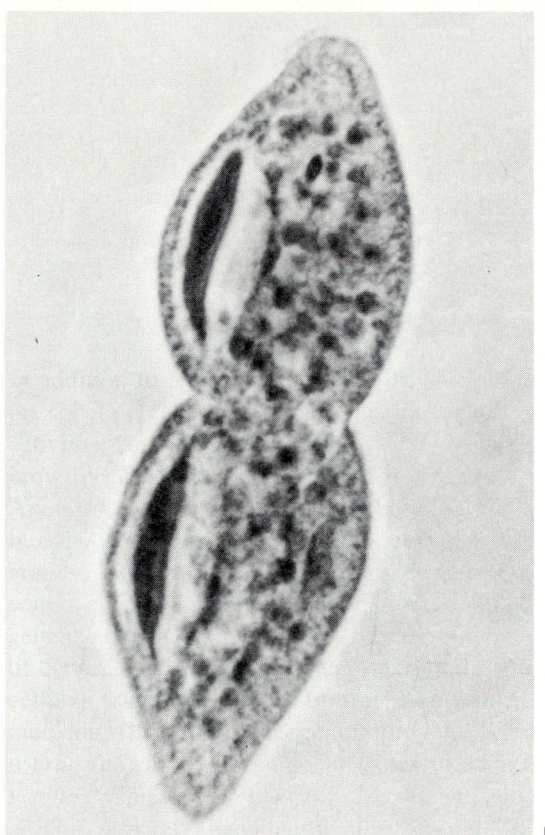

B

FIGURE 11.42
(A) Scanning electron micrograph of the ciliate protozoan *Paramecium*, showing the cytosome region. (Courtesy Eugene W. McArdle, Northeastern Illinois University). (B) Micrograph showing binary fission of a *Paramecium aurelia* (500×).

euglenoids, and other algae. The members of the Mastigophora here recognized as protozoa are heterotrophic and occur in the class Zoomastigophorea. Many members of the Mastigophora are plant and animal parasites. The genera *Trypanosoma* and *Leishmania* contain species that produce serious human diseases (Figure 11.41). *Trypanosoma gambiense*, for example, causes African sleeping sickness, and *T. cruzi* is the causative agent of Chagas' disease. Infections with the flagellate protozoan *Giardia* can cause severe diarrhea. *Leishmania donovani* is the causative agent of kalaazar disease, also known as *dum dum fever*. Human diseases caused by flagellate protozoa are normally transmitted by

arthropods, and control of many of these diseases depends on controlling the carrier rather than eliminating the disease-causing protozoa.

Ciliophora   **Ciliophora** are motile by means of **cilia**. Some members of the Ciliophora have a mouth-like region known as a **cytostome** (Figure 11.42). In addition to their role in locomotion, the cilia move food particles into the cytostome. The ciliate protozoa reproduce by various asexual and sexual means. Asexual reproduction is often by binary fission, and sexual reproduction is usually by conjugation. The Ciliophora normally contain two nuclei, a macronucleus and a micronucleus, both of which are diploid. The macronucleus is involved in asexual reproduction and the micronucleus in sexual reproduction. *Paramecium* is perhaps the best-known genus of ciliate protozoa. Other genera of Ciliophora include *Stentor*, *Vorticella*, *Tetrahymena*, and *Didinium*. Ciliate protozoa consume other microorganisms, including other protozoa, as their food source. For example, the genus *Didinium* can consume *Paramecium* species, providing a dramatic picture of the microbial world when viewed by scanning electron microscopy (Figure 11.43).

Sporozoa   The **Sporozoa** are parasites, exhibiting complex life cycles. The adult forms are nonmotile, but immature forms and gametes may be motile. The sporozoans derive nutrition by absorbing nutrients from the host cells they inhabit. The immature stages are referred to as **sporozoites**.

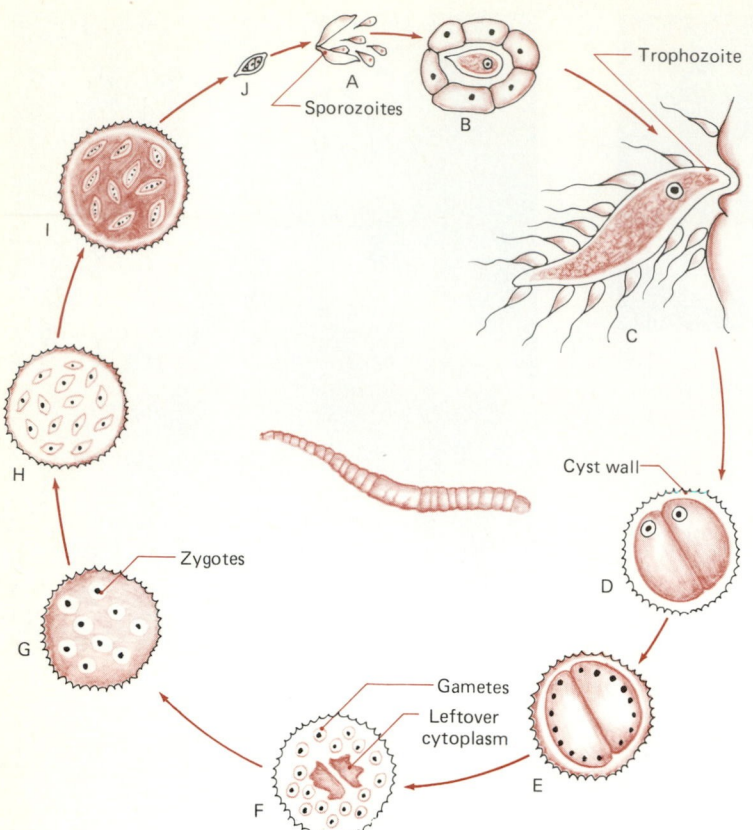

FIGURE 11.44 (left)

Diagram of the life cycle of the sporozoan *Monocystis*. *Monocystis* lives in the common earthworm. (A) Spores are eaten by the earthworm and sporozoites are released; (B) young trophozoite in a sperm morula; (C) the trophozoite grows into a worm, and earthworm sperm develop from the morula cells and attach to the trophozoite; (D) trophozoites associate in pairs and form a cyst wall (gametocytes); (E) nuclei of gametocytes divide; (F) gametes are produced; (G) gametes fuse, i.e., fertilization occurs to form zygotes; (H) zygotes secrete spore walls (sporocysts); (I) nuclei divide to form eight sporozoites in each spore; (J) spores are liberated when the earthworm dies. Spores may also be transferred to another worm during intercourse and infect testes of other worms. They may also be transferred to the worm cocoon, entering the ground when the cocoon disintegrates.

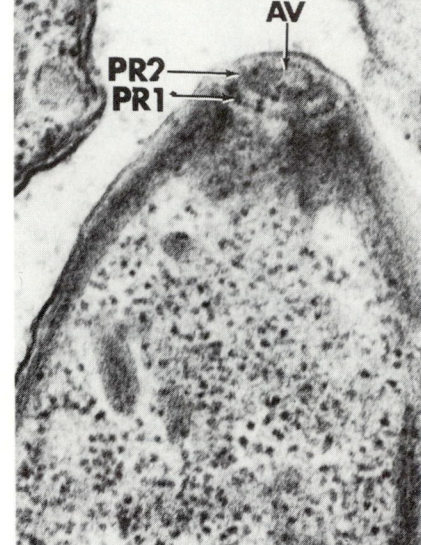

FIGURE 11.45 (right)

The apical structure of a member of the Apicomplexa. Micrograph of *Eimeria labbeana*, showing apical vesicles (AV) and preconoidal rings (PR1, PR2). (51,500×). (Courtesy T. Varghese; reprinted by permission of the Society of Protozoologists from T. Varghese, 1975, *Journal of Protozoology*, 22:68)

Reproduction of the **trophozoite**, the adult stage of a sporozoan, occurs asexually by multiple fission in which the mother cell divides into many daughter cells (Figure 11.44). The cells produced by multiple fission eventually mature into gametes that can be involved in sexual reproduction. The multiple fission process can result in the production of thousands of spores. *Plasmodium vivax* and various other species of *Plasmodium* cause malaria. During the course of a ma-

larial infection, life cycles of *Plasmodium* involve reproduction within human red blood cells, with the periodic release of large numbers of protozoa. As discussed earlier, members of the Apicomplexa, which include *Plasmodium* species, are characterized by the formation of an apical complex (Figure 11.45); it is interesting that such morphological observations, achievable only by using electron microscopy, are now employed for defining major taxonomic groupings.

Postlude

Eukaryotic microorganisms show a greater variety of reproductive strategies than bacteria, often exhibiting both asexual and sexual reproduction. This variety of reproductive strategies enhances the maintenance of a variety of genetic information. Meiosis and mitosis are essential processes for the redistribution of DNA during these life cycles. The fact that most eukaryotic microorganisms have several strategies for

reproduction enables them to establish relatively complex life cycles, in many cases involving an alternation between haploid and diploid phases. As part of their reproductive strategies, eukaryotic microorganisms produce various sexual and asexual spores. Fungi, for example, produce many sexual and asexual spores, which are used to define the major fungal taxonomic groups.

From the discussion of the systematics of the various groups of microorganisms, it is clear that each major group is classified on a totally different basis. In some cases, organisms have been classified in more than one group, such as the dinoflagellates that are treated by phycologists and protozoologists alike. Basing classification systems on phenotypic characteristics of necessity leads to some ambiguities. Without question the phylogenetic relationships among eukaryotic microorganisms will be clarified in future years, and the prevailing classification systems will be revised to include genetic analyses. For the most part, however, the taxonomy of eukaryotic microorganisms is still dominated by the classical approach based on field observations of the different morphological forms that occur during the life cycle of the organism.

Some fungi form macroscopic structures, which made it possible to observe them prior to the advent of the microscope. Fungi were used and studied from early times. The ancient Romans knew which fungi were epicurean delicacies, which were lethal, and which had hallucinogenic effects. Hooke's *Micrographia* (1665) included illustrations of microscopic fungi. Yeasts are recognizable in the drawings of Leeuwenhoek. The first book solely about fungi was the *Theatrum Fungorum* by Johannes Franciscus Van Starbeeck in 1675, using the drawings Charles de l'Escluse, also known as Clusius, prepared in 1601. The fungi *Rhizopus*, *Mucor*, and *Penicillium* are identifiable in the 1679 drawings of Marcello Malpighi.

The science of mycology, the study of fungi, however, probably owes its origins to Pier Antonio Micheli, an Italian botanist who, in 1729, published *Nova Plantarum Genera*, which included his studies on fungi. Almost half of the plants Micheli described were fungi, and many of the generic names still used today were first presented in this study. Micheli's most important contribution was the observation of the production of spores and the demonstration that the spores reproduced plants similar to their parent. Heinrich Anton deBary made major contributions to the field of mycology in his studies on plant pathology. He elucidated the life cycles of many rusts and defined the broad groups of Phycomycetes and Myxomycetes. In 1885 deBary proved that the blight that had caused the great Irish potato famine of the preceding decade was caused by a fungus; this was the first demonstration that a specific microorganism is the causative agent of disease.

A practical system of classification of the Fungi Imperfecti, those fungi that do not exhibit sexual reproductive phases according to spore groups, was developed by Pier Andrea Saccardo, who also collaborated on the 25-volume *Sylloge Fungorum* (1882–1925), which critically compiled most of the literature on fungal systematics published prior to 1920. In the early twentieth century, A. H. R. Buller also published a major monograph on fungal systematics. The first major compilation of yeast systematics was published in 1896 by Emil Hansen; this taxonomic system was greatly revised by A. Guilliermond between 1920 and 1928. The systematics proposed by Hansen and Guilliermond included the use of physiological, sexual, and phylogenetic relationships, as well as morphological observations, to determine classification. These characteristics, supplemented by direct analyses of fungal genetic information, form the basis of today's classification of the fungi.

Today, the fungi are classified largely on the basis of their modes of reproduction. The sexual spores of fungi are the most important features used in their classification and identification. Additionally, the asexual spores and vegetative structures of the fungi are used for the finer definition of taxa. Although fungi typically form filamentous mycelia, one group, the yeasts, are characteristically unicellular. The fungi include the slime molds, a taxonomic group with interesting life cycles, especially those that exhibit communication and coordinated behavior.

Like some fungi, many algae form macroscopic structures that are visible to the naked eye, and references to algae are found in early Eastern and Western literature. It was not until the mid-eighteenth century, though, that microscopic methods were used to examine algae. Once the reproductive phases of algae were recognized, life history studies could proceed and taxonomic systems of classification could be developed. The elucidation of algal sexual reproductive cycles was pioneered by Gustave Thuret, a wealthy Parisian amateur scientist, in studies conducted from 1840 to 1854 with *Fucus*. Nathaniel Pringsheim, working with *Vaucheria* during the same period, described the growth of algae and the development of sexual stages, allowing algal classification to be based on reproductive systems rather than just on superficial resemblances. During the early nineteenth century, many phycologists (algologists) published works that advanced the taxonomic classification of algae. At the end of the nineteenth century, phycologists employed their accumulated knowledge of algal morphology and reproduction to revise the classification schemes for the taxonomy of algae. Many of today's views on the systematics of the algae date from the late nineteenth and early twentieth centuries.

The algae are classified into seven groups largely on the basis of pigment production and the biochemical nature of the storage reserve materials. The seven groups of algae are the green algae, euglenoids, brown algae, golden and yellow-green algae, dinoflagellates, cryptomonads, and red algae. Some of these groups, such as the euglenoids, are closely related to protozoa. Others, such as the brown algae, are closely related to plants. Many algae produce complex macroscopic structures. The algae include the diatoms, organisms that produce frustules containing silica that are highly symmetric and quite beautiful.

In contrast to the algae, the protozoa are classified into major groups largely on the basis of their means of locomotion. Some protozoa form extensions of the cytoplasm known as *pseudopodia* or *false feet*, involved in both locomotion and the ingestion of food. The Ciliophora are protozoa that are motile by means of cilia. The Mastigophora are protozoa that are motile by means of flagella. The Sporozoa, all of which are parasites, are generally nonmotile. Members of this group of Protozoa were among the first microscopic organisms ever observed. The protozoa *Vorticella*, *Volvox*, and *Euglena* clearly are shown in the 1677 sketches of van Leeuwenhoek. *Paramecium*, as well as other protozoa, were described in 1678 by Christian Huygens. Leeuwenhoek continued to report his drawings of protozoa and in 1681 described what appears to be *Giardia intestinalis*, thus discovering parasitic protozoa. Louis Joblot, a professor of mathematics with an interest in optics that led him to microscopy, published the first treatise on protozoa in 1718. G. A. Goldfuss introduced the term *protozoan* in 1817, and a chapter about this group of organisms appeared in a book on the comparative anatomy of invertebrates in 1848 by Karl T. E. von Siebold. The term *protozoan* is derived from the Greek *protos*, meaning first, and *zoon*, meaning animal. In 1838 Christian Ehrenberg published a major monograph on the protozoa describing more than 500 species, including their digestive and reproductive systems. Felix Dujardin, a French professor of zoology, in 1841 published a classification system for the protozoa using primarily morphological features; Dujardin was an accurate observer, and his classification system was sounder than that of Ehrenberg.

Medical protozoology began in the mid-1800s. Pasteur, in 1870, reported that a protozoan was responsible for a disease of silkworms that had devastated the French silk industry during the 1800s. Also in 1870, T. R. Lewis observed *Amoeba* in the stools of individuals suffering from choleric symptoms and, shortly thereafter, F. Losch described *Entamoeba histolytica* as the causative agent of dysentery in man. Transmission of this disease by ingestion of *E. histolytica* was shown by E. L. Walker and A. W. Sellards in 1913. The discovery that disease-causing microorganisms could be transmitted by animal vectors, which was made by Theobold Smith, represented a major advance in medical protozoology and in our understanding of the mechanisms of disease transmission. Smith and co-workers (1893) proved that Texas cattle fever was caused by a protozoan and that transmission of the disease involved a tick vector. This proof opened the way for the discovery that a number of other diseases are transmitted by arthropod vectors. Alphonse Laveran and Camillo Golgi (1881–1886) showed that malaria was caused by a parasitic protozoan. Ronald Ross, in 1897–1898, found that the malarian parasite was transmitted in birds by a mosquito vector. The mode of transmission of the protozoan that caused malaria in humans was discovered by Battista Grassi, who disputed with Ross the priority of discovery of vector transmission of this disease. Joseph Dutton (1902) found that a trypanosome protozoan caused African sleeping sickness and was transmitted by the tsetse fly. William B. Leishman and C. Donovan (1903) discovered that kala azar disease was caused by a protozoan that subsequently was named *Leishmania donovani*; many microbial species names are derived from the names of the individuals who studied them.

Finally, a word of advice for those interested in the taxonomy of eukaryotic microorganisms. We have discussed only the basic characteristics of the more common organisms. These organisms exhibit a variety of forms and mycologists, phycologists, and protozoologists have developed separate and elaborate languages of specialized terminology (jargon) to describe the characteristics used in their classification. It is first necessary to learn the proper terminology before taxonomic keys and diagnostic tables become useful aids in identifying these microorganisms. Additionally, because many of the key features used in identifying eukaryotic microorganisms involve the observation of morphological forms, gaining practical laboratory and field experience with an expert taxonomist is critical in developing the skills needed to identify them. Introductory microbiology laboratory courses traditionally are heavily biased toward the bacteria and often omit the examination of eukaryotic microorganisms. With some practical experience, one can readily learn to recognize and properly identify many of these organisms in their natural habitats. In the ensuing chapters, we will consider the practical aspects of the microorganisms whose systematics, including physiology, morphology, and taxonomy, we have examined in this and previous chapters.

Suggested Supplementary Readings

AINSWORTH, G. C., and A. S. SUSSMAN (eds.). 1965–1973. *The Fungi: An Advanced Treatise* (4 volumes). Academic Press, New York.

ALEXOPOULOS, C. J., and C. W. MIMS. 1979. *Introductory Mycology*. John Wiley & Sons, Inc., New York.

BARNETT, J. A., R. W. PAYNE, and D. YARROW. 1979. *A Guide to Identifying and Classifying Yeasts*. Cambridge University Press, New York.

BOLD, H. C., and M. J. WYNNE. 1978. *Introduction to the Algae: Structure and Reproduction*. Prentice-Hall, Inc., Englewood Cliffs, N.J.

CHAPMAN, V. J., and D. J. CHAPMAN. 1975. *The Algae*. St. Martin's Press, New York.

COLE, G. T., and B. KENDRICK (eds.). 1981. *Biology of Conidial Fungi*. Academic Press, New York.

CORLISS, J. O. 1979. *The Ciliated Protozoa: Characterization, Classification and Guide to the Literature*. Pergamon Press, New York.

FARMER, J. N. 1980. *The Protozoa: Introduction to Protozoology*. C.V. Mosby Co., St. Louis.

GALL, J. G. (ed.). 1986. *The Molecular Biology of Ciliated Protozoa*. Academic Press, Orlando, Fla.

GRAY, W. D., and C. J. ALEXOPOULOS. 1968. *Biology of the Myxomycetes*. The Ronald Press Co., New York.

HARTWELL, L. L. 1974. *Saccharomyces cerevisiae* cell cycle. *Bacteriological Reviews* 38:164–198.

JAHN, T. L., E. C. BOVEE, and F. F. JAHN. 1979. *How to Know the Protozoa*. Wm. C. Brown Co., Dubuque, Iowa.

KUDO, R. R. 1977. *Protozoology*. Charles C. Thomas, Publisher, Springfield, Ill.

LASKIN, A., and H. A. LECHEVALIER. 1979. *Handbook of Microbiology: Fungi, Algae, Protozoa, and Viruses.* CRC Press, Inc., Boca Raton, Fla.

LEE, R. E. 1980. *Phycology.* Cambridge University Press, Cambridge, England.

LEVINE, N. D., J. O. CORLISS, F. E. G. COX, G. DEROUX, J. GRAIN, B. M. HONIGBERG, G. F. LEEDALE, A. R. LOEBLICH, J. LOM, D. LYNN, E. G. MERINGELD, F. C. PAGE, G. POLJANSKY, V. SPRAGUE, J. VAVRA, and F. G. WALLACE. 1980. A newly revised classification of the protozoa. *Journal of Protozoology* 27:37–58.

LODDER, J. 1984. *The Yeasts: A Taxonomic Study.* North Holland Publications, Amsterdam.

MILLER, O. K. 1979. *Mushrooms of North America.* E.P. Dutton Co., New York.

MOORE-LANDECKER, E. 1982. *Fundamentals of the Fungi.* Prentice-Hall Inc., Englewood Cliffs, N.J..

PHAFF, H. J., M. W. MILLER, and E. M. MRAK. 1978. *The Life of Yeasts.* Harvard University Press, Cambridge, Mass.

PICKETT-HEAPS, J. D. 1975. *Green Algae: Structure, Reproduction and Evolution in Selected Genera.* Sinauer Associates, Inc., Publishers, Sunderland, Mass.

ROSE, A. H., and J. S. HARRISON (eds.). 1986. *The Yeasts.* Academic Press, Orlando, Fla.

TRAINOR, F. R. 1978. *Introductory Phycology.* John Wiley & Sons, Inc., New York.

WESTPHAL, A. 1976. *Protozoa.* Blackie and Son Ltd., Glasgow, Scotland

Study Questions

1. On what basis are the major groups of fungi defined? What is the importance of spore formation in fungal classification?

2. On what basis are the major groups of algae defined? What is the importance of photosynthetic pigments and reserve materials in algal classification?

3. On what basis are the major groups of protozoa defined? What is the importance of mode of locomotion to protozoan taxonomy?

4. Why are dinoflagellates treated by both protozoan and algal taxonomists?

5. What are the differences between the ascomycetes and the basidiomycetes?

6. What is a diatom? What is unique about the structure of a diatom?

7. Should the brown algae be considered as plants or microbes?

8. Compare the reproduction of a yeast with that of a filamentous fungus. How are yeasts and filamentous fungi enumerated?

9. What is the role of sexual reproduction in eukaryotic microorganisms?

Situational Problem

Identifying Mushrooms

The ability to identify mushrooms properly can be a matter of life and death because some mushrooms are deadly poisonous. Mushrooms should never be eaten unless you are absolutely certain that they are not among the poisonous varieties. Many tragic stories appear in the news media when someone errs and eats a poisonous mushroom. In some cases, immigrants to the United States find and pick mushrooms that look just like the ones in their native country, only to discover that they are poisonous. Even knowledgeable people, such as the White House chef during the administration of President John Kennedy, have unfortunately mistaken the identity of a poisonous mushroom for one that they consider to be an edible delicacy. Hospitals and poison control centers have expert mycologists as consultants whom they contact in cases of suspected fungal poisoning to aid in the identification of the fungus and thus in the determination of the appropriate treatment process.

Suppose you want to collect mushrooms and serve them at a meal. You should compare the mushrooms you collect with an identification guide that is pertinent to your specific region. To determine what is involved in this task, assuming the season is appropriate, find and collect mushrooms in your local vicinity; otherwise, obtain several different types of mushrooms from your local supermarket or produce supplier. Then, using an identification guide, which you should be able to find in your library, try to identify the species of these mushrooms. If the mushrooms you identify are store bought, you can check your identification and actually eat the mushrooms. Do not eat any of the wild mushrooms you have collected.

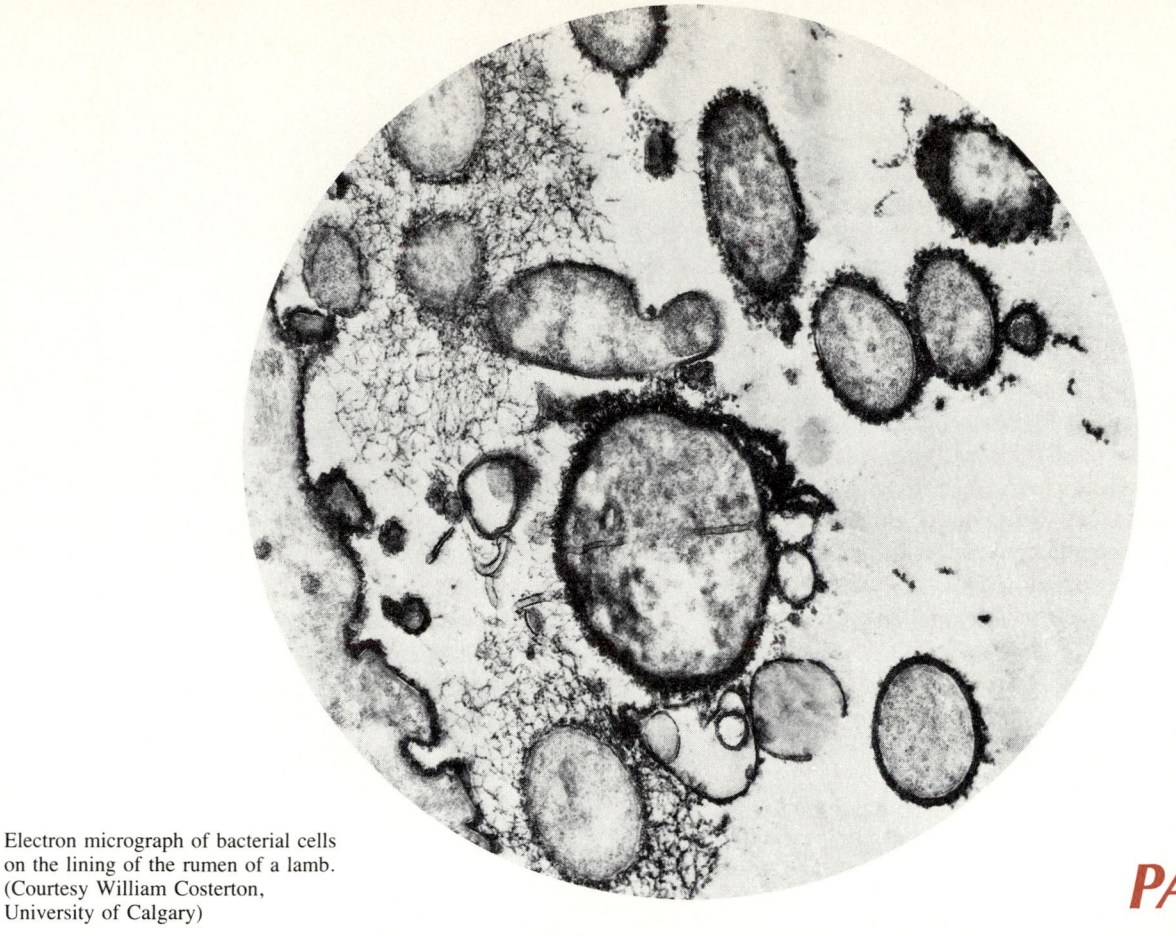

Electron micrograph of bacterial cells
on the lining of the rumen of a lamb.
(Courtesy William Costerton,
University of Calgary)

PART **6**

Environmental Microbiology

Influence of Environmental Factors on the Growth and Distribution of Microorganisms

Rates of microbial growth and death are greatly influenced by a number of environmental parameters, some environmental conditions favoring rapid microbial reproduction and others permitting no microbial growth. Conditions permitting the growth of one microorganism may preclude the growth of another. Not all microorganisms can grow under identical conditions. Each microorganism has a specific tolerance range for specific environmental parameters. Outside the range of environmental conditions under which a given microorganism can reproduce, it may either survive in a relatively dormant state or lose **viability**; that is, it may lose the ability to reproduce and will consequently die.

In both laboratory and natural situations some, environmental parameter or interaction of environmental parameters controls the rate of growth or death of a given microbial species. In nature, where conditions cannot be controlled and many species coexist, fluctuating environmental conditions favor population shifts because of the varying growth rates of individual microbial populations within the community at a given location. In the laboratory, it is possible to adjust conditions to achieve optimal growth rates for a given microorganism. Similarly, in industrial fermentors, conditions can be adjusted to optimize microbial growth rates, thereby maximizing the accumulation of desired microbial metabolic products. Many laboratory and industrial applications use pure cultures of microorganisms, facilitating the adjustment of the growth conditions so that they favor optimal growth of the particular microbial species.

By adjusting environmental conditions, one can also increase the death rate of microorganisms, an important consideration when trying to kill microorganisms, as in sterilization procedures. Sterilizing products is very important in many areas, including food processing, medical procedures, and the pharmaceutical industry. It is also possible to alter environmental conditions so that microorganisms do not die but do not reproduce, a method used for the preservation of microorganisms, as in culture collections and food preservation. Many times the conditions needed to sterilize a product would ruin it. Thus, for example, we preserve by freezing many foods whose taste and textural qualities would be destroyed if they were sterilized at high temperatures.

In this chapter, we will consider the influence of a number of environmental factors on microorganisms and their distribution in nature.

12.1 *Temperature*

Effects of Temperature on Microbial Growth Rates

Temperature is one of the most important environmental factors affecting the rates of both microbial growth and death. Temperature influences the rate of chemical reactions and the three-dimensional configuration of proteins, thereby affecting the rates of enzymatic activities (Figure 12.1). As long as the enzyme is not denatured, that is, as long as its three-dimensional structure is not disrupted, a rise of 10°C generally results in the approximate doubling of the rate of its reaction. The Q_{10} of a reaction describes the change in reaction rate that occurs when the temperature is increased by

10°C (Figure 12.2). At elevated temperatures, however, proteins are denatured, and enzymatic activities decline above the specific temperature that is characteristic of the heat stability of the particular enzyme. Enzymes have optimal temperatures, that is, at some temperature each enzyme exhibits maximal activity. Optimal temperatures vary among enzymes, and even the same enzyme from different organisms can have different optimal temperatures. At some temperature above optimal, denaturation occurs. Because of protein denaturation at elevated temperatures and the resultant change in membrane fluidity, there is an upper temperature limit for microbial growth. At temperatures above that limit, microorganisms are unable to survive because they cannot carry out their life-supporting metabolic activities.

The minimum and maximum temperatures at which a microorganism can grow establish the **temperature growth range** for that microorganism (Figure 12.3). Within this growth range there is an optimal growth temperature at which the highest rate of reproduction occurs. Because the generation time is the reciprocal of the instantaneous growth rate, the shortest generation time occurs at the optimal temperature. The **optimal growth temperature** is defined by the maximal growth rate, not the maximal cell yield. Sometimes greater cell or product yields are achieved at lower or higher temperatures. Some microorganisms, known as **stenothermophiles**, grow only at temperatures near their optimal growth temperature, whereas **eurythermal microorganisms** grow over a wider range of temperatures. In the laboratory, **incubators**, which are simply controlled-temperature chambers, are normally used to establish conditions that permit the growth of a microbial culture at temperatures favoring optimal growth rates.

Different microorganisms have different optimal growth temperatures (Figure 12.4). Some microorganisms grow best at low temperatures. Such organisms, known as **psychrophiles**, have optimal growth temperatures of under 20°C. Psychrophilic microorganisms may have enzymes that are inactivated at even moderate temperatures, of about 25°C, or may have proteases that are activated at these temperatures. Alternatively, the membrane fluidity of psychrophiles may be altered at moderate temperatures, restricting these organisms to low temperatures where they can maintain viability. Some psychrophilic microorganisms are capable of growing below 0°C as long as liquid water is available. Psychrophilic microorganisms are commonly found in the world's oceans and are also capable of growing in a household refrigerator, where they are important agents of food spoilage.

Other microorganisms, known as **mesophiles**, have optimal growth temperatures in the middle temperature range between 20 and 50°C. Most of the bacteria grown in introductory microbiology laboratory courses are mesophilic. Many mesophiles have an optimal temperature of about 37°C, which corresponds to human body temperature. Many of the normal resident microorganisms of the human body, such as *Escherichia coli*, are mesophiles. Similarly, most

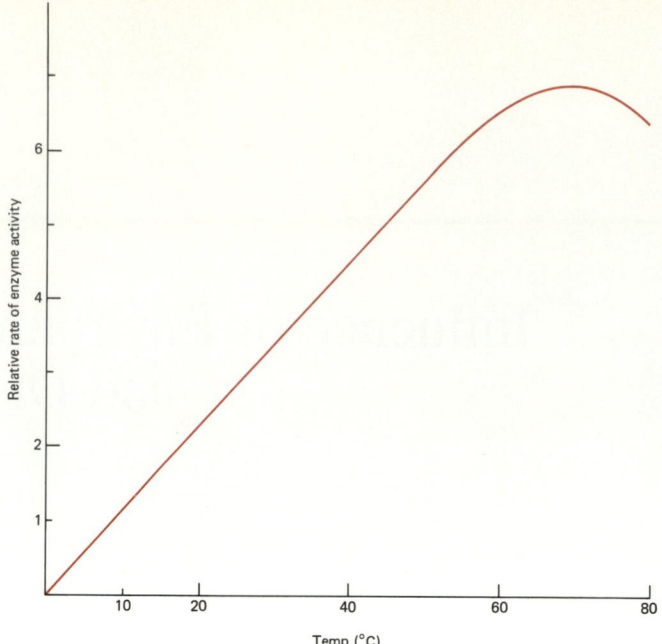

FIGURE 12.1
The effect of temperature on rates of enzyme activities is represented in this graph. As temperature increases, the rate of enzyme activity increases.

$$Q_{10} = \frac{\text{activity at temperature } T + 10°C}{\text{activity at temperature } T}$$

Enzyme	Q_{10} value (temp. range)
Catalase	2.2 (10–20°C)
Maltase	1.9 (10–20°C)
Urease	1.8 (20–30°C)

FIGURE 12.2
Formula for Q_{10} and some representative Q_{10} values. A rise of 10°C results in an approximate doubling of the rate of an enzymatic activity.

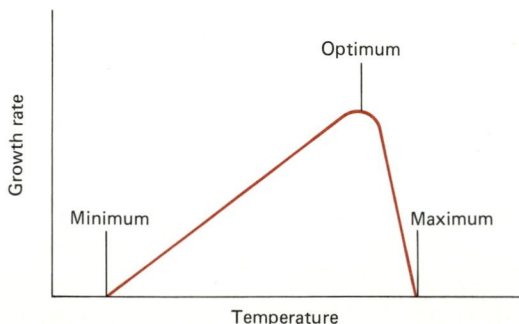

FIGURE 12.3
Effect of temperature on growth rate, showing growth ranges and optimal, minimum, and maximum temperatures for microorganisms.

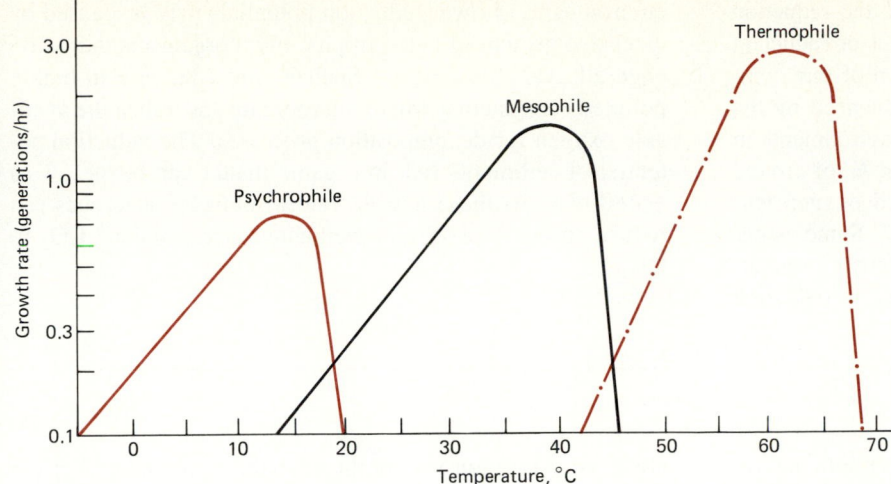

FIGURE 12.4
Temperature growth ranges and optima for psychrophilic, mesophilic, and thermophilic organisms.

human pathogens are mesophiles, and thus are able to grow rapidly and establish an infection within the human body. Although some microorganisms are restricted to growth near the optimal growth temperature, others can grow over a wide range of temperatures. Microorganisms generally can actively reproduce over a wider range of temperatures below the optimal growth temperature than above it. For example, psychrotrophic microorganisms are eurythermal mesophiles; their optimal growth temperature is between 20 and 50°C (making them mesophiles), but they are capable of growing at low temperatures (of about 5°C), such as inside a refrigerator.

Whereas mesophiles grow in the middle temperature range, **thermophiles** are organisms with high optimal growth temperatures. Thermophiles such as *Bacillus stearothermophilus* grow at relatively high temperatures, often only above 40°C.[1] The upper growth temperature for thermophilic microorganisms, though, is about 99°C. Many thermophilic microorganisms have optimal growth temperatures of about 55–60°C. Thermophiles are found in such exotic places as hot springs and effluents from laundromats. However, many thermophiles can survive very low temperatures, and viable thermophilic microorganisms are routinely found in frozen Antarctic soils. It should be made clear that the classification of an organism as a psychrophile, mesophile, or thermophile refers to the organism's optimal growth temperature and not the temperature range at which it can survive. Many *Bacillus* and *Clostridium* species, for example, are mesophiles, not thermophiles, even though their ability to produce endo-

spores permits them to survive in a dormant state at very high temperatures. The terms **psychroduric** and **thermoduric** are useful to describe those organisms that are capable of surviving at low and high temperatures, respectively, but that do not grow at these extremes.

As a rule, the maximal growth rates of thermophiles are greater than those of mesophiles, which in turn are greater than those of psychrophiles. The differences in optimal growth temperatures and temperature growth ranges among microorganisms result in a spatial separation of these different classes of organisms in nature. A microorganism can proliferate only when the environmental temperatures are restricted to the temperature growth range of that organism. The ability of a microorganism to compete for survival in a given system is increased when temperatures are near its optimal growth temperature.

The highest temperature at which microbial growth can occur is not known. The prevailing theory is that the critical determining factor is the availability of liquid water, rather than the actual temperature, and that as long as liquid water exists, temperatures up to some quite elevated limit need not preclude the existence of life. Bacteria isolated from hot areas of high pressure in the deep oceans, where water remains in liquid form despite temperatures above 300°C because of the high pressures, have been reported to grow at a temperature of 250°C, but these findings are very controversial and not widely accepted. However, bacteria definitely have been shown to grow under pressure at 110°C.

12.2 ˋReduction Potential

In addition to temperature, oxygen concentration and the reduction potential have a major influence on microbial activities. As we discussed in Chapter 5, many metabolic reac-

tions are oxidation-reduction reactions in which one compound is oxidized and another compound is reduced. The ability of an organism to carry out these oxidation-reduction reactions depends on the oxidation-reduction state of the environment. A positive reduction potential (E_h) value favors oxidation, whereas a negative E_h indicates a reducing envi-

[1]The term *cauldoactive* is used to describe extreme thermophiles that often fail to grow at temperatures below 50°C and are able to grow at temperatures above 100°C.

ronment. In a complex system, such as soil, the reduction potential is influenced by the strongest oxidant or reductant in that system, as well as by the concentration of that compound. The reduction potential is greatly influenced by the presence or absence of molecular oxygen. Environments in equilibrium with atmospheric oxygen have an E_h of around $+800$ mV, while environments with reduced oxygen tensions have reduction potentials below 800 mV. Some essential nutrient elements, such as iron and manganese, are soluble at low reduction potentials but precipitate in oxidizing

environments. Lower reduction potentials may be caused by extensive growth of heterotrophic microorganisms that scavenge all available oxygen. Such is often the case in highly polluted ecosystems, where microorganisms utilize the available oxygen for decomposition processes. The reduction potential of sediments rich in organic matter can be as low as -450 mV. At these low E_h values, obligate anaerobes can reduce sulfate to H_2S and methanogens can reduce CO_2 to CH_4.

12.3 Water Activity

Besides needing the right environmental temperature and reduction potential, all microorganisms require water for growth and reproduction. Water is an essential solvent and is needed for all biochemical reactions in living systems, and the availability of water has a marked influence on microbial growth rates (Figure 12.5). Pure distilled water has a **water activity (A_w)**[2] of 1.0. Adsorption and solution factors, however, can reduce the availability of water and thus lower the water activity. Water, for example, may be bound by a solute and hence unavailable to the microorganisms. Because solute concentrations also affect osmotic pressures, the growth of microorganisms at low water activities can also be viewed in terms of osmotic pressure (see the discussion later in this chapter). Adding high concentrations of sugars, such as su-

crose, to a solution lowers the availability of water. For example, maple syrup has a water activity of 0.9. Adding salt (NaCl) to a solution produces the same effect. A saturated solution of NaCl has an A_w of 0.8. Seawater, though, which has a salt concentration of only about 3 percent, has an A_w of 0.98.

In the atmosphere, the availability of water is expressed as **relative humidity (RH)**. RH $= 100 \times A_w$ and thus, a relative humidity of 90 percent corresponds to an A_w of 0.90. The relatively low availability of water in the atmosphere accounts for the inability of microorganisms to grow in the air. Microorganisms likewise are unable to grow on dry surfaces except when the relative humidity is high. Microbial growth on surfaces is a problem in tropical zones, where the available water in the atmosphere can support microbial growth, permitting microorganisms to grow on clothing, tents, and numerous other surfaces where this normally does not occur in temperate regions.

Water activity is an index of the water that is actually available for utilization by microorganisms. Most microorganisms require an A_w above 0.9 for active metabolism (Table 12.1). Some microorganisms, however, known as **xerotolerant** organisms, can grow at much lower water activities. Some yeasts grow on concentrated sugar solutions

[2]Water activity is an index of the amount of water that is free to react. It is equivalent to the atmospheric measure of water availability known as *relative humidity*.

FIGURE 12.5

The effect of water activity (A_w) on the growth rate of microorganisms. Normally, microorganisms grow best at high water activity levels and are severely inhibited by reduced water activity.

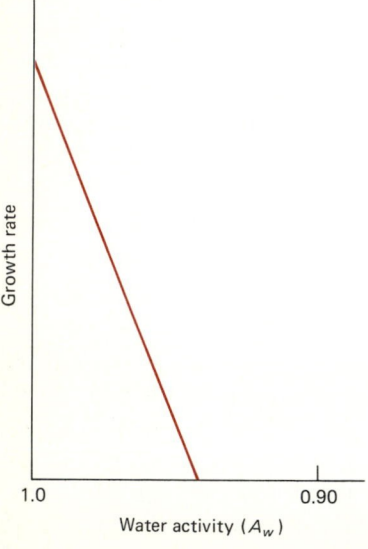

TABLE 12.1 Approximate Limiting Water Activities for Microbial Growth

Water Activity (A_w)	Bacteria	Fungi	Algae
1.00	Caulobacter Spirillum		
0.90	Lactobacillus Bacillus	Fusarium Mucor	
0.85	Staphylococcus	Debaromyces	
0.80		Penicillium	
0.75	Halobacterium Halococcus	Aspergillus Chrysosporum	Dunaliella
0.60		Saccharomyces rouxii Xeromyces bisporus	

with an A_w of 0.60. As a rule, fungi are able to grow at lower water activities than other microorganisms, such as bacteria. Fungi, therefore, grow on many surfaces where the available water will not support bacterial growth. This is why fungal, not bacterial, growth is commonly observed on the surface of bread.

The inability of microorganisms to grow at low water activities can be used for the preservation of many products. Salting was one of the early means of preserving foods and is still employed today. By adding high concentrations of salt, the water activity is lowered sufficiently to prevent the growth of most microorganisms. Many food products are also preserved by drying. This preservation method depends on maintaining the product in a dry state. Exposure to high humidity can negate the factor limiting microbial growth, promoting microbial spoilage of food products preserved in this manner.

The ability to withstand drying can have important path-ogenic implications. *Mycobacterium tuberculosis* is a classic example of an organism capable of withstanding severe desiccation and still remaining infective. This characteristic obviously has important public health implications. Whereas some microorganisms are relatively resistant to drying, others are unable to survive desiccating conditions for even a short period of time. For example, *Treponema pallidum*, the bacterium that causes syphilis, is extremely sensitive to drying and dies almost instantly in the air or on a dry surface. Many microorganisms produce specialized spores that can withstand the desiccating conditions of the atmosphere. Such spores generally have thick walls that retain moisture within the cell. Many fungal spores can be transmitted over long distances through the atmosphere; some spores even travel from one continent to another. The transmission of fungal spores through the air is a serious problem in agriculture because it permits the spread of fungal diseases of plants from one field to another.

12.4 Pressure

Osmotic Pressure

Changing the solute concentration alters not only the availability of water but also the **osmotic pressure**, that is, the force resulting from differences in solute concentration on opposite sides of a membrane. The cell-wall structures of bacteria and other microorganisms make them relatively resistant to changes in osmotic pressure, but extreme osmotic pressures can result in the death of microorganisms. In hypertonic solutions, microorganisms may shrink and become desiccated, and in hypotonic solutions the cell may burst. Organisms that can grow in solutions with high solute concentrations are called **osmotolerant** (Figure 12.6). These or-ganisms are able to withstand high osmotic pressures and also grow at low water activities. Some microorganisms are actually **osmophilic**, requiring a high solute concentration for growth. For example, the fungus *Xeromyces* is an osmophile, with an optimum A_w of approximately 0.9. Additionally, solutions with high sugar concentrations are used in laboratory procedures to protect protoplasts[3] and spheroplasts[4] against rupture due to osmotic pressure.

Salt Tolerance   Tolerance to salt can also be viewed in terms of water activity and osmotic pressure. A special term is used to describe salt tolerance. Some microorganisms, known as **halophiles**, require NaCl for growth (Figure 12.7). Moderate halophiles, which include many marine bacteria, grow best at salt concentrations of about 3 percent. Extreme halophiles exhibit maximal growth rates in saturated brine solutions. These organisms grow quite well at salt concentrations of greater than 15 percent and can grow in places like salt lakes and pickle barrels. High salt concentrations normally disrupt membrane transport systems and denature proteins, and extreme halophiles must possess physiological mechanisms for tolerating high salt concentrations. For example, the extreme halophilic bacterium *Halobacterium* possesses an unusual cytoplasmic membrane and many unusual enzymes that require a high salt concentration for activity.

[3]Protoplasts are cells with their cell walls completely removed.
[4]Spheroplasts are cells with their cell walls partially removed.

FIGURE 12.6

The growth of most microorganisms is restricted by high osmotic pressure, as occurs when sugar concentrations are high. However, osmotolerant and osmophilic organisms can grow at relatively high sugar concentrations.

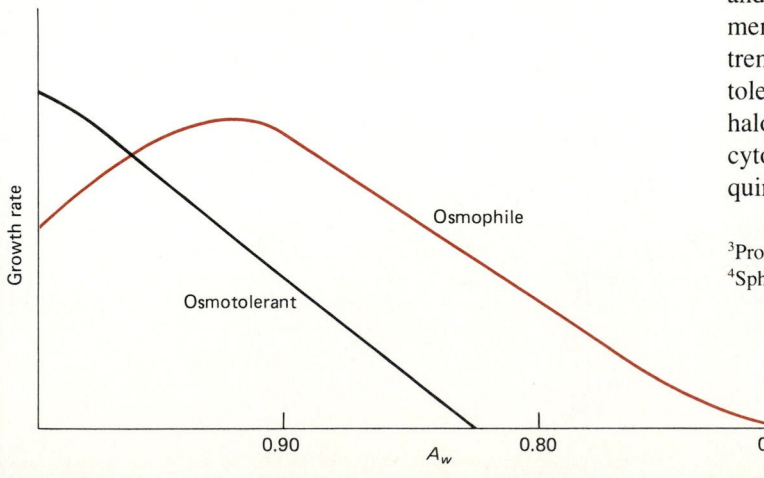

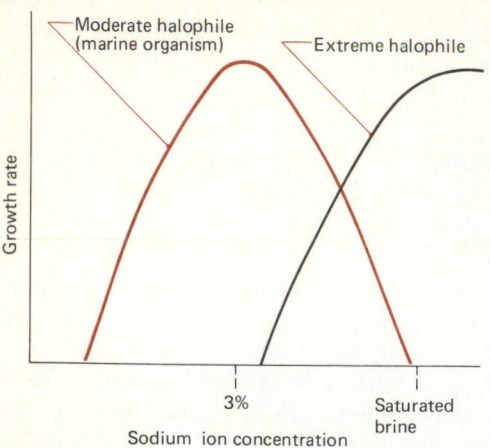

FIGURE 12.7
The effect of salt (sodium ion) concentration on the growth of microorganisms of varying salt tolerances. Halophiles require salt for growth.

Most microorganisms, however, do not possess these physiological adaptations and are not tolerant of high salt concentrations. The degree of sensitivity to salt varies among different microbial species. Many bacteria will not grow at

a salt concentration of 3 percent. Some strains of *Staphylococcus*, however, are salt tolerant and grow at salt concentrations greater than 10 percent. This physiological adaptation in *Staphylococcus* is important, as some members of this genus grow on skin surfaces where salt concentrations can be relatively high.

Hydrostatic Pressure

In addition to osmotic pressure, **hydrostatic pressure**, the pressure exerted by a water column as a result of the weight of the column, can influence microbial growth rates. Each 10 m of water depth is equivalent to approximately 1 atm. Most microorganisms are relatively tolerant of the hydrostatic pressures in most natural systems but cannot withstand the extremely high hydrostatic pressures that characterize deep ocean regions. Hydrostatic pressures of more than 200 atm generally inactivate enzymes and disrupt membrane transport processes. However, some microorganisms—referred to as **barotolerant**— can grow at high hydrostatic pressures, and there even appear to be some microorganisms—referred to as **barophiles**—that grow best at such pressures.

12.5 *Acidity and pH*

The **pH** of a solution describes the hydrogen ion concentration [H$^+$]. The pH is equal to $-\log$ [H$^+$] or $1/\log$ [H$^+$]. A neutral solution has a pH of 7.0; acidic solutions have pH values below 7; and alkaline or basic solutions have pH values greater than 7. Microbial growth rates are greatly influenced by pH values and are based largely on the nature of proteins. Because charge interactions within the amino acids of a polypeptide chain greatly influence the structure and function of proteins, enzymes are normally inactive at very high and very low pH values. The effects of pH and temperature on microbial growth and death rates are interactive (Figure 12.8). In general, microorganisms are less tolerant of higher temperatures at low pH values than they are at neutral pH values.

As noted in Chapter 5, many fermentation pathways result in the production of acids. In fermentors and batch cultures these acids can accumulate, drastically lowering the pH value and preventing continued microbial growth. In culture media and industrial fermentors, pH values must be controlled in order to achieve optimal growth rates. This is normally accomplished by buffering the solution. Buffers are used to maintain the pH value within a range, permitting continued microbial growth. **Buffers** are salts of weak acids or bases that keep the hydrogen ion concentration constant by maintaining an equilibrium with the hydrogen ions of the solution. Buffers thus dampen changes in pH. At neutral pH values, a phosphate buffer may be used; at alkaline pH values, borate buffers are often employed; and citrate buffers are frequently used for maintaining acidic conditions.

pH Tolerance Ranges

Microorganisms vary in their pH tolerance ranges (Table 12.2). Fungi generally exhibit a wider pH range, growing well over a range of 5–9, compared to most bacteria, which

FIGURE 12.8
The interactive effects of pH and temperature on microbial survival. The graph shows the influence of pH on the heat resistance of bacterial spores. The lower the pH, the lower the heat resistance of spore suspensions. (Based on N. W. Desrosier, 1970. *The Technology of Food Preservation*, AVI Publishing Co., Westport, Conn.)

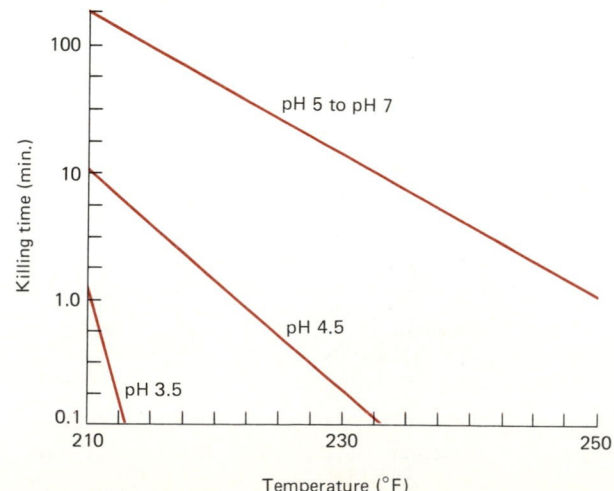

TABLE 12.2   Table of pH Tolerances of Various Bacteria

Organism	Minimum	pH Optimum	Maximum
Thiobacillus thiooxidans	1.0	2.0–2.8	4.0–6.0
Lactobacillus acidophilus	4.0–4.6	5.8–6.6	6.8
Escherichia coli	4.4	6.0–7.0	9.0
Proteus vulgaris	4.4	6.0–7.0	8.4
Enterobacter aerogenes	4.4	6.0–7.0	9.0
Clostridium sporogenes	5.0–5.8	6.0–7.6	8.5–9.0
Pseudomonas aeruginosa	5.6	6.6–7.0	8.0
Erwinia cartovora	5.6	7.1	9.3
Nitrobacter spp.	6.6	7.6–8.6	10.0
Nitrosomonas spp.	7.0–7.6	8.0–8.8	9.4

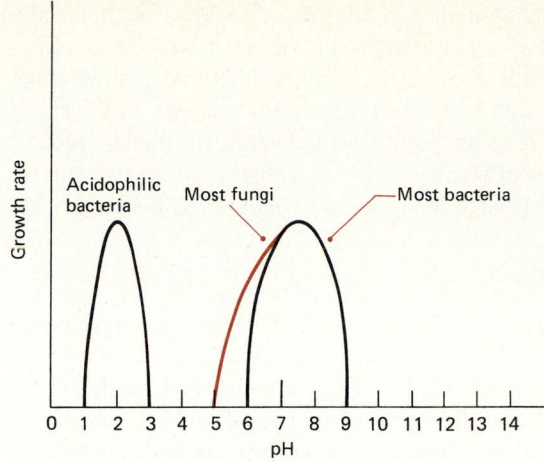

FIGURE 12.9

The pH ranges for the growth of fungi and bacteria. As a rule, fungi are more tolerant of low pH than bacteria, but some acidophilic bacteria are restricted to growth at very low pH.

grow well over a range of 6–9 (Figure 12.9). Similarly, some fungi grow well at lower pH values, as low as 0. Some other eukaryotic microorganisms, including protozoa and algae, are able to grow at low pH values; the lower limit for growth of some protozoa is approximately 2, and, for some algae, approximately 1. Although most bacteria are unable to grow at low pH values, there are exceptions. Certain bacteria tolerate pH values as low as 0.8. There are even some bacteria, called **acidophiles**, that are restricted to growth at low pH values. Some members of the genus *Thiobacillus* are acidophilic and grow only at pH values near 2. Acidophilic bacteria possess physiological adaptations that permit enzymatic and membrane transport activities at low pH. The cytoplasmic membrane of an acidophilic bacterium breaks down and cannot function at neutral pH values.

Differences in tolerance to acidic pH values can be used in designing selective growth media. A growth medium with a pH of 5.5 is favorable for the growth of most fungi but does not permit the growth of most bacteria. This factor is used in clinical isolation procedures; it is also employed in industry, where lowering the pH of a medium designed to support the growth of a fungus, such as *Saccharomyces*, can eliminate unwanted bacterial growth. Strong alkalinity can digest many cellular biochemicals. One way of killing microorganisms is to expose them to a strong base (alkaline digestion). The use of ammonia as a disinfectant in cleaning solutions is based on its ability to digest microorganisms.

12.6 *Radiation*

Another way of killing microorganisms is by exposure to certain forms of radiation. There is a continuous spectrum of electromagnetic radiation (Figure 12.10). We divide the **electromagnetic spectrum** into certain categories of radiation, including gamma rays (short wavelengths of 10^{-3} to 10^{-1} nm); x-rays (wavelengths of 10^{-3} to 10^2 nm); ultraviolet (uv) light (wavelengths of 4–400 nm); visible light (wavelengths or 400–800 nm); infrared radiation (wavelengths of 10^3 to 10^5 nm); and microwaves (wavelengths of greater than 10^6 nm). The radiant energy of the sun provides the driving force for the global ecosystem, and visible light radiation is required for photosynthesis. High-energy, short-wavelength radiation disrupts DNA molecules, and exposure to short-wavelength radiations may cause mutations, many of which are lethal. Exposure to gamma rays, x-rays, and uv radiation increases the death rate of microorganisms and is used in various sterilization procedures to kill microorganisms.

Ionizing Radiation

Gamma rays and x-rays have high penetrating power and are able to kill microorganisms within a sample by inducing or forming toxic free radicals. Free radicals are highly reactive chemical species that can lead to polymerization and other chemical reactions that disrupt the biochemical organization of microorganisms. Viruses as well as other microorganisms are inactivated by exposure to these **ionizing radiations** (Figure 12.11). Microorganisms, however, are generally more tolerant of ionizing radiation than are higher organisms.

Sensitivities to ionizing radiation vary (Table 12.3). Resistance to ionizing radiation is based on the biochemical constituents of a given microorganism. Nonreproducing (dormant) stages of microorganisms tend to be more resistant to radiation than growing organisms. For example, endospores are more resistant than the vegetative cells of many bacterial species. Exposure to 0.3–0.4 Mrad is necessary to cause a

10-fold reduction in the number of viable bacterial endospores. An exception is the bacterium *Micrococcus radiodurans*, which is particularly resistant to ionizing radiation. Vegetative cells of *M. radiodurans* tolerate as much as 1 Mrad of exposure with no reduction in viable count. It appears that efficient repair mechanisms are responsible for the high resistance to radiation exhibited by this bacterium.

Ultraviolet Light

Ultraviolet (uv) light does not have high penetrating power and is useful for killing microorganisms only on or near the surface of clear solutions. The strongest germicidal wavelength of 260 nm coincides with the absorption maxima of DNA, suggesting that a principal mechanism by which uv light exerts its lethal effect is through mutations, such as those resulting from the formation of thymine dimers in the DNA. Microorganisms have enzymes that can repair the alterations in the DNA caused by exposure to uv light, including the photoreactivation mechanism that excises thymine dimers. In addition to initiating the formation of thymine dimers, exposure to uv light interferes with other microbial biochemicals and functions. Exposure to 340-nm uv light has a powerful killing effect on microorganisms, although DNA does not strongly absorb light of this wavelength.

FIGURE 12.10

The spectrum of electromagnetic radiation.

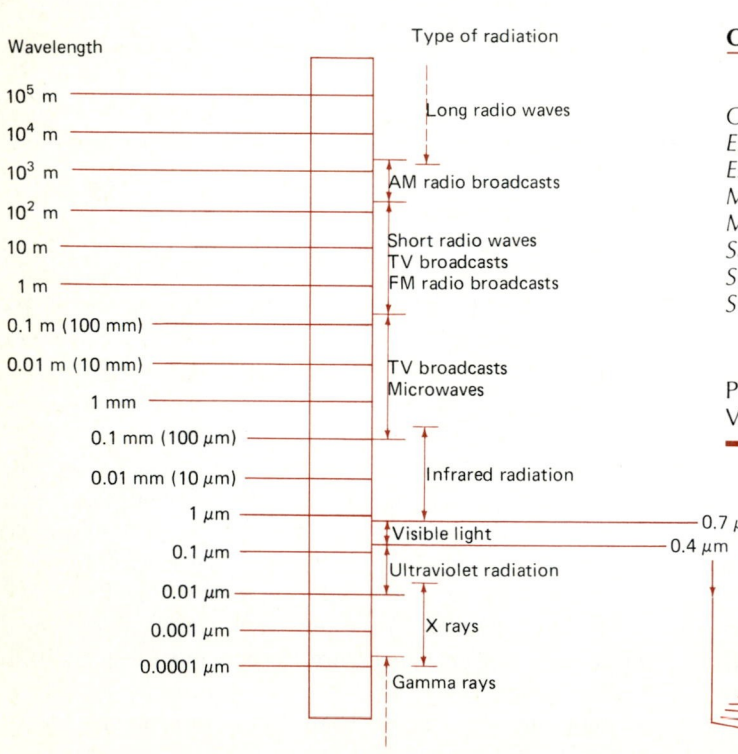

The Electromagnetic Spectrum

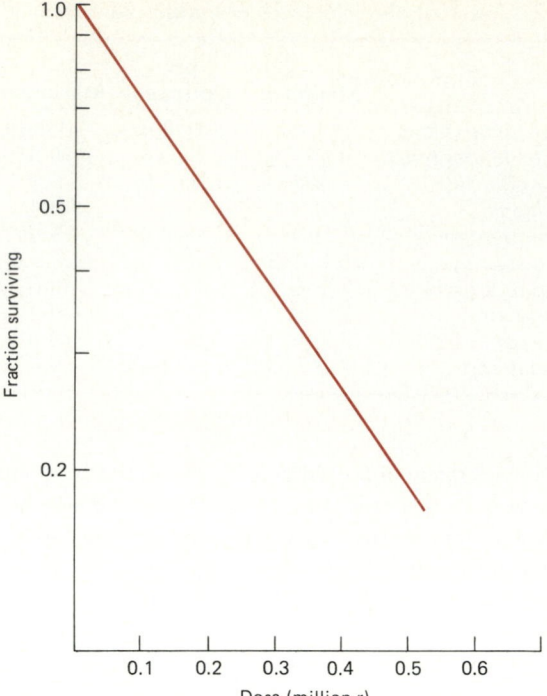

FIGURE 12.11

Viral inactivation by exposure to ionizing radiation. The graph summarizes the results of three experiments exposing Tobacco Mosaic virus to x-rays. (Based on W. Ginoza and A. Norman, 1957, *Nature*, 179:520–521)

TABLE 12.3   Radiation Tolerances for Various Microorganisms

Organism	Dose (Mrad)
Bacteria	
Clostridium botulinum (type E)	1.5
Enterobacter aerogenes	0.16
Escherichia coli	0.18
Micrococcus radiodurans	6.0
Mycobacterium tuberculosis	0.14
Salmonella typhimurium	0.33
Staphylococcus aureus	0.35
Streptococcus faecalis	0.38
Viruses	
Polio virus	3.8
Vaccinia virus	2.5

FIGURE 12.12

The graph illustrates the effect of sunlight on the viability of colorless and yellow-pigmented *Micrococcus lutea*. Note that death occurs in the white strain but not in the pigmented strain. (Based on M. M. Matthews and W. R. Sistrom, 1959, *Nature* 184:1892)

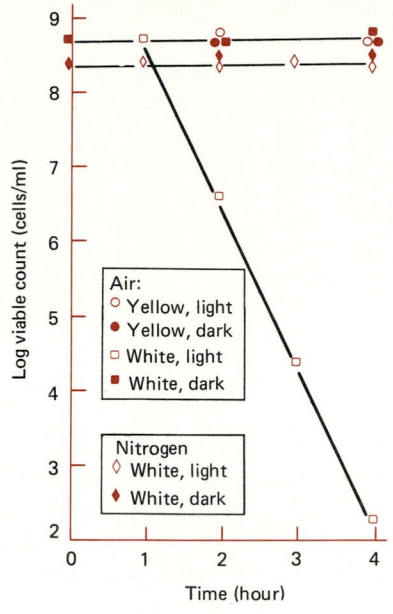

Discovery Process

Reasoning that carotenes might protect bacterial cells against the lethal effects of exposure to light, Matthews and Sistrom exposed a colorless mutant of *M. lutea* to light in the air. Assuming that the presence of carotenes protects the bacteria against the lethal action of light, the caroteneless mutants should be killed, whereas the yellow-pigmented wild type should survive exposure to light. This was indeed their finding. After an exposure to direct sunlight of 2 hours, more than 99 percent of the colorless mutants were killed compared to less than 1 percent of the yellow wild-type cells. From this they inferred that the yellow pigment confers protection from sunlight. Further experiments showed that killing does not occur anaerobically, but only in the presence of air. It is therefore properly deduced that light-induced killing is a photooxidation reaction requiring oxygen.

Visible Light

Exposure to **visible light** can also cause the death of microorganisms. Visible light may be absorbed by various biochemicals of microbial cells, resulting in changes in the energetic states of those biochemicals. In some cases, this process can lead to interference with metabolic activities or damage to cellular structures. Additionally, exposure to visible light can lead to the formation of singlet oxygen, which is chemically reactive and can result in the death of a microorganism. Some microorganisms produce **pigments** (colored compounds) that protect them against the lethal effects of exposure to light (Figure 12.12). For example, yellow, orange, or red carotenoid pigments interfere with the formation and action of singlet oxygen, preventing its lethal effect. Microorganisms possessing carotenoid pigments tolerate much higher levels of exposure to sunlight than nonpigmented microorganisms. Pigmented microorganisms are often found on surfaces that are exposed to direct sunlight, such as on leaves of trees. Many viable microorganisms found in the air produce colored pigments. Some microbial spores that are used principally for aerial dispersal are also pigmented.

Although exposure to high intensities of light can be lethal to some microorganisms, photosynthetic microorganisms require light in the visible spectrum to carry out oxidative photophosphorylation. The rate of photosynthesis is a function of light intensity. At some light intensities, the rate of photosynthesis reaches a maximum, and although light intensities above the optimal level do not result in further increases in the rate of photosynthesis, light intensities below this level result in lower rates. One interesting effect of lowered light intensity levels is the changeover from oxygen-evolving to anoxygenic photosynthesis exhibited by some cyanobacteria.

The wavelength of light also has a marked effect on the rate of photosynthesis. Different photosynthetic microorganisms are capable of using light of different wavelengths. For example, anaerobic photosynthetic bacteria use light of longer wavelengths than eukaryotic algae are capable of using. Many photosynthetic microorganisms have accessory pigments that enable them to use light of wavelengths other than the absorption wavelength for the primary photosynthetic pigments. The distribution of photosynthetic microorganisms in nature reflects the variations in the ability to use light of different wavelengths and the differential penetration of different colors of light into aquatic habitats.

Infrared and Microwave Radiations

Longer-wavelength infrared and microwave radiations have poor penetrating power and do not appear to kill microorganisms directly. Absorption of such long-wavelength radiation, however, results in increased temperature. Exposure to infrared or microwave radiation can thus indirectly kill microorganisms by exposing them to temperatures that are higher than their maximal growth temperatures. Exposure to microwave radiation produces the same effect. Because microwaves generally do not kill microorganisms directly, there is some concern in the food industry that cooking with microwave ovens may not adequately kill microorganisms contaminating food products.

12.7 Magnetism

Most bacteria are not affected by magnetic fields, but some bacteria exhibit **magnetotaxis**, that is, motility directed by a geomagnetic field. Magnetotactic bacteria contain dense inclusion bodies, called **magnetosomes**, that allow them to orient themselves in a magnetic field. Because the magnetotactic microorganisms that have been described are microaerophilic or anaerobic, the magnetotactic orientation may help these motile organisms to locate deeper, more reduced organic sediments in the aquatic environments in which they live.

12.8 Nutrients

Specific organic substrates often favor the growth of particular populations with specific catabolic activities. Some microorganisms require relatively high levels of organic matter, whereas others grow only at relatively low concentrations. The total amount of organic matter in a sample, regardless of its specific biochemical composition, can be described by the **chemical oxygen demand (COD)**, that is, the amount of oxygen needed to oxidize the organic matter completely to carbon dioxide. The COD is determined by measuring the amount of oxidizing reagent consumed during the oxidation of organic matter with dichromate or permanganate. The **biological oxygen demand (BOD)** yields an estimate of the readily usable concentrations of organic matter. The BOD is determined by allowing microorganisms to consume oxygen in solution during oxidation of the organic matter present in the sample. A high BOD indicates the abundance of organic substrates that can be used by microorganisms aerobically. When the BOD is high, however, microorganisms often use much of the available oxygen for the degradation of organic matter, creating a condition of oxygen depletion and resulting in the death of higher organisms, such as fish, that require O_2 for survival.

Most natural ecosystems are not characterized by high concentrations of usable organic matter; rather, microorganisms in many ecosystems, such as the oceans, live in environments with very low concentrations of nutrients. Periods of starvation are probably experienced by most free-living bacteria. Starvation survival is important for most bacteria in nature. **Oligotrophic** or **low-nutrient bacteria** possess physiological properties that permit them to use efficiently the limited nutrient resources available to them. Substrate uptake characteristics of oligotrophs permit acquisition of growth substrates against steep concentration gradients between the cell and its surrounding.

Many inorganic compounds are essential nutrients for microorganisms, but many of them are also toxic. It is generally necessary to know the chemical form of inorganic compounds when considering them as microbial nutrients or inhibitors. Inorganic compounds of importance in examining the ecology of microorganisms include the gases oxygen, carbon dioxide, carbon monoxide, hydrogen, nitrogen, nitrous oxide, and hydrogen sulfide; the elemental form of sulfur; the cations ammonium, ferrous iron, ferric iron, magnesium, calcium, silicon, sodium, potassium, boron, cobalt, copper, manganese, molybdenum, vanadium, nickel, zinc, mercury, cadmium, and lead; and the anions phosphate, carbonate, bicarbonate, sulfide, sulfate, nitrite, nitrate, chloride, chlorate, bromide, fluoride, silicate, selenite, and arsenate.

12.9 Microorganisms in Their Natural Habitats

Having considered the effects of various environmental parameters on microbial growth and survival, we can now examine the distribution of microorganisms in the environment. Microorganisms exist in numerous locations, known as **habitats**, characterized by vastly different environmental conditions. Microbial populations and activities vary greatly among habitats having vastly differing **abiotic parameters**, that is, nonliving environmental factors. The physical and chemical characteristics of a habitat influence the growth, activities, interactions, and survival of the microbial populations found there. Some microorganisms possess unique features that make them fitter for survival in a particular ecosystem. The diversity of populations in a given habitat contributes to the ability to change and maintain stability within the community of organisms in that habitat.

Some microorganisms are **autochthonous** (indigenous) to a particular habitat. These microorganisms are capable of survival, growth, and metabolic activity within that habitat.

Autochthonous microorganisms occupy the functional positions or **niches** of that ecosystem. In contrast to the indigenous members of the microbial community, some microorganisms in a given habitat may be foreign, or allochthonous. **Allochthonous** (nonindigenous) microorganisms typically have grown elsewhere and have been transported into a foreign ecosystem. Such foreign microorganisms do not occupy the niches of that ecosystem and are transient members of the microbial community. Generally, autochthonous microorganisms must exhibit adaptive features that make them physiologically compatible with the physical and chemical environments of the habitat. They must be functionally and competitively compatible with the other organisms living in that habitat. Thus, microorganisms occupying extreme environments, such as hot springs, desert soils, and ocean trenches, must possess physiological adaptations that permit them to survive and function under conditions that normally preclude biological activity.

Atmosphere

The **atmosphere** as a habitat is characterized by high light intensities, extreme temperature variations, low concentrations of organic matter, and a scarcity of available water, making it a hostile environment for microorganisms and a generally unsuitable habitat for microbial growth. Nevertheless, substantial numbers of microorganisms are found in the lower regions of the atmosphere (Table 12.4). These microorganisms do not grow within the atmosphere; rather, they represent allochthonous populations transported into the atmosphere from aquatic and terrestrial habitats. Transport through the atmosphere is important in the dispersal of many microorganisms, ensuring their continued survival by permitting **propagules**, the reproductive units of microbial populations, to reach more favorable habitats (Figure 12.13). Many plant pathogens are transported through the air from one field to the next, and the spread of various fungal diseases of agricultural crops can be predicted by measuring the concentrations of airborne fungal propagules. The spread of pathogenic bacteria in droplets is important in the transport of these microorganisms from an infected to a susceptible host.

Although some microorganisms become airborne as growing vegetative cells, more commonly microorganisms enter the atmosphere as spores. Metabolically, dormant spores are better adapted to survival in the atmosphere than actively growing vegetative cells. Some spores have pigments that protect them against exposure to damaging uv radiation in the atmosphere. Other spores have very thick walls that protect them against desiccation during aerial transport. As a rule, spores tend to be light and have aerodynamic shapes favorable for extended lateral travel through the atmosphere. Many microorganisms produce large numbers of spores and have developed adaptations for discharging their spores into the atmosphere. In some cases, spores are forcibly ejected into the atmosphere to facilitate aerial dissemination. Aero-

TABLE 12.4 Microorganisms Found in the Atmosphere*

Type of Organism	Percentage
Bacteria	
Gram-positive pleomorphic rods (e.g., *Corynebacterium*)	20
Gram-negative rods (e.g., *Achromobacter*, *Flavobacterium*)	5
Endospore formers (e.g., *Bacillus*)	35
Gram-positive cocci (e.g., *Micrococcus*)	40
Fungi	
Cladosporium	80
Alternaria	5
Penicillium	2
Other (e.g., *Aspergillus, Chaetomium, Dematium Fumago, Fusarium, Helminthosporium, Sclerotinia, Stachybotrys, Trichoderma Verticillium*)	13

*In tropospheres over North America.

sols are also important in the aerial transfer of various microbial populations that do not possess special adaptations for prolonged survival in the atmosphere.

Hydrosphere (Aquatic Habitats)

Freshwater Habitats **Freshwater habitats** include lakes, ponds, swamps, springs, streams, and rivers. There are a number of important chemical and physical parameters that make freshwater ecosystems more or less suitable habitats for microorganisms. Temperature, pH, nutrient concentrations, and oxygen levels are important parameters influencing the existence of particular microbial populations in freshwater habitats. **Oligotrophic lakes** are defined as those having low concentrations of nutrients and typically low rates of **primary productivity**, the photosynthetic production of organic matter. Autochthonous heterotrophic microorganisms in such low-nutrient habitats should be able to grow with low nutrient concentrations, and many freshwater bacteria exhibit unusual shapes that increase their surface area-to-volume ratio to permit more efficient uptake of nutrients. Stalked bacteria, such as *Caulobacter* and *Hyphomicrobium* species, are frequently found in lake water (Figure 12.14). In many situations, bacteria become attached to surfaces or form aggregates that permit the concentration of the available nutrients to levels that will support growth. In contrast to oligotrophic lakes, **eutrophic lakes** have high nutrient concentrations. These lakes generally have high rates of primary productivity, and their oxygen concentrations are generally low as a result of extensive microbial decomposition of organic nutrients.

Within aquatic habitats, there is a zonation in the distribution of microbial populations based on the physicochemical parameters of the habitat (Figure 12.15). Lakes, for example, are divided into a shallow **littoral zone** where light

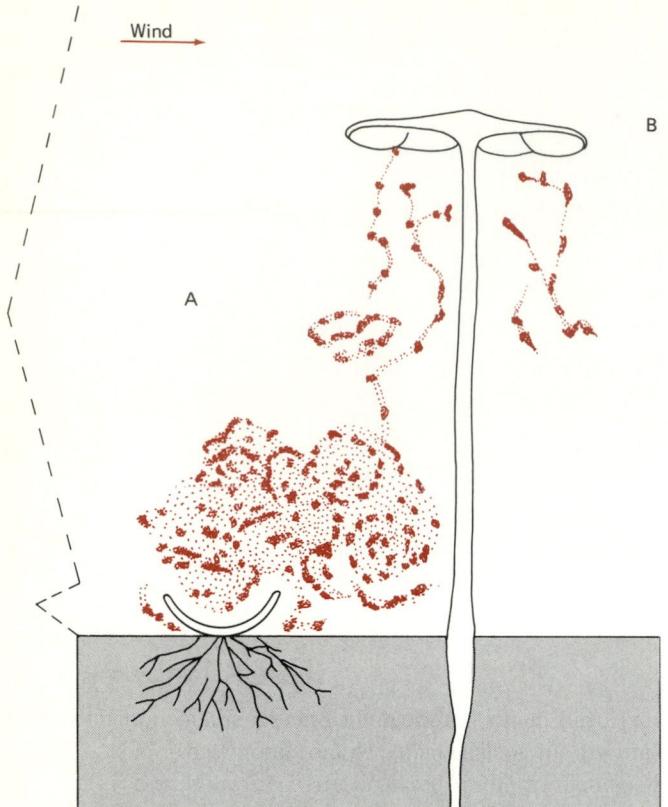

FIGURE 12.13 (*left*)
Many fungi depend on spore dispersal through the atmosphere to ensure survival of the species. This drawing shows the discharge of ascospores into the air from a cup fungus and an agaric. Laminar air flow above the ground (horizontal dashes) and turbulent air flow above this level (dashes in spirals) are illustrated. (A) The spores of the cup fungus have just been discharged as a puff into the turbulent air zone. (B) The agaric spores are steadily dropping into the turbulent air. (After C. T. Ingold, 1971, *Fungal Spores: Their Liberation and Dispersal*, Oxford University Press, New York.)

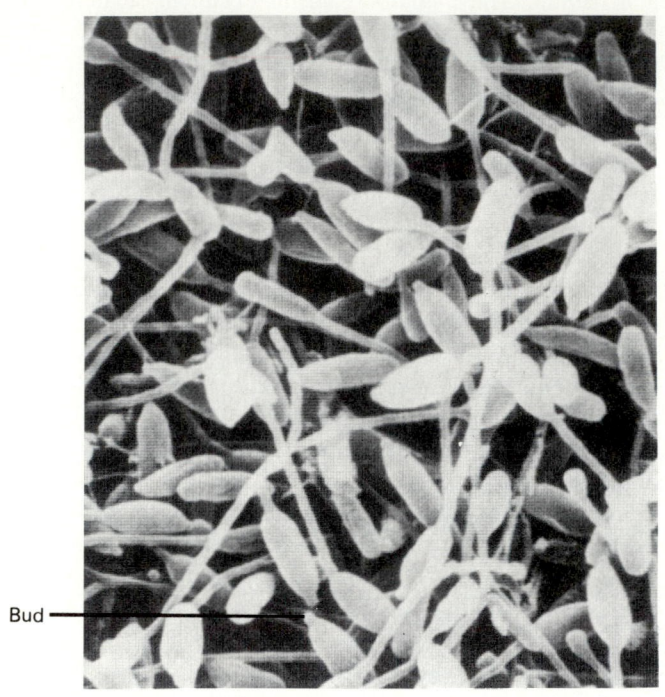

Bud

FIGURE 12.14
Scanning electron micrograph of *Hyphomicrobium*, showing accumulation of a large number of budding cells. The large surface area of a *Hyphomicrobium* cell permits this bacterium to obtain its required nutrients from dilute aqueous solutions and thus to grow in aquatic habitats with very low nutrient concentrations. (Courtesy Richard L. Moore, University of Calgary)

penetrates to the bottom (**benthos**), a **limnetic zone** where light penetration is sufficient so that primary productivity exceeds heterotrophic metabolism, and a **profundal zone** where consumption exceeds primary productivity (Figure 12.16). The **compensation depth**, separating the limnetic and profundal zones, is the depth of effective light penetration where photosynthetic activity balances respiratory activity. Cyanobacteria and algae are important populations in the community of the limnetic and littoral zones but decline in importance with depth in the profundal zone. The population levels of heterotrophic consumers depend on the concentrations of available nutrients. Normally, much higher populations of bacteria are found in sediment than in the water column because nutrients settle to the bottom of a water body and accumulate in the benthic sediment.

Temperature has a major influence on the distribution of microbial populations in freshwater habitats. Temperature influences the density of water, causing its cycling of water within a lake. Water is most dense at 4°C. Many temperate lakes exhibit **thermal stratification**, which undergoes seasonal changes (Figure 12.17). During the summer the upper layer, the **epilimnion**, is warm and oxygen-rich, while the lower layer, the **hypolimnion**, is characterized by low temperatures and low oxygen concentrations. The epilimnion is separated from the hypolimnion by the **thermocline**, a zone of rapid temperature change. Poor light penetration to the hypolimnion restricts oxygen-generating photosynthesis, and respiration depletes the existing oxygen. Mineral nutrients

are generally depleted in the epilimnion as a result of the extensive growth of photosynthetic and heterotrophic microbial populations.

The principal ecological functions of microorganisms in freshwater habitats include (1) the input of organic matter from primary production; (2) the decomposition of dead organic matter, liberating mineral nutrients for primary production; (3) the assimilation and reintroduction into the food web of dissolved organic compounds; and (4) the biogeochemical cycling of elements, which is the system of chem-

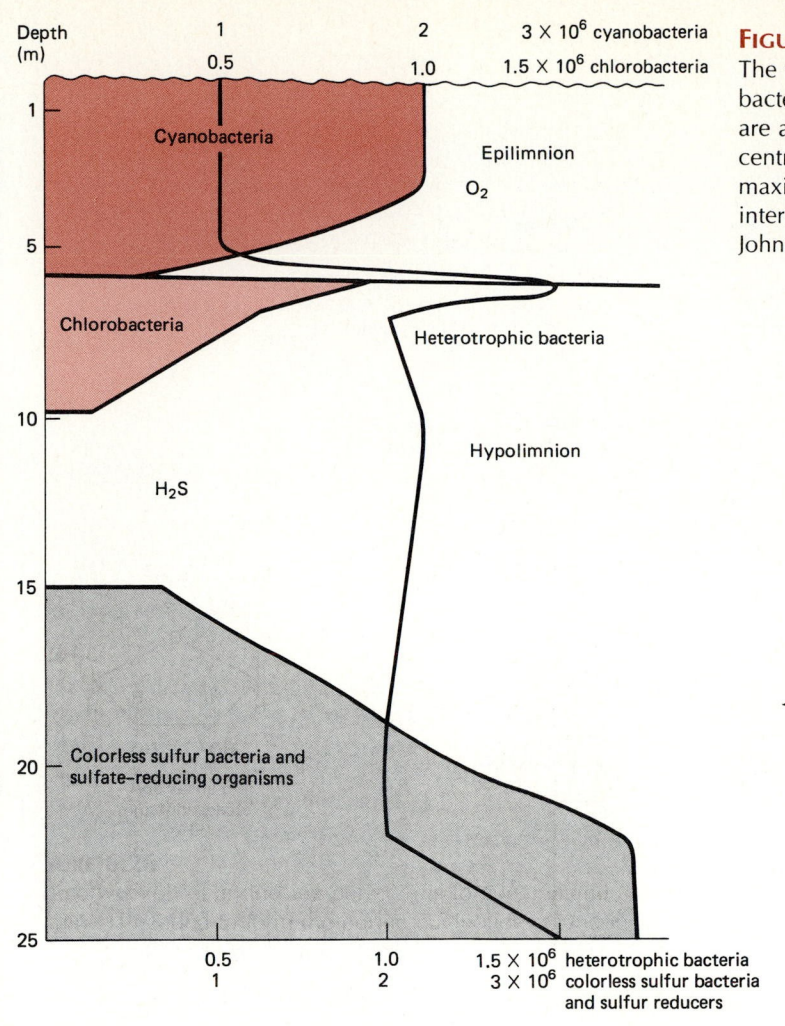

FIGURE 12.15 (*left*)
The vertical distribution of microorganisms in a lake. Cyanobacteria are abundant in the epilimnion; sulfate reducers are abundant in the lower hypolimnion; and maximal concentrations of heterotrophs occur just below the zone of maximal photosynthetic production and at the water–sediment interface. (After Rheinheimer, 1980, *Aquatic Microbiology*, John Wiley & Sons, Inc., New York.)

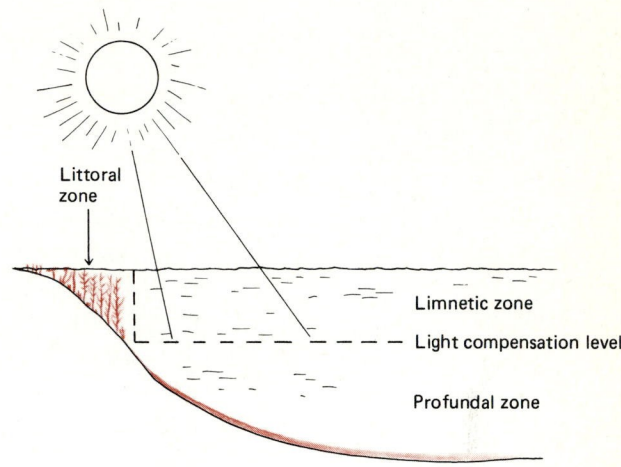

FIGURE 12.16 (*above*)
The divisions of a lake into zones in relation to light penetration.

ical transformations of elements by living systems (Figure 12.18). Autotrophic bacteria are autochthonous members of the microbial populations of lakes and generally play a very important role in the nutrient cycling of the lakes. The metabolic activities of chemolithotrophic bacteria are important in the cycling of nitrogen, sulfur, and iron within lakes. Members of the genera *Nitrosomonas*, *Nitrobacter*, and *Thiobacillus* are especially important in the cycling of nitrogen and sulfur within freshwater bodies.

Microorganisms in the sediments of freshwater habitats are generally different from those in overlying waters. In shallow ponds, anaerobic photosynthetic bacteria occur at the surface of the sediment, often conferring characteristic colors, such as purple, on such water bodies. Fungi and heterotrophic bacteria grow well on the debris that accumulates at the sediment–water interface. Within the sediment, anaerobic microorganisms become increasingly important. Denitrifying *Pseudomonas* species, hydrogen sulfide-producing *Desulfovibrio* species, and methanogens occur in sediments and are very important in the biogeochemical cycling of carbon, nitrogen, and sulfur within freshwater habitats. The metabolic activities of these bacterial populations produce gaseous compounds that can bubble up through the water column and transfer from the hydrosphere to the atmosphere.

Bacteria are able to use very low concentrations of organic compounds. This is very important in oligotrophic lakes because the metabolic activities of bacteria permit the concentration of dissolved nutrients within bacterial biomass. This ability of bacteria to concentrate organic matter permits the introduction of organic compounds into the **food web**, the integrated feeding relationships of the lake, that otherwise would be lost to the biological community. The growth of bacteria at such very low concentrations of nutrients is a form of **secondary productivity**, which is the heterotrophic recapture of dilute nutrients.

Some freshwater microorganisms have developed interesting adaptive features that permit them to function well in freshwater ecosystems. Species of the dinoflagellate *Ceratium*, for example, exhibit seasonal changes in cell shape. During summer when water is warm, the algae form elongated appendages that increase their buoyancy, and during winter when the water is cold and denser, they conserve energy by forming much shorter appendages. Some cyanobacteria possess gas vacuoles that allow them to adjust their

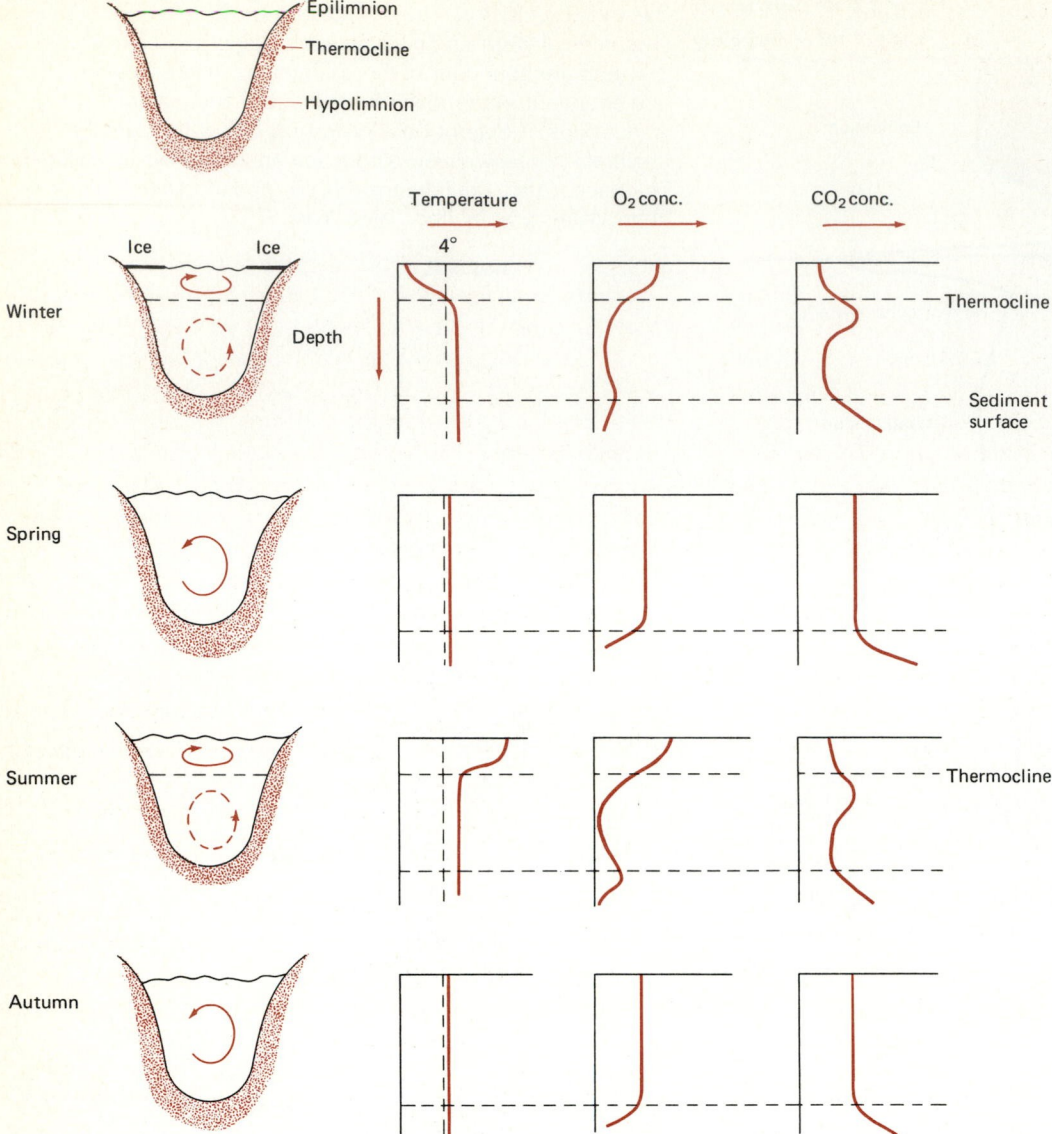

FIGURE 12.17

The divisions of a lake in relation to temperature and oxygen and carbon dioxide concentrations with depth at different times of the year under different systems of stratification. Winter stratification does not occur in some lakes or in most seas, in which case there is a continuous period of mixing from the autumn turnover to the onset of summer stratification.

buoyancy. By adjusting their height in the water column, depending on the intensity of light, these bacteria can maximize their use of light energy for photosynthesis (Figure 12.19).

Marine Habitats The **oceans** cover approximately 71 percent of the earth's surface, providing a sizable habitat for microbial populations. The oceans contain almost every naturally occurring chemical element, but most elements are present only in extremely low concentrations. Environmental conditions in the open ocean are relatively constant; normally, the salinity is in the range of 33–37 parts per thousand, the pH is near 8.3, and temperatures below 100 m in

depth are between -2 and 5°C. As with freshwater habitats, there is a surface **euphotic zone** where light penetrates to the compensation depth (Figure 12.20). Algae and cyanobacteria are important primary producers within this euphotic zone, especially in warm tropical waters. The **pelagic zone** or the open ocean extends from the surface to a depth of 6,000 m. Within the pelagic zone, recycling of mineral nutrients is extremely slow, and dead organic matter slowly sinks through the pelagic zone to the benthos. In the very deep ocean trenches, the rate of organic decomposition is very slow. In fact, lunches that accidentally sank with a deep-diving submarine remained intact until the submarine was recovered 10 months later.

FIGURE 12.18
Simplified diagram of nutrient cycling within a lake habitat.

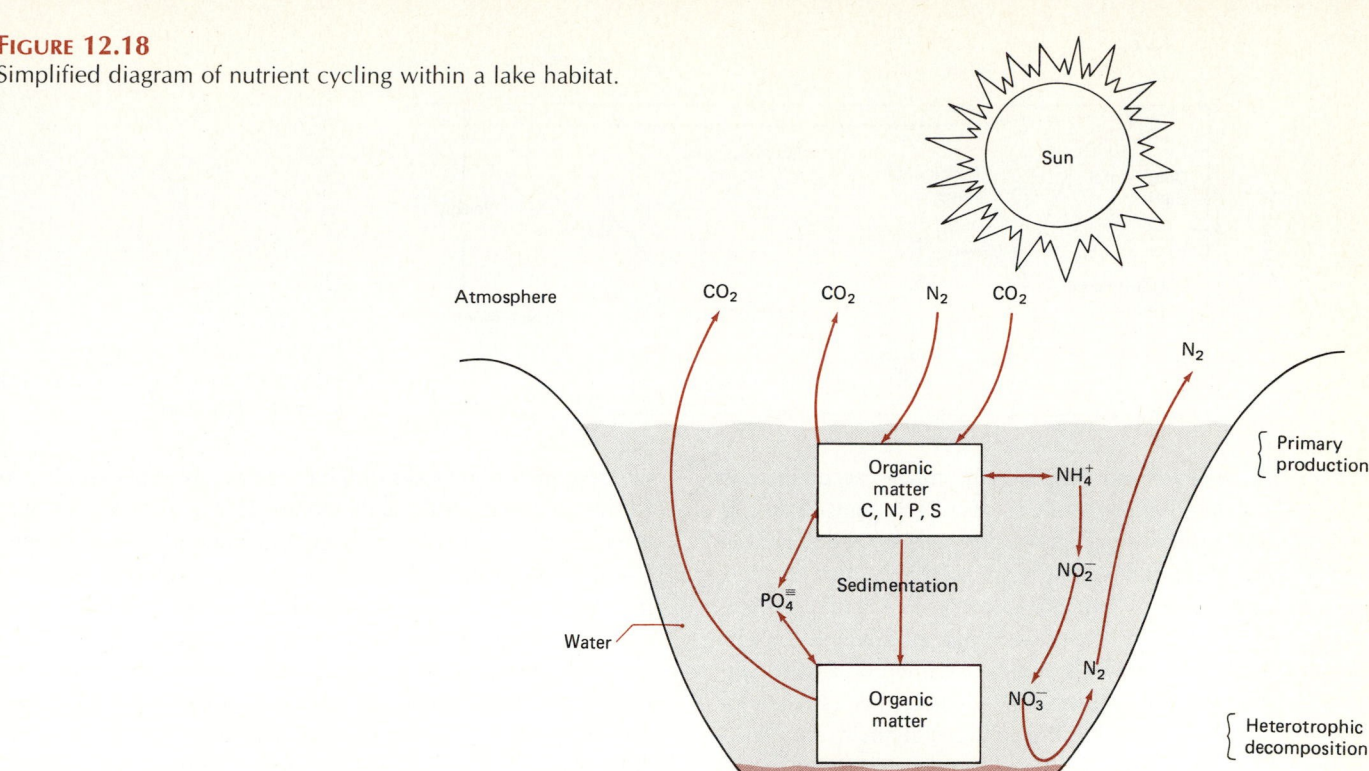

FIGURE 12.19
Gas vacuoles allow cyanobacteria to adjust their height in a water column so that they can locate themselves at a level of light intensity that will permit optimal photosynthetic activity. (A) Phase contrast micrograph of *Anaebaena* containing gas vacuoles. (B) Electron micrograph of a freeze-etched preparation of *Nostoc muscorum*, showing the cylindrical structure of the gas vacuoles. (From BPS—J. Robert Waaland, University of Washington)

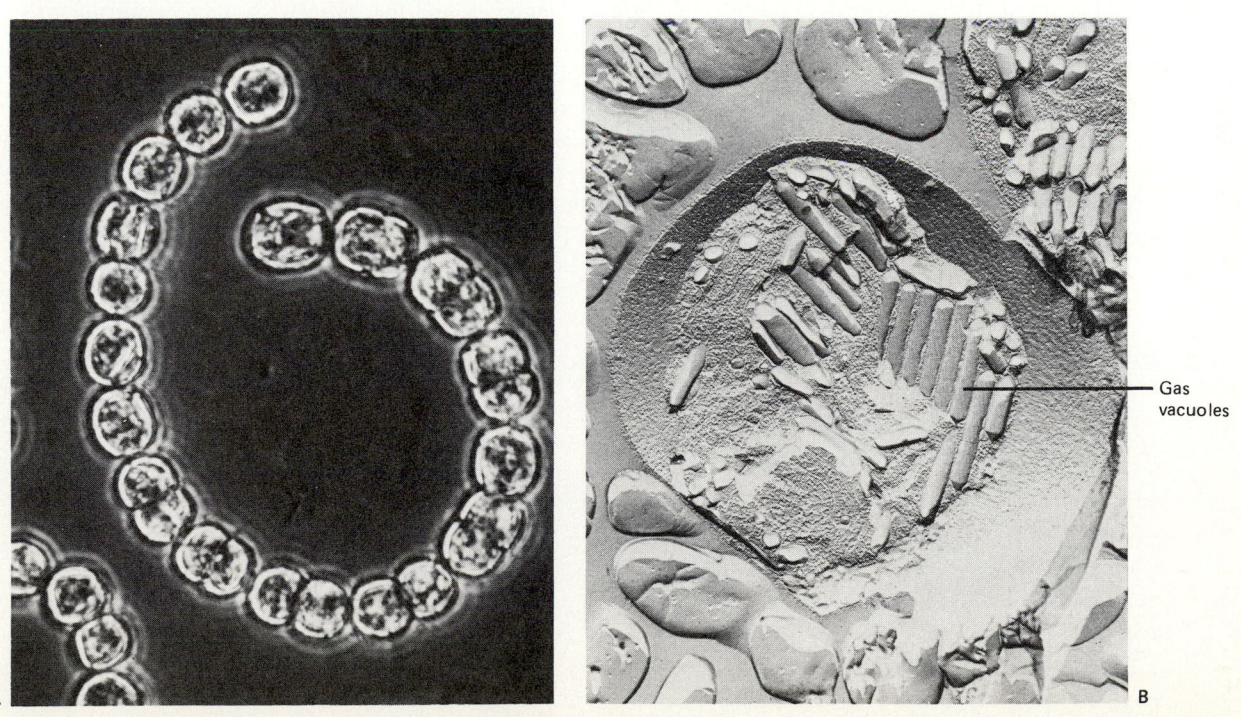

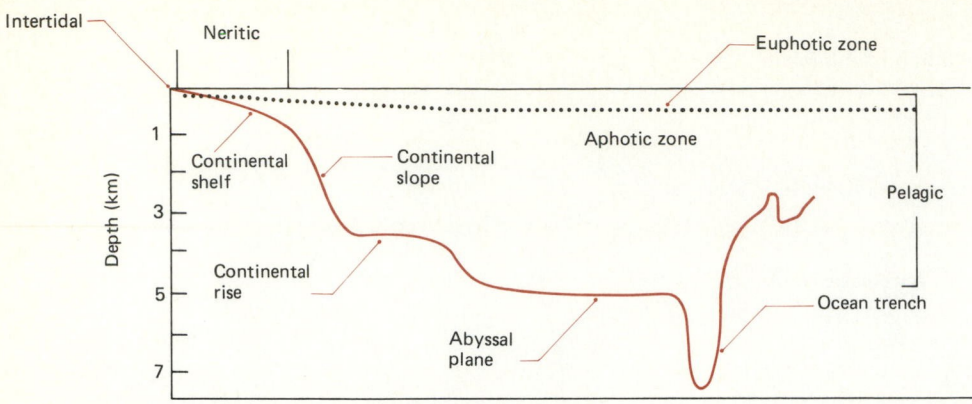

FIGURE 12.20
The divisions of the ocean, showing horizontal zonation.

Mineral nutrients and dead organic matter accumulate at the benthos. The highest biomass of microorganisms in marine habitats normally occurs near the surface and decreases with depth. Microbial numbers are relatively high in near-shore **estuarine waters** where freshwater and marine habitats interface, but are very low, 1–100 per milliliter, in pelagic waters. Within estuaries, the number of microorganisms varies with the tides. Upwelling zones, where this nutrient-rich water moves upward along the continental shelf, are particularly productive (Figure 12.21). Typical productivity values for open ocean, coastal, and upwelling areas of the ocean are 50, 100, and 300 g of organic carbon per square meter per year, respectively.

Primary production in pelagic marine habitats is extremely limited by a lack of mineral nutrients. Because of the expanse and depth of the oceans, planktonic microorganisms have an almost exclusive role in primary production, providing the main source of organic compounds for heterotrophic marine organisms. Higher plants and benthic macroalgae contribute significantly to primary production only in estuaries and littoral (nearshore) areas. Many algal species are exclusively marine, where the input of carbon to the marine ecosystem by algae is essential.

Autochthonous marine microorganisms should be capable of growth at salinities of 20–40 mg/L, and true marine bacteria generally have an optimal salt concentration of 33–35 mg/L. Marine bacteria require the ions in seawater to maintain proper membrane functions. Some marine bacteria have multiple membranes surrounding their cells, and exposure to fresh water disrupts these membrane layers. True marine bacteria must be capable of growth at nutrient concentrations found in the oceans. Because most of the ocean has a temperature below 5°C, psychrophilic and psychrotrophic bacterial populations are quite common. In the deep ocean trenches, barophilic and barotolerant bacteria are important indigenous members of the microbial community. Most marine bacteria are Gram negative and motile. *Pseudomonas* and *Vibrio* species often predominate in marine waters. These marine bacteria are generally nutritionally versatile, can metabolize a wide variety of organic substrates, and are capable of using almost any carbon source as a potential substrate.

Lithosphere (Soil Habitats)

Soil constitutes the major habitat of **terrestrial** microorganisms. Soil is a favorable habitat for the proliferation of microorganisms, and the number of microorganisms in soil habitats, typically 10^6–10^9/g, is usually much higher than that in freshwater or marine habitats. Microorganisms are not evenly distributed throughout the soil column. Higher numbers occur in the organically rich surface layers than in the underlying mineral soils (Figure 12.22). Particularly high numbers of microorganisms occur in association with plant roots (Table 12.5).

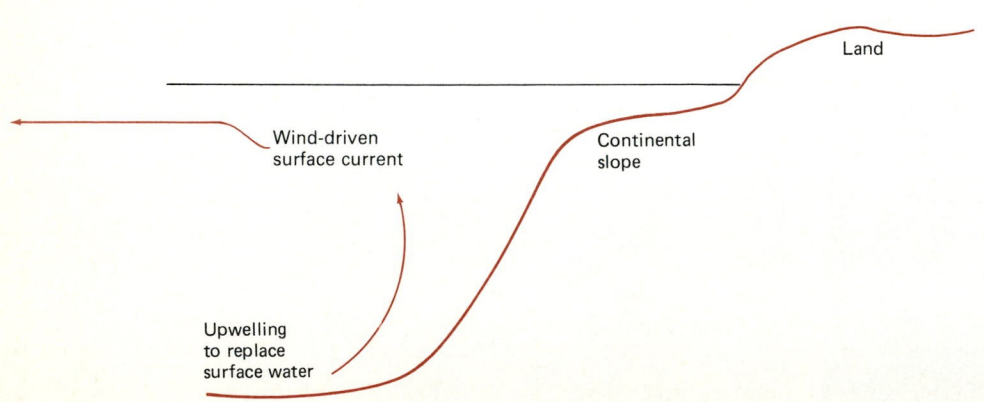

FIGURE 12.21
Upwelling of deep ocean water occurs along the continental slope and replaces surface waters driven offshore by the wind.

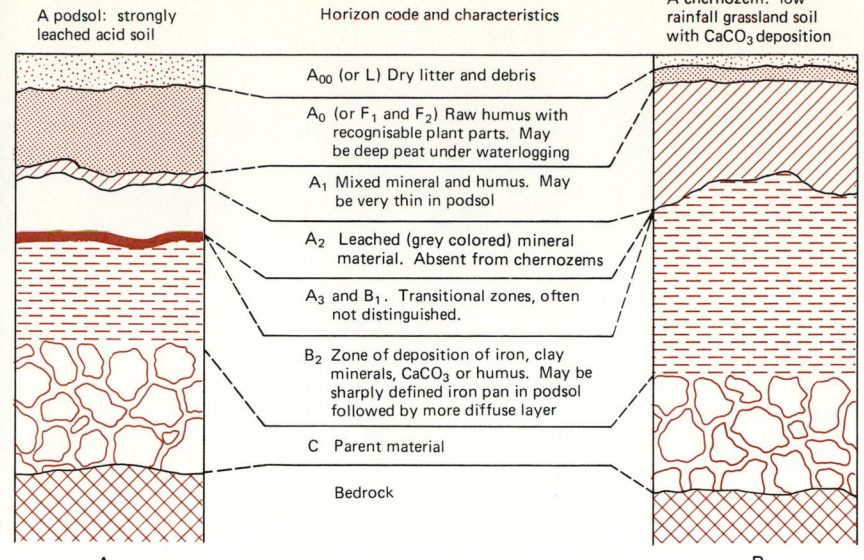

A podsol: strongly leached acid soil

Horizon code and characteristics

A chernozem: low rainfall grassland soil with CaCO₃ deposition

A_{00} (or L) Dry litter and debris

A_0 (or F_1 and F_2) Raw humus with recognisable plant parts. May be deep peat under waterlogging

A_1 Mixed mineral and humus. May be very thin in podsol

A_2 Leached (grey colored) mineral material. Absent from chernozems

A_3 and B_1. Transitional zones, often not distinguished.

B_2 Zone of deposition of iron, clay minerals, CaCO₃ or humus. May be sharply defined iron pan in podsol followed by more diffuse layer

C Parent material

Bedrock

A

B

FIGURE 12.22

Soil profiles of two soil types. (A) A podsol, which forms under conditions of acid leaching, and (B) a chernozem, which is characteristic of much drier conditions, such as grasslands.

There are major chemical and physical differences between soils in different climatic zones. Soils contain varying proportions of clay, silt, and sand particles (Figure 12.23). The soil texture is important because it determines the surface area available for microbial growth. The proportion of various clay minerals also affects the binding properties and chemistry of the soil. The physicochemical differences in temperature (varying from hot soils near the equator to cold polar soils), moisture (varying from dry desert soils to wet, partially submerged soils), pH (varying from extremely acid to strongly alkaline soils), and E_h (varying from oxidizing conditions of well-aerated soils to highly reducing conditions of waterlogged anaerobic soils) greatly influence which microorganisms occur within the indigenous microbial communities of different soil habitats. For example, fungal populations are favored in soils of low pH, and bacteria tend to occur in higher numbers in those of higher pH (Figure 12.24). It should be noted that soils are discontinuous habitats and that conditions between the microhabitats in which microorganisms actually live can vary. Soils, for example, are characterized by discontinuities in concentrations of oxygen, which create **microhabitats**, with aerobic and anoxic

(oxygen-free) zones occurring in relatively close proximity to each other (Figure 12.25). Thus, obligate aerobes and obligate anaerobes, as well as facultative microorganisms, usually are found living together in soils, each in its own microhabitat.

It is difficult to ascribe generalized adaptive features to most indigenous soil microorganisms. Soils have many different microhabitats with differing physicochemical properties, and even at a particular location there may be several different microenvironmental situations favoring the growth of different indigenous microbial populations. A higher pro-

FIGURE 12.23

A soil texture triangle. To define soil texture, the proportions of sand, silt, and clay are determined. The intersection of lines drawn perpendicular to the sides of the soil triangle indicates the soil texture.

TABLE 12.5. The Effect of Distance from a Plant Root on the Number of Microorganisms

Distance from root (mm)	Microorganisms Per Gram of Dry Weight of Soil		
	Nonfilamentous Bacteria	Streptomycetes	Fungi
	($\times 10^7$)	($\times 10^7$)	($\times 10^6$)
0	16.0	4.7	3.6
0–3	5.0	1.6	1.8
3–6	3.8	1.1	1.7
9–12	3.7	1.1	1.3
15–18	3.4	1.0	1.2

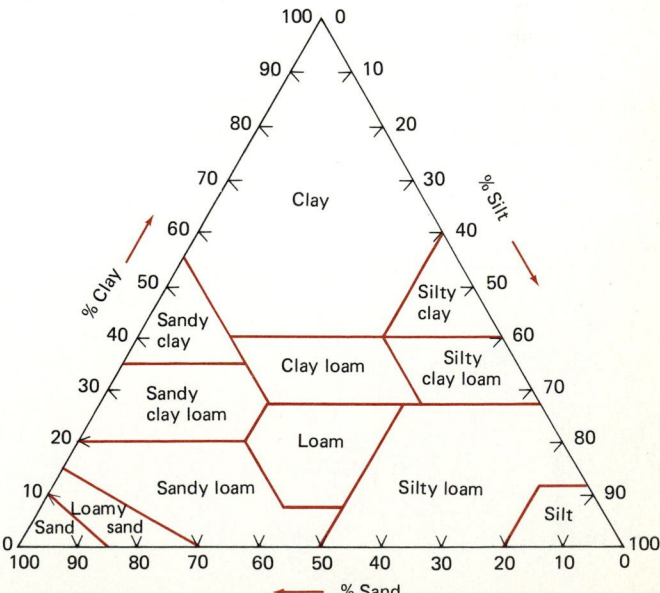

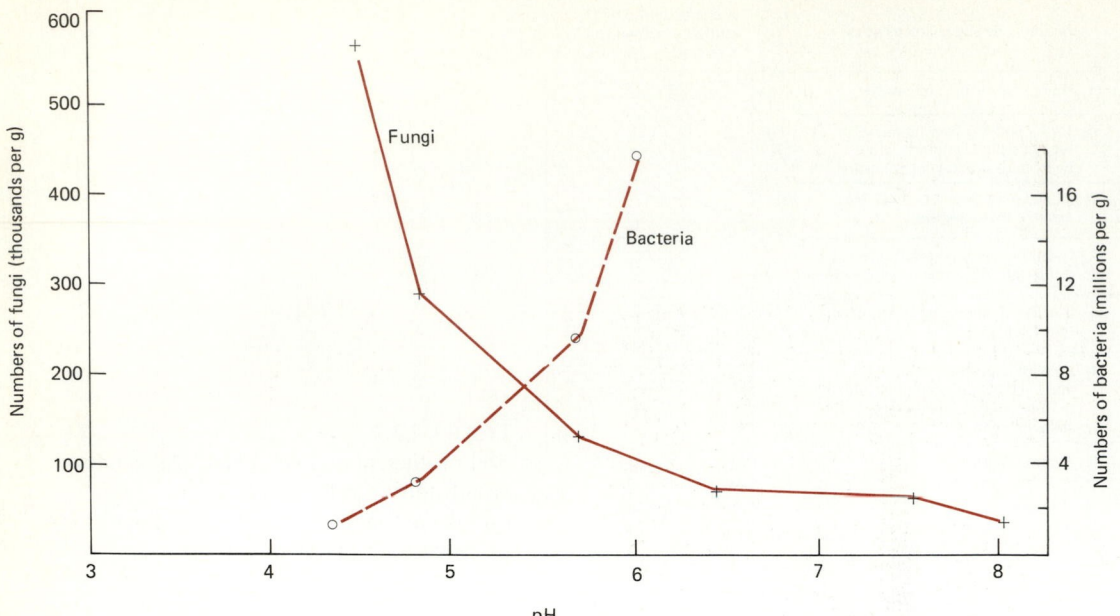

FIGURE 12.24
This graph shows the relationship between soil pH, the number of bacteria, and the number of fungal propagules.

portion of Gram-positive bacterial genera are found in soil than in freshwater and marine habitats. Many different bacterial genera are commonly found in soil (Table 12.6). Actinomycetes can comprise a significant proportion of the bacterial community in soil habitats, sometimes accounting for 10–33 percent of the total bacterial community. These organisms are relatively resistant to desiccation and can grow slowly at low nutrient concentrations. At least some soil bacteria are adapted to periods of starvation; *Arthrobacter* species can survive for over 80 days without nitrogen and carbon by reducing their endogenous rates of metabolism and relying

on reserve glycogen and polyhydroxybutyrate. Soil *Bacillus* and *Clostridium* species sporulate under adverse conditions.

Fungi constitute the major proportion of the microbial biomass in soils. Most types of fungi can be found in soil habitats, existing either as free-living organisms or in association with plant roots. The fungi most frequently isolated from soils are members of the Fungi Imperfecti, but numerous ascomycetes and basidiomycetes also occur. Most soil fungi are opportunistic, growing and carrying out active metabolism when conditions are favorable, with adequate moisture, aeration, and relatively high concentrations of usable sub-

FIGURE 12.25
(A) Microhabitats within soil particles as seen in a cross section of a soil crumb; this drawing also shows the uneven distribution of bacterial microcolonies and the occurrence of water and air within pore spaces. (B) Anoxic environments develop around plant roots and residues when voids between soil particles are sealed.

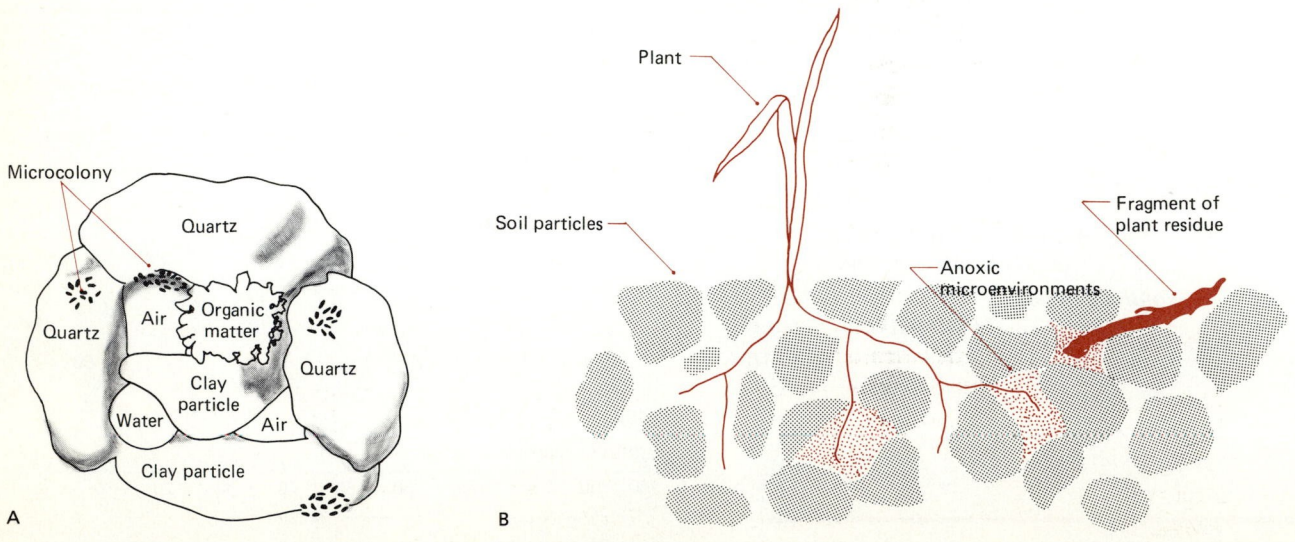

TABLE 12.6   Relative Proportions of Aerobic and Facultative Bacteria Commonly Found in Soil

Genus	Percentage of Total Number of Microorganisms
Arthrobacter	5–60
Bacillus	7–67
Actinomycetes	10–33
Pseudomonas	3–15
Agrobacterium	1–20
Alcaligenes	2–12
Flavobacterium	2–10
Corynebacterium	<5
Micrococcus	<5
Staphylococcus	<5
Xanthomonas	<5
Mycobacerium	<5

strates. Many soil fungi are able to metabolize polysaccharides, such as cellulose, that are the readily available substrates in soil. Numerous soil fungi also produce resting stages that permit them to remain dormant, even for decades, when conditions are not favorable for growth.

Organic matter in soil is largely composed of humus. Most microorganisms are not capable of using this very complex substance. Some of them, designated as autochthonous soil bacteria by Winogradsky, grow slowly in soil, utilizing refractory (nonreadily degradable) humic compounds. In contrast to these slow-growing organisms, most of which are Gram-negative rods and actinomycetes, **zymogenous microorganisms** are not able to utilize humic compounds but exhibit high levels of activity and rapid growth on readily metabolizable substrates available in the form of plant litter, animal droppings, and carcasses. Zymogenous organisms are intermittently active when such organic substrates enter soil.

As in aquatic habitats, microorganisms in soil are responsible for biodegradation and mineral cycling. Important plant polymers, such as cellulose and lignin, are degraded almost exclusively by microorganisms. Because soil is a nutrient-rich environment, the number and diversity of heterotrophic microorganisms, especially of bacteria and fungi, are characteristically high. Microorganisms in soils, however, play a relatively minor role in the input of organic carbon. Primary productivity in terrestrial habitats is dominated by higher plants. The important role of soil microorganisms in maintaining soil fertility will be discussed further when we consider agricultural microbiology.

Extreme Habitats

The conditions of **extreme environments** greatly restrict the range of microbial species that can grow in such habitats. The extremes of environmental conditions that microorganisms may have to tolerate include high temperatures approaching that of boiling water, low temperatures approaching freezing levels, low acidic pH values, high alkaline pH values, high salt concentrations, low water availability, high irradiation levels, low concentrations of usable nutrients, and high concentrations of toxic compounds. Many microorganisms that inhabit extreme environments, including hot springs, salt lakes, and Antarctic desert soils, possess specialized adaptive physiological features that permit them to survive and function within the physicochemical constraints of these ecosystems. The membranes and enzymes of microorganisms inhabiting extreme environments often have distinct modifications that permit them to function under conditions that would inhibit active transport and metabolic activities in organisms lacking these adaptive features.

Hot Springs and Thermal Vents   The most extreme and extensive high-temperature habitats are found in areas of volcanic activity. Steam vents in such areas may reach a temperature of 500°C. **Hot springs**, which occur throughout the world including Yellowstone National Park in the United States, have temperatures near 100°C (Figure 12.26). Microorganisms living in hot springs obviously must be adapted to function at high temperatures. The growth of microorganisms in hot springs is also often limited by low concentrations of organic matter, oxygen, and, depending on the particular hot spring, either acid or alkaline pH values. Despite these extreme conditions, several microorganisms do possess the adaptive features necessary to live in such habitats. For example, as water overflows the hot spring, it flows down channels, establishing a temperature gradient, with a clear zonation of microorganisms occupying habitats of differing maximal temperatures along this temperature gradient; bac-

FIGURE 12.26

Mammoth hot spring in Wyoming's Yellowstone National Park. The terraces are made of travertine, a form of calcium carbonate that has been dissolved from the limestone beneath the ground and carried to the surface by hot water. On the surface the calcium carbonate is no longer able to remain in solution and is deposited as travertine. The travertine itself is white but takes on the colors of the algae and bacteria that grow on it. (Courtesy Wyoming Travel Commission)

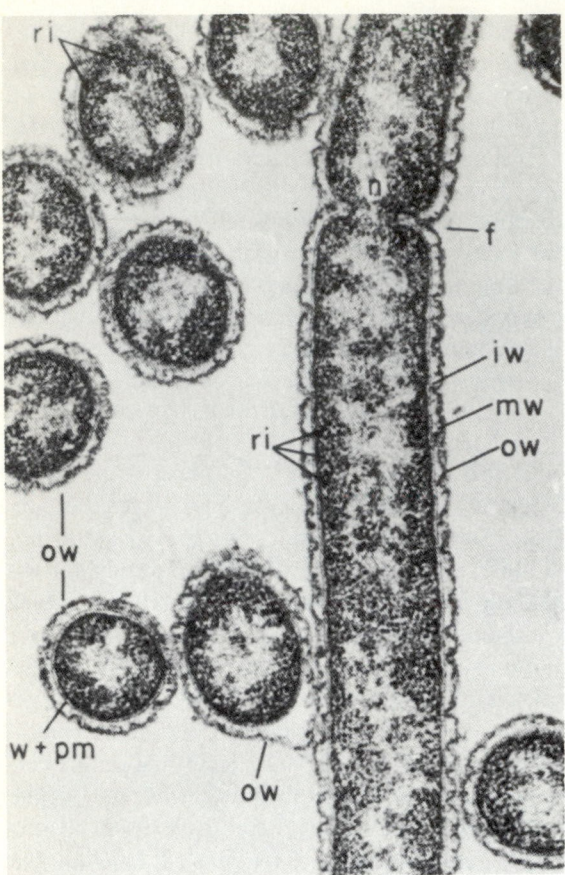

FIGURE 12.27

Micrograph of *Thermus aquaticus*, showing neoplasm (n) with dense, thin DNA fibrils surrounded by cytoplasm containing numerous ribosomes (ri). The cell envelope is composed of a plasma membrane (pm) and a wall exhibiting outer dense layers (ow), a middle light zone (mw), and an inner dense layer (iw). Note the cell division by furrowing (f). Where two cells are in contact, the external wall (ow) separates from the inner wall, which remains adherent to the plasma membrane. (Reprinted by permission of the American Society for Microbiology, Washington, D.C. from T. D. Brock, and M. R. Edwards, 1970, *Journal of Bacteriology*, 104:509–517)

teria are the most tolerant of elevated temperatures. At temperatures above 75°C, only a few bacterial species, including members of the genera *Thermus* and *Sulfolobus*, appear to grow (Figure 12.27). *Bacillus stearothermophilus* is often the dominant bacterial species in hot springs in temperature zones of 55–70°C, but many other microorganisms, including cyanobacteria and algae, also occur in such hot spring habitats. Cyanobacteria occur as layers of growth within specific zones of thermal ponds. The cyanobacteria grow in higher temperature zones than algae, which are restricted to growth below 55°C. Indeed, prokaryotes are often more tolerant of extreme environments than eukaryotes.

Thermal vent communities are located at depths of 800 to 1,000 m, where spreading of the sea floor allows seawater to percolate deeply into the crust and to react with hot core materials. These vent regions receive no sunlight and mini-

mal organic nutrient input from the low-productivity surface water above. Nevertheless, microbial growths cover all available surfaces on and near the vents, and high densities of unique clams, mussels, vestimentiferan worms, and other invertebrates cluster in the vicinity. The whole vent community is supported energetically by the chemoautotrophic oxidation of reduced sulfur, primarily by *Beggiatoa*, *Thiomicrospira*, and additional sulfide or sulfur oxidizers of great morphological diversity.

Many of the organisms living in hot spring and thermal rift habitats are **obligate thermophiles** and are restricted to growth at high temperatures. These thermophilic microorganisms have adaptive features that allow them to carry out active metabolism at temperatures of over 60°C. Many thermophilic microorganisms produce enzymes that are not readily denatured at high temperatures. Sometimes unusual amino acid sequences occur within the proteins of thermophiles, stabilizing these proteins at elevated temperatures. The membranes of thermophilic microorganisms possess a major proportion of high molecular weight and branched fatty acids that permit them to maintain their semipermeable properties at high temperatures. Thermophiles have relatively high proportions of guanine and cytosine in their DNA that raise the melting point and add stability to the DNA molecules of these organisms.

Salt Lakes   **Salt lakes** occur in arid regions where evaporation exceeds freshwater inflow or where a lake is fed by a salt spring. High concentrations of salt dehydrate cells and denature enzymes. Relatively few organisms can grow in highly saline waters, and often the biota of salt lakes is restricted to a few **halophiles** (salt-requiring) and **salt-tolerant bacteria**. Halophilic microorganisms have high internal concentrations of potassium chloride, and their enzymes must have a greater tolerance of salt than the enzymes of microorganisms that are not salt tolerant. In many cases, high concentrations of salt are required by halophiles to maintain their enzymatic activities. Many halophiles have unusual membranes, such as the purple bilayer membranes of *Halobacterium* (Figure 12.28). The role of the halobacterial membrane, which contains bacteriorhodopsin, in the phototrophic generation of ATP was discussed earlier. The cell wall of *Halobacterium* lacks murein and appears to be stabilized by sodium ions. The ribosomes of *Halobacterium* require high concentrations of potassium for stability. It is these types of adaptive features that permit halophiles to live in the saturated brine environments of salt lakes.

Desert Soils   **Desert soils**, by definition, receive less than 25 cm of rainfall per year. Many deserts are extremely dry and are subject to extreme diurnal variations in temperature because of the lack of moisture in the atmosphere overlying these areas to moderate the loss of heat. Microorganisms living in dry desert soils must be able to tolerate long periods of desiccation. In the dry valleys of Antarctica, microorganisms must also tolerate very low temperatures and, during

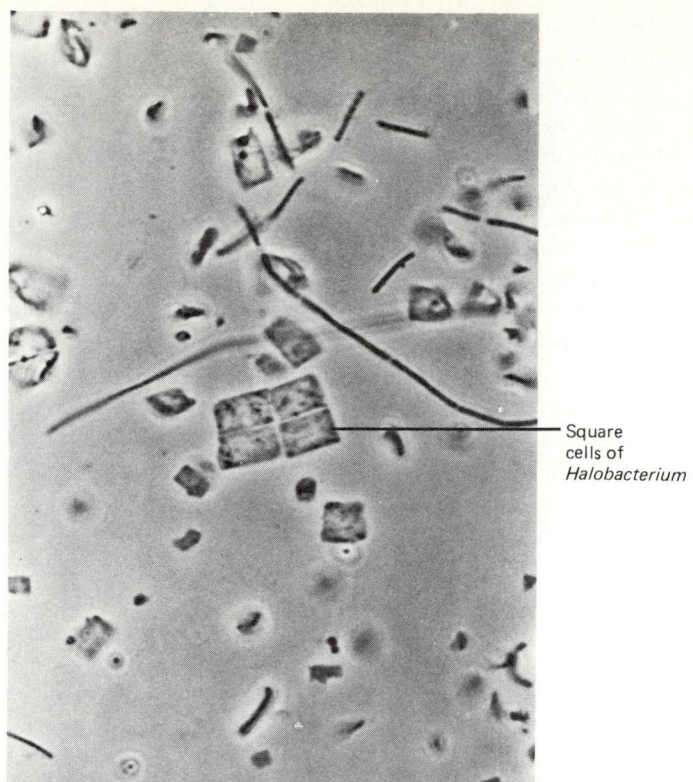

Square
cells of
Halobacterium

FIGURE 12.28
Micrograph of *Halobacterium* from a salt pond on the shore of the Red Sea. This particular species is most unusual in that it has square cells. (Courtesy Walther Stoeckenius, University of California)

part of the year, high irradiation levels. In such environments, many microorganisms have developed adaptations that allow them to survive in a dormant state during unfavorable conditions and to grow actively only during those brief periods of time when conditions are favorable, such as after a rainstorm. Many of the bacteria and fungi living in desert soils form spores that allow them to exist, if necessary, for decades between growth periods. When there is adequate

moisture, the spores germinate, and for a brief period the organisms can actively grow and reproduce. The lichen symbiosis, a mutually dependent association between fungi and photosynthetic microorganisms, is an adaptive association between microorganisms that permits growth under conditions of severe desiccation. Lichens are important in the dry habitats of Antarctica, where they can grow slowly during the relatively warm summer months. The slow growth rates and the ability to retain water permit lichens to exist in such extremely dry habitats.

Extraterrestrial Habitats   The planets in our solar system, other than Earth, are hostile habitats for living organisms, lacking water and organic carbon and having toxic concentrations of various gases in their atmospheres to make life as we know it impossible. The planet Mars, however, contains some water; therefore, experimental life detection systems were sent there as part of the United States National Aeronautics and Space Administration's Viking Mission. Martian microorganisms would have had to develop adaptations that would permit their existence under these harsh conditions; a Martian microbe would have to be hardy.

The life detection systems of the Viking Mars lander were designed to detect microbial life. More specifically, they were capable of detecting the increased turbidity that would be associated with microorganisms growing in solution, as well as the exchange of gases between microorganisms and the overlying atmosphere, including the production of volatile products from the degradation of organic matter and the fixation of carbon dioxide in organic compounds. Soils from the dry Antarctic valleys were used to test these life detection systems because conditions in these terrestrial Earth habitats are most similar to those of the dry, cold soils of Mars. The results of the Viking mission were initially confusing. They showed apparently positive test results indicating the presence of living microorganisms, but the results appear to have been due to strictly chemical reactions. The Viking exploration project scientists concluded that there are no living organisms in the Martian soils examined.

Postlude

Many factors influence microbial growth and death rates. Microorganisms exhibit ranges of tolerance to many abiotic factors. When a given parameter exceeds the tolerance range of a microorganism, the microorganism ceases to grow. In some cases, exceeding a tolerance range increases the death rate, and the microorganism is unable to survive; in other cases, the inability to grow does not result in the death of the microorganism, but rather in its preservation. Some microorganisms tolerate a wide range of values for a given parameter, whereas others are stenotolerant and are restricted to growth near their optimal value for that parameter.

For each of the parameters affecting microbial growth and death rates, microorganisms exhibit optimal growth rates at specific values of the given parameter. In laboratory and industrial situations, it is possible to adjust conditions to favor optimal growth rates or to preclude microbial growth. For example, cultures can be incubated at the optimal growth temperature if maximal growth rates are desired. If microbial growth is not desired, temperatures can be lowered to prevent microbial reproduction. Temperatures can also be raised to increase the death rate of microorganisms and to sterilize various materials. The chemical composition of a growth medium can be adjusted to favor selectively the growth of a particular microorganism. This can be accomplished by including inhibitory substances in the medium to prevent the growth of undesirable microorganisms.

A culture medium must contain the nutrients that are required by the organism for growth. In some cases, it is necessary to include a variety of growth factors, such as vitamins and amino acids. In other cases, microorganisms are capable of growing on simple media. In general, microorganisms require a source of carbon, nitrogen, phosphorus, iron, magnesium, sulfur, sodium, potassium, and chloride. Oxygen is also required by some microorganisms, but oxygen is toxic to others. Parameters that have a great influence on microbial growth and death rates include organic and inorganic nutrient concentrations, concentrations of organic and inorganic inhibitory substances, temperature, oxygen concentrations, water availability, radiation intensity, pressure, and pH. The combined effects of these factors generally determine the ability of a microorganism to survive and grow in a natural system. It is also these parameters that are manipulated in controlled situations to regulate the rates of microbial growth and death. A number of factors can be adjusted to increase microbial death rates and kill microorganisms. These include exposure to high heat and high irradiation levels. The modification of environmental parameters to control rates of mi-

crobial growth and death is applied in many areas of microbiology, and many of these applications will be discussed in later chapters. Understanding the environmental factors controlling microbial growth gives insight into the natural distribution of microorganisms in nature and provides the basis for developing methods to control microbial growth.

Microbial populations have developed various adaptations for functioning within diverse habitats. Microorganisms living in extreme environments possess adaptive features that permit them to survive and grow under relatively hostile conditions. In some cases, the adaptive features of microorganisms are aimed at survival until conditions are more favorable, such as the formation of spores that may be aerially transported. In other cases, microorganisms have developed modified structural proteins and membranes, such as those of the thermophilic and halophilic bacteria, or strategies for growth, such as those of lichens, that permit them to grow and carry out active metabolism under normally adverse conditions. The dynamic interactions between microbial populations and their abiotic surroundings contribute to the maintenance of integrated ecosystems in various habitats, the ecological activities of microorganisms being essential for supporting productivity and maintaining environmental quality within a habitat.

Suggested Supplementary Readings

ALEXANDER, M. 1971. *Microbial Ecology*. John Wiley & Sons, Inc., New York.

ALEXANDER, M. 1977. *Introduction to Soil Microbiology*. John Wiley & Sons, Inc., New York.

ATLAS, R. M., and R. BARTHA. 1987. *Microbial Ecology: Fundamentals and Applications*. Benjamin/Cummings Publishing Co., Inc., Menlo Park, Calif.

BAROSS, J. A., and J. W. DEMING. 1983. Growth of "black smoker" bacteria at temperatures of at least 250°C. *Nature* (London) 303:423–426.

BLAKEMAN, J. P. (ed.). 1981. *Microbial Ecology of the Phylloplane*. Academic Press, London.

BLAKEMORE, R. P. 1982. Magnetotactic bacteria. *Annual Review of Microbiology* 36:217–238.

BROCK, T. 1966. *Principles of Microbial Ecology*. Prentice-Hall, Inc., Englewood Cliffs, N.J.

BROCK, T. 1978. *Thermophilic Microorganisms and Life at High Temperatures*. Springer-Verlag, New York.

BROCK, T. (ed.). 1986. *Thermophiles: General Molecular and Applied Microbiology*. John Wiley & Sons, Inc., New York.

BROWN, A. D. 1976. Microbial water stress. *Bacteriological Reviews* 40:803–846.

BROWN, M., and P. WILLIAMS. 1985. The influence of environment on envelope properties affecting survival of bacteria in infections. *Annual Review of Microbiology* 49:527–556.

BURNS, R. G., and J. H. SLATER (eds.). 1982. *Experimental Mi-crobial Ecology*. Blackwell Scientific Publications, Oxford, England.

BUTTON, D. K. 1985. Kinetics of nutrient-limited transport and microbial growth. *Microbiological Reviews* 49:270–297

CAMPBELL, R. E. 1983. *Microbial Ecology*. Blackwell Scientific Publications, Oxford, England.

CODD, G. A. (ed.). 1984. *Aspects of Microbial Metabolism and Ecology*. Academic Press, London.

COLWELL, R. R., and R. Y. MORITA (eds.). 1974. *Effect of the Ocean Environment on Microbial Activities*. University Park Press, Baltimore.

DOETSCH, R. N., and T. M. COOK. 1973. *Introduction to Bacteria and Their Ecobiology*. University Park Press, Baltimore.

FLETCHER, M., and G. D. FLOODGATE (eds.). 1985. *Bacteria in Their Natural Environments*. Academic Press, London.

FRIDOVICH, I. 1977. Oxygen is toxic! *BioScience* 27:462–466.

FRIEDMAN, E. I. 1982. Endolithic microorganisms in the Antarctic cold desert. *Science* 215:1045–1053.

GOULD, G. W., and J. E. L. CORRY (eds.). 1980. *Microbial Growth and Survival in Extremes of Environment*. Academic Press, New York.

GRAY, T. R. G., and D. PARKINSON (eds.). 1968. *The Ecology of Soil Bacteria*. University of Toronto Press, Toronto.

GRAY, T. R. G., and J. R. POSTGATE (eds.). 1976. *The Survival of Vegetative Microbes*. Cambridge University Press, New York.

GREGORY, P. H. 1973. *The Microbiology of the Atmosphere.* John Wiley & Sons, Inc., New York.

GRIFFIN, D. M. 1981. Water and microbial stress. *Advances in Microbial Ecology* 5:91–136.

HERBERT, R. A., and G. A. CODD (eds.). 1986. *Microbes in Extreme Environments.* Academic Press, London.

HUGO, W. B. (ed.). 1971. *Inhibition and Destruction of the Microbial Cell.* Academic Press, London.

JANNASCH, H. W., and M. J. MOTTL. 1985. Geomicrobiology of deep-sea hydrothermal vents. *Science* 216:1315–1317.

JANNASCH, H. W., and C. W. WIRSEN. 1977. Microbial life in the deep sea. *Scientific American* 236(6):42–52.

KUSHNER, D. J. 1968. Halophilic bacteria. *Advances in Applied Microbiology* 10:73–97.

KUSHNER, D. J. (ed.). 1978. *Microbial Life in Extreme Environments.* Academic Press, London.

LARSEN, H. 1967. Biochemical aspects of extreme halophilism. *Advances in Microbial Physiology* 1:97–132.

LASKIN, A., and H. LECHEVALIER (eds.). 1974. *Microbial Ecology.* CRC Press, Inc., Boca Raton, Fla.

LYNCH, J. M., and N. J. POOLE (eds.). 1979. *Microbial Ecology: A Conceptual Approach.* Blackwell Scientific Publications, Oxford, England.

MacLEOD, R. A. 1985. Marine microbiology far from the sea. *Annual Review of Microbiology* 39:1–20.

MITCHELL, R. 1974. *Introduction to Environmental Microbiology.* Prentice-Hall, Inc., Englewood Cliffs, N.J.

MORITA, R. Y. 1976. Psychrophilic bacteria. *Bacteriological Reviews* 39:144–167.

MORRIS, J. G. 1975. The physiology of obligate anaerobiosis. *Advances in Microbial Physiology* 12:169–246.

RHEINHEIMER, G. 1986. *Aquatic Microbiology.* John Wiley & Sons, Inc., New York.

RUSSELL, A. D., W. B. HUGO, and G. A. J. AYLIFFE. 1982. *Principles and Practice of Disinfection, Preservation and Sterilisation.* Blackwell Scientific Publications, Oxford, England.

SIEBURTH, J. McN. 1979. *Sea Microbes.* Oxford University Press, New York.

SKERMAN, V. D. B. 1969. *Abstracts of Microbiological Methods.* John Wiley & Sons, Inc., New York.

SKUJINS, J. 1984. Microbial ecology of desert soils. *Advances in Microbial Ecology* 7:49–91.

SLATER, J. H., R. WHITTENBURY, and J. W. T. WIMPENNNY. 1983. *Microbes in Their Natural Environments, Thirty-Fourth Symposium of the Society for General Microbiology.* Cambridge University Press, Cambridge, England,

WOOD, E. J. F. 1965. *Marine Microbial Ecology.* Chapman and Hall, London.

Study Questions

1. What is the effect of temperature on microbial growth and death rates? What is the difference between optimal and maximal growth temperatures?

2. What is a mesophile? psychrophile? thermophile? psychrotroph?

3. What effect does a high salt concentration have on microorganisms? What is the difference between a halophile and a salt-tolerant organism?

4. What mechanisms have microorganisms evolved for protection against radiation exposure?

5. What is the difference between an autochthonous and an allochthonous microorganism in a particular habitat?

6. What properties are needed for the survival of microorganisms in the atmosphere?

7. What is a true marine bacterium?

8. Discuss the adaptation and zonal separation of microorganisms in a hot spring habitat.

9. Why were the Antarctic dry valleys used as models for the design of the Mars Viking lander?

Situational Problem 1

Designing a School Science Fair Project

Science fair projects have become a routine part of the elementary and high school curricula. Many times students grow microorganisms on various sugar-containing substances, such as jams and jellies, to demonstrate visible microbial growth and to describe the microorganisms that they observe with the aid of a microscope. Another common project is to observe the effects of various disinfectant substances, such as mouthwashes, on microbial growth. Some such projects involve the use of microorganisms. In most such cases, elementary and high schools lack the necessary facilities, and their teachers the needed expertise, to perform such projects safely. Every year students faced with the task of doing such projects turn to college students and professors for advice.

In advising such students, it is important to make sure that they understand basic microbiological methods, especially aseptic technique and how to transfer and dispose of microorganisms safely. Too often the students developing such elementary projects fail to recognize that the microbes they are working with are living organisms; they do not realize that microorganisms require specific growth conditions; and they ignore the fact that growth-supporting substances will dry out and that cultures must be repeatedly transferred to be maintained. They simply open the containers containing extensive microbial growth and hold the contents up to their faces to see and smell.

Suppose that you have been asked by a high school student to help with a project for the school science fair. This student is specifically interested in microorganisms and the effects of environmental factors on the distribution of microorganisms in nature. The interest stems from an article the student read in *National Geographic* on microorganisms growing in deep-sea vents at extremely high temperatures. The student would really like to study these thermal vent microbes but realizes that this is beyond the scope of his or her resources. He or she would like to do something similar. What project would you suggest?

Although the project must be the student's own work, you can help greatly with its design and can ensure that the right methods are used. Assuming that the student wants to do the project that you suggest, what are the next steps? What books would you recommend that the student consult as references? Develop a specific set of hypotheses and methods that would support the determination of their validity. Make sure that the methods are within the scope of the available resources and that the experiments can be concluded within 1–3 months.

Situational Problem 2

Searching for Life on Other Planets

If life exists at all on other planets, microorganisms are the most likely forms that would be present, even if higher forms of life also exist, because of the ubiquitous role of microorganisms in transforming elements into forms that can support the continued requirements of living organisms. An unmanned probe sent to Mars failed to detect living microorganisms, but new analyses of the composition of the Martian surface, and of other planets and their atmospheres, has raised new questions about where to search for extraterrestrial life. Assuming that you could help direct the search for extraterrestrial microbial life, what chemical and physical properties would you look for in the surface and atmospheric composition of potential planetary exploration sites? What conditions would you expect to favor life? What conditions would you view as precluding life? How would you actually go about detecting life on Mars?

CHAPTER 13

Biogeochemical Cycling and Interactions Among Microbial Populations

13.1 *Biogeochemical Cycling*

Microorganisms perform essential functions within **ecosystems**, the integrated systems of biological populations and abiotic factors that comprise the global **biosphere**.[1] The activities of microorganisms within the biosphere have a direct impact on the quality of human life. Without the essential biogeochemical cycling activities of microorganisms, higher forms of life, including humans, could not exist. **Biogeochemical cycling**, that is, the movement of materials via biochemical reactions through the global biosphere, is a consequence of the exchange of elements between the biotic and abiotic portions of the biosphere that occurs through characteristic cyclic pathways as a result of the metabolic reactions of living organisms. Just as there are central metabolic pathways within cells through which transformations of compounds occur, there are central pathways through which materials flow within ecosystems. All living organisms carry out chemical transformations that influence their environment. Many of these chemical changes are a consequence of oxidation-reduction reactions that occur during microbial metabolism. Changes in the chemical forms of various elements can lead to the physical translocations of materials, sometimes mediating transfers between the **atmosphere** (air), **hydrosphere** (water), and **lithosphere** (land). Physical movement, for example, between soluble and insoluble states, also forms an important part of biogeochemical cycling.

Because of their ubiquitous distribution and diverse enzymatic activities, microorganisms play a major role in bio-

geochemical cycling. The major elements of living biomass, carbon, hydrogen, oxygen, nitrogen, phosphorus, and sulfur are most intensively cycled by microorganisms; other elements are generally cycled to a lesser extent. The biogeochemical cycling activities of microorganisms are essential for the survival of plant and animal populations and determine, in large part, the potential productivity level of a given habitat. Human activities, such as the release of pollutants, can have a major influence on the rates of microbial cycling activities, leading to significant changes in the biochemical characteristics of a given habitat.

The Carbon Cycle

Carbon is actively cycled between inorganic carbon dioxide and the variety of organic compounds that compose living organisms. The **biogeochemical cycling of carbon** primarily involves the transfer of carbon dioxide and organic carbon between the atmosphere, where carbon occurs principally as inorganic CO_2, and the hydrosphere and lithosphere,[2] which contain varying concentrations of organic and inorganic carbon compounds (Figure 13.1). The autotrophic metabolism of photosynthetic and chemolithotrophic organisms is responsible for primary production, the conversion of inorganic carbon dioxide to organic carbon. Once carbon is fixed (reduced) into organic compounds, it can be transferred from

---

[1]The biosphere is that portion of the Earth and its atmosphere in which living organisms occur.

[2]In the lithosphere and hydrosphere, carbon dioxide reacts with water to form carbonate and bicarbonate, which are the principal inorganic forms of carbon found there.

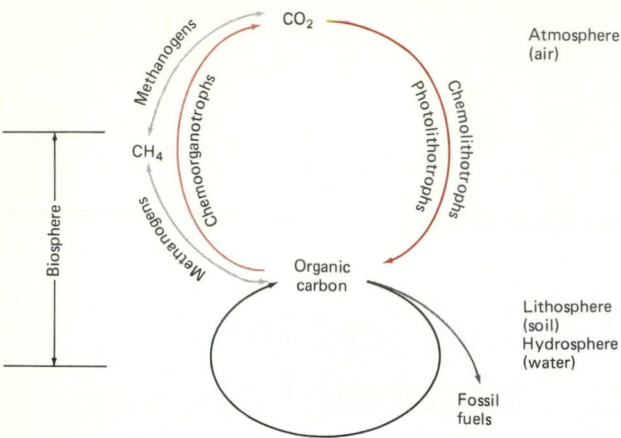

FIGURE 13.1
A simplified view of the carbon cycle.

population to population within the biological community, supporting the growth of a wide variety of heterotrophic organisms. The respiratory and fermentative metabolism of heterotrophic organisms returns inorganic carbon dioxide to the atmosphere, completing the carbon cycle.

The production of methane by a specialized group of methanogenic archaebacteria represents a shunt to the normal cycling of carbon because the methane that is produced cannot be used by most heterotrophic organisms and thus is lost from the biological community to the atmosphere. Normally, fossil fuels, such as coal and petroleum, are not actively cycled through the activities of microorganisms. Burning of fossil fuels also adds CO_2 to the atmosphere, which has led to a general rise in the concentration of atmospheric CO_2 and a resulting warming of global temperatures, a phenomenon known as the **greenhouse effect**

Trophic Relationships   The carbon dioxide converted into organic carbon by the **primary producers**, the autotrophs, in an ecosystem represents the **gross primary production** (total amount of organic matter produced) by the biological community in a given habitat. Part of the gross primary production is converted back to carbon dioxide by the respiration of the primary producers, and only the remaining organic carbon in the form of biomass and soluble metabolites—the **net primary production**—is available for heterotrophic consumers in terrestrial and aquatic habitats. The oxidative metabolism of the biological community removes organic carbon and the energy stored in such compounds from the ecosystem, and thus represents a decay of the energy stored within a given habitat. If the net primary production is greater than the community respiration, organic matter accumulates within the ecosystem. If, on the other hand, respiratory activities are greater than the net primary production, organic matter must be added from an external source or the community in that ecosystem will decline.

Most ecosystems are dependent on the fixation of carbon dioxide, that is, the input of organic matter by photosynthetic

organisms, including plants, algae, and photosynthetic bacteria. The thermal rift areas of the deep ocean regions near the Galapagos Islands represent an interesting exception because the ecosystems associated with these areas are based on the input of organic carbon by sulfur-oxidizing chemolithotrophic bacteria that grow in the warm, hydrogen sulfide-rich waters that enter the ocean through thermal vents (Figure 13.2). These organisms generate ATP and reduced coenzymes by oxidizing hydrogen sulfide, and use the ATP and reduced coenzymes to drive the reduction of CO_2 via the Calvin cycle.

Carbon and energy in organic compounds, formed by primary producers, move through the biological community of an ecosystem. The transfer of energy stored in organic compounds between the organisms in the community forms a **food web**, an integrated feeding structure (Figure 13.3). The feeding relationships between organisms establish the **trophic structure**, that is, the routes by which energy and materials are transferred out of an ecosystem. The movement of energy through an ecosystem occurs in steps from one trophic level to another. At the base of the food web are the primary producers that form the organic matter for the system. **Grazers** are organisms that feed upon primary producers. In phytoplankton-based food webs, algae and cyanobacteria are the primary food source for grazers. In detrital food webs, microbial biomass produced from growth on dead organic matter (detritus) serves as a primary food source for grazers. The grazers, in turn, are eaten by **predators**, which in turn may be preyed upon by larger predators.

The overall feeding relationships establish a pyramid of biological populations in the food web (Figure 13.4). The pyramid shape occurs because only a small portion of the energy stored in any trophic level is transferred to the next higher trophic level. Normally, 85–90 percent of the energy stored in the organic matter of a trophic level is consumed by respiration during transfer to the next trophic level and enters the decay portion of the food web. Consequently, the higher the trophic level, the smaller its biomass (Figure 13.4).

The decay portions of food webs are dominated by microorganisms (Figure 13.5, page 375). Microbial decomposition of dead plants and animals and partially digested organic matter is largely responsible for the conversion of organic matter to carbon dioxide and the reinjection of inorganic CO_2 into the atmosphere. The rates of organic matter mineralization depend on various factors, including environmental conditions—such as pH, temperature, and oxygen concentration—and the chemical nature of the organic matter. Some natural organic compounds, such as lignin, cellulose, and humic acids, are relatively resistant to attack and decay only slowly. Various synthetic compounds, such as DDT, may be **recalcitrant**, that is, completely resistant to enzymatic degradation. We depend on the activities of microorganisms to decompose organic wastes, and when microbial decomposition is ineffective, organic compounds accumulate. This is evidenced by the environmental accumulation of plastic ma-

FIGURE 13.2

Micrographs of bacteria growing in a thermal rift area off the Galapagos Islands. (A) In this scanning electron micrograph, note the dense microbial layer of prosthecate cells, resembling *Hyphomicrobium*, on the surface. (B) This transmission micrograph of the bacterial surface shows a loose layer of Gram-negative cells and metal deposits. Scale bar = 1 μm. These photomicrographs were made as part of a survey of the most commonly observed and morphologically conspicuous microorganisms found attached to natural surfaces or artificial deposits in the area of thermal vents at the Galapagos Rift ocean spreading zone at a depth of 2,550 m. (Courtesy Holger W. Jannasch, Woods Hole Oceanographic Institution, Woods Hole, Mass.; reprinted by permission of the American Society for Microbiology, Washington, D.C., from H. W. Jannasch and C. O. Wirsen, 1981, *Applied and Environmental Microbiology*, 41:528–538)

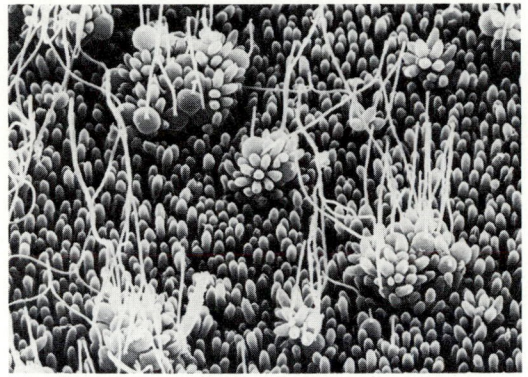
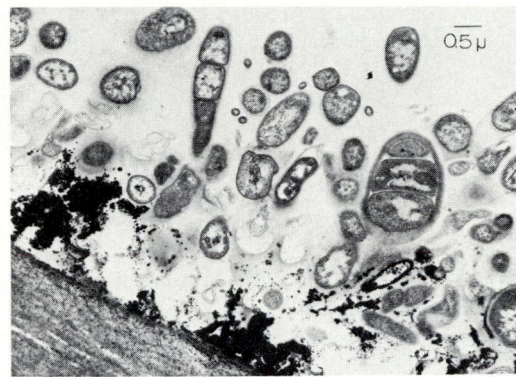

A   B

Discovery Process

The work of Holger Jannasch had established that microbial activities occur at very low rates in the deep oceans; deep ocean sediments, in effect, are biological deserts because of their low temperatures, high pressures, and low inputs of organic matter. These rates are so low that bologna sandwiches accidentally submerged inside the submersible Alvin were not decomposed during several months of exposure to microorganisms of the deep sea. What a surprise it was then when investigators found an area of extremely high biological productivity at a depth of 2,550 m in a region of thermal vents off the Galapagos Islands. The thermal vents warmed the waters, but what was the source of food supporting the growth of worms several feet long and clams several feet across? There was no light to support photosynthesis, and transport of organic matter from the surface was unlikely. The most likely explanation was chemolithotrophic primary production based on oxidation of hydrogen sulfide from the vents. Establishing that bacterial chemolithotrophy was the source of organic matter would be difficult; to reach the vents in the Alvin would take hours, time on the bottom to carry out experiments would be extremely limited, and working Alvin's mechanical arms would be

difficult. Nevertheless, this was the task undertaken by Holger Jannasch and Carl Wirsen of the Woods Hole Oceanographic Institution.

These investigators found that all surfaces intermittently exposed to H_2S-containing hydrothermal fluid were covered with mats composed of layers of prokaryotic, Gram-negative cells interspersed with amorphous manganese-iron metal deposits; although some of the cells were encased in dense metal deposits, there was little correlation between metal deposition and the occurrence of microbial mats; highly differentiated forms appeared to be analogues of certain cyanobacteria; isolates from massive mats of prosthecate bacteria were identified as *Hyphomicrobium* spp.; intracellular membrane systems similar to those found in methylotrophic and nitrifying bacteria were found in about 25 percent of the cells composing the mats; and thiosulfate enrichments made from mat material resulted in isolations of different types of sulfur-oxidizing bacteria, including the obligately chemolithotrophic genus *Thiomicospira*. These studies established that chemolithotrophic bacteria supported the productivity of the thermal rift region.

terials that are recalcitrant to microbial attack. Many modern problems relating to the accumulation of environmental pollutants reflect the inability of microorganisms to degrade rapidly enough the concentrated wastes of industrialized societies.

Microbial Mobilization and Immobilization of Carbon within the Biosphere The activities of microorganisms affect the accessibility of the biological community to the carbon and energy stored in organic compounds. Some microbial transformations of organic carbon, such as the

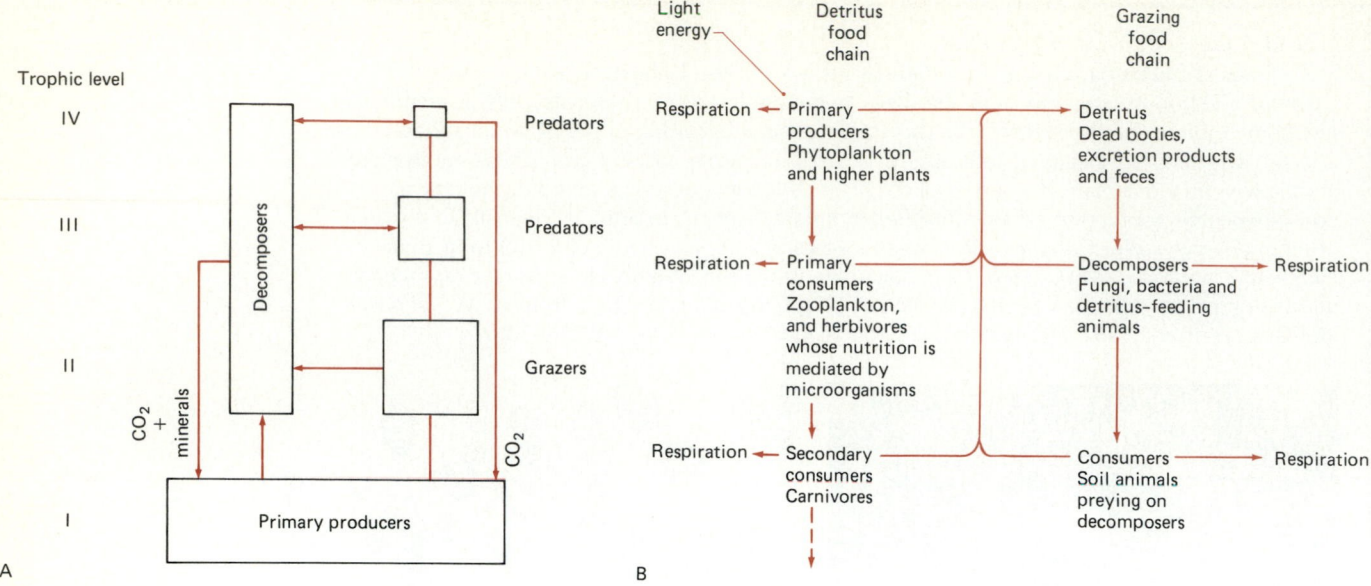

FIGURE 13.3

Two food webs. (A) This idealized food web shows the transfers between trophic levels. Organic carbon formed by primary producers is transferred to grazers and predators. CO_2 is returned to primary producers by decomposers and by respiration of grazers and predators. The population sizes that can be supported decline at higher trophic levels. (B) The grazing and detritus food chains are interlinked.

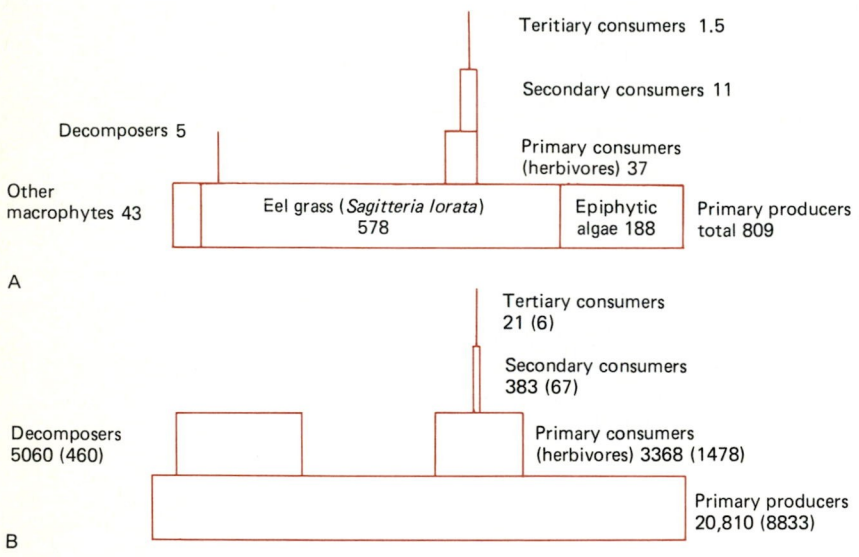

FIGURE 13.4

Food webs are structured in the shape of a pyramid. These pyramids represent Silver Springs, Florida. (A) Biomass pyramid: the figures represent grams of dry weight per square meter. The decomposers have a small biomass but high metabolic activity. (After H. T. Odum, 1957, *Ecological Monographs*, 27(1):55–112), (B) Energy pyramid: the figures represent the gross primary production or consumption in kilocalories per square meter per year. The portion of the total energy flow through the group of organisms that is actually fixed as organic biomass and that is potentially available as food for other populations at the next trophic level is indicated in brackets. Note the importance of the decomposers when their activity is expressed in energy terms, rather than in biomass, as in (A). (After E. P. Odum, 1959, *Fundamentals of Ecology*, W. B. Saunders Co., Philadelphia)

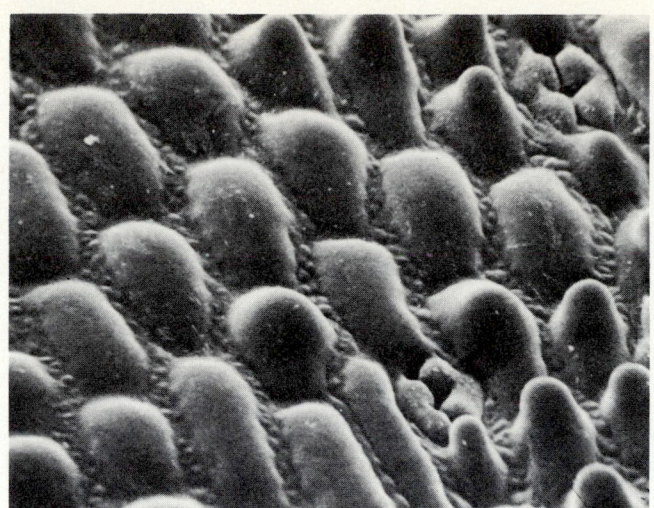
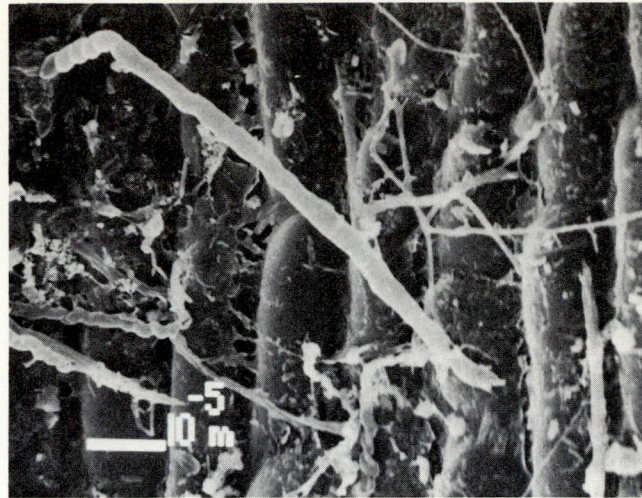

A   B

FIGURE 13.5

The decomposition of dead plant matter is seen in these scanning electron micrographs. (A) The uncolonized ventral surface of the aquatic plant *Carex* from Toolik Lake, Alaska; (B) the colonized and partially decomposed ventral surface of *Carex* litter exposed to sediment surface microbiota of the littoral zone of Toolik Lake for 13 days. Note the bacteria and fungal filaments enmeshed in a slime matrix. (Courtesy J. R. Vestal, University of Cincinnati)

production of humic acids in soil, tend to reduce the rate of cycling or to immobilize that portion of carbon and energy within the biosphere. **Humic substances** are defined as those portions of the soil that have undergone sufficient transformation to render the parent material unrecognizable. These are complex substances that are not readily metabolized. The formation of humic acids is a two-stage process involving the microbial degradation of organic polymers to their monomeric constituents, followed by chemical polymerization via a series of oxidation reactions, some of which are catalyzed by microbial enzymes, to produce the complex soil humic material.

Microbial transformations that alter the physical state of carbon-containing compounds, such as the production of gaseous compounds from water-soluble solid substances, also have a major effect on the availability of carbon to the biological community within a particular habitat. The production of methane from low molecular weight fatty acids by some bacterial species in anoxic habitats is an interesting microbial process that tends to remove carbon from the biological community. Relatively few organisms can oxidize methane and reintroduce it into the actively cycled carbon pool. Other transformations, such as the anaerobic degradation of lignocellulose, mobilize stored organic carbon by producing simpler biochemicals that can be utilized by a wide variety of organisms. In some cases, the growth of microorganisms effectively concentrates energy and nutrients within microbial biomass, making it possible for higher organisms to use the organic matter contained within the ecosystem on an energetically favorable basis; as examples, coprophagous (feces-eating) animals ingest fungal biomass growing on fecal matter as a major source of nutrition, and various shellfish are filter feeders, removing microorganisms suspended in water as their food source.

The Hydrogen and Oxygen Cycles

The movements of hydrogen and oxygen through the biosphere are closely tied to the carbon cycle (Figure 13.6). Hydrogen is an essential component of all organic compounds, and oxygen occurs in all organic compounds except hydrocarbons. Water represents a major reservoir of both hydrogen and oxygen. When water acts as an electron donor during photosynthesis, molecular oxygen is liberated and hydrogen and electrons are transferred to form reduced coenzymes that are used in biosynthesis. During respiration both oxygen and hydrogen are recycled, with the formation of water and carbon dioxide. Oxidation-reduction reactions of organic compounds, including the reduction of carbon dioxide to form organic matter, use hydrogen from reduced coenzymes for biosynthetic metabolism. The oxygen liberated by **photolysis**, the splitting of water that occurs during photosynthesis, is the major source of molecular oxygen upon which aerobic respiration depends. As discussed previously, the concentrations of molecular oxygen in a given habitat are critical in determining the members of the biological community and the potential metabolic activities of the microorganisms within that community.

The concentration of hydrogen ions, which determines the **pH**, is also extremely important in establishing the distribution of microorganisms within a given habitat. In addition to its direct effect on microorganisms and microbial enzymes, the pH also indirectly influences microbial activities by determining the degree of dissociation of many molecules.

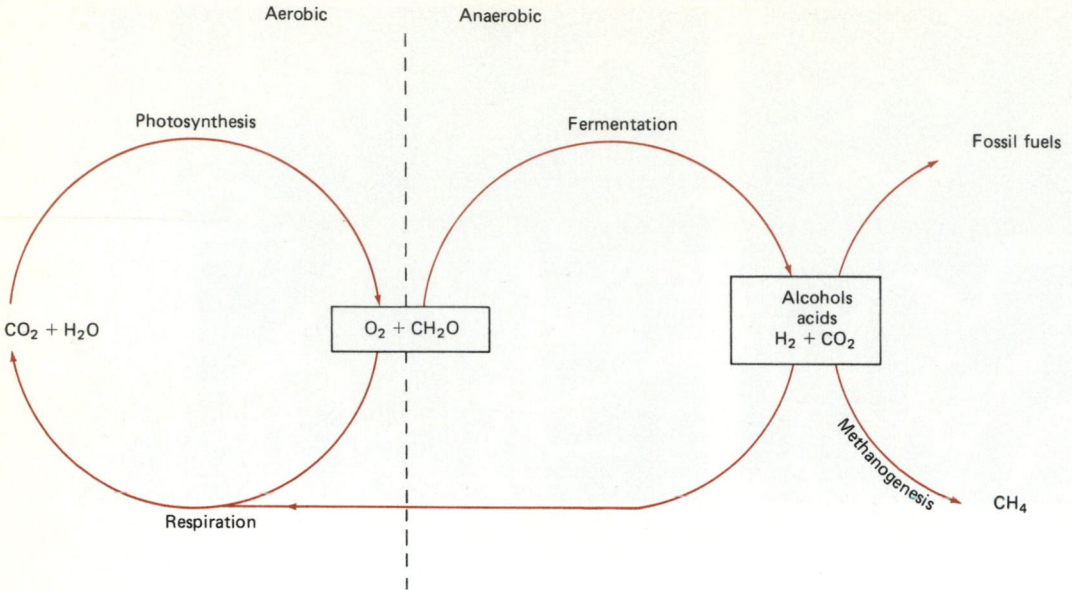

FIGURE 13.6

This diagram of integrated carbon, hydrogen, and oxygen cycles shows the involvement of oxygen and hydrogen in the aerobic and anaerobic oxidation of organic carbon and in the reduction of CO_2.

Of particular importance are the effects of pH on (1) the solubilities of gases, such as carbon dioxide, which is required for primary productivity; (2) the availability of dissolved nutrients, such as phosphorus, which often limits productivity within ecosystems; and (3) the concentrations of toxic substances, such as heavy metals, which may inhibit microbial activities within a given habitat.

The Nitrogen Cycle

Nitrogen Fixation   Whereas carbon, hydrogen, and oxygen are actively cycled by microorganisms, plants, and animals, the biogeochemical cycling of nitrogen is largely dependent on the metabolic activities of microorganisms alone (Figure 13.7). The productivity of many ecosystems is limited by the supply of fixed forms of nitrogen. Aside from the industrial chemical fixation of molecular nitrogen using the Haber-Bosch process to form nitrogen fertilizers, the process of **nitrogen fixation**, the conversion of N_2 to ammonia or organic nitrogen, is restricted almost exclusively to a limited number of bacterial species. Few microorganisms and no plants or animals are able to use atmospheric nitrogen directly; plants, animals, and most microorganisms depend on the availability of fixed forms of nitrogen for incorporation into their cellular biomass. Other than one exceptional case, where a green alga apparently fixes atmospheric nitrogen, this process is carried out strictly by bacteria.

It is estimated that microorganisms globally convert approximately 200 million metric tons of nitrogen to fixed forms of nitrogen per year compared to about 30 million metric tons produced by industrial production of nitrogen fertilizers. Ammonia is the first detectable product of nitro-

gen fixation. It is assimilated into amino acids and subsequently synthesized into proteins and nucleic acids. Proteins, amino acids, and inorganic ammonium ions are used as a source of nitrogen by many organisms that are unable to assimilate atmospheric nitrogen directly.

FIGURE 13.7

The nitrogen cycle: the chemical forms and key processes in the biogeochemical cycling of nitrogen. The critical steps of nitrogen fixation, nitrification, and denitrification are all mediated by bacteria.

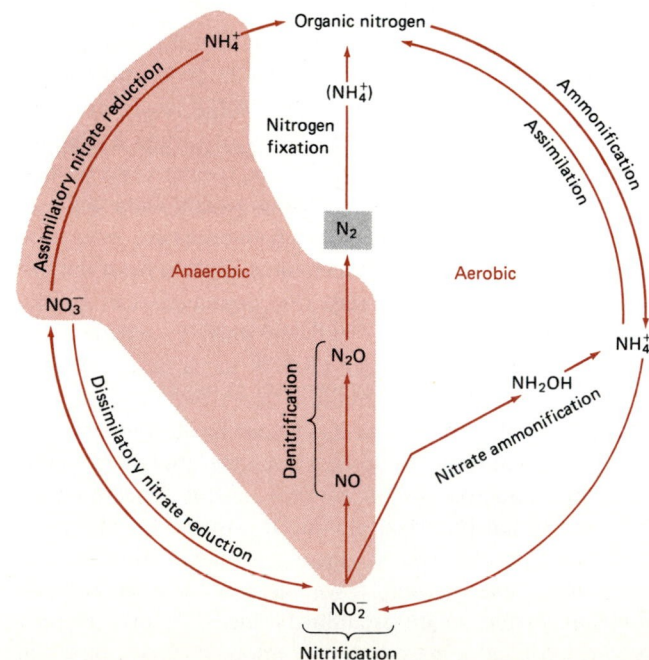

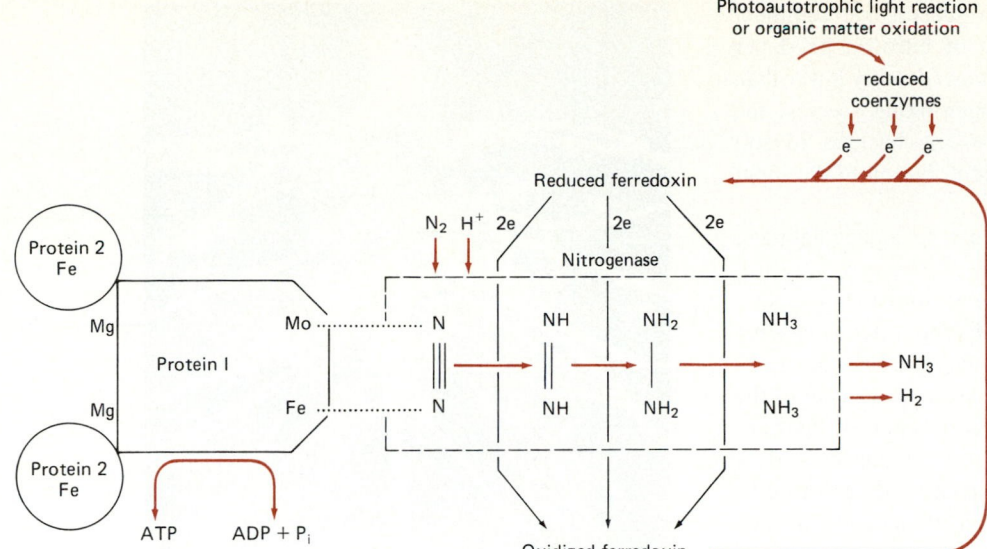

FIGURE 13.8

A generalized scheme for the action of the nitrogenase enzyme. In this process, electrons are donated to nitrogen, resulting in the formation of reduced ammonium. Initially, an electron is donated to the iron component of the nitrogenase, activating the enzyme. Interaction with ATP leads to the transfer of electrons and the binding of molecular nitrogen to the molybdenum atom of the nitrogenase. Additional electron transfers produce the reduced nitrogen compounds.

The fixation of atmospheric nitrogen depends on the **nitrogenase** enzyme system (Figure 13.8). In this enzyme system, composed of nitrogenase and nitrogenase reductase, electrons are transferred through either ferredoxin or flavodoxin to nitrogenase reductase and then to nitrogenase, where they are used to reduce N_2 and H^+ to NH_3 and H_2. The active site of nitrogenase, where reduction of nitrogen actually occurs, is associated with a molybdenum- and iron-containing cofactor. The production of H_2 that accompanies the reduction of nitrogen adds to the ATP requirements of nitrogen fixation. Some, but not all, nitrogen-fixing bacteria possess a hydrogenase and therefore are able to gain some energy by oxidizing hydrogen.

Nitrogenase is very sensitive to oxygen and is irreversibly inactivated upon exposure to even low concentrations. Nitrogen fixation, therefore, often is restricted to habitats where nitrogenase is protected from exposure to molecular oxygen; the nitrogenase enzyme is protected in the root nodule system by the red pigment leghemoglobin, which supplies oxygen to the organisms for respiration without denaturing the nitrogenase. Nitrogen is also inhibited by high concentrations of ATP, but large amounts of ATP are required to drive the electron transfer reactions catalyzed by the nitrogenase enzyme system. The fixation of atmospheric nitrogen requires a high energy input (approximately 30 ATP/N_2 fixed) and in terrestrial ecosystems is largely dependent on the availability of relatively high concentrations of organic matter for use in the respiratory generation of ATP.

Nitrogen Fixation in Soil. In terrestrial habitats, the microbial fixation of atmospheric nitrogen is carried out by free-living bacteria and by bacteria living in symbiotic (mutually dependent) association with plants. The relationships between microorganisms and plants involved in nitrogen fixation will be discussed later in this chapter. **Symbiotic nitrogen fixation** by *Rhizobium* or *Bradyrhizobium* is most important in agricultural fields, where these bacteria live in association with leguminous crop plants. In forests, other symbiotic nitrogen-fixing bacteria, including actinomycetes, live in association with various trees and make significant contributions to the soil nitrogen needed to support the growth of forests. When growing in association with plants, symbiotic nitrogen-fixing bacteria, such as *Rhizobium* and *Bradyrhizobium* species, generally exhibit rates of nitrogen fixation that are two to three orders of magnitude higher than those accomplished by free-living, nitrogen-fixing soil bacteria. *Rhizobium* species associated with alfalfa, for example, can account for an input of 250 kg of nitrogen fixed per hectare per year, as compared to 2.5 kg of nitrogen fixed per hectare per year for free-living, nitrogen-fixing *Azotobacter* species.

Azotobacter species have exceptionally high respiratory rates, far in excess of those of all other aerobic bacteria, and this may prevent molecular oxygen from reaching and inactivating the oxygen-sensitive nitrogenase. *Azotobacter* species also produce resting cells known as *cysts* that are quite resistant to desiccation but not to heat. Free-living, nitrogen-fixing members of the genera *Azotobacter*, *Azomonas*, and *Derxia* are common in temperate regions in neutral or alkaline soils and waters. These bacteria tend to be sensitive to low pH. In tropical regions, *Beijerinckia* species, which are more acid tolerant, are the prevalent nitrogen-fixing, free-living soil microorganisms.

The ability of microorganisms to fix nitrogen is now readily detected by the acetylene reduction assay. The assay is based on the fact that the nitrogenase system also catalyzes the reduction of acetylene—which, like molecular nitrogen, has a triple bond. The reduction of acetylene forms ethylene, which is easily detectable by gas chromatography. Consequently, many additional free-living bacteria have been shown to be capable of fixing atmospheric nitrogen. Most of these free-living, nitrogen-fixing bacteria exhibit nitrogen-fixing activities only at oxygen levels well below 0.2 atm.

Such conditions frequently occur in subsoil and sediment environments. Although the amount of nitrogen fixed per hectare by free-living soil bacteria is considerably lower than the amount fixed by symbiotic nitrogen-fixing species, the widespread distribution of the free-living bacteria in soil means that they make a significant contribution to the input of nitrogen to terrestrial habitats.

Nitrogen Fixation in Aquatic Habitats. In aquatic habitats, cyanobacteria, such as *Anabaena* and *Nostoc*, are very important in determining the rates of conversion of atmospheric nitrogen to fixed forms of nitrogen. Cyanobacteria, capable of nitrogen fixation, are distributed in both marine and freshwater habitats. These cyanobacteria are able to couple the ability to generate ATP through the conversion of light energy and organic matter, through the reduction of carbon dioxide, with the ability to fix atmospheric nitrogen to efficiently form nitrogen-containing organic compounds. In such organisms, the oxygen-sensitive nitrogenase enzyme is usually protected by thick-walled heterocysts where oxygen-evolving photosynthesis does not occur. Nonheterotrophic cyanobacteria fix nitrogen only under low oxygen tension. In low-nutrient aquatic environments, the use of light energy for generating ATP is critical in supplying sufficient ATP to drive the nitrogen fixation reactions. Rates of nitrogen fixation by cyanobacteria are typically 10 times higher than those shown by other free-living bacteria; thus, cyanobacteria form a very important component of aquatic food webs. Also, epiphytic cyanobacteria associated with the phyllosphere or leaf surfaces of Arctic mosses are the most important nitrogen fixers in the high Arctic ecosystem.

Nitrogen-fixing Symbiosis. The symbiotic relationship between members of the bacterial genera *Rhizobium* and *Bradyrhizobium* and leguminous plants is of extreme importance for maintaining soil fertility. Such species are able to invade the roots of suitable host plants, leading to the formation of **nodules** (tumorous growths), within which the bacteria are able to fix atmospheric nitrogen (Figure 13.9). *Bradyrhizobium* species nodulate soybeans, lupines, cowpeas, and a variety of tropical leguminous plants. *Rhizobium* species inoculate alfalfa, peas, clover, and a wide variety of temperate zone leguminous plants. The establishment of a symbiotic association between these microorganisms and a plant is very specific, and there is a mutual recognition between the bacteria and binding sites on the surfaces of the plant roots. The interaction between these microbes and a leguminous plant involves (1) the attraction of the bacteria to the plant roots by amino acids secreted by the plant; (2) the binding of the bacteria to receptors (lectins) on the plant root; (3) the activity of plant growth substances, leading to curling and branching of the rootlets; (4) the entry of bacteria into the root hairs; (5) the development of an infection thread; (6) the transformation of the plant cells to form a tumorous growth; (7) the multiplication of bacteria within the nodule; and (8) the transformation of the invading bacteria into distorted (pleomorphic) forms.

Within the nodule, metabolites are transferred between the

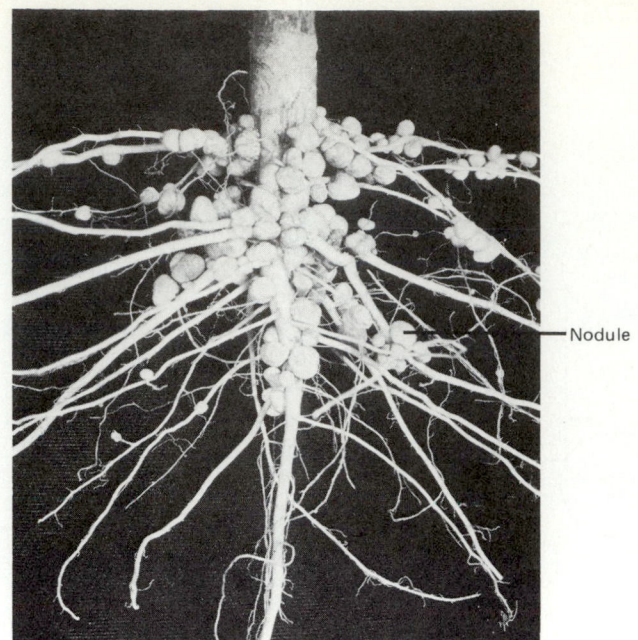

— Nodule

FIGURE 13.9
Nodules occur as tumorous growths on the roots of plants infected with nitrogen-fixing bacteria. It is within these tumor-like growths that *Rhizobium* fixes atmospheric nitrogen. Only a few types of plants can enter into a symbiotic relationship with nitrogen-fixing bacteria. Shown here is the extensive nodulation of the root system of a soybean plant. (Courtesy Nitragin Co., Milwaukee, Wis.)

bacteria and the plant. Within the infected plant tissue the root-nodule bacteria multiply, forming unusually shaped pleomorphic cells called **bacteroids**, which are no longer capable of independent reproduction (Figure 13.10). The bacteroid cells contain active nitrogenase enzymes, not generally found in free-living *Rhizobium* cells, that allow them to fix molecular nitrogen and provide their symbiotic plant partner with an available source of fixed nitrogen for growth. The plants provide organic compounds for the generation of required ATP by the symbiotic bacteria. Leghemoglobin in the nodule supplies oxygen to the bacteroids for their respiratory metabolism but also maintains a sufficiently low concentration of free oxygen so as not to inactivate the nitrogenase enzymes. The control of oxygen is therefore critical, for oxygen is both required and inhibitory for the nitrogen fixation process.

In addition to the symbiotic relationship between *Rhizobium* and *Bradyrhizobium* with leguminous plants, various other bacterial species, including cyanobacteria and actinomycetes, are able to enter into similar mutualistic relationships with a restricted number of other types of plants, resulting in the formation of nodules and the ability to fix atmospheric nitrogen. An equivalent sequence of events is involved in the formation of nodules resulting from the mutualistic relationships between bacteria and nonleguminous

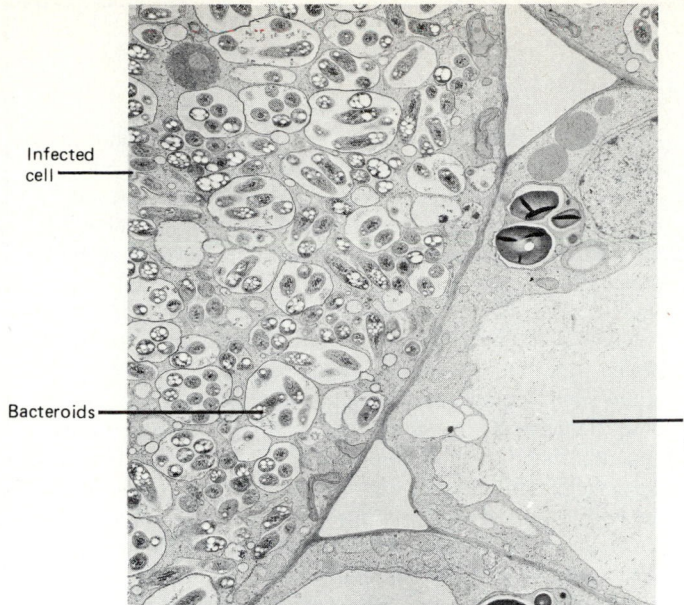

Infected cell

Bacteroids

Uninfected cell

FIGURE 13.10
Electron micrograph of *Rhizobium japonicum* bacteroids in an infected cell of a root nodule of soybean. The micrograph also shows an uninfected cell without bacteroids ($13,000\times$). (From BPS—E. H. Newcomb, University of Wisconsin)

plants, although leghoemoglobin is not the specific oxygen carrier in such relationships. *Rhizobium*, for example, can fix nitrogen in association with *Trema*, a tree found in tropical and subtropical regions. Likewise, actinomycetes establish specific mutualistic relationships that permit the bacteria to fix nitrogen, such as occurs when *Frankia alni* infects the roots of the alder tree, leading to the formation of nodules (Figure 13.11). Such an actinomycete-type nitrogen-fixing symbiosis is especially important with angiosperms.

Much higher crop yields and significant economic savings would be realized if plants could be grown without the need for adding artificially produced nitrogen fertilizer. The elimination of massive fertilizer applications to agricultural soils would also reduce problems associated with nitrification and groundwater contamination. For years scientists have been exploring the relationships between the root-nodule bacteria

and the plants with which these nitrogen-fixing bacteria can establish symbiotic relationships. Several researchers have tried to find especially effective nitrogen-fixing strains of *Rhizobium* that could increase crop yields. Using mutagens and screening procedures, Winston Brill and colleagues at the University of Wisconsin isolated strains of *Rhizobium* that were capable of very high rates of nitrogen fixation. However, field tests with these efficient *Rhizobium* strains did not increase crop yields; the superior nitrogen-fixing strains could not successfully compete with indigenous strains.

Ammonification   Microorganisms also perform important transformations of organic and inorganic fixed forms of nitrogen that alter the availability of needed nitrogen resources within aquatic and terrestrial ecosystems. Nitrogen in organic

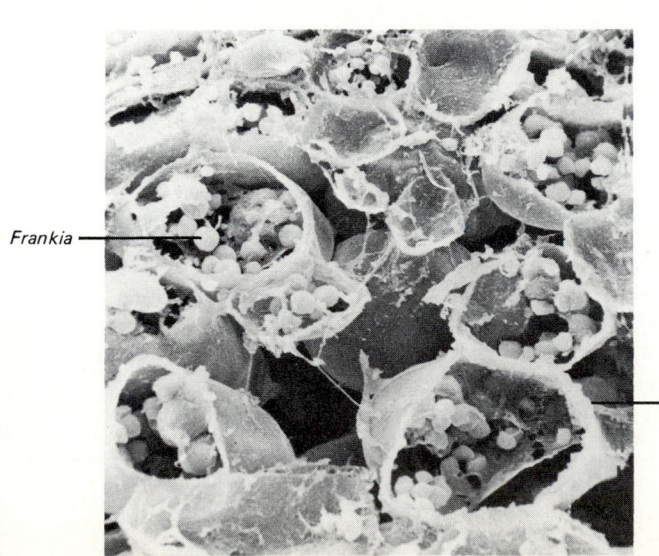

Frankia

Nodule

FIGURE 13.11
This micrograph shows a section through a root nodule of *Alnus glutinosa*; note the vesicular structures and hyphae of the nitrogen-fixing actinomycete *Frankia* ($1,750\times$). (Courtesy of Jan J. Becking, Institute for Atomic Sciences in Agriculture, Wageningen, The Netherlands)

ANALYTICAL PROCESS

Genetic Engineering to Create New Nitrogen-Fixing Organisms

One of the greatest benefits that may be realized through genetic engineering is the introduction of the capacity to fix nitrogen into plants, such as wheat, corn and rice, that are not able to utilize atmospheric nitrogen. Because of the inefficiency and lack of success with the mutation-screening approach, microbiologists are studying the genetics and biochemistry of infection by *Rhizobium* with the aim of employing recombinant DNA techniques to genetically engineer plants containing the bacterial genes for nitrogen fixation. Many research groups are carrying out these investigations. In one series of studies performed by Aladar Szalay and co-workers at Cornell University, the genes for nitrogen fixation are first inserted into the genome of a yeast; plasmids from *Escherichia coli* and from a yeast cell are cleaved and then fused to form a single hybrid plasmid, which can be recognized by the yeast cell and integrated into its chromosomal DNA (Figure 13.12). In the next step, the genes that will be introduced into the yeast are isolated from the chromosome of *Klebsiella pneumoniae*, a nitrogen fixer. The genes, collectively designated *nif*, code for some 17 proteins. Another *E. coli* plasmid is cleaved, and the isolated

nif genes are introduced to form a second hybrid plasmid. Because of the bacterial DNA already inserted into one of the yeast chromosomes, the yeast cell recognizes the hybrid *E. coli* plasmid. The plasmid is then integrated into the yeast chromosome. Although the insertion of the prokaryotic *nif* genes into the eukaryotic yeast cell demonstrates that genetic material can be transferred between different biological systems, the nitrogen-fixing proteins are not expressed in the yeast. More studies are needed to elucidate the factors controlling expression of the *nif* genes before success is obtained. It is increasingly apparent that the ability to engineer organisms, such as eukaryotic plant cells that can fix atmospheric nitrogen, depends on developing a thorough understanding of the molecular biology of gene expression and knowing how to create the environmental conditions for nitrogen-fixing activity. Once we understand the mechanisms of gene regulation in eukaryotes and the physiological requirements for nitrogen fixation, we will be able to apply this knowledge through genetic engineering to create organisms with novel properties.

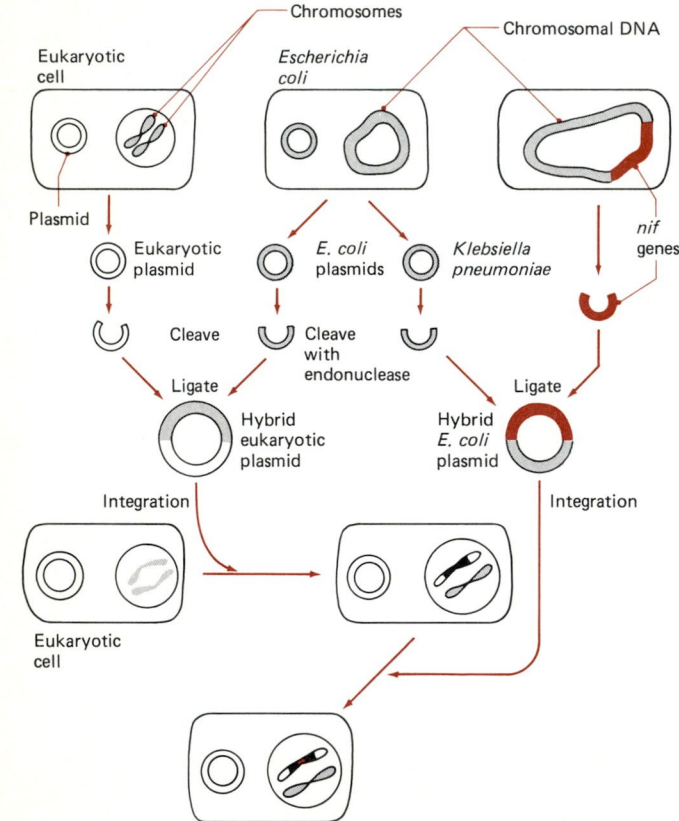

FIGURE 13.12

Genetic engineering techniques can be used to create plants that contain the bacterial *nif* gene for nitrogen fixation.

matter is found predominantly in the reduced amino form, such as occurs in amino acids. Many microorganisms, as well as plants and animals, are capable of converting organic amino nitrogen to ammonia; the process is known as **ammonification**. Deaminases play an important role in this process of ammonification, which transfers nitrogen from organic to inorganic forms. The microbial decomposition of urea, for example, results in the release of ammonia, which may be returned to the atmosphere or which may occur in neutral aqueous environments as ammonium ions. Ammonium ions can be assimilated by a number of organisms, continuing the transfer of nitrogen within the nitrogen cycle.

Nitrification   Although many organisms are capable of ammonification, relatively few are capable of **nitrification**, the process in which ammonium ions (oxidation level = -3) are initially oxidized to nitrite ions (oxidation level = $+3$) and subsequently to nitrate ions (oxidation level = $+5$). Nitrification is an example of aerobic respiration. Both the oxidation of ammonia to nitrite and the oxidation of nitrite to nitrate, the two steps of nitrification, are energy-yielding processes from which chemolithotrophic bacteria are able to derive needed energy. The metabolism of the chemolithotrophic nitrifying bacteria changes the oxidation levels of the ammonium and nitrite ions when these ions serve as electron donors for chemiosmotic generation of ATP.

Relatively low amounts of ATP, however, are generated by the oxidation of inorganic nitrogen compounds. Therefore, large amounts of inorganic nitrogen compounds must be transformed in order to generate sufficient ATP to support the growth of these chemolithotrophic bacteria. The oxidation of approximately 35 moles of ammonia is required to support the fixation of 1 mole of carbon dioxide, and approximately 100 moles of nitrite must be oxidized to support the fixation of 1 mole of carbon dioxide. As a consequence of the high amounts of nitrogen that must be transformed to support the growth of chemolithotrophic bacterial populations, the magnitude of the nitrification process is typically very high, whereas the growth rates of nitrifiers are generally relatively low compared to those of other bacteria.

The two steps of nitrification, the formation of nitrite from ammonium and the formation of nitrate from nitrite, are carried out by different microbial populations (Table 13.1). For the most part, the oxidative transformations of inorganic nitrogen compounds in the nitrification process are restricted to several species of autotrophic bacteria. In addition to the chemolithotrophic nitrifying bacteria, some heterotrophic bacteria and fungi are capable of oxidizing inorganic nitrogen compounds, but the rates of heterotrophic nitrification are normally four orders of magnitude lower than those of autotrophic nitrification. In soils *Nitrosomonas* is the dominant bacterial genus involved in the oxidation of ammonia to nitrite, and *Nitrobacter* is the dominant genus involved in the oxidation of nitrite to nitrate. Several other autotrophic bacteria, including ammonia-oxidizing members of the genera *Nitrosospira*, *Nitrosococcus*, and *Nitrosolobus*, and nitrite-

TABLE 13.1   Genera of Nitrifying Bacteria

Genus	Converts	Habitat
Nitrosomonas	Ammonia to nitrite	Soils, freshwater, marine
Nitrosospira	Ammonia to nitrite	Soils
Nitrosococcus	Ammonia to nitrite	Soils, freshwater, marine
Nitrosolobus	Ammonia to nitrite	Soils
Nitrobacter	Nitrite to nitrate	Soils, freshwater, marine
Nitrospina	Nitrite to nitrate	Marine
Nitrococcus	Nitrite to nitrate	Marine

oxidizing members of the genera *Nitrospira* and *Nitrococcus*, are also important nitrifiers in different ecosystems. Many of the nitrifying bacteria contain extensive internal membrane networks that are probably the sites of nitrogen oxidation (Figure 13.13).

Because relatively few microbial genera make significant contributions to the rates of nitrification, it is not surprising that this process is particularly sensitive to environmental stress. Toxic chemicals can block the nitrification process. Nitrification is an obligately aerobic process, and under anaerobic conditions, such as may exist when high concentrations of organic matter are added to soil or aquatic ecosystems, the nitrification process may cease. The process of nitrification is very important in soil habitats because the transformation of ammonium ions to nitrite and nitrate ions results in a change from a cation to an anion. Positively charged cations are bound by negatively charged soil clay particles and thus are retained in soils, but negatively charged anions, such as nitrate, are not absorbed by soil particles and are readily leached from the soil (Figure 13.14). Nitrification, therefore, represents a mobilization process in soils that results in the transfer of inorganic fixed forms of nitrogen from surface soils to subsurface groundwater reservoirs. In agriculture, inhibitors of nitrification, such as nitrapyrin, sometimes are intentionally added to soils to prevent the transformation of ammonium to nitrate, ensuring better fertilization of crops.

The transfer of nitrate and nitrite ions from surface soil to groundwater supplies is critical for two reasons: (1) it represents an important loss of nitrogen from the soil, where it is needed to support the growth of higher plants, and (2) high concentrations of nitrate and nitrite in drinking water supplies pose a serious human health hazard. Nitrite is toxic to humans because it can combine with blood hemoglobin to block the normal gas exchange with oxygen. Additionally, nitrites can react with amino compounds to form highly carcinogenic nitrosamines. Further, nitrate, although not highly toxic itself, can be reduced microbially in the gastrointestinal tracts of human infants to form nitrite, causing the "blue baby syndrome"; this reduction of nitrate does not occur in adults because of the low pH of the normal adult gastrointestinal tract. Nitrate and nitrite in groundwater is a particular problem in agricultural areas, such as the corn belt of the mid-

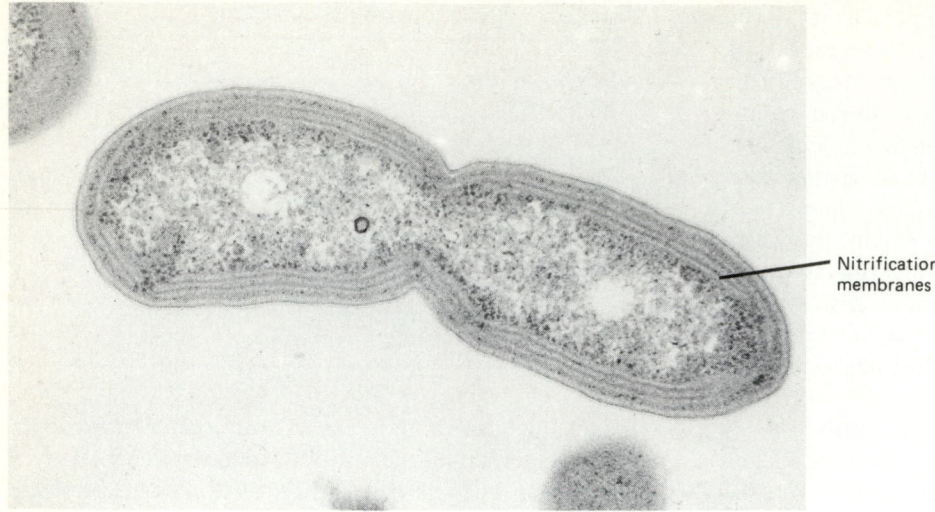

FIGURE 13.13
Micrograph of the internal membranes of the nitrifying bacteria *Nitrosomonas europaea* ($77,750\times$). (Courtesy Stan W. Watson, Woods Hole Oceanographic Institution, Woods Hole, Mass.; reprinted by permission of the American Society for Microbiology, from S. W. Watson and M. Mandel, 1971, *Journal of Bacteriology*, 107:563–569)

western United States, where high concentrations of nitrogen fertilizers are applied to soil. The use of nitrification inhibitors in combination with the application of ammonium nitrogen fertilizers can minimize the nitrate leaching problem, while at the same time supporting better soil fertility and increased plant productivity.

Denitrification   **Denitrification**, the conversion of fixed forms of nitrogen to molecular nitrogen, is another important, albeit undesirable, process in the biogeochemical cycling of nitrogen that is mediated by microorganisms. Some aerobic bacteria can use nitrate in place of oxygen as a final electron acceptor, reducing nitrate as a result of anaerobic respiration. Some bacteria, such as *E. coli*, are only able to reduce nitrate to nitrite, but a variety of other bacteria are able to carry out the two subsequent anaerobic respirations by which nitrite ion is reduced to nitrous oxide gas (N_2O) and subsequently to molecular (N_2). The process is called *denitrification* when N_2O or N_2 are produced. Some species of *Pseudomonas*, *Moraxella*, *Spirillum*, *Thiobacillus*, and *Bacillus* are capable of denitrification. Nitrous oxide formation occurs preferentially in habitats with high nitrate concentrations and/or low pH values. Formation of molecular nitrogen is favored when there is an adequate amount of organic matter to supply energy. Dissimilatory nitrate reductase, the enzyme involved in initiation of the denitrification process, is inhibited by oxygen, and denitrification generally occurs under anaerobic conditions. The return of nitrogen to the atmosphere by the denitrification process completes the nitrogen cycle.

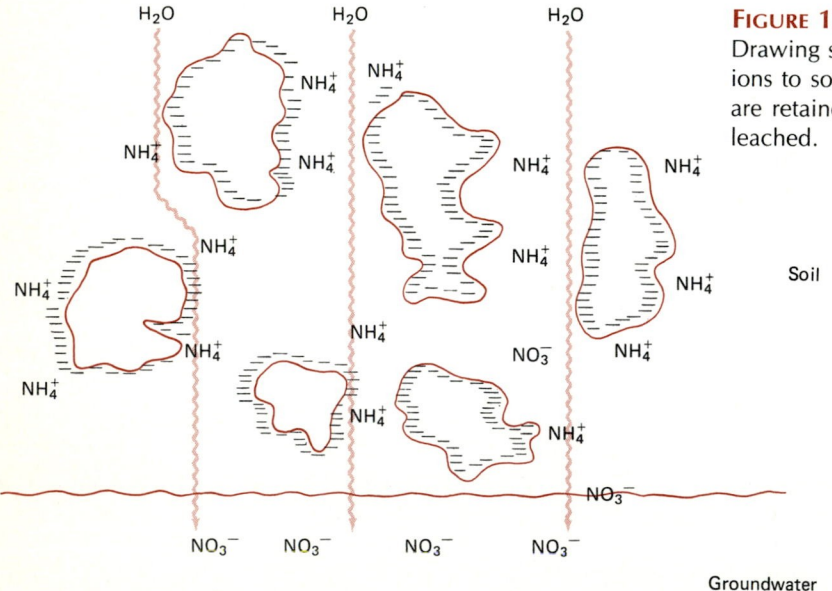

FIGURE 13.14
Drawing showing the relationship of ammonium and nitrate ions to soil particles; the positively charged ammonium ions are retained and the negatively charged nitrate ions are leached.

Nitrite Ammonification   Some bacteria, particularly *Clostridium* species, reduce nitrite to ammonium ions in a process called **nitrite ammonification**. Although involved in ATP generation, the process is not an example of anaerobic respiration. In nitrite ammonification, electrons from NADH are used to reduce nitrite rather than to reduce an organic compound. Consequently, the organic products of fermentation are more completely oxidized and the yield of ATP via substrate-level phosphorylation can be greater. Nitrite ammonification is competitive with denitrification. Unlike denitrification, nitrite ammonification does not remove nitrogen from the soil. In fact, much of the nitrate added to soils is reduced to ammonia by plants and fermentative bacteria rather than to N_2 by denitrifiers.

The Sulfur Cycle

Sulfur can exist in a variety of oxidation states within organic and inorganic compounds, and oxidation-reduction reactions—mediated by microorganisms—change the oxidation states of sulfur within various compounds, establishing the **sulfur cycle** (Figure 13.15). Microorganisms are capable of removing sulfur from organic compounds. Under aerobic conditions, the removal of sulfur (**desulfurization**) from organic compounds results in the formation of sulfate, whereas under anaerobic conditions hydrogen sulfide is normally produced from the mineralization of organic sulfur compounds (Figure 13.16). Hydrogen sulfide may also be formed by sulfate-reducing bacteria that utilize sulfate as the terminal electron acceptor during anaerobic respiration. Hydrogen sul-

fide can accumulate in toxic concentrations in areas of rapid protein decomposition, is highly reactive, and is very toxic to most biological systems. It can react with metals to form insoluble metallic sulfides.

The predominant source of hydrogen sulfide in different habitats varies. In organically rich soils, most of the hydrogen sulfide is generated from the decomposition of organic sulfur-containing compounds. In anaerobic sulfate-rich marine sediments, most of it is generated from the dissimilatory reduction of sulfate by sulfate-reducing bacteria, such as members of the genus *Desulfovibrio*. Anaerobic sulfate reduction is important in corrosion processes and in the biogeochemical cycling of sulfur.

Use of Hydrogen Sulfide by Autotrophic Microorganisms
Although hydrogen sulfide is toxic to many microorganisms, the photosynthetic sulfur bacteria use it as an electron donor for generating reduced coenzymes during their photosynthetic metabolism. The anaerobic photosynthetic bacteria often occur on the surface of sediments, where there is light to support their activities and a supply of hydrogen sulfide from dissimilatory sulfate reduction and anaerobic degradation of organic sulfur–containing compounds. Some photosynthetic bacteria deposit elemental sulfur as an oxidation product, whereas others form sulfate.

Some bacteria, including members of the genera *Beggiatoa* and *Thiothrix*, are capable of generating ATP by oxidizing hydrogen sulfide. These bacteria deposit elemental sulfur granules within the cell, which in the absence of hydrogen sulfide can be further oxidized to sulfate (Figure 13.17). *Beg-*

FIGURE 13.15
The sulfur cycle, showing various transformations of organic and inorganic compounds.

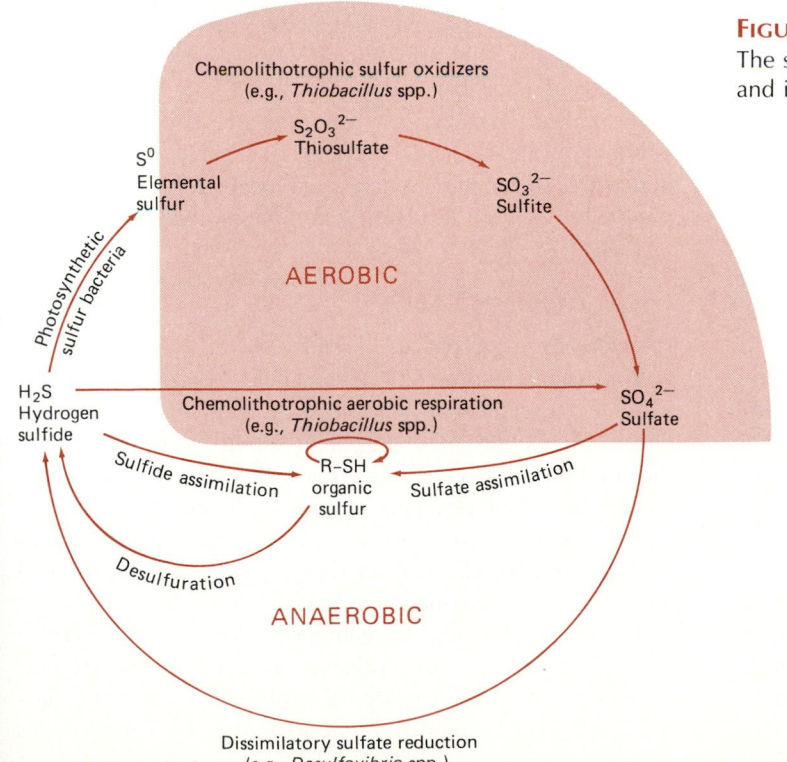

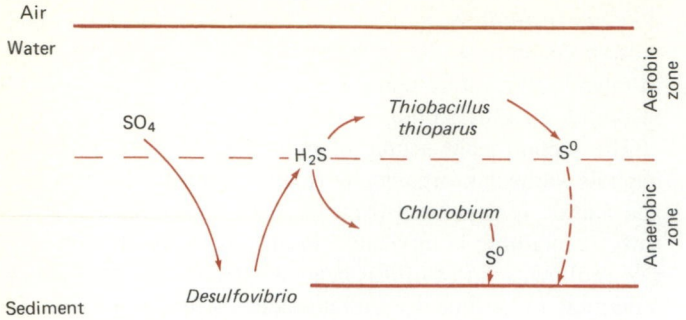

FIGURE 13.16
The products of sulfur transformations in aerobic and anaerobic habitats.

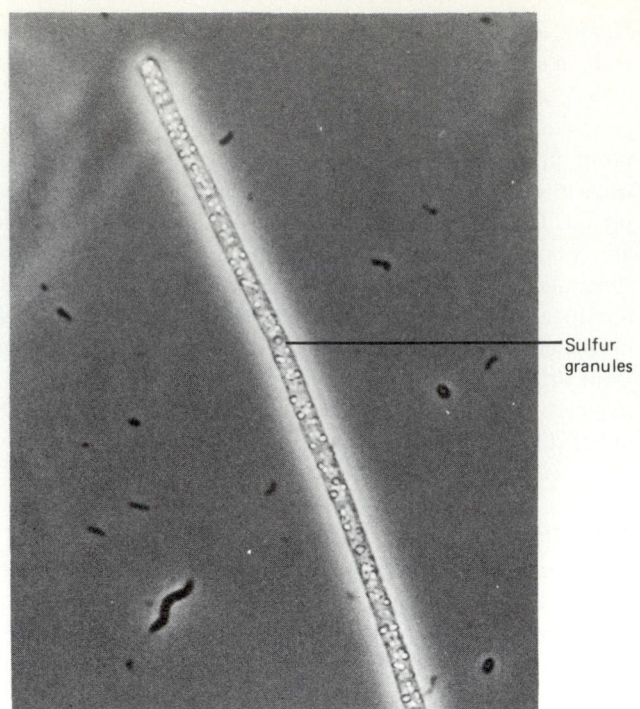

FIGURE 13.17
Micrograph of a *Beggiatoa* sp., showing the occurrence of intracellular sulfur granules (1,200×). (From BPS—Paul W. Johnson, University of Rhode Island)

giatoa and *Thiothrix* are not chemolithotrophs, and although energy is apparently derived from the oxidation of hydrogen sulfide, these organisms require organic carbon for growth. Chemolithotrophic members of the genus *Thiobacillus* oxidize sulfur as their source of energy. *Thiobacillus* species are used in bioleaching processes for mineral recovery. Some *Thiobacillus* species are acidophilic and grow well at a pH 2–3. The growth of such species can produce sulfate from the oxidation of elemental sulfur, leading to the environmental accumulation of sulfuric acid.

Acid Mine Drainage   **Acid mine drainage** is a consequence of the metabolism of sulfur and iron-oxidizing bacteria. Coal in geological deposits is often associated with pyrite (FeS_2), and when coal mining activities expose pyrite ores to atmospheric oxygen, the combination of autoxidation and microbial sulfur and iron oxidation produces large amounts of sulfuric acid. When pyrites are mined as part of an ore recovery operation, oxidation may produce large amounts of acid. The acid draining from mines kills aquatic life and renders the water it contaminates unsuitable for drinking or for recreational uses. At present, approximately 10,000 miles of U.S. waterways are affected in this manner, predominantly in the states of Pennsylvania, Virginia, Ohio, Kentucky, and Indiana. Strip mining is a particular problem with respect to acid mine drainage because this method of coal recovery removes the overlying soil and rock, leaving a porous rubble of tailings exposed to oxygen and percolating water. Oxidation of the reduced iron and sulfur in the tailings produces acidic products, causing the pH to drop rapidly and preventing the reestablishment of vegetation and a soil cover that would seal the rubble from oxygen. A strip-mined piece of land continues to produce acid mine drainage until most of the sulfide is oxidized and leached out; recovery of this land may take 50–150 years.

The overall reaction for the oxidation of pyrite can be summarized as $2 FeS_2 + 7.5 O_2 + 7 H_2O \rightarrow 2 Fe(OH)_3 + 4 H_2SO_4$. The sulfuric acid produced accounts for the high acidity and the precipitated ferric hydroxide for the deep brown color of the effluent. The mechanism of pyrite oxidation in acid mine drainage is quite complex. At neutral pH, oxidation by atmospheric oxygen occurs rapidly and spontaneously, but below pH 4.5, autoxidation is slowed drastically. In the pH range of 4.5–3.5, the stalked iron bacterium *Metallogenium* catalyzes the oxidation of iron. As the pH drops below 3.5, the acidophilic bacteria of the genus *Thiobacillus* oxidize the reduced iron sulfide in the pyrite. The rate of microbial oxidation of FeS is several hundred times greater than the rate of spontaneous oxidation, and although pyrite oxidation starts spontaneously, microbial oxidation of sulfur and iron is responsible for the continued production of high levels of acid mine drainage.

Other Element Cycles

Phosphorus   Phosphorus normally occurs as phosphates in both inorganic and organic compounds. Microorganisms assimilate inorganic phosphate and mineralize organic phosphorus compounds, and microbial activities are involved in the solubilization or mobilization of phosphate compounds. Unlike the other elements discussed, microorganisms normally do not oxidize or reduce phosphorus. The phosphorus cycle represents physical movement of phosphates without alteration of the oxidation level. In many habitats, phosphates are combined with calcium, rendering them insoluble and unavailable to most organisms. Various heterotrophic microorganisms are capable of solubilizing phosphates primarily through the production of organic acids. These actions of microorganisms mobilize phosphate, and activities of

FIGURE 13.18

An algal bloom in a small lake resulting from eutrophication. (Courtesy Ronald F. Lewis, U.S. Environmental Protection Agency, Cincinnati)

other microorganisms act to immobilize phosphorus. For example, microorganisms compete with plants for available phosphate resources because the assimilation of phosphates by microorganisms removes phosphates from the available nutrient pool required by plants.

In many habitats, productivity is limited by the availability of phosphate. When excess concentrations of phosphate enter phosphate-limited aquatic habitats, as for example when waste water–containing phosphate detergents are added to lakes, there can be a sudden increase in productivity, a process called **eutrophication**[3] (Figure 13.18). The blooms of algae and cyanobacteria associated with eutrophication can greatly increase the concentrations of organic matter in bodies of water. During the subsequent decomposition of this organic matter, the water column can be severely depleted of oxygen, causing major fish kills. The introduction of high concentrations of phosphate from phosphate laundry detergents created such serious eutrophication problems in many water bodies that some municipalities banned their use.

Although phosphate-reducing microorganisms have not been isolated, there are indications that some microorganisms are capable of using phosphate as a terminal electron acceptor in anaerobic respiration pathways under appropriate environmental conditions. Phosphate theoretically can serve as a terminal electron acceptor in the absence of sulfate, nitrate, and oxygen. The product of phosphate reduction would be phosphine (PH_3), which is volatile and spontaneously ignites upon contact with oxygen, producing a green glow. The production of phosphine has sometimes been observed near burial sites and swamps where there is extensive microbial de-

[3]Eutrophication is an increase in organic matter concentration that often occurs when a factor that normally limits primary productivity, which produces organic matter, no longer is a limiting factor. For example, adding phosphate to a lake where the concentration of phosphate is the key factor limiting primary productivity allows increased formation of organic matter. If phosphate, however, was not the principal factor limiting productivity, then adding it would not increase organic matter production and cause eutrophication.

composition of organic matter, giving rise to "ghostly" phenomena.

Iron   The cycling of **iron compounds** has a marked effect on the availability of this essential element for other organisms. Iron is transformed between the ferrous (Fe^{+2}) and ferric (Fe^{+3}) oxidation states by microorganisms (Figure 13.19). Ferric and ferrous ions have very different solubility properties, ferric compounds tending to be less soluble than ferrous compounds. Bacterial transformations of iron are important in corrosion processes and in the formation of acid mine drainage. Various bacteria, including members of the genera *Thiobacillus*, *Gallionella*, and *Leptothrix* are capable of oxidizing iron compounds. Some of these bacteria deposit

FIGURE 13.19

The iron cycle, showing interconversion of ferrous and ferric iron.

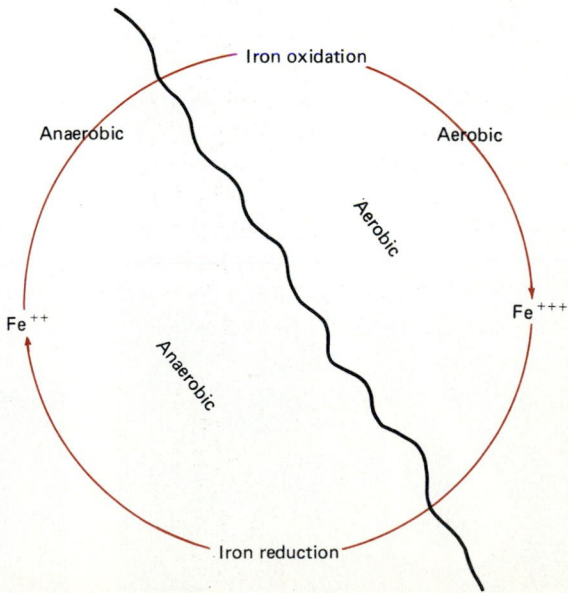

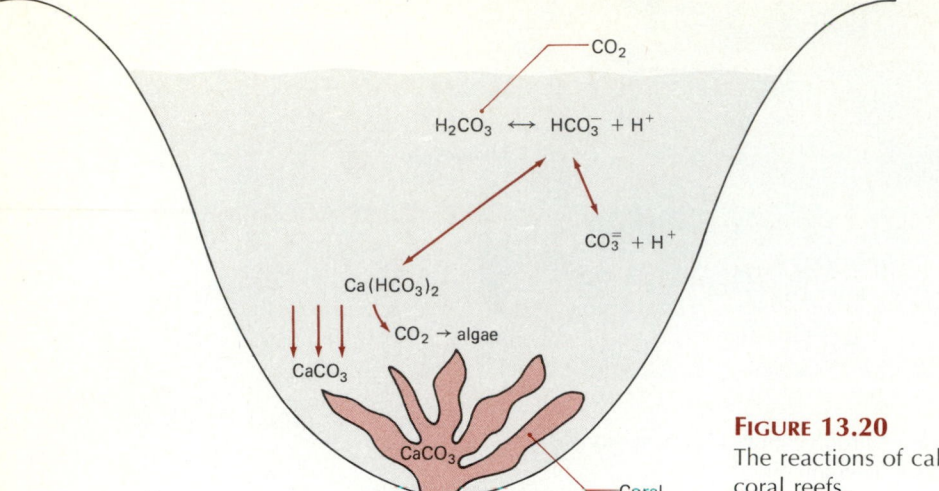

FIGURE 13.20
The reactions of calcium in seawater lead to the formation of coral reefs.

ferric hydroxide in an extracellular sheath. Over eons of time, the accumulation of iron-oxidizing bacterial sheath material can lead to the formation of substantial iron deposits.

Calcium Calcium also exhibits biogeochemical cycling between soluble and insoluble forms. Calcium bicarbonate is extremely soluble, calcium carbonate is much less so. The microbial production of acidic compounds solubilizes precipitated and immobilized calcium compounds. There is an interesting cycling of calcium in marine habitats where dissolved carbon dioxide reacts with available calcium, forming calcium bicarbonate and calcium carbonate (Figure 13.20).

During the formation of coral, calcium carbonate precipitates when carbon dioxide held in solution as calcium bicarbonate is removed by algal cells of the coral. This process results in the deposition of calcium carbonate and the formation of coral reefs. Calcium carbonate is also precipitated by various algae to form an outer frustule. Accumulation of algal frustules can lead to the formation of major limestone deposits, such as the famous white cliffs of Dover on the British coast of the English Channel.

Silicon Various algae, most notably the diatoms, form silicon-impregnated structures (Figure 13.21). These algae pre-

FIGURE 13.21
These micrographs of diatoms show the delicate beauty of their frustules. (A) Interior view of *Tabellaria flocculosa* (6,900×); (B) exterior view of *Stephanodiscus niagarae* (3,740×). (Courtesy Ed Theriot, University of Michigan)

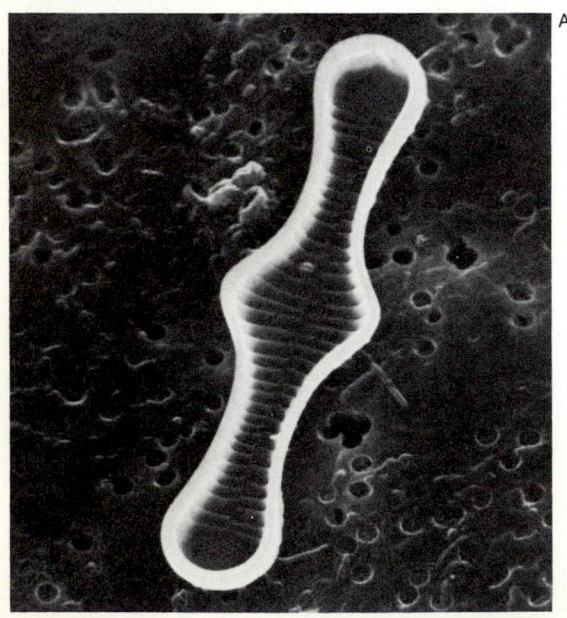

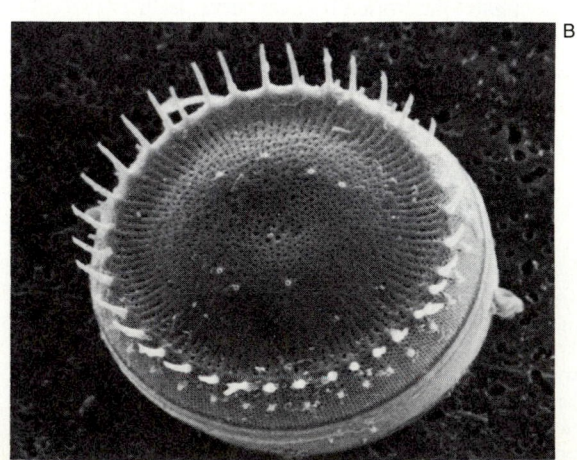

cipitate silicon dioxide to build their delicate, decorative shells. As much as 10 billion metric tons of silicon dioxide is precipitated by microorganisms in the oceans each year. The shells of these dead microorganisms accumulate and form silicon-rich oozes that later develop into deposits of diatomaceous or Fuller's earth.

Manganese   Manganese exists as a water-soluble divalent manganous ion and as a relatively insoluble tetravalent manganic ion. The microbial oxidation of manganous ions forms manganese oxides, which produce characteristic **manganese nodules** (Figure 13.22). The manganese for the nodules originates in anaerobic sediments; when the manganese enters aerobic habitats, it is oxidized and precipitates to form the nodules. The farming of manganese nodules in deep ocean sediments is considered a possible method of obtaining manganese for industrial usage.

Heavy Metals   Mercury, arsenic, and other heavy metals are also subject to microbial biogeochemical cycling. These transformations are important because they alter both the mobility and toxicity of the metals. For example, mercury is released into the environment largely as a consequence of its widespread use in industry and the burning of fossil fuels, although some mercury is also leached from rocks. The **methylation of mercury** causes increased toxicity and **biomagnification** (Figure 13.23). Mercury salts, though fairly toxic, are excreted efficiently; therefore, their release into the environment was not originally viewed with much concern. In anaerobic sediments, however, some microorganisms are capable of methylating mercury, that is, adding a methyl group to mercury. The product, methylmercury, is lipophilic and is readily concentrated in filter-feeding shellfish. Unlike inorganic and phenylmercury compounds, methylmercury is

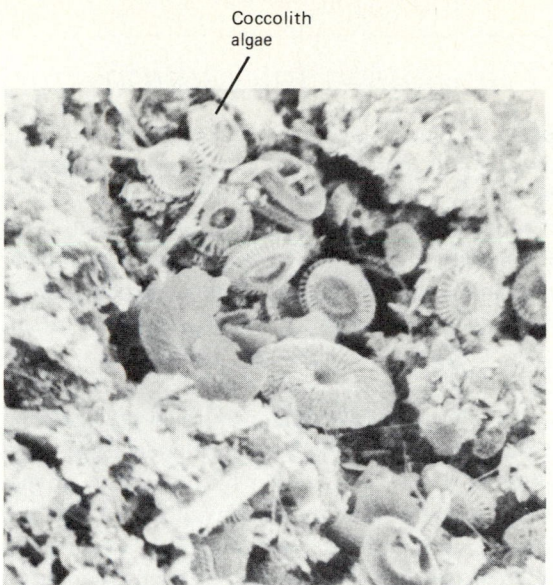

Coccolith algae

FIGURE 13.22
A manganese nodule, showing associated bacteria and algae. In this scanning electron micrograph, the nodule is covered with the debris of cocolith algae. (Courtesy Paul LaRock, Florida State University)

excreted by humans very slowly, having a half-life of 70 days, and it is highly neurotoxic. In Japan in the 1950s, the ingestion of shellfish containing methylmercury led to outbreaks of Minamata disease, a severe disturbance of the central nervous system associated with mercury poisoning. The buildup of methylmercury compounds in Scandinavian freshwater lakes and the U.S. Great Lakes forced large areas to be condemned for fishing.

FIGURE 13.23
The mercury cycle, showing the biological magnification of methylmercury in the aquatic food chain.

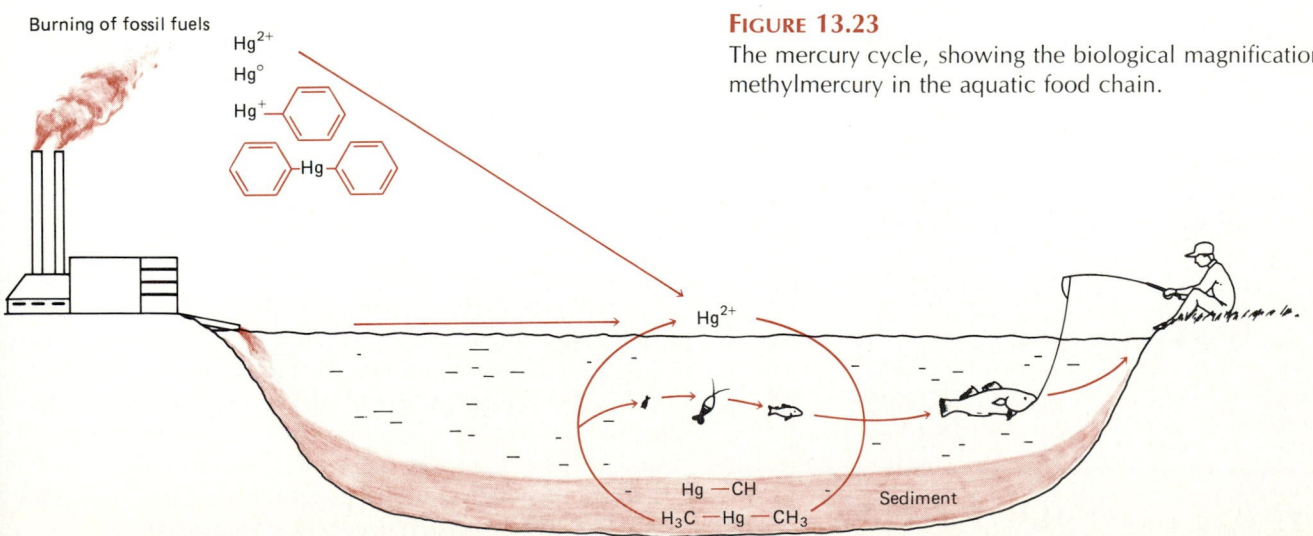

13.2 Population Interactions

Having considered the role of microorganisms in modifying their biochemical surroundings, we now turn to another aspect of microbial ecology: the interactions between diverse microbial populations. Interactions between two different biological populations can be classified according to whether both populations are unaffected by the interaction, one or both populations benefit, or one or both populations are adversely affected (Table 13.2). Within a biological community, various types of interactions can occur between diverse microbial populations, between microbial and plant populations, and between plant and animal populations. The interactions between populations within a community are dependent on the environmental conditions of the habitat, and under different environmental conditions the same populations can exhibit different interpopulation relationships.

The **positive interactions** between biological populations enhance the ability of the interacting populations to survive within the community of a particular habitat, sometimes permitting populations to co-exist in a habitat where individually they cannot exist alone. The development of positive interactions permits microorganisms to use available resources more efficiently in concert than can be accomplished by an individual microbial population growing alone. **Mutualistic or symbiotic relationships** between two microbial populations create essentially new organisms that are capable of occupying niches that cannot be occupied by either organism alone. For example, lichens, which represent a symbiotic assemblage of microorganisms, grow on rock surfaces where the individual populations cannot.

Negative interactions between populations act as feedback mechanisms that limit population densities. In some cases, negative interactions may result in the elimination of a population that is not well adapted for continued existence within the community of a given habitat. Within stable communities, negative interactions are themselves adaptive and ensure the maintenance of a balance between populations within the biological community. The negative feedback interactions limit population densities and provide a self-regulation mechanism that benefits the overall population in the long term because it prevents overpopulation and destruction of the habitat's resources. Negative interactions also tend to preclude the invasion of an established community composed of autochthonous (indigenous) populations by allochthonous (foreign) populations, and thus act to maintain community stability.

Neutralism

The first possible interaction that we will discuss, **neutralism**, actually represents a lack of interaction between two populations. Neutralism is rare but can occur between populations that are physically removed from each other. A lack of interaction is more likely at low population densities, when organisms are not likely to come into physical contact with each other, than when population densities are high. Dormant resting stages of microorganisms are more likely to exhibit neutralism toward other microbial populations than are actively growing vegetative cells. Low rates of metabolic activity, which characterize the resting stages of microorganisms, favor a lack of interaction. Low population densities and the formation of resting stages by some microbial populations allow temporal and spatial niches within a habitat and permit many populations to coexist there. Under these conditions, organisms coexist without competing for the same available resources in the habitat. Different microbial populations do not occupy the same niche within a community at the same time because competition between them for the same resources would lead to elimination of the least fit population and retention of the most adaptive population.

Commensalism

In a commensal relationship between two populations, one population benefits and the other one is unaffected. By definition, **commensalism** is a unidirectional relationship between populations. Commensal relationships between diverse microbial populations are quite common, often occurring when the unaffected population modifies the habitat in such a way that a second population benefits. For example, the removal of oxygen from a habitat, as a result of the metabolic activities of a population of facultative anaerobes, creates an environment that is favorable for the growth of obligately anaerobic populations. The lowered oxygen tension favors the anaerobic bacteria, and assuming that there is lack of competition for the same available substrates, the obligate anaerobes do not affect the existence of the facultative organisms. Various other chemical modifications of the environment of a given habitat by one microbial population likewise may benefit other populations without resulting in any negative or positive feedback interactions.

TABLE 13.2   Classification of Population Interactions

Name of Interaction	Effect of Interaction	
	Population A	**Population B**
Neutralism	0	0
Commensalism	0	+
Synergism (Protocooperation)	+	+
Mutualism (Symbiosis)	+	+
Competition	−	−
Amensalism	0 or +	−
Parasitism	+	−
Predation	+	−

0 = no effect.
+ = positive effect.
− = negative effect.

In some cases, a microbial population can physically alter a habitat, permitting a second population to exist. Production of a primary bacterial film on the hull of a ship, for example, permits secondary colonization by many other microorganisms that results in fouling. In a similar manner, a primary infection caused by one microbial species may allow opportunistic pathogens to establish secondary infections, with the opportunistic pathogens benefiting from their ability to invade and multiply within the host organism without adversely affecting the primary pathogen population.

Many commensal relationships between microbial populations are based on the production of **growth factors**. Some bacterial populations produce and excrete growth factors, such as vitamins, that can be utilized by other populations. As long as the growth factors are produced in excess and are excreted from an organism, a commensal interaction can occur. For example, fastidious microorganisms often depend on growth factors released from other organisms. Some marine bacteria, growing within the water column, depend on specific amino acids and/or vitamins produced by surface algae. Often it is difficult to culture such bacteria on defined media because of a lack of understanding of the organism's growth factor requirements.

Cometabolism, which occurs when an organism growing on a particular substrate gratuitously oxidizes a second substrate that it is unable to assimilate, forms the basis for many commensal relationships. Although the organism responsible for the transformation does not benefit, other populations may use the oxidation products. For example, *Mycobacterium vaccae*, growing on propane as a source of carbon and energy, will gratuitously oxidize cyclohexane to cyclohexanone (Figure 13.24). The cyclohexanone can be used by other microorganisms, such as populations of *Pseudomonas* species. In such a case, the *Pseudomonas* benefit because they are unable to metabolize cyclohexane; the *Mycobacterium* is unaffected because it does not assimilate cyclohexanone. In a similar sense, the waste products of one organism

may be a favorable substrate for the growth of another organism. Coprophagous fungi, for example, live on the fecal material of animal populations. The fungi benefit from the animals' deposition of fecal material, and the members of that animal population are unaffected by the relationship.

Still another basis for commensal relationships is the removal or neutralization of toxic materials. The oxidation of hydrogen sulfide by a microbial population, for example, can lower the concentration of this toxic material to a level at which other populations can exist. The existence of other microbial populations may not positively or negatively affect the hydrogen sulfide–oxidizing microorganisms.

A commensal relationship may also be established when one organism grows on the surface of another organism. For example, bacterial populations growing on the skin surface generally exhibit a commensal relationship with human beings. They benefit from being able to grow on a surface and from the nutrients and water provided in human sweat. Humans are not adversely affected, nor do they necessarily benefit directly from the growth of various microbial populations on the skin surface. Many animals have naturally occurring **epiphytic** (surface) bacterial populations (Figure 13.25). Analogously, epiphytic bacteria grow as commensal populations on many plant surfaces. They colonize the surfaces of algae, benefiting from the metabolic activities of the algae (Figure 13.26).

Synergism

Synergism or **proto-cooperation** between two populations indicates that both populations benefit from the relationship but that the association is not obligatory. Both populations are capable of surviving independently, although they both gain advantage from the synergistic relationship. Synergistic relationships are loose in that one member population can readily be replaced by another. In some cases, it is difficult to distinguish between commensalism and synergism. One

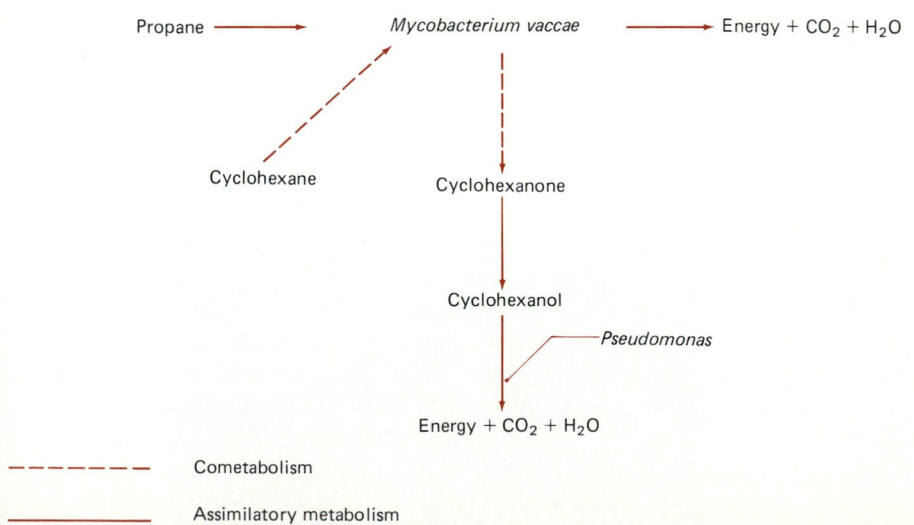

Propane ⟶ *Mycobacterium vaccae* ⟶ Energy + CO_2 + H_2O

Cyclohexane

Cyclohexanone

Cyclohexanol

Pseudomonas

Energy + CO_2 + H_2O

– – – – – – Cometabolism

———— Assimilatory metabolism yielding energy

FIGURE 13.24
The cometabolism of cyclohexane by *Mycobacterium vaccae* in the presence of propane is an example of commensalism based on cometabolism, which allows the growth of *Pseudomonas* on the cyclohexane.

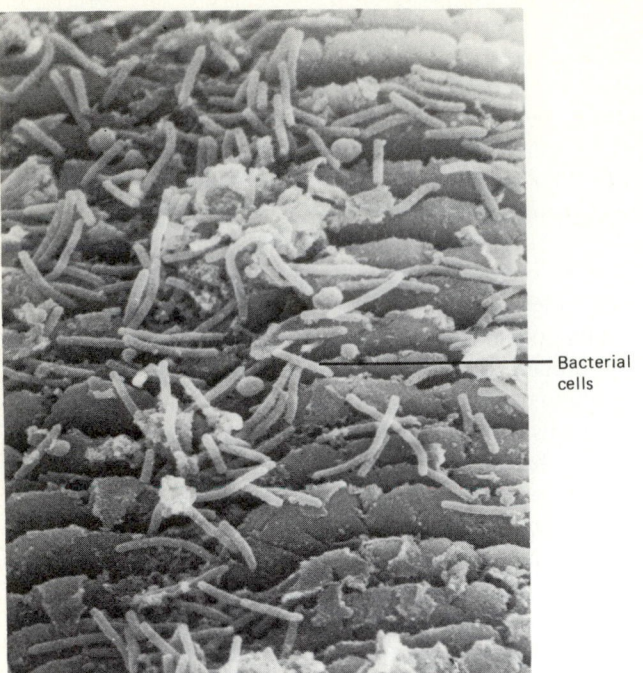

FIGURE 13.25 *(above)*

Scanning electron micrograph showing epiphytic bacteria growing on a tapeworm removed from the gut of a shark (2,000×). Bacteria can be found growing on the surfaces of most animals. (Courtesy Robert Apkarian, Yerkes Primate Research Center, Emory University)

form of synergism, **syntrophism**, occurs as a result of **cross-feeding**, in which the two populations supply each other's nutritional needs. Synergistic activities of two microbial populations may allow the completion of a metabolic pathway that neither organism is capable of carrying out alone. As a result of such cooperative metabolism, both organisms can derive carbon and energy from a substrate. In a theoretical example of cross-feeding (Figure 13.27), population 1 is able to metabolize compound A, forming compound B, but is unable to carry out the next step in a metabolic pathway to form compound C. Compound B accumulates in the cells of population 1 and is released into the surrounding medium, where it becomes available to population 2. Population 2 is unable to use compound A but can use compound B, forming compound C. Population 2 releases compound C into the medium, where it becomes available to population 1. Both populations 1 and 2 are able to carry out the subsequent metabolic steps of the pathway, producing needed energy and metabolic products, and by acting jointly the two populations are able to complete the pathway.

In a specific example of syntrophism, *Streptococcus faecalis* and *Escherichia coli* are able to convert arginine to putrescine together, although neither organism can carry out the transformation alone. *S. faecalis* is able to convert arginine to ornithine, which can then be used by a population of *E. coli* to produce putrescine; *E. coli* growing alone can transform arginine to produce agmatine, but cannot convert

FIGURE 13.26 *(below)*

Algae and plant surfaces are colonized by various epiphytic bacteria. (A) Epiphytic bacteria and diatoms on the filamentous surface of the alga *Pylaiella littoralis*. This micrograph was made from a Nomarski differential interference contrast microscope (650×). (B) Filaments of the bacterium *Leucothrix mucor* growing on the surface of the red alga *Polysiphonia* (63×). (From BPS—Paul W. Johnson, University of Rhode Island)

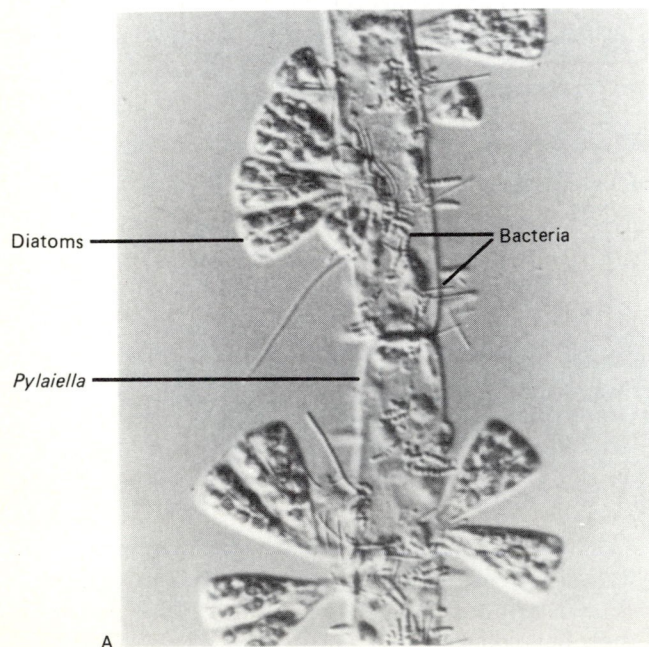

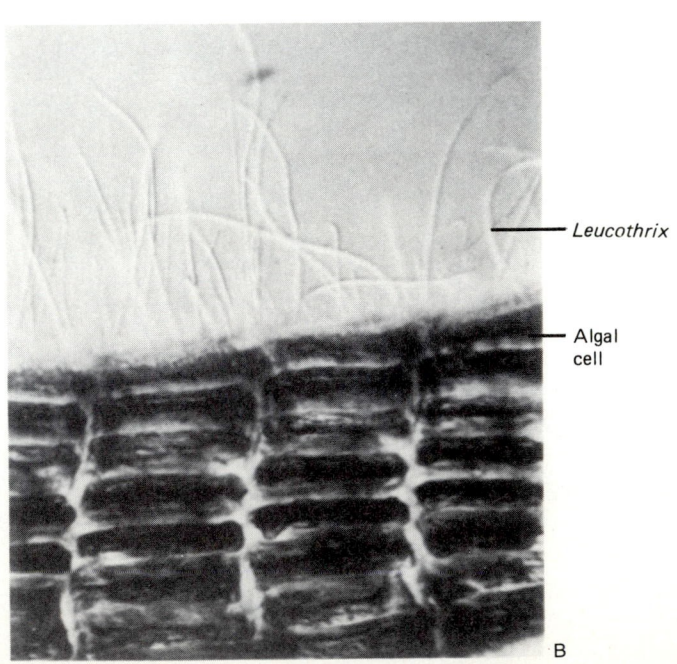

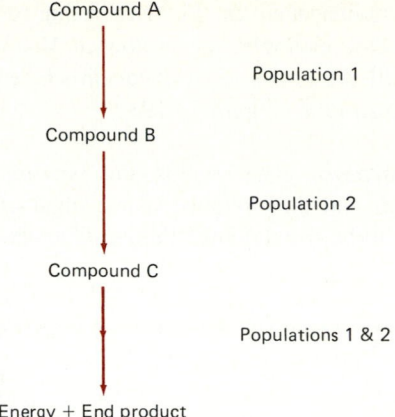

FIGURE 13.27
As shown in this theoretical example, cross-feeding establishes the basis for synergistic relationships.

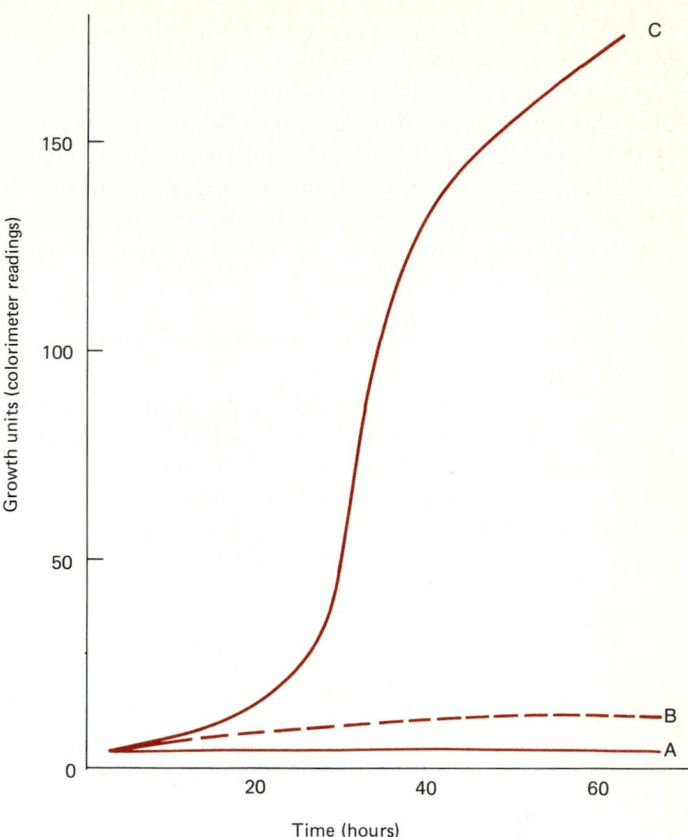

FIGURE 13.28
The synergistic relationship between *S. faecalis* and *L. arabinosus* in a medium lacking phenylalanine and folic acid. (A) *S. faecalis*, which requires folic acid; (B) *L. arabinosus*, which requires phenylalanine; (C) combined culture.

arginine to putrescine. *Lactobacillus arabinosus* and *S. faecalis* similarly can establish a synergistic relationship based on the mutual exchange of required growth factors (Figure 13.28). *L. arabinosus* requires phenylalanine for growth, which is produced by *S. faecalis*. *S. faecalis* requires folic acid, which is produced by *L. arabinosus*. In a minimal medium both populations can grow together, but neither population can grow alone.

Synergistic interactions between plants and microorganisms are important in providing the nutritional requirements of both members. Within the rhizosphere, the soil region in close contact with plant roots, the plant roots exert a direct influence on the soil bacteria, known as the **rhizosphere effect**. Likewise, the microbial populations in the rhizosphere have an important influence on the growth of the plant. As a consequence of these interactions, microbial populations reach much higher densities in the rhizosphere than in the free soil, and plants exhibit enhanced growth characteristics. The interactions of plant roots and rhizosphere microorganisms are based largely on interactive modification of the soil chemical environment by processes such as water uptake by the plant system; release of organic chemicals to the soil by the plant root; microbial production of plant growth factors; and microbially mediated availability of mineral nutrients.

Within the rhizosphere, the bacteria obtain required growth factors, such as amino acids, from the plant roots. In turn, the microorganisms in the rhizosphere have a marked influence on the growth of plants. In the absence of appropriate rhizosphere microbial populations, plant growth may be impaired. Microbial populations in the rhizosphere may benefit the plant by (1) removing hydrogen sulfide, which is toxic to the roots; (2) increasing solubilization of mineral nutrients; (3) synthesizing vitamins, amino acids, auxins, and gibberellins that stimulate plant growth; and (4) antagonizing potential plant pathogens through competition and developing amensal relationships based on the production of anti-

biotics. **Allelopathic substances**, which are substances formed by an organism that inhibit other organisms, produced by microorganisms in the rhizosphere may allow plants to enter amensal relationships with other plant populations. Allelopathic substances surrounding some plants prevent invasions of that habitat by other plants, and such amensal relationships between plant populations may actually be based on synergistic relationships between plant and microbe.

Mutualism

Mutualism or symbiosis is an obligatory interrelationship between two populations that benefits both of them. Because all relationships can be viewed as beneficial in the overall sense of maintaining ecological balance, the term *symbiosis* was originally used to denote any intimate relationship between two populations, whether beneficial or not. The present use of the term **symbiosis**, however, is restricted to situations where both organisms benefit and where the relationship is obligatory. **Mutualism** is an extension of synergism, allowing populations to unite and establish essentially a single unit population that can occupy habitats unfavorable for the existence of either population alone. Mutualistic relationships may lead to the evolution of new organisms, and various theories of evolution point to the structural similarities between mitochondria, chloroplasts, and prokaryotic cells to indicate that the development of eukaryotic organisms was based on the establishment of endosymbiotic relationships with bacteria, that is, where bacterial cells began to live within eukaryotic cells and both the bacteria and eu-

karyotes became mutually dependent on this relationship for survival. In a more specific example, the protozoan *Mixotricha paradoxa* is propelled by rows of attached spirochetes rather than by conventional cilia (Figure 13.29).

Microbe-microbe Interactions   The relationships between some heterotrophic fungi and their photosynthetic algal or cyanobacterial partners in the formation of **lichens** is an ex-

---

FIGURE 13.29

A section through the cell surface of the flagellate *Mixotricha paradoxa* (31,000×). Two bacteria (b) and parts of three brackets (br) are shown, together with numerous spirochetes. Note the long filaments (fl) on the posterior surfaces of the brackets, the dense granules (g) in their bases, the strands of the fibrous network (n), and the chains of small vacuoles (v). (Reprinted by permission of the Royal Society of London, from L. R. Cleveland and A. V. Grimstone, 1964, *Proceedings of the Royal Society of London (series B)*, 159:683)

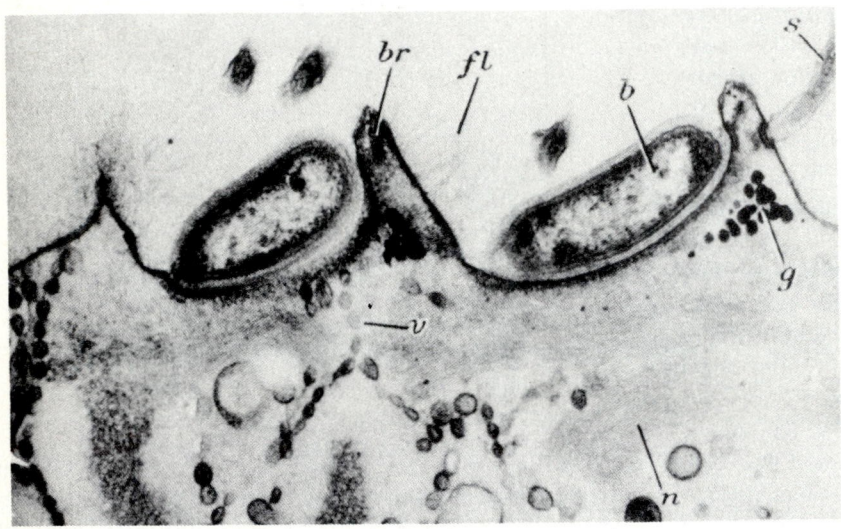

Discovery Process

Grimstone and Cleveland's careful microscopic studies showed that *M. paradoxa*, isolated from the intestines of termites, is a bizarre composite form; the host organism is a protozoan, a polymastigote flagellate with one forward and three trailing typical eukaryotic flagella. It moves in a weird straight motion propelled by spirochetes, which are symbionts arranged regularly on the cortex of the host. The four normal complex eukaryotic flagella seem to act as rudders, causing changes in the direction of the host movement. These cortical spirochetes are attached at specific sites, each connected to a raised portion of the host and simultaneously to a second extracellular symbiotic bacterium. The spirochete, host cortex, and associated cortical bacterium are patterned so specifically that this symbiosis must be the result of many years of evolution. This organism, which is normally subjected to the low oxygen levels of the termite gut, lacks mitochondria. Another symbiont, an endosymbiotic bacterium,

was observed to be surrounded by host endoplasmic reticulum and in a particular spatial arrangement with the cortical spirochetes.

Grimstone and Cleveland suggested that the endosymbiotic bacteria might be the functional replacement of the mitochondria and the source of ATP. Thus, *Mixotricha* normally harbors three independent microbial, prokaryotic symbionts: cortical spirochetes and their associated bacteria that provide the anomalous movement; endosymbiotic bacteria that perhaps replace mitochondria; and the *Mixotricha* itself, symbiotic in the intestine of its termite host. The discovery of the nature of *Mixotricha* lends support to those who hypothesize that major evolutionary steps resulted from endosymbiotic relationships that eventually led to the formation of new organisms in which the symbiotic partners had lost their individual identities.

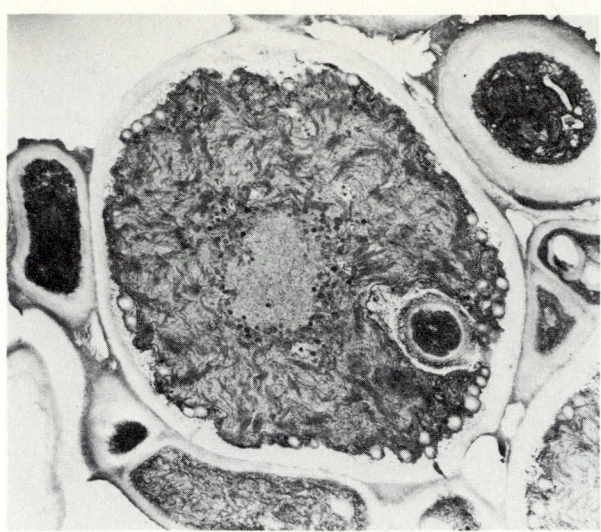

FIGURE 13.30

The ultrastructure of a lichen, showing the fungal and algal or cyanobacterial components.
(A) *Cladonia cristatella*, an algal cell (*Trebouxia erici*) with a large chloroplast that fills most of
the cell, and a central pyrenoid surrounded by lipid-containing pyrenoglobuli, which are small,
dark or black bodies. Note the haustorium, the penetrating hyphae, within the algal cell and
the appressoia, the hyphal cells, against the outer algal wall. In lichens with green algal symbionts,
most of the carbon fixed photosynthetically by the alga is excreted as some type of sugar alcohol
and made available to the fungus. (B) In this scanning electron micrograph, the fungal hyphae
can be seen growing into and around their photosynthetic partner (3000×). (A—courtesy J. B.
Jacobs and V. Ahamdjian, Woester Polytechical Institute; B—courtesy Robert Apkarian, Yerkes
Primate Research Center, Emory University)

cellent example of a mutualistic intermicrobial population relationship that results in the formation of an essentially new organism. Lichens are composed of a primary producer, the **phycobiont**, and a consumer, the **mycobiont** (Figure 13.30). The lichen has totally different physiological properties than either of the species of which it is composed. Lichens can grow in habitats, such as on rock surfaces, where neither algae nor fungi can exist alone. Most lichens are resistant to extremes of temperature and desiccation, a particular advantage on exposed surface habitats. The lichen is a very self-sufficient organism. The phycobiont utilizes light energy and atmospheric CO_2 to produce the organic matter consumed by

the mycobiont. In some lichens the cyanobacterial partner is also capable of fixing atmospheric nitrogen. The mycobiont provides physical protection for the lichen phycobiont and produces organic acids that can solubilize rock minerals, making essential nutrients available to the lichen.

Although the mutualistic relationships of algal and fungal populations in lichens are normally stable, they can be disrupted by environmental perturbations. Lichens are extremely sensitive to air pollution, and sulfur dioxide in the atmosphere is particularly inhibitory to lichens (Figure 13.31). Exposure to sulfur dioxide reduces the efficiency of the photosynthetic activity by the phycobiont, allowing the

FIGURE 13.31

The disappearance of lichens from urban areas is illustrated in this graph. By examining lichens on trees, it is possible to determine the extent of air pollution and sometimes even the direction of the source of air pollutants.

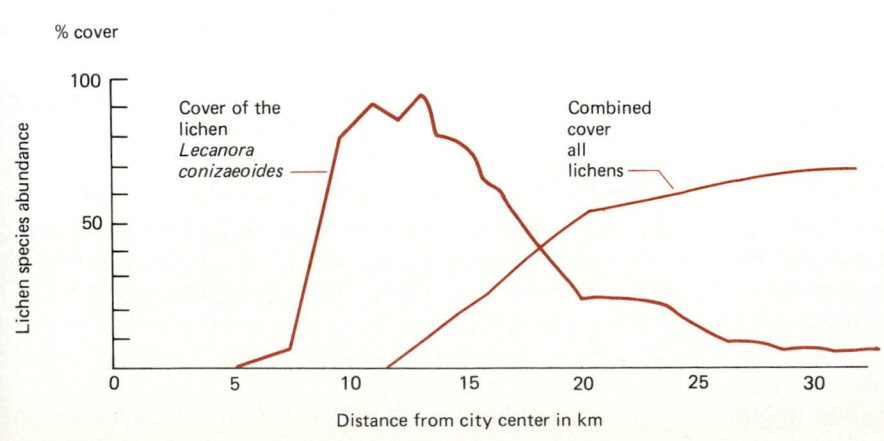

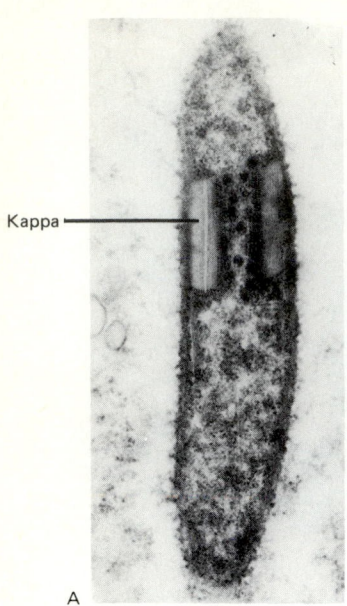
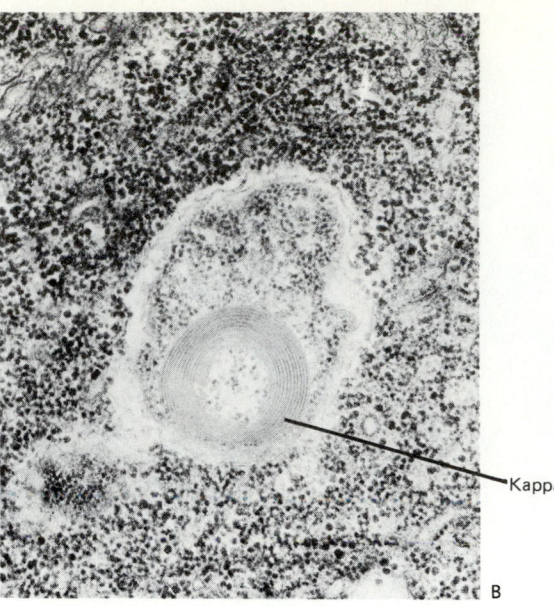

FIGURE 13.32
Kappa particles of *Paramecium aurelia*. (A) *P. aurelia*, showing the presence of endosymbiotic bacteria occurring as kappa particles. (B) An isolated kappa particle. (A—courtesy of John R. Preer, Jr., Indiana University; reprinted by permission of Cambridge University Press, New York, from J. R. Preer, Jr., and A. Jurand, 1968, *Genetic Research*, 12:331–340; B—courtesy Eugene W. McArdle, Northeastern Illinois University)

mycobiont to overgrow it and leading to the elimination of the symbiotic relationship. Once the careful metabolic balance is interrupted, the lichen and its member algal and fungal populations disappear from the habitat.

Another interesting mutualistic relationship occurs between the protozoan *Paramecium aurelia* and various bacterial species. The obligately **endosymbiotic bacteria** live within the protozoan, where they appear as structures such as kappa particles (Figure 13.32). There are two classes of populations of *P. aurelia*: killer strains that contain kappa particles and sensitive strains that lack the bacterial endosymbionts. The presence of kappa particles gives an important advantage to killer strains when they are in competition for available resources with sensitive strains because those with kappa particles can eliminate those that lack them. The endosymbiotic bacteria probably derive nutritional benefits from the protozoan.

Plant–Microbe Interactions

Mycorrhizae. Microorganisms establish important relationships with plants such as the nitrogen-fixing symbiosis between *Rhizobium* and leguminous plants discussed earlier. The formation of **mycorrhizae**, which are mutualistic relationships between fungi and plant roots, is another important symbiotic relationship between microorganisms and plants. The fungus derives nutritional benefits from the plant roots and contributes to the plant's nutrition. The establishment of mycorrhizal associations involves the integration of plant roots and fungal mycelia into a unified morphological unit. Some plants with mycorrhizal fungi are able to occupy habitats that they otherwise could not inhabit. The importance of this microbe–plant interaction is attested to by the fact that 95 percent of all plants have mycorrhizae.

There are several types of mycorrhizal associations differentiated on the basis of the degree of integration of the fungus into the root structure (Figure 13.33). **Ectomycorrhizae** are characterized by the formation of an external fungal sheath around the root and fungal penetration of the intercellular regions of the root. Such ectomycorrhizal associations occur in most oak, beech, birch, and coniferous trees. **Endomycorrhizae** involve fungal penetration of root cells. The **vesicular-arbuscular (VA)** type of mycorrhizal association in which the root cortex contains specialized inclusions, called **vesicles** and **arbuscules**, is the most common form of mycorrhiza. This association frequently goes unnoticed because it does not have a macroscopic effect on root morphology. Most major agricultural crop plants, including wheat, maize, potatoes, beans, soybeans, tomatoes, strawberries, apples, oranges, grapes, cotton, tobacco, tea, coffee, sugar cane, sugar maple, and others, form VA endomycorrhizal associations. It is apparent from this list that this type of mutualistic association is very common and widely distributed.

Animal–Microbe Interactions

Insects. There are some particularly interesting mutualistic relationships between microorganisms and animal populations. Some plant-eating insect populations, for example, actually cultivate microorganisms on plant tissues (Figure 13.34). The microorganisms degrade cellulosic plant residues, providing a digestible source of nutrition for the insects, which lack cellulase enzymes and cannot derive any nutritional benefit from simply eating plant material. The insects provide the microorganisms with a habitat in which they can proliferate. At the same time, the insects process the plant material, preparing a suitable medium for microbial growth and, in some cases, secreting substances that protect the growing microorganisms from invasion by other microbial species.

The **fungal gardens** of myrmicine ants, the attini, are an

A

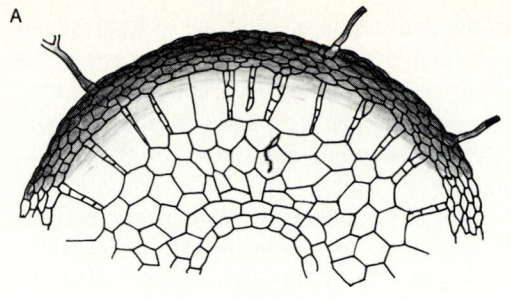

B

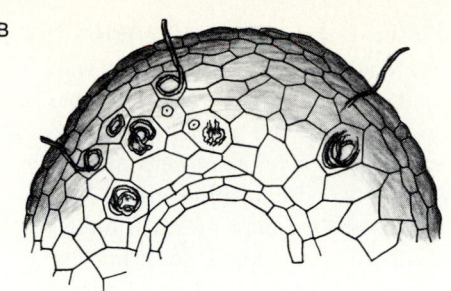

C

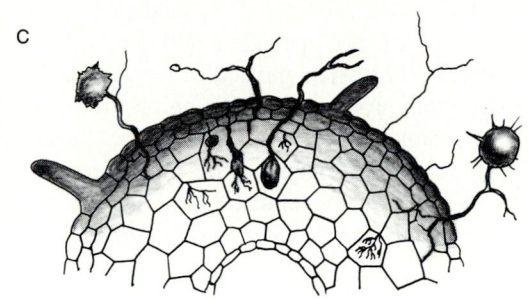

FIGURE 13.33

Structures of the different types of mycorrhizae. (A) Ectotrophic mycorrhizal rootlet, showing the fungal sheath and intercellular penetration; (B) entotrophic mycorrhizal rootlet showing penetration of hyphae; (C) rootlet with vesicular-arbuscular mycorrhiza showing penetration of hyphae and tree-like and vesicle-like hyphal structures.

FIGURE 13.34

Some insects cultivate fungi as their food source. Compare the appearance of uncultivated (left) and cultivated (right) fungal mycelia and their respective gardening insects (middle). (A) Ambrosia beetle; (B) termite; (C) Attini ant. (Based on S. W. T. Batra and L. R. Batra, 1967, *Scientific American* 217(5):112–120).

A

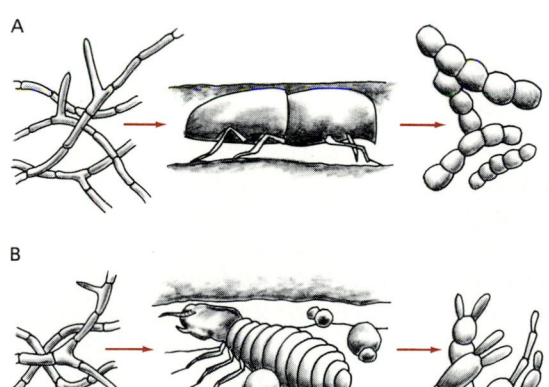

B

C

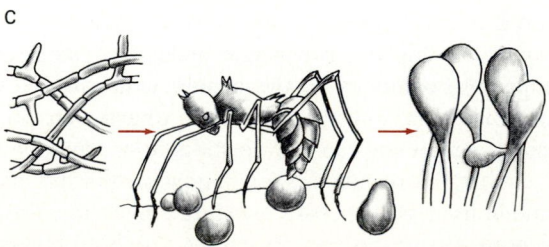

Discovery Process

It has been discovered that insects are unique among animals in having developed mutualistic relations with fungi. This may have come about because so many of them share the same small habitats. In addition, most insects are equipped to carry living spores of fungi either in their guts, in the folds between their joints that secrete waxy substances, or among their bristles. Elucidating the biochemical basis for the fungus–insect symbiotic relationships has involved various lines of investigation concerning the physiological ecology of both the insects and the fungi. Fungus-gardening insects are apparently able to produce and spread antibiotic substances that prevent the growth of alien fungi; substances secreted by the insects also transform the mutualistic fungi, causing either ambrosia, spherules, or bromatia (particles consisting of swollen tips of fungal filaments) to appear, rather than sexual fruiting bodies. Two distinct kinds of mutualistic associations between insects and fungi have been found. In the gardens of wood wasps, ambrosia beetles, termites, and ants, the fungus extracts nourishment from a substrate, and the insect feeds either on the fungus, the substrate predigested by the fungus, or both. The fungus is prevented from producing sexual fruiting structures but is supplied by its insect partner with a suitable habitat and a means of dispersal. In the colonies of the fungus *Septobasidium* and scale insects, the situation is reversed; the insect feeds on the fungus, and the fungus provides shelter for its castrated insect partner.

excellent example of an insect population growing fungi in pure culture. The ants macerate leaf material, mix it with saliva and fecal matter, and inoculate the prepared substrate with a pure fungal culture. After growth of the fungus, the ants harvest a portion of the fungal biomass and the by-products they ingest. Various wood-inhabiting insects, including ambrosia beetles and some termites, maintain similar mutualistic relationships with microbial populations. In these cases, the animals rely on the cellulolytic enzymes of microbial populations to convert plant residues into nutritional sources that they can use. The insect provides the microorganism with an optimal habitat for growth.

Ruminant Ecosystem. Ruminant animals, such as cows, llamas, and camels, establish similar mutualistic relationships with microbial populations. Although plants are the main food sources for these animals, ruminant animals do not produce cellulase enzymes themselves; instead, they depend on microbial populations for the degradation of the cellulosic materials they consume. The **rumen**, the large first chamber within the stomach of these animals, provides a stable, constant-temperature, anaerobic environment for the establishment of mutualistic associations with microbial populations. The plant material ingested by the animal provides a continuous source of nutrients for the microorganisms within the rumen, very much like what occurs in a continuous fermentor.

The overall equation for the fermentation that occurs in the rumen is $57.5 (C_6H_{12}O_6) \rightarrow 65$ acetate $+ 20$ propionate $+ 15$ butyrate $+ 60 CO_2 + 35 CH_4 + 25 H_2O$. Microbial populations within the rumen convert cellulose, starch, and other polysaccharides to carbon dioxide, hydrogen gas, methane, and low molecular weight organic acids (Figure 13.35). A portion of the low molecular weight fatty acids, carbon dioxide, and molecular hydrogen produced by various fermentative bacteria, such as *Ruminococcus*, are converted by methanogenic bacteria to methane. Although hydrogen is produced by many of the fermentative bacteria in the rumen, it does not accumulate because of its rapid utilization by methanogenic bacteria. Cows burp considerable amounts of the methane generated by these bacteria within the rumen. The organic acids produced by the microbial populations are absorbed into the bloodstream of the animal, where they are oxidized aerobically to produce the ATP needed to meet the animal's energy requirements. Because the rumen is anaerobic, most of the caloric content of the ingested plants is maintained in the fatty acids transferred to the bloodstream of the animal.

There are diverse bacterial and protozoan populations within the rumen. Some of these microbial populations are found only within the specialized environment of the rumen, and others also occur in analogous environments, making this a borderline case between synergism and mutualism. Clearly, though both animal and microbial populations benefit from this relationship, there is a delicate balance among the individual populations within the complex microbial community in the rumen, with each population contributing metabolically to the conversion of substrate to fermentation products. The population balances can be upset by sudden diet changes in ruminants, leading to a condition of bloat due to excessive gas formation. Restoration of the metabolic balance among microbial populations restores the healthy state of the animal.

Bioluminescence. The mutualistic relationship between some **luminescent bacteria** and marine invertebrates and fish is particularly interesting. The light emitted by these bacteria is blue-green and is emitted continuously, provided that oxygen is available. The production of light is mediated by the enzyme luciferase, and is based upon the reaction of reduced flavin mononucleotide ($FMNH_2$), molecular oxygen, and a long-chain aldehyde that produces FMN in an electronically excited state. The return of the excited FMN to its ground state results in the emission of light. Some fish have specific organs in which they maintain populations of luminescent bacteria, including members of the genera *Photobacterium* and *Beneckea* (Figure 13.36). Although the bacteria normally emit light continuously, the fish are able to manipulate the organs containing the luminescent bacteria so as to emit flashes of light. The fish supply the bacteria with nutrients and protection from competing microorganisms. The light emitted by the bacteria is used in various ways by different fish. In some cases, the pattern of light emission is used in sexual mating rituals. In deep-sea and nocturnal fish, such as the flashlight fish *Photoblepharon*, the light emitted by the bacteria aids the fish in finding food sources and warding off predators.

Competition

Competition occurs when two populations are striving for the same resource. Often it focuses on a nutrient present in limited concentrations, but it may also occur for other resources, including light and space. As a result of the competition, both populations achieve lower densities than would have been achieved by the individual populations in the absence of competition. Competitive interactions tend to bring about ecological separation of closely related populations, a fact known as the **competitive exclusion principle**. Competitive exclusion prevents two populations from occupying the same ecological niche. When two populations attempt to occupy the same niche, one will win the competition and the other will be excluded (Figure 13.37). Chemostats, which by definition have a growth-limiting substrate in the growth medium, are used frequently to study competition between populations under controlled conditions. As a rule, the population with the higher growth rate under the given set of environmental conditions in the habitat will win over the population with the lower growth rate. Fluctuations in environmental conditions can lead to shifts in competitive balances, resulting in population oscillations within the microbial community. Spatial separation allows microorganisms to escape competitive pressures, permitting coexistence of competitive populations.

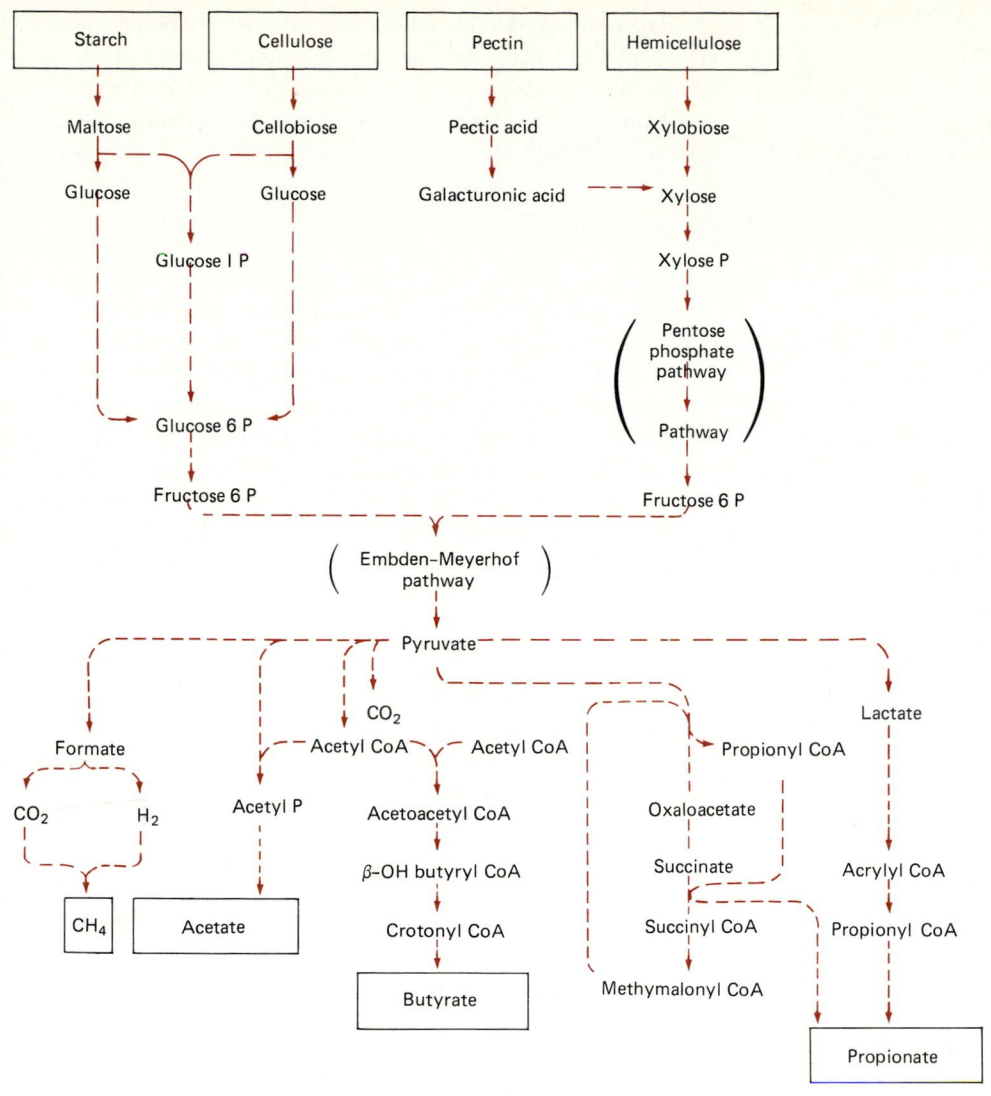

Substrate	Organism	Products	Substrate	Organism	Products
Starch	*Bacteroides amylophilus*	Formate, acetate, succinate	Cellulose (cont'd)	*Ruminococcus albus*	Acetate, formate, hydrogen, carbon dioxide
	Bacteroides ruminicola	Formate, acetate, succinate		*Clostridium lochheadii*	Acetate, formate, butyrate, hydrogen, carbon dioxide
	Selenomonas ruminantium	Acetate, propionate, lactate	Pectin	*Lachnospira multiparus*	Acetate, formate, lactate, hydrogen, carbon dioxide
	Succinomonas amylolytica	Acetate, propionate, lactate)			
	Streptococcus bovis	Lactate	Lactate	*Selenomonas lactilyitca*	Acetate, succinate
Cellulose	*Bacteroides succinogenes*	Succinate, acetate, formate		*Peptostreptococcus elsdenii*	Acetate, propionate, butyrate, valerate, hydrogen, carbon dioxide
	Butyrivibrio fibrisolvens	Acetate, formate, lactate, butyrate, hydrogen, carbon dioxide	Carbon dioxide + Hydrogen	*Methanobrevibacter ruminantium*	Methane

FIGURE 13.35

Many metabolic transformations occur within the rumen. This diagram shows the pathways involved in the rumen fermentations of the major insoluble carbohydrates present in plants. Compounds in boxes are major substrates or fermentation products. A number of bacterial species are involved in rumen fermentations.

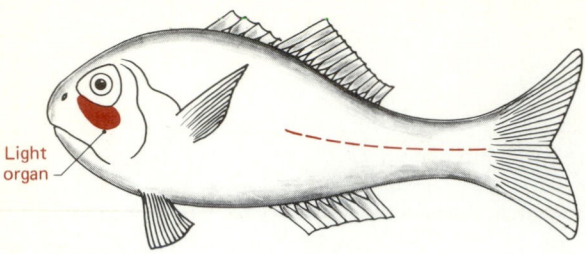

FIGURE 13.36
Various deep-sea fish have specialized organs containing luminous bacteria. The flashlight fish, *Anomalops katoptron*, has such luminous organs situated below the eye.

Amensalism

Amensalism, or **antagonism**, occurs when one population produces a substance inhibitory to another population. The first population gains a competitive edge as a result of its ability to inhibit the growth of competitive populations. The production of antibiotics, for example, can give the antibiotic-producing population an advantage over a sensitive strain when competing for the same nutrient resources. The production of lactic acid by *Streptococcus* and *Lactobacillus* species similarly eliminates competitors. The preemptive colonization of food products by lactic acid bacteria precludes the invasion of that food by other bacterial species. This fact is utilized in the production and preservation of food products by the addition of lactic acid as a preservative. Various other chemicals produced by microbial populations, including in-

organic compounds such as oxygen, ammonia, mineral acids, and hydrogen sulfide, and organic compounds, such as fatty acids, alcohols, and antibiotics, permit the establishment of commensal relationships between microbial populations.

Parasitism

In a relationship of **parasitism**, the parasite population is benefited and the host population is harmed. As a rule, parasitic relationships are characterized by a relatively long period of contact, and the parasite is smaller than the host. The parasite normally derives its nutritional requirements from the host cell, and in the process the host cell is damaged. The host–parasite relationship is typically quite specific. Some microorganisms are obligate parasites, their existence depending on the successful establishment of a parasitic relationship with a host organism. For example, viruses are obligate intracellular parasites, able to multiply only within suitable host cells. Similarly, rickettsiae are obligately parasitic bacteria and sporozoans are obligately parasitic protozoa. Many human diseases result from infections with microbial parasites, and some diseases of plants and animals will be considered in our discussion of agricultural microbiology. Such host–parasite relationships that cause disease syndromes clearly exert a negative influence on the susceptible host and benefit the parasite.

Parasitic relationships also exist between diverse microbial populations. Bacteriophage invade and multiply within bacterial cells, causing lysis and death of the bacteria; viruses

FIGURE 13.37
Competition between organisms eliminates the slower-growing species. Here two protozoans that have similar niches are competing. In a mixed culture the faster growing *Paramecium aurelia* replaces *P. caudatum*.

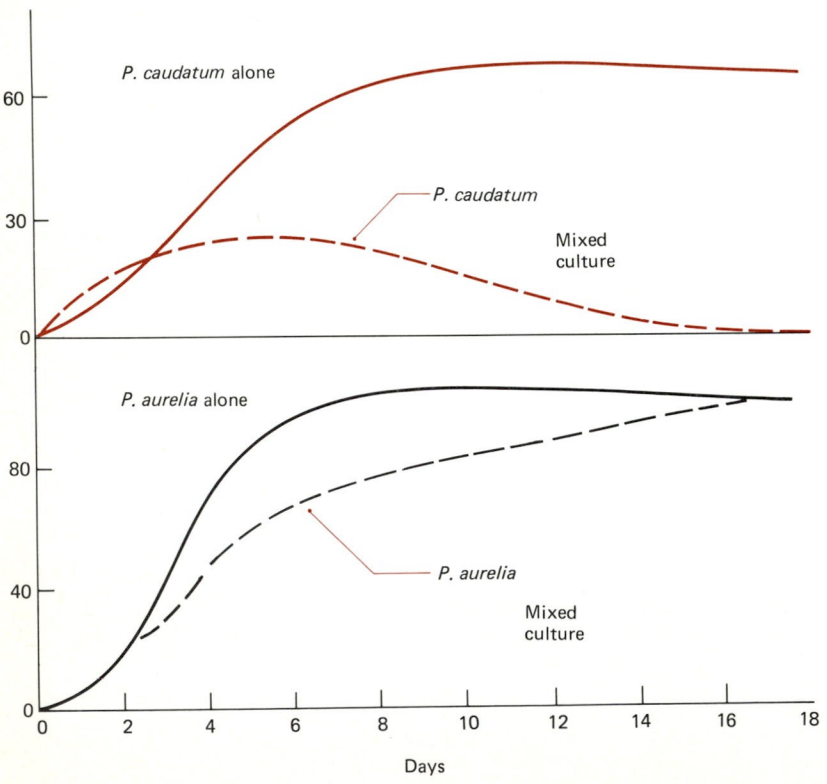

Days

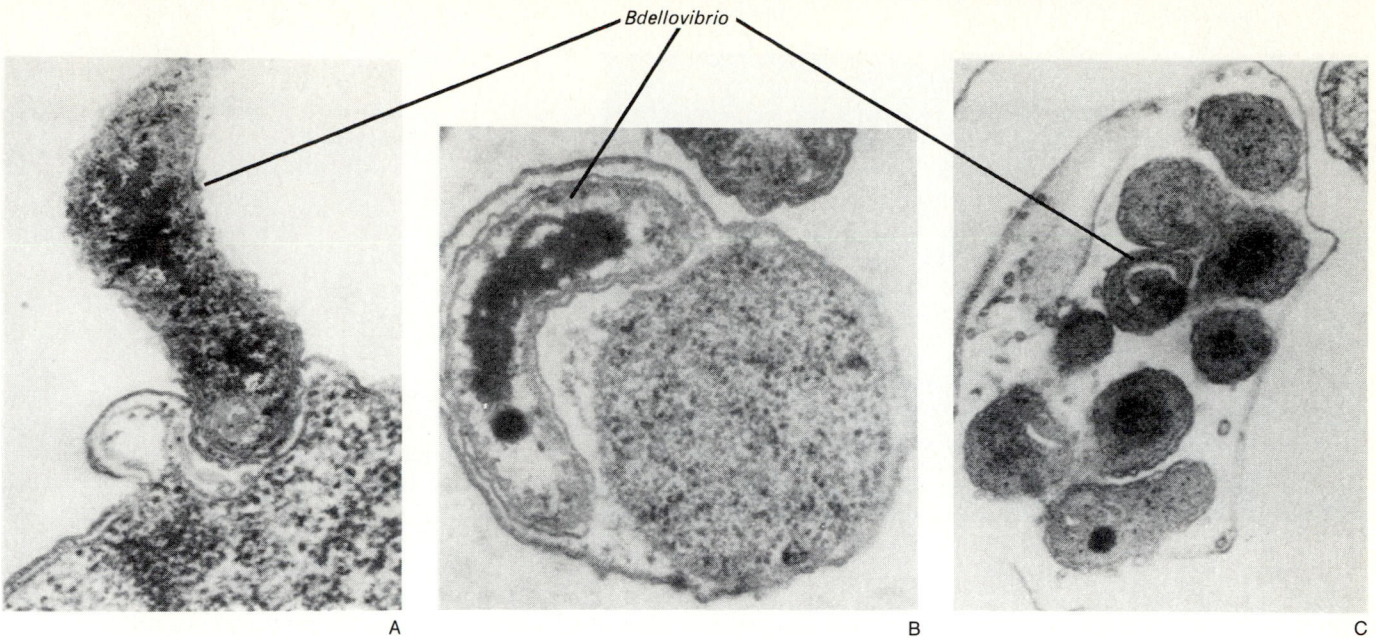

Bdellovibrio

A   B   C

FIGURE 13.38

(A) *Bdellovibrio bacteriovorus* entering an *Escherichia coli* host cell. (B, C) *B. bacteriovorus* within an *E. coli* host cell. (Courtesy Jeffrey C. Burnham, Medical College of Ohio, and Sam F. Conti, University of Massachusetts; reprinted by permission of the American Society for Microbiology, from J. C. Burnham, and S. F. Conti, 1968, *Journal of Bacteriology*, 96:1374)

invade fungi, algae, and protozoa. Some bacteria are parasites of other bacteria. For example, *Bdellovibrio* is parasitic on other bacterial populations, and is able to invade and multiply by binary fission within cells of *E. coli* (Figure 13.38). As a result of such parasitic interactions, populations of host cells generally decline. Parasitism as such acts as a mechanism for controlling population densities, which in an overall sense is beneficial in maintaining ecological stability.

Predation

Predation involves the consumption of a prey species by a predatory population. Normally, predator–prey interactions are of short duration and the predator is larger than the prey, but this is not always the case. The predatory populations derive nutrition from the prey species, and clearly, the predator population exerts a negative influence on the consumed prey population. Some microbial species are **predators** and others are **prey**. Many protozoa prey upon bacterial species (Figure 13.39), and the nondiscriminatory consumption of bacterial populations by protozoan predators is sometimes referred to as **grazing**. Similarly, protozoa and invertebrate animal populations graze on algal primary producers. Predation is an important process in establishing food webs to support the growth of higher organisms. Various filter-feeding animals are able to remove microorganisms from suspension. This grazing activity is important in transferring

FIGURE 13.39

This micrograph shows a partially lysed bacterium within the vacuole of an ameba that grazes on bacteria. The protozoan derives its nutrition from ingestion of bacteria (20,000×). (Courtesy O. Roger Anderson, Columbia University)

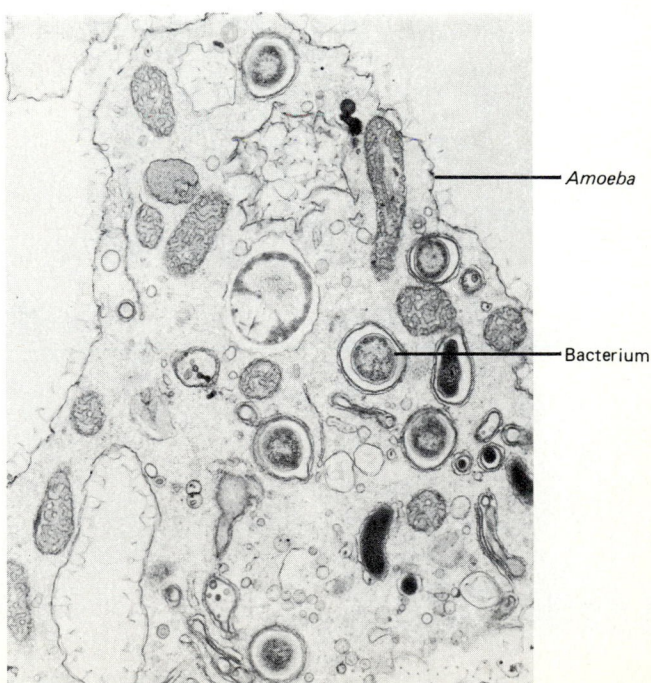

Amoeba

Bacterium

Paramecium   Didinium

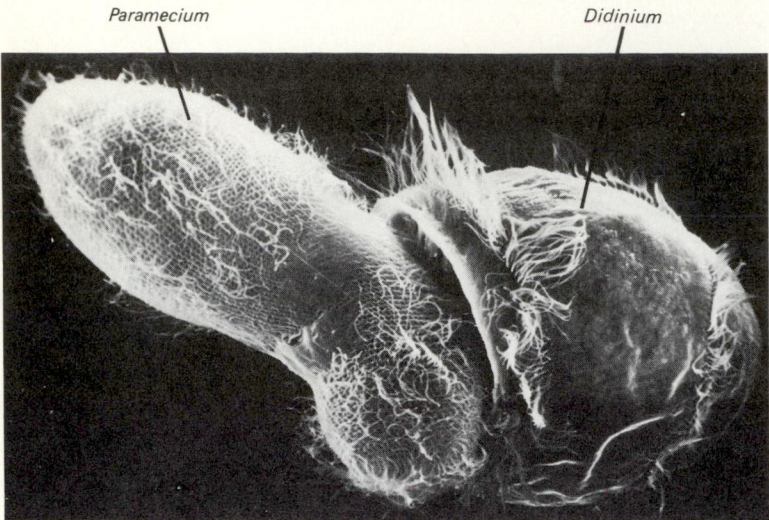

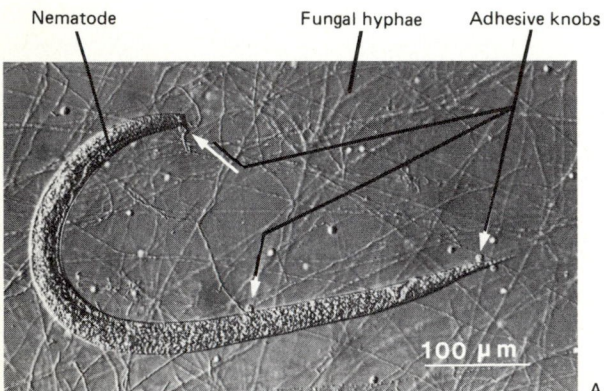

FIGURE 13.40
Didinium nasutum is seen here engulfing *Paramecium multimicronucleatum* (1,500×). (From BPS—H. S. Wessenberg and G. A. Antipa, 1970, *Journal of Protozoology* 17:250–270)

biomass from microorganisms to higher trophic levels in aquatic food webs.

Although the predator is normally larger than the prey, there are some interesting cases where a small microbial predator consumes a larger organism. For example, the protozoan *Didinium* can engulf and consume the larger protozoan *Paramecium* (Figure 13.40). Some fungi are able to trap and consume much larger nematodes (Figure 13.41). There are several mechanisms by which these fungi capture

nematode prey, including the production of networks of adhesive branches, stalk adhesive knobs, and adhesive or constrictive rings. When a nematode attempts to move past such a predatious fungus, the fungus traps it. Even violent movements by the nematode to escape the grasp of the fungus generally fail. The fungal hyphae penetrate and digest the nematode, consuming the animal.

Theoretically, interactions of predator and prey species could lead to regular cyclic fluctuations in the populations of

Nematode   Fungal hyphae   Adhesive knobs

FIGURE 13.41
Micrographs showing nematode-trapping fungi. Various types of nematode traps are produced by different fungal species. Once trapped, the nematode thrashes about, but the fungus does not let go. (A) Nomarski interference micrograph of *Dactylaria candida*, showing a nematode captured by several adhesive knobs. (Courtesy S. Olson and B. Norbring-Hertz, University of Lund), (B) Scanning electron micrograph of the adhesive network of *Arthrobytrys oligospora* (Courtesy B. Norbring-Hertz, University of Lund; reprinted from *Forum Mikrobiologie* by permission of G-I-T Verlag Ernst Giebler), (C) Micrograph showing a nematode trapped in a constricted ring trap of *A. dactyloides*. (Courtesy David Pramer, Rutgers, the State University)

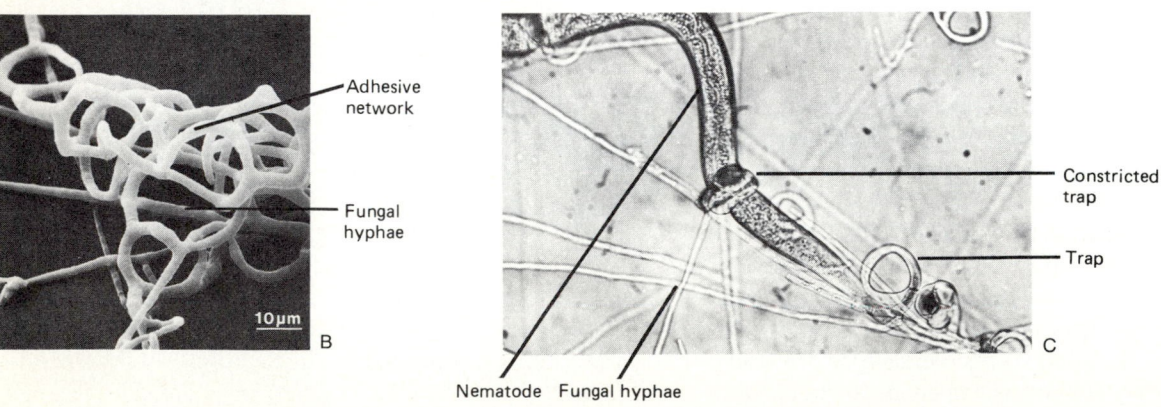

Adhesive network

Fungal hyphae

Constricted trap

Trap

Nematode Fungal hyphae

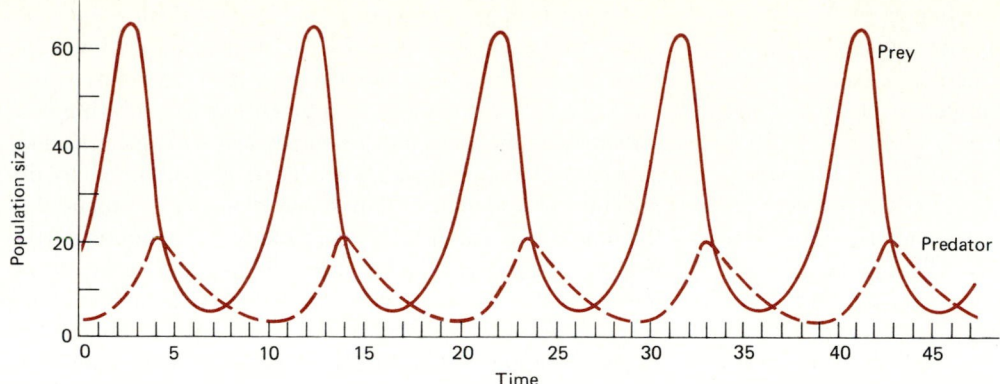

FIGURE 13.42
Theoretical population oscillations resulting from the interaction of predator and prey.

the two species (Figure 13.42). As the size of the prey population increases, it can support a larger predator population. The decline in the size of the prey population means that fewer predators can be supported, and therefore, the population of the predator population also declines. If either the predator or the prey were completely eliminated, the population of the other would be deleteriously affected. Without the prey as an available food resource, the predator would be eliminated, and without the control exerted by the predator, the prey population could grow too large, leading to the complete consumption of the available nutrient resources of the habitat. In reality this situation rarely occurs, as disturbances and other factors dampen the cyclic oscillations and generally prevent the elimination of either the predator or the prey population.

Postlude

The works of Sergei Winogradsky and Martinus Beijerinck largely established the basis for the field of **microbial ecology**, which examines the interactions of microorganisms with their biotic and abiotic surroundings. Beijerinck, a Dutchman, received his initial training in chemistry at the Delft Polytechnical School, to which he was later appointed professor in 1895. He was difficult to get along with and berated his students for even their smallest errors. Beijerinck was a keen observer with a strong general scientific background and liked to perform neat experiments with simple tools. Winogradsky was born in Russia but performed many of his studies in France and Switzerland, including work at the University of Strasbourg, while it was headed by the mycologist deBary, and at the Pasteur Institute.

Whereas Pasteur concentrated on the study of microbial heterotrophic growth using organic compounds, Winogradsky and Beijerinck made significant discoveries concerning microbial transformations of inorganic compounds and the autotrophic (self-feeding—not requiring preformed organic compounds) means of metabolism. In 1877 Theophile Schloesing and Achille Müntz proposed that nitrification, the biochemical transformations of inorganic nitrogen compounds, was a microbial process. Winogradsky isolated and described the autotrophic nitrifying bacteria, showing that they were responsible for transforming ammonium ions to nitrate ions in soil. He demonstrated that microorganisms could derive energy from inorganic chemical reactions and their carbon from carbon dioxide, and by doing so he defined chemoautotrophic metabolism. Winogradsky also described the microbial oxidation of sulfur, hydrogen sulfide, ferrous iron, and anaerobic nitrogen-fixing bacteria.

During the period when Winogradsky was working on the metabolism of inorganic substances, 1888 to 1901, Beijerinck was reporting on symbiotic and nonsymbiotic aerobic nitrogen fixation by bacteria, the process by which atmospheric nitrogen is combined with other elements, making this essential nutrient available to plants, animals, and other microorganisms. Beijerinck also isolated sulfate-reducing bacteria, which are important in the cycling of sulfur compounds in soil and sediment. All of these reactions form the basis of important transformations and movements of elements in soil ecosystems and determine the fertility of soil. Another particularly significant advance made by Beijerinck was the development of the technique of enrichment culture, which permits the isolation of a bacterium with a particular metabolic activity by the adjustment of incubation conditions.

The works of Beijerinck and Winogradsky were primarily concerned with soil processes, but the microbial transformations that they discovered formed the basis for an understanding of biogeochemical cycling reactions and the critical role of microorganisms in transforming elements on a global scale. These microbially mediated cycling reactions are essential for maintaining environmental quality and are necessary for supporting life on Earth. Microorganisms, because of their diversity and ubiquitous distribution, are extremely important in ecological processes. Critical steps in the major global biogeochemical cycles are mediated by microorganisms; for example, virtually all steps in the biogeochemical cycling of nitrogen involve microorganisms. The growth of higher organisms in various habitats relies on the biogeochemical cycling activities of microorganisms, which in many cases form crucial links in the food webs of an ecosystem, permitting the flow of energy and organic compounds to reach higher trophic levels. In some cases animal populations, such as ants, actually cultivate microorganisms in order to derive nutrition from plant sources they cannot digest themselves; cows and other ruminants depend on the degradation of cellulose and other plant polymers by microorganisms within the rumen for their nutrition.

Although many microbes are single-celled organisms, there are a variety of interpopulation interactions involving microorganisms. These interactions establish stability within the biological community of a given habitat, ensuring conservation of the available resources and ecological balance between cohabiting populations. There is no unnecessary duplication of function within stable biological communities; the most fit species fill the available niches, and other populations are competitively excluded. In some cases, an interpopulation relationship is beneficial to both interacting species, sometimes giving rise to an essentially new, unified "superorganism" with a greater functional capacity than that possessed by the individual organisms. Even the negative interactions between populations are beneficial to the balance of the overall community, preventing population imbalances and invasion of the indigenous autochthonous community by foreign allochthonous populations.

Suggested Supplementary Readings

ALEXANDER, M. 1971. *Microbial Ecology*. John Wiley & Sons, Inc., New York.

ALEXANDER, M. 1977. *Introduction to Soil Microbiology*. John Wiley & Sons, Inc., New York.

ALEXANDER, M. (ed.). 1984. *Biological Nitrogen Fixation: Ecology, Technology, and Physiology*. Plenum Press, New York.

ATLAS, R. M., and R. BARTHA. 1987. *Microbial Ecology: Fundamentals and Applications*. Benjamin/Cummings Publishing Co., Inc., Menlo Park, Calif.

BIRD, D. F., and J. KALFF. 1986. Bacterial grazing by planktonic lake algae. *Science* 231:493–495.

BLACKBURN, T. H. 1983. The microbial nitrogen cycle. In *Microbial Geochemistry* (C. W. E. Krumbein, ed.), pp. 63–89. Blackwell Scientific Publications, Oxford, England.

BLAKEMAN, J. P. (ed.). 1981. *Microbial Ecology of the Phylloplane*. Academic Press, London.

BRILL, W. 1980. Biochemical genetics of nitrogen fixation. *Microbiological Reviews* 44:449–467.

BRILL, W. 1981. Biological nitrogen fixation. *Scientific American* 245(3):68–81.

BROCK, T. 1966. *Principles of Microbial Ecology*. Prentice-Hall, Inc., Englewood Cliffs, N.J.

BROCK, T. 1978. *Thermophilic Microorganisms and Life at High Temperatures*. Springer-Verlag, New York.

BURNS, R. G., and J. H. SLATER (eds.). 1982. *Experimental Microbial Ecology*. Blackwell Scientific Publications, Oxford, England.

CAMPBELL, R. E. 1983. *Microbial Ecology*. Blackwell Scientific Publications, Oxford, England.

CODD, G. A. (ed.). 1984. *Aspects of Microbial Metabolism and Ecology*. Academic Press, London.

COLWELL, R. R., and R. Y. MORITA (eds.). 1974. *Effect of the Ocean Environment on Microbial Activities*. University Park Press, Baltimore.

DOETSCH, R. N., and T. M. COOK. 1973. *Introduction to Bacteria and Their Ecobiology*. University Park Press, Baltimore.

EHRLICH, H. L. 1981. *Geomicrobiology*. Marcel Dekker, New York.

FENCHEL, T. 1982. Ecology of heterotrophic microflagellates: Adaptations to heterogeneous environments. *Marine Ecology Progress Series* 9:25–33.

FENCHEL, T., and T. H. BLACKBURN. 1979. *Bacteria and Mineral Cycling*. Academic Press, London.

FERRY, B. W. 1982. Lichens. In *Experimental Microbial Ecology* (R. G. Burns and J. H. Slater, eds.), pp. 291–319. Blackwell Scientific Publications, Oxford, England.

FOCHT, D. D. 1982. Denitrification. In *Experimental Microbial Ecology* (R. G. Burns and J. H. Slater, eds.), pp. 194–211. Blackwell Scientific Publications, Oxford, England.

GOULD, G. W., and J. E. L. CORRY (eds.) 1980. *Microbial Growth and Survival in Extremes of Environment*. Academic Press, New York.

GRAY, T. R. G., and D. PARKINSON (eds.). 1968. *The Ecology of Soil Bacteria*. University of Toronto Press, Toronto.

GREGORY, P. H. 1973. *The Microbiology of the Atmosphere*. John Wiley & Sons, Inc., New York.

HOLLANDER, A. (ed.). 1977. *Genetic Engineering for Nitrogen Fixation*. Plenum Press, New York.

JANNASCH, H. W., and M. J. MOTTL. 1985. Geomicrobiology of deep-sea hydrothermal vents. *Science* 229:717–725.

JONES, J. G. 1986. Iron transformations by freshwater bacteria. *Advances in Microbial Ecology* 9:149–185.

JORGENSEN, B. B. 1980. Mineralization and the bacterial cycling of carbon, nitrogen, and sulphur in marine sediments. In *Contemporary Microbial Ecology* (D. C. Ellwood, J. N. Hedger, M. J. Latham, J. M. Lynch, and J. H. Slater, eds.), pp. 239–252. Academic Press, London.

LASKIN, A., and H. LECHEVALIER (eds.). 1974. *Microbial Ecology*. CRC Press, Inc., Boca Raton, Fla.

LJUNGDAHL, L. G., and K. E. ERICSSON. 1985. Ecology of microbial cellulose degradation. *Advances in Microbial Ecology* 8:237–299.

LOESCHE, W. J. 1986. Role of *Streptococcus mutans* in human dental decay. *Microbiological Reviews* 50:353–380.

LYNCH, J. M. 1982. The rhizosphere. In *Experimental Microbial Ecology* (R. G. Burns and J. H. Slater, eds.), pp. 395–411. Blackwell Scientific Publications, Oxford, England.

LYNCH, J. M., and N. J. POOLE (eds.). 1979. *Microbial Ecology: A Conceptual Approach*. Blackwell Scientific Publications, Oxford, England.

MARGULIS, L. 1981. *Symbiosis in Cell Evolution: Life and Its Environment on the Early Earth*. W.H. Freeman and Co., San Francisco.

MITCHELL, R. 1974. *Introduction to Environmental Microbiology*. Prentice-Hall, Inc., Englewood Cliffs, N.J.

POSTGATE, J. 1982. *Fundamentals of Nitrogen Fixation*. Cambridge University Press, New York.

POSTGATE, J. 1984. *The Sulphate-Reducing Bacteria*. Cambridge University Press, New York.

RHEINHEIMER, G. 1986. *Aquatic Microbiology*. John Wiley & Sons, Inc., New York.

SIEBURTH, J. McN. 1979. *Sea Microbes*. Oxford University Press, New York.

TIESSEN, H., and J. W. B. STEWART. 1985. The biogeochemicastry of soil phosphorus. In *Planetary Ecology* (D. C. Caldwell, J. A. Brierley, and C. L. Brierley, eds.), pp. 463–472. Van Nostrand Reinhold Co., New York.

WOOD, E. J. F. 1965. *Marine Microbial Ecology*. Chapman and Hall, London.

ZEIKUS, J. G. 1981. Lignin metabolism and the carbon cycle: Polymer biosynthesis, biodegradation and environmental recalcitrance. *Advances in Microbial Ecology* 5:211–243.

Study Questions

1. What is a food web? What are the trophic levels of a food web?

2. What functions do microorganisms play in the cycling of organic matter through food webs?

3. Discuss the role of microorganisms in the biogeochemical cycling of nitrogen; include the different processes and microbial populations involved in the global nitrogen cycle. Discuss the differences in nitrogen cycling in aquatic and soil habitats.

4. What are the problems associated with nitrification after fertilizer addition to agricultural soils?

5. How are microorganisms involved in the formation of acid mine drainage?

6. Why is neutralism favored at low population densities?

7. What is cometabolism? Why can this process be important in the degradation of complex organic pollutants?

8. How does commensalism differ from synergism?

9. What are the differences between predation and parasitism?

10. Discuss a mutualistic relationship between
    a. two microbial populations
    b. microorganisms and a plant
    c. microorganisms and an animal population

Situational Problem

Establishing the Role of Antibiotics in Nature

We all recognize the importance of antibiotics in modern medicine. Antibiotics are substances produced by microorganisms that selectively inhibit or kill other microbes. Many antibiotics are produced by soil actinomycetes; some are produced by fungi and other by bacteria. A natural assumption is that soil microbes produce antibiotics in their natural habitat and use them to gain advantage over their competitors; that is, antibiotics are presumed to be involved in naturally occurring amensal relationships in the soil. However, demonstrating that antibiotics play an important role, or even a minor role, in nature has been difficult.

Antibiotics are secondary metabolites, that is, they are not involved in the primary microbial metabolism that is concerned with energy generation and biosynthesis of cell constituents. As secondary metabolites, antibiotics are produced only when conditions are very favorable for microbial metabolism. The substrate concentrations supplied in fermentors for the commercial production of antibiotics are rarely found in nature. The concentrations of readily usable substrates in soil typically are near starvation levels and are not sufficient to support the production of secondary metabolites. Free antibiotics are not detectable in soils, at least not with the analytical instruments currently in use.

Additionally, while antibiotic resistance plasmids are

found in very high proportions of the microbial populations exposed to the antibiotics used in medicine—for example, in bacterial populations inhabiting hospitals, where large amounts of antibiotics are used—only small proportions of soil isolates are antibiotic resistant. Thus, whereas some soil microorganisms clearly have the potential for producing antibiotics, there is little indication that others need to respond to the antibiotics in their natural habitats in order to survive.

Suppose you were asked to determine whether antibiotics play a role in soil. What evidence would you consider necessary to resolve this issue, and how would you go about obtaining it? Try researching the approaches that have been used and the answers that have been proposed by searching *Biological Abstracts* to find relevant journal articles on this subject.

CHAPTER 14

Agricultural Microbiology

Microbial biogeochemical cycling activities and interpopulation relationships have a major influence on agricultural practices. As in other areas of economic microbiology, some microbial activities are beneficial and others detrimental, and it is the controlled balance of microbial activities that is important in determining agricultural success. Fertilizers and pesticides have become an integral part of modern agricultural practice because crop yields depend on maintaining high levels of soil fertility and limiting destruction due to infectious diseases and pest populations. The environmental fate and usefulness of these chemical fertilizers and pesticides are greatly influenced by the metabolic activities of microorganisms in soil. In some cases, the microbial populations themselves provide essential inputs of fixed forms of nitrogen to soil and effectively control pest populations. In other cases, microorganisms are responsible for the loss of valuable fertilizers and cause numerous harmful diseases of agricultural crops and farm animals.

14.1 Soil Fertility and Management of Agricultural Soils

Influence of Available Nitrogen on Soil Fertility

Microbial biogeochemical cycling activities are extremely important for the maintenance of **soil fertility**, that is, the ability of the soil to support plant growth. The nutrient in most limited supply normally is nitrogen, and thus, the concentration of fixed forms of nitrogen in soil usually determines the potential productivity of an agricultural field. The natural availability of fixed forms of nitrogen in agricultural soils is determined by the relative balance between the rates of microbial nitrogen fixation and denitrification. **Nitrogen-rich fertilizers** are widely applied to soils to support increased crop yields, but proper application of nitrogen fertilizers must take into consideration the solubility and leaching characteristics of the particular chemical form of the fertilizer and the rates of microbial biogeochemical cycling activities. In order to avoid the losses caused by leaching and denitrification, nitrogen fertilizer is commonly applied as an ammonium salt, free ammonia, or urea. When nitrification proceeds too quickly, as it does in some agricultural soils, wasteful losses of nitrogen fertilizer and groundwater contamination with nitrate occur. Nitrification of ammonium compounds also yields acidic products that may have to be neutralized by liming. To prevent the undesirable microbial transformation of nitrogen fertilizers, nitrification inhibitors, such as nitrapyrin, are often applied together with the nitrogen fertilizer. The use of nitrification inhibitors can increase crop yields by 10–15 percent for the same amount of nitrogen fertilizer applied. In addition, by decreasing the rate of nitrification, the problem of groundwater pollution by nitrate is prevented.

Crop Rotation **Crop rotation**, that is, alternating the types of crops planted in a field, is traditionally used to prevent the exhaustion of soil nitrogen and to reduce the cost of nitrogen fertilizer applications. Leguminous crops, such as soybeans, often are planted in rotation with other crops because of their symbiotic association with nitrogen-fixing bacteria, which reduce the soil's requirement for expensive nitrogen fertilizer. Leguminous plants produce more fixed nitrogen than they require, and the excess ammonium nitrogen is released to the soil. Of more importance in terms of soil fertility is the fact that most of the combined (fixed) nitrogen is released to the soil upon decomposition of the crop residues from leguminous plants that are plowed under (Table 14.1). Soybeans and corn are often rotated every few years in the midwestern United States because corn takes up nitrogen from the soil, substantially decreasing the concentration of soil nitrogen, but during the seasons when soybeans are grown, the level of fixed nitrogen in the soil increases. In some cases, nitrogen fixation can be enhanced by inoculation of legume seeds with appropriate *Rhizobium* strains, which increases the extent of nodule formation because of the increased numbers of rhizobia that effectively initiate the infective process that leads to nodule formation. Besides increasing the extent of nodule formation, it is possible to take steps to increase the rate of nitrogen fixation within the nodules. In molybdenum-deficient soils, a dramatic improvement in the rate of nitrogen fixation can be achieved by the application of small amounts of molybdenum because this element is a constituent of the nitrogenase enzyme complex, which is required for nitrogen-fixing activities. It is important that maximal rates of nitrogen fixation be achieved by rotation with leguminous crops to successfully replenish soil nitrogen.

Soil Management Practices

Besides nitrogen, soil organic matter (humus) is important in determining soil fertility. Humus acts as a nutrient reserve, increases the ion exchange capacity, and loosens the structure of the soil, which are all important determinants of the soil's ability to support plant growth. When virgin lands are put to agricultural use, their humus content decreases regularly over the next 40 to 50 years, eventually reaching a stable concentration at a much lower value. The probable causes for this phenomenon are losses due to wind and water erosion and increased aeration of the soil through tilling and the removal of most of the produced organic matter when the crop is harvested. The reduction in soil organic matter lowers the fertility of soil.

Soil management practices vary according to crop and soil characteristics but generally share several common features.

TABLE 14.1 Nitrogen Gains in Soils in the United States Obtained by Planting Leguminous Crops

Crop	Soil Nitrogen Increase (kg nitrogen fixed/hectare/year)
Alfalfa	100–280
Red clover	75–175
Pea	75–130
Soybean	60–100
Cowpea	60–120
Vetch	80–140

With the exception of rice grown in paddies, all major crops require aerobic soil conditions. The aeration status of soils is controlled principally by the soil moisture level. It is vital to provide adequate drainage for agricultural land in order to prevent waterlogging of soils that can lead to the formation of anaerobic soil conditions. In addition to the directly injurious effects of prolonged exposure to anaerobic soil conditions on plant roots, oxygen deprivation is indirectly deleterious to plants because of the microbial use of secondary electron acceptors, including nitrate, sulfate, and ferric iron, which results in the loss of vital nitrogen due to increased rates of denitrification, production of toxic H_2S, and deposition of sticky greenish ferrous iron in waterlogged soils.

The management of soil pH is often critical in agricultural production. The soil pH regulates solubility of plant nutrients, availability of potentially toxic heavy metals, and biodegradation of plant material. Some mineral cycling activities, such as sulfur oxidation and nitrification, result in acid production, and in some soils it is necessary to maintain the pH close to neutrality by liming.

Fields that are cleared of crop cover are subject to erosion of the topsoil, especially where the land slopes extensively. "No-till farming" is a relatively novel agricultural practice that provides a solution to the combined problems of humus loss, topsoil erosion, and soil compaction. In this type of soil management, plowing is eliminated or reduced to a minimum, planting is performed with drill-seeding machines, weeds are controlled with herbicides rather than by cultivation, and crop residues are left on the field as a mulch cover. In areas that are especially vulnerable to erosion, grass or other cover is planted for the winter and is killed with herbicides just before the next season's crop is planted. This practice builds up humus, loosens and conserves topsoil, and helps soil retain moisture. Besides its soil conservation benefits, no-till farming in some situations is economically competitive with, or even superior to, conventional farming methods.

14.2 Microbial Diseases of Crops

Whereas some microorganisms contribute to the soil fertility necessary for crop growth, some microbes cause plant diseases. There are tens of thousands of diseases of cultivated plants, and each agricultural crop is generally subject to over 100 different diseases. Diseases of agricultural crops cause serious economic losses, with annual worldwide crop losses

TABLE 14.2 Worldwide Agricultural Crop Losses Due to Plant Diseases

Crop	Loss (millions of tons)
Cereals	135
Potatoes	89
Sugar beets and sugar cane	232
Other vegetables	31
Fruits	33
Coffee, cocoa, and tobacco	3
Oil crops	14
Fiber crops and natural rubber	3

TABLE 14.4 Some Symptoms of Microbial Diseases of Plants

Symptom	Description
Necrosis (rots)	Death of plant cells; may appear as spots in localized areas
Canker	Localized necrosis resulting in lesions, usually on the stem
Wilt	Droopiness due to loss of turgor
Blight	Loss of foliage
Chlorosis	Loss of photosynthetic capability due to bleaching of chlorophyll
Hypoplasia	Stunted growth
Hyperplasia	Excessive growth
Gall	Tumorous growth
Scab	Localized lesions, usually slightly raised or sunken

due to diseases, insects, and weeds totaling about $200 billion (Table 14.2). Crop losses resulting from infectious diseases in the United States alone amount to over $4 billion per year, resulting in significant economic hardship to farmers and seriously reducing the world's supply of food. There are major regional differences in the value of crop losses due to diseases, insects, and weeds. The percentage losses differ considerably with the continent and are much greater in underdeveloped areas than in more developed regions (Table 14.3) because of the lack of available economic resources for pesticides, herbicides, fungicides, and fertilizers for maintaining healthy agricultural crops.

Outbreaks of plant diseases can cause immediate and long-lasting agricultural damage. The chestnut blight disease destroyed the native North American chestnut trees that had provided an important cash crop, especially in the Appalachian area. A leaf blight of maize in 1970 caused the destruction of more than 10 million acres of corn crops in the United States during that one year. Sometimes plant diseases even have far-reaching historical effects on masses of people. The potato blight in Ireland in 1845 resulted in mass starvation and widespread emigration from Ireland to North America.

Symptoms of Plant Diseases and Mechanisms of Microbial Pathogenicity

Microbial pathogens are a major cause of plant diseases. The mechanisms by which microbial pathogens cause plant diseases and the symptoms of different plant diseases vary with

TABLE 14.3 Regional Crop Losses Due to Plant Diseases, Insects, and Weeds

Region	Loss (% total crop)
Europe	25
Oceania	28
North and Central America	29
U.S.S.R. and China	30
South America	33
Africa	42
Asia	43

the causal agent and sometimes with the plant (Table 14.4). A plant is healthy when it can carry out its basic physiological functions, including (1) normal cell division, differentiation, and development; (2) absorption of water and minerals from the soil and translocation of these nutrients throughout the plant; (3) photosynthesis and translocation of the photosynthetic products to areas of utilization or storage; (4) metabolism of synthesized compounds; (5) reproduction; and (6) storage of food supplies for overwintering or reproduction. Plants become diseased whenever any of these functions are significantly disrupted.

Normally, **plant pathogens** (microorganisms that cause diseases of plants) weaken or destroy cells and tissues, reducing or eliminating the ability to perform their normal physiological functions and resulting in disease symptoms and reduced plant growth or death. The kinds of cells and tissues that become infected determine which physiological functions of the plant are initially impaired, as shown when (1) infection of the root, as occurs in root **rots**, interferes with absorption of water and nutrients from the soil; (2) infection of the xylem vessels, as occurs in vascular **wilts** and certain **cankers**, interferes with translocation of water and minerals to the crown of the plant; (3) infection of the foliage, as occurs in leaf spots, blights, and mosaics, interferes with photosynthesis; (4) infection of the cortex, as occurs in cortical canker and viral infections of phloem, interferes with the downward translocation of photosynthetic products; (5) infections of reproductive structures, as occur in bacterial and fungal blights and microbial infections of flowers, interfere with reproduction; and (6) infections of fruit, as occur in fruit rots, interfere with reproduction and/or storage of reserve foods for the new plant (Figure 14.1).

Infected plants can develop a variety of morphological abnormalities as a result of infection by a microbial pathogen. Invasion of plant cells by pathogenic microorganisms sometimes results in the rapid death of the plant. In other cases, the plants undergo slower changes. Pathogens that penetrate plants directly often elicit a morphological response from the plant that may result in the formation of abscission or gum layers (Figure 14.2). Formation of papillae may be an attempt by the plant's innate defense system to block the

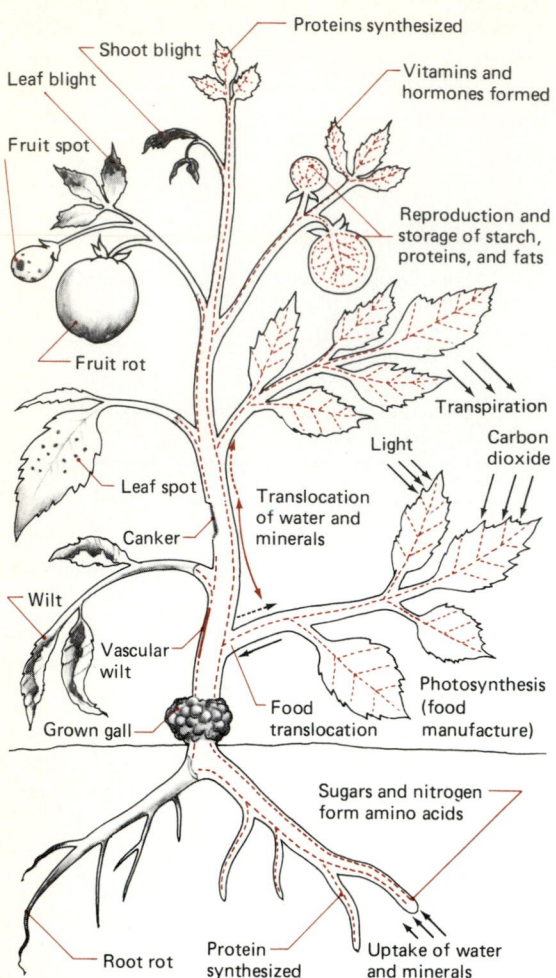

FIGURE 14.1
Basic plant functions can be interfered with by common types of plant diseases.

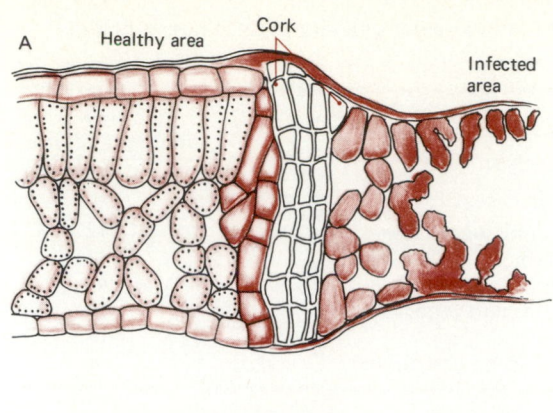

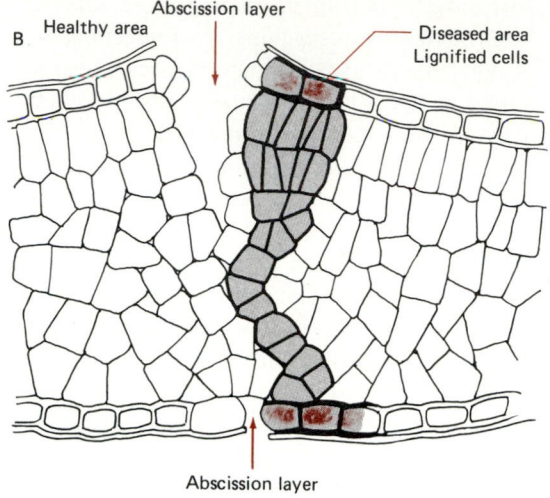

FIGURE 14.2
Cork (A) or abscission (B) layers are formed by leaves in response to microbial infections.

spread of the pathogens. The cell walls of infected plant tissues are often modified, resulting in swelling or other distortions of the cell. Cell-wall modifications may be due to the production of enzymes by the pathogenic microorganisms that degrade cell-wall components. Invasion by plant pathogens may disrupt cell permeability, leading to leakage and death of the plant cells. Changes in permeability may be caused by pectinase enzymes or by toxins produced by the plant pathogens.

Blockage of water transport in a plant can lead to desiccation and symptomatic wilt. *Fusarium*, which causes wilt in tomatoes, results in the reduction of water flow through the xylem. Stomatal dysfunction due to invading pathogens alters transpiration and water transport in the plant, and various bacterial pathogens cause wilting by blocking the stomata. *Erwinia stewartii* causes wilt of corn, *E. tracheiphila* wilt of cucumbers, *Pseudomonas solanacearum* wilt of tobacco and tomato, and *Corynebacterium insidiosum* wilt of alfalfa. These wilt diseases represent a serious cause of plant loss due to bacterial diseases.

Plant pathogens may alter the metabolic activities of the plant, and diseased plants sometimes show decreased growth as a result of changes in respiratory activity and rates of carbon dioxide fixation. Foliar pathogens sometimes produce **chlorosis** (bleaching of chlorophyll), which prevents the plant from carrying out photophosphorylation and producing the ATP needed for carbon dioxide fixation. Plant pathogens may also cause changes in protein synthesis. **Overgrowths** and **gall formation** involve alterations in the nucleic acid function controlling protein synthesis.

Microbial pathogens can disrupt normal plant functions by the production of degradative enzymes, toxins, and growth regulators. Because plant cells are held together by pectin substances, plant pathogens that produce pectinases cause the weakening of plant structures. For example, polygalacturonase enzymes degrade pectic substances, producing **soft rots** and other lesions of plant tissues. Other plant pathogens produce cellulase and/or hemicellulase enzymes that degrade the primary cell-wall components of plant cells. Some pathogenic microorganisms of plants produce toxins that interfere with the normal metabolic activities of the plant; for example, *Pseudomonas tabaci* produces an exotoxin, β-hydroxydiaminopimelic acid, which interferes with the metabolism of methionine and causes tobacco wildfire disease.

In some cases, plant diseases result from plant cells dividing too rapidly, causing **hyperplasia**, or becoming excessively enlarged, causing **hypertrophy**. Such hyperplastic or hypertrophied cells can result in the development of abnormally large, nonfunctional organs, an abnormal proliferation of organs, and/or the production of amorphous overgrowths on normal-looking organs. Overstimulated cells and tissues not only divert much needed carbon and energy away from the normal tissues but also, by their explosive growth, frequently crush adjacent normal tissues and interfere with the physiological functions of the plant. **Dwarfism** can result when plant pathogens degrade or inactivate plant growth substances. Auxin and indoleacetic acid production by some microorganisms is implicated in disease processes, such as production of galls. Gibberellins and cytokinins produced by some fungal plant pathogens result in excessive elongation of plant stems. Production of ethylene by some plant pathogens triggers metabolic changes in the plants, which lead to damage of plant tissues.

Transmission of Plant Pathogens

The development of plant diseases, which is initiated by the contact of pathogenic microorganisms with a plant surface, involves entry of the pathogen into the plant, growth of the infecting microorganisms, and finally, the appearance of disease symptoms. Pathogenic microorganisms may come into contact with the plant via its roots or aerial surfaces. Most fungal plant pathogens are dispersed through the air as spores, often coming to rest on the leaves or stems of plants. Most viral diseases of plants are transmitted by insect vectors, which bring these pathogens into contact with the plant's aerial surfaces. Some bacterial and fungal pathogens are also carried by insect vectors. Nematodes and other animals in soil can transmit pathogens to the plant root system. Motile pathogens in the soil, including plant pathogenic bacteria, such as *Pseudomonas* species, and fungi, such as *Oidium*, are attracted to plant roots through chemotaxis.

Typically, pathogenic microorganisms enter plants through natural openings and wounds caused by animals or injury from farm implements and equipment (Figure 14.3). Viruses normally enter plants through wounds caused by the vector carrying the virus. Some viruses enter plants through the roots with the groundwater taken up by the plant. Some plant pathogens, however, are able to penetrate the plant directly. Such penetration normally involves attachment of the pathogen to the plant surface, followed by formation of a penetration peg passing through the cuticle and the cell wall, accomplished mainly by physical forces. Penetration

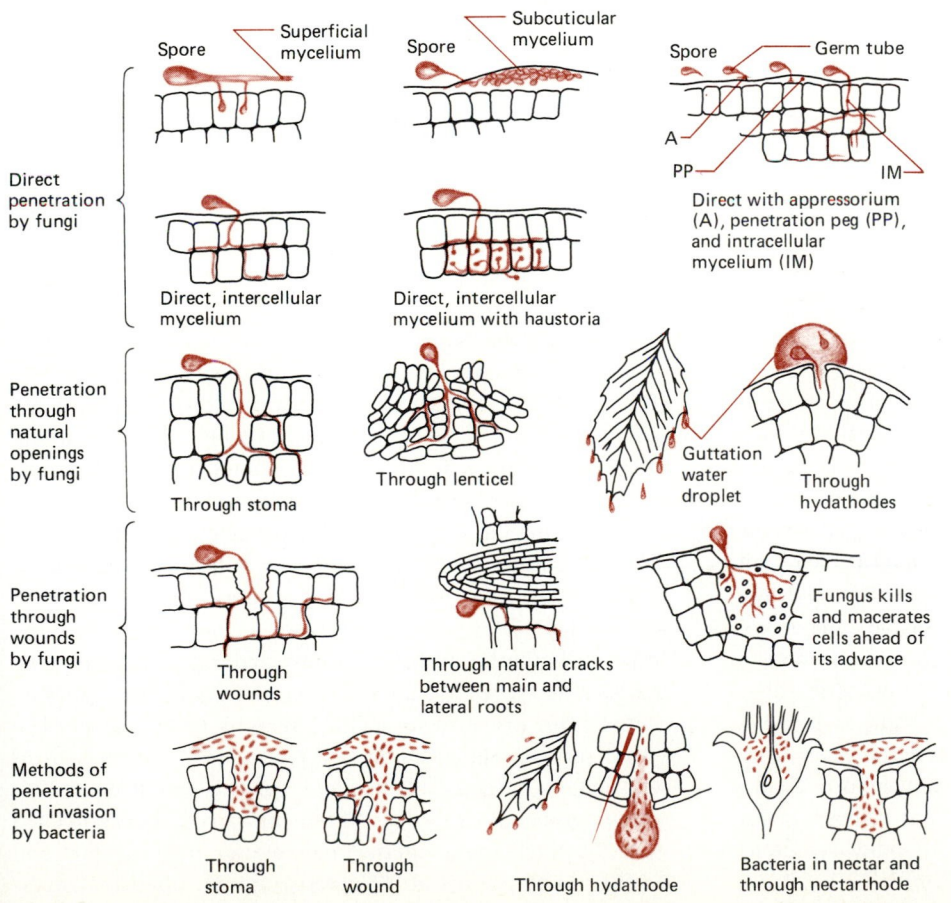

FIGURE 14.3

As shown in this diagram, microbial pathogens can enter plants in several ways. Fungal pathogens can penetrate plants directly through natural openings or through wounds. Bacterial pathogens can penetrate plants through stomata and wounds.

can also be enzymatic when plant tissues are subjected to enzymatic attack by the pathogen, which softens the plant tissue in the vicinity of penetration. In some cases the pathogen penetrates the cuticle, such as occurs with powdery mildews caused by *Erysiphe* species and soft rots caused by *Botrytis cinerea*.

Viral Diseases

A necessary attribute of most plant pathogens, especially viral pathogens, is the ability to survive outside host cells until susceptible plant hosts can be located. Plant pathogens must move from crop to crop and field to field for their continued existence, and they must also successfully overwinter if their crop hosts are seasonal. As obligate intracellular parasites, plant pathogenic viruses must find suitable plant cells for their replication. If free in soil, viruses would be subject to inactivation by soil microbial enzymes. Persistence within **vectors** (biological carriers such as insects) is one means by which viruses may survive in soil. The vector also normally provides the opening through which the virus enters the plant. The distribution of viral plant diseases often follows the spatial distribution pattern of the vectors. The patterns of dissemination of pathogenic viruses are fixed to a large extent by environmental factors that affect the survival and movement of the vector organisms, such as soil texture and moisture.

Many plant pathogenic viruses are transported to susceptible host plants by vectors that acquire these pathogens from soil or diseased plant tissues. Insects, such as aphids, leaf hoppers, mealy bugs, and nematodes, often act as vectors for viral diseases of plants. Even other microbes can serve as vectors for viral pathogens. *Olpidium brassicae*, a chytrid fungus, is the vector of tobacco necrosis virus and probably of several other plant pathogenic viruses. Pollen and plant seeds are also involved in the transmission of plant viruses. For example, tobacco rattle virus is detectable on the pollen of infected petunia plants and is disseminated through the air together with the pollen to susceptible plants. The spread of viruses on structures involved in the reproductive activities of the plant, such as pollen and seeds, ensures that viruses are maintained with susceptible host plant populations and that viral diseases are endemic to these populations.

Plant pathogenic viruses are named according to their ability to cause specific diseases (Table 14.5). Often the only symptom of a viral plant infection is a reduced growth rate that results in some degree of dwarfing. In the case of systemic viral diseases of plants, the most common symptoms are mosaics and ringspots (Figure 14.4). **Mosaics** are characterized by the formation of light-green, yellow, or white spots intermingled with the normal green aerial plant structures. **Ringspots** are characterized by the appearance of chlorotic or necrotic rings on the leaves. These primary symptoms may be accompanied by a variety of other symptoms in specific viral plant diseases (Figure 14.5).

TABLE 14.5   Some Plant Pathogenic Viruses

DNA Viruses

Caulimovirus group: cauliflower mosaic virus

RNA Viruses

Rod-shaped
  Tobavirus group: tobacco rattle virus, pea early browning virus
  Tobamovirus group: tobacco mosaic virus
Viruses with flexuosus or filamentous particles
  Potexvirus group: potato virus X
  Carlavirus group: carnation latent virus
  Potyvirus group: potato virus
  Beet yellows virus
  Festuca necrosis virus
  Citrus tristerza virus
Isometric viruses
  Cucomovirus group: cucumber mosaic virus
  Tymovirus group: turnip yellow mosaic virus
  Comovirus group: cowpea mosaic virus
  Nepovirus group: tobacco ringspot virus
  Bromovirus group: brome mosaic virus
  Tombushvirus group: tomato bushy stunt
  Alfalfa mosaic virus
  Pea enation mosaic virus
  Tobacco necrosis virus
  Wound tumor virus group: wound tumor virus, rice dwarf virus
  Tomato spotted wilt virus
Rhabdoviruses
  Lettuce necrotic yellows virus

Viroid Diseases

Recently, viroids have been identified as the causative agents of several plant diseases (Figure 14.6). **Viroids**, which are simply naked RNA molecules, have been implicated in potato spindle disease, chrysanthemum stunt, and citrus exocortis disease, as well as several other diseases. It is quite interesting that nucleic acid molecules, in contrast to more highly organized microbial pathogens, can be the source of disease.

Potato spindle tuber disease, one of the most destructive diseases of potatoes, is caused by potato spindle tuber viroid (PSTV). PSTV is spread primarily by knives used to cut potato seed tubers, as well as during handling and planting of the crop. Following inoculation of a tuber with PSTV, the viroid replicates itself and spreads systemically throughout the plant. Infected potato plants appear erect, spindly, and dwarfed, and the tubers are elongated. Citrus exocortis viroid (CEV), which is similar to PSTV, causes citrus exocortis, a disease that affects various citrus trees. Infected plants show splits in the bark and partially loosened outer bark that give the bark a cracked, scaly appearance. CEV is readily transmitted from diseased to healthy trees by cutting tools. The viroid survives on contaminated blades even after treatment with many chemical disinfectants but can be inactivated by sodium hypochlorite. The viroid appears to be associated with the nuclei and internal membranes of host cells and results in aberrations of the cytoplasmic membranes. Chrys-

FIGURE 14.5
Symptoms of viral diseases of plants.

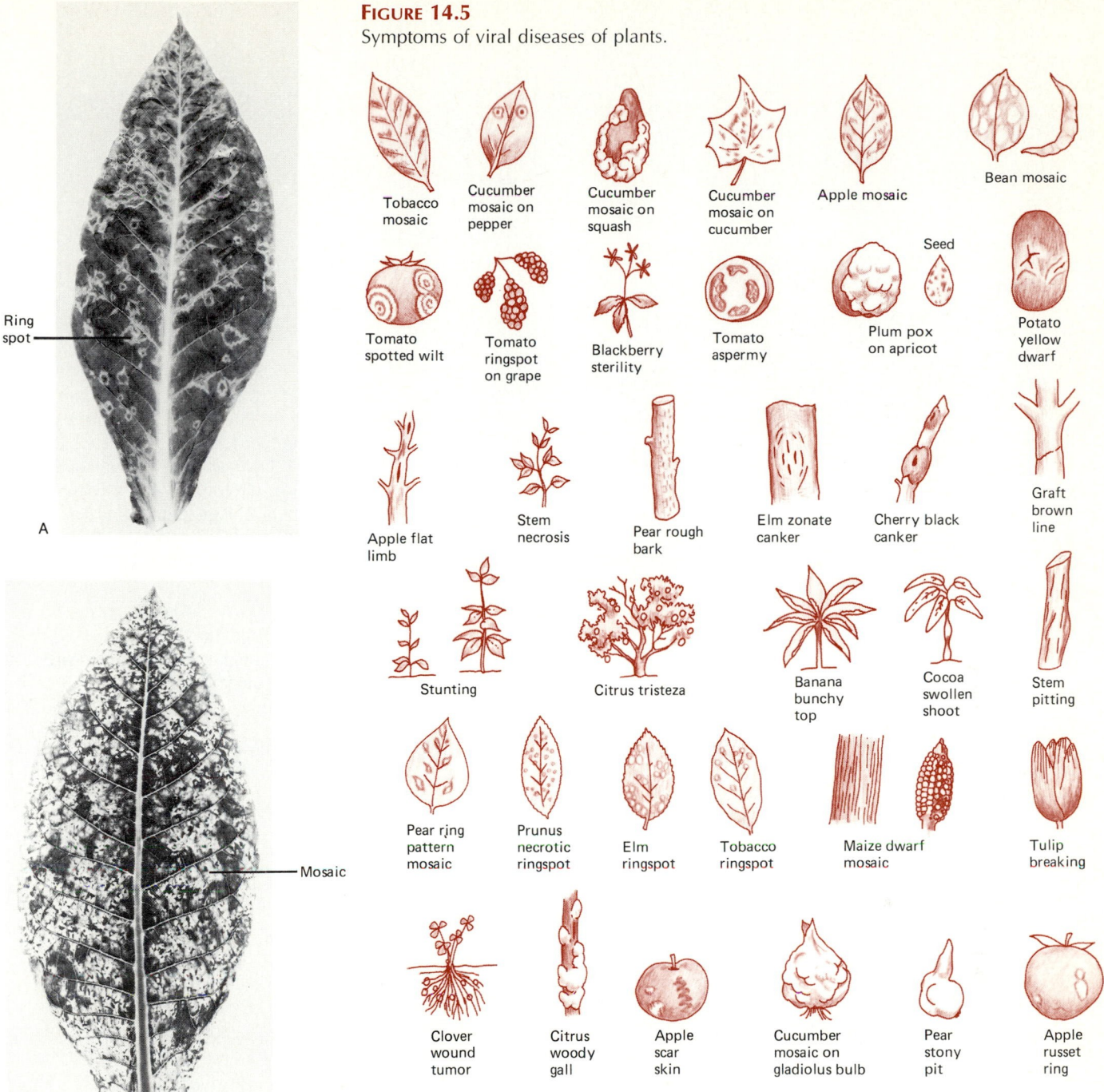

Ring spot

A

Mosaic

B

FIGURE 14.4
(A) Viral ring spot on a tobacco leaf. Note the peculiar line patterns, often spreading out on either side of the veins, which make identification of this disease easy. (B) This tobacco leaf shows the characteristic symptoms of mosaic disease. (Courtesy U.S. Department of Agriculture)

anthemum plants infected with chrysanthemum stunt viroid (ChSV) are smaller, paler, and inferior in quality to normal ones. ChSV moves slowly through a plant, often taking 5 to 6 weeks to move from an inoculated leaf into the stem, and new symptoms develop 3 to 4 months after inoculation.

Bacterial Diseases

Plant pathogenic bacterial species occur in the genera *Mycoplasma*, *Spiroplasma*, *Corynebacterium*, *Agrobacterium*, *Pseudomonas*, *Xanthomanas*, *Streptomyces*, and *Erwinia*. These bacteria are widely distributed and cause a large number of plant diseases, including hypertrophy, wilts, rots, blights, and galls (Table 14.6). Plant pathogenic bacteria cause many different disease symptoms, and most symptoms of plant disease can be caused by several different bacterial species (Figure 14.7).

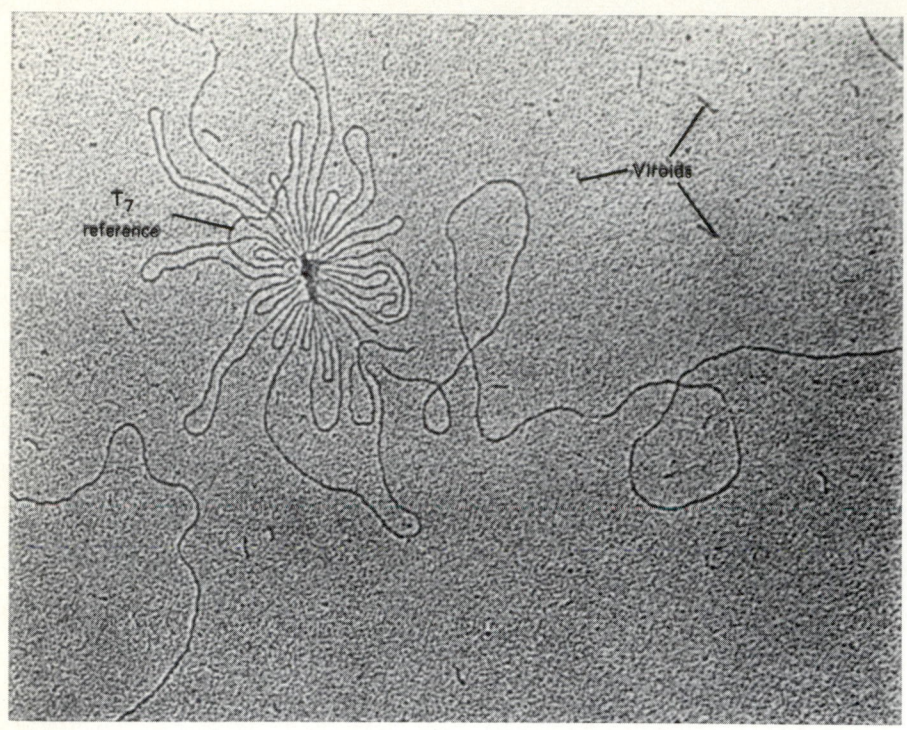

FIGURE 14.6
Photomicrograph of the potato spindle
tuber viroid (56,300×). Note the size
difference between the reference bacteriophage T7 DNA and the nucleic acid of
the viroids. (From BPS—T. O. Diener, U.S.
Department of Agriculture, and T. Koller
and J. S. Sago, Swiss Federal Institute
of Technology, Zurich)

FIGURE 14.7
Symptoms of bacterial diseases
of plants.

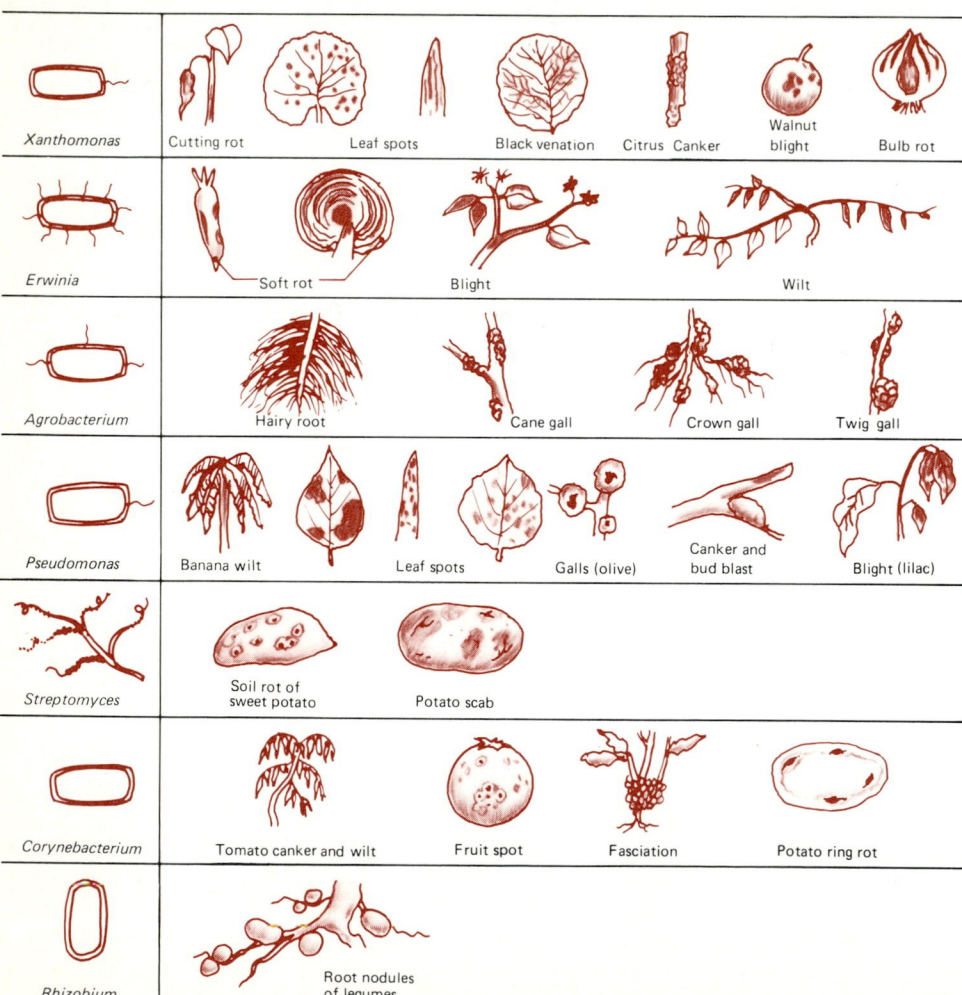

TABLE 14.6 Some Bacterial Diseases of Plants

Genus	Species	Disease
Pseudomonas	*tabaci*	Wildfire of tobacco
	angulata	Leaf spot of tobacco
	phaseolicola	Halo blight of beans
	pisi	Blight of peas
	glycinea	Blight of soybeans
	syringae	Blight of lilac
	solanacearum	Moko of banana
	caryophylli	Wilt of carnation
	cepacia	Sour skin of onion
	marginalis	Slippery skin of onion
	savastanoi	Olive knot disease
	marginata	Scab of gladiolus
Xanthomonas	*phaseoli*	Blight of beans
	oryzae	Blight of rice
	pruni	Leaf spot of fruits
	juglandis	Blight of walnut
	campestris	Black rot of crucifers, citrus canker
	vascularum	Gumming of sugar cane
Erwinia	*amylovora*	Fire blight of pears and apples
	tracheiphila	Wilt of cucurbits
	stewartii	Wilt of corn
	carotovora	Soft rot of fruit, black leg of potato, blight of chrysanthemum
Corynebacterium	*insidiosum*	Wilt of alfalfa
	michiganese	Wilt of tomato
	facians	Leafy gall of ornamentals
Streptomyces	*scabies*	Scab of potato
	ipomoeae	Pox of sweet potato
Agrobacterium	*tumefaciens*	Crown gall of various plants
	rubi	Cane gall of raspberries
	rhizogenes	Hairy root of apple
Myoplasma	sp.	Aster Yellows
	sp.	Peach X disease
	sp.	Peach Yellows
	sp.	Elm phloem necrosis
Spiroplasma	sp.	Cirtus stubborn disease
	sp.	Bermuda grass witches' broom
		Corn stunt

The relationship between bacterial plant pathogens and their host plant is greatly affected by the stationary nature of the plant, the periodicity of plant growth, and the protective surfaces of the plant. Bacterial plant pathogens must possess an independent mode of dispersal in order to reach new host plants and must have some mechanism for entering the plant. Because most plant pathogenic bacteria do not form resting stages, they must remain within the confines of the plant tissues. Even during times of plant dormancy, many plant pathogenic bacteria can remain viable on plant seeds and other plant tissues. Some bacterial populations exhibit no significant soil phase. For example, *Erwinia amylovora*, which causes fire blight in fruit trees, remains within infected tissues and is disseminated in plant exudates by insects or raindrops (Figure 14.8). During the winter, *E. amylovora* does not grow but remains dormant within infected tissues

of the stems and branches of trees, and in the spring the bacteria are distributed to susceptible plants.

Many bacterially caused plant diseases are seedborne. In these cases, the pathogenic bacteria must survive on the seeds for a transient period in the soil. Bacteria may be carried on seeds as a surface contaminant or within the seed. For example, *Pseudomonas phaseolicola* is carried in the micropyle and causes the halo blight of beans. *Xanthomonas malvacearum*, the causative organism of cotton blight, can be found on the external cotyledon margins during germination of the seed. Some bacterial diseases are caused by pathogens that have a permanent soil phase, for example, some fluorescent *Pseudomonas* species that cause soft rots in plants. Unlike these organisms, which grow on excreted organic compounds in the rhizosphere, most plant pathogenic bacteria are unable to survive for long periods of time in the soil.

Crown gall, which is caused by *Agrobacterium tumefaciens*, is a particularly interesting plant disease (Figure 14.9). Crown gall may occur on fruit trees, sugar beets, or other broad-leaved plants. The disease process is initiated when viable cells of *A. tumefaciens* enter wounded surfaces of susceptible dicotyledonous plants, usually at the soil–plant stem interface, either through the root or a wound. *A. tumefaciens* is able to transform host plant cells into tumorous cells, and the disease is manifested by the formation of a tumor growth, the crown gall. Once the disease is established, the tumor continues to grow even if viable *Agrobacterium* are eliminated. The tumor maintenance principle has been identified as a fragment of a large, **tumor-inducing (Ti) plasmid**. A fragment of this bacterial plasmid is transferred to the plant, where it is maintained in the tumor tissue. The Ti plasmid has great potential in recombinant DNA technology to introduce desired genetic information into a wide range of plants of agricultural significance (Figure 14.10). Thus, although *Agrobacterium* possessing the Ti plasmid cause great damage to crops, the Ti plasmid itself represents a vehicle for the creation of improved crops capable of disease resistance and increased yields with decreased management.

One of the more serious recent outbreaks of plant disease is citrus canker, caused by *Xanthomonas campestris* var. *citrii*. Infections with this plant pathogen produce lesions on the leaves. The outbreak of citrus canker in 1913 took 20 years and $6 million to control. The new strain of *X. campestris* that infected Florida citrus groves in 1984 is distinct from previous strains, having a high level of fatty acids that protect the pathogen. Quarantine procedures and the burning of infected seedlings, about 21 percent of the nursery stock of Florida, were used to suppress this outbreak of the disease.

One of the most interesting ecological relationships between bacteria and plants involves the role of certain phyllosphere bacteria in initiating ice crystal formation, which results in frost damage to the plant. Some strains of *Pseudomonas syringae* and *Erwinia herbicola* produce a surface protein that can initiate ice crystal formation. These bacteria

FIGURE 14.8
Life cycle of *Erwinia amylovora*, which causes fire blight in pears and apples. This bacterial pathogen causes serious economic losses.

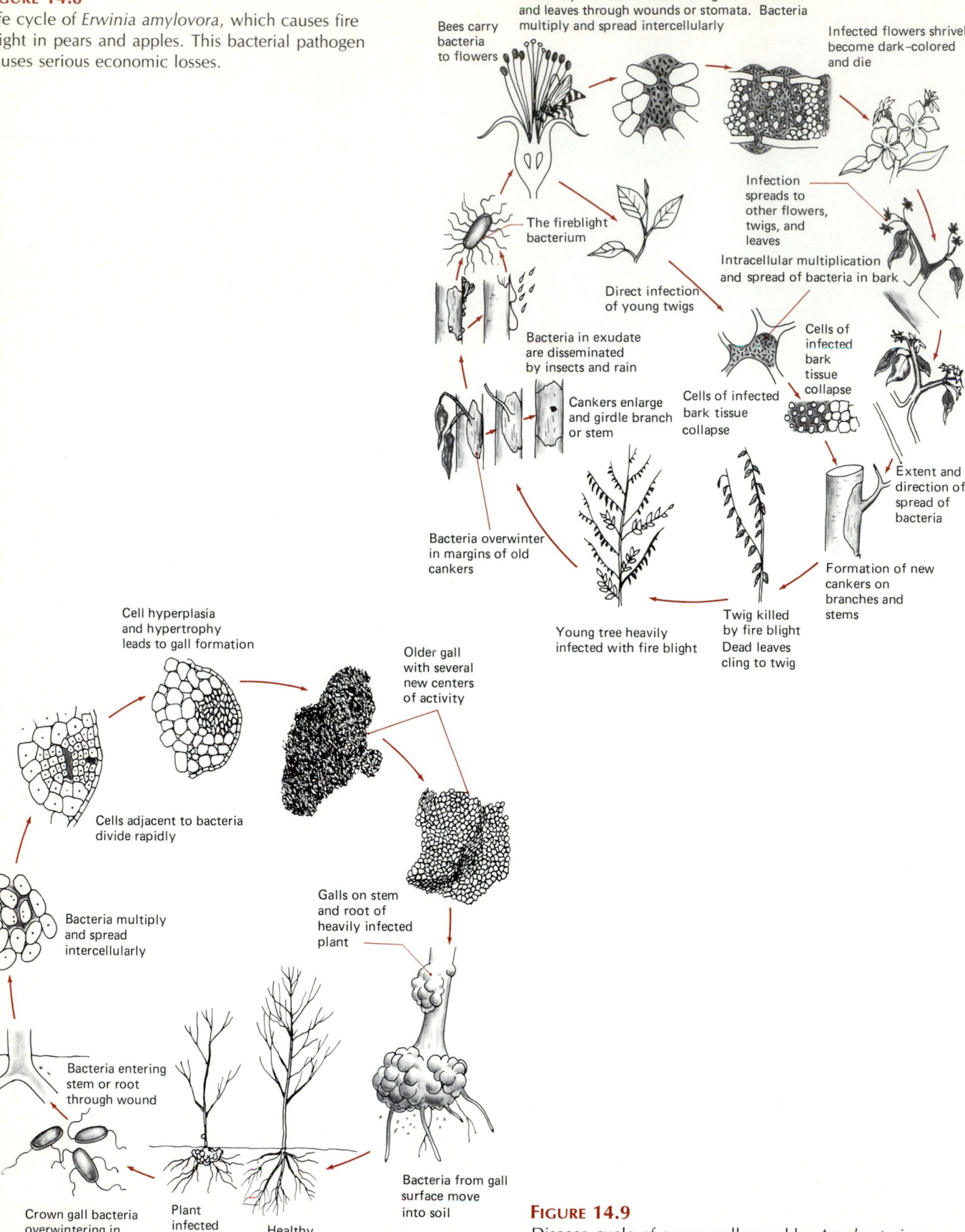

FIGURE 14.9
Disease cycle of crown gall causd by *Agrobacterium tumefaciens*.

FIGURE 14.10
The Ti plasmid has many potential uses in genetic engineering.

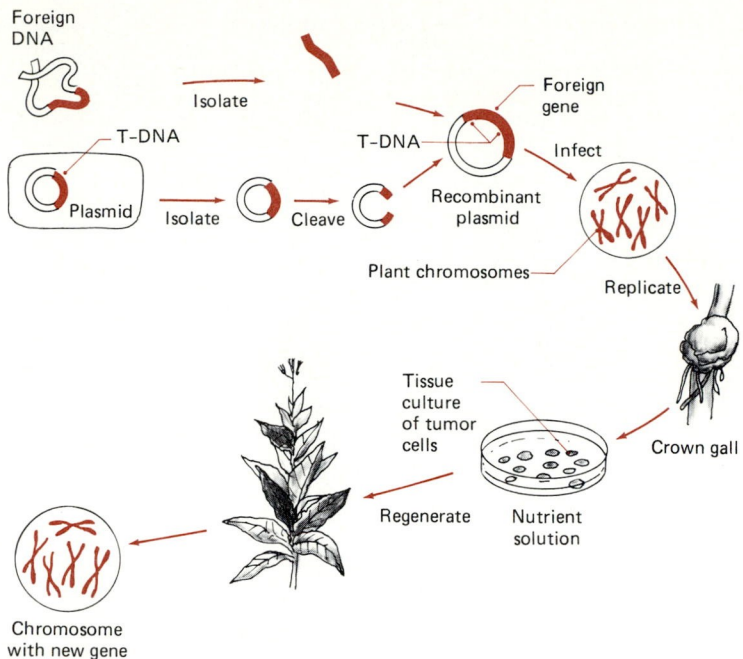

Discovery Process

The elucidation of how *Agrobacterium tumefaciens* causes crown gall has raised the possibility of exploiting the natural infectious process to introduce foreign DNA into plant cells. *A. tumefaciens*, which causes crown gall tumors in most dicotyledonous plants, carries a plasmid that induces the infected plant cells to synthesize nitrogen compounds called *opines*. During the normal infectious process, a section of the plasmid, called T-DNA, combines with chromosomal DNA in the nucleus of the plant cell. It may therefore be possible to employ the Ti plasmid as a vector for inserting foreign DNA into plant cells. To do so, the plasmid would be cut open at a site within the T-DNA, and the foreign gene would be spliced into it. When the tumor cells are grown in tissue culture, they continue to carry and replicate T-DNA during the normal divisional process. Investigators at the University of Leiden have been able to infect tobacco cells in culture with *A. tumefaciens*, and have found that tobacco plants regenerated from cultured tumor cells retain T-DNA and continue to make opine synthetase. Moreover, scientists at the Max Planck Institute for Plant Breeding in Cologne demonstrated that the gene carried by T-DNA that codes for the enzyme opine synthetase is passed on through the seed to succeeding generations as if it were an ordinary dominant gene. If foreign genes inserted into T-DNA are also transmitted to plant progeny, new plant strains could be genetically engineered.

are conditional plant pathogens, causing death due to frost damage only at temperatures that can initiate the freezing process. Laboratory experiments have demonstrated that ice crystals do not form until the temperature drops to -7 to $-9°C$, when ice-nucleation active populations are replaced with mutant strains that do not produce the ice-initiating proteins, thereby limiting frost damage. The development of genetically engineered strains of ice-minus *P. syringae*, that is, strains that do not form the ice-nucleating surface protein, and the proposal to apply such strains to field crops for frost protection have caused great public and scientific concern about the possible environmental consequences of such uses of genetically engineered microorganisms.

Fungal Diseases

Although bacteria and viruses cause many plant diseases, most plant diseases are caused by pathogenic fungi (Table 14.7). In addition to the large number of fungal plant pathogenic species, a wide variety of disease symptoms characterize fungal infections of plants (Figure 14.11). Many fungi are well adapted to act as plant pathogens. Fungal spore production permits aerial transmission between plants and allows plant pathogenic fungi to remain viable outside host plants. Survival and infectivity of most plant pathogenic fungi depend on prevailing environmental temperature and moisture conditions. Spores germinate and mycelia grow

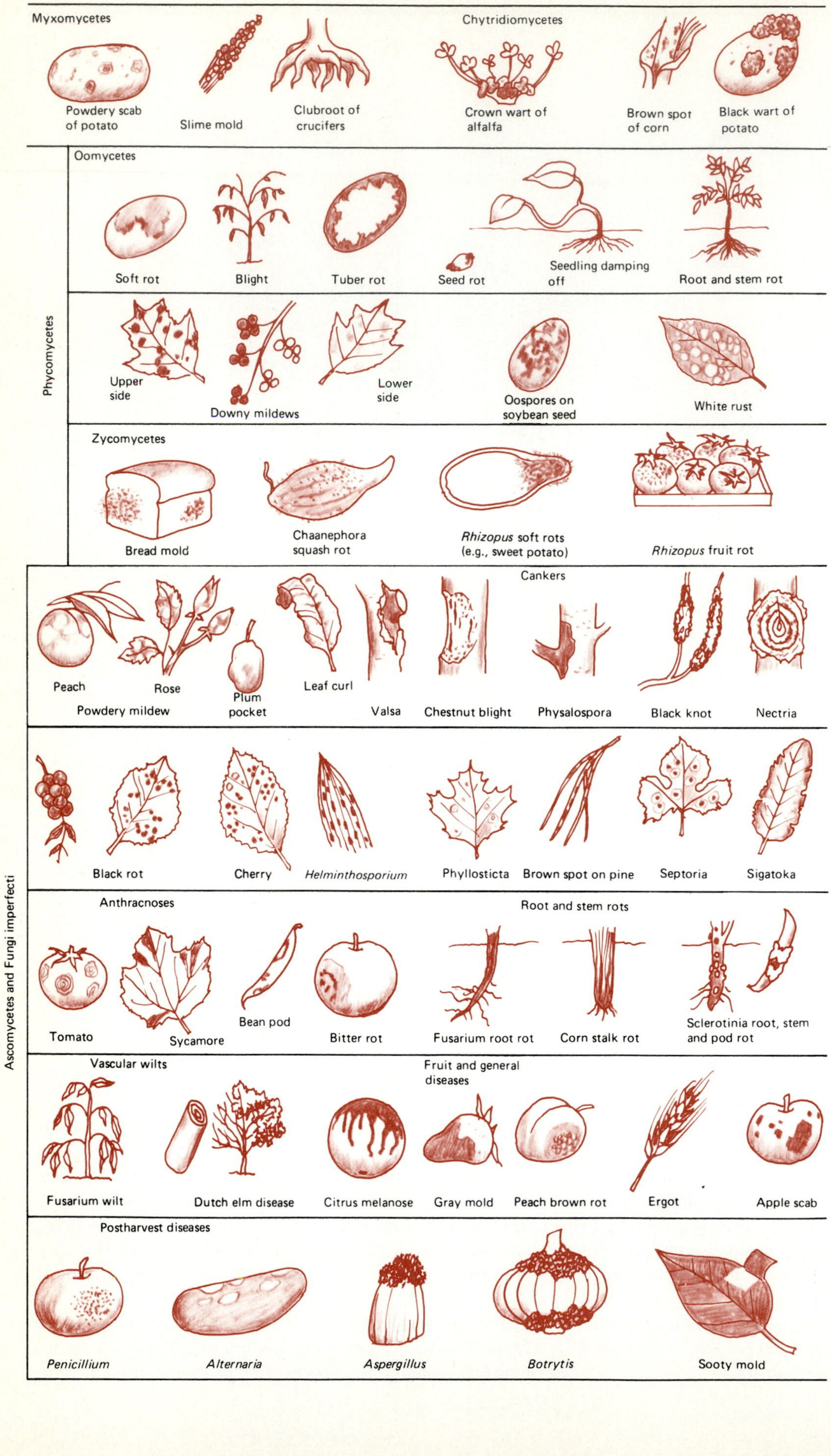

FIGURE 14.11
Symptoms of fungal diseases of plants.

TABLE 14.7   Some Diseases of Plants Caused by Fungi

Slime Molds

Plasmodiophora	Clubfoot of crucifers
Polymyxa	Root disease of cereals
Spongospora	Powdery scab of potato
Plasmopara	Downy mildew of grapes

Oomycetes

Albugo	White rust of crucifers
Phytophthora	Late blight of potato
Pythium	Seed decay, root rots

Chitridiomycetes

Olpidium	Root disease of various plants
Physoderma	Brown spot of corn
Synchytrium	Black wart of potato
Urophlyctis	Crown wart of alfalfa

Zygomycetes

Rhizopus	Soft rot of fruits

Ascomycetes

Ceratocystis	Dutch elm disease
Claviceps	Ergot of rye
Diaporthe	Bean pod blight
Dibotryon	Black knot of cherries
Diplocarpon	Black spot of roses
Endothia	Chestnut blight
Erysiphe	Powdery mildew of grasses
Lophodermium	Pine needle blight
Microsphaera	Powdery mildew of lilac
Mycosphaerella	Leaf spots of trees
Ophiobolus	Take all of wheat
Podosphaera	Powdery mildew of apple
Sclerotinia	Soft rot of vegetables
Taphrina	Peach leaf curl
Venturia	Apple scab

Basidiomycetes

Armillaria	Root rots of trees
Cronartium	Pine blister rust
Exobasidium	Stem galls of ornamentals
Fomes	Heart rot of trees
Marasmius	Fairy ring of turf grasses
Polyporus	Stem rot of trees
Puccinia	Rust of cereals
Sphacelotheca	Loose smut of sorghum
Tilletia	Stinking smut of wheat
Typhulai	Blight of turf grasses
Urocystis	Smut of onion
Uromyces	Rust of beans
Ustilago	Smut of corn, wheat, and others

Deuteromycetes

Alternaria	Leaf spots and blight of various plants
Aspergillus	Rots of seeds
Botrytis	Blights of various plants
Cladosporium	Leaf mold of tomato
Colletotrichum	Anthracnose of crops
Cylindrosporium	Leaf spots of various plants
Fusarium	Root rot of many plants
Helminthosporium	Blight of cereals
Penicillium	Blue mold rot of fruits
Phoma	Black leg of crucifers
Rhizoctonia	Root rot of various plants
Thielaviopsis	Black root rot of tobacco
Verticillium	Wilt of various plants

---

when the temperature is between -5 and $+45°C$ and there is an adequate supply of moisture. Spores, though, can retain viability for long periods of time during environmental conditions that do not allow germination. Plant pathogenic fungi generally exhibit a complex life cycle, spent in part in host plant infection and in part outside host plants in soil or on plant debris in the soil. As an example, the life cycle of *Rhizoctonia solani*, a fungus that causes a variety of diseases, is illustrated in Figure 14.12. Similarly, the fungus *Monilinia fructicola*, the causative agent of brown rot of stone fruits, overwinters as mycelia or conidia on plant materials in the ground; in the spring new conidia are produced, and the mycelia on mummified fruit in the ground produce ascospores; the ascospores are dispersed by wind, rain, and insects, reaching newly forming fruits and initiating their infection, which results in brown rot (Figure 14.13).

Fungal plant pathogens usually grow most rapidly and cause the most severe diseases of plants during the warmer months of the year because during winter many such plant pathogens are inactive. Some fungi, however, such as *Ty-phula* and *Fusarium* which cause snow mold of cereals and turf grasses, thrive only in cool seasons or regions. In some cases, the optimum temperature for disease development is different from the optimal growth temperature of either the pathogen or the host. For example, in the case of black root rot of tobacco, caused by the fungus *Thielaviopsis basicola*, the optimum temperature for establishing the disease is lower than the optimal growth temperature of the pathogen, and the host is less able to resist the pathogen at temperatures of 17–23°C. In other cases, such as root rots of wheat and corn caused by the fungus *Gibberella zeae*, the optimum temperature for development of disease is higher than the optimal growth temperature of both the pathogen and the wheat. Soil pH also has a marked effect on the infectivity of soil-borne plant pathogens. For example, *Plasmodiophora brassicae* causes club root of crucifers at pH values of approximately 5.7, but the disease is completely checked at pH 7.8.

Moisture, like temperature and pH, influences the initiation and development of infectious plant diseases. Some fungal pathogens are dispersed by rain droplets, initiating con-

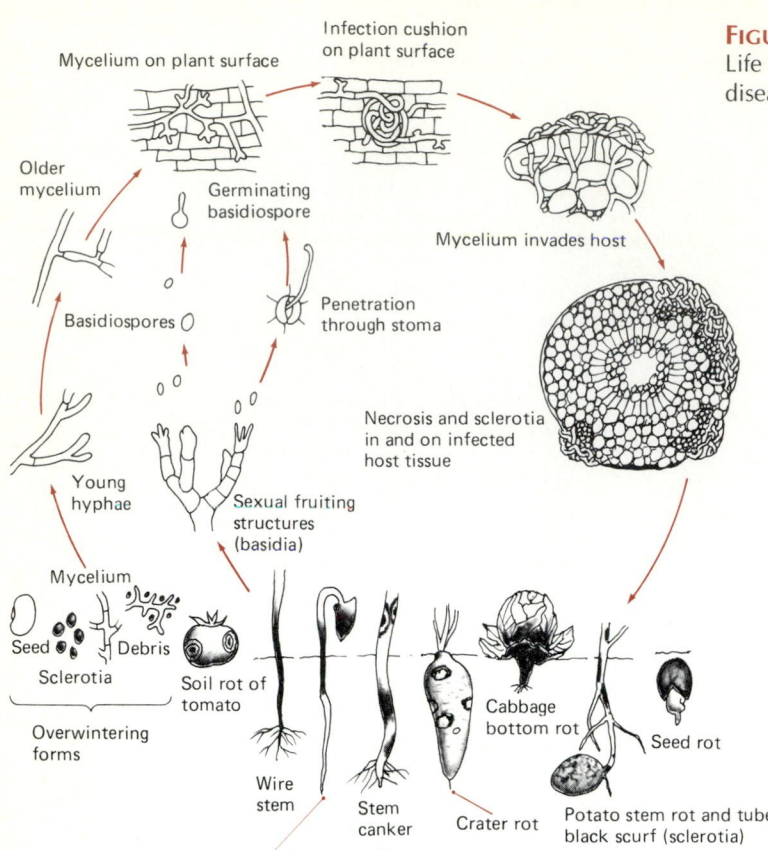

FIGURE 14.12
Life cycle of *Rhizoctonia solani*, which causes various plant diseases.

FIGURE 14.13
Peaches infected with brown rot. This disease also affects plums, prunes, and cherries. (Courtesy U.S. Department of Agriculture)

tact with the susceptible plant. The occurrence of plant diseases is closely correlated with the distribution of rainfall in some regions. For example, downy mildew of grapes and fire blight of pears are more severe when there is high rainfall or high relative humidity. Most fungal pathogens are dependent on high relative humidity for germination of spores in the phylloplane. Some diseases that affect the plant through the root system are most severe when soil moisture is near the saturation point. The increased moisture allows more rapid multiplication and dispersion of the pathogenic zoospores of *Pythium*.

Perhaps the most important economic fungal diseases of plants are caused by the **rusts** and **smuts**, which are Basidiomycota. There are over 20,000 species of rust fungi and over 1,000 species of smut fungi. The rust fungi require two unrelated hosts for the completion of their complex life cycle. Important plant diseases caused by rusts include black stem rusts of cereals, white pine blister rust, coffee rust, cedar-apple rust, and asparagus rust. As an example, the complex life cycle of the cedar-apple rust fungus is illustrated in Figure 14.14. The smuts are so named because they produce black, dusty spore masses resembling soot or smut. Important diseases caused by smut fungi include loose smut of oats, corn smut, bunt or stinking smut of wheat, and onion smut. Smut and rust fungi cause millions of dollars worth of crop damage annually.

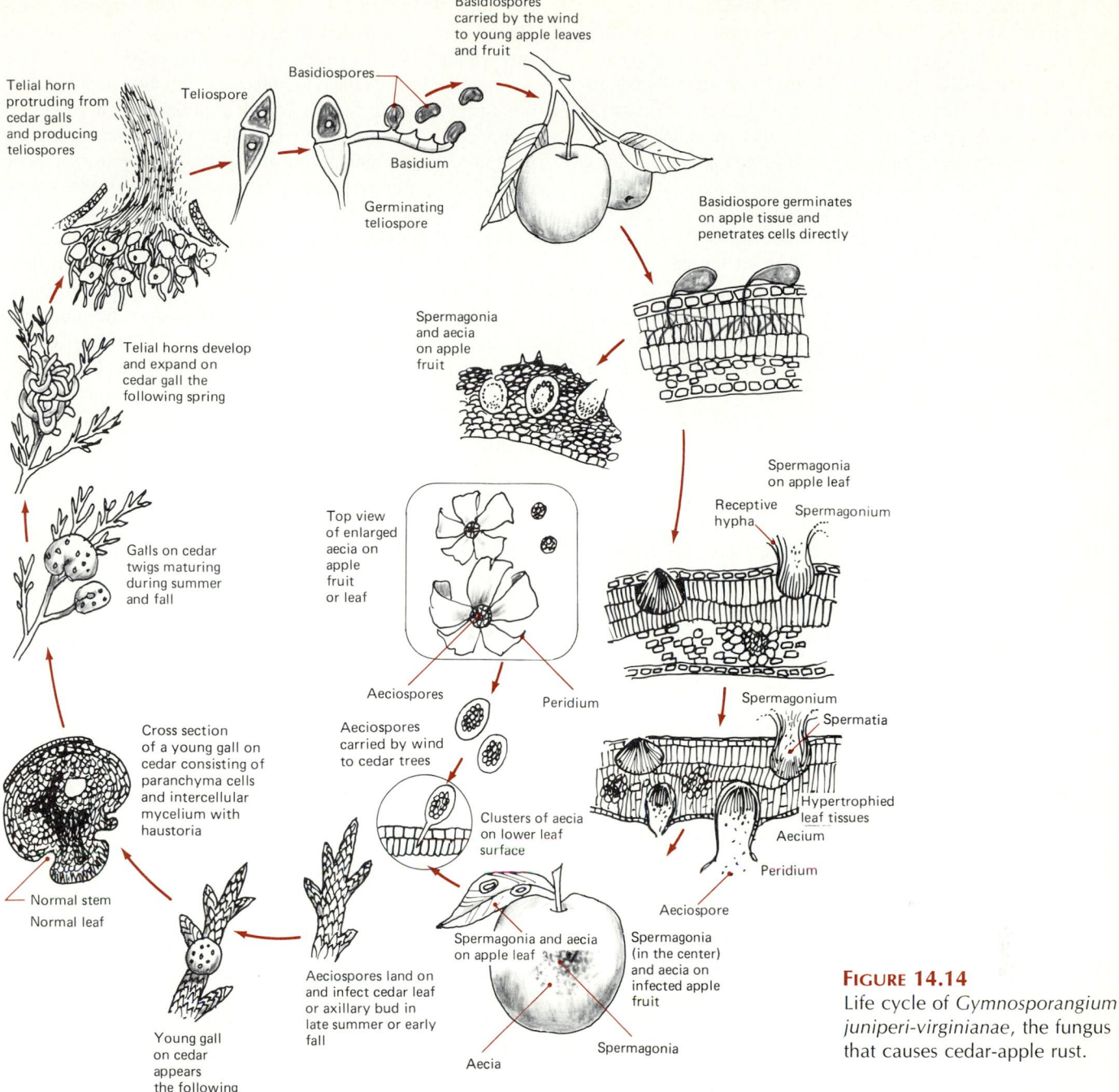

Basidiospores carried by the wind to young apple leaves and fruit

Basidiospores

Teliospore

Telial horn protruding from cedar galls and producing teliospores

Basidium

Germinating teliospore

Basidiospore germinates on apple tissue and penetrates cells directly

Telial horns develop and expand on cedar gall the following spring

Spermagonia and aecia on apple fruit

Spermagonia on apple leaf

Receptive hypha

Spermagonium

Galls on cedar twigs maturing during summer and fall

Top view of enlarged aecia on apple fruit or leaf

Spermagonium

Spermatia

Cross section of a young gall on cedar consisting of paranchyma cells and intercellular mycelium with haustoria

Aeciospores

Peridium

Aeciospores carried by wind to cedar trees

Hypertrophied leaf tissues

Aecium

Peridium

Normal stem
Normal leaf

Clusters of aecia on lower leaf surface

Aeciospore

Spermagonia and aecia on apple leaf

Spermagonia (in the center) and aecia on infected apple fruit

Young gall on cedar appears the following summer

Aeciospores land on and infect cedar leaf or axillary bud in late summer or early fall

Aecia

Spermagonia

FIGURE 14.14
Life cycle of *Gymnosporangium juniperi-virginianae*, the fungus that causes cedar-apple rust.

14.3 *Control of Crop Diseases*

Since obligate plant pathogens can remain viable for only a limited period of time outside host plant tissues, appropriate management procedures can be used to control plant pathogens of agricultural crops. The most important of these are **quarantine** (restriction of movement) procedures to restrict the spread of plant pathogens, **sanitary practices** to prevent infection of plants with contaminated soils and tools, the development and planting of **resistant crop varieties** that are not susceptible to particular plant pathogens, the use of **crop rotation practices** to limit contact between the pathogen and the host plant, and the use of **pesticides**, such as fungicides, to control populations of plant pathogens. Plant pathogens are normally specific for particular host plants. By the periodic sowing of new plant varieties, the plants that are susceptible to the diseases caused by the particular pathogenic microorganism established in that agricultural field are removed. Thus, rotating crops is an effective way of temporarily removing those plant varieties that are suitable hosts for specific microbial pathogens because the high populations of plant pathogens produced as a result of the infection of the susceptible crops are greatly reduced by this practice. The susceptible plant crop can be successfully reestablished when it is rotated back into that field in later years.

Agricultural management practices are frequently directed at avoiding plant disease through modification of host populations. Selective breeding methods have been used extensively in agriculture to develop plant populations that are genetically resistant to pathogens, and many new plant strains developed in this way are also successfully producing high crop yields. Pest populations, however, are subject to evolution, selection, and geographical spread, making it only a matter of time before newly developed plant varieties are subject to serious plant diseases. Thus, in agriculture there is a need for a continuous breeding program for new strains of resistant plants to replace strains of plants that are, or are becoming, susceptible to diseases caused by pathogens and pests. For example, since the 1930s, strains of resistant wheat have been continuously developed to resist infections with rust fungi and other pathogenic microorganisms.

Resistance to disease is an inherent feature of a plant designed to restrict the entry and/or subsequent deleterious effects of a pathogen. Plants with thicker cuticles or cork layers are more resistant to plant diseases. Insect vectors carrying plant pathogens may be unable to penetrate these thickened layers. Resistance of some species of barberry to penetration by basidiospores of *Puccinia graminis* has been attributed to the thickness of the cuticle in the epidermis of the leaves. The ability to close the stomata when conditions are favorable for infection is an adaptive feature of some plants that renders them relatively resistant to plant pathogens disseminated through the air. Plant-breeding programs are often aimed at developing disease-resistant varieties with such anatomical adaptations.

There are also several adaptive physiological features that make some plant varieties relatively resistant to strains of pathogenic microorganisms that can be selected in a plant breeding program. Altered metabolic pathways in plants may also render them unsusceptible to the enzymes or toxins produced by pathogenic microorganisms. Such changes in the plant's metabolism may effectively control plant diseases by making plant cells inhospitable habitats for the growth of pathogenic microorganisms. The production of inhibitory substances by the plant can prevent the establishment of an infection. Plants produce a variety of polycyclic and polyaromatic compounds, phytoalexins, that have antimicrobial

activities. Plants that have previously been exposed to invading microorganisms and have survived the infection retain such chemicals for a period of time, during which they have an increased resistance to infection by pathogenic microorganisms. Plants can be exposed to attenuated strains of pathogenic microorganisms to induce the formation of such inhibitory biochemicals, but plants do not exhibit the same extensive immune response observed in vertebrate animal populations.

Because many pathogenic microorganisms are transmitted from one plant to another, increased spacing between individual plants decreases the likelihood that a pathogenic microorganism will be successfully transmitted from infected to uninfected plants. Dense plantings of hemlock, for example, result in increased losses due to twig rust because the pathogen that causes this disease is spread from plant to plant. The density of host plant populations can also result in environmental modifications that affect the development of plant pathogens. Dense planting of tomatoes, for example, results in the retention of humidity, which favors the development of blight due to the fungus *Botrytis*.

Unfavorable environmental conditions during planting can increase the susceptibility of agricultural crops to microbial infections. At suboptimal temperatures, plants often produce increased amounts of exudates, containing amino acids and carbohydrates, which favor the rapid development of microbial populations in the soil surrounding the plant. Other suboptimal abiotic soil factors, such as lack of oxygen and excessive moisture content, also result in increased production of such exudates. The subsequent development of pathogenic *Pythium* and *Rhizoctonia* populations in the vicinity of seeds often results in fungal disease of developing seedlings.

Soil pH may be lowered deliberately to control soil-borne plant pathogens, such as *Streptomyces scabies* (Figure 14.15), which cause potato scab disease. The occurrence of this disease is greatly reduced when the soil pH is acidic, and because potatoes grow well in acid soil, it is easy to control this plant disease by acidifying the soil. As it would be dangerous and uneconomical to apply acid directly, soil acidification is conveniently accomplished by applying powdered sulfur to the field. The activity of *Thiobacillus thiooxidans* converts sulfur to H_2SO_4, producing the desired acidification.

Various control measures can be used to eliminate or diminish reservoirs of pathogenic microorganisms present at the time of planting. In horticultural practices, for example, soils are sometimes fumigated with fungicides or heat sterilized to remove plant pathogens. It is not practical, though, to sterilize soils in large agricultural fields. However, many plant pathogens contaminate plant seeds, and **disinfection of seeds** by chemical or hot water treatments can greatly reduce the incidence of disease within a plant population. Such seed disinfection practices are widely used for many crops, including corn.

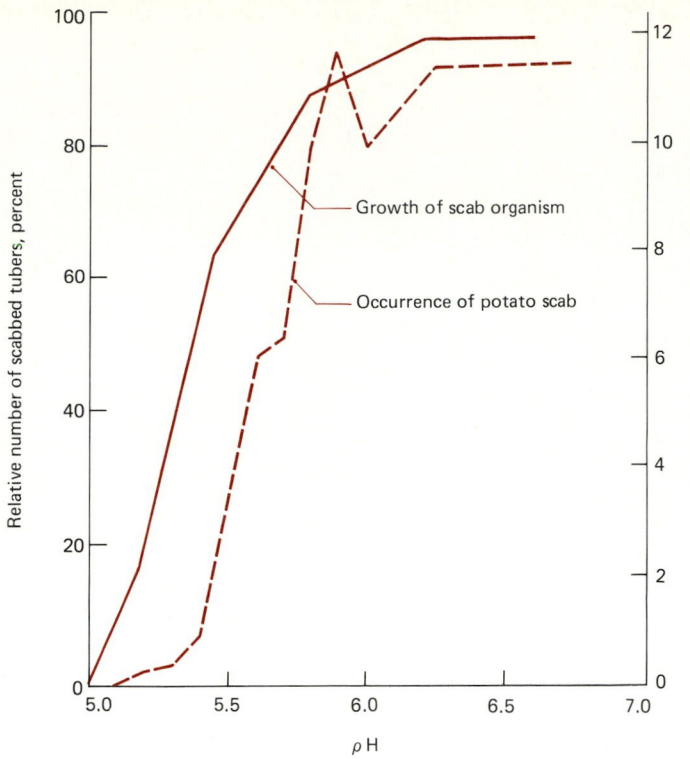

FIGURE 14.15

This graph illustrates the relationship between pH and *Streptomyces scabies*, the causative organism of potato scab disease. From the results shown here, it is apparent that soil acidification is an effective means of controlling this plant pathogen.

The physical removal of infected plant tissues is also an effective method for controlling the spread of disease. The removal of an infected crop from an agricultural field is an inherent consequence of the crop rotation method already discussed. The specificity of host–pathogen interactions precludes the establishment of infection by that pathogen in the other crops used in a system of rotation. The removal of diseased plants and debris following harvest also limits populations of plant pathogens.

In some plant diseases, control can be achieved by limiting the populations of plants that act as alternate hosts for the pathogens. For example, the rust fungus, *Puccinia graminis*, which infects wheat, uses the barberry shrub, *Berberis vulgaris*, as an alternate host (Figure 14.16). By eliminating or reducing the size of this alternate host population from wheat-growing regions, the transmission of disease caused by this rust fungus has been curtailed. Control of the insects and nematodes, which act as vectors for microbial plant pathogens, can be used to control plant diseases. Aphid control programs are especially important because these insects often act as vectors for plant diseases.

Microbial amensalism and parasitism can be used to control populations of pathogenic microorganisms. Negative interpopulation relationships protect many plants from infection with disease-causing microorganisms. The phenomenon of **soil fungistasis**, that is, the inhibition of fungi by soil that

is believed by some to be due to microbial activities, is widespread in soil. Some bacterial species produce antifungal substances, and the addition of *Bacillus* and *Streptomyces* species to soil has been shown to control damping off disease in cucumber, peas, and lettuce and several other diseases caused by the fungus *Rhizoctonia solani*. Damping off is the destruction of seedlings near the soil line that results in the seedlings falling to the ground. Adding organic compounds to soil can increase soil fungistasis, presumably by stimulating antibiotic-producing streptomycetes. It has also been suggested that cellulase- and chitinase-producing strains of microorganisms contribute to the control of populations of fungal pathogens that have cell walls containing cellulose or chitin. The addition of cellulase-producing myxobacteria to the rhizosphere of young seedlings has been found to control diseases caused by pathogenic fungi such as *Pythium*, *Rhizoctonia*, and *Fusarium*, which enter the plant through the soil. The treatment of soil with crab and lobster shells, which contain chitin, has been demonstrated to reduce markedly the severity of root-rot diseases caused by *Fusarium* species, and it has been suggested that the decreased disease incidence reflects higher levels of chitinase-producing microorganisms. However, the roles of chitinase and cellulase in controlling populations of plant pathogens have yet to be conclusively shown, and their use has not been developed into a commercially important method for controlling plant diseases.

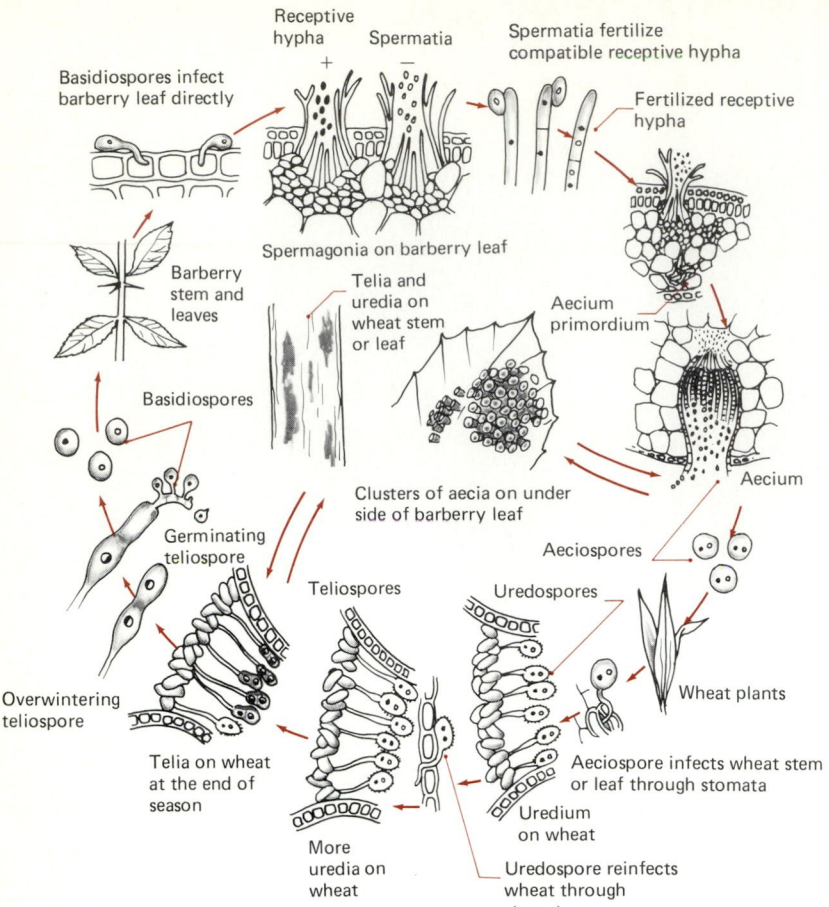

Receptive hypha Spermatia

Spermatia fertilize
compatible receptive hypha

Basidiospores infect
barberry leaf directly

Fertilized receptive
hypha

Spermagonia on barberry leaf

Barberry
stem and
leaves

Telia and
uredia on
wheat stem
or leaf

Aecium
primordium

Basidiospores

Aecium

Clusters of aecia on under
side of barberry leaf

Germinating
teliospore

Aeciospores

Teliospores Uredospores

Wheat plants

Overwintering
teliospore

Telia on wheat
at the end of
season

Aeciospore infects wheat stem
or leaf through stomata

Uredium
on wheat

More
uredia on
wheat

Uredospore reinfects
wheat through
stomata

FIGURE 14.16

Life cycle of stem rust of wheat, which is caused
by *Puccinia graminis*. Like other rusts, *Puccinia*
requires alternate hosts.

14.4 Pesticide Microbiology

The large-scale production of crops requires massive plant-ings of a single-plant species within an agricultural field. Such monocultures of plants are inherently unstable, and as a result, much agricultural production effort and expense goes toward the control of competing weeds, destructive in-sects, and microbial pathogens. The rising cost of manual labor and the increasingly large-scale agricultural operations in developed countries have promoted the growing use of chemical pest control. In 1948 there were only 3 insecticides in agricultural usage, but in 1975 1170 pesticides, 425 her-bicides, 410 fungicides, and 335 insecticides were registered for use in the United States, and 725 million kilograms of synthetic organic pesticides were produced. In terms of ef-fectiveness, economy, and quality control, pesticides have been an unqualified success in increasing crop production, but there are environmental dangers inherent in their wide-spread use. There have been many emotionally charged de-bates over the potential dangers of pesticide use, with sweep-ing endorsements and condemnations made by factions supporting conflicting interests and philosophies. The total abandonment of all pesticide use would unquestionably cause an international economic and public health crisis of major proportions, while continuing uncontrolled and indiscrimi-nate pesticide use would eventually cause irreversible envi-

ronmental damage. A responsible pesticide policy clearly de-mands a careful risk-benefit analysis for each compound and its various uses, with particular emphasis on its biodegrad-ability and its effects on nontarget organisms.

It is essential that pesticides used in agriculture not have an effect on nontarget organisms. Part of the licensing pro-cedure for new pesticides involves toxicity testing to dem-onstrate that wildlife, fish, and other nontarget organisms are not likely to be exposed to lethal concentrations of the pes-ticide when it is applied at recommended levels to agricul-tural fields. It is also important to show that a pesticide does not disrupt microbial biogeochemical cycling activities that could reduce soil fertility. As discussed earlier in this chap-ter, the metabolic activities of soil microorganisms, partic-ularly those involved in carbon and nitrogen transformations, are essential for achieving maximal crop yields. A pesticide, aimed at eliminating an agricultural pest that is diminishing crop yield, must not itself reduce plant productivity by in-terfering with the microbial maintenance of soil fertility.

Biomagnification

In addition to have no adverse effect on nontarget organisms, pesticides should be biodegradable in order to minimize any

TABLE 14.8   Environmental Persistence Times of Some Pesticides

Common Name	Chemical Structure	Persistence Time
Aldrin	1,2,3,4,10,10-1,2,4a,4,8,8a,hexahydro-endo-1,4-exo-5,8-dimethanonaphthalene	>15 years
Chlordane	1,2,3,4,6,7,8,8-octachloro-2,3,3a,4,7,7a-hexahydro-4,7-methanoindene	>15 years
DDT	1,1,1-trichloro-2,2-*bis*(*p*-chlorophenyl)-ethane	>15 years
Dicamba	3,6-dichloro-*o*-anisic acid	4 years
Diuron	3-(3,4-dichlorophenyl)-1,1-dimethylurea	>15 months
2-(2,4-DP)	2-(2,4-dichlorophenoxy)propionic acid	>103 days
Endrin	1,2,3,4,10,10-hexachloro-6,7-epoxy-1,4,4a,5,6,7,8,8a-octahydro-endo-1,4-endo-5,8-dimethanonaphthalene	>14 years
Fenac	2,3,6-trichlorophenylacetic acid	>18 months
Fluometuron	N'-(3-trifluoromethylphenyl)-N,N-dimethylurea	195 days
Heptachlor	1,4,5,6,7,8,8-heptachloro-3a,4,7,7a-tetrahydro-4,7-endomethanoindene	>14 years
Lindane	1,2,3,4,5,6-hexachlorocyclohexane	>15 years
Monuron	3-(*p*-chlorophenyl)-1,1-dimethylurea	3 years
Parathion	0,0-diethyl 0-*p*-nitrophenyl phosphorothioate	>16 years
PCP	pentachlorophenol	>5 years
Picloram	4-amino-3,5,6-trichloropicolinic acid	>5 years
Propazine	2-chloro-4,6-*bis*-(isopropylamino)-*s*-triazine	2–3 years
Simazine	2-chloro-4,6-*bis*-(ethylamino)-*s*-triazine	2 years
2,4,5-T	2,4,5-trichlorophenoxyacetic acid	>190 days
2,3,6-TBA	2,3,6-trichlorobenzoic acid	2 years
Toxaphene	chlorinated camphene	>14 years

possible harmful ecological side effects of their use in agriculture. Different pesticides exhibit vastly different residence times in the environment, with some persisting indefinitely (Table 14.8). What happens when a persistent organic compound, one that is not subject to biodegradation, is introduced into the biosphere? Chlorinated hydrocarbon insecticides have been detected in remote Arctic regions, thousands of miles away from their nearest possible application site. As a result, we have learned that it is insufficient to assess the effect of a persistent organic chemical solely on the environment to which it is to be directly applied. It may be dispersed and may exert unexpectedly detrimental effects on more fragile ecosystems far removed from the original application site.

Even though distribution tends to dilute organo-chlorines to the low parts per billion (ppb) range, these chemicals still cause concern. They do so because of a phenomenon called **biological magnification** or **biomagnification**, which is the increase in concentration of a chemical in biological organisms compared to its concentration in the environment. Biomagnification occurs when an environmental pollutant is both persistent and lipophilic. Because of their lipophilic character, such compounds are partitioned from the surrounding water into the lipids of both prokaryotic and eukaryotic microorganisms, and their concentrations in microbial cells may be one to three orders of magnitude higher than those in the surrounding environment (Figure 14.17). When microorganisms are ingested by members of the next higher trophic level in the food web, the pollutant is neither degraded nor excreted to any significant extent; in fact, its concentration is increased by almost another order of magnitude. Thus, such pollutants are concentrated as they are transferred to higher trophic levels. Consequently, their concentration is increased by almost an order of magnitude, so that the top trophic level organisms—such as birds of prey, carnivores, and large predatory fish—may carry a body burden of the environmental pollutant that exceeds the environmental concentration by a factor of 10^4–10^6.

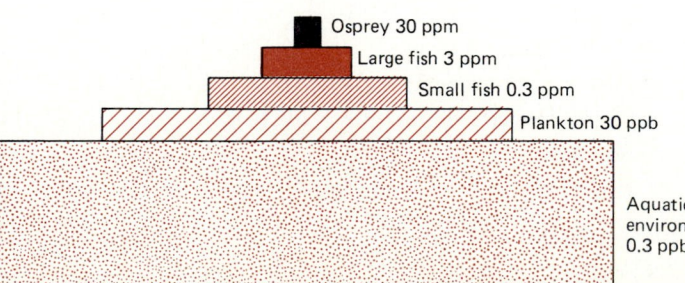

FIGURE 14.17

The biomagnification of DDT. Very small concentrations of dissolved DDT can move by physical partitioning into plankton. Further concentration of the pollutant takes place through the shrinkage of biomass between successive tropic levels.

A pesticide that is biomagnified may cause the death or serious debilitation of animals at the top of the food web. DDT and other chlorinated hydrocarbons were implicated in the death or reproductive failure of various birds of prey and other wildlife. Because human beings derive their food from various trophic levels, we are in a position of less exposure to biomagnified pesticides than a top-level carnivore. Nevertheless, at the time of unrestricted DDT use, the average American, with no occupational exposure, carried a body burden of 4–6 ppm DDT and its derivatives. Although this amount was not considered dangerous, the trend toward increasing contamination of the higher trophic levels of the biosphere became sufficiently clear and led to the ban on the use of DDT in the United States and several other countries for all but emergency situations.

Biodegradation

The majority of the currently used organic pesticides are subject to extensive **biodegradation**, which is decomposition by living organisms—generally microorganisms. To maintain environmental quality, pesticide biodegradation normally should occur within a single growing season. Synthetic pesticides show a bewildering variety of chemical structures, but most contain relatively simple hydrocarbon skeletons with a variety of substituents, such as halogens, amino, nitro, hydroxyl, and other functional groups. Aliphatic hydrocarbons are oxidized to fatty acids (Figure 14.18). The fatty acids are then degraded via the β-oxidation sequence, and the resulting C_2 fragments are further metabolized via the tricarboxylic acid cycle (see Chapter 6 for a discussion of

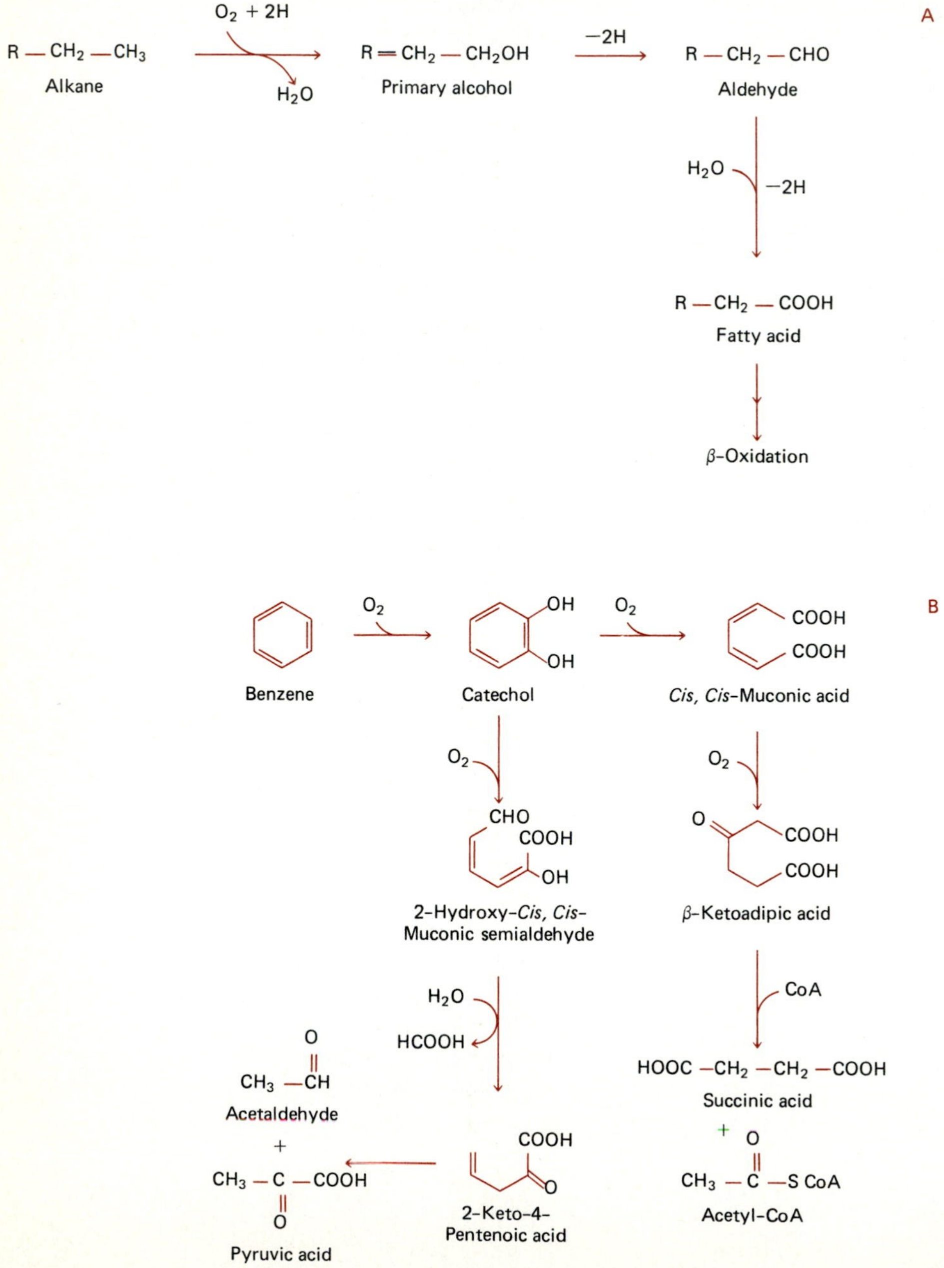

FIGURE 14.18
Degradation pathways for aliphatic (A) and aromatic (B) hydrocarbon pesticide moieties.

this process). Aromatic ring structures are metabolized by dihydroxylation and ring cleavage mechanisms. Prior to these transformations, substituents on the aromatic ring may be completely or partially removed. Substituents uncommon in natural compounds, such as halogens, nitro, and sulfonate groups, if situated so as to impede oxygenation, will frequently cause recalcitrance.

Often a simple change in the substituents of a pesticide may make the difference between **recalcitrance**, that is, complete resistance to biodegradation, and biodegradability. The chemical structures of some biodegradable and some recalcitrant pesticides are compared in Figure 14.19. The herbicide 2,4-D is biodegraded within days, but 2,4,5-T, which differs only by an additional chlorine substitution in the *meta*-position, persists for many months. The additional substitution interferes with the hydroxylation and cleavage of the aromatic ring. Propham is cleaved by microbial amidases so rapidly that for some applications the addition of amidase inhibitors becomes necessary, but propachlor, which has a tertiary amine group, is not subject to attack by such amidases and persists considerably longer. Methoxychlor is less persistent than DDT because the *p*-methoxy groups are subject to dealkylation, and the *p*-chloro substitution endows DDT with great biological and chemical stability.

As the organic molecules that enter the environment and must be degraded become more complex, and as they consist of aliphatic as well as aromatic, alicyclic, or heterocyclic portions, few generalizations can be made about their degradation patterns. If the moieties of the molecule are connected by ester, amide, or ether bonds that can be cleaved by microbial enzymes, the initial attack usually takes this form, and the resulting compounds are subsequently metabolized, as outlined before. If such an attack cannot occur, degradation will commonly be initiated at the aliphatic end of the molecule. However, if degradation at this end is blocked by extensive branching or by other substituents, the attack may start from the aromatic end. The site and mode of the initial attack are determined not only by molecular structure but also by the enzymatic capabilities of the microorganisms involved, as well as by the prevailing environmental conditions, including the availability of oxygen and soil pH. These factors modify not only the rate but also the pathway and the ultimate products of the degradation.

In some cases, one portion of the pesticide molecule is

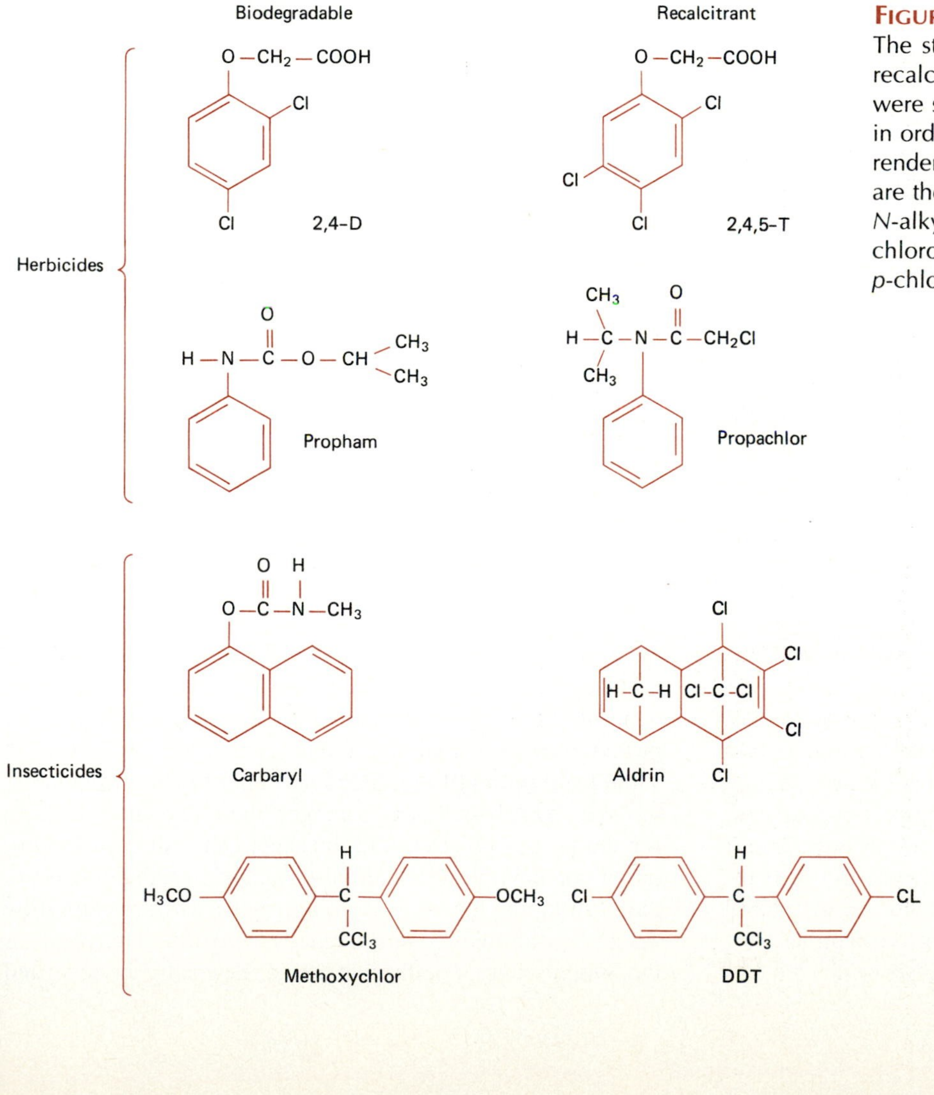

FIGURE 14.19

The structures of some biodegradable and recalcitrant pesticides. The pairs of compounds were selected for overall structural similarity in order to show the molecular features that render the compounds recalcitrant. These features are the 5-chloro-substitution in 2,4,5-T, the *N*-alkyl substitution in propachlor, the multiple chloro-substitutions in aldrin, and the two *p*-chloro-substitutions in DDT.

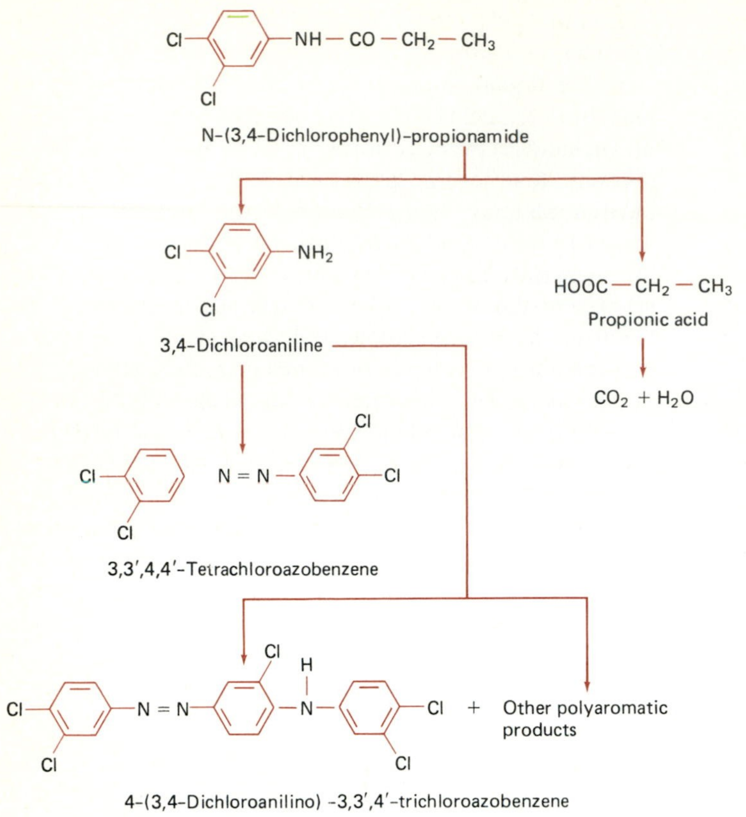

FIGURE 14.20
The biodegradation pathway for the herbicide propanil, N-(3,4-dichlorophenyl)-propinamide. The aliphatic portion of the molecule is degraded, but the aromatic portion is dimerized and polymerized to persistent residues. The transformation involves microbial synergism.

susceptible to degradation and another is recalcitrant. Some acylanilide herbicides are cleaved by microbial amidases, and the aliphatic moiety of the molecule is mineralized (converted to carbon dioxide and water). The aromatic moiety, stabilized by chlorine substitutions, resists mineralization, but the reactive primary amine group may participate in various biochemical and chemical reactions, leading to polymers and complexes that render the fate of such herbicide residues extremely complex. Figure 14.20 shows some of the transformations of the acylanilide herbicide propanil. Microbial acylamidases cleave the propionate moiety, which is subsequently mineralized. A portion of the released 3,4-dichloroaniline (DCA) is acted upon by microbial oxidases and

peroxidases, with the result that they dimerize and polymerize to highly stable residues, such as 3,3',4,4'-tetrachloroazobenzene (TCAB), and related azo compounds. The reasons for such transformations are still somewhat obscure. They may occur by chance when a microbial enzyme that has another metabolic function recognizes and acts upon the man-made residue. In some cases, the reaction seems to "detoxify" the residue from the microbe's point of view, but the overall persistence and environmental impact of the pesticide are increased by such synthetic transformations because some of the synthetic products may be toxic or carcinogenic to higher organisms.

14.5 Biological Control

Biological control using microbial pathogens offers a method that can augment the use of chemical pesticides in controlling pest populations. Microbial populations can be used directly for controlling plant and animal pest populations. Populations of pathogenic or predatory microorganisms that are antagonistic toward a particular pest population provide a natural means of controlling pest populations, and preparations of such antagonistic microbial populations are called **microbial pesticides**. The effective use of pathogenic microorganisms as pesticides depends on the ability to establish a disease epidemic among susceptible pest populations.

To be effective as microbial pesticides, the prospective microbial pathogen must be virulent and cause disease in the pest population when properly applied at the recommended concentration. The pathogen must not be sensitive to expected environmental variations, and after application it should survive until the infection within the pest population has been established. The pathogen should be rather specific for the pest population and must not cause disease in nontarget populations. It should also rapidly establish disease in the pest population so as to minimize destruction caused by the pest. Microbial control methods have been developed for the suppression of arthropod pests, especially insects, and several commercial microbial insecticides are available for agricultural use. Microbial suppression of animal populations is aimed at populations that cause crop and other plant damage and at those that act as vectors of disease-causing microorganisms. Controlling insect damage to crops through the use of chemical or microbial pesticides is critical for achieving high crop yields.

General Considerations

The development and use of microorganisms for the control of pest populations require considerations that go well beyond finding a pathogen for a particular weed or animal pest population. Questions of economics, production, quality control, application, side effects, and safety must all be satisfactorily answered. In many cases the economics of the development, production, and application of microbial pesticides does not permit commercial development of an otherwise promising biological control system. The effectiveness of the microbial pathogens must be carefully evaluated. Industry must be able to produce and stockpile large, stable populations of potential microbial pesticides. Quality control must ensure standardized batches of the microbial pesticides that have the same virulence so that recommended application rates may be used with confidence in the field. Application methods for microbial pesticides must permit the microorganisms to remain viable long enough to contact the pest population and to establish widespread disease. Resistant stages, such as spores or cysts, are best for this use. With some candidate microbial pathogens it is difficult to achieve repeatable results. Environmental conditions must be taken into account, as these affect the virulence of microbial pesticides. Persistence of microbial pesticides must also be considered. Some candidate microorganisms have very short survival times. Persistence is required for long-term control of pest populations.

Ideally, microbial pesticides should exhibit a high degree of host specificity. They should not adversely affect nontarget populations. The host specificity, though, should not be so narrow as to preclude its effectiveness against a simple genetic variance within the pest population. It is often difficult to predict whether a microbial pesticide can establish disease in nontarget populations because it is impossible to test the infectivity of the pesticide against all possible nontarget populations. Obviously, any microbial pesticide used should be harmless to humans and other valued plant and animal populations. The use of microbial pesticides is probably best when employed in an integrated program of management practices for agricultural crops and domestic animals that minimizes opportunities for infection or interaction, along with limited applications of appropriate chemical pesticides carefully timed for maximum effect.

Viral Pesticides

Pathogenic viruses possess the potential for use as pesticidal agents. Viruses cause diseases in insects and other arthropods. The specificity of the virus–host relationship makes viruses ideal candidates for use in the control of specific pest insect populations, with few or no deleterious effects on people and other animals. Insect pathogenic viruses frequently cause natural disease epidemics, known as **epizootics**, in insect populations. Viruses have been used in attempts to control outbreaks of a variety of insect pests including gypsy moths, Douglas fir tussock moths, pine processionary caterpillars, red-banded leaf rollers (a pest of apples), spruce budworms, codling moths (a pest of apples, walnuts, and other deciduous fruits), Great Basin tent caterpillars, alfalfa caterpillars, cabbage white butterflies, cabbage loopers, cotton bollworms, corn earworms, tobacco budworms, tomato worms, army worms, and wattle bagworms, among others.

The most thoroughly studied of these viruses are the nuclear polyhedrosis viruses, cytoplasmic polyhedrosis viruses, and granulosis viruses. The **nuclear polyhedrosis viruses** develop in the host-cell nuclei; the virions are occluded singly or in groups in polyhedral inclusion bodies. **Cytoplasmic polyhedrosis viruses** develop only in the cytoplasm of host midgut epithelial cells; the virions are occluded singly in polyhedral inclusion bodies. **Granulosis viruses** develop in either the nucleus or the cytoplasm of host fat, tracheal, or epidermal cells; the virions are occluded singly or, rarely, in pairs in small occlusion bodies called *capsules*.

Baculoviruses are perhaps the most carefully studied insect viruses. They include nuclear polyhedrosis and granulosis viruses. Pathogenic baculoviruses have been found principally for Lepidoptera, Hymenoptera, and Diptera. The infection is often transmitted by ingestion of contaminated food, with cell invasion probably beginning in the midgut. Several nuclear polyhedrosis viruses are produced in the United States on a large scale for control of insect pests. Inoculation of leaves with polyhedrosis viruses can initiate epizootics in Lepidoptera and Hymenoptera larvae that feed on plant leaves, causing significant population reductions in these insects.

Nuclear polyhedrosis viruses cause disease in sawflies. When the accidental introduction of the European sawfly into North America in the twentieth century threatened the spruce forests, introduction of nuclear polyhedrosis viruses into this pest caused a spectacular epizootic that reduced the sawfly populations and saved many spruce forests. Similarly, the European pine sawfly, which was causing serious damage to pines in New Jersey, Ohio, and Michigan, has been controlled by introducing insects containing nuclear polyhedrosis viruses into the population, resulting in epizootics.

An interesting example of the use of viral pesticides is the attempt to control the rabbit populations of Australia with myxoma virus. Rabbits were introduced into Australia from Europe in 1859, and because there were no natural enemies for the rabbit in Australia, their reproduction was unchecked. Myxoma virus, occurring naturally among South American rabbits, was found to be a virulent pathogen of the European rabbit, and in an effort to achieve control of the rabbit populations in Australia, the Myxoma virus was introduced. The virus rapidly spread among the rabbit populations, causing high seasonal morbidity and mortality. Initially 99 percent of the infected rabbits died, but after only a few years the virulence of the myxoma virus for the surviving rabbits declined. The survivors of the initial epidemics had been selected for their resistance to the virus. The resistance of the rabbits is innate and is not due to an immunological defense system. Thus, within only a few years, an equilibrium was achieved between the virus and the Australian rabbits. Myxomatosis was effective in lowering the Australian rabbit population to about 20 percent of its level before the introduction of the virus, and now that the virus is firmly established within the rabbit population, there is a pathogen that controls its level. However, the introduction of the myxoma virus did not completely eliminate the rabbits, as was originally envisaged.

Bacterial Pesticides

There are several bacterial pathogens of insects that currently are used, or have potential for use, as insecticides. They include endospore-forming *Bacillus* and *Clostridium* species, as well as non-endospore-forming species of *Pseudomonas*, *Enterobacter*, *Proteus*, *Serratia*, and *Xenorhabdus*. Of the potential bacterial pesticides, **Bacillus thuringiensis** has been most extensively exploited in the bacterial control of pest insect populations. Commercial preparations of *B. thuringiensis* are registered by at least 12 manufacturers in five countries for use on numerous agricultural crops, forest trees, and ornamentals for control of various insect pests (Table 14.9). *B. thuringiensis* has been tested successfully against more than 140 insect species, including members of the Lepidoptera, Hymenoptera, Diptera, and Coleoptera. It is a crystalliferous bacterium because, in addition to endospores, it produces discrete parasporal bodies within its cell (Figure 14.21). It is the proteinaceous parasporal crystal that is the toxic factor. Four separate toxic substances are produced by *B. thuringiensis*. Use of this bacterium brings about commercially acceptable levels of suppression of cabbage worms, cabbage loopers, and many other pests of vegetable crops. *B. thuringiensis* also readily suppresses populations of tent caterpillars, bagworms, and canker worms, which are pests of forest trees. Gypsy moth and spruce budworm can be suppressed by *B. thuringiensis*, but only when high application rates are used and uniform foliage coverage is attained.

Of particular interest is the potential use of *B. thuringiensis israelensis* (BTI) to control populations of the mosquito vectors of malaria. In fact, effective use of BTI now offers the best hope for controlling malaria because, unlike DDT, it is environmentally safe and because mosquitoes show no sign of developing resistance. Testing of BTI for control of malaria-carrying mosquitoes has begun only recently, but the results of tests in Africa against blackfly, the carrier of widespread river blindness, are excellent, according to the United Nations World Health Organization.

TABLE 14.9 Some Registered Uses for *Bacillus thuringiensis* Products in the United States

Pest	Crop
Vegetable and Field Crops	
Alfalfa caterpillar, *Colias eurytheme*	Alfalfa
Artichoke plume moth, *Platyptila carduidactyla*	Artichokes
Bollworm, *Heliothis zea*	Cotton
Cabbage looper, *Trichoplusia ni*	Beans, broccoli, cabbage, cauliflower, celery, collards, cotton, cucumbers, kale, lettuce, melons, potatoes, spinach, tobacco
Diamondback moth, *Plutella maculipennis*	Cabbage
European corn borer, *Ostrinia nubilalis*	Sweet corn
Imported cabbageworm, *Pieris rapae*	Broccoli, cabbage, cauliflower, collards, kale
Tobacco budworm, *Heliothis virescens*	Tobacco
Tobacco hornworm, *Manduca sexta*	Tobacco
Tomato hornworm, *Manduca quinquemaculata*	Tomatoes
Fruit Crops	
Fruit tree leaf roller, *Archips argyrospilus*	Oranges
Orange dog, *Papilio cresphontes*	Oranges
Grape leaf folder, *Desmia funeralis*	Grapes
Shade Trees, Ornamentals	
California oakworm, *Phryganidia californica*	
Fall webworm, *Hyphantria cunea*	
Fall cankerworm, *Alsophila pometaria*	
Great Basin tent catepillar, *Malacosoma fragile*	
Gypsy moth, *Lymantria (Porthetria) dispar*	
Linden looper, *Erannis tiliaria*	
Salt marsh caterpillar, *Estigeme acrea*	
Spring cankerworm, *Paleacrita vernata*	
Winter moth, *Operophtera brumata*	

Other *Bacillus* species cause milky disease of Japanese beetles. In Japan, where the beetle encounters natural antagonists, it is a relatively minor pest. In the United States, however, the beetle does not have any natural associated pathogens or other antagonists. The Japanese beetle feeds voraciously on some 300 species of plants and has been responsible for large economic losses. The greatest success in suppressing pest populations of Japanese beetles has been obtained by using bacteria that produce milky disease. For many years, a mixture of *Bacillus popilliae* and *B. lentimorbus* has been marketed under the trade name Doom (Figure 14.22). *B. lentimorbus*, which infects mainly first and second instar grubs, does not produce parasporal crystals; *B. popilliae* produces parasporal bodies and infects a high proportion of third instar grubs. The use of these *Bacillus*

Bacillus   Parasporal bodies

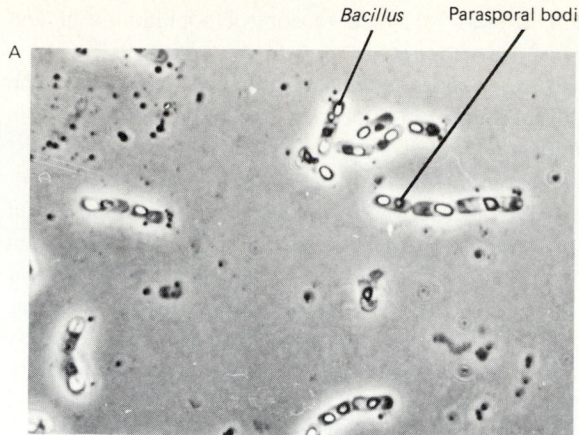

FIGURE 14.21

(A) Micrograph of *Bacillus thuringensis tolworthi* under phase contrast microscopy. Note the spores and parasporal inclusions within the cells. (Courtesy B. N. Herbert, Shell Research, Ltd., Kent, England), (B) Photograph showing commercial products containing *B. thuringensis*. (Courtesy Sandoz, Inc., San Diego, Calif.), (C) Scanning electron micrograph showing crystals of delta-endotoxin from Thuricide (24,000×). (Courtesy Dr. Guggenheim, University of Basel)

FIGURE 14.22

(A) A can of Doom, a biological pesticide that is effective in controlling Japanese beetles. (Courtesy Fairfax Laboratories, Clinton Corners, N.Y.), (B) Micrograph of *Bacillus popilliae*, one of the two active bacterial components of Doom. (Courtesy Michael Klein, U.S. Department of Agriculture, Wooster, Ohio)

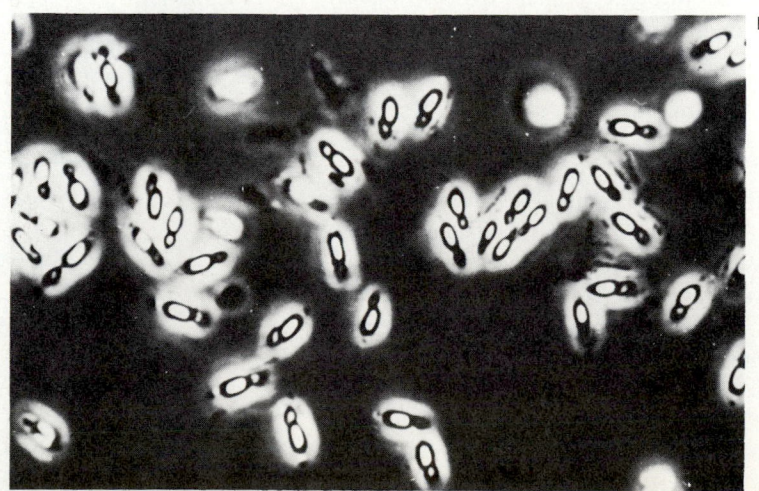

species to produce milky disease in grubs of Japanese beetles is probably responsible for the control of pest beetle populations, and although in the past there have been major infestations of Japanese beetles in the United States, today there are relatively few major outbreaks.

Fungal Pesticides

Pathogenic fungal infections, or **mycoses**, can cause epizootics; fungi are, therefore, potentially important in the control of pest populations. Insect mycoses are caused by Phycomyces, Ascomyces, Basidiomyces, and Deuteromyces. Most studies on entomogenous fungi have been concerned with members of the fungal genera *Beauveria*, *Metarrhizium*, *Entomophthora*, and *Coelomomyces*. *Beauveria bassiana* has been used extensively in the Soviet Union to control the eastward spread of Colorado beetle and against the codling moth; it also infects the corn earworm. Fungi of the genus *Aschersonia* have been used to control pests of citrus trees in the Soviet Union near the Black Sea. In Florida, *Hirsutella* has been used against the citrus rust mite. The fungus *Metarrhizium* is used in Brazil to control populations of leaf hoppers and frog hoppers.

The use of predaceous nematode-trapping fungi has been suggested for the control of pest nematode populations in soil. Some plants are subject to infection with nematodes. Nematode-trapping fungi have been used in attempts to control root knot disease of pineapples in Hawaii. Successful control of pest nematodes was achieved but was dependent on supplementing soil with organic matter. It is not clear whether control of the nematodes was due to the predaceous fungi or to the alteration in food-web relationships caused by addition of organic supplements. Subsequent attempts to demonstrate that nematode-trapping fungi, rather than soil supplements, were primarily responsible for controlling nematode populations in soil had ambiguous results. Under appropriate conditions, nematode-trapping fungi with supplemental organic matter appear to have the potential for diminishing the effects of nematodes on susceptible plant populations. Other possibilities for controlling pest nematodes include infection with parasitic fungi, protozoa, bacteria, and viruses.

14.6 Diseases of Farm Animals

Farm animals are subject to various infectious diseases that can spread through a herd, causing sizable economic losses to animal breeders and farmers. Diseases of animals other than humans are called **zoonoses**. Some of these diseases can also be transmitted to people and represent an occupational hazard to those working with animals. Cows, horses, swine, and fowl are among the barnyard animals susceptible to serious infectious diseases. As with human diseases, the common portals of entry for the microbial pathogens that cause zoonoses are the respiratory, gastrointestinal, and genitourinary tracts of these animals. Various pathogenic viruses, bacteria, fungi, and protozoa are responsible for disease outbreaks in populations of farm animals. Because farm animals are normally herded tightly together, these infectious diseases can easily spread through the entire herd, rapidly producing adverse economic effects. Some of the common infectious diseases of farm animals are described in Table 14.10.

Control of Animal Diseases

Methods of controlling infectious animal diseases are similar to those employed for preventing and treating human diseases. Vaccines have been developed for establishing immunity (resistance based upon the animals' physiological defense mechanisms) against many diseases, and extensive immunization programs are carried out on most farms. Some infectious diseases of farm animals, such as foot-and-mouth disease (Figure 14.23), can be prevented by vaccination, reducing economic losses. Genetic engineering has been employed to produce a foot-and-mouth vaccine, and recombi-

nant DNA technology has recently been used to produce a pseudorabies vaccine to protect swine. The new pseudorabies vaccine uses a virus with a genetic deletion, produced using recombinant DNA methods. The deletion prevents the virus

FIGURE 14.23
Symptoms of foot-and-mouth disease in cattle. This steer has excess saliva flowing from its mouth. The animal was experimentally infected 24 hours previously by inoculating the tongue with the virus of this disease. (Courtesy U.S. Department of Agriculture)

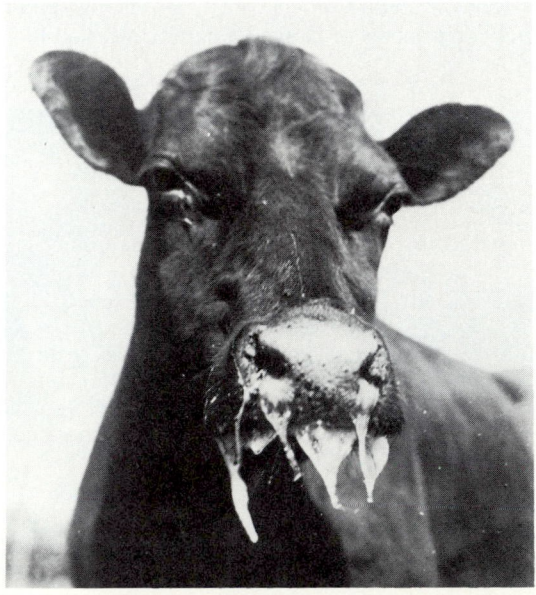

TABLE 14.10 Some Infectious Diseases of Farm Animals

Disease	Causative Agent	Description
Anaplasmosis	*Anaplasma marginale*	A protozoan disease of cattle. The protozoa are transmitted through tick, fly, and mosquito vectors. The disease is characterized by severe anemia caused by reproduction of the protozoa within the animal's red blood cells.
Anthrax	*Bacillus anthracis*	Anthrax affects cattle, sheep, horses, and swine. It is widespread in Mexico, South America, and Asia. *B. anthracis* occurs in soils and is transmitted to farm animals in contaminated feed and water. Infection results in septicemia, and there is a high mortality rate.
Avian tuberculosis	*Mycobacterium avium*	Chronic pulmonary infection occurring in domestic fowl.
Blackleg	*Clostridium chauvoei*	An acute infection of cattle, sheep, and other ruminant animals. The disease is characterized by severe swelling, and the infected subcutaneous tissues are permeated with blood and gas; the underlying muscles appear dark brown-black, which accounts for the name of this disease. Animals acquire this infection via the gastrointestinal tract by the consumption of contaminated feed or water.
Brucellosis	*Brucella* sp.	A serious infection occurring in cattle, swine, and goats. In cattle *B. abotus* causes abortion and infection of the udders.
Coccidiosis	*Eimeria* sp.	A protozoan disease of cattle, characterized by severe gastroenteritis and bloody diarrhea. Cysts are shed with fecal matter, and transmission is via the gastrointestinal route from ingestion of contaminated water or feed.
Contagious metritis	*Haemophilus* sp.	A sexually transmitted bacterial disease of horses occurring in the United States, France, and Argentina. The disease results in abortion in mares. Quarantining of breeding stock is practiced when outbreaks of this disease occur.
East Coast fever	*Thieleria parva*	A protozoan disease prevalent in cattle along the east coast of Africa. The protozoa are transmitted via a tick vector and reproduce within the red blood cells of infected cattle.
Erysipeloid	*Erysipelothrix rhusiopathiae*	Second only to swine influenza as an infectious disease of swine in the United States. Most common in North America during the summer. Transmission is via the gastrointestinal tract due to feed or water contaminated with fecal matter or urine containing the causative bacterium. The acute form of this disease can occur as septicemia, chronic endocarditis, and "diamond skin disease."
Foot-and-mouth disease	Rhinovirus	A highly contagious disease, especially among cattle, that also affects sheep, swine, and goats. The virus is transmitted through direct contact between infected animals and through ingestion of food and water containing the virus. The disease is characterized by formation of vesicular

TABLE 14.10 (Continued)

Disease	Causative Agent	Description
Foot-and-mouth disease (cont.)		lesions in the mouth, on the muzzle, and on the udders and teats. The lesions are painful and result in lameness and a reluctance of the animal to eat or move.
Fowl cholera	*Pasteurella multocida*	A highly communicable and often fatal disease of fowl.
Fowl typhoid	*Salmonella gallinarum*	A disease affecting chickens and turkeys that is transmitted from infected to susceptible fowl through fecal contamination of feed and water.
Glanders	*Pseudomonas mallei*	An infectious disease of horses and other equines. Transmitted primarily through the watering trough, feed, and direct contact. Glanders is characterized by formation of nodules and ulcerations of the mouth, respiratory tract, internal organs, and skin. Acute forms of the disease often are fatal.
Hog cholera	RNA virus (Togaviridae)	A highly communicable disease prevalent in the midwestern United States. It is characterized by a high fever, loss of appetite, lack of coordination, purulent conjunctivitis, and diarrhea. In the United States the mortality rate is approximately 90 percent.
Infectious coryza	*Haemophilus gillinarum*	A serious disease of fowl characterized by bloody and mucoid exudates of the nostrils and eyes.
Limberneck	*Clostridium botulinum*	Botulism of fowl resulting in paralysis of the neck muscles; other forms of poisoning that produce symptomatic paralysis of the neck are also classified as limberneck.
Newcastle disease	Newcastle virus (Paramyxovirus)	Also known as *avian pneumoencephalitis,* this disease causes extensive losses to poultry farms. The virus is spread through chicken coops via airborne droplets. The symptoms include coughing, sneezing, and gasping. A sharp decline in egg production occurs in egg-laying flocks. Nervous symptoms include characteristic paralysis of the legs or wings and twisting of the neck.
Piraplasmosis	*Piroplasma bigemina*	Also known as *tick fever,* this protozoan disease affects cattle, horses, and sheep. Transmission is through tick bites.
Pullorum disease	*Salmonella pullorum*	An economically important disease of chickens that is transmitted congenitally from hen to egg.
Swine influenza	Influenza virus	The disease is transmitted via the respiratory tract and can result in 100 percent infection fo susceptible animals in a herd. Swine flu is prevalent in the fall and winter. The symptoms appear suddenly and include fever, anorexia, prostration, cough, and abnormal breathing. Mortality rates are generally less than 2 percent.
Trypanosomiasis	*Trypanosoma brucei, T. congolense, T. vivax*	This protozoan disease affects large numbers of African cattle. It is transmitted by the tsetse fly.

from replicating and causing disease, but does not affect its ability to stimulate immunity in the vaccinated swine. Other vaccines will undoubtedly be produced by genetic engineers that will protect farm animals against disease. Antibiotics have been employed as prophylactic (preventive) measures to preclude the spread of infectious disease through animal populations. In fact, penicillin and tetracycline antibiotics are sometimes included in bird feed to prevent the spread of bacterial diseases. Antibiotics can also be used in conjunction with artificial infections to produce immunity. For example, oxytetracycline-controlled infections with *Thieleria parva* establish immunity in cattle against East Coast fever. There is a problem, however, with residual concentrations of antibiotics in the meat or milk of animals so treated. Therefore, the use of prophylactic use of antibiotics for farm animals is limited.

Sanitary measures in rearing animals are particularly effective in reducing the reservoirs of infectious microorga-

nisms. Proper ventilation in chicken coops is important for preventing the aerial spread of bacterial diseases. The development of resistant animal breeds has reduced the incidence of some diseases, such as tick-bite fever. Elimination of vector populations—particularly insect carriers of microbial pathogens—is an efficient means of controlling disease because one vector can transmit the disease to many animals. For example, in the case of East Coast fever, one infected tick can contain enough parasites to kill 100 cattle (Figure 14.24).

FIGURE 14.24

(A) Micrograph showing an infected cell and an uninfected cell of the tick vector of East Coast fever. An infected cell may contain up to 50,000 sporozoites of *Thieleria parva*. The protozoan is transmitted to cattle through tick bites. The *Thieleria* bind to lymphocytes and enter the host's white blood cells within about 5 minutes. (B–F) These micrographs show the uptake of a sporozoite by a lymphocyte. The protozoa later infect red blood cells to complete their life cycle. (Courtesy Don Fawcett, International Laboratory for Research in Animal Diseases, Nairobi, Kenya; reprinted by permission of the American Association for the Advancement of Science, 1982, *Science* 216:504)

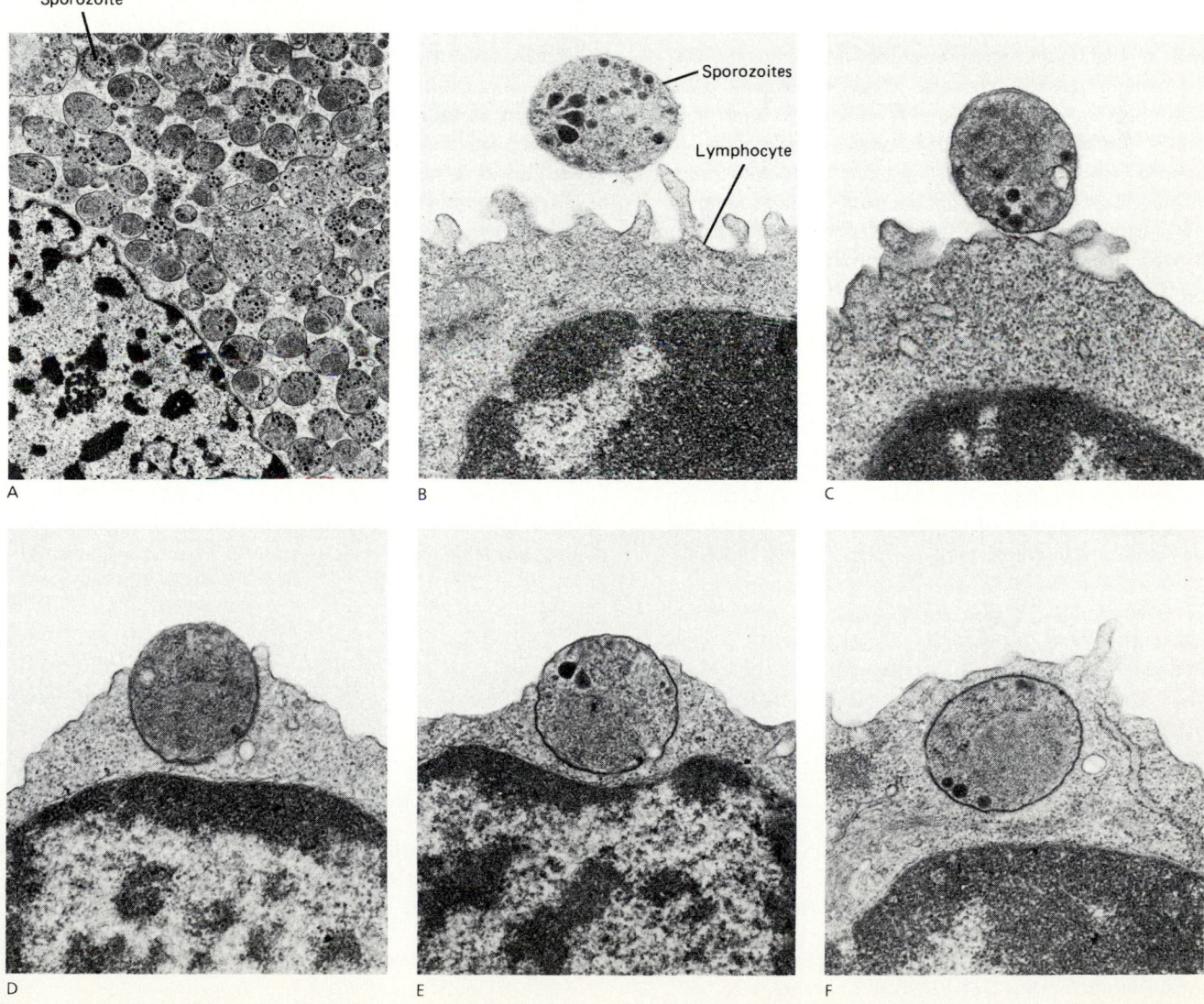

When prevention fails and an infectious disease becomes established in a herd, it is important to treat or eliminate the infected animals. It is essential that infected animals be removed from the herd, and in some cases it may be necessary to quarantine or even destroy an entire herd in order to eliminate the source of the infectious pathogens. When the disease is caused by a bacterial species, antibiotics can often be used to cure the infected animals. Treatment methods must be initiated promptly enough to prevent the spread of an infection through an entire herd and from one herd to another. It is critical that effective remedial measures be taken to control the transmission of disease among farm animals in order to reduce economic losses.

Postlude

The central ecological role of microorganisms in the biosphere is clear when agricultural practices are considered. The success of agricultural production of both crops and livestock depends on an integrated program of controlling microbial activities. The large-scale production of agricultural crops in monocultures creates an ecologically unstable ecosystem that requires major inputs of energy and intensive management to maintain soil fertility and prevent destruction of the crop by microbial and pest populations. Plants and farm animals, like human beings, are subject to many diseases, and limiting the spread of pathogenic microorganisms through crops and herds is a major task for farmers. Many agricultural practices, including breeding programs to develop hardier plant and animal varieties, are aimed at reducing losses due to microbial diseases. Harvests can be greatly reduced as a result of microbial plant diseases, causing great economic losses and sometimes creating serious food shortages.

Fertilizers are added to soils in order to overcome nitrogen limitations and increase crop production. The application of nitrogen fertilizers to soil must take into account the microbial nitrification process, lest the fertilizer be wastefully squandered and groundwater problems result. Some farm practices, most notably crop rotation, are aimed at taking advantage of microbial transformations of nitrogen that increase soil fertility and eliminating the need to add costly nitrogen fertilizers to the field in order to achieve high crop yields. The incorporation of the bacterial genes controlling nitrogen fixation into staple crops, such as wheat and corn, through the use of genetic engineering will go a long way toward ensuring that world food needs can be met when the proper technology is developed. Other political, rather than microbiological, factors—such as distribution and cultural preferences—are also involved in meeting world food needs.

Modern agriculture has become almost totally dependent on the use of chemical pesticides to control pest populations. Herbicides are applied to fields to prevent competition between the crop plant and weeds; fungicides are used to prevent fungal diseases, the most frequent infectious diseases of plants; and insecticides are used to control damage by insect pests. Biological control, through the controlled establishment of microbial diseases in pest populations, represents a useful alternative to conventional chemical pesticides. Here is a case where disease can be considered beneficial to humanity. The environmentally safe use of chemical pesticides depends on microbial biodegradation to prevent persistence and biomagnification of these toxic compounds. When microorganisms fail to degrade the pesticide, there may be serious environmental consequences far removed from the site of application, as seen in the case of DDT. The recognition that some compounds are recalcitrant to microbial attack violates the cardinal rule of microbiology—that microorganisms are infallible. As will be discussed in the next chapter, we rely so heavily on microorganisms to remove our wastes that it is difficult, even for microbiologists, to accept the truth that some organic compounds are resistant to biodegradation by microorganisms.

Suggested Supplementary Readings

AGRIOS, G. N. 1978. *Plant Pathology*. Academic Press, New York.

ALEXANDER, M. 1965. Biodegradation: Problems of molecular recalcitrance and microbial fallibility. *Advances in Applied Microbiology* 7:35–80.

ALEXANDER, M. 1977. *Introduction to Soil Microbiology*. John Wiley & Sons, Inc., New York.

ALEXANDER, M. 1981. Biodegradation of chemicals of environmental concern. *Science* 211:132–138.

ALEXANDER, M. 1985. Ecological constraints on nitrogen fixation in agricultural ecosystems. *Advances in Microbial Ecology* 8:163–183.

ARONSON, A. I., W. BECKMAN, and P. DUNN. 1986. *Bacillus thuringiensis* and related insect pathogens. *Microbiological Reviews* 50:1–24.

BARNES, E. H. 1979. *Atlas and Manual of Plant Pathology* Plenum Press, New York.

BRILL, W. J. 1981. Agricultural microbiology. Scientific American 245(3):198–215.

BURGES, H. D. (ed.). 1981. *Microbial Control of Pests and Plant Diseases, 1970–1980.* Academic Press, New York.

CAMPBELL, R. 1985. *Plant Microbiology* Edward Arnold, London.

CARSON, R. 1962. *Silent Spring.* Houghton-Mifflin, Boston.

COPPEL, H. C., and J. W. MERTINS. 1977. *Biological Insect Pest Suppression.* Springer-Verlag, Berlin.

DEACON, J. W. 1983. *Microbiological Control of Plant Pests and Disease.* Aspects of Microbiology No. 7. American Society for Microbiology, Washington, D.C.

DELWICHE, C. C. 1970. The nitrogen cycle. *Scientific American* 223(5):137–146.

DHINGRA, O. D., and J. B. SINCLAIR. 1985. *Basic Plant Pathology Methods.* CRC Press, Inc., Boca Raton, Fla.

DICKINSON, C. H., and J. A. LUCAS. 1982. *Plant Pathology and Plant Pathogens.* Blackwell Scientific Publications, Oxford, England.

DIENER, T. O. 1979. *Viroids and Viroid Diseases.* Wiley-Interscience, New York.

EATON, R. W., and K. N. TIMMIS. 1984. Genetics of xenobiotic degradation. In *Current Perspectives in Microbial Ecology* (M. J. Klug and C. A. Reddy, eds.), pp. 694–703. American Society for Microbiology, Washington, D.C.

GILLESPIE, J. H., and J. F. TIMONEY. 1981. *Hagan and Bruner's Infectious Diseases of Domestic Animals.* Cornell University Press, Ithaca, N.Y.

GRAY, T. R. G., and S. T. WILLIAMS (eds.). 1975. *Soil Bacteria.* Longman, Inc., New York.

HALVORSON, H. O., D. PRAMER, and M. ROGUL (eds.). 1985. *Engineered Organisms in the Environment: Scientific Issues.* American Society for Microbiology, Washington, D.C.

HARRIS, K., and K. MARAMOROSCH. 1981. *Pathogens, Vectors, and Plant Diseases.* Academic Press, New York.

HILL, I. R., and S. J. L. WRIGHT (eds.). 1978. *Pesticide Microbiology: Microbiological Aspects of Pesticide Behavior in the Environment.* Academic Press, London.

HUFFAKER, C. B., and P. S. MESSENGER (eds.). 1976. *Theory and Practice of Biological Control.* Academic Press, New York.

LINDOW, S. E. 1983. The role of bacterial ice nucleation in frost injury to plants. *Annual Reviews of Phytopathology* 21:363–384.

LYNCH, J. M. 1976. Products of soil microorganisms in relation to plant growth. *CRC Critical Reviews in Microbiology* 5:67–107.

LYNCH, J. M. 1982. *Soil Biotechnology: Microbiological Factors in Crop Productivity.* Blackwell Scientific Publications, Oxford, England.

MERCHANT, I. A., and R. A. PACKER. 1967. *Veterinary Bacteriology and Virology.* The Iowa State University Press, Ames.

MOHANTY, S. B., and S. K. DUTTA. 1981. *Veterinary Virology.* Lea & Febiger, Philadelphia.

PARKER, W. H. 1980. *Health and Disease in Farm Animals: An Introduction to Farm Animal Medicine.* Pergamon Press, New York.

SIEGMUND, O. H. (ed.). 1979. *The Merck Veterinary Manual.* Merck and Company, Inc., Rahway, N.J.

STEVENS, R. B. 1974. *Plant Disease.* John Wiley & Sons, Inc., New York.

WALKER, N. 1975. *Soil Microbiology.* John Wiley & Sons, Inc., New York.

WHEELER, H. 1975. *Plant Pathogenesis.* Springer-Verlag, New York.

WHIPPS, J. M., and J. M. LYNCH. 1986. The influence of the rhizosphere on crop productivity. *Advances in Microbial Ecology* 9:187–244.

Study Questions

1. How are agricultural soils managed to conserve soil nitrogen?

2. Why is the nitrogen fixation symbiotic relationship so important to soil fertility? How can genetic engineering extend the range of plants that can establish mutualistic relationships with nitrogen-fixing bacteria?

3. What are some of the symptoms of plant disease?

4. Discuss the transmission of plant diseases.

5. Discuss agricultural management practices aimed at controlling plant diseases.

6. What is biomagnification? What properties of a pesticide are important in determining its biomagnification through a food web?

7. Why must pesticides be biodegradable for their safe use in agriculture?

8. What is biological control? How are insect viruses used in the control of plant diseases?

Situational Problem

Safety Considerations for the Deliberate Release of Genetically Engineered Microorganisms in Agriculture

For many centuries, humans have selected genetically different varieties of plants and animals for various purposes. Horticulturalists select and breed plant varieties for their aes-thetic value. Agriculturists select and breed plants and animals for their nutritive value and resistance to disease. These practices are accepted as necessary for the production of the world's food supplies and go on without a great deal of public attention or scrutiny. However, the development of recombinant DNA technology, with its enormous potential for genetically engineering novel plants, animals, and microbes, has raised many scientific and ethical questions.

In addition to the development of novel microorganisms intended to be grown in contained fermentors in order to produce pharmaceuticals and other industrially valuable substances, much of the effort and potential of genetic engineering lies in the area of agriculture. Specific areas with great potential include the creation of new plants, particularly staple crops that, directly or in association with mutualistic microbes, can use atmospheric nitrogen instead of fertilizers; herbicide-resistant plants, produced perhaps by the incorporation of novel rhizosphere microbes; pesticide- and/or pest-resistant plants, again produced perhaps by the incorporation of novel rhizosphere microbes; novel pesticide-degrading microorganisms that can eliminate toxic levels of pesticides from contaminated areas; and frost-resistant plants covered with microorganisms that do not form ice-crystallization nuclei. The technical capability of producing these novel organisms is at hand and, in fact, several such organisms have already been created.

However, until recently these novel organisms had not been intentionally released into the environment, even on a test scale. The reason is concern over the ecological effects of introducing genetically engineered organisms, particularly microbes that cannot be readily seen and traced after their introduction into the environment. Many examples of the adverse effects of introducing new species, including the proliferation of imported rabbits in Australia and the progression of kudzu in the United States, make many environmentalists apprehensive about introducing newly created organisms. Yet we have had experience and success in introducing microbes to control pests, as exemplified by the use of *Bacillus thuringiensis* to control insect pests in North America and the introduction of Myxoma virus to control rabbits in Australia. Concern about the environmental risks versus the benefits of introducing organisms engineered for agricultural use has led to public and scientific debates, congressional legislation, and court rulings that make the outlook for the environmental use of any such organisms in agricultural practice uncertain. Editorials in local newpapers have called upon the public and local representative bodies to become active participants in determining the future uses of engineered microorganisms for environmental and agricultural purposes in their communities.

If your community were to hold a public hearing on this issue, as many are doing, you could speak as an informed layman. Consult news media, books, and journal articles on the issues related to the safety of deliberate release of genetically engineered microorganisms into the environment to determine the stand that you will take. Then prepare a 10-minute statement expressing your informed opinion. Be sure to include the critical issues as you see them and the ways in which uncertainties could be reduced so that the decision of the community representatives will be the correct one for your area.

CHAPTER 15

Environmental Quality: Biodegradation of Wastes and Pollutants

Human activities create vast amounts of various wastes and pollutants. The release of these materials into the environment sometimes causes serious health problems and may preclude desirable usage of our land and water resources. The use of rivers, for example, as a habitat for fish, as a source of irrigation and drinking water, and for the disposal of sewage depends on the careful management of the amounts of wastes entering the ecosystem and the level of pathogenic microorganisms associated with their release. Because water is now in short supply, it is frequently recycled and reused for various purposes. There are limits to the natural capacity of an ecosystem to cope with wastes and pollutants; that is, the microorganisms in the ecosystem have a limited capacity to biodegrade unwanted materials without causing serious deterioration of environmental quality or an increased incidence of disease. The level of wastes produced by dense human and domestic animal populations often exceeds the local ecosystem's biodegradative capacity, resulting in serious environmental pollution and epidemic outbreaks of disease. Each year more than 500 million people are afflicted with waterborne diseases, and more than 10 million of them die. Most of these disease outbreaks stem from fecal contamination of drinking water supplies. Proper treatment of wastes, employing microbial biodegradation, and disinfection of potable water supplies in order to kill contaminating pathogenic microorganisms can greatly improve the safety and quality of water supplies and the status of human health. In this chapter we will examine some of the methods involving microbiology that are employed to maintain environmental quality and safely dispose of waste and pollutant materials.

15.1 Solid Waste Disposal

Urban solid waste production in the United States amounts to roughly 150 million tons per year. Much of this material is inert, composed of glass, metal, and plastic, but the rest is decomposable organic waste, such as kitchen scraps, paper, and other household and industrial garbage. Sewage sludge derived from treatment of liquid wastes, animal waste from cattle feed lots, and large-scale poultry and swine farms are also major sources of solid organic waste. In traditional small family farm operations, most organic solid waste is recycled into the land as fertilizer. In highly populated urban centers and areas of large-scale agricultural production, the disposal of massive amounts of organic waste becomes a difficult and expensive problem.

There are several options for dealing with solid waste problems. Today, many of the inert components of solid waste, such as aluminum and glass, are recovered and recycled. Even paper, which is relatively resistant to microbial degradation, can be recovered from solid waste, and many books and newspapers are printed on recycled paper. The remaining bulk of the solid waste may be incinerated, creating potential air pollution problems, or the organic components can be subjected to microbial biodegradation in aquatic or terrestrial ecosystems. In many cases the solid waste is dumped at sea or discarded on land, allowing biodegradation to occur naturally without any special treatment, but excessive dumping of organic wastes into terrestrial and marine ecosystems can cause untoward problems unless the operation is carefully managed and monitored.

Sanitary Landfills

The simplest and least expensive way to dispose of solid waste is to place the material in **landfills** and to allow it to decompose. In landfill procedures, both organic and inorganic solid wastes are deposited together in low-lying land that has minimal real estate value. Because exposed waste can cause aesthetic and odor problems, attract insects and rodents, and pose a fire hazard, each day's waste deposit is covered over with a layer of soil, creating a **sanitary landfill** (Figure 15.1). When the landfill is full, the site can be used for recreational purposes and may eventually provide a foundation for construction. For 30 to 50 years after the establishment of a landfill, the organic content of the solid waste undergoes slow, anaerobic microbial decomposition. The products of anaerobic microbial metabolism include carbon dioxide, water, methane, various low molecular weight alcohols, and acids, which diffuse into the surrounding water and air, causing the landfill to settle slowly. Eventually, the decomposition slows greatly, signaling completion of the biodegradation of the solid waste, subsidence ceases, and the land is stabilized and suitable as a site for construction. Extensive amounts of methane are produced during this decomposition process, potentially providing a source of needed natural gas. At some solid waste disposal sites, such as the one at Palos Verdes, California, the methane that is produced is collected and sold to nearby power plants.

Although the use of sanitary landfills is simple and inexpensive, there are several problems associated with this waste disposal method. Premature construction on a still biologically active landfill site may result in structural damage to the buildings because of movement of the land base, and an explosion hazard may exist due to methane seepage into basements and other belowground structures. Aboveground plantings may also be damaged because of methane seepage. The number of suitable disposal sites available in urban areas is very limited, often necessitating long and expensive hauling of the solid waste to available sites. The possible seepage of anaerobic decomposition products, heavy metals, and a variety of recalcitrant hazardous pollutants from the landfill site into underground aquifers, which are used in many urban areas as water sources, has caused many municipalities to place severe restrictions on the location and operation of landfills and the types of materials that can be deposited in them. Thus, alternatives to the landfill technique for disposing of solid waste are being sought by many municipalities.

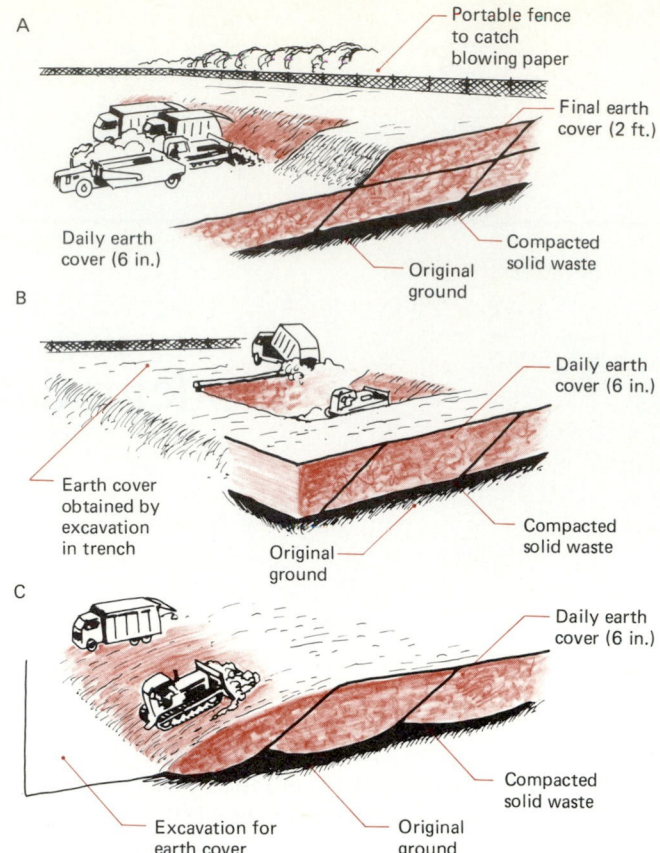

FIGURE 15.1
Sanitary landfills are used for the inexpensive disposal of solid waste. Several different methods, as shown in these drawings, can be employed. (A) The area method: a bulldozer spreads and compacts solid wastes. Cover material is hauled in and spread at the end of the day's operations. Note that a fence to catch any blowing debris is present in all landfill methods. (B) The trench method: the waste collection truck deposits its load in a trench, where it is spread and compacted. At the end of the day, soil is excavated from a future trench and used as the daily cover material. (C) Ramp variation: solid wastes are spread and compacted on a slope. The daily deposit is covered with earth from the base of the ramp. This variation is used with either the area or the trench method.

Composting

The organic portion of solid waste can be biodegraded by **composting**, the process by which solid heterogeneous organic matter is degraded by aerobic, mesophilic, and thermophilic microorganisms. Composting, like incineration, requires sorting of the solid waste into its organic and inorganic components. This can be accomplished either at the source, by the separate collections of garbage (organic waste) and trash (inorganic waste), or at the receiving facility, by using magnetic separators to remove ferrous metals and mechanical separators to remove glass, aluminum, and plastic materials. The remaining largely organic waste is ground up, mixed with sewage sludge and/or bulking agents, such as shredded newspaper or wood chips, and then composted. The addition of dehydrated sewage sludge to domestic garbage improves the carbon–nitrogen balance because sewage sludge is high in nitrogen and therefore enhances microbial biodegradation activities, as well as providing a means of disposing of some sewage sludge waste and a considerable number of decomposer microorganisms. The addition of 10 percent by weight sewage sludge to the material being composted improves its porosity, an important factor because 30 percent air space is needed to optimize the availability of oxygen for microbial respiration in the aerobic compost process and because water must drain out of the composting material to prevent waterlogging and the development of anaerobic conditions.

Composting is a microbial process that converts noxious organic waste materials into a stable, sanitary, humus-like product. Reduced in bulk, it can be used for soil improvement. The various composting methods are differentiated by the physical arrangement of the the solid waste; that is, composting can be accomplished in windrows, aerated piles, or continuous-feed reactors (Figure 15.2). The **windrow method**, in which the solid waste is arranged in long rows, is a simple but relatively slow process, typically requiring several months to achieve biodegradation of the metabolizable components and stabilization of the waste material. Odor and insect problems are controlled in this process by covering the windrows with a layer of soil or finished compost. Unless the decomposing material is turned several times during the process, the quality of the finished compost product is uneven. Because the process is so slow, large amounts of land must be used, causing the same problems as sanitary landfills in densely populated urban areas.

Composting rates can be enhanced in the **aerated pile method**, in which waste is arranged in piles and forced aeration is used to provide needed oxygen. Perforated pipes are buried inside the compost pile and air is pumped through the pile, both oxygenating and cooling it. The heat generated in the aerated pile process is used to evaporate water for the final drying of the product. The **continuous-feed composting process** uses a reactor that permits control of the environmental parameters (Figure 15.3). The reactor is analogous to an industrial fermentor and permits the production of a relatively uniform product. Compared to the other compost methods, the continuous-feed process requires a high initial financial investment. By optimizing conditions, composting in the reactor is accomplished in just 2 to 4 days, although the product requires additional curing for about a month prior to packaging and shipment.

In a compost of domestic garbage and sludge, numerous microbial species that come from soil, water, and human fecal matter are present. The relatively high moisture content of the compost material favors the development of bacterial rather than fungal populations. In the composting of solid organic wastes, the process is initiated by mesophilic heterotrophs, which, as the temperature rises, are replaced by thermophilic microorganisms. The initial temperature increase is probably due to the growth of mesophilic bacteria

FIGURE 15.2

These photographs show compost heaps used for decomposition of solid wastes. (A) Steam is rising from this compost heap because of the heat buildup that results from the rapid decomposition of organic matter, caused by thermophilic microorganisms. (B) In this heap a straight row is formed by a machine, which is also used to turn the heap periodically in order to maintain good aeration. (Courtesy Atal E. Eralp, USEPA, Cincinnatti) (C) This is a commercial compost facility owned by Paygro, Inc., in South Charlestown, Ohio. The T-shaped poles have thermocouples attached to them in order to monitor continously the temperature at various depths within the pile. Air is forced through the pile from underneath by fans to provide aeration. (Courtesy V. L. McKinley and J. R. Vestal, University of Cincinnatti)

in the interior portions of the composting material. Thermophilic microorganisms prominent in the composting process include the bacteria *Bacillus stearothermophilus*, *Thermomonospora* spp., *Thermoactinomyces* spp., and *Clos-*

COMPOST PLANT FLOW DIAGRAM

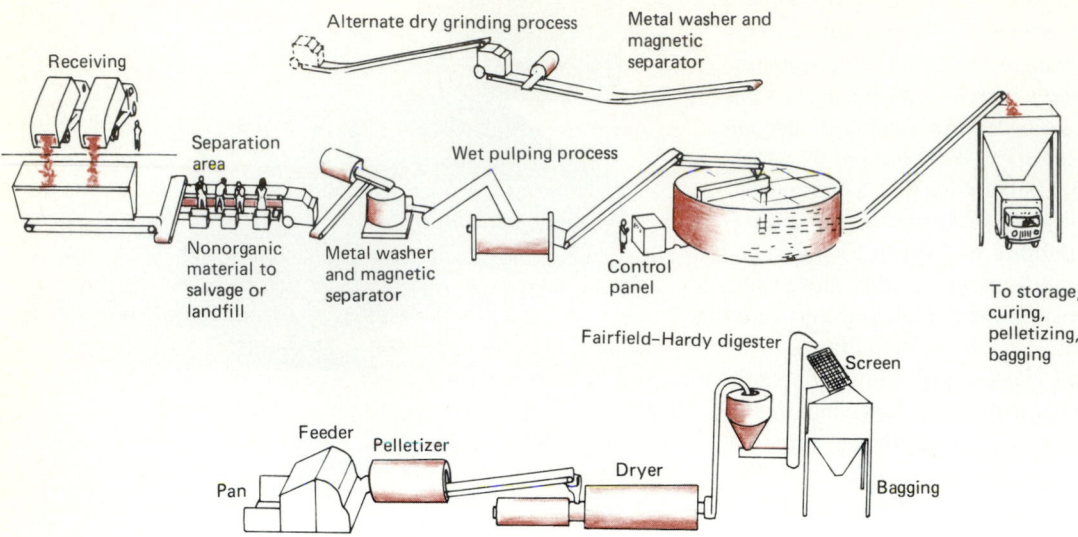

FIGURE 15.3

The design of this continuous-process compost plant permits the efficient and economical use of composting for solid waste disposal.

tridium thermocellum and the fungi *Geotrichum candidum*, *Aspergillus fumigatus*, *Mucor pusillus*, *Chaetomium thermophile*, *Thermoascus auranticus*, and *Torula thermophila*. In the continuous-reactor composting process, the reactor is maintained continuously at thermophilic temperatures by using the heat produced within the reactor by the biodegradation of the organic matter.

The control of several conditions is critical for achieving optimal composting. Temperatures needed to achieve maximal rates of organic matter decomposition are in the range of 50–60°C. Self-heating typically raises the temperature inside a static compost pile to 55–60°C or above in 2 to 3 days under favorable conditions, but after a few days at this optimal level, the temperature gradually declines unless the pile is turned to resupply oxygen and ensure that the thermophilic process occurs throughout the pile, instead of only at the core. Moisture must be adequate; 50–60 percent water content is optimal, but excess moisture—70 percent or above—interferes with aeration and lowers self-heating because of water's large heat capacity. The carbon-to-nitrogen ratio must not be greater than 40:1; a lower nitrogen content would preclude the formation of a sufficient microbial biomass, and

a greater nitrogen concentration, such as C:N = 25:1, would lead to volatilization of ammonia, causing odor problems and lowering the usefulness of the compost product as a fertilizer.

Although compost is a good soil conditioner and supplies some plant nutrients, it cannot compete with synthetic fertilizers for use in agricultural production. The sale of compost effectively reduces the cost of the waste disposal operation but generally does not render the waste disposal operation self-supporting. When sewage sludge is used as a major component of the original compost mixture, however, the finished product may contain relatively high concentrations of potentially toxic heavy metals, such as cadmium, chromium, and thallium. Because little is known about the behavior of these metals in agricultural soils, the use of sewage sludge–derived compost in agriculture is not widely practiced, but compost does find extensive applications in parks and gardens for ornamental plants, in land reclamation, particularly after strip mining, and as part of highway beautification projects. Even though landfill operations are less expensive than composting, the long-range environmental costs in terms of groundwater contamination favor the composting process.

15.2 *Treatment of Liquid Wastes*

Agricultural and industrial operations—along with everyday human activities—produce **liquid wastes**, including **domestic sewage**. These liquid waste discharges flow through natural drainage patterns or sewers, eventually entering natural bodies of water, such as groundwater, rivers, lakes, and

oceans. In theory the liquid wastes disappear when they are flushed into such water bodies, according to the adage "the solution to pollution is dilution." Bodies of water into which sewage flows must also serve local communities as the source of water for drinking, household use, industry, irrigation, fish and shellfish production, swimming, boating, and other recreational purposes, making the maintenance of the acceptable high quality of these natural waters essential. Fortunately, **self-purification** is an inherent capability of natural waters, based on the biogeochemical cycling activities and interpopulation relationships of the indigenous microbial populations. Organic nutrients in the water are metabolized and mineralized by autochthonous heterotrophic aquatic microorganisms. Ammonia is nitrified and, along with other inorganic nutrients, used and immobilized by algae and higher aquatic plants. Allochthonous populations of enteric and other pathogens that enter aquatic ecosystems are maintained at low levels and/or eliminated by the pressures of competition and predation of the autochthonous aquatic populations. Consequently, reasonably low amounts of raw sewage can be accepted by natural waters without causing a significant decline in the level of water quality.

Despite this fact, human demographic patterns of densely populated areas, large-scale agricultural operations, and major industrial activities result in the production of liquid wastes on a scale that routinely overwhelms the self-purification capacity of aquatic ecosystems, causing an unacceptable deterioration of water quality. The relative changes in some environmental parameters and populations in a river receiving sewage are illustrated in Figure 15.4. A prominent feature of river water receiving sewage effluents is the presence of the filamentous aerobic bacterium *Sphaerotilus natans*, known as the "sewage fungus" (Figure 15.5). A heterogeneous microbial community also develops amid the filaments of this bacterium below a sewage outfall. *S. natans* and the associated microbial community are efficient degraders of organic matter, consuming oxygen at a rate of two grams per hour per square meter. The bloom of the sewage fungus exemplifies the aesthetically displeasing results of excessive addition of organic matter to natural water bodies. Depending on the rate of sewage discharge, flow rate, water temperature, and other environmental factors, water may reestablish an acceptable quality level at some distance downstream from the sewage outfall, typically within 24–60 km. The maintenance of satisfactory water quality means that natural waters should not be overloaded with organic or inorganic nutrients or with toxic, noxious, or aesthetically unacceptable substances; that their oxygen, temperature, salinity, turbidity, or pH levels should not be altered so significantly that they lose their ability to support fish pro-

FIGURE 15.4

The effects of organic effluent on a stream and the changes that occur downstream of the outfall on materials, plants, and animals are represented in this graph. Water quality studies have found that near the outfall there is a great abundance of *Sphaerotilus natans*, the sewage fungus, high ammonium and nitrate concentrations, a very low oxygen concentration, and a specialized macrofauna not found in clean water. Depending on the rate of sewage discharge, water flow rate, temperature, and other environmental factors, the water quality may return to close to its original state at a typical distance ranging from a few to several dozen kilometers downstream from the sewage outfall.

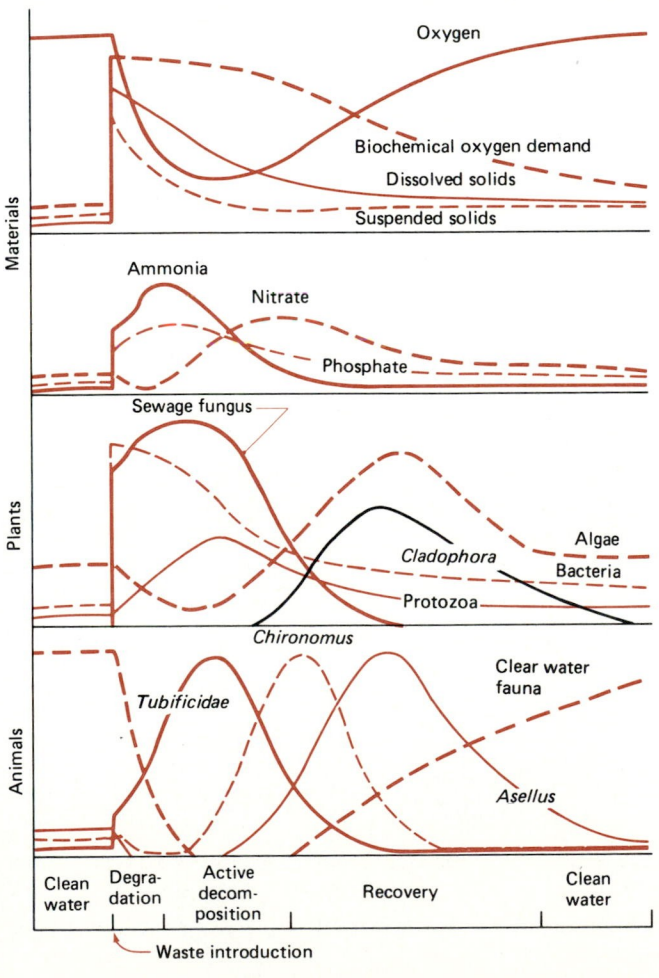

FIGURE 15.5

Electron micrograph of a thin section of *Sphaerotilus natans*, showing the sheath surrounding the cells. The cell width is approximately 500 nm. (From BPS—T. J. Beveridge, University of Guelph)

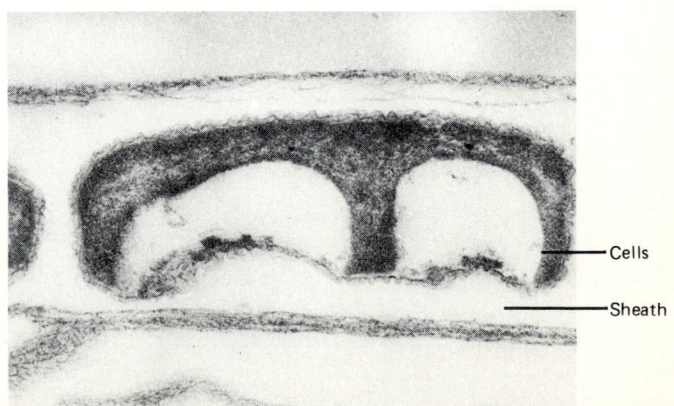

ANALYTICAL PROCESS

Biological Oxygen Demand (BOD)

We have developed several measures of water quality that help us manage aquatic ecosystems by indicating how much waste can safely be allowed to enter rivers and lakes without causing serious deterioration of water quality. One widely used measure of water quality, the **biological oxygen demand (BOD)**, represents the amount of oxygen required for the microbial decomposition of the organic matter in the water. The BOD procedure, which is used extensively in monitoring water quality and biodegradation of waste materials, is designed to determine how much oxygen is consumed by microorganisms during oxidation of the organic matter present in the sample.

The BOD can be easily determined in the laboratory by incubating a water sample and measuring the amount of oxygen consumed during a 5-day period (Figure 15.6). The procedure is based upon the consumption of oxygen by the microorganisms that are naturally present in the water sample. The oxygen remaining after 5 days of incubation can be determined chemically or, more commonly, with the use of oxygen electrodes. The difference between the starting concentration of oxygen and the residual oxygen represents the amount of oxygen consumed by the indigenous microorganisms in degrading the organic materials in the water sample, that is, the BOD.

Incubation at 20°C for 5 days is commonly used because the test was originally developed in Great Britain, where average water temperatures are near 20°C and where it takes a maximum of 5 days for anything entering a local river to reach the ocean. Once the organic matter reaches the ocean, it is no longer considered a threat to water quality. In the United States and other large countries, it may be necessary to modify the incubation period used in the standard 5-day BOD procedure to account for the extended residence time of organic matter in the waterways receiving organic pollutants. The development of appropriate modifications to the original procedure has been slow, in part because of a lack of understanding of the original assumptions used in establishing the standard 5-day incubation procedure. Appropriate modifications of the standard BOD procedure, based on actual residence times in inland waterways and desirable multiple uses of water, are presently being incorporated into water quality standards.

FIGURE 15.6

The BOD determination procedure and the specialized BOD bottle. Note that the shape of the neck of the bottle permits complete filling of the bottle to exclude air.

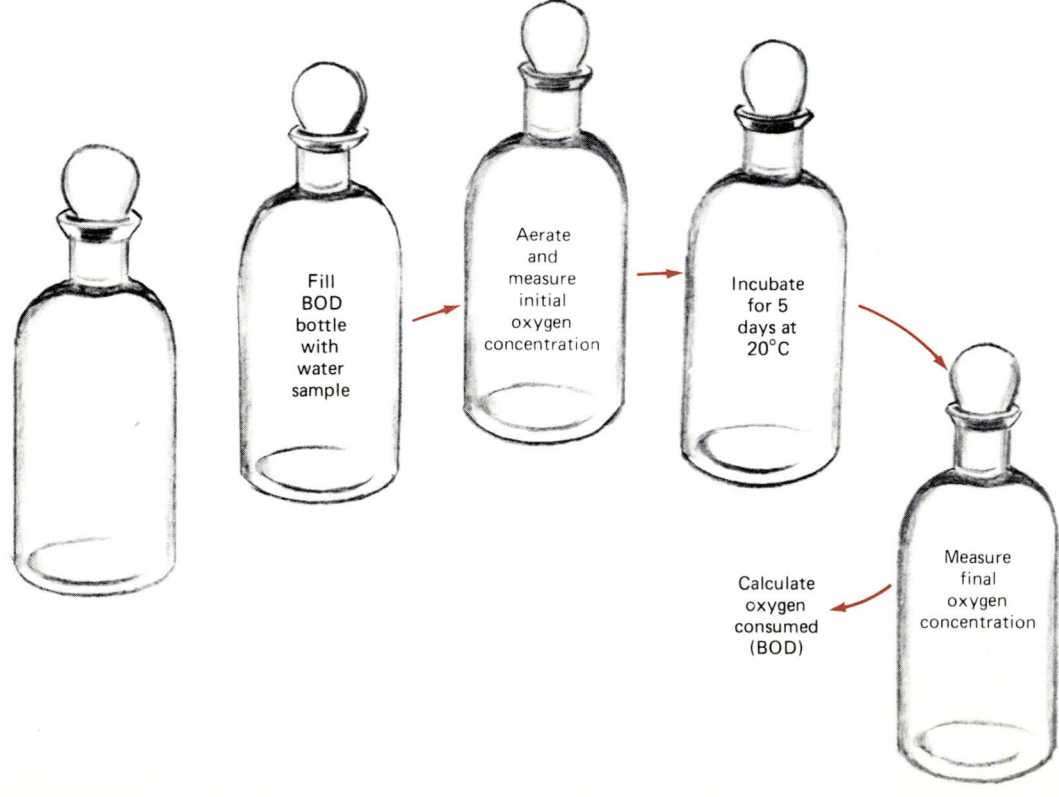

duction and recreational usage of the water body; and that they should not be allowed to become vehicles of disease transmission due to fecal contamination.

Sewage Treatment

One consequence of urbanization is the need to remove sewage and other organic wastes from concentrated population centers. Waterways that are normally used for waste removal under the premise that "the solution to pollution is dilution" can be overwhelmed by such concentrated inputs of organic matter. A high BOD generally indicates the presence of excessive amounts of organic carbon. The dissolved oxygen in natural waters seldom exceeds 8 mg/L because of its low solubility, and it is often considerably lower because of heterotrophic microbial activity, making oxygen depletion a likely consequence of adding wastes with high BOD values to aquatic ecosystems. The polluting power of different sources of wastes is reflected in the BOD of the material (Table 15.1). Exhaustion of the dissolved oxygen content is the principal result of a sewage overload on natural waters. Oxygen deprivation kills obligately aerobic organisms, including some microorganisms, fish, and invertebrates, and the decomposition of dead organisms within the water body creates an additional oxygen demand. Fermentation products and the reduction of the secondary electron sinks of nitrate and sulfate give rise to noxious odors, tastes, and colors, making the water putrid and septic.

Modern methods of liquid waste treatment are aimed at reducing the BOD of the waste before it is discharged into

TABLE 15.1   BOD Values for Different Types of Wastes

Type of Waste	BOD (mg/L)
Domestic sewage	200–600
Slaughterhouse wastes	1,000–4,000
Piggery effluents	25,000
Cattle shed effluents	20,000
Vegetable processing	200–5,000

a water body in order to maintain acceptable water quality. There are several different approaches to reducing the BOD, employing combinations of physical, chemical, and microbiological methods. Most communities in developed countries have facilities for treating **sewage,** which is the used water supply containing domestic waste together with human excrement and wash water and industrial waste, including acids, greases, oils, animal matter, vegetable matter, and storm waters. The use of household garbage disposal units also increases the organic content of domestic sewage. The treatment of liquid wastes is aimed at removing organic matter, human pathogens, and toxic chemicals. The treatment of domestic sewage reduces the BOD due to suspended or dissolved organics and the number of enteric pathogens so that the discharged sewage effluent will not cause unacceptable deterioration of environmental quality.

Sewage is subjected to different treatments, depending on the quality of the effluent deemed necessary to be achieved to permit the maintenance of acceptable water quality (Figure 15.7). **Primary treatments** rely on physical separation procedures to lower the BOD; **secondary treatments** rely on

FIGURE 15.7

Flow chart of the stages of sewage treatment. Primary treatment is principally physical, secondary treatment is principally biological, and tertiary treatment is principally chemical.

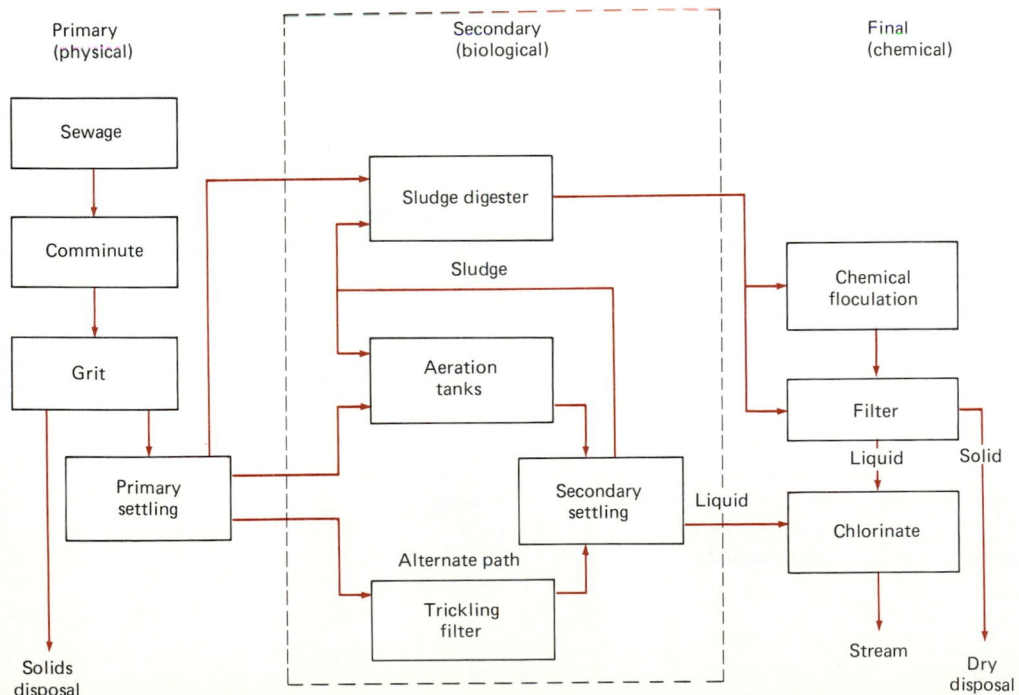

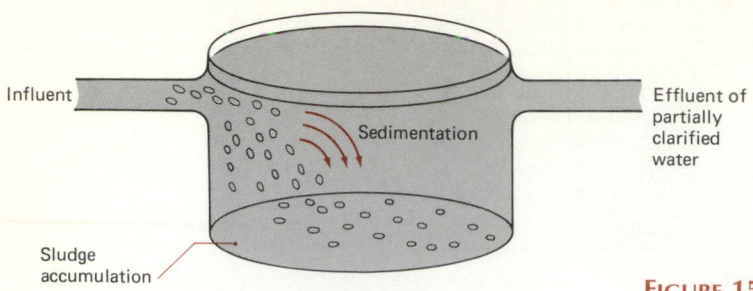

FIGURE 15.8

Diagram of a settling tank for primary sewage treatment.

microbial biodegradation to further reduce the concentration of organic compounds in the effluent; and **tertiary treatments** use chemical methods to remove inorganic compounds and pathogenic microorganisms. Municipal sewage treatment facilities are designed to handle organic wastes but are normally incapable of dealing with industrial wastes containing toxic chemicals, such as heavy metals. Industrial facilities frequently must operate their own treatment plants to deal with waste materials.

Primary Treatment   Primary sewage treatment removes suspended solids in settling tanks or basins (Figure 15.8). The

solids are drawn off from the bottom of the tank and may be subjected to anaerobic digestion and/or composting prior to final deposition in landfills or as soil conditioner. Only a low percentage of the suspended or dissolved organic material is actually mineralized during liquid waste treatment; most of it is removed by settling, and as a result the disposal problem is merely "displaced" to the solid waste area rather than being solved. Nevertheless, this displacement is essential because of the detrimental effects of discharging effluents with high BOD into aquatic ecosystems with naturally low dissolved oxygen contents. The liquid portion of the sewage, which contains dissolved organic matter, can be subjected to further treatment or discharged after primary treatment alone. Because liquid wastes vary in composition and may contain mainly solids and little dissolved organic matter, primary treatment may remove 70–80 percent of the BOD and may be sufficient. For typical domestic sewage (Table 15.2), however, primary treatment normally removes only 30–40 percent of the BOD.

Secondary Treatment   To achieve an acceptable reduction in the BOD, secondary treatment by a variety of means is necessary (Table 15.3). In secondary sewage treatment a small portion of the dissolved organic matter is mineralized, and the larger portion is converted to removable solids. The combination of primary and secondary treatment reduces the original sewage BOD by 80–90 percent. The secondary sewage treatment step relies on microbial activity, may be aerobic or anaerobic, and is conducted in a large variety of devices. A well-designed and efficiently operated secondary treatment unit should produce effluents with BOD and/or suspended solids of less that 20 mg/L. Because the secondary treatment of sewage is a microbial process, it is extremely sensitive to the introduction of toxic chemicals that may be contained in industrial waste effluents or that accidentally may contaminate the sewerage system. The accidental intro-

TABLE 15.2   Characteristics of Typical Municipal Waste Water

Component	Concentration (mg/L)
Solids	
Total	700
Dissolved	500
Fixed	300
Volatile	200
Suspended	200
Fixed	50
Volatile	150
BOD (biochemical oxygen demand)	300
TOC (total organic carbon)	200
COD (chemical oxygen demand)	400
Nitrogen (as N)	
Total	40
Organic	15
Ammonia	25
Nitrate	0
Phosphorus (as P)	
Total	10
Organic	3
Inorganic	7
Grease	100

TABLE 15.3   Efficiency of Various Types of Sewage Treatment

Treatment	BOD (% reduced)	Suspended Solids (% removed)	Bacteria (% reduced)
Sedimentation	30–75	40–95	40–75
Septic tank	25–65	40–75	40–75
Trickling filter	60–90	0–80	70–85
Activated sludge	70–96	70–97	95–99

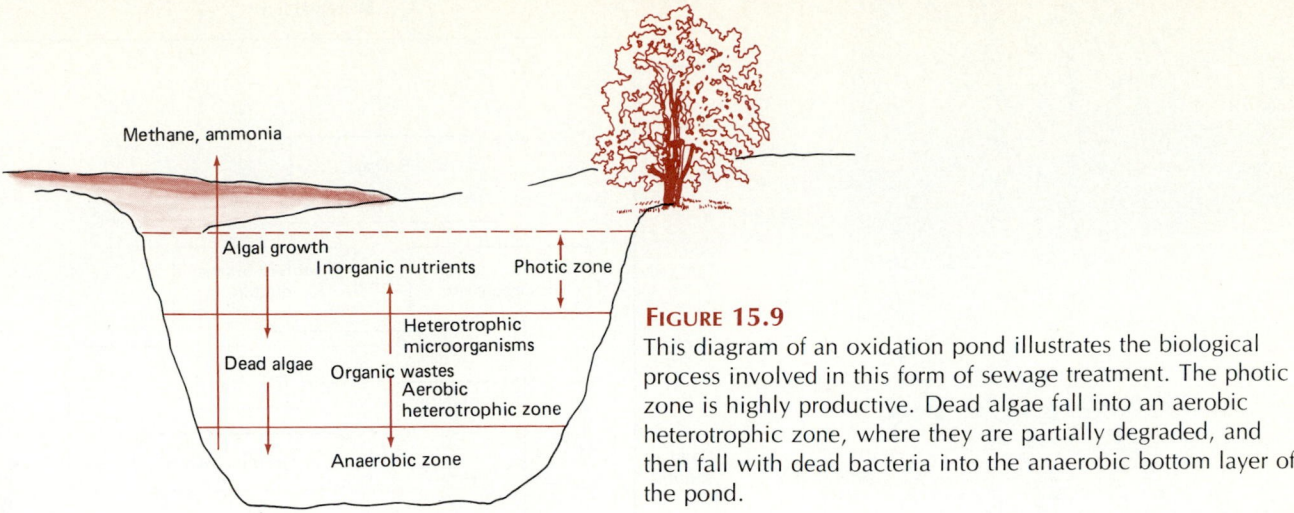

FIGURE 15.9

This diagram of an oxidation pond illustrates the biological process involved in this form of sewage treatment. The photic zone is highly productive. Dead algae fall into an aerobic heterotrophic zone, where they are partially degraded, and then fall with dead bacteria into the anaerobic bottom layer of the pond.

duction of the organic chemicals octachlorocyclopentene and hexachlorocyclopentadiene into the municipal sewerage system of Louisville, Kentucky, for example, poisoned the microorganisms in the sewage treatment facility, forcing the dumping of 7 billion gallons of untreated sewage into the Ohio River during a 3-month period before the toxic chemicals could be removed from the system. The accidental introduction of hexanes into the same sewerage system several years later caused a massive explosion and the disruption of normal sewage treatment for an extended period of time.

Oxidation Ponds. **Oxidation ponds**, also known as *stabilization ponds* and *lagoons*, are used for the simple secondary treatment of sewage effluents in rural communities and some industrial facilities (Figure 15.9). Heterotrophic bacteria degrade sewage organic matter within the ponds, producing cellular material and mineral products that support the growth of algae. The proliferation of algal populations in these lagoons produces oxygen that replenishes the oxygen depleted by the heterotrophic bacteria, permitting continued organic matter decomposition. As oxygenation is usually achieved by diffusion and by the photosynthetic activity of algae, such ponds need to be shallow. Typically, oxidation ponds are less than 10 feet deep, which maximizes the euphotic zone for algal growth. Oxygenation is usually incomplete, with consequent odor problems. The performance of oxidation ponds is strongly influenced by seasonal temperature fluctuations, and their usefulness, therefore, is largely restricted to warmer climatic regions. The bacterial and algal cells formed during the decomposition of the sewage settle to the bottom, eventually filling in the pond. Oxidation ponds generally are low-cost operations, but they tend to be inefficient and require large holding capacities and long retention times. The degradation of organic matter in these ponds is relatively slow, and the residence time for the treatment of domestic sewage may be as long as a week. The effluents containing oxidized products are periodically removed from the ponds, which are then refilled with raw sewage.

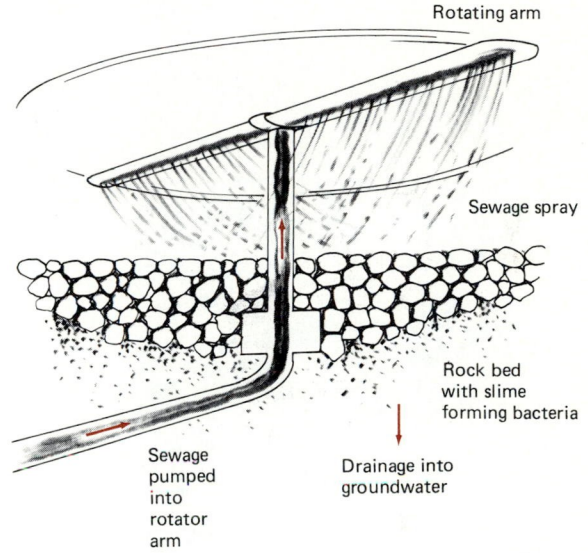

FIGURE 15.10

In a trickling filter the clarified sewage is sprayed over rocks. The bacterial film on the rocks aerobically decomposes the dissolved organic matter. The drainage may go through the soil column to the groundwater or drain into a nearby river or ocean.

Trickling Filter. The **trickling filter system** is a simple and relatively inexpensive film-flow type of aerobic sewage treatment method (Figure 15.10). The sewage is distributed by a revolving sprinkler suspended over a bed of porous material. The sewage slowly percolates through this porous bed, and the effluent is collected at the bottom. The porous material of the filter bed becomes coated with a dense, slimy bacterial growth, principally composed of *Zooglea ramigera* and similar slime-forming bacteria (Figure 15.11). The slime matrix thus generated accommodates a heterogeneous microbial community, including bacteria, fungi, protozoa, nematodes, and rotifers. The most frequently found

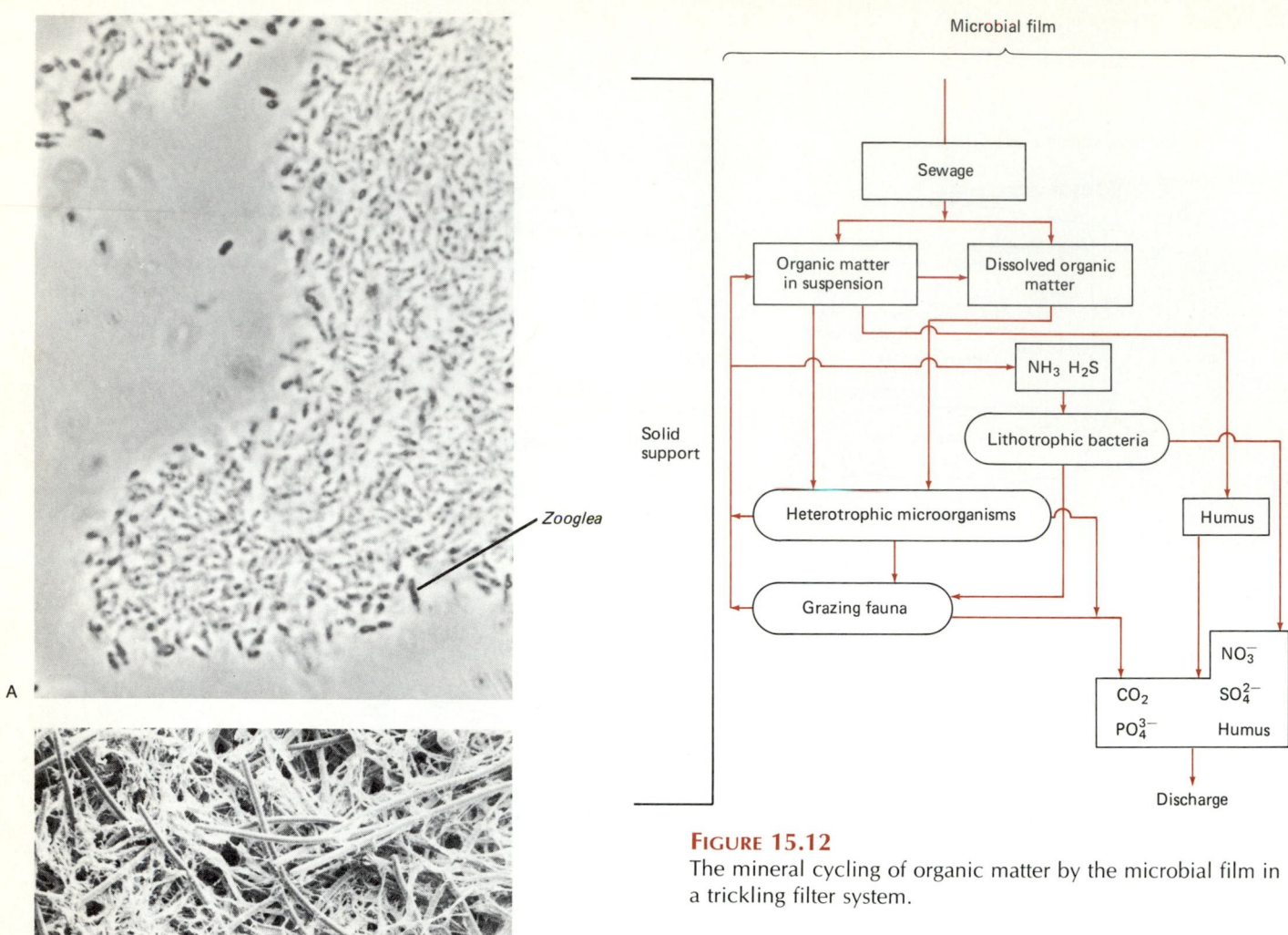

FIGURE 15.11
Photomicrograph of slime-forming bacteria on a trickling filter. (A) *Zooglea ramigera* surrounded by an extensive slime layer. (Courtesy Patrick Dugan, Ohio State University), (B) Scanning electron micrograph of a microbial community on the surface of a rock from a trickling filter; note the abundant filamentous algae and fungi that develop on the surface film (500×). (Courtesy Robert Apkarian, Yerkes Primate Research Center, Emory University)

FIGURE 15.12
The mineral cycling of organic matter by the microbial film in a trickling filter system.

bacteria are *Beggiatoa alba*, *Sphaerotilus natans*, *Achromobacter* spp., *Flavobacterium* spp., *Pseudomonas* spp., and *Zooglea* spp. This microbial community absorbs and mineralizes dissolved organic nutrients in the sewage, reducing the BOD of the effluent (Figure 15.12). Aeration occurs passively as a result of the movement of air through the porous material of the bed. The sewage may be passed through two or more trickling filters or may be recirculated several times through the same filter to reduce the BOD to acceptable levels. The effluent from the trickling filters may be clarified by allowing sloughed-off biomass to settle prior to discharge. A drawback of this otherwise simple and inexpensive treatment system is that a nutrient overload produces excess microbial slime, which reduces aeration and percolation rates, periodically necessitating renewal of the trickling filter bed. Also, cold winter temperatures strongly reduce the effectiveness of such outdoor treatment facilities.

Biodisc System. The **rotating biological contactor** or **biodisc system** is a more advanced type of aerobic film-flow treatment system. In the biodisc system, closely spaced discs, usually made of plastic, are rotated in a trough containing the sewage effluent (Figure 15.13). The discs are partially submerged and become coated with a microbial slime similar to the one that develops in trickling filters. Continuous rotation of the discs keeps the slime well aerated and in contact with the sewage. The thickness of the microbial slime layer in all film-flow processes is governed by the diffusion of nutrients through the film. Microbial growth on the surface of the discs is sloughed off gradually and is removed by subsequent settling. When the film becomes so

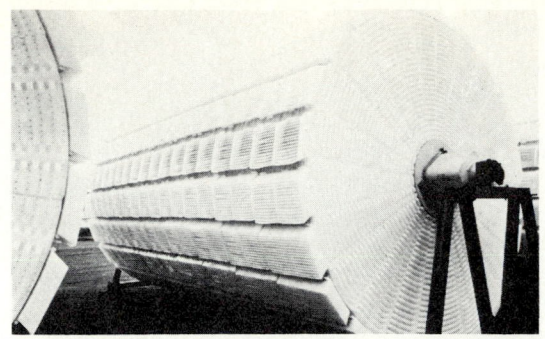

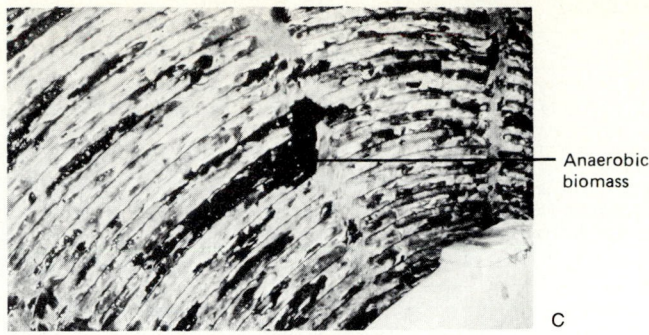

Anaerobic biomass

FIGURE 15.13

(A) A biodisc assembly; the discs rotate throughout the sewage treatment process, and microbes growing on the surfaces of the polyethylene cups degrade the organic matter in the sewage. (B) Close-up view of the normal appearance of the surface of the biodisc with a film of aerobic microorganisms. (C) When the system is overloaded, a heavy film of anaerobic microorganisms develops. Note the black area of anaerobic biomass and the occurrence of gelatinous, filamentous, sulfur-oxidizing growths on the outer layer. (Courtesy Autotrol Corp., Milwaukee, Wis.)

thick that oxygen and nutrients fail to reach the inner portions of the film, most of the innermost microorganisms die, causing detachment of the slime layer and temporary disruption of the process. The biodisc system is used in some communities for the treatment of both domestic and industrial sewage effluents. This system requires less space than trickling filters and is more efficient and stable in operation but needs a higher initial financial investment.

Activated Sludge. The **activated sludge process** is a very widely used aerobic suspension type of liquid waste treatment system (Figure 15.14). After primary settling, the sewage, containing dissolved organic compounds, is introduced into an aeration tank. Air injection and/or mechanical stirring provides the aeration. The rapid development of microorganisms is also stimulated by reintroduction of most of the settled sludge from a previous run (Figure 15.15), and the process derives its name from this inoculation with activated sludge.

During the holding period in the aeration tank, vigorous development of heterotrophic microorganisms has taken place. The heterogeneous nature of the organic substrates in sewage allows the development of diverse heterotrophic bacterial populations, including Gram-negative rods, predominantly *Escherichia, Enterobacter, Pseudomonas, Achromobacter, Flavobacterium,* and *Zooglea* spp.; other bacteria, including *Micrococcus, Arthrobacter,* various coryneforms and mycobacteria, *Sphaerotilus,* and other large filamentous bacteria; and low numbers of filamentous fungi, yeasts, and protozoa, mainly ciliates. The protozoa are important predators of the bacteria, along with rotifers. The bacteria in the

FIGURE 15.14

An activated sludge tank in a sewage treatment plant. Note the bubbling due to forced aeration. (From BPS—Carl May, Moss Beach, Calif.)

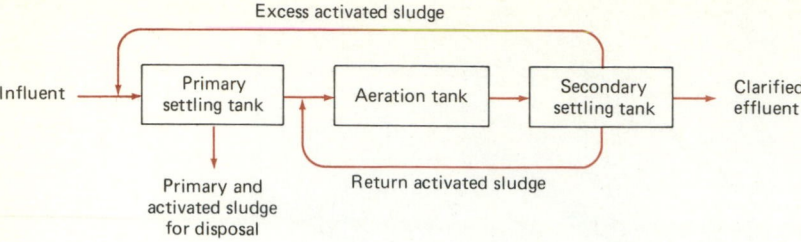

FIGURE 15.15
The flow of materials through an activated sludge secondary sewage treatment system. A portion of the sludge is recycled as inoculum for the incoming sewage.

activated sludge tank occur both in free suspension and as aggregates or flocs. The flocs are composed of microbial biomass held together by bacterial slimes, produced by *Zooglea ramigera* and similar organisms. Most of the ciliate protozoa, such as *Vorticella*, are of the attached filter-feeding type and adhere to the flocs, while feeding predominantly on the suspended bacteria. The floc is too large to be ingested by the ciliates and rotifers, and acts as a defense mechanism against predation. In the raw sewage, suspended bacteria predominate, but during the holding time in the aeration tank, their numbers decrease, and at the same time those bacteria associated with flocs greatly increase in number (Table 15.4).

As a consequence of extensive microbial metabolism of the organic compounds in sewage, a significant proportion of the dissolved organic substrates is mineralized, and another portion is converted to microbial biomass. In the advanced stage of aeration, most of the microbial biomass becomes associated with flocs that can be removed from suspension by settling. The settling characteristic of sewage sludge flocs is critical to their efficient removal. Poor settling

produces "bulking" of sewage sludge, caused by proliferation of such filamentous bacteria as *Sphaerotilus*, *Beggiatoa*, *Thiothrix*, and *Bacillus* and such filamentous fungi as *Geotrichum*, *Cephalosporium*, *Cladosporium*, and *Penicillium*. The causes of bulking are not always understood, but it is frequently associated with high C:N and C:P ratios and/or low dissolved oxygen concentrations. A portion of the settled sewage sludge is recycled for use as the inoculum for the incoming raw sewage, but the rest of the sludge requires additional treatment by composting or anaerobic digestion.

Combined with primary settling, the activated sludge process tends to reduce the BOD of the effluent to 10–15 percent of that of the raw sewage. The treatment also drastically reduces the number of intestinal pathogens in the sewage. This reduction is the result of the combined effects of competition, adsorption, predation, and settling. Predation by ciliates, rotifers, and *Bdellovibrio* is probably indiscriminate and affects pathogens as well as nonpathogenic heterotrophs. Also, pathogens tend to grow poorly or not at all under the environmental conditions of an aeration tank, and nonpathogenic heterotrophs proliferate vigorously. Therefore, whereas nonpathogenic heterotrophs reproduce to compensate for their predatory removal, the pathogens are continuously decimated (Table 15.5). Settling of the flocs removes additional pathogens, and the number of *Salmonella*, *Shigella*, and *Escherichia coli* typically is 90–99 percent lower in the effluent of the activated sludge treatment process than in the incoming raw sewage. Enteroviruses are removed to a similar degree, and the main removal mechanism appears to be adsorption of the virus particles onto the settling sewage sludge floc.

Septic Tank. The simplest anaerobic treatment system, the **septic tank**, is used extensively in rural areas that lack sew-

TABLE 15.4 Number of Bacteria at Different Stages of Sewage Treatment

Treatment	Total Bacteria (number/mL)	Viable Bacteria (number/mL)
Settled sewage	7×10^8	1×10^7
Activated sludge mixed liquor	7×10^9	6×10^7
Filter slimes	6×10^{10}	2×10^9
Secondary effluents	5×10^7	6×10^5
Tertiary effluents	3×10^7	4×10^4

TABLE 15.5 Percentage Reductions in the Numbers of Indicator Organisms in Different Types of Sewage Treatment Processes

Treatment	*Escherichia coli*	Coliforms	Fecal Streptococci	Viruses
Sedimentation	3–72	13–86	44–60	—
Activated sludge	61–100	13–83	84–93	79–100
Trickling filter	73–97	15–100	64–97	40–82
Lagoons	80–100	86–100	85–99	95

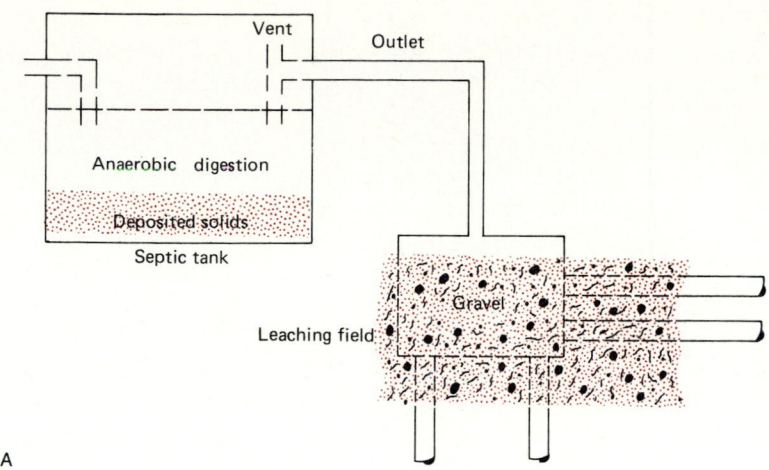

Cross-section
of septic tank

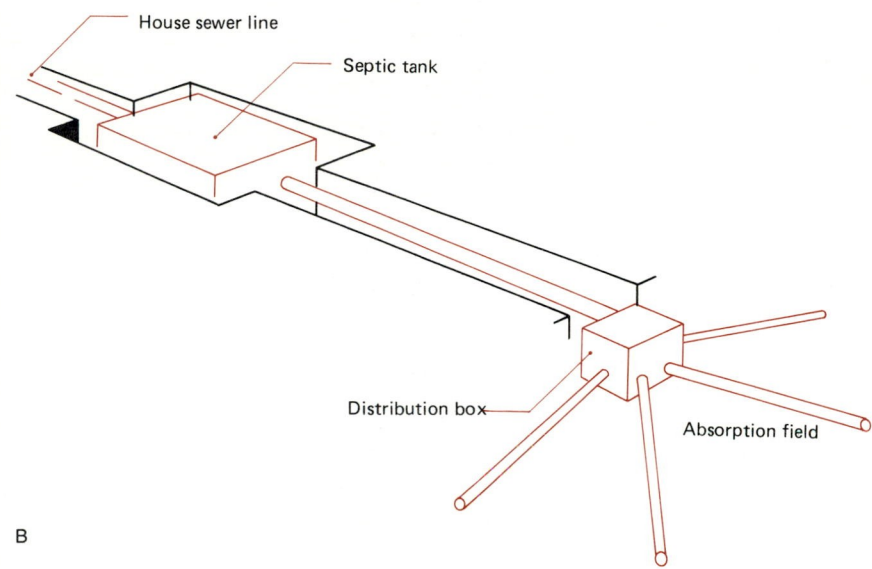

FIGURE 15.16

(A) A typical septic tank used for the disposal of sewage in a rural home is diagrammed. The sewage is degraded by anaerobic digestion. The treated effluent is disposed of in a leaching field of permeable soils, with the effluent pipes usually set in gravel. (B) Diagram of the installation of a septic tank for sewage disposal from a home. The sewage effluent is distributed into the surrounding soil.

erage systems (Figure 15.16). Many rural and suburban single-family dwellings use septic tanks. A septic tank acts largely as a settling tank, within which the organic components of the waste water undergo limited anaerobic digestion. The accumulated sludge is maintained under anaerobic conditions and is degraded by anaerobic microorganisms to organic acids and hydrogen sulfide. Residual solids settle to the bottom of the septic tank and the clarified effluent is allowed to percolate into the soil, where the dissolved organic compounds in the effluent undergo biodegradation. These products are distributed into the soil, along with the clarified sewage effluent. Septic tank treatment does not reliably destroy intestinal pathogens, and it is important that

the soils receiving the clarified effluents not be in close proximity to drinking wells to prevent contamination of drinking water with enteric pathogens.

Anaerobic Digesters. Large-scale anaerobic digesters are used for further processing of the sewage sludge produced by primary and secondary treatments (Figure 15.17). Although anaerobic digesters could be used for direct treatment of sewage, economic considerations favor aerobic processes for relatively dilute wastes, and the use of anaerobic digesters is restricted to treatment of concentrated organic wastes. Therefore, in practice, large-scale anaerobic digesters are used only for processing settled sewage sludge and the treatment of very high BOD industrial effluents. **Anaerobic**

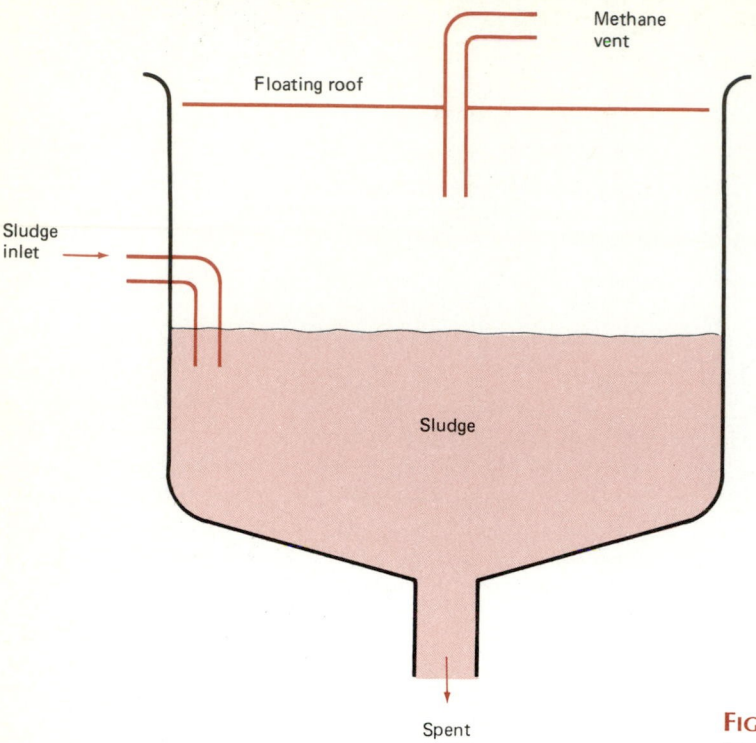

FIGURE 15.17
Diagram of an anaerobic sludge digester.

digesters are large fermentation tanks designed for continuous operation under anaerobic conditions. Provisions for mechanical mixing, heating, gas collection, sludge addition, and drawoff of stabilized sludge are incorporated into the design of a large-scale anaerobic digester to permit effective operation. Anaerobic digesters contain high amounts of suspended organic matter; between 20 and 100 g/L is considered favorable. Much of this suspended material is bacterial biomass, and viable counts can be as high as 10^9–10^{10} bacteria per milliliter. Fungi and protozoa are present in very low numbers and do not play a significant role in anaerobic digestion. A complex bacterial community is involved in the degradation of organic matter within an anaerobic digester, with the number of anaerobic microorganisms typically two to three orders of magnitude higher than the number of aerobes.

The anaerobic digestion of wastes is a two-step process in which a large variety of nonmethanogenic, obligately, or facultatively anaerobic bacteria participate. First, complex organic materials, including microbial biomass, are depolymerized and converted to fatty acids, CO_2 and H_2 (Figure 15.18). In the next step, methane is generated either by the direct reduction of methyl groups to methane or by the reduction of CO_2 to methane by molecular hydrogen or other reduced fermentation products, such as fatty acids. The final products obtained in an anaerobic digester are a gas mixture, approximately 70 percent methane and 30 percent carbon dioxide, microbial biomass, and a nonbiodegradable residue.

The optimal operation of anaerobic digesters requires good control of several parameters, such as retention time, temperature, pH, and C:N and C:P ratios. The optimal performance temperature is in the range of 35–37°C. The pH must remain in the range of 6.0–8.0, with 7.0 being optimal. Variations in pH and the inclusion of heavy metals or other toxic materials in the sludge can easily upset the operation of the anaerobic digester. In a "stuck" or "sour" digester, methane production is interrupted, fatty acids and other fermentation products accumulate, and it is difficult to restore normal operation. It is usually necessary to clean out the reactor and charge it with large volumes of anaerobic sludge from an operational unit, a costly and time-consuming exercise.

A properly operating anaerobic digester yields a greatly reduced volume of sludge compared to the starting material. The product obtained, however, still causes odor and water pollution problems and can be disposed of at only a few restricted landfill sites. Aerobic composting can be used to further consolidate the sludge, rendering it suitable for disposal in any landfill site or for use as a soil conditioner. Several gases are produced as a result of the anaerobic biodegradation of sludge, primarily methane and CO_2. The gas can be used within the treatment plant to drive the pumps and/or to provide heat for maintaining the temperature of the digester; after purification, it may be sold through natural gas distribution systems. Thus, in addition to its primary function in removing wastes, anaerobic digesters can produce needed fuel resources.

Tertiary Treatment   The aerobic and anaerobic biological liquid waste treatment processes just discussed are designed to reduce the BOD of biodegradable organic substrates and

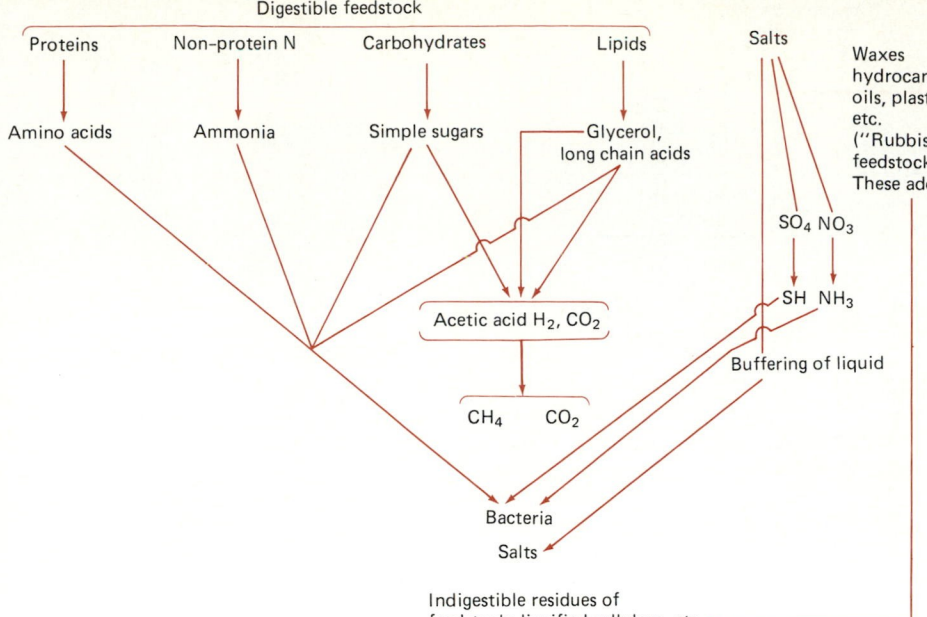

FIGURE 15.18
The metabolic reactions that take place within an anaerobic digester. Digestible feedstock polymers are hydrolyzed. The hydrolyzed material is converted to acid, hydrogen, and carbon dioxide, which then undergo methanogenesis. These reactions contribute monomers and energy for the formation of bacterial cells, which with indigestible residues of the feedstock form digested sludge.

oxidizable inorganic compounds, and all represent secondary treatment processes. **Tertiary treatment**, defined as any practice beyond a secondary one, is designed to remove non-biodegradable organic pollutants and mineral nutrients, especially nitrogen and phosphorus salts. Secondary treatment is still required to avoid overloading this expensive treatment stage with biodegradable materials that could have been removed in more economical ways. The removal of toxic, non-biodegradable organic pollutants, such as chlorophenols, polychlorinated biphenyls, and other synthetic pollutants, is necessary to reduce the toxicity of the sewage effluent to acceptable levels. Activated carbon filters are normally used in the removal of these materials from secondary-treated industrial effluents. Reverse osmosis is one way of eliminating organics and inorganics, but there are problems with this procedure because of microbial fouling of the filters.

The release of sewage effluents containing phosphates and fixed forms of nitrogen can cause serious **eutrophication**, that is, nutrient enrichment, in aquatic ecosystems. Sudden nutrient enrichment by sewage discharge or agricultural run-off triggers explosive algal blooms. Because of a variety of causes—some unknown but including mutual shading, exhaustion of micronutrients, and the presence of toxic products and/or antagonistic populations—the algal population usually "crashes," and the subsequent decomposition of the dead algal biomass by heterotrophic microorganisms exhausts the dissolved oxygen supply in the water, precipitating extensive fish kills and septic conditions. Even if the process does not proceed to this extreme, algal mats, turbidity, discoloration, and shifts in the fish population from valuable species to more tolerant but less valued forms represent undesirable changes due to eutrophication.

To prevent eutrophication, phosphate is commonly removed from sewage by precipitation as calcium, aluminum,

or iron phosphate. This can be accomplished as an integral part of primary or secondary settling or in a separate facility where the precipitating agent can be recycled. Nitrogen, present mainly as ammonia, can be removed by volatilization as NH_3 at a high pH. Some ammonia eliminated from the sewage in this manner, however, may return to the watershed in the form of precipitation and cause further eutrophication problems. **Breakpoint chlorination** is an alternative procedure for removing ammonia. The addition of hypochlorous acid (HOCl), in a 1:1 molar ratio, results in the formation of monochloramine (NH_2Cl), and further addition of HOCl to an approximate molar ratio of 2:1 results in nearly complete oxidation of the ammonia to molecular nitrogen. As chlorination of the sewage effluent is commonly practiced for disinfection purposes, chlorination to this "breakpoint" can be accomplished in the same process. The removal of ammonium nitrogen also lowers the BOD of the effluent because the ammonia undergoes nitrification in waters receiving the sewage effluent, which consumes oxygen dissolved in the receiving water.

A highly advanced tertiary water treatment system that integrates several of these tertiary treatment processes is currently in operation in South Tahoe, California (Figure 15.19). This advanced treatment system was installed to prevent the eutrophic deterioration of scenic Lake Tahoe, an oligotrophic crater lake. In this system, after conventional primary and secondary treatments, phosphate is precipitated by liming, and ammonia is removed by stripping the high-pH effluent in a stripping tower at elevated temperature with vigorous aeration. After ammonia stripping, the pH is adjusted to neutrality. After additional settling, aided by polyelectrolyte addition and further clarification, nonbiodegradable organics are removed by filtration through activated carbon. The result of this extensive treatment procedure is a

A

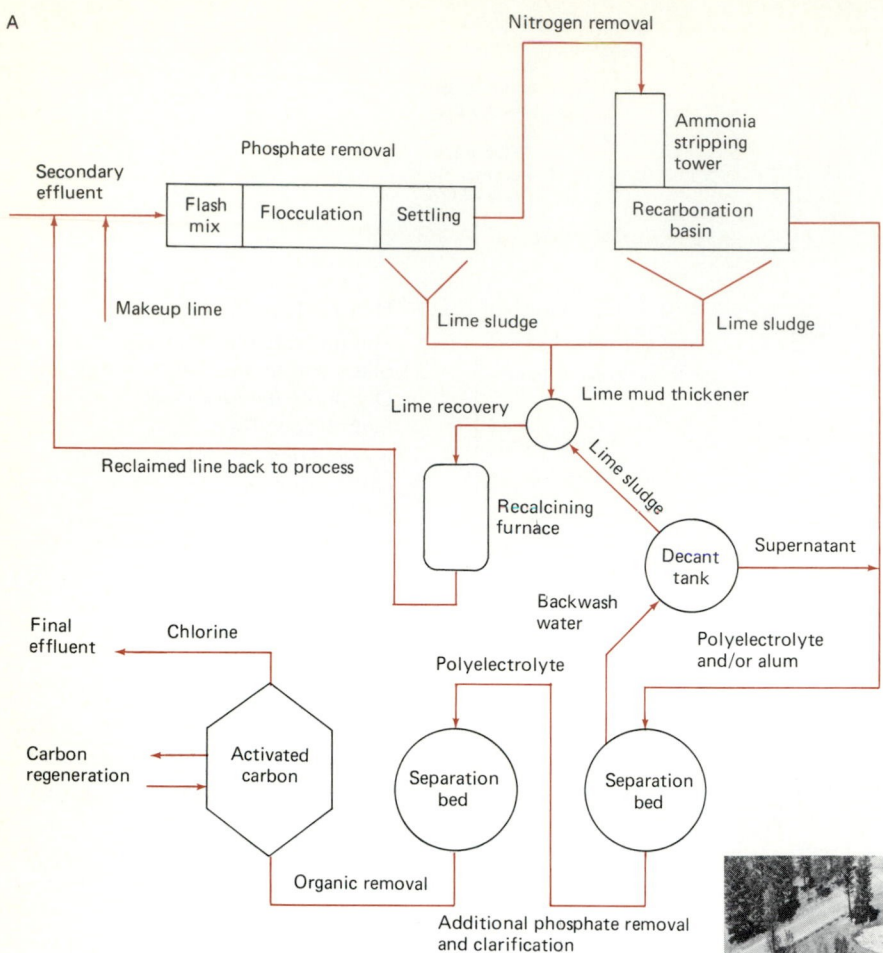

TABLE 15.6 Quality of Water After Tertiary Treatment at the South Tahoe, California, Sewage Treatment Facility

Parameter	Wastewater (mg/L)	Reclaimed Water (mg/L)
BOD	200–400	1
COD	400–600	3.25
TOC	—	1.0–7.25
Phosphate	25–30	0.2–1.0
Organic nitrogen	10–15	0.3–2.0
Ammonia nitrogen	25–35	0.3–1.5
Nitrate and nitrite	0	0

very high-quality effluent that can be released into Lake Tahoe without causing eutrophication (Table 15.6).

Disinfection The final step in the sewage treatment process is **disinfection**, designed to kill enteropathogenic bacteria and viruses that were not eliminated during the previous stages of sewage treatment. Disinfection is commonly accomplished by **chlorination**, using either chlorine gas (Cl_2) or hypochlorite ($CaOCl_2$ or $NaOCl$). Chlorine gas reacts with

B

FIGURE 15.19

In the advanced sewage treatment system of South Tahoe, California, phosphate, nitrogen, and recalcitrant organic compounds are removed. (A) The extensive treatment process illustrated in this diagram is operational at relatively few locations, where sewage effluents must cause minimal environmental perturbation. (B) Aerial photograph of the South Tahoe sewage treatment facitity. (From BPS—South Tahoe Municipal Water Facility)

water to yield hypochlorous and hydrochloric acids, the actual disinfectants. Hypochlorite is a strong oxidant, which is the basis of its antibacterial action. As an oxidant, it also reacts with residual dissolved or suspended organic matter, ammonia, reduced iron, manganese, and sulfur compounds. The oxidation of these compounds competes for available HOCl, reducing its disinfecting power. Amounts of hypochlorite sufficient to satisfy these reactions and to allow excess-free residual chlorine to remain in solution for disinfection would result in high salt concentrations in the effluent.

Therefore, it is desirable to remove nitrogen and other contaminants by alternative means and to use chlorination for disinfection only. A disadvantage of disinfection by chlorination is that the more resistant organic molecules, such as some lipids and hydrocarbons, are not completely oxidized but instead become partially chlorinated. Chlorinated hydrocarbons tend to be toxic and difficult to mineralize. Because alternative means of disinfection, such as ozonation, are more expensive, chlorination remains the principal means of sewage disinfection.

15.3 Treatment and Safety of Water Supplies

Providing safe water for drinking and other uses, free of pathogens and toxic substances, is closely related to the problem of safe disposal of liquid wastes. Fecal contamination of potable water supplies from untreated or inadequately treated sewage effluents entering lakes, rivers, or groundwaters that serve as municipal water supplies creates conditions for rapid dissemination of pathogens. The primary route of infection is direct ingestion of the pathogens in the drinking water, but additional infection opportunities arise when fruits, vegetables, and eating utensils are washed with contaminated water. The obvious remedy for this situation is to disrupt the transmission of enteropathogenic fecal organisms to water supplies. The pathogenic population is reduced considerably during sewage treatment, but this does not ensure adequate treatment of all sewage discharge. Therefore, major sanitation efforts are required to treat and safely distribute public water supplies. Such sanitation practices have led to the virtual elimination of waterborne infections in the developed countries, but such infections continue to be major causes of sickness and death in underdeveloped regions.

Disinfection of Potable Water Supplies

Most potable water supplies come from rivers and underground wells and springs. Water from underground sources is partially purified by filtration as it passes through the soil column, removing particulate matter and microorganisms. This does not, however, preclude the possibility of bacterial or viral contamination of the water supply, particularly if the source of the water is near a sewage effluent. In some rural areas, water is boiled or treated with antimicrobial chemicals to ensure its safety. In densely populated areas, municipal water treatment facilities are designed to ensure the safety of the drinking water supply.

The principal processes of a water treatment facility are sedimentation, filtration, and disinfection (Figure 15.20). **Sedimentation** is carried out in large reservoirs, where the water is held for a sufficient period of time to permit large particulate matter to settle out of the water. Sedimentation rates can be increased by the addition of aluminum sulfate, alum, which forms a floc that precipitates and carries with it microorganisms and suspended organic matter to the bottom

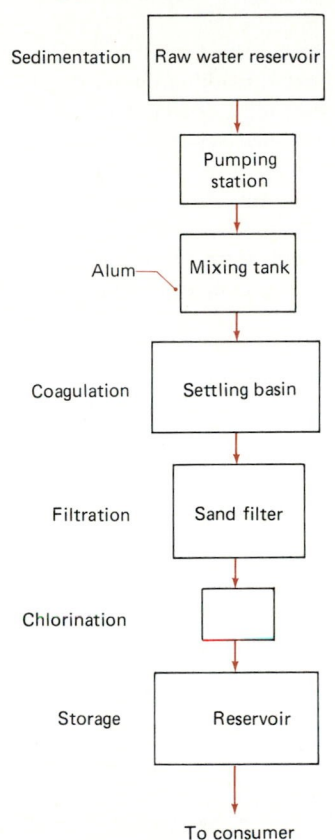

FIGURE 15.20
A flow diagram of the water purification process used by most municipalities.

of a settling basin. The water then undergoes **filtration** by passage through sand filter beds that removes up to 99 percent of the bacteria. The water may also be filtered through activated charcoal to remove potentially toxic organic compounds and organic compounds that impart undesirable color and/or taste to the water. The water is then subjected to disinfection to ensure that it does not contain any pathogens that could be a source of waterborne disease. The water may also be treated with fluoride to reduce dental caries in the community, and minerals may be removed to soften the water for washing.

ANALYTICAL PROCESS

Coliform Counts for Assessing Water Safety

The most frequently used indicator organism is the normally nonpathogenic coliform bacterium *Escherichia coli*. Positive tests for *E. coli* do not prove the presence of enteropathogenic organisms but do establish this possibility. Because *E. coli* is more numerous and easier to grow than the enteropathogens, the test has a built-in safety factor for detecting potentially dangerous fecal contamination. *E. coli* meets many of the criteria for an ideal indicator organism, but there are limitations to its use as such, and various other species have been proposed as additional or replacement indicators of water safety.

For *E. coli* to be a useful indicator organism of fecal pollution, it must be differentiated readily from nonfecal bacteria. The conventional test for the detection of fecal contamination involves a three-stage test procedure (Figure 15.21). In the first stage, lactose broth tubes are inoculated with undiluted or appropriately diluted water samples. The tubes showing gas formation are recorded as positive and are used to calculate the most probable number of coliform bacteria in the sample. Gas formation, detected in small inverted test tubes called *Durham tubes*, gives positive presumptive evidence of contamination by fecal coliforms; this is called a **presumptive test**. Gas formation in lactose broth at 37°C is characteristic not only of fecal *Salmonella*, *Shigella*, and *E. coli* strains but also of the nonfecal coliform *Enterobacter aerogenes* and some *Klebsiella* species. Therefore, in the second test stage of this procedure, the presence of enteric bacteria is confirmed by streaking samples from the positive lactose broth cultures onto a medium, such as eosin methylene blue (EMB) agar. Fecal coliform colonies on this medium acquire a characteristically greenish metallic sheen, *Enterobacter* species form reddish colonies, and nonlactose fermenters form colorless colonies, respectively; this is called a **confirmed test**. Alternatively, the confirmed test can be accomplished by using brilliant green lactose-bile broth (BGLB). If BGLB is used, it is then subcultured onto EMB. Subculturing colonies showing a green metallic sheen on EMB into lactose broth incubated at 35°C should produce gas formation, completing a positive test for fecal coliforms called a **completed test**.

FIGURE 15.21

An outline of the water quality testing procedures for determining the number of coliforms. Such tests are used for determining the safety of drinking and recreational waters.

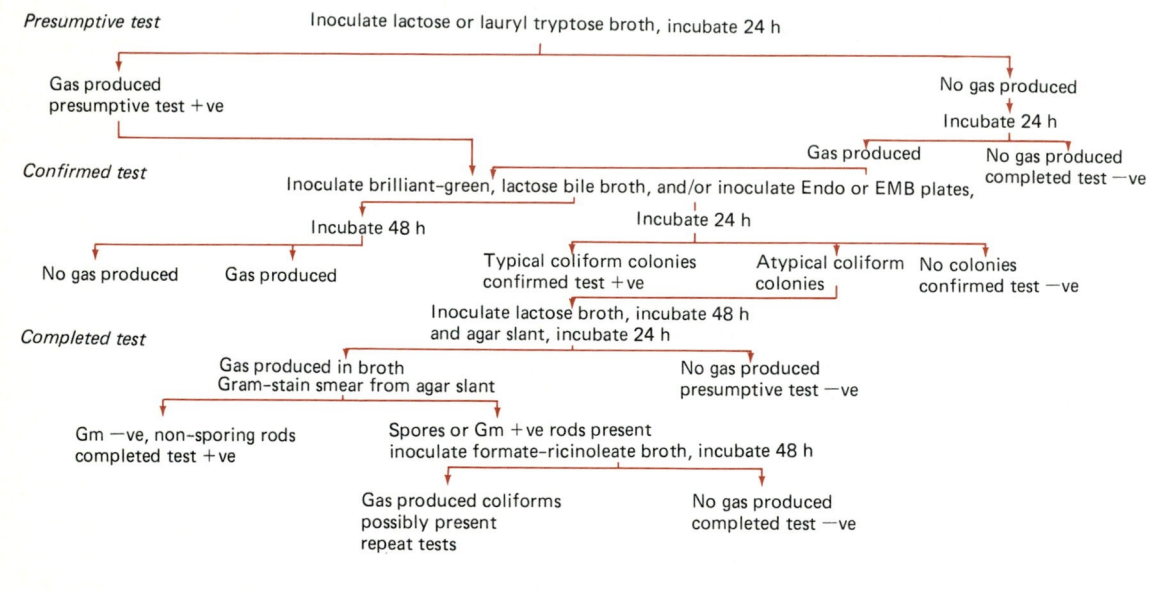

Chlorination has been the traditional method employed for disinfecting municipal water supplies. This treatment method is relatively inexpensive, and the free residual chlorine content of the treated water represents a built-in safety factor against pathogens surviving the actual treatment period and causing recontamination. The disadvantage of chlorination is the incidental production of trace amounts of organochlorine compounds, particularly trihalomethane (THM), a suspected carcinogen. The U.S. Environmental Protection Agency established in 1979 a maximal THM limit in drinking water of 100 μg/L. To stay within this limit using traditional chlorine disinfection, organic compounds would have to be removed by sand filtration or other methods that are impractical and too expensive. Fortunately, it was found that disinfection by monochloramine is effective and produces much lower amounts of THMs. As an example, traditional chlorination of Ohio River water produced 160 μg THM/L, while chloramine treatment produced THM levels consistently below 20 μg/L. Monochloramine may be generated right in the water to be disinfected by adding ammonium prior to or simultaneously with chlorine or hypochlorite. **Chloroamination**, the use of chloramines as drinking water disinfectants, is the least expensive way to reduce THM formation, and this practice is spreading rapidly.

Ozone (O_3) is a more expensive alternative for disinfecting water supplies that has sometimes been used for water disinfection, with good results in both Europe and the United States. Ozone treatment kills pathogens reliably and does not result in the synthesis of any undesirable trace organochlorine contaminants. However, because ozone is an unstable gas, water treated with it does not have any residual antimicrobial activity and is more prone to chance recontamination than chlorinated water. Ozone has to be generated from air on site in ozone reactors, using an electrical corona discharge. Only about 10 percent of the electricity is actually generating ozone; the rest is lost as heat, making disinfection by ozone considerably less cost effective than chlorination and chloramination.

Bacterial Indicators of Water Safety

The importance to public health of clean drinking water requires objective test methods to establish high standards of water safety and to evaluate the effectiveness of treatment procedures. To monitor water routinely for the detection of actual enteropathogens, such as *Salmonella* and *Shigella*, would be a difficult and uncertain undertaking. Instead, bacteriological tests of drinking water establish the degree of fecal contamination of a water sample by demonstrating the presence of **indicator organisms**. The ideal indicator organism should (1) be present whenever the pathogens concerned are present; (2) be present only when there is a real danger of pathogens being present; (3) occur in greater numbers than the pathogens to provide a safety margin; (4) survive in the environment as long as the potential pathogens; and (5) be easy to detect with a high degree of reliability of correctly identifying the indicator organism, regardless of what other organisms are present in the sample.

This three-stage test can be simplified. In the **Eijkman test**, suitable dilutions are incubated in lactose broth at 44.5°C, a temperature at which fecal coliforms still grow but nonfecal coliforms are inhibited. Gas formation constitutes a one-step positive test, but precise temperature control is mandatory because temperatures only a few degrees higher inhibit or kill the fecal coliforms. It is also possible to filter known volumes of diluted or undiluted water samples through 0.45–μm pore size bacteriological filters and incubate the filters directly on EMB agar, m-Endo agar, or other suitable media. Colonies of fecal coliforms appear with a

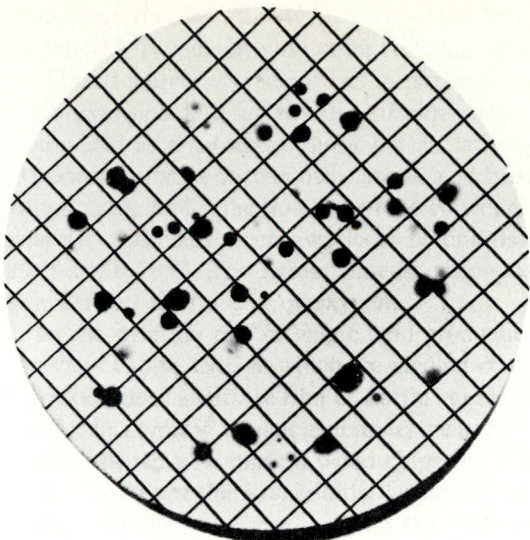

FIGURE 15.22
The membrane filtration system and colonies of coliform bacteria growing on the surface of a membrane filter. (From BPS—Millipore Corp.)

characteristic metallic sheen on EMB, for example, and can be easily counted (Figure 15.22).

Standards for Tolerable Levels of Fecal Contamination
The standards for tolerable limits of fecal contamination (Table 15.7) vary with the intended water use and are somewhat arbitrary, with large built-in safety margins, the usefulness of which has been borne out by long experience. The most stringent standards are imposed on the municipal water supplies to be used by many people. Somewhat higher coliform counts are sometimes tolerated in private wells used by only one family because such wells would not become a source of a widespread epidemic. Maintenance of a high drinking water standard does not absolutely exclude the possibility of ingesting enteropathogens with the water but helps keep this possibility to a statistically tolerable minimum. The built-in safety factors are twofold: (1) enteropathogens are very likely to be present in much lower numbers than fecal coliforms and (2) a few infective bacteria are unlikely to be able to overcome natural body defenses. A minimum infectious dose

TABLE 15.7 United States Water Standards for Coliform Contamination

Water Use	Maximal Permissible Coliform Count (number/100 mL)
Municipal drinking water	1
Waters used for shellfishing	70
Recreational waters	1,000

of several hundred to several thousand bacteria is usually necessary for an actual infection to be established. Drinking water supplies meeting the 1/100 mL coliform standard have never been demonstrated to be the source of a waterborne bacterial infection.

Fecal coliform counts are also used to establish the safety of water in shellfish harvesting and recreational areas. Because shellfish tend to concentrate bacteria and other particles acquired through their filter-feeding activity and are sometimes eaten raw, they can become a source of infection by waterborne pathogens. Therefore, there are relatively stringent standards for waters used for shellfishing. Clinical evidence for infection by enteropathogenic coliforms through recreational use of waters for bathing, wading, and swimming is unconvincing, but as a precaution, beaches are usually closed when fecal coliform counts exceed the recreational standard of 1,000/100 mL. Some regional standards require that disinfected sewage discharges not exceed this limit.

Water quality standards based on fecal coliform levels do not account for the possible transmission of viruses associated with fecal matter through municipal water supplies.

There is ample evidence, of course, for destructive epidemics by enteroviruses caused by untreated drinking water in various underdeveloped countries. Enteroviruses are somewhat more resistant to disinfection by chlorine or ozone than bacteria and, occasionally, active virus particles are recovered from treated water that meets fecal coliform standards. Thus, the possibility exists that water that meets accepted quality standards may still occasionally be a source of a viral infection. As many as 100 different viral types can be shed in human feces, but practical concern has been mainly with the viruses that cause infectious hepatitis, poliomyelitis, and viral gastroenteritis. Infectious hepatitis is sometimes spread by water supplies, though the more prevalent mode of infection is by the consumption of raw shellfish from fecally contaminated waters. Spread of polio infection through water supplies and/or recreational use of beaches has been suspected in many cases. The situation with regard to viral gastroenteritis is similar. At this point, we can only say that the possibility of an occasional sporadic viral infection through drinking water adequately treated by bacteriological standards cannot be excluded, but there is no hard evidence for any epidemics caused by such water.

15.4 Biodegradation of Environmental Pollutants

Human exploitation of fossil fuel reserves and the production of many novel synthetic compounds (**xenobiotics**) in the twentieth century have introduced into the environment many compounds that microorganisms normally do not encounter and thus are not prepared to biodegrade. Many of these compounds are toxic to living systems, and their presence in aquatic and terrestrial habitats often has serious ecological consequences, including major kills of indigenous biota. The disposal or accidental spillage of these compounds has created serious modern environmental pollution problems, particularly when microbial biodegradation activities fail to remove these pollutants quickly enough to prevent environmental damage. Sewage treatment and water purification systems are usually incapable of removing these substances if they enter municipal water supplies, where they pose a potential human health hazard.

The problems associated with pesticides and their biomagnification through food webs have been discussed in Chapter 14. Here we will discuss the problems associated with laundry detergents and oil spills. These are representative examples of the growing problems of toxic waste disposal and environmental pollution. Today there are many sites littered with decaying drums filled with toxic chemicals that are slowly seeping into the environment. The fate and the untoward effects of these chemicals remain to be determined in the never-ending battle between microbes and human wastes and pollutants.

Alkyl Benzyl Sulfonates

Alkyl benzyl sulfonates (ABS) are the major components of anionic laundry detergents. Cleaning occurs when ABS molecules form a monolayer around lipophilic droplets or particles that make up most stains or dirt on clothing, forming an emulsion that can be rinsed out of the fabric with water. The ABS molecule is a surface active molecule, having a polar sulfonate and a nonpolar alkyl end. During laundering, ABS molecules orient their nonpolar ends toward lipophilic substances and their sulfonate ends toward the surrounding water. The alkyl portion of the ABS molecule may be linear or branched (Figure 15.23). **Nonlinear ABS** is easier to manufacture and has slightly superior detergent properties than conventional soaps, but nonlinear ABS has proved to be resistant to biodegradation, causing extensive foaming of rivers receiving ABS-containing wastes. Some communities have banned the use of anionic detergents because of their persistence in groundwater supplies used as sources of potable water.

It is the methyl branching of the alkyl chain that interferes with biodegradation because the tertiary carbon atoms block the normal β-oxidation sequence. By changing the design of this synthetic molecule to that of a **linear ABS**, the blockage can be removed. The detergent industry has switched to linear ABS, which is free from this blockage and consequently more easily biodegraded. The ABS story is partic-

Nonlinear alkyl benzyl sulfonate

$$H_3 - CH - [CH_2CH]_2 - CH_2 - CH - \langle \rangle - SO_3Na$$
(with CH_3 branches)

Linear alkyl benzyl sulfonate

$$H_3C - CH_2 - [CH_2]_8 - CH_2 - CH_2 - \langle \rangle - SO_3Na$$

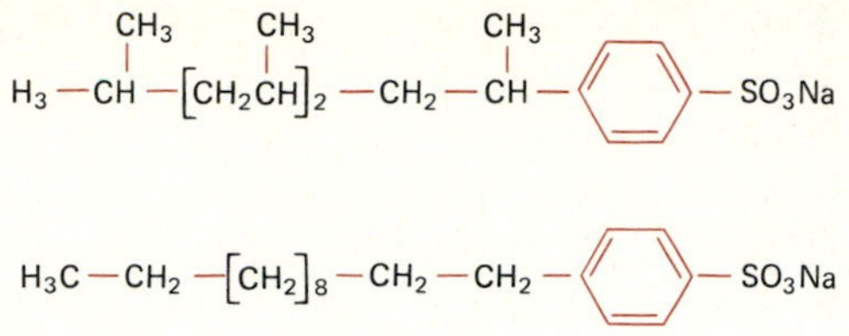

FIGURE 15.23
Linear and nonlinear ABS molecules. The methyl branches in the former group of compounds interfere with rapid biodegradation and cause foaming problems in contaminated waters. Linear ABS compounds do not share this problem.

ularly significant because it was one of the first instances in which a synthetic molecule was specifically redesigned to remove obstacles to biodegradation while preserving the useful characteristics of the compound.

Biodegradable polymers, for example, can be synthesized to replace or augment various plastics, including polyethylene, polystyrene, and polyvinylchloride, which are recalcitrant to microbial attack and therefore have been accumulating in the environment. By understanding the role of microbial biodegradation in maintaining environmental quality, human ingenuity has the potential to produce economically profitable synthetic compounds that are biodegradable and that can be safely disposed of in an environmentally safe manner.

Oil Pollution

Over 10 million metric tons of **oil pollutants** enter the marine environment each year as a result of accidental spillages and disposal of oily wastes. Periodically, pictures of dead birds floundering in a sea of oil after a major oil spillage occur on the front page of the daily newspaper, evoking images of impending ecological doom. Actually, only a small proportion of all marine oil pollutants comes from major oil spills; most of the oil pollution problem originates from minor spillages associated with routine operations.

Petroleum is a complex mixture composed primarily of aliphatic, alicyclic, and aromatic hydrocarbons (Figure 15.24). There are hundreds of individual compounds in every crude oil, the composition of each crude oil varying with its origin. As a result, the fate of petroleum pollutants in the environment is complex. The challenge for microorganisms to degrade all of the components of a petroleum mixture is immense. Nevertheless, microbial biodegradation of petroleum pollutants is a major process and is the reason that the oceans are not covered with oil today. As an example of the ability of microbes to degrade petroleum pollutants, measurements indicate that after the 1978 wreck of the supertanker *Amoco Cadiz* off the coast of France, microorganisms biodegraded 10 tons of oil per day in the affected area. Microbial biodegradation represented the major process responsible for the ecological recovery of the oiled coastal region.

The susceptibility of petroleum hydrocarbons to biodegradation is determined by the structure and molecular weight of the hydrocarbon molecule. *n*-Alkanes of intermediate chain length (C_{10}–C_{24}) are degraded most rapidly. Short chain alkanes (less than C_9) are toxic to many microorganisms, but they generally evaporate rapidly from oil slicks. As alkane chain length increases, so does resistance to biodegradation. Branching, in general, reduces the rate of biodegradation because tertiary and quaternary carbon atoms interfere with degradation mechanisms or can block degradation altogether. Aromatic compounds, especially of the condensed polynuclear type, are degraded more slowly than alkanes. Alicyclic compounds are frequently unable to serve as the sole carbon sources for microbial growth unless they have a sufficiently long aliphatic side chain, but they can be degraded via cometabolism by two or more cooperating microbial strains with complementary metabolic capabilities.

Petroleum has always entered the biosphere by natural seepage but at rates much slower than those of the forced recovery of petroleum by drilling, which is now estimated to be about 2 billion metric tons per year. The production, transportation, refining, and ultimately the disposal of used petroleum and petroleum products inevitably result in environmental pollution. Because the bulk of this load is, of course, heavily centered on offshore production sites, major shipping routes, and refineries, its input frequently exceeds the self-purification capacity of the receiving waters. Petroleum pollutants in the environment are destructive to birds and marine life and, when driven ashore, cause heavy economic losses due to aesthetic damage to recreational beaches.

In addition to killing birds, fish, shellfish, and other invertebrates, oil pollution can have more subtle effects on marine life. Even at a low parts per billion (ppb) concentration, dissolved aromatic components of petroleum can disrupt the chemoreception of some marine organisms. Because feeding and mating responses largely depend on chemoreception, such disruption can lead to elimination of many species from a polluted area even when the pollutant concentration is far below the lethal level. Another disturbing problem is the possibility that condensed polynuclear components of petroleum, some of which are carcinogenic and relatively resistant to biodegradation, may move up marine food chains and taint fish or shellfish. Polynuclear aromatic compounds are among the components of crude oil most resistant to microbial biodegradation and become a major component of the tarry residues left in the sea when oil biodegradative activities slow to a halt.

The successful biodegradative removal of petroleum hydrocarbons from the sea depends on the enzymatic capacities

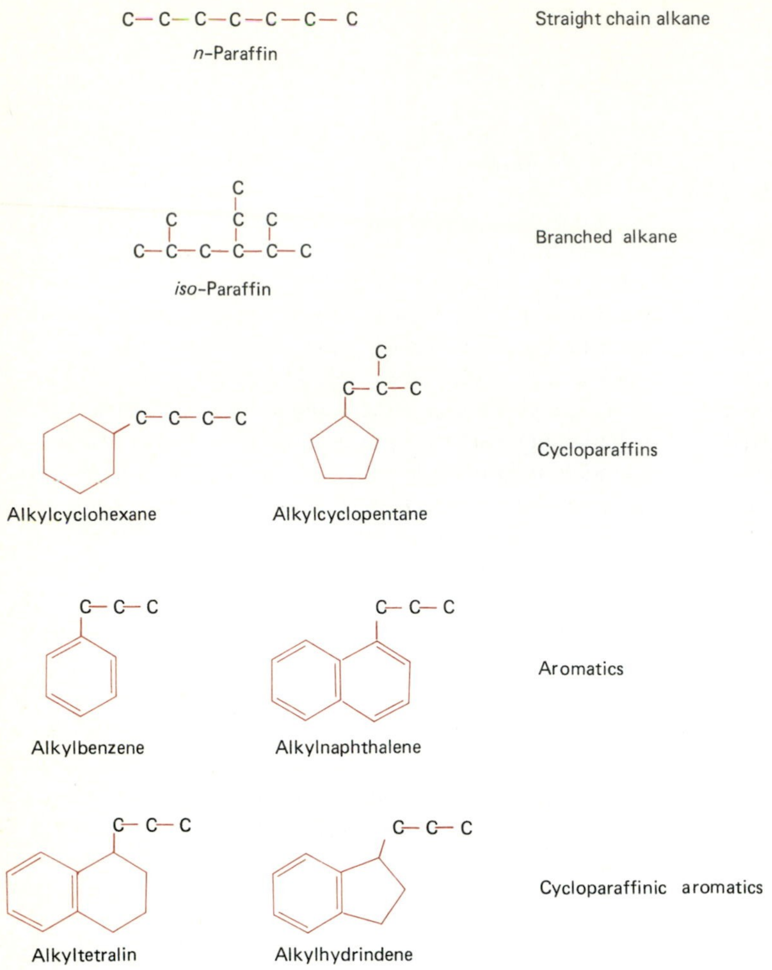

FIGURE 15.24

Hydrocarbon structures found in petroleum. Homologues, isomers, and combinations result in hundreds of individual hydrocarbon compounds in crude oil samples.

of microorganisms and various abiotic factors. Microbial hydrocarbon biodegradation requires suitable growth temperatures and available supplies of fixed forms of nitrogen, phosphate, and molecular oxygen. In the oceans, temperature and nutrient concentrations often limit the rates of petroleum biodegradation. The low concentrations of nitrate and phosphate in seawater are particularly limiting to hydrocarbon biodegradation because petroleum is primarily composed of hydrogen and carbon. For example, after the IXTOC I well blowout, which in 1980 created the largest known oil pollution incident, little biodegradation of the oil–water emulsion (mousse) occurred in the surface waters of the Gulf of Mexico because of severe nutrient limitations.

Although many microorganisms can metabolize petroleum hydrocarbons, no single microorganism possesses the enzymatic capability to degrade all, or even most, of the compounds in a petroleum mixture. More rapid rates of degradation occur when there is a mixed microbial community than can be accomplished by a single species. Apparently, the genetic information in more than one organism is required to produce the enzymes needed for extensive petroleum biodegradation. Recombinant DNA technology, however, permits the incorporation of the diverse types of genetic information extracted from several organisms into a single organism. Through genetic engineering, a ''superbug'' has been created that is capable of degrading many different hydrocarbon structures and that is potentially useful in oil pollution abatement programs. This hydrocarbon-degrading microorganism is the first organism for which a patent has been granted in the United States. The ruling by the Supreme Court that genetic engineering could in essence invent microorganisms has far-reaching consequences for the future use of recombinant DNA technology in the United States to develop microorganisms of economic significance.

Microorganisms created by microbiologists have the potential of helping to cleanse the environment of man-made pollutants. Despite the ability to create superbugs, the usefulness of such organisms in pollution abatement depends on the compatibility with their environment. In many cases, environmental factors, rather than the genetic capability of a microorganism, limit the biodegradation of pollutants. Thus, although genetically engineered organisms are a useful addition to the arsenal of antipollution measures, there is no panacea for solving human pollution problems.

Situational Problem

Designing the Waste Disposal System for a Self-Contained Residential Development

Suppose you are on the planning board for a resort retirement community in Arizona. The community is situated on a remote lake quite distant from oceans, rivers, and any other communities. The lake is planned as a center of recreational activities and as a major source of drinking water for the community. Additional water will come from an aquifer located 500 feet below the surface of the soil. The soil in this area is mostly sand. The community will have a golf course that requires water and nutrients. It is imperative that the disposal of wastes not destroy the aesthetic value of the lake or interfere with its use as a supply of potable water.

There are various options for the disposal of solid waste and sewage. The selection of a particular disposal method or methods depends in large part on the magnitude of the wastes generated, as well as the location and cost of the facility. In this case, the community is planned for a maximum of 400 residences. How would you go about selecting the appropriate waste disposal system? How would you monitor the effectiveness of the waste disposal system and foresee any deleterious effects on the multiple uses of the water supply? How could you economically link the disposal of wastes with the maintenance of the golf course? What could you do if, despite your best planning efforts, excessive waste entered the lake, leading to eutrophication and the potential introduction of enteric pathogens?

Photograph of koji fermentation tray showing growth of *Aspergillus kojae* on a mixture of soybeans and wheat. (Courtesy Kikkoman Foods, Walsworth, Wisconsin)

Food and Industrial Microbiology

CHAPTER 16 Food Microbiology

CHAPTER 17 Industrial Microbiology

Food Microbiology

Food products serve not only as sources of nutrition for humans and other animals but also as substrates for the growth of microorganisms. The uncontrolled growth of microorganisms in food causes spoilage, a serious problem accounting for sizable losses of food products that are critically needed to meet global nutritional requirements. A major concern in food microbiology is the development of an understanding of the processes involved in food spoilage and the appropriate measures for the preservation of food products. In particular, it is necessary to prevent contamination of food products with human pathogens and to control the potential proliferation of toxin-producing microorganisms, which can result in food poisoning and the transmission of foodborne pathogens.

Careful quality control in the food industry is essential for preventing outbreaks of foodborne disease. Additionally, the growth of microorganisms in foods can alter the quality of the product, causing sizable economic losses to the food industry. Although the uncontrolled growth of microorganisms in food products is deleterious, the controlled growth on or in foods is used to produce numerous foods and beverages that are considered gastronomic delights. For example, the growth of microorganisms in milk produces cheese; in fruit juices, wine; and in grain extracts, beer. In these cases, the growth of microorganisms and the accumulation of their metabolic products are desirable.

16.1 Food Spoilage

Food spoilage is defined as any change in a food that renders it undesirable or unsafe for human consumption, and the growth of microorganisms in a food represents only one process that may cause food to spoil. The growth of pathogenic microorganisms in a food is certainly undesirable, as it can make that food unsafe to eat, but other microbially induced changes in food, such as decreased nutritional content and altered taste, odor, color, and texture, can also make a food undesirable for human consumption. Such food spoilage is difficult to define because it depends on the culture of the consumer; for example, Eskimos bury fish to produce a stinky mess and in Britain meat should have the strong aroma of aging, but these same foods would likely be rejected as spoiled in most parts of the United States.

Foods are classified according to their susceptibility to microbial spoilage as nonperishable, semiperishable, or perishable. **Perishable foods**, such as meats, fish, poultry, most fruits and vegetables, eggs, and milk, readily spoil because of microbial activities and generally have short shelf lives unless steps are taken to remove, kill, or prevent the growth of associated microorganisms. **Semiperishable foods**, such as potatoes and apples, generally remain unspoiled for prolonged periods of time unless improperly handled. **Nonperishable foods**, such as sugar, flour, and numerous dry products, do normally not spoil, but even these foods can spoil under appropriate conditions. For example, if cereals, grains, and flours are stored under conditions of high moisture, various fungal and bacterial populations are able to grow and spoil the products. We have all observed the growth of fungi on bread that has been stored too long.

The changes that occur in a food during microbial spoilage depend on the particular microbial populations involved, their enzymatic activities, environmental conditions, and the nature of the food. The microorganisms involved in the spoilage process generally do not originate within the food, the inner tissues of most plants and animals being sterile, but rather come from the surface tissues or are a result of contamination during processing. Several factors controlling food spoilage are intrinsic to the food, but others are extrinsic environmental parameters. The most important extrinsic factors influencing food spoilage are (1) temperature of storage; (2) relative humidity; and (3) oxygen concentration. By controlling these parameters, it is possible to alter the rate of food spoilage. It is somewhat more difficult to control the inherent properties of animal and plant tissues that influence microbial spoilage of foods. The intrinsic moisture content, pH, and physical and chemical nature of the food, in part, control the numbers and types of microorganisms involved in the spoilage process. The biochemical composition of a food, in particular, has a marked influence on the microbial populations involved in the spoilage process and the microbial decomposition products associated with the spoilage of that food (Table 16.1).

Representative Spoilage Processes

Fruits, vegetables, meats, poultry, seafoods, milk and dairy products, and various other food products differ in their biochemical composition and therefore are subject to spoilage by differing microbial populations (Table 16.2).

Spoilage of Fruits and Vegetables   Fruits and vegetables are subject to rot because of the microbial degradation of pectin, the biochemical responsible for maintaining the firmness and texture of fruits and vegetables (Figure 16.1). Microbially produced pectinesterases and polygalacturonases hydrolyze pectins, resulting in the formation of **soft spots** in fruits and vegetables. About 20 percent of the harvested crops of fruits and vegetables are lost to spoilage primarily because of the activities of bacteria and fungi. Carbohy-

drates, which are present in high concentrations in fruits, vegetables, and other foods, are readily degraded by numerous microorganisms, resulting in the production of various degradation products, such as low molecular weight acids and alcohols. The accumulation of such products of microbial metabolism can alter the taste of a food, generally causing it to **sour**. The taste of sour milk, for example, is associated with the accumulation of lactic acid from the microbial transformation of carbohydrates. The accumulation of low molecular weight acids may also alter the smell of the product; sour milk often can be recognized just by smelling it. In canned foods, production of acid and no gas is referred to as **flat-sour spoilage** because the food becomes sour, but the can shows no evidence of food spoilage because no gas is produced; that is, the can remains flat. When gas is produced during spoilage of canned foods, the can swells, as can be seen by examining the lids, which are designed to push outward if pressure builds within the can because of microbial action.

TABLE 16.2   Some Spoilage Organisms of Important Foods

Class of Food Products*	Genera Dominating When Spoilage Occurs During Standard Conditions of Storage
Cereal grains (carbohydrate)	Aspergillus, Fusarium, Monilia, Penicillium, Rhizopus
Bread (carbohydrate)	Bacillus, Aspergillus, Endomyces, Neurospora, Rhizopus
Vegetables (carbohydrate)	Achromobacter, Pseudomonas, Flavobacterium, Lactobacillus, Bacillus
Fruits and juices (carbohydrate)	Acetobacter, Lactobacillus, Saccharomyces, Torulopsis
Fresh meat (protein and lipid)	Micrococcus, Cladosporium, Thamnidium, Achromobacter, Pseudomonas, Flavobacterium
Sausage, bacon, ham, etc. (protein and lipid)	Micrococcus, Lactobacillus, Streptococcus, Debaromyces, Penicillium
Whale meat (lipid and protein)	Streptococcus, Clostridium, Achromobacter, Pseudomonas, Flavobacterium
Poultry (protein and lipid)	Achromobacter, Pseudomonas, Flavobacterium, Micrococcus, Salmonella
Fish, shrimp (protein)	Achromobacter, Pseudomonas, Flavobacterium, Micrococcus, Vibrio
Shellfish (protein)	Achromobacter, Pseudomonas, Flavobacterium, Micrococcus, Vibrio
Milk and milk products (carbohydrate, lipid, and protein)	Streptococcus, Lactobacillus, Microbacterium, Achromobacter, Pseudomonas, Flavobacterium, Bacillus
Eggs (protein, and lipid)	Pseudomonas, Cladosporium, Penicillium, Sporotrichum

*Main biochemical constitutuents are shown in parentheses.

TABLE 16.1   Some Food Spoilage Processes

Process	Substrate	Products
Putrefaction	Proteins	Amino acids, amines, ammonia, and hydrogen sulfide
Souring	Carbohydrates	Acids, alcohols, and carbon dioxide
Rancidity	Lipids	Fatty acids and glycerol
Soft rot	Pectin	Methanol, galacturonic acid, and polygalacturonic acid

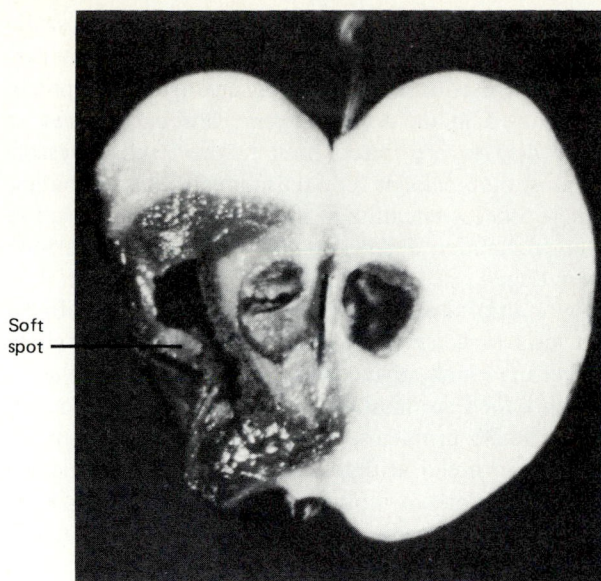

FIGURE 16.1
This rotting fruit shows soft spot formation. Such spoilage of fruits and vegetables causes great economic losses and limits the useful shipping times and shelf lives of many products. (From BPS—W. Rosenberg, Iona College)

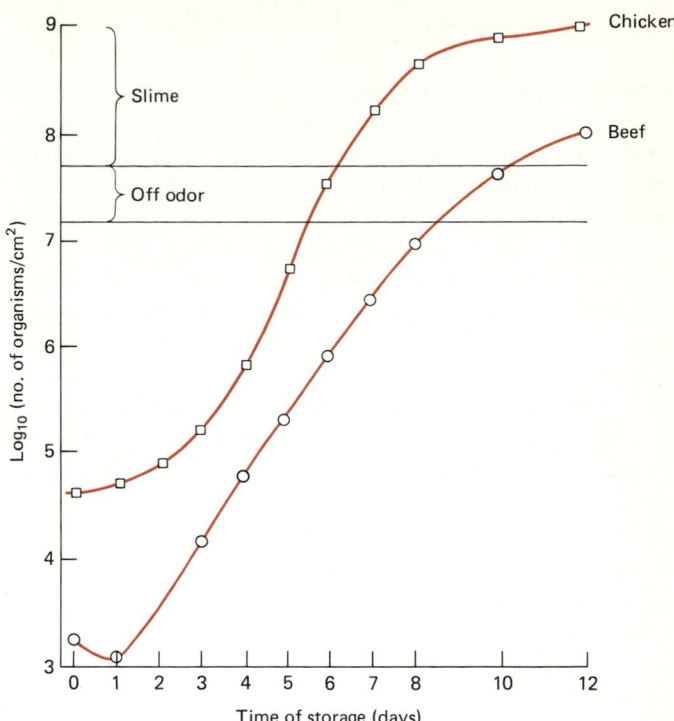

FIGURE 16.2
The increase in the numbers of microorganisms during meat spoilage, including the time of onset of off odor and slime formation, are shown in this graph. Storage was at 5°C. Note that although chicken initially has a considerably higher microbial content than beef, the proliferation of microorganisms in the beef begins after only 1 day, and the sharp increases in the chicken do not begin until day 3.

Spoilage of Meats Meats and other proteinaceous products can be decomposed by anaerobic bacteria, resulting in **putrefaction**. Putrefaction of meat is the result of the breakdown of proteins by proteinases. The subsequent degradation of amino acids produces foul-smelling, low molecular weight sulfur- and nitrogen-containing compounds, such as mercaptans, hydrogen sulfide, ammonia, and amines (Table 16.3). The evolution of noxious, odoriferous compounds from putrefying proteins renders food unacceptable for human consumption. The characteristic odor of hydrogen sulfide, for example, renders rotting eggs inedible, and the development of off odors and slime on poultry and beef reflects the presence of increased microbial populations and spoilage of the meat (Figure 16.2). In canned foods such spoilage is referred to as **sulfide stinkers**. Under aerobic conditions, the

decomposition of proteins generally does not result in the production of compounds with obnoxious odors. The spoilage of meat under aerobic conditions, though, can result in the accumulation of surface slime, generally because of the growth of *Pseudomonas*, *Achromobacter*, *Streptococcus*, *Leuconostoc*, *Bacillus*, *Micrococcus*, and *Lactobacillus* species. The physical state of the meat influences which microbial species may be involved in spoilage. The spoilage of fresh whole meats is normally associated with lactic acid bacteria, particularly *Lactobacillus*, *Leuconostoc*, and *Streptococcus* species, whereas the spoilage of ground beef is primarily due to the growth of *Pseudomonas*, *Acromobacter*, and *Micrococcus*.

Spoilage of Other Foods Spoilage by microbial degradation of fats and oils produces rancidity, caused by oxidation or hydrolysis of lipids. Spoiled butter, for example, becomes **rancid** because of the hydrolysis of butterfat, with the production of free fatty acids and glycerol and the accompanying development of undesirable flavors. Fish with high lipid levels, such as salmon and mackerel, may also become rancid because of the hydrolysis of fats and oils.

The growth of some microorganisms can alter the color of the food. For example, the growth of *Serratia marcescens*

TABLE 16.3 Products From the Microbial Decomposition of Amino Acids

Chemical Process	Products
Oxidative deamination	Keto acid + NH_3
Hydrolytic deamination	Hydroxy acid + NH_3
Reductive deamination	Saturated fatty acid + NH_3
Dehydrogenation deamination	Unsaturated fatty acid + NH_3
Decarboxylation	Amine + CO_2
Hydrolytic deamination + decarboxylation	Primary alcohol + NH_3 + CO_2
Reductive deamination + decarboxylation	Hydrocarbon + NH_3 + CO_2
Oxidative deamination + decarboxylation	Fatty acid + NH_3 + CO_2

can produce bright red bread; the combined growth of *Pseudomonas species and Streptococcus lactis* produces blue milk; the growth of fungi on meat surfaces can produce off colors, such as black spot caused by *Cladosporium herbarum*, white spot caused by *Sporotrichum carnis*, and green patches produced by *Penicillium* species; and the growth of heterolactic fermenters on cured meats, such as frankfurters, causes greening because of the action of peroxidases on the meat pigments. The development of unnatural colors in a food reduces its acceptability, regardless of whether there is any associated development of an off taste or smell. Microbial growth can also alter the texture of a food. For example, the growth of *Bacillus* species in products such as milk and dough can produce ropiness, a textural change resulting from the growth of encapsulated *Bacillus* species together with the hydrolysis of starch and proteins.

Indicators of Human Pathogens Associated with Foods

Perhaps the most serious food spoilage process involves the growth of pathogenic and toxin-producing microorganisms. The ingestion of foods containing toxins and human pathogens can cause a variety of gastrointestinal diseases. In some cases, the growth of pathogenic microorganisms in a food is accompanied by obvious signs of spoilage, such as the production of gas or foul-smelling compounds, which give a clear indication that the product should not be eaten. In other cases, there are no obvious signs of spoilage to indicate the presence of pathogens or toxins. It is therefore necessary to carry out inspection programs aimed at detecting the possible contamination of food products with pathogenic microorganisms. As a consequence of the high quality control procedures employed in the food industry, few occurrences of foodborne disease are traced back to the food processor. Rather, most outbreaks of food poisoning are caused by improper post process handling of the food, often as a result of improper preparation and/or handling at home or in a food service setting.

Quality control laboratories in the food industry routinely perform tests to detect pathogens in food products and to ensure that the numbers and types of microorganisms associated with a food are not likely to cause serious food spoilage and/or health problems. In some cases, quality control procedures are aimed at detecting the presence of specific microorganisms, but in most cases the tests consist of examining the food for indicator organisms. For example, coliform counts are routinely performed on representative samples of many food products as an indication of possible fecal contamination, since food contaminated with fecal material has a relatively high probability of containing human pathogens. Foods such as hamburger often have 10^5 total bacteria per gram, but as long as they do not contain any *Salmonella* or other pathogens, they are considered safe for human consumption. In the United States, agencies involved in the surveillance and regulation of foods are the Department of Agriculture (USDA), the Food and Drug Administration (FDA), and state boards of health. Standards based on the presence or absence of pathogens, such as those set by Congress and regulated by the USDA and FDA, do provide the safeguards needed for assurance of the safety of food products.

16.2 Food Preservation Methods

Food preservation methods are aimed at preventing the microbial spoilage of food products and the growth of foodborne pathogens. Thus, the two principal goals of food preservation methods are (1) increasing the shelf life of the food and (2) ensuring its safety for human consumption. There are a variety of food preservation methods (Table 16.4), all of which attempt to eliminate microorganisms from foods and/or to limit their growth rates. Although some of these preservation methods can increase the amount of time before spoilage occurs, that is, the **shelf life**, indefinitely, others extend the usefulness of the food for only a limited period of time.

The choice of a food preservation method depends on the nature of the food. It is critical that the procedures used for preserving foods not render the product unacceptable for human consumption. For example, although boiling milk for a long period of time could greatly extend its shelf life, the product would no longer be palatable. Food preservation methods must include the processing necessary to establish conditions that will prevent microbial growth during storage.

Understanding the principles employed in the preservation method is important so that subsequent storage and handling of the food do not negate the effectiveness of the preservation method. We will now consider some of the methods commonly used for food preservation.

TABLE 16.4   Principles and Principal Methods of Food Preservation

1. Prevention or delay of microbial decomposition
   a. by keeping out microorganisms (asepsis)
   b. by removal of microorganisms (e.g., filtration)
   c. by reducing the rate of microbial growth (e.g., by low temperature, drying, anaerobic conditions, and chemical inhibitors)
   d. by killing microorganisms (e.g., by heat or radiation)
2. Prevention or delay of self-decomposition of the food
   a. by inactivation of food enzymes (e.g., blanching)
   b. by prevention of chemical reactions (e.g., by using antioxidants)

Asepsis

Preventing the contamination of food products with pathogenic and spoilage microorganisms is an important part of food preservation methods. The use of **sanitary methods**, such as washing utensils that come in contact with food, minimizes microbial contamination and is critical whenever food products are handled. It should be remembered that anything that contacts the food, including hands and work surfaces, may be a source of microbial contamination and that once the food is inoculated, the microorganisms may reproduce rapidly under favorable conditions.

The use of appropriate methods for processing food products can reduce the levels of microbial contamination and prevent the spread of microorganisms through the food, but aseptic methods do not eliminate microorganisms from the food. For example, the careful processing of animal products in the slaughterhouse is critical in order to minimize microbial contamination of meats by microorganisms associated with the animal surfaces and gut contents, but even with aseptic handling, meat surfaces retain microbial contaminants. Trimming away spoiled portions of a food is an obvious way of reducing microbial contamination, but spread from the spoiled area to the remaining food may still occur unnoticed. Washing fruits and vegetables can be important for removing soil particles, animal fertilizers, and associated microorganisms that may cause food spoilage, but caution must be taken when washing products to make sure that the washing process itself does not add or spread spoilage microorganisms through the food. Chlorine is normally added to water to ensure that the wash water does not act as a source of microbial contaminants.

The **packaging** of food products is extremely important in preventing microbial contamination during transport and storage. Many foods have a natural protective covering that prevents or delays microbial decomposition of the sterile inner tissues, such as the outer skins of fruits and vegetables and the shells of eggs. Processing of foods, however, often removes or disrupts these protective layers, exposing the product to microbial spoilage. Most processed food products are wrapped or placed in sealed containers to preclude microbial contamination and spoilage of the food. Quality control in packaging is an important aspect of the food processing operation.

Filtration and Centrifugation

Filtration and centrifugation can be used to remove microorganisms from some food products. Filtration is used to eliminate completely (sterilize) or reduce the number of microorganisms in drinking water, beer, wine, soft drinks, and fruit juices. Industrial filtration methods employ various types of bacteriological filters, including sintered glass, cellulose, and nitrocellulose with pore sizes small enough to trap bacteria. Because of the clogging of filters, only clear liquids can be processed effectively by filtration. Centrifugation can be used to clarify a liquid prior to filtration. Generally, centrifugation will not remove all microorganisms from a food, but the reduction in microbial numbers that can be achieved by centrifugation can extend its shelf life.

High Temperatures

Killing foodborne microorganisms by exposure to high temperatures is a very effective means of preventing food spoilage. The ability to use high-temperature preservation methods for a given food is dependent on the temperature sensitivity of that food. If a food is destroyed by exposure to the high temperatures necessary to kill its associated microorganisms, high-temperature methods obviously cannot be used. High-temperature preservation methods may involve **sterilization**, exposing the food at a given elevated temperature for a period of time sufficient to kill all microorganisms associated with that food, followed by asepsis to prevent recontamination of the food. Canning, for example, uses heat for sterilizing the food and hermetic sealing under anaerobic conditions to prevent spoilage. In this preservation method the high-temperature exposure kills all of the microorganisms in the product, the can or jar acts as a physical barrier to prevent recontamination of the product, and the anaerobic conditions inside prevent oxidation of the biochemicals in the food (Figure 16.3).

Pasteurization Alternatively, exposure to high temperatures may be used merely to reduce the number of microorganisms associated with the food, as long as human pathogens are eliminated. **Pasteurization** uses relatively brief exposures to moderately high temperatures to reduce the number of viable microorganisms and to eliminate human pathogens, but a pasteurized food retains viable microorganisms, which means that additional preservation methods are needed to extend the shelf life of the product. Such a process is normally followed by other preservation methods, often refrigeration, to reduce the growth rates of the surviving microorganisms.

The temperatures and exposure times required for the effective preservation of foods depend on the nature of the food and the heat resistance properties of the microorganisms associated with that food. Pasteurization of milk, for example, is largely aimed at eliminating a limited number of non-spore-forming pathogenic bacteria, namely, *Brucella* species, *Coxiella burnetii*, and *Mycobacterium bovis*, which are associated with the transmission of disease via contaminated milk. These microorganisms are relatively sensitive to elevated temperatures, and the pasteurization of milk is therefore normally achieved by exposure to 62.8°C for 30 minutes (**low temperature–long time or LTH process**) or 71.7°C for 15 seconds (**high temperature–short time or HTST process**). Although milk in the United States is normally preserved by pasteurization and requires refrigeration to extend its shelf life, exposure to 141°C for 2 seconds can be used to sterilize it. Sterilized ultra high-temperature (UHT)

FIGURE 16.3

(A) Illustration showing the steps in a commercial canning procedure. (B) Photographs of a commercial canning operation; the cans are carried along a conveyor belt through the filling and sealing operations and into the retort; in modern automated systems, the speed of the conveyor determines the exposure time of the cans within the retort. (Courtesy American Can Company, Barrington, Ill.).

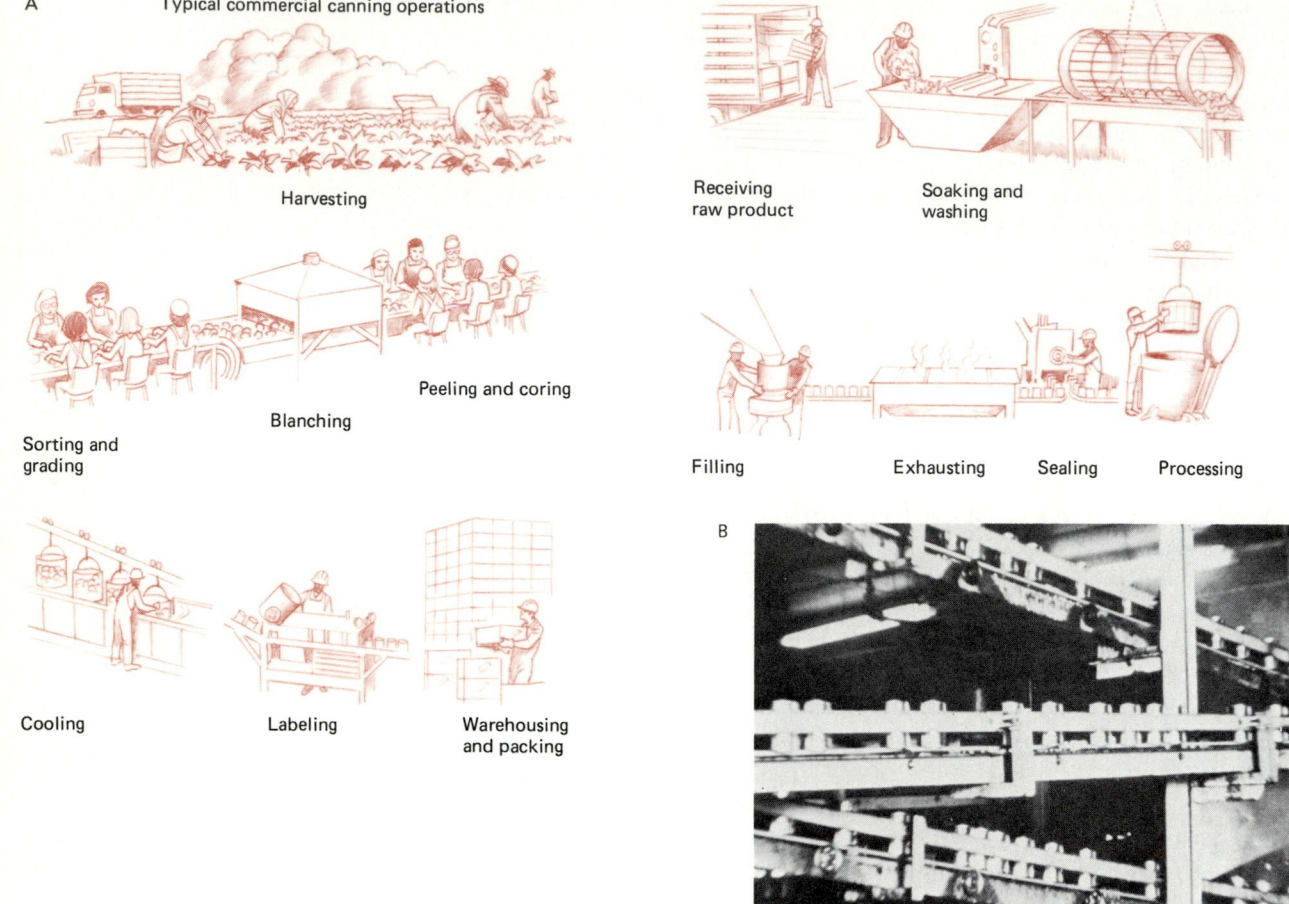

A   Typical commercial canning operations

Harvesting

Receiving raw product

Soaking and washing

Sorting and grading

Blanching

Peeling and coring

Filling   Exhausting   Sealing   Processing

Cooling   Labeling   Warehousing and packing

B

Discovery Process

The method of canning to preserve foods originated in 1795, when the French government offered a prize of 12,000 francs for the development of a practical method of food preservation. In 1809 Francois (Nicolas) Appert succeeded in preserving meats in glass bottles that had been kept in boiling water for varying periods of time. Appert was issued a patent for his process in 1810.

As early as 1820, the commercial production of canned foods was begun in the United States by W. Underwood and T. Kensett. Appert's discovery that foods could be preserved for prolonged periods of time when heated and stored in the absence of oxygen came at the same time that the questions of spontaneous generation and the role of microorganisms in fermentation and putrefaction were being debated by the premier scientists of the day. Spallanzani in 1765 had shown that beef broth that had

been boiled for an hour and sealed did not spoil. Appert applied to home economics the results of Spallanzani's experiments. Not being a scientist, Appert probably did not understand either why his method worked or its long-range significance. It was not until almost a half-century later that Pasteur, in disproving the theory of spontaneous generation, provided the scientific explanation of why Appert's canning method worked. Pasteur pointed out that Appert's method, even when modified by using temperatures below 100°C and relatively short incubation times, was practical for preventing undesirable ferments. Today the method of canning, begun over a century ago in ignorance of why it worked, that is, without the knowledge of its scientific rationale, is one of the most widely used methods for preserving foods.

milk has been marketed in several European countries for several years and has been introduced in the United States, where it has not met with great commercial success except for use as individual servings of coffee creamer. It has a very long shelf life provided that the container remains sealed. After the package is opened, other preservation methods, typically refrigeration, must be used to prevent spoilage.

Canning Unlike pasteurization, **canning** of foods normally involves exposure for longer periods of time to higher temperatures in order to kill endospore-forming microorganisms (Table 16.5). Some heat-canned foods do not have to be sterile because other factors, such as the food's low intrinsic pH or the removal of oxygen, which lower the oxidation/reduction potential, can prevent the growth of microorganisms. The foods should be "commercially sterile," which means that no viable microorganisms can be detected by conventional cultural methods or that the numbers of survivors is too low to be significant under conditions of storage. Exposure to 115°C for 15 minutes is generally considered necessary in home canning to ensure killing of endospore

formers in medium- (pH 4.5–5.3) to low-acid (pH>5.3) foods. Particular concern must be given to ensuring the sterility of such foods because of the possible growth of *Clostridium botulinum* and the seriousness of the disease botulism. Somewhat lower temperatures, for example, exposure to 100°C for 10 minutes, are often employed in home canning of acidic (pH<4.5) foods, in part because of the lowered thermal resistance of microorganisms under acidic conditions and because of the fact that *C. botulinum* is unable to grow at low pH values (Table 16.6). Several commercial canning operations that did not adhere to the necessary standards and whose operations resulted in outbreaks of botulism have been put out of business. Most problems occur with home canning, where lack of care can result in insufficient heating. A particular problem can occur when canning is performed without heating, a process called the **cold-pack method**. This method should not be used for low-acid foods such as meats, vegetables, or other foods in which dangerous microorganisms can grow and produce toxins. The cold-pack method, therefore, is reserved for high-acid foods such as citrus fruits, tomatoes, and peaches.

Determining the appropriate temperature and exposure time for a food preservation process involves establishing thermal death time curves for the most heat-resistant microorganisms that may be present in the food. The *D* value (**decimal reduction time**) is the exposure time at a given temperature needed to reduce the number of viable microorganisms by 90 percent (Figure 16.4). In the canning industry the relatively heat-resistant *Bacillus stearothermophilus* or *Clostridium* PA 3679 is used for determining an acceptable *D* value. The *D* values at different temperatures can be used to construct a thermal death time curve from which the *z* value can be calculated (Figure 16.5). The **thermal death time (TDT)** is the time required at a specific temperature to kill a specific number of microorganisms. The *z* **value** describes the temperature change, in degrees Fahrenheit, needed to change the *D* value (exposure time) by a

TABLE 16.5 Recommended Processing Times for Foods in Number 2 Size Cans

Food	Temperature (°F)		Time (min)
	Initial	Final	
Asparagus spears	120	240	25
Green beans	120	240	20
Lima beans	140	240	35
Baked beans in sauce	140	240	95
Beets in brine	140	240	30
Carrots in brine	140	240	30
Corn, cream style	140	240	105
Corn, whole kernel	140	240	50
Peas in brine	140	240	35
Spinach, whole leaf	140	252	50

TABLE 16.6 Classification of Canned Foods and Their Processing Requirements

Acidity Class	pH	Representative Foods	Spoilage Organisms	Processing
Low acid	7	Ripe olives, eggs, milk, poultry, beef, oysters	Mesophilic *Clostridium* spp.	High temperature (240–250°F)
	6	Beans, peas, carrots, beets, asparagus, potatoes	Thermophiles, plant enzymes	High temperature (240–250°F)
Medium acid	5	Figs, tomato soup, ravioli	Limit of growth of *C. botulinum*	High temperature 240–250°F
Acid	4	Potato salad, pears, peaches, oranges, apricots, tomatoes	Aciduric bacteria	Boiling water (212°F)
		Sauerkraut, apple, pineapple, grapefruit, strawberry	Plant enzymes	boiling water (212°F)
Highly acid	3	Pickles, relish, lemon juice, lime juice	Yeasts and other fungi	Boiling Water (212°F)

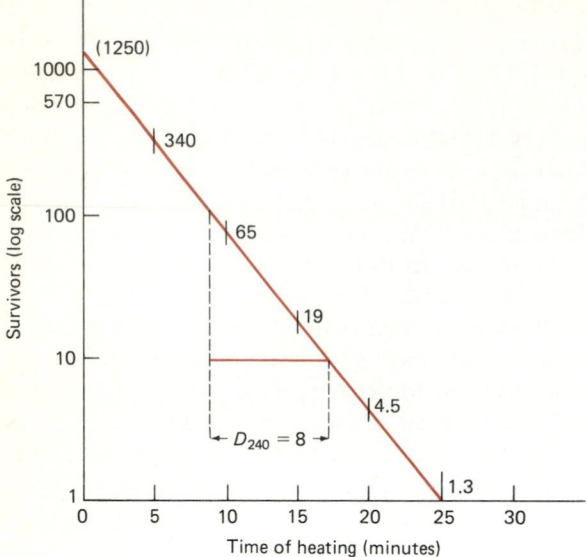

FIGURE 16.4

A survivor curve for endospores heated at 240°F in canned pea brine, pH 6.2. The D value, as determined from this curve, is 8 minutes. This curve establishes the rate of destruction of microorganisms at a given temperature. Obviously, the higher the initial microbial concentration in the food, the longer the heating time needed to reduce the number of survivors to safe levels.

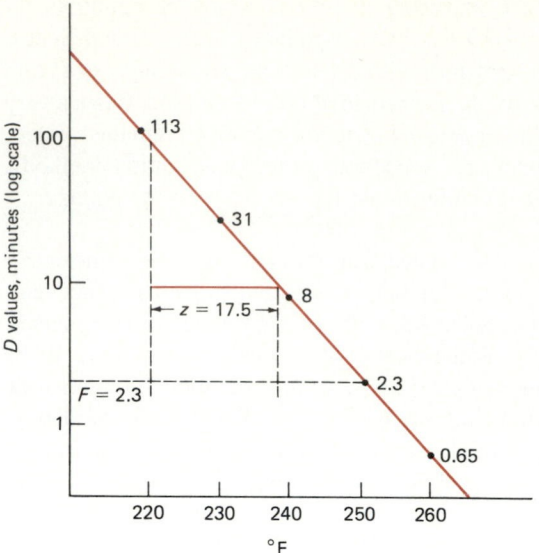

FIGURE 16.5

This diagram illustrates the z and F values for endospores heated at 240°F in canned pea brine, pH 6.2.

TABLE 16.7   Thermal Death Times of Bacterial Cells and Spores

Organism	Time (min)	Temperature (°C)
Vegetative Cells		
Salmonella typhosa	4	60
Staphylococcus aureus	19	60
Escherichia coli	25	60
Lactobacillus bulgaricus	30	71
Endospores		
Bacillus subtilis	18	100
Clostridium botulinum	200	100
Flat sour bacteria	Over 1,000	100

factor of 10 while still achieving the same effective kill of microorganisms. D values can also be used to calculate an **F value**, the exposure time in minutes at 250°F (121°C) needed to kill the spores or vegetative cells of a particular organism, that is, the D value at 250°F (Table 16.7). By knowing the D, F, and z values, the process time required for preservation of a food can be determined. For example, the endospores of *C. botulinum* have an F value of 0.21 minute. The canning industry must employ a $12D$ process for nonacid foods. Heating the food for 2.52 minutes, 12 times the F value, reduces the probability of the survival of *C. botulinum* endospores to 10^{-12}. Thus, if there were one spore in every can, the probability of contamination remaining after processing should be reduced to one in every million million cans. Heating at 121°C for 2.52 minutes therefore should ensure the safety of canned foods with respect to possible contamination with *C. botulinum*. Some endospore-forming microorganisms, such as flat sour bacteria (endospore-forming bacteria that produce acid but little or no gas in canned foods), have F values of 4 to 5 minutes; therefore, heating for longer periods of time is essential for killing flat sour and other resistant bacteria.

The exposure time needed to kill microorganisms varies for different food products. Differences in the required exposure times and temperatures depend on the intrinsic properties of the food, such as pH, and on the numbers and types of microorganisms that are normally associated with that food. For example, whereas milk is pasteurized at 71.7°C

for 15 seconds, pasteurization of an ice cream mix normally requires 82.2°C for 20 seconds. Thus it is necessary to determine the appropriate processing times for each food, even those that seem to be closely related. A general formula for calculating the process time needed to destroy a certain number of microorganisms in a food is $\log t/F = (250 - T)/z$, where t is the time in minutes needed to kill a specified number of test organisms, T is the temperature in degrees Fahrenheit, and z and F represent the thermal death relationship values for the given test microorganism. This formula can be used to determine the appropriate exposure times at a given temperature that are needed to preserve food products. It is also necessary to take into account the heating and cooling times in order to determine the appropriate processing time (Figure 16.6).

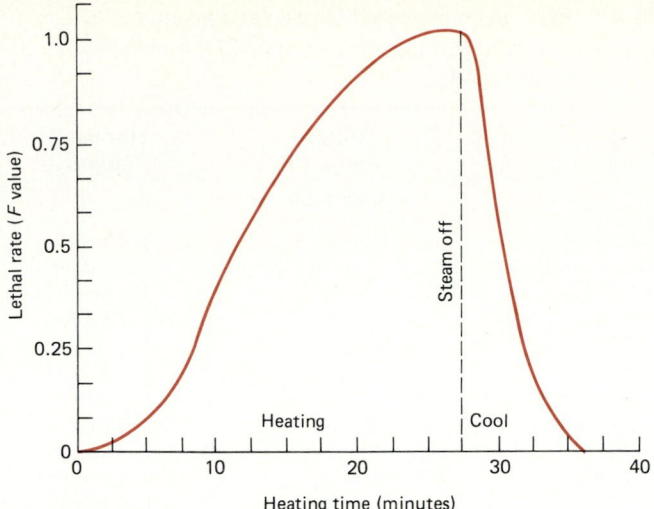

FIGURE 16.6

A lethality curve for a canned-food heat process. The lethal value of the process is described by the area under the curve. The entire heating and cooling time must be considered when establishing the proper process time.

Radiation

Ionizing radiation is used to pasteurize or sterilize some food products. The FDA has approved the use of gamma radiation for the preservation of several foods. Various other countries also employ food irradiation. Radiation exposure can be used to increase the shelf life of various foods, including seafoods, vegetables, and fruits (Table 16.8). In addition to killing spoilage organisms, radiation can inactivate enzymes involved in autocatalytic spoilage. Currently in the United States, spices, fruits, and pork can be preserved by irradiation, and fish and poultry are soon likely to be added to this list. Most sterilization procedures involving exposure

TABLE 16.8 Potential Useful Radiation Dosages for Extending the Shelf Life of Food Products

Product	Dose (krads)	Shelf Life (days)
Fishery Products		
Atlantic haddock fillets	100–250	30–37
Fresh shrimp	100–200	21–28
Pacific cod fillets	100	16–18
Pacific oysters	100	31
King crab meat	100	21
Meats and Poultry		
Fresh meat and poultry	50–100	21
Fruits		
Cherries	250	14–20
Oranges	200	90
Peaches, nectarines	200	14
Pineapples	300	14
Strawberries	200	14–18

to radiation employ gamma radiation from ^{60}Co or ^{137}Ce at levels of 100–200 krads. Exposure to radiation does not leave any residual radioactivity in the food, but the method is still controversial. Some public health authorities have expressed concern that odor-producing spoilage microorganisms may be more readily killed than some pathogenic microorganisms, removing a critical warning signal while leaving the dangerous pathogens that make the food unsafe to consume. Some concern has also been expressed that toxic, nonradioactive chemicals may be formed as a result of exposing foods to ionizing radiation even though the data to date indicate that irradiation can safely be used for the preservation of several foods.

As with thermal processing, D values can be calculated for radiation exposures in order to determine the appropriate radiation dosage level for preserving food products (Figure 16.7). But in contrast to the thermal process, only the total

FIGURE 16.7

D values for radiation treatments of (A) *Clostridium botulinum* endospores and (B) enzymes.

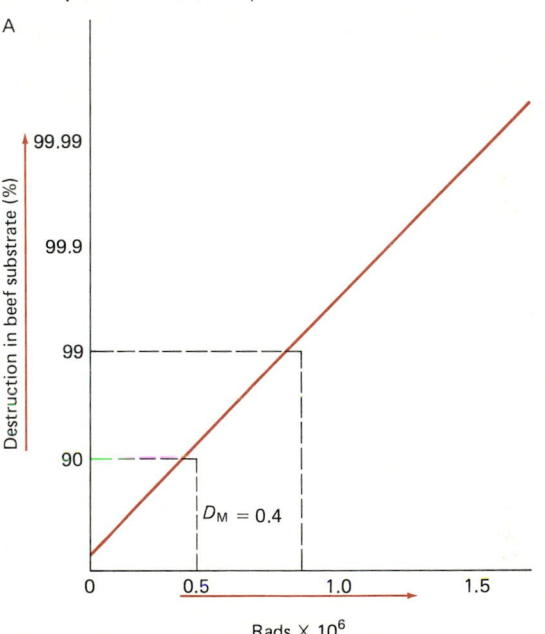

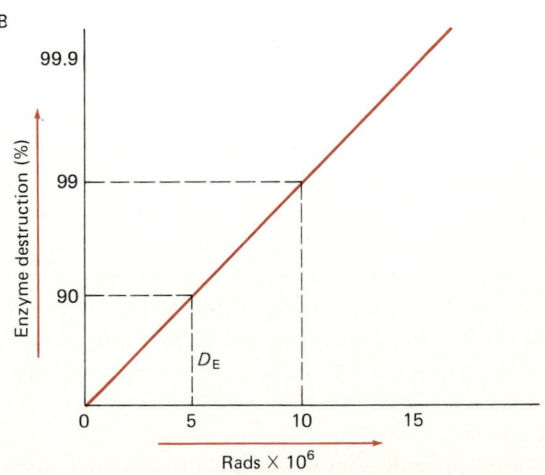

dosage is important, and the exposure time required to reach that dosage is irrelevant. In designing a preservation process that uses radiation, it is necessary to consider the types of microorganisms that may be present in the food and their radiation resistance. As a rule, bacterial spores are more resistant to radiation exposure than vegetative cells, with the exception of *Micrococcus radiodurans*, which is exceptionally resistant to radiation exposure. Developing a radiation preservation process by using *M. radiodurans* as a test organism ensures a margin of safety against spores, including those of *C. botulinum*.

Low Temperatures

Refrigeration and **freezing** are widely used for the preservation of foods (Table 16.9). Low temperatures restrict the rates of growth and enzymatic activities of microorganisms. Though most pathogenic microorganisms are mesophilic and thus unable to grow in refrigerated foods at 5°C, *C. botulinum* type E will grow and produce toxin (Figure 16.8). Psychrophilic and psychrotrophic microorganisms are able to grow slowly at 5°C. Thus, although refrigeration extends the shelf life of the product, it does not do so indefinitely. Also, delicate foods, like fish and chicken, spoil rapidly at refrigerator temperatures. Freezing at temperatures of −20°C or less precludes microbial growth entirely. Not all food products can be preserved by freezing because of the damage that may occur as a result of ice formation. Desiccation of frozen foods, commonly known as **freezer burn**, although not a microbial spoilage process, causes serious quality defects.

Blanching of fruits and vegetables, by scalding with hot water or steam prior to deep freezing, inactivates plant enzymes that may produce toughness, change in color, and loss of flavor and nutritive value. A brief scalding prior to freezing also reduces the number of microorganisms on the food surface by up to 99 percent, enhances the color of green vegetables, and displaces air trapped in food tissues. Commercially frozen foods are often prepared by quick freezing, a process that decreases damage to some food products. **Quick freezing** generally denotes a freezing time of 30 minutes or less, using small units of food. It can be accomplished (1) by direct immersion of food in a refrigerant, as in freezing fish in brine; (2) by indirect contact with the refrigerant, where the food or package is placed in a chamber at −17.8 to −45.6°C; or (3) by air-blast freezing, where frigid air at −17.8 to −34.4°C is blown across the materials being frozen. For **slow or sharp freezing**, where only natural air circulation is used, the temperature usually ranges between −15 and −29°C, and freezing takes 30 to 72 minutes. Quick freezing is usually the superior method because smaller ice crystals are formed, resulting in less destruction of food cells. The shorter solidification period means less diffusion of soluble materials and separation of ice. The prevention of microbial growth also is more prompt, and there is a more rapid slowing of enzyme action.

Freezing does not kill all microorganisms, although some

TABLE 16.9 Recommended Storage Temperatures and Approximate Storage Life of Various Fresh, Dried, and Processed Foods

Food	Storage Temp. (°F)	Storage Life (approx.)
Vegetables		
Asparagus	32	3–4 weeks
Beans (green)	45	8–10 days
Brussel sprouts	32	3–4 weeks
Carrots	32	10–14 days
Cauliflower	32	2–3 weeks
Celery	32	2–4 months
Corn, sweet	32	4–8 days
Cucumbers	47	10–14 days
Lettuce	32	3–4 weeks
Onions	32	6–8 months
Peas, green	32	1–2 weeks
Potatoes, sweet	58	4–6 months
Spinach	32	10–14 days
Squash	35	10–14 days
Tomatoes	32	7 days
Fruits		
Apples	32	2–7 months
Figs	30	5–7 days
Grapefruit	40	4–8 weeks
Lemons	50	1–4 months
Oranges	33	8–12 weeks
Peaches	32	2–4 weeks
Blueberries	32	3–6 weeks
Strawberries	32	7–10 days
Dairy Products		
Butter	34	2 months
Eggs, fresh	30	8–9 months
Eggs, dried	35	6–12 months
Meat, Poultry, Fish		
Beef	32	1–6 weeks
Ham	32	7–12 days
Poultry	32	1 week
Fish	35	5–20 days

microbial death occurs during freezing and during thawing as a result of ice crystal damage to microbial membranes. Therefore, when food is thawed, the surviving microorganisms can grow, leading to food spoilage and potential accumulation of microbial pathogens and toxins if the food is not promptly prepared or consumed. Once food products have been frozen, it is generally not advisable to thaw and refreeze them. Refreezing also alters the texture, flavor, and nutritional qualities of the food, and even though these changes are not related to microbial spoilage, they lower the acceptability of the food. Freezing, thawing, and refreezing generally disrupt the texture of the food, permitting invasion by microorganisms that are normally restricted to food surfaces. When thawed a second time, refrozen food products are more prone to microbial spoilage than are foods that are allowed to thaw only once. This problem can be avoided if proper care is taken during handling to prevent excessive microbial contamination and growth.

Desiccation

The **desiccation** (drying) of foods is a very effective means of preservation, provided that the food does not become moist during storage. The establishment of bacteriostatic environmental conditions is the general principle involved in using **dehydration** to preserve food products. Desiccation need not kill the microorganisms in food products, but microbial growth requires water. Bacteria generally will not grow below an A_w (water activity) level of 0.9, and fungi generally will not grow below an A_w value of 0.65 (Table

TABLE 16.10 Minimum A_w values for the Growth of Various Food Spoilage Fungi

Organism	Minimum A_w
Candida utilis	0.94
Botrytis cinerea	0.93
Rhizopus nigricans	0.93
Mucor spinosus	0.93
Candida scottii	0.92
Trichosporon pullulans	0.91
Candida zeylanoides	0.90
Endomyces vernalis	0.89
Alternaria citri	0.84
Aspergillus glaucus	0.70
Aspergillus echinulatus	0.64

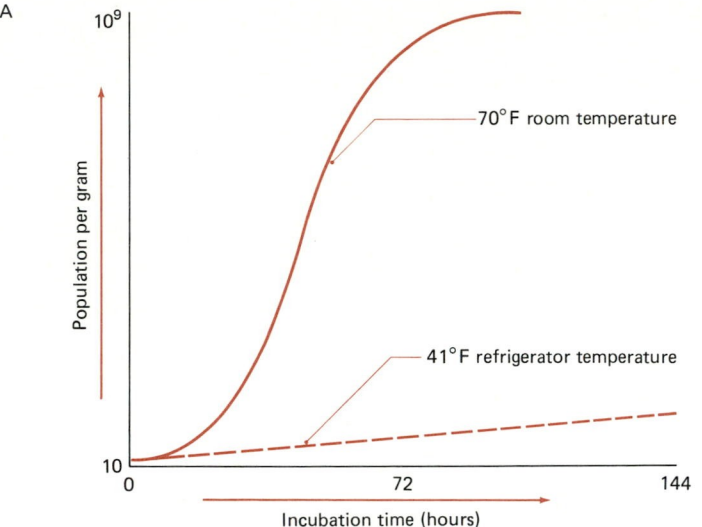

FIGURE 16.8

Temperature–growth relationships of microorganisms. Diagram (A) shows that the growth rate of microorganisms is much lower at refrigerator than at room temperatures; (B) shows that microorganisms grow over a wide range of temperatures. Knowing the temperature growth ranges for organisms, such as those capable of causing spoilage or disease, is important when selecting the proper temperature for the storage of different foods. The safe and growth-supportive temperatures for bacteria are indicated.

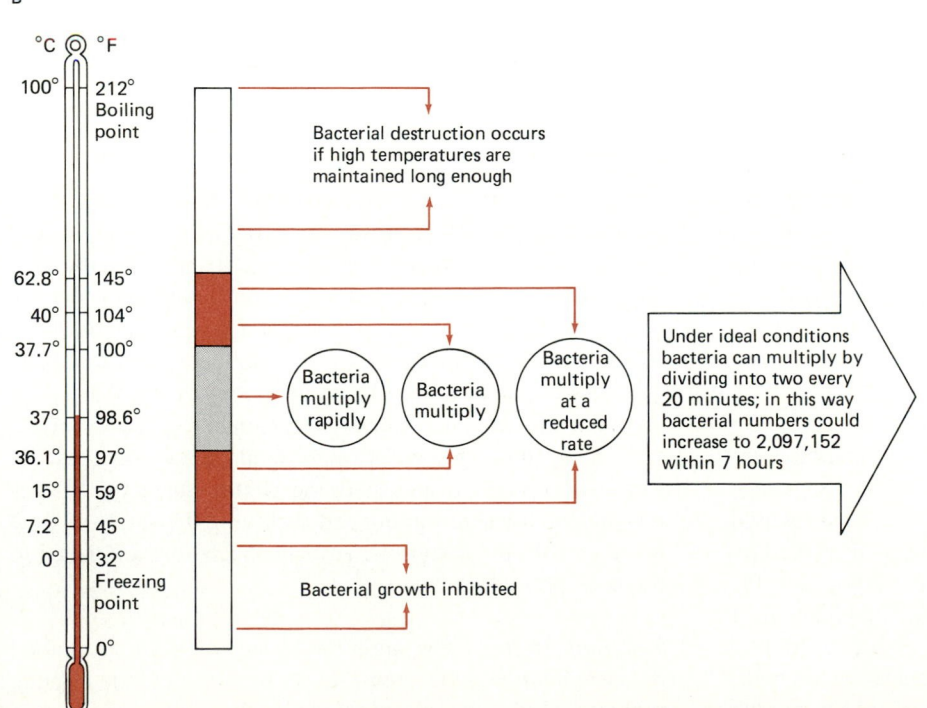

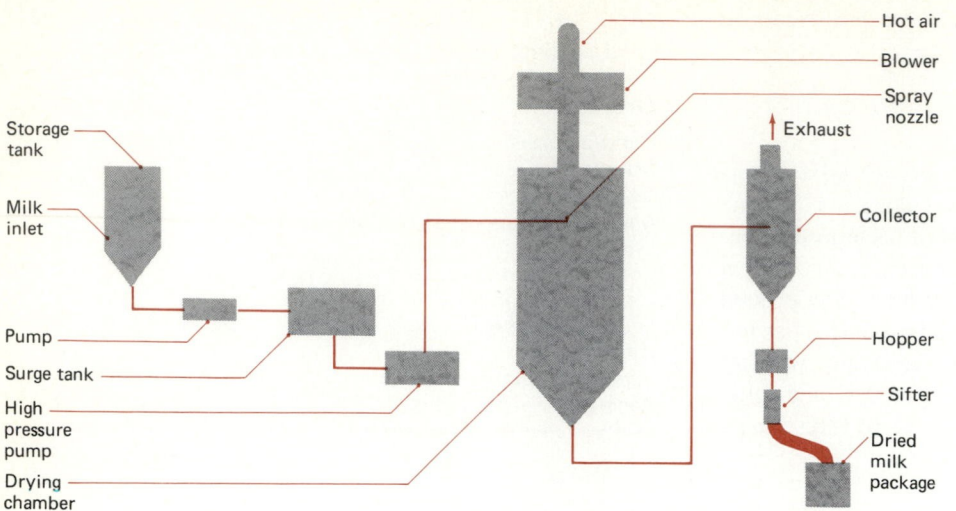

FIGURE 16.9
Diagram of a spray dryer system used in the milk industry to produce powdered milk by dehydration. Desiccated milk has a greatly extended shelf life compared to normal milk.

16.10). If food can be maintained at an A_w value of 0.65 or less, spoilage is unlikely to occur for several years. Products preserved by drying include fruits, vegetables, eggs, cereals, grains, meat, and milk. Drying can be accomplished by many different methods; one example is the ancient natural practice of sun drying, where food is placed on trays and put outdoors with no control of temperature or humidity. This method is limited to warm, dry climates and to fruits such as raisins, prunes, figs, apricots, peaches, and pears. Artificial drying can be accomplished by the passage of heated air with controlled humidity over the food.

Foods can also be dehydrated by the evaporation of their water by using a heated drum dryer, in which the product is passed over a heated drum, with or without a vacuum, or by using a spray dryer (Figure 16.9), in which the product is sprayed into dry, heated air. Spray drying is normally used for the preparation of dried eggs. Milk can be dehydrated by using either the spray drying or heated drum processes. Evaporated milk is prepared by removing 60 percent of the water from whole milk. In powdered milk, over 85 percent of the water is generally removed. Freeze drying is less destructive and yields higher-quality foods than drying at elevated temperatures but is much more expensive. Therefore, this process, which sublimes water directly from frozen foods under a high vacuum, is used only for high-value products, such as meats, camping rations, and coffee.

Anaerobiosis

Packaging of food products under anaerobic conditions, **anaerobiosis**, is effective in preventing aerobic spoilage processes. **Vacuum packing** in an airtight container is used to eliminate air; ground coffee, for example, is preserved by this method. The absence of oxygen prevents not only the growth of aerobic microorganisms but also autoxidation of the food as a result of intrinsic enzymatic activities, greatly increasing product shelf life. Establishing anaerobic conditions, however, also creates ideal conditions for the growth

of obligate anaerobes, such as *C. botulinum*, and therefore is normally used in conjunction with additional preservation methods. For example, in canning, heat sterilization precedes packaging under anaerobic conditions.

Controlled Atmospheres

Controlled atmospheres containing 10 percent CO_2 sometimes are used to preserve stored food products. This method is commonly employed in several countries for the preservation of fruits, such as apples and pears. Carbon dioxide in this concentration retards fungal rotting of fruits. In addition to CO_2, ozone can be added to storage chambers to preserve foods. Because of the high oxidizing activity of ozone, however, this procedure cannot be used with high-lipid foods because it would cause rancidity. These controlled-atmosphere preservation techniques are receiving increased use today in the food industry.

Chemical Preservatives

The use of chemical additives is widespread and represents an important means of preserving foods (Table 16.11). The chemical preservative used depends on the nature of the food and the likely spoilage microorganisms. Although there is great concern today over the addition of any chemicals to foods because of the finding that some chemicals that have been used as food additives, such as red dye number 2, are potential carcinogens, it must be remembered that the effective preservation of food prevents spoilage and the transmission of foodborne diseases. In the United States the FDA is responsible for determining and certifying the safety of food additives and must approve any chemicals that are added to foods as preservatives.

Salt and Sugar   The addition of salt or sugar to a food reduces the amount of available water and alters the osmotic pressure. High salt concentrations, such as occur in saturated

TABLE 16.11   Some Representative Chemical Food Preservatives

Preservatives	Maximum	Target Organisms	Foods
Propionic acid and propionates	0.32%	Fungi	Bread, cakes, some cheese
Sorbic acid and sorbates	0.2%	Fungi	Cheeses, syrups, jellies, cakes
Benzoic acid and benzoates	0.1%	Fungi	Margarine, cider, relishes, soft drinks, catsup
Sulfur dioxide, sulfites, bisulfites, metabisulfites	200–300 ppm	Microorganisms	Dried fruits, grapes, molasses
Ethylene and propylene oxides	700 ppm	Fungi	Spices
Sodium diacetate	0.32%	Fungi	Bread
Sodium nitrite	200 ppm	Bacteria	Cured meats, fish
NaCl	None	Microorganisms	Meats, fish
Sugar	None	Microorganisms	Preserves, jellies
Wood smoke	None	Microorganisms	Meats, fish

brine solutions, are bacteriostatic, and the shrinkage of microorganisms in brine solutions can cause loss of viability. Salting is effectively used for the preservation of fish, meat, and other foods. However, because of the association of high levels of salt in the diet with high blood pressure and heart disease, there is currently great interest in lowering the salt content of foods. Sugars, such as sucrose, also act as preservatives and are effective in preserving fruits, candies, condensed milk, and other foods. Some foods, including maple syrup and honey, are preserved naturally by their high sugar content.

Organic Acids   Various low molecular weight carboxylic acids are inhibitors of microbial growth. Lactic, acetic, propionic, citric, benzoic, and sorbic acids or their salts are effective food preservatives. An examination of the lists of food additives in the various foods in your pantry will rapidly convince you of the wide use of organic acids as preservatives. The effectiveness of a particular **acidulant** (acidic compound) depends on the pH of the food that determines the degree of dissociation of the acid. For example, at the same pH, the order of effectiveness as a preservative is HCl < citric acid < lactic acid < acetic acid.

Propionates are primarily effective against filamentous fungi. The calcium and sodium salts of propionic acid are used as preservatives in bread, cake, and various cheeses, and as propionates are effective inhibitors of rope formation in bread dough and milk. Besides their intentional addition to various foods, propionates form naturally during the production of Swiss cheese and act as a natural preservative.

Lactic and acetic acids also are effective preservatives that form naturally in some food products. Cheeses, pickles, and sauerkraut contain concentrations of lactic acid that normally protect the food against spoilage. Vinegar contains acetic acid, an effective inhibitor of bacterial and fungal growth. Acetic acid is used to pickle meat products and is added as a preservative to various other products, including mayon-naise and catsup. Both of these preservatives, however, will prevent surface fungal growth on a food only if molecular oxygen is excluded.

Benzoates, including sodium benzoate, methyl *p*-hydroxybenzoate (methylparaben), and propyl-*p*-hydroxybenzoate (propylparaben), are extensively used as food preservatives. Benzoates are primarily effective in the undissociated form, and thus, their use generally is restricted to highly acid foods. Benzoates are used as preservatives in such products as fruit juices, jams, jellies, soft drinks, salad dressings, fruit salads, relishes, tomato catsup, and margarine.

Sorbic acid, used primarily as calcium, sodium, or potassium salts (such as sodium sorbate), is most effective in acid foods. It is more effective as a preservative at pH 4–6 than the benzoates. Sorbates inhibit fungi and bacteria, such as *Salmonella*, *Staphylococcus*, and *Steptococcus* species. Sorbates are frequently added as preservatives to cheeses, baked goods, soft drinks, fruit juices, syrups, jellies, jams, dried fruits, margarine, and various other products. Citric acid is used as a preservative in some soft drinks.

Nitrates and Nitrites   Nitrates and nitrites are added to cured meats to preserve the red meat color and to protect against the growth of food spoilage and poisoning microorganisms. Nitrates also effectively inhibit *C. botulinum* in meat products such as bacon and ham. Recently, however, there has been great concern over the addition of nitrates and nitrites to meats because these salts can react with secondary and tertiary amines to form nitrosamines, which are highly carcinogenic.

Sulfur Dioxide, Ethylene Oxide, and Propylene Oxide   Sulfur dioxide and various sulfites have antimicrobial activities. Such fruit products as lemon juice, wine, and dried fruit can be preserved by fumigating with sulfur dioxide and by adding liquid sulfites. Ethylene and propylene oxides are microbicidal and can be used to sterilize food products.

These compounds are used as fumigating agents in the food industry and are primarily applied to dried fruits, nuts, and spices as antifungal agents.

Wood Smoke   Smoking of foods has been traditionally used both to impart a desirable flavor to the food and to deposit preservative chemicals from the smoke on it. This method of food preservation was used by the American Indians, who hung meat on the tops of the teepee poles while fires were lit inside. Smoking is still used today for preserving various meats, such as smoked hams. Wood smoke contains a large number of volatile compounds possessing bacteriostatic and bactericidal activities, including formaldehyde, phenol, cresols, and low molecular weight fatty acids.

Antibiotics   Antibiotics are among the chemicals that have been used as preservatives for certain food products in some countries. Tetracyclines, for example, have been used for preserving poultry and fish products. Concern over the increased occurrence of antibiotic-resistant bacterial strains has caused regulatory agencies in many countries to ban the use of antibiotics as food preservatives.

16.3 Microbiological Production of Food

Although microbial growth is a problem when it results in food spoilage, microorganisms are used beneficially in the food industry for food production. Many of the foods and beverages we commonly enjoy, such as wine and cheese, are the products of microbial enzymatic activity. For the most part, it is the fermentative metabolism of microorganisms that is exploited in the production of food products. The accumulation of fermentation products, such as ethanol and lactic acid, is desirable because of their characteristic flavors and other properties. Only a few processes, such as the production of vinegar, make use of microbial oxidative metabolism. The microbial production of foods can be viewed as an exercise in harnessing microbial biochemistry to produce desired, rather than adverse, changes in food products.

The production of fermented foods requires the proper substrates, microbial populations, and environmental conditions to obtain the desired end product. Quality control is essential in food fermentation to ensure that the product is of high quality. A fermented food may require additional preservation to prevent spoilage because further uncontrolled microbial growth could render it inedible. For example, once wine is produced, it must be maintained under anaerobic conditions in order to prevent its oxidation to vinegar.

The microbial processes used in food production traditionally employ microbial enzymatic activities to transform one food into another, with the microbially produced food product having properties vastly different from those of the starting material. In addition to the use of microorganisms to produce fermented food products, microbial biomass is now considered a potential source of protein for meeting the food needs of an expanding world population. Some microorganisms, such as mushrooms, have been used as food products for centuries. The growth of bacteria, algae, and fungi as proteinacious food, however, is a relatively new concept. Microbial biomass can be used as an animal feed supplement or may be developed as a direct source of protein for human consumption.

Fermented Dairy Products

Numerous products are made by the microbial fermentation of milk, including buttermilk, yogurt, and many cheeses. The fermentation of milk is primarily carried out by lactic acid bacteria. The lactic acid fermentation pathway and the accumulation of lactic acid from the metabolism of the milk sugar lactose are common to the production of fermented dairy products. The accumulated lactic acid in these products acts as a natural preservative. The differences in the flavor and aroma of the various fermented dairy products are due to additional fermentation products that may be present in only relatively low concentrations.

Buttermilk, Sour Cream, Kefir, and Koumis   Different fermented dairy products are produced by using different strains of lactic acid bacteria as starter cultures and different fractions of whole milk as the starting substrate (Table 16.12). Sour cream, for example, uses *Streptococcus cremoris* or *S. lactis* for the production of lactic acid, and *Leuconostoc cremoris* or *S. lactis diacetilactis* for the production of the characteristic flavor compounds. Cream is the starting substrate for this product. If skim milk is used as the starting material, cultured buttermilk is produced. Bulgarian buttermilk is made by using *Lactobacillus bulgaricus* for the production of both lactic acid and flavor compounds. Butter is normally made by churning cream that has been soured by lactic acid bacteria. *S. cremoris* or *S. lactis* is used to produce lactic acid rapidly, and *Leuconostoc citrovorum* produces the necessary flavor compounds. The *Leuconostoc* enzymes attack citrate in milk, producing diacetyl, which gives butter its characteristic flavor and aroma. Kefir and koumis, which are popular in some European countries, are fermentation products of *S. lactis*, *S. cremoris*, other *Lactobacillus* species, and yeasts. Lactic acid, ethanol, and carbon dioxide are formed during the fermentation and give these products their characteristic flavors.

TABLE 16.12   Some Foods Produced From Fermented Milks

Fermented Product	Microorganisms Responsible for Fermentation	Description
Sour cream	*Streptococcus* sp. *Leuconostoc* sp.	Cream is inoculated and incubated until the desired acidity develops.
Cultured buttermilk	*Streptococcus* sp. *Leuconostoc* sp.	Made with skimmed or partly skimmed pasteurized milk.
Bulgarian buttermilk	*Lactobacillus bulgaricus*	Product differs from commercial buttermilk in having higher acidity and lacking aroma.
Acidophilus milk	*Lactobacillus acidophilus*	Milk for propagation of *L. acidophilus* and the milk to be fermented are sterilized and then inoculated with *L. acidophilus*. This milk product is used for its medicinal therapeutic value.
Yogurt	*Streptococcus thermophilus,* *Lactobacillus bulgaricus*	Made from milk in which solids are concentrated by evaporation of some water and addition of skim milk solids. Product has consistency resembling custard.
Kefir	*Streptococcus lactis,* *Lactobacillus bulgaricus,* yeasts	A mixed lactic acid and alcoholic fermentation; bacteria produce acid, and yeasts produce alcohol.

Yogurt   Over 550,000 pounds of yogurt are produced annually in the United States. Yogurt is made by fermenting milk with a mixture of *L. bulgaricus* and *S. thermophilus*. Yogurt fermentation is carried out at 40°C. The characteristic flavor of yogurt is due to the accumulation of lactic acid and acetaldehyde produced by *L. bulgaricus*. Because of the tart taste of acetaldehyde, most yogurt produced in the United States is flavored by adding fruit.

Cheese   A wide variety of cheeses are produced by microbial fermentation. Cheeses consist of milk curds that have been separated from the liquid portion of the milk (whey). The curdling of milk is accomplished by using the enzyme rennin (casein coagulase or chymosin) and lactic acid bacterial starter cultures. Rennin is obtained from calf stomachs or by microbial production. Cheeses are classified as (1) soft if they have a high water content (50–80 percent); (2) semi-hard if the water content is about 45 percent; and (3) hard if they have a low water content (less than 40 percent). Cheeses are also classified as unripened if they are produced by single-step fermentation or as ripened if additional microbial growth is required during maturation of the cheese to achieve the desired taste, texture, and aroma (Table 16.13). Processed cheeses are made by blending various cheeses to achieve a desired product. If the water content is elevated during processing, thereby diluting the nutritive content of the product, the product is called a *processed food* rather than a cheese.

The natural production of cheeses involves lactic acid fermentation, with various mixtures of *Streptococcus* and *Lactobacillus* species used as starter cultures to initiate the fermentation. The flavors of different cheeses result from the use of different microbial starter cultures, varying incubation times and conditions, and the inclusion or omission of secondary microbial species late in the fermentation process.

Ripening of cheeses involves additional enzymatic transformations after the formation of the cheese curd, using enzymes produced by lactic acid bacteria or enzymes from other sources. Unripened cheeses do not require the additional enzymatic transformations. Cottage cheese and cream cheese are produced by using a starter culture similar to the one used for the production of cultured buttermilk and are soft cheeses that do not require ripening. Sometimes a cheese is soaked in brine to encourage the development of selected bacterial and fungal populations during ripening. Limburger is a soft cheese produced in this manner. During ripening the curds are softened by proteolytic and lipolytic enzymes, and the cheese acquires its characteristic aroma. The production of Parmesan cheese also involves brine curing.

Swiss cheese formation involves a late propionic acid fermentation, with ripening accomplished by *Propionibacterium shermanii* and *P. freudenreichii*. The propionic acid yields the characteristic aroma and flavor, and the carbon dioxide produced during this late fermentation forms the holes in Swiss cheese. Various fungi are also used in the ripening of different cheeses. The unripened cheese is normally inoculated with fungal spores and incubated in a warm, moist room to promote the growth of filamentous fungi. For example, blue cheeses are produced by using *Penicillium* species. Roquefort cheese is produced by using *P. roqueforti*, and camembert and brie by using *P. camemberti* and *P. candidum*.

Fermented Meats

Several types of sausage, such as Lebanon bologna, the salamis, and the dry and semidry summer sausages, are produced by allowing the meat to undergo heterolactic acid fermentation during curing. The fermentation has a preservative effect and also adds a tangy flavor to the meat. Various lactic

TABLE 16.13   The Classification of Some Cheeses

Cheese	Microorganisms		
Soft, unripened			
Cottage	*Streptococcus lactis*	*Leuconostoc citrovorum*	
Cream	*Streptococcus cremoris*		
Neufchatel	*Streptococcus diacetilactis*		
Soft, ripened, 1–5 months			
Brie	*Streptococcus lactis*	*Penicillium camemberti*	*Brevibacterium linens*
	Streptococcus cremoris	*Penicillium candidum*	
Camembert	*Streptococcus lactis*	*Penicillium camemberti*	
	Streptococcus cremoris	*Penicillium candidum*	
Limburger	*Streptococcus lactis*	*Brevibacterium linens*	
	Streptococcus cremoris		
Semisoft, ripened, 1–12 months			
Blue	*Streptococcus lactis*	*Penicillium roqueforti*	
	Streptococcus cremoris	*Penicillium glaucum*	
Brick	*Streptococcus lactis*	*Brevibacterium linens*	
	Streptococcus cremoris		
Gorgonzola	*Streptococcus lactis*	*Penicillium roqueforti*	
	Streptococcus cremoris	*Penicillium glaucum*	
Monterey	*Streptococcus lactis*		
	Streptococcus cremoris		
Muenster	*Streptococcus lactis*	*Brevibacterium linens*	
	Streptococcus cremoris		
Roquefort	*Streptococcus lactis*	*Penicillium roqueforti*	
	Streptococcus cremoris	*Penicillium glaucum*	
Hard, ripened, 3–12 months			
Cheddar	*Streptococcus lactis*	*Lactobacillus casei*	
	Streptococcus cremoris		
	Streptococcus durans		
Colby	*Streptococcus lactics*	*Lactobacillus casei*	
	Streptococcus cremoris		
	Streptococcus durans		

TABLE 16.13 (Continued)

Cheese	Microorganisms		
Edam	Streptococcus lactis Streptococcus cremoris		
Gouda	Streptococcus lactis Streptococcus cremoris		
Gruyere	Streptococcus lactis Steptococcus thermophilus	Lactobacillus helveticus	Propionibacterium shermanii or Lactobacillus bulgaricus and Propionibacterium freudenreichi
Swiss	Streptococcus lactis Streptococcus thermophilus	Lactobacillus helveticus	Propionibacterium shermanii or Lactobacillus bulgaricus and Propionibacterium freudenreichi
Very hard, ripened, 12–16 months			
Parmesan	Streptococcus lactis Streptotoccus cremoris Streptococcus thermophilus	Lactobacillus bulgaricus	
Romano	Lactobacillus bulgaricus	Streptococcus thermophilus	

acid bacteria are normally involved in the fermentation, but *Pediococcus cerevisiae* can be used for controlled production of these types of meats.

Leavening of Bread

Yeasts are added to bread dough to ferment the sugar, producing the carbon dioxide that leavens the dough and causes it to rise. The principal yeast used in bread baking is *Saccharomyces cerevisiae*, known as **bakers' yeast**. Bakers' yeast is produced in large quantities for the baking industry (Figure 16.10). The yeast is normally grown in a molasses-mineral salts medium at a pH of 4.3–4.5 and temperature of 30°C, with the molasses substrate added gradually to maintain a sugar concentration of 0.5–1.5 percent. The concentration of sugar in the fermentor is critical, as too high a concentration represses respiratory enzymes and alcohol production even under highly aerobic conditions. The yeasts are generally collected by centrifugation and pressed through a filter to remove excess liquid. For the baking industry, the yeasts are normally formed into cakes or are dried further to form active dry yeast containing less than 8 percent water and are frequently used for home baking purposes.

In the baking process, the yeast is used strictly as a source of enzymes to carry out alcoholic fermentation. The yeast does not grow during the first 2 hours after addition to the dough, by which time the leavening process is normally completed. Amylases in the dough convert starch to sugars, and the yeasts metabolize the sugars that are formed, producing carbon dioxide and ethanol. Besides *S. cerevisiae*, various other microorganisms, including coliform bacteria and *Clostridium* species, can be employed for leavening bread. The microorganisms used for bread leavening must produce carbon dioxide from the fermentation of sugars to be useful.

In modern home and commercial baking processes, excess amounts of yeast are normally added so that the fermentation time is very short. Older, more traditional bread-making processes use less yeast and longer fermentation times. However, when the processing time exceeds 2 hours, there can be an undesirable growth of fungi and bacteria. During fermentation the dough becomes conditioned as a result of the action of proteases on the flour protein, gluten. Enzymes are produced by the yeasts or may be added from other sources. As a result of this conditioning, the gluten matures, becoming elastic and capable of retaining the carbon dioxide gas produced by the yeasts during the fermentation. Sugar, or

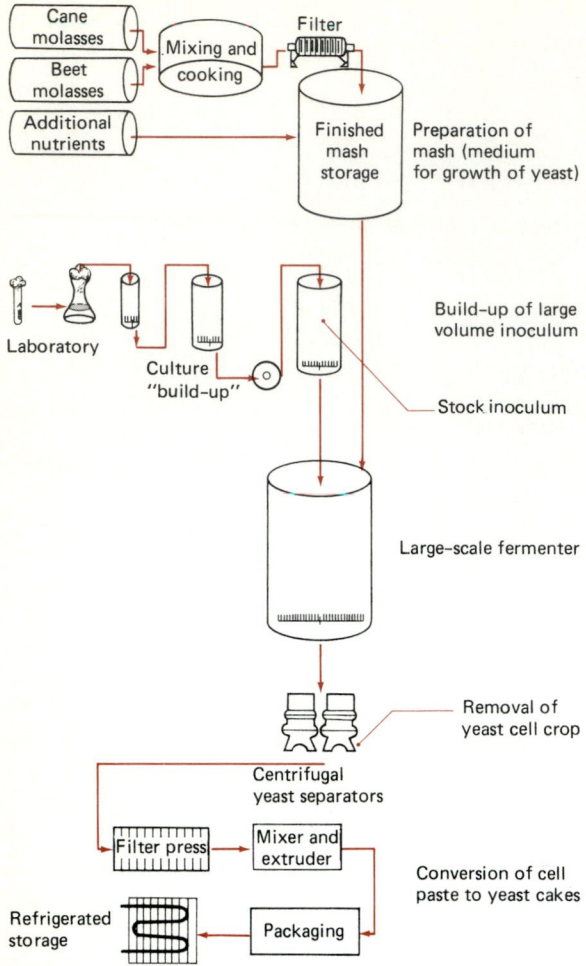

FIGURE 16.10
The steps in the commercial production of bakers' yeast.

amylase to convert starches to sugar, is normally added to the flour in order to increase the rate of gas production by the yeast. Addition of increased amounts of yeast and various salts to support yeast metabolism also increases the rate of gas production. The leavening process is normally carried out at 27°C, which is optimal for fermentation. Too high or too low a temperature can result in reduced rates of gas production.

After leavening, the bread is baked. Carbon dioxide bubbles are trapped in the dough and give rise to the honeycomb texture and increased volume of the baked bread. Although the interior of the bread does not reach 100°C, the heating is sufficient to kill the yeasts, inactivate their enzymes, expand the gas, evaporate the ethanol produced during the fermentation, and establish the structure of the bread loaf. During the baking there is also a gelatinization of the starch, which results in setting of the bread. In the dough the gluten gives structural support, but in the baked bread the structural support comes from the gelatinized starch.

In addition to leavening bread, microbes produce the characteristic flavors of some breads. For example, the production of San Francisco sour dough bread utilizes the yeast *Torulopsis holmii* and a heterofermentative *Lactobacillus* species to sour the dough and give this bread its characteristic sour flavor. Rye bread is also produced by initially souring the dough; cultures of *Lactobacillus plantarum*, *L. brevis*, *L. bulgaricus*, *Leuconoscoc mesenteriodes*, and *Streptococcus thermophilus* are employed as starter cultures in making different rye breads. The action of heterofermentative lactic acid bacteria produces the bread's characteristic flavor.

Alcoholic Beverages

Microorganisms, principally yeasts in the genus *Saccharomyces*, are used to produce various types of alcoholic beverages. The production of alcoholic beverages relies on **alcoholic fermentation**, that is, the conversion of sugar to alcohol by microbial enzymes. The flavor and other characteristic differences between various types of alcoholic beverages reflect differences in the starting substrates and the production process, rather than differences in the microbial culture or the primary fermentation pathways employed in the production of alcoholic beverages.

Beer and Ale   Beer is a very popular beverage with a high per capita consumption rate. The worldwide production of beer is over 18 billion gallons per year. Beer and ale are malt beverages, so named because the initial preparation of the substrate for microbial fermentation involves barley malt and the production of beer begins with the **malting** of the barley (Figure 16.11). Malt contains a mixture of amylases and proteinases prepared by germinating barley grains for about a week and crushing the grains to release the plant enzymes. Some beers, particularly those produced in Europe, are prepared entirely from malted barley. In the production of most beers, however, the malt is added to adjuncts in a process known as **mashing**. The malt **adjuncts**, such as corn, rice and wheat, provide carbohydrate substrates for ethanol production. During the mashing process, the amylases from the barley malt hydrolyze the starches and other polysaccharides, as well as the proteins in the malt adjunct. The mash is heated to reach temperatures of about 70°C, which facilitate the rapid enzymatic conversion of starch to sugars. The insoluble materials are allowed to settle from the mash and serve as a filter. The clear liquid that is produced in this process is called **wort**.

The wort is then cooked with hops, the dried flowers of the hop plant. This cooking concentrates the mixture, inactivates the enzymes, extracts soluble flavoring compounds from the hops, and greatly reduces the number of microorganisms prior to the fermentation process. Additionally, compounds in the hops extract, principally resins such as humulone, have antibacterial properties and protect the wort from the undesirable growth of Gram-positive bacteria that could sour the beer.

The fermentation of wort to produce beer in most countries is carried out by the yeast *Saccharomyces carlsbergensis*, a

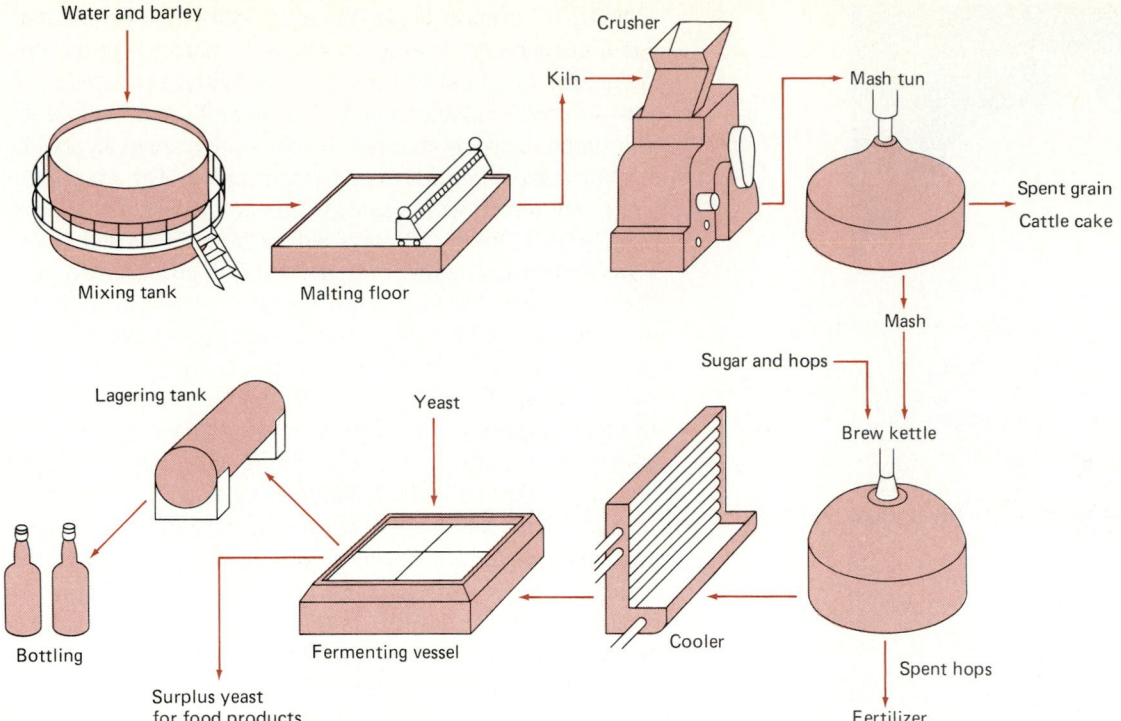

FIGURE 16.11

The steps in the brewing of beer. The production of beer begins with the malting of barley, in which the grain is induced to sprout, producing enzymes that will catalyze the breakdown of starch. The malt is ground and mixed with warm water, and often other cereals such as corn, before going to the mash tun, where over a period of a few hours enzymes break down the long chains of starch into smaller molecules of carbohydrate. The aqueous extract wort is separated from the mix and boiled with hops in a brew kettle. The boiling extracts flavors from the hops and stops the enzyme action in the wort. The hops are removed and the wort is put in a fermenting vessel, where it is pitched, or seeded, with yeast. After fermentation the beer may go to a lagering tank to mature; then it is pasteurized and bottled.

bottom fermenter. This means that at a late stage in fermentation, the yeasts flocculate or aggregate and settle, partially clarifying the beer. *S. cerevisiae* is also sometimes used in beer production, particularly in Great Britain and parts of the United States, but it is a top yeast and rises to the surface during fermentation. The inoculation of the yeast into the cooled wort, known as **pitching**, uses a heavy inoculum, of about 1 pound yeast per barrel of beer. The wort is initially aerated to facilitate reproduction of the yeast but is then allowed to become anaerobic, promoting the fermentative production of alcohol and carbon dioxide.

Usually the fermentation process is carried out at low temperatures and may take 1 to 2 weeks to reach completion. During fermentation the yeasts convert the sugars in the wort to alcohol and carbon dioxide and also produce small amounts of glycerol and acetic acid from the fermentation of the carbohydrates. Proteins and lipids are converted to small amounts of higher alcohols, acids, and esters, which contribute to the flavor of beer. The active fermentation process is accompanied by extensive foaming of the mixture because of the production of carbon dioxide (Figure 16.12). The

product is then known as a *green beer* and requires aging to achieve the characteristic flavor and aroma of the finished product.

The commercial production of beer is usually a **batch process**, in which the substrates and inoculum are added to a brewing kettle. When the fermentation is completed, the products are collected as a single batch. In some countries the production of beer is carried out in a **continuous flow-through process**, in which fresh substrate is continuously or periodically added to the fermentation and product is continuously collected. This production process is analogous to the operation of a chemostat.

During the aging process, precipitation of proteins, yeasts, and resins occurs, resulting in a mellowing of the flavor. The mature beer is removed and filtered. The finished product is carbonated to achieve a carbon dioxide content of 0.45–0.52 percent. In the commercial production of beer, the carbon dioxide is normally collected during the fermentation phase and reinjected during the finishing process. In home production of beer, a small additional amount of sugar is usually added to each bottle to permit limited additional fermentative

FIGURE 16.12
Photograph showing beer being produced. The foaming is due to the rapid action of the yeast, which produces carbon dioxide as well as alcohol. In the traditional European manner, the beer is carefully inspected by the brewmaster at all stages of fermentation to ensure production of a quality product. (Reprinted by permission of Quarto, Ltd., from M. Jackson, 1977, *The World Guide to Beer*)

production of CO_2, achieving carbonation within the bottle. Most bottled or canned beers are pasteurized at 60–61°C or filtered to remove viable yeasts. Commercially produced beer in the United States has an alcohol content of about 3.8 percent. In Canada it is 5 percent.

In addition to normal beer, there are a number of other malt beverages. **Light beers** are low carbohydrate beers produced by using a wort prehydrolyzed with fungal glucoamylases and amylases. The prehydrolysis of dextrin in the wort to maltose and glucose permits the yeasts to ferment the carbohydrates completely to alcohol and carbon dioxide, greatly reducing the concentration of residual carbohydrates in the beer. These low-calorie beers are particularly popular today for those who consume large amounts of beer and don't wish to develop a "beer belly."

Ale is produced by using *Saccharomyces cerevisiae*. The fermentation is carried out at temperatures of 12–25°C, permitting the fermentation to reach completion in only 5 to 7 days. The yeast cells are carried upward with the carbon dioxide, and excess cells are skimmed off the top during the fermentation period. A higher concentration of hops is used in the production of ale than in beer, contributing to the particularly tart taste of ale; some ales have higher alcohol concentrations than most beers.

Saki, a Japanese beverage, is a yellow rice beer. Its alcohol concentration normally is 14–17 percent. In the production of saki, a starter culture—normally *Aspergillus oryzae*—is used as a source of fungal enzymes to hydrolyze the rice starch to sugars that can then be converted to alcohol by *Saccharomyces* species during fermentation. The *Aspergillus* spores are mixed with steamed rice and incubated at approximately 35°C for 5 to 6 days before inoculation with yeast. The yeast fermentation of the rice mash to produce saki takes several weeks. **Sonti**, a similar product produced in India, uses the fungus *Rhizopus sonti* to convert the rice starch to sugars, which are subsequently fermented by yeasts.

Distilled Liquors   **Distilled liquors** or spirits are produced in a manner similar to that of beer, except that after the fermentation process the alcohol is collected by distillation, permitting the production of beverages with much higher alcohol concentrations than could be achieved during the fermentation process (Figure 16.13). The initial steps in the production of distilled spirits are analogous to those in beer production, beginning with a mashing process in which the polysaccharides and proteins in a starting plant material are converted to sugars and other simple organic compounds that can be readily fermented by yeasts to form alcohol.

Various starting plant materials are used for the production of different distilled liquor products. Rum is produced by using sugar cane syrup or molasses as the initial substrate; rye whiskey is produced from the fermentation of a rye mash; bourbon or corn whiskey uses corn mash; and brandy comes from the fermentation of grapes. The yeasts used in the production of distilled liquors typically are special distiller strains of *S. cerevisiae*, which yield relatively high concentrations of alcohol. The yeasts produced during fermentation are collected, dried, and used as animal feed. The mash is sometimes soured prior to the yeast fermentation process by allowing lactic acid fermentation to occur initially in order to prevent the growth of undesired microorganisms that might interfere with the fermentative action of the yeast.

The alcoholic product formed from the fermentation of wort, known as a *beer* or *wine*, is heated in a still and alcohol is collected. In addition to alcohol, various volatile organic compounds, fusel oils, are collected with the distillate and contribute to the characteristic flavors of the different distilled liquor products. Distilled products also differ from one another in the nature of the distillation process; Scotch whiskey, for example, is distilled in batches by using small pot stills, whereas many other distilled whiskeys are produced by using continuous distillation processes. The distilled alcohol product is normally aged to yield a mellow-tasting alcoholic beverage.

Wines   Wine is fermented primarily from grapes, although other fruits are sometimes used. Red wines are produced by using whole red grapes, whereas white wines are made from white grapes or from red grapes that have had their skins removed. The production of wine begins when the grapes

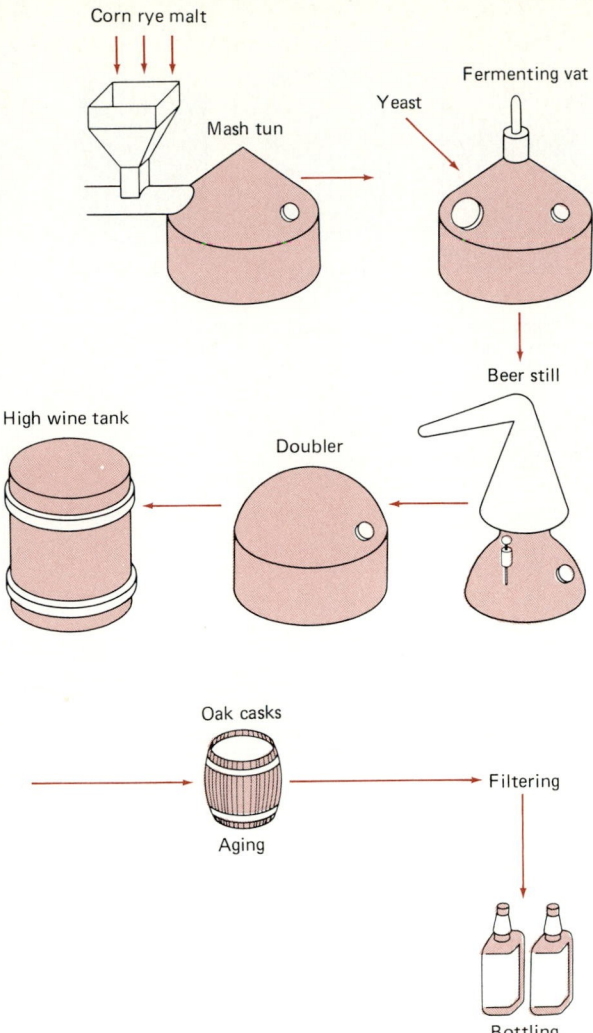

FIGURE 16.13

The process for producing distilled spirits. Distilled spirits, such as whiskey, are made in a process much like the one used to brew beer. Bourbon production is illustrated here. Grains of corn are mixed with smaller amounts of rye and malted barley, crushed, and mixed with warm water. The wort that emerges from the mash is transferred to a fermentor and pitched with yeast. After fermentation the beer is conveyed to a unit consisting of a beer still and doubler. The condensate is collected in a high wine tank and then matured for several years in oak casks before bottling.

a product of consistent quality can be produced. Initially, the grape must and yeasts are stirred to increase aeration and permit the proliferation of the yeasts. The mixing is later discontinued, permitting anaerobic conditions to occur that favor the production of alcohol. The sugar content of the grapes and the alcohol tolerance limit of the yeasts determine the final ethanol concentration. The sugar content of the grapes depends on the grape variety and its ripeness and varies from season to season, accounting in part for the fact that some years and vineyards are better than others for the production of quality wines.

The fermentation of red wines typically is carried out at 24–27°C for 3 to 5 days, and white wines take 7 to 14 days at 10–21°C. During fermentation wine is periodically **racked**, that is, it is filtered through the bottom sediments and added back to the top of the fermentation vat. Carbon dioxide produced during the fermentation process forces the skins and other debris to the surface. The color of red wine is due to extraction of the pigments from the grape skin by the alcohol produced during the fermentation. At the end of fermentation, wines typically have an alcohol content of 11–16 percent by volume. They are then aged to achieve their final bouquet and essence of flavor. During aging, some fermentation of the malic acid of grape juice is carried out by lactobacilli (malolactic fermentation), reducing the acidity of the wine.

By using similar processes, a variety of wines can be produced. Dry wines contain little or no sugar, whereas sweet wines contain some residual unfermented sugar. Distillation is used to achieve the high alcohol content—19–21 percent—of fortified or dessert wines. Normally, the carbon dioxide produced during alcoholic fermentation is allowed to escape and the wine is, therefore, still. In the case of champagne and other sparkling wines, however, the carbonation is essential. In some commercially produced champagne, carbon dioxide is reinjected into the wine after fermentation. In the classic French method of producing champagne, the wine is fermented in the bottle. After fermentation is complete, the bottles are inverted, and the yeast sediments into the neck of the specially shaped champagne bottles. The yeasts are frozen and removed as a plug without excessive loss of carbon dioxide. Wines stoppered with a cork must be stored on their side to prevent the cork from drying out, which would permit air to enter and allow the alcohol to be oxidized by bacteria to form acetic acid. The spoilage of wines, with the formation of vinegar, sour wine, is a serious problem. In the United States, most wine bottles are sealed with a plastic stopper and therefore need not be stored on their side to preclude the souring of the wine.

Vinegar

The production of **vinegar** involves an initial anaerobic fermentation to convert carbohydrates by *S. cerevisiae* to alcohol, followed by a secondary oxidative transformation of the alcohol to form acetic acid by *Acetobacter* and *Glucon-*

are crushed to form a juice or **must** (Figure 16.14). In the classic European method of wine production, wild yeasts from the surface skins of the grapes are the only inoculum for the fermentation. In modern wine production, however, the natural microbiota associated with the grapes are inactivated by sulfur dioxide fumigation or by the addition of metabisulfite so that the "wild microorganisms" do not compete with the defined yeast strains used to ferment the grapes in this process. The grape must is then inoculated with a specific strain of yeast, normally a variety of *S. cerevisiae*. By using specific yeast strains and controlled fermentation conditions,

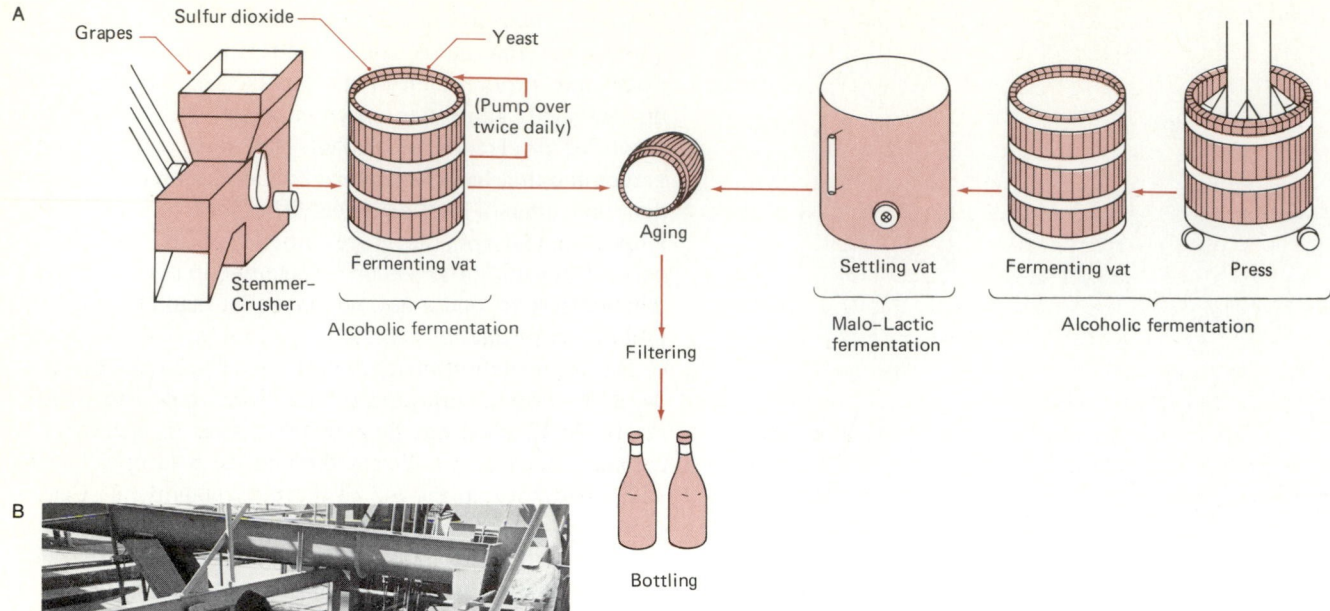

FIGURE 16.14

The wine production process. (A) Diagram of the batch process, the commonest way of making wine, although some of the cheaper wines are manufactured by a process of continuous fermentation in which grape juice is fed steadily into a fermenting stage and wine is steadily removed. The processes that produce red and white wines differ somewhat. (B) The commercial presses gently extract as much juice as possible without breaking the seeds. The corkscrew conveyor brings the grape must from the crusher. (C) Wine is aged in casks before bottling. (D) Bottles are filled mechanically under careful quality control. (B–D Courtesy The Wine Institute, San Francisco)

obacter. The starting materials for the production of vinegar may be fruits, such as grapes, oranges, apples, pears; vegetables, such as potatoes; malted cereals, such as barley, rye, wheat, and corn; and sugary syrups such as molasses, honey, and maple syrup. The type of vinegar is determined by the starting material. For example, wine vinegar comes from grapes and cider vinegar from other fruits. The history of the commercial production of vinegar shows an interesting progression in fermentor design to accomplish the necessary transfer of oxygen to the bacteria, and we will examine some of the types of vinegar generators that have been developed.

In slow methods for the production of vinegar—still used in some small European operations—an initial natural alcoholic fermentation achieves an alcohol concentration of 11–13 percent. After production of the alcoholic liquid, acetic bacteria are seeded into the solution and allowed to convert the alcohol slowly to acetic acid. In the **Orleans process** for producing vinegar, a barrel is filled about one-fourth full with raw vinegar from a previous run to provide the active inoculum. A wine, hard cider, or malt liquor is then added as a substrate. Sufficient air is left in the barrel to permit oxidative metabolism, acetic acid bacteria grow as a film on the top of the liquid, and the conversion of alcohol to acetic acid takes several weeks to several months to complete at 21–29°C. The rate of vinegar production is limited primarily by the transfer of oxygen.

To increase the rate of acetic acid production, a vinegar generator can be used in which the alcohol-containing liquid is trickled over a surface film of acetic acid bacteria (Figure 16.15). In a typical vinegar generator, the acetic acid bacteria are maintained as a film on wood chips. The alcohol liquid is sprinkled over the wood chips, and during the slow tric-

FIGURE 16.15

The problem in the production of vinegar consists of supplying the needed oxygen and substrate to the acetic acid bacteria. In the classic vinegar generator the bacteria develop as a film on wood chips, and the aerated liquid containing the ethanol substrate is dripped over the bacteria.

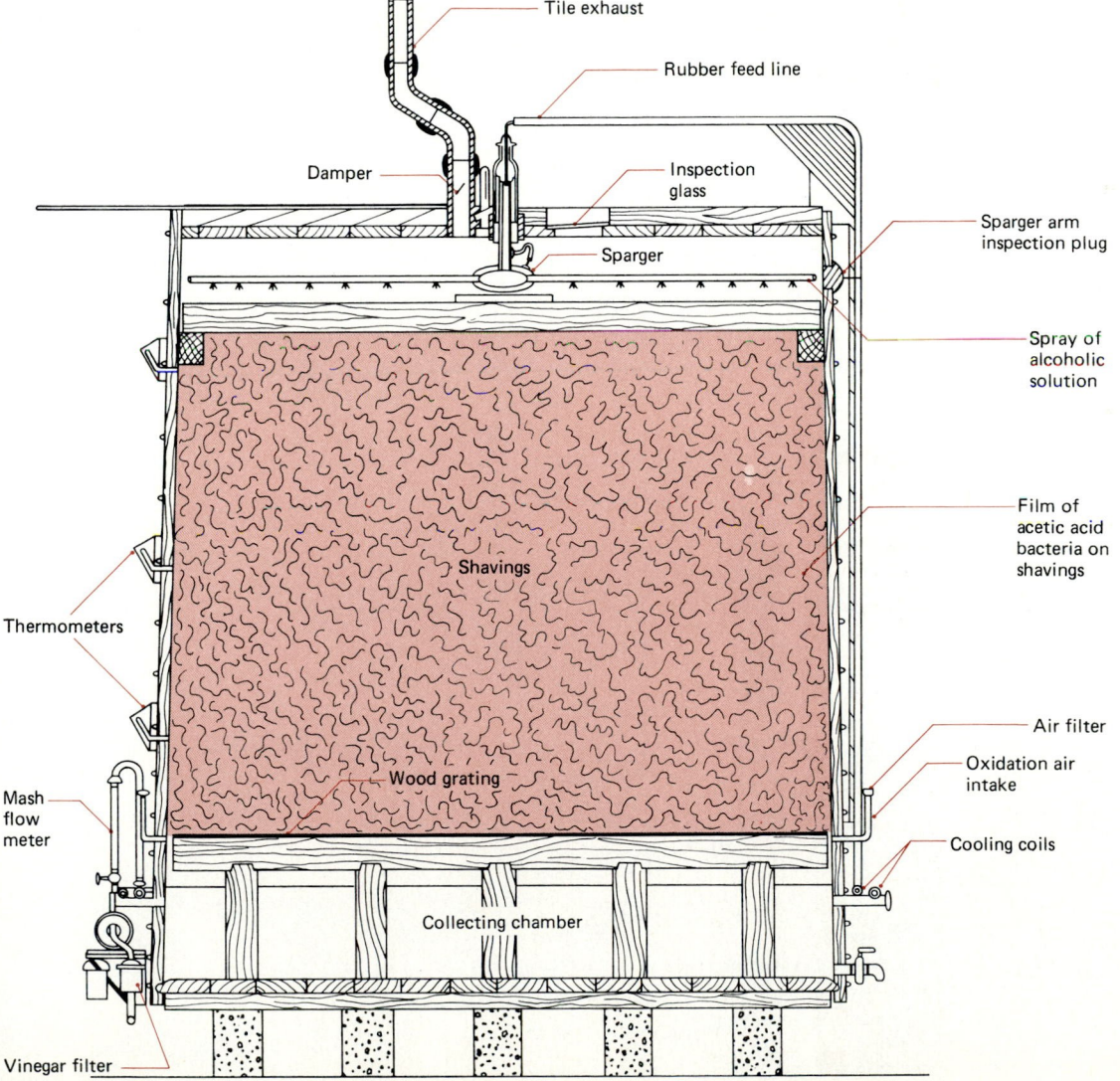

kling of the liquid down through the generator, the alcohol is converted to acetic acid. Air enters the generator from the bottom, facilitating the oxidative process. In order to control any excessive heat that may be generated, cooling coils are normally required. One or two runs of the alcoholic liquid through the generator is sufficient to produce high-quality vinegar.

Today, though, industrial producers of vinegar use **submerged culture reactors** (Figure 16.16). Forced aeration is used to maximize the rate of acetic acid production, and the bacteria grow in the fine suspension created by the air bubbles and the fermenting liquid. By using the submerged method for vinegar production, an 8–12 percent alcoholic liquid is inoculated with an *Acetobacter* species at 24–29°C with carefully controlled aeration. Using a 10 percent alcohol solution as substrate, the acetic acid yield can be 13 percent. Once the vinegar is formed, it is clarified by passage through a filter and allowed to age in order to achieve its final body, taste, and bouquet. The vinegar may be pasteurized at 60–66°C for a few seconds to inactivate any remaining viable bacteria.

Fermented Vegetables

Vegetables, such as cabbage, carrots, cucumbers, green tomatoes, leafy vegetables, greens, and olives, are fermented

FIGURE 16.16

In a modern submerged culture vinegar generator, a controlled level of oxygen is achieved. The alcohol in storage tank 1 is denatured with a little acetic acid and is diluted in denaturing tank 2. In mash tank 4 the substrate is prepared and readied from the denatured alcohol by the addition of water and nutrients. This readied mash is stored in feed tank 5. In the Vinegator (6) the mash is oxidized to vinegar in batches by *Acetobacter* under controlled aeration, constant temperature, and a controlled level of residual O_2. When the concentration of residual alcohol approaches zero, part of the vinegar is transferred to harvest receiver tank 8. Fresh substrate is then pumped into the Vinegator at a preset level. In clarifying tank 9, the new vinegar is mixed with the precipitant to reduce the present clouding. The precipitated cloud is removed by filtration and the clarified vinegar is stored in tank 13.

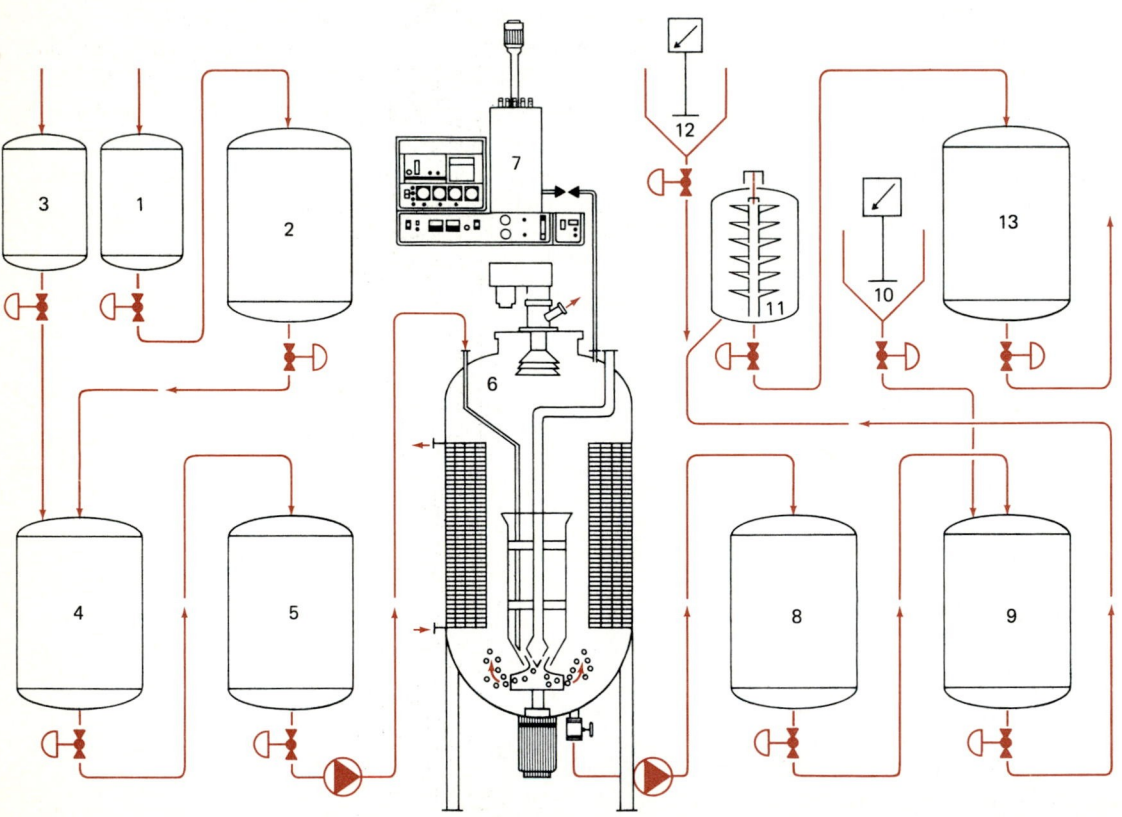

1 Alcohol storage tank	6 Vinegator	11 Filter
2 Denaturing tank	7 Laboratory-Vinegator	12 Precoat tank
3 Nutrient tank	8 Harvest receiver tank	13 Storage tank
4 Mash tank	9 Precipitation–clarifying tank	
5 Feed–tank	10 Precipitant tank	

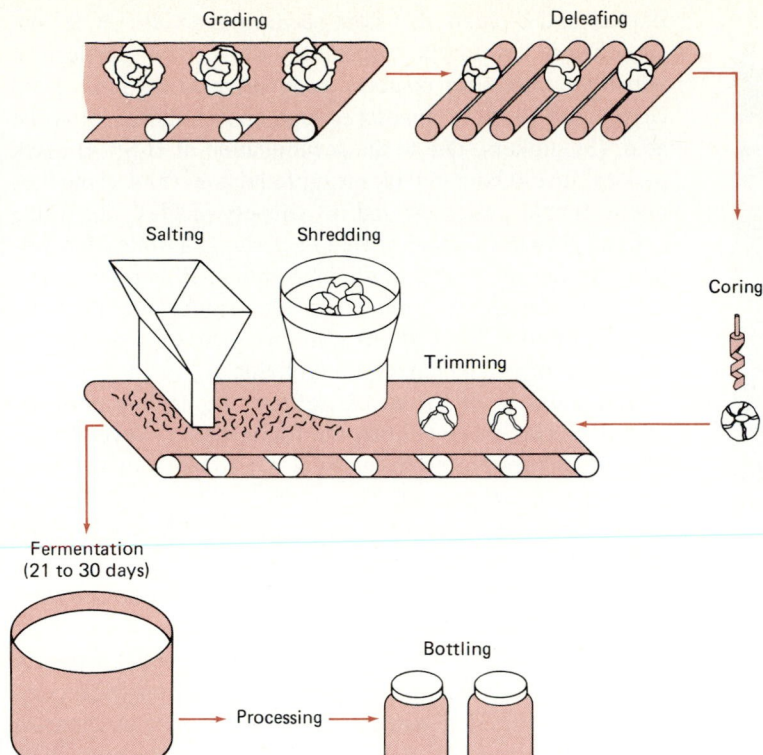

FIGURE 16.17

The production of sauerkraut in order to preserve cabbage is an application of fermentation that long predates canning and freezing and is still widely practiced on a commercial scale. In this method, brine is generated by osmotic gradients arising from the interaction of the salt and the natural fluids of the cabbage. In the brine, the lactic acid bacteria originating on the fresh cabbage become the dominant species in the extended process of fermentation.

by using lactic acid bacteria as a means of creating new food products that are not readily subject to spoilage. Other fermentations, particularly of soybeans, are carried out to produce specially desired flavors, aromas, and textures in food products.

Sauerkraut Sauerkraut is produced from a lactic acid fermentation of wilted, shredded cabbage (Figure 16.17). Salt, 2.25–2.5 percent, is added to shredded cabbage to help extract the plant juices, control the microbiota during the fermentation, and maintain an even dispersal of the bacteria. Anaerobic conditions develop in the salted, shredded cabbage and surrounding juice, primarily as a result of continued respiration of plant cells but also because of some bacterial metabolism.

The production of sauerkraut involves a succession of bacterial populations (Figure 16.18). Coliform bacteria, such as *Enterobacter cloacae*, are prominent in the initial mixed community, and produce gas and volatile acids as well as some lactic acid. The accumulating lactic acid exerts a selective pressure on the microbial community, causing population shifts and continued succession. As a result, after the initial fermentation there is a shift in the microbial community, and *Leuconostoc mesenteroides*, which grows well at 21°C and is not inhibited by 2.5 percent salt, becomes the dominant microbial population. Up to 1 percent lactic acid may accumulate—and yeasts and various bacteria may grow as a surface film—during this phase of the fermentation.

The continuing succession of bacterial populations next favors the development of *Lactobacillus plantarum*, which

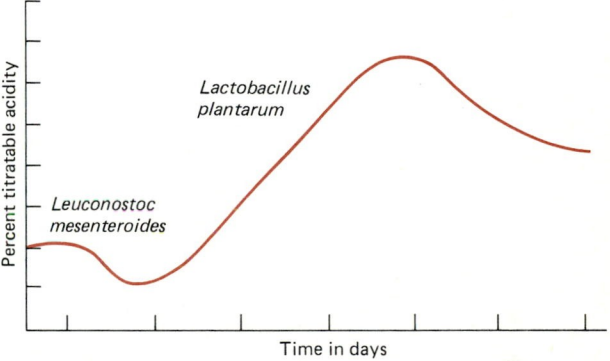

FIGURE 16.18

During sauerkraut production, there are successional changes in the bacterial populations in response to changing environmental conditions. The production of high-quality sauerkraut depends on the contributions of several different microbial populations.

produces acid but no gas. During this phase of the fermentation, the concentration of lactic acid reaches 1.5–2 percent. Growth of *L. plantarum* also removes mannitol, which is produced by *Leuconostoc* and has an undesirable bitter flavor. The fermentation can be stopped at this stage by canning or refrigerating the sauerkraut. If there is any residual sugar and mannitol after the action of *L. plantarum*, the successional process can continue with the development of *Lactobacillus brevis*, a gas-producing species. The growth of *L. brevis* can increase the lactic acid concentration to 2.4 percent and also imparts a bitter acid flavor to the sauerkraut.

FIGURE 16.19

Photograph showing pickle production. Problems occasionally arise during the fermentative production of pickles, leading to the formation of bloaters, stinkers, and other undesired products. (Courtesy Paramount Pickles, Louisville, Ky.)

High-quality sauerkraut has a lactic acid concentration of about 1.7 percent and a clean acid flavor, with low concentrations of diacetyl contributing to the aroma and flavor of the final product.

Pickles   The traditional method for producing pickles by fermenting cucumbers uses the natural microbiota associated with the cucumber, and controlled temperature and salt concentrations, to regulate the fermentation process (Figure 16.19). Controlled fermentation of pickles can also be achieved by inoculation with *L. plantarum* and *Pediococcus cerevisiae*. The traditional process takes 6 to 9 weeks to reach completion. During this period, the salt concentration is gradually increased to reach a final level of about 15.9 percent NaCl. At the beginning of the fermentation, when the salt concentration is low, many bacterial genera are able to grow, including *Pseudomonas*, *Flavobacterium*, and *Bacillus*. As the salt concentration is increased, the populations that become favored include the lactic acid bacteria *Leuconostoc mesenteroides*, *Streptococcus faecalis*, and *P. cerevisiae*. As the lactic acid and salt concentrations increase, *L. plantarum* becomes the dominant bacterium, beginning several days after the fermentation and continuing until the salt concentration surpasses 10 percent. The completion of the fermentation process involves yeasts that grow at high salt concentrations. During the final yeast fermentation stage, some carbohydrates are converted to alcohol. The growth of film-forming yeasts, such as *Debaryomyces*, *Pichia*, *Endomycopsis*, and *Candida*, lowers the lactic acid concentration.

Because of the complexity of changes in the microbial community during this natural fermentation, the process often goes awry and yields unmarketable pickles, such as (1) floaters and bloaters that float because of excessive gas accumulation within the cucumber; (2) hollow pickles, in which the cucumber contents have shriveled because of excessive salt or the formation of high concentrations of acetic acid; (3) stinkers, due to the accumulation of H_2S; (4) black pickles, due to bacterial pigment production; (5) soft pickles, due to fungal proteases; and (6) slippery pickles, due to the surface growth of encapsulated bacteria. Controlled fermentation conditions—and a pure inoculum of *P. cerevisiae* and *L. plantarum* after the removal of the natural microbiota by fumigation or chlorination—can be used to increase the likelihood of producing a quality pickle.

The sourness of the pickle reflects the amount of lactic acid that accumulates during the fermentation. Several varieties of pickles are produced by a modification of the basic fermentation process. In the production of dill pickles, a brine of 7.5–8.5 percent NaCl is used. The dill herb is added for flavoring, and vinegar also is normally added to prevent undesirable fermentation reactions. Because of the low concentration of salt, various indigenous soil bacteria on cucumber surfaces are able to grow during the initial stages of fermentation. As lactic acid accumulates, the bacterial community becomes dominated by *L. mesenteroides*, *S. faecalis*, *P. cerevisiae*, and *L. plantarum*. The final concentration of lactic acid in dill pickles is in the range of 1–1.5 percent.

Olives   The production of green olives involves lactic acid fermentation. The harvested olives are washed with a solution of sodium hydroxide that removes most of the oleuropein, a very bitter phenolic glucoside that gives unfermented olives a very undesirable flavor. The olives are then placed in a brine solution, and a lactic acid fermentation lasting for 60 days to 10 months is permitted to occur. During the first 2 weeks of the fermentation, the brine becomes stabilized as compounds are leached from the olives and microbial populations begin to multiply. At the intermediate stage, which occurs during the following 2 to 3 weeks, *Leuconostoc* is the dominant bacterial species and lactic acid accumulates. The final stage of fermentation is dominated by *L. plantarum* and *L. brevis*; yeasts and various bacteria also occur during this stage. The final acidity of the olives is approximately 7.1 percent lactic acid.

Soy Sauce   Several oriental foods are prepared by fermenting soybeans or rice. Soy sauce, a brown, salty, tangy sauce, which in Japanese is called *shoyu*, is produced from a mash consisting of soybeans, wheat, and wheat bran. Soy sauce is used as a condiment or as an ingredient in other sauces (Figure 16.20). The starter culture for the production of soy sauce is produced by **koji fermentation**, a dry fermentation in which a mixture of soybeans and wheat is inoculated with spores of *Aspergillus oryzae* (Figure 16.21). The mixture is moistened but is not submerged in liquid. The fungi grow on the surface of the soybeans and wheat, accumulating various enzymes including proteinases and amylases. Various bacterial populations, normally dominated by lactic acid bac-

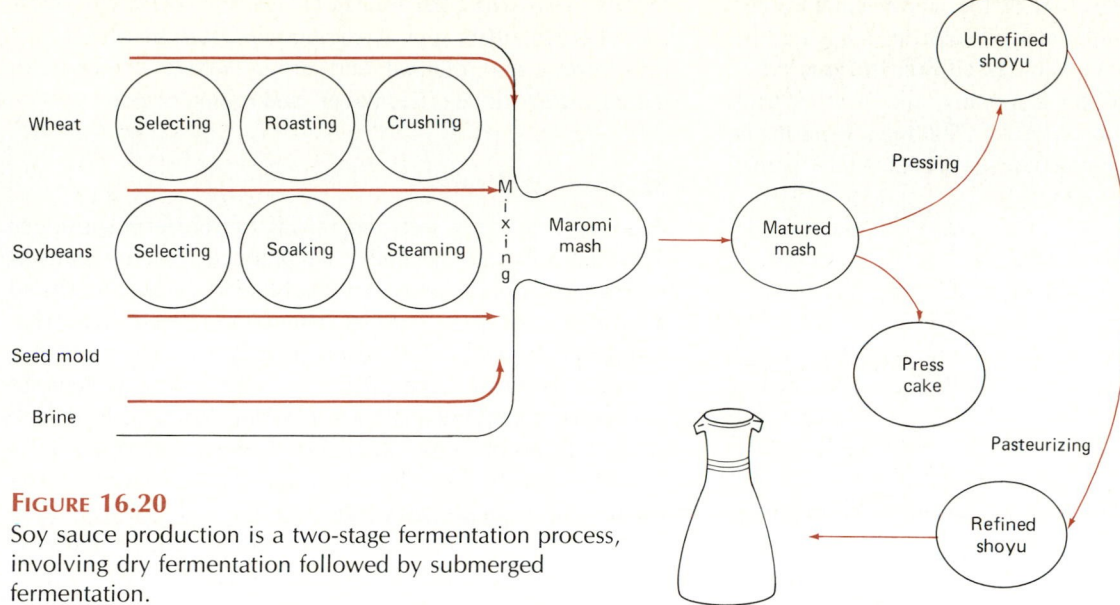

FIGURE 16.20
Soy sauce production is a two-stage fermentation process, involving dry fermentation followed by submerged fermentation.

FIGURE 16.21
Photograph of a koji fermentation tray, showing the growth of *Aspergillus sojae* on a mixture of soybeans and wheat. (Courtesy Kikkoman Foods, Walsworth, Wis.)

teria, also develop during this koji fermentation. After the starter culture develops, it is dried and extracted.

The extract is mixed with a mash consisting of autoclaved soybeans, autoclaved and crushed wheat, and steamed wheat bran. The mash with the koji is incubated in flat trays for several days at approximately 30°C and is then soaked with concentrated brine. The resulting mixture is called *maromi*. The mash is then incubated for a period ranging from 10 weeks to over 1 year, depending on the incubation temperature. During this incubation period the proteinases, amylases, and other enzymes of the koji are active, and there is a succession of microbial populations. The maturation begins with lactic acid bacteria, including lactic acid production by *Pediococcus soyae*, and later involves alcoholic fermentations by yeasts, such as *Saccharomyces rouxii*, *Zygosaccharomyces soyae*, and *Torulopsis* species. The most important organisms during the fermentation process are *A. oryzae*, which produces proteinases and amylases; *Lactobacillus* spe-

cies, which produce sufficient amounts of lactic acid to prevent spoilage by other microorganisms; and yeasts, which produce sufficient alcohol to increase the flavor.

An interesting problem was encountered when soy sauce production was begun in the United States. In Japan the maturation process is carried out in concrete tanks, and the necessary microbial populations are maintained in the porous concrete surface. In the United States, where sterilized stainless steel tanks are used for the secondary fermentation process, it was difficult to define, maintain, and add at the proper times the cultures that are needed for the successional process involved in producing quality soy sauce. Eventually, the process was perfected, and only a soy sauce connoisseur can tell the difference between the U.S. and Japanese products.

Miso **Miso** is also produced by using a koji fermentation with *A. oryzae*. Steamed polished rice, placed in shallow trays, is used in the production of the starter culture. The koji is mixed with a mash of steamed soybeans and after the addition of salt, the fermentation is allowed to proceed at 28°C for 1 week, at 35°C for 2 months, and at room temperature for several additional weeks. The miso is normally ground into a paste, to be combined with other food before eating.

Tempeh   **Tempeh** is an Indonesian food produced from soybeans. The soybeans are soaked at 25°C, dried, and inoculated with spores of various species of *Rhizopus*. The mash is incubated at 32°C for 20 hours, during which mycelial growth occurs. The product is then salted and fried before eating.

Tofu and Sofu   **Tofu** (Japanese) or **sofu** (Chinese) is a cheese-like product produced by fermenting soybeans with *Mucor* species. The soybeans are soaked, ground to a paste, and curdled by adding calcium or magnesium salts. The pressed curd blocks are placed in trays at 14°C and incubated for 1 month, during which time the fungal populations develop.

Natto   **Natto** also is produced from boiled soybeans and involves the incubation of *Bacillus subtilis* with soybeans for 1 to 2 days, during which time proteinase enzymes soften and add flavor compounds to the soybeans. Various other oriental foods are also produced by similar fermentations.

Poi   **Poi** is a fermented food product from the Hawaiian Islands. In the production of poi, the stems of the taro plant are steamed, ground, and subjected to fermentation for 1 to 6 days. During the first few hours, coliforms, *Pseudomonas*, and various other microorganisms predominate. Then a successional process occurs, with *Lactobacillus, Streptococcus*, and *Leuconostoc* becoming the dominant populations.

Finally, yeasts and the fungus *Geotrichum candidum* flourish. The fermentation products, principally lactic acid, acetic acid, formic acid, ethanol, and carbon dioxide, contribute to the characteristic texture, flavor, and aroma of poi.

Single Cell Protein

In addition to using microorganisms to transform substrates enzymatically into desired food products, microorganisms can be grown as a source of **single-cell protein (SCP)**, so named because the microorganisms are single-celled organisms rich in protein. Microorganisms grow rapidly and produce a high-yield, high protein food crop. The proteins of selected microorganisms contain all of the essential amino acids. Various bacteria, fungi, and algae are potential sources of large amounts of SCP. The algae *Scenedesmus* and *Spirulina*, for example, have been cultured in various warm ponds as a food source. The production of SCP from algae is advantageous because these organisms are able to utilize solar energy, greatly reducing the amount of fuel resources required to produce SCP. Some algae currently are harvested as a source of food.

Research on the concept of SCP production was begun during the 1960s by oil companies when petroleum was inexpensive and appeared to be an economically attractive substrate for growing SCP. The Imperial Chemical Works in Britain produces Pruteen, the SCP product of *Methylophilus methylotrophus*, a bacterium that grows on C_1 compounds. *M. methylotrophus* is grown on methanol, derived from methane, and the cell crop is harvested, centrifuged, dried, and sold in pellet or granular form (Figure 16.22). Because of dramatic increases in the price of oil, petroleum hydrocarbons are no longer considered as the primary substrates for producing SCP. The product simply could not be economically competitive with soybean and fish meal. Future less expensive sources of methanol, perhaps derived from

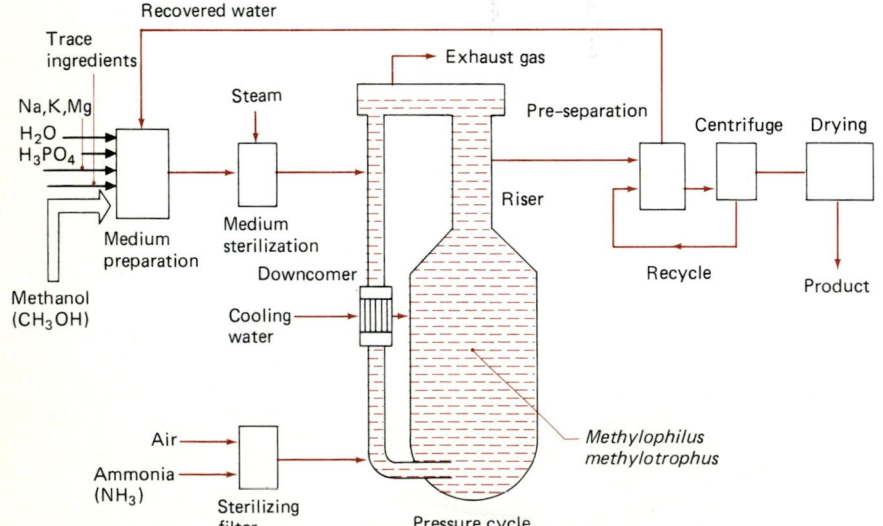

Production of bacterial protein from methanol

FIGURE 16.22
The production of bacterial protein from methanol. If an inexpensive source of methanol can be found, perhaps by microbial fermentation of waste products, the production of SCP may play a significant role in meeting world food needs.

cellulose, will likely revive the prospects for large-scale production of microbial SCP.

Yeasts are excellent candidates for development as commercial sources of SCP. Yeast-based SCP has a high vitamin content. Various species of yeast, including members of the genera *Saccharomyces, Candida*, and *Torulopsis*, can be grown on waste materials, recycling these substances into useful sources of food. The growth of yeasts on waste materials serves a dual function: the removal of the unwanted substances and the production of much needed protein-rich foods. In the USSR there is huge commercial production of *Candida* yeast protein from hydrolyzed peat. Approximately 1.1 million tons of yeast protein per year are being produced in a rapidly expanding Soviet industry that aims to reduce Soviet dependence on imported grain.

SCP is primarily produced as an animal feed. There are problems with using SCP for direct human consumption because of high concentrations, 6–11 percent, of nucleic acids. This may result in increased serum levels of uric acid, causing kidney stone formation or gout, possible allergic reactions, and possible gastrointestinal reactions including diarrhea and vomiting. Chickens and other animals, however, can be grown on SCP rather than on plant materials, helping to meet world food needs. Researchers are still trying to find the proper microorganism and set of production conditions to produce SCP that can be fed directly to humans.

Postlude

The field of food microbiology can be viewed as an application of the principles of controlling microbial growth to prevent food spoilage and the foodborne transmission of human diseases, together with the harnessing of microbial activities to produce food products for consumption. On the one hand, uncontrolled microbial food spoilage processes reduce the value of food products; on the other hand, controlled microbial spoilage processes are used to produce valuable food products. Various food preservation methods are employed to increase the shelf life of numerous food products. Several modern preservation methods, including canning and freezing, can extend the shelf life of the product indefinitely. Increasing the shelf life of products is extremely important because much of the world's population now lives in urban centers, requiring increased storage and transport and because many foods are grown seasonally but desired year round. The shipping and distribution of food to urban population centers, which are distant from agricultural areas of food production, necessitates increased storage times for foods, compared to agricultural societies, where the food can be consumed quickly after harvesting. Proper preservation of foods is important for public health reasons, as well as for saving needed resources to meet human nutritional needs.

The use of microorganisms to produce food products is also important to meet the food demands of the world's expanding population. Microbial fermentation products, such as cheeses, are part of the normal diet of most individuals. Various beverages, such as beer and champagne, are also frequently consumed. Many of these microbially produced foods are considered delicacies, adding enjoyment to eating. The use of microorganisms in producing SCP represents an important, potentially inexpensive source of protein-rich food for the future.

Developments in the field of food microbiology represent important ways in which our basic understanding of microbial processes can be applied to help overcome food shortages and to aid economic development in various parts of the world. Today's challenge to the food industry is to increase food production, to increase the shelf life of foods, and to respond to the public desire for foods that are nutritious, safe, and tasty.

Suggested Supplementary Readings

AYRES, J. C., J. O. MUNDT, and W. E. SANDINE. 1980. *Microbiology of Foods*. W.H. Freeman and Co., San Francisco.

DESROSIER, N. W. 1977. *The Technology of Food Preservation*. AVI Publishing Co., Westport, Conn.

FENNEMA, O. 1976–. *Principles of Food Science* (4 volumes). Marcel Dekker, New York.

FRAZIER, W. C., and D. C. WESTHOFF. 1978. *Food Microbiology*. McGraw-Hill Book Co., New York.

GAMAN, P. M., and K. B. SHERRINGTON. 1981. *The Science of Food: An Introduction to Food Science, Nutrition, and Microbiology*. Pergamon Press, New York.

HOBBS, B. C., and J. H. CHRISTIAN. 1974. *The Microbiological Safety of Foods*. Academic Press, New York.

INTERNATIONAL COMMISSION ON MICROBIAL SPECIFICATIONS FOR FOODS (ICMSF). 1974. *Microorganisms in Foods: Sampling for Microbiological Analysis—Principles and Specific Applications*. University of Toronto Press, Toronto.

INTERNATIONAL COMMISSION ON MICROBIAL SPECIFICATIONS FOR FOODS (ICMSF). 1978. *Microorganisms in Foods: Their Sig-*

nificance and Methods of Enumeration. University of Toronto Press, Toronto.

INTERNATIONAL COMMISSION ON MICROBIAL SPECIFICATIONS FOR FOODS (ICMSF). 1980. *Microbial Ecology of Foods* (2 volumes). Academic Press, New York.

JAY, J. M. 1986. *Modern Food Microbiology.* Van Nostrand Reinhold Co., New York.

KHARATYAN, S. G. 1978. Microbes as food for humans. *Annual Review of Microbiology* 32:301–327

MONTVILLE, T. J. 1987. *Food Microbiology* CRC Press, Inc., Boca Raton, Fla.

PEDERSON, C. S. 1979. *Microbiology of Food Fermentations.* AVI Publishing Co., Westport, Conn.

ROSE, A. H. 1981. The microbiological production of food and drink. *Scientific American* 245(3):126–139.

ROSE, A. H. (ed.). 1983. *Food Microbiology.* Academic Press, London.

SHARPE, A. N. 1980. *Food Microbiology.* Charles C. Thomas Publisher, Springfield, Ill.

WEISER, H. H., G. J. MOUNTNEY, and W. A. GOULD. 1971. *Practical Food Microbiology and Technology.* AVI Publishing Co., Westport, Conn.

Study Questions

1. What is food spoilage?

2. What is the difference between an intrinsic and an extrinsic factor with respect to food spoilage processes?

3. What is the difference between canning and pasteurization?

4. Discuss three processes for drying foods to prevent spoilage.

5. Why is adding salt to a food useful for preventing spoilage? Why are we trying to limit the use of this preservation method?

6. What are the differences in the methods used to produce beer, wine, and distilled liquors?

7. How is cheese made? What is ripening? When we consider the great variety of cheeses, how can they all be made from essentially the same starting material?

8. How is sauerkraut produced? Discuss the role of microbial succession in the production of sauerkraut.

9. What can go wrong in the production of pickles?

10. What is a chemical food preservative? What are the advantages and disadvantages of the use of chemical preservatives? Should they be added to food products?

11. What is single-cell protein (SCP)? What are the useful candidate substrates for SCP production? Can recombinant DNA technology help create a microbial strain that will solve world hunger? Discuss.

Situational Problem

Planning a Party with Foods and Beverages Produced by Microorganisms

Congratulations! You have been chosen as chairperson of the annual summer Biology Department picnic. Each year the party has a special biological theme. This year the theme is "The Fungi." In keeping with this theme, you decide that the foods and beverages served should be produced by fungi.

1. Plan the menu for this picnic. Assume that the food will be served outdoors and that there will be only minimal refrigeration available for several hours before and during the picnic.

2. You also feel that the picnickers should know how fungi contributed to the production of each of the items served. So the decorations will include posters that explain the role of the fungi. Prepare sketches of the posters to display with each of the foods and beverages at the picnic.

3. Because the picnic was a tremendous success, you have been asked to do an encore for the winter break party. This time the theme is "The Bacteria." Prepare a menu and sketches of appropriate posters illustrating the role of bacteria in the production of each of the foods and beverages that you would serve.

CHAPTER 17

Industrial Microbiology

17.1 *The Fermentation Industry*

Industrial microbiology, in its broadest sense, is concerned with all aspects of business that relate to microbiology. In a more restricted sense, industrial microbiology is concerned with employing microorganisms to produce a desired product and with preventing microorganisms from diminishing the economic value of various products. This duality of purpose is clearly seen in the food industry, a major area of industrial microbiology discussed in Chapter 16. Quality control to prevent microbial contamination of various industrial products and to limit microbial corrosion and biodeterioration are important concerns of industrial microbiology, as are producing and profiting from the sale of quality microbial products. Various commercial products of important economic value made by microorganisms are (1) pharmaceuticals, including antibiotics, steroids, human protein, vaccines, and vitamins; (2) organic acids; (3) amino acids; (4) enzymes; (5) organic solvents; and (6) synthetic fuels. Many of these products can be produced both microbially and by chemical synthesis. The choice of which process to employ generally depends on economics, and it is not surprising that some products that historically have been produced by microorganisms are now produced chemically and vice versa. The decision is dictated by variable costs; therefore, changes in the cost of raw materials and the market value of a particular product influence the feasibility of microbial production of that product.

In industrial microbiology the term **fermentation** is not used in its restricted scientific sense, referring to metabolic pathways that proceed by fermentation rather than respiration, but rather in a wider sense to include any chemical transformation of organic compounds carried out by using microorganisms and their enzymes. Industrial processes using microorganisms exploit the enzymatic activities of the microbe to produce substances of commercial value. Production methods in industrial microbiology bring together the raw materials (substrates), microorganisms (specific strains or microbial enzymes), and a controlled favorable environment (created in a fermentor) to produce the desired substance. The essence of an industrial process is to combine the right organism, an inexpensive substrate, and the proper environment to produce high yields of a desired product (Figure 17.1).

Critical activities of industrial microbiologists include (1) the search for microorganisms that carry out biotransformations of commercial importance, with the emphasis on finding or creating specific strains of microorganisms that will yield sufficient quantities of the desired product to permit commercial production on an economically favorable basis, and (2) the design of the optimal production process. Production process technology includes defining the substrate mixture—containing the least expensive components—that will produce the highest yield of the desired product, recognizing that many times the presence or absense of even trace amounts of a component will vastly alter the yield of the desired product; designing fermentors to optimize the environmental conditions in order to achieve maximal product yields; and developing recovery methods that achieve separation of the desired product from microbial cells, residual substrate, and other metabolic products in the most economical manner. The complexity of achieving even a simple sin-

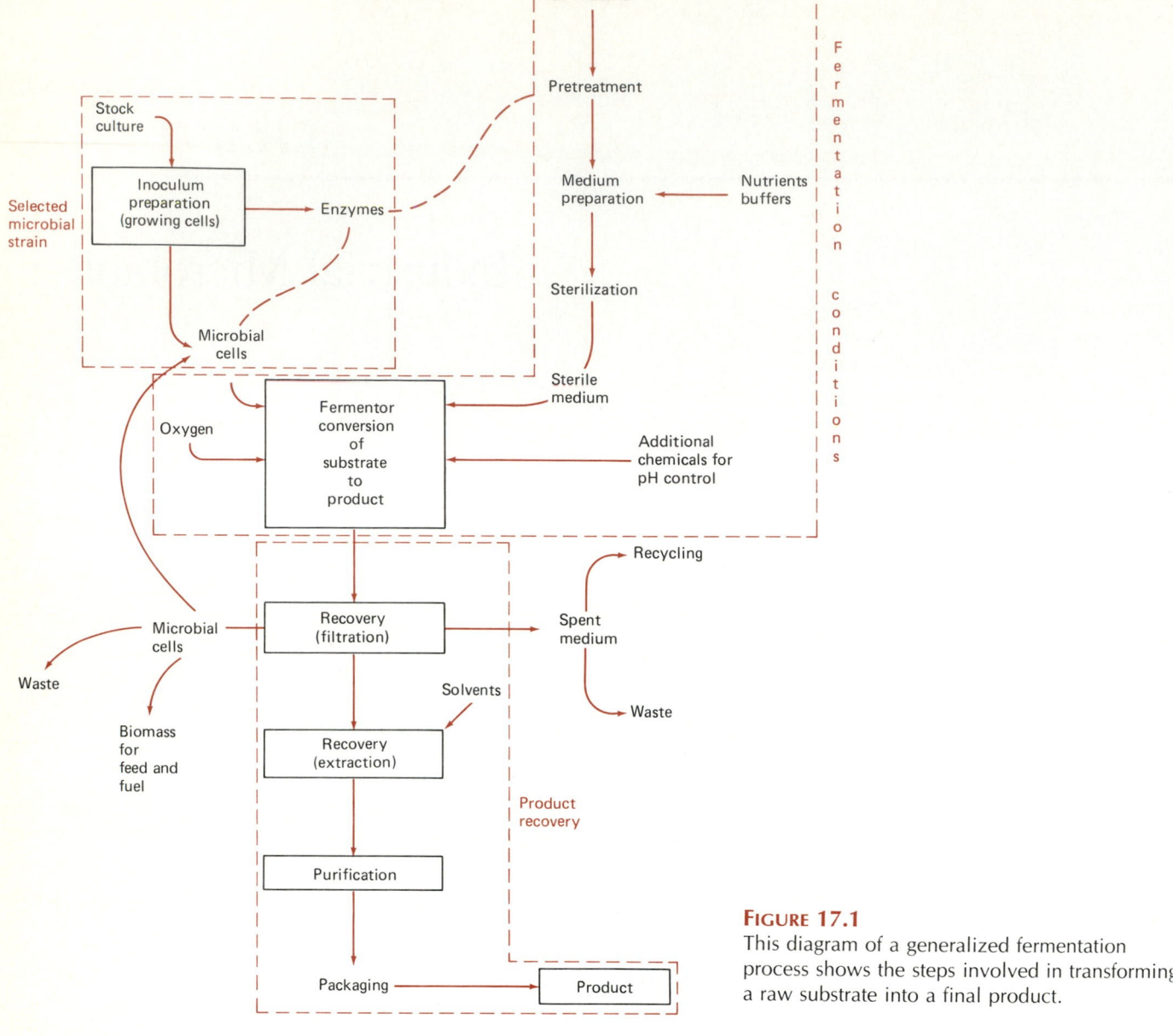

FIGURE 17.1
This diagram of a generalized fermentation process shows the steps involved in transforming a raw substrate into a final product.

gle-step transformation of a molecule is great when considered on a large industrial scale.

Selection of Industrial Microorganisms

The selection of microorganisms for use in the fermentation industry begins with screening to find the right microorganism. Of the many species of microorganisms, relatively few possess the genetic information needed to produce economically useful products (Table 17.1). The screening procedures employed in industry are designed to separate microorganisms that are potentially valuable in producing a commercially useful product from the rest. Large-scale industrial screening procedures incorporate assays that permit identification of these microorganisms.

Screening for microorganisms that possess the potential for producing industrially important substances includes both naturally occurring microorganisms and genetic variants. The classic approach used to find new antibiotic-producing strains has been to screen large numbers of isolates from soil samples for microorganisms that naturally produce antimicrobial substances. Additionally, mutations can be induced by exposure to radiation or mutagenic chemicals to increase genetic variability within the populations being screened, with the hope of isolating a unique microbial strain capable of producing a novel metabolite with the desired properties or a strain that produces large quantities of a valuable substance. Often, once a microorganism is identified as possessing the genetic information needed to produce a potentially useful substance, it is necessary to carry out successive stages of

TABLE 17.1   Some Microbial Species Used for Producing Commercial Products

Industrial Chemicals		Polysaccharides	
Saccharomyces cerevisiae	Ethanol (from glucose)	*Leuconostoc mesenteroides*	Dextran
Kluyveromuces fragilis	Ethanol (from lactose)	*Xanthomonas campestris*	Xanthan gum
Clostridium acetobutylicum	Acetone and Butanol	**Pharmaceuticals**	
Aspergillus niger	Citric acid	*Penicillum chrysogenum*	Penicillins
Amino Acids and Flavor-enhancing Nucleotides		*Cephalosporium acremonium*	Cephalosporins
Corynebacterium glutamicum	L-Lysine	*Streptomyces*	Amphotericin B, kanamycins, neomycins, streptomycin, tetracyclines and others
Corynebacterium glutamicum	5'-inosinic acid and 5'-guanylic acid	*Bacillus brevis*	Gramicidin S
Corynebacterium glutamicum	MSG	*Bacillus subtilis*	Bacitracin
Vitamins		*Bacillus polymyxa*	Polymyxin B
Ashbya gossypii	Riboflavin	*Rhizopus nigricans*	Steroid transformation
Eremothecium ashbyi	Riboflavin	*Arthrobacter simplex*	Steroid transformation
Pseudomonas denitrificans	Vitamin B_{12}	*Mycobacterium*	Steriod transformation
Propionibacterium shermanii	Vitamin B_{12}	*Escherichia coli* (via recombinant DNA technology)	Insulin, human growth hormone, somatostatin, interferon
Enzymes			
Aspergillus oryzae	Amylases		
Aspergillus niger	Glucamylase		
Trichoderma reesii	Cellulase		
Saccharomyces cerevisiae	Invertase		
Kluyveromyces fragilis	Lactase		
Saccharomycopsis lipolytica	Lipase		
Aspergillus	Pectinases and proteases		
Bacillus	Proteases		
Mucor pussilus	Microbial rennet		
Mucor meihei	Microbial rennet		

ANALYTICAL PROCESS

Screening of Antibiotic Producers

The search for antibiotics in the pharmaceutical industry presents a good example of how screening procedures are employed to select microorganisms for industrial applications. The discovery of new antibiotics results from laborious searches. Samples from many sources, including soils from around the world, are examined as potential sources of antibiotic-producing microorganisms; countless strains of microbial isolates are tested by pharmaceutical laboratories. One of the best penicillin-producing strains of *Penicillium* was isolated from an orange purchased at a roadside fruit stand; several antibiotic-producing actinomycetes were isolated from a manure-enriched pasture. Of the numerous investigations, few studies yield evidence of promising new compounds of potential clinical importance. Identification of compounds with antimicrobial activity is an essential step in the screening process.

A useful antibiotic-producing strain must produce metabolites that inhibit the growth or reproduction of pathogens. This essential property can be assayed by using test strains and examining whether the isolate being screened produces substances that inhibit the growth of these test organisms. If a suspension of the test organism is applied to the surface of an agar plate, the zone of inhibition around a colony may indicate that the organisms in that colony are producing an antibiotic. Alternatively, the crude filtrate of a broth-grown microbial culture can be added to a culture of a test organism to determine whether substances with antimicrobial activity are produced by the organism being screened.

A positive result in such a primary screening procedure in no sense ensures the discovery of an industrially useful antibiotic-producing strain. It simply identifies those strains of microorganisms that have the potential for further development. Secondary screening procedures are then carried out to determine whether the organism is indeed producing a substance of industrial interest that merits further investigation and development. These procedures may include both qualitative assays, aimed at identifying the nature of the substance being produced and deter-

A   B   C

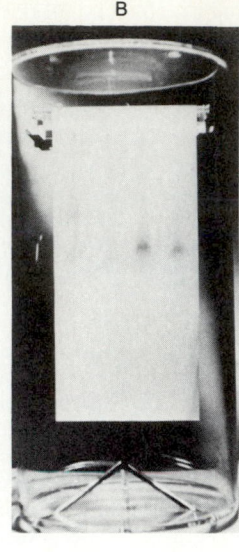
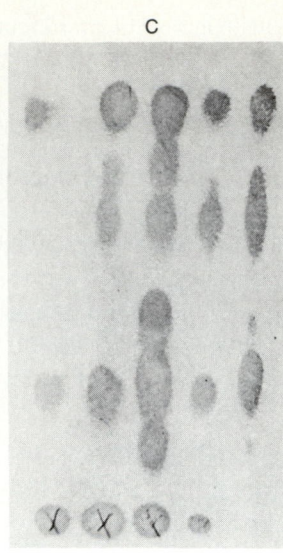

FIGURE 17.2

Paper chromatography, which permits the relatively easy separation of compounds from a complex mixture, is used during secondary screening procedures to identify active antibiotic compounds. Compounds are separated on the basis of their relative polarities in this form of affinity chromatography. When the paper chromatogram is placed directly on an agar plate, the resolved compounds are absorbed into the agar. Zones of inhibition indicate the presence and locations of antimicrobial compounds. Many compounds can be screened in this way to identify those few that may prove useful. Compounds showing antimicrobial activity can be isolated and tested to determine whether they are truly new compounds that merit further study. Tests include determining the effectiveness of a new compound against a wide variety of pathogenic microorganisms and elucidating the potential toxicity and other untoward side effects that may occur in mammals. Only a limited number of compounds identified in the original screening procedure as being of potential clinical use ever reach clinical trials, and fewer still are ever marketed. The development and successful marketing of antibiotics traditionally is the result of serendipity and hours of laborious screening, rather than of any great theoretical insight.

These photographs show some of the steps in screening for new antimicrobial compounds. (A) This room has many large tanks for paper chromatography, (B) An individual tank showing separation of a potentially active compound, (C) The resultant paper chromatogram is then placed on a large agar plate so that the separated product diffuses into the agar. A lawn of bacteria is applied to the agar, and after incubation zones of inhibition around isolated spots indicate the locations of potentially useful products. (Courtesy Marvin Hoehn and Bernard Abbott, Eli Lilly and Co., Indianapolis)

mining whether it is a new compound not previously considered for industrial production, and quantitative assays, aimed at determining how much of the substance is being produced.

In screening for antibiotic producers, the crude filtrate from a broth culture may be separated chromatographically and the antimicrobial activities of the separated components determined. In some cases, paper chromatography is used to separate compounds for testing (Figure 17.2); in other cases, high-pressure liquid chromatography (HPLC) is employed. The individual active components can then be isolated and used for further screening against additional test organisms to determine the microbial inhibition spectrum. This additional screening is useful in determining whether the substance has a broad or narrow range of activity and if it is particularly effective against

specific pathogens. If an organism is indeed found to possess the potential for creating a useful new antibiotic, many additional tests are required to determine whether sufficient quantities of the substance can be produced to permit industrial production.

The screening program should identify the optimal incubation conditions for maximal economic yield of the product. Usually toxicity testing must be performed to determine whether the product can selectively inhibit pathogens without causing severe side effects that would preclude its therapeutic use. The secondary screening procedure thus yields a great deal of information about potentially useful microorganisms, allowing emphasis to be placed on the development of processes employing microorganisms likely to produce economically valuable substances.

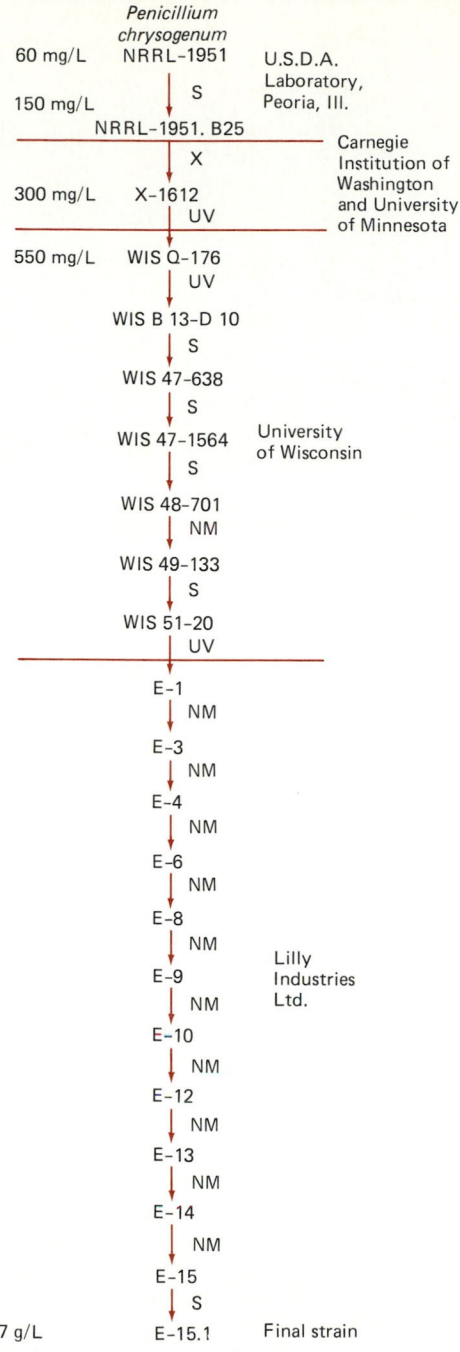

mutation before isolating a strain of that organism that can be employed for commercial production.

For example, the *Penicillium* species observed by Alexander Fleming to inhibit the growth of *Staphylococcus* had obvious potential for commercial development but did not produce sufficient quantities of penicillin to permit industrial production. Extensive screening of soil samples from around the world led to the isolation of a potentially useful strain from soil collected in Peoria, Illinois. Multiple successive mutations, though, were necessary to develop a strain of *Penicillium chrysogenum* capable of producing nearly 100 times the concentration of penicillin produced by the original strain, making production of penicillin commercially feasible (Figure 17.3). The mutation and screening approach has been important in the successful development of various strains of microorganisms currently used in the fermentation industry.

Genetic engineering has opened up many new possibilities for employing microorganisms to produce economically important substances. Whereas the mutation and selection approach is hit or miss, the use of recombinant DNA technology permits the purposeful manipulation of genetic information to engineer a microorganism that can produce high yields of a variety of products. Until the recent breakthroughs in the techniques of genetic engineering, a bacterium could produce only substances coded for in its bacterial genome. However, as discussed in Chapter 8, it is now possible to engineer bacterial strains that produce plant and animal gene products (Figure 17.4). Thus, bacteria now exist that produce human interferon, insulin, and other hormones.

The use of biotechnology has great social consequences. The ability to modify the genetic composition of all organisms—from microbes to humans— using genetic engineering raises serious ethical questions that society must now face. The development of microorganisms producing high yields of such substances promises to revolutionize the economics of the pharmaceutical industry. The seemingly unlimited potential for creating microorganisms capable of producing lucrative products has spawned a major new growth industry, genetic engineering. The ruling of the United States Supreme Court that genetically engineered microorganisms can be patented also adds economic incentive for industrial applications of recombinant DNA technology.

FIGURE 17.3

Induced mutations in a penicillin-producing strain were used to produce a new, commercially useful strain. Repeated mutations were necessary to create a strain of the mold *Penicillium chrysogenum* that synthesized enough penicillin to form the basis of a commercial process. Radiation and chemical agents were employed to induce mutations in the mold. (S = spontaneous mutation; X = x-radiation; UV = ultraviolet radiation; and NM = nitrogen mustard). Selection of the superior strains ultimately gave rise to strain E 15.1 that yielded 55 times as much penicillin as laboratory strains. Simultaneous improvements in fermentation techniques increased the yields still further; yield figures in this chart reflect both kinds of increase. Classic genetic techniques such as these are still important in the antibiotics industry, as the complexity of antibiotic synthesis makes it impractical to develop new strains by directly altering simple genes. Current fermentation methods yield more than 20 g/L.

FIGURE 17.4

A genetic engineering process for producing a bacterium capable of synthesizing human growth hormone.

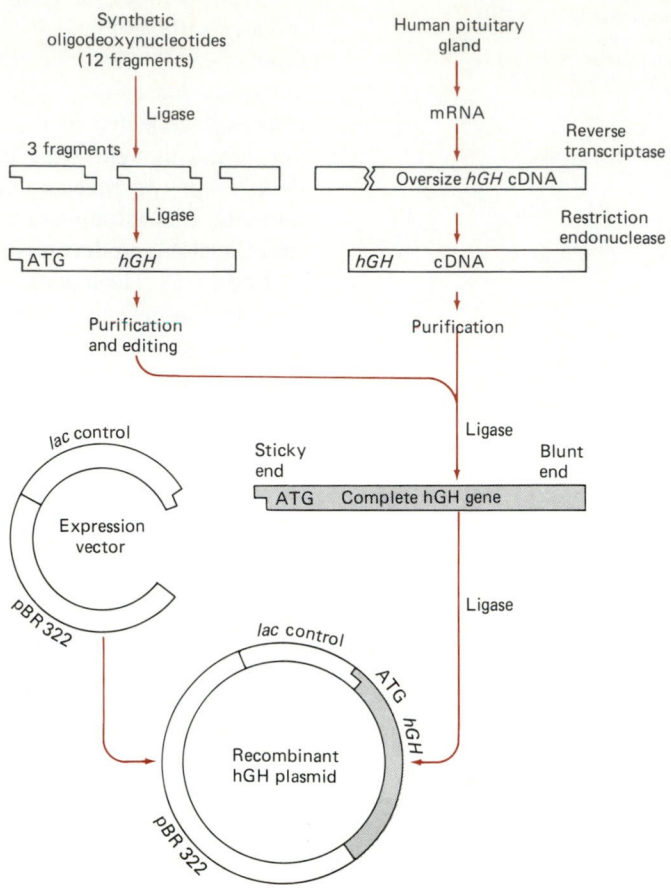

Discovery Process

The recognition that genes are split in eukaryotic cells has forced genetic engineers to develop special techniques for incorporating human genes into bacteria in order to produce useful gene products. Through a series of novel procedures involving a combination of chemical synthesis and isolation of the natural molecules, a gene for a human growth hormone has been created and placed into *Escherichia coli*. Human growth hormone is a polypeptide 191 amino acid units long elaborated by tissues of the pituitary gland; its absence leads to a form of dwarfism that can be cured by administration of the hormone. The segment of the gene that codes for the first 34 amino acids of the peptide was constructed chemically from blocks of nucleotides. To obtain the rest of the gene, a series of enzymes were used. Reverse transcriptase was employed to copy the gene for the hormone from mRNA obtained from human pituitary tissues. The use of reverse transcriptase simplifies the job of cutting and

splicing because the DNA produced in this way is colinear with the sequence of nucleotides in the mRNA. Restriction endonucleases cut out the needed fragment. DNA ligase was then used to join the natural and synthetic fragments. The complete gene produced in this manner has been inserted into a modified version of plasmid pBR322 incorporating the *lac* operon. The synthetic part of the growth hormone gene had been constructed earlier with its own initiation codon (ATG). The hormone could therefore be produced independently in bacterial cells without the need for attachment to a bacterial protein. Clinical trials of human growth hormone produced by genetically engineered bacteria are quite promising; children receiving injections of bacterially produced human growth hormone for a few months are approaching normal height for their age group, with no significant side effects.

It is critical to consider genetic regulatory mechanisms in both the mutation-screening and genetic engineering approaches to the development of microbial strains of industrial importance. The development of such strains often involves the need to overcome natural regulatory mechanisms that limit the amount of the gene product produced. In nature it is advantageous for a microorganism to produce only minimal amounts of a required substance because doing so gives that organism a competitive edge for survival. As a result, various genetic regulatory mechanisms have evolved to conserve available carbon and energy resources and to avoid production of excessive amounts of any product. In the fermentation industry, however, the valuable microbial strains are those that produce excessively high amounts of the desired product. Many mutant or genetically engineered strains used for industrial production no longer possess the genetic regulatory mechanisms for conserving their resources and producing limited amounts of a substance. Whereas such organisms would not do well in natural environments, where competition for available resources dictates which organisms survive, they do quite well in fermentors, where competition is eliminated and optimal conditions are created to favor the growth of that microbial strain in pure culture.

Because industry relies on specific microbial strains, it is important to maintain those specific genetic variants and protect against further spontaneous mutations that could alter the economics of the fermentation process. Industrial strains of microorganisms are therefore maintained in culture collections, generally in a dormant state where they are protected against mutational processes. The maintenance of these stock cultures is an essential, although sometimes boring, part of industrial microbiology. Checks are periodically run on production strains to ensure that they have retained their essential genetic capabilities. If undesirable alterations in the production strain are detected, new cultures are initiated from those maintained in the stock culture collection.

Production Process

After discovering or engineering a microorganism that produces a commercially valuable product, it is necessary to develop a fermentation process that optimizes conditions for the desired microbial activity and that yields maximal amounts of product with the highest economic profit. A balance must be achieved between production costs and the price of the product because excessive costs may render commercial production economically unfeasible.

The development of a commercial process occurs in a stepwise fashion, initially using small flasks, then small fermentors (under 10 gallons), intermediate-size fermentors (up to several hundred gallons), and finally, large-scale fermentors (thousands of gallons) (Figure 17.5). At each stage of production, development conditions are adjusted to produce maximal yields at minimal costs. The organic and inorganic composition of the medium, as well as the pH, temperature, and oxygen concentration, are the main factors that are varied

FIGURE 17.5

(A) A small-scale fermentor used in the design of an antibiotic fermentation process. In this dual-fermentor unit, the rates of aeration, pH, temperature, and other factors can be varied while growth and product formation are monitored until optimal conditions are determined. (Courtesy New Brunswick Scientific Co.) (B) Intermediate-size fermentors are used in a pilot plant to scale up the fermentation. (C) Large-scale fermentors are used for commercial antibiotic production. (Courtesy Marvin Hoehn and Bernard Abbott, Eli Lilly and Co., Indianapolis)

to maximize the efficiency of the production process. Even in a batch process, conditions are often varied during fermentation to achieve the maximal product yield, and conditions are monitored during the fermentation process to ensure that critical parameters remain within allowable limits.

It is necessary that the reaction chambers and substrate solutions be sterilized prior to the addition of the microbial strain being used in the production process. This is particularly important because the strains of microorganisms used in industrial fermentations are selected for their ability to produce the desired product in high yield, rather than for their ability to compete with other microorganisms. Infections of fermentation reactions with microbial contaminants can easily lead to a competitive displacement of the strain being employed to produce the product, with obviously deleterious results.

Fermentation Medium   The composition of the fermentation medium must include the nutrients essential to support the growth of the microbial strain and the formation of the desired product. Essential nutrients for microbial growth include sources of carbon, nitrogen, and phosphorus (Table 17.2). The choice of a particular nutritive source is made on economic as well as biological grounds. Depending on the nature of the fermentation process, all of the raw materials may be added at the beginning of the fermentation, or nutrients may be fed to the microorganisms gradually throughout the process. Often plant materials, such as molasses, are used as a carbon source. Some pretreatment of the raw material is frequently necessary to convert complex carbohydrate materials into relatively simple sugars that can be readily metabolized by microorganisms. Either organic nitrogen, sometimes in the form of cornsteep liquor, or inorganic nitrogen, such as ammonia, may be used to meet the nutritional needs of the microbial strain. Phosphorus is usually added as an inorganic salt.

Since crude raw materials are normally employed in the medium, many of the minor nutritional requirements of microorganisms are met because they naturally occur in appropriate concentrations in the raw material. In some fermentation processes, however, trace elements, such as heavy metals, must be present in specific concentrations to achieve acceptable yields of the desired product. The quality of the water used in the fermentation and the nature of the pipes used to supply solutions to the fermentation reaction can be exceptionally important. In some cases, metals leaching from pipes can inhibit microbial production of fermentation products, and in other cases such leached metals may be essential for achieving optimal yields of the desired product.

Aeration   Many industrial fermentations are aerobic processes, and therefore, it is important to achieve the **optimal oxygen concentration** to permit microbial growth with maximal product yield. The transfer of oxygen to microorganisms in large-scale fermentors is particularly difficult because the microorganisms must be well mixed and the oxygen dispersed to achieve relatively uniform concentrations in order to support maximal production rates. The development of fermentors for the growth of obligately aerobic microorganisms in a broth (submerged aerobic culture) requires careful design in order to achieve optimal oxygen concentrations throughout the solutions contained in high-volume fermentors. Many fermentor designs have mechanical stirrers to mix the solution, baffles to increase turbulence and ensure adequate mixing, and forced aeration to provide needed oxygen (Figure 17.6). It should be noted that a high concentration of microbial cells, as is achieved in a fermentor, can rapidly deplete the soluble oxygen in an aqueous solution, creating anaerobic conditions that may not be favorable to microbial production of the desired product. Forced aeration and mechanical mixing, though, are relatively expensive because of the high energy costs involved and must be economically justified for use in industrial fermentation processes.

pH   The pH of the reaction is also critical. The enzymes involved in forming the desired product all have optimal pH ranges for maximal activity and limited pH ranges in which activity is maintained. The rapid growth of microorganisms in a fermentor can quickly alter the pH of the reaction medium. For example, if the microorganisms accumulate acid, which in fact may be the desired product, the pH of a nonbuffered medium can decline precipitously, halting microbial production of the desired fermentation product. To prevent such changes, fermentation media are often buffered to dampen changes in the pH. Additionally, the pH of the reaction solution normally is continuously monitored, and acid or base is added as needed to maintain it within acceptable tolerance limits.

Temperature   The temperature of the reaction must also be carefully regulated to achieve optimal yields of product. Rapidly growing microorganisms can generate a large amount of heat that must be dissipated to prevent inactivation of enzymes. Cooling coils are often employed in fermentors to regulate temperature in order to maximize the rate of product accumulation. Heating coils are used in some fermentors

TABLE 17.2   Nutrient Sources for Industrial Fermentations

Nutrient	Raw Material
Carbon source	
Glucose	Corn sugar
	Molasses
	Starch
	Cellulose
Fats	Vegetable oils
Hydrocarbons	Petroleum fractions
Nitrogen source	
Protein	Soybean meal
	Cornsteep liquor (from corn milling)
	Distillers' solubles (from Alcoholic beverage manufacture)
Ammonia	Pure ammonia or ammonium salts
Nitrate	Nitrate salts
Nitrogen	Air (for nitrogen-fixing organisms)
Phosphorus source	Phosphate salts

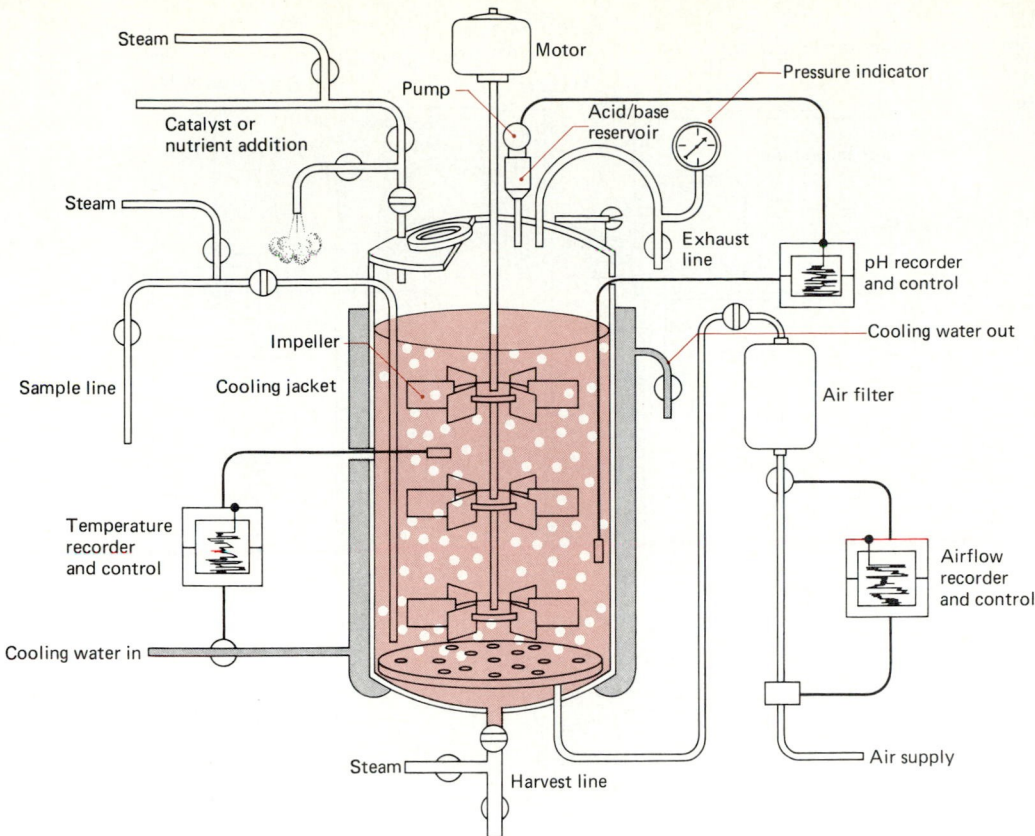

Steam
Motor
Pump
Acid/base reservoir
Pressure indicator
Catalyst or nutrient addition
Steam
Exhaust line
pH recorder and control
Impeller
Cooling water out
Sample line
Cooling jacket
Air filter
Temperature recorder and control
Airflow recorder and control
Cooling water in
Steam
Harvest line
Air supply

FIGURE 17.6

A batch reactor is employed for most current applications of industrial microbiology. In essence the reactor is a vessel in which a medium and microorganisms are mixed and then given an optimum environment in which the microbial enzymes can act. The temperature and pH are regulated. Filtered air, sometimes enriched with oxygen, is bubbled though the mixture. Samples are removed for chemical and biological assay. To prevent contamination, steam is directed through the various inlets to keep them sterilized, and the pressure inside the vessel is maintained at a value greater than atmospheric pressure. At the end of a specified period, ranging from hours to days, the batch is drained from the vessel and the product is isolated and purified.

where elevated temperatures are required to achieve optimal rates of product formation. These heating coils are also used for periodic sterilization of the fermentor chamber.

Batch versus Continuous Processes   In addition to its nutritional and environmental parameters, a fermentation process may be designed as a batch process, which is analogous to inoculating a flask containing a broth with a microbial culture, or as a continuous flow process, which is analogous to that of a chemostat. The choice of the process design depends on the economics of both production and recovery of the desired product. Compared to batch processes, flow-through fermentors are more prone to contamination with undesired microorganisms, making quality control difficult to maintain. The flow-through design, however, has the advantage of producing a continuous supply of product that can be recovered at a constant rate for commercial distribution (Figure 17.7). By their very nature, batch processes require significant startup times to initiate

the fermentation process, incubation times to allow fermentation products to accumulate, and recovery times during which the product is separated from the spent medium and microbial cells.

Immobilized Enzymes   The use of immobilized enzymes is an interesting alternative method for producing a desired product. In this process, microbial enzymes and/or microbial cells are adsorbed or bonded to a solid surface support, such as cellulose (Figure 17.8). The bonded and thus **immobilized enzymes** act as a solid-surface catalyst. A solution containing the biochemicals to be transformed by the enzymes is then passed across the solid surface. Temperature, pH, and oxygen concentration are set at optimal levels to achieve maximal rates of conversion. This type of process is very useful when the desired transformation involves a single metabolic step, but it is more complex when many different enzymatic activities are required to convert an initial substrate into a desired end product.

FIGURE 17.7 (right)

Flow-through fermentors are becoming increasingly popular in industry. Their use is economically favorable, although it is more difficult to maintain quality control of the product. In this diagram, a flow-through fermentor is shown. The design permits control of the same environmental factors as are controlled in batch processes. The main difference is that the substrates are continuously added and the products continuously harvested.

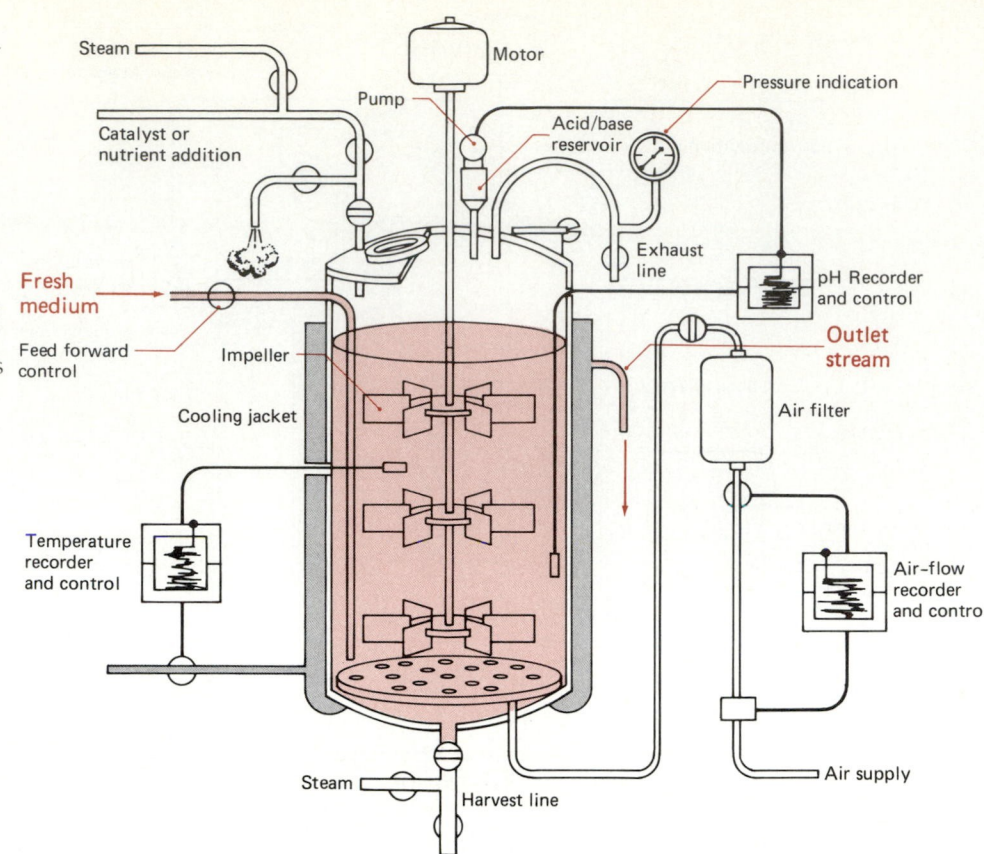

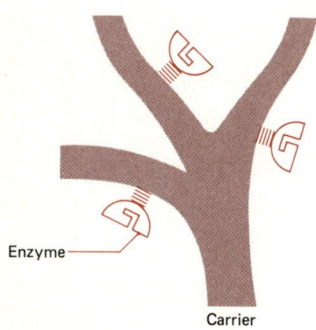

FIGURE 17.8 (left)

Immobilized enzymes are prepared by binding an enzyme to a carrier such as carboxymethyl cellulose. A solution of the substrate is allowed to flow past the bound enzymes, where the substrate molecules are converted to the product.

The use of immobilized enzymes makes an industrial process far more economical, avoiding the wasteful expense of continuously growing microorganisms and discarding the unwanted biomass. In such immobilized enzyme systems, it is essential to maintain enzymatic activity so that the enzymes are not washed off the surface or inactivated during the process. When whole cells, rather than cell-free enzymes, are employed in such immobilized systems, it is necessary to maintain viability of the microorganisms during the process. This generally involves adding necessary growth substrates.

Recovery Methods There are various methods employed in both continuous and batch culture processes for the recovery of fermentation products. The commonly used recovery methods include distillation, centrifugation, filtration, and chromatographic separation. The use of distillation for the recovery of ethanol, for example, was discussed in Chapter 16 when we considered the production of distilled liquors. Other recovery methods will be discussed when we consider the production of specific products later in this chapter. After a microbial fermentation product is successfully recovered it is packaged and marketed. Packaging is another important aspect of the production process because it must protect the product from contamination. Thus, every step in a fermentation process, from the isolation of the strains to the delivery of the product, must be carefully controlled, using basic microbiological principles to ensure the delivery of a high-quality product. Microbiologists, accordingly, play an important role in all stages of the process.

Quality Control of Industrial Products

The term **quality control** in industrial microbiology has several meanings. In the case of a production process, the concern is with the quality of the product. In microbial fermentation processes, quality control procedures must include checks on the fermentation medium and conditions, the microbial production strain, recovery efficiency, and packaging methods. Quality is determined by the yield and purity of the product. A low yield of product is economically unfavorable and may reflect a lack of proper control during the fermentation process, and a low-quality product may preclude the use of the product for its intended purpose. Environmental parameters must be maintained as close to optimal as possible to maximize the yield and quality of the product. The producing microbial strain, which in many processes is genetically unstable, must retain its essential genetic and physiological features in order to produce high yields of the desired product. Checks must be run and, when necessary, the strain must be replaced with a new stock culture. Extraction and packaging procedures must ensure a lack of excess contamination with substances other than the desired product.

Additionally, care often must be taken to prevent microbial contamination of the product. It is essential that many products be free of microbial contaminants even if they are not produced by microbial fermentation processes. Plastic and glass petri plates, pipettes, and many other items used in the microbiology laboratory must be sterile. Likewise, many pharmaceuticals, medical supplies, and surgical instruments must be free of microbial contaminants for their safe use. A variety of sterilization techniques are employed in different industrial processes to remove viable microorganisms from various products. The choice of a particular sterilization method depends on the nature of the product. Plastic products, for example, which cannot withstand exposure to high temperatures, are often sterilized by exposure to radiation or chemical-oxidizing agents, such as ethylene oxide. Liquid products, such as antibiotic solutions, may be filtered to remove microorganisms, and the bottling of many drugs is accomplished using aseptic techniques, often with the aid of sterile air laminar flow hoods (Figure 17.9). Laminar flow hoods use high-efficiency particulate air (HEPA) filters that trap microorganisms, producing clean but not necessarily sterile air. HEPA filters are constructed of cellulose acetate pleated around aluminum foil or glass fibers separated by

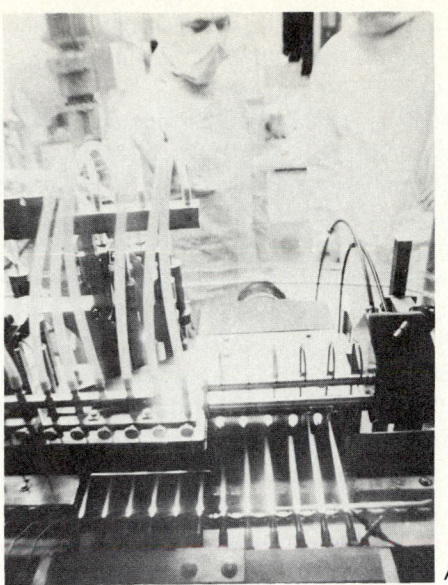

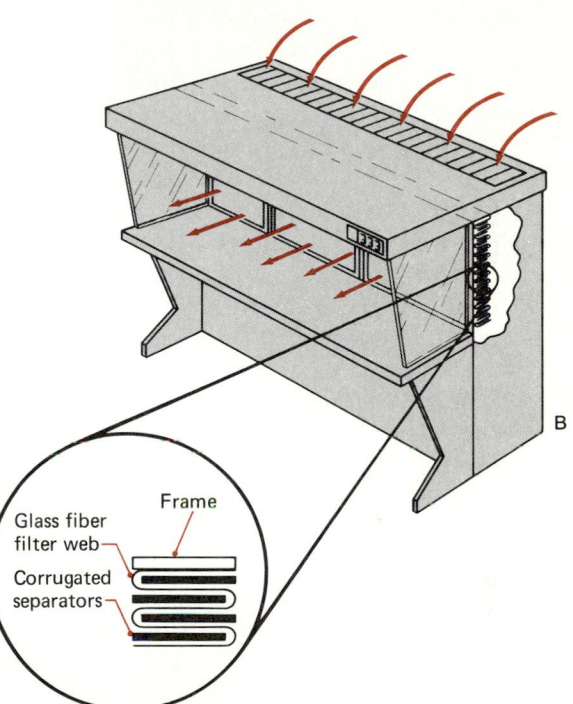

FIGURE 17.9

Photograph of a sterile drug-filling operation for producing medicinals in ampules. (A) Workers at a Giza, Egypt, manufacturing plant preparing pharmaceuticals in an area completely isolated and supplied with a change of filtered sterile air each minute. (Courtesy E. R. Squibb & Sons, Inc., Princeton, N. J.) (B) Diagram of a laminar flow hood used for aseptic transfer procedures.

spacers. The efficiency of HEPA filters is 99.97 percent for removal of particles that are 0.3 μm in diameter. In some industries, where any contamination could destroy the product, HEPA filters are used to supply the air to an entire work space. Industrial microbiologists in many industries are designing procedures to sterilize or reduce the number of microorganisms associated with the final product to acceptable levels.

It is especially critical that some products, such as surgical suturing material, be free of microbial contaminants. Representative samples of products must be checked frequently to monitor product quality. The discovery of a contaminated sample can lead to the recall and destruction of an entire production lot. Sometimes, notification of a product recall, such as for contaminated food items, is given through the mass media. The removal of contaminated products from use is costly but often essential. Quality control procedures must ensure that the number of microorganisms associated with a product is within the prescribed allowable limits for their safe use.

Production of Pharmaceuticals

The pharmaceutical manufacturing industry—a major source of employment for industrial microbiologists—is primarily concerned with disease processes, some of which are caused by microorganisms, and with making drugs, many of which are produced by microorganisms, to control disease processes. The world's supply of pharmaceuticals, including many antibiotics, steroids, vitamins, and vaccines, is produced in large part by microorganisms. The microbial production of pharmaceuticals is a major industry; antibiotic sales alone accounted for approximately $5 billion in worldwide sales in 1980. The role of microorganisms in producing these pharmaceuticals is economically important for industry and is essential for making these compounds available at a cost low enough to permit their wide use in preventing and treating numerous diseases. In this section, some representative examples will be discussed to illustrate the processes involved in the production of various pharmaceuticals.

Antibiotics   Of the thousands of different antibiotics, which are substances made in nature by various microorganisms that inhibit or kill other microorganisms, relatively few are produced commercially. The major antibiotics used in medicine and the microorganisms used for producing these antibiotics are shown in Table 17.3.

Penicillin. In a typical process for manufacturing **penicillin**, an inoculum of *Penicillium chrysogenum* is produced by inoculating a dense suspension of spores of the fungus onto a wheat bran-nutrient solution. The cultures are allowed to incubate for approximately 1 week at 24°C and are then transferred to an inoculum tank. In some cases, these spores are germinated to produce mycelia for inoculation into these tanks. The inoculum tanks are agitated with forced aeration for 1 to 2 days to provide a heavy mycelial growth for inoculation into a production tank. In some cases, additional step-up procedures are employed in which sequentially larger tanks are used to achieve larger amounts of mycelial inoculum for the production tanks.

TABLE 17.3   Some Antibiotics Produced by Microorganisms

Antibiotic	Produced by
Amphotericin B	*Streptomyces nodosus*
Bacitracin	*Bacillus licheniformis*
Carbomycin	*Streptomyces halstedii*
Chlorotetracycline	*Streptomyces aureofaciens*
Chloramphenicol	*Streptomyces venezuelae* or total chemical synthesis
Erythromycin	*Streptomyces erythreus*
Fumagillin	*Aspergillus fumigatus*
Griseofulvin	*Penicillium griseofulvin* *Penicillium nigricans* *Penicillium urticae*
Kanamycin	*Streptomyces kanamyceticus*
Neomycin	*Streptomyces fradiae*
Novobiocin	*Streptomyces niveus* *Streptomyces spheroides*
Nystatin	*Streptomyces noursei*
Oleandomycin	*Streptomyces antibioticus*
Oxytetracycline	*Streptomyces rimosus*
Penicillin	*Penicillium chrysogenum*
Polymyxin B	*Bacillus polymyxa*
Streptomycin	*Streptomyces griseus*
Tetracycline	Dechlorination and hydrogenation of chlorotetracycline; direct fermentation in dechlorinated medium
Vira A (adenine arabinoside)	*Streptomyces antibioticus*

The typical medium used for the production of penicillin has changed in the last few decades. Whereas in 1945 the typical medium contained 3.5 percent cornsteep liquor solids (waste product of starch manufacture), 3.5 percent lactose, 1 percent glucose, 1 percent calcium carbonate, 0.4 percent potassium phosphate, 0.25 percent vegetable oil, and a penicillin precursor, such as phenylacetic acid, the medium used today typically uses 10 percent total glucose or molasses by continuous feed, 4–5 percent cornsteep liquor solids, 0.5–0.8 percent total phenylacetic acid by continuous feed, and 0.5 percent total vegetable oil by continuous feed. The major change is the elimination of lactose from the medium and the use of continuous feed substrate addition to increase the efficiency of penicillin production. The phenylacetic acid is the precursor used to form the benzene ring side chain of the penicillin G molecule (see Figure 17.12 for the structure of the penicillin molecule). The addition of this precursor steers the fungal metabolic reactions to form increased amounts of penicillin.

The pH of the medium after sterilization is approximately 6, which is critical because penicillin is inactivated at both low and high pH values. The pH is maintained near neutrality during the course of the fermentation by the addition of alkali to the medium as needed. The incubation temperature for the fermentations is maintained at approximately 25–26°C, and aeration is provided during the production process.

The typical course of a penicillin fermentation takes 7 days, although longer times may be required when very large

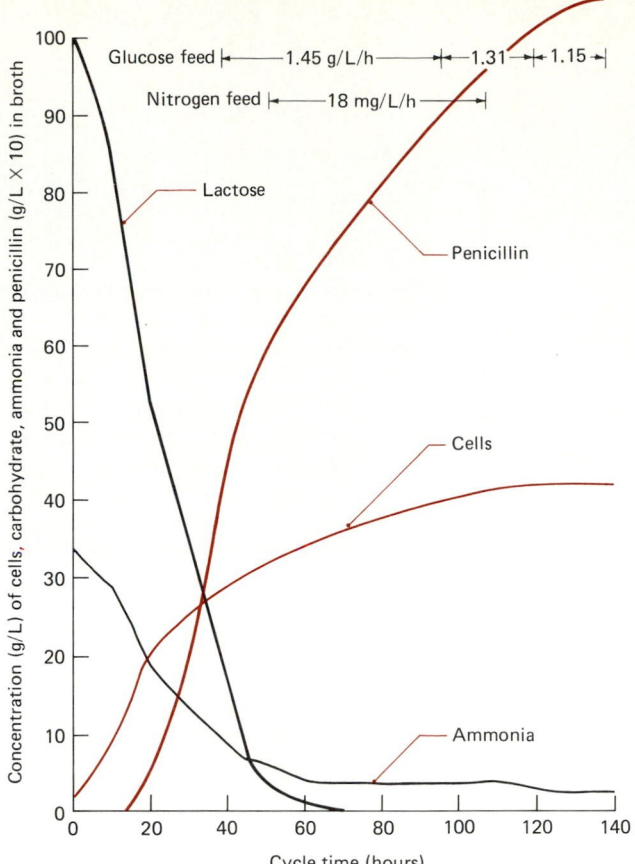

FIGURE 17.10

Time course changes in carbohydrate, nitrogen, penicillin, and biomass concentrations during penicillin fermentation. During the period of biomass accumulation, lactose and ammonia concentrations decline and there is little penicillin production. After 1 day of incubation, penicillin begins to accumulate. During this period, glucose and nitrogen are fed to the cells at rates that optimize penicillin production.

fermentors are used (Figure 17.10). During the first day of the fermentation, there is a large increase in the biomass of *Penicillium* mycelia. The carbohydrate substrate is rapidly used during this early phase, providing the necessary carbon and energy for the production of fungal mycelia. At a later stage, reduction of the carbohydrate concentration provides the necessary nutritional starvation conditions that favor penicillin production. The nitrogen required to support fungal growth comes from the cornsteep liquor. The production of penicillin, a secondary metabolite (**idiolite**) not required for the growth of the fungus, lags behind the accumulation of fungal biomass (**trophophase**). The accumulation of penicillin occurs in the (**idiophase**), which begins on the second day and reaches its maximal concentration a few days later (Figure 17.11).

When the fermentation is completed, the concentration of penicillin having reached maximal achievable levels, the liquid medium containing the penicillin is separated from the fungal cells, using a rotating vacuum filter. The fungal biomass is scraped from the surface of the filter drum, dried,

and marketed as an animal feed supplement. Penicillin is recovered from the filtrate, using various extraction procedures. It is extracted from the solution by using an organic solvent and then extracted back into aqueous solution. The exchange of penicillin back and forth between organic and aqueous solvents is accomplished by altering the pH and results in the partial purification of the antibiotic. Spent solvents used in the extraction of the penicillin are recycled. Potassium ions are then added to the aqueous solution, resulting in the formation of the crystalline potassium salt of penicillin G, which can be recovered by filtration or centrifugation. The filtered and dried penicillin salt is over 99.5 percent pure.

The penicillin G produced in this process can be further modified to form various penicillin derivatives (Figure 17.12). The modification of penicillin may be accomplished chemically or by using microbial enzymes. For example, 6-aminopenicillanic acid (6-APA) can be formed by fermentation, using bacterial acylase enzymes in an aqueous solution at 37°C. The same transformation of penicillin G to form 6-APA can also be accomplished chemically in three steps by using various chemical solvents, anhydrous conditions, and low temperatures. Similar transformations of the basic penicillin structure can also yield other penicillin derivatives. For example, in 1981 piperacillin was approved as a broad-spectrum antibiotic, and in 1982 azlocillin was also introduced for use against bacterial strains that are resistant to earlier-generation penicillins.

Cephalosporins. Similar semisynthetic approaches can be used for manufacturing other antibiotics. For example, cephalosporin C is made as the fermentation product of *Cephalosporium acremonium*, but this form of the antibiotic is not potent enough for clinical use. The cephalosporin C molecule, however, can be transformed by removal of an α-aminoadipic acid side chain to form 7-α-aminocephalosporanic acid, which can be further modified by adding side chains to form clinically useful products with relatively broad spectra of antibacterial action (Figure 17.13). Various side chains can be added to as well as removed from both 6-aminopenicillanic and 7-aminocephalosporanic acids to produce antibiotics with varying spectra of activities and varying degrees of resistance to inactivation by enzymes produced by pathogenic microorganisms. We are now into the so-called third-generation cephalosporins, such as moxalactam, which have been developed to combat bacteria that produce enzymes capable of degrading penicillins and cephalosporins.

Streptomycin. Streptomycin and various other antibiotics are produced using strains of *Streptomyces griseus*. As in penicillin fermention, spores of *S. griseus* are inoculated into a medium to establish a culture with a high mycelial biomass for introduction into an inoculum tank, with subsequent use of the mycelial inoculum to initiate the fermentation process in a production tank. The basic medium for the production of streptomycin contains soybean meal as the nitrogen source, glucose as the carbon source, and sodium chloride.

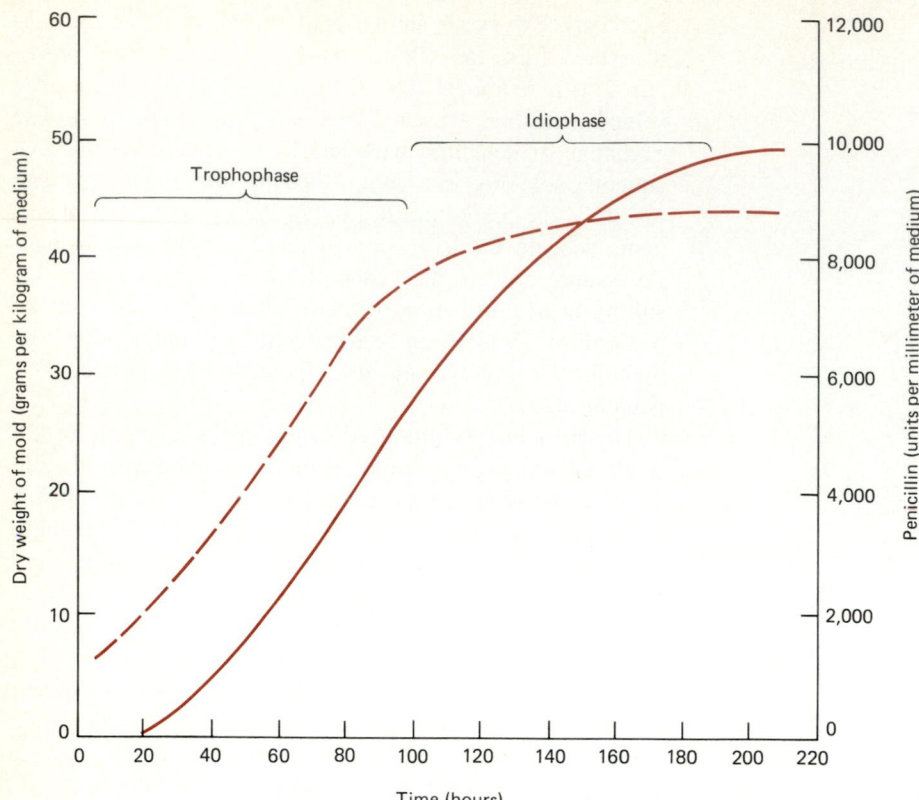

FIGURE 17.11

This graph shows how penicillin production lags behind production of fungal biomass. Penicillin, a secondary metabolite, is not formed as a direct result of the metabolism that keeps the cells alive. Hence, the accumulation of a secondary metabolite in a reactor vessel lags behind the growth of the cells that produce it. The graph shows the accumulation of mold cells (dotted line) and the subsequent accumulation of penicillin (solid line). The values of temperature and pH that are best for the growth of cells are seldom best for the synthesis of a secondary metabolite. In a batch process a compromise between the two sets of optimum conditions is sought.

FIGURE 17.12

The modification of penicillin G to 6-aminopenicillanic acid (6-APA) can be accomplished both chemically and microbiologically. The microbiological conversion (top reaction) has fewer steps and is less expensive. In the production of semisynthetic penicillins, 6-APA forms a chemical nucleus to which side chains can be attached, yielding new antibiotics. In the chemical conversion, three steps must be carried out at low temperatures and under strictly anhydrous conditions with a number of chemical solvents. The biological process relies on bacteria that make acylases, which are enzymes that cleave away the benzyl group, leaving 6-APA. The fermentation can be carried out in water at 37°C.

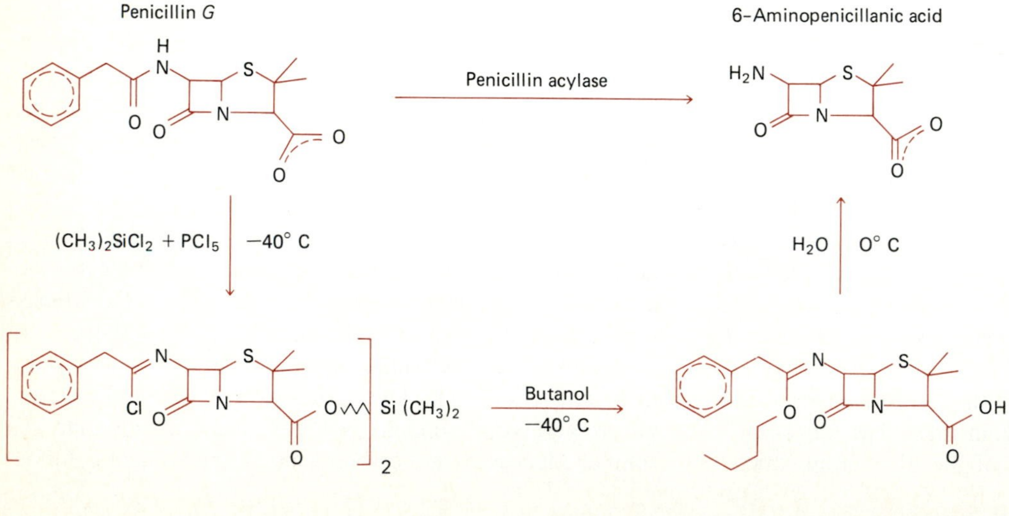

Penicillin *G*   6-Aminopenicillanic acid

Penicillin acylase

$(CH_3)_2SiCl_2 + PCl_5$ $-40°$ C

H_2O $0°$ C

Butanol
$-40°$ C

$Si(CH_3)_2$

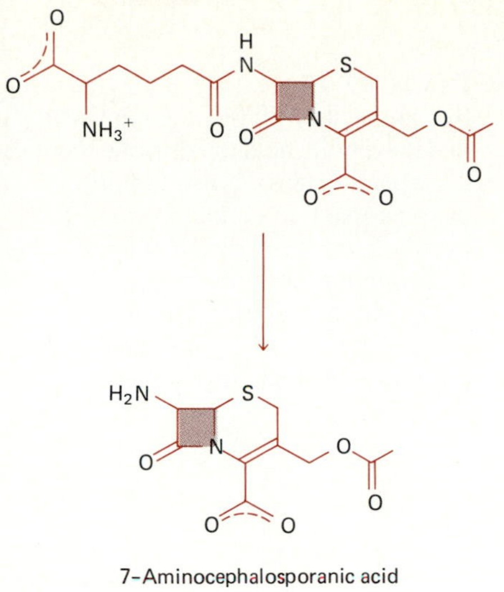

Cephalosporin *C*

7-Aminocephalosporanic acid

FIGURE 17.13

The conversion of cephalosporin C molecules to 7-aminocephalosporanic acid.

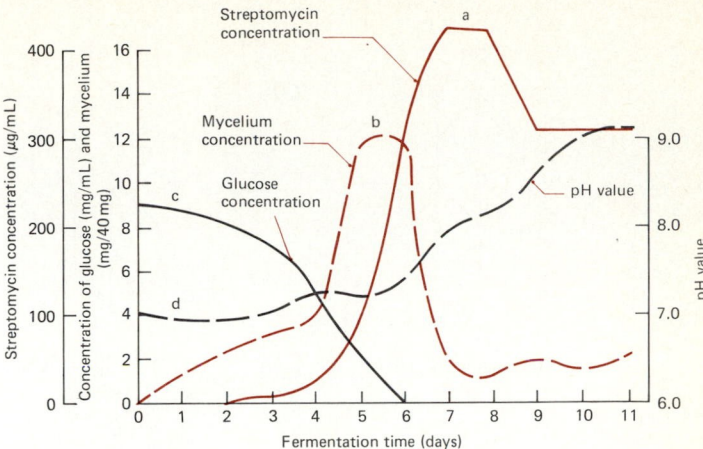

FIGURE 17.14

The time course of a streptomycin production process.
(a) streptomycin concentration; (b) mycelium concentration;
(c) glucose concentration; and (d) pH value.

The optimum temperature for this fermentation is approximately 28°C, and the maximal rate of streptomycin production is achieved in the pH range of 7.6–8.0. High rates of aeration and agitation are required to achieve maximal production of streptomycin. The fermentation process lasts for approximately 10 days and yields of streptomycin exceed 1 g/L.

The classic fermentation process for the production of streptomycin involves three phases (Figure 17.14). During the first phase there is rapid growth of *S. griseus*, with production of mycelial biomass. Proteolytic enzymatic activity of *S. griseus* releases ammonia to the medium from the soybean meal, causing a rise in pH. During this initial fermentation phase there is little production of streptomycin. During the second phase there is little additional production of mycelia, but the secondary metabolite streptomycin accumulates in the medium. The glucose added in the medium and the ammonia released from the soybean meal are consumed during this phase. The pH remains fairly constant, between 7.6 and 8.0. In the third and final phase of the fermentation, after depletion of carbohydrates from the medium, streptomycin production ceases and the bacterial cells begin to lyse. There is a rapid increase in pH because of the release of ammonia from the lysed cells, and the fermentation process normally is ended by the time the cells begin to lyse.

After completion of the fermentation, the mycelium is separated from the broth by filtration and the streptomycin is recovered. Streptomycin is a water-soluble basic substance and is insoluble in most organic solvents. One method of recovery and purification consists of adsorbing the streptomycin onto activated charcoal and eluting with acid alcohol. The antibiotic is then precipitated with acetone and further purified by using column chromatography. Several other chemical procedures can be employed for recovering and purifying streptomycin.

Steroids   The use of microorganisms to carry out biotransformations of steroids is very important in the pharmaceutical industry. Steroid hormones regulate various aspects of metabolism in animals, including humans. One such hormone, cortisone, has been found to relieve the pain associated with rheumatoid arthritis. Various cortisone derivatives are also useful in alleviating the symptoms associated with allergic and other undesired inflammatory responses of the human body. Additionally, various steroid hormones regulate human sexuality; some of them are manufactured as oral contraceptives. The physiological properties of a steroid depend on the nature and the exact position of the chemical constituents on the basic steroid ring structure. The chemical synthesis of steroids is very complex because of the requirement to achieve the necessary precision of substituent location.

For example, cortisone can be synthesized chemically from deoxycholic acid (Figure 17.15), but the process requires 37 steps, many of which must be carried out under extreme conditions of temperature and pressure, with the resulting product costing over $200 per gram. The major difficulty in chemically synthesizing cortisone is the need to introduce an oxygen atom at the number 11 position of the steroid ring, but this can be accomplished by microorganisms. The fungus *Rhizopus arrhizus*, for example, hydroxylates progesterone, forming another steroid with the introduction of oxygen at the number 11 position (Figure 17.16). The fungus *Cunninghamella blakesleeana* similarly can hydroxylate the steroid cortexolone to form hydrocortisone, with the introduction of oxygen at the number 11 position. Other transformations of the steroid nucleus carried

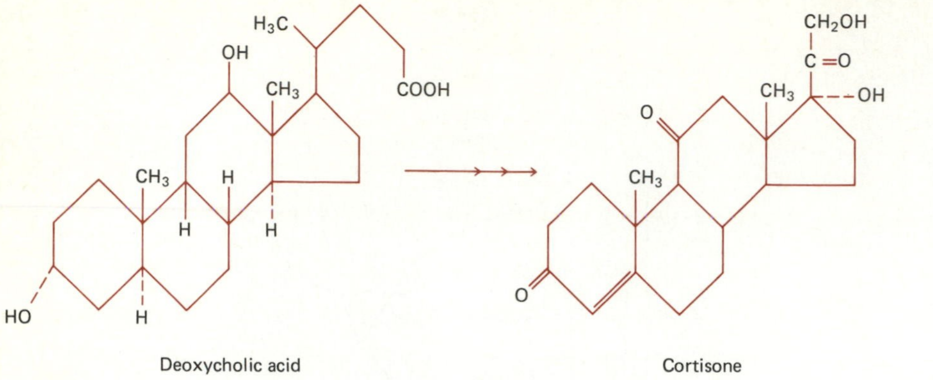

Deoxycholic acid   Cortisone

FIGURE 17.15

By chemical synthesis, the conversion of deoxycholic acid to cortisone requires 37 separate steps. It also requires extreme reaction conditions, and the need to introduce oxygen at the number 11 position represents a major difficulty. Microorganisms can introduce oxygen specifically at the number 11 position in a single reaction.

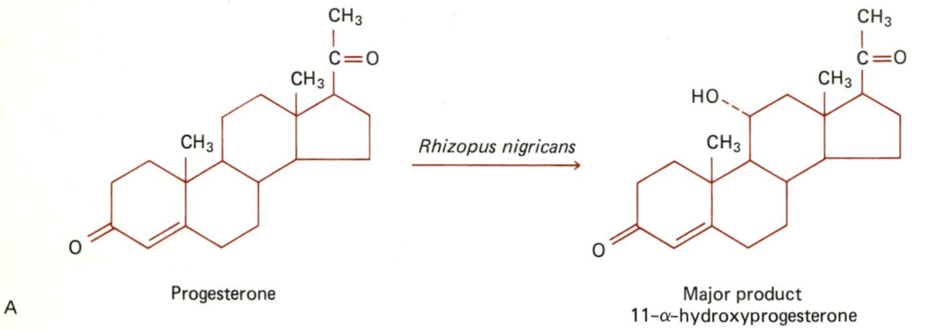

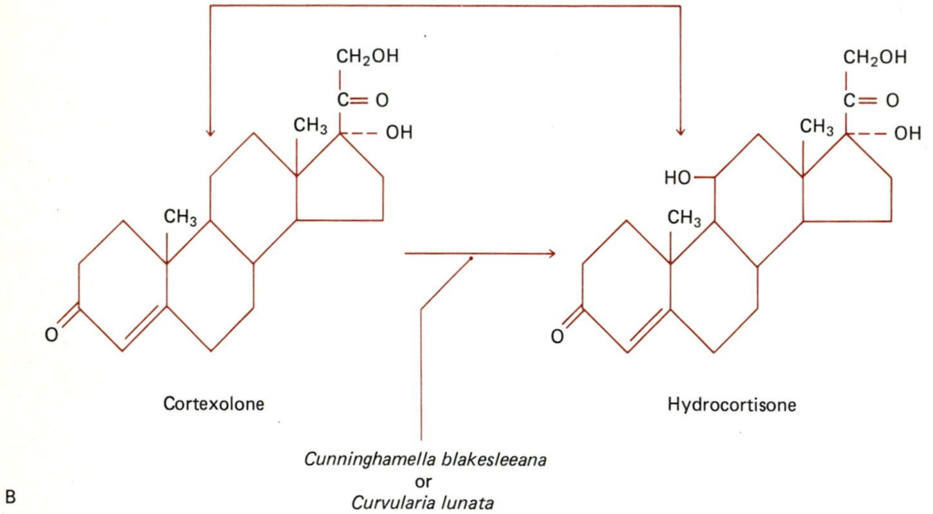

A

Progesterone   *Rhizopus nigricans*   Major product
11–α–hydroxyprogesterone

B

Cortexolone   Hydrocortisone

Cunninghamella blakesleeana
or
Curvularia lunata

FIGURE 17.16

The specificity of microbial enzymes allows them to be used for transformations of various cortisone derivatives. The hydroxylation of (A) progesterone by *Rhizopus arrhizus* and (B) cortexolone by *Cunninghamella blakesleeana* are reactions that illustrate the use of microorganisms to hydroxylate the cortisone ring in order to form new cortisone products.

out by microorganisms include hydrogenations, dehydrogenations, epoxidations, and removal and addition of side chains. The use of such microbial transformations in the formation of cortisone has lowered the original cost over 400-fold, so that in 1980 the price of cortisone in the United States was less than 50 cents per gram, compared to the original $200.

In a typical steroid transformation process, the microorganism, such as *Rhizopus nigricans*, is grown in a fermentation tank, using an appropriate growth medium and incubation

conditions to achieve a high biomass. In most cases, aeration and agitation are employed to achieve rapid growth. After the growth of the microorganisms, the steroid to be transformed is added, and so, for example, progesterone is added to a fermentor containing *R. nigricans* that has been growing for approximately 1 day, and the steroid is hydroxylated at the number 11 position to form 11-α-hydroxyprogesterone. The product is then recovered by extraction with methylene chloride or various other solvents, purified chromatographically, and recovered by crystallization. A large number of similar transformations are carried out to produce a great variety of steroid derivatives for different medicinal uses (Figure 17.17).

Human Proteins   In addition to its impact on many other fields of microbiology, genetic engineering has expanded the roles of microorganisms in the pharmaceutical industry to include the production of human proteins. By using recombinant DNA technology, human DNA sequences that code for various proteins have been incoporated into the genomes

of bacteria. By growing these recombinant bacteria in fermentors, human proteins can be produced commercially. Human insulin, for example is produced by a recombinant *Escherichia coli* strain and marketed as Humulin. Other strains are used to produce human growth hormone, tumor necrosis factor (TNF), interferon (human recombinant beta interferon—trade name, Betaseron), and interleukin-2 (human recombinant interleukin-2—trade name, Proleukin). Humulin is used to treat diabetes in cases where the individual is

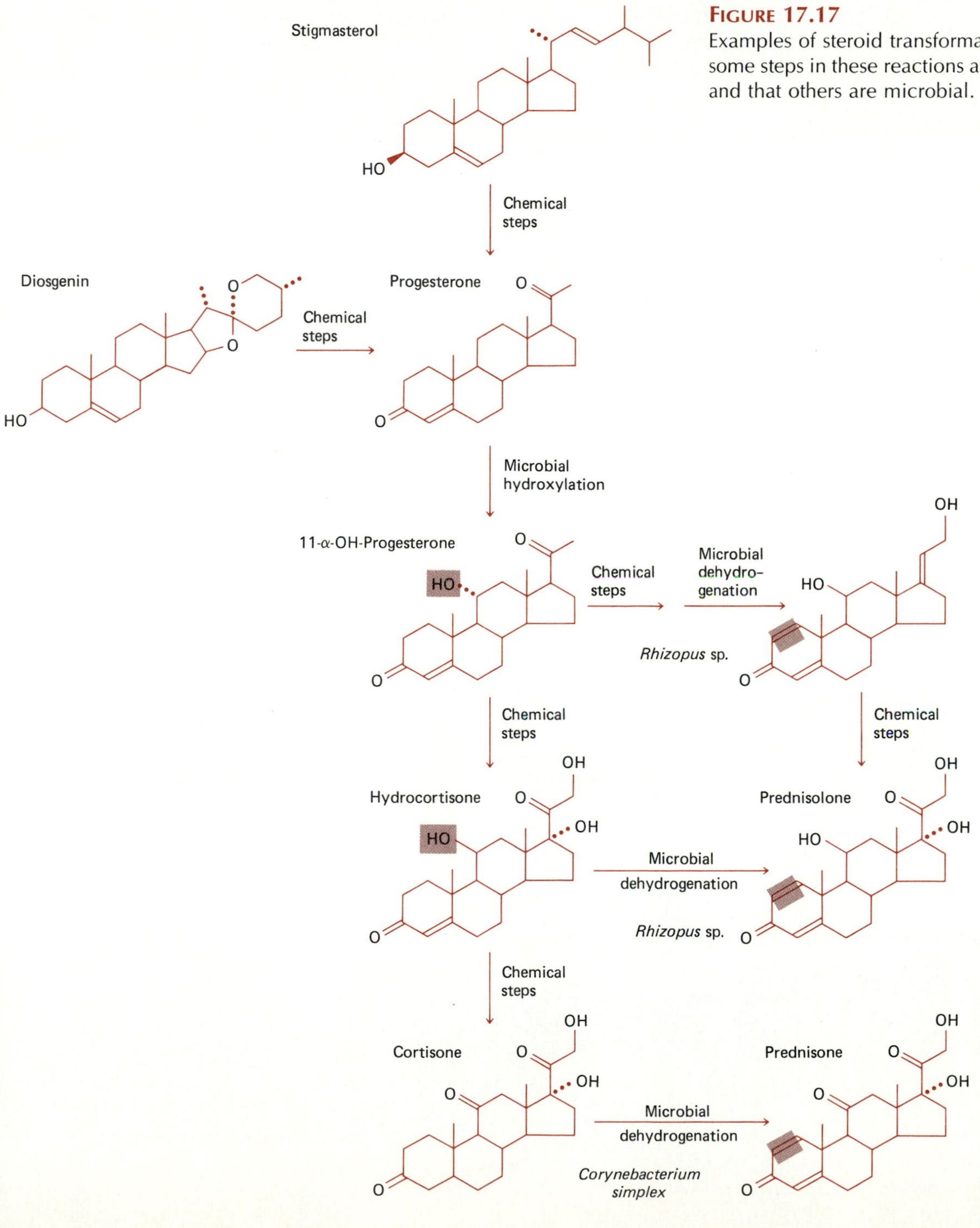

FIGURE 17.17
Examples of steroid transformations. Note that some steps in these reactions are chemically induced and that others are microbial.

allergic to insulin harvested from cattle. Human growth factor is used to treat diseases, such as dwarfism, resulting from a deficiency of this hormone. Interleukin-2, interferon, and TNF are important components of the natural human immune response, and their production may prove useful in treating some diseases where increased levels of these substances would be therapeutic. Interferon, for example, is important in the defense against viruses, and it may prove useful in treating viral infections. TNF is a natural substance produced in the body in small amounts by certain white blood cells called *macrophages* that appears to kill some cancer cells and infectious microorganisms without adversely affecting most normal cells. The production of large amounts of TNF by recombinant bacteria is aiding in the investigation of its potential use in the treatment of cancer. Undoubtedly, recombinant DNA technology will permit the production of other useful human proteins.

Vaccines   The use of **vaccines** is extremely important for preventing various serious diseases. The development and production of these vaccines constitute an important function of the pharmaceutical industry. The production of vaccines involves growing microorganisms possessing the antigenic properties needed to elicit a primary immune response. Vaccines are produced either by mutant strains of pathogens or by attenuating or inactivating virulent pathogens without removing the antigens necessary for eliciting the immune response.

For the production of vaccines against viral diseases, strains of the virus often are grown by using embryonated eggs (Figure 17.18). Individuals who are allergic to eggs cannot be given such vaccine preparations. Viral vaccines may also be produced by using **tissue culture**. For example, the older rabies vaccine, which was produced in embryonated

duck eggs and had painful side effects, has been replaced with a vaccine produced in human fibroblast tissue cultures that has far fewer side effects. The production of vaccines that are effective in preventing diseases caused by bacteria, fungi, and protozoa generally involves growing the microbial strain on an artificial medium, which minimizes problems with allergic responses. Commercially produced vaccines must be tested and standardized before use. It is critical that the vaccine not contain active forms of a virulent pathogen, lest the vaccine transmit the disease it aims to prevent. Unfortunately, there have been several outbreaks of disease associated with improperly prepared vaccines. High standards of quality control and appropriate safety test procedures can prevent such incidents. Various vaccines, including some new vaccines produced with recombinant microorganisms, will be discussed in Chapter 19.

Vitamins   Several **vitamins**, which are essential animal nutritional factors, can be produced by microbial fermentation (Table 17.4). Vitamin B_{12}, for example, can be produced as a by-product of *Streptomyces* antibiotic fermentations (Figure 17.19). A soluble cobalt salt is added to the fermentation reaction as a precursor to vitamin B_{12}. Relatively high amounts of this vitamin accumulate in the medium at concentrations that are not toxic to the *Streptomyces*. Vitamin B_{12} can also be produced commercially by direct fermentation, using *Propionibacterium shermanii* or *Pseudomonas denitrificans*, and these are the organisms used today for the production of this vitamin. *P. shermanii* can be grown in anaerobic culture for 3 days and in aerobic culture for 4 days to produce vitamin B_{12}. The growth medium for vitamin B_{12} production by these organisms contains glucose, cornsteep liquor (a waste product of starch manufacture), and cobalt chloride. The medium is maintained at pH 7 by using ammo-

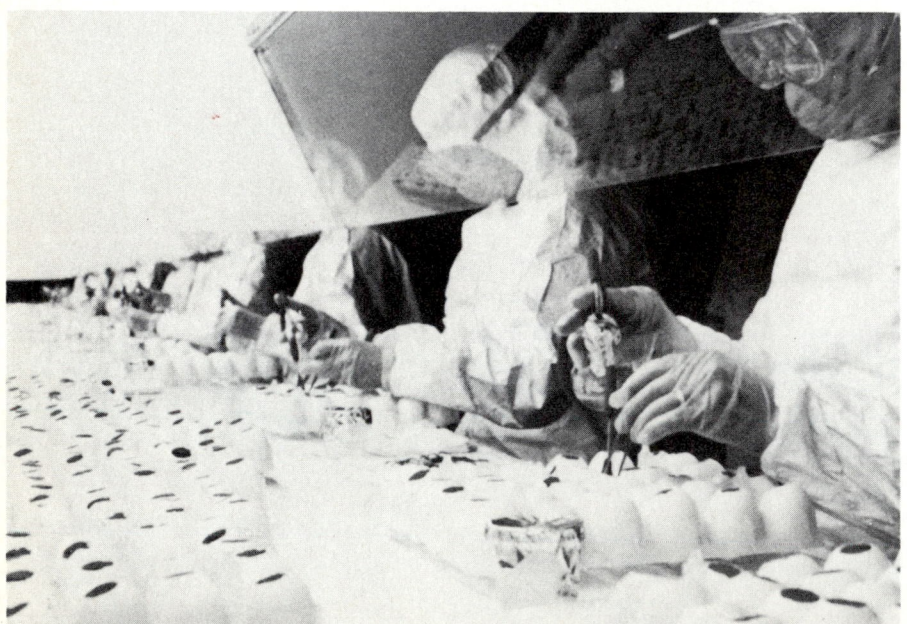

FIGURE 17.18
This photograph illustrates the large-scale production of vaccines, using inoculated embryonated eggs. (Courtesy Wyeth Laboratories, Philadelphia)

TABLE 17.4 Production of Some Vitamins Using Microorganisms

Vitamin	Culture	Medium	Fermentation Conditions	Yield
Riboflavin	*Ashbya gossypii*	Glucose, collagen, soya oil, glycine	6 days at 36°C, aerobic	4.25 g/L
L-Sorbose (in vitamin C synthesis)	*Gluconobacter oxidans* subsp. *suboxidans*	D-Sorbitol, 30% cornsteep	45 hours at 30°C, aerobic	70% based on substrate used
5-Keto gluconic acid (in vitamin C synthesis)	*Gluconobacter oxidans* subsp. *suboxidans*	Glucose, CaCO₃, cornsteep	33 hours at 30°C, aerobic	100% based on substrate used
Vitamin B₁₂	*Propionibacterium shermanii*	Glucose, cornsteep, ammonia, cobalt, pH 7.0	3 days at 30°C, anaerobic, + 4 days, aerobic	23 mg/L

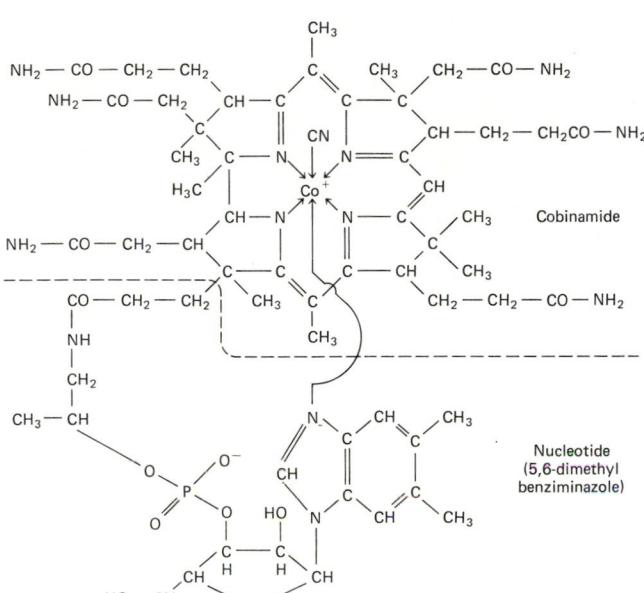

FIGURE 17.19

(A) The structure of vitamin B₁₂, cyanocobalamin, (B) Photograph showing column chromatography, which is used for the purification of vitamin B₁₂. (Courtesy Merck, Sharp and Dohme, Rahway, N.J.)

nium hydroxide. *P. denitrificans* is grown for 2 days in aerated culture for vitamin B₁₂ production, using a medium containing sucrose, betaine, glutamic acid, cobalt chloride, 5,6-dimethylbenzimidazole, and salts.

Riboflavin can also be produced as a fermentation product by using various microorganisms. Riboflavin is a by-product of acetone butanol fermentation and is produced by various *Clostridium* species. Commercial production of riboflavin by direct fermentation often uses the fungal species *Eremothecium ashbyii* or *Ashbya gossypii*. Riboflavin production using such fungi employs a medium containing glucose and/or corn oil. Corn oil may be added even when glucose is used as the primary growth substrate to increase yields of riboflavin. The

fermentation using *Ashbya gossypii* to produce riboflavin is normally carried out at 26–28°C, pH 6–7.5 for approximately 4–5 days. After growth of the yeast, the cells are recovered and used as a feed supplement for animals to supply needed riboflavin. Various other vitamins can also be produced by fermentation, but relatively low yields often limit their economic potential.

Production of Organic Acids

Several organic acids, including acetic, gluconic, citric, itaconic, gibberellic, and lactic acids, can be produced by microbial fermentation (Table 17.5).

TABLE 17.5   Some Organic Acids Produced by Fermentation

Product	Culture	Substrate (yield—%)	Process
Acetic acid	*Acetobacter* spp.	Ethanol (98–99)	Continuous aerated process using an alcoholic solution containing (%): glucose, 0.9; ammonium phosphate, 0.4; magnesium sulfate, 0.1; potassium citrate, 0.1; pantothenic acid, 0.00005. Extraction by filtration.
Lactic acid	*Lactobacillus delbrueckii*	Milk whey, molasses, pure sugars (90)	10–15% glucose, 5–6 days, 50°C in corrosion-resistant fermentor; pH 5.5–6.0 buffered with $CaCO_3$; no aeration, growth factors provided by malt. Extraction by precipitation after heating to 80°C and the addition of chalk (calcium lactate is formed); extraction with solvents; esterification with methanol followed by distillation.
Fumaric acid	*Rhizopus* spp.	Glucose (60)	3 days at 30°C with aeration; pH 5–6 maintained by the addition of NaOH. Extraction by acidification of media and crystallization.
Gluconic acid	*Aspergillus niger*	Glucose and cornsteep liquor (90)	36 hours at 30°C with aeration; pH 6.5. Extraction by filtration and purification using cation exchange column.

Acetic Acid The production of acetic acid or vinegar has been discussed in Chapter 16. Acetic acid can also be used for other commercial purposes, for example, as a stop bath in photographic processing. However, for economic reasons, most production for such purposes today is accomplished using chemical syntheses.

Gluconic Acid Gluconic acid also has a variety of commercial uses. Calcium gluconate, for example, is used as a pharmaceutical to supply calcium to the body; ferrous gluconate similarly is used to supply iron in the treatment of anemia; and gluconic acid in dishwasher detergents prevents spotting of glass surfaces due to the precipitation of calcium and magnesium salts. Gluconic acid is produced by various bacteria, including *Acetobacter* species, and by several fungi, including *Penicillium* and *Aspergillus* species. *Aspergillus niger*, for example, converts glucose to gluconic acid in a single enzymatic reaction (Figure 17.20).

The commercial production of gluconic acid, using *A. niger*, employs a submerged culture process. *A. niger* is initially grown to form a sufficient amount of mycelia, after which the conversion of glucose to gluconic acid, mediated by the fungal enzyme glucose oxidase, is purely an enzymatic reaction. A typical growth medium for the production of gluconic acid contains approximately 25 percent glucose, various salts, calcium carbonate, and a compound containing the element boron. The boron in the medium stabilizes calcium gluconate, maintaining this compound in solution and preventing its precipitation and permitting the use of excess calcium carbonate to neutralize most of the gluconic acid produced while keeping the pH within acceptable limits. The fermentation is conducted at 30°C with aeration and agitation. Cooling coils are used to control the heat generated in this oxidative process. The growth of fungal mycelia is limited by the concentration of nitrogen in the medium. The gluconic acid is recovered from the fermentation by addition

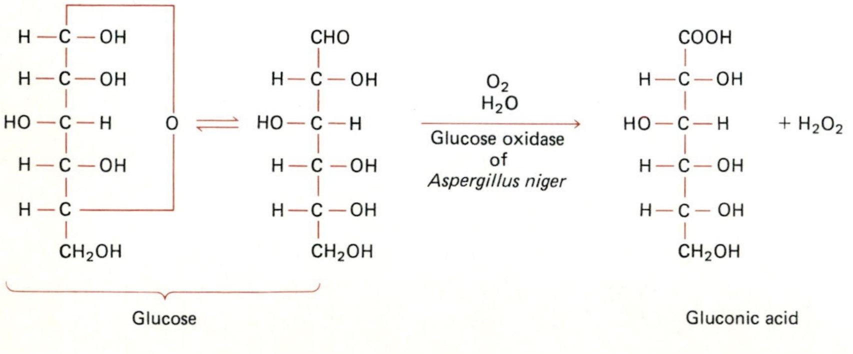

FIGURE 17.20
The conversion of glucose to gluconic acid by oxidation by the glucose oxidase of *Aspergillus niger*.

of calcium hydroxide to form crystalline calcium gluconate. Free gluconic acid can then be recovered by the addition of acid.

Citric Acid **Citric acid** is also produced by cultures of *A. niger*. Commercially produced citric acid is used in various ways, including as a food additive, especially in the production of soft drinks; as a metal chelating and sequestering agent; and as a plasticizer. The composition of the fermentation medium is critical for obtaining high yields of citric acid. It is essential to limit the growth of the fungus so that high levels of citric acid can accumulate; this can be accomplished by having a deficiency of trace metals or phosphate in the medium. A typical medium for the production of citric acid contains molasses, ammonium nitrate, magnesium sulfate, and potassium phosphate. Acid is added to achieve a low pH, and some of the metals in the medium are complexed with ferricyanide, removing them from solution; alternatively, metals are removed using cation exchange resins.

Itaconic Acid **Itaconic acid** is used as a resin in detergents. The transformation of citric acid by *Aspergillus terreus* can be used for the fermentative production of itaconic acid, although chemical procedures for producing this compound are also available (Figure 17.21). The fermentation process uses a well-aerated molasses–mineral salts medium at a very low pH, below 2.2. At higher pH values *A. terreus* degrades itaconic acid, and the desired product obviously would not accumulate. As citric acid fermentation, low levels of trace metals must be removed to achieve acceptable product yields (Table 17.6). The development of fungal mycelia in this fermentation is intentionally limited, often by using a low inoculum size, in order to produce high accumulations of itaconic acid. Recovery is accomplished by evaporation of the fermentation medium to crystallize the itaconic acid.

Gibberellic Acid **Gibberellic acid** and related gibberellins are plant hormones and are extensively used as growth-promoting substances to stimulate plant growth, flowering, and seed germination and to induce the formation of seedless fruit. The commercial production of gibberellins can be used to enhance agricultural productivity. Gibberellic acid is formed by the fungus *Gibberella fujikuroi* (= *Fusarium moniliforme*) and can be produced commercially, using aerated submerged culture. A glucose–mineral salts medium, an incubation temperature of approximately 25°C, and slightly acidic pH conditions are employed for the production of gib-

FIGURE 17.21

The conversion of citric acid to itaconic acid.

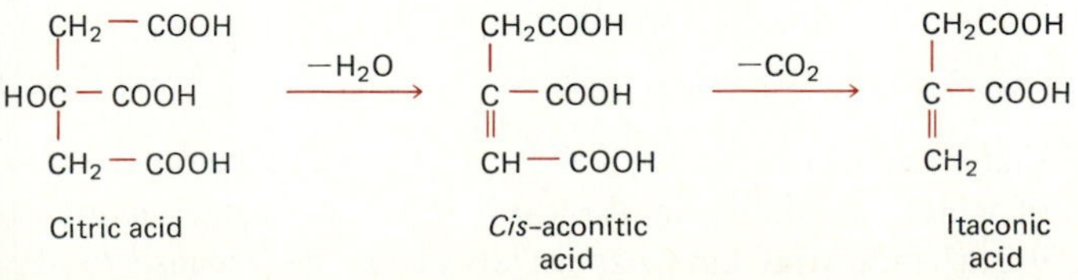

TABLE 17.6 Effects of the Concentrations of Some Metals in the Fermentation Medium on Itaconic Acid Production by a Mutant of *Aspergillus terreus*

Element (mg/L)	Yield (% conversion)
Zinc	
0	16
0.5	43
6	50
Copper	
0.5	55
1	52
3	53
6	55
Calcium	
0	9
337	43
2700	59
Iron	
0	57
1	25
2	17
4	17

berellic acid. Production normally takes 2 to 3 days, with accumulation of gibberellic acid lagging behind the growth of the fungus.

Lactic Acid **Lactic acid** has various commercial uses. It is used in foods as a preservative, in leather production for deliming hides, and in the textile industry for fabric treatment. Various forms of lactic acid are also used for other purposes—in resins as polylactic acid, in plastics as various derivatives, in electroplating as copper lactate, and in baking powder and animal feed supplements as calcium lactate. The formation of lactic acid in making various fermented dairy products has been discussed in Chapter 16. *Lactobacillus delbrueckii* is widely used in the commercial production of lactic acid, but various other *Lactobacillus*, *Streptococcus*, and *Leuconostoc* species also are of industrial importance for the production of this compound.

The typical medium for the production of lactic acid normally contains 10–15 percent glucose or another fermentable sugar, 10 percent calcium carbonate to neutralize the lactic acid formed, and ammonium phosphate and trace amounts of other nitrogen sources. Corn sugar, beet molasses, potato starch, and whey are often used as sources of carbohydrates for this fermentation. A typical production process for lactic acid uses an incubation temperature of 45–50°C and a pH of 5.5–6.5. The fermentor is agitated to suspend the calcium carbonate but is not aerated because this is an anaerobic process. The fermentation is normally completed within 5 to 7 days, with approximately 90 percent of the sugar converted to lactic acid. After the fermentation, calcium carbonate is

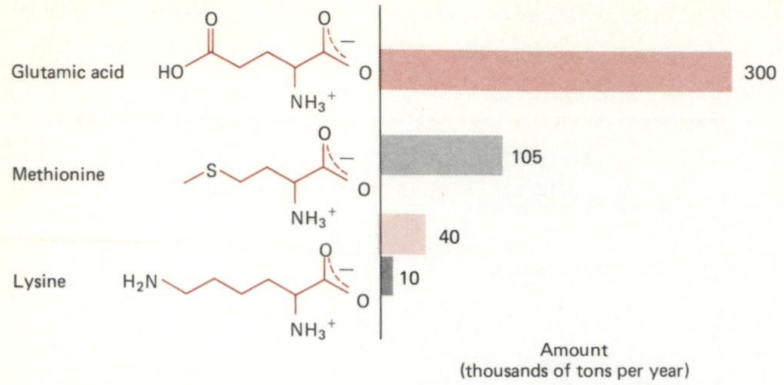

FIGURE 17.22

Amino acids are commercially important products. Worldwide sales of these products totaled $1.7 billion in 1980. Glutamic acid, methionine, and lysine are the ones now made in the largest quantities. Methionine and lysine are nutritionally essential amino acids used as animal feed additives. Glutamic acid and 80 percent of the lysine is made by fermentation. Methionine is manufactured by chemical synthesis. All amino acids have two isomers, only one of which can participate in biological reactions. Fermentation yields only the biologically active isomer. In chemical synthesis, half of the yield is the inactive one. This specificity makes the biological means more efficient, but it has not always been possible to exploit it. With increased understanding of cellular metabolism, all industrially valuable amino acids may soon be made by fermentation.

added to raise the pH to 10, and the medium is heated and filtered. This procedure kills the bacteria, coagulates proteins, removes excess calcium carbonate, and decomposes residual carbohydrates. The recovery of lactic acid of high enough purity for some applications is difficult to achieve, and the cost of recovery has forced the replacement of lactic acid with alternative chemicals for some commercial uses.

Production of Amino Acids

Microbial production of the amino acids lysine and glutamic acid presently accounts for over $1 billion in annual worldwide sales (Figure 17.22). Animals require various amino acids in their diets. Lysine and methionine are essential amino acids but are not present in sufficient concentrations in grains to meet animal nutritional needs. Lysine produced by microbial fermentation and methionine produced synthetically are used as animal feed supplements and as additives in cereals. Glutamic acid is principally made for use as monosodium glutamate (MSG), an ingredient in soup production widely used as a flavor enhancer. The flavoring industry in the United States consumed more than 30,000 tons of MSG in 1980, some of which was imported from Japan, a major producer of amino acids by fermentation, as well as some from Taiwan and South Korea.

In the microbial production of amino acids, only the desired L-isomer is formed, whereas their chemical synthesis produces a racemic mixture that requires costly separation procedures to remove the biologically inactive D-isomer half

of the mixture. The major problem in using microbial fermentation for commercial production of amino acids is overcoming the natural microbial regulatory control mechanisms that limit the amount of amino acid produced and released from the cells. Commercial amino acid production processes have successfully overcome these restrictions, and in the future, genetically engineered strains with defective control mechanisms and membranes will undoubtedly permit the economic production of a variety of amino acids by microbial fermentation.

Lysine   The direct production of L-lysine from carbohydrates uses a homoserine-requiring auxotroph of *Corynebacterium glutamicum*. The blocking of homoserine synthesis at the level of homoserine dehydrogenase results from feedback inhibition of that enzyme and leads to the accumulation of lysine (Figure 17.23A). Cane molasses is generally used as the substrate, and the pH is maintained near neutrality by adding ammonia or urea. As the sugar is metabolized, lysine accumulates in the growth medium. Through the use of the homoserine-requiring auxotroph, about 50 g/L of lysine can be produced in 2 to 3 days (Figure 17.23B).

Glutamic Acid   L-Glutamic acid and MSG can be produced by direct fermentation, using strains of *Brevibacterium*, *Arthrobacter*, and *Corynebacterium*. Cultures of *C. glutamicum* and *Brevibacterium flavum* are widely used for the large-scale production of MSG. The fermentation process employs a glucose–mineral salts medium and periodic additions of urea as a nitrogen source during the course of the fermentation; the pH is maintained at 6–8 and the temperature at about 30°C, and the medium is well aerated. The difficulty in the production of glutamic acid, as well as other amino acids, by direct fermentation is getting the cells to secrete sufficient quantities of the amino acid to permit commercial production. There are several methods for inducing leaky membranes that permit excretion of the amino acid product from the cell. One approach is to grow *C. glutamicum* in a medium with suboptimal concentrations of biotin (Figure 17.24). Without an adequate supply of biotin, the cells form membranes that are deficient in phospholipids, and the glutamic acid is secreted through these leaky membranes. Another approach is to add fatty acids or surface active agents (detergents) to disrupt the membranes and release the glutamic acid from the cells. Still another way of causing the cell to excrete amino acids is to add penicillin to the medium during the log phase of growth, causing the bacteria to become leaky and release glutamic acid to the surrounding medium. Adjusting the pH and adding sodium chloride can be used to convert glutamic acid to the desired MSG.

Production of Enzymes

Enzymes have a variety of commercial applications, some of which are shown in Table 17.7. Microbial production of useful industrial enzymes is advantageous because of the

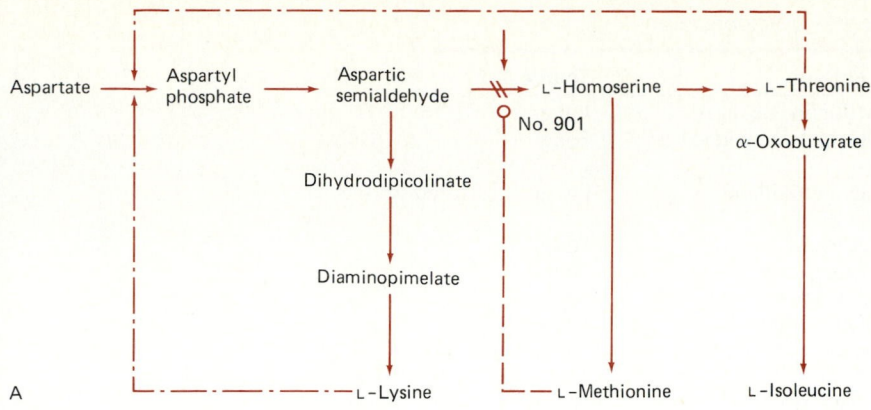

A

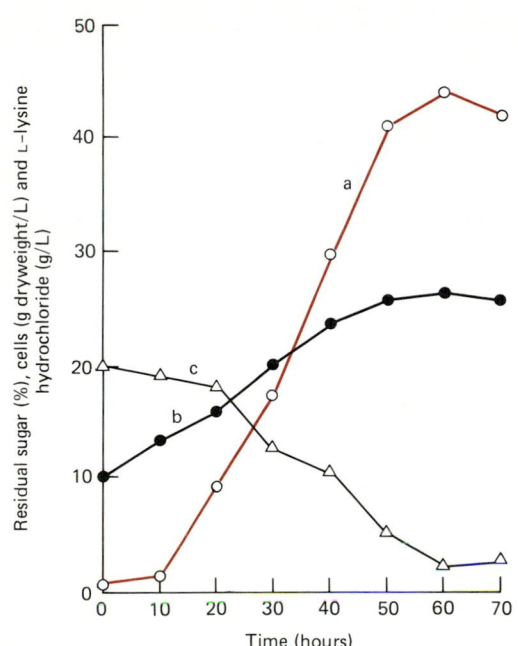

B

FIGURE 17.23

The production of L-lysine by *Corynebacterium glutamicum*. (A) The regulation of lysine biosynthesis by *C. glutamicum*. Dotted lines = operation of feedback inhibition; dashed lines = repression of enzyme synthesis. (B) A time course graph for this fermentation. Line a = changes in the concentration of L-lysine; line b = changes in the concentrations of cells; and line c = changes in the concentration of residual sugar.

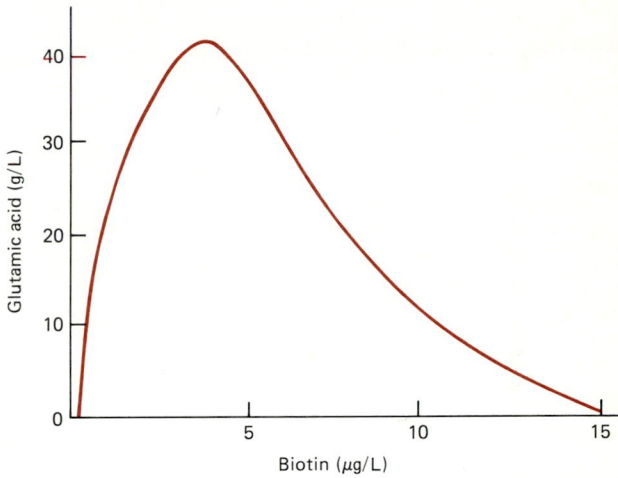

FIGURE 17.24
The effect of biotin concentration on the yield of glutamic acid.

large number of enzymes and the virtually unlimited supply that can be produced by microorganisms. A generalized scheme for the microbial production of commercial enzymes is shown in Figure 17.25. Enzymes produced for industrial processes include proteases, amylases, glucose isomerase, glucose oxidase, rennin, pectinases, and lipases. The four extensively produced microbial enzymes are protease, glucamylase, alpha-amylase, and glucose isomerase (Figure 17.26).

Proteases   **Proteases** are a class of enzymes that attack the peptide bonds of protein molecules, forming small peptides. Proteases produced by different bacterial species are used for different industrial purposes. The largest commercial application of bacterial alkaline proteases is in the laundry industry, principally in modern detergent formulations. The general trend toward the use of nonphosphate laundry detergents that function at lower wash temperatures has led to the increasing incorporation of enzymes into liquid and powdered detergents to improve their cleaning performance. In the United States, the proportion of detergents containing enzymes increased from 2.5 percent in 1979 to about 25 percent in 1985. In the cleaning industry proteases are used as spot removers in dry cleaning, as presoak treatments in laundering, and in laundry detergents. The action of the enzyme degrades various proteinaceous materials, such as milk and eggs, forming small peptide fragments that can be washed out readily. In dry cleaning, proteases are effective spot removers and are even useful in removing blood spots. These protease enzymes are relatively heat stable and are able to remain active in warm-hot water long enough for them to

TABLE 17.7   Important Uses for Enzymes Produced by Microorganisms

Industry	Application	Enzyme	Source
Analytical	Sugar determination	Glucose oxidase	Fungi
	Glycogen determination	Galactose oxidase	Fungi
	Uric acid determination	Urate oxidase	Fungi
Baking	Bread Baking	Amylase	Fungi
		Protease	Fungi
Brewing	Mashing-making beer	Amylase	Bacteria
		Glucamylase	Fungi
Carbonated beverages	Oxygen removal	Glucose oxidase	Fungi
Cereals	Breakfast foods	Amylase	Fungi
Chocolate, cocoa	Syrups	Amylase	Fungi, bacteria
Coffee	Coffee bean fermentation	Pectinase	Fungi
Confectionery	Soft-center candies	Invertase	Bacteria, fungi
Dairy	Cheese production	Rennin	Fungi
Dry cleaning	Spot removal	Protease, amylase	Bacteria, fungi
Eggs, dried	Glucose removal	Glucose oxidase	Fungi
Fruit juices	Clarification	Pectinases	Fungi
	Oxygen removal	Glucose oxidase	Fungi
	Debittering of citrus	Naringinase	Fungi
Laundry	Spot removal	Protease, amylase	Bacteria
	Cold-soluble laundry starch	Amylase	Bacteria
Leather	Bating	Protease	Bacteria, fungi
Meat	Meat tenderizing	Protease	Fungi, bacteria
Mayonnaise and salad dressings	Oxygen removal	Glucose oxidase	Fungi
Paper	Starch modification for paper coating	Amylase	Bacteria
Pharmaceutical and clinical	Digestive aids	Amylase	Fungi, bacteria
		Protease	Fungi, bacteria
		Lipase	Fungi
		Cellulase	Fungi
	Wound debridement (tissue removal)	Streptokinase-streptodornase	Bacteria
		Protease	
Photographic	Recovery of Silver from spent film	Protease	Bacteria
Plumbing	Drain opener	Keratinase (protease)	Bacteria
Starch and syrup	Corn syrups	Amylase, dextrinase	Fungi
		Glucose isomerase	Fungi, bacteria
	Production of glucose	Glucamylase, amylase	Fungi, bacteria
Textile	Desizing of fabrics	Amylase, protease	Bacteria
Wine	Clarification	Pectinases	Fungi

degrade the proteinaceous materials contaminating the fabric being washed. When used as a presoak, protease enzymes have sufficient time to act to degrade insoluble proteinaceous materials staining the fabric. Currently proteases for detergents are largely produced by *Bacillus licheniformis*. The enzymes produced by these *Bacillus* strains are active against protein molecules that make up common stains such as blood and grass.

Other alkaline proteases are being developed, using recombinant DNA technology, to function over a wide pH and temperature range, to remain stable under alkaline conditions

in the presence of detergent components such as sequestering agents, surfactants, and bleach, and to exhibit long shelf life stability. One recombinant strain, *Bacillus* sp. GX6644, secretes an alkaline protease that is highly active toward the milk protein casein, with highest activity occurring at pH 11 and at moderate temperatures of 40–55°C. Another recombinant strain, *Bacillus* sp. GX6638, produces several alkaline proteases, one of which remains active over a broad pH range (8–12), exhibits exceptional stability under highly alkaline conditions (88 percent of the initial activity at pH 12 after 25 hours), and functions in the presence of bleach.

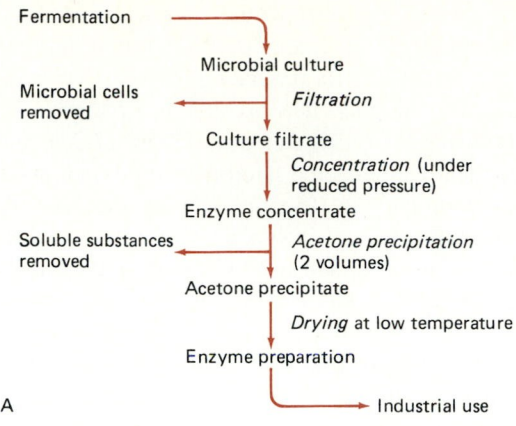

Fermentation
└→ Microbial culture

Microbial cells ←─── *Filtration*
removed

Culture filtrate
│ *Concentration* (under
│ reduced pressure)
Enzyme concentrate

Soluble substances ←─── *Acetone precipitation*
removed   (2 volumes)

Acetone precipitate
│ *Drying* at low temperature
Enzyme preparation
└─────→ Industrial use

A

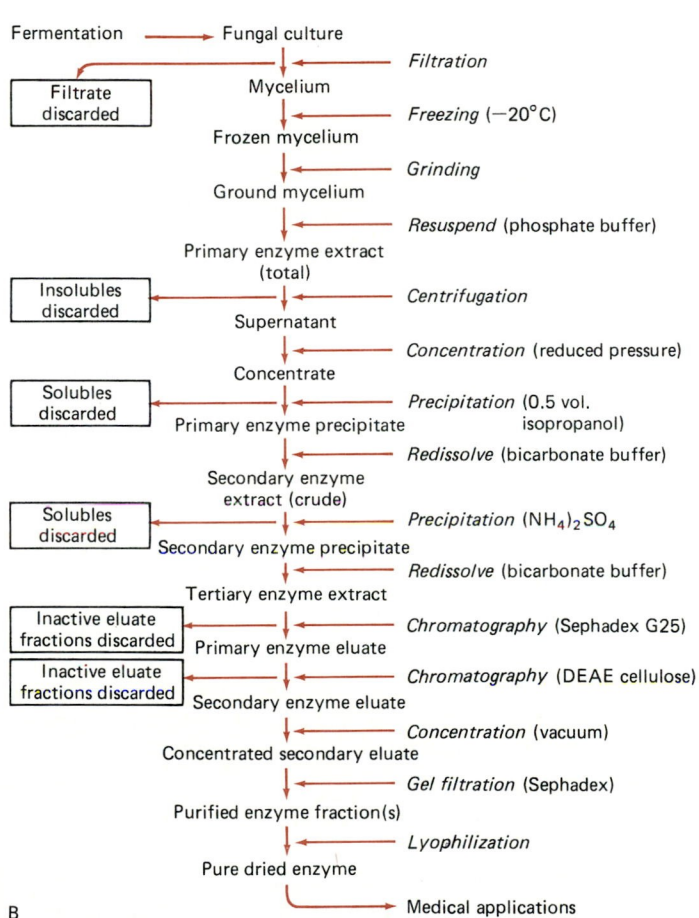

Fermentation ──→ Fungal culture
│ ←─── *Filtration*
Mycelium

Filtrate
discarded

│ ←─── *Freezing* (−20°C)
Frozen mycelium

│ ←─── *Grinding*
Ground mycelium

│ ←─── *Resuspend* (phosphate buffer)
Primary enzyme extract
(total)

Insolubles
discarded
│ ←─── *Centrifugation*
Supernatant

│ ←─── *Concentration* (reduced pressure)
Concentrate

Solubles
discarded
│ ←─── *Precipitation* (0.5 vol.
│   isopropanol)
Primary enzyme precipitate

│ ←─── *Redissolve* (bicarbonate buffer)
Secondary enzyme
extract (crude)

Solubles
discarded
│ ←─── *Precipitation* $(NH_4)_2SO_4$
Secondary enzyme precipitate

│ ←─── *Redissolve* (bicarbonate buffer)
Tertiary enzyme extract

Inactive eluate
fractions discarded
│ ←─── *Chromatography* (Sephadex G25)
Primary enzyme eluate

Inactive eluate
fractions discarded
│ ←─── *Chromatography* (DEAE cellulose)
Secondary enzyme eluate

│ ←─── *Concentration* (vacuum)
Concentrated secondary eluate

│ ←─── *Gel filtration* (Sephadex)
Purified enzyme fraction(s)

│ ←─── *Lyophilization*
Pure dried enzyme
└─────→ Medical applications

B

FIGURE 17.25

Generalized schemes for the production of (A) crude enzymes and (B) purified enzymes. As a rule, crude enzyme preparations involve removal of microbial cells and protein precipitation. For extracellular enzymes, cell breakage is not required, as illustrated for the preparation of crude enzymes. For intracellular enzymes, the cells initially are broken, as shown for the preparation of purified enzyme. Preparation of purified enzymes requires repetitive procedures that remove impurities, including reprecipitation and chromatographic separation.

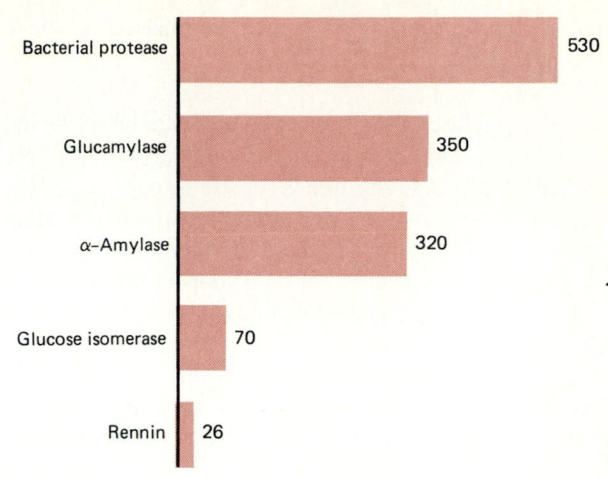

Amount (tons per year)

Bacterial protease — 530
Glucamylase — 350
α–Amylase — 320
Glucose isomerase — 70
Rennin — 26

FIGURE 17.26

Commercial sales of enzymes totaled $300 million in 1980. The five enzymes illustrated here are produced by microbiological methods. Bacterial protease, which degrades proteins by cleaving peptide bonds, is used as a cleaning aid. Glucamylase, alpha-amylase, and glucose isomerase convert starch into high-fructose corn syrup sweetener, which is replacing sucrose in soft drinks. Amylases break down starch to yield glucose; glucose isomerase converts glucose into fructose. Rennin, employed in making cheese, can be extracted from the fourth stomach of a cow or can be made biologically.

In addition to the development of enzymes for use in detergents, recombinant DNA technology has been employed to develop a bacterial strain that produces an enzyme, known as Kerazyme, that is used for dissolving hair and opening hair-clogged drains. Hair consists of the protein keratin, which is rather resistant to enzymatic attack. Agreements already guarantee over $3.8 million of product sales for Kerazyme over the next 5 years, making this the most successful product developed to date by genetic engineering.

Another major use of microbial proteases is in the baking industry. Proteases are used to alter the properties of the gluten proteins of flour. Fungal protease is added in the manufacture of most commercial bread in the United States to reduce mixing time and improve the quality of the loaf. Either fungal or bacterial protease is used in the manufacture of crackers, biscuits, and cookies. Fungal proteases are principally obtained from *Aspergillus* species, and bacterial proteases are primarily produced using *Bacillus* species. Fungal proteases have a wider range of pH tolerance than bacterial proteases.

Proteases are also used for various other products, including digestive aids. Adding proteases to beef can soften or tenderize it, making the meat more edible. A typical meat tenderizer contains 5 percent fungal protease, as well as MSG and other ingredients. In the leather industry, microbial proteases are used for bating of hides, which improves the quality by softening the leather. In the textile industry, protease enzymes are used for removing proteinaceous sizing and freeing silk fibers from the proteinaceous material in which they are embedded.

Amylases   Amylases are used for the preparation of sizing agents and the removal of starch sizing from woven cloth; preparation of starch sizing pastes for use in paper coatings; liquefaction of heavy starch pastes that form during heating steps in the manufacture of corn and chocolate syrups; production of bread; and removal of food spots in the dry cleaning industry, where amylase functions in conjunction with protease enzymes. Amylases are also sometimes used to replace or augment malt for starch hydrolysis in the brewing industry, as in the production of low-calorie beers.

There are various types of amylases, including (1) alpha-amylase, which converts starch to oligosaccharides and maltose; (2) beta-amylase, which converts starch to maltose and dextrins; and (3) glucamylase, which converts starch to glucose. All three enzymes are used in the production of syrup and dextrose manufacture from starch. Fungal production of amylases uses *Aspergillus* species. For example, *A. oryzae* is used to produce amylases from wheat bran in stationary culture, and *A. niger* is used to produce amylases in aerated submerged culture, using a starch–mineral salts medium. *Bacillus subtilis* and *B. diastaticus* are used for the commercial production of bacterial amylases.

The conversion of starch to a high-fructose corn syrup sweetener, using microbial enzymes, represents an economically significant and relatively new industrial process, producing over 2 million tons of high-grade sweetener per year. This sweetener, produced in a three-step process using the enzymes alpha-amylase, glucamylase, and glucose isomerase, is rapidly replacing sucrose as the primary sweetener in soft drinks. In the final step, glucose (approximately 50 percent) is converted into fructose by the enzyme glucose isomerase. The use of mutation-screening methods combined with genetic recombination techniques has permitted the development of strains of *B. subtilis* with greatly enhanced abilities to produce high yields of alpha-amylase. The development of such strains markedly increases the economic feasibility of producing sweeteners and similar products using microbial enzymes.

Other Enzymes   Various other microbial enzymes are produced for industrial applications. As indicated in Chapter 16, rennin is used in the production of cheese, and *Mucor pussilus* or *M. meihei* can be used for the commercial production of rennin for curdling milk in cheese production. Fungal pectinase enzymes are used in the clarification of fruit juices. Glucose oxidase, produced by fungi, is used for removing glucose from eggs prior to drying, since powdered dried eggs brown because of the chemical reaction of proteins with glucose, and removing the glucose stabilizes and prevents deterioration of the dried egg product. Glucose oxidase is also used to remove oxygen from various products, such as soft drinks, mayonnaise, and salad dressings, preventing oxidative color and flavor changes.

Microbial enzymes may also be used for the production of synthetic polymers. For example, the plastics industry now uses chemical methods for producing alkene oxides used in the production of plastics. It is possible to synthesize alkene oxides by using microbial enzymes (Figure 17.27), and the use of genetically engineered microbial strains can make such synthesis commercially feasible. The synthesis of alkene oxides from alkenes is accomplished by the sequential action of three enzymes: pyranose-2-oxidase from the fungus *Oudmansiella mucida*, a haloperoxidase from the fungus *Caldariomyces*, and an epoxidase from a *Flavobacterium* species. The production of propylene oxide using microbial enzymes could revolutionize the plastics industry, altering the cost of producing this widely used material.

Production of Solvents

Several organic solvents can be produced by using microbial fermentation, but generally organic solvents are produced today by chemical synthesis. For example, although ethanol is produced by fermentation for beverages and gasohol, industrial alcohol for use as a solvent is mostly produced by chemical synthesis. Fermentation processes have been important in the past in the industrial production of organic solvents and, as economic conditions change, will likely be used again in the future. The process for producing acetone and butanol was discovered by Chaim Weizmann (1874–1952), a Polish-born chemist then working in England. The discovery he made was influential in the British willingness to permit the establishment of a Jewish homeland in Palestine, and Weizmann later became the first president of of the new State of Israel. During World War I, the microbial production of acetone was very important for the production of the explosive cordite, and microbially produced butanol was converted to butadiene and used in making synthetic rubber. After the war the demand for acetone declined, but the need for *n*-butanol increased for its use in brake fluids, urea-formaldehyde resins, and the production of protective coatings, such as lacquers used on automobiles.

The microbial production of acetone and butanol uses anaerobic *Clostridium* species. The fermentation process discovered by Weizmann was based on the conversion of starch to acetone by *C. acetobutylicum*. Other species, such as *C. saccharoacetobutylicum*, are able to convert the carbohydrates in molasses to acetone and butanol. These *Clostridium* species synthesize butyric and acetic acids, which are then converted to butanol and acetone. The yields of these neutral solvents are typically low, approximately 2 percent by weight of the fermentation broth, representing a 30 percent conversion of carbohydrates to neutral solvents. The accumulation of higher concentrations of these solvents is limited by the toxicity of the compounds. The solvents produced by fermentation are recovered by distillation. In South Africa, because of the scarcity of petroleum and the abundance of plant residues as substrates for fermentation, butanol and acetone are produced by microbial fermentation

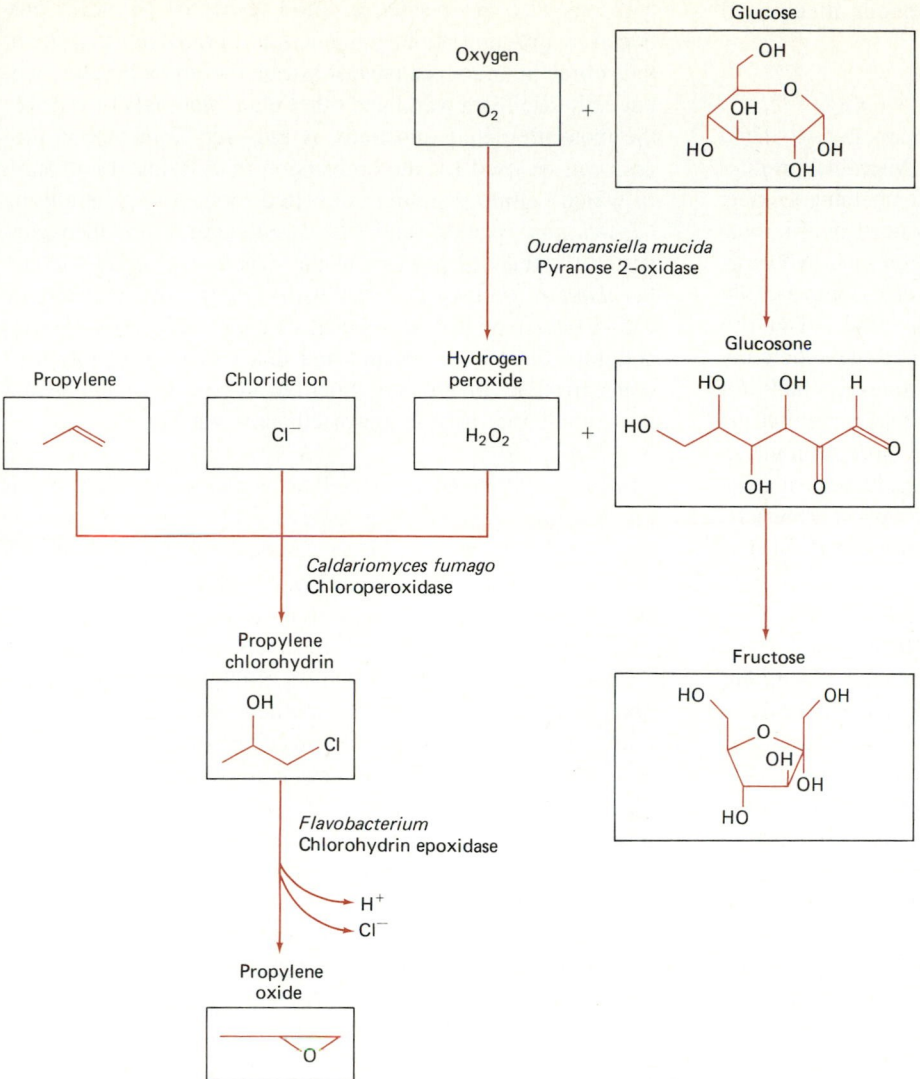

Glucose

Oxygen

O_2

+

Oudemansiella mucida
Pyranose 2-oxidase

Propylene   Chloride ion   Hydrogen
                                          peroxide   Glucosone

Cl^- H_2O_2 +

Caldariomyces fumago
Chloroperoxidase

Propylene
chlorohydrin   Fructose

OH

Cl

Flavobacterium
Chlorohydrin epoxidase

H^+
Cl^-

Propylene
oxide

O

FIGURE 17.27
Plastics, such as polypropylene and polyethylene, are made by the polymerization of
alkene oxides, which are now synthesized from petrochemicals. Their enzymatic
synthesis has been proposed by Saul L. Neidleman of the Cetus Corporation and relies
on three enzymes: pyranose-2-oxidase, a chloroperoxidase, and an epoxidase. The
fungal and bacterial sources of the enzymes are illustrated here. The system can also
generate valuable by-products, such as fructose.

employing this process. Elsewhere, these solvents are produced from petroleum; however, when the costs of petrochemicals increase, the feasibility of more extensive use of
fermentation in producing *n*-butanol is enhanced.

Today glycerol is also produced primarily by chemical
synthesis, based on the saponification of fats and the chemical oxidation of propane and propylene. Glycerol is used as
(1) a solvent in flavorings and food coloring agents; (2) a
lubricant in the manufacture of pet food, candy, cake icings,
toothpaste, glue, cellophane, and other products; and (3) an
emollient and demulcent in pharmaceuticals and cosmetics.
Glycerol is also used in the production of explosives and
propellants. The production of glycerol by fermentation in

Germany was an important factor during World War I because it was used in the production of munitions. The microbial production of glycerol is accomplished by adding
sodium sulfite to a yeast–ethanol fermentation process. The
sodium sulfite reacts with the carbon dioxide to produce sodium bisulfite, which prevents the reduction of acetaldehyde
to ethanol. This blockage results in a divergence of the metabolic pathway with the accumulation of glycerol. Glycerol
can be produced by using yeasts, such as *Saccharomyces
cerevisiae*, and bacteria, such as *Bacillus subtilis*. The microbial production of glycerol may be renewed as a result of
the finding that some yeasts can synthesize glycerol without
the need to add sodium sulfite, thus making the process eco-

nomically competitive with chemical methods of glycerol production.

Production of Fuels

Limited petroleum resources are forcing many industrialized nations to seek alternative fuel resources. Microbial production of **synthetic fuels** has the potential for helping to meet world energy demands. Useful fuels produced by microorganisms include ethanol, methane, hydrogen, and hydrocarbons. The use of microorganisms to produce commercially valuable fuels depends on finding the right strains of microorganisms that are able to produce the desired fuel efficiently and having an inexpensive supply of substrates available for the fermentation process. It is obviously imperative that the production of synthetic fuels not consume more natural fuel resources than are produced. Microbial production of fuels can be a particularly attractive process when waste materials, such as sewage and municipal garbage, are used as the fermentation substrate.

Ethanol   The microbial production of ethanol has become an important source of a valuable fuel, particularly in regions of the world that have abundant supplies of plant residues. Brazil produces and uses large amounts of ethanol as an automotive fuel and plans to replace gasoline with ethanol by the 1990s. **Gasohol**, a 9:1 blend of gasoline and ethanol, has become a popular fuel in the midwestern United States. At present about 100 million gallons of ethanol per year are used as a fuel, but 12 billion gallons per year would be required to completely replace gasoline use in the United States. There are three major limitations to the successful production of sufficient quantities of ethanol to serve as a major fuel source: (1) ethanol is relatively toxic to microorganisms, and therefore, only limited concentrations of ethanol can accumulate in a fermentation process; (2) carbohydrate substrates normally used for the production of ethanol in the food industry are relatively expensive, making the cost of fuel produced by fermentation high; and (3) distillation to recover ethanol requires a substantial energy input, reducing the net gain of fuel as an energy resource produced in this process.

Despite problems with the economics of ethanol production, several processes can be employed for the commercial production of ethanol as a fuel (Figure 17.28). The finding that the bacterium *Zymomonas mobilis* ferments carbohydrates, forming alcohol twice as rapidly as yeasts, appears to represent a significant advance in the search for a microbial strain for producing ethanol as a fuel. *Thermoanaerobacter ethanolicus*, a thermophilic bacterium, may be even more efficient than the organisms currently used for the fermentative production of ethanol. The use of thermophilic organisms would be particularly useful if the organism grew above the boiling point of ethanol, facilitating the recovery of the product. Corn sugar and plant starches are currently used as substrates for the production of ethanol, but the prices of these substrates vary greatly, depending on plant harvests,

and they also are needed as food resources. Biomass produced by growing photosynthetic microorganisms is a potential source of an inexpensive substrate for ethanol production, but cellulose from wood and other plant materials is probably the most promising substrate. A two-step fermentation process can be used for the conversion of cellulose to ethanol, in which cellulose is first converted to sugars, generally by *Clostridium* species, and the carbohydrates are then converted to ethanol by yeasts of the *Zymomonas* or *Thermoanaerobacter* species. It is very likely that genetic engineering can create a microbial species that can efficiently convert cellulose directly to ethanol and that will also tolerate high concentrations of ethanol. Such an organism should permit the commercial production of ethanol as a fuel.

Methane   Methane produced by methanogenic bacteria is another important potential energy source. Methane can be used for the generation of mechanical, electrical, and heat energy. Large amounts of methane can be produced by anaerobic decomposition of waste materials. Many sewage treatment plants are able to meet all or part of their own energy needs from the production of methane in their anaerobic sludge digesters. Excess methane produced in such facilities is sufficient to supply power for some municipalities. Efficient generation of methane can be achieved by using algal biomass grown in pond cultures, sewage sludge, municipal refuse, plant residue, and animal waste.

Methanogenic bacteria are members of the archaebacteria; they are obligate anaerobes and produce methane from the reduction of acetate and/or carbon dioxide. The production of methane generally requires a mixed microbial community, with some bacterial populations converting the available organic carbon into low molecular weight fatty acids that are substrates for methanogens.

Other Fuels   Hydrogen is another potential fuel source that can be produced by microorganisms. It is not currently employed as a major fuel but could be developed into one if an efficient production process were found. Photosynthetic microorganisms are capable of producing hydrogen and of growing using solar energy. Such organisms are able to convert solar energy into a fuel that can be stored and used to power electrical generators and provide a conventional source of energy. Such photoproduction of hydrogen is an intriguing, but as yet far from practical, idea.

Various microorganisms, including algae, also are capable of producing higher molecular weight hydrocarbons, but the potential use of such hydrocarbons as a fuel source has received relatively little attention. Although physicochemical processes are crucial, microorganisms are believed to play a role in the formation of petroleum deposits. A more thorough understanding of the basic mechanisms of microbial hydrocarbon formation and the formation of petroleum deposits should permit the development of genetically engineered microorganisms and fermentation processes to produce synthetic sources of petroleum hydrocarbons.

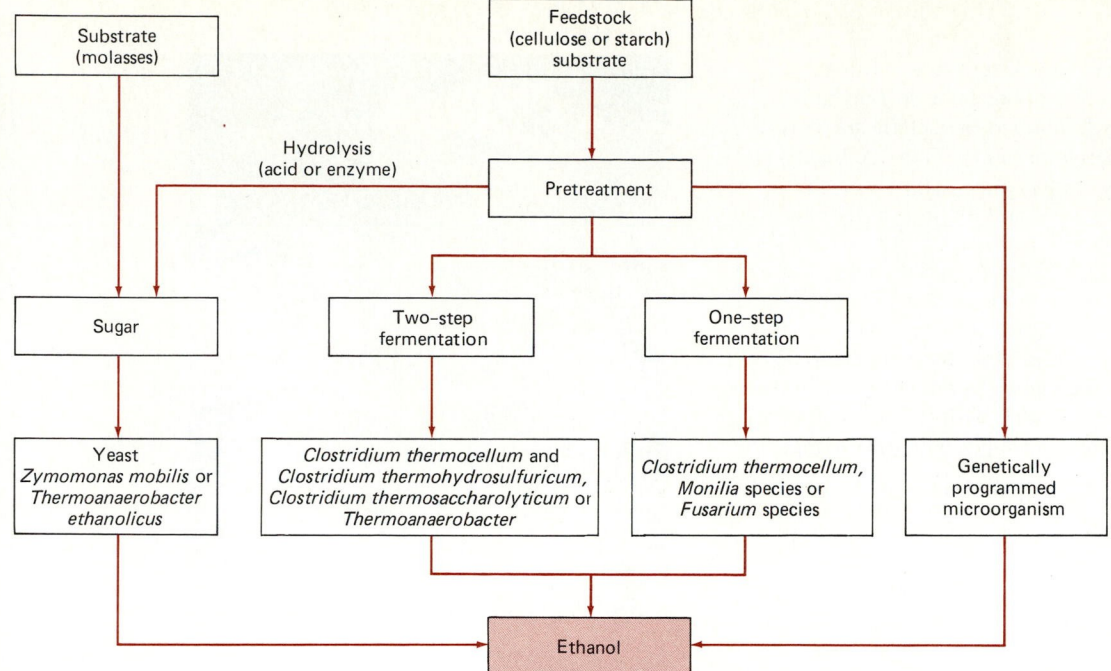

FIGURE 17.28

The biological production of ethanol as a fuel is accomplished by much the same method as the production of alcoholic beverages. The substrate is either crude sugar (from sugar cane or beet molasses) or starch (from corn, wheat, rye, or cassava) that has been converted into sugar. Yeasts are usually the fermenting organisms, but bacteria, such as *Zymomonas mobilis* and *Thermoanaerobacter ethanolicus*, may prove to be more efficent. Because the prices of crude sugar and starch vary so much, and because they are needed as foodstuffs, cellulose and related polymers of wood, which are abundant, renewable, and less expensive, are also used as substrate. The cellulose can be fermented in either one or two steps. In the two-step process, one microorganism breaks down the cellulose into its component sugar units, which are subsequently fermented into ethanol by another microorganism. In the one-step process, a single microorganism both breaks down the cellulose and ferments the resulting sugar into ethanol. Microorganisms that have this capability are rare, and so another strategy is to program a yeast or a bacterium genetically so that it converts cellulose into ethanol in one operation.

17.2 *Microbially Enhanced Recovery of Mineral Resources*

As the technological level of an ever-increasing world population rises, so does the need for industrially important minerals. As easily accessible, high-grade deposits of ores are depleted, it becomes increasingly important to find innovative and economical methods to recover such metals from lower-grade deposits, which for technical or economic reasons have not been used. Microorganisms can play an important role in recovering valuable minerals from low-grade ores, meeting some of the needs of industrialized society. Such recovery processes employ microbial metabolic activities to gain access to, rather than to actually produce, desired products.

Bioleaching of Metals

Microbial mining by the process of **bioleaching** recovers metals from ores that are not suitable for direct smelting because of their low metal content (Figure 17.29). Bioleaching uses microorganisms to alter the physical or chemical properties of a metallic ore so that the metal can be extracted. For example, metals can be extracted economically from low-grade sulfide or sulfide-containing ore by exploiting the metabolic activities of thiobacilli, particularly *Thiobacillus ferrooxidans*. Under optimal conditions in the laboratory, as much as 97 percent of the copper in low-grade ores has been

FIGURE 17.29

Bioleaching process for mineral recovery in a leaching dump of a large open-pit copper mine in Bingham Canyon, Utah. Very high numbers of bacteria are involved, more than a million per gram of ore. They are mainly members of the genus *Thiobacillus* and assist in the leaching operation by converting the iron in various compounds from the ferrous to the ferric form, which is an effective oxidizing agent. The bacteria oxidize pyrite to form sulfuric acid; to maintain the high acidity of the leaching solution; and to oxidize the insoluble copper-containing sulfide minerals in order to produce soluble copper sulfate, which migrates to the bottom of the dump, where it collects in catch basins. The solution is periodically pumped out to facilities where the copper is recovered. These photographs show large-scale bacteria-leaching operations. (A) Aerial overview of the mining operation at Bingham Canyon, Utah; (B) the leaching solution is sprayed on the field and recycled; (C) the leachate drains into a catch basin, from which the copper is recovered. (Courtesy K. C. Hochstetler, Salt Lake City)

A

B

C

Discovery Process

As early as 1000 B.C. copper was recovered from the drainage water of mines in the Mediterranean basin, but it was not until just 25 years ago that the role of bacteria in this traditional method of mineral extraction was understood. Ten percent of U.S. copper production today is a result of the deliberate exploitation of bacteria. In 1957 the relationship between the presence of *Thiobacillus ferrooxidans* and the dissolution of metals in copper-leaching operations was recognized. But even that short a time ago, the need for this biomining was not apparent, so the technology for exploiting bacterial leaching for the recovery of metals was not developed. The development of such technology is as much a function of economic need as of scientific capability. Now that energy is so costly and we have realized that the supply of accessible, high-grade metal ores is limited, we accept the need to extend the study of bacterial leaching beyond the work of those few investigators who have been examining these organisms and the reactions they carry out.

Increasing the rate of bacterial oxidation reactions through techniques such as genetic engineering would be an important improvement; the slowness of the biological process is one of its main drawbacks compared to chemical extraction methods. The greatest potential application of bacterial leaching for mineral recovery lies in the controlled treatment of vast quantities of solid materials, such as discarded waste rock, overburden, and tailings, which contain small quantities of widely dispersed, valuable metals. For example, the United States contains some 7 billion tons of sulfide deposit material, with an average nickel content of 0.2 percent, worth approximately $60 billion. This resource is not currently being exploited because of the inefficiency of current mining and extractive technologies and also because of possible detrimental effects on the environment. Advances in biomining technology would make it possible to recover this nickel, and along with it, some 400 million pounds of cobalt found in the same sulfide-rich material. Such a process must emphasize low capital investment, greater recovery of metal, and minimal environmental damage.

A

B

C

D

E

Plate 11 **(A)** Inspection of a gel following electrophoresis to determine whether foreign genes have been successfully incorporated by recombination. The DNA is stained with ethidium bromide and viewed with UV light. (From BPS—Agracetus) **(B)** A fermentor used for growing genetically engineered bacteria to produce human proteins. Human recombinant interferon and human recombinant interleukin-2 can be produced by bacteria in this fermentor. The color coded piping identifies the various solutions and gases supplied to and removed from the fermentor. (From Cetus Corporation/Chuck O'Rear) **(C)** Column chromatography is used to separate human proteins of interest from the cells and other products of genetically engineered microorganisms. In this process a highly purified product is collected that is free of traces of contaminating impurities. (From Cetus Corporation/Chuck O'Rear) **(D)** Aseptic filling of vials with recombinant beta interferon prior to freeze drying. The beta interferon produced by a genetically engineered bacteria is relatively stable, very pure and highly active. (From Cetus Corporation) **(E)** Crystals of Humulin® (humin insulin produced by a recombinant strain of *Escherichia coli*. These are the first crystals of humin insulin produced via recombinant DNA technology. (Courtesy Eli Lilly and Company)

A

B

C

D

E

F

Plate 12 **(A)** Technician adding DNA samples to a gel to assay by electrophoresis to determine whether foreign genes have been successfully transferred and incorporated by recombination. (From BPS—Agracetus) **(B)** Field test of tobacco plants genetically engineered for resistance to crown gall disease, a disease of plants caused by *Agrobacterium tumefaciens* characterized by tumorlike growth. (From BPS—Agracetus) **(C)** A tobacco plant that has been genetically engineered to include the luciferase gene that codes for luminescence. Light production from the plant occurs when the plant is watered with a solution containing luciferin—the substrate for luciferase. (From BPS—Marlene DeLuca, reprinted by permission of AAAS from *Science* 234:858) **(D)** In a carefully monitored first field test of the intentional environmental release of a genetically engineered microorganism conducted in California on April 24, 1987, ice-minus bacteria—so designated because they do not initiate ice crystal formation—were added to strawberry plants.

The organism produced under the name *Frostban* is a strain of *Pseudomonas syringae* that has a deletion so that it does not produce the protein that initiates ice nucleation on plant surfaces which leads to frost damage of the plant. Shortly after this test Steven Lindow of the University of California at Berkely conducted a similar test on potatoes. (From BPS—David J. Cross) **(E)** Extensive air and soil samples were collected during the field test of the deliberate release of *Pseudomonas syringae* to monitor the potential spread of bacteria from the point of application. (From BPS—David J. Cross) **(F)** Test tubes with water and strawberry blossoms at −2°C. Ice crystals have formed in the left tube where the strawberry blossoms have the wild type *Pseudomonas syringae* on their surfaces. No ice crystals have formed in the right tube where the strawberry blossoms have been inoculated with the genetically engineered ice-minus strain of *P. syringae*. (From BPS—Advanced Genetic Science, Oakland, California)

recovered by bioleaching, but such high yields are seldom achieved in actual mining operations. Even a 50–70 percent recovery of copper by bioleaching from an ore that would otherwise be completely unproductive would be an important achievement. The process is currently applied on a commercial scale to low-grade copper and uranium ores, and laboratory experiments indicate that it also has promise for the recovery of nickel, zinc, cobalt, tin, cadmium, molybdenum, lead, antimony, arsenic, and selenium from their low-grade sulfide-containing ores. The leaching process can also be used to separate the insoluble lead sulfate ($PbSO_4$) from other metals that occur in the same ore.

$MS + 2 O_2 \rightarrow MSO_4$ is the equation for the general process carried out by *T. ferrooxidans* and related species, where M represents a divalent metal. Because the metal sulfide (MS) is insoluble and the metal sulfate (MSO_4) is usually water soluble, this transformation produces a readily leachable form of the metal. *T. ferrooxidans* is a chemolithotrophic bacterium that derives energy through the oxidation of either a reduced sulfur compound or ferrous iron. It exerts its bioleaching action by oxidizing the metal sulfide being recovered directly, converting S^{2-} to SO_4^{2-}, and/or indirectly by oxidizing the ferrous iron content of the ore to ferric iron. The ferric iron, in turn, chemically oxidizes the metal to be recovered to a soluble form that can be leached from the ore.

If the ore formation is sufficiently porous and overlays a water-impermeable stratum, it is possible to leach the ore *in situ* without first mining it. An appropriate pattern of boreholes is established, with some of the holes used for the injection of the leaching liquor and others for the recovery of the leachate (Figure 17.30). More frequently, though, this bioleaching process is accomplished after the ore is mined, broken up, and piled in heaps on a water-impermeable formation or on a specially constructed apron (Figure 17.31). Water is then pumped to the top of the ore heap and trickles down through the ore to the apron. A continuous reactor leaching operation for recovery of copper from its low-grade sulfide ore is shown in Figure 17.32. The leaching water and ore usually supplies enough dissolved mineral nutrients to satisfy the needs of *T. ferrooxidans*, but in some cases mineral nutrients, such as ammonia and phosphate, must be added. In most of these bioleaching operations, the leached metal is then extracted with an organic solvent and subsequently removed from the solvent by stripping. Both the leaching liquor and the solvent are recycled.

Copper, which is in high demand for the electrical industry, is generally in short supply. A typical low-grade copper ore contains 0.1–0.4 percent copper. The "pregnant" leaching solution may contain 1 to 3 g of copper per liter. In copper leaching operations, the action of *Thiobacillus* involves both direct oxidation of copper sulfide (CuS) and indirect oxidation of CuS via generation of ferric ions from ferrous sulfide. Ferrous sulfide is present in most of the important copper ores, such as chalcopyrite ($CuFeS_2$). The copper is recovered either by solvent extraction or by using scrap iron. In the latter case, copper replaces iron according to the equation $CuSO_4 + Fe^\circ \rightarrow Cu^\circ + FeSO_4$ and is more advantageous for bioleaching because the organic solvent residues in the leaching liquor may inhibit the continued activity of *T. ferrooxidans*.

The recovery of uranium, a fuel required by the nuclear power generation industry, can also be enhanced by microbial activities. The microbial recovery of uranium from otherwise useless low-grade ores is helpful in overcoming the international energy shortage. Nuclear safety and waste disposal problems, as well as the limited supply of uranium, render current nuclear fission generators controversial; for all of these reasons, they may very well be only a stopgap solution to the international energy problem. Although bioleaching cannot influence safety considerations, this process can have an immediate and direct bearing on the economics of nuclear power production by providing a mechanism for commercial utilization of low-grade uranium deposits and for the recovery of uranium from low-grade nuclear wastes. Recovery of uranium from radioactive wastes is extremely important because it overcomes the problem of waste disposal, a major shortcoming of using nuclear power generators.

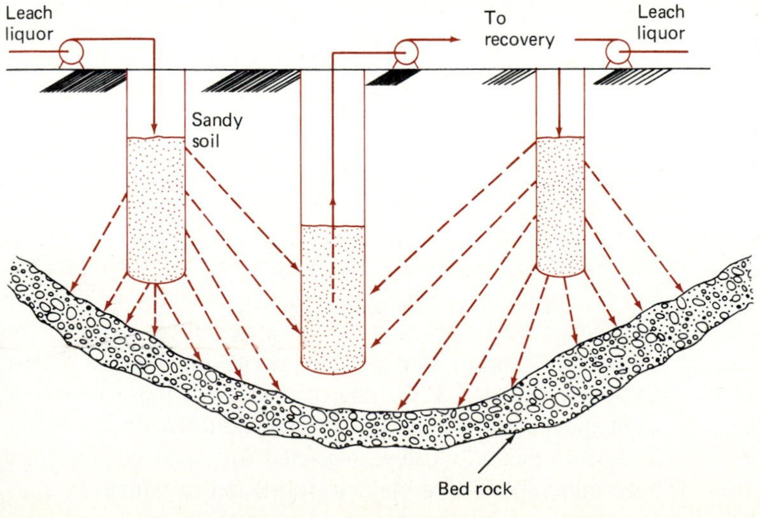

FIGURE 17.30

The hole-to-hole bioleaching process is used for low-grade ores. The process employs *Thiobacillus* species and is practiced where the ore is relatively porous and overlies an impermeable bedrock. Leaching is from multiple injection wells to a central collection well. (After J. Zajic, 1969, *Microbial Biogeochemistry*, Academic Press)

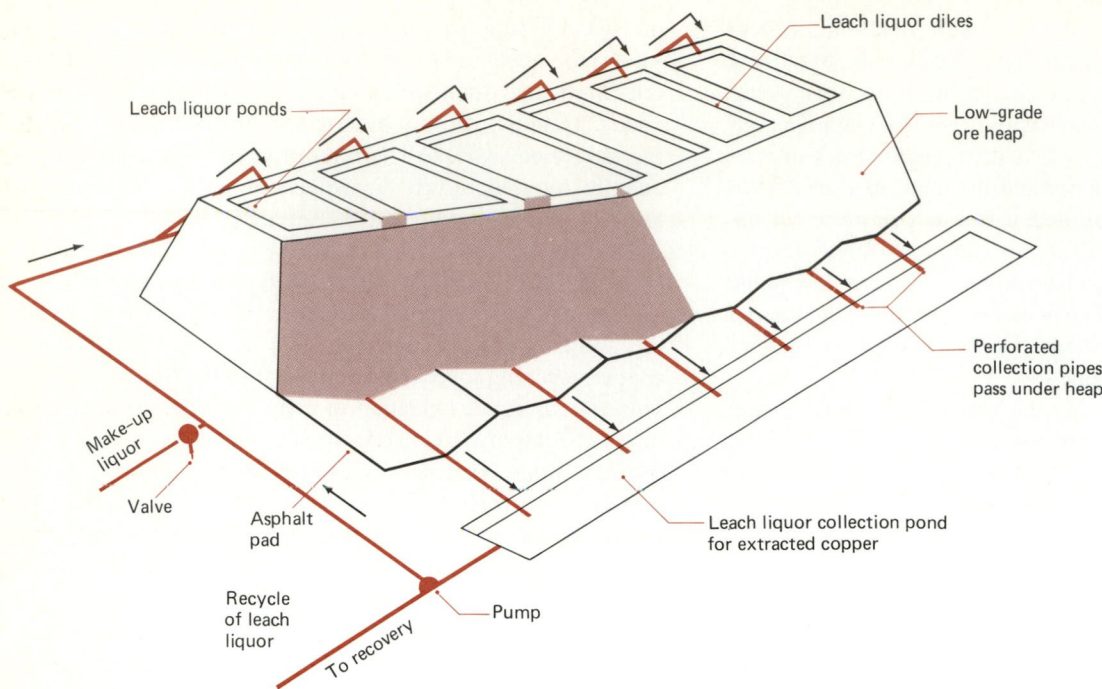

FIGURE 17.31

The heap bioleaching process is also used for the recovery of low grade ore. The ore is mined, crushed, and heaped into the shape of a cone, with its point cut off on an impervious asphalt pad. The leaching liquor is pumped to the top of the heap and percolates through the ore. The leachate is collected, processed, and recycled.

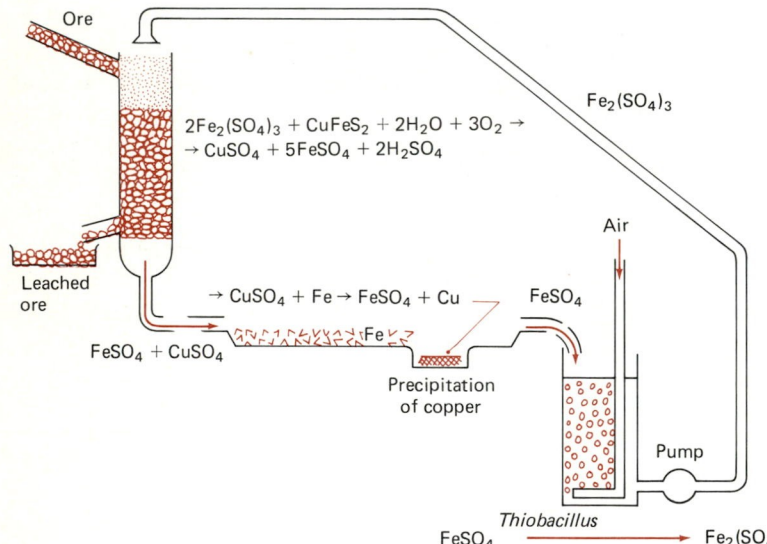

FIGURE 17.32

A continuous reactor leaching operation extracts copper from low-grade ore. The oxidation of sulfide and ferrous iron is carried out by *Thiobacillus ferrooxidans* which generates the acid for leaching. Copper is precipitated by exchange, using scrap iron. The Kennecott Copper Company holds the patent on this process.

Insoluble tetravalent uranium oxide (UO_2) occurs in low-grade ores. Although there is no evidence for the direct oxidation of UO_2 by *T. ferrooxidans*, UO_2 can be converted to the leachable hexavalent form (UO_2SO_4) indirectly by the action of this microorganism. *T. ferrooxidans* oxidizes the ferrous iron in pyrite (FeS), which often accompanies uranium ores, to ferric iron. The oxidized iron acts as an oxidant, converting UO_2 chemically to UO_2SO_4, which can be recovered by leaching. The technical and economic feasibility of employing *Thiobacillus* for the recovery of uranium and copper minerals depends on a variety of factors. The particular form of the naturally occurring mineral is very important. Bacterial leaching of uranium is most feasible in geological strata where the ore is in the tetravalent state and is pyritic, that is, closely associated with reduced sulfur and iron minerals. The geological formation in which the minerals occur is also important in determining the suitability of the bioleaching process. *In situ* bioleaching is ideal when there is a natural drainage system, as through a fault with an impermeable basin, that will permit economical recovery of the minerals. If these conditions do not exist, mining must precede the heap leaching process described previously.

The characteristics of the ore have an important effect on its susceptibility to bioleaching. The rate of leaching is determined in large part by the size of the mineral particles. Increasing the surface area, accomplished by crushing and/or grinding, generally increases production efficiency. Environmental factors must also be conducive for efficient bioleaching to occur. Optimal conditions for bioleaching using *T. ferrooxidans* area temperature of 30–50°C, a pH of 2.3–2.5, and an iron concentration of 2–4 g/L of leach liquor. Available oxygen and nutrients, such as ammonium, nitrogen, phosphorus, sulfate, and magnesium, are essential for the growth of *T. ferrooxidans*.

The oxidative activities of *Thiobacillus* can produce high temperatures in some mineral deposits, which may exceed the tolerance limits of the species being used. Obviously, this would lead to decreased bioleaching activity and mineral production. Because of these high temperatures, thermophilic sulfur-oxidizing microorganisms may be useful for some bioleaching processes. Members of the genus *Sulfolobus* are obligate thermophiles that can oxidize ferrous iron and sulfur in a manner similar to that of the members of the genus *Thiobacillus*. These acid-tolerant thermophilic bacteria can oxidize inorganic substrates and are used in the bioleaching of metallic sulfides. *Sulfoloblus* has been used for the bioleaching of molybdenite (molybdenum sulfide), whereas *Thiobacillus* is intolerant of high concentrations of this metal, as well as, of mercury and silver.

In addition to bioleaching to recover metals, some microorganisms are able to concentrate metals in their cells at levels several orders of magnitude higher than those occurring in the surrounding solution. Such bioconcentration has the potential for extracting rare metal ores from dilute solutions and for recovering metals from industrial effluents. In one such process, *Rhizopus* binds uranium from low-grade ores and nuclear wastes, contributing not only to the production of nuclear fuels but also to the solution of the nuclear waste disposal problem. Theoretically, microorganisms could be used to recover gold from the sea. It is not clear, though, whether such bioconcentration procedures will permit the development of economically feasible industrial processes for the recovery of valuable metals.

Oil Recovery

Besides their role in the recovery of metals, microorganisms can be used to increase the recovery of petroleum hydrocarbons. The **tertiary recovery of petroleum**, that, is the use of biological and chemical means to enhance oil recovery, and the enhanced recovery of hydrocarbons from oil shales are important because readily recoverable oil supplies have diminished. Tertiary recovery of oil employs solvents, surfactants, and polymers to dislodge oil from geological formations. The use of tertiary recovery methods has the potential for recovering 60–120 billion barrels of oil in United States reserves alone that otherwise could not be recovered. Xanthan gums produced by bacteria, such as *Xanthomonas campestris*, are promising compounds for the tertiary recovery of oil. These polymers have high viscosity and flow characteristics that allow them to pass through small pores in the rock layers containing oil deposits. When added during water flooding operations, that is, when water is pumped into petroleum reservoirs to force out the oil, xanthan gums help push the oil toward the production wells. These polymers are produced by conventional fermentation processes in which *X. campestris* is grown and the xanthan gums are recovered.

Bioleaching of oil shales also has the potential for greatly enhancing the recovery of hydrocarbons. Many oil shales contain large amounts of carbonates and pyrites, and the removal of these minerals increases the porosity of the shale, enhancing recovery of the oil. Acid dissolves the carbonates and can be produced by *Thiobacillus* species growing on the sulfur and iron in the pyrite. Such microbial leaching appears to have the potential for making recovery of hydrocarbons from oil shales economically feasible.

17.3 *Biodeterioration*

Several billion dollars in losses occur every year as a result of the microbial **biodeterioration** of materials, excluding food products. Microbial metabolic activities are involved in the corrosion of metals and the deterioration of various materials, including wood, paper, paint, and textile products. Various control measures are employed in different industries for reducing losses due to microbial biodeterioration. It is estimated that, without control measures, microbial biodeterioration would cause losses exceeding $50 billion per year. The magnitude of these figures attests to the significance of microbial biodeterioration in reducing the economic value of various products and the importance of controlling microbial activities to reduce these losses.

Paper

The production of paper is a major industry. In the United States, paper products are consumed at a rate in excess of 500 pounds per person per year. Losses to the pulp and paper industry through the activities of microorganisms are substantial. Microorganisms can affect all stages of the papermaking process; they are able to degrade wood, the raw material used for making paper. Growth of microorganisms during paper production reduces the quality of the final product, and microbial growth on paper products can preclude their intended use by consumers. In the paper-making process, pulpwood—broken into chips—is cooked under high pressure in a digester for a sufficient period of time to solubilize the lignins, pectins, and other adhesive biochemicals in the wood. Either acids, in the acid sulfite process, or bases, in the kraft sulfate process, are added during cook-

ing to aid in the digestion of the wood. When the cooking process is completed, the digester is suddenly opened, releasing the pressure and causing the wood chips literally to explode and separate into loosely bonded cellulose fibers. The cooking solution is recovered and used for various purposes. For example, sulfite liquor is processed and used as a nutrient for growing yeasts and as an animal feed supplement. The fibers recovered from the cooking process are often bleached to remove pigments. A slurry of cellulose fibers, 5–15 percent fiber in water, is milled to form sheets of paper.

Microbial biodeterioration can occur at various stages during this paper-making process. It is estimated that 10 percent of all pulpwood that is cut is lost to biodeterioration. Cellulolytic fungi are particularly important in the degradation of wood. Many different fungi cause wood rots, reducing the value of wood for the manufacture of paper (Figure 17.33). Microbial growth in pulpwood results in (1) loss of

Fungal growth

FIGURE 17.33

Photographs of wood showing fungal wood rot. The growth of the fungus causes biodeterioration of the wood and loss of integrity and usefulness of the wood for structural support. (A) Typical brown wood decay with white fungal mycelia on the surface; (B) brown rot caused by *Gloeophyllum trabeum* in a railroad cross-tie; (C) cross section of Douglas fir with brown rot. (Courtesy W. Eslyn, USDA, Forest Products Laboratory, Madison, Wis.)

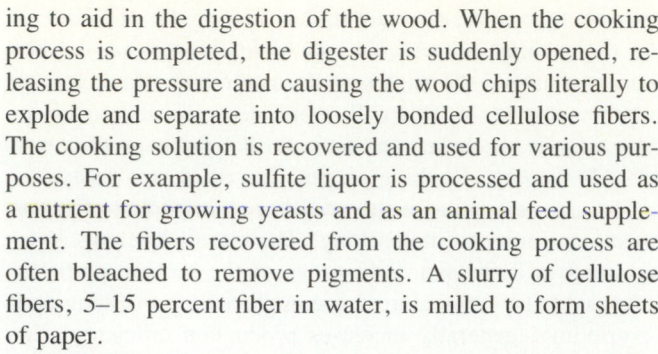

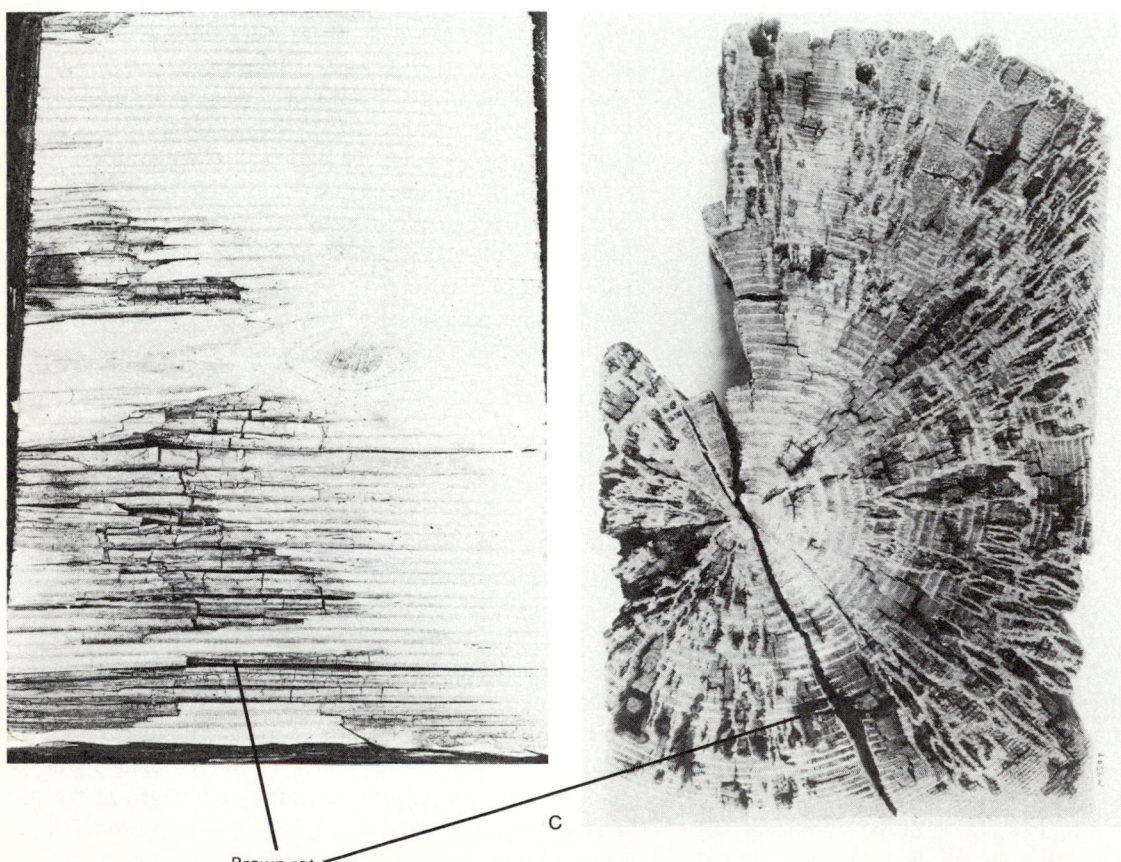

Brown rot

usable cellulose fibers; (2) loss of strength in the paper; (3) and discoloration and imperfections, including holes, spots, and various other blemishes in the finished product. The severity of these problems depends on the particular manufacturing process and the specific enzymatic activities of the contaminating microorganisms. Pulp stored for long periods of time for processing can be subject to very large losses because of microbial biodeterioration. Addition of copper sulfate or other antimicrobial agents to the pulp is used to minimize such losses.

During the paper-making process, growth of microorganisms on the surfaces of the equipment used in the paper mill can reduce the quality of the paper produced. Slimes can develop from the growth of masses of microorganisms on paper mill equipment and in the solutions used in making paper, causing significant losses because of the production of paper of unacceptable quality. It is estimated that microbial growth in the pulp and paper mill causes losses of $1–5 per ton of paper produced. The losses due to slime formation include production time lost to clean equipment, decreased performance of equipment, and the loss of finished products not acceptable for consumer use (Figure 17.34). Several environmental parameters can be modified to control slime deposits, including periodic washing of equipment with hot sodium hydroxide to kill contaminating microorganisms.

FIGURE 17.34
Photograph of paper, showing an imperfection (hole) due to slime formation resulting from microbial growth. (Courtesy Nalco Chemical Co., Oak Brook, Ill.)

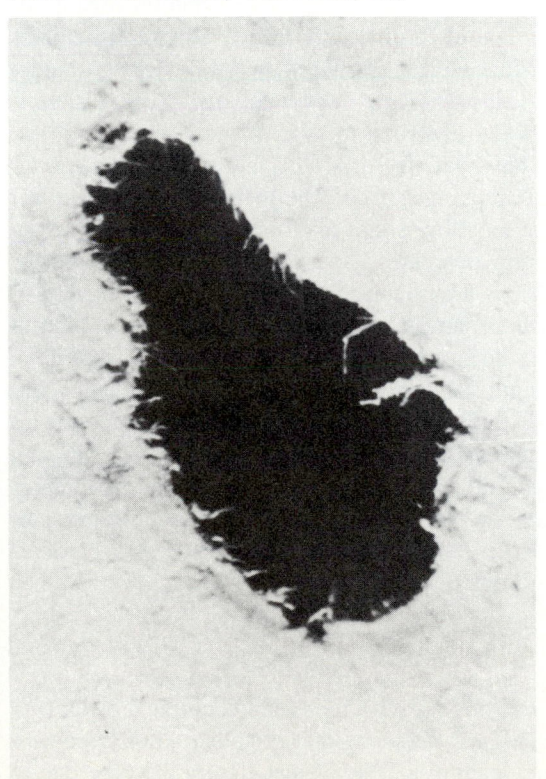

Various chemicals with antimicrobial activities can be used as microbicides in the paper industry. The use of chlorine is probably of greatest value in the paper industry for controlling slime. Chlorine is added to the water coming into the paper mill to control bacterial growth. This is the same chlorination process used to maintain acceptable levels of microorganisms in drinking water supplies. Various antifungal agents, including phenolic and organosulfite compounds, are also used in the paper mill to control undesired microbial growth. Mercurials have been used extensively in the paper-making industry but are now banned by the FDA for use in paper for food packaging.

Wood

In addition to paper, other wood products are subject to microbial biodeterioration. As indicated previously, many fungi produce cellulases that are able to decompose wood cellulose. To prevent biodeterioration, many wood products are preserved from microbial attack by coating the surface of the wood with paint or lacquer. Paint protects wood surfaces by preventing microorganisms and the water needed for their growth from reaching the surface of the wood. Various wood products are protected from biodeterioration by impregnation with antimicrobial agents. For example, railroad ties and telephone poles are normally impregnated with creosote, an antimicrobial phenolic compound, and marine wood pilings are treated with copper-naphthalene to prevent biodeterioration.

Paint

Although paints are used to protect wood surfaces against biodeterioration, paints themselves are organic compounds subject to microbial degradation. Paint components, including primary pigments, as well as finished paint products, are subject to microbial attack. The growth of microorganisms in a paint product can alter the color and binding properties of the paint (Figure 17.35). The growth of microorganisms in the paint can lead to peeling as a result of a loss of adhesion between the paint film and the coated surface. A large number of fungal and bacterial species have been isolated from paints. Therefore, various microbicidal agents are routinely added to paint products to prevent microbial growth and biodeterioration of the paint.

The only effective means of preventing microbial attack on liquid paints and paint films is through the use of chemical inhibitors. Relatively few microbicidal and microstatic agents have been found to be effective in paint products. The choice of paint preservative is dictated by (1) the effectiveness of the particular chemical in controlling the growth of contaminating microorganisms and (2) the particular application. Mercury and lead compounds have been widely used in paints to control microbial biodeterioration, but the human toxicity of many of these compounds has led to a severe curtailment of their use. Other antimicrobial agents used in

Discoloration

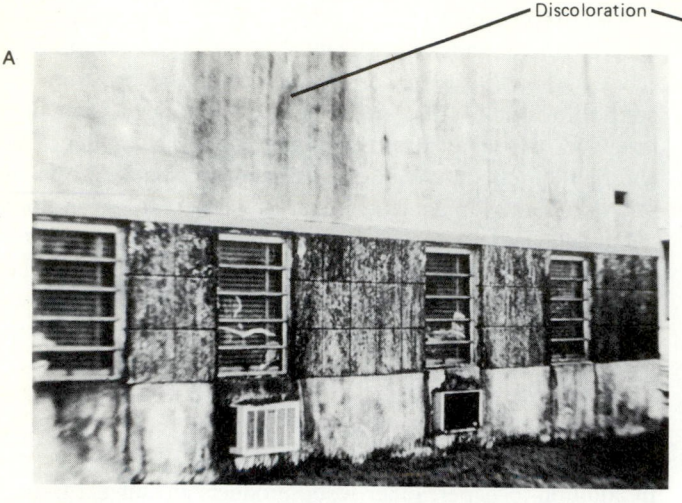

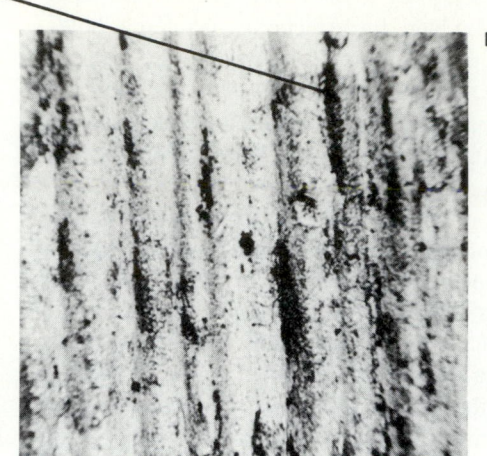

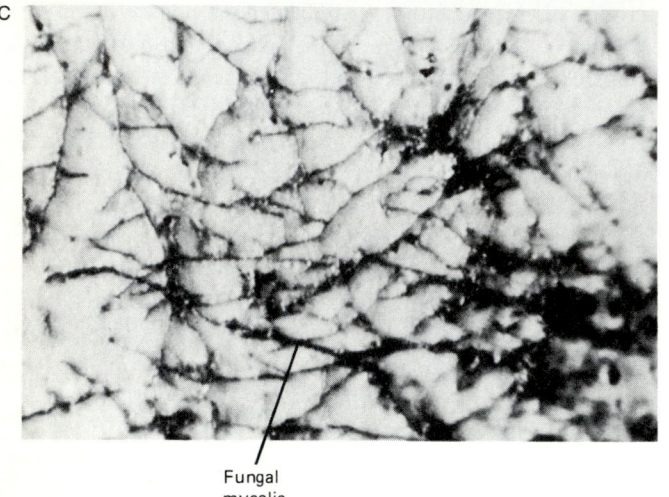

Fungal mycelia

FIGURE 17.35

These photographs show the detrimental effects of microbial growth on paint surfaces. (A) Fungal growth on the exterior of this building has caused discoloration of the painted surface. (B) A close-up view of the discolored area of the painted building. (C) Filaments of *Aureobasidium pullulans* that grow on paint film (20×). (Courtesy Amy Z. Leathers, PPG Industries, Coatings and Resins Division, Pittsburgh)

paints include chlorinated phenols, inorganic pigments such as barium metaborate, quaternary ammonium compounds, and various other microbicides.

Marine paints present a special problem for the prevention of biodeterioration. Such paint products must prevent microbial degradation of the wood and metal surfaces of ships and other marine structures. Marine paints should also resist marine fouling, the growth of microorganisms and higher plants and animals on the hulls of ships. The fouling of a ship's hull can greatly reduce the performance characteristics of the vessel, causing loss of speed and increased fuel consumption. Various biocides are routinely incorporated into marine paints to reduce microbial development of surface films that can support the attachment and growth of higher organisms,

such as barnacles. Copper oxides most frequently have been used as antibacterial, antialgal, and antifungal agents to prevent the development of slime on marine surfaces. Antifouling paints often contain 55 percent cuprous oxide, but such high levels of antifouling agents can cause corrosion problems. High concentrations of copper oxides, which slowly leach from the paint, are needed to control microbial growth over a long period of time. The use of very high concentrations of toxic substances, however, can create pollution problems when they leach into the surrounding waters.

Textiles

Textiles, particularly those composed of natural organic fibers, such as cotton and flax, are readily attacked by microorganisms. Most fabrics are not subject to extensive biodeterioration because they are dehydrated and because there is insufficient water available to support microbial growth. In the tropics, however, there is adequate moisture in the air (humidity) to support microbial growth on various textiles. Fungi are the most important microorganisms in such textile biodeterioration processes. **Mildew**, due to the growth of various fungi, is a particular problem in the deterioration of products such as canvas tents (Figure 17.36). The growth of fungi on a textile surface causes discoloration and weakening of fiber strength. The addition of waterproofing agents containing small quantities of phenolic compounds to textiles can reduce losses due to mildew formation. Storing camping equipment and other textile products in dry places is important in preventing biodeterioration of the fabric. Despite preventive measures, annual losses due to textile biodeterioration total millions of dollars.

Metal Corrosion

Corrosion of iron and steel pipes can occur as a result of a variety of chemical reactions that establish an electrochemical gradient, leading to loss of metal from the pipe due to

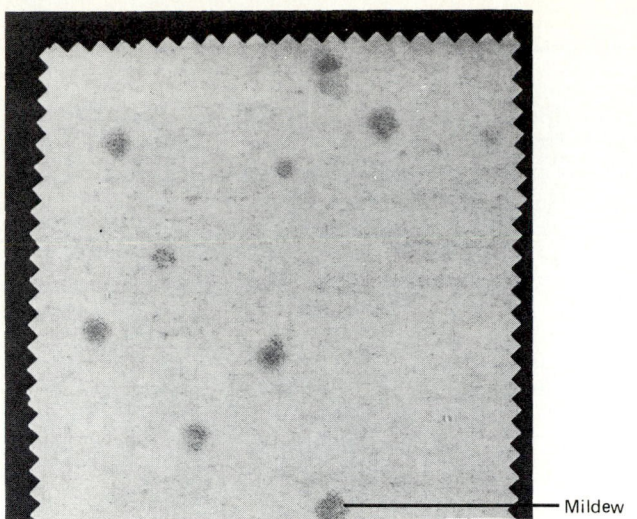

FIGURE 17.36

This cotton duck tent fabric has mildew spores (black spots) that are clearly visible after being weathered for 12 months. (Courtesy David A. Yeadon and Robert J. Harper, Jr., USDA, Southern Regional Research Center, New Orleans)

FIGURE 17.37

Corrosion of metal pipes. (A) These cast iron pipes show graphitization resulting from the activities of sulfate-reducing bacteria. The pipes that appear to be in good condition are actually the same pipes as those with the holes, except that the corrosion products have not been removed. (Courtesy Melvin Romanoff and Warren Iverson, National Bureau of Standards). (B) Scanning electron micrograph of the bacteria associated with an area of corroding pipe. (Courtesy Edwin E. Geldreich, USEPA, Cincinnatti)

electrolysis (Figure 17.37). The growth of several microbial species is also important in corrosion processes (Table 17.8). The growth of the sulfate-reducing bacterium *Desulfovibrio desulfuricans* is especially important in the corrosion of metals. Sulfate-reducing bacteria do not attack metal directly but accelerate the electrolytic corrosion process by promoting depolarization of the anodic and cathodic surfaces, thereby driving the corrosive reactions and causing anaerobic corrosion. The reduction of sulfate to hydrogen sulfide results in cathodic depolarization. The hydrogen sulfite formed from the reduction of sulfate reacts with ferrous iron to form iron sulfide; the effect of this reaction is anodic depolarization. Additionally, a very active hydrogenase associated with *Desulfovibrio* removes the protective layer of hydrogen that surrounds submerged iron pipes, exposing the underlying iron to corrosive attack. Also, the metabolic activities of anaerobic sulfate-reducing bacteria result in the formation of iron hydroxides, which are corrosion products. Sulfate-reducing bacteria are active within the pH range 6–8. Serious corrosion of iron and steel structures occurs in anaerobic waterlogged soils with a near-neutral pH. Anaerobic microbial corrosion of water mains, buried in land reclaimed from the sea north of Amsterdam, forced their replacement every 2 to 3 years. Anaerobic microbial corrosion of cast iron causes **graphitization**, a process in which a pipe loses much of its iron, becoming soft and brittle, and easily broken. This is what happened to the buried pipes in the Netherlands. Steel and aluminum pipes are also subject to anaerobic corrosion. Anaerobic microbial corrosion of steel results in more localized pitting, which sometimes causes perforation of the pipe.

Aerobic microbial corrosion is not as serious an economic problem as anaerobic corrosion. Aerobic corrosion processes are carried out by various bacteria, such as *Gallionella*, *Crenothrix*, and *Leptothrix* species, that oxidize metals and form metalic oxides as corrosion products. The gelatinous sheaths of these bacteria form a barrier to oxygen diffusion,

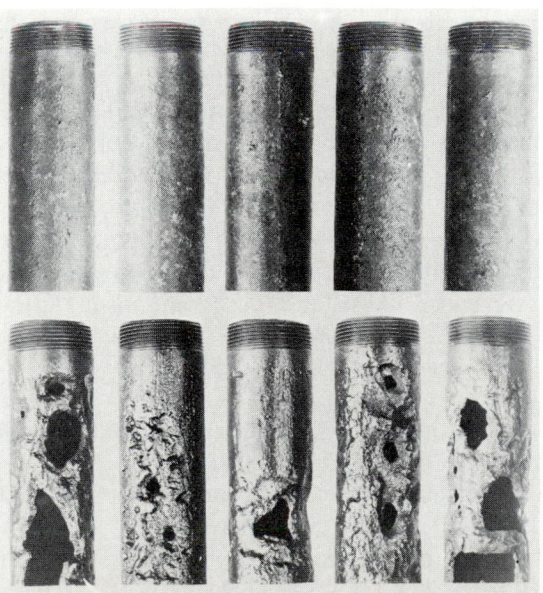

A
B

TABLE 17.8   Some Bacteria Involved in Corrosion Processes

Organism	Oxygen Requirement	Inorganic Components	Metabolic End Products	Habitat	Optimal Range	
					Temp.	pH
Sulfate-reducing *Desulfovibrio desulfuricans*	Anaerobic	Sulfate, thiosulfate	Hydrogen sulfide	Water, soil, mud, oil-reservoir	25–30	6–7.5
Sulfur-oxidizing *Thiobacillus thiooxidans*	Aerobic	Sulfur, thiosulfate	Sulfuric acid	Soil, water	28–30	2–4
Thiosulfate-oxidizing *Thiobacillus thioparus*	Aerobic	Thiosulfate, sulfur	Sulfur, sulfuric acid	Soil, water, mud, sewage	28–30	7
Iron bacteria *Crenothrix* *Leptothrix* *Gallionella*	Aerobic	Iron, manganese	Ferric or manganese oxides	Water	25	8
Nitrate-reducing *Thiobacillus denitrificans*	Facultative	Thiosulfate, sulfur, sulfide	Sulfate	Soil, mud, peat, water	30	7–9
Hydrogen-utilizing *Hydrogenomonas*	Microaerophilic	Hydrogen	Water	Soil, water	28–30	7

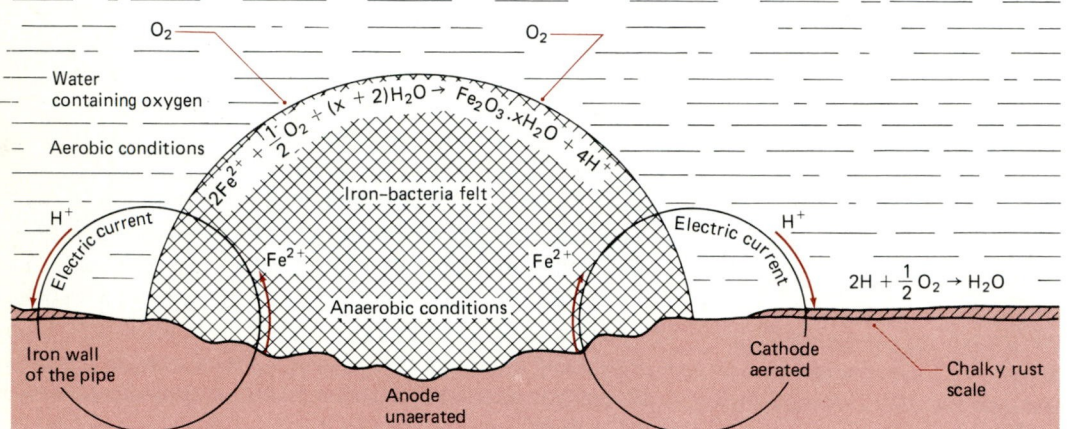

FIGURE 17.38

Diagram illustrating the corrosion process on the internal surface of a well-aerated water delivery pipe when a tubercle is formed by iron bacteria.

causing the development of an electrical potential difference between the surface covered by the bacteria (anodic region) and the exposed surface (cathodic region). This electrical differential can be extensive, leading to accelerated loss of metal and corrosion (Figure 17.38). Sulfur-oxidizing *Thiobacillus* species, which produce high concentrations of sulfuric acid, are also important in aerobic corrosion processes.

The microbially mediated corrosion of metal pipes is a serious problem in oil and gas delivery systems. One method of protecting pipelines against the action of sulfate-reducing microorganisms is to increase the pH above 9.5. Buried pipes can be coated to prevent contact between the metal surface, water, and soil microorganisms. Additionally, electric currents can be applied to the pipe to preclude corrosion processes. Various bacterial inhibitors are employed to control microbial corrosion. The most commonly used bacterial inhibitors are alkyl substituted amine and quaternary ammonium compounds. These chemical compounds are toxic to many bacteria, including *Desulfovibrio desulfuricans*, a bacterial species of major importance in corrosion processes.

Postlude

The application of technology to the control of microbial growth has great economic consequences. The metabolic activities of various microorganisms support several major industries, such as the manufacture of pharmaceuticals. Biodeterioration, on the other hand, results in large-scale economic losses because of uncontrolled microbial growth.

The production of biocides, aimed at preventing microbial degradation of various products, is an industry in itself. The emphasis in much of industrial microbiology is on harnessing microbial activities for beneficial purposes.

The success of using microorganisms in industrial processes, in the final analysis, is judged solely on economic grounds. The microbial fermentation industry is based on using the most efficient strain, the least expensive substrate, and an optimal environment in a fermentor to produce the highest possible concentrations of a substance that can be efficiently recovered and purified. Changes in the costs of raw materials and energy continuously shift the economic balance between using fermentation and chemical synthesis as the mode of production of a particular product. In some cases, using microorganisms is the only effective means of making a product, such as in the production of vaccines. In other cases, microbial activities and chemical syntheses are both used, as in the production of modified penicillin derivatives. In still other instances, chemical syntheses have replaced microbial fermentations, as in the production of organic solvents. However, when local economics are favorable, such as in Brazil and South Africa, microbial methods are still used for producing certain organic solvents.

There is a general maxim that microorganisms can be found or created to carry out any possible chemical transformation and that microorganisms can carry out chemical

modifications more efficiently than synthetic chemical methods. Using this philosophy, industrial microbiologists have developed screening procedures to seek and, where necessary, to modify—by purposefully increasing rates of mutation—specific strains of microorganisms capable of producing high yields of desired substances. Many antibiotics routinely used in medicine today were discovered, and the microbial strains producing these compounds isolated and developed, by the use of such screening procedures. The promise of recombinant DNA technology to permit the incorporation of foreign genetic information into microorganisms supports the belief that microorganisms will become increasingly important in the production of many goods not hitherto considered possible by microbial fermentation. The full benefits of genetic engineering have yet to be realized. We are already driving cars fueled in part by the products of microbial metabolic activities and eating foods produced with the aid of microorganisms. In the future, our use of such microbial products is certain to increase.

Suggested Supplementary Readings

AHARONOWITZ, Y., and G. COHEN. 1981. The microbiological production of pharmaceuticals. *Scientific American* 245(3):140–152.

ATLAS, R. M. (ed.). 1984. *Petroleum Microbiology*. Macmillan Publishing Co., Inc., New York.

BALL, C. (ed.). 1984. *Genetics and Breeding of Industrial Microorganisms*. CRC Press, Inc., Boca Raton, Fla.

BRIERLEY, C. L. 1982. Microbiological mining. *Scientific American* 247(2):44–53.

BU'LOCK, J. D., and B. KRISTIANSEN (eds.). 1987. *Basic Biotechnology*. Academic Press, London.

CASIDA, L. E. 1968. *Industrial Microbiology*. John Wiley & Sons, Inc., New York.

COOMBS, J. (ed.). 1985. *Dictionary of Biotechnology*. Elsevier Science Publishing Co., Inc., New York.

CRUEGER, W., and A. CRUEGER. 1984. *Biotechnology: A Textbook of Industrial Microorganisms*. Science Tech, Madison, Wisc.

DAVIS, J. B. 1967. *Petroleum Microbiology*. Elsevier Publishing Co., Amsterdam.

DEMAIN, A. L. 1981. Industrial microbiology. *Science* 214:987–995.

DEMAIN, A. L., and N. A. SOLOMON. 1981. Industrial microbiology. *Scientific American* 245(3):66–76.

DEMAIN, A. L., and N. A. SOLOMON (eds.). 1985. *Biology of Industrial Microorganisms*. Butterworth, Stoneham, Mass.

DEMAIN, A. L., and N. A. SOLOMON. 1986. *Manual of Industrial Microbiology and Biotechnology*. American Society for Microbiology, Washington, D.C.

EVELEIGH, D. E. 1981. The microbiological production of industrial chemicals. *Scientific American* 245(3):154–178.

HOPWOOD, D. A. 1981. The genetic programming of industrial microorganisms. *Scientific American* 245(3):91–102.

HUTCHINS, S., S. DAVIDSON, J. BRIERLY, and C. BRIERLY. 1986. Microorganisms in reclamation of metals. *Annual Review of Microbiology* 40:311–366.

JONES, D. T., and D. R. WOODS. 1986. Acetone-butanol fermentation revisited. *Microbiological Reviews* 50:484–524.

MILLER, B. M., and W. LITSKY. 1976. *Industrial Microbiology*. McGraw-Hill Book Co., New York.

PEPPLER, H. J., and D. PERLMAN (eds.). 1979. *Microbial Technology* (2 volumes). Academic Press, New York.

PHAFF, H. 1981. Industrial microorganisms. *Scientific American* 245(3):77–89.

PRIMROSE, S. B. 1987. *Modern Biotechnology*. Blackwell Scientific Publishers, Oxford, England.

REHM, H., and G. REED (eds.). 1981–. *Biotechnology: A Comprehensive Treatise* (8 volumes). Verlag Chemie International, Inc., Deerfield Beach, Fla.

RIVIERE, J. *Industrial Applications of Microbiology*. 1978. Halsted Press, New York.

ROSE, A. H. 1977–. *Economic Microbiology* (a multivolume treatise). Academic Press, New York.

SAUNDERS, V. A., and J. R. SAUNDERS. 1987. *Microbial Genetics Applied to Biotechnology*. Croom Helm, London.

SILVER, S. (ed.). 1986. *Biotechnology: Potentials and Limitations*. Springer-Verlag, Berlin.

VANEK, Z., and Z. HOSTALEK (eds.). 1986. *Overproduction of Microbial Metabolites: Strain Improvement and Process Control Strategies*. Butterworth, Stoneham, Mass.

Study Questions

1. How does the industrial use of the term *fermentation* differ from its use to describe microbial metabolism?

2. How are antibiotic-producing microorganisms found? How is it determined if they are producing substances of industrial importance?

3. Discuss the role of mutation and selection in the history of penicillin production. Why would a similar approach not be suitable for finding an insulin-producing strain of *E. coli*?

4. Why are microorganisms especially important in the production of steroids?

5. What is an immobilized enzyme? Discuss the role of such enzymes in industrial microbiology.

6. What is a fermentor? How are aeration and pH controlled in fermentors?

7. What are the essential properties of a substrate for an industrial fermentation?

8. How was the microbial production of butanol critical in determining the outcome of World War I?

9. How are microorganisms involved in the corrosion of metals?

10. What is bioleaching? How is this process used for the recovery of uranium?

11. Discuss several ways in which microorganisms can help meet the current fuel shortage.

12. What is biodeterioration? How do we control the biodeterioration of wooden ships? Of paints? Of paper?

Situational Problem

Entrepreneurial Advice on Biotechnology

Biotechnology is one of the more exciting applied fields of science and has major potential for economic growth that has yet to be realized. Several corporations have been formed to capitalize on recombinant DNA technology. Stock in these companies has been actively traded, but major product successes have been limited to date. Nevertheless, in an era of individualistic entrepreneurial enthusiasm, biotechnology is an appealing area of investment. Although genetic engineering clearly dominates the headlines in this field, biotechnology encompasses a broad field that combines the biological and engineering sciences for economic (applied) purposes. The application of microorganisms, whether created by genetic engineers or discovered in nature—to produce economically valuable products, such as antibiotics, or to control detrimental situations, such as environmental pollution—is the mainstay of biotechnology.

Suppose your friend unexpectedly inherited a large sum of money and asked you to join him in starting a biotechnological enterprise. What projects would you suggest? How would you know what work had already been done on that project? How could you realistically determine the economic investment needed and the potential profits that could be realized? Compose a proposal that could serve as a prospectus for additional investors.

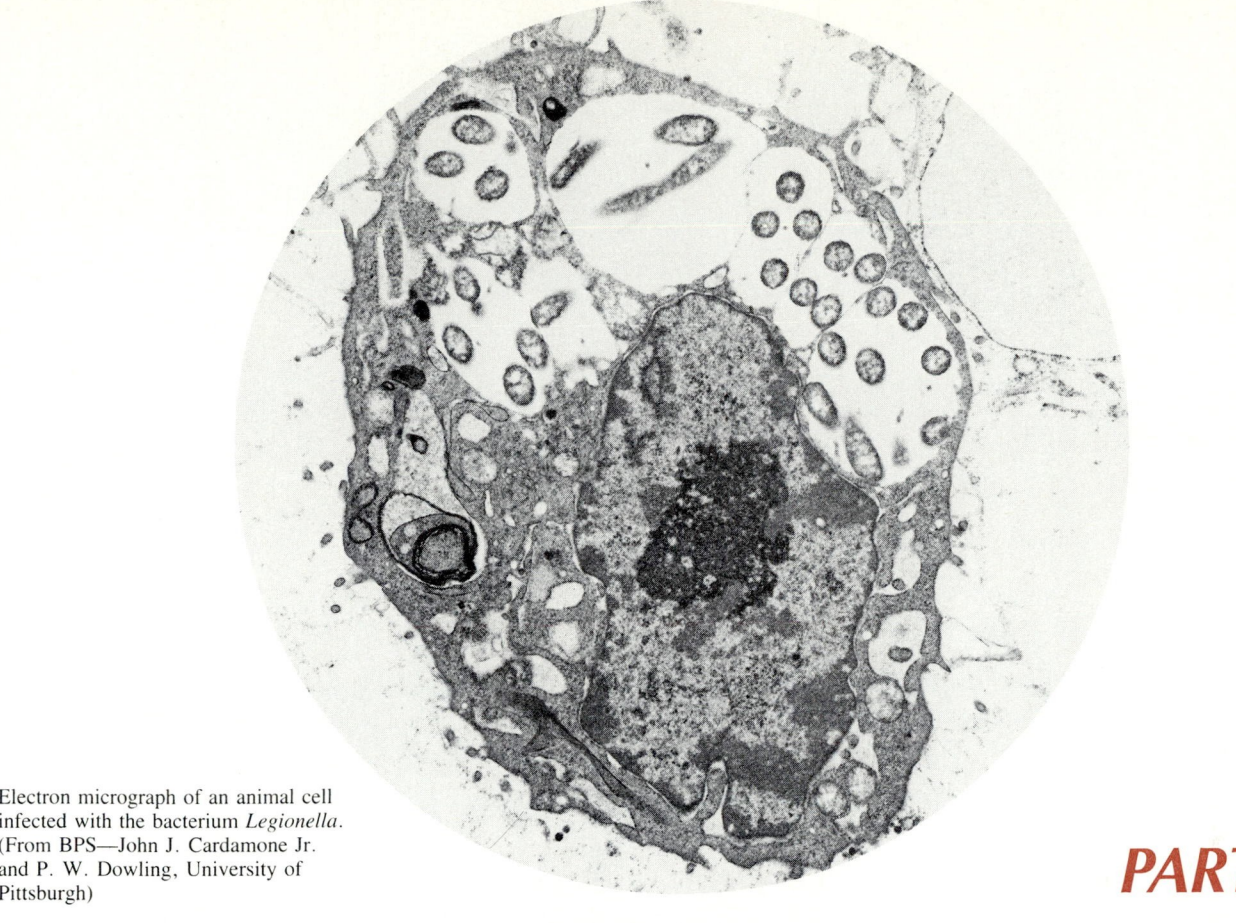

Electron micrograph of an animal cell infected with the bacterium *Legionella*. (From BPS—John J. Cardamone Jr. and P. W. Dowling, University of Pittsburgh)

PART *8*

Medical Microbiology

CHAPTER 18

Interactions Between Microorganisms and Humans

From the moment birth begins, humans are continuously exposed to microorganisms. Some of the interactions between humans and microbial populations are essential for the well-being of the individual, but others result in diseases. The nature of the interaction between a specific microorganism and a human host depends on the physiological properties of both the microorganism and the host. These relationships are classified using general ecological terms that apply to the interrelationships among any biological populations. Mutually beneficial relationships between microbial populations and human hosts are termed **synergistic**. A relationship is one of **commensalism** when the microbial population benefits without adversely or beneficially affecting the human host. **Parasitism** occurs when the microbial population benefits but the human host is adversely affected. Pathogenic microorganisms can establish a parasitic relationship with a host organism that results in disease of the host. The interactions of microorganisms and humans thus represent a continuum from those that are needed to maintain good health to those that cause human disease and even death.

18.1 *The Normal Human Microbiota*

Although we cannot see them, we are literally covered with microorganisms. In fact, the average adult human has 10^{13} eukaryotic animal cells (human cells) and 10^{14} associated prokaryotic and eukaryotic cells of microorganisms; stated another way, the normal human being is composed of just over 10^{14} cells, only 10 percent of which are human and the remaining 90 percent of which are microbial. This percentage is based on the number of cells, but because bacterial cells are far smaller than eukaryotic cells, the bacteria add little to the biomass of the human body. Most of these bacteria are associated with the gastrointestinal tract. The body surfaces of most animals, including humans, are populated by microorganisms, with distinct microbial populations inhabiting the surface tissues of the skin, oral cavity, respiratory tract, gastrointestinal tract, and genitourinary tract. The microbial populations most frequently found in association with particular tissues are referred to as *indigenous microbial populations*, **normal microflora**, or **normal microbiota**. Although the term *microflora* is used extensively, the term *microbiota* is preferable, as it avoids any inference that microorganisms are little plants. The concept of a normal microbiota does not imply a constant, identical association between particular microbial populations and humans. Rather, the normal microbiota qualitatively describes the species that are generally found within the stable mixture of microbial populations (microbial community) associated with particular body tissues, and within this microbial community the relative concentrations of individual populations can and do fluctuate throughout an individual's life and in response to a wide variety of external environmental influences.

Usually the normal microbiota of the human body are nonpathogenic, that is, they do not cause disease. However, the relationships between microorganisms, humans, and disease are dynamic, and microorganisms that are normally non-pathogenic can cause disease under appropriate conditions. Virtually any microorganism is a potential opportunistic pathogen capable of causing disease when the right set of conditions occurs. For example, a person's normal defense mechanisms are subject to failure, and a host with compromised defense mechanisms is susceptible to **infection** and **disease**.[1] The common adage "when you are tired, and run-down you are more prone to infection," though an obvious oversimplification, has much validity because infectious diseases are as much the result of the failure of the human physiological and immunological defense systems as of the special properties of pathogenic microorganisms. Stated another way, diseases caused by microorganisms occur when a pathogen is able to overcome host defenses because of the inherent properties of that microbe, known as **virulence factors**, that enable it to cause disease, or because there is a breakdown in the human host defenses against disease-causing microorganisms.

The acquisition of a resident microbiota by humans occurs in stages and therefore is termed a **successional process**. Although some parasites can migrate through the placenta, the human fetus is normally sterile, with colonization of body tissues beginning during the birth process. The different tissues of the body provide distinct habitats with varying environmental conditions for the growth of differing microbial populations. The growth of microorganisms on body tissue surfaces alters the local environmental conditions, leading to the successional changes in the populations of microorganisms associated with the tissues until a relatively stable, normal microbiota is established.

Not all body tissues, though, provide suitable habitats for the growth of microorganisms. For example, most of the urinary tract lacks a resident microbiota. Only the distal end of the urinary tract has a resident microbiota, and urine that does not contact this extremity is considered a sterile body fluid. Similarly, blood is considered a sterile body fluid because the circulatory system does not possess a resident microbiota. In reality, various microorganisms frequently enter the bloodstream but normally do not establish growing populations within the circulatory system. For example, a segment of the circulatory system associated with the liver, the hepatic portal system, normally contains low numbers of bacteria that pass through the intestinal wall as a result of abrasions in the lining of the intestinal tract caused by food particles. These bacteria are routinely eliminated from the circulatory system by specialized white blood cells, known as **Kupffer cells**, that occur in blood vessels of the liver. Such defensive cells and other antimicrobial factors in blood prevent the establishment of a resident microbiota within the circulatory system. Transient **bacteremia** (bacteria in the

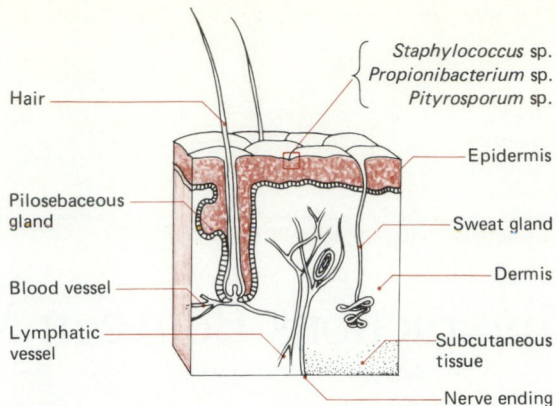

FIGURE 18.1

Drawing of the surface of the skin, indicating some of the indigenous microbiota found there.

bloodstream), however, can occur throughout the circulatory system, even in healthy individuals. For example, following some dental procedures, such as oral prophylaxis (cleaning of teeth) and tooth extractions, bleeding of the gums results in a transient systemic bacteremia, that is, spread of bacteria through the circulatory system to all parts of the body, which lasts for about 24 hours; chewing gum and food can also cause transient low-grade bacteremia.

Microorganisms Indigenous to the Skin

Factors Influencing the Growth of Microorganisms on the Skin   Human **skin surfaces** are not especially favorable habitats for the growth of microorganisms (Figure 18.1). Even though the skin is continuously exposed to microorganisms, most of them are unable to reproduce there. A major factor that determines the distribution of microbial populations on the skin surface is the microenvironment with its particular abiotic (environmental) properties. Environmental factors, such as temperature, water activity, pH, and salinity, represent a severe environmental stress that can prevent many microbial populations from colonizing the skin surface. The lack of available water is a major limiting factor controlling the extent of such microbial growth. Although sweat glands secrete fluids containing water on the skin surface, sweat contains high concentrations of salt and organic compounds that reduce the water activity (A_w) sufficiently to inhibit most microbial growth. Even if they are able to grow at low water activity levels, the microorganisms that become established within the resident microbiota of the skin must tolerate the osmotic stress caused by the high salt concentrations there.

Additionally, the presence of antimicrobial agents, including free fatty acids—some of which are unsaturated and/or of low molecular weight—produced by the host animal and antimicrobial compounds produced by those microorganisms that do successfully establish themselves as the normal microbiota of the skin, act to prevent foreign microorganisms

---

[1]Infection occurs when microorganisms reproduce within the body. In some cases, such microbial reproduction leads to disease, that is, to an altered and impaired physiological state of the human host.

from growing on the skin surface. Lipids on the skin surface are derived mainly from the secretion of **sebum** from the sebaceous glands. Sebum is secreted into hair follicles, and although some bacteria are able to colonize these depressions in the skin surface, many of the unsaturated free fatty acids in sebum have antimicrobial activity. The microorganisms inhabiting the skin surface must tolerate such natural antimicrobial biochemical secretions. Furthermore, many microorganisms living on the skin surface produce antimicrobial substances, some of which are low molecular weight fatty acids; those microbial populations successfully inhabiting the skin surface must resist these antimicrobial substances. Both the antimicrobial substances of human origin and those produced by the indigenous microbiota of the skin act to prevent the invasion of the skin surface by other microbial populations and thus to limit the diversity of the indigenous skin microbiota.

Resident Microbiota of the Skin   The dominant microbial populations on the skin surface are Gram-positive bacteria, which are normally relatively resistant to desiccation compared to Gram-negative bacteria. Members of the genera *Staphylococcus* and *Micrococcus* are frequently the most abundant microorganisms on the skin surface (Figure 18.2) because they are generally salt tolerant and can utilize the lipids present on the skin surface. Other Gram-positive bacteria usually found as part of the normal microbiota of the skin surface include *Corynebacterium*, *Brevibacterium*, and *Propionibacterium*. Gram-negative bacteria generally occur primarily in the moister regions of the skin surface, such as in the armpits and between the toes. Relatively few fungi are included in the normal microbiota of the skin surface. Two yeasts, *Pityrosporum ovale* and *P. orbiculare*, though, are able to metabolize the lipids found on the skin surface and normally occur on scalp tissues.

Effects of Washing on the Indigenous Microbiota of the Skin   Proper cleanliness habits and good hygienic practices tend to prevent the establishment of nonindigenous microorganisms among the natural skin microbiota by preventing the buildup of excessive concentrations of organic matter. The presence of high concentrations of organic matter can make the skin surface a more favorable environment for microorganisms, permitting the growth of many microorganisms that otherwise could not grow there. Washing the skin surface, however, removes excess organic matter and temporarily reduces the number of both resident and transient microorganisms (Figure 18.3). Shortly after the skin surface is washed, the populations of indigenous microbiota begin to return to their original numbers.

Oral Cavity

In contrast to the relatively few microbial populations that flourish on the skin surface, an abundance of microorganisms develops within the oral cavity. The oral cavity contains a

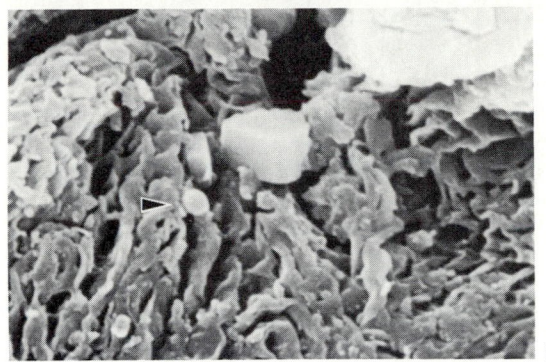

FIGURE 18.2

Scanning electron micrographs of bacteria on the surface of human skin. (A) Skin showing coccal-shaped bacteria (arrows) (5,000×). (B) Skin showing coccal-shaped bacteria and crystalline deposits from sweat (5,000×). (Courtesy Robert Apkarian, Yerkes Primate Research Center, Emory University)

great variety of surfaces, with differing environmental conditions providing varied habitats for colonization by diverse microbial populations. Within the oral cavity, microorganisms can grow on various surfaces, including the gums (gingiva) and teeth (Figure 18.4). The tooth surface is unique among body tissues in that it is a nonshedding hard surface. Teeth adsorb acidic glycoproteins from saliva. These glycoproteins have a net negative charge that initially repels bacteria, which also have a negative charge.

Resident Microbiota of the Oral Cavity   Some bacteria, however, are able to attach specifically to the glycoproteins covering surfaces in the oral cavity, enabling them to colonize those surfaces and to initiate the process of plaque formation. *Streptococcus* species typically constitute a high proportion of the normal microbiota throughout the oral cavity, and some of these species, including *S. mutans*, produce slime layers and adherence factors that allow them to stick to tooth surfaces. Interestingly, *S. mutans* is not found in

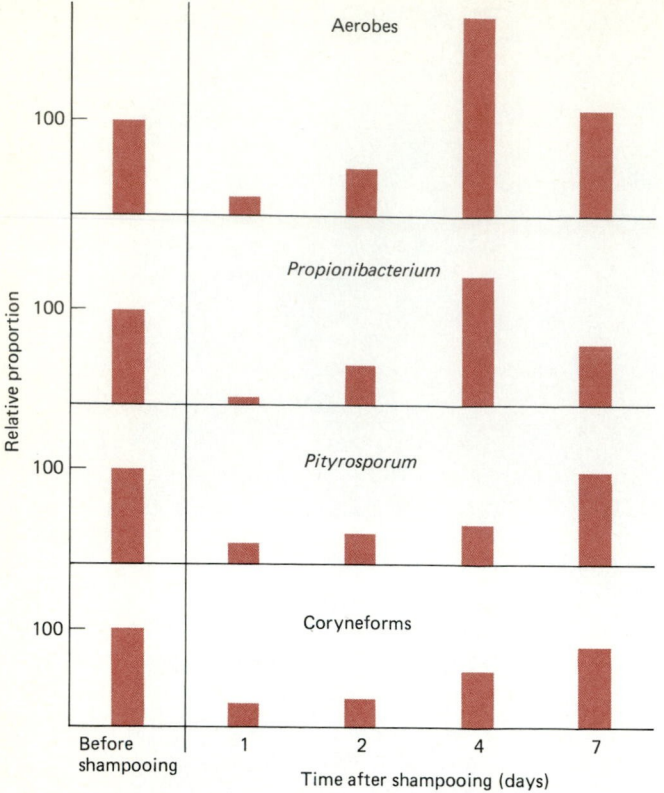

FIGURE 18.3 *(above)*

The changes in microbial populations on the human scalp after shampooing. Washing reduces the level of bacteria on the scalp, and it takes almost a week for prewash bacterial population levels to be established. Note that some populations, notably the aerobes and *Propionibacterium* populations, reacher higher levels 4 days after washing and then decline over the next few days to reach prewash levels. It is quite common to see such overshoots and oscillations of microbial populations after a disturbance; the final levels represent an interactive balance among the populations on the scalp. (Reprinted by permission of the Society for Applied Bacteriology from R. P. Marples, 1974, Effects of germicides on skin flora, in *The Normal Microbial Flora of Man*, F. A. Skinner and J. G. Carr, eds.)

toothless babies and elderly individuals. Other *Streptococcus* species, such as *S. sanguis*, colonize saliva. Some of these lactic acid bacteria form **dental plaque**[2] on the surfaces of teeth and are implicated in the formation of **dental caries** (cavities) (Figure 18.5). Plaque formation is enhanced by sucrose in the diet. Although *Lactobacillus* species are normally present in low numbers, they are frequently found in association with dental caries. Besides *Streptococcus* species, obligate anaerobic bacteria are found in high numbers in the oral cavity, including Gram-negative coccoid members of the genus *Veillonella* and Gram-positive species of *Bacteroides*, *Fusobacterium*, and *Peptostreptococcus*. Additional members of the normal microbiota of the oral cavity include species of the bacteria *Actinomyces*, *Neisseria*, *Treponema*, *Staphylococcus*, *Micrococcus*, *Vibrio*, *Leptotrichia*, and *Rothia*; fungi, such as the yeast *Candida albicans*; and protozoa, such as species of *Entamoeba* and *Trichomonas*. The somewhat surprisingly high incidence of obligate anaerobes within the resident microbiota of the oral cavity results from the high rates of metabolism by the facultatively anaerobic members of this microbial community that scavenge the free oxygen, producing conditions that favor the growth of obligate anaerobes.

Effects of Brushing   Whereas we routinely attempt to cleanse tooth surfaces of food particles and dental plaque, daily brushing represents only a temporary disturbance of the normal microbiota of the oral cavity. Within minutes of even the complete removal of microorganisms, populations of *Streptococcus sanguis*, *S. salivarius*, and *Actinomyces vis-*

---

[2]Dental plaque is a matrix of microbial cells and microbially produced extracellular polysaccharides (carbohydrates) that forms on the tooth surface. The initiation of plaque formation is mediated by glycoproteins produced by the human host. Plaque formation represents a succession of microbial populations, each of which alters the environmental conditions so that different populations are better able to survive there.

FIGURE 18.4

(A) The human oral cavity and (B) the cultivatable bacteria found in its various parts.

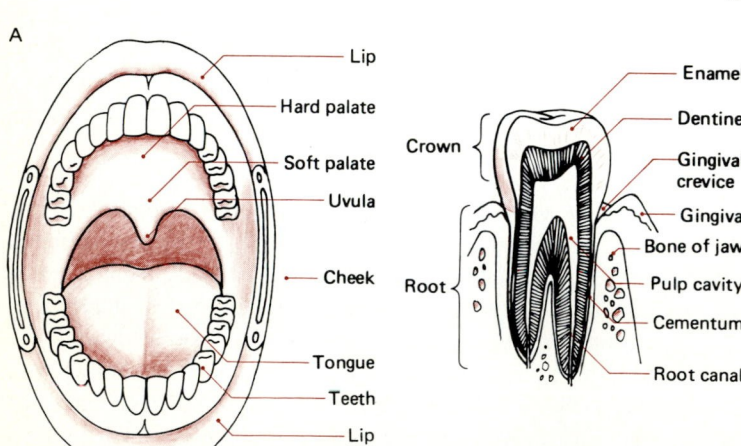

	% Cultivable bacteria		
	Tooth	Tongue	Gingival crevice
Streptococcus	25	45	25
Bacteroides	7	5	17
Fusobacterium	5	1	4
Corynebacterium	–	4	17
Veillonella	13	15	11
Actinomyces	33	2	9
Peptostreptococcus	11	5	9

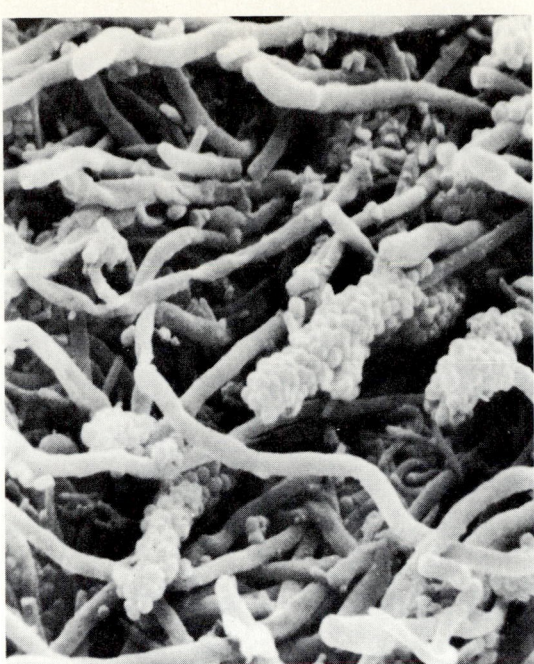

FIGURE 18.5
This scanning electron micrograph shows the formation of dental plaque (2,900×) in the human mouth just 3 days after the last cleansing (brushing) of teeth. (From BPS—Z. Skobe, Forsythe Dental Center)

cosus that colonize the tongue and saliva once again cover the tooth surface. Within a week or two after a thorough professional dental cleaning, the microbial community colonizing the tooth surface includes species of *Bacteroides*, *Veillonella*, *Neisseria*, *Fusobacterium*, and *Rothia*, in addition to *Streptococcus* and *Actinomyces* species. Although cleaning the teeth does not eliminate this resident microbiota, brushing does remove substrates that permit the overgrowth of microorganisms, the acidic products of microbial metabolism, and dental plaque, which are responsible for dental caries and periodontal disease.

Gastrointestinal Tract

Many microorganisms are washed from the oral cavity to the rest of the gastrointestinal tract. Also, we regularly ingest viable microorganisms with our food. There are marked differences between the upper and lower portions of the gastrointestinal tract with respect to their suitability for microbial growth and survival.

Although some microorganisms are able to reproduce within food particles in the stomach, this is a transitory phenomenon. Exposure to the gastric juices secreted into the stomach kills most microorganisms. The low pH of the stomach also precludes the existence of an indigenous microbiota associated with the lining of that organ. Similarly, enzymes in the upper portions of the small intestine do not favor the growth and survival of microorganisms. In contrast to the upper gastrointestinal tract, there is an abundant indigenous microbiota associated with the lower intestinal tract, that is, the lower regions of the small intestine and the colon or large intestine (Figure 18.6). The constant temperature, 37°C, and availability of water and nutrients render the large intestine a favorable habitat for the growth of a variety of microbial populations. Food, partially digested by the body's enzymes, is continuously supplied to the microorganisms inhabiting the intestinal tract, supporting the growth of a large resident microbiota. Consequently, the highest numbers of resident microbiota are associated with the large intestine (Figure 18.7). Most of the microorganisms in the large intestine are anaerobic. The facultative anaerobes there reduce the oxygen levels sufficiently so that strict anaerobes proliferate.

Resident Microbiota of the Gastrointestinal Tract   The initial residents of the intestinal tract in breast-fed infants are members of the genus *Bifidobacterium* compared to *Lactobacillus* species in bottle-fed infants. These initial colonizers of the intestinal tract are later displaced by other bacterial species, which include both obligate and facultative anaerobes, including members of the genera *Lactobacillus*, *Streptococcus*, *Clostridium*, *Veillonella*, *Bacteroides*, *Fusobacterium*, and coliform bacteria. The enteric bacteria, including members of the genera *Escherichia*, *Proteus*, *Klebsiella*, and *Enterobacter*, are facultative anaerobes. Although these facultative anaerobes were once thought to comprise the majority of the microorganisms inhabiting the intestinal tract, methodological improvements for the isolation and culture of obligate anaerobes have revealed that up to 99 percent of the intestinal microbiota may be obligate anaerobes of the genera *Bacteroides* and *Fusobacterium*. The actual proportions of the individual bacterial populations within the indigenous microbiota of the intestinal tract vary, depending in part on the diet of the host.

Respiratory Tract

The normal microbiota of the respiratory tract is quite different from that of the gastrointestinal tract. Even though the respiratory tract has various defense mechanisms for preventing microbial infections, the upper respiratory tract, including the nasal cavity and nasopharynx, is normally inhabited by various species of the genera *Streptococcus*, *Staphylococcus*, *Moraxella*, *Neisseria*, *Haemophilus*, *Bacteroides*, and *Fusobacterium*, as well as members of the spirochete and coryneform groups. The lower respiratory tract, though, lacks a normal resident microbiota. An abundance of white blood cells capable of engulfing and digesting microbial contaminants is present in the lower respiratory tract, preventing the establishment of an indigenous microbiota. The lower respiratory tract may be colonized temporarily by exogenous microbes—which may be pathogens— inhaled on dust particles or water droplets or by the same microorganisms that normally colonize the upper respiratory tract.

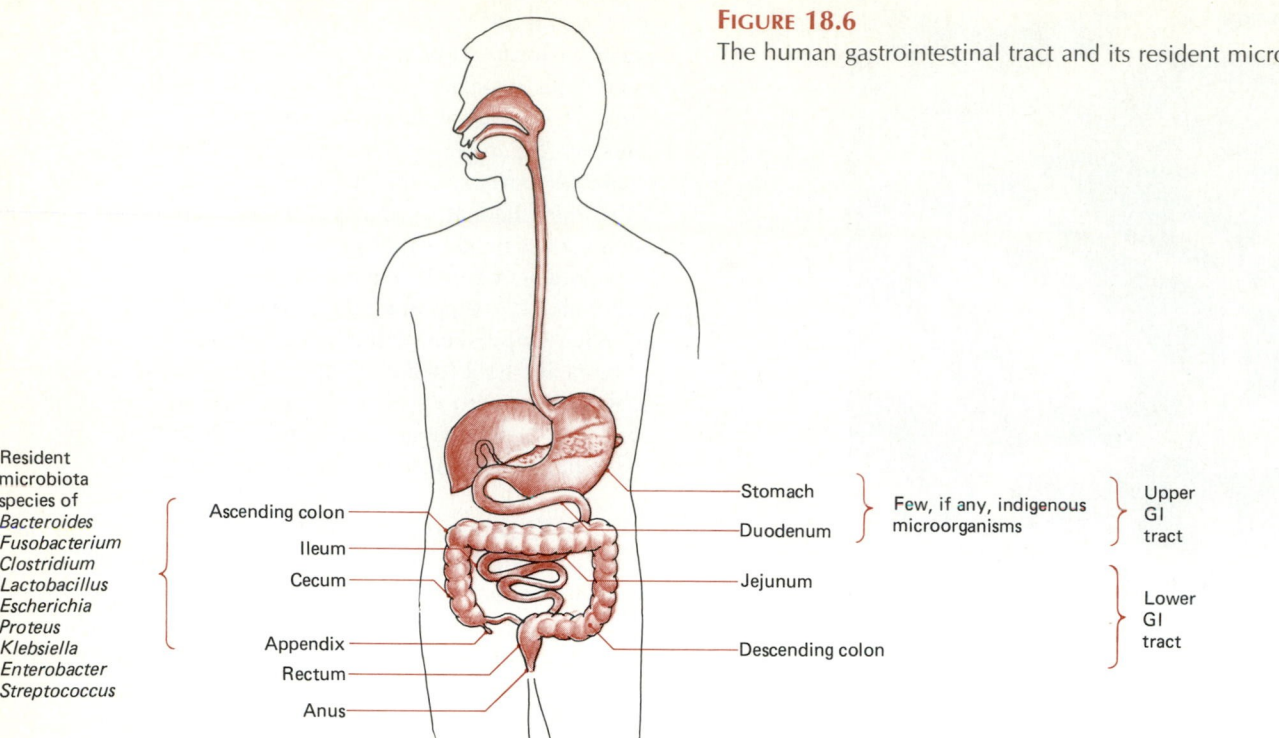

Resident
microbiota
species of
Bacteroides
Fusobacterium
Clostridium
Lactobacillus
Escherichia
Proteus
Klebsiella
Enterobacter
Streptococcus

Ascending colon
Ileum
Cecum
Appendix
Rectum
Anus

Stomach
Duodenum — Few, if any, indigenous microorganisms — Upper GI tract

Jejunum
Descending colon — Lower GI tract

Genitourinary Tract

Like the lower respiratory tract, most of the genitourinary tract, including the kidneys and urinary bladder, are normally free of microorganisms. The external genital regions of both sexes, though, contain various indigenous microbial populations. For example, the terminal areas of the urethra in both males and females are colonized by bacteria. Because of its relatively large surface area and the nature of its mucoidal secretions, the vaginal tract harbors an especially large variety of resident microbiota. The normal microbiota of the vaginal tract includes bacteria such as *Streptococcus*, *Lactobacillus*, *Bacteroides*, and *Clostridium*; coliforms; spirochetes; yeasts, including members of the genus *Candida*; and flagellated protozoa of the genus *Trichomonas* (Figure 18.8). The acidity of the vaginal tract (typically, pH 4.4–4.6), which is due in large part to the metabolic activities of the resident microbiota—particularly the lactic acid bacteria that contribute to maintenance of the low pH—limits the species of microorganisms that can survive and reproduce there. Thus, the microorganisms that multiply within the vaginal tract are normally acid tolerant. The resident microbiota of the vaginal tract is a carefully balanced, complex microbial community. Hormonal levels have an important influence on which populations occur within the vaginal tract. The indigenous prepubescent vaginal microbiota differs from the microbial community that follows sexual maturation, and these poulations, in turn, differ from postmenopausal ones. Also, there are large fluctuations in the resident microbiota

during the normal menstrual cycle. Occasionally, population imbalances occur, for instance, the proliferation of the yeast *Candida albicans* or the protozoan *Trichomonas vaginalis*, leading to a form of inflammation called *vaginitis*.

FIGURE 18.7
The microbiota of the intestine is shown in the mucous membrane of the large intestine. The bacteria (500×) are in crypts. (From BPS—D. C. Savage, University of Illinois, Champaign-Urbana)

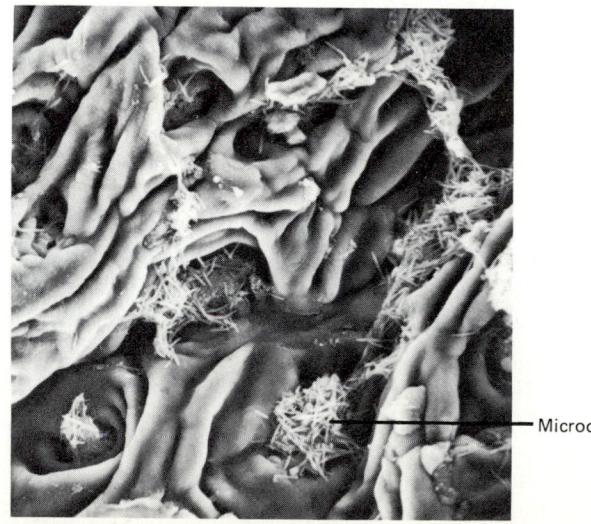

Microorganisms

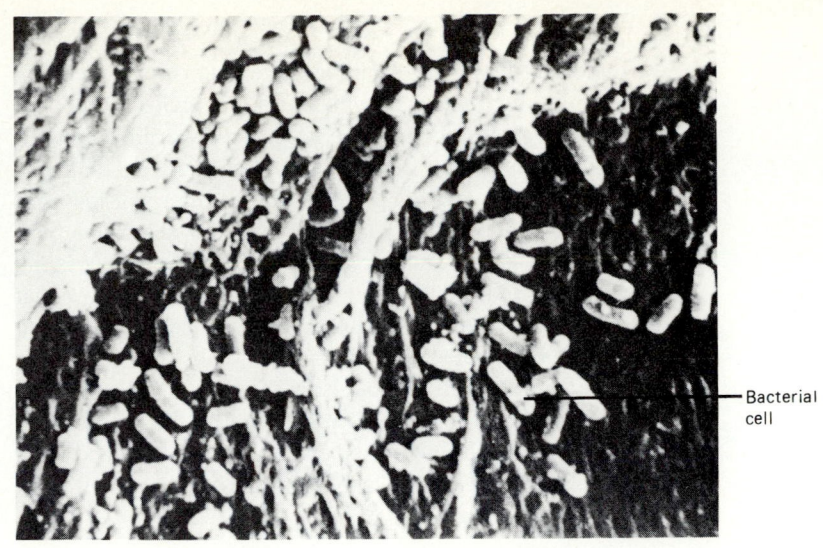

Bacterial
cell

FIGURE 18.8
The normal microbiota of the vaginal tract is represented by this bacterial microcolony (10,700×) at the intercellular borders on a rat vaginal epithelium. (Reprinted by permission of the American Society for Microbiology, Washington D.C., from B. Larsen, A. J. Markovetz, and R. P. Galask, 1977, *Applied and Environmental Microbiology*, 34:85)

18.2 Virulence Factors of Pathogenic Microorganisms

Having noted that microorganisms naturally occur in association with the surface tissues of healthy individuals, we now turn to the host–parasite interactions responsible for human diseases caused by microorganisms. What is it about some microorganisms that allows them to establish the infections that cause human disease? To answer this question, we need to examine the host defense mechanisms that normally preclude infectious diseases and the intrinsic properties of pathogenic microorganisms that allow them to overcome these defenses and initiate the physiological changes inherent in the disease process.

The simple presence of an organism does not equal disease. The invasion or infection of the body by a microorganism, even by a pathogen that typically causes disease, results in disease only when the potential of the microorganism to disrupt normal bodily functions is fully expressed. In many cases, though, infections with potentially pathogenic microorganisms do not lead to disease because the ability of the microbes to affect body functions adversely is not fully expressed and therefore the overt clinical manifestation of the disease does not occur. Many healthy individuals are carriers of potentially pathogenic microorganisms, that is, they are infected with the microbes but will not or have not yet developed a disease as a result of the infection. Although we now understand that **pathogenicity**, that is, the ability to cause disease, is not a property of the microorganism alone, it is still useful to examine the intrinsic properties of pathogenic microorganisms that contribute to their potential for causing human disease.

Disease-causing microorganisms possess properties, referred to as **virulence factors**, that enhance their pathogenicity and allow them to colonize and/or to invade human tissues and disrupt normal bodily functions. The **virulence** of pathogenic microorganisms, that is, their ability to induce human disease, depends in large part on two properties of microorganisms: invasiveness and toxigenicity.[3] **Invasiveness** refers to the ability of microorganisms to invade human cells and tissues and to multiply on or within them. Most of the microorganisms of the normal microbiota of humans do not invade the tissues on whose surfaces they grow. Microorganisms that possess invasive properties are able to establish infections within host cells and tissues. **Toxigenicity** refers to the ability of a microorganism to produce biochemicals, known as **toxins**, that disrupt the normal functions of cells or are generally destructive to human cells and tissues. Some toxin-producing microorganisms can grow outside of the host and still cause disease symptoms if the toxins enter human tissues. These toxin-producing strains need not establish an infection within the human body to cause disease.

The establishment of a microbially caused disease is a function of the virulence of the particular microorganism in terms of its invasiveness, its toxigenic properties, the dosage (numbers) of that microorganism, and the resistance of the host individual. It may therefore be expressed as follows:

$$\text{Infectious disease} = f\left(\frac{\text{microbial virulence} \times \text{dose}}{\text{host resistance}}\right)$$

Pathogens vary with regard to their invasiveness and toxigenicity. For example, *Streptococcus pneumoniae* is highly invasive but only mildly toxigenic. The source of its viru-

---

[3]The term *pathogenicity* refers to the qualitative ability of a microorganism to cause disease, whereas the term *virulence* quantitatively describes the extent of an organism's pathogenicity.

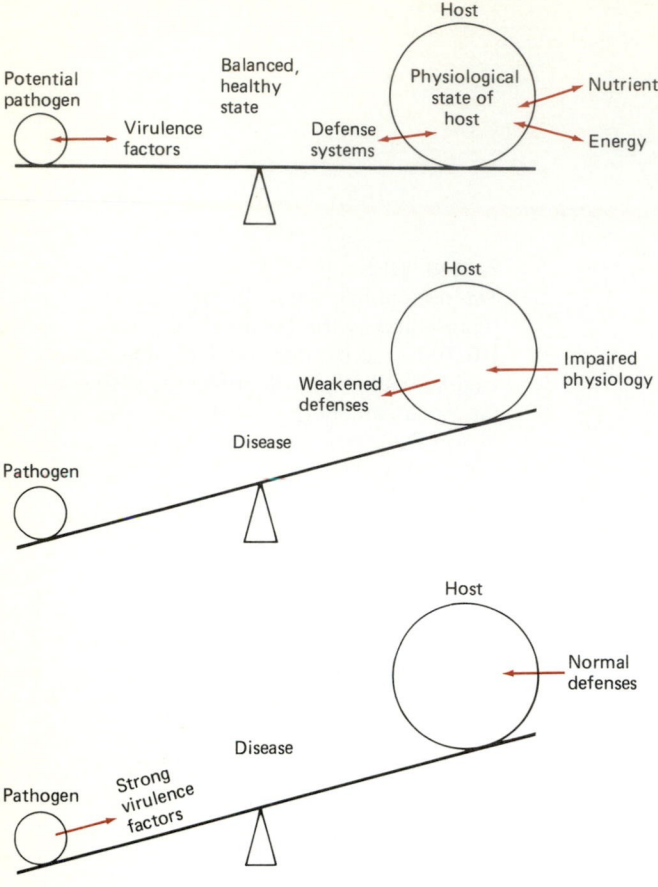

FIGURE 18.9

The interaction of a potential pathogen and a human host cell depends upon the intrinsic (virulence) factors of the microbe and the physiological state of the host with the environmental factors determined by that state.

lence is its ability to disseminate rapidly throughout the lung. In contrast, *Clostridium tetani*, which is only slightly invasive, moving little beyond the initial point of infection, is highly toxigenic, producing widespread effects that act at sites far removed from the initial site. The relationship between a pathogen or parasite and its host is dynamic and varies depending upon the physiological state of each (Figure 18.9). The extent of pathology varies greatly depending upon the microorganism and the physiological state of the host, that is, the properties of the specific microbial strain and the environment in which it may proliferate. Some microorganisms, such as influenza viruses, infect many individuals but the disease symptomatology generally is not too severe. However, influenza can be a fatal disease in the elderly and in those who are physiologically debilitated. Rabies virus, on the other hand, causes severe disease symptomatology and the death of infected individuals who do not receive preventive treatment because of the intrinsic properties of the virus, regardless of their age and general physiological condition.

Toxigenicity

An important factor determining the extent of pathogenicity is the ability and extent to which a microorganism produces toxins (biological poisons). Toxins cause discernible damage to human host systems and in some cases cause death (Table 18.1). In the past, toxins produced by microorganisms were classified as endotoxins if they comprised a heat-stable part of the microbial cell and as exotoxins if they were heat-labile proteins secreted by the cell. However, we now know that some exotoxins are not released until the cell is disrupted and that substances classified as endotoxins are sometimes released from the cell without lysis. Therefore, a better classification system for toxins is one based on the biochemical nature of the toxin, whereby endotoxins are equated with lipopolysaccharide toxins and exotoxins are equated with protein toxins.

Lipopolysaccharide Toxins   The term **bacterial endotoxin** is equated with the **lipopolysaccharide (LPS)** component of the Gram-negative eubacterial cell wall. Although all Gram-negative eubacteria have LPS in their cell walls, LPS is not toxic unless it is released from the outer layer of the cell. When Gram-negative bacteria die, their cell walls disintegrate, releasing the LPS toxin. Some growing Gram-negative bacteria also release LPS toxin; in these cases, the LPS can have a toxic effect on a host organism. Toxicity is associated with the lipid portion of the LPS molecule, termed lipid A, which does not have the same structure as membrane phospholipids. Rather, lipid A is composed of fatty acids, such as β-hydroxy myristic, attached by an ester or amide

FIGURE 18.10

Endotoxin is an LPS toxin. The toxicity is associated with the lipid portion of the molecule, lipid A. Lipid A contains β-hydroxymyristic acid (HMA)—a C14-saturated fatty acid—and various other long chain fatty acids (FA). The lipid A portion of the molecule is linked to a polysaccharide via ketodeoxy-octonate (KDO). The polysaccharide portion increases the water solubility of the molecule and hence its availability within the body to cause a toxic reaction.

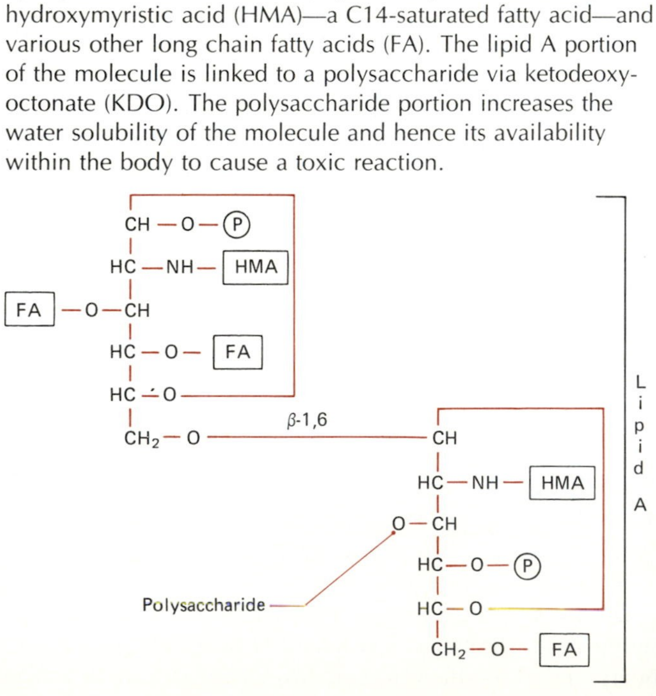

TABLE 18.1   Some Protein Toxins Produced by Microorganisms that Cause Disease in Humans

Microorganism	Toxin	Disease	Action
Clostridium botulinum	Several neurotoxins	Botulism	Paralysis; blocks neural transmission
Clostridium perfringens	α-Toxin	Gas gangrene	Lecithinase
	κ-Toxin		Collagenase
	θ-Toxin		Hemolysin
Clostridium tetani	Neurotoxin (tetanospasm)	Tetanus	Spastic paralysis interfers with motor neurons
	Tetanolysin		Hemolytic cardioxtoxin
Corynebacterium diphtheriae	Diphtheria toxin	Diphtheria	Blocks protein synthesis at level of translation
Streptococcus pyogenes	Streptolysin O	Scarlet fever	Hemolysin
	Streptolysin S		hemolysin
	Erythrogenic		Causes rash of scarlet fever
Shigella dysenteriae	Neurotoxin	Bacterial dysentery	Hemorrhage, paralysis
Staphylococcus aureus	Enterotoxin	Food poisoning	Intestinal inflammation
Aspergillus flavus	Aflatoxin B_1	Aflatoxicosis	Blocks protein synthesis at level of transcription
Amanita phalloides	α-Amanitin	Mushroom food poisoning	Blocks protein synthesis at level of transcription

linkage to a diglucosamine-β-1,6 disaccharide (Figure 18.10). This same structure is found in all eubacterial Gram-negative LPS. The physiological effects of LPS toxins include fever, circulatory changes, and other general symptoms, such as weakness and nonlocalized aches. The injury to the circulatory system by the LPS of the Gram-negative cell is basic to the action of this toxin, but its mechanism of action is not yet understood. The effects of LPS toxins are generally the same for all species of Gram-negative bacteria because of the common nature of lipid A. Thus, there is no specific characteristic disease symptomatology associated with the endotoxin of a particular bacterial species. LPS toxins of *Salmonella* and *Shigella* species are responsible in part for diseases, such as gastroenteritis, caused by these pathogens, but these pathogens also produce protein toxins (exotoxins) that are largely responsible for their pathogenicity; for example, *Shigella* produces protein toxins that act on nerve cells.

Protein Toxins   In contrast to LPS toxins, the effects of **protein toxins (exotoxins)** are specific to the microorganism producing the toxin, and these toxins cause distinctive clinical symptoms. Whereas LPS toxins are produced exclusively by Gram-negative bacteria, protein toxins are produced by both Gram-negative and Gram-positive bacteria. Most bacterial protein toxins are composed of a receptor

---

ANALYTICAL PROCESS

Detection and Quantitation of Endotoxin

Endotoxin (LPS) can be assayed using aqueous extracts from the blood cells (amoebocytes) of the horseshoe crab (*Limulus* amoebocyte lysate). The reaction of *Limulus* amoebocyte lysate with endotoxin is highly specific. The *Limulus* extract reacts with the lipid component of LPS, causing precipitation of the *Limulus* lysate, which increases the turbidity of the solution. The turbidity is proportional to the concentration of endotoxin and can be measured with a spectrophotometer. The assay is extremely sensitive and can detect less than 1000 bacteria per milliliter of sample. The *Limulus* amoebocyte assay is used to detect the presence of minute quantities of endotoxin in serum, cerebrospinal fluid, water, and other substances. The use of this assay is important in the diagnosis of Gram-negative bacterial infections of blood and cerebrospinal fluid. In aquatic microbiology the detection of LPS is used to quantitate the biomass of the indigenous microorganisms, most of which are Gram negative.

TABLE 18.2 Comparison of Selected Characteristics of Bacterial LPS Toxins (Endotoxins) and Protein Toxins (Exotoxins)

Characteristic	LPS Toxin	Protein Toxin
Chemical composition	LPS–protein complex	Protein
Source	Cell walls of Gram-negative bacteria; released upon death and autolysis of the bacteria	Gram-negative and Gram-positive bacteria; excretion products of growing cells or, in some cases, substances released upon autolysis and death of the bacteria
Effects on host	Nonspecific	Generally affects specific tissues
Thermostability	Relatively heat-stable (may resist 120°C for 1 hour)	Heat-labile; most are inactivated at 60–80°C
Toxoids	No	Yes
Lethal dose	Large	Small

protein component that attaches to a target cell and a toxic component that enters the cell and disrupts normal cell activity.

Protein toxins are more readily inactivated by heat than LPS toxins. A protein toxin can normally be inactivated by exposure to boiling water for 30 minutes, whereas LPS toxins can withstand autoclaving (Table 18.2). Some enterotoxins, however, are proteins or peptides that are relatively heat stable. Typically, protein toxins are excreted into the surrounding medium. For example, *Clostridium botulinum*, the causative organism of botulism, secretes a potent exotoxin into canned food products, the ingestion of even minute amounts of which is lethal. Protein toxins are generally more potent than LPS toxins, and far smaller amounts are needed to produce serious disease symptoms than are required for disease symptoms due to LPS. As examples of the potency of protein toxins, about 30 g of diphtheria toxin can kill 10 million

people and 1 g of botulinum toxin can kill everyone in the United States (over 225 million people).

Often protein toxins are referred to by the disease they cause, such as diphtheria toxin or botulinum toxin. They may also be categorized according to the symptoms they cause. As examples, neurotoxins affect the nervous system, enterotoxins cause an inflammation of the tissues of the gastrointestinal tract, and cytotoxins interfere with cellular functions.

Neurotoxins. **Neurotoxins** are toxins that interfere with the functioning of the nervous system, usually by blocking nerve cell transmission. Neurotoxins, even those produced by members of the same genus, differ markedly in their mode of action. The neurotoxins responsible for the symptomatology of botulism, a disease caused by *C. botulinum*, bind to nerve synapses, blocking the release of acetylcholine from nerve cells of the central nervous system and causing the loss of motor function (Figure 18.11). The inability to transmit impulses through motor neurons can cause respiratory failure, resulting in death.

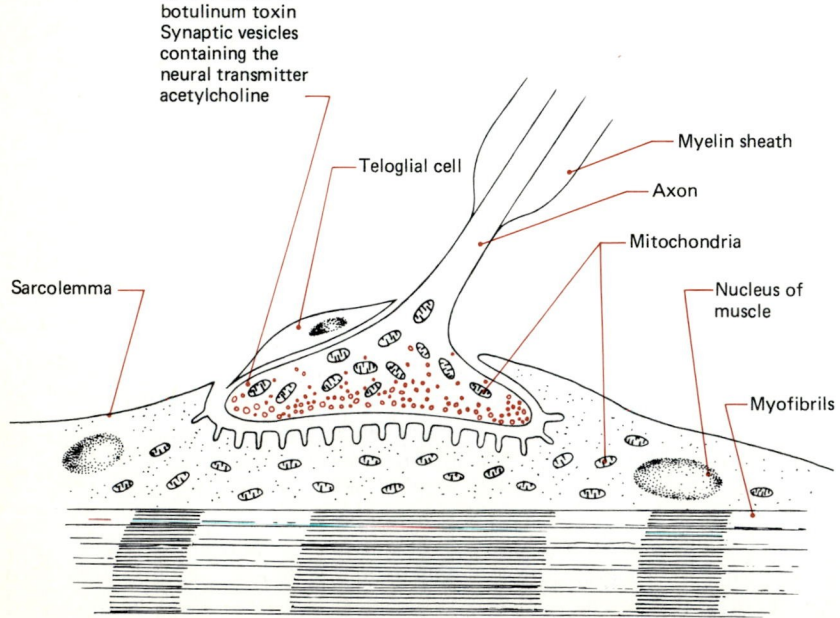

Site of action of botulinum toxin
Synaptic vesicles containing the neural transmitter acetylcholine

Teloglial cell

Sarcolemma

Myelin sheath

Axon

Mitochondria

Nucleus of muscle

Myofibrils

FIGURE 18.11

Diagram showing the action of botulinum toxin. The toxin blocks the transmission of nerve impulses across the synapses of the motor end plates, producing paralysis. The binding of the toxin to the myoneural (nerve–muscle) junction appears to be a two-step process, with toxin, binding first to one receptor and then to another. The binding of the toxin blocks conduction of the nerve impulse at or near the end of the point of final branching between the nerve filaments and the presynaptic part of the end plate complex of the muscle cell, preventing the proper transmission of the nerve impulse to the muscle. This results in flaccid paralysis.

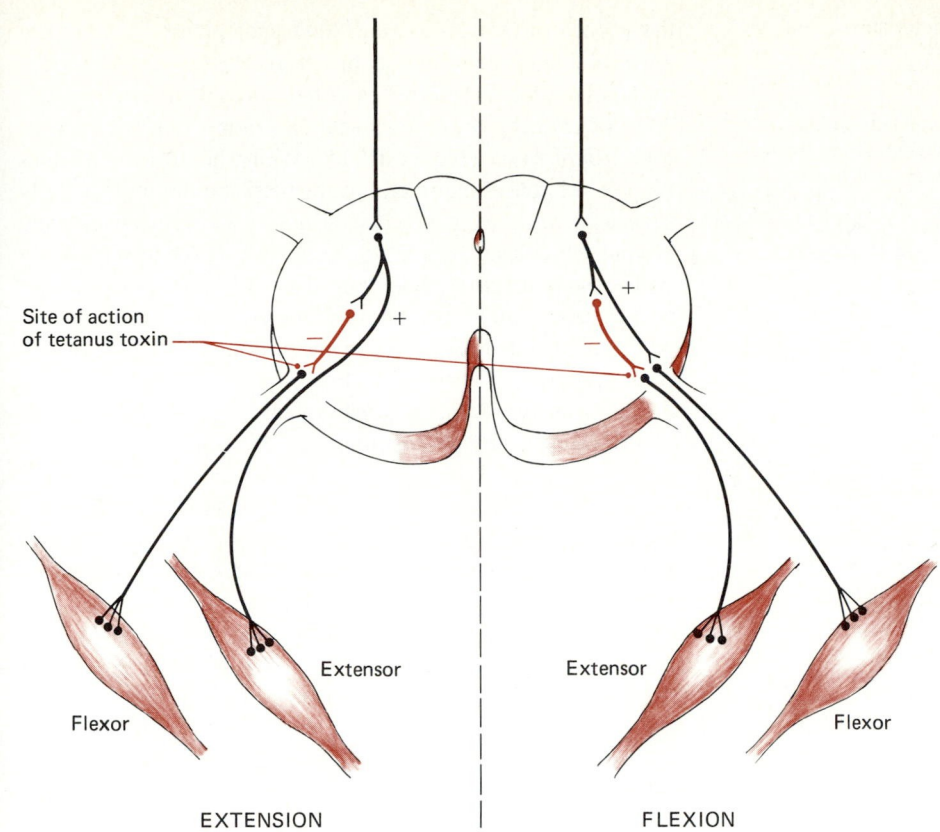

Site of action
of tetanus toxin

Flexor   Extensor   Extensor   Flexor

EXTENSION   FLEXION

FIGURE 18.12
Diagram showing the action of tetanus toxin. The excitation pathways for flexion and extension are reciprocally inhibited so that the excitation of one muscle action, such as flexion, causes inhibition of the antagonistic muscles, which is extension. Tetanus toxin acts at the inhibitory motor neurons so that this antagonism is blocked. The result is the spastic paralysis that characterizes tetanus.

The neurotoxin **tetanospasmin**, produced by *C. tetani*, the causative agent of tetanus, interferes with the peripheral nerves of the spinal cord (Figure 18.12). Tetanospasmin inhibits the ability of these nerve cells to transmit signals to the muscle cells properly, causing the symptomatic spastic paralysis of tetanus. Like the neurotoxin produced by *C. botulinum*, the neurotoxin of *C. tetani* paralyzes motor neurons, but unlike botulinum toxin, tetanospasmin acts only on the nerves of the cerebrospinal axis.[4]

The neurotoxin produced by *Shigella dysenteriae*, the so-called "shiga toxin," differs from the neurotoxins produced by *C. botulinum* and *C. tetani* in that it interferes with the circulatory vessels that supply blood to the central nervous system rather than affecting the nerve cells directly. The neurological effects of the shiga toxin are thus secondary to the primary action of the toxin on the vascular circulatory system.

Enterotoxins. **Enterotoxins** stimulate the cells of the gastrointestinal tract in an abnormal way. Various enteropathogenic bacteria, such as *Salmonella*, *Shigella*, and *Vibrio* species, produce enterotoxins. For example, the enterotoxin **choleragen** produced by *Vibrio cholerae*, the causative agent of cholera, inhibits the conversion of cyclic AMP to ATP by increasing the activity of adenylcyclase. The resulting elevated concentrations of cyclic AMP cause the release of inorganic ions, including chloride and bicarbonate ions, from the mucosal cells that line the intestine into the intestinal lumen (Figure 18.13). Although the exact mechanism of toxin action on adenylcyclase is not understood, the change in the ionic balance resulting from the action of this toxin causes the movement of large amounts of water into the lumen in an attempt to balance the osmotic pressure, leading to severe dehydration that sometimes results in the death of infected individuals.

Cytotoxins. **Cytotoxins** kill cells by enzymatic attack or by blocking essential cellular metabolism. Included among the cytotoxins are those proteins that cause lysis of blood cells and those that interfere with protein synthesis. For example, **diphtheria toxin**, produced by *Corynebacterium diphtheriae*, inhibits protein synthesis in mammalian cells. Diphtheria toxin preferentially affects cardiac and renal tissues. This toxin blocks transferase reactions during the translation of mRNA, preventing the addition of amino acids and thus the elongation of the peptide chain. The production of diphtheria toxin is particularly interesting because only lysogenized cells of *C. diphtheriae* produce diphtheria toxin proteins. This is a good example of how phage can convert an otherwise nonpathogenic bacterium into a pathogenic one.

---

[4]It is postulated that tetanus toxin inhibits the release of glycine from the inhibitory neurons (interneurons) in the anterior horn of the spinal cord. Because glycine is the inhibitory neurotransmitter in these interneurons, the result is convulsions similar to those produced by strychnine, which is known to compete with glycine for receptor sites.

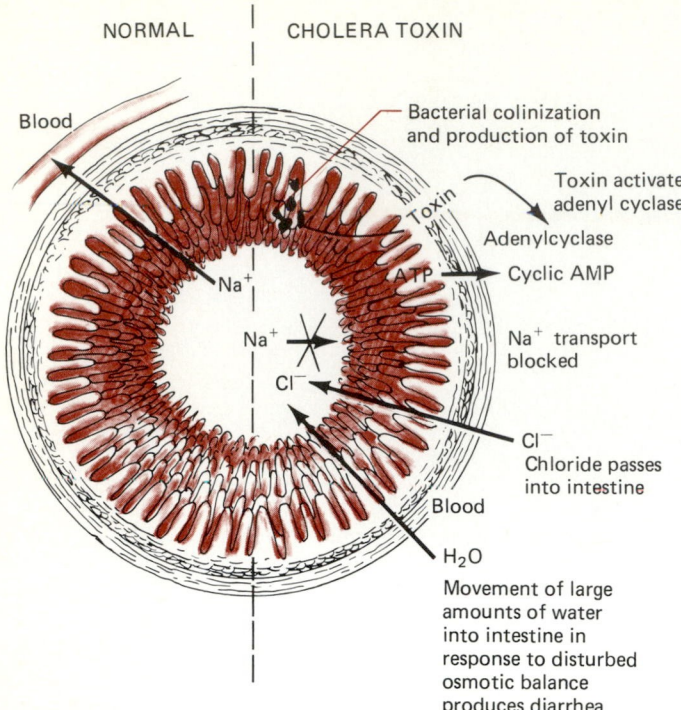

FIGURE 18.13

Diagram showing the action of cholera toxin. The action of the toxin at the lining of the intestine causes an osmotic imbalance and the flow of excessive amounts of water out of the body.

The protein toxin is coded for by the phage genome. Thus, only when *C. diphtheriae* is infected with a virus does a human infection with *C. diphtheriae* result in disease.

Some proteins produced by pathogenic microorganisms, called *cytolysins*, contribute to their virulence by causing the lysis of blood cells (Table 18.3). Cytolysins that cause the lysis of human erythrocytes are called **hemolysins** because their action results in the release of hemoglobin from these red blood cells. For example, *Streptococcus* species produce various hemolysins, including streptolysin O, an oxygen-labile and heat stable protein, and streptolysin S, an oxygen-stable, acid-sensitive, and heat-labile protein. The hemolytic action of these cytolysins produces zones of clearing when these bacteria are grown on blood agar plates. A complete zone of clearing around a bacterial colony growing on a blood agar plate is referred to as **beta hemolysis** and a partial zone of clearing around a bacterial colony is referred to as **alpha hemolysis** (Figure 18.14). Alpha hemolysis involves the conversion of hemoglobin to methemoglobin, generally seen as a zone of green discoloration with partial clearing around the colony. Hemolytic activity is associated not only with *Streptococcus* species but also with various other bacterial genera, including *Staphylococcus* and *Clostridium*. In addition to red blood cells, white blood cells are killed by some microbial cytotoxins. For example, leukocidin produced by *Staphylococcus aureus* causes lysis of leukocytes, contributing to the pathogenicity of this organism.

Toxins Produced by Eukaryotic Microorganisms   Bacteria are not the only microorganisms that produce toxins. Several eukaryotic microorganisms produce potent cytotoxins, some of which are nonproteins. For example, dinoflagellates produce highly potent toxins that cause paralytic shellfish poisoning. Many mushrooms are highly poisonous because of the potency of the **mycotoxins** (fungal toxins) they produce. The term *mycotoxin* actually encompasses a wide variety of biochemicals with varying modes of action. Some cause ultrastructural changes in the host, whereas others interfere with various metabolic activities of host cells. Although there is no generalized mechanism that applies to all mycotoxins, the mode of action of most mycotoxins appears to be based primarily on their ability to interact with macromolecules, subcellular organelles, and organs of animals. In the most infamous of the poisonous mushrooms, *Amanita*, the toxin alpha amanitin blocks transcription of DNA by interfering with RNA polymerase enzyme. Similarly, *Aspergillus* species produce aflatoxins that bind to DNA and prevent transcription of genetic information, resulting in a variety of adverse effects on humans and other animals.

Invasiveness: Enzymes as Virulence Factors

Some enzymes produced by microorganisms also interfere with normal mammalian functions, and some of these enzymes contribute to the virulence of microbial pathogens by

TABLE 18.3   Some Extracellular Enzymes Involved in Microbial Virulence

Enzyme	Action	Examples of Bacteria Producing Enzyme
Hyaluronidase (spreading factor)	Breaks down hyaluronic acid	*Streptococcus pyogenes*
Coagulase	Blood clots coagulation of plasma	*Staphylococcus aureus*
Phospholipase	Lyses red blood cells	*Staphylococcus aureus*
Lecithinase	Destroys red blood cells and other tissue cells	*Clostridium perfringens*
Collagenase	Breaks down collagen (connective tissue fiber)	*Clostridium perfringens*
Fibrinolysin (kinase)	Dissolves blood clots	*Streptococcus pyogenes*

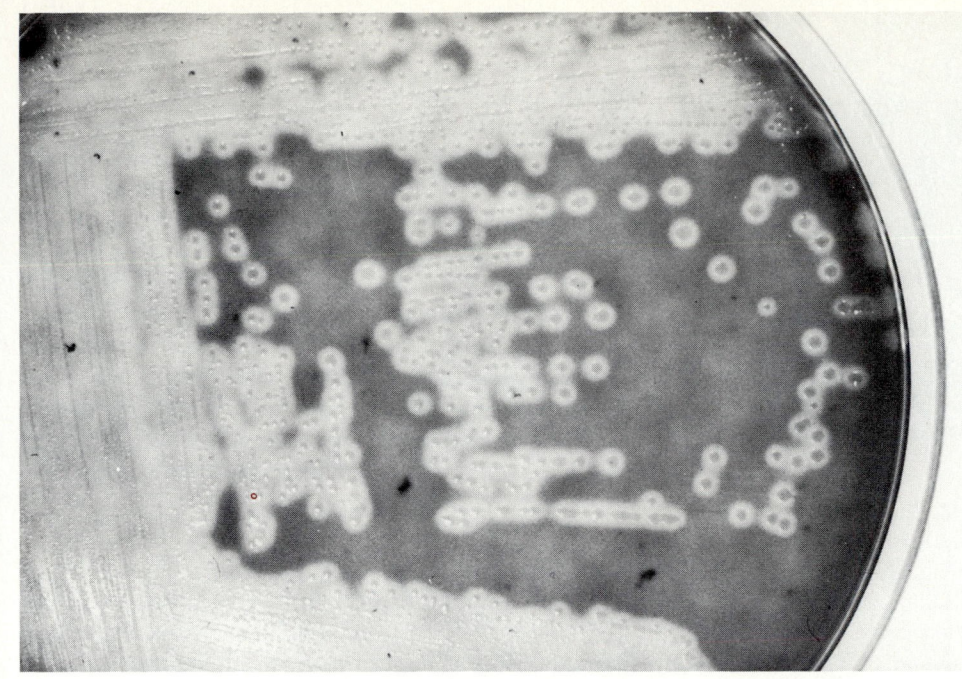

Figure 18.14
Beta-hemolytic growth of *Strep-tococcus pyogenes* on blood agar plates, showing zones of clearing where hemolysis has occurred around the colonies. (From BPS— L. M. Pope and D. R. Grote, University of Texas, Austin)

enabling the microorganisms to invade body tissues and cells. For example, various **phospholipase** enzymes produced by microorganisms can destroy animal cell membranes. Phospholipases can act as hemolysins, causing the lysis of red blood cells. Indeed, some substances that have been classified as toxins are now known to be toxic enzymes. For example, the alpha toxin of *Clostridium perfringens* is a **lecithinase**, also known as *phospholipase C* or *phosphatidylcholine phosphohydrolase*. This enzyme hydrolyzes lecithin, which is a lipid component of eukaryotic membranes. Lecithinase activity, associated with *C. perfringens*, is in part responsible for the ability of this organism to grow, to invade tissues, and to cause gas gangrene. Hydrolysis of lecithin by the alpha toxin of *C. perfringens* destroys the integrity of the cytoplasmic membranes of many cells and is the primary cause of the extensive tissue damage in this disease. Lecithinase also acts as a hemolysin, causing lysis of red blood cells in addition to destroying cells of various other tissues. The release of iron from the lysed blood cells allows this pathogen to grow in an environment that normally has a very low concentration of this essential growth nutrient.

Some *Staphylococcus* and *Streptococcus* species produce **fibrinolysin (kinase)**. The fibrinolytic enzymes staphylokinase and streptokinase catalyze the lysis of fibrin clots. The action of these two fibrinolytic enzymes may enhance the invasiveness of pathogenic strains of *Staphylococcus* and *Streptococcus* by preventing fibrin in the host from walling off the area of bacterial infection. Without the action of fibrin, the pathogens are free to spread to surrounding areas. In a somewhat different way, the production of **coagulase** enhances the virulence of some *Staphylococcus* species. On the other hand, the enzyme coagulase converts fibrinogen to fibrin. Some *Staphylococcus* species, such as *S. aureus*, pro-

duce this enzyme, and the deposition of fibrin around the staphylococcal cells presumably protects the cells against the circulatory defense mechanisms of the host. Coagulase-negative strains of *S. aureus*, however, still have been found to be virulent pathogens. It is thus difficult to associate virulence with the activity of a single enzyme, even though these enzymes appear to play a role in the virulence of a variety of pathogenic microorganisms.

Several other enzymes produced by microorganisms can destroy body tissues. For example, **hyaluronidase** breaks down hyaluronic acid, the substance that holds together the cells of connective tissues. Pathogens that produce hyaluronidases spread through body tissues, and therefore, hyaluronidase is referred to as the *spreading factor*. Various species of *Staphylococcus*, *Streptococcus*, and *Clostridium* produce hyaluronidase enzymes. Some *Clostridium* species also produce collagenase, an enzyme that breaks down the proteins of collagen tissues. The k toxin of *C. perfringens*, for example, is a collagenase that contributes to the spread of this organism through the human body. The breakdown of fibrous tissues enhances the invasiveness of pathogenic microorganisms. Thus, the actions of some microbial enzymes contribute to the virulence of pathogens by enhancing the ability of the microorganisms to proliferate within body tissues and by interfering with the normal defense mechanisms of the host organism.

Factors That Interfere with Phagocytosis

Several other factors contribute to the virulence of microorganisms, including the production of surface layers that interfere with the ability of **phagocytes** (phagocytic blood cells) to engulf and destroy bacteria that invade the human

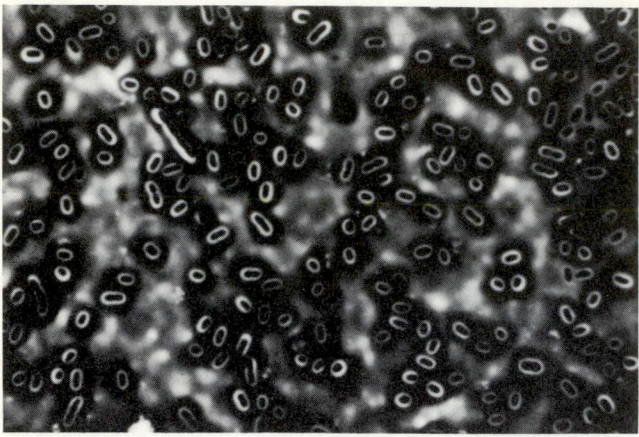

FIGURE 18.15
Micrograph showing encapsulated bacteria. The presence of a capsule increases the potential virulence of microbial pathogens because they are more resistant to host defense mechanisms. This is *Klebsiella pneumoniae* (1,000×). (From BPS—Leon LeBeau, University of Illinois Medical Center)

body. As discussed in earlier chapters, **capsules** protect some bacteria against the host defense mechanism of phagocytosis. Capsules surrounding the cells of strains of *Streptococcus pneumoniae*, for example, permit these bacteria to evade the normal defense mechanisms of the host, allowing them to reproduce and causing the symptomatology of pneumonia. The virulence of other bacteria, including *Haemophilus influenzae* and *Klebsiella pneumoniae*, is also enhanced by capsule production (Figure 18.15). Pili can also interfere with phagocytosis, and the pili of *Neisseria*, for example, can retard phagocytosis, increasing the persistence of the pathogenic *Neisseria* species.

Adhesion Factors

In addition to their role in avoiding phagocytosis, capsules and slime layers contribute to the ability of bacteria to attach or adhere to particular host cells or tissues; other cell surface carbohydrates, proteins, flagella, and fimbriae (pili) also play a role in the attachment of various pathogens to host cells and tissues. Many pathogenic bacteria must adhere to mucous membranes in order to establish an infection. Specific factors that enhance the ability of a microorganism to attach to the surfaces of mammalian cells are termed **adhesins**, and the production of such substances is another important factor that determines the virulence of particular pathogens. The pili of several pathogenic bacteria and their associated adhesins appear to play a key role in permitting the bacteria to adhere to host cells and establish infections. For example, enteropathogenic strains of *Escherichia coli* have particular adhesins associated with their pili that permit them to bind to the mucosal lining of the intestine. In a similar manner, *Vibrio cholerae* is able to adhere to the mucosal cells lining

the intestine, allowing the establishment of an infection (Figure 18.16).

Likewise, the adsorption of certain viruses onto specific receptor sites of human cells establishes the necessary prerequisite for the uptake of the viruses by those cells, leading to the reproduction of the viruses, the disruption of normal host cell function, and the production of disease symptoms by the invading viral pathogens. Some viruses, such as adenoviruses, have external spikes that aid in their attachment to host cells. Similarly, the spikes of orthomyxoviruses and paramyxoviruses attach to receptors of *N*-acetylneuraminic acid on the surfaces of human red blood cells. The ability of pathogenic microorganisms, including viruses, to attach to and invade particular cells and tissues establishes specific tissue affinities for pathogenic microorganisms.

Iron Uptake

Because most bacteria require an iron concentration above 10^{-8} M free iron, while that of most human tissue is less than 10^{-18} M, the ability of bacteria to grow within blood is limited by the lack of available iron. Some pathogens, however, are able to overcome this limitation and sequester the iron that they require from the blood. In order to supply themselves with the iron needed for reproduction, such pathogens produce low molecular weight compounds involved in iron transport, called **siderophores**, that bind iron tightly. Siderophores remove iron normally bound to transferrin or other iron-binding compounds found in blood. In the case of enteric bacteria that synthesize a siderophore called **enterobactin**, the siderophore–iron complex attaches to the outer membrane of the cell wall where a protein acts to dissociate

FIGURE 18.16
Micrograph showing *Vibrio cholerae* attached to the lining of the gut of this infant mouse at the tip of the villus (5,000×). The ability to attach to the intestine is an important characteristic of many pathogens that enter the body through the gastrointestinal tract. (From BPS—Garry T. Cole, University of Texas, Austin)

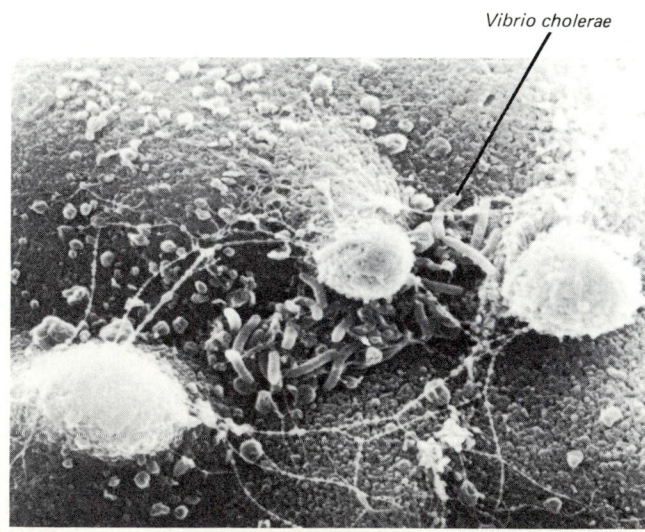

Vibrio cholerae

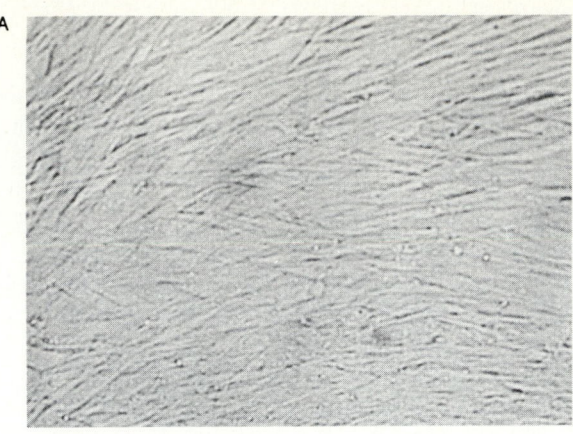
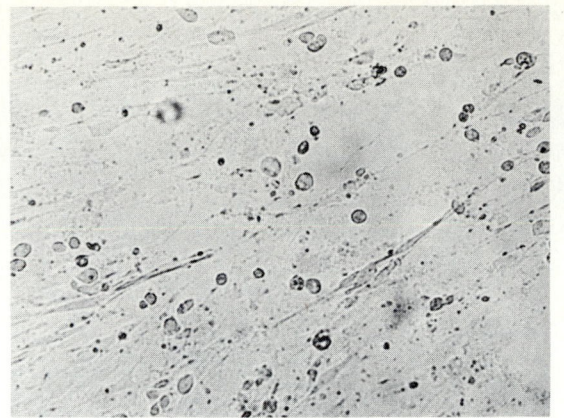

FIGURE 18.17

The characteristic cytopathic effect of Rhinovirus in tissue culture is indicated by cell shrinkage, nuclear pyknosis, and loss of adsorption to glass. Enteric viruses, such as Poliovirus and Coxsackie virus, produce similar cytopathic effects. (A) Micrograph of normal uninfected cells. (B) Micrograph showing cytopathic effects in cells after infection with Rhinovirus. (Courtesy Jack Schieble, California Department of Health Services.)

the iron from the siderophore, allowing the iron to be transported into the cell. Other pathogens, such as *Neisseria* and *Mycobacterium* species, sequester iron without producing siderophores by synthesizing an outer membrane protein that removes iron directly from transferrin. Thus, these bacteria have virulence properties with respect to their ability to acquire iron that enable them to overcome host resistance and initiate systemic infections.

Cytopathic Effects of Viruses

Like bacteria, pathogenic viruses have virulence factors that contribute to their ability to cause disease. When viruses reproduce within host cells, they can also produce substances that may destroy or interfere with the normal functioning of cells. The observable changes in the appearance of cells in-

fected with viruses are collectively known as **cytopathic effects (CPE)** (Figure 18.17). In some cases, human cells infected with viruses die. For example, polio viruses kill the human cells they infect. In other cases, infected cells develop nonlethal abnormalities. Inclusions sometimes occur within the nucleus or cytoplasm of infected cells. These inclusions may be stained with basic or acid dyes and viewed with a microscope. For example, cells infected with measles virus develop acidophilic inclusions in the nucleus and cytoplasm; cells infected with rabies virus develop acidophilic inclusions only within the cytoplasm; and cells infected with adenovirus develop basophilic inclusions within the nucleus. Some viruses, such as measles virus, cause infected cells to fuse, forming multinuclear giant cells. Additionally, some viruses possess genes, called **oncogenes**, that transform normal cells into malignant (cancerous) cells.

18.3   Host Defense Mechanisms

The human host has a number of different lines of defense against potentially pathogenic microorganisms. Some of these defenses are innate, and these are the topic of this section. Others are induced or learned responses, and these will be covered in Chapter 19 in the discussion of the specific immune response. The innate response is composed of physical, chemical, and cellular elements.

Physical Barriers

Intact body surfaces represent the first line of defense against microorganisms, physically blocking the entry of pathogens into the body (Figure 18.18). Most microorganisms, including the normal microbiota of humans, are generally nonin-

vasive and so do not penetrate the skin and mucous membranes. The outer surface of the skin layer is composed of **keratin**, a protein not readily degraded enzymatically by microorganisms. Keratin resists the penetration of water and thus proves a formidable external barrier to microorganisms. We have already discussed the contributions of saline sweat and sebum to the prevention of microbial growth on skin surfaces. Those microorganisms that are able to tolerate the environmental conditions of the skin and proliferate there are themselves producers of antimicrobial substances, also known as **allelopathic** substances,[5] that act to prevent the

---
[5]Allelopathic substances are chemicals produced by one organism that kill or inhibit other organisms. Allelopathic is a general term applied to substances produced by higher organisms as well as by microbes.

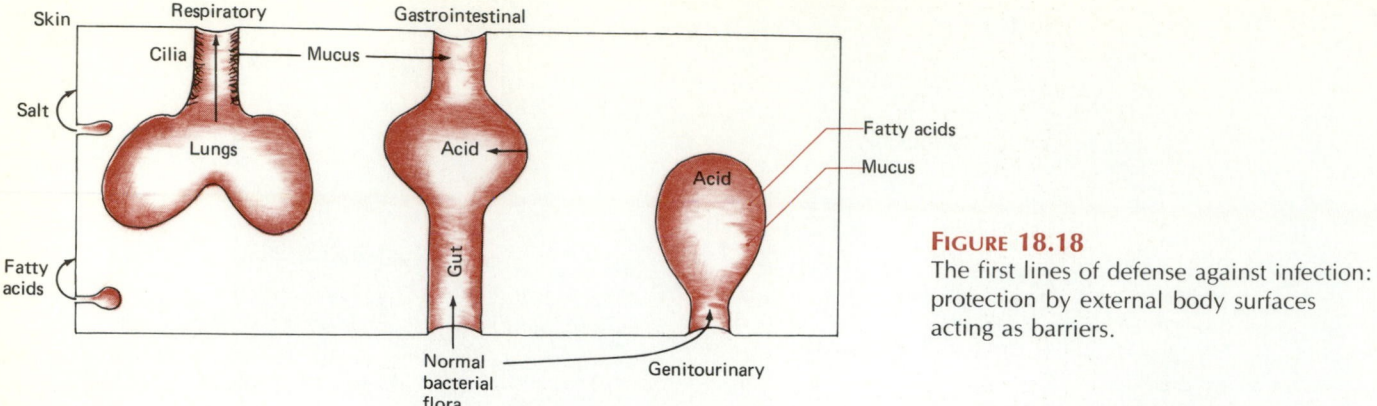

FIGURE 18.18
The first lines of defense against infection: protection by external body surfaces acting as barriers.

establishment of infection on or through the skin by pathogenic microorganisms.

The respiratory tract (Figure 18.19) is protected in part against the invasion of pathogenic microorganisms by secretions of **mucus**. Mucus is secreted from goblet cells and subepithelial glands. Microorganisms tend to become trapped in mucus and are swept out of the body by the **ciliated epithelial cells** that make up the lining of much of the respiratory tract. These cells effectively act as filters to prevent potential pathogens from penetrating the surface tissues of the respiratory tract, and their movement establishes an upward wave motion (Figure 18.20). Some of the mucus and microorganisms are swept out of the body through the oral and nasal cavities by the wave-like action of this so-called **mucociliary escalator system**. Sneezing and coughing also tend to remove many of these microorganisms from the respiratory tract (Figure 18.21). The gag reflex helps to remove postnasal drip and mucus swept up by the ciliated epithelium of the bronchi, with its associated microorganisms. Additionally, the swallowing reflex removes most remaining particulates from the respiratory tract, including microorganisms that become attached to mucus, moving the trapped microorganisms out of the respiratory tract and into the digestive tract.

In the digestive tract, mucous membranes make it difficult for pathogenic microorganisms to attach to and penetrate the gastrointestinal tract lining. Various body tissues are protected against accumulations of microorganisms by the movement of fluids across their surfaces. For example, tears continuously remove microorganisms from the eye, urine flushes them from the surfaces of the urinary tract, and perspiration washes them from the skin surface.

Chemical Defenses

Some of the fluids that wash body tissues also contain antimicrobial chemicals. The tissues of the eye, for example, are protected against the undesirable growth of microorganisms

FIGURE 18.19
The respiratory tract has several mechanisms of nonspecific resistance that protect against invasion by pathogenic microorganisms.

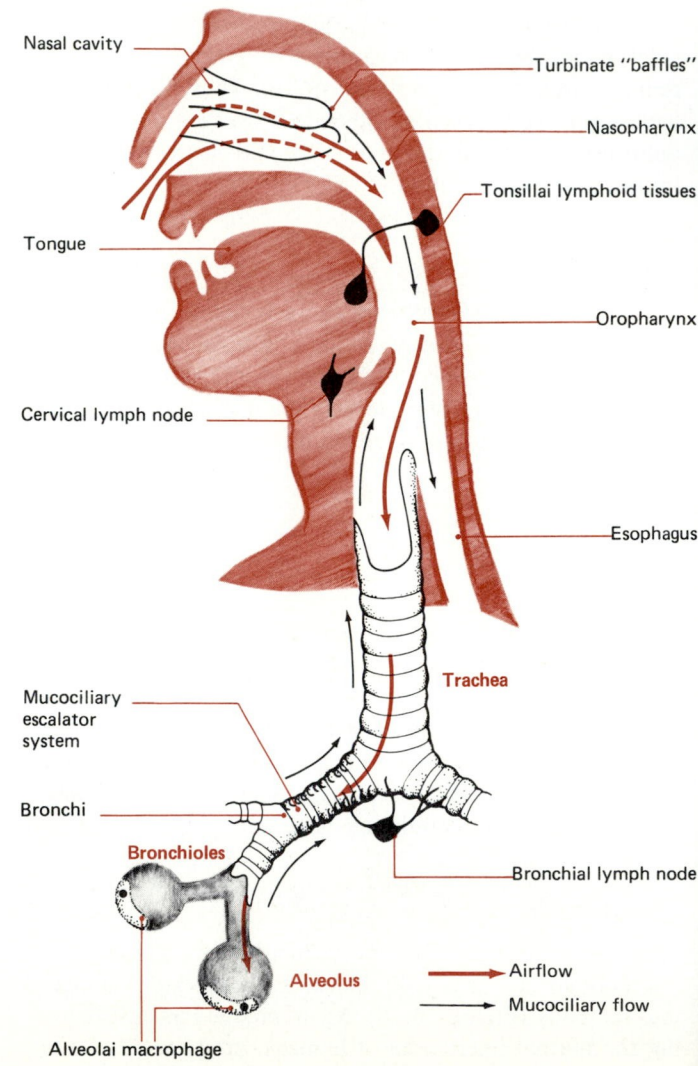

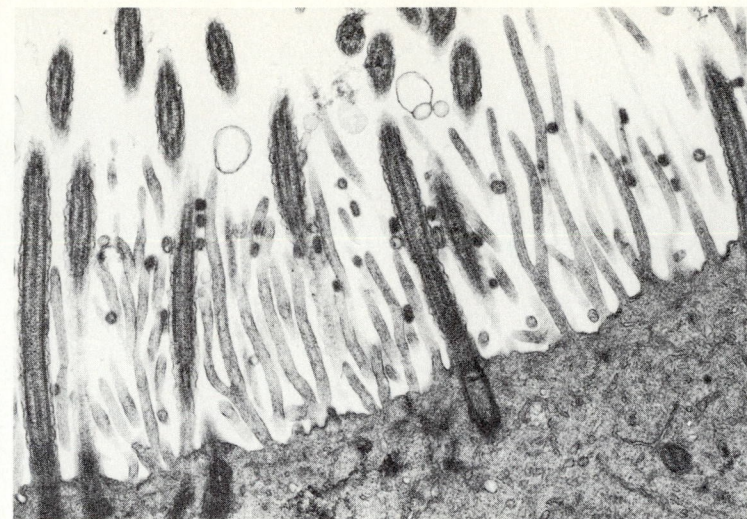

FIGURE 18.20
Electron micrograph showing the filtering of influenza viruses by the cilia lining the respiratory tract. The influenza viruses appear as small black particles. Many microorganisms are trapped on the surface mucus of the projecting cilia and fail to penetrate the respiratory tract, but in the case of influenza viruses, some virions may penetrate, leading to the onset of influenza. (Courtesy R. Dourmashkin, New York University)

by the presence of various enzymes in tears, including lysozyme. **Lysozyme**, which is also found in other body fluids, including saliva and mucus, degrades the cell walls of bacteria, conferring antimicrobial activity on these body fluids. The continuous washing of the eye with tears, which contain this antimicrobial substance, generally prevents the growth of microorganisms on the tissues of the eye. In a similar way, swallowing, coughing, and sneezing expose bacteria to body fluids with antimicrobial activity, thus reducing the number of potential pathogens.

Some chemicals within the body bind iron, thereby withholding this essential growth element from pathogenic microorganisms. By limiting the amount of available iron, these compounds limit the growth of pathogens. **Lactoferrin** and **transferrin** are examples of such iron-binding compounds. Lactoferrin is present in tears, semen, breast milk, bile, and nasopharyngeal, bronchial, cervical, and intestinal mucosal secretions. Transferrin is present in serum and the intercellular spaces of various tissues and organs. Transferrin transports iron from the small intestine, where it is absorbed, to tissues, where it is used. Excess iron is stored intracellularly in a form that is tightly bound. Consequently, the concentration of free iron in the blood and other tissues is normally less than 10^{-18} M, which is far lower than that required for growth by most microorganisms. The ability of some pathogens, such as those that produce siderophores, to overcome this iron limitation was discussed earlier.

The acid of the stomach provides another chemical barrier that acts to prevent microbial invasion of the body. Most microorganisms entering the digestive tract are unable to tolerate the low pH (normally 1–2) of the stomach. Thus, the number of viable microorganisms is greatly reduced during

FIGURE 18.21
A violent cough expels microorganisms from the respiratory tract. Unfortunately, this defense mechanism propels droplets into the air and thus potentially toward the respiratory tract of another individual. (A) Note the distant spread of droplets as this young boy coughs. (B) Note how the spread of droplets is inhibited by this little girl's placing her hand over her mouth.

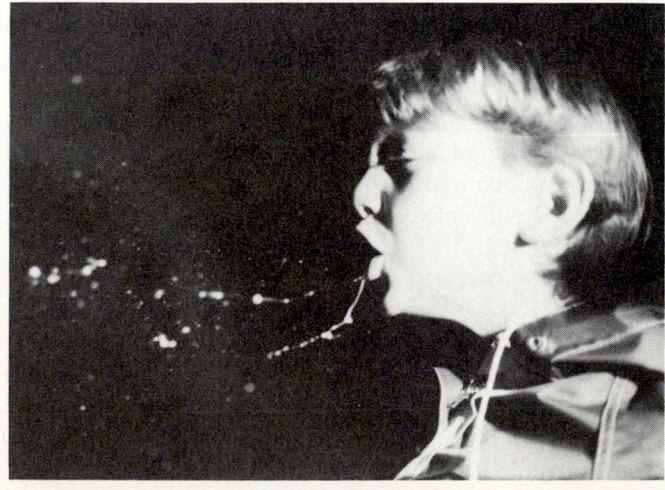
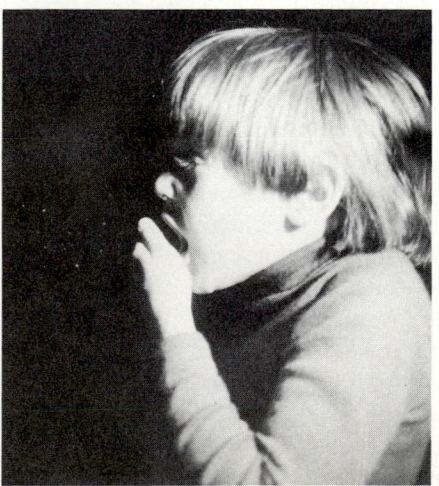

A

B

passage through the stomach. Bile and digestive enzymes in the intestines, in addition to the alkaline environment, further reduce the number of surviving microorganisms. Additionally, the large populations of indigenous microbiota in the lower intestinal tract protect the host against invasion by pathogens; the natural microbiota of the gastrointestinal tract form antagonistic relationships with nonindigenous microorganisms. As a result, most nonindigenous microorganisms entering the intestinal tract are degraded during passage through it or are removed, along with large numbers of indigenous microorganisms, in the passage of fecal material from the body. Bacterial biomass often comprises 75 percent of the dry weight of fecal matter.

Other body tissues are also protected against invasive microorganisms. The genitourinary tract is protected from pathogenic microorganisms by various mechanisms, including the outward flow of mucus, low pH, high salt concentrations, the presence of antimicrobial enzymes, and antagonistic relationships with the natural microbiota of the lower genitourinary tract.

Interferon Blood contains several chemical factors that defend against microbial infections. **Interferons**, for example, are a family of inducible glycoproteins produced by eukaryotic cells in response to viral infections and other stimuli that help eliminate such infections. Interferon production is also elicited by other microbial pathogens that reproduce within host cells, including bacteria, such as rickettsias and chlamydias, and protozoa, such as those causing malaria. The production of interferon occurs shortly after such infections (Figure 18.22). Interferon is produced both by infected tissue cells and by certain lymphocyte blood cells that are part of the lymphatic defense system. Interferon glycoproteins are of relatively low molecular weight and are normally produced only in low concentrations. These glycoproteins are released from infected cells and migrate to uninfected cells, protecting the cells from viral infections (Figure 18.23). Because interferon is produced in very limited quantities, only neighboring cells are immediately protected. Interferons do not block the entry of the virus into a cell but rather prevent the replication of viral pathogens within protected cells.

The synthesis of interferons is regulated at the level of transcription. It appears that an infection with a virus induces synthesis of interferon glycoproteins, with double-stranded RNA viruses being the most potent inducers of interferon synthesis. When interferons move to uninfected cells, the synthesis of specific proteins that inhibit translation is derepressed. It is hypothesized that these translational inhibitory proteins block the translation of viral mRNA without preventing the translation of host cell mRNA molecules. The interferon system thus involves a complex series of molecular events that induce an antiviral state. These events include recognition of an interferon-inducing molecule; derepression and synthesis of interferon proteins; modification and secretion of the interferon molecules; interaction of interferon with susceptible cells; activation and synthesis of previously re-

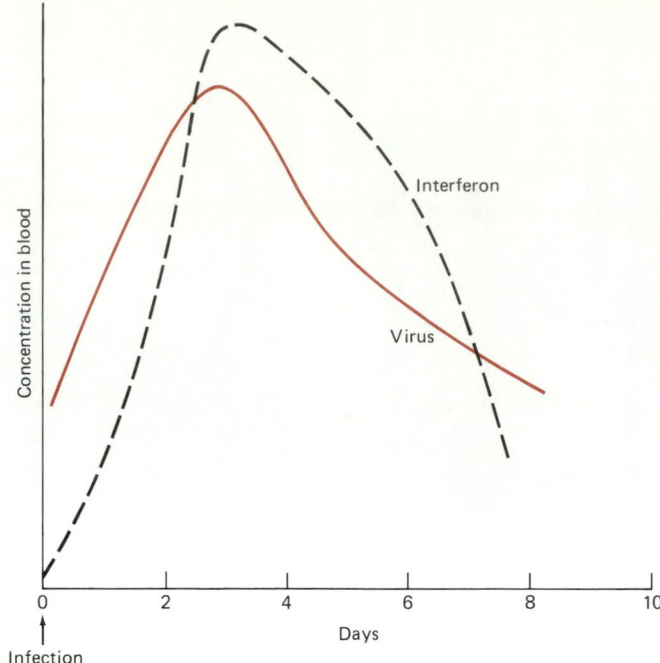

FIGURE 18.22
The time course for interferon production after a viral infection in the lungs of mice is illustrated in this graph.

pressed genetic information; and alteration of the cell's metabolism as expressed by some identifiable interferon interaction, such as resistance to viral infection. It should be emphasized, though, that interferons themselves are not antiviral substances. They have no direct effect on viruses, and their antiviral action is mediated by cells in which they induce an antiviral state.

Interferon production is considered a nonspecific resistance factor because interferon proteins do not exhibit specificity toward a particular pathogen, which means that interferon produced in response to one virus is also effective in preventing the replication of other viruses. Although interferon glycoproteins are not specific for a species of invading viruses or for other intracellular microbial parasites, they are specific for the host organism that produced them; that is, interferons produced by human cells are effective only in human cells and do not exert a protective effect against intracellular parasites in other animal species. Interferons appear to be an important component of the elaborate integrated defense system against viral infections, and their production plays a significant role in preventing viral infections like the common cold.

Besides its role in protecting against viral infections, interferon acts as a regulator of the complex defense network that protects the body against infections and the development of malignant cells. As such, interferon is involved in the control of various phagocytic blood cells that engulf and kill various pathogens (including bacteria) and abnormal or foreign mammalian cells (including cancer cells).

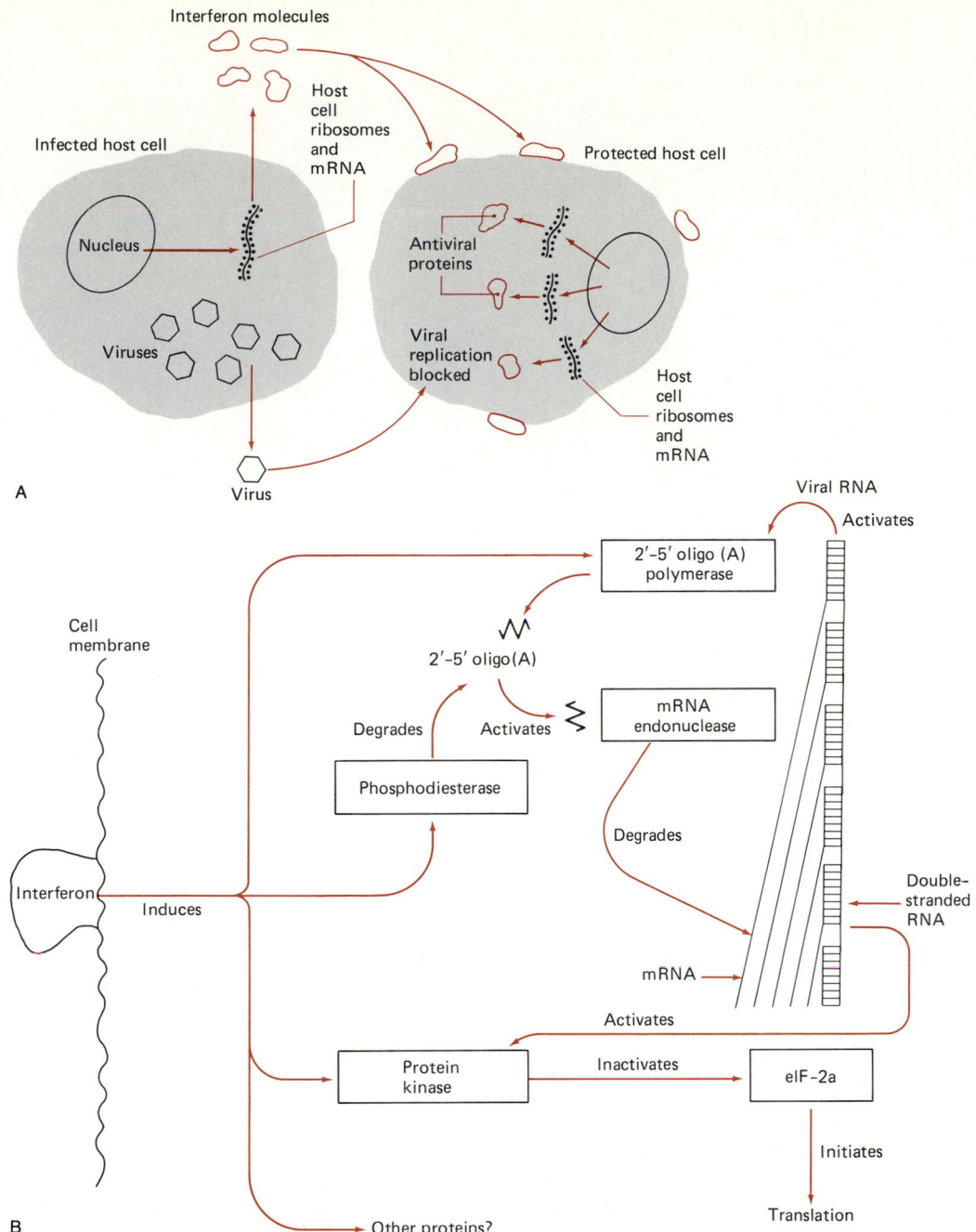

FIGURE 18.23

The production of interferon activates a system that prevents viral replication within host cells.
(A) General sequence of events showing how interferon induces protection of cells against viral
replication. Interferon is produced by a host cell that has been invaded by a virus. This particular
host cell is not protected, and the virus replicates within it. However, interferon released from
this infected host cell moves to other uninfected host cells, inducing an antiviral state that protects
those cells against viral infections. (B) Details of how interferon works reveal the complexity of
its action. Interferons bind to specific surface receptors of uninfected cells, stimulating them
to produce at least two enzymes. One of these, 2′,5′ oligoadenylate synthetase, catalyzes the
synthesis of an unusual polymer, 2′,5′ oligoadenylate, which activates an intracellular ribonuclease. The ribonuclease cleaves and thereby inactivates the viral RNA genome. The other
enzyme produced in response to interferon is a protein kinase that is activated only in the presence
of double-stranded RNA. The activated protein kinase catalyzes the phosphorylation of a factor,
eIF2a, that is required for the initiation of protein synthesis. The phosphorylated eIF2a is inactive,
and therefore protein synthesis, including synthesis of viral proteins, ceases.

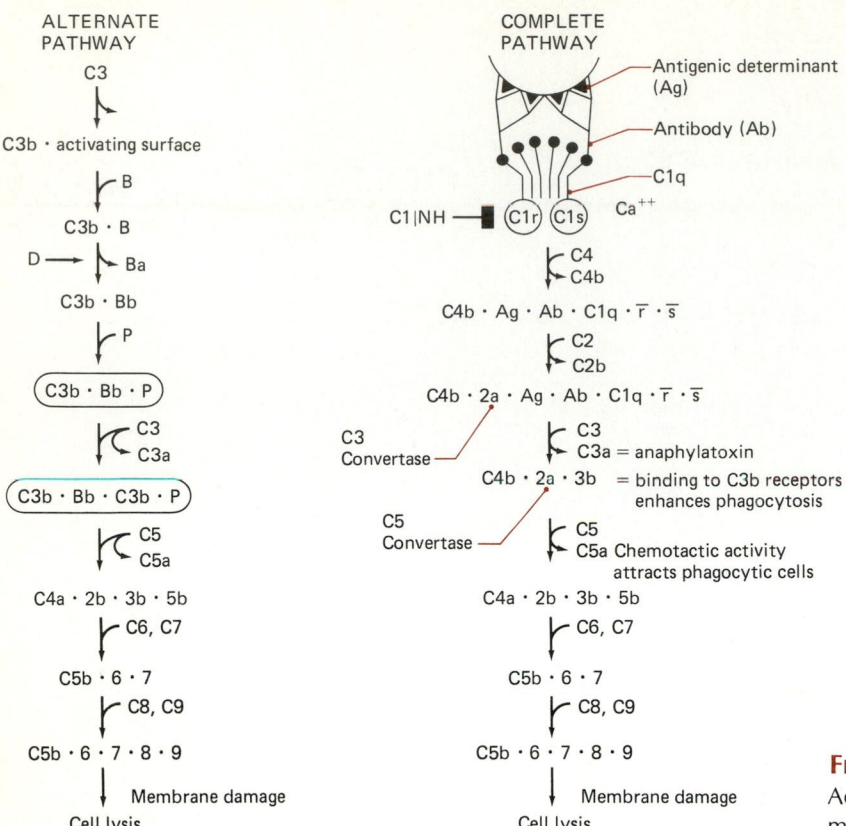

FIGURE 18.24
Activation pathways and functions of complement molecules.

Because of the importance of interferon in controlling viral infections and the proliferation of malignant cells, its commercial production is being developed with the expectation that interferon administration will prove useful in the treatment of various diseases. The human genes coding for the production of interferon have been cloned into *E. coli*, creating by genetic engineering a bacterial strain that produces this human protein. Such genetically engineered bacteria are able to produce sufficient quantities of interferon for therapeutic uses.

Complement   In addition to interferons, blood contains other glycoproteins called **complement** that play a role in the removal of invading pathogens. Complement glycoproteins include over a dozen different proteins in the blood serum. As the name implies, complement augments or complements other defenses that protect the body against microbial infections. The complement system normally acts in association with the specific immune response, as will be discussed in Chapter 19, but it can also act in an alternative pathway that does not require special molecules produced by the specific immune response system (Figure 18.24). Some biochemicals, such as the endotoxic lipid A component of LPS toxins, can trigger the nonspecific response of the complement system. The nonspecific initiation of the complement system involves the activation of the C_3 protein of the complement system. The C_3 complement molecule is split into C_{3a}, which acts as a chemotactic stimulus for certain white blood cells called *neutrophils*, and C_{3b}, which adheres

to the bacterial cell. The conversion of C_3 also permits the addition of other complement proteins in a cascade fashion. The successive addition of the various complement components results in altered permeability of the cytoplasmic membrane of the invading microorganism, leading to the death of that would-be pathogen.

Biochemicals, such as endotoxin of the Gram-negative cell wall, can trigger the nonspecific response of the complement system; thus, the presence of complement tends to prevent infections of the circulatory system by Gram-negative bacteria (Figure 18.26). Besides cascading onto the surfaces of bacterial cells, which results in cell lysis, complement molecules activate phagocytic blood cells. This phenomenon, known as **immune adherence** or **opsonization**, occurs when complement molecules bind to the surface of a microorganism and then interact with receptors on the surface of a phagocytic blood cell to enhance the efficiency with which that blood cell engulfs and destroys the invading microbe. Some of the complement molecules act as chemotactic agents to attract phagocytic cells to the site of infection. This is particularly important in the inflammatory response, which will be discussed later in this chapter.

Cellular Defenses

Phagocytosis   Should microorganisms penetrate the body's outer barriers and enter the circulatory system, they are subject to **phagocytosis** by a variety of cells of the blood, the mononuclear phagocyte system, and the lymphatic system.

ANALYTICAL PROCESS
Role of Resident Microbiota in the Host Defense System

In most cases, the relationships between animals and their normal microbiota are synergistic, that is, mutually beneficial. Germ-free animals develop abnormalities of the gastrointestinal tract and are more susceptible to disease than animals with normal associated microbiota. The normal associated microbiota of animals contributes in part to the normal defense mechanisms that protect animals against infection by pathogens. To determine the role of the indigenous microbiota, it is possible to deliver an animal by cesarean section (surgical removal of the fetus from the uterus via the abdomen) and raise that animal in the absence of microorganisms (Figure 18.25). Such **gnotobiotic** or **germ-free** animals provide suitable experimental models for investigating the interactions of animals and microorganisms. Comparing animals possessing normal associated microbiota with germ-free animals permits the elucidation of the complex relationships between microorganisms and host animals.

Germ-free experimentation extends the microbiologist's pure culture concept to *in vivo* studies. From the use of germ-free animals, it has been learned that an animal's lack of exposure to microbes results in a complex of deleterious effects. Germ-free animals differ from other members of their species. Their metabolic rate and cardiac output are reduced. Structures that are designed to defend against bacterial invasion—such as the lymphatic system, the antibody-forming system, and the mononuclear phagocyte system—are poorly developed in these animals. Some of the animal's organs that would normally have natural populations of bacteria are often reduced in size or capacity. It is important to note that some members of the normal indigenous microbiota exhibit antagonism toward potential pathogens, and their lack removes an important line of defense against pathogens. For example, acid production by the indigenous microbiota of the vaginal tract lessens the probability of infection with *Neisseria gonorrhoeae*, and the normal microbiota of this region also lessens the likelihood of overgrowth by *Candida* yeasts. Other mechanisms of antagonism by the indigenous microbiota that enhance host resistance to disease include alteration of the oxygen tension, production of antibiotics, and competition for available nutrients.

As would be expected, germ-free animals are more susceptible to bacterial infection. Organisms like *Bacillus subtilis* and *Micrococcus luteus*, which are harmless to other animals, cause disease in germ-free animals; more exotic pathogenic microorganisms, like *Vibrio cholerae* and *Shigella dysenteriae*, are able to establish infections far more readily where there are no normal microbiota that have a competitive advantage within the intestinal tract. When the normal microbiota are adversely affected—for example, by antibiotics that are used to treat a disease—an imbalance may occur that leads to the development of disease. For example, the use of antibiotics sometimes disrupts the balance of the microbial community of the gastrointestinal tract, permitting the growth

FIGURE 18.25
An isolette used for the rearing of germ-free animals. Sterile food and water can be fed to the animals so that they are never exposed to micoorganisms. (Courtesy Germ Free Laboratory, Miami, Fla.)

of *Clostridium difficile*, which causes a severe and sometimes fatal gastrointestinal tract infection (antibiotic-associated pseudomembranous enterocolitis). Similarly, women taking antibiotics sometimes develop vaginitis due to an overgrowth of the fungus *Candida albicans*, which is normally held in check by the indigenous bacteria of the vaginal tract.

At the same time, though, germ-free animals are resistant to *Entamoeba histolytica*, the causative organism of amoebic dysentery, because the protozoan requires the normal intestinal bacteria as a food source. Likewise, tooth decay is no problem to germ-free animals, even those on high-sugar diets, because of the lack of lactic acid bacteria in their oral cavities.

Besides their role in preventing certain diseases, the indigenous microbiota contribute to the nutrition of the animal by synthesizing nutrients essential to the welfare of the host. For example, germ-free animals require vitamin K, which normally is synthesized by the resident microbiota of the gastrointestinal tract. These microbiota also synthesize biotin, riboflavin, pantothenate, and pyridoxine, supplying these vitamins to the animal host. Thus, the maintenance of a healthy indigenous microbiota is essential to the maintenance of a healthy individual.

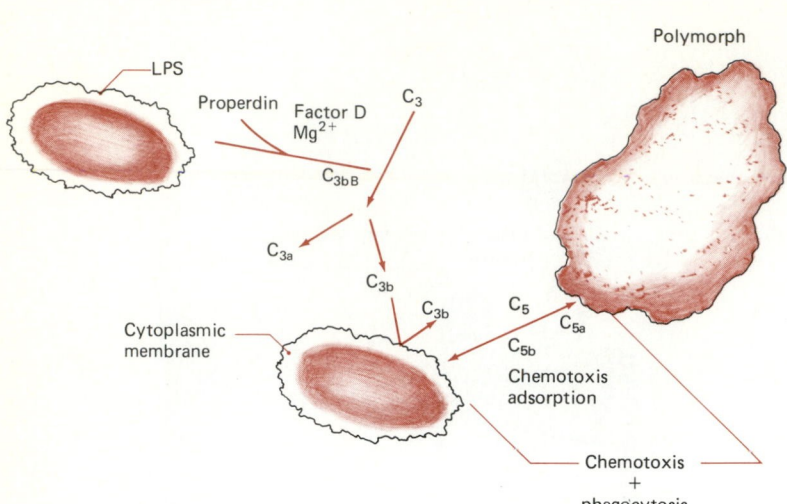

FIGURE 18.26
The alternative complement system involves a cascade of complement molecules to the surface of an invading microorganism, leading to enhanced phagocytosis and death of the pathogen by lysis.

Phagocytosis is a highly efficient host defense mechanism against the invasion of microorganisms. Such defense mechanisms are innate properties of the host organism that are active without prior exposure to microbial pathogens. They work independently of the specific immune response. As a rule, such innate defense mechanisms exhibit relatively low specificity toward particular species of microorganisms. Although phagocytic activities are part of the nonspecific defense mechanism against infection, they also play a role in the specific immune defense system that will be discussed in Chapter 19. In fact, phagocytosis is most efficient in removing bacterial cells when it is activated by components of the specific immune response system.

Blood is composed of various types of cells (Table 18.4). Several types of white blood cells (**leukocytes**) are involved in nonspecific resistance against pathogenic microorganisms (Figure 18.27). Some of the leukocytes, called *granulocytes*, contain cytoplasmic granules. These granulocytes are differentiated on the basis of staining reactions accomplished in the laboratory and include **basophils**, leukocytes that stain with basic dyes; **eosinophils**, leukocytes that react with acidic dyes, becoming red when stained with the dye eosin; and **neutrophils** (also called **polymorphonuclear neutrophils, polymorphs**, or **PMNs**) that contain granules that exhibit no preferential staining—that is, they are stained by neutral, acid, and basic dyes. The leukocytes that do not contain granular inclusions (agranulocytes) include the monocytes and lymphocytes. Monocytes are important in the nonspecific immune response, and lymphocytes are especially important in the specific immune response.

Neutrophils, the most abundant phagocytic cells in blood, are produced in the bone marrow and are continuously present in circulating blood, affording protection against the entry of foreign materials. These white blood cells exhibit chemotaxis and are attracted to foreign substances, including invading microorganisms, which they engulf and digest along with particulate matter that may be present, such as

TABLE 18.4   Normal Cellular Composition of Adult Human Blood

Cell Type	Number (per mL)
Leukocytes (white cells)	4,500–9,000
Granulocytes	
Neutrophils	3,000–6,750
Basophils	25–90
Eosinophils	100–360
Mononuclear cells	
Lymphocytes	1,000–2,700
Monocytes	150–170
Platelets	145,000–375,000
Erythrocytes (red cells)	3,600,000–5,400,00

cell debris (Figure 18.28). Neutrophils live for only a few days in the body but are replenished from the bone marrow in high numbers.

Phagocytic blood cells can have numerous lysosomes that contain hydrolytic enzymes capable of digesting microorganisms. During phagocytosis, the microorganism is engulfed by the pseudopods of the phagocytic cell and is transported by endocytosis across the cell membrane, where it is contained within a vacuole called a **phagosome**. The phagosome migrates to and fuses with a lysosome, thus exposing the microorganism to the enzymes within the lysosome, including degradative enzymes and enzymes that catalyze the production of biochemicals such as hydrogen peroxide and superoxide radicals (Figure 18.29). During phagocytosis there is an increase in oxygen consumption by the phagocytic cells associated with elevated rates of metabolic activities of the hexose monophosphate shunt. As a consequence, oxygen is converted to the superoxide anion, hydrogen peroxide, singlet oxygen, and hydroxyl radicals, all of which have antimicrobial activity. Phagocytosis involves a shift in metabolism from a respiratory to a fermentative process, with the consequent production of lactic acid. This leads to a

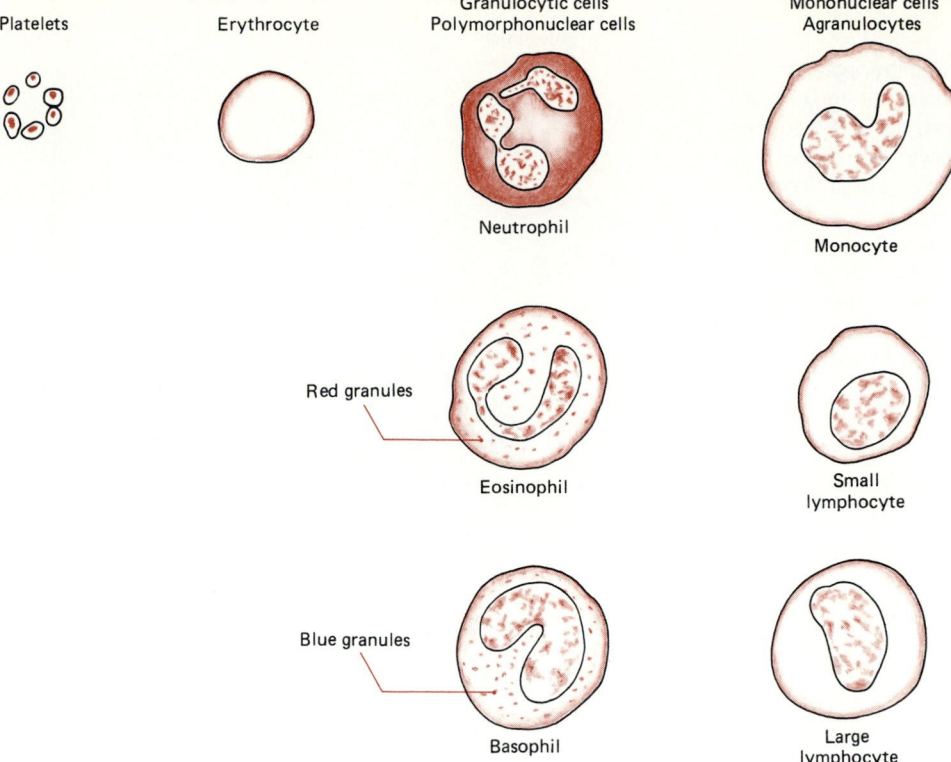

FIGURE 18.27

Blood contains different types of cells, many of which are involved in nonspecific protection against invasion by microorganisms. The different types of blood cells are illustrated here.

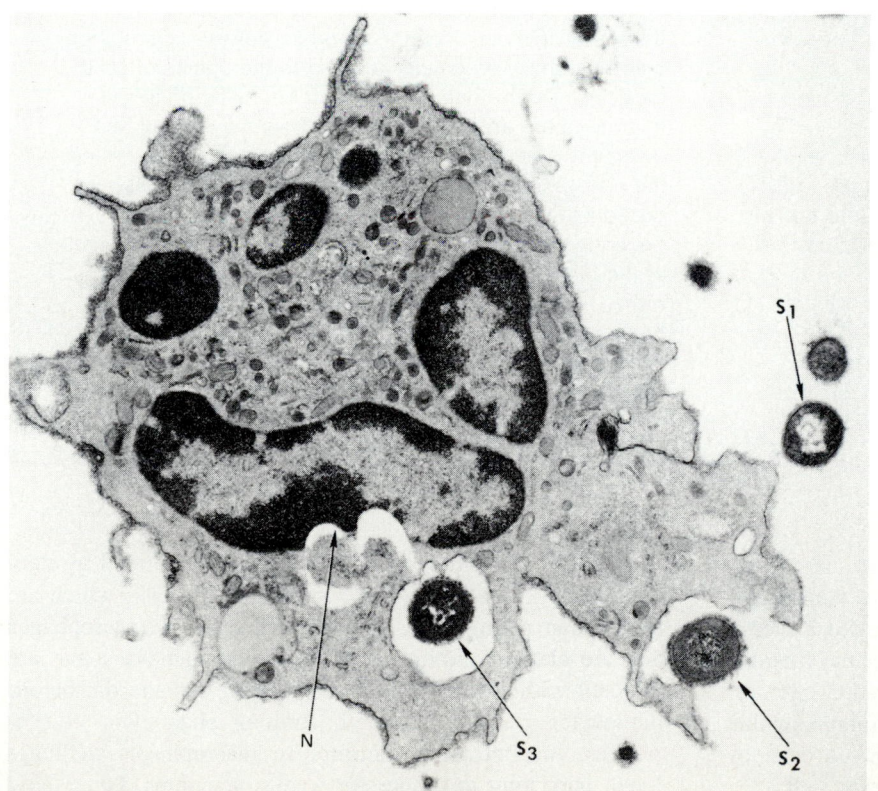

FIGURE 18.28

Micrograph of a polymorphonuclear leukocyte in the presence of *Streptococcus pyogenes*. S_1 is a free bacterial cell, S_2 is a bacterial cell being phagocytized, and S_3 is a bacterial cell that has been phagocytized. The nucleus is being digested in the region designated with the arrow labeled N. (From BPS—John G. Hadley, Battelle Pacific Northwest Laboratories, Richland, Wash.)

FIGURE 18.29

The phagocytic killing of pathogenic microorganisms involves the engulfment and digestion of the microorganisms by phagocytic blood cells. First, the prey becomes attached to the surface of the phagocyte membrane; the attachment is assisted by complement and/or antibodies (A). Numerous pseudopodia surround the prey, and the prey enters the cell by phagocytosis (B). The membranes of the pseudopodia fuse to enclose the prey within a phagosome (C). Enzymes from many granules are dumped into the phagosome, and the prey is often digested (D). The fluids and solutes are absorbed by the cell or removed by exocytosis (E).

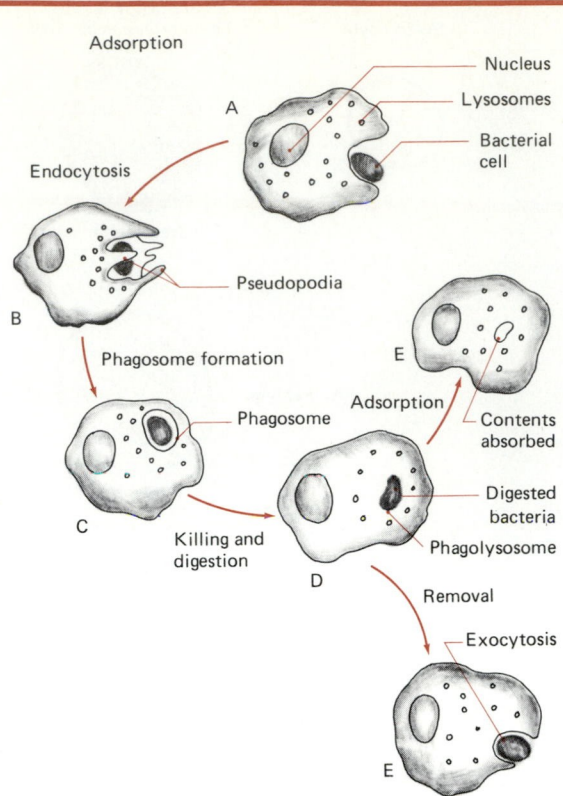

Discovery Process

Eli Metchnikoff in 1884 described phagocytosis in his studies on the infection of the common waterflea, *Daphnia*, by a fungus (*Monospora bicuspidata*). Metchnikoff was a moody, temperamental Russian who did his most creative scientific work in a manic state, applying detailed, detached observation to his studies. He meticulously reported his microscopic observations of how a microbe intruding into an organism is dealt with by that organism. He first examined starfish larvae stuck with thorns but then began to examine microbial infections of *Daphnia*. This was an ideal model system for use in his effort to determine how the bacterial cells that he saw inside white blood cells in the human body got there, because *Daphnia* is simple and transparent and the fungus *Monospora* is large and easily seen without staining. Metchnikoff was able to observe all stages of the *Monospora* in the abdominal cavity of the infected *Daphnia*.

There he saw spores penetrating the intestinal wall as a result of peristalsis, whereupon blood corpuscles immediately began to surround and attach themselves to the spores. He observed the blood corpuscles as circulating, colorless phagocytic cells adapted to the uptake of solid particles. The mobile cells, which he called *phagocytes*, migrate to the area of infection, where they engulf and digest the microbes. Metchnikoff observed and described the changes the fungal spores underwent until they were destroyed and separated into irregular grains. This pioneering work established the role of cellular components of the blood in destroying disease-causing microorganisms. Today we recognize the phagocytic activity of human white blood cells as a primary line of defense against invasion of the body by pathogenic microorganisms.

decrease in pH, which enhances the activity of many lysosomal enzymes. Additionally, phagocytic cells contain digestive enzymes that are involved in killing ingested bacteria, including some that kill bacteria by degrading the D-amino acids found in bacterial cell walls. The degraded microorganisms are transported to the cytoplasmic membrane within a vacuole and are removed from the phagocytic cells by exocytosis or are consumed within the phagocytic cell.

In addition to neutrophils, **monocytes** are formed by stem cells in bone marrow. These mononuclear cells, which are larger than neutrophils, are the precursors of macrophages and are able to move out of the blood to tissues that are infected with invading microorganisms. Outside the blood, monocytes become enlarged, forming phagocytic **macrophages**. In contrast to neutrophils, macrophages are long-lived, persisting in tissues for weeks or months. Once these differentiated macrophages are formed, they are capable of reproducing to form additional macrophages, whereas neutrophils are terminal cells that are nonreproductive and must be replenished from the bone marrow. Macrophages, like neutrophils, are able to engulf and digest microorganisms, but some of the lysosomal enzymes of macrophages are different from those of polymorphs. Some microorganisms are resistant to the enzymatic activities of neutrophils and/or macrophages. For example, *Mycobacterium tuberculosis* survives and even multiplies within macrophages. As a result, some microorganisms survive, continuing to grow and later to cause infection because of the failure of these phagocytic cells to kill the invading pathogens. This is one of the reasons that tuberculosis is a persistent disease and is difficult to treat.

Neutrophils are particularly important in blood, while macrophages are distributed throughout the body, including fixed sites within the mononuclear phagocyte system (Figure 18.30). The **mononuclear phagocyte system**, formerly called the *reticuloendothelial system*, refers to a systemic network of phagocytic cells distributed through a network of loose connective tissue and the endothelial lining of the capillaries and sinuses of the human body. The phagocytic cells associated with the lining of the blood vessels in bone marrow, liver, spleen, lymph nodes, and sinuses constitute this host defense system. Some of the macrophages in the mononuclear phagocyte system occur at fixed sites and are designated with particular names. For example, **microglia** are macrophages of the central nervous system; **Kupffer cells** are phagocytic cells that line the blood vessels of the liver; **dust cells** are macrophages fixed in the alveolar lining of the lungs; and **histiocytes** are fixed macrophages in connective tissues. Other macrophages of the mononuclear phagocytic system are called **wandering cells** because they move freely into tissues where foreign substances have entered. The wandering macrophages are attracted to these tissues through chemotaxis by chemical stimuli elicited by the foreign material. Wandering macrophages occur in the peritoneal lining of the abdomen and the alveolar lining of the lung, as well as in other tissues. The presence of relatively high numbers of macrophages in the respiratory tract is important in preventing the establishment of both pathogens and a normal indigenous microbiota within the tissues of the lower respiratory tract.

The lymphatic system also plays a major role in host resistance to pathogens through the lymphocytes, cells involved in the specific immune response, the role of which will be discussed in Chapter 19.

Inflammatory Response

The **inflammatory response** represents a generalized response to infection or tissue damage and is designed to localize invading microorganisms and arrest the spread of the infection. The inflammatory response is characterized by four symptoms: reddening of the localized area, swelling, pain, and elevated temperature. The **redness** results from capillary

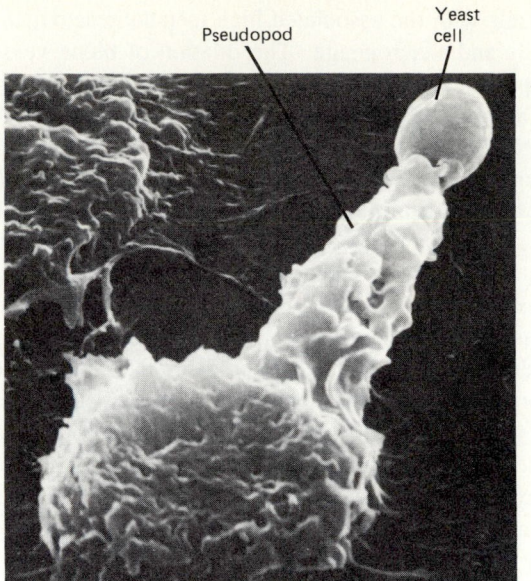

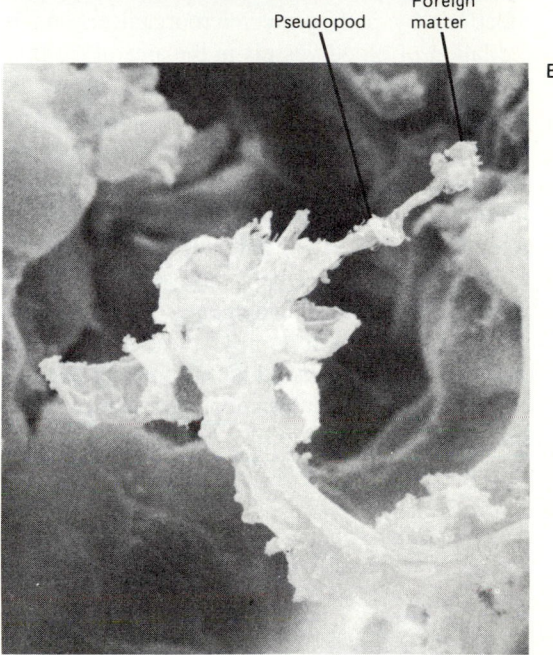

FIGURE 18.30

(A) This alveolar macrophage is utilizing an extended pseudopod for ingesting a yeast cell (4,300×). (From BPS—John G. Hadley, Battelle Pacific Northwest Laboratories, Richland, Wash.). (B) In this scanning electron micrograph of a Kupffer cell (3,000×) the pseudopodia of this fixed macrophage of the rat liver are visible. The topmost pseudopod appears to be attached to a foreign body. (Courtesy Robert Apkarian, Yerkes Primate Research Center, Emory University)

dilation that allows more blood to flow.[6] The **elevated temperature**, which is a localized phenomenon, also occurs because capillary dilation permits increased blood flow through

---

[6]The term *dilation* is a misnomer because the capillaries are no more dilated (expanded) than they normally are; during "dilation" there are simply fewer constricted capillaries, permitting increased blood circulation through more open capillaries.

these vessels with the associated high metabolic activities of neutrophils and macrophage. The dilation of blood vessels is accompanied by increased capillary permeability, causing **swelling** as fluids accumulate in the bases surrounding tissue cells. Actually, the swelling is due to increased permeability of the venules, but the term *increased capillary permeability* is entrenched in the clinical terminology used to describe this phenomenon. **Pain** is due to lysis of blood cells, triggering the production of bradykinin and prostaglandins, which are substances produced by human cells that alter the threshhold and intensity of the nervous system response to pain. Bradykinin decreases the firing threshold for pain nerve fibers and the prostaglandins, PGE_1 and PGE_2, intensify this effect. Aspirin, which is often used to decrease pain, antagonizes prostaglandin formation but has little or no effect on bradykinin formation. Thus, aspirin can decrease but not eliminate the pain associated with the inflammatory response.

Having considered the characteristic symptoms of the inflammatory response, let us discuss how it helps defend against infections by pathogenic microorganisms. As indicated, the dilation of blood vessels in the area of the inflammation increases blood circulation, allowing increased numbers of phagocytic blood cells to reach the affected area. Neutrophils are initially most abundant, but in the later stages of inflammation, monocytes and macrophages of the mononuclear phagocyte system predominate (Figure 18.31). The phagocytic cells are able to kill many of the ingested microorganisms. Phagocytic blood cells migrate to the affected tissues, passing between the endothelial cells of the blood vessel by a process known as **diapedesis**. The death of phagocytic blood cells involved in combatting the infection results in the release of histamine, prostaglandins, and bradykinins, which, in addition to their other effects, are vasodilators, that is, substances that increase the internal di-

ameter of blood vessels. Additionally, specialized cells that line connective tissues, called **mast cells**, react with complement, leading to the release of large amounts of histamine contained within these cells (Figure 18.32). These substances are biochemical mediators that alter the circulation during the inflammatory response (Figure 18.33). Thus, the death of some phagocytic cells and the release of certain biochemicals enhance the inflammatory response.

The area of the inflammation also becomes walled off as the result of the development of fibrinous clots. The deposition of fibrin isolates the inflamed area, cutting off normal circulation. The fluid that forms in the inflamed area is known as the **inflammatory exudate**, commonly called **pus**. This exudate contains dead cells and debris in addition to body fluids. After the removal of the exudate, the inflammation may terminate and the tissues may return to their normal function, which is why physicians, for example, often lance boils to remove the exudate.

The elevation of body temperature (**fever**) aids in the ability of the inflammatory response to protect against invading microorganisms, because enzymatic reactions occur more rapidly at higher temperatures and because some pathogens are very sensitive to even slightly elevated temperatures. Fever is a response to fever-inducing substances called **pyrogens**, including bacterial endotoxins, that enter the blood as a result of the death of microorganisms or are released by phagocytic blood cells. Elevated body temperatures enhance the rate of phagocytic activity and other aspects of the immune response. They also often reduce the rates of reproduction of pathogenic microorganisms, many of which have optimal temperatures coincident with the normal body temperature. Thus, the complex reactions of the inflammatory response work in an integrated fashion to contain and eliminate infecting pathogenic microorganisms.

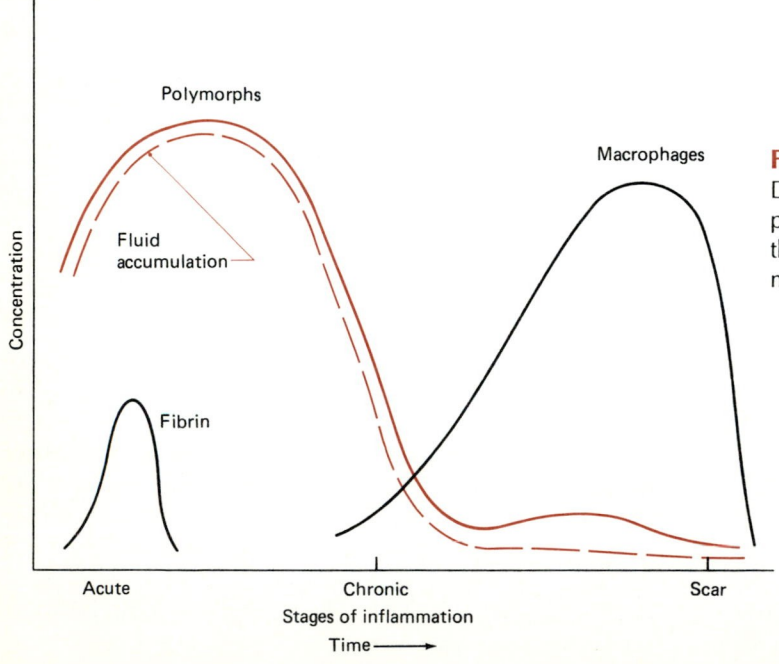

FIGURE 18.31
During the early stages of the inflammatory response, polymorphs are particularly important. In the later stages, the concentration of polymorphs declines and that of macrophages increases.

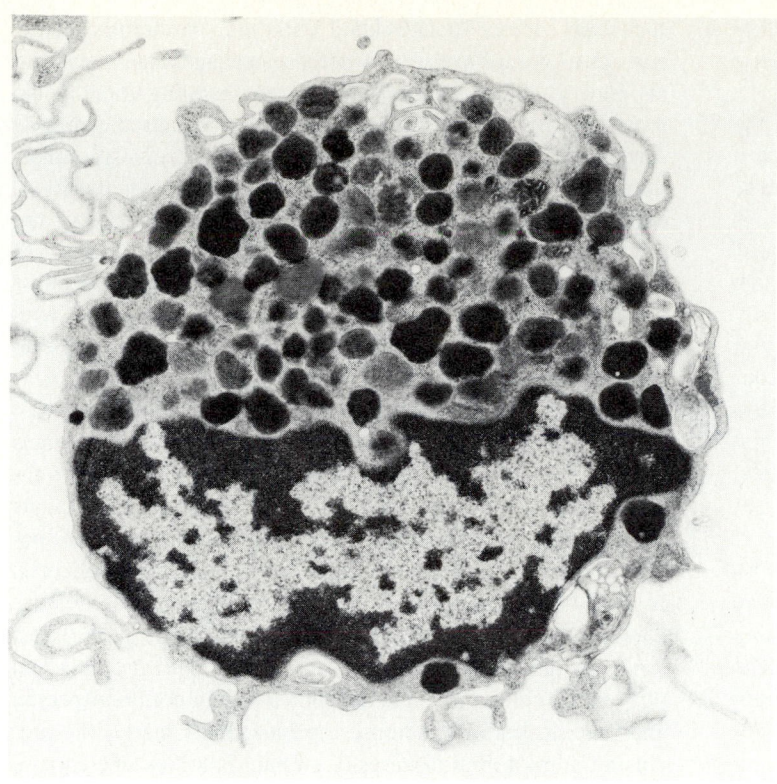

FIGURE 18.32
A human mast cell from the lung, showing granule inclusions of histamine. The release of histamine from these cells occurs as part of the immune response, leading to altered vascular permeability. (Courtesy Ann Dvorak, Beth Israel Hospital, Boston)

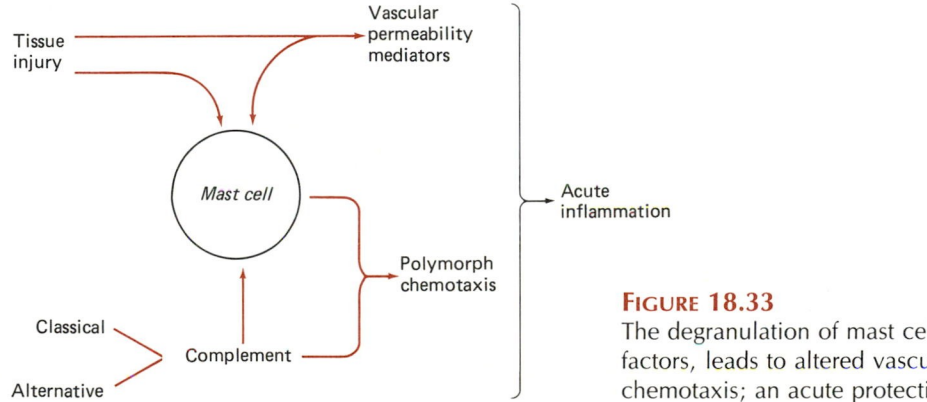

FIGURE 18.33
The degranulation of mast cells, which may be triggered by various factors, leads to altered vascular permeability and enhanced polymorph chemotaxis; an acute protective inflammatory reaction results.

Postlude

The interactions between microorganisms and humans represent a delicate balance between biological populations. The normal microbiota of humans play a role in maintaining a healthy condition, but various microorganisms can act as pathogens, causing disease. The normal human microbiota are usually indigenous to the surfaces of body tissues, most notably the oral cavity, gastrointestinal tract, vaginal tract, and skin surface. These resident microorganisms are generally noninvasive and limited to the surfaces of the human body. The environmental conditions of particular body tissues and the physiological properties of particular microbial populations determine the species composition of the resident microbial community of a given body tissue. In contrast to the normal human microbiota, pathogenic microorganisms that cause human disease generally are either invasive, toxigenic, or both. There are a variety of factors that confer virulence on microbial populations, including the ability to produce toxins, which disrupt the normal functioning of mammalian cells; invasiveness factors, such as enzymes that permit the microorganisms to spread through body tissues;

and adaptations, such as capsules, that permit the pathogenic microorganisms to evade the normal host defense mechanisms.

The ability of a microorganism to reproduce within the human body, producing an infection, requires that such a microorganism overcome or evade a variety of host defense mechanisms. The surfaces of body tissues are generally resistant to penetration by microorganisms. In addition to acting as a physical barrier, human surface tissues often contain biochemicals with antimicrobial activity, such as the fatty acids of skin tissues and the enzyme lysozyme in tears. Microorganisms that penetrate the surface barriers are confronted by several nonspecific resistance mechanisms. Several substances in the circulatory system, such as interferon and complement glycoproteins, have antimicrobial activities. Interferons, in particular, prevent the reproduction of viruses within human cells.

Additionally, various cells that circulate through the body, such as monocytes, neutrophils, and macrophages, have phagocytic activity. The ability to engulf and ingest invading microorganisms eliminates most would-be pathogens. The inflammatory response brings together several different mechanisms for preventing the proliferation and spread of infecting pathogenic microorganisms. The inflammatory response tends to limit the spread of an infection, to direct phagocytic cells to the site of infection, and to enhance the effectiveness of phagocytic removal of infecting pathogens. These host defense mechanisms are physiological and, as such, are subject to individual variation. Numerous factors can act to compromise host defense mechanisms, rendering an individual susceptible to disease-causing pathogenic microorganisms. For example, a wound breaks the normal surface barrier, permitting invasion by microorganisms; emotional and physiological stress, such as improper nutrition, exposure to extreme temperatures, and tension, lower the physiological potential for successful phagocytic removal of infecting microorganisms; and age may influence the physiological ability of an individual to resist infection, with young children and the elderly being particularly prone to various infectious diseases.

In addition to the mechanisms already described for maintaining a balance between microbial populations and humans without the occurrence of disease, there is a specific or acquired immune defense mechanism that protects individuals against diseases caused by microorganisms. The topic of specific acquired immunity will be discussed in the next chapter. It should be noted, though, that the specific immune defense mechanisms are interactive with several of the nonspecific host resistance mechanisms. The complement proteins and phagocytic cells, for example, play a role in both nonspecific and specific immune defense mechanisms. Clearly, the preclusion of microbial diseases of humans involves an elaborate and complex integrated system of host resistance mechanisms. Similar systems occur in other animals, making their study possible in the laboratory.

Suggested Supplementary Readings

ADAMS, D. O., and M. G. HANNA (eds.). *Contemporary Topics in Immunobiology*, Volume 13: *Macrophage Action*. C. V. Mosby, St. Louis.

ATASSI, M. Z., C. J. VANOSS, and D. R. ABSOLOM. 1984. *Molecular Immunology: A Textbook*. Marcel Dekker, New York.

BARON, S. 1985. *Medical Microbiology*. Addison-Wesley Publishing Co., Reading, Mass.

BEAMAN, L., and B. L. BEAMAN. 1984. The role of oxygen and its derivatives in microbial pathogenesis and host-defense. *Annual Review of Microbiology* 38:27–48.

BOWDEN, G. H. W., D. C. ELWOOD, and I. R. HAMILTON. 1979. Microbial ecology of the oral cavity. *Advances in Microbial Ecology* 3:135–218.

BROWN, M. R. W., and P. WILLIAMS. The influence of environment on envelope properties affecting survival of bacteria in infections. *Annual Review of Microbiology* 39:527–556.

BRUBAKER, R. R. 1985. Mechanisms of bacterial virulence. *Annual Review of Microbiology* 39:21–50.

BULLEN, J. J., and E. GRIFFITHS. 1987. *Iron and Infection: Molecular, Physiological and Clinical Aspects*. John Wiley & Sons, Inc., New York.

CHU, F. S. 1978. Mode of action of mycotoxins and related compounds. *Advances in Applied Microbiology* 22:83–143.

CLARKE, R. T. J., and T. BAUCHOP (eds.). 1978. *Microbial Ecology of the Gut*. Academic Press, London.

COLLEE, J. G. 1981. *Applied Medical Microbiology*. Blackwell Scientific Publications, Oxford, England.

CROSA, J. H. 1984. The relationship of plasma-mediated iron transport and bacterial virulence. *Annual Review of Microbiology* 38:69–89.

CUATRECASAS, P. (ed.). 1977. *The Specificity and Action of Animal, Bacterial and Plant Toxins*. Chapman and Hall, London.

FALCONE, G., et al. 1984. *Bacterial and Viral Inhibition and Modulation of Host Defenses*. FEMS Symposia, 1982. Academic Press, New York.

FRIEDMAN, R. M. 1981. *Interferons: A Primer*. Academic Press, New York.

GARVEY, J. S., N. E. CREMER, and D. H. SUSSDORF. 1977. *Methods in Immunology: A Laboratory Text for Instruction and Research*. W.A. Benjamin, Reading, Mass.

GILMAN, A. G., L. S. GOODMAN, and A. GILMAN (eds.). 1980. *Goodman and Gilman's The Pharmacological Basis of Therapeutics*. Macmillan Publishing Co., Inc., New York.

GODING, J. W. 1983. *Monoclonal Antibodies: Principles and Practices*. Academic Press, London.

GOLUB, E. S. 1981. *The Cellular Basis of the Immune Response*. Sinauer Associates, Sunderland, Mass.

GOREN, M. 1977. Phagocyte lysosomes: Interactions with infectious agents, phagosomes, and experimental perturbations in function. *Annual Review of Microbiology* 31:507–533.

GOTZE, O., and H. J. MULLER-EBERHARD. 1976. The alternative pathway of complement activation. *Advances in Immunology* 24:1–35.

HILDEMANN, W. H. 1984. *Essentials of Immunology.* Elsevier, New York.

HOOD, L. E. 1984. *Immunology.* Benjamin/Cummings Publishing Co., Menlo Park, Calif.

Immunology: Readings from Scientific American. 1976. W.H. Freeman and Co., San Francisco.

ISENBERG, H. D., and A. BALOWS. 1981. Bacterial pathogenicity in man and animals. In *The Prokaryotes* (M. P. Starr, H. Stolp, H. G. Truper, A. Balows, and H. G. Schlegel, eds.), Volume 2, pp. 83–122. Springer-Verlag, Berlin.

JOKLIK, W., H. WILLETT, and D. B. AMOS. 1984. *Zinsser's Microbiology.* Appleton-Century-Crofts, Chicago.

KASS, E. H., and S. M. WOOLF. 1973. *Bacterial Lipopolysaccharides: The Chemistry, Biology and Clinical Significance of Endotoxins.* University of Chicago Press, Chicago.

KIMBALL, J. W. 1986. *Introduction to Immunology.* Macmillan Publishing Co., New York.

KIRKWOOD, E. M., and C. J. LEWIS. 1984 *Understanding Medical Microbiology.* John Wiley & Sons, Inc., New York.

LINTON A. H. 1982. *Microbes, Man and Animals: The Natural History of Microbial Interactions.* John Wiley & Sons, Inc., Chichester, England.

MACARIO, C. L., and E. C. deMACARIO (eds.). 1985. *Monoclonal Antibodies Against Bacteria.* Academic Press, Orlando, Fla.

MARPLES, M. J. 1969. Life on the human skin. *Scientific American* 220(1):108–129.

McNABB, P. C. and T. B. TOMASI. 1981. Host defense mechanisms at mucosal surfaces. *Annual Review of Microbiology* 35:477–496.

MIMS, A. A., and D. O. WHITE. 1984. *Viral Pathogenesis and Immunology.* Blackwell Scientific Publications, Oxford, England.

ROBBINS, S. L., and R. S. COTRAN. 1979. *The Pathological Basis of Disease.* W.B. Saunders Co., Philadelphia.

ROSEBURY, T. 1962. *Microorganisms Indigenous to Man.* McGraw-Hill Book Co., New York.

SAVAGE, D. C. 1977. Microbial ecology of the gastrointestinal tract. *Annual Review of Microbiology* 31:107–133.

SCHLESINGER, R. B. 1982. Defense mechanisms of the respiratory system. *BioScience* 32(1):45–50.

SKINNER, F. A., and J. G. CARR (eds.). 1974. *The Normal Microflora of Man.* Academic Press, London.

SMITH, J. R., R. J. LAUDICINA, and R. D. RUFO. 1986. *Learning Guides for the Medical Microbiology Laboratory.* John Wiley & Sons, Inc., New York.

STEWART, W. E. (ed.). *Interferons and Their Actions.* 1977. CRC Press, Inc., Boca Raton, Fla.

WATSON. 1984. *Nutrition, Disease Resistance, and Immune Function.* Marcel Dekker, New York.

WILKINSON, P. C. 1974. *Chemotaxis and Inflammation.* Churchill Livingstone, Edinburgh, Great Britain.

WOGAN, G. N. 1975. Mycotoxins. *Annual Review of Pharmacology* 15:436–451.

Study Questions

1. What is meant by the normal human microbiota?

2. What body surfaces are normally colonized by microorganisms?

3. What body fluids normally are free of bacteria? Why?

4. How does the resident microbiota of the skin differ from that of the gastrointestinal tract? What are some of the major bacterial genera occurring in these two habitats?

5. What factors influence which bacterial genera can establish themselves within the indigenous microbial community of the skin?

6. How does the presence of an indigenous microbiota benefit humans?

7. What attributes contribute to the virulence of pathogens? What is the difference between toxigenicity and invasiveness?

8. What is a toxin? What is the difference between an endotoxin and an exotoxin?

9. What is phagocytosis? How does it contribute to our resistance to pathogenic microorganisms?

10. What is an inflammatory response? How does inflammation act to prevent the spread of pathogens throughout the body?

11. How is the respiratory tract protected against invasion by pathogenic microorganisms?

12. How is interferon involved in protection against viral infections?

Situational Problem

Defining a Pathogen

United States federal regulations concerning the deliberate release of genetically engineered microorganisms into the environment established one set of guidelines for pathogens and another for nonpathogens. Other regulations concern the shipment and transport of pathogens. The guidelines fail to deal specifically with opportunistic pathogens, which are organisms that normally do not cause disease but that can do so under certain conditions. For example, *Escherichia coli*, which is part of the normal intestinal microbiota of a healthy individual, can cause serious urinary tract and spinal column infections under certain circumstances. Therefore, it may be considered by some as a nonpathogen and by others as a pathogen.

Typically, when governmental regulations and/or guidelines are proposed, a time period is made available for public comment. Assume that we are still in that period when public comment is requested to help shape the final regulations. What definition would you propose for pathogenic microorganisms that could be applied universally for regulatory purposes? Justify your position in a cogent letter that could be sent to your congressional representative.

CHAPTER 19

The Immune Response

The **immune response system** of humans and other higher animals provides a mechanism for a specific response to the invasion of particular pathogenic microorganisms and other foreign substances. It is largely this specific physiological response that protects us against disease. In contrast to non-specific defense mechanisms, the immune response is characterized by specificity, memory, and the acquired ability to detect foreign substances. The human immune response is able to recognize macromolecules that are different in some way from the normal macromolecules of the body. The ability to differentiate "self" from "nonself" at the molecular level is necessary for the development of the specific immune response. The specificity of the immune response permits the recognition of even very slight biochemical differences between molecules. Consequently, the macromolecules of one microbial strain can elicit a different response from those of even a very closely related strain of the same species.

The specific immune response is adaptive or acquired, in that once a response to a particular macromolecule, called an *antigen*, has occurred, a memory system is established

that permits a rapid and specific secondary response upon reexposure to that same substance (Figure 19.1). The ability to recognize and respond rapidly to pathogenic microorganisms establishes a state of immunity that precludes infection with those specific pathogens. The ability to recognize the microorganisms that previously elicited an immune response forms the basis for acquiring or developing immunity to specific diseases. As a consequence of **acquired immunity**, we usually suffer from many diseases, such as chicken pox, only once. We can also intentionally expose ourselves to specific foreign macromolecules, through the use of vaccines, to artificially establish a state of immunity. The use of vaccination to prevent disease will be discussed in Chapter 20 after we establish the basis for its use in this chapter.

Examination of the basis of immunity reveals that there are two different forms of the immune response. In one response mode, called **antibody-mediated immunity**, specific proteins called (**antibodies**) are made when foreign antigens are detected. By definition, an **antigen** is any macromolecule that elicits the formation of an antibody and that can subse-

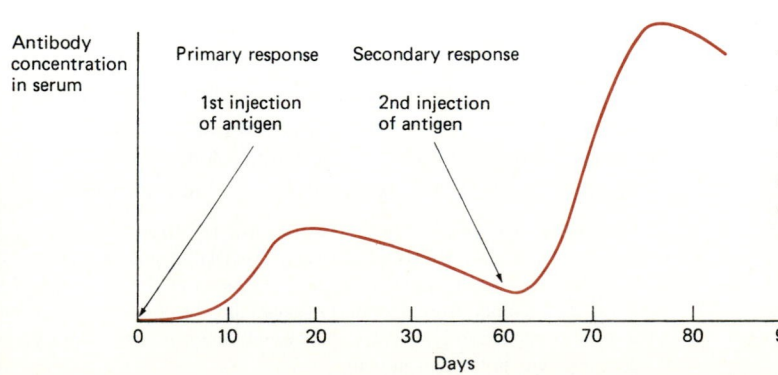

FIGURE 19.1
The primary and secondary immune responses. A rabbit was injected on two separate occasions with staphylococcal antigens. In this case, the antigens were injected as denatured protein toxins or toxoids. The antigen elicited the formation of antibody. The antibody response on the second contact with antigen was more rapid and intense.

567

quently react with an antibody. In antibody-mediated immunity, plasma cells derived from certain white blood cells (**B lymphocytes**) synthesize antibodies in response to the detection of a foreign macromolecule with antigenic properties.[1] Antibodies, which are also called **immunoglobulins**, are made in response to specific antigens and react with those antigens.[2] In the second form of the immune response, called **cell-mediated immunity**, certain cells of the body acquire the ability to destroy other cells that are recognized as foreign or abnormal. In contrast to antibody-mediated immunity, cell-mediated immunity depends on another class of lymphocyte cells, known as **T lymphocytes**. T lymphocytes interact with "foreign" cells to bring about their destruction. There are also interactions between B and T lymphocytes and between these cells and other blood cells that establish an integrated defense network.

The B and T lymphocytes, upon which immunity depends, are differentiated in the lymphatic system (Figure 19.2). Both types of lymphocytes originate from bone stem cells and become differentiated during maturation. The precursors of T cells pass through the liver and spleen before reaching the thymus gland, where they are processed. Although differentiated in the thymus gland, T cells are inactive until they mature later in other lymphoid tissues. T cells are processed within specialized T-cell domains of lymphoid tissues. The thymus-dependent differentiation of **T cells** or **thymocytes** occurs during childhood, and by adolescence the secondary lymphoid organs of the body generally contain their full complement of T cells. The T cells then generally circulate throughout the body. Human B lymphocytes appear to develop in the bone marrow.[3] Human B cells are found predominantly in lymphatic tissues. Like T cells, B cells undergo secondary processing within lympatic tissues, including the spleen, tonsils, and lymph nodes. The B cells are processed within T-cell–independent regions of the lymphoid tissues. Within lymphatic tissues, B lymphocytes give rise to antibody-secreting plasma cells in response to antigenic stimulation.

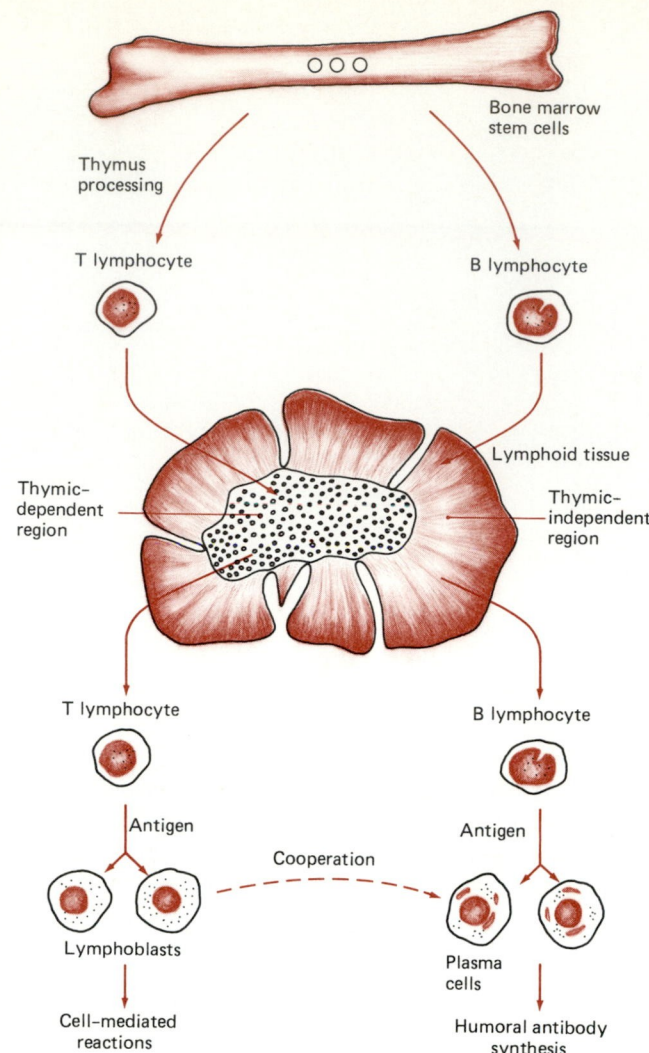

FIGURE 19.2
The differentiation of T and B cells forms the basis for the two arms of the immune defense network.

19.1 Antibody-Mediated Immunity

Antigens

A wide variety of macromolecules can act as antigens, including all proteins, most polysaccharides, nucleoproteins, lipoproteins, and various small biochemicals if they are attached to proteins or polypeptides. The two essential properties of an antigen are its **immunogenicity** (ability to stimulate antibody formation) and its specific reactivity with antibody molecules. The antigen molecule consists of a reactive portion, the **hapten**, that reacts chemically with the antibody molecule and a carrier portion that is necessary for the stimulation of antibody production. A hapten can react with antibody molecules but is unable to elicit the formation

---

[1] Blood is composed of fluid and various types of blood cells. The cellular component consists of red blood cells and several types of white blood cells. Antibodies are found in serum, which is the fluid portion of coagulated (clotted) blood. In contrast, plasma is the fluid portion of blood in which the cells are still suspended.

[2] In old medical terminology, blood and other vital body fluids were considered as "humors" after the Greek word for fluids. Thus, antibody-mediated immunity is often referred to as *humoral immunity* because the antibody molecules flow extracellularly through the body fluids.

[3] The term *B lymphocyte* actually refers to bursa-dependent lymphocytes, so named because these lymphocytes are differentiated in chickens and other birds in the lymphoid organ known as the *bursa of Fabricius*. Even though humans do not possess a bursa of Fabricius, the designation *B lymphocyte* is applied to lymphocytes that can differentiate into antibody-synthesizing cells in the human body.

of antibody, whereas a complete antigen molecule both reacts and elicits the production of specific antibodies. Antigenic molecules may be multivalent, having multiple reactive sites, or monovalent, having only one reactive site. Generally, multivalent antigens elicit a stronger immune response than monovalent antigens. In some cases, a multivalent antigen, variously called a **heterophile antigen**, **heterologous antigen**, or **Forssman antigen**, can react with antibodies produced in response to a different antigen.

In many cases, antigens are associated with cell surfaces and are therefore called **surface antigens**. Human cells have specific surface antigens; for example, human blood types are determined by the presence or absence of antigens designated A and B on the surfaces of red blood cells. Microorganisms, including viruses and bacteria, also have many surface antigens, some of which may be associated with particular anatomical structures. For example, strains of *Salmonella* have specific antigens associated with the the proteins of their flagella, called **flagellar** or **H antigens**, and other specific antigens associated with the surface lipopolysaccharides (LPS) of the cell wall, called **somatic** or **O antigens**.

One of the reasons that the immune response is effective in preventing disease is that the toxins contributing to the virulence of pathogenic microorganisms usually have antigenic properties. Antibodies produced against toxins are referred to as **antitoxins**. Most protein toxins are highly antigenic, eliciting the synthesis of high titers (concentrations) of antibody. Similarly, bacterial LPS toxins are antigenic and are responsible for the initiation of antibody production against many Gram-negative bacteria. Denatured proteins often retain their antigenic properties. Of particular importance are denatured proteins called **toxoids**, which retain their antigenic properties but do not cause the onset of disease symptoms; they are antigenically active macromolecules but are no longer toxins. Toxoids are useful for eliciting antibody-mediated immune responses and are used as vaccines for protecting individuals against diphtheria, tetanus, and various other diseases caused by pathogenic microorganisms that produce protein toxins.

Antibody (Immunoglobulin) Molecules

Immunoglobulins (antibodies) are proteins found in the serum fraction of blood. There are five classes of immunoglobulins: IgG, IgA, IgM, IgD, and IgE. The characteristics of these five major classes of immunoglobulin molecules, all of which are globular proteins, are summarized in Table 19.1. Each class of immunoglobulin serves a different function in the immune response.

Immunoglobulins all have the same basic structure, consisting of four peptide chains, two identical heavy chains, and two identical light chains,[4] which are joined by disulfide

bridges linking the chains (Figure 19.3, Table 19.1). Differences in the heavy chains are responsible for the five major classes of immunoglobulins. There are five types of heavy chains, referred to as alpha (α), gamma (γ), delta (δ), epsilon (ϵ) and mu (μ). The light chains of the immunoglobulin molecules may be either kappa (κ) or lambda (λ).

The chains of the immunoglobulin can be split enzymatically. Papain cleaves the immunoglobulin to form two identical **Fab (antigen-binding) fragments** and an additional **Fc (crystallizable) fragment**, whereas pepsin cleaves the immunoglobulin molecule in another location, forming a divalent antibody-binding fragment but not forming Fc fragments (Figure 19.3). It is the Fab fragment that actually binds to antigen molecules, whereas the Fc fragment augments the action of the immunoglobulin molecule by binding to complement molecules and/or phagocytic cells.

The ends of the Fab fragments are said to be hypervariable, containing different amino acid sequences that give rise to varying three-dimensional structures for different immunoglobulin molecules (Figure 19.4). The variable portions at the ends of the peptide chains of the Fab fragments contain about 110 amino acids. The constant regions of the heavy chains, coded for by genes in the C region, are involved in secondary biological functions after the binding of antigen to the antibody molecule. The flexibility and variability of the immunoglobulin molecules permit the specificity of their reactions with antigenic biochemicals. Allosteric changes in the immunoglobulin macromolecule result in a conformation having a three-dimensional structure that allows the antigen and antibody molecules to fit properly together.

Types of Immunoglobulins **IgG** is the largest immunoglobulin fraction, generally comprising approximately 80 percent of the body's immunoglobulins. It is the predominant circulating antibody and readily passes through the walls of small vessels (venules) into extracellular body spaces, where it reacts with antigen and stimulates the attraction of phagocytic cells to invading microorganisms. Reactions of IgG with surface antigens on bacteria activate the complement system and attract additional neutrophils to the site of the infection. IgG also crosses the placenta and confers immunity upon the fetus that lasts for the first months after birth. It also plays a major role in preventing the systemic spread of infection through the body and in facilitating recovery from many infectious diseases.

IgA occurs in mucus and in secretions such as saliva, tears, and sweat. It is important in the respiratory, gastrointestinal, and genitourinary tracts, where it plays a major role in protecting surface tissues against invasion by pathogenic microorganisms. IgA also is secreted into human breast milk and plays a role in protecting nursing newborns against infectious diseases. The IgA molecules bind with surface antigens of microorganisms, preventing the adherence of such antibody-coated microorganisms to the mucosal cells lining the respiratory, gastrointestinal, and genitourinary tracts. The IgA molecules do not initiate the classical com-

---

[4]The terms *heavy* and *light* refer to the relative molecular weights of the polypeptide chains. There are more amino acids in the heavy chain; hence, it has a greater moleuclar weight than the light chain.

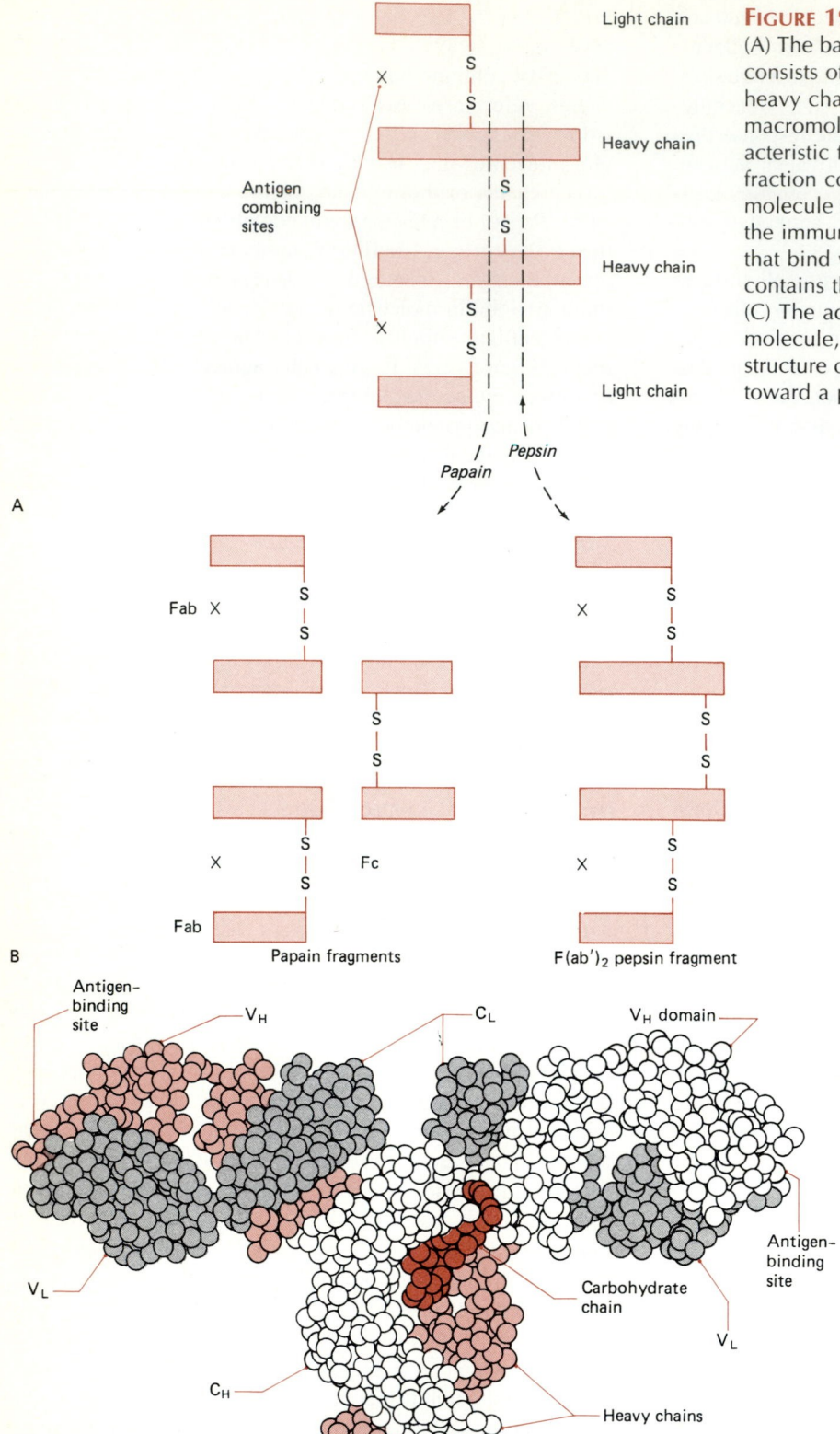

A

B   Papain fragments   F(ab')₂ pepsin fragment

C

FIGURE 19.3
(A) The basic structure of the immunoglobulin molecule consists of two identical light chains and two identical heavy chains. (B) Splitting of the immunoglobulin macromolecule by papain and pepsin produces characteristic fractions with distinct functions. The Fab fraction contains the ends of the immunoglobulin molecule that are variable and that bind with ends of the immunoglobulin molecule that are variable and that bind with antigen. The Fc or crystallizable fraction contains the constant portion of the immunoglobulin. (C) The actual immunoglobulin is a three-dimensional molecule, and it is the specific orientation of the structure of the molecule that determines the specificity toward a particular antigen.

TABLE 19.1   Properties of the Five Classes of Immunoglobulins

Property	Immunoglobulin				
	IgG	*IgA*	*IgM*	*IgD*	*IgE*
Molecular weight	150,000	160,000 and dimer	900,000	185,000	200,000
Number of basic four-peptide units	1	1, 2	5	1	1
Heavy chains	γ	α	μ	δ	ε
Light chains	κ + λ	κ + λ	κ + λ	κ + λ	κ + λ
Valency for antigen binding	2	2, 4	10	2	2
Concentration range in normal serum	8–16 mg/mL	1.4–4 mg/mL	0.5–2 mg/mL	0–0.4 mg/mL	17–450 μg/mL
Percentage of total immunoglobulin	80	13	6	0–1	0.002
Complement fixation					
Classical	+ +	−	+ + +	−	
Alternative	−	±	−	−	
Crosses the placenta	+	−	−	−	−
Fixes to homologous mast cells and basophils	−	−	−	−	+
Binds to macrophages and neutrophils	+	±	−	−	−
Major characteristics	Most abundant Ig of body fluids; combats infecting bacteria and toxins	Major Ig in seromucous secretions; protects external body surfaces	Effective agglutinator produced early in immune response	Mostly present on lymphocyte surface	Protects external body surfaces; responsible for atopic allergies
Structure					

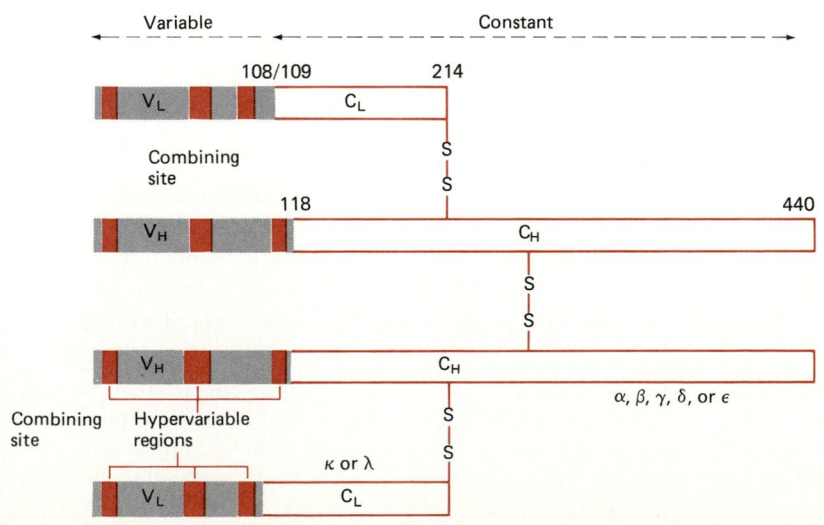

FIGURE 19.4

Illustration of an immunoglobulin, showing constant (C_L = constant light chain and C_H = constant heavy chain) and variable (V_L = variable light chain and V_H = variable heavy chain) amino acid composition regions. The amino acid residues are numbered, starting from the N-terminal end. C_L starts at residue 108 for kappa types and 109 for lambda. The high degree of variation in amino acid residues can be seen in the hypervariable region.

ANALYTICAL PROCESS

Electrophoretic Separation of Immunoglobulins

Immunoglobulins are difficult to separate but can be differentiated by using electrophoresis, a method that separates proteins on the basis of the charge, their size, and the shape of the macromolecule (Figure 19.5). Electrophoresis is the process of separating charged particles by migration in an electric field. A protein sample is usually placed on a solid support medium, such as agarose or polyacrylamide, which in turn is placed between two electrodes, a positively charged anode and a negatively charged cathode. When the current is turned on, protein migration takes place at a characteristic rate for each protein determined by its net charge, which is a function of pH, and its molecular weight. The classes of immunoglobulins form relatively broad electrophoretic bands, indicating the heterogeneity of molecules within each class and making it impossible to use routine electrophoresis for the quantitation of individual classes of immunoglobulins. Such analyses are useful in clinical medicine in diagnosing certain diseases, such as one associated with deficiencies of the immune response.

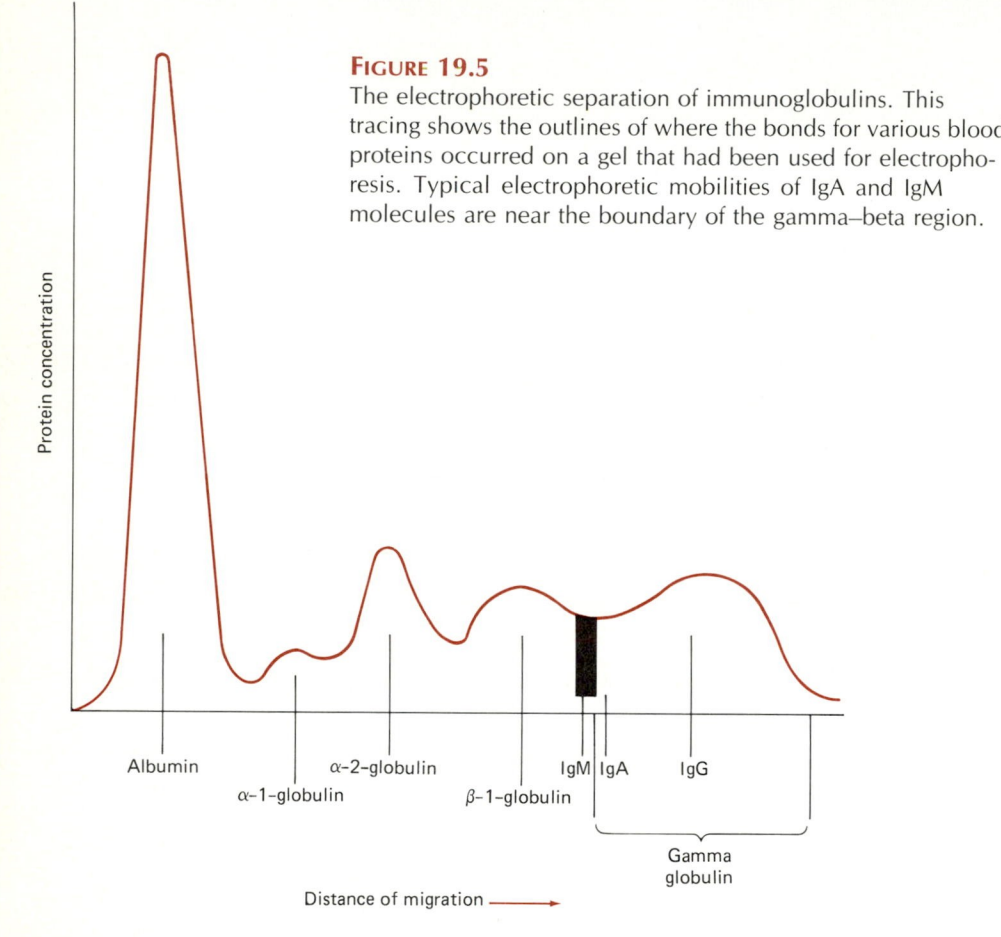

FIGURE 19.5
The electrophoretic separation of immunoglobulins. This tracing shows the outlines of where the bonds for various blood proteins occurred on a gel that had been used for electrophoresis. Typical electrophoretic mobilities of IgA and IgM molecules are near the boundary of the gamma–beta region.

plement pathway but can play a role in activating the alternative complement pathway discussed in Chapter 18. Plasma contains relatively high concentrations of monomeric IgA molecules, but it is the dimers of IgA that bind to receptors on the surface of secretory cells, leading to their secretion into body fluids (Figure 19.6). The IgA picks up secretory protein that protects it against proteases. In mucus secretions, IgA is the major immunoglobulin molecule involved in the immune response that protects external body surfaces. Individuals with asthma tend to lack IgA and to have excessively high levels of another immunoglobulin, IgE.

IgM is a high molecular weight immunoglobulin, occurring as a pentamer; that is, IgM contains five monomeric units of the basic four-peptide chain immunoglobulin molecule. IgM molecules are formed prior to IgG molecules in response to exposure to an antigen. Because of its high number of antigen-binding sites, the IgM molecule is effective in attaching to multiple cells that have the same surface antigens. As such, it is important in the initial response to a bacterial infection and in the activation of complement. During the later stages of infection, IgG molecules are more important. IgM molecules occur primarily in the blood serum

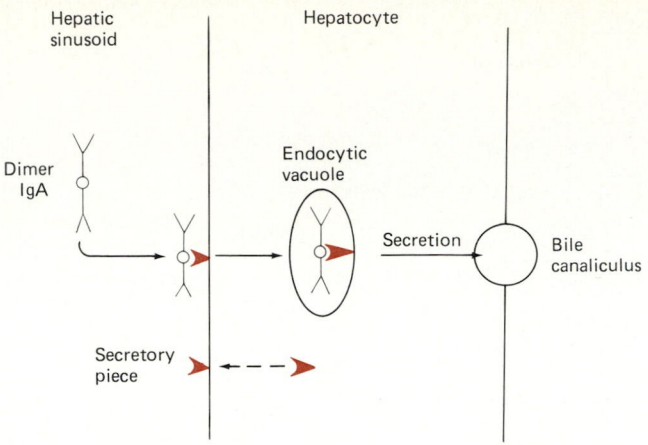

FIGURE 19.6
The mechanism of IgA secretion is exemplified by the transfer of circulating IgA into the bile. Dimeric IgA in the sinusoid bonds to a surface secretory piece, thereby activating the uptake and secretion of the immunoglobulin. There is perhaps an analogy with the uptake of IgG into macrophages through stimulation of the surface IgG receptors.

and, together with IgG molecules, are important in preventing the circulation of infectious microorganisms through the circulatory system.

IgD antibody molecules, together with IgM, are present on the surface of some lymphocyte cells. Although the precise role of IgD remains to be fully defined, it appears to play a role as an antigen receptor in lymphocyte activation and suppression. Within blood plasma, IgD molecules are short-lived, being particularly susceptible to proteolytic degradation.

IgE molecules are normally present in the blood serum as a very low proportion of the immunoglobulins. The ratio of IgG to IgE is normally 50,000:1. IgE serum levels, though, are elevated in individuals with allergic reactions, such as hay fever, and in some persons with chronic parasitic infections. The main role of IgE appears to be the protection of external mucosal surfaces by mediating the attraction of phagocytic cells and the initiation of the inflammatory response. IgE molecules are important because they bind to mast cells and basosphils, where they play a role in mediating immune reactions, including, unfortunately, hypersensitivity reactions such as hay fever and other allergic responses. The tight binding of IgE to cell membranes in circulating mast cells and leukocytes, based on the characteristics of the Fc portion of the IgE molecule, is an unusual property of IgE.

Within each of the major classes of immunoglobulins, there are variants. For example, IgG can be grouped into four subclasses, IgG1, IgG2, IgG3, and IgG4; each subclass has differences in the heavy chains of the immunoglobulin molecule. These variations can be classified as **isotypes**, referring to variations in the heavy chain constant regions associated with different classes and subclasses of immunoglobulins; **allotypes**, referring to genetically controlled allelic forms of the immunoglobulin molecule; and **idiotypes**, referring to the individual specific immunoglobulin molecules that differ in the hypervariable regions of the Fab fragments. These many variations in immunoglobulins establish the needed diversity of macromolecules for an effective immune response.

Genetic Basis for Immunoglobulin Diversity

The genetic variability coding for this diversity of antibody molecules provides the basis for the immune response to many different antigens. Before the recognition that eukaryotic cells had split genes—a topic discussed in Chapter 7—it was difficult to understand how the genome encoded the information for the enormous diversity of antibodies. The human immune system has a virtually unlimited capacity to generate different antibodies, which recognize and bind to millions of potential antigens. One early theory proposed that each human cell contains a separate gene encoding each antibody chain that an individual is capable of synthesizing. However, there is not enough DNA in human cells to accomplish this and still have genes that code for other functions; the human genome contains perhaps a million genes, and only a small fraction of these can specify antibodies. Another proposal was that each cell contains a small number of genes for encoding antibody chains but that these genes are so susceptible to mutation that multiple mutations accumulating in mature B cells confer on the organism the ability to produce a variety of different antibodies. This theory also cannot adequately account for the diversity of antibodies. Thus, the theory held for many years that there was a one-to-one correspondence between genes and polypeptides cannot account for the diversity of antibodies, and another theoretical explanation is necessary.

The essence of the explanation of how a limited number of genes can generate the great diversity of antibodies is that the genes ultimately specifying the structure of each antibody are not present as such in germ cells (the male sperm and the female egg) or in the cells of the early embryo. Rather, the currently accepted theory is that there are variable and constant regions of antibodies that are encoded by separate groups of genes, and that recombination permits shuffling and joining of the components so that billions of different combinations can occur. Polypeptides thus can be synthesized from information contained in several gene fragments scattered over the genome. The reshuffling results in numerous combinations of varieties of the light chains (kappa and lambda) and heavy chains (alpha, delta, epsilon, gamma, and mu) that make up the complete immunoglobulin molecules. There are three clusters of genes, one for heavy chain synthesis and two for light chain synthesis. Within each cluster, there are a number of gene fragments that contain the information for the complete immunoglobulin molecules.

Light Chain Diversity   The active gene for a light chain is assembled and expressed by a process of somatic recombi-

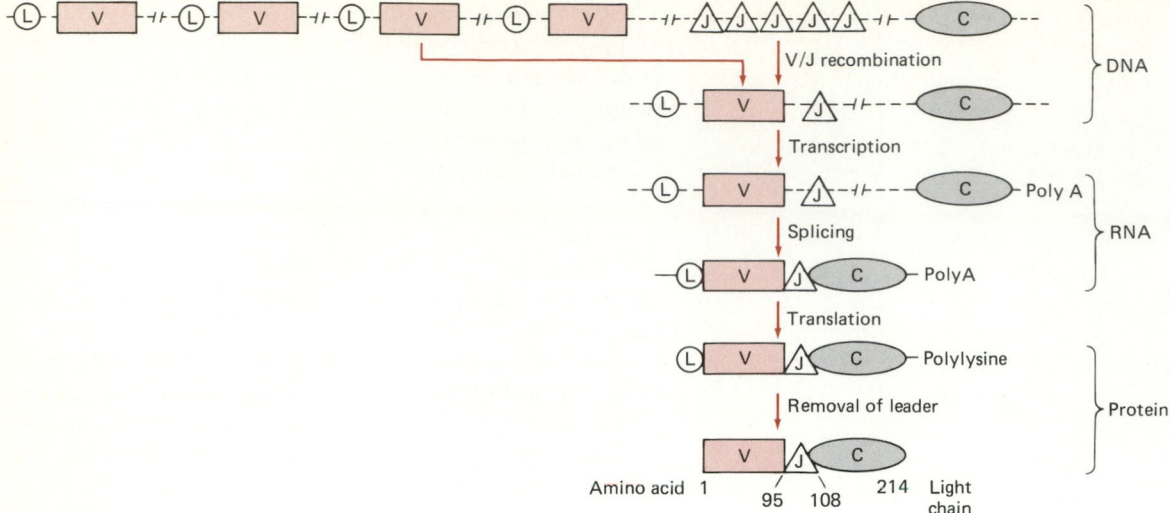

FIGURE 19.7

The variable region of the chain is encoded by V and J genetic sequences, the constant region by a C gene. Each V sequence is separated from a leader sequence (L) by a short intron. The L/V segments are separated from five J sequences by a long intron. In the human kappa system, the J's, in turn, are separated from a single C gene by an intron. In the human lambda-chain system there are six C genes, each linked to its own J sequence. During lymphocyte development, one V gene, with its L sequence, is recombined with one of the J sequences to form, along with the single C gene, an active kappa or lambda gene. The entire gene is transcribed to form hnRNA. The introns and any extra J's in the hnRNA are spliced out to produce mRNA. The mRNA is translated to form the light-chain precursor. The leader is cleaved away as the mature chain passes across the cytoplasmic membrane to be assembled into an antibody molecule.

nation and RNA splicing (Figure 19.7). In this system, the components of the active gene are present in the germ line in the cells of the embryo in multiple versions. Each complete light chain genetic region includes a variable region and a constant region. For the light chain, the variable region is encoded by variable (V) and joining (J) sequences and the constant region by a C gene. A large number of variable genes, with closely related nucleotide sequences, have been identified in embryonic DNA. There may be 150 alternative V sequences, each separated from a leader (L) sequence by a short intervening sequence. The L sequence specifies a hydrophobic leader that is 17 to 20 amino acids long. The other coding region of the V gene specifies most but not all of the variable region. The L/V segments are separated from five joining sequences by a long noncoding sequence of DNA. The J sequence, which codes for the remaining portion of the variable region, is a short sequence that is repeated several times, with slight but significant variations, at intervals of about 300 nucleotides. In the human light kappa-chain system, the J sequences, in turn, are separated from a C gene by another intervening sequence. In the human light lambda-chain system, the arrangement is somewhat different in having six C genes, each one apparently linked to its own J sequence.

The joining of one of 150 V genes to one of 5 J genes can generate 150 × 5, or 750, different active genes for a light-chain variable region. However, even greater light chain di-

versity occurs because the V/J recombination site is not precisely defined. A V gene and a J gene can apparently be joined at different crossover points, and if there are 10 alternative joining sites, there would be a 10-fold increase in diversity, giving rise to 7,500 different possible combinations.

During lymphocyte development, one V gene with its L sequence is recombined with one of the J sequences to form, along with the single C gene, an active kappa gene. The entire gene is transcribed into an hnRNA transcript. The hnRNA is converted to mRNA by removing the intervening sequences (introns), including extra joining segments that may be present in the hnRNA. The mRNA is translated to form a light-chain precursor containing a leader that acts as a signal sequence to initiate the transport of the protein across the cytoplasmic membrane. The leader is cleaved away as the light-chain precursor moves across the cytoplasmic membrane to produce the mature light chain. Light chains fold up in such a way that the hypervariable region forms the antibody-antigen combining site.

Heavy Chain Diversity   In the heavy chain, the formation of the variable region is governed by the same principles that apply to the light chain, but the potential for diversity is even greater. The additional diversity comes from a sequence of embryonic DNA called the *diversity (D) gene*. The D gene accounts for a significant portion of the hypervariable region of the heavy chain. The heavy chain is thus a recombinational product of the V, J, D, and C genes (Figure 19.8). Active heavy-chain genes are assembled by recombination of one of the L/V sequences, a D gene, and a J gene to code for the variable region of the chain and a C gene to code for the constant region. In the heavy chain, there are eight separate C sequences, each one coding for a different constant region. Assuming that there are 50 D sequences, 80 V sequences, 6 J sequences, and 100 recombinational variations, for embryonic human cells there are therefore 2.4 million possible different heavy chains. Taken together with the 7,500 possible combinations available to the human kappa light chain, the 2.4 million heavy chains yield a total of some 18 billion possible antibodies coded for by only about 300 separate genetic segments in the embryonic DNA. Mutations within the genes encoding the variable region of the immunoglobulin also contribute to immunoglobulin diversity increasing the number of potential immunoglobulins even further.

Clonal Selection

The complexity of the system in generating the genetic diversity of immunoglobulins is matched by the maturation process of lymphocytes. Even greater immunoglobulin diversity is generated during the maturation of B lymphocytes. A precursor of the antibody-producing cells, the pre-B lymphocytes, makes a mu heavy-chain constant region linked to a specific variable region (a product of V/D/J recombination). This heavy chain at first remains inside the pre-B cell, but after the onset of light-chain and delta heavy-chain synthesis, both the mu and the delta heavy chains combine with the light chains to form complete IgM and IgD molecules, with the concurrent appearance of both IgM and IgD on the cell surface. Both antibodies have the same variable regions, and so both are directed against the same antigen.

The subsequent steps in lymphocyte maturation occur in response to interactions with specific antigens. An antigen binds to a receptor at the best-fitting antigen-combining site. By this interaction, the cell displaying the selected immunoglobulin is driven further along its developmental pathway. In the course of B-cell maturation, the IgD and IgM disappear from the cell surface and either IgM, IgG, IgE, or IgA is instead secreted by the cell. Because each heavy chain gives the antibody a different effector function, the same combining site can take part in different immune reactions. The process by which the same variable region appears in association with different heavy-chain constant regions is called **heavy-chain class switching**

The lymphocyte depends on two mechanisms to carry out class switching. One mechanism is based on differential RNA transcription and splicing, the other on a version of DNA recombination. RNA transcription and splicing account for the successive appearance of membrane-bound and secreted IgM and for the simultaneous appearance of IgM and IgD (Figure 19.9). A heavy-chain class switch is accomplished by DNA recombination in which a switching signal

(S) precedes each constant region. The switching signal can modify the recombination that joins a V/D/J sequence to one of the downstream constant-region sequences by changing the particular constant region that is united to the the V/D/J sequence (Figure 19.10).

Events During Fetal Development   How, though, can specific immunoglobulins be synthesized in response to specific foreign antigens, and how can we avoid synthesizing immunoglobulins that react with our own antigens? To explain this, we hypothesize that interactions between B lymphocytes and antigens during fetal development are quite different from those that occur after birth. The **clonal selection theory** depends on the development and differentiation of a large population of B lymphocytes during fetal development. It assumes that during the development of the fetus, a complete set of lymphocytes is produced, with each lymphocyte cell containing the genetic information for initiating an immune response to a single specific antigen, with receptors for each antigen located within the cytoplasmic membranes of the differentiated B lymphocytes. When a B lymphocyte combines the variable genes for the light and heavy chains, it irrevocably establishes the antigen-binding specificity of that cell. According to the clonal selection theory, an individual lymphocyte responds to only one antigen because it has only one type of receptor, and the correspondence between the lymphocyte receptor and the antigenic determinant accounts for the specificity of the antibody-mediated immune response. However, in terms of the constant region, a given cell may change its receptor and may even express more than one type of receptor. By heavy chain switching, the B cell may turn out to express IgA, IgE, IgG, IgM, or a combination of immunoglobulin types.

To explain the development of tolerance to self antigens, the clonal selection theory says that during fetal development, surface receptors of certain B cells react with self antigens. Those lymphocytes that have reacted with self antigens are unable to divide to form a clone of cells; rather, the reaction of B lymphocytes with antigens during fetal development leads to the destruction of those lymphocytes (Figure 19.11). As such, there is a negative selection for lymphocytes capable of reacting with the antigens of the human body, and only lymphocytes that fail to react with antigens during the fetal development period survive. Theoretically, any antigen that is detected and reacts with the appropriate fetal lymphocyte is recognized as self. A tolerance of self antigens develops because all of the B lymphocytes that can produce antibody to self antigens are destroyed during development of the fetus. Thus, at birth, the remaining genetically differentiated B lymphocytes should be competent to react only with foreign antigens, that is, with nonself antigens.

Accordingly, one should not exhibit **autoimmunity**; that is, one should not show an immune response against one's own antigens. Autoimmunity can occur, though, if B cells are not exposed to particular human antigens during fetal

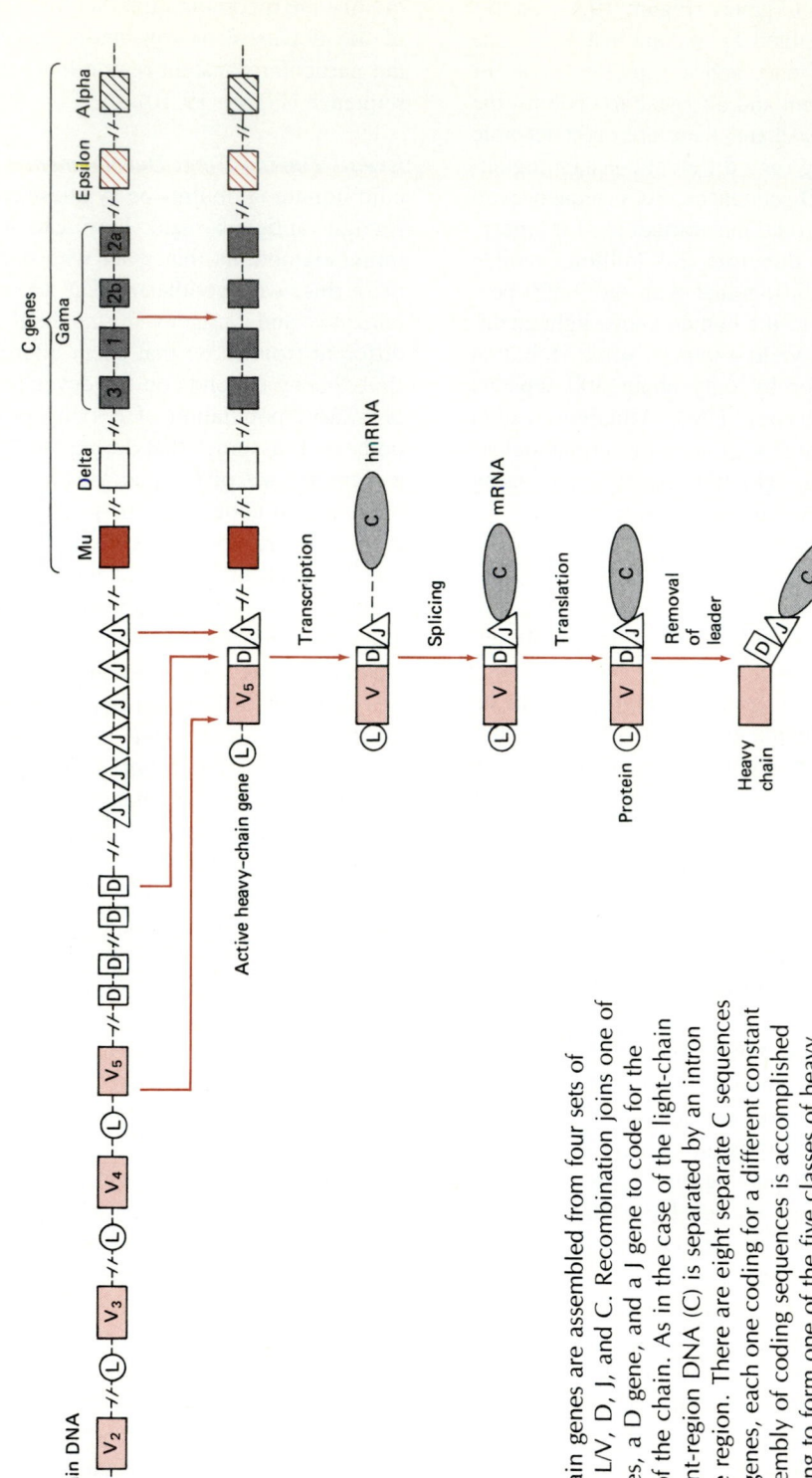

FIGURE 19.8

Active heavy-chain genes are assembled from four sets of gene sequences: L/V, D, J, and C. Recombination joins one of the L/V sequences, a D gene, and a J gene to code for the variable region of the chain. As in the case of the light-chain gene, the constant-region DNA (C) is separated by an intron from the variable region. There are eight separate C sequences for heavy-chain genes, each one coding for a different constant region. Final assembly of coding sequences is accomplished by RNA processing to form one of the five classes of heavy chains that determine the class of immunoglobulin.

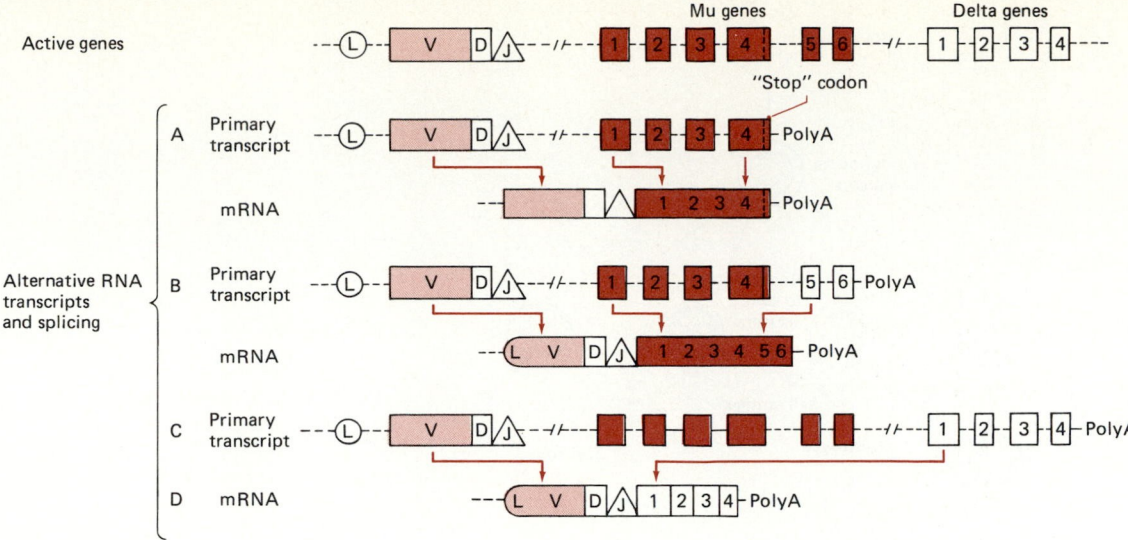

FIGURE 19.9

Regulatory mechanisms permit differential transcription and splicing of the hnRNA to form different mRNAs that account for the successive appearance of membrane-bound and secreted IgM and for the simultaneous appearance of IgM and IgD. (A) The complete configuration of the heavy-chain genes for the variable region and for the mu and delta constant regions in an antibody-producing cell. (B) Transcription can terminate, forming an hnRNA with none of the delta genes and with only some of the mu genes; if this occurs, the RNA containing only mu genes is spliced to produce an mRNA lacking the amino acid sequence that would bind the IgM to the cell, and the IgM is therefore secreted. (C) Transcription can form an hnRNA with all of the mu genes and none of the delta genes. In this case, splicing produces an mRNA with a sequence for attachment of the IgM to the cell, and the IgM is therefore membrane bound. (D) Transcription can form an hnRNA with the delta genes. Splicing can eliminate the mu genes and connect the delta genes directly to the variable-region sequence, making mRNA that encodes a delta chain and thus IgD.

FIGURE 19.10

Each constant-region gene, which is actually a split gene with several introns, is preceded by a switching signal (S). The switching signals mediate a recombination that joins a V/D/J sequence to one of the downstream constant-region sequences. The switched DNA is transcribed and the RNA is spliced to make an mRNA encoding whichever constant region is joined to the variable sequence.

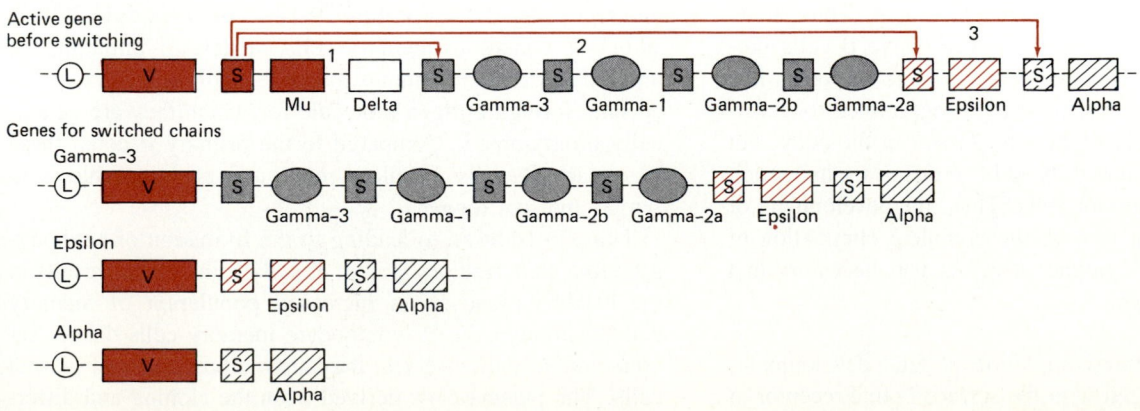

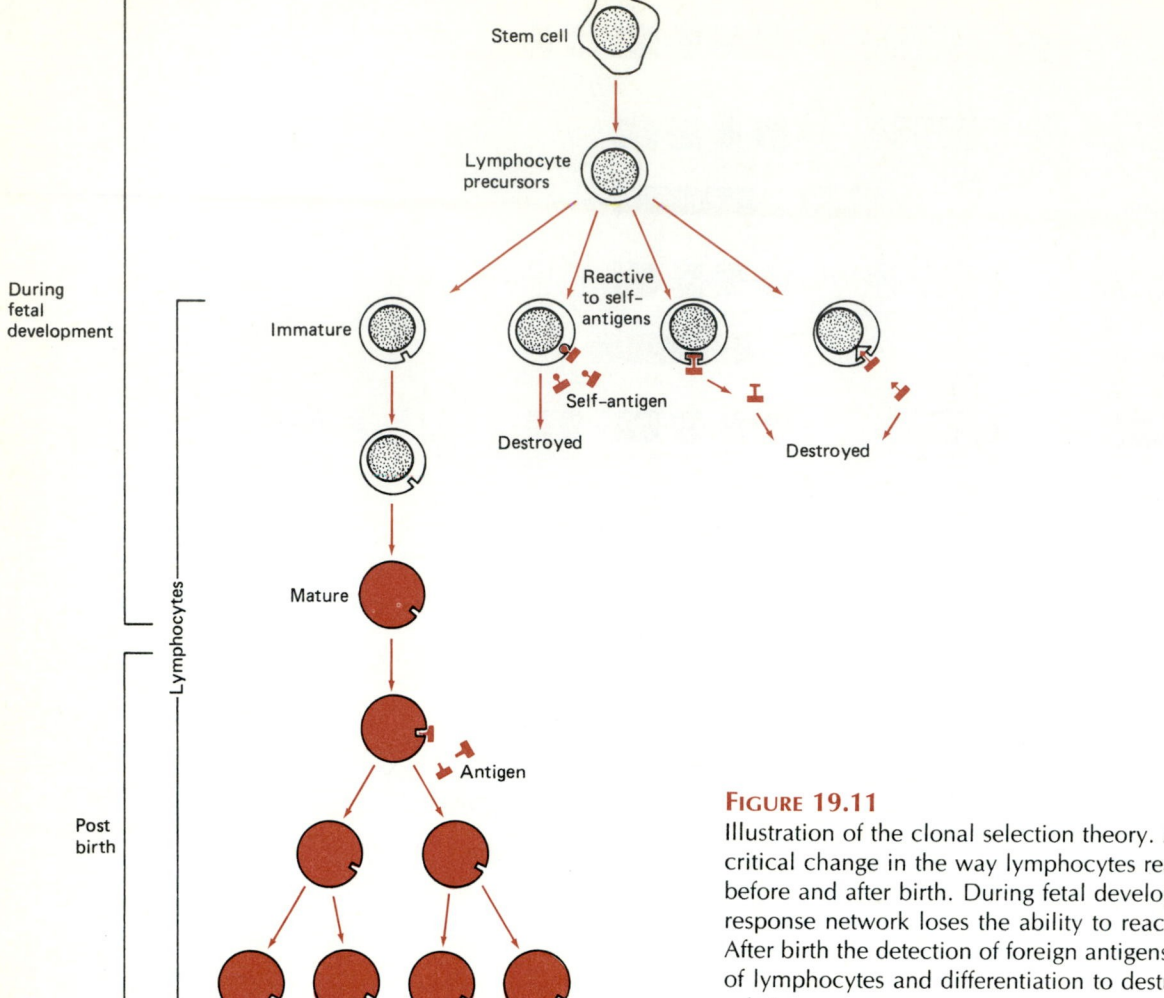

FIGURE 19.11
Illustration of the clonal selection theory. As shown, there is a critical change in the way lymphocytes react with antigen before and after birth. During fetal development the immune response network loses the ability to react with self-antigens. After birth the detection of foreign antigens leads to the cloning of lymphocytes and differentiation to destroy the foreign substance.

development. In such cases, some B cells programmed for reacting with self antigens survive fetal development. For example, if B cells are not exposed during fetal development to the specific antigens that later occur on male sperm cells, male infertility can result because the body reacts to such antigens as foreign antigens. In fact, such autoimmunity is an important cause of male infertility. Additionally, autoimmunity occurs if mutations reestablish B cell lines that were properly eliminated during fetal development. Thus, there are several ways in which the body can have B cells programmed to react with self antigens. In fact, it appears that some B cell lines that are genetically programmed for reacting with self antigens are typically present in the body, but that these B cells are normally held in check by the action of T-lymphocyte suppressor cells. Thus, the development of self-tolerance does not require the complete elimination of self-reactive precursor lymphocytes, but it is necessary that suppression be dominant.

Postnatal Events After completion of fetal development, the reaction of an antigen with the surface-bound receptor of

a lymphocyte cell initiates a totally different response than occurs during the fetal development period. According to the clonal selection theory, after fetal development is complete, the interaction of an antigen with a specific immunoglobulin molecule that is integrated into the cytoplasmic membrane of a B lymphocyte initiates the differentiation and multiplication of that cell to form two different cell populations: **plasma cells** that are able to synthesize a specific antibody and secondary B cells (Figure 19.12). The **secondary B lymphocytes** constitute memory cells capable of initiating the antibody-mediated immune response upon detection of the specific foreign antigen molecule for which they are genetically programmed. Compared to the primary B cells, these secondary B cells circulate more actively from blood to lymph and live longer.

Thus, in addition to leading to the formation of antibody-secreting cells, the reaction of B cells with antigen results in the establishment of an increased population of memory cells. Although the B-lymphocyte memory cells do not secrete antibody themselves, they are the precursors of plasma cells. The plasma cells derived from the cloning and differ-

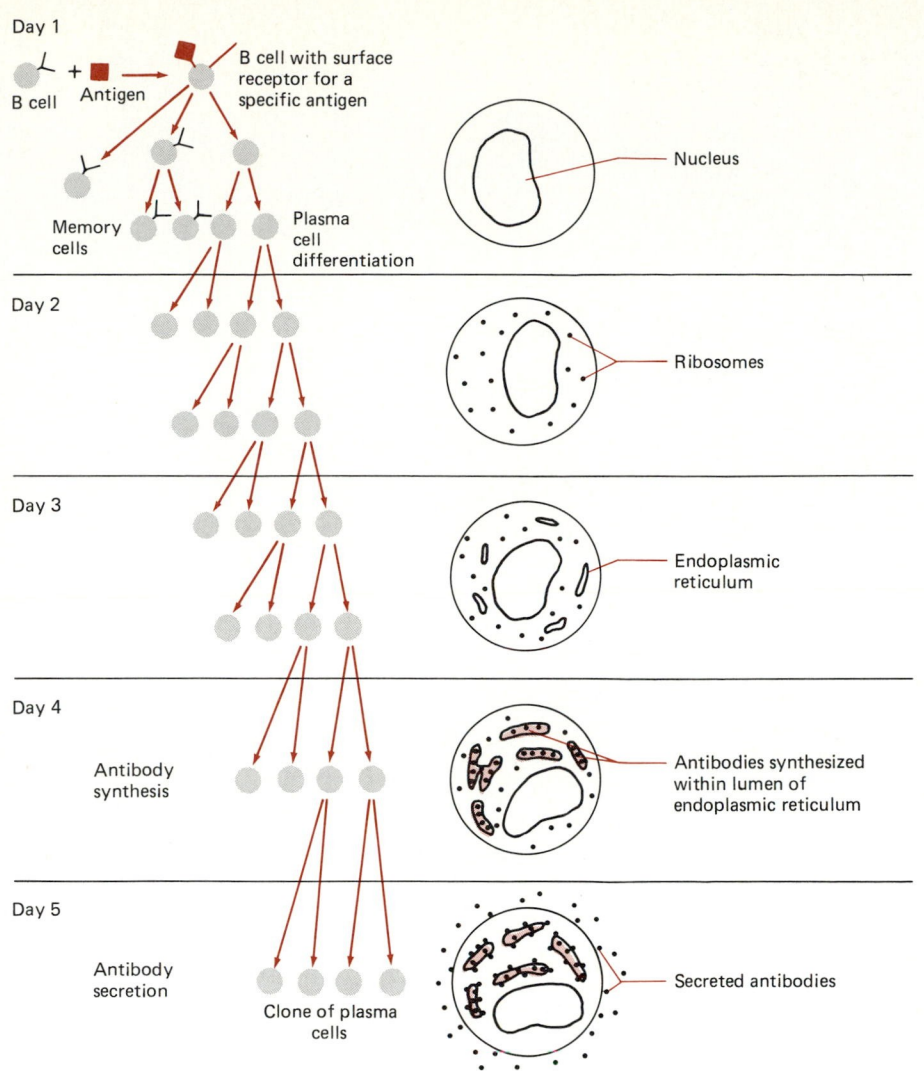

Day 1

B cell + Antigen → B cell with surface receptor for a specific antigen

Memory cells

Plasma cell differentiation

Nucleus

Day 2

Ribosomes

Day 3

Endoplasmic reticulum

Day 4

Antibody synthesis

Antibodies synthesized within lumen of endoplasmic reticulum

Day 5

Antibody secretion

Clone of plasma cells

Secreted antibodies

FIGURE 19.12
The response of B cells to antigen results in the production of a memory system and the development of antibody-synthesizing plasma cells.

entiation of B lymphocytes are responsible for the secretion of extracellular antibodies into the blood serum. Each clone of a plasma cell line secretes a single specific antibody molecule. Because relatively few B cells with the appropriate receptors are present at the first exposure to a given antigen, the primary response is characteristically slow, producing relatively low yields of antibody. There is a long lag period in the primary response, during which selection, differentiation, and cloning of appropriate B cell lines must occur.

After an antigen has elicited an antibody-mediated immune response, there is an increase in the number of B lymphocytes capable of reacting with that antigen. Because antigen binds to and selects for cells having receptors of the highest affinity, this process results in an increase in the number of lymphocyte cells with receptors of high affinity for the particular antigenic molecule. The cloning of these cells establishes a bank of **memory cells**, so that upon subsequent exposure to the same antigen, perhaps years later, there is a larger population of B cells that can initiate the secondary immune response rapidly and efficiently.

In addition to the activation of B lymphocytes by antigen,

a cooperative interaction between B and T cells is necessary to stimulate antibody production effectively (Figure 19.13). Cooperation between B and T cells requires an antigen with at least two different antigenic determinants, suggesting that the antigen functions as a bridge between the interacting T and B cells. Macrophage are also involved in the cooperative interactions of the various cells of the immune response. As part of the inflammatory response, macrophage process antigens that become attached to the surfaces of the macrophage cells. When these macrophage migrate to lymphatic tissues, they interact with **helper T cells**, which respond to antigens only when they are presented by macrophages. Helper T cells are required before an activated B cell can initiate cloning and extensive antibody synthesis. Usually a B cell cannot be triggered to initiate antibody synthesis without interacting with helper T cells. These helper T cells provide a necessary second signal for the initiation of cloning; antigen-stimulated T helper cells release interleukin-2 that stimulates multiplication of T helper cells and also the release of B-cell growth factor, which in turn enhances the division of B cells. Antigen-presenting cells (APCs), which include macro-

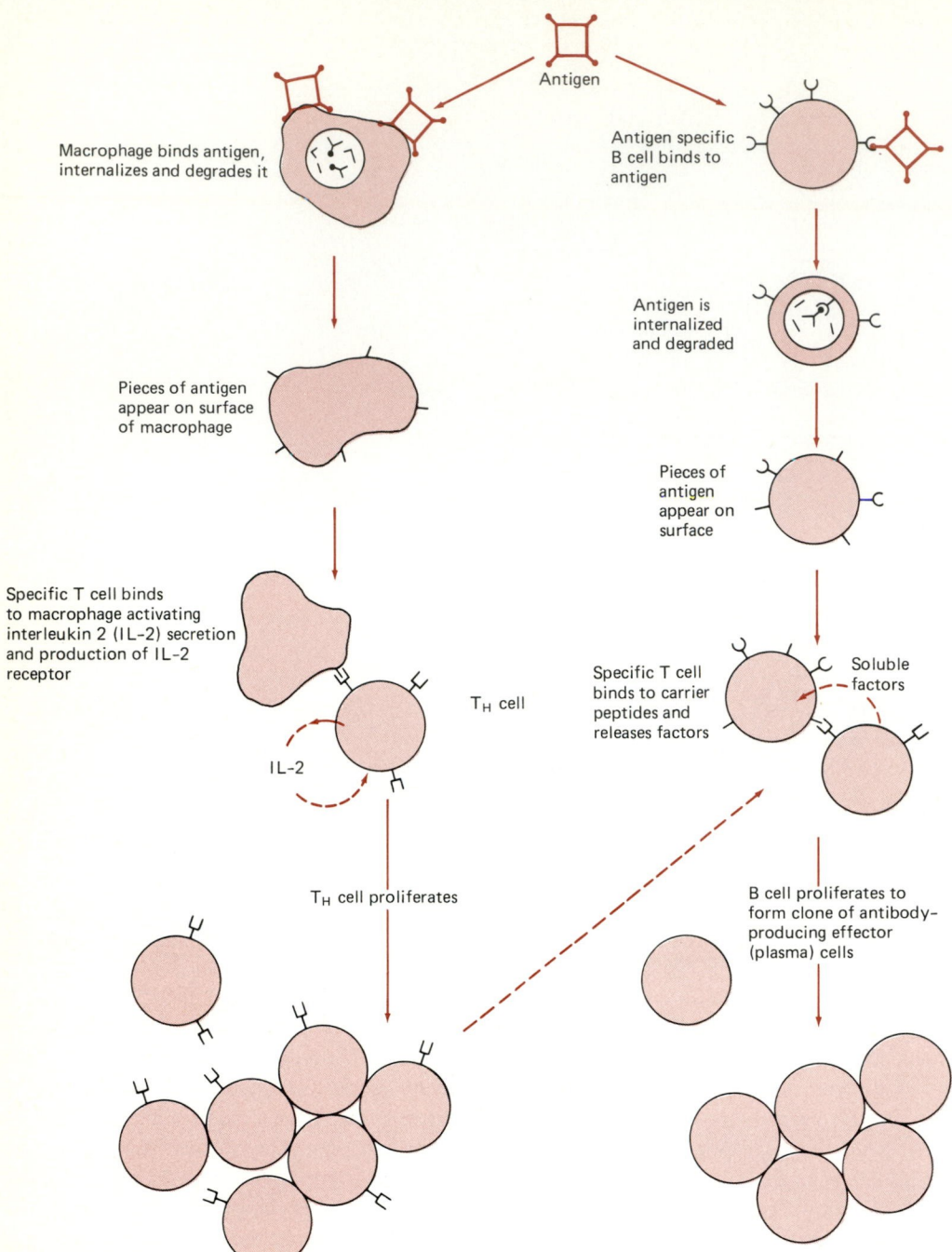

Antigen

Macrophage binds antigen, internalizes and degrades it

Antigen specific B cell binds to antigen

Antigen is internalized and degraded

Pieces of antigen appear on surface of macrophage

Pieces of antigen appear on surface

Specific T cell binds to macrophage activating interleukin 2 (IL-2) secretion and production of IL-2 receptor

T_H cell

Specific T cell binds to carrier peptides and releases factors

Soluble factors

IL-2

T_H cell proliferates

B cell proliferates to form clone of antibody-producing effector (plasma) cells

FIGURE 19.13
Once thought to be totally independent systems, B cells and T cells are now known to engage in extensive cooperative interations in response to an antigen. Stage 1 is antigen specific; helper T cells are activated by macrophage-processed antigen to stimulate the resting B cell, which has bound antigen through its hapten-specific immunoglobulin receptors, and after a small number of divisions to transform to a blast cell with receptors for soluble growth factors. Stage 2 is nonantigen specific; these soluble growth factors are produced by interactions between the T cell and macrophages, possibly different from the original type initially representing the antigen, and stimulate the blast cells to repeated division and maturation, to become antibody-producing effector cells.

phages and other specialized cells found in skin and lymphoid organs, are also involved in the cooperative antibody response, playing an essential role by binding antigen on their surfaces so that T helper cells with receptors for the antigen bind to them. APCs present the antigen to the T lymphocytes in the proper orientation so that the T cells can recognize the antigen. Once an antigen binds to a T helper cell, interleukin-1 is released, further stimulating activities of T cells, B cells, and macrophages.

The other class of T cells, T suppressor cells, interfere with the binding of antigens to T helper cells, thereby limiting the immune response. T suppressor cells specifically inhibit antibody production. These T cells are generated by antigenic stimulation, responding to antigens with repeating identical antigenic determinants, and are antigen specific in their suppression. Suppressor T cells play an important role in regulating the immune response.

Antigen–Antibody Reactions

The basis of antibody-mediated immunity, which plays an important role in preventing and eliminating microbial infections, depends on the reactions of antigen with antibody molecules. The reactions of IgA molecules with bacteria and

ANALYTICAL PROCESS
Culture of Hybridomas

The fact that each specific antibody is synthesized by a different cell line of B lymphocytes and their derived plasma cells makes it possible to culture cells producing **monoclonal antibodies**. This can be accomplished by fusing myeloma cells—tumor cells of the immune system that produce large amounts of unusual immunoglobulins and can be readily cultured—with lymphocyte cells programmed for the synthesis of a single antibody. The resulting hybrid cell, known as a **hybridoma**, can be cultured and can synthesize specific antibodies (Figure 19.14).

Hybridomas are produced by fusing B cells, which are genetically programmed to produce a particular (single) antibody, with a cancer cell, which is capable of proliferating rapidly under laboratory conditions. The technique used for making hybridomas was designed in 1975 by Cesar Milstein and Georges Kohler in England. They injected antigen into a mouse, whose B cells began to produce antibodies against that antigen. They then removed the mouse's spleen, which contained many B cells. These cells were incubated with cancerous tumor cells derived from myeloma tumors of bone marrow; some of the B cells fused with the tumor cells to produce the hybrid cells. The hybrids are screened to select those that produce a desired antibody. Grown in culture, the descendants of clones of a single hybridoma cell continue to produce only the type of antibody characteristic of the parent B cell, hence the name *monoclonal* antibody. The hybridoma formed is a permanent union; hybridoma cells are uniform, highly specific, and can readily be produced in large quantities.

These highly specific monoclonal antibodies are useful in clinical procedures and may prove useful in the treatment of some diseases. If, for example, monoclonal antibodies could be made from antigens that are unique to particular pathogens, these antibodies could be used to treat specific diseases. Monoclonal antibodies are already used for the diagnosis of allergies and infectious diseases, such as hepatitis, rabies, and some venereal diseases. Early stages of cancer may also be detectable with monoclonal antibodies because certain types of cancer cells have surface antigens that differ from those of normal

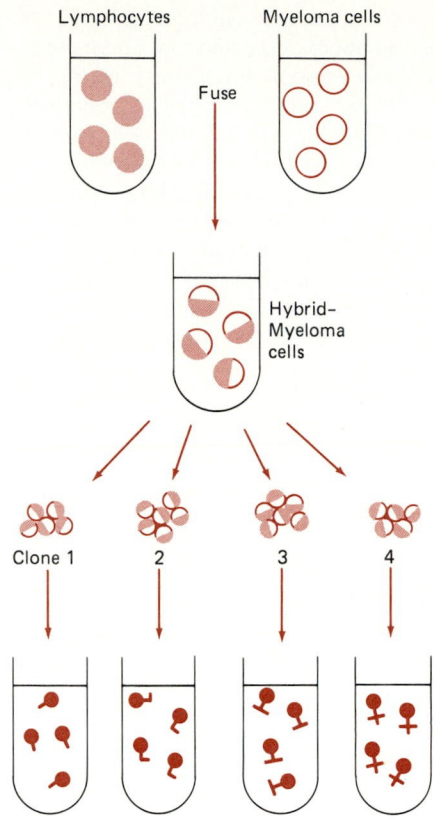

FIGURE 19.14
Monoclonal antibody synthesis involves fusion of a B cell with a myeloma cell.

cells. It is possible that, in the future, monoclonal antibodies will be used to treat cancer and infectious diseases. Monoclonal antibodies, used alone or chemically attached to drugs, could locate and destroy a cancer cell or pathogenic microorganism without damaging the healthy tissue surrounding it.

---

viruses in the fluids surrounding surface tissues prevent the adsorption of many potential pathogens onto these surface barriers. In this way, IgA antibody molecules prevent the establishment of infections. IgG antibody molecules, acting as antitoxins, are able to neutralize toxin molecules by combining with them, thus blocking their reactions and preventing the onset of disease symptomatology. Even poisonous cobra venom can be neutralized by reaction with appropriate antibody molecules. The interactions of antibody with surface antigens of bacterial cells render many pathogenic bacteria susceptible to phagocytosis. In fact, the ingestive

phagocytic attack on most bacteria requires an initial antigen–antibody reaction before phagocytic blood cells can engulf the invading bacteria. The increased phagocytosis associated with antibody-bound antigen, called **opsonization**, is important in the destruction of pathogenic bacteria. Antigen–antibody reactions are required to overcome infections by bacteria, such as *Haemophilus influenzae*, that are inherently resistant to phagocytosis. These *in vivo* reactions between antigen and antibody molecules constitute a major line of defense against invading bacteria and other microorganisms.

ANALYTICAL PROCESS

In Vitro *Immunological Reactions*

In addition to their importance *in vivo*, the reactions between antigen and antibody molecules are important in many laboratory procedures, including many used for the *in vitro* diagnosis of disease. Immunological reactions performed *in vitro* are called *serological reactions*. Serology is based upon the ability to observe or otherwise detect specific antigen–antibody reactions. It is important that only the specified antigen-antibody reaction be detected. False (improper) results can occur if there is cross reactivity, that is, if an antigen reacts with an antibody that was made for a reaction with a different antigen. This is a potential problem inherent in all serological reactions.

A variety of procedures are used to detect antigen–antibody reactions for diagnostic purposes. Each method has particular advantages and limitations. The critical factors involved in selecting a methodological approach are sensitivity, specificity, reliability, and cost.

Precipitin Reactions

When antigen and antibody molecules react, they sometimes form large polymeric macromolecules whose solubility properties result in their **precipitation**. Precipitation of the antigen–antibody complex can be readily visualized by allowing the antigen and antibody molecules to mix in a fluid or an agar gel. The formation of insoluble complexes that aggregate and precipitate is dependent on the relative concentrations of the antigen and antibody molecules (Figure 19.15). Thus, to achieve maximal precipitation of the antigen–antibody complex, the react-

ing antigen and antibody molecules must be present in the proper relative concentrations.

In order to establish the appropriate relative concentrations for precipitation reactions, the antigen and antibody molecules are generally allowed to diffuse together, establishing a concentration gradient. A characteristic precipitation band occurs at the appropriate relative proportions of the antigen and antibody. Some precipitin tests employ a single diffusion gradient. In **single diffusion methods**, an antigen is allowed to diffuse unidirectionally into a tube containing a uniform concentration of soluble antibody so that the antigen establishes a concentration gradient through the tube (Figure 19.16). If the antigen and antibody react to form an insoluble complex, a band of precipitation will occur at a particular level in the tube. The formation of a precipitant requires that the antigen and antibody match; that is, the antibody must be reactive specifically with the antigenic determinants of the particular biochemical applied to the tube.

In **double diffusion methods**, such as the Ouchterlony method, both the antigen and the antibody are allowed to diffuse toward each other (Figure 19.17). In this method, antigen and antibody molecules are placed in separate wells cut into an agar gel. These molecules diffuse toward each other; if they react, they form a precipitation band in the region of optimal relative proportions of the antigen and antibody molecules. This method can be used both to identify antigens and to compare the immunological relationships between them. In order to assess the relationship between antigens,

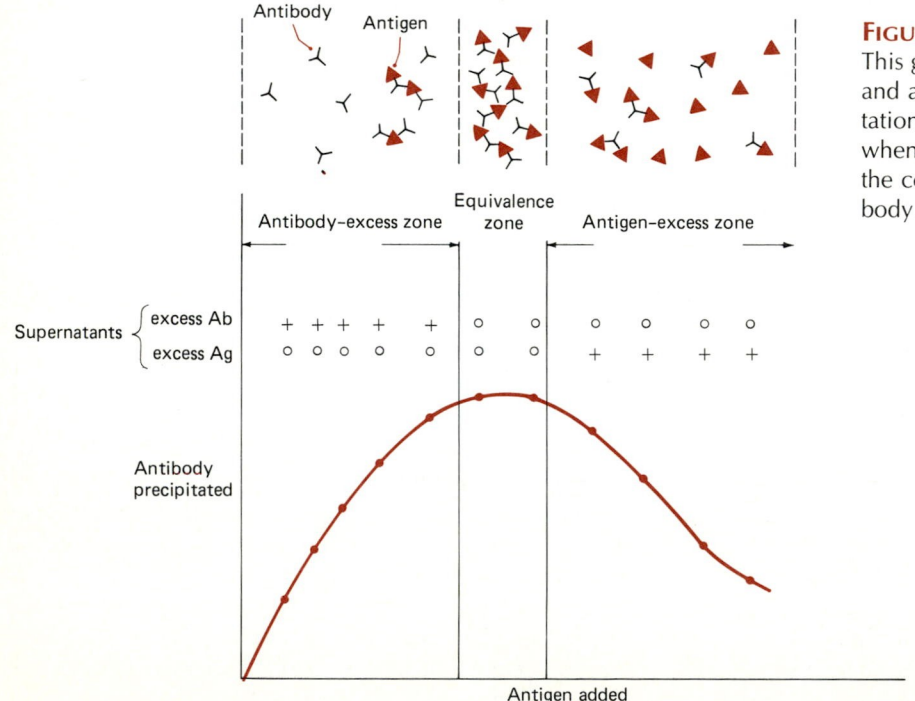

FIGURE 19.15
This graph illustrates the effects of antigen and antibody concentrations on precipitation. Maximal precipitation occurs when there is an optimal ratio between the concentrations of antigen and antibody.

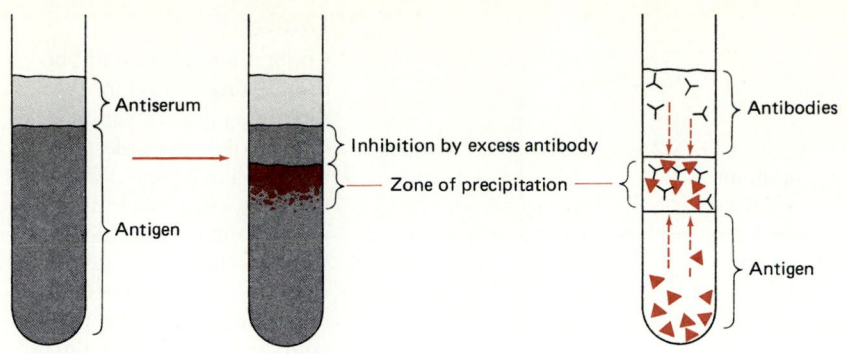

FIGURE 19.16
In the single diffusion method, antibody is allowed to diffuse into a tube containing dispersed antigen, leading to a zone of precipitation in the region where the antigen and antibody achieve the proper relative concentrations.

multiple adjacent wells are filled with the antigens and allowed to diffuse toward a single well filled with antibody molecules. The formation of confluent precipitation bands indicates identity; the formation of a spur indicates partial identity; and a lack of confluent precipitation indicates a lack of identity between the antigen molecules. The arced shape of the precipitation band formed is characteristic of the antigen molecule and is generally concave to the well containing the reactant of higher molecular weight because the large molecules tend to diffuse more slowly through an agar gel.

Immunoelectrophoresis combines the efficiency of electrophoretic separation with the selectivity and sensitivity of detection of the immunological reaction between antigen and antibody molecules. To improve the resolution of separation of the components of a complex mixture, two separate electrophoretic runs can be made in different directions. The antigens or antibodies are separated by electrophoresis and react with their counterparts to form a visible precipitate. **Counter immunoelectrophoresis** is used in a number of diagnostic procedures to detect various microorganisms in body fluids such as cerebrospinal fluid, urine, and blood (Figure 19.18). This procedure depends on a precipitin reaction to identify homology between the antigen and the antibody, relying

on immunodiffusion with electrophoresis driving the antigen and antibody toward each other, with precipitation occurring when matching antigens and antibodies meet in the right concentrations so that antigen–antibody reactions result in visible precipitation.

Radioimmunoassay

Radioimmunoassay techniques also permit the sensitive quantitation of antigen molecules (Figure 19.19). The radioimmunoassay method was initially developed by using ^{125}I radioactive iodine for the detection of very low levels of hormones, such as insulin. The steps of radioimmunoassay are (1) reaction of a specific antibody with a sample containing an unknown quantity of antigen; (2) addition of radiolabeled antigen, which combines with any unreacted antibody; (3) separation of the radiolabeled antigen–antibody complex from uncombined radiolabeled antigen; (4) determination of the amount of radiolabeled antigen–antibody complex formed; and (5) calculation of the concentration of the unknown antigen. Quantitation of the concentration of the antigenic compound is based on the fact that the unknown antigen competes with the radiolabeled antigen for antibody binding sites, and thus the quantity of radiolabeled antigen–antibody complex is diminished in direct proportion to the amount

FIGURE 19.17
In the double diffusion method, the antibody and antigen diffuse toward each other. (A) The line of confluence obtained with two antigens that cannot be distinguished by the antiserum used, that is, they appear identical. (B) Spur formation by partially related antigens having a common determinant, a, but individual determinants, b and c, reacting with a mixture of antibodies directed against a and b. The antigen with determinants a and c can only precipitate antibodies directed to a. The remaining antibodies (Ab) cross the precipitin line to react with the antigen from the adjacent well that has determinant b, giving rise to a "spur" over the precipitin line. (C) Crossing over of lines formed with unrelated antigens.

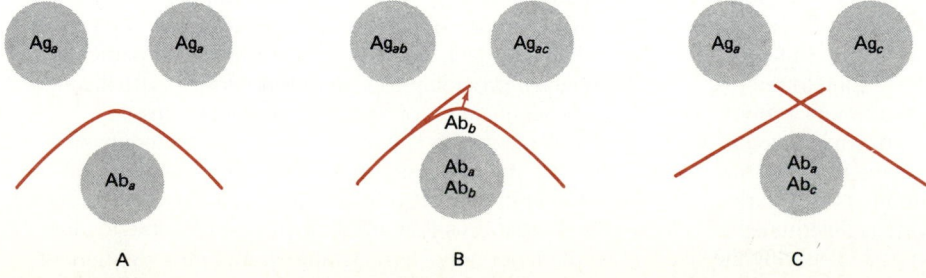

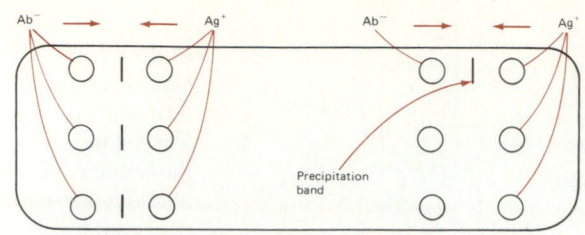

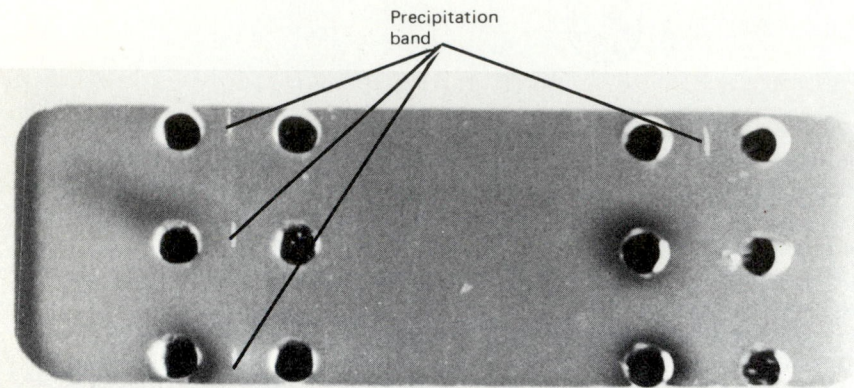

Precipitation band

FIGURE 19.18
Counter immunoelectrophoresis testing is used for the rapid diagnosis of various infections. For example, it is used to screen spinal fluid for bacteria, as shown in this photograph of actual test results. At the left are positive controls, showing precipitation bands betweeen wells filled with antibodies and antigens for *Haemophilus influenzae* (top), *Neisseria meningitidis* (middle), and *Streptococcus pneumoniae* (bottom). At the right are reactions with the patient's serum, showing a positive test for the presence of *H. influenzae* and negative results for the other two antigens.

of unlabeled antigen–antibody complex formed. The ability to detect and measure low levels of radioactivity, coupled with the specificity of the antigen–antibody reaction, makes the radioimmunoassay a very sensitive procedure for the quantitation of a variety of antigenic compounds.

Enzyme-linked Immunosorbent Assay

Because of the problems inherent in working with radioactive chemicals, several other related procedures employing nonradioactive labeled antigens have been developed. These include the **enzyme-linked immunosorbent assay (ELISA)**, in which an antibody typically is bound to an enzyme, such as peroxidase or phosphatase, which can produce colored reaction products from an appropriate substrate (Figure 19.20). The ELISA is similar to the radioimmunoassay in that it is based on competition between a tagged antigen molecule of known quantity and an unlabeled antigen of unknown quantity. The ELISA assays can be used for the detection of antibodies, such as those produced in response to infections with *Salmonella*, *Yersinia*, *Brucella*, *Rickettsia*, *Treponema*, *Legionella*, *Mycobacterium*, and *Streptococcus* species. The ELISA assay can also be used to detect the toxins (antigens) produced by *Vibrio cholerae*, *Escherichia coli*, and *Staphylococcus aureus*.

In this assay, samples thought to contain antibodies against a particular antigen are incubated with an excess of the antigen that has been attached to the surface of a shallow plastic reaction vessel. After rinsing to remove any unbound antibody, the amount of bound antibody is quantitated by adding a solution of enzyme-conjugated antibody that binds to constant domains of antibodies in the sample. Excess conjugated antibody is rinsed away, and the activity of the bound enzyme is determined by adding the substrate of the reaction and measuring the

formation of products. Because the products of the reactions used in ELISA procedures are colored, the amount of product formed can be readily determined by the intensity of the color that has developed, using a spectrophotometer. The activity of the bound enzyme is proportional to the amount of antigen-binding antibody in the sample; therefore, the original concentration of such antibodies can be estimated from a series of control assays employing known concentrations of specific antibody. The procedure can be switched by attaching antibody to the plastic reaction vessels in order to measure the concentration of antigen.

Immunofluorescence

Similar assay procedures may also employ dyes to label the antigen molecules. Among the dyes that can be used to tag antibodies are fluorescent dyes, such as fluorescein and rhodamine, which can be coupled readily to antibody molecules to form **conjugated fluorescent dyes**. Such dyes can be used specifically to stain the antigenic reactive sites of microorganisms (Figure 19.21). The stained preparations are visualized by using a fluorescence microscope with an excitation wavelength appropriate for the dye moiety of the conjugated fluorescent antibody stain. The conjugated fluorescent antibody dye molecules retain the specificity of the antigen-binding site of the antibody molecule. Thus, these dyes can be used to react specifically with microorganisms possessing particular antigens, providing a powerful tool for the visualization and identification of specific microorganisms.

This technique is particularly useful in identifying specific strains of microorganisms within a mixed microbial community; it is also invaluable in identifying pathogenic microorganisms that are difficult or impossible to culture in the laboratory. **Direct fluorescent antibody staining**

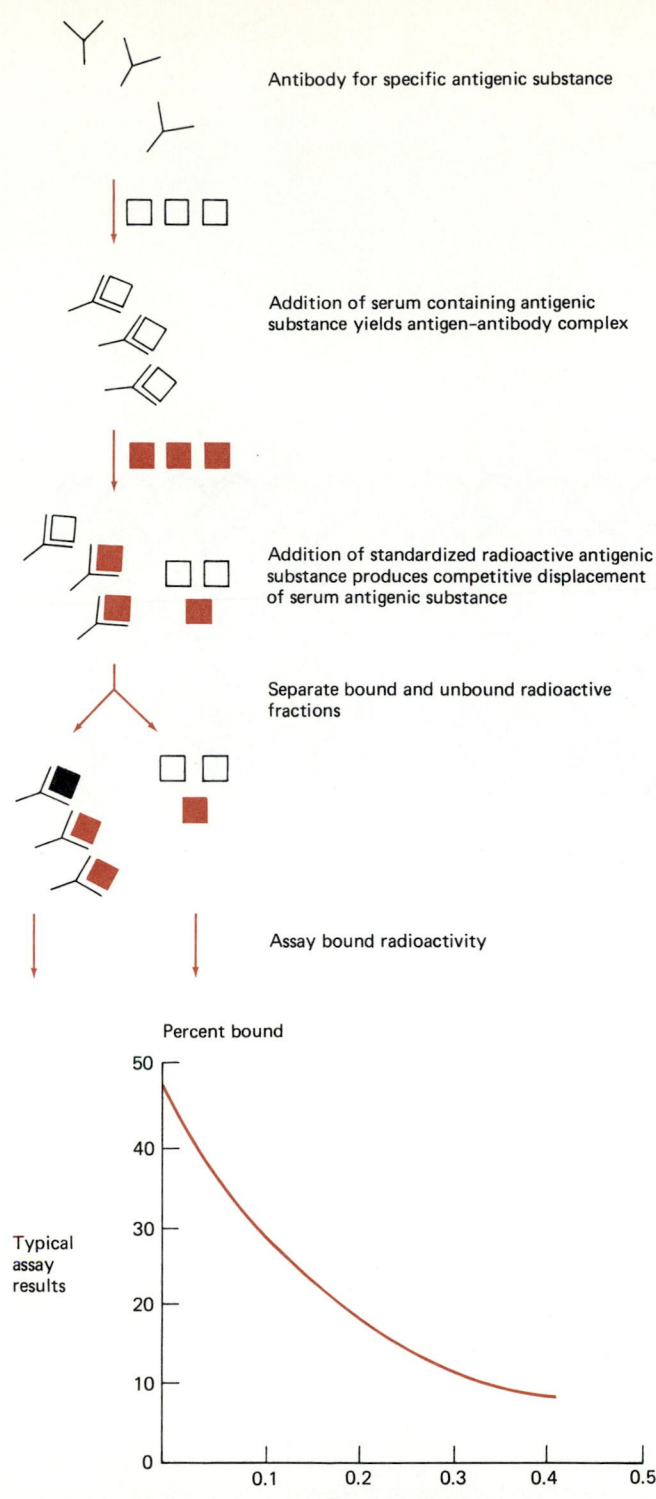

Antibody for specific antigenic substance

Addition of serum containing antigenic substance yields antigen–antibody complex

Addition of standardized radioactive antigenic substance produces competitive displacement of serum antigenic substance

Separate bound and unbound radioactive fractions

Assay bound radioactivity

Percent bound

Typical assay results

Nanograms in serum

FIGURE 19.19

Radioimmunoassay has become an important technique for quantitating polypeptides, steroids, thyroid hormones, and vitamin B_{12}. The assay is based upon competition between a known standardized amount of an antigenic substance that is radiolabeled and an unknown amount of that same antigenic substance in a body fluid that is nonradioactive for an antibody. The relative amounts of radiolabeled and nonlabeled antigen that bind with the antibody indicate the levels of the antigens in the body fluid sample. High levels of radioactivity in the antibody-bound fraction indicate a low level of the antigenic substance in the body-fluid sample. Low levels of radioactivity in the antibody-bound fraction indicate a high level of the antigenic substance in the body-fluid sample.

(**FAB**) uses a defined antibody conjugated with a fluorescent dye to stain unknown microorganisms. In this procedure, microorganisms are stained only if the antibody reacts with the surface antigens of the microorganisms on the slide. Viewing with a fluorescence microscope permits the specific detection of microorganisms that react with the fluorescent antibody stain. In this procedure, specific pathogenic microorganisms can be identified even in the presence of other microorganisms. The fluorescence antibody staining method is useful both for identifying clinical isolates and for detecting pathogens, such as *Treponema pallidum* or *Legionella pneumophila*, within exudates from infected tissues. **Indirect immunofluorescence tests** are also employed in identifying bacteria, such as *T. pallidum*. In the fluorescent test for treponemal antibodies (FTA), dead *T. pallidum* cells are fixed to a slide, and a patient's serum is added. The slide is washed and a fluorescent anti-immunoglobulin is added. If the antibodies in the patient's serum react with the *T. pallidum* on the slide, the bacteria will be stained (positive test), whereas if no reaction occurs, the fluorescent antibody is washed off the slide and no fluorescing bacteria are visible (negative test).

Agglutination Reactions

When the antigen is located on a cell surface, the reaction of antigen and antibody can produce clumping or **agglutination** of the cells (Figure 19.22). The agglutination reaction is probably best known for its use in blood typing. Human red blood cells contain antigens on their surface. The membranes of human red blood cells may possess either type A or type B polysaccharide antigens, both type A and type B antigens, or neither of these two antigens. Blood types are determined by mixing known **antisera** (anti-A and anti-B antibodies) with a blood sample (Figure 19.23). An agglutination reaction indicates the presence of the corresponding antigen. Type A blood has type A but not type B surface antigens on the red blood cells. Type B blood has type B but not type A antigens. Type AB blood has both type A and type B red blood cell surface antigens. People with type O blood have neither A nor B antigens on their red blood cell surfaces.

Various other antigens, including M, N, and Rh, can also be identified on the red blood cell surface. Some of these antigens can create incompatibilities between blood types; thus, before transfusions, the blood of the donor and the recipient is screened for about 8–10 of the antigens most commonly involved in transfusion reactions. Transfusion incompatibility must be avoided because it

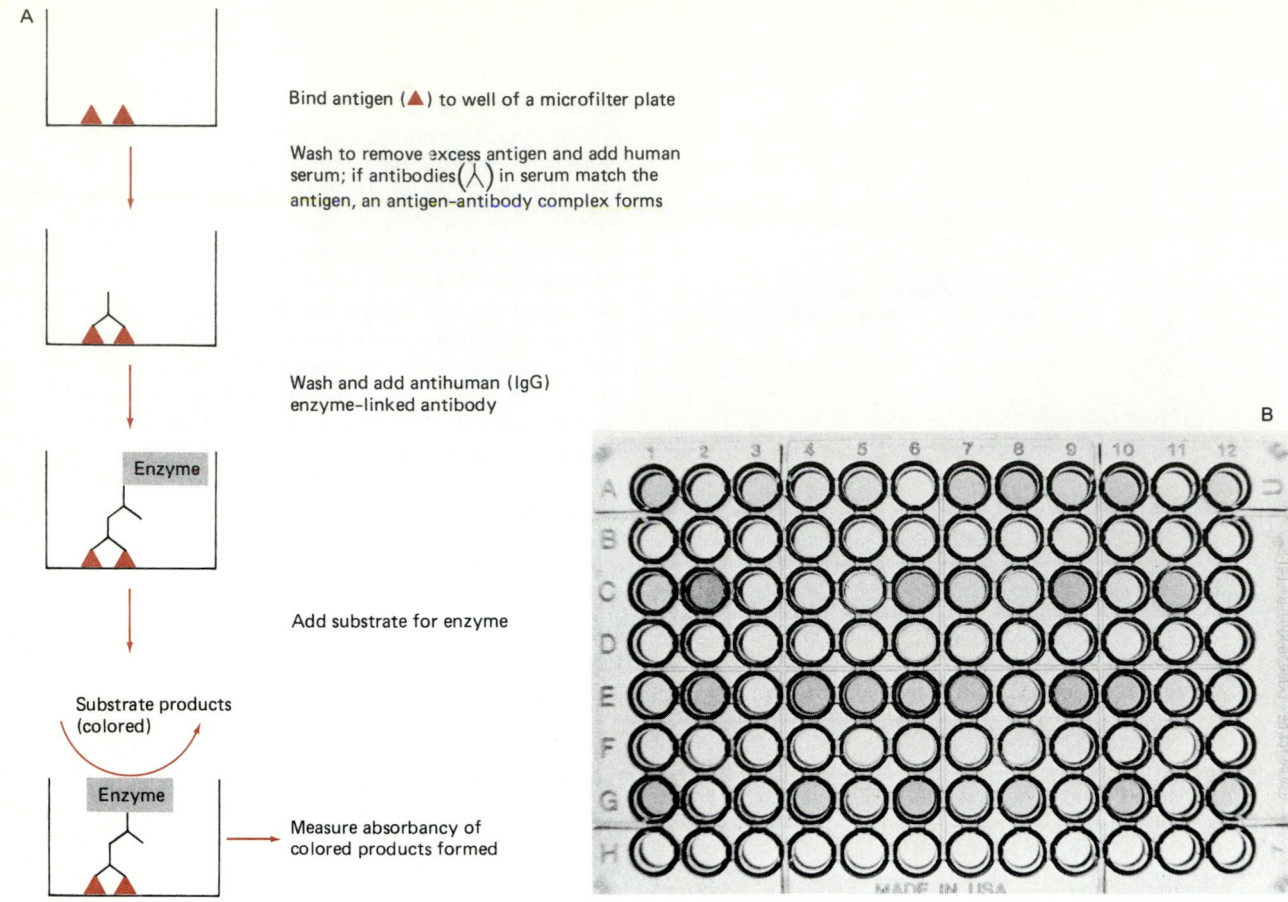

A

Bind antigen (▲) to well of a microfilter plate

Wash to remove excess antigen and add human serum; if antibodies (⋀) in serum match the antigen, an antigen-antibody complex forms

Wash and add antihuman (IgG) enzyme-linked antibody

Enzyme

Add substrate for enzyme

Substrate products (colored)

Enzyme

Measure absorbancy of colored products formed

B

FIGURE 19.20

ELISA procedures are used for the detection of various pathogens. (A) Diagram of indirect antibody detection by ELISA. (B) Results of the ELISA procedure for the detection of Rubella virus. Rubella antigen is attached to the surfaces of the wells of a plastic microtiter plate. The patient's serum is added to the well. Specific antibody, if present in the patient's serum, binds to the attached antigen. Phosphatase-conjugated antihuman IgG is added next. It binds to the antibody–antigen complex. Then p-nitrophenol phosphate, a substrate for phosphatase, is added. The p-nitrophenol phosphate is hydrolyzed by the bound enzyme conjugate, producing the colored product p-nitrophenol. After a specified time the enzyme substrate reaction is stopped with sodium hydroxide. The absorbance of the hydrolyzed substrate is measured at 405 nm with a spectro-photometer and is directly proportional to the amount of antibody in the patient's serum. In this photograph, the darker the well, the greater the amount of antibody in the patient's serum. (Courtesy Karen Cost, Norton's Hospital, Louisville, Ky.) Alternatively, horseradish peroxidase could have been used in conjunction with its substrate, hydrogen peroxide. Peroxidase activity is measured in these assays by adding an electron donor, such as benzidine, which polymerizes to form an insoluble brown substance.

can cause death. The detailed serotyping of red blood cell antigens can also be used for other purposes, such as determining paternity. Because the antigens of the red blood cells are genetically determined, the genetic information is heritable, and the pattern of minor antigens on the red blood cell surface, while not unique, is sufficiently characteristic to reflect genetic linkage.

A number of diseases can be rapidly diagnosed by employing agglutination tests. For example, infectious mononucleosis, caused by the Epstein-Barr virus, is routinely diagnosed through the use of serological agglu-tination tests. The blood serum of individuals infected with this virus contains antibodies that will agglutinate

sheep red blood cells. Such antibodies are called **hetero-phile antibodies** because they cross-react with antigens other than the ones that elicited their formation. Hetero-phile antibodies may be present in normal sera, but in individuals with infectious mononucleosis, the concentra-tion of such antibodies is greatly elevated. By testing the ability of serial dilutions of blood serum to cause agglutination of sheep red blood cells (Figure 19.24), the titer[5] of heterophile antibodies can be determined. In performing these tests, the patient's serum is heated to

[5]The titer is a measure of concentration; it is the reciprocal of the dilution.

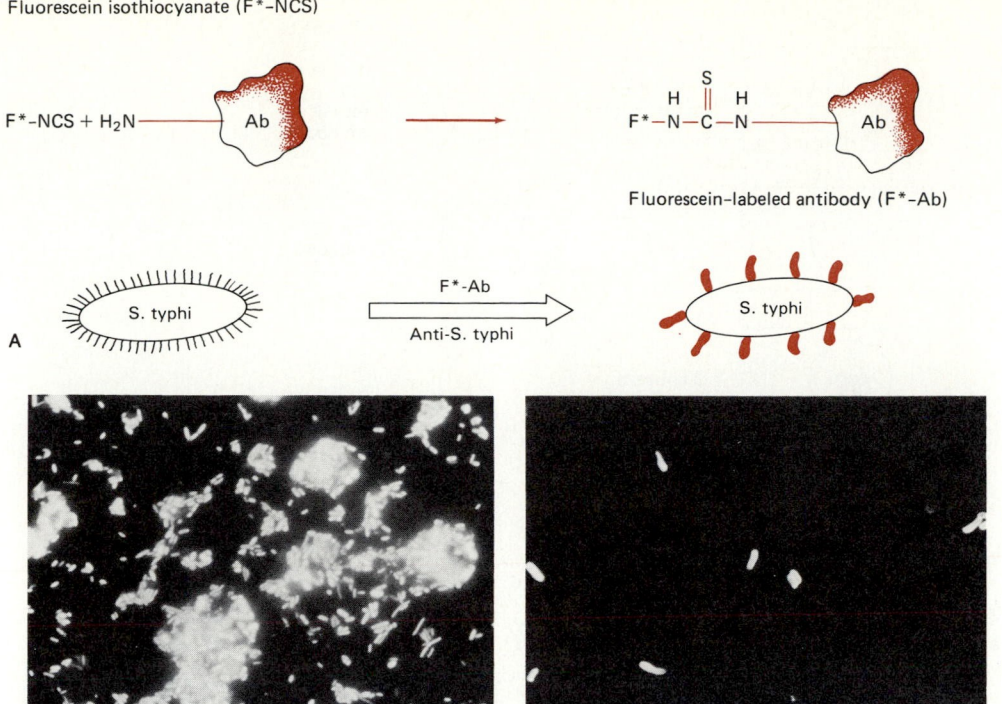

Fluorescein isothiocyanate (F*-NCS)

Fluorescein-labeled antibody (F*-Ab)

FIGURE 19.21

Fluorescent antibody staining allows the detection of specific microorganisms. As such, this method is useful in the clinical identification of pathogenic microorganisms. (A) Diagram showing how conjugated fluorescent antibodies are prepared and used in the staining of bacteria. (B) Photomicrograph of a sample of human fecal material stained with acridine orange, showing the total microbial content. (C) Photomicrograph of the same human fecal sample, showing conjugated antibody fluorescent staining of *Salmonella typhimurium*. Comparing (B) and (C), the specificity of immunofluorescence is obvious. (Courtesy B. Bohlool; reprinted by permission of Plenum Press from B. B. Bohlool and E. L. Schmidt, 1980, The immunofluorescence approach in microbial ecology. *Advances in Microbial Ecology* 4:203–241)

56°C for 30 minutes to destroy complement that could result in lysis rather than agglutination. A titer of 1:224 or greater, that is, the detection of antibodies in serum that has been diluted by a factor of at least 1:224, is considered presumptive evidence of infectious mononucleosis. Additional differential tests for infectious mononucleosis can be carried out to confirm the diagnosis. In these tests, the patient's serum is mixed with guinea pig kidney tissue in one tube and beef erythrocyte antigen in another. The guinea pig tissue does not absorb heterophile antibodies produced in response to the Epstein-Barr virus and does not reduce the ability of the serum to agglutinate sheep cells, whereas erythrocyte antigens do absorb these heterophile antibodies, lowering the ability of the serum to agglutinate red blood cells.

The detection of heterophile antibodies, also called **cross-reactive antibodies**, is useful in diagnosing several other diseases as well. For example, the **Weil-Felix test** is used to diagnose some diseases caused by *Rickettsia* species. In these tests, antibodies produced in response to a particular rickettsial infection agglutinate bacterial

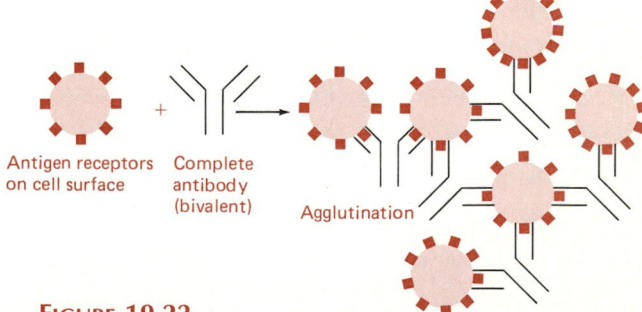

Antigen receptors on cell surface   Complete antibody (bivalent)   Agglutination

FIGURE 19.22

Agglutination occurs when antigens are located on the surface of a cell; the reaction of antibody with the surface antigens causes the cells to agglutinate.

strains of *Proteus*, designated OX-19, OX-2, and OX-K. Febrile agglutinins, which are antibodies produced in response to various fever-producing bacteria, can be similarly detected by agglutination tests. The **Widal test** uses antigens from *Salmonella* species for these tests,

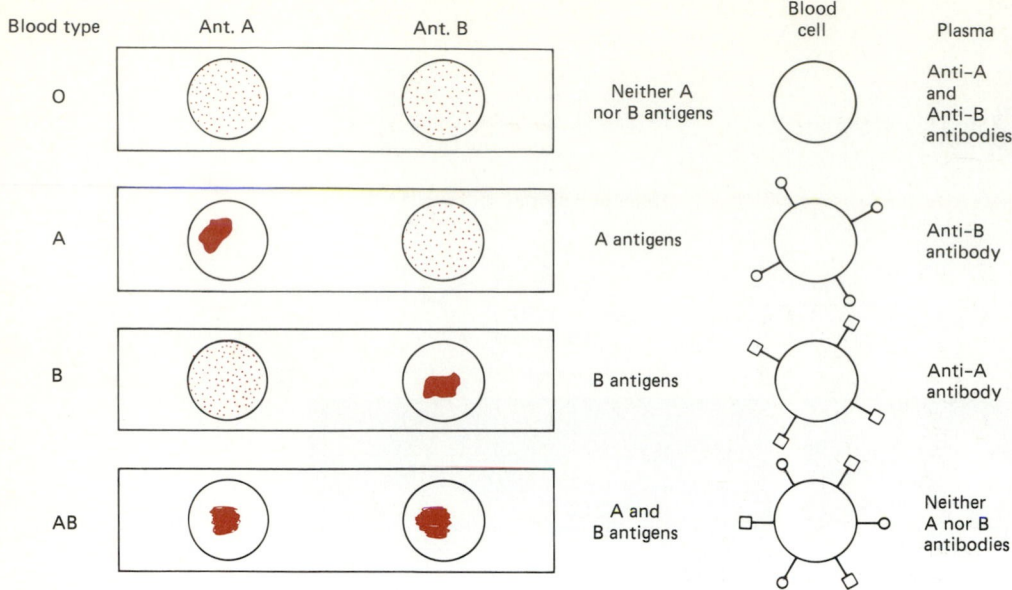

FIGURE 19.23

Illustration of blood typing to determine the major blood groups (A, B, AB, and O) based on agglutination reactions.

permitting the diagnosis of typhoid and paratyphoid fevers.

Cases of atypical pneumonia caused by *Mycoplasma* can also be detected by several different types of agglutination tests. In one test, the patient's own red blood cells are used to test for agglutination. Antibodies produced in response to pneumonia caused by *Mycoplasma pneumoniae* cause agglutination of blood cells (**hemagglutination**) at 4°C but not at normal body temperature. The detection of cold agglutinins, antibodies that cause agglutination at refrigerator temperatures but not at body temperature, can indicate several diseases other than atypical pneumonia, but the symptoms normally permit determination of the etiology of the disease.

The use of *in vitro* antigen–antibody reactions to identify a microorganism, known as **serotyping**, provides essential information for the identification of many pathogenic bacterial species. *Salmonella*—including *S. typhi*, the causative organism of typhoid fever—and *Neisseria gonorrhoeae*, the causative organism of gonorrhea, as

well as many other pathogenic bacteria—can be identified rapidly using agglutination reactions. Over 2,000 serotypes in the genus *Salmonella* can be defined by the O (somatic cell) and H antigens (flagella), with each serotype defined by a grouping of O and H antigens.

Streptococcus species can also be serotyped by using **coagglutination** procedures. This approach is used in the Streptex and Phadabac systems. In coagglutination, dead *Staphylococcus* cells are coated with IgG and mixed with the streptococci and antistreptococcal antibodies. The reaction forms a lattice matrix of agglutinated cells. The larger clumps of cells produced by coagglutination provide greater test sensitivity than simple agglutination tests.

Passive Agglutination

Because the agglutination test requires that the antigens be on particle surfaces, soluble antigens are not suitable for direct agglutination tests. In cases where the antigen to be detected is soluble and not associated with a cell or other particle, **passive agglutination** tests can be em-

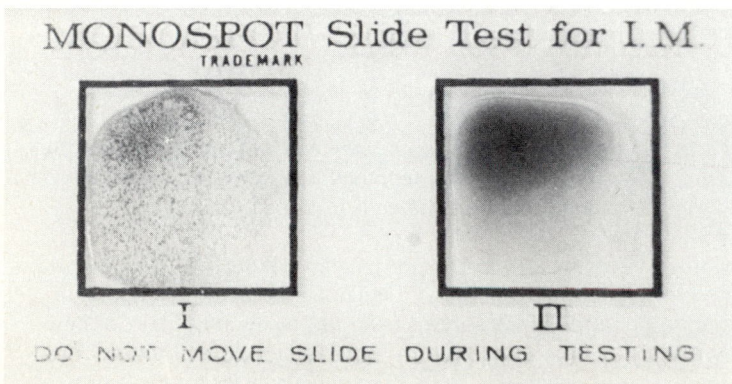

FIGURE 19.24

An agglutination slide test for the diagnosis of infectious mononucleosis provides a simple, rapid, and accurate method for detecting the presence of antibodies formed in response to infection with the Epstein-Barr virus. In this photograph, I is a positive reaction and II is a negative reaction.

ployed. In such tests, the antigen is attached to a particle surface before an agglutination test is run. The particle may be a cell or, more commonly, a latex bead. Very small quantities of antigen-coated beads are required for these tests. Some rapid screening tests for detecting pregnancy use antigen-coated latex beads. These tests detect the antibodies produced in response to the condition of pregnancy.

Passive agglutination, using latex beads, is rapidly replacing various older procedures and is now employed for detecting *Haemophilus influenzae*, *Streptococcus pneumoniae*, *Neisseria meningitidis*, *Staphylococcus aureus*, and rubella virus. Latex beads may also be coated with soluble antibodies and used in agglutination tests. Pneumonia caused by *S. pneumoniae* can be detected in this manner. Blood serum from patients with pneumococcal pneumonia possesses a protein called *C reactive protein (CRP)*. The CRP reacts with the C polysaccharide of the capsular material of *S. pneumoniae*. Anti-CRP attached to latex beads will react with CRP, causing readily observed agglutination. A positive agglutination test of the patient's serum with anti-CRP coated latex beads is presumptive evidence of pneumococcal pneumonia. The inflammatory response is monitored by measuring the CRP and is considered more reliable than other methods.

Passive agglutination can also be used to detect *Treponema pallidum* in the diagnosis of syphilis. The antigens from *T. pallidum* can be conjugated with red blood cells in passive agglutination tests to detect the presence of antibodies against *T. pallidum* in a patient's serum.

Another test used for the diagnosis of syphilis is designed to detect the presence of an IgM antibody produced in individuals infected with *T. pallidum*. This IgM antibody, known as *reagin*, reacts with phospholipids and can be detected by using cardiolipin, an alcoholic extract of beef hearts. The reaction of reagin with cardiolipin results in agglutination or flocculation, which can be visualized by using low-power microscopy. The Venereal Disease Research Laboratory (VDRL) and rapid plasma reagin (RPR) tests, which are conveniently performed with the use of a prepared kit, are widely employed for the detection of syphilitic reagin.

Hemagglutination Inhibition

It is also possible to assess the inhibition of agglutination in order to determine the quantities of certain antigens. Some viruses cause hemagglutination of red blood cells, a fact used in **hemagglutination inhibition (HI)** tests to determine whether a patient has been exposed to a specific virus. In the HI test, the patient's serum is incubated with viral antigens and the mixture is then added to red blood cells. If the patient's serum contains antibody for the specific virus, the viral antigens are neutralized or inactivated, and agglutination of the red blood cells is reduced or absent. Agglutination occurs if the patient does not have antibodies to the antigens of the particular virus. The degree of hemagglutination can thus be employed to diagnose viral diseases. Dilutions of the antigen-treated sera are used to determine the endpoint of HI (Figure 19.25), a useful procedure for diagnosing diseases such as rubella and influenza.

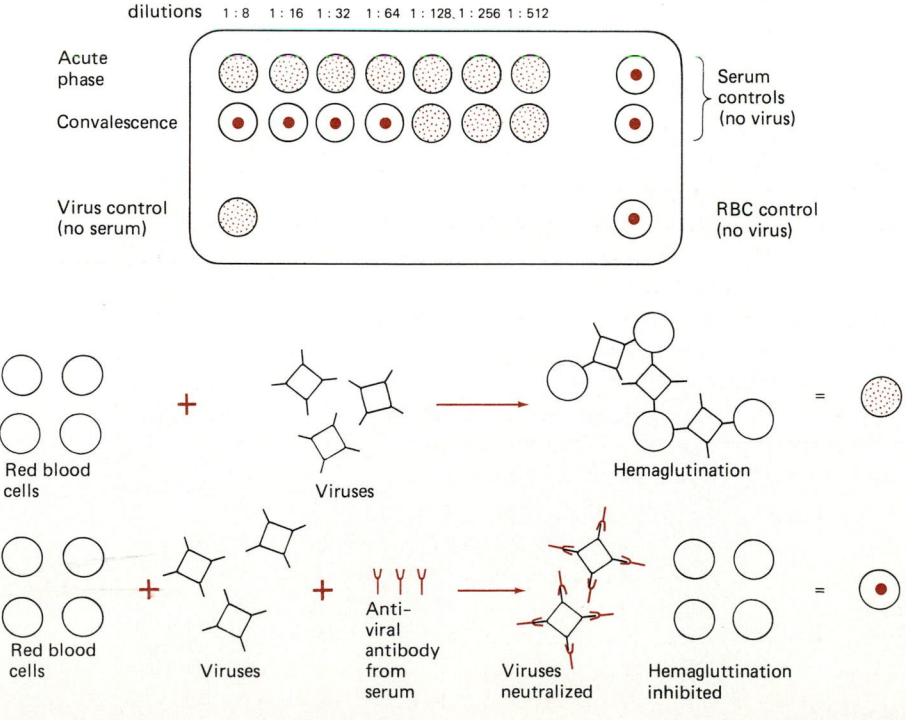

FIGURE 19.25

This is a diagram of a hemagglutination test for a virus in a microtiter plate. During the acute phase, hemagglutination occurs at all serum dilutions shown. The titer of virus hemagglutination during convalescence here is 128 because this is the lowest dilution (1:128) at which hemagglutination occurs; at lesser dilutions there is hemagglutination inhibition. This quantitative test can detect acute and convalescent cases of viral diseases such as hepatitis.

Complement

In addition to their role in the alternative complement pathway discussed in Chapter 18, complement molecules can react with antigen–antibody complexes to initiate the **classical complement pathway** (Figure 19.26). Complement action is a complex biochemical process in which a small number of events at the molecular level result in massive reactions at the cellular level of an infecting microbe and also at the physiological level of the human host. This complex series of complement reactions eventually leads to the lysis of microbial cells or the enhanced ability of phagocytic blood cells to eliminate such cells. Complement represents a family of glycoprotein molecules designated C_1, C_2, C_3, and so forth, with the numbers assigned based on the order of their discovery. In the alternative complement pathway, the sequence of complement reactions leading to cell lysis begins with C_3 (see Figure 18.24). Additional complement molecules are involved in the initiation of the classical complement pathway.

The action of complement via either the alternative or the classical pathway results from a number of glycoprotein molecules that are present in blood in an inactive form. In the classical pathway, the process is initiated by the formation of a complex between an antigen and an antibody. The initial involvement of complement is in the reaction of a complement molecule with the constant region of an antibody molecule. This reaction occurs only after the variable region of the antibody has reacted with an antigen; that is, the classical complement pathway is not initiated unless an antigen–antibody complex has already been formed. Once initiated, however, the complement pathway proceeds in an autocatalytic and cascading fashion, with one complement molecule after another adding to the reaction complex. As a result of the interaction of complement with the antigen–antibody complex, reactions can occur that otherwise

FIGURE 19.26

The sequence of classical complement activation by membrane-bound antibody, showing the formation of C_{3a} and C_{5a} fragments chemotactic for polymorphs and with anaphylatoxin activity causing histamine release. Immune adherence through C_3 to macrophages, platelets, or red cells facilitates phagocytosis. Fixation of C_8 and C_9 generates cytolytic activity. Fragments released during the activation of C_4 and C_2 are thought to have chemotactic and kinin-like activity, respectively.

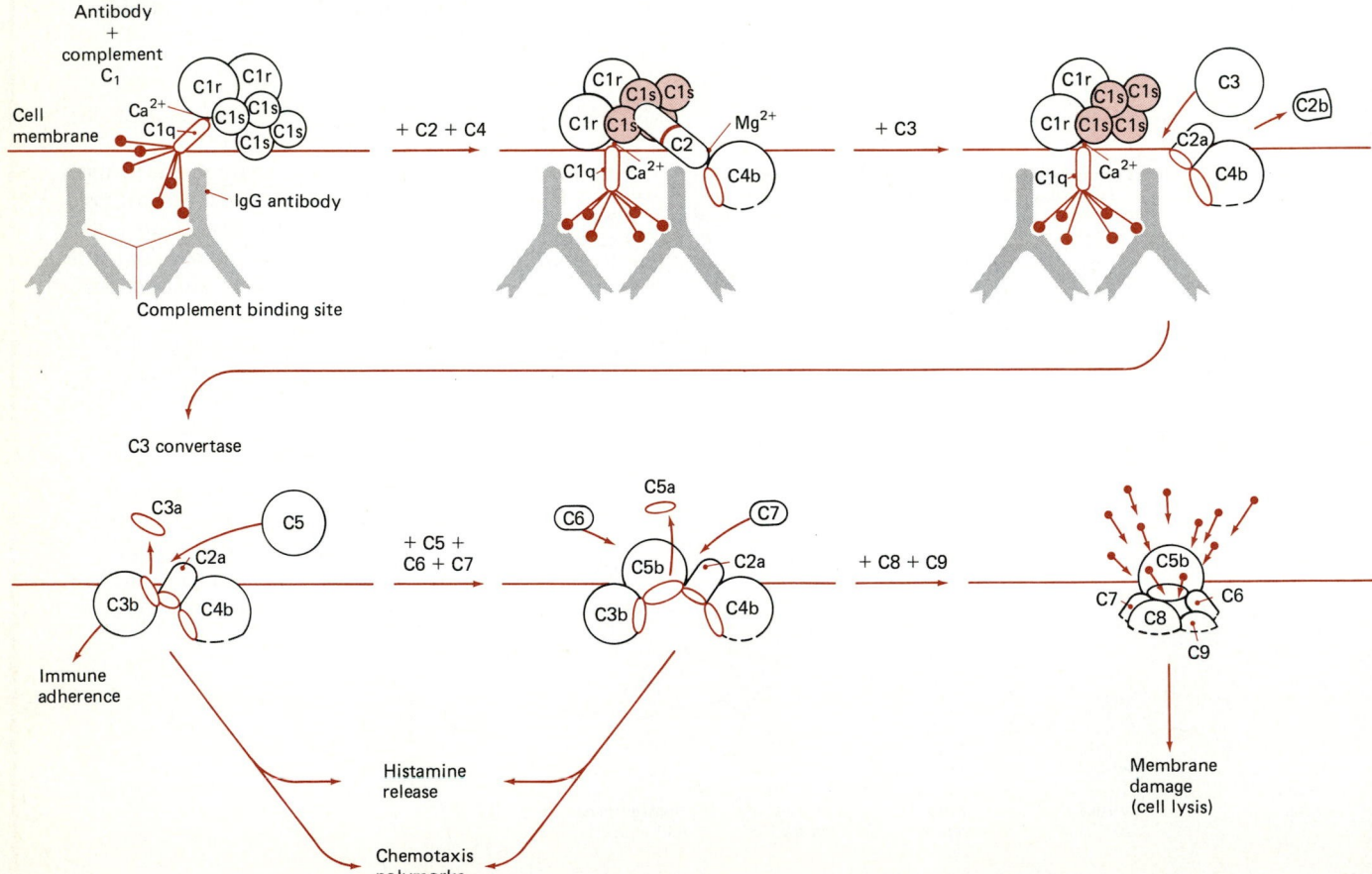

ANALYTICAL PROCESS

Complement Fixation

Complement molecules are fixed as a consequence of their reaction in both the classical and alternative complement pathways, and it is possible to measure the degree of **complement fixation** for diagnostic procedures (Figure 19.27). In a typical complement fixation test, an individual's serum, containing an unknown amount of antibody and free complement, is mixed with a known antigen, such as a specific virus. The reactant mixture is then added to a test indicator system, such as sensitized sheep red blood cells, which have been agglutinated by reacting with anti-sheep RBC antibody. If, after reaction with antigenic compounds, there is free complement available

from the individual's serum, the sheep red blood cells undergo hemolysis. Thus, in this system, free human complement augments the antigen–antibody reaction that involves sheep red blood cells, resulting in lysis of the sheep red blood cells. This reaction can be easily visualized and indicates that the individual is probably not suffering from a disease caused by that virus. Had a virus been eliciting an immune response, the antibody and complement molecules available in the individual's serum would have reacted with the viral antigens, resulting in the fixation of the complement molecules. In the absence of free complement, the sheep red blood cells in the

FIGURE 19.27

Illustration of the complement fixation test procedure. (A) Typical positive serum, containing antibody; (B) typical negative serum, antibody absent.

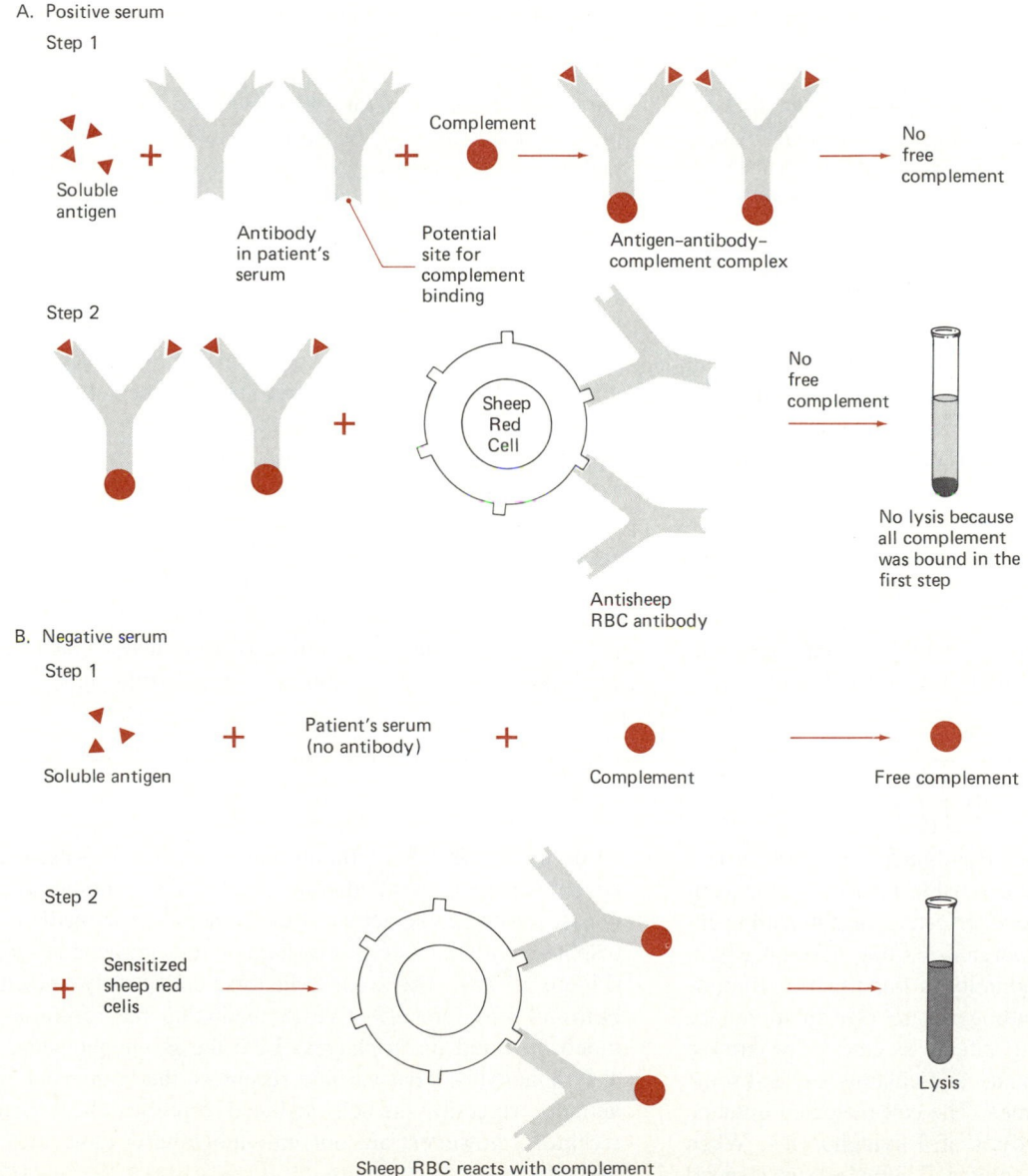

indicator system would have remained agglutinated, and no lysis would have occurred. A positive reaction for complement fixation (the failure to observe lysis of the sensitized sheep red blood cells) is indicative of an infection in the individual being examined.

Measurement of complement fixation is another serological approach to diagnosing the etiological agents of disease. Although it has been replaced with more reliable methods for diagnosing syphilis, the **Wassermann test** still represents a classical example of a diagnostic **complement fixation test**. The complement fixation test employs an indicator system, normally consisting of sheep red blood cells, and homologous antibodies for the red blood cell antigens. The patient's serum is heated to 56°C for 30 minutes to inactivate any available free complement. The serum is then mixed with a known antigen and free complement, and the mixture is added to the indicator system. Hemolysis of the sheep red blood cells indicates that there is free complement, that an antigen–antibody–complement complex did not form

when the patient's serum was added to the known antigen and complement. This constitutes a negative test for the disease caused by the organism from which the known antigen was derived. If, on the other hand, lysis of the sheep red blood cells does not occur, it indicates that complement was fixed in the reaction; thus, there is presumptive evidence that the patient has the given infection.

In the Wassermann test, it was found that the IgM antibody, reagin, was responsible for fixing complement. Reagin is released from damaged liver cells during syphilis infection, and the test was therefore diagnosing a symptom of the disease rather than specific antibodies that are produced for *Treponema pallidum*. Although no longer used for the diagnosis of syphilis, complement fixation tests are used clinically for the diagnosis of various other diseases, including a number of fungal diseases such as *Histoplasmosis* and *Coccidioidomycosis*. Complement fixation tests can also be used for the detection of viral and rickettsial antigens and antibodies.

---

would not happen; thus, the initiation of the complement sequence by the antigen–antibody complex augments the specific immune response.

Specifically in the classical pathway, when IgM or IgG antibodies combine with an antigen on a cell surface, complement C_1 can bind to the Fc region of the antibody molecule, forming an antigen–antibody–complement complex. This complex initiates the cascade of complement molecules, with several consequences. At each stage of the complement cascade, different complement molecules are activated, so that they consequently activate the next complement molecule in the pathway. The initiation of the classical complement pathway involves C_1, C_2, and C_4. Three proteins (C_1q, C_{1r}, C_{1s}), comprising the C_1 complex, are triggered by the antigen–antibody complex to become active proteases. The proteolytic action of the activated C_1 complex, in turn, activates C_2 and C_4.

The complement cascade has several consequences, including the chemotactic attraction of neutrophils and the ini-

tiation of the inflammatory response. The adherence of complement molecules to the surfaces of microorganisms also facilitates the adherence of macrophages and neutrophils to the microbial cells, enhancing phagocytosis. Enhanced phagocytosis occurs because both neutrophils and macrophages have receptors on their surfaces for complement C_{3b}. The binding of C_{3b} to a pathogen permits the establishment of a bridge between that pathogen and a neutrophil or macrophage so that the phagocytic blood cell remains in contact with the pathogen. Individuals lacking C_3 complement have inadequate phagocytic activities and are particularly susceptible to bacterial infections. Additionally, the complement sequence results in damage to the cytoplasmic membrane of a pathogen, resulting in the lysis and death of the cell on which the complement cascade occurs. The full complement system leads to membrane damage and can cause lysis in Gram-negative bacteria by permitting lysozyme to reach the cytoplasmic membrane. The ability to bring about cell lysis is important for the destruction of bacterial pathogens.

19.2 Cell-Mediated Immune Response

Whereas the antibody-mediated immune response system recognizes substances that are outside host cells, the **cell-mediated immune response** is effective in eliminating infections by pathogenic microorganisms that develop within host cells. Cell-mediated immunity is important in controlling infections where the pathogens are able to reproduce within human cells, including infections caused by viruses; some bacteria, such as rickettsias and chlamydias; and some protozoa, such as trypanosomes. The cell-mediated immune response depends on the actions of T lymphocytes. When stimulated by an appropriate antigen, T lymphocytes respond

by dividing and differentiating into **cytotoxic T cells**, which are sometimes also called *killer T cells*. Various other T cells release biologically active soluble factors that mediate the responses of other cells involved in the immune response (Figure 19.28). The soluble factors, collectively known as **lymphokines**, are effective in mediating the responses of monocytes and macrophages. Like the B lymphocytes, the T lymphocytes have surface receptors that can react with antigen, triggering the cell-mediated response. These surface receptors, however, are not immunoglobulin molecules, as they are in B lymphocytes.

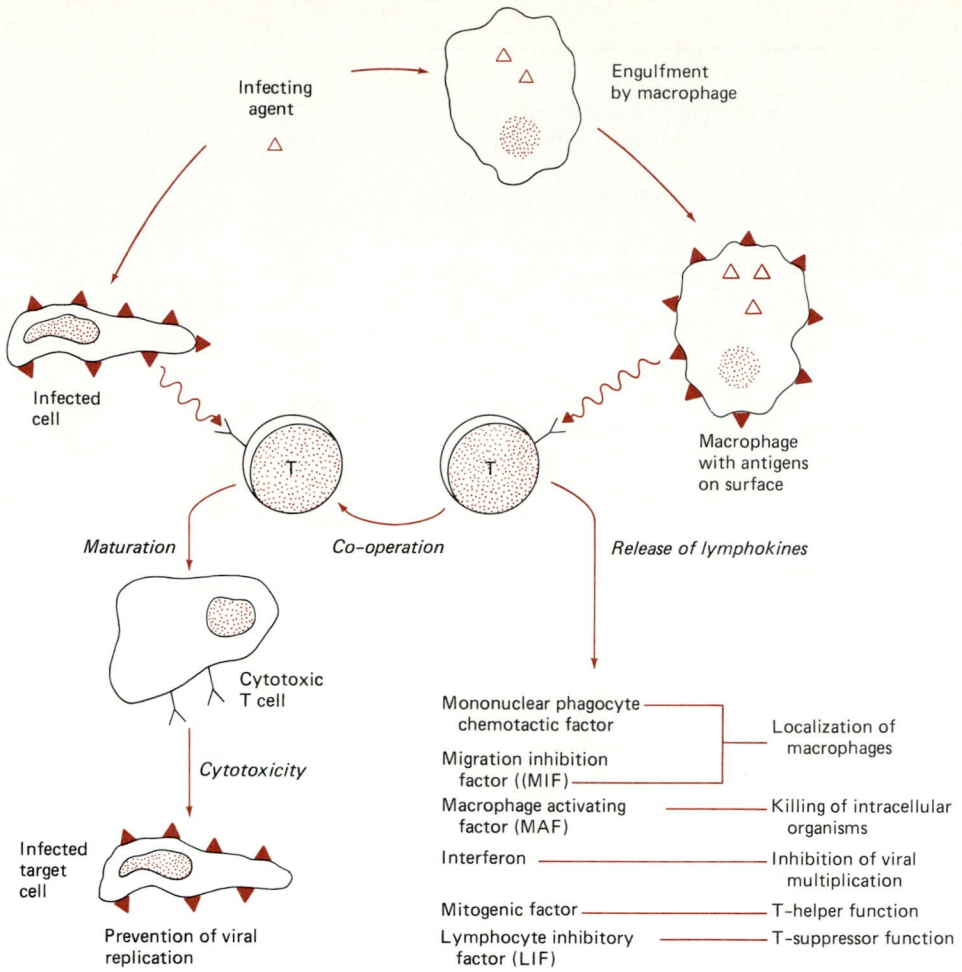

Infecting agent

Engulfment by macrophage

Macrophage with antigens on surface

Infected cell

T

T

Maturation

Co-operation

Release of lymphokines

Cytotoxic T cell

Cytotoxicity

Infected target cell

Prevention of viral replication

Mononuclear phagocyte chemotactic factor

Migration inhibition factor ((MIF)

Localization of macrophages

Macrophage activating factor (MAF)

Killing of intracellular organisms

Interferon

Inhibition of viral multiplication

Mitogenic factor

T-helper function

Lymphocyte inhibitory factor (LIF)

T-suppressor function

FIGURE 19.28
The cell-mediated immune response, the response of T cells to antigen, showing lymphokines and cytotoxic cell formation. This response operates through the generation of cytotoxic T cells and the release of lymphokines by the stimulation of two distinct T-cell subpopulations. Different lymphokines may be produced by different lymphocyte subsets. The intense proliferation induced by antigenic stimulation has not been shown but is an essential factor in the amplification of the response.

Lymphokines

As part of the cell-mediated immune response, **lymphokines**—the biochemical mediators of cellular immunity—are released by T lymphocytes after antigenic stimulation. These soluble factors exhibit different activities (Table 19.2). The lymphokines produced by T cells include (1) macrophage chemotactic factor, a lymphokine that causes the attraction and accumulation of macrophages to the site of lymphokine release; (2) migration inhibition factor, which inhibits macrophage cells from migrating farther once they have reached the site of lymphokine attraction; (3) macrophage activating factor (MAF), which results in an alteration of macrophage cells that increases their lysosomal activities and thus their ability to kill and ingest organisms; (4) skin reactive factor, which enhances capillary permeability and thus the movement of monocytes across the vascular spaces; and (5) immune interferon, which activates antiviral proteins, preventing further intracellular multiplication of viruses.[6] In addition to these lymphokines, T helper cells are involved in stimulating B cells as part of the antibody-mediated immune re-

sponse, and T suppressor cells are involved in shutting off the antibody-mediated response by suppressing the synthesis of antibody. It is thus apparent that lymphokines and T lymphocytes play a major role in regulating the activities of other cells involved in the immune response.

One of the specific lymphokines, **immune or gamma (γ)-interferon** is secreted by lymphocytes in response to a specific antigen to which they have been sensitized or stimulated to divide. In some of its physiological effects, this immune interferon is different from other interferon molecules and may kill tumor cells. Like other interferons, immune interferon molecules have antiviral activities, stimulating the synthesis of antiviral proteins, including 2,5 adenylate polymerase. This polymerase, when bound to double-stranded RNA, a viral replicative intermediate, activates an endonuclease that can cleave viral RNA. The primary function of immune interferon, though, appears to be different from that of other interferons. Immune interferon may regulate the proliferation of the lymphoid cells that are stimulated to divide in response to interactions with antigenic biochemicals. Immune interferon may also enhance phagocytosis by macrophages, as well as enhance the cytotoxicity of lymphocytes and the activities of killer T cells.

[6]Interferon and MAF probably represent different activities of the same protein.

TABLE 19.2   Effects of Lymphokines on Various Cells

Lymphokine	Action
Macrophage chemotactic factor	Attracts macrophages to site of lymphokine release, typically causing an accumulation of macrophages at a site of inflammation
Migration inhibitory factor	Inhibits migration of macrophages away from the site of inflammation
Macrophage activating factor (MAF)	Stimulates phagocytic activity of macrophages (γ-interferon) by enhancing their lysosomal activities
Immune interferon	Activates antiviral proteins, T cells, and macrophages
Interleukin-1	Stimulates activities of T cells, B cells, and macrophages
Interleukin-2	Stimulates antigen activated T-helper cells
B-cell growth factor	Stimulates production of B cells
Leukocyte inhibitory factor	Inhibits migration of phagocytic neutrophils away from the site of infection
Leukocyte chemotactic factor	Attracts phagocytic neutrophils to the site of infection
Skin reactive factor	Enhances capillary permeability and movement of monocytes across vascular spaces
Platelet-activating factor	Activates platelets to aggregate
Eosinophil chemotactic factor	Attracts phagocytic neutrophils to the site of parasitic infection

Cytotoxic T Cells

In addition to the antigen-induced release of lymphokines from T cells, intracellular infections, such as viral infections, can result in the generation of T cells that are specifically cytotoxic for intracellularly infected host cells (Figure 19.29). The populations of T cells that include cytotoxic T cells and T aggressor cells appear to direct their activity against antigens of the tissue cell surface called the *major histocompatibility antigens*. These histocompatibility antigens are genetically controlled by the **major histocompatibility complex (MHC)** region of the genome. The MHC antigens are highly polymorphic because each allelic gene codes for a different antigen, creating large numbers of distinct antigens. These MHC antigens play an important role in the recognition of self and nonself tissues and are especially important in the immune response to tumors and tissue transplants.

The ability of cytotoxic T cells to recognize and to react with antigens on cell surfaces depends on the gene products of the MHC. Small variations in the MHC antigens, which occur in tumor cells and cells infected with viruses, can be detected by T cells, leading to the appropriate cytotoxic response by these T cells. The ability to recognize antigens on the surfaces of cytoplasmic membranes permits T cells to monitor and regulate the proliferation of abnormal types of cells. This appears to be extremely important not only for preventing the unchecked development of cells harboring intracellular pathogens but also for inhibiting tumor development. As a result of the ability of cytotoxic T cells to recognize transformed human cells, many would-be tumor cells do not become malignant growths.

In a number of intracellular viral infections, antibodies are ineffective. In such cases, cell-mediated immunity augments the antigen-mediated immune response. Although antibody molecules can neutralize free viruses, antibodies are unable to penetrate and attack viruses multiplying within host cells. It is the cell-mediated immune response that has the capability of eliminating cells infected with viruses. The sensitized T aggressor cells are able to recognize the presence of

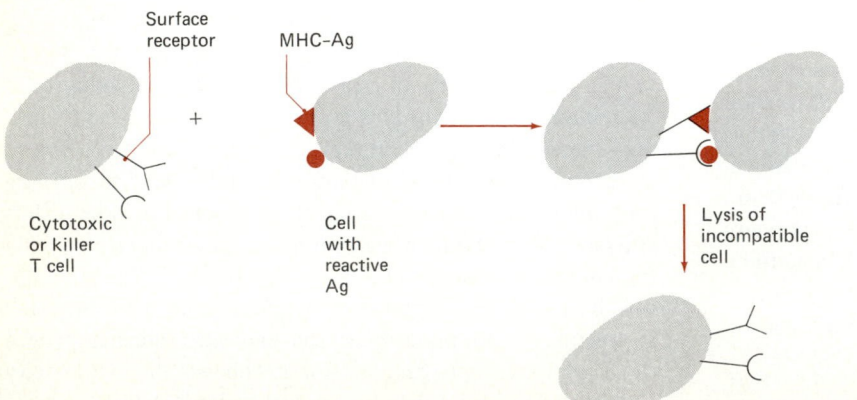

FIGURE 19.29
The interaction of cytotoxic T cells with cells having foreign MHC antigens. The MHC and surface antigens in this region of the genome play an important role in the recognition of incompatible cells.

histocompatibility antigens that have been modified by the presence of a viral infection in order to direct the cytotoxic attack of these lymphocytes against only those cells that are infected with a virus. Additionally, the T helper and T suppressor cells, which interact with the B cells to regulate antibody-mediated immunity, depend on their ability to recognize antigens coded for by the genes of the MHC region.

The MHC gene products, in association with T lymphocytes, thus seem to provide an integrated mechanism for regulation of the immune response. Indeed, as the complexity of the specific immune response is revealed, it is becoming increasingly clear that humans have evolved an interactive network of various cells and macromolecules formed by these cells to regulate the immune response in a functional manner. It is also clear that antigens and receptors on cell surfaces permit molecular-level communication between cells so that cells can act in an interactive and cooperative fashion.

19.3 Dysfunctional Immunity

Whereas the specific immune response plays a critical role in protecting humans and other higher vertebrates from numerous infectious diseases, failures of the immune response can compromise the ability of humans to resist infection, leaving an individual susceptible to a variety of diseases. Indeed, many infectious diseases are believed to result from failures of the immune response to protect the individual adequately against the invasion of and/or toxicity associated with pathogenic microorganisms. Failures of the immune response can also result in autoimmune diseases in which the inability to differentiate properly between self and nonself antigens results in reactions with self antigens and the killing of some of one's own cells. The development of tumor cells can also be viewed as a failure of the immune response, but in this case the failure to recognize and to respond properly to inappropriate cells within the body allows malignant cells to proliferate in an uncontrolled manner. The normal active immune response can also be undesirable in some cases, such as in transplants where the immunological recognition of and response to the foreign antigens of a donor results in tissue rejection. Additionally, allergies are the result of physiological changes caused by certain types of immune responses substances present in foods or dust that do not normally activate the immune system. Individuals suffering from allergies know too well that immune responses may occasionally be dysfunctional.

Immunodeficiencies

There are several types of deficiencies that can occur within the immune system, resulting in the failure of the system to recognize and respond properly to the antigens of pathogenic microorganisms. Individuals with **immunodeficiencies** are more prone to infection than those who are capable of a complete and active immune response. Immunodeficiencies can affect the variety of cells that are interactive in the immune response system (Table 19.3).

The most severe type of immunodeficiency, **severe combined deficiency**, results from a failure of stem cells to differentiate properly. Individuals with severe combined deficiency, having neither B nor T lymphocytes, are incapable of any immunological response. Any exposure of such individuals to microorganisms can result in the unchecked growth of the microorganisms within the body, resulting in certain death. Individuals suffering from severe combined deficiency can be kept alive in sterile environments, where they are protected from any exposure to microorganisms (Figure 19.30). Bone marrow grafts may be employed to establish normal immune functions, but the grafts must come from siblings with histocompatible bone marrow.

Less severe immunodeficiencies occur when only B-cell or only T-cell functions are lacking. **DiGeorge syndrome** results from a failure of the thymus to develop correctly, so

TABLE 19.3 Types of Immunodeficiencies

Deficiency	Example	Immune Response		Common Infections with This Immunodeficiency	Treatment
		Humoral	Cellular		
Complement	C_3 deficiency	Normal	Normal	Pyogenic bacteria	Antibiotics
Myeloid cell	Chronic granulomatous disease	Normal	Normal	Catalase-positive bacteria	Antibiotics
B-cell	Infantile sex-linked α-γ-globulinemia (Bruton)	Absent	Normal	Pyogenic bacteria; *Pneumocystis carinii*	γ-globulin
T cells	Thymic hypoplasia (DiGeorge syndrome)	Lower	Absent	Certain viruses; *Candida*	Thymus graft
Stem cell marrow	Severe combined deficiency (Swiss type)	Absent	Absent	All of the above	Bone graft

FIGURE 19.30
David was born in 1971 with severe combined immune deficiency. He survived for 13 years because he was isolated from the microbe-laden environment. He is shown here just before his ninth birthday playing with a fish tank inside the isolation bubble where he has lived since birth. He also had a sterile isolator "space suit" that gave him very limited mobility outside of this bubble, which protected him from infections that his body could not combat. David died in 1984 after an attempt to treat his immunodeficiency by injecting bone marrow from a sibling into his body. This treatment has been successfully used to treat other children with immunodeficiencies. The bone marrow from the immunologically compatible sibling provided the missing B- and T-cell lines that are essential for establishing immunity. Unfortunately, after the bone marrow transplant, there was an uncontrolled proliferation of B cells in David's body that apparently attacked his own tissues, resulting in his tragic death. Much has been learned from David that will help treat future cases of severe combined immune deficiency. (Courtesy United Press International Photo)

that T lymphocytes do not become properly differentiated. Individuals suffering from this condition do not exhibit cell-mediated immunity and thus are prone to viral and other intracellular infections. Additionally, because T helper cells are involved in enhancing antibody production by B cells, the antigen-mediated or humoral response is depressed in individuals suffering from DiGeorge syndrome. The complete absence of the thymus is rare, and partial DiGeorge syndrome—in which some T cells are produced, although in lower numbers than in individuals with fully functional thymus glands—is more common.

Bruton's congenital a-γ-globulinemia results in the failure of B cells to differentiate and produce antibodies, but the cell-mediated response is normal. This immunodeficiency

disease is sex limited and affects only males. Boys with Bruton's globulinemia are particularly subject to bacterial infections, including those by pyrogenic (fever-inducing) bacteria, such as *Staphylococcus aureus*, *Streptococcus pyogenes*, *S. pneumoniae*, *Neisseria meningitidis*, and *Haemophilus influenzae*. The treatment of this disease involves the repeated administration of IgG to maintain adequate levels of antibody in the circulatory system.

The most common form of immunodeficiency also involves a failure of B cells. This immunodeficiency is known as **late-onset hypogammaglobulinemia**. In this condition there is a deficiency of circulating B cells and/or B cells with IgG surface receptors. Such individuals are unable to respond adequately to antigen through the normal differentiation of B cells into antibody-secreting plasma cells. Other immunodeficiencies may affect the synthesis of specific classes of antibodies. For example, some individuals exhibit IgA deficiencies, producing depressed levels of IgA antibodies. Such individuals are prone to infections of the respiratory tract and body surfaces normally protected by mucosal cells that secrete IgA.

Immunodeficiencies may also affect the complement system. People who fail to produce sufficient C_3 complement are unable to respond properly to bacterial infections. The lack of an active complement system limits the inflammatory response and lytic killing of pathogenic bacterial cells. Immunodeficiencies may also affect the functioning of monocytes, neutrophils, and macrophages. Phagocytic cells lacking enzymes that produce hydrogen peroxide and other antimicrobial forms of oxygen do not have proper lysosomal functions that kill bacteria, and pathogenic bacteria are able to multiply within such metabolically deficient phagocytic cells. Antibiotics can be used to protect individuals from who are deficient in both complement and active phagocytic cells invading pathogenic bacteria.

Acquired Immune Deficiency Syndrome   In 1979, a new disease—**acquired immune deficiency syndrome** or **AIDS**—started to appear. Individuals with this disease exhibit immunosuppression. Specifically, they have depressed levels of T helper cells, causing them to have more T suppressor than T helper cells, which effectively shuts off the immune response network. As a result of the immunodeficiency, individuals with AIDS are subject to infection by a wide variety of disease-causing microorganisms and to the development of a form of cancer (Kaposi's sarcoma). Onset of the disease is sudden; early symptoms include low-grade fever, swollen glands, and general malaise. Generally, one infection follows another in victims of this disease until death occurs.

Since its discovery, a great deal has been learned about AIDS, and some treatments and possible vaccines for prevention hold promise in checking the future spread of this disease. However, for the near future, no cure or vaccine for preventing this disease is likely to be found. It is now known that AIDS is caused by a retrovirus named *human immuno-*

deficiency virus (HIV).[7] The virus is transmissible primarily by sexual contact, both homosexual and heterosexual. In the United States the homosexual male population is the most widely infected group, but in other countries heterosexual transmission appears to be a significant route of spread, and both males and females are equally infected. Besides sexual transmission, the virus is transmitted through contaminated blood and blood products and can cross the placenta from mother to fetus. Steps have been instituted to screen the blood supply in the United States used for transfusions to ensure its safety. Intravenous drug abusers are at high risk of contracting this disease through contaminated syringes. Although the virus is found in the saliva of infected individuals, there is no evidence for transmission by even prolonged casual contact; only direct sexual contact or exchange of blood are known to cause AIDS transmission.

Not all individuals exposed to the virus, as evidenced by the development of antibodies against HIV in their blood, will develop AIDS, although the percentages are as yet unknown. Some develop AIDS-related complexes, such as lymphadenopathy syndrome. It is not clear what factors lead to AIDS in some cases and not in others. It is known that the HIV virus reproduces in T cells, leading to a decline in T helper cells that renders the immune response ineffective. Although only a few health care workers have been infected with HIV to date, those cases arising from accidental self-injection with contaminated syringes or other direct exposure of contaminated blood, there is great concern among surgeons, dentists, paramedics, and other health care workers, as well as fire and police personnel who are at risk of con-

[7]As noted in Chapter 9, the name of the virus that is popularly used is *human T cell leukemia virus #3 (HTLV-3)*, but the formal taxonomic name for this virus is now *human immunodeficiency virus (HIV)*. The formal scientific journals use the latter name, but the former continues to be used in the lay media.

tamination in the course of their work. Many such workers have adopted a policy of wearing masks and gloves when dealing with situations in which they may be exposed to blood containing HIV.

Recent studies indicate that the drug azidothymidine (AZT) is useful in limiting replication of the virus and alleviates some of the effects of infection with HIV. AZT blocks the reverse transcription needed by this retrovirus for replication. The drug has been recommended for individuals who have developed *Pneumocystis* pneumonia and impairment of the immune system as a result of AIDS. Additional drugs, including ribavirin, have also shown promise in limiting the progression of infection with HIV that leads to AIDS. Very preliminary test results indicate that ribavirin blocks the development of AIDS in individuals infected with HIV who have developed pre-AIDS syndrome but who have not yet contracted AIDS.

Hypersensitivity Reactions

Whereas immunodeficiencies cause diseases, an excessive immunological response to an antigen can also result in tissue damage and a physiological state known as **hypersensitivity**. Hypersensitivity reactions occur when an individual is sensitized to an antigen, so that further contact with that antigen results in an elevated immune response. The hypersensitivity reaction may be immediate, occurring shortly after exposure to the antigen, or delayed, occurring a day or more afterward. There are several types of hypersensitivity reactions, each mediated by different aspects of the immune response (Table 19.4).

Anaphylactic Hypersensitivity   **Anaphylactic hypersensitivity (type 1 hypersensitivity)**—a systemic, potentially life-threatening condition—occurs when an antigen reacts with antibody bound to mast or basophil blood cells, leading to

TABLE 19.4   Comparison of the Characteristics of the Four Types of Hypersensitivity Reactions

Hypersensitivity Example/Reaction	Alternative Name	Description
Type 1 atopic allergies; asthma	Anaphylactic hypersensitivity	IgE attached to mast cell or basophil reacts with antigen, causing lysis of the blood cell and release of biochemicals that are potent physiological mediators, such as histamine
Type 2 transfusion incompability; Rh incompatability	Antibody-dependent cytotoxic hypersensitivity	Antigen on cell surface combines with antibody and cell dies
Type 3 serum sickness	Complex-mediated hypersensitivity; Arthus reaction	Formation of immune complex involving antigen, antibody, and complement triggers an inflammatory response, and immune complex is deposited in tissues
Type 4 contact dermatitis	Cell-mediated hypersensitivity; delayed hypersensitivity	Hypersensitivity that occurs only after a delay and involves T cells

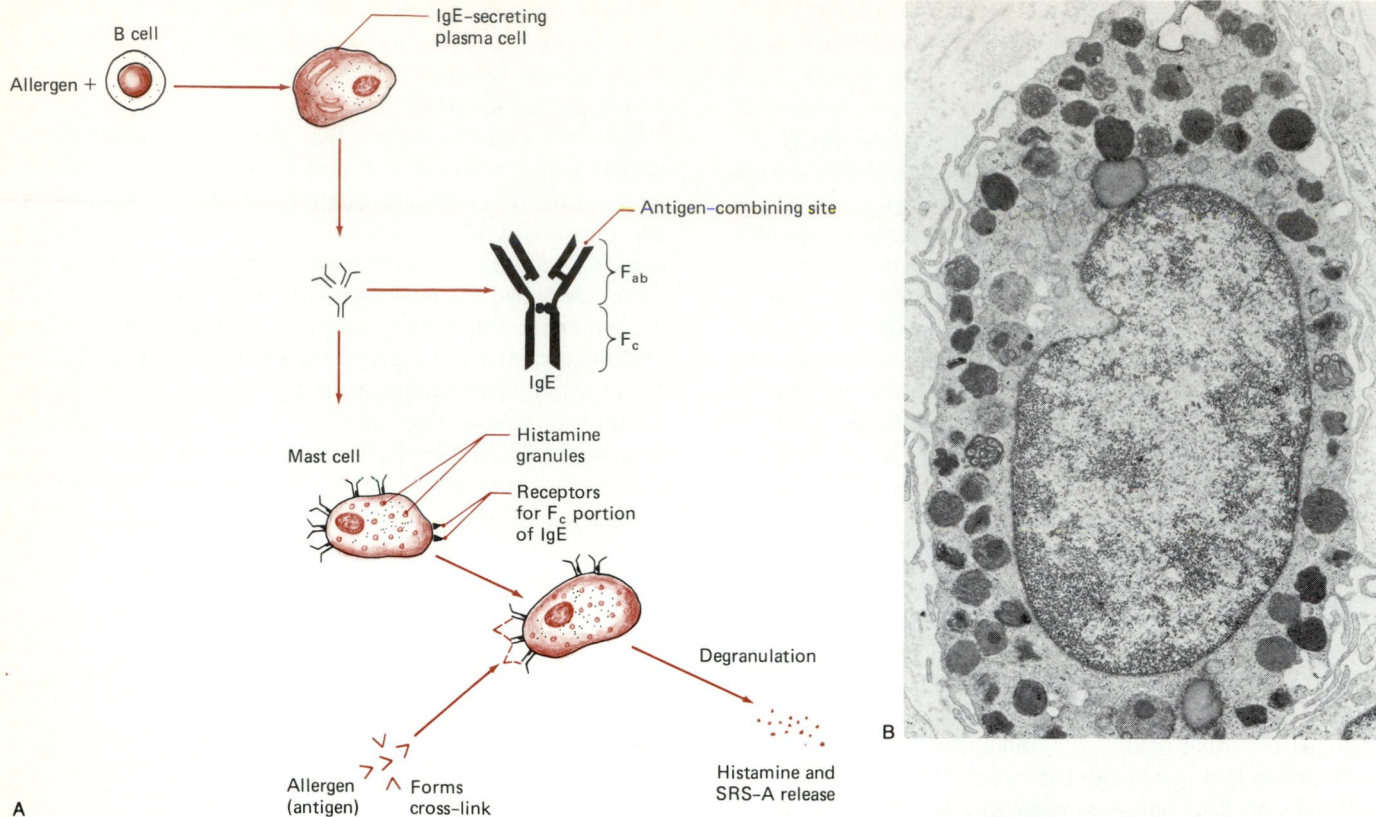

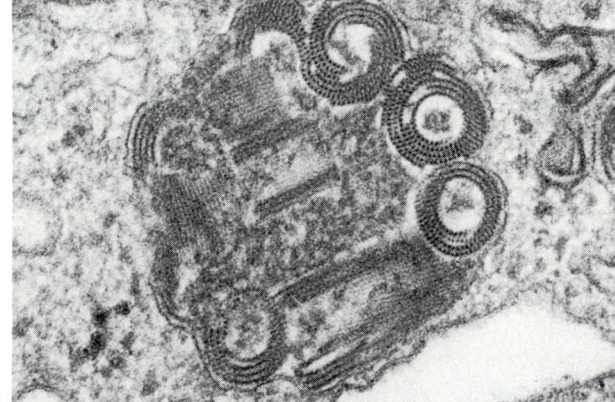

FIGURE 19.31

(A) Illustration of the type 1 hypersensitivity reaction, an immune system response to an allergen. Allergens come into contact with specific IgE antibody molecules on the surface of a basophil or a mast cell. A subsequent membrane response causes the release of chemical mediators from granules in the basophil or mast cell. The mast cell or basophil will degranulate when a cross-linkage forms between receptors on the surface of the cell. This may occur upon exposure of IgE-sensitized basophils or mast cells to an allergen with which the IgE will react (antigen-mediated degranulation) or upon reaction of IgE-sensitized cells with anti-IgE antibodies (antibody-mediated degranulation). (B) Electron micrograph of a typical mast cell, showing extensive granular inclusions of the potent vasoactive agent, histamine. (14,000×). (C) Higher magnification of granular inclusions of a mast cell, showing their characteristic "scrolls", which consist of coils of a crystalline or fibrillar structure (150,000×). (Courtesy D. Zucker-Franklin, New York University Medical Center; reprinted by permission of Grune and Stratton, Inc., from D. Zucker-Franklin, 1980, Ultrastructural evidence for the common origin of human mast cells and blastophils. *Blood* 56:536)

disruption of these cells with the release of vasoactive mediators, such as histamine (Figure 19.31). This condition is also known as *immediate hypersensitivity* because it occurs shortly after exposure to the antigen that triggers this response. The antigens that initiate this response are called **allergens**. In the absence of proper modulation by T cells, a clone of B lymphocytes is transformed by the binding of an allergen that would not normally elicit a response, leading to the formation of antibody-secreting plasma cells. The

plasma cells make IgE antibodies against the allergen. The reason that IgE is made preferentially is not yet known, but some individuals inherit the genetic trait for producing high levels of IgE and are prone to develop type 1 hypersensitivities. Allergies are specific because IgE is specific. The Fc region of the IgE binds to specific sites on the surfaces of mast and basophil cells, sensitizing the individual against the allergen. The surface of a mast cell can be covered with as many as 500,000 IgE receptors. When specific IgE antibodies are synthesized in response to an allergen, they move through the bloodstream to mast cells in connective tissue and become firmly fixed to the receptors, a process known as **sensitization**. The next time the individual is exposed to the same allergen, that allergen can react directly with the IgE fixed to mast cells rather than causing B lymphocytes to initiate antibody synthesis.

When two adjacent IgE molecules on the surface of a mast cell are bridged by two reactive sites on the allergen molecule, a sequence of events causes the cytoplasmic membrane of the mast cell to become permeable to calcium ions. The calcium ions activate enzymes that promote ATP generation, the assembly of microtubules, and the contraction of microfilaments. These events cause the granular contents of the mast cell to migrate to the cytoplasmic membrane, with the subsequent release of histamine, heparin, serotonin, and chemical factors that activate blood platelets and attract eosinophils and phagocytic white blood cells. Each of these chemical mediators contributes in its own way to the allergic reaction.

The bridging of the mast cells by an allergen also leads to the formation of prostaglandins and leukotrienes.[8] These mediators of hypersensitivity are produced by different pathways that involve arachidonic acid. Some of the prostaglandins that are produced cause constriction of smooth muscle in the bronchi of the lungs; one prostaglandin causes dilation of the bronchi. Members of the prostaglandin family also affect the activity of mucous glands and the stickiness of blood platelets. The alternative leukotriene pathway involves the conversion of arachidonic acid to a mixture of leukotrienes known as **slow-reacting substance of anaphylaxis**, or **SRS-A**. The leukotrienes are 100 to 1,000 times as potent as histamine or the prostaglandins in constricting bronchi.

The release of the contents of basophil or mast cells establishes the basis for a severe physiological response. The sudden release of a large amount of histamine (a potent vasodilator) and other pharmacologically active compounds—such as heparin, platelet-activating factors (PSFs), SRS-A, and serotonin—into the bloodstream can produce anaphylactic shock, causing respiratory or cardiac failure. Bee stings and the administration of antibiotics, such as penicillin, can trigger such anaphylactic hypersensitivity reactions[9] (Figure 19.32). Symptoms may include hives, abdominal cramps, diarrhea, nausea, vomiting, respiratory difficulties, and rapid death. Prompt administration of adrenalin (epinephrine) is used to counter anaphylactic hypersensitivity reactions. Epinephrine raises the blood pressure, thereby reversing the action of the vasodilators released as a result of the hypersensitivity reaction.

Atopic allergies result from a localized expression of type 1 hypersensitivity reactions. The interaction of antigens (allergens) with cell-bound IgE on the mucosal membranes of

---

[8]The interaction of an allergen with a mast cell activates serine esterase, initiating a series of reactions that generate phosphatidyl choline. When the calcium ions enter the cell, the enzyme phospholipase A_2 is activated, which promotes the conversion of phosphatidyl choline to lysophosphatidyl choline and arachidonic acid. Prostaglandins and leukotrienes are produced from arachidonic acid by two different enzymatic pathways. The prostaglandin pathway is initiated when the enzyme cyclooxygenase converts arachidonic acid to prostaglandins G_2 and H_2, which are subsequently converted to the active prostaglandins D_2, E_2, F_{2a}, I_2, and thromboxane A_2. Prostaglandins F_{2a} and thromboxane A_2 are potent but short-lived constrictors of smooth muscle in the bronchi of the lungs. Prostaglandin E_2 has the opposite effect, dilating the bronchi.

[9]Many atopic allergies occur as a result of low molecular weight compounds, such as penicillin, that act as haptens.

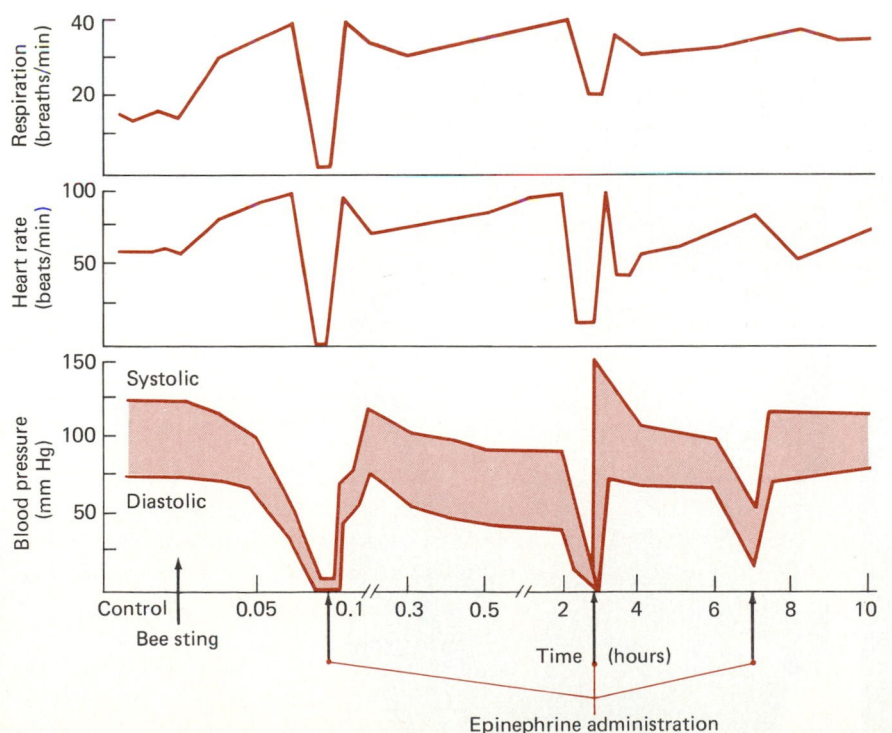

FIGURE 19.32
Systemic anaphylaxis triggered by the sting of a single honeybee and its treatment with epinephrine. (Based upon J. Vick and L. M. Lichtenstein, 1977, Allergic responses to airborne allergens and insect venoms *Federation Proceedings* 36:1727)

the upper respiratory tract and conjunctival tissues initiates a localized type 1 hypersensitivity reaction. Hay fever and allergies to certain foods are examples of such atopic allergies. When the allergen interacts with sensitized cells of the upper respiratory tract, the symptoms often include coughing, sneezing, congestion, tearing eyes, and respiratory difficulties. In cases where the allergen enters the body through the gastrointestinal tract, the symptoms often include vomiting, diarrhea, or hives. Antihistamines are useful in treating many such allergic reactions because they neutralize the main mediator of the physiological response; the antihistamine blocks the vasoactive action of the histamine release from sensitized mast and basophil cells.

In some cases, the allergic reaction primarily affects the lower respiratory tract, producing a condition known as **asthma**. Asthma is characterized by shortness of breath and wheezing. These symptoms occur because the allergic reaction causes a constriction of the bronchial tubes, producing spasms. The primary mediator of asthma is not histamine but SRS-A. Therefore, antihistamines are not of therapeutic value in treating this condition, and the treatment of asthma

generally involves administration of epinephrine or aminophylline.

Atopic allergies can be diagnosed by skin tests (Figure 19.33). Subcutaneous or intradermal injection of antigens results in a localized inflammation reaction if the individual is allergic to that antigen, that is, if that individual exhibits a type 1 hypersensitivity reaction. In this way, allergens for a particular individual are identified. The symptoms of atopic allergies can be controlled, at least in part, by avoiding the identified allergens and by using antihistamines. The antihistamines combine with the histamines that are released from mast cells as part of the allergic response, thereby blocking the action of a principal chemical mediator of the allergic reaction.

In addition to treating the immediate symptoms of an allergic reaction, attempts can be made to desensitize the individual. Desensitization usually is achieved by identifying and then administering repeated doses of the allergen. The procedure generally is time-consuming and costly. Over a period of time, however, the allergic response can be reduced or eliminated. The mechanism by which desensitization

FIGURE 19.33

Photographs showing skin testing for allergies; the positive reactions show inflammation and the development of pustules; results are rated on a scale of 1–4 +. (A) Results of intradermal inoculation into the arm of several allergens; each test shows erythema and a wheal; the pseudopods on the bottom right indicate a clear 4 + reaction to intradermal antigens. (B) The scratch test is performed on the back or forearm; the scratches should be 1 cm long and 2.5 cm apart, and the allergens should then be applied to the torn skin surface. As shown in this photograph, a large number of tests against different allergens (about 50) can be performed simultaneously on the back of the patient; in the series of tests shown, positive scratch tests—with erythema, wheal, and pseudopods—are evident in row 9, and negative tests are exemplified in row 1 and column 1. (Courtesy B. Issacs, Norton's-Kosair-Children's Hospital, Louisville, Ky.)

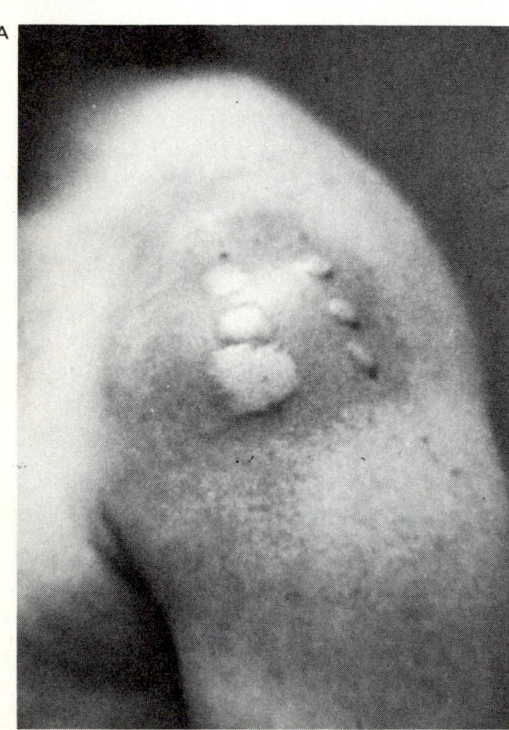
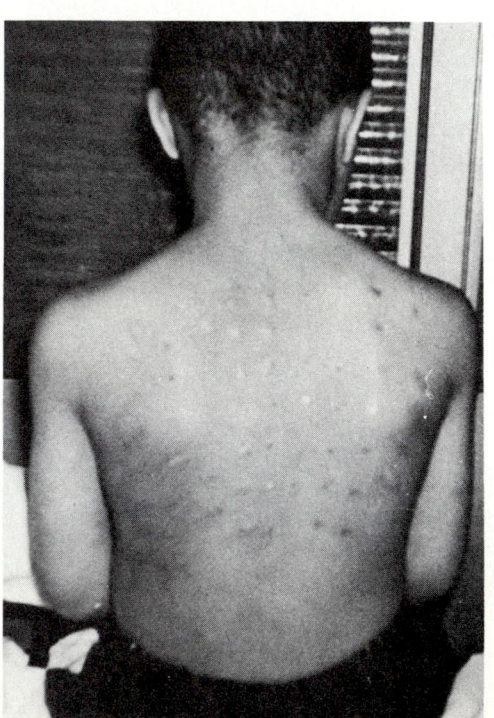

works may consist of directing immunoglobulin production in the direction of IgG, which is not involved in the allergic response, and away from IgE, which is a critical mediator of atopic allergies. Once made, IgG molecules can react with the allergen molecules before they reach IgE, preventing degranulation of mast cells and basophils and thereby blocking the allergic reaction.

Antibody-Dependent Cytotoxic Hypersensitivity **Antibody-dependent cytotoxic hypersensitivity** reactions (**type 2 hypersensitivity**) occur by a different mechanism. In type 2 hypersensitivity reactions, an antigen present on the surface of the cell combines with an antibody, resulting in the death of that cell by stimulating phagocytic attack or by initiating the sequence of the complement pathway that results in cell lysis. This type of antibody-dependent cytotoxic response occurs after transfusions with incompatible blood types. An individual's blood serum also contains antibody to any antigens that do not occur in the cytoplasmic membranes of the red blood cells of that individual. If, for example, a person with type A blood (antigen A on blood cell surfaces and antibody B circulating) were given a transfusion with type B blood (antigen B on blood cell surfaces and antibody A in serum), the circulating antibodies in the recipient would react with the surface antigens of the donor cells, initiating the addition of complement molecules and the lysis of the donated cells. Symptoms of such incompatible transfusions include fever, chills, chest pain, nausea, vomiting, jaundice, and sometimes death. It is, therefore, essential that blood transfusions be made with compatible blood types.

With respect to compatible blood types, persons with type O blood are sometimes called *universal donors* and individuals with type AB blood are sometimes called *universal recipients*. The reason for this is that type O blood cells lack both A and B antigens on their surfaces and therefore lack the antigens generally associated with transfusion incompatibility; regardless of the circulating antibodies in the recipient, the donated blood cells do not have the antigens to react, and the anti-A and anti-B antibodies in the donated blood are rapidly diluted when introduced into the larger volume of blood in the recipient. Similarly, persons with type AB blood do not have circulating antibodies against either A or B antigens and therefore lack antibodies that would react with the A and B antigens on blood cells that are introduced regardless of the cell type. However, the concepts of the universal donor and the universal recipient refer only to the major A and B antigens. There are various other antigens on blood cell surfaces, including the Rh antigen, that can cause incompatibility reactions. Therefore, except in emergencies, transfusions are given only after adequate analysis of cell antigens and only with matching blood types.

Rh incompatibility between mother and fetus is another example of type 2 hypersensitivity (Figure 19.34). Rh incompatibility occurs when the father is Rh positive, the mother is Rh negative, and the fetus is Rh positive. In this case, the mother develops Rh antibodies in response to exposure to the Rh antigens of the fetus. Generally, the mother is exposed to the fetal Rh antigens at the time of birth, so that she does not develop an immune response until after the birth of the first child. In subsequent pregnancies, however, the anti-Rh antibodies (IgG) circulating through the mother's body can cross the placenta and attack the cells of the fetus, causing anemia. During development of the fetus, fetal blood is purified by the mother's liver (Figure 19.35). At birth, the fetal blood is no longer purified by the maternal circulatory system and the infant develops jaundice. This disease, **hemolytic disease of the newborn** (previously called *erythroblastosis fetalis*), can be treated by removal of the fetal Rh-positive blood and replacement by transfusion with Rh-negative blood that will not be attacked by the anti-Rh antibodies that crossed the placenta and now are circulating within the newborn. At a later time, when the anti-Rh antibodies passively acquired (passive natural immunity)[10] from the mother have been diluted and eliminated, these transfused cells are later replaced by Rh-positive cells produced by the infant. Cell destruction does not occur at this time because the maternal antibodies have been eliminated from the infant's circulatory system.

To prevent hemolytic disease of the newborn, passive artificial immunization of the Rh-negative mother with Rhogam (anti-Rh antibodies) is used at the time of birth. The anti-Rh antibodies react with the fetal Rh-positive cells that enter the mother at the time of birth through traumatized tissue. The reaction of anti-Rh antibodies with Rh-positive cells limits the development of an anamnestic (memory) immune response (active natural acquired immunity) in the mother by binding to the Rh antigens that have been introduced, thereby preventing their recognition by the immune system of the mother. Thus, artificial passive immunization is used to prevent the development of active natural immunity. This treatment is repeated at each birth when the baby is Rh-positive and the mother is Rh-negative. As a result of this treatment, a serious antibody-dependent cytotoxic hypersensitivity reaction can be prevented.

Complex-mediated Hypersensitivity **Complex-mediated hypersensitivity** (**type 3 hypersensitivity**) reactions involve antigens, antibodies, and complement, which initiate an inflammatory response. These reactions occur when the formation of antibody–antigen complexes triggers the onset of an inflammatory response (Figure 19.36). Such an inflammatory response is part of the normal immune response, but if there are large excesses of antigen, the antigen–antibody–complement complexes may circulate and become deposited in various tissues. Inflammatory reactions from

[10]The term passive indicates that antibody production occurred outside the individual's body. In contrast, the term *active-acquired* indicates production of antibody within the individual's body as a result of the learned (acquired) ability to recognize antigens. *Artificial* indicates medical treatment as opposed to natural means of acquiring passive or active immunity. Thus, natural passive immunity occurs when a fetus acquires antibodies from its mother, whereas artificial passive immunity occurs when an individual is injected with serum containing antibodies.

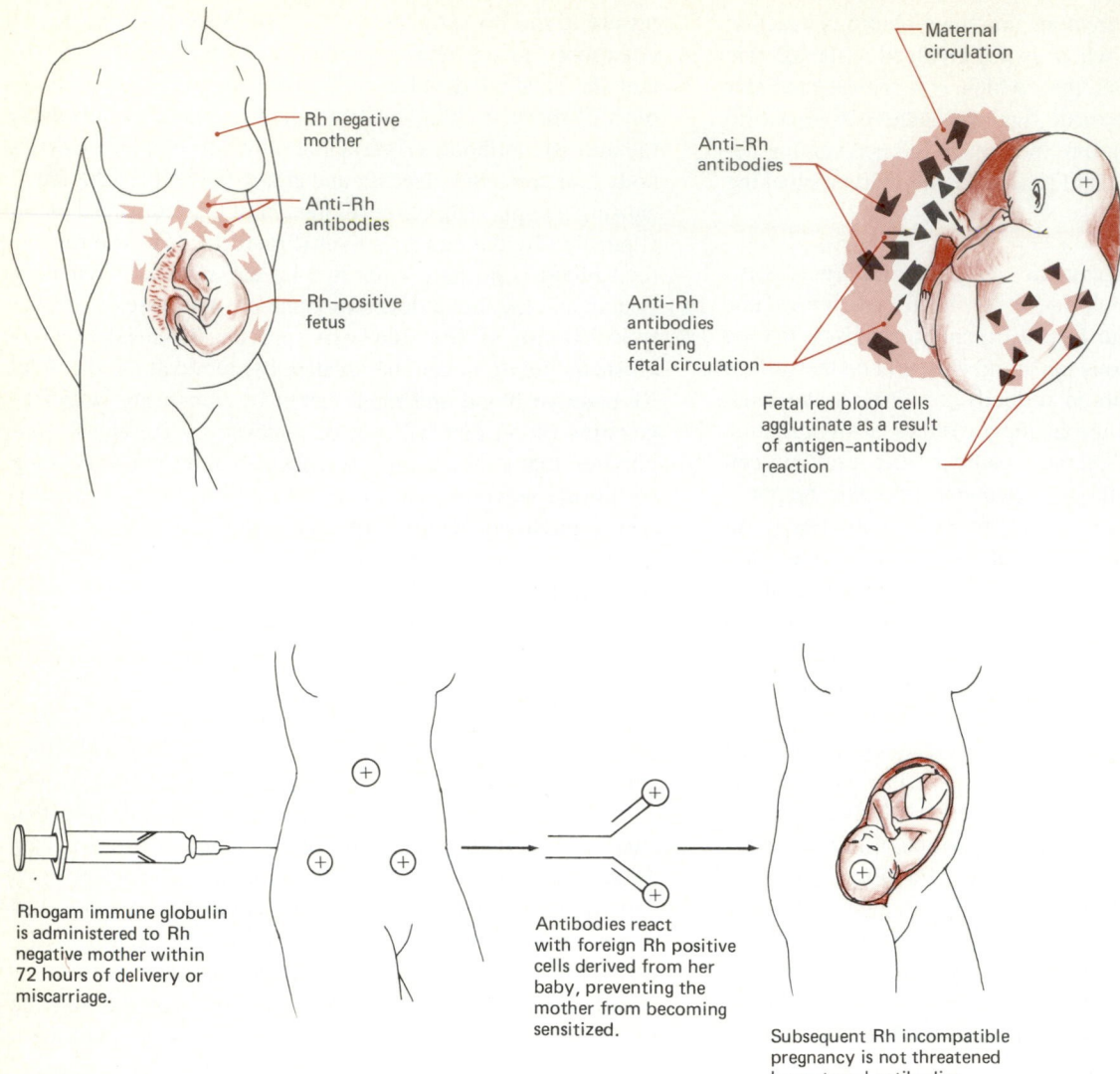

FIGURE 19.34

Illustration of the mechanism for the development of hemolytic disease of the newborn due to Rh incompatibility and the treatment for its prevention.

such deposition of immune complexes can cause physiological damage to kidneys, joints, and skin. Various examples of type 3 hypersensitivity or **Arthus immune-complex reactions** are listed in Table 19.5.

In the Arthus reaction, the site becomes infiltrated with neutrophils, and there is extensive injury to the walls of the local blood vessels. This sometimes occurs in the lungs because of repeated exposure to antigens on the surfaces of inhaled particulate matter, causing the development of extrinsic allergic alveolites. When an antigen–antibody complex initiates such a reaction in the alveoli, the symptoms generally include cough, fever, and difficulty in breathing. These symptoms typically develop over a period of 4 to 6 hours, and the attack usually subsides within a few days after the removal of the source of the antigen. Persons in various occupations have a high risk of developing this condition. For example, farmers often develop this reaction because of

TABLE 19.5   Examples of Type 3 Hypersensitivity Reactions

Disease Condition	Caused by Exposure to:
Farmer's lung	Actinomycete spores
Cheese washer's disease	*Penicillium casei* spores
Furrier's lung	Fox fur protein
Maple bark stripper's disease	*Cryptostroma* spores
Pigeon fancier's disease	Pigeon antigens
Serum sickness	Foreign blood serum

repeated exposure to the airborne spores of actinomycetes growing on hay. Sugarcane workers, mushroom growers, cheesemakers, and pigeon fanciers are also prone to this condition because of exposure to airborne antigens associated with their activities.

Serum sickness is another type of immune complex disorder. This disease results when patients are given large

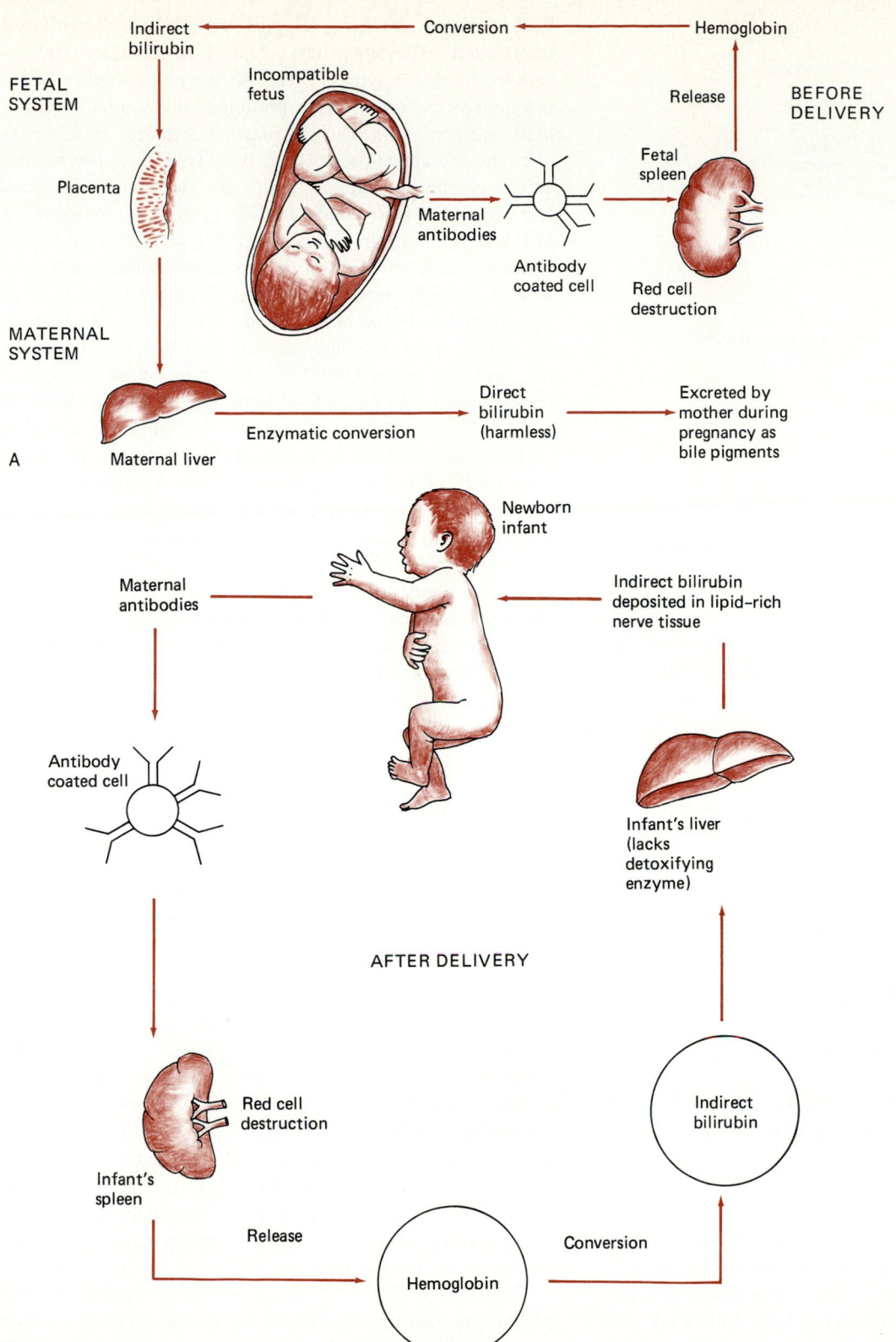

FIGURE 19.35
(A) Steps in the formation and excretion of hemoglobin derivatives during fetal development.
(B) Events after birth of an Rh-incompatible fetus that result in hemolytic disease of the newborn.

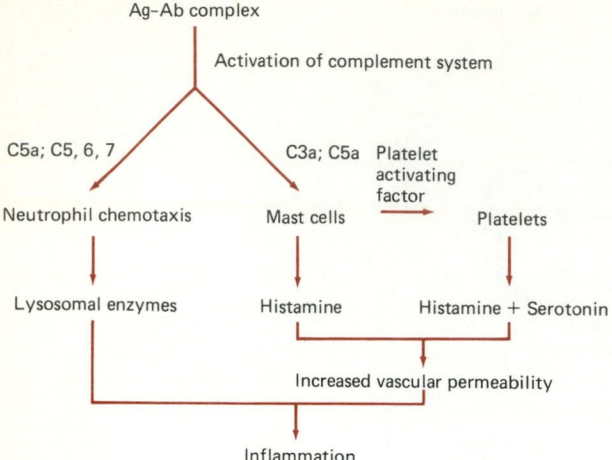

FIGURE 19.36

Pathways to inflammation triggered by antigen–antibody complexes in type 3 hypersensitivity reactions. (After J. W. Kimball, 1986, *Introduction to Immunology*, Macmillan Co.)

doses of foreign serums, such as horse serum antitoxins to protect against tetanus and diphtheria—a once widely used practice—and antilymphocyte serum for immunosuppression to protect against rejection of transplanted tissues. The antigens in these foreign serums stimulate an immune response. Because large infusions of serum are given, these antigens have not been degraded and cleared from the body by the time circulating antibodies appear. Immune complexes form between the residual antigens and the circulating antibodies. These antigen–antibody complexes are deposited at various body sites including the joints, kidneys, and blood vessel walls. Symptoms of serum sickness generally appear 7–10 days after injection of the foreign serum and include fever, nausea, vomiting, malaise, hives, and pain in muscles and joints. In many cases of serum sickness, the immune complexes are carried to the kidneys and cause nephritis (inflammatory disease of the kidneys).

This condition of **glomerulonephritis** can also be brought about by persistent infections resulting in the formation of antigen–antibody complexes that are deposited within the glomeruli of the kidneys. Immune complexes—formed by antibody reactions with antigens produced by *Streptococcus pyogenes* (the causative agent of strep throat, which produces protein toxins that may circulate through the body and cause other diseases), hepatitis B virus (the cause of serum hepatitis), *Plasmodium* species (protozoa that cause malaria), and *Shistosoma* (helminthic worms that cause shistosomiasis)—may lead to this condition. The persistence of these infections provides a continuing supply of antigen to react with circulating antibodies produced by the infected individual. The immune complex that forms accumulates in the kidneys, eventually causing nephritis due to complex-mediated hypersensitivity.

Cell-mediated (Delayed) Hypersensitivity **Cell-mediated** or **delayed hypersensitivity** (**type 4 hypersensitivity**) reactions involve T lymphocytes. As the name implies, these reactions occur only after a prolonged delay after exposure to the antigen, often reaching maximal intensity 24–72 hours after the initial exposure (Figure 9.37). Delayed hypersensitivity reactions occur as allergies to various microorganisms and chemicals. **Contact dermatitis**, resulting from exposure of the skin to various chemicals, is a typical delayed hypersensitivity reaction. Poison ivy is one of the best-known examples of such contact dermatitis (Figure 19.38). Contact with catechols in the leaves of the poison ivy plant leads to the development of a characteristic rash with itching, swelling, and blistering. Catechols appear to act as **haptens**,[11] reacting with skin proteins to form active antigens; lipids in the skin help to retain the catechols. By combining with skin proteins, catechols bring about a cell-mediated response involving T cells. Upon primary exposure to poison ivy, no dermatitis (skin rash) occurs, but subsequent exposure of sensitized individuals to the oils of the poison ivy plant results in dermatitis after an initial delay of several days. Various other agents—including metals, soaps, cosmetics, and biological materials—can also cause contact dermatitis. The treatment of contact dermatitis often involves the administration of corticosteroids which depress cell-mediated immune reactions.

Transplantation and Immunosuppression

Sometimes it is necessary to suppress the immune response, as in cases where organs or tissues are transplanted from one individual to another. Tissues contain surface antigens that are coded for by the MHC and reactions to these antigens generally preclude successful transplantation of tissues unless the transplant tissues come from a compatible donor, that is, one with an identical histocompatibility complex, or unless the normal immune response is suppressed. Since finding a compatible donor is virtually impossible, for even siblings—except identical twins—do not share many histocompatibility antigens, suppression of the immune response is usually necessary in organ transplant cases. In a sense, the normal immune defense mechanism is dysfunctional in transplantation and grafting because in these cases it does not serve the desired or useful function. The ability to transplant major organs, such as kidneys and hearts, is dependent on the ability to control the normal immune response and prevent rejection of the transplanted tissues.

There are a number of drugs, including cyclosporin and prednisone, that may be used to suppress the normal immune response. The most widely used immunosuppressant for blocking rejection of major organ transplants is cyclosporin. This drug is produced by microorganisms. It blocks rejection of the transplanted tissue by suppressing the cell-mediated

[11]Haptens are small molecules that elicit an immune response when bound to larger molecules.

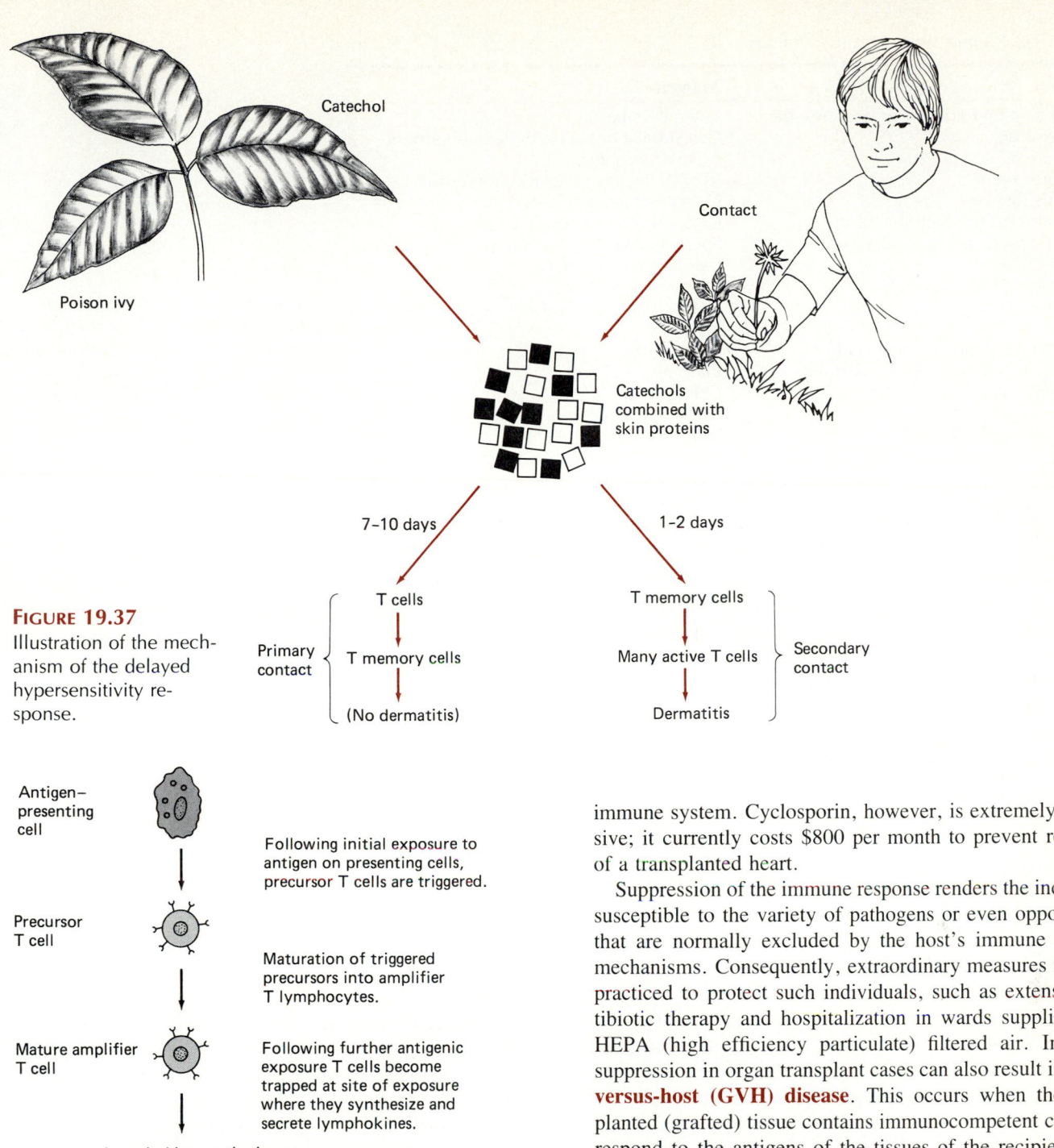

FIGURE 19.37
Illustration of the mechanism of the delayed hypersensitivity response.

FIGURE 19.38
Sequence of events in the development of contact dermatitis due to exposure to poison ivy.

immune system. Cyclosporin, however, is extremely expensive; it currently costs $800 per month to prevent rejection of a transplanted heart.

Suppression of the immune response renders the individual susceptible to the variety of pathogens or even opportunists that are normally excluded by the host's immune defense mechanisms. Consequently, extraordinary measures must be practiced to protect such individuals, such as extensive antibiotic therapy and hospitalization in wards supplied with HEPA (high efficiency particulate) filtered air. Immunosuppression in organ transplant cases can also result in **graft-versus-host (GVH) disease**. This occurs when the transplanted (grafted) tissue contains immunocompetent cells that respond to the antigens of the tissues of the recipient. The depressed immune system of the recipient is unable to control the transplanted tissue, and the reaction can be fatal. This commonly is a problem in bone marrow transplants because bone marrow contains large numbers of B and T cells that can initiate an immune response against the immunosuppressed recipient.

Autoimmune Diseases

There are a number of **autoimmune diseases** that result from the failure of the immune response to recognize self antigens. Such autoimmunity diseases often result in the progressive

TABLE 19.6   Some Types of Autoimmunity

Disease	Antigen
Hashimoto's thyroiditis; primary myxedema	Thyroglobulin
Thyrotoxicosis	Cell surface thyroid-stimulating hormone (TSH) receptors
Pernicious anemia	Intrinsic factor; parietal cell microsomes
Addison's disease	Cytoplasm adrenal cells
Premature onset of menopause	Cytoplasm steroid-producing cells
Male infertility (some)	Spermatozoa
Juvenile diabetes	Cytoplasm and surface of islet cells
Goodpasture's syndrome	Glomerular and lung basement membrane
Myasthenia gravis	Skeletal and heart muscle; acetyl choline receptor
Autoimmune hemolytic anemia	Erythrocytes
Idiopathic thrombocytopenic purpura	Platelets
Ulcerative colitis	Colon "LPS"
Sjögren's syndrome	Ducts, mitochondria, nuclei, thyroid; IgG
Rheumatoid arthritis	IgG; collagen
Systemic lupus erythematosus	DNA; nucleoprotein; cytoplasmic soluble Ag; array of other Ag, including elements of blood-clotting factors

degeneration of tissues. A number of such diseases are summarized in Table 19.6. In **systemic lupus erythematosus**, numerous autoantibodies are produced that react with self antigens, including some directed at DNA molecules. In this disease, antigen–antibody complexes often circulate and settle in the glomeruli of the kidney, causing kidney failure. In cases of **myasthenia gravis**, antibodies react with nerve–muscle junctions. In autoimmune hemolytic anemia, antibodies react with red blood cells, causing anemia. Various other disease conditions may reflect the failure of the immune system to recognize self antigens. These diseases can be treated by using immunosuppressive substances to prevent the self-destruction of body tissues by the body's own immune response.

Postlude

The defense mechanisms of a host against pathogenic microorganisms involve a complex network of interactions between various host cells. The specific immune response represents a major line of protection against invading microorganisms and other foreign biochemicals. The immune response is activated by exposure to certain biochemical determinants known as *antigens*. Immunity depends on the ability to detect and respond rapidly to such antigens.

Two types of lymphocyte cells, B lymphocytes and T lymphocytes, are both involved in conferring immunity but differ in their modes of action. B cells are primarily involved in the antibody-mediated immune response. Upon exposure to an appropriate antigen, B cells undergo differentiation, forming plasma cells that are able to secrete antibodies. There is a diversity of B lymphocytes, each containing different surface receptors, and each B cell can establish clones of plasma cells capable of responding to a particular antigenic determinant. The lymphocytes contain the genetic information for producing specific antibodies, and lymphocyte cell lines maintain a memory system for recognizing and responding to foreign antigens. The clonal selection theory explains how they come to possess the lymphocytes that are able to distinguish self from nonself antigens. Failures of this recognition process can render one susceptible to infectious diseases or may result in autoimmunity.

Five major classes of antibodies are secreted, each serving a somewhat different function in the immune response. IgG antibodies are the major circulating antibodies reacting with antigens, neutralizing toxins, and preventing the spread of pathogenic microorganisms through the circulatory system. IgA antibodies involved in the protection of surface cells from microbial attachment are secreted from mucosal cells. The reaction of antigen and antibody forms a complex that can react further with complement molecules to initiate the classical complement pathway, resulting in the lysis or death of invading microorganisms. The classical complement pathway also enhances phagocytosis and initiates the inflammatory response, both of which contribute to the elimination of invading microorganisms.

There are various possible reactions between antigen and antibody molecules, many of which are used *in vitro* for diagnostic purposes. Precipitation reactions, tagged-antibody or tagged-antigen assays, radioimmunoassays, and agglutination reactions are employed frequently in the clinical laboratory for diagnostic purposes. Many pathogenic microorganisms can be identified by their specific reactions with defined antibodies. Agglutination reactions are used most commonly in blood typing and in other procedures where antigens are bound to particle surfaces.

Within the human body, antibody-mediated immunity constitutes a major line of defense against toxins and the development of bacterial infections outside host cells. The immunological defense against intracellular pathogens, such as viruses, depends in large part on cell-mediated immunity. The cell-mediated immune response involves T cells that may release lymphokines, which initiate a variety of defense mechanisms, including increased antibody synthesis, enhanced phagocytosis, and inflammation. Cell-mediated immunity protects the body against a variety of intracellular parasites, most notably viruses. In addition to releasing other lymphokines, activated T cells release interferon, preventing the multiplication of viruses in uninfected cells. T cells may also be differentiated into cell lines with cytotoxic properties that are able to recognize and kill cells harboring viruses. Cytotoxic T cells eliminate infected cells, and this appears to be the main mechanism by which viral infections are elimi-

nated. The ability of T cells to recognize altered host cells depends on their interaction with surface antigens of the cell that differ from the antigens coded for by the MHC. MHC antigens are also important in the ability of cytotoxic T cells to recognize tumor cells, preventing the formation of malignancies, and the ability of T cells to recognize foreign tissues, determining the compatibility of transplanting graft tissues.

Failures of the immune response to produce the necessary differentiated tissue lines, such as macrophages, B lymphocytes, and T lymphocytes, result in immunodeficiency diseases in which an individual is unable to respond properly and thus to defend against invasion by microorganisms. Many human infections occur when the host defense mechanisms are compromised and the immune response fails to respond optimally to potential pathogens. The regulation and expression of the immune response is an extremely complex system that is developed only in higher vertebrates, such as humans. The primary function of this highly developed and integrated system seems to be the protection of self from invasion by disease-causing foreign agents.

Suggested Supplementary Readings

ALLEN, J. C. 1980. *Infection and the Compromised Host*. Williams & Wilkins Co., Baltimore.

BACH, M. K. 1982. Mediators of anaphylaxis and inflammation. *Annual Review of Microbiology* 36:371–413.

BAGLIONI, C., and T. W. NILSEN. 1981. The action of interferon at the molecular level. *American Scientist* 69:392–399.

BARRET, J. T. 1983. *Textbook of Immunology: An Introduction to Immunochemistry and Immunobiology*. C.V. Mosby, St. Louis.

BELLANTI, J. (ed.). 1985. *Immunology III*. W.B. Saunders Co., Philadelphia.

BENACERRAF, B. 1981. Role of MNC gene products in immune regulation. *Science* 212:1229–1238.

BENACERRAF, B., and E. R. UNANUE. 1984. *Textbook of Immunology*. Williams & Wilkins Co., Baltimore.

BIER, O. G., W. D. daSILVA, D. GOTZ, and I. MOTA. 1981. *Fundamentals of Immunology*. Springer-Verlag, New York.

BIGLEY, N. J. 1980. *Immunologic Fundamentals*. Year Book Medical Publishers, Inc., Chicago.

BOWRY, T. R. 1984. *Immunology Simplified*. Oxford University Press, New York.

BURKE, D. C., and A. G. MORRIS (eds.). 1983. *Interferons: From Molecular Biology to Clinical Applications*. Thirty-Fifth Symposium of the Society for General Microbiology. Cambridge University Press, Cambridge, England.

COOPER, E. L. 1982. *General Immunology*. Pergamon Press, New York.

DAVIES, D. R., and H. METZGER. 1983. Structural basis of antibody function. *Annual Review of Immunology* 1:87–117.

DODD, R. Y., and L. F. BARKER. 1985. *Infection, Immunity and Blood Transfusions*. A.R. Liss, New York.

DORIA, G., and A. ESHKOL. 1980. *The Immune System*. Academic Press, New York.

EISEN, H. N. 1980. *Immunology*. Harper & Row Publishers, Inc., Hagerstown, Md.

FUDENBERG, H., D. SITES, J. CALDWELL, and J. WELLS (eds.). 1980. *Basic and Clinical Immunology*. Lange Medical Publications, Los Altos, Calif.

GALLO, R. 1987. The AIDS virus. *Scientific American* 256(1):46–72.

GLUCKMAN, J. C., D. KLATZMAN, and L. MONTAGNIER. 1986. Lymphoadenopathy-associated virus infection and acquired immunodeficiency syndrome. *Annual Review of Immunology* 4:97–118.

GOLUB, E. S. 1981. *The Cellular Basis of the Immune Response*. Sinauer Associates, Sunderland, Mass.

HABERMEHL, K. O. (ed.). 1985. *Rapid Methods and Automation in Microbiology and Immunology*. Springer-Verlag, New York.

HAMILTON, H., and M. B. ROSE (eds.). 1985. *Immune Disorders*. Springhouse Publishers, Springhouse, Penn.

HILDEMANN, W. H. 1984. *Essentials of Immunology*. Elsevier Publishing Co., New York.

HONJO, T. 1983. Immunoglobulin genes. *Annual Review of Immunology* 1:499–528.

HOOD, L. E., I. L. WEISMANN, W. B. WOOD, and J. H. WILSON. 1984. *Immunology*. Benjamin/Cummings Publishing Co., Inc., Menlo Park, Calif.

HUFFER, T., et al. 1986. *Introduction to Human Immunology*. Jones and Barlett, Boston.

Immunology: Readings from Scientific American. 1976. W.H. Freeman and Sons San Francisco.

KIMBALL, J. W. 1986. *Introduction to Immunology*. Macmillan Publishing Co., Inc., New York.

KIRKWOOD, E. M., and C. J. LEWIS. 1984. *Understanding Medical Microbiology*. John Wiley & Sons, Inc., New York.

KLEIN, J. 1982. *Immunology: The Science of Self-Nonself Discrimination*. John Wiley & Sons, Inc., New York.

MILLER, K., and S. NICKLIN (eds.). 1987. *Immunology of the Gastrointestinal Tract*. CRC Press, Inc., Boca Raton, Fla.

MILSTEIN, C. 1980. Monoclonal antibodies. *Scientific American* 243(4):66–74.

MIMS, C. A., and D. O. WHITE. 1984. *Viral Pathogenesis and Immunology*. Blackwell Scientific Publications, Oxford, England.

MIZEL, S. B., and P. JARET. 1986. *The Human Immune System: The New Frontier in Medicine*. Simon & Schuster, New York.

MYRVIK, Q. N. (ed.). 1984. *Fundamentals of Immunology*. Lea & Febiger, Philadelphia.

NAHMIAS, A. J., and R. J. O'REILLY (eds.). 1981. *Immunology of Human Infection*. Plenum Publishing Co., New York.

NEWBY, T. J., and C. R. STOKES (eds.). 1984. *Local Immune Responses of the Gut*. CRC Press, Inc., Boca Raton, Fla.

NGO, T. T., and H. M. LENHOFF (eds.). 1985. *Enzyme-Mediated Immunoassay*. Plenum Publishing Co., New York.

POTTS, E., and M. MORRA. 1986. *Understanding Your Immune System*. Avon Books, New York.

ROITT, I. M., J. BRISTOFF, and D. MALE. 1986. *Immunology*. Blackwell Scientific Publications, Oxford, England.

ROSE, N. R. 1981. Autoimmune diseases. *Scientific American* 244(2):80–103.

SIKORA, K., and H. SMEDLEY. 1984. *Monoclonal Antibodies*. Blackwell Scientific Publications, Oxford, England.

SINGHAL, S. K., and T. L. DELOVITCH (eds.). 1986. *Mediators of Immune Regulation and Immunotherapy*. Elsevier, New York.

SOULSBY, E. J. L. (ed.). 1987. *Immune Responses in Parasitic Infections: Immunology, Immunopathology, and Immunoprophylaxis*. CRC Press, Inc., Boca Raton, Fla.

STANSFIELD, W. D. 1981. *Serology and Immunology*. Macmillan Publishing Co., Inc., New York.

STEELE, R. W. 1983. *Immunology for the Practicing Physician*. Appleton and Lange, New York.

STEFFEN, C., and H. LUDWIG (eds.). 1981. *Clinical Immunology and Allergology*. Elsevier Publishing Co., New York.

WEIR, D. M., C. C. BLACKWELL, L. A HERZENBERG, and L. A. HERZENBERG (eds.). 1986. *Handbook of Experimental Immunology*. 4 volumes. Blackwell Scientific Publications, Oxford, England.

YAGUE, J. J. WHITE, C. COLECLOUGH, J. KAPPLER, E. PALMER, and P. MARRACK. 1985. The T cell receptor: The alpha and beta chains define idiotype, and antigen and MHC specificity. *Cell* 42(1):81-88.

Study Questions

1. What is an antigen?
2. What is an antibody?
3. How does the clonal selection theory explain the ability of the immune response network to distinguish between self and nonself antigens?
4. Discuss the differences between a primary and a secondary (memory) immune response.
5. Discuss the differences between the antibody-mediated and cell-mediated immune response systems. Why do we need two such elaborate defense systems?
6. What is an agglutination reaction? How is this reaction used in blood typing? What is meant by compatible blood for transfusion purposes?
7. How does an agglutination reaction differ from a precipitin reaction?
8. How can agglutination reactions be carried out with soluble antigens?
9. What is complement? How is it involved in the immune response to a bacterial infection?
10. What are autoimmune diseases? How can such diseases occur?
11. What is an allergy? How are allergies related to the immune response?
12. What is a lymphokine? What functions do lymphokines have in the immune response?
13. How is the MHC region of the genome related to tissue compatibility?
14. What are the five major classes of immunoglobulins? Compare the role of each immunoglobulin class in the immune response.

Situational Problem

Should a Child with AIDS Be Allowed to Attend School?

Acquired immune deficiency syndrome (AIDS) is perhaps the greatest threat to human health in history. The news media report stories about AIDS almost daily. The disease weakens the immune defense system, leaving the individual susceptible to numerous infections and to a relatively rare form of cancer. AIDS most commonly is acquired by sexual contact but can also be contracted through contaminated blood or acquired congenitally. Children have contracted

AIDS by these latter routes of transmission. The fear of AIDS has led to hotly contested debates about whether children with this disease should be allowed to attend school along with healthy children. Particular concern has been expressed that younger children with AIDS will bite other children and that, because the virus that causes AIDS has been found in the saliva of individuals with this disease, biting will lead to transmission of AIDS.

Parents have organized school boycotts to prevent the exposure of their children to individuals known to have AIDS. School boards and other school officials have been divided in their belief as to whether children with AIDS should be allowed to attend classes. The courts have vacillated on the issue. Public health authorities have repeatedly indicated that AIDS does not appear to be transmissible by even prolonged casual contact.

1. What would you do if you had a healthy child whose classmate was diagnosed as having AIDS? Would you allow your child to continue attending that class or that school? Would you request the school board to prevent the child with AIDS from attending that school? What reasons would you give to defend your decision?

2. What would you do if your child's teacher was diagnosed as having AIDS? Would you allow your child to continue to attend that school? Would you request that the school board replace the teacher? What reasons would you give to defend your decision?

3. What would you do if your child was diagnosed as having AIDS? Would you allow your child to continue attending school? What reasons would you give to defend your decision?

CHAPTER 20

Prevention, Diagnosis, and Treatment of Human Diseases

Despite the elaborate defense mechanisms aimed at preventing the invasion of the human body and the establishment of diseases by pathogenic microorganisms, humans are subject to a variety of infectious diseases. The development of a basic understanding of the interrelationships between humans and microorganisms, particularly the immune defense system and the virulence of specific microbial pathogens, has led to practices that prevent or diminish the incidence of human diseases caused by microorganisms. Soon after the germ theory of disease was accepted, hygienic and aseptic practices were instituted, greatly reducing the incidence of disease. Later, antimicrobial agents were added to the weapons used to control infectious diseases. Many infectious diseases, when they do occur, are relatively easily treated with antibiotics. We need only think about the number of times a physician has prescribed an antibiotic for us or someone in our family to appreciate the widespread use of antibiotics in

medical practice. Physicians also take great precautions regarding hygienic practices to prevent infection. Childbirth, for example, once involved a very high risk of infection to the mother from the midwife's dirty hands, but today's midwives and obstetricians thoroughly wash their hands and use clean equipment during delivery, thereby greatly reducing the risk of infection. Indeed, the control of microorganisms pathogenic to humans is fundamental to the practice of modern medicine. Many once widespread deadly diseases, such as cholera and whooping cough, are rare today because of: the institution of hygienic practices; the development of vaccines and immunization programs that have drastically reduced the incidence of certain diseases; and the effective use of antibiotics to control various infectious diseases. The use of vaccines for preventing disease and of antimicrobial agents for treating infectious diseases has led to greatly increased life expectancies.

20.1 Epidemiology

In order to control the spread of infectious diseases, it is necessary to understand the nature of the pathogens and their routes of transmission. The examination of disease transmission is part of the field of epidemiology. Epidemiologists consider the **etiology** (the cause of the disease) and the factors involved in the transmission of infectious agents, especially in relation to populations. With this information the epidemiologist attempts to determine how a disease outbreak in a population can be effectively controlled.

Epidemiology, the study of the factors and mechanisms that govern the spread of disease within a population, is based on the statistical probability that a susceptible individual will be exposed to a particular pathogen and that such exposure will result in disease transmission. The likelihood of disease transmission depends on the concentration and virulence of the pathogen, the distribution of susceptible individuals, and the potential sources of exposure to the pathogenic microorganisms. In many cases, a disease outbreak can be traced to a single source of exposure. The epidemiologist often acts as a detective to locate the origin of a disease outbreak, in some cases searching for a source of tainted food, in others for direct contact with infected individuals, and so forth. The epidemiologist is aided by the knowledge that disease transmission frequently occurs via the air, the ingestion of contaminated food or water, and/or direct contact with infected individuals or contaminated inanimate objects (fomites). By determining where and what people have eaten, where they have been, and with whom they have been in contact, the epidemiologist can establish a pattern of disease transmission.

The number of cases reported each day and the locations of disease occurrences enable epidemiologists to distinguish between a **common source outbreak**, which is characterized by a sharp rise and rapid decline in the number of cases, and a **person-to-person epidemic**, which is characterized by a relatively slow, prolonged rise and decline in the number of cases[1] (Figure 20.1). In the United States, the Centers for Disease Control (CDC) in Atlanta, Georgia, compiles the statistics necessary for such determinations. When common sources (**reservoirs**) or carriers (**vectors**) of pathogens are identified, action can be taken to break the chain of disease transmission, for instance, by recalling potentially contaminated foods from the marketplace. When person-to-person **(propagated) transmission** is responsible for disease outbreaks in a population, steps can be taken to reduce the number of susceptible individuals, for example, by immu-

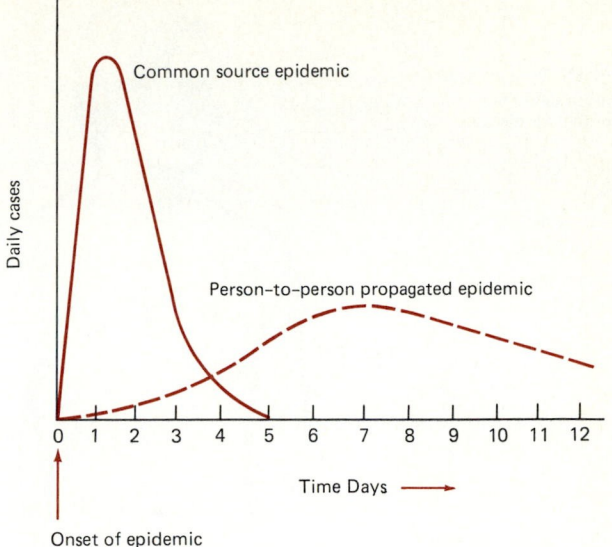

FIGURE 20.1
The shape of the curve of reported cases of a disease is indicative of whether the epidemic originates from a common source or is propagated from person to person.

nization, thereby breaking the chain of transmission. It is the aim of the epidemiologist to identify the sources of disease outbreaks and to advise public health officials regarding the steps that should be taken to prevent them.

20.2 *Preventing Disease by Avoiding Exposure to Pathogenic Microorganisms*

Perhaps the most effective way of preventing diseases caused by microorganisms is to avoid exposure to pathogenic microorganisms. Total avoidance of microorganisms, though, is not practical because we are continuously exposed to microorganisms carried through the air, in water and foods, and on the surfaces of virtually all objects that we contact. Only in the rarest of cases, when the immune system is totally nonfunctional, is absolute avoidance of contact with microorganisms practiced. It is possible, though, to control microbial populations and our interactions with them in ways that reduce the probability of encountering pathogenic microorganisms, thus reducing the incidence and spread of infectious diseases. A greatly diminished incidence of many diseases caused by microorganisms is the consequence of an understanding of the modes of transmission of pathogenic micro-

organisms and of preventive measures to reduce exposure to disease-causing microorganisms. Many modern sanitary practices are aimed at reducing the incidence of diseases by preventing the spread of pathogenic microorganisms or by reducing their number to concentrations that are insufficient to cause disease.

The methods employed for preventing exposure to specific disease-causing microorganisms vary depending on the particular route of transmission (Figure 20.2). Proper sewage treatment and drinking water disinfection programs reduce the likelihood of contracting a disease through contaminated water. Recognition of the fact that many serious diseases, such as typhoid, are transmitted through water contaminated with fecal material is the basis for enforcement of strict water quality control standards. Quality control measures are also applied throughout the food industry to prevent the transmission of disease-causing microorganisms through food products. Methods used to preserve food products and prevent the growth of microorganisms that spoil food and may cause human disease were discussed in Chapter 16. Pasteurization of milk is a good example of a process designed to reduce exposure to pathogenic microorganisms that occur

---

[1]In a common source epidemic, many individuals simultaneously acquire the infectious agent from the same source, as occurs in outbreaks of cholera in the Far East, where the pathogen is acquired from contaminated drinking water. In a person-to-person epidemic, there is a chain of transmission from one infected individual to another, as occurs in sexually transmitted diseases such as acquired immune deficiency syndrome (AIDS).

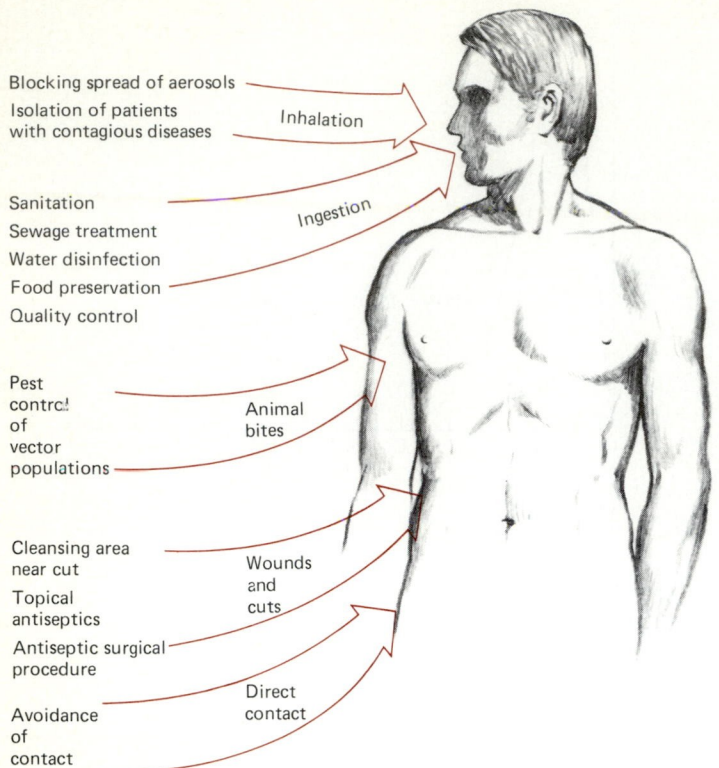

Blocking spread of aerosols

Isolation of patients with contagious diseases

Inhalation

Sanitation

Sewage treatment

Water disinfection

Food preservation

Quality control

Ingestion

Pest control of vector populations

Animal bites

Cleansing area near cut

Topical antiseptics

Antiseptic surgical procedure

Wounds and cuts

Avoidance of contact

Direct contact

FIGURE 20.2

Routes of disease transmission and steps that may be taken to prevent exposure to pathogenic microorganisms.

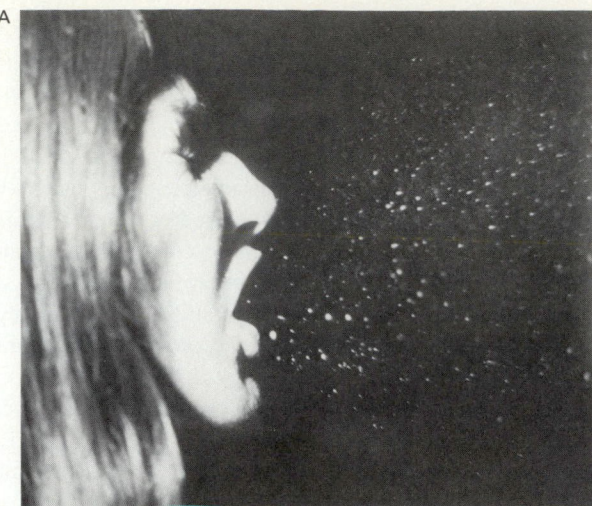

FIGURE 20.3

These photographs show droplet spread by sneezing and the effect of covering one's nose and mouth with a handkerchief when sneezing. (A) An unstifled sneeze; (B) the use of a handkerchief blocks the spread of droplets.

and proliferate in untreated milk. Chloramination of municipal water supplies, that is, treatment with chloramines, is widely used to prevent exposure to the pathogenic microorganisms that occur in water supplies and thus to ensure the safety of drinking water. The aim of washing one's hands before eating is to avoid the accidental contamination of one's food with soil or other substances that may harbor populations of disease-causing microorganisms.

Failure to maintain quality control of water and food supplies often results in outbreaks of disease; for example, cholera outbreaks often occur when sewage is allowed to mix with drinking water supplies, such as frequently occurs in the Far East when monsoon rains cause flooding, resulting in contamination of drinking water supplies. Outbreaks of botulism are associated with improperly canned food products, that is, with food products that have not been heated long enough to kill contaminating spores of *Clostridium botulinum*. Growth of *C. botulinum* in a canned food results in the exposure of the individuals who ingest the food to the lethal toxins produced by this bacterium. Extensive quality control testing is required in most countries to prevent the outbreaks of disease associated with contaminated water and food supplies.

It is more difficult to control the transmission of disease-causing microorganisms through the air than it is to ensure water and food quality control. Several steps, though, can be taken to reduce the likelihood of exposure to airborne

pathogenic microorganisms. Covering one's nose and mouth with a handkerchief when sneezing and coughing reduces the number of potential pathogens that may become airborne in aerosols and that therefore may be transmitted from an infected individual to a susceptible host (Figure 20.3). The use of surgical masks when visiting individuals who are particularly susceptible to infections, such as newborn infants, is an important precaution in preventing the spread of infectious diseases. Likewise, preventing the exposure of individuals whose immunological defense mechanisms are compromised by a variety of conditions—such as treatment for cancer or a recent organ transplant—to airborne pathogenic microorganisms is an important aspect of patient management practice. Similarly, masks should be worn in the presence of patients with tuberculosis or other pulmonary diseases, or

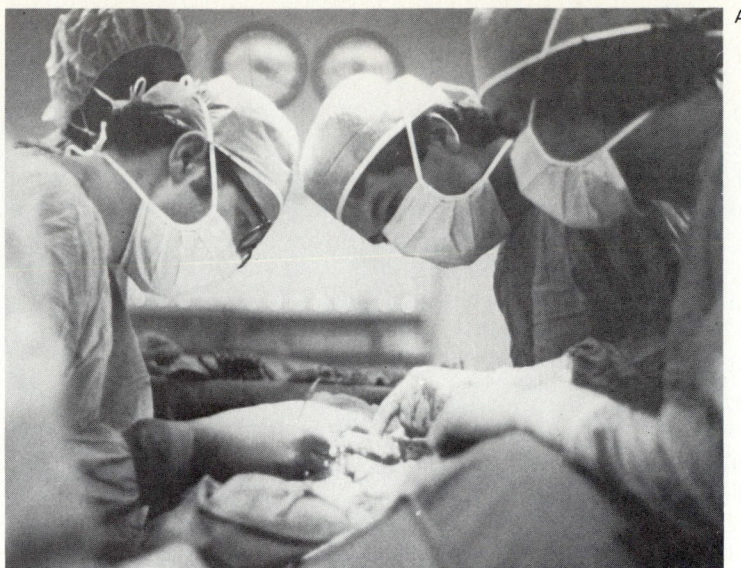

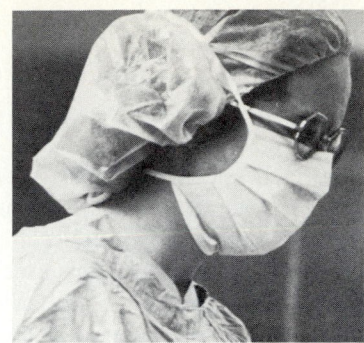

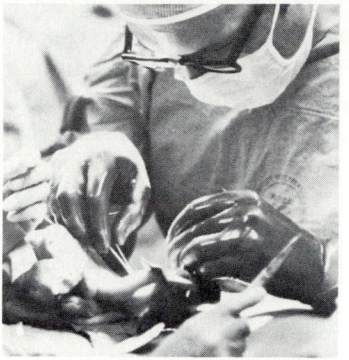

FIGURE 20.4

Operating room surgical staff take many precautions to preclude microbial contamination of tissue exposed during the surgical procedure. (A) Everyone in the room is masked, gloved, capped, and gowned. (B) Note how much of the face is covered by the mask. (C) Note that the parts of the patient not involved in the operation are also covered. (Courtesy Michelle Ising, University of Louisville, Ky.)

other precautions, such as remaining a safe distance from infected individuals, should be taken to minimize exposure. Isolation of individuals with contagious microbial diseases, in which the infectious agent is airborne, is often practiced. For example, children with measles, chicken pox, or mumps are often kept away (isolated) from other children who are not immune to these diseases. Such practices decrease the probability of exposure to pathogenic organisms and prevent, or at least reduce, the transmission of disease.

Avoiding direct contact with infected individuals is also important in preventing the spread of diseases when the pathogen is transmitted by direct contact. Historically, the isolation of leprosy patients in remote colonies is an example of the extreme steps that have been taken to prevent contact of such individuals with the general population. Today this extreme practice is not needed because of the use of antimicrobial agents and because we recognize that the disease is not as infectious as was once thought. However, avoiding sexual contact with individuals suffering from sexually transmitted diseases, such as syphilis and gonorrhea, interrupts the transmission of the pathogens that cause these diseases; avoidance of sexual contact with infected individuals and proper use of prophylactic condoms are essential for and will undoubtedly remain the main method for controlling the spread of sexually transmitted diseases.

In the case of pathogens that enter the body through breaks in the skin surface, a variety of procedures are employed to reduce the probability of exposure. Great care, for example, is taken during surgical procedures to prevent contamination by accidental introduction of microorganisms into the exposed tissues. Clean operating rooms and sterile instruments, garments, gloves, and masks are used by a hospital surgical staff (Figure 20.4; see also Figure 20.6). Wounds are cleansed to prevent the introduction of foreign material that may harbor potential pathogens, and antiseptics are applied to skin surfaces to minimize the entry of pathogenic microorganisms into tissues normally protected by an intact skin covering.

Practices are normally employed to control insect and other animal populations that act as vectors for the transmission of diseases caused by pathogenic microorganisms and to control the populations or nonbiological sources that may act as reservoirs of pathogens. **Vectors** are carriers of pathogenic microorganisms involved in the transmission of disease. The most notable vectors of pathogenic microorganisms are mosquitoes, lice, ticks, and fleas. Some public health measures, such as mosquito control programs, are aimed at reducing the sizes of these vector populations and thus lowering the probability of exposure to the pathogenic microorganisms capable of causing diseases such as plague, typhus fever, yellow fever, malaria, and a variety of other diseases transmitted by insect vectors. Although it was possible to eliminate smallpox because humans were the only reservoirs of the pathogens, it is not possible to eliminate bubonic plague because various wild animal populations serve as reservoirs for the bacterial pathogen *Yersinia pestis*, the causative agent of plague, which is transmitted to humans by flea vectors.

20.3 Preventing Disease with Chemical Antimicrobial Agents

A number of inhibitory chemicals are employed for the control of microbial growth. Chemicals that kill microorganisms or prevent their growth are called **antimicrobial agents**. Some common types of antimicrobial agents used to control microbial growth and prevent infections are listed in Table 20.1. Chemical inhibitors are widely used as food preservatives to prevent pathogens from accumulating in food and thereby prevent disease when the food is ingested. Concentration and contact time are critical factors that determine the effectiveness of an antimicrobial agent against a particular microorganism (Figure 20.5). Microorganisms vary in their sensitivity to particular antimicrobial agents. Generally, growing microorganisms are more sensitive than organisms in dormant stages, such as spores. Many antimicrobial agents are aimed at blocking active metabolism and preventing the organism from generating the macromolecular constituents needed for reproduction. Because resting stages are metabolically dormant and are not reproducing, they are not affected by such antimicrobial agents. Similarly, viruses are more resistant than other microorganisms to antimicrobial agents because they are metabolically dormant outside host cells.

Antimicrobial agents are used in a wide variety of applications. They are classified according to their application and spectrum of action. **Germicides** are chemical agents that kill microorganisms, but not necessarily bacterial endospores. Such chemicals may exhibit selective toxicity and, depending on their **spectrum of action**, may act as **viricides** (killing viruses), **bactericides** (killing bacteria), **algicides** (killing algae), or **fungicides** (killing fungi). Whereas germicides kill growing microorganisms, **microbiostatic agents** merely inhibit the growth. When the microstatic agent is removed, microorganisms resume their growth. **Disinfectants** can be either germicides or microstatic agents that kill or prevent the growth of pathogenic microorganisms. Household cleaning agents often contain disinfectants to control the growth of microorganisms. Ammonia and bleach (hypochlorite) are widely used disinfectants. In general, agents that oxidize biological macromolecules, such as hypochlorite, are effective disinfectants. Disinfectants are not, however, considered safe for use on living tissue and are applied only to inanimate objects.

Antiseptics are similar to disinfectants but may be applied safely to living tissues. The use of antiseptics in surgical practice was introduced by Joseph Lister (Figure 20.6). Antiseptics are used for topical (surface) applications and are not necessarily safe for ingestion. Although not as effective as many of the other antiseptic agents, soap and water do reduce the number of microorganisms on the skin. The term **sanitizing agent** is often used to describe a compound that reduces the number of microbes without necessarily killing

them or inhibiting their growth. Alcohol is far more effective than soap and water in reducing the number of microorganisms on the skin surface (Figure 20.7). It is probably the

TABLE 20.1   Summary of Chemical Agents Used to Control Microbial Growth

Antimicrobial Agent	Description
Phenolics	Phenol is no longer used as a disinfectant or antiseptic because of its toxicity to tissues. Derivatives of phenol such as 0-phenylphenol, hexylresorcinol, and hexachlorophene are used as disinfectants and antiseptics.
Halogens	Chlorination is extensively used to disinfect water; drinking water, swimming pools, and waste treatment plant effluent are disinfected by chlorination. Iodine is an effective antiseptic; iodophors are used as disinfectants and antiseptics; the soaps used for surgical scrubs often contain iodophors.
Alcohols	Alcohols are bactericidal and fungicidal, but are not effective against endospores and some viruses; ethanol and isopropanol are commonly used as disinfectants and antiseptics. Thermometers and other instruments are disinfected with alcohol, and swabbing of the skin with alcohol is done before injections.
Heavy metals	Heavy metals such as silver, copper, mercury, and zinc have antimicrobial properties and are used in disinfectant and antiseptic formulations. Silver nitrate was used to prevent gonococcal eye infections. Mercurochrome and merthiolate are applied to skin after minor wounds. Zinc is used in antifungal antiseptics. Copper sulfate is used as an algicide.
Dyes	Several dyes, such as gentian violet, inhibit microorganisms and are used as antiseptics for treating minor wounds.
Surface-active agents	Soaps and detergents are used to remove microbes mechanically from the skin surface. Anionic detergents (laundry powders) remove microbes mechanically; cationic detergents, which include quaternary ammonium compounds, have antimicrobial activities. Quaternary compounds (quats) are used as disinfectants and antiseptics.
Acids and alkalies	Organic acids can control microbial growth and are frequently used as preservatives. Sorbic, benzoic, lactic, and propionic acids are used to preserve foods and pharmaceuticals. Benzoic, salicylic, and undecylenic acids are used to control fungi that cause diseases such as athlete's foot.

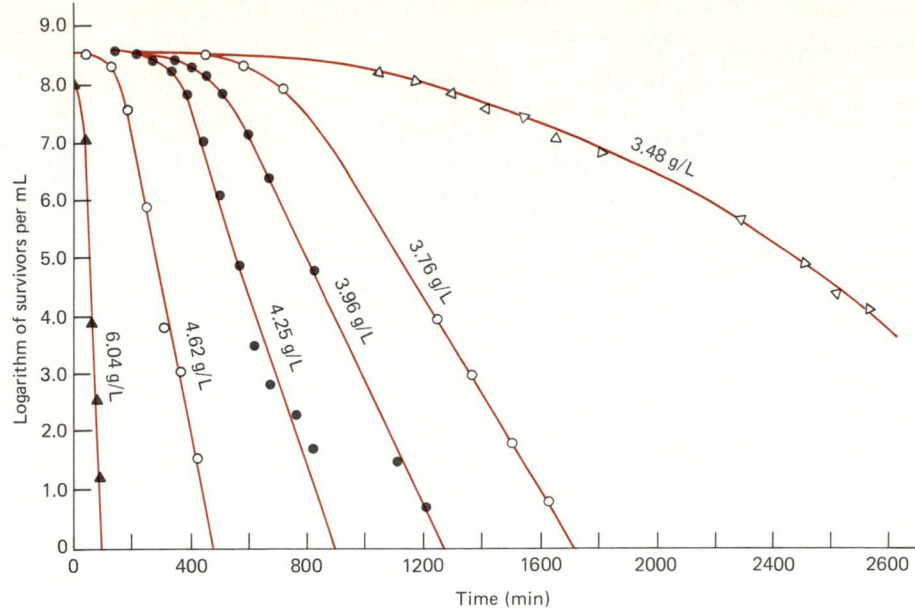

FIGURE 20.5

This graph charts the results of exposure of *Escherichia coli* to various concentrations of phenol at 35°C. The number of survivors, expressed logarithmically, is plotted against time. The concentrations of phenol used were (A) 6.04 g/L; (B) 4.62 g/L; (C) 4.25 g/L; (D) 3.96 g/L; (E) 3.76 g/L; and (F) 3.48 g/L. Increasing the concentration of the disinfectant increases the speed with which bacterial cells are killed. (Based on R. C. Jordan and S. E. Jacobs, 1945, *Journal of Hygiene*, 44:210. Cambridge University Press)

FIGURE 20.6

Joseph Lister (1827–1912) recognized the importance of preventing the contamination of wounds in order to curtail the development of infection. Here Lister is shown spraying carbolic acid over a patient undergoing an operation (ca. 1867). (Bettmann Archive)

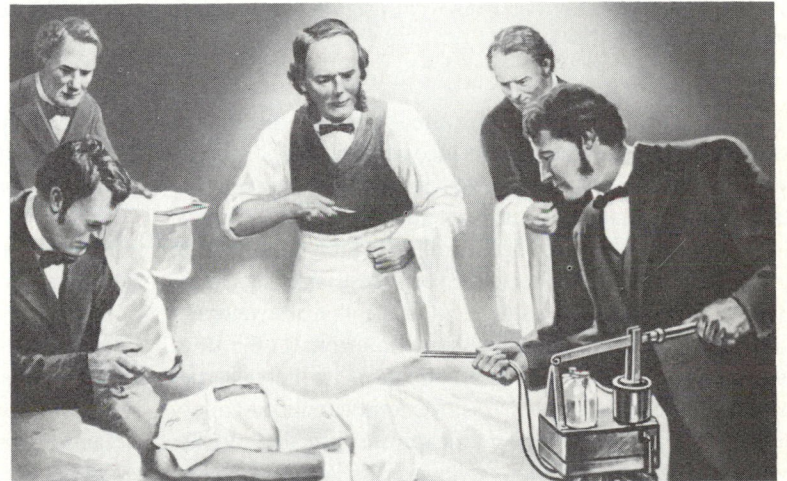

Discovery Process

The use of antiseptics to prevent infections was introduced by Joseph Lister, an English Quaker and physician, who revolutionized surgical practice in 1867 by introducing antiseptic principles. The discovery in the early 1850s of anesthesia and its administration to patients made surgery much easier but, of course, did nothing to reduce the incidence of post surgical disease, which was often as high as 90 percent, especially in military hospitals. Lister knew that in the 1840s Ignaz Semmelweis, a Hungarian physician who worked in maternity wards in Vienna, had shown that physicians who went from one patient to another without washing their hands were responsible for transmitting childbed fever (puerperal fever). He was also aware that Pasteur in the 1860s had demonstrated that microorganisms are present in the air. Lister used carbolic acid (phenol) as an antiseptic during surgery. He first used bandages soaked in carbolic acid to dress wounds due to compound fractures in order to diminish the likelihood of infection. Later he used a carbolic acid spray in addition to direct application of this compound during surgical procedures. He eventually discarded the practice of spraying after 17 years of trials as unnecessary, but he retained the use of direct application.

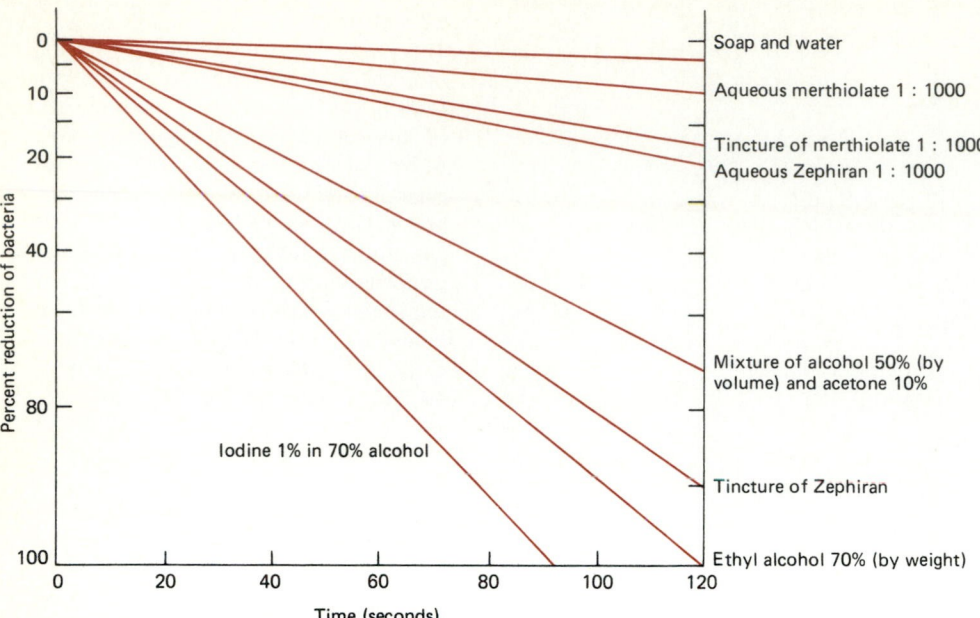

FIGURE 20.7

This graph illustrates the comparative effectiveness of various antiseptic agents. In each test, the bacterial microbiota before antiseptic application was considered to be 100 percent. The residual biota immediately after the use of the antiseptic is shown as a proportion of the original one. The steeper the curve, the greater the effect. Note that if the original concentration was $10^6/cm^2$, even the greatest reduction shown would have a resident population of about $10^4/cm^2$ (Based on P. B. Price, 1957, Skin antisepsis, in *Lectures on Sterilization*, (J. H. Brewer, ed., Duke University Press, Durham, N.C.)

most widely used antiseptic and is used to reduce the number of microorganisms on the skin surface in the area of a wound, as well as for the disinfection of various contaminated objects. Alcohol denatures proteins, extracts membrane lipids, and acts as a dehydrating agent, all of which contribute to its effectiveness as an antiseptic. Even viruses are inactivated by alcohol. The drawbacks of alcohol are that it evaporates too quickly and that it dries and sometimes cracks the skin.

Iodine is another effective antiseptic agent, killing all types of bacteria, including spores. It is frequently applied to minor wounds to kill microorganisms contaminating the surface, preventing infection[2]. Various dyes used in selective media, such as crystal violet, are similarly used as antiseptic agents. Such stains are normally effective bactericidal agents in concentrations of less than 1:10,000. Heavy metals are also used

in antiseptic formulations. Mercuric chloride, copper sulfate, and silver nitrate are a few examples of heavy metal–containing compounds that are used to kill microorganisms. Silver nitrate, for example, was once applied to the eyes of newborn human infants to kill possible microbial contaminants in order to preclude the transmission of gonococcal infections from an infected mother to the infant's eyes. It has been replaced with the antibiotic erythromycin. Phenolics and related compounds are often used as surgical scrubs. Both cationic and anionic detergents are also used as antiseptics; various detergents are quite effective, particularly those containing quaternary ammonium salts. A check of your pharmacy's shelves will illustrate the number of different agents and chemical formulations that are marketed as antiseptics.

20.4 Preventing Disease Using the Body's Immune Response

Before considering how the body's immune response can be used to prevent disease, we should discuss how an epidemic occurs and how disease is transmitted among susceptible individuals (Figure 20.8). **Epidemics** are outbreaks of disease in which unusually high numbers of individuals in a population contract a disease. As these individuals recover from the disease, they become immune, that is, unsusceptible, and thus no longer participate in the chain of disease transmission. The number of individuals who must be immune to

[2]Iodine is not used by diabetics because it causes localized tissue damage.

ANALYTICAL PROCESS

Evaluation of the Effectiveness of Disinfectants and Antiseptics

Several standardized test procedures have been employed for evaluating the effectiveness of disinfectants. The classic test procedure, used until a few decades ago, is the phenol coefficient. The phenol coefficient test compares the activity of a given product with the killing power of phenol under the same test conditions. To determine the phenol coefficient, dilutions of phenol and the test product are added separately to test cultures of *Staphylococcus aureus* or *Salmonella typhi*. The tests are run in liquid culture. After exposure for 5, 10, and 15 minutes, a sample from each tube is collected and transferred to a nutrient broth medium. After incubation for 2 days, the tubes from the different disinfectant dilutions are examined for visible evidence of growth. The phenol coefficient is defined as the ratio of the highest dilution of a test germicide that kills the test bacteria in 10 minutes, but not in 5 minutes, to the dilution of phenol that has the same killing effect. For example, if the greatest dilution of a test disinfectant producing a killing effect is 1:100, and the greatest dilution of phenol showing the same result is 1:50, the phenol coefficient is 100/50 or 2.0.

The phenol coefficient indicates the relative antimicrobial activity of various disinfectants but does not establish the appropriate concentration that should be used for disinfecting surfaces. The Association of Official Analytical Chemists' (AOAC) use-dilution method, which has replaced the phenol coefficient as the standard method for evaluating the effectiveness of disinfectants, establishes appropriate dilutions of a germicide for actual conditions. In this procedure, disinfectants are tested against *S. aureus* strain ATCC 6538, *Salmonella cholerasuis* strain ATCC 10708, and *Pseudomonas aeruginosa* strain ATCC 15442. Small stainless steel cylinders are contaminated with specified numbers of the test bacteria. After the cylinders are dried, they are placed in a series of specified dilutions of the test disinfectant. At least 10 replicates of each organism at the test dilutions of the disinfectant

are used. The cylinders are exposed to the disinfectant for 10 minutes, allowed to drain, transferred to appropriate culture media, and incubated for 2 days. After incubation the tubes are examined for growth of the test bacteria. No growth should occur if the disinfectant was effective at the test concentration. An acceptable use dilution is one that kills all test organisms at least 95 percent of the time.

In addition to the quantitative techniques described previously, potential antimicrobial activity can be qualitatively evaluated by adding pieces of filter paper saturated with the test agent to agar media seeded with test bacteria. The lack of growth in the vicinity of the filter paper containing the test agar indicates the antimicrobial effect of the agent. Standardized test procedures, such as the use-dilution test, are also useful for comparing the effectiveness of different antiseptics.

Two factors must be evaluated in determining the effectiveness of antiseptics: the antimicrobial activity of the agent and the lack of toxicity to living tissues. A particularly meaningful approach for comparing antiseptics that encompasses both of these factors is the generation of a toxicity index. In the tissue toxicity test, germicides are tested for their ability to kill bacteria and their toxicity to chick-heart tissue cells. The toxicity index is defined as the ratio of the greatest dilution of the product that can kill the animal cells in 10 minutes to the dilution that can kill the bacterial cells in the same period of time under identical conditions. For example, if a substance is toxic to chick-heart tissue at a dilution of 1:1,000 and is bactericidal for *S. aureus* at a dilution of 1:10,000, the toxicity index would be 1,000/10,000 or 0.1. Typical toxicity values for tincture of iodine solution and tincture of merthiolate are 0.2 and 3.3, respectively. Ideally, an antiseptic should have a toxicity index of less than 1.0, that is, it should be more toxic to bacteria than to human tissue.

---

prevent an epidemic outbreak of disease is a function of the infectivity of the disease (I), the duration of the disease (D), and the proportion of susceptible individuals in the population (s). When the triple product, $s \times I \times D$, is low because of a high proportion of immune individuals, that is, when approximately 70 percent of the population is immune, the whole population generally is protected, a concept known as **herd immunity**. Although immunity in 70 percent of the population usually prevents propagation of a pathogen through the population, the proportion of the population that must be immune to prevent an epidemic varies depending on the effectiveness with which the pathogen is transmitted and its virulence. Herd immunity can be established by artificially stimulating the immune response system through the use of

vaccines, rendering individuals unsusceptible to a particular disease and thereby protecting the entire population.

Immunization

Vaccines are preparations of antigens designed to stimulate the normal primary immune response, resulting in a proliferation of the memory cells and the ability to exhibit a secondary memory or anamnestic response upon subsequent exposure to the same antigens. **Immunization**, that is, the intentional exposure to antigens to elicit an immune response, was first introduced in the 1700s to control smallpox; it has since been used to prevent various other diseases (Figure 20.9). The antigens within the vaccine need not be as-

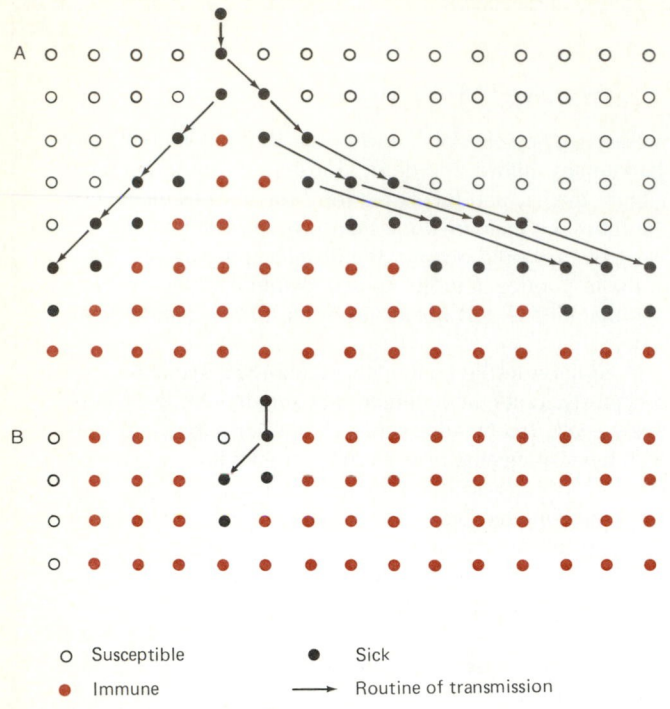

Day 1
Day 2
Day 3
Day 4
Day 5
Day 6
Day 7
Day 8

Day 1
Day 2
Day 3
Day 4

o Susceptible   ● Sick
● Immune   ⟶ Routine of transmission

FIGURE 20.8

The kinetics of the spread of an infectious disease and the effect of increasing the number of immune individuals in the population in limiting epidemic outbreaks of disease. (A) The individuals in this population are all susceptible. The introduction of a sick individual initiates an epidemic outbreak of the disease. At the height of the epidemic, 50 percent of the population are ill. Eventually, all individuals develop the disease and subsequently become immune. (B) Eighty percent of the population are immune to the disease when a sick individual enters the population. No epidemic occurs because susceptible individuals fail to contact an infected individual and to contract the disease.

sociated with active virulent pathogens. They need only elicit an immune response, with the production of antibodies possessing the ability to cross-react with the critical antigens associated with the pathogens against which the vaccine is designed to protect. Vaccines are useful because they confer immunity; that is, they render an individual unsusceptible to a disease, without actually producing the disease, or at least not a serious form of it.

Some of the vaccines that are useful in preventing diseases caused by a variety of microorganisms are listed in Table 20.2. Vaccines may contain antigens prepared by killing or inactivating pathogenic microorganisms, or they may use attenuated (weakened) strains that are unable to cause severe disease symptoms. Some vaccines are prepared by denaturing microbial exotoxins; the denatured proteins produced are called **toxoids**. Protein exotoxins, such as those involved in tetanus and diphtheria, are suitable for toxoid preparation, and the vaccines for preventing these diseases employ toxins inactivated by treatment with formaldehyde. These toxoids retain the antigenicity of the protein molecules; that is, they elicit the formation of antibody and react with antibody molecules, but because the proteins are denatured, they are unable to initiate the biochemical reactions associated with the active toxins that cause disease conditions. In some cases, whole microorganisms rather than individual protein toxins are used for preparing vaccines. When microorganisms are killed by treatment with chemicals, radiation, or heat, the antigenic properties of the pathogen are retained without the risk that exposure to the vaccine could cause the onset of the disease associated with the virulent live pathogens. The vaccines used for the prevention of whooping cough (pertussis) and influenza are representative of the preparations containing

antigens that are prepared by inactivating pathogenic microorganisms.

Even when the vaccines are killed cells, problems can occur in some cases. A small percentage of children, for example, have allergic reactions to the pertussis component of the standard diphtheria-pertussis-tetanus (DPT) vaccine, leading some to question the wisdom of government-mandated administration of this vaccine. Most manufacturers of this vaccine have ceased its production rather than face the liability lawsuits associated with such reactions, leaving the supply of DPT vaccine dangerously short. Enhanced quality control programs by the major remaining producer

TABLE 20.2 Some Vaccines Useful in Preventing Microbial Diseases

Disease	Types of Vaccine
Smallpox	Attenuated live virus
Yellow fever	Attenuated live virus
Hepatitis B	Attenuated live virus
Measles	Attenuated live virus
Mumps	Attenuated live virus
Rubella	Attenuated live virus
Polio	Attenuated live virus (Sabin)
Polio	Inactivated virus (Salk)
Influenza	Inactivated virus
Rabies	Inactivated virus
Tuberculosis	Attenuated live bacteria
Pertusis	Inactivated bacteria
Cholera	Inactivated bacteria
Diphtheria	Toxoid
Tetanus	Toxoid
Haemophilus meningitis	Capsular material
Pneumococcal pneumonia	Capsular material

FIGURE 20.9

(A) Edward Jenner (1749–1823) vaccinated James Phipps in about 1800 with cowpox material, resulting in the development of resistance to smallpox infection by the boy and establishing the scientific credibility of vaccination to prevent disease. (Culver Pictures)
(B) The vaccination of a child against rabies conducted under the direction of Louis Pasteur. (Bettmann Archive)

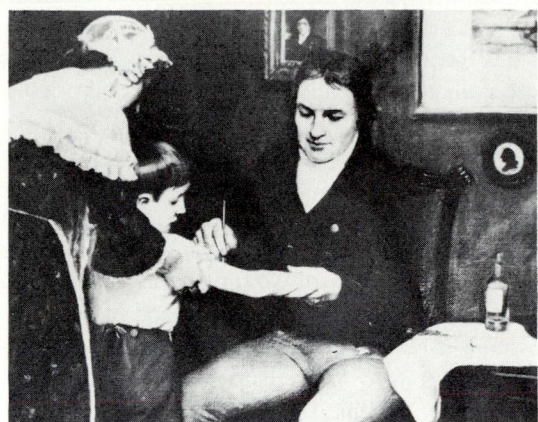

A

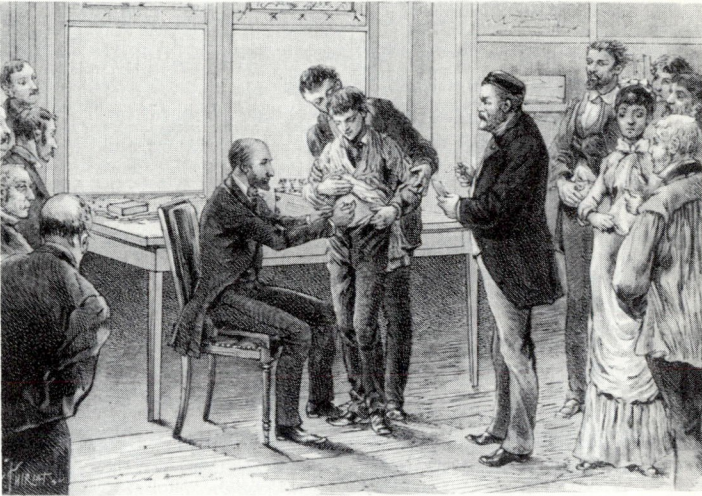

B

Discovery Process

The practice of immunization was used in the Far East for centuries before it was introduced into England in 1718 by Lady Mary Montagu, whose husband had been the British ambassador to Turkey. Lady Mary used her considerable influence in the court of King George I to gain publicity for the increased use of immunization. She even arranged for testing of her idea on prisoners and orphans, then a common practice, though she had no explanation of how or why it worked. Despite her efforts, immunization was not accepted by the scientists and physicians of the time as a useful practice for preventing disease. It was not until the report by Edward Jenner to the Royal Society in London 80 years later that credence was given to this practice. Jenner's 1798 report on the value of vaccination with cowpox as a means of protecting against smallpox established the basis for the immunological prevention of disease. Jenner was a middle-class country doctor, whose interest in science, like that of Leeuwenhoek, was typical of his position: scholarly but amateurish. The work, begun with Jenner's discovery of the effectiveness of vaccination in preventing smallpox, culminated in the 1970s with the eradication of smallpox from the face of the earth.

Much work was needed in addition to Jenner's report to achieve the elimination of smallpox and to use vaccines for preventing many other diseases. Pasteur greatly furthered the development of vaccines when in 1880 he reported that attenuated microorganisms could be used to develop effective vaccines against chicken cholera. The production of these vaccines depended on prolonging the time between transfers of the cultures, a fact accidentally discovered through an error by Charles Chamberland,

who used an old culture during one of the experiments he was conducting with Pasteur. Following his work on chicken cholera, Pasteur directed his attention to the study of anthrax. Because he enjoyed being the center of attention and controversy, Pasteur staged a dramatic public demonstration to test the effectiveness of his anthrax vaccine. Witnesses were amazed to see that the 24 sheep, 1 goat, and 6 cows that had received the attenuated vaccine were in good health, whereas all of the animals that had not been vaccinated were dead of anthrax.

In 1885 Pasteur announced to the French Academy of Sciences that he had developed a vaccine for preventing another dread disease, rabies. Although he did not understand the nature of the causative organism, Pasteur had developed a vaccine that worked. Pasteur's motto was "Seek the microbe," but the microorganism responsible for rabies is a virus, which could not be seen under the microscopes of the 1880s. Pasteur, nevertheless, was able to weaken the rabies virus by drying the spinal cords of infected rabbits and allowing oxygen to penetrate the cords. Thirteen inoculations of successively more virulent pieces of rabbit spinal cord were injected over a period of 2 weeks during the summer of 1885 into Joseph Meister, a 9-year-old boy who had been bitten by a rabid dog. "Since the death of the child was almost certain, I decided in spite of my deep concern to try on Joseph Meister the method which had served me so well with dogs. . . . I decided to give a total of 13 inoculations in ten days. Fewer inoculations would have been sufficient, but one will understand that I was extremely cautious in this first case. Joseph Meister escaped not only the rabies that he might have received from his bites, but also the rabies which I inoculated into him." (Louis Pasteur, 1885, *Comptes Rendus. Academic des Sciences*). With the successful development of a vaccine for preventing rabies, crowds flocked to Pasteur's laboratory. The development of the rabies vaccine crowned Pasteur's distinguished career. While Pasteur used attenuated microorganisms for his vaccines, Daniel Salmon and Theobold Smith (1886) demonstrated that it was not necessary to use a live attenuated microbial strain in order to achieve artificial immunity. They developed the first killed vaccine for the control of hog cholera, a significant advance in the creation of vaccines to prevent disease.

and the development of a new form of the vaccine promise to reduce the incidence of adverse reactions.

In contrast to these vaccines, other vaccine preparations contain living but attenuated strains of microorganisms. Pathogens are **attenuated** by a variety of procedures, including moderate use of heat, chemicals, desiccation, and growth in tissues other than the normal host. The Sabin vaccine for poliomyelitis, for example, uses viable polioviruses attenuated by growth in tissue culture. These viruses are capable of multiplication within the digestive tract and the salivary glands but are unable to invade the nerve tissues and thus do not produce the symptoms of polio. The vaccines for measles, mumps, rubella, and yellow fever similarly utilize viable but attenuated viral strains. Attenuated strains of rabies virus can be prepared by desiccating the virus after growth in the central nervous system tissues of a rabbit or following growth in a chick or duck embryo. Vaccines containing viable attenuated strains require relatively low amounts of the antigens because the microorganism is able to replicate after administration of the vaccine, resulting in a large increase in the amount of antigen available within the host to trigger the immune response mechanism.

Quality control is extremely important in preparing all vaccines, particularly those using attenuated strains. Some people given swine flu vaccine during the 1976 scare about an impending outbreak of this disease actually contracted flu because of the inadequate inactivation (killing) of the viruses in hastily prepared vaccines. Others developed a neurological disorder called Guillain-Barré syndrome after vaccination against swine flu. In the 1950s, several tragic cases of polio occurred in children given the Salk polio vaccine, which was prepared with inactivated polioviruses, because of the failure to fully inactivate some batches of the vaccine. Because the Salk vaccine is prepared from a particularly virulent strain of poliovirus, replication of the virus in those inoculated with the problem batches caused paralytic polio.

The failure of the quality control program for the Salk vaccine was partly responsible for the general switch to the live attenuated Sabin polio vaccine. The Sabin vaccine is prepared with attenuated viral strains that are not particularly virulent and that do not invade the nervous system, causing paralysis. It uses strains of poliovirus that have the three predominant antigens of the major polioviruses, designated type 1, 2, and 3 antigens. The Sabin vaccine is administered orally, and the virus multiplies within the gastrointestinal tract. Although the virus is attenuated, mutations and recombinations are possible during replication. Some recent cases

of polio have been reported with the Sabin vaccine, causing some to reevaluate the relative merits of the Salk versus the Sabin vaccine.

One way to avoid the problems associated with both attenuated (live) and inactivated (killed) vaccines, such as the Salk and Sabin vaccines, is to use only individual components of the microorganism to elicit an immune response. For example, the capsule of *Streptococcus pneumoniae* is used to make a vaccine against pneumococcus pneumonia. This vaccine is used in high-risk patients, such as individuals over 50 years old who have chronic diseases such as emphysema. Another vaccine has been produced from the capsular polysaccharide of *Haemophilus influenzae* type b, a bacterium that frequently causes meningitis in children 2–5 years old. The *Hib vaccine*, as it is called, is being widely administered to children in the United States. While this vaccine is not always effective in establishing protection in children under 2 years, it is administered to children between 18 and 24 months old who attend day care centers because they have a greater risk of contracting *H. influenzae* infections.

It is not always easy to find antigens associated with pathogens that confer long-term, active immunity. Desperate efforts are now underway to formulate a vaccine that will prevent AIDS. Years of research, however, have failed to produce vaccines against other sexually transmitted diseases—such as syphilis—as well as other prevalent diseases—such as malaria and tooth decay. Attempts were made to make a vaccine against gonorrhea using pili from *Neisseria gonorrhoeae*, the bacterium that causes this disease, but were not successful because long-lasting immunity against *N. gonorrhoeae* does not develop; the vaccine, though, has been used by the military to achieve short-term immunity. Other vaccines are in development that use ribosomes instead of surface components of the cell. Additionally, synthetic proteins are being considered as potential antigens for protection against various diseases, and recombinant DNA technology is being used to create **vector vaccines** containing the genes for the surface antigens for various pathogens. A vector vaccine is one that acts as a carrier for antigens associated with pathogens other than the one from which the vaccine was derived. The attenuated virus used to eliminate smallpox is a likely vector for simultaneously introducing multiple antigens associated with different pathogens, such as the chicken pox virus; several prototype vaccines using the smallpox vaccine as a vector have been made.

The effectiveness of a vaccine depends on a number of factors, including the antigens in the vaccine, the other chem-

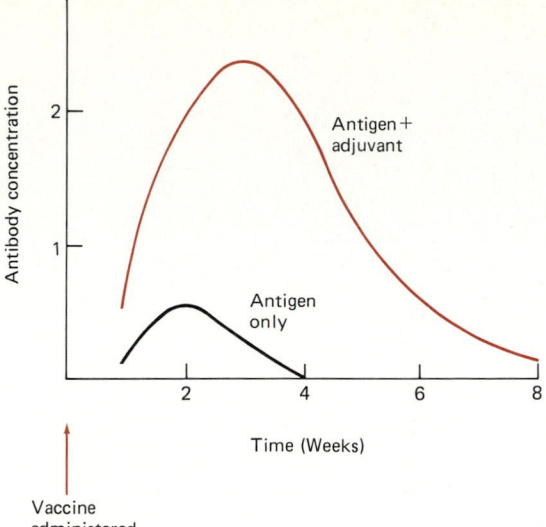

FIGURE 20.10
Adjuvants enhance antigenicity, as illustrated by this graph.

TABLE 20.3 Recommended Schedule for Vaccine Adminstration

Disease	Primary Immunization	Booster Doses
Diphtheria, pertussis, tetanus	Intramuscular DPT injections at 2, 4, 6, and 15 months	One intramuscular booster at 3–6 years (tetanus every 10 years)
Influenza	Seasonally for high-risk elderly and chronically ill	
Measles, mumps, rubella	One subcutaneous MMR injection at 15 months	None
Polio	Sabine vaccine: oral at 2, 4, and 15 months	One oral dose at 4–6 years
	Salk vaccine: intramuscular injections at 2, 3, 4, and 16 months	Intramuscular injections every few years

icals in the vaccine, and the route of administration. Some chemicals, known as **adjuvants**, greatly enhance the antigenicity of other biochemicals (Figure 20.10). The inclusion of an adjuvant therefore can greatly increase the effectiveness of the vaccine. When protein antigens are mixed with aluminum compounds, for example, a precipitate is formed that is more useful for establishing immunity than are the proteins alone. Alum-precipitated antigens are released slowly in the human body, enhancing the stimulation of the immune response. The use of adjuvants eliminates the need for repeated booster doses of the antigen—which increase the intracellular exposure to antigens to establish immunity—and permits the use of smaller doses of the antigen in the vaccine.

The antigens in the vaccine may be introduced into the body by a number of routes: **intradermally** (into the skin), **subcutaneously** (under the skin), **intramuscularly** (into the muscle), **intravenously** (into the bloodstream), into the mucosal cells lining the respiratory tract through inhalation, or orally into the gastrointestinal tracts. The effectiveness of a given vaccine depends in part on the normal route of entry for the particular pathogen. For example, polioviruses normally enter via the mucosal cells of the upper respiratory or gastrointestinal tract; therefore, the Sabin polio vaccine is administered orally, enabling the attenuated viruses to enter the mucosal cells of the gastrointestinal tract directly. It is likely that vaccines administered this way stimulate secretory antibodies of the IgA class in addition to other immunoglobulins. Intramuscular administration of vaccines, like the Salk polio vaccine, is more likely to stimulate IgG production, which is particularly effective in precluding the spread of pathogenic microorganisms and toxins produced by such organisms through the circulatory system.

Multiple exposures to antigens are sometimes needed to ensure the establishment and continuance of a memory re-

sponse. Periodic **booster vaccinations** are necessary, for example, to maintain immunity against tetanus. Several administrations of the Sabin vaccine are needed during childhood to establish immunity against poliomyelitis. No booster vaccinations, though, are needed to establish permanent immunity against measles, mumps, and rubella. The recommended schedule for the administration of some vaccines is shown in Table 20.3. Although vaccines are normally administered prior to exposure to antigens associated with pathogenic microorganisms, some vaccines are administered after suspected exposure to a given infectious microorganism. In these cases, the purpose of vaccination is to elicit an immune response before the onset of disease symptomatology. For example, tetanus vaccine is administered after puncture wounds may have introduced *Clostridium tetani* into deep tissues, and rabies vaccine is administered after animal bites may have introduced rabies virus. The effectiveness of vaccines administered after the introduction of the pathogenic microorganisms depends on the relatively slow development of the infecting pathogen prior to the onset of disease symptoms and the ability of the vaccine to initiate antibody production before the active toxins are produced and released to the site where they can cause serious disease symptoms.

Not all diseases can be prevented by using vaccines, and some antigens confer immunity that lasts for only weeks or months. Such short-lived immunity may be effective in preventing disease if there is a known likelihood of exposure to a given pathogen, but it is not feasible to attempt the large-scale use of vaccines that confer only short-term immunity.

The greatest success in preventing disease through the use of vaccines can be seen in the case of smallpox (Figure 20.11). The vaccine used to prevent smallpox contains a live strain of pox virus. The vaccine most commonly used is prepared from scrapings of lesions from cows or sheep. The

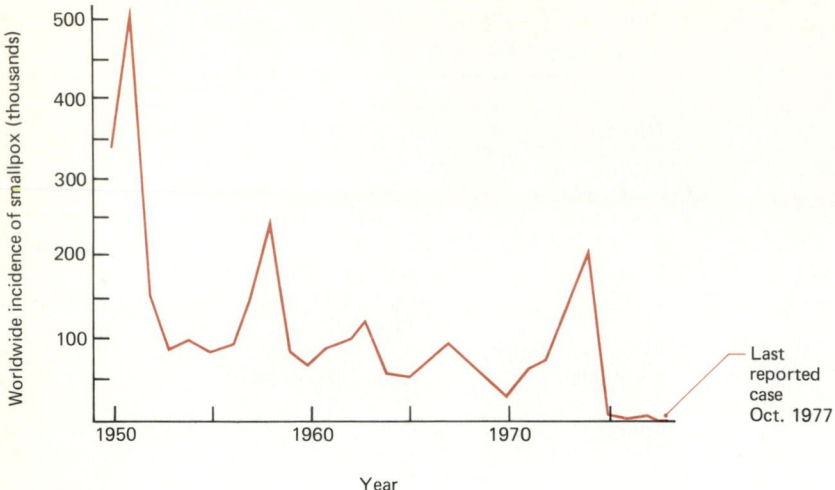

FIGURE 20.11
This graph shows the decline in the incidence of smallpox as a result of an effective international vaccination program.

scrapings are treated with 1 percent formaldehyde to kill bacterial contaminants and 40 percent glycerol to stabilize the viral antigens. These antigens are quite labile, which is why "live" viral preparations are required for successful vaccination to achieve immunity. Various commercial viral strains have been used for the production of commercial vaccines. Although these strains were presumed to have been derived from cowpox virus, it now appears, based on its antigenic properties, that an attenuated strain of smallpox virus may have been inadvertently used. Because of the length of time that this virus has been cultivated, it is difficult to identify its original source positively, but the pox virus used for vaccine preparation clearly differs from the cowpox viruses found in nature.

Regardless of the origins of the viral strain used in the vaccines, smallpox, a once dreaded disease, has been completely eliminated through an extensive worldwide immunization program conducted under the auspices of the World Health Organization (WHO). The success of the WHO program depended on the use of lyophilized vaccines to overcome the problem of inactivation of the viral antigens in hot climates. The program was not without risks because the virus used for vaccination was virulent enough to cause a fatality rate of 1 in 1 million vaccinations. By immunizing a sufficient portion of the world's population against smallpox, though, it was possible to interrupt the normal transmission of smallpox virus from infected individuals to susceptible hosts. A consequence of the success of this immunization program is that it is no longer necessary to vaccinate against smallpox. The successful elimination of smallpox through a vaccination program was dependent on the fact that humans are the only known host for the smallpox virus and that the virus has a relatively short survival time outside human host tissues. Smallpox presumably has been eliminated permanently and, as such, is the only infectious human disease known to have been eliminated through human intervention, ingenuity, and cooperation.

Other Immunological Procedures

There are also several other immunological procedures that may be used to prevent disease. For example, a variety of **antitoxins** (antibodies that neutralize toxins) can be used to prevent toxins of microbial or other origin from causing disease symptomatology. The administration of antitoxins establishes passive artificial immunity. Antitoxins are used to neutralize the toxins in snake venom, saving the victims of snake bites. The toxins in poisonous mushrooms can also be neutralized by administration of appropriate antitoxins. The administration of antitoxins and immunoglobulins to prevent disease occurs after exposure to a toxin and/or an infectious microorganism.

It is also possible to establish passive immunity by the administration of IgG obtained from another individual. Passive immunity lasts for a limited period of time because IgG molecules have a finite lifetime in the body and because the administration of IgG does not involve the establishment of a memory immune response capability. Such passive immunity is conferred naturally upon an infant by the passage of IgG molecules across the placenta during fetal development. IgG and IgA are found in the colostrum and milk of nursing mothers, protecting newborns against infectious diseases during the early period of life. The administration of IgG is also particularly useful therapeutically in preventing disease in persons with immunodeficiencies and other high-risk individuals.

20.5 Diagnosis of Disease

Pattern of Disease

Despite preventive measures and host defenses, infectious diseases occur and their diagnosis is important for instituting appropriate treatment. In many cases, the reproduction of pathogenic microorganisms within the body produces specific diseases that are associated with characteristic signs and symptoms. **Signs** are objective changes, such as a rash or fever that a physician can observe. **Symptoms** are subjective changes in body function, such as pain or loss of appetite, that are experienced by the patient. A characteristic group of signs and symptoms constitutes a **disease syndrome**. Often the physician is able to diagnose a disease exclusively on the basis of the symptoms reported by the patient and the signs observed. In other cases, more elaborate laboratory tests are necessary to identify the cause of the disease.

In **acute** diseases the symptoms and signs develop rapidly, reaching a height of intensity, and end fairly quickly. Measles, cholera, and influenza are all examples of acute diseases. In **chronic** diseases the symptoms persist for a prolonged period of time. The persistent cough of chronic bronchitis is typical of the long-term signs associated with a chronic disease.

Whether a disease is acute or chronic, when it is caused by an infectious agent, it produces a characteristic pattern. The progress of any infectious disease in a given patient can be divided into several stages: incubation, prodromal, period of illness (also called *onset of the acute phase*), period of decline (also called *progression of the acute phase*), and convalescence.

In the normal course of disease, there is an **incubation period** after the pathogen enters the body and before any signs or symptoms appear. The incubation period varies in different diseases. During this period the microorganism has invaded the host, is migrating to various tissues, but has not yet begun to increase to sufficient numbers to cause discomfort or infectivity. The onset of symptoms marks the end of the incubation period and the start of the **prodromal stage**. Now the patient is aware of discomfort but does not yet have precise sufficiently symptoms to permit the clinician to make a diagnosis. However, sufficient replication of the pathogen has occurred to render the patient contagious to others. Moreover, the nonspecific inflammatory defenses have become operative.

The **period of illness** occurs next, during which time the disease is most severe. The various signs and symptoms that characterize the particular disease occur in this period. During the period of illness (acute stage), the patient often is sufficiently ill to alter his or her normal work or school activities. Clones of B or T cells are being selected to initiate the immune defense. This phase of the disease progresses toward either death or convalescence. Recovery depends upon whether the immune systems or medical treatments are adequate. Assuming the disease is not fatal or chronic, the signs and symptoms begin to disappear during the **period of decline**. **Convalescence** progresses either to a carrier stage or to freedom from the pathogen. In some cases, the immune memory system may protect the person from recurrence of the infection for several months, several years, or life. Full **recovery** marks the end of the disease syndrome.

Indicators That Diseases Are of Microbial Etiology

The characteristics of the immune response system provide the basis for determining whether a disease is of probable microbial etiology. In most cases, a microbial infection elicits an inflammatory response characterized by fever, pain, swelling, and redness. Although an inflammatory response does not necessarily reflect an infectious disease, an elevated body temperature (fever) is often considered presumptive evidence of a microbial infection. The physician observing a patient with a red, sore throat and fever assumes that the symptoms are the result of a microbial infection. In many such cases, when the presumptive evidence strongly indicates pharyngitis (infection of the pharynx), treatment is usually administered without rigorous clinical diagnosis and confirmation of the cause, even though it would have been appropriate to identify the etiological agent by laboratory testing.

Differential Blood Counts In other cases, where the identification of a microbial infection is not clear-cut, additional presumptive evidence of an infectious process can be obtained by performing a differential blood count, in which the relative concentrations of different types of blood cells are determined. This clinical procedure can provide a general indication of the nature of the infecting agent, that is, if the disease is mediated by a virus, bacterium, fungus, or protozoan. Changes in the composition of the blood usually occur as a consequence of a microbial infection. Such changes generally result from the immune response and are reflected in shifts in the relative quantities and types of white blood cells. An elevated white blood cell count (**leukocytosis**) is characteristic of many systemic infections. A systemic bacterial infection, for example, is normally characterized by a progressive **neutrophilia (neutrophilic leukocytosis)**, an increase in neutrophil cells, particularly an increase in young neutrophil cells known as **stab** or **band cells** (Figure 20.12). Compared to mature neutrophils, stab cells have a U-shaped nucleus that is slightly indented but not segmented. The increase in stab cells, indicative of neutrophilia, is known as a *shift to the left*, referring to a blood cell classification system in which immature blood cells are positioned on the left side of a standard reference chart and

| | Total leukocytes | Baso-philes | Eosino-philes | Normal Neutrophils | | | | Lympho-cytes | Mono-cytes |
				Myelo-cytes	Juveniles	Stabs	Segments		
Normal	7500	0–1	2–4	0	0–1	3–5	58–66	21–30	4–8
Scarlet fever	16,680	2	0	84	1	15	58	18	7
Appendicitis	13,800	0	0	0	10	59	20	10	0
Staphylococcus septicemia	34,950	0	0	0	12	31	46	8	3
Tularemia	19,550	1	0	0	6	53	23	12	5

FIGURE 20.12

Some representative differential white blood cell counts for various infections.

mature blood cells are placed on the right. The recovery phase of an infection is characterized by a reduction in fever, a decrease in the total number of leukocytes, and an increase in the number of monocytes. Gradually, the relative numbers of the various white blood cells return to their respective normal ranges.

In addition to systemic infections, some localized infections, like abdominal abscesses, result in neutrophila. Not all bacterial infections, though, show this characteristic leukocytosis. Some, such as typhoid fever, paratyphoid fever, and brucellosis, actually result in a persistent depression in the number of neutrophil cells (**neutropenia**). Many viral infections similarly result in lowered numbers of white blood cells (**leukopenia**). A general indication of whether a disease is of bacterial or viral origin, therefore, may be obtained by performing a white blood cell count and determining whether there is a significant shift in the quantity of neutrophils.

Changes in the quantities of eosinophils may also indicate the nature of the infection. The number of eosinophil cells generally declines during systemic bacterial infections. **Eosinophilia** (increased numbers of eosinophils) is symptomatic of allergic diseases and parasitic infections, including those mediated by protozoans. Thus, the observation of elevated numbers of eosinophils is useful in the preliminary diagnosis of such diseases.

In some diseases, such as infectious mononucleosis, there are characteristic changes in the white blood cells (Figure 20.13). In this disease, there is a transient leukocytosis because of an increase in B lymphocytes, which characteristically are enlarged—making them appear like monocytes—and show obvious changes in the nucleus, including the shape, size, and density of the nuclear region. These changes are useful in the diagnosis of this disease. A few microbial infections, malaria for example, result in decreased numbers of red blood cells (**anemia**). Thus, a simple examination of the blood often gives a preliminary indication of the etiology

of a disease condition, establishing the direction of additional test procedures for positively identifying the causative agent.

Skin Testing   **Skin testing**, based upon delayed hypersensitivity reactions, can be another useful procedure in the presumptive diagnosis of several infectious diseases. These tests, however, are merely screening methods used as diagnostic aids with regard to prior exposure to an infectious agent or antigen; they cannot be used for positive diagnosis of a disease. In skin testing, antigens derived from a test organism are injected intradermally. The development of redness in 24–72 hours is evidence of a delayed hypersensitivity reaction, indicating that the patient had previously been exposed and become sensitized to that specific antigen. A positive skin test may indicate an active infection caused by the organism from which the antigens are derived, but it usually reflects an earlier exposure to that organism.

The classic skin test for a microbial infection is the **tuberculin reaction** for detecting probable cases of tubercu-

FIGURE 20.13

There are various characteristic changes in white blood cells that occur as a result of infectious mononucleosis. (A) A normal B lymphocyte. (B) An atypical B cell indicative of infectious mononucleosis; the lymphocyte is enlarged, pleomorphic, and vacuolated.

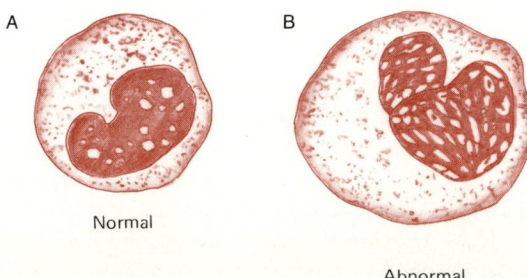

A   Normal

B   Abnormal

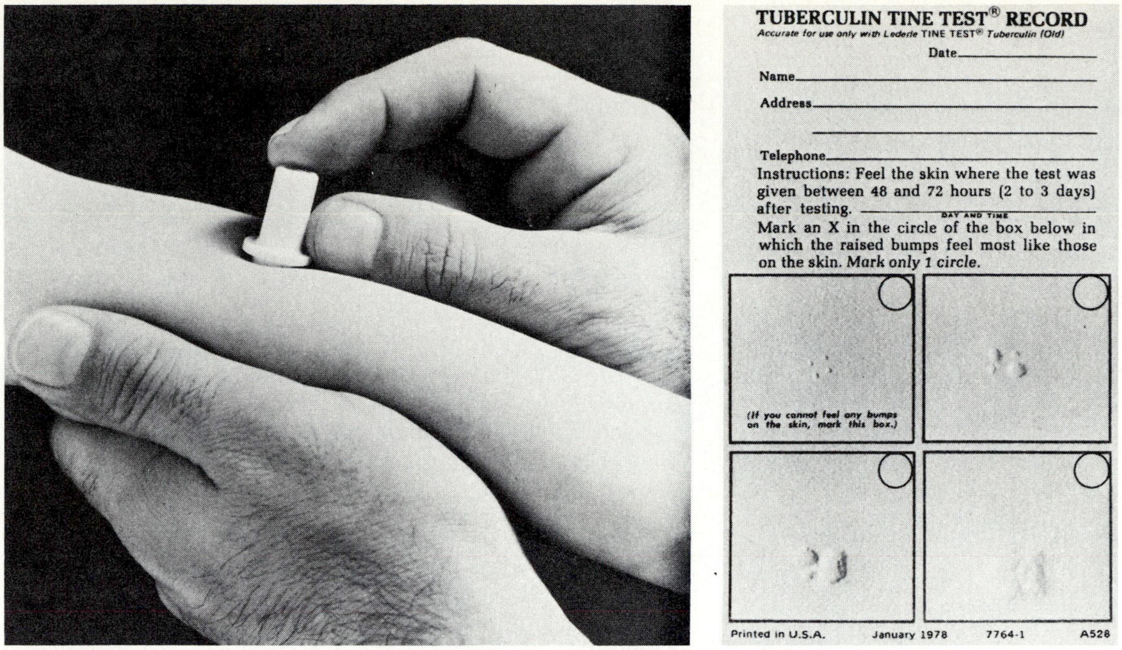

FIGURE 20.14

The results of a skin test, showing a positive tuberculin reaction. Development of inflammation after a delay is indicative of a positive reaction. (A) Application of the antigen to the skin in the Tine test. (B) Record card for the Tine test, showing development of positive reaction. (Courtesy Lederle Laboratories, Wayne, N.J.)

losis (Figure 20.14). A purified protein derivative (PPD) extract from *Mycobacterium tuberculosis* is injected subcutaneously, and the area near the injection is observed for evidence of a delayed hypersensitivity reaction. A positive test results in **erythema** (reddening) and **induration** (hardening) of the skin, with the peak reaction occurring in 48–72 hours. The reliability of the tuberculin test, though, depends on how the antigen is administered. In the **Montoux test**, commonly used in the United States, an appropriate dilution of PPD is injected intradermally into the superficial layers of the skin of the forearm. Other test procedures, including the once widely used **Tine test**, employ various mechanical devices and multiple punctures to expose the individual to the antigen. They are not as reliable as the Montoux PPD procedure.

Similar skin tests are available for the diagnosis of coccidiomycosis, using coccidiodin, an antigen derived from *Coccidioides immitis*; histoplasmosis, using histoplasmin, a crude filtrate from *Histoplasma capsulatum*; leprosy, using lepromin derived from *Mycobacterium leprae*; brucellosis, using brucellergen obtained from a *Brucella* species; and the venereal disease lymphogranuloma venereum, using lygranum from *Chlamydia* species. In many cases, skin tests are used to screen a population for individuals who are infected with a pathogen but who have not developed clinical symptoms; this procedure identifies persons for which additional rigorous test procedures should be performed. For example, tuberculosis testing is routinely carried out on schoolchildren to identify possible carriers of *M. tuberculosis*.

Screening and Isolation Procedures

Whereas disease symptomatology, changes in blood composition, and skin testing can serve as indicators of microbial infection and possibly of the nature of the microorganisms causing the disease, many other nonmicrobiological factors may produce similar symptoms and clinical findings. The positive diagnosis of an infectious disease, therefore, requires the isolation and identification of the pathogenic microorganism or the identification of antigens specifically associated with a given microbial pathogen.

A variety of procedures are employed for the isolation, collection, and identification of pathogenic microorganisms from different tissues (Table 20.4). Different procedures are required for the isolation of different types of microorganisms, and clinical procedures are designed to screen and facilitate the recovery of those etiological agents of disease that predominate within specific tissues. When the symptomatology suggests that the disease may be caused by a rare pathogen and/or routine screening fails to detect a probable causative microorganism, additional specialized isolation procedures may be required.

Identification of Pathogenic Microorganisms

Determining that a disease is caused by microorganisms and isolating those microorganisms are only part of the job of diagnosing a disease. It is also necessary to identify the specific causative organism. When the cause of a disease is first

TABLE 20.4 Some Procedures Used for the Diagnosis of Various Diseases, Indicating the Collection Method, Culture Medium, and Organisms Detected

Body Part	Collection Method	Culture Media	Organism	Result	Disease
Upper respiratory tract: throat and nasopharyngeal cultures	Sterile cotton swabs	Blood agar	*Streptococcus pyogenes*	Beta-hemolysis	Pharyngitis, rheumatic fever
		Chocolate agar	*Haemophilus influenzae,* *Neisseria meningitidis*		Epiglottitis Meningitis
		Thayer-Martin medium	*N. gonorrhoeae*		Gonorrhea
		Bordet-Gengou	*Bordetella pertussis*		Whooping cough
		Tellurite serum agar	*Corynebacterium diphtheriae*	Smooth, glistening gray-black colonies	Diphtheria
Lower respiratory tract	Transtracheal aspiration of sputum	Blood agar	*Streptococcus pneumoniae,* *Staphylococcus aureus*		Pneumonia
		Chocolate agar	*Streptococcus pyogenes*		
		MacConkey's agar	*Klebsiella pneumoniae,* *Haemophilus influenzae*		
		Stained smears	*Histoplasma capsulatum*		
		Sabouraud's agar	*Coccidioides immitis, Candida albicans*		
		Lowenstein-Jensen	*Mycobacterium tuberculosis*	Acid-fast, red colonies, increased turbidity	Tuberculosis
Central nervous system	Lumbar puncture for cerebrospinal fluid	Liquid enrichment media Blood agar Chocolate agar	*Streptococcus pneumoniae,* *Neisseria meningitidis,* *Haemophilus influenzae*		Meningitis
Circulatory tract blood	Renal puncture	Radiolabeled glucose medium Roll-tube streak anaerobic culture	Various		Septicemia
Urinary tract	Midstream catch of voided urine	Blood agar	*Escherichia coli*	Less than 10^5 bacteria/mL	Urinary tract infections
		Cysteine lactose Electrolyte-deficient agar MacConkey's agar, EMB agar	*Klebsiella, Proteus, Pseudomonas, Salmonella Serratia, E. coli,* and other Gram-negative rods		
Genital tract	Urethral exudate (males) Swabs from cervix, vagina and anal canal (females)	Thayer-Martin medium and chocolate agar	*Neisseria gonorrhoeae*	Gram-negative kidney bean shaped diplococci	Gonorrhea

TABLE 20.4   *(continued)*

Body Part	Collection Method	Culture Media	Organism	Result	Disease
Intestinal tract	Stool samples	*Salmonella-Shigella* Hektoen enteric media, xylase-lysine-desoxycholate media, brilliant green, EMB, Endo and MacConkey's agar			
Eyes and ears	Fluids	Blood, chocolate, MacConkey's agar, Gram stain			
Skin	Swabs, aspirates, or washings from lesions	Aerobic and anaerobic culture techniques	*Clostridium tetani, C. perfringens*		

investigated, one must identify the etiologic agent using an unambiguous procedure. For infectious diseases, Koch's postulates are used; these postulates, discussed in Chapter 1, permit the establishment of a cause-and-effect relationship between a specific pathogen and a particular disease. Once Koch's postulates are fulfilled and the causative microbial agent of a disease is known, it is possible to diagnose a disease based upon symptomatology and the positive identification of that pathogen in the patient.

A wide range of biochemical, serological, and gene probe procedures are available for the definitive identification of microbial isolates of clinical significance. Some of these procedures were discussed in Chapter 10. Accuracy, reliability, and speed are important factors governing the selection of clinical identification protocols. The selection of the specific procedures to be employed for the identification of pathogenic isolates is guided by the presumptive identification of the organism at the genus or family level, based on the observation of colonial morphology and other growth characteristics on the primary isolation medium, and on the microscopic observation of stained specimens. Some of the protocols and criteria used for the identification of various clinical isolates will be discussed in the following pages.

Both classical dichotomous key and computerized probabilistic matrix approaches are used for the identification of pathogenic microorganisms. Pathogenic filamentous fungi and protozoa are generally identified on the basis of the morphological characteristics of the organism, the growth appearance, and a limited number of biochemical tests. Various other morphological and biochemical characteristics are used for identifying other pathogens. Conventional identification schemes for bacteria and yeasts rely on the determination of a variety of biochemical features exhibited by growing isolates. In general, fewer than 20 tests are required to identify clinical bacterial isolates at the species level. The purpose is to distinguish the isolates present in the specimen, using the minimal number of tests to define distinct taxa accurately.

In addition to using growth and biochemical characteristics for the identification of pathogenic microorganisms, a variety of serological test procedures are employed, as discussed in Chapter 19. Serological tests are particularly useful in identifying pathogens that are difficult or impossible to isolate on conventional media and in identifying many varieties of pathogenic strains not easily distinguished by biochemical testing. For example, over 2,000 serotypes in the genus *Salmonella* are defined by the O (somatic cell) and H antigens (flagella), with each serotype defined by a constellation of O and H antigens. The identification of pathogenic viruses and nonculturable bacteria, such as *Treponema pallidum*, generally depends on serological testing.

20.6 *Treating Diseases with Antimicrobial Agents*

When normal host defense mechanisms and preventive measures fail to protect an individual against the establishment of a particular pathogenic microorganism, there are a large number of antimicrobial agents available for treating the resulting diseases. Such drugs have become an essential part of modern medical practice. The antimicrobial agents used in medical practice are aimed at eliminating infecting microorganisms or preventing the establishment of an infection. **Antibiotics**, which are defined as antimicrobial substances produced by microorganisms, were discovered by Sir Alexander Fleming (Figure 20.15). They have been used in medicine only since the mid-1940s.

Although many of the antimicrobial compounds used today are in fact produced by microorganisms, and therefore

FIGURE 20.15

The discovery and action of antibiotics are shown in these photographs, where the inhibitory effects of anti-biotic-producing strains on sensitive strains are demonstrated. (A) Sir Alexander Fleming (1881–1955), shown here developing penicillin in his laboratory, had the insight to recognize the significance of the inhibition of bacterial growth in the vicinity of a fungal contaminant when most other scientists probably would have simply discarded the contaminated plates. (Central Office of Information, London). (B) The original culture plate of Alexander Fleming, showing the growth of a *Penicillium* colony and the lack of colonies of *Staphylococcus* near the fungal colony. The colonies developing closest to the *Penicillium* colony show evidence of bacteriolysis. (From BPS—E. Chain and H. W. Florey, 1944, *Endeavor* 3:9). (C) Photograph showing zones of inhibited bacterial growth of a normal antibiotic-sensitive bacterium surrounding agar plugs selected from six different *Streptomyces* strains. (Courtesy Boyd Woodruff, Merck, Sharpe and Dohme, Rahway, N.J., reprinted by permission of the Society of General Microbiology from H. B. Woodruff and L. E. McDaniel, 1958, The antibiotic approach, in *Symposia of the Society of General Microbiology*, No. VIII, *The Strategy of Chemotherapy*, p. 49)

A

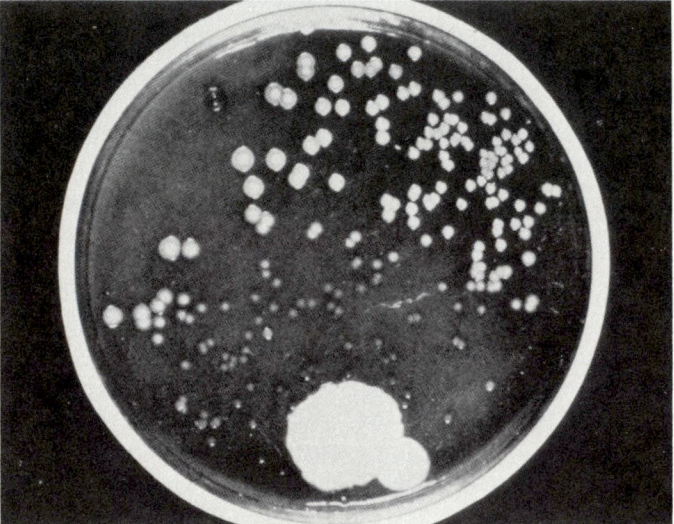

B

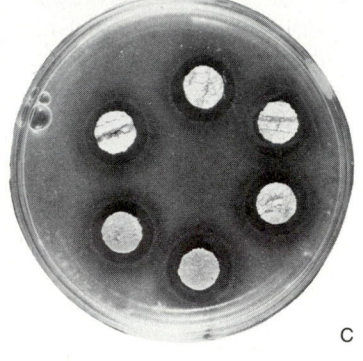

C

Discovery Process

Various chemical formulations for preventing microbial growth and infection were described by Robert Koch and his disciples, including Paul Ehrlich, who, like Pasteur, had been trained as a chemist. From 1880 to 1896, Ehrlich worked in Koch's laboratory, and in 1896 he became director of the first of his own institutes, which he dedicated to finding "substances which have their origin in the chemist's retort," that is, substances produced by chemical synthesis, to cure infectious diseases. Ehrlich's research between 1880 and 1910 established the early basis for modern chemotherapy. He established the correct formula for atoxyl, an arsenical, which was

being considered for use in treating sleeping sickness, and developed almost 1,000 new derivatives of this compound. Compound 606, salvarsan, proved to be effective in treating syphilis. Sahachiro Hata, a Japanese expert on spirochetes, tested the atoxyl derivatives, using syphilitic rabbits in Ehrlich's laboratory. He found compound 914, neosalvarsan, to be a reliable drug for curing syphilis and relapsing fever, both caused by spirochetes. The use of neosalvarsan in 1912 represents the first widespread use of synthetic drugs; these drugs became known as *magic bullets* and were portrayed as being able to find and kill disease-causing germs.

A major breakthrough in chemotherapy occurred in 1929, when the Scottish bacteriologist Alexander Fleming, working in a London teaching hospital, reported on the antibacterial action of cultures of a *Penicillium* species. Fleming observed that the mold *Penicillium notatum* killed his cultures of the bacterium *Staphylococcus aureus* when the fungus accidentally contaminated the culture dishes. It is likely that the fungal contaminant of Fleming's cultures, which was to bring medical practice into the modern era of drug therapy, blew into his laboratory from the floor below, where an Irish mycologist was working with strains of *Penicillium*. Such a serendipitous event can change history, but in science it takes a special individual like Fleming to recognize the significance of the observation. As Pasteur said, "Chance favors the prepared mind." Fleming's chance discovery of the effect of mold contamination on a bacterial culture plate was possible only because Fleming's background and knowledge had been enriched by the growth of general scientific awareness such that what he saw made sense to him and fitted a historical pattern of scientific investigation.

After growing the fungus in a liquid medium and separating the fluid from the cells, Fleming discovered that the cell-free liquid was an inhibitor of many bacterial species. His publication on the active ingredient, which he called *penicillin*, was the first report of the production of an antibiotic. However, Fleming did not isolate pure penicillin, nor did he demonstrate its chemotherapeutic effects. Fleming himself pointed out that the pioneering work on antibiotics was thwarted because of the failure of microbiologists to pursue the chemical investigations necessary to separate and purify the active agents in their extracts. By the 1930s, however, microbiology had become far more chemistry oriented, and this approach culminated in the preparation of solid penicillin. In 1940, 10 years after Fleming's initial report, Howard Florey and Ernst Chain successfully isolated and purified penicillin. Other scientists established the therapeutic value of penicillin, and this antibiotic remains the cornerstone of the modern medical treatment of many infectious diseases.

Major advances in the development of chemotherapeutic agents continued to be made in the 1930s. Gerhard Domagk, a German physician employed as a chemist by a dye works company, the forerunner of modern pharmaceutical companies, developed prontosil, the first sulfa drug, effective against streptococcal infections. At the Pasteur Institute, a husband-and-wife research team, the Trefouëls, discovered the active constituent of prontosil, sulfanilamide, the first real wonder drug, so called because of its amazing ability to cure serious diseases. In the early 1940s, the Russian immigrant soil microbiologist, Selman Waksman, and co-workers at Rutgers University in New Jersey found that various bacteria of the actinomycetes group produced antibacterial agents. Streptomycin, produced by *Streptomyces griseus*, became the best-known of the new antibiotic wonder drugs. The antibiotics produced by actinomycetes generally have a broader spectrum of action than penicillin and thus can be used to treat a number of diseases for which penicillin is ineffective. Most antibiotics in current use are produced by actinomycetes. The importance of penicillin and the subsequently discovered antibiotics in treating diseases of microbial origin cannot be overestimated.

are actually antibiotics, some are produced partly or entirely by chemical synthesis. Even though not of microbial origin, similar compounds synthesized by organic chemists are usually also called *antibiotics*. To avoid problems in terminology the all-inclusive term **antimicrobic** often is used. In this section, the terms *antibiotic* and *antimicrobic* will be used interchangeably. These chemicals should not be confused with the large number of drugs used in medical practice for alleviating the symptoms of disease or for treating diseases not caused by microorganisms.

To be of therapeutic use, an antimicrobial agent must exhibit **selective toxicity**. A therapeutically useful antimicrobial agent must inhibit infecting microorganisms and exhibit greater toxicity to the infecting pathogens than to the host organism. A drug that kills the patient is of no use in treating infectious diseases, whether or not it also kills the pathogens! Even selective, therapeutically useful antimicrobics, though, can produce side effects (Table 20.5). As a rule, antimicrobics are of most use in medicine when their mode of action involves biochemical features of the invading pathogens not possessed by normal host cells.

In addition to selective toxicity, consideration must be given to the target organ. Because of differential solubilities, antimicrobics exhibit specific distribution patterns within the body, which must be recognized when choosing the proper agent (Table 20.6).

Classification and Selection of Antimicrobial Agents

Some antimicrobial agents are **microbiocidal**, killing microorganisms, and others are **microbiostatic**, inhibiting the growth of microorganisms but not actually killing them (Figure 20.16). Microbiostatic agents prevent the proliferation of infecting microorganisms, holding populations of pathogens in check until the normal immune defense mechanisms eliminate the invading pathogens. Antibiotics represent a major class of antimicrobial agents. By definition, antibiotics are biochemicals produced by microorganisms that inhibit the growth of, or kill, other microorganisms. By their very nature, antibiotics and synthetic antimicrobics must exhibit selective toxicity because they are produced by one microorganism and exert varying degrees of toxicity against others. The discovery and use of antibiotics have revolutionized medical practice in the twentieth century.

The selection of a particular antimicrobial agent for treating a given disease depends on several factors, including (1) the sensitivity of the infecting microorganism to the par-

TABLE 20.5   Major Toxicities of Selected Antimicrobics

Antimicrobic Agent	Mechanism	Signs
Aminoglycosides	Binds hair cells of organ of Corti	Deafness
	Binds vestibular cells	Vertigo
	Competitive neuromuscular blockage	Respiratory paralysis
	Tubular necrosis	Nephrotoxicity
Amphotericin	Distal tubular damage	Nephrotoxicity
	Renal tubular acidosis	Nephrotoxicity
Carbenicillin	Inhibition of platelet aggregation	Bleeding
Cephalosporins	Cortical stimulation	Myoclonic seizures
Cephaloridine	Proximal tubular damage	Nephrotoxicity
Chloramphenicol	Damages stem cell	Aplastic anemia
	Inhibits protein synthesis	Reversible anemia
Clindamycin	Proliferation of *Clostridium difficile*	Diarrhea
Emetine	Permeability changes	Hypotension
Isoniazid	Liver cell damage	Hepatitis
Neomycin	Villous damage	Malabsorption
Penicillins	Cortical stimulation	Myoclonic seizures
Polymyxins	Noncompetitive neuromuscular blockage	Respiratory paralysis
	Tubular necrosis	Nephrotoxicity
Rifampin	Liver cell damage	Hepatitis
Sulfonamides	Glucose 6-phosphate deficiency	Hemolytic anemia
	Collecting duct obstruction	Nephrotoxicity
Tetracyclines	Liver cell damage	Hepatitis
	Degradation products	Fanconi syndrome

TABLE 20.6   Distribution of Antimicrobics to Specific Body Areas

Bone

Penicillins, tetracyclines, cephalosporins, lincomycin, and clindamycin antimicrobics penetrate bone and bone marrow; levels are higher in infected bone than in normal bone.

Central Nervous System

Only lipid-soluble antimicrobics cross the blood-brain barrier and reach brain tissues. In the presence of inflammation, such as brain abscess, various penicillins achieve appreciable concentrations in the brain. Levels of most antimicrobics in the cerebrospinal fluid (CSF) are low. Penicillin G and ampicillin can achieve adequate CSF levels in the presence of inflammation; oxacillin, naficillin, and methicillin can be used to treat staphylococcal meningitis. CSF levels of chloramphenicol are adequate to treat *Streptococcus, Neisseria,* and *Haemophilus* but not most Gram-negative bacteria. Cefoxamine, moxalactam, and cefoperazone enter CSF in the presence of inflammation in concentrations that are adequate to treat *Streptococcus, Neisseria, Haemophilus, Klebsiella,* and *Escherichia coli* infections.

Ears and Sinuses

Most of the penicillins reach levels in the middle ear fluid in sufficient concentrations for the treatment of otitis media. Concentrations of antimicrobics in sinuses are adequate for ampicillin, amoxicillin, tetracyclines, erythromycin, sulfonamides, and trimethoprim to treat infections.

Eyes

Few antimicrobics penetrate the eye well. Levels of penicillins and cephalosporins in the aqueous humor are less than 10 percent of the peak serum levels and inhibit only highly sensitive bacteria.

Pleural and Pericardial Fluids

Most of the penicillins, cephalosporins, sulfonamides, macrolides, clindamycin, chloramphenicol, and antituberculosis drugs diffuse into serus cavities.

Pulmonary

Concentrations of most antibiotics within the lung are satisfactory, provided there is sufficient blood flow. Penicillins and tetracyclines show variable sputum concentrations. Antituberculosis agents, such as isoniazid and rifampin, achieve appreciable levels in pulmonary tissue.

Skin

Tetracyclines and clindamycin concentrate in skin tissue and are effective in treatment of acne.

Synovial Fluid

Most antibiotics used in the treatment of joint infections reach inflamed joints in adequate concentrations.

Urinary Tract

Treatment of kidney and other urinary tract infections depends largely upon the concentrations in the urine rather than upon serum levels. Nalidixic acid and nitrofurantoins are effective in treating urinary tract infections.

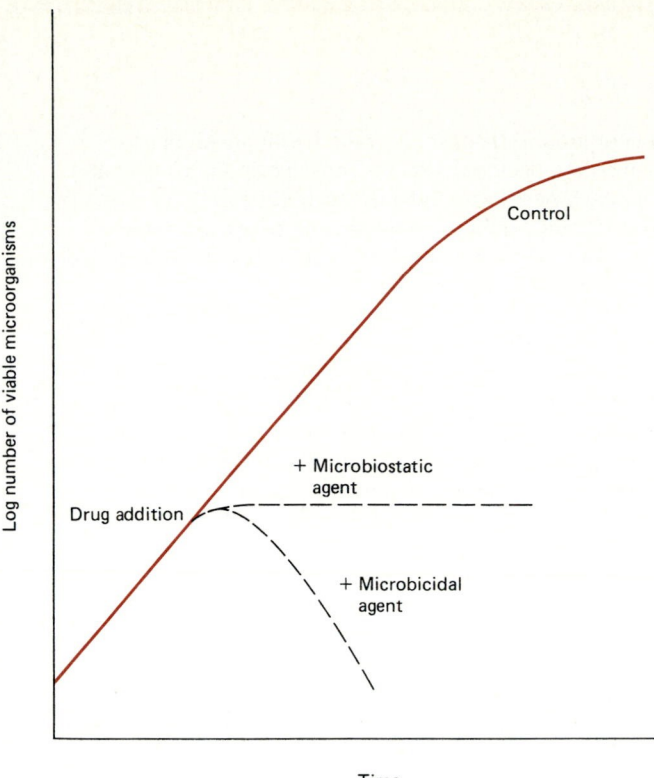

FIGURE 20.16
This graph shows the effects of microbicidal and microbiostatic agents on the number of viable microorganisms.

ticular antimicrobial agent; (2) the side effects of the antimicrobial agent with regard to direct toxicity to mammalian cells and to the microbiota normally associated with human tissues; (3) the biotransformations of the antimicrobial agent that occur *in vivo*, relative to whether the agent will remain in its active form for a sufficient period of time to be selectively toxic to the infecting pathogens; and (4) the chemical properties of the antimicrobial agent that determine its distribution within the body, relative to whether or not adequate concentrations of the active antimicrobial chemical will be able to reach the site of infection in order to inhibit or kill the pathogenic microorganisms causing the infection. For example, although many antibiotics possess antimicrobial activities that are effective against the pathogenic bacteria that cause urinary tract infections, only a limited number of antibiotics are effective in treating these infections because relatively few can reach and be concentrated in the tissues of the urinary tract in their active form. Additionally, one antimicrobial agent can influence the effects of another antimicrobial agent. In some cases, the use of two drugs enhances the effectiveness of the treatment, whereas in other cases one drug interferes with the inhibitory effects of the other (Figure 20.17).

Antibiotics Effective Against Bacterial Infections

The biochemical differences in the cell structures of bacterial cells and eukaryotic cells form the basis for the effective use of antibiotics against bacterial infections. The bacterial cell wall, with its unique peptidoglycan, and the 70S ribosome represent two major sites against which antibacterial agents may be directed. Most of the common antibiotics used in medicine for treating bacterial infections are inhibitors of cell-wall or protein synthesis. Some antibiotics are more selective than others with respect to the bacterial species that they inhibit. A **narrow-spectrum antibiotic** may be targeted at a particular pathogen, for instance at Gram-positive cocci, or at a particular bacterial species. In contrast is the **broad spectrum antibiotic**, inhibiting a relatively wide range of bacterial species, including both Gram-positive and Gram-negative types. The choice of a particular antibiotic depends in part on the biochemical properties of the infecting bacterial strain. In most cases, physicians make an educated guess as to which antibiotic is appropriate for treating a particular infection, and the selection of the antibiotic is based on the

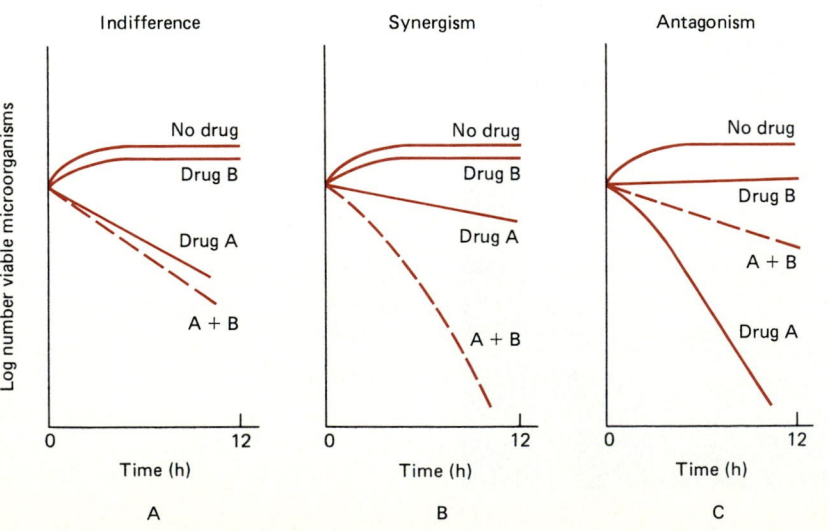

FIGURE 20.17
This diagram illustrates the possible interactions between antimicrobial agents. (A) Indifference; (B) synergism; (C) antagonism. Interactions are reflected in the log numbers of microorganisms that are still viable after application of these agents.

ANALYTICAL PROCESS

Antimicrobial Susceptibility Testing

As soon as antibiotics became commercially available to medical practitioners, the need for antimicrobial susceptibility testing immediately became apparent. Determination of the antimicrobial susceptibility of a pathogen is important in aiding the clinician to select the most appropriate agent for treating that disease. It is pointless to prescribe an antibiotic that is ineffective against the microorganism causing the disease. Additionally, physicians want to avoid indiscriminate administration of antibiotics because the selective pressures of excessive antibiotic usage can and have led to the evolution of antibiotic-resistant strains of pathogens that become problems when they cause infections that do not respond to the antibiotics routinely used to treat specific diseases. The clinical microbiology laboratory provides information, through standardized *in vitro* testing, with regard to the activities of antimicrobial agents against microorganisms that have been isolated and identified as the probable etiological agents of disease. Antibiotic susceptibility testing, which relies on the observation of antibiotics inhibiting the growth and/or killing cultures of microorganisms *in vitro*, provides the physician with the information needed to prescribe the proper antibiotics for treating infectious diseases.

The first method used for antibiotic susceptibility testing was developed by Alexander Fleming. In Fleming's *ditch plate* technique, a strip of agar in the form of a ditch is removed from a petri plate and replaced with medium containing the mold extract penicillin. Multiple streak inocula of the organisms to be tested are made at right angles to the ditch. Strains that grow up to the ditch are resistant to penicillin; strains that do not are sensitive. As numerous new antibiotics were discovered and became available for use by physicians, more rapid and reliable methods were needed for determining which antibiotic to use in the treatment of a particular patient. In 1943 J. W. Foster and H. B. Woodruff first reported the use of antibiotic-impregnated filter paper strips on agar plates inoculated with the test organism. Zones of inhibition

occur around strips impregnated with antibiotics to which the bacterial strain is sensitive. J. G. Vincent and H. W. Vincent improved the technique that same year by introducing antibiotic-impregnated paper disks, thus increasing the number of antibiotics that could be tested simultaneously against a bacterial isolate in one petri dish. In 1944 D. C. Morely devised a method for drying the paper disks after soaking them in an antibiotic solution. These early methods for testing antibiotic susceptibility were not designed to produce quantitative results.

At the end of the 1950s, antimicrobial susceptibility testing was marked by lack of an acceptable standardized procedure, which led to variable results depending on which laboratory was used. In response to this situation, a WHO committee was formed to evaluate antibiotic susceptibility testing; the decisions of this committee formed the groundwork for the development of the Bauer-Kirby standardized technique. In 1961 T. G. Anderson developed the steps that Bauer and Kirby incorporated into a standardized method. In the **Bauer-Kirby procedure** the following items are standardized: the amount of each antibiotic impregnated into the disks used in the test procedure; the composition of the test medium; the nature of the inoculum; the incubation time; and the measurement of the size of the zone of inhibition needed to establish antibiotic susceptibility. The qualitative susceptibility of microorganisms to antimicrobial agents can be determined on agar plates by using filter paper disks impregnated with antimicrobial agents.

The Bauer-Kirby test system is a standardized antimicrobial susceptibility procedure in which a culture is inoculated onto the surface of Meuller-Hinton agar, followed by the addition of antibiotic impregnated disks to the agar surface. In such agar diffusion methods the antibiotics diffuse into the agar, establishing a concentration gradient. Inhibition of microbial growth is indicated by a clear area (zone of inhibition) around the antibiotic disk (Figure 20.18). The diameter of the zone of inhibition reflects the solubility properties of the particular anti-

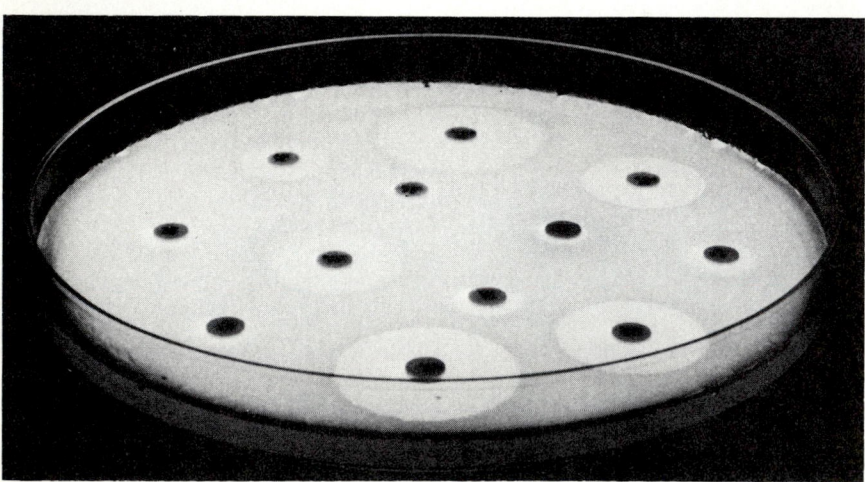

FIGURE 20.18
This photograph shows an antibiotic sensitivity testing plate, revealing the sensitivity of a bacterial isolate to various antibiotics, as indicated by the sizes of the zones of inhibition around the antibiotic disks on the agar plate. (Courtesy Bernard Abbott, Eli Lilly Research Laboratories, Indianapolis)

TABLE 20.7   Intepretation of Zones of Inhibition for Bauer-Kirby
Antibiotic Susceptibility Testing

Antibiotic	Disc Conc.	Resistant	Inhibition Zone Diameter (mm) Intermediate	Susceptible
Amikacin	0.01 mg	13 or less	12–13	14 or more
Ampicillin	0.01 mg	11 or less	12–13	14 or more
Bacitracin	10 units	8 or less	9–11	13 or more
Cephalothin	0.03 mg	14 or less	15–17	18 or more
Chloramphenicol	0.03 mg	12 or less	13–17	18 or more
Erythromycin	0.015 mg	13 or less	14–17	18 or more
Gentamicin	0.01 mg			13 or more
Kanamycin	0.03 mg	13 or less	14–17	18 or more
Lincomycin	0.002 mg	9 or less	10–14	15 or more
Methicillin	0.005 mg	9 or less	10–13	14 or more
Nalidixic acid	0.03 mg	13 or less	14–18	19 or more
Neomycin	0.03 mg	12 or less	13–16	17 or more
Nitrofurantoin	0.3 mg	14 or less	15–16	17 or more
Penicillin G staphylococci	10 units	20 or less	21–28	29 or more
Penicillin, other organisms	10 units	11 or less	12–21	22 or more
Polymyxin	300 units	8 or less	9–11	12 or more
Streptomycin	0.01 mg	11 or less	12–14	15 or more
Sulfonamides	0.3 mg	12 or less	13–16	17 or more
Tetracycline	0.03 mg	14 or less	15–18	19 or more
Vancomycin	0.03 mg	9 or less	10–11	12 or more

biotic—that is, the concentration gradient established by diffusion of the antibiotic into the agar—and the sensitivity of the given microorganism to the specific antibiotic. Standardized zones for each antibiotic disk have been established to determine whether the microorganism is sensitive (S), intermediately sensitive (I), or resistant (R) to the particular antibiotic (Table 20.7). The results of Bauer-Kirby testing indicate whether a particular antibiotic has the potential for effectively controlling an infection caused by a particular pathogen.

The Bauer-Kirby agar diffusion test procedure is designed for use with rapidly growing bacteria. It is not directly applicable to filamentous fungi, anaerobes, or slow-growing bacteria, although modifications of the media composition and incubation conditions can be made for testing the antibiotic susceptibility of such microorganisms. Different standardized systems are used for performing antibiotic sensitivity testing in these cases. For example, prereduced Wilkins-Chalgren agar, anaerobic transfer techniques, and anaerobic incubation can be used for determining the antibiotic sensitivities of anaerobic bacteria.

Besides Bauer-Kirby and other agar diffusion tests, many clinical laboratories use the Autobac system for measuring microbial growth, based on light scattering or equivalent automated liquid diffusion methods for antibiotic sensitivity testing. This system uses Eugonic broth and/or low-thymidine broth culture and standardized antibiotic disks, but the antibiotics in the disks elute into the broth rather than into agar, as in the Bauer-Kirby procedure. The concentrations of the antibiotics and the density and growth phase of the cultures are adjusted so that uniform interpretive guidelines can be used for assessing antibiotic sensitivities. A normalized light

scattering index is generated to determine S, I, and R; for R this index is 0.00–0.50; for I, 0.51–0.60; and for S, 0.61–1.00, except for penicillin G, where it is 0.00–0.90. Other automated systems are also available for performing this procedure, including the Microscan, BBL sceptre, Vitek AMS, and Abott MSII. These automated systems simplify and enhance the reliability of antimicrobial susceptibility testing, making it likely that they will be used more frequently, thereby reducing the excessive use of inappropriate antimicrobics by some physicians.

Another approach to antimicrobial susceptibility testing is to determine the **minimum inhibitory concentration (MIC)** using tube dilution procedures. This test determines the concentration of an antibiotic that is effective in preventing the growth of the pathogen and gives an indication of the dosage that should be effective in controlling the infection. A standardized microbial inoculum is added to tubes containing serial dilutions of an antibiotic, and the growth of the microorganism is monitored as a change in turbidity. In this way, the breakpoint or MIC of the antibiotic, that is, the lowest concentration that prevents growth of the microorganism in vitro, can be determined (Figure 20.19). The MIC indicates the minimal concentration of the antibiotic that must be achieved at the site of infection to inhibit the growth of the microorganism being tested. By knowing the MIC and the theoretical level of the antibiotic that may be achieved in body fluids, such as blood and urine, the physician can select the appropriate antibiotic, the dosage schedule, and the route of administration (Table 20.8). Generally, a margin of safety of 10 times the MIC is desirable to ensure successful treatment of the disease.

The MIC can be used for determining the antibiotic sensitivity of both aerobic and anaerobic microorganisms.

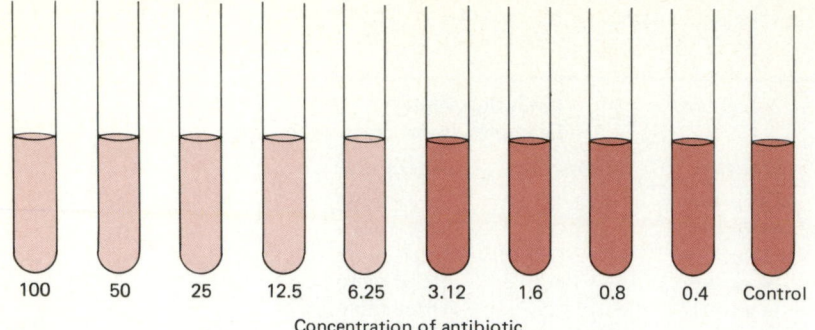

100 50 25 12.5 6.25 3.12 1.6 0.8 0.4 Control

Concentration of antibiotic
μg/mL

FIGURE 20.19
Tube dilution procedure for determining the breakpoint in MIC determination. The MIC for the test illustrated here is 6.25 μg/mL. Compared with the Bauer-Kirby procedure, the MIC provides quantitative information that can be used to determine the dose rate and administration schedule for an antibiotic.

The use of microtiter plates, which require only a few hundred microliters per sample well, and automated inoculation and reading systems makes the determination of MIC feasible for use in the clinical laboratory (Figure 20.20). MIC test can even be performed on normally sterile body fluids without isolating and identifying the pathogenic microorganisms. For example, blood or cerebrospinal fluid containing an infecting microorganism can be added to tubes containing various dilutions of an antibiotic and a suitable growth medium. An increase in turbidity would indicate the growth of microorganisms and the fact that the antibiotic at that concentration was ineffective in inhibiting microbial growth, whereas a lack of growth would indicate that the pathogenic microorganisms were susceptible to the antibiotic at the given concentration. By determining the MIC, the appropriate dosage as well as the right antibiotic can be selected to treat an infectious disease.

MIC tests are used to establish the concentration of an antibiotic that will inhibit growth, but not to determine whether the antibiotic is microbicidal. It is, however, also possible to determine the **minimal bactericidal concentration (MBC)**, also known as the *minimal lethal concentration (MLC)*. The MBC is the lowest concentration of an antibiotic that will kill a defined proportion of viable organisms in a bacterial suspension during a specified exposure period. Generally, a 99.9 percent kill of bacteria at an initial concentration of 10^5–10^6 cells/mL during an 18–24-hour exposure period is used to define the MBC.

In order to determine the minimal bactericidal concentration, it is necessary to plate the tube suspensions showing no growth in tube dilution (MIC) tests onto an agar growth medium in order to determine whether the bacteria are indeed killed or whether they survive exposure to the antibiotic at the concentration being tested. Although determination of the MIC is adequate for establishing the appropriate concentration of an antibiotic that should be administered to control an infection in patients with normal immune response levels, the MBC is essential in cases of endocarditis and is particularly useful in determining the appropriate concentration of an antibiotic for use in treating patients with lowered immune

TABLE 20.8   Achievable Levels of Some Common Antibiotics in Various Body Fluids

Antibiotic	Achievable Peak Blood Levels (μg/mL)	Achievable Urine Levels (μg/mL)	Dose	
Clindamycin	1–4	>20	Oral	150–300 mg
	6–10	>60	IV	300–600 mg
Erythromycin	1–2		Oral	250–500 mg
	10–20		IV	300 mg
Penicillin	2–3	>300	Oral	500 mg
	6–8	>300	IM	500 mg
	4–7	>300	IV	500 mg
Ampicillin	1–3	>50	Oral	250–500 mg
	2–6	>20	IM	250–500 mg
	10–25	>100	IV	1,000–1,500 mg
Cephalothin	3–18	>300	Oral	250–500 mg
	9–24	>1,000	IM	500–1,000 mg
	30–85	>1,000	IV	1,000–2,000 mg
Gentamicin	2–10	>20	IM/IV	1–2 mg
Tetracycline	1–2	>200	Oral	250–500 mg
	10–20	>200	IV	500 mg
Chloramphenicol	10–12	>100	Oral	1,000 mg
	20–30	>200	IV	1,000 mg
Nitrofurantoin		>100	Oral	50–100 mg

IV = intravenous.
IM = intramuscular.

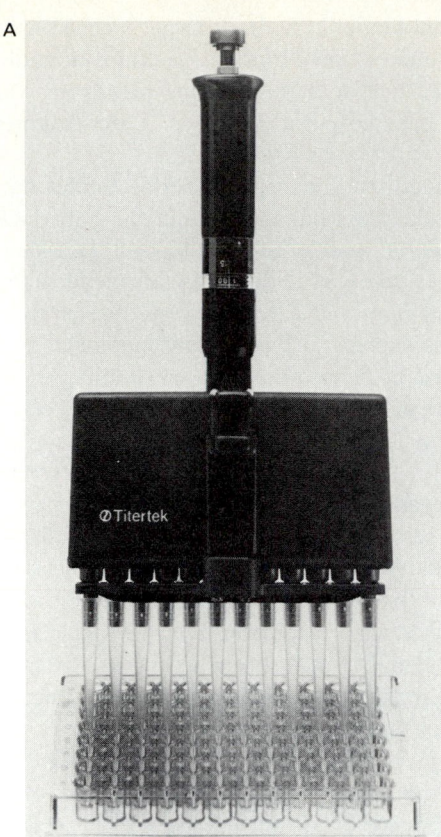
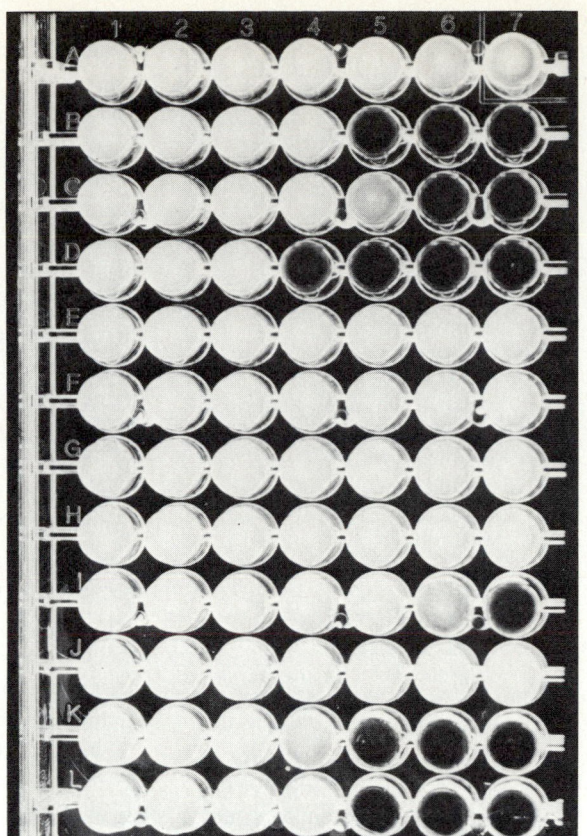

FIGURE 20.20

Test plates used for automated MIC determination. (A) This Titertek automatically fills the microtiter plates. (Courtesy Flow Laboratories, McLean, Va.) (B) In this photograph, the results of MIC testing are shown. Lack of growth, that is, inhibition, appears as clear wells. (From BPS—Leon J. LeBeau, University of Illinois Medical Center, Chicago)

defense responses, such as those receiving chemotherapy for cancer.

Another way of determining the effectiveness of an antibiotic is to measure serum killing power using the Schlicter test. Instead of adding dilutions of an antibiotic to suspensions of bacteria in a growth medium, a bacterial suspension is added to dilutions of the patient's serum. The ability of the bacteria to grow in the patient's blood is assessed by measuring changes in turbidity. Assuming that the patient is being treated with an antibiotic, no bacterial growth should occur. The breakpoint in the dilutions where bacterial growth occurs reflects the concentration of the antibiotic in the patient's blood and the *in vivo* effectiveness of the antibiotic in controlling the infection. Inhibition at dilutions of the patient's serum of to 1:8 or more is considered as an acceptable level.

most likely pathogen causing the disease symptomatology and the antibiotics generally known to be effective against this pathogen. Many times a physician will select a broad-spectrum antibiotic in order to ensure timely and effective treatment. Only in special cases, such as when a patient fails to respond to a particular antibiotic and an infection persists, is an attempt normally made to isolate the pathogenic bacterium and to determine its range of antibiotic sensitivity.

However, concern is mounting in the medical field about the overuse of antibiotics because the undesired side effect is the selection for disease-causing antibiotic-resistant strains. It is now considered proper medical practice to perform culture and sensitivity studies to determine the proper antibiotic for treating a patient. Only in cases of life-threatening infections should antibiotics be used without such testing to avoid selective pressure for the development of antibiotic-resistant pathogens. The importance of this problem was underscored when the American Medical Association called upon physicians to avoid unnecessary use of antibiotics.

The reason for concern about how we use antibiotics is that numerous bacterial strains have acquired the ability to resist the effects of some antibiotics, with some bacterial strains, generally those containing R plasmids, having **multiple antibiotic resistance**. The basis of resistance in some cases is the ability of the particular strain to produce enzymes

TABLE 20.9 Generic and Trade Names of Some Common Antibiotics

Generic Name	Trade Name
Tetracycline	Acromycin, Panmycin, Tetracyn, Tetrachel, Rexamycin
Oxytetracycline	Teramycin
Chlorotetracycline	Aureomycin
Demeclocycline	Declomycin
Methacycline	Rondomycin
Doxycycline	Vibramycin
Minocycline	Minocin, Vectrin
Penicillin G	Crysticillin, Duracillin
Ampicillin	Amcill, Omnipin, Penbritin, Polycillin
Cephalothin	Keflin
Cephalexin	Keflex
Chloramphenicol	Chloromycetin, Mychel
Gentamicin	Garamycin
Kanamycin	Kantrex
Erythromycin	Ilotycin
Nystatin	Mycostatin, Nilstat
Trimethoprim-sulfamethoxazole	Bactrin, Septra
Chloroquine	Aralen, Avloclor, Resochin

that degrade the antibiotic, preventing the active form of the antibiotic from reaching the bacterial cells where they could be inhibitory. For example, some bacterial strains produce penicillinases (β-lactamases) that are able to degrade the antibiotic penicillin, making such strains resistant to penicillin. Resistance may also be due to decreased drug uptake, decreased transformation of the drug to its active form, and/or decreased sensitivity of the microbial structure against which the drug is directed.

Modes of Action of Antibacterial Agents In this section, we will consider the properties and modes of action of many of the common antibiotics currently in use. It should be noted that only the generic names of the antibiotics will be used, although the same compound may be produced by several companies under several different trade names (Table 20.9). Many municipalities now require pharmacies to fill a prescription with its generic brand to encourage competitive pricing, unless a specific brand name is required. Such laws are controversial because the pharmaceutical manufacturers lose the incentive to develop new antibiotics and to maintain high and costly quality control when a less stringently regulated and less expensive product will be used to fill a prescription. The public wants cost-effective but high-quality antibiotics.

Cell-wall Inhibitors. The **penicillins** and **cephalosporins** are two widely used classes of antibiotics that inhibit the formation of bacterial cell-wall structures. Penicillins are synthesized by strains of the fungus *Penicillium*. The cephalosporins are produced by members of the fungal genus *Cephalosporium*. The penicillins and cephalosporins both contain a β-lactam ring and thus have related biochemical structures (Figure 20.21). There are various penicillin and cephalosporin antibiotics contain differing biochemical substituent groups and exhibit differing spectrums of antibacterial activity. Because of these differing properties, various penicillins and cephalosporins are used in the treatment of specific diseases (Table 20.10).

Both the penicillins and cephalosporins inhibit the formation of peptide cross-linkages within the peptidoglycan backbone of the cell wall. These antibiotics specifically inhibit the enzymes involved in the cross-linkage for transpeptidase reactions (Figure 20.22). It appears that the β-lactam portion of cephalosporin and penicillin antibiotics binds to the transpeptidase enzyme, preventing the binding of the enzyme to the normal substrate, D-alanine. Bacterial cell walls lacking the normal cross-linking peptide chains are subject to attack by **autolysins**, which are autolytic enzymes produced by the organism that degrade the cell's own cell-wall structures. The result is that, in the presence of cephalosporins or penicillins, growing bacterial cells are subject to lysis because, without functional cell-wall structures, the bacterial cell is not protected against osmotic shock. It should be noted that the penicillin and cephalosporin antibiotics do not themselves remove intact cell walls and thus are ineffective against resting or dormant cells.

Many of the penicillins, such as penicillin G, have a relatively narrow spectrum of activity, being most effective against Gram-positive cocci, including *Staphylococcus* species. Others, such as ampicillin, have a broader spectrum of activity, inhibiting some Gram-negative as well as Gram-positive bacteria. Ampicillin, an amino-substituted penicillin, is active against many Gram-negative rods, including *Escherichia coli*, *Haemophilus influenzae*, *Shigella* species, and *Proteus* species. To inhibit peptidoglycan synthesis effectively in Gram-negative bacteria, the antibiotic must pass through the outer lipopolysaccharide (LPS) layers to reach the peptidoglycan located at the inner portion of the cell wall. The broad spectrum activity of ampicillin appears to be based on its ability to penetrate to the site of action of the transpeptidase enzyme, whereas narrow-spectrum penicillins, such as penicillin G, are relatively inefficient at reaching this site.

Penicillin G and various other β-lactam antibiotics are subject to inactivation by penicillinase enzymes (β-lactamases). Penicillinase-producing bacterial strains are able to degrade the β-lactam ring structure of many penicillins, rendering them ineffective in treating such bacterial strains. For example, penicillin G is normally effective against *Neisseria gonorrhoeae*, a Gram-negative coccus that causes gonorrhea, but some penicillinase-producing strains of *N. gonorrhoeae* have now been found, requiring the use of antibiotics other than penicillin G in the treatment of cases of gonorrhea caused by these strains. There may also be other causes for the penicillin resistance of *N. gonorrhoeae*. About 1 in 10^9 cells of *N. gonorrhoeae* are resistant to penicillin; thus, high enough antibiotic concentrations must be given for a long enough time to allow the natural body defense mechanisms to eliminate all of the infecting bacteria. Structural modifications of penicillin G, such as occur in methicillin, can

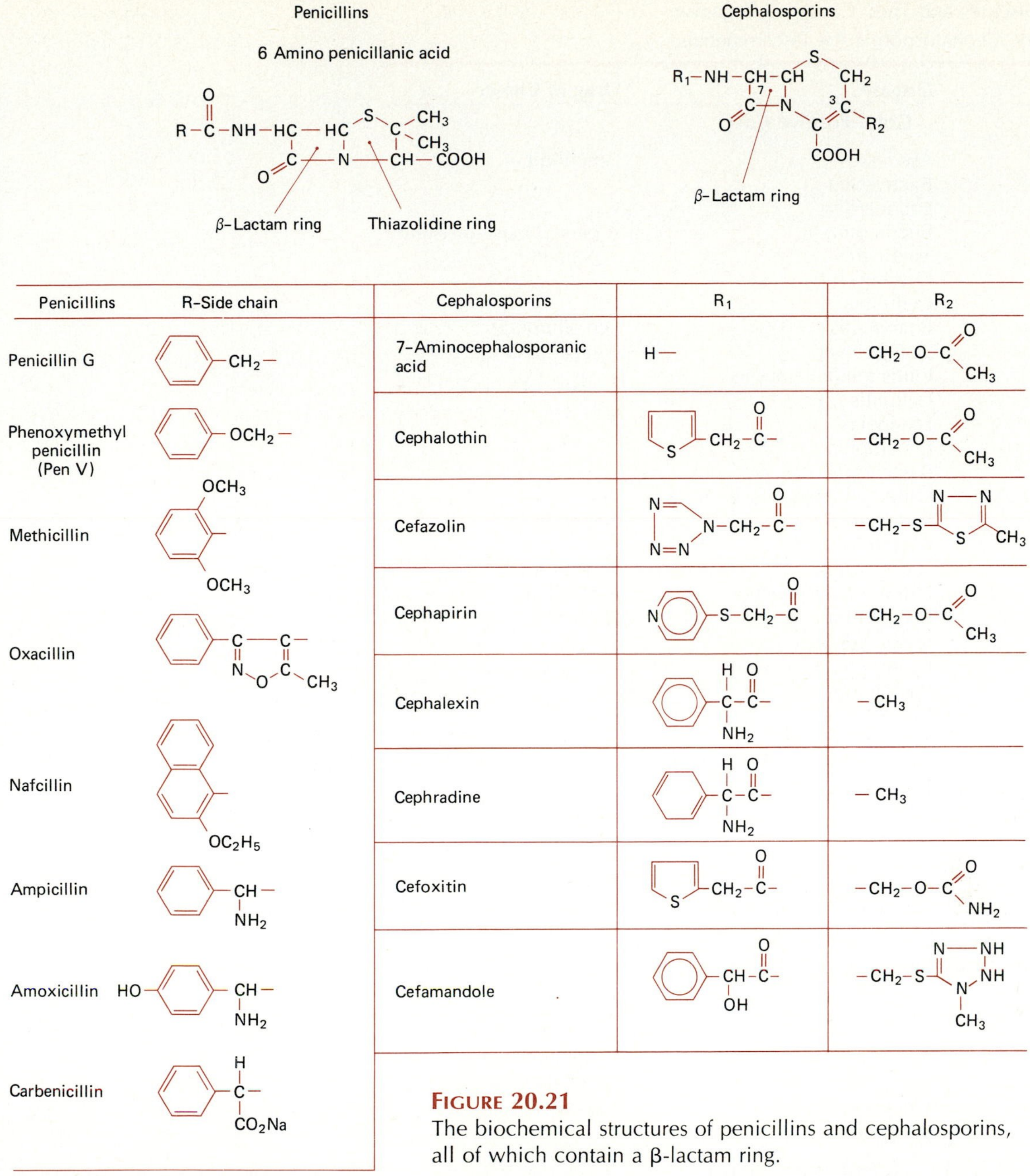

Penicillins

6 Amino penicillanic acid

Cephalosporins

β-Lactam ring   Thiazolidine ring

β-Lactam ring

Penicillins	R-Side chain	Cephalosporins	R_1	R_2
Penicillin G		7-Aminocephalosporanic acid		
Phenoxymethyl penicillin (Pen V)		Cephalothin		
Methicillin		Cefazolin		
Oxacillin		Cephapirin		
Nafcillin		Cephalexin		
Ampicillin		Cephradine		
Amoxicillin		Cefoxitin		
Carbenicillin		Cefamandole		

FIGURE 20.21

The biochemical structures of penicillins and cephalosporins, all of which contain a β-lactam ring.

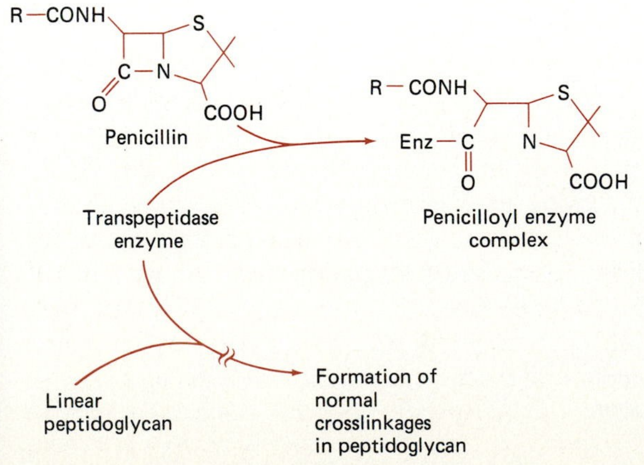

Penicillin

Transpeptidase enzyme

Penicilloyl enzyme complex

Linear peptidoglycan

Formation of normal crosslinkages in peptidoglycan

FIGURE 20.22

The mode of action of penicillin and cephalosporin antibiotics involves blockage of the normal cross-linkages in the peptidoglycan layer of the bacterial cell wall by forming an inactive complex with a key enzyme in cell-wall synthesis.

TABLE 20.10 Some Diseases and Their Causative Organisms for Which Penicillins and Cephalosporins Are Recommended

Causative Organism	Disease	Drug of Choice
Gram-Positive Cocci		
Staphylococcus aureus	Abscesses	Pencillin G
	Bacteremia	
	Endocarditis	
	Pneumonia	A penicillinase-resistant
	Meningitis	penicillin
	Osteomyelitis	
	Cellulitis	
Streptococcus pyogenes	Pharyngitis	Penicillin G
	Scarlet fever	Penicillin V
	Otitis media, sinusitis	
	Cellulitis	
	Erysipelas	
	Pneumonia	
	Bacteremia	
	Other systemic infections	
Streptococcus (viridans group)	Endocarditis	Penicillin G
	Bacteremia	
Streptococcus faecalis (enterococcus)	Endocarditis	Penicillin G
	Urinary tract infection	Ampicillin
	Bacteremia	Penicillin G
Streptococcus bovis	Endocarditis	Penicillin G
	Urinary tract infection	
	Bacteremia	
Streptococcus (anaerobic species)	Bacteremia	Penicillin G
	Endocarditis	
	Brain and other abscesses	
	Sinusitis	
Streptococcus pneumoniae (pneumonococcus)	Pneumonia	Penicillin G
	Meningitis	
	Endocarditis	
	Arthritis	
	Sinusitis	
	Otitis	
Gram-Negative Cocci		
Neisseria gonorrhoeae (gonococcus)	Genital infections	Ampicillin or amoxicillin Penicillin G
	Arthritis-dermatitis syndrome	Ampicillin or amoxicillin Penicillin G
Neisseria meningitidis (meningococcus)	Meningitis	Penicillin G
	Bacteremia	
Gram-Positive Rods		
Bacillus anthracis	"Malignant pustule" pneumonia	Penicillin G
Corynebacterium diphtheriae	Pharyngitis	Penicillin G
	Laryngotracheitis	
	Pneumonia	
	Other local lesions	
Erysipelothrix rhusiopathiae	Erysipeloid	Penicillin G
Clostridium perfringens	Gas gangrene	Penicillin G
Clostridium tetani	Tetanus	Penicillin G
Gram-Negative Rods		
Haemophilus influenzae	Otitis	Amoxicillin
	Sinusitis	Ampicillin
	Bronchitis	
	Epiglottitis	
Enterobacter aerogenes	Urinary tract infection	Cephamandole
Klebsiella pneumoniae	Urinary tract infection	Cephalosporin
	Pneumonia	

TABLE 20.10   (continued)

Causative Organism	Disease	Drug of Choice
Pasteurella multocida	Wound infection Abscesses Bacteremia Meningitis	Penicillin G
Bacteroides spp.	Oral disease Sinusitis Brain abscess Lung abscess	Penicillin G
Fuscobacterium nucleatum	Ulcerative pharyngitis Lung abscess Genital infections Gingivitis	Penicillin G
Streptobacillus moniliformis	Bacteremia Arthritis Endocarditis Abscesses	Penicillin G
Spirochetes		
Treponema pallidum	Syphilis	Penicillin G
Treponema pertenue	Yaws	Penicillin G
Leptospira	Weil's disease Meningitis	Penicillin G
Actinomycetes		
Actinomyces israelii	Cervical, facial, abdominal, thoracic and other lesions	Penicillin G

render the molecule resistant to penicillinases but may also narrow the spectrum of action, limiting the primary use of such antibiotics to the treatment of infections caused by penicillinase-producing *Staphylococcus* species.

In contrast to the penicillins, the cephalosporins generally have a broad spectrum of action, and many of them, such as cefoxitin and cephalothin, are relatively resistant to penicillinase. As such, the cephalosporins are useful in treating a variety of infections caused by Gram-positive and Gram-negative bacteria. Many physicians are now using broad-spectrum cephalosporins where the use of narrow-range and more specifically directed penicillins would be adequate. Cephalosporins are most prudently used as alternatives to penicillins for patients who are allergic to penicillin and for those pathogens that are not penicillin sensitive. Cephalothin is often the antibiotic of choice for treating severe staphylococcal infections, such as endocarditis, to avoid complications in cases where the infecting *Staphylococcus* species produces β-lactamases. Cefamandole, another of the cephalosporins, is widely used in treating pneumonia, as it is active against *Haemophilus influenzae*, *Staphylococcus aureus*, and *Klebsiella pneumoniae*, which are frequently the causative agents of respiratory tract infections resulting in pneumonia. The cephalosporins may also be used in place of penicillins for the prophylaxis of infection by Gram-positive cocci following surgical procedures.

In addition to the penicillins and cephalosporins, several other antibiotics inhibit cell-wall synthesis, including **vancomycin, bacitracin**, and **cycloserine** (Figure 20.23). These antibiotics do not block the enzymes involved in the formation of peptide cross-linkages in the murein component of the wall, but rather block other reactions involved in the synthesis of the bacterial cell wall. Cycloserine is a structural analogue of D-alanine and can prevent the incorporation of D-alanine into the peptide units of the cell wall (Figure 20.24). In the presence of D-cycloserine, the subunits that are necessary for cell-wall synthesis cannot be adequately synthesized. Cycloserine is a broad-spectrum antibiotic produced by *Streptomyces orchidaceus*. Its therapeutic use is limited by its toxic reactions involving the central nervous system. Cycloserine is inhibitory for *Mycobacterium tuberculosis* and has been used in conjunction with other antibiotics in the treatment of tuberculosis.

Vancomycin and bacitracin prevent the linkage of the *N*-acetylglucosamine and *N*-acetylmuramic acid moieties that compose the peptidoglycan molecule. Bacitracin is produced by strains of *Bacillus subtilis*. The use of bacitracin is restricted to topical application because this antibiotic causes severe toxicity reactions. Vancomycin is produced by *Streptomyces orientalis* and is especially effective against strains of *S. aureus*. Vancomycin is used to treat only serious infections caused by penicillin-resistant strains of *Staphylococcus* and/or when the patient exhibits allergic reactions to penicillins and cephalosporins.

Inhibitors of Protein Synthesis. The antibiotics streptomycin, gentamicin, neomycin, kanamycin, tobramycin, and amikacin are inhibitors of bacterial protein synthesis (Figure 20.25). These **aminoglycoside** antibiotics contain amino

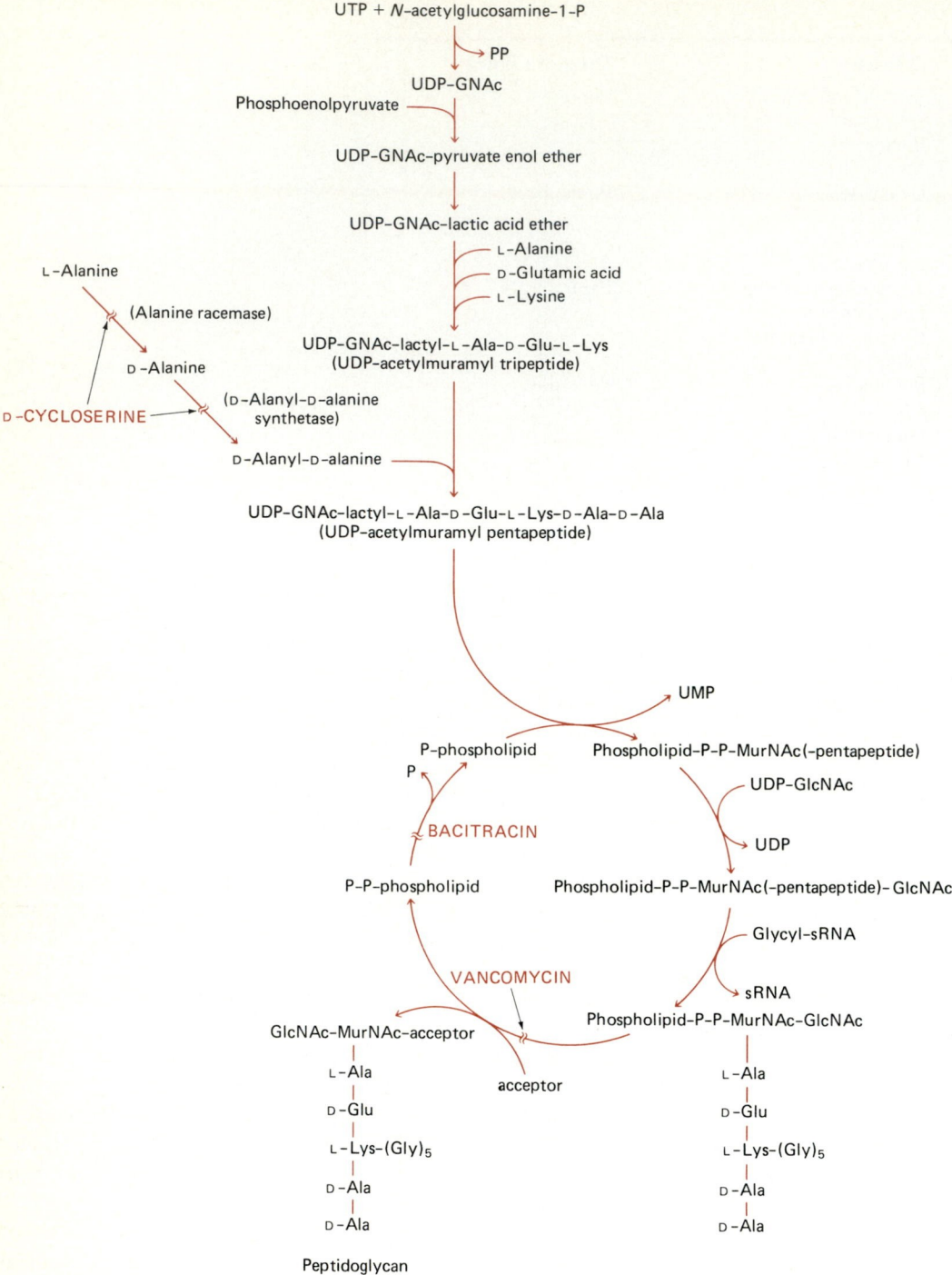

FIGURE 20.23

The modes of action of vancomycin, bacitracin, and cycloserine in inhibiting cell-wall synthesis.

FIGURE 20.24

The structures of D-cycloserine and D-alanine. One can easily see how cycloserine acts as an analogue of the normal amino acid needed for proper cell-wall formation.

D-Cycloserine   D-Alanine

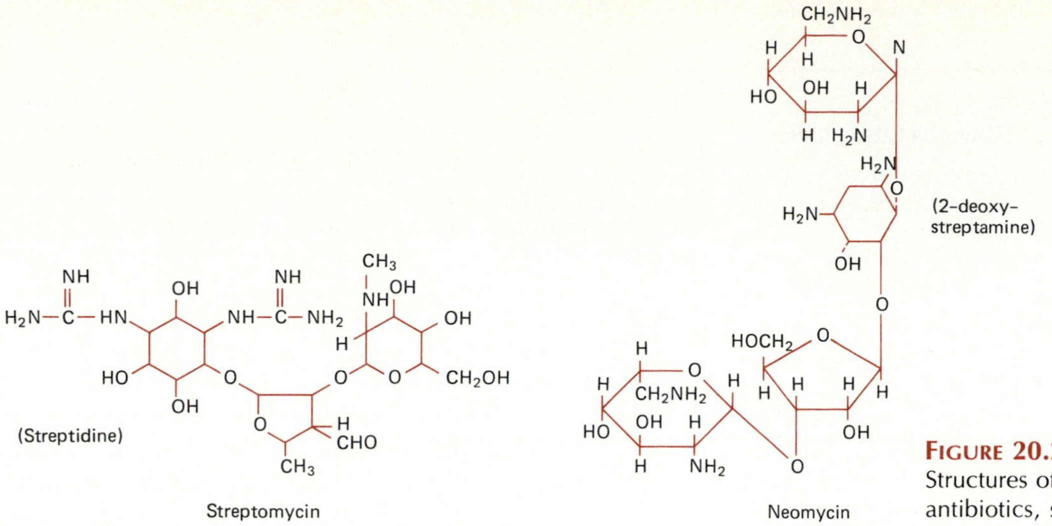

FIGURE 20.25
Structures of representative aminoglycoside antibiotics, streptomycin and neomycin.

sugars linked by glycosidic bonds. The aminoglycosides are used almost exclusively in the treatment of infections caused by Gram-negative bacteria. These antibiotics are relatively ineffective against anaerobic bacteria and facultative anaerobes growing under anaerobic conditions, and their action against Gram-positive bacteria is also limited. The aminoglycoside antibiotics are produced by actinomycetes. For example, streptomycin is produced by *Streptomyces griseus*, neomycin by *S. fradiae*, kanamycin by *S. kanamyceticus*, and gentamicin by *Micromonospora purpurea*; amikacin is a semisynthetic derivative of kanamycin.

The aminoglycoside antibiotics bind to the 30S ribosomal subunit of the 70S prokaryotic ribosome, blocking protein synthesis and decreasing the fidelity of translation of the genetic code. Aminoglycosides disrupt the normal functioning of the ribosomes by interfering with the formation of initiation complexes, the first step of protein synthesis that occurs during translation. Additionally, aminoglycosides induce misreading of the mRNA molecules, leading to the formation of nonfunctional enzymes. The interference of protein synthesis results in the death of the bacterium. Various mutations, though, can occur that reduce the effect of misreading some mRNA molecules, in some cases even leading to a dependence on streptomycin-induced misreading of the genetic information.

To be effective, the aminoglycoside antibiotics must be transported across the cytoplasmic membrane. Although sensitive bacteria transport the aminoglycosides across the membrane, accumulating these antibiotics intracellularly, resistant strains may lack a mechanism for aminoglycoside transport across the membrane. Resistant strains may also produce enzymes that degrade or transform the aminoglycoside molecules. For example, various enzymes associated with the cytoplasmic membranes of some bacterial strains are capable of adenylating, acetylating, or phosphorylating aminoglycoside antibiotics. Additionally, mutations can occur that alter the site at which the aminoglycosides normally bind to the bacterial ribosomes. Some *Pseudomonas aeruginosa*

strains, for example, possess ribosomes to which streptomycin is unable to bind.

The aminoglycoside antibiotics are useful in treating a variety of diseases (Table 20.11). Because of its serious side effect on the seventh cranial nerve that can cause deafness with prolonged usage, streptomycin is used in the treatment of only a limited number of bacterial infections. It is sometimes used in the treatment of brucellosis, tularemia, endocarditis, plague, and tuberculosis. Gentamicin is effective in treating urinary tract infections, pneumonia, and meningitis. Gentamicin is, however, extremely toxic and thus is used only in severe infections that may prove lethal if unchecked, particularly when the infecting bacteria are not sufficiently sensitive to other less toxic antibiotics. Tobramycin has properties similar to those of gentamicin, but *P. aeruginosa* is particularly sensitive to tobramycin. Thus, this antibiotic is sometimes used for the treatment of pneumonia and other infections caused by *Pseudomonas* species. Neomycin, which is active against a broad spectrum of Gram-negative bacteria, is primarily used in topical application for various infections of the skin and mucous membranes. Kanamycin, a narrow-spectrum antibiotic, is frequently used by pediatricians for infections due to *Klebsiella*, *Enterobacter*, *Proteus*, and *E. coli*. Amikacin, which has the broadest spectrum of activity of the aminoglycosides, is the antibiotic of choice for treating serious infections caused by Gram-negative infections acquired in hospitals because such infections are often due to bacterial strains that are resistant to multiple antibiotics, including other aminoglycosides.

In addition to the aminoglycoside antibiotics, a number of other antibiotics inhibit bacterial protein synthesis. These antibiotics include the tetracyclines, chloramphenicol, erythromycin, lincomycin, clindamycin, and spectinomycin. Some recommended therapeutic uses of these antibiotics are shown in Table 20.12. Unlike the aminoglycoside antibiotics, which are bacteriocidal, these inhibitors of bacterial protein synthesis are generally bacteriostatic.

The **tetracyclines**, like the aminoglycosides, bind specif-

TABLE 20.11 Some Diseases and Their Causative Organisms for Which Aminoglycoside Antibiotics Are Recommended

Causative Organism	Disease	Drug of Choice
	Gram-Negative Rods	
Enterobacter aerogenes	Urinary tract; other infections	Gentamicin; tobramycin
Proteus	Urinary tract; other infections	Gentamicin; tobramycin
Pseudomonas aeruginosa	Bacteremia	Gentamicin; tobramycin
Acinetobacter	Various nosocomial infections; bacteremia	Gentamicin
Yersinia pestis	Plague	Streptomycin ± tetracycline
Serratia	Variety of nosocomial and opportunistic infections	Gentamicin
Mycobacterium tuberculosis	Tuberculosis	Stereptomycin + other antibiotics

TABLE 20.12 Some Therapeutic Uses of Tetacyclines, Chloramphenicol, Erythromycin, and Clindamycin

Causative Organism	Disease	Drug of Choice
	Gram-Negative Rods	
Salmonella	Typhoid fiver Paratyphoid fever Bacteremia	Chloramphenicol
Haemophilus influenzae	Pneumonia Meningitis	Chloramphenicol
Haemophilus ducreyi	Chancroid	A tetracycline
Brucella	Brucellosis	A tetracycline ± streptomycin
Vibrio cholerae	Cholera	A tetracycline
Flavobacterium meningosepticium	Meningitis	Erythromycin
Pseudomonas mallei	Glanders	Streptomycin + a tetracycline
Pseudomonas pseudomallei	Melioidosis	A tetracycline ± chloramphenicol
Campylobacter fetus	Enteritis	No treatment or erythromycin
	Bacteremia	Chloramphenicol
Bacterioides fragilis	Brain abscess	Chloramphenicol
	Lung abscess Intra-abdominal abscess Bacteremia Endocarditis	Clindamycin
Legionella pneumophila	Legionnaire's disease	Erythromycin
	Spirochetes	
Borrelia recurrentis	Relapsing fever	A tetracycline
	Miscellaneous Agents	
Mycoplasma pneumoniae	Atypical pneumonia	Erythromycin
Rickettsia	Typhus fever	Chloramphenicol
	Murine typhus Brill's disease Rocky Mountain spotted fever	A tetracycline
Chlamydia trachomatis	Trachoma	A sulfonamide + a tetracycline
	Inclusion conjuctivitis	A tetracycline
	Nonspecific urethritis	A tetracycline

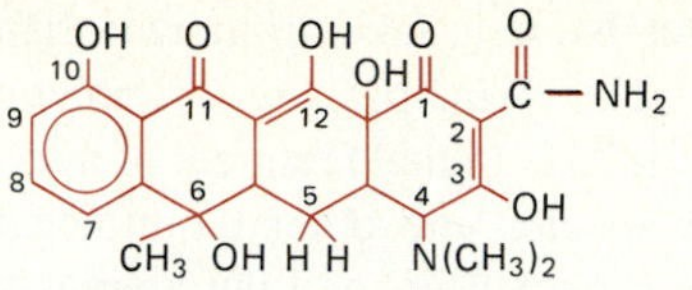

FIGURE 20.26
The structures of tetracycline antibiotics.

Compound	Substituent(s)	Position(s)
Chlortetracycline	—Cl	(7)
Oxytetracycline	—OH, —H	(5)
Demeclocycline	—OH, —H; —Cl	(6; 7)
Methacycline	—OH, —H; =CH₂	(5; 6)
Doxycycline	—OH, —H; —CH₃, —H	(5; 6)
Minocycline	—H, —H; —N(CH₃)₂	(6; 7)

ically to the 30S ribosomal subunit, apparently blocking the receptor site for the attachment of aminoacyl tRNA to the mRNA ribosome complex and thus preventing the addition of amino acids to a growing peptide chain. The sensitivity to tetracyclines depends on the transport of the tetracycline molecules across the cytoplasmic membrane. Some tetracyclines, such as doxycycline, appear to pass directly across the membrane, whereas others enter the cell only by active transport. Resistance to tetracyclines develops because of the movement of a transposon between a plasmid and the bacterial chromosome, and involves an alteration of the mechanisms of membrane transport of the tetracycline molecules.

There are a variety of tetracycline antibiotics that have the same basic four-ring structure, with some variations in the substituents (Figure 20.26). The tetracyclines are produced by various *Streptomyces* species. For example, chlortetracycline (aureomycin) is produced by *S. aureofaciens*, oxytetracycline by *S. rimosus*, and demeclocycline by *S. aereofaciens*; methacycline, doxyclycline, minocycline, and tetracycline are all semisynthetic derivatives. The tetracyclines are effective against a variety of pathogenic bacteria, including rickettsia and chlamydia species. Tetracylines, for example, are used therapeutically in treating the rickettsial infections of Rocky Mountain spotted fever, typhus fever, and Q fever and the chlamydial diseases of lymphogranuloma venereum, psittacosis, inclusion conjunctivitis, and trachoma. Tetracyclines are also useful in treating a variety of other bacterial infections, including pneumonia caused by *Mycoplasma pneumoniae*, brucellosis, tularemia, and cholera.

Unlike the antibiotics discussed so far that inhibit bacterial protein synthesis, **chloramphenicol** acts primarily by binding to the 50S ribosomal subunit, preventing the binding of tRNA molecules to both the aminoacyl and peptidyl binding sites of the ribosome. Consequently, peptide bonds are not formed when chloramphenicol is present in association with

the bacterial ribosome. It is used in the laboratory as a specific inhibitor of protein synthesis. Chloramphenicol, which is produced by *Streptomyces venezuelae*, is a fairly broad-spectrum antibiotic active against many species of Gram-negative bacteria. Resistance to chloramphenicol is generally associated with the presence of an R plasmid that codes for enzymes able to transform the chloramphenicol molecule. The production of an acetyl transferase enzyme can inactivate the chloramphenicol molecule because acetylated derivatives of chloramphenicol do not bind to bacterial ribosomes. This appears to be the main mechanism by which resistance to chloramphenicol occurs. Chloramphenicol has a number of toxic effects, including aplastic anemia, that limit its therapeutic uses to those where the benefits outweigh the dangers associated with toxic reactions. Chloramphenicol is used for treating typhoid fever, as well as various other infections caused by *Salmonella* species; it is also effective against anaerobic pathogens and can be used in treating diseases, such as brain abscesses, normally caused by anaerobic bacteria.

Like chloramphenicol, **erythromycin** acts by binding to 50S ribosomal subunits, blocking protein synthesis. Erythromycin, produced by *Streptomyces erythreus*, is a macrolide antibiotic, so named because it contains a multimembered lactone ring attached to deoxy sugar moieties (Figure 20.27). This antibiotic is most effective against Gram-positive cocci, such as *Streptococcus pyogenes*. Erythromycin is not active against most aerobic Gram-negative rods but does exhibit antibacterial activity against some Gram-negative organisms, such as *Pasteurella multocida*, *Bordetella pertussis*, and *Legionella pneumophilia*. Therapeutically, erythromycin is recommended for the treatment of Legionnaire's disease and is also effective in treating diphtheria, whooping cough, and the type of pneumonia caused by *M. pneumoniae*. Erythromycin may also be used as an alternative to penicillin in treating staphylococcal infections, streptococcal infections, tetanus, syphilis, and gonorrhea.

Like erythromycin, the antibiotics lincomycin and clindamycin bind to the 50S ribosomal subunit, blocking protein synthesis. Lincomycin is produced by *Streptomyces lincolnensis*, and clindamycin is a semisynthetic derivative of lincomycin. The use of these antibiotics is restricted by their side effects, such as severe diarrhea. Clindamycin is partic-

FIGURE 20.27
The biochemical structure of erythromycin.

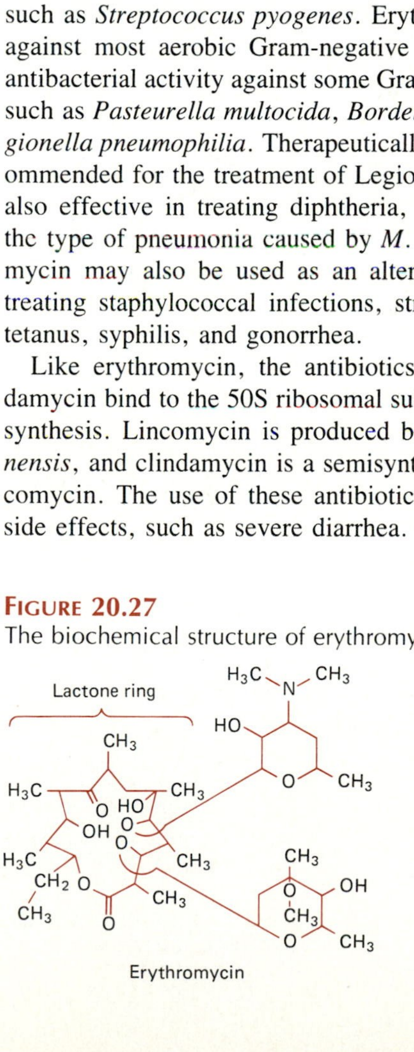

Erythromycin

ularly effective against Gram-positive bacteria, including anaerobes, and in the treatment of infections due to *Bacteroides* and *Fusobacterium* species.

Several other antibiotics that inhibit protein synthesis are not useful in treating bacterial infections because they inhibit protein synthesis in mammalian cells to the same extent that they do in bacterial cells. If the mode of action of these antibiotics is not specific for bacteria, they are not therapeutic antibacterial agents. For example, puromycin is an analogue of tRNA molecules and can compete with them in binding to ribosomes. The mode of action of this antibiotic does not distinguish between inhibiting eukaryotic and prokaryotic protein synthesis. Similarly, dactinomycin (actinomycin D) blocks protein synthesis in both bacterial and eukaryotic cells; this antibiotic binds to double-stranded DNA, preventing transcription of the genetic information to form an mRNA molecule. Although not useful in treating bacterial infections, dactinomycin does have a therapeutic role in treating some malignancies when it is desirable to block the rapid division of cancer cells. Rifampin, a semisynthetic derivative of rifamycin B, also blocks protein synthesis at the level of transcription. Rifampin inhibits DNA-dependent RNA polymerase enzymes and thus can block transcription; this antibiotic is more effective against bacterial RNA polymerase enzymes than mammalian RNA polymerases and therefore can be used therapeutically in treating some bacterial diseases. Rifampin is used in combination with other antibiotics in the treatment of mycobacterial diseases, such as tuberculosis.

Inhibitors of Membrane Transport The cytoplasmic membrane is the site of action of some antimicrobial agents. The polymyxins, such as polymyxin B, interact with the cytoplasmic membrane, causing changes in the structure of the bacterial cell membrane and leakage of cell contents. Polymyxin B is bactericidal, and its effectiveness is restricted to Gram-negative bacteria. The action of polymyxin B is related to the phospholipid content of the cell wall and the membrane

complex. Sensitive bacteria take up more polymyxin B than resistant strains. The principal use of polymyxin B and colistin (polymyxin E) is in the treatment of infections caused by *Pseudomonas* species and other Gram-negative bacteria that are resistant to penicillins and the aminoglycoside antibiotics. Both polymyxin B and colistin are useful in treating severe urinary tract infections caused by Gram-negative bacteria that are resistant to other antibiotics.

DNA Inhibitors Some antimicrobial agents act by blocking normal DNA replication. In particular, the **quinolones** interfere with DNA gyrase, preventing the establishment of a replication fork and the replication of DNA needed for cell multiplication. Although DNA synthesis is blocked, transcription and translation (protein synthesis) can still occur. Bacteria exposed to quinolones elongate rather than divide normally. The quinolones include nalidixic acid, ciprofloxacin, norfloxacin, amifloxacin, and enoxacin. These antimicrobics are effective against a broad range of Gram-positive and Gram-negative bacteria, including some—such as mycobacteria—that are resistant to many other compounds.

Inhibitors with Other Modes of Action Sulfonamides, sulfones, and para-aminosalicylic acid are structural analogues of the vitamin para-aminobenzoic acid, which makes them useful antibacterial agents (Figure 20.28). A cell mistakenly using an analogue, such as sulfonamide, in place of the normal substance, para-aminobenzoic acid in this case, results in the formation of molecules that are unable to perform their essential metabolic functions; in this case, there is a failure of critical coenzyme functions. Folic acid is an essential coenzyme composed in part of para-aminobenzoic acid. Mammalian cells are unable to synthesize folic acid; they require an intake of folic acid as part of their diet and cellular uptake via an active transport system. Bacterial cells, in contrast, normally synthesize their required folic acid and are unable to transport it across their cytoplasmic membranes (Figure 20.29). The analogues of para-aminobenzoic acid are

FIGURE 20.28

Sulfonamide, showing its similarity to *para*-aminobenzoic acid.

p-Aminobenzoic acid (PABA)

Sulfonamides

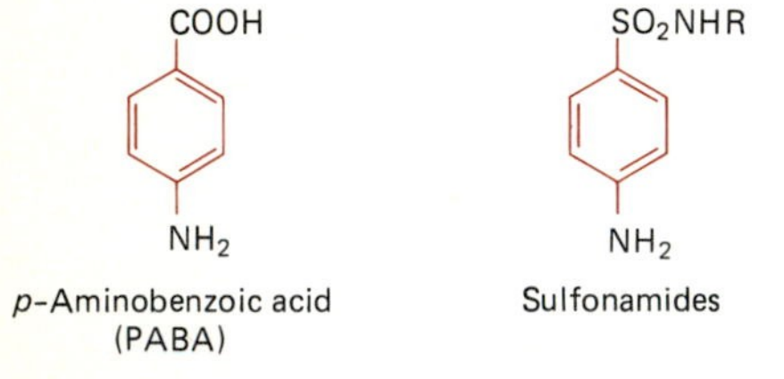

FIGURE 20.29

Folic acid metabolism, showing inhibition by analogues of *para*-aminobenzoic acid.

Sulfanilamide

p-Aminobenzoic acid (PABA)

(Pteridine)   (PABA)   (Glutamic acid)

Folic acid

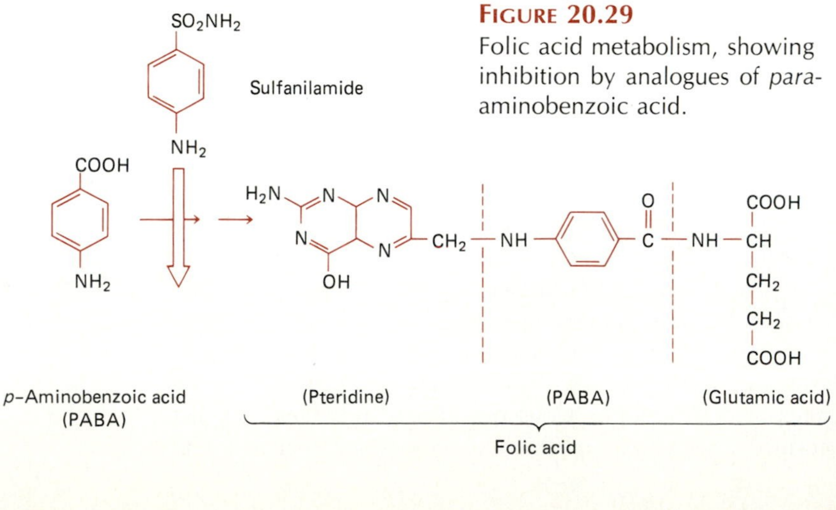

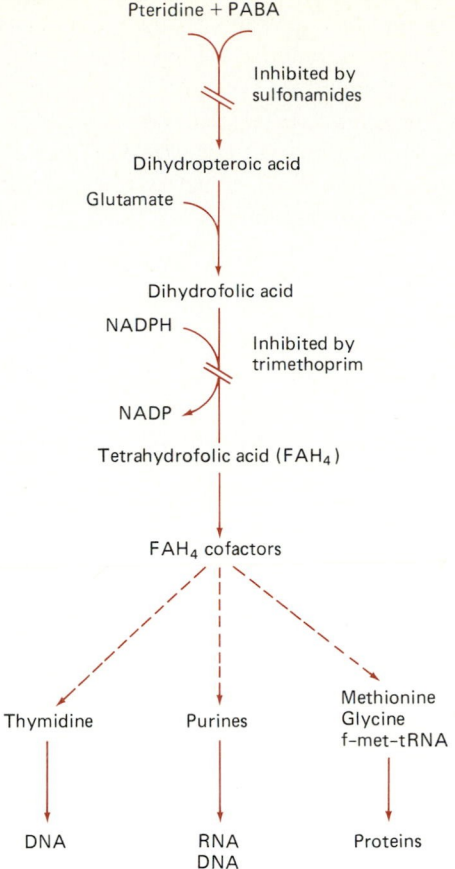

FIGURE 20.30
The effects of trimethoprim and sulfomethoxazole on thymidine, purine, and protein synthesis.

effective competitors with the natural substrate for the enzymes involved in the synthesis of folic acid and, as such, are able to inhibit the formation of this required coenzyme, causing a bacteriostatic effect. The sulfones are useful in treating leprosy. The use of sulfonamides and para-aminosalicylic acid has declined as a result of the development of resistant strains and of more effective antibiotics with less toxic side effects.

Trimethoprim is an inhibitor of dihydrofolate reductase, especially in bacteria. Dihydrofolic acid is a coenzyme required for 1-carbon transfers, such as those that occur in the synthesis of thymidine and purines (Figure 20.30). Trimethoprim is effective in blocking bacterial growth by preventing the formation of the active form of the required coenzyme. Its effectiveness is enhanced when is coupled with sulfamethoxazole. Trimethoprim is a broad-spectrum antibacterial agent and is effective in the treatment of many urinary and intestinal tract infections. It is used primarily for the treatment of urinary infections due to *E. coli*, *Proteus*, *Klebsiella*, and *Enterobacter*.

In addition to the use of trimethoprim-sulfamethoxazole, several other compounds are used as antiseptics in treating

urinary tract infections.[3] These compounds, which inhibit the growth of many bacterial species, include methenamine, nalidixic acid, oxolinic acid, and nitrofurantoin. The usefulness of these drugs, though, depends on the fact that they are concentrated in the urinary tract tissues and thus can act as antiseptics at this location. Nalidixic acid is effective against most Gram-negative bacteria that cause urinary tract infections. Nitrofurantoin inhibits several bacterial enzymes, but its mode of action is unknown. It is used only in a limited number of cases because it is generally not as effective as other antibiotics, including sulfanomides.

Although isoniazid is not particularly effective against *Mycobacterium tuberculosis* when used alone, it is generally used in association with other antibiotics in treating tuberculosis. The specific mechanism of isoniazid is not known, but its primary action appears to involve the inhibition of mycolic acid biosynthesis. Mycolic acids are unique components of the cell walls of mycobacteria, and blockage of the biosynthesis of these compounds could specifically inhibit mycobacteria.

Antimicrobial Agents Effective Against Infections Caused by Eukaryotic Microorganisms

In contrast to the relative ease with which bacterial infections can be treated, the fact that fungi, protozoa, and humans all have eukaryotic cells limits the sites against which antimicrobial agents can be selectively directed to control eukaryotic pathogens of human beings. Consequently, there are relatively few antimicrobial agents of therapeutic value against infections caused by eukaryotic microorganisms, particularly in treating systemic infections.

Antifungal Agents   There are sufficient differences between a fungal cell and a human cell so that some therapeutically useful compounds with antifungal activity have been discovered. Some of the therapeutic uses of **antifungal agents** are listed in Table 20.13. The **polyene antibiotics**, amphotericin

[3]The term *antiseptic* is used to indicate that these substances actually wash the surface of the urinary tract.

TABLE 20.13   Some Therapeutic Uses of Antifungal Agents

Causative Organism	Disease	Drug of Choice
Candida albicans	Skin and superficial mucous membrane lesons	Amphotericin B Nystatin
Cryptococcus neoformans	Meningitis	Amphotericin B + flucytosine
Candida albicans	Pneumonia	Amphotericin B
Aspergillus	Meningitis	
Mucor	Skin lesions	
Histoplasma capsulatum	Lung lesions Histoplasmosis	Amphotericin B
Coccidioides immitis	Coccidiomycosis	Amphotericin B
Blastomyces dermatitidis	Blastomycosis	Amphotericin B

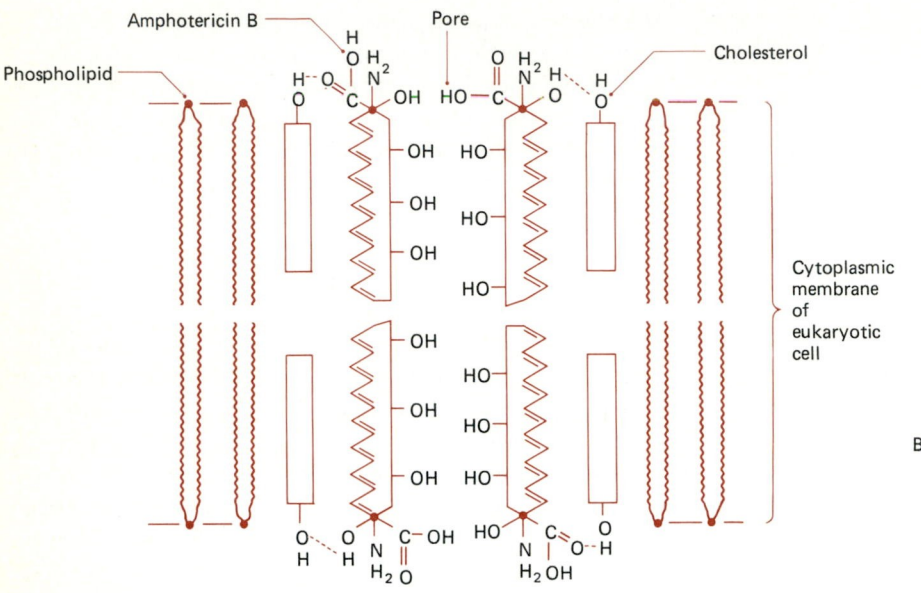

FIGURE 20.31

(A) The structure of the polyene antibiotic, amphotericin B, and (B) its effect on the cytoplasmic membrane of a eukaryotic cell.

B, and nystatin are used in treating a variety of fungal diseases. The polyene antibiotics act by altering the permeability properties of the cytoplasmic membrane, leading to the death of the affected cells. Interactions of polyenes with the sterols in the cytoplasmic membranes of eukaryotic cells appear to form channels or pores in the membrane, allowing leakage of small molecules (Figure 20.31). Differences in the sensitivity of various organisms are determined by the concentrations of sterols in the membrane. Because mammalian cells, like fungi, contain sterols in their cytoplasmic membranes, it is not surprising that polyene antibiotics also cause alterations in the membrane permeability of mammalian cells and toxicity to mammalian tissue, as well as the death of fungal pathogens.

Nystatin, which is produced by *Streptomyces noursei*, is primarily used in the treatment of topical infections caused by members of the fungal genus *Candida*. Vaginitis, thrush, and *Candida* infections of the gastrointestinal tract are effectively treated by using nystatin. **Amphotericin B**, which is produced by *Streptomyces nodosus*, has a relatively broad spectrum of activity and is used in the treatment of systemic fungal infections. It is the most effective therapeutic agent for treating systemic infections due to yeast and fungi. The potential toxic side effects of amphotericin B usage, however, such as kidney damage, require careful supervision of its administration. Patients requiring amphotericin B must be hospitalized so that the initial reaction to the therapy can be carefully supervised. Patients who have received amphotericin B almost invariably exhibit some toxic side effects, but without this drug, systemic fungal infections are almost invariably fatal. Amphotericin B is used in the treatment of cryptococcosis, histoplasmosis, coccidioidomycosis, blastomycosis, sporotrichosis, and candidiasis.

In addition to the polyene antibiotics, imidazole derivatives such as miconazole and clotrimazole have a broad spectrum of antifungal activities and are used in the topical treatment of superficial mycotic infections. These two antimicrobial agents appear to alter membrane permeability, leading to the inhibition and/or death of selected fungal species. Flucytosine, a fluorinated pyrimidine (Figure 20.32), is also

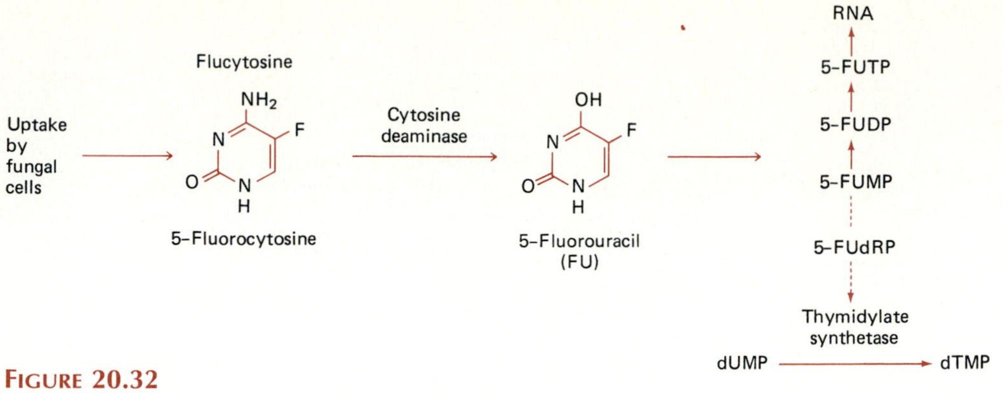

FIGURE 20.32

The structure of flucytosine and its transformation to fluorouracil.

effective in treating systemic fungal infections. Flucytosine is less effective, but also less toxic, than amphotericin B and is primarily used in combination with it. Within fungal cells, flucytosine is converted to fluorouracil and is further metabolized to form an inhibitor of thymidylate synthetase, causing an inhibition of normal nucleic acid synthesis. Mammalian cells do not convert as much flucytosine to fluorouracil as do fungal cells, accounting for the selective toxicity of this antifungal agent.

Griseofulvim is another antibiotic that is effective against some fungal infections. This antibiotic is produced by *Pencillium griseofulvum* and causes a disruption of mitotic spindles, inhibiting fungal mitosis. Griseofulvim is used in the treatment of fungal diseases of the skin, hair, and nails caused by various species of dermatophytic fungi, like *Microsporum*, *Epidermophyton*, and *Trichophyton*. These dermatophytes concentrate griseofulvim by an active uptake process, and their sensitivity is correlated directly with their ability to concentrate the antibiotic.

Antiprotozoan Agents   Treatment of human protozoan diseases with antimicrobial agents presents a special problem because many of the pathogenic protozoa exhibit a complex life cycle, often including stages that develop within mammalian cells. Different antimicrobial agents are generally needed for use against different forms of the same pathogenic protozoan, depending on the stage of the life cycle and the involved tissues. For example, the protozoan species of the genus *Plasmodium* that cause malaria exhibit complex life cycles, part of which occur in the liver and blood of human beings (Figure 20.33). The erythrocytic stage of the *Plasmodium* life cycle that occurs within human blood cells is the most sensitive to **antimalarial drugs**. The life stages that occur within the liver are difficult to treat, and the sporozoites injected into the bloodstream by mosquitoes are not affected by antimalarial drugs. The antimalarials effective against the erythrocytic forms of the protozoan include chloroquine and amodiaquine, neither of which is effective against the stages of the *Plasmodium* that occur in the liver (Figure 20.34). These antimalarial agents appear to interfere with DNA rep-

lication. The effect of these drugs is a rapid **schizontocidal action**, that is, the rapid interruption of schizogony or multiple division that occurs within red blood cells. The sensitivity of malarial protozoa to these drugs depends on the active transport of these compounds into the protozoa and their selective accumulation intracellularly.

Chloroguanide is also used in the suppression of malaria. This drug is transformed within the body to a triazine derivative that inhibits the enzyme dihydrofolate reductase and thus interferes with the essential metabolic reactions involving this coenzyme, which are required for the proliferation of the malaria protozoa. Chloroguanide is sometimes used concurrently with sulfonamide compounds that also interfere with folate metabolism. It binds more strongly to the plasmodial enzyme than to the comparable mammalian dihydrofolate reductase, accounting for its selective inhibition. In addition to affecting the schizont stage, chloroguanide influences the sterilizing action of gametocytes. Because resistance to the synthetic antimalarial drugs is increasing, quinine, one of the early drugs used for the treatment of malaria, is once again being used.

For the radical cure of malaria, that is, the eradication of both the erythrocytic and liver stages of the protozoan, primaquine is normally used. This drug is used in conjunction with chloroquine and chloroguanide. The precise mode of action of primaquine has not been elucidated. Because of its toxic side effects, it is primarily used in the treatment of relapsing malarial infections. Pyromethamine, which also inhibits folic acid metabolism, has also been used in the treatment of malaria. Many plasmodium strains, however, have developed resistance to this drug, limiting its usefulness in treating malaria.

Several other drugs are used in the treatment of various other protozoan infections (Table 20.14). As with malaria, the life cycle of the particular protozoan determines which agents will be effective in controlling the infection. Only a few of these antiprotozoan agents will be discussed further here. Quinacrine hydrochloride is used to treat *Giardia lamblia*, a protozoan disease spread through contaminated water that has become a major problem in the United States. Me-

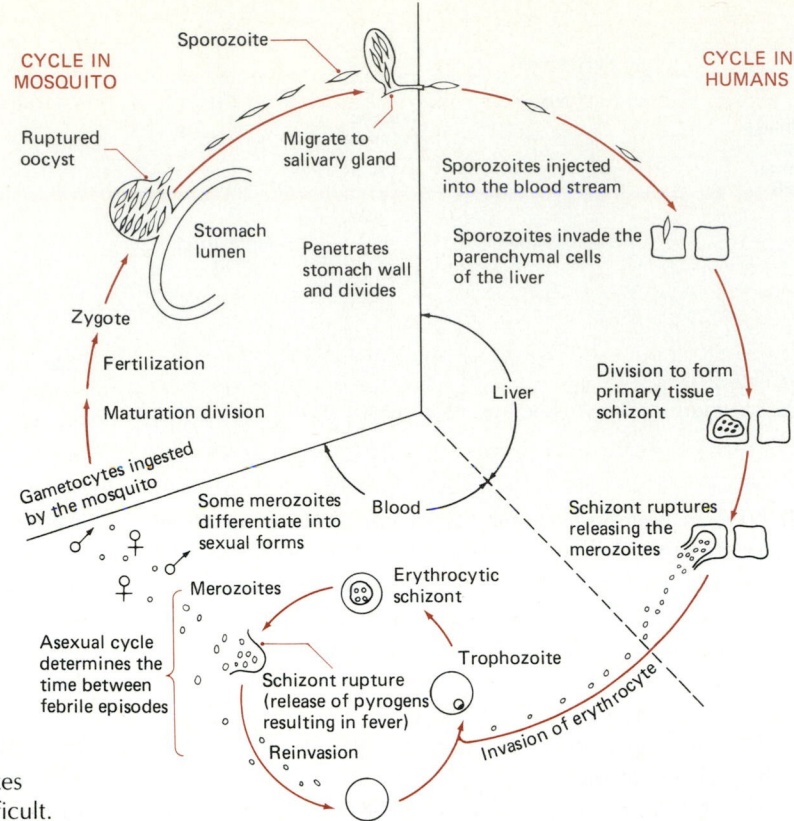

FIGURE 20.33
The complex life cycle of the malarial protozoan makes treatment of this disease with antibiotics extremely difficult.

tronidazole is also used in the treatment *Giardia* infections, as well as in cases of dysentery caused by the protozoan *Entamoeba histolytica*. Metronidazole interferes with hydrogen transfer reactions, specifically inhibiting the growth of anaerobic microorganisms, including anaerobic protozoa. Pentamidine and related diamidine compounds are useful in treating infections by members of the protozoan genus *Trypanosoma*. Compounds of this type interfere with DNA metabolism.

Melarsoprol is an arsenical, that is, an arsenic-containing compound useful in treating some stages of human trypanosomiasis, particularly because of its ability to penetrate

cerebrospinal fluid (Figure 20.35). Arsenicals react with the sulfhydryl groups of proteins, inactivating a large number of enzymes. It appears that mammalian cells can metabolize these compounds to nontoxic forms more rapidly than protozoan cells, accounting for the selective toxicity of melarsoprol to trypanosome protozoans. Sodium stibogluconate, an antimony-containing compound, is useful in treating diseases caused by members of the protozoan genus *Leishmania*. Antimony compounds of this type inhibit the enzyme phosphofructokinase in some life history stages of the leishmanias, accounting for its inhibitory effects. Other antiprotozoan agents useful in the chemotherapy of protozoan diseases include suramin, a nonmetalic compound that inhibits a wide variety of enzymes, and nifurtimox, which is effective against *Trypanosoma cruzi*, the causative organism of Chagas' disease.

FIGURE 20.34
The biochemical structures of chloroquine and amodiaquine, which are used to treat malaria.

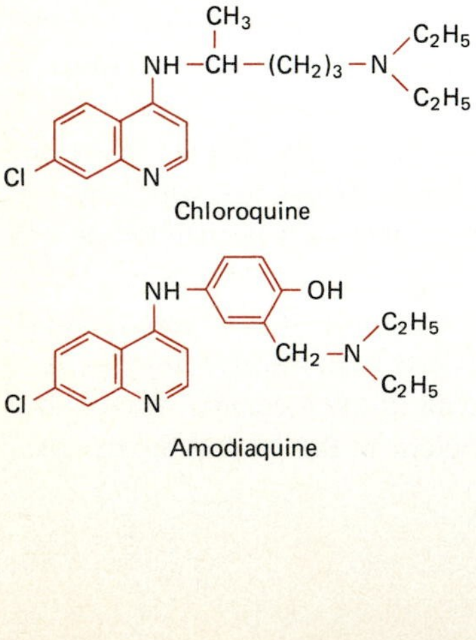

Chloroquine

Amodiaquine

FIGURE 20.35
The structure of melarsoprol, an effective antiprotozoan agent.

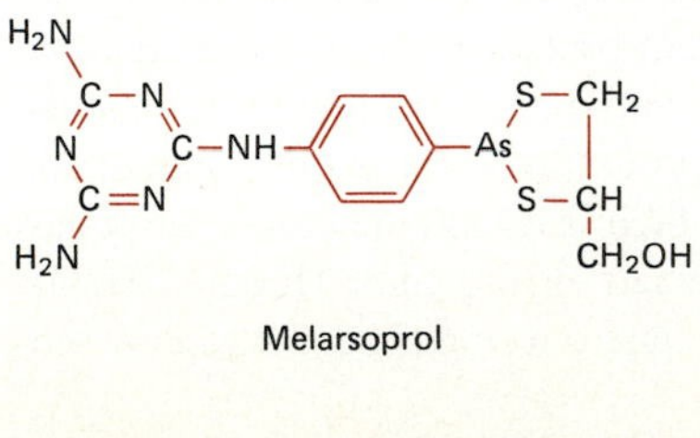

Melarsoprol

TABLE 20.14   Some Drugs Used in the Treatment of Diseases Caused by Protozoan Pathogens

Infecting Organism Disease	Drug of Choice
Entamoeba histolytica	
Asymptomatic cyst passer	Diiodohydroxyquin
Mild intestinal disease	Metronidazole
Severe intestinal disease	Metronidazole
Hepatic abscess	Metronidazole
Giardia lamblia	Quinacrine hydrochloride
Balantidium coli	Oxytetracycline
Dientamoeba fragilis	Diiodohydroxyquin
Trichomonas vaginalis	Metronidazole
Pneumocystis carinii	Trimethoprim-sulfamethoxazole
Toxoplasma gondii	Pyrimenthamine plus trisulfapyrimidines
Leishmaniasis	Sodium stibogluconate
Leishmania donovania (kala azar, visceral leishmaniasis)	
Leishmania tropica (oriental sore, cutaneous leishmaniasis)	Sodium stibogluconate
Leishmania braziliensis (American mucocutaneous leishmanisasis)	Sodium stibogluconate
African trypanosomiasis	
Trypanosoma gambiense	Pentamidine
Trypanosoma rhodesiense	Suramin
Either *T. gambiense* or *T. rhodesiense* in late disease with central nervous system involvement	Melarsoprol
South American trypanosomiasis (Chagas' disease)	
Trypanosoma cruzi	Nifurtimox

TABLE 20.15   Some Antiviral Agents and Their Therapeutic Uses

Causative Organism	Disease	Drug of Choice
Herpes simplex virus	Keratoconjunctivitis	Vidarabine
		Trifluridine
		Acyclovir
	Encephalitis	Vidarabine
	Cold sores	Acyclovir
	Genital herpes	Acyclovir
Influenza virus A	Influenza	Amantadine
HIV	AIDS	Azidothytmidine

Antiviral Agents

The search for antiviral drugs comparable to the antibiotics used to control bacterial infections has largely been fruitless. There are no broad-spectrum antiviral agents currently in clinical use, and most viral infections cannot be treated effectively by using antiviral chemicals. The integral role of the mammalian host cell in the process of viral replication complicates the difficulty of finding compounds that specifically inhibit viral replication. Most compounds that prevent the reproduction of viruses also interfere with mammalian cell metabolism, resulting in adverse effects on human cells precluding the therapeutic use of such agents. Very few antiviral agents have been found with clinical applicability, and these generally have a narrow spectrum of antiviral activity.

The best treatment for many viral diseases is prevention through the appropriate use of vaccines and the controll of vectors that act as carriers for viruses. Generally, recovery from a viral infection is dependent on the natural immune defense response of the body. For most viral infections, treat-

ment is aimed at maintaining a physiological state in which an effective immune response can be ensured by following the sage advice, "rest and drink plenty of fluids." The discovery of the role of interferons in the natural immune response to viruses holds promise for the future if interferons can be produced commercially and if their administration proves to be of therapeutic value for specific diseases. Genetic engineering would seem to provide the greatest hope for the commercial production of interferons, making these antiviral substances available for medical treatment of viral infections. Because, as we will see in Chapters 21 and 22, viruses cause many human diseases, the development of antiviral agents of therapeutic value is very important for improving our ability to treat numerous diseases.

Only a few chemical agents have been developed so far for therapeutic use in treating specific viral infections (Table 20.15). Among these antiviral compounds, idoxuridine, an analogue of thymidine that interferes with DNA metabolism in both viral and mammalian cells, has clinical uses in the treatment of herpes simplex keratitis. Amantadine is recommended for the prophylaxis of high-risk patients in cases of documented influenza A virus epidemics. Amantadine appears to interfere with the absorption of influenza A viruses by human cells and thus with their entry into cells to establish an infection. There are several compounds that have antiviral activity against herpes viruses. Trifluridine is a nucleic acid base analogue that inhibits DNA synthesis. This fluorinated pyrimidine nucleoside is effective against epithelial keratitis of the eye caused by Herpes simplex virus. Vidarabine was originally developed for the treatment of leukemia but has proven to be more effective in treating herpes simplex encephalitis and keratoconjuntivitis. Vidarabine (vira-A) is an adenine arabinoside, which is phosphorylated within mammalian cells to the corresponding nucleotide that acts as an inhibitor of viral DNA polymerase. The selectivity of vidarabine is due to its inhibition of mammalian DNA synthesis to a lesser extent than viral DNA replication.

Acyclovir (9-[2-hydroxymethyl]guanine) has proven to be the best antiherpes drug so far discovered. Acyclovir is more effective than idoxuridine and vidarabine for the treatment of herpetic ocular disease. It is also effective in treating cold sores and genital herpes caused by Herpes simplex. It may also be of value in treating chicken pox and shingles. Acy-

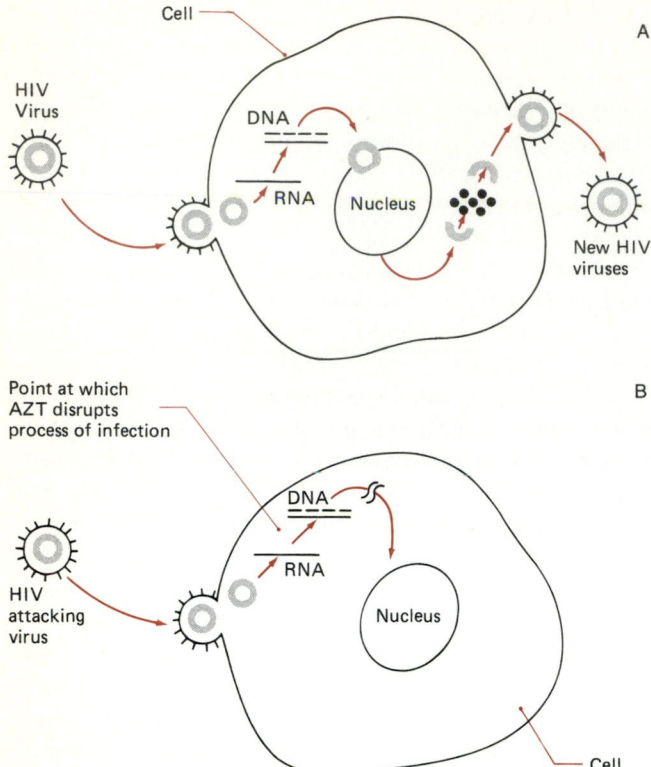

FIGURE 20.36
AZT, an analogue of thymidine, is effective in blocking the
replication of the virus that causes AIDS. (A) Normally, when
HIV (HTLV-3) invades a T-helper cell, it uses reverse transcriptase to form a virus-directed DNA. The DNA made with
the viral RNA as a template moves to the cell nucleus and
directs the synthesis of viral components needed for the
successful replication of the virus. (B) In the presence of AZT,
the virus is unable to synthesize the needed DNA by reverse
transcription and hence is unable to replicate. The drug
appears to be useful in reversing some of the symptoms of
AIDS but is not considered a cure for the disease.

clovir, a nucleoside analogue, is converted *in vivo* to an
acylguanosine triphosphate that inhibits herpes simplex viral
DNA polymerase, thus blocking viral DNA replication. The
activation of acyclovir is initiated by a viral-directed thymidine kinase enzyme that converts this compound to an
acycloguanosine monophosphate, which is subsequently converted to the acycloguanosine di- and triphosphates. In an
uninfected cell, there is only very limited conversion of acyclovir to the phosphorylated acylguanosines. Because an enzyme coded for by the herpes virus is required to activate
acyclovir, this compound exhibits selective antiviral activity,
making it therapeutically valuable.

Recently, the drug azidothymidine (AZT) has been found
to be effective in the treatment of AIDS. AZT is a analogue
of the DNA base thymidine. It was first synthesized as a
possible anticancer drug but proved ineffective against tumors. HIV, which causes AIDS, is a retrovirus, and AZT is
effective in preventing the reverse transcriptase of this virus
from forming viral-directed DNA that is needed for the successful replication of the virus (Figure 20.36). The drug is
still experimental, and the full range of its possible side effects unknown. Nevertheless, because untreated AIDS is fatal and the initial test results with AZT have been so promising, the United States Food and Drug Administration has
approved its use for the treatment of most individuals with
AIDS.

Postlude

The prevention of infectious diseases is clearly desirable, and
various measures can be employed. The probability of contracting such a disease can be greatly reduced by using
measures that control the populations of pathogenic microorganisms and the probability of exposure to such organisms.
These measures often rely on proper sanitation, food handling, and preservation; prevention of wound contamination;
and minimizing the spread of aerosols that contain microorganisms. Public health measures aimed at controlling infectious diseases are also frequently aimed at controlling vec-

tor populations, such as mosquitoes, fleas, ticks, and lice,
that can act as carriers of pathogenic microorganisms.

Many vaccines have been developed, and their use in rendering hosts unsusceptible to specific pathogenic microorganisms is important in controlling numerous infectious diseases. Smallpox, for example, has been eliminated as a
consequence of the extensive use of vaccination. Polio, measles, mumps, rubella, yellow fever, tetanus, and rabies can
be prevented by the appropriate use of vaccines. The effectiveness of vaccines depends on the normal memory response

of the human immune system and on establishing active immunity through exposure to antigens that elicit the desired immune response without causing disease. Both inactivated and attenuated microorganisms are employed in different vaccines used for immunization purposes.

The clinical identification of the etiological agents of disease relies on both biochemical and serological testing. The culture methods and identification systems employed depend on the preliminary diagnosis of the disease and on the microorganisms presumed to be the causative agents based on preliminary screening procedures. The procedures used in clinical microbiology and immunology laboratories are subject to change as new and better screening procedures to isolate pathogenic microorganisms are developed. Speed and accuracy in determining the nature of the disease, the disease etiology, and the appropriate antimicrobial agents that may be effective in controlling it are of the essence because an error or delay in diagnosing a disease can prove fatal. Automation has greatly improved speed and reliability in the clinical laboratory, and pathogenic microorganisms can now be isolated and identified within hours. The development of various serological procedures has greatly improved the accuracy and speed of many diagnoses. Similarly, antibiotic susceptibility testing can be accomplished rapidly. At present, hospital clinical laboratories attempt to provide a complete diagnostic report, identifying the pathogen and its antibiotic sensitivities, within 24 hours. It has been predicted that by the year 2000 virtually all diagnoses will be made at bedside by serological methods, providing the physician with immediate information needed to initiate proper therapy, and improved antibiotic susceptibility testing may permit *in vivo* monitoring of the effectiveness of antibiotic treatment.

When host defense mechanisms fail and exposure to a pathogenic microorganism results in the development of an infection, there are a variety of antimicrobial agents that can be employed for the treatment of disease. The bacterial diseases are perhaps the easiest infectious diseases to treat because the differences between prokaryotic and eukaryotic cells provide sites against which chemicals can exert selective inhibition against bacteria without killing or severely inhibiting host mammalian cells. The discovery and use of antibiotics, which are antimicrobial agents produced by microorganisms, have revolutionized medical practice. The choice of a particular antibiotic for treating a disease rests on the nature of the infecting microorganism and the tissues in which the infection occurs. In general, the murein component of the bacterial cell wall and the 70S ribosomes of bacteria provide targets against which many of the therapeutically useful antibiotics are aimed. Prudent medical practice requires that antibiotics be specifically aimed at the infecting microorganisms, that is, wherever possible, specific or narrow-spectrum antibiotics should be employed to avoid the selective pressure for the development of antibiotic-resistant strains of pathogenic microorganisms. Such strains are becoming a very serious problem in treating infectious diseases.

The treatment of infections by viruses and eukaryotic microorganisms presents a greater problem than bacterial infections for finding specific inhibitors that are not also directed at mammalian cells. Systemic mycotic and protozoan infections are particularly difficult to treat, often persisting for long periods of time. Many of the drugs used in the treatment of such infections also cause side effects toxic to human tissues.

Microbiologists associated with the pharmaceutical industry are continuously screening for new antibiotics or derivatives of known antimicrobial agents that might be of therapeutic value in treating infectious diseases. The activities of the pharmaceutical industry in discovering and producing antimicrobial agents were considered in Chapter 17. Advances continue to be made in our ability to treat diseases caused by microorganisms as new antimicrobial agents are discovered and developed, holding out the promise of achieving a higher standard of human health in the future.

Suggested Supplementary Readings

BERDY, J. (ed.). 1980–1982. *Handbook of Antibiotic Compounds.* CRC Press, Inc., Boca Raton, Fla.

BROWN, F. 1984. Synthetic viral vaccines. *Annual Review of Microbiology* 38:221–236.

BRYAN, L. E. 1982. *Bacterial Resistance and Susceptibility to Chemotherapeutic Agents.* Cambridge University Press, New York.

COHEN, M. L., and R. V. TAUXE. 1986. Drug-resistant *Salmonella* in the United States: An epidemiological perspective. *Science* 234:964–969.

CURRAN, J. W., W. M. MORGAN, A. M. HARDY, H. W. JAFFE, and W. R. DOWDLE. 1985. The epidemiology of AIDS: Current status and future prospects. *Science* 229:1352–1357.

DALTON, H. P., and H. C. NOTTEBART, JR. (eds.). 1986. *In-terpretive Medical Microbiology.* Churchill Livingstone, New York.

EWING, W. H. 1985. *Edwards' and Ewing's Indentification of Enterobacteriaceae.* Elsevier Publishing Co., New York.

FINEGOLD, S. M., and E. BARON. 1986. *Diagnostic Microbiology.* C.V. Mosby Co., St. Louis.

FUNDENBERG, H. H., D. P. SITES, J. L. CALDWELL, and J. V. WELLS (eds.). 1980. *Basic and Clinical Immunology.* Lange Medical Publications, Los Altos, Calif.

GILMAN, A. G., L. S. GOODMAN, and A. GILMAN (eds.). 1980. *Goodman and Gilman's Pharmacological Basis of Therapeutics.* Macmillan Publishing Co., Inc., New York.

GLASS, R. I. 1986. New prospects for epidemiological investigations. *Science* 234:951–955.

HUGO, W. B., and A. D. RUSSELL. 1983. *Pharmaceutical Microbiology*. Blackwell Scientific Publications, Oxford, England.

KAGAN, B. M. 1974. *Antimicrobial Therapy*. W.B. Saunders, Co., Philadelphia.

KONEMAN, E. W., S. D. ALLEN, V. R. DOWELL, JR., and H. M. SOMMERS (eds.). 1983. *Color Atlas and Textbook of Diagnostic Microbiology*. J.B. Lippincott Co., Philadelphia.

LANCINI, G., and F. PARENTI. 1982. *Antibiotics: An Integrated View*. Springer-Verlag, New York.

LENNETTE, E. H., A. BALOWS, W. J. HAUSLER, JR., and J. P. TRUANT (eds.). 1986. *Manual of Clinical Microbiology*. American Society for Microbiology, Washington, D.C.

LORIAN, V. (ed.). 1986. *Antibiotics in Laboratory Medicine*. Williams & Wilkins, Baltimore.

LYNN, M., and M. SOLOTOROVSKY (eds.). 1981. *Chemotherapeutic Agents for Bacterial Infections* (Benchmark Papers in Microbiology). Academic Press, New York.

MANDELL, G. L., R. G. DOUGLAS, JR., and J. E. BENNETT. 1985. *Anti-Infective Therapy*. John Wiley & Sons, Inc., New York.

MILGROM, F., C. J. ABEYOUNIS, and K. KANO (eds.). 1981. *Principles of Immunological Diagnosis in Medicine*. Lea & Febiger, Philadelphia.

Physician's Desk Reference. Published annually. Medical Economics Co., Oradell, N.J.

PRATT, W. B. 1977. *Chemotherapy of Infection*. Oxford University Press, New York.

PULVERER, G., and J. JELJASZEWICZ (eds.). 1985. *Chemotherapy and Immunity*. VCH Publishers, New York.

RAVEL, R. 1978. *Clinical Laboratory Medicine: Clinical Application of Laboratory Data*. Year Book Medical Publishers, Inc., Chicago.

ROITT, I. M., J. BROSTOFF, and D. MALE. 1986. *Immunology*. Blackwell Scientific Publications, Oxford, England.

ROSE, N. R., and H. FRIEDMAN (eds.). 1986. *Manual of Clinical Immunology*. American Society for Microbiology, Washington, D.C.

RYTEL, M. J. 1979. *Rapid Diagnosis in Infectious Diseases*. CRC Press, Inc., Boca Raton, Fla.

SMITH, J. R., R. J. LAUDICINA, and R. D. RUFO. 1986. *Learning Guides for the Medical Microbiology Laboratory*. John Wiley & Sons, Inc., New York.

SOULSBY, E. J. L. (ed.). 1987. *Immune Responses in Parasitic Infections: Immunology, Immunopathology, and Immunoprophylaxis*. CRC Press, Inc., Boca Raton, Fla.

STANSFIELD, W. D. 1981. *Serology and Immunology*. Macmillan Publishing Co., Inc., New York.

VOLK, W. A. 1982. *Essentials of Medical Microbiology*. J.B. Lippincott, Co., Philadelphia.

ZUCKERMAN, A. J., J. E. BANATVALA, and J. R. PATTISON. 1987. *Principles and Practice of Clinical Virology*. John Wiley & Sons, Inc., New York.

Study Questions

1. How should you avoid exposure to airborne pathogens?

2. What is a vaccine? How is vaccination used to prevent disease?

3. How are differential blood counts used to determine if a disease is caused by a microorganism?

4. Why is it important to identify cultures sent to the clinical microbiology laboratory quickly and accurately?

5. What are serological tests? How are they used in the identification of pathogens?

6. How is immunofluorescence used to identify *Treponema pallidum*? Why are serological methods critical for identifying this pathogenic bacterium?

7. Why is it essential to perform antimicrobial susceptibility testing on pathogenic isolates?

8. How does a physician select an antibiotic for treating an infectious disease? Describe several approaches used to determine the sensitivity of a pathogen to antibiotics. Is antimicrobial sensitivity the sole criterion for selecting an antibiotic? Discuss.

9. What is an MIC? Why is this an increasingly common test in clinical microbiology laboratories?

10. Is penicillin useful in treating the common cold? Explain.

11. Why is it easier to find antibacterial agents than to discover useful antifungal agents?

12. Why is it so difficult to find antimicrobial agents for treating viral diseases?

13. What antibiotics should you prescribe for each of the following conditions?
    a. urinary tract infection
    b. upper respiratory tract bacterial infection
    c. fungal infection of the vaginal tract
    d. herpes encephalitis
    e. malaria

14. Discuss the mode of action of penicillin.

15. Why is penicillin ineffective against bacteria that produce β-lactamases?

16. Why is an inhibitor of transcription not useful in treating bacterial infections of humans?

17. Discuss the mode of action of streptomycin.

18. Discuss the mode of action of acyclovir.

19. Describe the stages of an acute infection. Comment on the number of pathogens, components of the immune response, and symptoms of the patient at each stage.

Situational Problem 1

Expressing an Opinion on Government-Mandated Vaccinations

The introduction of vaccines, which started in the 1700s, was initially welcomed by the public because of the promise of the eradication of feared diseases such as smallpox. By the twentieth century, the vaccination of children against once deadly diseases—including tetanus, diphtheria, and whooping cough (pertussis)—became routine practice in developed nations. Through the efforts of WHO, these vaccines were also introduced into developing nations, leading to the worldwide elimination of smallpox and better control of several other diseases. Within the past few decades, vaccines have been introduced that prevent various other diseases, including measles, mumps, rubella, influenza, polio, and others. The control of measles and polio in North America has been particularly effective but requires constant vigilance to ensure that children continue to be immunized.

The effectiveness of a vaccination program depends upon reducing the number of susceptible (nonimmune) individuals to such a low level that even if a case of the disease occurs, it cannot spread because of the statistical improbability of a viable pathogen reaching a susceptible individual. To ensure that an adequate proportion of the population is immune to specified diseases, the governments of developed nations have instituted mandatory vaccination programs. Typically, proof of vaccination is required to attend school, making vaccination of children necessary.

In recent years, however, the wisdom of such mandated vaccination has come under severe attack and public scrutiny. The problem arises from the fact that some children exhibit adverse reactions to vaccines. Particular problems have been encountered with the pertussis vaccine; a small proportion of children receiving the vaccine exhibit very severe side effects, including mental retardation and even death. Requiring parents to have their children immunized with this vaccine, despite the knowledge that a small proportion will die or suffer severe illness as a result, raises serious ethical questions for the medical profession and the general public. These questions became frequent topics of debate on television talk shows and in the halls of Congress. Issues discussed included how to balance the interest of the health and welfare of the general public through the use of mass immunizations with the legitimate needs of individuals to whom immunization may be a threat or a violation of religious principles and the role of government in this aspect of public and personal health and safety.

In our litigious society, it is not surprising that several lawsuits and large cash settlements have resulted when children have suffered ill effects following required vaccinations. These financial settlements and the increased cost of their liability insurance forced several vaccine manufacturers to halt production of pertussis vaccine in the mid-1980s, jeopardizing the adequacy of the world supply of this vaccine. After extensive debate, the U.S. government decided to help underwrite the cost of this insurance.

Compose a letter to the editor of your local newspaper or to your congressional representative expressing your support of or dissatifaction with this decision. Consider the medical, ethical, and financial ramifications of your position.

Situational Problem 2

When Should AIDS Testing Be Required?

For many years, a blood test for the diagnosis of syphilis (the Wassermann test) was required in most states before a marriage license could be obtained. This requirement was based on the recognition that syphilis is a sexually transmitted disease, the presumption that protecting married couples would lead to the control of this feared disease, and the lack of adequate treatment methods. Unfortunately, the prevalence of extramarital sexual activity and the inadequacy of the most frequently used, now outdated test failed to control the disease. With the introduction of a penicillin treatment that could cure the disease and the recognition that there are many other sexually transmitted diseases that were not being diagnosed, most states dropped the required blood test to obtain a marriage license.

Today the fear of AIDS has caused some to propose that mandatory blood testing be required for various situations, including as a prerequisite for applying for a marriage license. The military services of the United States now require

AIDS testing of new recruits. Various employers also are requiring such tests, and predictions are that over 5 percent of all new job positions will require such tests for all applicants. There have been proposals to require AIDS testing of all immigrants to the United States and all those seeking a marriage license. The American Civil Liberties Union and various other groups have expressed concern that such required testing is an unwarranted infringement on personal rights.

Under what conditions or situations do you feel AIDS testing should be required? What justification can you provide for your position?

CHAPTER 21

Human Diseases Part 1: The Respiratory, Gastrointestinal, and Genitourinary Tracts as Portals of Entry

Despite the extensive defense mechanisms of the human body and the preventive measures that can be taken to avoid infections with microbial pathogens, humans remain susceptible to a variety of infectious diseases. The dynamics of the interactions between microorganisms and humans preclude the total elimination of all, or even most, **infectious diseases** (diseases caused as a result of microbial growth within the body). Our increased understanding of the interrelationships between pathogenic microorganisms and their human hosts,

however, has resulted in the reduced incidence of many infectious diseases, better management of sick individuals, and decreased mortality from many once deadly diseases. In this and the following chapter we will consider some of the diseases of humans caused by microorganisms. Emphasis will be placed on the epidemiology of the disease, considering the characteristics of various diseases including the interrelationships between pathogenic microorganisms, human host populations, and disease symptomology.

21.1 Portals of Entry and Transmission of Infectious Agents

Pathogenic microorganisms gain access to the body through a limited number of routes known as **portals of entry**. The routes of entry are the respiratory tract, gastrointestinal tract, genitourinary tract, skin, and wounds. Most pathogenic microbes cause disease only if they enter the body via a specific route. For example, depositing *Clostridium tetani* on the intact skin surface has no effect, but depositing in deep wounds results in the deadly disease tetanus. Pathogens can become established within the body in only a limited number of ways because of the nonspecific and immune defenses associated with different body tissues and the inherent properties of the microorganism.

The restrictive nature of the portals of entry also means that a sufficient number of microorganisms is necessary to initiate an infective process. The number of pathogens needed to establish a disease is known as the **infectious dose**. For some pathogens the infectious dose is one, but for others, hundreds of thousands of microbes may be necessary to over-

whelm the host defenses and allow the invading microorganisms to reproduce within the body. Various factors influence the infectious dose required to initiate a disease, including the nature of the pathogen, the portal of entry, and the state of the host defenses. In many cases, diminished host defenses permit relatively low numbers of potential pathogens to establish an infection. Malnutrition, for example, results in lowered amounts of antimicrobial body fluids and inadequate host defenses to protect against infectious microbes.

The transmission of infectious agents involves the movement of pathogens from a source to the appropriate portal of entry (Figure 21.1). The source of an infectious agent is known as the **reservoir**. In some cases, the reservoirs of human pathogens are nonliving sources such as soil and water. For example, tetanus is generally acquired when spores of *C. tetani*, which are widely distributed in soil, contaminate a wound. Often diseases acquired from such sources are noncommunicable, that is, they are singular

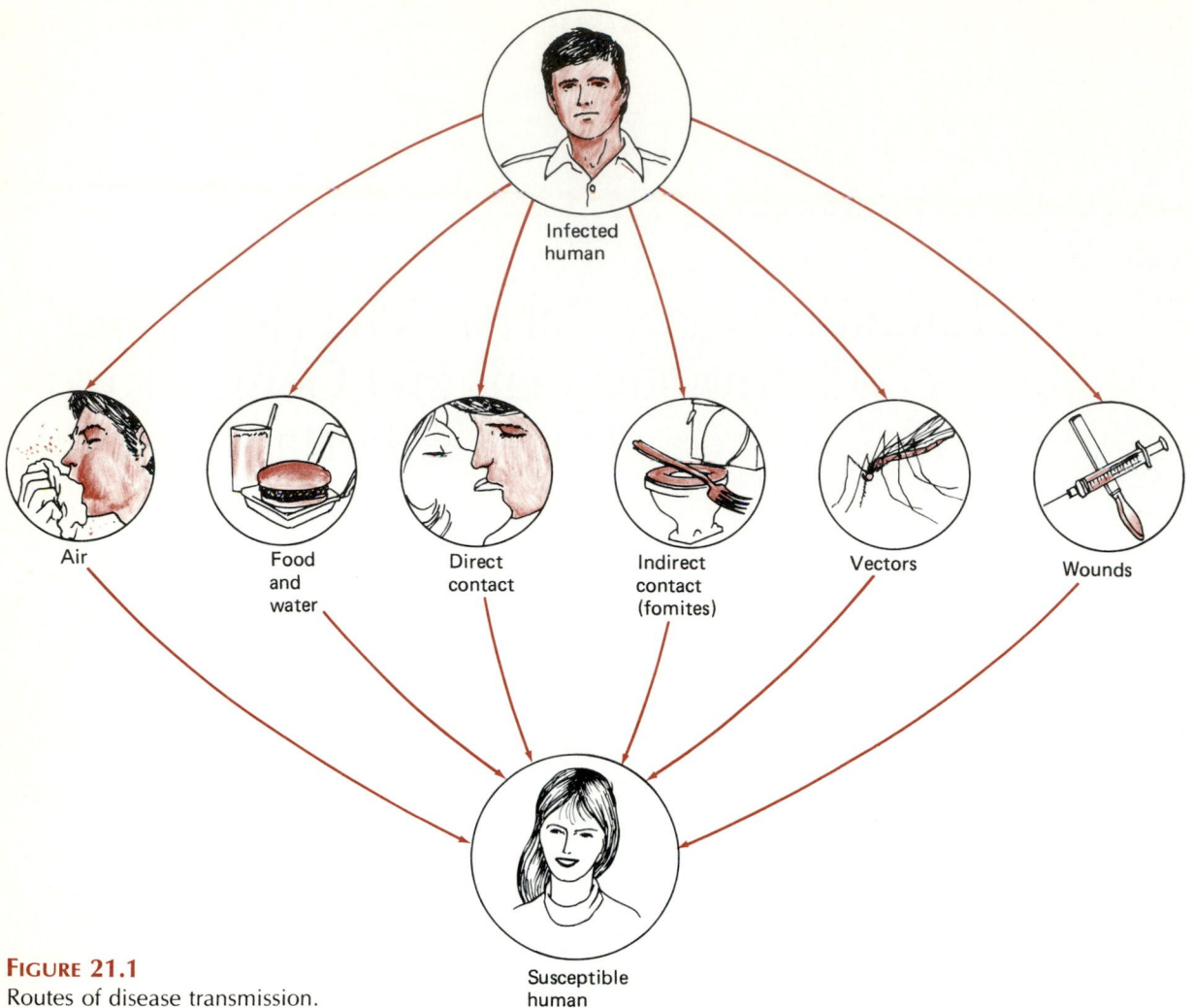

FIGURE 21.1
Routes of disease transmission.

events and are not normally transmitted from one infected individual to the next. So it is that, for example, health care workers treating a patient with tetanus are at no greater risk of contracting this disease than is the rest of the population.

Although water and soil are sometimes the reservoirs of human pathogens, they are more frequently only indirectly involved in the transmission of pathogens from an infected individual to the next susceptible host. Humans are the principal reservoirs for microorganisms that cause human diseases. People infected with a pathogen act as a source of contagion for others. The term *contagious disease* indicates that a pathogen will move with ease from one infected individual to the next. People who come in contact with someone suffering from a contagious disease are at risk of contracting that disease unless they are immune.

In some cases, infected individuals do not develop disease symptomatology. Such individuals are called *asymptomatic carriers* or simply **carriers**. Although they do not become sick themselves, carriers are important reservoirs of infectious agents. The classic case of disease transmission by such

a carrier occurred in the early 1900s when a cook, Mary Mallon, spread typhoid fever from one community to another. In 10 years Mary worked in seven different households in which 50 cases of typhoid fever occurred with 3 deaths. She would quit and disappear each time there was a case of typhoid in the household of her employer. When she was finally tracked down in 1907, she was diagnosed as a carrier of *Salmonella typhi*. She refused to undergo the surgical removal of infected tissues and was imprisoned. She was released in 1910 due to public outcry protesting her confinement, only to continue spreading typhoid to other communities, including a sanitorium in New Jersey and a hospital in New York. In 1915, she was recaptured and forcibly hospitalized for the rest of her life. Typhoid Mary, as she became known, died in 1938 at the age of 70.

Humans, both carriers and those showing overt signs of disease, are not the only sources of infectious agents. In some cases, nonhuman animal populations are the source. Diseases that primarily affect wild and domestic animals are known as **zoonoses**. Some zoonoses can be transmitted to humans by direct contact with infected animals, by ingesting contaminated meat, or more frequently by vectors. **Vectors** are carriers of disease agents. The vector need not develop disease; it need only transmit it from a reservoir to a susceptible individual. Arthropods, such as mosquitoes, are frequently the vectors of human disease.

Because the transmission of pathogens occurs via restricted routes, it is possible to control our interactions with microbial populations in ways that reduce the probability of contracting infectious diseases. The methods employed for preventing exposure to specific disease-causing microorganisms vary depending on the particular route of transmission. Many modern sanitary practices are aimed at reducing the incidence of diseases by preventing the spread of pathogenic microorganisms or by reducing their populations to concentrations that are insufficient to cause disease. Mosquito and rodent control, sanitary waste disposal, sewage treatment, chlorination of water supplies, pasteurization, and various other methods are used to restrict the spread of pathogens. The greatly diminished incidence of many diseases caused by microorganisms is the consequence of an adequate understanding of the modes of transmission of pathogenic

microorganisms and preventive measures that reduce exposure to disease-causing microorganisms.

Our discussion of human diseases caused by microorganisms is organized according to the portal of entry of the pathogen because this approach focuses on the route of transmission and the factors governing the spread of disease. In many cases, pathogenic microorganisms establish localized infections in the region of the portal of entry, but in other cases pathogens are able to spread systemically and establish infections involving other body tissues. Many pathogens possess adaptive features, known as **invasive properties**, that permit them to penetrate the body's defense mechanisms through a particular portal of entry. The normal body openings that serve as portals of entry for pathogenic microorganisms are the respiratory, gastrointestinal, and genitourinary tracts, and the diseases caused by pathogens that enter via these routes will be discussed in this chapter. In Chapter 22, we will consider those infective processes where pathogens establish infections in surface areas of the body, such as the skin and eye, or enter the body when the normal surface barriers are physically disrupted as the result of surgery, wounds, or animal bites.

21.2 Diseases Caused by Pathogens That Enter Via the Respiratory Tract

It should not be surprising, as we inhale 10,000–20,000 L of air per day that usually contains between 10,000 and 1,000,000 microorganisms, that the respiratory tract provides a portal of entry for many human pathogens. Potential pathogens freely enter the respiratory tract through the normal inhalation of air and when substances are placed in the mouth. Various viruses, bacteria, and fungi are able to multiply within the tissues of the respiratory tract, sometimes causing localized infections and sometimes entering the circulatory system through the numerous blood vessels associated with the respiratory tract and spreading through the bloodstream to other sites in the body. In order to establish an infection via the respiratory tract, a pathogen must overcome the natural immunological defense mechanisms that are particularly extensive in the lower respiratory tract, where there are numerous phagocytic cells. The microorganisms that generally establish infections in the upper respiratory tract are different from those that are able to move past the cilia and mucus secretions designed to restrict the movement of particles, including microorganisms, to the lower respiratory tract. Several factors contribute to the fact that the respiratory tract is a major portal of entry for pathogenic microorganisms. The upper respiratory tract is in continuous contact with air that contains many microorganisms. The respiratory system has a very large surface area to facilitate gas exchange, and there is a great deal of interaction between the respiratory tract and the circulatory system to permit reoxygenation of blood in the lower respiratory tract. Thus,

the potential for respiratory infection is great, but fortunately the actual rate of disease is low.

Transmission through the air is undoubtedly the main route of transmission of pathogens that enter via the respiratory tract. **Airborne transmission** often occurs when droplets containing pathogenic microorganisms are transferred from an infected to a susceptible individual. Droplets regularly become airborne during normal breathing, but the coughing and sneezing associated with respiratory tract infections are primarily responsible for the spread of pathogens in aerosols and thus for the airborne transmission of disease. As discussed in Chapter 20, the incidence of these diseases can be reduced by covering one's nose and mouth while coughing and sneezing and by avoiding contact with contagious individuals, a practice we are taught to follow at an early age.

Diseases Caused by Viral Pathogens

The Common Cold   The common cold is the most frequent infectious human disease, and it is safe to assume that we have all had a cold at some time. More than 200 million work and school days are lost each year in the United States alone because of colds. The term **common cold** actually refers to a cluster of diseases primarily caused by viruses, characterized by a localized inflammation of the upper respiratory tract and the release of mucus secretions, and generally accompanied by sneezing and sometimes coughing. The etiological agents responsible for the majority of cases of the common cold have yet to be identified, but we do know that in adults approximately 25 percent of all colds are caused by **rhinoviruses**, compared to only about 10 percent of colds in children. Rhinoviruses are members of the Picornaviridae, distinguished from other members of this family, which includes poliovirus, by their inactivation at low pH and their ability to maintain infectivity at 50°C. There are over 100 immunologically distinct types of rhinoviruses capable of causing the common cold; hence, it is not surprising that immunity does not offer continuous protection against all of the antigenically distinct viruses capable of causing it.

Viruses causing the common cold infect the cells lining the nasal passages and pharynx, producing an inflammatory response with associated tissue damage in the infected region. The initial viral infection can be followed by a secondary bacterial infection as the normal microbiota of the upper respiratory tract invade the damaged tissues. Symptoms include nasal stuffiness, sneezing, coughing, headache, malaise (a vague feeling of discomfort), sore throat, and sometimes a slight fever. There is no specific treatment for the common cold, which is a self-limiting clinical syndrome. Recovery usually occurs within 1 week without complications as a result of the natural immune defense response. Like many other respiratory diseases, colds occur primarily during the winter months, in part because of the physiological stress posed by exposure to cold temperatures and in part because of increased contact of individuals during indoor winter activities.

Influenza   Compared to the common cold, **influenza** is often a more serious and debilitating disease. Influenza is transmitted by inhalation of droplets containing influenza viruses, which are released into the air as droplets originating from the respiratory tracts of infected individuals. The outer envelope of an influenza virus has numerous protruding spikes that affect both the pathogenicity and antigenicity of the particular viral strain (Figure 21.2). Changes in the combinations of genes that cause antigenic shifts and the production of new strains of influenza viruses generally are associated with changes in the biochemistry of these spikes. There are two types of spikes, designated **H (hemagglutinin)** and **N (neuraminidase) spikes**. The H spikes cause clumping (agglutination) of red blood cells; presumably they are important in increasing the ability of the influenza virus to attach to human cells during the establishment of an infection and are also a valuable aid in the serological identification of the particular strain of influenza virus. Antibodies against the H spikes are very important in the body's resistance against infection by that particular strain of influenza virus. The N spikes are less important in decreasing resistance to influenza infections and appear to be involved in the release of viruses from infected cells following viral replication.

Major groups of influenza viruses are designated according to the antigens associated with their capsids. There are three major groups of influenza viruses, designated types A, B, and C. Specific strains of influenza virus are further designated by variations in the protein composition of their H and N spikes. Outbreaks of influenza are cyclical, with major outbreaks caused by type A virus occurring every 2 to 4 years, those caused by type B virus occurring every 4 to 6 years, and outbreaks caused by type C virus occurring only rarely.

Within each of the major types of influenza virus there are various antigenic subtypes that are responsible for different outbreaks of influenza (Figure 21.3). Major antigenic changes, known as *antigenic drift*, occur because of accumulated genetic mutations and recombinations that can even cause gene reassortment between an animal and a human strain. An **antigenic shift**, resulting from the addition of new genes, produces new strains of influenza virus. Strains of influenza viruses are often described by the location where

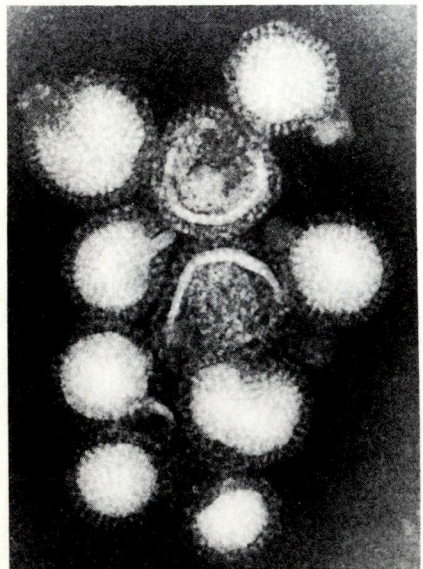

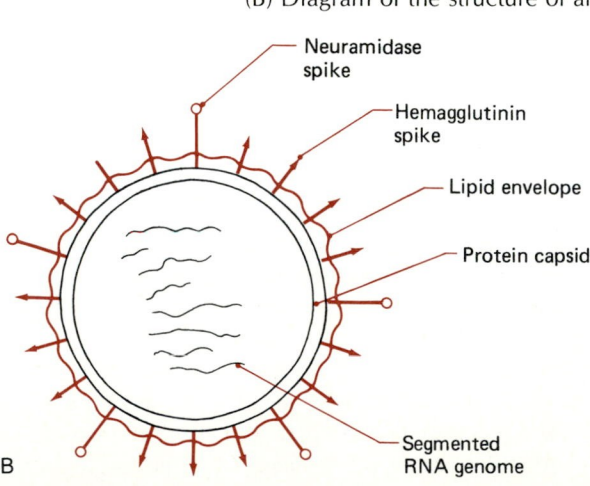

A   B

FIGURE 21.2

(A) Electron micrograph of the influenza A Hong Kong virus. (From BPS: F. A. Murphy, Centers for Disease Control, Atlanta) (B) Diagram of the structure of an influenza virus.

Neuramidase spike

Hemagglutinin spike

Lipid envelope

Protein capsid

Segmented RNA genome

FIGURE 21.3

Influenza outbreaks show regular cyclic fluctuations due to antigenic changes in the virus and associated variations in the susceptibility of the population. The highest incidence of influenza occurs during the winter, as shown in this graph of reported deaths in 121 United States cities compared with influenza isolates reported to the World Health Organization (WHO) Collaborating Centers in the United States from September 1972 to August 1979.

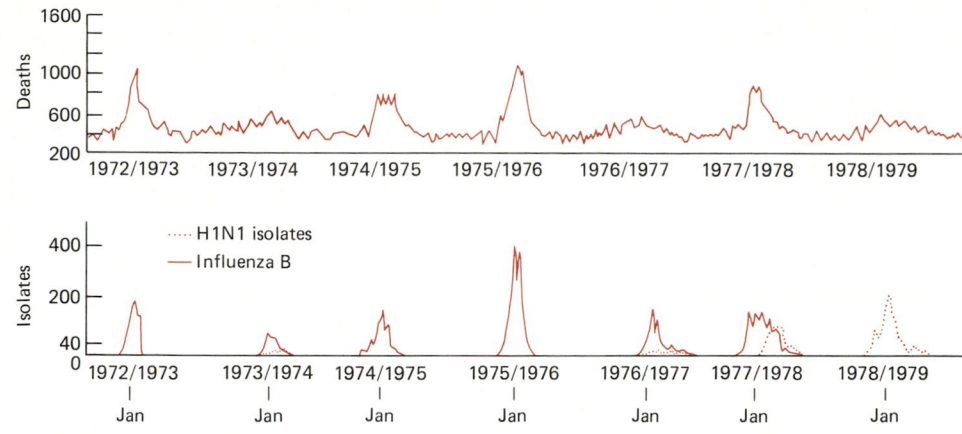

Discovery Process

Outbreaks of influenza occur each year. In some years, their severity and incidence are greater than in others. It has been discovered that major outbreaks of influenza occur when new strains of influenza viruses arise through genetic modification and when a large proportion of the human population lacks immunity to those particular strains. Epidemics depend on the emergence of these new strains. Influenza viruses display a marked tendency to mutate, and so new variants are constantly arising against which the majority of people do not possess the required immune mechanisms. Therefore, epidemics are always a threat. Epidemics of influenza are usually short and subside quickly, followed several weeks later by a secondary wave, a free period, and then a tertiary wave. In secondary and tertiary outbreaks, the number of people attacked is less than in the primary outbreak, but the disease tends to be more severe, complications are more common, and the mortality rate is higher. By detecting the primary outbreak and identifying the particular strain of influenza virus responsible for it, steps can be taken to minimize the outbreak. For example, the Asian flu epidemic of 1957 was the most severe since the great pandemic of 1918–1919, but far fewer people died in 1957 because the outbreak did not come as a surprise.

A new and potent strain of type A influenza virus appeared in southern China early in 1957. It reached Amsterdam, Sydney, and San Francisco within weeks, but even before it spread from the Far East, a worldwide network of scientists was mobilized. Researchers in about 30 laboratories worked to identify the new strain and to trace the pattern of influenza outbreaks. Epidemiologists at WHO centers in the United States and Great Britain organized the information and mapped the worldwide course of the disease. The early detection of this outbreak of a new form of influenza, Asian flu, allowed time for the preparation of an effective vaccine. Within 10 months of detection of the first cases of Asian flu, pharmaceutical manufacturers had produced 10 million doses of vaccine. The use of the vaccine to immunize large numbers of individuals who otherwise would have been susceptible to influenza, and the availability of antibiotics to combat secondary bacterial infections, minimized mortality due to this serious health threat. The Surgeon General of the United States announced that "for the first time in history the medical community was ahead of an impending epidemic." The ability to predict impending influenza outbreaks, to identify the causative viral stain, and to produce effective vaccines allows us to quickly minimize the effects of influenza outbreaks.

outbreaks of the disease associated with that particular antigenic variety of the flu virus were first detected. For example, the Taiwan strain of influenza virus, first seen in the Orient in 1986, is a type A influenza virus designated H_1N_1. The antigenic designation is important because it indicates to epidemiologists whether there has been a substantial change in the antigenic properties of the virus and whether a sufficient proportion of the population will be susceptible to that strain such that an epidemic outbreak is likely.

Influenza is characterized by the sudden onset of a fever, with temperatures abruptly reaching 102–104°F approximately 1 to 3 days after the actual exposure, and infection.

The disease is further characterized by malaise, headache, and muscle ache. In uncomplicated cases of influenza, the viral infection is self-limiting and recovery occurs within a week. However, influenza can lead to complications, such as a secondary bacterial infection, causing pneumonia. Complications associated with influenza infections are prevalent among the elderly and individuals with compromised host defense responses. Such individuals should be immunized against the prevalent strain of influenza virus prior to the outbreak of influenza epidemics because complications can result in death.

One serious complication associated with outbreaks of influenza is the development of **Reye's syndrome**, an acute pathological condition affecting the central nervous system. Reye's syndrome also occurs after infections with other viruses, and the specific relationship to influenza virus is not clear. Occurrences of Reye's syndrome are highly but inexplicably correlated with outbreaks of influenza B virus. Reye's syndrome is associated principally with children. For reasons that have yet to be elucidated, there is a greater incidence of Reye's syndrome when aspirin is used to treat the symptoms of a viral infection. Consequently, pediatricians warn against the use of aspirin for children with influenza and other viral infections of the respiratory tract. Although a direct cause-and-effect relationship between aspirin use and viral infections has not been established, aspirin manufacturers have placed warning labels on the bottles, especially on children's aspirin.

Another condition involving the central nervous system associated with influenza infections is Guillain-Barré syndrome. In 1976 there was an increased incidence of this syndrome associated with an active immunization program against a predicted outbreak of swine flu, which did not occur. Swine flu is caused by an influenza virus that normally occurs in pigs but that can be transmitted to humans. The outbreak of Guillain-Barré syndrome appears to have resulted from contamination of some batches of vaccine with viable influenza viruses.

There have been several major or **pandemic**[1] outbreaks of influenza during the nineteenth and twentieth centuries. During 1918–1919, outbreaks of influenza resulted in the death of over 20 million people. Many of these deaths may have been the result of secondary infections with *Streptococcus pneumoniae*, rather than of the primary influenza infection. Continued genetic drift, with major antigenic shifts in the virus resulting from genetic recombination, permits these periodic epidemic outbreaks. The introduction of a virus into a population, most of whose members are susceptible to infection, creates the conditions needed for the establishment of an epidemic. Influenza outbreaks spread worldwide, with person-to-person propagated transmission from the site of an initial outbreak with a new strain. It is possible to watch the

disease spread from one area to another (Figure 21.4). Each year epidemiologists make predictions about the severity of influenza outbreaks, and public health officials take the necessary steps of immunizing high-risk individuals with the correct antigenic type and warning the public about the dangers of this disease. Even in a nonepidemic year, influenza causes a significant number of deaths; for example, the death rate due to influenza in 1980—a nonepidemic year—in the United States was 0.3 per 100,000 population.

The clinical diagnosis of influenza depends on the isolation of an influenza virus, using tissue culture with a cell line such as monkey kidney cells, and the serological detection of increased titers of antibody in the patient's serum that are reactive with Myxovirus antigens. As with most other viral diseases, treatment of uncomplicated cases of influenza centers on treating the symptoms, with recovery from the disease dependent upon the immune response of the infected individual. The antiviral drug amantadine is effective against influenza A viruses, but only during the incubation period. Once the symptoms of the disease have appeared, amantadine is ineffective; thus, this drug is of limited clinical use. Primary control of influenza is achieved by vaccinating individuals who are prone to the complications resulting from this disease, leaving others unprotected to suffer periodically from influenza.

Measles   Unlike influenza, which occurs at any age, **measles (rubeola)** is a highly contagious disease occurring almost exclusively among children. The measles virus is readily transmitted from an infected child to a susceptible host, as illustrated by the fact that there is a greater than 90 percent incidence of an acute infection after exposure to measles virus by susceptible children. Measles can be prevented by childhood immunization; therefore, the rate of infection in the United States, at least where immunization is practiced routinely, has been declining regularly in recent years, although in other countries measles has not declined in the same way (Figure 21.5). It had been predicted that measles would be eliminated from the United States by 1982, but there have been several outbreaks among college-age individuals who had not been immunized and who had not contracted this disease as children. Some college campuses were quarantined in the early 1980s during severe outbreaks.

After initial viral multiplication in the mucosal lining of the upper respiratory tract, measle viruses appear to be disseminated to lymphoid tissues, where further multiplication occurs. Before the onset of symptoms, large numbers of measles viruses are shed in secretions of the respiratory tract and eye, and in urine, promoting the rapid epidemic spread of this disease. Infection with measles viruses can involve a number of organs, and there is a high rate of mortality associated with measles in regions of the world where malnutrition and limited medical treatment facilities predominate. The disease is associated with 2–3 million deaths per year in developing nations. When measles is fatal, the virus generally invades the central nervous system.

---

[1]A pandemic is an outbreak of disease that affects large numbers of people in a major geographical region or that has reached epidemic proportions simultaneously in different parts of the world.

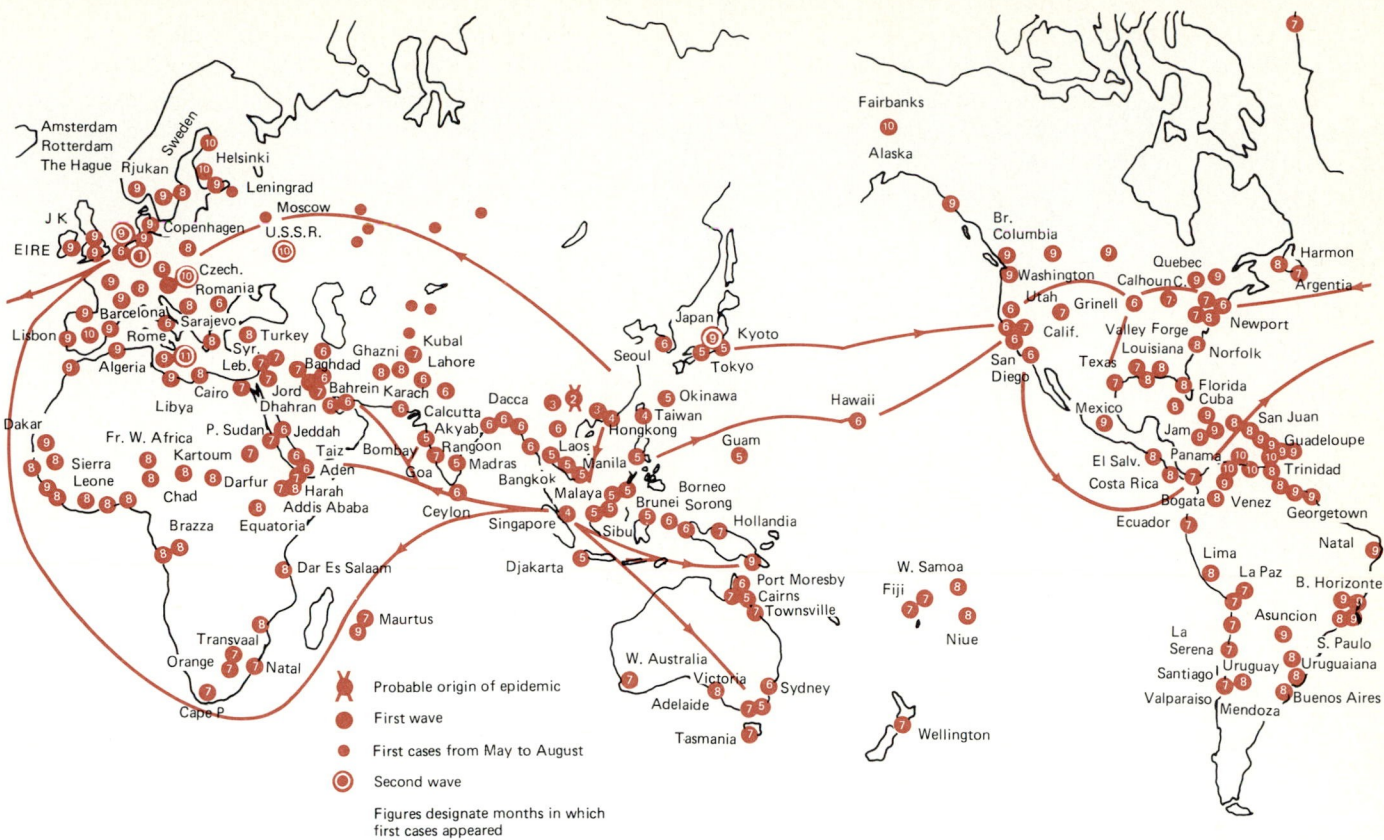

FIGURE 21.4

Map showing the worldwide spread of influenza during an epidemic outbreak. This epidemic originated in Southeast Asia and spread in waves to all other population centers of the world.

Measles is characterized by the eruption of a **skin rash** approximately 14 days after exposure to the measles virus. The rash generally appears initially behind the ears, spreading rapidly to other areas of the body during the next 3 days. Disease symptomatology often begins a few days before the onset of the characteristic measles rash. These initial symptoms include high fever, coughing, sensitivity to light, and the appearance of **Koplik's spots** (red spots with a white dot in the center that occur in the oral cavity, generally appearing first on the inner lip). Treatment is normally supportive, including rest and the intake of sufficient fluids. In uncomplicated cases, the fever disappears within 2 days, and the individual returns to normal activities a few days later. If the fever persists for more than 2 days after the eruption of the rash, it is likely that a complication, such as bronchitis or pneumonia, has developed. In these cases, additional treatment is needed to cure the secondary infection.

German Measles   Like measles, transmission of **rubella virus**, the causative agent of **German measles (rubella)**, appears to be via droplet spread, with the initial infection occurring in the upper respiratory tract. In contrast to the measles virus, however, the rubella virus exhibits a relatively low rate of infectivity; thus, prolonged exposure appears to be needed for the establishment of infection. After multipli-

cation in the mucosal cells of the upper respiratory tract, rubella viruses appear to be disseminated systemically through the blood. Approximately 18 days after initiation of the infection, a characteristic rash, appearing as flat pink spots, occurs on the face and subsequently spreads to other parts of the body. Enlarged, tender lymph nodes and a low-grade fever characteristically precede the occurrence of the German measles rash. In children and adolescents, rubella is usually a mild disease. If it is acquired during pregnancy, however, the fetus can become infected with the rubella virus, resulting in congenital rubella syndrome, characterized by the development of multiple abnormalities in the infant. There is a very high rate of mortality, exceeding 25 percent, in cases of congenital rubella syndrome. Vaccination has greatly reduced the incidence of rubella (Figure 21.6) in children and is also used to confer immunity on women of childbearing age who had not contracted the disease at an earlier age.

Mumps   **Mumps** is another disease occurring primarily in childhood. The ribonucleic acid–protein core of the mumps virus comprises a complement-fixing antigen, and the envelope has spikes with neuraminidase, hemagglutinating, and hemolytic activities. The mumps virus is transmitted via contaminated droplets of saliva. The initial infection appears to

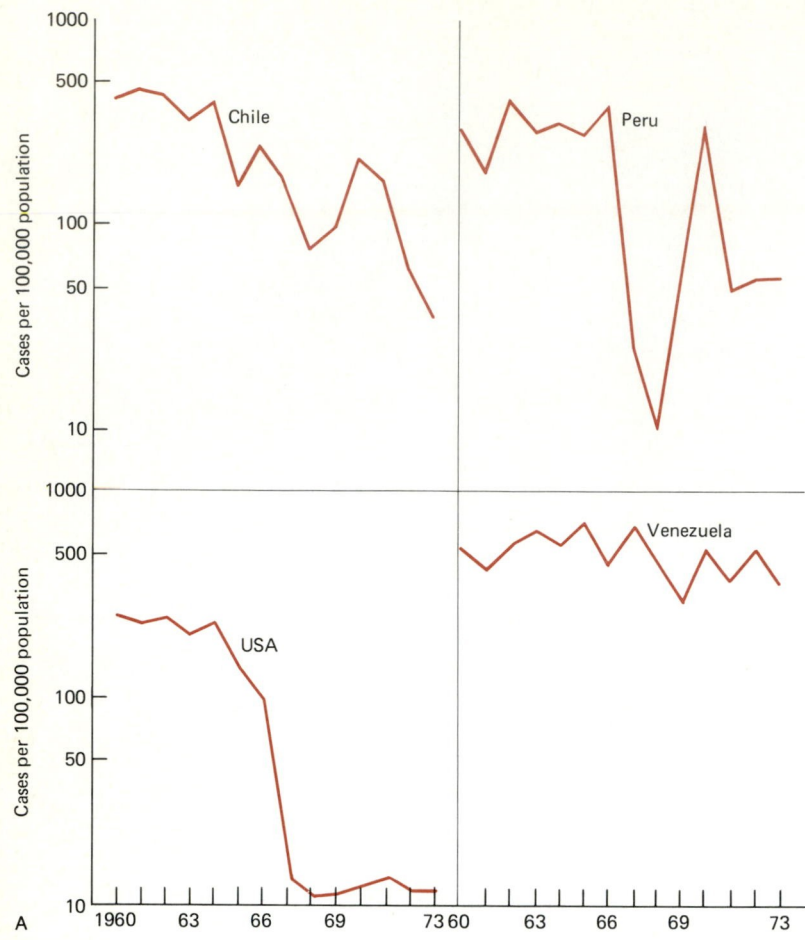

FIGURE 21.5

(A) Reported cases of measles per 100,000 population in four countries. Although the incidence of this disease has been greatly reduced in the United States through immunization practices, it remains a major problem in other countries. (B) The number of cases of measles in the United States was declining steadily and was predicted to reach zero around 1984. However, outbreaks of measles among college-age individuals who had not been vaccinated and had escaped measles during childhood reversed this trend in the mid-1980s.

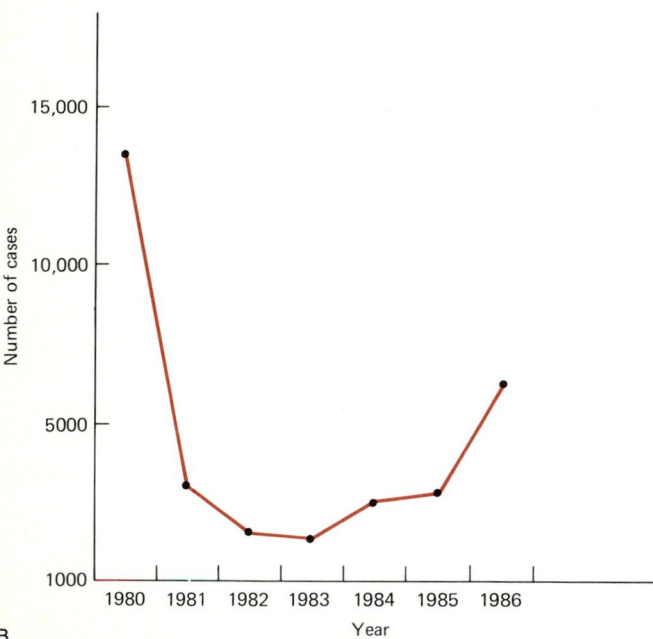

in about 75 percent of patients, with swelling in one gland usually preceding swelling on the other side by about 5 days. The average incubation period for mumps is 18 days, and the swelling of the salivary glands generally persists for less than 2 weeks. The mumps virus may spread to various body sites, and although the effects of the disease are normally not long-lasting, there can be several complications; for example, mumps is a major cause of deafness in childhood. In males past puberty, the mumps virus can cause **orchitis** (inflammation of the testes), but old wives' tales to the contrary, mumps rarely results in male sterility.

Chicken Pox and Shingles **Chicken pox (varicella)**[2] is caused by the **varicella-zoster virus**, a member of the herpesvirus group, which also causes shingles (herpes zoster). Ninety percent of all cases of chicken pox occur in children under 9 years of age. In children, chicken pox is generally a relatively mild disease, but when it occurs in adults the symptoms are characteristically severe. Chicken pox, a highly contagious disease, is probably transmitted via contaminated droplets and direct contact with vesicle fluid containing varicella-zoster virus. The initial site of viral repli-

occur in the upper respiratory tract, with subsequent dissemination to the salivary glands and other organs. Mumps is characterized by the enlargement of one or more of the salivary glands, resulting from viral replication within those glands. Swelling on both sides (**bilateral parotitis**) occurs

[2]The disease has nothing to do with chickens. The name is derived from the old English *gicken*, meaning itching.

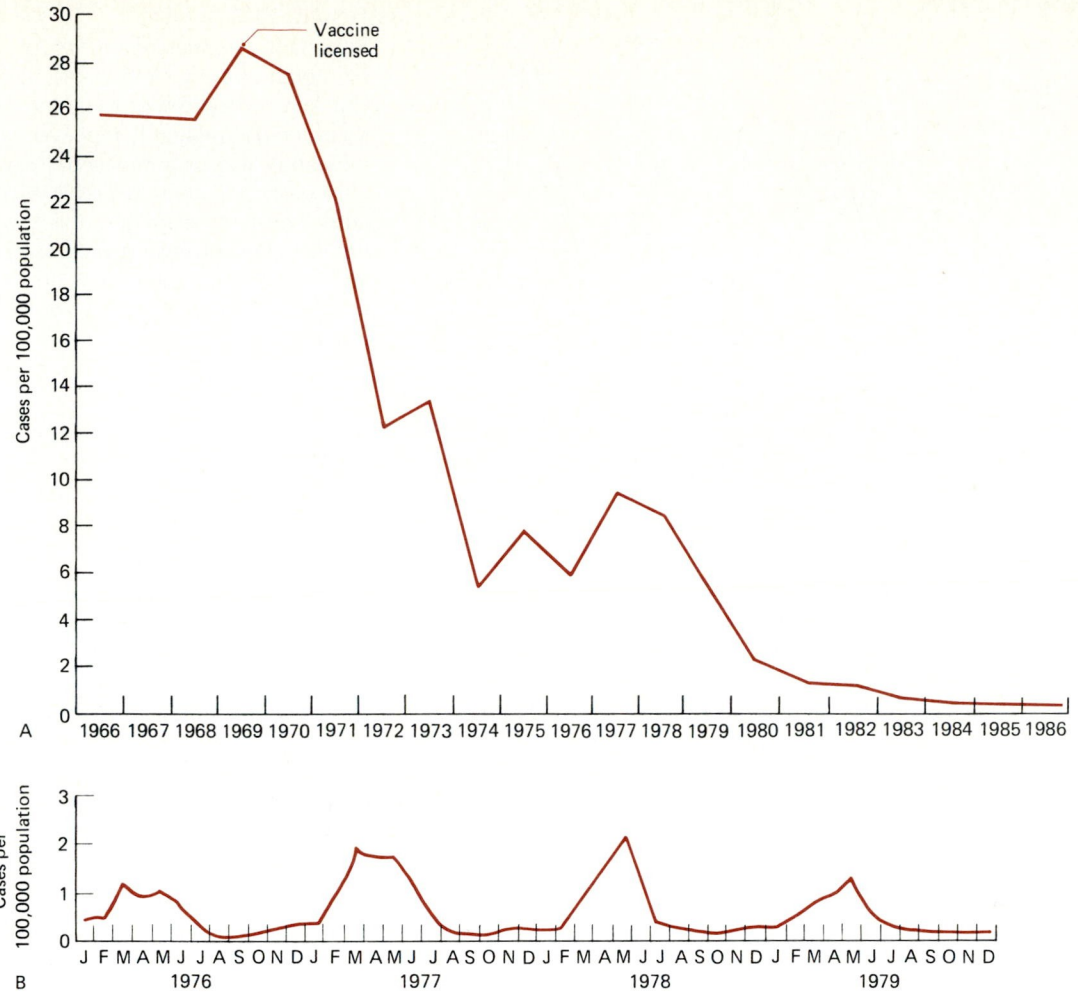

FIGURE 21.6

Incidence of rubella (German measles) in the United States. (A) Reported case rates by year, showing the great decline in the incidence of this disease after the introduction of a vaccine for its prevention. (B) Reported case rates by month, showing that the disease rate peaks in the spring of each year, from March to May.

cation has not been positively established but appears to be in the upper respiratory tract. Local lesions (vesicles) occur in the skin after dissemination of the virus through the body. These skin lesions become encrusted, and the crusts fall off in about 1 week. Vesicles also occur on mucous membranes, especially in the mouth. In some cases, the varicella-zoster virus spreads to the lower respiratory tract, resulting in pneumonia; in this way, several other tissues, including the central nervous system, can also be involved in complicated cases of chicken pox. Unlike the situation with other childhood viral diseases, vaccination practices have not yet been introduced, and outbreaks of chicken pox continue to show regular seasonal cycles of the same magnitude (Figure 21.7).

In adults, **shingles** is the principal disease resulting from infections with the varicella-zoster virus. It is this virus, acquired in childhood, that can remain within the body, perhaps as a provirus—a state similar to lysogeny—or associated with nerve endings. Shingles is the result of reactivation of a latent virus. This disease affects individuals who have developed circulating antibodies in response to infection with this virus. It appears to be the reaction of antibodies with body sites, such as nerve endings, associated with the emergence of the latent virus that causes the symptoms of shingles, which are therefore quite different from the original case of chicken pox associated with active viral infection. In cases of shingles, the virus reaches the sensory ganglia of the spinal or cranial nerves, producing an inflammation. There is usually an acute onset of pain and tenderness along the affected sensory nerves. A rash also develops along these nerves, usually lasting for 2 to 4 weeks, but the pain may last for weeks or months.

Infectious Mononucleosis   **Infectious mononucleosis** is caused by the **Epstein-Barr (EB) virus**, a member of the herpesvirus group. The EB virus occurs in oropharyngeal secretions of infected individuals and appears to be trans-

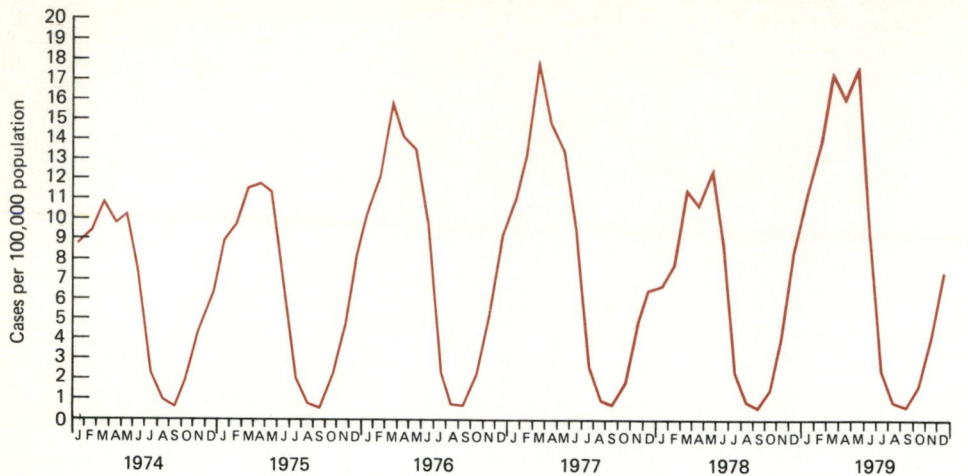

FIGURE 21.7
Reported case rates of chicken pox by month in the United States. Chicken pox continues to be the second most frequently reported infectious disease in this country. The seasonal pattern has remained fairly constant, with the peak incidence occurring between March and May. At present there is no licensed vaccine for preventing this disease.

mitted by exchange of oropharyngeal secretions containing this virus, although transmission can also occur by droplet spread. Infectious mononucleosis most commonly occurs in young adults 15–25 years of age, a fact that may be explained by the exchange of saliva during kissing, a prevalent activity often involving more partners in this age group than others. In the course of the infection, B lymphocytes are infected, leading to the characteristic changes in the white blood cells used in diagnosing the disease. The symptoms of this disease include a sore throat, low-grade fever that generally peaks in the early evening, enlarged and tender lymph nodes, general fatigue, and weakness. In most cases of infectious mononucleosis, the symptoms are relatively mild and the acute stage of the illness lasts for less than 3 weeks. However, the EB virus has been associated with both Burkitt's lymphoma and nosopharyngeal carcinoma, two forms of cancer.

Diseases Caused by Bacterial Pathogens

Pneumonia   **Pneumonia**, an inflammation of the lungs involving the alveoli, can be caused by a number of viral and bacterial agents. Pneumonia is often a complication that occurs when the host defense mechanisms are compromised as a result of other diseases. Frequently, pneumonia is a nosocomial (hospital-acquired) infection occurring after surgery or during treatment for another disease, when patients are ''run down'' and their physiologically impaired state reduces the effectiveness of the immune response system. The lack of movement and deep breathing in postsurgical patients reduce the efficiency of the normal defense mechanisms in clearing the lungs of mucus and bacteria, and the accumulation of fluids favors the establishment of a microbial infection. There is a high rate of mortality in cases of pneumonia (Figure 21.8); more than half of the cases of pneumonia are caused by bacteria, and this disease ranks among the top causes of death from infectious diseases. Bacteria that cause pneumonia most frequently enter the lungs via the air, although transport of pathogens to the lungs through the bloodstream can also occur.

The most frequent etiological agent of bacterial pneumonia in adults is *Streptococcus pneumoniae* (pneumococcus), a Gram-positive, capsule-forming coccus (Table 21.1). Often pneumococcal pneumonia is an endogenous disease, originating in the individual's normal throat microbiota. The symptoms of pneumococcal pneumonia, which is most prevalent in adult males, include the sudden onset of a high fever, production of colored, purulent sputum, and congestion. In most patients, an upper respiratory tract infection with the

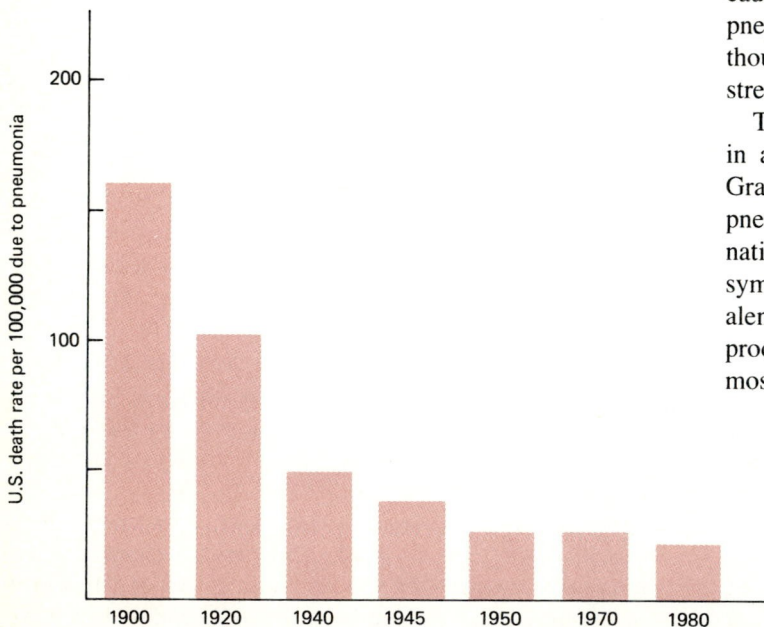

FIGURE 21.8
Death rate due to pnuemonia in the United States in the twentieth century. Note that the dramatic decreases in the first half of the century have now leveled off.

TABLE 21.1   Frequency of Major Types of Bacterial Pneumonia

Clinical Entity	Etiological Agent	Percent of Cases	Indicated Antimicrobial Drug
Pneumococcal lobar pneumonia	*Streptococcus pneumoniae*	Over 90	Penicillin
Klebsiella (Friedlander's) pneumonia	*Klebsiella pneumoniae*	1–5	Gentamicin
"Flu" pneumonia	*Haemophilus influenzae* type b	1–5	Ampicillin

characteristic symptom of a sore throat precedes the development of pneumococcal pneumonia. Several other bacteria, including *Staphylococcus aureus*, *Haemophilus influenzae*, and *Klebsiella pneumoniae*, are also responsible for a significant number of cases of pneumonia. In children *H. influenzae* type B frequently is the cause of pneumonia. Vaccines have been developed using polysaccharides from purified capsules against both *S. pneumoniae* and *H. influenzae*.

During the development of pneumonia, bacteria reproduce in the lung tissue, forming a lesion. The phagocytic portion of this inflammatory response results in decreased numbers of bacteria within the lesion. Bacteria spread through the alveoli and into the pulmonary system. The exudate that develops during pneumonia interferes with gas exchange in the lungs. Without treatment the death rate from pneumococcal pneumonia is about 30 percent. Antibiotic treatment cures bacterial pneumonia, with penicillin the antibiotic of choice for treating pneumonia caused by *S. pneumoniae*. The specific antibiotic treatment for pneumonias caused by bacteria other than *S. pneumoniae* varies with their specific antibiotic sensitivities.

Atypical Pneumonia   Several bacteria cause **atypical pneumonias** requiring special treatments. Identification of the pathogen and determination of its antibiotic susceptibility are essential in selecting the best antimicrobial agents in cases of atypical pneumonia.

Primary Atypical Pneumonia. Mycoplasma pneumoniae causes an atypical self-limiting pneumonia (primary atypical pneumonia) that has a low death rate. During World War II this disease became known as *walking pneumonia*. It is often the cause of pneumonia among children of school age. *M. pneumoniae* lacks a cell-wall structure and is therefore not sensitive to penicillin. This organism, however, is sensitive to tetracycline and erythromycin, which can be used effectively to treat this type of atypical pneumonia. Unlike other mycoplasmas, *M. pneumoniae* can attach to the epithelial surface of the respiratory tract. This bacterium does not penetrate the epithelial cells, nor does it produce a protein toxin, but the hydrogen peroxide released by the bacterium causes cell damage, including loss of the cilia lining the respiratory tract and death of surface endothelial cells.

Legionnaire's Disease. Another atypical form of pneu-
monia is caused by *Legionella pneumophila* and related species in this genus. In addition to the typical symptoms of pneumonia, **Legionnaire's disease** is often characterized by kidney and liver involvement and by an unusually high incidence of associated gastrointestinal symptoms. The fever associated with this disease starts low but then typically reaches 104–105°F. If untreated, the fatality rate is about one in six. *L. pneumophila* produces β-lactamase enzymes and is not sensitive to most penicillins and cephalosporins, but it is sensitive to other antibiotics, such as erythromycin and tetracycline. Erythromycin is the antibiotic of choice when Legionnaire's disease is diagnosed. *L. pneumophila* is a Gram-negative, fastidious, rod-shaped organism whose nutritional requirements for growth complicated early isolation attempts, initially confounding epidemiologists who were trying to discover the etiological agent of Legionnaire's disease.

The disease syndrome caused by these organisms is referred to as *Legionnaire's disease* because the first detected outbreak of this disease occurred during a convention of the American Legion in Philadelphia during July 1976. In the investigation of this outbreak, the first 90,000 man-hours of investigation, costing over $2 million and employing virtually all conventional isolation procedures, failed to reveal the causative agent of this disease. Finally, the breakthrough, revealing that this disease is of bacterial etiology, involved the use of indirect immunofluorescent staining with antibodies from the sera of affected individuals. Later, it was discovered that the bacterium, subsequently named *Legionella pneumophila* (lung-loving), could be grown on a chocolate agar medium, made with heated blood that looks as though it contains chocolate, if iron and cysteine are included as growth factors. It was also later found, by examining stored blood sera, that a 1968 outbreak of a disease in Pontiac, Michigan, the etiology of which had not been identified, had been caused by a different strain of *L. pneumophila*. Various other outbreaks of this disease have since been identified.

Species of *Legionella* appear to be natural inhabitants of bodies of water. During periods of rapid evaporation, such as occur during summer, the bacteria can become airborne in aerosols, and inhalation of contaminated aerosols can lead to the onset of illness. In some cases, outbreaks of Legionnaire's disease have been traced to air-conditioning cooling systems. These bacteria multiply in the cooling system waters, which are rapidly evaporated to provide cooling, and inadvertently become airborne and circulate through the air-conditioning system.

Psittacosis. **Psittacosis**, which is also known as *ornithosis* or *parrot fever*, is another type of atypical pneumonia. This disease is caused by an obligate intracellular bacterium, *Chlamydia psittaci*. Birds act as a reservoir for *C. psittaci*, from which the name of the disease is derived. This infection is contracted by inhalation of *C. psittaci*; thus, the bacteria enter the body through the respiratory tract. Even though *C. psittaci* is rather fastidious, the primary route of transmission of psittacosis is from birds to human beings via aerosol dispersal of droplets and contaminated dust particles. Parakeets, canaries, other pet birds, and domestic fowl are frequently the sources of human infection. *C. psittaci* multiplies in the cells of the mononuclear phagocyte system prior to systemic dissemination through the bloodstream. The symptoms of psittacosis include fever, headache, malaise, and coughing. Psittacosis is generally a mild disease, and in uncomplicated cases recovery normally occurs within 1 week, aided by the use of tetracyclines.

Bronchitis **Bronchitis**, an inflammatory disease involving the bronchial tree that does not extend into the pulmonary alveoli, can be caused by several microorganisms, including *Streptococcus* species, *Staphylococcus* species, *Haemophilus influenzae*, *Mycoplasma pneumoniae*, and various types of viruses. *M. pneumoniae* and various viruses appear to be the most frequent causative organisms of bronchitis, but it is difficult to define the specific etiologic agent responsible for this disease because bronchitis almost always occurs as a complication of another disease condition, such as pharyngitis (sore throat). The symptoms of bronchitis are normally preceded by those associated with a normal upper respiratory tract infection, such as malaise, headache, and sore throat. The onset of bronchitis is marked by the development of a cough that eventually yields mucopurulent sputum, that is, sputum containing mucus, reflective of the development of bronchial congestion. Acute bronchitis can be treated effectively with antibiotics, such as penicillin and tetracycline. Chronic bronchitis does not develop because of microbial infection alone. Rather, it appears to depend on irritation of the bronchi by repeated microbial infections and/or the inhalation of irritants, such as cigarette smoke. These irritations compromise the normal secretory and ciliary function of the bronchial mucosa, and the resulting excessive mucus secretion in the bronchi favors bacterial growth and the establishment of infection.

Rheumatic Fever *Streptococcus* species, which cause a variety of diseases, are normally transmitted through the air in contaminated droplets and establish the primary infection in the tissues of the upper respiratory tract. In some cases, the infection is limited to these tissues, causing conditions such as pharyngitis and tonsillitis. In other cases, the streptococci

or protein exotoxins produced by streptococci enter the circulatory system and spread systemically. In **scarlet fever**, for example, the systemic spread of hemolysins produced by *S. pyogenes* is manifest as a rash of pinhead red spots, and in rheumatic fever the systemic spread of *S. pyogenes* involves multiple body sites.

Rheumatic fever is generally the most serious consequence of *S. pyogenes* infections. In rheumatic fever the systemic spread of *S. pyogenes* toxins affects multiple body sites. The symptoms of this disease vary, but characteristically there is a high fever, painful swelling of various body joints, and cardiac involvement, including subsequent development of heart murmurs from childhood occurrences. The symptoms of rheumatic fever normally begin to occur a little over 2 weeks after a characteristic sore throat, associated with an upper respiratory tract infection with *S. pyogenes*. Because of the serious manifestations of rheumatic fever, it is important to diagnose the etiological agents of sore throats in children. Throat swabs plated on blood agar can readily be screened for the presence of β-hemolytic streptococci, and when they are detected, serological and/or biochemical tests are carried out to determine if group A streptococci, the group that includes *S. pyogenes*, are present. Penicillin is effective in treating group A streptococcal infections, and its use in treating streptococcal pharyngitis can prevent the occurrence of rheumatic fever.

The specific causal relationship between *S. pyogenes* and the symptoms of rheumatic fever has not been established. It is likely that antibodies produced in response to group A streptococcal cell-wall antigens are cross-reactive with cardiac antigens and that it is an autoimmune response that actually results in cardiac damage. The treatment of rheumatic fever, therefore, includes the use of anti-inflammatory drugs to reduce tissue damage, as well as antibiotics to remove the infecting streptococci.

Carditis In addition to causing cardiac involvement as one manifestation of rheumatic fever, streptococcal species are among the most prevalent bacteria responsible for inflammation of the heart muscle, **carditis**. For example, over 50 percent of the reported cases of endocarditis in the United States during the 1960s were attributed to streptococcal infections, although rickettsia and fungi also cause carditis. **Endocarditis**, an inflammation of the endocardium—a specialized membrane of epithelial and connective tissue that lines the cardiac chambers and forms much of the heart valve structure—most frequently occurs when normally nonpathogenic members of the microbiota associated with body surfaces enter the circulatory system and attach to cardiac tissues. Viridans streptococci, present in high numbers in the oral cavity, is the most frequently identified etiological agent of endocarditis. Dental treatments are often implicated in the mobilization of this bacterium, initiating its entry into the bloodstream through the tissues of the oropharyngeal cavity. **Myocarditis** (inflammation of the myocardium, which is principally cardiac muscle) can also result from microbial

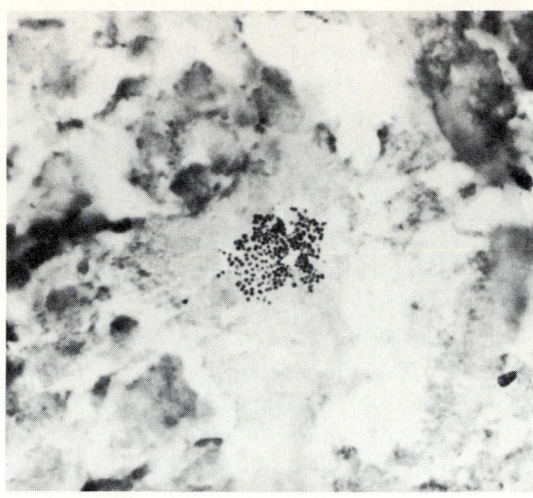

FIGURE 21.9
This micrograph shows a section of heart tissue infected with viridans streptococci. (Courtesy John J. Bochino, Norton's Kosair Children's Hospital, Louisville, Ky.)

infections, most commonly by *Streptococcus* species. Microbial growth on the heart tissues causes serious abnormalities, and, if untreated, is often fatal (Figure 21.9). Treatment of infections of the heart tissues usually employs broad-spectrum antibiotics that are not inactivated by β-lactamases and often requires prolonged administration of high doses of antibiotics to ensure removal of the infecting bacteria.

Diphtheria   Diphtheria, caused by strains of *Corynebacterium diphtheriae*, results from a protein toxin produced by strains of *C. diphtheriae* harboring a temperate phage. Diphtheria toxin is a potent protein exhibiting toxicity against almost all mammalian cells. The bacteria generally do not invade the tissues of the respiratory tract; rather, it is the dissemination of the toxin through the body that causes the severe symptoms of this disease. *C. diphtheriae* is normally transmitted via droplets from an infected individual to a susceptible host, establishing a localized infection on the surface of the mucosal lining of the upper respiratory tract.

There is generally a localized inflammatory response, pharyngitis, in the vicinity of bacterial multiplication in the upper respiratory tract. In severe infections with *C. diphtheriae*, symptoms include low-grade fever, cough, sore throat, difficulty in swallowing, and swelling of the lymph glands. Complications from diphtheria can block respiratory gas exchange and result in death due to suffocation. The extensive use of vaccine to prevent diphtheria has greatly reduced the incidence of this disease but has not altered the fatality ratio (Figure 21.10).

As noted in Chapter 20, some children immunized with this vaccine have had serious adverse reactions that can even cause death, leading to questions about whether mandatory vaccination should continue. If immunization is discontinued, however, cases of this disease would increase. In immunized individuals, infection with toxicogenic strains of

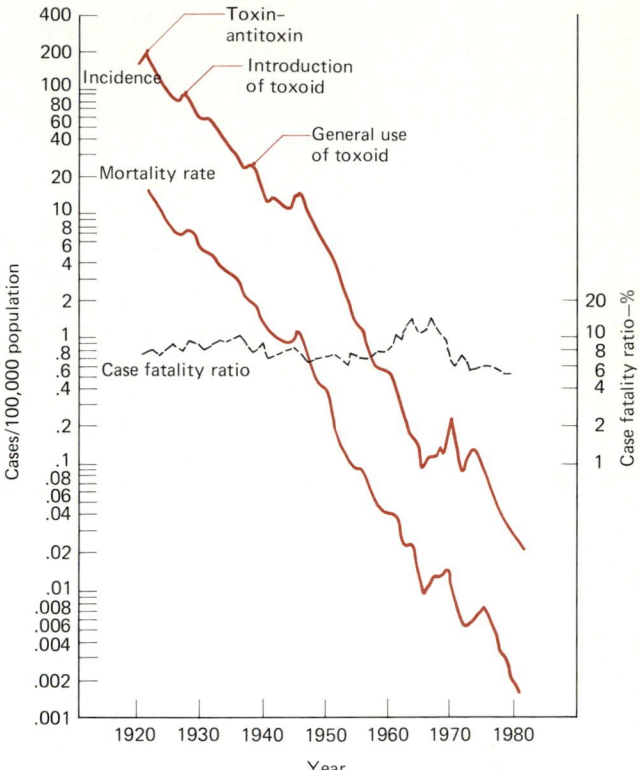

Diphtheria—reported annual incidence and mortality rates, and case fatality ratio, United States — 1920-1980

FIGURE 21.10
Incidence, mortality rates, and case fatality ratio for diphtheria in the United States since 1920. The rates of this disease have declined dramatically since effective vaccination programs were begun. However, the case fatality ratio has remained relatively constant, underscoring the importance of preventing this disease and the difficulty in treating it.

C. diphtheriae is generally restricted to a localized pharyngitis with no serious complications. Diphtheria, however, remains a serious problem in socioeconomically depressed regions of the world, where extensive immunization is not practiced. Treatment of diphtheria involves the use of antitoxin to block the cytopathic effects of diphtheria toxin, which prevents the occurrence of serious symptoms associated with this toxin. This immunological treatment is augmented by the use of antibiotics, like erythromycin, to eliminate the bacterial infection.

Whooping Cough   Whooping cough or pertussis derives its name from the distinctive symptomatic cough associated with this disease. Other symptoms resemble those of the common cold, although vomiting often occurs after severe coughing episodes. Whooping cough is caused by *Bordetella pertussis*, a Gram-negative coccobacillus, which exhibits fastidious nutritional requirements. *B. pertussis* is capable of reproducing within the respiratory tract, and high numbers of this bacterium are found on the surface tissues of the bronchi and trachea. *B. pertussis* produces several toxins that establish the pathogenicity of this organism. Erythromycin and tetracyclines are effective in eliminating the infecting bacteria, although the treatment of whooping cough primarily involves maintenance of an adequate oxygen supply. The administration of pertussis vaccine has greatly reduced the occurrence of whooping cough, and the disease is prevented by routine immunization of infants.

Tuberculosis   **Tuberculosis**, caused by *Mycobacterium tuberculosis* and related mycobacterial species, is primarily transmitted via droplets from an infected to a susceptible individual, although it can also be transmitted by the ingestion of contaminated food. Before the extensive use of pasteurization, milk contaminated with *M. tuberculosis* was associated with outbreaks of this disease. The principal portal of entry for *M. tuberculosis*, however, is through the respiratory tract because much lower numbers of bacteria are required to establish an infection compared to transmission through the gastrointestinal system.

The common form of tuberculosis involves an infection of the pulmonary system, with multiplication of *M. tuberculosis* occurring in the lower respiratory tract despite the phagocytic activity of macrophage that protect this area from infection by most potential bacterial pathogens. The pulmonary form of tuberculosis involves inflammation and lesions of lung tissue, which can be detected by chest x-rays (Figure 21.11).

FIGURE 21.11

X-ray showing the development of tuberculosis in the lungs of an adult. The left lung is normal, but the right lung shows a large, calcified tubercle, as well smaller tubercles and infiltration in the apex of the right lung. (From BPS—R.B. Morrison, M.D., Austin, Tex.)

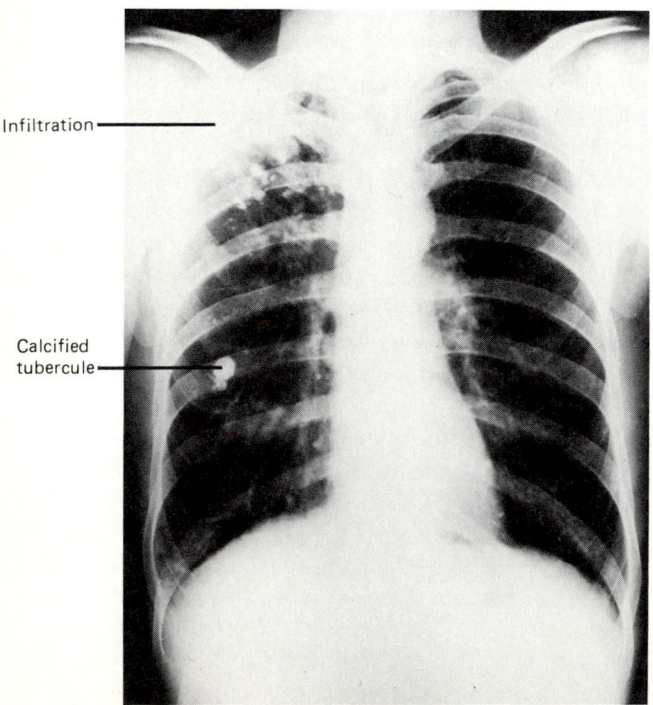

Infiltration

Calcified tubercule

The bacteria spread from the primary lesions to the draining lymph and then through lymph and blood to other parts of the body. Infection with *M. tuberculosis* elicits a cellular immune response because the bacteria are able to reproduce within phagocytic cells, and a delayed hypersensitivity reaction is typical of infection with *M. tuberculosis*. Dormant mycobacteria can remain within the body, and the infectious process can be reactivated at a later time, with various physiological factors probably contributing to recurrence of the disease.

The course of tuberculosis varies greatly among infected individuals. In some cases the infection is restricted to the area of primary lesions, and in others it spreads into various other tissues. Disease symptoms, including fatigue, weight loss, and fever, generally do not appear until extensive lesions have developed in the lung tissues. As a result of the slow growth rate of *M. tuberculosis* and the ineffectiveness of phagocytic cells in killing this bacterial species, tuberculosis is generally a persistent and progressive infection. Without treatment, it is often fatal. Effective treatment of tuberculosis is generally prolonged and involves the use of multiple antibiotics such as streptomycin, rifampin, and isoniazid. Malnutrition and stress are important factors relating to the resistance to tuberculosis and the course of the disease. **Extrapulmonary tuberculosis** commonly develops as a result of reactivation of dormant lesions established during the primary infection. The common locations of extrapulmonary involvement are the genitourinary system, bones and joints, lymph nodes, pleura, and peritoneum. In some cases, the central nervous system is also involved, as in the case of novelist Thomas Wolfe, who, after a severe case of influenza or pneumonia, developed a reactivation of tuberculosis that spread through the bloodstream, leading to death from meningeal tuberculosis. It is interesting to note that the disease was not clinically confirmed, and some mycologists claim that coccidioidomycosis rather than tuberculosis was responsible for Wolfe's death.

Meningitis   **Meningitis**, an inflammation of the meninges (the membrane surrounding the brain and spinal cord), can be caused by various viruses and bacteria. The typical form of transmission of bacteria causing meningitis is via droplet spread, with the initial infection occurring in the respiratory tract followed by transmission via the bloodstream to the meninges. Injuries that expose the central nervous system to bacterial contaminants provide an alternative portal of entry. *Neisseria meningitidis*, *Haemophilus influenzae*, *Streptococcus pneumoniae*, and *Escherichia coli* are the most common etiological agents of bacterial meningitis. There is a high degree of correlation between the age of the patient and the specific etiological agent (Table 21.2). For example, *N. meningitidis*, also known as the *meningococcus*, is often the causative agent of bacterial meningitis in patients between 5 and 40 years of age but is rarely found in cases of meningitis in younger children. Similarly, *E. coli* often causes meningitis in infants but not in adults.

TABLE 21.2   Correlation of Age with the Etiological Agents of Meningitis

	Percent of Isolates Found in Patients			
Etiological Agent	Under 2 months	2–60 months	5–40 years	Over 40 years
Neisseria meningititis	—	20	45	10
Haemophilus influenzae	—	60	5	2
Escherichia coli and other Enterobacteriaceae	55	—	—	10
Pseudomonas aeruginosa	2	—	—	—
Streptococcus pneumoniae and other *Streptococcus* spp.	28	12	25	55
Staphylococcus spp.	5	—	10	13
Other	10	8	10	10

Meningitis is characterized by sudden fever, severe headache, painful rigidity of the neck, nausea, vomiting, and frequently convulsions, delirium, and coma. If untreated, bacterial meningitis is usually fatal. Because death may occur within hours of recognition of the infection, accurate and swift diagnosis and speedy initiation of treatment are essential. The diagnosis of the etiological agent of meningitis depends on the isolation and identification of the pathogens from the cerebrospinal fluid. A number of antibiotics are used in the treatment of meningitis, and the antibiotic of choice is determined by the expected antibiotic susceptibility of the causative agent of the disease; in many cases, it is one of the penicillin derivatives.

Otitis Media   **Otitis media** is an inflammatory disease of the mucosal lining of the middle ear. Bacterial infections of the middle ear normally originate from an upper respiratory tract infection, with the bacteria entering the ear through the auditory (eustachian) tube, the principal portal of entry to the ear. *Streptococcus pneumoniae* is the etiological agent in over 50 percent of the cases of otitis media, and ampicillin is effective in treating such infections. *S. pyogenes* and *Haemophilus influenzae* are also frequently the causative agents. Manifestations of middle ear infections normally include severe pain and fever; in cases caused by *Streptococcus* species, the tympanic membrane is usually fiery red.

Q Fever   **Q fever**, caused by the rickettsia *Coxiella burnetii*, is a unique rickettsial disease because it is not manifested itself as a rash and is the only one normally transmitted via the respiratory tract. The disease can also be transmitted through direct contact of the bacteria with the eyes and via the gastrointestinal tract. *C. burnetii* is normally found in nonhuman animal populations, such as cattle and sheep, where it is transmitted via tick vectors. The bacteria can become airborne on **fomites** (inanimate objects), such as hair and dust particles, and establish human infections by invading the lower respiratory tract, leading to a systemic infection. The symptoms of Q fever often include fever, headache, chest pain, nausea, and vomiting. Tetracyclines and chloramphenicol are normally used to treat this disease.

Diseases Caused by Fungal Pathogens

Histoplasmosis   **Histoplasmosis**, one of several fungal diseases of humans, is caused by the fungus *Histoplasma capsulatum* (*Emmonsiella capsulata*), which enters the respiratory tract through the inhalation of spores that are then deposited in the lungs. *H. capsulatum* is a dimorphic fungus exhibiting yeast-like growth at 37°C and filamentous growth at 25°C. Normally, histoplasmosis is a self-limiting disease in which symptoms may be absent or resemble a mild cold. In some cases, however, the systemic distribution of the fungus to different organs of the body may prove fatal. The fungus persist in the lung and in tissues and cells of the mononuclear phagocyte system. Amphotericin B is used in the treatment of cases of progressive disseminated histoplasmosis.

Histoplasmosis is endemic to certain regions of the world, such as the Ohio and Mississippi river valleys of the United States (Figure 21.12). The fungus is found in soils contaminated with bird droppings, and dust particles released from abandoned bird roosts appear to be involved in some outbreaks of this disease. The apparent association of bird roosts with histoplasmosis has been used as the justification for large-scale kills of blackbirds, but the usefulness of this procedure has not been conclusively demonstrated.

Coccidioidomycosis   **Coccidioidomycosis** is caused by *Coccidioides immitis*, which—like *Histoplasma capsulatum*—exhibits dimorphism. Coccidioidomycosis is also referred to as **valley or San Joaquin fever** because of the geographic distribution of *C. immitis* and the associated areas of occurrence of this disease. Because of the association of the spores of *C. immitis* with the arid soils of the southwestern United States, these soils must be disinfected before being shipped to other regions. Many individuals living in this region have a positive skin test, indicating that they have been infected at some time with this fungus. Visitors to the states of Nevada, California, Utah, Arizona, and New Mexico often develop symptoms of a mild cold because of infection with *C. immitis*. Normally, *C. immitis* occurs in soil, and transmission of coccidioidomycosis involves inhalation

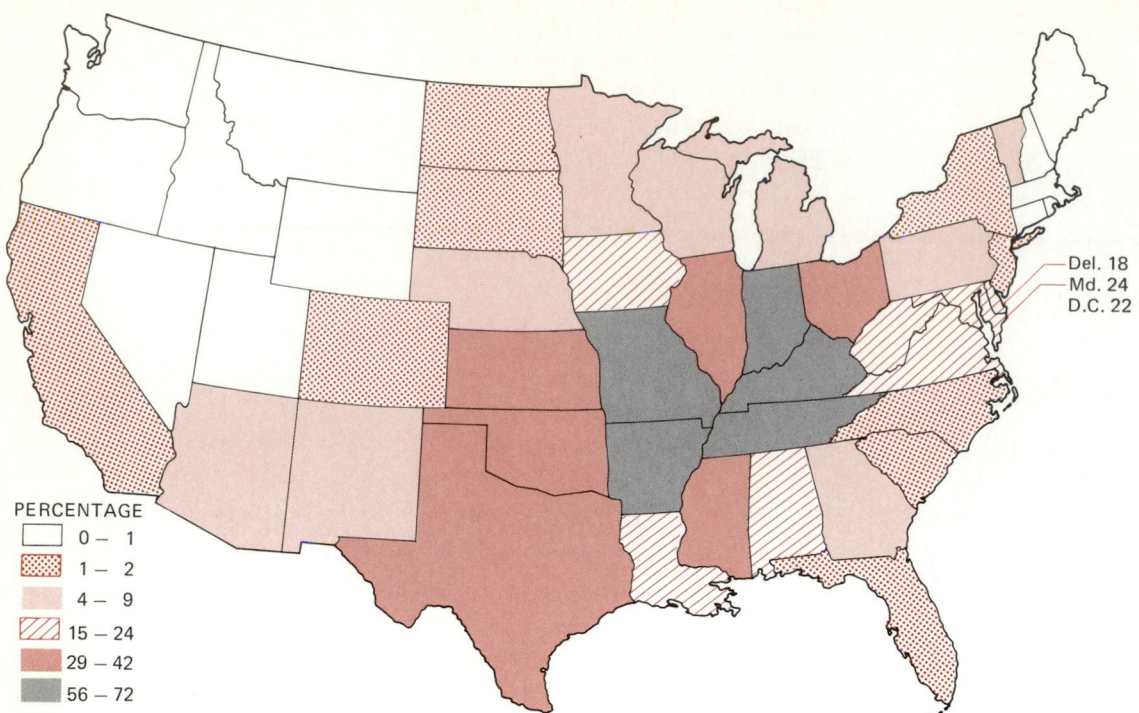

PERCENTAGE
- 0 – 1
- 1 – 2
- 4 – 9
- 15 – 24
- 29 – 42
- 56 – 72

Del. 18
Md. 24
D.C. 22

FIGURE 21.12

This map, showing the incidence of skin reactivity to histoplasmin among naval recruits, indicates that histoplasmosis in the United States occurs primarily in the Mississippi and Ohio valleys. In southern Kentucky, middle Tennessee, and southern Missouri, the incidence of skin reactivity is as high as 90 to 95 percent. (From L. B. Edwards et al., 1969, American Review of Respiratory Diseases, 99:1)

of dust particles containing conidia of this fungus. When deposited in the bronchi or alveoli, the conidia of *C. immitis* elicit an inflammatory response. Within host tissues, *C. immitis* appears as spherules containing multiple spores. In some cases *C. immitis* remains localized in the area of the primary lesion, but the organism can be distributed to other parts of the body. Symptoms of coccidioidomycosis include chest pain, fever, malaise, and a dry cough. In most cases, no special treatment is required for the cure of localized coccidioidomycosis, and upon recovery the individual is immune to this disease. However, when there is evidence of systemic dissemination of *C. immitis*, treatment

with amphotericin B is normally used in effectively treating the disease.

Blastomycosis **North American blastomycosis** is a systemic disease caused by *Blastomyces dermatidis*. The primary site of infection is the lungs, from which the fungus can be disseminated to many other body tissues. *B. dermatidis* is a dimorphic fungus that normally inhabits soils. Blastomycosis occurs most frequently in the southeastern United States. Skin lesions are common. Like other systemic mycoses, blastomycosis can be treated with amphotericin B.

21.3 Diseases Caused by Pathogens That Enter Via the Gastrointestinal Tract

Microorganisms routinely enter the gastrointestinal tract in association with ingested food and water. The large resident microbiota that develop in the human intestinal tract after birth are important for the maintenance of good health and is usually not involved in disease processes. These microbiota are normally noninvasive and are associated with the surface tissues and ingested food material. Some pathogenic microorganisms, however, possess toxigenic or invasive

properties that permit them to cause disease when they enter the gastrointestinal tract.

There are two distinct processes that can initiate disease through the gastrointestinal tract. In the first type, microorganisms growing in food or water can produce toxins, and their ingestion initiates a disease process. Such diseases are classified as **food poisoning** or **intoxication** because the etiological agents of the disease need not grow within the body; that is, there is no true infectious process. Toxins absorbed through the gastrointestinal tract can cause neural damage and death in some cases, as well as localized inflammation and gastrointestinal upset in others. In the second type of disease-causing process, invasive pathogens establish an initial infection through the gastrointestinal tract and cause localized gastrointestinal upset or systemic disease symptomatology. Generally, the establishment of infection through the gastrointestinal tract requires a relatively large infectious dose; that is, a relatively large number of pathogenic microorganisms are required to successfully overcome the inherent defense mechanisms of the gastrointestinal tract. Quite different measures are required to prevent and treat infectious gastrointestinal diseases compared to those for specific microorganisms responsible for food poisoning.

Diseases Caused by Viral Pathogens

Gastroenteritis   Gastroenteritis involves an inflammation of the lining of the gastrointestinal tract. In most cases viral gastroenteritis is a self-limiting disease, often referred to as the *24-hour* or *intestinal flu*. **Viral gastroenteritis** is not caused by an influenza virus and is not related to true cases

of flu; rather, it is due to several different viruses, including adenoviruses, coxsackieviruses, polioviruses, and members of the **ECHO virus group**. The **Norwalk agent**, a small DNA virus identified as being responsible for an outbreak of "winter vomiting disease" that occurred in Norwalk, Ohio, in 1968, appears to be an important etiological agent of various viral gastroenteritis outbreaks. **Rotavirus**, a large RNA virus, also appears to be a very common etiological agent of diarrhea in infants, particularly in socioeconomically depressed regions of the world.

Viruses causing gastroenteritis normally replicate within cells lining the gastrointestinal tract, and large numbers of viruses are released in fecal matter. Contamination of food with fecal matter is an important route of transmission of viral gastroenteritis, as well as many other diseases caused by microorganisms that enter via the gastrointestinal tract. The characteristic symptoms of viral gastroenteritis include sudden gastrointestinal pain, vomiting, and/or diarrhea (Figure 21.13). Recovery normally occurs within 12 to 24 hours of the onset of disease symptomatology. As a result of the vomiting and diarrhea, there can be a severe loss of body fluids and dehydration. The loss of water and the resultant imbalance in electrolytes can have serious consequences,

FIGURE 21.13

The time course of the symptoms of viral gastroenteritis is shown in this illustration of the response of two volunteers to oral administration of stool filtrate derived from a volunteer who received the original Norwalk rectal-swab specimen. The height of the shaded curve is roughly proportional to the severity of the sign or symptom. Essentially, this experiment both confirms the cause of the disease by Koch's postulates and demonstrates the variability of symptomatology; in viral diseases such as this, human subjects rather than experimental animals are used because of the specificity of the virus–host relationship. (Reprinted by permission of the University of Chicago Press, from R. Dolin et al., 1971, *Journal of Infectious Diseases*, 123:307)

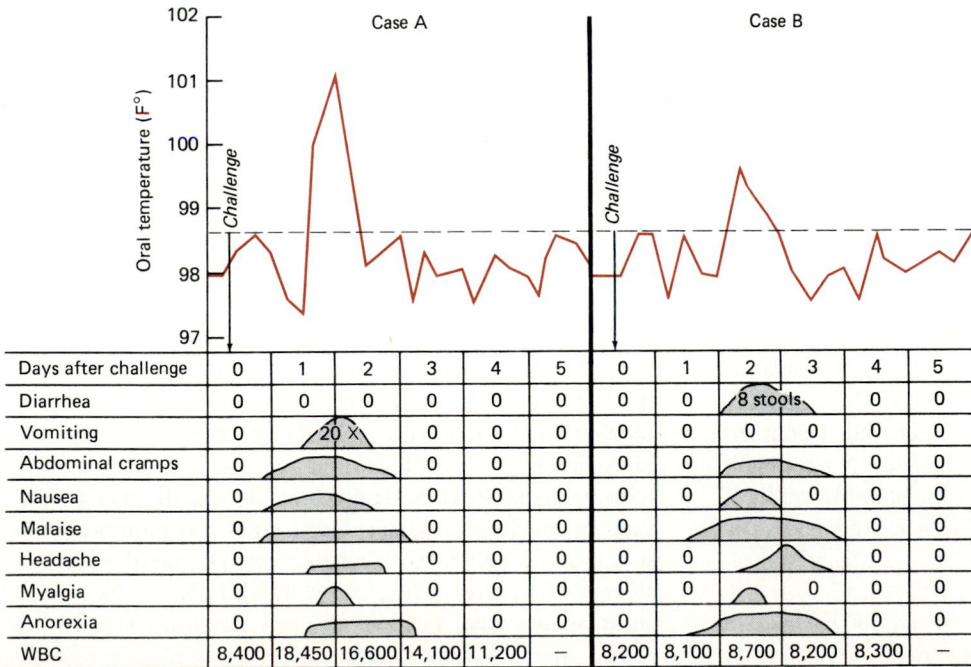

Days after challenge	0	1	2	3	4	5	0	1	2	3	4	5
Diarrhea	0	0	0	0	0	0	0	0	8 stools		0	0
Vomiting	0	20 X	0	0	0	0	0	0	0	0	0	0
Abdominal cramps	0			0	0	0	0	0			0	0
Nausea	0			0	0	0	0	0		0	0	0
Malaise	0			0	0	0	0				0	0
Headache	0			0	0	0	0	0		0	0	0
Myalgia	0			0	0	0	0	0		0	0	0
Anorexia	0				0	0	0				0	0
WBC	8,400	18,450	16,600	14,100	11,200	—	8,200	8,100	8,700	8,200	8,300	—

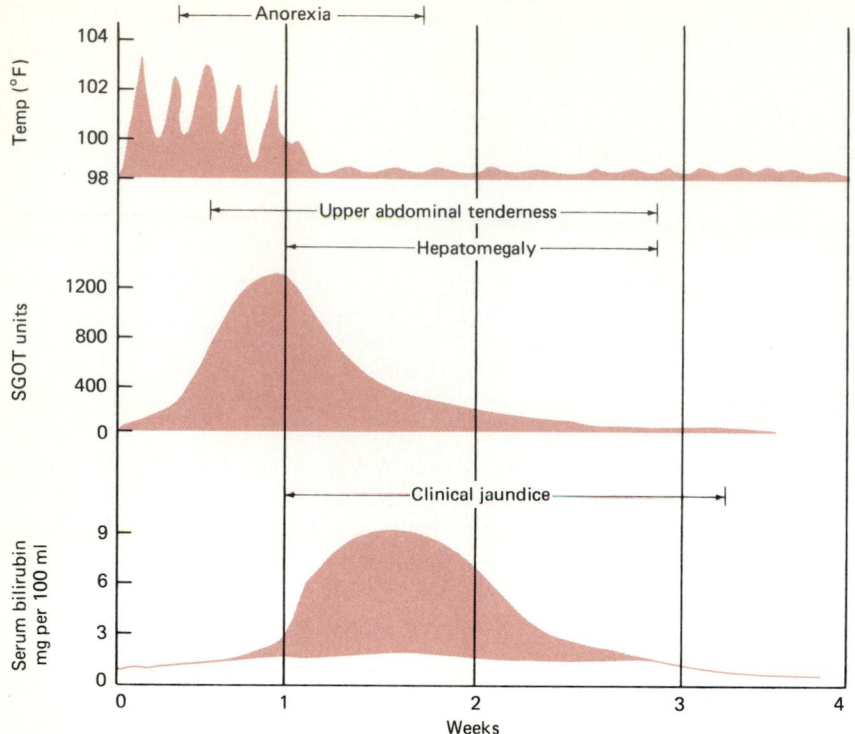

FIGURE 21.14

The time course of the symptoms of infectious hepatitis shows that jaundice begins 1 week after the initial fever. (After W. P. Havens, Jr., in *Viral and Rickettsial Infections of Man,* F. L. Horsfall, Jr. and I. Tamm, eds., J.B. Lippincott Co., Philadelphia)

particularly in infants, where viral gastroenteritis is sometimes fatal.

Hepatitis   **Hepatitis** is a systemic viral infection that primarily affects the liver. There are several types of hepatitis viruses, designated types A, B, D, and non-A, non-B. Type A hepatitis virus normally enters the body via the gastrointestinal tract and causes infectious hepatitis. Although there are some documented cases of foodborne hepatitis B, types B, D and non-A, non-B hepatitis viruses, which will be discussed in Chapter 22, normally enter the body through skin punctures and cause serum hepatitis. Hepatitis type A virus is usually transmitted by the fecal-oral route and is prevalent in areas with inadequate sewage treatment. Several outbreaks of viral hepatitis have been associated with contaminated shellfish that have concentrated viruses from sewage effluents.

The initial symptoms of infectious hepatitis include fever, abdominal pain, and nausea, followed by jaundice, the yellowing of the skin indicative of liver impairment caused by the virus (Figure 21.14). Damage to liver cells also results in increased serum levels of enzymes, such as transaminases, normally active in liver cells. The detection of increased serum levels of these enzymes is used in diagnosing this disease. Infectious hepatitis caused by the type A virus tends to be less serious than serum hepatitis. In most cases of infectious hepatitis, the infection is self-limiting, and recovery occurs within 4 months.

Poliomyelitis   **Polioviruses** may enter the body through either the gastrointestinal or respiratory tracts. Transmission through the ingestion of food and water containing polioviruses is considered very important. Polioviruses are able to multiply within the tissues of the oropharynx and intestines. Viruses entering the bloodstream are disseminated, and further viral replication occurs within lymphatic tissues. Polioviruses can cross the blood-brain barrier, where they continue to multiply within neural tissues and cause varying degrees of damage to the nervous system.

The initial symptoms of poliomyelitis, commonly referred to as **polio**, include headache, vomiting, constipation, and sore throat. In many cases, these early symptoms are followed by obvious neural involvement, including paralysis due to the injury of motor neurons. Although the paralysis can affect any motor function, in over half of the cases of paralytic poliomyelitis the arms and/or legs are involved. Fortunately, paralytic symptoms are 1,000 times less frequent than nonparalytic infections, and many cases of poliovirus infection fail to show any evidence of clinical symptomatology.

Poliomyelitis is prevalent in children and as such is also called **infantile paralysis**. The disease also strikes adults; in fact, the fatality rate in adults is much higher than that in children (Table 21.3). The use of the **Salk and Sabin polio vaccines** has dramatically reduced the incidence of this disease (Figure 21.15). It is important that preschool children be immunized because major outbreaks of poliomyelitis have traditionally been associated with transmission among children in close contact in a schoolroom. Despite the ability to prevent this serious disease, many children are not immunized voluntarily, even in affluent countries such as the United States. Many school systems now require evidence

TABLE 21.3   Age-Specific Case Fatality Rates of Paralytic Poliomyelitis in the United States, 1960–1968

Age Group	Cases	Deaths	Case Fatality Rate (%)
0–4	1899	92	4.8
5–9	959	65	6.8
10–14	379	33	8.7
15–19	205	18	8.8
20–29	462	71	15.4
30–39	275	65	23.6
40 and over	136	41	30.1

of polio vaccination before a child can be enrolled. This is essential because the reduced incidence of this disease means that there are fewer paralyzed individuals to serve as visible reminders of its seriousness. Constant efforts to reinforce parental awareness of the importance and success of vaccination against poliomyelitis are worthwhile.

A postpolio syndrome has surfaced in recent years among survivors of polio, occurring about 40 years after the initial attack. The symptoms are similar to those of the original disease, but there is no trace of viral involvement. It has been theorized that the neurological cells that took over the functions of polio-damaged tissues are now suffering the after effects of 40 years of overwork.

Diseases Caused by Bacterial Pathogens

Botulism   **Botulism**, the most serious form of bacterial food poisoning, is caused by neurotoxins produced by *Clostridium botulinum*. The toxins are absorbed from the intestinal tract and transported via the circulatory system to motor nerve synapses, where their action blocks normal neural transmissions. Various strains of *C. botulinum* elaborate different toxins. Types A, B, and E toxins cause food poisoning of humans. Type E toxins are associated with the growth of *C. botulinum* in fish or fish products, and most outbreaks of botulism in Japan are caused by type E toxins because large amounts of fish are consumed there. Type A is the predominant toxin in cases of botulism in the United States, and type B toxin is most prevalent in Europe.

Over 90 percent of the cases of botulism involve improperly home-canned food. Of 236 outbreaks of this disease in the United States between 1899 and 1974, 57 percent were caused by contaminated vegetables, 15 percent by contaminated fish, and 12 percent by contaminated fruit. The endospores of *C. botulinum* are heat resistant and can survive prolonged exposure at 100°C. Certain canned foods provide an optimal anaerobic environment for the growth of *C. botulinum* that results in the release of toxin into the food. *C. botulinum*, though, cannot grow and produce toxin at low pH and thus is not a problem in acidic food products. Symptoms of botulism can appear 8–48 hours after ingestion of

FIGURE 21.15

(A) Photomicrograph of poliovirus. (From BPS—J. J. Cardamone, Jr. and B. A. Phillips, University of Pittsburgh).
(B) Incidence of poliomyelitis per 100,000 population in the Americas. Note that although the incidence rates were initially similar, the occurrence of this disease in North America dramatically declined after the introduction of vaccines; no

similar decline has occurred in Central and South America, where extensive vaccination has not been carried out. Since 1973 the incidence of poliomyelitis has remained near zero in North America. Declines in the incidence of polio have also occurred in some South American countries where extensive vaccination programs have been carried out. Today there are still countries with high rates of polio while nearly no cases are occurring in the United States and Canada.

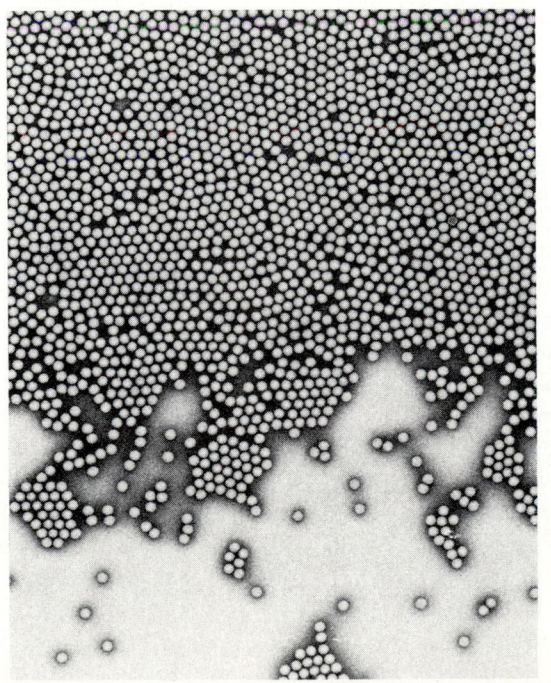

A

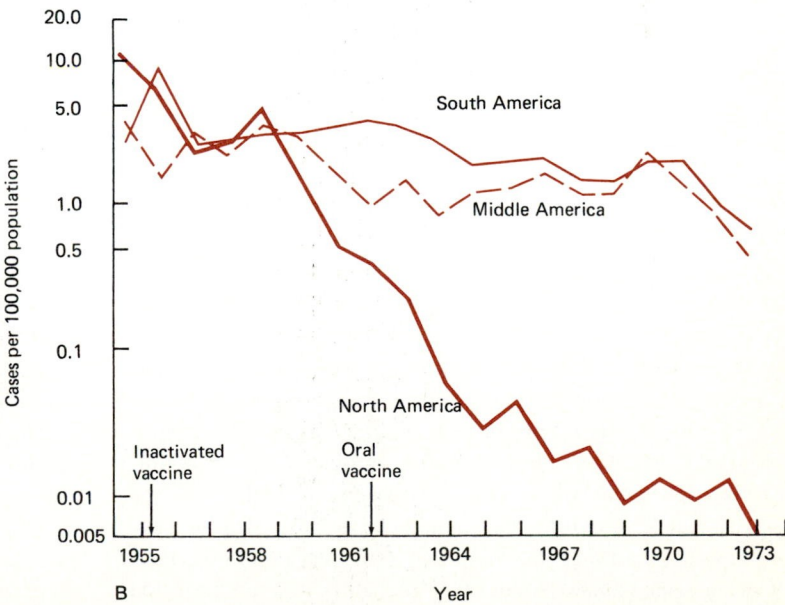

B

the toxin, and their early onset normally indicates that the disease will be severe. Type A toxin botulism is generally more severe than the disease caused by other types of toxin. In severe cases of botulism there is paralysis of the respiratory muscles, and despite improved medical treatment, the mortality rate is still about 25 percent. The use of trivalent ABE antibodies is useful in treating this disease, but it is of paramount importance to ensure continued respiratory functioning.

C. botulinum is normally incapable of establishing an infection in adults because of the low pH of the stomach and the upper end of the small intestine. However, in infants, prior to the colonization of the intestinal tract by *Lactobacillus* species, *C. botulinum* can reproduce and elaborate neurotoxin from the gastrointestinal tract tissues. There is evidence that some cases of **sudden infant death syndrome**, or crib death, can be attributed to *C. botulinum*. Accordingly, additional concern is being given to food products, particularly honey, that infants consume with respect to the possible ingestion of *C. botulinum* endospores.

Perfringens Food Poisoning
Another clostridial species, *Clostridium perfringens*, is a major cause of a less severe form of food poisoning. *C. perfringens* generally accounts for over 10 percent of the outbreaks of foodborne disease in the United States. The ingestion of food containing toxin produced by *C. perfringens* and the adsorption of the toxin into the cells lining the gastrointestinal tract initiate this disease. Toxin type A of *C. perfringens* is associated with most cases of clostridial food poisoning, particularly with cooked meats if a gravy is prepared with the meat. The spores of *C. perfringens* type A can survive the temperatures used in cooking many meats, and if incubated in a warm gravy, there is sufficient time for the spores to germinate and the growing bacteria to produce enough toxin to cause this disease.

The symptoms of food poisoning associated with *C. perfringens* generally appear within 10–24 hours after ingestion of food containing the toxin. They include abdominal pain and diarrhea, but vomiting, headache, and fever normally do not occur. Unlike botulism, recovery from food poisoning caused by *C. perfringens* generally occurs within 24 hours.

Staphylococcal Food Poisoning
Strains of *Staphylococcus aureus*, which cause food poisoning, produce an exotoxin that is also an enterotoxin. The release of toxin from the bacterial cells causes an inflammation of the lining of the gastrointestinal tract. The staphylococcal enterotoxins do not exert a direct local effect; they must be absorbed through the bloodstream and reach the digestive tract in order to initiate a food poisoning syndrome. *S. aureus* can reproduce within many different types of food products. Enterotoxin-producing strains of *S. aureus* often enter foods from the skin surfaces of people who handle food. Custard-filled bakery goods, dairy products, processed meats, potato salad, and various canned foods are frequently found to be the source of the toxin. Salads prepared for a summer picnic can easily be contaminated (inoculated) with *S. aureus*, and when they are left in the sun in a wicker basket (incubated), the bacteria can multiply, producing a sufficient amount of enterotoxin to provide an unexpected nighttime encore to the day's fun.

The symptoms of staphylococcal food poisoning occur relatively rapidly after ingestion of toxin-containing food, usually within 1–6 hours. The symptoms generally include nausea, vomiting, abdominal pain, and diarrhea. They usually subside within 8 hours of onset, and complete recovery usually occurs within a day or two. The prevention of staphylococcal food poisoning depends on proper handling and preservation of food products to prevent contamination and subsequent growth of enterotoxin-producing strains of *Staphylococcus*.

Gastroenterocolitis
Gastroenterocolitis, an inflammation of the intestinal lining, can result from various bacterial infections. *Clostridium perfringens* and *Staphylococcus aureus*, for example, frequently involved in cases of food poisoning, can also establish infections in the gastrointestinal tract and cause gastroenterocolitis.

Enterotoxin-producing strains of *Escherichia coli* are also capable of causing both mild and severe forms of enterocolitis. In most cases, enterotoxin-producing strains of *E. coli* do not invade the body through the gastrointestinal tract; rather, toxin released by cells growing on the surface lining of the gastrointestinal tract causes diarrhea. Aside from diarrhea, abdominal cramps are normally the only other clinical symptom of this disease. Travelers from the United States to Mexico often suffer severe diarrhea as a result of ingestion of strains of *E. coli* foreign to their own microbiota and therefore generally avoid drinking the water. Many cases of severe diarrhea in children are caused by noninvasive, enterotoxin-producing strains of *E. coli*.

In some cases, enteropathogenic strains of *E. coli* invade the body through the mucosa of the large intestine to cause a serious form of dysentery. Invasive strains of *E. coli* are primarily associated with contaminated food and water in Southeast Asia and South America. The ability to invade the mucosa of the large intestine depends on the presence of a specific K antigen in enteropathogenic serotypes of *E. coli*. The enterotoxins produced by *E. coli* cause a loss of fluids from intestinal tissues. With proper replacement of body fluids and maintenance of the essential electrolyte balance, infections with enterotoxic *E. coli* normally are not fatal.

Salmonellosis. **Salmonellosis**, caused by members of the genus *Salmonella*, which are Gram-negative short rods, is commonly manifested as gastroenterocolitis. Various *Salmonella* species, especially the numerous serotypes of *S. enteritidis*, are commonly the etiological agents of salmonellosis. Like many enteropathogenic bacteria, *Salmonella* species have pili that enable them to adhere to the lining of the gastrointestinal tract. Although *Salmonella* species are able to reproduce within the intestines, causing inflammation, they do not normally penetrate the mucosal lining and enter the bloodstream; in some cases, however, *Salmonella* species can gain access to the circulatory system, causing bacteremia. For example, paratyphoid fever, which is caused by strains of *S. paratyphi* and *S. typhimurium*, is characterized by gastroenteritis and a relatively high rate of bacteremia (Figure 21.16). *Salmonella* species causing gastroenteritis are normally transmitted by ingestion of contaminated food. Birds and domestic fowl, especially ducks, turkeys, and chickens, including their eggs, are commonly identified as the sources of *Salmonella* infections. Inadequate cooking of large turkeys and the ingestion of raw eggs cause a significant number of cases of salmonellosis.

Enterocolitis from *Salmonella* infections is normally characterized by abdominal pain, fever, and diarrhea that lasts for 3–5 days. The onset of disease symptoms normally occurs 8–24 hours after ingestion of contaminated food. Nausea and vomiting may be the initial symptoms, but they usually do not persist once pain and diarrhea begin. The feces may contain mucus and blood. Generally, the disease is self-limiting, with recovery occurring within 1 week. During acute salmonellosis the feces may contain 1 billion *Salmonella* cells per gram. Fecal contamination of water and food supplies can contribute to the transmission of this disease. Except for children and the aged, the treatment of salmonellosis normally does not require the use of antibiotics because relatively rapid recovery normally occurs.

Shigellosis. **Shigellosis**, or **bacterial dysentery**, is an acute inflammation of the intestinal tract caused by species of the Gram-negative genus *Shigella*, including *S. flexneri*, *S. sonnei*, and *S. dysenteriae*. The transmission of *Shigella* species normally occurs by the direct anal-oral route, although water and food supplies are involved in some outbreaks of bacterial dysentery. *Shigella* species are able to penetrate the mucosal cells of the large intestine and multiply in the submucosa. Areas of intense inflammation develop around the multiplying bacteria and microabscesses form and spread, leading to bleeding ulceration (Figure 21.17).

The symptoms of *Shigella* infections include abdominal pain, fever, and diarrhea, with mucus and blood in the excreta. Bacterial dysentery normally is a self-limiting disease, with recovery occurring 2–7 days after onset. The severe dehydration associated with this disease can cause shock and lead to death in children, in whom the incidence of bacterial dysentery is highest. In cases of childhood shigellosis, antibiotics are used to treat the disease. However, many *Shigella* strains now contain antibiotic-resistant plasmids; as a result, the spectrum of antibiotics needed to combat shigellosis has increased. Tetracycline, ampicillin, and nalidixic acid are now employed to treat bacterial dysentery caused by different *Shigella* species.

Other Forms of Gastroenterocolitis. Campylobacter fetus var. *jejuni* has been found to be the causative agent of many cases of gastroenteritis in infants. In fact, *C. fetus* may be more important in juvenile gastroenteritis than *Salmonella* species. The transmission of *C. fetus* appears to be via con-

taminated food or water. *C. fetus* is a Gram-negative, motile, spiral-shaped bacterium, formerly known as *Vibrio fetus*, which also causes fetal abortion in cattle and sheep.

Vibrio parahaemolyticus is responsible for many cases of gastroenteritis in Japan and perhaps in the United States. It occurs in marine environments, and the ingestion of contaminated seafood, particularly the eating of raw fish, is the main route of transmission. Gastroenteritis caused by *V. parahaemolyticus* requires the establishment of an infection within the gastrointestinal tract, rather than simple ingestion of an enterotoxin. The symptoms generally appear 12 hours after ingestion of contaminated food and include abdominal pain, diarrhea, nausea, and vomiting. Recovery from this form of gastroenteritis normally occurs in 2–5 days, and the mortality rate is very low.

Bacillus cereus is responsible for a small proportion of gastoenterocolitis cases. Strains of this bacterial species cause a relatively mild form of gastroenteritis, and recovery normally occurs in less than a day. The occurrence of gastroenteritis due to *B. cereus* requires the ingestion of a large number of spores. The symptoms include abdominal pain, profuse diarrhea, and nausea.

Yersinia enterocolitica and related species produce a severe form of enterocolitis. The symptoms of infection resemble those of appendicitis and include abdominal pain, fever, diarrhea, vomiting, and leucocytosis. Often an appendectomy is performed before this disease is properly diagnosed as **yersiniosis**. Outbreaks of yersiniosis are most common in Western Europe but have also been confirmed in the United States. In an outbreak of yersiniosis in New York involving over 200 school children, the infection was traced to a common source of chocolate milk. Ten children underwent unnecessary appendectomies before the true etiology of the disease was established. *Y. enterocolitica* is widely distributed and has been found in water, milk, fruits, vegetables, and seafoods. This organism is psychrotrophic and thus is able to reproduce within refrigerated foods, where it can multiply and reach an infectious dose. In fact, *Y. enterocolitica* grows better at 25°C than at 37°C.

Typhoid Fever   Outbreaks of **typhoid fever**, a systemic infection caused by *Salmonella typhi*, are associated with contaminated water supplies and the handling of food products by individuals infected with this bacterium. Although the portal of entry for *S. typhi* normally is the gastrointestinal tract, infections with *S. typhi* do not initially cause gastroenteritis; rather, the bacteria simply enter the body via this route and cause infections at other sites. In the course of the disease, however, the intestines become involved, along with various other organs. A relatively low infectious dose is required for *S. typhi* to establish an infection. The infecting bacteria rapidly enter the lymphatic system and are disseminated through the circulatory system. Phagocytosis by neutrophil cells does not kill *S. typhi*, and the bacteria continue to multiply within phagocytic blood cells. The surface Vi antigen of *S. typhi* apparently interferes with phagocytosis,

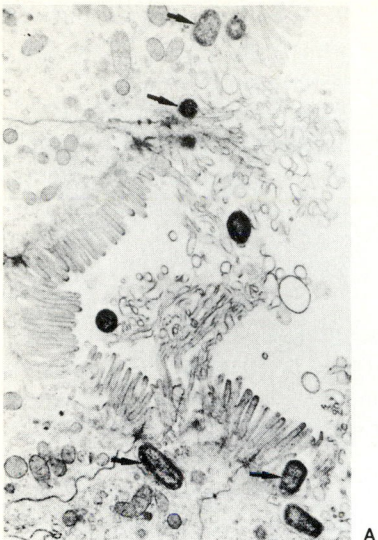

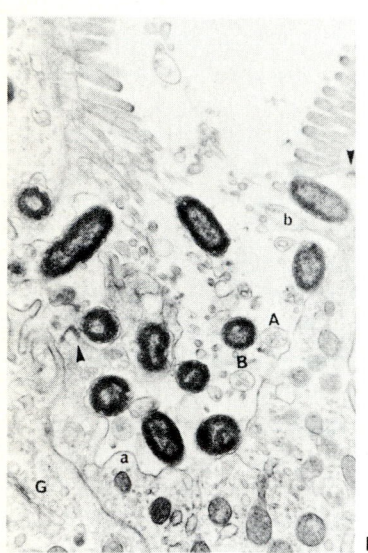

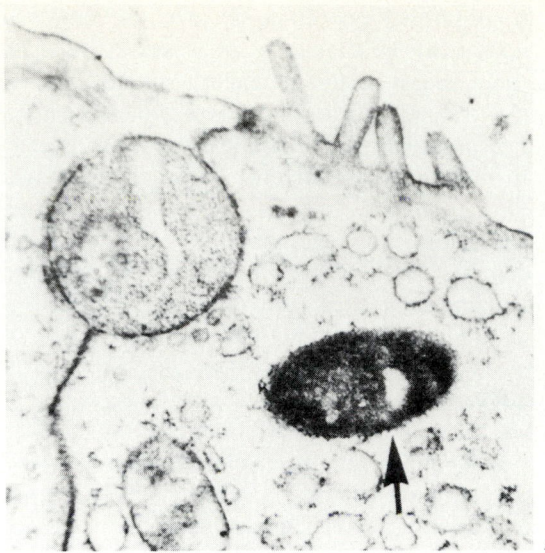

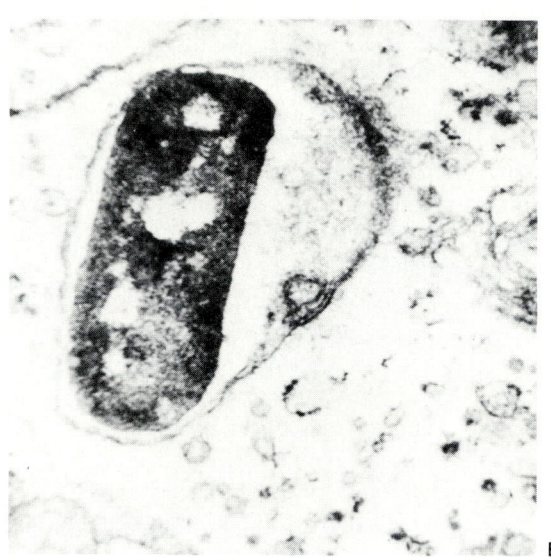

FIGURE 21.16

Micrographs showing invasion of the intestine by *Salmonella*. (A) Note the extensive degeneration of the microvilli terminal cell and apical cytoplasm that is localized at site of bacterial penetration (arrows). Other cytoplasmic components and adjacent cells are unaltered (11,000×). (B) Note the projections (referred to as *blebs*) with (A) or without (a) vesicles that arise from the host cell cytoplasm and are pinched off (B,C and b,c) into a cavity that also contains degenerating microvilli. There are increasing numbers of small vesicles around the Golgi apparatus (G). An intercellular junctional complex is laterally displaced (27,000×). (A reprinted by permission of the American Society for Microbiology, Washington, D.C., from A. Takeuchi, 1975, in: *Microbiology—1975*, p. 176; B D.C., from A. Takeuchi, 1975, in: *Microbiology—1975*, p. 176; B reprinted by permission of the Hoeber Medical Division of Harper & Row Publishers, Inc., Hagerstown, Md., from A. Takeuchi, 1967, *American Journal of Pathology*, 50:125)

FIGURE 21.17

Micrographs showing intestine infected with *Shigella*. (A) Apical portion of the intestinal epithelium 24 hours after infection. A *Shigella* organism (arrow) lies free in the cytoplasm near the lumen surface. The microvilli and terminal web have undergone regressive changes. The endoplasmic reticulum is swollen. Membrane-bound intracytoplasmic inclusions contain osmophilic granular material resembling ribsosomes and a mitochondrion. A similar structure is present in the intracellular space (8000×). (B) A composite of intraepithelial membrane-enclosed *Shigella* organisms (10,000×). The membranes frequently enclose cytoplasmic components. G = Golgi apparatus. membranes frequently enclose cytoplasmic components. G = Golgi apparatus. (Reprinted by permission of the Hoeber Medical Division of Harper & Row Publishers, Inc., Hagerstown, Md., from A. Takeuchi et al., 1965, *American Journal of Pathology*, 47:1037, 1041)

and elimination of infecting cells depends on the antibody-mediated immune response.

After invasion of the mononuclear phagocyte system, the *S. typhi* infection becomes localized in lymphatic tissues, particularly in Peyer's patches of the intestine, where ulcers can develop. Localized infections always develop and cause damage to the liver and gallbladder, and sometimes also to the kidneys, spleen, and lungs.

The symptoms of typhoid fever include fever (104°F), headache, apathy, weakness, abdominal pain, and a rash with rose-colored spots. The symptoms develop in a stepwise fashion over a 3-week period. If no complications occur, the fever begins to decline at the end of the third week. However, if it remains untreated, the mortality rate averages 10 percent. Chloramphenicol is effective in the treatment of typhoid fever, and its use and that of other antibiotics has reduced the death rate to approximately 1 percent.

Cholera   Cholera is caused by the Gram-negative, curved rod *Vibrio cholerae*, serotypes *cholerae* and *El Tor*. Although we typically associate cholera with Asia, sometimes referring to the disease as *Asiatic cholera*, it also occurs in the United States, primarily in the Gulf Coast region, where cases have been traced to contaminated shellfish. Cholera is a particular problem in socioeconomically depressed countries, where there is poor sanitation and inadequate sewage treatment and where medical facilities have only a limited capacity to deal with outbreaks. This disease is endemic in the Ganges delta, and there are annual epidemic outbreaks of cholera in India and Bangladesh. In these endemic areas of Asia the death rate is normally 5–15 percent. Seasonal outbreaks of cholera often occur in Southeast Asia when monsoon rains wash sewage material into drinking water supplies. During sudden epidemics, the mortality rate may reach 75 percent.

V. cholerae is able to multiply within the small intestine and produces the enterotoxin responsible for the symptoms of cholera, which occur suddenly and include nausea, vomiting, abdominal pain, diarrhea with "rice water stools," and severe dehydration, followed by collapse, shock, and in many cases death. *V. cholerae* itself does not invade the body and is not disseminated to other tissues. It is the enterotoxin produced by *V. cholerae* that binds irreversibly to the epithelial cells of the small intestine, stimulating the formation of cyclic AMP within the mucosal cells there. The accumulation of cyclic AMP initiates secretion of water and electrolytes into the lumen of the small intestine because of changes in the membrane permeability of the mucosal cells. The rapid loss of fluid from the cells of the gastrointestinal tract associated with this disease often produces shock, and if it remains untreated, the mortality rate is high. The initial diarrhea that results from infection with *V. cholerae* can cause the loss of several liters of fluid within a few hours. The treatment of cholera centers on replacing fluids and maintaining the electrolyte balance, that is, on combating

shock. Treatment with tetracycline generally reduces the duration of the disease.

Appendicitis   There are more than 200,000 cases per year in the United States of **appendicitis**, an inflammation of the appendix occurring when there is an obstruction of the lumen of this organ. Appendicitis is caused by a mixture of bacterial populations that constitute the normal microbiota of the intestinal tract, and virtually all members of this normal microbiota can contribute to appendicitis. Symptoms normally include abdominal pain, localized tenderness, fever, nausea, vomiting, and leukocytosis. Treatment usually involves the surgical removal of the appendix. Serious complications from appendicitis arise if the infection is permitted to progress and the appendix ruptures. In such cases, systemic bacteremia ensues, and there is a high rate of mortality. Antibiotics are not effective in treating uncomplicated acute appendicitis and are unable to stop the development of infection before rupture of the appendix can occur. In cases where appendicitis is complicated by rupture of the appendix, however, antibiotic treatment is, essential in controlling the spread of infection to other tissues and the use of antibiotics has greatly reduced the number of deaths from this complication.

Fungal Food Poisoning

Although fungi normally do not cause infections via the gastrointestinal system, **mycotoxins** produced by some fungi are responsible for serious cases of food poisoning. Many mycotoxins are potent neurotoxins. Various species of mushrooms contain toxins that can be absorbed through the gastrointestinal tract, and the ingestion of poisonous mushrooms, such as *Amanita phalloides*, is normally fatal. The amatoxins and phallotoxins produced by *A. phalloides* and other species of *Amanita* cause symptoms of food poisoning 8 to 24 after their ingestion. Initial symptoms include vomiting and diarrhea; later, degenerative changes occur in liver and kidney cells, and death may ensure within a few days of ingesting as little as 5–10 mg of toxin.

Some filamentous fungi, other than mushrooms, also produce toxins that can cause human disease. *Aspergillus* species growing on peanuts and grains produce **aflatoxins**, which are potent carcinogens, as well as toxic. They are known to cause death in sheep and cattle and may be involved in some human disease conditions. Aflatoxins are the only known carcinogens for which the United States government has set permissible levels; all other products with carcinogenic activity are banned outright.

Ergotism, another disease caused by fungi, results from ingesting grain containing ergot alkaloids produced by *Claviceps purpurea*. The toxins of *C. purpurea* cause degeneration of the capillary blood vessels, and this type of food poisoning has a relatively high mortality rate. Symptoms of ergotism may include vomiting, diarrhea, thirst, hallucinations, convulsions, and lesions of the extremities. Various outbreaks of mass hallucinations have been traced to contamination of food with ergot alkaloids, and there are even theories that the Salem witch hunts in colonial Massachusetts were related to grain contamination and widespread ergotism.

Algal Food Poisoning

Algae are rarely considered as the etiological agents of disease, but **paralytic shellfish poisoning** is caused by toxins produced by dinoflagellate *Gonyaulax*. Blooms of *Gonyaulax* cause red tides in coastal marine environments. During such algal blooms, the algae and the toxins they produce can be concentrated in bivalve shellfish, such as clams and oysters. The ingestion of shellfish containing algal toxins can lead to symptoms that resemble those of botulism. Shellfishing is banned in areas of *Gonyaulax* blooms to prevent this form of food poisoning.

Diseases Caused by Protozoan Pathogens

Amebic Dysentery   **Amebic dysentery** or **amebiasis** is caused by the protozoan *Entamoeba histolytica*. Infections with *E. histolytica* may be asymptomatic or may involve mild or severe diarrhea and abdominal pain. Amebic dysentery occurs as a result of inadequate sewage treatment and contamination of water with *E. histolytica*, whose cysts are not killed by the chlorination methods normally used to treat municipal drinking water. Infection is acquired by ingesting contaminated food or water containing cysts of *E. histolytica*. Infestation occurs in the small intestine without causing any disease symptoms.

Trophozoites (motile flagellate forms) multiply within the colon, where they appear to ingest red blood cells, yeasts, and bacteria. When immune surface defense mechanisms are lowered, trophozoites are able to invade the epithelial cells of the colon. The invasion of the colon lining results in the formation of ulcers. The trophozoites spread through the submucosa, cutting off the blood supply through the mucosal lining, which leads to sluffing off of mucosal cells and enlargement of ulcers. Mucus and blood from these lesions characteristically appear in the feces. Cysts are shed with the fecal matter and can enter and contaminate water supplies. In some cases, *E. histolytica* can spread to the liver, lung, or skin, causing ulcer or abscess formation in these tissues. Several antiprotozoan drugs, such as metronidazole, are effective in treating amebiasis, and the choice of which one to use depends on whether the infection is restricted to the intestinal tract or whether other organs, such as the liver, are involved.

Giardiasis   *Giardia lamblia*, a flagellated protozoan, is responsible for most cases of diarrhea infection caused by protozoa. *G. lamblia* forms both motile trophozoites and nonmotile cysts, and the cysts are the infective form. The cysts of *G. lamblia* can enter the gastrointestinal tract through contaminated water. A high incidence of **giardiasis** occurred among groups touring Leningrad during the 1970s as a result of contaminated water supplies. In 1973 a major outbreak of giardiasis occurred in upstate New York, with an estimated 4,800 individuals developing symptoms of the disease. The following year, giardiasis was the most common waterborne disease in the United States. *G. lamblia* can live saprophytically within the small intestine without causing any symptoms of giardiasis, and in the United States almost 4 percent of the population appears to be infected by this organism. Excessive growth of the organism, however, can cause disease symptoms that include diarrhea, dehydration, mucus secretion, and flatulence. Metronidazole is generally used in the treatment of this disease.

21.4 Diseases Caused by Pathogens That Enter Via the Genitourinary Tract

The extremities of the genitourinary tract in both males and females contain an indigenous microbiota, but the inner tissues, including the bladder and kidneys, are normally sterile. These tissues are subject to infection with **opportunistic pathogens** if they become contaminated by microorganisms from the indigenous microbiota of the extremities of the genitourinary tract or the gastrointestinal tract. Additionally, the genitourinary tract provides the portal of entry for pathogens that are directly transmitted during sexual intercourse. Infections with such pathogens are known as **venereal** or **sexually transmitted diseases**

Urinary Tract Infections

Urinary tract infections can be caused by a wide variety of microorganisms. The urethra and urinary bladder are most frequently the sites of infection within the urinary tract, with the resulting infections referred to as **urethritis** and **cystitis**, respectively. Although many microorganisms can cause urinary tract infections, the most common etiological agents are Gram-negative bacteria, particularly those normally occurring in the gastrointestinal tract. Accidental contamination of the urinary tract with fecal matter appears to be one of the

TABLE 21.4 Representative Relative Proportions of Bacterial Isolates from Hospital-Acquired and Community-Acquired Urinary Tract Infections

Community-Acquired Organisms	Percent	Hospital-Acquired Organisms	Percent
Escherichia coli	57	Escherichia coli	39
Klebsiella pneumoniae	13	Proteus mirabilis	18
Proteus mirabilis	8	Pseudomonas aeruginosa	17
Streptococcus sp.	8	Streptococcus spp.	11
Pseudomonas aeruginosa	7	Klebsiella pneumoniae	10
Proteus rettgeri	5	Serratia marcescens	8
Others	2		

most important means of transmission of such infections. Although *Escherichia coli* probably is the most common cause of urinary tract infections, members of the genera *Klebsiella, Enterobacter, Serratia, Proteus,* and *Pseudomonas* are also isolated relatively frequently (Table 21.4). Infections with *Serratia* species and *Pseudomonas aeruginosa* most often occur after catheterization procedures because a catheter can carry microorganisms from the extremities of the genitourinary tract, contaminating the inner tissues and resulting in urethritis and cystitis.

The symptoms of urethritis normally include pain and a burning sensation during urination. Cystitis is characterized by suprapubic pain and the urge to urinate frequently. Growth of bacteria in the urinary tract can cause an obstruction. Various antibiotics, such as nitrofurantoin and nalidixic acid, are used in treating infections of the lower urinary tract. Depending on whether there is an obstruction or another underlying cause of the infection, the prognosis is normally good.

Infections of the urethra may spread to the kidney, causing serious, life-threatening disease. **Pyelonephritis** can cause kidney failure and death, but as with other urinary tract infections, the use of antibiotics is generally effective in curing this disease. *E. coli* is the most frequent etiological agent of pyelonephritis in patients who develop infections outside of hospital settings. *Proteus mirabilis, Enterobacter aerogenes, Klebsiella pneumoniae, P. aeruginosa, Streptococcus* species, and *Staphylococcus* species are additional causative agents. Infections with *Proteus* species are especially significant because they have the enzymatic capacity to decompose urea in the urine to ammonia and carbon dioxide. Urgent and frequent urination is normally symptomatic of pyelonephritis. Pyelonephritis and other urinary tract infections are more frequent in adult women than in men or young girls, in part because of anatomical differences in the structure of the urinary tract of females, and also because of the physiological changes that occur during menstruation and pregnancy that favor the establishment of microbial infections.

Female Genital Tract Infections

Vaginal Tract Infections The female genital tract is subject to infection by a variety of microorganisms. The trauma associated with sexual intercourse, menstrual bleeding, normal changes during pregnancy, and childbirth contributes to infections of the female genital tract. Infections of the vulva and vagina are common and generally not serious. **Vulvovaginitis** is an inflammation of both the vulva and the vagina, and can be caused by viruses, bacteria, fungi, and protozoa. In about 20 percent of the cases of vulvovaginitis, the flagellate protozoan *Trichomonas vaginalis* is the etiological agent. The fungus *Candida albicans* is also frequently implicated. Bacterial and fungal overgrowth of the vaginal tract is common during pregnancy, when the pH of the vaginal tract increases. The use of birth control pills also tends to result in higher pH values in the vagina, and the administration of antibacterial antibiotics, which can adversely affect the indigenous bacterial populations, also favors the development of *Candida* infections of the vaginal tract.

The excessive growth of *T. vaginalis, C. albicans,* and various bacteria causes changes in the mucosal cells lining the vaginal tract. The symptoms of vulvovaginitis include increased vaginal discharge and burning. In cases where *T. vaginalis* is the causative agent, there is normally a profuse, greenish, odorous discharge. When *Candida* infections are involved, the discharge is normally thicker and cheesier, and is released in lesser quantity. Antibiotic treatment is effective in controlling vulvovaginitis, with the choice of drug depending on the etiological agent. Metronidazole can be used in cases of protozoan infections, and nystatin generally is used for *Candida* infections.

Infections of the lower parts of the female genital tract may also spread to higher structures, including the cervix and fallopian tubes. Infections of the upper regions of the genital tract are particularly serious during pregnancy and can lead to septic abortion. Some infections of the gestating female genital tract, such as with the Gram-positive, acid-fast bacterium *Listeria monocytogenes*, can also be transferred to the fetus, resulting in congenital abnormalities. Microbial infection with *L. monocytogenes* can occur either through the placenta or during childbirth. In some cases of female genital tract infections, it is necessary to perform a caesarean section delivery in order to protect the infant from contamination.

Toxic Shock Syndrome **Toxic shock syndrome** is caused by *Staphylococcus aureus*. This bacterium can enter the body via the genital tract, and elaboration of its toxins causes high

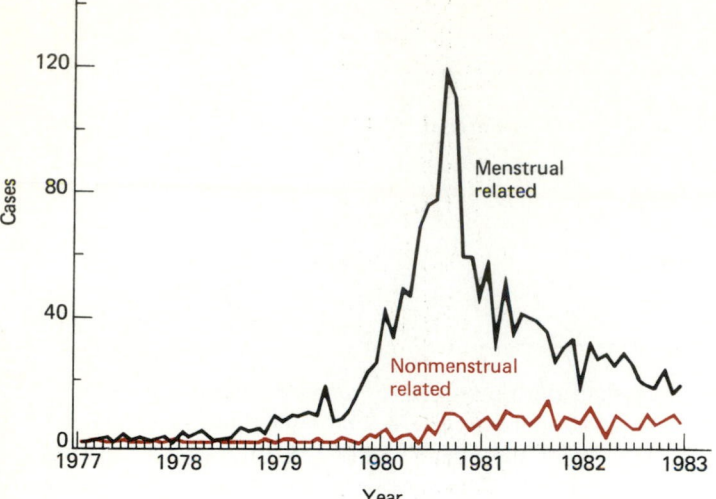

FIGURE 21.18
Reported cases of toxic shock syndrome, showing the high incidence of menstruation-related cases of this disease in 1980. (Based upon *Morbidity and Mortality Weekly Reports*, Centers for Disease Control, Atlanta)

fever, nausea, vomiting, and in many cases death. This disease is not restricted to women and can occur after the introduction of *S. aurens* via other portals of entry, including surgical wounds. The occurrence of toxic shock syndrome, though, is especially correlated with the use of tampons during menstruation, particularly if these devices are left in place for a long period of time (Figure 21.18). The association of this disease with the use of tampons received a great deal of publicity in the early 1980s, forcing one major manufacturer to remove its product from the market. The fibers of some tampons absorb magnesium, permitting the proliferation of *Staphylococcus,* with the production of large amounts of toxins. The hormonal changes that occur during menstruation also favor the proliferation of bacteria in the vaginal tract.

Puerperal Fever **Puerperal** or **childbed fever** is a systemic bacterial infection that may be acquired via the genital tract during childbirth or abortion. The most frequent etiological agents of postpartum sepsis are β-hemolytic group A and B *Streptococcus* species. *Staphylococcus*, *Pseudomonas*, *Bacteroides*, *Peptococcus*, *Peptostreptococcus*, and *Clostridium* species, as well as other bacteria, can also cause this disease. The source of infection is normally the obstetrician, obstetrical instruments, or bedding. The bacteria causing puerperal fever are not normally part of the resident microbiota of the vaginal tract. Prior to the introduction of aseptic procedures, puerperal fever was often a fatal complication after childbirth and remains an important complication following childbirth and abortion procedures. It was the leading cause of maternal death in Massachusetts in the mid-1960s. Penicillin is usually effective in treating postpartum sepsis. The use of proper obstetric procedures generally prevents this disease.

Sexually Transmitted Diseases

Sexually transmitted diseases are contracted by direct sexual contact with an infected individual, generally during sexual intercourse. The physiological properties of the patho-

gens causing these diseases restrict their transmission, for the most part, to direct physical contact because the etiological agents of sexually transmitted diseases have very limited natural survival times outside infected tissues. The incidence of sexually transmitted diseases reflects contemporary sexual behavioral patterns, but in part may also reflect changes in reporting and recording cases of these diseases. At present, outbreaks of some sexually transmitted diseases are considered to be reaching epidemic proportions. The social implications of their transmission often overshadow the fact that these are infectious diseases and must be treated as medical problems, with the emphasis on curing the patient and reducing the incidence of disease by preventing the spread of the infectious agents. The overall control of sexually transmitted diseases depends on breaking the network of transmission, which necessitates public health practices that seek to identify and treat all sexual partners of anyone diagnosed as having one of the sexually transmitted diseases.

Gonorrhea **Gonorrhea** is a sexually transmitted disease caused by the Gram-negative diplococcus *Neisseria gonorrhoeae*, often referred to as *gonococcus*. *N. gonorrhoeae* is a fastidious organism readily killed by drying and exposure to metals. The sensitivity of *N. gonorrhoeae* to desiccation makes the chances of transmission of gonorrhea through inanimate objects, such as toilet seats in public restrooms, negligible. The adherence of *N. gonorrhoeae* to the lining of the genitourinary tract during sexual transmission and the ensuing spread of the infection depend on the pili of this bacterial species. *N. gonorrhoeae* is able to penetrate to the subepithelial connective tissue during spread of the infection (Figure 21.19). *N. gonorrhoeae* infects the mucosal cells lining the epithelium. This bacterium is able to infect the urethra, cervix, anal canal, pharynx, and conjunctivae.

After an alarming increase in the number of cases of gonorrhea in the United States between 1960 and 1975, there has been a slow decline in the incidence of this disease; however, the frequency of antibiotic resistance among iso-

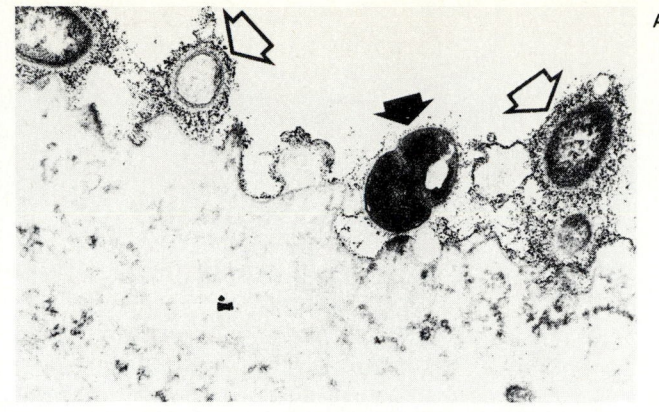

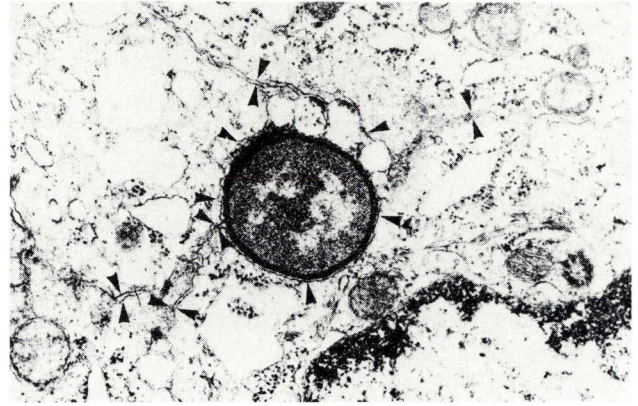

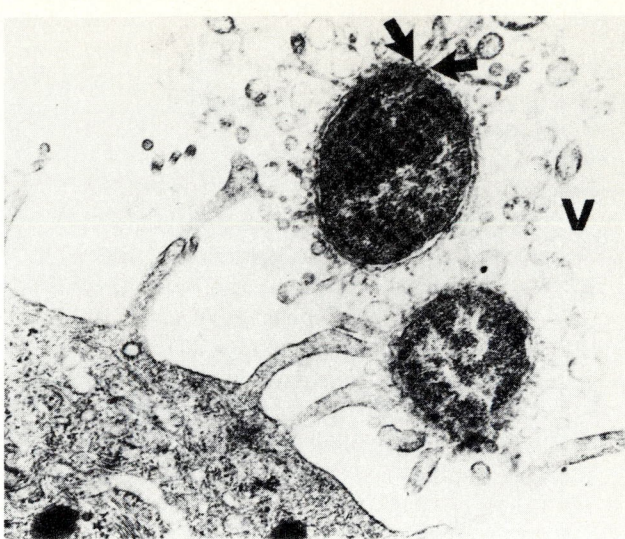

FIGURE 21.19

Electron micrographs of gonococci in the genital tracts of patients with gonorrhea. (A) Gonococcus (solid arrow) and commensal bacteria lying on the cervical epithelium of a female (25,650×). Colloidal thorium treatment reveals well-defined acidic capsules on the commensal bacteria (hollow arrows). Glycoproteins of the epithelial cell coat and in cervical mucus have also been stained. By contrast, there is no evidence of a capsule in the gonococcal surface. Light staining with colloidal thorium can be attributed to contaminating host cell material. (B) Gonococcus lying between epithelial cells in a human fallopian tube organ culture 8 hours after its introduction (21,900×). The gonococcus is tightly enclosed by the surrounding host cell surfaces (arrows), which are linked in places by desmosomes. (C) Gonococci approaching the surfaces of a urethral epithelial cell from a male with early symptomatic gonorrhea (29,200×). Microvilli from the host cell have made contact with the bacteria. The vesicles (V) are characteristically associated with host-grown gonococci and appear at some points (arrows) to be budded off from the bacterial surface. (Reprinted by permission of the American Society for Microbiology, Washington, D.C., from M. E. Ward et al, 1975, in: *Microbiology—1975*, D. Schlessinger, ed., pp.188–199)

lates has been increasing since 1975 (Figure 21.20). This increase coincides with the widespread introduction of birth control pills, which contain hormones that cause pH values in the vaginal tract to increase, removing the normal protection against infections by acid-sensitive bacteria, including *N. gonorrhoeae*. Gonorrhea is normally contracted from

someone who is asymptomatic or who has symptoms but does not seek treatment. The rate of gonorrhea acquisition among males is about 35 percent after a single exposure to an infected female and rises to 75 percent after multiple sexual contacts with the same individual.

In most cases, gonorrhea is a self-limiting disease, but in both sexes the infection may spread to contiguous parts of the genital urinary tract, and *N. gonorrhoeae* may be disseminated to other parts of the body. For example, infections with *N. gonorrhoeae* can spread to the joints, causing gonorrheal arthritis; to the heart, causing gonorrheal endocarditis; and to the central nervous system, causing gonorrheal meningitis.

Often in women, the early stages of gonorrhea are not associated with any overt symptoms, and many women with gonorrhea remain asymptomatic carriers. The cervix often is the site of gonococcal infection, but various other tissues may be involved if the infection spreads. If the infection spreads to the uterus, it causes a chronic infection of the uterine tubes called *salpingitis*. This condition typically is characterized by abdominal pain. Salpingitis can cause infertility. It is also the cause of fertilization to outside the uterus, a life-threatening situation called *ectopic pregnancy*. A gonococcal infection may spread to the urethra, causing an inflammation called *gonococcal urethritis*. Gonorrhea also can lead to **pelvic inflammatory disease (PID)**,[3] which results from a generalized bacterial infection of the uterus, pelvic organs, uterine tubes, and ovaries (Figure 21.21). PID may occur without overt symptoms of gonorrhea but nevertheless may cause infertility. Occlusions of the fallopian tubes due to scarring produces sterility in some cases of gonorrhea.

Female infections with *N. gonorrhoeae* exhibit a wide variety of symptoms. Most commonly, symptomatic cases exhibit inflammation with some pain and swelling, abnormal

---

[3]PID can also be caused by other pathogens. *Chlamydia*, which cause nongonococcal PID, are most commonly the source of this disease.

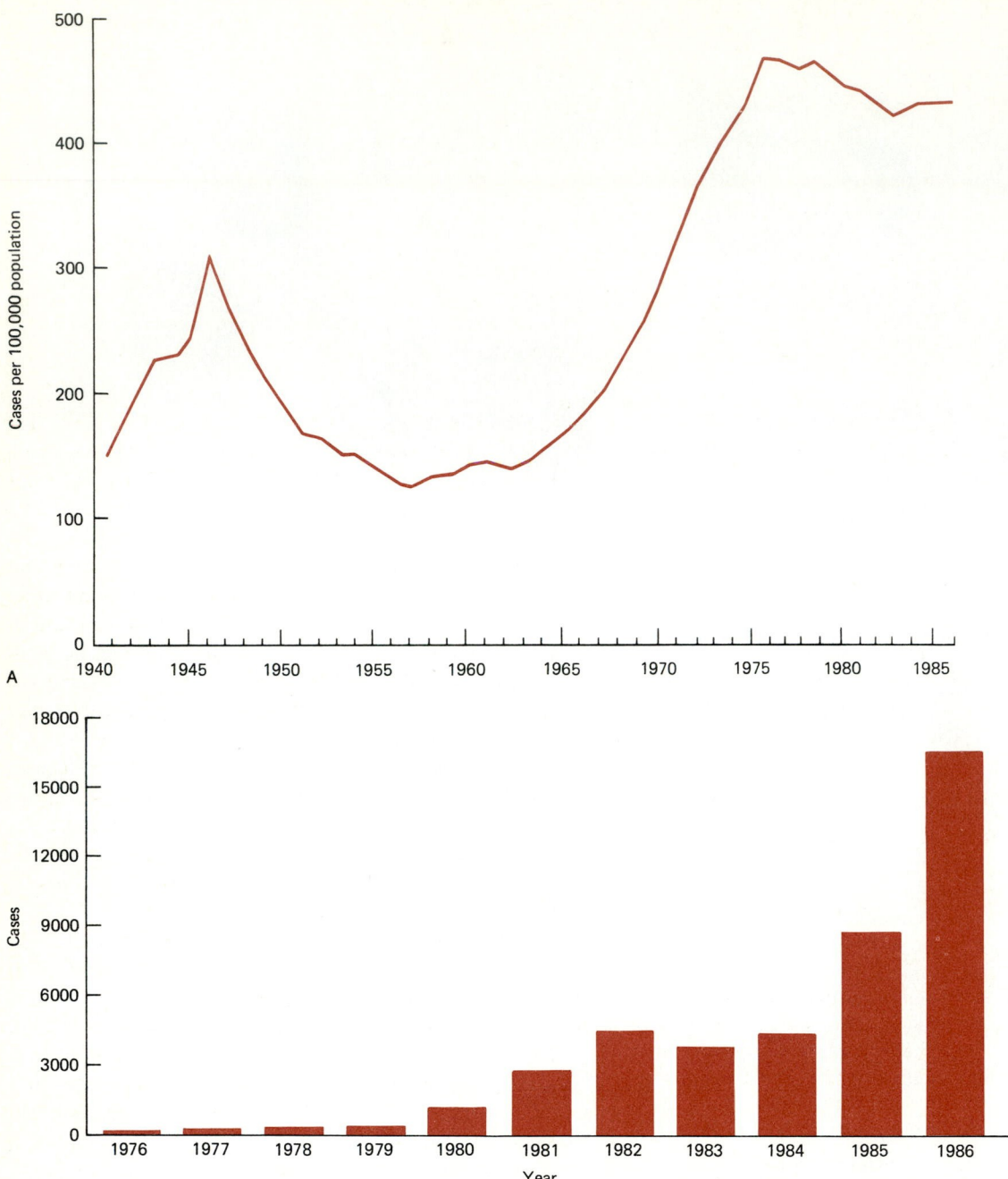

FIGURE 21.20

(A) The incidence of gonorrhea shows an alarming epidemic increase in recent years, as demonstrated by these reported civilian cases in the United States. Note the sharp decline from 1946 to 1956, followed by the dramatic increase since then, particularly between 1960 and 1975. The precise cause for this rise is unknown; some point to increased sexual activity in a permissive society; it seems likely that the use of oral contraceptives that lead to altered vaginal pH has played a role in the increased ability of *Neisseria gonorrhoeae* to survive, resulting in increased transmission rates of this disease. Since 1975 there has been a gradual decline in the incidence of gonorrhea in the United States. (B) Although the number of cases of gonorrhea has declined, the incidence of antibiotic resistant isolates, of *N. gonorrhoeae* has increased dramatically since 1975.

vaginal discharge, and abnormal menstrual bleeding. Because many women with gonorrhea are asymptomatic, the eyes of all infants are routinely washed immediately after birth with erythromycin sulfate to prevent infections of the eye, which could result from the transmission of *N. gonorrhoeae* from mother to infant during passage through the vaginal tract. *N. gonorrhoeae* can also be transmitted to the eyes of adults by rubbing them with hands contaminated with *N. gonorrhoeae*.

In men, gonorrhea results in a characteristic painful, purulent urethral discharge. The pus results from the migration of phagocytic leukocytes to the site of infection. Symptoms

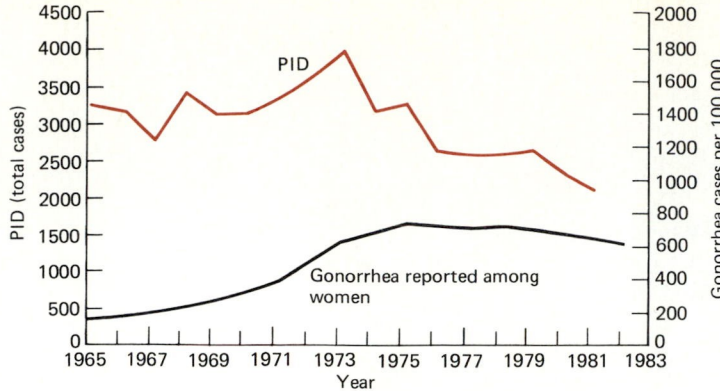

FIGURE 21.21

Total number of consultations for pelvic inflammatory disease (PID) and reported cases of gonorrhea among women per 100,000 population. (Based upon *Morbidity and Mortality Weekly Reports,* Centers for Disease Control, Atlanta)

of gonorrhea in males are usually apparent less than 1 week after infection. If the disease is untreated, occlusion of the vas deferens due to scarring produces sterility in some males.

Gonorrhea is readily treated with antibiotics, with penicillin being the drug of choice. Other antibiotics, such as tetracycline, are also effective. In recent years there has been an increase in the tolerance of *N. gonorrhoeae* to antibiotics, creating a major concern in the treatment of gonorrhea (Figure 21.22). The recent identification of β-lactamase–producing strains of *N. gonorrhoeae* in some cases may lead to a movement away from of penicillin in treating gonorrhea because such strains are resistant to most penicillins. With penicillin-resistant strains of *N. gonorrhaeae*, a third-generation cephalosporin is the antibiotic of choice. Because of the increased incidence of penicillin-resistant gonorrhea, the Centers for Disease Control in Atlanta may recommend that relatively high doses of ceftriaxone and equivalent third-generation cephalosporins become the standard treatment for most cases of gonorrhea.

Syphilis   **Syphilis** is a sexually transmitted disease caused by *Treponema pallidum.* This organism is a bacterial spiro-

FIGURE 21.22

Histogram showing the increased resistance of *Neisseria gonorrhoeae* to penicillin. The increasing occurrence of resistant strains coincides with increased use of antibiotics and reflects selective pressure that allows increased survival of the resistant strains at the expense of the susceptible strains.

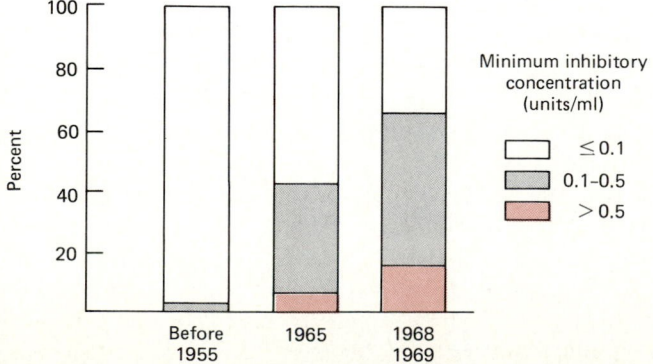

chete, fastidious in its growth requirements and readily killed by drying, heat, and disinfectants such as soap, arsenicals, and mercurial compounds. Historically, hot bath spas and arsenic- and mercury-containing compounds have been used in the treatment of syphilis. The inability of *T. pallidum* to survive for long outside the body makes transmission through inanimate objects (fomites) virtually nonexistent. Transmission depends on direct contact with the infective syphilitic lesions containing *T. pallidum*. *T. pallidum* enters the body via abrasions of the epithelium and by penetrating mucous membranes. The bacteria migrate to the lymphatic system shortly after penetrating the dermal layers.

Syphilis manifests itself in three distinct stages. During the primary stage, a **chancre** develops at the site of *Treponema* inoculation. Primary lesions generally occur on the genitalia. The average incubation period for the manifestation of primary syphilis is 21 days after infection. The primary lesions typically heal within 3–6 weeks, often giving the individual the false impression that the disease has been cured. The secondary stage of syphilis normally begins 6–8 weeks after the appearance of the primary chancre. During this stage, there are cutaneous lesions and lesions of the mucous membranes that contain infective *T. pallidum*. Lesions may appear on the lips, tongue, throat, penis, vagina, and numerous other body surfaces (Figure 21.23). There may be additional symptoms of systemic disease during this stage, such as headache, low-grade fever, and enlargement of the lymph nodes. After the secondary stage, syphilis enters a characteristic latent period during which there are no clinical symptoms of the disease. This phase can be detected by serological means because the Wassermann and similar tests are still positive. The latent phase marks the end of the infectious period of syphilis.

The tertiary phase of syphilis, also known as **late syphilis**, usually does not occur until years after the initial infection. During tertiary syphilis, damage can occur to any organ of the body. In about 10 percent of the cases of untreated syphilis, this phase involves the aorta, and damage to this major blood vessel can result in death. In approximately 8 percent of the cases of untreated syphilis, there is central nervous system involvement with a variety of neurological manifestations, including personality changes and paralysis.

If untreated, approximately 25 percent of the individuals

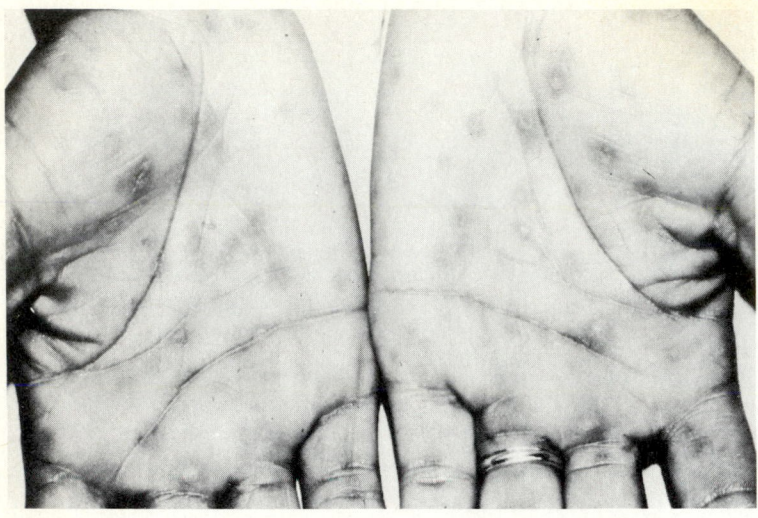

FIGURE 21.23
Secondary syphilis lesions on the palms of hands resulting from syphilis infection. (From BPS—Leonard Winograd, Stanford University)

who contract syphilis will suffer one or more relapses of the secondary stage during the first 4 years of illness, 15 percent will develop tertiary benign lesions, 10 percent cardiovascular lesions, and 8 percent central nervous system lesions. The risks of debilitating symptoms and death make this a very serious form of venereal disease. Fortunately, syphilis can be treated with penicillin and other antibiotics, particularly during the early stages. Unlike gonorrhea, the number of cases of syphilis in the United States has been relatively constant (Figure 21.24). There has been a large increase in reported cases of syphilis among homosexual men; for example, over 60 percent of the cases of syphilis reported in the state of Washington during 1971 involved homosexual transmission. As with other sexually transmitted diseases, the control of syphilis, depends on finding and treating all sexual contacts who may have contracted this disease and may be involved in its further transmission. No long-term immunity develops after infections with *T. pallidum*, and individuals who are cured by treatment with antibiotics remain susceptible.

In addition to sexually transmitted syphilis, *T. pallidum* can be transmitted across the placenta of pregnant women with syphilis, infecting the fetus and causing stillbirth or congenital syphilis in the newborn. Stillbirth is likely if pregnancy occurs during the primary or secondary stages of syphilis. Congenital syphilis is most likely during the latent period of the disease. Congenital syphilis has very serious consequences, usually resulting in mental retardation and

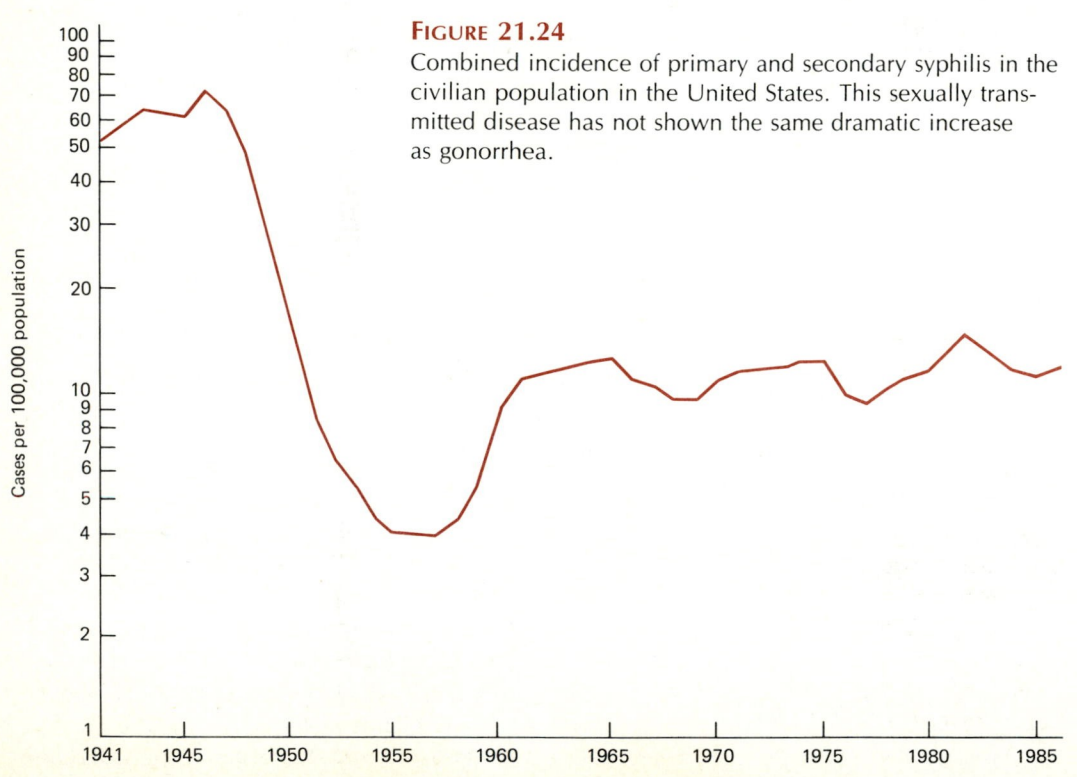

FIGURE 21.24
Combined incidence of primary and secondary syphilis in the civilian population in the United States. This sexually transmitted disease has not shown the same dramatic increase as gonorrhea.

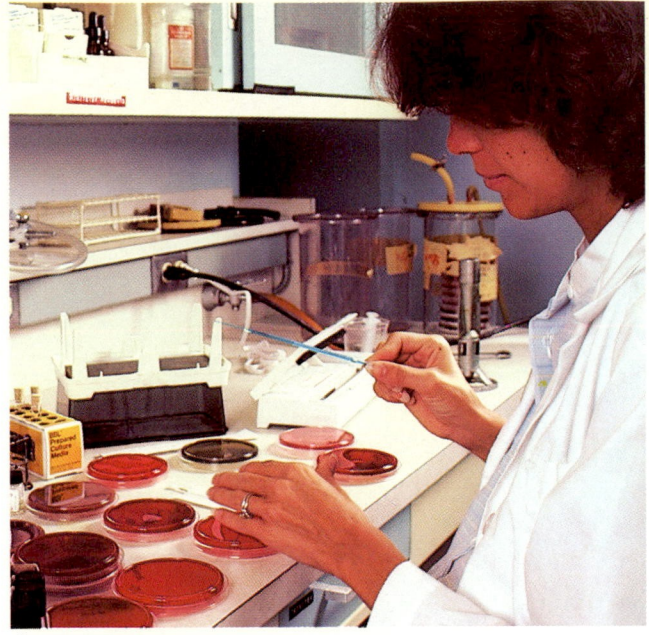

A

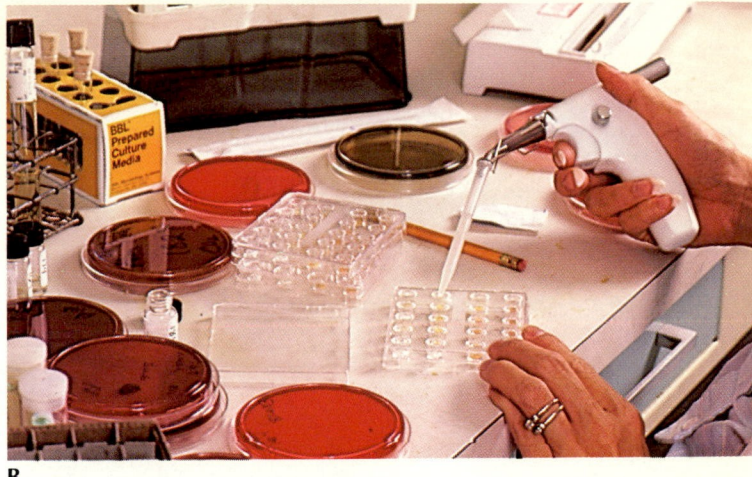

B

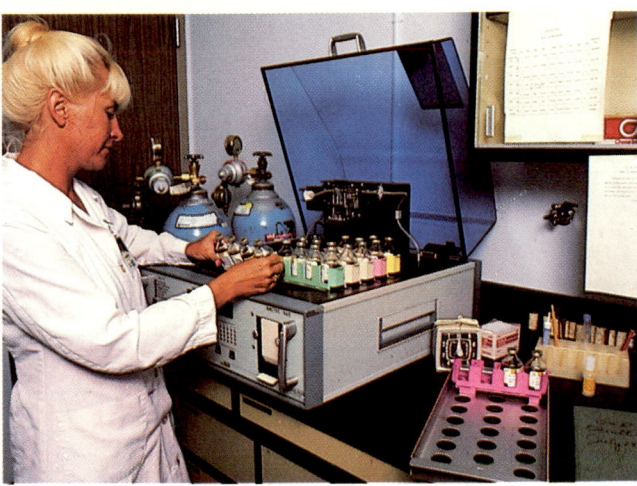

C

D

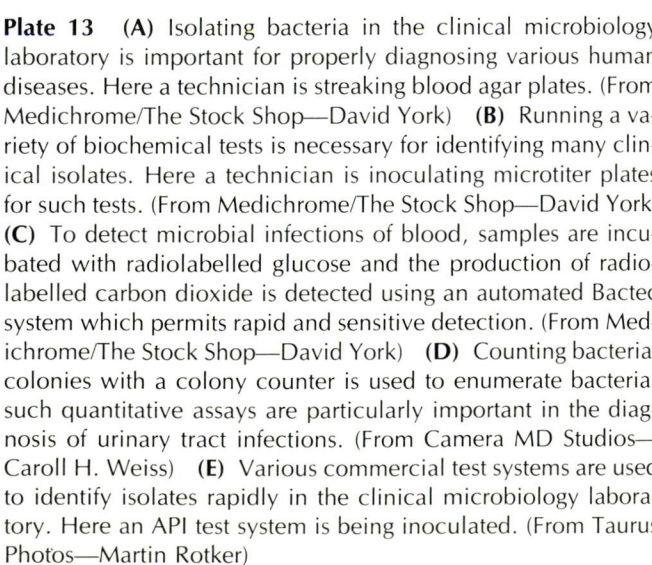

E

Plate 13 **(A)** Isolating bacteria in the clinical microbiology laboratory is important for properly diagnosing various human diseases. Here a technician is streaking blood agar plates. (From Medichrome/The Stock Shop—David York) **(B)** Running a variety of biochemical tests is necessary for identifying many clinical isolates. Here a technician is inoculating microtiter plates for such tests. (From Medichrome/The Stock Shop—David York) **(C)** To detect microbial infections of blood, samples are incubated with radiolabelled glucose and the production of radiolabelled carbon dioxide is detected using an automated Bactec system which permits rapid and sensitive detection. (From Medichrome/The Stock Shop—David York) **(D)** Counting bacterial colonies with a colony counter is used to enumerate bacteria; such quantitative assays are particularly important in the diagnosis of urinary tract infections. (From Camera MD Studios—Caroll H. Weiss) **(E)** Various commercial test systems are used to identify isolates rapidly in the clinical microbiology laboratory. Here an API test system is being inoculated. (From Taurus Photos—Martin Rotker)

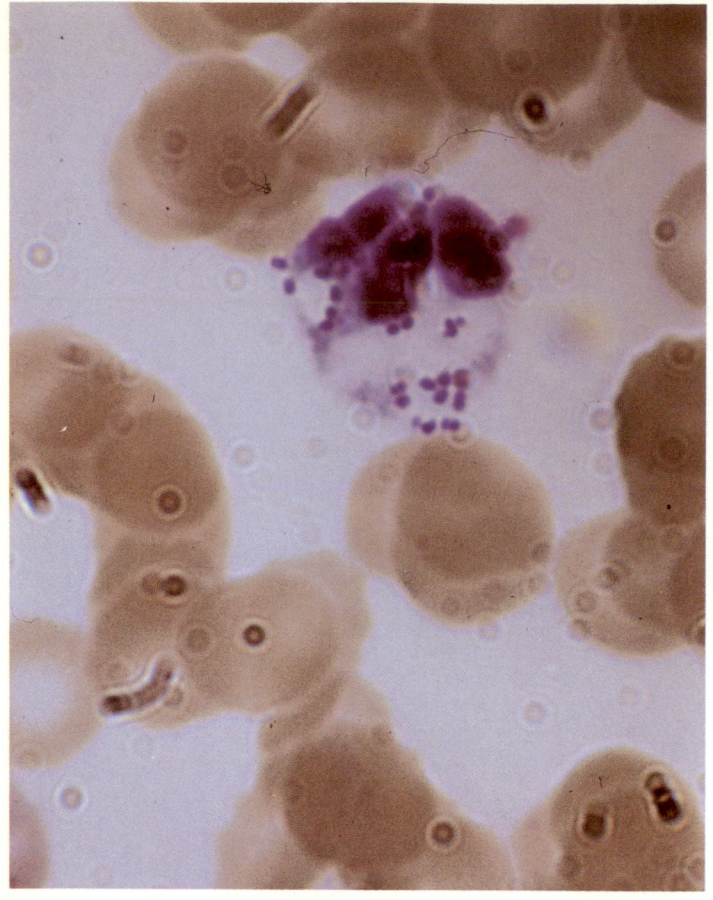

A

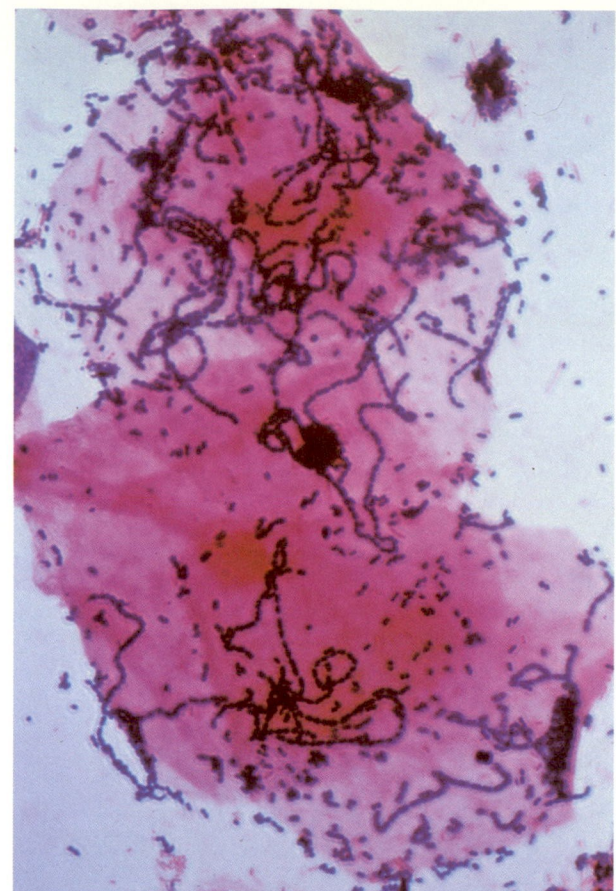

B

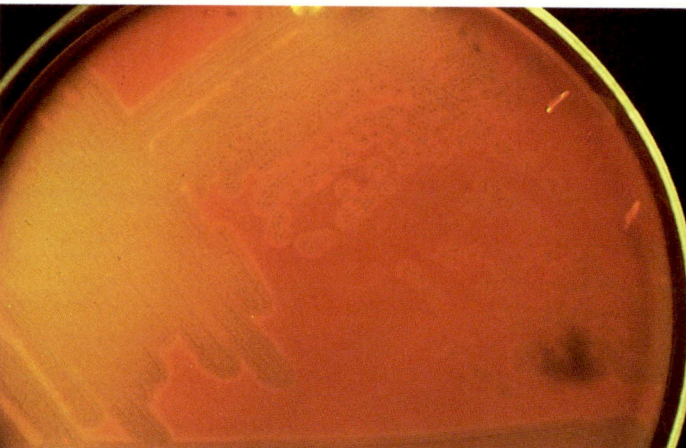

D

C

Plate 14 **(A)** Micrograph showing phagocytosis of *Streptococcus pneumoniae* by white blood cells. (From BPS—Leon LeBeau, University of Illinois Medical Center) **(B)** Micrograph of a throat smear showing streptococci (blue) and two epithelial cells (pink). (From BPS—Leon LeBeau, University of Illinois Medical Center) **(C)** A blood agar plate inoculated with a throat swab showing growth of a β-hemolytic streptococcus (small colonies surrounded by clear zones along the streak lines). (From BPS—R. L. Moore, BioTechniques Laboratories) **(D)** A blood agar plate inoculated with a culture of *Streptococcus pneumoniae* showing partial clearing (greening) around colonies indicative of α-hemolysis.

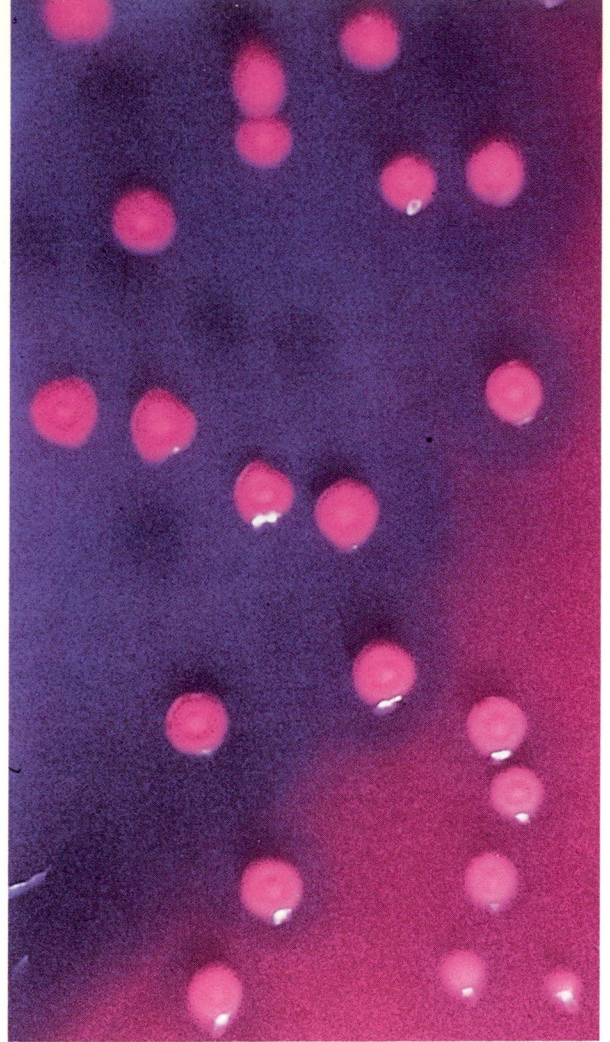

A

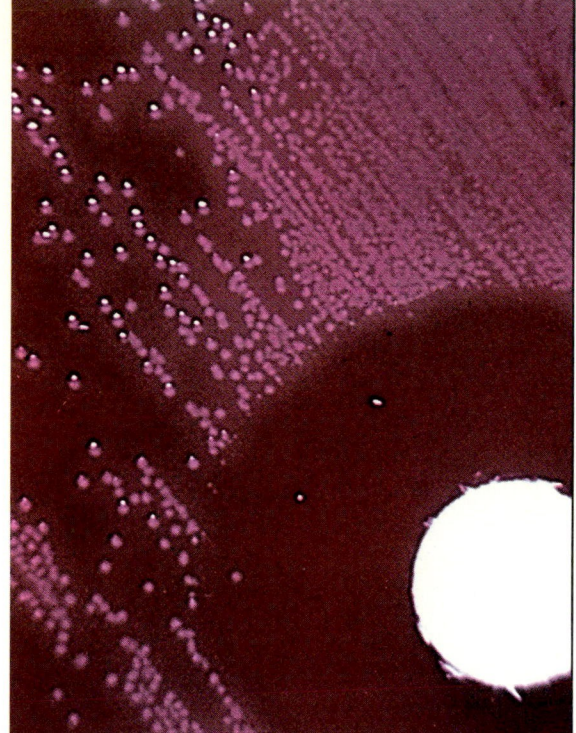

B

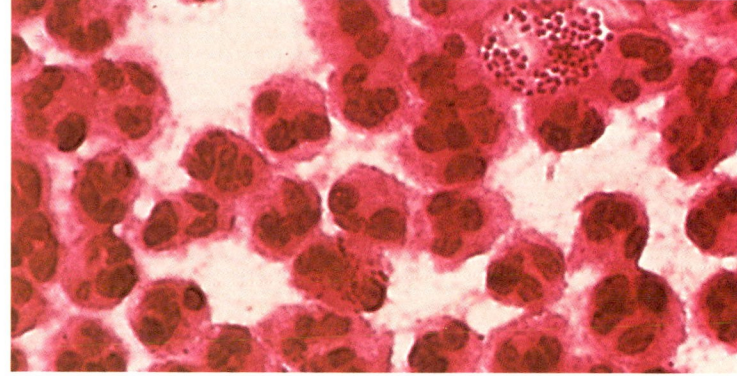

C

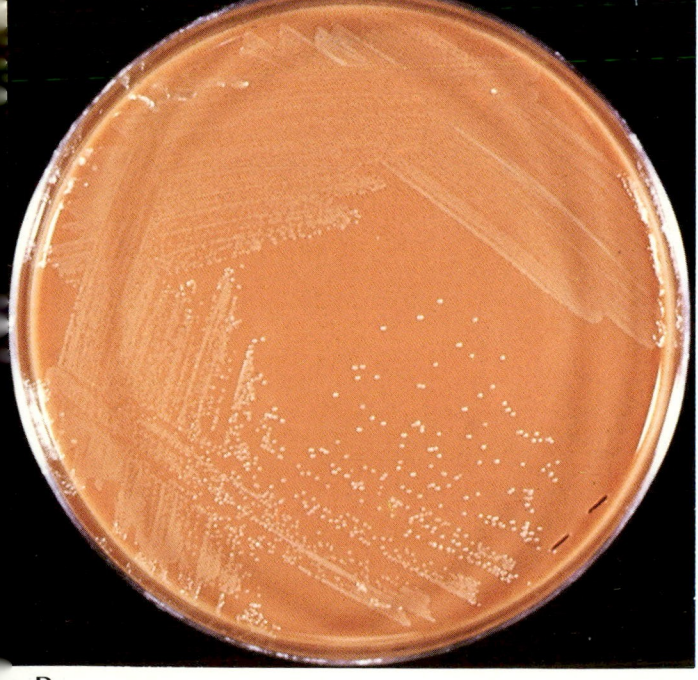

D

Plate 15 **(A)** Colonies of *Escherichia coli* growing on Mac-Conkey agar showing characteristic colony morphology and coloration. *E. Coli* frequently causes urinary tract infections and also is the most common cause of bacterial meningitis in children. (Courtesy James Snyder, Humana Hospital-University, Louisville, Kentucky) **(B)** Colonies of *Streptococcus pneumoniae* growing on blood agar from the cerebrospinal fluid in a case of bacterial meningitis. The bacterium shows α-hemolysis (partial clearing around colonies) and is inhibited by optochin (clear zone around white filter disk impregnated with optochin). (Courtesy James Snyder, Humana Hospital—University, Louisville, Kentucky) **(C)** *Neisseria gonorrhoeae* and white blood cells in pus from a case of gonorrhea. The bacteria can be seen as diplococci within one of the white blood cells. (Courtesy James Snyder, Humana Hospital—University, Louisville, Kentucky) **(D)** Culture of *Neisseria gonorrhoeae* growing on chocolate agar. (Courtesy James Snyder, Humana Hospital—University, Louisville, Kentucky.)

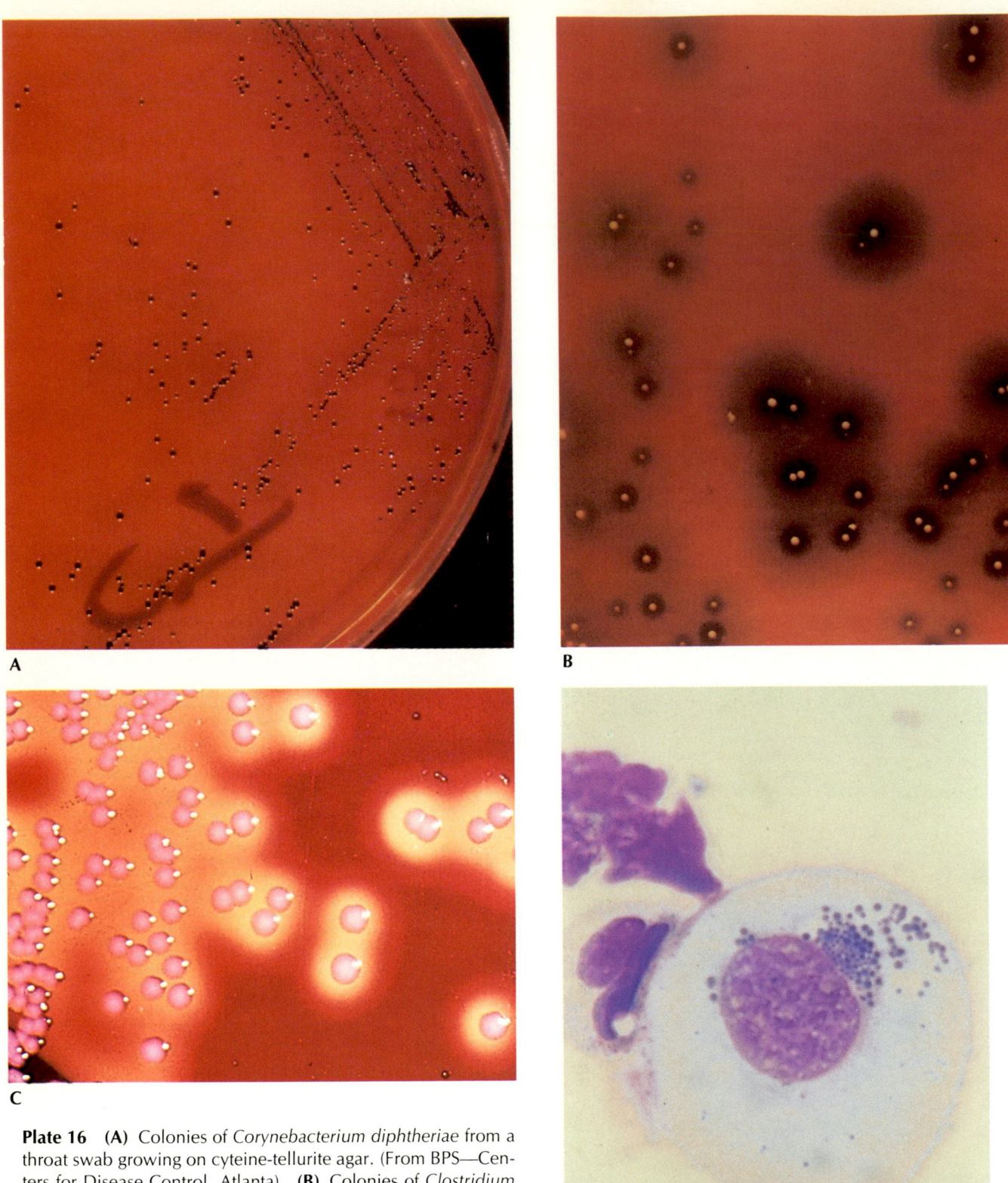

Plate 16 **(A)** Colonies of *Corynebacterium diphtheriae* from a throat swab growing on cyteine-tellurite agar. (From BPS—Centers for Disease Control, Atlanta) **(B)** Colonies of *Clostridium perfringens* growing on blood agar. (From BPS—Leon Lebeau, University of Illinois Medical Center) **(C)** Colonies of *Staphylococcus aureus* growing on blood agar showing characteristic morphology and β-hemolysis (clearing around colonies). (Courtesy James Snyder, Humana Hospital—University, Louisville, Kentucky) **(D)** Micrograph showing an inclusion body of *Chlamydia trachomatis* from a conjunctival scraping. (Courtesy James Snyder, Humana Hospital—University, Louisville, Kentucky)

neurological abnormalities in the infant; the probability of survival in such infants depends on the specific nature of the neurological impairment.

Nongonococcal Urethritis (NGU)

Nongonococcal urethritis is the term used to describe sexually transmitted diseases that result in inflammation of the urethra caused by bacteria other than *N. gonorrhoeae*. This disease is also called **nonspecific urethritis (NSU)**. It is estimated that between 4 and 9 million people in the United States have contracted it. Most cases are mild. Females often are asymptomatic and males usually notice some pain and discharge during urination. In serious cases, the inflammation associated with this condition can cause infertility. In males the epididymis may become inflamed, and in females the cervix or fallopian tubes may become blocked. Most cases of NGU appear to be caused by *Chlamydia trachomatis*, a small, obligately intracellular, parasitic bacterium. *Mycoplasma hominis* and *Ureaplasma urealyticum* also have frequently been reported to cause NGU. These two bacterial species lack cell walls. It has been difficult to diagnose the causes of NGU because of problems with culturing and identifying these bacteria. All of these bacteria are inhibited by antibiotics such as tetracyclines but not by penicillin. Compared to gonococcal urethritis, NGU has a longer incubation period. Individuals treated with penicillin for cases diagnosed as gonorrhea may later develop NGU if multiple bacteria were associated with the sexually transmitted disease.

C. trachomatis, like *N. gonorrhoeae*, can be transmitted during birth from the mother to the eyes of the newborn. *Chlamydia* infections of the eye can be serious. Because many females are asymptomatic, it is difficult to take selective measures to prevent this occurrence. Therefore, erythromycin is applied to the eyes of newborns shortly after birth to protect them against both *N. gonorrhoeae* and *C. trachomatis*. Before the prevalence of *C. trachomatis* was recognized, silver nitrate was used to treat the eyes of newborns. Silver nitrate inhibits *N. gonorrhoeae* but not *C. trachomatis*, whereas erythromycin is effective against both of these infectious agents.

Lymphogranuloma Venereum

Lymphogranuloma venereum is a more serious sexually transmitted disease caused by *Chlamydia trachomatis*. This disease is more common in the tropics than in the United States, where only a few hundred cases a year occur. In lymphogranuloma venereum, *C. trachomatis* enters the body through small abrasions in the genitourinary tract. The chlamydial cells are phagocytized and carried to lymph nodes. Lymph nodes are the primary sites affected in this disease, with swelling and tenderness usually occurring 5–21 days after healing of the primary lesions in the area of initial infection. In about 25 percent of the cases, there is a genitoanorectal syndrome characterized by a bloody, mucopurulent rectal discharge. Lymphogranuloma venereum can cause serious consequences, but these rarely occur. Although treatment with tetracycline and other antibiotics is usually effective in preventing the onset of late

and serious symptoms, it does not seem to shorten the usual time of required for enlarged lymph nodes to heal (4–6 weeks).

Chancroid

Chancroid is a relatively rare sexually transmitted disease caused by *Haemophilus ducreyi*. Chancroid occurs most frequently in the underdeveloped nations of Africa, the Caribbean, and Southeast Asia, but the incidence of this disease has been increasing in the United States. Soft chancres develop 3–5 days after sexual exposure, and untreated lesions may persist for months. Chancroidal ulcers heal quickly but often leave deep scars. Sulfanilamide and tetracycline are used in the treatment of this disease.

Granuloma Inguinale

Granuloma inguinale, also known as *granuloma venereum*, is caused by a Gram-negative, rod-shaped bacterium, *Calymmatobacterium granulomatis*. Granuloma inguinale most frequently occurs in India, the west coast of Africa, islands of the South Pacific, and some South American countries. The disease is relatively rare in the United States. It appears to be sexually transmitted, as evidenced by the fact that the initial lesions occur on the genitalia and appear 9–50 days after sexual intercourse with an infected individual. There appears to be a very low rate of infection with this sexually transmitted disease, and in many cases sexual partners do not contract it. The genitalia develop characteristic ulcers, older portions of which exhibit loss of pigmentation. Chloramphenicol, erythromycin, and tetracycline, as well as other antibiotics, are used in the treatment of this disease.

Genital Herpes

Genital herpes is usually caused by a herpes simplex type 2 virus.[4] Herpes simplex 2 is most frequently transmitted by sexual contact and causes infection of the genitalia. It is estimated that 20 million Americans now have genital herpes and that there will be at least half a million new cases per year unless effective means of controlling this disease are found. In women the primary site of herpes infection is the cervix, but it may also involve the vulva and vagina. In men the herpes implex virus frequently infects the penis. The primary infection includes symptoms of genital soreness and ulcers in the infected areas. The virus and manifestations of infection may be transmitted to other areas of the body, most notably the mouth and anus. Genital herpes may have particularly serious repercussions in pregnant women because the virus can be transmitted to the infant during vaginal delivery, causing damage to the infant's central nervous system and/or eyes. Herpes is lethal in up to 60 percent of infected newborns, and in surviving babies there is a 50 percent risk of blindness or neurological damage.

The ulcers produced by herpes simplex type 2 infection generally heal spontaneously in 10–14 days, but because of

---

[4]Diseases caused by herpes simplex type 1 virus will be discussed in the next chapter because herpes simplex type 1 does not appear to be as commonly transmitted during sexual intercourse, and the genitourinary tract need not be the primary portal of entry.

the budding mode of reproduction of herpes viruses, the infection is not eliminated when the ulcers heal; rather, a reservoir of infected cells remains within the nerve cells of the body. Later multiplication of the viruses can produce new ulcers, even in the absence of additional sexual activity. It is not known exactly what initiates subsequent attacks of herpes, but such recurrences may be triggered by sunlight, sexual activity, menstruation, and stress. The disease remains transmissible, which interferes with the establishment of stable sexual relationships; there are also many adverse psychological effects associated with genital herpes. Genital herpes disrupts marital relationships, and the epidemic outbreak may contribute to a reversal of the sexual revolution. Several of the newly developed antiviral drugs should be useful in the treatment of herpes viral infections. In particular, acyclovir is used to reduce the longevity of genital herpes lesions.

Genital Warts   Warts are benign tumors caused by infections with papilloma viruses. These viruses are transmitted by direct contact, normally infecting the skin and mucous membranes. Direct sexual contact usually is the source of the infecting papilloma viruses when **genital warts** occur. Warts generally do not appear for several weeks after infection. Chemical (e.g., acid) and physical (e.g., freezing) methods can be used to remove warts, but these benign tumors also disappear without treatment. There is increasing evidence that some of the papilloma viruses that cause genital warts can cause cancer. Previously it was thought that herpes viruses might lead to cervical cancer, but it now appears that adolescent females who have had extensive sexual contacts and have developed genital warts have an elevated rate of cervical cancer. Frequent Pap smears are suggested for such women.

Acquired Immune Deficiency Syndrome (AIDS)   We have already discussed **acquired immune deficiency syndrome** or **AIDS** in Chapter 19 when considering the diseases of the immune system. However, because of the importance of this disease and the fact that it is sexually transmissible, we will expand upon that earlier discussion.

HIV binds to a specific and selective subset of T cells, the T-4 helper/inducer cells, which are phenotypically defined by the presence of CD-4 molecules along their surfaces. The CD-4 molecule which is used by cells to recognize antigens in association with the major histocompatibility complex, appears also to be the receptor for the virus. HIV also invades other human cells, including monocytes and cells of the central nervous system, which may serve as reservoirs of the virus within the body.

HIV has eight identified genes, most of whose functions have been delineated. There is a gene coding for the core proteins, a gene for reverse transcriptase, an envelope gene for the outer coat, and also a gene for augmenting replication, called the *tat* gene. The *tat* gene product, which induces the expression of the genes both for interleukin-2 and its receptor,

is necessary for cell death, but it is unclear how HIV actually kills T4 cells. The control regions for expression of the viral genes are located at the ends of the genome in the long terminal repeats (LTRs), that is, long repetitive sequences of nucleotides. An early consequence of T-cell activation is the turning on of the gene coding for interleukin-2, a protein that stimulates T-cell division. The nucleotide sequences of the LTR control regions of HIV are similar to sequences in the control region of the interleukin-2 gene and also that of the gamma-interferon gene. Conceivably, the same factors that turn on interleukin-2 production when T cells are activated also activate HIV genomes. This could explain why repeated exposure to HIV or even infections with other microorganisms can trigger onset of AIDS since stimulating the cell-mediated immune response, that is, activating T cells, may also stimulate viral replication. Additional studies will reveal the precise mechanism by which HIV causes AIDS and will lead to better treatment and eventual prevention of this disease.

The incidence of AIDS has been increasing in epidemic fashion in the last few years (Figure 21.25). The scope of human immunodeficiency virus (HIV)-associated diseases includes the asymptomatic carrier state and AIDS-related complex (ARC), an infection with the virus without full-blown disease, but with symptoms including fatigue, malaise, anorexia, wasting, and in certain patients, lymphadenopathy. These symtoms can lead to severe ARC and even to death without the occurrence of the characteristic opportunistic infections of AIDS.

Certain groups have exhibited a particularly high incidence of AIDS (Table 21.5). These high-risk groups include homosexual men, drug abusers, and hemophiliacs.[5] Intravenous drug users are at risk because the virus can be transmitted via blood-contaminated hypodermic needles. The disease can also be transmitted by sexual contact, which is the main means of transmission among homosexuals within infected communities. Heterosexual transmission also occurs, and if AIDS develops during pregnancy, it can be transmitted through the placenta to the fetus. Unlike other sexually transmitted diseaes, it appears that prolonged or repeated exposure is necessary to cause AIDS. Although the virus has been detected in the saliva of infected individuals, direct transmission of the disease by kissing, airborne droplets, and eating utensils has not been demonstrated. AIDS itself is incurable at this time, but treatment with azidothymidine limits replication of the HIV virus and currently is the most hopeful, albeit experimental, treatment. Patients with this disease can be treated with passive immunization and antibiotics to protect them against life-threatening secondary infections.

Clearly, AIDS is a serious disease, and there have been some excessive public reactions. In some hospitals, staff members have refused to treat patients with AIDS. Establishments in San Francisco catering to the homosexual com-

---

[5]Haitians were initially considered a high-risk group, but this is no longer the case. It is not clear why Haitians exhibited a high incidence of AIDS when this disease was first discovered.

munity have been closed on the grounds that they constitute a public health hazard. These reactions are reminiscent of those in our dark past when people with leprosy were labeled lepers and driven from society. Until we fully understand how to treat and prevent the spread of AIDS, unfounded public reactions are likely to continue. By early 1987, there were approximately 31,000 cases of full-blown AIDS in the United States and according to the World Health Organization (WHO) about 40,000 to 55,000 cases worldwide, but these are clearly underestimates of the true incidence of AIDS. By 1987 there likely were at least 150,000 cases of ARC and, more importantly, probably between 1 and 2 million Americans who, although infected with HIV, were completely asymptomatic. This number has important implications when considering strategies to block progression from an asymptomatic carrier state to full-blown disease and in considering the future impact of the disease. If up to 2 million Americans are infected with HIV, 20 to 30% will likely develop the disease after 5 years, giving us 270,000 cumulative cases of AIDS by the end of 1991, with 74,000 cases in 1991 alone with a health care cost by that time of $8-20 billion.

FIGURE 21.25

Graph showing the incidence of AIDS in the United States. In 1986, approximately 86 percent of all reported cases of AIDS in the world were in the United States. Based upon the current rate of increase of this disease, it is predicted that by 1990, 1 out of every 10 adults in the United States will have been infected with the AIDS virus if the spread of this disease is not checked.

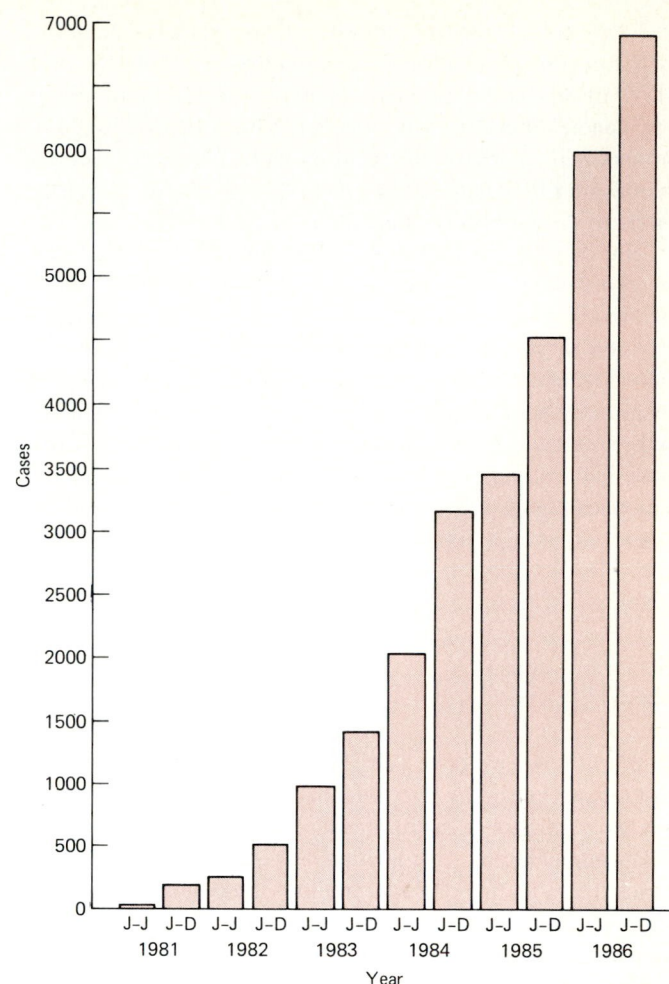

TABLE 21.5   AIDS Patients in the United States*

Patient Group	Before 1983	Total Number of Cases by Year (Percent of Total Cases for Each Year)			
		1983	1984	1985	1986
Adult Male					
Homosexual/bisexual	562 (66.9)	1,252 (60.6)	2,720 (64.2)	5,306 (65.3)	8,322 (64.8)
IV drug user	98 (11.5)	295 (14.2)	561 (13.2)	1,132 (13.9)	1,674 (13.0)
Both homosexual and IV drug user	74 (8.7)	194 (9.3)	396 (9.3)	576 (7.0)	925 (7.2)
Hemophilia—coagulation disorder	6 (0.7)	11 (0.5)	31 (0.7)	66 (0.8)	119 (0.9)
Heterosexual contacts	41 (4.8)	69 (3.3)	106 (2.5)	131 (1.6)	195 (1.5)
Transfusion recipients	1 (0.1)	14 (0.7)	28 (0.6)	96 (1.2)	185 (1.4)
Undetermined	16 (1.9)	51 (2.4)	81 (1.9)	158 (1.9)	342 (2.6)
Adult Female					
IV drug user	26 (3.0)	79 (3.8)	152 (3.6)	276 (3.4)	430 (3.3)
Hemophilia—coagulation disorder	0 (0.0)	0 (0.0)	2 (0.1)	2 (0.1)	3 (0.1)
Heterosexual contacts	16 (1.9)	32 (1.5)	60 (1.4)	131 (1.6)	275 (2.1)
Transfusion recipients	2 (0.2)	12 (0.6)	20 (0.5)	57 (0.7)	90 (0.7)
Undetermined	7 (0.8)	17 (0.8)	24 (0.5)	65 (0.8)	92 (0.7)
Pediatric	1 (0.1)	41 (1.9)	50 (1.2)	124 (1.5)	178 (1.4)

*Based on data from *Morbidity and Mortality Weekly Reports* (MMWR) issued by the Centers for Disease Control in Atlanta.

Even worse is the recognition that whereas less than 0.03% of the general population in the United States is infected with HIV, in central Africa approximately 10% and up to 18% of the general population is infected. Thus, 10 to 20% of all infants will be born to infected mothers. Moreover, because blood often is not screened, 10 to 20% of the daily transfusions in Africa will be contaminated with HIV.

The devastating impact of AIDS around the world, both in economic and human terms, has lead to major public health efforts to educate the public about AIDS and to identify and to publicize ways to halt the spread of this disease. Blood screening has been instituted for certain groups, such as military personnel and prison inmates, to identify individuals who have been exposed to HIV and quarantine procedures have sometimes been used to limit others from coming in contact with individuals who have a positive reaction. Some individuals now stockpile their own blood for use in later surgical procedures rather than risk exposure to blood that may contain HIV. The Surgeon General of the United States has recommended the institution of an extensive program of AIDS education in the public schools and that condoms be widely used to block the sexual spread of HIV. Advertisements now appear on television and in the lay media with the message that the proper use of condoms, that is, the practice of ''safe sex,'' is the way to limit the spread of AIDS. Either there will have to be a drastic change in sexual activities or a vaccine that protects against AIDS will have to be developed to halt the epidemic rise in deaths attributable to AIDS. We will undoubtedly eventually find methods of preventing and better ways of treating AIDS, as we have for most other infectious diseases but in the meantime curbing the spread of AIDS depends upon education, awareness, and self-protection.

Postlude

In this chapter we have considered a number of the diseases of humans caused by microbial pathogens that enter the body via the respiratory, gastrointestinal, and genitourinary tracts. The pathogens that cause human diseases are generally restricted to a particular portal of entry through which they can enter the body and establish an infection.

Airborne transmission, particularly in the form of droplet spread of aerosols containing pathogenic microorganisms from infected to susceptible individuals, is important in outbreaks of diseases caused by pathogenic microorganisms that enter through the respiratory tract. Crowded conditions favor airborne transmission of pathogens that enter the body this way. As a result, outbreaks of diseases caused by such pathogens often affect individuals who work or live together, such as children attending the same school.

Contaminated food and water supplies play an important role in disease transmission in cases where the pathogens or toxins produced by microorganisms enter the body through the gastrointestinal tract. Food poisoning often affects a number of individuals who have eaten together. Many times all members of a family or all patrons of a restaurant will be stricken after ingestion of contaminated food. There have been occasional outbreaks of foodborne infection or poisoning among the passengers on a ship or airplane who have been served the same meal. Major outbreaks of foodborne disease represent a frequent problem in institutions that serve food prepared in large quantities, such as prisons and nursing homes.

In the case of pathogens that enter the body through the genitourinary tract, contamination with microorganisms comprising the normal microbiota of the gastrointestinal tract and sexually transmitted pathogens are principally involved in such infections. The control of sexually transmitted diseases requires public health measures to identify the source of infection in order to interrupt the chain of transmission. Recognition that these diseases represent a medical, rather than a social, problem is important in effectively treating and reducing the incidence of these diseases.

Generally, infections that are localized near the point of entry are less serious than those that spread systemically. In many cases, microbial infections are self-limiting, and recovery usually occurs without serious complications. Some diseases, however, are associated with a high mortality rate if left untreated, but fortunately, antibiotics are effective in treating many of them. Control of microbial diseases depends on lowering the incidence of infection wherever possible by taking preventive measures. By knowing the etiological agents and the routes of transmission of specific diseases, public health measures can be taken to control many of them. These measures include vaccination, water treatment, sanitation improvement, and educational programs, as well as treatment of infected individuals to reduce the reservoirs of pathogenic microorganisms. The effective elimination of smallpox as a disease of humans can be repeated for several other diseases, and in almost all cases more effective treatment and preventive measures can reduce the incidence of infectious diseases. We should remember that the rational treatment of infectious diseases has existed for only a few decades. We have made tremendous advances in the post–World War II era in reducing the incidence of infectious diseases and the fatalities due to pathogenic microorganisms. We have the medical means to reduce the destructiveness of many diseases in many areas of the world but lack the determination to mobilize the resources necessary to do so. We need to apply our efforts to preventing and/or developing effective treatment methods for the various remaining infectious diseases.

Suggested Supplementary Readings

BOYD, R. F., and B. G. HOERL. 1981. *Basic Medical Microbiology*. Little, Brown and Co., Boston.

BOYD, R. F., and J. J. MARR. 1980. *Medical Microbiology*. Little, Brown and Co., Boston.

BRAUDE, A. I. (ed.). 1981. *Medical Microbiology and Infectious Disease*. W.B. Saunders Co., Philadelphia.

CHESNEY, P. J., M. S. BERGDOLL, J. P. DAVIS, and J. M. VERGERONT. 1984. The disease spectrum, epidemiology, and etiology of toxic-shock syndrome. *Annual Review of Microbiology* 38:315–338.

CURRAN, J. W., W. M. MORGAN, A. M. HARDY, H. W. JAFFE, and W. R. DOWDLE. 1985. The epidemiology of AIDS: Current status and future prospects. *Science* 229:1352–1357.

DAVIS, B. D., R. DULBECCO, H. N. EISEN, H. S. GINSBERG, and W. B. WOOD. 1981. *Microbiology*. Harper & Row, Publishers, Inc., Hagerstown, Md.

DUGUID, J. P. (ed.). 1979. *Medical Microbiology*. Churchill Livingstone Inc., New York.

EBBERSON, P., R. J. BIGGAR, and M. MELBYE. 1984. *AIDS: A Basic Guide for Clinitians*. Munksgaard, Copenhagen.

EMMONS, C. W., C. H. BINFORD, J. P. ULZ, and K. J. KWON-CHUNG. 1977. *Medical Mycology*. Lea & Febiger, Philadelphia.

FITZGERALD, T. J. 1981. Pathogenesis and immunology of *Treponema pallidum*. *Annual Review of Microbiology* 35:29–54.

GALLO, R. C. 1986. The first human retrovirus. *Scientific American* 255(6):88–101.

GALLO, R. C. 1987. The AIDS virus. *Scientific American* 256(1):46–57.

GILBERT, D. N., and J. P. SANFORD (eds.). 1979. *Infectious Diseases*. Grune and Stratton, New York.

HOEPRICH, P. D. (ed.). 1983. *Infectious Diseases: A Modern Treatise of Infection Processes*. Harper & Row, Publishers, Inc., Hagerstown, Md.

HOLMES, K. K., P-A, MARDH, P. F. SPARLING, and P. J. WIESNER. 1984. *Sexually Transmitted Diseases*. McGraw-Hill Book Co., New York.

JAWETZ, E., J. L. MELNICK, and E. A. ADELBERG. 1986. *A Review of Medical Microbiology*. Appleton and Lange, New York.

JOKLIK, W. K., H. P. WILLETT, and D. B. AMOS (eds.). 1984. *Zinsser Microbiology*. Appleton-Century-Crofts, New York.

MANDELL, G., R. G. DOUGLAS, JR., and J. E. BENNETT. 1985. *Principles and Practice of Infectious Diseases*. John Wiley & Sons, Inc., New York.

PECHERE, J. C. (ed.). 1984. *Infections: Recognition, Understanding, Treatment*. Lea & Febiger, Philadelphia.

REESE, R. E., and R. G. DOUGLAS, Jr. 1986. *A Practical Approach to Infectious Diseases*. Little, Brown and Co., Boston.

RIEMANN, H., and F. L. BRYAN (eds.). 1979. *Food-Borne Infections and Intoxications*. Academic Press, New York.

ROBERTS, R. B. 1986. *Infectious Diseases: Pathogenesis, Diagnosis, and Therapy*. Year Book Medical Publishers, Inc., Chicago.

STUART-HARRIS, C. 1981. The epidemiology and prevention of influenza. *American Scientist* 69:166–172.

TOP, F. H., and P. WEHRLE. 1976. *Communicable and Infectious Diseases*. C.V. Mosby Co., St. Louis.

VOLK, W. A., D. C. BENJAMIN, R. J. KADNER, and J. T. PARSONS. 1986. *Essentials of Medical Microbiology*. J.B. Lippincott Co., Philadelphia.

WILSON, G. S., and A. A. MILES (eds.). 1975. *Topley and Wilson's Principles of Bacteriology and Immunity*. Williams & Wilkins, Baltimore.

YOUMANS, G. P., P. Y. PATERSON, and H. M. SOMMERS. 1980. *The Biological and Clinical Basis of Infectious Disease*. W.B. Saunders Co., Philadelphia.

Study Questions

1. What is a portal of entry?
2. How are pathogens that enter the body through the gastrointestinal tract generally transmitted? How do we control the transmission of these pathogens?
3. How are pathogens that enter the body through the respiratory tract generally transmitted? How do we control the transmission of these pathogens?
4. What is the difference between a foodborne infection and food poisoning (intoxication)? How is this difference reflected in the way we control different types of foodborne disease?
5. Name three viral, three bacterial, and three protozoan disease-causing organisms that enter the body through the respiratory tract.
6. For each of the following diseases, what are the causative organisms and characteristic symptoms:
   a. Influenza
   b. Tuberculosis
   c. Rheumatic fever
   d. Histoplasmosis
   e. Poliomyelitis
   f. Legionnaire's disease
   g. Whooping cough
   h. Typhoid fever
   i. Botulism
   j. Cholera
7. What organisms cause meningitis?
8. What organisms cause urinary tract infections?
9. What organisms normally cause vaginitis?
10. What is a sexually transmitted disease? Give some examples of these diseases and their causative organisms. Why is it difficult to control the spread of these diseases, even though many of them are readily treated with antibiotics?
11. Name five diseases that are routinely prevented by the use of prophylactic immunization.

Situational Problem

Trying to Diagnose Diseases

In order to help finance your college education, you have a part-time job in a physician's office. Your job includes responsibility for gathering information about the patient that may aid in making a diagnosis. Because you are a premed major, you decide to find out whether you can diagnose the disease based on the information you acquire. You keep a private record of your presumed diagnoses and later check them against the diagnoses made by the physician.

1. The first patient of the day is a 12-year-old boy who suddenly developed localized severe pain on the right side of the abdomen. At a party the previous night, he ate 12 hot dogs and various other foods. He is currently exhibiting nausea and vomiting. He weighs 130 pounds. He has a temperature of 101°F. Examination of a blood sample shows an elevated white blood cell count. Based on this information, what disease would you suspect?

2. The second patient is a 30-year-old woman who has been experiencing a series of upper respiratory tract infections. She is diabetic. Last month she had a case of pneumonia that was diagnosed as caused by *Streptococcus pneumoniae*. She was treated with a third-generation penicillin and recovered fully. Now she again has pneumonia, but this time the clinical laboratory has diagnosed the causative agent as *Pneumocystis carinii*. She is being treated with metronidazole. What underlying disease do you suspect?

3. Next, you answer a phone call from one of the doctor's regular patients. The patient informs you that his entire family had gone to the beach for a summer picnic. After swimming for some time they had lunch, which included chicken salad, potato salad, lemonade, and apple pie. A few hours later, they dug up some clams, which they ate raw. By the time they reached home, they all had abdominal pain and the children were vomiting. No member of the group had a fever. What disease would you suspect?

4. The next patient to enter the office is an 18-year-old freshman at Wisconsin University who has just come home for summer vacation. Shortly before finals for the spring semester, she noticed that she was tired and had little energy. She had been having an active social life but still had a high B average. She has blond hair, blue eyes, is 5 feet 5 inches tall, and weighs 124 pounds. She tells you that she has been feeling slightly feverish each evening but that the feeling always disappears by morning. Just before she left school for vacation, the university's health clinic took a blood sample and told her to have her family physician call for the test results. You call the clinic and are told that the test revealed an elevated white blood cell count and the presence of abnormal white blood cells. What disease would you suspect?

5. A female patient has several painful lesions on her genitals. She has been sexually active. She tells you that this is not the first time she has had such lesions. At each occurrence, they have healed within a few weeks, but shortly thereafter new lesions have appeared in the same region. She does not now, nor has she previously had, lesions elsewhere on her body. The process of healing and recurrence has occurred every few weeks over the past year, and she has finally decided to see a physician about this condition. What disease would you suspect?

CHAPTER 22

Human Diseases Part 2: The Skin as a Portal of Entry and Diseases of Superficial Body Tissues

Relatively few microorganisms are able to invade the body through the skin; therefore, the skin normally acts as an effective barrier to microbial infection. If, however, the skin tissues are mechanically interrupted as a result of wounds, burns, animal bites, or surgical procedures, a portal of entry is opened through which pathogenic microorganisms may enter the body. The process of breaking the skin surface not only provides a portal of entry for potentially pathogenic microorganisms but also often inoculates microorganisms directly into the circulatory system and inner body tissues. For example, when a child scrapes a hand or knee on the ground, the wound is frequently contaminated with dirt and associated microorganisms. Even a puncture wound with a sterile hypodermic syringe can pick up microorganisms from the vicinity of the puncture and carry them through the skin surface. The routine cleansing of wounds and the use of topical antiseptics after minor skin punctures and abrasions are accepted prophylactic measures taken to prevent the establishment of infections. Many animals are carriers of microorganisms, and their bites simultaneously disrupt the skin barrier and inoculate the wound with microorganisms whose pathogenic potential may be life-threatening. Arthropods, in particular, commonly act as vectors of some very dangerous human pathogens, and their bites can establish serious infections.

Paradoxically, although the superficial tissues of the body serve as barriers to penetration by pathogenic microorganisms, sometimes they themselves are also subject to infection. The eyes, for example, are normally protected from infections by the antimicrobial activities of tears, but some microorganisms can establish infections of the eye. The skin is extremely resistant to microbial infection, but some microorganisms nevertheless manage to establish cutaneous infections. The oral cavity, including the tooth surfaces and gingiva, is colonized by a large resident microbiota that, under particular conditions, cause diseases within the oral cavity, such as dental caries and periodontal diseases. Additionally, a limited number of microorganisms are able to penetrate the skin, and direct contact between these microorganisms and the skin surface can initiate a disease syndrome.

22.1 Diseases Transmitted Through Animal Bites

Many human infections that are transmitted via animal bites have lower animals as reservoirs, that is, infections in lower animal populations maintain the pathogen population. These diseases, termed **zoonoses**, are defined as infectious diseases of lower animals that are transmissible to humans. In these instances, the lower animal populations and humans act as alternative hosts for the proliferation of the pathogens. In considering the transmission of such diseases, it is important to examine the **reservoirs** and **vectors** of the infectious agents, as well as the nature of the specific etiological agent (Table 22.1). The prevention of infectious diseases that enter the body through animal bites often involves controlling in-

TABLE 22.1   Representative Diseases of Humans Transmitted by Arthropod Bites

Disease	Etiological Agent	Biological Vector	Reservoir
Yellow fever	Yellow fever virus	Mosquito (*Aedes aegypti*, *Haemagogus* spp.)	Humans, monkeys
Dengue fever	Dengue fever virus	Mosquito (*Aedes* spp., *Armigeres obturbans*)	Humans
Eastern equine encephalitis	Encephalitis viruses	Mosquito (*Aedes* spp., *Culex* spp., *Mansonia titillans*)	Humans, horses, birds
Colorado tick fever	Colorado tick fever virus	Wood ticks (*Dermacentor andersoni*)	Golden mantle, ground squirrel
Plague	*Yersinia pestis*	Rodent fleas (*Xenopsylla cheopis*), Human fleas (*Pulex irritans*)	Rodents (rats)
Tularemia	*Francisella tularensis*	Ticks (*Dermacentor* spp., *Amblyomma* spp.), Deerflies (*Chrysops discalis*)	Rodents, ticks
Rocky Mountain spotted fever	*Rickettsia rickettsii*	Ticks (*Dermacentor* spp., *Amblyomma* spp., *Ornithodoros* spp., etc.)	Rodents
Endemic typhus fever	*Rickettsia typhi*	Fleas (*Xenopsylla cheopis* and others)	Humans
Relapsing fever	*Borrelia recurrentis* and other species	Body louse (*Pediculus humanus*)	Humans, ticks
Chagas' disease	*Trypanosoma cruzi*	Cone-nosed bugs (*Triatoma* spp., *Panstronglyus* spp., *Rhodnius* spp.)	Dogs, cats, oppossums, rats, armadillos
African trypanosomiasis (sleeping sickness)	*Trypanosoma gambiense*; *T. rhodesiense*	Tsetse flies (*Glossina* spp.)	Humans, wild mammals
Malaria	*Plasmodium vivax*, *P. malariae*, *P. falciparum*, *P. ovale*	Mosquito	Humans
Leishmaniasis	*Leishmania donovani*; *L. tropica*; *L. braziliensis*	Sandflies (*Phlebotomus* spp.)	Dogs, foxes, rats, mice, two-toed sloth, gerbils, humans

fected reservoir and carrier animal populations. For example, mosquito control programs are employed to prevent the spread of many arthropod-borne diseases such as yellow fever and malaria. It should be remembered, however, that even though animals play a critical role in the transmission of such diseases, it is the viruses, bacteria, or protozoa that are the actual etiological agents.

Diseases Caused by Viral Pathogens

Rabies   **Rabies** is principally a disease of carnivorous animals other than humans. The rabies virus, a bullet-shaped, single-stranded RNA virus, can be transmitted to people through the bite of an infected animal (Figure 22.1). In urban settings, dogs are most frequently involved in the transmission of rabies. Wild animals that transmit rabies include foxes, skunks, jackals, mongooses, squirrels, raccoons, coyotes, badgers, and vampire bats (Figure 22.2). A new live vaccine has been developed in France for the introduction of rabies viral antigens. Hence, it may be possible to establish immunity in wild animal populations. The new vaccine is genetically engineered using the smallpox vaccine (vaccinia virus) as a vector. Rabies viruses multiply within the salivary glands of infected animals and normally enter humans in the animal's saliva through the portal of entry established by the

animal's bite. The rabies virus is not able to penetrate the skin by itself, and deposition of infected saliva on intact skin does not necessarily result in transmission of the disease. Transmission from bats can also occur via the respiratory tract by the inhalation of aerosols formed in the atmosphere around dense populations of infected bats.

When rabies viruses enter through an animal bite, they are normally deposited within muscle tissues, where they subsequently multiply. The rabies viruses reach peripheral nerve endings and migrate to the central nervous system. Cytoplasmic inclusion bodies, known as **Negri bodies**, develop within the neurons of the brain. Multiplication of rabies viruses within the nervous system causes a number of abnormalities, which manifest as the symptoms of this disease. The initial symptoms of rabies include anxiety, irritability, depression, and sensitivity to light and sound. These symptoms are followed by the development of hydrophobia (fear of water) because of difficulty in swallowing. As the infection progresses, there is paralysis, coma, and death.

Once the clinical symptoms of rabies begin, the disease is considered to be invariably fatal; therefore, treatment of rabies requires vaccination before the symptoms become manifest. The rabies vaccine stimulates antibody synthesis that prevents proliferation of the virus before it can cause irreversible damage to the central nervous system. It is critical

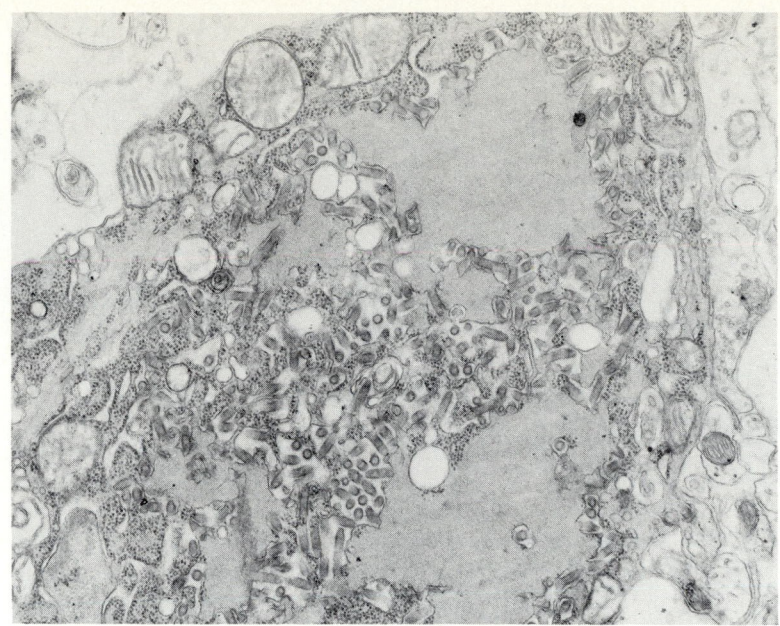

FIGURE 22.1

Micrograph of rabies virus from the neuron of an inoculated mouse (30,500×). Note how the cell is crowded with virus particles. (From BPS— A. Harrison, Centers for Disease Control, Atlanta)

to examine the animal that has bitten a human for the presence of rabies virus, to diagnose the disease in the animal, and to determine if vaccination is necessary. Frequently, the public is asked to help identify and locate a dog that has bitten someone in order to determine if it is necessary to carry out the vaccination procedure. Only if the animal is conclusively shown not to be infected with rabies can the immunization procedure be safely omitted for the bitten individual.

Rabies is one of the few diseases in which active immunization is used as a treatment after suspected infection. The vaccination procedure employed for many years used a vaccine prepared from rabies virus propagated in embryonated duck eggs and inactivated by β-propiolactone. The treatment involved 21 daily injections followed by booster inoculations 10 and 20 days later. Today the vaccine is produced in tissue culture. The new vaccine requires only a few intramuscular injections, which are administered together with rabies im-

mune globulin. The new rabies vaccine has a higher concentration of the necessary antigens for eliciting an immune response, and only three injections over a 7-day period are required to establish immunity.

Yellow Fever   Yellow fever, caused by a small RNA *Flavivirus* of the family Togaviridae, is transmitted by mosquito vectors, predominantly *Aedes aegypti*. There are two epidemiological patterns of transmission. Urban transmission involves vector transfer by *A. aegypti* from an infected to a susceptible individual, and jungle yellow fever normally involves transmission by mosquito vectors among monkeys, with transfer via mosquito vectors to humans representing an occasional deviation from the normal transmission cycle (Figure 22.3). Outbreaks of yellow fever were a major problem in the construction of the Panama Canal, leading to Walter Reed's instrumental work in establishing the relationship between yellow fever and mosquito vectors in 1901. Today

FIGURE 22.2

Incidence of rabies in wild and domestic animals in the United States.

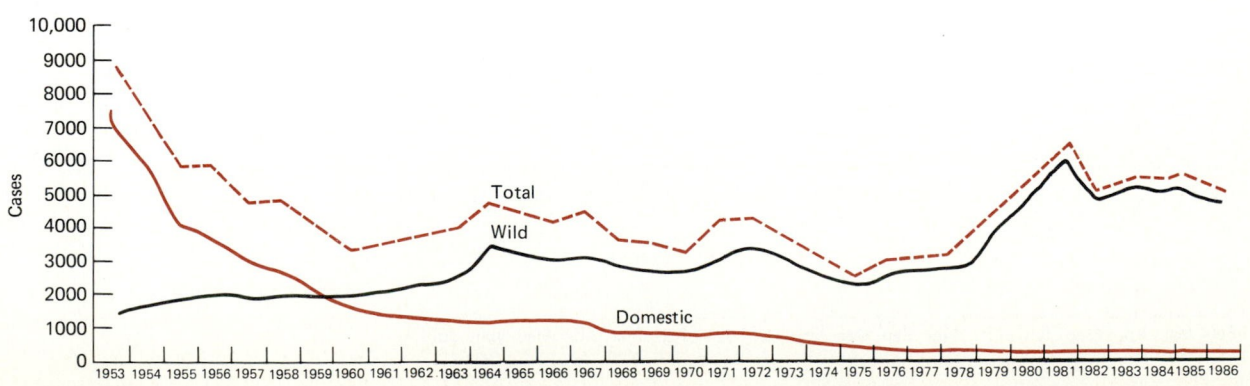

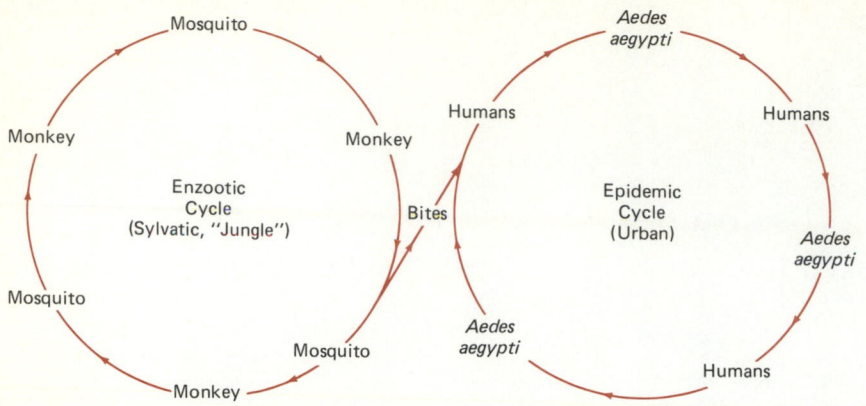

FIGURE 22.3
Transmission cycle patterns for yellow fever, showing the relationship between enzootic and epidemic transmission cycles of the disease.

yellow fever occurs primarily in remote tropical regions, and current outbreaks take place primarily in Central America, South America, the Caribbean, and Africa.

The onset of yellow fever is marked by **anorexia** (loss of appetite), nausea, vomiting, and fever. The multiplication of the virus results in liver damage, causing the **jaundice** from which the disease derives its name. The symptoms generally last for 1 week, after which either recovery begins or death occurs. The mortality rate for yellow fever is about 5 percent. There is no effective antiviral drug at present for treating the disease; however, it can be prevented by vaccination, using the 17D strain of yellow fever virus. The urban form of transmission has been largely controlled by effective mosquito eradication programs, but the jungle form of transmission cannot easily be interrupted because of the large natural reservoir of yellow fever viruses maintained within monkey populations.

Dengue Fever   **Dengue fever** also is caused by a *Flavivirus* transmitted by the mosquito *Aedes aegypti*. Like other members of the family Togaviridae, the dengue virus multiplies within the cytoplasm. Outbreaks of dengue fever occur in the Caribbean, South Pacific, and Southeast Asia (Figure 22.4). There are several different serotypes of dengue virus and three distinct disease syndromes: classic, hemorrhagic, and shock. A characteristic rash and fever develop during all forms of this disease. The classic form of dengue is a self-limiting disease, but in the hemorrhagic and shock syndromes the mortality rate can reach 8 percent in children.

The dengue virus replicates within cells of the circulatory system, causing **viremia** (viral infection of the bloodstream), which persists for 1–3 days during the febrile period. Multiplication of the dengue virus within the circulatory system causes vascular damage. It has been postulated that immune reactions contribute to the formation of complexes that ini-

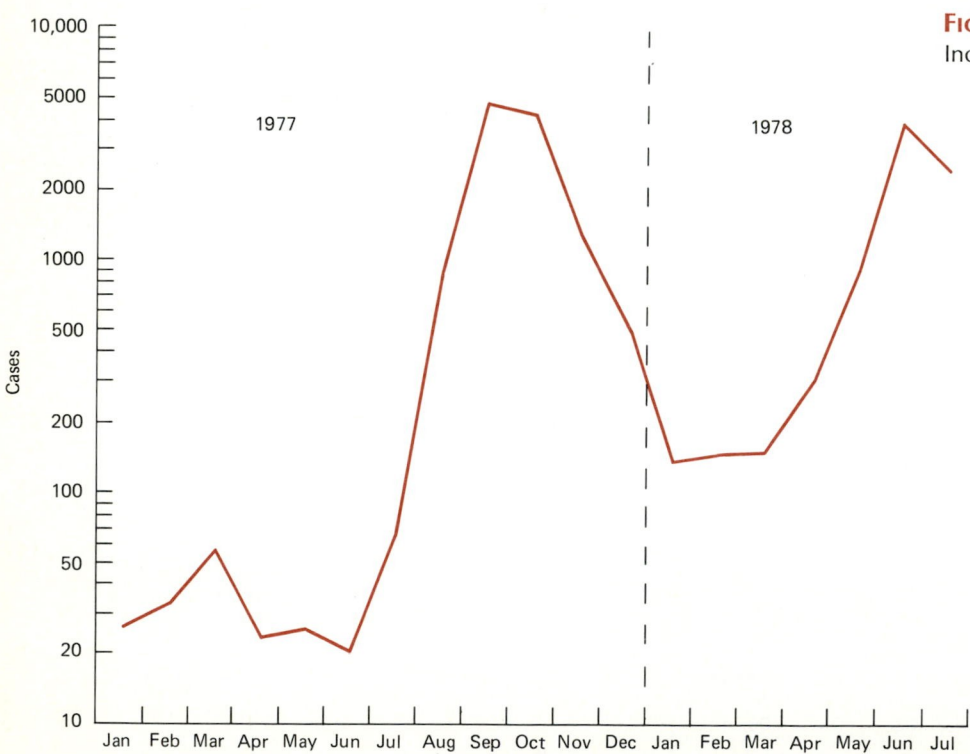

FIGURE 22.4
Incidence of dengue fever in Puerto Rico.

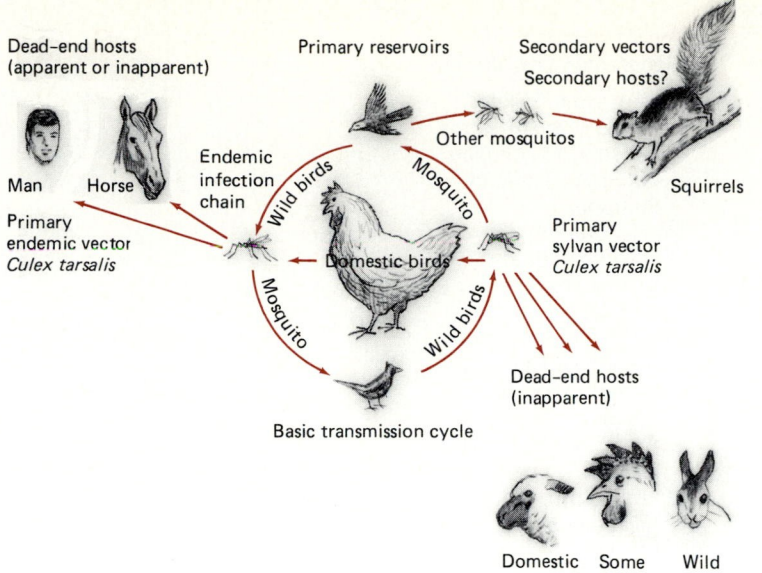

FIGURE 22.5
Epidemiological transmission of equine encephalitis. The chains of transmission for rural St. Louis encephalitis are similar, except that horses are inapparent, rather than apparent, hosts. Equine encephalitis infections also have a similar summer infection chain, but a few significant differences exist: the identity of the vector infecting humans is unknown; domestic birds do not appear to be a significant link in the chain; and it has a bird-to-bird secondary cycle in pheasants, whose role is unclear.

tiate intravascular coagulation (blood clotting within the blood vessels) or hemorrhagic lesions. Previous exposure to dengue virus and the presence of a cross-reacting antibody seem to be important in determining the severity of the disease symptomatology. Damage to the circulatory vessels appears to occur when antigen–antibody complexes activate the complement system, with the release of vasoactive compounds.

Encephalitis   **Encephalitis**, a disease syndrome defined by an inflammation of the brain, can be caused by various viruses, many of which are arthropod borne and belong to the family Togaviridae. Viruses capable of causing encephalitis

in humans are maintained in populations of various vertebrates, particularly birds and rodents, as well as populations of arthropods. Transmission to humans, via an arthropod vector in which the virus has multiplied, represents a dead end in the transmission cycle (Figure 22.5). Viral multiplication within the arthropod initially occurs in the gut, followed by dissemination through the hemolymph and multiplication within the salivary glands. Viruses accumulate in the saliva of the arthropod, which facilitates transfer to a person bitten by the arthropod. Outbreaks of viral encephalitis exhibit seasonal cycles, with increased numbers of cases occurring during the summer, when the vector populations are at their peak (Figure 22.6).

The different forms of viral encephalitis include **eastern equine, western equine, Venezuelan equine, St. Louis, Japanese B, Murray valley, California**, and **tick-borne encephalitis** (Table 22.2). The specific viral etiological agent, arthropod vector, and geographic distribution are different for each of these forms. Infections with encephalitis-causing viruses begin with viremia, followed by localization of the viral infection within the central nervous system, where lesions develop. The locations of the lesions within the brain are characteristic for each type. With the exception of St. Louis encephalitis, in which kidney damage also occurs, the pathological changes are normally restricted to the central nervous system.

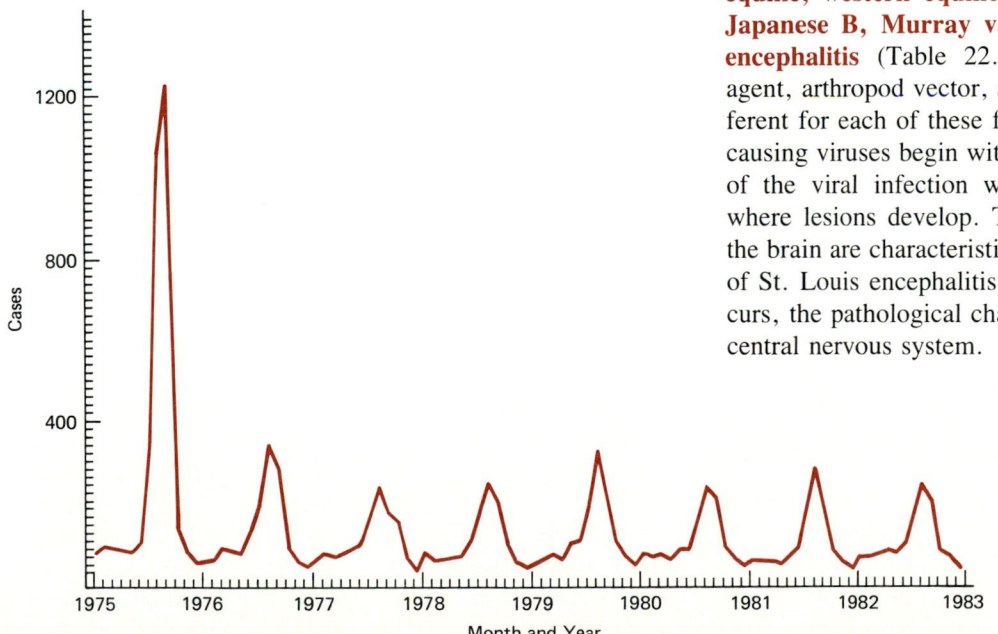

FIGURE 22.6
Incidence of encephalitis in the United States. The high number of cases for 1975 was due to an epidemic of St. Louis encephalitis.

TABLE 22.2   Togaviridae that Cause Encephalitis in Humans

Genus	Subgroup	Viral Species	Vector	Geographic Distribution
Alphavirus	I	Eastern equine encephalitis	Mosquito	Eastern United States, Canada, Brazil, Cuba, Panama, Dominican Republic, Trinidad, Philippines
		Venezuelan equine encephalitis	Mosquito	Brazil, Colombia, Ecuador, Trinidad, Venezuela, Mexico, Florida, Texas
		Western equine encephalitis	Mosquito	Western United States, Argentina, Canada, Mexico, Guyana, Brazil
Flavivirus	I	St. Louis encephalitis	Mosquito	United States, Trinidad, Panama
		Japanese B encephalitis	Mosquito	Japan, Guam, Eastern Asian mainland, India, Malaya
		Murray Valley encephalitis	Mosquito	Australia, New Guinea
		Ilheus	Mosquito	Brazil, Guatemala, Honduras, Trinidad
	IV	Tick-borne group (Russian spring-summer encephalitis group)	Tick	USSR, Canada, Malaya, United States, Central Europe, Finland, Japan, India, Great Britain
		Rio Bravo (bat salivary gland)	Unknown	California, Texas

Encephalitis symptoms are often subclinical. When the illness is symptomatic, encephalitis begins with fever, headache, and vomiting, followed by stiffness and then paralysis, convulsions, psychoses, and coma. Different forms of viral encephalitis have different outcomes. For example, in symptomatic cases of eastern equine encephalitis the mortality rate is approximately 80 percent, whereas in western equine encephalitis it is less than 15 percent. Individuals who recover from symptomatic encephalitis may have permanent neurological damage.

No antiviral drug has been developed for the treatment of encephalitis caused by arthropod-borne viruses. Vaccines have been developed, though, that establish immunity against the specific viruses that cause the disease. As a rule, control of arthropod-borne viral encephalitis must depend on the control of vector populations. Insecticides have been widely used in public health programs to reduce the populations of mosquitoes and ticks, the vectors of viral encephalitis. Mosquito eradication programs are often intensified when diagnoses of encephalitis raise the possibility of widespread outbreaks.

Not all forms of encephalitis are arthropod borne. For example, herpes simplex virus, which is not normally transmitted through vectors, can cause both meningitis and encephalitis. Vidarabine, an adenine arabinoside, has been demonstrated to be effective in treating herpes simplex virus

encephalitis, but this drug has not been proven effective against other encephalitis-causing viruses.

Another form of encephalitis, **kuru**, has a very different mode of transmission. Transmission occurs through the ingestion of human brain tissue during ritual cannibalism in New Guinea. The etiological agent of kuru is proposed to be a prion, but the existence of prions has yet to be definitively established. It is clear, though, that it may take years for symptoms to occur after the initial infection. In kuru there is a progressive vacuolation of the neurons that produces a progressive inability to control voluntary muscular movement (ataxia). The symptoms are similar to those of several other degenerative nervous diseases, including scrapie and Creutzfeldt-Jakob diseases, and prions have also been proposed as the etiological agents of these diseases.

Colorado Tick Fever   **Colorado tick fever** is caused by a double-stranded RNA virus. The virus is maintained largely within populations of the golden mantle ground squirrel, which acts as a reservoir, and is transmitted to humans through a tick vector (*Dermacentor andersoni*). Colorado tick fever occurs mainly in the western United States. Characteristically, there are two episodes of fever, separated by a few days. Malaise, muscle aches, and vomiting are also symptomatic manifestations. The disease is self-limiting, and recovery normally occurs within 1–2 weeks without treatment. Prevention of Colorado tick fever involves avoiding tick bites, for example, by wearing protective clothing in tick-infested areas and by using arthropod repellents.

Diseases Caused by Bacterial Pathogens

Plague   **Plague** is caused by *Yersinia pestis*, a Gram-negative, nonmotile, pleomorphic rod. *Y. pestis* is normally maintained within populations of wild rodents and is transferred from infected to susceptible rodents by fleas. *Y. pestis* is able to multiply within the gut of the flea, which blocks normal digestion, causing the flea to increase the frequency of feeding attempts and so to bite more animals, increasing the probability of disease transmission. Plague is endemic to many rodent populations; for example, *Y. pestis* is permanently established in rodent populations from the Rocky Mountains to the West Coast of the United States.

The transmission of plague was extremely widespread during the Middle Ages because of poor sanitary conditions and the abundance of infected rat populations in areas of dense human habitation. The development of rat control programs and improved sanitation methods in urban areas have greatly reduced the incidence of this disease. It is not possible, however, to completely eliminate plague in humans because of the large number of alternative hosts in which *Y. pestis* is maintained. In rural environments, for example, *Y. pestis* is found in ground squirrels, prairie dogs, chipmunks, rabbits, mice, rats, and other animals.

The introduction of *Y. pestis* into humans through flea bites initiates a progressive infection that can involve any organ or tissue of the body. Phagocytosis is effective in killing many of the invading bacteria, but some cells of *Y. pestis* are resistant and continue to multiply and spread through the circulatory system. In **bubonic plague**, *Y. pestis* becomes localized and causes inflammation of the regional lymph nodes. The enlarged lymph nodes are called *buboes*, from whence the name of the disease is derived. The symptoms of bubonic plague include malaise, fever, and pain in the area of the infected regional lymph nodes. Severe tissue necrosis can occur in various areas of the body, and the skin appears blackened. It was this symptom that gave the name *black death* to the disease in the Middle Ages. As the infection progresses, the symptoms become quite severe, and without treatment the mortality rate is 60 to 100 percent. The disease can also progress in humans to involve the pulmonary system, leading to pneumonic plague.

Transfer of *Y. pestis* can then occur through droplet spread, establishing outbreaks of **primary pneumonic plague**. The pneumonic form has been very important in plague epidemics. If the bacteria invade the lungs, as in primary pneumonic plague, the disease often progresses rapidly, manifested by severe prostration, respiratory difficulties, and death within a few hours of onset. Plague can be effectively treated with antibiotics. Streptomycin generally is the drug of choice against *Y. pestis*, although other antibiotics, such as chloramphenicol and tetracycline, can also be used.

Relapsing Fever   **Relapsing fever** is caused by various species of *Borrelia*, a Gram-negative spirochete (Figure 22.7). These pathogenic bacteria are normally transmitted to humans through a bite by the body louse, *Pediculus humanus*. Epidemics of louse-borne relapsing fever occur under conditions that favor the proliferation of body lice, such as crowding with relatively poor sanitation. The lice act strictly as vectors and do not develop the disease symptoms when they acquire *Borrelia* from an infected human. Crushing an infected louse can release *Borrelia* into the break in the skin caused by the louse bite. *Borrelia* species can also be transmitted through tick vectors, accounting for periodic outbreaks of relapsing fever in the western United States.

The clinical manifestations of relapsing fever occur intermittently. There is normally a sudden onset of fever, approaching 105°F, that lasts for 3–6 days. The fever then falls rapidly and the temperature remains normal for 5–10 days, followed by a second onset of fever that generally lasts 2–3 days. Additional relapses frequently occur in the tick-borne disease, but not usually when the disease is louse borne. During the course of relapsing fever, *Borrelia* species multiply in the blood and various other body tissues. The natural removal of *Borrelia* from the body depends on an antibody-mediated immune response rather than on phagocytosis. Anatomical abnormalities develop in the spleen, and lesions may also occur in various other body organs. Tetracycline and chloramphenicol are effective against *Borrelia* and are used in treating this disease. With proper treatment, over 95 percent of patients with relapsing fever recover. Control of this disease depends on avoiding contact with vectors carrying *Borrelia*, particularly body lice and ticks. In many areas of the world, pesticides are useful in maintaining low populations of rodents and other animals that act as nonhuman vertebrate hosts of vector ticks.

Rocky Mountain Spotted Fever   **Rocky Mountain spotted fever** is caused by a rickettsia and is the first of several diseases to be discussed in this section that are caused by rickettsias transmitted to humans via biting arthropod vectors (Table 22.3). In Rocky Mountain spotted fever, *Rickettsia rickettsii* is transmitted to humans through the bite of a tick. *R. rickettsii* is normally maintained within various tick populations, such as the wood and dog ticks (Figure 22.8). The bacteria multiply within the midgut of the tick and are passed congenitally from one generation of ticks to the next. Humans are accidental hosts of *R. rickettsii* as a result of occasional bites of infected ticks that allow the transfer of *R. rickettsii* to them.

Rocky Mountain spotted fever occurs in areas of North and South America, most commonly in the spring and summer, when ticks and humans are most likely to come in contact, and normally occur in well-defined localized re-

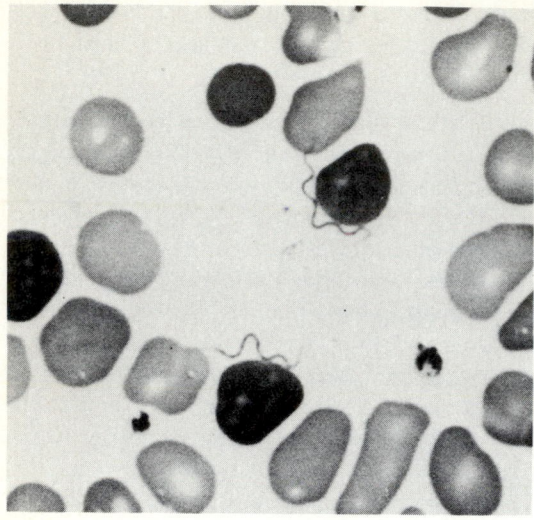

FIGURE 22.7
Micrograph of *Borrelia hermsi* from a smear of mouse blood stained by the Giemsa technique, followed by counterstaining with crystal violet (2200×). The characteristic morphology of the bacteria is shown in this preparation. (Courtesy Richard T. Kelly, Baptist Memorial Hospital, Memphis)

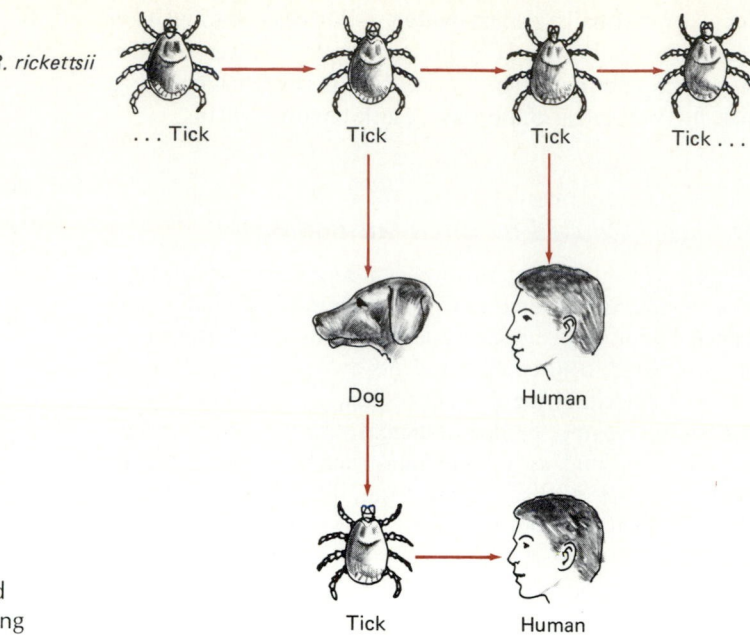

FIGURE 22.8
Transmission pattern of Rocky Mountain spotted fever.

TABLE 22.3   Summary of Certain Important Epidemiological and Clinical Characteristics of Rickettsial Diseases

| Disease | Epidemiological Features | | |
	Usual Mode of Transmission to Humans	Reservoir	Georgraphic Occurrence
Spotted Fever Group			
Rocky Mountain spotted fever	Tick bite	Ticks, rodents	Western Hemisphere
Tick typhus	Tick bite	Ticks, rodents	Mediterranean littoral, Africa, Asia
Rickettsialpox	House-mouse mite bite	Mites, mice	United States, Russia, Korea
Typhus Group			
Primary louse-borne typhus	Infected louse feces rubbed into broken skin or as aerosol to mucous membranes	Man	Worldwide
Brill-Zinsser disease	Recrudescence months or years after primary attack of louse-borne typhus	–	Worldwide
Murine typhus	Infected flea feces rubbed into broken skin or as aerosols to mucous membranes	Rodents	Worldwide (scattered pockets)
Scrub typhus	Mite bite	Mites, rodents	Southeast Asia, Western and Southestern Pacific, Japan

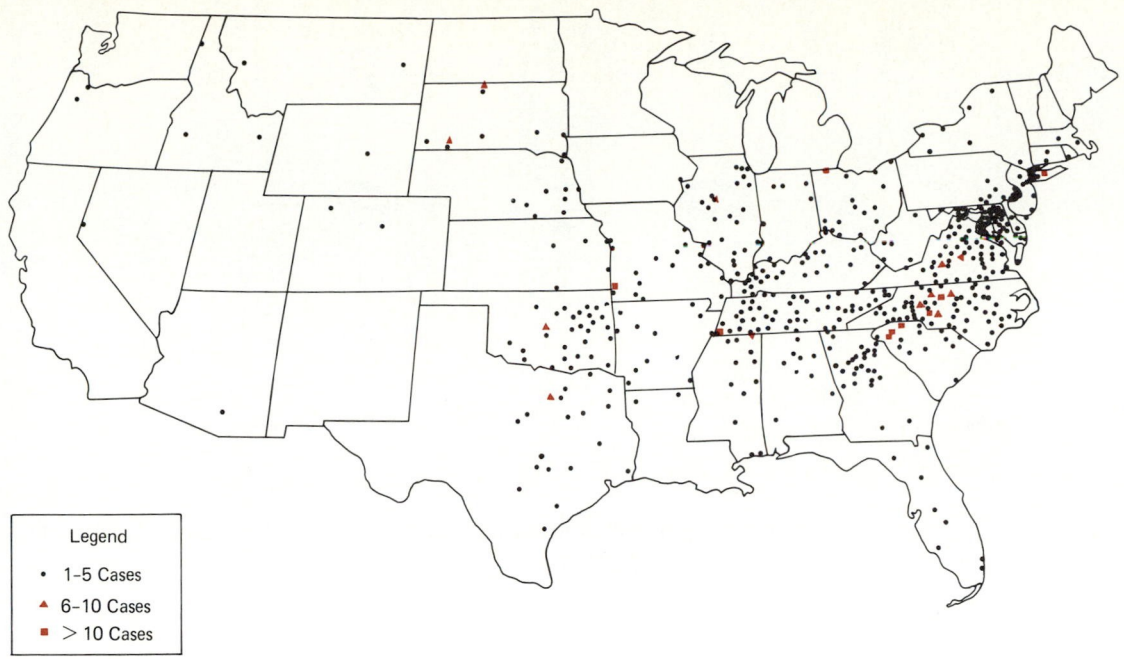

Legend
- • 1–5 Cases
- ▲ 6–10 Cases
- ■ > 10 Cases

FIGURE 22.9

Map showing the incidence of Rocky Mountain spotted fever in the continental United States.
Note that the highest rate of incidence is in the eastern half of the country.

gions. During the mid-twentieth century, many cases of Rocky Mountain spotted fever occurred in the United States in the region of the Rocky Mountains, but relatively few cases have been reported there in recent years. On the other hand, outbreaks of this disease have risen dramatically in the eastern United States, where most cases now occur (Figure 22.9).

When injected into humans, *R. rickettsii* multiply within the endothelial cells lining the blood vessels. Vascular lesions occur and account for the typical production of the characteristic skin rash associated with this disease. The rash is most prevalent on the extremities, particularly the palms of the hands and soles of the feet. In approximately 19 percent of cases of Rocky Mountain spotted fever there is no rash, making these cases especially hard to diagnose. Lesions probably also occur in the meninges, causing severe headaches and mental confusion. If treated with antibiotics, such as chloramphenicol and tetracycline, the disease is rarely fatal, but if untreated, the overall mortality rate is probably greater than 20 percent. Prevention of Rocky Mountain spotted fever primarily involves control of populations of infected ticks and avoidance of tick bites. However, control is difficult to achieve, and in 1985 there were almost 900 cases of Rocky Mountain spotted fever in the United States.

Typhus Fever   There are several types of **typhus fever**, all of which are caused by rickettsias transmitted to humans via biting arthropod vectors (Figure 22.10). **Infectious** or **classic typhus fever** is caused by *Rickettsia prowazekii* and is transmitted to humans via the body louse. The chain of transmis-

sion is restricted to humans and lice. Lice contract the disease from infected humans and, in turn, pass the rickettsia on to susceptible human hosts. *R. prowazekii* multiplies within the epithelium of the midgut of the louse. When an infected body louse bites another human it defecates at the same time, depositing feces containing *R. prowazekii*, which enter through the wound created by the bite. The name **epidemic typhus** is derived from the fact that this form of typhus is transmitted from person to person only by the body louse. Under crowded conditions with lice infestation, the disease can spread easily and cause large numbers of cases. For example, millions of cases occurred in World War I and in the concentration camps during World War II. The onset of epidemic typhus involves fever, headache, and rash. The heart and kidneys are frequently the site of vascular lesions. If untreated, the mortality rate in persons 10–30 years old is approximately 50 percent. The Weil-Felix test is useful in diagnosing classic typhus fever (Table 22.4). Chloramphenicol, tetracycline, and doxyclycline are effective in treating epidemic typhus.

Relapses louse-borne typhus can occur years after the primary attack. Such recurrences are referred to as **Brill-Zinsser disease**. *R. prowazekii* apparently can survive in some cells of the body for a prolonged period of time and give rise to this later infective phase. The factors that initiate recurrence of typhus are unknown, and there is no known way of preventing these episodes. Prevention of louse-borne typhus fever involves the control of louse infestation, and delousing infected patients effectively interrupts the chain of transmission.

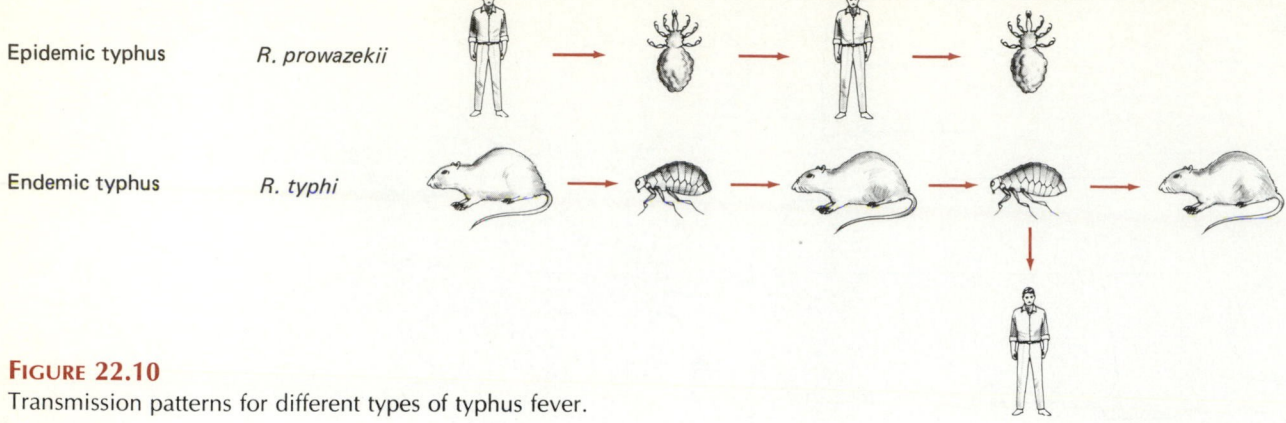

FIGURE 22.10
Transmission patterns for different types of typhus fever.

Murine or **endemic typhus fever** is caused by *Rickettsia typhi* and is transmitted to humans by rat fleas. Murine typhus is normally maintained in rat populations endemically through transmission by rat fleas. Occasionally, rat fleas attack humans, and if they are infected with *R. typhi*, the disease can be transmitted. As with louse-borne typhus, the flea deposits pathogenic bacteria in the fecal matter, which is rubbed into the flea bite by the host because of the local irritation caused by the bite. The symptoms of murine typhus are similar to those of classic typhus fever but are generally milder. Chloramphenicol and tetracycline are effective in treating this disease, and there is a relatively low mortality rate. Prevention of murine typhus depends on limiting rat

populations, which also limits the size of the vector rat flea population.

Diseases Caused by Protozoan Pathogens

Malaria   On a worldwide basis, **malaria** is one of the most common human infectious diseases. The annual incidence is about 150 million cases. Malaria has been largely eliminated from North America and Europe but remains the most serious infectious disease in tropical and subtropical regions of the world (Figure 22.11). It is caused by four species of *Plasmodium* (Table 22.5). *P. vivax* and *P. falciparum* are most frequently involved in human infections. The *Anopheles*

TABLE 22.4   Some Typical Weil-Felix Reactions

Causative Organism	Agglutination with Proteus of Strain*			Disease
	OX-19	OX-2	OX-K	
Rickettsia prowazekii	+ + + +	+	−	Epidemic typhus
R. mooseri	+ + + +	+	−	Murine typhus
R. rickettsii	+/+ + + +	+/+ + + +	−	Rocky Mountain spotted fever
R. tsutsugamushi	−	−	+ +	Scrub typhus
R. akari	−	−	−	Rickettsialpox
Rochalimaea quintana	−	−	−	Trench fever
Coxiella burnetii	−	−	−	Q fever

*Number of pluses indicates strength of reaction; minus indicates no reaction.

TABLE 22.5   Summary of Important Characteristics of Human Malarias

	P. falciparum	*P. vivax*	*P. ovale*	*P. malariae*
Incidence	Common	Common	Uncommon	Uncommon
Cell increase during primary hepatic schizogony	1–40,000 in 5.5–7 days	1–10,000 in 6–8 days	1–15,000 in 9 days	1–2,000 in 13–16 days
Cell increase during secondary hepatic schizogamy	1–8 to 24 (avg., 16) in 48 hr	1–12 to 24 (avg., 16) in 48 hr.	1–6 to 16 in 48 hr.	1–6 to 12 (avg., 8) in 72 hr
Incubation period	8–27 days (avg., 12)	8–27 days (avg., 14) (rarely months)	9–17 days (avg., 15)	15–30 days
Mortality	High in nonimmune persons	Uncommon	Rarely fatal	Rarely fatal

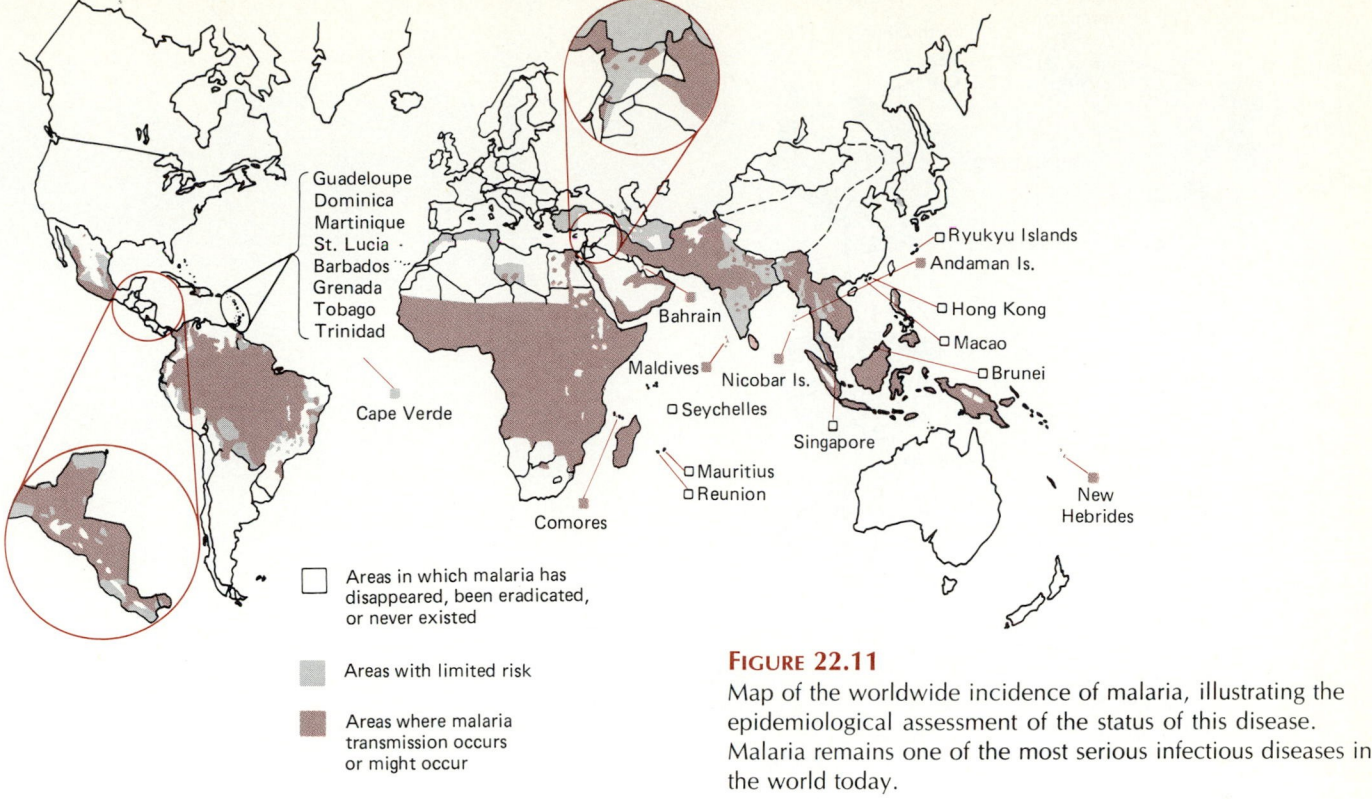

Guadeloupe
Dominica
Martinique
St. Lucia ·
Barbados
Grenada
Tobago
Trinidad

Cape Verde

Bahrain

Maldives

Nicobar Is.

□ Seychelles

□ Mauritius
□ Reunion

Comores

☐ Ryukyu Islands
■ Andaman Is.

□ Hong Kong
□ Macao
□ Brunei

Singapore

New
Hebrides

☐ Areas in which malaria has
disappeared, been eradicated,
or never existed

▨ Areas with limited risk

■ Areas where malaria
transmission occurs
or might occur

FIGURE 22.11

Map of the worldwide incidence of malaria, illustrating the epidemiological assessment of the status of this disease. Malaria remains one of the most serious infectious diseases in the world today.

mosquito is the vector responsible for transmitting malaria to humans. After inoculation into the body, the sporozoites of *Plasmodium* begin to reproduce within liver cells (Figure 22.12). Multiplication of the *Plasmodium* sporozoites occurs by schizogony, in which a single sporozoite can produce as many as 40,000 merozoites. The invasion of erythrocytes by the hepatic merozoites begins the erythrocytic phase of malaria, causing anemia and other severe manifestations.

Symptoms of malaria begin approximately 2 weeks after the infection is established by the mosquito bite and include chills, fever, headache, and muscle aches. These symptoms appear periodically, generally lasting for less than 6 hours. Schizogony occurs every 48 hours with *P. vivax* and *P. ovale* and every 72 hours with *P. malariae*, resulting in a synchronous rupture of infected erythrocytes that triggers the onset of disease symptoms.

Malarial infections persist for long periods of time and are rarely fatal, except when the disease is caused by *P. falciparum*. There is no vaccine for malaria, but attempts are being made to engineer one genetically. The disease can be prevented by drug prophylaxis. Individuals traveling to areas with high rates of malaria, such as Southeast Asia and Africa, often use antimalarial drugs, such as chloroquine, to avoid contracting this disease. The use of insect netting and other measures to prevent being bitten by an infected mosquito is extremely important. In the United States, control measures have been effective, but periodic morbidity increases have occurred after overseas military ventures (Figure 22.13).

Leishmaniasis   **Leishmaniasis**, caused by infections with members of the protozoan genus *Leishmania*, is transmitted to humans by sand fly vectors. *Leishmania* species reproduce in humans and other animals intracellularly as a nonmotile form, the amastigote. In sand flies the protozoa exist in a flagellated form, the promastigote. *L. mexicana* and *L. tropica*, sometimes referred to as *Old World cutaneous leishmaniasis* or *oriental sore*, cause infections limited to the skin. *L. braziliensis* causes infections of the skin and mucocutaneous junctions. The most serious form of leishmaniasis is caused by *L. donovani*, which multiplies throughout the mononuclear phagocyte system and causes kala-azar.

Leishmaniasis is geographically restricted to regions where sand flies can reproduce and acquire *Leishmania* species from infected canines and rodents. The protozoa are able to reproduce within the sand flies and are present in the saliva. The use of DDT has been effective in some regions in eliminating the sand fly vector. In other cases, infected rodent populations have been controlled, greatly reducing the incidence of leishmaniasis.

Trypanosomiasis   Trypanosomiasis is caused by infections with species of the protozoan genus *Trypanosoma*. **American trypanosomiasis**, or **Chagas' disease**, occurs in Latin America and is caused by *T. cruzi*, which is usually transmitted to humans by infected triatomid (cone-nosed) bugs. *T. cruzi* is a flagellate protozoan, but in vertebrate hosts it forms a nonflagellate form, the amastigote (Figure 22.14).

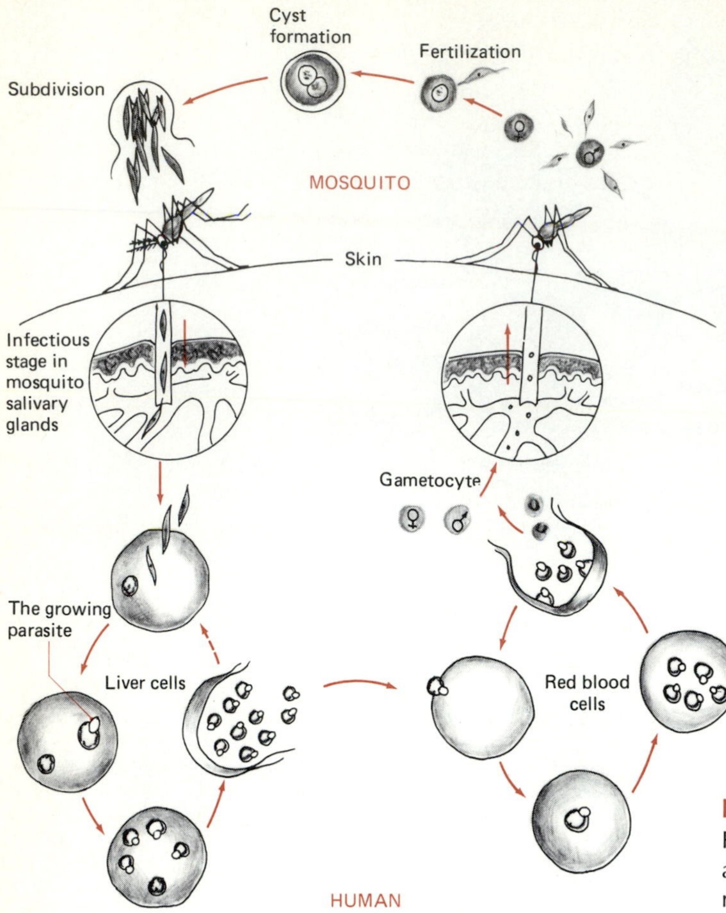

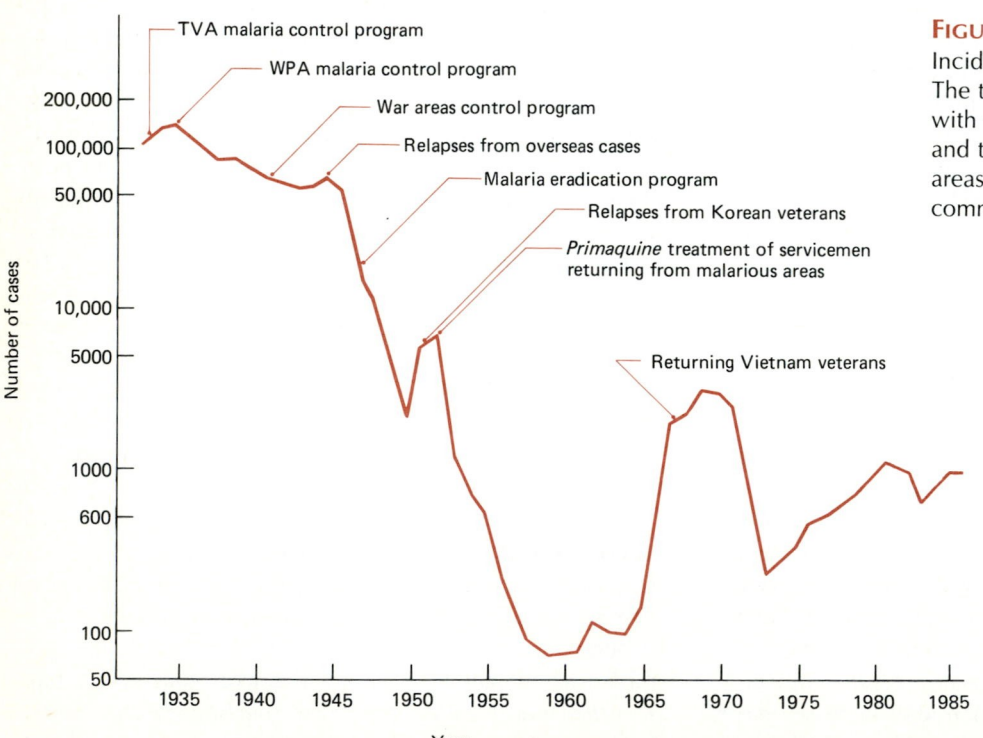

FIGURE 22.12

Reproduction of a malaria parasite within the mosquito vector and the human body. The protozoan exhibits two distinct reproductive phases.

FIGURE 22.13

Incidence of malaria in the United States. The total number of cases has fluctuated with the application of control measures and the return of military personnel from areas of the world where malaria is common.

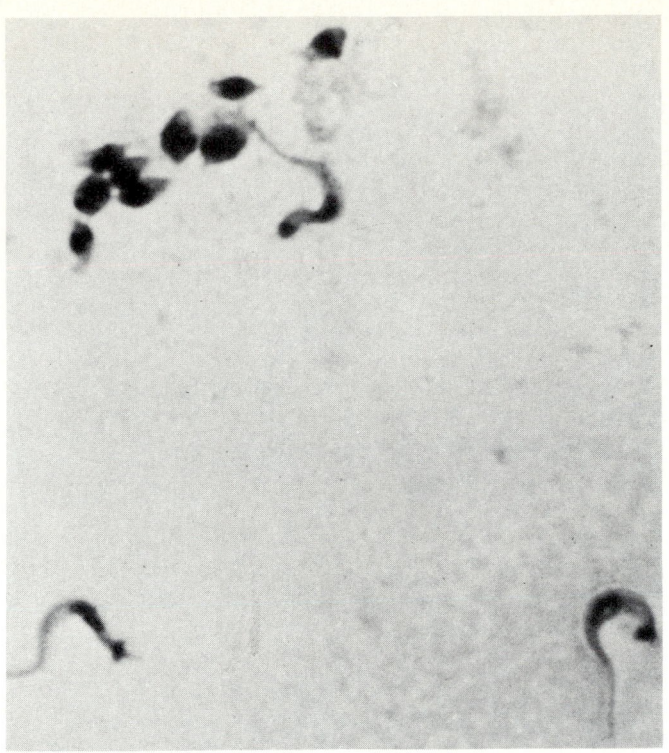

FIGURE 22.14
Micrograph showing flagellate (trypomastigote) and nonflagel-late (spheromastigote) forms of *Trypanosoma cruzi*. The trypomastigote, formerly known as the *trypanosomal stage*, occurs within the blood and within the insect vector. The spheromastigote, formerly known as the *amastigote stage* and prior to that as the *Leishmanial stage*, occurs within tissues, such as heart muscle and brain. (Courtesy Centers for Disease Control, Atlanta)

Dogs and cats are reservoirs of *T. cruzi*. The vectors of Chagas' disease normally live in the mud and wood houses of South America, and construction of better housing elimi-nates the habitat for vector populations that brings them into close contact with humans.

When *T. cruzi* infects human hosts, the protozoa initially multiply within the mononuclear phagocyte system. Later, the myocardium and nervous system are invaded. Damage to the heart tissue occurs as a result of this infection. In 90 percent of the cases there is spontaneous remission, but 10 percent of the hospitalized patients die during the acute phase of the disease because of myocardial failure. Death due to heart disease as a result of Chagas' disease may also occur well after recovery from the acute phase. Chagas' disease is the leading cause of cardiovascular death in South America, and the incidence of this disease in Brazil is extraordinarily high. Several antiprotozoan drugs, such as aminoquinoline, are effective in treating Chagas' disease if the symptoms are recognized early, but once the progressive stages have be-gun, treatment is supportive rather than aimed at eliminating the infecting agent.

African trypanosomiasis, also known as **African sleep-ing sickness**, is caused by infections with *T. gambiense* and *T. rhodesiense*. There is some debate over the taxonomic position of these two species, and it is possible that both are actually strains of *T. brucei*. The etiological agents of Afri-can trypanosomiasis are transmitted to humans through the tsetse fly vector. Tsetse flies acquire *Trypanosoma* species from various vertebrate animals, such as cows, which act as reservoirs of the pathogenic protozoa. In humans, infec-tions with *T. gambiense* or *T. rhodesiense* are disseminated through the mononuclear phagocyte system, and there is evi-dence of localization within regional lymph nodes. Multipli-cation of the protozoa can damage heart and nerve tissues. Progression through the central nervous system takes months to years. If untreated, the initially mild symptoms, which include headaches, increase in severity and lead to fatal men-ingoencephalitis. If the disease is diagnosed before there is central nervous system involvement, it can be successfully treated with antiprotozoan agents, such as suramin. If there is central nervous system involvement, melarsoprol, an ar-senical, is used. Prevention of African trypanosomiasis in-volves the control of the tsetse fly population, which is ac-complished by clearing vegetation to destroy the natural habitats of the tsetse fly.

22.2 *Diseases Transmitted Through Direct Contact*

In some cases, the deposition of pathogenic microorganisms on the skin surface can lead to an infectious disease. Some diseases transmitted in this manner are restricted to superfi-cial skin infections, but in other cases the pathogens are able to enter the body and spread systemically. Although rela-tively few microorganisms possess the enzymatic capability to establish infections through the skin surface, some micro-organisms are able to enter the subcutaneous layers through the channels provided by hair follicles. The transmission of some **contact diseases** may follow minor abrasions that al-low the pathogens to circumvent the normal skin barrier.

Many human contact diseases can be transmitted in other ways, such as through the respiratory or gastrointestinal tract or via animal bites. They are grouped together here on the basis of the relative frequency and/or importance of this mode of transmission.

Diseases Caused by Viral Pathogens

Warts   Warts are benign tumors of the skin caused by **pap-illomaviruses**, which are small icosahedral DNA viruses. Transmission of wart viruses appears to occur primarily by

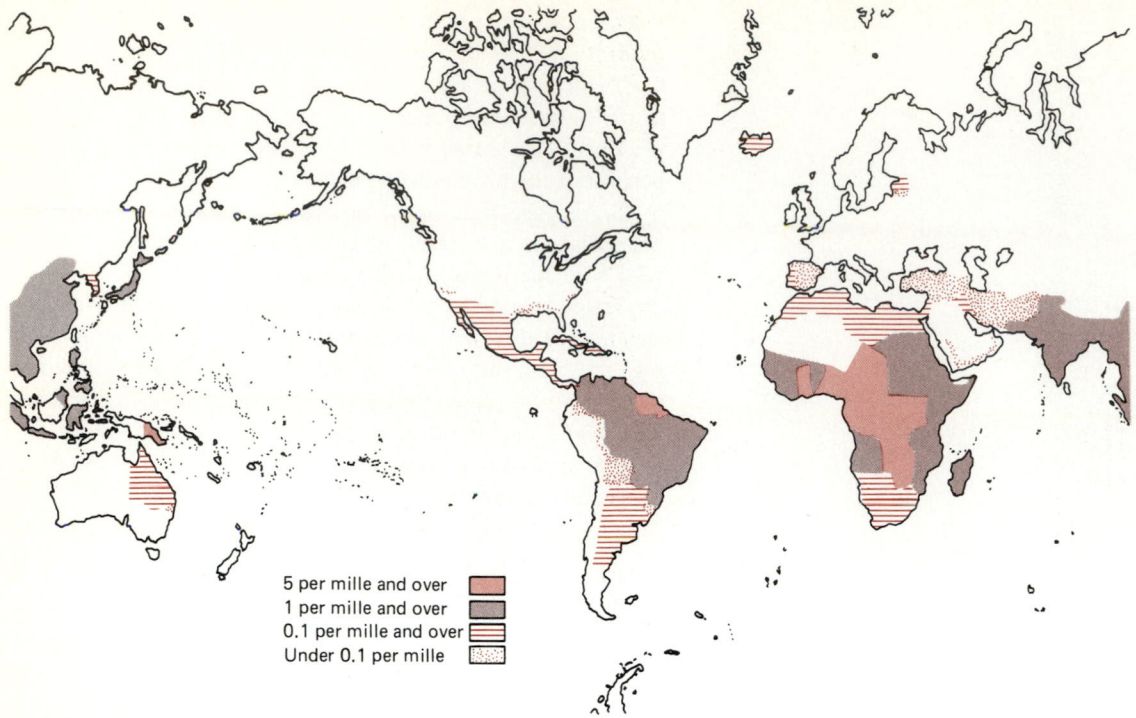

5 per mille and over
1 per mille and over
0.1 per mille and over
Under 0.1 per mille

FIGURE 22.15
Map showing the worldwide incidence of leprosy or Hansen's disease. Today this disease is
concentrated in subequatorial zones.

direct contact of the skin with wart viruses from an infected individual, although indirect transfer also may occur through fomites. The human papillomaviruses appear to infect only humans and no other animals, with children developing warts more frequently than adults. Warts can occur on any of the body surfaces; their appearance varies, depending on their location. At present there is no effective antiviral treatment for human warts, and therapy often involves destruction of infected tissues by applying acid or freezing. In general, warts are a self-limiting and recovery can be expected without treatment within 2 years.

Herpes Simplex Infections   As indicated in Chapter 21, there are two types of herpes simplex viruses. **Herpes simplex 1 virus** is most frequently involved in nongenital herpes infections. Infections with this virus most commonly involve the skin above the waist, lips, and mouth, and the virus appears to be transmitted through direct contact of the surface epithelial tissues with the virus. A focal infection develops around the site of inoculation, and dissemination of the virus occurs from the primary focal lesion. In most cases, the lesions are limited to the epidermis and surface mucous membranes; however, in some cases, the virus can spread systemically, causing central nervous system infections.

Herpes viruses in the mouth and on the lips result in the development of lesions known as *cold sores* or *fever blisters*. Similar lesions also develop in infected regions of the skin. The lesions generally heal within several days of their ap-

pearance, but new lesions develop periodically, indicating the persistence of herpes simplex infections. Several of the new antiviral drugs, such as acyclovir, are effective against herpes viruses and are useful in treating cutaneous herpes infections.

Diseases Caused by Bacterial Pathogens

Leprosy   In 1980 there were about 3 million cases of leprosy, or Hansen's disease worldwide, caused by *Mycobacterium leprae* (Figure 22.15). There are two forms of leprosy. **Lepromatous** or **cutaneous leprosy** is characterized by granulomatous nodules in the skin, whereas **tubercular** or **neural leprosy** involves lesions around the peripheral nerves. Bacteria present in the lesions are infective and can be transmitted through the skin or mucous membranes of susceptible individuals. There may be 1 billion viable cells of *M. leprae* per gram of skin in advanced cases of lepromatous leprosy, and direct skin contact appears to be very important in the transmission of this disease.

Leprosy may also be spread through droplets, which can have as many as 10^7 cells per milliliter of *M. leprae*, from infected individuals to susceptible hosts when the primary infection occurs within the lung tissues. There is an extremely long incubation period for leprosy, usually 3–5 years, before the onset of disease symptomatology. The symptoms of leprosy vary, but the earliest detectable ones generally involve skin lesions. Unlike other mycobacterial

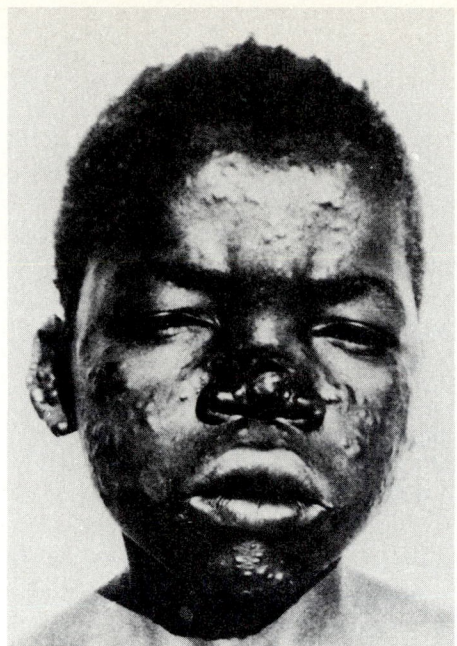
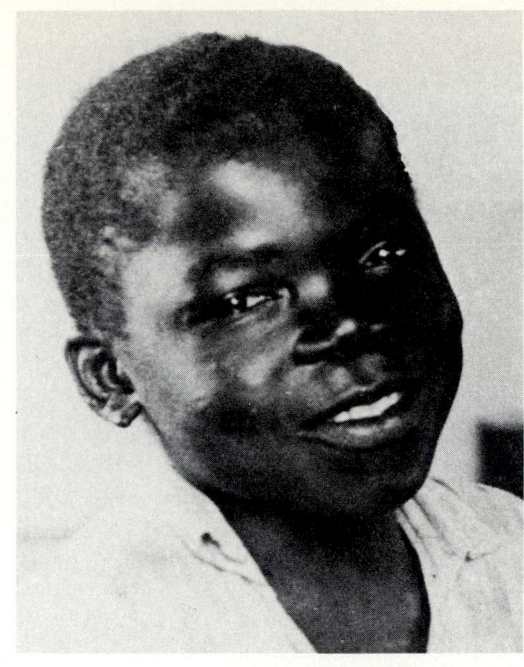

FIGURE 22.16
Throughout history, leprosy has been treated as a dreaded disease, and individuals who have contracted it have been outcast from societies. Although recovery is slow, this disease can be effectively treated, as shown in these photographs. (A) A young boy with borderline leptomatous leprosy. (B) The same child 2 years later, after treatment with appropriate medication. (Courtesy American Leprosy Missions, Bloomfield, N.J.)

species, *M. leprae* is able to reproduce within nerve tissues, damaging to the nervous system by reproducing within certain nerve cells (Schwann cells).

During the course of leprosy, many organs and tissues of the body may be infected in addition to the infection of nerve cells characteristic of all forms of leprosy. In the tubercular form relatively few nerves and skin areas are involved, but in lepromatous leprosy multiplication of *M. leprae* is not contained by the immune defense mechanisms, and the bacteria are disseminated through many tissues. Leprosy can be treated with dapsone, which is bacteriostatic, or rifampin, which is bactericidal. Prolonged treatment with antimicrobial agents is needed to control leprosy infections. Leprosy is rarely fatal, and complete recovery occurs after treatment in many cases (Figure 22.16).

Anthrax Anthrax is primarily a disease of animals other than humans, but it can occasionally be transmitted to people. The disease is caused by *Bacillus anthracis*, a Gram-positive, endospore-forming rod. Transmission to humans can occur by direct contact of the skin with the endospores of *B. anthracis*, via the respiratory tract through inhalation of spores, and via the gastrointestinal tract through the ingestion of spores. The cutaneous route of transmission accounts for 95 percent of the cases of anthrax in the United States. Contact with animal hair, wool, and hides containing spores

of *B. anthracis* is often implicated in transmission of anthrax, and the disease is therefore known as *wool sorter's disease*. Deposition of spores of *B. anthracis* under the epidermis permits germination, with subsequent production of toxin by the growing bacteria.

The localized accumulation of toxin causes necrosis of the tissue, with the formation of a blackened lesion. The development of cutaneous anthrax can initiate a systemic infection, and untreated cutaneous anthrax has a mortality rate of 10–20 percent. Cutaneous anthrax can be treated with penicillin and other antibiotics, reducing the death rate to less than 1 percent. Avoiding contact with infected animals and preventing the development of anthrax in farm animals through the use of anthrax vaccine have effectively reduced the incidence of this disease.

Tularemia Tularemia is caused by *Francisella tularensis*, a Gram-negative, fastidious coccobacillus. *F. tularensis* causes infections in many mammals and arthropod vectors. Transmission to humans can occur by ingesting contaminated material and handling infected animals. *F. tularensis* may gain entry to the body directly through the skin, particularly through minute openings such as at hair follicles and near the fingernails. Tularemia is an occupational problem for individuals who handle potentially contaminated animals, such as hunters, butchers, cooks, and agricultural workers.

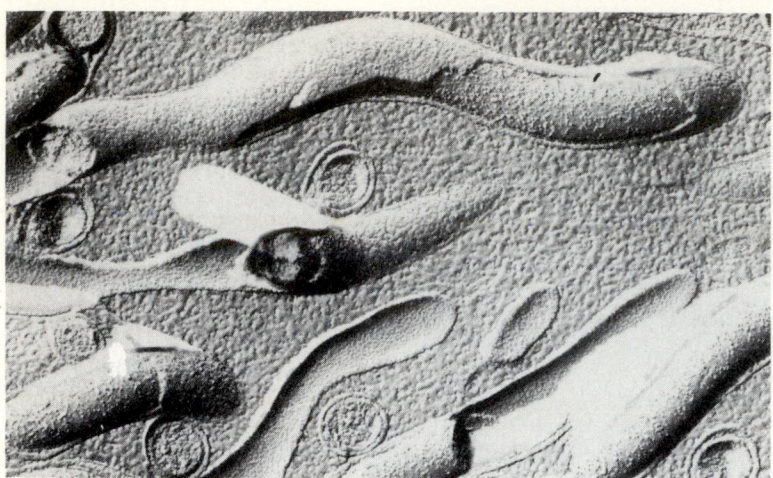

FIGURE 22.17
A freeze-eched micrograph of the spirochete *Leptospira interrogans*. (From BPS—Stanley C. Holt, University of Massachusetts)

Often ulcers of the fingers are symptomatic of tularemia that is contracted through the handling of infected animals. The localized ulcers result from the multiplication of *F. tularensis*, as does a systemic infection with localization within the regional lymph nodes. In most cases there is fever, skin ulceration, and enlargement of the lymph nodes. The death rate in untreated tularemia acquired through the skin is approximately 5 percent, but rates approaching 30 percent occur when the disease is contracted through inhalation. Streptomycin, tetracycline, and chloramphenicol are effective in controlling this disease. Prevention of tularemia can be achieved by avoiding contaminated animals. Because rabbits have an especially high rate of infection with tularemia, special precautions should be taken in their handling.

Brucellosis   **Brucellosis** is caused by several *Brucella* species, all of which are Gram-negative, small, nonmotile, aerobic rods. This is an infectious disease of nonhuman animals that can be transmitted to humans through ingestion of contaminated milk and handling of infected animals. The use of pasteurization has greatly reduced the transmission of brucellosis via the gastrointestinal tract, and most human infections today result from direct contact with infected animals. Brucellosis is an occupational hazard in farming, veterinary practice, and meat packing, where *Brucella* species can enter through the skin, particularly in areas of minor abrasions.

Once *Brucella* species enter the body, the bacteria spread rapidly through the mononuclear phagocyte system and multiply within phagocytic cells. Infecting *Brucella* species normally become localized in regional lymph nodes. There may be enlargement of the spleen and liver; other symptoms include weakness, chills, malaise, headache, backache, and fever. The fever may rise and fall in some cases, and thus this disease is often referred to as **undulant fever**. The death rate in untreated cases of brucellosis is about 3 percent. The use of tetracyclines, however, is effective in treating this disease and reduces the death rate to nearly zero. Prevention of brucellosis involves eliminating the disease in animals such as cattle, sheep, and goats, the reservoirs of *Brucella*.

It is important to segregate and treat infected animals in order to prevent the spread of this disease through animal herds. Vaccines are effective in limiting the spread of brucellosis through some animal populations, such as cattle, but there are no effective vaccines for others, such as hogs.

Leptospirosis   **Leptospirosis**, or **Weil's disease**, is principally a disease of nonhuman animals caused by *Leptospira interrogans* (Figure 22.17). Leptospires are long spirochetes that can enter the human body through the skin, especially if there are small abrasions. The disease can also be transmitted through the respiratory tract. Infected animals excrete leptospires in their urine, and humans generally contract the disease through contact with contaminated material harboring viable *Leptospira* species.

During the acute phase of this illness, the symptoms normally include a high spiking fever, chills, headache, muscle aches, malaise, abdominal pain, nausea, and vomiting. Various body organs can be involved, and most fatal cases result from infection of the kidney and subsequent renal failure. In most cases, leptospirosis is subclinical, and the prognosis for complete recovery is excellent. When severe symptoms develop, however, the death rate is about 10–40 percent. Penicillin may provide a useful treatment and may shorten the duration and severity of the illness. Avoidance of potentially contaminated environments, such as sewers and contaminated streams, is effective in preventing outbreaks of this disease.

Pyoderma and Impetigo   **Pyoderma** is an infection of the skin caused primarily by *Staphylococcus* and *Streptococcus* species. **Impetigo**, which occurs almost exclusively in children during warm weather, is a type of pyoderma caused by infections with *Staphylococcus aureus* and/or *Streptococcus pyogenes*. Direct contact with infected material appears to be important in the transmission of impetigo. Strains of *S. pyogenes* are deposited and remain viable on normal skin surfaces, and minor trauma to the skin permits invasion by the streptococci. *S. pyogenes* produces hyaluronidase, which permits spread of the bacteria. The infection results in the formation of a lesion that can be invaded by *S. aureus* from the skin surface, establishing a secondary infection. Typical manifestations of impetigo are localized lesions that progress to form pustules. Pyoderma caused by *Streptococcus* and *Staphylococcus* species tends to be benign, although the pustules may persist and spread, forming satellite infections. The disease normally begins on the extremities, and intense itching is associated with the lesions that develop. Scratching facilitates spread to other parts of the body.

Glomerulonephritis   In some cases, systemic infections can result from bacterial pyoderma. The spread of streptococcal skin infections is especially likely if there are wounds that open a portal of entry for systemic dissemination. The best evidence now suggests that **glomerulonephritis** is an autoimmune disease— directed against a protein of *Streptococcus pyogenes*, which spreads to the kidneys during a streptococcal infection—and is not due to an active kidney infection. Streptococcal and staphylococcal skin infections can be treated by the application of topical antimicrobial drugs, such as bacitracin, if there are relatively few lesions. In cases where the pyoderma is widespread, systemic administration of antimicrobial agents, such as penicillin or erythromycin, is effective in treating the disease. Pyoderma caused by streptococcal and staphylococcal skin infections can generally be prevented by washing the skin to prevent the accumulation of large numbers of nonresident microbiota and by cleansing and applying antiseptics to minor skin abrasions.

Acne   **Acne** is a common problem during adolescence. The disease occurs when microbial invasion at the base of hair follicles initiates excessive secretions by the sebaceous glands. Various bacteria, including *Propionibacterium acnes (Corynebacterium acnes)* and *Staphylococcus epidermidis*, have been identified as the etiological agents of acne. Outbreaks of acne most commonly occur on the face, upper back, and chest. The disease is characterized by inflammatory papules, pustules, and cysts. Acne most frequently occurs during puberty because of the hormonal changes that take place during this period. Once adaptation to mature levels of sex hormones occurs, adults are normally not susceptible to this inflammatory disease.

Mastitis   **Mastitis** is an infection of the mammary glands. This disease is normally caused by infections with *Staphylococcus aureus*, although other bacteria may also be involved. The bacteria are able to penetrate the skin surface through the nipples of nursing mothers and multiply in the mammary glands. Treatment of mastitis involves administration of penicillin or other antibiotics that are effective against *Staphylococcus* and other Gram-positive bacteria. The advisability of continuing to nurse during mastitis is currently being debated among physicians, some advising cessation of nursing and others considering that the worst thing to to.

Mycetoma   **Mycetoma**, an infectious disease of the skin, can progress to involve subcutaneous tissues and bones. This disease is caused by various species of actinomycetes, including *Nocardia madurae*, and by various fungal species. One form of mycetoma, known as **madura foot**, is important in the Madura Province of India. The feet are the most common sites of involvement because the actinomycetes and fungi causing mycetoma most frequently enter the skin through abrasions acquired while walking barefoot on soil, although the infectious agents may penetrate the skin in other ways. Multiplication of the bacteria or fungi within the infected area results in swelling, which is frequently grotesque. It is necessary to identify the specific etiological agent of mycetoma in order to determine the appropriate treatment. When the causative organism is a fungus, amphotericin B can be used. If a *Nocardia* species or another actinomycyte is responsible, sulfonamides are normally an effective treatment. The remaining worldwide incidence of this disease is largely a function of socioeconomic conditions because its occurrence can be greatly reduced by wearing shoes.

Diseases Caused by Fungal Pathogens

As just discussed, fungi as well as bacteria can cause mycetoma. Additionally, various fungal species are responsible for a number of superficial infections of the skin. Many of the fungi that cause superficial skin infections are **dermatophytes**; that is, they infect only the skin and its appendages, such as hair and nails. Even though the growth of dermatophytic fungi is restricted to the skin, the host–parasite interaction can cause serious manifestations. Contact of the skin with spores of a dermatophyte can initiate infection, with the dermatophytic fungi growing filamentously within the dead kerratin-containing layers of the skin. The colonization of the skin by dermatophytic fungi initiates a cell-mediated immune response that generally occurs 10–35 days after infection. This response causes inflammatory damage to the skin tissue; however, it also prevents further lateral spread of the fungus.

Diseases caused by dermatophytic fungi are manifested as **tinea** or **ringworm**. These diseases are normally well localized and never fatal. The identification of the disease depends on which regions of the body are infected (Table 22.6). Most dermatophytic fungi are members of the genera *Microsporum* and *Trichophytum*. Transmission of dermatophytic fungi is enhanced by conditions of high moisture and sweating, and retention of moisture increases the probability of contracting superficial infections of the skin. The transmission of **athlete's foot**, for example, is often associated with the high moisture levels and bare feet of athletes in a locker room, although it is now known that this disease is not acquired unless the individual has skin abrasions that the fungus can infect. Drying the feet and using antifungal agents, however, can reduce the spread of this disease. It is virtually impossible to protect all body areas against potential infection with

TABLE 22.6   Epidemiology of Dermatomycoses

Disease	Causative Agent	Transmission	Examples of Sources
Tinea capitis (ringworm of the scalp)	*Microsporum* spp., *Trichophyton* spp.	Direct or indirect contacts	Lesions, combs, toilet articles, headrests
Tinea corporis (ringworm of the body)	*Epidermophyton*, *Microsporum* spp., *Trichophyton* spp.	Direct or indirect contacts	Lesions, floors, shower stalls, clothing
Tinea pedis (ringworm of the feet [athlete's foot])	*Epidermophyton*, *Trichophyton* spp.	Direct or indirect contacts	Lesions, floors, shoes and socks, shower stalls
Tinea unguium (ringworm of the nails)	*Trichophyton* spp.	Direct contact	Lesions
Tinea cruris (ringworm of the groin [jock itch])	*Trichophyton* spp., *Epidermophyton*	Direct or indirect contacts	Lesions, athletic supports

superficial dermatophytic fungi, resulting in a high incidence of dermatomycoses.

Sporotrichosis   In contrast to the superficial fungal diseases just discussed, **sporotrichosis** is a subcutaneous mycosis caused by the fungus *Sporothrix schenckii*. Inoculation of *S. schenckii* into the skin as a result of a minor injury initiates this infection. Sporotrichosis occurs worldwide and is distributed especially widely in South Africa, France, and Mexico. The infection begins with the formation of a subcuta-neous nodule, with secondary nodules developing later as the infection spreads through the lymphatic system. Lesions due to sporotrichosis normally occur on the extremities and spread to other parts of the body. The spread of the disease occurs slowly, permitting time for diagnosis and therapy. As a result, the death rate is very low. The cutaneous form of sporotrichosis can be treated with a solution of potassium iodine, but if the fungus is widely disseminated, administration of amphotericin B is necessary.

22.3 Infections Associated with Wounds and Burns

Infections After Wounds

Wounds disrupt the protective barrier of the skin and provide a portal of entry through which microorganisms can enter the circulatory system and deep body tissues. Microorganisms on the skin surface can readily pass through the opening of a wound. As a result, many infections associated with wounds are caused by opportunistic pathogens derived from the normal microbiota of the skin. To avoid entry of bacteria into a wound, the area is usually covered with gauze to protect against contamination.

Staphylococcus aureus is the most frequent etiological agent of wound infections, and *Streptococcus pyogenes* is also often involved. In severe wounds, where the integrity of the gastrointestinal tract is disrupted, enteric bacteria are frequently the causative agents of wound infections (Table 22.7). In many cases, wound infections are localized at the site of the wound, but infections established in this manner can spread systemically and may involve many body tissues and organs. For example, infections with *S. aureus* established through skin wounds can spread and form abscesses in bone marrow (osteomyelitis) and other body tissues, including the spine and brain. Superficial wounds can generally be treated with topical antiseptics or antibiotics to prevent the establishment of infections. Serious deep wounds, how-

TABLE 22.7   Bacterial Causes of Wound Infections Recorded at an Urban Hospital

Staphylococcus aureus	48%
Enteric and other bacteria associated with the gastrointestinal tract	49%
Escherichia coli	
Proteus spp.	
Klebsiella spp.	
Enterobacter spp.	
Bacteroides spp.	
Streptococcus faecalis	
Streptococcus pyogenes, group A	3%

ever, may require the prophylactic use of systemic antibiotics to prevent the onset of severe infections. Infections of deep wounds may involve anaerobic bacteria of the genera *Clostridium*, *Bacteroides*, and *Fusobacterium*, as well as *Staphylococcus* and *Streptococcus* species.

Gas Gangrene   Deep wounds not only provide a portal of entry for microorganisms, but the tissue damage often interrupts circulation to the area, creating conditions that permit the growth of obligately anaerobic bacteria. **Gas gangrene** is a serious infection that may result from the growth of

Clostridium perfringens and other *Clostridium* species. The development of gas gangrene is dependent on the deposition of endospores of *Clostridium* in the wound tissue and the occurrence of anaerobic conditions that permit the germination and multiplication of these obligately anaerobic bacteria.

The *Clostridium* species that cause gas gangrene produce toxins, the diffusion of which extends the area of dead and anaerobic tissues. The exotoxins produced by these species are tissue necrosins and hemolysins that account, in part, for the rapid spread of infection. The growing *Clostridium* species produce carbon dioxide and hydrogen gases, as well as odoriferous low molecular weight metabolic products. The gas that accumulates is primarily hydrogen because it is less soluble than CO_2. In most cases, the onset of gas gangrene occurs within 72 hours of the occurrence of the wound; if untreated, the disease is fatal. Even with antimicrobial treatment, there is a high rate of mortality; therefore, radical surgery—amputation—is often employed to prevent the spread of infection. If treated rapidly enough, localized areas of necrotic tissue can be excised and high doses of penicillin administered to block the spread of the infection. The prevention of gas gangrene depends on ensuring that wounds do not provide a suitable environment for the growth of the anaerobic *Clostridium* species. This requires adequate drainage of wounds to prevent the establishment of anaerobic conditions and the removal of foreign material and dead tissue.

Tetanus   **Tetanus** is caused by *Clostridium tetani*. This organism produces a neurotoxin that can cause severe muscle spasms. Tetanus is sometimes referred to as **lockjaw** because the muscles of the jaw and neck contract convulsively so that

FIGURE 22.18
This painting of a soldier dying of tetanus, by Charles Bell, shows the characteristic position of the jaw and neck associated with this disease. (By the kind permission of the President and Council of the Royal College of Surgeons of Edinburgh)

the mouth remains locked closed, making swallowing difficult (Figure 22.18). *C. tetani* is widely distributed in soil. Transmission to humans normally occurs as a result of a puncture wound that inoculates the body with spores of *C. tetani*. If anaerobic conditions develop at the site of the wound, the endospores of *C. tetani* germinate, and the multiplying bacteria produce neurotoxin. *C. tetani* is noninvasive and multiplies only at the site of inoculation. The neurotoxin it produces however, spreads systemically, causing the symptoms of this disease.

Virtually any type of wound into which foreign material is introduced may carry spores of *C. tetani* and lead to the development of tetanus (Table 22.8). Tales of the association of rusty nails with this disease probably originated because farmers often developed tetanus after stepping on such nails that were contaminated with soil and endospores of *C. tetani*, but clearly the rusty nails themselves are not the cause of this disease. If untreated, tetanus is frequently fatal, but if recovery does occur, there are no lasting effects. Tetanus can be treated by the administration of tetanus antitoxin to block the action of the neurotoxin. The disease can be prevented by immunization with tetanus toxoid, and tetanus booster vaccinations are frequently given after wound injuries to ensure immunity against this disease.

Infections After Burns

Burns remove the protective skin layer, exposing the body to numerous potential pathogens. Microbial infection after extensive burns, where a large portion of the skin is damaged, is a very serious complication that often results in the death of the patient. *Staphylococcus aureus*, *Streptococcus pyogenes*, *Pseudomonas aeruginosa*, *Clostridium tetani*, and various fungi often cause infections in burn victims. It is important to avoid contamination of the burn area that can

TABLE 22.8   Portals of Entry of *Clostridium tetani* in Cases of Tetanus

Portal of Entry	Frequency (%)
Wounds	31
Punctures	27
Lacerations	8
Abrasions	3
Crushes	3
Surgical and obstetric	3
Injections	4
Ulcers	3
Other wounds	14
Unknown	7

A
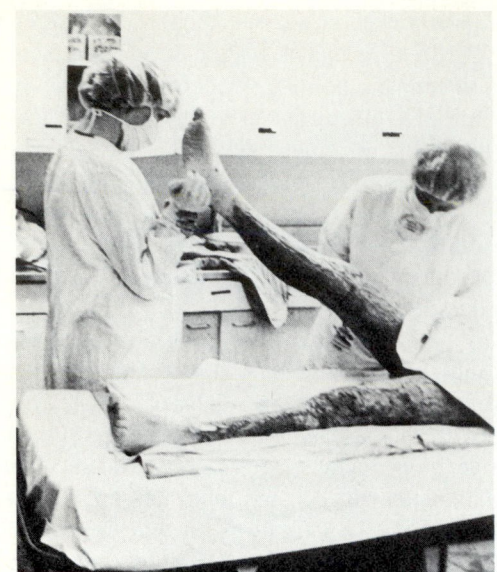
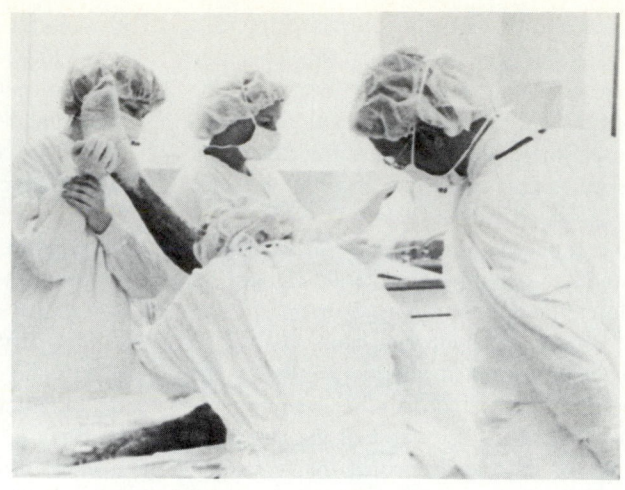
B

FIGURE 22.19

Burn victims are especially vulnerable to infection because of the destruction of the surface skin barrier. The open-air procedure is currently the method of choice at many burn units; in this procedure the burn wound is left open to the air, except when dressings are applied. As a result, extreme precautions are taken to minimize exposure to microorganisms, including keeping burn victims in isolation rooms and ensuring that individuals in contact with such patients are properly attired to prevent the transmission of potentially pathogenic microorganisms. (A) The vulnerability of burn tissues to infection is apparent in this photograph, showing the leg of a man with severe burns over most of his body. (B) When applying dressings, all staff must be masked, gloved, and gowned. (Courtesy Norton, Kosairs, Childrens Hospital, Louisville, Ky.)

introduce opportunistic pathogens into exposed tissue (Figure 22.19). To ensure that infections are detected in time to permit treatment with antimicrobial agents, microbiological tests are frequently performed; typically, two times a week a moist swab is run over a 1 cm² area and cultured for quantitation of bacteria and fungi; a count of greater than 10^5 is indicative of colonization and pending invasion. For prophylaxis, silver-containing antimicrobials, such as silvidene, are frequently used. When infections are detected, specific topical agents are selected on the basis of the sensitivity of the infecting agent. If infections do develop, the prognosis depends on the size of the burn, the extent of infection, and the physiological state of the patient.

22.4 Infections Associated with Medical Procedures

Medical procedures are designed to cure diseases, but some procedures used in the treatment of disease can inadvertently introduce pathogenic microorganisms into the body and initiate an infectious process. Several types of **nosocomial infections** (hospital-acquired infections)—including pneumonia acquired in hospitals, urinary tract infections that develop as a result of the insertion of a catheter, and puerpereal fever that develops from gynecological procedures—have been discussed in Chapter 21. In this section, some additional problems will be considered concerning the development of infections that develop as a result of medical procedures that penetrate the skin, as for example, in association with surgical incisions.

Infections After Surgical Procedures

Surgical procedures often expose deep body tissues to potentially pathogenic microorganisms. A surgical incision circumvents the normal body defense mechanisms. Great care is therefore taken in modern surgical practice to minimize microbial contamination of exposed tissues (Figure 22.20). These practices include the use of clean operating rooms with minimal numbers of airborne microorganisms, sterile instruments, masks, and gowns, all of which prevent the spread of microorganisms from the surgical staff to the patient, and the application of topical antiseptics prior to making the incision in order to prevent accidental contamination of the

FIGURE 22.20

In operating theaters, walls are disinfected, instruments sterilized, and staff appropriately garbed to prevent infection of exposed tissues. Note that the surgical staff are all attired so as to minimize the chances of any microorganisms associated with the surgical staff contaminating the surgical wound. Surgical staff wear hats, masks, gloves, gowns, and even shoe covers, and scrub (wash) their skin extensively before entering the operating room. (Courtesy National Institutes of Health, Washington, D.C.)

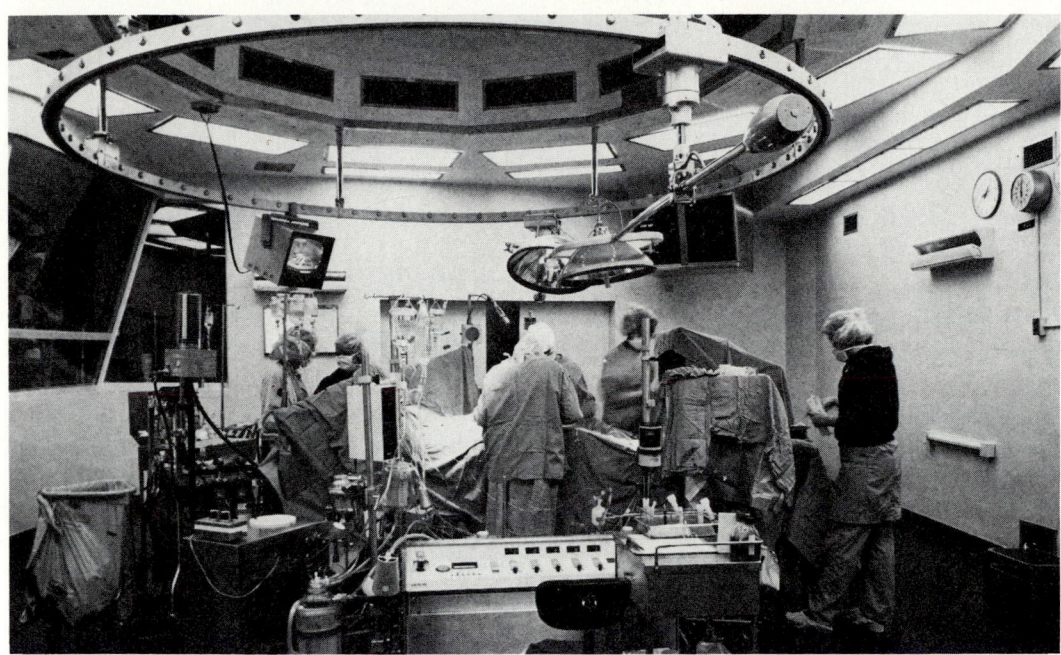

Discovery Process

In the 1850s, when Joseph Lister was beginning his career as a surgeon, the infamous hospital diseases of erysipelas, pyanemia, septicemia, and gas gangrene made surgical wards nightmares pervaded by death rather than places of healing. Although it was clear that dirt and overcrowding increased the incidence of disease, there was no adequate understanding of the role of microorganisms in the spread of disease or the particular susceptibility of surgical wounds to infection. Anesthetics, which were introduced to surgical practice at the time Lister began his career, made the pain of surgery more tolerable, and hence, more people were willing to submit to the surgeon's blade. However, anesthetics did not lower the mortality rate attributable to postoperative infections. Even in a successful year, one-third of Lister's patients who had undergone major surgery, such as amputation, died from infections. When the work of Pasteur—which demonstrated that microorganisms carried in the air are responsible for putrefaction of animal flesh—became known to Lister, it made sense to him that airborne microbes were responsible for postoperative infections. Lister knew that Pasteur killed germs by heating, but this did not solve the problem of how to prevent airborne contamination of open wounds. Lister reasoned that if he covered

a wound with a dressing that did not exclude air but that killed the microorganisms floating in the air, he might be able to reduce the death rate. He remembered a newspaper article describing the use of carbolic acid to treat sewage and recalled that the acid also killed certain parasites on cattle. In 1865 Lister decided to try out his theory on a patient with a compound fracture; complications from such cases frequently resulted in infection and death. Although his first trial was a failure, Lister continued his experimental trials on the use of carbolic acid to protect wounds against infection; these experiments soon met with success. "Though hardly expecting success, I tried the application of carbolic acid to the wound, to prevent decomposition of the blood, and so avoid the fearful mischief of suppuration throughout the limb. Well, it is now 8 days since the accident, and the patient has been going as if there was no external wound." Lister extended his work to the use of aerosol sprays of carbolic acid to prevent infection during surgery. He published his work, ensuring that others knew of his discoveries. Lister's work marked the beginning of the era of antiseptic surgery. His success depended on having the right background, steadfast determination, and scientific reasoning.

TABLE 22.9   Some Factors in the Development of Surgical Wound Infections

Situation	*Staphylococcus aureus* (%)	Enteric bacteria (%)	*Streptococcus pyogenes* (%)
Emergency operation	10	21	1
In wound of second operation	11	12	0
First operation of day	17	16	0
Other	10	0	2
Total	48	49	3

wound with the indigenous skin microbiota of the patient. After many surgical procedures antibiotics are given for several days as a prophylactic measure.

Despite all of these precautions, infections still sometimes occur after surgery (Table 22.9). These infections can be quite serious, as the patient is already in a debilitated state. The onset of such infections is generally marked by a rise in temperature. A purulent lesion may develop around the wound. Particularly serious complications may follow open heart surgery if the patient develops endocarditis, caused by *Staphylococcus* or *Streptococcus* species. In surgical procedures involving cutting of the intestines, the normal gut microbiota may contaminate other body tissues, causing **peritonitis** and infections of other tissues caused by enteric bacteria, unless great care is taken to minimize such contamination, and antibiotics are used to prevent microbial growth (see Table 22.9). The specific microorganisms causing infections of surgical wounds and the tissues that may be involved depend on the nature of the surgery and the tissues that are exposed to potential contamination with opportunistic pathogens.

Serum Hepatitis

Serum hepatitis is caused by hepatitis B virus and hepatitis non-A–non-B virus. An estimated 80,000–100,000 new cases of serum hepatitis occur in the United States each year. The incidence is much higher in Africa and Asia. Although the principal means of transmission of serum hepatitis involves transfusions with contaminated blood, this virus may also be transmitted by various other routes. There is a high rate of transmission of serum hepatitis among drug addicts, who frequently use contaminated syringe needles. Perhaps 10 percent of those infected with hepatitis B and hepatitis non-A–non-B virus become chronic carriers. The carrier rate

of these viruses in blood donors in the United States appears to be between 0.5 and 1 percent, but in other countries it may be as high as 5 percent. It is estimated that there are 2 million carriers of hepatitis viruses in the world. The blood of an infected individual may remain infective for months or years.

The clinical manifestations of serum hepatitis may include the development of jaundice and generally are very similar to those described for hepatitis A infections in Chapter 21. When the viral infection is transmitted through blood transfusions, the mortality rate associated with serum hepatitis is about 10 percent, reflecting the high dose of viruses normally transmitted via this route and the fact that the patient receiving the transfusion is in a debilitated state. One of the most important ways of preventing the transmission of this disease is to screen blood donors and to eliminate contaminated blood from blood banks. Avoiding the reuse of syringe needles among drug addicts also reduces the spread of this disease.

The United States Food and Drug Administration has approved a new vaccine for hepatitis B produced from viral particles isolated from the blood of human carriers of the disease. This vaccine, made available in 1982, was the first completely new viral vaccine in a decade and the first ever licensed in the United States made directly from human blood. The production of this vaccine involves a 65-week cycle of purification and safety testing to preclude inclusion of intact infectious viruses or other undesired factors from the blood of hepatitis B carriers. The cost is about $100 for the three doses needed to establish immunity. Several recent advances in genetic engineering, however, indicate that a second-generation hepatitis B vaccine may be available soon at a considerably lower price. At present the approved hepatitis B vaccine is intended for high-risk individuals, including health care workers and drug addicts.

22.5 Infections of the Eye

Conjunctivitis

Conjunctivitis is an inflammation of the conjunctiva (mucous membrane) covering the cornea and under the eyelids and results from bacterial or viral infections of the eye. Bacterial conjunctivitis can be caused by a variety of bacteria, including *Haemophilus* species, *Moraxella* species, *Staphy-* lococcus aureus, *Streptococcus pneumoniae*, *Pseudomonas aeruginosa*, *Corynebacterium* species, *Neiserria gonorrhoeae*, and *Chlamydia* species. *Haemophilus aegyptius* is the common etiological agent of **epidemic conjunctivitis** or **pink eye** among school children. The establishment of an infection of the eye requires not only inoculation with opportunistic pathogens but also that the bacteria overcome the natural defenses of the eye. Thus, bacteria causing conjunctivitis must resist being washed out by tears and inactivation by lysozyme and antibodies. Contamination of the eye with foreign material and injuries of the eye tissue favor the establishment of infections.

The manifestations of conjunctivitis and the seriousness of this disease depend on the particular pathogenic microorganisms. The symptoms of conjunctivitis generally include swelling and reddening of the eyelids and the formation of a purulent discharge. The eye normally becomes red and itchy because of extensive dilation of the capillaries; consequently, the disease is referred to as *pink eye*. There may also be some photophobia and blurring of vision. Many cases of bacterial conjunctivitis are self-limiting, and the disease symptoms disappear within 1 week after onset. In some cases, however, such as when the infection is caused by *P. aeruginosa*, the disease may progress to involve the cornea. When the cornea is infected with *P. aeruginosa*, perforation may result within 24 hours after the onset of infection. Other bacterial and viral species may also cause ulceration and damage of the cornea.

Some forms of conjunctivitis can be transmitted from an infected mother to a child during birth. Both *N. gonorrhoeae* and *Chlamydia trachomatis*, the causative agents of trachoma and inclusion conjunctivitis, respectively, can be transferred in this manner. The use of silver nitrate to treat the eyes of newborns can prevent the transmission of *N. gonorrhoeae* but not of *C. trachomatis*. **Inclusion conjunctivitis** caused by *C. trachomatis* can also be acquired by adults and children by direct contact with infected individuals and from swimming in nonchlorinated waters contaminated with this bacterial species from genital sources. In cases of bacterial conjunctivitis, the use of antimicrobial agents limits the severity of the disease. Sulfacetamide, neomycin, and gentamicin are frequently used in treating infections of the eye.

In addition to bacteria, several viruses cause acute con-junctivitis. Adenoviruses are the etiological agents of **epidemic keratoconjunctivitis** or **shipyard eye**. The disease derives its name from the fact that outbreaks have occurred in shipyards and industrial plants, where trauma to the eye may occur as an occupational hazard. In this form of conjunctivitis, the conjunctivae become inflamed and the cornea may become keratinized (covered with dead keratin protein).

Epidemic hemorrhagic conjunctivitis is caused by an enterovirus. This disease is associated with epidemic outbreaks in Africa, Asia, and Europe. It is transmitted by direct or indirect contact with discharges from infected eyes and by droplets containing the infective enterovirus. This form of conjunctivitis is characterized by subconjunctival hemorrhages. The disease is self-limiting, and antimicrobial agents are not useful in treating it. Viral conjunctivitis can also be caused by herpes simplex virus type 1. Ulcerations caused by infection with this virus can cause blindness in some cases. The use of antiviral drugs is effective in controlling herpes simplex virus but not other viral infections of the eye.

Trachoma

Trachoma is a form of keratoconjunctivitis caused by *Chlamydia trachomatis*. Trachoma occurs worldwide but is most prevalent in dry regions, including the southwestern United States, where this disease is a particular problem on Indian reservations. Transmission occurs through direct contact of the eye with infectious material. The development of the chlamydial infection initially involves a localized infection of the conjunctivae, but this is followed by spread to other areas of the eye, including the cornea. In *C. trachomatis* infections vascular papillae and lymphoid follicles are formed, eye tissues can become deformed and scarred, and a chronic infection can lead to partial or total loss of vision. In fact, trachoma is the chief cause of blindness in the world. Several antibiotics, including sulfonamides, tetracyclines, and erythromicin, are useful in treating trachoma.

22.6 *Infections of the Oral Cavity*

Dental Caries

The surfaces of the oral cavity are heavily colonized by microorganisms. Excessive growth of microorganisms in the mouth can cause diseases of the tissues of the oral cavity. One of the most common human diseases caused by microorganisms is **dental caries**. Caries is initiated at the tooth surface as a result of the growth of *Streptococcus* species. These streptococci can initiate caries because they have the following essential properties: (1) they can adhere to the tooth surface; (2) they produce lactic acid as a result of fermentative metabolism, thereby dissolving the dental enamel

surface of the tooth; and (3) they produce a polymeric substance, which causes the acid to remain in contact with the tooth surface. *S. mutans, S. sanguis,* and *S. salivarius* are implicated as the causative agents of dental caries. These bacteria produce dextran sucrase, which catalyzes the formation of extracellular glucans from sucrose, leading to the formation of dental plaque (Figure 22.21). Dental plaque is an accumulation of a mixed bacterial community in a dextran matrix; it may be as many as 500 cells thick. There is a high degree of structure within plaque, indicative of sequential

FIGURE 22.21

Streptococcus mutans in dental plaque, the bacterial accumulation on the tooth, is seen in this carious lesion of tooth enamel (3600×). These bacteria produce polymers and an enzyme, dextran sucrase, that creates a sticky film on the tooth surface, providing a favorable environment for bacteria to grow and attack tooth enamel; it also restricts the movement of their metabolic product, lactic acid, away from the tooth surface, resulting in tooth decay. Of particular importance is sucrose in the diet, which is necessary for the formation of the plaque polymer. Reducing the intake of sucrose and cleaning the tooth surface to remove plaque buildup reduces dental caries. (From BPS—Z. Skobe, Forsythe Dental Center)

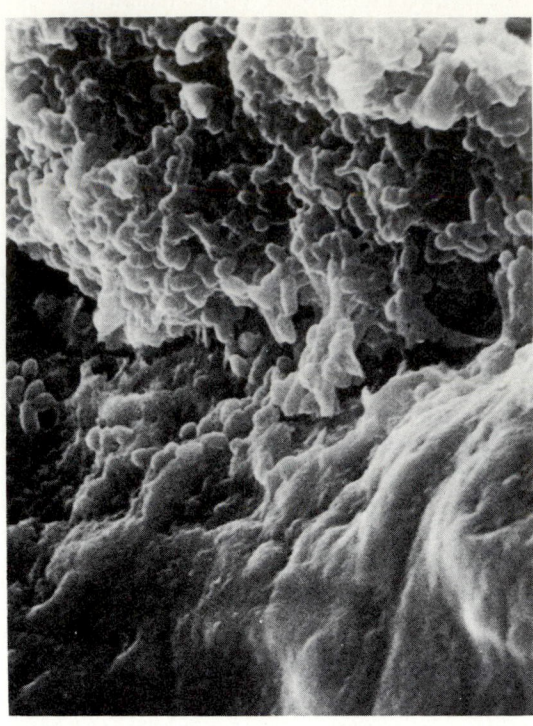

Discovery Process

The ecological principles that have evolved from studies of the adherence of bacteria to surfaces of the mouth, particularly the teeth, appear to have broad applicability to host–parasite relationships in other areas of the body and may provide a basis for understanding the natural resistance or susceptibility of hosts and tissues to pathogenic agents. Although a variety of bacterial types are indigenous to the mouth, probably because of the diversity of surfaces available for colonization and its relatively unselective conditions, streptococci and Gram-positive, filamentous organisms are most prevalent. *Streptococcus mutans* is suspected to be of etiological importance in dental caries. R. J. Gibbons and J. van Houte found that highly pathogenic strains of *S. mutans* characteristically form large bacterial masses on the surfaces of teeth, whereas most of the other nonpathogenic species they tested did not. The formation of dental plaques by *S. mutans* was significantly enhanced by adding sucrose to the diets of test animals. The addition of other carbohydrates, such as glucose, maltose, and fructose, did not have this effect. *S. mutans* can synthesize extracellular glucans and fructans from sucrose, which in turn allow it to attach and accumulate on the walls of culture vessels, immersed teeth, and other solid objects.

The mouth is subjected to fluid flow. Therefore, colonization requires either that bacteria attach themselves to the surface or that bacterial multiplication occur at a rate faster than the dilution rate caused by the flowing secretions. The available evidence does not support the latter theory. Until recently, though, little has been known about the mechanisms responsible for bacterial attachment to surfaces. Other studies by Gibbons, van Houte, and co-workers suggest that reversible sorption occurs and that only a very small percentage of the cells that do sorb to surfaces in the mouth become sufficiently firmly bound to initiate colonization. In the case of *S. mutans*, this firm attachment to a tooth is promoted by the *de novo* synthesis of extracellular polysaccharides from sucrose while the organisms are weakly associated with the surface. This appears to be the mechanism by which dietary sucrose fosters the implantation of *S. mutans* into the mouths of rodents and humans, because *in vitro* experiments and experiments with animals have shown that the presence of preformed glucans on the bacterial surface prior to sorption does not promote firm attachment. Electron microscopic studies have shown that oral *Streptococcus* species possess a fibrillar coating; in *S. salvarius*, *S. mitis*, and *S. pyogenes* these fibrils mediate the attachment to epithelial surfaces. It also appears that substances in saliva mediate the establishment of the primary film that leads to the formation of dental plaque, with subsequent bacterial growth resulting in the formation of dental caries. It is hoped that once the mechanism of attachment is fully understood, ways will be found to prevent plaque formation, with concomitant elimination of dental caries.

colonization by different bacterial populations and the different positions of each population within this complex bacterial community.

Dental caries can be prevented by limiting dietary sugar substrates, from which the bacteria produce acid and plaque, and by removing accumulated food particles and dental plaque by periodic brushing and flossing of the teeth. Such hygienic practices are particularly effective if performed after eating meals and snacks. The development of aspartame-sweetened sugarless gums is an aid to some people in limiting exposure of teeth to sugars. Additionally, tooth surfaces can be rendered more resistant to microbial attack by including calcium in the diet, such as by drinking milk, and by fluoride treatments of the tooth surface. The administration of fluorides in the diet, such as by consumption of fluoridated water, during the period of tooth formation can reduce dental caries by as much as 50 percent. New methods of plaque removal, including the use of enzymes in mouthwashes and slightly acidic toothpastes, should aid in limiting dental caries. Attempts are also being made to limit colonization of the tooth surface by using antibodies to block the sites of attachment of oral streptococci. One related new approach to preventing dental caries involves the use of low-acid–producing streptococci to preempt colonization of the oral cavity by the normal lactic acid–producing strains of *Streptococcus*; this approach is still highly experimental.

Periodontal Disease

In addition to causing dental caries, microorganisms growing in the oral cavity can cause diseases involving the supporting tissues of teeth (**periodontal disease**). Periodontal disease can be manifested as **gingivitis** (inflammation of the gingiva), **necrotizing ulcerative gingivitis** (**trench mouth** or **Vincent's disease**), **periodontitis** (inflammation of the periodontium, also known as **pyorrhea**), and **juvenile periodontitis** (noninflammatory degeneration of the periodontium leading to bone regression) (Figure 22.22). Periodontal disease develops when dental plaque accumulates between the tooth and the surrounding tissues. Bacteria associated with plaque clearly play a role in the development of periodontal disease. Antibiotics that suppress microbial populations in the oral cavity limit the the advance of periodontal disease, indicating the role of bacteria in the etiology of gingivitis and periodontitis. Although it is unlikely that Koch's postulates can ever be satisfied to identify the specific etiological agents of periodontal disease, it appears that different bacterial populations are involved in different stages of the development of this disease.

Healthy periodontal tissues in humans appear to be associated with relatively few indigenous microorganisms. Most of these organisms are Gram-positive cocci that are located supragingivally on the tooth surface. Microorganisms commonly associated with these tissues include *Streptococcus mitis*, *S. sanguis*, *Staphylococcus epidermidis*, *Rothia dentocariosa*, *Actinomyces viscosus*, *A. naeslundii*, and occa-

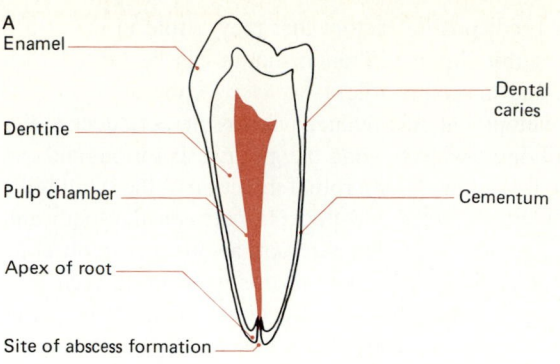

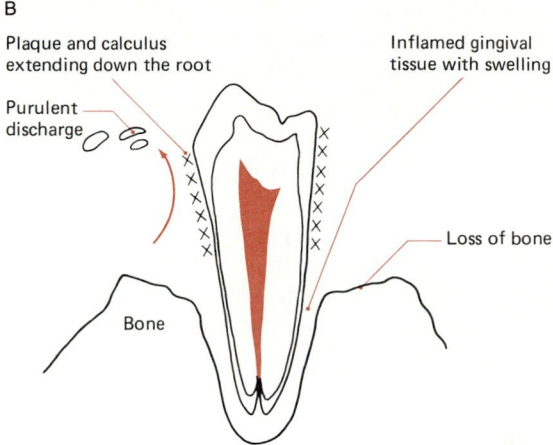

A
- Enamel
- Dentine
- Pulp chamber
- Apex of root
- Site of abscess formation
- Dental caries
- Cementum

B
- Plaque and calculus extending down the root
- Purulent discharge
- Bone
- Inflamed gingival tissue with swelling
- Loss of bone

FIGURE 22.22
(A) The anatomy of a tooth, showing the sites of microbial attack. (B) Diagram showing that the development of peridontal disease involves the tissues surrounding the tooth.

sionally, species of *Neisseria* and *Veillonella*, among others. The onset of gingivitis is marked by characteristic changes in the bacterial populations associated with gingival plaque. Initially there appears to be an overgrowth of the normal supragingival plaque accompanied by a large increase in the proportion of *Actinomyces* species, such as *A. vicosus*. These *Actinomyces* species may be important in the attachment of other bacterial populations. As the inflammation of the gingiva progresses, anaerobic Gram-negative bacteria, such as species of *Veillonella*, *Campylobacter*, and *Fusobacterium*, become prevalent. These Gram-negative bacteria appear to be located primarily on the surface of the plaque in subgingival sites.

The development of acute necrotizing ulcerative gingivitis is the only form of periodontal disease in which invasion of microorganisms into the tissues occurs. Clinical manifestations of trench mouth include fever, swelling, ulceration, and bleeding of the gums with tissue necrosis, causing ulcers between the teeth. Ultrastructural studies have shown that an unknown spirochete infiltrates the tissues and is the likely cause of this disease. This infection occurs mostly in adolescents and young adults. The name *trench mouth* is derived from the prevalence of this condition in the trenches of World War I. Fatigue, poor diet, poor oral hygiene, and anxiety are important predisposing factors that play a role in the establishment of this disease. Trench mouth can be treated with penicillin and metronidazole.

The development of advanced forms of periodontal disease, involving tissues beyond the gingiva, is serious and can lead to the loss of teeth. Microbial infection of the subgingiva can establish abscesses and pockets between the tooth and supporting tissues in which microorganisms can proliferate (**peridontal pocket**). Microorganisms growing in such protected pockets erode the alveolar bone and destroy periodontal membranes, causing loosening of the teeth that can lead to tooth loss. The control of periodontal disease can be achieved by frequent removal of dental plaque, by daily flossing, and by periodic professional dental cleanings. Flossing removes plaque, which plays an important role in the development of periodontal disease; prophylactic dental treatment also removes calcified plaque (calculus), which may help promote periodontal disease.

In some cases, there are large accumulations of plaque for years without destruction of the supporting structures. In contrast, some individuals develop dramatic loss of the periodontium with little evidence of plaque accumulation or gingival inflammation. It may be that local factors or the development of specific bacterial populations within the plaque determine the course of periodontal disease. The progression of inflammation into the periodontal membrane involves the development of a mixed community that is dominated by Gram-negative bacteria.

The specific bacterial populations vary in different syndromes of periodontitis. For example, up to 75 percent of the isolates in cases of advanced periodontitis have been found to be Gram-negative rods, predominantly *Bacteroides melaninogenicus* subspecies *asaccharolyticus* and *Fusobacterium nucleatum*. *Eikenella corrodens*, spirochetes, and various unidentifed motile bacteria also occur in relatively high numbers in cases of periodontitis. Juvenile periodontitis (**periodontosis**) is a distinct disease associated with different Gram-negative bacterial populations. It involves a rapidly progressing periodontal breakdown that affects young patients. The periodontal lesions are localized to permanent first molars and incisors or are generalized. The deep pocket bacteria in cases of juvenile periodontitis are comprised of approximately 65 percent Gram-negative bacteria. *Capnocytophaga* species (formerly *Bacteroides* species) and *Actinobacillus actinomycetemcomitans* are the predominant isolates, suggesting that these baceria play an important role in the etiology of this disease.

Postlude

In this chapter, we have examined a number of human diseases caused by microbial pathogens that enter the body through the skin and cause diseases of superficial body tissues. Trauma, especially if associated with the introduction of foreign material, is important in establishing infections of surface body tissues. Direct contact and fomites contaminated with pathogenic microorganisms are often involved in the transmission of infections of the eye and skin, and proper hygienic practices, which reduce the likelihood of contact with pathogenic microorganisms, greatly reduce the incidence of these diseases.

In the case of diseases of the oral cavity caused by microorganisms indigenous to the mouth, it is not possible to avoid contact with the pathogens. Consequently, the prevention of dental caries and periodontal disease depends on the effective removal of dental plaque with its associated microbial populations. Essentially, the prevention of these diseases depends on the elimination of microbial metabolic products before they can accumulate in concentrations sufficient to cause diseases involving the tissues of the oral cavity. Inclusion of elements in the diet, such as fluoride, that render the host tissues more resistant to microbial attack, and the reduced intake of foods containing high carbohydrate levels, which microorganisms can readily utilize as substrates to proliferate and form dental plaque, can greatly reduce the incidence of dental caries and periodontal disease.

Wounds and animals bites that disrupt the protective skin layers provide the main portals of entry by which pathogenic microorganisms penetrate the skin and gain entry to the body. Minimizing contamination of wounds with foreign matter and associated microorganisms is important in controlling infections that may occur when the skin barrier is broken. Surgical practices utilize elaborate aseptic procedures to minimize potential infection, but nevertheless, infections sometimes occur after surgery, attesting to the vulnerability of the body to microbial infection when the skin barrier is disrupted and host defense mechanisms are impaired.

Animal bites are extremely important in the transmission of various infectious diseases. Many arthropod populations act as vectors, providing both the microbial inoculum and the portal of entry for initiating an infectious process. Reducing the populations of vectors, such as mosquitos, ticks, fleas, and lice, is important in controlling the morbidity rates of diseases transmitted in this manner. Additionally, because many diseases transmitted via arthropod vectors are maintained in nonhuman animal populations, it is important to recognize and control the reservoir populations.

In the preceding six chapters, the principles of interaction between microorganisms and humans that produce human disease have been considered. The establishment of infectious disease processes in humans involves dynamic interactions between microbial pathogens and human host cells and tissues. The human body possesses an elaborate defense network that acts to prevent microbial infections, including the antibody-mediated and cellular immune defense mechanisms, which generally preclude the growth of microorganisms in body tissues. However, as we have seen, pathogenic microorganisms possess metabolic capabilities that allow them to overcome or escape these defense mechanisms and produce disease. The manifestations of disease depend on the dynamics of the interaction between the host and parasite.

In this and the preceding chapter, the epidemiology of infectious diseases has been emphasized. Microbial pathogens are generally restricted to specific portals of entry that determine their potential routes of transmission. Understanding the transmission processes permits the development of effective means of preventing or at least reducing the incidence of many diseases. When prevention fails and an infectious disease occurs, the clinical microbiology laboratory employs various methods for diagnosing the etiology of the disease and determining the applicability of the various antimicrobial agents available to treat it.

At least in developed countries, the prevention and control of infectious diseases have greatly reduced morbidity and mortality rates. Similar advances have yet to be fully realized in socioeconomically depressed regions of the world, where poor nutrition, poor sanitation, and lack of sufficient medical facilities allow the continuation of high rates of morbidity and mortality from infectious diseases. Advances in our understanding of host–parasite relationships and improvements in the delivery of health services based on our current understanding of disease processes can alleviate much of the suffering and death associated with infectious diseases.

Suggested Supplementary Readings

BOYD, R. F., and B. G. HOERL. 1981. *Basic Medical Microbiology*. Little, Brown and Co., Boston.

BOYD, R. F., and J. J. MARR. 1980. *Medical Microbiology*. Little, Brown and Co., Boston.

BRAUDE, A. I. (ed.). 1981. *Medical Microbiology and Infectious Disease*. W.B. Saunders Co., Philadelphia.

BURGDORFER, W., and R. ANACKER. 1982. *Rickettsiae and Rickettsial Diseases*. Academic Press, New York.

BURNETT, G. W., H. W. SCHERP, and G. S. SCHUSTER. 1976. *Oral Microbiology and Infectious Disease*. Wiliams & Wilkins Co., Baltimore.

DAVIS, B. D., R. DULBECCO, H. N. EISEN, H. S. GINSBERG, and W. B. WOOD. 1981. *Microbiology*. Harper & Row, Publishers, Inc., Hagerstown, Md.

DIXON, R. E. 1981. *Nosocomial Infections*. Yorke Medical Books, New York.

DUGUID, J. P. (ed.). 1979. *Medical Microbiology*. Churchill Livingstone Inc., New York.

EMMONS, C. W., C. H. BINFORD, J. P. ULZ, and K. J. KWON-CHUNG. 1977. *Medical Mycology*. Lea & Febiger, Philadelphia.

FAUST, E. C., P. C. BEAVER, and R. C. JUNG. 1975. *Animal Agents and Vectors of Human Disease*. Lea & Febiger, Philadelphia.

HOEPRICH, P. D. (ed.). 1983. *Infectious Diseases: A Modern Treatise of Infection Processes*. Harper & Row, Publishers, Inc., Hagerstown, Md.

JAWETZ, E., J. L. MELNICK, and E. A. ADELBERG. 1986. *A Review of Medical Microbiology*. Appleton and Lange, New York.

JOKLIK, W. K., H. P. WILLETT, and D. B. AMOS (eds.). 1984. *Zinsser Microbiology*. Appleton-Century-Crofts, New York.

LOESCHE, W. J. 1986. Role of *Streptococcus mutans* in human dental decay. *Microbiological Reviews* 50:353–380.

MANDELL, G., R. G. DOUGLAS, JR., and J. E. BENNETT. 1985. *Principles and Practice of Infectious Diseases*. John Wiley & Sons, Inc., New York.

NOLTE, W. A. (ed.). 1980. *Oral Microbiology*. C.V. Mosby Co., St. Louis.

PECHERE, J. C. (ed.). 1984. *Infections: Recognition, Understanding, Treatment*. Lea & Febiger, Philadelphia.

REESE, R. E., and R. G. DOUGLAS, JR. 1986. *A Practical Approach to Infectious Diseases*. Little, Brown and Co., Boston.

ROBERTS, R. B. 1986. *Infectious Disease: Pathogenesis, Diagnosis, and Therapy*. Year Book Medical Publishers, Inc., Chicago.

TOP, F. H., and P. WEHRLE. 1976. *Communicable and Infectious Diseases*. C.V. Mosby Co., St. Louis.

VOLK, W. A., D. C. BENJAMIN, R. J. KADNER, and J. T. PARSONS, 1986. *Essentials of Medical Microbiology*. J.B. Lippincott Co., Philadelphia.

WILSON, G. S., and A. A. MILES (eds.). 1975. *Topley and Wilson's Principles of Bacteriology and Immunity*. Williams & Wilkins Co., Baltimore.

YAMAGUCHI, T. (ed.). 1981. *Color Atlas of Clinical Parasitology*. Lea and Febiger, Philadelphia.

YOUMANS, G. P., P. Y. PATERSON, and H. M. SOMMERS. 1980. *The Biological and Clinical Basis of Infectious Disease*. W.B. Saunders Co., Philadelphia.

Study Questions

1. What is a vector? What are some common vectors of infectious diseases?
2. For each of the following diseases, what is the causative organism, the reservoir for the etiological agent, and the vector for disease transmission?

   a. Plaque
   b. Rocky Mountain Spotted Fever
   c. Malaria
   d. Smallpox
   e. Endemic typhus

3. How is rabies treated?

4. What is the etiological agent of malaria, and why is this disease difficult to treat?

5. Why are serious *Clostridium* infections associated with deep wounds?

6. What are the differences between the toxins produced by *Clostridium botulinum* and *C. tetani*?

7. Why do urinary tract infections frequently occur after catherization procedures?

8. What is the cause of serum hepatitis? Why is this disease associated with drug addicts?

9. Why are burn victims prone to fatal microbial infections?

10. Discuss the etiology of dental caries. Why should removing sucrose from the diet lower the rate of this disease?

Situational Problem

Trying to Help Find the Source of Diseases

If you have a job with the Department of Public Health in Peoria, Illinois, you might be in charge of recording the cases of infectious diseases as reported by local hospitals. These reports are forwarded to the Centers for Disease Control (CDC) in Atlanta. Although it is not specifically part of your job, you decide to try to identify the sources of the etiological agents for each of the diseases you report to CDC and to alert your supervisor to any cases where you feel steps should be taken on the local level to prevent additional cases of the disease.

1. One week after a family of four returned home after traveling to Africa, where they were on safari for 10 days, one of the children became quite ill, exhibiting periodic severe chills that recur every 24 hours, serious muscle pain, and severe headache. When the pain was not severe the child slept a great deal and, when awakened, was quite tired. None of the other family members exhibited any of these symptoms. The physician has reported this as a probable case of malaria. What source of the disease would you suspect, and would you recommend any local precautionary steps to preclude the development of additional cases?

2. Five men from an investment group traveled to Panama on a business trip. They stayed at a five-star hotel that had excellent restaurants and a chlorinated water supply. They all ate exactly the same meals. Shortly after their return, one of the men found that he had little appetite. After eating very little for a day, he developed the symptoms of nausea, vomiting, and fever. The next day he developed jaundice. The symptoms lasted for about a week, after which he exhibited a full recovery. The physician has reported this as a probable case of yellow fever. What source of the disease would you suspect, and would you recommend any local precautionary steps to preclude the development of additional cases?

3. Because of injuries sustained in an automobile accident, a woman required transfusions with 2 units of blood. During her recovery, she developed a fever that oscillated daily for a week and then disappeared. She showed no sign of pneumonia. Just before her body temperature returned to normal, she started to experience abdominal pain and nausea. A few days later, she showed yellowing of the skin. Elevated levels of transaminases were detected in her blood at this time. Five days later, she died. The physician has reported this as a probable case of hepatitis. What source of the disease would you suspect, and would you recommend any local precautionary steps to preclude the development of additional cases?

Appendices

The Metric System and Some Useful Conversion Factors

Metric System Prefixes

pico (p) $= 10^{-12}$
nano (n) $= 10^{-9}$
micro (μ) $= 10^{-6}$
milli (m) $= 10^{-3}$
centi (c) $= 10^{-2}$
deci (d) $= 10^{-1}$
deka (da) $= 10$
hecto (h) $= 10^2$
kilo (k) $= 10^3$
mega (M) $= 10^6$

Length

1 kilometer (km) $= 0.62$ mile (mi)
1 mi $= 1.609$ km
1 meter (m) $= 3.28$ feet (ft) $= 1.09$ yards (yd)
1 ft $= 0.305$ m
1 yd $= 0.914$ m
1 centimeter (cm) $= 0.394$ inch (in)
1 in $= 2.54$ cm
1 ft $= 30.5$ cm
1 millimeter (mm) $= 0.039$ in
1 Angstrom (Å) $= 10^{-10}$ m

Area

1 hectare (ha) $= 10,000$ m^2 $= 2.471$ acres
1 acre $= 0.4047$ ha
1 sq km (km^2) $= 100$ hectares $= 0.3861$ sq mi
1 sq mi $= 2.590$ km^2
1 sq m (m^2) $= 10,000$ cm^2 $= 1.1960$ sq yd $=$ 10.764 sq ft
1 sq yd $= 0.8361$ m^2
1 sq ft $= 0.0929$ m^2
1 sq cm (cm^2) $= 100$ sq. mm $= 0.155$ sq in
1 sq in $= 6.4516$ cm^2

Mass

1 metric ton (t) $= 1,000$ kilograms $= 1.10$ short tons
1 short ton $= 0.91$ t
1 kilogram (kg) $= 1,000$ grams $= 2.205$ pounds
1 pound $= 453.60$ g
1 gram (g) $= 1,000$ milligrams $= 0.0353$ ounce
1 ounce $= 28.35$ g
1 milligram (mg) $= 10^{-3}$ g $= 0.02$ grains
1 microgram (μg) $= 10^{-6}$ g
1 nanogram (ng) $= 10^{-9}$ g
1 picogram (pg) $= 10^{-12}$ g

Volume (Solids)

1 cu m (m^3) = 1,000,000 cm^3 = 35.315 cu ft =
1.3080 cu yd
1 cu ft = 0.0283 m^3
1 cu yd = 0.7646 m^3
1 cu cm (cm^3) = 10^3 mm^3 = 0.0610 cu in
1 cu in = 16.387 cm^3

Volume (Liquids)

1 liter (L) = 10^3 milliliters = 1.06 quart
1 gallon (gal) = 3.785 L
1 kiloliter (kL) = 1,000 L = 264.17 gallons
1 qt = 0.94 L
1 milliliter (mL) = 10^{-3} L = 0.034 fluid oz
1 fl oz = 29.57 mL
1 microliter (μL) = 10^{-6} L

Temperature

degrees Fahrenheit (°F) = 9/5°C + 32
degrees centigrade (°C) = 5/9 (°F − 32)
0°C = 32°F (freezing point of water)
100°C = 212°F (boiling point of water)

Review of Basic Chemistry

A2.1 *Nature of Organic Molecules*

Molecules are specific combinations of chemical elements formed when atoms establish chemical bonds by sharing or exchanging electrons in their outer orbitals. The electrons in an atom are arranged in a defined order, located at specified energy levels or shells (Figure A2.1). The electron shell, or orbital of electrons, represents a volume occupied by an electron cloud at some distance from the central nucleus of the atom. The maximum number of electrons that can occupy a particular energy level increases with the distance from the nucleus; for example, the first orbital holds a maximum of two electrons; the second orbital, eight electrons. The electrons in the outermost valence shell normally form chemical

bonds, with stable bonds occurring when atoms fill their valence shells. The number of bonds that an element can form depends on the number of electrons required by that atom to fill its outer electron shell; carbon, the central atom in organic molecules, for example, can establish four bonds because it has four electrons in its outer electron shell that can hold a maximum of eight electrons.

Carbon-containing molecules—with the exception of carbon dioxide, carbon monoxide, and a few others, which contain carbon but do not hydrogen—are organic; all other molecules are inorganic. In addition to carbon, hydrogen, oxygen, and nitrogen, the elements phosphorus, sulfur, sodium, potassium, magnesium, calcium, iron, and chlorine also occur in relatively high abundance within the biochemicals of microorganisms. To fill their outer electron shells, nitrogen requires three electrons, oxygen requires two, and hydrogen requires one; therefore, nitrogen can establish three chemical bonds, oxygen two, and hydrogen one. Both inorganic and organic molecules are essential components of microorganisms, and although life is based on organic carbon, it is likewise totally dependent on inorganic water.

H

First electron shell

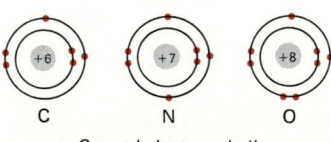

C   N   O

Second electron shell

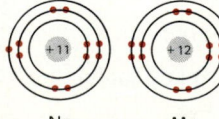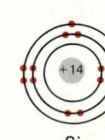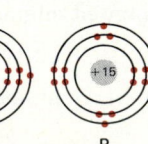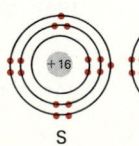

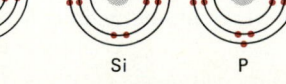

Na   Mg   Si   P   S   Cl

Third electron shell

FIGURE A2.1

The electrons of an atom are arranged in orbitals. Each orbital represents a different energy level.

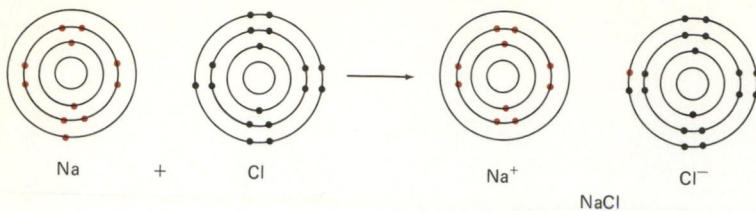

FIGURE A2.2

The formation of ionic bonds involves the loss and gain of electrons. In the formation of NaCl, the chlorine atom gains an electron to fill its outer electron shell, and the sodium atom loses an electron so that all of the remaining electron shells are filled. After the formation of the ionic bond, the sodium has a positive charge and the chloride a negative charge.

Types of Chemical Bonds

Two basic types of chemical bonds hold molecules together. Ionic bonds are based on charge differences between chemical elements that develop when atoms donate or acquire electrons to fill their outer electron shells. For example, the salt sodium chloride represents a molecule formed by an ionic bond between the elements sodium and chlorine (Figure A2.2). The sodium atom donates an electron, forming a positively charged sodium ion (cation); the chlorine accepts the electron, forming a negatively charged chloride ion (anion); the positively charged sodium ions and negatively charged chloride ions are held together by electrostatic forces. Ionic bonds are relatively weak, and in aqueous solution they are readily broken with the dissociation of ions. Consequently, ionic bonds are not strong enough to hold together the macromolecules of living systems.

In contrast to the ionic bond, where an electron is completely transferred from one element to another, covalent bonds form when elements share electrons (Figure A2.3). The number of covalent bonds that an element is capable of forming depend on its electronic structure. Each covalent bond involves the sharing of a pair of electrons, with each atom contributing one electron. In some cases, atoms share two pairs of electrons, giving rise to a double bond. Once formed, covalent bonds require a relatively high amount of energy to break them; they provide the stability needed to establish the macromolecules required for the existence of microbes and other living organisms.

Covalent bonds exhibit differences in relative strength, depending on which elements share electrons (Table A2.1).

FIGURE A2.3

The formation of the covalent bond involves the sharing of electrons. In molecular hydrogen (H_2) each atom shares its first orbital electron with the other, effectively completing the first orbital of each atom.

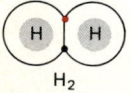

H_2

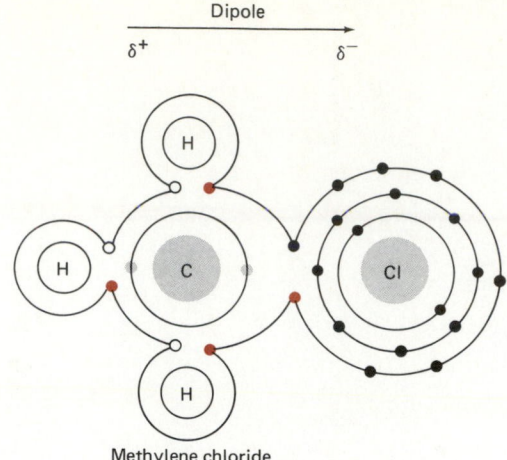

Methylene chloride

FIGURE A2.4

The covalent bonds of methylene chloride show the formation of dipole moment due to the uneven sharing of electrons. The shared electron draws closer to the chlorine (Cl) than to the carbon (C), thus establishing a charge separation.

There are also differences in the distribution of electrons between the atoms forming covalent bonds. If two atoms of the same element share electrons to form a molecule, the electrons are evenly distributed between the atoms. This occurs in a molecule such as hydrogen (H_2). However, when a covalent bond forms between the atoms of two different elements, the electrons are shared unevenly, with the electrons exhibiting a greater affinity for one of the atoms and being drawn closer to that atom (Figure A2.4). This gives rise to polarity in which one atom has a greater positive charge and the other atom a greater negative charge. The molecule is said to have a dipole resulting from the separation of charge, with the polarity of a covalent bond depending on the specific elemental atoms involved in establishing the bond. The polarity of covalent bonds establishes the basis for charge interactions and the formation of additional weak bonds between organic molecules.

TABLE A2.1   Relative Strengths of Some Chemical Bonds

Bond	Relative Strength
Covalent (energy of interaction, 30–100 kcal/mole)	
H-H	1.0
C-C	0.8
C-H	1.0
C-O	0.8
C-N	0.7
Ionic (energy of interaction, 10–20 kcal/mole)	
Na · · · Cl	0.3
Hydrogen (energy of interaction, 2–10 kcal/mole)	
H-O · · · O	0.1
H-N · · · N	0.1

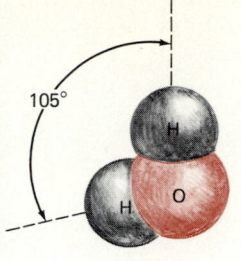

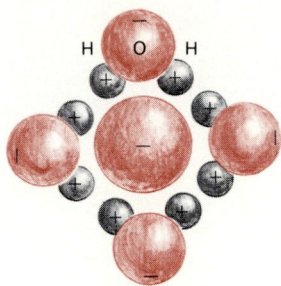

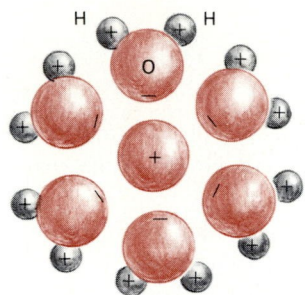

FIGURE A2.5

The spatial arrangement of atoms in a water molecule results in a dipole moment due to the unequal distribution of electrons between hydrogen and water. As a result, water is a good polar solvent because it can surround both positively and negatively charged ions.

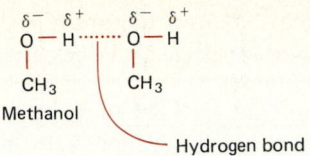

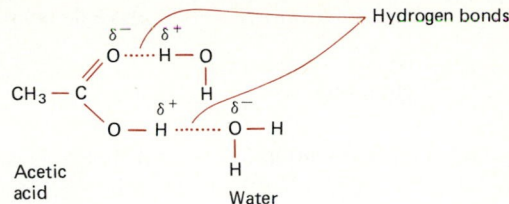

FIGURE A2.6

Hydrogen bonding occurs between molecules as a result of charge interactions stemming from the dipole moments of the molecules.

Water, for example, is an excellent polar solvent because it has a large dipole moment, that is, a large charge separation due to the unequal sharing of electrons between the oxygen and hydrogen atoms (Figure A2.5). The internal separation of charge in the water molecule permits intermolecular associations with other polar molecules. The positive regions of the solute molecule associate with the negative oxygen atom of water, and the negative regions of the solute associate with the positive hydrogen atoms of water. These electrostatic interactions are responsible for separating polar solute molecules, enhancing their dissolution in aqueous solution. The solvent properties of water are essential for the occurrence of biochemical reactions within microbial cells. The interactions between polar groups also lead to the exclusion of nonpolar molecules, or nonpolar portions of large molecules, creating hydrophobic interactions between nonpolar molecules. The separation of the polar (*hydrophilic* means water-loving) and nonpolar (*hydrophobic* means water-fearing) ends of phospholipids establishes the basis for the selective permeability of biological membranes, an essential function of living systems.

Weak intermolecular bonds may occur between neighboring molecules. A hydrogen atom that is covalently bonded

to oxygen or nitrogen within a molecule may be simultaneously attracted to a nitrogen or oxygen atom of a neighboring molecule, forming a weak link known as a *hydrogen bond* between the two molecules (Figure A2.6). The strength of the hydrogen bond is only about one-tenth that of the covalent bond, but such weak bonds have very important functions in living systems. Hydrogen bonds are essential for maintaining the double helical structure of DNA. Hydrogen bonding accounts for the fluid nature of water and its ability to act as a good polar solvent. The capillary action of water, its ability to move upward through narrow pores, is dependent on the hydrogen bonding of the water molecules to the solid surfaces of a porous material, such as glass. In proteins, besides the strong covalent bonds that link the carbon skeleton of the molecule, intramolecular hydrogen bonds between the functional groups of different amino acids act to stabilize and determine the configuration of these large molecules. It is the fact that the weak hydrogen bonds can be readily formed and broken that allows the flexibility needed for these various biochemical functions.

Functional Groups and Structure of Organic Molecules

Organic molecules have vastly differing properties that are determined by their chemical composition, size, and the arrangement of the atoms within them, that is, the way in which the specific elements of the molecule are bonded together. Organic molecules consisting of only carbon and hydrogen, hydrocarbons, are relatively nonreactive, and although such compounds are important in fossil fuels, they do not comprise the major biochemicals of microorganisms. Rather, the major biochemical constituents of microorganisms include oxygen, nitrogen, sulfur, and other elements in addition to carbon and hydrogen. As a rule, the organic molecules that enter into biochemical reactions of microor-

TABLE A2.2 Common Functional Groups of Organic Molecules and Some Representative Classes of Molecules in Which They Occur

Group	Name	Class of Molecules
R—OH	Hydroxyl group	Alcohols, carbohydrates
R—C(OH)=O	Carboxyl group	Carboxylic acids, fatty acids, amino acids
R—C(R)=O	Keto group	Ketones, sugars
R—C(H)=O	Aldehyde group	Aldehydes, sugars
R—C(H)(H)—H	Methyl group	—
R—N(H)—H	Amino group	Amines, amino acids
R—P(OH)(OH)=O	Phosphate group	Organic phosphates, phospholipids, nucleotides
R—SH	Sulfhydryl group	Mercaptans, a few amino acids

ganisms possess functional groups, and the major classes of organic molecules comprising living systems are based on the presence of particular functional groups (Table A2.2). The substitution of a functional group for one of the hydrogens on the carbon skeleton of an organic molecule renders the molecule more reactive, establishing the basis for microbial metabolism.

An important property of a functional group is its ability to react with a different functional group of another molecule. The interactions between functional groups involve the formation of weak bonds. The nature of these weak bonds ensures that only complementary functional groups can react to form a bond, establishing a specificity of chemical reactions. The reactions between organic acids and bases, which form the basis for many fundamental biochemical reactions of microorganisms, involve such specific interactions between functional groups and the formation of weak ionic bonds between complementary groups. The ability of biochemicals to dissociate and form acids, which in water solution produce hydrogen ions (H^+), and bases, which in water solution produce hydroxyl ions (OH^-), is important in many of the biochemical reations of microorganisms. The concentration of hydrogen ions determines the pH of a solution, which is defined as $-\log[H^+]$. Pure water contains 10^{-7} moles of hydrogen ions per liter and thus has a pH of 7. At pH 7 the concentrations of H^+ and OH^- are equal; water therefore has a neutral pH. pH values below 7 are acidic, and those above 7 are basic. The pH affects the degree of dissociation of acids and bases, which has a marked influence on the properties of biochemical molecules. For example, the enzymatic activities of microorganisms are deter-

mined in large part by the degree of dissociation of their functional amino and carboxylic acid groups of the enzyme molecules.

The location of a functional group within an organic molecule can be quite important in determining the biochemical properties of that molecule. In order to keep track of the location of functional groups, a convention has been adopted for numbering the carbon atoms in an organic molecule. The specific numbering systems vary for different classes of organic molecules. An example of the numbering of a simple hydrocarbon and a carboxylic acid is shown in Figure A2.7. In some molecules, in addition to the formal numbering system, the term α-*carbon* is used to describe the carbon atom to which the functional group that typifies that class of compounds is directly attached, and the succeeding carbons are also designated by subsequent letters of the Greek alphabet. By establishing formal conventions the exact arrangement of atoms within the biochemical can be specified, permitting

FIGURE A2.7
An example of the numbering of a carbon chain of a hydrocarbon and a carboxylic acid and the use of Greek letters to designate the position relative to the functional group.

$$\overset{6}{H_3C} - \overset{5}{CH_2} - \overset{4}{CH_2} - \overset{3}{CH_2} - \overset{2}{CH_2} - \overset{1}{CH_3}$$
Hexane

$$\overset{6}{H_3C} - \overset{5}{\underset{\epsilon}{CH_2}} - \overset{4}{\underset{\delta}{CH_2}} - \overset{3}{\underset{\gamma}{CH_2}} - \overset{2}{\underset{\beta}{CH_2}} - \overset{1}{\underset{\alpha}{COOH}}$$
Hexanoic acid

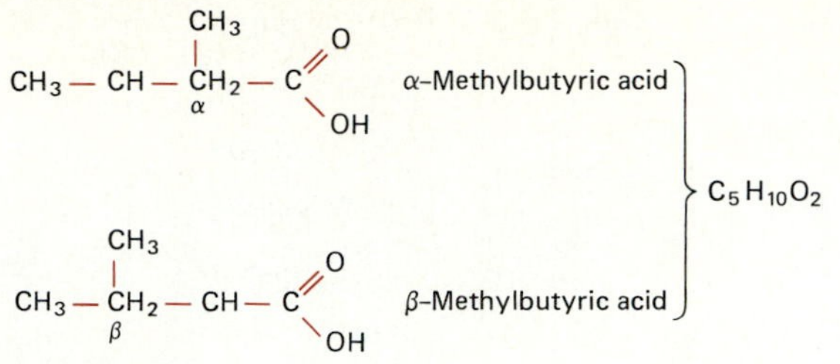

CH₃ — CH — CH₂ — C ⟨=O, OH⟩   α-Methylbutyric acid

CH₃ (above CH), α (below CH)

CH₃ — CH₂ — CH — C ⟨=O, OH⟩   β-Methylbutyric acid

CH₃ (above CH), β (below CH₂)

$C_5H_{10}O_2$

FIGURE A2.8
Structural isomers have identical molecular formulas, but the substituents are located at different positions. In this example of methylbutyric acid, the only difference is the position of the methyl group. Nevertheless, these structural isomers have different physicochemical properties.

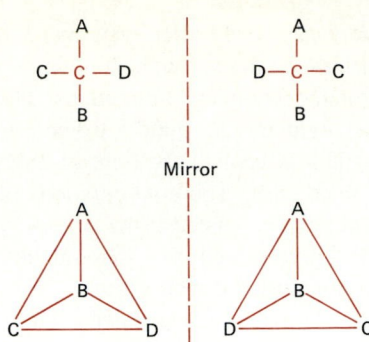

FIGURE A2.10
An asymmetric carbon allows molecules to exist in two different forms that are mirror images of each other.

discussion of the properties and function of that specific molecule in the biochemistry of microorganisms.

Isomerism

Compounds can have the same chemical composition, that is, the same molecular formula but still have very different chemical and physical properties, depending on the arrangement of the molecule. Such molecules with identical molecular formulas, but different spatial arrangements, are called *isomers*. The carbon atoms of isomers can be bonded at different positions, producing structures with different spatial arrangements (structural isomers) (Figure A2.8). Additionally, because organic molecules are three-dimensional, compounds that have the same structural formulas can still have different spatial arrangements. The spatial arrangements of

the atoms in the molecule are defined by the configuration and conformation of the molecule. *Configuration* describes the relative arrangement in space of the atoms that make up the molecule, and *conformation* describes the different spatial arrangements of the molecule that result from rotation about single covalent bonds (Figure A2.9).

Stereoisomers are compounds with the same molecular formula and the same skeletal structure but different three-dimensional arrangements of the atoms in space. Because a carbon atom can form bonds with four different groups, creating an asymmetric center, many organic compounds occur as stereoisomers. Organic molecules can have two different forms (*configurations*) about an asymmetric carbon atom that represent mirror images of the molecule (Figure A2.10). The mirror images of the molecule are not superimposable and represent truly different configurations. The different config-

FIGURE A2.9
Organic molecules vary in their configuration and conformation, as shown in this illustration. The stabilities of the differing configurations and conformations vary, and in each case there is a preferred (most stable) form. (A) The chair and boat conformations of cyclohexane. (B) The two possible configurations of the chair conformation of D-glucose.

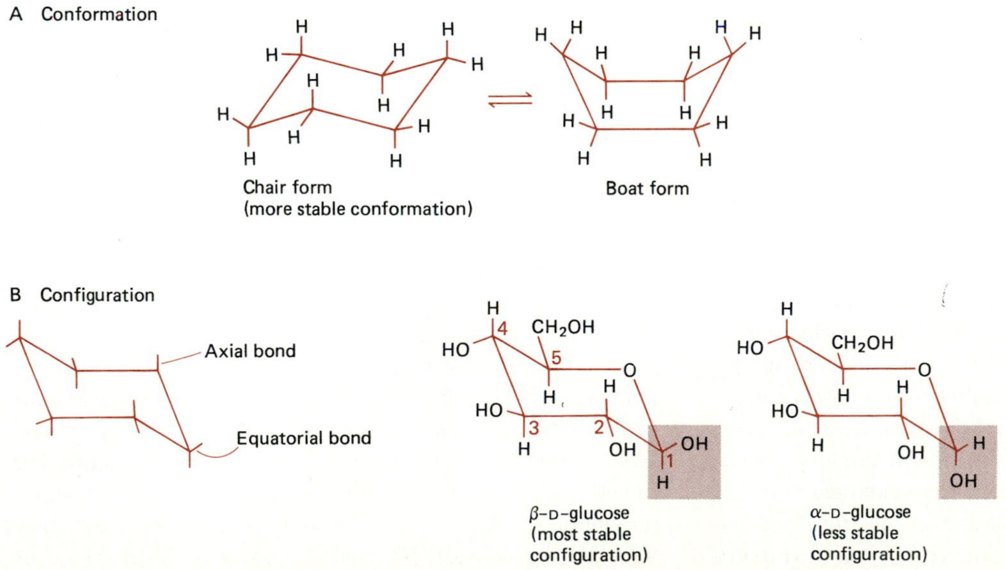

A   Conformation

Chair form
(more stable conformation)

Boat form

B   Configuration

Axial bond

Equatorial bond

β-D-glucose
(most stable
configuration)

α-D-glucose
(less stable
configuration)

urations of the stereoisomers can have optical activity, and when a plane of polarized light shines through the two different isomers, it will be rotated in opposite directions. The direction of rotation of the light by the optical isomers is termed *dextrorotatory* (+) if it is bent to the right or *levorotatory* (−) if it is bent to the left. The configurations of all optically active compounds can be related to the arbitrarily chosen reference molecule glyceraldehyde, the smallest sugar that has an asymmetric carbon atom. The two stereoisomers of glyceraldehyde are designated D and L (Figure A2.11). All stereoisomers that are related to L-glyceraldehyde are designated L, and those related to D-glyceraldehyde are designated D, regardless of the direction of rotation of polarized light given by the isomer. The designations D and L therefore denote the absolute configuration of a molecule. Many biochemicals found in microorganisms contain either D or L forms of the molecule, but not both. Proteins contain only L amino acids, although D amino acids occur as constituents of other macromolecules. Stereoisomerism is important because the different stereoisomers of a molecule exhibit different reactivities in chemical reactions. Generally, only one of the optical isomers is capable of participating in a given biochemical reaction.

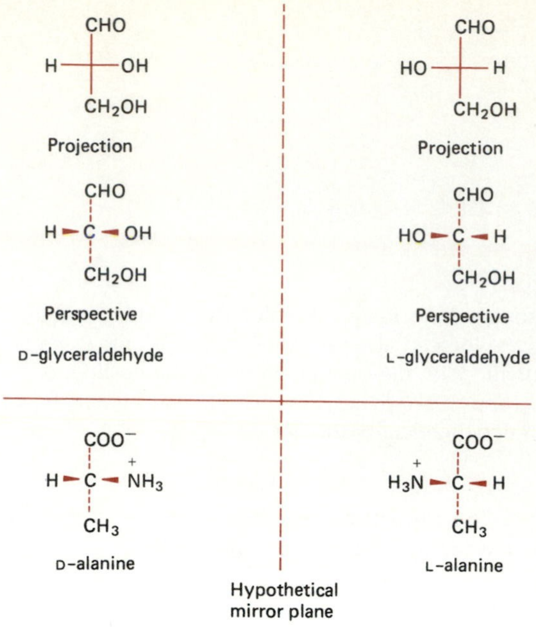

FIGURE A2.11

The D and L forms of a glyceraldehyde molecule and its relation to L alanine.

A2.2 *Macromolecules Comprising Microorganisms*

In the biochemical molecules of living systems, elements are linked together by covalent bonds to form relatively high molecular weight macromolecules with differing functional groups and spatial arrangements. The major classes of macromolecules that make up the structural and functional units of microorganisms are the carbohydrates, lipids, proteins, and nucleic acids.

Carbohydrates

Carbohydrates have the basic chemical formula $C_n(H_2O)_n$ and include simple sugars, such as glucose, and the macromolecules formed from such simple monosaccharides. By definition, carbohydrates are either polyhydroxy-aldehydes, or polyhydroxy-ketones, or molecules that yield polyhydroxy-aldehydes or polyhydroxy-ketones upon hydrolysis. Those carbohydrates that cannot be hydrolyzed are called *monosaccharides* (Figure A2.12). Individual monosaccharides may be linked to form larger polymeric carbohydrate molecules. A disaccharide contains 2 monosaccharide units, an oligosaccharide contains 2 to 10 monosaccharide units, and a polysaccharide contains more than 10 monosaccharide units.

In aqueous solution the five- and six-membered monosaccharides can form ring structures, and the structures of these carbohydrates frequently are illustrated in the ring form as opposed to the simpler straight-chain representation (Figure

A2.13). As a result of ring formation, monosaccharides act as if they have an extra asymmetric carbon; therefore, in addition to being designated as D or L, these compounds are designated as α or β.

The bond formed between monosaccharides to create disaccharides and larger polysaccharides is called a *glycosidic bond*. The glycosidic bond forms between the aldehyde group of one of the monosaccharide units and one of the alcohol groups of the other monosaccharide unit, with the elimination of water (Figure A2.14). When a glycosidic bond forms, the functional aldehyde or ketone group of one of the monosaccharide units is generally left free and therefore remains reactive. The disaccharides sucrose and trehalose are exceptions in which the reactive groups of both monosaccharides are linked. The glycosidic bond is specified by the position numbers of the carbons that are linked by the bond and by the three-dimensional orientation of the bond. Glycosidic bonds are termed α or β, depending on their three-dimensional orientation. The same monosaccharides can form various disaccharides and polysaccharides by forming glycosidic bonds with different orientations. For example, maltose is formed by an α-1-4 linkage of two glucose units, and cellobiose is formed by a β-1-4 linkage of two glucose units. Similarly, starch represents a polymer of glucose units linked by α-1-4 glycosidic bonds, and cellulose represents a polymeric chain of glucose-monosaccharides that are linked with β-1-4 glycosidic linkages (Figure A2.15). Starch occurs

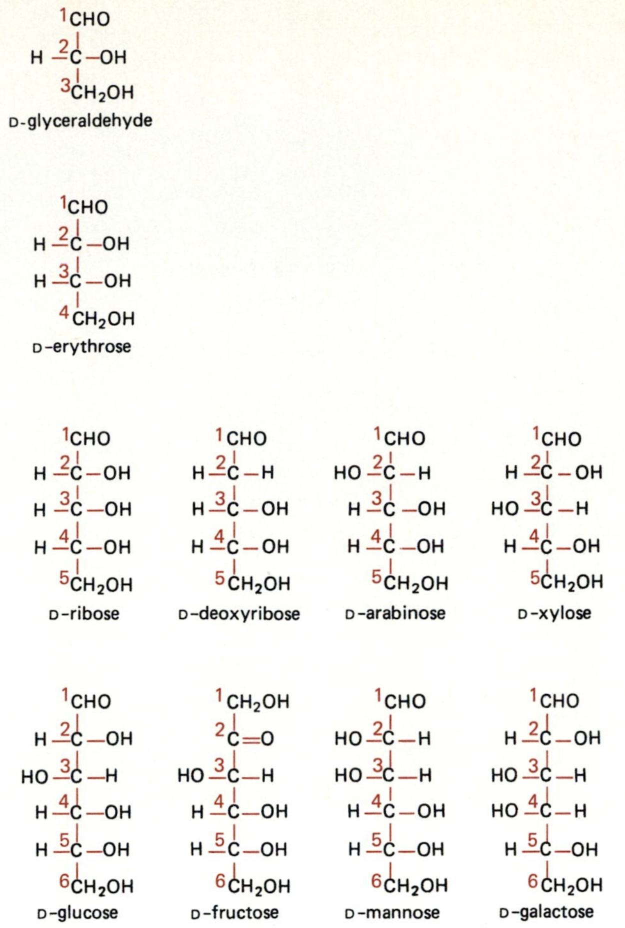

¹CHO
H —²C—OH
³CH₂OH
D-glyceraldehyde

¹CHO
H —²C—OH
H —³C—OH
⁴CH₂OH
D-erythrose

¹CHO
H —²C—OH
H —³C—OH
H —⁴C—OH
⁵CH₂OH
D-ribose

¹CHO
H —²C—H
H —³C—OH
H —⁴C—OH
⁵CH₂OH
D-deoxyribose

¹CHO
HO —²C—H
H —³C—OH
H —⁴C—OH
⁵CH₂OH
D-arabinose

¹CHO
H —²C—OH
HO —³C—H
H —⁴C—OH
⁵CH₂OH
D-xylose

¹CHO
H —²C—OH
HO —³C—H
H —⁴C—OH
H —⁵C—OH
⁶CH₂OH
D-glucose

¹CH₂OH
²C=O
HO —³C—H
H —⁴C—OH
H —⁵C—OH
⁶CH₂OH
D-fructose

¹CHO
HO —²C—H
HO —³C—H
H —⁴C—OH
H —⁵C—OH
⁶CH₂OH
D-mannose

¹CHO
H —²C—OH
HO —³C—H
HO —⁴C—H
H —⁵C—OH
⁶CH₂OH
D-galactose

FIGURE A2.12
Structural formulas of common monosaccharides.

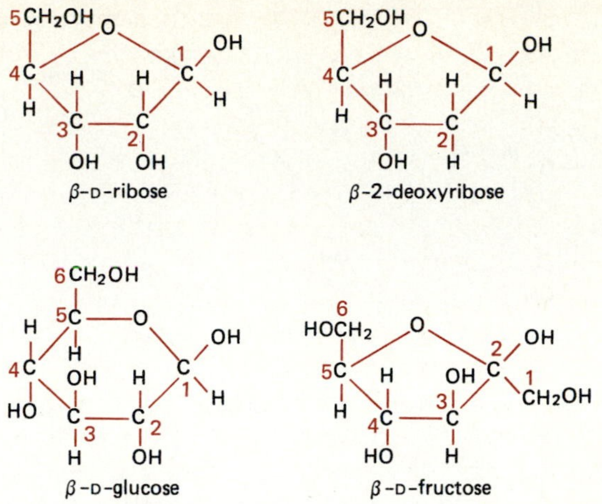

β-D-ribose

β-2-deoxyribose

β-D-glucose

β-D-fructose

FIGURE A2.13
The ring structure of some representative carbohydrates.

in two forms: amylose, which is an unbranched polymer of glucose with α-1-4 glycosidic bonds; and amylopectin, which has branches formed by α-1-6 glycosidic bonds, in addition to the normal α-1-4 bonds found in amylose. The branches normally occur every 20–30 glucose units. Glycogen resembles amylopectin but has even greater branching.

Carbohydrate molecules also may establish bonds with other molecules to form carbohydrate derivatives. For example, phosphate may bond with glucose to form glucose 6-phosphate (Figure A2.16). The 6 refers to the position of

FIGURE A2.14
The formation of a glycosidic bond, showing the α and β orientations.

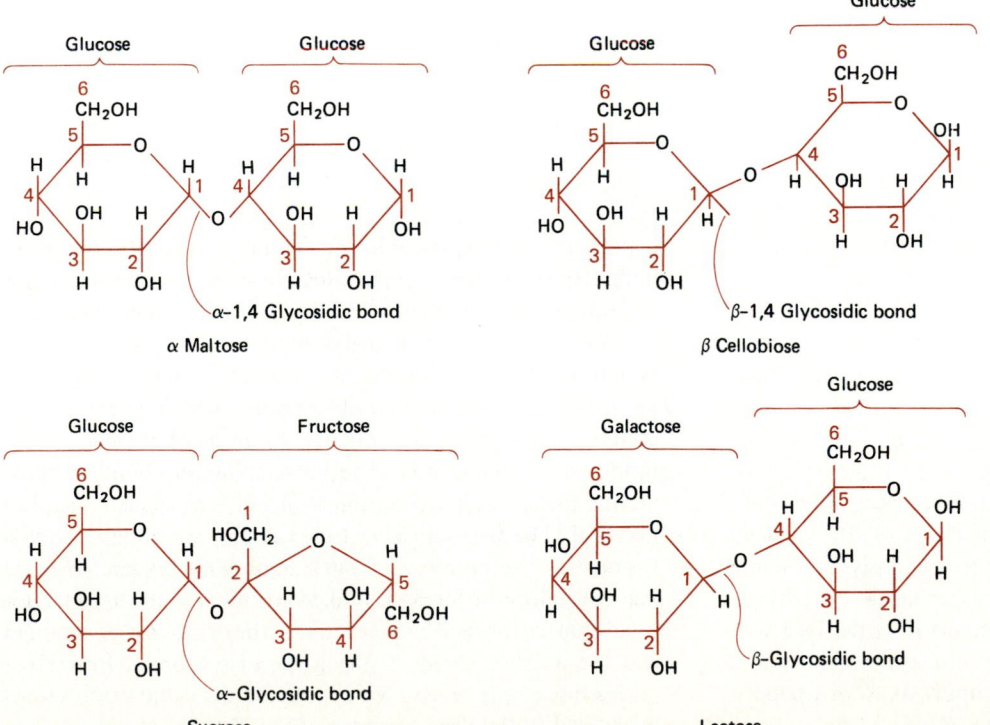

α-1,4 Glycosidic bond
α Maltose

β-1,4 Glycosidic bond
β Cellobiose

α-Glycosidic bond
Sucrose

β-Glycosidic bond
Lactose

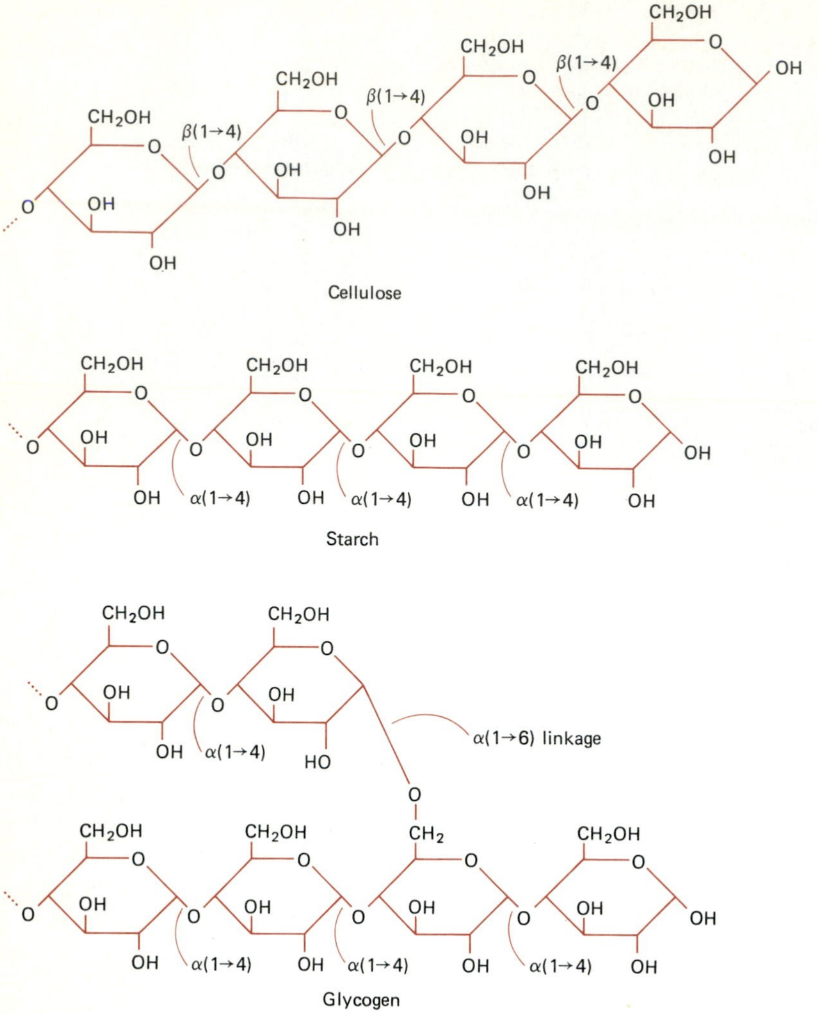

FIGURE A2.15

Several polymers of glucose—cellulose, starch, and glycogen.

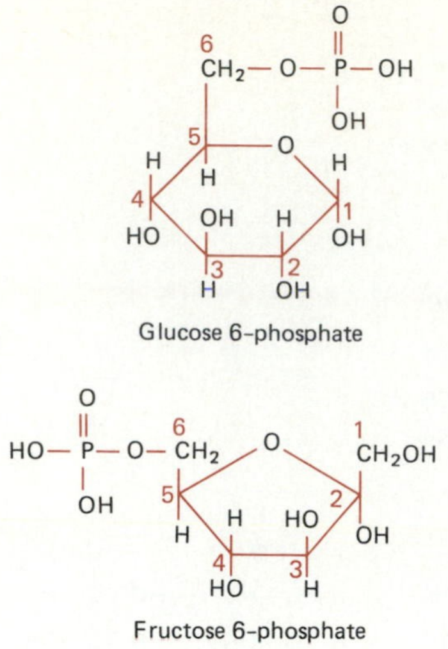

FIGURE A2.16

The structural formulas of glucose 6-phosphate and fructose 6-phosphate.

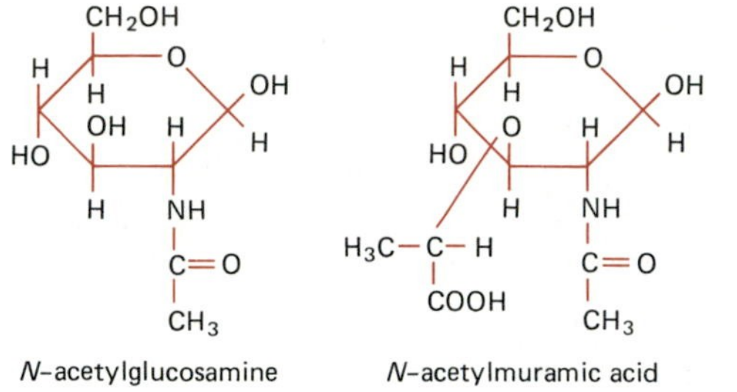

FIGURE A2.17

The structural formulas of N-acetylglucosamine and N-acetylmuramic acid.

the phosphate on the glucose. The bond formed between the monosaccharide and phosphate is an ester linkage. Phosphate ester linkages are important in linking the monosaccharide units that form the backbone of the molecules DNA and RNA. In addition to bonding with phosphate, carbohydrates may form derivatives containing nitrogen molecules, such as amino sugars. Two such carbohydrate derivatives are N-acetylglucosamine and N-acetylmuramic acid, which occur in the cell walls of bacteria (Figure A2.17). Carbohydrates also may bond with peptides and protein molecules to form peptidoglycans and glycoproteins. In a peptidoglycan molecule, a polysaccharide forms the backbone of the molecule and peptides form the side chains, whereas a glycoprotein is a protein macromolecule with oligosaccharide or polysaccharide side chains; that is, the protein forms the backbone of the molecule and the carbohydrate the minor side chains. Carbohydrates may also bind with lipids to form lipopolysaccharides.

Lipids

Lipids are water-insoluble biochemicals that are soluble in nonpolar solvents, such as chloroform. Lipids are the major constituent of biological membranes, and the solubility properties of lipid molecules enable them to function as the components of these semipermeable barriers. There are two major classes of lipids: complex lipids, which normally are composed of fatty acids bonded to an alcohol, and simple lipids, such as steroids. The most common complex lipids are the fats, which are combinations of fatty acids bonded to glycerol. The triglycerides, for example, are complex lipids formed by the bonding of fatty acids with glycerol (Figure A2.18). The fatty acid is linked to the alcohol by an ester bond that is formed between the carboxyl group of the acid and the alcohol group of the glycerol molecule. In triglycerides three fatty acids, which may be the same or different, are linked to the three carbons of the glycerol molecule. The

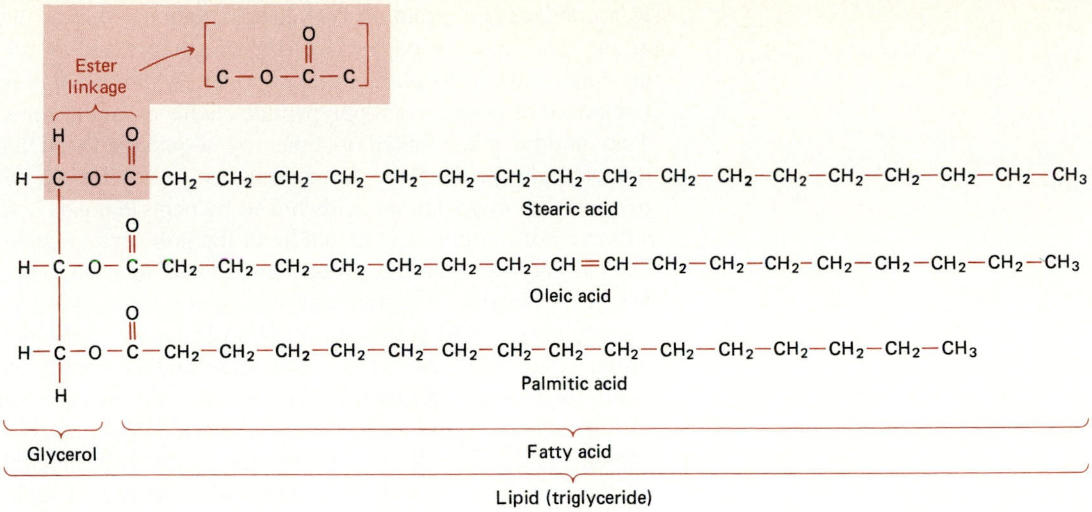

Ester linkage → $\left[\begin{array}{c} \overset{\displaystyle O}{\overset{\|}{C} - O - C - C} \end{array}\right]$

$$H - \overset{\overset{\displaystyle H}{|}}{C} - O - \overset{\overset{\displaystyle O}{\|}}{C} - CH_2 - CH_2 - CH_2 - CH_2 - CH_2 - CH_2 - CH_2 - CH_2 - CH_2 - CH_2 - CH_2 - CH_2 - CH_2 - CH_2 - CH_2 - CH_2 - CH_3$$

Stearic acid

$$H - \overset{|}{C} - O - \overset{\overset{\displaystyle O}{\|}}{C} - CH_2 - CH_2 - CH_2 - CH_2 - CH_2 - CH_2 - CH_2 - CH = CH - CH_2 - CH_2 - CH_2 - CH_2 - CH_2 - CH_2 - CH_2 - CH_3$$

Oleic acid

$$H - \overset{|}{C} - O - \overset{\overset{\displaystyle O}{\|}}{C} - CH_2 - CH_2 - CH_2 - CH_2 - CH_2 - CH_2 - CH_2 - CH_2 - CH_2 - CH_2 - CH_2 - CH_2 - CH_2 - CH_2 - CH_3$$

Palmitic acid

$$\overset{|}{H}$$

Glycerol   Fatty acid

Lipid (triglyceride)

FIGURE A2.18

The structural formula of a triglyceride with palmitic acid, stearic acid, and oleic acid constituents.

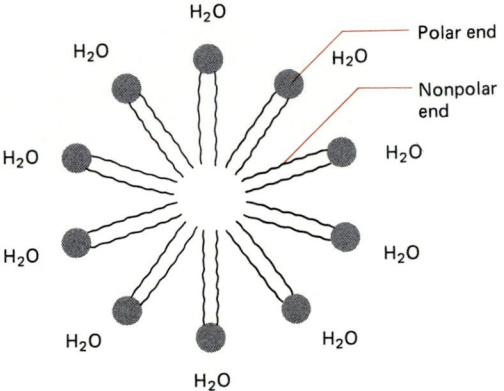

FIGURE A2.19

Because one part of the molecule is more polar than the other nonpolar part, lipids form clusters, called *micelles*, in water. The nonpolar portions point inward and the polar sections point toward the water.

FIGURE A2.20

The structural formula of a phospholipid, showing the hydrophilic and hydrophobic nature of this class of compound.

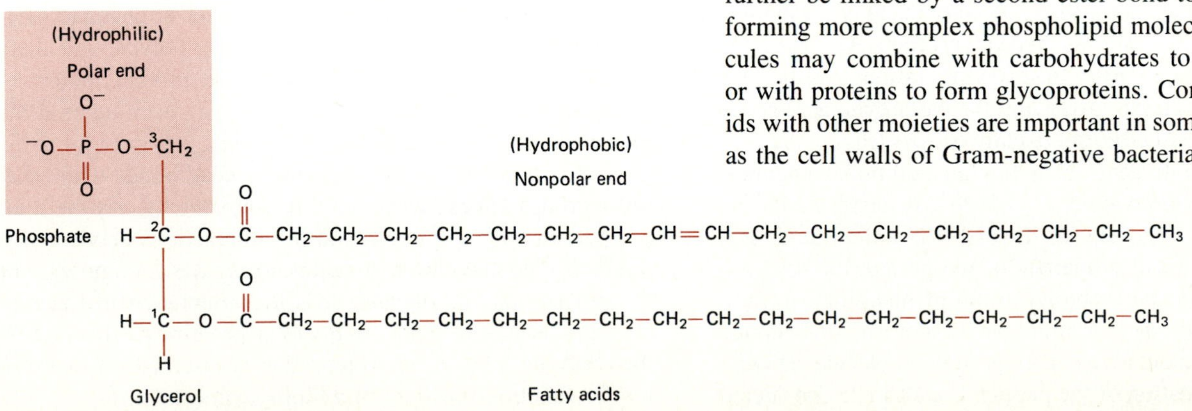

(Hydrophilic)

Polar end

$$\overset{\overset{\displaystyle O^-}{|}}{^-O - P - O - ^3CH_2}$$
$$\overset{\|}{O}$$

Phosphate

most common fatty acids found in complex cellular lipids are palmitic acid, a 16-carbon saturated fatty acid, stearic acid; an 18-carbon saturated fatty acid; and oleic acid, an 18-carbon unsaturated fatty acid having one double bond. The nonpolar nature of these relatively long-chain fatty acids causes lipids to exhibit hydrophobic reactions with water. The salts of C_{16}–C_{18} are soaps and form micelles in water that are stabilized by hydrophobic interactions; in micelles the negatively charged carboxyl groups of the fatty acids all point outward toward the water, and the nonpolar hydrocarbon portions are buried within the micelle (Figure A2.19). Similarly, in biological membranes the negatively charged portion of the lipid interfaces with water, and the nonpolar portion is buried within the membrane structure.

The principal lipids of biological membranes are phospholipids. These phospholipids are derivatives of glycerol in which fatty acids are bonded to only two of the carbons of the glycerol molecule (Figure A2.20). The third carbon is linked by an ester bond to phosphate. The negatively charged phosphate acts as a hydrophilic head of the lipid molecule and is attracted to water, and the hydrocarbon portion of the fatty acid acts as a hydrophobic tail. The phosphate may further be linked by a second ester bond to another alcohol, forming more complex phospholipid molecules. Lipid molecules may combine with carbohydrates to form glycolipids or with proteins to form glycoproteins. Combinations of lipids with other moieties are important in some structures, such as the cell walls of Gram-negative bacteria.

(Hydrophobic)

Nonpolar end

$$H - ^2\overset{|}{C} - O - \overset{\overset{\displaystyle O}{\|}}{C} - CH_2 - CH_2 - CH_2 - CH_2 - CH_2 - CH_2 - CH_2 - CH = CH - CH_2 - CH_2 - CH_2 - CH_2 - CH_2 - CH_2 - CH_2 - CH_3$$

$$H - ^1\overset{|}{C} - O - \overset{\overset{\displaystyle O}{\|}}{C} - CH_2 - CH_2 - CH_2 - CH_2 - CH_2 - CH_2 - CH_2 - CH_2 - CH_2 - CH_2 - CH_2 - CH_2 - CH_2 - CH_2 - CH_2 - CH_3$$

$$\overset{|}{H}$$

Glycerol   Fatty acids

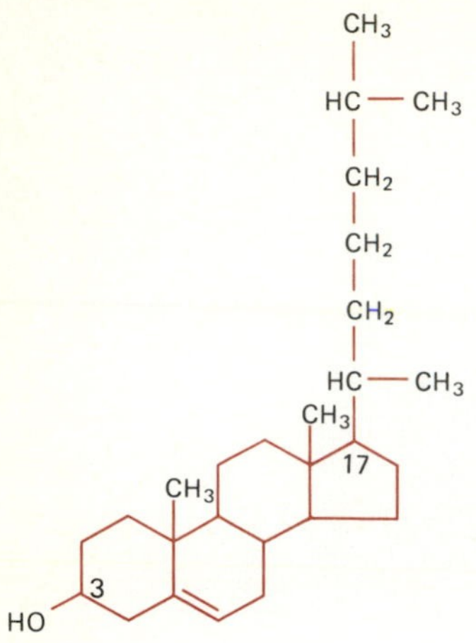

CH₃
|
HC — CH₃
|
CH₂
|
CH₂
|
CH₂
|
HC — CH₃
CH₃
17
CH₃
3
HO

Cholesterol

FIGURE A2.21

The structural formula of the sterol cholesterol.

Most simple lipids are unsaturated hydrocarbon molecules. The steroid molecules are four-ringed compounds that can form a variety of derivatives. The sterol cholesterol (Figure A2.21) is an important steroid derivative in animal tissues, and the accumulation of cholesterols is important in cardiovascular disease; other steroid derivatives in mammalian systems include the bile acids, adrenocortical hormones, and both male and female sex hormones. Different sterols occur in plants, fungi, algae, and protozoa, but bacteria generally lack sterols in their membranes.

Proteins

Proteins are large macromolecules of considerably higher molecular weight than lipids; the molecular weights of proteins range from 6,000 to several million. Proteins are composed of long chains of amino acids linked by peptide bonds and are extremely important molecules in biological systems, often composing 50 percent of the cell's dry weight. The amino acid subunits of proteins possess two functional groups, an amino group and a carboxylic acid group. Only 20 amino acids normally are found in the protein macromolecules. In some cases, these 20 amino acids are modified to form derivatives such as hydroxyproline. All 20 of the essential amino acids have an L configuration, and in each of them both the amino group and the carboxylic acid group are linked to the same central α-carbon atom (Figure A2.22). Each of the amino acids also has an additional chemical group bonded to the α-carbon atom that is designated as an *R group*. These R groups distinguish the amino acids and establish the chemical properties of the protein molecule.

The bonding of the carboxyl group of one amino acid to the amino group of another amino acid establishes the peptide linkage that characterizes the protein molecule (Figure A2.23). The formation of the peptide bond can be considered as a condensation reaction between the amino group of one amino acid and the carboxyl group of another amino acid accompanied by the elimination of water. A protein may be composed of one or more polypeptide chains of amino acids. Two amino acids linked together by a peptide bond are termed a *dipeptide*; three amino acids so linked form a *tripeptide*; and many amino acids linked by peptide bonds constitute a *polypeptide*. In the middle of the polypeptide chain, the carboxyl and amino groups are linked to each other. When a polypeptide forms, one end of the molecule has a free carboxylic acid group and the other end has a free amino group (Figure A2.24). The amino free end is termed the *amino* or *N-terminal* and the carboxyl free end the *carboxyl* or *C-terminal*. These two ends of a peptide chain are biochemically distinguishable and impart an ordered direction on the protein molecule. The ability of biochemical molecules to exhibit directionality is important in their functioning in biological systems.

Protein molecules also have a three-dimensional structure that determines their biological properties (Figure A2.25). The sequence of amino acids in the polypeptide forms the primary structure of a protein. It is the number and order of specific amino acids within the peptide chains of the protein—that is, the primary structure—that establish the properties of the protein macromolecule. The long polypeptide chain of a protein molecule is twisted, often forming a helix. The specific conformational pattern of a protein, stabilized by hydrogen bonding, is known as the *secondary structure*. Protein molecules also exhibit folding of the polypeptide chains that establishes a tertiary structure. In globular proteins the polypeptide chains are tightly folded into a tight spherical or globular shape. Folding of the polypeptide molecule is also stabilized by interactions of the sulfhydryl groups of the sulfur-containing amino acids. The interactions of sulfhydryl groups of the amino acid cysteine form disulfide bridges that contribute to the protein molecule's tertiary structure. Some proteins are composed of more than one polypeptide molecule (Figure A2.26). The quaternary structure of a protein reflects the interactions of two or more polypeptide chains and describes how the peptide chains are arranged or clustered in space.

Ultimately, it is the primary structure that determines the higher-order structures that describe the three-dimensional nature of the protein molecule. The distinctive characteristics of the R groups of the amino acids in the polypeptide chain impose constraints on the shape of the protein molecule. Hydrophobic groups tend to associate with each other at the interior of folded chains, and hydrophilic groups tend to assume positions where they can form weak bonds with other polar groups. These weak bond interactions specify the folding patterns of the protein and the interactions between different polypeptide chains in determining the configuration of the protein macromolecule. It is the orientation in space of the various amino acid groups of a protein, established by the configuration of the protein, that is essential for that protein to function properly in a biological system.

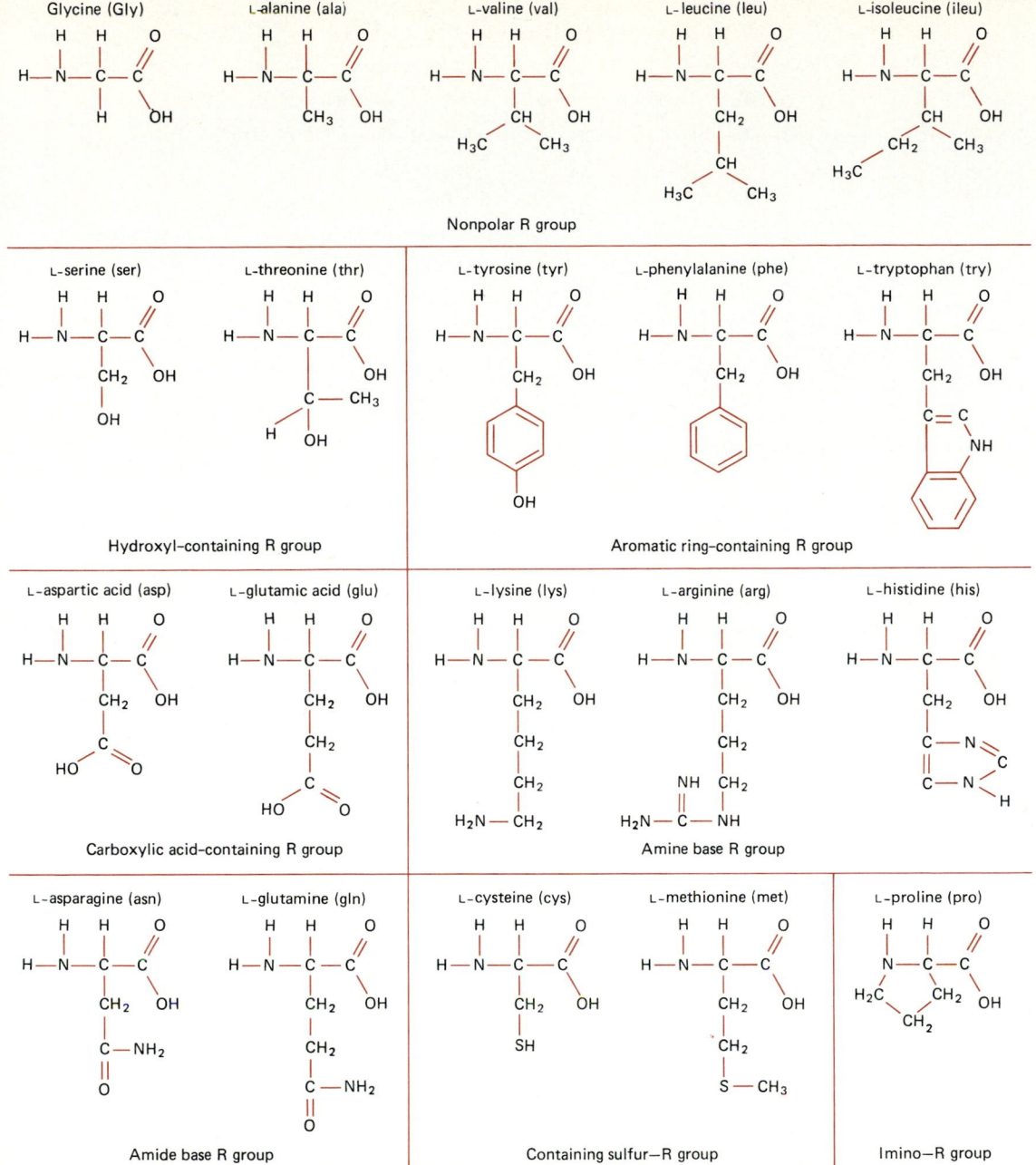

FIGURE A2.22

The structural formulas of 20 common amino acids. Each is an L-α-amino acid. The structures differ in the other constituents.

One of the most important functions of proteins is their role as enzymes. Although not all proteins are enzymes, all enzymes are proteins. Enzymes are responsible for catalyzing biochemical reactions in living systems. Each microbial cell must possess many enzymes, thousands in fact, in order to carry out the essential metabolic activities involved in its growth and reproduction. The order of the amino acids at the substrate-binding and active sites of the enzyme is especially critical in determining the specificity of action of that en-

FIGURE A2.23

Peptide bonds link amino acids into polymeric units, binding together the units of proteins.

Free amino end

Free carboxyl end

$H_3N-CH-C-N-CH-C-N-CH-C-N-CH-C-N-CH-C-N-CH-C-N-CH-C-N-CH-COO^-$

Ser | Cys | Glu | Leu | Ala | Lys | Pro | Phe

FIGURE A2.24

A polypeptide has a free amino end and a free carboxyl end.

zyme, that is, which specific biochemical reactions the enzyme is capable of catalyzing. Enzymes with different three-dimensional conformations at their active sites catalyze different metabolic reactions. The high degree of specificity exhibited by enzymes is based on their biochemical composition, which establishes their proper three-dimensional configuration, and even slight changes, such as a change in the order of relatively few amino acids in a long polypeptide chain, can alter the activity of an enzyme.

Active enzyme molecules must have the proper three-dimensional conformation to exhibit catalytic activity. Denaturation of the protein molecule, that is, disruption of

the three-dimensional shape, eliminates the activity of an enzyme molecule. Enzymes are globular proteins, and their tight folding pattern determines the active sites for the catalytic activities demonstrated by these macromolecules. The critical three-dimensional configuration of a protein can be easily disrupted without breaking the primary polypeptide chain. Changes in the ionic strength of a solution, such as occur when salt is added to the solution, alter the weak bond interactions stabilizing the secondary and tertiary structures of the protein. Similarly, changes in pH alter the weak bond interactions that maintain the structure of the protein macromolecule (Figure A2.27). As a result, the protein unfolds

FIGURE A2.25

Proteins have primary, secondary, tertiary, and quaternary structures.

Primary

Secondary

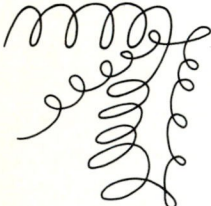

Tertiary

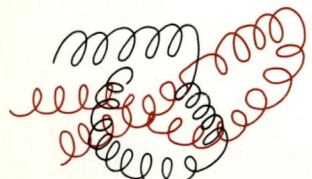

Quaternary

FIGURE A2.26

The immunoglobulin G (IgG) molecule (A) is composed of four polypeptide chains held together by disulfide bridges; (B) sulfhydryl bonds also can establish the structure of a single peptide chain.

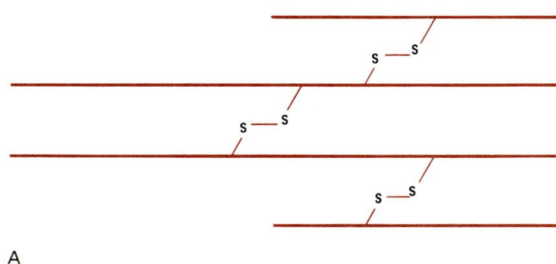

A

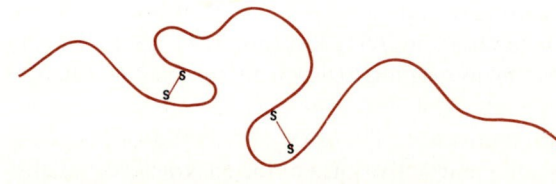

B

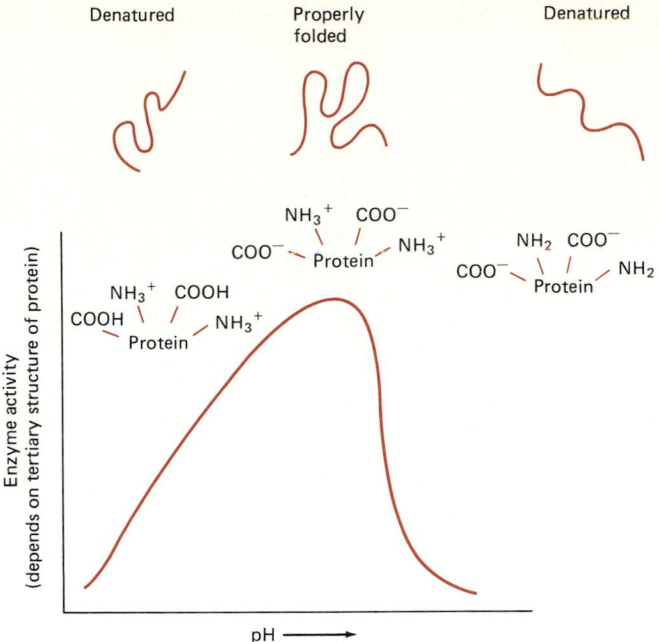

FIGURE A2.27
The effects of pH on the dissociation of the amino acids in a protein and its effect on the tertiary structure is shown in this graph.

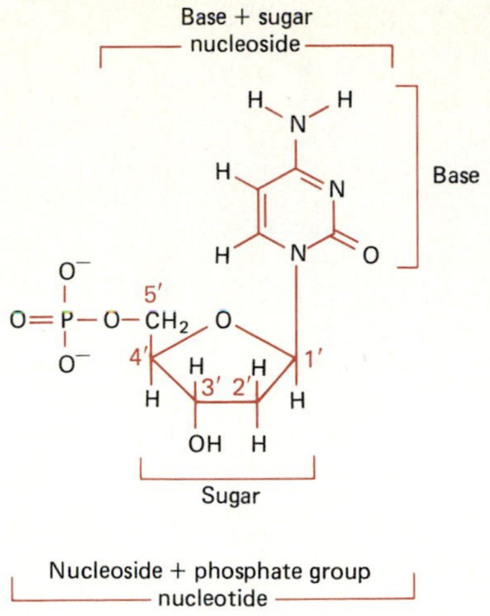

and loses its biological activity. In most cases such denaturation of proteins is irreversible, but in some cases a denatured protein can spontaneously refold, reestablishing its proper three-dimensional shape and its capability of exhibiting biological activity. At high temperatures, changes in the shapes of protein molecules also occur; hence, enzymes are inactivated at elevated temperatures. We use this property to heat-kill microorganisms.

Nucleic Acids

Nucleic acids, like proteins, are large macromolecules, but instead of being composed of amino acids, the subunits of nucleic acid macromolecules are nucleotides linked together to form a polymer. The nucleotide subunits of nucleic acids are themselves composed of smaller units. A nucleotide is composed of three covalently linked individual units: a nitrogenous base (a heterocyclic ring structure containing nitrogen, as well as carbon, sometimes referred to as a *nucleic acid base* or simply as a *base*); a 5-carbon monosaccharide (either ribose or deoxyribose); and a phosphate group (Figure A2.28). The nitrogenous base is linked to the 1-carbon of the monosaccharide by an *N*-glycosidic bond. The nitrogenous base plus the monosaccharide, without the phosphate group, is called a *nucleoside*. The phosphate group forms a diester bond between the 3-carbon of one monosaccharide and the 5-carbon of another monosaccharide unit, and it is this bond that links the molecule together (Figure A2.29).

Two types of nucleic acids occur in biological systems; deoxyribonucleic acid (DNA) and ribonucleic acid (RNA)

FIGURE A2.28
The structural formula of a nucleotide, the basic unit of the informational macromolecules of living systems.

(Figure A2.30). In most organisms, DNA stores the genetic information of the organism. There are several types of RNA molecules that serve different functions, all of which are involved in converting the genetic information into proteins in order to carry out the biological functions of the organism. In DNA the monosaccharide that occurs is deoxyribose; ribose is the monosaccharide found in RNA. Four nitrogenous bases occur in DNA: adenine, guanine, cytosine, and thymine. Adenine (A) and guanine (G) are substituted purine bases that have two rings. Cytosine (C) and thymine (T) are substituted pyrimidine bases and have only one ring. The

FIGURE A2.29
Nucleotides are linked by phosphate diester bonds to form dimers and polymeric units.

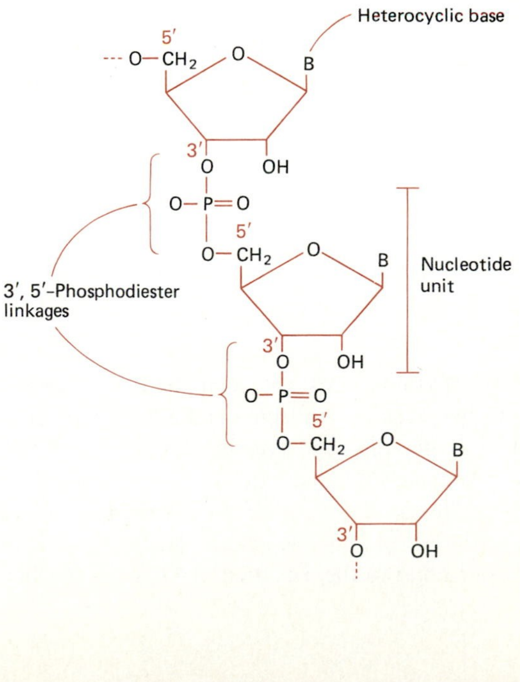

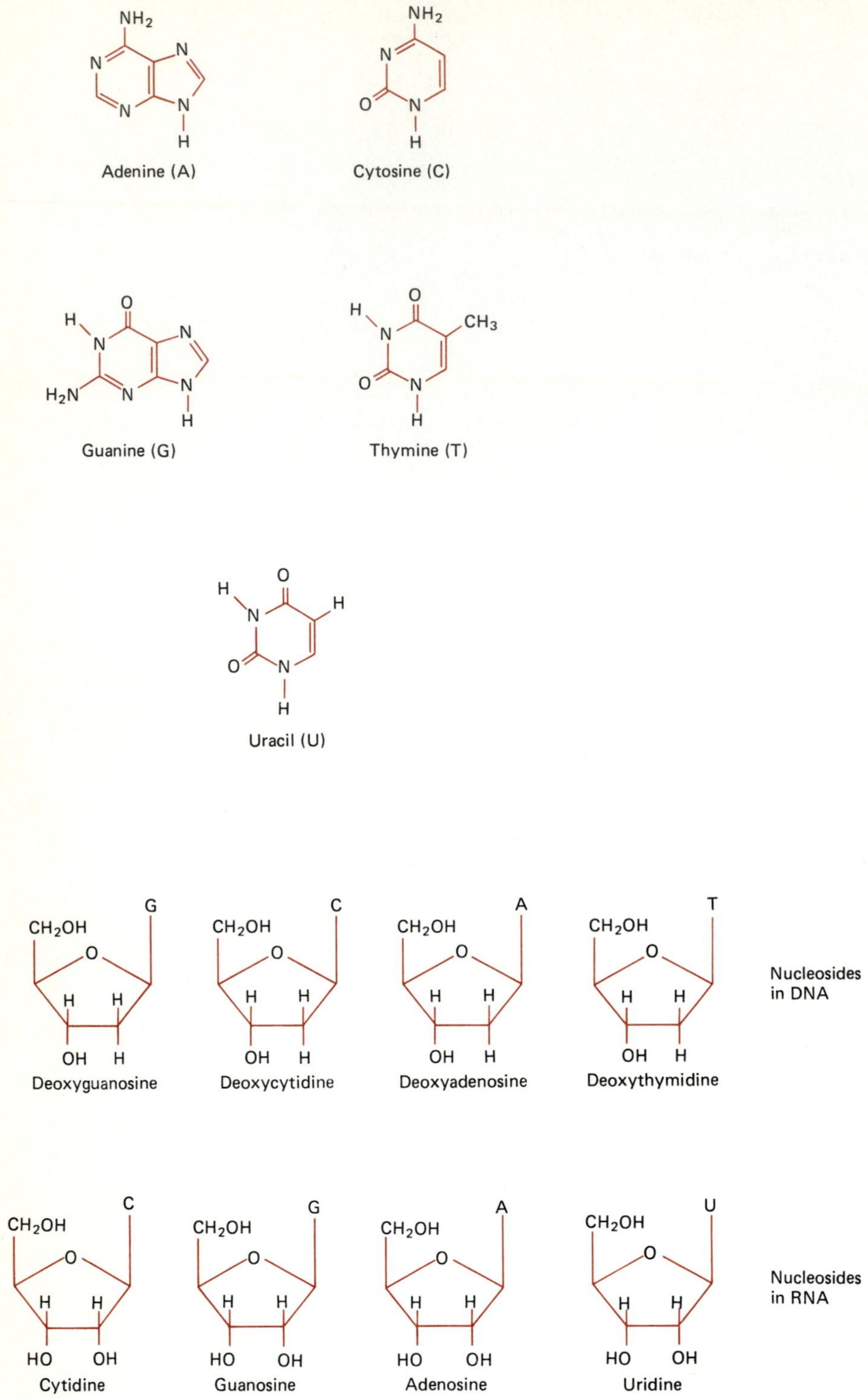

FIGURE A2.30
The structural formulas of bases and sugars in DNA and RNA.

pyrimidine base thymine does not occur in RNA. Rather, another pyrimidine, uracil (U), occurs in the RNA molecule. The bases adenine, guanine, and cytosine also occur in RNA.

In both RNA and DNA the nucleotides are linked by 3′–5′phosphate diester linkages (Figure A2.31). Consequently, at one end of the nucleic acid molecule, there is no phosphate diester bond to the 3-carbon of the monosacchar-

ide, and thus there is a free hydroxyl group at the 3-carbon position (3′OH free end); at the other end of the molecule, the 5-carbon is not involved in forming a phosphate diester linkage, and thus there is a free phosphate hydroxyl group at the 5-carbon position (5′P free end).

The DNA double helix is composed of two primary polynucleotide chains held together by hydrogen bonding be-

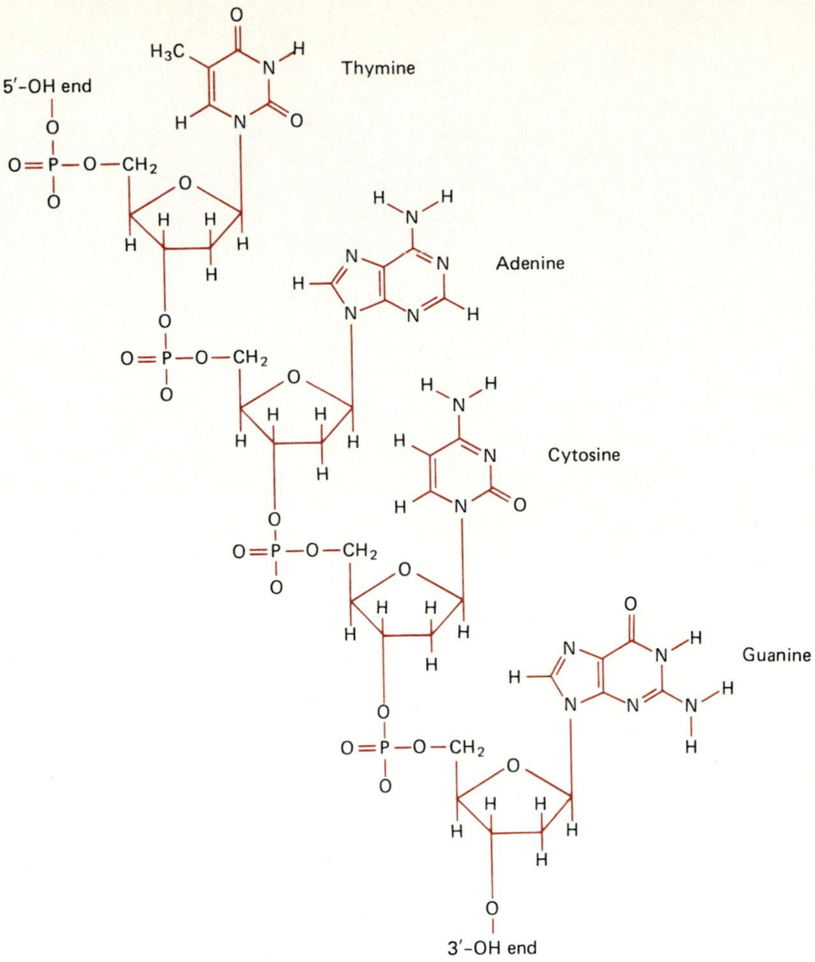

FIGURE A2.31
Because nucleic acids are linked by phosphodiester bonds between the 3-carbon of one nucleic acid base and the 5-carbon of the other, nucleic acids have 3'OH and 5'OH free ends at opposite ends of polymeric molecules.

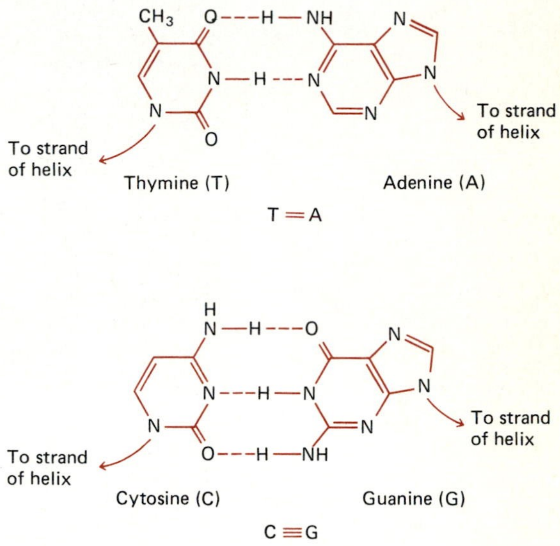

FIGURE A2.32
The two strands of DNA are held together by hydrogen bonding between complementary base pairs.

tween complementary nucleotide bases (Figure A2.32). Within the double helical DNA, adenine always pairs with thymine, and guanine always pairs with cytosine. There are two hydrogen bonds established between the adenine-thymine base pairs and three hydrogen bonds established between the guanine-cytosine base pairs. The sequence of nucleotides within the DNA molecule codes for the sequence of bases in RNA and, ultimately, the sequence of amino acids in proteins.

The directional nature of nucleic acid molecules is critical in establishing the necessary direction of reading the genetic information. Within the double helical DNA macromolecule the two polynucleotide chains run in opposite directions; that is, one chain runs from the 3'OH to the 5'P free end, and the complementary chain runs in an antiparallel direction from the 5'P to the 3'OH free end. A consequence of the antiparallel nature of the DNA molecule is that different information is stored in each of the chains. The genetic code, based on only the few letters (nucleotides) in its "alphabet,"

provides the necessary biochemical basis for encoding the genetic information of the great diversity of living organisms.

In addition to their involvement in the nucleic acids DNA and RNA, several nucleotides form derivatives that have extremely important functions within microorganisms (Figure A2.33). Adenosine triphosphate (ATP) is a molecule produced in cells that is used to store chemical energy until it is needed to drive various energy-requiring biochemical reactions. The production and utilization of ATP are essential to the bioenergetics of the cell, and all of the metabolic pathways of microorganisms are involved in either producing or consuming ATP. Nicotinamide adenine dinucleotide (NAD) is another nucleotide derivative that plays a critical role in the metabolism of microorganisms. NAD and related compounds act as coenzymes and are involved in oxidation-reduction reactions. Much of the metabolism of microorganisms involves oxidation-reduction reactions requiring the use of coenzymes, accounting for the importance of these nucleotide molecules in biochemical reactions.

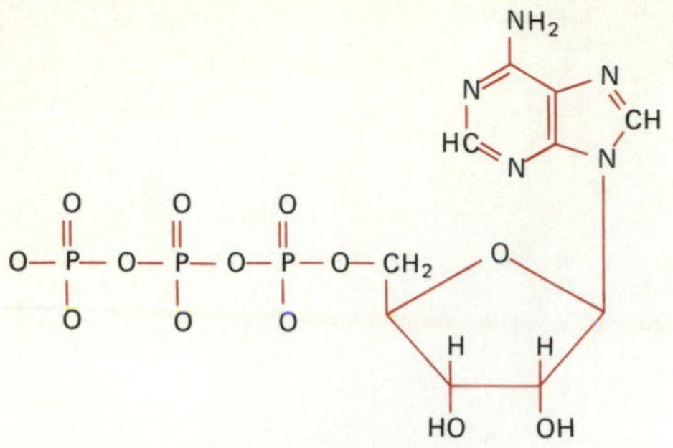
Adenosine triphosphate (ATP)

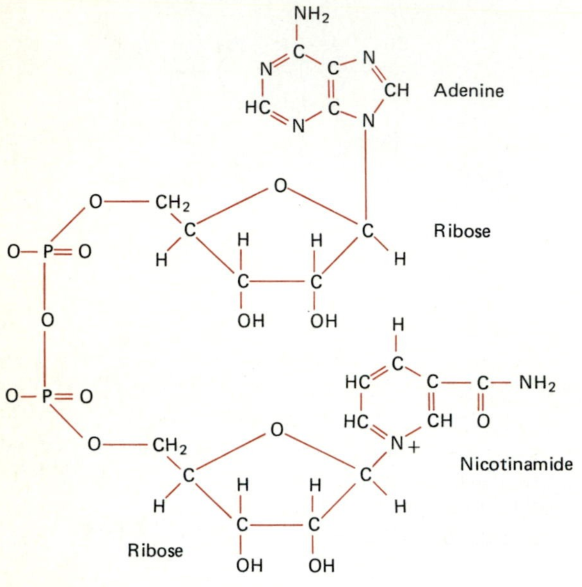
Adenine
Ribose
Nicotinamide
Ribose
NAD

FIGURE A2.33
The structural formulas of ATP, the key compound involved in biological energy transfers, and NAD, a key coenzyme involved in hydrogen transfers within a cell.

Glossary of Microbiological Terms

abiotic referring to the absence of living organisms.

abrasion an area denuded of skin, mucous membrane, or superficial epithelium by rubbing or scraping.

abscess a localized accumulation of pus.

absorption the uptake, drinking in, or imbibing of a substance; the movement of substances into a cell; the transfer of substances from one medium to another, e.g., the dissolution of a gas in a liquid; the transfer of energy from electromagnetic waves to chemical bond and/or kinetic energy, e.g., the transfer of light energy to chlorophyll.

accessory pigments pigments that harvest light energy and transfer it to the primary photosynthetic reaction centers.

acellular lacking cellular organization; not having a delimiting cytoplasmic membrane; organizational description of viruses, viroids, and prions.

acetyl a two-carbon organic radical containing a methyl group and a carbonyl group.

acetyl CoA acetyl coenzyme A; a condensation product of coenzyme A and acetic acid; an intermediate in the transfer of 2-carbon fragments, notably in their entrance into the tricarboxylic acid cycle.

acetylene reduction assay assay for nitrogen fixation based upon the conversion of acetylene to ethylene by nitrogenase, the enzyme responsible for nitrogen fixation.

achromatic lens an objective lens in which chromatic aberration has been corrected for two colors and spherical aberration for one color.

acid fast the property of those bacteria, such as mycobacteria, that retain their initial stain and do not decolorize after washing with dilute acid-alcohol.

acid foods foods with a pH value less than 4.5.

acid mine drainage consequence of the metabolism of sulfur- and iron-oxidizing bacteria when coal mining exposes pyrite to atmospheric oxygen and the combination of autoxidation and microbial sulfur and iron oxidation produces large amounts of sulfuric acid, which kills aquatic life and contaminates water.

acidic a compound that releases hydrogen (H^+) ions when dissolved in water; a compound that yields positive ions upon dissolution; a solution with a pH value less than 7.0.

acidic stains stains with a positively charged chromophore (colored portion of the dye) that are attracted to negatively charged cells.

acidophiles microorganisms that show a preference for growth at low pH, e.g., bacteria that grow only at very low pH values, ca. 2.0.

acidulant acidic compound used as a chemical food preservative.

acne an inflammatory disease involving the oil glands and hair follicles of the skin, found chiefly in adolescents and marked by papules or pustules, especially about the face.

acquired immune deficiency syndrome (AIDS) an infectious disease syndrome caused by HIV retrovirus, characterized by the loss of normal immune response system functions, followed by various opportunistic infections.

acquired immunity the ability of an individual to produce specific antibodies in response to antigens to which the body has been previously exposed based on the development of a memory response.

acrasin substance secreted by a slime mold that initiates aggregation to form a fruiting body identified as 3'5' cyclic AMP.

Acrasiogymnomycotina cellular slime molds; subdivision of Gymnomycota, which are fungi that feed largely on bacteria and form sporocarps.

Actinomycetales order of bacteria characterized by the formation of branching filaments whose families are distinguished on the basis of the nature of their mycelia and spores.

actinomycetes members of an order of bacteria in which species are characterized by the formation of branching and/or true filaments.

activated sludge process an aerobic secondary sewage treatment process using sewage sludge containing active complex populations of aerobic microorganisms to break down organic matter in sewage.

activation energy the energy in excess of the ground state that must be added to a molecular system to allow a chemical reaction to start.

active immunity immunity acquired as a result of the individual's own reactions to pathogenic microorganisms or their antigens; attributable to the presence of antibody or immune lymphoid cells formed in response to an antigenic stimulus.

active transport the movement of materials across cell membranes from regions of lower to regions of higher concentration, requiring the expenditure of metabolic energy.

acute referring to a disease of rapid onset, short duration, and pronounced symptoms.

acute stage stage of a disease when symptoms appear and the disease is most severe.

acyclovir antiviral agent used in the treatment of diseases caused by *Herpes simplex*.

adaptive enzymes enzymes produced by an organism in response to the presence of a substrate or a related substance; also called *inducible enzymes*.

adenine a purine base component of nucleotides, nucleosides, and nucleic acids.

adenosine a mononucleoside consisting of adenine and D-ribose.

adenosine diphosphate (ADP) a high-energy derivative of adenosine containing two phosphate groups, one less than ATP; formed upon hydrolysis of ATP.

adenosine monophosphate (AMP) a compound composed of adenosine and one phosphate group formed upon the hydrolysis of ADP.

adenosine triphosphatase an enzyme that catalyzes the reversible hydrolysis of ATP; the membrane-bound form of this enzyme is important in catalyzing the formation of ATP from ADP and inorganic phosphate.

adenosine triphosphate (ATP) a major carrier of phosphate and energy in biological systems, composed of adenosine and three phosphate groups; the free energy released from the hydrolysis of ATP is used to drive many energy-requiring reactions in biological systems.

adhesins substances involved in the attachment of microorganisms to solid surfaces; factors that increase adsorption.

adhesion factors substances involved in the attachment of microorganisms to solid surfaces; factors that increase adsorption.

adjuncts starchy substrates, such as corn, wheat, and rice, that provide carbohydrates for ethanol production and are added to malt during the mashing process in the production of beer.

adjuvants substances that increase the immunological response to a vaccine and, for example, can be added to vaccines to slow down absorption and increase effectiveness; substances that enhance the action of a drug or antigen.

ADP see *adenosine diphosphate*.

adsorption a surface phenomenon involving the retention of solid, liquid, or gaseous molecules at an interface.

aer- combining form meaning air or atmosphere.

aerated pile method method of composting for the decomposition of organic waste material where the wastes are heaped in separate piles and forced aeration provides oxygen.

aerial mycelia a mass of hyphae occurring above the surface of a substrate.

aerobes microorganisms whose growth requires the presence of air or free oxygen.

aerobic having molecular oxygen present; growing in the presence of air.

aerobic bacteria bacteria requiring oxygen for growth.

aerobic respiration metabolism involving a respiration pathway in which molecular oxygen serves as the terminal electron acceptor.

aerosol a fine suspension of particles or liquid droplets sprayed into the air.

aflatoxin a carcinogenic poison produced by some strains of the fungus *Aspergillus flavus*.

African sleeping sickness also known as *African trypanosomiasis*; a protozoan disease that affects the nervous system, caused by *Trypanosoma*, a flagellate injected by the bite of the tsetse fly.

agar a dried polysaccharide extract of red algae used as a solidifying agent in various microbiological media.

Agaricales order of basidiomycetes; basidium-producing fungi that includes the mushrooms and boletes.

agglutin an antibody capable of causing the clumping or agglutination of bacteria or other cells.

agglutination the visible clumping or aggregation of cells or particles due to the reaction of surface-bound antigens with homologous antibodies.

agricultural microbiology the study of the role of microorganisms in agriculture.

AIDS see *acquired immune deficiency syndrome*.

airborne transmission route by which pathogens are transported to a susceptible host via the air; main route of transmission of pathogens that enter via the respiratory tract through the air.

akinetes thick-walled resting spores of cyanobacteria and green algae.

alcoholic fermentation conversion of sugar to alcohol by microbial enzymes; fermentation that produces alcohol (ethanol) and carbon dioxide from glucose; also known as *ethanolic fermentation*.

alcohols organic compounds characterized by one or more -OH (hydroxyl) groups.

aldehydes a class of substances derived by oxidation from primary alcohols and characterized by the presence of a -CHO group.

ale alcoholic beverage produced with top-fermenting *Saccharomyces cerevesiae* and a high concentration of hops to produce a tart taste and a high alcohol concentration.

algae a heterogeneous group of eukaryotic, photosynthetic, unicellular and multicellular organisms lacking true tissue differentiation.

algicides chemical agents that kill algae.

alkaline a condition in which hydroxyl (OH^-) ions are in abundance; solutions with a pH of greater than 7.0 are alkaline or basic.

allele one or more alternative forms of a given gene concerned with the same trait or characteristic; one of a pair or multiple forms of a gene located at the same locus of homologous chromosomes.

allelopathic a substance produced by one organism that adversely affects another organism.

allergen an antigen that induces an allergic response, i.e., a hypersensitivity reaction.

allergy an immunological hypersensitivity reaction; an antigen–antibody reaction marked by an exaggerated physiological response to a substance in sensitive individuals.

allochthonous an organism or substance foreign to a given ecosystem.

allosteric effector substance that can bind to the regulatory site of an allosteric enzyme, resulting in the alteration of the rate of activity of that enzyme.

allosteric enzymes enzymes with a binding and catalytic site for the substrate and a different site where a modulator (allosteric effector) acts.

allosteric inhibitor an allosteric effector that results in reduced rates of activity of an allosteric enzyme.

allotypes antigenically different forms of a given type of immunoglobulin that occur in different individuals of the same species.

alpha-hemolysis α-hemolysis; partial hemolysis of red blood cells, as evidenced by the formation of a zone of partial clearing (greening) around certain bacterial colonies growing on blood agar.

Amastigomycota division of the fungi; the subdivisions are Zygomycotina, Ascomycotina, Basidiomycotina, and Deuteromycotina; they do not produce motile cells and vary from single cells to mycelia that may be coenocytic or may have extensive septation.

amastigotes rounded protozoan cells lacking flagella; a form assumed by some species of Trypanosomatidae, e.g., *Plasmodium*, during a particular stage of development.

amebiasis see *amebic dysentery*.

amebic dysentery an inflammation of the colon caused by *Entamoeba histolytica*; also known as *amebiasis*.

amensalism an interactive association between two populations that is detrimental to one while not adversely affecting the other.

Ames test test for the detection of chemical mutagens and potential carcinogens.

amino an -NH_2 group.

amino acids a class of organic compounds containing an amino (-NH_2) group and a carboxyl (-COOH) group.

amino end the end of a peptide chain or protein with a free amino group, i.e., an alpha amino group not involved in forming the peptide bond.

aminoacyl site site on a ribosome where a tRNA molecule attached to a single amino acid initially binds during translation.

aminoglycosides broad-spectrum antibiotics containing an aminosugar, an amino- or guanido-inositol ring, and residues of other sugars that inhibit protein synthesis, e.g., kanamycin, neomycin, and streptomycin.

ammonification the release of ammonia from nitrogenous organic matter by microbial action.

ammonium ion the cation NH_4^+.

amoebida protozoa having no distinct shape that form lobopodia and lack a skeletal structure.

AMP see *adenosine monophosphate*.

amphibolic pathway a metabolic pathway that has both catabolic and anabolic functions.

amphotericin B broad-spectrum antifungal agent used for treating systemic infections.

amplifying gene expression activity by which eukaryotes can increase the amount of rRNA and thus the number of ribosomes that can be used to translate the information in a stable mRNA molecule, thus producing large amounts of the enzyme coded for by a given gene.

amylases enzymes that hydrolyze starch.

anabolism biosynthesis; the process of synthesizing cell constituents from simpler molecules, usually requiring the expenditure of energy.

anaerobes organisms that grow in the absence of air or oxygen; organisms that do not use molecular oxygen in respiration.

anaerobic the absence of oxygen; able to live or grow in the absence of free oxygen.

anaerobic culture chamber enclosure designed to exclude oxygen from the atmosphere, generally by generating hydrogen, which reacts with available oxygen as a catalyst to produce water.

anaerobic digester a secondary sewage treatment facility used for the degradation of sludge and solid waste.

anaerobic life life in the absence of air.

anaerobic photosynthetic bacteria bacteria that carry out the reactions of photosystem I.

anaerobic respiration the use of inorganic electron acceptors other than oxygen as terminal electron acceptors for energy-yielding oxidative metabolism.

anaerobiosis the state or condition characterized by the absence of air; anaerobic; lack or removal of oxygen from the atmosphere.

analogously similar phenotypically similar microorganisms.

anamnestic response a heightened immunological response in persons or animals to the second or subsequent administration of a particular antigen given some time after the initial administration; a secondary or memory immune response, i.e., the rapid reappearance of antibody in the blood after exposure to an antigen to which the subject had previously shown a primary immune response.

anaphylactic hypersensitivity an exaggerated immune response to foreign protein or other substances, involving the degranulation of mast cells and the release of histamine.

anaphylactic shock physiological shock resulting from an anaphylactic hypersensitivity reaction, e.g., to penicillin or bee bites; in severe cases, death can result within minutes.

anemia a condition characterized by having less than the normal amount of hemoglobin, reflecting a reduced number of circulating red blood cells.

animal viruses viruses that multiply within animal cells.

anions negatively charged ions.

anisogametes gametes differing in shape, size, and/or behavior.

annulus a ring-shaped structure; a transverse groove in the cellular envelope of dinoflagellates.

anode the positive terminal of an electrolytic cell.

anorexia absence of appetite.

anoxic absence of oxygen; anaerobic.

anoxygenic photosynthesis photosynthesis that takes place in the absence of oxygen and during which oxygen is not produced; photosynthesis that does not split water and evolve oxygen.

anoxyphotobacteria bacteria that can carry out only anoxygenic photosynthesis; a group of photosynthetic, phototrophic bacteria occurring in aquatic habitats that do not evolve oxygen.

antagonism the inhibition, injury, or killing of one species of microorganism by another; an interpopulation relationship in which one population has a deleterious (negative) effect on another.

anthrax an infectious disease of animals, including humans, cattle, sheep, and pigs, caused by *Bacillus anthracis*.

anti- combining form meaning opposing in effect or activity.

antibacterial agents agents that kill or inhibit the growth of bacteria, e.g., antibiotics, antiseptics, and disinfectants.

antibiotics substances of microbial origin that in very small amounts have antimicrobial activity; current usage of the term extends to synthetic and semisynthetic substances that are closely related to naturally occurring antibiotics and that have antimicrobial activity.

antibodies glycoprotein molecules produced in the body in response to the introduction of an antigen or hapten that can specifically react with that antigen; also known as *immunoglobulins*, which are part of the serum fraction of the blood formed in response to antigenic stimulation and which react with antigens with great specificity.

antibody-dependent cytotoxic hypersensitivity type 2 hypersensitivity; reaction in which an antigen present on the surface of a cell combines with an antibody, resulting in the death of that cell by stimulating phagocytic attack or initiating the complement pathway; examples include blood transfusions between incompatible types and Rh incompatibility.

antibody-mediated immunity immunity produced by the activation of the B-lymphocyte population, leading to the production of several classes of immunoglobulins.

anticodon a sequence of three nucleotides in a tRNA molecule that is complementary to the codon triplet in mRNA.

antifungal agents agents that kill or inhibit the growth and reproduction of fungi; may be fungicidal or fungistatic.

antigen any agent that initiates antibody formation and/or induces a state of active immunological hypersensitivity and that can react with the immunoglobulins that are formed.

antigen–antibody complex the molecular combination that results from the reaction between antigen and complementary antibody molecules.

antigenic shift genetic change caused by the addition of new genes e.g. those that produce new strains of influenza viruses.

antihistamines compounds used for treating allergic reactions and cold symptoms that work by inactivating histamine that is released as part of the immune response.

antimalarial drugs agents effective against the erythrocytic stage of the plasmodium life cycle.

antimicrobial agents chemical or biological agents that kill or inhibit the growth of microorganisms.

antimicrobics antimicrobial agents; chemicals that inhibit or kill microorganisms and can be safely introduced into the human body; include synthetic and naturally produced antibiotics; term generally used instead of *antibiotic*.

antiseptics chemical agents used to treat human or animal tissues, usually skin, in order to kill or inactivate microorganisms capable of causing infection; not generally considered safe for internal consumption.

antisera blood sera that contain antibodies.

antitoxin antibody to a toxin capable of reacting with that poison and neutralizing the toxin.

antiviral agents substances capable of destroying or inhibiting the reproduction of viruses.

Aphyllophorales saprophytic order of Basidiomycetes that contains the shelf or bracket fungi, cantharelles, coral, toothed, and pore fungi.

API-20E system see *miniaturized commercial identification systems*.

aplanospores nonmotile, sexual spores.

apochromatic lens an objective microscope lens in which chromatic aberration has been corrected for three colors and spherical aberration for two colors.

appendaged bacteria see *budding bacteria*.

appendicitis inflammation of the appendix; a condition generally caused by an obstruction of the appendix followed by an overgrowth of indigenous microbes, requiring surgical removal of the appendix.

aquatic growing, living in, or frequenting water; a habitat composed primarily of water.

aqueous of, relating to, or resembling water; made from, with, or by water; solutions in which water is the solvent.

arbuscules specialized inclusions in root cortex in the vesicular-arbuscular type of mycorrhizal association.

archaebacteria prokaryotes with cell walls that lack murein, having ether bonds in their membrane phospholipids; analysis of rRNA indicates that the archaebacteria represent a primary biological kingdom related to both eubacteria and eukaryotes; considered to be a primitive group of organisms that were among the earliest living forms on Earth.

arthropods animals of the invertebrate phylum Arthropoda, many of which are capable of acting as vectors of infectious diseases.

arthrospores spores formed by the fragmentation of hyphae of certain fungi, algae, and cyanobacteria.

Arthus immune-complex reaction see *complex-mediated hypersensitivity*.

artifact the appearance of something in an image or micrograph of a specimen that is due to causes within the optical system or preparation of the specimen and is not a true representation of the features of the specimen.

asci plural of *ascus*.

ascocarp structure on true ascomycetes that produces asci.

ascomycetes members of a class of fungi distinguished by the presence of an ascus, a sac-like structure containing sexually produced ascospores.

Ascomycotina sac fungi; fungal subdivision of Amastigomycota that produces sexual ascospores within an ascus; asexual reproduction is carried out by fission, fragmentation of hyphae, formation of chlamydospores, and production of conidia, mycelia composed of septate hyphae and hyphal cell walls containing chitin.

ascospores sexual spores characteristic of ascomycetes, produced in the ascus after the union of two nuclei.

ascus the sporangium or spore case of fungi, consisting of a single terminal cell.

-ase suffix denoting an enzyme.

asepsis state in which potentially harmful microorganisms are absent; free of pathogens.

aseptic techniques precautionary measures taken in microbiological work and clinical practice to prevent the contamination of cultures, sterile media, etc., and/or infection of persons, animals, or plants by extraneous microorganisms.

asexual lacking sex or functional sexual organs.

asexual reproduction reproduction without union of gametes; formation of new individuals from a single individual.

assay analysis to determine the presence, absence, or quantity of one or more components.

assembly stage of viral replication during which packaging of a nucleic acid genome within a protein capsid occurs.

assimilation the incorporation of nutrients into the biomass of an organism.

asthma type 1 hypersensitivity reaction that primarily affects the lower respiratory tract; the condition is characterized by shortness of breath and wheezing.

ataxia the inability to coordinate muscular action.

athlete's foot disease caused by dermatophytic fungi affecting chronically wet feet with skin abrasions; tinea or ringworm of the feet.

atmosphere the whole mass of air surrounding the Earth; a unit of pressure approximating 1×10^6 dynes/cm^2.

ATP see *adenosine triphosphate*.

ATPase see *adenosine triphosphatase*.

attenuation any procedure in which the pathogenicity of a given organism is reduced or abolished; reduction in the virulence of a pathogen; control of protein synthesis involving the translation process.

attenuator site the site between the operator region and the first structural gene of the operon where transcription can be interrupted, as in the *trp* operon.

attractants chemicals that cause bacteria to move toward them; chemicals that cause phagocytic white blood cells to move toward them.

atypical pneumonia bronchopneumonia (infection of the lower respiratory tract) of unknown etiology, not secondary to any other acute infectious disease.

autochthonous microorganisms and/or substances indigenous to a given ecosystem; the true inhabitants of an ecosystem; term often used to refer to the common microbiota of the body or those species of soil microorganisms that tend to remain constant despite fluctuations in the quantity of fermentable organic matter in the soil.

autoclave apparatus in which objects or materials may be sterilized by air-free saturated steam under pressure at temperatures in excess of 100°C.

autoimmunity immunity or hypersensitivity to some constituent in one's own body; immune reactions with self antigens.

autolysins endogenous enzymes involved in the breakdown of certain structural components of the cell during particular phases of cellular growth and development.

autolysis the breakdown of the components of a cell or tissues by endogenous enzymes, usually after the death of the cell or tissue.

autospores sexually formed, nonmotile spores resembling the parent cell morphologically.

autotrophs organisms whose growth and reproduction are independent of external sources of organic compounds, the required cellular carbon being supplied by the reduction of CO_2 and the needed cellular energy being supplied by the conversion of light energy to ATP or the oxidation of inorganic compounds to provide the free energy for the formation of ATP.

autotrophy mode of metabolism that does not use preformed organic matter; able to use carbon dioxide as source of carbon and light or inorganic chemicals as source of energy.

autoxidation the oxidation of a substance upon its exposure to air.

auxotrophs nutritional mutants that require growth factors not needed by the parental strain.

avirulent lacking virulence; a microorganism lacking the properties that normally cause disease.

axopodia semipermanent pseudopodia, e.g., the pseudopodia that emanate radially from the spherical cells of heliozoans and some radiolarian species.

Azobacteraceae family of Gram-negative bacteria that exhibit pleomorphic morphology and fix molecular nitrogen.

B cell a differentiated lymphocyte involved in antibody-mediated immunity.

B lymphocytes white blood cells that are able to produce specific immunoglobulins; their surfaces carry specific immunoglobulin antigen-binding receptor sites.

B memory cells specifically stimulated B lymphocytes not actively multiplying but capable of multiplication and production of plasma cells upon subsequent antigenic stimulation.

Bacillaceae family of Gram-positive, endospore-forming rod- and coccal-shaped bacteria.

bacilli bacteria in the shape of cylinders.

Bacillus thuringiensis a bacterial species used as a pesticide on numerous agricultural crops, forest trees, and ornamentals for the control of various insect pests.

bacitracin antibiotic that inhibits bacterial cell wall synthesis.

bacteremia condition in which viable bacteria are present in the blood.

bacteria members of a group of diverse and ubiquitous prokaryotic, single-celled organisms; organisms with prokaryotic cells, i.e., cells lacking a nucleus.

bacterial chlorophyll photosynthetic pigment of green and purple anaerobic photosynthetic bacteria.

bacterial chromosomes the single DNA macromolecule that contains the genetic information of bacterial cells.

bacterial dysentery dysentery caused by *Shigella* infections; also called *shigellosis*.

bacterial endospores see *endospores*.

bacterial endotoxin see *endotoxin*.

bactericidal any physical or chemical agent able to kill some types of bacteria.

bacteriological filter a filter with pores small enough to trap bacteria, about 0.45 μm or smaller, used to sterilize solutions by removing microorganisms during filtration.

bacteriology the field of science dealing with bacteria, including their relation to medicine, industry, and agriculture.

bacteriophage a virus whose host is a bacterium; a virus that replicates within bacterial cells.

bacteriostatic an agent that inhibits the growth and reproduction of some types of bacteria but need not kill the bacteria.

Bacteroidaceae family of Gram-negative anaerobic bacteria, many of which are important in the normal microbiota of humans.

bacteroids irregularly shaped (pleomorphic) forms that some bacteria can assume under certain conditions, e.g., *Rhizobium* in root nodules.

baeocytes reproductive cells of pleurocapsalean cyanobacteria; also called *endospores*.

baker's yeast *Saccharomyces cerevisiae*; yeast used in the baking industry.

band cells see *stab cells*.

barophiles organisms that grow best and/or only under conditions of high pressure, e.g., in the ocean's depths.

barotolerant organisms that can grow under conditions of high pressure but do not exhibit a preference for growth under such conditions.

bartonellosis carrion's disease; a bacterial infection of humans endemic to the Andes, caused by *Bartonella bacilliformis*, which attacks red blood cells.

base analogs chemicals that structurally resemble the DNA nucleotides and therefore may substitute for them, but do not function in the same manner.

base substitutions mutations that occur when one pair of nucleotide bases in the DNA is replaced by another pair.

basic stains dyes whose active staining parts consist of a cationic, negatively charged group that may be combined with an acid, usually inorganic, that has affinity for nucleic acids.

basidiocarps the fruiting bodies of basidiomycetes.

basidiomycetes a group of fungi distinguished by the formation of sexual basidiospores on a basidium.

Basidiomycotina club fungi; fungal subdivision of Amastigomycota; includes the smuts, rusts, jelly fungi, shelf fungi, stinkhorns, bird's nest fungi, puffballs, and mushrooms; produce sexual basidiospores on the surfaces of basidia.

basidiospores sexual spores formed on basidiocarps by basidiomycetes.

basidium club-like structure of basidiomycetes on which basidiospores are borne.

basophils white blood cells containing granules (granulocyctes) that readily take up basic dye.

batch culture see *batch process*.

batch process common simple form of culture in which a fixed volume of liquid medium is inoculated and incubated for an appropriate period of time; cells grown this way are exposed to a continually changing environment; when used in industrial processes, the culture and products are harvested as a batch at appropriate times.

beer beverage produced by microbial alcoholic fermentation and brewing of cereal grains.

benthos the bottom region of aquatic habitats; collective term for the organisms living at the bottom of oceans and lakes.

Bergey's Manual reference book describing the established status of bacterial taxonomy; describes bacterial taxa and provides keys and tables for their identification.

beta-galactosidase an enzyme catalyzing the hydrolysis of β-linked galactose within dimers or polymers.

beta-hemolysis complete lysis of red blood cells, as shown by the presence of a sharply defined zone of clearing surrounding certain bacterial colonies growing on blood agar.

beta-lactamase an enzyme that attacks a β-lactam ring, such as a penicillinase that attacks the lactam ring in the penicillin antimicrobials, inactivating such antibiotics.

beta-oxidation metabolic pathway for the oxidation of fatty acids resulting in the formation of acetate and a new fatty acid that is two carbon atoms shorter than the parent fatty acid.

binary fission a process in which two similarly sized and shaped cells are formed by the division of one cell; process by which most bacteria reproduce.

binding proteins chemosensors in the cell envelope that bind specifically and tightly to substances in the membrane transport process, detect certain chemicals, and signal the flagella to respond.

binomial nomenclature the scientific method of naming plants, animals, and microorganisms composed of two names consisting of the species and genus.

bioassay the use of a living organism to determine the amount of a substance based on the growth or activity of the test organism under controlled conditions.

biochemicals substances produced by and/or involved in the metabolic reactions of living organisms.

biodegradation the process of chemical breakdown of a substance to smaller products caused by microorganisms or their enzymes.

biodeterioration the chemical or physical alteration of a product that decreases the usefulness of that product for its intended purpose caused by microorganisms or their enzymes.

biodisc system a secondary sewage treatment system employing a film of active microorganisms rotated on a disc through sewage.

bioenergetics the transfer of energy through living systems; energy transformations in living systems.

biogeochemical cycling the biologically mediated transformations of elements that result in their global cycling, including transfer between the atmosphere, hydrosphere, and lithosphere.

bioleaching the use of microorganisms to transform elements so that the elements can be extracted from a material when water is filtered through it.

biological control the deliberate use of one species of organism to control or eliminate populations of other organisms; used in the control of pest populations.

biological oxygen demand (BOD) the amount of dissolved oxygen required by aerobic and facultative microorganisms to stabilize organic matter in sewage or water; also known as *biochemical oxygen demand*.

bioluminescence the generation of light by certain microorganisms.

biomagnification an increase in the concentration of a chemical substance, such as a pesticide, as the substance is passed to higher members of a food chain.

biomass the dry-weight, volume, or other quantitative estimation of organisms; the total mass of living organisms in an ecosystem.

biosphere the part of the Earth in which life can exist; all living things together with their environment.

biosynthesis the production (synthesis) of chemical substances by the metabolic activities of living organisms.

biotic of or relating to living organisms, caused by living things.

biotype a variant form of a given species or serotype, distinguishable by biochemical or other means.

biphasic growth curve growth curve reflecting diauxie, i.e., the preferential utilization of one substrate at a given rate before another substrate is metabolized at a different rate.

blanch to scald briefly, steam, or parboil prior to deep freezing to inactivate plant enzymes that might produce toughness, change in color, and loss of flavor and nutritional value during storage.

blastomycosis a chronic mycosis caused by *Blastomyces* in which lesions develop, e.g., in the lungs, bones, and skin.

blight any plant disease or injury that results in general withering and death of the plant without rotting.

blood plasma the fluid portion of the blood minus all blood corpuscles.

blood serum the fluid expressed from clotted blood or clotted blood plasma.

blood type an immunologically distinct, genetically determined set of antigens on the surfaces of erythrocytes (red blood cells), defined as A, B, AB, and O.

bloodstream the flowing blood in a circulatory system.

bloom a visible abundance of microorganisms, generally referring to the excessive growth of algae or cyanobacteria at the surface of a body of water.

BOD see *biological oxygen demand*.

booster vaccines vaccine antigens administered to elicit an anamnestic response and to maintain extended active immunity.

botany the study of plants, including their origin, development, structure, function, and distribution.

botulism food intoxication or poisoning that is severe and often fatal, caused by *Clostridium botulinum*.

bracket fungi see *shelf fungi*.

bradykinin a peptide formed by enzymatic modification of an α-globulin (kininogen) acting as a vasodilator and playing a role in neural recognition of noxious stimuli.

breakpoint chlorination procedure for the removal and oxidation of ammonia from sewage to molecular nitrogen by the addition of hypochlorous acid.

bright-field microscope a microscope that uses visible light transmitted through a specimen to illuminate that specimen.

Brill-Zinsser disease recurrent louse-borne typhus.

broad spectrum antibiotic antibiotic capable of inhibiting a relatively wide range of bacterial species, including both Gram negative and Gram positive types.

bronchitis inflammation occurring in the mucous membranes of the bronchi, often caused by *Streptococcus pneumoniae, Haemophilus influenzae*, and certain viruses.

brown algae members of the division Phaeophycophyta; see also *Phaeophycophyta*.

brucellosis a remittent febrile disease caused by infection with bacteria of the genus *Brucella*.

Bruton's congenital γ-globulinemia chromosomal linked B-cell immunodeficiency disease affecting males in which all immunological classes are totally or partially absent.

bubonic plague form of plague involving inflammation of regional lymph nodes caused by *Yersinia pestis*.

bud scar site on a yeast cell produced by the process of fungal budding, which limits the number of progeny that can be derived from a mother cell.

budding a form of asexual reproduction in which a daughter cell develops from a small outgrowth or protrusion of the parent cell; the daughter cell is smaller than the parent cell.

budding bacteria heterogeneous group of bacteria that form extensions or protrusions from the cell; these bacteria have reproductive or physiological functions; some reproduce by budding, others by binary fission.

buffer a solution that tends to resist the change in pH when acid or alkali is added.

burst size the average number of infectious viral units released from a single cell.

butanediol fermentation pathway metabolic sequence during which acetoin is produced, carbon dioxide is released and NADH is reoxidized to NAD^+; the end product is butanediol.

butanol fermentation pathway metabolic sequence carried out by certain *Clostridium* species, with pyruvate converted to either acetone and carbon dioxide, isopropanol and carbon dioxide, butyrate, or butanol.

butyric acid pathway see *butanol fermentation pathway*.

C₄ pathway a carbon dioxide fixation pathway in both heterotrophs and autotrophs that produces oxaloacetate.

calcium dipicolinate chemical component of bacterial endospores contained within the core and involved in conferring heat resistance on endospores.

Calvin cycle the primary pathway for carbon dioxide fixation (conversion of carbon dioxide to organic matter) in photoautotrophs and chemolithotrophs.

cankers plant diseases, or conditions of those diseases, that interfere with the translocation of water and minerals to the crown of the plant.

canning method for the preservation of foodstuffs in which suitably prepared foods are placed in metal containers that are heated, exhausted, and hermetically sealed.

capillary one of a network of tiny hair-like blood vessels connecting the arteries to the veins.

capsid a protein coat of a virus enclosing the naked nucleic acid.

capsomere the individual protein units that form the capsid of a virus.

capsule a mucoid envelope composed of polypeptides and/or carbohydrates surrounding certain microorganisms; a gelatinous or slimy layer external to the bacterial cell wall.

carbohydrates a class of organic compounds consisting of many hydroxyl (-OH) groups and containing either a ketone or an aldehyde.

carbolic acid phenol.

carboxyl end the terminus of a polypeptide chain with a free alpha carboxyl group not involved in forming a peptide linkage; also known as the *C terminal end*.

carboxylic acid an organic chemical having a -COOH functional group.

carcinogen cancer-causing agent.

carditis inflammation of the heart.

caries bone or tooth decay with the formation of ulceration; also known as *dental caries* or *cavities*.

carotenoid pigments a class of pigments, usually yellow, orange, red, or purple, that are widely distributed among microorganisms.

carpogonia the basal bodies bearing female gametes in some red algae.

carpospore red algal spore produced during fertilization.

carriers individuals who harbor pathogens but do not exhibit any signs of illness.

catabolic pathway a degradative metabolic pathway; a metabolic pathway in which large compounds are broken down into smaller ones.

catabolism metabolic reactions involving the enzymatic degradation of organic compounds to simpler organic or inorganic compounds with the release of free energy.

catabolite repression repression of the transcription of genes coding for certain inducible enzyme systems by glucose or other readily utilizable carbon sources.

catalases enzymes that catalyze the decomposition of hydrogen peroxide (H_2O_2) into water and oxygen and the oxidation of alcohols to aldehydes by hydrogen peroxide.

catalyst any substance that accelerates a chemical reaction but itself remains unaltered in form and amount.

catalyze to subject to modification, especially an increase in the rate of a chemical reaction.

catheterization insertion of a hollow tubular device (a catheter) into a cavity, duct, or vessel to permit injection or withdrawal of fluids.

cathode the electrode at which reduction takes place in an electrolytic cell; a negatively charged electrode.

cations positively charged ions.

cDNA see *complementary DNA*.

cell the functional and structural subunit of living organisms, separated from its surroundings by a delimiting membrane.

cell envelope structure found only in Gram-negative cell walls, extending outward from the cytoplasmic membrane to the outer membrane.

cell lysis see *lysis*.

cell-mediated hypersensitivity type 4 hypersensitivity or delayed hypersensitivity; reaction involving T lymphocytes and occurring 24–72 hours after exposure to the antigen; contact dermatitis, including poison ivy, is an example.

cell-mediated immune response see *cell-mediated immunity*.

cell-mediated immunity specific acquired immunity involving T cells, primarily responsible for resistance to infectious diseases caused by certain bacteria and viruses that reproduce within host cells.

cell wall structure outside of and protecting the cell membrane, generally containing murein in prokaryotes and composed chiefly of various other polymeric substances, e.g., cellulose or chitin, in eukaryotic microorganisms.

cellulase an extracellular enzyme that hydrolyzes cellulose.

cellulose a linear polysaccharide of β-D-glucose.

central metabolic pathways metabolic sequences that play key roles in both catabolism and biosynthesis.

centrifugation process in which particulate matter is sedimented from a fluid, or fluids of different densities are separated, using a centrifuge.

centrifuge an apparatus used to separate by sedimentation particulate matter suspended in a liquid by a centrifugal force.

cephalosporins a heterogeneous group of natural and semisynthetic antibiotics that act against a range of Gram-positive and Gram-negative bacteria by inhibiting the formation of cross-links in peptidoglycan.

cerebrospinal fluid the fluid contained within the four ventricles of the brain, the subarachnoid space, and the central canal of the spinal chord.

chanchre the lesion formed at the site of primary inoculation by an infecting microorganism, usually an ulcer.

chancroid a lesion produced by an infection with *Haemophilus ducreyi* involving the genitalia; a sexually transmitted disease caused by *H. ducreyi*.

charging the attachment of an amino acid to its specific tRNA molecule.

chemical mutagens chemical substances that can modify nucleotide bases; chemicals that increase the rate of mutation.

chemical oxygen demand the amount of oxygen required to oxidize completely the organic matter in a water sample.

chemical preservative chemical substances added to prevent the spoilage of a food or the biodeterioration of any substance by inhibiting microbial growth and/or activity.

chemiosmosis the generation of ATP by the movement of hydrogen ions into pores in the cytoplasmic membrane that are associated with the ATPase system.

chemiosmotic hypothesis the theory that the living cell establishes a proton and electrical gradient across a membrane and that, by controlled reentry of protons into the region contained by that membrane, the energy to carry out several different types of endergonic processes may be obtained, including the ability to drive the formation of ATP.

chemoautotrophs microorganisms that obtain energy from the oxidation of inorganic compounds and carbon from inorganic carbon dioxide; organisms that obtain energy through chemical oxidation and use inorganic compounds as electron donors; also known as *chemolithotrophs*.

chemoheterotrophs see *heterotrophs*.

chemolithotrophs microorganisms that obtain energy through chemical oxidation and use inorganic compounds as electron donors and cellular carbon through the reduction of carbon dioxide; also known as *chemoautotrophs*.

chemoorganotrophs organisms that obtain energy from the oxidation of organic compounds and cellular carbon from preformed organic compounds.

chemostat an apparatus used for continuous-flow culture to maintain bacterial cultures in the log phase of growth, based on maintaining a continuous supply of a solution containing a nutrient in limiting quantities that controls the growth rate of the culture.

chemotaxis a locomotive response in which the stimulus is a chemical concentration gradient; movement of microorganisms toward or away from a chemical stimulus.

chemotherapy the use of chemical agents for the treatment of disease, including the use of antibiotics to eliminate infecting agents.

chicken pox common, acute, and highly contagious infection caused by the herpesvirus varicella-zoster, occurring most frequently in children and producing a distinctive rash.

chitin a polysaccharide composed of repeating *N*-acetylglucosamine residues, abundant in arthropod exoskeletons and fungal cell walls.

chlamydias obligate intracellular parasites whose reproduction is characterized by a change of the small, rigid-walled, infectious form of the organism into a larger, thin-walled, noninfectious form that divides by fission.

chlamydospores thick-walled, typically spherical or ovoid resting spores produced asexually by certain types of fungi from cells of the somatic hyphae.

chlor- combining form indicating that chlorine is substituted for hydrogen.

chloramination the use of chloramines to disinfect water.

chloramphenicol aminoglycoside antibiotic that inhibits bacterial protein synthesis; it acts by binding to the 50S ribosomal subunit, preventing the binding of tRNA.

chlorination the process of treating with chlorine, as in disinfecting drinking water or sewage.

Chlorobiaceae green sulfur bacteria; family of nonmotile, obligately phototrophic bacteria that produce green or green-brown carotenoid pigments.

Chloroflexaceae family of anaerobic phototrophic bacteria; its members have flexible walls, form filaments, and exhibit gliding motility.

Chlorophycophyta green algae; may be unicellular, colonial, or filamentous; most cells are uninucleate; some form coenocytic filaments, contain contractile vacuoles, or store starch as reserve material; their cell walls are composed of cellulose, mannans, xylans, or protein.

chlorophyll the green pigment responsible for photosynthesis in plants; the primary photosynthetic pigment of algae and cyanobacteria.

chloroplasts membrane-bound organelles of photosynthetic eukaryotes where the biochemical conversion of light energy to ATP occurs; the sites of photosynthesis in eukaryotic organisms.

chlorosis the yellowing of leaves and/or plant components due to bleaching of chlorophyll, often symptomatic of microbial disease.

cholera an acute infectious disease caused by *Vibrio cholerae* characterized by diarrhea, delirium, stupor, and coma.

choleragen enterotoxin produced by *Vibrio cholerae* that blocks the conversion of cyclic AMP to ATP.

Chromatiaceae purple sulfur bacteria; family of phototrophic bacteria that produce carotenoid pigments, appear orange-brown, purple-red, or purple-violet, and deposit elemental sulfur.

chromatic aberration an optical lens defect causing distortion of the image because light of differing wavelengths is focused at differing points instead of at a single focal point.

chromatids fibrils formed from a eukaryotic chromosome when it replicates prior to meiosis or mitosis.

chromatin the deoxyribonucleic acid–protein complex that constitutes a chromosome; the readily stainable protoplasmic substance in the nuclei of cells.

chromosomes structures that contain the nuclear DNA of a cell.

chronic a disease condition in which the symptoms persist for a long time.

chroococcacean cyanobacteria subgroup of cyanobacteria, unicellular rods or cocci that reproduce by either binary fission or budding; generally nonmotile.

Chrysophycophyta division of algae that includes the yellow-green algae, golden algae, and diatoms; all produce chrysolaminarin; most are unicellular and some are colonial.

Chytridiomycetes class of Mastigomycota; fungi that produce zoospores; motile with a single posterior whiplash flagellum; many are parasitic on plants, algae, and other fungi.

chytrids members of the Chytridiales, which are mainly aquatic fungi that produce zoospores with a single posterior flagellum.

-cide suffix signifying a killer or destroyer, as in a chemical that kills microbes.

cilia thread-like appendages, having a 9 + 2 arrangement of microtubules occurring as projections from certain cells, that beat rhythmically, causing locomotion or propelling fluid over surfaces.

ciliated epithelial cells cells that line the respiratory tract and act as filters by sweeping microorganisms out of the body with a wave-like motion.

Ciliophora members of one subphylum of protozoa that possess simple to compound ciliary organelles in at least one stage of their life cycle; these protozoa are motile by means of cilia.

circadian rhythm daily cyclical changes that occur in an organism even when it is isolated from the natural daily fluctuations of the environment.

circulatory system the vessels and organs comprising the cardiovascular and lymphatic systems of animals.

cistron the functional unit of genetic inheritance; a segment of genetic nucleic acid that codes for a specific polypeptide chain; synonym for *gene*.

citric acid intermediary metabolite in the Krebs cycle; an organic acid, produced by *Aspergillus niger*, used as a food additive, a metal chelating and sequestering agent, and a plasticizer.

citric acid cycle see *Krebs cycle*.

clamp cell connections hyphal structures in many basidiomycetes formed during cell division by dikaryotic hyphal cells, i.e., formed by hyphal cells containing two nuclei of different mating types.

classical complement pathway series of reactions initiated by the formation of a complex between an antigen and an antibody that lead to the lysis of microbial cells or the enhanced ability of phagocytic blood cells to eliminate such cells.

classification the systematic arrangement of organisms in groups or categories according to established criteria.

clonal selection theory a theory that accounts for antibody formation by supposing that during fetal development a complete set of lymphocytes are developed, with each lymphocyte containing the genetic information for initiating an immune response to a single specific antigen for which it has only one type of receptor; B cells that react with self antigens during this period are destroyed.

clone a population of cells derived asexually from a single cell, often assumed to be genetically homologous; a population of genetically identical individuals; to replicate cells of uniform genetic composition; to replicate a specific gene sequence.

cloning vector segment of DNA used for the replication of foreign DNA fragments.

club fungi see *Basidiomycotina*.

coagglutination an enhanced agglutination reaction based on using antibody molecules whose Fc fragments are attached to cells so that a larger matrix is formed when the Fab portion reacts with other cells for which the antibody is specific.

coagulase an enzyme produced by pathogenic staphylococci, causing coagulation of blood plasma.

cocci spherical or nearly spherical bacterial cells, varying in size and sometimes occurring singly, in pairs, in regular groups of four or more, in chains, or in irregular clusters.

coccidioidomycosis a disease of humans and domestic animals caused by *Coccidioides immitis*, usually occurring via the respiratory tract.

codominance the partial expression of the genetic information contained in both the recessive and dominant alleles of a gene.

codon a triplet of adjacent bases in a polynucleotide chain of an mRNA molecule that codes for a specific amino acid; the basic unit of the genetic code specifying an amino acid for incorporation into a polypeptide chain.

coenocytic referring to any multinucleate cell, structure, or organism formed by the division of an existing multinucleate entity or when nuclear divisions are not accompanied by the formation of dividing walls or septa; multinucleate hyphae.

coenzymes the nonprotein portions of enzymes; small, nonprotein organic chemicals that are not tightly bound to the enzymes with which they function and that act as acceptors or donors of electrons or functional groups during enzymatic reactions.

cofactors inorganic substances, such as minerals, required for enzymatic activity.

cold pack method a method of canning uncooked high-acid food by placing it in hot jars or cans and sterilizing in a bath of boiling water or steam.

colicinogenic plasmids plasmids that code for colicins.

colicins proteins produced by some bacteria that inhibit closely related bacteria.

coliforms Gram-negative, lactose-fermenting, enteric rods, e.g., *Escherichia coli*.

colinear two related linear information sequences arranged so that the unit may be moved from one to the other without rearrangement; RNA and DNA molecules with precisely matching base pairs.

coliphage a virus that infects *Escherichia coli*.

colonization the establishment of a site of microbial reproduction on a material, animal, or person without necessarily resulting in tissue invasion or damage.

colony the macroscopically visible growth of microorganisms on a solid culture medium.

colony forming units number of microbes that can replicate to form colonies, as determined by the number of colonies that develop.

Colorado tick fever the only recognized tick-borne viral disease in the United States; an acute febrile disease characterized by sudden onset of fever, headache, and severe muscle pain.

coma a state of unconsciousness.

cometabolism the gratuitous metabolic transformation of a substance by a microorganism growing on another substrate; the cometabolized substance is not incorporated into an organism's biomass, and the organism does not derive energy from the transformation of that substance.

commensalism an interactive association between two populations of different species living together in which one population benefits from the association, while the other is not affected.

common cold an acute, self-limiting inflammation of the upper respiratory tract due to a viral infection.

common source outbreak disease outbreak characterized by a sharp rise and a rapid decline in the number of cases.

compensation depth the depth of an aquatic habitat at which photosynthetic activity balances respiratory activity; in lakes, the depth of effective light penetration, separating the limnetic and profundal zones.

competent in transformation, the state of a recipient cell in which DNA can pass across its membrane, depending upon environmental conditions and the cell growth phase.

competition an interactive association between two species, both of which need some limited environmental factor for growth and thus grow at suboptimal rates because they must share the growth-limiting resource.

competitive exclusion principle the statement that competitive interactions tend to bring about the ecological separation of closely related populations and precludes two populations from occupying the same ecological niche.

competitive inhibition the inhibition of enzyme activity caused by the competition of an inhibitor with a substrate for the active (catalytic) site on the enzyme; impairment of the function of an enzyme due to its reaction with a substance chemically related to its normal substrate.

complement group of proteins normally present in plasma and tissue fluids that participates in antigen–antibody reactions, allowing reactions such as cell lysis to occur.

complement fixation the binding of complement to an antigen–antibody complex so that the complement is unavailable for subsequent reactions.

complement fixation test test that measures the degree of complement fixation for diagnostic purposes.

complementary DNA (cDNA) in cloning eukaryotic genes in bacteria, a single-stranded DNA molecule that is complementary to the complete mRNA.

completed test in assays for assessing water safety, gas formation by subcultured colonies showing a greenish metalic sheen on EMB agar grown on lactose broth incubated at 35°C; positive test for fecal coliforms.

complex-mediated hypersensitivity type 3 hypersensitivity; reaction that occurs when excess antigens are produced during a normal inflammatory response and antibody–antigen–complement complexes are deposited in tissues.

complex medium a medium made with constituents whose compositions are not fully known and may vary.

composting the decomposition of organic matter in a heap by microorganisms; a method of solid waste disposal.

concentration gradient condition established by the difference in concentration on opposite sides of a membrane.

condenser lenses the lenses on a microscope used for focusing or directing light from the light source onto the object.

conditionally lethal mutations mutations that cause the loss of microbial viability only under certain environmental conditions.

confirmed test in assays for assessing water safety, the formation of greenish, metallic colonies of fecal coliforms on EMB agar or brilliant green lactose–bile broth.

conidia thin-walled, asexually derived spores, borne singly or in groups or clusters in specialized hyphae.

conidiophores branches of mycelia-bearing conidia.

conjugated fluorescent dyes fluorescent dyes coupled with antibody molecules used to tag antibodies.

conjugation the process in which genetic material is transferred from one microorganism to another, involving a physical connection or union between the two cells; a parasexual form of reproduction sometimes referred to as *mating*.

conjugative plasmids F and other plasmids that encode for self-transfer from one cell to another.

conjunctivitis inflammation of the mucous membranes covering the eye, the conjunctiva.

consortium an interactive association between microorganisms that generally results in combined metabolic activities.

constipation a condition in which the bowels are evacuated at long intervals or with difficulty; the passage of hard, dry stools.

constitutive enzymes enzymes whose synthesis is not altered in response to changes in the environment but rather are continuously synthesized.

contact dermatitis delayed hypersensitivity reaction resulting from exposure of the skin to chemicals; poison ivy is an example.

contact diseases diseases caused by agents that are able to enter the subcutaneous layers of the skin through hair follicles.

contagion the process by which disease spreads from one individual to another.

contagious disease an infectious disease that is communicable to healthy, susceptible individuals by physical contact with someone suffering from that disease, contact with bodily discharges from that individual, or contact with inanimate objects contaminated by that individual.

contamination the process of allowing the uncontrolled addition of microorganisms to an area or substance.

continuous culture see *continuous flow-through process*.

continuous feed composting process a composting process that uses a reactor to establish the environmental parameters that maximize the degradation process.

continuous flow-through process a process for growing microorganisms without interruption by continual addition of substrates and recovery of products.

continuous strand of DNA the strand of DNA that can be synthesized continuously because it runs in the appropriate direction for the continuous addition of new free nucleotide bases; also referred to as the *leading strand of DNA*.

contractile vacuoles pulsating vacuoles in certain protozoa used for the excretion of wastes and the exclusion of water for the maintenance of proper osmotic balance.

controlled atmosphere storage method of preserving foods by altering the normal atmospheric concentrations of various gases, usually by increasing the CO_2 concentration and decreasing the O_2 concentration to prevent the spoilage of fruit.

convalescence recovery period of a disease during which signs and symptoms disappear.

copraphagous capable of growth on fecal matter; feeding on dung or excrement.

cornsteep liquor the concentrated water extract by-product resulting from the steeping of corn during the production of cornstarch; used as a medium adjunct to supply nitrogen and vitamins in industrial fermentations.

corrosion the eating away of a metal resulting from changes in the oxidative state.

cortex a layer of a bacterial endospore important in conferring heat resistance on that structure.

coryneform group bacterial group of Gram-positive, irregularly shaped rods with a tendency to show incomplete separation following cell division and to exhibit pleomorphic morphology.

coryza an inflammation of the mucous membranes of the nose, usually marked by sneezing and the discharge of watery mucus.

cosmid phage–plasmid artificial hybrids.

counter current immunoelectrophoresis a technique based on the immunological detection of substances that relies on the movement of an antibody and an antigen toward each other in an electric field, resulting in the rapid formation of a detectable antigen–antibody precipitate.

counter immunoelectrophoresis see *countercurrent immunoelectrophoresis*.

covalent bond a strong chemical bond formed by the sharing of electrons.

cowpox a mild, self-limiting disease caused by a vaccinia virus, involving the formation of vesicular lesions on the hands and arms of humans and the udders of cows.

CPE see *cytopathic effects*.

cristae convolutions of the inner membrane that extend into the interior of the mitochondria of eukaryotic cells.

critical point drying a method for removing liquids from a microbiological specimen by adjusting the temperature and pressure so that the liquid and gas phases of the liquid are in equilibrium with each other; used to minimize disruption of biological structures for viewing by scanning electron microscopy.

crop rotation the alternation of the types of crops planted in a field.

cross-feeding the phenomena in which the growth of an organism is dependent on the provision of one or more metabolic factors or nutrients by another organism growing in the vicinity; also termed *syntrophism*.

cross-reactive antibodies see *heterophile antibodies*.

cross walls see *septae*.

crossing over the process in which, in effect, a break occurs in each of the two adjacent DNA strands, and the exposed 5′P and 3′OH ends unite with the exposed 5′P and 3′OH ends of the adjacent strands so that there is an exchange of homologous regions of DNA.

crown gall plant disease caused by *Agrobacterium tumefaciens*, which infects fruit trees, sugar beets, and other broad-leafed plants, manifested by the formation of a tumor growth.

Cryptophycophyta group of unicellular brown algae that reproduce by longitudinal division, producing two flagella of equal length.

culture to encourage the growth of particular microorganisms under controlled conditions; the growth of particular types of microorganisms on or within a medium as a result of inoculation and incubation.

curvature of field distortion of a microscopic field of view in which specimens in the center of the field are in clear focus, while those in the peripheral region are out of focus.

cutaneous pertaining to the skin.

cyanobacteria prokaryotic, photosynthetic organisms containing chlorophyll a, capable of producing oxygen by splitting water; formerly known as *blue-green algae*.

Cyanobacteriales order of Oxyphotobacteria whose primary bacterial photosynthetic pigment is chlorophyll a; the blue-green algae or cyanobacteria.

cyclic oxidative photophosphorylation see *cyclic photophosphorylation*.

cyclic photophosphorylation a metabolic pathway involved in the conversion of light energy to chemical energy, with the generation of ATP that does not produce the reduced coenzyme, NADPH.

cycloserine antibiotic that inhibits bacterial cell wall synthesis.

cyst a dormant form assumed by some microorganisms during specific stages in their life cycles, or assumed as a response to particular environmental conditions in which the organism becomes enclosed in a thin- or thick-walled membranous structure, the function of which is either protective or reproductive; a normal or pathological sac with a distinct wall containing fluid.

cystitis an inflammation of the urinary bladder.

cytochromes reversible oxidation-reduction carriers in respiration.

cytokinesis the division of cytoplasm following nuclear division.

cytolysis the dissolution or disintegration of a cell.

cytopathic effects (CPE) generalized degenerative changes or abnormalities in the cells of a monolayer tissue culture due to infection by a virus.

Cytophagales gliding bacteria exhibiting widely differing morphological forms and modes of metabolism that do not form fruiting bodies; however, some form filaments and others are chemolithotrophs.

cytoplasm the living substance of a cell, exclusive of the nucleus.

cytoplasmic membrane the selectively permeable membrane that forms the outer limit of the protoplast, bordered externally by the cell wall in most bacteria.

cytoplasmic polyhedrosis virus a type of virus used as a viral pesticide that develops in the cytoplasm of host midgut epithelial cells.

cytoplast the unified structure that provides the rigidity needed to hold the various structures of the eukaryotic cell in their appropriate locations.

cytosine a pyrimidine base found in nucleic acids.

cytosis the movement of materials into or out of a cell, involving the engulfment and formation of a membrane-bound structure rather than passage through a membrane.

cytoskeleton protein fibers composing the structural support framework of a eukaryotic cell.

cytostomes mouth-like openings of some protozoa, particularly ciliates.

cytotoxic T cells specialized class of T lymphocytes that are able to kill cells as part of the cell-mediated immune response.

cytotoxins substances capable of injuring certain cells without causing cell lysis.

D value see *decimal reduction time*.

dark-field microscope a microscope in which the only light seen in the field of view is reflected from the object under examination, resulting in a light object on a dark background.

deaminase an enzyme involved in the removal of an amino group from a molecule, liberating ammonia.

deamination the removal of an amino group from a molecule, especially an amino acid.

death phase the part of the normal growth curve that represents the inability of microorganisms to reproduce.

decarboxylase an enzyme that liberates carbon dioxide from the carboxyl group of a molecule by hydrolysis.

decarboxylation the splitting off of one or more molecules of carbon dioxide from organic acids, especially amino acids.

decimal reduction time the time required at a given temperature to heat inactivate or kill 90 percent of a given population of cells or spores; the time needed to reduce the number of viable microbes under a specified set of conditions by an order of magnitude.

decomposers organisms, often bacteria or fungi, in a community that convert dead organic matter into inorganic nutrients.

deficiencies deletions of large numbers of base pairs that can result in the loss of genetic information for one or more complete genes.

defined medium the material supporting microbial growth in which all of the constituents, including trace substances, are quantitatively known; a mixture of known composition used for culturing microorganisms.

degenerate describes the redundancy inherent in the genetic code that occurs because there are several codons coding for the insertion of the same amino acid into the polypeptide chain.

dehydration removal of water; drying.

dehydrogenase an enzyme that catalyzes the oxidation of a substrate by removing hydrogen.

delayed hypersensitivity see *cell-mediated hypersensitivity*.

deletion mutations mutations caused by the removal of one or more nucleotide base pairs from the DNA.

denaturation the alteration in the characteristics of an organic substance, especially a protein, by physical or chemical action; the loss of enzymatic activity due to modification of the tertiary protein structure.

dendrograms graphic representations of taxonomic analyses, showing the relationships between the organisms examined.

dengue fever a human disease caused by togavirus and transmitted by mosquitoes, characterized by fever, rash, and severe pain in joints and muscles.

denitrification the formation of gaseous nitrogen or gaseous nitrogen oxides from nitrate or nitrite by microorganisms.

dental caries tooth decay.

dental plaque matrix of microbial cells and microbially produced extracellular polysaccharides that forms on the tooth surface and can be removed by brushing and flossing.

deoxyribonucleic acid (DNA) the carrier of genetic information; a type of nucleic acid occurring in cells, containing adenine, guanine, cytosine, and thymine, and D-2-deoxyribose linked by phosphodiester bonds.

deoxyribose a 5-carbon sugar having one oxygen less than the parent sugar ribose; a component of DNA.

derepress the regulation of transcription by reversibly inactivating a repressor protein.

dermatitis an inflammation of the skin.

dermatophytes fungi characterized by their ability to metabolize keratin and capable of growing on the skin surface, causing disease.

desert a region of low rainfall; a dry region; a region of low biological productivity.

desiccation removal of water; drying.

desulfurization removal of sulfur from organic compounds.

detergent a synthetic cleaning agent containing surface-active agents that do not precipitate in hard water; a surface-active agent having a hydrophilic and a hydrophobic portion.

detrital food chain a food chain based on the biomass of decomposers rather than on that of primary producers.

detritus waste matter and biomass produced from decompositional processes.

deuteromycetes fungi with no known sexual stage; also known as *Fungi Imperfecti*.

Deuteromycotina see *Fungi Imperfecti*.

diagnostic table a table of distinguishing features used as an aid in the identification of unknown organisms.

diapedesis the process by which leukocytes move out of blood vessels.

diarrhea a common symptom of gastrointestinal disease, characterized by increased frequency and fluid consistency of stools.

diatomaceous earth a silicaceous material composed largely of fossil diatoms, used in microbiological filters and industrial processes.

diatoms unicellular algae having a cell wall composed of silica, the skeleton of which persists after the death of the organism.

diauxie the phenomenon in which, given two carbon sources, an organism preferentially metabolizes one completely before utilizing the other.

dichotomous key a key for the identification of organisms, using steps with opposing choices until a final identification is achieved.

dictysomes the individual stacks of membranes in a Golgi apparatus.

differential blood count procedure for finding the ratios of various types of blood cells, used to determine the relative numbers of white blood cells as a diagnostic indication of an infectious process.

differential media bacteriological media on which the growth of specific types of organisms leads to readily visible changes in the appearance of the media so that the presence of these organisms can be determined.

differentially permeable membrane a membrane that selectively restricts the movement of molecules.

diffraction the breaking up of a beam of light into bands of differing wavelength due to interference.

diffusion the movement of molecules across a concentration gradient from the area of higher concentration to the area of lower concentration.

DiGeorge syndrome immune disorder caused by the partial or total absence of cellular immunity, resulting from a deficiency of T lymphocytes because of incomplete fetal development of the thymus.

digestive vacuoles membrane-bound organelles formed when a eukaryotic cell engulfs a food source and then fuses with lysosomes, permitting digestion of the contents.

dikaryotes cells with two different nuclei resulting from the fusion of two cells.

dimorphism the property of existing in two distinct structural forms, e.g., fungi that occur in filamentous and yeast-like forms under different conditions.

dinoflagellates algae of the class Pyrrophycophyta, primarily unicellular marine organisms, possessing flagella.

diphtheria an acute, communicable human disease caused by strains of *Corynebacterium diphtheriae*.

diphtheria toxin cytotoxin produced by *Corynebacterium diphtheriae* that inhibits protein synthesis in mammalian cells by blocking transferase reactions during translation.

diplococci cocci occurring in pairs.

diploid having double the haploid number of chromosomes; having a duplication of genes.

dipole moment the polarity resulting from the separation of electric charges; in chemical bonds, resulting from an unequal distribution of electrons.

direct counting procedures methods for the enumeration of bacteria and other microbes that do not require the growth of cells in culture but rather rely upon direct observation or other detection methods by which the undivided microbial cells can be counted.

direct fluorescent antibody staining (FAB) method used to detect the presence of an antigen by staining with a specific antibody linked with a fluorescent dye; the conjugated fluorescent antibody reacts directly with the antigens.

disaccharides carbohydrates formed by the condensation of two monosaccharide sugars.

discontinuous strand of DNA the strand of DNA that lags behind the replication of the continuous strand because DNA polymerases can add nucleotides in only one direction; therefore, synthesis of this strand can begin only after some unwinding of the double strand has occurred and takes place via the synthesis of short segments that run in the opposite direction to the overall direction of synthesis; also referred to as the *lagging strand of DNA*.

disease a condition of an organ, part, structure, or system of the body in which there is incorrect functioning due to the effect of heredity, infection, diet, or environment; a physiologically impaired state of a plant or animal resulting from microbial infection, microbial products, or microbial activities; a physiological condition that occurs when microorganisms overcome host defense systems.

disease syndrome stages in the course of a disease.

disinfectants chemical agents used for disinfection.

disinfection the destruction, inactivation, or removal of microorganisms likely to cause infection or produce other undesirable effects.

dispersal breaking up and spreading in various directions, e.g., the spread of microorganisms from one place to another.

dissemination the scattering or dispersion of microorganisms or disease, e.g., the spread of disease associated with the dispersal of pathogens.

dissociation separation of a molecule into two or more stable fragments; a change in colony form often occurring in a new environment, associated with modified growth or virulence.

distilled liquor alcoholic beverage produced by microbial alcoholic fermentation followed by chemical distillation to achieve a high alcohol concentration.

DNA see *deoxyribonucleic acid*.

DNA double helix the two primary polynucleotide chains of DNA held together by hydrogen bonding between complementary nucleotide bases.

DNA homology the degree of similarity of base sequences in DNA from different organisms.

DNA polymerases enzymes that catalyze the phosphodiester bonds in the formation of DNA.

dolipore septae the thick internal transverse openings between cell walls of basidiomycetes.

domestic sewage liquid household wastes.

dominant allele the allelic form of a gene whose information is preferentially expressed.

donor any cell that contributes genetic information to another cell.

dormant an organism or spore that exhibits minimal physical and chemical change over an extended period of time but remains alive.

double diffusion method precipitin reaction technique in which an antigen and an antibody diffuse toward each other from separate wells cut into an agar gel.

doubling time see *generation time*.

drugs substances used in medicine for the treatment of disease.

dust cells macrophage cells fixed in the alveolar lining of the lungs.

dwarfism plant condition resulting from degradation or inactivation of plant growth substances by pathogens.

dysentery an infectious disease marked by inflammation and ulceration of the lower part of the bowels, with diarrhea that becomes mucous and hemorrhagic; disease condition characterized by diarrhea.

dysfunctional immunity an immune response that produces an undesirable physiological state, e.g., an allergic reaction, or the lack of an immune response resulting in a failure to protect the body against infectious or toxic agents.

early proteins proteins that are made early in viral replication.

ECHO virus group group of viruses frequently found as causative agents of gastroenteritis.

eclipse period the period in the lytic reproduction cycle in which complete infective viruses are not present.

ecology the study of the interrelationships between organisms and their environments.

ecosystem a functional self-supporting system that includes the organisms in a natural community and their environment.

ectomycorrhizae a stable, mutually beneficial (symbiotic) association between a fungus and the root of a plant where the fungal hyphae occur outside the root and between the cortical cells of the root.

effluent the liquid discharge from sewage treatment and industrial plants.

Eijkman test in assays for assessing water safety, gas formation from dilutions of water samples incubated in lactose broth at 45°C demonstrates the presence of fecal coliforms.

electromagnetic spectrum a range of energy in the form of waves of differing lengths that produce varying electric and magnetic fields as it travels through space from its source to a receiver.

electron a negatively charged subatomic particle that orbits the positively charged nucleus of an atom.

electron acceptors substances that accept electrons during oxidation-reduction reactions.

electron donors substances that give up electrons during oxidation-reduction reactions.

electron transport chain a series of oxidation-reduction reactions in which electrons are transported from a substrate through a series of intermediate electron carriers to a final acceptor establishing an electrochemical gradient across a membrane that results in the formation of ATP.

electrophoresis the movement of charged particles suspended in a liquid under the influence of an applied electron field.

elementary body small, rigid-walled, infectious form of chlamydias.

elevated temperature higher than normal temperature; characteristic symptom of the inflammatory response associated with the high metabolic activities of neutrophils and macrophages.

EMB agar see *eosin methylene blue agar*.

Embden-Meyerhof pathway a specific glycolytic pathway; a sequence of reactions in which glucose is broken down to pyruvate.

Embden-Myerhof-Parnas Pathway see *Embden-Myerhof Pathway*.

embryonated eggs hen or duck eggs containing live embryos, used for culturing viruses and preparing tissue cultures.

encephalitis an inflammation of the brain.

end- combining form indicating within.

end product the chemical compound that is the final product in a particular metabolic pathway.

end product inhibition see *feedback inhibition*.

end product repression the process of shutting off transcription when a by-product of the metabolism coded for by the genes in that transcription region accumulates.

endemic peculiar to a certain region, e.g., a disease that occurs regularly in an area.

endergonic a chemical reaction with a positive ΔG; a chemical reaction requiring input of free energy.

endocarditis infection of the endocardium or heart valves caused by bacteria or, in the cases of intravenous drug abusers, fungi.

endocardium the membrane lining the interior of the heart.

endocytosis the movement of materials into a cell by cytosis.

endogenous produced within; due to internal causes; pertaining to the metabolism of internal reserve materials.

endomycorrhizal associations mycorrhizal associations in which there is fungal penetration of plant root cells.

endonuclease an enzyme that catalyzes the cleavage of DNA, normally cutting it at specific sites.

endoparasitic slime molds see *Plasmodiophoromycetes.*

endophytic a photosynthetic organism living within another organism.

endoplasmic reticulum the extensive array of internal membranes in a eukaryotic cell involved in coordinating protein synthesis.

endospores thick-walled spores formed within a parent cell; in bacteria, heat-resistant spores.

endosymbiotic a symbiotic (mutually dependent) association in which one organism penetrates and lives within the cells or tissues of another organism.

endosymbiotic evolution theory that bacteria living as endosymbionts within eukaryotic cells gradually evolved into organelle structures.

endothelial a single layer of thin cells lining internal body cavities; the inner layer of the seed coat of some plants.

endothermic a chemical reaction in which energy is consumed; a chemical reaction requiring an input of heat energy.

endotoxins toxic substances found as part of some bacterial cells; the lipopolysaccharide component of the cell wall of Gram-negative bacteria.

energy charge measure of the energy status of a cell, describing its relative proportions of ATP, ADP, and AMP.

enrichment culture any form of culture in a liquid medium that results in an increase in a given type of organism while minimizing the growth of any other organism present.

enter- combining form meaning the intestine.

enteric of or pertaining to the intestines.

enteric bacteria bacteria that live within the intestinal tract.

Enterobacteriaceae family of Gram-negative, facultatively anaerobic rods, motile by means of peritrichous or polar flagella, divided into five tribes.

enterobactin siderophore synthesized by enteric bacteria.

enterotoxins toxins specific for cells of the intestine, causing intestinal inflammation and producing the symptoms of food poisoning.

enthalpy the total heat of a system; ΔH.

Entner-Doudoroff pathway glycolytic pathway that results in the net production of only one ATP molecule per molecule of glucose substrate metabolized.

entomogenous fungi fungi living on insects; fungal pathogens of insects.

entropy that portion of the energy of a system that cannot be converted to work; ΔS.

enumeration determination of the number of microorganisms.

envelope the outer covering surrounding the capsid of some viruses.

enzymatic reactions chemical reactions catalyzed by enzymes.

enzyme-linked immunosorbent assay (ELISA) a technique used for detecting and quantifying specific serum antibodies based upon tagging the antigen–antibody complex with a substrate that can be enzymatically converted to a readily quantifiable product by a specific enzyme.

enzymes proteins that function as efficient biological catalysts, in-creasing the rate of a reaction without altering the equilibrium constant by lowering the energy of activation.

eosin methylene blue agar a medium used for the detection of coliform bacteria; the growth of Gram-positive bacteria is inhibited on this medium, and lactose fermenters produce colonies with a green metallic sheen.

eosinophil a white blood cell having an affinity for eosin or any acid stain.

eosinophilia an increase above normal in the number of eosinophils in the peripheral blood.

epi- prefix meaning upon, beside, among, above, or outside.

epidemic an outbreak of infectious disease among a human population in which, for a limited time, a high proportion of the population exhibits overt disease symptoms.

epidemiology the study of the factors and mechanisms that govern the spread of disease within a population, including the interrelationships between a given pathogenic organism, the environment, and populations of relevant hosts.

epifluorescence microscopy a form of microscopy employing stains that fluoresce when excited by light of a given wavelength, emitting light of a different wavelength; exciter filters are used to produce the proper excitation wavelength, and barrier filters are used so that only the fluorescing specimens are visible.

epigenetic direct products derived from an organism's genome, e.g., ribosomal RNA.

epilimnion the warm upper surface layer of an aquatic environment.

epiphytes organisms growing on the surfaces of other organisms, e.g., bacteria growing on the surface of an algal cell.

episomes segments of DNA capable of existing in two alternate forms, one replicating autonomously in the cytoplasm, the other replicating as part of the bacterial chromosome.

epitheca the larger of the two parts of the cell wall (frustule) of a diatom.

epizootic an epidemic outbreak of infectious disease among animals other than humans.

Epstein-Barr virus a member of the herpesvirus group; the causative agent of infectious mononucleosis.

equilibrium a state of balance, a condition in which opposing forces equalize with one another so that no movement occurs; in a chemical reaction, the condition where forward and reverse reactions take place at equal rates so that no net change occurs; when a reaction is at equilibrium, the amounts of reactants and products remain constant.

equilibrium constant the relationship among concentrations of the substances within an equilibrium system regardless of how the equilibrium condition is achieved.

ergotism a condition of intoxication that results from the ingestion of grain contaminated by ergot alkaloids produced by the fungus *Claviceps purpurea.*

erythema abnormal reddening of the skin due to local congestion, symptomatic of inflammation.

erythrocytes red blood cells.

erythromycin an antibiotic produced by a strain of *Streptomyces* that inhibits protein synthesis.

estuary a water passage where the ocean tide meets a river current; an arm of the sea at the lower end of a river.

ethanolic fermentation a type of fermentation in which glucose is converted to ethanol and carbon dioxide.

etiological agent an agent, such as a microbe, that causes a disease.

etiology the study of the causation of disease.

Euascomycetidae true ascomycetes; fungi that produce asci in ascocarps that develop from dikaryotic hyphae.

eubacteria prokaryotes other than archaebacteria.

eugenotes theoretical primitive versions of prokaryotes.

Euglenophycophyta unicellular division of algae that contain chlorophylls a and b and appear green, lack a cell wall, and are surrounded by a pellicle; they store paramylon as reserve material and reproduce by longitudinal division; they are widely distributed in aquatic and soil habitats.

eukaryotes cellular organisms having a membrane-bound nucleus within which the genome of the cell is stored as chromosomes composed of DNA; eukaryotic organisms include algae, fungi, protozoa, plants, and animals.

eukaryotic cells cells having a true nucleus containing the genome and various additional membrane-bound organelles.

euphotic the top layer of water, through which sufficient light penetrates to support the growth of photosynthetic organisms.

eurythermal microorganisms that grow over a wide range of temperatures.

eutrophication the enrichment of natural waters with inorganic materials, especially nitrogen and phosphorus compounds, that support the excessive growth of photosynthetic organisms.

evolution the directional process of change of organisms by which descendants become distinct in form and/or function from their ancestors.

exergonic a reaction accompanied by a liberation of free energy.

exo- prefix indicating outside, an outside layer, or out of.

exocytosis movement of materials into a cell.

exoenzymes enzymes that occur either attached to the outer surface of the cell membrane or in the periplasmic space; enzymes released into the medium surrounding a cell, including enzymes that attack extracellular polymers by sequentially removing units from one end of a polymer chain.

exogenous due to an external cause; not arising from within the organism.

exon the region of a eukaryotic genome that encodes the information for protein or RNA macromolecules or regulates gene expression; a segment of eukaryotic DNA that codes for a region of RNA that is not excised during posttranscriptional processing.

exonucleases enzymes that progressively remove the terminal nucleotides of a polynucleotide chain.

exothermic a chemical reaction that produces heat.

exotoxins protein toxins secreted by living microorganisms into the surrounding environment.

exponential growth phase the period during the growth cycle of a microbial population when growth is maximal and constant and there is a logarithmic increase in population size.

expression vector in gene cloning, a genetic vector that contains not only the desired gene but also the necessary regulatory sequences that permit control of the expression of that gene.

extracellular external to the cells of an organism.

extraterrestrial originating or existing outside of the Earth or its atmosphere.

extreme environments environments characterized by extremes in growth conditions, including temperature, salinity, pH, and water availability, among others.

exudate viscous fluid containing blood cells and debris that accumulate at the site of an inflammation or lesion.

F pilus attachment structure that projects from cells of certain bacteria involved in mating, found on cells that donate DNA.

F plasmid fertility plasmid coding for the donor strain that includes genes for the formation of the F pilus.

F value the number of minutes required to heat inactivate or kill an entire population of cells or spores in an aqueous solution at 121°C.

Fab (antigen-binding fragment) either of two identical fragments produced when an immunoglobulin is cleaved by papain; the antigen-binding portion of an antibody, including the hypervariable region.

facilitated diffusion diffusion at an enhanced rate; movement from a region of high concentration to one of low concentration that occurs more rapidly than it would on the basis of the concentration gradient.

facultative anaerobes microorganisms capable of growth under either aerobic or anaerobic conditions; bacteria capable of both fermentative and respiratory metabolism.

FAD see *flavin adenine dinucleotide*.

FADH$_2$ reduced flavin adenine dinucleotide.

family a taxonomic group; the principal division of an order; the classification group above a genus.

fastidious an organism difficult to isolate or culture on ordinary media because of its need for special nutritional factors; an organism with stringent physiological requirements for growth and survival.

fatty acids straight chains of carbon atoms with a COOH at one end in which most of the carbons are attached to hydrogen atoms.

Fc (crystallizable) fragment the remainder of the molecule when an immunoglobulin is cleaved and the Fab fragment separated; the crystallizable portion of an immunoglobulin molecule containing the constant region; the end of an immunoglobulin that binds with complement.

feedback activation the binding of a substance to an allosteric site, thus activating the enzyme and increasing its activity.

feedback inhibition a cellular control mechanism by which the end product of a series of metabolic reactions inhibits the activity of an earlier enzyme in the sequence of metabolic transformations; thus, when the end product accumulates, its further production ceases.

ferment to cause fermentation in; that which causes fermentation.

fermentation a mode of energy-yielding metabolism that involves a sequence of oxidation-reduction reactions in which an organic substrate and the organic compounds derived from that substrate serve as the primary electron donor and the terminal electron acceptor, respectively; in contrast to respiration, there is no requirement for an external electron acceptor to terminate the metabolic sequence.

fermentation pathways metabolic sequences for the oxidation of organic compounds to release free energy to drive the formation of ATP in which the organic substrate acts as electron donor and a product of that substrate acts as an electron acceptor.

fermented food food product of microbial fermentation.

fermenter an organism that carries out fermentation.

fermentor a reaction chamber in which a fermentation reaction is carried out; a reaction chamber for growing microorganisms used in industry for a batch process.

fertility fruitfulness; the reproductive rate of a population; the ability to support life; the ability to reproduce.

fertility plasmid see *F plasmid*.

fever the elevation of body temperature above normal.

fibrin the insoluble protein formed from fibrogen by the proteolytic action of thrombin during normal blood clotting.

fibrinogen a protein in human plasma synthesized in the liver; it is the precursor of fibrin, which is used to increase the coagulability of blood.

fibrolysin a proteolytic enzyme capable of dissolving or preventing the formation of a fibrin clot.

filament any elongated, thread-like bacterial cell.

filamentous fungi fungi that develop hyphae and mycelia; also called *molds*.

filterable virus an obsolete term used to describe infectious agents that are able to pass through bacteriological filters.

filtration the separation of microorganisms from the medium in which they are suspended by passage of a fluid through a filter with pores small enough to trap the microbes.

fire algae see *Pyrrophycophyta*.

fission a type of asexual reproduction in which a cell divides to form two or more daughter cells.

fixation of carbon dioxide the conversion of inorganic carbon dioxide to organic compounds.

flagella flexible, relatively long appendages on cells used for locomotion.

flagellates organisms having flagella; one of the major divisions of protozoans, characterized by the presence of flagella.

flagellin soluble, globular proteins constituting the subunits of bacterial flagella.

flat-field objective a microscope lens that provides an image in which all parts of the field are simultaneously in focus; an objective lens with minimal curvature of field.

flat-sour spoilage a type of microbially caused spoilage that occurs in canned foods in which acid but no gas is produced.

flavin adenine dinucleotide (FAD) a coenzyme involved in transfers of electrons during oxidation-reduction reactions of the Krebs cycle and oxidative phosphorylation.

floc a mass of microorganisms caught together in a slime produced by certain bacteria, usually found in waste treatment plants.

Floridean starch primary reserve material of Rhodophycophyta.

fluid mosaic model currently accepted model of the structure of the cytoplasic membrane that describes this membrane as a bi-lipid layer of proteins distributed in a mosaic-like pattern both on the surface and in the interior of the membrane, with lateral as well as transverse movement of proteins occurring through the structure.

fluorescence the emission of light by certain substances upon absorption of an exciting radiation; the wavelength of the emitted light is different from that of the excitation radiation.

fluorescence microscope a microscope in which the microorganisms are stained with a fluorescent dye and observed by illumination with short-wavelength light, e.g., ultraviolet light.

fomes inanimate objects that can act as carriers of infectious agents.

fomites objects and materials that have been associated with infected persons or animals and that potentially harbor pathogenic microorganisms.

food additive a substance or mixture of substances other than the basic foodstuff, which is intentionally present in food as a result of any aspect of production, processing, storage, or packaging.

food infection disease resulting from the ingestion of food or water containing viable pathogens that can establish an infectious disease, e.g., gastroenteritis, from ingestion of food containing *Salmonella*.

food intoxication disease resulting from the ingestion of toxins produced by microorganisms that have grown in a food.

food poisoning a general term applied to all stomach or intestinal disorders due to food contaminated with certain microorganisms, their toxins, chemicals, or poisonous plant materials; disease resulting from the ingestion of toxins produced by microorganisms that have grown in a food.

food preservation the prevention or delay of microbial decomposition, or self-decomposition of food and prevention of damage due to insects, animals, mechanical causes, etc.; the delay or prevention of food spoilage.

food spoilage the deterioration of a food that lessens its nutritional value or desirability, often due to the growth of microorganisms that alter the taste, smell, or appearance of the food or the safety of ingesting it.

food web an interrelationship among organisms in which energy is transferred from one organism to another; each organism consumes the preceding one and in turn is eaten by the following member in the sequence.

Foraminiferida marine members of the protozoan class Sarcodina that form one or more chambers composed of silicareous or calcareous tests.

formalin a 40 percent solution of formaldehyde, a pungent-smelling, colorless gas used for fixation and preservation of biological specimens and as a disinfectant.

Forssman antigen a heat-stable glycolipid; a heterophile antigen, an immunologically related antigen found in unrelated species.

frame shift mutation a type of mutation that causes a change in the three base sequences read as codons, i.e., a change in the phase of transcription arising from the addition or deletion of nucleotides in numbers other than three or multiples of three.

free energy the energy available to do work, particularly in causing chemical reactions; ΔG.

freeze etching a technique used to examine the topography of a surface exposed by fracturing or cutting a deep-frozen cell, making a replica, and removing the biological material; used in transmission electron microscopy.

freeze-drying see *lyophilization*.

freezer burn light colored spots that appear on frozen food, caused by loss of surface moisture due to faulty packaging or improper freezing methods.

freezing conversion of a liquid to a solid by reducing the temperature; a method used for the preservation of food by storage at $-20°C$, based on the fact that low temperatures restrict the rate of growth and enzymatic activities of microorganisms.

freshwater habitats lakes, ponds, swamps, springs, streams, and rivers.

fruiting body a specialized microbial structure that bears sexually or asexually derived spores.

frustules the silicaceous cell walls of a diatom.

fungal gardens fungi grown in pure culture by insects.

fungi a group of diverse, widespread unicellular and multicellular eukaryotic organisms, lacking chlorophyll and usually bearing spores and often filaments.

Fungi Imperfecti fungi with septate hyphae that reproduce only by means of conidia, lacking a known sexual stage; Deuteromycetes.

fungicides agents that kill fungi.

fungistasis the active prevention or hindrance of fungal growth by a chemical or physical agent.

galls abnormal plant structures formed in response to parasitic attack by certain insects or microorganisms; tumor-like growths on plants in response to an infection.

gametangium a structure that gives rise to gametes or that in its entirety functions as a gamete.

gametes haploid reproductive cells or nuclei, the fusion of which during fertilization leads to the formation of a zygote.

gamma interferon see *immune interferon*.

gamma globulin any of the serum proteins with antibody activity.

gas gangrene a disease involving tissue death that develops when certain species of toxin-producing bacteria grow in anaerobic wounds or necrotic tissues.

gas vacuoles membrane-limited, gas-filled vacuoles that occur commonly in groups in the cells of a number of cyanobacteria and certain other bacteria.

gasohol a mixture of gasoline and ethanol used as a fuel.

gastroenteritis an inflammation of the stomach and intestine.

gastroenterocolitis an inflammation of the gastrointestinal tract accompanied by the formation of pus and blood in the stools.

gastrointestinal syndrome gastroenteritis associated with nausea, vomiting, and/or diarrhea.

gastrointestinal tract the stomach, intestines, and accessory organs.

gastromycetes basidiomycete group that includes the puffballs, earthstars, stinkhorns, and bird's nest fungi.

gelatin a protein obtained from skin, hair, bones, tendons, etc.; used in culture media for the determination of a specific proteolytic activity of microorganisms.

gelatinase a hydrolytic enzyme capable of liquefying gelatin.

gene a sequence of nucleotides that specifies a particular polypeptide chain or RNA sequence or that regulates the expression of other genes.

gene cloning replication of foreign DNA inserted by recombinant DNA technology.

gene pool set of genes of an organism.

generalized transduction a form of recombination in which a phage carries bacterial DNA from a donor to a recipient cell, resulting in the exchange of many homologous genes.

generation time the time required for the cell population or biomass to double.

genetic code code for specific amino acids formed by three sequential nucleotides in mRNA; the 64 codons formed by sequences of three nucleotides that specify the genetic information of all organisms.

genetic engineering the deliberate modification of the genetic properties of an organism either through the selection of desirable traits, the introduction of new information on DNA, or both; the application of recombinant DNA technology.

genetic mapping determination of the relative positions of genes in DNA or RNA.

genetics the science dealing with inheritance.

genital herpes a sexually transmitted disease caused by a herpes virus; an infection by herpes simplex virus marked by the eruption of groups of vesicles.

genital warts a disease characterized by benign tumor development, caused by human papilloma virus transmitted by sexual contact; warts most commonly develop on the moist areas of the genitalia.

genitourinary tract the combined urinary and genital systems; the combined reproductive system and urine excretion system, including the kidneys, ureters, urinary bladder, urethra, penis, prostate, testes, vagina, fallopian tubes, and uterus.

genome the complete set of genetic information contained in a haploid set of chromosomes.

genotype the genetic information contained in the entire complement of alleles.

genus a taxonomic group directly above the species level, forming the principal subdivisions of the family.

geometric isomers nonequivalent structures based on the positions at which ligands are attached to a central atom.

germ any microorganism, especially any of the pathogenic bacteria.

germ free animal an animal with no normal microbiota; all of its surfaces and tissues are sterile, and it is maintained in that condition by being housed and fed in a sterile environment.

germ theory of disease theory that infectious and contagious diseases are caused and transmitted by the activity of microorganisms.

German measles rubella, an acute systemic infectious disease of humans caused by rubella viruses invading via the mouth or nose, characterized by a rash.

germicide a microbicidal disinfectant; a chemical that kills microorganisms.

germination a degradative process in which an activated spore becomes metabolically active, involving hydrolysis and depolymerization.

giardiasis an infection with *Giardia* protozoa in the human intestine.

gibberellic acid organic acid used as a plant growth hormone, formed by the fungus *Gibberella fujikuroi* in aerated submerged culture.

gingivitis inflammation of the gums.

gliding motility movement that occurs when some bacteria are in contact with solid surfaces.

globular protein the general name for a group of water-soluble proteins.

glomerulonephritis an inflammation of the filtration region of the kidneys.

gluconeogenesis the biosynthesis of glucose from noncarbohydrate substrates.

gluconic acid organic acid produced by a submerged culture process from mycelia of *Aspergillus niger*.

glucose the monosaccharide sugar $C_6H_{12}O_6$.

glutamine synthetase/glutamate synthase pathway pathway for the formation of L-glutamate used when ammonium concentrations are low.

glycocalyx specialized bacterial structure with an attachment function composed of a mass of tangled fibers of polysaccharides or branching sugar molecules surrounding a cell or colony of cells.

glycogen a nonreducing polysaccharide of glucose found in many tissues and stored in the liver, where it is converted when needed into sugar.

glycolysis an anaerobic process of glucose dissemination by a sequence of enzyme-catalyzed reactions to a pyruvic acid.

glycoproteins a group of conjugated proteins that, upon decomposition, yield a protein and a carbohydrate.

glycosidic bonds bonds in disaccharides and polysaccharides formed by the elimination of water.

glyoxylate cycle a metabolic shunt within the Krebs cycle involving the intermediate glyoxylate.

Golgi apparatus a membranous organelle of eukaryotic organisms involved in the formation of secretory vesicles and the synthesis of complex polysaccharides.

gonidia reproductive cells of unicelluar green algae *Volvox* that lack flagella.

gonorrhea a sexually transmitted disease caused by *Neisseria gonorrhoeae*; infectious inflammation of the mucous membrane of the urethra and adjacent cavities caused by *N. gonorrhoeae*.

graft-versus-host (GVH) disease disease that occurs when transplanted or grafted tissue contains immunocompetent cells that respond to the antigens of the recipient's tissues.

Gram-negative cell wall bacterial cell wall composed of a thin peptidoglycan layer, lipoproteins, lipopolysaccharides, phospholipids, and proteins.

Gram-positive cell wall bacterial cell wall composed of a relatively thick peptidoglycan layer and teichoic acids.

Gram stain differential staining procedure in which bacteria are classified as Gram-negative or Gram-positive, depending on whether they retain or lose the primary stain when subject to treatment with a decolorizing agent; the staining procedure reflects the underlying structural differences in the cell walls of Gram-negative and Gram-positive bacteria.

grana a membranous unit formed by stacks of thylakoids.

granules small intracellular particles that usually stain selectively.

granuloma inguinale the chronic destructive ulceration of the external genitalia due to *Donovania granulomatis*.

granulosis virus viral pesticide that develops in either the nucleus or the cytoplasm of host fat, tracheal, or epidermal cells.

graphitization anaerobic microbial corrosion that causes cast iron pipes to lose their iron, resulting in the formation of soft, brittle pipes that are easily broken.

grazers organisms that prey upon primary producers; protozoan predators that consume bacteria indiscriminately; filter-feeding zooplanktons.

green algae see *Chlorophycophyta*.

greenhouse effect rise in the concentrations of atmospheric CO_2 and a resulting warming of global temperatures.

gross primary production total amount of organic matter produced in an ecosystem.

groundwater all subsurface water.

group translocation see *group transport*.

group transport the transfer of materials across the cytoplasmic membrane of a bacterial cell that results in chemical modification of the substance as it moves across the membrane.

growth any increase in the amount of actively metabolic protoplasm accompanied by an increase in cell number, cell size, or both.

growth curve a curve obtained by plotting the increase in the size or number of microorganisms against the elapsed time.

growth factors any compound, other than the carbon and energy source, that an organism requires and cannot synthesize.

growth rate increase in the number of microorganisms per unit time.

guanine a purine base that occurs naturally as a fundamental component of nucleic acids.

Guillain-Barré syndrome acute febrile polyneuritis; a diffuse neuron paralysis.

Gymnomycota slime molds; a fungal group that borders on the protozoa; their vegetative cells lack cell walls and exhibit a phagotrophic mode of nutrition.

H$^+$ hydrogen ion or proton; a hydrogen ion concentration described by the pH that is the logarithm of the hydrogen ion concentration.

H antigen a flagellar antigen found in certain bacteria.

habitat a location where living organisms occur.

halophiles organisms requiring NaCl for growth; extreme halophiles grow in concentrated brines.

haploid a single set of homologous chromosomes; having half of the normal diploid number of chromosomes.

hapten a substance that elicits antibody formation only when combined with other molecules or particles but that can react with preformed antibodies.

heavy chain class switching in lymphocyte maturation, the process in which the same variable region of antibodies appears in association with different heavy-chain constant regions.

heliozoida protozoa that produce numerous radiating axopodia found in fresh water.

helix a spiral structure.

helper T cells see *T helper cells*.

hemagglutination the agglutination or clumping of red blood cells.

hemagglutination inhibition (HI) the inhibition of hemagglutination (antibody-mediated clumping of red blood cells), usually by means of specific immunoglobulins or enzymes, used to determine whether a patient has been exposed to a specific virus.

hemagglutinin spikes projections from surfaces of influenza viruses that cause agglutination of red blood cells; they increase the ability of influenza viruses to attach to human cells.

heme an iron-containing porphyrin ring occurring in hemoglobin.

hemocytometer a counting chamber used for estimating the number of blood cells.

hemoglobin the iron-containing, oxygen-carrying molecule of red blood cells, containing four polypeptides in the heme group.

hemolysin a substance that lyses erythrocytes (red blood cells).

hemolysis the lytic destruction of red blood cells and the resultant escape of hemoglobin.

hemolytic disease of the newborn disease that stems from an incompatibility of fetal (Rh-positive) and maternal (Rh-negative) blood, resulting in maternal antibody activity against fetal blood cells; also known as *erythroblastosis fetalis*.

hemorrhagic showing evidence of bleeding; the tissue becomes reddened by the accumulation of blood that has escaped from capillaries into the tissue.

hepatitis inflammation of the liver.

herbicides chemicals used to kill weeds.

herd immunity the concept that an entire population is protected against a particular pathogen when 70 percent of the population is immune to that pathogen.

heritable any characteristic that is genetically transmissible.

herpes simplex infection localized blistery skin rash caused by herpes simplex virus, usually on the lip or the genitalia.

hetero- combining form meaning other, other than usual, different.

heterocysts cells that occur in the trichomes of some filamentous cyanobacteria that are the sites of nitrogen fixation.

heterocystous cyanobacteria cyanobacteria that form heterocysts.

heteroduplex an intermediate form of DNA occurring during homologous recombination.

heterogamy the conjugation of unlike gametes.

heterogeneous composed of different substances; not homologous.

heterogeneous nuclear RNA (hnRNA) see *heterogeneous RNA*.

heterogeneous RNA high molecular weight RNA formed by direct transcription in eukaryotes that is then processed enzymatically to form mRNA.

heterokaryon a cell containing genetically different nuclei, occurring in some fungal hyphal cells.

heterolactic acid fermentation fermentation of glucose that produces lactic acid, acetic acid, and/or ethanol, and carbon dioxide, carried out by *Leuconostoc* and some *Lactobacillus* species.

heterologous antigen multivalent antigen; see also *Forssman antigen*.

heterophile antibodies antibodies that react with heterophile antigens; commonly found in sera of individuals with infectious mononucleosis.

heterophile antigens immunologically related antigens found in unrelated species; multivalent antigen; see also *Forssman antigen*.

heterotrophs organisms requiring organic compounds for growth and reproduction, the organic compounds serve as sources of carbon and energy.

heterotrophy mode of metabolism requiring preformed organic matter.

heterozygous a microorganism whose allelic forms of a gene differ.

Hfr see *high frequency recombinant*.

high copy number in gene cloning, a large number of repetitive copies of a gene that are produced.

high frequency recombinant a bacterial strain that exhibits a high rate of gene transfer and recombination during mating; the F plasmid is integrated into the bacterial chromosome.

histamine a physiologically active amine that plays a role in the inflammatory response.

histiocytes macrophages that are located at a fixed site in a certain organ or tissue.

histocompatibility antigens genetically determined isoantigens present on the lipoprotein membranes of nucleated cells of most tissues that cause an immune response when grafted onto a genetically disparate individual and thus determine the compatibility of tissues in transplantation.

histones basic proteins rich in arginine and lysine that occur in close association with the nuclear DNA of most eukaryotic organisms.

histoplasmosis a disease of humans and animals caused by the fungus *Histoplasma capsulatum*, characterized by fever, anemia, leukopenia, and emaciation, primarily involving the mononuclear phagocyte system.

hnRNA see *heterogeneous RNA*.

holdfast a structure that allows certain algae and bacteria to remain attached to the substratum.

homo- combining form denoting like, common, or same.

homolactic fermentation the fermentation of glucose that produces lactic acid as the sole product, carried out by certain species of *Lactobacillus*.

homologous pertaining to the structural relation between parts of different organisms due to evolutionary development of the same or a corresponding part; a substance of identical form or function.

homologous recombination recombination of regions of DNA containing alleles of the same genes.

homologously dissimilar genetically dissimilar.

homology genetic relatedness.

homozygous microorganism whose allelic forms of a gene are identical.

host a cell or organism that acts as the habitat for the growth of another organism; the cell or organism upon or in which parasitic organisms live.

host-range mutation viral mutation that alters the host that the virus can infect.

hot springs thermal springs with a temperature above 98°C.

HTST process high temperature–short time pasteurization process at a temperature of at least 71.5°C for at least 15 seconds; the most widely used form of commercial pasteurization.

humic acids any of the various organic acids obtained from humus, the soil matter whose origin is no longer identifiable; complex polynuclear aromatic compounds comprising the soil organic matter.

humoral referring to the body fluids.

humoral immune defense system see *antibody-mediated immunity*.

humus the organic portion of the soil remaining after microbial decomposition.

hyaluronidase enzyme that catalyzes the breakdown of hyaluronic acid.

hybridization of nucleic acids artificial construction of a double-stranded nucleic acid by complementary base pairing of two single-stranded nucleic acids.

hybridomas cells formed by fusion of lymphocytes (antibody precursors) with myeloma (tumor) cells that produce rapidly growing cells that secrete monoclonal antibodies.

hydr- combining form meaning water.

hydrocarbons compounds composed only of hydrogen and carbon.

hydrogen bond a weak attraction between an atom that has a strong attraction for electrons and a hydrogen atom that is covalently bonded to another atom that attracts the electron of the hydrogen atom.

hydrolysis the chemical process of decomposition involving the splitting of a bond and the addition of the elements of water.

hydrophilic a substance having an affinity for water.

hydrophobia fear of water, one of the symptoms of rabies.

hydrophobic a substance lacking an affinity for water; not soluble in water.

hydrosphere the aqueous envelope of the Earth, including bodies of water and aqueous vapor in the atmosphere.

hydrostatic pressure pressure exerted by the weight of a water column; it increases approximately 1 atm with every 10 m in depth.

hyperchromatic shift the change in absorption of light exhibited by DNA when it is melted, forming two strands from the double helix.

hyperplasia the abnormal proliferation of tissue cells, resulting in the formation of a tumor or gall.

hypersensitivity an exaggerated immunological response upon reexposure to a specific antigen.

hypertonic a solution whose osmotic pressure is greater than that of a standard solution.

hypertrophy an increase in the size of an organ, independent of natural growth, due to enlargement or multiplication of its constituent cells.

hypervariable region a region of immunoglobulins that accounts for the specificity of antigen–antibody reactions; genetically specified terminal regions of the Fab fragments.

hyphae branched or unbranched filaments that constitute the vegetative form of an organism, occurring in filamentous fungi, algae, and bacteria.

Hyphochytridiomycetes class of Mastigomycota; fungi that produce uniflagellate zoospores of the tinsel type.

hypolimnion the deeper, colder layer of an aquatic environment; the zone below the thermocline.

hypotheca the smaller of the two parts of the cell wall of a diatom.

hypotonic a solution whose osmotic pressure is less than that of a standard solution.

icosahedral virus a virus having cubical symmetry and a complex,

20-sided capsid structure.

icosahedron a solid figure contained by 20 plane faces.

identification the process of determining the closest relationship of an unknown organism to a group that has already been defined.

identification key a series of questions that leads to the unambiguous identification of an organism.

idiolite a secondary metabolite; for example, penicillin is an idiolite because it is not required for the growth of the fungus.

idiophase the phase of metabolism in batch culture in which secondary metabolism is dominant over primary growth-directed metabolism; the phase of antibiotic or other secondary product accumulation.

idiotypes immunoglobulin molecules with distinct variable regions determining the specificity of the antigen–antibody reaction.

IgA immunoglobulin A; an antibody that occurs primarily in mucus, semen, and secretions such as saliva, tears, and sweat; an immunoglobulin that plays a major role in protecting mucous membrane surface tissues against microbial infection.

IgD immunoglobulin D; an immunoglobulin that is present on the surface of some lymphocytes, along with IgM, and appears to play a regulatory role in lymphocyte activity.

IgE immunoglobulin E; an immunoglobulin that normally is present in blood serum in very low concentrations but that becomes elevated in individuals with allergies; an immunoglobulin that attaches to mast and basophil cells and triggers an allergic response when it reacts with allergens in sensitized individuals.

IgG immunoglobulin G; the largest fraction of the body's immunoglobulins and a major antibody that circulates through the body; an immunoglobulin that plays a major role in protecting the body against systemic microbial infections.

IgM immunoglobulin M; a high molecular weight immunoglobulin occurring as a pentamer that is formed prior to IgG in response to exposure to an antigen; an immunoglobulin that is important in the early response to a microbial infection.

immobilization the binding of a substance so that it is no longer reactive or able to circulate freely.

immobilized enzyme an enzyme bound to a solid support.

immune the condition following initial contact with a given antigen in which antibodies specific for that antigen are present in the body; the innate or acquired resistance to disease.

immune adherence see *opsonization.*

immune interferon a lymphokine having antiviral properties secreted by lymphocytes in response to a specific antigen to which they have been sensitized.

immune response system the integrated mechanisms for responding to the invasion of the body by particular pathogenic microorganisms and other foreign substances; characterized by specificity, memory, and the acquired ability to detect foreign substances.

immunity the relative unsusceptibility of a person or animal to active infection by pathogenic microorganisms or the harmful effects of certain toxins; resistance to disease by a living organism.

immunization any procedure in which an antigen is introduced into the body in order to produce a specific immune response.

immunodeficiency the lack of an adequate immune response due to inadequate B- or T-cell recognition and/or response to foreign antigens; a lack of antibody production.

immunoelectrophoresis a two-stage procedure used for the analysis of materials containing mixtures of distinguishable proteins, e.g., serum, using electrophoretic separation and immunological detection.

immunofluorescence any of a variety of techniques used to detect a specific antigen or antibody by means of homologous antibodies or antigens that have been conjugated with a fluorescent dye.

immunogenicity the ability of a substance to elicit an immune response.

immunoglobulins (antibodies) a varied class of proteins found in plasma and other body fluids, including all known antibodies; the antibody fraction of serum; the five classes of antibodies IgA, IgD, IgE, IgG, and IgM.

immunological referring to the immune response.

immunology the study of immunity.

immunosuppressant a drug that depresses the immune response.

impetigo an acute inflammatory skin disease caused by bacteria, characterized by small blisters, weeping fluid, and crusts.

IMViC tests a group of tests (indole, methyl red, Voges-Proskauer, citrate) used in the identification of bacteria of the Enterobacteriaceae family.

in situ in the natural location or environment.

in vitro in glass; a process or reaction carried out in a culture dish or test tube.

in vivo within the living organism.

inclusion bodies accumulations of reserve materials in bacteria.

incompatibility group incompatible plasmids that do not coexist in the same cell.

incompatible plasmids pairs of plasmids that cannot be replicated with stability in the same cell.

incubation the maintenance of controlled conditions to achieve the optimal growth of microorganisms; the period of time between the establishment of an infection and the onset of disease symptoms.

incubation period the period of time between the establishment of an infection and the onset of disease symptoms.

indicator organism an organism used to indicate a particular condition, commonly applied to coliform bacteria, e.g., *Escherichia coli* or *Streptococcus faecalis*, when their presence is used to indicate the degree of water pollution due to fecal contamination.

indigenous native to a particular habitat.

indirect immunofluoresence test used in identifying bacteria such as *Treponema pallidum* by adding dead cells to the patient's serum and adding fluorescent anti-immunoglobulin; if the bacteria stain, the test is positive; the test is indirect because the fluorescent antibody reacts with human IgG (acting in this case as an antigen) after the IgG has reacted with the dead bacterial cells.

inducers substances responsible for activating certain genes, resulting in the synthesis of new proteins.

inducible enzymes enzymes that are synthesized only in response to a particular substance in the environment.

induction an increase in the rate of synthesis of an enzyme; the turning on of enzyme synthesis in response to environmental conditions.

induration hardening of the skin, a positive hypersensitivity reaction.

infantile paralysis see *poliomyelitis.*

infection a condition in which pathogenic microorganisms have become established in the tissues of a host organism.

infectious a disease-causing agent or disease that can be transmitted from one person, animal, or plant to another.

infectious dose the number of pathogens that are needed to overwhelm host defense mechanisms and establish an infection.

infectious mononucleosis glandular fever, an acute infectious disease that primarily affects the lymphoid tissues; caused by Epstein-Barr virus, which enters the body via the respiratory tract.

inflammation the reaction of tissues to injury characterized by localized heat, swelling, redness, and pain.

inflammatory exudate pussy material from blood vessels deposited in tissues or on tissue surfaces as a defensive response to injury or irritation.

inflammatory response a nonspecific immune response to injury characterized by redness, heat, swelling, and pain in the affected area.

influenza an acute, highly communicable disease that tends to occur in epidemic form, caused by a myxovirus and characterized by malaise, headache, and fever.

infusoria archaic term for microorganisms.

inhibition prevention of growth or multiplication of microorganisms; reduction in the rate of enzymatic activity; repression of chemical or physical activity.

inhibitors substances that repress or stop a chemical action.

initial body larger, thin-walled, noninfectious form of chlamydias.

initiation in protein synthesis, the stage at which the translating complex of mRNA, ribosome, and tRNA first assembles.

inoculate to deposit material, an inoculum, onto medium to initiate a culture, carried out with an aseptic technique; to introduce microorganisms into an environment that will support their growth.

inoculum the material containing viable microorganisms used to inoculate a medium.

insecticides substances destructive to insects; chemicals used to control insect populations.

insertion a type of mutation in which a nucleotide or two or more contiguous nucleotides are added to DNA.

insertion mutations a mutation in which one or more nucleotides are inserted into a gene.

insertion sequence a transposable genetic element that can move around bacterial chromosomes, occurring at different locations on the chromosome.

insertional inactivation in gene cloning, insertion of foreign DNA at an antibiotic-resistant site, causing loss of resistance because the nucleotide sequence of the antibiotic resistance gene is disrupted.

interference microscope a microscope that relies on destructive and/or additive interference of light waves to achieve contrast.

interferons glycoproteins produced by animal cells that act to prevent the replication of a wide range of viruses by inducing resistance.

intergenic mutations mutations within a single gene that affect other genes.

intermediary metabolism intermediate steps in the cellular synthesis and breakdown of substances.

intoxication poisoning, as by a drug, serum, alcohol, or any poison.

intracellular within a cell.

intradermal within the skin.

intragenic suppressor mutations a second mutation that negates the effect of the first mutation within a single gene.

intramuscular within the substance of a muscle.

intravenous within or into the vein.

intron an intervening region of the DNA of eukaryotes that does not code for a known protein or a regulatory function.

invasiveness the ability of a pathogen to spread through a host's tissues.

ion an atom that has lost or gained one or more orbital electrons and is thus capable of conducting electricity.

ionic bond a chemical bond resulting from the transfer of electrons between metallic and nonmetallic atoms; positive and negative ions are formed and held together by electrostatic attraction.

ionization the process that produces ions.

ionizing radiation radiation, such as gamma and x-radiation, that induces or forms toxic free radicals, which cause chemical reactions disruptive to the biochemical organization of microorganisms.

iso- combining form meaning for or from different individuals of the same species.

isogamete a reproductive cell similar in form and size to the cell with which it unites; found in certain protozoas, fungi, and algae.

isogamy fertilization in which the gametes are similar in appearance and behavior.

isolation methods aseptic procedures used for the establishment of pure cultures, usually involving the separation of microorganisms on a solid medium into individual cells that are then allowed to reproduce to form clones of single microorganisms.

isomer one of two or more compounds having the same chemical composition but differing in the relative positions of the atoms within the molecules.

isotope an element that has the same atomic number as another element but a different atomic weight.

isotypes antibodies differing in heavy chain constant regions associated with different classes and subclasses of immunoglobulins.

itaconic acid an organic acid used as a resin in detergents; made by the transformation of citric acid by *Aspergillus terreus*.

-itis suffix denoting a disease; specifically, an inflammatory disease of a specified part of the body.

Jaccard coefficient a measure of similarity used in cluster analysis to show the relationship between individuals; it does not consider negative matches.

jaundice yellowness of the skin, mucous membranes, and secretions resulting from liver malfunction.

kappa particles bacterial particles that occur in the cytoplasm of certain strains of *Paramecium aurelia*; such strains have a competitive advantage over other strains of *Paramecium* and are known as *killer strains*.

karyogamy the fusion of nuclei, as of gametes in fertilization.

kelp brown algae with vegetative structures consisting of a holdfast, stem, and blade; it can form large macroscopic structures.

keratin a highly insoluble protein that occurs in hair, wool, horn, and skin.

ketone an organic compound derived by oxidation from a secondary alcohol, containing a characteristic -C=O group.

killer T cells see *cytotoxic T cells*.

kinase see *fibrinolysin*.

kingdom a major taxonomic category consisting of several phyla or divisions; the primary divisions of living organisms.

K_m the Michaelis constant; describes the affinity of an enzyme for a substrate; the substrate concentration at half of the maximal velocity of an enzyme.

Koch, Robert a German physician and bacteriologist; responsible for major advances in medical microbiology and basic microbiological (pure culture) procedures; see also *Koch's postulates*.

Koch's postulates a process for elucidating the etiological (causative) agent of an infectious disease.

koji fermentation dry fermentation in which a mixture of soybeans and wheat is inoculated with spores of *Aspergillus oryzae* and moistened, not submerged in liquid, so that fungi grow on the surface; used in the production of soy sauce.

Koplik's spots small red spots surrounded by white areas occurring on the mucous membranes of the mouth during the early stages of measles.

Krebs cycle the tricarboxylic acid cycle; the citric acid cycle; the metabolic pathway in which acetate derived from pyruvic acid is converted to carbon dioxide and reduced coenzymes are produced.

Kupffer cells macrophages lining the sinusoids of the liver.

kuru disease caused by a prion affecting the central nervous system; observed among cannibals in New Guinea.

-labile unstable, readily changed by physical, chemical, or biological processes.

lac operon inducible enzyme system of *Escherichia coli* for the utilization of lactose.

lactam an organic compound containing an -NH-CO- group in ring form.

lactamase an enzyme that breaks a lactam ring.

lactic acid organic acid with antimicrobial activity produced by lactic acid bacteria (*Lactobacillus*, *Streptococcus*, *Leuconostoc*) involved in antagonistic relationships among microorganisms and used as a preservative.

lactic acid fermentation fermentation that produces lactic acid as the primary product.

Lactobacillaceae family of Gram-positive, asporogenous, regularly shaped rods that produce lactic acid as a major fermentation product.

lactoferrin an iron-containing compound that binds the iron necessary for microbial growth, resulting in a slight antimicrobial action.

lactose a disaccharide in milk; when hydrolyzed, it yields glucose and galactose.

lag phase a period following inoculation of a medium during which the number of microorganisms does not increase.

lagging strand of DNA see *discontinuous strand of DNA*.

laminar flow the flow of air currents in which streams do not intermingle; the air moves along parallel flow lines; used in a laminar flow hood to provide air free of microbes over a work area.

landfill a site where solid waste is dumped and allowed to decompose; a process in which solid waste containing both organic and inorganic material is added to soil and allowed to decompose.

late-onset hypogammaglobulinemia immunodeficiency disorder characterized by a shortage of circulating B cells and/or B cells with IgG surface receptors.

late proteins proteins coded for late in the developmental sequence of a virus.

late syphilis tertiary phase of syphilis that can damage any body organ, occurring several years after the initial infection.

latent potential; not manifest; present but not visible or active.

latent period the period of time following infection of a cell by a virus before new viruses are assembled.

leach to wash or extract soluble constituents from insoluble materials.

leader sequence the beginning sequence of nucleotides in an mRNA molecule involved in the initiation of protein synthesis at the ribosomes.

leading strand of DNA see *continuous strand of DNA*.

leaf spots plant diseases in which infection of the foliage interferes with photosynthesis.

leavening substance used to produce fermentation in dough or liquid; the production of CO_2 that results in the rising of dough.

lecithinases extracellular phospholipid-splitting enzymes.

Legionellaceae family of Gram-negative, fermentative, rod-shaped bacteria that require iron and cysteine as growth factors.

Legionnaire's disease a form of pneumonia caused by *Legionella pneumophila*.

leishmaniasis a disease caused by protozoa of the genus *Leishmania*.

leprosy Hansen's disease, a chronic contagious disease affecting humans and armadillos, caused by *Mycobacterium leprae*.

leptospirosis disease of humans or animals caused by *Leptospira*.

lesion a region of tissue mechanically damaged or altered by any pathological process.

lethal dose the amount of a toxin that results in the death of an organism.

lethal mutations mutations that result in the death of a microorganism or its inability to reproduce.

leukocidin an extracellular bacterial product that can kill leukocytes.

leukocyte a type of white blood cell characterized by a beaded, elongated nucleus.

leukocytosis an increase above the normal upper limits of the leukocyte count.

leukopenia a decrease below the normal lower limit of the leukocyte count.

lichens a large group of composite organisms, each consisting of a fungus in symbiotic association with an alga or cyanobacterium.

life a state that characterizes living systems, encompassing the complex series of physicochemical processes essential for maintaining the organization of the system and the ability to reproduce that organization.

ligases a group of enzymes that catalyze reactions in which a bond is formed between two substrate molecules using energy obtained from the cleavage of a pyrophosphate bond.

light beer beer with a low calorie content produced with fungal enzymes to ensure that simple substrates are available for alcoholic fermentation.

light microscope a microscope in which visible light is used to illuminate the specimen; often referred to as a *bright-field microscope*.

light scattering dispersion of light when it strikes particles; used in some instruments for estimating quantities of suspended particles, including microorganisms.

lignins a class of complex polymers in the woody material of higher plants.

limnetic zone in lakes, the area where light penetration is sufficient so that primary production exceeds heterotrophic metabolism.

linear alkyl benzyl sulfonate (ABS) synthetic molecule with a straight hydrocarbon chain, benzene ring, and, sulfate group designed as a component of anionic laundry detergent that is easily biodegraded.

lipases fat-splitting enzymes; enzymes that break down lipids.

lipids fats or fat-like substances that are insoluble in water and soluble in nonpolar solvents.

lipophilic preferentially soluble in lipids or nonpolar solvents.

lipopolysaccharide toxin see *endotoxin*.

lipopolysaccharides molecules consisting of covalently linked lipids and polysaccharides; a component of Gram-negative bacterial cell walls.

liquid diffusion method a method for detecting a substance based upon the diffusion of that substance into a medium to achieve the appropriate concentration for a reaction to occur; a method used in serology for detection of antigens and antibodies based upon its ability to achieve a zone of equivalence in which antigen–antibody reactions can occur.

liquid wastes waste material in liquid form, the result of agricultural, industrial, and all other human activities.

liter a metric unit of volume equal to 1000 milliliters; approximately equal in volume to a quart.

lithosphere the solid part of the Earth.

lithotrophs microorganisms that live in and obtain energy from the oxidation of inorganic matter; autotrophs.

litmus plant extract dye used as an indicator of pH and of oxidation or reduction potential.

littoral situated or growing on or near the shore; the region between the high and low tide marks.

living system a system separated from its surroundings by a semipermeable barrier; composed of macromolecules, including proteins and nucleic acids, having lower entropy than its surroundings, and thus requiring inputs of energy to maintain its high degree of organization, capable of self-replication and normally based on cells as the primary functional and structural units.

lobopodia false feet that are extensions of ectoplasm, which includes the flow of endoplasm.

lockjaw see *tetanus*.

locus the point on a chromosome occupied by a gene.

logarithmic phase (log phase) see *exponential growth phase*.

low acid food food with a pH above 4.5 and below 7.0.

LPS see *lipopolysaccharides*.

LTH process low temperature–long-time pasteurization process, e.g., at 63°C for 30 minutes.

luminescence the emission of light without production of heat sufficient to cause incandescence, produced by physiological processes or by friction, chemical, or electrical action.

ly-, lys-, lyt- combining forms meaning to loosen or dissolve.

lymph a plasma filtrate that circulates through the body.

lymph nodes an aggregation of lymphoid tissues surrounded by a fibrous capsule found along the course of the lymphatic system.

lymphocytes lymph cells.

lymphogranuloma venereum a sexually transmitted disease caused by a *Chlamydia*, characterized by an initial lesion, usually on the genitalia, followed by regional lymph node enlargement and systemic involvement.

lymphokines a varied group of biologically active extracellular proteins formed by activated T lymphocytes involved in cell-mediated immunity; the chemical mediators of cellular immunity.

lyophilization the process of rapidly freezing a substance at low temperature and then dehydrating the frozen mass in a high vacuum; a process in which water is removed by sublimation, moving from the solid to the gaseous phase.

lysins antibodies or other entities that under appropriate conditions are capable of causing the lysis of cells.

lysis the rupture of cells.

lysogeny the nondisruptive infection of a bacterium by a bacteriophage.

lysosomes organelles containing hydrolytic enzymes involved in autolytic and digestive processes.

lysozymes enzymes that hydrolyze peptidoglycan; they act as bactericidal agents when they degrade the bacterial cell walls.

lytic of or relating to lysis or a lysin.

lytic phage viruses that cause lysis of cells within which they reproduce.

MacConkey's agar a solid medium used for the growth of enteric bacteria.

macro- combining form meaning long or large.

macromolecules very large organic molecules having polymeric chain structures, as in proteins, polysaccharides, and other natural and synthetic polymers.

macrophages mononuclear phagocytes; large, actively phagocytic cells found in spleen, liver, lymph nodes, and blood; important factors in nonspecific immunity.

macroscopic of a size visible to the naked eye.

madura foot a form of mycetoma, i.e., an infectious disease of the skin that can progress to involve subcutaneous tissues and bone, caused by an antinomycete that is important in the Madura province of India.

magnetosomes dense inclusion bodies within bacterial cells that contain iron granules and act as magnetic compasses, permitting bacteria to move in response to the Earth's magnetic field.

magnetotaxis motility directed by a geomagnetic field.

magnification the extent to which the image of an object is larger than the object itself.

major histocompatibility complex (MHC) the genetic region in human beings that controls not only tissue compatibility but also the development and activation of part of the immune response.

malaise a general feeling of illness, accompanied by restlessness and discomfort.

malaria an infectious chronic disease caused by the *Plasmodium* protozoa; transmitted by mosquitoes and characterized by intermittent fever, anemia, chills, and sweating.

maltase an enzyme that converts maltose to glucose.

malting enzymatic conversion of barley by plant amylases and proteases that is used to prepare grain for microbial alcoholic fermentation.

maltose a disaccharide formed upon hydrolysis of starch or glycogen and metabolized by a wide range of fungi and bacteria.

manganese nodules nodules (round, irregular mineral masses) produced by microbial oxidation of manganese oxides.

marine of or relating to the oceans.

mash the crushed malt or grain meal steeped and stirred in hot water with amylases to produce wort as a substrate for microorganisms.

mast cells cells that contain granules of histamine, serotonin, and heparin, especially in connective tissues involved in hypersensitivity reactions.

Mastigomycota true fungi; some are unicellular, whereas others form extensive filamentous, coenocytic mycelia to produce motile cells with flagella; asexual reproduction involves zoospores, nutrition provided by the absorption of nutrients.

Mastigophora a subclass of protozoans characterized by the presence of flagella.

mastitis inflammation of the breast.

mating the meeting of individuals for sexual reproduction.

measles an acute, contagious systemic human disease caused by a paramyxovirus that enters via the oral and nasal routes, characterized by the presence of Koplik spots and a characteristic rash.

medical microbiology the study of medical science as it relates to microorganisms.

medium the material that supports the growth/reproduction of microorganisms.

meiosis cell division that results in a reduction of the state of ploidy, normally from diploid to haploid during the formation of the germ cells.

melting temperature of DNA the midpoint temperature of a denaturation curve used in the analysis of DNA composition in which DNA is heated and the double-stranded helix is converted to single-stranded DNA.

membrane filter a cellulose–ester membrane used for microbiological filtrations.

memory cells clones of lymphocytes with receptors of high affinity for a particular antigenic molecule.

meninges the membranes covering the brain and spinal cord.

meningitis inflammation of the membranes of the brain or spinal cord.

mesophiles organisms whose optimum growth is in the temperature range of 20–45°C.

mesosomes intracellular membranous structures observed as infoldings of bacterial cell membranes in electron microscopy; their function is unknown, and in fact they now appear to be artifacts of specimen preparation.

messenger RNA (mRNA) the RNA that specifies the amino acid sequence for a particular polypeptide chain.

metabolic pathway a sequence of biochemical reactions that transforms a substrate into a useful product for carbon assimilation or energy transfer.

metabolism the total of all chemical reactions by which energy is provided for the vital processes and new cell substances are assimilated.

metabolites chemicals participating in metabolism; nutrients.

metabolize to transform by means of metabolism.

metachromatic granules cytoplasmic granules of polyphosphate occurring in the cells of certain bacteria that stain intensively with basic dyes but appear a different color.

Methanobacteriaceae family of methane-producing bacteria.

methanogens methane-producing prokaryotes; a group of archaebacteria capable of reducing carbon dioxide or low molecular weight fatty acids to produce methane.

methylation the process of substituting a methyl group for a hydrogen atom.

Methylomonadaceae a family of Gram-negative bacteria that can utilize carbon monoxide, methane, or methanol as the sole source of carbon; they also utilize respiratory metabolism.

MIC see *minimum inhibitory concentration*.

Michaelis-Menten equation mathematical description of the relationship between the rate of an enzymatic reaction and the substrate concentration.

micro- combining form meaning small.

Micro-ID system a miniaturized commercial identification system.

microaerophiles aerobic organisms that grow best in an environment with less than atmospheric oxygen levels; oxygen-requiring microorganisms that grow only at reduced oxygen concentrations.

microbes microscopic organisms; microorganisms.

microbial ecology the field of study that examines the interactions of microorganisms with their biotic and abiotic surroundings.

microbial mining a mineral recovery method that uses bioleaching to recover metals from ores not suitable for direct smelting.

microbial pesticides preparations of populations of pathogenic or predatory microorganisms that are antagonistic toward a particular pest population.

microbicidal any agent capable of destroying, killing, or inactivating microorganisms so that they cannot replicate.

microbiology the study of microorganisms and their interactions with other organisms and the environment.

microbiota the totality of microorganisms associated with a given environment.

microbiostatic chemical agents that inhibit the growth of microorganisms but do not kill them; when the agent is removed, growth is resumed.

microbodies organelles within a cell containing specialized enzymes whose functions involve hydrogen peroxide.

microcidal see *microbicial*.

Micrococcaceae family of Gram-positive cocci whose cells occur singly or as irregular clusters.

microcysts refractile and encapsulated myxospores.

microfibrils thread-like structures found in the cell walls of filamentous fungi, consisting of chitin.

microfilament an elongated structure composed of protein subunits.

microglia macrophages of the central nervous system.

micrometer one millionth (10^{-6}) of a meter; one thousandth (10^{-3}) of a millimeter.

microorganisms microscopic organisms, including algae, bacteria, fungi, protozoa, and viruses.

microscope an optical or electronic instrument for viewing objects too small to be visible to the naked eye.

microtome an instrument used for cutting thin sheets or sections of tissues or individual cells for examination by light or electron microscopy.

microtubules cylindrical protein tubes that occur within all eukaryotic organisms; they aid in maintaining cell shape, comprise the structure of organelles of cilia and flagella, and serve as spindle fibers in mitosis.

mildew any of a variety of plant diseases in which the mycelium of the parasitic fungus is visible on the affected plant; biodeterioration of a fabric due to fungal growth.

mineralization the microbial breakdown of organic materials into inorganic materials brought about mainly by microorganisms.

miniaturized commercial identification systems small devices containing multicompartmentalized chambers that each perform separate biochemical tests, used for the identification of bacterial species.

minimum inhibitory concentration (MIC) the concentration of an antimicrobial drug necessary to inhibit the growth of a particular strain of microorganism.

Minitek system a miniaturized commercial identification systems.

miso product of koji fermentation of rice with *Aspergillus oryzae*, it is ground into a paste and combined with other foods being eaten.

missense mutations type of base substitution that results in the change in the amino acid inserted into the polypeptide chain specified by the gene in which the mutation occurs.

mitochondrion a semiautonomous organelle found in eukaryotic cells, the site of respiration and other cellular processes; consists of an outer membrane and an inner one that is convoluted; site of ATP generation in eukaryotic cells.

mitosis the sequence of events resulting in the division of the nucleus into two genetically identical cells during asexual cell division; each of the daughter nuclei has the same number of chromosomes as the parent cell.

mixed acid fermentation a type of fermentation carried out by members of the Enterobacteriaceae that converts glucose to acetic, lactic, succinic, and formic acids.

mixed amino acid fermentation pathway metabolism of amino acids resulting in their deamination and decarboxylation.

mixotrophs organisms capable of utilizing both autotrophic and heterotrophic metabolic processes, e.g., the concomitant use of organic compounds as sources of carbon and light as a source of energy.

modification the methylation of nucleotide residues in DNA; modification of newly synthesized DNA by specific enzymes in a manner characteristic of the particular bacterial strain.

moiety a part of a molecule having a characteristic chemical property.

mold a type of fungus having a filamentous structure.

mole % G + C the proportion of guanine and cytosine in a DNA macromolecule.

mollicutes a class of prokaryotic organisms that do not form cell walls, e.g., *Mycoplasma*.

monera prokaryotic protists with a unicellular, simple colonial organization; bacteria.

mono- combining form meaning single, one, or alone.

monocistronic mRNA that contains the information for only one polypeptide sequence.

monoclonal antibody an antibody produced from a clone of cells making only that antibody.

monocytes ameboid, agranular, phagocytic white blood cells derived from the bone marrow.

mononuclear having only one nucleus.

mononuclear phagocyte system the macrophage system of the body, including all phagocytic white blood cells except granular white blood cells; the reticuloendothelial system.

monosaccharide any carbohydrate whose molecule cannot be split into simpler carbohydrates; a simple sugar.

Montoux test test for tuberculosis in which an appropriate dilution of purified protein derivative is injected intradermally into the superficial layers of the skin of the forearm.

morbidity the state of being diseased; the ratio of the number of sick individuals to the total population of the community; the conditions inducing disease.

mordant a substance that fixes the dyes used in staining tissues or bacteria; a substance that increases the affinity of a stain for a biological specimen.

morphogenesis morphological changes, including growth and differentiation of cells and tissues during development; the transformations involved in the growth and differentiation of cells and tissues.

morphology the study of the shape and structure of microorganisms.

mortality death; the proportion of deaths within a population.

mortality rate death rate; number of deaths per unit population per unit time.

most probable number (MPN) the statistical estimate of the size of a bacterial population through the use of dilution and multiple tube inoculations.

motility the capacity for independent locomotion.

mRNA see *messenger RNA*.

mucociliary escalator system defense system that lines the upper respiratory tract and protects it against pathogens; the system consists of mucous membranes and cilia; mucous secretions trap microbes and cilia beat with an upward wave-like motion to expel microorganisms from the respiratory tract.

mucopeptide peptidoglycan component of bacterial cell walls.

mucosa a mucous membrane, the lining of body cavities that communicate with the exterior.

mucous membrane the type of membrane lining body cavities and canals that have communication with air.

mucus a viscid fluid secreted by mucous glands consisting of mucin, water, inorganic salts, epithelial cells, and leukocytes.

multilateral budding in fungi, budding that occurs all around the mother cell.

multiple antibiotic resistance the ability to resist the effects of two or more unrelated antibiotics by bacterial strains generally containing R plasmids.

mumps an acute infectious disease caused by a virus, characterized by swelling of the salivary glands.

murein peptidoglycan; the repeating polysaccharide unit comprising the backbone of the cell walls of eubacteria.

mushrooms fungi that are members of the Agaricales; the basidiocarps of basidiomycetes.

must the fluid extracted from crushed grapes; the ingredients, e.g., fruit pulp or juice, used as substrate for fermentation in wine making.

mutagen any chemical or physical agent that promotes the occurrence of mutation; a substance that increases the rate of mutation above the spontaneous rate.

mutant any organism that differs from the naturally occurring type because its base DNA has been modified, resulting in an altered protein that gives the cell properties different from those of its parent.

mutation a stable, heritable change in the nucleotide sequence of the genetic nucleic acid, resulting in an alteration in the products coded for by the gene.

mutation rate the average number of mutations per cell generation.

mutualism a stable condition in which two organisms of different species live in close physical association, each organism deriving some benefit from the association; symbiosis.

myasthenia gravis autoimmune disease resulting from the failure of the immune response to recognize self antigens; antibodies react with nerve–muscle junctions.

myc- combining form meaning fungus.

mycelia the interwoven mass of discrete fungal hyphae.

mycetoma a chronic infection usually involving the foot, characterized by the presence of pus-filled nodules and caused by a wide variety of fungi or bacteria; also known as *madura foot*.

mycobiont the fungal partner in a lichen.

mycolic acids fatty acids found in the cell walls of *Mycobacterium* and several other bacteria related to the actinomycetes.

mycology the study of fungi.

mycoplasmas members of the group of bacteria composed of cells lacking cell walls, bounded by a single triple-layered membrane, exhibiting a variety of shapes; the smallest organisms capable of self-reproduction.

mycorrhizae a stable, symbiotic association between a fungus and the root of a plant; the term also refers to the root–fungus structure itself.

mycosis any disease in which the causal agent is a fungus.

mycotoxins toxic substances produced by fungi, including aflatoxin, amatoxin, and ergot alkaloids.

mycovirus a virus that infects fungi.

myocarditis infection of the myocardium; can result from viral, bacterial, helminthic, or parasitic infections, hypersensitivity immune reactions, radiation therapy, or chemical poisoning.

myocardium the muscular tissue of the heart wall.

myx- combining form meaning mucus.

myxamoebae: nonflagellated ameboid cells that occur in the life cycle of the slime molds and members of the Plasmodiophorales.

Myxobacterales fruiting myxobacteria; gliding, small, rod-shaped bacteria normally embedded in a slime layer; under appropriate conditions they aggregate to form fruiting bodies.

myxobacteria see *Myxobacterales*.

myxomycetes true slime molds, class of Plasmodiogymnomycotina; some form myxamoebae, others swarm cells; their classification is based on the structure of the fruiting body.

myxospores resting cells in the fruiting bodies of members of the Myxobacteriales.

NAD⁺ see *nicotinamide adenine dinucleotide*.

NADH reduced nicotinamide adenine dinucleotide.

NADP⁺ see *nicotinamide adenine dinucleotide phosphate*.

NADPH reduced nicotinamide adenine dinucleotide phosphate.

nasopharynx the upper part of the pharynx continuous with the nasal passages.

natto food product from the Orient; the fermentation product of boiled soybeans and *Bacillus subtilis*.

necrosis the pathological death of a cell or group of cells in contact with living cells.

negative interactions interactions between populations that act as feedback mechanisms and limit population densities.

negative stain a stain with a positively charged chromophore.

negative staining the treatment of cells with dye so that the background, rather than the cell itself, is made opaque; used to demonstrate bacterial capsules or the presence of parasitic cysts in fecal samples.

Negri bodies acidophilic, intracytoplasmic inclusion bodies that develop in cells of the central nervous system in cases of rabies.

Neisseriaceae family of Gram-negative cocci and coccobacilli, including the genera *Neisseria*, *Branhamella*, *Moraxella*, and *Acinetobacter*.

nematodes worms of the class Nematoda.

neoplasm the result of the abnormal and excessive proliferation of the cells of a tissue; if the progeny cells remain localized, the resulting mass is called a *tumor*.

net primary production amount of organic carbon in the form of biomass and soluble metabolites available for heterotrophic consumers in terrestrial and aquatic habitats.

neuramidase spikes projections from surfaces of influenza viruses that are involved in the release of viruses from infected cells following viral replication.

neurotoxin a toxin capable of destroying nerve tissue or interfering with neural transmission.

neutralism the relationship between two different microbial populations characterized by the lack of a recognizable interaction.

neutralization of toxic materials conversion of toxic materials to nontoxic forms.

neutropenia a decrease below the normal standard in the number of neutrophils in the peripheral blood.

neutrophilia (neutrophilic leukocytosis) an increase above the normal standard in the number of neutrophils in the peripheral blood.

neutrophils large granular leukocytes with highly variable nuclei consisting of three to five lobes and cytoplasmic granules that stain with neutral dyes and eosin; phagocytic white blood cells.

niche the functional role of an organism within an ecosystem; the combined description of the physical habitat, functional role, and interactions of the microorganisms occurring at a given location.

nicotinamide adenine dinucleotide (NAD⁺) a coenzyme used as an electron acceptor in oxidation-reduction reactions.

nicotinamide adenine dinucleotide phosphate (NADP⁺) the phosphorylated form of NAD⁺ formed when NADPH serves as an electron donor in oxidation-reduction reactions.

nine + two (9 + 2) system the arrangement of microtubules in eukaryotic flagella and cilia, consisting of nine peripheral pairs of microtubules surrounding two single central microtubules.

nitrate a salt of nitric acid; NO_3^-.

nitrate reduction the reduction of nitrate to reduced forms; for example, under anaerobic and microaerophilic conditions, bacteria use nitrate as a terminal electron acceptor for respiratory metabolism, producing N_2O and N_2.

nitrification the process in which ammonia is oxidized to nitrite and nitrite to nitrate; a process primarily carried out by the strictly aerobic, chemolithotrophic bacteria of the family Nitrobacteraceae.

nitrifying bacteria Nitrobacteraceae; Gram-negative, obligately aerobic, chemolithotrophic bacteria occurring in fresh and marine waters and in soil that oxidize ammonia to nitrite or nitrite to nitrate.

nitrite a salt of nitrous acid; NO_2^-; nitrites of sodium and potassium are used as food additives and preservatives.

nitrite ammonification reduction of nitrite to ammonium ions by bacteria; does not remove nitrogen from the soil.

Nitrobacteraceae nitrifying bacteria; family of chemolithotrophic bacteria that oxidize nitrite or ammonia to generate ATP; found in soil, fresh water, and seawater.

nitrogen fixation the reduction of gaseous nitrogen to ammonia, carried out by certain prokaryotes.

nitrogen-rich fertilizers products containing fixed forms of nitrogen that can serve as plant nutrients when applied to crop fields to support increased production.

nitrogenase the enzyme that catalyzes biological nitrogen fixation.

nitrogenous containing nitrogen.

nodules tumor-like growths formed by plants in response to infections with specific bacteria within which the infecting bacteria fix atmospheric nitrogen; a rounded, irregularly shaped mineral mass.

Nomarski differential interference microscope a specialized type of interference microscope that produces high-contrast images of unstained specimens with a three-dimensional appearance; its special features are a polarizing filter, an interference contrast condenser, and a prism analyzer plate.

nomenclature the naming of organisms, a function of taxonomy governed by codes, rules, and priorities laid down by committees.

noncompetitive inhibition inhibition of enzyme activity by a substance that does not compete with the normal substrate for the active site and thus cannot be reduced by increasing the substrates concentration.

noncyclic photophosphorylation a metabolic pathway involved in the conversion of light energy for the generation of ATP in which an electron is transferred from an electron donor, normally water, by a series of electron carriers, with the eventual formation of a reduced coenzyme, normally $NADPH_2$.

nongonococcal urethritis any inflammation of the urethra not caused by *Neisseria gonorrhoeae*.

nonhomologous recombination recombination involving little or no homology between the donor DNA and the region of the DNA in the recipient where insertion occurs.

nonlinear alkyl benzyl sulfonate (ABS) a component of anionic laundry detergent that contains a braided alkane chain, is resistant to biodegradation, and causes foaming of receiving waters; banned because of its persistence in groundwater.

nonperishable foods food products that are not subject to spoilage by microorganisms under normal storage conditions and consequently have an extended shelf life as long as those conditions are maintained.

nonreciprocal recombination see *nonhomologous recombination*.

nonsense codon a codon that does not specify an amino acid but acts as a punctuator of mRNA.

nonsense mutation a mutation in which a codon specifying an amino acid is altered to a nonsense codon.

normal microbiota microbial populations most frequently found in association with particular tissues that typically do not cause disease; also known as indigenous microbial populations.

normal microflora see *normal microbiota*.

Norwalk agent small DNA virus responsible for an outbreak of winter vomiting disease in Norwalk, Ohio, in 1968.

nosocomial infection an infection acquired while in the hospital.

nuclear membrane a double layer with a distinct space between the two membranes surrounding the genomes of eukaryotic cells.

nuclear polyhedrosis virus viral pesticide that develops in host-cell nuclei.

nuclease an enzyme capable of splitting nucleic acids to nucleotides, nucleosides, or their components.

nucleic acid a large, acidic, chain-like macromolecule containing phosphoric acid, sugar, and purine and pyrimidine bases; the nucleotide polymers RNA and DNA.

nucleoid region the region of a prokaryotic cell in which the genome occurs.

nucleolus an RNA-rich intranuclear body not bounded by a limiting membrane that is the site of rRNA synthesis in eukaryotes.

nucleoprotein a conjugated protein closely associated with nucleic acid.

nucleosome the fundamental structural unit of DNA in eukaryotes, having approximately 190 base pairs folded and held together by histones.

nucleotide the combinination of a purine or pyrimidine base with a sugar and phosphoric acid; the basic structural unit of nucleic acid.

nucleus an organelle of eukaryotes in which the cell's genome occurs; the differentiated protoplasm of a cell surrounded by a membrane that is rich in nucleic acids.

numerical aperture the property of a lens that describes the amount of light that can enter it.

numerical taxonomy a system that uses overall degrees of similarity and large numbers of characteristics to determine the taxonomic position of an organism; allows organisms of unknown affiliation to be identified as members of established taxa.

nutrient a growth-supporting substance.

nutritional mutations mutations that alter the nutritional requirements of the progeny of a microorganism.

nutritional requirements the essential growth substances needed for metabolism and reproduction.

nystatin polyene antibiotic used in the treatment of topical *Candida* infections.

O antigens lipopolysaccharide–protein antigens occurring in the cell walls of Gram-negative bacteria.

objective lens the microscope lens closest to the object.

obligate aerobes organisms that grow only under aerobic conditions, i.e., in the presence of air or oxygen.

obligate anaerobes organisms that cannot use molecular oxygen; organisms that grow only under anaerobic conditions, i.e., in the absence of air or oxygen; organisms that cannot carry out respiratory metabolism.

obligate intracellular parasites organisms that can live and reproduce only within the cells of other organisms, such as viruses, all of which must find suitable host cells for their replication.

obligate thermophiles organisms restricted to growth at high temperatures.

occluded closed or shut up.

oceans the whole body of salt water that covers nearly three-fourths of the Earth's surface.

ocular lens the eyepiece of a microscope; the lens closest to the eye.

3'-OH free end unattached hydroxyl group at the 3-carbon position at one end of a nucleic acid molecule.

-oid combining form meaning resembling.

oil immersion lens a high-power objective lens of a microscope designed to work with the space between the objective and the specimen, filled with oil to enhance resolution.

oil pollutants petroleum hydrocarbons that contaminate the environment.

Okazaki fragments the short segments of newly synthesized DNA along the trailing or discontinuous strand that are linked by a ligase to form the completed DNA.

oligotrophic lakes and other bodies of water that are poor in those nutrients that support the growth of aerobic, photosynthetic organisms; microorganisms that grow at very low nutrient concentrations.

oncogenes genes that can lead to malignant transformations of animal cells.

oncogenic viruses viruses capable of inducing tumor formation, i.e., animal cell transformations.

one-step growth curve describes the lytic reproduction cycle that releases a large number of phage simultaneously.

oogamy a form of fertilization that involves either a motile male gamete and a relatively large, nonmotile female gamete or gametangial contact in which the gametangia are morphologically different.

oomycetes water molds, class of Mastigomycota; fungi that reproduce using flagellated zoospores.

oospores thick-walled, resting spores of fungi.

operator region a section of an operon involved in the control of the synthesis of the gene products encoded within that region of DNA; a regulatory gene that binds with a regulatory protein to turn on and off transcription of a specified region of DNA.

operon a group or cluster of structural genes whose coordinated expression is controlled by a regulator gene.

operon model a model that explains the control of the expression of structural genes, such as for lactose metabolism, by the regulation of the transcription of the mRNA directing the synthesis of the products of those structural genes.

opportunistic pathogens organisms that exist as part of the normal body microbiota but that may become pathogenic under certain conditions, e.g., when the normal antimicrobial body defense mechanisms have been impaired; organisms that are not normally considered pathogens but that cause disease under some conditions.

opsonization the process by which a cell becomes more susceptible to phagocytosis and lytic digestion when a surface antigen combines with an antibody and/or other serum component.

optical isomers compounds having the same number and kind of atoms and grouping of atoms but differing in their configurations or arrangements in space; specifically, their structures are not superimposable.

optimal growth temperature the temperature at which microbes exhibit the maximal growth rate.

optimal oxygen concentration the oxygen concentration at which microbes exhibit the maximal growth rate with maximal product yield.

orchitis inflammation of the testes.

organelle a membrane-bound structure that forms part of a microorganism and that performs a specialized function.

Orleans process method for the production of vinegar in which raw vinegar from a previous run provides the active inoculum; classic slow process for producing vinegar that relies on a microbial surface film.

oscillatorian cyanobacteria subgroup of cyanobacteria that form filamentous structures composed of straight or helical vegetative cells.

-ose combining form denoting a sugar.

osmophiles organisms that grow best or only in or on media of relatively high osmotic pressure.

osmosis the passage of a solvent through a membrane from a dilute solution into a more concentrated one.

osmotic pressure the force resulting from differences in solute concentrations on opposite sides of a semipermeable membrane.

osmotic shock any disturbance or disruption in a cell or subcellular organelle that occurs when it is transferred to a significantly hypertonic or hypotonic medium, with lysis of cells resulting from osmotic pressure.

osmotolerant organisms that can withstand high osmotic pressures and grow in solutions of high solute concentrations.

otitis media inflammation of the inner ear.

outer membrane a structure found in Gram-negative cell walls that acts as a coarse molecular sieve and allows the diffusion of hydrophilic and hydrophobic molecules.

oxidase an enzyme (oxidoreductase) that catalyzes a reaction in which electrons removed from a substrate are donated directly to molecular oxygen.

oxidation an increase in the positive valence or a decrease in the negative valence of an element resulting from the loss of electrons that are taken on by some other element.

oxidation pond a method of aerobic waste disposal employing biodegradation by aerobic and facultative microorganisms growing in a standing water body.

oxidation-reduction potential a measure of the tendency of a given oxidation-reduction system to donate elections, i.e., to behave as a reducing agent, or to accept electrons, i.e., to act as an oxidizing agent; determined by measuring the electrical potential difference between the given system and a standard system.

oxidative phosphorylation a metabolic sequence of reactions occurring within a membrane in which an electron is transferred from a reduced coenzyme by a series of electron carriers, establishing an electrochemical gradient across the membrane that drives the formation of ATP from ADP and inorganic phosphate by chemiosmosis.

oxidative photophosphorylation a metabolic sequence of reactions occurring within a membrane in which light initiates the transfer of an electron by a series of electron carriers, establishing an electrochemical gradient across the membrane that drives the formation of ATP from ADP and inorganic phosphate by chemiosmosis.

oxidize to produce an increase in the positive valence through the loss of electrons.

Oxyphotobacteria subclass of Photobacteria; bacteria capable of splitting water to form oxygen as part of photosynthetic metabolism; bacteria capable of producing oxygen during photosynthesis.

ozonation the killing of microorganisms by exposure to ozone.

5′-P free end unattached phosphate ester group at the 5-carbon position at one end of a nucleic acid molecule.

packaging natural or artificial wrapping or covering designed to protect or delay microbial decomposition of sterile interior tissues.

pain characteristic of the inflammatory response, an unpleasant sensation due to lysis of blood cells, triggering the production of bradykinin and prostaglandins that alter the threshold and intensity of the nervous system's response to pain.

palindrome a word reading the same backward and forward.

pallindromic sequence a base sequence the complement of which has the same sequence; a nucleotide sequence that is the same when read in the antiparallel direction.

pandemic an outbreak of disease that affects large numbers of people in a major geographical region or that has reached epidemic proportions simultaneously in different parts of the world.

papilloma viruses small, icosahedral DNA viruses that cause warts.

paralytic shellfish poisoning disease caused by toxins produced by the dinoflagellate *Gonyaulax*, which concentrates in shellfish such as oysters and clams.

parasites organisms that live on or in the tissues of another living organism, the host, from which they derive their nutrients.

parasitism an interactive relationship between two organisms or populations in which one is harmed and the other benefits; generally, the population that benefits, the parasite, is smaller than the population that is harmed.

parfocal pertaining to microscopic oculars and objectives that are so constructed or so mounted that in changing from one to another, the image remains in focus.

parotitis inflammation of the parotid gland, as in mumps.

passive agglutination a procedure in which the combination of an antibody with a soluble antigen is made readily detectable by the prior adsorption of the antigen to erythrocytes or to minute particles of organic or inorganic materials.

passive diffusion unassisted movement of molecules from areas of high concentration to areas of low concentration.

passive immunity short-term immunity brought about by the transfer of preformed antibody from an immune subject to a nonimmune subject.

Pasteur effect the slower rate of glucose utilization by a microorganism growing aerobically by respiratory metabolism than by the same organism growing anaerobically, reflecting feedback inhibition; in those organisms capable of both fermentative and respiratory metabolism, the inhibition of glucose utilization in anaerobically grown cells upon exposure to oxygen.

Pasteur, Louis French chemist and bacteriologist; responsible for major advances in microbiology.

pasteurization reduction in the number of microorganisms by exposure to elevated temperatures but not necessarily the killing of all microorganisms in a sample; a form of heat treatment that is lethal for the causal agents of a number of milk-transferable diseases, as well as for a proportion of normal milk microbiota, which also inactivates certain bacterial enzymes that may cause deterioration in milk.

pathogenicity the ability of an organism to cause disease in the host it infects.

pathogens organisms capable of causing disease in animals, plants, or microorganisms.

pathology the study of the nature of disease through the study of its causes, processes, and effects, along with the associated alterations of structure and function.

pelagic zone the region of an aquatic environment that comprises the entire body of water, excluding the mud or sand that forms the bed or bottom of that environment.

pellicle a thin protective membrane occurring around some protozoa, also known as a *periplast*; a continuous or fragmentary film that sometimes forms at the surface of a liquid culture; it consists entirely of cells or may be largely extracellular products of the cultured organisms.

pelvic inflammatory disease (PID) any acute, subacute, recurrent, or chronic infection of the oviducts and ovaries, with adjacent tissue involvement; most commonly caused by *Neisseria gonorrhoeae*.

penicillins a group of natural and semisynthetic antibiotics with a β-lactam ring that are active against Gram-positive bacteria inhibiting the formation of cross-links in the peptidoglycan of growing bacteria.

pentose a class of carbohydrates containing five atoms of carbon.

pentose phosphate pathway a metabolic pathway that involves the oxidative decarboxylation of glucose 6-phosphate to ribulose 5-phosphate, followed by a series of reversible, nonoxidative sugar interconversions.

pepsin a proteolytic enzyme.

peptidase an enzyme that splits peptides to form amino acids.

peptide bond a bond in which the carboxyl group of one amino acid is condensed with the amino group of another amino acid.

peptides compounds of two or more amino acids containing one or more peptide bonds.

peptidoglycan the rigid component of the cell wall in most bacteria, consisting of a glycan (sugar) backbone of repetitively alternating *N*-acetylglucosamine and *N*-acetylmuramic acid with short, attached, cross-linked peptide chains containing unusual amino acids; also called *murein*.

peptidyl site the site on the ribosome where the growing peptide chain is moved during protein synthesis.

Peptococcaceae family of Gram-positive cocci with complex nutritional requirements whose cells occur singly or in pairs, or in regular or irregular masses; they are obligately anaerobic, producing low molecular weight fatty acids, carbon dioxide, hydrogen, and ammonia.

peptones a water-soluble mixture of proteoses and amino acids produced by the hydrolysis of natural proteins either by an enzyme or by an acid.

perfringens food poisoning food poisoning by the ingestion of *Clostridium perfringens* type A, a self-limiting condition characterized by abdominal pain and diarrhea.

period of decline stage of disease after the period of illness during which the signs and symptoms of the disease disappear.

period of illness the acute phase of a disease during which the patient experiences characteristic symptoms.

periodontal disease disease of the tissues surrounding the teeth.

periodontal pockets holes in the gums deepened by periodontal disease.

periodontosis juvenile periodontitis, noninflammatory degeneration of the periodontium leading to bone regression.

periodontitis inflammation of the periodontium, the tissues surrounding a tooth.

periplasm the region between the cytoplasmic membrane and the outer cell wall membrane of Gram-negative bacteria.

periplasmic space in Gram-negative bacterial cells, the area between the outer cell wall membrane and the cytoplasmic membrane.

periplast see *pellicle*.

perishable foods food products that are readily subject to spoilage by microorganisms and consequently have a short shelf life.

peritonitis inflammation of the peritoneum.

peritrichous referring to the arrangement of a cell's flagella in a more or less uniform distribution over the surface of the cell.

permeability the property of cell membranes that permits transport of molecules and ions in solution across the membrane.

permease an enzyme that increases the rate of transport of a substance across a membrane.

peroxidase an oxidoreductase that catalyzes a reaction in which electrons removed from a substrate are donated to hydrogen peroxide.

peroxide the anion O_2^- or HO_2^-, or a compound containing one of these anions.

peroxisomes microbodies that contain D-amino acid oxidase, α-hydroxy acid oxidase, catalase, and other enzymes, found in yeasts and certain protozoa.

person-to-person epidemic epidemiological disease pattern characterized by a relatively slow, prolonged rise and decline in the number of cases.

pertussis see *whooping cough*.

pest a population that is an annoyance for economic, health, or aesthetic reasons.

pesticides substances destructive to pests, especially insects.

petri dish a round, shallow, flat-bottomed dish with a vertical edge together with a similar, slightly larger structure that forms a loosely fitting lid, made of glass or plastic, widely used as receptacles for various types of solid media.

pH the symbol used to express the hydrogen ion concentration, signifying the logarithm to the base 10 of the reciprocal of the hydrogen ion concentration; $-\log[H^+]$.

Phaeophycophyta brown algae that produce xanthophylls; algae where the primary reserve materials are laminarin and mannitol and the cell wall is two-layered and composed of alginic acid.

phage see *bacteriophage*.

phagocytes any of a variety of cells that ingest and break down certain categories of particulate matter.

phagocytosis the process in which particulate matter is ingested by a cell, involving the engulfment of that matter by the cell's membrane.

phagosomes membrane-bound vesicles in phagocytes formed by the invagination of the cell membrane and the phagocytized material.

phagotrophic referring to the ingestion of nutrients in particulate form by phagocytosis.

pharmaceutical a drug used in the treatment of disease.

pharyngitis inflammation of the pharynx.

phase contrast microscope a microscope that achieves enhanced contrast of the specimen by altering the phase of light that passes through the specimen relative to the phase of light that passes through the background, eliminating the need for staining in order to view microorganisms and making the viewing of live specimens possible.

phenetic pertaining to the physical characteristic of an individual without consideration of its genetic makeup; in taxonomy, a classification system that does not take evolutionary relationships into consideration; a classification system that assesses similarity based upon appearance.

phenol coefficient a number that expresses the antibacterial power of a substance relative to that of the disinfectant phenol.

phenotype the totality of observable structural and functional characteristics of an individual organism, determined jointly by its genotype and the environment.

-phile combining form meaning similar to or having an affinity for.

philopodia false feet that are filamentous projections composed entirely of ectoplasm.

-phobic combining form meaning having an aversion for or lacking affinity for.

phosphatases enzymes that hydrolyze esters of phosphoric acid.

phosphodiester bond the bonding of two moieties by a phosphate group; each moiety is held to the phosphate by an ester linkage.

phosphofructokinase an enzyme that mediates the addition of a phosphate group to glucose 6-phosphate, with the formation of glucose-1,6-diphosphate, a key step during glycolysis.

phospholipase an enzyme that catalyzes the hydrolysis of a phospholipid.

phospholipid a lipid compound that is an ester of phosphoric acid and also contains one or two molecules of fatty acid, an alcohol, and sometimes a nitrogenous base.

phosphorylation the esterification of compounds with phosphoric acid; the conversion of an organic compound into an organic phosphate.

phosphotransferase system a type of group translocation in which a phosphate group is added to a sugar as it passes through the membrane of a bacterium, such as *Escherichia coli*.

photo- combining form meaning light.

photoautotrophs organisms whose source of energy is light and whose source of carbon is carbon dioxide; characteristic of algae and some prokaryotes.

photoheterotrophs organisms that obtain energy from light but require exogenous organic compounds for growth.

photolithotrophs see *photoautotrophs*.

photolysis liberation of oxygen by splitting of water during photosynthesis.

photophosphorylation a metabolic sequence by which light energy is trapped and converted to chemical energy, with the formation of ATP.

photoreactivation a mechanism whereby the effects of ultraviolet radiation on DNA may be reversed by exposure to radiation of wavelengths in the range 320–500 nm; an enzymatic repair mechanism of DNA present in many microorganisms.

photosynthesis the process in which radiant (light) energy is absorbed by specialized pigments of a cell and is subsequently converted to chemical energy; the ATP formed in the light reactions is used to drive the fixation of carbon dioxide, with the production of organic matter.

photosynthetic membranes specialized membranes in photosynthetic bacteria that are the anatomical sites where light energy is converted to chemical energy in the form of ATP during photosynthesis.

photosystem I see *cyclic photophosphorylation*.

photosystem II see *noncyclic photophosphorylation*.

photosystems pathways of electron transfer initiated by light energy; pathways of ATP synthesis in photosynthetic bacteria used to convert light energy to chemical energy.

phototaxis the ability of bacteria to detect and respond to differences in light intensity, moving toward or away from light.

phototrophs organisms whose sole or principal primary source of energy is light; organisms capable of photophosphorylation.

phycobilisomes granules found in cyanobacteria and some algae on the surface of their thylakoids.

phycobiont the algal partner of a lichen.

phycocyanin type of pigment in cyanobacteria and some algae that confers blue color.

phycoerythrin type of pigment in cyanobacteria and red algae that confers red color.

phycology the study of algae.

phycomycete a group of true fungi that lack regularly spaced septae in the actively growing portions of the fungus and produce sporangiospores by cleavage as the primary method of asexual reproduction.

phycovirus any virus whose host cell, within which it replicates, is a cyanobacterium or alga.

phylogenetic referring to the evolution of a species from the simpler forms; in taxonomy, a classification based on evolutionary relationships.

phylogeny evolutionary relatedness.

phylum a taxonomic group composed of groups of related classes.

physiology the study of the functions of living organisms and their physicochemical parts and metabolic reactions.

phytoalexin polyaromatic antimicrobial substances produced by higher plants in response to a microbial infection.

phytoplankton passively floating or weakly motile photosynthetic aquatic organisms, primarily cyanobacteria and algae.

phytoplankton food chain a food chain in aquatic habitats based on the grazing of primary producers.

pickles cucumbers that have been fermented in brine, vinegar, or the like, using natural microbiota or *Lactobacillus plantarum* or *Pediococcus cerevisiae*.

pigments colored compounds.

pili filamentous appendages that project from the cell surface of certain Gram-negative bacteria apparently involved in adsorption phenomena; filamentous appendages involved in bacterial mating.

pilin a chain of proteins, the subunits of pili.

pink eye infection of the eye caused by *Haemophilus aegyptius*, characterized by swelling of the eyelids, discharge from the eye, and bleeding within the conjunctiva, as well as redness and itching; characteristic of many eye inflammations.

pitching the inoculation of yeast into cooled wort or grape must during the production of beer or wine, respectively.

plague a contagious disease often occurring as an epidemic; an acute infectious disease of humans and other animals, especially rodents, caused by *Yersinia pestis* that is transmitted by fleas.

planapochromatic lens a flat field apochromatic objective microscope lens.

plankton collectively, all microorganisms that passively drift in the pelagic zone of lakes and other bodies of water, chiefly microalgae and protozoans.

plant pathogens microorganisms that cause plant diseases.

plant pathology the study of the diseases of plants.

plant viruses viruses that replicate within plant cells.

plaque the accumulation of bacterial cells within a polysaccharide matrix on the surfaces of teeth; also known as *dental plaque*.

plaques clearings in areas of bacterial growth due to lysis by phage.

plasma cells cells that are able to synthesize a specific antibody and secondary B cells.

plasma membrane see *cytoplasmic membrane*.

plasmids extrachromosomal genetic structures that can replicate independently within a bacterial cell.

Plasmodiogymnomycotina subdivision of Gymnomycota; includes two classes, Protostetliomycetes and Myxomycetes.

Plasmodiophoromycetes endoparasitic slime molds, class of Mastigomycota; fungi that are obligate parasites of plants, algae, and other fungi, forming a plasmodium within host cells.

plasmodium malaria-causing protozoa; the life stage of acellular slime molds, characterized by a motile, multinucleate body.

plasmogamy fusion of cells without nuclear fusion to form a multinucleate mass.

plastids a class of membrane-bound organelles found within cells of higher plants and algae, containing pigments and/or certain products of the cell, e.g., chloroplasts.

plate counting method of estimating numbers of microorganisms by diluting samples, culturing on solid media, and counting the colonies that develop to estimate the number of viable microorganisms in the sample.

pleomorphism the variation in size and form among cells in a clone or a pure culture.

pleurocapsalean cyanobacteria unicellular subgroup of cyanobacteria, exhibiting multiple fission to produce coccoid reproductive cells that fail to separate completely following binary fission, forming multicellular aggregates.

ploidy in a eukaryotic nucleus or cell, the number of complete sets of chromosomes.

PMNs see *neutrophils*.

pneumonia inflammation of the lungs.

poi Hawaiian fermented food product made from the stems of the taro plant.

polar budding in fungi, budding that occurs at only one end of the mother cell.

polar flagella flagella emanating from one or both polar ends of a cell.

polar mutations mutations that prevent the translation of subsequent polypeptides coded for in the same mRNA molecule.

polarized light light vibrating in a defined pattern.

poliomyelitis inflammation of the gray matter of the spinal cord, caused by a picornavirus.

poliovirus a picornavirus that causes an inflammation of the gray matter of the spinal cord, i.e., the disease polio.

pollutant a material that contaminates air, soil, or water; substances—often harmful—that foul water or soil, reducing their purity and usefulness.

poly-β-hydroxybutyric acid a polymeric storage product formed by some bacteria.

polycistronic coding for multiple cistrons; mRNA molecules that code for the synthesis of several proteins, often the proteins are functionally related and under the control of a specific operon.

polyene antibiotics used to treat fungal diseases; they act by al-

tering the permeability properties of cytoplasmic membranes.

polymerase an enzyme that catalyzes the formation of a polymer.

polymers the products of the combination of two or more molecules of the same substance.

polymorph a leukocyte with granules in the cytoplasm; also known as a *polymorphonuclear leukocyte (PMN)*.

polymorphonuclear having a nucleus that resembles lobes connected by thin strands of nuclear substance.

polymorphonuclear neutrophils see *neutrophils*.

polypeptide a chain of amino acids linked together by peptide bonds, but of lower molecular weight than a protein.

polyphosphate reserves of organic phosphate that can be used in the synthesis of ATP.

polysaccharides carbohydrates formed by the condensation of monosaccharides, e.g., starch and cellulose, that have multiple monosaccharide subunits.

polysomes complexes of ribosomes bound together by a single mRNA molecule; also known as *polyribosomes*.

porins proteins found in the outer membranes of Gram-negative cells in groups of three, they form cross-membrane channels through which small molecules can diffuse.

portals of entry the sites through which pathogens can gain access and entry to the body.

positive interactions between biological populations, interactions that enhance the ability of the interacting populations to survive within the community a particular habitat.

positive stain a stain with a positively charged chromophore.

post-transcriptional modification action on hnRNA within the nucleus to form mRNA.

potable fit to drink.

pour plate a method of culture in which the inoculum is dispersed uniformly in molten agar or other medium in a petri dish; the medium is allowed to set and is then incubated.

pitching the inoculation of mash with yeast during the production of alcoholic beverages; the inoculation of a substrate with microorganisms.

precipitation separation of a substance in solid form from a solution, as by means of a reagent.

precipitin reaction a serological test in which the interaction of antibodies with soluble antigens is detected by the formation of a precipitate.

predation a mode of life in which food is primarily obtained by killing and consuming animals; an interaction between organisms in which one benefits and one is harmed, based on the ingestion of the smaller organism, the prey, by the larger organism, the predator.

predators organisms that practice predation.

preemptive colonization alteration of environmental conditions by pioneer organisms in a way that discourages further succession.

presumptive test in assays for assessing water safety, gas formation in Durham tubes containing lactose broth and water samples is positive evidence of fecal contamination.

prey an animal taken by a predator for food.

Pribnow sequence a sequence within nucleotide bases in DNA that determines the site of transcription initiation.

primary atypical pneumonia pneumonia caused by *Mycoplasma pneumoniae*.

primary immune response the first immune response to a particular antigen that has a characteristically long lag period and a relatively low titer of antibody production.

primary pneumonic plague form of plague caused by invasion of the lungs by *Yersinia pestis*.

primary producers organisms capable of converting carbon dioxide to organic carbon, including photoautotrophs and chemoautotrophs.

primary sewage treatment the removal of suspended solids from sewage by physical settling in tanks or basins.

prions infectious proteins; substances that are infectious and reproduce within living systems but appear to be proteinaceous, based on degradation by proteases, and to lack nucleic acids based on resistance to digestion by nucleases.

probabilistic identification matrices combinations of characteristics of organisms used to characterize large numbers of strains of a taxonomic group in order to establish the variability of a particular feature within a group; data matrices used to allow organisms of unknown affiliation to be identified as members of established taxa.

processed foods cheese products to which water has been added, thereby diluting their nutritional value.

Prochlorales order of Oxyphotobacteria; the primary photosynthetic pigments are chlorophyll a and b, only members of the genus *Prochloron* occur as green, single-celled, extracellular symbionts of marine invertebrates.

prodromal stage time period in the infectious process following incubation when the symptoms of the illness begin to appear.

profundal zone in lakes, the area where consumption exceeds primary productivity.

progenotes theoretical primitive, self-replicating, protein-containing, cell-like structures.

progeny offspring.

projector lens the lens of an electron microscope that focuses the beam on the film or viewing screen.

prokaryotes cells whose genomes are not contained within a nucleus; the bacteria.

prokaryotic cells bacterial cells.

promastigote an elongated, flagellated form assumed by many species of the Trypanosomatidae during a particular stage of development.

promoter specific initiation site of DNA where the RNA polymerase enzyme binds for transcription on the DNA.

propagated transmission see *person-to-person transmission*.

propagules the reproductive units of microorganisms.

prophage the integrated phage genome formed when this genome becomes integrated with the host's chromosome and is replicated as part of the bacterial chromosome during subsequent cell division.

prophylaxis the measures taken to prevent the occurrence of disease.

Propionibacteriaceae family of Gram-positive rods that produce propionic acid, acetic acid, or mixtures of organic acids by fermentation; consists of the genera *Propionibacterium* and *Eubacterium*.

propionic acid fermentation pathway metabolic sequence carried out by the propionic bacteria that produces propionic acid.

prosthecae a cell wall–limited appendage forming a narrow extension of a prokaryotic cell.

proteases exoenzymes that break down proteins into their component amino acids.

protein toxins proteins secreted by bacteria that act as poisons.

proteinase one of the subgroups of proteases or proteolytic enzymes that act directly on native proteins in the first step of their conversion to simpler substances.

proteins a class of high molecular weight polymers composed of amino acids joined by peptide linkages.

proteolytic enzymes enzymes that break down proteins.

protista in one proposed classification system, a kingdom of organisms lacking true tissue differentiation, i.e., the microbes; in another classification system, a kingdom that includes many of the algae and protozoa.

proto-cooperation synergism; a nonobligatory relationship between two microbial populations in which both populations benefit.

protobionts see *progenotes*.

protonmotive force potential chemical energy in a gradient of hydrogen ions and electrical energy across the bacterial cytoplasmic membrane.

protoplasm the viscid material constituting the essential substance of living cells upon which all the vital functions of nutrition, secretion, growth, reproduction, irritability, and locomotion depend.

protoplasts spherical, osmotically sensitive structures formed when cells are suspended in an isotonic medium and their cell walls are completely removed; a bacterial protoplast consists of an intact cell membrane and the cytoplasm it contains.

Protostetliomycetes class of Plasmodiogymnomycotina.

prototrophs parental strains of microorganisms that give rise to nutritional mutants known as *auxotrophs*.

protozoa diverse eukaryotic, typically unicellular, nonphotosynthetic microorganisms generally lacking a rigid cell wall.

protozoology the study of protozoa.

Pseudomonadaceae family of Gram-negative, straight or curved rods that are motile by means of polar flagella; most strains carry out obligately aerobic respiration, unable to fix atmospheric nitrogen; nutritionally versatile; some produce characteristic fluorescent pigments; widely distributed in soil and water.

pseudomurein component of the cell walls of archaebacteria.

pseudoplasmodium formed by swarming together or aggregation of myxamebae; undergoes a developmental sequence to form a sporocarp.

pseudopodia false feet formed by protoplasmic streaming in protozoa; used for locomotion and the capture of food.

psittacosis an infectious disease of parrots, other birds, and humans, caused by *Chlamydia psittaci*.

psychro- combining form meaning cold.

psychroduric microorganisms capable of surviving but not of growing at low temperatures.

psychrophile an organism that has an optimum growth temperature below 20°C.

psychrotroph a mesophile that can grow at low temperatures.

puerperal fever an acute febrile condition following childbirth, caused by infection of the uterus and/or adjacent regions by streptococci.

pure culture a culture that contains cells of one kind; the progeny of a single cell.

purine $C_5H_4N_4$, a cyclic nitrogenous compound, the parent of several nucleic acid bases.

purple membrane the portion of the cytoplasmic membrane that contains bacteriorhodopsin, found in *Halobacterium*.

pus a semifluid, creamy yellow or greenish-yellow product of inflammation composed mainly of leukocytes and serum.

putrefaction the microbial breakdown of protein under anaerobic conditions.

pyelonephritis inflammation of the kidneys.

pyknosis a condition in which the nucleus is contracted.

pyoderma a pus-producing skin lesion.

pyogenic pus producing.

pyorrhea see *periodontitis*.

pyrimidine a six-membered cyclic compound containing four carbon and two nitrogen atoms in a ring; the parent compound of several nucleotide bases.

pyrite a common mineral containing iron disulfite.

pyrogenic fever producing.

pyrogens fever-producing substances.

Pyrrophycophyta fire algae; generally brown or red because of xanthophyll pigments; unicellular and biflagellate; store starch or oils as the reserve material; the cell walls contain cellulose.

Q_{10} describes the actual change in the rate at which a reaction proceeds when the temperature is increased by 10°C; for enzymatic reactions the Q_{10} usually is about 2.

Q Fever an acute disease in humans characterized by sudden onset of headache, malaise, fever, and muscular pain, caused by *Coxiella burnetii*; the reservoirs of infection are cattle, sheep, and ticks.

quality control a system for verifying and maintaining a desired level of quality in a product or process by careful planning, use of proper equipment, continued inspection, and corrective action when required; in fermentation processes, quality is determined by the yield and purity of the product.

quarantine the isolation of persons or animals suffering from an infectious disease in order to prevent transmission of the disease to others.

quick freezing subjecting cooked or uncooked foods to rapid refrigeration, permitting them to be stored almost indefinitely at freezing temperatures.

quinolones antimicrobial agents that act by blocking normal DNA replication by interfering with DNA gyrase.

R plasmid a plasmid encoding for antibiotic resistance.

rabies an acute and usually fatal disease of humans, dogs, cats, bats, and other animals, caused by the rabies virus and commonly transmitted in saliva by the bite of a rabid animal.

racking a step in the fermentation of wine in which the wine is filtered through the bottom sediments and added back to the top of the fermentation vat.

radappertization reduction in the number of microorganisms by exposure to ionizing radiation.

radioimmunoassay a highly sensitive serological technique used to assay specific antibodies or antigens, employing a radioactive label to tag the reaction.

radioisotopes radioactive isotopes; isotopes emitting radioactivity.

radiolaria free-living protozoa occurring almost exclusively in marine habitats; they contain axopodia, with a skeleton of silicon or strontium sulfate.

radurization sterilization by exposure to ionizing radiation.

rancid having the characteristic odor of decomposing fat, chiefly due to the liberation of butyric and other volatile fatty acids.

raphe a slit or pore in the cell wall of a diatom.

rDNA recombinant DNA.

reading frame groups of three nucleotide sequences.

reagins a group of antibodies in serum that react with the allergens responsible for the specific manifestations of human hypersensitivity; a heterophile antibody formed during syphilis infections.

reaneal to reestablish double-stranded DNA.

rearrangement of genes change in the relative positions of genes within the chromosome, thus altering the expression of the information contained in those genes.

recalcitrant a chemical that is totally resistant to microbial attack.

rec (recombination) genes genes that code for enzymes involved in heteroduplex formation during homologous recombination.

recessive allele the allelic form of a gene whose information is not expressed.

recipient strain any strain that receives genetic information from another strain.

reciprocal recombination recombination that occurs as a result of crossing-over in which a symmetrical exchange of genetic material takes place, i.e., the genes lost by one chromosome are gained by the other, and vice versa.

recombinant any organism whose genotype has arisen as a result of recombination; also, any nucleic acid that has arisen as a result of recombination.

recombinant DNA technology see *genetic engineering*.

recombination the exchange and incorporation of genetic information into a single genome, resulting in the formation of new combinations of alleles.

recovery the end of a disease syndrome.

red algae see *Rhodophycophyta*.

red tides aquatic phenomenon caused by toxic blooms of *Gonyaulax* and other dinoflagellates that color the water and kill invertebrate organisms; the toxins concentrate in the tissues of filter-feeding mollusks, causing food poisoning if the mollusks are ingested.

redness characteristic of the inflammatory response resulting from capillary dilation.

reducing power the capacity to bring about reduction.

reduction an increase in the negative valence or a decrease in the positive valence of an element resulting from the gain of electrons.

reduction potential the relative susceptibility of a substrate to oxidation or reduction.

reductive amination the reaction of an α-carboxylic acid with ammonia to produce an amino acid.

refraction the deviation of a ray of light from a straight line in passing obliquely from one transparent medium to another of different density.

refractive index an index of the change in velocity of light when it passes through a substance causing a deviation in the path of the light.

refrigeration method used for the preservation of food by storage at 5°C, based on the fact that low temperatures restrict the rates of growth and enzymatic activities of microorganisms.

regulatory genes genes that serve a regulatory function; genes that do not code for specific peptides but instead regulate the expression of structural genes.

relapsing fever a human disease characterized by recurrent fever, caused by a *Borrelia* species and transmitted by ticks and lice.

relative humidity the availability of water in the atmosphere.

renin enzyme obtained from a calf's stomach that can hydrolyze proteins.

repellents chemicals that push substances away from them; chemicals that cause microbes to move away from them.

replica plating a technique by which various types of mutants can be isolated from a population of bacteria grown under nonselective conditions, based on plating cells from each colony onto multiple plates and noting the positions of inoculation.

replication multiplication of a microorganism; duplication of a nucleic acid from a template; the formation of a replica mold for viewing by electron microscopy.

replication fork the Y-shaped region of a chromosome that is the growing point during replication of DNA.

replicative RNA strands templates for the synthesis of new viral genomes produced by RNA polymerase.

replicon a nucleic acid molecule that possesses an origin and is therefore capable of initiating its own replication.

repressible a characteristic of enzymes that allows them to be made unless stopped by the presence of a specific repression substance.

repression the blockage of gene expression.

repressor protein a protein that binds to the operator and inhibits the transcription of structural genes.

reproduction a fundamental property of living systems by which organisms give rise to other organisms of the same kind.

reservoirs the constant sources of infectious agents found in nature.

resistance plasmids see *R plasmid*.

resistant crop varieties species of agricultural plants that are not susceptible to particular plant pathogens.

resolution the fineness of detail observable in the image of a specimen.

resolving power a quantitative measure of the closest distance between two points that can still be seen as distinct points when viewed in a microscope field; depends largely on the characteristics of the microscope's objective lens and the optimal illumination of the specimen.

respiration a mode of energy-yielding metabolism requiring a terminal electron acceptor for substrate oxidation; oxygen is frequently used as the terminal electron acceptor.

respiration pathways metabolic sequences for the oxidation of organic compounds to release free energy in order to drive the formation of ATP that require an external electron acceptor.

respiratory tract the structures and passages involved in the intake of oxygen and the expulsion of carbon dioxide in animals.

restriction enzymes enzymes capable of cutting DNA macromolecules; endonucleases.

reticuloendothelial system see *mononuclear phagocyte system*.

retroviruses family of enveloped RNA animal viruses that use reverse transcriptase to form a DNA macromolecule needed for their replication.

reverse transcription mechanism for RNA synthesis in which the RNA viruses use their RNA genome as a template for an RNA-directed DNA polymerase; RNA-directed synthesis of DNA that is the reversal of normal informational flow within a cell.

Reye's syndrome a neurological disease that sometimes occurs after a viral infection.

Rh incompatibility type 2 hypersensitivity reaction that occurs when a mother is Rh negative and the father and fetus are Rh positive; the mother develops Rh antibodies during the birth of such an infant that may cross the placenta and cause anemia in her next Rh-positive fetus.

rheumatic fever a febrile disease characterized by painful migratory arthritis and a predilection to heart damage leading to chronic valvular disease; the cause is unknown.

rhinoviruses causal agents of 25 percent of all common colds in adults.

Rhizobiaceae Gram-negative family of rod-shaped bacteria capable of fixing atmospheric nitrogen.

rhizopod a root-like pseudopodium of protozoa.

rhizosphere an ecological niche that comprises the surfaces of plant roots and the region of the surrounding soil in which the microbial populations are affected by the presence of the roots.

rhizosphere effect evidence of the direct influence of plant roots on bacteria, demonstrated by the fact that microbial populations usually are higher within the rhizosphere (the region directly influenced by plant roots) than in root free soil; elevation of levels of microbial populations directly due to the influence of plant roots.

Rho protein protein required to interrupt transcription.

Rhodophycophyta red algae that occur in marine habitats and contain phycocyanin, phycoerythrin, and chlorophyll pigments; the primary reserve material is Floridean starch; exhibit a specialized type of oogamous sexual reproduction; some produce tetraspores and have a bilayered cell wall.

Rhodospirillaceae purple, nonsulfur bacteria; family of phototrophic bacteria that produce red-purple carotenoid pigments; consist of the genera *Rhodospirillum*, *Rhodopseudomonas*, and *Rhodomicrobium*; carry out photoheterotrophic metabolism, converting carbon dioxide to organic matter by the Calvin cycle.

ribonucleic acid (RNA) a linear polymer of ribonucleotides in which the ribose residues are linked by $3',5'$-phosphodiester bridges; the nitrogenous bases attached to each ribose residue may be adenine, guanine, uracil, or cytosine.

ribosomal RNA (rRNA) RNA of various sizes that make up part of the ribosomes, constituting up to 90 percent of the total RNA of a cell; single-stranded RNA, but with helical regions formed by base pairing between complementary regions within the strand.

ribosomes cellular structures composed of rRNA and protein; the sites where protein synthesis occurs within cells.

70S ribosomes sites of protein synthesis in bacterial cells, mitochondria, and chloroplasts.

80S ribosomes sites of protein synthesis in the cytoplasm of eukaryotic cells.

ribulose 1,5-bisphosphate carboxylase enzyme that determines the rates of the Calvin cycle; enzyme that catalyzes the reaction between carbon dioxide and ribulose 1,5-bisphosphate.

rickettsialpox a disease caused by *Rickettsia akari*, characterized by enlargement of the lymph nodes, fever, chills, headache, secondary rash, and leukopenia following the formation of the initial papule at the location of the bite of a mite.

rickettsias members of the family Rickettsiaceae; Gram-negative bacterial parasites or pathogens of vertebrates and arthropods that reproduce within host cells by binary fission.

ringspots symptom of viral plant disease characterized by the appearance of chlorotic or necrotic rings on the leaves.

ringworm any mycosis of the skin, hair, or nails in humans or other animals in which the causal agent is a dermatophyte; also called *tinea*.

ripen to bring to completeness or perfection; to age or cure, as in cheese; to develop a characteristic flavor, odor, texture, and color.

RNA see *ribonucleic acid*.

RNA polymerase an enzyme that catalyzes the formation of RNA macromolecules.

Rocky Mountain spotted fever a tick-borne human rickettsial disease that occurs in parts of North America, caused by *Rickettsia rickettsii*.

rods bacteria in the shape of cylinders.

roll tube method technique used to create anaerobic conditions in which a prereduced, sterilized medium is rolled during cooling so that it covers the inside of the test tube and inoculation is accomplished under a stream of carbon dioxide or nitrogen.

rolling circle model replication pattern of viral DNA in which a circular DNA molecule is used to spin off unidirectionally a linear DNA molecule.

rotating biological contactor see *biodisc system*.

rotavirus a large DNA virus, the common etiological agent for diarrhea in infants.

rots plant diseases characterized by the breakdown of tissue caused by any of a variety of fungi or bacteria.

rRNA see *ribosomal ribonucleic acid*.

rubella see *German measles*.

rumen one of the four compartments that form the stomach of a ruminant animal where anaerobic microbial degradation of plant residues occurs, producing nutrients that can be metabolized by the animal.

runs straight-line movements by motile bacteria.

rusts plant diseases caused by fungi of the order Uredianales, so called because of the rust-colored spores formed by many of the causal agents on the surfaces of the infected plants.

Sabin vaccine attenuated live viral antigenic preparation administered for the prevention of polio.

sac fungi see *Ascomycotina*.

saki yellow rice beer made in Japan.

salinity the concentration of salts dissolved in a solution.

Salk vaccine inactivated viral antigenic preparation administered for the prevention of polio.

salmonellosis any disease of humans or animals in which the causal agent is a species of *Salmonella*, including typhoid and paratyphoid fevers, but most frequently referring to a gastroenteritis.

salt lake an inland water body with a high salt concentration normally approaching saturation.

salt-tolerant bacteria bacteria that can grow at concentrations of NaCl of 3–15 percent, which most bacteria cannot tolerate.

San Joaquin fever see *coccidioidomycosis*.

sanitary engineering the science dealing with the removal of waste materials.

sanitary landfill a method for disposal of solid wastes in low-lying areas, with wastes covered with a layer of soil each day.

sanitary methods techniques that prevent contamination of food or objects with pathogenic and spoilage organisms, including washing, sanitizing, and packaging.

sanitary practices any practice that produces sanitary conditions, such as by cleaning and/or sterilizing, or removes microorganisms and/or the substances that support microbial growth.

sanitize to make sanitary, as by cleaning or sterilizing.

sanitizing agents compounds that reduce the number of microbes without necessarily killing them or inhibiting their growth.

saprophytes organisms, e.g., bacteria and fungi, whose nutrients are obtained from dead and decaying plant or animal matter in the form of organic compounds in solution.

Sarcodina a major taxonomic group of protozoa characterized by the formation of pseudopodia.

saturation phenomenon in enzyme kinetics in which raising the concentration of a substrate does not continue to increase the rate of the reaction; the maximal concentration of a substance that will dissolve in a given solvent.

scanning electron microscope (SEM) an electron microscope in which a beam of electrons systematically sweeps over the specimen, and the intensity of secondary electrons generated at the specimen's surface where the beam's impact is measured and the resulting signal is used to determine the intensity of a signal viewed on a cathode ray tube that is scanned in synchrony with the scanning of the specimen.

scanning electron microscopy a form of electron microscopy in which the image is formed by a beam of electrons that has been reflected from the surface of a specimen.

scarlet fever infection caused by *Streptococcus pyogenes* transmitted by inhalation and direct contact; most common in children 2–10 year of age; characterized by sore throat, nausea, vomiting, fever, rash, and strawberry-colored tongue; treated with penicillin or erythromycin.

schizogamy a form of asexual reproduction characteristic of certain groups of protozoa; coincident with cell growth; nuclear division occurs several or numerous times, producing a schiziont that then further segments into other cells.

schizontocidal action effect of antimalarial drugs that rapidly interrupts schizogony within red blood cells.

sclerotia hard resting bodies that are resistant to unfavorable conditions and may remain dormant for prolonged periods.

SCP see *single cell protein*.

sebum the secretion of the sebaceous gland containing unsaturated free fatty acids that act as antimicrobics.

secondary B lymphocytes memory B-lymphocyte cells capable of initiating the antibody-mediated immune response for which they are genetically programmed.

secondary immune response the response of an individual to the second or subsequent contact with a specific antigen, characterized by a short lag period and the production of a high antibody titer.

secondary productivity the heterotrophic recapture of dilute nutrients; formation of bacterial biomass from utilization of nutrients at low concentrations.

secondary sewage treatment the treatment of the liquid portion of sewage containing dissolved organic matter, using microorganisms to degrade the organic matter that is mineralized or converted to removable solids.

secretory pertaining to the act of exporting a fluid from a cell or organism.

sedimentation the process of settling, commonly of solid particles from a liquid.

seedborne method of transmission of a bacterial plant disease in which the pathogens survive on seeds for a transient period in soil.

segmented genome a viral genome composed of several separate RNA molecules.

selective medium an inhibitory medium or one designed to encourage the growth of certain types of microorganisms in preference to any others that may be present.

selective toxicity the toxic effect of some antimicrobial agents on some microorganisms but not on others; inhibitory or killing effect against certain cells (e.g., microbes) but not against others (e.g., human cells).

self-limiting a disease that normally does not result in mortality even without medical intervention; an infection that is eliminated by natural host immune defenses prior to mortality and without the need for antimicrobics to curtail progression of the infection.

self-purification inherent capability of natural waters to cleanse themselves of pollutants based on biogeochemical cycling activities and interpopulation relationships of indigenous microbial populations.

semiconservative replication the production of double-stranded DNA containing one new strand and one parental strand.

semiperishable foods food products that are not readily subject to spoilage by microorganisms and consequently have a long shelf life unless improperly handled.

sense strand the strand of DNA that codes for the synthesis of RNA.

sensitization a process in which specific IgE antibodies are synthesized in response to an allergen, move through the bloodstream to mast cells in connective tissue, and become firmly fixed to receptors so that the next time the individual is exposed to the same allergen, that allergen can react directly with the IgE fixed to the mast cells.

septa a wall structure separating two cells.

septa formation in binary fission, the inward movement of the cytoplasmic membrane and cell wall; the separation of the two complete bacterial chromosomes.

septae plural of *septa*.

septate separated by cross walls.

septic tank a simple anaerobic treatment system for waste water where residual solids settle to the bottom of the tank and the clarified effluent is distributed over a leaching field.

septicemia a condition in which an infectious agent is distributed throughout the body via the bloodstream; blood poisoning, the condition attended by severe symptoms in which the blood contains large numbers of bacteria.

septum in bacteria, the partition or cross wall formed during cell division that divides the parent cell into two daughter cells; in filamentous organisms, e.g., fungi, one of a number of internal transverse cross walls that occur at intervals within each hypha.

serology the *in vitro* study of antigens and antibodies and their interactions; immunological (antigen–antibody) reactions carried out *in vitro*.

serotypes the antigenically distinguishable members of a single species.

serotyping tests to identify microorganisms based upon serological procedures that detect the presence of specific characteristic antigens.

serum the fluid fraction of coagulated blood.

serum hepatitis a form of viral hepatitis transmitted by the parenteral injection of human blood or blood products contaminated by the causal agent.

serum killing power the antimicrobial activity of the serum of a patient receiving antibiotics; an *in vivo* measure of antibiotic activity.

serum sickness a hypersensitivity reaction that occurs 8–12 days after exposure to a foreign antigen; symptoms caused by the formation of immune complexes include a rash, joint pain, and fever.

severe combined deficiency a genetically determined type of immunodeficiency caused by the failure of stem cells to differentiate properly; victims are incapable of any immunological response.

sewage the refuse liquids or waste matter carried by sewers.

sewage treatment the treatment of sewage to reduce its biological oxygen demand and to inactivate the pathogenic microorganisms present.

sex pilus see *F pilus*.

sexual reproduction reproduction involving the union of gametes from two individuals.

sexual spore a spore resulting from the conjugation of gametes or nuclei from individuals of different mating type or sex.

sexually transmitted diseases (STDs) diseases whose transmission occurs primarily or exclusively by direct contact during sexual intercourse.

sharp freezing a method of food preservation in which natural air between −15 and −29°C is circulated; freezing takes between 30 and 72 hours.

sheath a tubular structure formed around a filament or bundle of filaments, occurring in some bacteria.

sheathed bacteria bacteria whose cells occur within a filamentous sheath that permits attachment to solid surfaces and affords protection.

shelf fungi members of the order Aphyllophorales; fungi that grow on trees with tough leathery fruiting bodies.

shelf life the period of time during which a stored product remains effective, useful, or suitable for consumption.

shigellosis bacillary dysentery caused by bacteria of the genus *Shigella*.

shingles an acute inflammation of the peripheral nerves caused by reactivation of an infection with the herpes varicella virus, which has remained latent after causing chicken pox in that individual; characterized by painful, small, red, nodular skin lesions.

shipyard eye epidemic keratoconjunctivitis; acute, self-limiting viral infection of the eyes characterized by redness, edema, swelling, and discomfort; derives its name from the fact that it commonly occurs among shipyard workers, who have a high rate of eye injury from metal shavings.

shunt a diversion from the normal path as an alternative pathway in metabolism.

Siderocapsaceae unicellular family of chemolithotrophic bacteria that oxidize iron or manganese, depositing iron and/or manganese oxides in capsules or extracellular material.

siderophores iron chelators that solubilize ferric hydroxide making soluble iron available.

sigma unit a subunit of RNA polymerase that helps to recognize the promoter site.

signal sequence a region of nucleotides at the beginning of an mRNA molecule and the corresponding sequence of amino acids in the synthesized protein that indicates that the protein is an exoprotein and is responsible for initiating the export of that protein across the cytoplasmic membrane.

signs observable and measurable changes in a patient caused by a disease.

silent mutations changes in the genotype that do not alter the phenotype; changes in the sequence of nucleotides that do not alter the amino acids in the gene product.

simple matching coefficient a similarity measure used in taxonomic analysis that includes both negative and positive matches in its calculation.

single cell protein (SCP) protein produced by microorganisms and primarily composed of microbial cells; sources of this protein include bacteria, fungi, and algae.

single diffusion method precipitin reaction technique in which an antigen is allowed to diffuse unidirectionally into a tube containing a uniform concentration of soluble antibody so that the antigen establishes a concentration gradient through the tube.

singlet oxygen form of oxygen in which two of the electrons in the valence shell have antiparallel spins that is chemically reactive with and lethal to microorganisms.

site-specific recombination see *nonhomologous recombination*.

skin rash cutaneous eruption; sign of a disease condition.

skin surfaces an environment characterized by lack of available water, high salt concentrations, low water activity, and the presence of antimicrobial agents; generally an unfavorable habitat for microbial growth.

skin testing testing procedure based on delayed hypersensitivity reactions useful in the presumptive diagnosis of some diseases.

slime layer a capsular layer surrounding microbial cells composed of diffuse secretions that adhere loosely to the cell surface.

slime molds see *Gymnomycota*.

slow freezing method of food preservation in which natural air between −15 and −29°C is circulated; freezing takes between 30 and 72 hours.

slow-reacting substance of anaphylaxis (SRS-A) a mixture of leukotrienes that acts as a potent bronchial constrictor.

sludge the solid portion of sewage.

smallpox an extinct disease caused by the variola virus, which caused an acute, highly communicable disease in humans characterized by cutaneous lesions on the face and limbs.

smuts plant diseases caused by fungi of the order Ustilaginales; typically involve the formation of masses of dark-colored teliospores on or within the tissues of the host plant.

snapping division after binary fission, cells do not completely separate; they appear to form groups resembling Chinese ideographs.

sneeze a sudden, noisy, spasmodic expiration through the nose, caused by the irritation of nasal nerves.

sodium a metallic metal of the alkali group.

soft spots evidence of microbial spoilage of fruits and vegetables resulting from the action of microbially produced pectinesterases and polygalacturanases.

sofu Chinese word for tofu; see *tofu*.

solid waste refuse, waste material composed largely of inert materials—glass, plastic, and metal—and some decomposable organic wastes, including paper and kitchen scraps.

somatic antigens anitgens that form part of the main body of a cell, usually at the cell surface; distinguishable from antigens that occur on the flagella or capsule.

somatic cells any cell of the body of an organism except the specialized reproductive germ cell.

sonti Indian rice beer made with *Rhizopus sonti*.

SOS system radical, complex, multifunctional system for repairing DNA damage.

sour tart taste due to the presence of acid in foods; can be evidence of microbial spoilage or the desired result of controlled fermentaion to produce a food product; caused by accumulation of acidic products of microbial metabolism.

soy sauce brown, salty, tangy sauce made in Japan from soybeans, wheat, and wheat bran fermented with *Aspergillus oryzae*.

specialized transduction form of gene transfer and recombination accomplished by the transmission of bacterial DNA from a donor to a recipient cell by a temperate phage in which only a small amount of genetic information is transferred; the transferred genes occur at specific locations.

species a taxonomic category ranking just below a genus; includes individuals that display a high degree of mutual similarity and that actually or potentially inbreed.

specificity the restrictiveness of interaction; of an antibody, refers the range of antigens with which an antibody may combine; of an enzyme, refers to the substrate that is acted upon by that enzyme; of a pathogen or parasite, refers to the range of hosts.

spectrophotometer an instrument that measures the transmission of light as a function of wavelength, allowing quantitative measure of the intensity of two sources or wavelengths.

spectrum a range, e.g., of frequencies within which radiation has some specified characteristic, such as the visible light spectrum.

spectrum of action the range of bacteria against which an antibiotic may be targeted; may be narrow or broad.

spermatia in certain ascomycetes and basidiomycetes, nonmotile, male reproductive cells.

spherical aberration a form of distortion of a microscope lens based on the differential refraction of light passing through the thick central portion of a convex-convex lens and the light passing through the thin peripheral regions of the lens.

spheroplasts spherical structures formed from bacteria, yeasts, and other cells by weakening or partially removing the rigid component of the cell wall.

spirilli bacteria in the shape of spirals.

spirochetes bacterial group characterized by the presence of helically coiled rods wound around one or more central axial filaments; mobile by a flexing motion of the cell.

split genes genes coded for by noncontiguous segments of the DNA so that the mRNA and the DNA for the protein product of that gene are not colinear; genes with intervening nucleotide sequences not involved in coding for the gene product.

spontaneous generation formation of living organisms from nonliving entities by natural processes, now proven an impossibility.

sporangiospores asexual fungal spores formed within a sporangium.

sporangium a sac-like structure within which numbers of motile or nonmotile, asexually derived spores are formed.

spore an asexual reproductive or resting body that is resistant to unfavorable environmental conditions capable of generating viable vegetative cells when conditions are favorable; resistant and/or disseminative forms produced asexually by certain types of bacteria by a process that involves differentiation of vegetative cells or structures; characteristically formed in response to adverse environmental conditions.

sporocarp special type of fruiting body that bears a mucoid droplet at the tip of each branch containing spores with cell walls.

sporotrichosis chronic disease caused by the fungus *Sporotrix schenckii* which usually enters the body through breaks in the skin, producing characteristic lesions on the fingers and hands; treated with application of potassium iodide solution.

sporozoa a subphylum of parasitic protozoa in which mature organisms lack cilia and flagella, characterized by the formation of spores.

sporozoite the cells produced by the division of the zygote of a sporozoan.

sporulation the process of spore formation.

spread plate technique a method of microbial inoculation whereby a small volume of liquid inoculum is dispersed with a glass spreader over the entire surface of an agar plate.

sputum the material discharged from the surface of the air passages, throat, or mouth, consisting of saliva, mucus, pus, microorganisms, fibrin, and/or blood.

stab cells immature lymphocytes.

stain a substance use to treat cells or tissues in order to enhance contrast so that specimens and their details may be detected by microscopy.

stalks relatively wide bacterial appendages that can attach to a substrate or to other cells; may serve to increase the efficiency of nutrient acquisition.

staphylococcal food poisoning an acute, nonfebrile condition caused by the enterotoxins of certain strains of *Staphylococcus*.

stationary growth phase a growth phase during which the death rate equals the rate of reproduction, resulting in a zero growth rate in batch cultures.

statospore a resting spore of some algae, consisting of two pieces.

stem cell a formative cell; a blood cell capable of giving rise to various differentiated types of blood cells.

stenothermophiles microorganisms that grow only at temperatures near their optimal growth temperature.

sterilization process that results in a condition totally free of microorganisms and all other living forms.

sterilize to render incapable of reproducing or free from microorganisms.

stigma red eyespot, a pigmented region in the chloroplasts of many unicellular green algae.

stock culture a culture that is maintained as a source of authentic subcultures; a culture whose purity is ensured and from which working cultures are derived.

storage vacuoles membrane-bound organelles involved in maintaining accumulated reserve materials segregated from the cytoplasm within eukaryotic cells.

strain a cell or population of cells that has the general characteristics of a given type of organism, e.g., a bacterium or fungus, or of a particular genus, species, and serotype.

streak plate technique a method of microbial inoculation whereby a loopful of culture is scratched across the surface of a solid culture medium so that single cells are deposited at a given location.

Streptococcaceae family of Gram-positive cocci whose cells occur as pairs or chains, exhibiting facultative, anaerobic, fermentative metabolism.

streptomycin an aminoglycoside antibiotic produced by *Streptomyces griseus*, affecting protein synthesis by inhibiting polypeptide chain initiation.

Strickland reaction see *mixed amino acid fermentation pathways*.

strict anaerobes microorganisms that cannot tolerate molecular oxygen and are inhibited or killed in its presence; microorganisms that cannot use oxygen or survive in its presence.

stroma the interior compartment of the chloroplast where carbon dioxide fixation occurs during photosynthesis.

structural gene a gene whose product is an enzyme, structural protein, tRNA, or rRNA, as opposed to a regulator gene whose product regulates the transcription of structural genes.

structural RNA see *ribosomal RNA*.

subcutaneous beneath the skin.

submerged culture reactors fermentors used for the commercial production of vinegar, using forced aeration to maximize the rate of acetic acid production, with bacteria growing in a fine suspension created by the air bubbles and the fermenting liquid.

subspecies division of species that describes a specific clone of cells.

substrate a substance upon which an enzyme acts.

substrate-level phosphorylation reaction in which ATP is formed from ADP by the direct transfer of a high-energy phosphate group from an intermediate substrate in a metabolic pathway, as opposed to chemiosmotic generation of ATP.

substrate specificity a characteristic of enzymes reflecting the fact that the enzyme and substrate must fit together in a specific way for the enzyme to lower the activation energy.

succession the replacement of populations by other populations better adapted to fill the ecological niche.

sudden infant death syndrome crib death; some cases may be caused by *Clostridium botulinum*.

sulfide a compound of sulfur with an element or basic radical.

sulfide stinker a type of microbially caused spoilage that occurs in canned foods, producing the noxious odor of hydrogen sulfide from putrefying proteins.

sulfur cycle biogeochemical cycle mediated by microorganisms that changes the oxidation state of sulfur within various compounds.

superoxide dismutase an enzyme that catalyzes the reaction between superoxide anions and protons, the products being hydrogen peroxide and oxygen.

superoxide radical a toxic free radical of oxygen (O_2^-).

suppressor mutation a mutation that alleviates the effects of an earlier mutation at a different locus.

surface antigens antigens associated with cell surfaces.

surfactant a surface-active agent.

susceptibility the likelihood that an individual will acquire a disease if exposed to the causative agent.

Svedberg unit the unit in which the sedimentation coefficient of a particle is commonly expressed; when values are given in seconds, the basic unit is 10^{-13} seconds.

swan-necked flasks flasks whose necks were curved by Pasteur for use in his experiments disproving the theory of spontaneous generation.

swarm cells flagellated cells of Myxomycetes that fuse together to form a true plasmodium.

swelling characteristic of the inflammatory response associated with the accumulation of fluids in the bases surrounding tissue cells.

symbiosis an obligatory interactive association between members of two populations, producing a stable condition in which the two organisms live together in close physical proximity to their mutual advantage.

symbiotic nitrogen fixation fixation of atmospheric nitrogen by bacteria living in mutually dependent associations with plants.

symptom a physiological disorder that results in a detectable deviation from the normal healthy state and is usually indicated by complaints from a patient.

symptomatology the symptoms of disease taken together.

synchrony a state or condition of a culture in which all cells are dividing at the same time.

synergism in antibiotic action, when two or more antibiotics are acting together, the production of inhibitory effects on a given organism that are greater than the additive effects of those antibiotics acting independently; an interactive but nonobligatory association between two populations in which each population benefits.

syngamy the union of gametes to form a zygote.

synthetic fuels fuels, such as ethanol, methane, hydrogen, and hydrocarbons, produced by microorganisms; see also *gasohol*.

syntrophism a phenomenon in which the extent of growth of an organism is dependent on the provision of one or more metabolic factors or nutrients by another organism growing in the vicinity.

syphilis a chronic, communicable, sexually transmitted disease peculiar to humans, caused by *Treponema pallidum* and characterized by a variety of lesions.

systematics a system of taxonomy; the range of theoretical and practical studies involved in the classification of organisms.

systemic infections infections that are disseminated throughout the body via the circulatory system.

systemic lupus erythematosus autoimmune disease resulting from the failure of the immune response to recognize self antigens; results in kidney failure.

T aggressor cells see *cytotoxic T cells*.

T cells lymphocyte cells that are differentiated in the thymus and are important in cell-mediated immunity, as well as in the modulation of antibody-mediated immunity.

T helper cells a class of T cells that enhance the activities of B cells in antibody-mediated immunity.

T lymphocytes see *T cells*.

T suppressor cells a class of T cells that depress the activities of B cells in antibody-mediated immunity.

taxis a directional locomotive response to a given stimulus exhibited by certain motile organisms or cells.

taxon a taxonomic group, e.g., genus, family, or order.

taxonomic hierarchy an organizational levels used to group living things; the levels are kingdom, phylum, class, order, family, genus, and species.

taxonomy the science of biological classification; the grouping of organisms according to their mutual affinities or similarities.

teichoic acids polymers of ribitol or glycerol phosphate found in the cell walls of some bacteria.

teliospores thick-walled, binucleate resting spores of rusts and smuts.

tempeh food from Indonesia made from soybeans fermented with spores of *Rhizopus*.

temperate bacteriophage bacteriophage with the ability to form a stable, nondisruptive relationship within a bacterium; a prophage in which the phage DNA is incorporated into the bacterial chromosome.

temperature degree of heat or coldness of a body or substance, as measured by a thermometer or other graduated scale; environmental parameter that influences the rates of chemical reactions and the three-dimensional configuration of proteins.

temperature growth range the range between the maximum and minimum temperatures at which a microorganism can grow.

temperature sensitive mutations mutations that alter the range of temperatures over which a microorganism may grow, using specific substrates.

template a pattern that acts as a guide for directing the synthesis of new macromolecules.

termination sites sequences of nucleotides in the DNA that act as signals to stop transcription.

terrestrial relating to or consisting of land, as distinct from water or air.

tertiary recovery of petroleum the use of biological and chemical means to enhance oil recovery.

tertiary sewage treatment a sewage treatment process that follows a secondary process, aimed at removing nonbiodegradable organic pollutants and mineral nutrients.

test algal cell wall structure containing calcium or silicon; the outer protective covering or shell formed by some protozoa.

tetanospasmin neurotoxin produced by *Clostridium tetani* that interferes with the ability of peripheral nerves of the spinal column to transmit signals to the muscle cells properly.

tetanus lockjaw, a disease of humans and other animals in which the symptoms are due to a powerful neurotoxin formed by the causal agent, *Clostridium tetani*, present in an anaerobic wound or other lesion, characterized by sustained involuntary contraction of the muscles of the jaw and neck.

tetracyclines a group of natural and semisynthetic antibiotics that have in common a modified naphthalene ring; bacteriostatic-agents with a broad spectrum of activity.

theca a layer of flattened, membranous vesicles beneath the external membrane of a dinoflagellate; an open or perforated shell-like structure that houses part or all of a cell.

theory of spontaneous generation nonscientific theory that held that living organisms could arise without external cause from nonliving matter.

thermal death time the time required at a given temperature for the thermal inactivation or killing of a specified number of microorganisms.

thermal stratification division of temperate lakes into an epilimnion, thermocline, and hypolimnion, subject to seasonal change; zonation of lakes based on temperature where warm and cold water masses do not mix.

thermal vents hot areas located at depths of 800–1000 m on the sea floor, where spreading allows seawater to percolate deeply into the crust and react with hot core materials; life around the vents is supported energetically by the chemoautotrophic oxidation of reduced sulfur.

thermocline zone of water characterized by a rapid decrease in temperature, with little mixing of water across it.

thermoduric microorganisms capable of surviving but not growing at high temperatures.

thermodynamics the basic relationships between properties of matter, especially those affected by changes in temperature, and a description of the conversion of energy from one form to another.

thermophiles organisms having an optimum growth temperature above 45°C.

thylakoids flattened, membranous vesicles that occur in the photosynthetic apparatus of cyanobacteria and algae; the thylakoid membrane contains chlorophylls, accessory pigments, and electron carriers and is the site of light reaction in photosynthesis.

thymine a pyrimidine component of DNA.

thymine dimers cannot act as templates for DNA polymerase and so prevent the proper functioning of polymerases; formed by base substitutions creating covalent linkages between pyrimidine bases on the same strand of the DNA, caused by exposure to ultraviolet light.

thymocytes see *T cells*.

tine test test for tuberculosis in which a mechanical device makes multiple punctures in the skin to expose the individual to the antigen.

tinea the lesions of dermatophytosis; also called *ringworm*.

tinsel flagellum flagellum of eukaryotic organisms that bear fine, filamentous appendages along their lengths.

tissue culture the maintenance or culture of isolated tissues and of plant or animal cell lines *in vitro*.

tissues in plants and animals, a group of similar cells performing the same function.

titer the concentration in a solution of a dissolved substance.

tofu Japanese cheese-like food product made by fermenting soybeans with *Mucor* species.

tonsilitis inflammation of the tonsils, commonly caused by *Streptococcus pyogenes*.

toxic shock syndrome a disease caused by the release of toxins from *Staphylococcus* species, resulting in a physiological state of shock; major outbreaks of the disease have been associated with the use of tampons during menstruation.

toxicity the quality of being toxic; the kind and quantity of a poison produced by a microorganism or possessed by a nonbiological chemical.

toxigenicity the ability to produce toxins.

toxin any organic microbial product or substance that is harmful or lethal to cells, tissue cultures, or organisms; a poison.

toxoid a modified protein exotoxin that has lost its toxicity but has retained its specific antigenicity.

toxoplasmosis an acute or chronic disease of humans and other animals caused by the intracellular pathogen *Toxoplasma gondii*; transmission occurs by ingestion of insufficiently cooked meats containing tissue cysts.

trachoma a communicable disease of the eye caused by *Chlamydia trachomatis*.

transamination the transfer of one or more amino groups from one compound to another; the formation of a new amino acid by the transfer of an amino group from another amino acid.

transcription the synthesis of mRNA, rRNA, and tRNA from a DNA template.

transduction the transfer of bacterial genes from one bacterium to another by bacteriophage.

transfer RNA (tRNA) a type of RNA involved in carrying amino acids to the ribosomes during translation; for each amino acid there are one or more corresponding tRNAs that can bind it specifically.

transferrin serum beta-globulin that binds and transports iron.

transferring a culture the aseptic process by which microbial specimens are taken from one culture tube or plate and transferred to another.

transformation a mode of genetic transfer in which a naked DNA fragment derived from one bacterial cell is taken up by another and subsequently undergoes recombination with the recipient's chromosome; in tissue culture, the conversion of normal cells to cells that exhibit some or all of the properties typical of tumor cells; morphological and other changes that occur in both B and T lymphocytes on exposure to antigens to which they are specifically reactive.

transformed cells cells produced *in vitro* that have altered surface properties and continue to grow even when they contact a neighboring cell.

transition a point mutation in which one purine or one pyrimidine is replaced by another.

translation the assembly of polypeptide chains with mRNA serving as the template, a process that occurs at the ribosomes.

translocation nonhomologous recombination.

transmission electron microscope (TEM) an electron microscope in which the specimen transmits an electron beam focused on it; image contrasts are formed by the scattering of electrons out of the beam, and various magnetic lenses perform functions analogous to those of ordinary lenses in a light microscope.

transposable genetic elements specific segments of DNA that can undergo nonreciprocal recombination and thus move from one location to another.

transposons translocatable genetic elements; genetic elements that move from one locus to another by nonhomologous recombination, allowing them to move around a genome.

transversion a point mutation in which a purine is replaced by a pyrimidine or a pyrimidine by a purine.

trench mouth also known as *Vincent's angina* and *necrotizing ulcerative gingivitis*; caused by fusiform bacillus or spirochete infection, characterized by painful, superficial, bleeding gingival ulcers; treated with antibiotics, analgesics, and possibly removal of tissue.

tricarboxylic acid cycle see *Krebs cycle*.

trichome a chain or filament of cells that may or may not include one or more resting spores.

trickling filter system a simple, film-flow aerobic sewage treatment system; the sewage is distributed over a porous bed coated with bacterial growth that mineralizes the dissolved organic nutrients.

triplet code describes the genetic code because three sequential nucleotides in mRNA are needed to code for a specific amino acid.

tRNA see *transfer RNA*.

-troph combining form indicating a relation to nutrition or nourishment.

trophic level steps in the transfer of energy stored in organic compounds from one organism to another.

trophic structure steps in the transfer of energy stored in organic compounds from one organism to another.

trophophase during batch culture that phase in which growth-directed metabolism is dominant over secondary metabolism.

trophozoite a vegetative or feeding stage in the life cycle of certain protozoa.

trp operon region of DNA contains the structural genes that code for the enzymes required for the biosynthesis of the amino acid tryptophan and the regulatory genes that control the expression of these structural genes.

true fungi see *Mastigomycota*.

true slime molds see *Myxomycetes*.

trypanosomiasis any of a group of human and animal diseases in which the causal organism is a member of the genus *Trypanosoma*.

tuberculin reaction classic skin test for detecting probable cases of tuberculosis in which a purified protein derivative of *Mycobacterium tuberculosis* is injected subcutaneously and the area near the injection site is observed for evidence of a delayed hypersensitivity reaction.

tuberculosis an infectious disease of humans and other animals, caused in humans by *Mycobacterium tuberculosis*; may affect any organ or tissue of the body, but usually the lungs.

tularemia an acute or chronic systemic disease characterized by malaise, fever, and an ulcerative granuloma at the site of infection, caused by *Francisella tularensis*.

tumbles turning movements that occur when bacteria stop traveling in a straight line.

tumor-inducing (Ti) plasmid plasmid found in *Agrobacterium tumefaciens* that codes for tumorous plant growths (galls) when this bacterium infects plants.

turbidity cloudiness or opacity of a solution.

turbidostat a system in which an optical sensing device measures the turbidity of the culture in a growth vessel and generates an electrical signal that regulates the flow of fresh medium into the vessel and the release of spent medium and cells.

twiddles see *tumbles*.

tyndallization a sterilization process designed to eliminate endospore formers in which the material is heated to 80–100°C for several minutes on each of 3 successive days and incubated at 37°C during the intervening periods.

type culture collections centralized storage depositories for the preservation of all microbial species.

type 1 hypersensitivity see *anaphylactic hypersensitivity*.

type 2 hypersensitivity see *antibody-dependent hypersensitivity*.

type 3 hypersensitivity see *complex-mediated hypersensitivity*.

type 4 hypersensitivity see *cell-mediated hypersensitivity*.

types specific microbial strains; see also *subspecies*.

typhoid fever an acute infectious disease of humans caused by *Salmonella typhi* that invades via the oral route; the symptoms include fever and skin, intestinal, and lymphoid lesions.

typhus fever an acute infectious disease of humans characterized by a rash, high fever, and marked nervous symptoms, transmitted by the body louse and rat flea infected with *Rickettsia prowazekii*.

ultracentrifuge a high-speed centrifuge that produces centrifugal fields up to several hundred thousand times the force of gravity; used for the study of proteins and viruses, for the sedimentation of macromolecules, and for the determination of molecular weights.

ultraviolet light short wavelength electromagnetic radiation in the range 100–400 nm.

undulent fever see *brucellosis*.

unicellular having the form and characteristics of a single cell.

uracil a pyrimidine base, a component of nucleic acids.

urea $CO(NH_2)_2$, a product of protein degradation.

ureases enzymes that split urea into carbon dioxide and ammonia.

urethra the canal through which urine is discharged.

urethritis inflammation of the urethra.

urinary tract the system that functions in the elaboration and excretion of urine.

urkaryote the proposed progenitor of prokaryotic and eukaryotic cells; the primordial living cell.

uv see *ultraviolet light*.

V_{max} the maximal velocity of an enzymatic reaction occurring when the enzyme is saturated with substrate.

VA mycorrhizae see *vesicular-arbuscular mycorrhizae*.

vaccine any antigenic preparation administered to stimulate the recipient's immune defense mechanisms with respect to given pathogens or toxic agents.

vacuole a membrane-bound cavity within a cell that may function in digestion, secretion, storage, or excretion.

vacuum packing use of anaerobiosis as a method of food preservation in which air is excluded from the food package.

vaginal tract a region of the female genital tract, the canal that leads from the uterus to the external orifice of the genital canal.

vaginitis inflammation of the vagina.

valley fever see *coccidioidomycosis*.

valves see *frustules*.

vancomycin antibiotic that inhibits bacterial cell wall synthesis.

variant a strain that differs in some way from a particular organism.

varicella-zoster virus member of the herpesvirus group; the causal agent of chicken pox and shingles.

vector vaccines vaccines that act as carriers for antigens associated with pathogens other than the one from which the vaccine was derived; created through recombinant DNA technology.

vectors organisms that act as carriers of pathogens and are involved in the spread of disease from one individual to another; genetic elements that can carry foreign DNA.

vegetative cells cells that are engaged in nutrition and growth; they do not act as specialized reproductive or dormant forms.

vegetative growth production of a new organism from a portion of an existing organism exclusive of sexual reproduction.

venereal disease see *sexually transmitted diseases*.

vesicles specialized inclusions in root cortex in the vesicular-arbuscular type of mycorrhizal association.

vesicular arbuscular mycorrhizae a common type of mycorrhizae characterized by the formation of vesicles and arbuscules.

viability the ability to grow and reproduce.

viable plate count method for the enumeration of bacteria whereby serial dilutions of a suspension of bacteria are plated onto a suitable solid grown medium, the plates are incubated, and the number of colony-forming units is counted.

Vibrionaceae family of Gram-negative, facultatively anaerobic rods consisting of the genera *Vibrio, Aeromonas, Plesiomonas,* and *Photobacterium*.

vinegar a condiment prepared by the microbial oxidation of ethanol to acetic acid.

viral of or pertaining to a virus.

viremia viral infection of the bloodstream.

viricides chemicals capable of inactivating viruses so that they lose their ability to replicate.

virion a single, structurally complete, mature virus.

viroids the causal agents of certain diseases, resembling viruses in many ways but differing in their apparent lack of a virus-like structural organization and their resistance to a wide variety of treatments to which viruses are sensitive; naked infective RNA.

virology the study of viruses and viral diseases.

virulence the capacity of a pathogen to cause disease, broadly defined in terms of the severity of the disease in the host.

virulence factors special inherent properties of disease-causing microorganisms that enhance their pathogenicity, allowing them to invade human tissue and disrupt normal body functions.

virulent pathogen an organism with specialized properties that enhance its ability to cause disease.

virus a noncellular entity that consists minimally of protein and nucleic acid and that can replicate only after entry into specific types of living cells; it has no intrinsic metabolism, and its replication is dependent on the direction of cellular metabolism by the viral genome; within the host cell, viral components are synthesized separately and are assembled intracellularly to form mature, infectious viruses.

visible light radiation in the wavelength range of 400–800 nm that is required for photosynthesis but can be lethal to nonphotosynthetic microorganisms.

vitamins a group of unrelated organic compounds, some or all of which are necessary in small quantities for the normal metabolism and growth of microorganisms.

volutin see *metachromatic granules*.

volva a cup-shaped remnant of the universal veil that surrounds the base of the stalk in mature fruiting bodies of certain fungi.

vomiting the forcible ejection of the contents of the stomach through the mouth.

vulvovaginitis inflammation of the vulva and vagina, usually caused by *Candida albicans*, herpes viruses, *Trichomonas vaginalis*, or *Neisseria gonorrhoeae*.

wandering cells cells capable of ameboid movement, including free macrophages, lymphocytes, mast cells, and plasma cells.

warts small, benign tumors of the skin, caused in humans by the human papilloma virus.

Wassermann test outdated classical complement test for the diagnosis of syphilis.

water activity (A_w) a measure of the amount of reactive water available, equivalent to the relative humidity; the percentage of water saturation of the atmosphere.

water molds see *oomycetes*.

Weil-Felix test serological test for the diagnosis of some diseases caused by *Rickettsia* species, especially typhus fever, using heterophile antibodies.

Weil's disease see *leptospirosis*.

whiplash flagella smooth flagella of algae and fungi.

whooping cough pertussis, an acute respiratory tract disease occurring mainly in children, caused by *Bordetella pertussis* and characterized by paroxysms of coughing that usually end in loud whooping inspirations.

Widal test agglutination test for the diagnosis of typhoid fever, using antigens from *Salmonella* species.

wilts plant diseases characterized by a reduction in host tissue turgidity, commonly affecting the vascular system; common causal agents are species of the fungi *Fusarium* and *Verticillium* and the bacteria *Erwinia* and *Pseudomonas*.

windrow method a slow composting process that requires turning and covering with soil or compost.

wine an alcoholic beverage produced by microbial fermentation of grapes and other fruit.

wobble hypothesis proposed by Frances Crick, this hypothesis accounts for the observed pattern of degeneracy in the third base of a codon and says that this base can undergo unusual base pairing with the corresponding first base in the anticodon.

wort in brewing, the liquor that results from the mixture of mash and water held at 40–65°C for 1 to 2 hours, during which the starch is broken down by amylases to glucose, maltose, and dextrins, and proteins are degraded to amino acids and polypeptides.

xanthophyll a pigment containing oxygen and derived from carotenes.

xenobiotic a synthetic product not formed by natural biosynthetic processes; a foreign substance or poison.

xerotolerant able to withstand dryness; an organism capable of growth at low water activity.

yeasts a category of fungi defined in terms of morphological and physiological criteria; typically, unicellular, saprophytic organism that characteristically ferment a range of carbohydrates and in which asexual reproduction occurs by budding.

yellow fever an acute, systemic disease that affects humans and other primates; caused by a togavirus and transmitted to humans by mosquitoes.

yersiniosis infection caused by *Yersinia enterolytica* whose symptoms resemble those of appendicitis.

Z pathway the combination of the cyclic and noncyclic photophosphorylation pathways in oxygenic photosynthetic organisms describing the metabolic reactions accounting for the trapping of light energy, and the generation of ATP, oxygen, and NADPH during photosynthesis.

z value the number of degrees fahrenheit required to reduce the thermal death time 10-fold.

zone of greening area of green discoloration with partial clearing around the colony resulting from alpha hemolysis.

zoology the study of animal life, including its origin, development, structure, function, and classification.

zoonoses diseases of lower animals.

zoospores motile, flagellated spores.

Zygomycotina fungal subdivision of Amastigomycota; its members have coenocytic mycelia and form zygospores, exhibit sexual reproduction, or produce asexual sporangiospores.

zygospores thick-walled resting spores formed after gametangial fusion by members of the zygomycetes.

zygote a single diploid cell formed from two haploid parental cells during fertilization.

zymogenous term used to describe soil microorganisms that grow rapidly on exogenous substrates.

Glossary of Bacterial Genera

Acetobacter Cells, ellipsoidal to rod-shaped; motile by peritrichous flagella or nonmotile; endospores not formed; young cells Gram negative, although in older cultures some strains become Gram variable; metabolism respiratory, never fermentative; oxidize ethanol to acetic acid in neutral and acid reactions (pH 4.5); strict aerobes; optimal growth at ca. 30°C; G + C 55–64 moles %.

Acinetobacter Rods, usually very short and plump, approaching coccus shape in stationary phase, predominantly in pairs and short chains; no spores formed; flagella not present; Gram negative; oxidative metabolism; oxidase negative, catalase positive; optimal growth at 30–32°C; G + C 40–47 moles %.

Actinomyces Gram positive, irregularly staining bacteria; non-acid-fast, non-spore-forming, and non-motile; filaments with true branching may predominate and are particularly evident in 18- to 48-hour microcolonies; carbohydrates are fermented, with the production of acid but no gas; some species may show greening or complete lysis of rabbit red blood cells; facultative anaerobes; most are preferentially anaerobic, and one species grows well aerobically; CO_2 is required for maximum growth.

Agrobacterium Motile, Gram-negative rods; metabolism respiratory; optimal growth at 25–30°C; G + C 59.6–62.8 moles %.

Alcaligenes Cells, rods, coccal rods, or cocci, usually occurring singly; motile with peritrichous flagella; Gram negative; metabolism respiratory, never fermentative; do not fix gaseous nitrogen; oxidase positive; optimal growth at 20–37°C; G + C 57.9–70 moles %.

Arthrobacter Cells that in complex media undergo a marked change in form during the growth cycle; older cultures (generally 2–7 days old) are composed entirely or largely of coccoid cells; on transfer to fresh complex medium, growth occurs by enlargement (swelling) of the coccoid cells followed by elongation; do not form endospores; Gram positive, but the rods may be readily decolorized and may show only Gram-positive granules in otherwise Gram-negative cells; not acid fast; metabolism respiratory, never fermentative; strict aerobes; optimal growth at 20–30°C; G + C 60–72 moles %.

Azotobacter Large ovoid cells with marked pleomorphism; Gram negative, with marked variability; fix atmospheric nitrogen; grow well aerobically but can also grow under reduced oxygen tension; optimal growth at between 20–30°C; G + C 63–66 moles %.

Bacillus Cells rod-shaped, straight, or nearly so; majority are motile; heat-resistant endospores formed; not more than one in a sporangial cell; Gram reaction: positive, or positive only in the early stages of growth; metabolism strictly respiratory, fermentative, or both, using various substrates; strict aerobes or facultative anaerobes; G + C 32–62 moles %.

Bacteroides Gram-negative, non-spore-forming rods; metabolize carbohydrates or peptone; fermentation products of sugar-utilizing species include combinations of succinic, lactic, acetic, formic, or propionic acids, sometimes with short-chain alcohols; obligately anaerobic; optimal growth at 37°C; G + C 40–55 moles %.

Bartonella In stained blood films the organisms appear as rounded or ellipsoidal forms or as slender, straight, curved, or bent rods occurring either singly or in groups within erythrocytes; within tissues they are situated within the cytoplasm of endothelial cells as isolated elements or are grouped in rounded masses; Gram negative; not acid fast; stain poorly or not at all with many aniline dyes but satisfactorily with Romanowsky's and Giemsa's stains; optimal growth at 37°C.

Bdellovibrio Cells are single, small, curved, motile rods in the parasitic state; motile; Gram negative; parasitic strains attach to and penetrate bacterial host cells; metabolism respiratory, not fermentative; optimal growth at 30°C; G + C 45.5–51.3 moles %.

Bifidobacterium Rods highly variable in appearance; Gram positive; not acid fast; non-spore-forming; nonmotile; anaerobic; optimal growth at 36–38°C; G + C 57.2–64.5 moles %.

Bordetella Minute coccobacilli arranged singly or in pairs, more rarely in short chains; Gram negative, bipolar; colonies on potato-glycerol-blood agar medium are smooth, convex, pearly, glistening, nearly transparent, surrounded by a zone of hemolysis without a definite periphery; metabolism respiratory, never fermentative; strict aerobes; optimal growth at 35–37°C.

Borrelia Cells helical, with 3–10 or more coarse, uneven, often irregular coils; Gram negative; fermentative metabolism; strict anaerobes; optimal growth at 28–30°C.

Brucella Coccobacilli or short rods; mammalian parasites and pathogens; no capsules; nonmotile; do not form endospores; Gram negative; metabolism respiratory; strict aerobes; some require 5–10 percent added CO_2 for growth, especially on initial isolation; optimal growth at 20–40°C; G + C 56–58 moles %.

Campylobacter Slender non-spore-forming, spirally curved rods; motile; respiratory; methyl red and Voges-Proskauer negative; found in the reproductive organs, intestinal tract, and oral cavity of animals and humans; G + C 30–35 moles %.

Chlamydia Nonmotile, spheroidal cells; obligately intracellular growth cycle; the principal developmental stages are (1) the elementary body, which is a small, electron-dense spherule containing a nucleus and numerous ribosomes surrounded by a multilaminated wall, (2) the initial body, which is a large, thin-walled, reticulated spheroid containing nuclear fibrils and ribosomal elements, and (3) an intermediate body representative of a transitional stage between the initial and elementary bodies; the elementary body is the infectious form of the organism; the initial body is the vegetative form that divides by fission intracellularly but is apparently noninfectious when separated from the host cells; Gram negative; because these bacteria are unable to synthesize their own high-energy compounds, they have been described as "energy parasites"; optimal growth at 33–41°C; G + C 39–45 moles %.

Citrobacter Motile, peritrichously flagellated rods; not encapsulated; Gram negative; can use citrate as their sole carbon source.

Clostridium Rods, usually motile by means of peritrichous flagella; form endospores; Gram positive but may appear Gram negative in the late stages of growth; fermentative; most strains are strictly anaerobic; G + C 23–43 moles %.

Corynebacterium Straight to slightly curved rods with irregularly stained segments and sometimes granules; frequently show club-shaped swellings; snapping division produces angular and palisade (picket fence) arrangements of cells; generally nonmotile; Gram positive, although some species (e.g., *C. diphtheriae*) decolorize easily, especially in old cultures; not acid fast; carbohydrate metabolism fermentative and respiratory; grow best aerobically; G + C 52–68 moles %.

Coxiella Short rods resembling organisms of the genus *Rickettsia* in their staining properties, dependent on host cells for growth, and close natural association with arthropod and vertebrate hosts; rods occur preferentially in the vacuoles of the host cell rather than in cytoplasm or nucleus, as do the species of *Rickettsia*.

Desulfovibrio Curved rods; motile; Gram negative; obtain energy by anaerobic respiration, reducing sulfates or other reducible sulfur compounds to H_2S; strict anaerobes; optimal growth at 25–30°C.

Eikenella Rods to coccobacilli; Gram negative; acid and gas not produced from carbohydrates; facultative anaerobes; G + C 56 moles %.

Erwinia Cells predominantly single, straight rods; motile by peritrichous flagella; Gram negative; fermentative; facultative anaerobes; pathogenic for plants; optimal growth at 27–30°C; G + C 50–58 moles %.

Escherichia Straight rods; Gram negative; motile by peritrichous flagella or nonmotile; citrate cannot be used as the sole carbon source; glucose and other carbohydrates are fermented, with the production of lactic, acetic, and formic acids; the formic acid is split into equal amounts of CO_2 and H_2; lactose is fermented by most strains; facultative anaerobes; indole positive; no H_2S produced from TSI agar; methyl red positive; Voges-Proskauer negative; optimal growth at 35–40°C; G + C 50–51 moles %.

Francisella Very small coccoid to ellipsoidal pleomorphic rods; nonmotile; Gram negative; strictly aerobic; optimal growth at 37°C.

Haemophilus Coccobacillary; nonmotile; Gram negative; strict parasites, requiring growth factors present in blood; aerobic to facultatively anaerobic; optimal growth at 37°C; G + C 38–42 moles %.

Klebsiella Nonmotile, encapsulated rods; can use citrate and glucose as their sole carbon sources; glucose is fermented, with the production of acid and gas; Voges-Proskauer positive; methyl red negative; H_2S not produced from TSI agar; catalase positive, oxidase negative; optimal growth at 35–37°C; G + C 52–56 moles %.

Lactobacillus Rods; chain formation common; do not produce spores; Gram positive but become Gram negative with increasing age; metabolism fermentative even though growth generally occurs in the presence of air; some are strict anaerobes; lactic acid is the major end product of fermentation; optimal growth at 30–40°C; G + C 33.3–53.9 moles %.

Legionella Cells rod-shaped; Gram negative; weakly oxidase positive, catalase positive; nonmotile; fastidious, with narrow optimal temperature and pH ranges; do not utilize carbohydrates; urea not utilized; aerobic; G + C 39 moles %.

Listeria Small, coccoid, Gram-positive rods; do not produce spores or capsules; not acid fast; motile by peritrichous flagella; acid but no gas produced from glucose and several other carbohydrates; esculin is hydrolyzed; aerobic to microaerophilic; optimal growth at 20–30°C; G + C 38 moles %, except one species with G + C 56 moles %.

Micrococcus Cells spherical, occurring singly or in pairs and characteristically dividing in more than one plane to form a regular cluster, tetrads, or cubical packets; no resting stages known; Gram positive; metabolism strictly respiratory; aerobes; optimal growth at 25–30°C; G + C 66–75 moles %.

Moraxella Rods, usually very short and plump (coccobacilli); Gram negative; oxidative metabolism; a limited number of organic acids, alcohols, and amino acids serve as carbon and energy sources; carbohydrates not utilized; oxidase positive, catalase usually positive; H_2S not produced; strict aerobes; optimal growth at 32–35°C; G + C 40–46 moles %.

Mycobacterium Slightly curved or straight rods; filamentous or mycelium-like growth may occur; acid-alcohol fast at some stage of growth; Gram positive, but not readily stained by Gram's method; nonmotile; no endospores, conidia, or capsules; growth slow to very slow; optimal growth at ca. 40°C; G + C 62–70 moles %.

Neisseria Cocci, occurring singly but often in pairs, with adjacent sides flattened; endospores not produced; nonmotile; capsules may be present; Gram negative; complex growth requirements; few carbohydrates utilized; aerobic or facultatively anaerobic; catalase positive; oxidase positive; optimal growth at ca. 37°C; G + C 47–52 moles %.

Nitrobacter Cells or short rods; Gram negative; chemolithotrophs that oxidize nitrite to nitrate and fix CO_2; strictly aerobic; temperature range for growth 5–40°C; G + C 60.7–61.7 moles %.

Nitrosomonas Cells ellipsoidal or short rods; motile or nonmotile; occur singly, in pairs, or in short chains; Gram negative; chemolithotrophic; oxidize ammonia to nitrite and fix CO_2; strictly aerobic; optimal growth at 5–30°C; G + C 47.4–51 moles %.

Nocardia Produce true mycelia, but mycelia production may be rudimentary; Gram positive; some species acid-fast to partially acid fast; obligate aerobes; nonmotile; pigments are produced by several species; G + C 60–72 moles %.

Pasteurella Cells ovoid or rod-shaped; nonmotile; do not produce endospores; Gram negative, but bipolar staining is common; metabolism fermentative; methyl red negative; Voges-Proskauer negative; aerobic to facultatively anaerobic; optimal growth at 37°C; G + C 36.5–43.0 moles %.

Propionibacterium Gram positive, non-spore-forming, nonmotile rods. Usually pleomorphic, diphtheroid, or club-shaped; metabolize carbohydrates, peptone, pyruvate, or lactate; fermentation products include combinations of propionic and acetic acids; anaerobic to aerotolerant; optimal growth at 30–37°C; G + C 59–66 moles %.

Proteus Straight rods; motile by peritrichous flagella; nonpigmented; acid produced from glucose; methyl red test usually positive; Gram negative; optimal growth at 37°C; G + C 38–42 moles %.

Pseudomonas Cells single, straight, or curved rods; motile by polar flagella; monotrichous or multitrichous; no resting stages known; Gram negative; metabolism respiratory, never fermentative; some are facultative chemolithotrophs; G + C 58–70 moles %.

Rhizobium Rods, commonly pleomorphic; Gram negative; metabolism respiratory; characteristically able to invade root hairs of leguminous plants and initiate production of root nodules; within nodules bacteria are pleomorphic (bacteroids); nodule bacteroids characteristically involved in fixing molecular nitrogen; optimal growth at 25–30°C; G + C 59.1–65.5 moles %.

Rickettsia Parasitic bacteria occurring intracellularly or intimately in association with tissue cells other than erythrocytes or with certain organs in arthropods; growth generally occurs in the cytoplasm of host cells; transmitted by arthropod vectors; short rods; no flagella; Gram negative; G + C 30–32.5 moles %.

Salmonella Rods, usually motile by peritrichous flagella; can utilize citrate as their carbon source; Gram negative; indole negative; H_2S produced on TSI agar; methyl red positive; Voges-Proskauer negative; catalase positive; oxidase negative; nitrate reduced; optimal growth at 37°C; G + C 50–53 moles %.

Serratia Motile, peritrichously flagellated rods; many strains produce pink, red, or magenta pigments; glucose is fermented, with or without the production of a small volume of gas; methyl red negative; Voges-Proskauer positive; Gram negative; facultatively anaerobic; catalase positive; oxidase negative; G + C 53–59 moles %.

Shigella Nonmotile rods; not encapsulated; cannot use citrate or malonate as their sole carbon source; H_2S is not produced; glucose and other carbohydrates are fermented, with the production of acid but not gas; Gram negative; generally catalase positive; oxidase negative.

Sphaerotilus Straight rods, occurring in chains, with a sheath of uniform width, which may be attached by means of a holdfast;

Gram negative; metabolism respiratory, never fermentative; aerobic but can grow at reduced oxygen temperatures; optimal growth at 25–30°C; G + C 65.5–70.5 moles %.

Spirillum Rigid, helical cells; motile by means of polar, multitrichous flagella; generally exhibit bipolar flagellation; Gram negative; strictly respiratory metabolism; aerobic-microaerophilic; optimal growth at 30°C; G + C 38–65 moles %.

Staphylococcus Cells spherical, occurring singly, in pairs, and characteristically dividing in more than one plane to form irregular clusters; nonmotile; no resting stages known; Gram positive; metabolism respiratory and fermentative; catalase positive; a wide range of carbohydrates may be utilized, with the production of acid; facultative anaerobes; optimal growth at 35–40°C; G + C 30–40 moles %.

Streptococcus Cells spherical, occurring in pairs or chains; Gram positive; metabolism fermentative; predominant end product of glucose fermentation is lactic acid; catalase negative; facultative anaerobes; minimal nutritional requirements generally complex; optimal growth at ca. 37°C; G + C 33–42 moles %.

Streptomyces Produces true mycelia; slender hyphae; the aerial mycelium at maturity forms chains of three to many spores; reproduction by germination of the aerial spores, sometimes by growth of fragments of the vegetative mycelium; Gram positive; produce a wide variety of pigments; highly oxidative heterotrophs; aerobes; optimal growth at 25–35°C; G + C 69–73 moles %.

Thiobacillus Small, rod-shaped cells; motile by means of a single polar flagellum; no resting stages known; Gram negative; energy derived from the oxidation of one or more reduced or partially reduced sulfur compounds; final oxidation product is sulfate; obligate aerobes; optimal growth at 28–30°C; G + C 50–68 moles %.

Treponema Unicellular, helical rods with tight regular or irregular spirals; cells have one or more axial fibrils inserted at each end of the protoplasmic cylinder; motile; Gram negative; metabolism fermentative, using amino acids and/or carbohydrates; strict anaerobes; G + C 32–50 moles %.

Ureaplasma Cells coccoid; lack a true cell wall; cells bounded by a triple-layered membrane; colonies exhibit a cauliflower head or fried egg appearance; urea hydrolyzed, with simultaneous production of carbon dioxide and ammonia; catalase negative; glucose fermented aerobically and anaerobically; optimal growth at 35–37°C; G + C 26.9–29.8 moles %.

Veillonella Small cocci; nonmotile; Gram negative; carbohydrates and alcohols not fermented; produce acetate, propionate, CO_2, and H_2 from lactate; oxidase negative; complex nutritional requirements; carbon dioxide required; anaerobic; optimal growth at 30–37°C; G + C 40–44 moles %.

Vibrio Short rods, axis curved or straight; motile by a single polar flagellum; Gram negative; not acid-fast; endospores not produced; nonencapsulated; metabolism both respiratory and fermentative; fermentation of carbohydrates produces mixed products but no CO_2 or H_2; oxidase positive; facultative anaerobes; G + C 40–50 moles %.

Yersinia Ovoid cells or rods; nonmotile or motile with peritrichous flagella; nonencapsulated; various carbohydrates fermented, without production of gas; methyl red positive; Voges-Proskauer negative; optimal growth at 30–37°C; G + C 45.8–46.8 moles %.

Page numbers in boldface type refer to illustrations.
The letter *n* following page numbers refers to footnotes.

Deficiencies
  and genetic mutations, 201
  immunologic, 595–597
Defined culture medium, 96
Dehydration
  of food products, 475–476, **476**
  of specimens, 38
DeLatour, Charles Cagniard, 8
Delbruck, Max, 239, 266
Delesserica, 334
  decipiens, **335**
Deletion mutation, 201
Demeclocycline, **643**
Denatured protein, 348, 735
  antigenic properties of, 569
Dendrogram, 272, **272**
Dengue fever, 261, 692, **694,** 694–695
Denitrification, 283, 382
Density gradient ultracentrifugation, 177
  DNA homology determined by, 274–275
Dental caries, 77, 291, 540, 713–715
Dental plaque, 77, **77,** 132, 540n, **541,** 713, **714,** 716
Deoxyadenosine, **736**
Deoxycytidine, **736**
Deoxyguanosine, **736**
Deoxyribonucleic acid. *See* DNA
Deoxyribose, **729,** 735
Deoxythymidine, **736**
Dermacentor andersoni, as vector for Colorado
    tick fever, 696
Dermatitis, contact, 604, **605**
Dermatophilaceae, 305
Dermatophytes, 707
Derxia, 377
Desensitization procedures, 600–601
Desert soils, as microbial habitats, 366–367
Desiccation, as food preservation method, 475–476
Desoxycholate-citrate agar, **100**
Desulforomaculum, 291
Desulfovibrio, 287, 359, 383
  desulfuricans, 287
    in corrosion of metals, 531, 532
Desulfurization, 383
Detergents, 456–457
  as antimicrobial agents, 614, 616
  enzymes in, 517–518
Deterioration of materials, microbial activities in,
    527–532
Deuteromycetes, 314, 328–329, 430
  representative genera, **329**
Deuteromycotina, 321, **322,** 328–329
Dextran production with microbial species, 497
Dextran sucrase production by streptococci, 713, 714
Dextrorotation of isomers, 728, **728**
D'Herella, Felix, 265
DHU loop of tRNA molecule, **185**
Diagnosis of disease, 623–627
  blood counts in, differential, 623–624
  identification of pathogen in, 625–627
  procedures used in, 626–627
  skin testing in, 624–625
Diagnostic tables, 276–277
Diaminopimelic acid, 56, 159
Diapedesis, 562
Diatomaceous earth, 333
Diatoms, **13,** 331–333, **332, 386,** 386–387
Diauxie, 190, **192**
Dichotomous key, 276
Dictyosomes, **70**
Dictyostellium discoideum, 318, 319
  life cycle of, **319**
Didinium, 339, 400
  nasutum, **339, 400**
Dientamoeba fragilis, antiprotozoan agents for, 649
Differential blood counts, 623–624
Differential culture media, 99
Differentially permeable membrane, 48
Differentiated tissues, **12**
Diffusion, 50, **51,** 52–53
  facilitated, 52–53, **53**
  passive, 52, 53, **53**
  in precipitin reactions, 582, **583**
DiGeorge syndrome, 595–596
Digestion, in eukaryotic cells, 70

Digestive vacuole, 70
Diiodohydroxyquin, 649
Dilation, capillary, 561n
Dimorphism, 313, 322
Dinoflagellates, 63n, **329,** 333–334, 337
  toxins produced by, 548, 678
Dipeptides, 732
Diphtheria, 214, 667
  diagnosis of, 626
  vaccine for, 618, 667
    recommended schedule for, 621
Diphtheria toxin, 546, 547
Dipicolinic acid, 159
Diploid cells, 208–209
Diptera, 427, 428
Direct count procedures, 102–103, **103**
Disaccharides, 125, 728
Discomycetes, 323, 324
Discontinuous strand of DNA, 178
Disease, 538n
  acute, 623
  bacterial. *See* Bacterial diseases
  caused by parasites, 398
  caused by protozoa, 339, 342, **702**
  chronic, 623
  of crops, 406–421
  diagnosis of, 623–627
  early methods in treatment of, 3, 5
  of farm animals, 430–434
  and food spoilage, 468
  fungal infections, 415–418. *See also* Fungal
      infections
  germ theory of, 10
  incubation period of, 623
  Koch's postulates of, 11–12
  mechanisms in, 538
  periods of illness and decline in, 623
  portals of entry in, 655–657
    gastrointestinal tract, 670–678
    genitourinary tract, 678–688
    respiratory tract, 657–670
    skin, 691–718
  prevention of, 611–622, **613, 615**
  prodromal stage of, 623
  signs of, 623
  symptoms of, 623
  transmission of infectious agents in, **553,** 611,
      612, 655–657, **656**
  treatment of, 627–651
  viruses in. *See* Virus diseases
Disinfectants, 614
  evaluation of, 617
Disinfection
  of potable water supplies, 453–455
  of seeds, 420
  in sewage treatment process, 452–453
Dissimilatory sulfate- or sulfur-reducing bacteria,
    286–287, **287**
Distillation, for ethanol recovery, 504
Distilled liquors, 484, **485**
DNA, 17, 20, 171, **172,** 735–737, **736, 737**
  analysis of composition, 274
  in bacterial chromosome, 61, **62**
  and binary fission, 100, **101**
  chi form of, **213**
  in chloroplasts, 67
  complementary (cDNA), 228
  discovery as genetic material, 216
  double helix of, 63, 173, **174**
  in eukaryotic cells, 62, 63, 63n, **64**
  and F pilus, 78, 78n
  homology of, 274–275
  hybridization of, **275**
  inhibition by antimicrobials, 644
  melting curve for, **274**
  in mitochondria, 66
  and mutations, 199–201, 204, 205, 207, 354
  and protein synthesis, 175–176
  recombinant. *See* Genetic engineering
  replication of, 176–180, **177, 178, 179, 180**
  structure of, 171, **172,** 173
    discovery of, 175, **175**
  synthesis compared to RNA, 180–181
  transfer in prokaryotes, 214–220

DNA gyrase, 176
DNA polymerases, 176, **178**
DNA viruses, 238, 245, 248, 249, 250, 259
  animal, 255, 259–261
  growth curves of, **253**
  plant, **410**
Dolipore septa, 325, **326**
Dollond, John, 7
Domagk, Gerhard, 629
Domestic sewage, 440
Dominant allele, 209
Donovan, C., 342
Doolittle, W. F., 84
Doom, as biological pesticide, **429**
Double helix of DNA, 63, 173, **174**
Doubling time, 106
Doxycycline, **643**
Drugs
  antimicrobial. *See* Antimicrobial agents
  manufacturing of. *See* Pharmaceutical
      manufacturing
Drying of food, 351, 475–476, **476**
Dujardin, Felix, 342
Dum dum fever, 339
Dust cells, 561
Dutch elm disease, 324
Dutton, Joseph, 342
Dwarfism, 409
Dyes, as antimicrobial agents, 614
Dysentery, 342
  amebic, 336, 459, 678
  bacterial, 675

Early proteins, 248
Ears
  antimicrobial distribution in, 630
  detection of pathogens in, 627
  otitis media, 669
Earthstars, 328
East Coast fever, 433, **433**
Eastern equine encephalitis, 696
Echovirus, 261
  in gastroenteritis, 671
Eclipse period in growth curve for bacteriophage,
    249, **250**
Ecosystems, 371
Ectomycorrhize, 394
Edwardsiella, 284
Ehrenberg, Christian, 342
Ehrlich, Paul, 8, 628
Eijkman test, 455
Eikenella corrodens, in periodontitis, 716
Eimeria labbeana, 340
Electromagnetic spectrum, 353, **354**
Electron arrangement in atom, 723, **723**
Electron microscope, 25, 27, 35–42
  characteristics of, **25**
  scanning, **25,** 35, 38, 41, **41,** 42
  transmission, **25,** 35–38, **39, 40**
  viewing chromatin with, 64
Electron transport chain, 127, **129,** 129
Electronic optics, 36
Electrophoresis, 17
  gel, 209
  immunoelectrophoresis, 583
  of immunoglobulins, **572**
Elementary body of chlamydias, 289
ELISA procedure, 584, **586**
Elphidium crispa, **337**
Embden-Meyerhof-Parnas pathway, 119
Embden-Meyerhof pathway, 119, 121, **121,** 123, **123**
Emetine, toxicity of, 630
Emission wavelength, 31
Emmonsiella capsulata, 324, 324n
Encephalitis
  equine, **695**
    Eastern, 692, 696
    Venezuelan, 696
    Western, 696
  herpetic, antiviral agents in, 649
  Ilheus, 696
  Japanese B, 696
  Murray Valley, 696
  Rio Bravo, 696

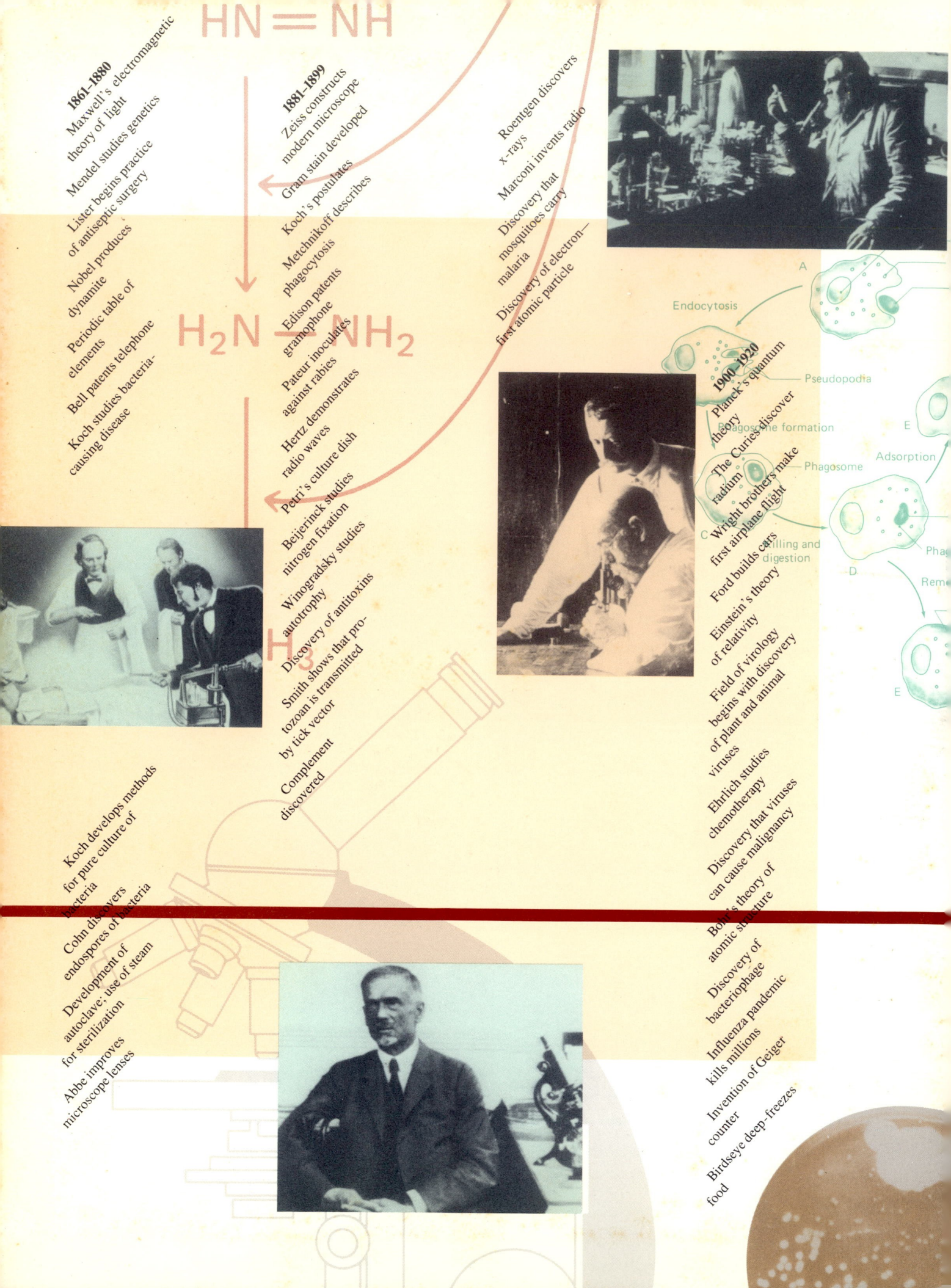

HN=NH

H₂N—NH₂

H₃

1861–1880
Maxwell's electromagnetic theory of light
Mendel studies genetics
Lister begins practice of antiseptic surgery
Nobel produces dynamite
Periodic table of elements
Bell patents telephone
Koch studies bacteria-causing disease

1881–1899
Zeiss constructs modern microscope
Gram stain developed
Koch's postulates
Metchnikoff describes phagocytosis
Edison patents gramophone
Pasteur inoculates against rabies
Hertz demonstrates radio waves
Petri's culture dish
Beijerinck studies nitrogen fixation
Winogradsky studies autotrophy
Discovery of antitoxins
Smith shows that protozoan is transmitted by tick vector
Complement discovered

Roentgen discovers x-rays
Marconi invents radio
Discovery that mosquitoes carry malaria
Discovery of electron—first atomic particle

Koch develops methods for pure culture of bacteria
Cohn discovers endospores of bacteria
Development of autoclave; use of steam for sterilization
Abbe improves microscope lenses

Endocytosis
Pseudopodia

1900–1920
Planck's quantum theory
Phagosome formation
The Curies discover radium
Phagosome
Wright brothers make first airplane flight
Adsorption
Ford builds cars
Killing and digestion
Einstein's theory of relativity
Field of virology begins with discovery of plant and animal viruses
Ehrlich studies chemotherapy
Discovery that viruses can cause malignancy
Bohr's theory of atomic structure
Discovery of bacteriophage
Influenza pandemic kills millions
Invention of Geiger counter
Birdseye deep-freezes food

A
C
D
E